The Cambridge
Encyclopedia of

Space

Missions, Applications and Exploration

**Fernand Verger,
Isabelle Sourbès-Verger,
Raymond Ghirardi**

with contributions by

Xavier Pasco

Foreword by **John M. Logsdon**

Translated by **Stephen Lyle** and **Paul Reilly**

PUBLISHED BY THE PRESS SYNDICATE OF THE UNIVERSITY OF CAMBRIDGE
The Pitt Building, Trumpington Street, Cambridge, United Kingdom

CAMBRIDGE UNIVERSITY PRESS
The Edinburgh Building, Cambridge, CB2 2RU, UK
40 West 20th Street, New York, NY 10011-4211, USA
477 Williamstown Road, Port Melbourne, VIC 3207, Australia
Ruiz de Alarcón 13, 28014 Madrid, Spain
Dock House, The Waterfront, Cape Town 8001, South Africa

http://www.cambridge.org

Previously published in French as *Atlas de géographie de l'espace*
First published in English 2003

Printed in the United Kingdom at the University Press, Cambridge

Typeset in Joanna 10.25/12.5 pt, in QuarkXpress™ [DS]

A catalogue record for this book is available from the British Library

Library of Congress Cataloguing in Publication data
Verger, Fernand.
[Atlas de la géographie de l'espace. English]
The Cambridge encyclopedia of space: missions, applications, and exploration/by Fernand
Verger, Isabelle Sourbès-Verger, Raymond Ghirardi; with contributions by Xavier Pasco.
p.   cm.
Includes bibliographical references and index.
ISBN 0 521 77300 8
1. Astronautics–Encyclopedias. 2. Rocketry–Encyclopedias. 3. Outer space–Exploration–
Space Policy–Encyclopedias. I. Sourbès-Verger, Isabelle. II. Ghirardi, Raymond. III. Title.
TL788.V48 2002
629.4'03–dc21   2002067408

ISBN 0 521 77300 8 hardback

Contents

Foreword

This is an exciting volume, hard to put down. Its coverage is literally cosmic in scope. No area of space activity goes unexamined. The book's visualizations of various aspects of space activities and capabilities are unique, provide new perspectives on what actually goes on in the region beyond the atmosphere. Just to pick one example, Figure 4.3 is a remarkable achievement. It summarizes in one chart the whole history of space activity in a clear and immediately understandable fashion. To see the clustering of satellites in various Earth orbits, and then the relatively few space probes that have explored the Solar System away from Earth, charts the path of space development to date in a fashion that dramatically improves upon what can be communicated by words alone. There are many, many similar standout depictions of complex information throughout the volume.

Most of us who have spent long careers working in the space sector are wont to say 'space is just a place,' then ignore the implications of that reality as we discuss what happens in orbit and beyond. Not so Fernand Verger and his colleagues. Professor Verger is one of the most distinguished geographers in France, and his influence on this volume is evident. The *Cambridge Encyclopedia of Space* takes a geographical perspective whenever possible. It first of all describes outer space in physical terms, as an environment with its own natural characteristics that both facilitate and limit what can be done there. This unique perspective sets the stage for the rest of the work.

The first artificial Earth satellite went into orbit less than a half-century ago. This volume sets out in both words and images humanity's achievements, benefits, and aspirations since that historic step towards *homo sapiens* a space-faring species. It provides an understanding of the physical, economic, and political realities that must be taken into account as next steps are planned. It depicts the many uses that have already been made of the capability to put people and machines into space, and suggest next steps in space development.

As the volume discusses the building blocks of space activity – spaceports, launch vehicles, and various space missions themselves, its words are complemented throughout by innovative visual and graphical presentations and by well-chosen photographs. The text is of course an essential element of any encyclopedia, and the text here both provides comprehensive and reliable information and offers penetrating insights regarding the factors that shape activity in space. That said, it is its visual material that sets this volume apart from any previous attempts to capture in one place the complexity of space activity. Professor Verger and his associates have spent many years perfecting their depictions of space activity, and they have made a real contribution to our appreciation of how far we have come in opening up the space frontier, and to the increasingly global character of space exploration and exploitation. The *Cambridge Encyclopedia of Space* will be an essential reference work for every space professional and a boon for those just learning about this new arena for human activity.

John M. Logsdon
Director, Space Policy Institute
The George Washington University
Washington, DC, USA

Preface

Over the last forty years, circumterrestrial space has been gradually occupied as unmanned satellites have been put into orbit to carry out a range of different functions, and in a more limited way, as human beings have also increased their presence, annexing the closer regions of the cosmos to the inhabited world. It has hence become possible to build up a geography of space, and it is this idea that lies at the heart of the present work.

Designed along the lines of an explanatory atlas, this encyclopedia allows the reader to understand the extent to which space has been occupied and to follow the main motivations underlying its development. To begin with, it provides a cartographical view of this occupation, briefly specifying the conditions that prevail in the medium and the physical laws that hold sway over the use of circumterrestrial space. Many constraints must be faced in space development. These constraints explain the unequal distribution of satellites and probes gravitating in a number of different orbits, nearby or distant, circular or eccentric, equatorial, polar or other, depending on their mission, whether it be for exploration of our cosmic neighbourhood or further afield, civilian Earth observation, telecommunications, military surveillance, or human occupation.

However, space-based activities can also be considered in terms of their relationship to Earth. The successive passages of satellites criss-cross the whole surface of our planet, their tracks winding around it like the thread around a ball of wool. Satellites supply a new image of the globe and encourage links between different peoples. At the same time, the complexity of space technology creates a genuine hierarchy amongst the countries of the world, reasserting the traditional balance of power on Earth, yet introducing new features. The main steps in space conquest have led to the steady constitution of what appears today to be an exclusive club of space powers. However, the different activities have been mastered to quite varying degrees. Almost all countries around the planet now use space systems. Many are those who operate the satellites that only a much more restricted group of countries are able to put together. On the other hand, very few nations can provide their own launch capacity, and even fewer can claim to master the whole range of manned and unmanned, civilian and military space resources.

The present book aims to describe and account for space endeavour around the world and to provide a careful analysis of the policies that guide the great space powers. Apart from the chapter specifically devoted to space policy, the means of access and main areas of application are presented to show how the various programmes express different national preferences and their consequences for world affairs. The geopolitical aspect of the space phenomenon is indeed a key feature, since satellites procure for us a new vision of our planet and a clearer picture of its resources. Hence, remote sensing which is so important for cartographic applications raises the problem of how data should be made available, for it is as relevant to national independence and international security as it is to territorial development. In the same way, the flow of information, by telephone or television, for positioning or other purposes, provides the subject for a cartographic representation which illustrates the main areas of exchange, the weakened and transformed notion of border, and the appearance of ever sharper international features heavily dominated by the United States. Finally, the navigation programmes are closely linked to questions of strategic independence in a field where applications are still emergent.

Space activities thus have many repercussions and, on a global level, increase the weight of the dominant powers, whether they be military or civilian. These manifest themselves on an economic level through the development of new systems made possible by state-of-the-art technology and answer to a growing need to dominate the markets. Space bestows an undeniable advantage upon those that lay claim to it, not only by the information it offers, but also by the possibilities for direct intervention which it opens up. Finally, the occupation of space by human beings and projects to set up long-term space outposts lead to new prospects, although sensitive to the vacillation of political commitment.

Going beyond a simple description of the way current projects attempt to occupy space, this work aims to provide a conceptual basis for a genuine geography of space, without which it would be difficult to comprehend its development or the growing number of related issues in today's world.

Acknowledgements

Images have been supplied by CNES, SpotImage, DLR, ESA, GRGS, ImageSat, ISAS, ISRO, NASA, NASDA, NRSA, Radarsat, Space Imaging.

CHAPTER ONE

The environment of outer space

- **DEEP SPACE**

- **NEAR SPACE**

Deep space

In the Solar System distance measurements are commonly expressed in *astronomical units* (AU). One AU is the length of the semi-minor axis of the Earth's orbit, that is, 149 597 870 km. In the *distant universe*, which extends beyond the range of today's probes, they are expressed in light years (1 light year = 9.461×10^{12} km) or parsecs (1 parsec = 3.26 light years).

Deep space refers to the central part of the Solar System (fig. 1.1). This contains:

■ one star, the Sun, of radius 696 000 km,

■ the nine *principal planets* whose mean distances from the Sun, or mean heliocentric distances, vary between 0.39 AU for Mercury and 39.44 AU for Pluto (see table 1.1). The motion of the planets against the celestial sphere of fixed stars has been observed since the earliest times, earning them the name of wandering bodies (from the Greek πλανητης).

The planets are divided into two groups according to their physical properties. The *terrestrial planets*, Mercury, Venus, Earth and Mars, are relatively small, with a solid surface and an atmosphere. The *Jovian* or *giant planets*, Jupiter, Saturn, Uranus and Neptune, are distinctly larger and much less dense. Finally, Pluto falls into neither of these two families, being a small planet of low density.

The Solar System also includes:

■ Natural satellites (moons) and rings orbiting certain planets (table 1.1).

■ The *asteroids*, numbering several thousand, and grouped into families with orbits showing a wide range of eccentricities, sizes and inclinations to the ecliptic. The majority follow quasi-circular orbits at heliocentric distances between 2 and 3.5 AU, where they make up the main belt. Further out, the Hildas gravitate around 4 AU and the Trojans around 5 AU (see fig. 8.30).

■ Several hundred other objects, similar to asteroids, which gravitate beyond the orbit of Neptune. These form the Kuiper Belt. Pluto and its moon Charon are now considered as members of this group.

■ The *comets*, which have eccentric orbits, and primitive trajectories with semi-major axes measuring several tens of thousands of AU.

In addition, like the rest of the interplanetary medium, deep space is filled with:

■ a flow of ionised particles originating in the Sun and known as the *solar wind*,

■ *interplanetary dust*.

Certain parameters are used to locate positions of celestial bodies in the Solar System relative to Earth.

The *geocentric distance* depends on both the orbit of the celestial body and the orbit of Earth, and also on the position of these bodies on their orbits at the relevant time. Neglecting the different inclinations of the orbits relative to the ecliptic, it depends on the *phase angle*, that is, the angle between the planet, the Sun and Earth. At its maximum, the geocentric distance is close to the sum of one AU and one radius of the planetary orbit in question. This occurs at conjunction, in the configuration planet–Sun–Earth. Its minimum is close to the absolute value of the difference between one AU and the radius of the planetary orbit at opposition or inferior conjunction.

The *synodic period* is the time required for the system planet–Sun–Earth to come back to the same configuration as viewed from Earth. This period varies from 115.9 days in the case of Mercury, to 2 years 49.5 days for Mars. The synodic period marks the return of certain special conditions, such as a favourable configuration for sending out probes.

	Name	Symbol	Equatorial diameter in km	Average density	Escape velocity in km / s	Synodic period		Heliocentric distance (AU)	Inclination on the ecliptic	Existence of satellites	Existence of rings
terrestrial planets	Mercury	☿	4 878	5.44	4.25		115.9 days	0.387	7°00'		
	Venus	♀	12 104	5.25	10.36	1 year	218.7 days	0.723	3°24'		
	Earth	♁	12 756	5.52	11.18			1	0°	○	
	Mars	♂	6 794	3.94	5.02	2 years	49.5 days	1.524	1°51'	○	
giant planets	Jupiter	♃	142 800	1.31	59.64	1 year	33.6 days	5.203	1°19'	○	○
	Saturn	♄	120 660	0.69	35.41	1 year	12.8 days	9.555	2°30'	○	○
	Uranus	♅	50 800	1.21	21.41	1 year	4.4 days	19.218	0°46'	○	○
	Neptune	♆	48 600	1.67	23.52	1 year	2.2 days	30.110	1°47'	○	○
	Pluto	♇	3 000	1 ?	?	1 year	1.5 days	39.439	17°10'	○	

Table 1.1. Planetary characteristics.

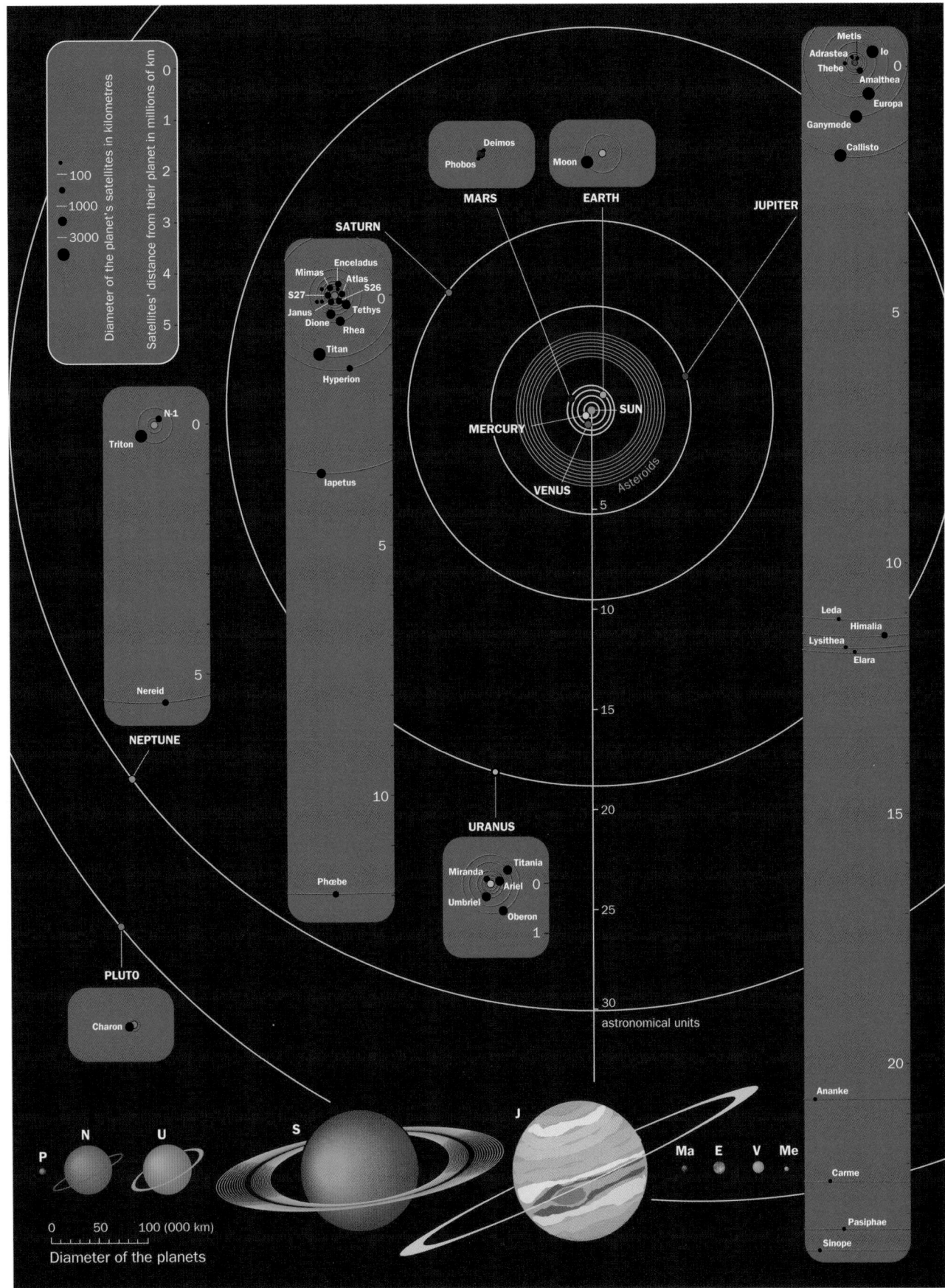

Figure 1.1. **The Solar System.**

Near space

The near space medium is dominated by the presence of a gaseous envelope, the *atmosphere*, held around Earth by gravity, and by the existence of a magnetic field, the *magnetosphere*, generated by the outer part of the terrestrial core.

Above the atmosphere, a plasma (ionised gas, composed of positively and negatively charged particles and ions) is trapped by the lines of force of the magnetosphere. This plasma constitutes the ionosphere and the various plasmatic regions of the magnetosphere.

The atmosphere

The atmosphere can be subdivided in different ways (fig. 1.2):

- By its composition, which remains constant in the lower convection zone or *homosphere*, up as far as the *homopause* (90 km). Ozone is most abundant between 40 and 50 km. Light elements predominate more and more in the *heterosphere* and, photoionised by solar ultraviolet radiation, give rise to the ionospheric plasma. Beyond the *exobase* (500 or 600 km), only the lightest atmospheric components remains (helium and hydro-

gen). These are liable to escape, although slowly, from Earth's gravitational attraction. This region is called the *exosphere*.

- By pressures and densities, which decrease more and more slowly with altitude. 50% of the mass of the atmosphere is located below 5 km, 90% below 16 km and 99.9% below 60 km. Even the most tenuous atmosphere exerts a braking effect on satellites, out as far as 600 km, thereby limiting their lifespans.

- By temperatures, which decrease with altitude by 5 °C to 10 °C per kilometre in the *troposphere* down to a first minimum in the *tropopause*. This is situated at 9 km altitude near the poles and about twice that near the equator. Aeroplanes fly mainly in the troposphere. Above the tropopause, temperatures increase with altitude under the effects of ozone dissociation by solar UV radiation, reaching 80 °C at around 50 km altitude, at the top of the *stratosphere*. This is called the *stratopause*. They then decrease to −70 °C in the *mesosphere*. Finally, beyond the *mesopause* (90 km), they increase rapidly in the *thermosphere* (which coincides

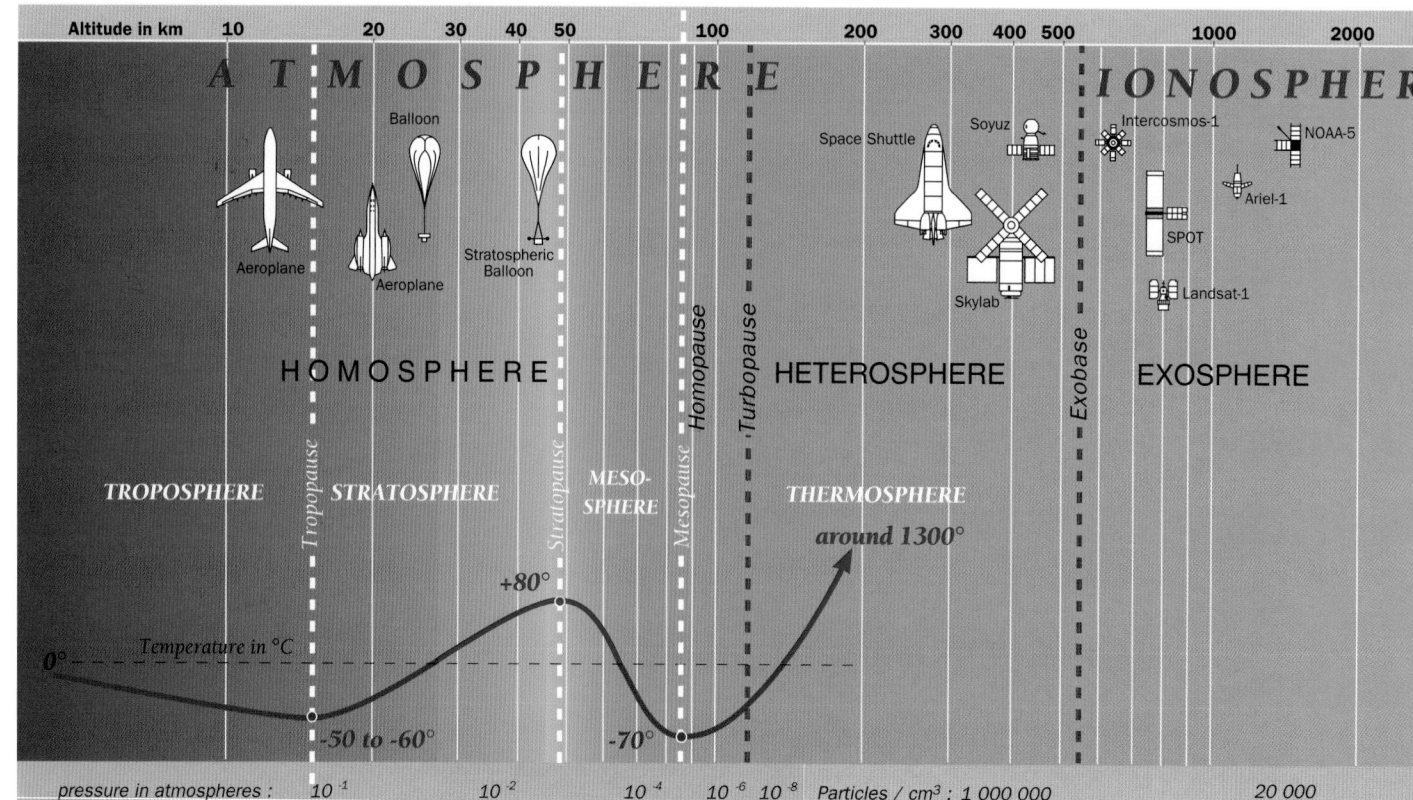

Figure 1.2. **Main divisions of near space showing airborne and space vehicles that visit them.**

4

Polar cleft
Earth
Line of magnetic poles

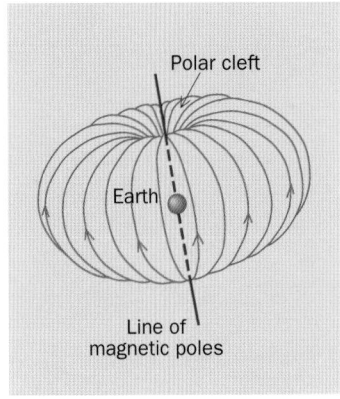

Figure 1.3. **Terrestrial magnetic field without the effects of the solar wind.**

Line of force
Earth
Solar wind
Shock wave
Magnetosheath
Neutral sheet

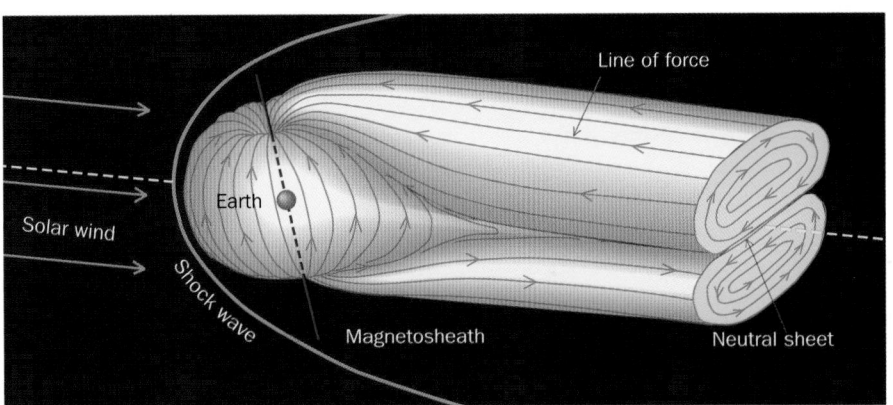

Figure 1.4. **Terrestrial magnetic field under the influence of the solar wind.** The magnetic field extends on the right, beyond the cross-section represented here to show the structure.

with the heterosphere) to reach the *exospheric temperature* of 1200 °C to 1300 °C, corresponding to the absorption of extreme UV solar radiation by photoionisation.

The magnetosphere

The magnetosphere forms a magnetic cavity in the solar wind, resulting from the interaction between the terrestrial dipolar magnetic field (fig. 1.3) and the interplanetary magnetic field 'frozen' into the solar wind (fig. 1.4). The surface on which the two fields balance is called the *magnetopause*. Beyond the magnetopause is a shock wave due to the supersonic flow of the solar wind. Between the two is the *magnetosheath*, where the magnetic field varies greatly in intensity and direction. This is dominated by particles from the interplanetary plasma (fig. 1.5).

Plasmas in the magnetosphere

The *ionosphere* begins in the outer thermosphere and is due to ionisation of the components of the terrestrial atmosphere carried to high altitude. This ionisation is caused by solar UV radiation and also by X-rays and primary cosmic radiation. Also to be found here are particles from the interplanetary plasma and

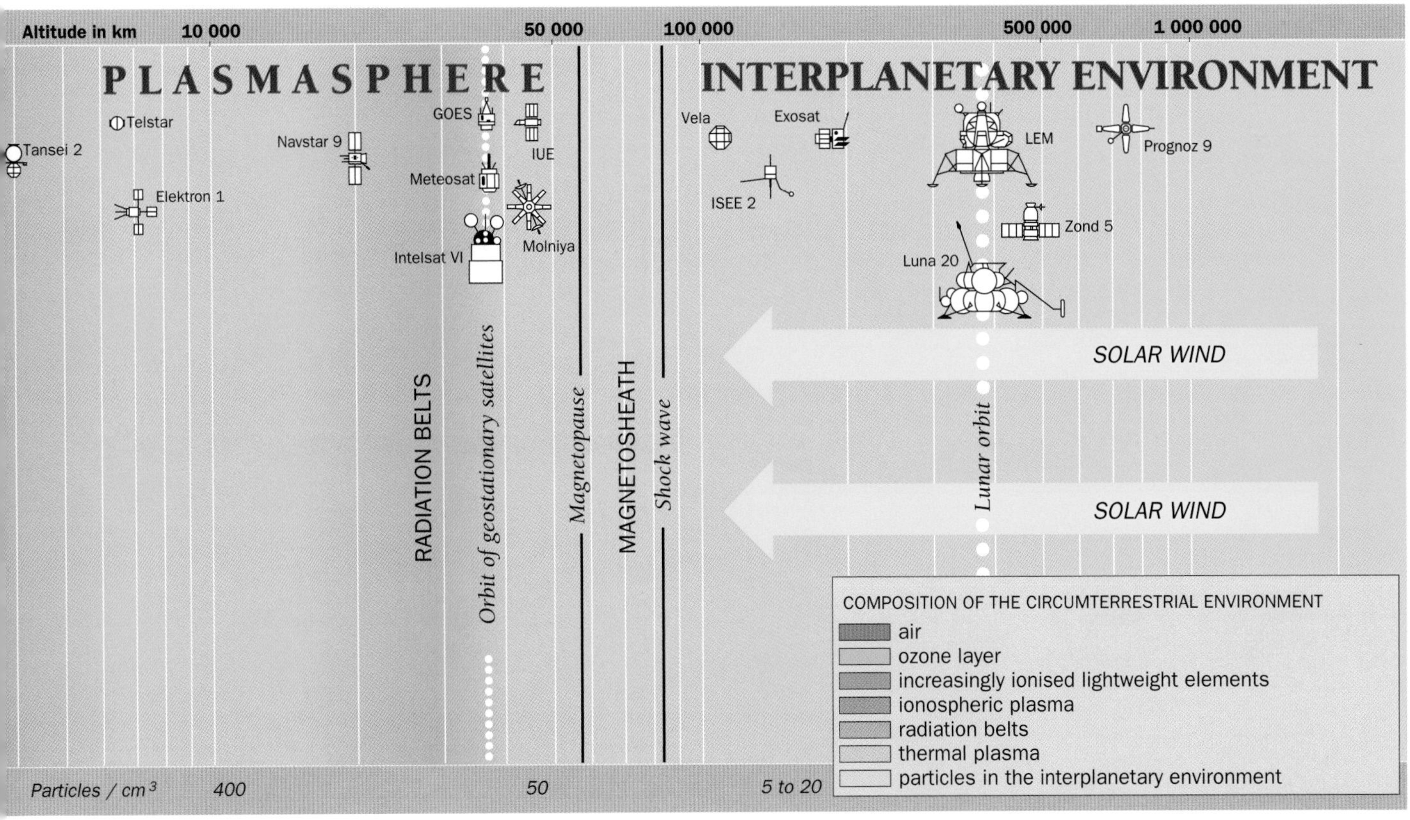

Altitude in km
10 000
50 000
100 000
500 000
1 000 000
PLASMASPHERE
INTERPLANETARY ENVIRONMENT
Tansei 2
Telstar
Elektron 1
Navstar 9
GOES
IUE
Meteosat
Intelsat VI
Molniya
Vela
Exosat
ISEE 2
LEM
Prognoz 9
Zond 5
Luna 20
SOLAR WIND
SOLAR WIND
RADIATION BELTS
Orbit of geostationary satellites
Magnetopause
MAGNETOSHEATH
Shock wave
Lunar orbit
COMPOSITION OF THE CIRCUMTERRESTRIAL ENVIRONMENT
air
ozone layer
increasingly ionised lightweight elements
ionospheric plasma
radiation belts
thermal plasma
particles in the interplanetary environment
Particles / cm 3
400
50
5 to 20

disintegration products from meteoritic and cometary grains in the upper atmosphere (source of metallic ions).

Above the terrestrial ionosphere proper, lines of force in the magnetosphere channel ionised particles in the direction of the magnetic field, thereby organising their distribution. The polar clefts represent the path of least resistance for charged particles to enter Earth's atmosphere. In periods of increased solar activity, accelerated particles (particularly in the magnetotail) are able to penetrate as far as the upper atmosphere along this route. There they excite atmospheric molecules, causing the polar auroras, borealis and australis.

Different zones can be distinguished in the magnetospheric plasma, in particular:

- The *Van Allen belts*, stable trapping zones in which particles (e.g., very high energy protons and electrons originating in cosmic rays, low energy electrons) bounce back and forth along the lines of force (taking between 0.1 and 2 seconds from one pole to the other, depending on the particle), whilst rotating about the Earth. Electrons complete this rotation in times from 1 to 10 hours in the sense of Earth's rotation, whilst protons move in the retrograde sense with a period of 5 seconds to 30 minutes. High energy protons are generally located between 2000 and 6000 km, but may fall much lower (down to 400 km) with the help of a magnetic anomaly between Brazil and South Africa. The radiation belts are filled with high energy particles which damage the silicon cells making up solar arrays, making them poor locations for satellites.
- The *plasmasphere* is made up of a very low energy plasma of the same composition and apparent origin as the ionospheric plasma, but at lower density (50 particles/cm^3).

Thermal conditions

Bodies in space are subject to different temperatures that depend less and less on the air temperature as the air becomes more rarified. In the mesosphere and beyond, heat exchanges are effected mainly by radiation, the principal sources in near space being the Sun (1200 kcal/m^2/h) and Earth (187 kcal/m^2/h by its own radiation, 430 kcal/m^2/h by reflection of solar radiation).

Solar radiation supplies the energy for most satellites operating in near space. In deep space, the greater the distance from the Sun, the lower the available energy and the more likely it becomes that other energy sources will be necessary, such as nuclear energy. Furthermore, the predominance of solar radiation in the thermal analysis of space vehicles means that, when there is no atmosphere, the surface exposed to the Sun heats to high temperature whilst the opposite surface cools by radiating into space. In order for satellites to function properly, it is important to establish a certain homogeneity of temperature (see table 1.2). One way of doing so is to use reflective cladding on the parts exposed to the Sun and absorbent cladding on the parts which need to be warmed up. During eclipses, in which the Earth prevents the satellite from receiving solar radiation, problems arise both from the reduction in radiation received and the thermal shock to which the various components of the satellite are exposed.

Dust and debris

Like the rest of interplanetary space, near space is filled with dust of cometary or other origins. Some of this dust may reach Earth's surface in the form of meteorites (the larger fragments) or micrometeorites (tiny fragments, less than 10 micrometres across). Fragments of intermediate dimen-

	DENSITY OF THE ATMOSPHERE	THERMAL EFFECT OF INFRARED RADIATION	ULTRAVIOLET RADIATION
POSITION	0 to 800 km	From 90 km ; stable from 500 km (inverse to the square of the distance from the Sun)	From 50 km stable at around 500 km
EFFECTS ON SATELLITES	Deceleration proportional to atmospheric density (= $1/2\,\rho v^2$ x ballistic coefficient, ρ being the specific density, v the velocity). Negligible at 800 km, the loss of altitude is on the order of metres per revolution at 600 km, tens of metres at 400 km and hundreds of metres at 200 km. Overheating is linked with deceleration.	Flux of 1200 kcal/m^2 per h facing the Sun, none in the shade, variation of temperature on the order of $+150\,^{\circ}$C to $-150\,^{\circ}$C	Only affects certain sensitive materials (e.g. film)
EFFECTS ON HUMANS	Aeraemia at 5 000 m Anoxia at 15 000 m Boiling of body fluids at 19 000 m	Depend on the emissive power and reflectivity of module walls	Destruction of exposed tissue
COUNTER-MEASURES	Reacceleration; heat shield; pressurisation	Insulating materials; temperature exchangers; louvres, anti-infrared visors; air conditioning	Materials and visors containing filters

Table 1.2. **Environmental influences on space flights.**

sions are called meteors, which may disintegrate in the terrestrial atmosphere and be observed as shooting stars.

Near space is also cluttered with more and more debris from rocket launches, former satellites no longer in use, or satellites that have been destroyed. The disintegration of one satellite alone, Kosmos 1275, produced 242 detected fragments and millions of particles from the same source which are now moving through circumterrestrial space

(see pages 54–56). Impacts from interplanetary dust and debris represent a real threat to satellites. Impact craters due to tiny particles have been observed on the Space Shuttle.

Figure 1.5. **Cross-section of the magnetosphere.** The Van Allen belts constitute a dangerous zone for satellites. Fortunately, the orbits of many Low Earth Orbit (LEO) satellites are located below them, whilst those of the Geostationary Earth Orbit (GEO) satellites lie well above them (from T. Encrenaz and J.-P. Bibring, 1987).

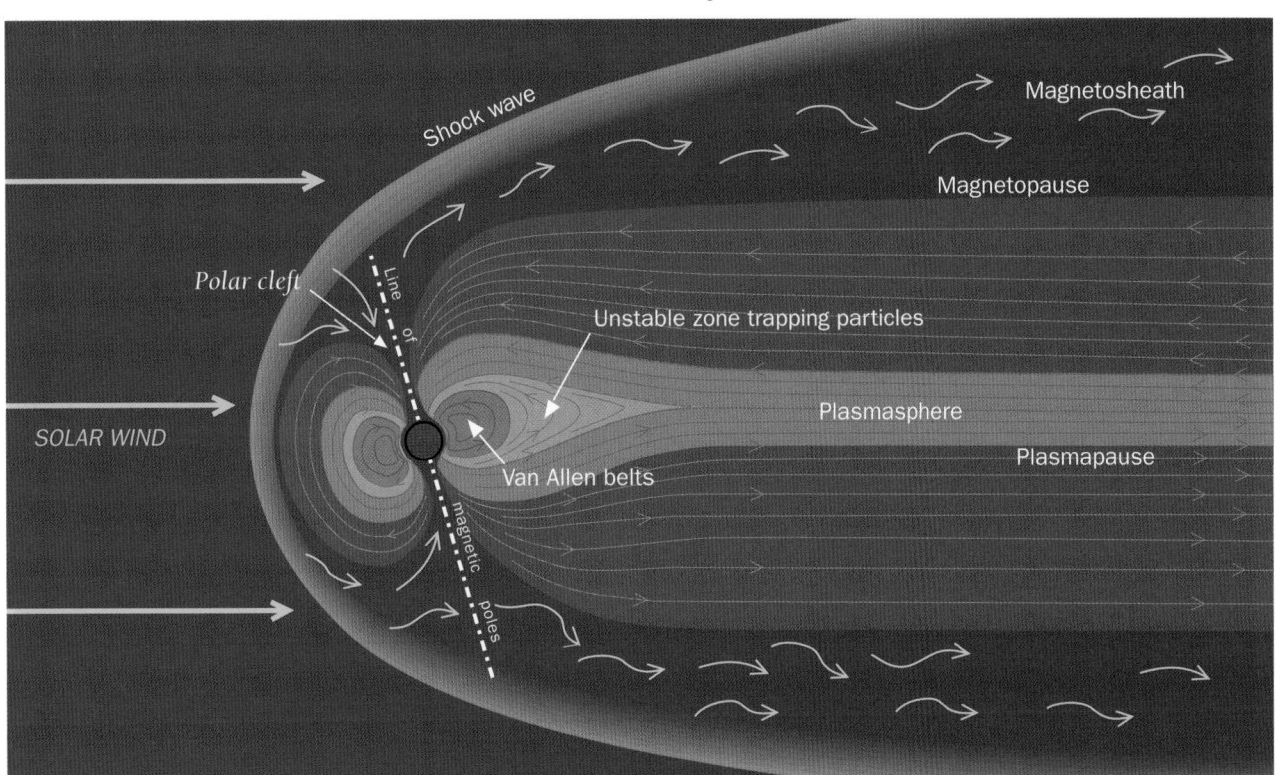

COSMIC RADIATION from solar eruptions	COSMIC GALACTIC RADIATION	RADIATION from the proton belt	METEORITES	WEIGHTLESSNESS
From 10 km, with maximum secondary cosmic radiation (the effect of primary cosmic radiation on air molecules) occurring between 10 and 15 km.		Between 2000 and 6000 km, with a max. at about 3000 km, drops to 400 km above southern Atlantic	Above the atmosphere	Not linked to altitude but to placing satellites in orbit
Increased braking through reheating/expansion of the upper atmosphere. Electrical breakdowns caused by magnetic storms created by solar eruptions. Destruction of cells on solar panels.	Destruction of solar panels		Impacts capable of unbalancing, damaging or destroying satellites	Mechanistic conditions under which gravity is not exerted (a favourable factor permitting the use of microgravity)
Severe irradiation (150 to 270 rads with protection of 2 g /cm^2). Eruptions (two or three a year) do not normally last longer than two hours.	Constant but weak irradiation, from 30 to 40 millirads per day	Irradiation varies with the altitude and as a function of the trajectory of the particles in relation to the space module (from 0.4 to 80 rads per day from 400 to 2400 km with protection of 2g /cm^2)		Circulatory troubles; joint troubles
Thickness and type of coating: but if the payload becomes too heavy it may be incompatible with current launch capacities.			Coatings and resistant shields	Artificial gravity produced by rotation is foreseen for future space stations

Orbits

General principles

Celestial mechanics and artificial satellites

The motion of natural and artificial satellites is governed by the law of universal gravitation. *Newton's law* can be stated as follows: two material points attract one another with a force proportional to each of their masses and inversely proportional to the square of the distance separating them. In symbolic form,

$$F = GM_1M_2/r^2$$

where G is the universal constant of gravitation and M_1, M_2 are the masses of the material points. The force on a satellite in orbit around Earth is equal to its mass multiplied by GM/r^2, where M is the mass of the Earth. Such a constant GM exists for any celestial body and for Earth has been accurately measured to be $\mu = 398\,603$ km^3/s^2. This is known as the geocentric gravitational constant. *Kepler's laws* determine the characteristics of satellite orbits:

■ The trajectory of a satellite in Earth orbit is an ellipse with one focus at the centre of the Earth (Fig. 2.1)
■ The radius vector d joining the satellite to the centre of Earth sweeps out equal areas in equal times (Fig. 2.2).

At each point of the orbit, there is therefore a constant relationship between the speed *V* of the satellite, its distance from the centre of Earth and the angle γ between its direction of motion and the local horizontal (the plane perpendicular to the radius vector): d.V.cos γ is constant.

The product dV is the same at apogee and at perigee, since in both cases $\gamma = 0$ and cos $\gamma = 1$. The shorter the radius at perigee, the greater the speed attained there (and conversely,

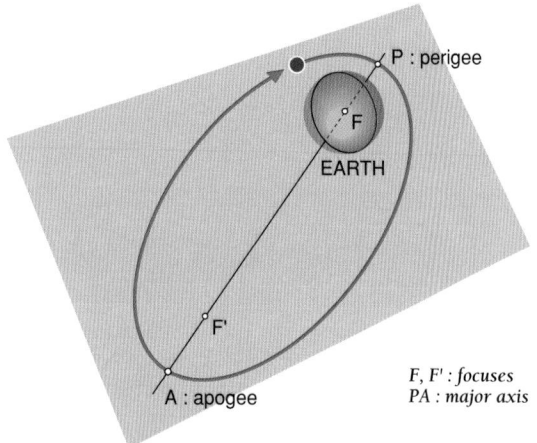

Figure 2.1. **Kepler's first law.** The case shown here is that of an artificial Earth satellite. The closest approach to Earth is called the perigee (P), and the furthest point the apogee (A).

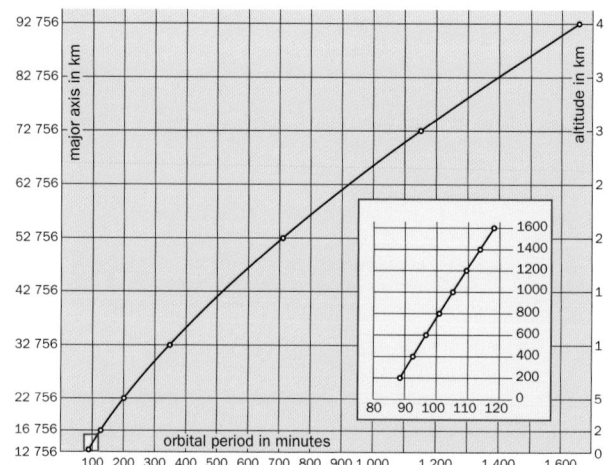

Figure 2.3. **Kepler's third law applied to Earth satellites.** Relation between major axis and period, detailed in the case of low orbits (inset).

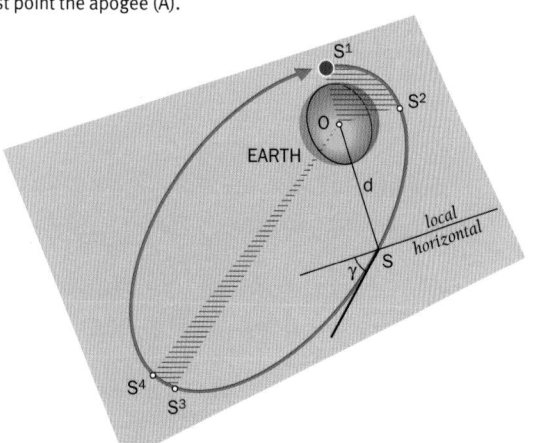

Figure 2.2. **Kepler's second law.** The two shaded areas are equal and the satellite takes the same time in going from S_1 to S_2 as it does in going from S_3 to S_4.

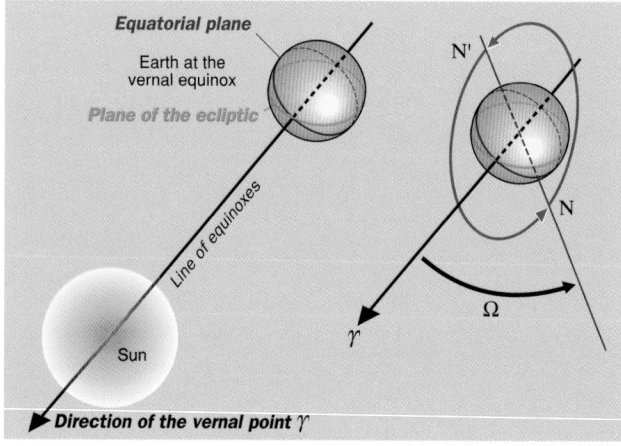

Figure 2.4. **Direction of the vernal equinox.**

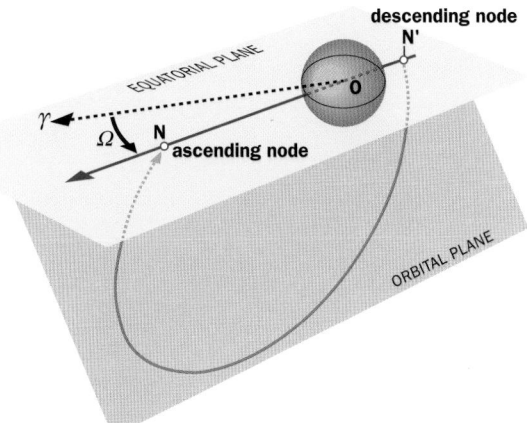

Figure 2.5. **Right ascension of ascending node.**

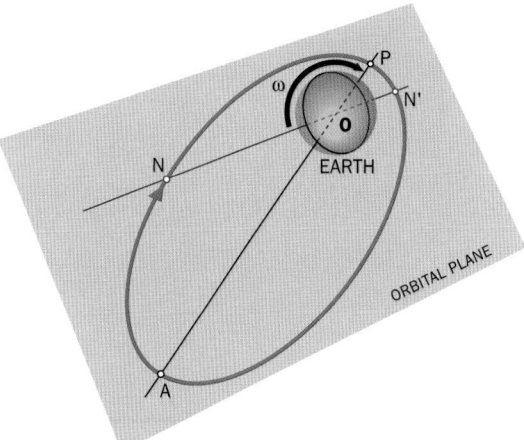

Figure 2.7. **Inclination.** Points on the surface of Earth located vertically below descending and ascending nodes are called *n'* and *n*, respectively.

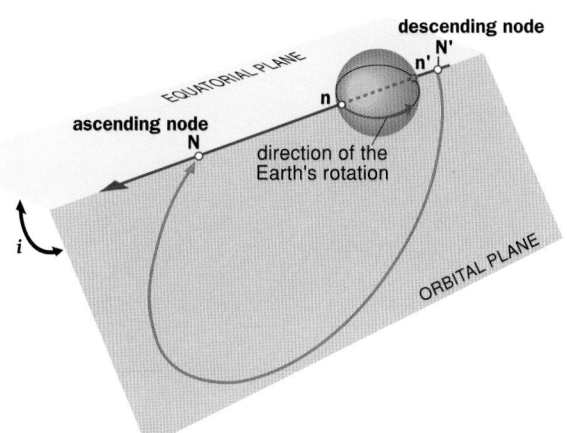

Figure 2.6. **Argument of the perigee.**

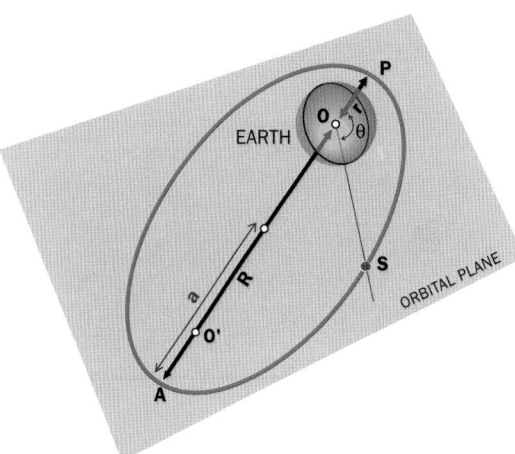

Figure 2.8. **Orbital elements and true anomaly of a satellite.**

the longer the radius at apogee, the lower the speed attained there). For a circular orbit, the speed is constant.

- The time T for one revolution (the period) is proportional to the length of the major axis: $T^2 = PA^3 (\pi^2/2\mu)$ (Fig. 2.3). $\pi^2/2\mu$ is constant for all Earth satellites.

Coordinate axes
The orbit cuts the plane of the equator at two points. The point where the satellite passes from the northern hemisphere to the southern hemisphere is called the descending node N′ and the point where it passes from the southern hemisphere to the northern hemisphere is called the ascending node N. The line NN′ is the line of nodes.

The plane of the terrestrial equator intersects the plane of the ecliptic along a line passing through the centre of Earth called the line of equinoxes.

The direction of this line towards the Sun on 21 March determines a point at infinity known as the *vernal equinox* Y. This direction remains fixed in the Solar System, independently of the Earth's motion. It is used as a reference point for astronomical coordinates (Fig. 2.4). Strictly speaking, this direction is not fixed relative to the stars. A very slow preces-

sional motion of Earth causes it to move around, making a complete rotation in 27 800 years.

Orbital elements
Five parameters define the orbits of artificial satellites around Earth.

Ω Every satellite follows an orbit lying within a fixed plane relative to astronomical coordinates.

The angle Ω between the line of nodes and the direction of the vernal equinox in the equatorial plane defines the orientation of the line of nodes. It is the right ascension of the ascending node (Fig. 2.5), measured from 0° to 360° from the vernal equinox towards the ascending node, in the direction of Earth's rotation.

i The line of nodes N′N divides the equatorial plane and the orbital plane into two parts. The angle i between the half-plane of the orbit containing the trajectory of the satellite from N′ to N and the half-plane of the equator containing the trajectory of a point on the equator from n′ to n is called the inclination. It takes values from 0° to 180° (Fig. 2.6). The orbit is said to be prograde or direct when i is less than 90° and retrograde when i is greater than 90°.

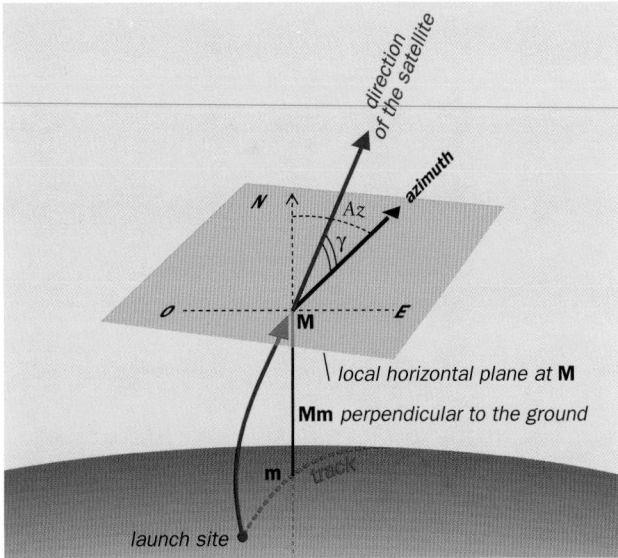

Figure 2.9. **Angles specifying the direction of a satellite.**

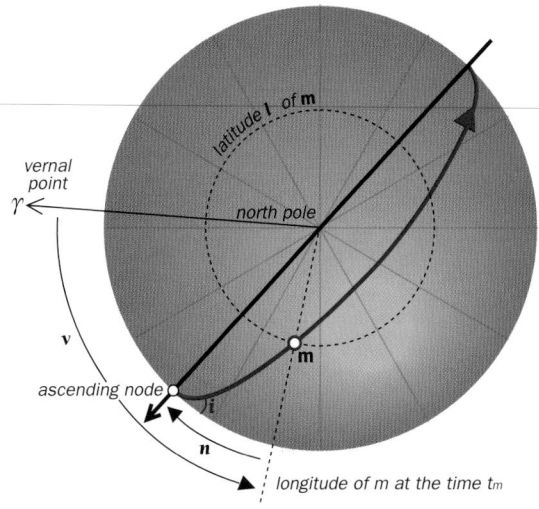

Figure 2.11. **Determining the right ascension of the ascending node.**

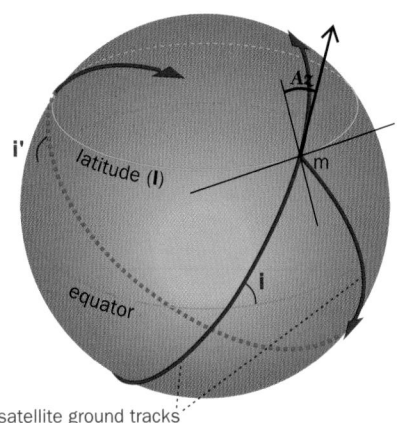

Figure 2.10. **Determining the inclination.** The inclination is the same for two tracks with azimuths at supplementary angles (totalling 180°): $i = i'$.

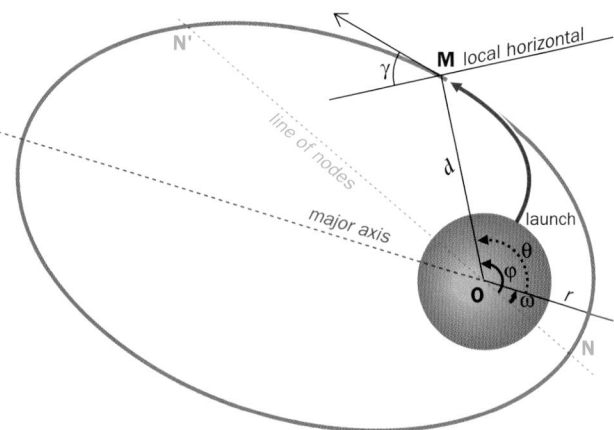

Figure 2.12. **Argument of the perigee when γ differs from 0.** In this case, the perigee, which is not the injection point, is at lower altitude than the latter. By Kepler's second law, energy is lost. For this reason, γ is usually close to 0.

ω The position of the major axis in the orbital plane is defined by the angle ω between the line joining the centre of Earth to the perigee, on the one hand, and the line of nodes on the other. This is the argument of the perigee (Fig. 2.7), taking values between $0°$ and $360°$, and measured from the ascending node towards the perigee in the same direction as the orbit of the satellite.

a, e, R, r The shape of an ellipse is completely specified by a pair of numbers. For calculational convenience, the semi-major axis a and eccentricity e (the ratio of the distance between the foci to the length of the major axis) are used. However, the radius at apogee R and the radius at perigee r can also be used to specify the orbital shape (Fig. 2.8). Since they are closer to the values commonly mentioned in satellite directories, they will be used in the following.

t A sixth parameter is used to define the position of a satellite on its orbit. This can be the time t elapsed between the time t_0 when the satellite passed a certain reference point, usually the perigee.

The position can also be given by the true anomaly of the satellite, that is, the angle θ between the perigee, the centre of the Earth and the satellite (Fig. 2.8). The Box on page 15 explains how t can be calculated from θ.

Conditions for satellite orbit acquisition
All orbital elements can be deduced from certain data given at an arbitrary point M on the orbit.

These data are:

■ speed V,
■ distance to the attractive focus d,
■ time t_m of passage at M,
■ geographic coordinates L and l,
■ direction defined by two angles measured from the projection of the velocity vector onto the local horizontal plane (Fig. 2.9):
 – vertical angle γ between this projection and the direction of the satellite,
 – horizontal angle Az between this projection and the north.

An orbit is acquired by establishing the right conditions at the initial point of the orbit, that is, at the point where propulsion ceases, also called the injection point.

i The inclination i of the orbit is determined by the azimuth Az and latitude l at the injection point (Fig. 2.10).

$$\cos i = \sin Az \cdot \cos l$$

Ω The right ascension of the ascending node Ω is determined by the injection time t_m and the longitude L (Fig. 2.11).

In order to ensure that the satellite orbits within a given plane, a particular time must be chosen for orbit acquisition, depending on the longitude, at which the line of nodes forms the required angle with the direction of the vernal equinox.

The direction of the vernal equinox **Y** and the longitude of the injection point form an angle v at time t_m.

The angle **Ω** is the difference between angles v and n, where n is the angle between the longitude of the injection point and the line of nodes.

$$\sin n = \cos i \sin l / \cos l \sin i$$

ω The argument of the perigee ω is determined by the angle **γ** between the direction of the satellite and the local horizontal (Figs. 2.12 and 2.13).

The radius d makes an angle φ with the direction of the ascending node N, such that $\sin \varphi = \sin l / \sin i$, and an angle θ with the direction of the perigee, called the true anomaly of

the point M. The angle ω is the difference between angles φ and θ (see Box on page 15 for the calculation of θ).

r The radius at perigee r is determined by the distance d to the focus and the angle **γ** (see Box on page 15 for the calculation of r).

When γ = 0, r = d.

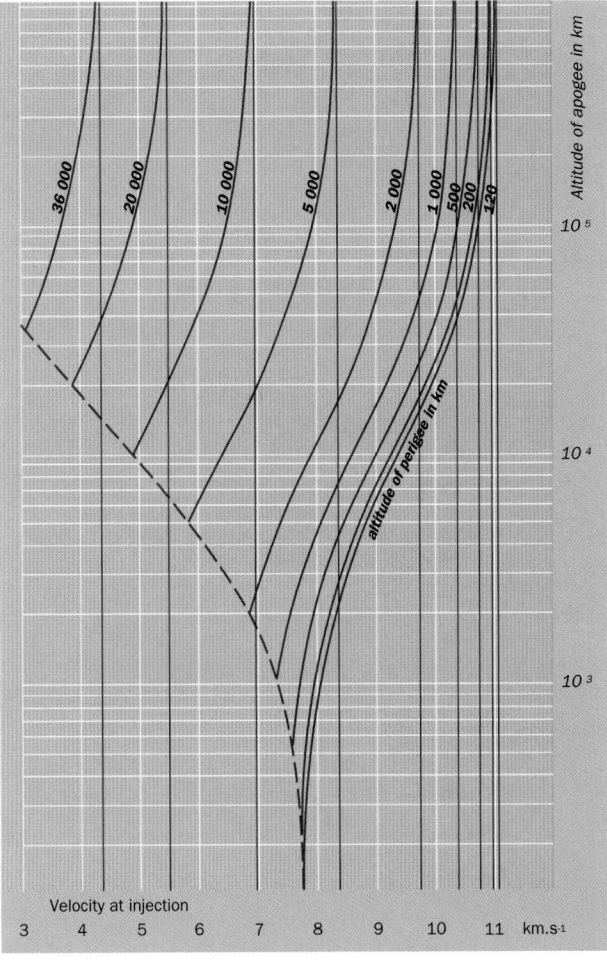

Figure 2.15. **Relation between apogee height and injection speed.** The injection point coincides with the perigee and the vertical lines indicate the value of the second cosmic velocity for nine different perigee heights. The dashed line shows the relation between injection speed and altitude for a circular orbit. The higher the injection speed, the lower the speed corresponding to the circular orbit.

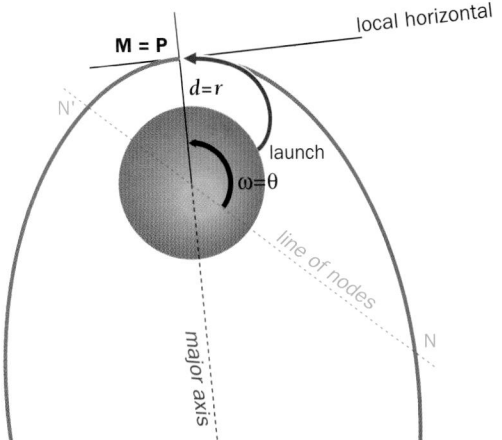

Figure 2.13. **Argument of the perigee when γ is 0.** Perigee and injection point **M** coincide and **sin** ω = **sin l/sin i.**

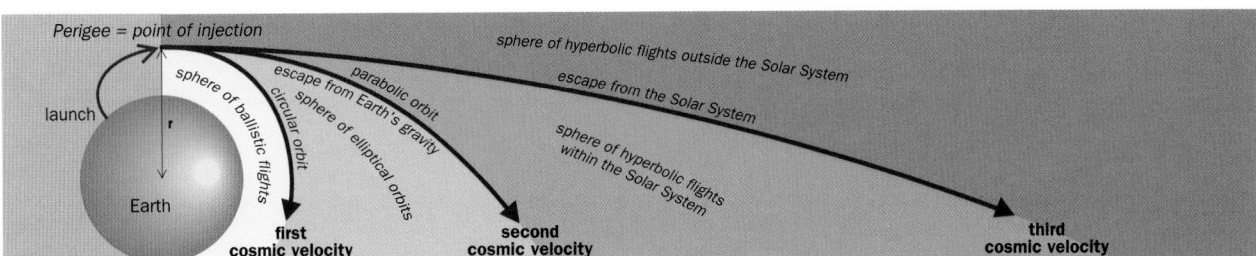

Figure 2.14. **The three cosmic velocities and four main types of trajectory.** When the injection velocity is less than the first cosmic velocity, the trajectory is an ellipse. The apogee rather than the perigee now coincides with the injection point and the perigee is situated at a lower altitude. If the perigee lies within the atmosphere, or exists only virtually, below the Earth's surface, the satellite accomplishes a ballistic flight and falls back to Earth.

R The radius at apogee R is determined for a given radius at perigee r by the speed at perigee Vp, which is obtained from V by the relation $Vp = V (d \cos \gamma/r)$.

$$Vp = \sqrt{(2\mu/r) - (2\mu/(R+r))}.$$

- The first cosmic velocity is the one for which R = r, and Vp takes the value $\sqrt{\mu/r}$. The orbit is then circular, whatever the altitude. At lower speeds, the vehicle follows a ballistic path and falls back to Earth (Fig. 2.14).
- The second cosmic velocity is the one for which R is infinite, and Vp takes the value $\sqrt{2\mu/r}$. The orbit is then a parabola.

Between the first and second cosmic velocities, orbits are elliptical (Fig. 2.15). Beyond the second cosmic velocity, the satellite escapes from Earth's attraction and the orbit is a hyperbola.

- The third cosmic velocity, related to the speed Vt of Earth in its orbit about the Sun, is the one at which the satellite succeeds in escaping from the Solar System ($Vp = 16.85$ km.s^{-1} in low orbit):

$$Vp = \sqrt{(2\mu/r) - Vt^2(3 - 2\sqrt{2})}.$$

t The time of passage at perigee t_0 is separated from the time t_m by a lapse of time t (see Box on page 15 for the calculation of t).

When $\gamma = 0$, t_m and t_0 are equal.

Modifications and perturbations to orbits

Each of the six parameters relating to an orbit can be modified, either by natural perturbations, or by manœuvres resulting from commands transmitted to the satellite motors.

Ω The orientation of the orbital plane is naturally perturbed by *nodal precession*, due to Earth's equatorial bulge. The closer the satellite is to Earth and the further its inclination is from 90°, the faster is the rotation of the plane (Fig. 2.16).

The rotation during one orbit is:

$$\Delta\Omega = -0.58 \, (D/(R+r))^2 \, (1/(1-e^2)^2) \cos i.$$

D is the diameter of Earth, and $\Delta\Omega$ is given in degrees.

For a circular orbit:

$$\Delta\Omega = -0.58 \, (D/(R+r))^2 \cos i.$$

Any manœuvre which modifies Ω is energetically costly and it is preferable to take advantage of natural precession.

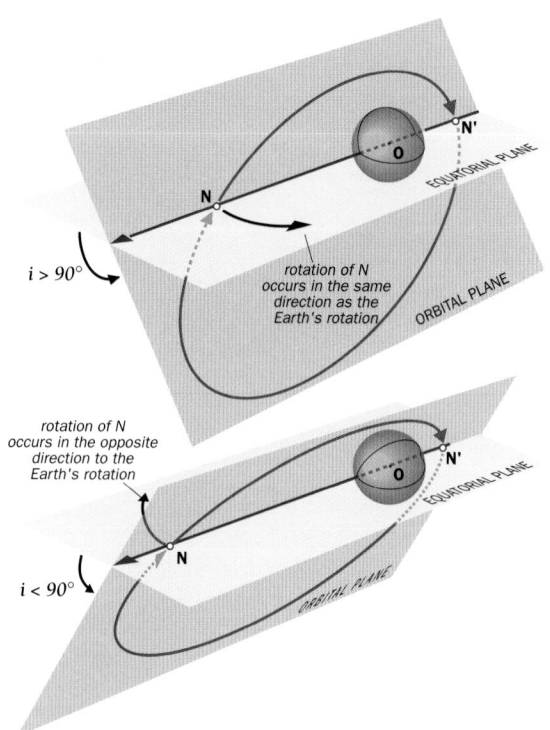

Figure 2.16. **Rotation of the orbital plane.** This occurs in the same direction as Earth's rotation if the inclination is greater than 90° and in the retrograde direction if the inclination is less than 90°. It is zero for an inclination of 90°.

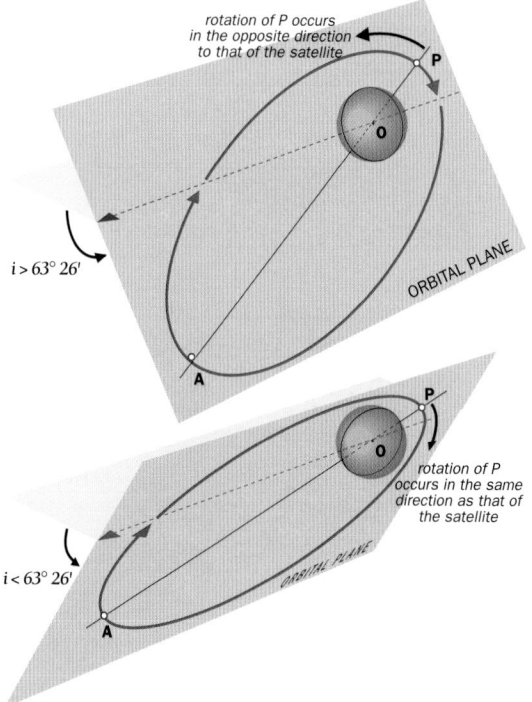

Figure 2.18. **Rotation of the perigee.** This occurs in the opposite direction to the satellite motion if the inclination is greater than 63° 26′ and in the same direction as the satellite motion if the inclination is less than 63° 26′. It is zero at inclination 63° 26′.

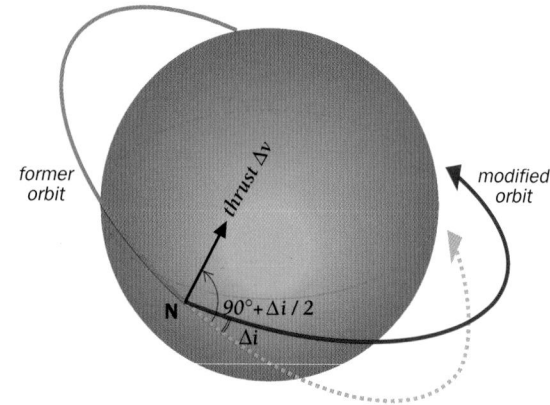

Figure 2.17. **Modifying the inclination.**

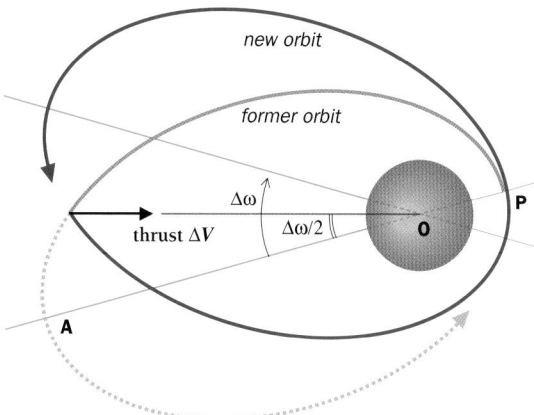

Figure 2.19. **Modifying the argument of the perigee.**

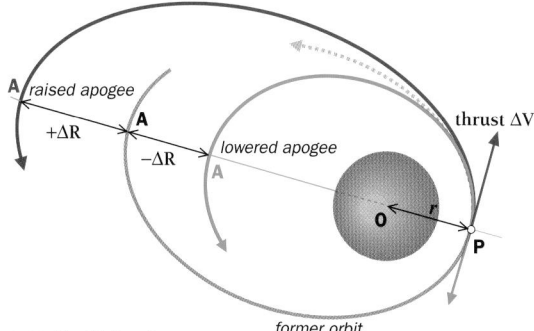

Figure 2.20. **Modifying the apogee.**

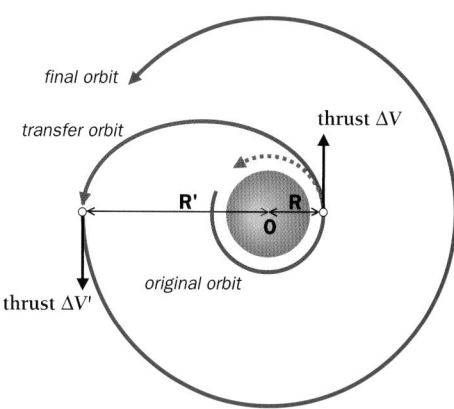

Figure 2.21. **Hohmann transfer path.**

i There is a natural perturbation of the inclination, due to the Sun–Moon attraction. Negligible for low orbits and tiny at 40 000 km, it increases with the apogee height.

To obtain a change Δi in i, a thrust ΔV is applied at an angle $90° + \Delta i/2$ to the direction of the satellite, at one of the nodes of the orbit (Fig. 2.17).

$$\Delta V = 2V \sin(\Delta i/2).$$

V is the speed of the satellite at the chosen point.

ω A natural perturbation of the argument of the perigee is *apsidale precession*, due to Earth's equatorial bulge. The closer the satellite is to Earth and the further its inclination is from $63° \, 26'$, the faster is the rotation of the perigee (Fig. 2.18).

The rotation during one orbit is:

$$\Delta\omega = 0.29 \, ((4-5\sin^2 i)/(1-e^2)^2) \, (D/(R+r))^2.$$

D is the diameter of Earth and $\Delta\omega$ is given in degrees.
For a circular orbit:

$$\Delta\omega = 0.29 \, (4-5\sin^2 i) \, (D/(R+r))^2.$$

To obtain a change $\Delta\omega$ in ω, a thrust of ΔV is applied at the point of the orbit where the radius makes an angle $\Delta\omega/2$ with the major axis, and towards the centre of Earth (Fig. 2.19).

$$\Delta V = 2\sqrt{\mu/r}(e\sqrt{1+e})\sin(\omega/2).$$

r is the radius at perigee and e the eccentricity of the orbit.

R Atmospheric drag, and to a much lesser extent solar radiation pressure, lower the apogee without changing the perigee height. Every elliptical orbit thus tends towards circularity at the height of the perigee. Once circularisation has been achieved, the orbit continues to fall, whilst remaining circular. The probable lifetime of a satellite in circular orbit is:
– a few days at 200 km,   – one century at 800 km,
– a few weeks at 300 km,   – several centuries at 1000 km,
– a few years at 600 km,   – a million years at 30 000 km.

CONDITIONS AT A SPECIFIC POINT AND PARAMETERS OF THE TRAJECTORY

if the following are known at a point M of the orbit (injection point or otherwise):
■ speed at $M = V$
■ length of radius vector $OM = d$
■ angle between direction of motion of satellite at M and local horizontal γ
the values of orbital parameters e, a, R, r can be obtained by the relations:

$$e^2 = \sin^2\gamma + (1-(dV^2/\mu))^2\cos^2\gamma \qquad a = md \, /(^2m-dV^2)$$

$$R = V^2d^2\cos^2\gamma/(\mu(1-e)) \qquad r = V^2d^2\cos^2\gamma/(\mu(1+e))$$

If the time t_m at which the satellite passes through M is known, the time t_o of its last passage at perigee is obtained by subtracting $t = (T/2\pi)(u-e.\sin u)$ from t_m, where T is the period and u the eccentric anomaly of M in radians.
u can be calculated as a function of the true anomaly of M, θ:

$$\cos\theta = (V^2d.\cos^2\gamma-\mu)/e\mu \qquad tg(\theta/2) = \sqrt{R/r}tg(u/2)$$

or as a function of d:

$$d = (R+r)(1-e.\cos u).$$

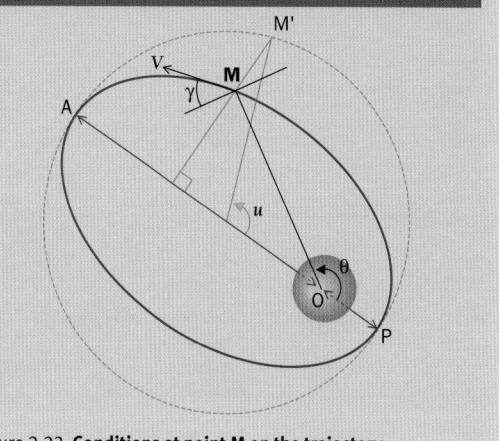

Figure 2.22. **Conditions at point M on the trajectory.**

To obtain a change ΔR in R, a thrust ΔV is applied at perigee, in the direction of the satellite if R is to be increased, and in the opposite direction if it is to be reduced (Fig. 2.20).

$$\Delta V = \Delta R\mu/(Vp(R+r)^2).$$

Vp is the speed at perigee and r the radius at perigee.

r To obtain a change Δr in r, a thrust ΔV is applied at apogee, in the direction of the satellite if r is to be increased, and in the opposite direction if it is to be reduced.

$$\Delta V = \Delta r\mu/(Va(R+r)^2).$$

Va is the speed at apogee.

A *Hohmann transfer* is the most economical manœuvre for changing from one circular orbit to another. It uses an ellipse that is tangent to both of the circular orbits and requires two thrusts ΔV and $\Delta V'$, in the direction of motion of the satellite if the new circular orbit is to have greater radius, and in the opposite direction if it is to have smaller radius (Fig. 2.21).

$$\Delta V = \sqrt{2\mu R'/(R(R+R'))} - \sqrt{\mu/r}$$

$$\Delta V' = \sqrt{\mu/R'} - \sqrt{2\mu R/(R'(R+R'))}$$

R is the radius of the initial orbit, R' the radius of the final orbit.

t All perturbations affecting R and r also affect the period of the satellite, and hence the time t_0 of its passage at perigee. t_0 can be modified by increasing R or r to slow the satellite down, then reducing them to return to the initial orbit. The opposite manœuvre brings about the same result. The longitude of a geostationary satellite is modified via a similar manœuvre.

Sun-synchronous satellite orbits

A *Sun-synchronous* or *heliosynchronous* orbit is one lying in a plane at a fixed angle with the Earth–Sun direction (Fig. 2.23). This property amounts to giving the orbital plane a nodal precession (see Fig. 2.16) equal to 360° per year, or 0.986° per day.

Such a precession then corresponds to the mean displacement of the Earth around the Sun. It is obtained by a retrograde inclination whose value depends on the altitude of the satellite (Fig. 2.24).

Sun-synchronous orbits ensure:

■ constant local solar time for passage through a given location, thereby guaranteeing an almost constant illumination, varying only with the seasons;

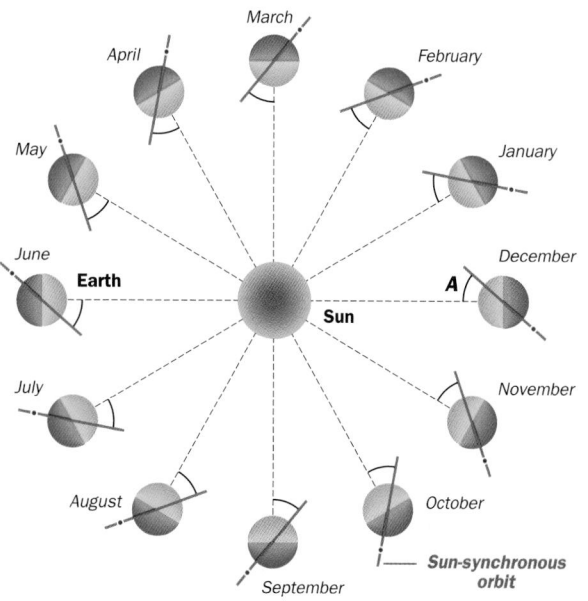

Figure 2.23. **Definition of Sun-synchronous orbit.** The angle A between the orbital plane and the Earth–Sun line remains constant throughout the year.

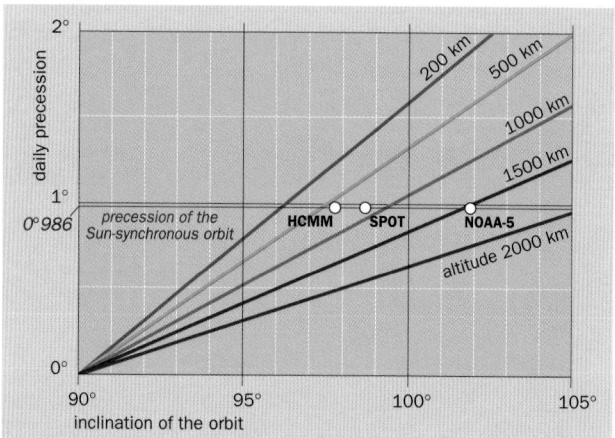

Figure 2.24. **Relations between daily precession, altitude and inclination for circular orbits.** Positions of Sun-synchronous Earth observation satellites HCMM, SPOT 3 and NOAA 5 are given as examples.

- coverage of almost the whole surface of the globe, since the orbit is quasi-polar.

Satellites are launched into Sun-synchronous orbit by choosing:

- the time at which the satellite must pass a given point, determined by the launch time,
- the altitude and inclination of the orbit, factors which combine to determine the required nodal precession. The higher the orbit, the greater the inclination.

Given the conditions satisfied by satellites placed in circular and Sun-synchronous orbits, this type of orbit is particularly suitable for Earth observation. Indeed, they guarantee a constant altitude relative to the

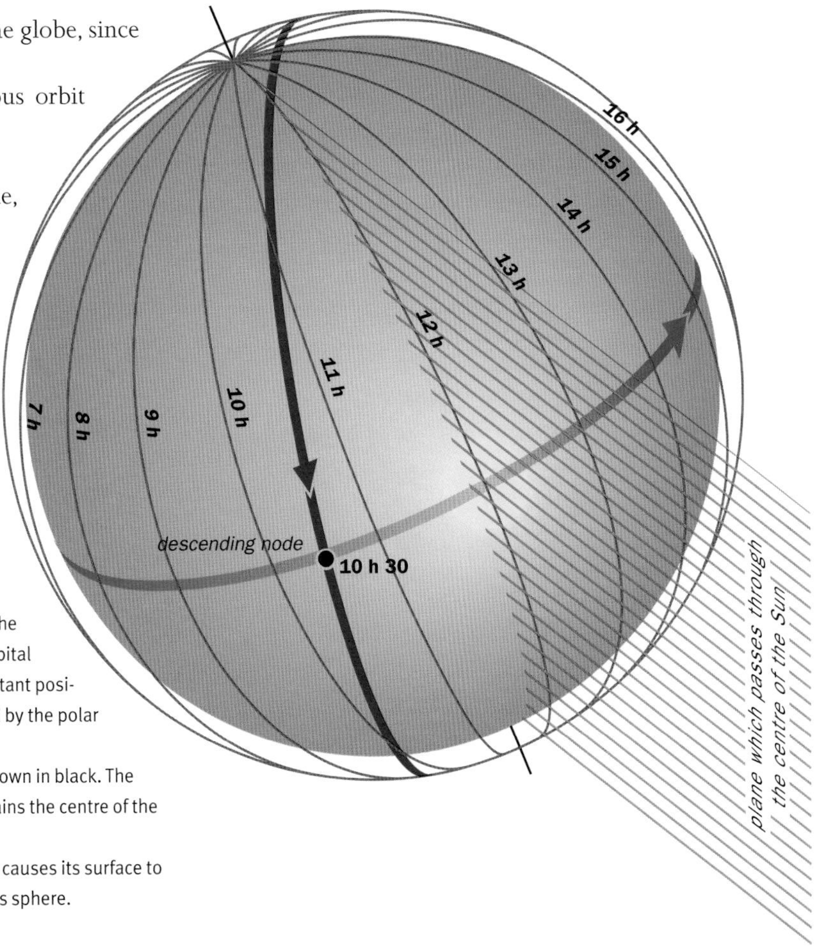

Figure 2.25. Times of passage of SPOT 3 on its descending pass.

The satellite orbit is shown in red. The plane containing the orbit cuts the polar axis at the centre of the Earth. The orbital plane of the Sun-synchronous satellite remains in a constant position, making an angle of 27.3° with the plane determined by the polar axis and the direction of the Sun.

Meridians of a sphere corresponding to solar time are shown in black. The plane determined by noon and midnight meridians contains the centre of the Sun, by definition.

The rotation of Earth (shown in colour inside the sphere) causes its surface to pass beneath the different time meridians of the previous sphere.

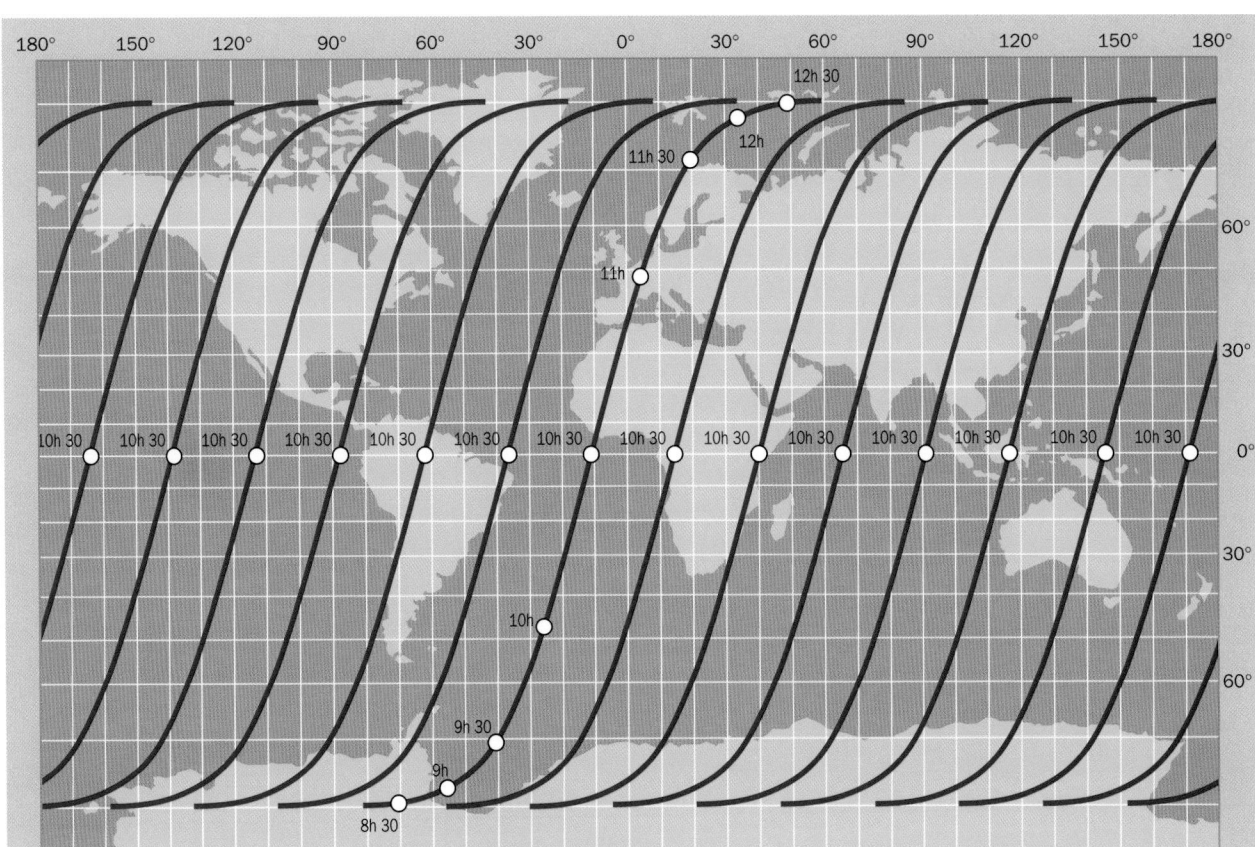

Figure 2.26. **Times of passage of SPOT 3 in local civil time during its descending passes.**

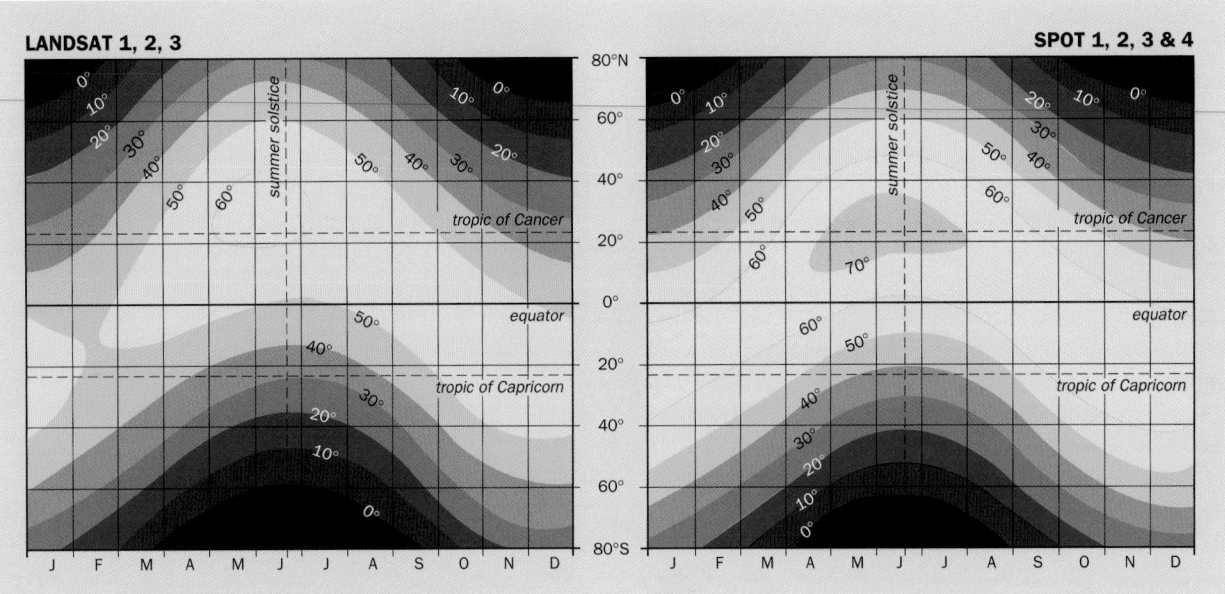

LANDSAT 1, 2, 3　　　　　　　　　　　　　　　　　　**SPOT 1, 2, 3 & 4**

Figure 2.27. **Solar elevation at the point beneath the satellite.** Charts give the value of the solar elevation at the point beneath the satellite at the time of passage of Landsat 1, 2 and 3 and the time of passage of SPOT 1, 2 and 3, during the year and at different latitudes. The descending node crossing of SPOT 1, 2 and 3 (10 h 30 m) is later than that of Landsat 1, 2 and 3 (9 h 30 m). This difference implies a higher solar elevation for SPOT than for Landsat. The elevation is maximal north of the tropic of Cancer at the summer solstice, and it is maximal at the equator at the equinoxes.

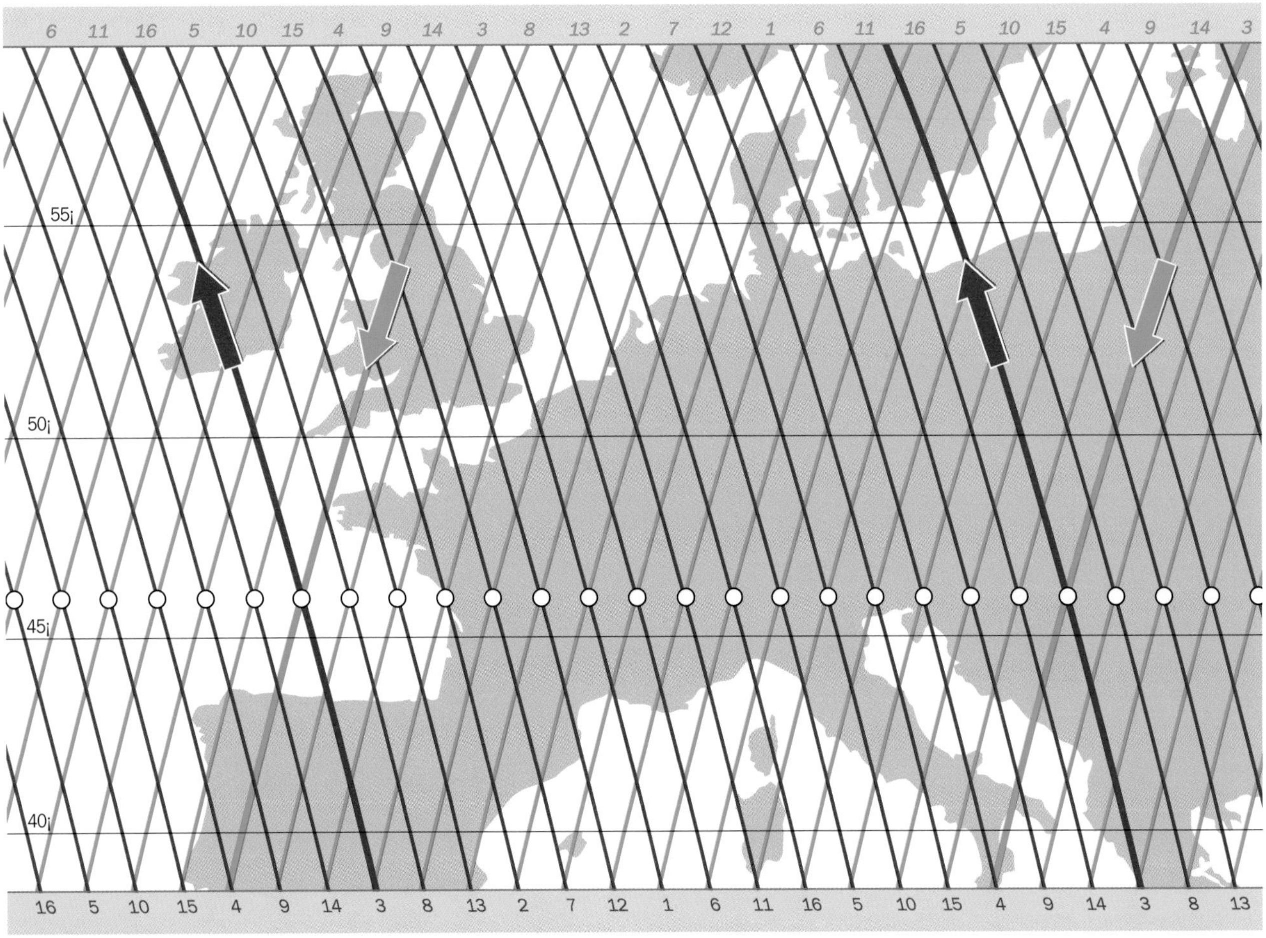

Figure 2.28. **Periodicity of the crossover of ascending and descending passes for Sun-synchronous satellites.** The crossover of ascending daytime passes, in red, and descending nighttime passes, in green, occurs at half-day intervals and at precisely defined latitudes. The example of the HCMM satellite shows the crossover of these passes at half-day intervals at latitude 46° north. Days of the month are numbered in the margin.

observed zone and a solar illumination that varies only with the season. They are particularly well suited to passive remote sensing, meteorological and atmospheric studies, and acquisition of thermal inertia data. Several programmes use Sun-synchronous orbits. Their altitudes, and hence inclinations, cover a wide range, as do their times of passage at different latitudes (Fig. 2.25). Parameters defining these missions must therefore include crossing time at the descending node in local civil time (see Table 2.1). Crossing times at different latitudes can be calculated from this parameter (Figs. 2.25 and 2.26).

Crossing times of Sun-synchronous satellites are determined according to the mission. Passive remote sensing satellites use visible and near infrared spectral bands and their orbits are thus fixed in such a way that they overfly highly illuminated regions during either their ascending trajectory or their descending trajectory.

The illumination of a point on Earth's surface overflown by a Sun-synchronous satellite depends on the height of the Sun above the horizon at that point. This height is the *solar elevation* and depends on the season. Above the tropics, it has a maximum at one of the solstices and a minimum at the other. At the equator, on the other hand, the elevation is maximal at the equinoxes (Fig. 2.27).

As a consequence of the different times of passage at successive latitudes, the distribution of equal elevation curves is not symmetric relative to the equator. The maximum elevation is located in the northern hemisphere for the majority of remote sensing Sun-synchronous satellites. This is because

the early morning is preferred for the descending node crossing (see Table 2.1).

Certain satellites, like the Landsat or the SPOT series, move along their descending trajectory above the sunward face and their ascending trajectory at night. Others, such as HCMM, do the opposite. The crossover points of ascending and descending passes for satellites in Sun-synchronous orbit occur at constant latitudes and with fixed periodicities.

Satellite	Hours	Minutes	Satellite	Hours	Minutes
LANDSAT 1	9	30	IRS 1-B	10	30
NOAA 2	8	30	NOAA 12	7	25
NOAA 3	8	30	FUYO 1	10	15
NOAA 4	8	30	SPOT 3	10	30
LANDSAT 2	9	30	NOAA 14	7	30
NOAA 5	8	30	IRS P-2	10	40
HCMM	2	0	ERS 2	10	30
LANDSAT 3	9	30	RADARSAT 1	6	0
NIMBUS 7	23	50	IRS 1C	10	30
TIROS N	3	0	IRS P3	10	30
NOAA 6	7	30	ADEOS 1	10	30
NOAA 7	2	30	SEASTAR	12	0
LANDSAT 4	9	37	SPOT 4	10	30
NOAA 8	7	30	RESURS-01 4	10	15
LANDSAT 5	9	37	IRS 1D	10	30
NOAA 9	2	30	LANDSAT 7	10	0
NOAA 10	7	30	FENGYUN 1C	9	0
SPOT 1	10	30	ORBVIEW 3	10	30
MOS 1A	10	25	IKONOS 2	10	30
IRS 1-A	10	25	CBERS 1	10	30
MOS 1B	10	30	EO 1	10	00
SPOT 2	10	30	CLOUDSAT	1	31
FENGYUN 1B	7	55	ENVISAT	10	0
ERS 1	10	15	SPOT 5	10	30

Table 2.1. **Descending node crossing times for the main Sun-synchronous satellites.** Satellites are classified in chronological order of launch.

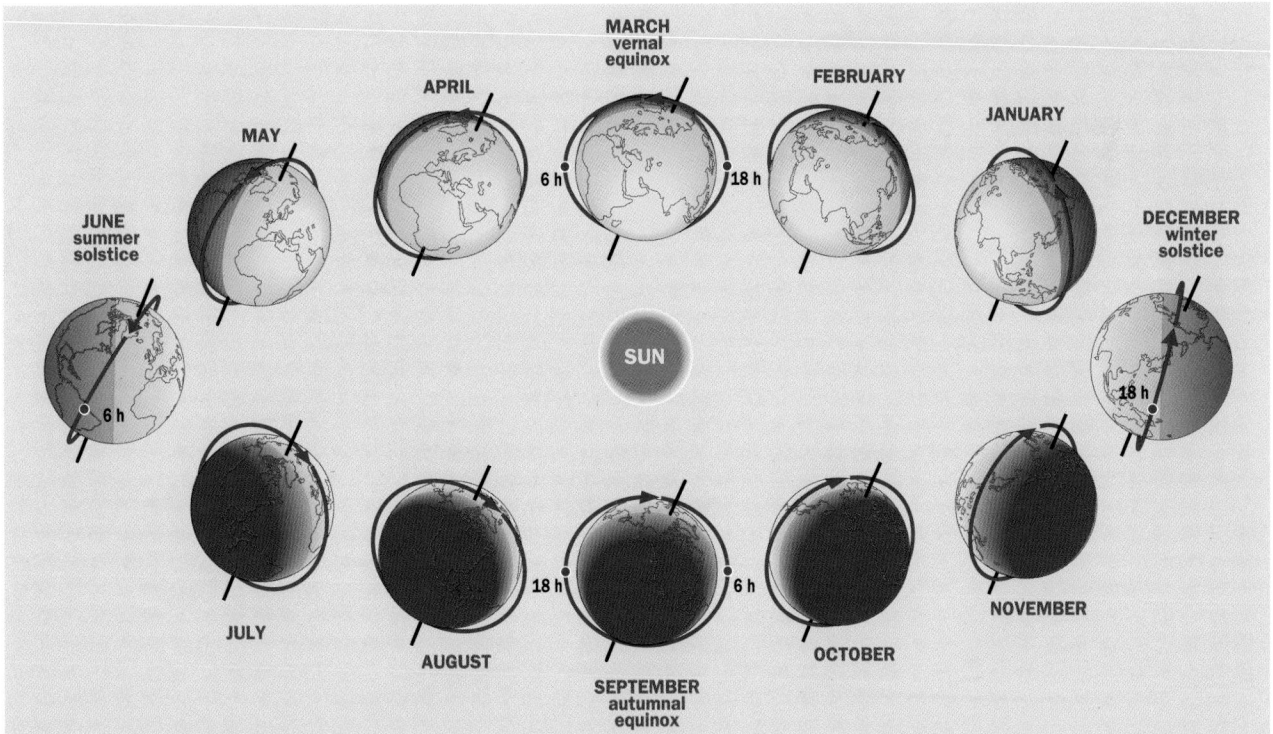

Figure 2.29. **Trajectory of a satellite in a dawn–dusk Sun-synchronous orbit during the twelve months of the year.** Radarsat 1 has been given as an example. It crosses the descending node at 6 a.m. The orbit is so disposed that its line of nodes remains perpendicular to the Earth–Sun line, thus allowing the satellite to receive solar energy continuously, except during short eclipses around the summer solstice. The angle between the orbital plane and the polar axis adds to the inclination of the Earth's polar axis at the summer solstice and subtracts from it at the winter solstice.

Hence, the daytime ascending passes and nighttime descending passes of the HCMM satellite crossed over at half-day intervals at latitude 46° north (Fig. 2.28).

This periodicity was used to evaluate the surface temperature difference of the regions overflown during daytime and nighttime passages. In this way, the thermal inertia of the ground could be determined from HCMM satellite data.

For the NOAA satellites, crossovers are used to the same end. However, the choice of different descending node crossing times for odd and even rank satellites makes it possible to use shifted lines of sight over the whole length of the pass.

Dawn–dusk orbits

Dawn–dusk orbits constitute a special class of Sun-synchronous orbit. They are given this name because the satellite crosses the nodes roughly during dawn or dusk for the regions overflown, and the line of nodes is perpendicular to the Earth–Sun line. The node crossing times are thus close to 6 o'clock in the morning and evening. This means that the satellite is almost permanently illuminated by the Sun. Its solar panels intercept the Sun's radiation and it has an almost constant supply of energy.

However, as a consequence of the inclination of these Sun-synchronous satellites, short eclipses occur during each cycle over a period of several weeks around the summer solstice.

Such orbits are used for certain active remote sensing satellites, such as Radarsat. Their radars require a lot of energy and it is important to avoid eclipses, which deprive them of solar radiation. Radarsat is eclipsed for 18 minutes during the summer solstice (Fig. 2.29).

Geostationary satellite orbits

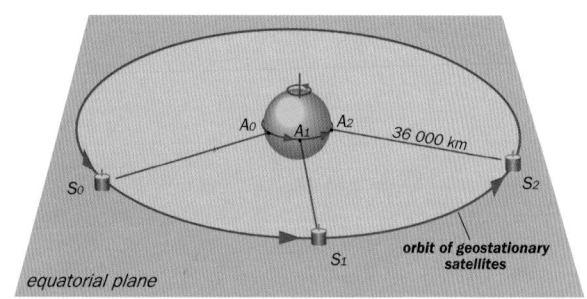

Figure 2.30. **Orbits of geostationary satellites.** On such an orbit, when the satellite **S** transits from **S₀** to **S₁** and then to **S₂**, the point **A** on the equator vertically below **S** moves with the same angular velocity and transits from **A₀** to **A₁** and then to **A₂**.

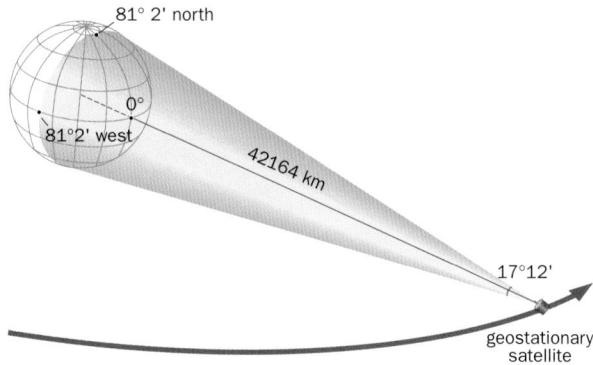

Figure 2.31. **Coverage by a geostationary satellite.** A geostationary satellite views 42% of the Earth's surface from the top of a cone. The angle of the cone at the apex is 17° 12′ and it is tangent to the Earth along a circle lying between latitudes 81° 2′ north and south, and for a satellite located at longitude 0°, between longitudes 81° 2′ west and east.

Certain satellites appear to hover above the same point of Earth's surface. These are called geostationary satellites. They follow a special orbit, first given an accurate description in 1945 by British engineer and science fiction writer, Arthur C. Clarke. However, the first geostationary satellite was not launched until 1963 because of the technical complexity of the operation.

Today, these satellites play an essential role in relaying communications around the Earth. They also carry out meteorological observations and military surveillance.

Determination of the orbit

In order to remain vertically above the same point on the Earth's surface, a geostationary satellite must:

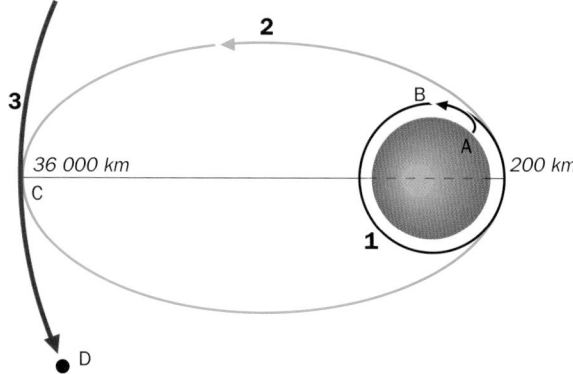

Figure 2.32. **Launch and station acquisition for a geostationary satellite.** Once the satellite has separated from its launch vehicle at **B**, deployment consists in placing it, after one or two revolutions along intermediate orbits (1 and 2), on the operational orbit (3) at a precisely defined longitude **D**.

- have constant latitude, which is only possible at latitude 0°, and thus follow an orbit with zero inclination according to Kepler's law (page 10);
- have constant longitude, and thus have the same angular velocity as Earth (Fig. 2.30), which is only possible in a circular orbit, by Kepler's second law;

- have a period equal to Earth's rotational period (23 h 56 m for 360°), which means that the satellite must fly at altitude 35 786 km. This figure is normally rounded up to 36 000 km. At this altitude, 42% of the surface of the globe is visible to the geostationary satellite (Fig. 2.31).

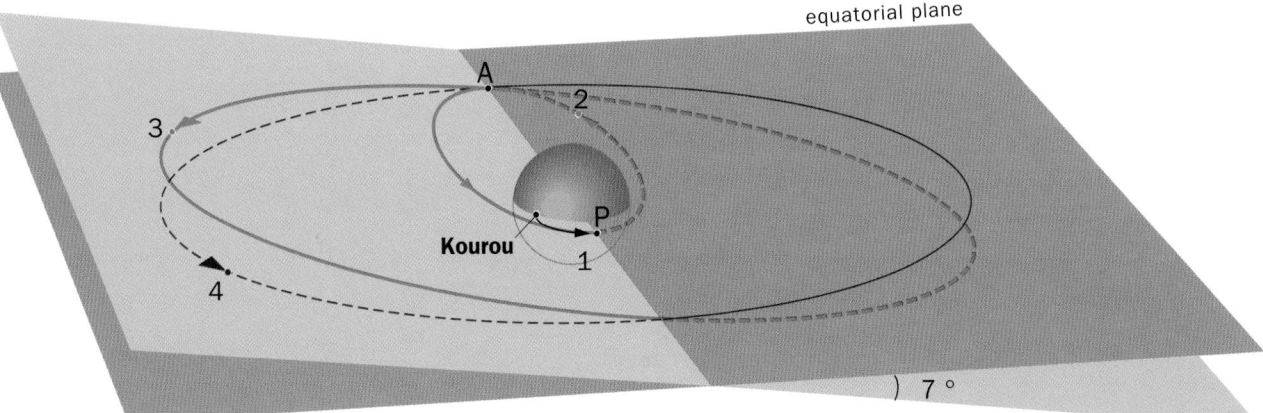

Figure 2.33. **Launch by Ariane from the Kourou base, French Guiana.**
- Phase one: transfer orbit. – The satellite and its apogee kick motor separate from the Ariane launch vehicle at an altitude of about 200 km altitude, before crossing the equatorial plane (1) at the perigee **P** and embarking directly upon a highly eccentric transfer orbit. The latter is inclined at 7° and its major axis already lies in the equatorial plane. Its apogee **A** is 36 000 km from Earth and its period is then 11 hours.
- Phase two: circularising the orbit and modifying the inclination. – After several revolutions (2), and during a passage at **A**, the apogee motor is fired.

The resulting thrust changes the path of the satellite and it gradually acquires a circular orbit (3), called a drift orbit, at the apogee height and still in a plane inclined at 7°. A further thrust applied at A changes the inclination to a value close to 0° (4), thereby placing the orbit in the equatorial plane. These are called apogee manœuvres. The two thrusts can be executed simultaneously, the line of apsides being in coincidence with the line of nodes.
- Phase three: satellite drift and attitude acquisition. – In the last manœuvre the satellite drifts along the orbit to its final longitudinal position, during which attitude manœuvres and fine adjustments are carried out. Such orbit trim manœuvres, executed at the right longitude, guarantee a circular and equatorial orbit, together with the required period of 23 h 56 m. They are made by thrusts tangential or normal to the orbit (see page 000).

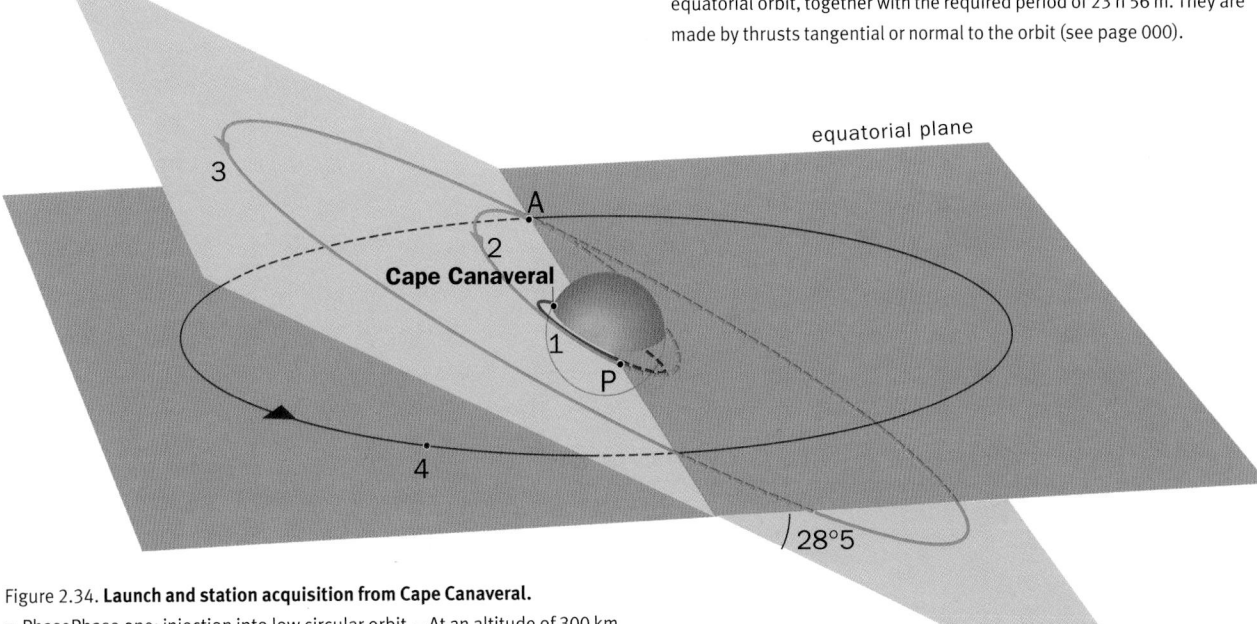

Figure 2.34. **Launch and station acquisition from Cape Canaveral.**
- PhasePhase one: injection into low circular orbit. – At an altitude of 300 km and an inclination of 28.5°, the satellite is said to be in parking orbit (1).
- Phase two: thrust given at **P** by the satellite's perigee kick motor. – The satellite then follows a transfer orbit with apogee at height 36 000 km. This is called a perigee manœuvre. The inclination is still 28.5° (2).
- Phase three: circularising the orbit and changing its inclination. – This is achieved by apogee manœuvres similar to those required in a Kourou

launch, but with a second, more powerful thrust, since the inclination must be swung through an extra 20° (3).
- PhasePhase four: the satellite drifts to its final longitudinal position and fine adjustments are made (4).

Launch and stationing

Continuous attitude and orbit trim manœuvres are required to place a satellite in its operating station.

The complexity of launch and station acquisition manœuvres depends on the geographical position of the launch pad. The most cost effective bases for inserting a satellite in zero inclination orbit are those located near the equator (Fig. 2.32). The latitude of a base is therefore a key factor in determining the thrust needed to

place the satellite in the equatorial plane (see modification of orbit on pages 14–16).

■ Launch from the Guianan Space Centre in Kourou, a base very close to the equator (latitude 5° 10′ north), allows simpler manœuvres. They comprise three phases (Fig. 2.33). The first two phases usually require two days, whilst the exact deployment of the satellite at the chosen longitude, followed by stabilisation in a suitable attitude, takes much longer, sometimes as long as several weeks.

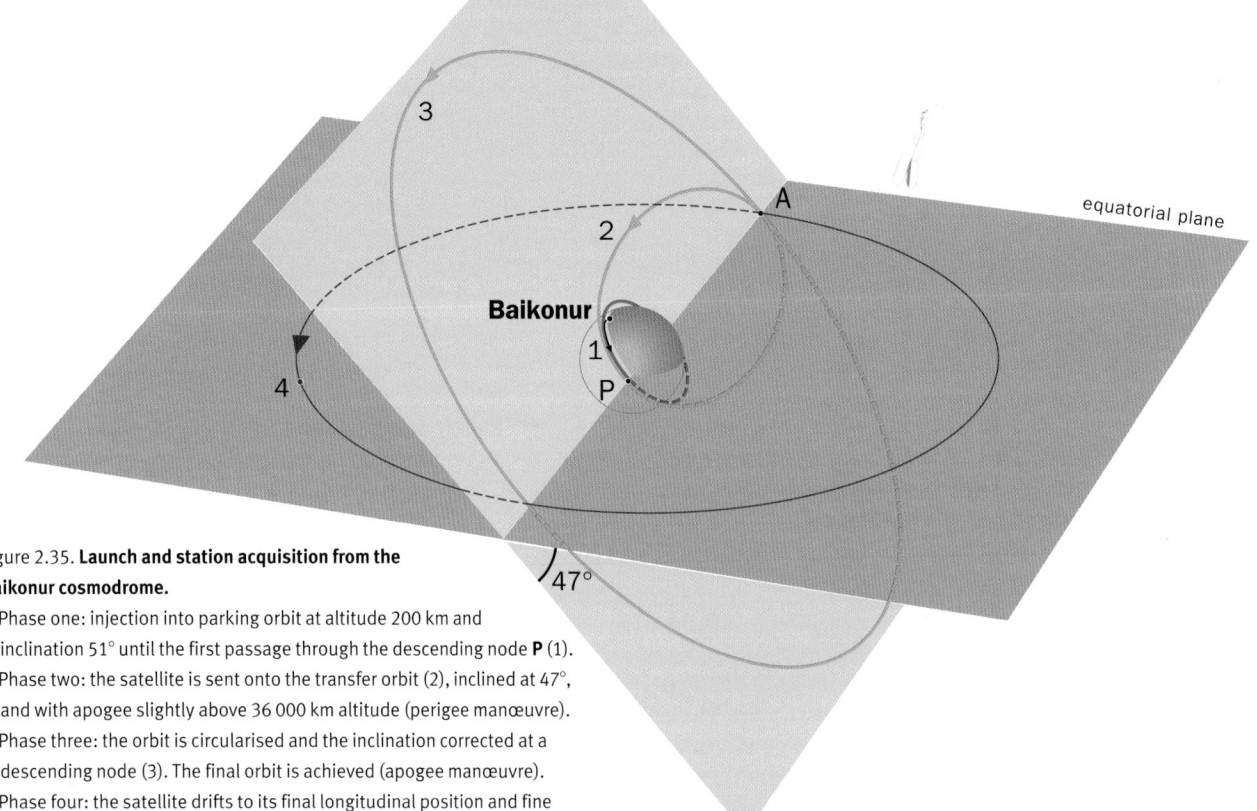

Figure 2.35. **Launch and station acquisition from the Baikonur cosmodrome.**

■ Phase one: injection into parking orbit at altitude 200 km and inclination 51° until the first passage through the descending node **P** (1).

■ Phase two: the satellite is sent onto the transfer orbit (2), inclined at 47°, and with apogee slightly above 36 000 km altitude (perigee manœuvre).

■ Phase three: the orbit is circularised and the inclination corrected at a descending node (3). The final orbit is achieved (apogee manœuvre).

■ Phase four: the satellite drifts to its final longitudinal position and fine adjustments are made (4).

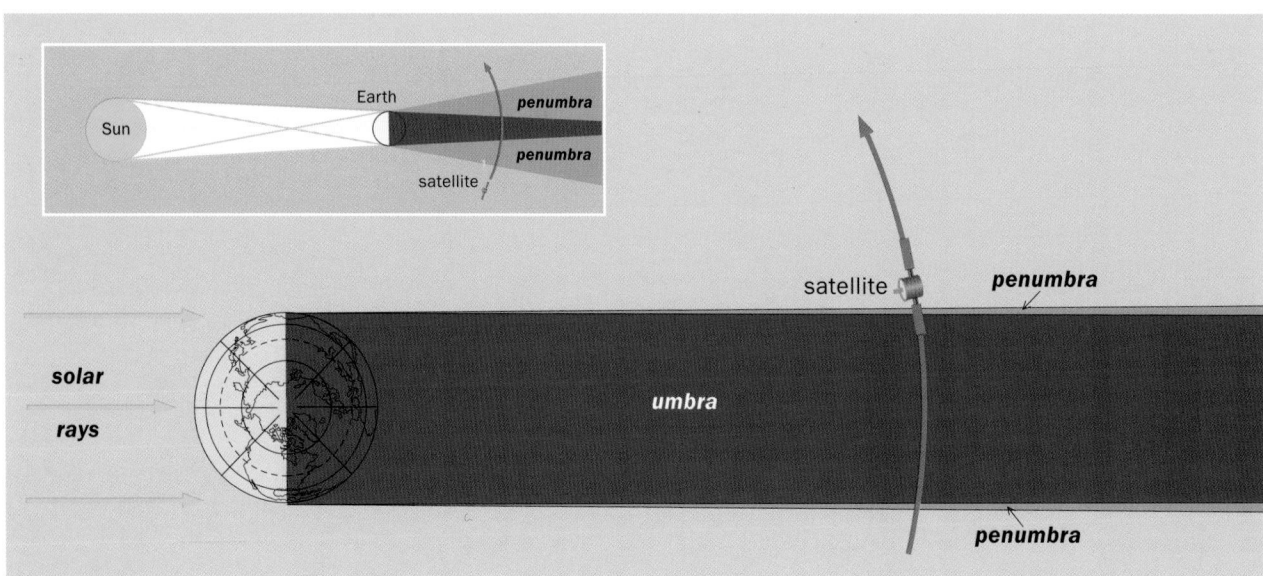

Figure 2.36. **Umbral and penumbral crossings by geostationary satellites at the equinoxes.** Scales are not respected on the overall view in the inset.

- Launch from Cape Canaveral in the United States, at latitude 28° 30′ north when carried out by the Space Shuttle, requires in extra manoeuvre compared with launch from Kourou by Ariane, owing to the limited capabilities of the Shuttle with regard to altitude (Fig. 2.34).
- Launch by the Proton rocket from Baikonur (latitude 45° 9′ north) in Kazakhstan involves similar manœuvres to launch from Cape Canaveral. The latitude at Baikonur, despite its being the southernmost base in the ex-USSR, requires more powerful thrust motors (Fig. 2.35).

Eclipses

The equatorial plane, in which geostationary satellites gravitate, makes a constant angle of 23° 27′ with the plane of the ecliptic. There are thus several days of the year, near the equinoxes, in which satellites may receive no solar light at all, if they pass through the Earth shadow, or only a part, if they pass through the penumbra (Fig. 2.36). Other eclipses are also caused by the Moon passing between Sun and satellite. Continued operation of the satellite therefore assumes some kind of energy supply to replace what is temporarily

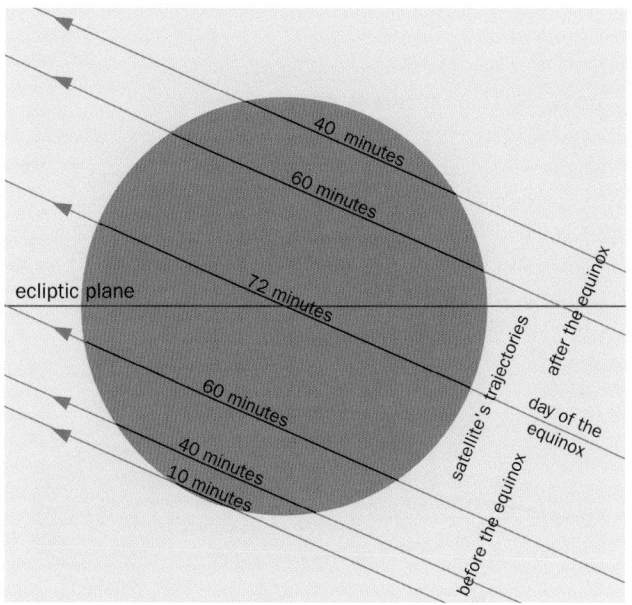

Figure 2.37. **Cross section of shadow cone at 36 000 km from Earth.** On the day of the equinox, the satellite crosses the shadow cone along a diameter of its cross section.

Figure 2.38. **Determining eclipse periods for a geostationary satellite.** The path of the satellite in the equatorial plane of the Earth is shown in orange, during solstices $S_1 S_2$ and equinoxes $S_3 S_4$.

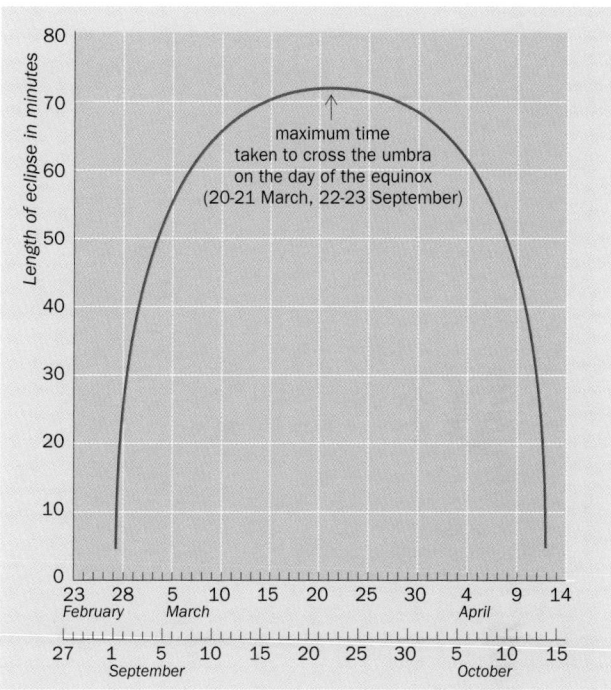

Figure 2.39. **Duration of total eclipse near the equinoxes.** The period of total eclipse gradually increases from 0 to 72 minutes during the 21 days preceding the equinox, then decreases during the 21 days following it.

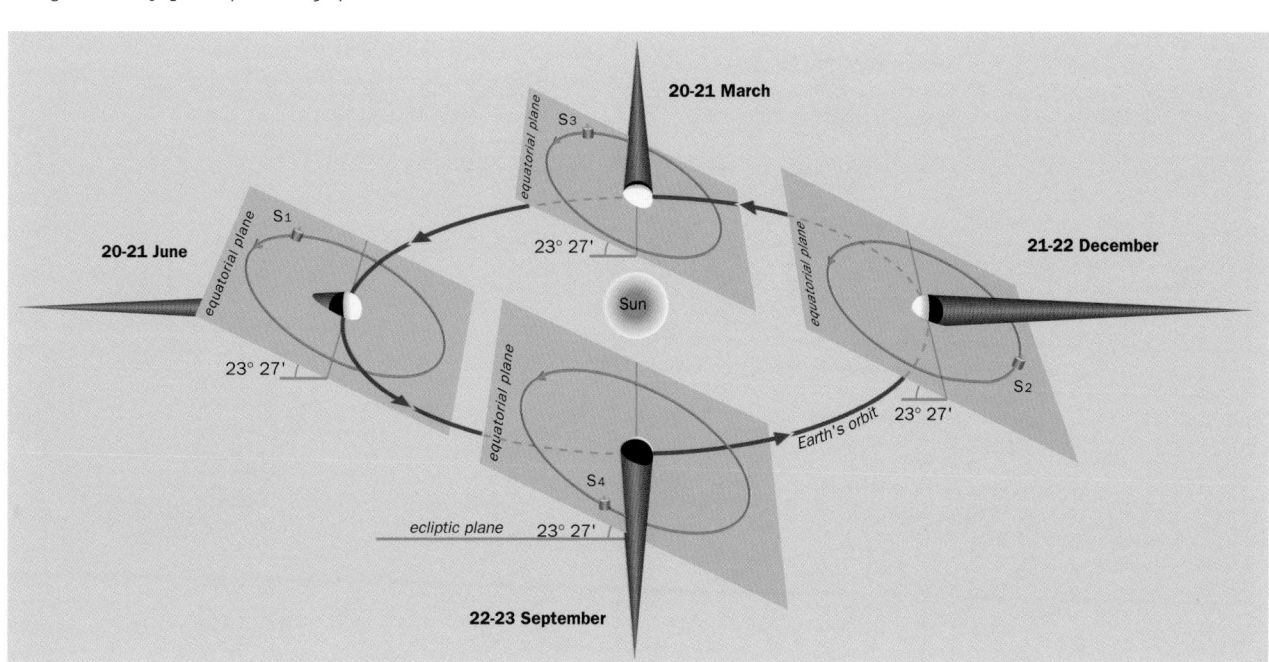

unavailable from the solar panels. The satellite may not resume normal service until some time after the return of full illumination, since its various appendages require a certain time to recover their operating temperatures.

At the two equinoxes in March and September, the satellite, the Earth and the Sun are aligned at midnight in local civil time. On the day of the equinox, the satellite spends 72 minutes in darkness, and this represents the maximum daily eclipse time. From 21 days before and up to 21 days after, the satellite crosses the umbral cone each day (Fig. 2.37). During the rest of the year, the orbits of geostationary satellites pass above or below the umbral cone, reaching maximal distance from them at the solstices (Fig. 2.38). Equinoctial eclipses are centred on the stroke of midnight for the meridian vertically below the satellite (Fig. 2.39).

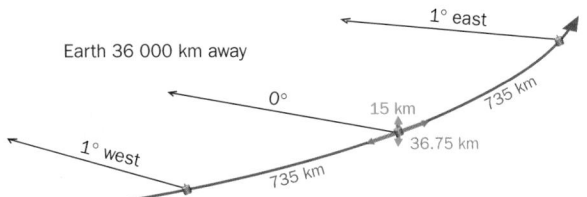

Figure 2.40. Position of a geostationary satellite on its orbit. The total length of the orbit is 264 390 km and one degree along it corresponds to 735 km. Tolerance is theoretically 73.5 km in longitude and 30 km in latitude.

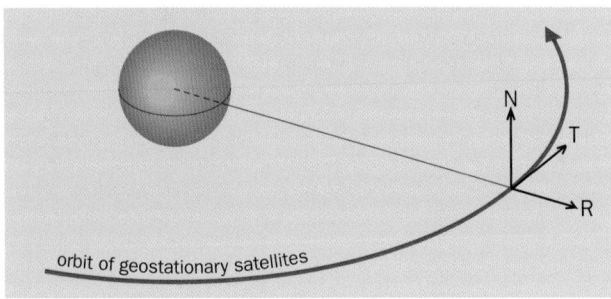

Figure 2.41. Positional corrections. Corrections to inclination are executed by thrust along the vector **N**, whilst corrections to eccentricity are made by thrusts along vectors **T** and **R** corrections to longitude by thrusts along vectors **R** and **T**.

Figure 2.42. Projection suited to the coverage of a geostationary satellite.

Stationkeeping

Perturbing forces, such as terrestrial gravitational forces, atmospheric drag or lunar and solar attractions, can modify a satellite's position, so that continual correction of longitude and latitude is necessary. This is achieved by commands from ground-based control centres, which ensure that the satellite remains within certain tolerance windows (Fig. 2.40).

Stationkeeping also includes stabilising the satellite, with the solar panel axis perpendicular to the orbital plane, and correct pointing of antennas. Stationkeeping thus consists in acting on orbital parameters via thrusts supplied by the satellite guide motors (Fig. 2.41).

Expedient cartographic projection

The satellite views Earth from the tip of a cone situated at an altitude of 36 000 km. The relevant regions of the Earth's surface are represented by constructing a special cartographic projection adjusted to this viewpoint (Fig. 2.42). This is a cylindrical projection tangent to Earth at the equator. Meridians are parallel and equally spaced, whilst lines of latitude are straight lines perpendicular to the meridians, but more and more closely spaced as the latitude increases (Fig. 2.43).

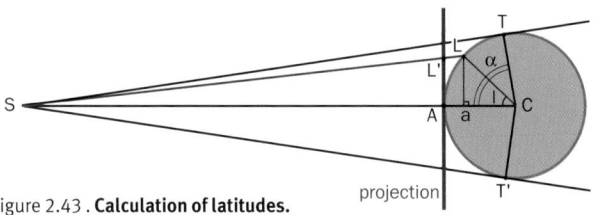

Figure 2.43 . Calculation of latitudes.

■ Calculation of limiting latitudes:
$\cos \alpha = TC/SC$
TC = Earth radius = 6 378 km
SC = 35 786 km + 6 378 km = 42 164 km
so that $\alpha = 81° 2'$

■ Calculating the position **L′** on the map of the parallel at latitude **L**:
The meridian arc **LA** is represented by the straight line **L′A** on the projection. Hence, **L′A/La = SA/Sa** and **L′A = SA.La/Sa**
where **La = R.sin l**, **SA** = 35 786 and **Sa** = 35 786 + (R. R.cos l)
so that **L′A = 35 786.R.sin l/35 786 + (R−R.cos l)**, expressed in kilometres.

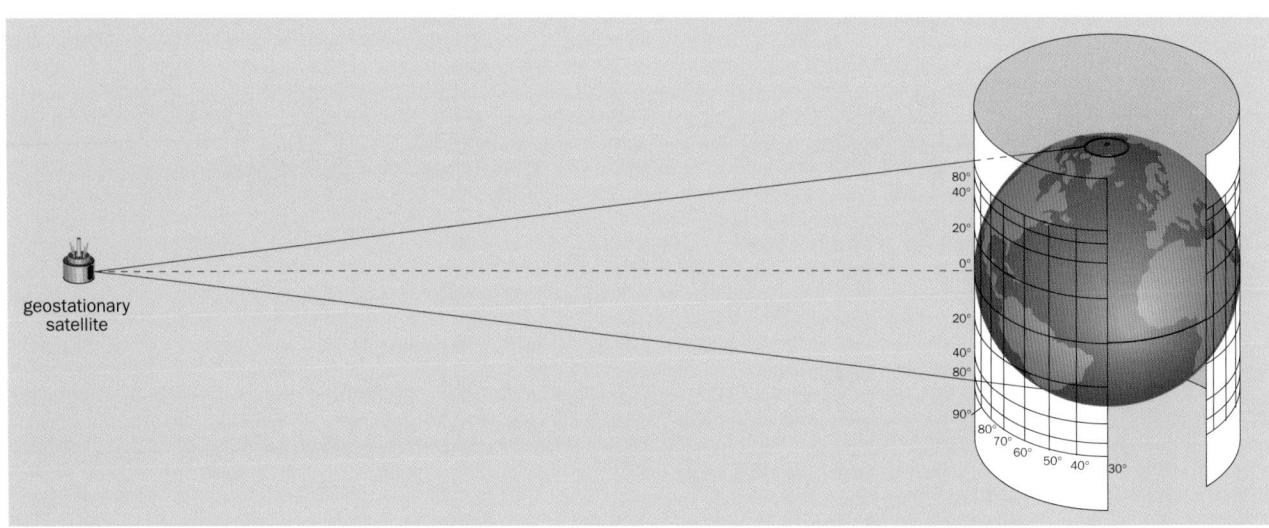

Lagrange points and associated orbits

A system comprising two celestial bodies (such as the Earth and the Moon, or a planet and the Sun), creates a conjugate gravitational field that can be represented in cross section by curves joining points of equal gravity. It can be seen (Fig. 2.44) that some of these curves intersect. Such intersections represent points of zero relative acceleration and are known as *Lagrange points*, named after the astronomer who first established their existence and theoretical positions at the end of the eighteenth century. An object placed at one of these points remains stationary relative to the system formed by the two bodies. Taking the Earth–Moon system as an example (Fig. 2.45), a satellite would orbit around Earth with the same angular velocity as the Moon, and its distance from either object would remain constant.

The Lagrange points are only approximately stable in practice, since other celestial bodies generally perturb the system. For example, the Sun perturbs the Earth–Moon system. A satellite therefore tends to oscillate about the equilibrium point, describing a more or less regular orbit called a halo orbit. Orbits around Lagrange points can be useful for several reasons.

In the Earth–Sun system, the point L_1 was used to maintain the ISEE-3 satellite on a semi-stable orbit between the Earth and the Sun from 1978 to 1982. Since it was located outside the magnetosphere, it was well placed to observe the solar wind (Fig. 2.46). The SOHO probes (*SOlar and Heliospheric Observatory*) and also Wind, since November 1996, also occupy such positions. The point L_2 will be occupied by future astronomical satellites FIRST, PLANCK, GAIA and NGST. This position, on the other side of the Earth to the Sun, allows an unperturbed view of the sky.

In the future, Lagrange points could be exploited for setting up space colonies at constant distance from Earth without involving any energy expenditure (points L_4 and L_5 of the Earth–Moon system). They might also be used for space ports serving as relay stations for large scale interplanetary transportation, since launch from a Lagrange point involves only a very low energy expenditure.

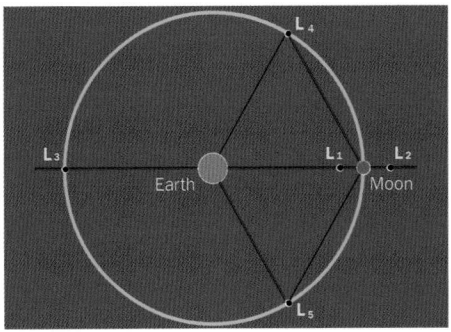

Figure 2.45. **Lagrange points in the Earth–Moon system.** The two points **L₄** and **L₅**, forming an equilaterial triangle with the Earth–Moon line, are stable: if an object moves away slightly, it is brought back to its original position. The three points **L₁**, **L₂** and **L₃** are unstable: the slightest movement away from the point will cause an object to go off into space.

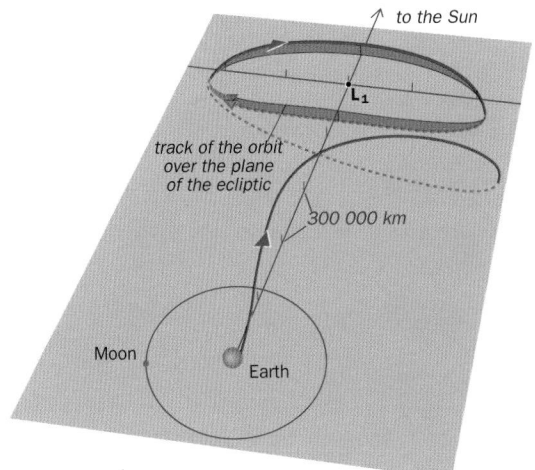

Figure 2.46. **The halo orbit about point L₁ of the Earth–Sun system.** Placed between Sun and Earth, the satellite records solar activity before crossing the magnetopause. In this way, it can distinguish solar phenomena from those caused by internal processes within the magnetosphere. This position has the advantage of avoiding configurations in which the Earth eclipses the Sun, and thereby allows continuous observation of the Sun. In addition, the orbital plane is inclined relative to the plane of the ecliptic, thereby precluding interference between radio links and solar radiation.

Figure 2.44. **Curves of equal gravitational potential in a system comprising two celestial bodies.** Points labelled **L₁** to **L₅** are the Lagrange points.

Space probe orbits

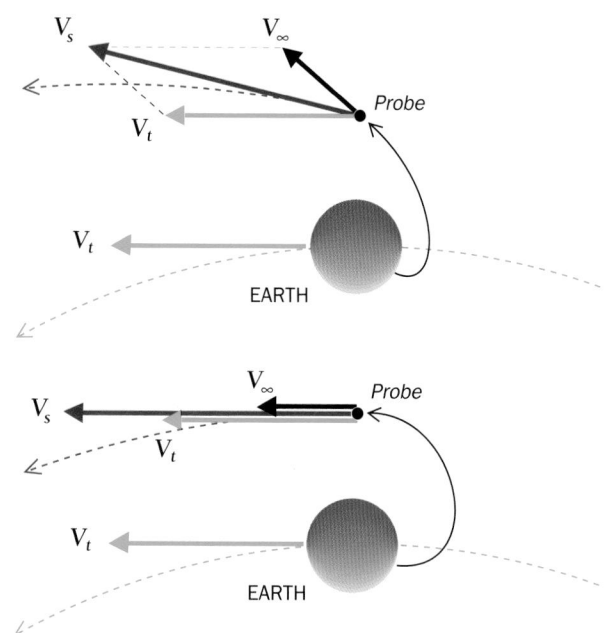

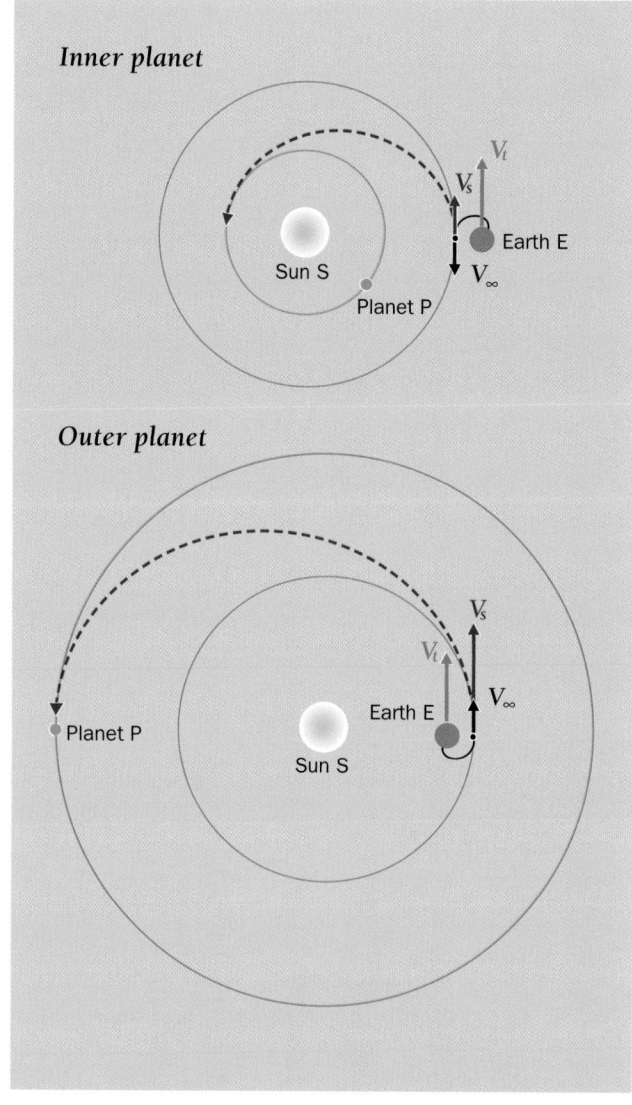

Figure 2.47. **Components of the velocity of a space probe.** If the probe is launched parallel to the orbit of Earth (below), its speed is the arithmetic sum of V_t and $V\infty$.

Figure 2.48. **Hohmann transfer trajectory between Earth and a planet.** Using the relation on page 16, the required hyperbolic excess velocity can be calculated as a function of the distances from the Sun to Earth and the target planet:

$$V\infty = \sqrt{2\mu.SP/(SE(SP+SE))} - \sqrt{\mu/SE}$$

μ is the heliocentric gravitational constant, equal to 1.32×10^{11} km^3/s^2. The travel time is equal to half the period of the orbit:
$$t^2 = PE^3.\pi^2/8\mu$$

Table 2.2. **Velocities and times required to reach a planet from Earth via Hohmann transfer orbit.** The last column gives the best time yet obtained by a more direct trajectory.

PLANET	Hyperbolic excess velocity (in km/s)	Launch velocity (in km/s⁻¹)	Time taken (in days)	Secant trajectories best real time (in days)
Mercury	7.59	13.5	105	-
Venus	2.5	11.4	146	94
Mars	2.99	11.5	259	131
Jupiter	8.85	14.2	997	546
Saturn	10.4	15.0	2 214	1 164
Uranus	11.3	15.8	5 834	3 078
Neptune	11.7	16.2	11 200	4 388
Pluto	12.3	16.2	17 000	-

Heliocentric orbit and hyperbolic excess velocity

When a probe exceeds the second cosmic velocity V_e (see page 14), it moves onto a heliocentric trajectory with a certain velocity V_s. This velocity is the vector sum of the velocity of Earth V_t (29.8 km.s^{-1}), and a part $V\infty$ of the velocity V that the launch vehicle has transmitted to the probe. It can be shown that $V\infty = \sqrt{V^2 - Ve^2}$. $V\infty$ is called the hyperbolic excess velocity of the probe (Fig. 2.47).

In order to attain a planet in the Solar System, the path requiring the lowest hyperbolic excess velocity is the Hohmann trajectory (Fig. 2.48). It is also the slowest, and in practice more direct paths are used depending on the amount of energy available to the probe (see Table 2.2 and Fig. 2.49). Furthermore, it is exceptional for the target planet to lie exactly in the plane of the ecliptic (although the deviation is generally only very slight). Consequently, the velocity vector must almost always have some component orthogonal to this plane.

Synodic period and space rendezvous

In order to achieve a rendezvous between probe and planet at some point P_2, the planet must be in a position P_1 at the time of launching such that it arrives at the point P_2 at the anticipated date (Fig. 2.49). This configuration is only repeated at the end of each synodic period. The latter depends on the respective orbital periods of the Earth and the planet in question (see pages 9–10):

$$1 / \text{synodic period} = 1 / \text{terrestrial period} - 1 / \text{planetary period}$$

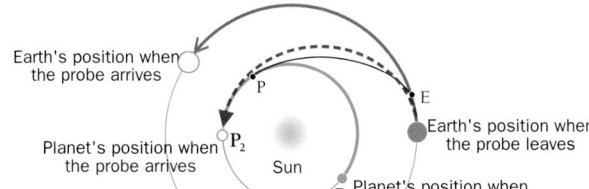

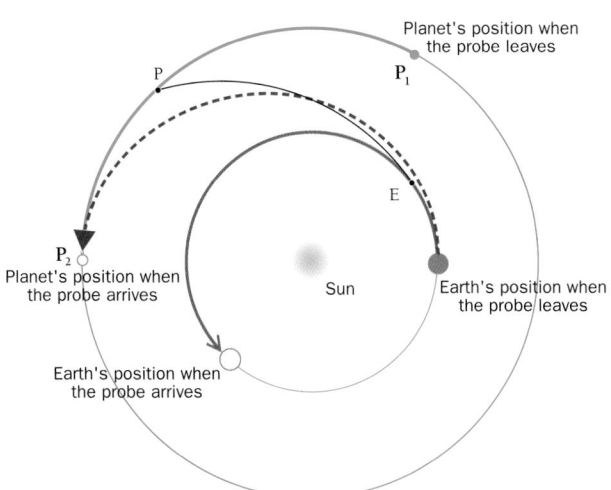

Figure 2.49. **Planetary positions at the launch of a space probe.** Inferior planet (above) and superior planet (below). The direct trajectory EP is more costly in energy terms, but quicker.

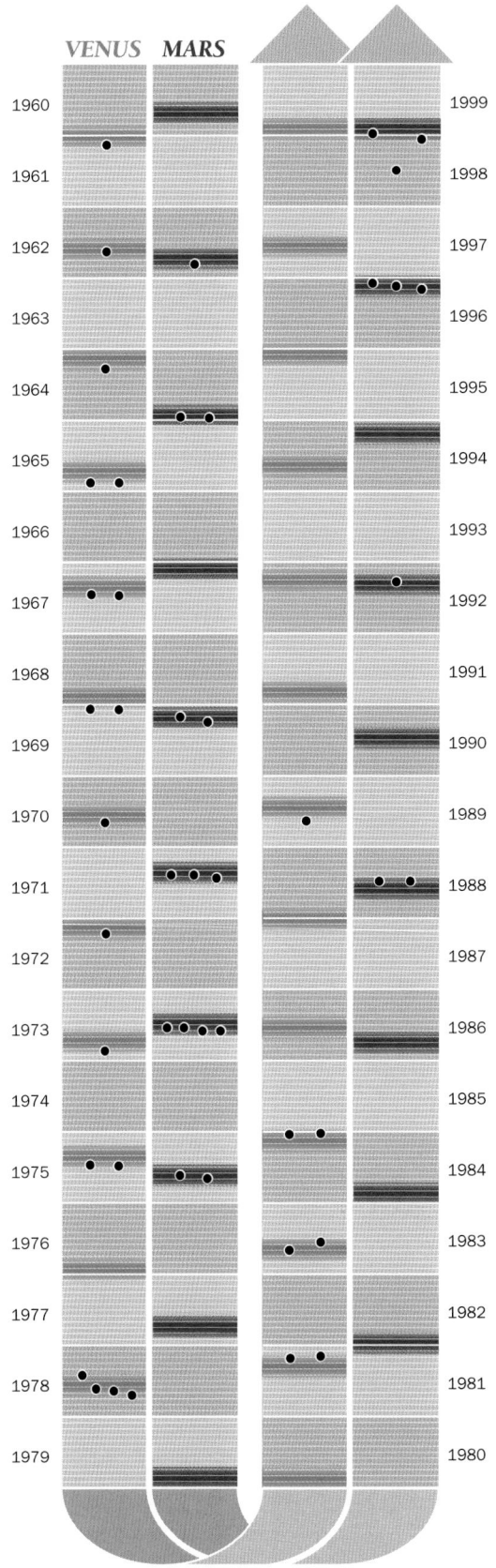

Figure 2.50. Launch windows for Venus and Mars from 1960 to 2000. Black spots indicate use of these windows for launches up to 1 January 2000. The launch of Planet B outside the window on 2 July 1998 was made possible by gravity-assisted manœuvres around the Moon, making up for the insufficient thrust of the M-5 launch vehicle.

Launch windows

In practice, the use of shorter and more direct trajectories extends the range of available times for launch, referred to as a launch window. This gain in flexibility depends on how much power the launch vehicle can provide to reach the desired speed, and also on the need for braking given the higher speed upon arrival in the neighbourhood of the planet, if the probe must land or go into orbit around it (limit due to probe power capability and the weight of propellant to be transported). The launch window can thus be viewed as a compromise based on the power of the launch vehicle and the characteristics of the probe (Fig. 2.50).

Sphere of gravitational influence

A simplified way of viewing a complex interplanetary trajectory is to assume that around each planet there exists a sphere of gravitational influence where the path can be considered as a conic section with the planet at one of its foci. As soon as the probe leaves this zone, its trajectory is then treated as a conic section with the Sun at one of its foci (Fig. 2.51 and Table 2.3).

When a probe passes into the sphere of gravitational influence of a planet, its trajectory in the frame of reference of that planet will thus be a conic section, in fact, a hyperbola (since the probe arrives from a point at infinity relative to the planet). The parameters determining this hyperbola can be deduced from the characteristics at the entry point into the sphere of gravitational influence (Fig. 2.52): velocity V_1, angle with local horizontal γ, distance to focus d (see page 15 for calculation of orbital elements). The speed of the probe increases under the effects of the planet's attraction until periapsis is reached, then diminishes. However, when it leaves the sphere of gravitational influence, its final speed is

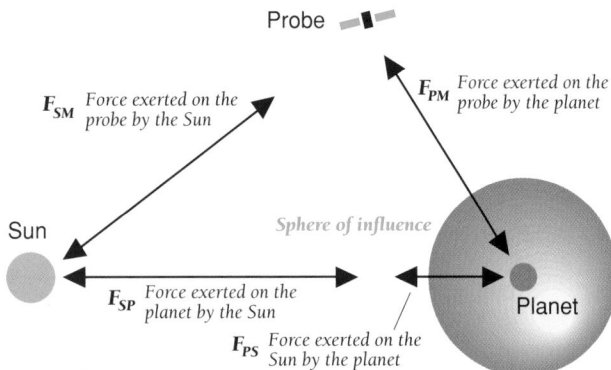

Figure 2.51. **Definition of the sphere of gravitational influence of a planet.** Considering the set of attractive forces between Sun, planet and space probe, the sphere of gravitational influence is defined as the region where

$$(F_{SM}-F_{SP})/F_{PM}<(F_{PM}-F_{PS})/F_{SM}$$

The zone is roughly a sphere centred on the planet, out to a distance R given by

$$R = \sqrt[5]{1/2.SP\,(\mu_p/\mu_s)^{2/5}}$$

SP is the distance of the planet from the Sun, μ_p and μ_s are the planetary and heliocentric gravitational constants.

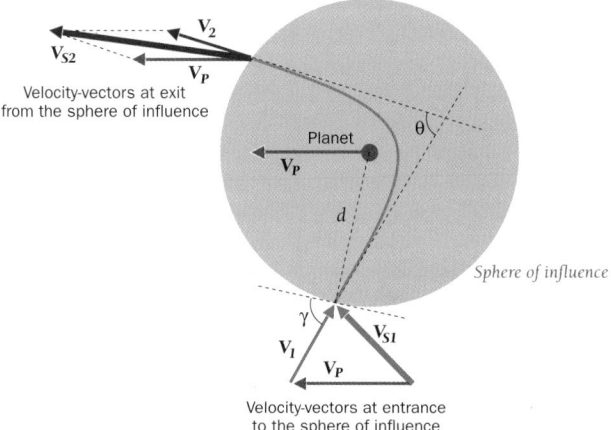

Figure 2.52. **Speed and direction of a space probe on entering and leaving the sphere of influence of a planet.**

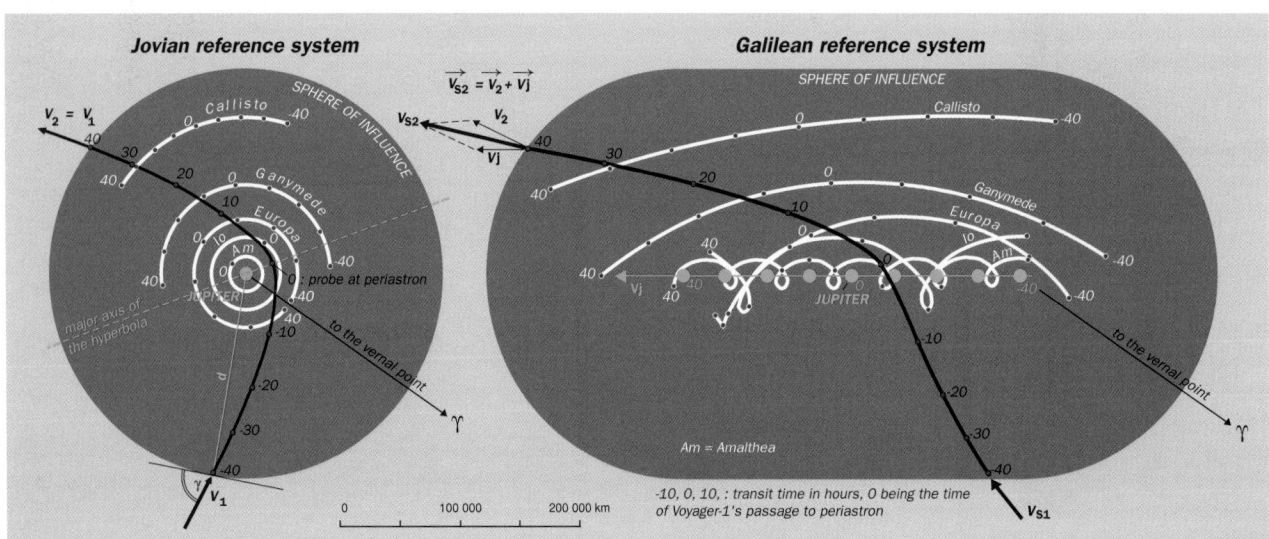

Figure 2.53. **Gravity-assisted manœuvre.** Voyager 1 swing-by in the sphere of gravitational influence of Jupiter.

the same as its entry speed ($V_2 = V_1$); only its direction is changed through an angle θ such that $\cos(\theta/2) = 1/e$ (where e is the eccentricity of the hyperbola).

But relative to the Solar System (Fig 2.53), the velocity V_{S2} on leaving the sphere of gravitational influence is the vector sum of the exit velocity V_2 and the velocity V_P of the planet around the Sun. Relative to the Sun, a probe passing through the gravitational influence of a planet therefore leaves the sphere of influence with both speed and direction modified.

Gravity-assisted manœuvres

The close passage of a probe in the neighbourhood of a planet is often exploited as a way of modifying speed, direction and inclination relative to the ecliptic plane, with very low energy expenditure. Indeed, a quite minimal correction to the velocity suffices, if executed at a suitable moment before entry into the sphere of gravitational influence, to alter the angle γ by 0 to 90° and to select the point of entry into the zone. Of course, an angle too close to 90° would lead to collision with the planet, or passage through any atmosphere that might be present. This manœuvre is referred to as a gravity-assisted manœuvre or *swing-by*. It was used for the first time in 1974, when Mariner 10 was deviated towards Mercury by Venus. It has also been used for the trajectories of Pioneer 11, Voyagers 1 and 2 (Fig. 2.54), Vega 1 and 2, Galileo and Cassini.

Galilean reference frame

In order to represent probe trajectories, a frame of reference is required. Such a frame is the set of coordinate systems rel-ative to which measurements are specified. For this purpose, the axes of these coordinate systems are taken as fixed. In this frame, a surface must also be chosen, upon which the coordinates of the relevant objects can be projected.

The most convenient frame for representing trajectories in the Solar System is the so-called Galilean frame. This takes the centre of the Sun as a fixed point and the vernal equinox (see page 10) as one coordinate axis. Since the planets and most space probes move in planes very close to the ecliptic, it is this plane that is usually chosen for projection.

The coordinates used in the encyclopedia for locating planets are those established by the French Bureau of Longitudes. The x-axis is the direction of the vernal equinox in the year 2000.

PLANET	Radius of the sphere of influence (in km)	Attraction potential (in km³s²)
Mercury	97 300	21 900
Venus	537 000	351 000
Earth	805 000	398 603
Mars	503 000	59 800
Jupiter	41 900 000	524 800 000
Saturn	47 700 000	301 100 000
Uranus	83 200 000	26 720 000
Neptune	122 000 000	22 730 000

Table 2.3. **Sphere of gravitational influence and planetary gravitational constants (product of the planetary mass with the universal constant of gravitation G)**

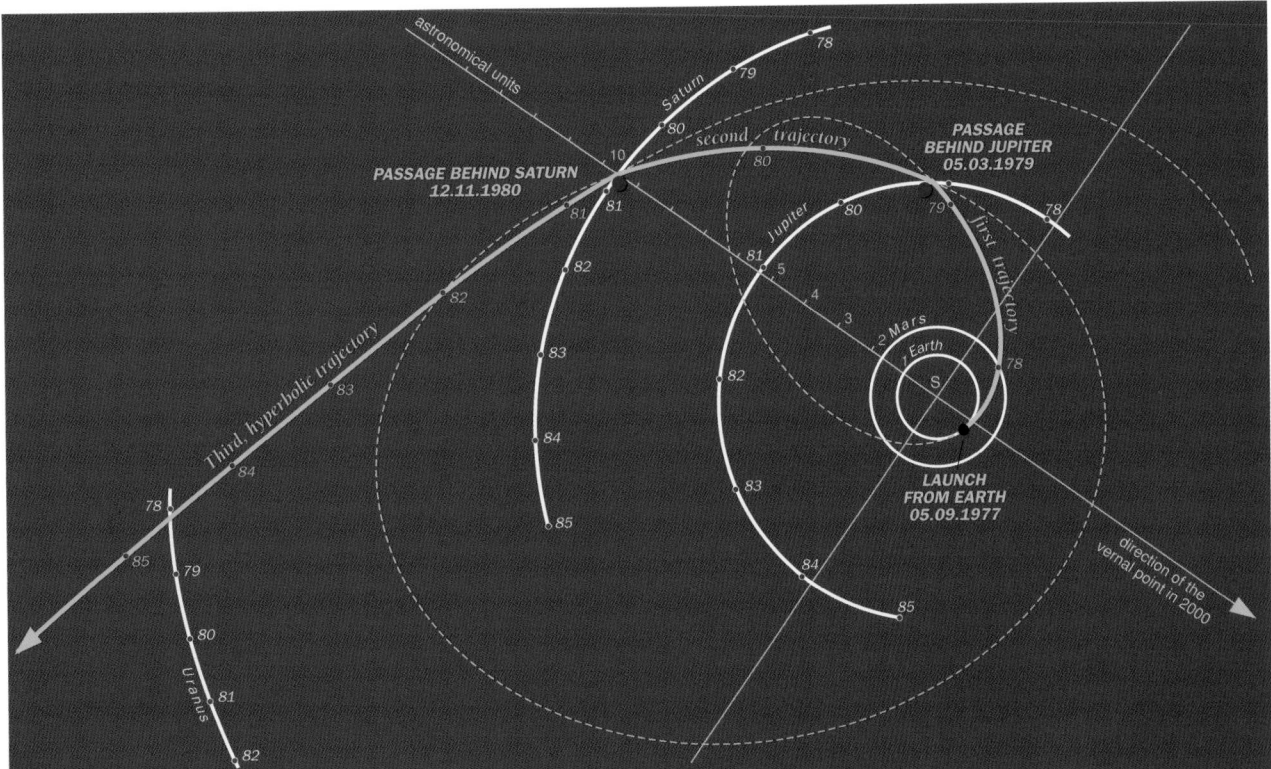

Figure 2.54. **Example of a complex trajectory.** The path of Voyager 1 in a Galilean reference frame comprises three successive heliocentric orbits of increasing amplitude (two ellipses and a hyperbola). Transfer from one curve to the next is obtained by swing-by around Jupiter and Saturn.

CHAPTER THREE

Ground tracks

- OVERVIEW

- FROM ORBIT TO EARTH: MAPPING THE GROUND TRACK

- TYPES OF GROUND TRACK

- GROUND TRACKS AND ORBITAL CYCLES

Overview

The ground track is an imaginary line described by the intersection of the Earth's surface with the nadir of the orbiting satellite. It is formed by a combination of the spacecraft's motion and the Earth's rotation. It offers a means of visualising on the surface of the globe the satellite's movement relative to the Earth.

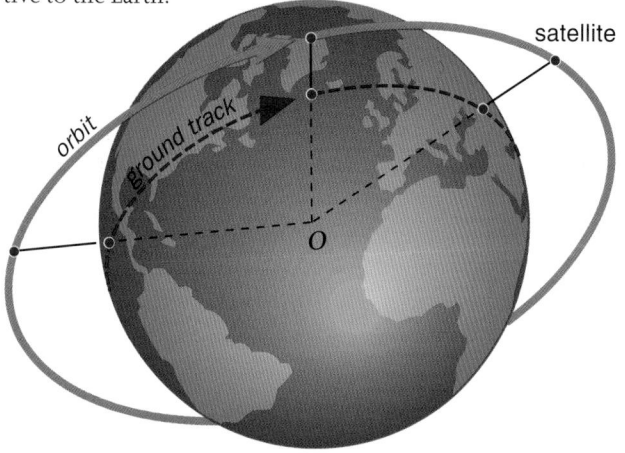

Figure 3.1. **Satellite ground track assuming the Earth does not rotate.**

The orbital plane bisects the Earth globe in a great circle which by definition passes through the centre of the Earth. If the Earth were assumed not to rotate, the track would coincide with the circumference of this circle (fig. 3.1).

Relative to this great circle around a hypothetically non-rotating Earth, the real ground track is modified by the Earth's rotation (fig. 3.2). The nature of the modification is determined by the satellite's angular velocity (and hence by its altitude) and by the Earth rotation rate at the latitude over which the satellite is located.

Influence of altitude and latitude

The lower the satellite's angular velocity the greater the impact on ground track orientation.

A satellite moving in a circular orbit exhibits constant angular velocity. As orbit altitude rises angular velocity falls. The higher the orbit the more the ground track is shifted to the west by the Earth's rotation (fig. 3.3).

In an eccentric orbit, velocity varies, rising from the apogee to the perigee. It follows that the nearer the satellite comes to apogee, the more the track is shifted westward (fig. 3.4).

The Earth relative rotation rate diminishes as latitude increases, falling to zero at the poles. The ground track is therefore displaced to a diminishing extent as the latitudes overflown increase (fig. 3.5).

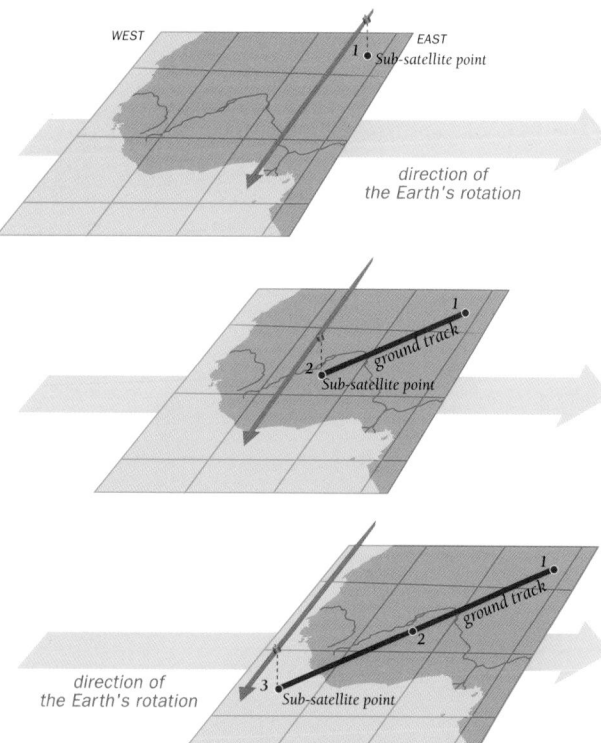

Figure 3.2. **Ground track over a region with account taken of the Earth's rotation.** While the satellite (shown here in polar orbit) progresses southward, the Earth rotates to the east and the ground track shifts in relation to the orbital plane.

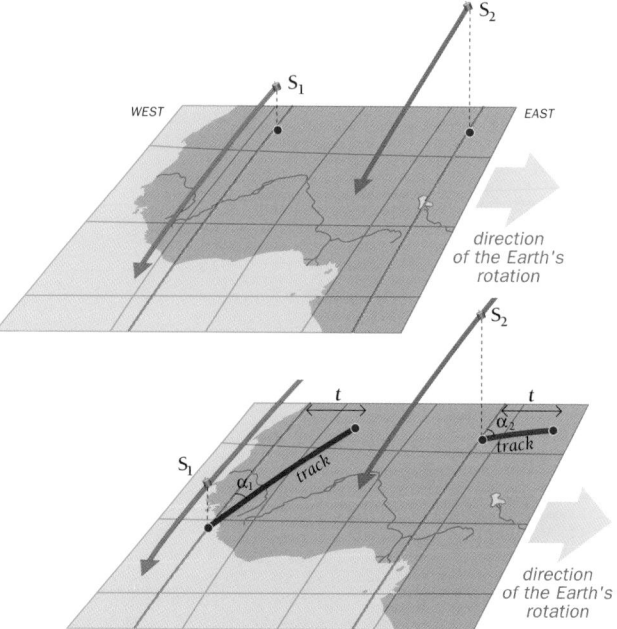

Figure 3.3. **The influence of altitude on the ground tracks of satellites in circular orbit.** Over the same time interval t, the track of satellite S_2, orbiting at a higher altitude than S_1, shifts further westward ($\alpha_2 > \alpha_1$).

Main characteristics of ground tracks

Orbit inclination determines the extreme northern and southern latitudes of the terrestrial segment swept by the satellite track. The extent of this zone from latitude to latitude – which is symmetrical either side of the Equator – is known as the *latitude coverage* (fig. 3.6).

As it progresses, the ground track of a satellite in retrograde orbit intersects increasingly westerly meridians.

The ground track of a satellite in direct orbit

- remains at the same meridian if the west–east component of its velocity vector is equal to the Earth rotation rate, which is to say if its ecliptic longitude advances at a rate of 15° 2′ 30″ per hour;
- intersects increasingly easterly meridians if its progress exceeds that rate;
- intersects increasingly westerly meridians if its progress falls short of that rate.

These last two classes of track are sometimes referred to, respectively, as direct and retrograde. A track is said to be "ascending" where it crosses the Equator from south to north and descending where the crossing is from north to south. The ground track has a similar shape either side of the Equator in the case of circular orbits and a different shape where the orbit is eccentric (fig. 3.7).

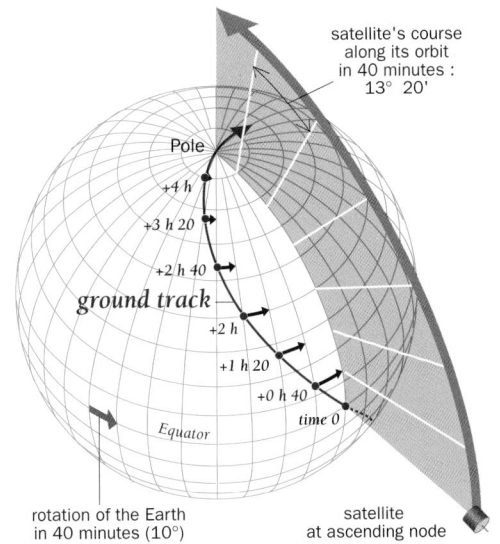

Figure 3.5. **The influence of latitude.** In this example, the satellite is in circular polar orbit and its angular velocity is 13° 20′ in 40 minutes. The Earth turns by 10° every 40 minutes and the track shifts 10° westward, representing a linear deviation of 1111 km at the Equator and 555 km at 60° (black arrows). Deviation is zero at the pole.

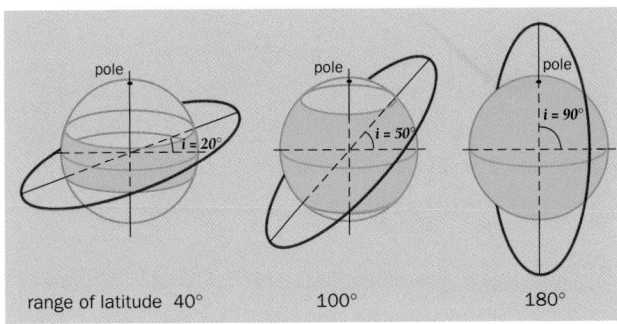

Figure 3.6. **Latitude coverage and inclination.** Only a satellite inclined at 90° can fly over all latitudes.

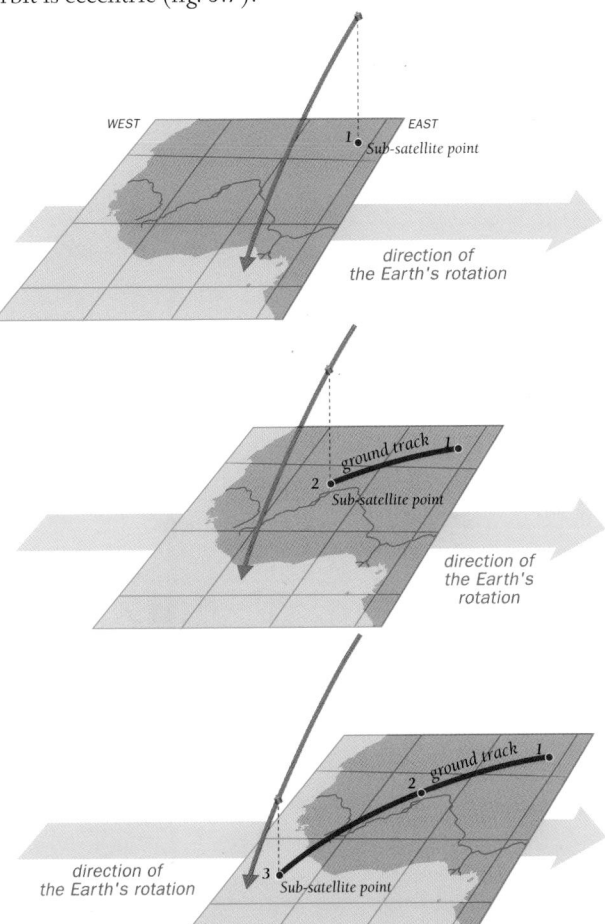

Figure 3.4. **The influence of altitude on the ground tracks of satellites in eccentric orbit.** As the satellite approaches Earth, its velocity increases and the westward shift in the track decreases.

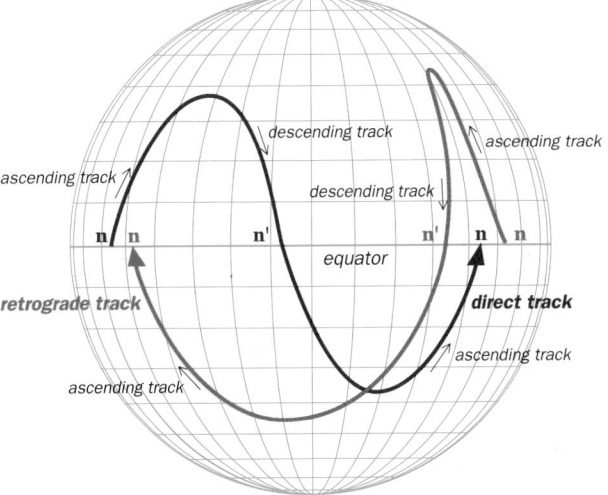

Figure 3.7. **Ground track geometry.** The track of the satellite in circular orbit (shown in red) is symmetrical in relation to point n′ situated on the Equator. The track of the satellite in eccentric orbit (shown in blue) is non-symmetrical in relation to the same point, except where the major axis of the orbit is in the plane of the Equator.

From orbit to Earth: mapping the ground track

Transposing the track onto a map

The process required to go from the trajectory of an orbiting satellite to the plotting on a map of the corresponding ground track can be broken down into a series of operations.

The first step is to establish the five orbital elements together with a positioning quantity for the satellite. For the latter it is convenient to take the point in time t at which the satellite crosses the ascending node N, which is taken to be the starting point for the track. The starting point is thus on the Equator, its longitude being that of the vernal equinox at t increased by angle Ω, the right ascension of the ascending node.

After a period of time Δt the satellite reaches a position S_1 corresponding to track position s_1. The coordinates of this position can be derived from the orbital inclination and the

Figure 3.8. **The coordinates of a point on the track.** The latitude *l* and longitude *L* of point P are calculated from inclination *i* and the angle, φ, described by the satellite's radius vector since the ascending node crossing N:

Assuming a fixed Earth:

$\sin l = \sin i \, \sin \varphi$

$\sin L = \sin \varphi \, \cos i / \cos l$

Assuming a rotating Earth, latitude is the same as for a fixed Earth, while longitude has to be adjusted by the angle through which the Earth has turned during the time, *t*, taken by the satellite to describe the arc NP, i.e. 360 t/1436 (for times in minutes and angles in degrees).

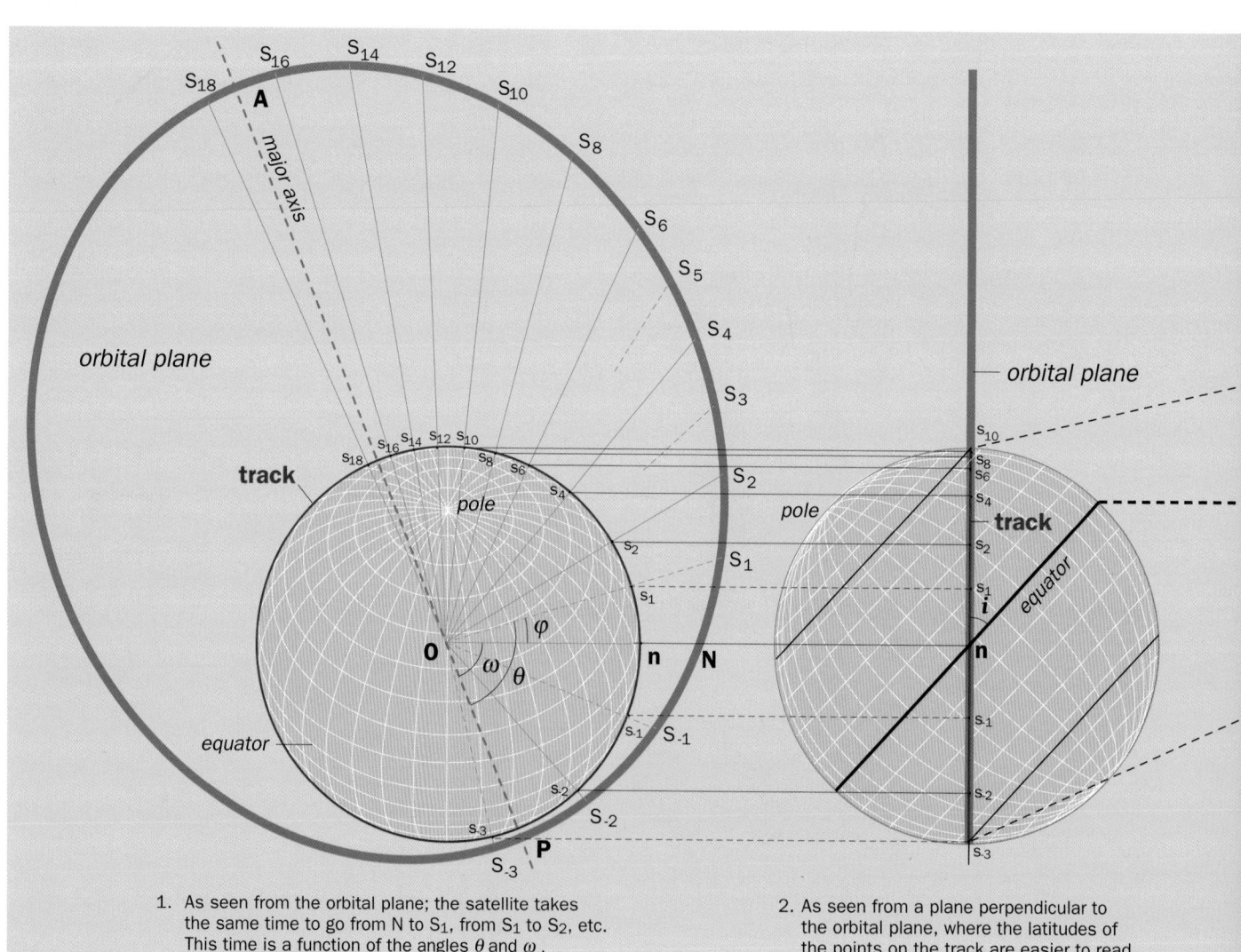

1. As seen from the orbital plane; the satellite takes the same time to go from N to S_1, from S_1 to S_2, etc. This time is a function of the angles θ and ω.

2. As seen from a plane perpendicular to the orbital plane, where the latitudes of the points on the track are easier to read.

Figure 3.9. **Plotting a ground track from an orbit.**

angle φ described by the satellite's radius vector since the ascending node crossing (fig. 3.8). The angle φ is the difference between θ, the true anomaly of S_1 and ω, the orbit's argument of perigee (calculation of θ as a function of Δt is discussed on page 15).

Once the coordinates of S_1 have been established, this point on the track can be plotted on a planisphere. The track can in this way be constituted point by point over a complete orbital revolution and then extended in like manner from the next ascending node crossing. These operations are summarised graphically in figure 3.9 for a number of satellite crossing points effected at regular time intervals.

Impact of orbit modifications

The ground track being determined by the five orbital elements, it is necessarily affected by changes in any of these, whether intentional or attributable to natural causes (see pages 14–16).

Modifying Ω, the right ascension of the ascending node, which is to say nodal precession (see page 14) has the effect of adding the direct or retrograde rotation of the orbital plane

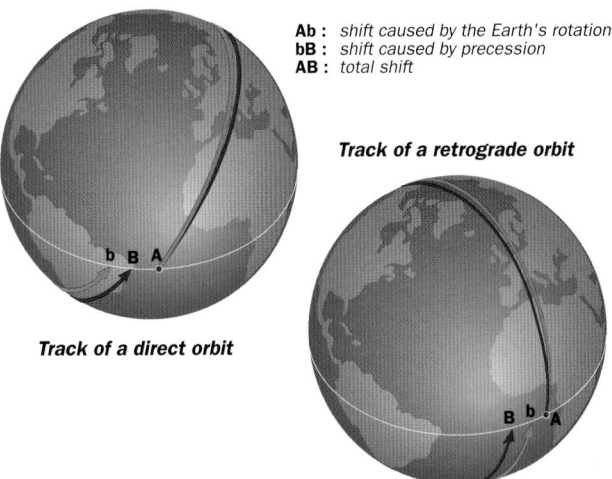

Ab : *shift caused by the Earth's rotation*
bB : *shift caused by precession*
AB : *total shift*

Track of a retrograde orbit

Track of a direct orbit

Figure 3.10. **Modification of a track by nodal precession.** The distance AB between two successive Equator crossings of an ascending track depends solely on the orbital period if nodal precession is left out of account. The latter does however modify distance AB, increasing it if the orbit is direct, decreasing it if it is retrograde. There is no precession at 90° orbital inclination.

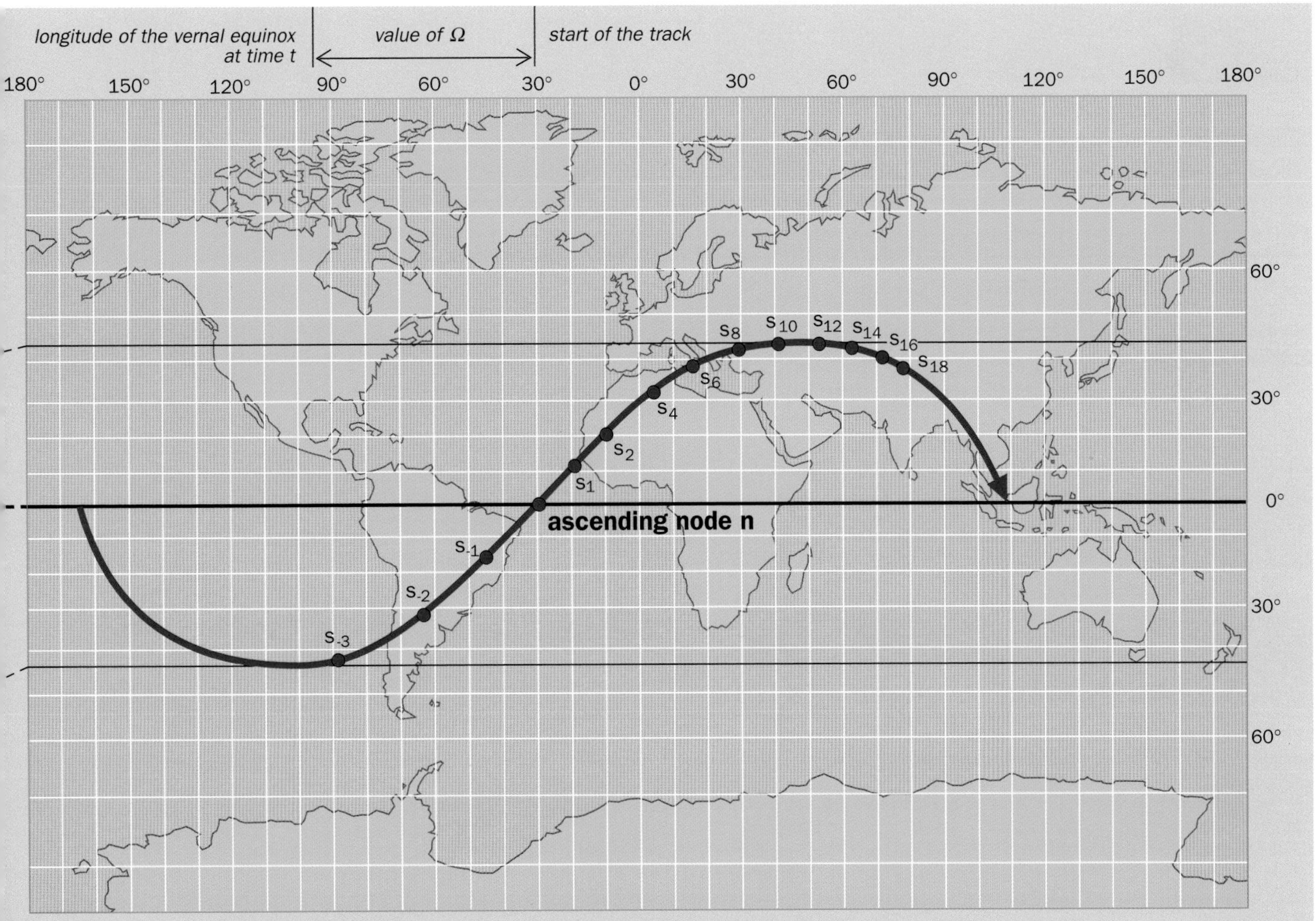

3. The track is transferred to a map. The position of the ascending node, which serves as the starting point for the track, is obtained by adding the value of Ω to the position of the vernal equinox at the moment when the satellite crosses to the ascending node.

to that of the Earth. At each successive revolution, the track is shifted westward if precession is negative and eastward if it is positive (fig. 3.10). The gap between two consecutive tracks is thus modified without any change in orbital period.

The altitude at which the satellite is flying affects the value of the precession to which it is subject: assuming inclination remains constant, precession is greater the lower the satellite's altitude (see fig. 2.24).

Modifying the argument of perigee ω, the rotation of the major axis in the orbital plane, alters the shape of the track but not the gap between two successive tracks: for a given orbital position, the corresponding point on the track changes latitude and the westward shift is modified (fig. 3.11). Only an inclination of 63° 26′ will cancel this rotation – which is why this orbit was chosen for the Soviet Molniya telecommunications satellites, whose apogee has always to be over the same point on Earth.

Modifying the inclination i not only modifies Ω and ω but also changes the latitude coverage value.

Modifying the orbit shape parameters R and r modifies velocity and hence the form of the track and also modifies the orbital period and hence the gap between two successive tracks, known as the *tracking interval* (fig. 3.12).

Projection options

How tracks are mapped does of course depend on the projection method adopted. In choosing a particular method, a number of criteria have to be considered: the size of the area to be represented, uniformity of scale along the track, maintenance of conformality or equal area. Track coordinates are plotted using the system of relations – proper to the projection concerned – by which the longitude and latitude of a point on Earth are assigned to horizontal and vertical coordinates representing the same point on a map.

The Miller projection (see fig. 3.9) adopted for the purposes of the present chapter was designed to attenuate the

excessive variations in scale for latitudes characteristic of the Mercator projection. It is neither conformal nor equal area and its system of equations is:

$$x = L$$
$$y = 1.25 \log \mathrm{tg} \, (\pi/4 + 0.4\,l)$$

In addition to the projections commonly used in geography, a number of projections have been developed with properties which lend themselves to the representation of ground tracks, the prime mover in this area being the US Geological Survey, the leading cartographic organisation in the United States. It has in particular developed an oblique Mercator projection (the Space Oblique Mercator, SOM) specially designed for the very precise continuous mapping of the tracks of satellites in circular orbit. The SOM projection preserves conformality and maintains the same linear scale along the entire track (fig. 3.13).

The Geological Survey has also proposed various types of projection designed for straight-line representation of the tracks of satellites in circular orbit. These projections, known collectively as satellite-tracking projections, are neither equal area nor equidistant. Conformality and scale are preserved at certain latitudes only, along one or two parallels of latitude.

These projections have been developed in direct cylindrical and conic form.

Cylindrical projections may be tangential, in which case conformality and scale are preserved at the Equator, or they may be secant, in which case conformality and scale are preserved along two symmetrical parallels either side of the Equator (fig. 3.14).

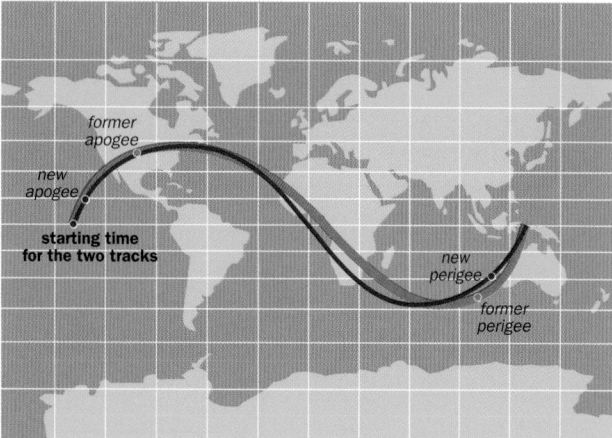

Figure 3.11. **Modification of a track by apsidal precession.** The satellite, as it passes in the vicinity of the apogee – the part of the orbit in which velocity values are at their lowest – is displaced towards latitudes where the Earth's rotation rate is higher. It follows that the track is subject to greater shift in this area.

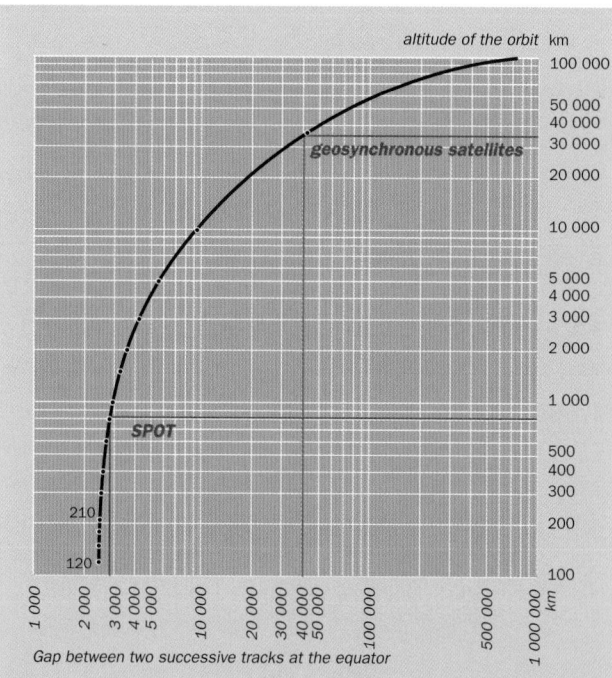

Figure 3.12. **Tracking interval as a function of altitude for circular orbits.** The gap between two successive ascending or descending tracks is the same.

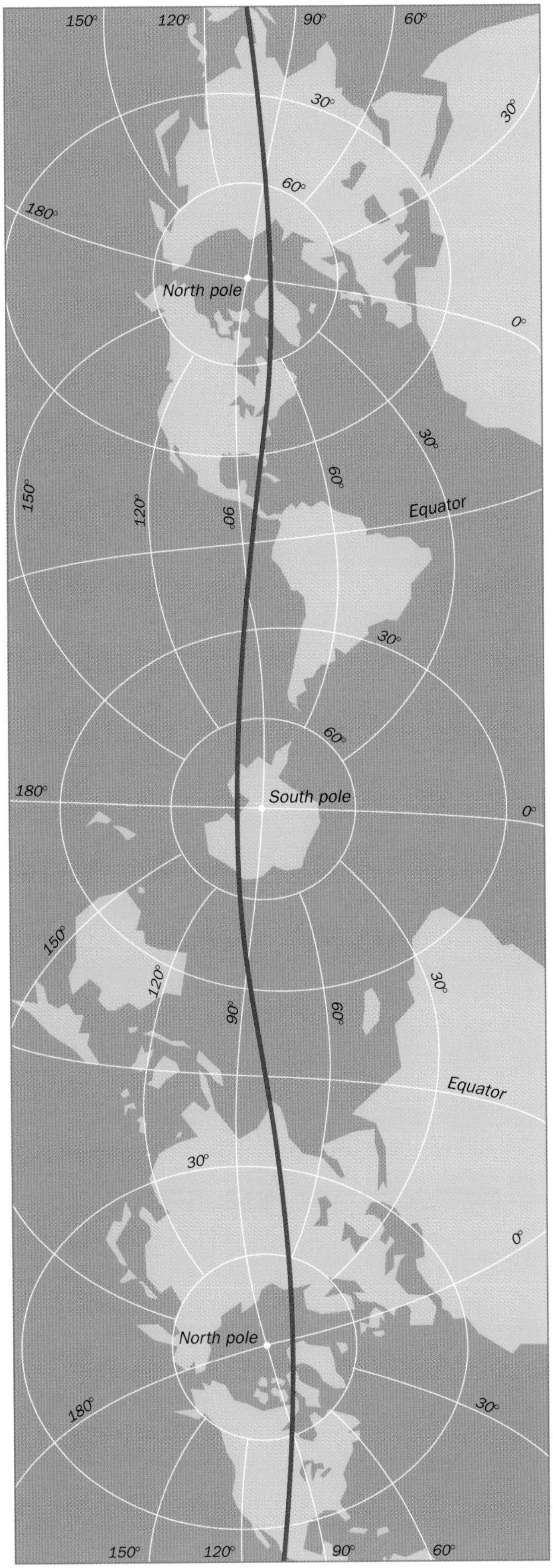

These two options also exist for conic projections. The projection shown in fig. 3.15 preserves conformality along the parallels at 45° and 70° N, while in fig. 3.16 conformality is preserved at 45° and 80.9° N, the northern tracking limit. With both these projections, scale is preserved along only one of the two basic parallels.

The conic projection portrayed in fig. 3.17 is tangential at the limit of latitude coverage (80.9° N). Conformality and scale are preserved along this parallel only.

Figure 3.13. **Space Oblique Mercator projection.** The figure shows the track described over one-and-a-half Landsat-5 revolutions. The orbit is inclined at 98.2°. The central line chosen for the projection (great circle of tangency) is not a straight line as with the standard oblique Mercator projection but rather the track itself. Owing to the deviation resulting from the Earth's rotation, the track is a highly flattened sinusoid (it forms an angle of only 8° with its *x* axis) passing at a maximum latitude of 81.8°. The meridians and parallels are curved lines except for the meridian corresponding to the northernmost pass on the track. The track described in the course of each revolution succeeds the previous track on a modified representation of the Earth's surface, whose rotation can be observed.

This projection was designed to produce minimum distortion in the swath scanned by the satellite's sensor. As a result, the projection is not strictly conformal – except along the track – as would be the case with an unmodified oblique Mercator projection (see J.P. Snyder, 1987, page 216).

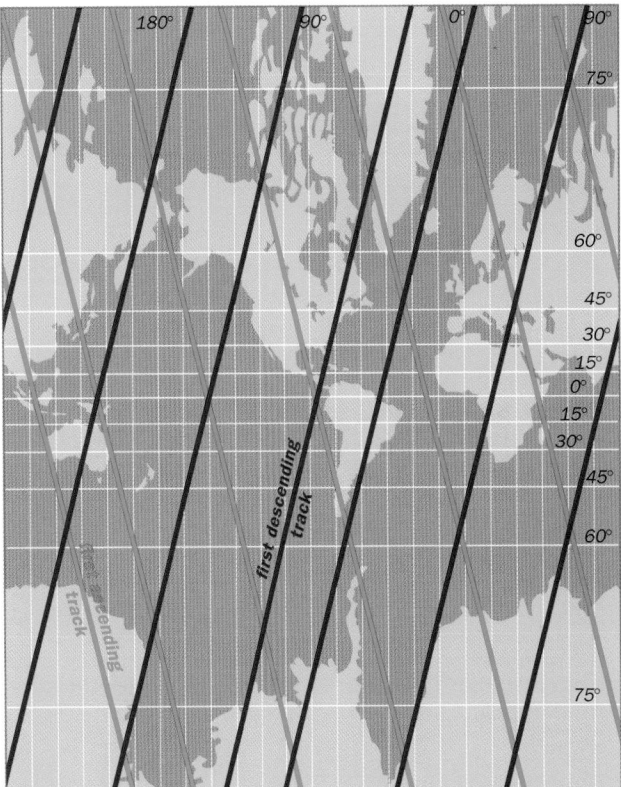

Figure 3.14. **Direct secant cylindrical projection at 30° N and S.** The figure shows Landsat-1, -2 and -3 tracks every 45 passes. The descending tracks form an angle with the succeeding ascending tracks at the latitude which marks the limit of coverage (not shown here). The bisector of this angle is perpendicular to the parallels (see J.P. Snyder, 1987, page 232).

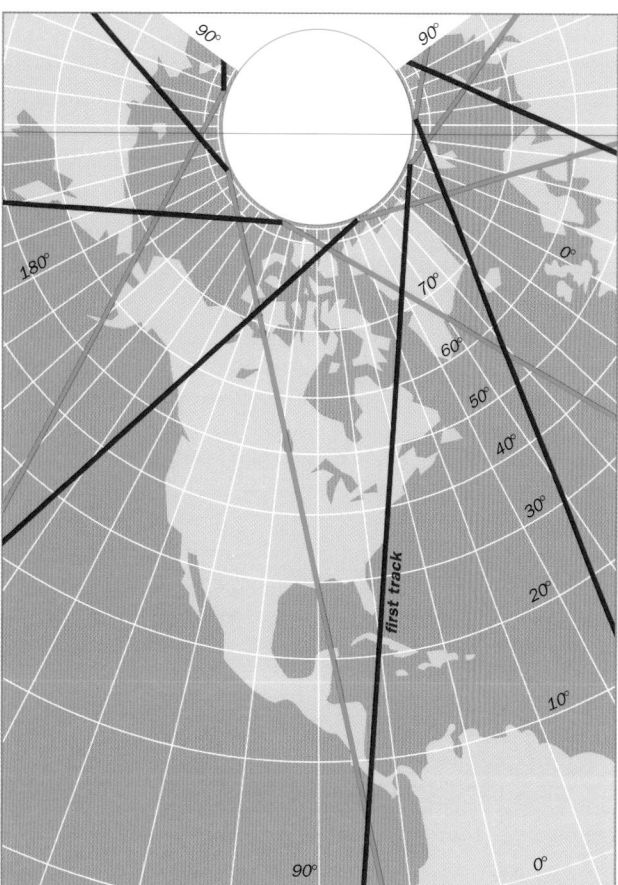

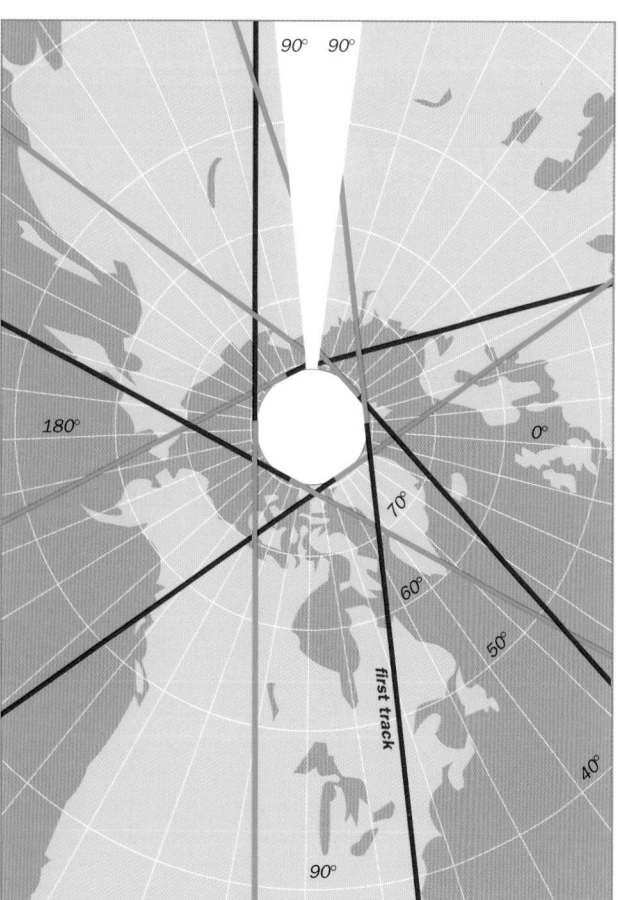

Figure 3.15. **Conic projection preserving conformality at 45° and 70° N.** The figure shows Landsat-1, -2 and -3 tracks every 45 passes. An ascending track (green line on the map) forms an angle with the succeeding descending track (red line) at the latitude which marks the northern tracking limit. The bisector of this angle is perpendicular to the projection parallels (see J. P. Snyder, 1987, page 233).

Figure 3.16. **Conic projection preserving conformality at 45° and 80.9° N.** The figure shows Landsat-1, -2 and -3 tracks every 45 passes. The ascending (green line) and descending (red line) track are continuous along a straight line tangential to parallel 80.9° N, which marks the northern tracking limit (see J. P. Snyder, 1987, page 234).

Figure 3.17. **Tangential conic projection at the limit of latitude coverage.** The tracks are the same as in figure 3.16 and exhibit the same continuity between ascending and descending tracks. The projection shown here is very similar to a polar azimuthal projection (see J. P. Snyder, 1987, page 235).

Types of ground track

Circular orbits

Figure 3.18 shows the various shapes described by the ground tracks associated with satellites in circular orbit at a number of inclinations and altitudes. Some of the orbits shown are not used in practice and have been included solely in order to present the variations as a continuum. Taking zero inclination as an example, the only real-case altitude is 36 000 km. Two types of ground track associated with commonly used orbits receive

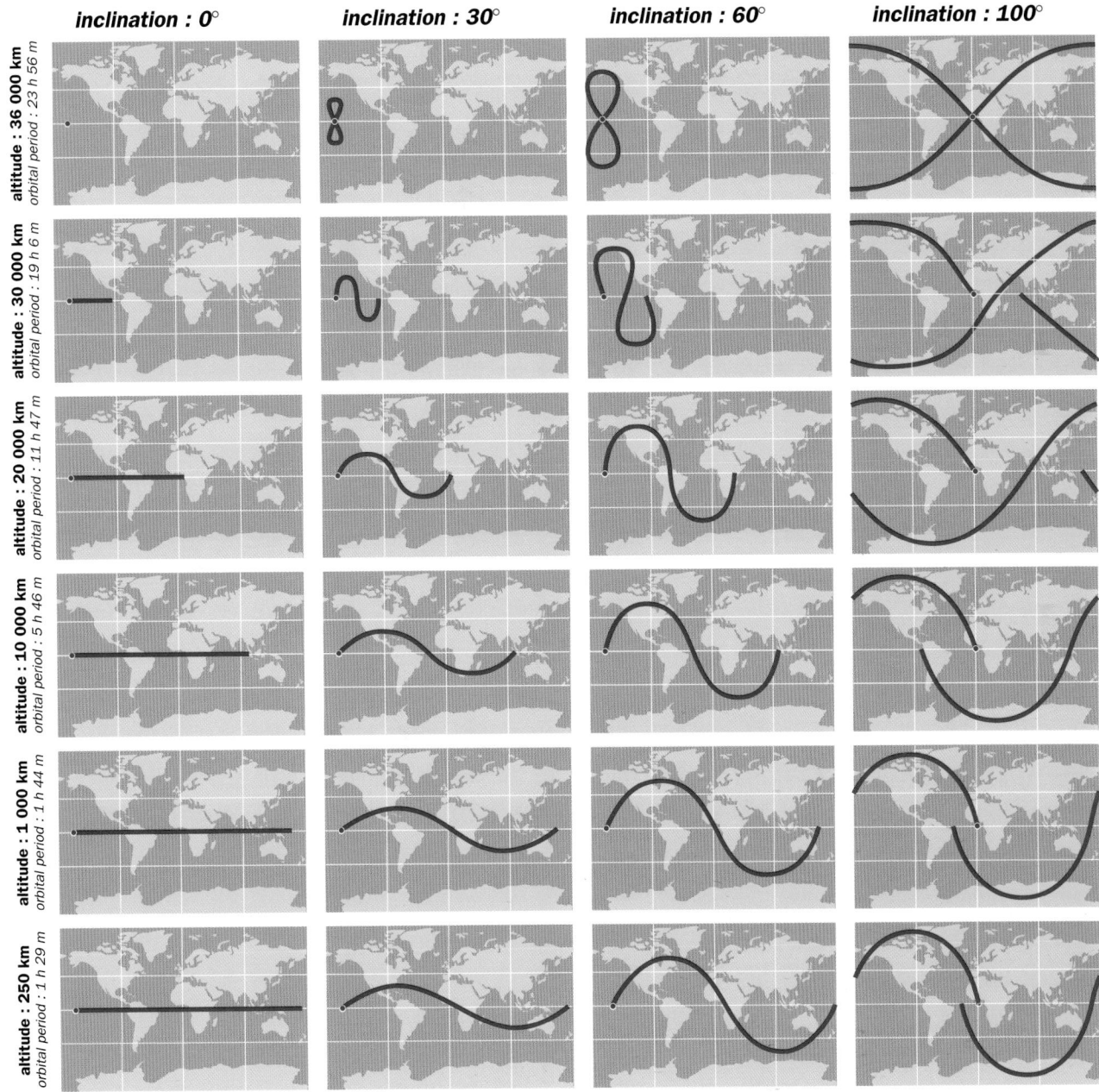

Figure 3.18. Types of track depending on altitude and inclination. The theoretical track corresponding to a single orbital revolution starting from an ascending node crossing has been transposed onto each planisphere. The first planisphere (top left) shows the track of a geostationary satellite. It will be seen from the other planispheres in the same row that a satellite placed at the same altitude as the geostationary orbit but inclined at an angle other than 0°, passes over the same point on the Equator once every orbital revolution. The track is shaped like a figure 8, the height of which depends on the angle of inclination. The satellites whose tracks appear in the first row are referred to as geosynchronous. The tracks of satellites flying at 0° inclination but at altitudes lower than that of the geostationary orbit extend in a straight line coincident with the Equator; the length of the track is a function of altitude (left-hand column).

more detailed treatment, in figs. 3.19 and 3.20. The first shows the ground tracks of a satellite in a medium-inclination prograde low Earth orbit – the class of orbit occupied by most manned spacecraft and space stations. The example given here is that of Skylab.

The second represents the ground tracks described by a satellite in retrograde orbit at an altitude of 600 km to 900 km. This class of orbit is typically that of sun-synchronous remote-sensing satellites whose tracks are of particular importance since it is these which determine the ground area observed.

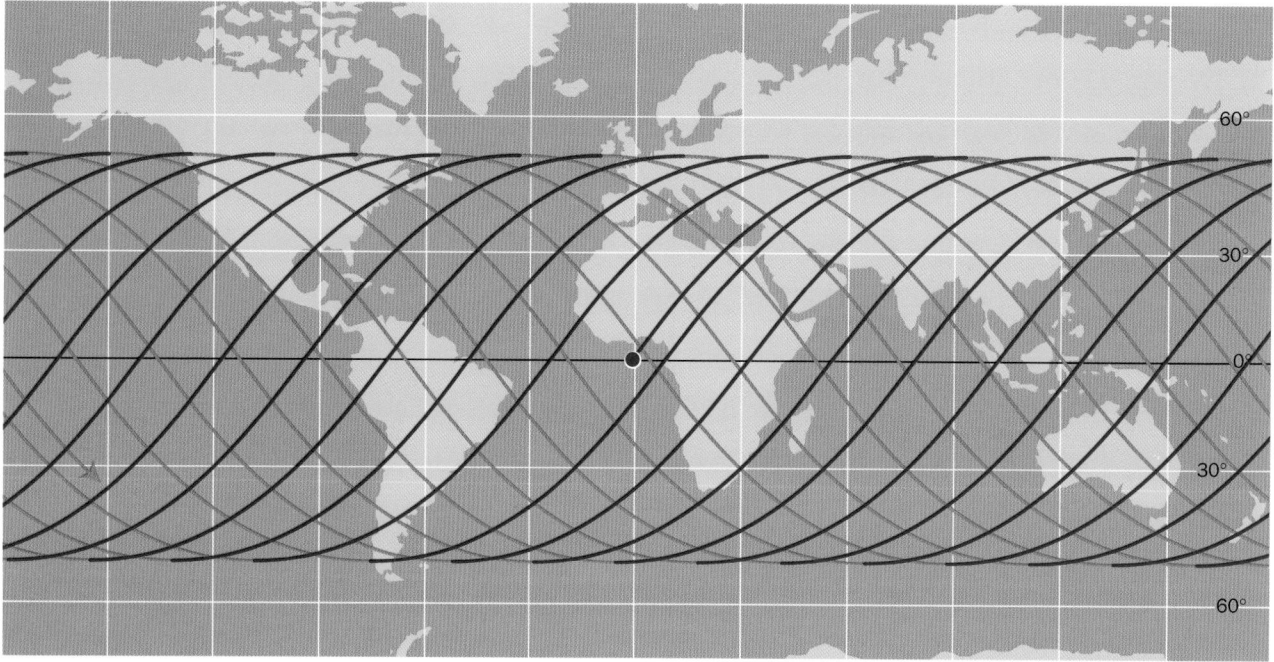

Figure 3.19. **Tracks described by International Space Station over a 24-hour period.** Orbit altitude is 385 km and inclination 51.6°. The orbital period is 92.18 minutes and the distance at the Equator between two successive tracks in the same direction is 2608.4 km.

Figure 3.20. **Tracks described by HCMM over a 24-hour period.** Orbit altitude is 630 km and inclination 97.5°. The orbital period is 97.2 minutes and the distance at the Equator between two successive tracks in the same direction is 2712 km.

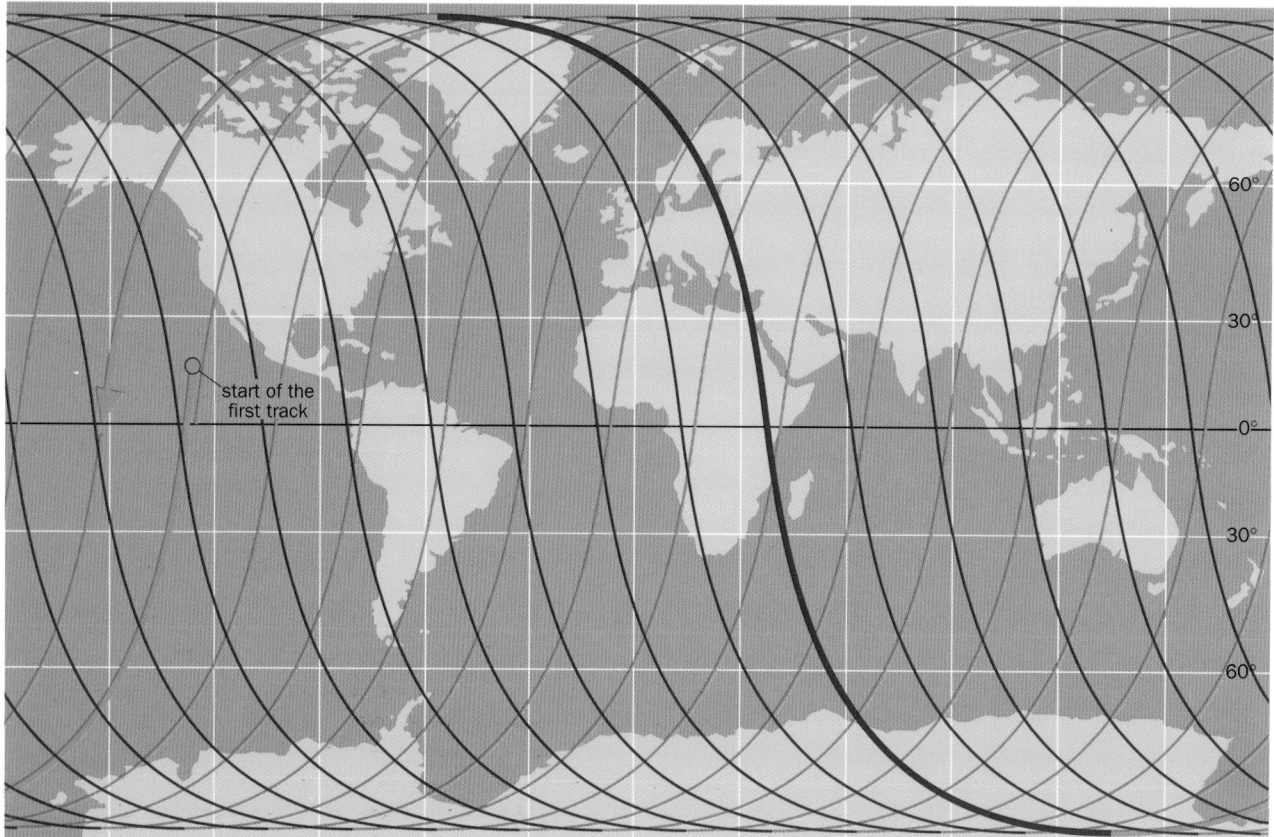

start of the
first track

Eccentric orbits

Eccentric orbits are associated with a wide variety of ground tracks. The four examples presented here concern Soviet satellites. The first (fig. 3.21) relates to an orbit with a medium-altitude apogee. The remaining figures (3.22, 3.23 and 3.24) illustrate the various shapes which may be assumed by the ground tracks of satellites in high-apogee orbits, depending on the argument of perigee.

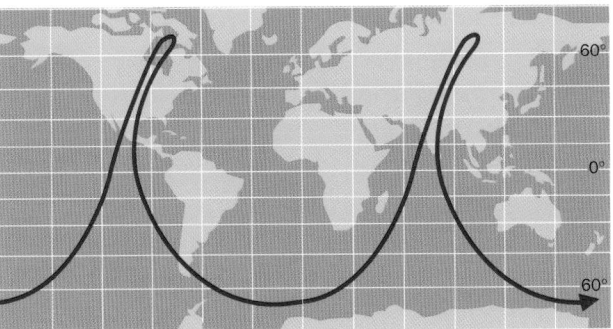

Figure 3.23. **Another example of a Molniya satellite track.** The difference in argument of perigee – 280° in place of 270° – explains the difference in track compared with fig. 3.22.

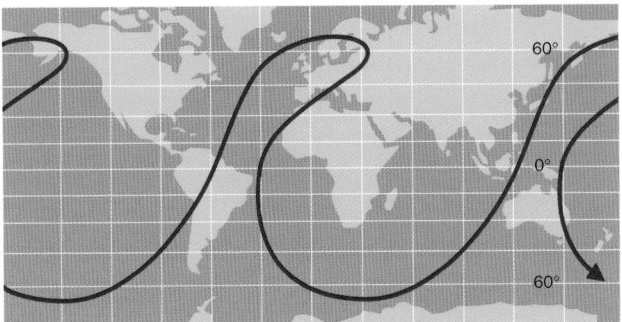

Figure 3.24. **Track described by a military early warning satellite.** The orbit is of the Molniya type but with a 316° argument of perigee. The descending track becomes retrograde between 60° N and 20° S. Orbital positioning is calculated to give apogee passes over areas of strategic interest in the Atlantic (Greenland) and the Pacific (Alaska).

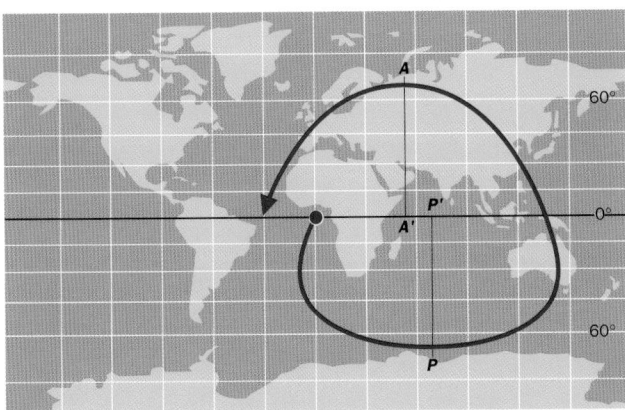

Figure 3.21. **Track described by the Elektron 4 scientific satellite over one orbital revolution.** Apogee and perigee are at 66 300 km and 447 km respectively and orbital inclination is 61°. The orbital period is 21 hours 54 minutes. The track segment located in the northern hemisphere exhibits a different shape than the segment in the southern hemisphere but both take a meridian as axis of symmetry (AA′ to the north, PP′ to the south). This is the case when the orbit's major axis is perpendicular to the nodal line ($\omega = 90°$ or 270°), i.e. when the perigee and apogee correspond to the highest latitudes on the satellite's trajectory.

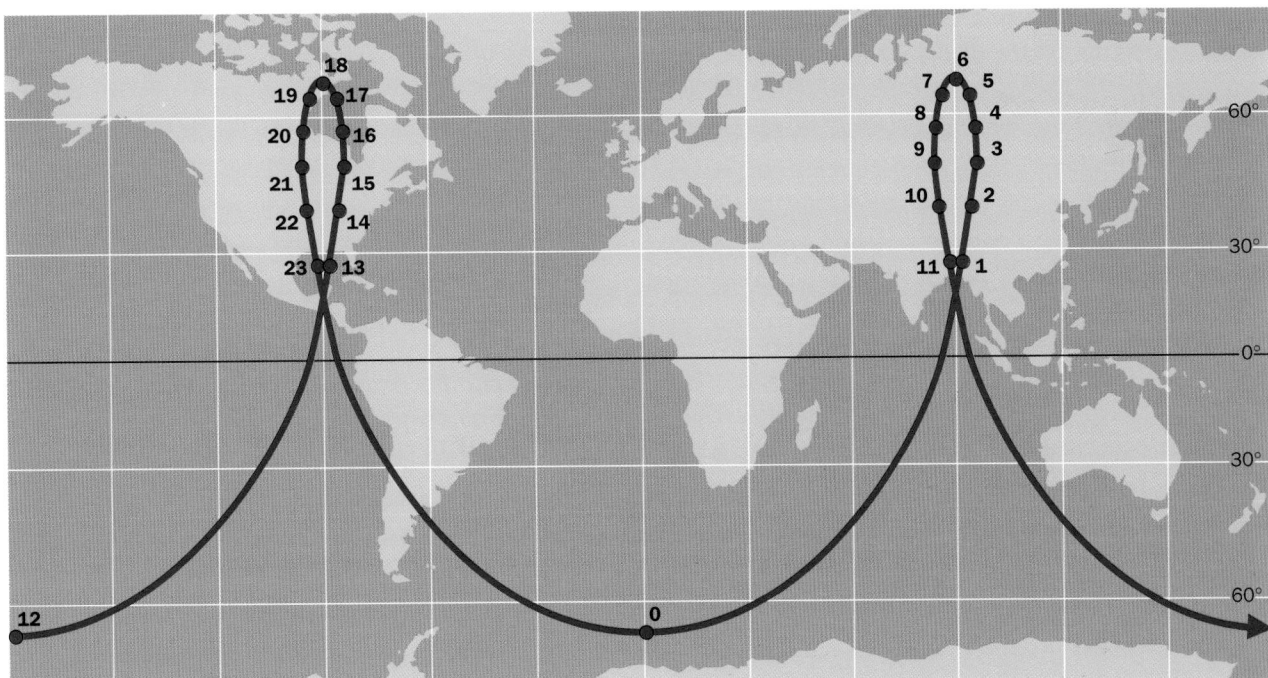

Figure 3.22. **Track described by a Molniya telecommunications satellite over a 24-hour period.** Perigee and apogee are at 400 km and 40 000 km respectively and orbital inclination is 65° with an argument of perigee of 270°. At about 50° latitude the satellite's eastward progress becomes slower than the Earth's rotation; the track becomes retrograde and describes a loop. It can be seen from the points overflown by the satellite hour by hour that an orbit of this kind will allow it to stay over Russian territory for more than 8 hours. Thus only three satellites succeeding each other at 8-hour intervals are needed for one to be in sight at any time.

Ground tracks and orbital cycles

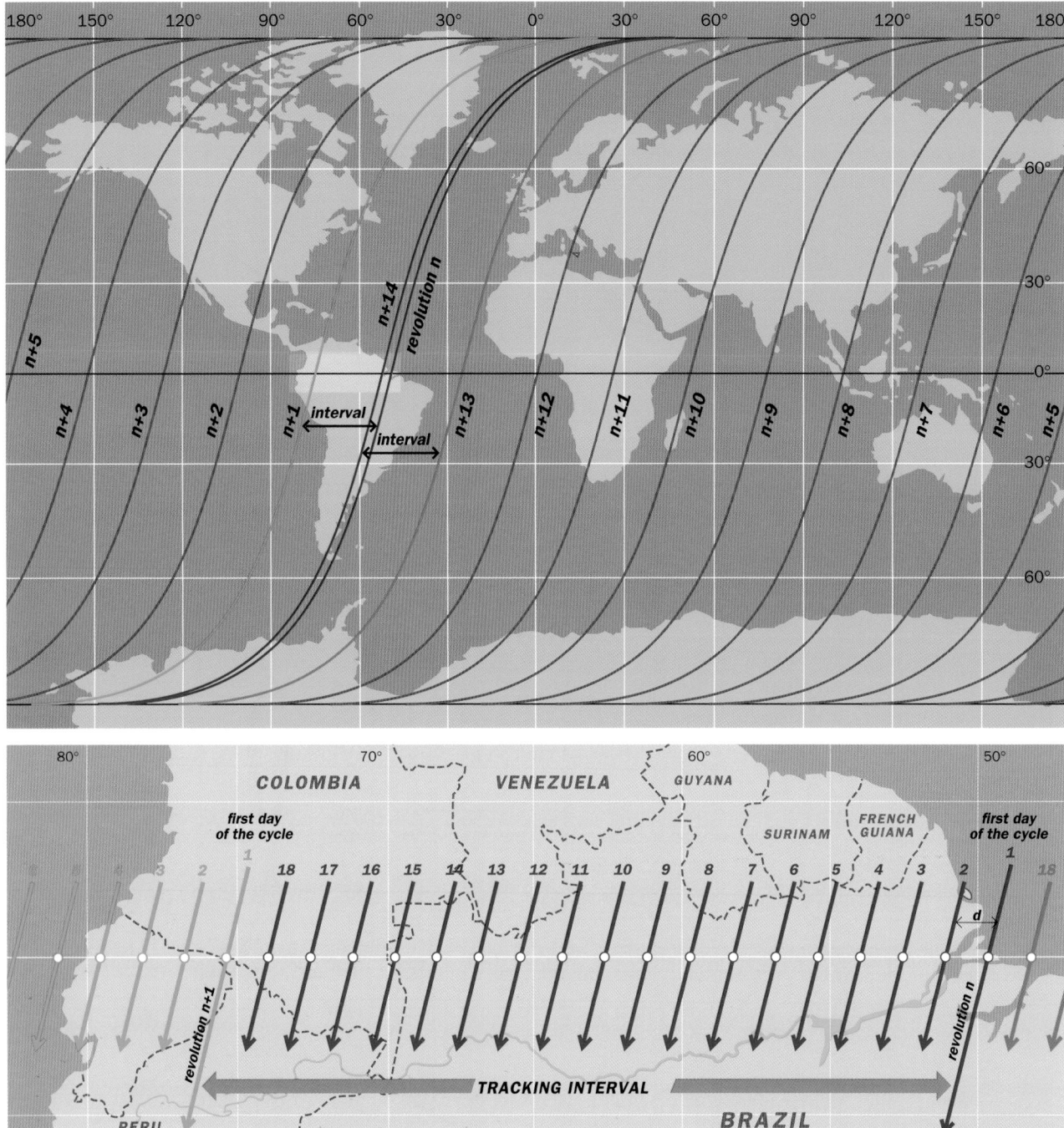

Figure 3.25. **Succession of descending tracks over a Landsat-1, -2 and -3 orbital cycle.** Above: descending tracks during the first day of the cycle. The distance between the track described in one revolution and the track described in the subsequent revolution is known as the tracking interval. On completion of 14 revolutions, track $n+14$ is located to the west of track n since the combined length of 14 tracking intervals exceeds that of the Equator. Track $n+14$ is executed on day two of the cycle. Below: descending tracks over the 18-day orbital cycle. On track $n+14$ the descending node pass occurs at a point on the Equator located at distance d to the west of track n. This is because 14 tracking intervals exceed by distance d the length of the Equator. The cumulative effect of these daily offsets over the 18-day cycle is to leave the track on day 18 again coincident with the track on the first day – since d divides exactly 18 times into tracking interval P.

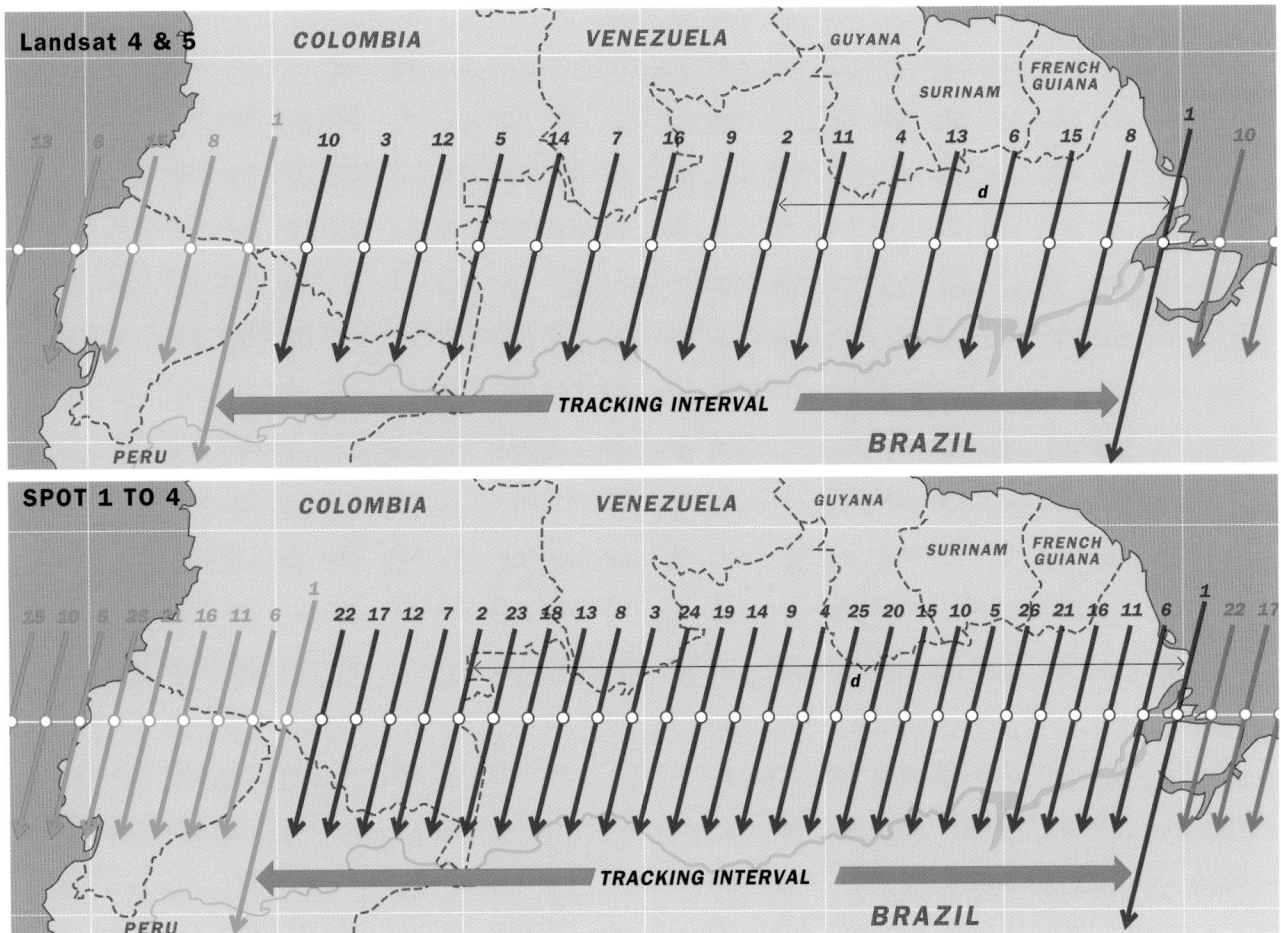

Figure 3.26. **The succession of descending tracks over a Landsat 4 and 5 and SPOT 1 to 4 orbital cycle.** Value **d** does not divide exactly into value **P**.

By *orbital cycle* is meant the whole number of orbital revolutions which a satellite must describe in order once again to be flying in the same direction over the same point on the Earth's surface. This concept is of special importance for remote-sensing satellites in sun-synchronous circular orbit with their requirement to provide full observation coverage of the Earth with the exception of areas at very high latitudes which their inclination prevents them overflying.

In the case of Landsat-1, -2 and -3 (fig. 3.25), the orbital cycle can be analysed as follows. In the course of orbital revolution n, the satellite crosses the equatorial plane at the descending node above a given point on Earth. At the following pass, the track shifts to the west by distance P – the tracking interval (see fig. 3.12) – which at the Equator is 2874 km. The value of P is determined by the Earth's rotation and the nodal precession for the orbit concerned (see page 14) over one revolution.

Fourteen revolutions – i.e. 24 hours and a few minutes – later the ground track is now located to the west of the track of the first revolution. The cumulative effect of these daily offsets is for the satellite to overfly the same point on Earth in the same direction once every 251 revolutions: $(18 \times 14) - 1 = 251$. This is the duration of the satellite's orbital cycle.

In the case described above of the first three Landsat satellites, the offset value d is an exact divisor of tracking interval P.

With some satellites, the track adjacent to the initial track is offset to the east after approximately one day. Here the tracks are still numbered monotonically but in descending order towards the west (as for example the three-day cycle of the European ERS 1 satellite). In yet other cases, d is not a sub-multiple of P and the tracks are not ordered monotonically, examples being Landsat-4 and -5 (fig. 3.26) and HCMM. In the latter instance, the track sequence is as follows: 1, 12, 7, 2, 13, 8, 3, 14, 9, 4, 15, 10, 5, 16, 11, 6.

In the case of SPOT 1 to 4 (fig. 3.26) the sequencing has the advantage of allowing the image of a given scene to be acquired in a period of five days at the Equator (and in shorter periods at the highest latitudes). This is because the sensors carried by SPOT are adjustable for oblique viewing and can thus be directed at regions located either side of the satellite nadir within a 950 km observation corridor – wider therefore than seven adjacent tracks at the Equator.

Regularity of cycle is a highly valued feature for Earth observation purposes and the orbital elements of sun-synchronous remote-sensing satellites are monitored and adjusted to ensure that cycles coincide as closely as possible. It is this feature which also allows image catalogues to be produced in which an orbit number is assigned to each track in a cycle.

CHAPTER FOUR

Occupation of space

- **THE GEOGRAPHY OF SPACEBORNE OBJECTS**

- **SATELLITES AND PROBES**

- **CIVILIAN AND MILITARY APPLICATIONS**

The geography of spaceborne objects

The space development process

The conquest of space is of universal interest for humankind. However, it cannot be undertaken without the necessary technical, scientific and financial capabilities. These impose a high level of selection upon potential candidates. In this regard, the simple presence in space of a national satellite, even one launched by another nation, appears essential to many countries, at least as far as their image is concerned.

The idea of power inspired by a mastery of space technology was indeed considerably strengthened in the particular historical context of the Cold War, which held sway when the conquest of space was in its infancy. It was knowhow acquired from intercontinental missile programmes that finally made it possible to gain access to space. Likewise, the first satellite programmes, whether aimed at science, Earth observation, telecommunications or early warning, were essentially related to security issues. From the very start, different nations have nevertheless established different ways of developing the potential that space offers. These approaches depend as much on the internal political situation of the country in question as on their role in international politics. Besides those countries for which security has played an overriding role (the United States and ex-USSR), others particularly in Europe have chosen to favour a synergy between civilian and military applications. Such applications are often referred to as dual-use or dual-purpose. Still others, like Japan or India, have opted for a purely civilian approach, although this situation may change at a later date.

A brief glance at the way space programmes have been set up in the main countries reveals from the very outset fundamental differences in the structure of the space sector. On the American side, a clear attempt was quickly made to differentiate between civilian and military activities, whilst in the Soviet Union, consequent to the special status of new technologies, space development was integrated into a military–industrial complex. Symbolically, the Vanguard launch vehicle developed by the US Navy was chosen to launch the first American satellite, nicknamed 'the grapefruit'. This launch vehicle belonged to a tradition of cooperation with the scientific community for experiments carried aboard probes. For their part, the Russians were intent on optimising their skills in the use of a single launch vehicle, the R7, directly derived from their intercontinental missile programme. In the end, after the failure of Vanguard, Explorer 1 was launched by the Jupiter C rocket, a product of the US Army's intercontinental missile programme. The respective masses of Sputnik 1 and Explorer 1 thus resulted from the

technical dictates of availability and knowhow (fig. 4.1). The mass of the first Russian atomic bomb, which required a very powerful launch vehicle, gave them the means to place heavy loads into orbit, whilst American ability to miniaturise systems, which proved to be the more fruitful in the longer term, looked on the face of it like an affirmation of weakness.

The growing use of civilian applications, such as those in telecommunications, combined with the gradual spread of technological and industrial capabilities, explains the ever-increasing presence of other nations in space. Notwithstanding, the original objectives, closely tied to prestige and image on the international scene, are still a relevant factor. Even if certain applications, such as military activities or manned flights, are still almost exclusively controlled by the United States and Russia, the ideas of sovereignty and independence remain key factors in the overall exploitation of space (see chapter 5).

Most of the steps taken towards the conquest of space (fig. 4.2) are marked by the appearance of new satellite families, each corresponding to some particular programme. There was a short period, mainly involving scientific satellites with apogees lying between 1000 and 1500 km, which was primarily aimed at acquiring a better understanding of the medium but also knowhow in the problem of launching probes beyond Earth orbit. The occupation of space subsequently intensified and diversified throughout the 1960s. This new exploratory phase saw the first programmes for human space flight, which symbolise the notion of space conquest. New vehicles were designed, culminating in the development of the Soyuz module and Apollo cabin. In parallel, the first commercial applications began to take shape, especially in the telecommunications area. Almost at the same time, the first satellites belonging to European countries, rather than the USSR or the United States, were making a timid appearance. These were French and British experimental programmes. At this time, France became the third space power, possessing its own launch capability in the form of its Diamant rockets.

In the 1970s, a second phase of developments began. The lunar exploration programme came to a close, although it had fallen far short of a complete conquest of the Moon. This meant that new priorities could be set up, turned towards satellites with specific applications and the commercialisation of space-based services. Manned programmes no longer offered any genuine novelty but, in the meantime, the space club had already considerably expanded. Many countries now possessed their own satellites, and some

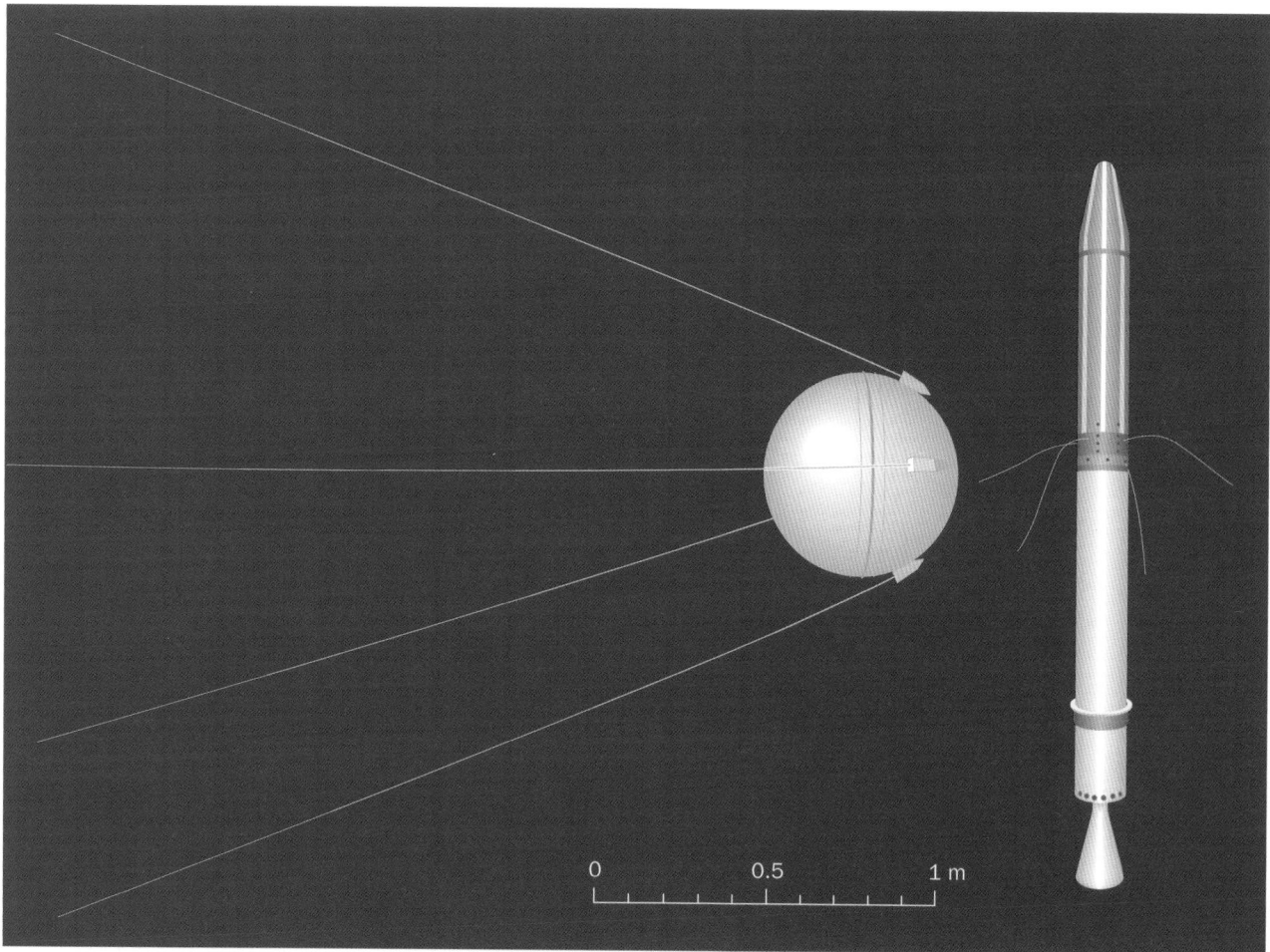

Figure 4.1. **The first two artificial satellites, from the Soviet Union and the United States.**

Left: Sputnik 1, with diameter 58 cm and mass 83.6 kg, circled around the Earth 1400 times between 4 October 1957 and 4 January 1958, on an orbit with perigee at 228 km and apogee at 947 km.

Right: the first American satellite Explorer 1 comprised a payload (the striped cylinder at the top) of length 75 cm, and a solid propellant rocket motor. The whole unit constituted the fourth stage of a Jupiter rocket, 2.04 m long and weighing 13.9 kg. It was launched into an orbit at 356–2540 km on 31 January 1958. Emissions failed on 23 May 1958 and it finally fell to Earth on 31 March 1970.

countries such as Japan, China and India were even capable of carrying out their own launches. At the same time, the increasing military presence in space remained the prerogative of the two major powers.

The 1980s can be characterised by the arrival of new technologies. The Space Shuttle in the area of manned space flight perhaps represents the best illustration. At the same time, the USSR was able to maintain a continued human presence in space aboard their space station Mir. Whilst satellites with specific applications were becoming more and more diverse, countries and groups of countries began to enter into competition with the United States. The SPOT satellites are witness to this new trend. Their images were sold for the first time by a private company, Spot Image. The deregulation of telecommunications induced similar effects and even brought in private operators. Finally, in the domain of military applications, the American 'Star Wars' programme arose from a more open militarisation, at precisely the moment when the United States and the Soviet Union were stressing the role played by reconnaissance satellites in the context of disarmament talks.

The 1990s, in their turn, were marked by the consequences of the collapse of the Soviet Union. This left the United States in first place, especially with regard to military activities. They endeavoured to preserve their advantage through a kind of cooperation closer to subcontracting than well balanced international relations. Competition became tougher in the commercial, telecommunications and launch sectors, not to mention the area of Earth observation, whilst the number of space powers continued to grow. In other areas, savings were the order of the day, with serious consequences for a great many scientific programmes that were either shelved or simply cancelled. Competition and collaboration both progressed without becoming mutually exclusive. And finally, the dual-use approach with its emphasis on combined military and civilian financing, took on a new dimension.

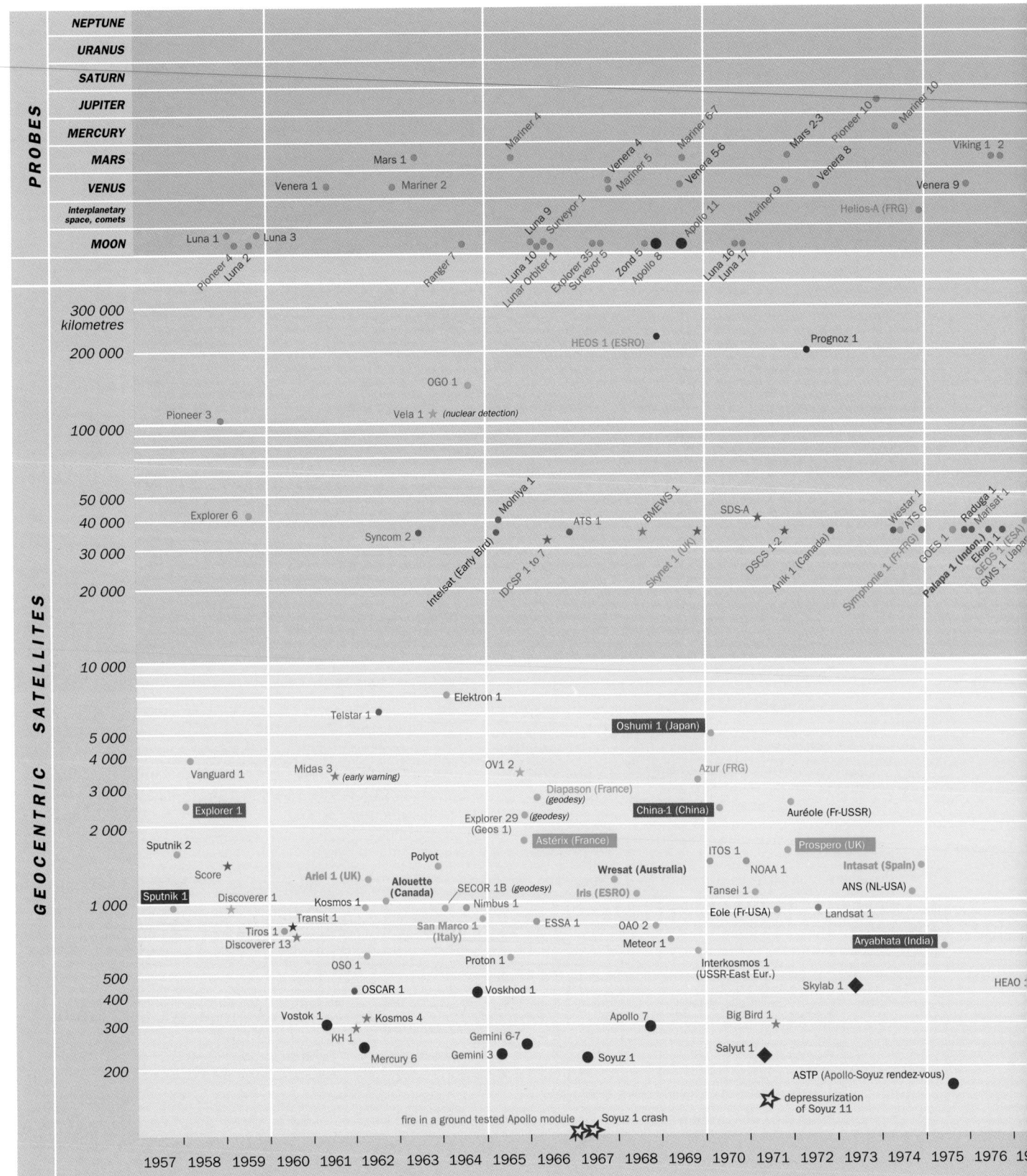

Figure 4.2. **Summary of the main events in the history of space between 1957 and 2000.** Satellites are represented by their altitude and apogee. In the case of probes, dates indicate the time of closest passage to the explored planet, rather than the time of launching.

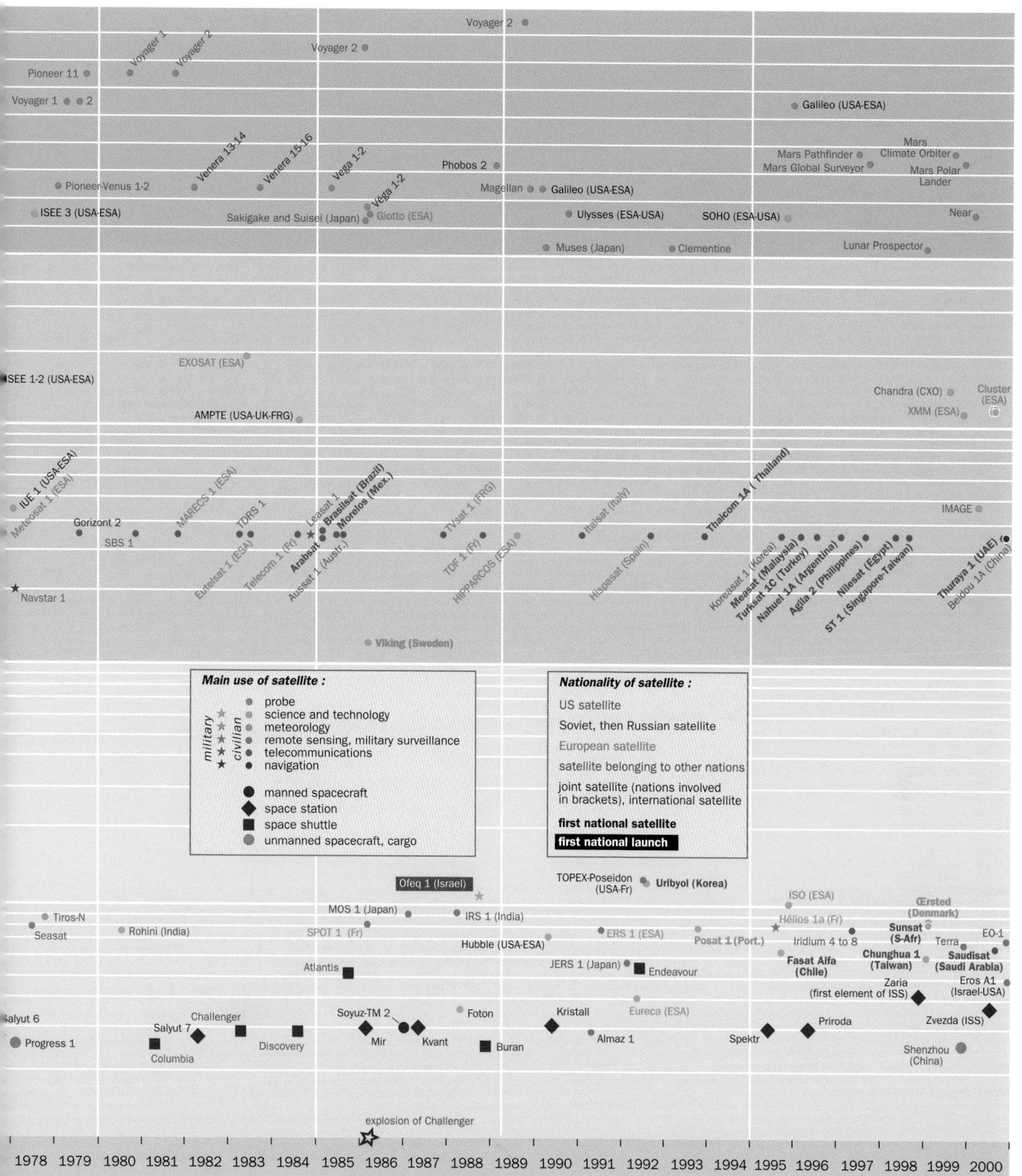

Voyager 2
Voyager 2
Pioneer 11
Voyager 1 ● ● 2
Galileo (USA-ESA)

Mars Climate Orbiter
Mars Pathfinder
Mars Global Surveyor
Mars Polar Lander

Venera 13-14
Venera 15-16
Vega 1-2
Phobos 2
Pioneer-Venus 1-2
Vega 1-2
Magellan   Galileo (USA-ESA)
Near
ISEE 3 (USA-ESA)
Sakigake and Suisei (Japan)   Giotto (ESA)   Ulysses (ESA-USA)   SOHO (ESA-USA)

Muses (Japan)   Clementine   Lunar Prospector

EXOSAT (ESA)
ISEE 1-2 (USA-ESA)
Chandra (CXO)   Cluster (ESA)
AMPTE (USA-UK-FRG)   XMM (ESA)

IUE 1 (USA-ESA)
Meteosat-1 (ESA)
Thalcom 1A (Thailand)
IMAGE
MARECS 1 (ESA)
Gorizont 2   TDRS 1   Leasat 1
Brasilsat (Brazil)
SBS 1   Morelos (Mex.)   TVsat 1 (FRG)   Italsat (Italy)   Koreasat (Korea)
Eutelsat 1 (ESA)   Arabsat   Measat (Malaysia)   Thuraya 1 (UAE)
Telecom 1 (Fr)   TDF 1 (Fr)   Turksat 1C (Turkey)   Nilesat (Egypt)   Beidou 1A (China)
Aussat 1 (Austr.)   HIPPARCOS (ESA)   Hispasat (Spain)   Nahuel 1A (Argentina)
Navstar 1   Aglia 2 (Philippines)
ST 1 (Singapore-Taiwan)
Viking (Sweden)

Main use of satellite :

military / civilian

★ / ● probe
★ / ● science and technology
★ / ● meteorology
★ / ● remote sensing, military surveillance
★ / ● telecommunications
★ / ● navigation

● manned spacecraft
◆ space station
■ space shuttle
● unmanned spacecraft, cargo

Nationality of satellite :

US satellite

Soviet, then Russian satellite

European satellite

satellite belonging to other nations

joint satellite (nations involved in brackets), international satellite

first national satellite

first national launch

Ofeq 1 (Israel)
TOPEX-Poseidon (USA-Fr)   Uribyol (Korea)
ISO (ESA)   Œrsted (Denmark)
MOS 1 (Japan)   IRS 1 (India)   Hélios 1a (Fr)
Tiros-N   Sunsat (S-Afr)
Seasat   Rohini (India)   SPOT 1 (Fr)   ERS 1 (ESA)   Iridium 4 to 8   Terra   EO-1
Posat 1 (Port.)
Hubble (USA-ESA)   Chunghua 1 (Taiwan)   Saudisat (Saudi Arabia)
Fasat Alfa (Chile)
JERS 1 (Japan)   Endeavour   Zaria (first element of ISS)   Eros A1 (Israel-USA)
Atlantis
Salyut 6   Soyuz-TM 2   Foton   Kristall   Eureca (ESA)   Priroda   Zvezda (ISS)
Challenger
Salyut 7   Mir   Kvant   Almaz 1   Spektr   Shenzhou (China)
Progress 1   Columbia   Discovery   Buran

explosion of Challenger
☆

1978 1979 1980 1981 1982 1983 1984 1985 1986 1987 1988 1989 1990 1991 1992 1993 1994 1995 1996 1997 1998 1999 2000

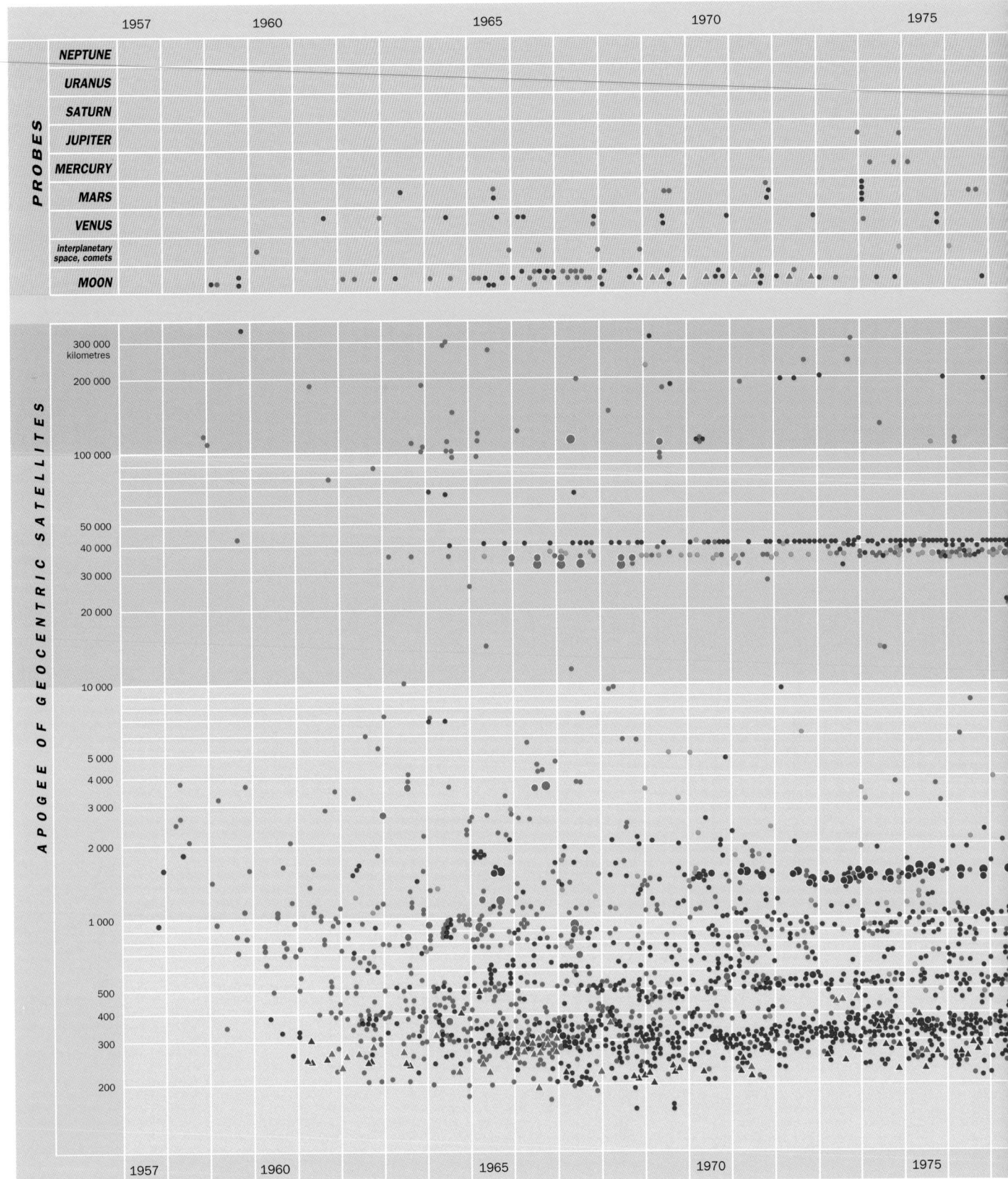

Figure 4.3. **Chronology of probes and satellites up to 1 January 2001.**
Satellites have been represented by their altitude at apogee, and probes by date of arrival at, or closest flyby of their objective.

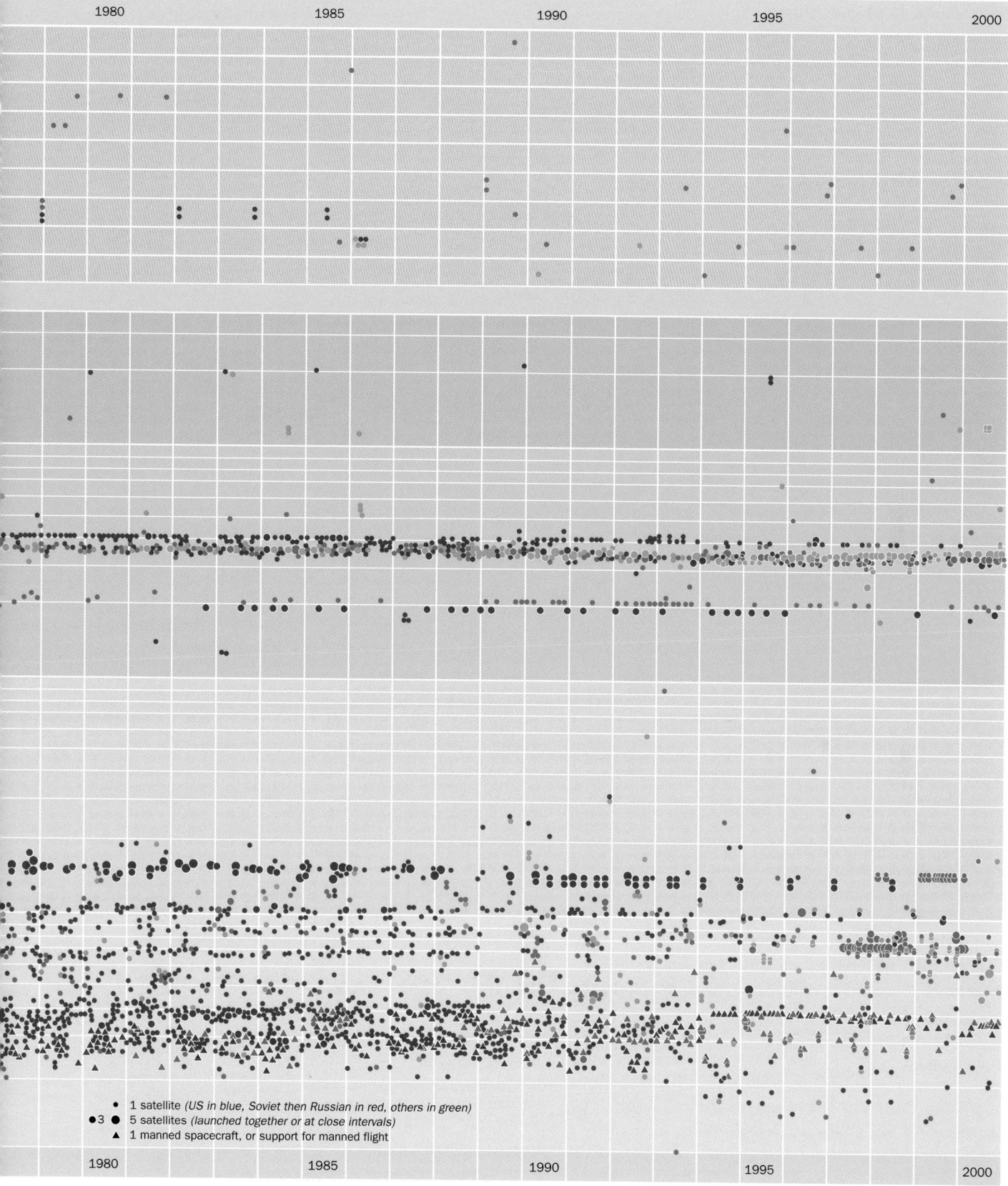

1980 1985 1990 1995 2000

- 1 satellite *(US in blue, Soviet then Russian in red, others in green)*
- 3 ● 5 satellites *(launched together or at close intervals)*
- ▲ 1 manned spacecraft, or support for manned flight

1980 1985 1990 1995 2000

Distribution of satellites at apogee

When the whole cross-section of satellite nationalities is taken into account, the occupation of space has tended to favour two zones (fig. 4.3). The first is located between altitudes of 300 and 1500 km, where many satellites use low orbits for military and civilian Earth observation. Since the end of the 1990s, these have been joined by telecommunications constellations. The second zone, which is narrower and more distant, lies at an altitude of around 36 000 km and is used by geostationary satellites, mainly for telecommunications and television. These satellites are characterised by the fact that they remain immobile relative to the ground stations they serve. Other regions of space are only sporadically occupied, although the race to the Moon led to a wide range of different launches for the purpose at hand.

Different space strategies emerge when satellites are grouped according to their nationality. At low altitudes, the significant presence of Soviet, then Russian satellites up until the end of the 1980s is dominated by frequent launches of rather unsophisticated devices whose lifetime was all the more limited due to their low orbiting altitude. The Kosmos series, covering a wide range of different functions, is strongly represented. However, in terms of apogee, different types of satellite can be distinguished. Large numbers of Kosmos satellites orbiting between 250 and 400 km are mainly concerned with the task of photo-reconnaissance. Some of these are launched at regular intervals to ensure permanent coverage of the whole planet. Others are sent into orbit at times of international crisis to provide fast and accurate information from high-risk areas. The short lifetime of these satellites, ranging from a few days to several weeks, is due to the fact that they must return to Earth with their stocks of photographs.

In the 1980s, the trend was towards longer periods in orbit, sometimes reaching several months, but without reducing the launch rate. This number would drop significantly during the 1990s. This was not only a consequence of longer satellite lifetimes combined with a technical ability to transmit data electronically; it can also be put down to drastic cuts in the Russian military budget.

Other satellites, still known as Kosmos but gravitating at apogees of around 1500 km, and often launched in clusters of eight satellites, are designed for telecommunications activities in the military domain. They complement heavier systems orbiting at altitudes of around 800 km. Scientific and meteorological satellites such as Interkosmos, Meteor and others, form the rest of the occupation of low orbits between 500 and 1000 km, whilst manned flights such as Vostok and Soyuz, launches of space stations Salyut and Mir and access to these take place at the lowest altitudes, between 250 and 300 km.

American presence at the same altitudes is much more limited. Manned flights, less regular than their Soviet counterparts, are likewise located around 200 km, except for the Skylab space station launched at over 400 km in 1973 and Shuttle flights around 450 km up until the explosion of Challenger in 1986. US satellites with apogee between 500 and 1000 km are far less common than the corresponding Soviet satellites and are mainly scientific and military satellites (Discoverer), meteorological satellites (Nimbus, ESSA, TIROS), Earth observation satellites (Landsat) or ocean observation satellites (Seasat). As time goes by, the number of launches decreases. This is explained by a steady reduction in NASA's financial support. Likewise, there are few satellites between 1000 and 1500 km. Apart from several experimental telecommunications systems (Score, Telstar), the main occupants are a few meteorological (NOAA) and navigation satellites (Transit). The first constellations of telecommunications satellites for mobile phones (Iridium and Globalstar) led to a clear densification in the occupation of low orbits between 1997 and 2000. However, the financial problems encountered in the commercial running of these systems have led to other projects of the same type being postponed (Teledesic or Skybridge). As a result, the sudden increase in American launches to these altitudes has become rather unpredictable.

Other countries owning satellites are grouped together in figure 4.3, whether or not they have their own launch capability. In addition, the occasional participation of such countries in collaboration with the United States or the ex-USSR is not featured when they are merely contributing on-board equipment for satellites or space stations. At low altitudes, they have relatively few satellites. These are mainly experimental capsules, and scientific or Earth observation satellites. Increasing use of small satellites, such as the UoSAT platforms of the University of Surrey in the UK, nevertheless reflects an increased presence. This phenomenon is likely to continue insofar as it provides a cheaper way of carrying out national space projects, especially those of developing countries. The absence of military spaceborne activities, despite the occasional exception (China, United Kingdom, France–Italy–Spain, Israel, and others), is clearly visible in the occupation diagrams.

At an altitude of 36 000 km, geostationary satellites constitute the second most densely occupied zone. The main features here are quite different. Soviet, then Russian presence arrived later and is less significant, since part of their telecommunications requirement is provided by satellites in highly eccentric Molniya orbits (400 km, 40 000 km). These provide links for eight hours a day. Owing to their rather unusual trajectories (see fig. 3.22), more of them are needed for the same coverage. Geostationary satellites give permanent coverage of the same region and they also have much longer lifetimes. Today, the system combines geostationary Gorizont or Raduga satellites, capable of relaying television programmes and telephone calls, with Molniya-type internal links.

Geostationary satellites are thus mainly American, all the more so in that the Intelsat satellites have been declared to have international status and therefore feature among the satellites of other countries, although the organisation was principally American until 1982.

The geostationary satellite launches of other nationalities comprise an important share of the international launch market. There is a steady increase in the number of systems belonging to international groups in addition to national telecommunications systems such as Anik (Canada), Télécom (France), Nahuel (Argentina), Nilesat (Egypt), and others, or regional telecommunications systems such as Arabsat, Eutelsat, Asiasat, and others.

The highest altitudes were attained with the conquest of the Moon. The rather symbolic value of the Apollo programme is witnessed by the sudden drop in American missions following the success of the Apollo flights. The Soviets for their part also brought their unmanned lunar exploration to a close just three years later, in 1976. Arriving on the scene at a later date, with a rather more restricted form of exploration, Japan remains the only other country to be represented at this altitude.

The distribution of satellites in terms of apogee thus reveals different space occupation strategies and mirrors the financial and technical vicissitudes of the various space programmes. Notwithstanding, any interpretation must bear in mind other factors. Since satellite lifetimes are not featured, it would be wrong to associate number and efficacy, particularly as far as military applications are concerned.

The fact remains that space offers different resources at different altitudes and that these are unequally exploited by the various countries depending on their means and national priorities.

Distribution in the Clarke orbit

Geostationary satellites are technically more difficult to launch and this explains why this region of space was only occupied later (fig. 4.4). NASA successfully achieved a first launch in 1963 and managed to launch another geostationary satellite each year up until 1966. The first Soviet geostationary satellite only made its appearance in 1974, and until 1979, only one or two further devices were launched each year. Europe has only been able to carry out its own launches since 1981. Since then, the general and systematic use of these satellites for telecommunications purposes has meant that twenty or so have been launched at regular intervals. The time allowed for reservations has thus had to be limited so as to avoid a situation in which the most popular positions are blocked for excessively long periods. Such positions include the Asia–Pacific zone or European zones, bitterly disputed for television services (fig. 4.5).

This increased density has produced its own problems. A geostationary satellite can be maintained in orbit almost indefinitely, whereas it can only be kept in operation for a limited time owing to wear and tear on the system and depletion of the propellant required for stationkeeping. In fact, the concerns regularly expressed about future use of this orbit during international conferences generally relate to the risk of interference rather than the physical proximity of the objects in question. Since the orbit measures about 265 000 km in length, two satellites separated by a longitude of 1° are actually 735 km apart.

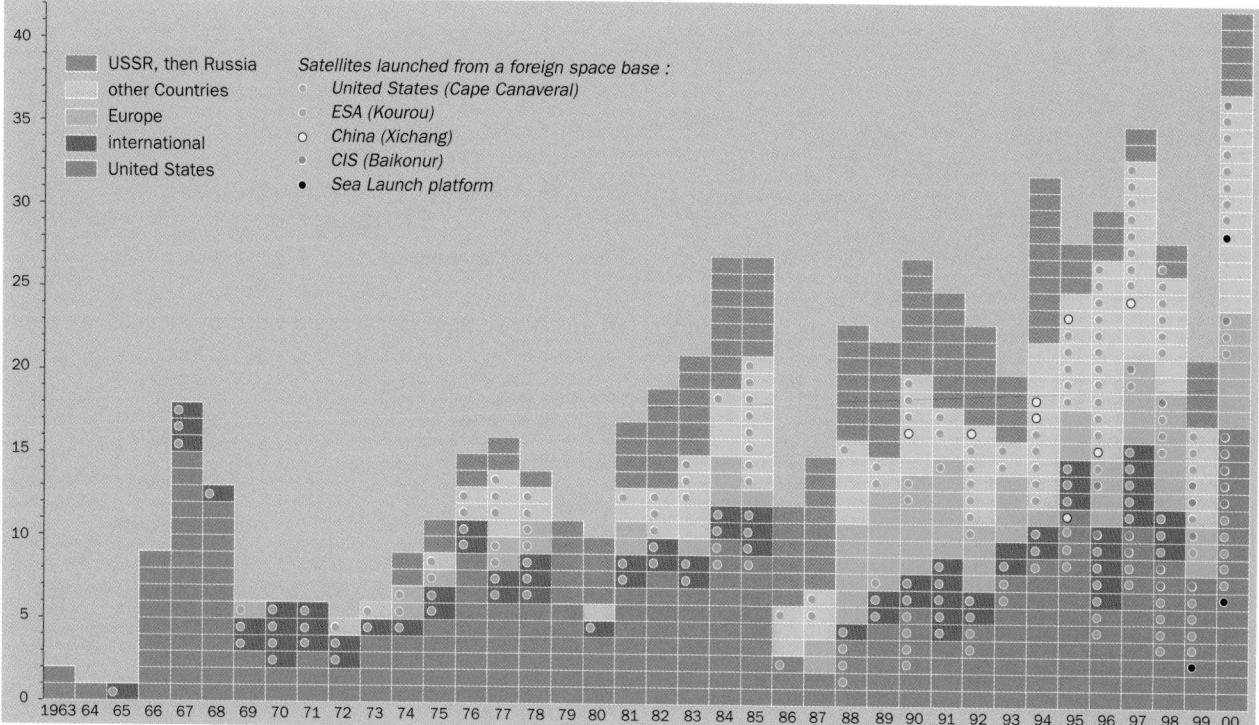

Figure 4.4. **Number of geostationary satellites sent into orbit each year according to country or group of countries.**

Space debris

The gradual crowding of space by dead satellites and various kinds of debris is a growing problem for both geostationary satellites and future low-Earth orbiting constellations. This problem is dealt with by international organisations such as the COmmittee on Peaceful Uses of Outer Space (COPUOS) or the International Telecommunication Union (ITU), but also by more specialised committees. One such is the Committee on Space Debris, made up of experts of different nationalities and set up in 1993 on the initiative of the American Academy of Sciences to assess the medium and long term consequences of this congestion. Another is the Inter-Agency Space Debris Coordination

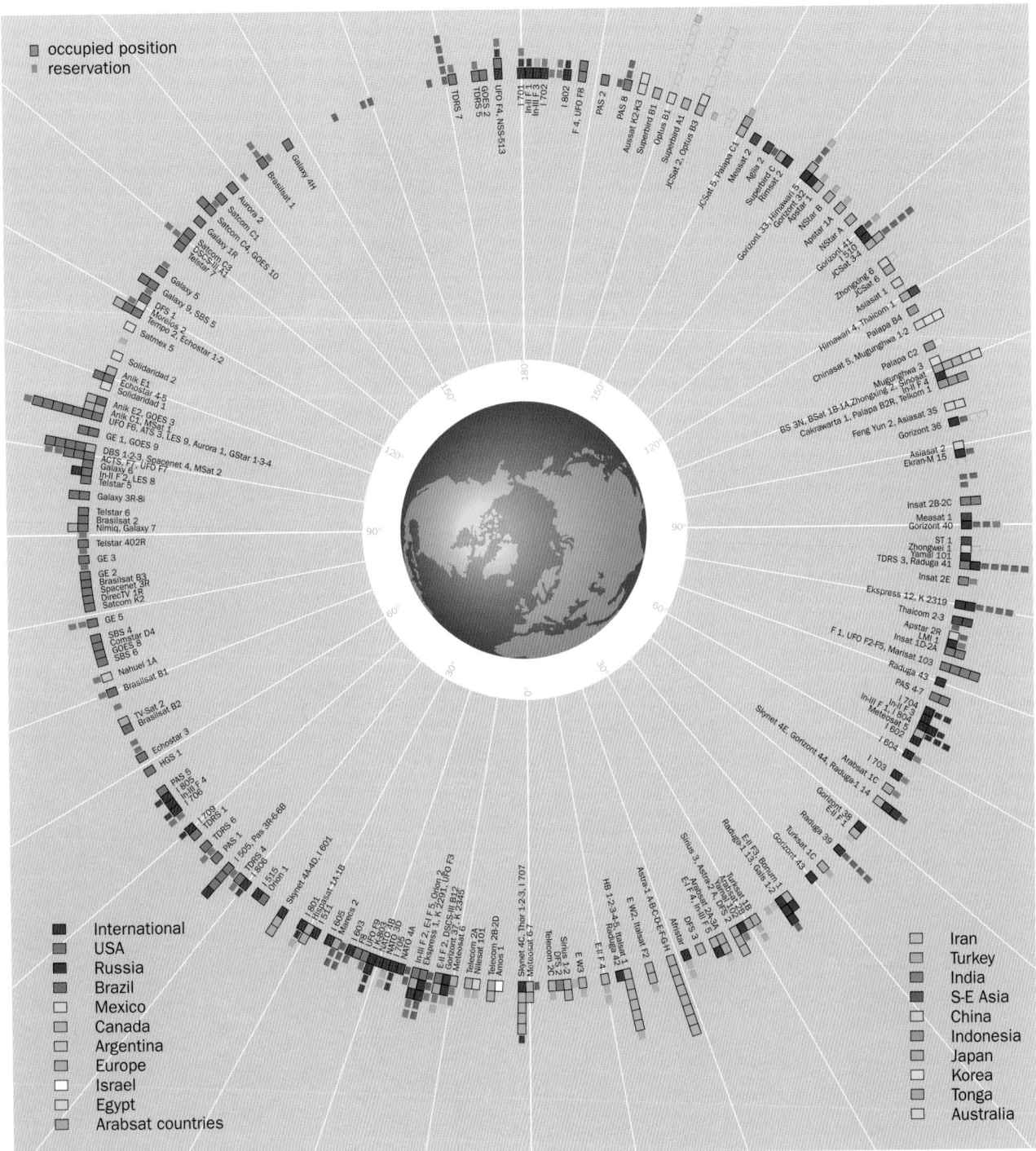

Figure 4.5. Orbital positions occupied as of 18 October 1999.
Squares edged in black represent satellites in service or in reserve at the given date. The positions of certain US military satellites are not known and they are not represented (telecommunications satellites DSCS 3A2, B4, 5, 7, 9, 10, 13, 14, early warning satellites DSP 5R 14, 15, 16, 17, 20 and ELINT Vortex 6, Magnum 1 and 2, Mercury 1 and 2, Advanced Orion 1 and 2).
Squares with no border represent notifications delivered by the ITU but not yet occupied by a satellite at the date the chart was compiled.
Abbreviations : E Eutelsat, F Fleetsatcom, HB Hot Bird, I Intelsat, In Inmarsat, K Kosmos.

Committee (IADC) which groups together the world's space agencies: NASA, RKA, ESA, CNES, BNSC, NASDA, CNSA, ISRO, ASI and DLR.

Only the United States and Russia monitor satellites and space debris, the latter with the means at the disposal of the republics making up the CIS. No other countries possess this capability for space surveillance. Spaceborne objects are monitored, using various techniques such as radar and telescope, by the *North American Air Command Defense* (NORAD) and its Russian counterpart the *Sistema Kontrolia Kosmitcheskogo Prostranstva* (SKKP). The need for global coverage requires a genuine network involving the cooperation of allied countries and also ships with specialised equipment, especially important in the case of the ex-USSR. In the present context, the break-up of the USSR with its serious budgetary difficulties means that Russian capabilities have undoubtedly diminished, although it continues to produce ephemerides and a precise catalogue of orbiting objects.

The limiting detection capacity of NORAD is 10 cm for low orbits and 1 m at 36 000 km. A total of 24 000 manmade objects have thereby been detected and catalogued. In 2000, of the 9000 objects monitored in space, more than 2700 correspond to satellites and of these, only about 500 are still active. Debris has many origins (fig. 4.6). Almost half is made up of fragments from launch vehicles, arising from the normal process of separation between the payload and the injection stage and also from accidental explosions or the deliberate destruction of satellites, especially military satellites. The latter may be caused by antisatellite tests or the deliberate explosion of reconnaissance satellites that have gone out of control.

In 1999, 51% of the objects still in orbit were of American origin and 42% were of Russian origin. This may seem surprising given that, since 1957, the Russians alone have been responsible for putting more than two thirds of the objects into orbit (fig. 4.7). This situation can be explained by certain characteristics of the launch vehicles and especially by the different altitudes at which the Russians generally station their satellites. Debris orbiting at low altitude burns up quickly in the atmosphere, whilst material circulating at 36 000 km can be considered to have an almost unlimited lifetime.

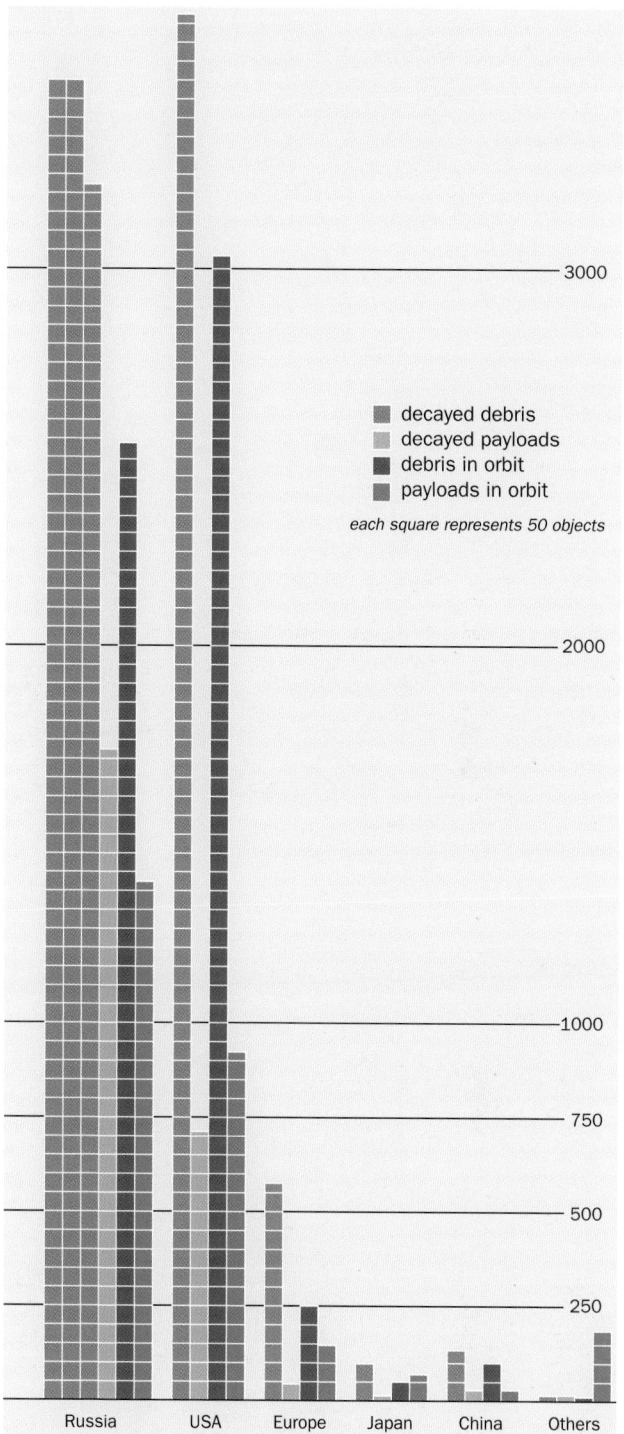

Figure 4.7. **Objects placed in orbit since 1957 according to country or group of countries.**

Debris resulting from a launch is attributed to the launching country. This explains why almost no debris falls under the heading 'other countries', since this group includes all countries or organisations that do not possess their own launch vehicle. The graph represents the state of objects in orbit as of 31 December 1999, as recorded by the *United States Space Command*. Only debris measuring more than 10 cm for low orbits and 1 m for geostationary orbits has been taken into account.

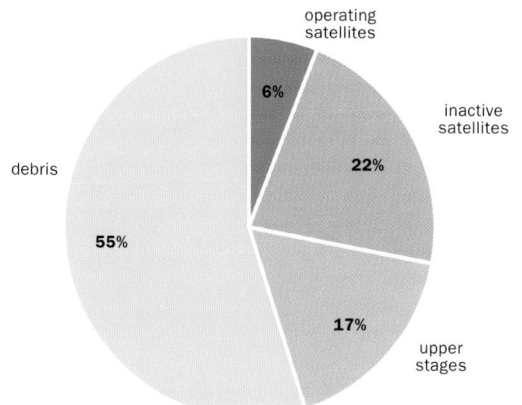

Figure 4.6. **Distribution of debris of size 10 cm or greater in 1999.**

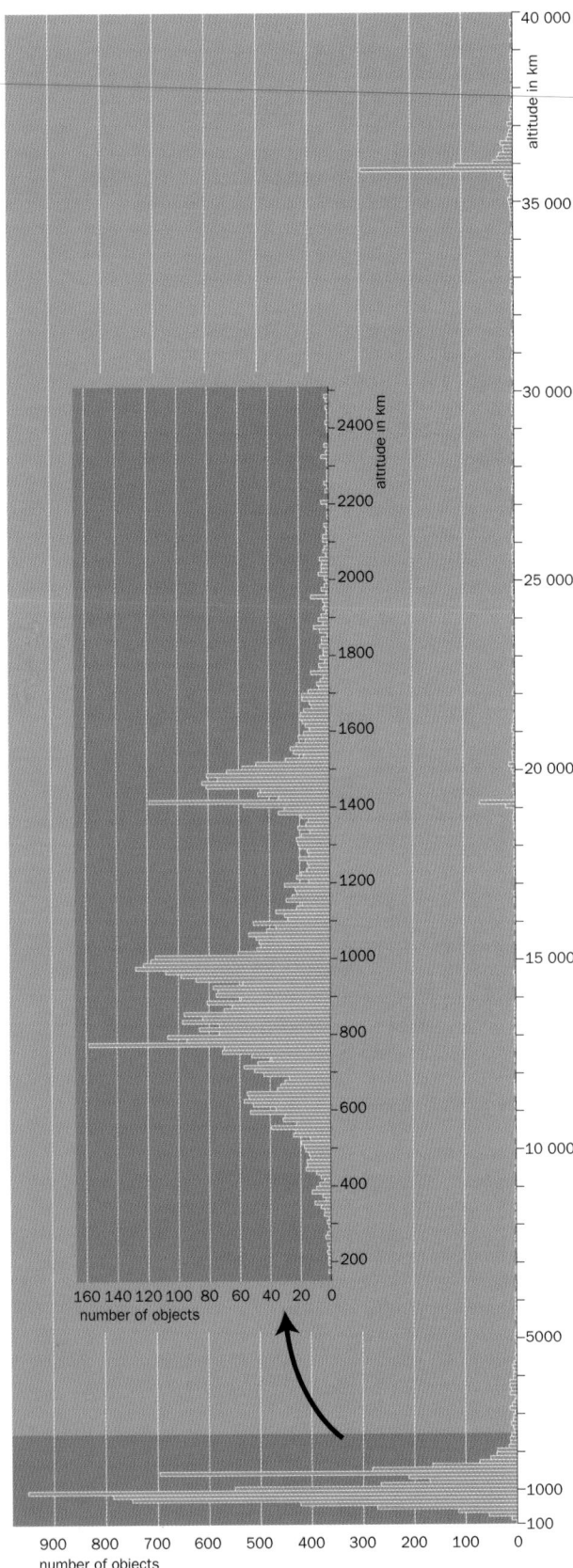

The nature of space debris also differs according to the mean altitude at which it orbits. For the main part, this debris orbits below 1600 km, that is, at altitudes typically used by unmanned research and application satellites but also by manned missions (fig. 4.8). Statistics concerning the probability of collision show that, although the threat is small at the present time, it remains a serious issue. Today, a satellite is more likely to collide with a piece of orbital debris than with a small meteorite. The risks due to these two sources are, in fact, rather different. Unlike space debris, meteorites arrive in more or less constant numbers and are fairly evenly distributed throughout space (fig. 4.9).

The increasing number of incidents, or even accidents, on various manned or unmanned satellites (Mir, the Shuttle, a great number of Kosmos satellites, etc.) demonstrates the reality of the danger. Experiments carried out at the *Long Duration Exposure Facility* (LDEF) have revealed a greater number of impacts than predicted. Several difficulties are involved concerning the detection threshold currently attained and also the localisation of debris, not least because the size, direction and eccentricity of orbit may differ even for fragments having a common origin. Present estimates suggest that there may be as many as several billion pieces of debris with size less than 10 cm. However, the risks remain limited to surface impacts, at least as long as the density of such debris is not too high. The relative velocity of debris is a crucial factor in assessing the consequences of a collision.

For the moment, very few technical solutions are available for removing existing debris. Many proposals have been put forward to improve detection of orbiting objects. With regard to this monitoring programme, European countries have been cooperating on a large scale within the framework provided by ESA, making use of European and American data. However, given the present level of technology, most of the effort aims at prevention, through launch vehicle and satellite design. Indeed, some small launch vehicles such as the American Pegasus or missile derivatives such as the Russian Rockot have proved to be particularly harmful to the environment.

Figure 4.8. **Distribution of debris for altitudes up to 40 000 km in 1999.** The inset records debris measuring more than 10 cm, in 10 km altitude bands. The altitude at which debris circulates relates to its origin. The main concentrations correspond to the altitudes of those satellites which generated them (courtesy of CNES, France).

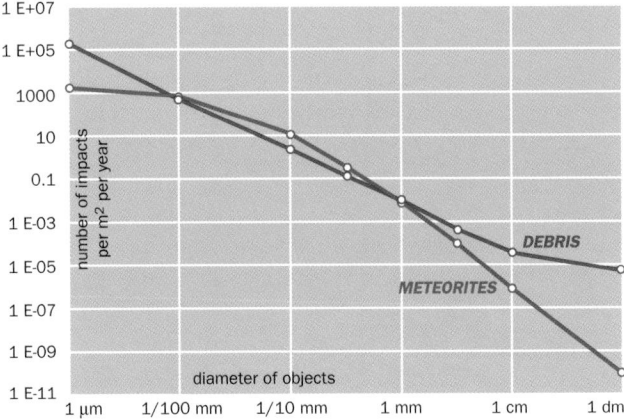

Figure 4.9. **Comparison between the flux of debris and meteorites at altitude 940 km.** In almost every case, there is a greater risk of impacting debris than impacting meteorites.

Satellites and probes

Satellites and probes alike comprise a platform and a payload. Platforms are configured according to function, but also reflect design factors introduced by the manufacturer. Hence, even for their unmanned satellites, the Soviets have almost always built in relatively heavy-duty airtight housing, allowing their instruments to operate under pressure conditions close to those prevailing on Earth. This has meant the need for powerful rockets to launch them. For their part, at least for civilian requirements, the Americans, Europeans and Japanese prefer to use much more sophisticated and lighter unmanned spacecraft, without pressurised housing. On the other hand, certain military satellites are protected against possible neutralising actions, such as nuclear or laser radiation.

Whereas there has long been a trend towards increasing satellite masses, new approaches favouring smaller, cheaper and more quickly assembled satellites are beginning to receive attention. The new type of spacecraft often provides more limited capabilities, whilst being more flexible to use. Applications have recently been developed across the board, although the earlier type of system has not yet been superseded (fig. 4.12).

Platforms include a certain range of subsystems, designed to carry out the satellite mission as determined by its payload. These subsystems are many and varied, with a high level of interdependence.

The correct attitude for the satellite is maintained by a special purpose subsystem. Satellites can be stabilised by a gyroscopic effect, that is, by the satellite spinning rapidly on its own axis. This is the case for the European meteorological satellite Meteosat, for example. In this case, a specially designed device steers the satellite antennas towards Earth stations at all times, despite the satellite rotation. Such antennas are fixed on rotating bearings and are said to be counter-rotating or despun. Other satellites, such as the Landsat satellites, are stabilised by a so-called three-axis system in such a way that they always present the same orientation relative to the Earth. This uses three mutually perpendicular flywheels or momentum wheels which act on the satellite attitude through variations in their angular speed. The correct attitude is then maintained on the basis of astronomical reference observations, made by star trackers, and solar and terrestrial horizon sensors.

Another system is designed to propel the satellite, for example, transforming an eccentric orbit into a circular orbit (apogee motor), modifying the orbit on a temporary basis to carry out some specific mission, changing orbital slot along the Clarke orbit, or even leaving that orbit at the end of the satellite lifetime in the case of the higher orbits.

The energy supply of the satellite is generally provided by a solar generator. Solar panel arrays are a typical feature in the outline of many space vehicles. Fuel cells and nuclear generators are also used. The latter are used in particular for space probes which may find themselves deprived of sufficient solar energy when crossing regions remote from the Sun.

A special purpose subsystem is devoted to the delicate problem of thermal regulation, particularly restrictive in the case of manned satellites. Indeed, a satellite receives solar radiation on the sunward face and reradiates in all directions. In addition, solar illumination is interrupted during eclipses. These present a serious problem for low-orbiting satellites, such as Sun-synchronous remote-sensing satellites whose ascending trajectories cross a significant part of the Earth's umbral cone during each revolution. Exceptions are the dawn–dusk satellites (see fig. 2.29). Geostationary satellites are likewise affected by this phenomenon near the equinoxes (see fig 2.37). In addition to temperature variations between variously exposed faces of the satellite and those due to eclipses, there are also variations due to the operation of the satellite itself. These include heat produced by telecommunications transmitters and other sources. Finally, space probes are subject to a specific thermal variation due to their varying distance from the Sun. At perihelion, the Helios probes thus receive almost nine times as much solar energy as they did at their departure from Earth, whereas Pioneer 10, now beyond the orbit of Pluto, receives less than two parts in a thousand. Various means are used to regulate the temperature in the different parts of a satellite (see table 1.2).

The technical platform can be used in different ways depending on the nature of the payload and the latter is determined by the aims of the mission: telecommunications, remote sensing of Earth resources, military reconnaissance, manned flight, and so on.

The payload on board telecommunications satellites includes a receiver, a certain number of transponders, that is, selective power amplifiers, and also antennas and reflectors. Antennas may have semi-global (hemispheric) or regional coverage, narrow steerable (spot) beams, or omnidirectionality for ranging (fig. 4.10). They are controlled by radio steering devices.

The payload on board remote-sensing satellites comprises a whole range of sensors, including photographic equipment, scanning radiometers, television cameras, and so on (fig. 4.11). In addition, terrestrial and solar horizon sensors are used to maintain a strict control over the attitude of the satellite, since an accurately determined line of sight is required for imaging. Remote control and ranging measurements use communication systems with Earth stations or relay satellites.

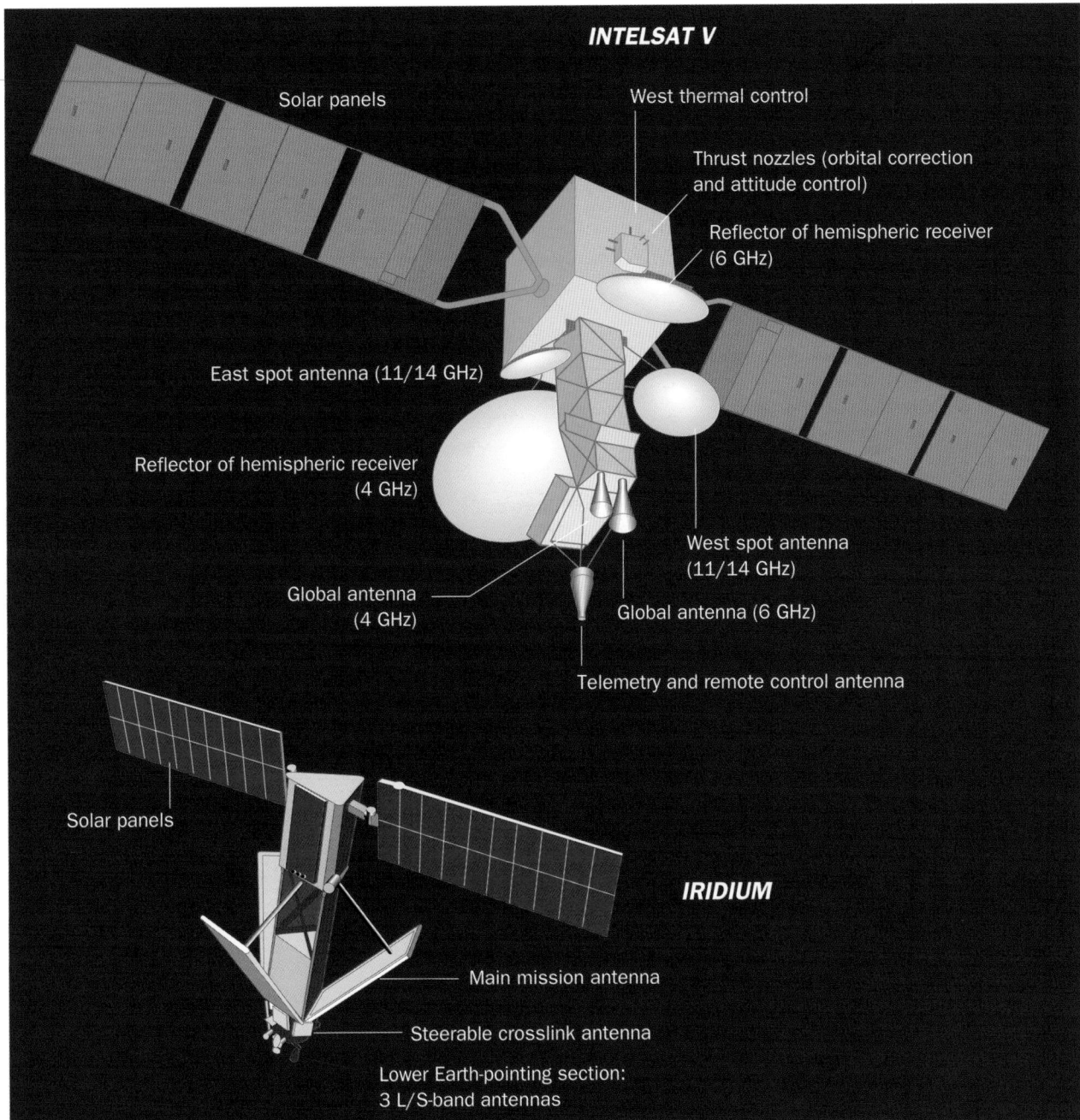

Figure 4.10. **Two telecommunications satellites: Intelsat V and Iridium.** The total wingspan of Intelsat V and Iridium is respectively 15.6 m and 7.9 m.

Finally, for economic reasons, certain satellites are designed to fulfill a range of different missions. Hence, the recoverable Chinese satellites in the FSW series, like the Russian Foton capsules, can just as well serve the purpose of military photo-reconnaissance as carry out civilian microgravity experiments.

Space stations must satisfy very specific requirements, allowing people to live and work in space. Their on-board systems are subject to more stringent safety standards, especially with regard to systems for attitude control and stabilisation, thermoregulation, survival and communications. Today, space stations are provided with several docking ports for unmanned and manned spacecraft and for permanent modules that are gradually added on to the central element. Such operations involve high-performance rendezvous procedures and must be achieved without in any way perturbing the overall stability of the space station.

Space probes are unmanned craft which leave geocentric orbit in order to explore the interplanetary medium or other celestial bodies (fig. 4.13). Thrust and maneuver subsystems are of the utmost importance, but all equipment must be particularly long-lived in order to achieve long term objectives in an extremely inhospitable cosmic environment. Finally, links with Earth are often difficult. As the craft moves further away, response times can become long.

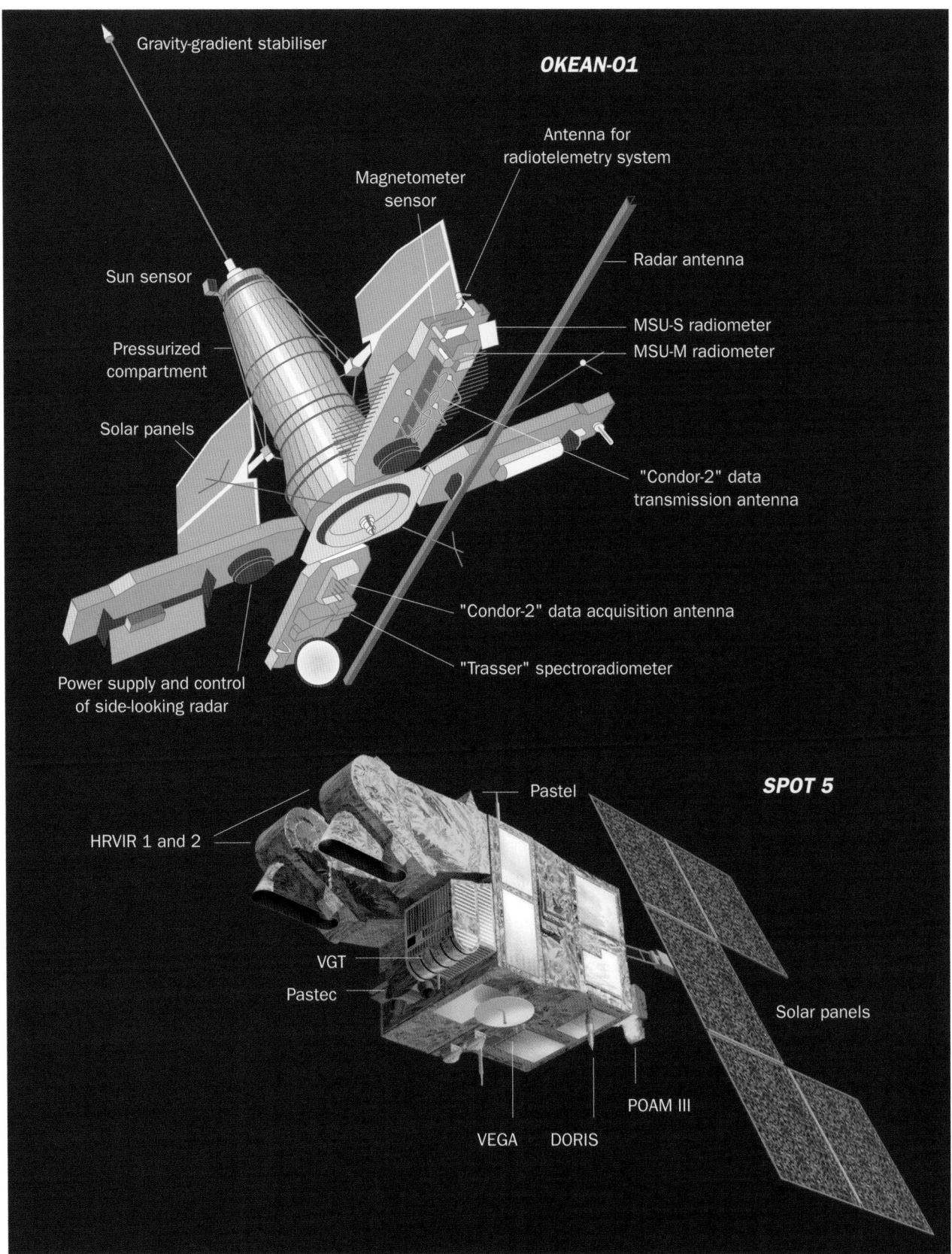

Gravity-gradient stabiliser

OKEAN-01

Antenna for
radiotelemetry system

Magnetometer
sensor

Sun sensor

Radar antenna

MSU-S radiometer
MSU-M radiometer

Pressurized
compartment

Solar panels

"Condor-2" data
transmission antenna

"Condor-2" data acquisition antenna

"Trasser" spectroradiometer

Power supply and control
of side-looking radar

SPOT 5

Pastel

HRVIR 1 and 2

VGT

Pastec

Solar panels

POAM III

VEGA   DORIS

Figure 4.11. **Two remote sensing satellites: Okean and SPOT 5.**
Condor 2 is a digital system for data collection and transmission. Multispectral scanning radiometers MSU-S and MSU-M operate in the visible and near infrared. Besides the main sensors, SPOT 5 also carries Pastel (laser communications passenger), Pastec (technology demonstration passenger), DORIS and POAM. (Artist's view of the SPOT 5 satellite, © David Ducros, CNES 2000.)

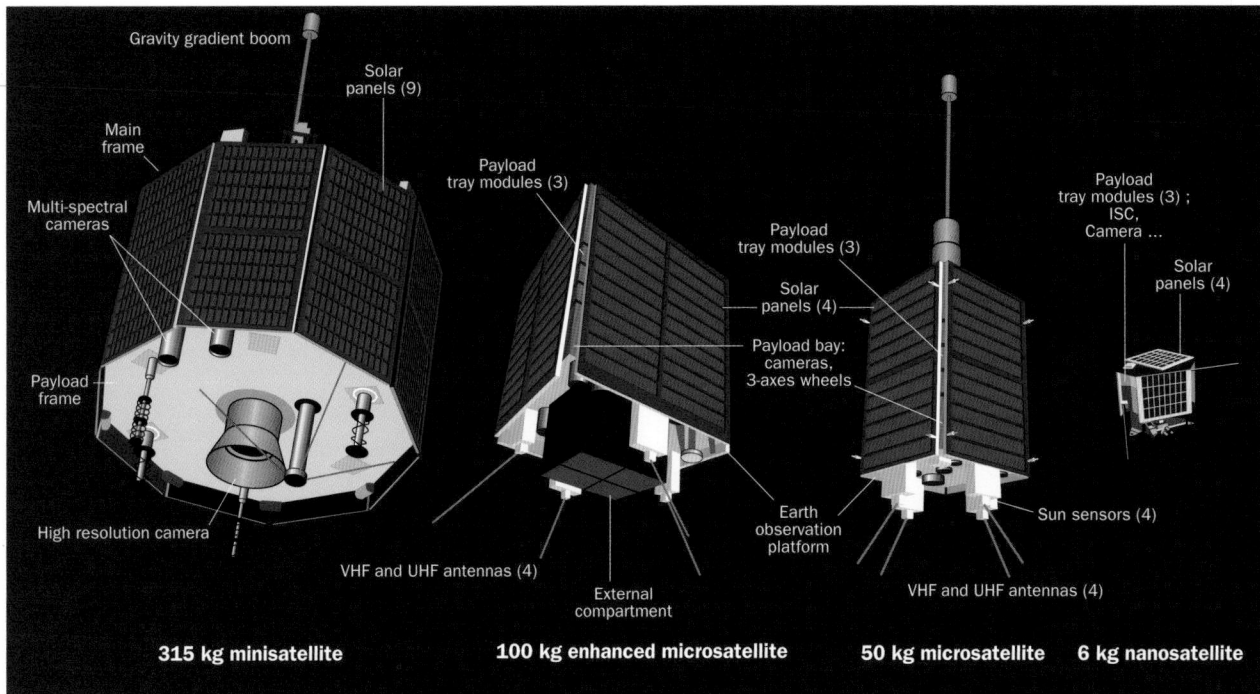

Figure 4.12. **The UoSat series of satellites from SSTL (Surrey Satellite Technology Ltd).**

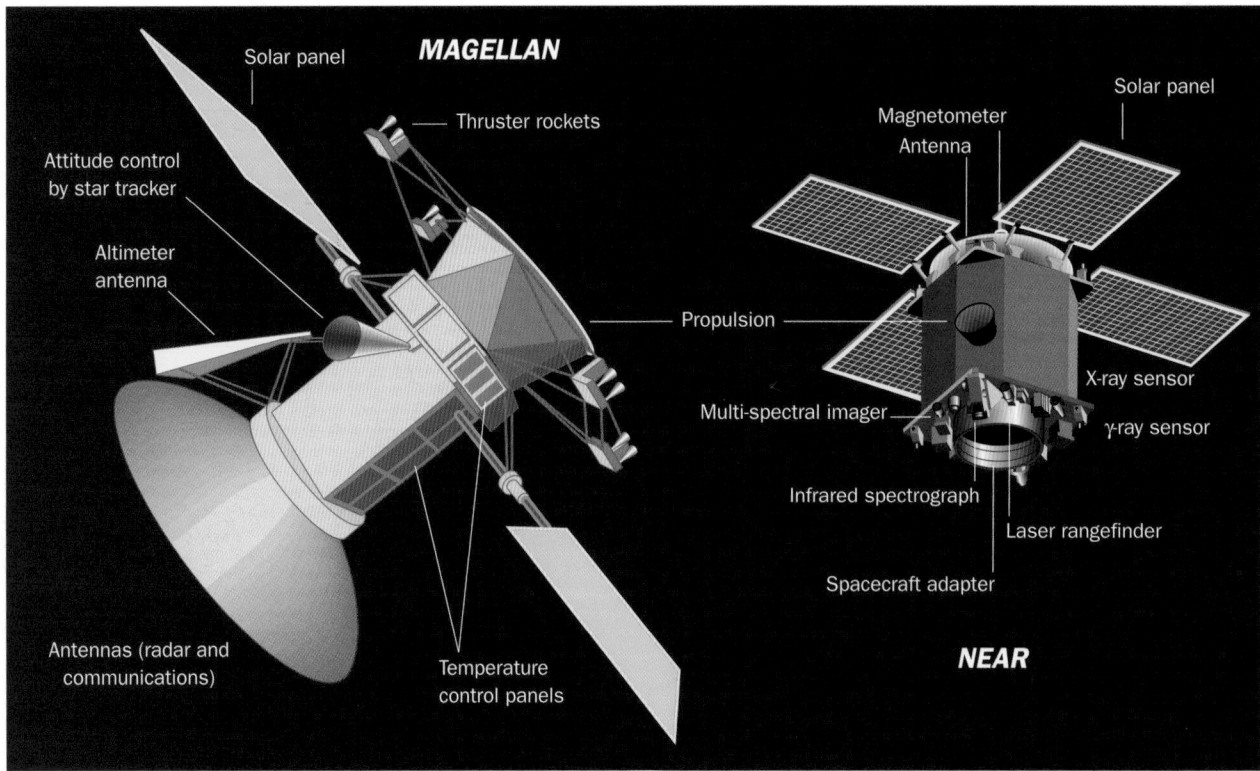

Figure 4.13. **Two space probes: Magellan and NEAR.**

Civilian and military applications

More than 2700 satellites are currently orbiting the Earth, according to the *Directorate of Public Affairs of the US Space Command* (fig. 4.14). Although it gives some idea of the relative importance of the various countries, this type of information merely provides a rough indicator insofar as active satellites represent only a small part of the total, and lifetimes in certain high altitude circular orbits may be extremely long. Even though some countries would thus appear to be over-represented, as in the case of Japan with several rather long-lived satellites, certain overriding features are nevertheless relevant, in particular the imbalance between the two first space powers and the rest of the world. Moreover, if the nationality of the satellite is taken as the criterion, then many countries whose role is in fact limited to that of a user, are featured, on a par with the small group of countries capable of building and launching their own satellites. However, the unequal occupation of space by different nations remains clearly apparent.

If satellites are analysed according to their missions (figs. 4.15 to 4.19), a key factor distinguishing between nations is whether or not they develop military space applications. This factor contributes to further widening the gulf between the United States and, to a lesser extent, the ex-USSR on the one hand, and the rest of the world on the other. The recent, low-profile appearance of European countries, China and Israel clearly shows that any ambition on the international scene must include the acquisition of military space capabilities, among other things. At the same time, such acquisitions are also an undeniable demonstration of autonomy in the fields of information and action. There is a direct connection between the weakening of Russian capabilities and its reduced presence in international relations.

The association of military and civilian technologies no doubt favours the development of military capabilities in more and more countries. Indeed, countries like Japan and India now possess the potential to use space for limited military purposes. Today, in terms of the satellites used or programmes under implementation, the United States falls into a category of its own, whilst Russia often finds its ambitions are thwarted (see chapter 5).

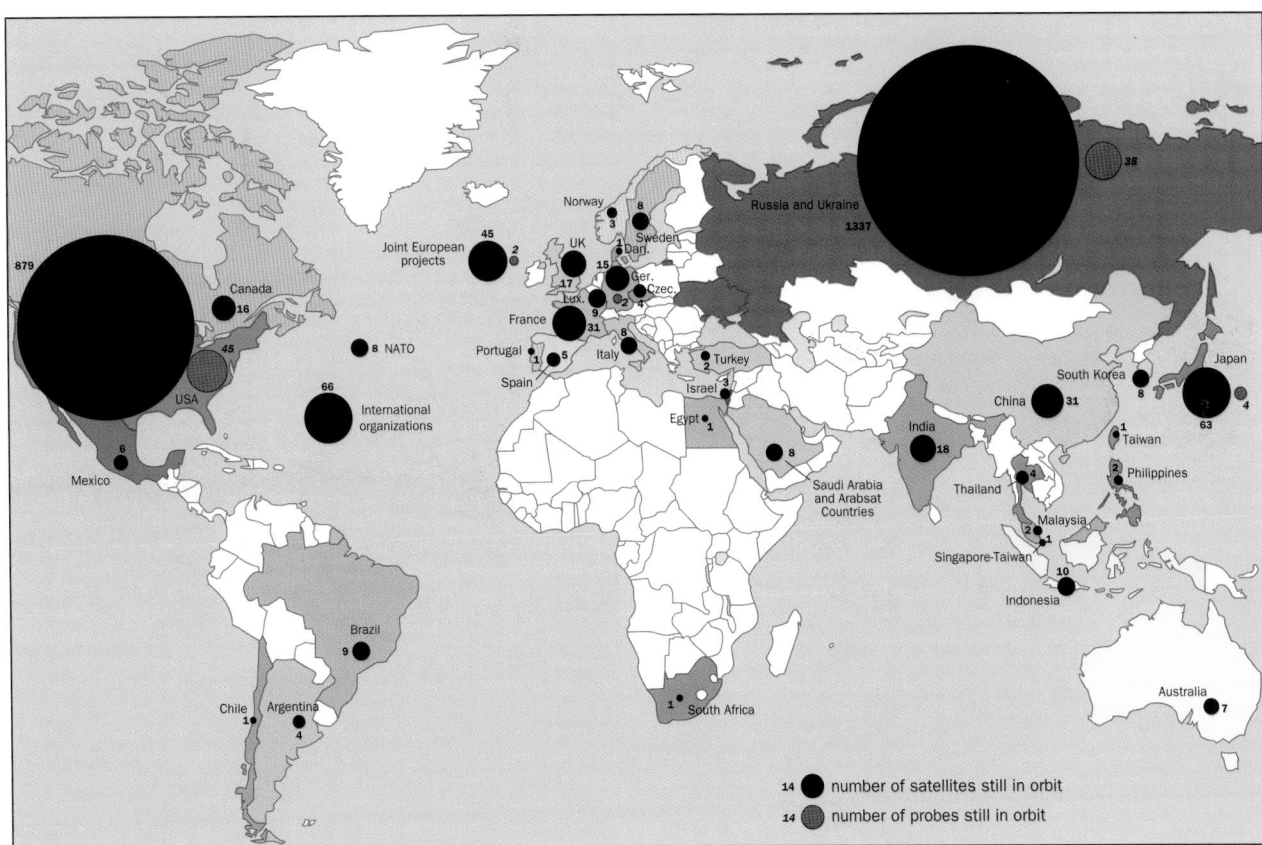

Figure 4.14. **Satellites and probes in orbit on 1 January 2000.**

Certain programmes or missions can satisfy both civilian and military needs, at least up to a certain point. These include in particular the areas of scientific research, Earth observation, telecommunications and even space station development. In the case of the USSR, this logic had been pushed almost as far as it would go. All undertakings in space were integrated into the military–industrial complex in such a way that rather undifferentiated systems could easily be produced, thus well suited to civilian as well as military needs. This trend towards standardisation is becoming more and more common. The aim is, of course, to obtain better returns on space expenditure. Having once been the only recourse of countries with limited resources, which gradually developed military capabilities from civilian knowhow, this approach can now be observed in the United States, but operating in the opposite direction. In other words, civilian means are brought in to aid military systems. The reality of this trend is exemplified by the US DoD's takeover and operation of the Iridium system, which was originally designed as a commercial concern. Their original logic was to use it to complement their own systems. When Iridium went bankrupt, the DoD was compelled to maintain operational capabilities for its own purposes and the original logic went somewhat astray.

Other applications are designed to meet specific requirements. Space probes, for instance, have purely civilian financial backing, even though state-of-the-art technology originally destined for military missions has sometimes been recycled and even though results are not only of interest to the civilian world. At the opposite extreme, some special military applications such as photo-reconnaissance or electronic intelligence, together with certain experiments, whether they be presented as defensive or offensive, are of a purely military nature, even if the concept of war in space remains confined to the realms of science fiction (see chapter 12).

Among the main areas of space development, telecommunications satellites occupy first place in terms of numbers. This numerical superiority is still more significant insofar as the lifetimes of telecommunications satellites are much longer than those of Earth observation satellites, the second most commonly encountered category of spaceborne object. In fact, the high number of photo-reconnaissance satellites rather distorts the true importance of this category since most Soviet spacecraft carried out short-lived missions. Taking the count since the beginning of the space age, science and technology satellites arrive in third place, grouping together tests across several different areas of activity. Finally, human space

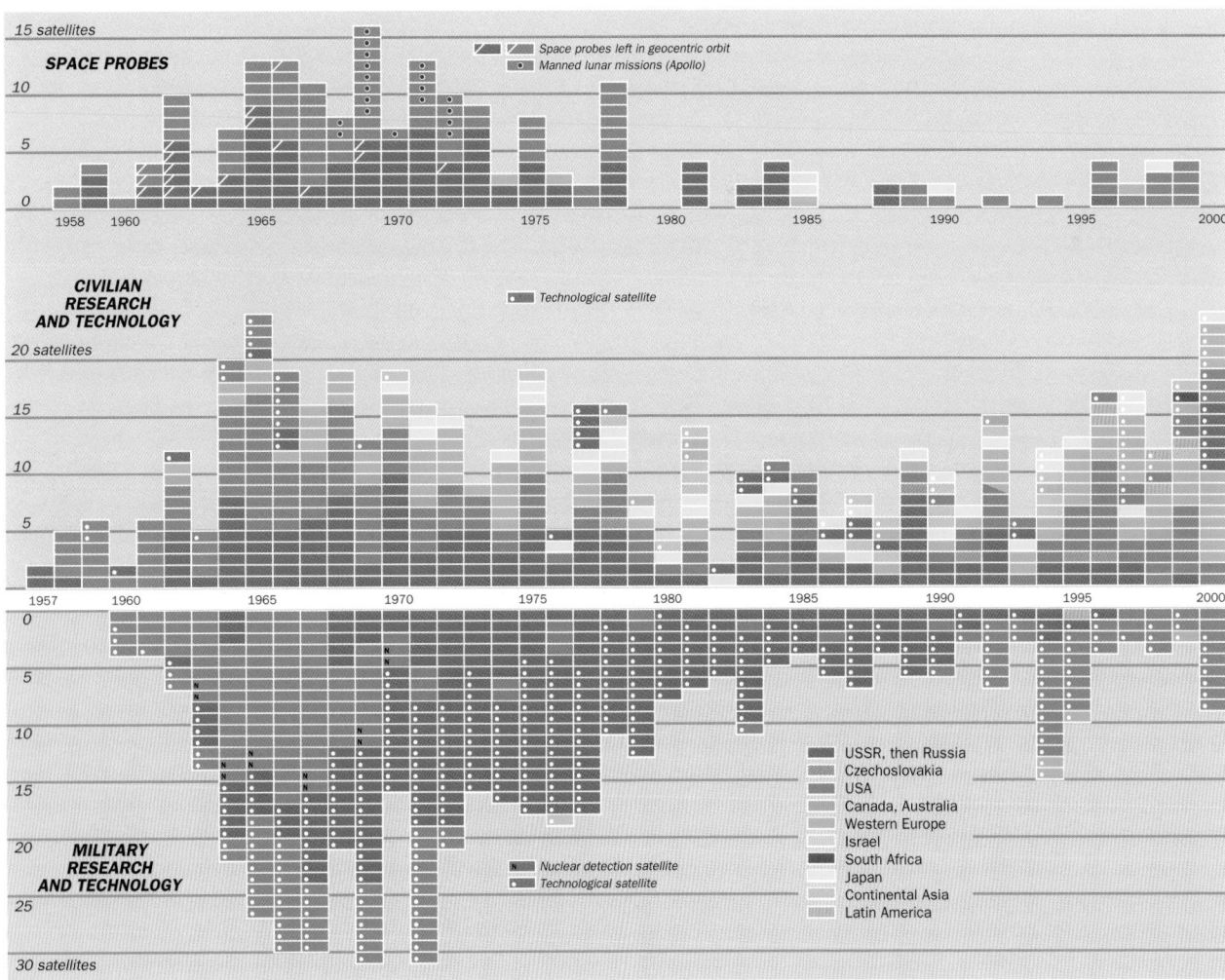

Figure 4.15. **Probes, scientific satellites and technology satellites by year of launch and country to which they belong.**

flight represents the smallest number of launches. In the eyes of the public, this category nevertheless corresponds to the most prestigious activity, even if high budgets and lack of immediate applications are often a subject of controversy.

Historically, the first satellites were scientific, opening the way to a wide range of space applications by revealing the circumterrestrial medium and its potential. With this in mind, all countries intent on developing space activities are represented in this domain of activity. Needless to say, the means at their disposal are wildly disparate, ranging from UoSAT type satellites of a few kilograms to Russian satellites weighing in at more than 7 tonnes (fig. 4.15). Many scientific projects are the result of collaboration between research centres, sometimes very wide-ranging. Consequently, the distinction between civilian and military activities is more often a question of organisation than of handling specific needs, reflecting the origins of financial backing and affiliation of scientific teams – with the exception of a few rather special experiments – at least in the beginning of the space age. Scientific satellites have sparked off several new fields of research and their widely disseminated results are of interest to almost the whole of the international scientific community (see chapter 7).

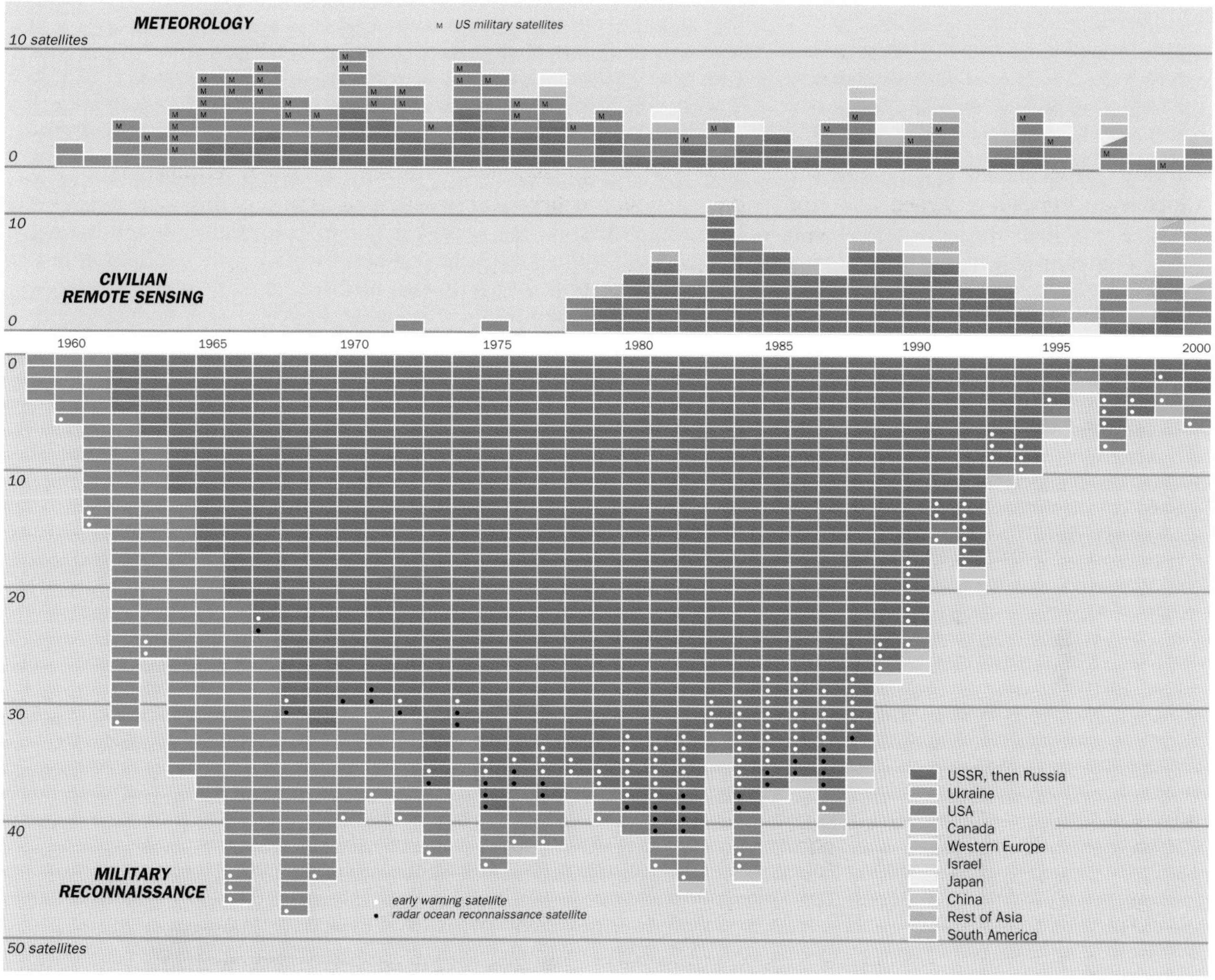

Figure 4.16. **Earth observation satellites by year of launch and country to which they belong.** Earth observation here includes meteorology, remote sensing of Earth resources and military reconnaissance. Early warning satellites and radar ocean reconnaissance satellites are also represented on the graph.

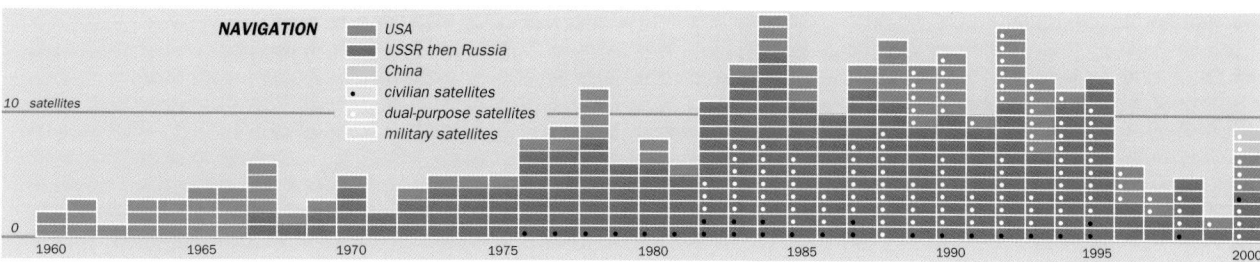

Figure 4.17. **Navigation satellites by year of launch and country to which they belong.**

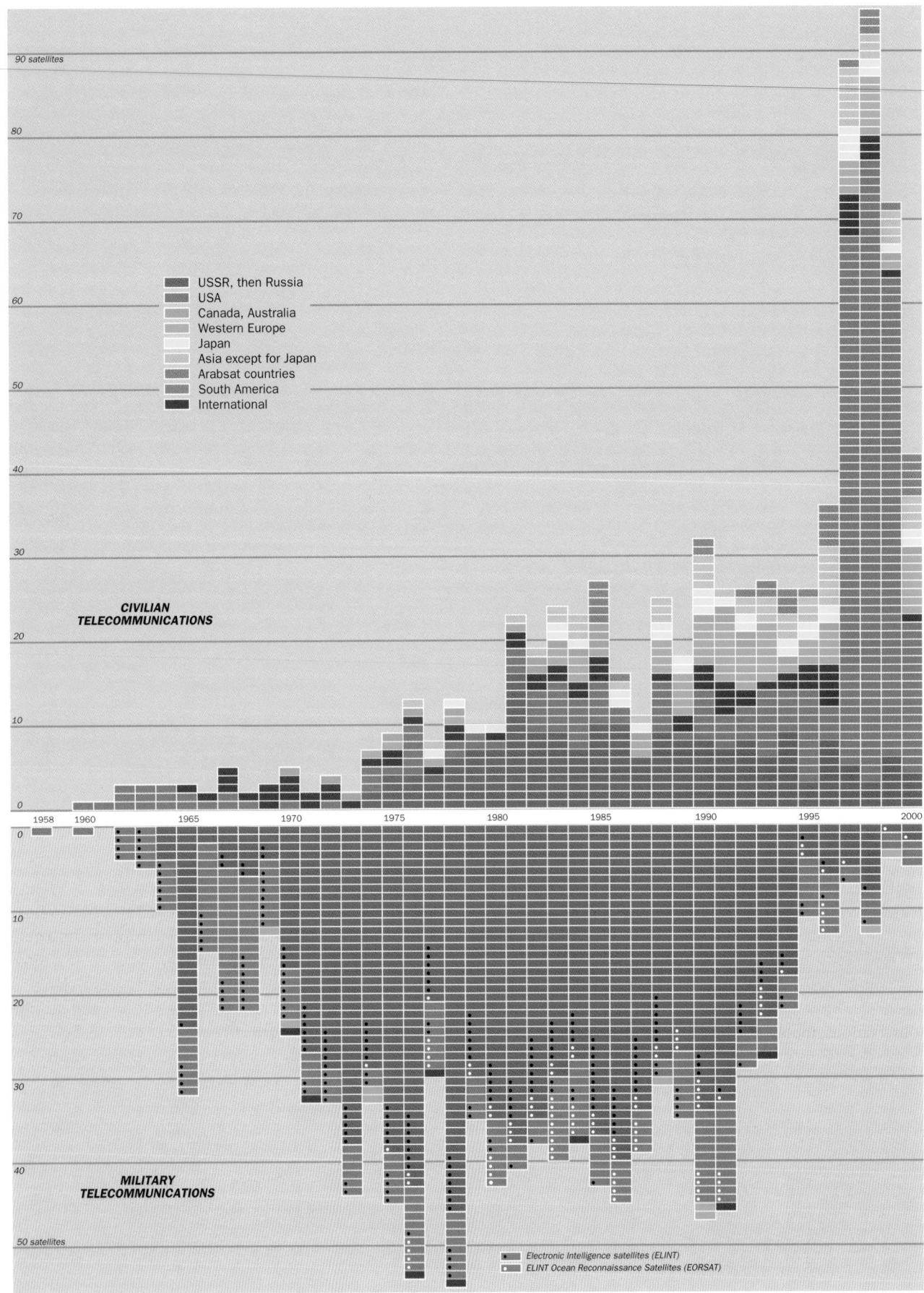

Figure 4.18. **Telecommunications satellites by year of launch and country to which they belong.**

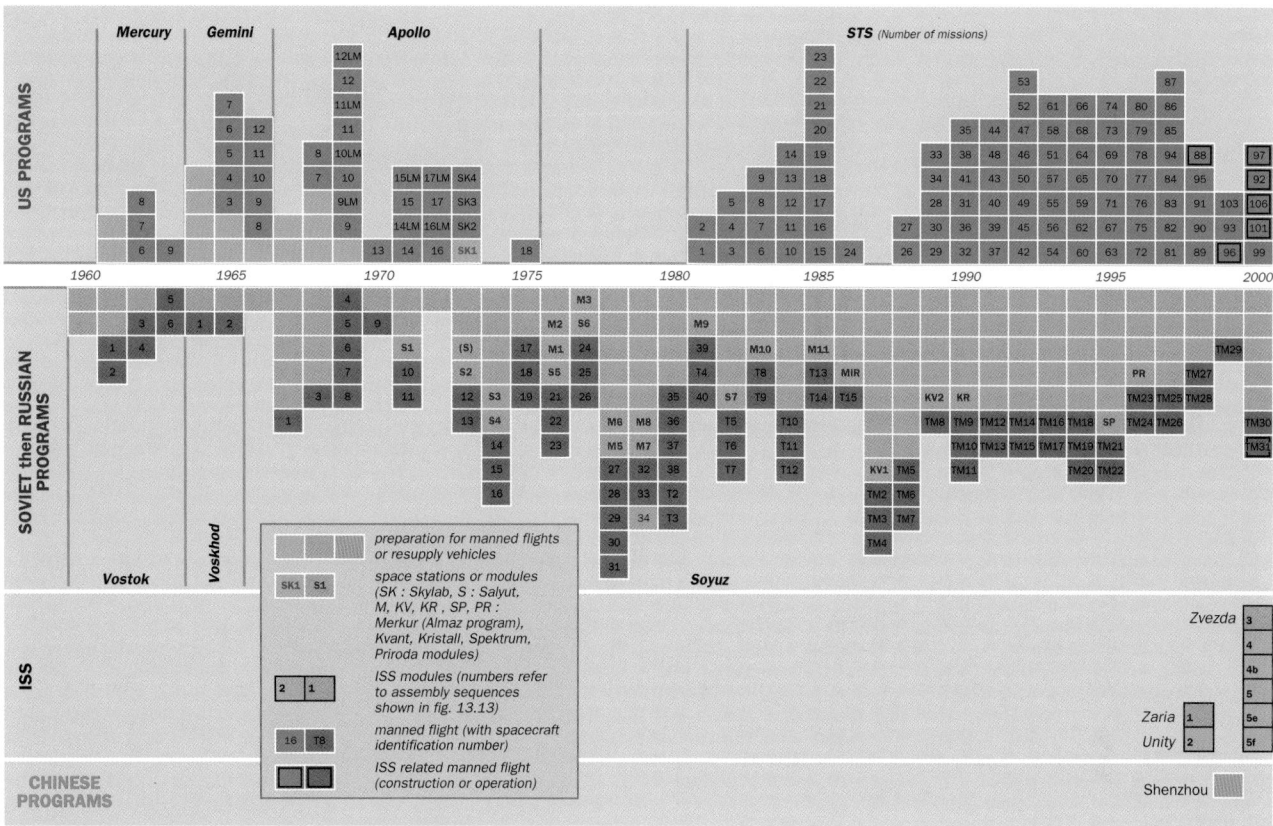

Figure 4.19. **Manned spacecraft, space stations and modules, and preparation or supply satellites for manned flights by year of launch and country to which they belong.**

The particularity of Earth observation is the high proportion of military missions compared with civilian programmes (fig. 4.16). In fact, from a technical standpoint, the systems used are fundamentally the same in each case (see chapter 9). In meteorology, for example, the distinction is largely a matter of the status of the satellite operator.

Some missions nevertheless use special sensors, such as early warning satellites looking for missile launches, or intelligence satellites analysing the ranging parameters of enemy systems. Globally speaking, military satellites must meet special operating requirements with regard to both ground-based receiving stations and space-based systems: a high level of confidentiality for links, faster transmission of observations, mobile Earth stations, and so on.

The Soviet Union and, to a lesser extent, the United States occupy unique positions due to their high proportion of military satellites with short lifetimes. Although this situation barely evolved for a period of 35 years, two factors eventually caused a change to take place: the break-up of the Soviet Union, which led to a reduction in the number of military satellite launches, and the setting up of spaceborne military surveillance programmes by European countries such as France. Finally, the current establishment of national programmes, some of which are now operational, expresses an increasing integration of space-based observation capabilities into overall territorial development.

In navigation (fig. 4.17), the increasing number of civilian users has led to a novel phenomenon whereby originally military systems have been opened up to an international community of users. Considering the fact that only the United States and Russia actually possess such navigation and positioning systems, a debate is currently underway, particularly in Europe and Japan, on ways of limiting this civilian and military dependence (see chapter 11).

Telecommunications is the only domain to be governed by largely commercial dictates and as such operates in a quite different way. Civilian systems have overtaken their military counterparts in number and a large part of the backing has private origins (fig. 4.18). The regularity of satellite launches is witness to the global growth of this type of use, bearing in mind that geostationary satellites have long life expectancy and ever increasing capacity. National, regional and international systems exist side-by-side and every country on the planet can now use space telecommunications, via the international networks if need be (see chapter 10).

Human occupation of space remains in a category of its own, for which only the United States and Russia have the technical knowhow (fig. 4.19). In this area, there is no longer a great divide between the military and civilian sectors, since both have always been closely associated in the training of astronauts, and indeed also in the financial backing of certain missions.

However, due to the human presence, and the fact that there are no direct applications, there is a clear problem of cost-effectiveness, considering the high investment required for such programmes. Maintaining and using the Space Shuttle, a cross between launch vehicle and space station with a mass of almost 100 tonnes, is an extremely complex exercise, exemplifying the difficulties involved in any ambitious human spaceflight programme. In fact, even the idea of the space-based workman, so widely publicised when the Hubble telescope was repaired in orbit, is actually of limited application given the special conditions prevailing in these regions. Indeed, this lack of practical justification partly explains the sporadic nature of the American presence in space. Programmes often reflect special circumstances related to domestic and foreign political problems, as in the case of Apollo, or else they owe their existence to technological and commercial considerations, generally in the medium or long term, as in the case of the Space Transportation System (STS, see fig. 6.27).

The almost continuous Soviet activity in this domain reflects a different context, namely one in which the prestige of such programmes provided a way of demonstrating technological parity with the United States, a theme for which finances were readily forthcoming. Since perestroika, internal difficulties have combined with the former ideological exploitation of cosmonauts to generate a serious questioning of such programmes. The transformation of the American space station Freedom into another, the Alpha programme, based on a mainly Russian–American cooperation (later to become the International Space Station (ISS)), meant that Russian manned activities could be pursued with largely external financial backing, as in the case of the space station Mir since 1991. The main rational arguments now used to support such pursuits are two-fold. Firstly, on a technological level, this experience is unique, involving as it does a complete infrastructure which includes among other things unmanned supply systems. Secondly, in physiological terms, it provides for experimentation that could be obtained in no other way.

Today, the more sentimental argument which appeals to the greatness of a human presence in space is nevertheless undiminished, although not so loudly proclaimed. By allowing friendly nations the possibility of sending one of their astronauts into space, the Soviet Union did a great deal to improve its international image. Trips on board the American Space Shuttles correspond to much the same approach. The end of Mir in 2001 and the beginnings of the ISS mark a new phase of international cooperation with regard to human presence in space, this time under American leadership.

CHAPTER FIVE

Space policy and budgets

- **BUDGETS AND SPACE ACTIVITIES THROUGHOUT THE WORLD**

- **RUSSIA AND THE CIS REPUBLICS**

- **THE UNITED STATES**

- **EUROPE**

- **JAPAN**

- **CHINA**

- **INDIA**

- **ISRAEL, CANADA, BRAZIL AND AUSTRALIA**

Budgets and space activities throughout the world

A political concern for independence and a high level of technological knowhow are prerequisites for belonging to the select club of space powers. There is nevertheless a great inequality between those countries whose space activities constitute a genuine area of national activity and a source of political and economic power, and those countries where space is merely a sideline with a marginal role in the economy, although it nevertheless allows them a limited and intermittent presence on the international scene.

Space budgets around the world
One way of assessing a country's interest in space is to compare its budget with spending in other countries, particularly if it is first expressed as a percentage of the Gross National Product (GNP). In 2001, the US budget distinguished itself, being a staggering five times the second largest space budget (fig. 5.1). The latter is that of European countries, calculated from the total of all national expenditure in addition to the budget of the European Space Agency. Japan stood in third position followed by China, India and Russia, in that order. Israel, Canada and Brazil represent the last significant category in the sense that, with a budget below 300 million dollars, other states devote these resources mainly to applications, usually in telecommunications.

Figure 5.1. **Space budgets around the world in 2001.**

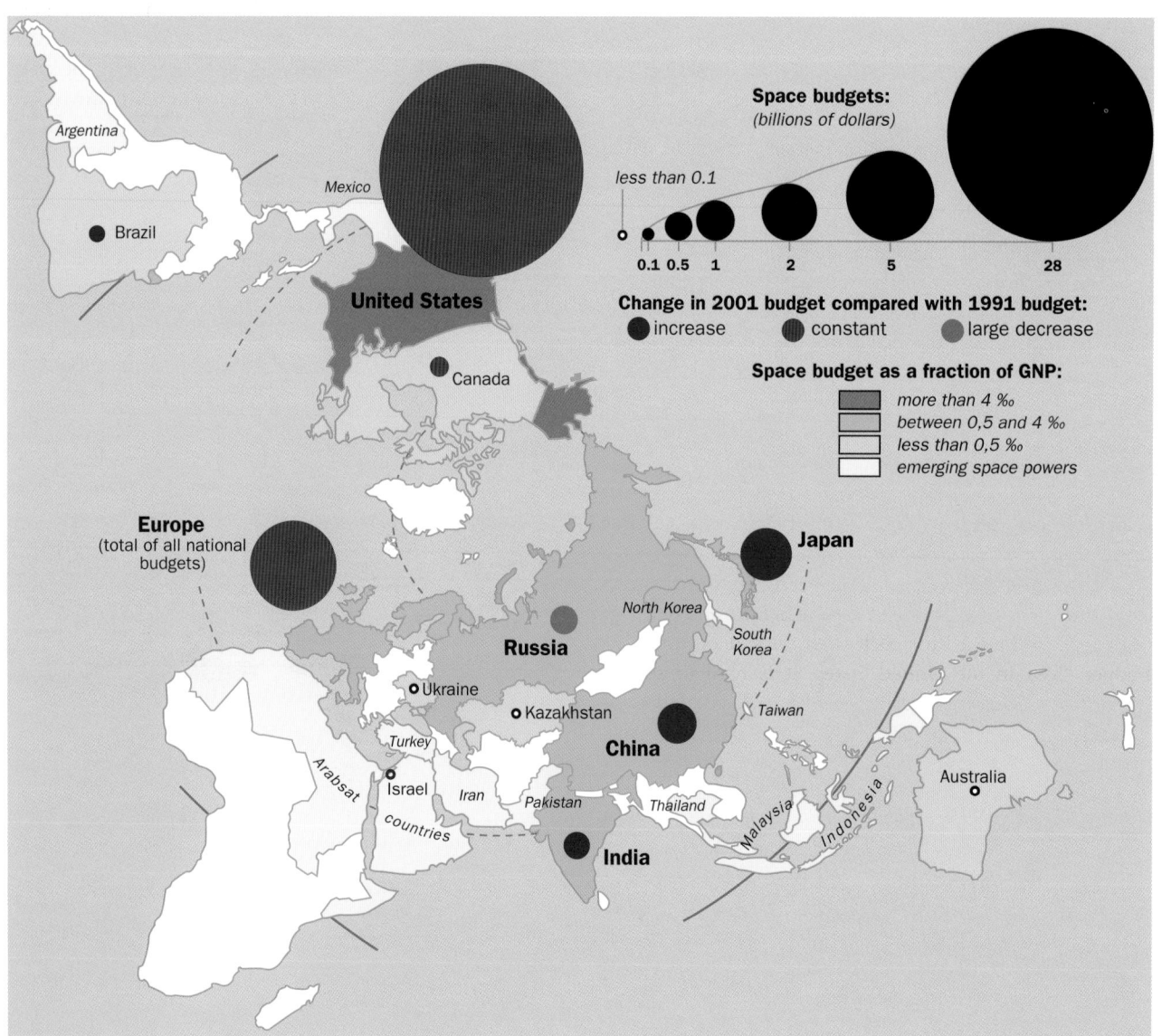

It is interesting to compare this situation with that prevailing ten years earlier when the Soviet Union was beginning to break up. At that time, the annual civilian and military resources devoted to space clearly distinguished the United States (26 billion dollars) and the Soviet Union (whose budget is estimated to have been around 20 billion dollars) from the rest of the world. This appears to be perfectly consistent with the level of activities they maintained in terms of programmes and launches. The other space powers followed the same order, that is, Europe, Japan, China and India, but with a lower level of investment. The main change thus concerns Russia's position. In 2001, Russia spends around 300 million dollars, placing it after India, in sixth position. When the various protagonists are compared in terms of the number of launches, for example (see figs. 5.4 to 5.6), a different picture is revealed, showing that a country's influence in space cannot be assessed purely in terms of its budget, especially where comparative analysis is concerned.

One of many reservations on this count is the reliability of declared budgets. For instance the Russian budget voted by the Duma would not be sufficient to explain observed activity levels. In this case, it must be assumed that there are other sources of revenue, such as sales of equipment either directly or through joint ventures (launches or propulsion technology), partnerships with financial backing from abroad (as in the case of manned flights), and so on.

Another obviously dubious case is the Chinese budget, characterised by its astonishing stability. This is belied by the great diversity of programmes in the area of launch vehicles and manned flight among other things. The most commonly quoted figure, mentioned during speeches, is one billion dollars. No official figure is available, and this explains how impossibly low values of 200 million dollars can sometimes be found in the literature. Such figures may correspond to the civilian space budget, if indeed such a distinction is meaningful in the political and economic context of Chinese space activities.

In a wider context, not only concerning space activities, assessment of military expenditure should always be made with caution. In the Russian case, it may be asked whether quoted figures refer to civilian space activities and hence the budget of RASA (in Russian RAKA), or indeed the whole budget. Even in the United States, it is hard to separate space-based military programmes from military activities in general and the same is almost impossible when R and D is considered.

Moreover, the sometimes significant differences in production costs from one country to another are completely ignored when budgets are merely expressed in dollars and no reference is made to the local standard of living. Even though space programmes presuppose skilled labour and expensive technology the world over, Russian, Chinese or Indian equipment can be manufactured at a much lower unit cost in dollars than it could be in the West. This question has often been disregarded and has certainly led to overestimates of the resources attributed to space in the Soviet Union. When attempts were first made to sell Russian launch vehicles abroad, it was immediately clear that prices were much lower than in the West. It may be that drastic price cuts can be explained by the desperate situation of the NPOs, at the risk of being accused of dumping. In addition, they may have had an inadequate awareness of the true production costs insofar as they operated on the attribution of resources in manpower and equipment rather than in rubles. In any case returns were nevertheless sufficient to keep the sector in business. The Chinese approach with its highly competitive prices also tends to demonstrate much lower production costs in that country. The same can be said of Indian activity levels when compared with the amount of their dollar budget.

Whatever reservations must be made concerning the figures, Russia remains an instructive case because of the sudden collapse and the genuine difficulties experienced by the Russian space sector over the past ten years. The ruble is undervalued relative to the dollar and this means that the budget is certainly greater in real terms. However, comparison with investment in terms of the GNP, also characterised by undervaluation of the dollar, reflects a genuine loss of political interest which is widely manifested, even in public speeches. The current capacity of the space sector, benefiting as it has from previous investments, is thus heavily dependent on

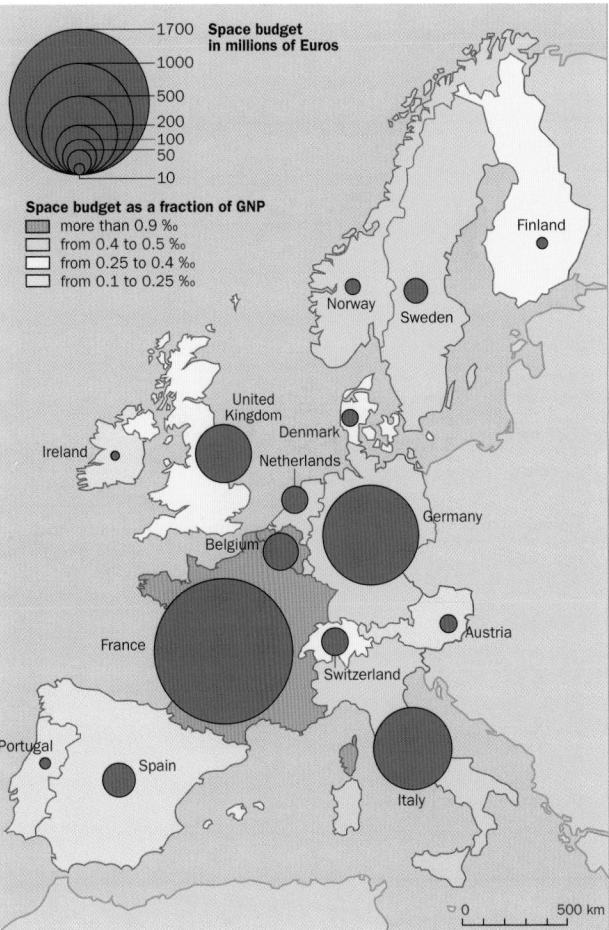

Figure 5.2. **Space budgets of European countries in 2001.**

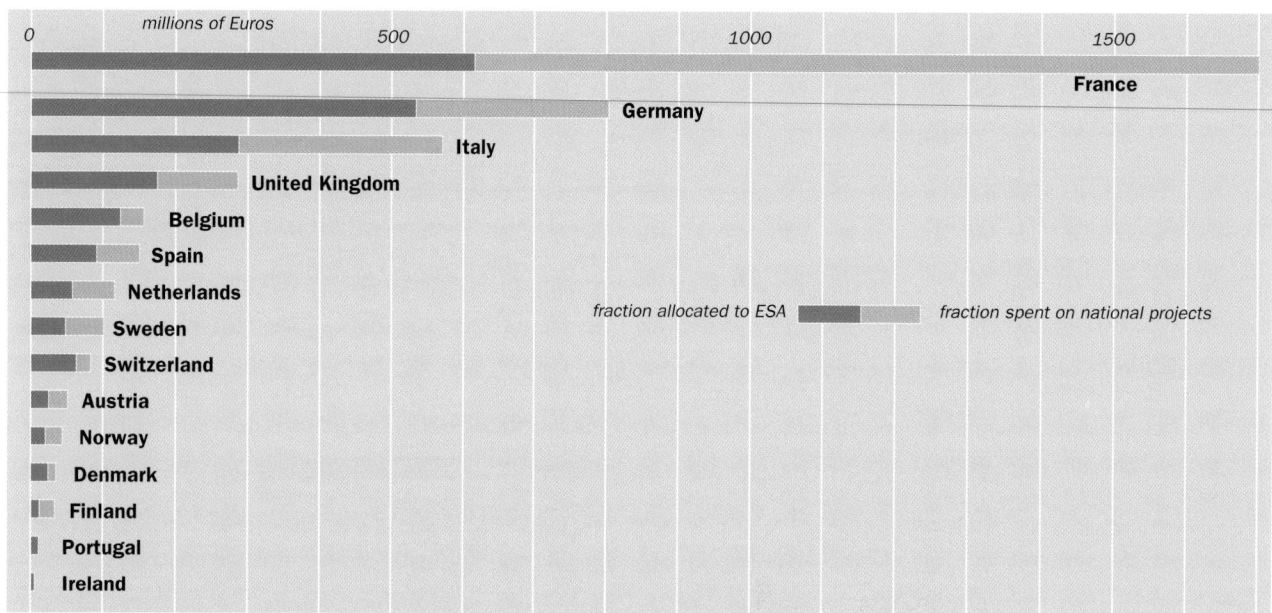

Figure 5.3. **Share of the budget of European countries devoted to the European Space Agency in 2001.**

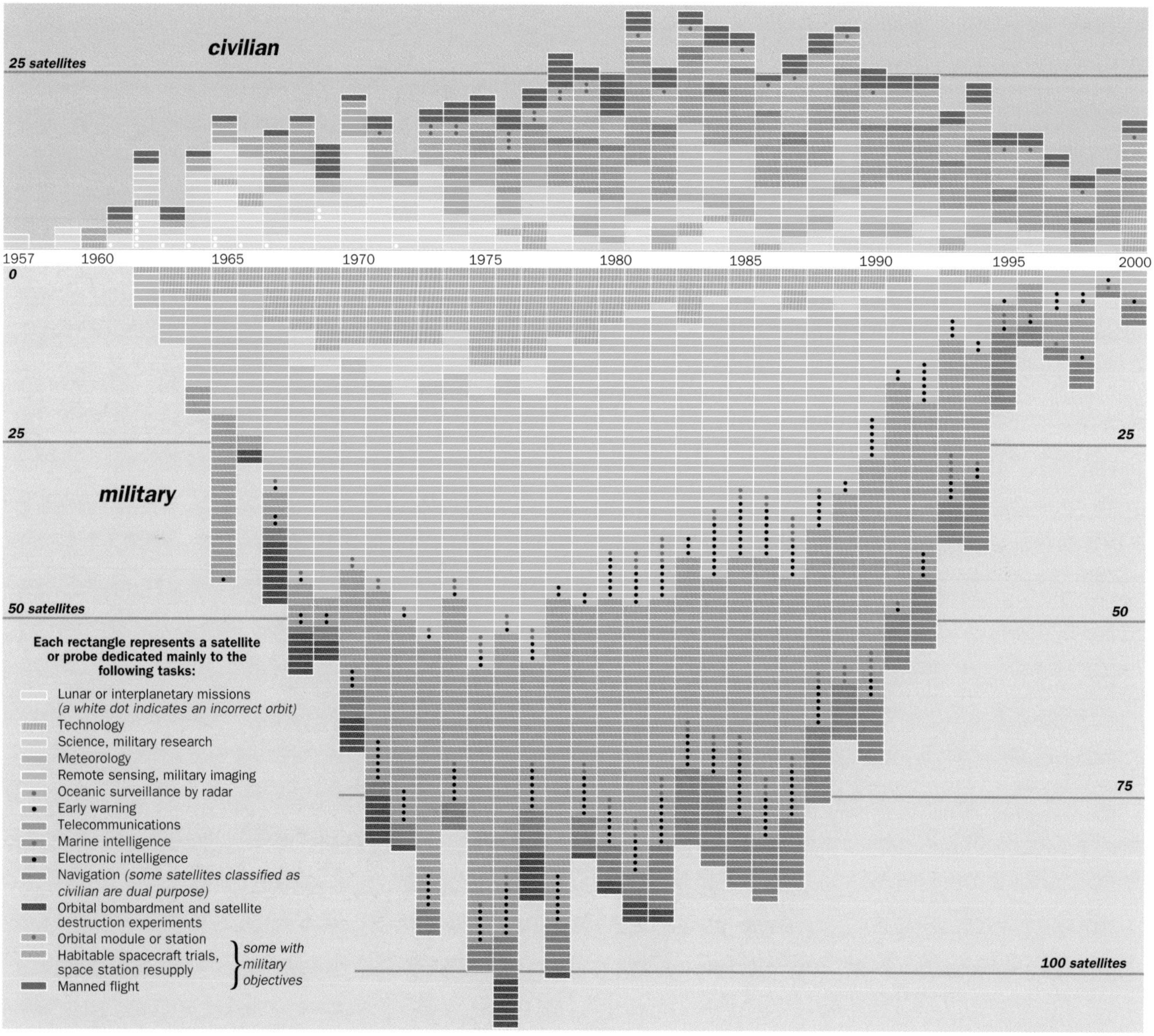

Figure 5.4. **Missions of Soviet or Russian satellites launched between 1957 and 2001.**

revenues from 'partnership' (or subcontracting) with foreign countries, in particular the United States. The amount of such revenues is very likely greater than the Russian budget itself.

Analysing budgets expressed relative to the GNP, and the same methodological reservations hold concerning GNP figures, especially for Russia and China, the following general conclusions can be drawn.

The United States remains the first space power by a considerable margin with a budget of 28 billion dollars representing 0.4% of the GNP.

The second space power today in budgetary terms is Europe with an expenditure of 6 billion dollars. Certain novel features arise from the ambiguities and the current limits of European construction. Different countries show very different levels of interest in space (fig. 5.2) and since activities at the European Space Agency are exclusively devoted to specific areas, the main contributors also run their own programmes. It is instructive to observe that those countries most heavily committed to space activities are allocating an increasing share to the ESA whilst national expenditure holds steady (fig. 5.3). Another striking feature is the appearance of national activities in those countries whose space budget was originally devoted entirely to ESA programmes. This doubtless results from the development of space knowhow on a national level, combined with a desire to carry out some programmes independently. When added

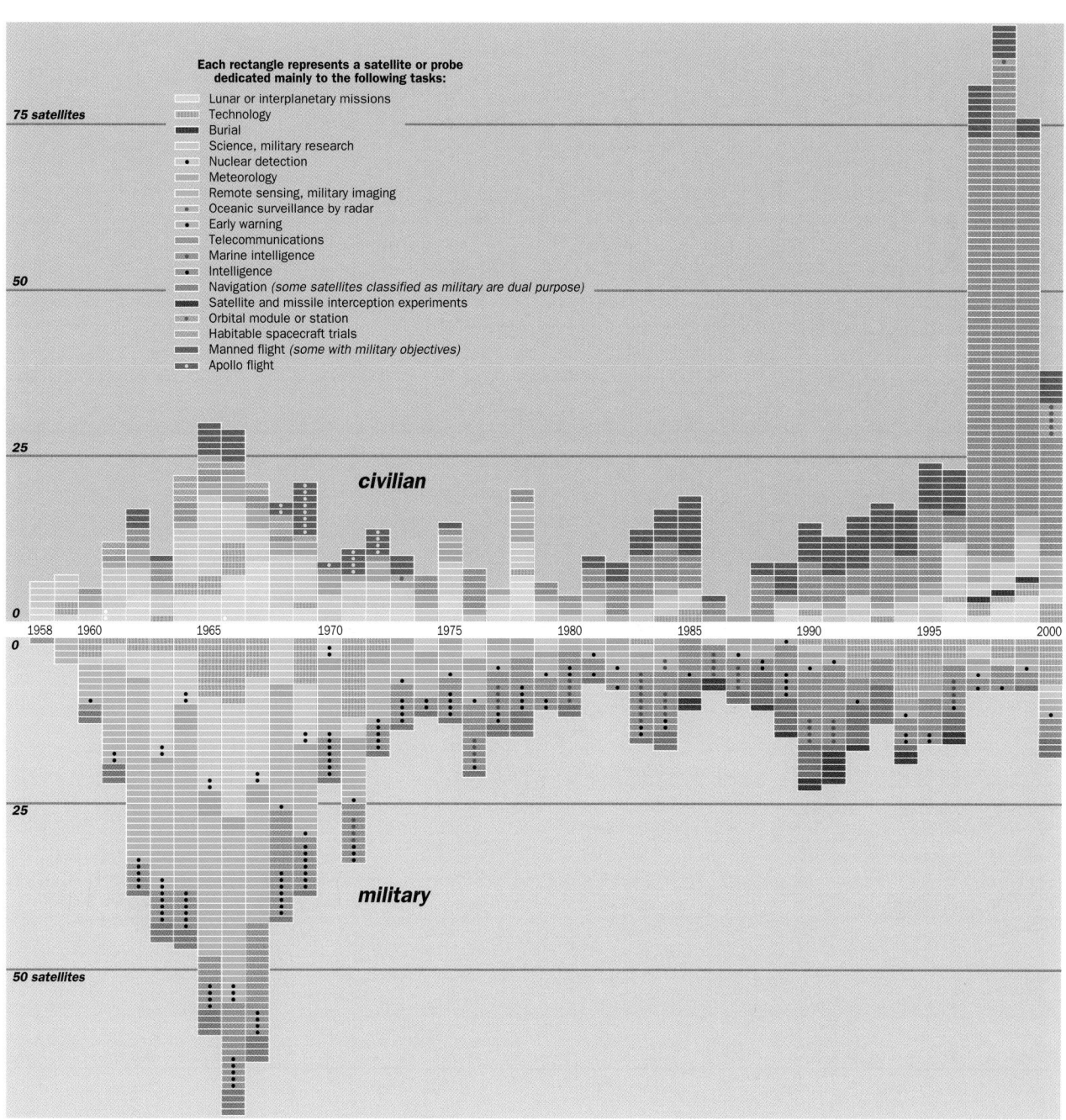

Figure 5.5. **Missions of US satellites launched between 1958 and 2001.**

together, the budgets of the European countries are relatively high, but the existence of differences from country to country limits spending efficiency.

Japan takes third place in terms of budget with a steady increase in its resources, which have doubled since 1991, reaching 2.5 billion dollars today. This nevertheless represents only a very small part of its GNP. Although investment in space programmes is increasing, they hold a relatively modest position amongst national priorities.

China now seems to be spending more on space activities than Russia. Interest in this area is still closely linked to political considerations of prestige, as witnessed by manned space projects and commercial issues.

India is positioned somewhere between China and Russia on the one hand, and Europe and Japan on the other, considering its 330 million dollar budget in relation to its revenues and the steady advance of its credits.

With official budgets of the order of 300–100 million dollars, Canada, Brazil and Israel maintain a steady but restricted involvement in space.

Although the Ukraine and Australia stand together at the bottom of this budgetary hierarchy, the situation in the two countries is quite different. The Ukraine is attempting to maintain the level of knowhow acquired during the days of the Soviet Union, whilst Australian activities represent only a tiny part of the national economy.

Satellite functions and nationality

Comparing the number of launches or the way they are distributed in terms of missions, the two leading protagonists in space, the ex-USSR and the USA, are once again clearly set apart (figs. 5.4, 5.5 and 5.6). A wide range of activities and involvement in certain specific areas are good indicators of a country's maturity in this domain. The United States and Russia continue to be atypical members of the space club in this respect.

Moreover, the number of launches is less revealing than the nature of the missions themselves when assessing a country's mastery of space resources. For instance, the large number of Soviet military satellites arises from a deliberate decision to favour mass production of robust and reliable equipment with a relatively short lifetime. Conversely, the reduction in the number of American satellites dedicated to military missions since the 1970s can be explained by the use of elaborate and costly equipment and in no way reveals a lower level of concern (see chapter 12).

Other countries can be clearly distinguished from the two main space powers by the fact that their military activities are restricted to a few European satellites and a limited level of Chinese experimentation. The main satellite missions in these countries concern scientific research, applications with an economic objective such as telecommunications, and actions that are often conceived in a partnership context.

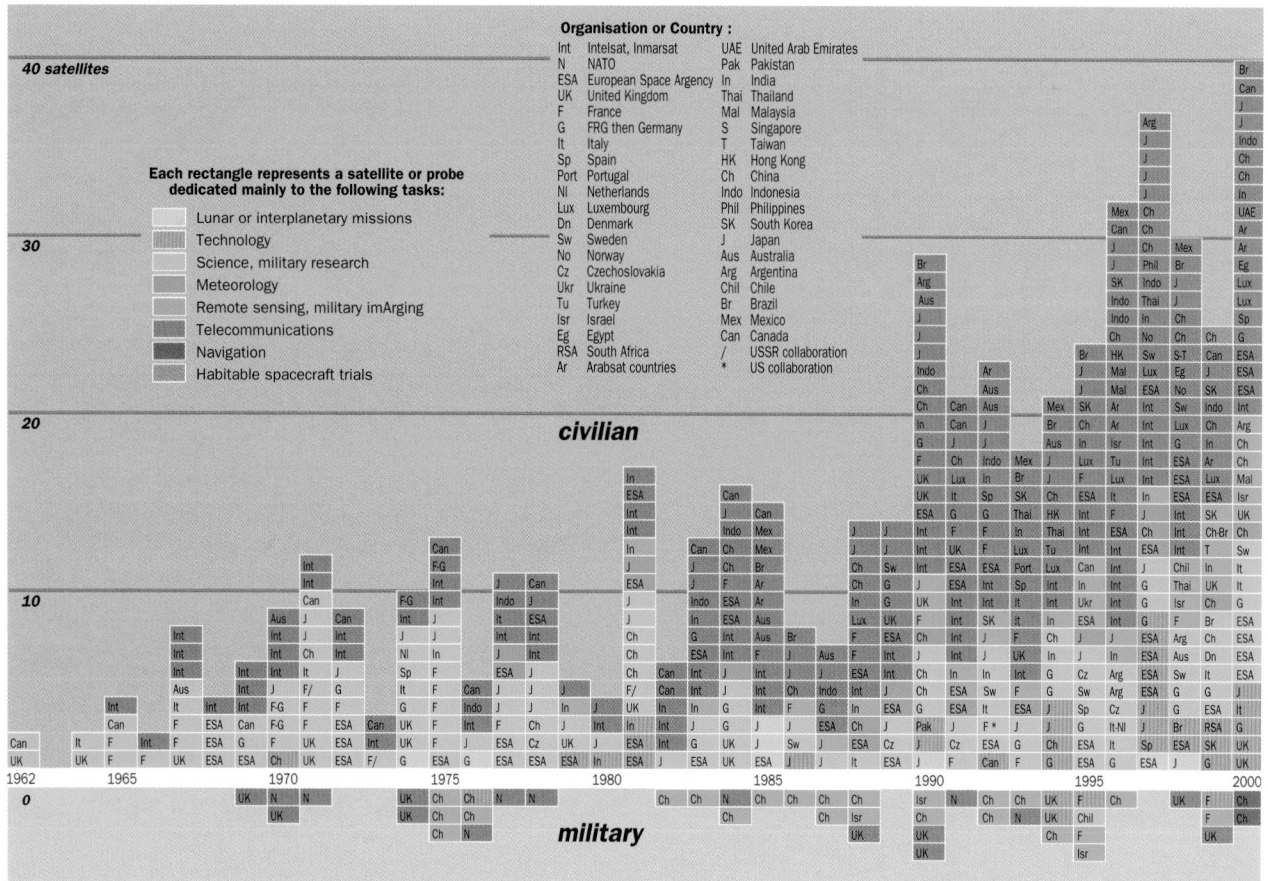

Figure 5.6. **Missions of satellites from other countries launched between 1962 and 2001.**

Russia and the CIS Republics

Successful space conquest gave the Soviet Union the opportunity, heavily exploited by the political powers, to appear as a highly advanced technological state, capable of taking humanity into a new era of its history. Like the nuclear industry, but with the added advantage of creating a more positive image, space provided a way for the Soviet Union to show that it could match the prowess of the United States. It even seemed able to take the lead, at least until faced with the accomplishments of the Apollo lunar programme. In fact, Soviet space programmes responded to global strategic concerns arising in the Cold War context and from US–Soviet rivalry, features which formed the basis for international relations at the time. The break-up of the USSR brought this situation to a sudden end. Russia inherited the greater part of the USSR legacy, exploiting its potential as best it could to prevent collapse but without including space issues amongst its new political projects. The Ukraine also took advantage of previously acquired skills through partnerships with Western nations, whilst Kazakhstan maintained activities at its space base via leasing agreements with Russia, its sole user. After ten years of difficult adaptation to the new national situation, the Russian space industry has secured its position by opening up to the rest of the world and promoting clearly defined specialities that have left their mark on the way it has restructured its activities.

Main features of the Soviet space system

The structure of the Soviet space industry long remained obscure, being integrated into the Military Industrial Commission (MIC). Perestroika and the ensuing more open attitude in Russia have made it possible to form a clearer idea of internal organisation. The specificity of the Soviet way of working may nevertheless remain difficult for a Westerner to comprehend.

Space flight research predates the 1917 revolution but it was under the Bolshevik regime that the first great research centres were set up and K. Tsiolkovsky was elected to the Academy of Sciences. After a reorganisation in 1928, the Academy was assigned the task of developing research in a new area, that of technology. Under Stalin, a tighter structuring of activities in the field of 'interplanetary communications' took place. The term referred to space flight, with special interest in rockets written into the first five-year plan. Hence, by the middle of the 1930s, the new generation of Soviet space scientists had already found their place. Both S. Korolev and V.P. Glushko directed teams working on missiles and, at the end of the Second World War, this research was integrated into the weapons production industry. It was thus given absolute priority. The Kremlin was particularly interested in developing intercontinental missiles in a tense international situation. At the beginning of the 1950s, the first Soviet rockets were tested at Kapustin Yar and Baikonur by two teams, one led by V.P. Glushko and S. Korolev and the other by M. Yangel and V. Chelomey. When Stalin died, the space programme run by L. Beria was cut back to launch vehicle activities. Research and industrial centres taking part came from the aeronautic sector and fell under the authority of the VPK (Military Industrial Commission). The Communist Party agencies were the main decision-making bodies. At the Politburo, the person in charge of the defence industry was also responsible for the space sector, and the defence committee was supervised by the Central Committee's secretary for defence issues.

With the arrival of N. Khrushchev, the political leadership became interested in satellites. A special committee was set up under the auspices of the Academy of Sciences and it was announced that the Soviet Union would take part in the International Geophysical Year. The space programme attracted interest from military officials and scientists alike, and the secretary general accepted S. Korolev's proposal to merge the satellite launch with the missile programme. The extraordinary impact of Sputnik on the USSR's international image was used to make the space programme a symbol of communist superiority over capitalism. In 1959, the organisation of the space programme was reviewed in order to optimise its efficiency. Responsibility for ballistic missiles was entrusted to the Strategic Rocket Forces (RVSN) whilst cosmonaut training was assigned to the Air Force in 1961. A new government committee was created to coordinate scientific research, replacing the Academy of Sciences, which handed over control of its technical institutes.

The lack of a central structure to run the space programme as a whole was compensated by a personalisation of responsibility. Engineers and scientists heading 'design bureaus' were directly involved in the space coordination committee and could have a significant influence on members of the Politburo and the Central Committee. S. Korolev, in particular, held a great many official responsibilities and was involved in both technological and scientific aspects of the various programmes.

When L. Brezhnev came to power, so also did a whole team of people raised in the traditions of the Military Industrial Commission, with intimate knowledge of the workings of the space industry. In order to bring greater cohesion to the sector, a specialised ministry was created in 1965. This was the Ministry for General Machine Building (MOM) whose mission was to coordinate space activities as a whole, from applied research to manufacturing. Notwithstanding, some important institutes remained under the authority of the Ministry of Aviation Industry (MAP) and the fact that personal relations were maintained between programme directors and political leaders often meant that MOM was required to ratify decisions taken at the highest level. S. Korolev died in 1966, depriving the sector of a unifying personality, and the increased rivalry that arose between General Designers did nothing to improve the situation. At the same time, MOM was unable to rationalise the management of space activities.

On the other hand, the Brezhnev era brought about a certain degree of institutionalisation in the space sector. Decisions were taken by the highest authorities of the Communist Party, whilst management was carried out on a ministerial level and the Academy of Sciences dealt with fundamental research and foreign relations.

From 1985 to 1991, M. Gorbachev's reforms began with the appearance or revitalisation of new authoritative bodies. Directives applied by the Council of Ministers were no longer issued solely by Communist Party agencies, even though these maintained a strong hold. At least in theory, they were also released by legislative bodies including in particular the Supreme Soviet. Moreover, Gorbachev's Union Treaty project presupposed that republican and federal structures would be integrated into the definition and realisation of programmes. In fact, these reforms never saw the light of day. Despite new political prospects concerning the role of MIC and in particular the desire to convert companies in the space sector, the overall appearance of the system was barely altered. The Soviet space industry continued to operate via the same channels with regard to decision-making and allocation of resources.

Power basically remained in the hands of the ministries since they applied decisions taken on the political level and allocated finances as well as manpower and equipment. Some ministries involved in the space sector also fulfilled a manufacturing role. These included MOM, by far the most important, MAP, the Ministry of Defence Industry, the Ministry of Electronics Industry and the Ministry of Radio Industry. Some like the Ministry of Defence or the Ministry of Telecommunications played more of a service role, whilst still others were essentially users, such as the State Committees for hydrometeorology or geodesy and cartography.

Unlike the ministries and State Committees, the Academy of Sciences only had access to research institutes. In order to realise its projects, it had to go through design bureaus depending on the ministries, and in particular MOM. Soviet scientific research was organised by various bodies: the Academy of Sciences for theoretical work, and the State Committee for Higher Education or the various industrial ministries for applied work. The Academy of Sciences has always been responsible for representing space activities outside the Soviet Union, although space was never an activity in its own right, and the main institute for space research,

the IKI created in 1965, was eventually absorbed into the astronomy section. More specialised in planetary geochemistry, the Vernadsky Institute founded in 1966 sometimes found itself in direct competition with IKI. Many other institutes of the Academy were involved in space research and space featured in a number of the eighteen main research programmes.

Apart from the institutes making up the Academy, the institutes of further education such as the Moscow Aviation Institute (MAI), and the central research institutes of the ministries, such as the TsNIIMash for MOM or TsAGI for MAP, played a key role not only as a result of their specialisation and technological knowhow, but also through their intermediate position between the ministries and industrial units. In theory, links between civilian and military institutes were handled by the Committee for Science and Technology (GKNT) although there was some debate about how effective it was.

At the end of the Soviet era, the ministries in charge of space activities still depended on the VPK and hence on the military industrial sector. The role of the Ministry of Defence and the existence of one branch in particular, the Strategic Rocket Forces, strengthened the commonly held view in the West that Soviet space activities were heavily militarised. In reality, although it was true that space held a priority role in the economy, military influence was by no means domonant. The industrial sector was also given a great deal of independence. What characterised the space sector was rather the heavy bureaucracy and many reforms were put forward to improve the way it worked. The disappearance of the Soviet Union would change everything from the very foundations, even bringing the political interest of space activities into question.

Sharing the legacy

The break-up of the Soviet Union gave rise to two types of problem which had to be faced simultaneously. One was the independence of the republics, which led to the loss of certain facilities, and the other was the complete disarray of state executive bodies that had previously managed space activities. The geographical distribution of Soviet space potential created special difficulties. For example, Kazakhstan asserted its ownership rights over the Baikonur cosmodrome, immediately rechristening it Tyuratam in 1992, before it once again became Baikonur. Likewise, republics possessing ground-based installations such as receiving stations or tracking stations announced their intention to take total control over them. In the same vein, several republics took the line that they actually owned a share in the space achievements of the ex-USSR, among them the Mir space station. They argued that these programmes had been financed on a federal basis.

In fact, the states making up the CIS were quite aware of their interdependence and rapidly set about establishing an organisation that would guarantee shared use of such infrastructures. Space was one of the first items in the cooperation treaty signed, with the exception of the Ukraine at Minsk in 1991. In this agreement, Russia maintained its hold over the running of space activities. Indeed the treaty confirmed Russia's dominant position in this respect, and it would have been difficult for any country to challenge. Russia controlled more than 80% of ex-USSR space potential and replaced the Soviet Union as its successor state in international organisations. In particular, by covering almost all military expenditure, it kept its strong-hold over CIS armed forces, including the Strategic Rocket Forces dependent on them at that time.

Any initial illusions about the potential profits from space activities were soon dispelled and most new states were dissuaded from demanding a common space sector when they discovered the costs involved. The Baikonur cosmodrome was leased to Russia and the Space Forces, responsible for satellite launches, were placed under the authority of the Russian Ministry of Defence. Practical questions ranging from the running of ground stations to the sharing out of general interest space systems were dealt with by specific agreements, some less official than others.

For its part, Russia had to cope with the poorly organised links between government, military and industry. Serious problems were posed for a sector accustomed to regular state revenues. The established decision-making authorities (the Communist Party executive bodies) and management organisations (MOM) disappeared from the scene, and the Military Industrial Complex remained largely in the background.

Russia

The new organisation of the Russian space sector featured a key innovation: the creation of a space agency on 25 February 1992 (fig. 5.7). This was the *Rossyskoe Kosmicheskoe Agenstvo* (RKA), an extremely lightweight structure comprising only a few hundred people responsible for government policy and the federal space programme. In comparison, MOM, for which space was only one field of activity among many others, employed around two million people. RKA inherited some existing facilities, such as TsNIIMash, the institute responsible for elaborating space research programmes, whilst other entities assumed a separate existence, such as the Glavkosmos agency created in 1985 to bring together all space activities and watch over their development. Despite internal rivalry, RKA soon imposed its authority through the experience of its personnel, most of whom came from former political and management structures. There was also increasing support from space companies keen to be represented at government level and to avoid the ascendency of NPO Energia, the largest company of the space sector. Moreover, RKA gained the support of its foreign counterparts who used it as a landmark within a generally disorientating global organisation. The new space agency thus began to play a key role in international partnerships, including the establishment of commercial agreements.

In July 1994, a further decree gave RKA responsibility for 38 space companies depending on the Goskomoboronprom (State Committee of the Russian Federation on Defence

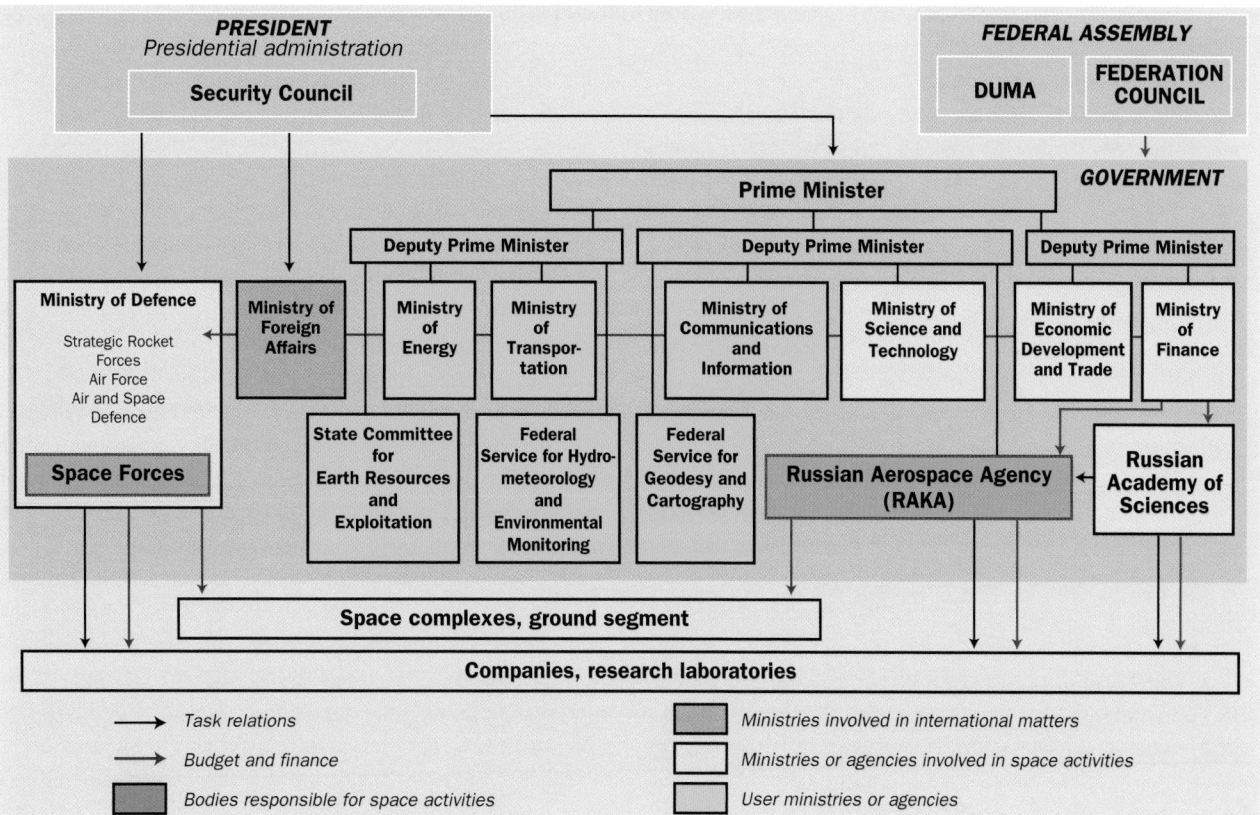

Figure 5.7. **Organisation chart of Russian space policy in 2001.**

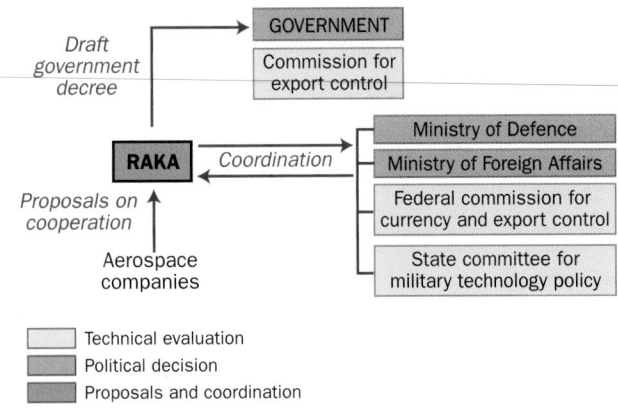

Figure 5.8. **Approval process on export of space technologies in Russia.**

Branches of Industry), thus broadening the organisation's range of activities. When yet another decree set out its new powers in May 1995, the agency officially became the central element. It defines Russian policy in the field of space activities and has responsibility for elaborating and carrying out national civilian programmes. It represents government interests on a national and international level. For all these activities, decrees were issued to ensure that the agency act in association with the Ministry of Defence and the body in charge of defence industries (fig. 5.8).

In March 1999, it was announced that authority over 350 companies in the aviation industry was to be transferred to RKA. At the time, all these companies depended on the Ministry of Defence. In July 1999, by governmental decree, RKA became RAKA (in English, Russian Aviation and Space Agency, RASA) or Rosaviakosmos. There were several reasons for grouping space and aviation industries together under the same regulatory authority. RAKA's range of activities was extending into the field of marketing, which it was to coordinate and support, a role made official by decree in October 1999. This attests to an increased willingness on the part of the Russian government to develop international cooperation. In addition, a synergic effect was expected given that Western partners were themselves gathered into large groups (US Lockheed, European EADS, and so on). Finally, although the agency had no formal regulatory role over the aeronautic companies, some of its leaders were already on their boards of directors and the idea of creating a large group encompassing the whole of the aeronautic industry was not a new one. In practice, the process seems rather slow. Even the official name of the agency only changed in the middle of the year 2000, and by the beginning of 2001, there had still been no transfer announcements.

The second important executive body is the Ministry of Defence. It retains the role of defining military space policy and it finances purely military activities and companies. The Russian Space Forces (VKS) are in charge of launches and space base infrastructure. After reintegrating the Strategic Rocket Forces (RVSN) between 1997 and 2000, they offi-

cially became independent once more in June 2001 when military reforms restructured the whole of the armed forces. The Air Force continued to run the Gagarin Cosmonaut Training Centre.

There are relations at different levels between the Ministry of Defence and RAKA. Within the Ministry, the Russian Space Forces handle launches and operate the military space segment, whilst the Space Defence Forces, which also became independent of the RVSN under the military reform, are responsible for monitoring space and air space and deal with ballistic missile defence. In addition, the Ministry of Defence sets up its own programmes. However, the civilian–military duality of space applications and difficulties of financing mean that the RAKA is often brought in on these projects, thus limiting the share of investment for both bodies. Certain Earth observation programmes (Orlets and Kometa see chapter 12) serve as examples.

As far as users are concerned, the role of the Ministry of Communications should not be underestimated. It negotiates frequency allocations and now elaborates its own space telecommunications programmes with independent financial backing. In contrast, the State Committees and Federal Services (hydrometeorology and environment, or geodesy and cartography) are clients with strictly limited means.

Beyond this, the main ministries also feature from time to time, insofar as space activities overlap their specific fields of interest (budget, privatisation, international agreements, and so on).

The transformation of the former State Committee, the GKNT, into the Ministry of Science and Technology together with a renewed insistence on the importance of its mission signal a general reassessment of the situation. Given the inadequate budget, this concern has little practical effect.

Situated outside the government, the Russian Academy of Sciences remains the principal entity defining Russian space science, in which international partnership, although limited, plays the main role in view of the fragility of the Russian budget. The fact that relations with the Academy are mentioned amongst the missions of one of the first Deputy Prime Ministers strongly implies a desire for coordination. It also shows, at least on the level of public declarations, a certain awareness of the need to preserve research activities.

A further innovation in the way the Russian space sector is run stems from the existence of a legislative power. The Duma votes the budget and is likely to intervene in certain aspects such as the regulation of space-related business. In fact, its powers are somewhat restricted and the vote for the budget is a largely formal affair. Members are only partly informed and there is no guarantee that the agreed sums will ever be paid over. Furthermore, space activities, which incite no special interest as such, are only discussed by committees devoted to other areas, such as defence, education, or foreign relations.

The new organisation of the Russian space sector closely resembles that in Western countries, with space activities growing less specific. Management difficulties, highlighted by all concerned, stem largely from a low level of financial support and lack of any national policy. From this standpoint, President Vladimir Putin's declarations in January 2001 demonstrate a new official awareness of the potential of the space sector. Speaking before the Security Council on state policy with regard to the space industry, his main emphasis was on the high technology role of space as a basis for economic competitivity and national security. The tone of argument may well have changed since the days of President Yeltsin, but in practice the share of the budget allocated to such activities shows that these ambitions are to be accomplished rather through international partnership. State funding is only just sufficient to preserve what has already been achieved in the sector.

The industrial fabric

Russian companies have a rather unusual internal organisation directly inherited from the Soviet system. Design bureaus denoted KB (*Konstruktorskoe Byuro*) or OKB (*Osoboe Konstruktorskoe Byuro*) are the key feature. In the Soviet era, they sometimes employed thousands of people. Their task was to design and develop systems to the prototype stage, or even to the level of a pilot production run. The most important descended from the pioneering organisations of space conquest. Originally headed by prestigious and powerful General Designers such as S. Korolev, V. Chelomey, M. Yangel or V.P. Glushko, they often continue to carry their names and long maintained their initial specialisations. In 1974, these Design Bureaus were integrated into Science and Production Associations (or NPOs in Russian) as part of a global reform of the Soviet economy which aimed to strengthen links between research institutes and manufacturing units.

Since 1989 political priority has been given to conversion of the military industrial sector. In itself this did not cause any particular difficulties to the space industry insofar as a great many space systems could easily be adapted to fulfill civilian requirements. The basic problem was rather how to finance these activities. Drastic budgetary cuts and such a low level of national investment were catastrophic for a sector which, in every country around the world, operates basically under governmental funding. Moreover, companies were handicapped by their numbers and size, not to mention the complete lack of financial and social support for restructuring and a geographical distribution across Soviet territory that was designed to satisfy anything but economic constraints (fig. 5.9). In addition, the possibilities for commercialisation, even though they seemed at first glance to be relatively straightforward to companies who could be so proud of the diversity and quality of their products, especially in the launch vehicle sector, turned out to be limited in certain obvious ways. The main problem was a lack of business experience in companies where the previous system for allocating resources paid little attention to the notion of production costs. They were also faced with a relatively closed international market in which the United States and Europe were firmly committed to preserving their market share.

In fact space activities were only kept going because OKBs in this sector were able to function more or less independently. Even though they suffered from uncertainties in the supply of raw materials, the OKBs at least possessed roughly the whole production line. In particular, they could lay claim to vast stocks referred to as strategic reserves which allowed them to mitigate the sudden drop in production. However, since 1995, it seems that these reserves have almost been exhausted.

The number of companies and the workforce of the Russian space industry was considerably reduced and it had somehow to obtain a new lease of life. It was faced with a slow domestic market, owing to the lack of both civilian and military financing. In addition, it suffered from a relatively ineffective internal organisation that can be put down to competition between the NPOs and the drain of skilled labour, in particular from certain subcontractors. It thus became an absolute priority for companies to penetrate the international market. Independent attempts to commercialise space products were viewed with some reluctance by an American and European establishment that had no desire to see a new competitor in a market where they were for the moment the only players. The development of joint ventures, in the launch market for example, was one way of getting round the obstacle. The reliability and low cost of Russian technology in certain fields such as propulsion gradually led to increased interest, not only from the established space powers, but also from countries new to the sector. Several prescriptions proved to be successful. The first kind, well known in the context of manned space flight, proceeds through international projects backed up by intergovernmental agreements with specific agreements between the relevant space agencies. The second kind consists in joint ventures with foreign companies, as is now common practice in the launch area. A third approach involves contracts for developing specific technology, as often happens for propulsion or telecommunications. The last category is the marketing of services, where launches once again constitute a prime example.

Today, private investment makes up the major share of the budget in Russian companies. With an estimated fifteen times fewer orders than at the end of the Soviet era, the space industry earns most of its profits from international activities. Results may be in the neighbourhood of 800 million dollars, representing almost three times the official budget. According to various experts, the sector may have climbed back to its 1993 level, but the reinvestment of this money supply is not a straightforward matter. The RAKA executive does not conceal the fact that these funds are essential if the sector is to survive. However, the more affluent firms are more and more reluctant to participate, arguing that they must maintain the same level of investment as their Western competitors and partners if they are to remain in existence.

MOSCOW REGION

Legend:
- □ ⬚ NPO and OKB
- ○ ○ Factories
- △ Institutes
- ■ Research and services
- ▨ Launch vehicles and engines
- ▨ Satellites and subsystems
- □ Components
- ▨ Ground systems
- ☆ ☆ Launch bases

0 ___ 500 km

NPO, OKB and factories

1 NPO, KB, Holding Company Arsenal
2 KB Biofizpribor
3 TsKB Mashinostroyeniya
4 NPO PM (applied mechanics)
5 KB Khimavtomatiki
6 VMZ (Voronezh Mechanical Plant)
7 Barrikady (plant)
8 Frunze MZ (plant)
9 NPO Polyot
10 PO KMZ (optomechanics plant)
11 TsSKB, Samara GKNPTs
12 KB Foton
13 Progress (plant)
14 Votkinsk plant
15 NPO Iskra
16 Kirov plant
17 KB Makeyev
18 UKVZ (Ust-Katav wagon-building plant)
19 AKO Polyot (plant)
20 Krasmash (plant)
21 NPO Almaz
22 NPO Geofizika
23 KB OM (Barmin)
24 KBTM
25 KB Kometa
26 NPO Kvant
27 OKB MEI

28 NPO Molniya
29 KB Motor
30 NPO AP
31 NPO TP
32 KB Salyut
33 NPO Tekhnomash
34 NPO Tekhnopribor
35 TsKB TM (Rubin)
36 NPO Vega
37 NPO VIAM
38 NPO VILS
39 OKB Vympel Corp.
40 GKNPTs Khrunichev
41 NPP Nauka (plant)
42 NPAOZT Elas
43 NPO Lavochkin
44 NPO Energomash
45 Krasnogorsk plant
46 NPO Planeta
47 RSC, NPO Energia
48 KB Khimmash (Isayev)
49 OKB Fakel
50 NPO Kompozit
51 NPO Mashinostroyeniya
52 NPO Soyuz
53 NPO, NPP, OKB Zvezda
54 KB Myasishchev
55 Tekhnopribor (plant)

56 Kalinin Machine Building
57 KB Start
58 OKB Nadiradze
59 Biisk Special Chemical Plant
60 NPO Source
61 NPO Elektronika
62 Orenburg Machine Building Plant
63 PO Strela
64 Avia Dvigatel
65 NPO Troud
66 NPO Almaz
67 NPO Integral
68 NPO Yuzhnoe
69 Pavlogradski Mechanical Plant
70 NPO Orbita JSC
71 NPO Khardron
72 Kharkov Electrical Equipement Plant
73 Kommunar Production Association
74 Chernigov Radioinstrument Plant
75 NPP Obriy
76 NPO Ukrkosmos
77 NPO Priroda
78 NPO Radiant
79 Almaz JSC
80 NPO Kurs
81 Kiev Radio Plant
82 Kievpribor Production Association

Research institutes

83 Vavilov institute
84 RIRT (Russian Institute of Radionavigation and Timing)
85 NII Germes
86 NII Mashinostroyeniya
87 GPO Agat
88 Babakin Research Centre
89 IMBP
90 NTTs Kompleks
91 MNII RS
92 MNIRTI (Glavnoye NIRTII)
93 NII PM
94 NII TP (Keldysh Research Center)
95 RNII KP
96 Tupolev ANTK
97 VNIIEM NPP
98 NII EM
99 TsNII Mashinostroyeniya, export
100 NII Khimmash
101 GNTs NII
102 NII Kompozit
103 IEM Institute
104 Institute of Atmospheric Optics
105 Dnieprovski State Design Institute
106 S. & R. Institute for Radio Measurements
107 Instrument Building Tech. Research Institute
108 Soyuz Research & Design Institute

Figure 5.9. **Geographical distribution of the main CIS establishments known to be involved in the space industry.**

The Ukraine and Kazakhstan

From a geographical standpoint, the OKBs and NPOs are very unevenly distributed across the ex-USSR. Almost all of the largest and best known companies are located on Russian territory. The Ukraine is an exception to this rule, since it plays host to a significant number of space companies, well integrated into the Russian manufacturing circuit, as well as ground-based facilities (see fig. 5.9). The creation of the National Space Agency of the Ukraine (NSAU) by presidential decree in March 1992 was intended as an assertion of sovereignty, seen as essential in the face of Russian hegemony. However, adequate financing was not forthcoming. In such a context, the main Ukrainian space manufacturer NPO Yuzhnoye chose to maintain its former links with Russian companies, in particular for producing the Zenit launch vehicle. Potential competition with Russia on the launch market subsequently led them to seek a wider range of partnerships. Specific agreements were signed with the United States to fix quotas of Ukrainian launches until the year 2000, but the situation remained relatively ambiguous since part of the production was handled in Russia. In the same way, the Sea Launch project involves Russian partners among others (see chapter 6).

In fact, the situation in the Ukrainian space sector greatly resembles that in its Russian counterpart. Once again space is not a political priority and at the same time defence conversion, self-financing and restructuring problems are rife. To make things worse, the space sector is heavily dependent on state authority. Efforts to preserve acquired knowhow and gain returns on earlier investment have not been free of difficulties. After a long period of hardship, the sector seems today to be in a better position. Partnerships have brought in sufficient resources to maintain activities and companies are beginning to diversify. For example, the conversion of ballistic missiles into small launch vehicles, sometimes marketed through partnership agreements, is now bringing results even though profits are still limited.

In September 1991, Kazakhstan staked its claim as a space power, at least in a symbolic sense, by announcing the creation of its own space agency. However, it has no other income for this purpose than the moneys received from leasing out the Baikonur cosmodrome to Russia. Since then, several agreements have been signed with Russia without establishing any real consensus between the two parties. This explains the recurrent debate at certain levels of Russian administration about the need for space bases on its own national territory. In fact, Baikonur remains an essential feature of Russia's manned space activities and it is heavily used for commercial launches into both geostationary and non-geosynchronous orbits and manned space flight to ISS. In 2000, the officially quoted rent represents in theory one quarter of the Russian space budget. This shows the artificial nature of the situation, although the two partners claim that relations are stabilised in a context where both feel the agreement is to their advantage.

The United States

There can be no doubt that Sputnik roused the interest of the American authorities in space. However, more directly military considerations had already defined its main uses. Related to the growing importance of ballistic missiles, the purpose of the first space programmes was twofold. One aim was to keep watch over Soviet arms stocks and another was to identify potential targets. This priority, attributed in the mid-1950s to observation and surveillance satellites, was to firmly fix the dividing line between sustained military activity and rather less regular civilian programmes. Recent political transformations are not without consequence for this long-standing arrangement, still the hallmark of the way American programmes are generally organised.

The main distinguishing feature of American space policy is certainly the sustained overall level of financing, set well above the kind of budget allocated elsewhere in the world. In such conditions, the development of a significant industrial base has further reinforced this favourable tendency. The United States is the first military space power by a very wide margin and it also leads the field in the market for applications such as telecommunications and Earth observation, whilst its launch activities are currently under restructuration. Today it is an absolute priority of American space policy to maintain and even enhance US supremacy. As far as civilian activities are concerned, the political powers are moving towards a deregulation of space telecommunications and the end of the large international organisations. At the same time, they hope to introduce services such as launch services into WTO negotiations. On the military side, the desire to dominate is made even clearer with notions like Space Control. At the turn of the third millennium, the United States views space as a basic component of American power which should be reinforced.

Organisation of the space sector

Since its inception, the American space sector has always been officially divided into civilian and military responsibilities. Civilian research and experimentation activities have been run by NASA since 1958, whilst military space interests were finally grouped together under the authority of the US Space Command in 1984. Scientists and manufacturers constitute two key categories in the accomplishment of space programmes. The philosophy behind the overall organisation is largely based on a clear awareness of their specific roles, as is shown by a short historical review of the US space sector and its organisation.

The United States hardly provided a favourable context for the beginnings of the space era, except possibly on a cultural level with a wealth of books and magazines devoted to science fiction. Traditionally, activities such as scientific research could only be developed in private institutions like universities, foundations or specialised societies, and never directly by the government. Specific contracts could be issued by the administration once the worth of a certain area of research had been established.

The Second World War marked a turning point in that a new notion of federal government responsibility came to the fore. Scientists and manufacturers were mobilised by pressure of circumstances and new government structures set up. In this context, military authorities carried the most weight, but after the war, when the benefits of effective government support had been observed, there was no longer any question of returning to a research system based entirely upon private initiative. Even though it was now recognised as a necessary step, the establishment of a civilian organisation was still no simple matter and it was under the authority of the US Army that Werner von Braun's team first pursued their research in the United States.

In parallel, the conversion of the war industries brought out the specific problem of the aeronautic industry for which a purely civilian market was insufficient. The government was forced once again to abandon its principles and guarantee the development of activities that were crucial for national security by placing its own orders for aircraft.

At the beginning of the Cold War, launch activities were carried out under direct military control. The German V2 bombers had demonstrated the effectiveness of these new weapons, but their application to space conquest incited little enthusiasm, except amongst science-fiction readers and members of groups like the American Rocket Society. Even the latter had had to change its name (from the American Interplanetary Society) in order to increase its credibility.

In fact several problems had already arisen. The first concerned global strategy. The United States was convinced that it must assume the role of defending the free world in a now permanent manner. From a financial standpoint, they hoped to contain the military budget, whilst at the same time maintaining a large number of bases around the world. They also hoped to promote a mastery of the air and the atom. Despite the fact that a report by the Rand Corporation stressed the potential of space conquest for strategic purposes as early as 1946, missile research did not initially stand out as a priority. Rocket manufacturers already integrated into the military industrial sector saw their credit drastically cut. Moreover, the US Navy and Army were each developing their own specific programmes. The US Navy was pursuing the Viking programme in a scientific study of the upper atmosphere and the US Army was occupied with developing its Redstone launch vehicle. For its part, the US Air Force had obtained a privileged position in the development of long range vehicles but had opted for bombers. However, in 1951, the Air Force initiated the Atlas long range intercontinental missile project which soon became a political priority. Space conquest represented a further stage that had already been discussed in another report by the Rand Corporation. This report stressed the technical feasibility and also a great many political implications. In particular, it discussed the risk that a satellite launch might be viewed by the Soviet Union as a direct threat to its sovereignty. In such circumstances, the already long-standing proposal to launch a satellite within the framework of the International Geophysical Year offered an interesting opportunity. The choice of launch vehicle was based on the same reasoning. Although the preference for the Viking rocket in 1955 was not the best from a purely technical point of view, it had been designed for scientific research and this strengthened the civilian character of the project.

When Sputnik was launched, it caused a great stir in the media which undoubtedly marked the main turning point for the United States. What counted from that point on was to reestablish the American image in world affairs and in the opinion of its own public. The need to develop innovative technology opened up a new era in which the federal government was compelled to increase its power in order to compete effectively against its communist rival. The new need was met by reorganising the Department of Defense and creating congressional committees to deal with space-related matters. NASA itself was founded partly to encourage growth in the space sector, but partly also to subordinate military space activities under civilian authority, as attested by the fact that W. von Braun's development team and that of the Jet Propulsion Laboratory (JPL) were transferred from army control to NASA.

By specifying the Moon as the great challenge that the United States was to take up, J.F. Kennedy gave space a new symbolic value. On the one hand, the basic principles of US space policy, those defended at the United Nations or the idea that civilian authority should prevail over military, remained unchanged. On the other, space was now to play a leading role in mobilising the country's intellectual and industrial resources. D. Eisenhower wanted to protect America from the ambitions of the military industrial complex and the technological and scientific elite who were, in his view, particularly tempted to impose their lobbying in the space sector. In the end, space became more an arena for symbolic exploits motivated by political necessities. In

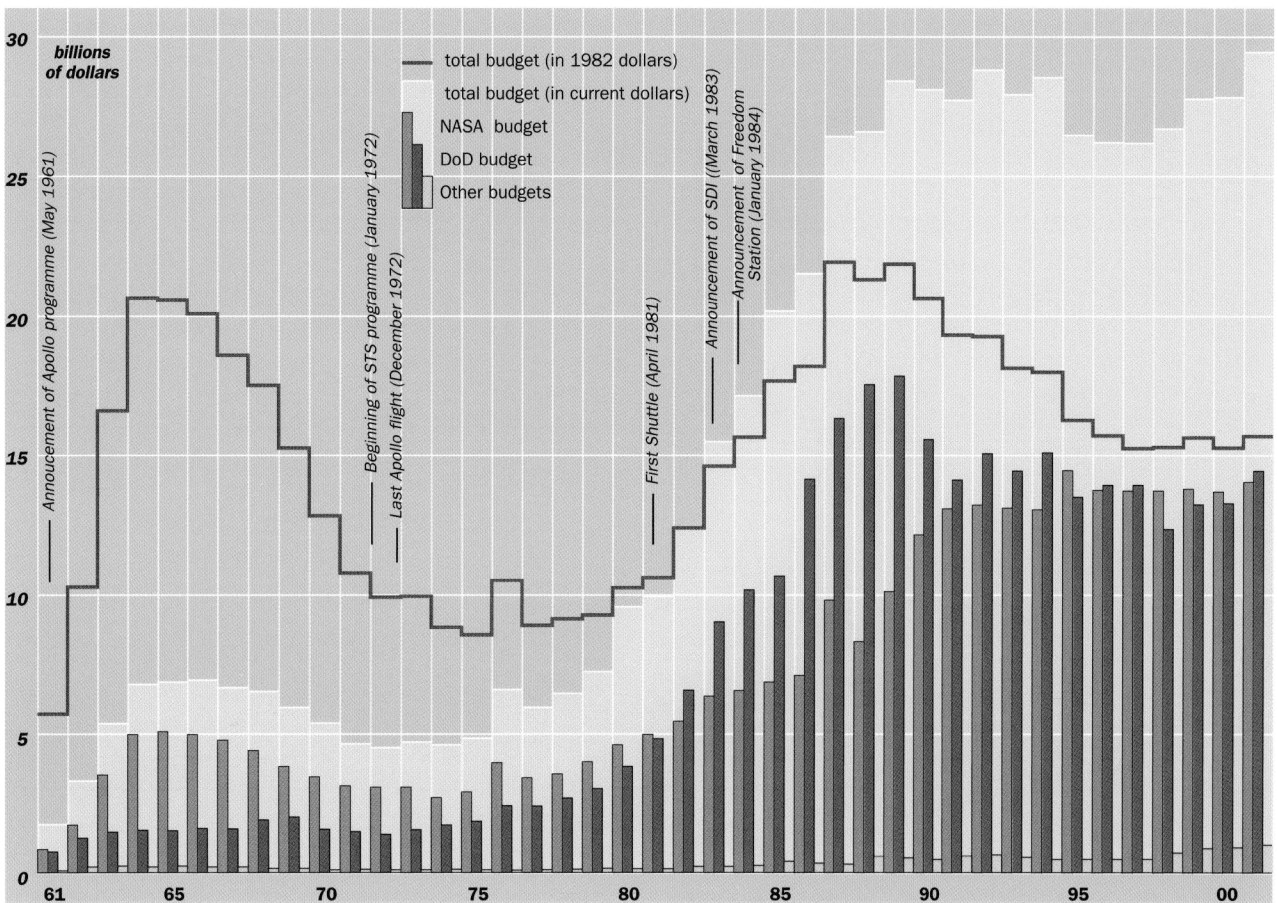

Figure 5.10. **Evolution of the US space budget according to administration.** Administrations other than NASA and the DoD are the Ministries of Energy, Commerce and Agriculture, the Department of the Interior and the National Science Foundation.

another respect, the structure of the space sector and the separation of civilian and military power led to a specific kind of bureaucratic response. NASA's strategy was to preserve its resources, whilst the Department of Defense obtained a substantial increase in investment for the military space sector.

The way the US space budget has evolved clearly illustrates this combination of political constraints and internal stipulations (fig. 5.10). Space endeavour reached its peak with the Apollo programme. This was followed by a sudden but then more gradual decline until the end of the 1970s. The Space Shuttle programme was unable to supplant the lunar programme, the latter being motivated by far more than a pure interest in space. Political indifference led to a series of budgetary reductions and in the end the Space Shuttle fell a long way short of initial projects to build a fully reusable vehicle. Concerning the space station, which represented the next logical step, its ups and downs soon proved how difficult it would be to justify investment in a large scale manned programme. Initial ambitions for the Freedom international space station in 1984 were significantly revised downwards. Whilst a first mutation of the project, known as Alpha, was built on American–Russian collaboration in 1993, successive adjustments to the budget and technical specifications eventually led to the current International Space Station programme. In 1997, it was finally decided to finance the construction of this new station over five years with a total construction cost estimated at 5.2 billion dollars. By 2000, costs had risen to 9 billion dol-

lars, an amount further increased by requests for budget extensions from the general contractor, Boeing Co. At the same time, the company announced that they would have to push back the deadline, and NASA finally scheduled the fully operational station for April 2006. The commitment by Congress back in 1997 to an annual budget of 2.5 billion dollars to set up the station by April 2002 thus falls seriously short of the target and could be extended for a further four years at an additional cost of slightly more than 6 billion dollars. Despite these difficulties, which caused NASA to spend more than 5.4 billion dollars on manned flight activities in 2001, investment has gone ahead in other sectors. The share devoted to space sciences and technology remains particularly significant. At 6.1 billion dollars, it represents the agency's largest budget, corresponding to 43% of annual expenditure.

American space policy is established by three main institutions: the Executive Office of the President, the legislative authorities and NASA (fig. 5.11). Choices made in the area of space policy largely reflect the prevailing balance between their various approaches.

When a proposal goes for consideration by the executive authorities, it traditionally transits via certain key figures belonging to the government or the presidential cabinet.

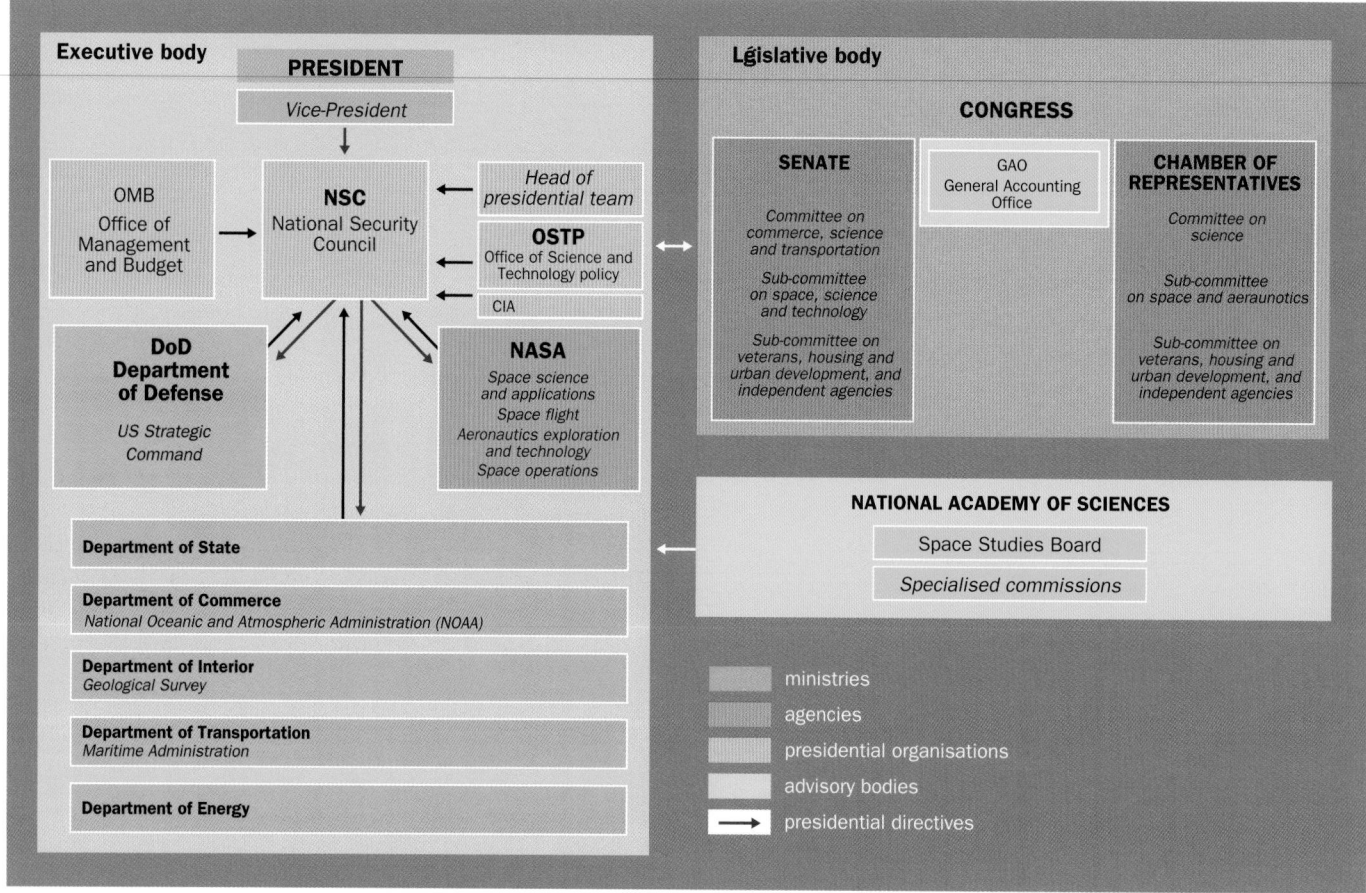

Figure 5.11. **Organisation of American space policy in 2002.**

Within the government, the Ministry of Defense and the State Department are the two ministries most closely concerned with decisions relating to the space sector. The Ministry of Defense intervenes for programmes raising questions of national security and for the development of the various military systems. Among the different representatives, the Air Force Secretary holds a privileged position since the Air Force is in charge of space matters at the Pentagon and its involvement in programmes is an important element of its own strategy. The State Department is especially concerned with international repercussions of the major programmes. Other ministries such as the Department of Commerce or the Department of the Interior may also be interested for various reasons, usually for specific applications like cartography or management of natural resources, for example.

At the White House, two bodies depending on the presidential cabinet are particularly involved: the Office of Management and Budget (OMB) and the Office of Science and Technology Policy (OSTP). The task of the OMB is to monitor public programmes of all kinds and fix a budget ceiling. Relations between the OMB and NASA are often difficult. The OMB is close to the main concerns of the President in office and has a considerable power of intervention. It represents the main political counterweight to the power and demands of the space agency.

Since the National Space Council was abolished in February 1993, the OSTP has been the presidential body in charge of space matters. The role of this authority extends to preparing the major decisions, cutting across both civilian and military considerations, as was the case for the presidential directive of 29 March 1996 concerning the restricted opening of the GPS system to commercial use. The OSTP acts through the National Science and Technology Council (NSTC), created in November 1993 to coordinate all federal investment in research and development. Apart from the President, others on the council are the Vice President, the scientific affairs advisor, the budget director, and roughly fifteen other members of the presidential cabinet and federal agencies responsible for scientific and technological programmes. Since 1994, five presidential directives on space have been prepared within the NSTC as part of the Presidential Review Directives (PRD).

In 2001, a significant part of the functions of the OSTP is set to be taken over by the National Security Council which will then be responsible for coordinating space-related matters where they are considered to affect national security. This coordination task should be implemented through a Policy Coordinating Committee for Space (PCCS) set up within the NSC. Its members will include the main representatives of the relevant parts of the government.

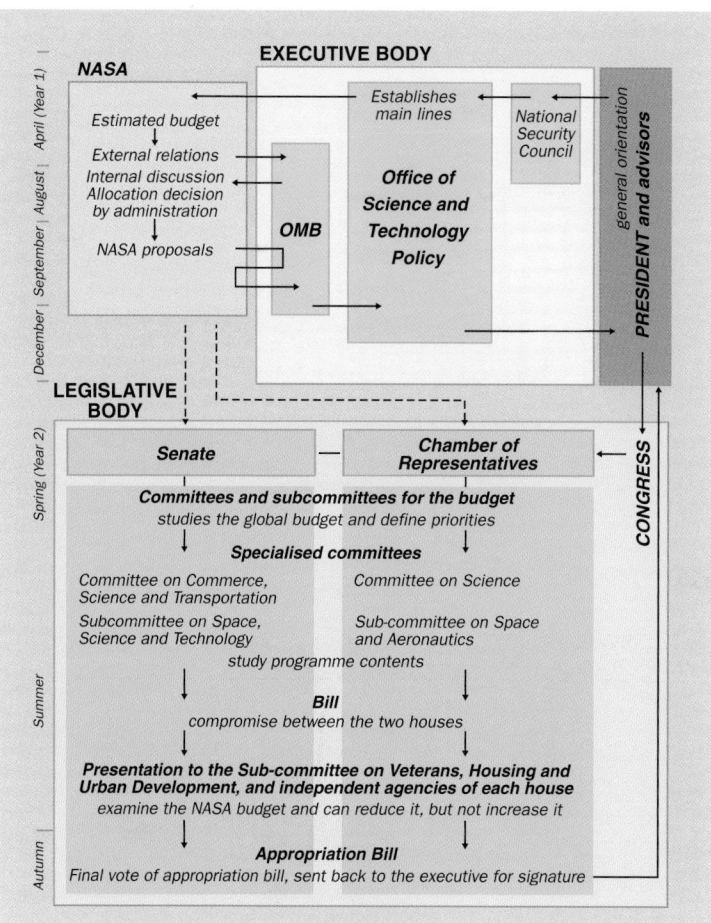

Figure 5.12. **Elaboration of NASA's budget.**

NASA's funding is finally determined by the Appropriation Committees of the Chamber of Representatives and the Senate. These propose the budget that Congress must vote on. They are highly politicised bodies and constitute one of the key centres of legislative decision-making as far as space programmes are concerned, confronting them, in the same budget line, with welfare and health programmes, not to mention the war veterans budget.

At the end of this process, the funds actually voted by Congress often differ widely from those originally proposed by the White House. However, space policy has never given rise to direct confrontation between the executive and legislative bodies. In fact, the long delay between proposals and final decisions also explains how priorities may sometimes shift.

Decompartmentalisation of military activities

For a few years now the United States has been keen to decompartmentalise space-related activities and structures left over from the Cold War. Following government directives, NASA and the Pentagon have been entertaining ever closer relations, in particular, by the increase in the last few years of horizontal structures like the Joint Program Offices. After 1995, this tendency was consolidated by several restructurations of Pentagon departments. One of the principal aims was to instill the use of space more deeply into military practice, so that it might eventually become a crucial element in future weapons systems and the running of military operations. It fell upon the C3I organisation (Command, Control, Communications and Intelligence) to make the best use of space programmes and set out the main features, including those involving international partnership. The choice of this department to house space activities at the Pentagon clearly demonstrates the importance assumed by space in the so-called Information Dominance Strategy, through the development of the main military information networks. In parallel, these reorganisations also sought to give users a more important share in laying down guidelines and defining programmes. The role of Space Architect was thus created in 1994 to collect requests from the different departments of the Pentagon and organise a concerted development of programmes. In 1998, this role was broadened with the appointment of a National Space Security Architect whose coordinating task now extended to space-based intelligence. This realignment of existing space activities in the Pentagon was carried out within the framework of the National Space Security Steering Group, a sort of interdepartmental forum devoted to space activities.

The creation of the National Imagery and Mapping Agency (NIMA) in October 1996 stemmed from the same urge to decompartmentalise, associating purely military missions traditionally handed down to the National Reconnaissance Office (NRO) with missions of a more civilian character, in accordance with the new strategic guidelines announced on several occasions by the Executive. NIMA took over all the image handling tasks formerly carried out by the NRO and

The President's close advisors represent a third distinct component of the decision-making process. Since space programmes involve large investments, over periods sometimes longer than a decade, assessments made by the President's direct entourage are decisive. A trend towards major technological programmes aimed at reinvigorating the scientific and technological components of the education system has emerged. However, it is still a rather general orientation that has not really been put into practice yet.

Once the executive has made its choice, the weight of the legislative authorities is decisive in the channel followed by proposals for programmes (fig. 5.12). At the beginning of the year, the executive sets out a detailed budget proposal and at the same time, the President justifies it in his message to Congress. These measures in the spring serve as a basis for successive discussions and hearings that take place over several months and correspond to a double check. To begin with, programmes pass through Authorization Committees which examine them in terms of their scientific and technological achievement. These committees are composed of specialised members who establish the specifications and make budget proposals. Generally speaking, those states most actively involved in aerospace activities are well represented and there is a clear attempt to push forward projects based on existing industrial capacities.

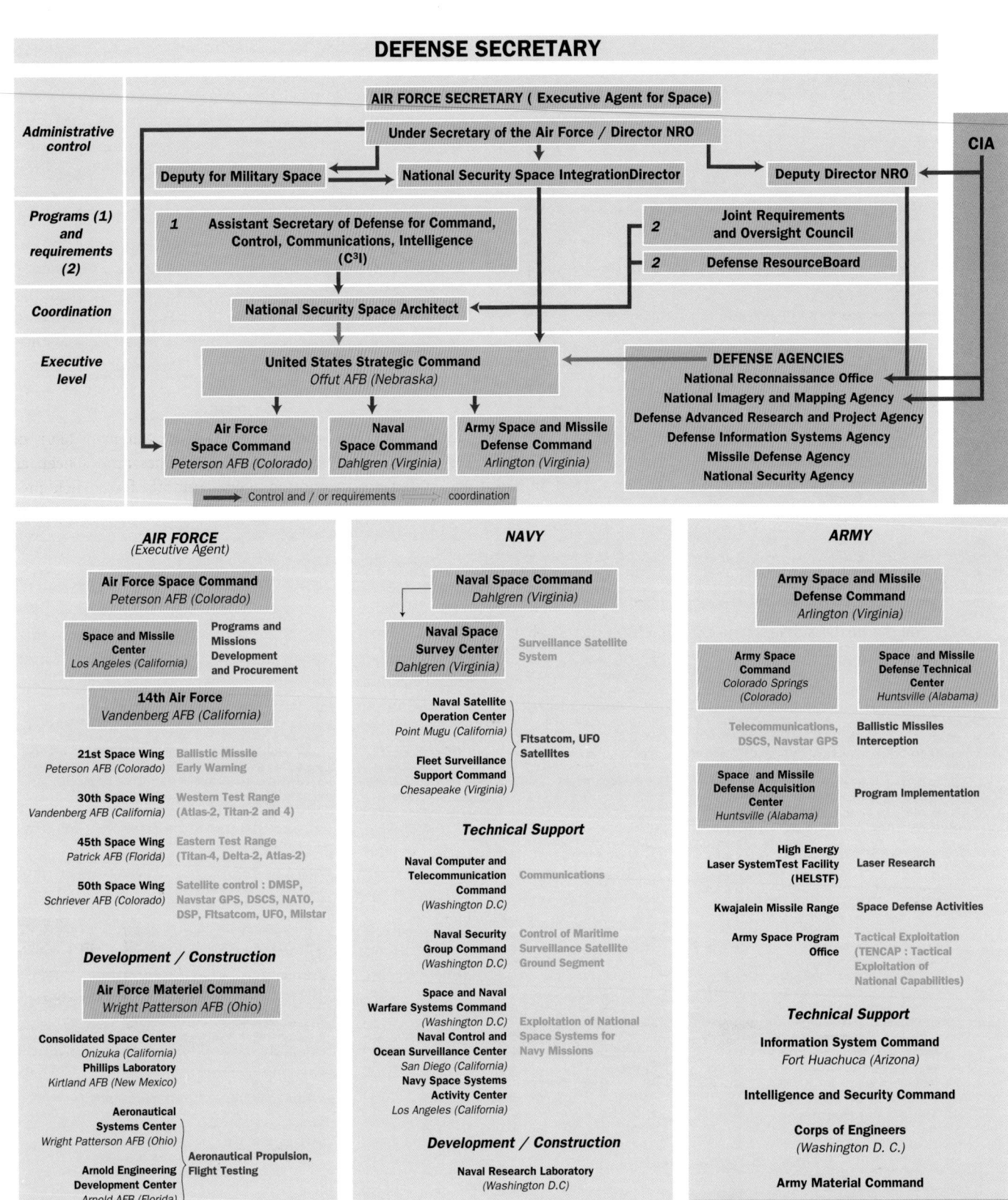

Figure 5.13. **Organisation of the DoD in 2002.** The US Air Force receives about 80% of the American military budget and is the executive agent for space activities at the Pentagon. Symmetrically, the US Navy and Army count for 80% in the use of military space systems.

the National Photographic Interpretation Center (NPIC) and also the whole range of cartographic missions carried out by the Defense Mapping Agency whose staff was transferred to the new agency. Another mission is to stimulate manufacturing and business activity and extend the use of commercial systems for governmental purposes.

In 2001, the Defense Secretary Donald Rumsfeld gave new impetus to the Pentagon's space activities. The latest reorganisations of the Air Force should be noted, aimed at better accounting for space activities from both a staffing and a budgetary standpoint. Also worth noting are efforts, coordinated by the under secretary to the Air Force, to improve coherence between Air Force space activities and the NRO, the body in charge of military observation satellites since 1960.

Research activities

American scientists involved in the space programme may belong to research centres run directly by NASA or the DoD, or they may work in universities like MIT or CalTech in traditional forms of collaboration. The extent to which individual scientists can affect the definition and accomplishment of programmes has always directly depended on their personal influence in the case of well known figures like James Van Allen or Carl Sagan. Otherwise they must express their ideas through the President's various scientific advisory bodies. Among these, the National Academy of Sciences currently has the dominant consultative role. Generally speaking, the scientific community is particularly concerned with the efficiency of space missions and optimal use of finances in terms of results. In the first instance, scientists are thus rather opposed to major manned programmes such as the Shuttle or the space station. On the other hand, they have learnt to be pragmatic and take advantage of backing wherever it presents itself, in the common instance where their basic opposition has not been heard. Moreover, biologists and some planetary scientists are tempted by the prospects of exploring the Moon or Mars and have shown interest in new unmanned planetary exploration projects put forward by NASA. Programmes like Mars Sample Return, intended to prepare for possible human exploration in the 21st century, have raised renewed interest and at the same time NASA has announced that traces of life may have been discovered on the Red Planet.

Space is often perceived as a stimulating area for science, offering research opportunities in a wide range of fields. The technological and scientific driving force provided by previous programmes like Global Change, the Space Exploration Initiative (SEI) or the Strategic Defense Initiative (SDI), or today's New Millennium Programme has always been an argument for promoting them. Spinoffs from such programmes are considerable, especially since space has from the beginning been characterised by the complementarity of civilian and military research (fig. 5.13).

Manufacturing companies

This dual nature of space activities makes it difficult to discover the exact involvement of industry in these programmes. Although the names of the main contractors used

Figure 5.14. **Geographical distribution of the main space-related companies and research institutes in the United States.** Companies have been located at the site of their head office.

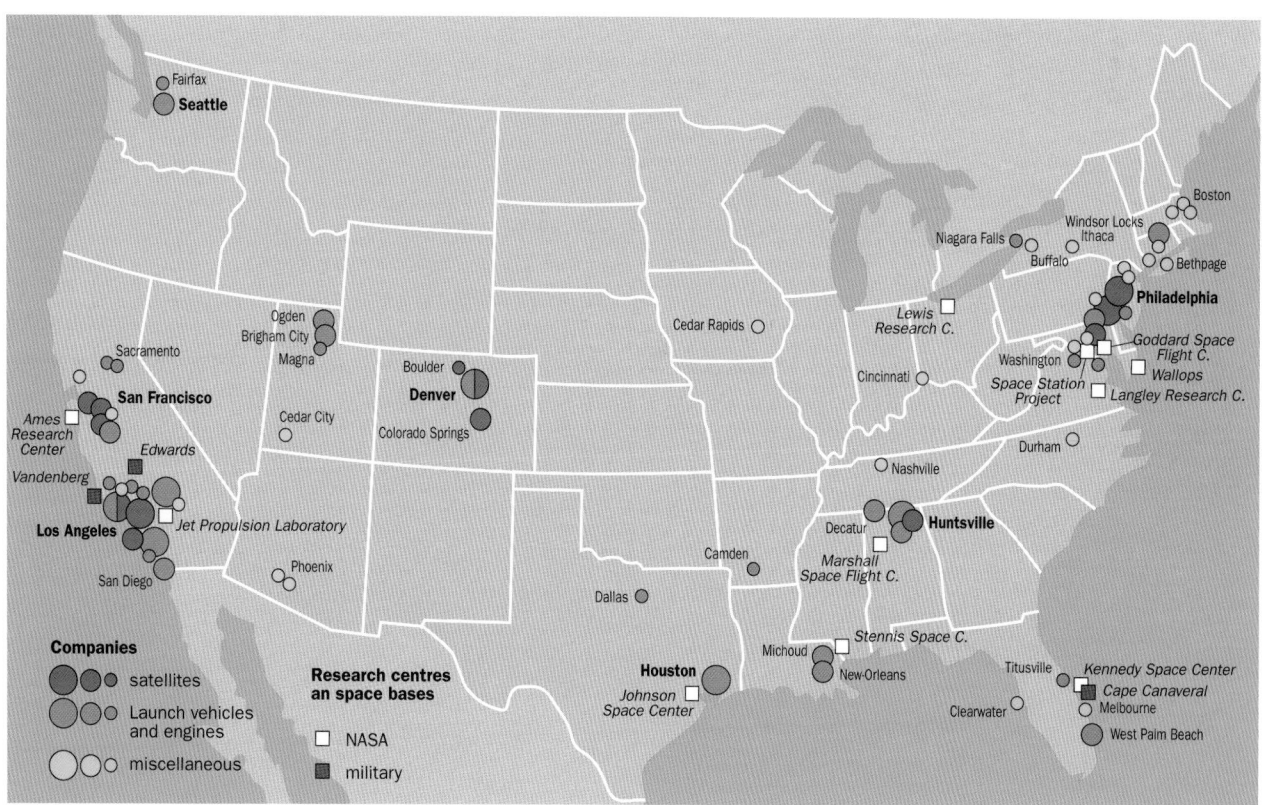

by the DoD and NASA are known, it remains difficult to determine precisely how much of their effort is devoted to space, especially since, more often than not, no distinction is made between space and missile activities. Furthermore, almost half of the military space budget appears to finance programmes for which absolutely no official information is available. The location of the biggest firms (fig. 5.14) clearly shows the importance of certain states favoured by the aerospace industry for a very long time now, where NASA and DoD research institutes are also present. For manufacturers, space programmes have often represented an important supplementary income to their other activities, particularly during recessionary periods caused by cuts in military spending or growing international competition. Space-related contracts may concern launch vehicles, satellites or subsystems whose manufacture requires specific technological knowhow. Such contracts provide an opportunity for the companies in question to gain experience in areas that will be essential for the future. In addition, funds allocated to the major space programmes guarantee a stable income over several years. However, such resources seem today to be devoted in preference to research and development of new technologies. This attitude can be explained by manufacturers' awareness that major programmes may be reoriented. Indeed, these programmes are decided in specific political contexts and are thus liable to subsequent modification. Investment in state-of-the-art technology does not exclude economies of scale, as demonstrated by the industrial recombinations occurring over the past few years. The rapid development of the two main groups has been achieved by establishing coherent combinations enabling them to take full advantage of 'portfolio effects' (increased concentrations in the ownership of product lines). Hence, during 1995, Boeing Co, already experienced in satellite manufacturing, took over the activities of MacDonnell Douglas and thereby extended its range of expertise to space launching via the Delta rocket. This merger followed on from a similar strategy first used slightly earlier by Lockheed. The latter had broadened its horizons by integrating the launch activities at General Dynamics (Atlas rockets), and those of Martin Marietta (Titan launch family). Since then, it seems that the priority of the two companies Boeing Co. and Lockheed Martin has been to cover the whole spectrum of space activities. Over the years, this has involved a growing control over all satellite-related activities, from the manufacture of platforms and payloads to the operation of satellites once in orbit. It should be noted in particular that Boeing Co. took over the space activities of Hughes Electronics (Hughes Space and Communications) in October 2000. This includes the production line for the HS platforms, a series which gave Hughes Electronics first place on the market for telecommunications satellites. This takeover was particularly significant because it struck a blow to one of the last bastions of resistance to the mergers and acquisitions strategy of the two main groups. It should nevertheless be noted that Hughes had already been pursuing the same strategy for a number of years in the running of satellite telecommunications services (for example, the takeover of Panamsat Corp.). This part of Hughes' activities was turned into a subsidiary and sold off to Rupert Murdoch's media group, whereupon it escaped from Boeing Co.

For its part, Lockheed Martin finally absorbed Comsat, the US representative at the international telecommunications organisation Intelsat, when Intelsat's activities where privatised. This takeover gave Lockheed, already a leader in satellite manufacture, an important position in the new private international company Intelsat, and a foot in the satellite operating business.

Two large groups now constitute the main industrial component. Boeing Co. is the leader with a turnover of 12 billion dollars in 2000 for its space activities alone, followed by Lockheed Martin which achieved a turnover of 7 billion dollars for the same period.

The main activities of the two big American companies have been gradually shifting. This evolution clearly reflects a transformation of space activities as a whole, particularly as regards their public component, with growing uncertainty about whether the major traditional governmental programmes will come to fruition. Indeed, these large scale manoeuvres were originally triggered by government calls for streamlining of industrial activities in this area. The financial performance of Lockheed Martin has recently been in decline. This shows that the balance between a refocusing of activities on the space business, which could for example have been justified by a possible stimulation of military space activities in the United States, and an unavoidable broadening of experience remains elusive.

Europe

The very notion of a European space industry is in itself rather complex. This is partly because space developments have been carried out independently of the general process of European construction. In addition, different civilian and military bodies, either exclusively national or acting through various kinds of partnership, have contributed to defining space policy and developing industrial activities. The European Space Agency has become the main authority in the European space industry. However, the growing role of the European Union, the development of military space activities, and internal changes in the industrial sector are new features that should be taken into account along with the internal evolution of the national space sectors in individual European member countries. Europe comes third as a world space power, judging by budget and achievements, and the European space industry is determined to pursue its efforts in this direction. However, the varying interpretations of national interest made in different European countries must now converge towards a more integrated approach to truly European questions. The economic aspect, and especially the role of the space sector in the information society, has been clearly identified. A concern for strategic independence and a strong position on the international scene as regards key activities such as navigation or launch has also been clearly formulated. Today, debate has turned to two weaker elements of the European space industry: the military aspect and, to a lesser extent, the human occupation of space, a purely spatial activity with fewer spinoffs outside its role as a symbol.

The beginnings of partnership

During the 1920s and 1930s, astronautics raised renewed interest across Europe. In Germany, Austria, the United Kingdom and France many pioneers studied the problem of rocket propulsion. The first trials involved both solid and liquid propellants and in 1929, three years after R. Goddard, the Austrian F. Sänger secretly flew a first rocket model. In 1931, in Germany, the Space Travel Society (*Verein für Raumschiffahrt*) also successfully carried out a first rocket flight using liquid propellant. Its president was H. Oberth and W. von Braun was a member. In France, R. Esnault-Pelterie was pursuing theoretical research on interplanetary flight and was especially interested in the practical problems of rocket propulsion. Finally, in the United Kingdom, the British Interplanetary Society was also studying space flight but did not carry out experimentation.

By the time of the Second World War, Germany was certainly the most active in the area of rocket research. The army was particularly encouraging given the potential of this new weapon, which did not feature on the list of arms restrictions imposed by the Versailles Treaty. Progress achieved during the war and the final development of the V2 marked a turning point in the history of astronautics. At the end of the war, the United States and the Soviet Union were the first to turn their attention to the development of ballistic missiles. When Sputnik was launched, France and the UK immediately undertook their own national programmes, hoping eventually to obtain access to space. Financial considerations led the main European states to envisage partnerships, although this does not mean that they abandoned their own national ambitions.

By 1960, the importance of space had been understood by people from a wide range of different backgrounds, in particular, those scientists already involved in European nuclear research. The European Preparatory Commission for Space Research (COPERS, from the French COmmission *Préparatoire Européenne pour la Recherche Spatiale*) was thus set up. This gave rise to two distinct organisations in 1962, although their creation was not officially ratified until 1964. The European Space Research Organization (ESRO) was devoted to scientific questions and the construction of satellites. Among member states, those interested in a launch vehicle programme belonged to another organisation, the European Launcher Development Organisation (ELDO). Handicapped by French and British strategic concerns, ELDO had difficulty overcoming the weight of national ambitions. In addition, it was more concerned about reconciling existing resources than elaborating a unifying project. Hence, the three stages of the Europa rocket were to be manufactured in the UK, France and Germany, respectively, whilst the guidance system was entrusted to Benelux. For its part, Italy was to develop an experimental satellite.

In parallel with the political organisation of European space activities, manufacturers proved to be particularly motivated. The official aim of their Eurospace partnership, also founded in 1960, was to promote the development of space-related companies by alerting national and European authorities to the new potential. The early expression of interest on the part of European manufacturers can be explained by the fact that they belonged to a sector that was already very dynamic, that is, the aeronautic industry. This was combined with hopes of obtaining new government orders in significant quantities and often accompanied by subsidies.

The arguments developed by Eurospace took into account three main motivations. The first was political and referred to the implications of such technical progress for national security and influence on the international scene. The second was economic, stressing the indirect benefits to be obtained from space activities, whether it was the development of new methods based on automation and reliability, or in the longer term, the manufacture of new materials in space. Finally, the third argument asserted the importance of observing the Earth from space in order to better manage our resources. European space activities had therefore to be organised along three lines: a scientific programme, a satellite application programme, and a launch programme.

In fact, the applications programme was only weakly represented by the European Conference on Telecommunications by Satellite (CETS, from the French *Conférence Européenne des Télécommunications par Satellites*) which had served as representative at the Intelsat organisation. The justification for its existence disappeared at the end of the negotiations, just at the

moment when the commercial interest in this type of activity was gradually coming to light.

General recommendations were also supported by several organisations within Europe, particularly after 1967. For example, the European Council, the Western European Union and the European Community, aware of what was at stake, advocated a European presence in this new field of activity.

At the beginning of the 1970s, European cooperation in the space sector presented a rather disconnected front, in the shape of three organisations bringing together different numbers of countries (fig. 5.15). They had different administrative structures, running procedures and budgets but were interdependent, which greatly complicated the situation. For each major decision, suitable political representatives of the member countries had to come together. Interministerial meetings were decisive but the annual space conferences held from 1966 were not sufficient for effective organisation. In addition, although scientific activities at ESRO were promising, ELDO had run into a great many problems. After the successive failures of Europa 2, all parties agreed that it was essential to create a single organisation.

The European Space Agency (ESA)

A complete reorganisation of Europe's space policy was essential if Europe was to rival the Soviet Union and the United States as a world space power. The European Space Agency was officially created in 1975 and officially ratified by the different parliaments in 1980.

At this time, the United Kingdom was vacillating on the question of Europe and had been the cause of several political crises within ELDO. It was thus keen on a rapid unification of space organisations. However, the appointment of a Director General was postponed several times because the two main contributors, France and Germany, each had their own candidate. In the end, the problem was solved by appointing a British director. A compromise was also found for a second sensitive question, namely, who would pay for what in the Kourou base infrastructure. The final text of the ESA convention reproduced the main points in the ESRO convention but extended space activities to cover applications satellites. Further powers were assigned to the Agency's executive, the Ariane programme being directly handled by the general management.

The European Space Agency is built around two bodies: the directorate comprising a Director General and an Assistant Director General, and the Council which is the main decision-making authority. Each state has one vote, a two-thirds majority being required to adopt programmes and budget. The Council can meet at a ministerial level for the more political decisions, whilst the major part of everyday decisions are dealt with at the level of top civil servants representing the different member countries.

The agency was originally conceived as a research and development organisation, deprived of commercial capabilities and denied any military leanings. Its aim was to

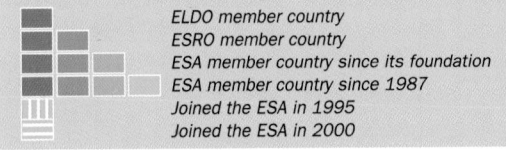

ELDO member country
ESRO member country
ESA member country since its foundation
ESA member country since 1987
Joined the ESA in 1995
Joined the ESA in 2000

Figure 5.15. **Countries and organisations making up the European Space Agency.** The head office is in Paris and the organisations are ESTEC (European Space Research and Technology) with headquarters at Noordwijk in the Netherlands, ESOC (European Space Operations Centre) at Darmstadt Germany, ESRIN (European Space Research Institute) at Frascati in Italy, and EAC (European Astronaut Centre) at Porz Wahn in Germany.

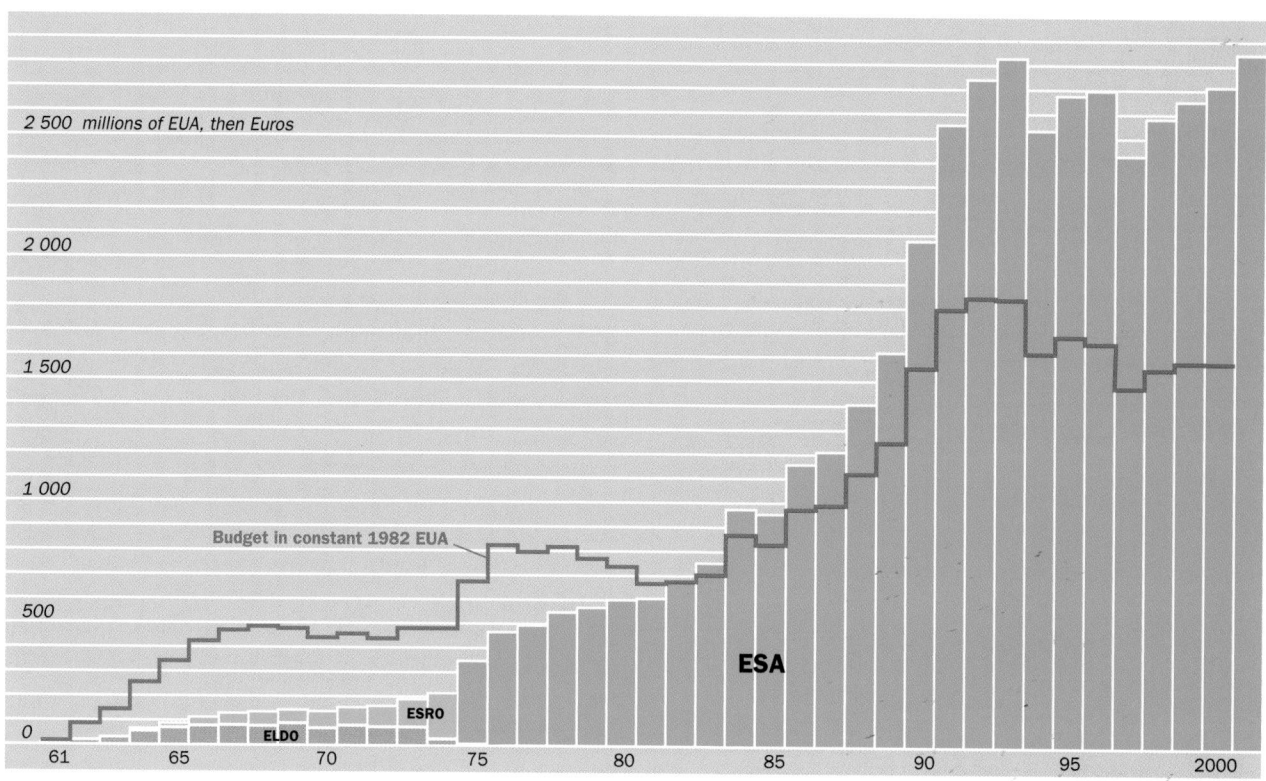

2 500 millions of EUA, then Euros

2 000

1 500

Budget in constant 1982 EUA

1 000

500

ESA

ESRO

0   ELDO

61   65   70   75   80   85   90   95   2000

Figure 5.16. **Evolution in the budget of the European Space Agency and organisations that preceded it.**

rationalise space activities in the different European countries and thereby create the world's third great space organisation. In practice, the basic working principles of ESA, that is, one country one vote and an ever stricter application of the principle of fair industrial returns, have led to a drift away from initial objectives. Agency policy has more and more often been reduced to a quest for compromise between member countries with differing national strategies. Besides obligatory scientific programmes, the flexibility of the system allows the development of optional activities. This has meant that the main stakeholders have specialised in areas of activity where the size of their contribution guarantees them a dominant role.

In accordance with choices made on a national level, France has thus placed itself in the lead for launch programmes and manned flights, symbols of European independence. Germany, the second main contributor and traditionally more favourable towards cooperation with the United States, has built up acknowledged skills in the field of manned flight. Italy is in an unusual situation since manufacturers have introduced a wide range of contributions to ESA programmes, despite national budgetary difficulties and limited industrial returns. In contrast, the United Kingdom, with very modest ambitions lying mainly in the area of Earth observation, has clearly benefited from ESA's principle of fair returns.

The funding of European space activities increased steadily until 1990 but then entered a lean period in 1994, pursuing the main programmes, Ariane 5 and participation in the International Space Station, without commitment to major new projects. It then grew slowly until the ESA budget recovered its 1993 level in 2001 (fig. 5.16). This situation resulted largely from the climate of economic austerity in member countries. Many experts were concerned that Europe would not invest sufficiently to develop expertise in new fields of activity which might in the longer term prove to be essential. Today, new impetus has been given to programmes considered to have strategic significance, such as the Galileo navigation system. At the same time launch activities are being pursued. Further crucial decisions must nevertheless be taken concerning telecommunications and reusable launch vehicles in the framework of the agency's new plan. A first debate was scheduled at the interministerial council meeting in November 2001.

ESA has proven its ability both in managing major programmes and in carrying out original space science. However, the existence of new features, whether they concern the evolution of technology, changes in national space preferences or developments in the general framework of the European community, all require a redefinition of objectives and ambitions for the future European space policy.

Space activities in European member states
One of ESA's missions (Article II of the Convention) was to coordinate the European space programme and national programmes with a view to gradually Europeanising the latter. In practice, European space programmes have not supplanted purely national activities. This is sometimes because a consensus has not been reached and sometimes because the national programmes embody military concerns. In fact,

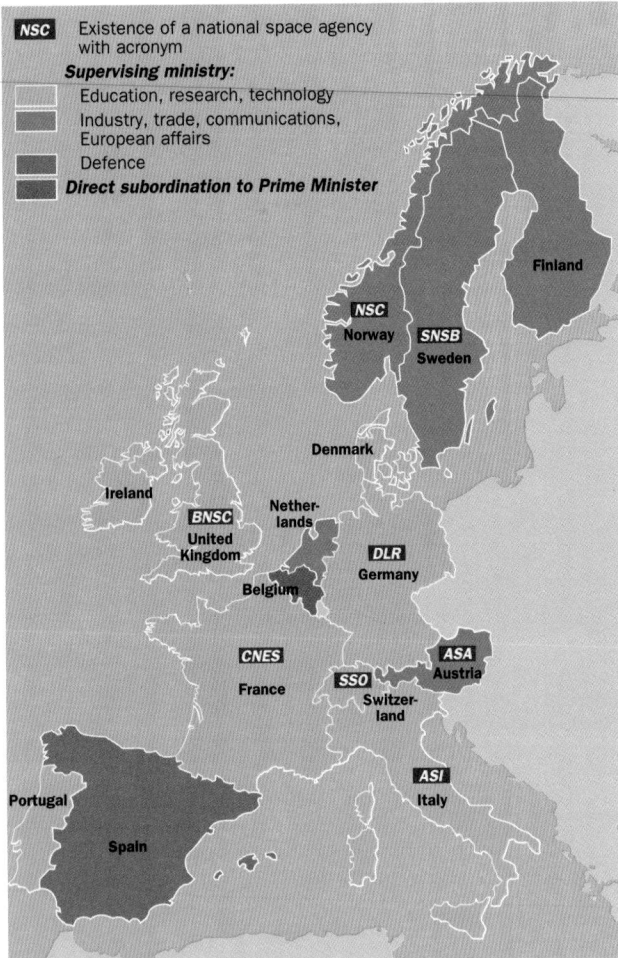

Figure 5.17. **Organisations responsible for space-related matters in the different member countries of the European Space Agency in 2001.**

Hence, the existence of a national space agency does not necessarily prove that space plays a key role for that country. Apart from France, where the CNES does in fact play a central role, other agencies exist in Austria, Italy, the United Kingdom, Sweden and Spain. These agencies have different purposes. Some are mainly responsible for civilian activities, like the British National Space Centre (BNSC) in the United Kingdom, whilst military activities exist in parallel even if they are limited to telecommunications and observation from space. In The Netherlands the agency responsible for space activities also deals with aeronautic affairs and in Ireland, space matters are dealt with by the science and technology agency. In Germany, the space agency has been integrated into a larger ensemble.

Depending on the case, the ministries supervising space matters are, under various appellations, those responsible for science, research, technology and education (Austria, Denmark, Italy), trade and industry (Finland, Ireland, Norway, United Kingdom, Sweden), or the economy (Netherlands). In one case, space even depends directly on the prime minister (Belgium). In France, the space agency CNES came under the supervision of three ministries, industry, research and defence, from 1993 to 1997. In June 1997 it was transferred to the authority of just two ministries, the Ministry of Education and Research and the Ministry of Defence. In Germany in 1993, the *Deutsche Agentur für Raumfahrtangelegenheiten* (DARA) was integrated into the *Deutsche Forschungsanstalt für Luft- und Raumfahrt* (DLR), whose responsibilities and name were slightly modified (DLR becoming *Deutsche Forschungs- und Managementzentrum für Luft- und Raumfahrt*). The result is that the Ministry of Research and the Ministry of Defence have an overseeing role related to their budgetary contribution. In other cases, space may depend on interministerial bodies, as in Switzerland. This generally corresponds to a rather low level of activity. However, the interministerial approach, whether institutionalised or not, is adopted in the majority of countries.

The complexity of the space question is clearly shown by the internal deliberations that take place at national level concerning the best ways to organise space-related structures, and also the switching of ministerial supervision when new governments are installed. In Germany, the merging of the DARA with a technical organisation, to the benefit of the latter, no doubt represents an attempt to streamline, but it spells the end of a purely spatial speciality. The main trend today favours synergism. The idea of partnership with manufacturers, described in the plan of action which the CNES set out in 1997 to present the main lines of its future activities, also features amongst ideas discussed at ESA. The tasks of the space agencies are up for reappraisal in every country. This reflects the gradually changing relations between the various protagonists and a certain maturity in the sector after more than 35 years of practice. Such redefinitions must take into account the way the various European space authorities are to fit together as well as their specific relationship with ESA.

national organisation of space activities and the weight of national budgets, which differ from country to country (see figs. 5.18 and 5.19), show that both attitude and degree of involvement are far from uniform across Europe.

The national authorities responsible for space matters vary widely (fig. 5.17). A first category is composed of countries with their own agencies devoted more or less exclusively to space. In a second category, space questions are directly handled by a ministry. In yet other cases, a simple interministerial entity may deal with these matters. Civilian ministries, with varying degrees of authority, can be divided into two main categories revealing quite different approaches. Depending on the country, space may be classed with research and technology, or it may be associated with industry and foreign trade. As far as the military space sector is concerned, defence ministries are responsible for activities specific to them, and relations with civilian activities are generally rather restricted. Interministerial coordination is a useful way of taking occasional users into account, such as those dealing with the environment.

In fact, the way space activities are organised does not necessarily reveal the importance they have for a given country.

France

In 1961, when the first European organisations were beginning to appear, France founded its own agency, the *Centre National d'Études Spatiales* (CNES). Apart from developing the capacity to reach space, the aim was to display a policy of independence and a presence on the international scene. One of the initial tasks at CNES was to concentrate expertise acquired in missile development and build the Diamant launch vehicle, whose first two stages made use of ballistic launch technology. Another task was to develop scientific satellites along similar lines to ESRO. However, CNES was also called upon to handle application satellites. The existence of a specialised structure of this kind, inspired by its American counterpart NASA, was unique in Europe until 1985.

CNES had yet to find its place in relation to French military organisations, and also in relation to European partnerships. Concerning this second aspect, the fact that funds transited through CNES and that its administrators held significant positions within the national delegation facilitated a good working relationship. CNES involvement in application satellites even played a decisive role in extending ESRO's missions. On the military side, the much-debated decision to use liquid propellants for Diamant B made it possible to develop a civilian capacity that used a different technology to military missiles, a feature which limited any conflict of interests concerning knowhow. The Diamant programme was finally abandoned in 1975, but CNES had already learned how to manage major programmes and cutting-edge technology. This experience played a key role in the development of the Ariane launch vehicle when ELDO was dissolved. It also affected the allocation of tasks within the ESA, where CNES insisted on taking the role of project manager.

Another original feature of French space involvement was the existence of extremely varied bilateral arrangements with European, American and Soviet partners. These concerned every aspect from telecommunications through scientific experiments to manned flights. The gradual appearance of military space programmes, growing more and more highly organised since the 1980s, represents a new line of development. The arrival of representatives from the Ministry of Defence Armament Office (*Délégation Générale pour l'Armement*) and the Military Command (*État-Major*) increased the need for coordination groups both to define objectives and to deal with more technical aspects. Globally speaking, the current trend in space affairs is towards a closer relationship with user needs, be they civilian or military, with greater consideration for industrial and sales aspects.

The CNES budget (fig. 5.18) is evolving in a favourable way for European partnership, even though the national share is holding steady. The basic question is how to harmonise relations with ESA given that the two agencies are of comparable size. This directly affects the status of technical

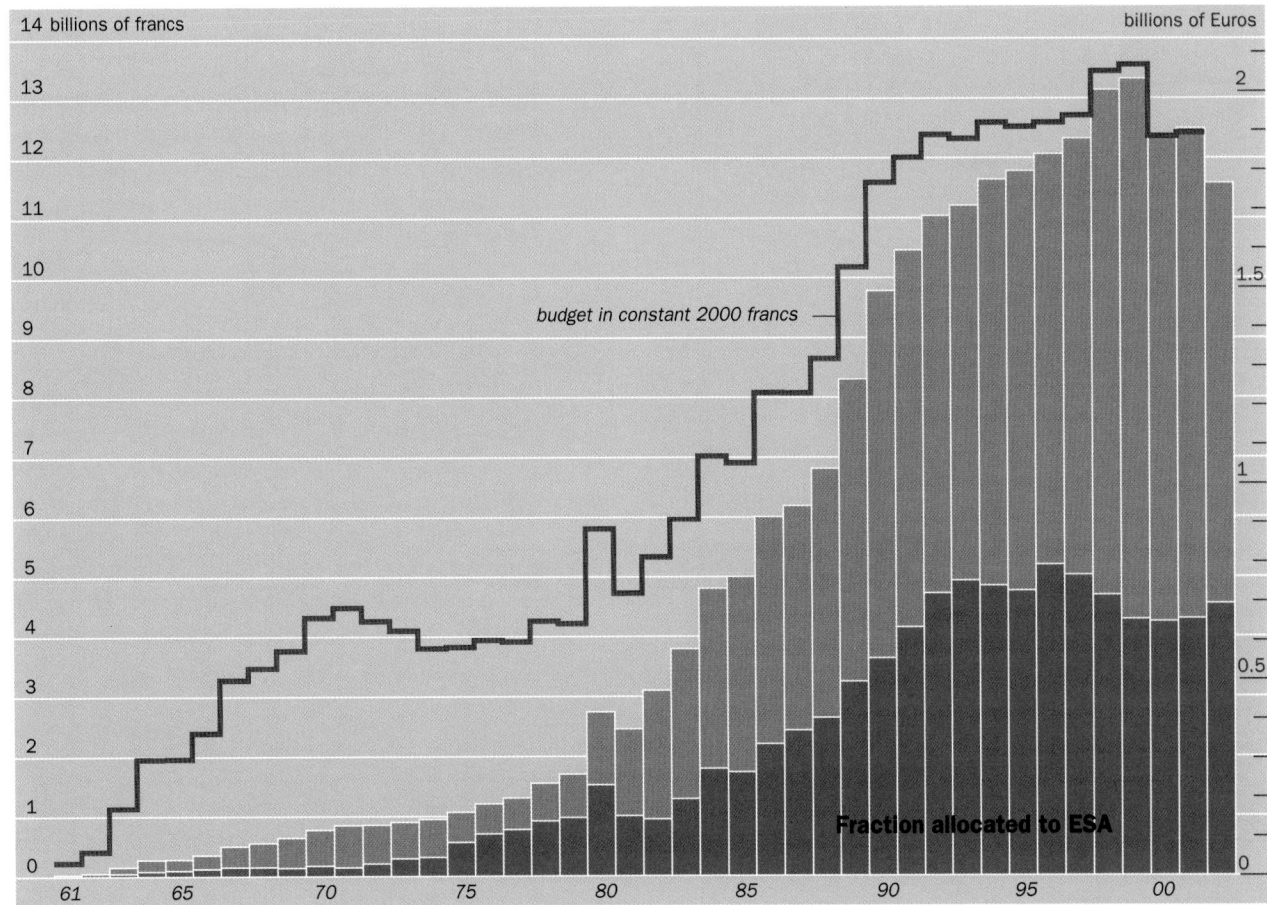

Figure 5.18. Evolution of the CNES budget from 1961 to 2001.

institutes, including the Toulouse centre (CST, *Centre spatial de Toulouse*) which runs satellite programmes, the Evry centre which is in charge of the Ariane programme, and the Guiana Space Centre (GSC). Although the GSC is automatically integrated into European space affairs by the fact that it is considered as the European space base, the case of the Evry centre is much less clear. On the one hand, the project management of the Ariane programme is based there, delegated to France by Europe, whilst on the other, it is run as a national centre. The centre could be Europeanised through a closer arrangement with Germany and a redefinition of its relationship with industry, in particular, with the new European company EADS which is largely responsible for Ariane production. The future of the CST remains the most complex problem because it possesses a similar type of expertise to ESTEC.

Germany, Italy and the United Kingdom

Germany is the second biggest contributor to ESA. It has always tended to favour programmes developed in a context of European or bilateral partnership. The share of its space budget paid over to ESA is sometimes as much as two thirds

of the total (fig. 5.19). Between 1982 and 1993, national programmes also benefited from an increase in space expenditure, and the creation in June 1990 of a federal space agency known as the *Deutsche Agentur für Raumfahrtangelegenheiten* (DARA) demonstrated an institutionalisation of space activities. The agency was set up in a quite different way to its French counterpart, since it had the status of a GmbH (*Gesellschaft mit beschränkter Haftung*), the equivalent of a limited liability company, under the supervision of the Ministry of Research and Technology (*BundesMinisterium für Forschung und Technologie*, or BMFT). This organisation was overseen by a supervisory council and had an advisory board open to industry and research representatives, features implemented so that it could manage programmes in a more flexible way. However, in 1997 DARA merged with the DLR as part of a streamlining campaign. Financial restrictions from 1997 raised questions about the pursuit of specifically national ambitions, but are also reflected in a lesser contribution to ESA. In addition, German participation in the ISS represents 41% of their space budget, a further destabilising factor. In 2000, annual budget forecasts for the period 2000–2004 show a high level of stability. Although the ISS still represents 28% of Germany's resources and space exploration a further 18%, a readjustment has been made to benefit Earth observation and space transportation programmes in which German companies play a lead role.

Italy is the third space power within the European Space Agency, despite the fact that it suffers from severe budgetary problems (see fig. 5.19). Italy's space agency, the *Agenzia Spaziale Italiana* (ASI), created in 1989, has little weight compared with the involvement of manufacturing companies. However, it seems that space is the subject of a genuine political interest. The existence of a five-year national programme and the steady growth of Italy's contribution to the ESA attest to a fairly ambitious space policy. After a slightly hesitant period from 1994, the published strategy for 1995–2004 and the reestablishment of a five-year plan for the period 1998–2003 suggest that this policy is turning towards a development of national space projects, with a longer term aim to devote half the overall space budget to such projects.

There is a clear concern to develop synergism between national and European ambitions. This is manifested in the context of the ISS, to which Italy is making a significant contribution, by virtue of the strong bilateral relations it maintains with the United States. It is also visible in the Europeanisation of the small launch vehicle Vega, initially developed under national funding. The same type of policy is being set up in the area of Earth observation where the COSMO/Skymed programme is already the subject of bilateral partnerships.

The United Kingdom operates a much less ambitious policy. Its budget remains limited (see fig. 5.19) with a dominant part allocated to ESA, averaging 68% since 1994. It mainly promotes the development of applications such as Earth observation. The new British space plan for the period

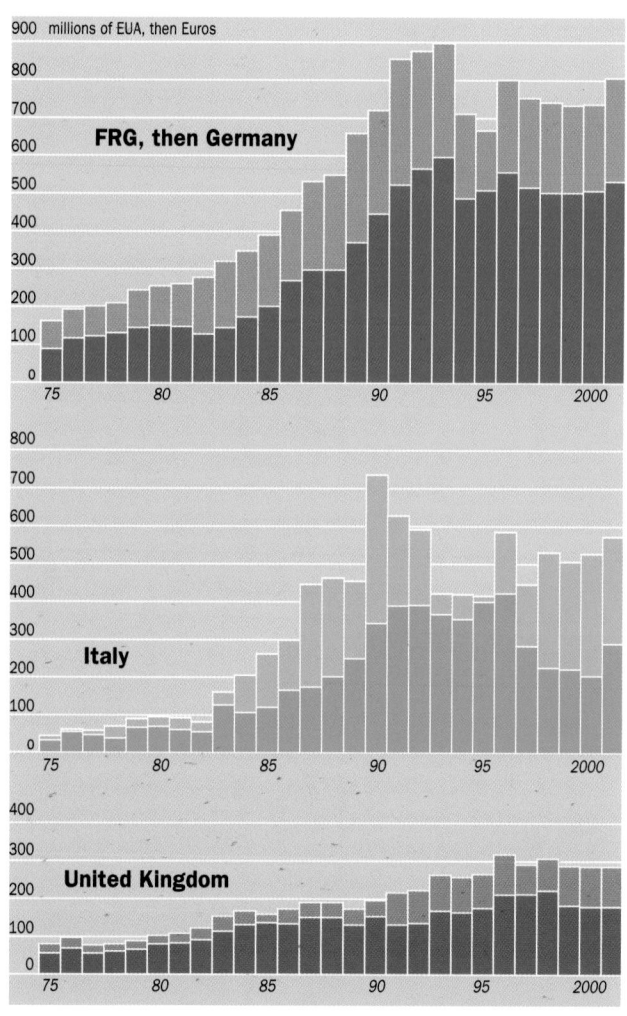

Figure 5.19. Space budgets in Germany, Italy and the United Kingdom from 1975 to 2001. The share devoted to the European Space Agency is more darkly shaded in each graph.

1999–2002 mentions support for the British space industry in the field of applications and also in the development of innovative technologies as its main objective. Given the low budget at its disposal, the British National Space Centre is very concerned about cutting the cost of programmes, particularly those at ESA. The United Kingdom also favours coordination between European organisations and in particular improved synergism between research financed by the European Union and space research. On a national level, BNSC is seeking to strengthen the involvement of private investors right at the outset of programmes, even though users are mainly state-run organisations, as in the case of navigation and to a lesser extent Earth observation. The idea is that users should commit themselves to purchasing data in the medium term.

European industrial activities

The development of industrial activity in the space sector has resulted directly from major European space programmes in the area of launch vehicles and also in telecommunications, scientific satellites and Earth observation satellites. Although more distant, manned space projects also offer European companies an opportunity in state-of-the-art technology with possibilities for development in the near future.

The spread of space activities across ESA member countries (fig. 5.20) has been actively promoted by the agency's principle of fair returns. This principle ensures that each participating country receives industrial contracts in proportion to the amount it has contributed to programmes validated by the agency. This system has some undesirable consequences.

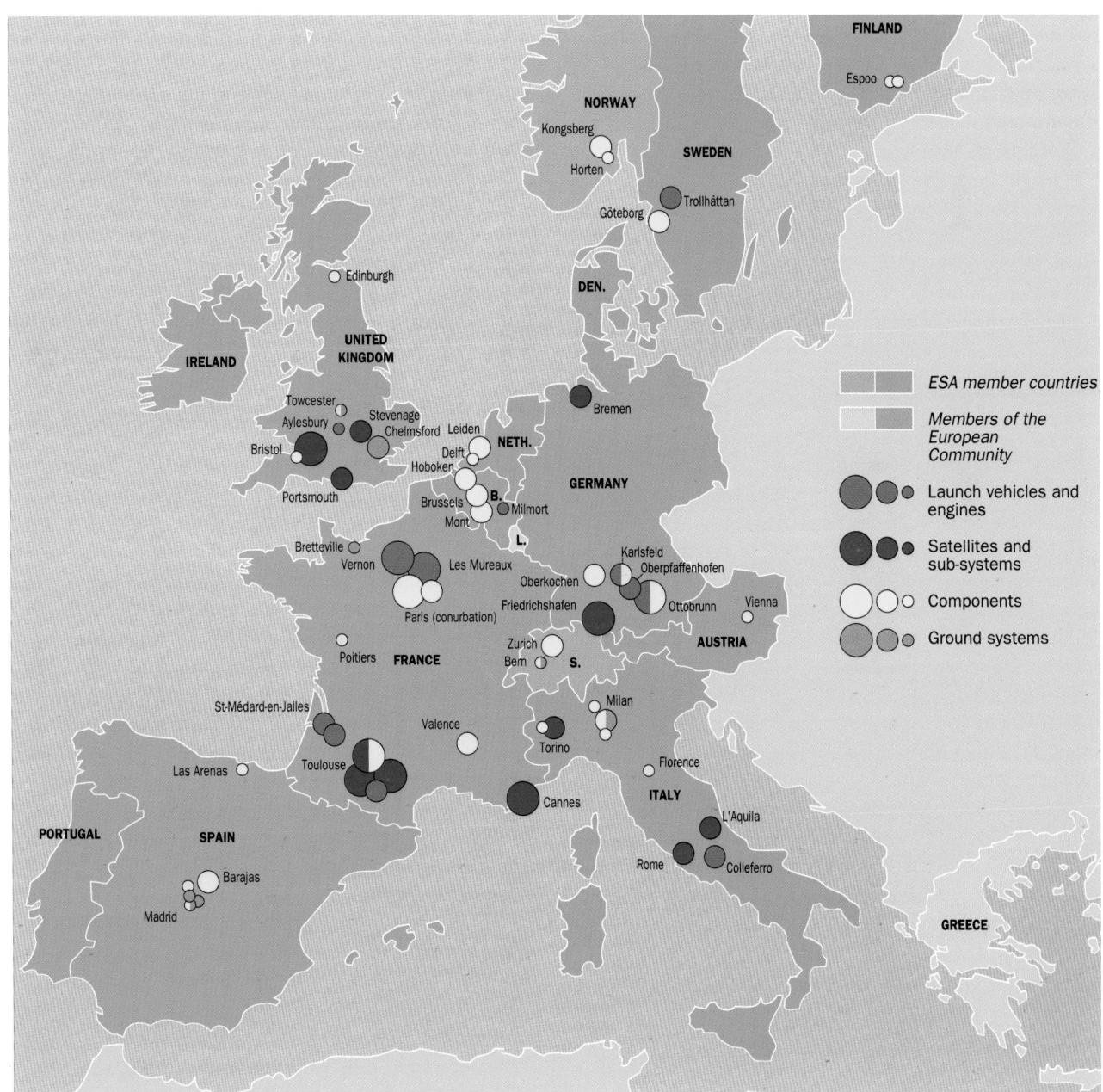

Figure 5.20. **Location of space industries within ESA member countries.**

For example, small countries tend to develop specific projects, adapted to suit their own capabilities, thus favouring the maintenance of national programmes. In addition, the better performing manufacturing companies regret the fact that competition is no longer given free rein. They argue that this increases the cost of European programmes and weakens Europe's position in the international marketplace. One proposed solution is to extend the principle of fair returns to sectors other than space within the European Community. This presupposes a greater level of integration within Europe as a whole.

The industrial space sector has a certain number of special features related to the very high level of public funding, with very limited commercial inclination. Moreover, the small size of the European market compared with the US market tends to give space manufacturing the image of a prototype-building industry, at least in the area of satellite construction. A transition towards producing longer series can be envisaged as certain new projects appear on the scene. One example is provided by low-Earth orbiting satellite constellations and another by efforts to develop joint programmes that take full advantage of comparable facilities, whether the systems are exclusively civilian or result from a dual approach (both civilian and military). It must nevertheless be stressed that the duplication of skills remains a severe handicap in Europe at the present time.

The restructuring of companies on a European scale aims to improve the global competitiveness of European industry by concentrating not only research and development spending but also the means of production. This trend first occurred on a national level. It is very clear in Germany with the gradual concentration of companies around DASA from the end of the 1980s, and also in Italy where the Alenia Aerospace group was formed in 1996. The creation of transnational centres of excellence (e.g., Matra Marconi Space) then required a prior acceptance of dependent relationships that goes against the very notion of national independence so long proclaimed in areas of such technological and symbolic importance. On the contrary, it forms part of a desire to establish a truly European policy, justified precisely by the notion of mutual dependence. In 2001, industrial knowhow has been divided very much according to the American scheme into satellites and launch vehicles with the appearance of two large groups: Astrium, a subsidiary of EADS (Aérospatiale Matra, DASA, CASA) and Alcatel Space (Alcatel espace, Aérospatiale Cannes and the space activities of Thomson–CSF). However, the size of the domestic market remains a critical factor for the international competitiveness of these companies.

Space and European construction

When the rivalry between America and Russia ended at the beginning of the 1990s, the general context for space activities was modified by a reduced need for demonstrations of prestige. At the same time, Europe had reached a new level of political and economic maturity by the second half of the

1990s, and from this standpoint achievements in space had lost much of their demonstrative value. The common European entity had become the world's third space power. Beyond this, the development of commercial space applications profoundly changed the approach to space-related matters. The industrial basis, precondition for longer term European independence in the areas of high technology and access to information, was now a priority.

All these external conditions and the limits of the current process for elaborating space policy and handling international industrial competition seem today to be on a strong footing, even though solutions remain to be clarified. Certain main lines of action have been proposed to establish a coherent framework for European space activities. These concern research and technology, telecommunications, Earth observation, industrial development, space legislation and education. At the same time, the European Union is the established partner in international negotiations and space is likely to feature more and more often in trade agreements.

The European Single Act established in 1988 conferred a broader mission upon the European Community with regard to political and economic aspects of security. The document referring to the Community and space, drawn up at this time at the request of the European Parliament, shows that space was henceforth a topic in its own right amongst EC considerations. The specificity of the European Space Agency's technical role was not put in doubt, but gaps in the decision-making process, concerning both general policy and industrial and commercial policy, led to reflection on how to better integrate space activities into the rest of European activities.

The European Union is interested in space for several reasons. For one thing, space techniques play a role in more and more sectors of community policy, including agriculture, environmental issues, fishing and many others. For another, the growing relevance of space applications in international negotiations, such as GATT which aims to establish rules for world trade, led the European Union to specify its objectives. In addition, the existence of the second mainstay of European policy, the Common Foreign and Security Policy (CFSP), which came into being with the Maastricht Treaty, means that the whole range of possibilities offered by space can be better taken into consideration.

In 1999, European ministers called upon the European Commission and the ESA executive to set up a 'Coherent European Strategy for Space', an initiative that was supported by the European Parliament. In 2001, the European Commission and ESA produced a common document entitled 'Europe and Space: Turning a new chapter'. This reasserted the general principles: to support the basic requirements of space activity (access to space and industrial capability), scientific knowledge and the development of benefits for society as a whole. In this context, the Commission's role was more precisely to determine the regulatory framework for space activities, to coordinate space research with European-wide research potential and to bring the relevant organisa-

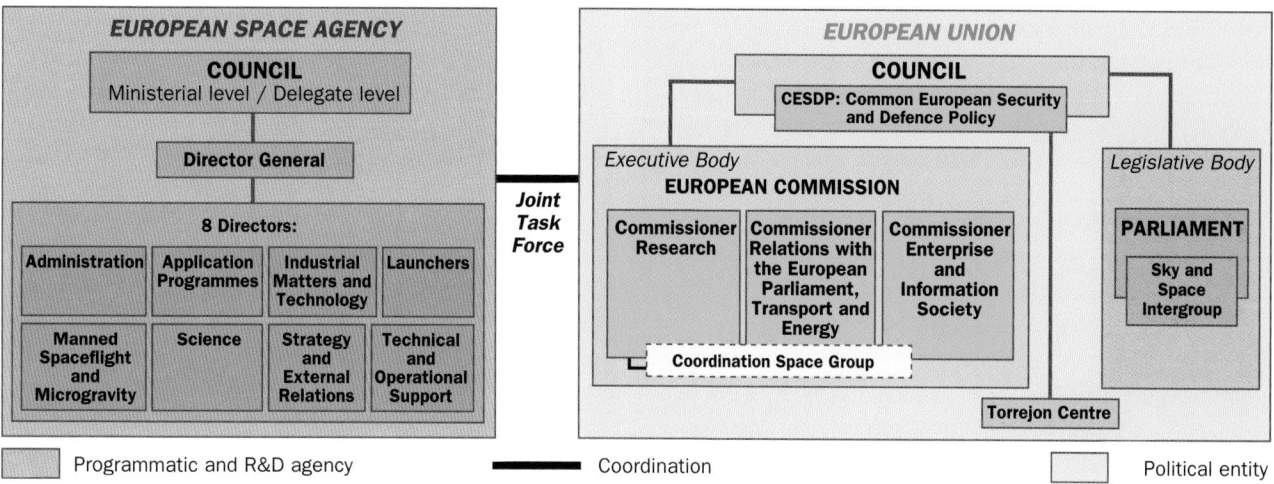

Figure 5.21. **General organisation of European space policy.**

tions and capabilities together for common projects, as in the case of Galileo and possibly the GMES programme (Global Monitoring for Environment and Security). The setting up of a joint task force between ESA and the Commission should lead to proposals for a permanent joint structure that would come into being at the end of 2001 (fig. 5.21).

European military space activities

The European military space sector first came into existence within the framework of the Western European Union (WEU), which has the job of defining conditions for European security, including related technological and industrial problems. To begin with, the WEU initiated several reports and colloquia on space. These approached the subject through a variety of themes, concerned first with the scope of European space activities and then more precisely, the management of a European space system designed to improve security. They then tackled the question of observation satellites as a European instrument for checking the

application of arms control treaties, particularly the Conventional Forces in Europe (CFE) Treaty. In 1991, the Western European Union Satellite Centre for satellite data interpretation was set up in Torrejon, Spain, marking the conclusion of a long process of reflection. Five years later, the appraisal carried out by the WEU of activities at the Torrejon centre during its experimental stages showed that maximal efficiency had not yet been achieved. One of the main problems was to implement genuine cooperation in sensitive areas like intelligence. More globally, the WEU had to face the basic dissimilarity between member countries, in terms of financial resources as well as political and strategic approach. However, the decision in May 1997 to support and strengthen activities at the Torrejon centre shows that, at least on a political level, the importance of space methods is officially recognised, even though most current programmes are still being developed in the context of direct bilateral or multilateral cooperation between the relevant countries.

Japan

In 2001, following the integration of the WEU in the European Union, the centre was designated a permanent military organisation reporting to the Council of the European Union, demonstrating that it plays a recognised role and that its missions do indeed belong to the development of the Common European Security and Defence Policy (CESDP).

For historical reasons space activities in Japan have a rather peculiar status. In the Japanese case it was quite out of the

question to follow the conventional approach of the major space powers, using experience built up in their aeronautic industries to develop aerospace expertise and manifesting their interest in space as a way of asserting national sovereignty. Having developed an approach that was very largely based upon technological motivations, the Japanese space sector now seems better prepared to take on the challenge of applications. The idea of developing a commercially competitive launch vehicle, as witnessed by the H 2A programme, and Japanese industry's first successful commercial satellite construction, built by Melco for Australian telecommunications, indicates a new perception of space activities. In 2001, the state of Japanese space programmes, whether they concern scientific or applications satellites, launch vehicles or manned flight, give Japan the image of a rather average space power with a low level of political involvement. At the end of 1998, after North Korea had launched its missile Taepo Dong, Japan decided to equip itself with surveillance satellites. This was the Information Gathering Satellite (IGS) programme, intended to be dual purpose, that is, meeting both military and civilian needs. It may lead to a new perception of space even if the notion of 'strategic' applications is likely to refer, at least for a certain time, more to economic than to security considerations.

Japanese space policy has three characterising features: a desire for consensus between all parties, a bottom-up decision-making process amongst the agencies, and respect for international commitments. These three components are present across the board in Japanese acquisition of space capabilities.

Japanese research was first carried out using funds from the Ministry of Education. A specific budget for the development of sounding rockets was allocated to the Institute of Industrial Sciences at the University of Tokyo in 1955. Traditionally interested in technological activities, Japan also expressed the desire to develop specific research activities in

the area of propulsion. A year later, the Science and Technology Agency (STA) was created, confirming this ambition but opting for a more technological than scientific approach. Then at the beginning of the 1960s, a further step was taken when the National Space Activities Council (NSAC) was set up. This group founded the Institute of Space and Aeronautical Science (ISAS) within the University of Tokyo and also a centre specifically dedicated to promoting space activities, the National Space Development Centre (NSDC). A third step was taken in this series of developments when NSAC was replaced by the Space Activities Commission (SAC), reporting to the Prime Minister and presided over by the Ministry of Science and Technology, and NSDC was replaced by the NAtional Space Development Agency (NASDA). No choice was made between university research and applied research, so that the two branches were pursued simultaneously. In addition, the original objective of acquiring homegrown expertise was somewhat modified because the request for technical assistance made by Prime Minister Sato during negotiations with the United States over return of the Okinawa Islands was favourably received by the Americans. An agreement signed in July 1969, two months before NASDA was set up, was based on the idea of a transfer of American technology and hence a slightly different orientation.

At the end of the 1970s, Japanese space activities took on their final shape, in fact only slightly modified by the administrative reform of 2001, which streamlined the sector under the authority of the Ministry of Education, global concern

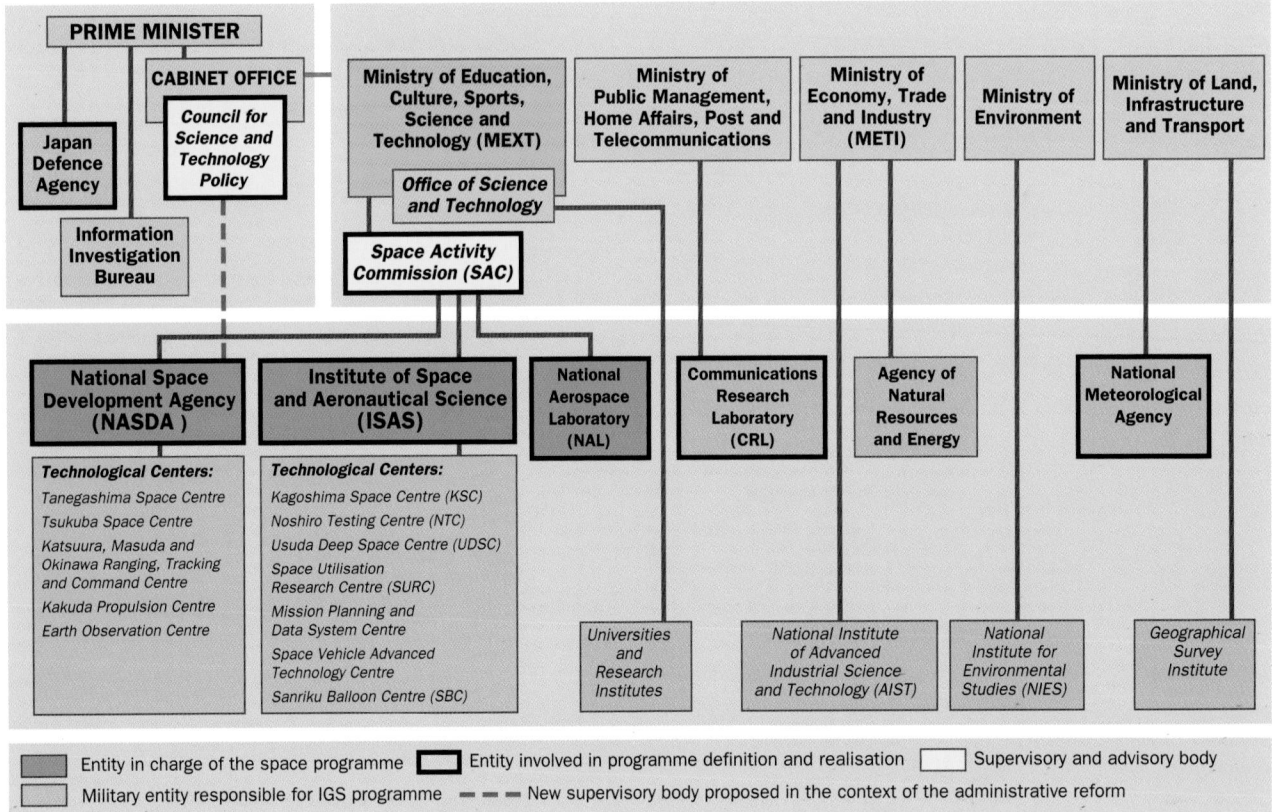

Figure 5.22. **Organisation chart of Japanese space activities in 2001.**

for technology being attested by creation of the Council for Science and Technology Policy under direct supervision by the Prime Minister (fig. 5.22). In 1978, SAC published the first main outline of Japanese space policy for the coming fifteen years. This document, entitled Outline of Japan's Space Development Policy, set out what it claimed was a global and coherent programme including both scientific research and satellite and launch vehicle construction, not forgetting the longer term aim of manned space flight. It was administered by STA which supervised both NASDA and the main institute for space technology, the National Aerospace Laboratory (NAL). Two separate agencies managed space programmes. The predominantly scientific ISAS obtained independence from its university origins in 1981, whilst NASDA was left to develop applications programmes. This division of tasks even concerned launch vehicle development and the running of launch bases. ISAS thus launched the first Japanese satellite, Oshumi, in 1970, whilst NASDA achieved its first launch with the N 1 rocket in 1975. The complementarity between the two organisations was taken for granted. They were not viewed as being in direct competition, and today efforts are being made to generate synergism.

The basic principle of elaborating a space development plan is a quite standard approach in Japan. It has the advantage of guaranteeing a certain level of dependability in the programming. However, it also amounts to letting professionals control the main tendencies and this in turn reveals a relative lack of interest in space matters on the part of the political powers. In 1996, a new fifteen-year plan was published, entitled Fundamental Policy of Japan's Space Activities. This advocated pursuing space policy in the following directions: developing expertise, giving an institutional definition of the respective roles of ISAS and NASDA, promoting international cooperation, encouraging private sector interest in space, and supporting a wide dissemination of space activities.

Two ministries are particularly important amongst government bodies consulted when the plan is revised, which happens on an annual basis, with a more thorough updating every five years. One of these is the Ministry of International Trade and Industry (MITI), which became the Ministry of Economy and Industry (MEXI) after the administrative reform of 2001. Indeed it has a very wide range of involvement, partly through its power of incentive over Japanese companies but also through its global responsibility for Japanese trade policy. From this point of view, however, the space industry does not stand out as a major field of activity and its activities are severely hampered by conditions laid down in the US specific regulations Super 301. For example, space activities were grouped together with the arms industry when the new ministry departments were set up. The Ministry of Post and Telecommunications (MPT) plays a more significant direct role. It regulates telecommunications and satellite operating activities and has considerable influence over the definition of programmes in which its own institute, the Communication Research Laboratory (CRL) is

involved. Such programmes are treated as experimental for the purpose of respecting Super 301. The economic and industrial considerations at stake in telecommunications, and Japan's special interest in systems that are not vulnerable to natural disasters such as typhoons, earthquakes and floods, mean that the MPT's involvement in defining programmes has sometimes led to confrontation with STA and NASDA. Finally, the Ministry of Transport is concerned with the development of weather forecasting systems and navigation activities. Various agencies thus contribute to the elaboration and application of space programmes, depending on their specific field of interest.

A special place is reserved for the private sector. Indeed, Japanese business people are directly interested in the development of space activities. This is demonstrated by the presence of the Federation of Economic Organisations (*Keidanren*) on a council for the promotion of space activities. For its part, the Society of Japanese Aerospace Companies has the task of coordinating sales promotion activities. In addition, the traditionally strong relations between manufacturers and research institutes and universities, which are partly financed by business, leads to greater coherence in space activities.

The overall coherence of Japanese space activities has allowed them to achieve quite significant results. The space programme comprises a wide range of civilian activities whilst the budget remains relatively modest (fig. 5.23). Indeed, after a large increase between 1972 and 1982, then again since 1987, the combined pursuit of launch vehicle, satellite and manned flight programmes presupposes a significant increase in expenditure. In its 1996 report, SAC recommended a doubling of government spending on space activities, and at the same time asked for greater private investment. However, a spate of technical problems with both satellites and launch vehicles from 1997 forced SAC to streamline programmes and seek out more cost effective approaches, under pressure from the Ministry of Finance. This led to the abandonment of the J 1 launch vehicle. The same tendency was further strengthened by administrative reforms in 2001, aimed at making the governmental machine more efficient. In April 2001, the three main space organisations, ISAS, NASDA and NAL, now grouped together under the supervision of the Ministry of Education, Culture, Sport, Science and Technology (MEXT), announced that joint projects would be set up after the fiscal year 2001.

The Japanese space industry has several unusual features. Founded upon the experience of the big electronics firms and in close cooperation with the United States, it is now largely autonomous. This industrial parentage, unprecedented amongst the other space powers, has led to an original model for the development of space activities. The Japanese approach to industrial management favours flexibly run teams, capable of working on different types of project. This enhances the possibility of integrating technologies developed in connected areas. It should be said, however, that in 2000, only seven Japanese companies featured in the

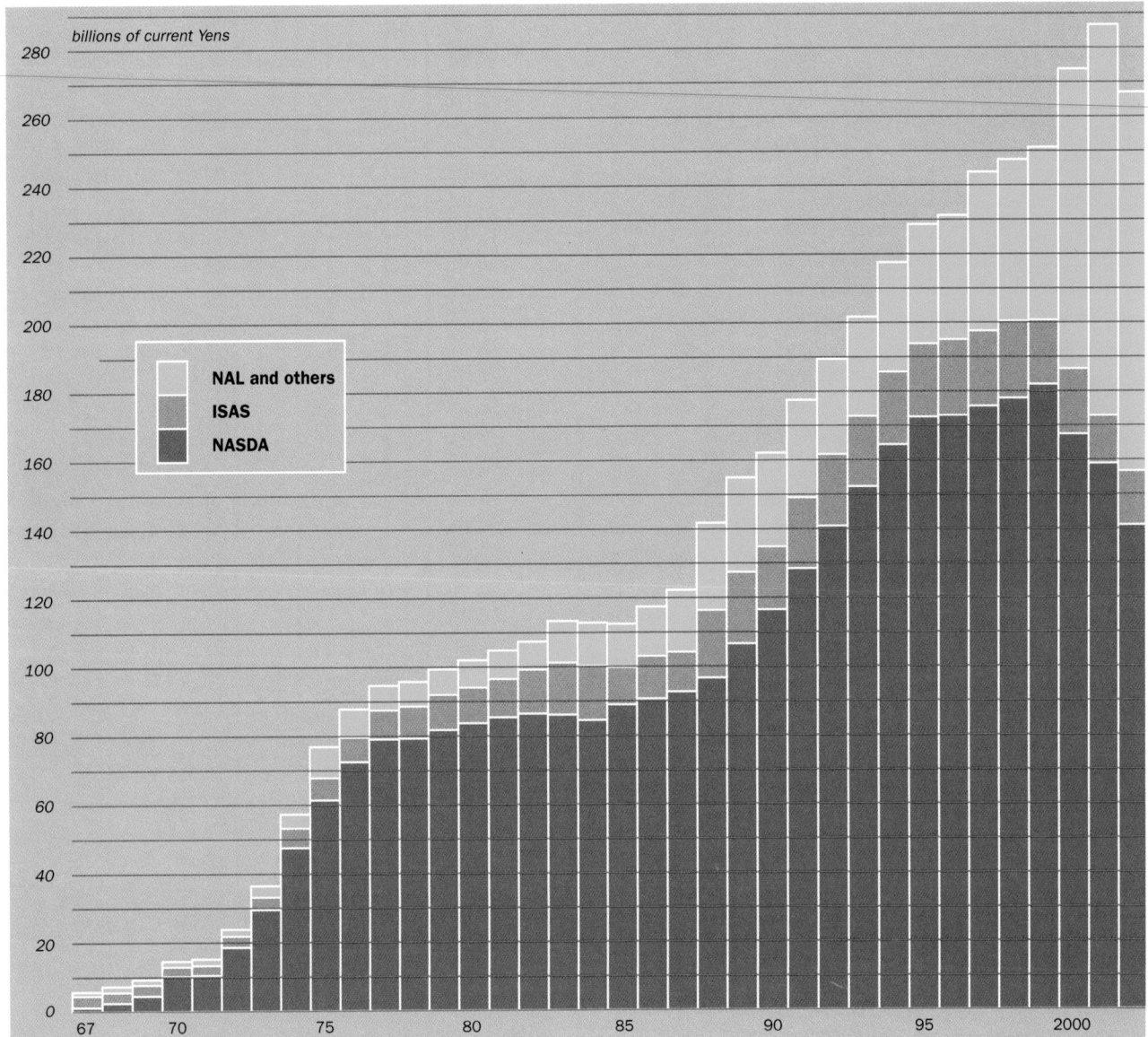

Figure 5.23. **Japanese space funding according to administration from 1966 to 2001.** The budget of the Ministry of Education, grouping together the different agencies, greatly increased after the administrative reform of 2001.

world's top fifty space-related firms and the first of these, NEC Corp, only reached the fourteenth place.

Depending mainly on the general electronics branch, the space divisions of the big Japanese firms function in a very different context to the American or European space industries. Hence, in a period of recession, their low profitability makes them prime targets for recovery packages. The end of the 1990s was very difficult in this context. The merger of Melco, NEC and Toshiba to form the MHI consortium, which then became IHI (Ishikawajima-Harima Heavy Industries) with the addition of Nissan, marked a thorough shake-up in the sector. This restructuring will affect the way relations evolve between NASDA and the industrial base.

Today, Japan has all the means to acquire independence in this domain, but future developments depend to a large extent on the international environment. This is true for both the commercialisation of space products and the possible implementation of military programmes. The evolution of its international relations, particularly with the United States, would appear to be a determining factor in its medium term attitude towards space. However, more diverse partnerships, particularly in Europe, show that Japan is determined to assert itself as a fully-fledged member of the international space community. The pursuit of technological excellence remains a significant feature in the definition of programmes, as attested by Earth observation capabilities and research into reusable vehicles. The desire to be a faultless partner is another important factor and the conditions imposed by partnership can help to consolidate the financial support for a programme. Manufacturers' more purposeful involvement and a clearer perception of economic questions on the part of newly arriving institutional players should bring the Japanese space sector still closer to its European counterpart.

China

When China proposed in 1986 to carry out commercial launches, it set its standard for space activities on the international scene. By proposing to compete with the United States and Europe, China demonstrated its complete independence and also its expertise in the manipulation of modern technology. The effects of this commercial policy were limited by Western attempts to set up launch quotas and by technical difficulties encountered by the Long March launch vehicles. A decisive turning point was nevertheless reached and China is now one of the major space powers, all the more so in that it is aiming towards the full panoply of civilian and military space systems, both manned and unmanned.

Chinese space activities date from the very beginning of the space age. In 1956, Tsien Hsue-shen, accused of being a communist in 1950, left the United States and proposed a plan to the State Council for developing ballistic missiles. Propulsion research thus featured in the science and technology development plan decided in 1958. In parallel with missile research, confirmed in 1957, Mao Tse-tung decided in 1958 to develop a space programme under the name of Mission 581. The first task was to design a launch vehicle, an objective which appeared in the twelve-year science and technology development plan.

Various civilian and military authorities were then involved. For strategic reasons, China was carrying out research into missile technology. The first Chinese sounding rocket was launched in 1960 in a project run by the Academy of Sciences, whilst in the same period, the first Chinese missile was also developed by a Ministry of Defence organisation known as Military Academy 5. Soviet technical assistance soon came to an end as relations between the two countries broke down and China could only count on its own resources. In 1962, a special committee responsible for defence and space technologies was set up and the seventh Ministry of Machine Building was given the supervision of space activities. In this context, rendered even more unstable by the Cultural Revolution, the establishment within the ministry of a military committee responsible for space did something to preserve space activities. In fact, the main political powers of the day agreed that China should be present in space. This is attested by the existence of two competing institutes. One was located symbolically in Shanghai where the Gang of Four held power, and the other in Beijing. Each succeeded in placing their first satellite in orbit using their own launch vehicle within a year (see chapter 6).

At the end of the Cultural Revolution during the 1970s, a relative return to stability was reflected in the space sector by a confirmation of national capabilities, particularly in industry, but also by a new openness towards cooperation and technology. The so-called Four Modernisations programme, which aimed to revitalise agriculture, industry, science and technology, and defence, was begun when Deng Xiaoping came to power. A strengthened interest in space was viewed as an essential feature of a powerful nation.

In the middle of the 1980s, China demonstrated its wide range of space capabilities. It had mastered the launch and recovery of unmanned capsules in the FSW programme and could reach geostationary orbits using the Long March 3 launch vehicles. The commercial policy then promoted in a more politically open international context was double-edged. It sought to obtain some financial returns on space activities largely developed using military funding, and at the same time it showed a desire to ease international relations. It then became possible to form a general idea of how the space sector was organised (fig. 5.24).

Although information concerning the political and military aspects of space activities remained somewhat disjointed and secret, the main tendencies were determined by the Central Military Commission of the Chinese Communist Party. The Commission on Science, Technology and Industry for National Defence (COSTIND) was put forward as a major protagonist in budget allocation and global coordination of space activities. The military's loss of control over space activities, albeit incomplete, was accomplished in several stages. To begin with, civilian activities aimed towards business applications were entrusted to the China Great Wall Industry Corporation (CGWIC) which appeared in 1986 and depended on a new ministry set up at the same time, the Ministry of Astronautics Industry. This ministry disappeared in 1993 and the China AeroSpace Corporation (CASC) was set up to deal with research and development and industrial production. A second major step was taken in 1998 with the creation of a new COmmission on Science, Technology and Industry for National Defence (COSTIND) as a department of the State Council. This occurred in the context of the 'Institutional Restructuring program of the State Council' adopted at the Ninth People's Congress. The stated aim was to streamline management in a certain number of companies in order to procure more efficient production of defence goods and enhance their role in the national economy. At the same time this structure lost its purely military dimension, which now fell under the control of the General Equipment Department. Bases and launches run via the China Satellite Launch and Tracking Control General Administration (CLTC) remained under the authority of the army, which supervised the various specialised institutes. The desire to open up the space

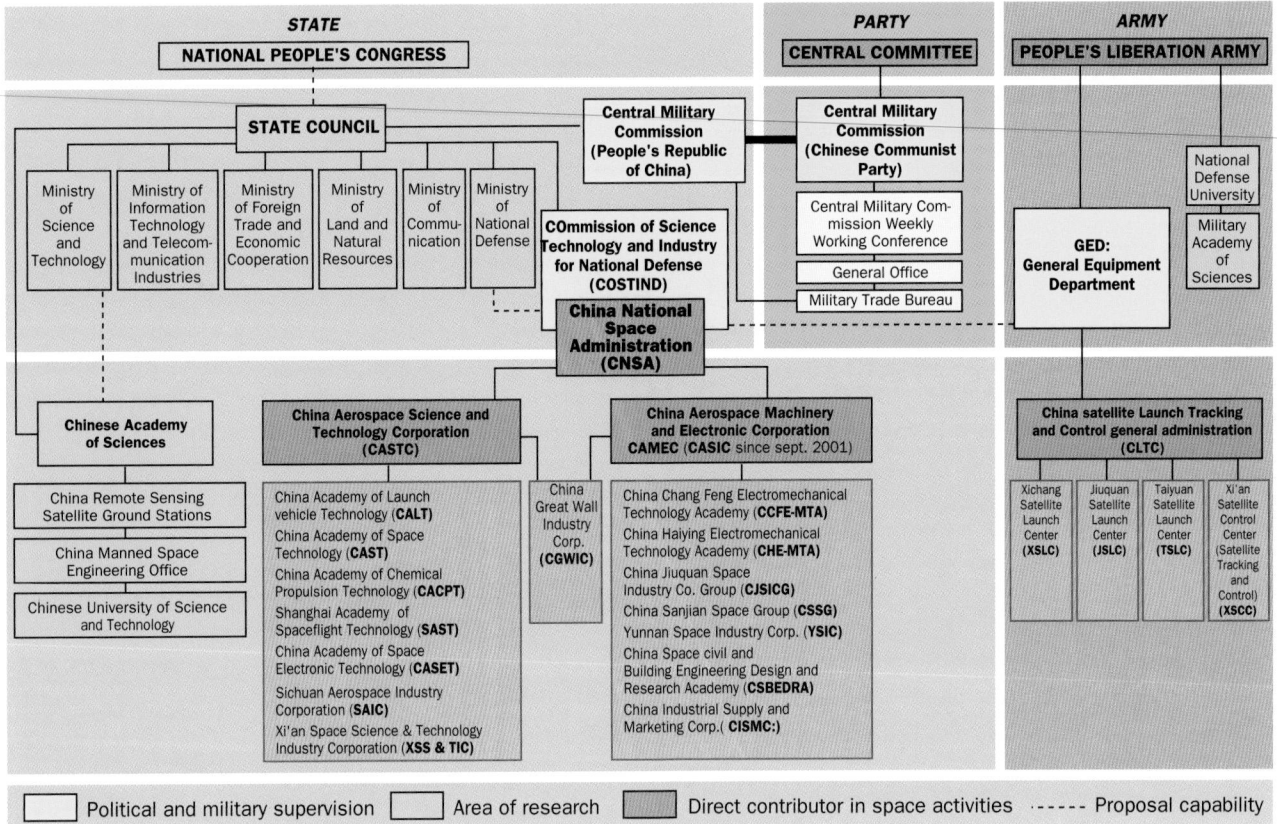

Figure 5.24. **Organisation chart of Chinese space policy in 2001.**

sector led China to set up a specific body, the China National Space Administration (CNSA), responsible for negotiating international contracts as well as administrating civilian space applications. By bringing together different activities, CNSA signalled an attempt at streamlining. It may also represent a desire to come closer to the institutional model adopted by other space powers. In parallel, and pursuing the same idea of improving industrial performance, CASC was itself divided into two specialised entities, CASTC and CAMEC. The latter became the China Aerospace Science and Industry Corporation (CASIC) in September 2001.

Even though certain organisational aspects of the Chinese space sector remain obscure, a much greater mystery surrounds the size of the budget. According to different estimates, none official, it oscillates between 175 million and 1.5 billion dollars. The lower estimate probably refers only to civilian activities, and the second should be treated with caution since it is difficult to assess local production costs. The CNSA has announced that investments for the next generation of civilian satellites in the tenth five-year plan (2001–2005) will be considerably greater than they were in the previous plan, without giving further details about the amount.

However, apart from budgetary questions, China seems to be making an effort to be more transparent with regard to space activities as a whole. Whilst still subordinate to COSTIND, CNSA confirmed its role as the Chinese space agency by publishing in 2000 a White Paper on space policy which is available on its Web site. This change of attitude

corresponds to the fact that the Chinese space programme has reached a certain level of maturity. It also reflects a political desire on the part of its leaders to take their proper place on the international stage, through a more active participation in worldwide partnerships and international regulatory authorities.

At the turn of the century, China can boast a space programme covering the whole range of satellite activities and technologies and, according to the White Paper, it has taken its place among the most advanced countries in the world in many important areas of technology, such as satellite recovery, launch vehicles, applications satellites and manned vehicle tests. The manned flight programme, Project 921 known as Shenzhou, or Magic Vessel, began in November 1999 when Shenzhou 1 was launched, followed on 9 January 2001 by Shenzhou 2. The six-day flight of the orbital module was carried out as a prestige operation and should soon be followed by the first mission of two or four *yuhangyuan*, or astronauts, announced by the CASC for the end of 2002.

The White Paper stresses certain fundamental principles of Chinese policy: independence, self-reliance and self-renovation. In practice, they are combined via an open attitude towards cooperation and international exchange which China is promoting across a very wide base including developing countries. Having already established space relations with the United States and Europe, China is now intent on further extending its space programme to other international partnerships with Russia for manned flights and with Brazil in the area of space applications. The bilateral programme CBERS 1, the China–Brazil Earth Resources Satellite, also called Ziyuan 1 by the Chinese, is presented to developing countries as a good example of cooperation in a high technology sector.

This does not mean that China underestimates the challenges still to be taken up in order to develop technologies and space applications that are changing so rapidly around the world, particularly those related to space industrialisation and commercialisation. The director of CNSA, also deputy commissioner of COSTIND, presented his organisation's new development plan (211 Plan) during the United Nations World Space Week, held for the first time in Beijing in October 2000. Scheduled for the 21st century are a common satellite platform, a new generation launch vehicle, a combined space applications system and a space research and exploration project.

These are the ambitions of a major space power, the international status that China would like to acquire. Its space policies thus aim to achieve civilian and military objectives of national importance defined by the state, introducing the idea of commercialisation only in a restricted form. Chinese organisations are authorised to develop international commercial activities which seem likely to bring in further funds, such as the marketing of launch vehicles.

India

Among the more recently emerged space powers, India is the only country apart from China with a real need for space technology to develop its territory, forecast meteorological threats, and improve communications. Today, having used space technologies on many occasions in the context of international cooperation and then developed its own knowhow, even commercialising some of its own achievements, India has built up the image of a country able to master cutting-edge technology. National pride in these successes and the geographical distribution of space centres in the various Indian provinces clearly demonstrate the symbolic importance of space as a federating agent. At the same time, the growing maturity of the space sector, and especially of its industrial foundations, signals entry into a new era. Whilst pursuing its policy of cooperation with the major space powers, India is building up more and more contacts with newer arrivals on the space scene and could envisage extending activities to space exploration and manned flight.

Indian space activities first appeared at the beginning of the 1960s under the control of the Department of Atomic Energy, when the Indian National Committee for Space Research (INCOSPAR) was set up. In 1963, the first sounding rocket was launched from the Thumba Equatorial Rocket Launching Station (TERLS) and the experience gained led to the founding of the Space Science and Technology Centre in Thumba in 1965. Four years later, the Indian Space Research Organisation (ISRO) was created to bring together all space activities, although still under the supervision of the Department of Atomic Energy.

Three years later, space activities were given an organisational foundation with the establishment of the Indian Space Commission and a Department of Space, both directly subordinated to the Prime Minister (fig. 5.25). This new field of activity was soon under full development, with the creation in 1974 of the National Remote Sensing Agency, then the attribution of government agency status to ISRO in 1975, and Soviet launch of the first Indian satellite Aryabhata on 19 April 1975. India's successes with both satellites and launch vehicles during the 1980s confirmed this trend. A state-run company Antrix was set up in 1992 to market Indian space products, marking a decisive turning point. India was now a fully fledged actor on the international space scene.

The organisation of the Indian space sector shows a clear desire to develop useful space activities through the presence of two committees at the Space Department. One, INSAT, deals with meteorology and telecommunications programmes, whilst the other deals with the management of natural resources. The main lines of Indian space policy are laid out for a ten-year period and updated at the end of each five-year plan. The smooth running of the system is ensured by appointing the same person as president of the Space Commission, the Space Department and ISRO.

The direct subordination of space activities to the Prime Minister also attests to a highly technical appreciation of the importance of space rather than some politically motivated involvement. Space is not attributed to cabinet ministers, the most influential members of Indian government, and its status as a department places it on a purely administrative level, like any other ministry department. The desire for national independence and sovereignty is undeniable as in other areas of technology, but international cooperation was originally the main vector for progress. The state of development of

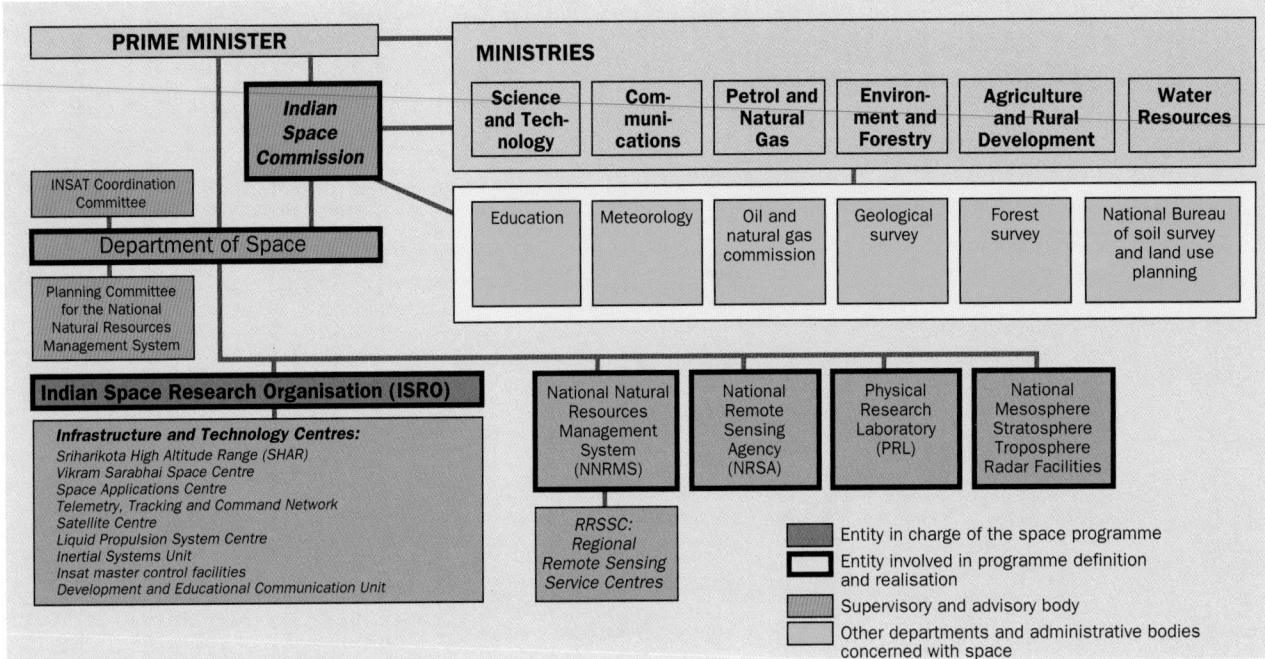

Figure 5.25. **Organisation chart of Indian space activities in 2001.**

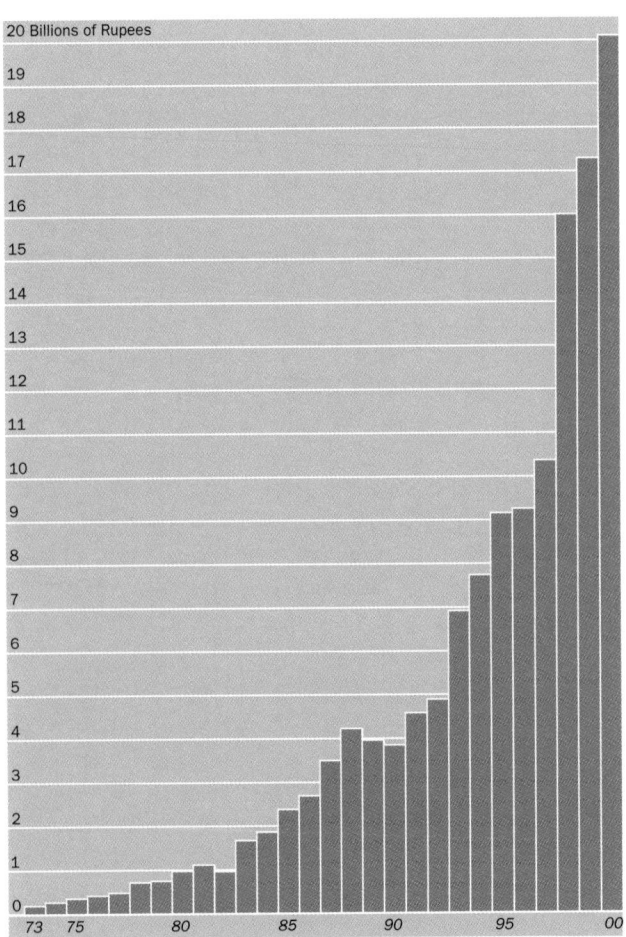

Figure 5.26. **The Indian space budget from 1973 to 2001.** ISRO is the main organisation responsible for carrying out space programmes. The National Remote Sensing Agency (NRSA) and the Physical Research Laboratory (PRL) have much lesser responsibilities.

Indian industry could otherwise only have procured very limited results.

Internal cooperation was enshrined as a basic principle, as for the INSAT system, a joint undertaking of the Departments of Space, Telecommunications, Meteorology and Radio. Apart from this, most programmes were developed in a framework of international partnerships, suitably diversified to balance out the risks of political dependence. The Soviet Union thus carried out several Indian satellite launches in the 1970s and also took an Indian cosmonaut into space in 1984. The United States made several geostationary telecommunications satellites available in 1975–1976, allowing India to test education programmes (the SITE programme). Germany and France also took part in joint experiments. Furthermore, with a view to promoting its image as a leading non-aligned country, India proposed in the context of the United Nations to share its space experience with other developing countries in the framework of the SHARES programme.

Since the beginning of the 1990s, India has proven its ability to move towards self-reliance. The steady increase in government funding attests to this resolution. New programmes have been implemented and new generations of satellites and launch vehicles have been introduced, leading to some major advances (fig. 5.26). In fact, most of the budget is spent on contracts with Indian manufacturers, technological transfer being considered one of ISRO's main missions.

India presents its space activities as exclusively civilian. However, the way the space authorities were set up historically reveals that a clear link has always existed with strategic questions. Neither can this aspect have totally disappeared. India's achievements with launch vehicles have undoubtedly gone hand in hand with a similar acquisition of expertise in the field of ballistic missiles. The main manufacturing firm in this area is Hindustan Aeronautics Limited, also a contracting party with military authorities such as the Defence Research and Development Laboratory. In the same way, capabilities acquired in Earth observation through operation of the IRS 1C satellite, whose sensors provide 5 metre resolution, could easily give rise to development of a military programme.

Space development is certainly not completely detached from the country's security concerns. However, the current organisation of the sector, the relatively low budget, and the still weak industrial base, which therefore necessitates continued international cooperation, are factors that can no doubt explain official insistence on the civilian status of the Indian space sector.

Israel, Canada, Brazil and Australia

Although these countries display quite different space capabilities, it is their position within the overall hierarchy of space powers that unites them. As a group they do possess their own internal hierarchy ranging from a low level of space power to an incomplete mastery of space technology. However, they share the characteristic of being rather borderline space powers with the common aim of acquiring a minimal level of independence.

Israel is the most accomplished amongst them. When the Ofeq 1 satellite was launched in 1988, Israel joined the restricted group of eight space powers possessing all the necessary means to access space. Indeed, it has both launch vehicles and space base, and has the capacity to develop its own satellites. Because of the strategic nature of space activities and the specific regional situation faced by Israel, information is almost non-existent regarding the decision-making process and the respective roles of the various authorities. Space activities are run by the Israeli Space Agency, created in 1982 and placed under the authority of the Ministry of Science, Culture and Sport. It operates mainly as a coordinating body, responsible for policy and programmes, and actively seeks to promote synergism between academic research in science and technology and national R and D resources in industry. The founding of the National Committee for Space Research as early as 1963 attested to university interest, whilst the strategic potential of space in terms of international image and operational needs also contributed to its development. In this context it is characteristic of Israel's approach to maintain a presence across the board, in science, telecommunications, observation and manned flight, although its budget remains extremely limited. However, the Ministry of Defence must also be considered as an important source of financial backing, particularly in the development of launch vehicles and Earth observation, despite a high level of discretion displayed over the precise nature of its involve-

ment. Israel Aircraft Industries (IAI), the main contractor, shares the aim of maximising return on investment. Since the end of the Cold War, it has made every effort to integrate the development of new technologies necessary to Israel's security with potential commercial applications.

Israel collaborates in diverse ways with many other nations, including the United States, Europe, Russia, Brazil and China. This is a consequence of the specific geographical, geopolitical and economic factors which preclude any extensive development of space activities within Israel itself. The aim is not only to achieve more effective results, but also to improve Israel's image as a fully participating space power on the international scene. This is exemplified by the flight of an Israeli astronaut aboard the Space Shuttle in 2001.

Canada illustrates a completely different approach. It revealed an early interest in space as a way of developing its extensive territory. Space telecommunications represented the first step with the launch of Anik 1 in 1972 and ten other satellites by 1991. The objective was to provide for a vast region with a very low population density and extreme climatic conditions. Whilst developing this successful satellite family, specific Canadian expertise was built up, as symbolised by the company Telesat and the establishment of the first commercial telecommunications system. Canada's involvement in new space platforms has kept the country on the cutting edge in certain fields of modern technology. The more recent Radarsat Earth observation programme is part of the same approach since the objectives were to contribute to the development of national territory and subsequently establish a business aspect.

The Canadian Space Agency was created rather late on, in 1989, as a federal organisation with a relatively modest and steadily increasing budget (fig. 5.27). It is dedicated to acquisition and support of Canadian experience and expertise, and thus tends to favour applications programmes with genuine industrial consequences. Canada has thereby acquired very specific areas of knowhow which allow its companies, although relatively small, a high level of export for their products on the international market. By cultivating its policy of filling specialised technological gaps on the market, Canada has been involved in many projects with the United States, and also with Europe. Indeed, Canada has had a strong cooperation agreement with the European Space Agency since 1978. It has diversified its activities in telecommunications, Earth observation, robotics and manned flight, thus maintaining a significant role in international risk management and development aid programmes. Prestige considerations nevertheless coexist with a desire for efficiency. Having guaranteed technological excellence in the robotics field, Canada has not neglected manned activities, proving itself to be a loyal partner in the International Space Station programme. The increased maturity of the space sector seems to be a growth factor and, through its novel system of private–public partnership, Canada is gradually widening its field of expertise. Indeed Akjuit Aerospace Company has even reactivated a base (Churchill Research Range) for small launch vehicles. However, Canada would not seek to set up major independent facilities. Indeed, this would go against its general philosophy, since the country would soon come up against tough competition in an already saturated market. Canada thus tends to perceive the strategic importance of space through its desire to promote research and development in science and technology, thereby enhancing the country's economic and industrial potential and guaranteeing it a significant position on the international scene.

Brazil is one of the main emerging space powers, putting forward launch autonomy as its overriding objective. However, its economic potential is limited and it maintains a peculiar relationship with the United States which means that it can only move forward in gradual stages.

The Grupo de Organizaçao da Comissao Nacional de Atividades Espaciais (GOCNAE) was created in 1961, witness to the longevity of Brazilian space research. GOCNAE mainly focused on atmospheric phenomena. Ten years on, the Instituto de Pesquisas Espaciais was set up to develop space applications that would help the country to develop its territory, through various projects integrating foreign satellite data. Experience acquired in this context brought awareness of the potential of space technologies and the country began to move towards autonomous development in the 1980s. Various programmes were established. These included climate studies and monitoring of terrestrial resources, generally carried out in partnership with other countries. The administrative and financial independence of the IPE was strengthened in 1985 with the creation of the Ministry of Science and Technology. It became a national institute (INPE) in 1990, and in 1995 obtained full autonomy by government decree. Apart from the pursuit of applications, Brazil seeks to achieve independence by implementing a launch vehicle programme and developing launch facilities at the Alcantara space base. This involves both direct and indirect involvement of the military.

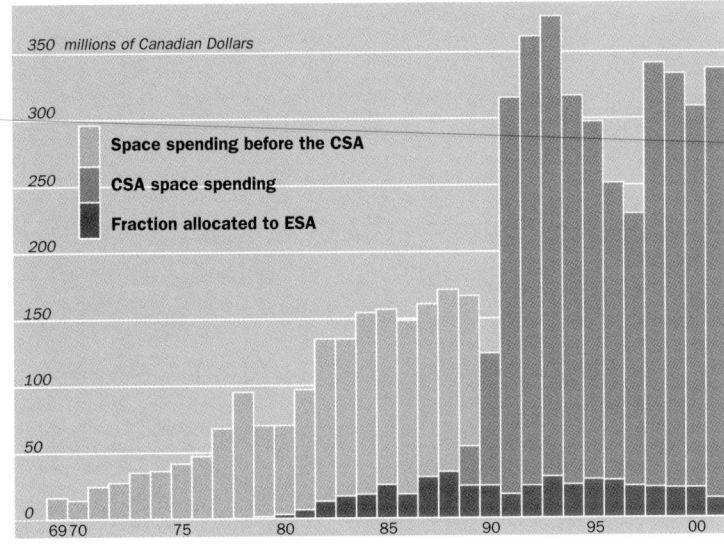

Figure 5.27. **The Canadian space budget from 1969 to 2001.**

The relationship with the United States is a key factor in such development, given American reluctance to propagate propulsion knowhow connected in any way with missile technology and the consequent strict control of sensitive technologies. In the same context, the Alcantara base was the subject of a specific agreement with the United States concerning conditions of use.

A borderline space power, Australia possesses several potential resources that it seems reluctant to develop. A member of the Commonwealth, Australia hosted the first British Blue Streak and Black Arrow tests before Woomera was selected in 1961 as launch base for the future Europa launch vehicle. Using the American Redstone launch vehicle, Australia was the third country to launch a satellite into space. Among its claims to fame in this area, it relayed Neil Armstrong's first steps on the Moon in 1969 via the Parkes radio telescope in New South Wales. The end of the Europa programme and the absence of any national space policy resulted in abandonment of space activities. Despite its geographical advantages, being situated in a semi-desert area of low population the Woomera base has not been reactivated since the last launch there in 1971. Projects for bases elsewhere have never really materialised, even though the prospects seem to be improving since the adoption of new legislation in 1998. Space nevertheless remains a dormant area of national policy and internal debate continues over the need to create a national space agency that could coordinate initiatives and strengthen both research potential and industrial knowhow. One potential development factor is Australia's open attitude towards partnership, not only with Western nations but also with Russia and Japan. However, this feature is not a clear component of any determined federal policy and although one Australian-born astronaut has actually flown aboard the Shuttle, it was in his capacity as an American national. Australia provides a good illustration of how difficult it is for a country to host space activities when there is no genuine governmental support.

CHAPTER SIX

Access to space

- OVERVIEW

- INTERNATIONAL COMPARISON

- RUSSIA AND THE CIS REPUBLICS

- THE UNITED STATES

- EUROPE

- JAPAN

- CHINA AND INDIA

- ISRAEL, BRAZIL AND OTHER COUNTRIES

Overview*

The possession of rocket launch technology and space bases is a key element in a country's independence. The number and capacity of available launch vehicles, together with levels of activity at the various space bases, also indicate the extent to which a nation can claim involvement in space or a role as a provider of services on the international scene. The situation in different countries varies widely. At one end of the scale are nations intent on maintaining autonomy through a certain level of launch capability, whilst at the other are countries whose commercial activities complement a significant level of national activity.

Since the beginning of the space age, 18 bases and almost 40 types of launch vehicle have sent more than 5 200 satellites into orbit. In 2001, although there has been a steady widening of the oligopoly that tends to prevail in space, only eight countries possess the means to launch their own satellites from their own bases, and there is already a considerable imbalance in their relative importance. Moreover, only two other countries seem likely to join this select group in the near future. Amongst the earliest and most significant space powers, the United States is on a par with Russia in terms of the number of launches. Previously, Russia would have been classed as the world's first space power according to the launch rate criterion. In the meantime, European activities have reached a stable level. Although some bases have been dormant, sometimes for a very long time, their recent reactivation in certain cases proves that new countries and new users are beginning to turn their attention to establishing autonomous access to space, possibly even with a view to selling this commodity.

New launch vehicles have primarily been developed by the major space powers in order to rationalise their launch capabilities. The aim has been to develop more efficient launch vehicles and so reduce costs, whether it be for their own satellites or to obtain a better place on the international market. They also seek to gain financial returns on the means at their disposal. They have thus attempted to extend the range of possibilities derivable from strategic missiles, in accordance with their need for small satellites, LEO constellations, and so on.

In this area of space activities as in many others, the United States stands alone thanks to its reusable launch vehicle, the Space Shuttle, and a range of innovative programmes.

In contrast, through varying approaches, Russia has aimed rather to commercialise its vast experience in the domain of cheap and reliable launch systems. Europe is faced with a low domestic demand and must carve out a niche on the international market, defining the development of its launch vehicle in terms of commercial satellite requirements and carefully gauging the levels of competitiveness and complementarity posed by foreign launch capacities. In Japan, the H 2 launch vehicle was first developed in order to demonstrate a certain technological and industrial self-sufficiency, but it proved to be too costly. It was thus replaced by the more cost-effective H 2A which recycled foreign technology. For their part, the Chinese spent some time improving the reliability of their launch systems and consolidated their position by offering launches at attractive rates. They remain victim to strict control over the transfer of Western technologies, particularly American, which severely limit their international clientele.

Other countries such as India and to a lesser degree Brazil have been steadily developing their launch capacities, starting with small launch vehicles capable of attaining low orbits and moving on to systems capable of injecting payloads into geostationary orbit. India has had geostationary capacities since April 2001. In contrast, Israel will have great difficulty following suit given the very limited extent of its territory, and must content itself with more restricted means.

There is a final category of newly emerging powers such as North Korea, Iran or Iraq which are following the original logic of the first space powers, using missile launch technology to acquire basic knowhow. Needless to say, this has led to controversy in the light of arms control agreements such as the Missile Technology Control Regime (MCTR).

Today, once they have reached a certain level of maturity, bases and launch vehicles developed for the purposes of sovereignty and strategic independence must face up to the international law of supply and demand. Although the integration of launch services into the jurisdiction of the World Trade Organisation is still under debate, and despite the high level of government subsidies, launch activities tend to be considered as a business concern. In this context, the private sector is expected to play an increasing role, even though direct economic and industrial benefits remain limited.

---

* Numbers of launches do not include suborbital flights. A launch is counted as successful if the payload is placed in orbit, even if the launch vehicle or the satellite malfunctions.

International comparison

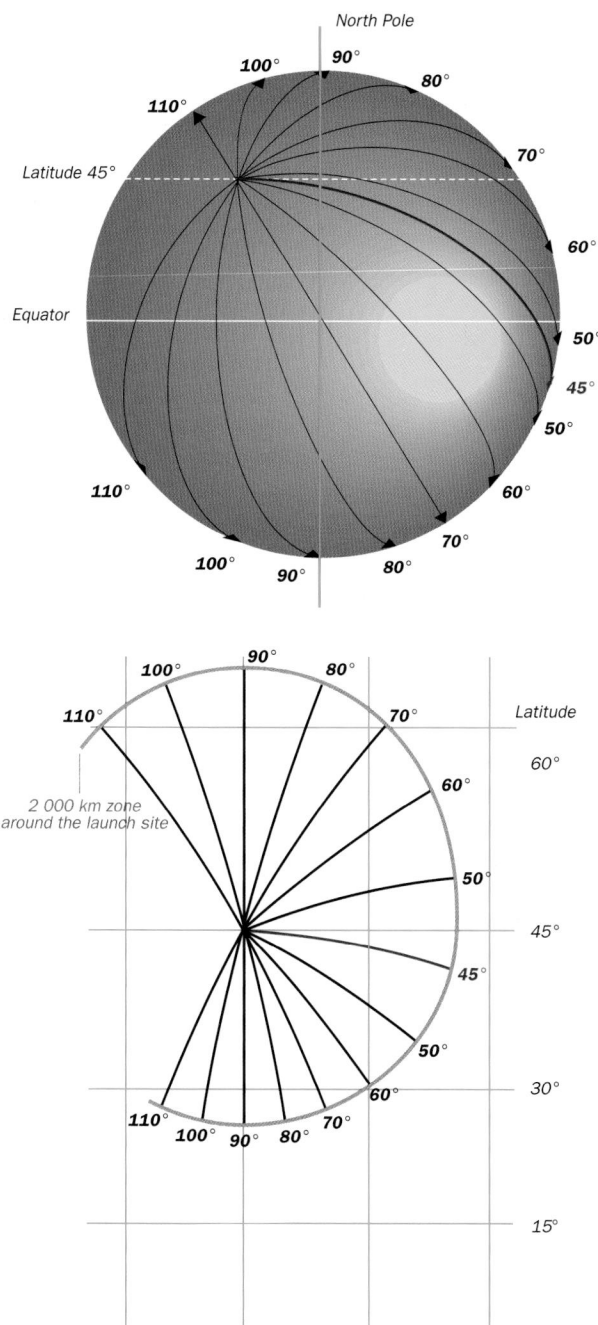

Figure 6.1. **Ground tracks from a base at 45° latitude.** An eastward launch projects a ground track corresponding to an inclination of 45°, as shown in red on the globe at the top of the figure. Other tracks (thin black lines) show that, whatever the launch direction, the inclination is always greater than the value of the latitude at the base. Tracks are shown here supposing the Earth to be fixed. In the lower picture, the same ground tracks are represented on a Mercator projection, each being cut off at a length of 2 000 km. The same approach has been used to show tracks on the planisphere of fig. 6.3.

Space bases constitute a record of space activities on the Earth's surface in the sense that they dictate the way space occupation can be achieved. Geographical constraints imposed by the site and its situation determine launch vehicle characteristics and conditions of orbital insertion. However, the places where launches are carried out on the Earth's surface also correspond to geopolitical and technological factors which determine their localisation within national territorial regions. The development of rocket launchers, on the other hand, depends on specific national policies arising from respective governmental priorities, be they strategic or industrial and commercial.

Situation of bases

The latitude at which bases are sited is a geographical factor of prime importance. When satellite paths are projected onto the Earth, supposing the latter to be fixed, the track is a great circle (see fig. 3.1). If the track is not itself the equator, it must intersect the equator. In such conditions (fig. 6.1), a satellite launched from a point of given latitude l begins by following a path in a plane that intersects the equatorial plane at an angle greater than or equal to l. This is why a base located at one of the poles could only directly access polar orbits (inclination = 90°). The cost of subsequent alterations to orbital planes increases with the value of the correction angle. It is easy to understand why low latitude locations are preferred.

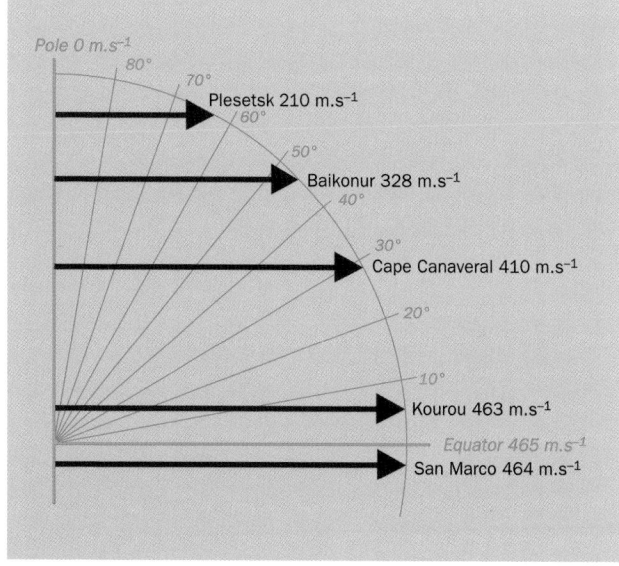

Figure 6.2. **Tangential speed provided by the Earth's rotation at the latitudes of various launch facilities.**

Conditions unfavourable to the siting of launch bases

Reduction, in kilometres per hour, of assistance
from the Earth's rotation on either side of the Equator

Densely populated areas threatened by falling debris

Satellite launches

geostationary and non-geosynchronous satellites	non-geosynchronous satellites only	
●	○	heavy and super heavy launchers
●	○	medium-lift launchers
•	○	micro and small launchers
★	☆	projected launch site

▼ launch pad on a floating platform
▲ home port for ships servicing the platform

Figure 6.3. **Planisphere showing satellite launch facilities.** All bases that launched satellites between 1957 and 2000 are shown on this planisphere, together with the main projects. The tangential speed due to rotation of the Earth and population levels in the surrounding areas are key factors when siting bases within a country's national territory.

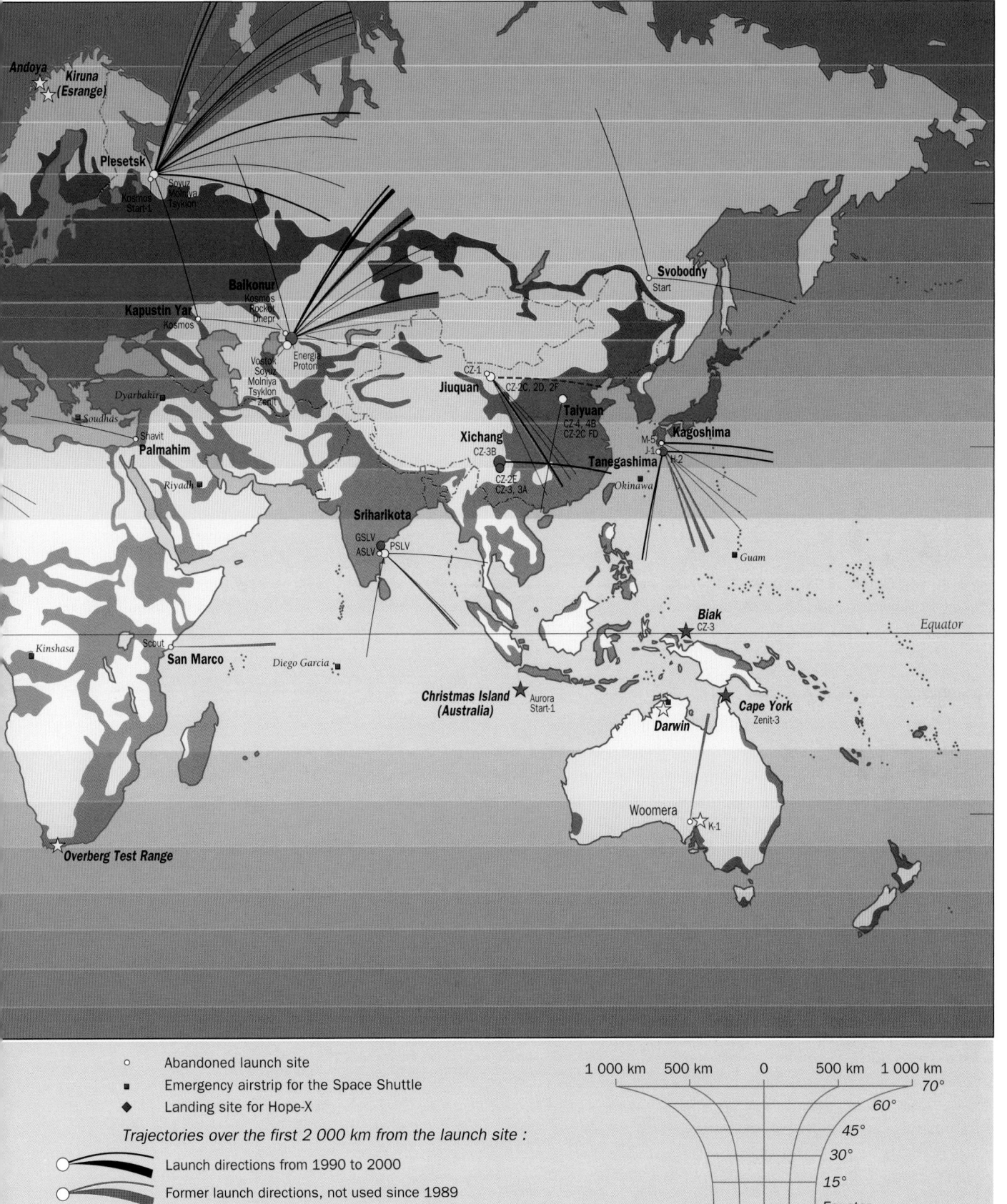

Andoya
Kiruna
(Esrange)

Plesetsk
Soyuz
Kosmos Molniya
Start-1 Tsyklon

Svobodny
Start

Baikonur
Kapustin Yar Kosmos
Rockot
Kosmos Dnepr

Vostok Energia
Dyarbakir Soyuz Proton
Molniya
Tsyklon
Soudhas Zenit CZ-1
Shavit Jiuquan CZ-2C, 2D, 2F
Palmahim
Taiyuan
CZ-4, 4B
CZ-2C FD
Riyadh Xichang M-5 Kagoshima
CZ-3B J-1
Tanegashima H-2
CZ-2E
CZ-3, 3A Okinawa

Sriharikota
GSLV Guam
ASLV PSLV

Equator

Biak
CZ-3
Kinshasa
Scout
San Marco
Diego Garcia

Christmas Island Aurora
(Australia) Start-1 Cape York
Zenit-3
Darwin

Woomera
K-1

Overberg Test Range

○ Abandoned launch site
■ Emergency airstrip for the Space Shuttle
◆ Landing site for Hope-X

Trajectories over the first 2 000 km from the launch site :

○ Launch directions from 1990 to 2000
○ Former launch directions, not used since 1989
○ Launch directions of manned flights (dashed line: planned flights)

1 000 km   500 km   0   500 km   1 000 km
70°
60°
45°
30°
15°
Equator

Moreover, the speed at which a satellite eventually goes into orbit depends on the speed it initially obtains from the rotational motion of the launch site about the Earth's axis, as well as the speed transferred to it by the rocket itself. The rotational speed of a launch site at latitude l is equal to cos l multiplied by the rotational speed of the Earth's surface at the equator. It is therefore maximal at the equator and zero at the poles (fig. 6.2). However, this extra velocity is only favourable for launches with positive azimuth, that is, towards the east. It must be subtracted for launches with negative azimuth, towards the west. This explains why there are many more prograde launches than retrograde launches in the case of non-geostationary satellites. As far as geostationary satellites are concerned, they can only orbit in the same direction as the Earth's rotation, by definition (see fig. 2.30).

For the costly launches required to put geostationary satellites into their distant equatorial orbits, there is a great advantage to be gained by using low latitude launch bases, since these guarantee the best possible contribution from the rotational speed of the base, as well as an inclination close to the final value of 0°.

A second geographical factor of key importance is the population level in neighbouring regions. Launch pads must be located in such a way that when spent stage's or the rocket itself in the case of malfunctioning, fall back to Earth, only marine zones or low population areas are affected.

This feature, combined with the predominance of eastbound launches, clearly favours the eastern edges of the continents (Kourou, Cape Canaveral, Sriharikota, Tanegashima and Kagoshima, San Marco, Alcantara) or sites located to the west of continental desert regions (Hammaguir, Baikonur and Volgograd). Only the three Chinese bases seem to escape this rule, in that highly populated regions lie beneath launch trajectories (fig. 6.3). The need for tracking stations along the line of flight, to monitor the satellite during ascent to orbit, may have played some role in these choices. Ascension Island serves this purpose for low inclination launches from the Guiana Space Centre (see fig. 6.29) and seaborne tracking remains a possibility as in the American or even Russian cases. Israel is compelled to fire towards the west, over the Mediterranean, in order to avoid overflying its Arab neighbours, a quite exceptional case.

Characteristics of sites

Permanent installations at launch complexes include areas reserved for vehicle assembly, propellant processing, satellite preparation, and rocket launch. Launch pads comprise launch tables and platforms, together with launch integration buildings which protect rockets and their umbilical masts on the tables. The various installations are reached by an internal communications system. Equipment and personnel must also reach bases by convenient external means of transport: airport, sea port, road or rail. Sites should be located some distance from built-up areas for safety reasons and require a great deal of open space for development, whilst retaining a certain level of isolation.

Bases by country

Launch sites are distributed over the territories of a very limited number of countries. For various geographical reasons, or because of choices related to internal organisation, some nations are home to several bases. However, numbers do not necessarily correspond to the level of space activity in these countries (fig. 6.3).

The Soviet Union created the first base at Tyuratam, some 300 km from the town of Baikonur, in Kazakhstan. It also developed the Plesetsk cosmodrome in the north of Russia, mainly for military purposes, and a smaller, little-used base at Kapustin Yar near the Caspian Sea. Russia inherited the last two because they happened to lie within its boundaries, and negotiated agreements with Kazakhstan for continued use of Baikonur. Since 1997, Russia has also used the Svobodny missile launch site in Russian Siberia, transforming it into a space base for small ICBM-derived launch vehicles.

The United States has two main bases, at Cape Canaveral in Florida and at Vandenberg in California. It also operates a secondary base at Wallops Island.

The European Space Agency has only one launch complex, the Guiana Space Centre near Kourou in French Guiana, built on the coast very near the equator. The second base belonging to a European member state is the San Marco base, also very near the equator. It is built on platforms in the Indian Ocean, less than five kilometres from the coast of Kenya. The level of activity is very low, with no launches since 1988.

In the Far East, India has a base on the small island of Sriharikota off the eastern coast of the Indian peninsula, whilst Japan has bases on the eastern side of the Japanese islands of Kagoshima and Tanegashima. All three Chinese bases are inland. The oldest is Jiuquan, followed by Xichang and then Taiyuan, which was first used in 1988.

In the Mediterranean region, the Israeli base of Palmahim, just south of Tel Aviv, can only be used by small rocket launchers firing towards the west for security reasons. As mentioned above, this means that they leave the surface in the opposite direction to the Earth's rotation and hence lose speed in the final prograde orbit.

Finally, Brazil opened the Alcantara launch complex in 1997 for its lightweight VLS launch vehicle. This base is ideally situated, almost on the equator.

With the increasing desire to commercialise space resources, projects are being considered for new launch bases in several countries. These are mainly proposed by the private sector. SpacePort Canada was the first project of this type, as early as 1994, when Akjuit Aerospace Inc. took out a thirty year lease on the Churchill Research Range in Manitoba. This base, originally designed for launching suborbital sounding rockets, closed in 1984. The company intends to build two launch complexes for satellites weighing between 1000 and 2000 kg. The spaceport was inaugurated in April 1998 by the launch of the Black Brant 9 sounding rocket. It carried a scientific payload for the Canadian Space Agency (ACTIVE mission).

The Kodiak Launch Complex in Alaska falls into the same category of development, encouraged by the Commercial Space Act. The base will cater for small- and medium-sized launch vehicles aiming at polar and retrograde orbits. The first launch took place in September 2001.

At the same time, several countries which for various reasons are not developing sufficiently big national programmes to justify large-scale investment are offering national sites for joint ventures.

As an example, Australia has long been pursuing this policy from a base in Woomera that was founded in 1946. It was here that the United Kingdom tested the Blue Streak stage of the Europa rocket and launched one satellite in 1971. The base is remote, located in the middle of a large semi-desert area, and the Kistler Aerospace Corporation hopes to take advantage of this in order to recover the two stages of its reusable launch vehicle, the Kistler K1. For its part, Russia is thinking of installing its small Start launch vehicle. Since 1986, there have been regular proposals to create another base at Cape York on the northern coast of Queensland. At latitude 12° south, this would be more propitious for geostationary satellite launches. However, the Cape York Space Agency consortium would appear to be somewhat lacking in impetus. Since 1993, the possibility of a base close to Darwin has theoretically been under investigation for small PacAstro launch vehicles in association with KITComm Pty. Ltd. More recently, in June 2001, the Australian government announced credit facilities for the development of the Christmas Island Spaceport within the framework of a bilateral agreement with Russia for the development of the new Aurora launch vehicle.

Likewise, South Africa has been offering the Overberg Test Range near Bredasdorp for commercial use for some time now. It was previously used to test the Republic of South Africa's launch vehicle RSA 3, abandoned in 1994, and may have been used to test the Israeli launch vehicle Shavit in June 1989.

Other projects merely extend existing activities. For instance, Norway and Sweden are candidates for launching small polar satellites on behalf of ESA, using their respective bases at Andoya and Esrange.

Off-shore, submarine and airborne launches

The restrictions imposed by a fixed launch site can be avoided. One way is to use launch vehicles light enough to be fired from a large aircraft, which thus plays the role of a first stage. The Orbital Sciences Corporation (OSC) has been sending small payloads into orbit since 1990 with its Pegasus launch vehicle carried aboard a Hercules, then a Lockheed L 1011 Tristar (Stargazer). In 2001, more than thirty missions were thereby accomplished from six different sites in the United States, Europe and the Marshall Islands. The Russian company Air Launch Aerospace Corporation (Kompomash, Polyot, KB Makeyev, KB Khimavtomatiki and others) is planning to do the same using a two-stage liquid oxygen–natural gas rocket, launched from the fuselage of an Antonov 124. Other Russian launch vehicles (Space Clipper, Surf) may also be put forward. It is convenient to use a landing strip close to a space base which is already equipped to process payloads and propellant. OSC flights generally leave from the Vandenberg, Cape Canaveral and Wallops Island bases. However, any long enough strip may be used and one flight has taken off from the Spanish base at Torrejón near Madrid. Indeed, the main advantage of this type of launch is its flexibility.

Other projects are currently under investigation, such as Airborne Launch put forward by the Israeli IAI–MLM Division. It is proposed to launch three microsatellites from Hercules C 130 cargo aircraft carrying one-, two-, three- or even four-stage Leolink launch vehicles. Another project put forward by the Israeli company Raphael would be very similar to this air-launched microsatellite system, but using an F15 type aircraft carrying a three-stage launch rocket of 5 to 6 tonnes capable of injecting payloads of between 50 and 100 kg into orbits of 300 to 1 000 km.

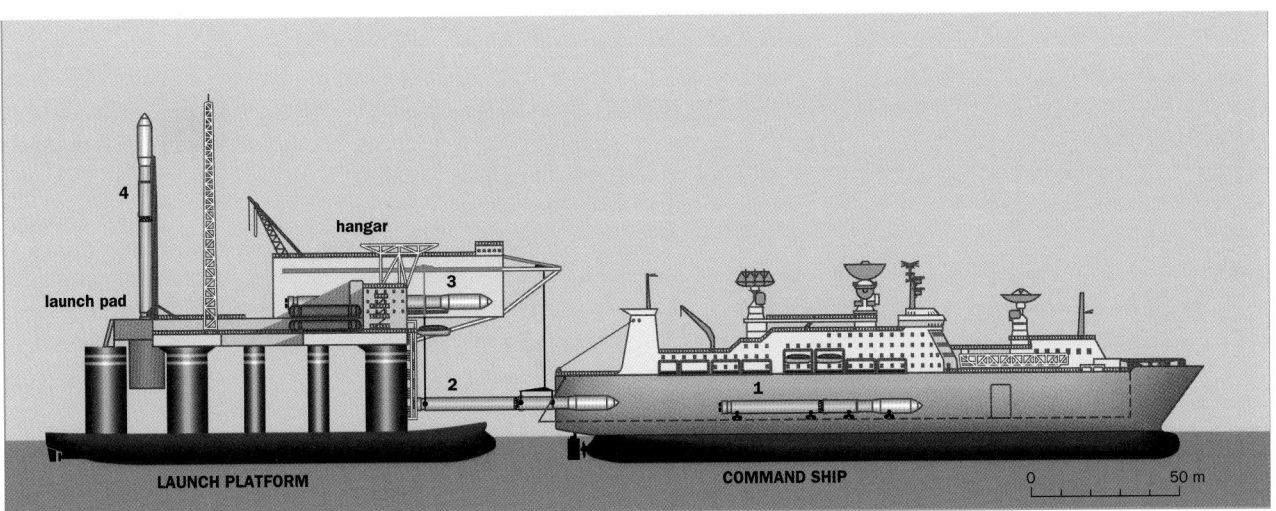

Figure 6.4. **The Sea Launch floating launch system.**

Ships and submarines can also serve as platforms for sending spacecraft into orbit. The Russians have carried out a single launch from a submarine in the Barents Sea and it was successful. They used a decommissioned Shtil SLBM (submarine-launched ballistic missile). KB Makeyev offers a wide range of SLBMs for submarine, ship or even aircraft launch, capable of putting small loads into low orbit.

The floating launch system Sea Launch Service is innovative in many ways. It is a joint venture between the American firm Boeing, the Anglo-Norwegian Kvaerner Group, the Ukrainian SDO Yuzhnoye/PO Yushmash and the Russian RSC Energia. A 28 000 tonne floating launch platform Odyssey operates with a 30 000 tonne command and transportation ship called the Sea Launch Commander, both manufactured by Kvaerner. Rockets are brought from a base at Long Beach, California, down to equatorial latitudes in the vicinity of Christmas Island in order to launch payloads into geostationary orbit (fig. 6.4). Stages one and two are a Yuzhnoye-produced Zenit 3, and the third stage is a Block DM built by RSC Energia. Boeing handles mission control and sales, as well as payload processing and integration. After seven successful launches between March 1999 and June 2001 and only one failure (in March 2000), the feasibility of this novel approach can be considered proven and it can be expected to incite emulation.

Level of activity in bases

Comparing bases from one country to another brings out the unequal levels of space activity in different nations, as well as the varying levels between bases within the same country (fig. 6.5).

This analysis reveals certain persistent factors, such as the predominance of Russia and the United States, with Russia being responsible for almost two-thirds of all launches since the beginning of the space age and other space powers accumulating between them only about 10% of the total. The imbalance can be put down to the relative importance of national governmental needs and technical choices concerning satellite lifetimes. Both these factors help to explain Russia's predominant position. Kourou is an interesting case because it is mainly concerned with commercial launches, covering almost half the market requirements but only 14% of the total number of launches. The low relative level of activity is an indirect indication of just how limited the market is, particularly as it has been more and more sharply contested by Russia and the United States since the mid-1990s. The idea of launching a vehicle like Soyuz from Kourou, the subject of lively debate in 2000, could mark a major turning point. Among arguments favouring stronger partnership with Russia (Soyuz is commercialised by the European–Russian company Starsem), it would more significantly represent a new commercial attitude towards use of the Kourou base as a commodity in itself. Up to now Europe's main objective has rather been to maintain Ariane's position on the launch market.

Comparing levels of activity country by country, Russia can be singled out not only by the fact that it has had a base located outside its national territory since the break-up of the Soviet Union, but also by its recent development of a new launch site, Svobodny. The Baikonur base, which was the Soviet Union's only space base until 1965, specialises in the launch of manned flights and geostationary satellites, activities which have guaranteed its regular use. In 1995, however, its annual level of activity dropped for the first time below that of its American counterpart at Cape Canaveral. Commercial launches of foreign satellites remain limited and are not sufficient to compensate for the reduced level of national activities. Moreover, even though the agreement with Kazakhstan would not appear to pose any major difficulties, it remains a thorn in the side of Russian space policy. The high cost of the lease, representing anywhere between 10 and 17% of the Russian space budget according to differing estimates, and in particular the loss of sovereignty over a key component of the country's space activities raise the question of a longer term alternative. The opening of a new cosmodrome has sometimes been explained in this light. However, the very restricted use of the new base and the characteristics of the launch vehicles used there mean that Svobodny is still a long way from becoming a serious competitor.

Over more than 25 years, it was Plesetsk that sent the greatest number of spacecraft into orbit. The gradual decrease in its level of activity, clearly visible since the end of the 1980s, reflects the slump in Russian space involvement. In 1996, for the first time, Russia launched fewer satellites than the United States. The decline at Plesetsk can be explained by the fact that this base has always specialised in the launch of low-Earth orbiting military satellites, either for telecommunications or surveillance, with varying but generally rather limited lifetimes. The steady fall in the number of launches is thus partly due to the increased orbital lifetime of reconnaissance satellites and the general reduction in military resources. The Plesetsk cosmodrome does nevertheless have some advantages. It is located on Russian territory, guaranteeing a degree of independence that cannot be claimed for Baikonur. For this reason, it has recently benefited from renewed interest as materialised by a series of new installations (see page 122). In addition, launches of low-Earth orbiting civilian satellites are on the increase and this may provide a further opening.

The Kapustin Yar base near Volgograd, long considered to be top secret, spent eleven years dormant after a hundred or so launches between 1961 and 1988. The launch of foreign satellites by two Kosmos 3M in 1999 and 2000 seems to open up new commercial possibilities. However, support for such use of the base is not unanimous. The Russian Space Forces (VKS) have drawn attention to two disadvantages: firstly, the area is highly populated and secondly, launches overfly Kazakhstan. For these reasons, they advocate the base at Svobodny.

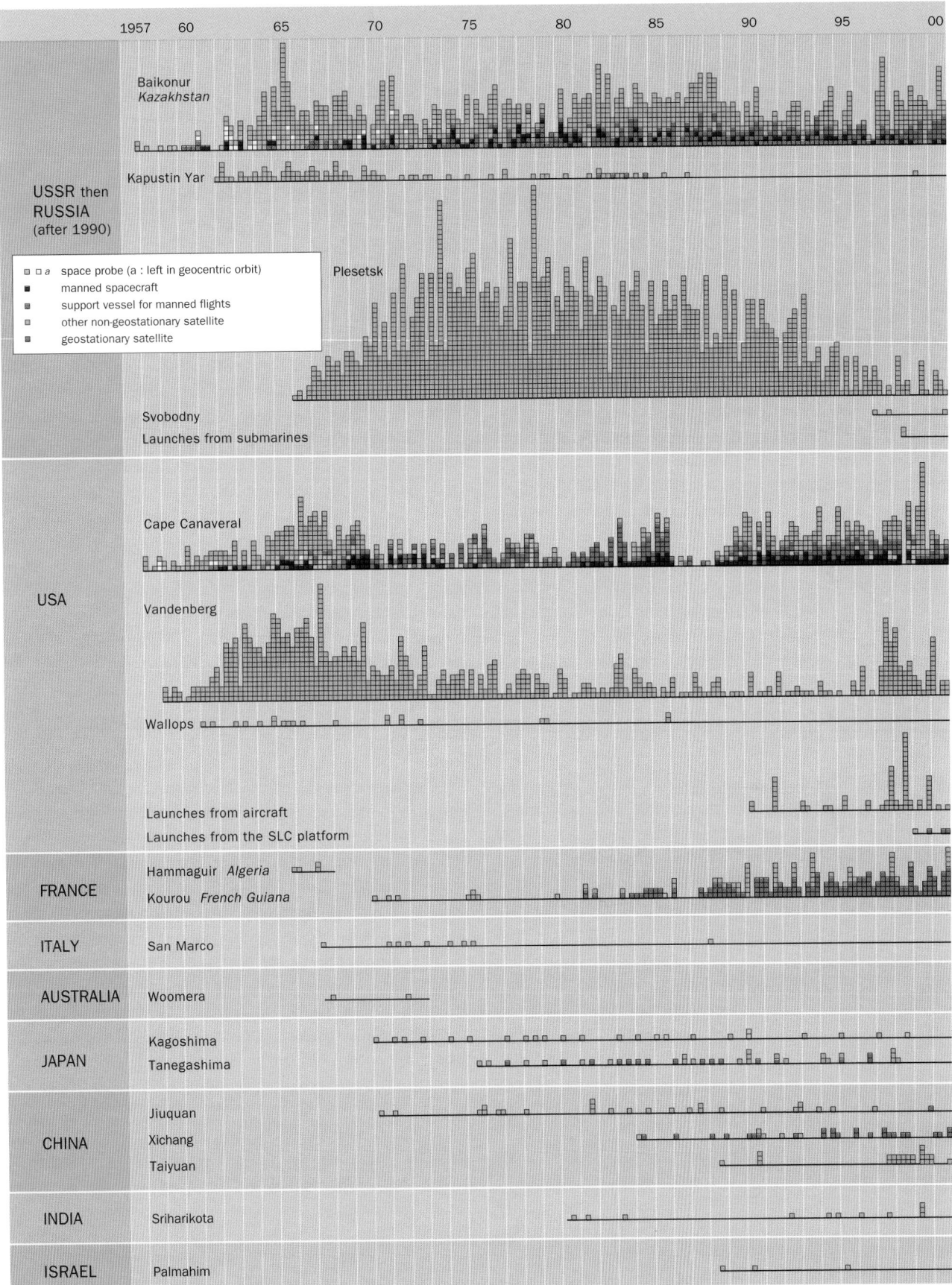

Figure 6.5. **Number of satellites or probes placed in orbit at different launch bases, by quarter from 1957 through to 2000.** Each square represents a satellite or probe placed in orbit.

In the United States, intensive use of the Space Shuttle and a significant level of geostationary satellite launches from Cape Canaveral guarantee a lead role for America's oldest launch facility. In 1996, there were thirty launches and it became the world's busiest space base, with almost as many launches as Plesetsk and Baikonur put together. Comparing with a rate of 120 Russian satellite launches per year in the 1970s, it is clear that far fewer satellites are now being launched.

The space base at Vandenberg is in many respects similar to Plesetsk. Its steady decline is also connected with a reduction in launches of low-Earth orbiting reconnaissance satellites, especially due to the fact that their orbital lifetimes were being extended before those in the USSR. There was at one time a plan to use the base to inject the Shuttle into highly inclined orbits, but this was dropped. However, civilian constellation launches have led to a significant revival, even more so than at Plesetsk, given the primarily American origin of these projects. The leasing of certain installations at Vandenberg to the private sector signalled a trend towards low-orbiting light-weight satellite launches. Commercial expectations were then revised downwards and this led to suspension of financial backing. The Wallops base, counterpart to Kapustin Yar, has not been used for rocket launches since 1986. There have been regular launches of small satellites from aircraft since 1990, although this activity remains modest.

Far behind the main Russian and American bases but well ahead of those belonging to other countries, the Kourou launch complex used by the European Space Agency occupies an unusual position insofar as it cannot lay claim to any large captive market. In particular the military sector is remarkable by its absence. The base thus concentrates mainly on commercial launches. For this reason, there is a preponderance of geostationary satellites, a characteristic shared by the Chinese base of Xichang which also focuses on commercial satellites.

There are two space bases in Japan, each of the space agencies ISAS and NASDA possessing its own launch site and facilities. Since both agencies are severely handicapped by constraints related to the sites and their use, this situation is likely to persist for some time to come, although there have been attempts to rationalise by reconciling the activities of the two agencies. The oldest base is run by ISAS at Kagoshima. It is devoted to scientific satellite launches, whilst the base at Tanegashima carries out all other satellite launches under the control of NASDA.

In China, the Taiyuan base was first used for missile launches, then for the small number of Chinese Sun-synchronous satellites. In 1998–99 the launch of Iridium satellites allowed it to reach a significant level of activity. The Jiuquan and Xichang bases have suffered from difficulties encountered with the Long March launch vehicles, but also from American restrictions on exportation of satellites and of licenses which seriously curtail their commercial use. For these reasons activity at these bases remains rather limited.

Finally, whilst development of the Indian GSLV launch vehicles will undoubtedly increase activity at Sriharikota, use of the Israeli base at Palmahim is restricted to the launch of small LEO satellites relating to the question of national independence.

Hence, even though activity levels in the various bases are gradually moving towards a better balance, the degree to which they are used still directly reflects the hierarchy of states in their capacity as space powers.

Launch vehicles

Access to space is a decisive factor in achieving space power. By controlling its own launch technology, a country is able to pursue its space ambitions with greater independence. The difficulties encountered in launching the French–German telecommunications satellite Symphonie in 1978, through lack of a European launch vehicle, revealed the power in the hands of the launch country, in this case the United States, particularly when the satellite is a potential competitor on the commercial scene. This desire to possess its own space facilities was what motivated Europe to build Ariane. Likewise for the Long March programme in China and other national programmes in Japan, India and Israel, or more recently in Brazil and Korea, despite the fact that the development of launch vehicles requires a high level of manufacturing skills and reliable long term financial backing. Today, political and strategic prerequisites still hold sway over commercial considerations. To speak, as is sometimes done, of competitiveness in the launch business amounts to ignoring the high level of public investment which remains a decisive factor. For the same reason, the cost of launches depends on features that are not by any means exclusively economic. Finally, the common technological basis for launch vehicles and long range missiles has strengthened a certain perception of launch programmes as affirming national sovereignty.

There are many technological constraints involved in developing a launch vehicle. The rocket must reach a speed greater than or equal to the first cosmic velocity (see page 14), at a high enough altitude to avoid excessive atmospheric braking. It must operate for the main part outside the atmosphere. This clearly distinguishes it from vehicles using the atmosphere as a source of oxidising agent. Launch vehicles are all faced with the problem of optimising the weight ratio between payload and propulsion system. This explains the difficulty in reaching an orbit with a one-stage rocket, as sought in the Single Stage to Orbit (SSTO) project. As a result, rockets use either superposed or juxtaposed stages which are jettisoned once their propellant has been used up.

The first launch vehicles and the missiles that preceded them used three types of propellant in their various stages. The combination of kerosene and liquid oxygen, then the more easily stored combinations of nitrogen tetroxide and hydrazine derivatives, so-called hypergolic propellants, are derived from propellants developed for the first generations of ballistic missiles. The same is true of solid propellants, sometimes referred to as powders with reference to the Chinese black powder, although they actually have a rather hard, rubbery consistency.

Cryogenic propulsion based on liquid hydrogen and liquid oxygen does not have the same origins. It is used to supply the huge thrust required by the Space Shuttle at the instant of blast-off and in the upper stages of rockets. However, the extreme temperature requirements of cryogenics mean that they are much more difficult to store and this is not compatible with the military need for missiles that stand permanently ready for launch. In the ongoing arms proliferation debate, this places a bound on the sometimes over-hasty correlation between certain launch vehicle programmes and the development of intercontinental ballistic capabilities.

There is another, non-polluting type of propellant using liquid oxygen and natural gas whose importance may well increase, at least for the development of lightweight launch vehicles.

Almost all space powers used knowhow acquired from missile design to develop their first launch vehicles. German V2 technology was thus largely taken up by the United States and the first successful American space launch was the work of Werner von Braun and his team. On the Soviet side, it was to a large extent the capability of the R7 intercontinental missile that determined the main lines of the Soviet space programme. To a much lesser degree, France's place as the third space power as of 1965 can also be attributed to the existence of a military missile programme. The process has been similar in both China and Israel. It is interesting to note the fears raised when Iraq attempted to launch the Tamouz 1 in December 1989 and North Korea the Taepo Dong in 1998. These attest to the continuing possibility of likening certain launch vehicles to intercontinental missiles, particularly in the area of guidance. In contrast, the case of the Japanese H 2 launch vehicle and its supposed intercontinental missile potential is more difficult to interpret, given that Japan's military efforts have been restricted to self-defence since the Second World War.

One of the decisive consequences of using military technology for the basic design of launch vehicles was the decision to develop non-reusable vehicles. It was not until the 1970s when the United States designed the Space Shuttle that the idea of a reusable engine was finally turned into a reality, with various studies also being carried out in Europe and Japan. There is some ambiguity in the use of the word 'shuttle'. It sometimes refers to the space plane alone, and sometimes to America's Space Transportation System (STS). This almost certainly arises because it is difficult to distinguish between conventional launch vehicles and a spacecraft which, although it represents a completely new way of inserting satellites into orbit, uses standard rocket engines to blast off. A second generation of recoverable spacecraft has been a recurrent subject of study at NASA. Such projects include the X 33 programme, an SSTO with seven cryogenic engines having the same capacity as the STS, and the X 34 programme, twenty times less powerful, having two stages one of which could return and land. These programmes were abandoned in 2000 but further initiatives have been planned within the framework of NASA's Space Launch Initiative. There have also been projects for small recoverable private launch vehicles such as those developed by Kelly Aerospace or the Kistler Aerospace Corp.

From the mid-1980s, new lines of research began to investigate the idea of a hypersonic aircraft without rocket engines. Such a vehicle would take off and land in the same way as an aircraft, thereby removing the constraints imposed by the significant ground installations involved in vertical take-off. It would reach orbital speeds by consuming onboard oxygen supplies as soon as it left the denser regions of the atmosphere. These projects are plagued with technical problems, high costs and conflicting assessments of the potential for this type of system, especially when it comes to civilian and commercial applications.

Launch capacity

For over 20 years, the Soviet Union and the United States carried out almost all launches, creating a monopolistic regime within their respective blocs. Replacing the USSR by Russia, they remain the only countries with full access to space across the whole range of possible missions, including manned flights. At the other extreme in the hierarchy, newly emerging space powers such as Israel and Brazil only possess light lift launch vehicles and are unable to reach geostationary orbits (fig. 6.6).

When the European launch vehicle Ariane was developed in the middle of the 1980s, it was done with the rapidly expanding commercial launch market in mind. Indeed, a growing number of countries and companies sought to have their own satellites launched, particularly into geostationary orbits. Arianespace was by definition more readily available than American operators who were already swamped by the domestic demand from the US government and military. Moreover, Ariane's position was consolidated by the United States' decision to make the Space Shuttle its only launch facility, a choice made for purely internal reasons. When technical problems caused a significant reduction in Shuttle flights, whilst American needs, particularly those of the military, were still on the increase, this led to a proportional growth in the number of Ariane's foreign customers. The Challenger accident in 1986 marked the high point in this trend and as the Ariane programme was also experiencing difficulties at this time, many satellites had to be added to the waiting list by the mid-1980s.

The Chinese government, followed by the Soviets, thus began to offer their own launch vehicles to the Western market. The Chinese had just launched their first geostationary satellite using their CZ 3 launch vehicle in 1984, whilst the Soviets could offer the Proton launcher, whose reliability was well established, and the new, more modern Zenit, inaugurated in 1985.

These new services were greeted with some reluctance on the part of the United States and Europe. With some success,

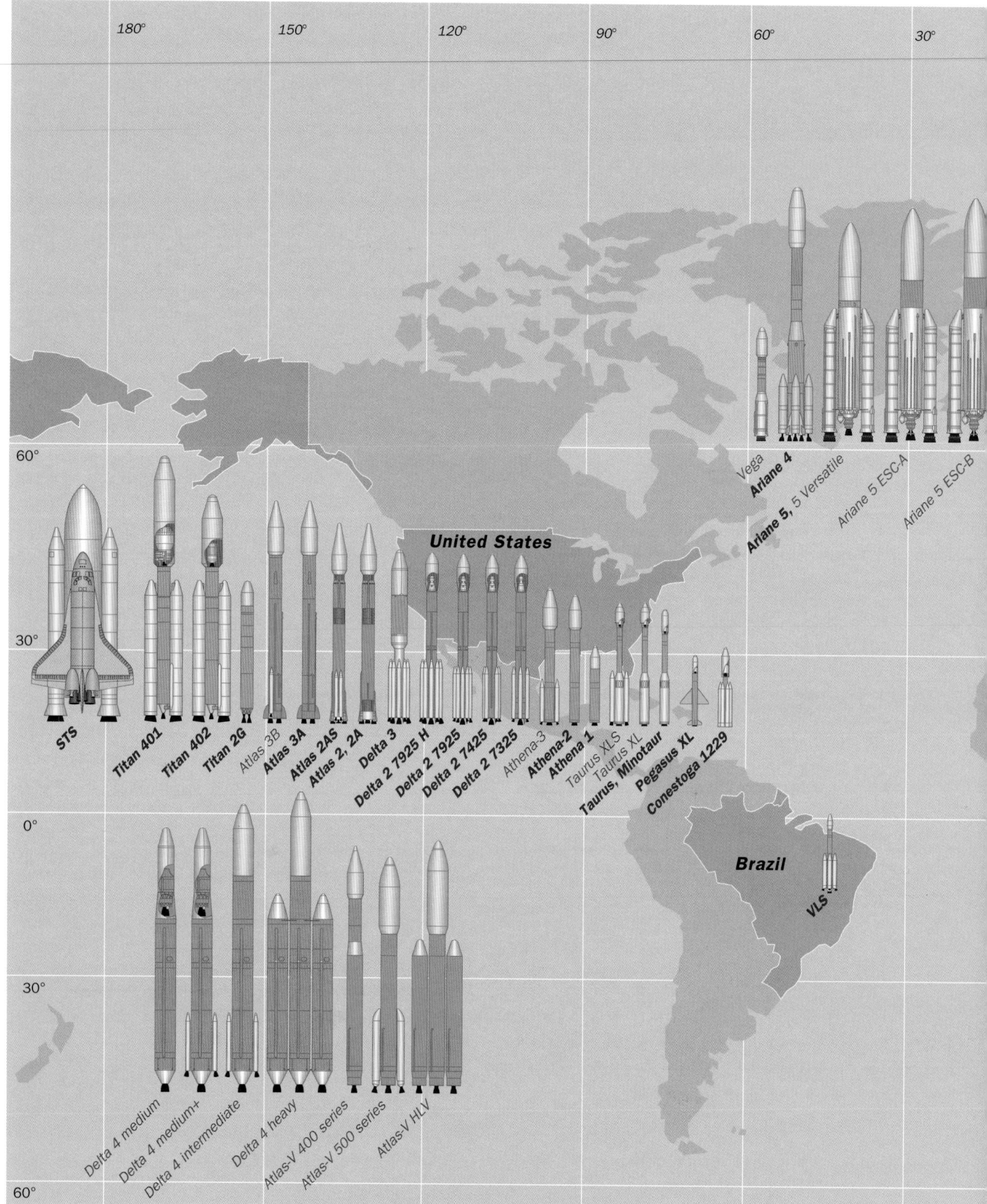

Figure 6.6. **Launch vehicles in service or under development throughout the world at the end of the year 2000.** Names of launch vehicles currently in use are shown in red.

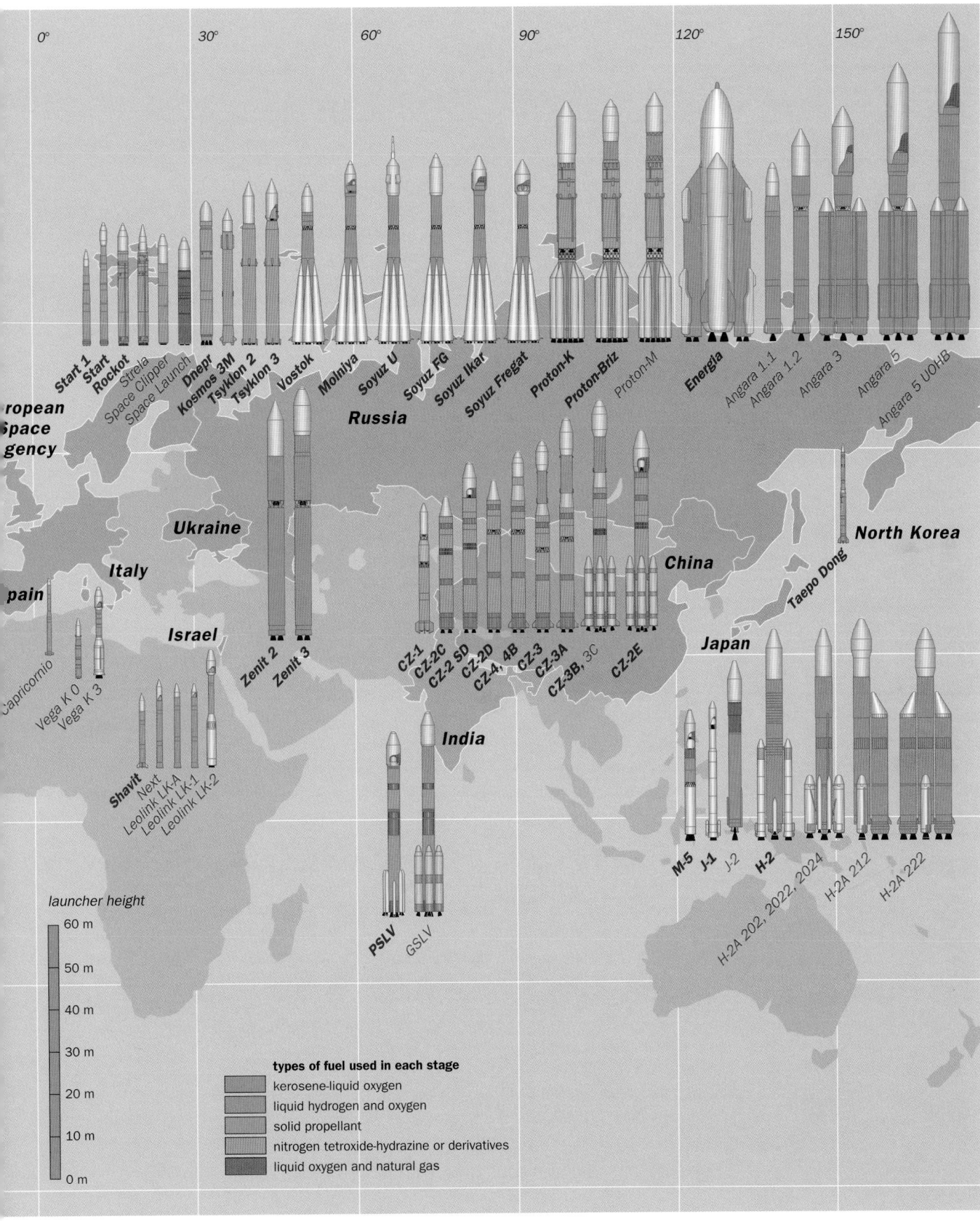

0° 30° 60° 90° 120° 150°

Start 1 Start Rockot Strela Space Clipper Space Launch Dnepr Kosmos 3M Tsyklon 2 Tsyklon 3 Vostok Molniya Soyuz U Soyuz FG Soyuz Ikar Soyuz Fregat Proton-K Proton-Briz Proton-M Energia Angara 1.1 Angara 1.2 Angara 3 Angara 5 Angara 5 UOHB

European Space Agency

Russia

Ukraine

Italy

Spain

Israel

North Korea

Taepo Dong

China

Japan

Capricornio Vega K 0 Vega K 3

Zenit 2 Zenit 3

CZ-1 CZ-2C CZ-2 SD CZ-2D CZ-4, 4B CZ-3 CZ-3A CZ-3B, 3C CZ-2E

Shavit Next Leolink LK-A Leolink LK-1 Leolink LK-2

India

PSLV GSLV

M-5 J-1 J-2 H-2 H-2A 202, 2022, 2024 H-2A 212 H-2A 222

launcher height

60 m
50 m
40 m
30 m
20 m
10 m
0 m

types of fuel used in each stage
- kerosene-liquid oxygen
- liquid hydrogen and oxygen
- solid propellant
- nitrogen tetroxide-hydrazine or derivatives
- liquid oxygen and natural gas

they applied measures laid down by the COCOM (COordinating COMmittee for export control) which limit the export of sensitive technology. It should be remembered that satellites were often developed on the basis of American licences. International agreements subsequently restricted the number of launches conceded to Chinese facilities – 9 geostationary satellites between 1989 and 1994, followed by 11 more between the end of 1995 and the end of 2000 – whilst a series of failures dealt a serious blow to China's credibility as a player on the international market.

The collapse of the USSR introduced a new feature. In 1993 quotas were granted to Russia by the United States – eight geostationary satellites between 1996 and the end of 2000 – at the same time as it was decided to formalise Russian–American cooperation in space. These steps were politically institutionalised by the Gore–Chernomerdin agreements following discussions between the two presidents Yeltsin and Clinton and led in 1993 to the creation of the Lockheed–Khrunichev–Energia (LKE) joint venture. Its objective was to market the Proton launch vehicle. Two years later, the merger of American firms Lockheed and Martin Marietta transformed LKE into a new joint venture known as International Launch Services (ILS). This organisation could offer its customers Proton and Atlas launches from Baikonur and Cape Canaveral. Its ambition at the time was to provide 50% of the world's commercial launches by the year 2000. In the meantime, in 1994 another joint venture called Sea Launch was created to sell and operate Zenit 3 launch capacities (see page 126). Within this new context which associated the interests of American and Russian companies, the initial quota policy was relaxed to allow 5 geostationary satellites for the Ukraine, 11 for Sea Launch, and 16 for Russia, scheduled between the end of 1995 and the end of 2001. In 1996, the Khrunichev centre which manufactured the Proton launch vehicle found itself with orders for 20 rockets up until 2001. The United States finally abandoned the idea of quotas in 2000, provided that the contracting parties, Russia and Ukraine, accepted conditions imposed by the Missile Technology Control Regime (MTCR) on the proliferation of ballistic weapons. Although the situation in China is more complex in the context of its rather special relationship with the United States, the trend is nevertheless towards an end to the quota policy. American restrictions tend to be brought to bear on the granting of satellite exportation licences.

The second half of the 1990s thus marked the beginning of growing competition after the shortage of launch vehicles in the 1980s. Europe was not involved in the 1993 negotiations between the Americans and Russians and had to review its own strategy. In 1996 the Starsem joint venture was set up to handle international sales of Soyuz launch vehicles. It comprises a 35% holding by the European Aeronautic Defence and Space Company (EADS) (former Aerospatiale), a 15% holding by another European company Arianespace, a 25% participation by the Samara Space Centre (TsSKB) and a further 25% by the Russian Aviation and Space Agency.

Although similar to the LKE agreement, this joint venture was set up in a slightly different perspective, since it aimed to complement Ariane 5 by specialising in the orbital insertion of small satellites making up constellations. Another European agreement was signed in 1995 with the creation of the Russian–German company Eurockot – DaimlerChrysler Aerospace (today part of EADS) 51%, Khrunichev 49% – to market the light lift Rockot launch vehicle.

The hiatus left by the shuttle accident, and the shuttle was America's only remaining launch vehicle, did not affect the commercial launch market alone. The United States found itself without launch capabilities for its military and governmental satellites. The huge task of realising alternatives was entrusted to the US Department of Defense. The Delta and Atlas production lines were reactivated, 40 Titan IV rockets were ordered between 1985 and 1997, the Pegasus, Taurus, Athena and other light lift launch programmes were subsidised, and the Evolved Expendable Launch Vehicle (EELV) programme was set in motion by a Presidential Directive of August 1994.

The EELV programme aims to develop a new family of conventional launch vehicles that would be more flexible to use than current launch services and 20 to 25% cheaper. At first it provided a source of financial backing for the modernisation of the Delta and Atlas launchers (Delta 3, Atlas 2AR and Atlas 2ARS), but then a choice had to made between the two competitors. In January 1999 it was finally decided to retain two 'competing partners', with Boeing receiving two-thirds of the subsidies for 19 launches and Lockheed the remaining third for 9 launches. The two future modular ranges Delta 4 and Atlas 5, each with light, medium and heavy lift variants, will have to cover all conceivable types of geostationary satellite. Official American reports and declarations leave little room for doubt about the scope of sales ambitions.

At the beginning of the new millennium, the world launch market has become extremely competitive. The relationship between the demand for satellite launches and the supply of launch facilities has been reversed by comparison with the situation in the 1980s. The competition is also more complex. This has arisen because of the greater number of tried and tested rocket facilities still in service, whilst new launch vehicles continue to appear yet the satellite market itself is actually expanding more slowly than was originally forecast. Companies must find strategies that match the different time scales involved in launch vehicle development and satellite evolution, especially when their activities are by their very nature limited to a dozen or so launches per year in the best of cases. Given that the newly emerging space powers are concerned about achieving independence in this domain, a trend which may reduce the size of the satellite market, prospects would appear to be relatively bleak. Such a situation will encourage existing space powers to become still more competitive. Current sales forecasts for 2005 suggest a rate of between 12 000 and 14 000 dollars per kilogram lofted into orbit.

Russia and the CIS Republics

With the collapse of the USSR, Russia was the main heir to Soviet space potential, with the Ukraine retaining some knowhow and Kazakhstan inheriting the major base at Baikonur. The space sector suffered considerably from drastic cuts in public funding, but a high level of launch knowhow has made it possible for private enterprise to maintain activities via joint ventures.

Baikonur

The first space base ever realised covers an area of 1560 km² and possesses a further 46000 km² of steppe, north-east of the Aral Sea, in the Republic of Kazakhstan. Its climate is typical of inland areas, with a wide diurnal and annual temperature range (from +45 °C to −35 °C) and low rainfall (less than 250 mm a year). The scarcity of cloud formation is a considerable advantage for launch activities. Water is provided by the Syr Daria and access by the Moscow–Tashkent railroad (figs. 6.7 and 6.8).

This base was long shrouded in the greatest secrecy by the Soviets, who referred to it as the Baikonur Cosmodrome, whereas it is actually located 370 km from the town of Baikonur. The nearest town is in fact Tyuratam on the right bank

Russian name	Sheldon Code	DoD Code	First launch to orbit	Payload capability (kg) *LEO: low Earth orbit, GTO: geostationary transfer orbit, GEO: geostationary orbit, SSO: sun-synchronous orbit. (300 km): height of circular orbit, (300-4 000 km): apogee and perigee of eccentric orbits. (65°): inclination. P, Tt, Sv, Eq (Plesetsk, Tyuratam, Svobodny, equatorial site): launch site*	Launches up to 01/01/2001 ( ) successful launches
Vostok	A-1	SL-3	23 09 1958	Tt: 4730 (LEO) 1150 (920 km)	158 (142)
Molniya-M	A-2-e	SL-6	10 10 1960	Tt: 000 (400-40000 km, 65°) 900 (400-200000 km, 65°) 1620 (Moon) 1180 (Venus) 950 (Mars)	285 (279)
Soyuz	A-2	SL-4	23 11 1963	Tt: 6855 (220 km, 51.6°)	789 (769)
Soyuz-U Ikar			09 02 1999	Tt: 4100 (450 km, 51.8°) 3300 (1400 km, 51.8°)	6 (6)
Soyuz-ST				Tt: 5 390 (400 km, 51.8°) 4600 (1 400 km, 51.8°) 3610 (800 km, 98.6°) 1630 (GTO 31°) *with Fregat*	0
Soyuz-U Fregat			08 02 2000	Tt: 5 300 (400 km, 51.8°) 4200 (1400 km, 51.8°) 3 460 (800 km, 98.6°) 1530 (GTO 31°)	4 (4)
Avrora				12000 (LEO) 9200 (SSO) 6000 (1500 km)	0
Avrora-Corvet				Eq: 4500 (GTO) 2100 (GEO)	0
Kosmos-3M	C-1	SL-8	18 08 1964	P: 1400 (400 km, 51°) 1140 (400 km, 83°) 830 (1600 km, 51°) 640 (1600 km, 83°)	421 (406)
Tsyklon-2	F-1-m	SL-11	16 12 1965	P: 1500 (200 km, 82.5°)	102 (101)
Tsyklon-3	F-2	SL-14	24 06 1977	P: 3500 (200 km, 82.5°) 2500 (1000 km, 82.5°)	119 (114)
Proton-K	D-1	SL-13	16 11 1968	Tt: 19760 (200 km, 48°) 2800 (800 km, 98.6°)	30 (27)
Proton-K 4 st.	D-1-e	SL-12	10 03 1967	Tt: 4350 (GTO) 1880 (GEO) 5700 (Moon) 5300 (Venus) 600 (Mars) *with DM-2M block*	252 (242)
with Breeze-M			05 07 1999	Tt: 14000 (2000 km, 51.6°) 5500 (GTO) 3200 (GEO)	2 (1)
Proton-M			07 04 2001	Tt: 6220 (GTO) 4000 (GEO) *with Breeze M*	0
Zenit-2	J-1	SL-16	13 04 1985	Tt: 13740 (200 km, 51.6°) 11380 (200 km, 99°) Eq: 15700 (200 km, 0°)	34 (28)
Zenit-3SL			27 03 1999	Eq: 5250 (GTO, 0°) 2000 (GEO, 0°)	5 (4)
Energia	K-1	SL-17	15 05 1987	Tt: 65000 (LEO) *200000 with 8 boosters*	2 (2)
Start-1	L-1	SL-18	25 03 1993	Sv: 420 (300 km, 90°) 300 (500 km, 90°) 110 (1000 km, 90°)	4 (4)
Start			28 03 1995	Sv: 645 (300 km, 90°) 530 (500 km, 90°) 275 (1000 km, 90°)	1 (0)
Rokot	M-1	SS-19	26 12 1994	P: 1810 (200 km, 63°) 1600 (200 km, 86°) 1100 (1800 km, 63°) 880 (1800 km, 86°) 800 (800 km, 98.6°)	3 (2)
Strela	M-1	SS-19		Sv: 1430 (250 km, 63°) 1180 (250 km, 90°) 520 (1800 km, 63°) 300 (1800 km, 90°)	0
Space Clipper		SS-24		2500 (200 km, 65°) 820 (1500-20000 km, 63.5°) 570 (400-40000 km, 65°)	0
Dnepr-1		SS-18	21 04 1999	Tt: 4400 (200 km, 65°) 4100 (200, 90°) 3700 (200-1000 km, 65°) 3500 (200-1000 km, 90°)	2 (2)
Priboy				Eq: 2400 (200 km, 0°) 1840 (200 km, 90°) 1850 (800 km, 0°) 1360 (800 km, 90°)	0
Shtil-1N		SSN-23	07 07 1998	670 (200 km) 410 (700 km)	1 (1)
Vysota		SSN-8		Eq: 130 (250 km, 0°) 130 (220-1500 km, 0°)	0
Volna		SSN-18		Eq: 130 (250 km, 0°) 130 (220-1500 km, 0°)	0
Air Launch				2600 (200 km, 0°) 2000 (200 km, 90°) 1400 (800 km, 98.6°) 1300 (1600 km, 63°)	0
Angara				P: 26200 (200 km, 62.7°)	0
(with Breeze)				P: 24700 (300 km, 62.7°) 23000 (300 km, 82.5°) 22500 (800 km, 62.7°) 3500 (GEO)	0
(with KVRB)				P: 4500 (GEO)	0
Angara 1.1				P: 2000 (LEO, 62.7°)	0
Angara 1.2				P: 3700 (LEO, 62.7°) Eq: 1800 (GTO)	0
Angara 3				P: 24500 (LEO) 6800 (GTO) 4000 (GEO) Tt: 25200 (LEO) 7500 (GTO) 4700 (GEO)	0
Angara 5				P: 30000 (LEO) Tt: 31000 (LEO) 7800 (GTO) 5100 (GEO) Eq: 8800 (GTO) 5900 (GEO)	0
Angara 5 UOHB				Tt: 11200 (GTO)	0

Table 6.1. **CIS launch vehicles in use or under development in 2001.** Owing to the lack of official information concerning Soviet launch facilities, Western organisations produced their own classifications, such as those of the US Department of Defense or the US Congress (Science Policy Research Division) in 1967. The latter, called the Sheldon code, attributes an upper case letter to the lower stage and a number to the upper stage. A lower case letter defines a specific characteristic. For example, an 'e' indicates the inclusion of an escape stage for interplanetary missions and highly eccentric orbits.

of the Syr Daria. A second town, Leninsk, soon appeared on the base, to house workers involved in space activities. In 1996, Kazakhstan changed its name from Leninsk to Baikonur, and in 2000, it officially had a population of 60 000 inhabitants.

The base at Baikonur attracts a lot of public interest, since it was from here that the first satellite was launched, as well as the first manned flight into orbit. There are several sites of historical interest, such as the residences occupied by Yuri Gagarin and Sergei Korolev and a park in which each astronaut plants a tree before departing for space.

The Baikonur base was the only one of the three space bases in the former Soviet empire to be located outside the territory of the Russian Federation when the Soviet Union broke up (see p. 74). Given that all manned spacecraft launches are made here, and that it is the best of the three for launching geostationary satellites, it would have been unthinkable for Russia to stop using it. On 30 December 1991, the Minsk agreement stipulated free access for all members of the CIS. On 25 May 1992, an agreement between Russia and Kazakhstan fixed the Russian share of running costs at 94%. Further agreements on 28 March and 10 December 1994 guaranteed Russian sovereignty over the cosmodrome and the town of Leninsk for a period of 20 years, with the possibility of renewal for a further 10 years, in return for an annual rent of 115 million dollars. The question of sovereignty and the Russian role in the base still seem

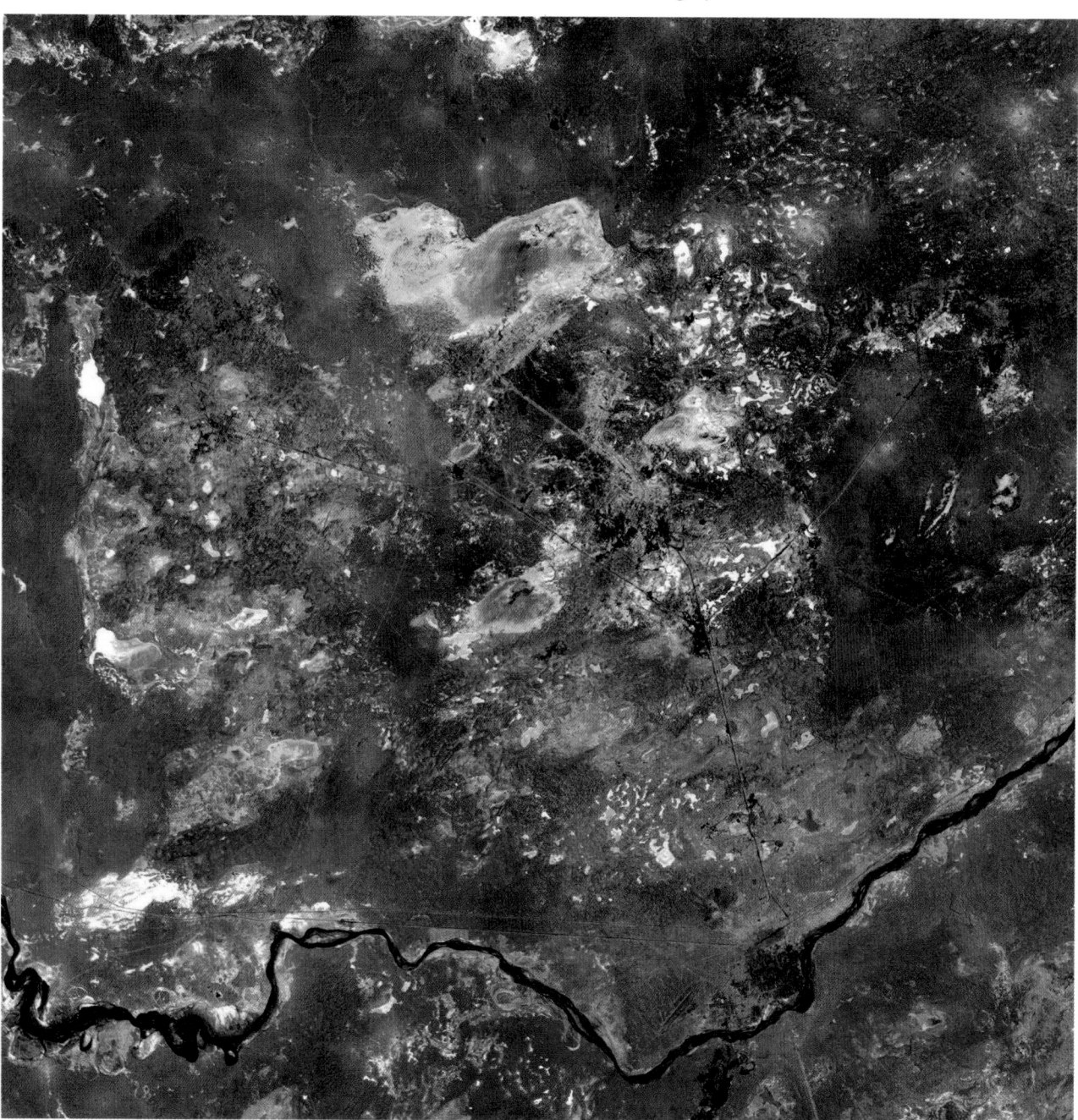

Figure 6.7. **SPOT image of Baikonour.** A mosaic of four SPOT 4 images taken on 05 June 2001 and 10 April 2001 (west) and 22 June 2000 and 11 May 2001 (east). © CNES 2000, 2001, Distribution SpotImage.

to be debated. Various agreements have been signed in the light of Russia's strong military presence (28 000 soldiers) and its inability to settle payments within the required time. Whatever contentions there may be, the desire for compromise has always won through. Kazakhstan has no use for such a base and would be unable to maintain it, whilst it remains essential to Russia. In any case, the two countries have preserved a close relationship in other areas.

The problem of financial backing has been partially solved by opening Baikonur to commercial launches. These have been made possible via joint ventures with the Americans using the Proton rocket and the Europeans using Soyuz launch vehicles. The respective companies, International Launch Services (ILS) and Starsem, use and contribute to improving existing equipment. Proton payloads arrive by the Jubilee track built for the Buran shuttle, whilst Starsem uses a large building on the premises of the Energia complex, completely renovated for payload processing. In the same spirit, test firings of the Rockot launch vehicle (in collaboration with Germany) took place at Baikonur, although commercial launches for the Eurockot company have been scheduled from Plesetsk. At the beginning of 2000, as a consequence of the relative internationalisation of the cosmodrome, the Russian Aviation and Space Agency Rosaviakosmos (RAKA, formerly RKA) officially took over its running from the Russian Space Forces (VKS). However, the VKS is still respon-

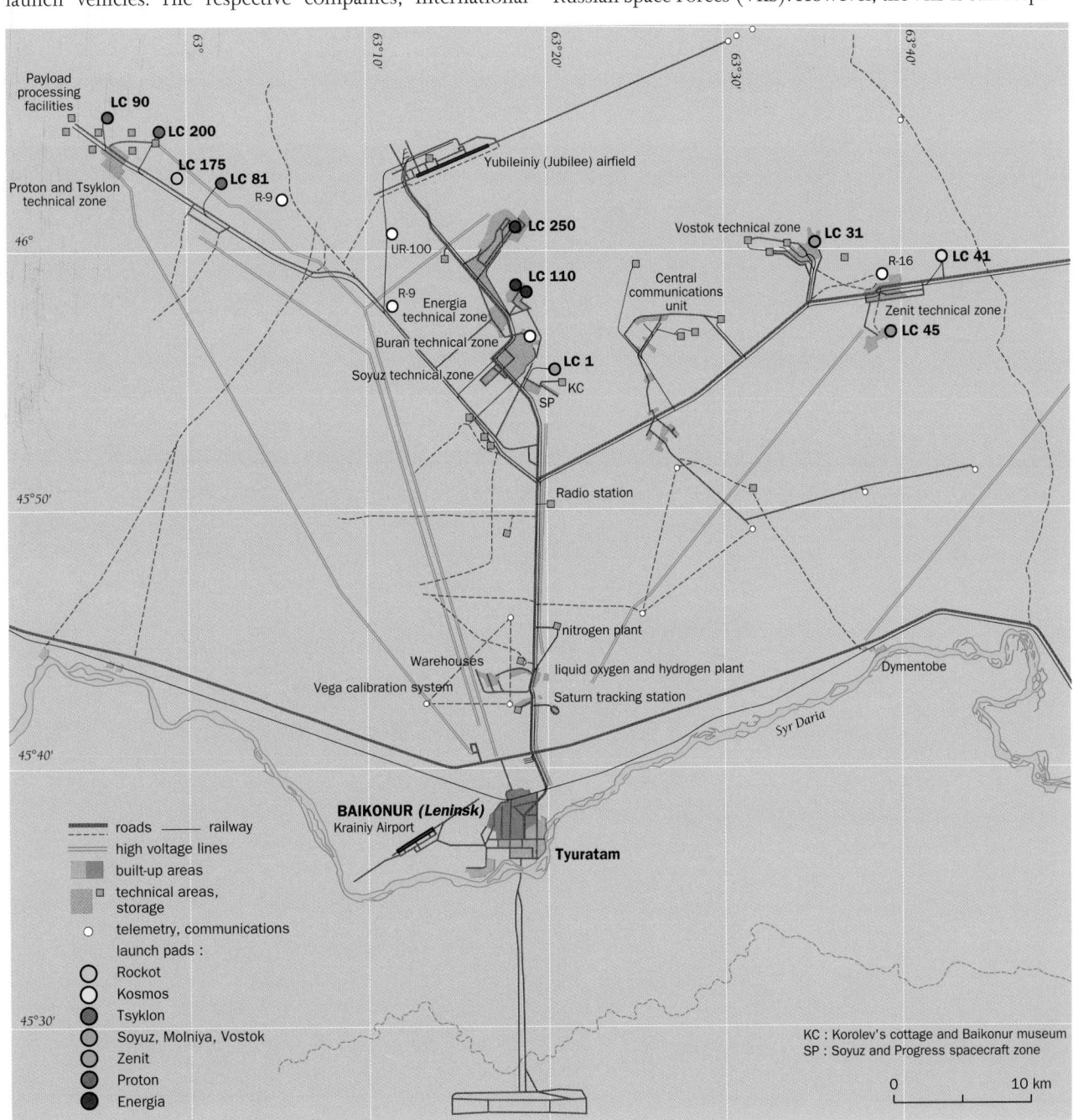

Figure 6.8. **Interpretative sketch of the Baikonur SPOT image.**

sible for launch complex development and lack of credit is becoming a serious problem in the face of a growing need to modernise the whole base.

Plesetsk

The Plesetsk cosmodrome is the main launch facility for military satellites and was originally a missile base. It was created for that purpose in 1960 and long kept secret. Kosmos 112 was the first satellite to be launched there in 1966. The official name of Northern Cosmodrome was attributed to it later, when the base was opened to civilian launch activities in the context of the Interkosmos organisation.

The base is sited in a rather featureless zone in the taiga south of Arkhangel'sk, on the glacial formations of the central Russian plateau, in the middle of a vast extent of subarctic pine and birch forest. The whole military and space complex taken together covers an area of 1762 km². This includes ballistic missile bases, space launch complexes and associated facilities. It is easy to pick out internal and external infrastructure on the SPOT image in figure 6.9: power lines, railroads (the Arkhangel'sk–Moscow line and branch lines serving the cosmodrome itself), roads and landing strips. It is more difficult to identify installations specific to the cosmodrome and the missile base visible on the photograph. Fig. 6.10 gives a reasonable interpretation of certain features.

Soyuz and Molniya launch vehicles use 4 pads in the Korolev zone, whilst Kosmos uses 2 pads, Tsyklon 2 pads and the Rockot an adapted Kosmos pad in the Yangel zone. The light lift launchers Start 1 and Start were also first fired from Plesetsk. Insofar as Plesetsk is located on Russian national territory, a certain level of development is programmed, such as the construction of two pads for Zenit to the north of the Yangel zone, and a launch complex for the future Angara launch vehicle.

Figure 6.9. **SPOT image of Plesetsk.** A mosaic of four SPOT 4 images taken on 26 June 2000 (west) and 15 July 1998 (east). © CNES 1998, 2000, Distribution SpotImage.

Kapustin Yar

The little used Kapustin Yar Space Centre developed from a former missile base, created in 1947 for medium range missiles and sounding rockets. It is located at latitude 48.4° north and longitude 45.8° east, on the north-east bank of the lower Volga in semi-arid steppe scattered with dried up lakes. It is by far the least active base. The Kosmos launch vehicles were the first to make their appearance here from March 1962, followed by the Kosmos 3M used to launch light Kosmos and also Interkosmos. Six tests of a minishuttle prototype between 1982 and 1987, with four orbital insertions, constitute the last large-scale space programme to be undertaken at this cosmodrome. The complete lack of launches since 1987 seemed to indicate that orbital flights had been abandoned, but the base was reactivated in April 1999 for the first commercial launch of the new Polyot-Assured Space Access Inc. A second launch of small German and Italian scientific satellites also occurred in 2000.

Svobodny

This former missile base is a quite different case insofar as it has only recently been transformed into a satellite launch centre, first for the light lift Rockot and Start 1 launchers (one example of the latter was successfully launched in March 1997), using existing silos, then possibly for the future heavy lift launch vehicle Angara.

The base is located in eastern Siberia on the banks of the Zeya, a tributary of the Amur, between the towns of Svobodny and Belogorsk, not far from the Trans-Siberian railroad (fig. 6.11). Its latitude and longitude are roughly 51° north and 128.5° east. Neither the climate, which is continental with very cold winters, nor the latitude are particu-

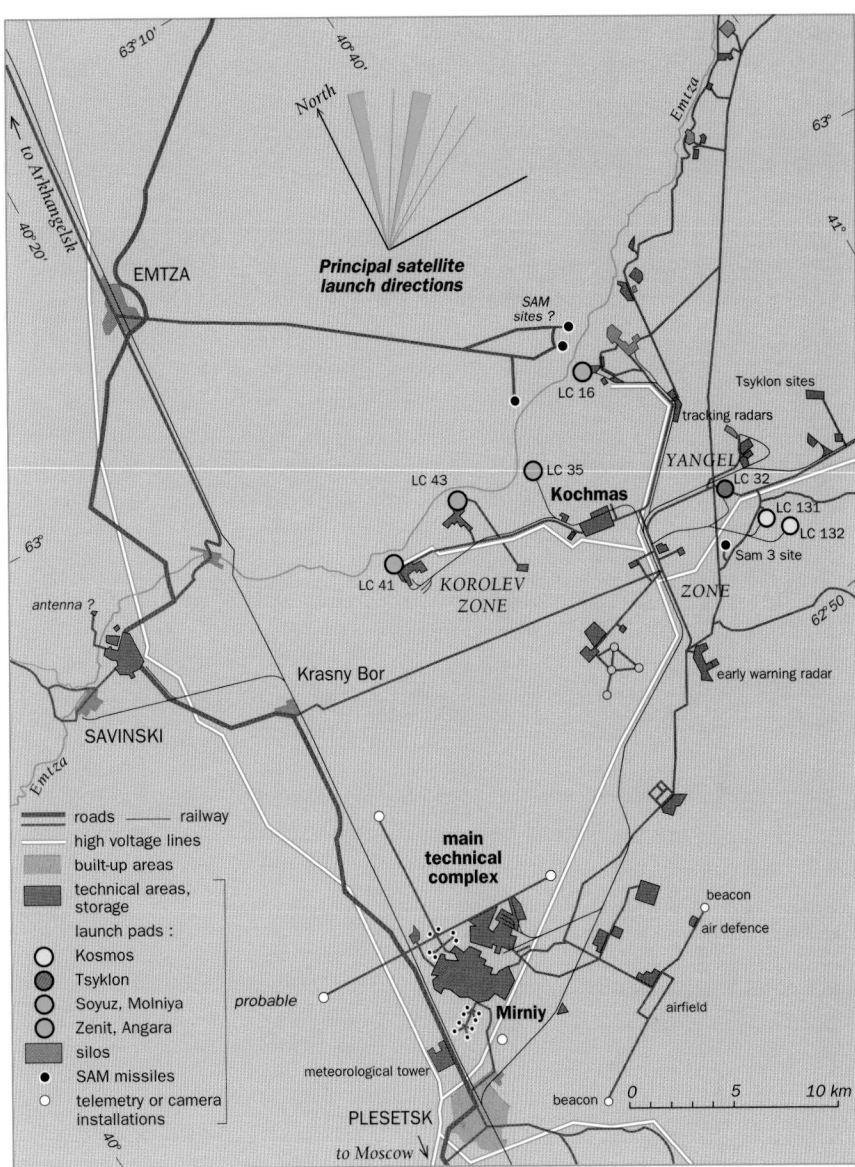

Figure 6.10. **Interpretative sketch of the SPOT image of Plesetsk.**

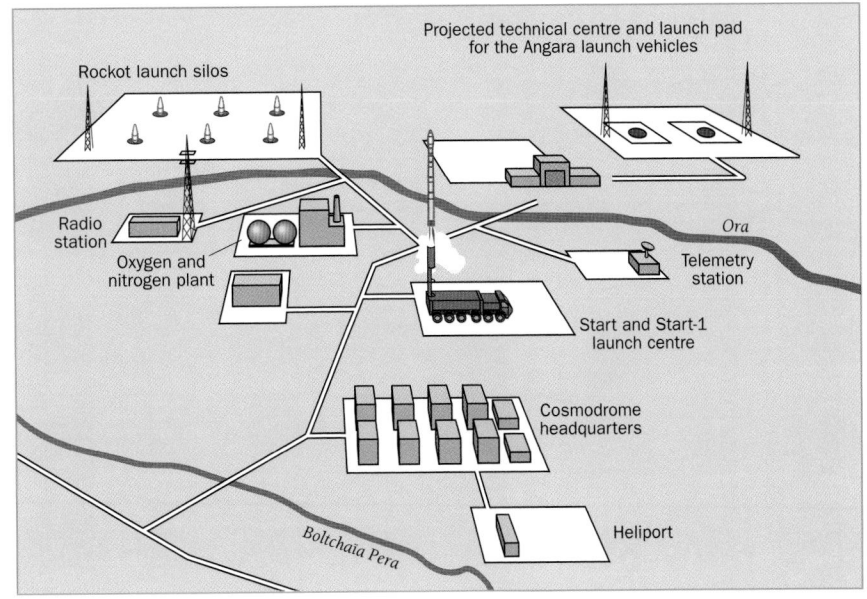

Figure 6.11. **Sketch of the Svobodny launch facility.** Based on information from the Coordinational Scientific Information Centre, Moscow.

larly favourable for setting up a space base. However, uncertainty over the status and possible future use of Baikonur, combined with the presence of equipment ideally suited for converting long range strategic missiles into launch vehicles, are convincing reasons for developing new launch facilities on Russian territory.

Launch vehicles from Semyorka to Energia

The history of Soviet launch vehicles begins in 1956 when the first intercontinental ballistic missile (ICBM) was built, giving birth to the R7 or Semyorka, a one-stage rocket with four strap-ons which sent the first Sputnik into orbit on 4 October 1957. This rocket constitutes the backbone of the so-called Korolev launchers, named after the man who designed them. Three of these, Vostok, Molniya, and Soyuz U, are still in use (fig. 6.12). Vostok derives from the Semyorka by adding a small supplementary stage. The Voskhod launch vehicle is made by building in an extra stage which is three times as powerful and Molniya is a Voskhod carrying a third stage. Soyuz appeared in 1966 and comprises a Voskhod with upgraded second stage. In 1973, this evolved into the Soyuz U, and then between 1982 and 1985, the Soyuz U2, which was just a Soyuz burning Sintin, a synthetic kerosene. (Since 1995, Sintin has proved too expensive and is no longer used.)

Until Soyuz 2 is ready, Starsem has been using the Soyuz U for its first few contracts (Globalstar). The launcher is supplemented with the Ikar on-orbit manoeuvring stage, borrowed from fifth and sixth generation military surveillance satellites. An upgraded version, the Soyuz ST, was tested in 2000. In addition to Ikar, it has a fourth stage, namely the Fregat module, developed by NPO Lavotchkin for the Phobos and Mars 96 probes, and available since test flights in February and March 2000. The same engine could equip Soyuz 2, allowing increased payloads or geostationary launches.

A modernised Soyuz has often been discussed over the last ten years or so, first referred to as Rus, then Soyuz FG, Soyuz/ST, Soyuz 2 and finally Aurora (or Avrora in Russian). The Soyuz FG, which flew in 2001, is a Soyuz U with modernised boosters and first stage engines. The Soyuz/ST should be an FG endowed with a new second stage and a digital guidance system borrowed from Zenit, and the Soyuz 2 is an ST with a new RD 120 M engine on the first stage. Finally, Aurora is a Soyuz 2 extended by a Corvet third stage, derived from the Block D of the Proton and Zenit, enabling it to reach geostationary orbit. In addition, the first stage is equipped with an NK 33 engine. This launch vehicle should be available from 2003.

By 1 January 2001, 175 Vostok and precursors, 305 Voskhod and 813 Soyuz had been launched, as well as 308 Molniyas. This includes 19 failures for Vostok, 14 for Voskhod and 20 for Soyuz, whilst 32 failures occurred for the Molniyas, 22 being due to Block L.

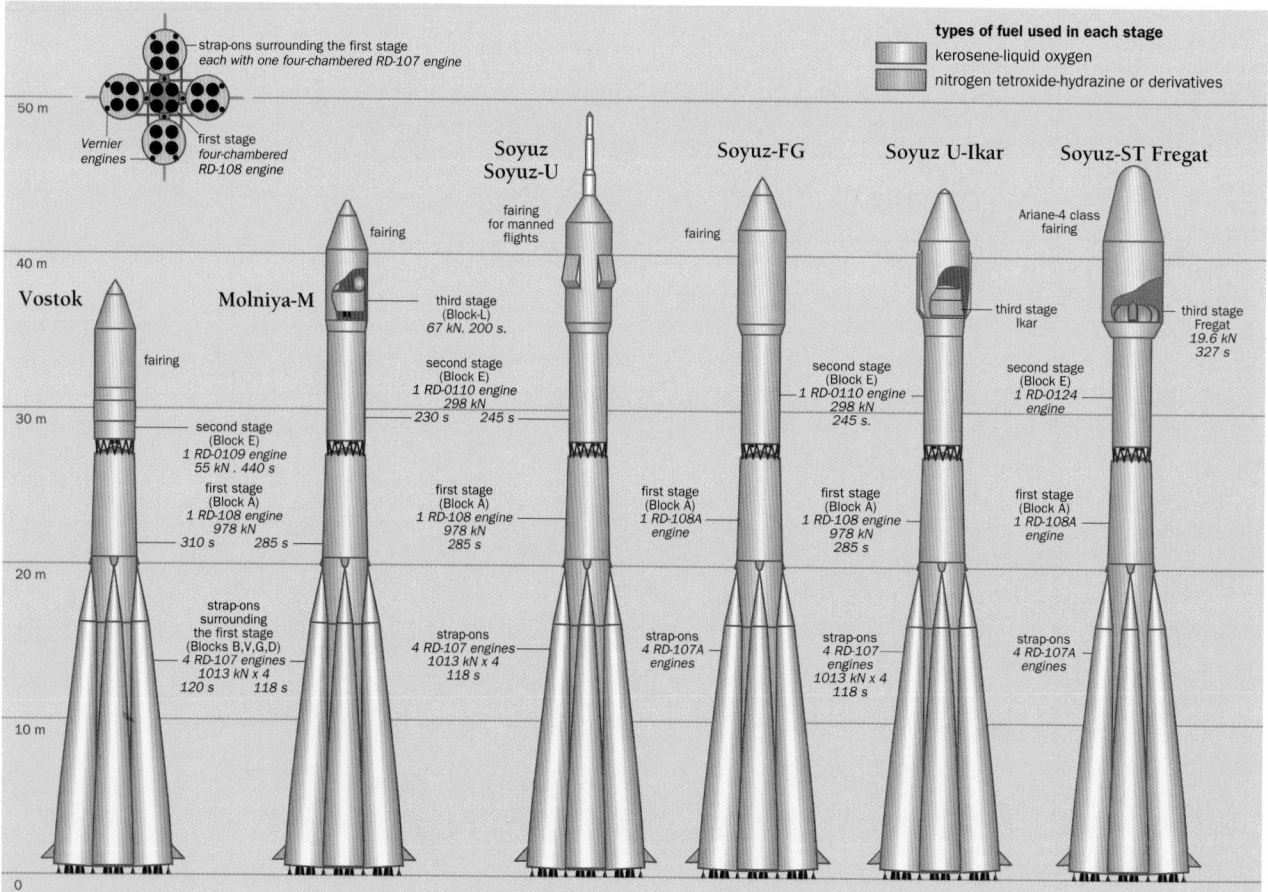

Figure 6.12. **Korolev or type A launch vehicles.**

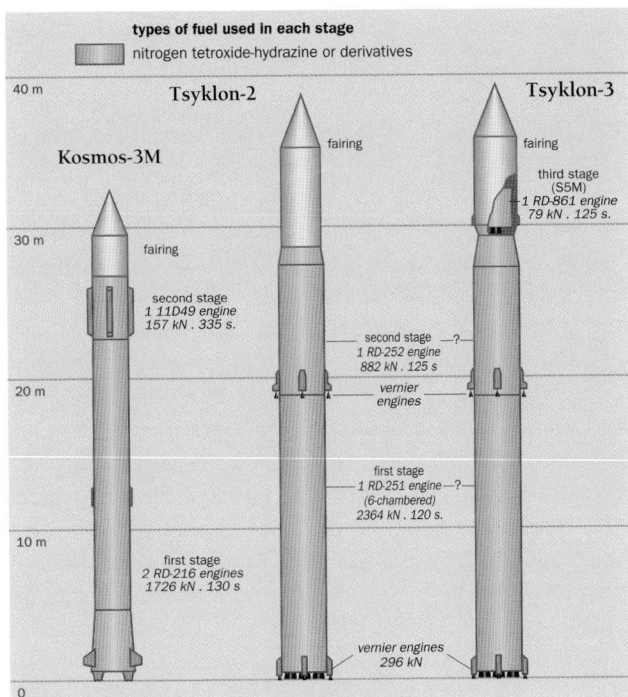

Figure 6.13. **Yangel launch vehicles, types C and F.**

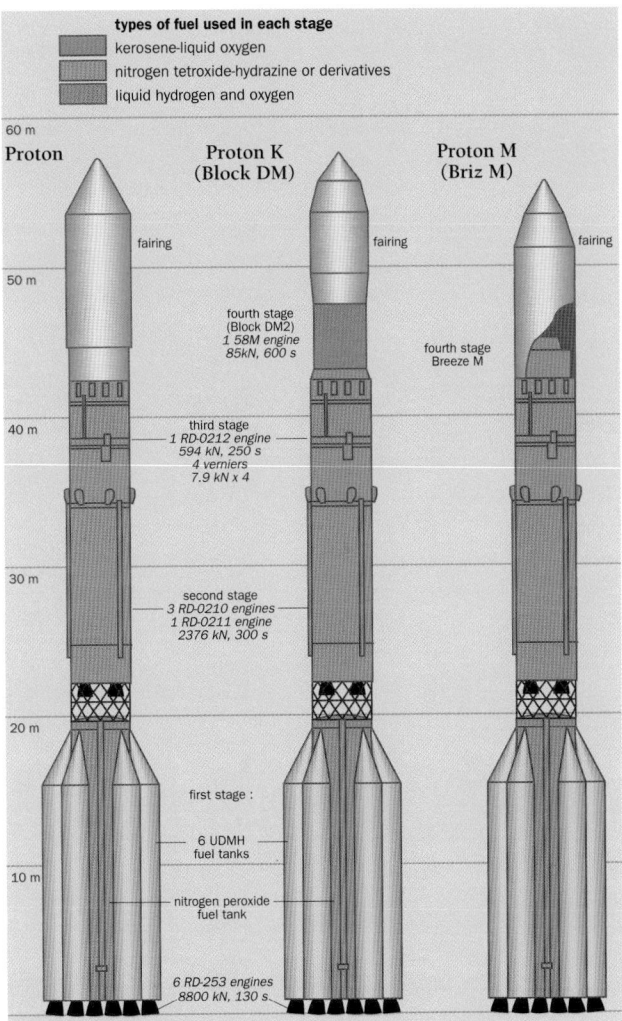

Figure 6.14. **The Proton rocket.** Proton is the only Russian rocket used to launch geostationary satellites, apart from the Zenit 3 chosen by the Sea Launch Corporation.

Another family of launchers, the Yangel launch vehicles, derived from ballistic missiles, has carried out the great majority of military missions. The light lift launcher Kosmos was used between 1962 and 1977. After 151 successful launches and 22 known failures, it gave way to a more powerful version, Kosmos 3M, with two stages, the second of which can be reignited, and the Tsyklon 2 and 3, with two or three stages depending on the mission (fig. 6.13). Kosmos 3M is built at Omsk by Polyot AKO, and Tsyklon by the Ukrainian company NPO Yuzhnoye.

In parallel, the Soviets developed a novel heavy lift launch vehicle under the direction of Vladimir Chelomey. This is the Proton rocket, currently referred to as Proton K. It first appeared in June 1965 in a two-stage version, with four launches including one failure. The four-stage version appeared in 1967, and the three-stage version in 1968. Current designs result from modifications made in 1986 and 1991. The main advantage of this now famous launch vehicle is its flexibility, due in part to a modular conception. The fourth stage of the version designed for high orbits and probes can use kerosene and liquid oxygen (Block D, DM) or Sintin (Block DM2, DM-2M, projected DM3). An upgraded version, the Proton M, will use a new fourth stage manufactured by Khrunichev. This is either the cryogenic KVRB, which will equip the Angara launch vehicle and which already equips the Indian GSLV launch vehicle, or the UDMH–nitrogen tetroxide Breeze M upper stage installed inside the fairing. A fifth stage, Fregat, will also be available for the Proton. The second stage could also be equipped with the cryogenic engine RD-0210 as originally planned for the Angara (fig. 6.14).

The Soviets were proposing the Proton as a commercial launch vehicle as early as 1983, aiming particularly at geostationary satellites. However, they had to await the American change of position related to the dismantling of the USSR in 1993 before the first contract could be signed, with Inmarsat. ILS handles international sales.

Almost 20 years after the Proton first appeared, and ignoring development of the Tsyklon system, it was not until 1987 that the Energia launch vehicle marked the appearance of a new family of heavy duty launch facilities. The long gestation period can be put down to the failure of the giant N1 launch vehicle which, like the American Saturn rocket, was intended for conquest of the Moon. Tests for the N1 were officially discontinued in 1969. Energia is an extremely powerful launch vehicle comprising a central cryogenic core flanked by two to eight kerosene–liquid oxygen boosters (fig. 6.15). It is designed to place payloads of between 65 and 200 tonnes into low-Earth orbit, as would be required to construct large space stations or to assemble interplanetary spacecraft, or

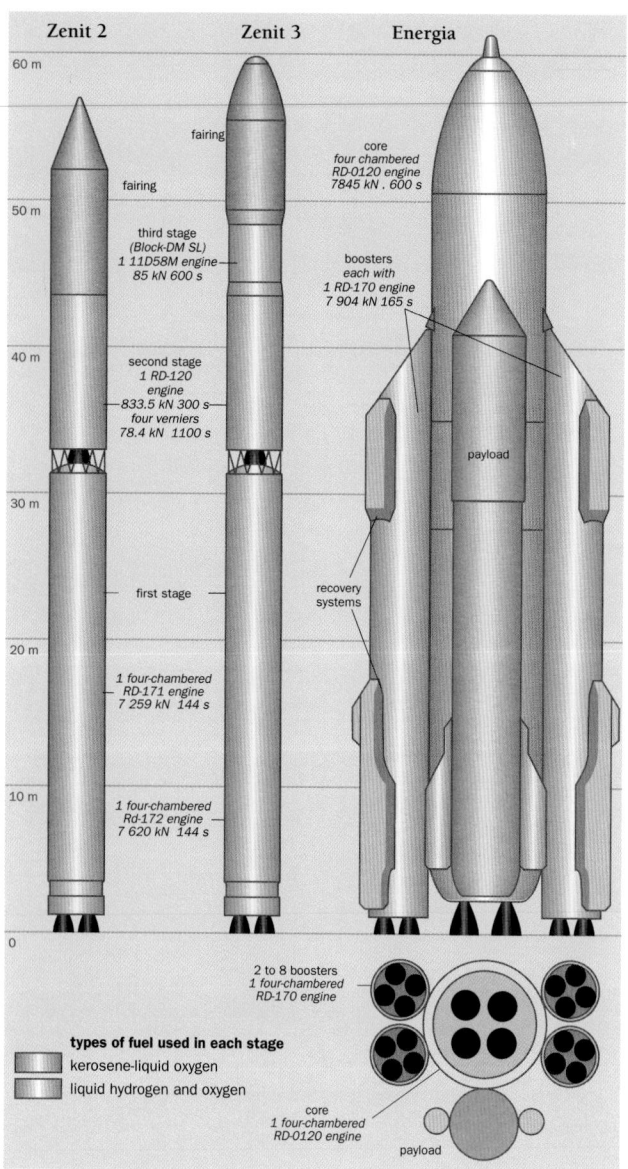

Figure 6.15. **Energia and Zenit.**

indeed with four boosters, to launch the Buran space shuttle (see chapter 13). Further developments are possible, by adjoining a second stage to the central core and a stage designed for injection into high orbits.

The problem of financing combined with the obligation for profitability in post-Soviet Russia have left Energia without immediate applications. No commercial payload currently requires such a powerful launch vehicle. Although two successful test launches were carried out in 1987 and 1988, this programme was subsequently shelved, awaiting an unlikely agreement with the United States to use Energia for the construction of the International Space Station, or the appearance of sufficiently heavy commercial satellites. Not only has a technical team been kept in service, but maintenance of the two launch pads at Tyuratam and the rockets already built has not officially been discontinued. However, future applications seem more and more doubtful.

In 1985, before the first Energia launch, the medium lift launch vehicle Zenit was revealed (fig. 6.15). It was built in the Ukraine by NPO Yuzhnoye using one of Energia's boosters equipped with a second stage. Zenit is the most modern of CIS launch facilities. The first foreign contract was signed in May 1995, for the launch of 36 satellites in the Globalstar project. There is a two-stage version and a three-stage version is under development. The Ukrainians propose both as the launch vehicle for the Australian space base project at Cape York, and this technology is already being used by Sea Launch. The Zenit 3 were endowed with a third stage Block DM for the occasion and a fairing supplied by Boeing (see fig. 6.15). Zenit can be developed in many ways, for example, by adding a reignitable fourth stage, or launching from an Antonov An-225. A heavy lift Zenit can be created by grouping together several first stages around a central tank.

Figure 6.16. **New light launchers and converted missiles.**

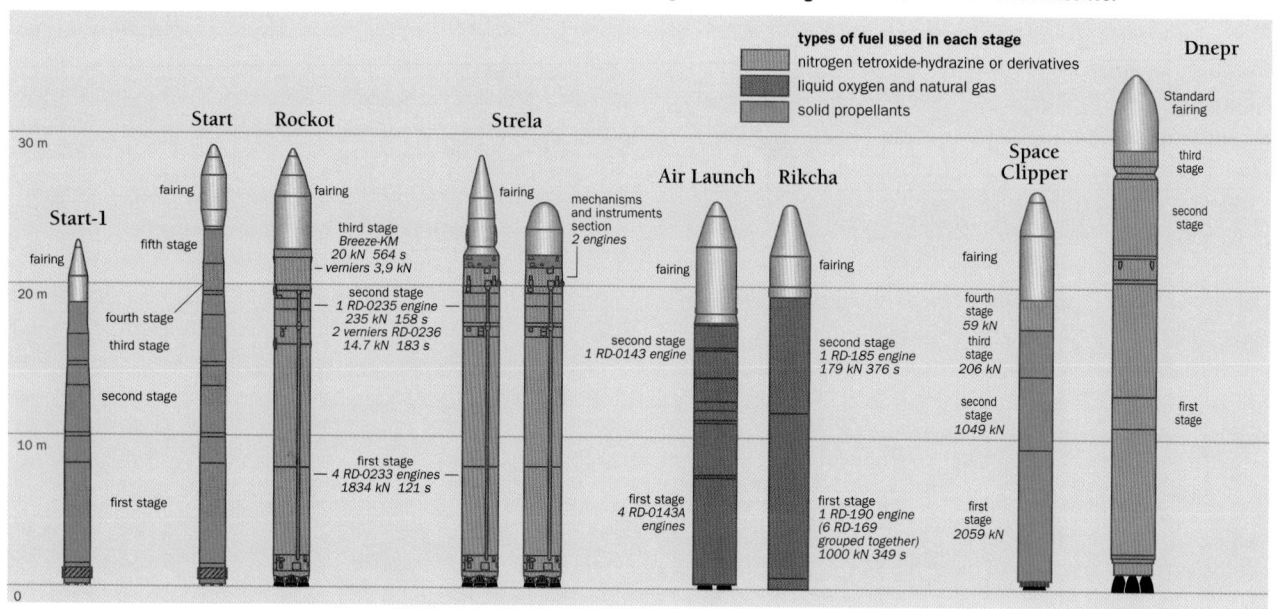

Ballistic missile conversion

This chapter in the history of Russian launch capabilities is directly related to disarmament agreements concerning hundreds of strategic intercontinental and submarine-launched ballistic missiles (ICBM and SLBM). Like the United States, Russia sought to transform its decommissioned missiles into light or medium lift launch vehicles with a view to selling them on the international market. Western users could thus benefit from very low prices for lifting small satellites into orbit.

The Topol missile was converted into the Start 1 launch vehicle at the Votkinsk plant by adding a fourth stage (first launch on 25 March 1993). The first launch of the five-stage version Start (fig. 6.16) on 28 March 1995 was a failure. Start is marketed by the Kompleks Scientific and Technological Centre of the Moscow Institute of Thermal Technology (now Kompleks-MIT) which is currently negotiating with Australia, Canada and the Brazilian base of Alcantara for Start launches outside Russia. In March 1997, the Svobodny base was inaugurated by a Start 1 launch and it now hosts the occasional commercial launch, such as that of the Israeli satellite Eros in 2000.

In parallel, Khrunichev developed the Rockot space launcher (fig. 6.16) from the two-stage Stilleto intercontinental missile by adding the liquid-propellant third stage Breeze. The first launch took place on 26 December 1994. Eurockot was formed by a joint venture between Khrunichev and the German company DASA, which has since become DaimlerChrysler Aerospace. It hopes to supply the European small satellite market, with prices only half those proposed by new US lightweight launch vehicles. For the moment, it has no obvious competitor in Europe itself, except for the Kosmos 3M which is slightly out of date.

NPO Mashinostroyeniya also offers an adaptation of the same ICBM going by the name of Strela (fig. 6.16).

NPO Yuzhnoye is working on the ICBM Scalpel and the missile R36M2 (SS 18 Satan). Under the name Space Clipper, the first should give birth to a modular series of six light to medium lift launch vehicles with three or four solid rocket stages, launched from an Antonov An-124 (fig. 6.16). Topped by a third stage – Fregat or S5M from

the Tsyklon – the R36M2 missile becomes the three-stage Dnepr medium lift launch vehicle, capable of lifting 4 400 kg into low orbit.

KB Makeyev is investigating the conversion of three Russian SLBMs, the four-stage Shtil (SSN 23 Skiff), the three-stage Volna (SSN 18 Stingray) and the two-stage Vysota (SSN 8), into lightweight liquid-propellant launch vehicles. Shtil was successfully test fired from a submarine on 5 July 1998 and is now regularly used. A medium lift launch vehicle, Priboy/Surf, could also be built by adding the first solid-propellant stage of another SLBM, Rif, to the four stages of Shtil. All could be launched from ships or the Antonov An-225.

New-design space launchers

As in the United States and to satisfy the same need for improved performance on the international launch market, there exist several projects for small launch vehicles that are

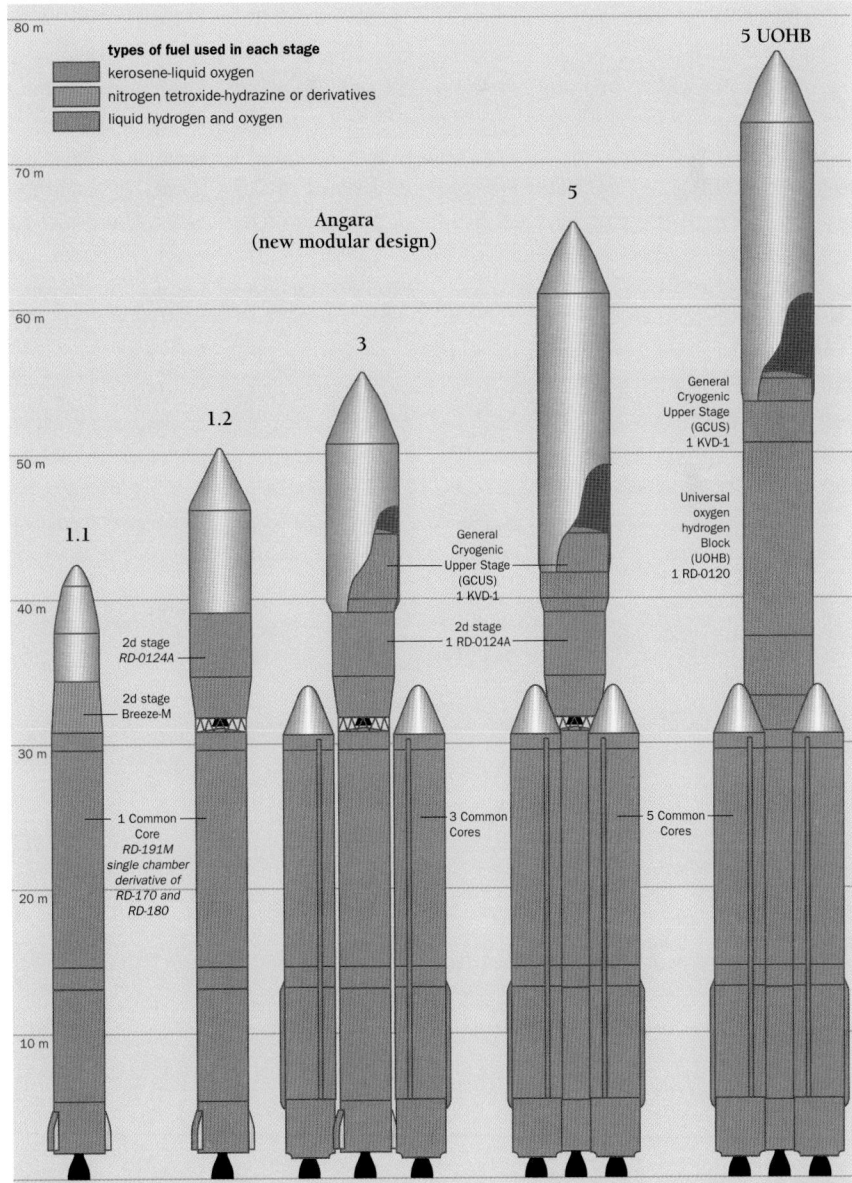

Figure 6.17. **The Angara projects.**

not directly derived from conventional ballistic missiles. Other more powerful launchers are being modernised at the initiative of private enterprise, sometimes in partnership with foreign companies called in to organise international sales.

Besides these SLBM conversion projects, KB Makeyev joined forces with NPO Energomash in the Kompomash Corporation to seek foreign partnerships for the commercialisation of a new two-stage lightweight launch vehicle equipped with liquid oxygen and liquid methane engines developed by NPO Energomash. A first project christened Rikcha seems to have been overshadowed by the creation of the Air Launch Aerospace Corporation in 1997. The launch vehicle (fig. 6.16) is carried aboard an Antonov 124-100 and cast from the rear cargo bay at an altitude of 11 000 metres just six seconds before firing. The market here concerns satellites of less than 2 500 kg in low orbits.

The most ambitious project is the Angara family of modular space launchers. Design and manufacture are being handled by KB Khrunichev and international sales by ILS.

A heavy duty launch vehicle was envisaged to begin with. The first stage used an upgraded version of the Zenit RD-174 engine comprising four RD-171 chambers. The second stage used Energia's cryogenic RD-120 engine. Liquid oxygen tanks were located on either side of the two stages. It was possible to add the Breeze M or the cryogenic KVRB stage of the Proton M as a third stage when lifting satellites of 4 500 to 5 000 kg into geostationary orbit or satellites of 28 to 30 tonnes into low Earth orbit.

A second project was presented in 1999, consisting of a family of five launch vehicles using the same core module, following the design of the American EELV programme (fig. 6.17). This common core will use the RD-191M engine, developed from the RD-171 and RD-180 (Zenit, Atlas III).

Angara 1.1 has one common core and a Breeze M second stage, whilst Angara 1.2 has one common core and a Block I second stage derived from that of the Zenit and the same as the one which should equip Soyuz 2 and Aurora. These are scheduled to replace current Kosmos M and Tsyklon vehicles. Angara 3 and 5 have three and five common cores sur-

mounted by a Block I and the General Cryogenic Upper Stage derived from the KVRB of the Proton M. They will be able to launch between 14 and 24.5 tonnes into low Earth orbit and will gradually replace the Zenit 2 and Proton vehicles. A super-heavy version is also planned, replacing the second stage of version 5 by the Universal Oxygen Hydrogen Block equipped with the RD-0120 engine. This version will be able to insert 11.2 tonnes into GTO.

A launch had originally been scheduled from Plesetsk in 2001 but is now expected for 2003. In the end, the launch pad may be built at Baikonur, even though the original deadline has not been respected. The great flexibility of the Angara modular system means that it is likely to be one of the main protagonists on the launch market.

Certain innovations such as the recovery of boosters may also lead to lower costs, testing new technologies in the field of semi-reusable launch vehicles. In this context, the Khrunichev centre is studying the future Baikal launcher (Angara IVA). It has a reusable first stage in the form of a winged URM which comes back and lands using two jet engines borrowed from the Yak 130.

Today launch vehicles constitute a strong point in Russian space activities. Their reliability and low production cost are attractive features for many potential customers. For a few years, the existence of stocks built up in the context of the strategic reserves policy made it possible to reap rewards from earlier investments. Since 1997, however, the exhaustion of these stocks, combined with a lack of government financial backing, has held back the development of new designs. This has compelled companies to enter partnership agreements with foreign firms whose real concerns may not prove to be the same as those of the Russian space sector. A turning point seems to have been reached in 2001. There is greater awareness of the real potential of these launch vehicles on the international market. Russian leaders are seeking agreements with countries such as Australia, for launches from the Christmas Islands, and with newcomers on the space launch market, confirming a desire for greater diversity in their partnerships.

The United States

Although America's launch bases are still considered essential for strategic independence, the question of modernising and privatising them has today been raised in the country, with the aim of transferring part of the burden of activity to the private sector. Following the debate in Congress in 1999 concerning the commercialisation of launch activities, the US administration made proposals in 2000 which suggested various types of private–public partnership for the management and ownership of space bases. Some solutions envisage a straightforward sharing out of tasks between public and private partners, whilst others advocate the gradual transfer of part of federal responsibility to new entities (Spaceport). The latter would be created on a regional basis or fall under the responsibility of the federated states, and might bring together industry and administration on a contractual basis. The idea of a completely business-run launch sector has also been invoked but does not seem to be seriously considered. In fact, for reasons of national security, the administration intends to maintain a right of preemption both for defining required facilities and establishing launch calendars. In parallel it remains to define the level at which private enterprise will contribute, with all the consequences attached to the sharing out of activities and the obligations which follow from them, and this in a launch market where sales prospects are somewhat diminished today. In this context, new arrangements envisaged by Congress between NASA and industrial partners, such as those introduced during the 106th Congress which proposed tax relief to stimulate economic activity at launch centres, have not yet been put into practice.

Cape Canaveral

Cape Canaveral is situated on the east coast of Florida, a shoreline made up of off-shore sandbars and low-lying dunes separated by marshland. The climate in this region is subtropical, prone to cyclones and salt-bearing winds. The latter has caused such corrosion in some installations, including pad 14 which saw the first American astronauts Shepard, Grissom and Glenn leave for space, that they have had to be demolished. Cape Canaveral, from which men really did leave our planet to walk on the Moon, is strangely close to the site imagined by Jules Verne for his Columbiad launch. The commonly used term Cape Canaveral actually covers a whole range of launch facilities (figs. 6.18 and 6.19): the Cape Canaveral Air Station (CCAS), under US Air Force control (45th Space Wing), the Kennedy Space Center (KSC) on Merrit Island, created by NASA, and a manufacturing centre (Boeing and Lockheed Martin) run by the USAF with some intervention from NASA.

The 45th Space Wing also controls the Eastern Range Tracking Network, which extends from the Cape as far as the Indian Ocean where data is relayed by the Western Range Tracking Network run from Vandenberg. It manages the Titan, Atlas and Delta launch pads (although the latter two were run by NASA until 1989). It is working with the Florida Spaceport Authority to adapt the LC 20 and LC 46 pads to the new lightweight Athena and Taurus launch vehicles. The CCAS is an ageing space base, where several installations date from the 1950s and the pads themselves constitute something of a bottleneck. The USAF's industrial partners are currently seeking ways of rejuvenating the site.

The Kennedy Space Center is entirely devoted to Space Shuttle launches and construction of the ISS. NASA has preferred to use the longer and better prepared landing strips at the Edwards Air Force Base in California's Mojave Desert for the return of shuttle flights, but new facilities have now been set up to improve shuttle operations.

Vandenberg

The Vandenberg base is situated on the Pacific coast in California, to the north of Point Arguello and north-west of Los Angeles in a rather mountainous region (figs. 6.20 and 6.21). Launches are all high inclination and take place southwards, over the Pacific as far as the edge of the Mexican coastline towards the south-east. Launches have occasionally been made towards the west in satellite interception experiments.

Vandenberg AFB is used for missiles, sounding rockets and space launchers, accessible to NASA and to private operators for Atlas and Delta launches. It was originally hoped to launch the Space Shuttle from Vandenberg as well as from Cape Canaveral, but the project was dropped even though the SLC 6 had already been built. The site was used to launch the LMLV 1 in August 1995.

An agreement between the California Spaceport Authority and the USAF led to the establishment of Spaceport West. This is run by Spaceport System International on a ten hectare site leased for 25 years, to the south of SLC 6. The future privately-run base will use the Payload Preparation Room planned for the Space Shuttle and will build a launch pad for Delta 2, Taurus and Athena 1 and 2 launches.

Wallops Island

The Wallops Flight Facility is located on an island off the alluvial coastline of Virginia (fig. 6.22). It is a secondary base administered by NASA and as yet has only seen the launch of about twenty small satellites from its single launch pad for the Scout launch vehicle. The last launch was in 1985 and the

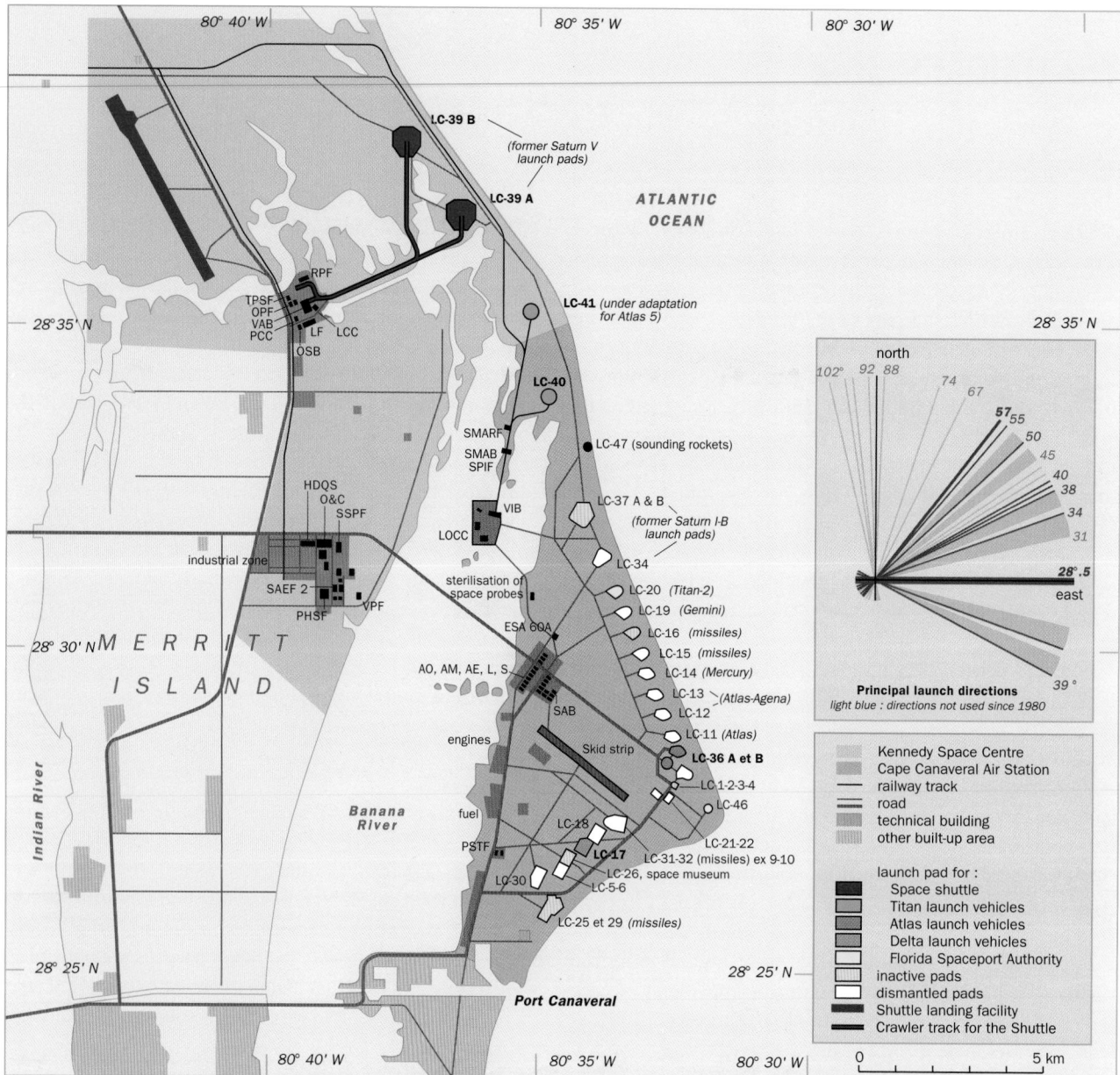

Figure 6.18. **Interpretative sketch of the SPOT image of Cape Canaveral.**
The Kennedy Space Centre (KSC) was created by NASA for the Apollo pro-gramme. Today it has become the centre of operations for the STS and the International Space Station.

The Space Shuttle technical zone contains the two launch pads LC-39A and LC-39B, a landing strip, the Vehicle Assembly Building (VAB) where the STS is assembled, the Launch Control Center (LCC), the Logistic Facility (LF) to stock and manage spare parts for the STS, the Operation Support Building (OSB) with library and reference material, the Processing Control Center (PCC) for orbiter tests and crew training, the Orbital Processing Facility (OPF) to pre-pare orbiters, the Thermal Protection System Facility (TPSF) for manufacture and repair of protective tiles, and the Rotation/Processing Facility (RPF) to prepare solid rocket boosters.

The industrial zone at KSC houses the Headquarters (HDQS), the Operation & Checkout (O&C) for reception, assembly and testing of horizontal payloads, and crew quarters, the Vertical Processing Facilities (VPF) for reception, assembly and testing of vertical payloads, including upper stages, the Space-craft Assembly & Encapsulation Facilities 2 (SAEF 2) for fuelling and fitting

payload pyrotechnic equipment, the Payload Hazardous Servicing Facilities (PHSF) for fuelling and fitting pyrotechnic equipment associated with heavy payloads and solid-propellant stages, and finally, the Space Station Process-ing Facilities (SSPF) which can process other payloads.

The Cape Canaveral Air Station (CCAS) is the USAF base which runs the Titan launch pads and, since 1989, the Delta and Atlas launch pads. Apart from KSC facilities such as SAEF 2, PHSF, etc., operators based at CCAS use hangars AO, AM, AE, L, S (also used by NASA) for satellite and payload processing, the Payload Spin Test Facility (PSTF), and the Explosive Safe Area 60A (ESA 60A) for fuelling and fitting satellite pyrotechnic equipment.

The US Air Force is more specifically involved with the Titan technical zone, with its Vertical Integration Building (VIB) for assembling the central core of the Centaur stage, the Solid Motor Assembly Building (SMAB) for assembling the IUS stage boosters, the Spacecraft Processing Integration Facility (SPIF), the equivalent of the VPF for classified payloads, the Solid Motor Assembly and Readiness Facility (SMARF) for booster preparation, the Launch Opera-tion Control Center (LOCC), and the Satellite Assembly Building (SAB).

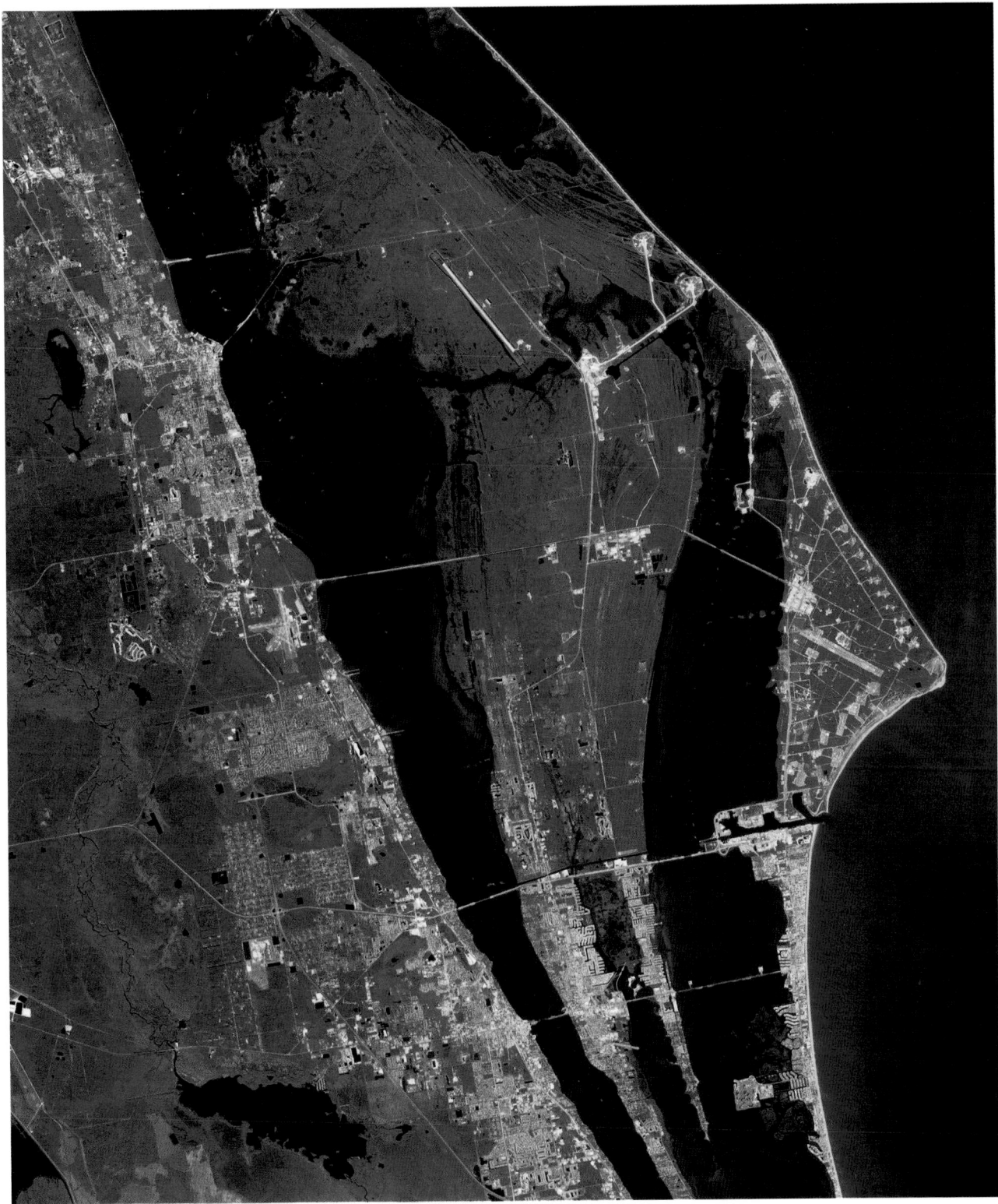

Figure 6.19. **SPOT image of Cape Canaveral.** SPOT 1 image taken on 17 May 2001. © CNES 2001, Distribution SpotImage.

Figure 6.20. **Interpretative sketch of the SPOT image of Vandenberg.** Both the north and south bases are run by the 30th Space Wing. North Vandenberg is still an important missile facility, but only retains a single Delta launch pad. Each Space Launch Complex (SLC) at Vandenberg possesses all installations needed to prepare launch vehicles and satellites. SSI installations are more widely dispersed and include among other things the Solid Rocket Motor Facilities (SRMF), the Solid Rocket Motor Storage Facility (SRMSF), the Integrated Processing Facility (IPF, former Payload Preparation Room) and the Operation Support Building (OSB).

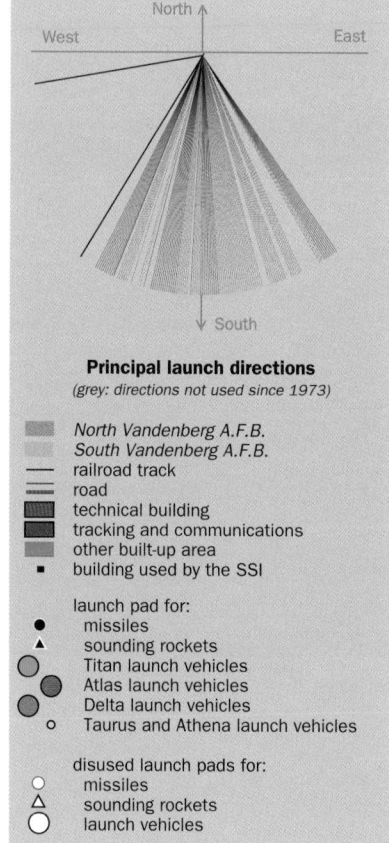

North ↑

West   East

Principal launch directions
(grey: directions not used since 1973)

South ↓

- North Vandenberg A.F.B.
- South Vandenberg A.F.B.
- railroad track
- road
- technical building
- tracking and communications
- other built-up area
- ▪ building used by the SSI

launch pad for:
- ● missiles
- ▲ sounding rockets
- ● Titan launch vehicles
- ● Atlas launch vehicles
- ● Delta launch vehicles
- ○ Taurus and Athena launch vehicles

disused launch pads for:
- ○ missiles
- △ sounding rockets
- ○ launch vehicles

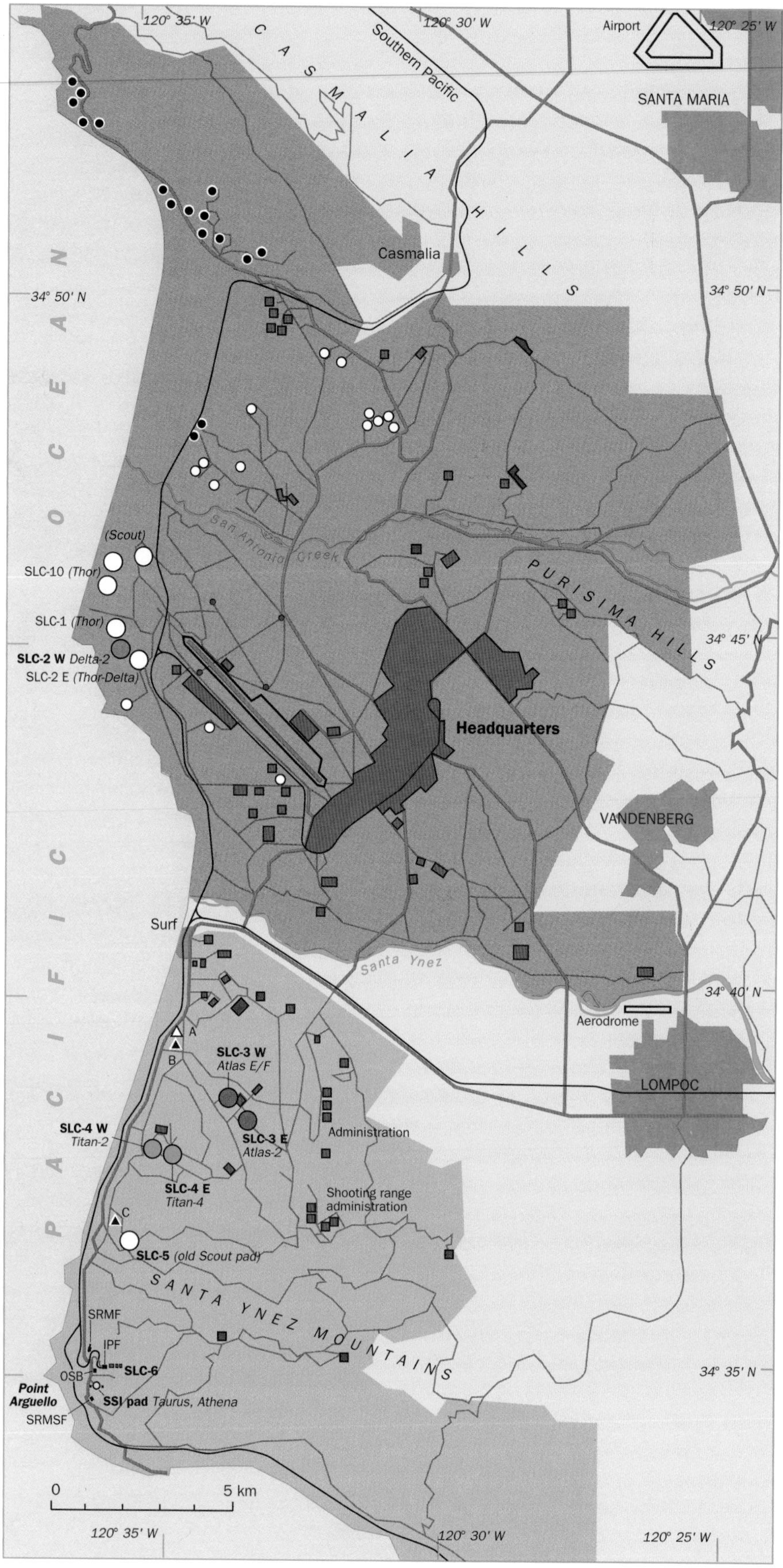

Figure 6.21. **SPOT image of the Vandenberg launch base.** A mosaic of two SPOT 4 images taken on 23 July 2000. © CNES 2000 Distribution SpotImage.

Scout has not been manufactured since 1994. The future of EER International's consortium project, with installation of a new launch pad for the Conestoga launch vehicle, seems uncertain given the failure of the first launch in October 1995 and the absence of further attempted launches since then. Wallops Island is also home to a NASA tracking station used for a great many satellites, in particular northbound launches from the Guiana Space Centre.

Kodiak Launch Complex

The Kodiak Launch Complex, promoted by the Alaska Aerospace Development Corp., is situated about 350 km south of Anchorage in Alaska. It is the first commercial spaceport to be set up in its own right, away from any pre-existing federal facilities. The Florida-based Command and Control Technologies Corporation (CCT), which holds a NASA licence, was in charge of computer hardware and software installations at the base in order to guarantee conformity with the minimal requirements for a non-governmental launch complex. The first mission to be carried out by Lockheed Martin with an Athena 1 rocket should launch NASA's Starshine 3 satellite and three experimental satellites for the US Department of Defense.

The flexible management and simplified facilities of the Kodiak spaceport are designed to provide more affordable launch services. The base is devoted to polar and high inclination satellites at a rate of three or four launches per year. Proposals to use the base for missile tests scheduled as part of the National Missile Defense programme may be viewed as an attempt to guarantee an alternative income.

Launch vehicles from Jupiter to Saturn

Apart from a small number of lightweight launch vehicles developed in the 1990s, all conventional US space launchers have sprung from intermediate range ballistic missiles (IRBMs) such as Jupiter (US Army), Redstone (US Navy), and Thor (US Air Force), or intercontinental ballistic missiles (ICBMs) such as Atlas and Titan, designed in the 1950s.

Jupiter C lifted America's first satellite into orbit in February 1958, but was not further developed. The Vanguard launch vehicle, derived from Redstone, led to the Scout light launch family as early as 1960. These were built by Vought Aircraft Industries for NASA and the US Department of Defense. After 110 orbital launches (including 11 failures) from Vandenberg, Wallops and San Marco, and a long series of upgrades designed to increase the power output of its four solid-propellant stages, the last Scout was launched on 9 May 1994.

The IRBM Thor gave rise to the Thor launch vehicle in 1960, built by the US company Douglas for NASA. Its development capacity is attested by the fact that more than 20 versions were used between 1958 and 1980 for 238 launches to Earth orbit, with 30 failures. It also gave rise to the Delta range of rockets in 1960. This range itself led to a great many variants, in fact more than 30, before being superseded since 1989 by the Delta 2 generation (fig. 6.23). The Delta 2 7920 has nine solid-propellant boosters and two liquid-propellant stages, whilst the Delta 2 7925 carries a third solid-propellant stage as well, the Payload Assist Module PAM-D. A light launcher programme (Delta-Lite, now Boeing Corp) has

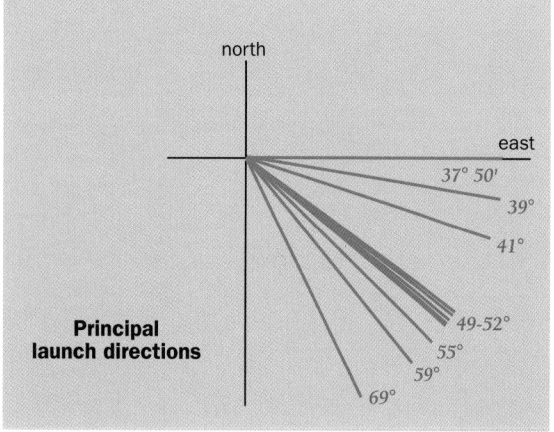

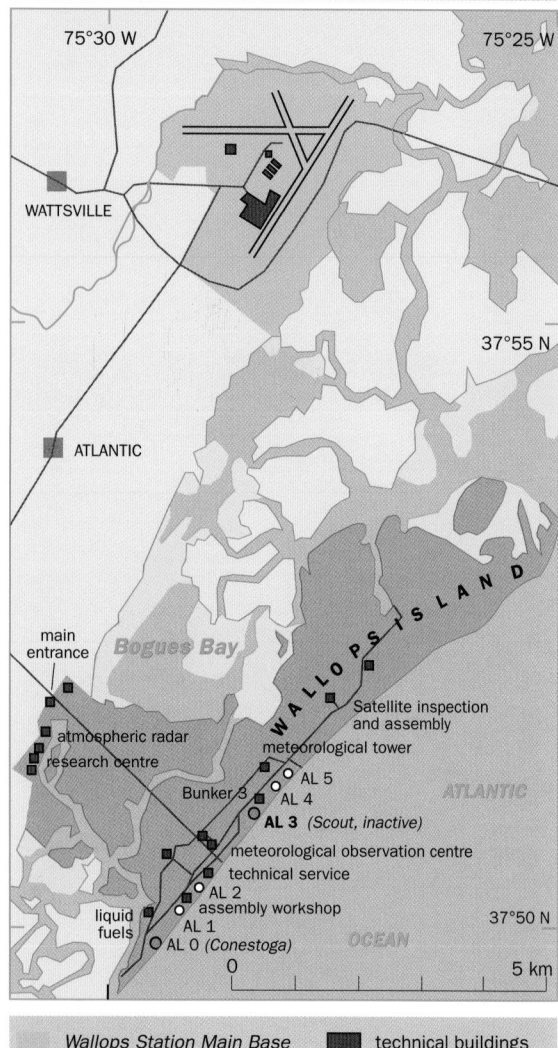

Figure 6.22. **Wallops Space Center.**

existed for some time, but it seems that in this class MDD prefers to offer variants of the Delta 2 with fewer boosters, such as the Delta 7425 (three heliocentric launches), the Delta 7426 (one cometary launch) or the Delta 7420 (five launches for Globalstar).

Delta 3 serves as a link between Delta 2 and the EELV. It uses a widened Delta first stage, shortened at the top, longer boosters and a cryogenic second stage that will also be used by the Atlas EELV (fig. 6.23).

A total of 282 Delta launches were carried out up until 1 January 2001, with only 10 failures, including the first Delta 3 launch.

Atlas derives from the ICBM of the same name built by General Dynamics (later absorbed into Lockheed Martin). It is a one-and-a-half stage launch vehicle: the kerosene and liquid oxygen contained in the core of the rocket supply three engines, two of which are jettisoned after running for 157 seconds. The US Air Force has used ICBM-derived launch

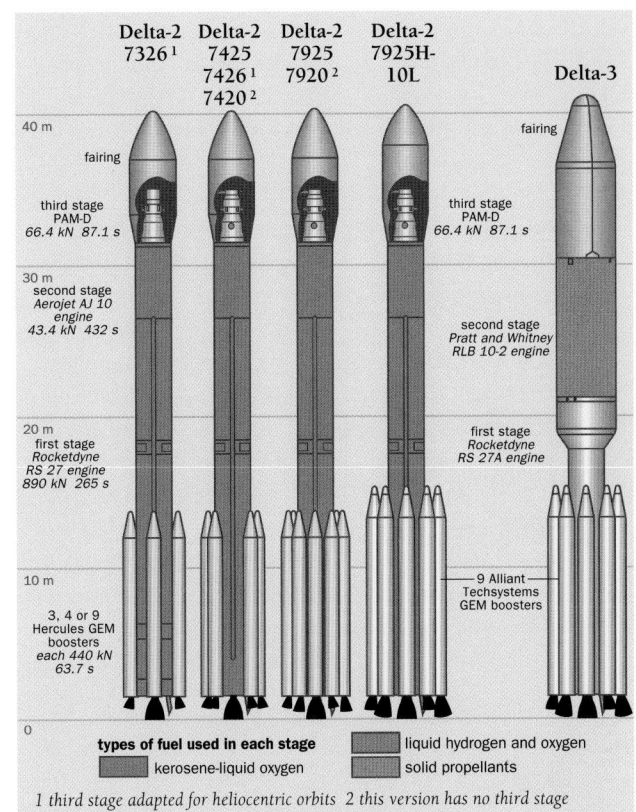

Figure 6.23. **Conventional US launch vehicles from McDonnell Douglas: Delta 2 and 3.**

Table 6.2. **American launch vehicles in use or under development in 2001.**

Name	First launch to orbit	Payload capability (kg) LEO: low-Earth orbit, GTO: geostationary transfer orbit, GEO: geostationary orbit, SSO: sun-synchronous orbit, EE: Earth escape. (300 km): height of circular orbit, (300-4 000 km): apogee and perigee of eccentric orbits. (65°): inclination. CC, V, Wo, Eq (Cape Canaveral, Vandenberg, Woomera, equatorial site): launch site	Launches up to 01/01/2001 ( ) successful launches
Delta-2 7320	24 06 1999	V: 2865(LEO)	2 (2)
Delta-2 7326	24 10 1998	CC: 2865 (185 km, 28.7°) 2095 (185 km, 90°) 1004 (GTO, 27°) 594 (EE)	2 (2)
Delta-2 7425/26	11 12 1998	CC: 1102 (GTO, 27°)	3 (3)
Delta-2 7420	14 02 1998	CC: 1800 (950 km, 52°)	7 (7)
Delta-2 7925/25 H	26 11 1990	CC: 1882 (GTO, 27°) 2 141 (20 000 km, 55°) 25 H 2 030 (GTO, 28.7°)	45 (44)
Delta-2 7920	04 11 1995	CC: 5039 (LEO, 28.7°) 3220 (SSO)	17 (17)
Delta-3	27 08 1998	CC: 8350 (LEO, 28.7°) 3 810 (GTO, 27°)	3 (2)
Delta-4 medium		CC: 4173 (GTO, 27°)	0
Delta-4 medium +		CC: 4853, 5 760, 6 668 (GTO, 27°)	0
Delta-4 heavy		CC: 13200 (GTO, 27°)	0
Atlas-2	07 12 1991	CC: 6780 (185 km, 28.5°) 2 950 (GTO, 27°) 2 000 (EE) V: 5 510 (185 km, 90°)	10 (10)
Atlas-2A	10 06 1992	CC: 7316 (185 km, 28.5°) 3 180 (GTO, 27°) 2 160 (EE) V: 6 190 (185 km, 90°)	20 (20)
Atlas-2AS	15 12 1993	CC: 8610 (185 km, 28.5°) 3 830 (GTO, 27°) 2 680 (EE) V: 7 210 (185 km, 90°)	21 (21)
Atlas-3A	24 05 2000	CC: 8690 (185 km, 28.5°) 4 037 (GTO, 27°)	1 (1)
Atlas-3B		CC: 10720 (185 km, 28,5°) 4 500 (GTO, 27°)	0
Atlas-5 400 series		CC: 8573 (LEO) 5 262 (GTO, 27°)	0
Atlas-5 500 series		CC: 10290 to 20 490 (LEO) 3960 to 8660 (GTO, 27°)	0
Atlas-5 HLV		CC: 19030 (LEO) 9390 (GTO, 27°) 6 577 (GEO)	0
Titan-2	19 01 1965	CC: 3175 (185 km, 28.6°) V: 2 177 (185 km, 90°) 2 360 (185 km, 63.5°) 3 028 (546 km 97.8° with kick stage)	21 (20)
Titan-401	07 02 1994	CC: 5773 (GEO)	12 (11)
Titan-402	14 06 1989	CC: 2860 (GEO)	6 (6)
Titan-403, 4, 5	08 06 1990	CC: 21900 (LEO, 28.6°) V: 14 090 (680 km, 98.2°)	12 (11)
STS	12 04 1981	CC: 24950 (204 km, 28.45°) 18 600 (204 km, 57°) 3 563 (925-39 450 km, 63°) 2270 (GEO with IUS)	101 (100)
Pegasus, -H, -XL	05 04 1990	200 XL 279 (463 km, 90°) 288 XL 382 (463 km, 0°) 130 XL 200 (800 km, 98°)	29 (27)
Taurus, Taurus-XL	13 03 1994	CC: 1770 XL 2180 (185 km, 28.7°) 445 XL 640 (GTO, 28.7°) V: 1435 XL 1750 (185 km, 90°) 230 XL 275 (EE)	5 (5)
Taurus-XLS		CC: 736 (GTO, 28.7°)	0
Minotaur	27 01 2000	V: 1750 (185 km, 90°)	2 (2)
Athena-1	15 08 1995	CC: 795 (200 km, 28.5°) 370 (1200 km, 57°) V: 515 (200 km, 90°) 225 (1200 km, 90°) 295 (800 km, 99°)	3 (2)
Athena-2	07 01 1998	CC: 1980 (200 km, 28.5°) 1190 (1200 km, 57°) 450 (Moon) V: 1490 (200 km, 90°) 960 (800 km, 99°)	3 (2)
Athena-3		CC: 3655 (LEO) 1136 (GTO)	0
Conestoga-1229		363 (180 km, 38°) 295 (180 km, 90°)	0
Conestoga-1620	23 10 1995	Wa: 1180 (180 km, 38°) 945 (463 km, 40°) 1000 (180 km, 90°)	1 (0)
Conestoga-1679		1500 (180 km, 38°) 1250 (180 km, 90°)	0
PA-2		225 (750 km, 90°)	0
K-1		Wo: 4600 (200 km, 45°) 3250 (200 km, 86°) 1200 (800 km, 98°) 1700 (780 km, 86°) 1400 (1200 km, 52°)	0

vehicles since 1965 to launch small military satellites (Atlas E and F). The last of this type (an Atlas E) was launched on 24 March 1995.

The most powerful versions of Atlas included a second stage:

■ The oldest form is the Agena stage, manufactured by Lockheed since 1959. It doubled as a liquid-propellant stage (using UDMH and IRFNA) and as a structure capable of remaining in orbit where it could serve as a base for military reconnaissance satellites, a precursor of Big Bird and KH 11. It was also used as a docking target in preparation for the Gemini missions. Its restartable engine allowed orbital adjustment and rendezvous manoeuvres. It constituted the second stage of most of the Thor rockets, Atlas-Agena from 1960, and the third stage of Titan 3B. The Agena stage has not been used since 1985.

■ The most recent form is the cryogenic Centaur stage, the world's first liquid hydrogen/liquid oxygen stage, produced in 1966 by General Dynamics' Convair division. Centaur is not only used as the second stage of Atlas. One

of its versions is an option with the Titan 4, whilst another was designed to equip the Space Shuttle but was dropped in 1986, and yet another constitutes the second stage of the Delta 3.

Current Atlas Centaurs are versions of Atlas II and III, with longer first and second stages compared with the previous version Atlas I. The latter was used for the last time on 25 April 1997. Atlas II A has a more powerful Centaur stage with two engines and Atlas II AS adds four Castor IV A solid boosters to the Atlas IIA. Atlas III A and III B (or 2AR and 2ARS) make the transition towards the EELV. The launch of Atlas III in May 2000 marked the appearance of a new first stage equipped with the Russian RD-180 engine. This will be the Common Core Booster of the Atlas V EELV. Atlas III B will be used to test the Common Element Centaur, an extended Centaur second stage, in a new single-engine version as well as in the two-engine version (fig. 6.24). Altogether, 310 Atlas rockets had been launched by 2001 with 29 failures.

The last series of conventional launch vehicles in service is Lockheed Martin's Titan family (fig. 6.24). It began with the adaptation of the Titan 2 ICBM for the Gemini missions. Launches began in April 1964 and ended in November 1966. However, the decommissioned Titan 2 missiles have

Figure 6.24. **Conventional US launch vehicles from Lockheed Martin: Atlas and Titan.** Titan 403, 404 or 405 are the same as Titan 402 without the upper stage. 403 designates flights from Vandenberg, 404 indicates the use of a short fairing (15.25 metres), and 405 designates flights from Cape Canaveral.

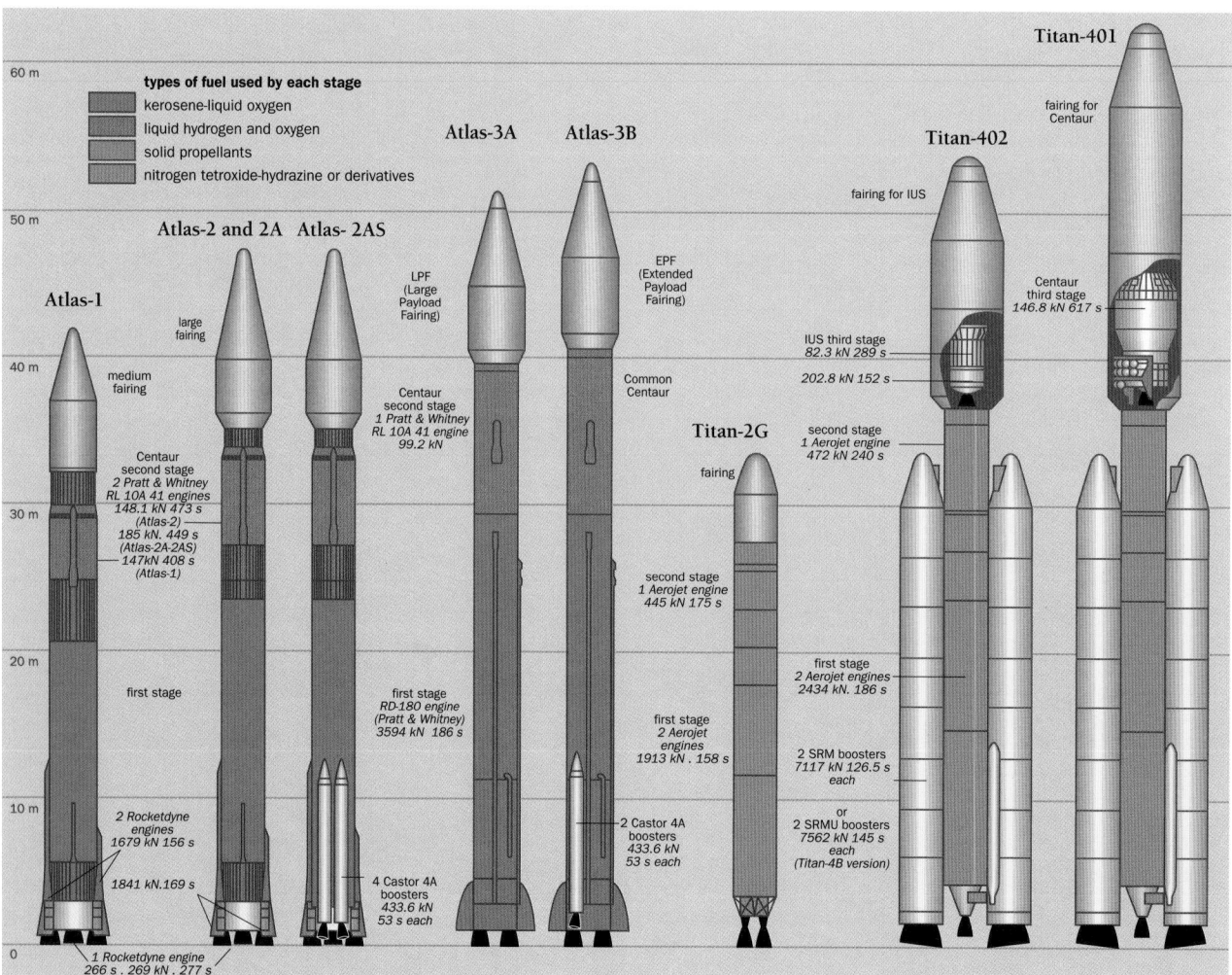

been reworked since 1988 to provide military launches. The basic version is the Titan 2G, but more powerful models are envisaged with an added central core of two to eight Castor 4A boosters (Titan 2S) or two liquid-propellant boosters (Titan 2L).

The modular series of Titan 3 launch vehicles appeared in September 1964. The two-stage central core consuming liquid propellants (UDMH and nitrogen tetroxide) is a longer and more powerful version of the Titan 2 core. There are optional solid boosters, a wide choice of third stage (apart from Agena and Centaur already mentioned, there is the restartable Transtage which consumes the same propellants as the central core), and four types of fairing allowing single or multiple launches. The latest representative of this series is the Titan 34D (1982–1989). With extended second stage and alterations to the fairing, enlarged by four metres, this became the commercial Titan 3, which made its first flight on 1 January 1990. Its third stage was the Payload Assist Module (PAM), Transfer Orbit Stage (TOS) or Inertial Upper Stage (IUS) for low or transfer orbits, and Transtage for geostationary orbits. Because of its high cost, Titan 3 is poorly suited to commercial launches and has not been used since 25 September 1992. Finally, there is Titan 4, an exclusively military launch vehicle, which had its first flight on 14 June 1989. It derives from the Titan 34D by extending the boosters and also the first and second stages. It can be used without the third stage (Titan 403, 404 or 405), with IUS (Titan 402) or with Centaur (Titan 401). Titan 4B, a modernised version with more powerful boosters, appeared in 1997.

The challenging Apollo programme required the construction of a very powerful rocket. This was achieved by NASA with the help of W. von Braun's team between 1959 and 1967. The Saturn family (Saturn I, IB and V) was the first that did not derive from a military missile. The most powerful version, Saturn V, measured 111 m in height. Its first stages were 10.06 m in diameter and developed a total thrust of 38 000 kN, whilst the third cryogenic stage had a diameter of 6.6 m and a thrust of 890 kN. When it was decided in 1972 to develop a recoverable space transportation system, the Saturn launch vehicles were abandoned. The last flight was by Saturn IB for the Apollo–Soyuz rendezvous in 1975. The various Saturn rockets carried out 26 missions with no failures.

Small launch vehicles

There has been a proliferation of light launch vehicles either developed or still under development by private companies, often with the support and financial backing of the US Air Force (fig. 6.25). This is due to a combination of factors: a shortage of US launch vehicles as of 1986, the steady miniaturisation of satellites, and governmental complication of missile conversions.

Pegasus is a small launch facility released from a Hercules aircraft. It was developed between 1987 and 1990 by Orbital Sciences Corporation and Hercules Aerospace (which has since become Alliant Techsystems). The original version has

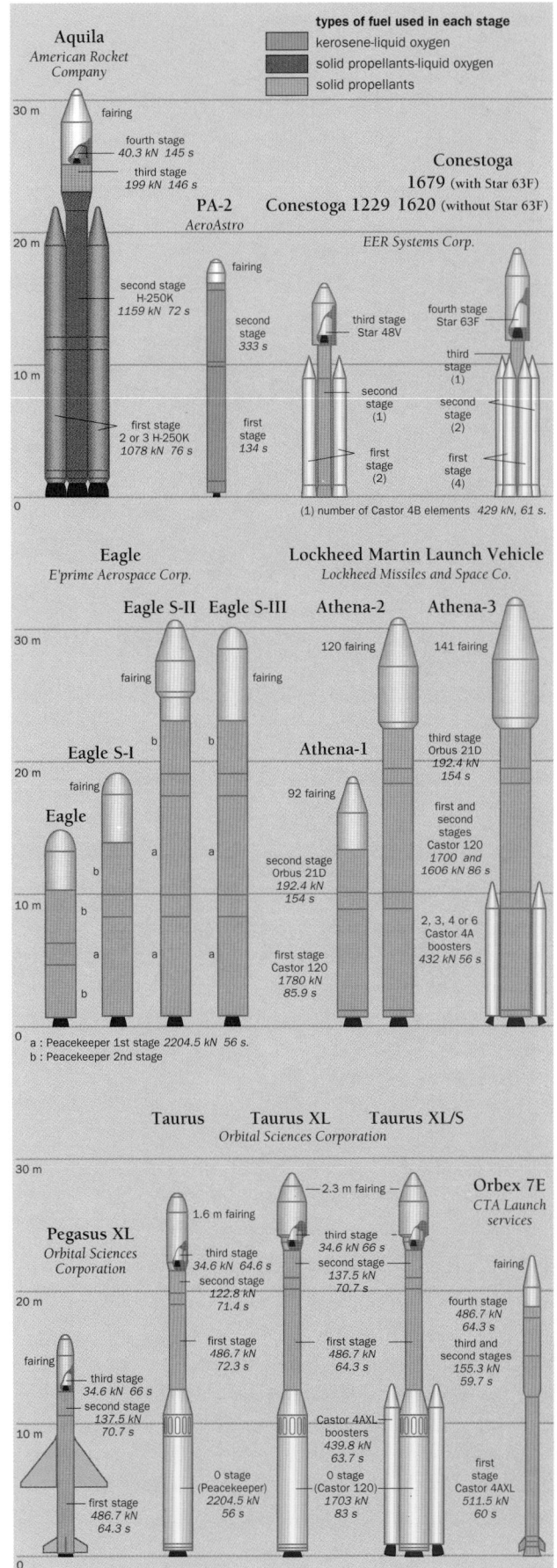

Figure 6.25. **New light lift launchers.**

been successfully used six times since its first flight on 5 April 1990. The lengthened XL version had a difficult beginning with two failures and a payload separation problem in November 1996, but has since functioned without mishap.

The same launch vehicle can be ground launched when a further stage, stage zero, is added. It then becomes the Taurus launch vehicle which first flew on 13 March 1994, without incident. The XL and XL/S versions with two Alliant GEM or Castor 4A XL strap-on boosters are also available, as is the Minotaur version since January 2000, with the two first stages of a Minuteman 2.

The Taurus stage zero is adapted from the solid-propellant first stage of the Peacekeeper ballistic missile (military version) or the Castor 120 (commercial version), both being manufactured by Thiokol. Several light to medium lift launch vehicles have been developed or are under development using these elements and others from the same manufacturer, such as the Castor 4A, 4A XL, 4B boosters, and small apogee motors called Orbus produced by United Technologies Chemical Systems Division.

Lockheed Martin Missiles and Space is thus developing the Lockheed Martin Launch Vehicle family (LMLV), renamed Athena, by combining Castor 120, Castor 4A and Orbus 21D. The first launch of LMLV 1 failed in August 1995 but it was followed by a success in 1997. Athena 2 successfully accomplished its first flight on 7 January 1998. There has been a significant effort to diversify launch sites. Having been launched from Vandenberg and Cape Canaveral, Athena was launched from the Kodiak spaceport in 2001, and it transported US DoD and NASA experimental satellites into orbit.

In parallel, the E Prime Aerospace Corporation has been developing its Eagle range from upgraded first and second stages of Peacekeeper in order to get round restrictions placed by the Strategic Arms Reduction Treaty (START) and produce a purely commercial launch vehicle. Following the first technical developments in 1999, the main task now is to obtain financial backing.

The case of the Conestoga launch vehicle illustrates the difficulties involved in financing this type of project. The project to construct the Conestoga light launch family from Castor 4B assemblages was originated by Space Services Inc. In 1990, EER Systems Corporation announced its intention to purchase Space Services Inc., but a launch failure on 23 October 1995 led to the end of the programme.

Another project suspended for lack of financing was that of CTA Inc. Their lightweight launch vehicle Orbex 7E consisted of a Castor 4A XL first stage and two Orbus 7S stages. Two further projects, which do not use standardised elements, have also met with difficulties to secure financing. These are Aquila of the American Rocket Company and PA 2 from AeroAstro. Aquila is a novel project using solid-fuel elements with liquid oxygen as oxidiser. Tests carried out between 1986 and 1993 were conclusive and two versions planned, Aquila 21 and 31, with two or three elements constituting the first stage. The PA 2 project, a light launcher with two liquid-propellant stages, is currently seeking backing or contracts, particularly in Sweden and Australia, where it might be launched.

This proliferation of lightweight launch vehicles is directly related to the boom in small communications and Earth observation satellites. With advertised prices of about 25 million dollars per launch for the most powerful among them, this new class of launch vehicle can cut costs by a factor of three compared with conventional competitors. Orbital Sciences Corporation and Lockheed Martin alike can already count on well-filled order books, if necessary by launching their own satellites. Orbital Sciences uses its Taurus and Pegasus launch vehicles to lift its Orbcomm satellites into orbit, whilst Lockheed Martin has a clientele for small experimental observation satellites and has already launched Lewis, Rocsat and Ikonos 1 and 2 with varying degrees of success (the first and third being lost). Neither are the activities of these launch vehicles restricted to the civilian sector. An agreement has in principle been reached to launch new generations of low-Earth orbiting early warning satellites using Athena.

Despite this growth, it should not be forgotten that programmes related to small satellite projects are still subject to a great deal of uncertainty. Whether they be military or civilian, only a small number are likely to subsist. The role of government involvement should not be underestimated either. It tends to subject such projects to long term constraints inspired by public research and development guidelines or new concepts of national security rather than help them to adopt a purely commercial perspective. Furthermore, given the delays expected, competition from missile-derived vehicles may prove to be significant, especially since opposition to their use by the political powers could soon be completely lifted. Early Bird, the first civilian observation satellite with one metre resolution, was launched in December 1997 by a converted Russian SS 25 missile, Start 1.

The EELV programme

In the context of the new Evolved Expendable Launch Vehicle programme (EELV), the main American manufacturers have set up their various projects in competition with one another. Two of them, Lockheed Martin and Boeing Co. were jointly selected by the Pentagon in January 1999 with respectively 7 and 21 launches already ordered. In both cases, the product consists of a modular family including a medium lift launch vehicle built around a basic module common to the whole range, one or more intermediate launchers provided with a cryogenic second stage, and one or more heavy launchers obtained by adding two to four strap-on boosters which are in fact two first stages.

Lockheed Martin has opted for a gradual transformation of its Atlas line. Within the framework of an agreement with the Russian company Energomash, these launch vehicles will incorporate the RD-180 variable throttling engine as the Common Core Booster (CCB). This was originally developed for the Russian N 1 lunar launcher. Already tested on Atlas III, it is currently produced in Russia by Energomash but will

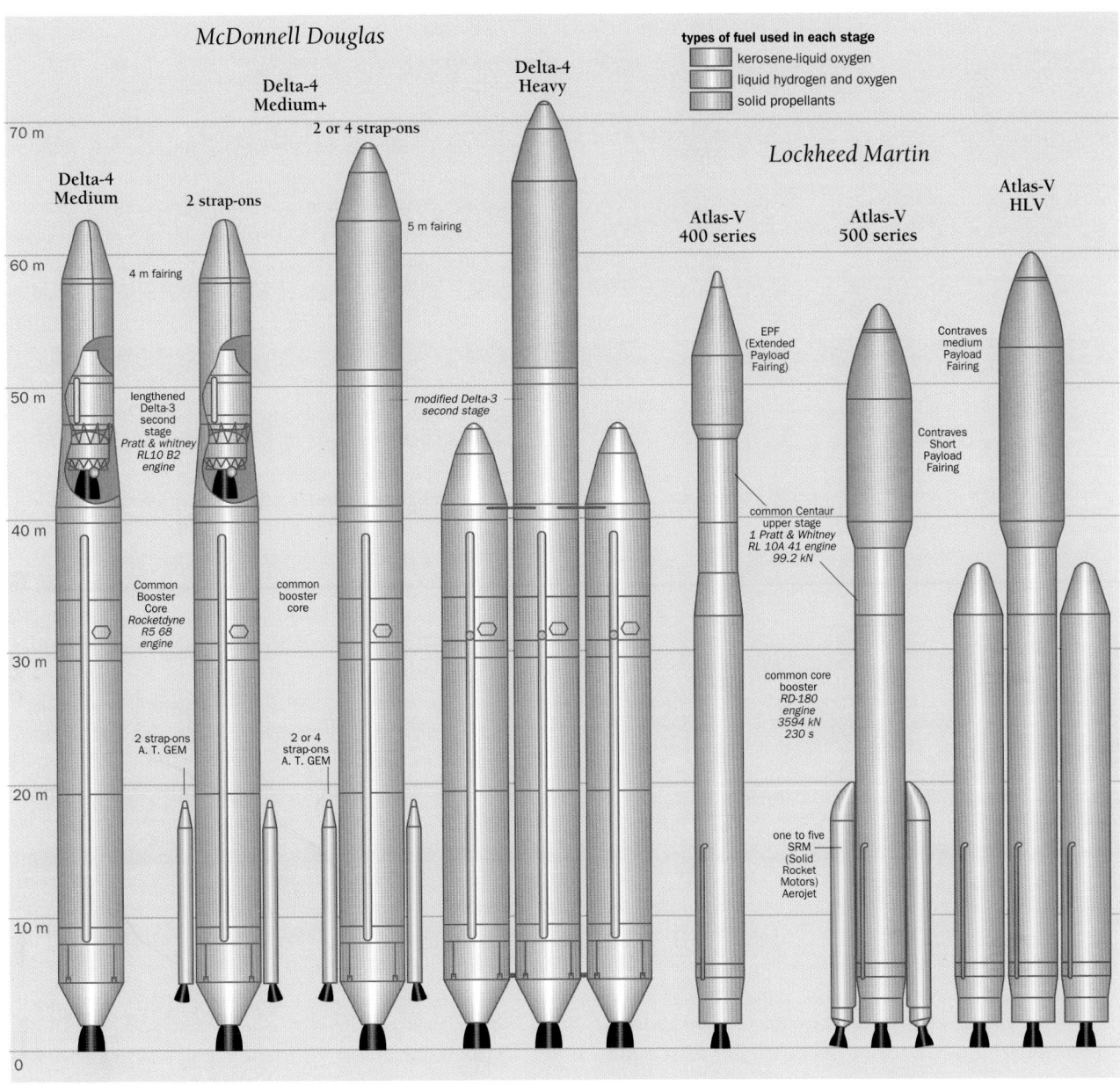

McDonnell Douglas

Delta-4
Medium

Delta-4
Medium+

Delta-4
Heavy

types of fuel used in each stage
- kerosene-liquid oxygen
- liquid hydrogen and oxygen
- solid propellants

Lockheed Martin

Atlas-V
400 series

Atlas-V
500 series

Atlas-V
HLV

2 strap-ons

2 or 4 strap-ons

70 m

60 m

50 m

40 m

30 m

20 m

10 m

0

4 m fairing

lengthened
Delta-3
second
stage
Pratt & whitney
RL10 B2
engine

Common
Booster
Core
Rocketdyne
R5 68
engine

2 strap-ons
A. T. GEM

common
booster
core

2 or 4
strap-ons
A. T. GEM

5 m fairing

modified Delta-3
second stage

EPF
(Extended
Payload
Fairing)

common Centaur
upper stage
1 Pratt & Whitney
RL 10A 41 engine
99.2 kN

common core
booster
RD-180
engine
3594 kN
230 s

one to five
SRM
(Solid
Rocket
Motors)
Aerojet

Contraves
medium
Payload
Fairing

Contraves
Short
Payload
Fairing

Figure 6.26. **Launch vehicles in the EELV programme.**

subsequently be manufactured by Pratt & Whitney at West Palm Beach. A joint venture called RD Amross between the two companies has been set up for this purpose. Atlas V 400 comprises the CCB and the Common Element Centaur as second stage, whilst the Atlas V 500 series carries one to five strap-ons in addition to this configuration. The latter are Aerojet's Solid Rocket Motors (SRMs). Finally, Atlas V Heavy uses three CCBs (fig. 6. 26).

For its Delta 4 range, Boeing Co. will use a slightly modified Delta 3 cryogenic stage. The basic module is completely new and will also be cryogenic. Two or four strap-ons will be used to diversify the medium and intermediate versions (fig. 6.26).

Reusable launch vehicles

Although the Space Transportation System (STS) was originally conceived as a completely recoverable launch vehicle, the system eventually designed in 1981 was only partially recoverable (fig. 6.27). Financial constraints led to a design consisting of three separable elements, two of which were jettisoned in flight, so that only the actual shuttle vehicle itself was completely recoverable. The three components of the STS are:

- The Space Shuttle or orbiter, built by Rockwell. It is propelled by three cryogenic engines and thus constitutes an integral part of the launch system, unlike the Soviet shuttle, which is just a payload carried by Energia. Six different Shuttles have been built. Enterprise served only to carry out non-orbital tests and Challenger was destroyed just after lift-off on 28 January 1986, leaving Columbia, Discovery, Atlantis and Endeavor, which replaced Challenger. The Shuttle can transport a crew of two to eight astronauts and has a cargo bay measuring 4.5 m by 18 m to carry payloads. The latter can remain in the cargo bay

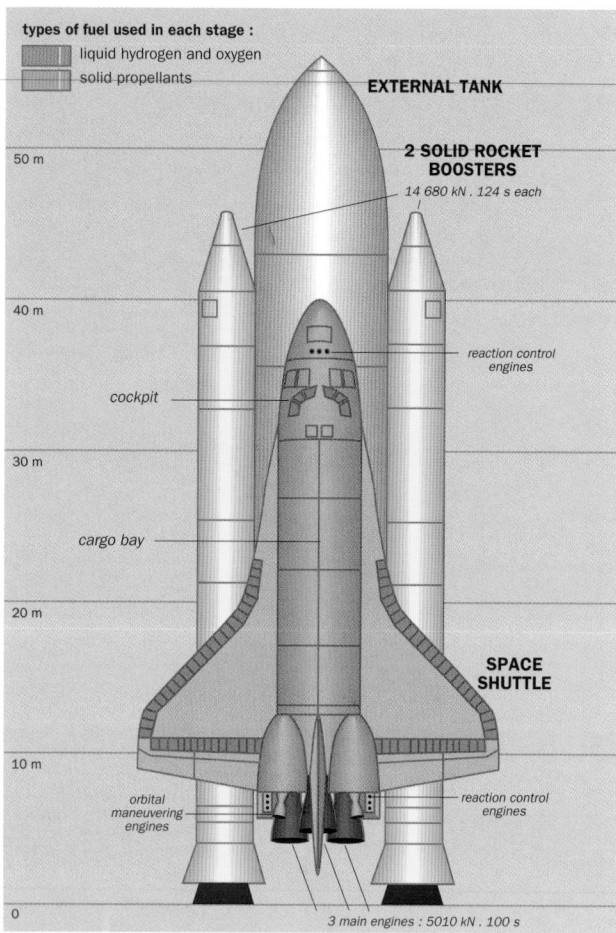

types of fuel used in each stage :
▦ liquid hydrogen and oxygen
▦ solid propellants

EXTERNAL TANK

50 m

2 SOLID ROCKET BOOSTERS

14 680 kN . 124 s each

40 m

reaction control engines

cockpit

30 m

cargo bay

20 m

SPACE SHUTTLE

10 m

orbital maneuvering engines

reaction control engines

0

3 main engines : 5010 kN . 100 s

Figure 6.27. **The Space Transportation System.**

during the flight or they can be placed in orbit in the immediate vicinity of the Shuttle and recovered at the end of the mission or during a later mission. They can also be inserted into other orbits using propulsion modules which serve the same purpose as the last stage of a conventional launch vehicle. (In fact, some such modules, such as IUS, PAM-D or the Italian IRIS, can play both roles.) The Shuttle is equipped with a 15 m long articulated arm, the Remote Manipulator System (RMS) of Canadian manufacture. It is used to recover or manoeuvre satellites.

- The External Tank (ET), built by Martin–Marietta, which contains the 2 000 m³ of liquid hydrogen and oxygen required by the Shuttle's main engines. It is jettisoned at an altitude of 114 km and cannot be recovered.

- Two Solid Rocket Boosters (SRBs), built by Morton–Thiokol. These are dropped into the sea with parachutes after a burn time of 120 seconds and can in principle be reused about 20 times. For security reasons, they must be almost completely dismantled and components recovered separately.

The Space Shuttle resumed flights on 29 September 1988 after two years of absence, but at a slower rate than before the accident. NASA's budgetary difficulties (one Shuttle flight is estimated to cost 350 million dollars), severe safety restrictions and the length of between-flight maintenance (repair-

ing protective tiles, corrosion in the engines, and so on) mean that only about ten flights can be managed per year. This leaves almost no room for the launch of commercial satellites.

Today, deprived of the launches it was to carry out for the IDS, the Space Shuttle has become a burden that NASA is finding more and more difficult to bear. In fact, almost a quarter of NASA's budget goes on the Shuttle. The space agency hopes to cut costs by entrusting Shuttle operations to a private company United Space Alliance, created for the purpose. This simple subcontracting agreement does not include any genuine marketing of Shuttle flights, at least not in the first phase. This prospect is only mentioned by NASA as a potential source of revenue insofar as it is feasible alongside frequent government flights.

Following the Presidential Directive of 5 August 1994, entitled the Space Transportation Policy, a new generation of reusable vehicles went under investigation at NASA. Two distinct types of programme were envisaged. The first, called X 33, explored the feasibility of a completely reusable vehicle requiring only one stage to reach low orbit (Single Stage to Orbit, abbreviated to SSTO). After inviting tenders for several projects, NASA retained Lockheed Martin's Lifting Body idea. The vehicle was to be propelled by two linear Aerospike engines built by Rocketdyne. These consume both atmospheric and onboard oxygen. It was to take off vertically like a conventional rocket but land like a glider. NASA was planning to invest just under a billion dollars to finance a series of suborbital test flights for the X 33 Flying Demonstrator, and the first of these took place at the beginning of 1999. For its part, Lockheed Martin was to devote 220 million dollars to the project. At the end of this research programme, Lockheed Martin hoped to develop commercial applications for this type of vehicle (the Venture Star project). However, in 2001, the US government announced that the project was to be dropped and funds redirected towards improving the Shuttle.

The aim of the X 34 programme was to realise a prototype for a two-stage launch vehicle with reusable first stage weighing 54 tonnes and measuring 27 metres in length, capable of lifting payloads of 1000 to 1200 kg into low orbit. To begin with, it was to use a Russian RD-180 engine until hybrid engines could be developed. One novelty of this programme was the high level of involvement by manufacturing companies like Orbital Sciences and Rockwell, both for finance and exploitation. NASA was limited to a technical support role. Herein lay the whole difficulty of the undertaking. By February 1996, doubts and delays were rife. After an initial withdrawal by its industrial partners, NASA invited them to join a new partnership in a less ambitious programme. The 18-metre long prototype built by Orbital Sciences is ready for material and structural tests, with two airborne launch trials scheduled. However, in 2001, NASA finally decided to withdraw from the programme, leaving Orbital Sciences to choose a new partner.

The space agency's investments in next-generation reusable launch vehicles are now being channelled into a new programme, the Space Launch Initiative (SLI), set up in 2001. The aim of this programme is a reappraisal of all existing industrial projects, awarding multiple contracts to design and develop new launch vehicles. The final choices for full-scale development will be made around 2005. With a total budget of more than five billion dollars until 2005 (including not only the financing of new programmes but also Shuttle safety upgrades), SLI will allow NASA to pursue all its current commitments.

Among those benefiting from this backing, it is worth mentioning the Kistler Aerospace Corporation which is developing the recoverable K1 launch vehicle. The first of these should be launched from Woomera in Australia, where a pad is already under construction, awaiting the authorisation of a launch from the Nevada Test Site. The K1 launch vehicle uses Russian NK 33 and NK 44 engines developed for the lunar programme. The two stages, LAP and OV, should return to Earth intact with the help of parachutes and airbags. Amongst the seven subcontractors involved in this programme, there are several large companies: Lockheed Martin Space Systems is making the tanks, and Aerojet the engines (fig. 6.28). To go beyond low orbits, the launch vehi-

cle could use a so-called active dispenser, the equivalent of a third stage but non-recoverable, which would actually be part of the satellite. This would bring light geostationary or planetary missions within reach. The K1 would also be available to NASA for cargo missions to the ISS.

Other privately run projects are under study, such as the Astroliner project devised by Kelly Space and Technology in partnership with Vought Aircraft Industries. Like Pegasus, these small reusable launch vehicles would use an aircraft as launch platform.

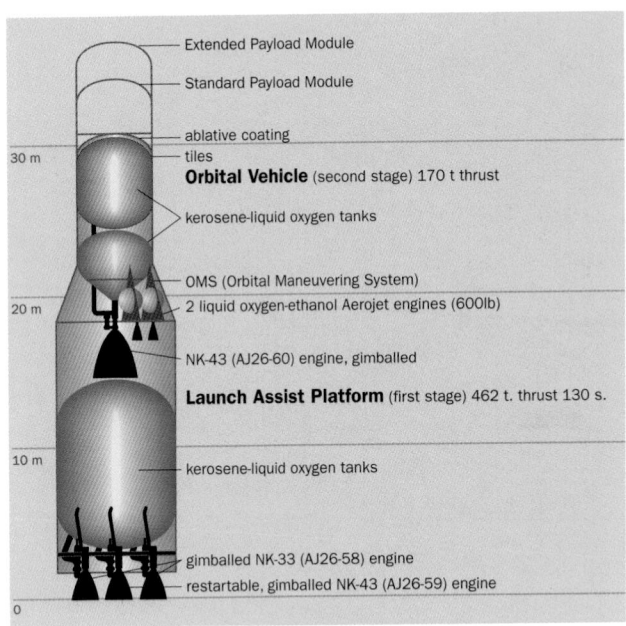

Figure 6.28. **The K 1 launch vehicle.** The first stage is built around three Russian NK 33 engines designed by Dvigatel, from whom Aerojet acquired the licence. The central engine reignites after separation to give the stage a ballistic return trajectory. The second stage has a single modified NK 33 and two engines for orbital manoeuvring. These get the payload onto a circular trajectory and reignite one day later to deorbit and bring it back to the vicinity of the base. The two stages are equipped with parachutes and airbags. The second also has thermal protection.

Europe

Independent access to space is a key feature of European space policy. The latter is largely dominated by the position of France, which has invested heavily in what it considers to be a sector of strategic importance. Europe possesses a single launch base on French territory overseas, and a single launch vehicle, Ariane, although it does possess variants. Today it must face up to growing international competition whilst in the past, it benefited from an extremely favourable international context, allowing it to take a dominant share in the commercial launch market. A crucial question today is the way the European family of launch vehicles should evolve. The potential of European industry should be taken into

account, whether it follows the American example by restructuring through mergers, or forms international alliances, in particular with Russian companies. In this context, a clear policy must be determined for the future of Europe's space bases bearing in mind the requirements of strategic independence as well as those of commercial viability.

Kourou

The French government decided to build the Guiana Space Centre (GSC) near Kourou in April 1964 when Algeria became independent and France lost its primary base at Hammaguir. The latter was used to launch the first French rockets in programmes going by the name of precious stones (Rubis, Topaze, Émeraude, and so on). The GSC was set up under the responsibility of the French space agency, the Centre National d'Études Spatiales (CNES), and became operational with the launch of the sounding rocket Véronique in 1968. Seven satellites were then sent into orbit by Diamant rockets (see fig. 6.5) before this first launch complex was abandoned in January 1976. From this point, exploitation of the GSC was tightly linked to the development of the Ariane launch vehicle in the context of an agreement with the European Space Agency. As early as 1975, France offered to share the GSC with ESA. For its part, ESA agreed to finance new facilities for Ariane and contributed to the annual budget of the base. In this

way, it was possible to finance the operations and investments needed for upcoming launch campaigns as well as launch complexes and manufacturing units for the launch vehicles.

The base is located on the Guianan coast between Kourou and Sinnamary, in a hot and tropical region with high mean cloud cover. It is built on coastal sandbars bordered by mangrove swamps (fig. 6.30). The low population in the region made it possible to reserve an area of 850 km^2 with 52 km of coastline, providing various launch directions without threat to people and property. Moreover, the region is a low risk area for hurricanes and earthquakes.

Apart from the advantages of being situated almost on the equator, ideal for geostationary launches, the Guiana Space

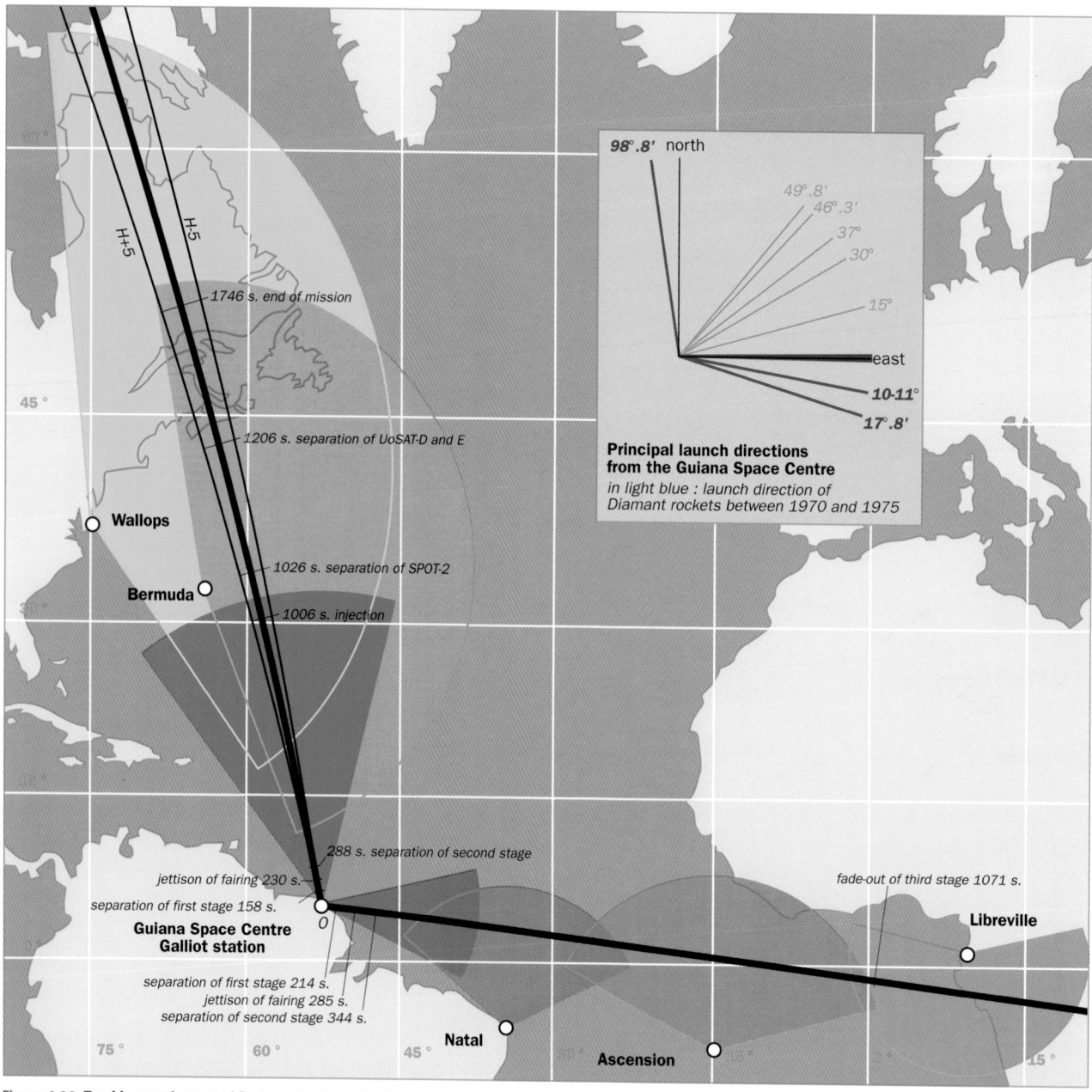

Figure 6.29. **Tracking stations used for launches from the Guiana Space Centre.** Coloured zones show the coverage of each tracking station along the ground track of the two main firing directions from Kourou. The timing given as an example for the polar launch is that of the SPOT 2 launch on 22 January 1990.

Centre is also well located for polar orbits. In both cases, vehicles overfly the Atlantic Ocean (fig. 6.29). For equatorial launches, tracking stations carrying out ranging measurements to determine the exact position of the launch vehicle and satellites are located at Natal in Brazil, then in the middle of the Atlantic Ocean where a NASA/DoD station operates from the British island of Ascension and finally on the African coast at Libreville in Gabon. For polar launches, two NASA tracking stations are used, one being on an island in the British archipelago of Bermuda and the other at Wallops in the United States.

The GSC has all the facilities of a fully equipped space base with radar tracking stations, laser ranging stations on neighbouring hilltops, a weather station, remote destruction facility, and so on (figs. 6.31 and 6.32). Its development has led to growth of the urban centre in Kourou. There is a port at the mouth of the river Kourou and an airport. Since the socioeconomic equilibrium of the French department of Guiana is an important factor if European investments are to be perpetuated and the base used in favourable conditions, every effort is made in French Guiana to encourage a parallel economic growth in

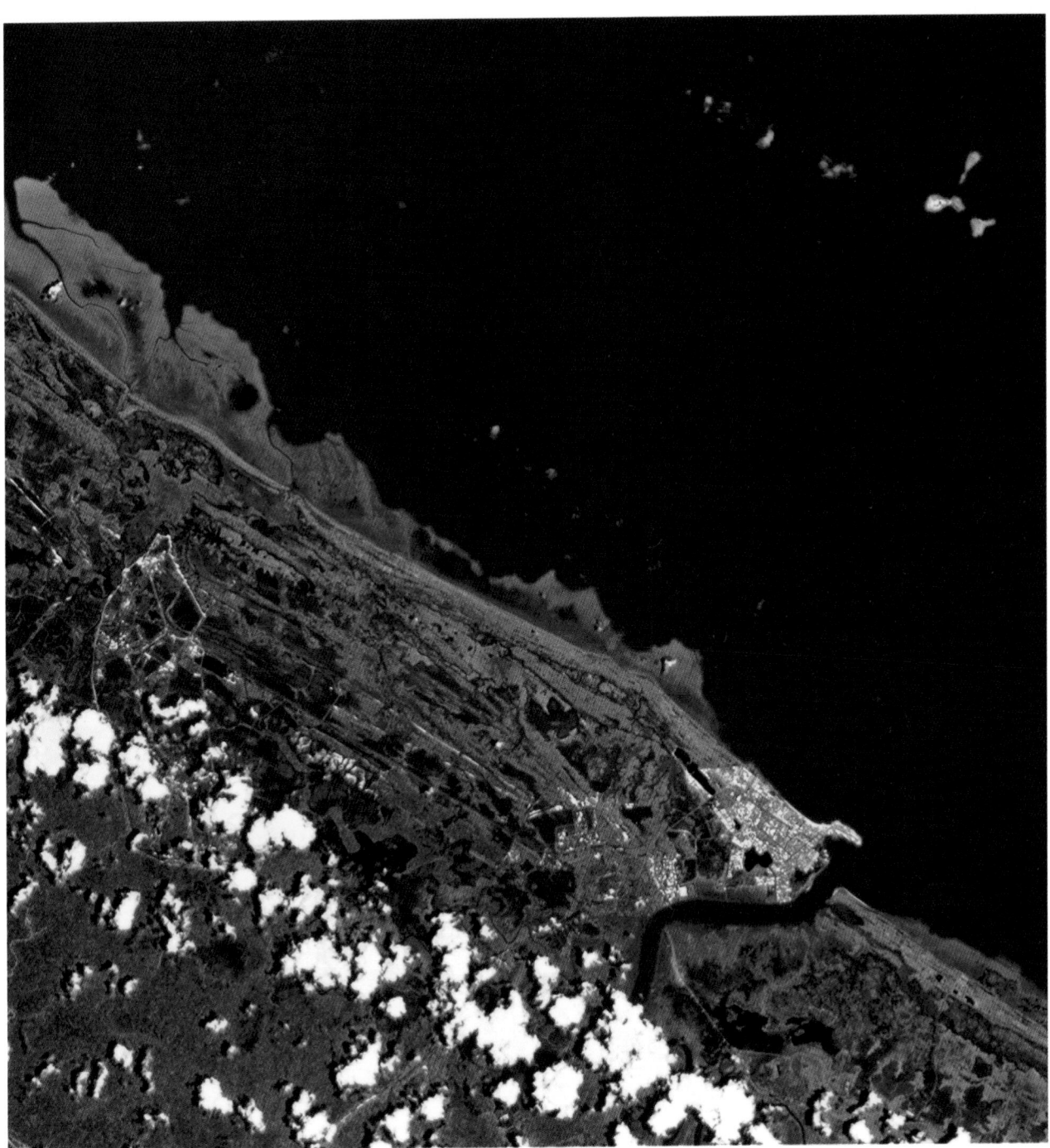

Figure 6.30. **SPOT image of Kourou.** SPOT image taken on 28 September 1996. The Salut islands are visible about fifteen kilometres north-east of Kourou.
© CNES 1996 Distribution SpotImage.

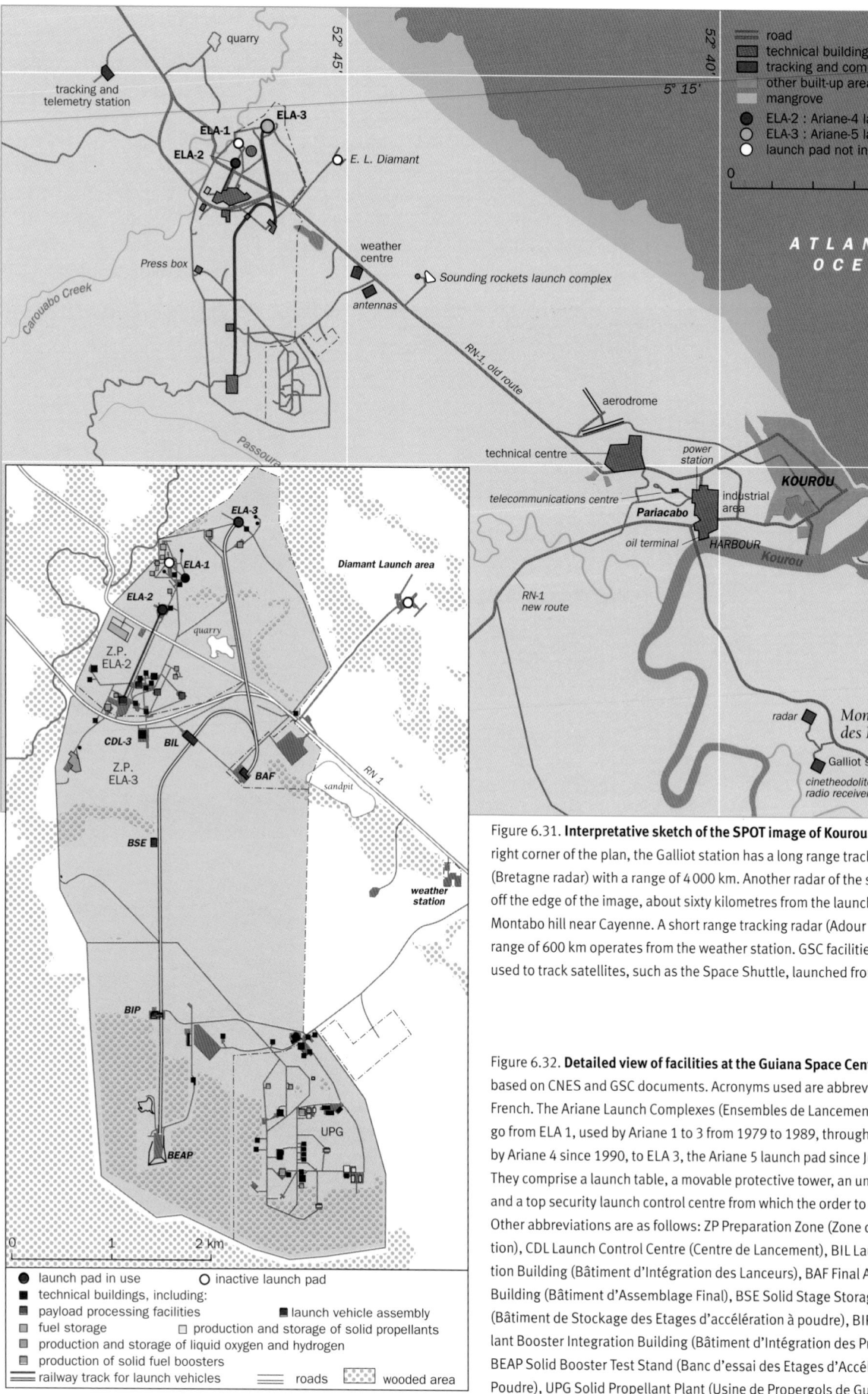

Figure 6.31. **Interpretative sketch of the SPOT image of Kourou.** In the lower right corner of the plan, the Galliot station has a long range tracking radar (Bretagne radar) with a range of 4 000 km. Another radar of the same type lies off the edge of the image, about sixty kilometres from the launch pads on the Montabo hill near Cayenne. A short range tracking radar (Adour radar) with a range of 600 km operates from the weather station. GSC facilities are also used to track satellites, such as the Space Shuttle, launched from other bases.

Figure 6.32. **Detailed view of facilities at the Guiana Space Centre.** The plan is based on CNES and GSC documents. Acronyms used are abbreviations of the French. The Ariane Launch Complexes (Ensembles de Lancement Ariane ELA) go from ELA 1, used by Ariane 1 to 3 from 1979 to 1989, through ELA 2, used by Ariane 4 since 1990, to ELA 3, the Ariane 5 launch pad since June 1996. They comprise a launch table, a movable protective tower, an umbilical mast and a top security launch control centre from which the order to fire is issued. Other abbreviations are as follows: ZP Preparation Zone (Zone de Préparation), CDL Launch Control Centre (Centre de Lancement), BIL Launch Integration Building (Bâtiment d'Intégration des Lanceurs), BAF Final Assembly Building (Bâtiment d'Assemblage Final), BSE Solid Stage Storage Building (Bâtiment de Stockage des Etages d'accélération à poudre), BIP Solid-Propellant Booster Integration Building (Bâtiment d'Intégration des Propulseurs), BEAP Solid Booster Test Stand (Banc d'essai des Etages d'Accélération à Poudre), UPG Solid Propellant Plant (Usine de Propergols de Guyane).

domains outside the space industry. The latter nevertheless remains the dominant activity.

The Ariane launch complexes (*Ensembles de Lancement Ariane*, ELA 1, 2 and 3) and payload preparation area (*Ensemble de Préparation des Charges Utiles EPCU*) belong to the European Space Agency. They are made available to the company Arianespace, which operates and maintains them. The latter has undertaken the construction of a new EPCU (S5) designed for future Ariane 5 missions and doubling the capacity of existing facilities.

Europe is keen to maintain its position in the face of new American competition and must therefore consider a broader use of the Kourou base. The request by the Russian government to launch Soyuz from Kourou illustrates this new political and strategic dimension. It is not a simple decision to take insofar as Soyuz would become a formidable competitor for Ariane itself in the area of geostationary launches if it could be launched from an equatorial site. However, opening up to a foreign launch vehicle marketed by Starsem, the advantages of the base could be put to better financial gain and such a move could even provide some control over a potential source of competition. This second approach seems to be favoured on the basis of announcements made in the summer of 2001. The final decision should be taken in 2002.

The Europa programme

European launch vehicle development has a long and tortuous history. After the Second World War, the only countries in Europe capable of setting up ballistic missile programmes, with the longer term aim of developing space launchers, were the United Kingdom and France. The British developed the light Black Arrow launch vehicle using their Black Knight missile for the two upper, liquid-propellant stages and their Skylark sounding rocket for the first, solid-propellant stage. Black Arrow only launched two satellites, from the Australian base at Woomera, in 1970 (failure) and 1971, before it was abandoned in 1973.

For its part, France produced a series of experimental missiles (Rubis, Topaze, Émeraude) which led up to the three-stage Diamant launch vehicle. The first stage used liquid propellant, whilst the two upper stages used solid propellants. Twelve launches were carried out from 1962, first at Hammaguir, then at Kourou, before the programme was abandoned in 1975.

A first attempt to concentrate national efforts into a European programme was made in 1962, leading on 29 February 1964 to the creation of the European Launcher Development Organisation (ELDO). This brought together the United Kingdom, France, Germany, Italy, Belgium and The Netherlands. The UK was the project manager for the Europa 1 programme using the launch base at Woomera. The first stage derived from the Blue Streak missile, manufactured under the Atlas licence and declassified in 1960. The other two stages were French (Coralie) and German (Astris). In 1970,

the second phase of the programme, Europa 2, brought a certain number of changes. The launch site was changed from Woomera to Kourou and a fourth stage was added for launching geostationary satellites. However, ten attempts at Woomera and one at Kourou all ended in failure. Dissension between the partners led to the programme being dropped in 1973. ELDO was dissolved and the European Space Agency was set up in its place.

The European launch vehicle Ariane

For its launch vehicle development, the European Space Agency (ESA) adopted the Ariane programme, undertaken by France in 1972, with the Centre National d'Études Spatiales (CNES) as the general contractor. France held a 63.87% stake and was thus the main driving force behind the development of an independent European launch facility. The Ariane programme resulted in the first Ariane 1 launch in 1979. It was followed by Ariane 2 and 3 (first flight in 1984), Ariane 4 (first flight in 1988) and Ariane 5 (first test flight in 1996). The economic situation was very favourable since the United States had opted for exclusive development of its Space Shuttle, whilst the USSR was not yet marketing its launch vehicles. Europe quickly climbed to first position and the Ariane launch vehicle became one of the most symbolic features of European space endeavour. Commercial exploitation of the launch vehicle was in principle assigned to the European company Arianespace in 1980 and became effective as of 1984. In 2001, the company has 43 shareholders from 12 different countries, including the CNES, several space and electronics manufacturers, and a number of banks. Arianespace is in charge of production, marketing and launch, but it is also involved to some extent in maintaining launch infrastructures and the site as a whole.

The first four generations of Ariane derive from one another by successive extensions to a stage or addition of boosters. The first two stages consume liquid propellants: UDMH and nitrogen tetroxide for Ariane 1, UH25 (a mixture of UDMH and hydrazine) and nitrogen peroxide for Ariane 2, 3 and 4. The third stage is cryogenic. The currently used version, Ariane 4, can be adapted to the payload through a choice of boosters and fairings. Six configurations are possible through choice of boosters (fig. 6.33). Payload processing combines three models for the fairing (short, medium or long) and a multiple launch external bearing structure called the SPELDA (French abbreviation for *Structure Porteuse Externe pour Lancement Double Ariane*) that can be inserted between the long or medium fairings and the third stage and which also exists in three versions (mini, short or long). A slightly longer version of the mini-SPELDA was developed in 1996. Medium and long fairings can be equipped with the Ariane dual launch system (SYLDA, *SYstème de Lancement Double Ariane*), inherited from Ariane 3, or the ASAP (Ariane Structure for Auxiliary Payloads), making it possible to launch six satellites of less than 50 kg as well as a large satellite. An extension to the third stage in 1992 (H 10+), and further

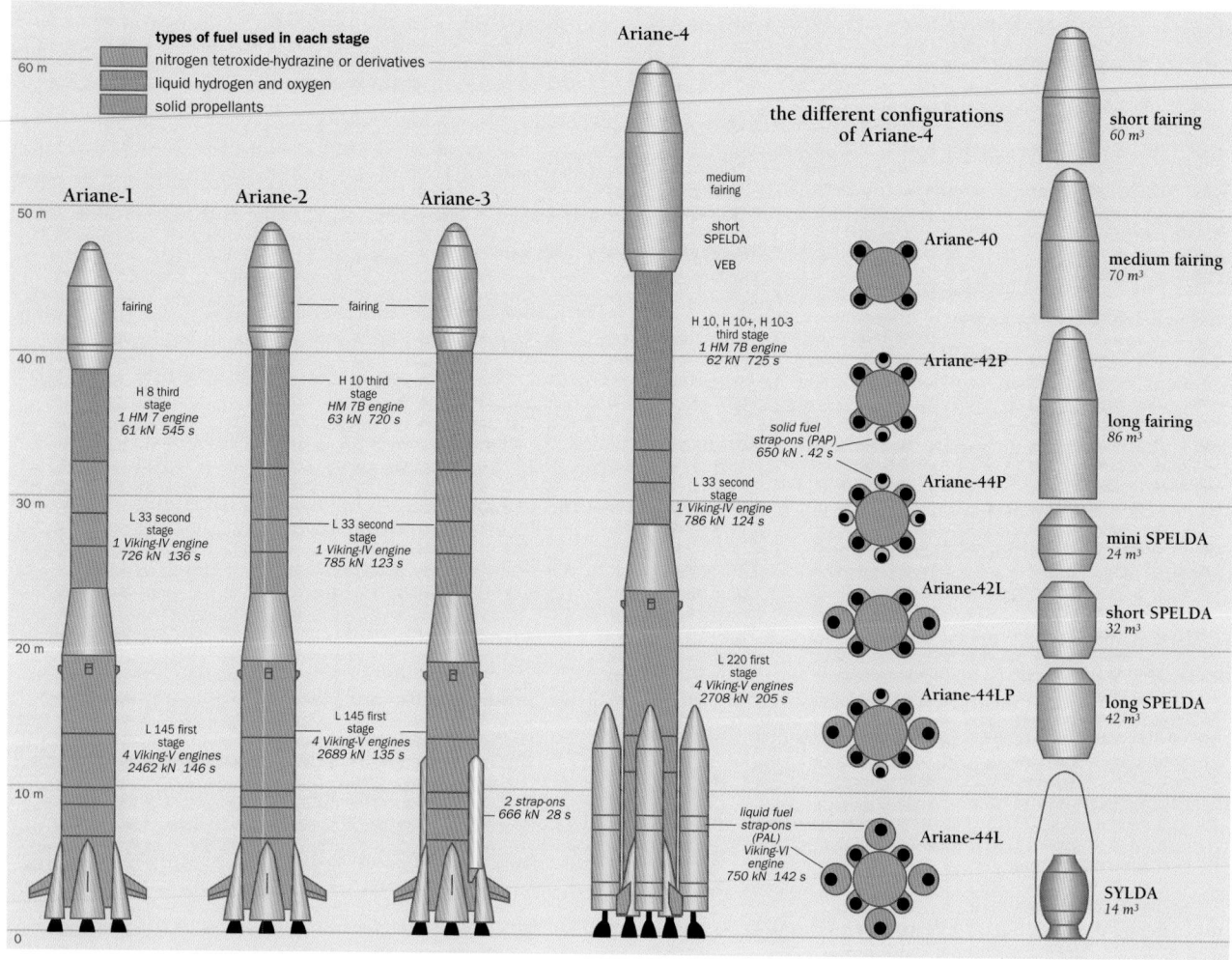

Figure 6.33. **The four generations of the Ariane launch vehicle.**

modifications in 1994 (H 10-3), increased the freight capacity of the launch vehicle by 9 to 15%.

Ariane 4 will be withdrawn from service in 2003, and Ariane 5 will carry out all launches sold by Arianespace.

In 1985 it was decided to implement the Ariane 5 programme. The first flight failed in June 1996, but the following two qualification flights were a success and the vehicle was commercialised in 1999. The European Space Agency now has at its disposal a heavy lift launch vehicle of a very different design to the first four versions of Ariane. It comprises a cryogenic first stage flanked by two large solid boosters, and a reignitable liquid-propellant second stage. Three fairings are available, short, medium and long, and there are six possible versions of the SYLDA (from 4.9 to 6.4 metres long). The length of the fairings can be adjusted using a set of four collars measuring between 0.5 and 2 metres. The SPELTRA (*Structure Porteuse Externe pour Lancement TRiple Ariane*) is available in two versions of lengths 5.5 and 7 metres. It is inserted between the second stage and the fairing for multiple launches to geostationary orbit. An ASAP 5 platform is also available, as are dispensers for deploying satellite clusters.

Arianespace aims at a wide market ranging from constellations to heavy vehicles. Apart from its other functions, Ari-

ane 5 was originally designed to launch the Hermes shuttle which would provide access to the International Space Station. After several years of wavering over Europe's means of access to the ISS, it was decided in 2000 to launch nine unmanned transfer vehicles (ATV) via Ariane 5 between 2003 and 2014.

In order to maintain Ariane's position on the market and make it more competitive and better suited to future payloads, several developments have been planned. Ariane 5 Evolution, scheduled for 2002, enhances capacity with a new generation of Vulcain engine, whilst Ariane 5 Versatile, also scheduled for 2002, has a modified upper stage with the Vulcain 2 engine. The latter has multiple restart capability to allow more flexibility when injecting satellites into orbit, both geostationary and otherwise. Looking even further ahead, the Ariane 5 Plus programme has two phases. The first, ESC-A, planned for 2002 will reuse the lower stage of Ariane 5 Evolution and add a cryogenic upper stage with the engine of the Ariane 4 third stage. The second phase, ESC-B, expected for the end of 2005, will be the most powerful version, with a capacity of 12 tonnes to geostationary transfer orbit. It will use the new Vinci engine which is more powerful and capable of multiple restarts in flight (fig. 6.34).

To provide a longer term view, the CNES has proposed the Ariane 2010 initiative. It aims to bring together research and development activities around a launch vehicle project with two-fold objective: to reach a capacity of 15 tonnes for dual launch to geostationary transfer orbit and/or reduce by 30% the cost of a 12 tonne launch via Ariane 5 ESC-B. Ariane 2010 may also be the starting point for the Future Launcher Technology Programme (FLTP). It is planned to demonstrate the reusable launch technology of this programme between 2002 and 2007. Since one of the main objectives is to significantly cut the cost of lifting one kilo into orbit, comparisons will have to be made between partially or totally reusable launch vehicles and conventional systems.

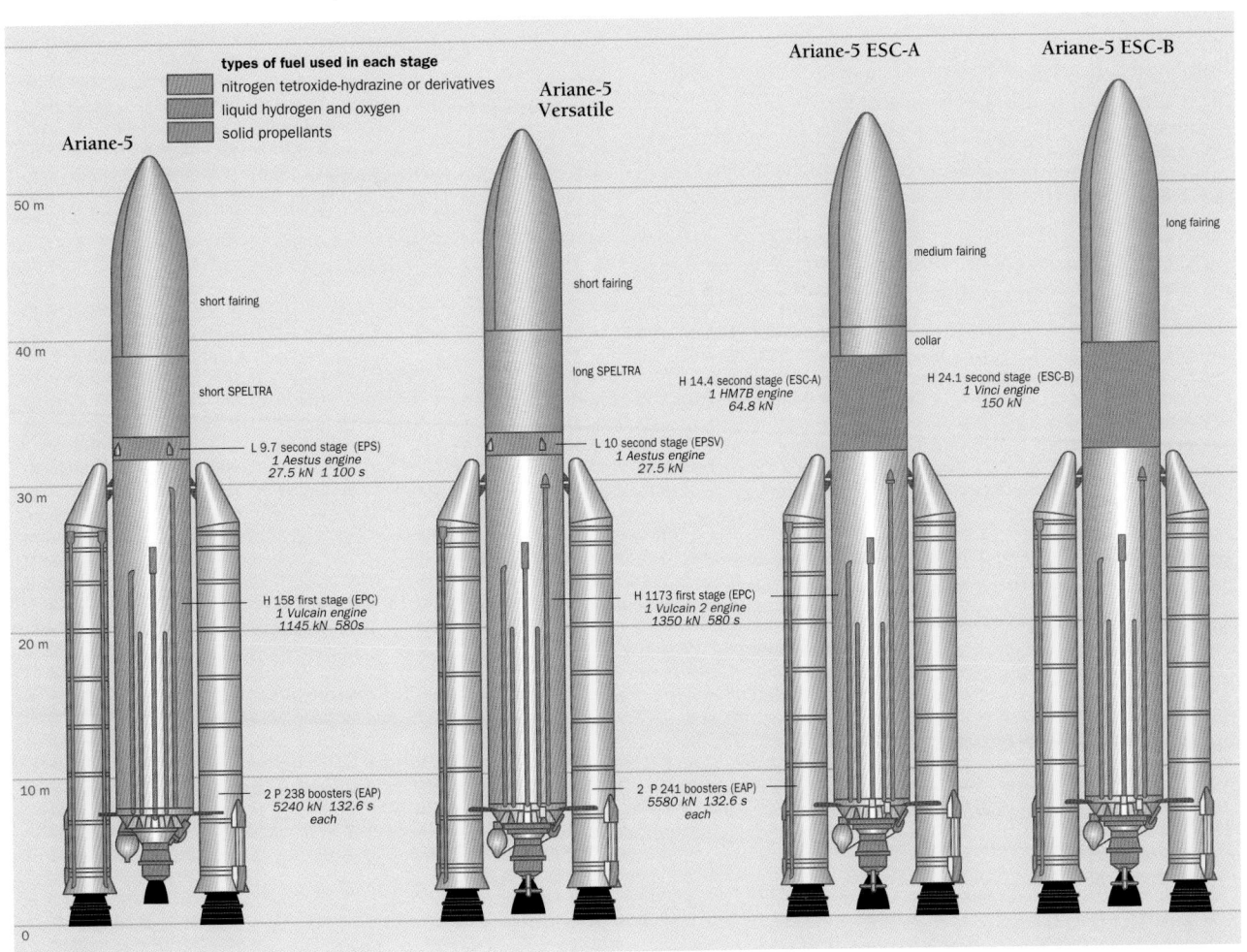

Figure 6.34. **The Ariane 5 heavy launch vehicle.**

Name		First launch to orbit	Final launch	Payload capability (kg) *LEO: low Earth orbit, GTO: geostationary transfer orbit, GEO: geostationary orbit. (300 km): height of circular orbit, (300-4 000 km): apogee and perigee of eccentric orbits. (65°): inclination. All launches are from Kourou*	Launches up to 01/01/2001 ( ) successful launches
Ariane-1		24 12 1979	22 02 1986	1700 (GTO, 7°)	11 (9)
Ariane-2		03 05 1986	02 04 1989	2065 (GTO, 7°)	6 (5)
Ariane-3		04 08 1984	12 07 1989	2580 (GTO, 7°)	11 (10)
Ariane-4	-40	22 01 1990		1900 (GTO, 7°), 2740 (800 km, 98.6°) with H 10-3: 2105 (GTO, 7°)	7 (7)
	-42P	20 11 1990		2600 (GTO, 7°), with H 10-3: 2930 (GTO, 7°)	14 (13)
	-44P	04 04 1991		3000 (GTO, 7°), with H 10-3: 3465 (GTO, 7°)	13 (13)
	-42L	12 05 1993		3200 (GTO, 7°), with H 10-3: 3480 (GTO, 7°)	12 (12)
	-44LP	15 06 1988		3700 (GTO, 7°), with H 10-3: 4220 (GTO, 7°)	25 (24)
	-44L	04 06 1989		4200 (GTO, 7°), with H 10-3: 4720 (GTO, 7°)	30 (29)
Ariane-5		25 05 1996		6800 (GTO, 7°), 5900 double (GTO, 7°), 10000 (800 km, 98.6°), 18000 (70-300 km, 51.6°)	8 (7)
Ariane-5	Versatile			8000 (GTO, 7°)	0
Ariane-5	ESC-A			10500 (GTO, 7°)	0
Ariane-5	ESC-B			12000 (GTO, 7°)	0
Vega K0				640 (200 km, 0°) 400 (1000 km, 0°) 470 (200 km, 90°) 310 (800 km, 98°)	0
Vega K3				1600 (200 km, 0°) 600 (1000 km, 0°) 1200 (200 km, 90°) 550 (800 km, 98°)	0

Table 6.3. **The European launch family in 2001.**

Small European launchers

The question of a complete range of European launch facilities was raised by the Ministerial Council of ESA in June 2000. A resolution was voted concerning European space launch strategy, stipulating that Ariane 5 Plus should be complemented by small and medium launch vehicles manufactured in Europe. These launch vehicles will be built up from common elements (stages, subsystems, technology, production units and operational infrastructure). They may benefit from new solid-propellant propulsion techniques such as those developed in the P 80 programme.

For the small launch vehicle, ESA officially opted at the end of 2000 to develop Vega, a lightweight launcher advocated by Italy since 1996 (fig. 6.35). Italian knowhow lies mainly in solid propellants, especially at BPD, a subsidiary of Fiat Avio since 1996. Moreover, since 1966 Italy has had its own small space base for launching Scout rockets. It comprises two converted oil exploration platforms: one, San Marco, is used as a launch pad whilst the other, Santa Rita, is the control centre (fig. 6.36). It is located in the Bay of Formosa, 5 km from the Kenyan coast at latitude 2.9° south. Eight scientific satellites were launched between 1966 and 1976, followed by a ninth in 1988; five of these were Italian, three American and one British. Since 1988, the base has been dormant and no longer possesses a suitable launch vehicle since the Scout was abandoned in 1994. For this reason, since 1988, BPD had been proposing projects like Vega K0 that might make up for the disappearance of the Scout launcher. The Vega launch vehicle finally adopted by ESA uses the principle of technological building blocks and reuses an adapted version of the Ariane 5 booster as first stage. It is named after the second brightest star of the northern hemisphere.

In 2000, 85% of the financial backing had been found for the Vega programme, with 20% coming from ESA member states (Sweden, Spain, The Netherlands, Switzerland and Belgium) and 65% from Italy in cooperation with France. The prime contractor will be a joint venture between Aerospatiale and Fiat Avio and marketing will be carried out by Arianespace. Vega's first launch is scheduled for 2005. It will handle polar and low-Earth orbiting satellites weighing between 300 kg and 2 tonnes. Four launches should be possible each year, lifting scientific or Earth observation satellites into orbit from Kourou's ELA 1 site. For the time being, this task is carried out from Plesetsk by the Rockot launch vehicle marketed by the Russian–German company Eurockot, which concluded a partnership agreement with Starsem in June 2000.

With regard to the medium launch facility, Starsem currently proposes Soyuz with attractive low cost and proven reliability. The ESA commission of June 2000 suggested using components derived from Ariane 5 (P 230) and the small launch vehicle (P 80 and third stage) to develop a purely European launch vehicle that would aim at the constellation market. However, Soyuz performs particularly well in this area and the prospects seem limited.

Reusable launch vehicles

The ESA Council decided, in November 2001, on a second phase for the Future Launcher Technology Programme, instigated in May 1999, without German participation. It mainly involves research and development. However, long term European aims are still obscure. Indeed, two areas require reflection: the design of a reusable launch vehicle aimed at cutting the cost of reaching orbit and the development of a manned vehicle (CRV) for accessing the ISS. There is a general consensus that Europe cannot bypass reusable technologies, but several competing approaches exist, even with regard to cooperation.

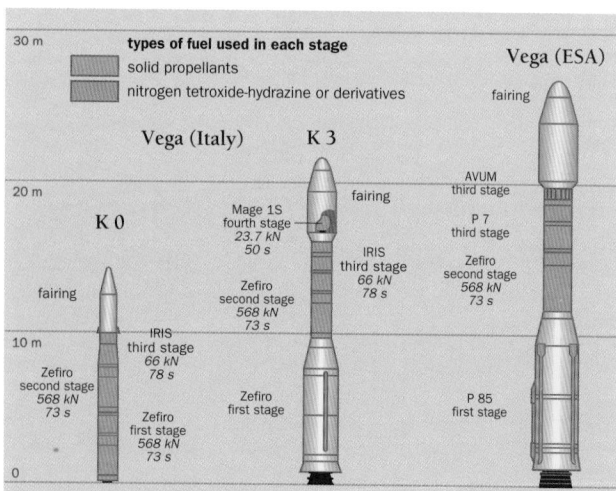

Figure 6.35. **The Vega light launch vehicle.** The launch vehicle finally chosen by ESA uses the P 230 booster from Ariane 5 as a first stage. The second and third stages will be adapted and derived from the Italian Zefiro engine intended for the original versions of the Italian projects (K0 and K3). The Attitude and Vernier Upper Module will play the role of a fourth stage and orbital corrector once the third stage has been fired.

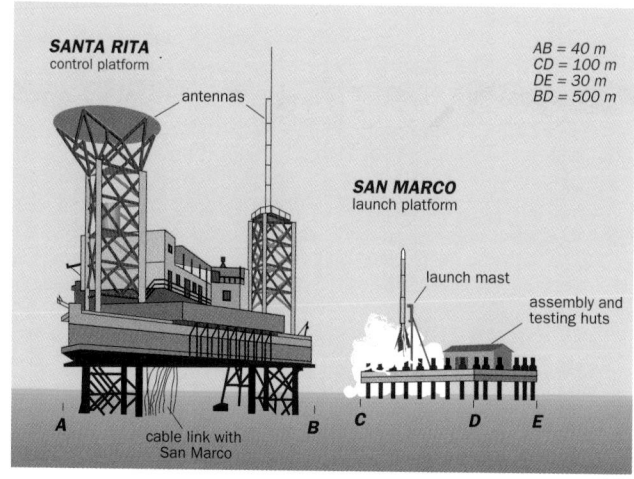

Figure 6.36. **The Italian San Marco launch facility.** The tide in the Indian Ocean has a mean amplitude of two metres in this area.

Japan

Unlike other members of the space club, Japan developed its launch programme in response to scientific and technological considerations, as would be expected for a university-run programme. From the Second World War until the 1952 San Francisco peace treaty, it was forbidden for Japan to undertake any research and development in aeronautics. However, the first studies in rocketry began at the beginning of the 1950s. In 1955, Tokyo University started to build light rockets in order to be able to take part in the International Geophysical Year. In fact, the development of launch vehicles within the university soon became highly controversial. Heavy investment was involved and some financial return had to be gained on space activities, whilst the question of applications fell by definition outside the competence of the academic world. From the very start there was opposition from the Science and Technology Agency (STA), not to men-

tion the Minister of Finance, and this resulted in the creation in 1964 of the Office of Space Development Promotion (precursor of NASDA) and the National Aeronautical Laboratory (NAL) under the supervision of the STA. However, strong scientific opposition led to the simultaneous creation in 1964 of the Institute of Space and Astronautical Science (ISAS). The latter became a relatively autonomous part of the Ministry of Education in 1981 in view of its scientific vocation. Clear limits were set to this autonomy since it was forbidden for ISAS to develop launch vehicles with diameter greater than 1.4 metres.

In 1969, in a context of strong economic growth, Japan was anxious to appear as a fully fledged member of the exclusive club of highly industrialised countries. Since space activities are particularly symbolic in this context, it thus founded the National Space Development Agency (NASDA). This

Figure 6.37. **Aerial view of the Kagoshima Space Centre (photo ISAS).**

organisation is responsible for technological and industrial applications. However, ministries requiring their own space systems, especially for telecommunications, prefer to keep some room for manoeuvre, independently of NASDA. The latter is thus not involved in the decision-making process when satellites with government applications are at issue. The rather special relationship with the United States led at the same date to a political agreement whereby the US would provide technological assistance for the development of launch systems. In this context, NASDA focuses on technological development and the acquisition of homegrown knowhow, which was originally intended as its primary mission.

Within this unusual framework, Japan has developed two branches of space activity. ISAS and NASDA each possess their own launch base and launch vehicles, with very different technology and capabilities. There have been some signs of reconciliation, however, as with the J-1 programme put forward by ISAS, even though the results have been limited.

In 1990, the Rocket Systems Corporation was created along the same lines as Arianespace to develop and market the H 2 launch vehicle and also to prepare for launches from Tanegashima. This step attests to a certain desire to introduce a commercial dimension to their activities. In fact, the target of technological independence on the one hand and the constraints imposed by business relations with the United States, anxious to respect the rules of fair competition on the other, explains why a business aspect has only appeared so recently in the Japanese space programme. The decision in 2000 to replace the H 2, built entirely from Japanese technology and too expensive compared with its competitors, by the more competitive H 2A launch vehicle is witness to this change of approach. The first launch of the H 2A took place successfully in August 2001. The purchase of foreign technology such as Russian engines or the development of European collaboration, as in the HOPE-X programme of atmospheric reentry tests, also confirms a new level of maturity and an attempt to diversify the acquisition of technological knowhow.

ISAS base and launch vehicles

The ISAS base is the Kagoshima Space Centre (fig. 6.37), built in 1963 on the steep hills south of Kyushu, at various heights between 220 m and 320 m above sea level. Launches are lim-

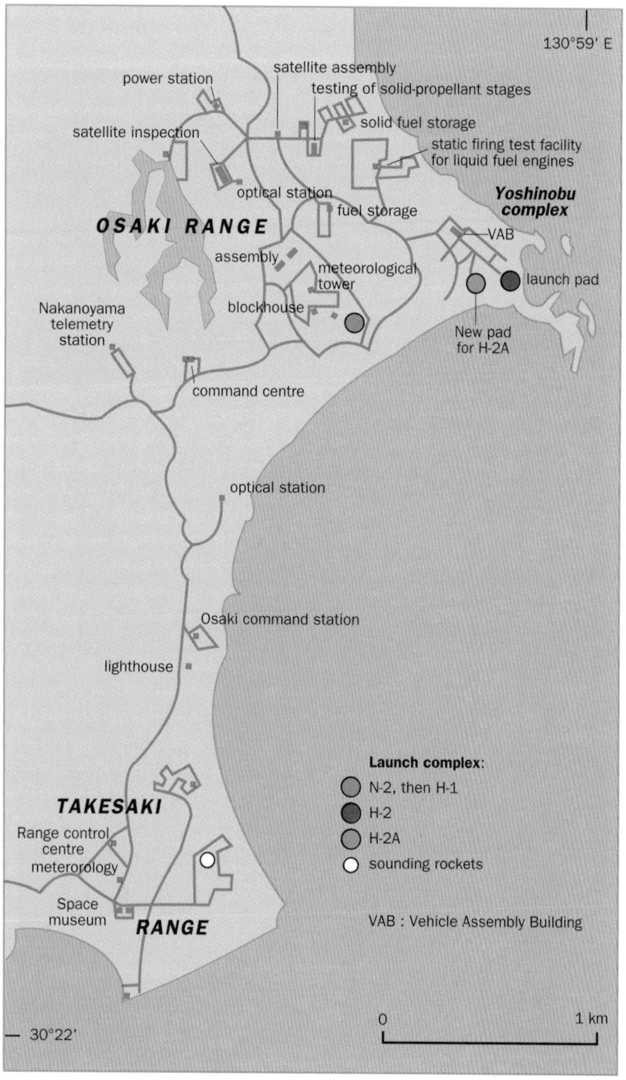

Figure 6.38. **The Tanegashima Space Centre.**

Figure 6.39. **The island of Tanegashima viewed by the OPS sensor on JERS 1.**
The space centre is located in the south-east of the island as indicated.
© NASDA.

ited to two periods in the year: January to February when cloud cover is at its lowest and August to September when cloud cover is actually rather heavy. This is to protect fishing activities which are forbidden over a wide area of the ocean to the east of the base when launches are programmed.

ISAS launch vehicles are three- or four-stage solid-propellant rockets built by Nissan and developed from Kappa and Lambda sounding rockets. In 1970, after four failures, Lambda 4S had one and only one successful launch, making Japan the fourth nation to reach space. It was replaced by the Mu launch vehicles which went through four versions: M 4S used between 1970 and 1972, M 3C between 1974 and 1979, M 3H between 1977 and 1978, M 3S between 1980 and 1984, and its extended version M-3S2 between 1985 and 1995. The new M 5 launch vehicle is the most powerful ever developed by ISAS. It launched its first scientific satellite in 1997, followed by the Nozomi Mars probe in 1998 (see fig. 6.41).

NASDA base and launch vehicles

The NASDA launch base is the Tanegashima Space Centre (figs. 6.38 and 6.39), built in 1968 on the island of the same name, 1000 kilometres south of Tokyo. The base is sited on low granite hills. Climatic and operational restrictions are similar to those at Kagoshima although the agreement has been extended from 90 to 130 days (71 days between 22 July and 30 September and 59 days between 1 January and 28 February). In order to cater for exceptional situations, to meet the requirements of an international collaboration agreement or to benefit from launch windows for interplanetary missions, a further period of 60 days has been accorded between 26 June and 15 July and also between November and December. Launch periods can thus reach a maximum of 190 days per year. Satellite launches are confined to the Osaki range where a second launch pad is under construction in the Yoshinobu complex, to the west of the first H 2 launch pad (fig. 6.40).

NASDA developed its first N 1 launch vehicles 1 (N for Nippon) from the first stage of the Thor rocket built by Mitsubishi under license from McDonnell-Douglas. The second stage was made in Japan with an American engine and also used liquid propellants. The launch vehicle was completed by a third stage using solid propellant and three strap-on

boosters, also based on American technology. It successfully launched seven satellites between 1975 and 1982.

N 2 is a more powerful version of N 1, with a stretched second stage and nine strap-on boosters. Eight launches were carried out between 11 February 1981 and 19 February 1989.

With the H 1 launch vehicle, Japan began to free itself from American technology, the original aim, by developing a cryogenic second stage. The other two stages were barely changed in relation to the N 2 except that the nine solid boosters were produced by Nissan. Nine launches took place between 13 August 1986 and 1992, all successful.

In 1994, the H 2 went into service. It was composed of two cryogenic stages built by Mitsubishi and two large solid boosters produced by Nissan (fig. 6.41), and marked the advent of a wholly Japanese launch vehicle. Thus, after long-standing cooperation with the United States, Japanese launch policy achieved its aim of technological self-sufficiency.

NASDA was planning to use the H 2 as the basis for a modular series and reduce production costs, which had already gone down by 30% since the first H 2 vehicles were built. When the launcher failed on 15 November 1999, the development plan was revised and efforts shifted onto the H 2A version. This series includes the 202 models with two, four or six boosters, and also the 212 and 222 which couple two

Figure 6.40. **The Yoshinobu complex.** The new launch pad for the H-2A launch vehicle, showing the Mobile Launcher (ML) at the centre. This moves launch vehicles between the Vehicle Assembly Building (VAB) and the pad. NASDA Document.

Name	First orbital launch	Final launch	Payload capability (kg) LEO : low Earth orbit, GTO : geostationary transfer orbit, GEO : geostationary orbit, EE : Earth escape (300 km) : height of circular orbit, (300-4 000 km) : apogee and perigee of eccentric orbits. (65°) inclination. K, Tan (Kagoshima, Tanegashima) : launch site	Launches up to 01/01/2001 ( ) successful launches
M-3S2	08 01 1985	15 01 1995	K: 770 (250 km, 31°) 170 (EE)	8 (7)
M-5	12 02 1997		K: 2000 (200 km, 31°) 800 (800 km, 98°) 520 (Moon) 450 (Mars, Venus)	3 (2)
H-1	13 08 1986	11 02 1992	Tan: 1400 (800 km, 98°) 550 (GEO) 800 (EE)	9 (9)
H-2	03 02 1994	15 11 1999	Tan: 10000 (250 km, 30°) 4000 (GTO, 30°) 2000 (GEO) 4300 (800 km, 98°)	7 (6)
H-2A 202			Tan: 4100 (GTO, 30°)	0
H-2A 2022			Tan: 4500 (GTO, 30°)	0
H-2A 2024			Tan: 5000 (GTO, 30°)	0
H-2A 212			Tan: 7500 (GTO, 30°)	0
H-2A 222			Tan: 9500 (GTO, 30°)	0
J-1			Tan: 870 (250 km, 30°)	0
J-2			Tan: 2750 (200 km, 30°) 1070 (800 km, 98.6°)	0

Table 6.4. **Japanese launch vehicles in use or under development in 2001.**

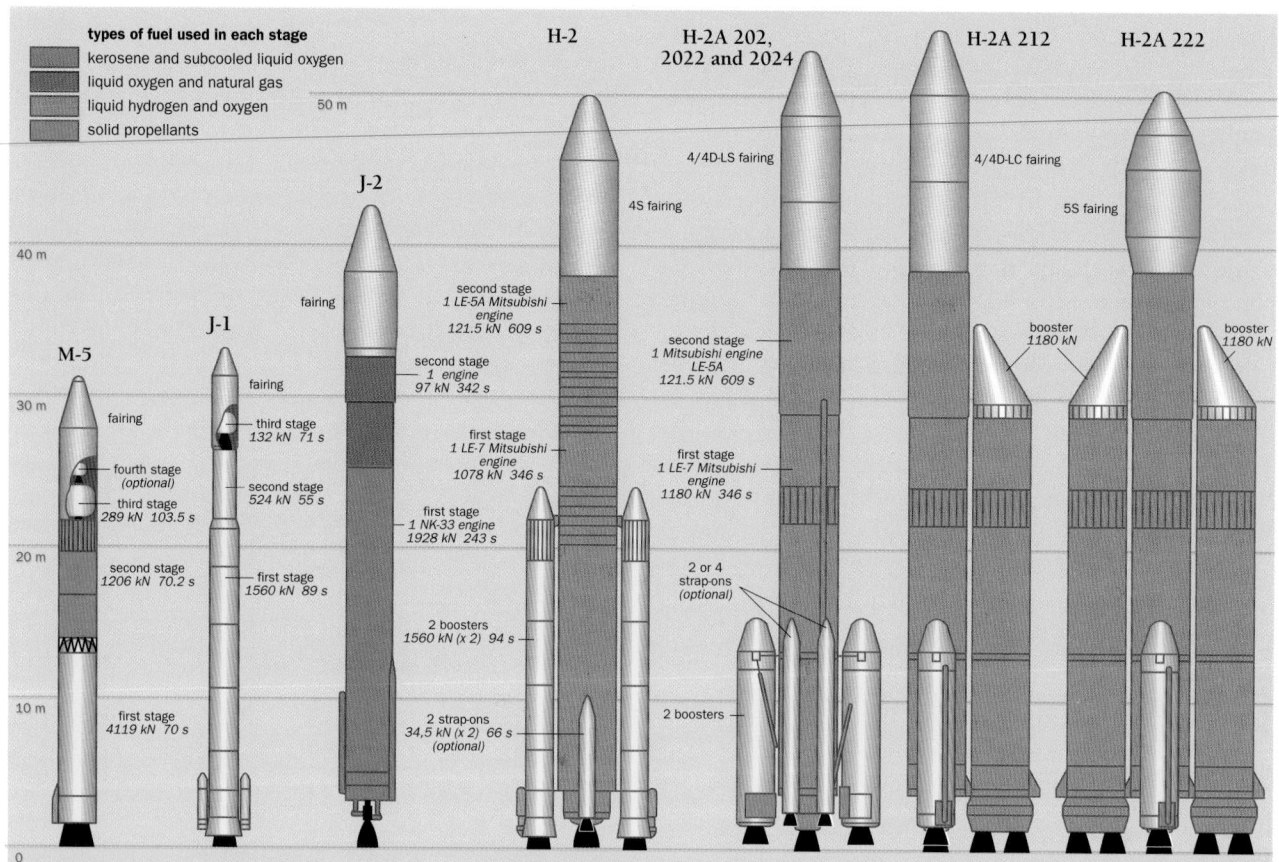

Figure 6.41. **Japanese launch vehicles.** The four available fairings for the H 2 and the H 2A are interchangeable. Model 4/4d is the 4S adapted for dual launches.

and three first stages. When completed, the programme should be able to loft 8 tonne payloads into GTO. There is also a variant of the H 2A with no second stage, designed to launch the unmanned minishuttle HOPE (H 2 Orbiting space PlanE, see chapter 13).

Prospects for commercialising the H 2A family depend on a significant reduction in launch costs. The H 2 was the most expensive launch facility on the market and NASDA's aim is to cut the price for the H 2A by 55%. Although still vague, these prospects led to successful negotiations with fishermen's associations for the extension of launch periods.

NASDA has also shown its willingness to cover the small satellite class, as with the J 1 launcher built from an H 2 booster and the second and third stages of the M-3S2. Although it successfully carried out a suborbital flight in 1996 and it is scheduled to launch the OICETS satellite in 2003, this will be its second and last flight. An SAC assessment report in 1998 recommended the end of this programme. NASDA plans to replace it by a cheaper J 1 Upgrade (or J 2). The first stage will use the technology of the Atlas 3 first stage and the Russian NK 33 engine, whilst the second stage will have a new

liquid oxygen–liquefied methane engine. The J 2 should make it possible to experiment with technology for the reusable launch programme and if necessary provide a launch capability for LEO satellites. In parallel, IHI created the joint venture company Galaxy Express with MHI and five other Japanese companies to develop and market the J 1U as of 2006.

For future space transportation systems, Japan intends to follow the recommendations of a committee comprising NAL, ISAS and NASDA who put forward a 15-year plan divided into three five-year phases. In the first phase, ISAS will develop a combined ramjet/turbojet air-breathing engine to be tested on a NAL flight demonstrator whilst NASDA develops a reusable rocket engine. The High Speed Flight Demonstrator (HSFD) will also serve to carry out trials. In the second phase, engineers will produce a scaled-down model of the first stage with its air-breathing engine and a full-scale model of the second stage based on a small experimental reusable rocket. In the third phase, a fully operational first stage will be built, onto which the second stage orbiter will be mounted in about 2015. The whole vehicle will be able to take off and land horizontally.

China and India

Chinese and Indian launch activities result from very different strategies. Chinese programmes reflect a typical power and independence scenario using knowhow obtained in the field of ballistic missiles. In 1970, China launched its first satellite and became the fifth space power. In contrast, India chose to develop its launch programme separately in order to acquire its own technology, which was not the simplest option.

When China offered two Long March launch vehicles, the CZ 3 and the CZ 2C, on the international market in 1985, it thus affirmed the need to earn some financial return from this sector. Such a step also leads to a certain degree of demil-itarisation. However, technical difficulties and American reluctance, motivated more by political than by economic doubts, have limited this new approach, even though it remains a reality. In 2001, the Great Wall Company can thus offer its range of launch vehicles, but it is held back by the need for authorisation from the US administration when satellites are manufactured under American licence. Needless to say, this represents a considerable share of the market.

Figure 6.42. **The Jiuquan space centre viewed by SPOT.** A mosaic of two SPOT 4 images taken on 9 September 1998 . The space centre is located on either side of the Ruoshui river. © CNES 1998 Distribution SpotImage.

Chinese bases

The East Wind Launch Facility, or Dong Feng in Chinese, was China's first space base. It handles launches into orbits with inclinations lying between 57° and 70°. It was originally a large military base built at the end of the 1950s for ballistic missile tests (figs. 6.42 and 6.43). It is located in Kansu Province on the southern edge of the Gobi desert (40.6° N, 100.2° E), near the town of Jiuquan from which it takes its name. It was equipped with two launch pads between 1968 and 1970, one for the FB 1 and CZ 2 launch vehicles, the other for the CZ 1. Two other pads were subsequently added, one for the CZ 2C in 1975 and one for the CZ 2D in 1992. A further launch pad has been under construction for the CZ 2F since 1999. This aims to satisfy the needs of a manned flight programme. However, it seems that because of accessibility problems a new base may be developed in the Hainan region for recovery of the Shenzhou capsules.

As this base is poorly situated for geostationary satellite launches, being too far north, a second launch facility was built in 1980, 65 km from the town of Xichang (28.25° N, 102° E), at an altitude of 1830 m in the Anning valley, in the south of Sichuan Province. Despite large scale transfers of the rural population, the zone remains densely populated. The base possesses three launch pads, the first for the CZ 3 (1984), the second for the CZ 2E (1990), and the third for the CZ 3A (1994) and then the CZ 3B. All commercial geostationary launches are carried out from Xichang and it is the only one with facilities for cryogenic stages.

A third base is located in the north-west of Shanxi Province. It goes by the name of Taiyuan although it is 280 km from the town of Taiyuan (37.5° N, 112.6° E). It was revealed when the CZ 4 vehicle was launched in 1988. The area has a temperate climate, remaining cool in summer. Although originally reserved for launching Chinese Sun-synchronous meteorological and Earth observation satellites (two launches between 1988 and 1997), it saw increased activity in 1997 and 1998 thanks to a number of Iridium launch contracts. It has facilities to launch the CZ 2C/SD, an adapted CZ 2C designed for Iridium following the 1993 contract with Motorola and which became available in 1997, as well as the CZ 4A (1988) and CZ 4B (1999).

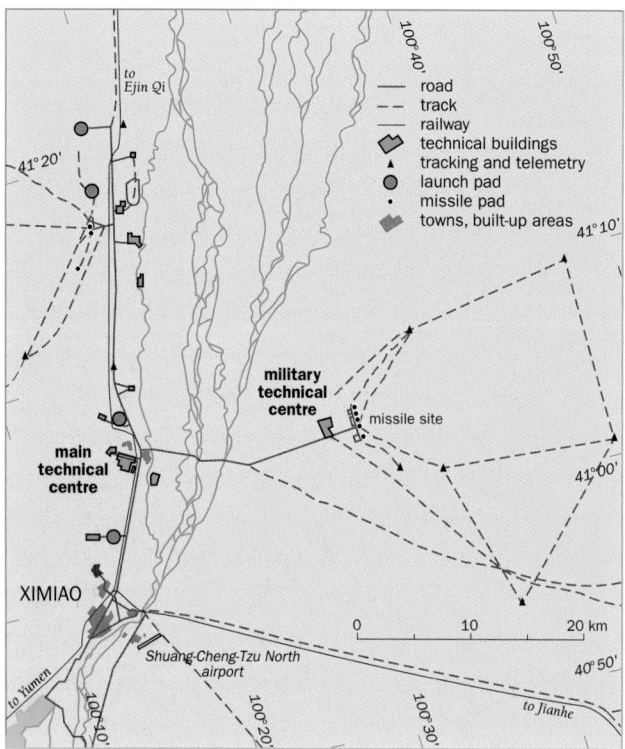

Figure 6.43. **Interpretative sketch of the SPOT image of the Jiuquan space centre.**

Name	Nationality	First launch to orbit	Payload capability (kg) LEO: low-Earth orbit, GTO: geostationary transfer orbit, GEO: geostationary orbit, EE: Earth escape (300 km): height of circular orbit, (300-4 000 km): apogee and perigee of eccentric orbits. (65°:) inclination. J, TY, X, Sri, Pal (Jiuquan, Taiyuan, Xichang, Sriharikota, Palmahim): launch site	Launches up to 01/01/2001 ( ) successful launches
CZ-1	Chinese	24 04 1970	J: 900 (300 km, 57°)	2 (2)
CZ-2C	Chinese	26 11 1975	J: 2000 (300 km, 57°) 750 (800 km, 98°)	14 (14)
CZ-2C SD	Chinese	01 09 1997	TY:	7 (7)
CZ-2D	Chinese	09 08 1992	J: 3400 (200 km, 57°) 850 (800 km, 98°)	3 (3)
CZ-4	Chinese	06 09 1988	TY: 1500 (800 km, 98°)	2 (2)
CZ-4B	Chinese	10 05 1999	TY:	3 (3)
CZ-3	Chinese	29 01 1984	X: 1340 (GTO, 28°)	13 (13)
CZ-3A	Chinese	08 02 1994	X: 2300 (GTO, 28°)	6 (6)
CZ-3B	Chinese	17 02 1996	X: 4850 (GTO, 28°)	5 (4)
CZ-3C	Chinese		J: 8800 (200 km, 57°) X: 3 460 (GTO, 28°)	0
CZ-2E (or F)	Chinese	16 07 1990		8 (7)
ASLV	Indian	24 03 1987	Sri: 150 (400 km, 46,5°)	4 (2)
PSLV	Indian	20 09 1993	Sri: 3000 (400 km, 13°) 1 000 (800 km, 98,6°) 450 (GTO, 13°)	5 (4)
GSLV	Indian	18 04 2001	Sri: 2500 (GTO, 13°)	0
Capricornio	Spanish		140 (400 km, 27°) 100 (700 km, 27°) 100 (400 km, 90°) 92 (400 km, 97°) 55 (700 km, 98°)	0
Shavit	Israeli	19 08 1988	Pal: 160 (250-1200 km, 143°)	6 (3)
Next	Israeli		600 (400 km, 28°) 350 (1600 km, 28°) 450 (400 km, 90°) 250 (1600 km, 90°)	0
Leolink LK-2	Israeli		1000 (LEO)	0
VLS	Brazilian	02 11 1997	200 (750 km, 2°)	2 (0)

Table 6.5. **Other launch vehicles in use or under development in 2001.**

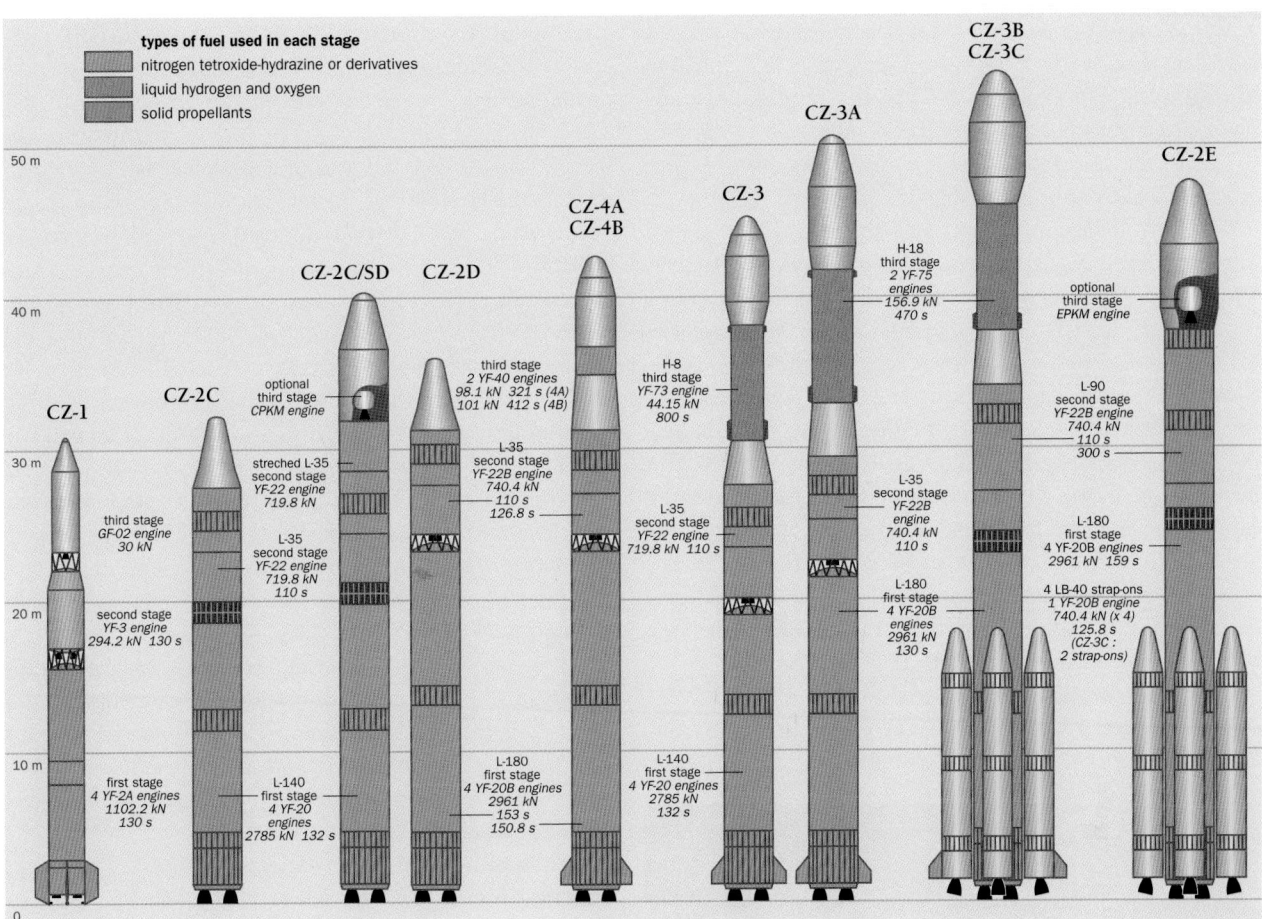

types of fuel used in each stage
- nitrogen tetroxide-hydrazine or derivatives
- liquid hydrogen and oxygen
- solid propellants

Figure 6.44. **Versions of the Long March launch vehicle in use or under development in 2001.**

Chinese launch vehicles

The CZ launch family (Chang Zheng in Chinese, also known as Long March) initially derived from the IRBMs and ICBMs DF 2, 3 and 4 built in the 1950s using Soviet technology (fig. 6.44).

In its first version, CZ 1 had two stages and used kerosene and liquid oxygen as propellants. It made two launches in 1970 and 1971 before being temporarily abandoned. However, a modernised version operating on UDMH and nitric acid, and with a Chinese- or Italian-made solid propellant third stage (IRIS) is currently offered for commercial launches.

In 1973, the FB 1 launch vehicle (Feng Bao, Storm) and in 1975, the CZ 2C were capable of placing recoverable FSW observation satellites in low-Earth orbit. The two vehicles were very similar. The FB 1 seemed to be a CZ 2 with simplified electronics. The first stage (L 140) like the second (L 35) consumed UDMH and nitrogen tetroxide. The FB 1 disappeared from the scene in 1981. The CZ 2C, capable of lofting 2.8 tonnes into a 200 km circular orbit, is still in use. It can be launched from Jiuquan or Taiyuan.

In 1984, the CZ 3 was the first to be able to carry geostationary satellites. The first two stages were those of the CZ 2C and a cryogenic third stage (H 8) was added. At this point, China became the third space power to possess the technology for liquid hydrogen–liquid oxygen engines.

In 1988, the three-stage CZ 4 launch vehicle was designed to launch satellites into Sun-synchronous orbit from the new base at Taiyuan, specially built for this purpose. CZ 4 was the first to use the L 180 stage, a stretched L 140 stage. The second stage was still the L 35.

China then produced five new launch vehicles, combining a new cryogenic stage H 18, a second stage L 90 that was more powerful than the L 35, and the liquid-propellant LB 40 booster to each of the previously mentioned stages. The following thus appeared in rapid succession: the CZ 2E (4 LB 40, L 180, L 90), the CZ 2D (L 180, L 35), the CZ 3A (L 180, L 35, H 18), the CZ 3B (4 LB 40, L 180, L 35, H 18), an extended CZ 2C also called CZ 2C/SD (L 140, L 90 or extended L 35) designed for the constellation market and finally, the CZ 4B which is a development of the CZ 4. A final model, the CZ 3C, should complete the CZ 3 family of geostationary satellite launchers, and further developments are planned for the CZ 2E and 3B in the near future (CZ 2E/A, CZ 3B/A), since the boosters will be stretched to the length of the first stage. The objective for 2003 is to be able to lift 11.8 tonnes into low-Earth orbit using the CZ 2E and 7 tonnes into GTO using the CZ 3B. Finally, the CZ 2F is a 2E that has been modified to cater for manned flights with a very similar fairing to the Russian Soyuz.

China has also announced that it is considering a heavy launcher, Long March X, capable of launching 23 tonnes into low orbit and 11 tonnes into GTO. It also claims to be studying reusable launch technology. The idea they are working on is a semi-reusable vehicle made up of a first stage derived from Long March and a reusable orbiter.

The great diversity of the Long March launch family has arisen from China's desire to penetrate the international market by providing for a wide range of needs. The evolving capacity of the CZ 3B (4.5 tonnes, then 5 tonnes and 5.2 tonnes to GTO) shows that it aims to meet the requirements of the telecommunications satellite market by entering the category of the Atlas and Delta launch vehicles available at the same periods. The development of CZ 2E fits the same pattern, aiming to cater for the increasing mass of telecommunications satellites, especially since the Space Shuttle was out of action following the Challenger accident in 1986 so that opportunities were encouraging. The same can be said of future versions although China has become aware of the great difficulty in penetrating this market.

Out of 47 satellites launched between 1970 and 2000, China launched 27 foreign satellites for Pakistan, Australia, Sweden, the United States, the Philippines and Brazil. In fact, commercial telecommunications satellites manufactured under American licence, which represent the major share of the market, are few and far between. The last was launched in 1998. The Long March family met with technical difficulties in 1995 and 1996, but loss of confidence can only partly explain the decline. Of greater relevance is the American quota policy, and especially the suspension of licences for telecommunications satellite exports which had previously been renewed on a regular basis.

Indian bases and launch vehicles

Officially, India runs a completely civilian space programme. It has undertaken to build up its knowhow in the launch area through a range of international collaborations. The results of this enterprise are clearly demonstrated by the successful launch of the Geostationary SLV in April 2001, the latest in their range of launch vehicles.

India built the Sriharikota space base (SHAR, Sriharikota High Altitude Range) on an island in the Bay of Bengal (fig. 6.45). It had originally been one of India's many sounding rocket launch complexes. The monsoon climate here is characterised by a rather long cloudy season and very heavy rain in October to November. There are also a great many depressions which form in the Bay of Bengal and reach the coast with the all the characteristics of a tropical cyclone.

India developed its SLV 3 (Satellite Launch Vehicle) from sounding rockets and placed its first satellite in orbit in

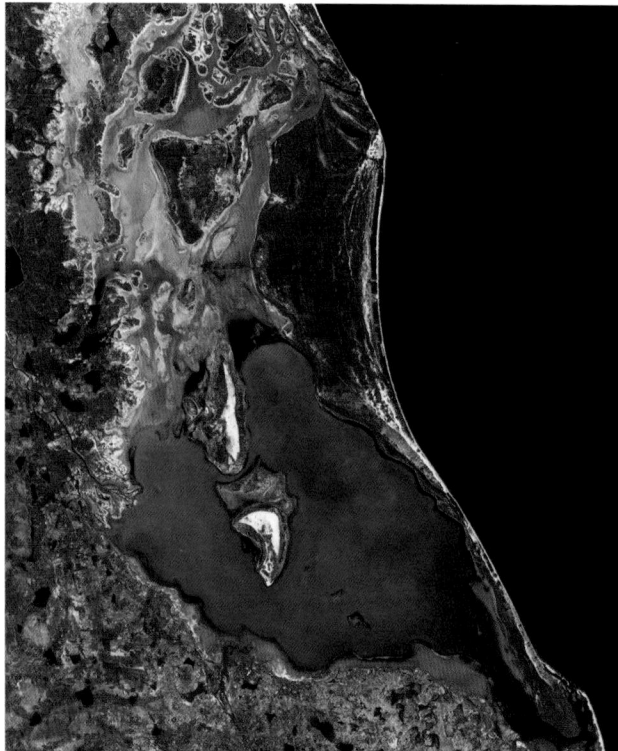

Figure 6.45. **The Sriharikota space centre viewed by SPOT.** SPOT 4 image taken on 30 January 1999. The island of Sriharikota lies between the Pulicat lake on the west, and the Bay of Bengal. The shallow waters of the lake are shown light blue on this false colour image, whilst the deeper waters of the Bay of Bengal are shown a darker blue. The island has a similar natural environment to Cape Canaveral and the Guiana Space Centre, with offshore sandbanks and low dunes covered with vegetation. The situation is ideal due to the low population in the region and its location on the Coromandel coast to the west of a wide stretch of sea. © CNES 1999 Distribution SpotImage.

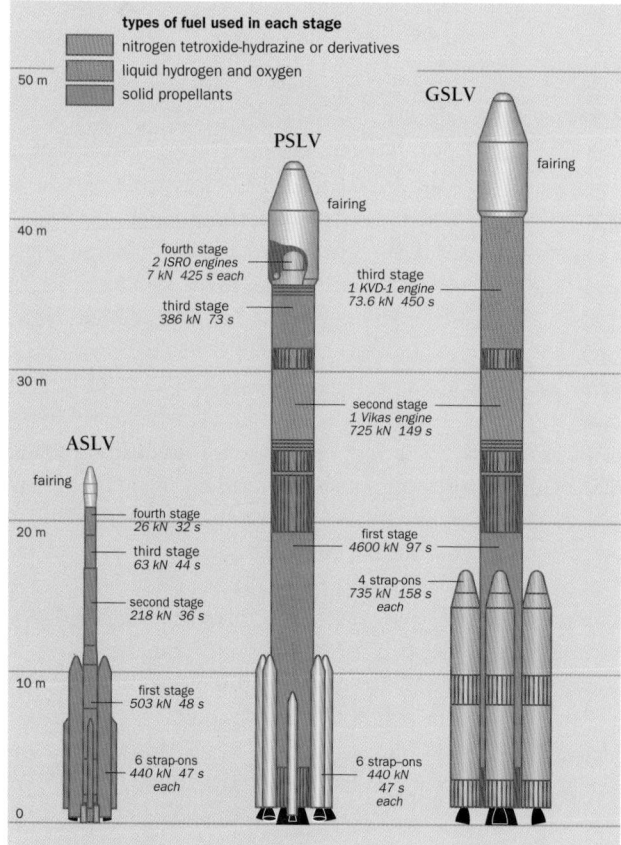

Figure 6.46. **Indian launch vehicles.**

1980. The SLV 3 had four solid-propellant stages and was quite similar to the American Scout rocket. After one failure and three successful launches, it was superseded by the more powerful version ASLV (Advanced SLV). This failed twice in 1987 and 1988 before it succeeded in lifting two satellites to orbit in 1992 and 1994. Today, India possesses the PSLV (Polar SLV), with liquid-propellant second and fourth stages, and since April 2001, the GSLV (Geostationary SLV) developed on the basis of Russian KVD 1 engines from KB Khimmash. The agreement with Khrunichev in 1992 to supply this engine was contested by the United States who considered that it involved a transfer of technology which violated the MTCR. For this reason, an Indian engine was developed with the same characteristics and an amended agreement established for a three-year period.

The PSLV suffered a failure in 1993, but succeeded in placing its first satellite in orbit on 20 September 1994 (fig. 6.46).

India would like to commercialise this product, taking advantage of the low cost of local manufacturing. However, there are several serious drawbacks. It is very heavy, two and a half times the mass of its competitors for the same payload. In addition it is complex to operate and can only be manufactured at a rate of one unit per year.

The GSLV is scheduled to make its second qualifying flight in late 2002. New versions are planned for the end of 2003, using a locally built engine (Cusp), and for 2006, with a Mark III version. Its current capacity of 2 tonnes to GTO must be raised to 4 or 5 tonnes if it is to launch upcoming Indian Insat satellites and if it is ever to be opened to commercialisation. India's launch programmes reveal a gradual move towards autonomy although a wide range of technology from Russia, the United States and Europe has been used. This is clearly visible in the horizontal or vertical assembly techniques used for the various vehicles.

Israel, Brazil and other countries

The Palmahim space base south of Tel Aviv (31.52° north) has served to launch three Offeq satellites. However, in order to avoid overflying the Arab states on its eastern border, launches have to be made towards the west. The national launch vehicle Shavit (Comet) has three solid-propellant stages and derives from the Jericho 2 missile, which itself used the same technology as the md 660 missile from the French company Marcel Dassault. Israeli knowhow also arises from cooperation with the United States (fig. 6.47). Although it can guarantee Israel's launch autonomy, Shavit's limited capacity, combined with the handicap of firing towards the west, will only allow it to loft payloads of a few hundred kilograms. Its manufacturer, Israel Aircraft Industries, proposed it as a micro launch vehicle for NASA, but Pegasus was selected instead. A stretched commercial version of Shavit called NEXT was also put forward for exploitation in the United States and was even retained by NASA for the SELV programme provided that a joint venture company could be set up.

The French-based company Leolink created by Astrium and Israel Aircraft Industries. offers three small launch vehicles derived from NEXT: LK A will be wholly manufactured in Israel, whilst LK 1 and LK 2 will be built using components from Thiokol (the first three stages), from Astrium (the fourth stage) and from IAI itself. The company

may be joined by a subsidiary of the American company L-3 Communications, Coleman Aerospace, which would manufacture and market the LK launch vehicles on the US market. LK A launched Ofeq 5 from Palmahim in May 2002 and the first commercial flight is planned from Alcantara in Brazil.

Brazilian projects

Brazil is preparing to enter the select club of satellite-launching countries, building on its long experience of sounding rockets. The Alcantara launch complex was inaugurated in 1990. It lies almost on the equator (2.17° south) and conditions are ideal for launching and tracking geostationary satellites. However, the VLS launch vehicle developed by the Instituto de Actividades Espaciais (IAE) is a lightweight launcher with four solid-propellant stages (fig. 6.47) which could not lift satellites to geostationary orbit. In addition, it has failed on two occasions, in November 1997 and in 1999. For this reason, Brazil has offered its future space base for international cooperation. Such a move should provide some return on investment as well as improving its image as a space power.

Many countries have shown interest. An example is the Ukraine which signed a partnership agreement to launch its Tsyklon rocket in November 1999. Another is Israel which hopes to use facilities at the base for its LK launch vehicle,

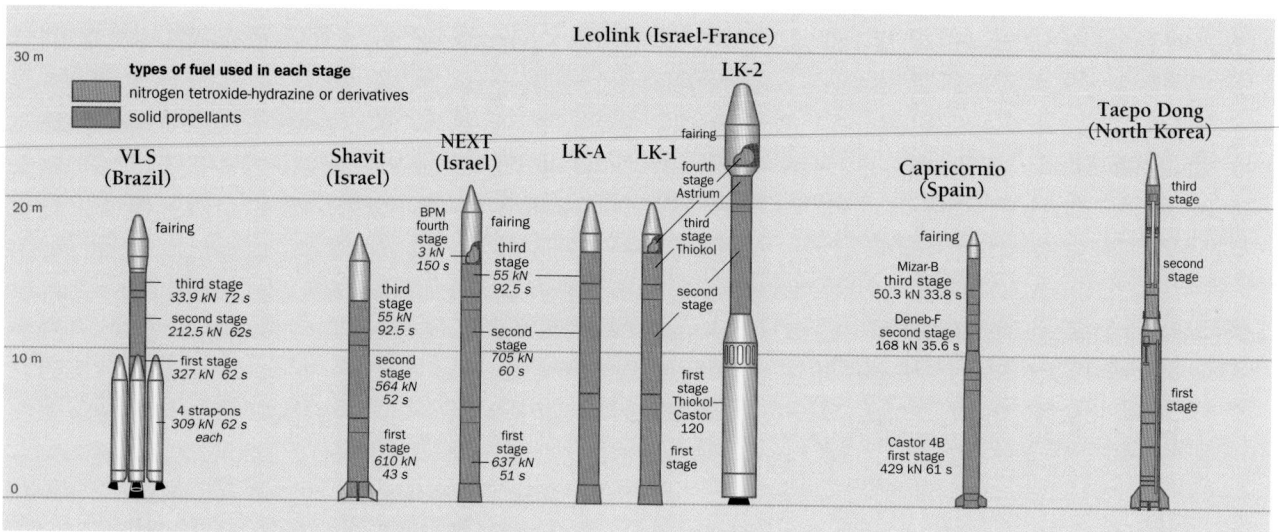

Figure 6.47. **Israeli, Brazilian and other launch vehicles in use or under development in 2000.** The micro launch vehicle Capricornio is a project from the Instituto Nacional de Técnica Aeroespacial, in association with the Span-ish defence ministry. It aims to launch scientific and military payloads with mass less than 150 kg from Hierro island.

not to mention both China and Russia. Brazil's desire to acquire rocket technology could facilitate partnership agreements but must be weighed up against its privileged relationship with the United States, which remains the main source of cooperation. The signing of the Technical Safeguards Agreement in April 2000 should make it easier to exploit the base commercially by allowing American companies to take part in its development. However, this does not mean that it authorises the transfer of launch vehicle technology. Investments made during 2001 attest to a revitalisation of the base.

New space powers

The relationship between missiles and launch vehicles is particularly well illustrated by the cases of Iraq and North Korea. Their respective tests in December 1989 and August 1998 arose from political and strategic motivations and were perceived by Western countries as a threat to international security. In neither case was a satellite observed by Western or Russian stations, even though it was claimed that satellites had been launched. Both launch vehicles derive from Scud missiles (fig. 6.47). Iraq's Tamouz was probably composed of four Scud missiles coupled side by side for the first stage and a single Scud for the second stage. The case of North Korea's Taepo Dong 1 is more controversial insofar as it may be made up of two or three stages. The first stage was probably propelled by a cluster of four Scud C engines and the second by a single Scud C whilst the whole thing seems to have been topped by a solid-propellant third stage. The situation prevailing in Iraq today may mean that this type of programme lies dormant, but the North Korean case is more complex. In terms of international diplomacy, there may have been an advantage in announcing a rocket launch rather than a missile test, for the purpose of calming international reactions and giving an impression of technological competence. At the same time, the programme could actually give rise to a primitive launch vehicle. The Taepo Dong seems in many ways similar to Jupiter C.

The Musudan-ri facility used to launch Kwangmyong-son 1, according to official claims, is located in the north-east of the country in the north of Hamgyong Province on the coast of the Sea of Japan.

CHAPTER SEVEN

Circumterrestrial scientific missions

- SCIENTIFIC RESEARCH

- STUDY OF THE EARTH

- OBSERVATION OF THE CIRCUMTERRESTRIAL ENVIRONMENT

- ASTRONOMICAL OBSERVATION

- OTHER FIELDS OF INVESTIGATION

Scientific research

In 1950 the world's scientists returned to the idea of the International Polar Year which had focused the efforts of several nations on the study of the poles in 1882, and again fifty years later in 1932. The aim this time was to organise a further international year during a phase of heightened solar activity predicted for 1957–58. The International Council of Scientific Unions decided to extend the project to the whole of the Earth within the framework of an International Geophysical Year (IGY), lasting from 1 July 1957 to 31 December 1958.

During a meeting in Rome in 1954, the organising committee recommended that artificial satellites should be launched in the context of the IGY. Scientists thus initiated and encouraged a project which seemed to be technically feasible according to those units of the armed services involved in missile development. Indeed, the latter had long been trying to persuade decision-makers that their missiles had potential for space conquest. Moreover, the symbolic significance of getting the first artificial satellite into Earth orbit was already recognised. In the United States, a study by the Rand Corporation, in its role as a military think tank, had already stressed the psychological impact of such an undertaking in 1950. In 1952, the scientist A.V. Grosse, known for his participation in the Manhattan Project which developed the first atomic bomb, had submitted a report to President Truman emphasising the various military and scientific advantages of such an endeavour, not to mention its prestige value. In the Soviet Union, S. Korolev and M. Keldysh, president of the Academy of Sciences, had also tried to convince J. Stalin that a rocket programme could be combined with the development of intercontinental missiles. They were unsuccessful, since there was no immediate strategic advantage in launching a satellite. N. Khrushchev, on the other hand, proved to be more open to the prestige arguments and a special committee for space development was set up in 1954 within the Academy of Sciences.

The proposal made by the Special Committee for the International Geophysical Year revived the idea, with results this time. In July 1955, the American and Soviet governments each announced, within a day of one another, that they intended to launch a satellite as part of the IGY. Their motivations were rather different. N. Khrushchev took it as a means of proclaiming on the international level the Soviet Union's accomplishments in the field of intercontinental rockets, whereas the Americans, who had already considered the possibility of developing satellite programmes for military reconnaissance, decided to take advantage of the IGY scientific satellite by creating a legal precedent for freedom of movement in space.

Even though strategic considerations were placed at the fore in opening up the conquest of space, scientific satellites nevertheless played a pioneering role in exploiting it further. They were aimed at a better understanding of the Earth environment, satisfying the requirements of fundamental research whilst at the same time contributing to the advancement of both civilian and military space technology. Their primary mission was naturally to investigate their immediate surroundings, including the atmosphere, ionosphere, magnetosphere, auroral zones and micrometeorites. Effective environmental characterisation was an essential step towards developing future generations of satellites. Later, they would often be used at the inception of national space programmes, provided that costs could be kept to a reasonable level and they could be launched without the need for heavy launch facilities. Moreover, international programmes make it possible for countries with sufficient scientific potential to cooperate over space research (fig. 7.1).

The main space agencies provide the framework for almost all undertakings in space science, bringing together the relevant scientific communities to define long term programmes. One such is the Earth Observing Enterprise of NASA's Earth Observing System (EOS) which involves satellites like Terra, EO 1 and Landsat 7. Others are Cornerstones or Earth Explorer for the ESA, and the interagency programme IASTP (Inter Agency Solar Terrestrial Physics). They support purely scientific missions by ongoing technology programmes (e.g., the Earth Science Technology Programme, the New Millennium Programme (NMP) for NASA, SMART for ESA), and design and finance platforms like Explorer, SMall EXplorer (SMEX), Medium Explorer (MIDEX), managed by NASA, or the minisatellites PROTEUS (Plateforme Reconfigurable pour l'Observation, les TElécommunications et les Usages Scientifiques), managed by the CNES (Centre National d'Études Spatiales, France). They also handle satellite tracking and command as well as data transfer, using TTC networks (Tracking, Telemetry and Command), such as ESA's Estrack network or ISRO's Istrac network. NASA is the only agency, apart from the Russian agency RAKA, to a much lesser degree, to have set up a network for distant satellites and probes. This is the Deep Space Network (DSN). It comprises three steerable, high-gain, parabolic reflector antennas located in Goldstone (California), Madrid (Spain) and Canberra (Australia). Apart from sending out commands and receiving telemetric data, the network can pinpoint probes and identify their trajectories, make VLBI observations, and carry out radio science experiments. Probes and satellites in far-ranging orbits can also be tracked by the Russian TsDUC

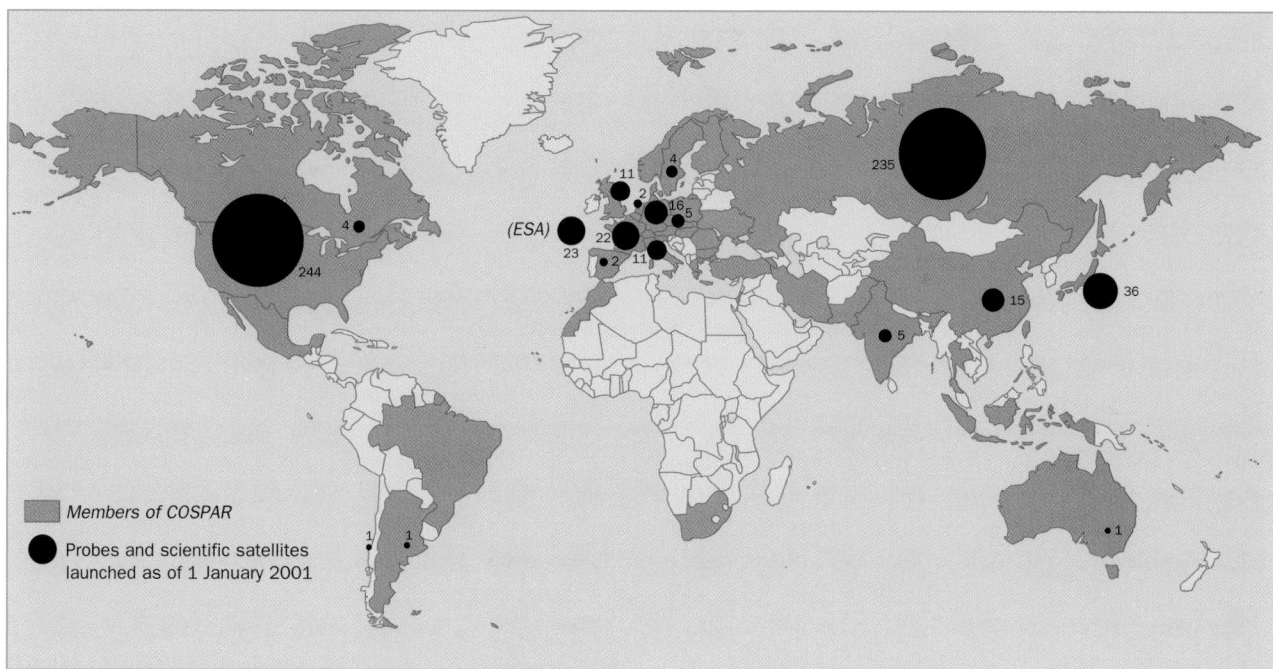

Figure 7.1. **Countries involved in space-based research as of 1 January 2002.**
Iraq has been suspended as a member of COSPAR (Committee of SPAce Research) and has not been shown.

Long Range Space Communications System, which is centred at Yevpatoria in the Ukraine and connected into a network with five other Earth stations (Medvezhi Ozera, Pushino, Simiez, Ulan Ude, and Ussurisk).

Branches of space science

Scientific research can benefit in many ways from access to space. It is thereby freed from conditions imposed on Earth-based observers before satellites came into being, although attempts had been made with rocket probes and balloons to escape from the various terrestrial constraints in a very limited number of areas, such as aeronomy. Many constraints can be lifted or at least attenuated by going into space: gravity, opaqueness of the atmosphere to certain bands of the electromagnetic spectrum, the impossibility of making in situ or close-range studies, and the necessarily fragmented and non-global character of studies carried out from the Earth's surface.

In the general context of studies of the Earth, space geodesy has opened a wide field of research in geodynamics through the measurement of lithospheric deformations. It is also used as a reference when determining sea levels or studying continental topography. It provides one of the best examples of a science with two-fold interest, both on a fundamental level and in immediate civilian and military applications.

The study of the Earth and its environment has long represented an important aspect of the space sciences, although perhaps not the most spectacular amongst them. Moreover, it is not always easy to separate exploratory missions aimed at research from observational missions directed towards applications, in the way attempted by the European Space Agency. Related to the first are investigations into the dynamics and physicochemistry of the stratosphere and the mesosphere. These are concerned with assessing the depletion of the ozone layer, or establishing the energy balance in the atmosphere on a regional or global scale, combined with simultaneous efforts to understand the mechanisms behind the evolution of the biosphere. The second type of mission is mainly concerned with understanding atmospheric and oceanographic phenomena, together with remote sensing of terrestrial resources (see Chapter 9). These missions have recently come to the fore with a view to making a global model of the Earth as a system and explaining phenomena of global change which jointly affect the atmosphere, lithosphere, cryosphere, hydrosphere and biosphere. They are the subject of large-scale programmes such as Living Planet run by ESA, or Earth Observing System (EOS) coordinated by NASA.

Exploration of the Solar System has also benefited from removal of the atmospheric screen. But it is by reducing distances and making observations that were previously impossible for geometrical reasons that space probes have proved themselves such a valuable tool. They have placed landers on the Moon, Mars and Venus, thereby providing a basis for comparative planetology, a rich resource for the history of our own planet as well as that of the Solar System. Space probes have already observed the Sun from nearby and made close approaches to several comets and asteroids, as well as all the giant planets, witnesses to the presolar nebula (see Chapter 8).

Stellar astronomy has been transformed by space-based observation, despite the high standards of generations of astronomers who carried out their observations over the ages, and the considerable advances made by ground-based observatories during the last century. Apart from astrometry, it is the new branches of astrophysics which have been opened up bypassing the atmospheric filter in the millimetre, submillimetre, infrared and ultraviolet wavelengths, as well as in the high energy region of X-rays and γ-rays, which mark the greatest progress towards improved understanding of the distant Universe.

Investigations in gravitational physics, a newly explored area of fundamental physics, aim to test the equivalence of gravitational and inertial mass, and also the theory of general relativity, and to detect low frequency gravitational waves. It is essential to carry out experiments in space, since it provides weightless conditions, large distances for interferometry, and significant variations in gravitational potential.

The possibility of experimenting in weightless conditions has had a significant impact on research in the life sciences, using plant, animal and human materials to observe the effect of gravity on the organisation and functioning of living beings and on the ability of organisms to adapt to weightlessness. Current studies are concerned with cell metabolism and mechanisms for perceiving gravity on the cellular level, in order to explain, among other things, the role played by weight in plant gravitropism. This research in physiology is of particular importance in preparing astronauts for life in space. Much work is being carried out on the regulation of cardiovascular fluids, the evolution of bone tissues and muscle fibres in weightless conditions, neurosensorial adaptation mechanisms, and also on the biology of early embryonic development. Furthermore, the effects of cosmic radiation on living cells must be studied as part of the preparation for future manned flights. Research into weightlessness has been a particular area of interest on board satellites with recoverable capsules. Examples are the Russian satellites Foton and Biokosmos, followed by Bion, or the Chinese FSW satellites. It has also been a feature on board American and Russian space stations and shuttles, often involving international cooperation.

Microgravity conditions open the way to new experimental possibilities in condensed matter physics. Research carried out in the science of materials, notably concerning crystal growth in weightless conditions, may well have industrial spinoffs.

Study of the Earth

Space geodesy

The purpose of geodesy is to determine the shape of the Earth and geographical variations in its gravitational field. It serves as a basis for our geometrical knowledge of the Earth, itself the foundation for any rigorous approach to cartography, whether it be for civilian or military purposes. Through study of gravitational fields, it also provides up-to-date knowledge of the matter distribution within the Earth, that is, in the crust, mantle and core. Before the space age, geodesy was confined to distances and angles measured from the ground to other points on the ground or to the celestial bodies. The first satellites were used as a means of viewing the Earth from the outside, thus providing distance measurements between base points that were not derived from ground measurements. The first experiments consisted in using the star background behind the satellite to calculate simultaneous sighting angles from several ground stations. By such a process, it became possible to link up networks separated by sea or ocean. This was the idea used to establish a link between France and Algeria in 1963, thanks to the Echo 1 satellite. A worldwide geodetic network was thereby set up by the US Coast and Geodetic Survey with a precision of 20 to 40 m. Sightings used the American satellite PAGEOS (Passive Geodetic Earth Orbiting Satellite), a highly reflecting inflatable sphere, 41 m in diameter.

Another approach to space geodesy involves the positioning of stations through the Doppler–Fizeau effect. To this end, satellite radial velocities are measured along the line of sight. Theoretically, it only requires a continuous emission at some fixed wavelength. However, two distinct wavelengths are generally used in order to eliminate perturbations due to ionospheric transmissions. This method, the most economical, has been applied since 1962 on satellites in the Transit network, then on French satellites Diapason and Diadème, and quite generally on all navigation satellites, with precision of the order of 50 cm to 1 m. This precision is greater on the Navstar satellites of the GPS system, and even more so with the French system DORIS. DORIS (Doppler Orbitography and Radiopositioning Integrated by Satellite) uses a world network of 50 so-called orbitography beacons with accurately known coordinates. It can position to within 10 cm in the three dimensions and can measure the ground distance between two positioning beacons with a precision of 1 cm over a measured length of 100 km. It has equipped the SPOT satellites since SPOT 2, Helios 1 A and B, and the Franco-American satellite TOPEX–POSEIDON, and its successor JASON and ENVISAT. The German PRARE (Precise Range And Rate Equipment) system carried on board ERS 1 would have given more precise results but it did not function correctly. However, an improved PRARE has been operating on ERS 2 since 1995.

Since 1964, space geodesy has also used laser ranging. Satellites such as Starlette and Lageos, and more recently EGP, Étalon 1 and 2, Lageos 2, Stella, GFZ 1, are equipped with catadioptric reflectors. These return a part of the emitted beam back towards a laser emitting station. Other satellites not specialised in geodesy, such as ERS 1, fulfill the same role. Catadioptric reflectors were also placed on the Moon during the Apollo 11 and Lunakhod missions. By measuring the time for a return signal, stations can currently be positioned with an accuracy of the order of 2 to 5 cm. However, ground stations are costly and the method is reserved for high-precision measurements.

Radar altimetry, first introduced with Skylab in 1973, followed by GEOS 3 in 1974, is of interest to geodesists because it allows measurements to be made above the oceans. A radar on board a satellite emits pulses vertically and at a high rate towards the ocean surface. The time taken for the return trip yields the distance to the satellite several times per second. Precise knowledge of the satellite trajectory can then be used to deduce the position of the ocean surface along the orbit (fig. 7.2). The main altimetry missions have been carried out by Seasat, Geosat, ERS 1, ERS 2, TOPEX–POSEIDON and Geosat Follow-on. They are pursued by the first Proteus mission of the CNES (France), Jason, by polar satellites ENVISAT (Europe)

and in the future Japanese ALOS (Advanced Land Observation Satellite), as well as by satellites in the US programme EOS.

Very Long Baseline Interferometry (VLBI) is used in radioastronomy to locate celestial radio sources, in particular quasars. It determines the direction of the source by comparing the phases of the same emission observed at several bases. Its main applications in the space domain are to satellite-based radioastronomy and also to planetology via space probes. By reversing the operating principle, that is, considering the relative position of radio sources observed by two ground bases to be known, the distance between these bases can be calculated to great accuracy (within a few centimetres), even when they are very widely separated.

Dynamical geodesy is concerned with perturbations to satellite orbits, the aim being to describe dynamical forces, and in particular the gravitational field at the surface of the Earth. A very detailed analysis of satellite trajectories is required. For example, they can be significantly perturbed by radiation pressure, particularly as the first geodesy satellites were very large inflatable spheres. These trajectories are also affected by tides on Earth and lunar gravitational attraction. However, dynamical geodesy aims to study variations in the Earth's gravity (fig. 7.3). Satellite trajectories can be followed very closely when a large number of precise measurements are available to relate them to ground stations. This explains why many programmes designed to acquire such data are referred to as trajectography programmes. In practice, computations are complicated by the need to account at the same time for the satellite equations of motion, in which gravitational force parameters are unknowns, and ground station coordinates, in which geocentric position parameters constitute another series of unknowns.

If the weight (combining gravity and centrifugal force) is known at each point, thereby defining a normal to the local

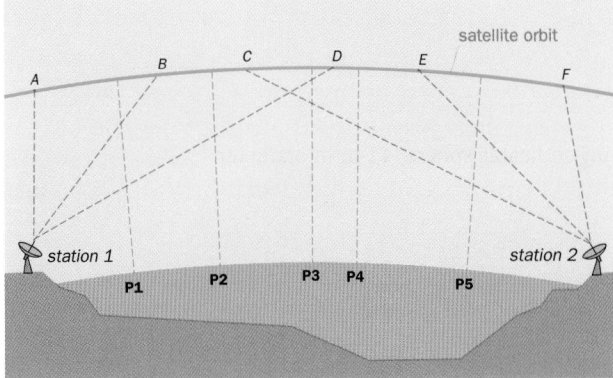

Figure 7.2. **Use of altimetry in geodesy.** The satellite trajectory is established relative to stations 1 and 2 by measurements towards A, B, C, D, E, F, and so on. Distances between points P1, P2, P3, P4, P5 and the trajectory are found using the satellite altimeter so that these points can be situated in the geodetic system to which the stations belong. Before the space age, it was impossible to include the ocean surface in geodetic networks, but it is in these regions that measurements are now the most common and the most precise.

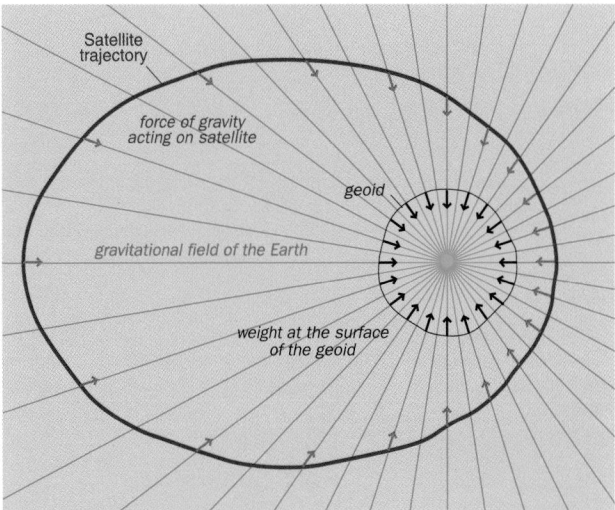

Figure 7.3. **Dynamical geodesy.** Blue lines show the Earth's gravitational field. By observing the satellite trajectory, the value of this field can be calculated along the orbit (red arrows). It is then possible to deduce weight values at the surface of the Earth (black arrows). The geoid is the equal weight surface coinciding with the mean sea level.

horizontal, gravitational equipotential surfaces can be constructed. The equipotential surface corresponding to mean sea level, called the reference equipotential, is the geoid.

In the construction of geodesy networks, geoid models replace the convenient but approximate ellipsoids of reference that were used before the space age. The setting up of a precise geoid is a long term undertaking, involving millions of observations. It is always susceptible to global refinements made by ground-based laser, radio sounding, altimetric or gravimetric observations, and local refinements resulting from concentrated series of altimetric measurements above the oceans. There are several geoid models. Global models are based upon theories in which the potential is expanded in spherical harmonics. The best known are the so-called Joint Gravity Model (JGM) established in 1996 by NASA and the University of Texas, and the GRIM 5 model established in 1999 by the Groupe de Recherche de Géodésie Spatiale in France and the Geoforschungszentrum in Germany (fig. 7.4).

More accurate determination of weight at different locations could only be obtained through new methods for direct measurement of gravity. One such technique consists in measuring to great accuracy the difference in radial velocity of two satellites moving very close to one another along the same low orbit. Relative differences in the gravitational field could then be assessed. The American project GRM (Geopotential Research Mission) was intended to provide an extremely accurate geoid by this approach, and thereby improve possibilities for exploiting oceanographic measurements, but it has since been abandoned. In March 2002, the US–German GRACE project (GRAvity recovery and Climate Experiment), which should measure the gravitational field and its long wavelength variations was launched. Satellites will be linked together by microwaves and move along the same orbit at a distance of 170–270 km from one another. It will also be able to evaluate the ground water content, a basic piece of data for understanding exchanges with the atmosphere.

Another method known as gradiometry uses two or more ultra-sensitive accelerometers placed within the same satellite to measure the gravity gradient between them. A micro-accelerometer (Cactus) has already functioned for three years on board the French satellite Castor. Measurements were designed to investigate the upper atmosphere and solar radiation pressure. Application to geodesy was the subject of ESA's Aristoteles project, abandoned in 1992. It has been replaced by the GOCE project (Gravity field and steady state Ocean Circulation Explorer), one of the four core missions of ESA's Earth Explorer programme, retained as a priority. GOCE uses a three-axis gravity gradiometer to measure the gravitational field to great accuracy (1 mGal) and the local level of the geoid to the nearest centimetre. Applications range from geodynamics and tectonics to oceanography and glaciology.

Military interest in accurate knowledge of the geoid and gravitational fields is clear, both for rigorous cartography and fine adjustment of missile launches, as well as for the return of de-orbited space capsules. It is no surprise that the very first true geodesy satellites were launched by the US Department of Defense, viz., Anna 1B, launched in 1962 following

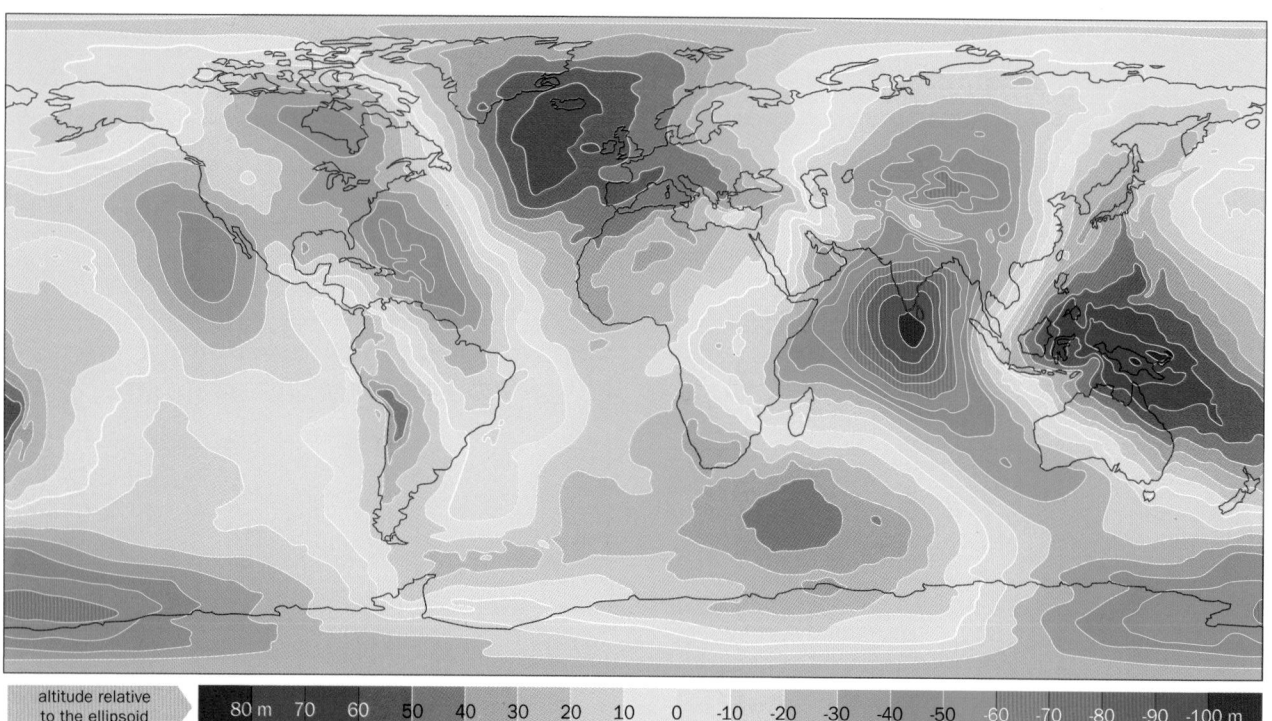

altitude relative to the ellipsoid   80 m 70 60 50 40 30 20 10 0 -10 -20 -30 -40 -50 -60 -70 -80 -90 -100 m

Figure 7.4. **The GRIM 5-C1 model for the Earth's gravitational potential (1999).** Curves represent the difference in altitude of the geoid with respect to an ellipsoid of reference (semi-major axis 6 378 136 m, oblateness 1/298.25781). The Antarctic, Turkestan and especially the region to the south of Sri Lanka lie below the surface of the ellipsoid, whilst the regions around Iceland and New Guinea lie above it.

the failure of Anna 1A, with the first optical geodesy beacons. Other military satellites were to come, including the GGSE series (Gravity Gradient Stabilization Experiment), SECOR (SEquential COllation of Range, also called Experimental Geodetic Research Satellite) and GEOS (Geodetic Earth Orbiting Satellite) which belongs to the Explorer programme. Launched by the US Navy in 1985, Geosat led to progress in positioning and a significant improvement in the precision of submarine missile launches. Its successor, Geosat Follow-on, was launched in February 1998.

The Soviets have also built up considerable knowhow in the field of geodesy for military purposes. Their first series of geodesy satellites, Sfera, appeared in 1968. A second generation, known as Musson and considered as dual purpose (i.e., both civilian and military), continued along the same lines from 1981 with the launch of one Kosmos per year. This rate declined to one every three years after the collapse of the Soviet Union, with the launch of Geo IK 1 in 1994 and Zeya in 1997, the first satellite to be launched from Svobodny (figs. 7.5 and 7.6).

Tectonics and internal geodynamics

In tectonics, precise distance measurements can be used to detect the very slight motions of the tectonic plates, fault systems and landslides. Applications to the study and prediction of earthquakes and volcanic eruptions require a dense network of beacons whose positions can be identified by GPS. Another measurement technique has turned out to be more effective, being more continuous and homogeneous. This is radar interferometry, whereby interferograms are constructed from radar images taken by operational satellites or from images stored in the archives. This method should be refined when SAR (Synthetic Aperture Radar) is used in the L band, since the latter is less sensitive to interference by vegetation. Another line of research might exploit possible perturbations in magnetic and electric fields several hours before earthquakes or eruptions. (Anomalous signals were observed in the readout of the Soviet satellite Aureol 3 (Arcad) when it flew close by a seismically active zone prior to a tremor.) This will be the mission of the CNES microsatellite project Demeter.

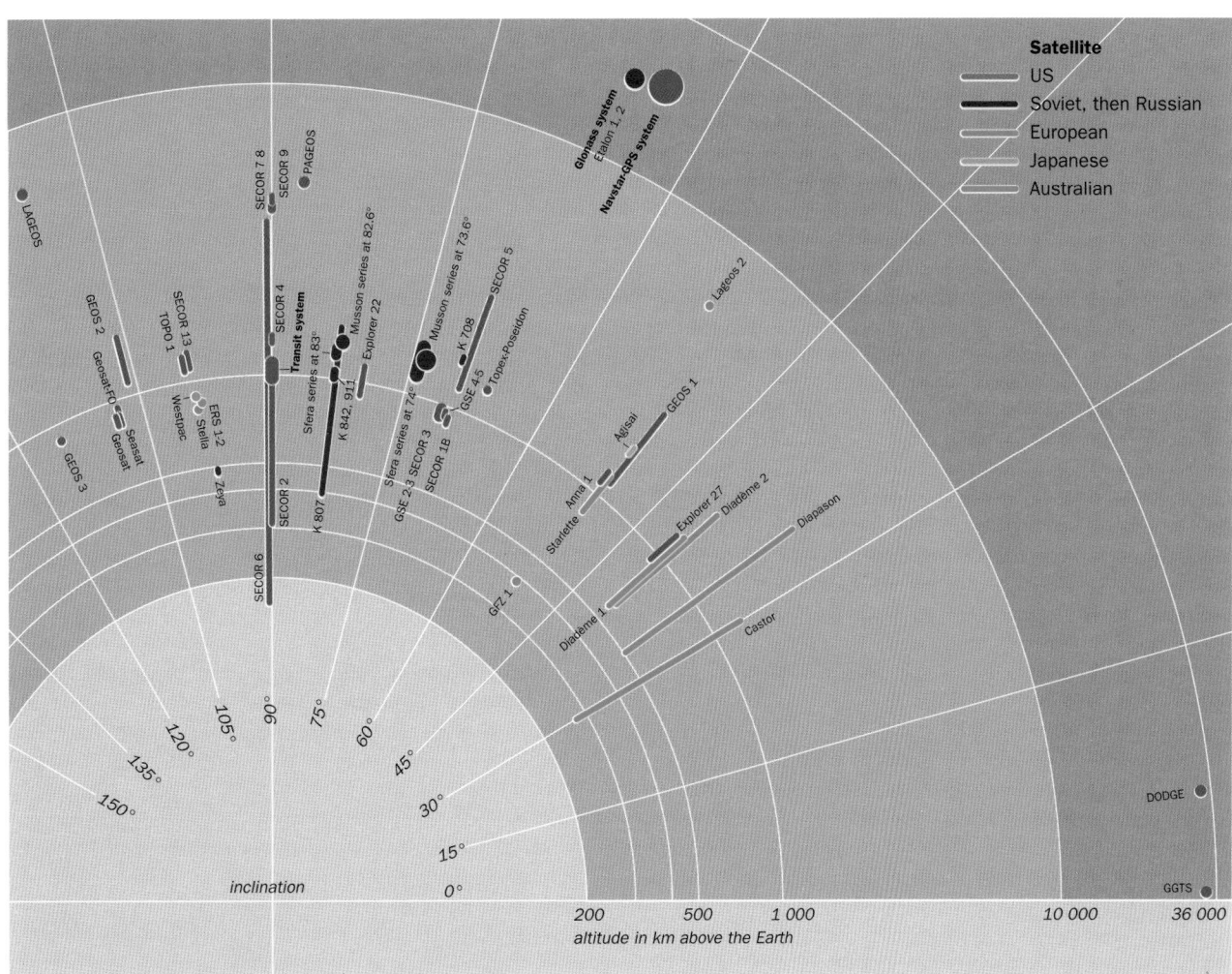

Figure 7.5. Inclination, apogee and perigee of geodesy satellites. Geodesy satellites generally have perigees above 700 km in order to minimise atmospheric braking. Orbits with eccentricities of order 0.2 are preferred for their favourable perigee motion. Apart from these constraints, a wide range of inclinations is used in order to cover the Earth with the most complete series of measurements possible, involving a predominance of high inclinations and even some polar orbits.

In geodynamics, refined and detailed knowledge of the Earth's combined gravitational and centrifugal force field serves to study slow or deep motions of the planet. These include rising land masses and variations in sea level, subduction of oceanic plates, and convection cells in the mantle. There is little doubt that it will also throw light upon the history of past continental motions. In addition, DORIS has revealed vertical movements of ground stations arising from internal geodynamic phenomena. Its results will be useful in quantifying deformations of the geoid. Furthermore, by observing displacements of Earth stations in a Galilean frame, either through distance measurements to a satellite in very high orbit, or through radio interferometry, it becomes possible to measure fluctuations (precession and nutation) in the Earth's axis of rotation and angular velocity. These reflect, in particular, changes in the Earth's internal density and the elasticity of its various layers.

Another range of observable phenomena related to the internal structure of the Earth concerns the magnetic field and its (secular) time variations. NASA's Magsat satellite, in operation from 1979 to 1981, produced the first complete instantaneous survey of the Earth's magnetic field. In addition, by recording its variations, maps could be constructed to show the fluid circulation at the surface of the Earth's core

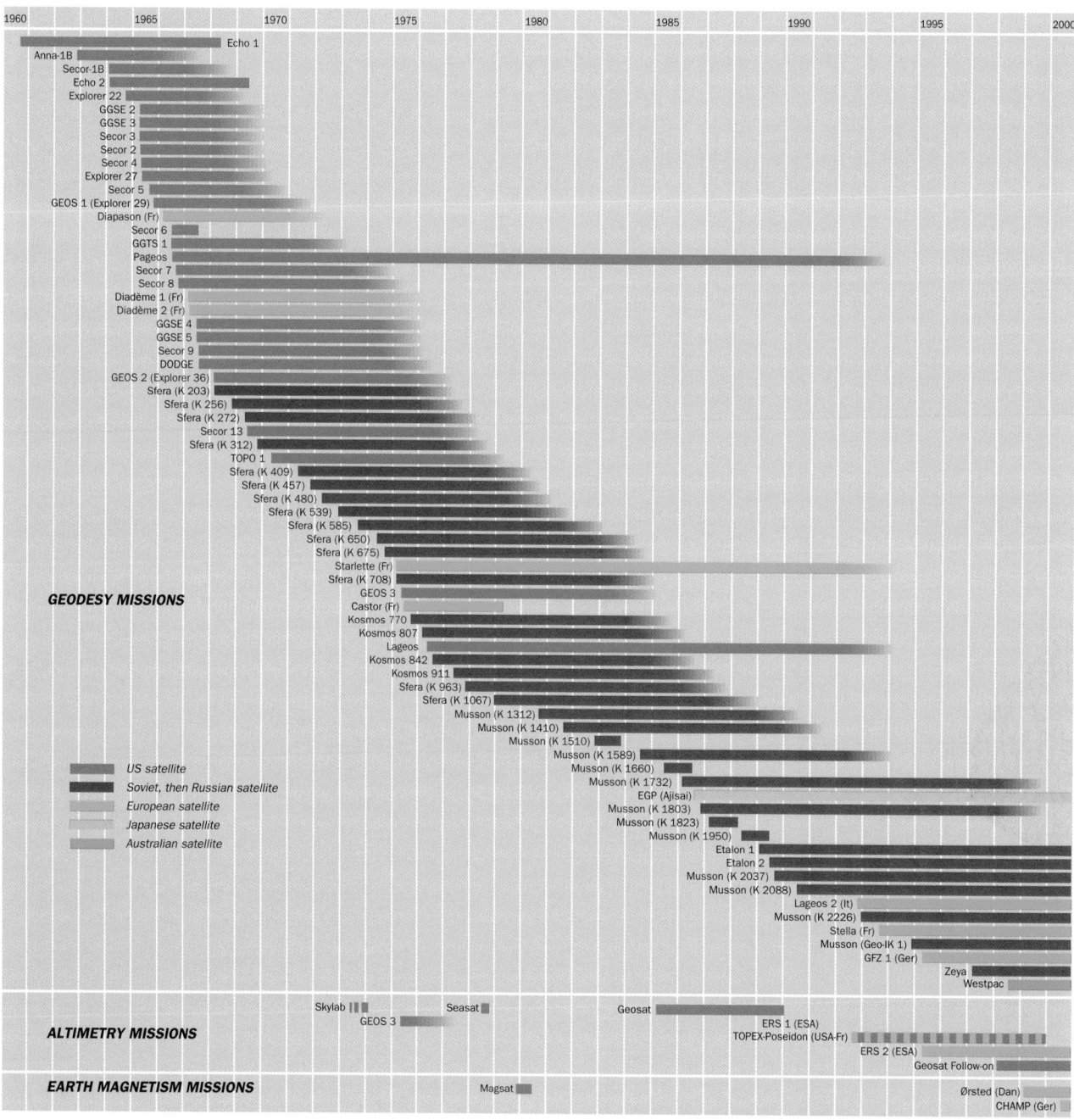

Figure 7.6. **Chronology of geophysical satellites.** Many satellites not specifically used for geodesy do carry out some geodetic missions and are not shown in this chronology. However, it does include the two passive telecommunications satellites Echo 1 and Echo 2 for which geodetic applications were of prime importance.

(fig. 7.7). The Danish satellite Oersted, launched in February 1999, is the first component of a permanent observatory for terrestrial magnetism. In July 2000, the CHAMP programme (CHAllenging Minisatellite Payload) for geoscientific research and applications carried two magnetometers and an accelerometer for mapping the Earth's global long to medium wavelength gravity field and temporal variations. Similar data is provided by one of the instruments on board the Argentinian scientific and Earth observation satellite SAC-C, launched in November 2000.

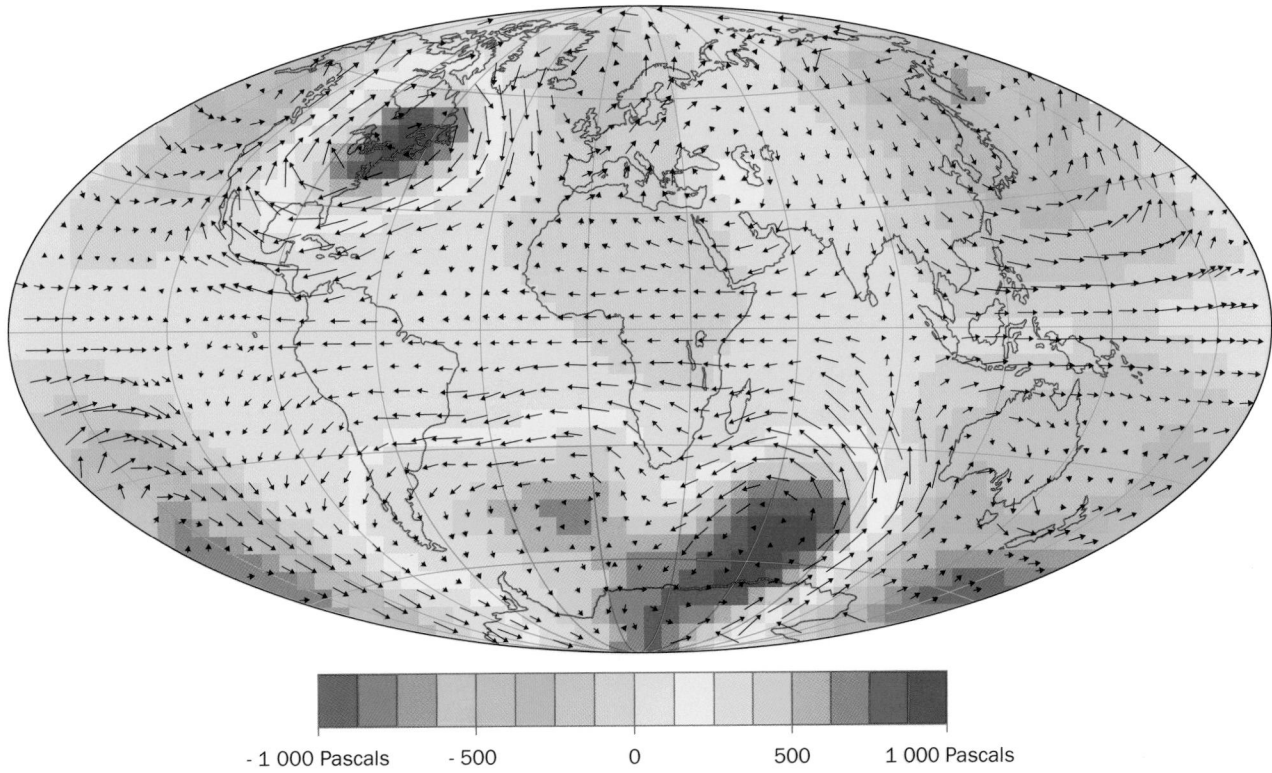

- 1 000 Pascals   - 500   0   500   1 000 Pascals

Figure 7.7. **Fluid circulation at the surface of the Earth's core.** The map shows pressure variations (in colours) and geostrophic circulation (arrows). It was put together from instantaneous data concerning the magnetic field and its secular variations as collected by Magsat. It is assumed that motions at the core surface are geostrophic, i.e., determined by the Coriolis force. Courtesy of CNES.

Observation of the circumterrestrial environment

Phenomena on the Earth's surface (cryosphere, hydrosphere, lithosphere and biosphere) are very closely linked to atmospheric phenomena. In the past, observations tended to be made separately, but today the trend is towards a more global model of the Earth as a system, controlled by the Earth's radiation budget. The magnetosphere, ionosphere and upper atmosphere are studied as a whole in order to understand the large-scale flow of plasmas and energy transfers at all altitudes and latitudes.

In order to model the Earth as a system, the radiation budget must be established at the top of the atmosphere, and combined with long periods of data concerning the field of precipitations and atmospheric flow. This data must be acquired throughout the day and in every season of the year. All types of surface must be investigated, including ice, water, and land surfaces covered with vegetation or otherwise, and all types of phenomenon must be observed, including volcanic eruptions, fires and the like. Finally, each

Figure 7.8. **MODIS image.** This image, taken on 28 February 2000 by the Terra satellite, covers Egypt, the Sudan, Israel, Jordan and Saudi Arabia. Across the middle of the desert, the Nile appears as a green ribbon which eventually fans out in the delta. The link between the Red Sea and the Mediterranean provided by the Suez canal is clearly visible. Scientific satellites with environmental objectives often gather images of the Earth. Data analysis must synthesise imaging data and other measurements. Courtesy of NASA.

In addition, the measurements used to make these models require more and more specialised equipment, designed specifically for surface phenomena, cloud cover, aerosols, the evolution of the ozone layer, atmospheric circulation, wind activity at different altitudes, or evaluation of the radiation budget. These instruments may be carried as passengers on satellites whose primary mission is quite different (in particular, remote-sensing or meteorology satellites, but sometimes also military satellites). They may be carried by dedicated satellites (e.g., TOMS, Quickscat, ACRIMSat), with the advantage that the mission profile, in terms of orbital characteristics, bus subsystems, transmission and so on, is optimised for the instrument in question. Finally, they may be found in varying numbers on board large satellites where they operate in synergy to collect synchronised data concerning disparate but complementary phenomena (e.g., Terra, Envisat).

These systems serve to follow climatic change and the evolution of greenhouse gases in the context of the United Nations Framework Convention on Climate Change (UNFCCC) adopted in 1992 and implemented in 1994. The Third Conference of the Parties to the Convention adopted the Kyoto Protocol in December 1997. This aims to stabilise the concentration of greenhouse gases. The use of satellites for continuous global observation of the land surface, and ocean and atmospheric movements is clearly essential for sustainable management of planet Earth. The American Earth Enterprise programme and ESA's Living Planet programme attempt to fulfill these requirements, among other things.

The European programme entitled Global Monitoring for Environment and Security (GMES) was initiated in 2000. It aims to synthesise data acquired by scientific satellites, often with very tightly constrained targets, and meteorological and Earth observation satellites. This type of programme, not yet precisely specified, reflects a growing awareness of the need to define and pursue complementary actions on a global scale.

Earth's surface and interface with the atmosphere

Atmospheric physics and land surface studies bring together Earth observation by remote sensing and scientific programmes with the vast ambition of modelling the way planet Earth works as a system. The boundary between the two areas of investigation is becoming ever harder to define. Such models also incorporate a great many measurements obtained by non-spaceborne techniques, e.g., balloon, sounding rocket, and ground-based measurements.

Ground cover and changes in land use play a major role in general models of the Earth system. This explains why so many remote sensing systems, like the SPOT 4 vegetation passenger (see fig. 9.43) and POLDER described in Chapter 9, are used to observe biomass changes in forest ecosystems and wetlands in order to assess general levels of carbon dioxide. Landsat's TM sensor or SPOT's HRV sensor provide the higher resolutions required to observe regions of methane (CH_4) and nitrous oxide (N_2O) emissions. This research on the interface between Earth's surface and atmosphere associ-

level of the atmosphere and its envelopes must be involved. Consequently, many and varied satellites are required for such an undertaking. These studies are necessarily part of large-scale programmes such as the Earth Observing Enterprise, Earth Observing System (NASA), or Earth Explorer and Cornerstones (ESA).

These models make abundant use of data collected by the optical sensors of remote-sensing and meteorological satellites, discussed in chapter 9, especially for plant cover, ices, material in suspension and others. They also use Earth observation measurements made by the instruments described in the first part of this chapter. In particular, by combined use of altimetry and a precise geoid, a whole range of minor perturbations due to winds, eddies, temperature differences and salinity can be measured, quite apart from the amplitudes of tides, perturbations which only disturb the mean sea surface from the surface of the geoid by a few decimeters. Such measurements are essential when establishing dynamical models for the circulation of ocean waters and exchanges with the atmosphere. With regard to continental land surfaces, gravimetric instruments are used to evaluate ground moisture content, basic data for understanding exchanges with the atmosphere.

ates large multipurpose satellites like Terra and Envisat with small satellites like EO 1.

EOS AM 1 Terra. The AM 1 satellite, subsequently renamed Terra, was launched on 18 December 1999 (see fig. 9.20). It has been in operation since 24 February 2000. With a mass of more than 5000 kg, it carries five instruments including three that are either wholly or partially devoted to land surface observation. The MODerate-resolution Imaging Spectroradiometer (MODIS) measures radiation from the Earth's surface and clouds in 36 discrete spectral bands (see fig. 9.13). Its wide viewing swath spans 2330 km, so that it can cover a large part of the planet each day (fig. 7.8). The Advanced Spaceborne Thermal Emission and Reflection radiometer (ASTER) has 14 channels ranging from the visible to the infrared (see fig. 9.13), with spectral resolution between 15 and 90 m depending on the channel and variable orientation of the sensors. The Multiangle Imaging SpectroRadiometer (MISR) observes various types of cloud, as well as aerosols and smoke plumes, with a resolution between 275 and 1100 m, and has a stereoscopic view of the Earth's surface thanks to images taken from nine different angles.

Earth Observing 1 (EO 1) was launched on 21 November 2000 as part of the New Millennium Program. The general philosophy behind the programme is to use satellites of much lower mass and energy consumption than their predecessors. Following the Landsat 7 trajectory with a time lag of one minute, this small satellite, weighing only 425 kg, uses its Advanced Land Imager (ALI) to take the same images in the same lighting conditions as Landsat 7, which it thereby complements. This sensor covers the following spectral bands: MS-1': 0.433–0.453 μm; MS-1: 0.450–0.515 μm; MS-2: 0.525–0.605 μm; MS-3: 0.630–0.690 μm; MS-4: 0.775–0.805 μm; MS-4': 0.845–0.890 μm; MS-5': 1.20–1.30 μm; MS-5: 1.55–1.75 μm; MS-7: 2.08–2.35 μm. It has a ground resolution of 30 m and a panchromatic band 0.480–0.690 μm, with a ground sampling distance of 10 m. Another EO 1 sensor, Hyperion, derived from technology used in the HyperSpectral Imaging instrument (HSI) of the Lewis satellite, tests the hyperspectral technology with 220 spectral bands between 0.4 and 2.5 μm and a resolution of 30 m (see fig. 9.13). Finally, the Atmospheric Corrector (AC) has high spectral resolution across the range from 0.850 to 1.50 μm, but a low spatial resolution of 250 m. It corrects surface imagery for atmospheric variability related mainly to the presence of water vapour.

EOS PM Aqua. Two of the six instruments aboard this satellite, which weighs over 3000 kg, will be concerned with the Earth's surface: the first is the Humidity Sounder for Brazil (HSB) and MODIS, already carried on Terra. Brazil's presence in the project marks an increasing desire in many countries to gain better control of their development using space technology.

EnviSat. The ESA's multimission Polar PlatForm project (PPF) was originally designed as part of an ambitious European programme, symbolised in 1985 by the Columbus

manned space programme. When Europe gave up the idea of an independent presence in space, the original Polar Orbiting Earth observation Mission (POEM 1), which was to have combined scientific research and operational meteorology objectives, was eventually replaced by Envisat 1 and METOP 1. Envisat 1 was launched into a sun-synchronous orbit at 800 km altitude in 2002, and pursues environmental objectives, as its name suggests. It has a much higher mass (8140 kg) than the ERS satellites (ERS 2 = 2610 kg), in keeping with the heavy multimission platforms used formerly. Ten instruments work in synergy to investigate the fundamental problems of environmental science: global warming, climate change, ozone depletion, ice and ocean monitoring. One aim is to continue to acquire the data previously supplied by ERS 1 and ERS 2 (see chapter 9, p. 266). Hence, the Advanced Along Track Scanning Radiometer (AATSR) continues measurements made by ATSR 1 and 2 (Along Track Scanning Radiometer) on board ERS 1 and 2. Likewise, the Advanced Synthetic Aperture Radar (ASAR) (see fig. 9.38) carries on and improves the functions of AMI aboard the ERS satellites, whilst RA 2 (Radar Altimeter) extends data from the ERS radar altimeters. Another onboard detector is the MicroWave Radiometer (MWR), which measures the water vapour content of the atmosphere and the liquid water content of clouds in order to correct measurements made by the radar altimeter RA 2. In a secondary role, MWR will also provide information about ground moisture. In addition to three atmospheric instruments described below, the Envisat 1 payload is completed by a DORIS system.

A more distant project is Land Surface Processes and Interactions, one of ESA's four core missions, planned to study biogeographic phenomena such as the carbon cycle and surface–atmosphere exchanges, with the help of a great many experimental ground stations. This is a very complex mission and has not yet been fully defined.

ALOS. The payload of the Japanese satellite ALOS to be launched in 2004 includes three instruments destined to make an important contribution to environmental surveillance. These are the Panchromatic Remote-sensing Instrument for Stereo Mapping (PRISM) for digital elevation mapping, the Advanced Visible and Near Infrared Radiometer type 2 (AVNIR-2) for precise land coverage observation, and the Phased Array type L-band Synthetic Aperture Radar (PALSAR) for day-and-night and all-weather land observation. PRISM has a resolution of 2.5 m and the stereoscopic view is obtained by forward, nadir and backward looking telescopes. It operates in a single spectral band between 0.52 and 0.77 μm. The AVNIR-2 instrument operates in four spectral bands (0.42–0.50 μm, 0.52–0.60 μm, 0.61–0.69 μm, and 0.76–0.89 μm) with a ground resolution of 2.50 m (see fig. 9.13). In its ScanSAR mode, PALSAR can obtain a wider swath than conventional SARs.

Other satellites are used for the same purposes. Through accurate gravimetric measurements, the American satellite GRACE (see p. 164) participates to evaluate ground moisture content and its seasonal variations as a sideline to its main mission. The ESA's SMOS project (Soil Moisture and Ocean Salinity Mission) will pursue the same purpose with an L-band two-dimensional interferometric radiometer.

Ocean circulation will be a key study objective for the future GOCE satellite, one of the four new core missions of ESA's Earth Explorer programme (already mentioned in the context of gravimetry). In the area of glaciology, apart from GOCE, ESA is also preparing Cryosat, a specialised satellite equipped with a Ku-band radar altimeter, whilst NASA has developed ICESat (Ice, Clouds and land Elevation Satellite), whose lidar will also be able to monitor cloud cover in three dimensions. The lidar technique for measuring cloud structure and altitude was tested aboard the Mir space station between 1996 and 1999, using the French lidar ALISSA (Atmosphère par LIdar Sur Saliout), which has rather low resolution. NASA's Vegetation Canopy Lidar (VCL) will use lidar to chart plant cover and topography in three dimensions and ESA plans to use lidar for its future mission, WAter vapour Lidar Experiment in Space (WALES).

Atmosphere

Many instruments designed for ground- or balloon-based aeronomy studies have also been made part of specialised satellites, just as they have been carried on board meteorological satellites such as Nimbus, NOAA (National Oceanographic and Atmospheric Administration) and the Russian Meteors, or even on board other satellites, usually in sun-synchronous orbit. Many specialised satellites of various nationalities have thus contributed for over 35 years to our understanding of the upper atmosphere, carrying a number of detectors for rather diverse experimental measurements (figs. 7.9 and 7.10). Among various precursors, the Air Density Explorers (ADE) Ex 9, 19, 24 and 39, very light inflatable spheres, evaluated the density gradient at the point of contact between atmosphere and ionosphere, as did the Italian San Marco satellite series. The density, temperature, pressure and chemical composition of the upper atmosphere were studied by the Aeronomy Explorers (AE), later to become Atmosphere Explorers, Ex 17, 32, 51, 54 and 55, the last three of which had an initial perigee of 150 km. The ozone layer and aerosols were observed by the American AD Explorers, the British Ariel 2, 3 and 4 satellites launched in 1964, 1967 and 1971, then the US Solar Mesosphere Explorer (SME) launched in 1981. The French–American satellite Eole, pick-

Figure 7.9. **Inclination, apogee and perigee of satellites observing the upper atmosphere and micrometeorites.** Variations in aeronomic parameters are measured by observation from different altitudes.

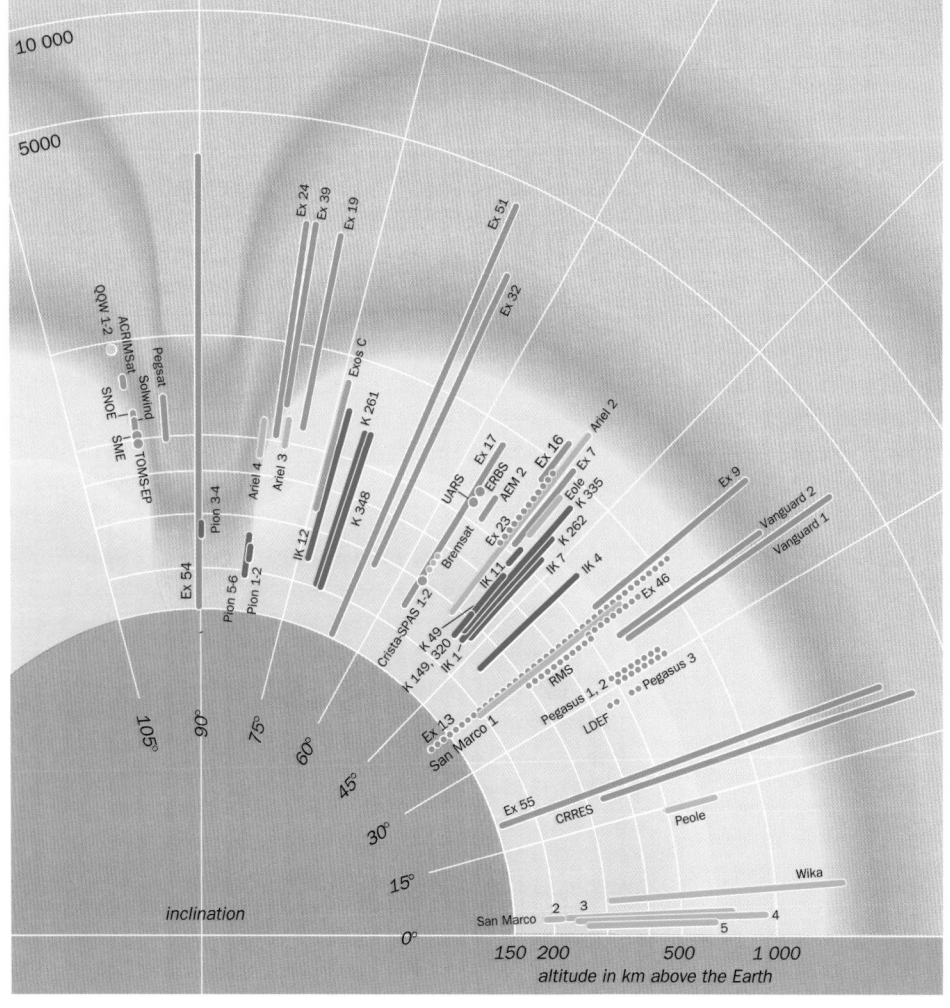
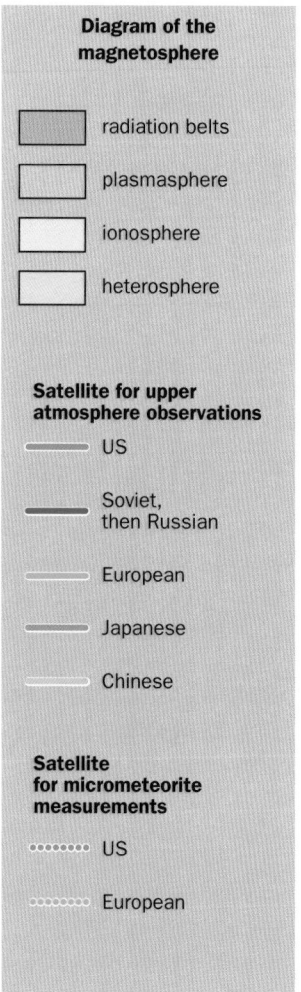

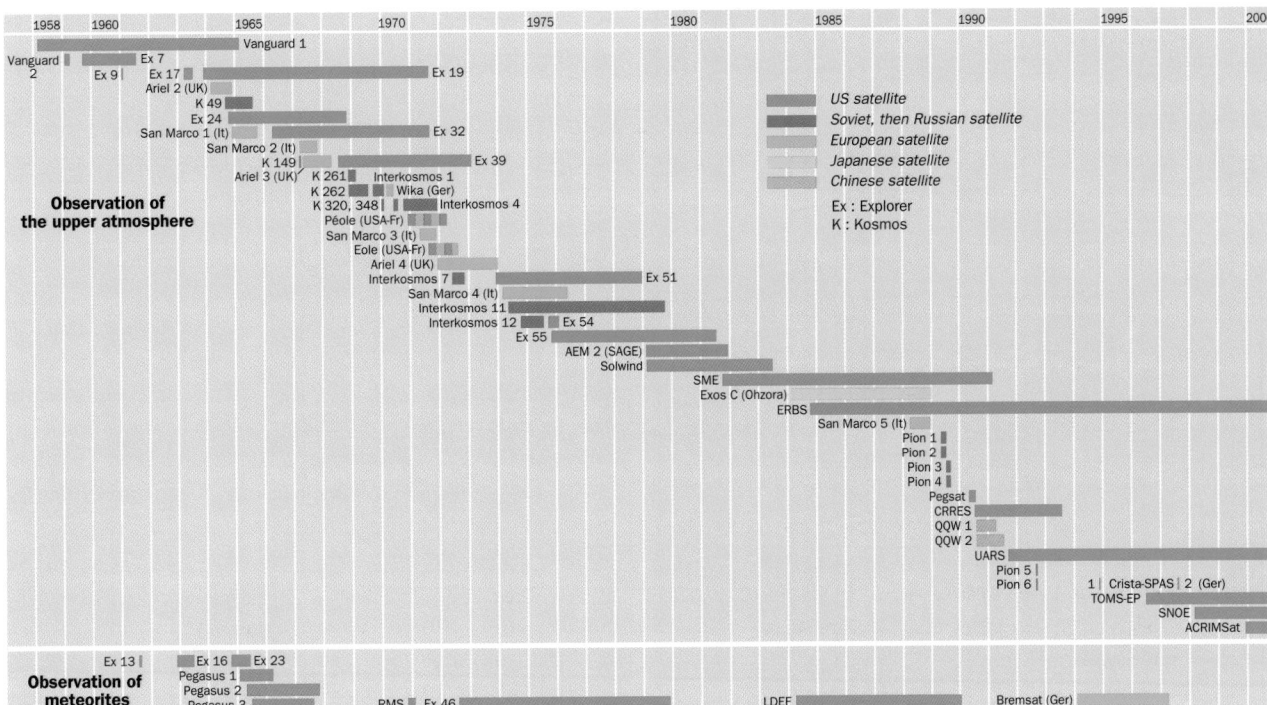

Figure 7.10. **Chronology of satellites observing the upper atmosphere and micrometeorite observation satellites.** Many Earth observation and meteorological satellites are involved in the study of atmospheric physics and land surface via the images they gather and also in most cases through measurements obtained using non-imaging sensors. These satellites are shown in figures 9.16 and 9.20.

ing up from the preparatory Peole satellite in 1971, monitored balloon probes in an innovative attempt to study high altitude winds in the southern hemisphere. Interkosmos 1, 4, 7, 11, 12, following Kosmos 261 and 262, carried out a parallel study of the influence of solar radiation on the upper atmosphere.

Many observations are made from manned spacecraft, which provide an opportunity for some experimentation. For example, Spacelab 1 carried out research on the upper atmosphere using the Imaging Spectrometric Observatory (ISO). This made spectral measurements across profiles of the mesosphere and thermosphere. Another example is provided by the MAPS programme (Measurement of Air Pollution from Satellites) undertaken in conjunction with four Shuttle flights (STS 2, 41G, 59 and, in 1994, STS 68). The German cryogenic spectrometer CRISTA (CRyogenic Infrared Spectrometers and Telescopes for the Atmosphere) analysed atmospheric gases by limb sounding in 1994 (STS 66) and 1997 (STS 85). It operated from a free flying platform SPAS (Shuttle PAllet Satellite) released at the beginning of the flight and retrieved at the end.

Ozone

A combination of space-based and ground-based measurements have revealed an increase in average temperatures and a depletion of the ozone layer. It is likely that human activities are partly responsible for this evolution, through emission of greenhouse gases and chlorofluorocarbons (CFCs). Studies of the ozone layer distribution and the factors affect-

ing it have given rise to many space programmes using a range of different detectors. Following experimental measurements made by the AEs and Ariel satellites described above, NASA's Nimbus 7 satellite, launched in October 1978, is a recognisable precursor in several fields of investigation, including study of the ozone layer, for which it provided measurements up until May 1993. It carried two specialised detectors: the Solar Backscatter UltraViolet spectral radiometer (SBUV) and the Total Ozone Mapping Spectrometer (TOMS). These acquired complementary data concerning the upper atmosphere. SBUV supplies a vertical profile of the ozone layer by means of 12 narrow spectral bands between 0.25 and 0.34 μm, whilst TOMS operates in 6 spectral bands between 0.312 and 0.380 μm, with three possible orientations around the nadir. SBUV flew on NOAA 9, 11, 13 and 16, and TOMS has flown on Meteor 3-05 from 1991 to 1994, ADEOS 1 (ADvanced Earth Observation Satellite, fig. 7.11) from November 1996 to June 1997, and TOMS–Earth Probe launched in 1996. TOMS will be used to equip specific (TOMS) satellites and others (ADEOS 2). The evolution of the ozone layer can be monitored around the whole globe by means of these two types of detector.

Since 1984, another type of detector has come on the scene which no longer views the Earth's surface through the atmosphere, but rather, views the Sun twice during each orbit of the satellite, once during its transit from night to day, and once during its transit from day to night. For example, the Stratospheric Aerosol and Gas Experiment (SAGE) flew on the Earth Radiation Budget Satellite (ERBS) and will equip

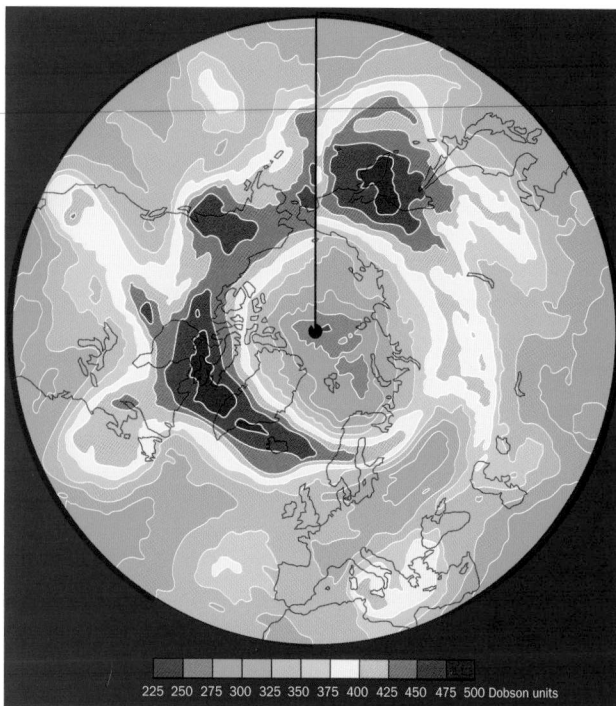

Figure 7.11. **Survey of ozone distribution in the northern hemisphere in April 1997.** The map was established using data from the ADEOS 1 satellite. Document kindly provided by NASDA–NASA.

a future Meteor 3. The Global Ozone Monitoring Experiment (GOME) was carried on board the ERS 2 satellite launched in 1995 and supplied a detailed history of the ozone layer over a period of several years. It also studied the atmospheric consequences of major forest fires. Finally, an improved version of GOME known as GOME 2 will be carried on board the European METOP satellites (see p. 243).

The POAM experiments (Polar Ozone and Aerosol Measurements) have been carried out on board the SPOT satellites, and POAM 3 is currently operating on board SPOT 4. This experiment measures the reduction of solar radiation as it passes through the atmosphere between altitudes 10 and 40 km above the terrestrial poles. Measurements are made twice during each orbit of the satellite, once as it transits from night to day, and once as it transits from day to night. Seasonal and long term changes in atmospheric components such as water, ozone, nitrogen dioxide and others can thereby be monitored. A GOMOS detector that views the stars rather than the Sun is carried on board Envisat (see above) and studies the vertical profile of the atmosphere. Finally, the Italian Triana satellite planned for 2004 will be placed at Lagrange point 1 in order to study ozone and the global Earth radiation budget in the atmosphere using its EPIC spectrometer (Earth Polychromatic Imaging Camera).

Earth radiation budget

The aim here is to compare visible and UV solar radiation with the radiation reflected by the Earth (particularly by clouds and snow) and the IR–TIR radiation of the atmos-

phere. The ERBE radiometers (Earth Radiation Budget Experiment) operated simultaneously on the Earth Radiation Budget Satellite (ERBS) and NOAA 9 and 10. The American Cloud and Earth Radiant Energy Sensor (CERES), comprising two scanning radiometers that measure radiation at the Earth's surface and at the upper limit of the atmosphere, is present on the TRMM satellites and Terra, and will equip the future satellite Aqua. Finally, the French SCARAB detector (SCAnner for RAdiation Budget) flew on Meteor 3-06 (1994) and on Resurs O1-04 (1998) before being used on Resurs 3M-01.

In 1984, the satellite ERBS began to establish the Earth's radiation budget between the 57° parallels north and south with its eight radiometers, each including a scanning and a non-scanning instrument. Its SAGE sensor made the first chemical analyses of the upper atmosphere. Since February 1990, the instruments have been operating in non-scanning mode only.

The Upper Atmosphere Research Satellite (UARS), launched in September 1991, is the first satellite of the Mission To Planet Earth (MTPE). It carries ten instruments: four measure the energy input from the Sun, three record structure and changes in the upper atmosphere, establishing temperature profiles and the ozone, methane and water vapour content, one detects traces of chlorofluorocarbons (CFCs) and other key molecules, and the last two, the High Resolution Doppler Imager (HRDI) and the Wind Imaging Interferometer (WIND II), map winds in the upper atmosphere.

The Tropical Rainfall Measuring Mission (TRMM), a US–Japanese collaboration, measures precipitations in the zone between the 35° parallels north and south. It follows a relatively low orbit (350 km altitude) and is not Sun-synchronous so that it can pass at a range of local times. It carries five instruments. The most innovative is the Precipitation Radar (PR). This measures the vertical distribution of precipitations by the radar reflectivity of cloud systems and the weakening of a signal as it passes through the precipitations (fig. 7.12). The TRMM Microwave Imager (TMI) and Visible and InfraRed Scanner (VIRS) complete the satellite's rain package. It also carries a Lightning Imaging Sensor (LIS) and the CERES experiment (Cloud and Earth Radiant Energy Sensor). TRMM data will be used to evaluate the rainfall consequences of the phenomenon known as El Niño. This effect, occurring in the equatorial regions of the Pacific Ocean close to the American continent, is manifested through an increased sea level recorded by altimetry satellites (e.g., TOPEX–POSEIDON, ERS 1 and 2 or Envisat with the RA2) and increased temperature measured by infrared thermography (e.g., AATSR on Envisat 1).

Three instruments are dedicated to atmospheric observations on EOS AM-1 Terra. The MOPITT experiment (Measurements Of Pollution In The Troposphere) measures the concentrations of methane and carbon monoxide in the

Figure 7.12. **Monthly averages of daily rainfall in mm per day.** Established using data from the TRMM satellite for the months of January and July 1998, 1999 and 2000. Document kindly communicated by NASDA–NASA.

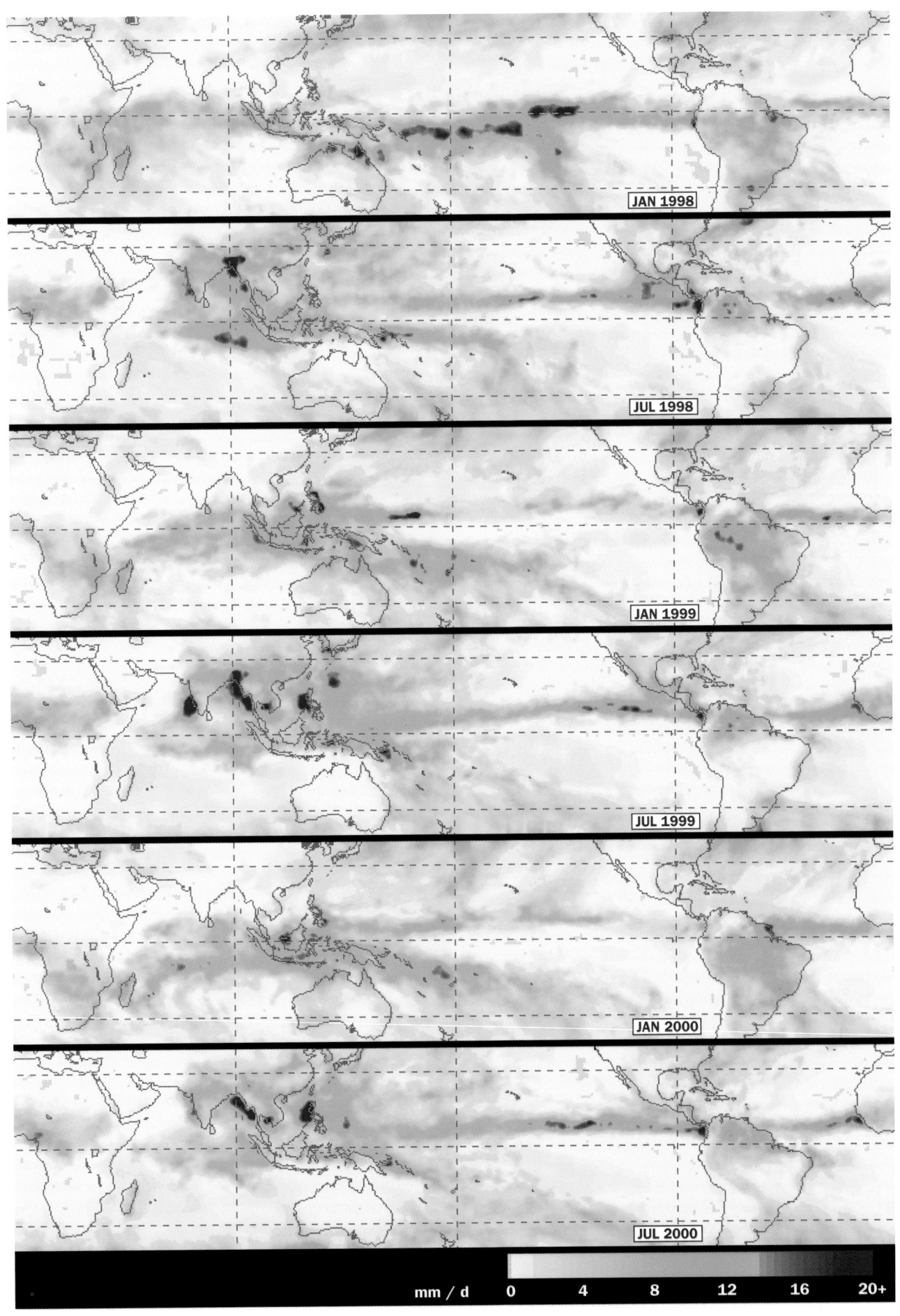

JAN 1998

JUL 1998

JAN 1999

JUL 1999

JAN 2000

JUL 2000

mm / d   0   4   8   12   16   20+

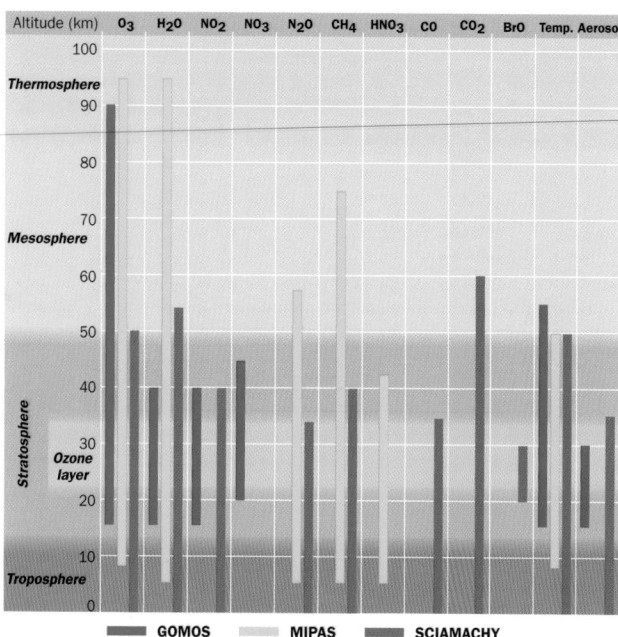

Figure 7.13. **Different altitude ranges in which measurements are made by the GOMOS, MIPAS and SCIAMACHY sensors on board Envisat 1.** (Based on ESA data.)

lower atmospheric layers with a resolution of 22 km. It is able to distinguish these gases from others such as carbon dioxide or water vapour. These measurements make a major contribution to understanding the causes of the greenhouse effect. The Multiangle Imaging SpectroRadiometer (MISR) observes the Earth's surface stereoscopically and also the various cloud types, aerosols and smoke plumes, with a resolution of 275 to 1100 m, taking images along nine different lines of sight. Finally, CERES measures the Earth's radiation budget (see above).

Four of the six instruments carried by EOS PM Aqua have atmospheric objectives: AIRS (Atmospheric Infrared Sounder, AMSR-E (Advanced Microwave Scanning Radiometer–EOS), AMSU-A (Advanced Microwave Sounding Unit), and CERES mentioned earlier.

Envisat instruments devoted to atmospheric observations measure composition at a range of altitudes. The GOMOS instrument (Global Ozone Monitoring by Occultation of Stars) measures the temperature and the ozone, NO_2, NO_3 and water vapour content across sections of the atmosphere up to altitudes of 100 km. It operates both night and day with vertical resolution better than 1.7 km. The Scanning Imaging Absorption Spectrometer for Atmospheric Cartography (SCIAMACHY) and the Michelson Interferometry Passive Atmospheric Sounder (MIPAS) also study atmospheric constituents (fig. 7.13).

EOS Chem, a US satellite weighing in at 3 tonnes, is designed to measure ozone and aerosols, and generally to study atmospheric chemistry and dynamics below 30 km altitude. Planned for 2002, its main instruments are the High Resolution Dynamics Limb Sounder (HIRDLS), the Micro-

wave Limb Sounder (MLS), the Tropospheric Emission Spectrometer (TES) and the Ozone Monitoring Instrument (OMI). ESA also envisages a mission of similar type, the Atmospheric Chemistry Explorer (ACE), as one of its future core missions.

The Atmospheric Dynamics Mission (ADM) is another of the ESA's four core missions, planned for around 2005. It will carry a Doppler wind lidar on a heliocentric orbit at 400 km for direct global observation of wind activity at twenty different atmospheric levels. The last of ESA's core missions, the Earth Radiation Mission (ERM), should study the divergence of radiative energy in the atmosphere and aerosol–cloud–radiation interactions, as well as the vertical distribution of water and ice and their transport by clouds.

Several small satellites are also dedicated to understanding the atmosphere. The SNOE minisatellite (Student Nitric Oxide Explorer) has been studying variations in the density of nitric oxide in the lower thermosphere since February 1998. Another minisatellite, ACRIMSat, was launched in December 1999 to complement data from the first four UARS sensors with measurements made by its EOS ACRIM 3 instrument. Quickscat, launched by NASA in June 1999, carries the Seawinds scatterometer designed to map winds at the ocean surface (see fig. 9.35). The same instrument will also be deployed aboard ADEOS 2.

The US Cloudsat, French Parasol, and European PRANUA-ATLASS projects will involve small and medium-sized satellites designed to improve understanding of aerosol–cloud–radiation interactions. The Franco-American satellite Calipso (formerly Picasso–Cena) which will fly in formation with EOS PM and Cloudsat, and NASA and the University of Colorado's more complex Solar Radiation and Climate Experiment (SORCE) will have the same objectives but carrying four instruments each. Finally, although the main mission of the European Odin project will be in astronomy, its submillimeter capabilities will also be used to study ozone depletion mechanisms.

It is worth mentioning a new technique for exploration of the upper atmosphere to be implemented in the Tethered Satellite System (TSS). In this US–Italian project, the satellite is deployed under the Space Shuttle, to which it is attached by a long cable. The first TSS flight, TSS 1, with a cable about 20 km long, was intended to demonstrate that a current could be generated in the cable as it cut through the Earth's magnetic field lines. Simultaneous measurements were then envisaged using a large number of probes fixed along the length of the cable and hence at different altitudes, some as low as a hundred or so kilometers above the Earth's surface. Using this technique, it should be possible to orbit at altitudes where atmospheric braking would severely limit the duration of the experiment, effectively rendering it impossible, were it not for the presence of another body gravitating at higher altitude. Other applications, apart from study of the upper atmosphere, are likely to see the light of day over the next few decades. However, the great technical difficulty involved in the operation meant that it was only possible to deploy 256 m of cable during flight STS 46 in 1992. The experiment was attempted a second time in 1996, from flight STS 75, but the cable once again failed to deploy correctly.

Ionosphere and magnetosphere

As a result of radio sounding, it has been known since the beginning of the 20th century that Earth has an ionised environment. However, waves emitted from the ground are blocked at around 300 km by the reflection peak of the ionosphere, the layer with maximal electronic density. The only way to investigate the Earth environment above 300 km altitude is through sounding rockets and satellites (fig. 7.14).

The very first satellites (Sputnik and Explorer) discovered and studied the Van Allen belts. Later investigations have focused on the composition of the ionospheric plasma and plasma waves, the polar auroras, interactions with solar and cosmic radiation, and others. The main programmes have been carried out since the 1960s by a large number of Explorer satellites (Explorer 8, 20, 22, 27), particularly the four Ionosphere Explorers, Canadian Alouette and Isis satellites, Soviet Proton satellites, the Interkosmos satellites (scientific cooperation between socialist countries), and the series of scientific research satellites of the Japanese ISAS and European ESRO (Iris, Aurora, Boreas, ESRO 4). On a national scale, European countries have had a limited presence in this field, with the British Ariel satellites, the French FR 1, the German Azur and Aeros series, the Spanish Intasat, and the Swedish Viking (fig. 7.15).

Almost all our present knowledge of the magnetosphere has been acquired by satellite. The relationship between the magnetosphere and the solar wind, the distribution of magnetic field lines, as well as the composition and electromagnetic radiation of the plasmasphere have been the subject of

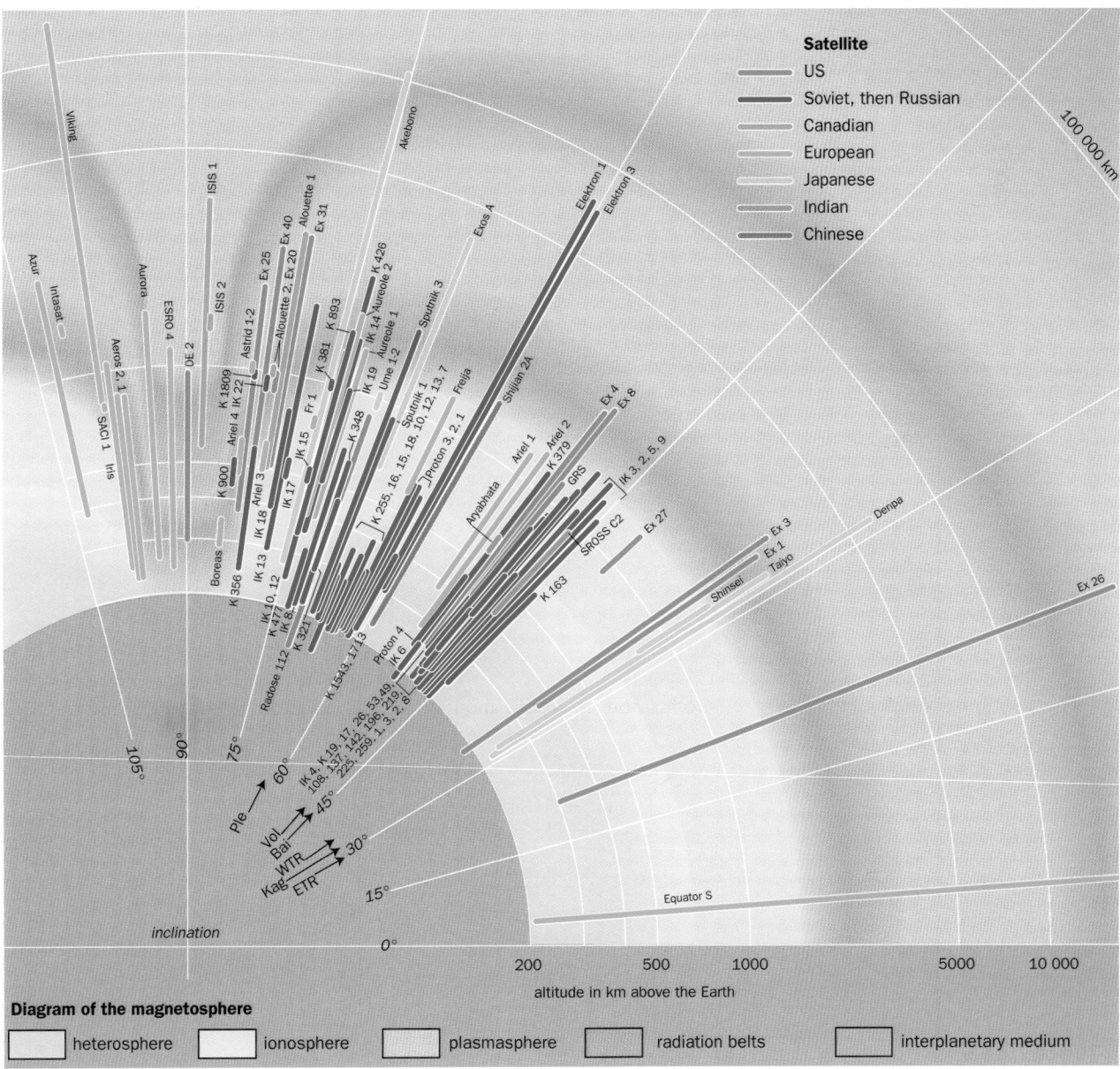

Figure 7.14. **Inclination, apogee and perigee of satellites observing the ionosphere and radiation belts.**

major investigations: NASA's Interplanetary Monitoring Platforms (IMP), the joint NASA–ESA International Sun–Earth Explorers (ISEE), and the Soviet Prognoz series. The lower magnetosphere has been explored by the Orbital Geophysical Observatories (OGO), which deployed six satellites in orbits with different inclinations, as well as the USSR's Elektron series and the US Dynamics Explorer series (DE), which deployed pairs of satellites, simultaneously launching them to different altitudes in order to determine spatio-temporal fluctuations so significant in the study of magnetospheric phenomena. The polar clefts have been studied by many magnetospheric satellites: IMP 7, OGO, DE, HEOS 2 (Highly Eccentric Orbit Satellite), the three Auréole satellites contributing to the Arcad programme (ARCtic Auroral Density), two Interkosmos satellites and Viking, but also by ionospheric satellites which have observed the polar auroras. Satellites

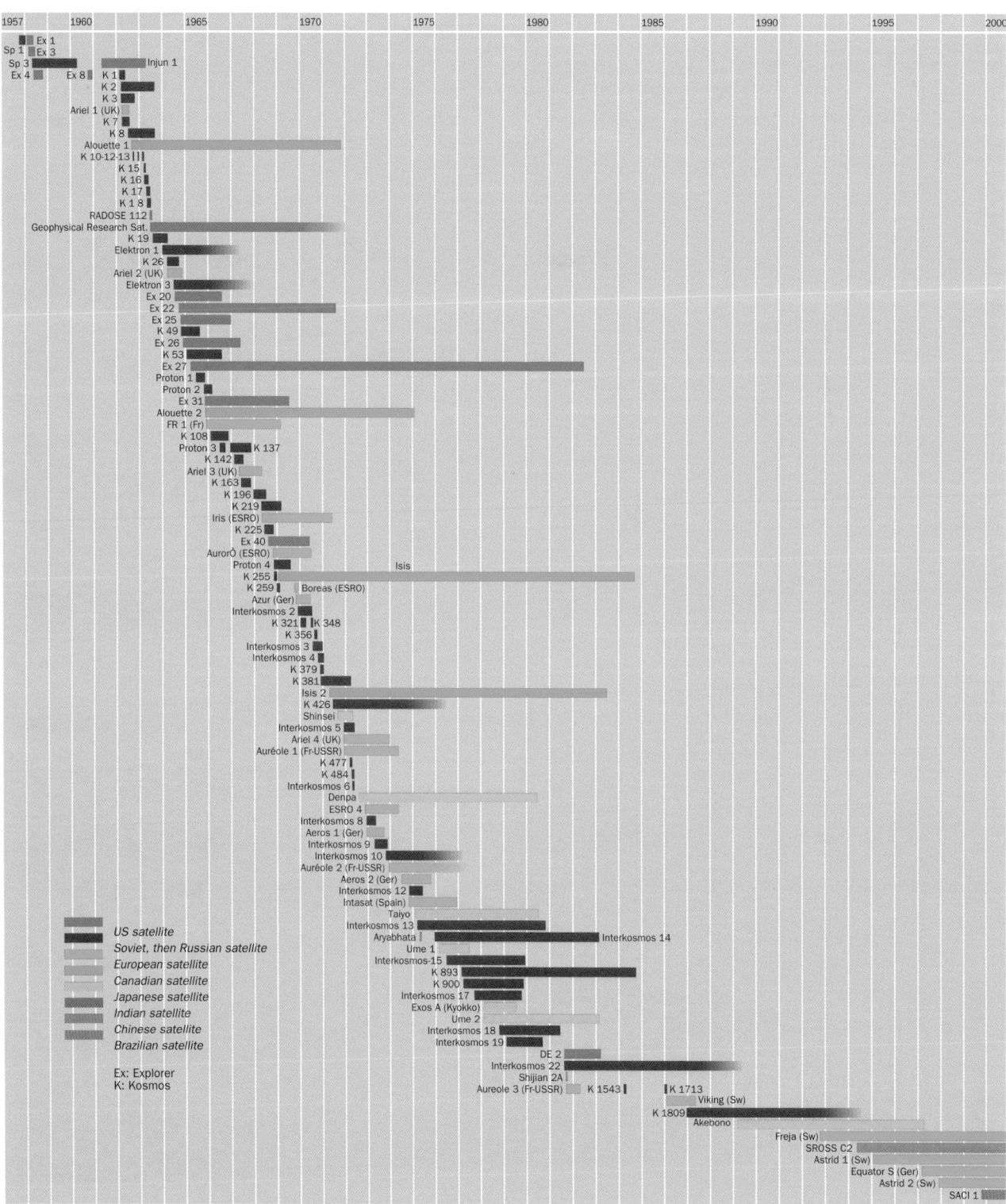

Figure 7.15. **Chronology of satellites observing the ionosphere.**

belonging to the AMPTE programme (Active Magnetospheric Particle Tracer Explorers), launched simultaneously, released clouds of lithium and barium into the magnetosphere and solar wind in order to observe their movements (figs. 7.16 and 7.17).

This first exploratory phase revealed the main systems making up the Earth's environment. They appear as a juxtapo-

sition of large scale cells: the interplanetary medium, with largely solar origins but also a galactic component, the magnetosphere with its magnetotail and auroral zone, the ionosphere and the upper atmosphere (fig. 1.5). These meet at relatively thin interfaces in which critical zones (where coupling dominates) play a crucial role through the turbulent microscopic processes occurring there. The main critical

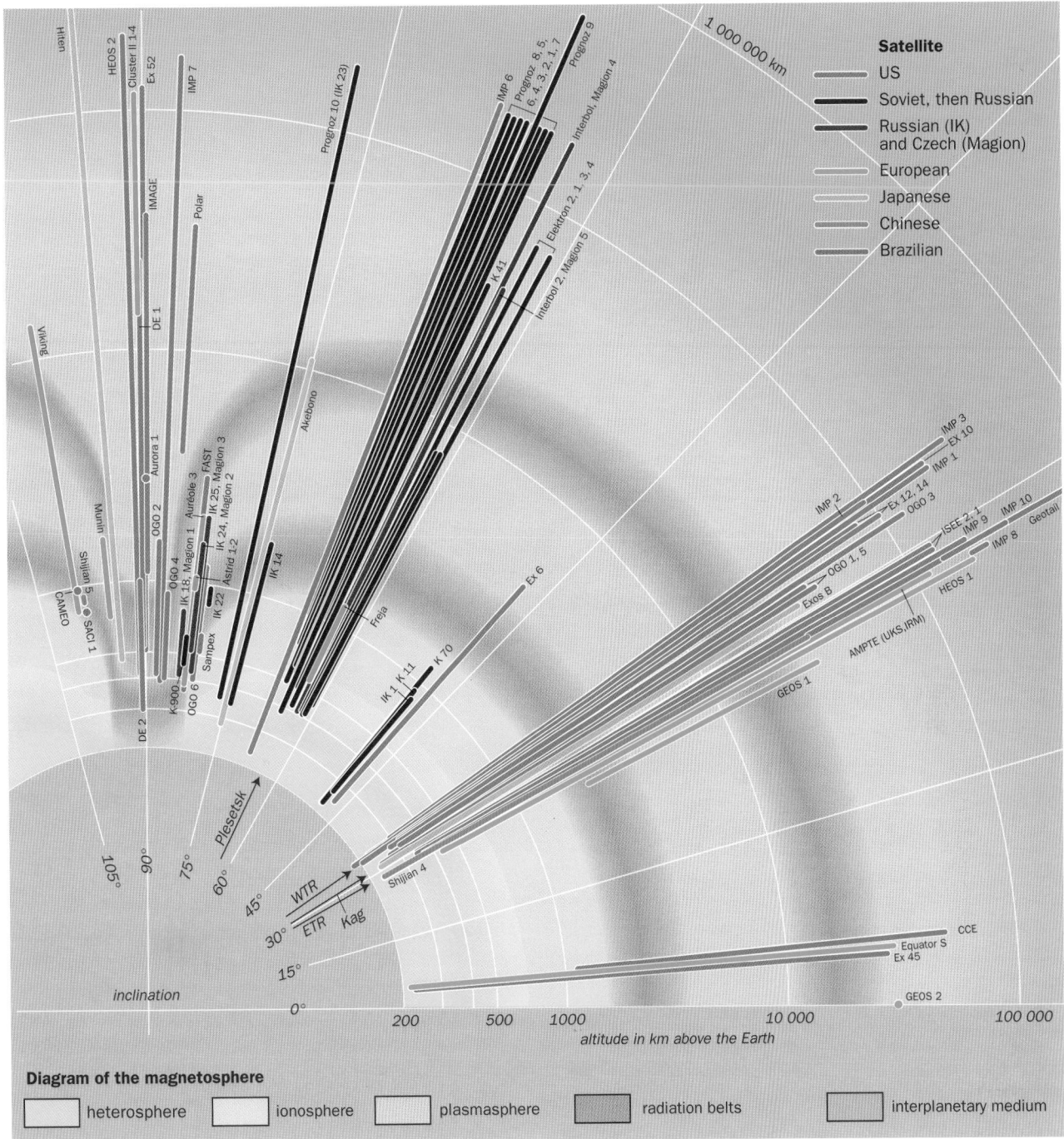

Figure 7.16. **Inclination, apogee and perigee of satellites observing the magnetosphere.** Such satellites generally have very high apogee. Exceptions are satellites observing the polar clefts or the lower magnetosphere. Inclinations are concentrated around three values, namely, 90° for the study of the polar clefts, 61–65° for satellites launched from the Russian base Plesetsk, and 28–34° corresponding to the latitudes of Japanese and US bases. The European Space Agency chose to give its Geostationary Earth Orbit Satellites (GEOS) a geostationary orbit. However, a failure in the 3rd stage of the launch vehicle left GEOS 1 in its transfer orbit. In fact, inclination is not the most significant orbital element as far as such measurements are concerned. The aim is rather to take into account the spatial structure of the magnetosphere.

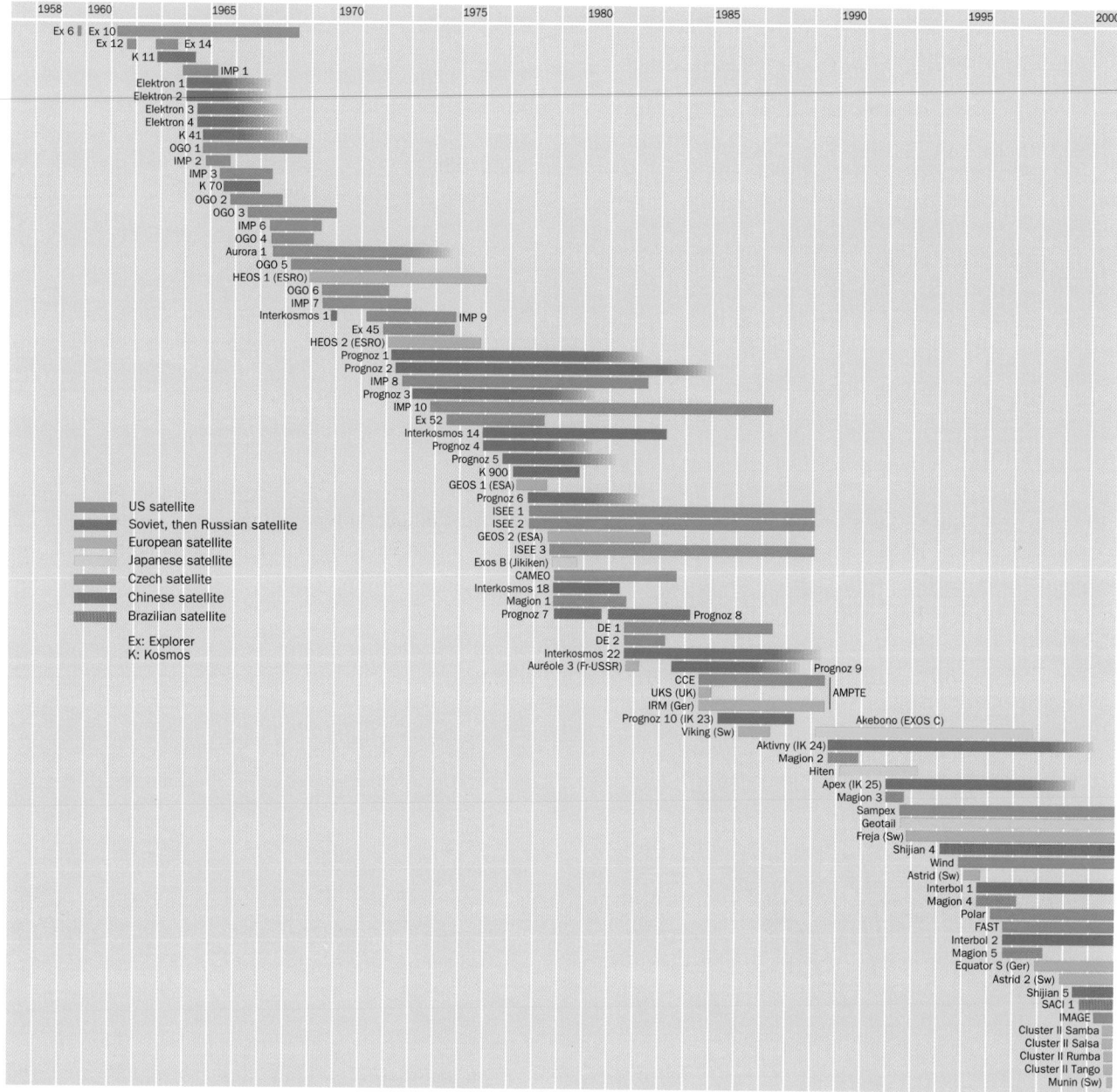

Figure 7.17. **Chronology of satellites observing the magnetosphere.**

zones are the bow shock wave, the magnetopause, the boundary layers of the plasmasphere, the sheets of the magnetotail where magnetic substorms are generated during solar flares, and the auroral acceleration regions.

The main coupling regions are located between the zone dominated by continuous solar wind–sporadic coronal mass ejections (CMEs) and the magnetospheric system, between the magnetospheric plasmas and the auroral ionosphere (or directly between the combined solar wind–CME and the auroral ionosphere), and between the ionosphere and the upper atmosphere. There are two prerequisites for establishing a 'meteorological' model of the general dynamics of these media: the first is a detailed understanding of coupling in critical zones; the second is frequent and rapid monitoring of solar events. The first condition requires a combina-

tion of macroscopic remote sensing observations and synchronised microscopic measurements made in situ, a project best carried out by constellations of satellites. The second prerequisite is or will be met by solar observatories.

These considerations led in 1977 to the creation of the international programme known as InterAgency Solar Terrestrial Physics (IASTP), which brings together ESA, NASA, ISAS and IKI as well as more than 100 universities and research centres in 16 countries. It was not until 1992 that the first satellite of the programme was launched. This was the Japanese satellite Geotail, prepared by Japanese missions Exos D (Akebono, dawn) and Muses A (Hiten, celestial virgin). It studied energy flow in the tail of the magnetosphere. This was followed in 1994 by NASA's Wind satellite (solar wind studies), ESA's SOHO in 1995 (solar observatory), and NASA's Polar in 1996 (studying auroral regions). The constellation Cluster was scheduled for launch in 1996. The aim was to overcome the difficulties involved in interpreting data gathered at a single point in space. A group of four satellites were to be deployed in a tetrahedral arrangement on highly eccentric geocentric orbits, with apogees at 119 000 km and perigees at 19 000 km. These measurements would then have been correlated with SOHO observations of solar activity. Unfortunately, these satellites were destroyed with Ariane 5 on its first flight. However, a new cluster experiment was decided in April 1997 and the four satellites Salsa, Samba, Rumba and Tango, were successfully launched in July and August 2000.

In parallel with the IASTP programme, Sweden has specialised in studies of the polar regions, with Freja (1992), Astrid 1 (1995) and Astrid 2 (1998). The international programme Interbol, involving 20 countries in collaboration with Russia, launched Interbol 1 in 1995, with its Czech subsatellite Magion 4. This studied the magnetotail. It was followed in 1996 by Interbol 2 and Magion 5 which studied the auroral region. The US small explorers SAMPEX (Solar Anomalous and Magnetospheric Particle Explorer) and FAST (Fast Auroral Snapshot explorer) were deployed in 1992 and 1996, respectively. The first studied particles in outer space and the second examined the auroral region with high spatial and temporal resolution. The US–German satellite Equator S (1997) completed the IASTP programme's magnetospheric investigations with data from the equatorial regions, whilst the Indian satellite SROSS C2 (1992) studied the thermosphere–upper atmosphere, and the Brazilian SACI 1 (1999) studied plasma bubbles and the ionospheric anomaly above Brazil and the South Atlantic. IMAGE (Imager for Magnetopause-to-Aurora Global Exploration), launched by NASA in March 2000, is the first to replace temporally discrete in situ observations of magnetospheric ions and electrons by a pair of imagers, one observing energetic neutral atoms (ENAs) and the other the geocorona, thus providing the first global view of these plasmas. It will observe ENA energy arising from charge exchange between cold neutral hydrogen in the geocorona and local energetic ions.

Apart from the American TIMED satellite (Thermosphere, Ionosphere Mesosphere Energetics and Dynamics mission), IASTP should be further supplemented by satellite constellations. After Cluster 2, Europe should launch IBIZA (Investigation BIsatellitaire des Zones Aurorales). A pair of manoeuvrable satellites will explore a limited region of the magnetosphere–ionosphere interface in great detail in order to seek out the fine structure of its electric and magnetic fields. It will also attempt to understand the processes whereby charged particles are accelerated and measure the associated electromagnetic radiation. Rather later, the Quattro projects will involve two to four satellites in the regions were magnetic substorms are thought to form, whilst IMAG-Stereo will send two imaging microsatellites to the same regions. The US projects are Magnetospheric Multiscale (six satellites, plasma boundaries), Magnetospheric Constellation

(several tens of microsatellites, dynamics of the magnetosphere) and Global Electrodynamics (two satellites, outer space–magnetosphere interface and outer space–atmosphere interface). The same pair of imagers used for IMAGE should also operate as part of a constellation via the TWINS mission (Two Wide-angle Imaging Neutral-atom Spectrometers), aboard two American satellites (probably military) in Molniya orbit. TIMED, another monosatellite launched in 2001, will also use remote sensing to study the coupling between the upper atmosphere and the ionosphere.

A good understanding of the Earth environment is essential for military space activities. In parallel with civilian programmes, which often include instruments from a range of different countries, military programmes are more finely tuned to meet specific needs, such as the protection of satellites against radiation of natural or artificial origin, knowledge of the various layers and perturbations occurring in them, identification of infrared signatures, and others.

In the United States, military scientific enterprise was considerable up until the beginning of the 1970s, dealing as it did with the first investigations of the medium. It began with the Discoverer series, originally launched to develop the recoverable capsule technique. From Discoverer 17, data was obtained concerning the ionosphere, radiation belts, cosmic radiation and terrestrial infrared emissions. The later ERS series (Environmental Research Satellites) and OV series (Orbiting Vehicles) studied mechanical, electrical, electronic and biological consequences of natural radiation. Between 1965 and 1971, a great deal of data was gathered on ions, protons, and electrons in the ionosphere, as well as X-rays and magnetism. The Calsphere series studied the density of the upper atmosphere and atmospheric braking. Telecommunications research satellites such as Lofti 1 and 2, P11 AS, Orbis, Radsat, P76-5, Solwind, and Polar Bear studied radiation at various wavelengths, LOGACS 1 the upper atmosphere, and SCATHA the influence of the solar wind and particles on satellites. The US Navy's Solrad series complemented NASA's three Solrad satellites. The most recent environmental satellites were FORTE and STEP 4, launched in 1997. STEP 4 carried three experiments on the effects of ozone, the ionosphere and magnetic ion fluxes, but was unable to deploy its solar panels. The technology satellite ARGOS (1999) carried a plasmasphere and magnetic storm monitoring experiment, among other things.

In addition, military satellite series pursuing various operational missions have collected data on the upper atmosphere and polar auroras (DMSP satellites, see chapter 9), solar and galactic radiation (VELA series), and terrestrial infrared radiation (MIDAS series). Some of this data has been made available to the scientific community. The Starfish experiment was designed to test the effects of high altitude nuclear explosions on natural radiation in the ionosphere and upper atmosphere, and caused some damage to several satellites. The Starad satellites (USAF, 1962), Injun 3 (USN, 1962) and Hitchhiker (USAF, 1963) collected data, as did NASA's Explorer 15 (1962) and the Soviet satellite Kosmos 5 (1962).

In the Soviet Union, it is difficult to identify missions with a specifically military vocation, since there is a tendency to amalgamate civilian and military space activities. Since the needs are globally the same, similar missions have doubtless been accomplished by Kosmos satellites with basically undifferentiated status. The Academy of Sciences would then deal with disseminating authorised results.

Micrometeorite studies

A great many scientific satellites and all space probes have been equipped with micrometeorite counters or detectors. However, only nine, including eight American, have exclusively carried out micrometeorite studies (figs. 7.9 and 7.10). The Long Duration Exposure Facility (LDEF) was launched in 1984 by the eleventh flight of the Space Shuttle and recovered during the 32nd flight in January 1990. It carried mainly technological experiments concerning the exposure of various materials to micrometeorites and radiation. Most of the results are internationally available, although some experiments are classified.

The German satellite Bremsat was launched by the Shuttle in February 1994. The objective was to observe dust and assess the micrometeorite population in circumterrestrial space, placing particular attention on the study of anthropogenic pollution. When it fell to Earth in February 1995, it gathered data down to an altitude of 110 km, before its destruction at 60 km.

Astronomical observation

It is in astronomy that space-based techniques have led to the greatest progress, despite the initial scepticism of some commentators. Indeed, satellites make it possible to observe wavelengths that are either partially or totally blocked out by the atmosphere: γ-rays, X-rays, ultraviolet and infrared. These same techniques also improve observation previously carried out via atmospheric windows, for example, by increasing luminosity and eliminating scintillation phenomena. In radioastronomy, the baseline for interferometry can be extended.

Following the short experiments that marked the beginnings of spaceborne astronomy, recent programmes have tended to involve longer missions, lasting between 10 and 20 years, with wide coverage of the sky (figs. 7.18 and 7.19). They require more and more sensitive detectors, able to exclude all sources of outside interference, together with more accurate pointing and better stabilisation. Although the latter are generally difficult to achieve on board manned spacecraft, astronomical and solar observation is nevertheless carried out from such vessels. They are less suitable than satellites because, apart from difficulties in obtaining stable pointing, there can be interference from neighbouring experiments. In addition, time is generally rather limited in manned missions, except for those involving permanent space stations. Astronomical observations have constituted an important element in missions carried out by Salyut 1, 3, 4, 5, 6, 7, Skylab, Soyuz 13, ASTP, STS-51F (Spacelab 2) and Mir (Kvant module). The Astro 1 mission (ultraviolet and X-ray observations) was carried out by STS 35. It is planned to install an alpha magnetic spectrometer (AMS mission) on the shuttle or the ISS to search for dark matter and antimatter.

Observation of the Solar System

The Sun is the closest star to Earth and hence of great interest for astronomical observation. At the same time, it has a direct influence on conditions in the Earth environment. Indeed, it is the main source of energy on Earth. Study of solar radiation and magnetism is therefore an essential factor in understanding the structure of the terrestrial magnetosphere.

Today, observation of the Sun involves simultaneous long term observations over a range of different wavelengths, together with measurements of the magnetic field and Doppler–Fizeau effect. During successive orbits, satellites provide better coverage over time than Earth-based observation. They must be scheduled according to the major solar events, as for example was the Solar Maximum Mission (SMM) in February 1980.

When ESA first established its Space Science programme in 1984, the Solar Terrestrial Physics Programme (STP) was one of the first Cornerstones to be envisaged. This included SOHO and the Cluster missions. SOHO (SOlar and Heliospheric Observatory) was launched in December 1995 and deployed at Lagrange point 1 two and a half months later. It is the most comprehensive of the solar observatories. Three of its twelve experiments are devoted to helioseismology: GOLF (Global Oscillation at Low Frequencies) studies radial and tangential vibrations, MDI/SOI (Michelson Doppler Imager/Solar Oscillations Investigation) the structure of the convection zone, and VIRGO (Variability of Solar Irradiance and Gravity Oscillations) variations in the solar wind. Six others are involved in remote sensing the solar atmosphere and three carry out in situ analyses. These are EIT (Extreme ultra violet Imaging Telescope), CDS (Coronal Diagnostic Spectrometer), UVCS (Ultra Violet Coronagraph Spectrometer), LASCO (Large Angle and Spectrometric COronagraph, fig. 7.20) and SUMER (Solar Ultraviolet Measurements of Emitted Radiation) which study magnetism and magnetic reconnection phenomena, whilst the solar wind and high energy particle emissions are studied by SWAN (Solar Wind ANisotropies), CELIAS (Charge, ELement and Isotope Analysis System), ERNE (Energetic and Relativistic Nuclei and Electron experiment), and COSTEP (COmprehensive SupraThermal and Energetic Particle analyser).

Launched in November 1994, Wind began by investigating the solar wind and plasmas in the near-Earth region from a series of high elliptical orbits. It was only in November 1996 that it became a genuine solar observatory when it was stationed at Lagrange point 1. It has eight instruments including gamma ray and high energy particle detectors. Ulysses extended magnetism and solar wind measurements to high

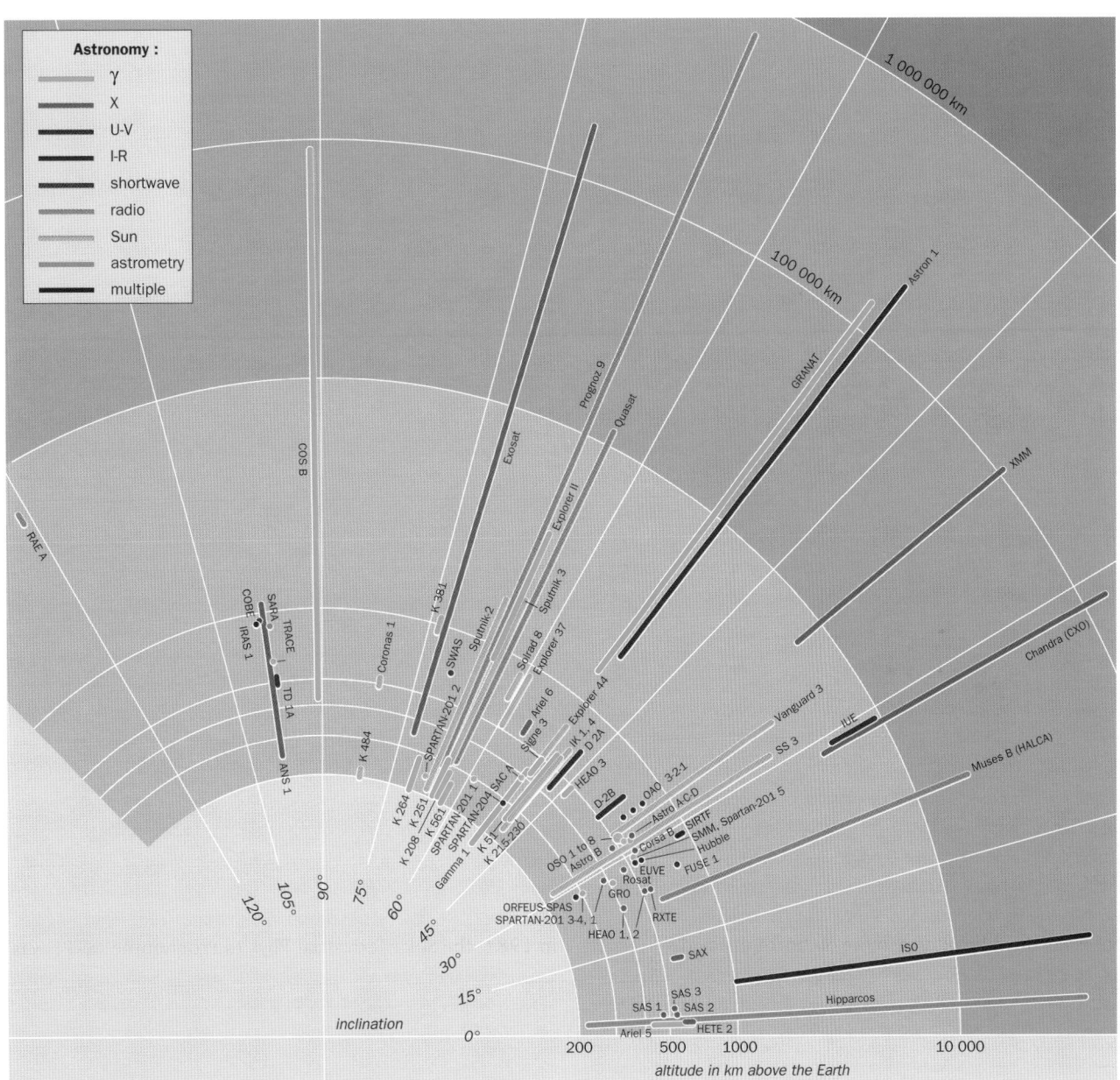

Figure 7.18. Inclination, apogee and perigee of astronomy satellites. Early on, most stellar astronomy satellites followed low orbits with inclinations depending on the location of the launch base. Later, these orbits were diversified. High apogee orbits increase observation time outside the Van Allen belts and reduce time lost through occultation by the Earth (e.g., COS B, Exosat, Astron 1, IUE, ISO, Granat). In radioastronomy, higher apogees increase measurement accuracy, a function of the distance from the Earth (e.g., Prognoz 9, located at almost a million kilometres). The geosynchronous orbit of the international satellite IUE allowed regular links with receiving stations. The sun-synchronous orbits of IRAS 1 and COBE were chosen to minimise solar infrared emissions. The orbit of the Hubble Space Telescope (HST) at an altitude of 600 km and with an inclination of 28°5 allows it to be maintained, repaired or recovered by the Space Shuttle. Finally, orbits about the Lagrange point 1 of the Earth–Sun system are privileged positions for observing the Sun, since the Earth and Moon remain outside the field of observation (e.g., Wind, SOHO and ACE), whilst orbits about Lagrange point 2 are ideal for stellar observations, since the Sun, Moon and Earth all lie outside the field of observation (e.g., future observatories FIRST, Planck and Gaia).

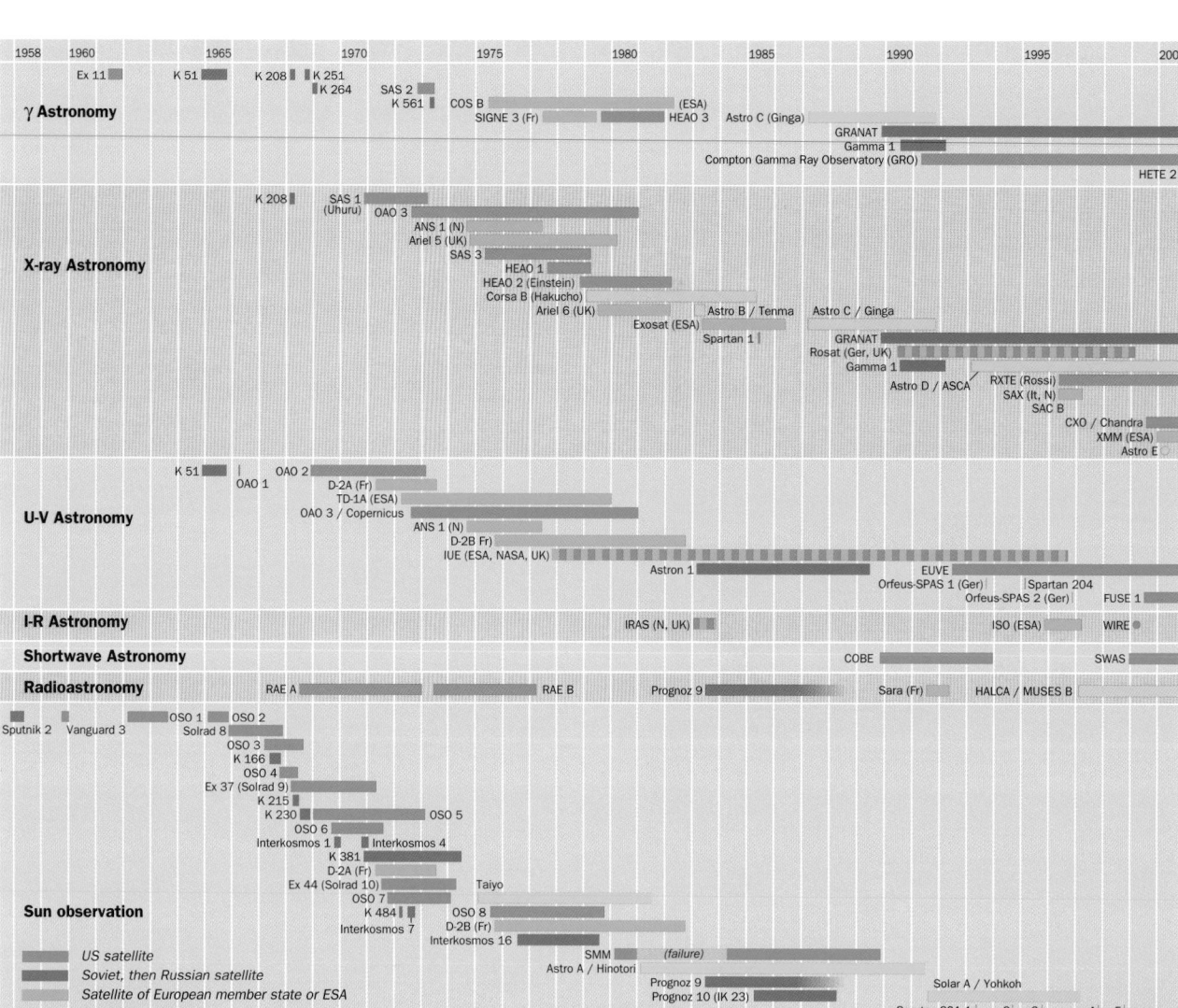

Figure 7.19. **Chronology of astronomy satellites in geocentric orbit.**

solar latitudes (see fig. 8.8). ACE (Advanced Composition Explorer) is the last of the big Explorer projects. Launched towards Lagrange point 1 in August 1997, it carries ten instruments, including eight designed to observe solar and cosmic radiation, a magnetometer and a real time solar wind monitor. Finally, TRACE (Transition Region And Coronal Explorer), launched in April 1998, is a small explorer equipped with an accurate UV and EUV telescope, whilst the Japanese satellite Yohkoh (August 1991) and the Argentinian satellite SAC-A (December 1998) carry X-ray telescopes.

At the beginning of the year 2000, spaceborne solar physics experiments were being carried out by Yohkoh (Solar A), Coronas I, SOHO, Wind, Ulysses, ACE, TRACE, and SAC-A. The free-flying retrievable SPARTAN 201 platforms are periodically deployed from the Space Shuttle to make short observations of the corona in UV and white light. Other scheduled projects are Solar B run by ISAS (accurate

global observation in the visible and X-rays), the Picard microsatellite from the CNES (accurate measurement of variations in the Sun's diameter, differential radiometer to evaluate the solar constant, UV photometers), the Small Explorer HESSI (High Energy Solar Spectroscopic Imager), which is an X-ray and γ-ray imager and spectrometer, cooled to −198 °C to study explosive energy released in solar flares, and above all STEREO (Solar TErrestrial RElation Observatory) from NASA.

Two satellites following the Earth's heliocentric orbit at angles of a few tens of degrees ahead and behind the Earth will make observations in three dimensions. In particular, they will observe coronal mass ejections (CMEs) and corresponding force lines of the interplanetary field using X-ray and UV imaging spectrometers, a visible light coronagraph and a radio direction finder (radio goniometer). Finally, a solar probe able to make the necessary in situ

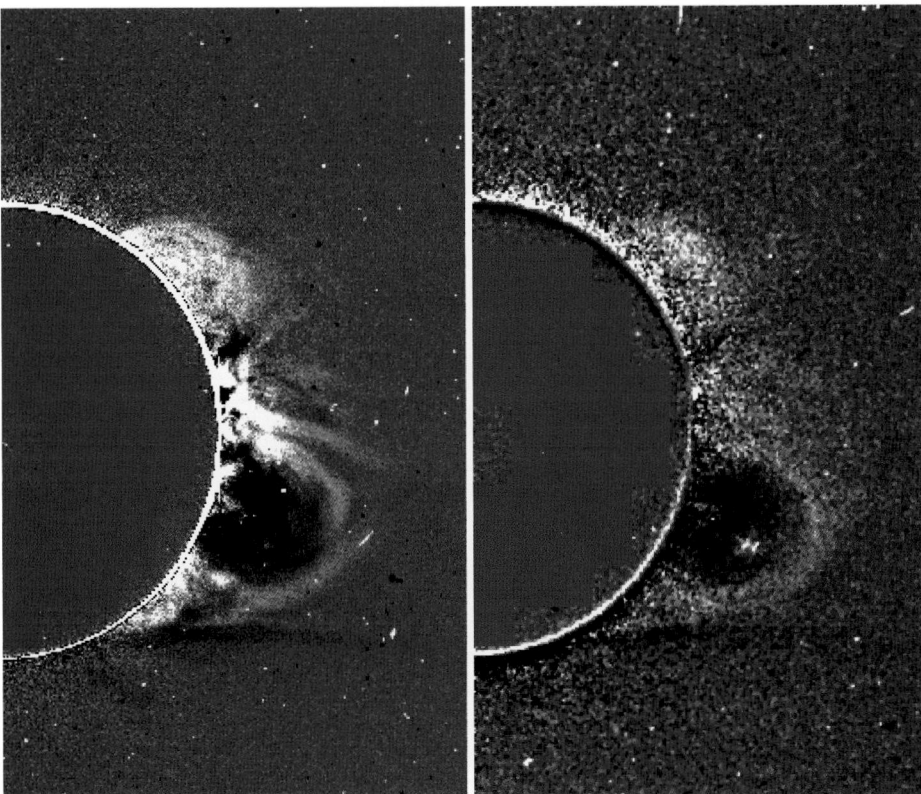

Figure 7.20. **Image taken by SOHO**. Onset of a coronal mass ejection as observed by the LASCO C1 coronagraph on 23 December 1996 in the forbidden line Fe XIV 5303 Å at 21:04 UT (left), and the K corona at 21:07 UT (right). Courtesy of NASA.

studies, and thereby resume the investigations of the German Helios probe series between 1974 and 1976, has been awaited for over twenty years. The latest project from JPL and NASA plans to make a single flyby at high solar latitudes some time around 2008–10, at a minimum distance of 3 solar radii, carrying out in situ measurements and using imaging devices.

Apart from the Sun, other bodies in the Solar System have benefited from observations made from geocentric orbits. This is the case for many lunar and planetary observations carried out by the Hubble Space Telescope, which also studies space beyond the Solar System (fig. 7.21).

Observation beyond the Solar System

■ Astrometry seeks to measure the positions of heavenly bodies as accurately as possible and lies at the foundation of much research on the origin of the Universe. It has been used to improve experimental determination of the post-Newtonian parameter γ, and these improvements will undoubtedly progress in the future. The satellite Hipparcos, launched in 1989, derives from the acronym HIgh Precision PARallax COllecting Satellite and celebrates the Greek astronomer who first established an accurate star map. Although unable to acquire geostationary orbit due to failure of its apogee kick motor, it nevertheless fulfilled its mission in an eccentric orbit. Between 1989 and 1993, it established the position coordinates

and components of proper motion of 120 000 stars. Other space telescopes, the Middle Explorer FAME (Full sky Astrometric Mapping Explorer) and DIVA (Deutsches Interferometer für Vielkanalphotometrie und Astrometry) should be launched by NASA and the German DLR in 2004. They should each be able to determine the position and brightness of close to 40 million stars, far more than was possible with Hipparcos.

Gaia, planned by ESA for 2012, will improve colour photometric measurements for over a billion stars, reaching out over much greater regions of our galaxy and even as far as the Large and Small Magellanic Clouds. Accurate observations of extragalactic Cepheids in the Magellanic Clouds will be of great importance in taking the measure of the Universe. These are variable stars whose period of variation is proportional to their absolute magnitude, so that their distance can be estimated by comparing with their apparent magnitudes.

■ Radioastronomy is used to study the structure of radio galaxies, jets from active galactic nuclei (AGNs), supernova remnants, and others. The Japanese satellite HALCA (High Altitude Laboratory for Communications and Astrophysics) launched in May 1997, and known at various stages in the project as Muses B or VSOP, will test the feasibility of coupling an orbiting radio telescope with a ground-based antenna to form a telemetric network.

■ The Cosmic Background Explorer (COBE) observed millimetre wavelengths to study the 2.7 K background interpreted as a fossil left over from the primordial Big Bang. It should be succeeded by NASA's MIDEX Microwave Anisotropy Probe (MAP), and the European satellite Planck.

■ Submillimetre wavelengths are absorbed by water and oxygen molecules. They can thus be used to measure the abundance of these molecules in dense interstellar clouds, where they correspond to radiation at 15 to 30 K. When there is compression due to gravitational effects, heating occurs, causing a thermal pressure which tends to oppose further compression. It is thought that water and oxygen play an important role in dissipating this thermal energy during star formation in molecular clouds. The small explorer SWAS, launched in December 1998, will seek out these water and oxygen molecules and associated objects in the Galaxy. It should be followed by the Odin minisatellite, run by Sweden with the collaboration of

Figure 7.21. Image of Saturn taken on 4 January 1998 by the Hubble Space Telescope. This false-colour image shows the planet's reflected infrared light. The view provides information on the clouds and hazes in Saturn's atmosphere. Two of Saturn's moons are visible: Dione at bottom left and Tethys at top right, in front of the planetary limb. (Photo no. STScI-PRC98-18. Courtesy of Erich Karkoschka, University of Arizona, and NASA.)

France, Finland and Canada. Odin will carry sensors in three spectral bands around 0.5 mm and another at 3 mm. Planetary atmospheres and gases ejected by several comets should also be studied. Towards 2007, the FIRST satellite (Far InfraRed and Submillimetre Telescope), renamed the Herschell Space Observatory in December 2000, is one of the ESA Cornerstones, but involves NASA participation. It will operate in the infrared and in sub-millimetre bands between 80 and 670 μm. Its main mission will be to study galaxies and young molecular clouds (less than a billion years old).

■ Only infrared observation from space can overcome the atmospheric screen and interference due to environmental emissions. Telescopes are cooled to a few degrees kelvin in order to reduce interfering radiation from the satellites themselves. Infrared observations provide a good way of observing the birth of stars and galaxies, and the early stages in their evolution. Following the infrared observations made by the two Voyager probes, the British–Dutch–American InfraRed Astronomy Satellite (IRAS) was equipped with detectors in four bands at 12, 25, 60 and 100 mm. Despite its rather poor angular resolution, it was the first to survey the whole sky, in 1983, cataloguing 250 000 infrared sources in the Universe. It also discovered five comets. Its successor was ESA's Infrared Space Observatory (ISO). This had a cryostat which cooled the telescope with liquid helium at −270 °C, evaporated slowly and vented to space. It made measurements in the infrared between 2.5 mm and 200 mm. Between 1995 and 1998, it observed specifi-cally selected regions rather than the whole sky, with a high spectral resolution, far superior to that of IRAS. Launched in March 1999, the Small Explorer WIRE (Widefield InfraRed Explorer) failed to operate. ISO should be succeeded by the Space Infrared Telescope Facility (SIRTF), a Great Observatory run by NASA, a thousand times more sensitive than IRAS, and FIRST mentioned earlier. For their part, the Japanese plan to launch the InfraRed Imaging Surveyor (IRIS or ASTRO F) around 2003. Its two instruments, the FIS photometer (Far Infrared Surveyor) and the three-channel camera IRC (InfraRed Camera, 2 to 25 μm) are cooled to 6 K by liquid helium, although the second will be able to operate without cooling.

Other projects are NASA's Terrestrial Planet Finder (TPF) and ESA's Darwin, designed for direct detection of extra-solar planets using space-based infrared interferometry.

■ The visible–near infrared wavelength range is accessible to ground-based astronomy. In space, it is represented by NASA's Hubble Space Telescope (HST), in collaboration with ESA. The HST perhaps represents the most prestigious contribution of space to astronomy, but has encountered many difficulties. Studies began in 1972. This very large telescope has an aperture of $^1/_{24}$ and an effective focal length of 57.6 m. It orbits at 610–620 km, within the range of the Space Shuttle, and maintenance visits were originally scheduled every 30 months. This would allow solar panels or gyroscopes to be replaced, among other things. It was also planned to return the telescope to the ground every 6 years. However, NASA later announced that visits would be less frequent and return to base avoided if possible. Data is transmitted via the TDRS network which also handles communications with the Shuttle and military satellites, giving priority to the latter. Launch was originally planned for 1983, but finally took place in May 1990. On 6 June, a defect was discovered in the concavity of the primary mirror. The five original optical instruments were rendered inoperative: the Wide Field/Planetary Camera (WF/PC), Faint Object Camera (FOC), Faint Object Spectrometer (FOS), High-Resolution Spectrometer (HRS) and High-Speed Photometer (HSP). It was impossible to change the primary mirror in flight but two solutions were found. The first involved replacing each instrument by new ones that had been corrected optically. The second would

exploit the fact that the last four of the instruments mentioned use the same optical axis. A correcting device could thus be inserted upstream and it would only remain to replace the WF/PC. It was this second solution that was adopted during the first service mission (STS 61) in December 1993. The HSP was dropped to make room for the correcting device COSTAR (Corrective Optics Space Telescope Axial Replacement). This improved the resolution of the other 3 instruments by 70%. In February 1997, the second service mission (STS 82) replaced the HRS by the Space Telescope Imaging Spectrograph (STIS) and the FOS by the Near Infrared Camera and Multi-Object Spectrometer (NICMOS). COSTAR is now only needed for the FOC, which ESA preferred not to replace for budgetary

Figure 7.22. **Image taken by the Hubble telescope in May 2000.**
An exceptional alignment shows a spiral galaxy in front of another spiral galaxy. This alignment allows us to observe the dark material in the foreground galaxy, visible only by its silhouette in front of the more distant galaxy. The pair of galaxies is called NGC 3314. (Courtesy of NASA and the Hubble Heritage Team.)

reasons. Fourteen years after the date originally scheduled, the HST was finally operating according to plan. Despite progress in ground-based observation over the same period, HST still slightly outperforms telescopes on Earth: it has a resolution of 0.01 arcsec compared with a limit of 0.3 arcsec at ground level, and it can operate in the ultraviolet (fig. 7.22).

Four instruments are thus still collecting data today. WF/PC 2 operates from the far ultraviolet to the near infrared (120 to 1100 nm). It can function in wide field mode (0.1 arcsec per pixel) and in planetary mode (0.043 arcsec per pixel). STIS can study objects across a spectral range from the UV (115 nm) through the visible red and near-infrared (1000 nm). NICMOS provides capability for infrared imaging and spectroscopic observations of astronomical targets. It detects light with wavelengths between 0.8 and 2.5 μm. FOC was built by the European Space Agency. There are two complete detector systems for the FOC. Each operates between 115 and 650 nm and uses an image intensifier tube to produce an image on a phosphor screen that is 100 000 times brighter than the light received. This system is so sensitive that objects brighter than 21st magnitude must be dimmed by the camera's filter systems to avoid saturating the detectors. Even with a broad-band filter, the brightest object which can be accurately measured is 20th magnitude.

Hubble's successor, the Next Generation Space Telescope (NGST) under development at NASA is planned for launch around 2007. It will have spectrometers covering a region ten times greater than Hubble in the near infrared. Although it has proven extremely useful to be able to access Hubble with the Space Shuttle, the NGST will probably be placed at Lagrange point 2 in order to keep the Sun, Earth and Moon out of its field of view, since these would perturb its observations.

In the same wavelengths, the minisatellite project Picard designed by the French CNES should open the way to research in asteroseismology in about 2002. By observing luminosity variations in certain stars, it will be possible to obtain information about their mass, age, rotation and magnetism, as well as detecting any planetary transits in front of the star.

■ The study of ultraviolet radiation was first seriously taken up by the International Ultraviolet Explorer (IUE), launched in 1978, following a few preliminary experiments carried out for the main part by NASA and ESA. A complete sky survey was pursued by the Extreme Ultra-Violet Explorer (EUVE) using EUV spectroscopy. This provides data on the physical and chemical properties of hot gases and plasmas. The Far Ultraviolet Spectroscopic Explorer (FUSE) launched in June 1999 for high resolution spectroscopy of faint sources precedes the GALEX project, a SMEX equipped with a UV imaging spectrometer to study the evolution of galaxies. HETE failed to deploy in 1996, but the French–Japanese High Energy Transient Explorer (HETE 2) should detect X-ray and UV emissions accompanying gamma bursts at sufficiently short notice to be able to point ground-based optical observatories.

■ High energy astronomy includes X-ray and γ-ray observation, studying radiation with wavelength shorter than that of ultraviolet light. This type of radiation is usually specified by energy rather than wavelength. For example, the rather loosely defined boundary between X-rays and γ-rays can be situated around 500 000 eV.

■ X-ray astronomy deals with the following kinds of source: stellar coronas, supernova remnants, active galactic nuclei, binary systems, quasars, and accretion phenomena near black holes. A general survey of sources and structural study of galactic nuclei was carried out by Rosat (Rœntgen Satellite, Germany–UK) from 1990 through to the beginning of 1999. Astro D (or ASCA, the Advanced Satellite for Cosmology and Astrophysics), Rossi XTE (X-ray Timing Explorer) and SAX (Satellite Astronomia Raggi-X) continue the study of rapid fluctuations in specific sources. Astro E was to take over from Astro D by carrying the first quantum X-ray calorimeter into space, but it was destroyed at launch on 10 February 2000. The ESA satellite XMM (X-ray Multiple-mirror Mission), subsequently known as Newton, is the second Cornerstone of the ESA's science programme. It was launched in December 1999 into an eccentric orbit inclined at 40°, with a period of 47.86 h corresponding to twice the Earth's period of rotation. Equipped with the European Photon Imaging Camera (EPIC), the Reflection Grating Spectrometer (RGS) and the Optical Monitor (OM), one of its main subjects of investigation is the spectroscopy of heavy elements such as iron and silicon in stars and binary systems. Launched on 23 July 1999, the AXAF satellite (Advanced X-ray Astrophysics Facility), renamed the Chandra X-ray Observatory (CXO) after the astrophysicist Chandrasekhar, is one of NASA's Great Observatories. It increases resolution of all X-ray sources by a factor of 100, making it possible to study the central black hole in the Galaxy. It will be followed by a constellation of four satellites to be deployed around the libration point L2 in 2007–8. This is the Constellation X-ray Mission (previously HTXS). It should obtain detailed spectral signatures of hot plasmas and thereby determine the composition, temperature and velocity of the emitting material. This should bring a better understanding of black holes. Finally, Russia intends to launch the Spectrum XG satellite using a Proton rocket. This will succeed Granat (1989–1997) in X-ray and γ-ray observation. Spectrum X (Spectrum-Roentgen-Gamma, SRG, SXG) will follow an orbit inclined at 51.5° with apogee 200 000 km and perigee 10 000 km. It will have a total mass of 5700 kg, with 2750 kg of scientific instruments supplied by various countries, including Denmark, the UK, Germany,

Italy, the USA, Finland, Switzerland, Israel, Hungary, Kyrgyzstan, Canada and Turkey.

■ Gamma astronomy observes objects emitting in the highest energy region of the electromagnetic spectrum. It is therefore particularly relevant to violent processes occurring in stellar and galactic evolution, revealed as gamma bursts. Gamma astronomy requires sophisticated instruments. It was first developed with the launch of the American SAS 2 in 1972 and European COS B in 1975. The Soviet satellite Granat launched in 1989 studied gamma radiation, in particular using the French telescope SIGMA. The very large Compton Gamma-Ray Observatory (CGRO), one of NASA's Great Observatories, was launched in April 1991 and deorbited in June 2000. It was equipped with four detectors covering a vast range of energies from 20 keV to 30 GeV. The Spectrum XG satellite described above will also study gamma radiation.

The SGRBE mission (Swift Gamma Ray Burst Explorer) has been scheduled by NASA to study gamma bursts, within the framework of the MIDEX programme. These satellites should be superseded by the INTErnational Gamma-RAy Laboratory (INTEGRAL), a joint project between ESA and Russia, and the Gamma-ray Large Area Space Telescope (GLAST), a joint project bringing together NASA, ESA and ISAS.

Other fields of investigation

Fundamental physics

In space, there are new possibilities for experimenting with the two main theories of modern physics, general relativity and quantum mechanics. All experiments planned involve measuring or detecting gravity.

■ A first type of experiment seeks to check the principle of equivalence between (passive) gravitational and inertial mass. The idea is to measure the relative acceleration of two freely moving masses with different compositions in a drag-free satellite orbiting in the Earth's gravitational field. Examples are NASA's MiniSTEP project (Satellite Test of the Equivalence Principle), with ESA participation, and the MicroSCOPE project from the CNES. The first measures accelerations by superconducting quantum interference devices developed at Stanford University, and the second uses cryogenic electrostatic accelerometers developed at the French Office National d'Etudes et de Recherches Aérospatiales (ONERA).

■ A second type of experiment seeks to measure the post-Newtonian parameter γ already estimated from naturally occurring relativistic effects: precession of the major axis of Mercury's orbit, bending and time delay of a light signal as it passes through the Sun's gravitational field, and deviation in the angle of reception of radio signals from quasars. In this category, NASA's GP 2 project (Gravity Probe) will compare the precession of cryogenic gyroscopes in polar orbit with the values predicted by theory. A more distant project, ESA's Solar Orbit Relativity Test (SORT) will measure the angle between two satellites following the same heliocentric orbit and the delay in the return of light pulses, using a ground-based laser and an ultrastable onboard atomic clock. A preparatory mission for SORT is ACES (Atomic Clock Ensemble in Space), scheduled for 2003 aboard the International Space Station. This will test two clocks and the laser synchronisation link between Earth and the onboard clocks.

■ A third type of experiment aims to detect gravitational waves directly using devices carried by widely separated satellites, the whole setup operating as an interferometer in the frequencies 0.1 to 0.0001 Hz. The satellites will be linked by ultrastable laser and carry ion propulsion engines for drag compensation (drag-free satellites). The NASA project Omega comprises 6 MIDEX minisatellites in very high geocentric orbits. They will move in pairs at the corners of an equilateral triangle with side one million kilometres. ESA's LISA project (Laser Interferometer Space Antenna) comprises 3 satellites forming an equilateral triangle of side 5 million kilometres, all following the same heliocentric orbit as the Earth and at 20° from it. Targets will be very powerful sources such as close binary systems, active galactic nuclei, coalescence of two giant black holes during galaxy collisions, and the formation of black holes. Both projects are very difficult to realise and may be merged.

Effects of microgravity and cosmic rays

In this domain, a great many experiments are carried out aboard space vehicles. They have three main aims. A first objective is to understand the effects of microgravity and cosmic rays on the lives of human beings in the spacecraft. A second objective is to compare biological processes on Earth, in the presence of gravity, with those occurring in space, in

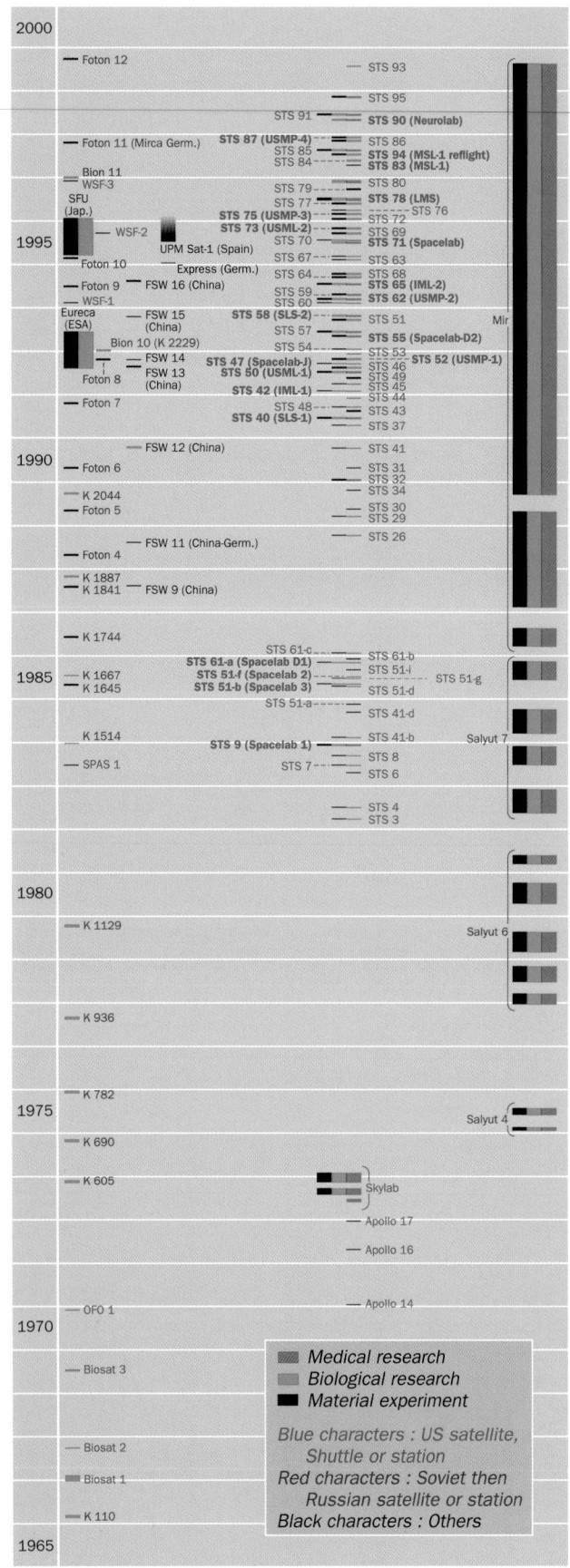

microgravity conditions, in order to understand the specific role played by gravity. A third objective is to carry out experimentation and production of materials that cannot be done in the weight conditions on Earth.

The first experiments naturally focused on biological questions, in order later to tackle the medical aspects of human presence in space. The first living creatures to experience microgravity were the dog Laika on board Sputnik 2 in 1957, and the dogs travelling aboard the Vostok spacecraft prior to Gagarin's inaugural flight. Experiments were carried out on other animals in space, including a chimpanzee on Mercury, a macaque monkey on Biosatellite III, and bullfrogs on OFO 1 (Orbiting Frog Otolith). The latter mission chose bullfrogs because they have a very similar vestibular organ (balance system) to human beings.

Although many observations were made of astronauts on both American (Mercury, Gemini, Apollo) and Soviet spaceflights (Vostok, Voskhod), they were limited to simple measurements of heart rate and breathing. The most detailed research compared the state of astronauts before and after flight. The rudimentary nature of equipment carried on board the satellites only hindered the development of biological and medical research in space. Experiments became more frequent and more sophisticated when full scale space stations had been set up.

Aboard Skylab, experiments were carried out on animals, e.g., the physiological cycles of mice, circadian cycles of vinegar flies, the weaving of spider's webs, and also on plants, e.g., growth of rice. However, most experiments concerned humans, in particular metabolic changes during spaceflight. Research in materials science was also developed on Skylab. This domain had also witnessed some previous experimentation, in particular aboard Apollo 14, 16, and 17. In the physics of materials, almost zero apparent gravity modifies any system in the fluid phase with strongly varying density distribution. The lack of convection or sedimentation in such conditions opens the way to new industrial applications, particularly in the field of alloys and crystals.

In addition to the development of recoverable capsules such as the Russian Foton or Chinese FSW, the launch of Salyut 5, and later Salyut 6, followed by Mir on the Soviet side and the successive American Shuttle flights marked the beginning of a new era in the biological, medical and materials sciences. Each gained in some specific way from the space context, often as part of international collaboration (fig. 7.23). Likewise, exobiological experiments carried out over

Figure 7.23. **Chronology of satellites carrying out research on the physics of materials and the life sciences in microgravity conditions (1965–2000).** Shuttle flights whose principal mission concerns microgravity are indicated in bold type. Preparations for the International Space Station explain the reduction in space-based experimentation in 1999 and 2000.

three months outside Mir paved the way to a new understanding of the capacity of molecules to survive in the space environment, since repeated observation of organic matter on micrometeorites collected in the Antarctic had raised fundamental questions about the origins of life on Earth.

In biology, apart from medical studies, microgravity experiments have been carried out on both plants and animals, and in particular rats. Animal osteoporosis was studied on Russian recoverable capsules Bion 10 and 11, and also aboard Spacelab 2. The fertilisation and development of embryos has been widely studied for various animals. The first fertilisation of xenopus frog's eggs was achieved in 1992, aboard the IML laboratory (STS 42). The experiment was repeated in 1994, aboard STS 65 in the context of the second International Microgravity Laboratory (IML 2). The first embryonic development of the xenopus in space from egg-laying to hatching was achieved during the SL-J mission. The first in vivo fertilisation was realised with newts in July 1994 as part of a Japanese experiment on board the American Space Shuttle, in the context of IML 2. The French experiment FERTILE in 1996 and 1998 (Fécondation et Embryogenèse Réalisée chez un Triton In vivo dans L'Espace), carried out aboard Mir and completed by Génésis in 1999, obtained similar results with the Spanish ribbed newt (pleurodeles waltl). It was shown that alterations tend to correct themselves as time goes by, without anatomic or functional abnormalities, and that subsequent reproduction is normal.

Medical research specifically aims at improving the safety of human beings in space, as they are required to stay there for ever longer periods. The effects of microgravity on the skeleton have been the subject of a great many missions, giving priority to long stay missions. Research into bone density was mainly carried out on cosmonauts aboard Mir. Some of these stayed on the space station for very long periods, up to 437 days (see chapter 13), a significant advantage for such investigations. Systematic analysis of weight-bearing and non-weight-bearing bones revealed that a lack of mechanical stresses plays an important role in space osteoporosis (bone loss) of the weight-bearing bones when missions last longer than six months.

It is particuarly important to study cardiovascular physiology given the particular problems caused by cardiovascular deconditioning when astronauts experience long term exposure to microgravity. A particular problem is orthostatic hypotension when they return to Earth, which can lead to loss of consciousness. These problems were studied during the flight of Spacelab D1 aboard STS 61 and also during the Spacelab Life Sciences mission (SLS 2) on STS 58, but particularly through the various missions aboard manned Russian space stations.

Neuroscientific investigations have been carried out to determine the influence of microgravity on the central nervous system. NASA's Neurolab mission was part of a 26 experiment international collaboration largely concerned with the human body. Likewise, experimentation aboard Mir identified a certain number of novel mechanisms involving synaptic and neuronal plasticity phenomena in the nervous system. Examples are the role of the visual system relative to other neurosensory organs in maintaining a certain posture, or the role played by weight in memorising complex shapes.

Apart from the problems caused by weightlessness, space radiobiology is another area of importance for the life of astronauts in space. Naturally, these questions were raised at the very beginning of the space age, with specialised satellites such as Biosat 2 following later. As space stations were developed and space walks became more common, problems caused by radiation from space received even greater attention. A quantitative and qualitative evaluation of irradiation was carried out using various physical and biological radiation dosimetry systems, such as Nausicaa or CIRCE aboard Mir. These studies revealed the influence of spatial and temporal variations in the magnetic field and in particular the effects of solar flares.

In the area of materials science, many experiments have concerned crystal growth, e.g., aboard the various Foton satellites, the Kristall module set up on Mir in 1990, and also many Shuttle missions. Among a large number of other examples, it is worth mentioning the U.S. Microgravity Laboratory (USML) which flew with STS 50 in 1992, then with STS 73 in 1995, and Microgravity Science Laboratory 1 (MSL 1) on STS 83, then on STS 94 in April and July 1997. Flight STS 87 carried USMP 4 (U.S. Microgravity Payload) which focused on materials science: the MEPHISTO experiment concerned crystal growth, whilst PEP (Particle Engulfment and Pushing by a solid/liquid interface) investigated the solidification of liquid metal alloys. Regular research has also been carried out on combustion in the absence of convection, in order to ensure fire safety aboard space vehicles, particularly during the above USMP 4 operation, the ELF experiment (Enclosed Laminar Flames) and aboard Mir (see fig. 7.23). Other experiments have taken place on unmanned platforms, unperturbed by human presence. These platforms are placed in orbit and recovered by the Space Shuttle during the same flight, as in the case of the Wake Shield Facility (WLF), or recovered during the next flight, as in the case of Eureca. They can also be launched by rocket and recovered by the Shuttle, as happened with the Japanese Space Flyer Unit (SFU), launched by an H2 rocket.

Aboard Mir, the hydrodynamics of critical fluids has received particular attention amongst various research topics concerning the physics of matter. In materials science, the fact that there is no convection has made it possible to measure mass diffusion coefficients to great accuracy, a boon for the modelling of industrial metallurgical processes.

CHAPTER EIGHT

Exploration beyond geocentric orbit

- **EXPLORATION AND GEOGRAPHY**

- **THE MOON**

- **SOLAR SYSTEM OBSERVATION MISSIONS**

Exploration and geography

Traditionally, geography is concerned with the journeys and itineraries of navigators and explorers. It was the desire to understand and explain the various features discovered in new territories that supplied the driving force for 19th century geography.

Before perhaps creating in the very long term an extension to inhabitable places, exploration of other planets in the Solar System is useful for two reasons: firstly, it opens the way to a geography of completely new paths, and secondly, it provides a rich field of comparison for physical studies of Earth itself (meteorology, internal and external geodynamics, and others).

Whilst all previous voyages were made from one point on Earth towards another region that always remained in the same relative position with regard to the starting point, trajectories in space involve much more original spatio-temporal coordinate systems. The path is calculated in such a way that the direction followed by the space probe actually aims at an unoccupied point in space at the time of departure, and in fact this point will remain unoccupied until the probe meets its target there at the very end of the journey. In addition, the fact that the target is moving means that certain periods of time are much more propitious for certain types of launch. Such periods are called launch windows (see figs. 2.49 and 2.50) and repeat themselves with the synodic period.

Exploration should result in a better understanding of the cosmic environment and celestial bodies within the Solar System. The first step in obtaining such knowledge involves cartographic survey of extraterrestrial geomorphology. Naturally, this geomorphology is interpreted in terms of our experience of terrestrial geomorphology. On the other hand, the latter also gains from study of planetary environments, which differ so widely in their histories, whilst at the same time retaining many similarities.

The advent of space exploration has meant that geomorphology on a planetary level can only be a comparative geomorphology today. Impact craters, for example, are poorly understood on Earth, where erosion and biotic changes tend to heal the wounds of meteoritic collisions, blurring their scar out of all recognition. Studying craters on the Moon, Mercury or Mars can only bring a better understanding of Earth's own ancient history.

Even meteorology finds comparison in the study of atmospheric phenomena on other planets. An example is the greenhouse effect observed on Venus, or the large-scale motions of the atmospheres of Mars and Venus. Without doubt, terrestrial oceanography can itself benefit from investigations of fluid circulation on planets where the Coriolis force is weak, such as Venus, which rotates so much more slowly about its axis than Earth.

For various reasons, the Moon and planets do not catch the attention of the general public and the scientific community in equal measures. Our natural satellite, the Moon, is the closest celestial body to Earth, and it has excited the imagination of humankind since the earliest times. Six centuries before Jesus Christ, philosophers like Thales of Miletus and Pythagoras had already judged the Moon to be similar in nature to Earth. The question as to whether life could exist on the Moon was therefore posed. In his work *De Facie in orbe lunae* (On the face of the Moon's disk), Plutarch once again exposed the arguments for and against in the first century AD. But it was Galileo's telescope that marked the beginning of a new era in the geography of the imagined occupation of space. Observations became more frequent and astronomers discovered the topography of the lunar surface, even producing the first maps. In *Somnium* (The Dream), J. Kepler made use of this new curiosity to familiarise the reader with astronomy's most recent discoveries: the different lengths of day and night and the extent of temperature variations. The existence of life on the Moon was not excluded, but it would have to manifest very different characteristics to the human life form. Tales of imaginary space travel began to appear with the works of J. Wilkins, F. Godwin and Cyrano de Bergerac, and then the science-fiction writings of E.A. Poe and J. Verne who made the Moon the objective of their travellers' whim. Even today, the Moon is the only celestial body apart from Earth to have felt the tread of human foot.

After the Moon, Mars and Venus share our fascination. Venus was perhaps the first to catch our attention by its brilliance, although only observable in the morning or evening sky, since it never moves more than 48° from the direction of the Sun. But progress in observational techniques have surely favoured Mars, the Red Planet, whose highly contrasted aspect appears particularly clearly at opposition. Mars was mapped as early as 1659 by the Dutch astronomer C. Huygens, following the first sketches by F. Fontana. This would have been impossible for Venus whose great brightness is due to a thick and all-obscuring cloud cover. Observation of Mars and the question of life forms on this planet raised many controversies, however. Following the famous novel by H.G. Wells, the Martians literally invaded science-fiction literature.

The other planets, such as Mercury or the giant planets, have never aroused the same interest on the part of the general public. Only comets have fascinated, and as often as not, terrified the imagination of human beings. Probes have advanced our understanding of comets to some extent, in particular with the study of comet Halley during its passage through the plane of the ecliptic in 1986.

Space exploration is the prerogative of a small number of countries, just as the great discoveries of the 15th and 16th centuries can be attributed to the efforts of a very few nations. In those days, the prerequisites were access to the oceans, financial resources and a certain level of culture and technology. Likewise, with its own special requirements, planetary exploration was initially accessible only to the Soviet Union and the United States of America, although Europe and Japan have since launched their own space probes (see fig. 8.6). Finally, exploration programmes are tightly bound to the political and budgetary context. The 1960s were witness to the first unmanned probes and also manned missions to the Moon, whilst the 1970s really saw space exploration taking wing, with the Viking missions to Mars, Mariner 10, the success of the Pioneer craft to Jupiter, Venus and Saturn, and the departure of Voyager. During the 1980s and 1990s, ever more sophisticated and expensive probes were launched. Their exceptionally wide range of scientific instruments (18 on Galileo, 20 on Cassini–Huygens) meant longer and longer periods of design and development.

This has appeared as a sensitive issue with the failure of two of these probes, loss of communication with Mars Observer before it went into Mars orbit in 1993 and the launch failure of Mars 96, misadventures which were to signal significant changes in the organisation of planetary exploration. Without sufficient financial backing, the Russian programmes were shelved one after the other and would not appear today to be any nearer to revival without strong cooperation incentive. Beginning in 1994 with the launch of Clementine, a 500 kg class probe jointly developed with the BMDO, NASA adopted a 'faster, better, cheaper' approach. Missions of the future should carry miniaturised, light-weight payloads (20 to 30 kg) and involve a small number of experiments (2 or 3) and a short development period (2 to 3 years) whilst costing no more than 250 million dollars, i.e., under one third the cost of Mars Observer. The Discovery strategy led in 1996 to the launch of NEAR (Near Earth Asteroid Rendezvous), closely followed by Mars Pathfinder, Lunar Prospector and Stardust. After Mars Pathfinder, exploration of Mars was soon to benefit from a special programme (Mars Surveyor) with similar features. Then came the successive failures of Mars Climate Orbiter and Mars Polar Lander in 1999, arousing strong criticism of the 'faster, better, cheaper' doctrine.

At the same time, the European scientific community had been hard hit by the failure of Mars 96. (Most of the experiments on board had originated in European laboratories.) With the thinning out and disappearance of Russian missions, an easy and relatively inexpensive means for Europeans to gain access to space was under threat. Moreover, scientists in Europe were beginning to fear that NASA missions might also become less accessible than in the past, since each probe carried few instruments and a single principal investigator was appointed to them at the beginning of every project. They therefore held high expectations of ESA, whose planetary programme had to take account of this new situation.

ESA had been preparing two substantial missions since 1985, namely Cassini–Huygens and Rosetta. However, faster and lighter missions such as Mars Express will soon characterise its future orientation. In the same vein, the SMART programme of small hi-tech probes will be applied to exploration and microsatellite missions may be launched via the Ariane 5 ASAP system.

The Moon

From the very beginning of the space age, and even before the first flight of a human being in space, the Moon has been viewed as a prestigious target. American and Soviet scientists soon undertook to reveal its secrets and the first photographs of the far side of the Moon, taken by Luna 3 in October 1959, were widely circulated as a demonstration of the high quality of Soviet science. Coming as it did in the middle of the Cold War, such an event was bound to have a profound impact. Conquest of the Moon was finally to become symbolic when J.F. Kennedy decided to make it the primary objective of American activities in space until the end of the 1960s.

After several fruitless attempts at approach (Pioneer) and orbit (Pioneer Orbiter), NASA acquired the basic knowhow involved in manned flights through three complementary programmes (fig. 8.1).

The Ranger programme began with six failures, before Ranger 7, 8 and 9 succeeded in transmitting some 17 000 pictures with their four RBV TV cameras on board (two wide

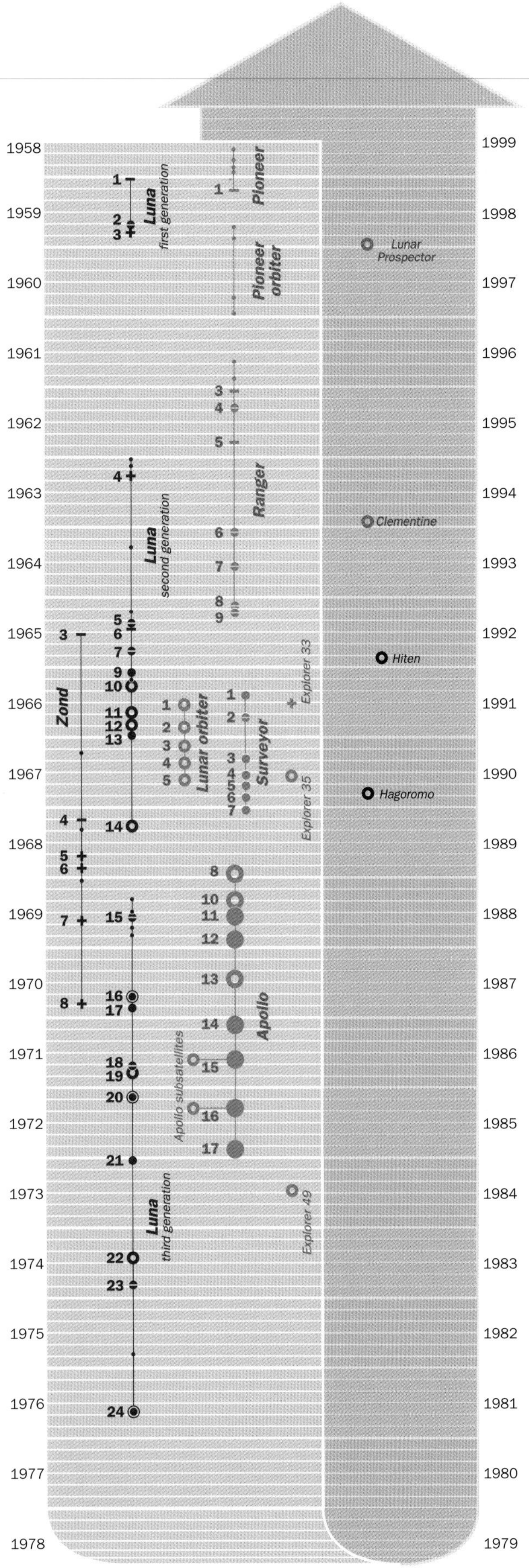

Figure 8.1. **Chronology of missions to the Moon.**

The approach to the Moon

- – – Pass-by, in heliocentric orbit
- + + Pass-by, then return to Earth
- ○ ○ ○ Into lunar orbit
- ⇌ ⇌ Crash on the Moon
- ● ● Soft landing
- ◉ Soft landing, then return to Earth
- ○ Manned mission into lunar orbit
- ● Manned mission with landing
- · · Failure

American spacecraft are shown in blue, Soviet in red and Japanese in black

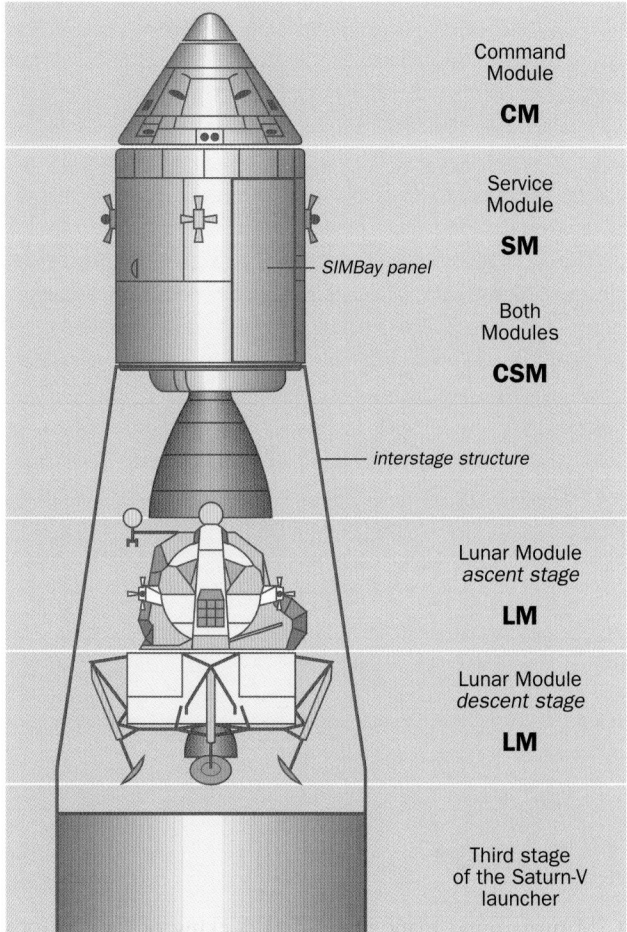

Figure 8.2. **Components of an Apollo mission at liftoff.**

Apollo vessel itself included a command module (CM) for three crew members, protected by a heat shield, and a service module (SM), which was jettisoned upon return to Earth, containing the Service Propulsion System (SPS), attitude adjustment thrusters, propellants, supplies for the crew, and the radio antenna for communications with Earth. The lunar module (LM) comprised a descent stage with its engine, and an ascent stage, with its liftoff engine, for two people. At the top of the CM and the LM was the link-up device that would enable the LM to redock after the landing. For missions including a lunar rover, the latter was fixed onto one side of the LM descent stage. The combined CM–SM was designated by the acronym CSM.

The CM and LM each received a code name after liftoff (for example, Columbia and Eagle for Apollo 11, or Odyssey and Aquarius for Apollo 13).

Command Module CM

Service Module SM

SIMBay panel

Both Modules CSM

interstage structure

Lunar Module *ascent stage* LM

Lunar Module *descent stage* LM

Third stage of the Saturn-V launcher

angle and two telescopic). The pictures, with varying degrees of resolution, were taken during the 14 minutes just prior to collision with the lunar surface.

The Lunar Orbiter programme mapped 99% of the lunar surface at medium resolution using a wide angle camera, whilst also seeking out possible landing sites and sending back high resolution images taken with a telescopic camera. The films were developed on board, optically scanned and transmitted down to Earth by radio. Further data was gathered concerning radiation intensity, micrometeorites and gravity (revealing the so-called mascons, positive gravitational anomalies often located in the middle of lunar maria). The probes were then deliberately crashed onto the surface so that they would not obstruct later missions.

The probes of the Surveyor programme went on to supply more accurate pictures of future sites during their approach, before soft landing to gather data concerning the strength and composition of the lunar soil. They also tested techniques for radar controlled landing.

Only two of these missions were unsuccessful: Surveyor 2 which crashed, and Surveyor 4 which suffered a transmission failure.

At the same time as these preliminary missions, the manned explorations announced in J.F. Kennedy's speech of 25 May 1961 became the objective of the Apollo programme, first proposed by NASA in July 1960. Three basic options were identified: direct flight, module assembly in terrestrial orbit, or rendezvous in lunar orbit. It was the last choice that was selected by the middle of 1961. The command module (fig. 8.2) and the Saturn 1 launch vehicle were tested in Earth orbit from 28 May 1964, the command and service module ensemble (CSM) going by the name of Apollo 4, and Saturn 5 was tested on 9 November 1967. The two stages of the lunar module were tested in flight in January 1968 (Apollo 5). In the meantime, on 27 January 1967, a very serious accident on the ground led to the death of three astronauts and the schedule for manned flights was frozen. It was only in October 1968 that men were first carried into low orbit (Apollo 7). Things then began to move more quickly. In December 1968, Apollo 8 put three men into orbit around the Moon for the first time, during a period of 20 hours. In March 1969 the general rehearsal of CSM turnaround manoeuvres and docking of the LM took place in low-Earth orbit (Apollo 9). Then in May 1969, a final circumlunar test was made of the separation and docking of the manned LM, with descent to just 14.5 km above the lunar surface (Apollo 10). At last, on 16 July 1969 the first man set foot on the Moon (Apollo 11).

Of the six ensuing flights (fig. 8.3), only Apollo 13 failed, owing to an accident that almost cost the lives of its crew. After Apollo 17, NASA ended the programme, although the Houston Space Centre continued to receive and process data from ALSEP (Apollo Lunar Scientific Experiment Packages) until as late as October 1977. The scientific results have been considerable, primarily through the return of samples (fig.

8.4). Their analysis has given rise to literally thousands of publications. Experiments were carried out on the lunar surface, detecting particles, analysing the composition of the atmosphere, determining electrical and physical properties of the ground material, and setting up laser reflectors, among other things. The ALSEP instruments were deployed by Apollo 12, 14, 15, 16 and 17. These instruments, which varied over the five missions, were supplied with energy from a nuclear reactor, allowing them to return data for a period of several years. They included the Cold Cathode Ion Gauge, measuring atmospheric pressure, the Passive Seismic Experiment, the Lunar Surface Magnetometer, the Lunar Surface Gravimeter, the Heat Flow Experiment, and the Lunar Seismic Profiling Experiment, which supplied information on geological characteristics down to depths of three kilometres.

In orbit, cameras and spectrometers were placed behind a movable panel in the SIMBay (Scientific Instrument Module Bay, see fig. 8.2) on the service modules of Apollo 15, 16 and 17. A spacewalk by an astronaut was required to recover the rolls of film before the service module was abandoned. The same bay was used to eject two subsatellites designed to study the Moon's gravity (mascons), as well as magnetic fields and plasmas.

For their part, Soviet scientists went through the same stages. They began with a significant lead, as demonstrated by the success of Luna 3, but then fell ever further behind, owing to lack of coordination and a serious underestimation of the budget that would be required (see chapter 13). The second generation Luna probes covered the same missions as the Lunar Orbiters (Luna 10, 11, 12, 14) or the Surveyors (Luna 5, 7, 9, 13). The manned flight programme only just got off the ground with Zond 3 to 8 before being abandoned, following repeated failure of the N-1 rocket.

The Soviets subsequently turned to an alternative programme in which unmanned vehicles were developed to bring back samples and explore the lunar soil (fig. 8.4). This programme was carried out by the third generation of Luna probes which, after several failures, succeeded in transferring 100 g (Luna 16), 50 g (Luna 20) and 170 g (Luna 24) of lunar soil and rock into a capsule, by means of a remotely controlled arm. The capsule, 50 cm in diameter and weighing 39 kg, was then placed by the ascent stage into a direct return orbit towards Earth, where it was set down by parachute. On two occasions, a Lunokhod vehicle (equivalent of the Lunar Rover) was placed on the lunar soil (Luna 17 and 21). Lunokhod 1 was in operation over a period of 11 months and travelled 10 540 m, whilst Lunokhod 2 functioned for 5 months and covered 37 km. Both transmitted a great number of photos and panoramic views. They also made physical and chemical analyses of the soil and carried a laser reflector that is still used to determine the Earth–Moon distance.

After this successful first phase of human exploration, the Moon was somewhat neglected by both Americans and Soviets alike. It was the Japanese who first attempted a further circumlunar mission. On 18 March 1990 the magnetospheric

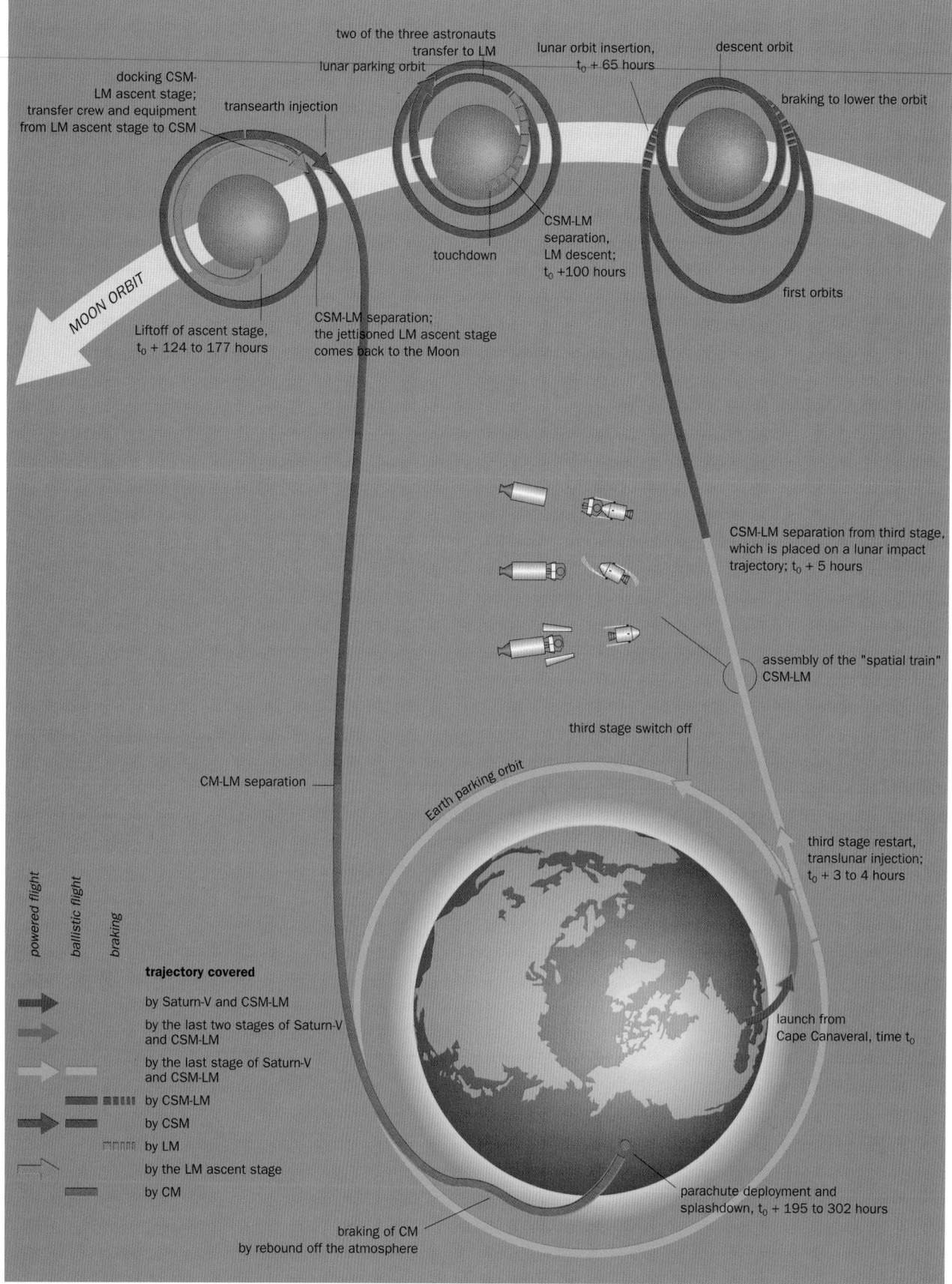

two of the three astronauts
transfer to LM
lunar parking orbit

lunar orbit insertion,
t_0 + 65 hours

descent orbit

docking CSM-
LM ascent stage;
transfer crew and equipment
from LM ascent stage to CSM

transearth injection

braking to lower the orbit

MOON ORBIT

CSM-LM
separation,
LM descent;
t_0 +100 hours

touchdown

first orbits

Liftoff of ascent stage,
t_0 + 124 to 177 hours

CSM-LM separation;
the jettisoned LM ascent stage
comes back to the Moon

CSM-LM separation from third stage,
which is placed on a lunar impact
trajectory; t_0 + 5 hours

assembly of the "spatial train"
CSM-LM

third stage switch off

CM-LM separation

Earth parking orbit

third stage restart,
translunar injection;
t_0 + 3 to 4 hours

powered flight

ballistic flight

braking

trajectory covered

by Saturn-V and CSM-LM

by the last two stages of Saturn-V
and CSM-LM

by the last stage of Saturn-V
and CSM-LM

by CSM-LM

by CSM

by LM

by the LM ascent stage

by CM

launch from
Cape Canaveral, time t_0

parachute deployment and
splashdown, t_0 + 195 to 302 hours

braking of CM
by rebound off the atmosphere

Figure 8.3. **Main phases of an Apollo mission.** The times in hours are given as an example, varying from one mission to another.

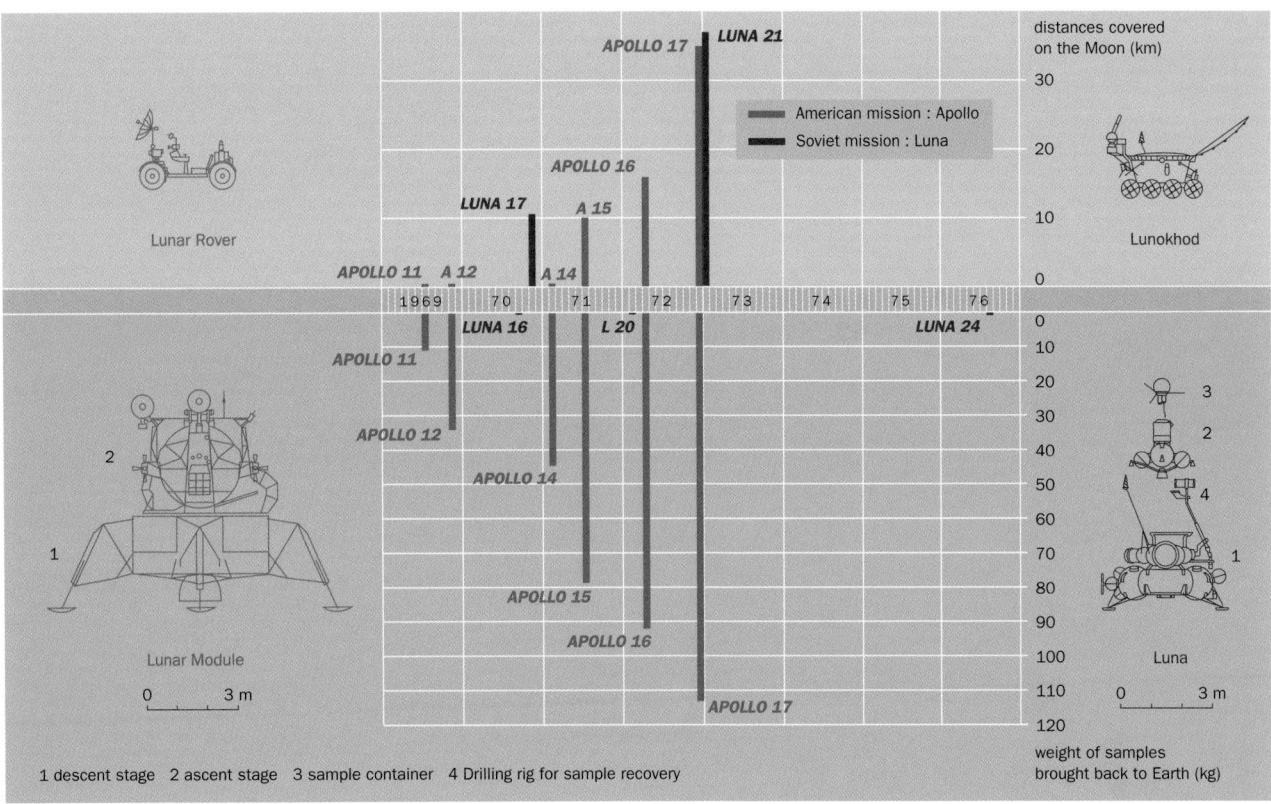

Figure 8.4. **Men and robots on the Moon: taking samples and travelling across the surface.** Soviet abandonment of a manned programme, together with a political and strategic concern to participate in the acquisition of new knowledge, resulted in their programme for unmanned exploration of the Moon. The only purpose of the American vehicles was to facilitate the astronauts' mission. Of the equipment shown in the picture, only the Soviet sample capsules were intended for return to Earth. The ascent stage of the American module served to reunite with the Apollo command module which had remained in circumlunar orbit.

satellite Hiten, precursor of Geotail, inserted a 12 kg subsatellite Hagoromo (Robe of Heaven) into a 7400–20000 km lunar orbit. Contact was lost even before the orbit was acquired. The probe Hiten itself also ended up in lunar orbit on 15 February 1992, finally impacting the lunar surface on 10 April 1993. Japan is planning to send another probe to the Moon in 2004. Lunar A will carry the Lunar Imaging Camera (LIC) and should launch two surface penetrators into the lunar soil, one on the visible side, and the other on the far side, which is permanently hidden from Earth. They will be equipped with seismometers and heat flow probes. Lunar A is intended to open the way one year later to a shared ISAS–NASDA project named SELENE (SELenological and ENgineering Explorer). It will include an orbiter and satellite relay. The orbiter will be equipped with a magnetometer, X-ray and γ-ray spectrometers, a multiband imager, a mapping camera, a laser altimeter, and a radar sounder. The gravitational field will be determined by interferometry with the relay satellite. A plasma imager will be directed towards Earth's plasmasphere, and the lunar environment will be studied by a plasma analyser and a charged particle spectrometer. Radio science experiments are also planned. At the end of the mission, the propulsion stage will detach from the probe to attempt a soft landing.

In January 1994, more than twenty years after the end of American and Soviet competition in the race to the Moon, the United States launched the probe Clementine 1. Its aim was to space qualify certain new technologies, some American (miniaturised subsystems developed as part of the American Strategic Defense Initiative), some European (on-

board JPEG data compression), to carry out scientific mapping of the lunar surface, and to observe the asteroid Geographos. Whilst moving in lunar orbit between 26 February and 3 May 1994 it transmitted 1200000 images back to Earth, without any loss of information. Clementine's later journey towards Geographos had to be abandoned following a computer breakdown.

The programme known as Space Exploration Initiative (SEI), announced by President G. Bush on the twentieth anniversary of the first manned lunar landing, seemed to demonstrate renewed interest in our natural satellite, with different aims to those motivating its initial conquest. However, financing was later refused for this programme. The Lunar Prospector mission originally considered as the first step in the SEI was integrated into NASA's *Discovery* strategy in February 1995. The Lunar Prospector probe was finally launched in July 1998. The main aims were geochemical studies by gamma spectrometry, gravimetric and magnetic field measurements, and the search for water ice in the permanently shaded recesses of the polar craters. Concerning the latter, the presence of water would naturally be of great

NEAR SIDE

◑ ◑ ◑ Crash		● ● Landing		◉ Landing and return to Earth

of the Japanese probe Hiten, of a Soviet probe Luna,

of an American spacecraft Surveyor (**S**), Ranger (**R**), of a manned Apollo LM (**A**), of Lunar Prospector (**LP**).

Figure 8.5 **Near side and far side of the Moon.** The Moon's period of rotation about its axis is equal to its orbital period about Earth, creating a fundamental difference between its two faces. Up until 1959, only the near side was known to astronomers, using their telescopes. With the space age, the far side was discovered for the first time.

This illustration contrasts the visible side of the Moon, used for landings, with the hidden side, which has not yet been attained by human vehicles because of the difficulty of communications with Earth. Future plans to use lunar sites are governed by the same dichotomy. Projects to build astronomical observatories thus opt for the far side, without earthshine or terrestrial radio

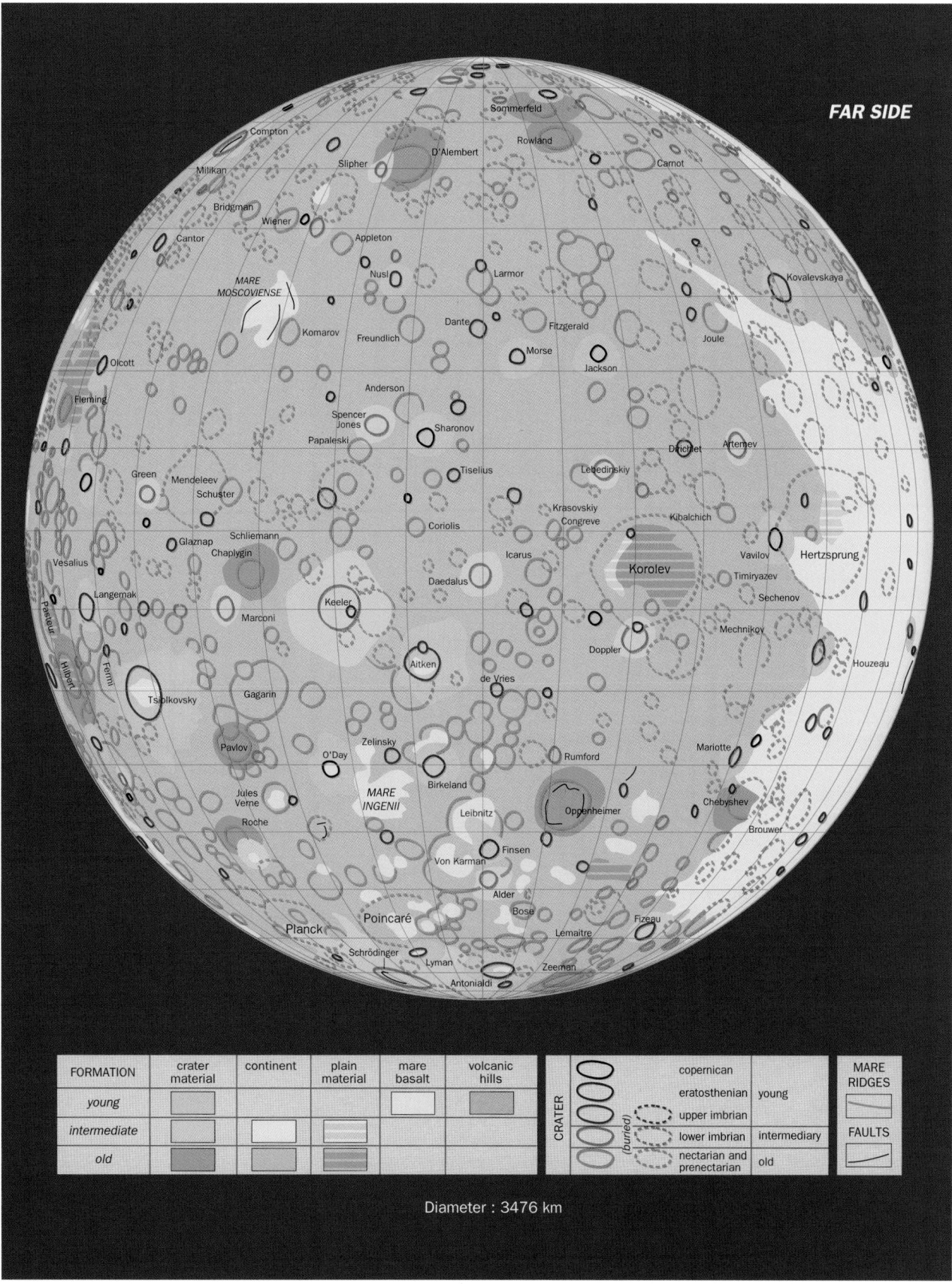

FAR SIDE

FORMATION	crater material	continent	plain material	mare basalt	volcanic hills
young					
intermediate					
old					

CRATER		
	copernican	
	eratosthenian	young
	upper imbrian	
	lower imbrian	intermediary
(buried)	nectarian and prenectarian	old

MARE RIDGES

FAULTS

Diameter : 3476 km

interference. Mining operations, on the other hand, would mainly be concerned with the earthward side for reasons of communications, and are poorly compatible with observatories anyway. The toponymy of the two lunar faces reveals the historical time lag between their geographical descriptions.

Place names on the longer known of the two have Latin origins. Those on the far side owe their Russian toponymy to the first Soviet missions, whilst American missions used the names of famous people from the world over as eponyms.

importance for any future human occupation of the Moon. Significant levels of hydrogen were detected by the neutron spectrometer at the lunar north pole and even higher levels in the south. This is interpreted as indicating that there is water ice mixed in with the lunar regolith fairly close to the surface. On 31 July 1999, at the end of its mission, the path of Lunar Prospector was modified so that it would impact the floor of a crater at the lunar south pole. At the same time, the Hubble Space Telescope and a great number of Earth-based observatories attempted spectrometric examination of the dusts raised by the collision. Although no trace of water was detected, this in no way proves its absence.

Hence, as a result of this long series of Soviet and American missions, the Moon is the best known celestial body in the Solar System, after Earth itself (fig. 8.5). It is also the only one from which rock samples have been procured.

As had already been observed from Earth with regard to its visible face, the lunar surface is composed of dark regions and bright regions. Although water is not actually visible anywhere in the lunar landscape, the dark regions are called *mare*, plural *maria*, from the Latin word meaning 'sea', following a convention adopted by the International Astronomical Union (IAU). They constitute only 2% of the Moon's far side, but more than 30% of its near side, although the southern region of each face exhibits the opposite asymmetry. The maria are low-lying and rather uniform areas, generally made up of extensive volcanic lava flows, composed of basalts that are richer in ferrous irons than terrestrial basalts. The uniformity of the seas is often disrupted by craters.

The light regions are called *terra*, plural *terrae*, or more commonly, highlands. They cover 84% of the whole surface and are made of a material very different from that of the seas. The main component is anorthositic breccias with a little norite and the occasional piece of basalt. Geomorphologically, the highlands are constituted by the juxtaposition or combination of a very great number of craters. Following a long debate as to the endogenous or exogenous origin of these craters, it has now been established that they are in fact impact craters caused by the fall of meteorites onto the Moon's surface, at least in the vast majority of cases. They fall into three classes according to size.

The smallest and simplest craters, measuring less than 10 km across and 2 km deep, generally have a rather flat internal topography. The crater Linné, with diameter 2500 m, is a good example from this category.

Larger craters, with diameters from 20 to 200 km, are more complex. Rebound phenomena create a central rocky peak with concentric terracing formed by landslip along the inner wall of the crater. The crater Tycho, with diameter 85 km, is a good example from this second category, with a central peak and terracing.

The biggest craters, measuring upwards of 200 km across, are commonly called basins. An example is the Imbrium basin, 1100 km in diameter. The central rocky peak is here replaced by more complex ring-shaped features.

Although the origin of the Moon remains controversial, its subsequent history is now much better understood. Relative chronologies can be based on geomorphological observations. The density of impacts, whose frequency has been falling throughout lunar history, reveals the age of the various regions of the surface. The most recent features are superimposed on the older ones. More recent craters might occur on a basaltic rock layer, whilst more recent basaltic flows might swallow up the outer wall of an older crater, and so on. However, in the Moon's case, a chronological stratigraphy has been established by means of absolute dating techniques such as uranium–lead or potassium–argon dating (see fig. 8.5).

Taken as a whole, the lunar landscape must be interpreted without reference to any directed flow of material, in stark contrast to what is observed on Earth. On our own planet, orientated erosion tends to wear down unsubmerged regions, accumulating eroded material in the oceans. On the Moon, no such processes have ever existed, owing to the lack of liquid flow and the absence of any atmosphere.

Solar System observation missions

Escaping from Earth's gravitational attraction, probes follow novel trajectories and widen the horizons of human observation to quite unknown proportions (fig. 8.6).

Probes observing the Sun and the interplanetary medium
Not all space vehicles placed in heliocentric orbit are destined for planetary exploration. Some aim to analyse the interplanetary medium and observe phenomena on the surface of the Sun (fig. 8.7).

Pioneer 5 was the first such mission, carrying instruments to study magnetic fields, solar plasma and cosmic rays. The following probes, Pioneer 6 to 9, formed a network for solar surveillance. They were better equipped and were able to make more detailed plasma studies, as well as observing the

Sun's surface and solar flares by telescope. It was these probes that discovered the Earth's magnetotail and made the first accurate measurements of the density of the solar plasma, in addition to observing comet Kohoutek and making related measurements.

They were followed by the two German Helios probes. With perihelia at 48 and 45 million kilometres from the Sun, they became the closest artificial objects ever to approach the Sun. Apart from studying the magnetic field, the solar plasma and cosmic rays, their mission involved detection of micrometeorites and photometry of zodiacal light.

The Ulysses mission was designed to go into polar orbit around the Sun and make observations which had until then only been carried out in the plane of the ecliptic. This was a joint project between the European Space Agency and NASA, initially described in 1979 as an international mission for the study of the solar poles. Two probes were to be launched in opposite directions. However, NASA had to abandon the construction of their vehicle in 1984. This meant that a certain number of instruments and the possibility of simultaneous

stereoscopic observations were also abandoned. After several postponements, launch was finally achieved in 1990 by the Space Shuttle Discovery. Following a first flyby of the Sun's south pole in 1994 and then the north pole in 1996, the Ulysses mission was extended until 2001, for a second flyby of the poles in 2000 and 2001 (see fig. 8.8).

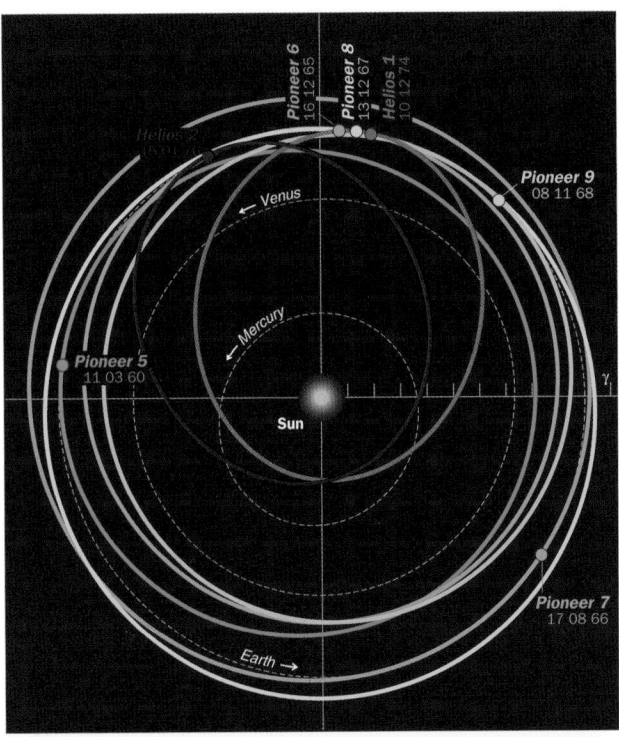

Figure 8.7. **Orbits of heliocentric probes.** Graduations on the horizontal axis correspond to 10 astronomical units and ϒ is the direction of the vernal equinox in 2000. Orbits of ISEE 3, Wind and SOHO are shown in figure 2.46.

Figure 8.8. **The trajectory of Ulysses.** The probe passed above the north pole of Jupiter, whose gravitational attraction pulled it into an orbital plane inclined at 80°. A launch speed of 42 kilometres per second would have been required to inject Ulysses into the same orbit without the gravity-assist manœuvre. Such a feat lies well beyond the capabilities of today's launch vehicles.

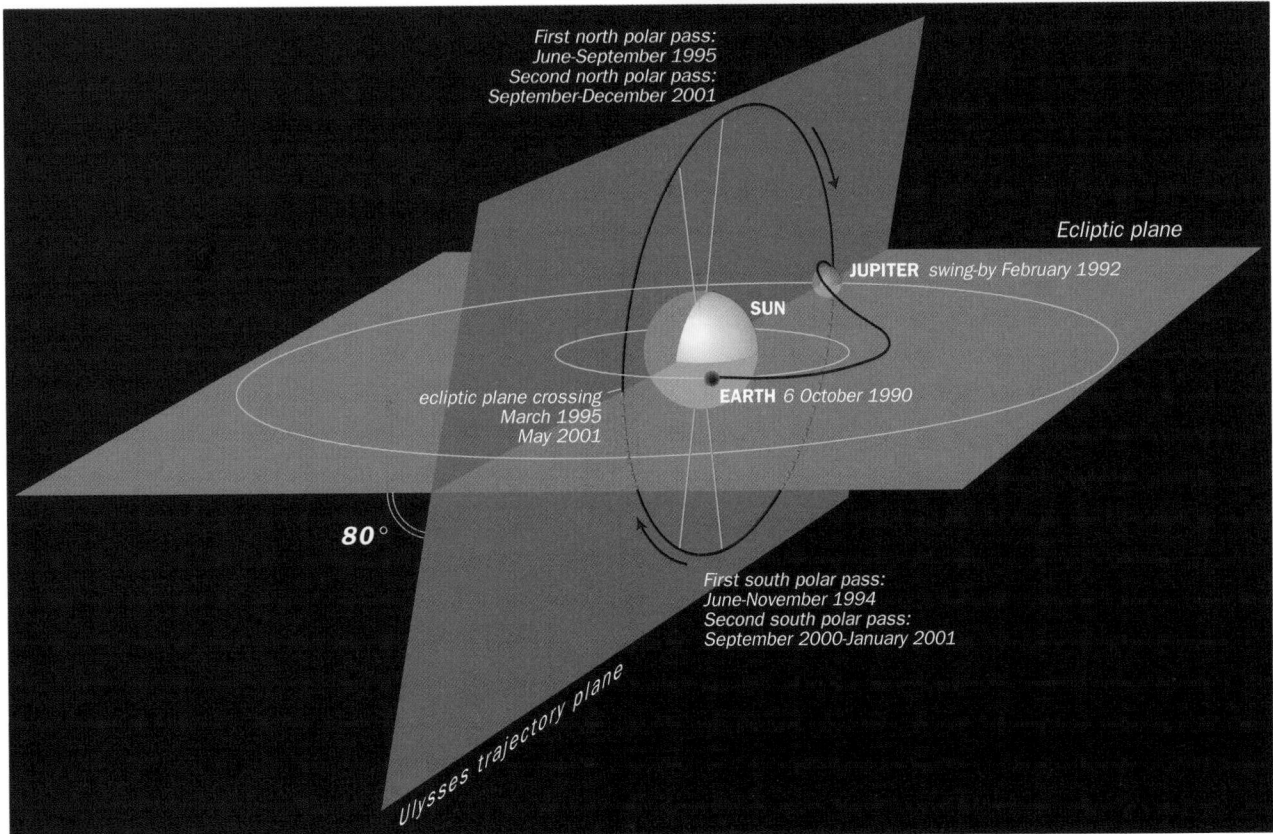

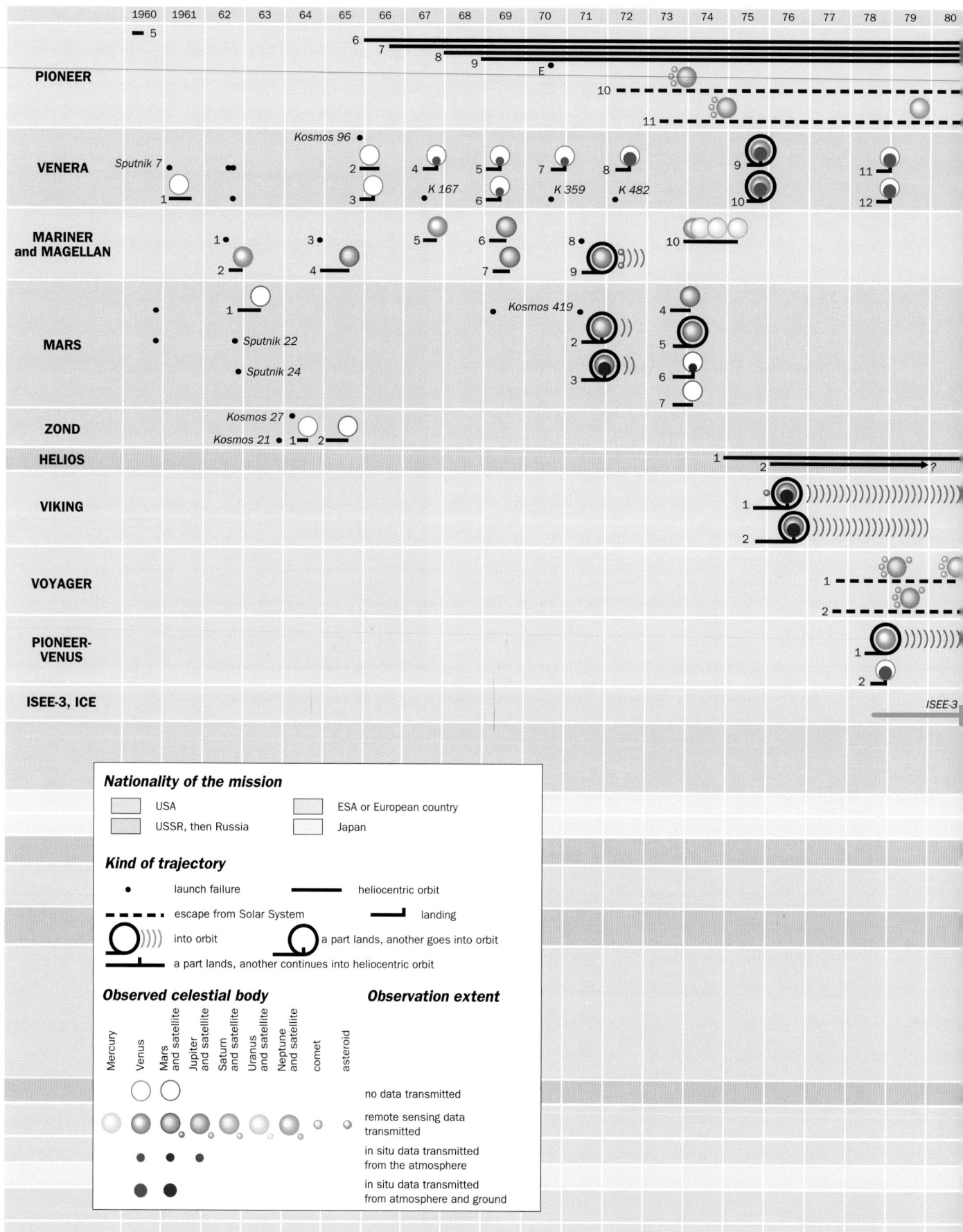

Figure 8.6. **Chronology of missions to explore the Solar System.** Lunar missions are given on figure 8.1.

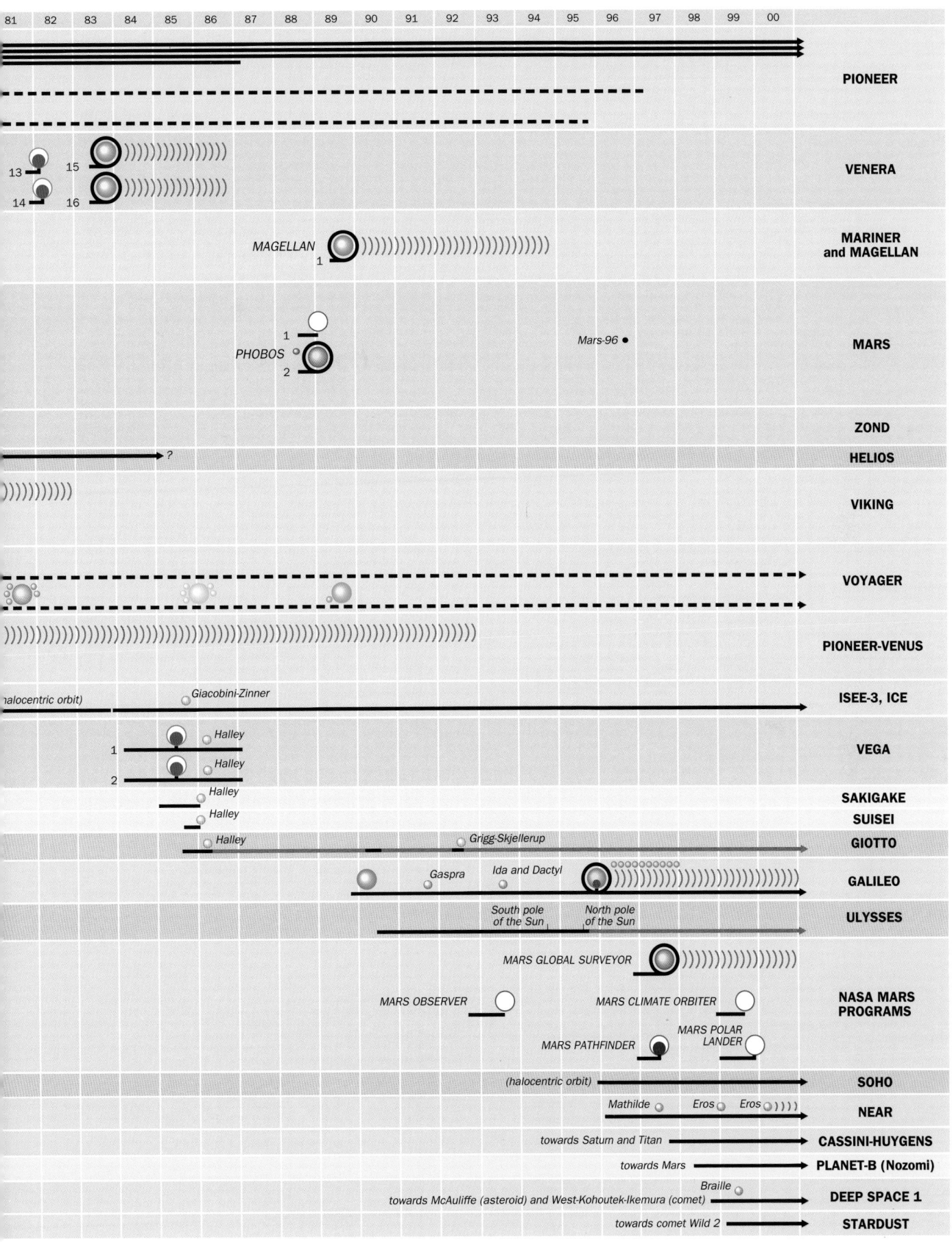

One of NASA's Discovery projects, called Genesis, launched in August 2001, will orbit in the Sun's vicinity to gather particles of solar matter and bring them back to Earth. This project complements the quest for samples of asteroidal and cometary material which should lead to a better understanding of the protosolar nebula.

Mars

Soviet probes Mars 1 and Zond 2 lost contact before their Mars flyby (fig. 8.10), so American probe Mariner 4 was the first to supply in situ data concerning this planet, in the form of video pictures, and measurements of the magnetic field and atmospheric pressure. Mariner 6 and 7 followed, transmitting 75 and 126 pictures of Mars, respectively, as well as the first images of Phobos (Mariner 7).

The Soviets launched two probes in the 1971 window and four others in the 1973 window, for little return: a few pictures from the Mars 2 and 3 orbiters and the Mars 3 descent module, rendered worthless by the 1971 sandstorm, 70 high resolution images of part of the southern hemisphere, taken by Mars 5, and in situ atmospheric measurements by the Mars 6 descent module which revealed a high concentration of argon. No further Soviet attempts were made before the 1988 Phobos missions.

In contrast, American missions between 1971 and 1975 were a complete success. Most of our current knowledge of Mars can be attributed to them. Mariner 9 was launched after Mars 2 but was first to acquire its orbit and, thanks to a greater flexibility in its programme, was able to await the end of the sandstorm before transmitting 7323 images covering 90% of the planet's surface. It also procured images of Phobos and Deimos, together with atmospheric pressure and ground temperature measurements.

With the Viking mission, the Americans landed their first descent modules on Mars. They thus obtained a whole series of data on the molecular composition of the atmosphere, the densities, temperatures and pressures from an altitude of 90 km down to ground level, the ground winds and a chemical analysis of the surface. They searched in vain for traces of biological activity. It later transpired that physical and chemical conditions on the surface would have wiped out any evidence of past life forms, and the only chance of locating any such traces therefore lies in searching deeper down. Apart from imaging Deimos, the orbiters covered the whole planetary surface with a mosaic of 2 500 images as well producing a complete thermal survey.

Exploration of Mars was taken up once more in 1988 but met with a series of failures. The first was the Soviet mission Phobos. Each of the two Phobos probes comprised an orbiter, a descent module (the DAS) that was supposed to land on Phobos and attach itself by means of a penetrator, and a second descent module (hopper) intended to move across the surface of Phobos in 20 metre bounds. Unfortunately, Phobos 1 failed through a telemetry message error and Phobos 2 lost contact with Earth when it reached the Mars neighbourhood. The only results obtained were magnetospheric measurements and infrared spectrophotometric observations of certain regions of the Martian surface.

The American probe Mars Observer was launched in 1992 but communications broke off before it went into Martian orbit in August 1993. The main aim of this costly mission (980 million dollars) was accurate mapping of the planet. A Russian mission to Mars, in which most ESA member countries were participating, was long delayed by the break-up of the Soviet Union. Mars 96 was finally launched, in a less ambitious form than the original conception, but came to grief in the Pacific soon after launch in November 1996.

Following the failure of Mars Observer, NASA reorientated its Mars exploration programme towards the lightweight probes of the Mars Surveyor programme. In November 1996, Mars Global Surveyor was launched, going into Mars polar orbit on 12 September 1997. A further American probe, Mars Pathfinder, launched in December 1996, succeeded in landing on the surface of Mars on 4 July 1997, using parachutes and airbags, following a faster outward journey.

Mars Global Surveyor began to take data by the middle of May 1999, after an aerobraking phase that was longer than expected due to the partial opening of a solar panel. The

Figure 8.9. **Mars Pathfinder, presidential panorama in Ares Vallis (see figure 8.11).** Panorama of the landing site from the Mars Pathfinder camera. Image NSSDC, courtesy of NASA.

probe carried four main instruments. The MOC (Mars Observer Camera) produced a daily wide angle image and images with resolution 1.5 m, in order to draw up the accurate map that the aborted Mars Observer mission had been unable to provide. The MOLA (Mars Orbiter Laser Altimeter) supplied hypsometric readings. The thermal emission spectrometer produced a thermal infrared survey. The magnetometer/electron reflectometer measured the magnetic field at ground level and also in the Martian ionosphere. It revealed a remnant crustal field which may well be witness to an ancient dipolar field produced by the planetary core.

Mars Pathfinder landed a small six-wheeled robot called Sojourner on the Martian surface. It was equipped with a stereoscopic pair of cameras and an alpha–proton spectrometer, both German designed, the latter being associated with a US made X-ray spectrometer (fig. 8.9). The lander carried the data transmission antenna, together with temperature, pressure and wind sensors.

In contrast, the campaign associated with the 1998 window was quite disastrous, with the loss of Mars Climate Orbiter when it crashed onto the planet following a manoeuvring error, and the still unexplained radio silence of Mars Polar Lander.

Finally, on 2 July 1998, Japan launched a satellite for observation of the Martian upper atmosphere, its ionised environment and its magnetic field. This was the Nozomi (Planet B) probe. Following difficulties orientating Nozomi towards its escape trajectory at the end of 1998, an alternative solution was found which pushed the Mars rendezvous back from October 1999 to December 2003 or January 2004.

In October 2000, following the two failures in 1998, NASA redefined its Mars exploration projects within the framework of the Mars Explorer Program. This comprises:

- Mars Odyssey to follow a 2 hour orbit, launched 7 April 2001. Instruments are THEMIS (THermal EMission Imaging System), a high resolution camera and spectrometer in the thermal infrared designed to study geomorphology and mineralogy; GRS (Gamma Ray Spectrometer) to identify chemical elements, in particular subsurface hydrogen, and MARIE to study radiation in the vicinity of the planet.
- Two Mars Explorer Rovers in 2003.

- A Mars Reconnaissance Orbiter in 2005, for very high resolution observation (20–30 cm).
- A mobile science laboratory in 2007, for prolonged, long range missions, and in particular for preparing sample-return missions.
- Two sample-return missions in 2014 and 2016. The first might be brought forward to the 2011 window, through collaboration with CNES and ASI, which are involved in sample acquisition and return and the communications network, and could provide an Ariane 5 launch vehicle.

Apart from the main missions, NASA is awaiting proposals for 'scout' missions that would place sensors in small rovers or aeroplanes. The European project NetLander may be built in at this level.

NASA has also planned a Mars–Earth communications network based on satellites in areostationary orbit (word deriving from Ares, the Greek name for Mars).

ESA has decided upon a mission for 2003 that will complement these projects. This is Mars Express which will be equipped with seven instruments: ASPERA 3, to study the interaction between solar wind and atmosphere using the energetic neutral atom technique; HRSC (High Resolution Stereo Colour imager) with a resolution of 1 metre; OMEGA, an infrared mapping spectrometer for surface minerals and atmospheric composition; MaRS, to measure variations in gravity, pressure and temperature using radio waves; an atmospheric Fourier spectrometer to measure the distribution of water vapour; SPICAM, an ultraviolet atmospheric spectrometer for atmospheric composition and ozone measurements; and MARSIS, a subsurface sounding radar altimeter to identify ice, water or dry ground down to a depth of two or three kilometres. A small lander, Beagle 2, will seek out traces of biological activity through analysis of samples excavated by a crawling mole. It will be equipped with a camera and atmospheric sensors.

Another, as yet undefined, European mission (NetLander) may use microsatellites to set up a network of geophysical sensors (seismometers) and geochemical sensors.

And finally, another objective lives on, namely the possibility of manned flight to Mars, although it may have to be postponed to a much later date.

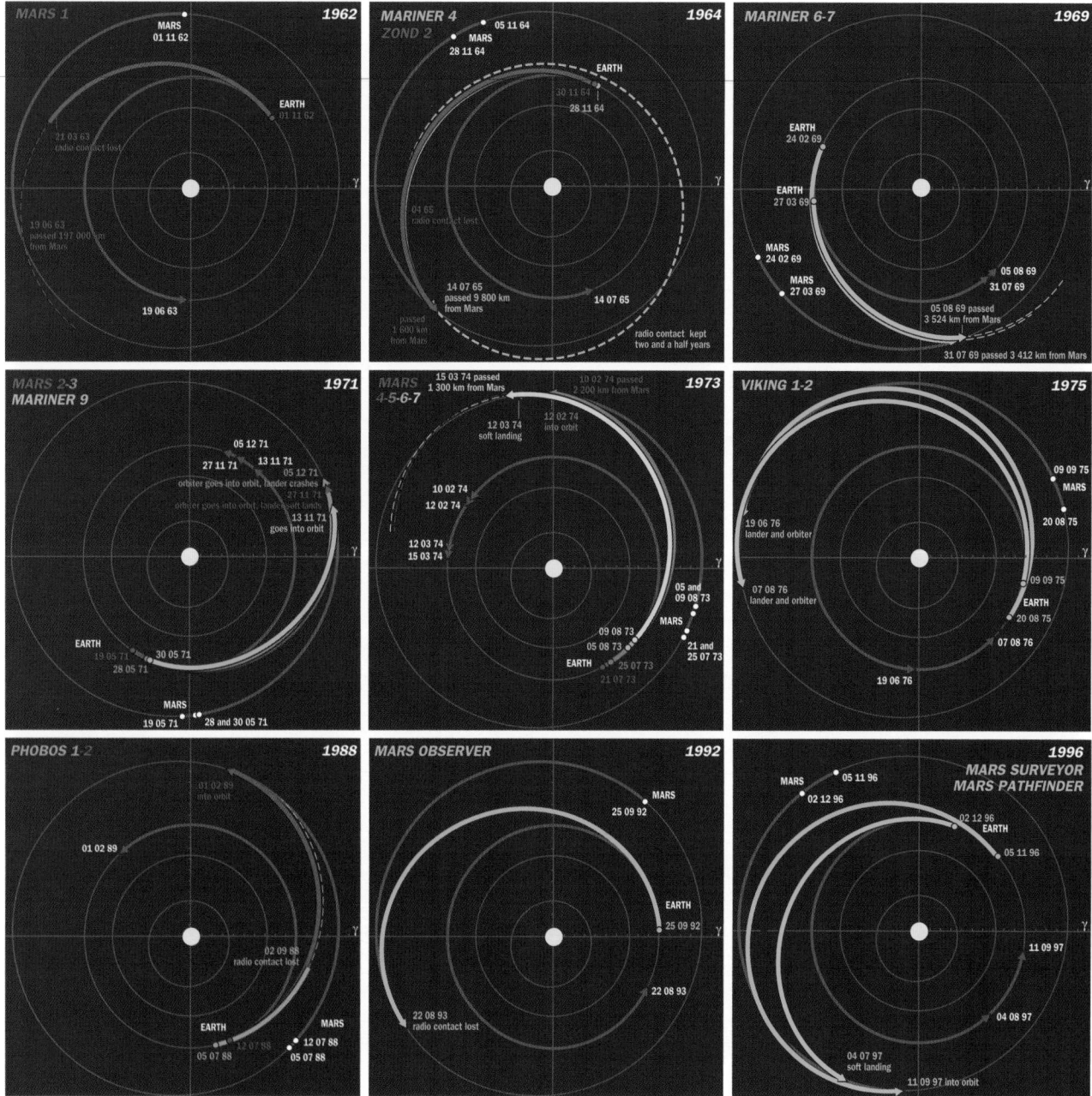

Figure 8.10. **Paths of missions to Mars.** All windows used until 2000 have been shown.

Like the Moon, Mars features a great number of impact craters. The frequency of these impacts was at its highest during the post-accretional phase, and then gradually fell off. The density of craters can be used to date geological formations. A parallel has even been attempted between the age of lunar surface regions, established in an absolute manner by radioactivity measurements, and the ages of Martian geological features. However, this method assumes some relationship between the meteoritic bombardment of the two celestial bodies, and this has not yet been established. When no absolute chronology is available, a relative chronology can nevertheless sort out formations of different ages. The most densely cratered regions are extensive in the highlands of the southern hemisphere. They correspond to the Noachian era. The cratered plains,

such as the hummocky plains and certain areas of Acidalia Planitia date from an intermediate period known as the Hesperian. The varied features of the volcanic plains, polar caps, and the stratified regions around the poles are almost crater-free and correspond to the most recent or Amazonian era.

There is no natural zero level for altitudes on Mars, so they are defined relative to an ellipsoid of revolution with an equatorial radius of 3393.4 km. Measured from this reference level, differences in altitude vary between -4 km and $+25$ km, which is a considerable range, one and a half times the maximal height difference observed on Earth.

The most typical of the Martian landscapes is perhaps represented by the very large centres of volcanic activity. A notable example is the Tharsis Rise, with an average altitude of 10 km

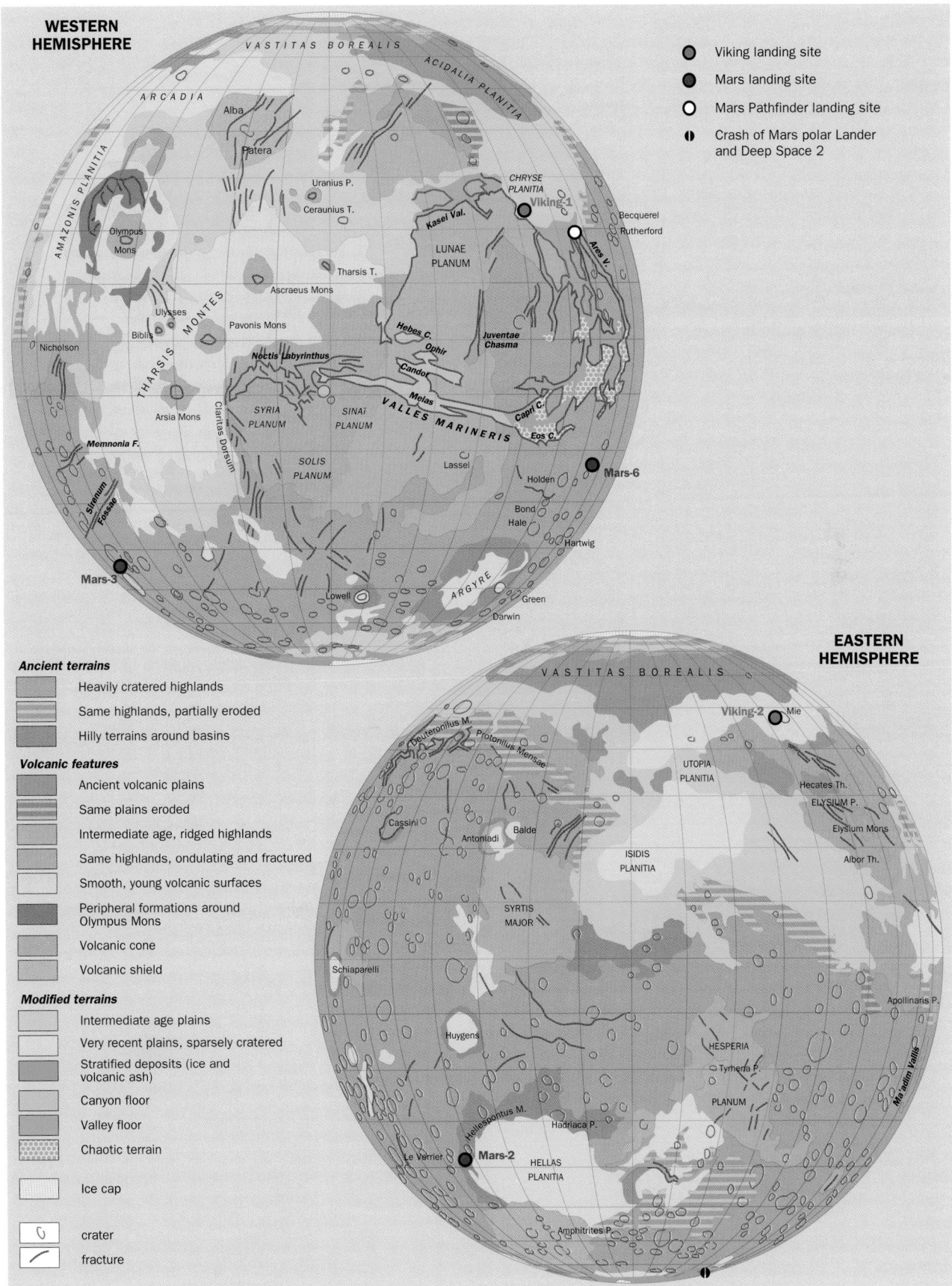

WESTERN HEMISPHERE

VASTITAS BOREALIS
ACIDALIA PLANITIA
ARCADIA
Alba
Patera
Uranius P.
CHRYSE PLANITIA
AMAZONIS PLANITIA
Ceraunius T.
Kasei Val.
Viking-1
Becquerel
Rutherford
Olympus Mons
Tharsis T.
LUNAE PLANUM
Ares V.
Ascraeus Mons
Ulysses
Bibli
Pavonis Mons
Hebes C.
Ophir
Juventae Chasma
Nicholson
Noctis Labyrinthus
Candor
Melas
Capri C.
THARSIS MONTES
Claritas Dorsum
SYRIA PLANUM
SINAÏ PLANUM
VALLES MARINERIS
Eos C.
Arsia Mons
Mars-6
Memnonia F.
SOLIS PLANUM
Lassel
Holden
Bond
Hale
Sirenum Fossae
Hartwig
Mars-3
Lowell
ARGYRE
Green
Darwin

● Viking landing site
● Mars landing site
○ Mars Pathfinder landing site
◐ Crash of Mars polar Lander and Deep Space 2

EASTERN HEMISPHERE

VASTITAS BOREALIS
Viking-2
Mie
Deuteronilus M.
Protonilus Mensae
UTOPIA PLANITIA
Hecates Th.
Cassini
Balde
ELYSIUM P.
Antoniadi
ISIDIS PLANITIA
Elysium Mons
Albor Th.
SYRTIS MAJOR
Schiaparelli
Apollinaris P.
Huygens
HESPERIA
Maadim Vallis
Tyrrhena P.
PLANUM
Hellespontus M.
Hadriaca P.
Le Verrier
Mars-2
HELLAS PLANITIA
Amphitrites P.

Ancient terrains
Heavily cratered highlands
Same highlands, partially eroded
Hilly terrains around basins

Volcanic features
Ancient volcanic plains
Same plains eroded
Intermediate age, ridged highlands
Same highlands, ondulating and fractured
Smooth, young volcanic surfaces
Peripheral formations around Olympus Mons
Volcanic cone
Volcanic shield

Modified terrains
Intermediate age plains
Very recent plains, sparsely cratered
Stratified deposits (ice and volcanic ash)
Canyon floor
Valley floor
Chaotic terrain
Ice cap

▽ crater
／ fracture

Figure 8.11. The main geomorphological regions of Mars. The two polar caps are shown, composed of water ice in the north and dry ice (frozen carbon dioxide) in the south, which is colder. Mars has a diameter of 6794 km.

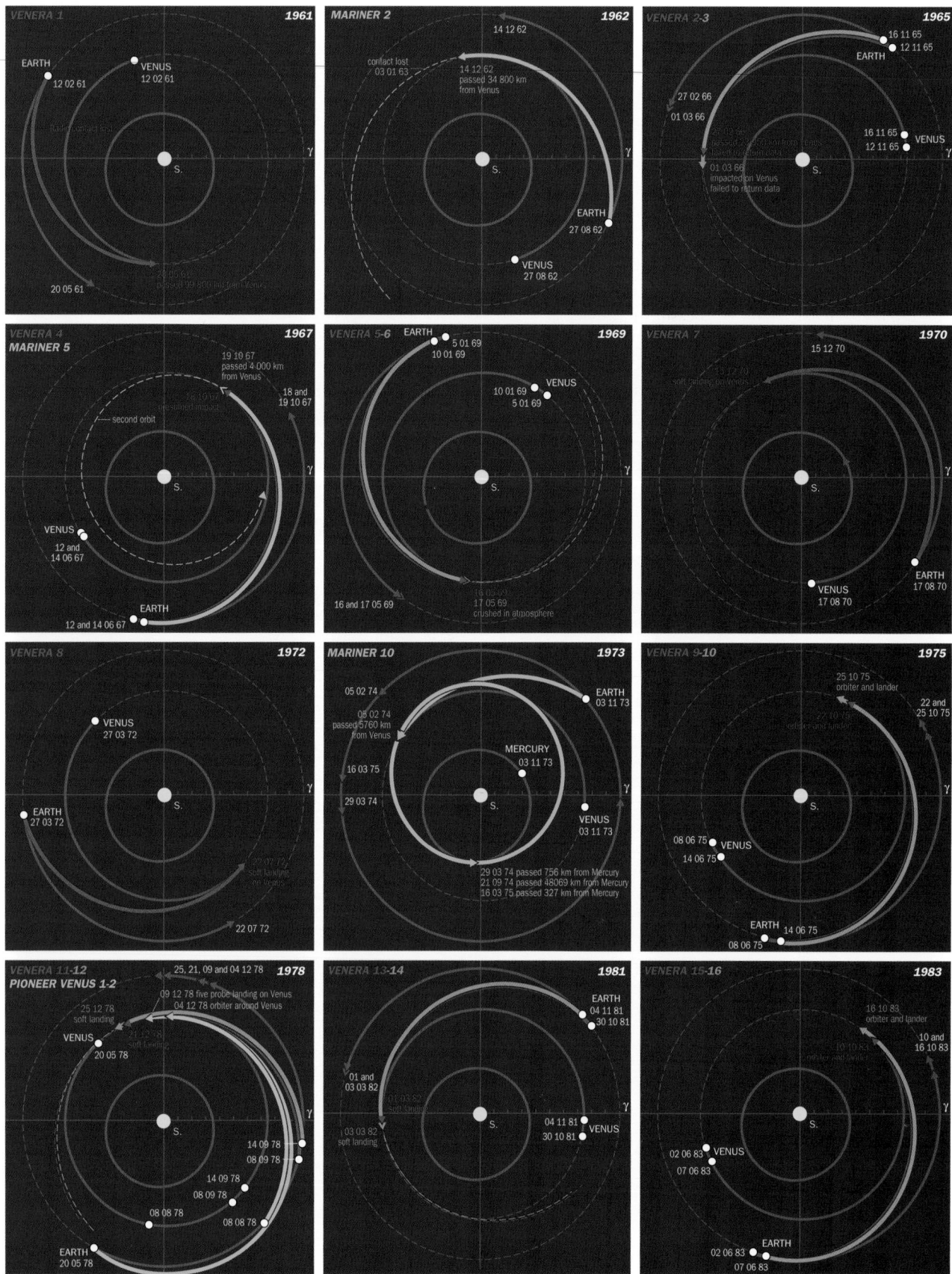

Figure 8.12. **Paths of missions towards Venus and Mercury.** Only Mariner 10 had a trajectory taking it near Mercury. All windows used up until 1999 have been shown, except for the 1984 window (Vega mission, see Fig. 8.34) and the 1989 window (Magellan).

and diameter 6000 km. The largest volcanoes are located here, including Olympus Mons which reaches a height of 25 km. To the south-east lies Valles Marineris, a gigantic canyon stretching out over 6000 km and descending to a depth of 6 km. The other high plateau, the Elysium region, also contains a group of very large volcanoes, such as Hecates Tholus.

There is a second type of landscape whose main representative lies in the southern hemisphere. This is the plateau densely covered with rather shallow craters of many different sizes and crisscrossed with hierarchical networks of paleochannels, a feature unique to the Martian environment. Large impact basins occur on this heavily cratered plateau. An example is Hellas Planitia, measuring 2000 km across, whose floor lies at −4 km (fig. 8.11).

In the smooth lowland plains, less affected by craters, great valleys some 25 km across originate in the plateau and stretch out as far as 1500 km. The zone of contact between the plateau and the lowland plains follows a roughly circular arc inclined at about 35° to the Martian equator. It reveals various features ranging from gentle slopes, through broken and hummocky escarpments, to straightforward steep rises climbing to commanding heights of over 2 km.

The general disposition of the Martian surface reveals a fundamental dichotomy, which basically opposes the two hemispheres and is explained by both endogeneous and exogeneous causes. On a smaller scale than this great division, Martian geomorphology is much more varied than lunar geomorphology. The presence of liquid water on the surface of Mars is impossible today, because the mean temperature is of the order of −60 °C and atmospheric pressure a mere 6 millibars. It must therefore be assumed that those features resembling landscapes formed by running water on Earth are inherited from a distant past in which the atmosphere was denser and the climate less cold. Liquid flows could then have carved out the networks of valleys observed on the surface of the Red Planet. Water may nevertheless exist in solid form, buried deep beneath the surface, where it would constitute a permafrost.

Venus

With Venera 1, the Soviets were the first to make a close flyby of Venus, although it lost contact before reaching its objective, just like the four following Soviet probes. Consequently, it was the Americans with Mariner 2 who made the first observations (fig. 8.12). They also devised and put into practice the occultation method whereby temperature and density data can be obtained concerning the atmosphere of a planet from the distortion of a signal emitted by the probe when it passes behind the atmosphere. This method has been applied systematically ever since.

With Venera 4, Soviet scientists initiated a new approach which made use of a flyby probe and a descent module, thereby obtaining their first results. They discovered the hydrogen corona around Venus and obtained the first temperature and pressure profiles of the atmosphere, as well as carrying out

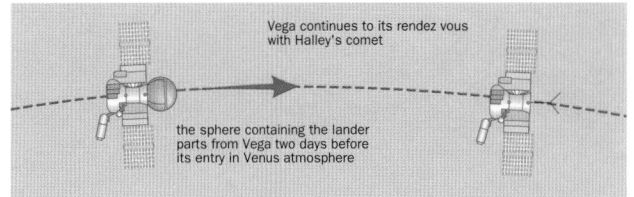

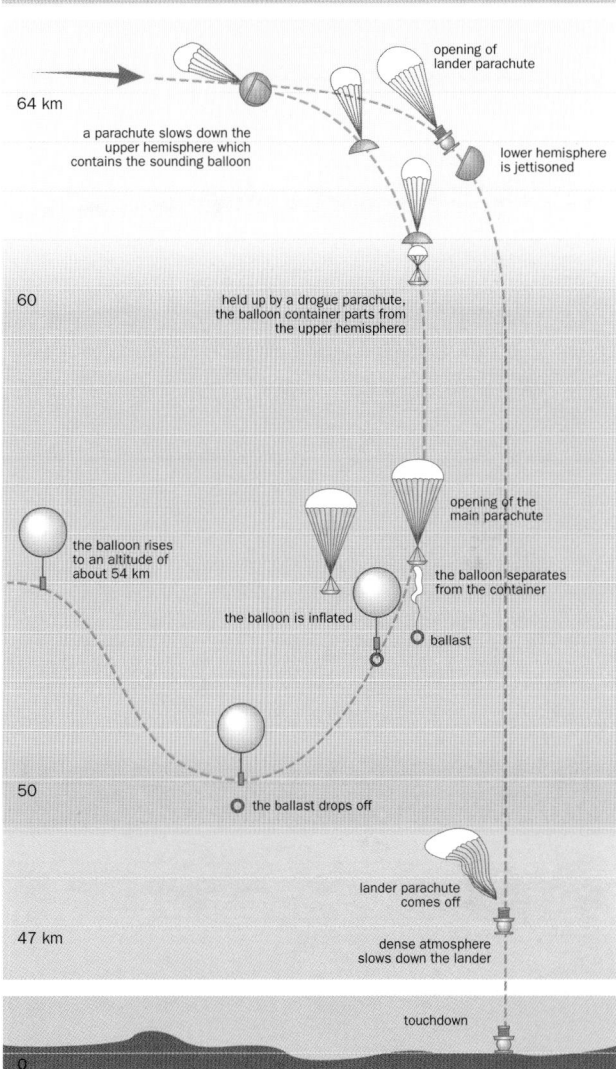

Figure 8.13. **Vega 1 and 2 descent modules: phases of descent into the Venusian atmosphere.** Released on 11 and 15 June 1985, the balloon probes drifted across the cloud layer transmitting temperature, pressure and solar illumination data.

chemical analysis of the atmosphere. The descent module was improved from mission to mission, until with Venera 8 it could resist the temperatures and pressures at the surface (450 °C, 100 atmospheres) and send back the first soil analyses.

The latest Soviet model appeared with the launch of Venera 9. It included a flyby probe and descent module, replaced on Venera 15 and 16 by a radar system. The module was equipped with a robotic arm for taking samples in the case of Venera 13 and 14, and also on the Vega probes. The latter disposed an atmospheric balloon developed by the French Centre National d'Etudes Spatiales (CNES) (fig. 8.13). The main achievements were analysis of atmospheric gases and

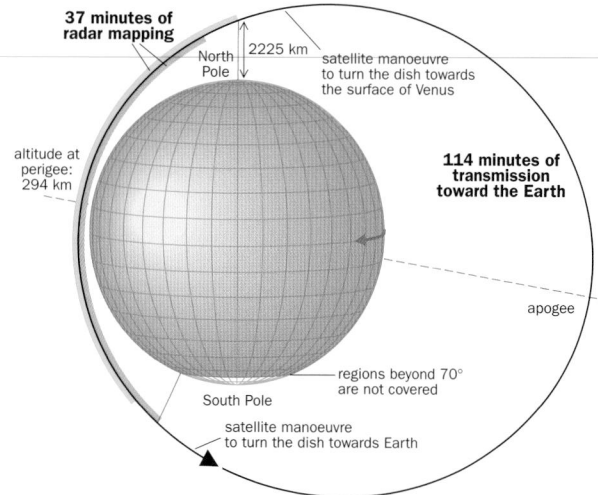

Figure 8.14. Orbit of the Magellan probe around Venus. The near-polar plane of Magellan's orbit meant that it could view almost the whole planet passing beneath it during the Venusian day, which lasts 243 terrestrial days. Venus rotates in a retrograde direction, so the succession of passes occurred in a prograde direction. Magellan's orbital period was 3h 15m. During this time, 37 minutes were used to image the surface of Venus. However, by starting imaging at different times, the detected swaths could be distributed over different latitudes. The swath ended alternately at latitudes 56° and 70° south, thereby covering a bigger area in the same operating time, whilst reducing the degree of overlap. The elliptical polar orbit was chosen to achieve close-up detection of the planet at periapsis and transmission to Earth at apoapsis. The argument of the periapsis was chosen near latitude 10° north to give good coverage of one of the poles. The north pole was preferred on the basis of previous observations. The orbital eccentricity was reduced in May 1993 so that high resolution measurements could be made of the gravitational field.

aerosols, the possible detection of lightning, data concerning movements of the atmosphere, chemical composition of the Venusian regolith (Venera 13 and 14, Vega 2) and its density (Venera 9 and 10), seismic activity (Venera 13 and 14), pictures of the surface (Venera 9, 10, 13, 14), a radar survey (resolution 1.5 to 2 km) and thermal mapping of half of the northern hemisphere of the planet (Venera 15 and 16).

Between 1962 and 1974, the United States pursued their own observations of Venus. Mariner 5 observations were used to calculate the equatorial radius, whilst ionospheric data together with the observed lack of intrinsic magnetism led to an original model for the interaction between ionosphere and solar wind. Then Mariner 10, although primarily dedicated to observation of Mercury, transmitted 3500 ultraviolet images on its Venus flyby, and thereby greatly improved our understanding of large-scale motions in the upper atmosphere of the planet.

After a break of about ten years (see fig. 8.6), the United States returned to Venus with two missions: Pioneer Venus and Magellan which supplied some of the most useful data yet acquired from this planet.

The Pioneer Venus 1 orbiter, launched on 20 May 1978, made a radar map of 83% of the planet's surface, vertically

Figure 8.15. Topography of Venus from Magellan radar coverage. The main continent, named Aphrodite after the Greek name for Venus, lies south of the equator. The second continent, named Ishtar after the Babylonian name for Venus, lies in the northern hemisphere and contains Maxwell Montes, the highest peak on the planet at an altitude of 10 800 metres. The reference level for altitudes is a sphere of radius 6051 km that corresponds to the mean Venusian land level. © NASA-JPL.

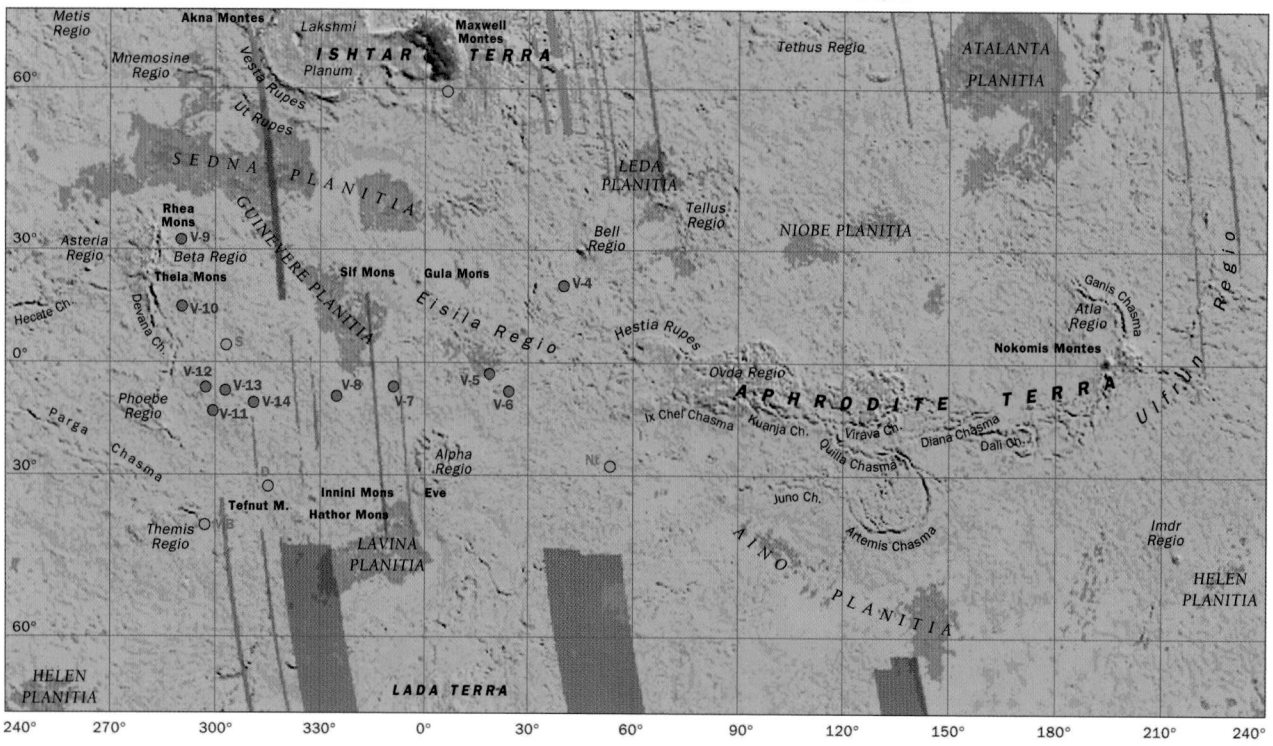

● **V-6** Venera landing

○ **VP** Pioneer landing (*MB: Multiprobe bus S: Sounder probe D: Day probe Nt: Night probe N: North probe*)

accurate to within about 200 m, as well as imaging clouds in the upper atmosphere, and making estimates of density and gravity. In addition, it complemented Mariner 5 and 10 observations of the interaction between the Venusian ionosphere and the solar wind.

Pioneer Venus 2, launched on 8 August 1978, comprised a transport vehicle (or bus) and four entry probes. The first analysed the atmosphere between altitudes of 150 and 115 km before breaking up. The four probes studied the temperature, pressure and composition of cloud and gas in the lower atmosphere.

Magellan was launched on 4 May 1989 and injected into orbit around Venus on 10 August 1990 (fig. 8.14). Carrying only one instrument, Magellan nevertheless achieved several important results. It covered almost 98% of the Venusian surface by a radar mapping, with a resolution of 100 metres at periapsis and 360 metres at the pole, as well as making an altimetric radar survey to a vertical resolution of 30 metres and evaluating the surface temperature by studying the ground response to radar waves (fig. 8.15).

The very similar size and density of Venus and Earth have given rise to many comparisons between the tectonic and volcanic phenomena occurring on the two planets. In particular, images returned by the Magellan probe were used to seek out signs of tectonic motions. Certain rather striking circular structures, known as *coronæ*, are observed on the surface of Venus. Measuring several hundred kilometres across, they are bordered by deep depressions, themselves skirted by significant topographical features. These rather unusual formations are generally interpreted as a surface manifestation of rising magma from the mantle. The coronal lithosphere would then be pushing beneath the surrounding lithosphere in some kind of subduction process. In order to acquire more data concerning this hypothesis, NASA reduced the eccentricity of Magellan's orbit at the end of its mission.

Although Venus remains enigmatic in many respects, no further mission has yet been clearly defined by the various space agencies, apart from NASA's Venus Laboratory Mission, rather loosely scheduled for 2005–2010. However, it should be emphasised that the American Pioneer Venus and Magellan missions have provided a considerable harvest of data that will take some time to exploit to the full. The detailed mapping of Venus currently underway at the US Geological Survey will require a great deal of interpretation and can be expected to proceed for years to come.

Mercury

Only one space probe, Mariner 10, has yet observed Mercury, the planet closest to the Sun. This probe was launched in the direction of Venus and deflected towards Mercury by gravity assist, in order to save energy. Without this ingenious trajectory, so much propellant would be needed for the trip that no launch vehicle would ever be able to place the probe into orbit. Favorable astronomical conditions for such a mission are extremely restricted, only recurring about once

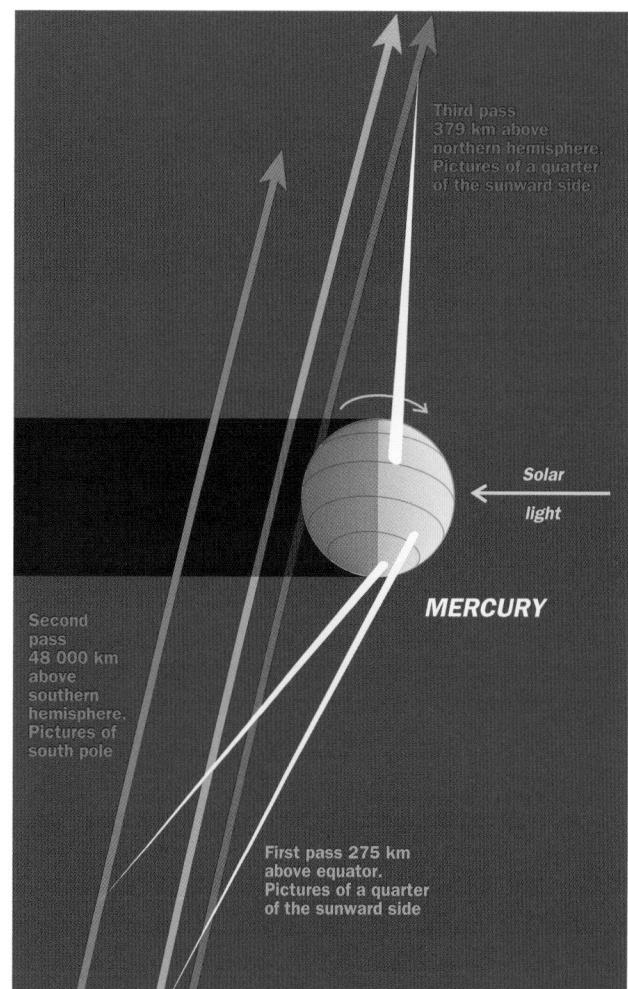

Figure 8.16. **The three Mariner 10 flybys of Mercury.** The same sunlit hemisphere was observed during the Mercurian afternoon whilst the probe approached the planet and during the Mercurian morning as the probe drew away.

every 10 years. The 1973 window was used to launch Mariner 10. The probe was injected into an eccentric solar orbit with period exactly twice the period of Mercury. In this way, the planet was overflown every 176 days (see fig. 8.12). Since Mercury completes three rotations about its axis for every revolution about the Sun, the probe always observes the same position (fig. 8.16).

Communications were maintained with the probe during the first three close passes. In order to collect magnetic field data, the probe had to fly past the night side, whereas it would have been easier to take images whilst flying over the sunward face. On the other hand, it is crucially important for comparative planetology to observe the magnetic fields around Earth and Mercury, since they are so strong in these two cases, whilst being virtually absent for Mars and the Moon. Mariner 10 also made measurements of the surface temperature on Mercury and demonstrated that there was no atmosphere, although there do exist some traces of helium.

Despite the severe restrictions due to the trajectory, the probe was able to gather 5500 images using a high resolution television camera, of which 1838 could be used for topographical purposes (the best resolutions being of the order of 100 m), covering 57% of the sunward surface.

The topographic relief of Mercury has thus been partially mapped with some degree of detail. A completely new toponymy was required since nothing was previously known of the Mercurian surface. Astronomers gave the features the names of famous people from many different disciplines,

historical periods and countries, such as Goya, Beethoven or Zola, and not just the names of astronomers, even if the major faults go by the names of Schiaparelli and Antoniadi. The very significant scarps on Mercury are named after the sailing boats of famous explorers (the Victoria scarp takes its name from Magellan's ship) whilst many of the basins carry the same name as the planet itself, but in different languages.

The landscape on Mercury is strikingly similar to that on the Moon. It is made up of smooth, dark plains which lie between crater-covered uplands. The craters here are more numerous as their size decreases, thus representing the size distribution of the interplanetary bodies that produced them. Some are up to several hundred kilometres across, the largest being the Caloris basin, which measures 1300 km in diameter. The name comes from the Latin word for 'hot', since the Sun lies directly above it during alternating perihelion passages, and the temperature rises drastically on such occasions. This enormous crater was caused by a very large meteoroid impact, and the surrounding region is wrinkled and cracked as a consequence. The same chaotic features can be observed on the opposite side of the planet and it is believed that seismic waves generated by the impact were focused to exactly this antipodal point on passing through the planet. For comparable diameters, craters are shallower than those on the Moon, and secondary craters are on average closer to the main craters than they are on the Moon. These two features must somehow be related to the stronger gravity on Mercury. Geomorphologically, Mercury is notable for its mighty tectonic scarps, reaching up more than 1000 metres above the surrounding plains and extending in straight lines over hundreds of kilometres.

The different terrains can be dated by the density, the degree of overlap and the worn appearance of the craters located on them. Their careful study leads to a geological history of Mercury, showing that it first underwent a period of intense bombardment, followed by a period of relative calm that has lasted for 3800 million years or so. This chronology compares well with what is known about the Moon and shows that the main phases of meteoritic bombardment may have occurred simultaneously throughout the Solar System.

Both NASA and ESA have projects to revisit Mercury, although it is a priority in neither case. Messenger (MErcury: Surface Space, ENvironment, GEochemistry and Ranging) is a Discovery project with departure planned for March 2004 and arrival in September 2009, after swingbys around Venus and Earth. The ESA mission to Mercury is also on the drawing board, but its exact date and content have not yet been finalised.

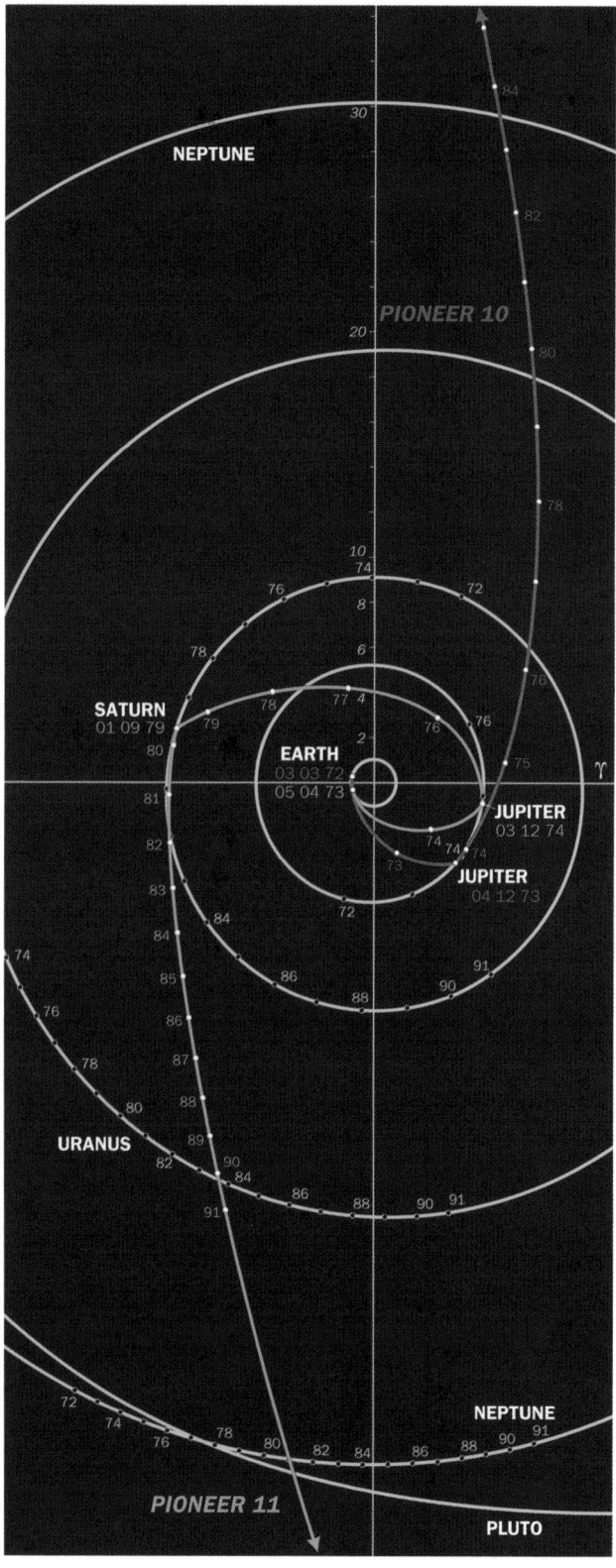

Figure 8.17. **Paths of Pioneer 10 and Pioneer 11.** Both probes were launched in 1972 from Cape Canaveral by Atlas Centaur rockets, the first on 3 March, and the second on 5 April. Pioneer 10 passed close to Jupiter, and Pioneer 11 flew by both Jupiter and Saturn. The two probes will leave the Solar System in opposite directions, orthogonal to the direction of the vernal equinox γ. Pioneer 10 went on transmitting data until NASA broke off reception on 31 March 1997. As in the next figure, dots on the probe trajectories and planetary orbits indicate positions at the beginning of each year. Distances on the vertical axis are in astronomical units.

The giant planets: a Grand Tour

It is an arduous task to explore the more distant reaches of the Solar System, beyond Mars. Advanced technology and considerable financing are essential. Upheld by enthusiastic and devoted scientists, these projects must nevertheless face the questions raised by the political powers that be.

During the 1960s it was noticed that a rare astronomical opportunity was about to occur, one which would only be repeated 175 years later. A shortlived alignment would make it possible for a space vehicle launched towards Jupiter to continue its path out to Saturn, Uranus and Neptune, before leaving the Solar System. As the launch had to be made in

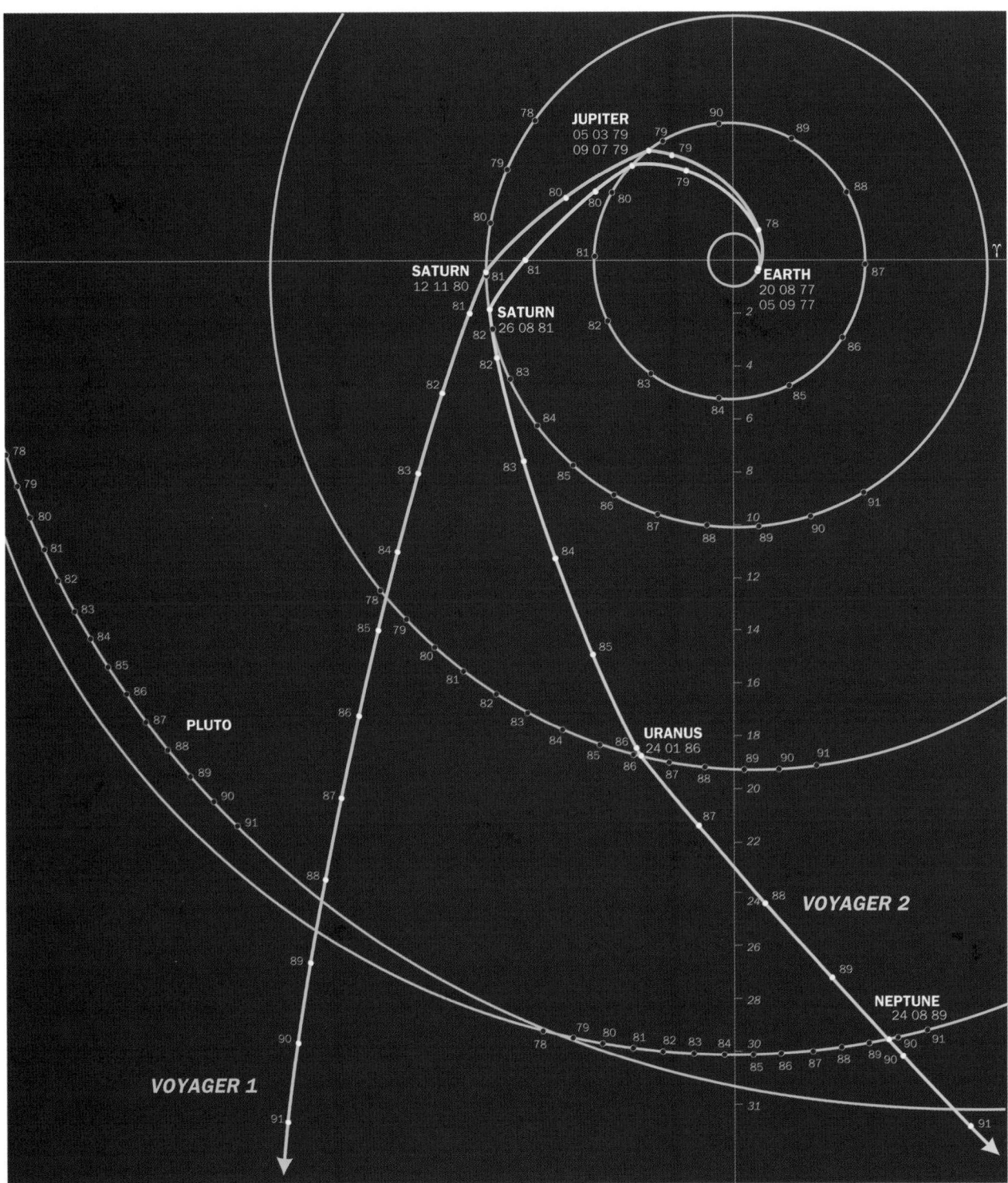

Figure 8.18. **Paths of Voyager 1 and Voyager 2.** Both probes were launched in 1977 from Cape Canaveral by Titan III Centaur rockets. Voyager 2 left on 20 August, before Voyager 1, which was launched on 5 September. The configuration of the four giant planets in 1977 was a rare opportunity for a single probe to fly past each one in succession, with the help of gravity-assist manœuvres. This series of rendezvous was successfully accomplished by Voyager 2, whilst Voyager 1 encountered only Jupiter and Saturn.

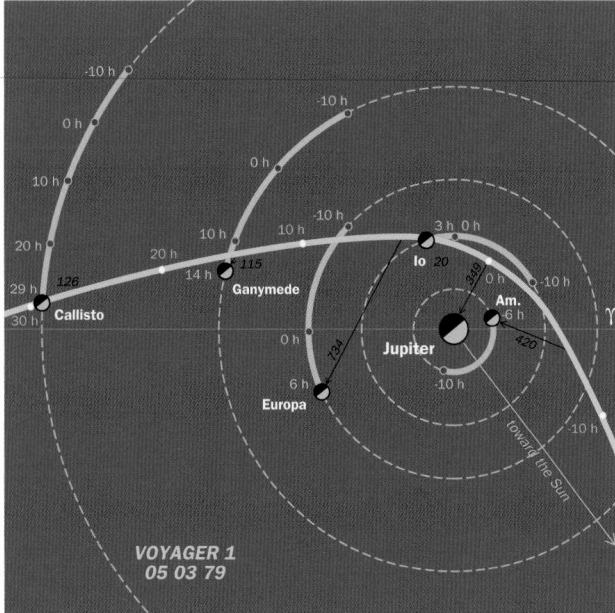

Figure 8.19. **Voyager 1 in the gravitational sphere of influence of Jupiter.** For most of the moons, flyby took place when the probe was moving away from Jupiter. Io, Ganymede and Callisto were the most closely approached satellites. As in the following figures, dots on the trajectories indicate time in hours, with 0 being the time of passage of the probe at its closest approach to the planet. Numbers in black are the minimal distances in thousands of kilometres between the probe and the centre of the planet or moon. Paths are represented in the reference frame of the planet and on its equatorial plane. Figure 2.53 shows the difference between this type of representation and one shown in the Galilean frame.

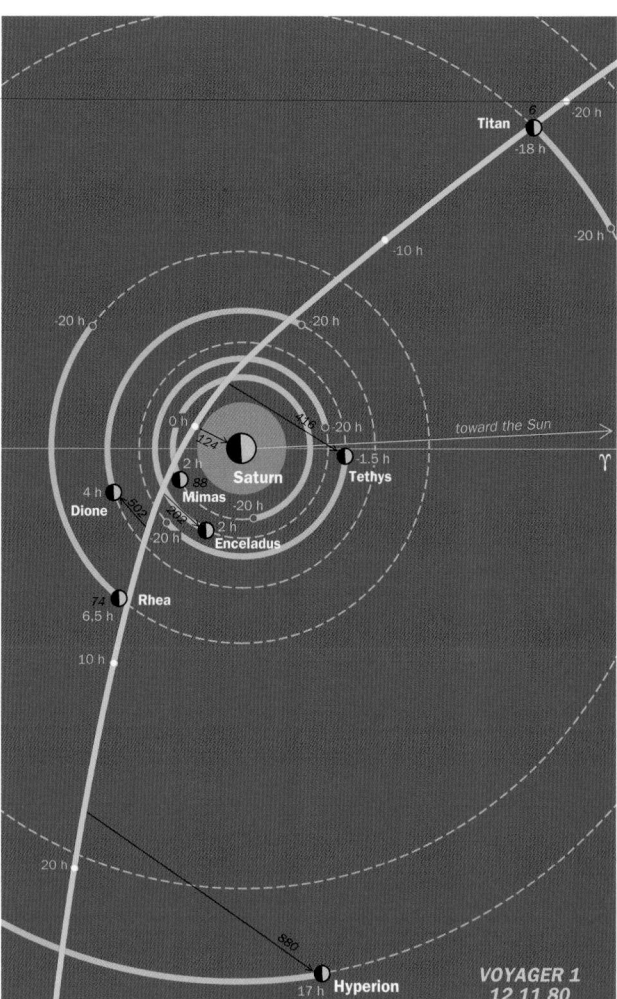

Figure 8.20. **Voyager 1 in the gravitational sphere of influence of Saturn.** Titan, Mimas and Rhea were the most closely approached moons.

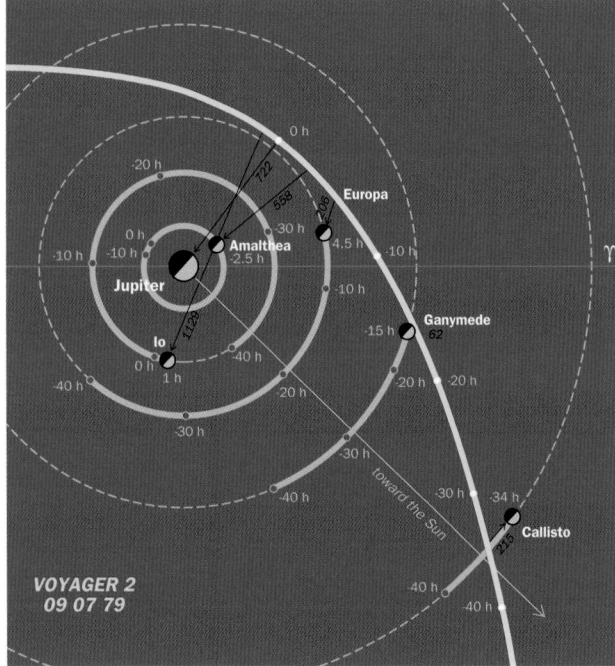

Figure 8.21. **Voyager 2 in the gravitational sphere of influence of Jupiter.** Most close approaches to moons took place when the probe was moving towards Jupiter. As the moons always present the same face towards the planet, their sunlit sides observed by this probe were opposite to those observed by Voyager 1. Callisto, Ganymede and Europa were the most closely approached moons.

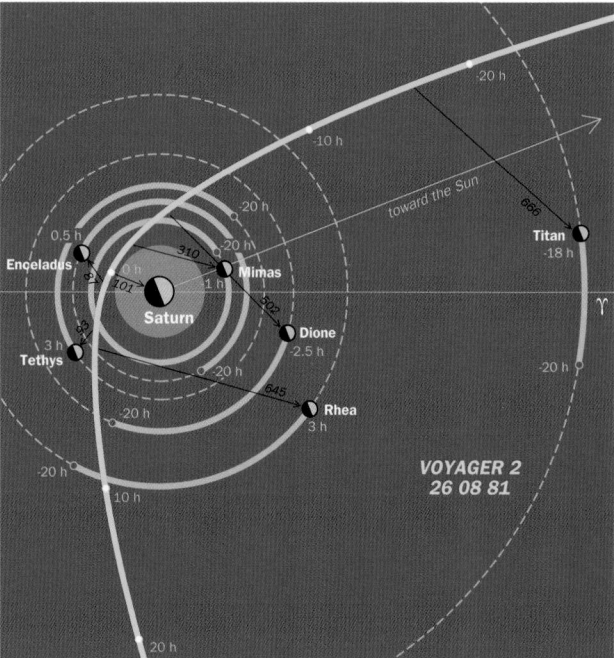

Figure 8.22. **Voyager 2 in the gravitational sphere of influence of Saturn.** Enceladus and Tethys were the most closely approached moons.

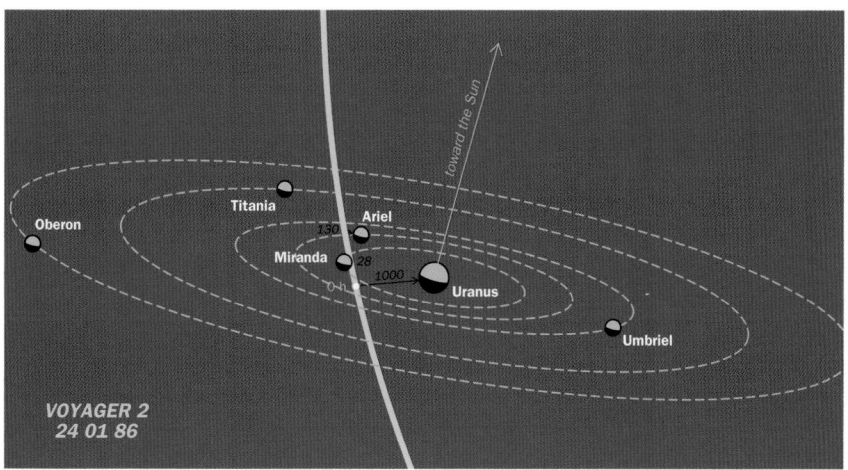

Figure 8.23. **Voyager 2 in the gravitational sphere of influence of Uranus.** This representation in the planet's frame of reference is projected onto the plane of the probe's trajectory. The equatorial plane of Uranus is inclined at 98° to the orbital plane, which is itself very close to the plane of the ecliptic (0.46°). Its moons move in the equatorial plane, except for Miranda which has a small inclination (4°) relative to this plane. Hence, the same pole of the planet and its satellites was sunlit throughout. The closest passage occurred on 24 January 1986, with Uranus at 71 000 km, Ariel at 120 000 km and Miranda at 28 000 km from Voyager 2.

1977, an ambitious project was devised by NASA, entitled the Grand Tour. In its original form, with notable innovations in the area of artificial intelligence and protection against radiation and impact, it was a highly sophisticated and hence costly undertaking. The Pioneer 10 and 11 missions were programmed as forerunners for the Grand Tour.

In 1972, the decision was made to undertake the Space Shuttle programme, thereby putting an end to the priority for space conquest along the lines of the Apollo programme. The ambitious Grand Tour was abandoned. It was to be replaced by a less expensive and also less prestigious mission, Mariner Jupiter–Saturn, which was to become Voyager 1 and 2. Later on, this project was also called the Grand Tour,

witnessing NASA's concern to conceal any sacrifices made for specific programmes and maintain a presence in space exploration. Thus, up until 1990, only two programmes, each limited to a couple of flyby probes, were exploring the Solar System beyond Mars. The Soviets for their part have doubtless never attempted any operation towards the giant planets, lacking the computing and electronic capabilities that a mission of this type would require.

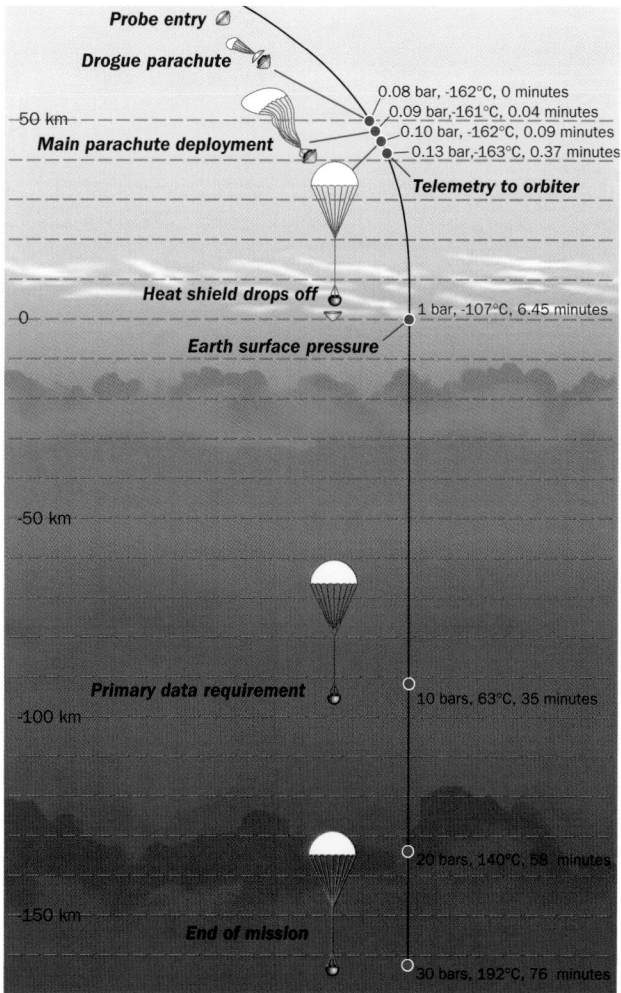

Figure 8.24. **Path of Galileo towards Jupiter.**

Figure 8.25. **Galileo descent probe: investigation of the Jovian atmosphere.**

The probes Pioneer 10 and Pioneer 11 carried only a photopolarimeter, being stabilised by rotation about their own axes and hence poorly suited to imaging activities. The other instruments, however, sent back accurate data on the low density of dusts and debris in the asteroid belt, and also on the magnetospheres of Jupiter and Saturn. It was discovered that Jupiter had a huge magnetic field. In addition, the chemical composition of the Jovian atmosphere was analysed (fig.8.17). These two probes also carried a metal plate engraved with the silhouette of a man and a woman, and symbolic indications with which the scientists of another world, if they exist, should be able to locate the geographical origin and launch date of the probe.

Pioneer 10 transmitted 300 images of Callisto, Ganymede, Europa and Jupiter, including 40 close-ups, and discovered a thirteenth natural satellite in orbit around Jupiter. Pioneer 11 transmitted 130 images of the same three satellites and also the north pole of Jupiter, before going on to return pictures of Saturn, which revealed a new ring and two hitherto unknown moons.

The Pioneer 11 mission was able to check that the tiny gateway leading towards Uranus and Neptune would not be blocked by some unknown ring of particles, which might have seriously damaged the future Grand Tour explorer, at least as it was intended during the development of the Pioneer probes, and as it was finally realised with the Voyager 2 probe.

The Voyager 1 and 2 probes were stabilised about three axes and in each case comprised a steerable platform carrying two television cameras (with fields of view 3° and 0.4°), two infrared spectrometers and a photopolarimeter. The six other instruments were magnetometers, three detectors for cosmic rays, plasma and low energy charged particles, a planetary radioastronomy instrument and a plasma wave subsystem. Imaging equipment sent back a spectacular collection of high resolution pictures (fig.8.18).

Voyager 1 returned 19000 images of Jupiter and its satellites Io, Europa, Ganymede and Callisto (fig. 8.19) as well as 17000 images of Saturn, Titan, Rhea, Dione, Tethys, Mimas and Enceladus (fig.8.20).

Voyager 2 sent back 15000 images of Jupiter, Amalthea, Europa, Ganymede and Callisto (fig. 8.21), 17500 images of Saturn, Phoebe, Iapetus, Hyperion, Tethys and Enceladus (fig. 8.22), 6000 images of Uranus and its satellites Miranda, Ariel, Umbriel, Titania and Oberon (fig. 8.23). These pictures revealed two new rings and three new satellites gravitating around Jupiter. The same region has since been revisited by the American probe Galileo, in December 1995 (see below). The Voyager 2 mission also revealed intense volcanic activity on Jupiter's moon Io, further rings of ice and dust around Saturn, and ten moons and two further rings around Uranus. It also demonstrated the uniqueness of Miranda which appears to be a kind of planetary conglomerate made up of several former moons, destroyed in some long-forgotten cataclysm. The geomorphology of this tiny satellite is certainly extremely complex, with giant scarp cliffs, fault canyons, cratered plains and grooved regions.

Other rewards from this programme concern the motion, chemical composition and temperature of the various atmospheres encountered, the material make-up of the moons, and also the extent of the magnetospheres and the plasma distributions within them. In addition, the Voyager 2 mission revealed something of the originality of Uranus, whose magnetic field lies at an angle of 60° to the planet's axis of rotation.

Voyager 2 passed within 1300 km of Neptune on 25 August 1989 and also within 6000 km of its satellite Triton.

Following the Neptune flyby, at which point both the Voyager probes were still in good condition, NASA was able to initiate an extension to its programme, called the Voyager Interstellar Mission (VIM). The aim is to study conditions in the most distant confines of the heliosphere. The two probes should pass one after the other through the termination shock and the heliosheath, thereby obtaining their first taste of the interstellar medium. At the time of writing (February 2000), the probes are still within the heliosphere, at 11499 (Voyager 1) and 9106 (Voyager 2) billion kilometres from Earth, moving with speeds of 17.281 and 15.807 km/s, respectively. In order to save energy, the instruments on the scan platform, where only the UV spectrometer was still in operation, have not been heated since 1998 in the case of Voyager 2 and 2000 in the case of Voyager 1. The magnetometers will no longer function from 2010 and 2011, respectively, so that the other instruments (the last 5 mentioned above) will be able to operate until at least 2020.

The giant planets: Galileo and Cassini

Exploration of the giant planets was resumed with the launch of American probe Galileo. The programme was delayed for a certain time by the Challenger disaster. Galileo was launched by the Space Shuttle Atlantis in October 1989. After two asteroid flybys, it flew out to Jupiter and released a descent probe which fell towards the planet, before itself entering into circumjovian orbit (fig. 8.24).

It is equipped in a similar way to Voyager. A scan platform without spin (the orbiter being stabilised by rotation) carries a high resolution SSI (solid state imaging) camera with a field of view of 0.4°, a Near Infrared Mapping Spectrometer (NIMS), and UltraViolet Spectrometer (UVS, with an extreme UV detector attached), and a photopolarimeter radiometer (PPR). Other instruments are magnetometers and detectors for high energy particles (EPD), plasma (PLS), and dust (DDS), the plasma wave subsystem, a heavy ion counter (HIC), a radio astronomy instrument and a celestial mechanics instrument (to determine masses of celestial bodies and atmospheric structures).

For its part, the descent probe carried an atmospheric structure instrument (ASI) giving temperatures, pressures, density and molecular weight of the layers it crossed, a Neutral Mass Spectrometer (NMS), a Helium Abundance Detector (HAD), a Net Flux Radiometer (NFR) and a detector of lightning and high energy particles.

The probe was released in August 1995 and its descent was crowned with success, despite the considerable difficulty of the undertaking. Data transmitted via Galileo over a period of 57 minutes greatly improved our knowledge of the composition and motions of the planet's atmosphere. The measured deceleration showed that the atmosphere was denser than expected, whilst composition analyses proved it to be less rich in helium and neon than had previously been supposed. These observations forced scientists to reconsider the generally held view that Jupiter might be representative of the protosolar nebula (fig. 8.25).

The Galileo orbiter went into circumjovian orbit on 7 December 1995. It studied the Jovian magnetotail and took excellent pictures of Jupiter, notably of the enigmatic Great Red Spot, which measures more than twice the Earth diameter in the east–west direction. It also observed the wind circulation which rotates counterclockwise around the spot at speeds of 400 km/h. The orbiter was initially injected into a highly eccentric Jovian orbit of period 230 days. After its first flyby of Ganymede, it then went into a series of swing-by manœuvres. The aim was to modify its period and inclination and finally achieve 11 orbits disposed like the petals of a flower, as had originally been programmed. These would allow a close flyby of all the Galilean moons (fig. 8.26 and table 8.1), as well as the most distant of the small inner satellites, and observation of the rings. The twelfth orbit marked

the beginning of an extension to the programme christened GEM (Galileo Europa Mission). The first eight orbits of the new mission led to a low altitude flyby of Europa, whilst the next four used the gravitational assistance of Callisto to lower the perijove, and the last two (orbits 24 and 25) brought it close enough to observe Io. This should have been the end of the mission, since it was expected that radiation and electric currents in the Io torus would destroy its circuits. In fact, the probe survived these two passages and a further extension of the mission (Galileo Millenium Mission) began in 2000 with flybys of Europa and Io (fig. 8.28).

Io is the closest Galilean moon to Jupiter and cuts through the planet's magnetic field lines as it follows its orbit. This means that it develops a strong electric charge, and this sets up a closed electric circuit between the satellite and its mother planet. Galileo made interesting observations of the geomorphological evolution of this satellite. The series of pictures taken by the Voyager probes had already allowed a kinematic analysis of the Io landscape. Galileo was able to attest to the rapid changes that had occurred over the 17 year period separating the Voyager and Galileo missions. These few years were witness to an intense geomorphological evolution

Figure 8.26. **Orbits described by Galileo around Jupiter during the initial stages of the mission.** Blue circles around Jupiter represent orbits of the Galilean moons: Callisto (C), Ganymede (G), Europa (E) and Io (I).

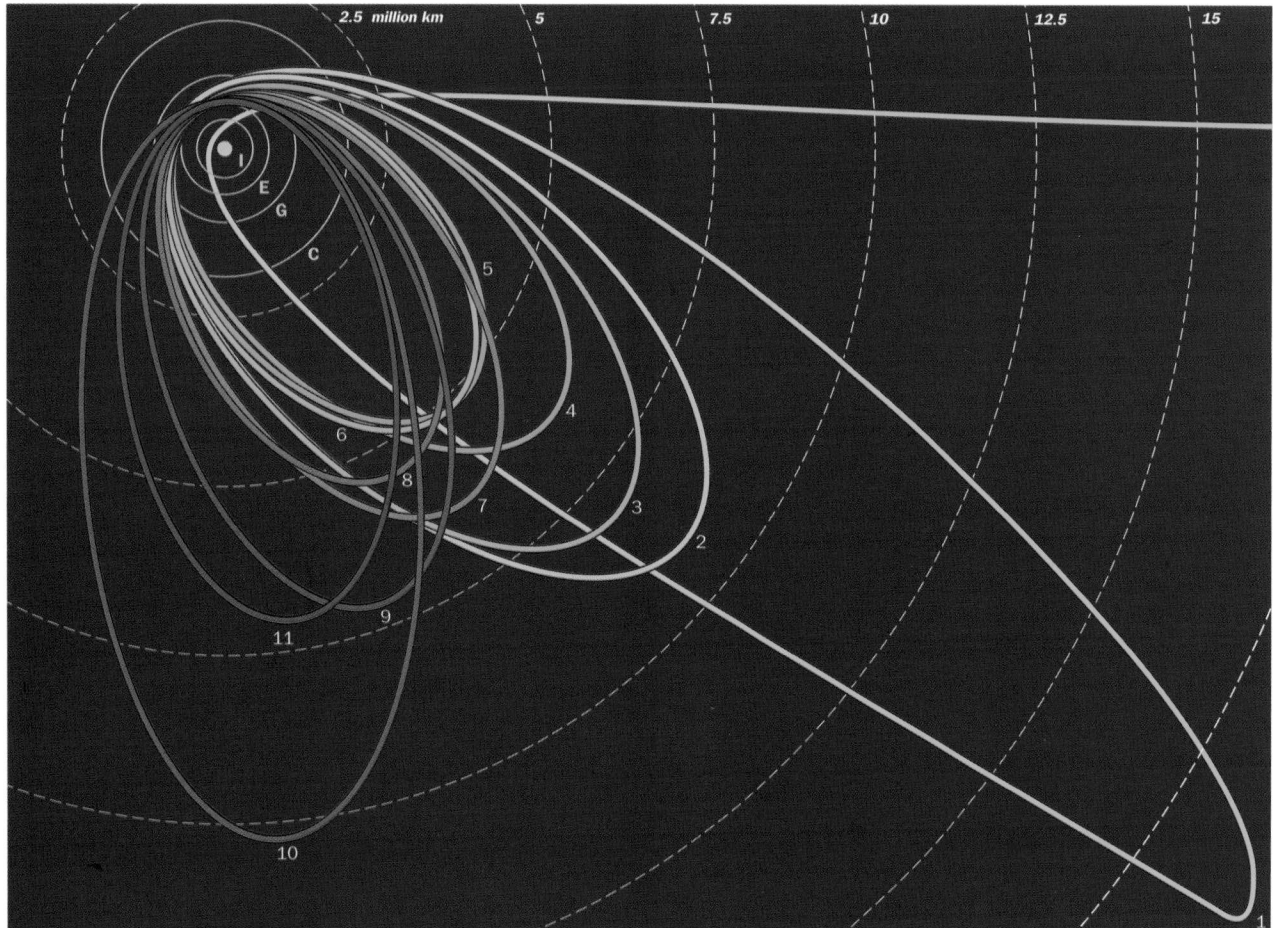

ORBIT	CALLISTO closest approach date	altitude	best resolution	GANYMEDE closest approach date	altitude	best resolution	EUROPA closest approach date	altitude	best resolution	IO closest approach date	altitude	best resolution	Amalthea	Thebe	Adrastea	Metis	Elara	Himalia
INITIAL PROGRAM																		
1				27 06 96	844	10	27 06 96	154000	1600	28 06 96	695000	8800						
2	09 09 96	422000	4300	06 09 96	250	47	07 09 96	671000	4900	07 09 96	439000	4900	•					
3	04 11 96	1100	29				06 11 96	31947	418	06 11 96	242000	2500						
4				19 12 96	789000	-	19 12 96	692	26	18 12 96	319000	5800	•	•	•			
5	21 01 97	598000	-				20 01 97	27419	-									
6	22 02 97	277000	-	02 21 97	315000	7300	20 02 97	587	12	19 02 97	399000	4000	•	•				
7	02 04 97	634000	6400	05 04 97	3059	136	04 04 97	23244	387	03 04 97	529000	5400						
8	06 05 97	33499	670	07 05 97	1585	140				07 05 97	948000	10000					•	•
9	25 06 97	416	125	26 06 97	79961	830				27 06 97	605000	-	•	•	•	•		
10	17 09 97	524	64				19 09 97	621000	7300	18 09 97	317000	3800	•	•	•	•		•
11	04 11 97	673000	-				06 11 97	1125	11	07 11 97	778000	-						
GALILEO EUROPA MISSION																		
12				15 12 97	14300	-	16 12 97	201	-	16 12 97	483000	-						
13				10 02 98	627300	-	10 02 98	3562	-	10 02 98	436000	-						
14	29 03 98	204000	-	29 03 98	917000	-	29 03 98	1649	-	29 03 98	250000	-						
15				31 05 98	325000	-	31 05 98	2521	-									
16				21 07 98	146000	-	21 07 98	1837	-									
17							26 09 98	3582	-	26 09 98	798000	-						
18							22 11 98	2281	-	23 11 98	994000	-						
19	01 02 99	897000	-				01 02 99	1495	-	01 02 99	854000	-						
20	05 05 99	1311	-	05 05 99	634000	-				02 05 99	787000	1000						
21	30 06 99	1047	-							01 07 99	125000	1400						
22	14 08 99	2296	-	12 08 99	834000	-	12 08 99	215000	-	12 08 99	735000	-				•		
23	16 09 99	1057	-	14 09 99	585000	-				14 09 99	446000	-						
24				10 10 99	925500	-	10 10 99	223000	-	11 10 99	611	-						
25				25 11 99	608000	-	25 11 99	11500	-	25 11 99	300	-						
GALILEO MILLENNIUM MISSION																		
26							03 01 00	351	-				•	•		•		
27										22 02 00	200	-						

Table 8.1 . **Galileo flybys of the Jovian planets.** Dates are given in days, months and years, flyby altitudes are in kilometres, and the best resolutions indicate pixel size in metres. Black spots in the columns of the small satellites indicate images taken by at least one of Galileo's four imaging instruments. Dashes indicate that data is lacking.

in Io's volcanic landscapes, dominated by sulphur dioxide. Apart from Earth, Io is the only other planet known to possess active volcanoes, such as Mount Pele or the volcano Masubi. Many of these volcanoes spew out plumes of material, sometimes reaching altitudes of several hundred kilometres. They cover the surface of Io with debris composed primarily of sulphur compounds whose varying temperatures deck them out in a whole range of colours, from red, through orange and yellow to white.

Galileo approached to within 201 km of Europa on one of its passages. Approximately the same size as the Moon, it revealed itself as an icy world (fig. 8.27). The numerous streaks and cracklike features crisscrossing the surface, already discovered during the Voyager missions in 1979, were resolved into triple bands formed by dark outer edges and a central band of lighter material. They can be interpreted through tectonics, and also by a new explanation involving dusty geysers. The latter would carry a mixture of ice and dark silicate dusts up to the surface, the bright central ray being formed by the rise of very pure water ice.

The scarcity of impact craters is generally explained by assuming that these features are erased in a continuous process, as surface ice layers are renewed. Europa also pro-

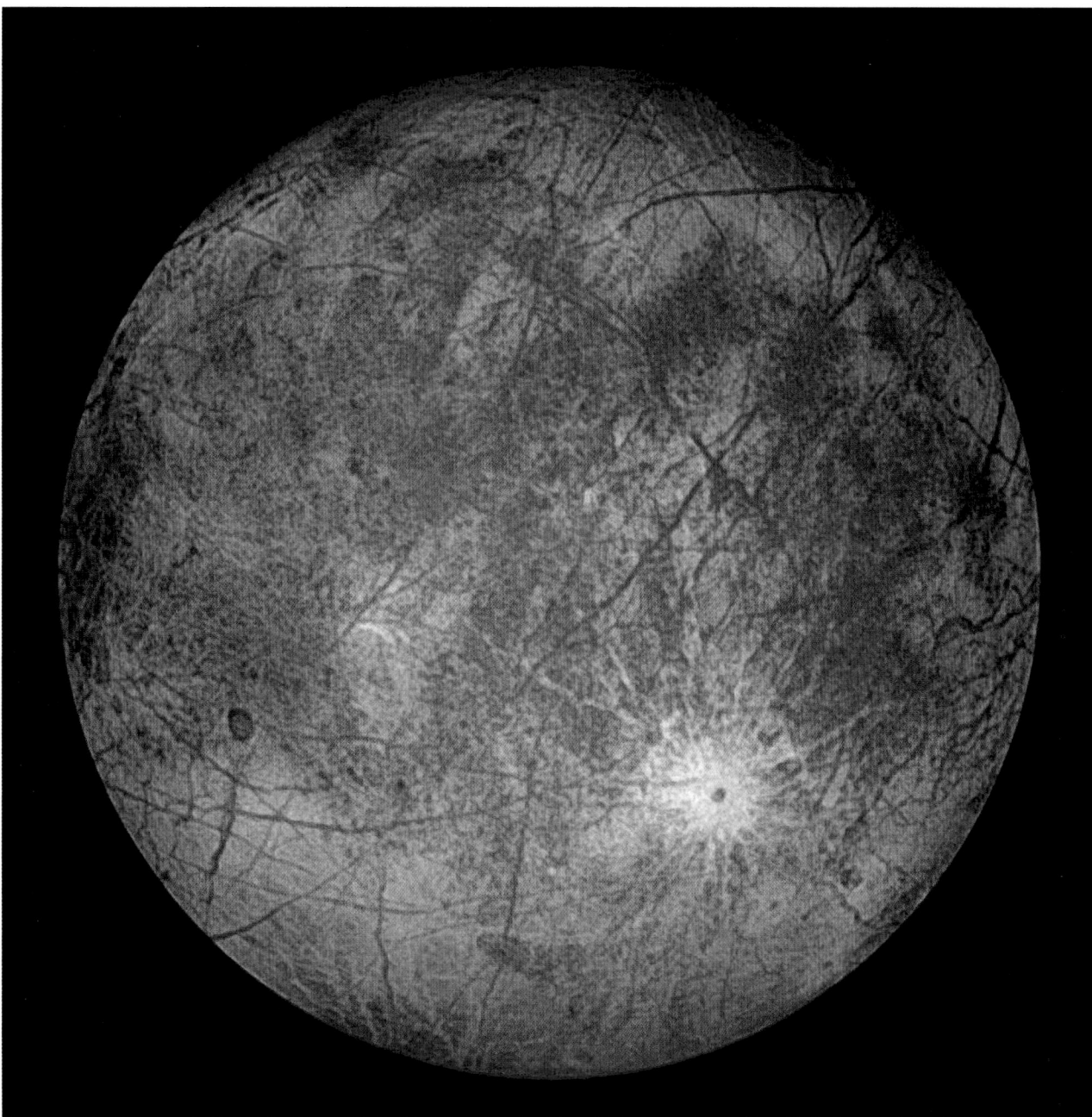

Figure 8.27. **Image of Europa by Galileo.** Image of the trailing hemisphere of Jupiter's ice-covered satellite Europa, taken on 7 September 1996 at a range of 677 000 km (see Table 8.1). The image is a false colour composite version of violet, green and infrared bands. Europa is about 3160 km in diameter. The bright feature in the lower third of the image is a young impact crater. Image processed by Deutsche Forschungsanstalt für Luft- und Raum-fahrt, JPL-NASA, courtesy of NASA.

vides a particularly interesting field of investigation for exo-biologists. The presence of water and an internal heat source may mean that there is a liquid ocean under the surface ice. The January 2000 flyby revealed directional changes in the magnetic field which could be related to movements of a conducting fluid medium like salt water.

Ganymede is the largest moon in the Solar System and possesses a magnetic field that Galileo was able to measure thanks to its varying acceleration during passage close by the planet. The density of Ganymede – only one third that of Earth – implies that there may be a considerable quantity of water involved in the satellite's make-up. This water forms an icy crust, beneath which there is a rocky mantle, itself enclosing a metallic core. The surface of Ganymede includes two very different kinds of terrain. Dark regions with a high density of impact craters are separated by light-coloured zones composed of younger ices.

Finally, Callisto has an icy crust incorporating silicates and is uniformly pockmarked with a great number of impact craters. The latter implies a much older surface than the other Galilean moons, without surface renewal via ice movements.

Another ambitious programme – the fourth of the projects recommended in 1983 by the Solar System exploration

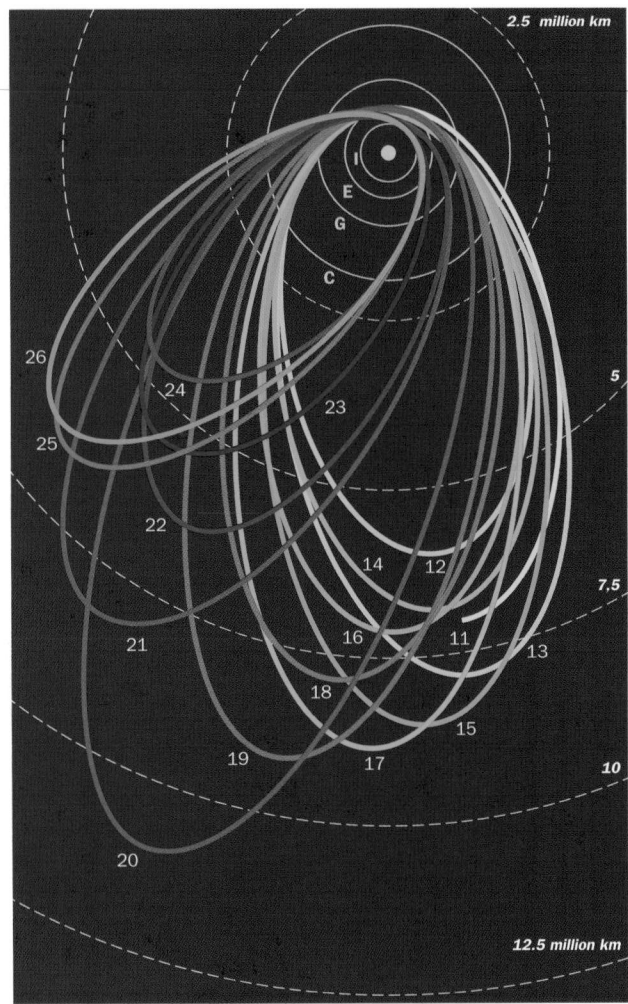

Figure 8.28. Galileo Europa Mission and Galileo Millenium Mission.

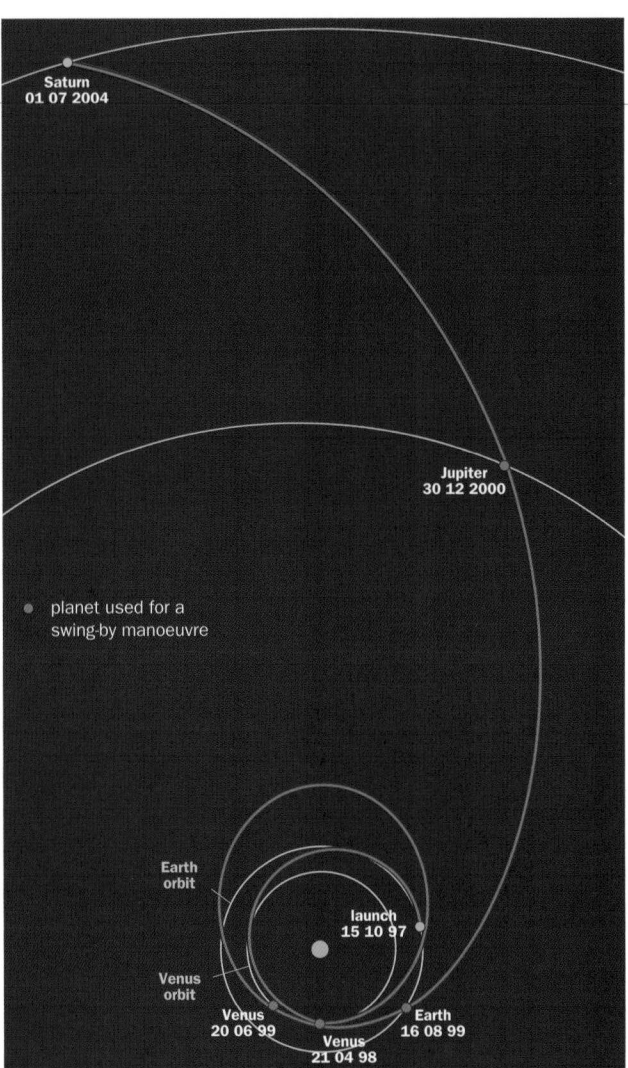

Figure 8.29. The path of Cassini towards Saturn.

committee at NASA – associates the American backed orbiter Cassini with the European Space Agency's Huygens descent probe. The mission was selected in 1985, but delayed for budgetary reasons before finally getting off the ground on 15 October 1997 with the help of a Titan 4 Centaur rocket. It made two gravity-assisted swing-bys of Venus, then one past Earth before encountering Jupiter in 2000 and going on to the Saturn neighbourhood in 2004. On 11 July of that year it will fly past Saturn's most distant moon Phoebe at a distance of 52 000 km (fig. 8.29).

The Cassini orbiter (named after French astronomer Jean-Dominique Cassini who studied Saturn in the 17th century) is stabilised around 3 axes. It carries a Visual Infrared Mapping Spectrometer (VIMS), an ISS camera (Imaging Science Subsystem) for visible wavelengths, with focal lengths of 250 and 2000 mm, a 13.8 GHz radar, an ion neutral mass spectrometer, a cosmic dust analyser, a plasma/radio wave spectrometer, a spectrometer for Plasma Science Investigation (PSI), an ultraviolet imaging spectrograph, a Magnetospheric IMaging Instrument (MIMI), a dual technic magnetometer, a Radio Science Subsystem (RSS, planetary gravitational fields) and a composite infrared spectrometer. Over the 4 years of its nominal mission, it will orbit Saturn 63 times. Titan will be over-

flown during 33 of these orbits and radar mapped, the closest scheduled approach being 950 km. Close-up observations are also planned for Iapetus, Enceladus, Dione and Rhea, whilst Tethys, Mimas and Hyperion will be observed from afar.

During the first Titan flyby, Cassini will release the Huygens probe. This is named after Dutch astronomer Christiaan Huygens who discovered Titan in 1655. The probe will begin by studying Titan's atmosphere as it falls through it, slowing its descent by means of two successively deployed parachutes. It is just possible that, after landing, it may still function for several minutes. It will then transmit information to Earth via the Cassini orbiter. The Huygens instruments are an Atmospheric Structure Instrument (ASI), a gas chromatograph mass spectrometer (GCMS), an Aerosol Collector/Pyrolyser (ACP), a Descent Imager/Spectral Radiometer (DISR), a Surface Science Package (SSP), and a Doppler Wind Experiment (DWE).

Titan, which is bigger than Mercury, has already been overflown at 6500 km by the Voyager 1 probe. Images showed a rather uniform ball of orangish fog. Its very dense atmosphere is 80% nitrogen and about 6% methane, whilst containing other hydrocarbons and molecular hydrogen. It is possible that Titan's atmosphere and surface have a similar composition to those on Earth at the very beginning of its history, with carbon-containing compounds very conducive to exobiological speculation.

Galileo and Cassini are the last two major exploration missions to the outer planets. Future missions remain to be defined but it seems doubtful whether they could exist on this scale within the framework of the Discovery programme.

Two possibilities, the Europa Orbiter and the Pluto–Kuiper Express are under study. Europa Orbiter will study Jupiter's satellite and its putative ocean with instruments unavailable to Galileo: a radar, a high resolution laser altimeter and probes to be placed on the Europan surface to carry out a seismic study.

The Pluto–Charon system seems more and more likely to be an element in a belt of small bodies (35 have so far been recorded with diameters between 200 and 2000 km), gravitating beyond the outer planets. They may well represent the residue from the process in which the outer planets were accreted, just like the main belt asteroids for the inner planets. It has been called the Kuiper belt. The Pluto–Kuiper Express project would send out two miniaturised probes towards Pluto, each one flying past a different hemisphere of the planet. The probes would then be guided towards some of the bodies in the Kuiper belt.

Asteroids and comets

Asteroids were first noticed in the 19th century. They were found to occupy a belt between Mars and Jupiter, at just the distance assigned by Bode's law concerning the heliocentric distances of planets in the Solar System. However, it was only with the close approaches made possible by space probes that decisive progress could be made in understanding them (fig. 8.30). Many asteroid observation and sample return missions were thought out in the 1980s. The Vesta mission, that was to associate the Soviet Union and Europe with the launch of four probes out to the main asteroid belt, was cancelled in 1989 by the European Space Agency, which preferred to back Cluster and Cassini. The American programme CRAF (Comet Rendezvous and Asteroid Flyby) was to study, among other things, the C-type asteroid (499) Hamburga and comet Kopff, but it was cancelled at the beginning of 1992 for budgetary reasons. In a first version of its mission, Cassini was to make observations of main belt asteroids, but NASA decided against, since it would have slowed the probe's progress out to more important issues.

In the end, it was Galileo, on its way to Jupiter, which took the first flyby pictures of an asteroid, on 29 October 1991. The subject was (951) Gaspra, an asteroid measuring 18 km across. The Galileo probe then went towards (243) Ida and discovered its satellite Dactyl on a flyby in August 1993.

Following the success of Galileo, the beginning of the American Discovery programme was marked by the NEAR mission (Near Earth Asteroid Rendezvous). It achieved the first orbit around an asteroid. Launched on 17 February 1996, it first passed at 1212 km from Mathilde (June 97), then at

3828 km from Eros (December 98), without being able to enter into orbit around it. It finally made a successful orbital approach at 323 km on 14 February 2000. This distance should gradually be reduced to a mere 15 km (fig. 8.31). Its instruments (a CCD camera, γ-ray, X-ray and near infrared spectrometers, a magnetometer, and a laser altimeter) should allow a detailed physical, chemical and mineralogical description of the asteroid (fig. 8.32).

Another probe, Deep Space 1, the first New Millennium mission, set off in October 1998 towards asteroid McAuliffe and comet West–Kohoutek–Ikemura. Its main aim is to

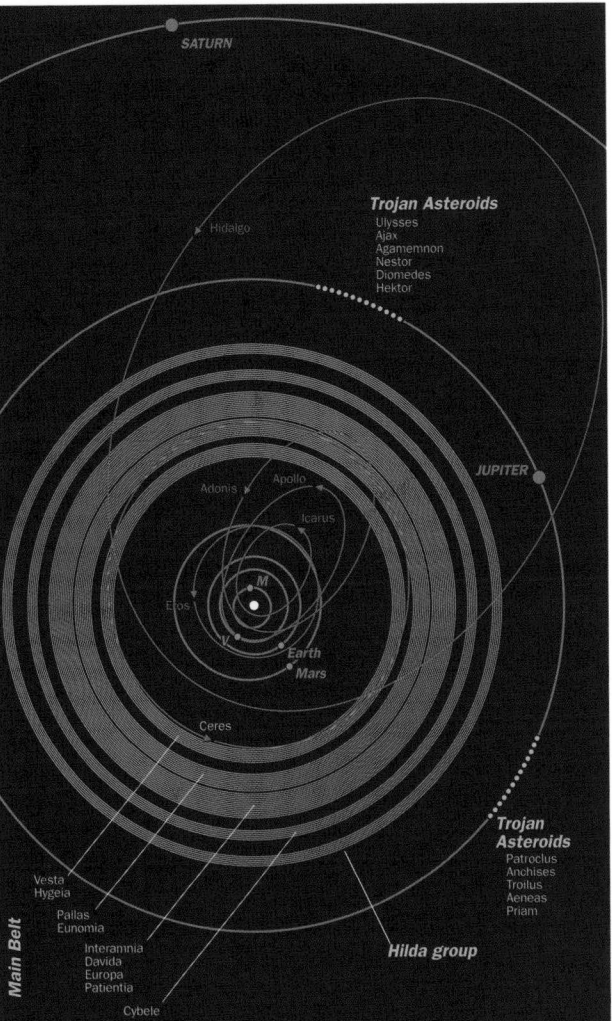

Figure 8.30. **Asteroid distribution in the Solar System.** Most asteroids are located in the main belt where they constitute several rings separated by the Kirkwood gaps. The most significant of these gaps also separates the main belt from the Hilda family. Orbital periods corresponding to the unoccupied locations of the Kirkwood gaps are simple fractions (1/3, 2/5, 3/7, 1/2) of the period of Jupiter, and they probably result from Jupiter's gravitational perturbations. The Trojan family move on the same orbit as Jupiter, at Lagrange points L4 and L5 of the Sun–Jupiter system. A hundred or so rather small asteroids have highly eccentric orbits, and often a significant inclination relative to the ecliptic plane. Several of these orbits are shown in red as an example. The number of asteroids measuring more than one kilometre across is estimated at several hundred thousand. The largest is Ceres, measuring 1025 km. The names of the ten largest asteroids are shown in green.

space qualify new miniaturised technologies (the probe weighs a mere 365 kg), and in particular to test out an ion propulsion motor.

Muses C is a sample-return mission to an asteroid. It is planned for launch by ISAS in 2002, and should arrive at asteroid 1998 FR36 in September 2005. Samples would be returned in June 2007. The probe will carry a CCD camera, X-ray and NIR spectrometers, a lidar, and the SSV (small separable vehicle), a micro rover developed by the JPL which should be placed on the surface. The probe will take samples of matter by firing a small projectile and then funneling the fragments into a sample holder. The latter, equipped with an efficient thermal protection, will be carried into a return orbit towards Earth when the probe passes nearby. It will have to withstand a temperature 30 times higher than the Space Shuttle on reentry.

However, observation, analysis and the possible return of samples from comets cause even greater enthusiasm amongst the scientific community. If the Solar System did originate, as is commonly affirmed, from the gravitational collapse of a protosolar nebula of ice and dust some 4 billion years ago, and if indeed comets are the most ancient celestial bodies, those which have undergone the least modification, then it is clear that presolar grains and traces of primitive condensates should be found in cometary material. By comparing these grains with material from the Sun, asteroids and various planets, a better understanding should be obtained of the phenomena that led to the formation of the Solar System.

Many Earth-orbiting satellites like IUE, Hubble and ISO have already supplied spectral data on comets. However, scientists believe that observations must be made from closer at hand, so that samples can be taken and analysed *in situ*. Several important missions are devoted to *in situ* exploration of periodic comets

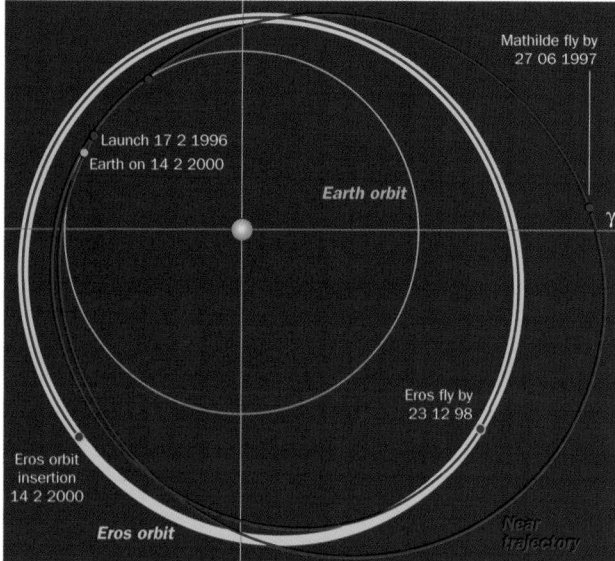

Figure 8.31. **The path of NEAR towards Eros.**

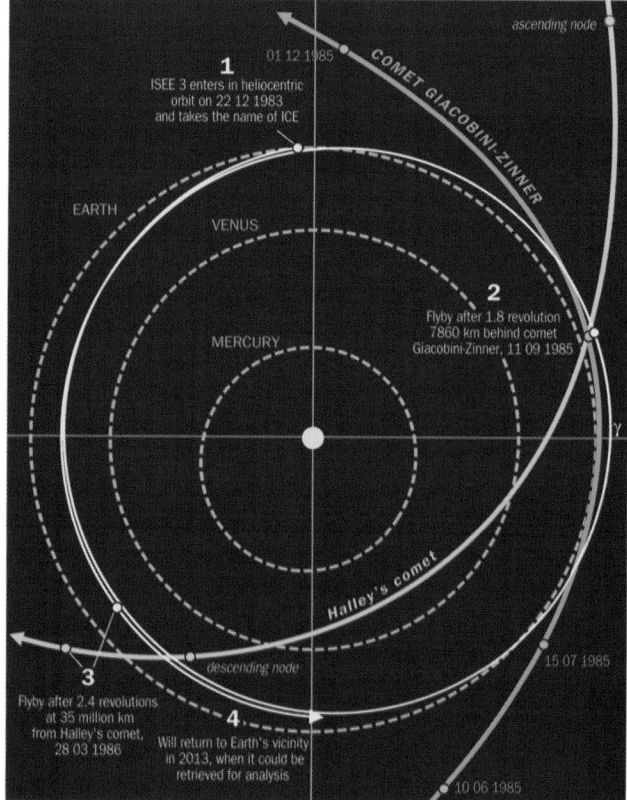

Figure 8.33. **The ICE cometary observation missions.** Using lunar gravitational assistance, ICE was directed towards two successive encounters, one with comet Giacobini–Zinner on 11 September 1985, and the other with comet Halley on 28 March 1986.

Figure 8.32. **Mosaic of the northern hemisphere of asteroid Eros.**
View constructed from six images taken on 29 February 2000 by the NEAR–Shoemaker spacecraft. Courtesy of NASA.

by means of probes. Unexpected comets, although often the most interesting, could only be reached with great difficulty, since much time is required to prepare missions.

The joint NASA–ESA satellite ISEE 3 (International Sun Earth Explorer) operated in connection with ISEE 2 and ISEE 1 to measure the solar wind. This satellite remained in a halo orbit for four years, but still had sufficient reserves to become a space probe. It was removed from its original orbit by means of a lunar gravity assist and sent off to a rendezvous with an old comet discovered in France by Giacobini in 1900 and again in Germany by Zinner in 1931. The satellite subsequently changed its name from ISEE 3 to ICE (International Cometary Explorer). On 8 September 1985, ICE flew behind the nucleus of comet Giacobini–Zinner at a distance of just 7800 km. Although it was not carrying a camera, the probe was nevertheless able to transmit completely new data concerning electrical and magnetic effects associated with the passage of the comet, and this to distances fifty times greater than those initially planned (fig. 8.33).

The mysterious comet Halley, whose regular passage has been observed every 76 years since 240 BC, has a sufficiently accurately known path for astronomers to organise several missions in its direction when it passed close to the Sun in 1986. A total of five space probes with complementary objectives were sent off on this occasion (fig. 8.34), although NASA dropped its plans to send a probe towards Halley, to the great disappointment of many American scientists.

The Soviets joined together with France and other nations to launch the probes Vega 1 and Vega 2 on 15 and 21 December 1984, respectively. The two syllables of the name Vega signify the double objective of this programme, which was devoted partly to exploration of Venus, and partly to study of Galleia, the Russian name for Halley. After flying by Venus, where they released descent modules (see fig. 8.13) in June 1985, the probes continued their route towards comet Halley, approaching to within 10 000 km in March 1986. For its part, the European Space Agency had launched the Giotto probe, named after the Italian painter who painted comet Halley in one of his frescos in 1304. Giotto reached its rendezvous with the comet eight months later, on 13 and 14 March 1986. The comet was then moving at 8 kilometres per second and a very great flight accuracy was needed to place Giotto at a distance of only 596 km from the cometary nucleus. From this vantage point, it sent back 2000 pictures showing the nucleus to be a solid object covered with black dust. The two other probes, Sakigake and Sui-

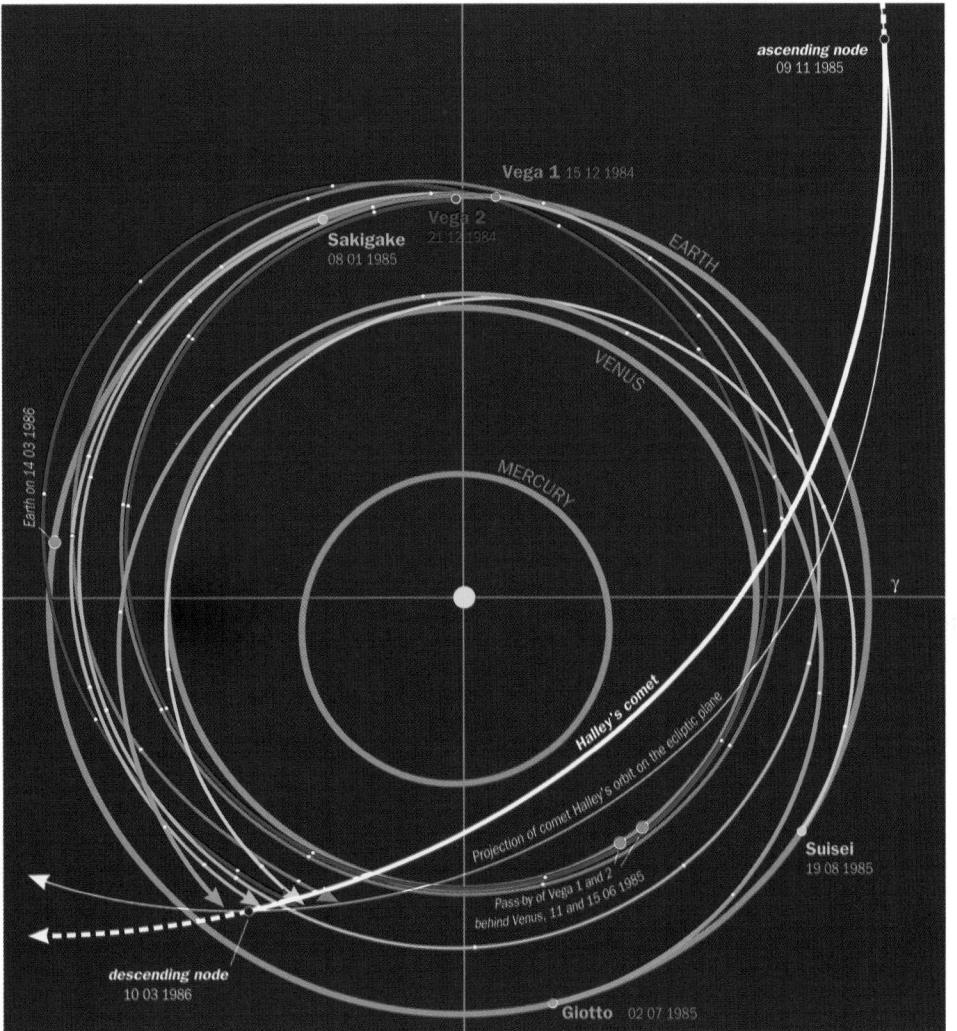

Figure 8.34. **Observational missions to comet Halley.** Setting off at various dates spread between December 1984 and July 1985, the five probes all arrived in the vicinity of comet Halley at about the same time, just a few days before or after 10 March 1986. At this particular date, the comet was passing through the descending node of its orbit, a propitious moment for rendezvous with a vehicle moving in the plane of the ecliptic. Rendezvous would also have been possible at the ascending node, but offered fewer advantages. For example, the Earth–probe distance was greater (implying slower communications), and the Sun–comet distance was also greater (implying less cometary activity and only half the energy available via solar arrays).
White dots on probe paths indicate their positions at the beginning of each month.

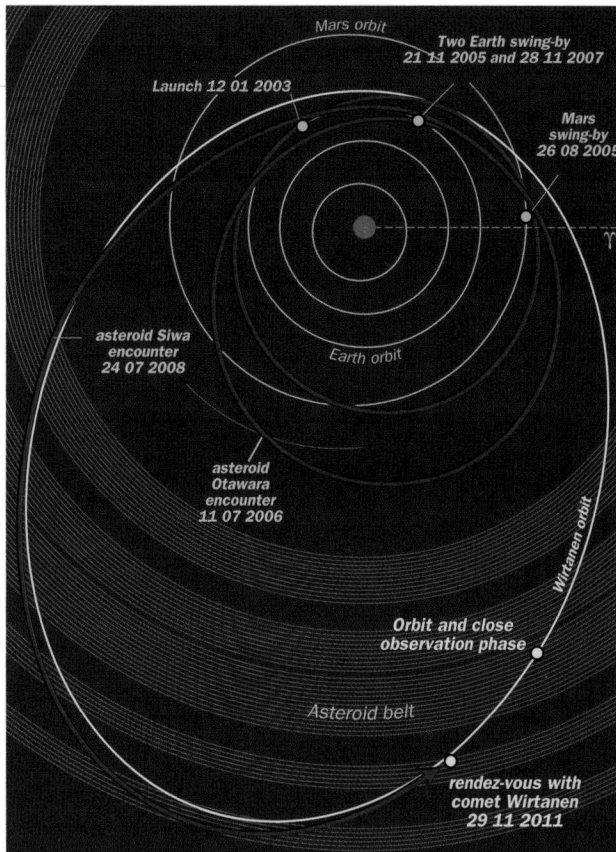

Figure 8.35. **The orbit of Rosetta.** The Rosetta probe was due for launch at the beginning of 2003. It will subsequently engage in gravity-assisted manoeuvres with Mars in August 2005 and then Earth in November 2005, before flying within 1000 km of the asteroid otawara in 2006 and 1000 km of Siwa in 2008. Having reached its aphelion in September 2010, Rosetta will go into orbit around comet Wirtanen in 2011, accompanying it right through until 2013.

sei, were Japanese. Their names mean 'reconnaissance' and 'comet', respectively, in Japanese. They were present at the same rendezvous in March 1986, but at much greater distances from the comet.

Lunar rock samples brought back to Earth for analysis have led to considerable progress in our understanding. The idea of a Comet Nucleus Sample Return mission has been under consideration by the main space agencies since 1984. There exist two modi operandi. The simplest consists in crossing the coma and gathering dust torn from the nucleus. The other option involves landing one or more penetrators on the nucleus itself, with the task of sampling the surface, or better still, the interior. In both cases, samples are brought back to Earth in a hermetically sealed capsule provided with thermal protection for reentry.

The first solution gave rise to the CAESAR projects (Comet Atmosphere and Earth SAmple Return) at ESA, and SOCCER at ISAS, before NASA finally retained the probe Stardust put forward by the JPL. Stardust departed in February 1999 in the hope of collecting sufficiently large cometary particles (greater than 15 microns) for analysis. Rendezvous with comet Wild 2

is scheduled for January 2004. Particles of interstellar dust will also be gathered. The capsule containing the samples will be dropped on the next close passage to Earth in 2006.

The second solution was originally the subject of the first joint NASA–ESA project, in 1984. This project was christened Rosetta in 1988, after the Rosetta stone which led to the deciphering of Egyptian hieroglyphs. NASA was to provide a Mariner Mark 2 bus, and ESA the lander and return capsule. However, the project was not retained by NASA in 1996. Since ESA had made this mission the third cornerstone of its Horizon 2000 programme when it was devised in 1993, it was maintained as part of the programme with some modifications, but keeping the same name. The mission was due for launch in January 2003. It includes an orbiter that will go into orbit around comet Wirtanen (fig. 8.35) and a lander that will make in situ analyses, since the idea of returning samples has since been abandoned.

The orbiter will carry four imagers: OSRIS (two cameras in visible wavelengths with high and medium resolution), ALICE (UV spectrometry), VIRTIS (visible and IR mapping spectrometry), and MIRO (microwave spectrometry). It will also dispose four composition analyzers: ROSINA (neutral gas and ion mass spectrometry), MODULUS Berenice (isotopic ratios of light elements by gas chromatography), COSIMA (dust mass spectrometer) and MIDAS (grain morphology). In addition there will be radio science experiments (RSI), cometary plasma environment and solar wind interaction studies (RPC), dust studies (GIADA), and a tomographic study of the nucleus by radio sounding (CONSERT, on both orbiter and lander).

The lander is equipped to take samples down to depths of one metre and make detailed analyses of cometary material. Techniques include elemental analysis by X-ray, α particle and proton spectrography (APX), pyrolysis and analysis by mass spectrometry and gas chromatography (COSAC), and gas and isotopic composition analysis (MODULUS Ptolemy). The CIVA instrument is an optical stereomicroscope associated with an IR spectrometer, SESAME (surface electrical and acoustic monitoring experiment) is a dust impact monitor, MUPUS is a MUlti-PUrpose Sensor for surface and subsurface science, and ROMAP is a magnetometer and plasma monitor.

Two other cometary projects fall within the framework of the Discovery programme. The CONTOUR probe (COmet Nucleus TOUR) will set off in 2002 for three cometary rendezvous in 2003, 2006 and 2008, and has a mission similar to NEAR. This time the probe could be rapidly reprogrammed if a new comet were detected. An encounter with a comet on its first passage through the inner Solar System would indeed be preferable, since its constituents would be less likely to have suffered alterations. Finally the Deep Impact probe, scheduled for 2004, will send a 500 kg mass into a high speed collision with comet P/Tempel 1. Its high resolution camera and IR spectrometer will then observe the impact crater and the debris raised from the nucleus, with a closest passage of 700 km.

Earth observation

- OVERVIEW

- SENSORS

- IMAGES OF THE EARTH

- METEOROLOGY

- REMOTE SENSING OF TERRESTRIAL RESOURCES

- OPTICAL REMOTE-SENSING SYSTEMS

- SAR-EQUIPPED EARTH RESOURCES SYSTEMS

- REMOTE SENSING – THE WAY AHEAD

Earth observation

Overview

Earth observation is one of the areas in which space activity offers the broadest range of applications, thanks to the ability of satellites to acquire, in the course of repeat visits, overviews of broad areas which when juxtaposed cover the entire planet without the sort of constraints which go with political frontiers.

Earth observation satellites fall into three main families – meteorological satellites, medium resolution remote-sensing satellites and high resolution satellites, the latter restricted initially to military reconnaissance activity but today being used increasingly for civil applications.

Satellites belonging to the first of these families typically supply images covering very large areas but offering relatively poor resolution and relying on multiple passes. They are geostationary or placed at the upper end of the low circular orbits at altitudes of around 900 km–1800 km. In either case, they may be deployed in groups of satellites belonging to a single programme, as with the Russian Meteor-II craft or the American NOAA satellites.

Satellites in the second family generally operate from circular orbits at lower altitudes – between 600 km and 1000 km – and provide, at a lower revisit frequency, images with higher spatial resolution. Examples are the American Landsat, French SPOT and Indian IRS series.

Satellites in the third family commonly describe eccentric orbits, performing their observations close to the low altitude perigee – which can be as low as about 160 km in the case of many Kosmos satellites; this makes for very fine ground resolution. A number of recent satellites, designed for a longer operational life, do however occupy circular orbits, at altitudes below 300 km in the case of those offering the highest resolution. Previously confined to defence duties, this family of satellites is now also being used to meet civil requirements. The launch in 1999 of the Ikonos 2 satellite, belonging to a private company, SpaceImaging, marked the emergence of a civil capability for metre-range resolution from circular orbits below 700 km, orbits not that distant from those occupied by certain military systems offering comparable resolution.

These three major families also differ from each other in terms of status. Meteorological satellites are operated by government bodies in the framework of programmes which may sometimes be international, such as the World Weather Watch. And while there is something of a trend towards commercialising some products, provision of a public service is still the guiding concern.

Remote-sensing satellites relying solely on optical technology and those carrying radar instrumentation, are used for cartography or for the study of continental and marine

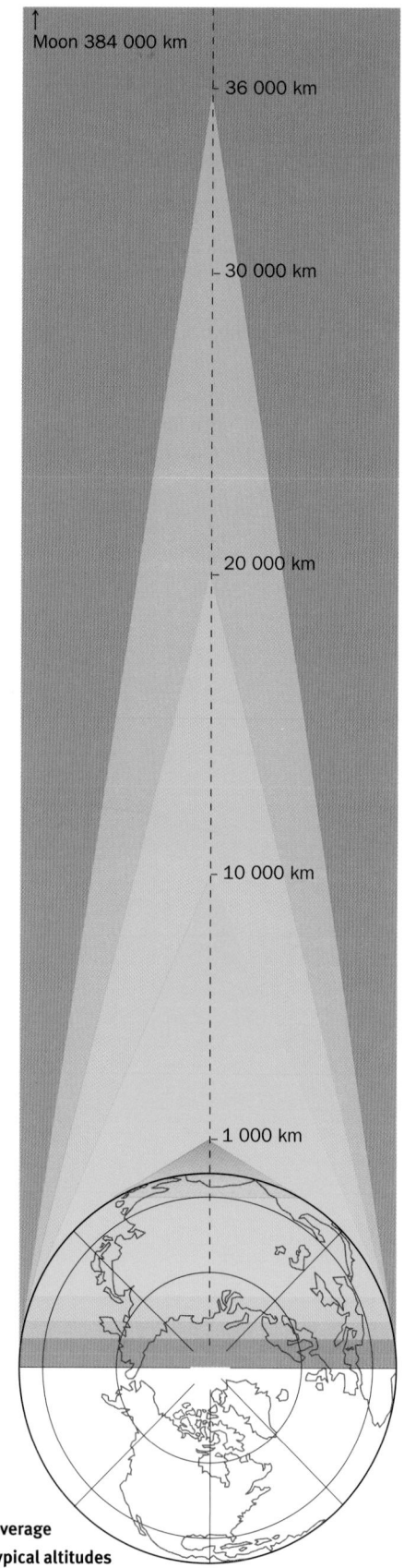

Figure 9.1. **Earth coverage from a number of typical altitudes**

resources. Exploitation of these systems, following initial funding by government agencies, is increasingly being handled by government-supported commercial outfits such as SPOT Image in France and Eosat in the United States. Since 1995 private firms have also been showing considerable interest in this area of activity.

High resolution satellites are an established feature of military reconnaissance activity (see Chapter 12) but are still something of a rarity in civil Earth observation. This new category of civil programme relies on private funding and is governed at least in theory by industrial and commercial considerations.

In terms of how data is used, it is not always that easy to distinguish between the various families of satellite, especially where the mission addressed − environmental management or risk monitoring for example − is very wide-ranging. NOAA images are a case in point: though intended primarily for meteorology, considerable use is also made of them in the remote sensing of plant resources. Similarly, data col-

lected by civil remote-sensing satellites such as SPOT has potential military uses and degraded data from Russian military satellites or again declassified images acquired by US military satellites may be used and commercialised for civil purposes. The emergence of metre-range resolution civil commercial satellites is confusing the picture still further but points to a growth in the range of potential users.

Altitudes and satellite coverage

The area viewed from satellites expands with altitude, eventually encompassing an entire hemisphere at infinite altitude. At lower altitudes, the variation in coverage is very marked: at 200 km, Apollo 9 could see 1.5% of the Earth's surface at a 151° angle, whereas NOAA 11 surveys more than 10% from its 1600 km vantage point. Conversely, at high altitudes very little variation occurs: at 20 000 km Navstar looks down on 38% of our planet's surface at an angle of 30°, while Meteosat, stationed at 36 000 km, improves that figure only slightly to 42% at an angle of 17° (figs. 9.1, 9.2 and 9.3).

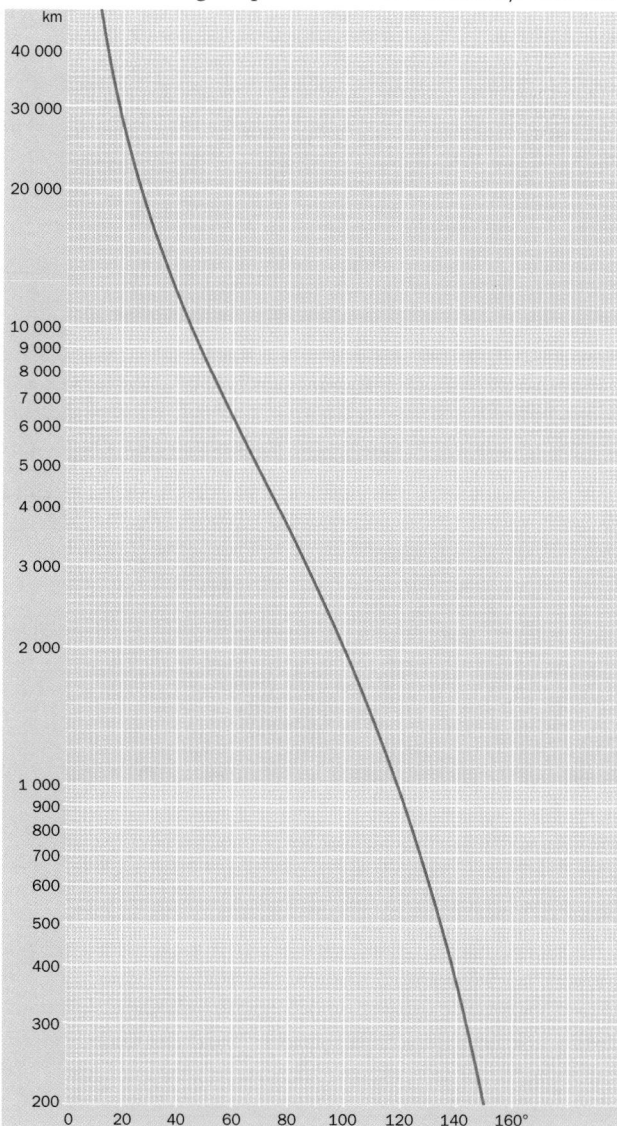

Figure 9.2. **Angle at which the Earth is viewed in relation to altitude**

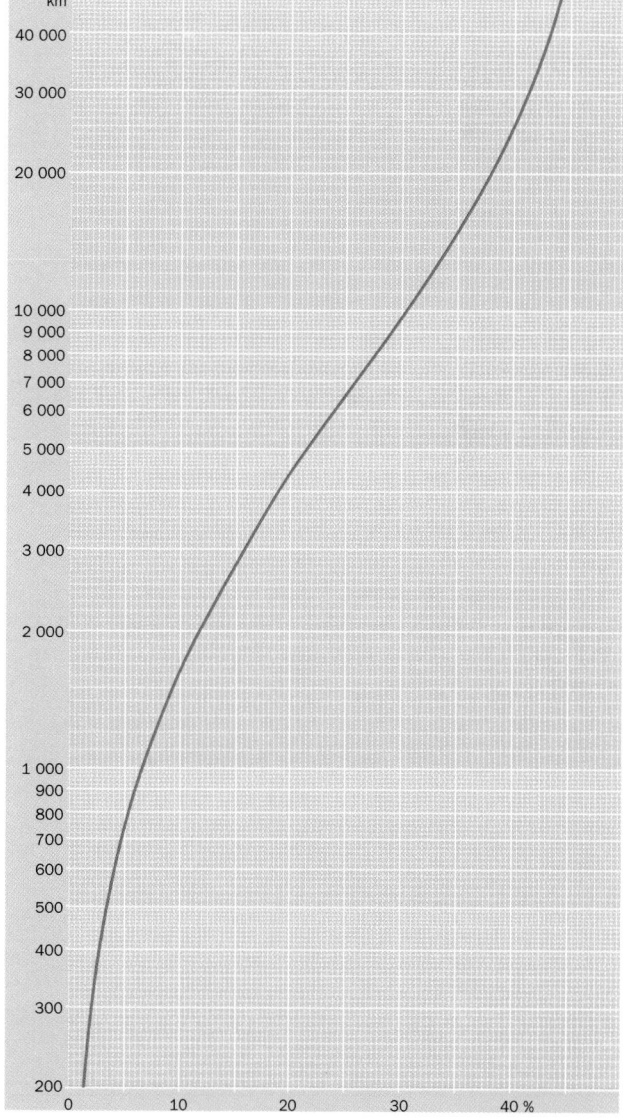

Figure 9.3. **Percentage of the Earth's surface that is visible in relation to altitude**

Sensors

Sensors measure the electromagnetic radiation emanating from a geometrically defined field. The dimensions of the field are determined by the sensor's optics. The entire field may be explored in one take, with simultaneous acquisition of all points in the image, as with still or TV cameras. The field may also be scanned sequentially, complete images being built up from a juxtaposition of individual readings, as is the case with scanning systems.

The sensor's overall field of view corresponds to the size of the geographical area selected for observation. The *Instantaneous Field of View* (IFOV) is defined by the solid angle from which the electromagnetic radiation measured by the sensor at a given point in time emanates. While the sensor's overall field of view and the IFOV coincide in the case of still cameras and to a certain extent when CCDs are used, this does not apply in the case of scanners, whose overall field of view corresponds to the continuous movement of an instantaneous field.

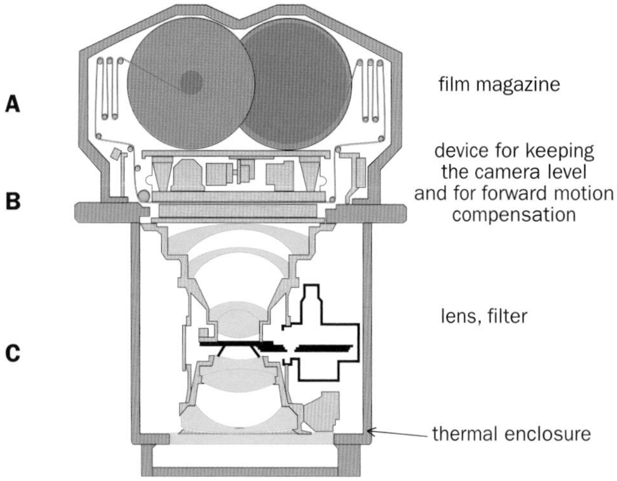

Still cameras

Satellite-borne still cameras support instant acquisition of an image of the area of the Earth's surface over which the satellite is flying (fig. 9.4). The field of view on the ground is determined by the lens used and the altitude at which the satellite is flying. Photographs are generally centred on the nadir and the field dimensions are defined either by angular apertures independent of flight altitude or by altitude-dependent linear measurements on the ground (length of sides or of diagonal) (table 9.1 and fig. 9.5).

The quality of some images can be enhanced by a technique (known as forward motion compensation, or FMC) which counters the change in the satellite's position while the photograph is being taken.

Figure 9.4. Example of a spaceborne still camera: the Large Format Camera (LFC)

The LFC is a very powerful multispectral imaging system used on the US Space Shuttle Challenger on its October 1984 mission. It consists of:

A – a magazine containing 1220 m of high-resolution film configured for 2400 images in a 460 mm along track x 230 mm across track format. These rather unusual dimensions support acquisition of images covering ground areas of 153 181 km² (553 km × 277 km) from an altitude of 370 km.

B – devices for keeping the camera perfectly level during exposure and for compensating forward motion.

C – a lens with a focal length of 305 mm and an angular aperture of 74°.

In addition, the Altitude Reference System (ARS), comprising two cameras with 152 mm focal lengths, records the star field at the point in time of each acquisition. Simultaneous collection of data concerning the satellite's altitude and position makes for highly precise mapping.

Labels on Figure 9.4:
- film magazine
- device for keeping the camera level and for forward motion compensation
- lens, filter
- thermal enclosure

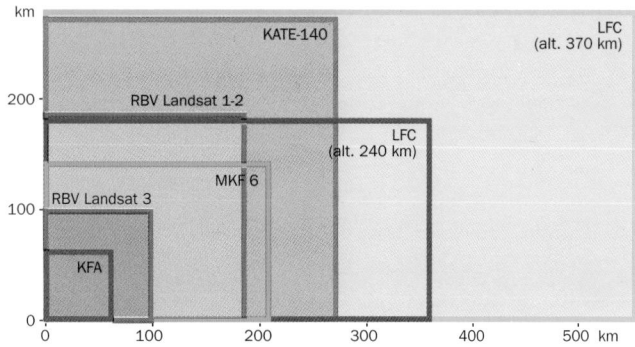

Figure 9.5. **Overall coverage area of some still and TV cameras**

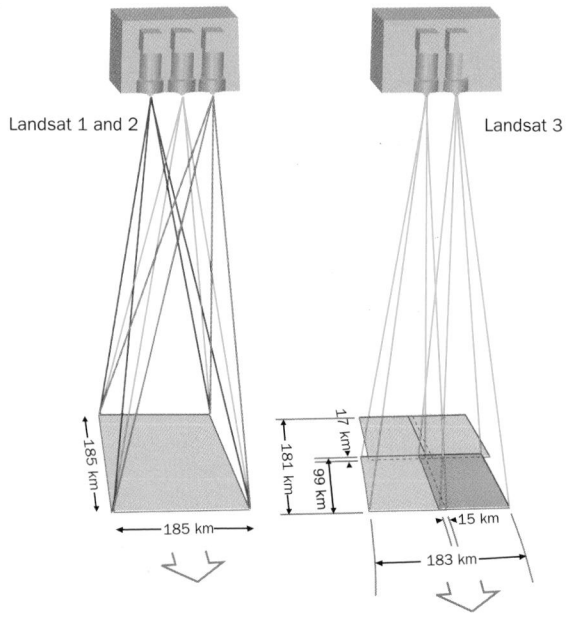

Figure 9.6. **The RBV systems on Landsat 1 and 2 and on Landsat 3**

camera	satellite	altitude (km)	number of lenses	focal length in mm	aperture	viewing angle	ground field in km	scale of the image	frame size in mm	exposure time (seconds)	forward motion compensation	resolution (metres)
Hasselblad	Gemini 4 to 7, 1965 Apollo 9 (S.065), 1969 Apollo 12, 1969	200	4	80	4-16	55°	70 x 70		57 x 57	1/125 - 1/250	—	125 70 70
Multiband Camera	Skylab (S.190 A), 1973	435	6	152	2.8-16	21°2	163 x 163		57 x 57	1/500	+	99
Earth Terrain Camera (ETC)	Skylab (S.190 B),1973	435	1	460		14°2	109 x 109		114 x 114	1/100 - 1/200	+	38
MKF 6	Soyuz 22, 1976 Salyut 7, 1982	250 275	6	125	4-13.5	41°	209 x 140	1/2 500 000	80 x 56		+	25 25
KATE 140	Soyuz 35, 1980 Salyut 7, 1982	250 275	1	140	6.8-16	85°	270 x 270	1/1 500 000	180 x 180	1/10 - 1/250	—	30 30
Metric Camera (MC)	STS 9 (Spacelab), 1983	250	1	305	5.6-11.	41°	180 x 180	1/820 000	230 x 230	1/250 - 1/500	—	30
Large Format Camera (LFC)		370 240	1	305	8	74° x 41°	553 x 277 358 x 179	1/1 200 000	460 x 230	1/30 - 1/250	+	
ARS	STS-41 G, 1984	240	2	152	2,8	stellar coverage			70 x 70	1/5	—	
KFA 1000	Resurs F	250	1	1 000		16°2	75 x 75	1/250 000	300 x 300		—	5 (down to 2)
KVR 1000	Kometa, 1983	220	1	1 000	5	11°4	37 x 165	1/220 000	758 x 40		+	0.75
TK 350	Kometa, 1983	220		350	5.6		200 x 300	1/630 000	284 x 189			10
KATE 200	Resurs F1, 1986	260		200			180 x 180	1/1 000 000	180 x 180			15
KFA 3000	Resurs F3, 1986	260		3 000			21 x 21	1/70 000	300 x 300			2
MK 4	Resurs F2, 1988	220		300			130 x 130	1/730 000	180 x 180			5
Return Beam Vidicon (RBV)	Landsat 1-2, 1972 et 1975	900	3	126		8°	185 x 185			1/62 - 1/83 - 1/125		80
Return Beam Vidicon (RBV)	Landsat 3, 1978	900	2	250		4°	98 x 98					80

Table 9.1. **Technical characteristics of spaceborne still and TV cameras**

Some still cameras used on military reconnaissance satellites feature a very narrow field and hence a very long focal distance. A case in point is the KFA 3000 high-resolution camera equipping various Russian satellites which comprises a telescope with a focal length of 3 m.

Television cameras

Though very different from still cameras, TV cameras nevertheless also take the complete image immediately and their overall field of view coincides with the instantaneous field. When the shutter has closed the images are stored in a charge storage layer. The images are explored subsequently by means of scanning. The Return Beam Vidicon (RBV) systems on Landsat 1 to 3 are a good example of this (fig. 9.6).

Scanners and push-broom sensors

Scanners take measurements in the instantaneous field of view as it moves along the scanlines. The ground width of the IFOV, perpendicular to the scanning direction, widens as the distance between the sensor and the ground increases; in other words the field expands away from the satellite's nadir point and towards the limb. The lower the satellite altitude the greater the impact of this variation in strip width. Moreover, the ground length of the instantaneous field of view in the scanning plane increases with its skew from the viewing angle. It follows that with sensors which scan the Earth's surface from one limb to the other, the ground area encompassed by the instantaneous field of view varies considerably (fig. 9.7).

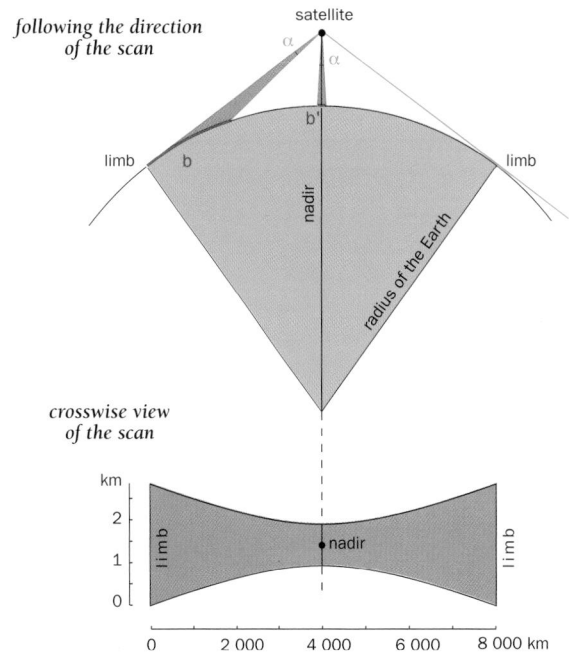

Figure 9.7. **Ground dimensions of the instantaneous field of view with limb-to-limb scanning.** The example shown here is the VHRR sensor on the NOAA 2 to 5 satellites.

Surfaces sensed by scanners are scanned in successive lines. The scanning process and scanlines result from a combination of the satellite's movements and those of the scanner itself.

Where geostationary satellites are concerned, the succession of scanlines is obtained by a mechanical device which orients the telescope. Each new line of images scanned by the

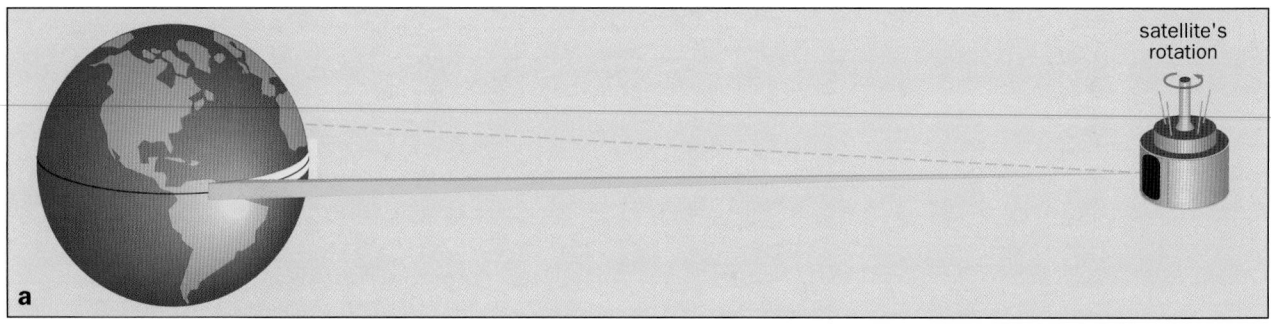

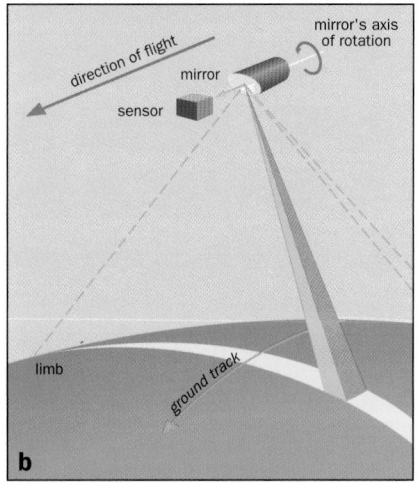

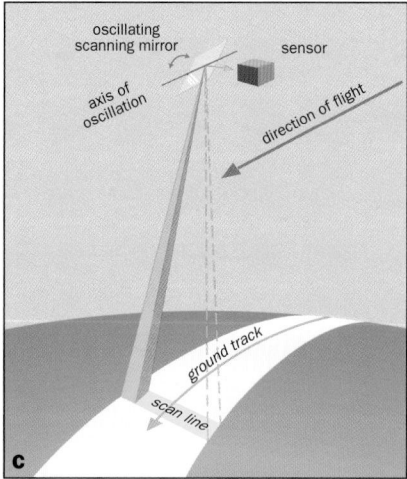

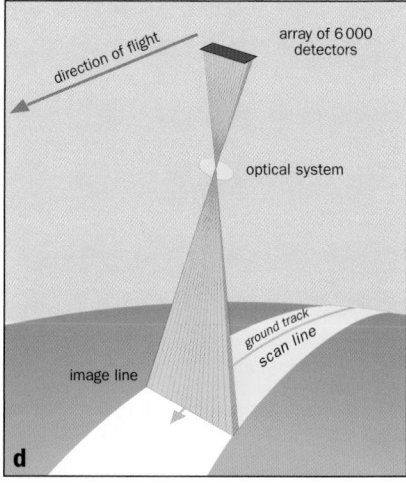

Figure 9.8. **The main scanning systems.**

a) Limb-to-limb scanning by satellite rotation. The example shown is the geostationary satellite Meteosat.

b) Scanning by mirror rotation.

c) Scanning by mirror oscillation.

d) Push-broom system. An array comprising a large number of detectors sweeps the ground in the direction of flight.

telescope is offset by a constant step from the line before. The Meteosat telescope thus moves through 18° in the course of 2500 passes in a north–south direction. Line scanning is effected by satellite rotation.

In the case of non-geostationary satellites, the general succession of scanlines is obtained by the motion of the satellite along its orbit. Scanlines may go from one terrestrial limb to the other or be restricted to much shorter distances corresponding to only part of our planet's apparent arcs.

Scanners performing limb-to-limb imaging explore a complete circle, of which the Earth's surface occupies only a small arc. In the case of geostationary satellites, this represents at most 17° of the total 360° sweep. Such scanners use various types of rotating system. In some cases – Meteosat being one example – a 360° scan is obtained by rotation of the satellite itself (fig. 9.8a). Another type of scanner uses a mirror revolving round an axis parallel to the satellite's velocity vector, a case in point being the Advanced Very High Resolution Radiometer (AVHRR) fitted to satellites NOAA 6 to 11 (fig. 9.8b). Some sensors do not sweep a complete circle, scanning only a limited arc on the visible part of the Earth. Scanning may be effected by oscillation of a mirror about an axis parallel to the satellite's velocity vector, as with

the MSS and TM sensors carried by the Landsat series (fig. 9.8c). Finally, some mechanical systems perform conical scanning, the advantage of this approach being that a constant distance is maintained between the sensor and the ground throughout the scanned area: this is the case with the OCTS sensor which equipped ADEOS 1.

Detector cells arranged in linear array, with no mechanical scanning mechanism, can also be used to collect data along ground scanlines, using the platform's motion to sweep the ground in the direction of flight (push-broom concept). The HRV and HRVIR sensors on SPOT satellites, which measure energy using detectors configured in a Charge Coupled Device (CCD) array, are an example of this kind of arrangement (fig. 9.8d). CCD arrays of this kind can be replaced by Active Pixel Sensors (APS) which are intended to offer simplified sensing electronics through incorporation of many functions in the sensor itself.

The median line corresponding to the centre of the field of view scanlines is known as the instrument track. This coincides with the satellite ground track where viewing and scanning are centred on the nadir but is offset from it in the case of oblique viewing (fig. 9.9a). Some sensors have a capability of inclining their viewing axis either side of nadir. The High Resolution Visible (HRV) instrument on SPOT 1 to 3, for example, can adjust to 90 different inclinations in 0.6° steps within a total angle of 54° (fig. 9.9b).

Where oblique viewing angles are programmable, imaging at large distances from the satellite track becomes possible, which makes for considerable flexibility. By acquiring

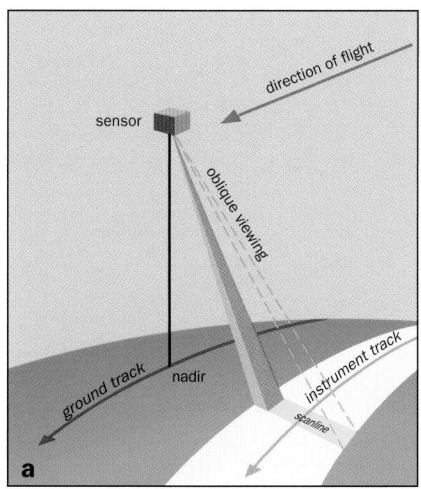 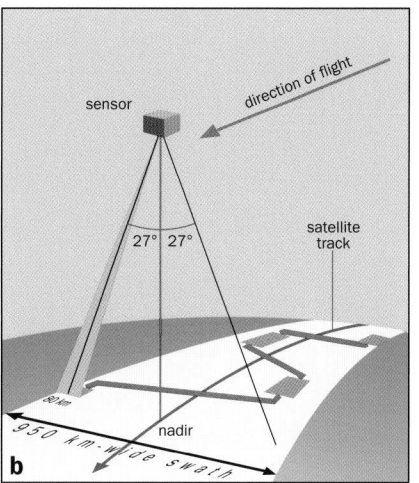 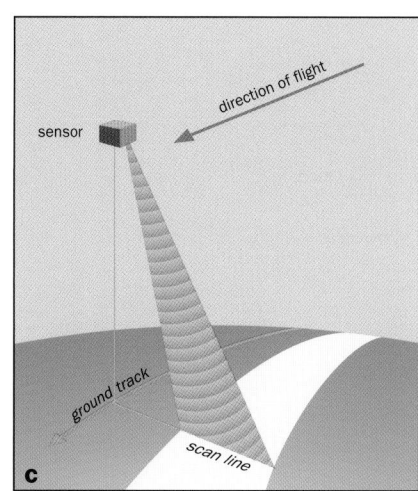

Figure 9.9. **Oblique viewing**
a) Instrument track and satellite track
b) SPOT oblique viewing. The viewing angle may be remotely adjusted from the ground. The areas observed must be situated within a 950 km wide viewing corridor straddling the satellite track. Scene widths range from 60 km near nadir to 80 km with extreme sideways viewing.
c) Off-nadir swath of the Side Aperture Radar.

images of the same ground area from different viewpoints, the terrain can be mapped stereoscopically. This can be achieved through oblique viewing, as with SPOT 1 to 4, but it is also possible to build up stereoscopic images from series of observations taken from different viewpoints on the same orbit; this involves combining forward and nadir observations, as in the case of JERS 1.

The value most frequently used in defining the qualities of a remote-sensing instrument is the instantaneous field of view (IFOV). This is determined by the sensor's optics and is defined by a plane angle measured in radians and a solid angle measured in steradians. The instantaneous field of view is constrained by the radiometric sensitivity of the detector, which has to be exposed long enough to receive sufficient energy to operate effectively. The use of detector arrays, making it possible to obtain longer exposure times than with mechanical scanning systems, has done much to reduce the instantaneous field of view and this form of detection is now used on all fine-resolution sensors. The need for a minimum quantity of energy to ensure satisfactory observation quality also explains why higher spatial resolutions are incompatible with finer spectral resolutions. To take the example of the HRV on SPOT 1 to 3: panchromatic operation can be associated with 10 m resolution, whereas multispectral mode operation with its narrower spectral bands is consistent with 20 m resolution.

The lower the satellite's altitude and the nearer the angle formed by the viewing axis and the ground to 90°, the smaller the ground area intercepted by the solid angle. At the sub-satellite point, that area is taken to be a square, the length of whose sides can be seen from the graph in fig. 9.10.

Sensors which collect data in the thermal infrared band have lower resolving power than those operating in the visible or near infrared spectral regions. IFOV angles are generally wider in the thermal infrared than in either the visible or the near infrared.

Imaging radars

Sensors which measure naturally emitted radiation in the form of light or heat are passive remote-sensing instruments.

Remote sensing is active, on the other hand, where the sensor itself emits radiation and then measures the radiation returned to the sensor, one example being radar instruments. Given the altitudes at which the satellites operate and the oblique sensing often used, enough energy must be emitted for the system to pick up the echo at considerable distances. Providing such quantities of energy on board is however a difficult matter, which explains why the use of space radar is lagging so far behind that of airborne radar.

Radar instruments have the advantage over optical sensors that they can operate round the clock – since they themselves emit the signal whose return values they then measure. And since they use microwaves, they can penetrate cloud cover with no effect on imagery.

Radars emit signals and receive the returning echoes, which are sent back by the ground in a variable length of time depending on the distance separating sensor and target. Target objects are located by measuring the varying distances between them and the sensor and not, as with other types of sensor, by the sensing angle. Synthetic Aperture Radars (SARs) use the frequency variation caused by the Doppler effect to produce high-resolution images.

Viewing is necessarily oblique and is generally restricted to one side of the satellite trajectory, either left or right (fig. 9.9c). In the case of Radarsat, viewing is normally to the right of the flight track but the look direction can be reoriented to left of the track by 180° rotation manoeuvres around the yaw axis (fig 9.11).

Radars can use vertical (V) or horizontal (H) polarisation for emission and reception. Four types of image can thus be obtained, two with parallel polarisation (HH and VV) and two with cross-polarisation (HV and VH). These various permutations were all operational together for the first time in

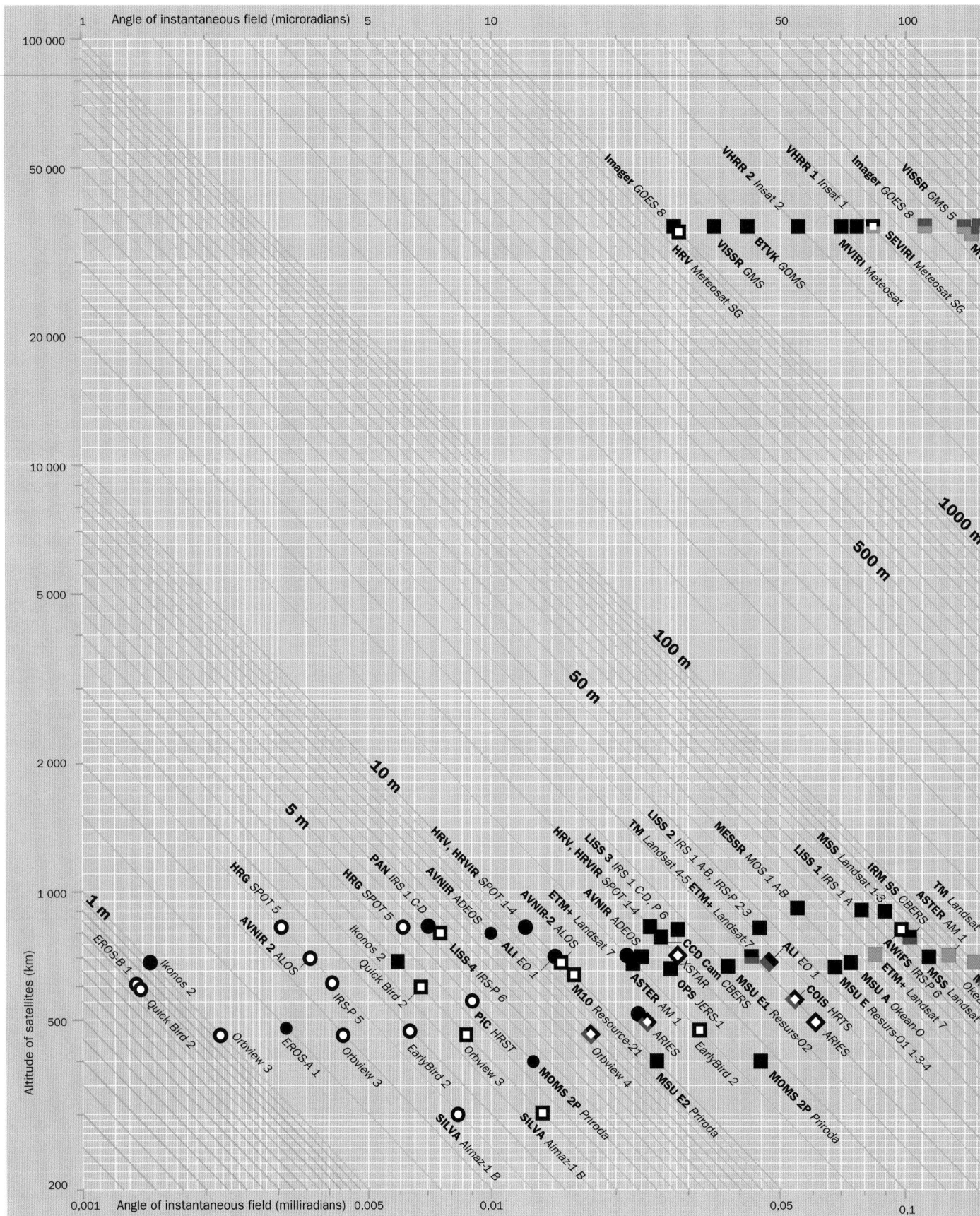

Figure 9.10. **Instantaneous fields of view.** Relationship between the satellite altitude (vertical scale), the angle of instantaneous field of view (horizontal scale) and the IFOV at the sub-satellite point (oblique scale).

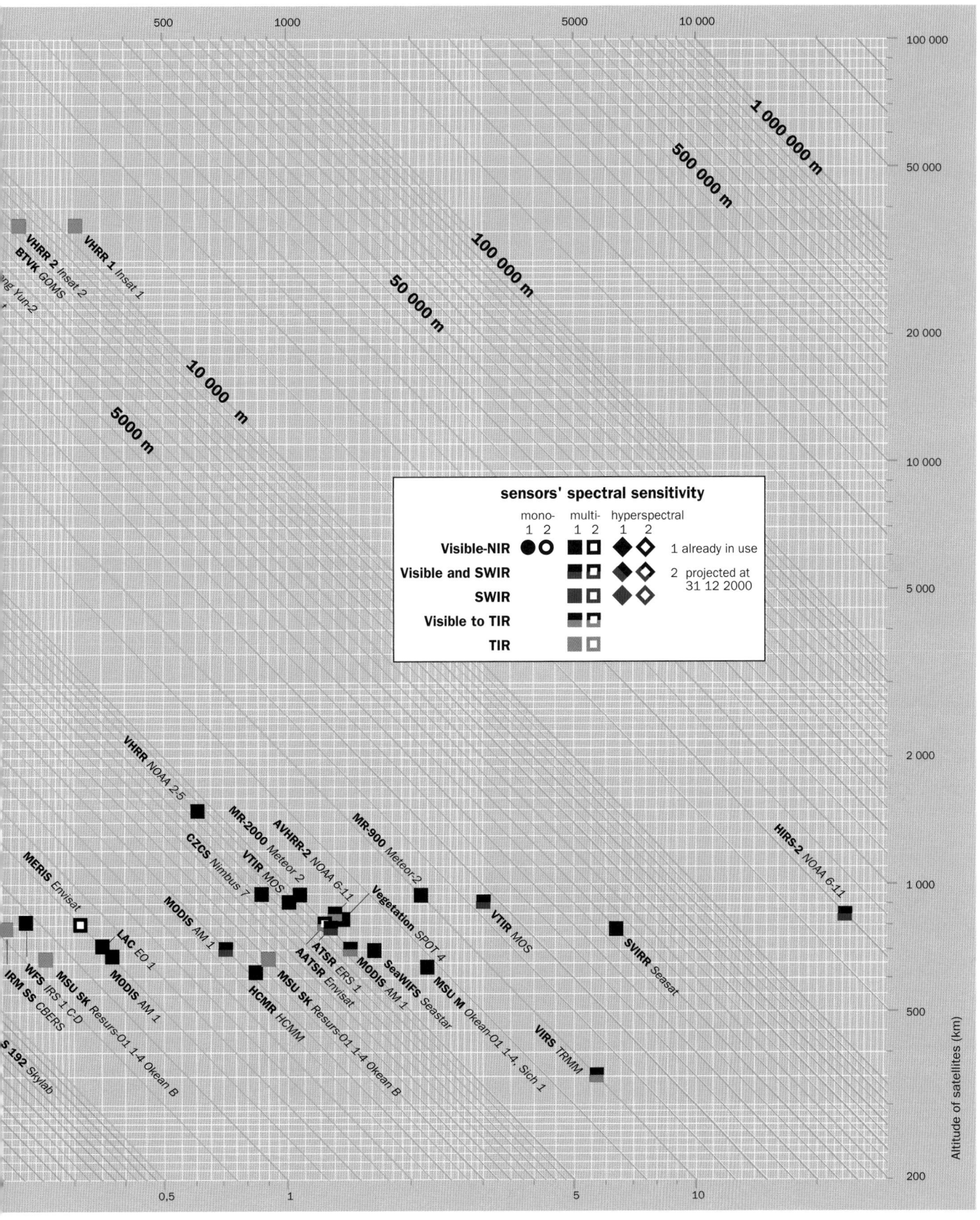

500   1000   5000   10 000   100 000

1 000 000 m

500 000 m

100 000 m

50 000 m

10 000 m

5000 m

50 000

20 000

10 000

5 000

sensors' spectral sensitivity

	mono-	multi-	hyperspectral	
	1 2	1 2	1 2	
Visible-NIR	●○	■□	◆◇	1 already in use
Visible and SWIR		■□	◆◇	2 projected at
SWIR		■□	◆◇	31 12 2000
Visible to TIR		■□		
TIR		■□		

VHRR 2 Insat 2
VHRR 1 Insat 1
BTVK GOMS
ng Yun-2

VHRR NOAA 2-5

MR-2000 Meteor 2
CZCS Nimbus 7
VTIR MOS
AVHRR-2 NOAA 6-11

MR-900 Meteor-2

HIRS-2 NOAA 6-11

MERIS Envisat

LAC EO 1

MODIS AM 1

Vegetation SPOT 4

ATSR ERS 1
AATSR Envisat
MODIS AM 1
SeaWIFS Seastar

VTIR MOS

SVIRR Seasat

MODIS AM 1

HCMR HCMM

MSU SK Resurs-O1 1-4 Okean B

MSU M Okean-O1 1-4, Sich 1

WFS IRS 1 C-D

MSU SK Resurs-O1 1-4 Okean B

IRM SS CBERS

VIRS TRMM

S 192 Skylab

2 000

1 000

5 000

500

Altitude of satellites (km)

200

0,5   1   5   10

Normal mode 0° **'Antarctic' mode** 0°

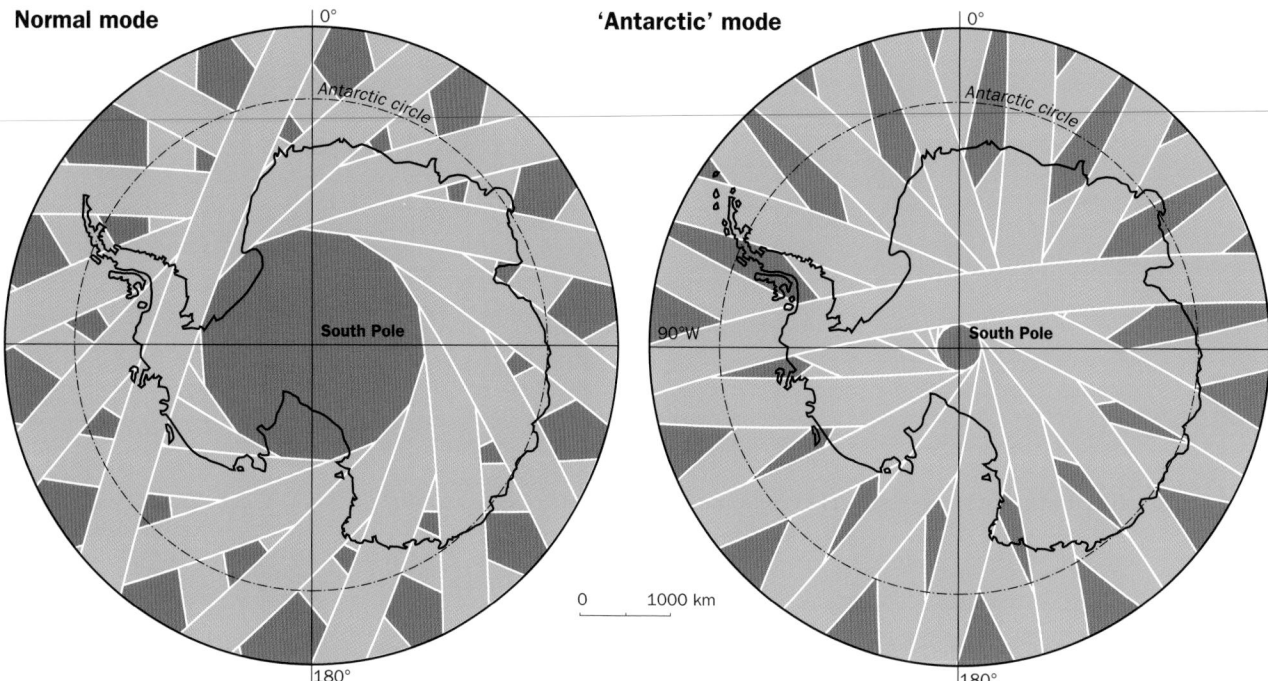

Figure 9.11. **Radarsat swaths over the Antarctic.** The strips represent the first day of synthetic aperture radar coverage with a 500 km swath. Viewing is to the right in normal operating mode and to the left in 'Antarctic' mode.

SIR C. By using a range of polarisations the radiometric qual-ity of imagery can be improved without a resolution penalty.

Finally, it is also possible to use the phase difference between two radar echoes returned by the same terrain and received along two trajectories very close together, a tech-nique known as interferometry. This technique, use of which is becoming increasingly widespread, offers a means of reconstituting terrain relief, in order in particular to evaluate digital elevation models. The critical limit – the separation beyond which the two radar beams become incoherent and interferometry can no longer be performed – increases as the wavelength used gets longer. The critical distance for SAR data collected on two trajectories by the ERS satellites operat-ing in the C-band – something like 800 m – is shorter than in the case of Seasat operating in the L-band.

Non-imaging systems
Earth observation also makes use of sensors collecting data which do not build up to an image of the Earth in any literal sense. There are many different kinds of non-imaging sensor. They may take measurements relating to the entire visible part of the Earth – this is the case with some wide instanta-neous field-of-view sensors on the Nimbus 7 Earth Radiation Budget (ERB) instrument – or they may concentrate on very limited areas of the Earth's surface, as in the case of the ERS 1 radar altimeter. Scatterometers are another form of non-imaging sensor with a very limited IFOV. They measure backscatter off the sea surface at different viewing angles, readings then used to calculate wind speed and direction (fig. 9.12). Scatterometers have been flown on Seasat, ERS 1

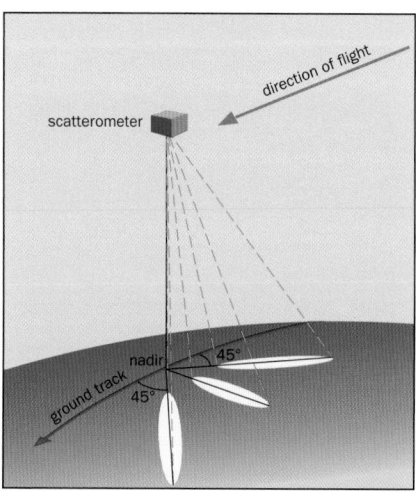

Figure 9.12. **Scatterometer viewing angles.** This is the arrangement on the AMI on ERS 1 and 2.

and 2 and Quikscat. Though not designed for imaging pur-poses, these sensors can be used to produce, by interpola-tion, cartographic images of the phenomena observed.

Spectral bands
All satellite remote-sensing systems designed to observe the Earth's surface use those regions of the electromagnetic spec-trum in respect of which the atmosphere is transparent – hence the name 'atmospheric windows' (fig. 9.13). Trans-mittance depends on many different factors, such as atmos-pheric composition (concentrations of water vapour, carbon dioxide, ozone etc.) or again aerosol content. It depends also on the density of the atmosphere through which incident energy and reflected radiation has to pass. Measurements are affected by the illumination angle, the viewing angle and the altitude of the region being observed.

Sensors responding to the same spectral bands as the human eye have been very widely used for observation from space. Interpretation of images acquired at wavelengths in the visible (450 nm to 710 nm) relies on criteria with which we are familiar from our everyday experience. Vegetation has low reflectance in the visible, solar radiation being absorbed by chlorophyll in the blue and red wavebands. Maximum reflectance is obtained in the yellow-green band at about 550 nm. Generally speaking the reflectance of minerals, higher than that of vegetation, increases at longer wavelengths. Differences in mineralogical composition, in particle sizes and in water concentration do however produce considerable variation at the more detailed level.

Where liquid water is concerned, the important role played by specular reflection must be stressed. Here reflectance is reasonably strong only in the short wavelengths of the visible range, which explains why expanses of pure water are blue. Generally speaking, the reflectance of water depends primarily on the matter suspended or dissolved in it, although the nature of the bed may also be a factor. Gelbstoff, dissolved organic matter resulting primarily from the decomposition of terrestrial vegetation and carried by waterways fed by landwash, is characterised by very high absorptance in the blue. Chlorophyll-a exhibits two maximum absorptance peaks at around 430 nm and 680 nm. Minerals in suspension strongly absorb energy in the blue but show very weak absorptance between 500 nm and 600 nm.

The situation is very different where solid water is concerned. Snow is highly reflective in the visible with some decline in the near infrared. Ice has lower reflectance, on a diminishing scale from granular ice, through pure ice to slush.

Moving outside the visible spectrum, the near infrared (NIR) band, from 0.7 μm to 1.1 μm – already regularly used in aerial photography – is highly valued for the weakness of the atmospheric scatter occurring in that range and for the sharply contrasting levels of reflectance it offers. Chlorophyll for example is strongly reflective in the NIR while water is strongly absorbent. And indeed on many satellites the responsivity of the sensors is limited to the visible and near infrared bands. These are also the spectral bands with the longest history of use on military photoreconnaissance satellites.

Beyond the NIR the terms used to refer to the different parts of the infrared vary according to specialists and languages. Short Wave Infrared (SWIR), from 1.4 μm to 2.5 μm, is referred to in the SPOT 4 context by the French acronym MIR, standing for *Moyen InfraRouge*. In this spectral region, absorption of radiation by water is particularly marked at wavelengths close to 1.45 μm, 1.95 μm and 2.5 μm, while vegetation shows reflectance maxima in the 1.65 μm and 2.2 μm bands. The first of these bands was selected for SPOT 4's HRVIR sensor and, on Landsat 4 and 5, for the TM sensor – which also featured a channel centred on the second band.

The Thermal Infrared (TIR) is another much used spectral range, by weather satellites in particular. The data concerned is used to determine surface temperatures of the oceans, landmasses and clouds.

In the TIR atmospheric window, water is strongly emissive; the emissivity of minerals, though weaker, varies considerably from one type to another. The sensor packages on many satellites combine measurements in this spectral range with readings in the visible, examples being the VHRR and later the AVHRR instruments on NOAA satellites. Sensors carried by military reconnaissance satellites also rely on this band, to acquire night images in particular, as has been the case with the American KH 7 series since 1966. The data readings, taken at repeat intervals of about 12 hours, provide information concerning surface inertia to day-time warming and night-time cooling. This repeat acquisition may be achieved locally by intersection of ascending and descending trajectories, as was the case with HCMM (fig. 2.28), or by combining data from two satellites with overflights 12 hours apart, the arrangement adopted for NOAA satellites.

Hitherto tested only on aeroplanes, hyperspectral sensors are now being looked at seriously as spaceborne apparatus. Their use will support very fine spectral resolution of signatures, significantly widening the range of remote-sensing techniques. Hyperspectral systems break the continuous spectrum of the incoming signal down into a large number of narrow bands. The Warfighter 1 instrument to be flown on Orbview 4 will for example analyse the spectral range between 0.45 μm and 5 μm into some 280 bands. The very fine spectral resolution offered by hyperspectral systems analysing not only the visible but also the NIR, SWIR and part of the TIR is opening up new perspectives, particularly for the study of seawater (chlorophyll), minerals and vegetation cover.

Other parts of the spectrum, such as microwave wavelengths, are being used increasingly widely. Satellite radar imagery was first used for civil purposes on the 1978 Seasat 1 mission, which featured a sidelooking L-band synthetic aperture radar. Radar sensing modes were also tested on the Space Shuttle in the course of three experiments, SIR A, SIR B and SIR C.

Many different bands are used, from the largest to the smallest (L, S, C, X). The largest offer greater penetration of the ground and vegetation while the shortest tell the user more about the roughness of surface minerals. The water content of the observed area greatly affects the penetrative power of radar waves. Microwave radiation penetrates dry ground far more deeply than ground which is waterlogged. It is thus possible using this technique to detect objects covered by dry sand in desert regions.

The number of spaceborne radars can be expected to grow rapidly, and these will operate at various incidence angles in several wavebands (the L, C and X bands under the EOS programme). They will deliver fresh insights into the surface roughness of continental landmasses and the nature of their subsurface and provide a basis for marine status mapping. Radar imaging is of course also of great interest to the military, a point borne out by the prominent roles played by Kosmos 1870 and the Lacrosse satellite.

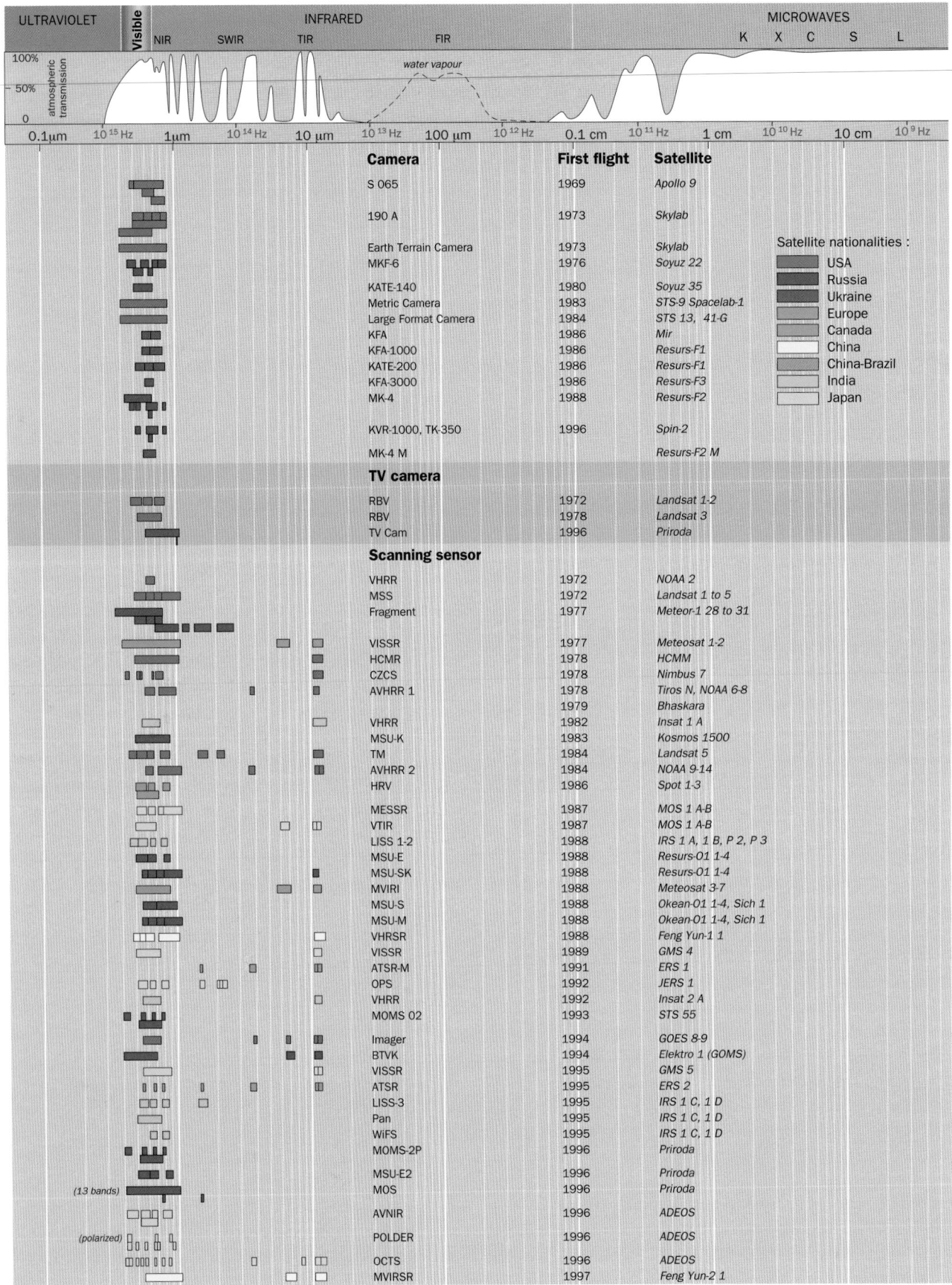

Figure 9.13. **The spectral bands used by the main imaging sensors launched and planned in 2002.**

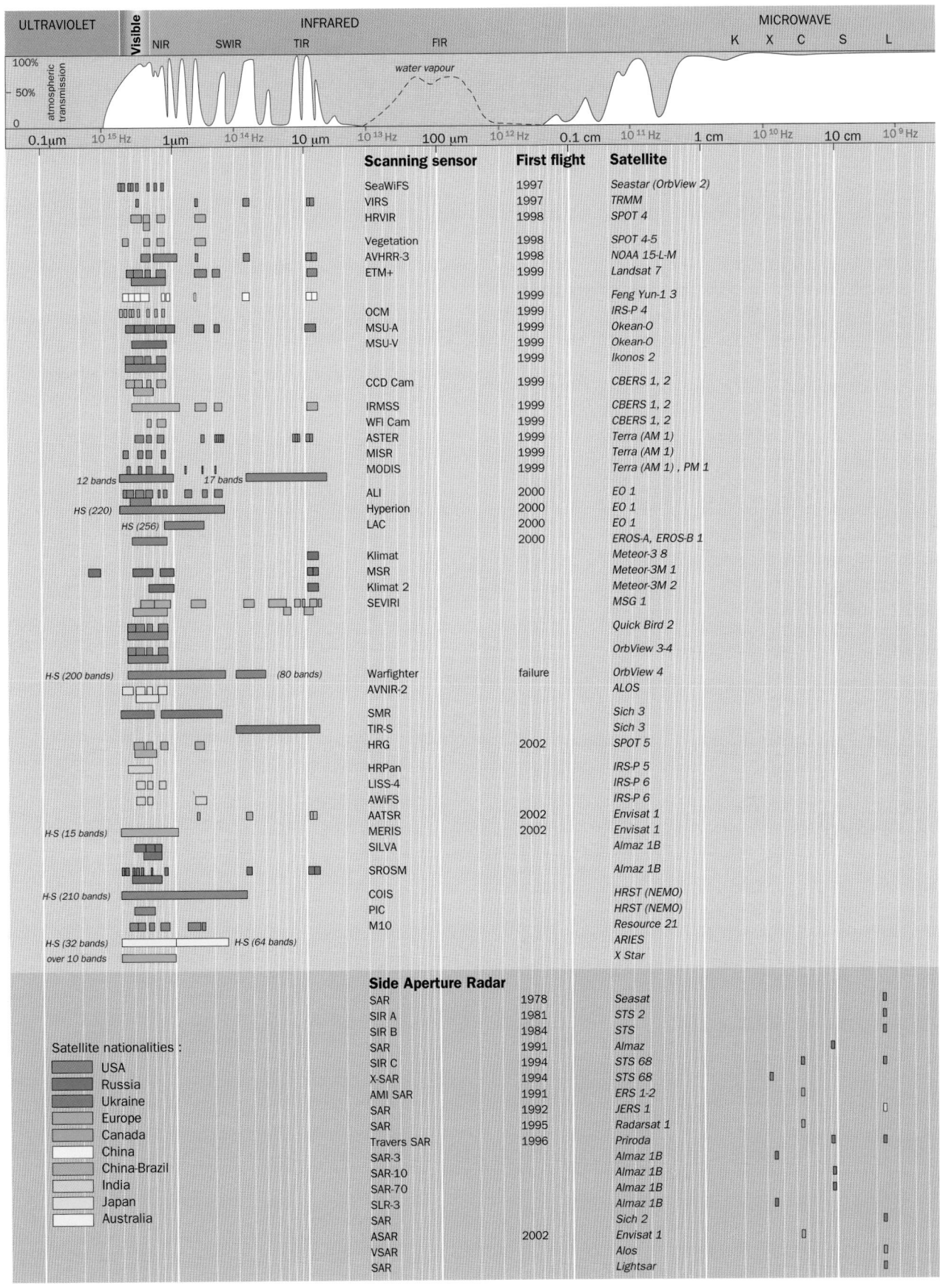

Scanning sensor	First flight	Satellite
SeaWiFS	1997	Seastar (OrbView 2)
VIRS	1997	TRMM
HRVIR	1998	SPOT 4
Vegetation	1998	SPOT 4-5
AVHRR-3	1998	NOAA 15-L-M
ETM+	1999	Landsat 7
	1999	Feng Yun-1 3
OCM	1999	IRS-P 4
MSU-A	1999	Okean-O
MSU-V	1999	Okean-O
	1999	Ikonos 2
CCD Cam	1999	CBERS 1, 2
IRMSS	1999	CBERS 1, 2
WFI Cam	1999	CBERS 1, 2
ASTER	1999	Terra (AM 1)
MISR	1999	Terra (AM 1)
MODIS	1999	Terra (AM 1) , PM 1
ALI	2000	EO 1
Hyperion	2000	EO 1
LAC	2000	EO 1
	2000	EROS-A, EROS-B 1
Klimat		Meteor-3 8
MSR		Meteor-3M 1
Klimat 2		Meteor-3M 2
SEVIRI		MSG 1
		Quick Bird 2
		OrbView 3-4
Warfighter	failure	OrbView 4
AVNIR-2		ALOS
SMR		Sich 3
TIR-S		Sich 3
HRG	2002	SPOT 5
HRPan		IRS-P 5
LISS-4		IRS-P 6
AWiFS		IRS-P 6
AATSR	2002	Envisat 1
MERIS	2002	Envisat 1
SILVA		Almaz 1B
SROSM		Almaz 1B
COIS		HRST (NEMO)
PIC		HRST (NEMO)
M10		Resource 21
		ARIES
		X Star

Labels in the spectral chart area:
12 bands · 17 bands · HS (220) · HS (256) · H-S (200 bands) · (80 bands) · H-S (15 bands) · H-S (210 bands) · H-S (32 bands) · H-S (64 bands) · over 10 bands

Side Aperture Radar		
SAR	1978	Seasat
SIR A	1981	STS 2
SIR B	1984	STS
SAR	1991	Almaz
SIR C	1994	STS 68
X-SAR	1994	STS 68
AMI SAR	1991	ERS 1-2
SAR	1992	JERS 1
SAR	1995	Radarsat 1
Travers SAR	1996	Priroda
SAR-3		Almaz 1B
SAR-10		Almaz 1B
SAR-70		Almaz 1B
SLR-3		Almaz 1B
SAR		Sich 2
ASAR	2002	Envisat 1
VSAR		Alos
SAR		Lightsar

Satellite nationalities :
- USA
- Russia
- Ukraine
- Europe
- Canada
- China
- China-Brazil
- India
- Japan
- Australia

Images of the Earth

The mass of images supplied by the various sensing systems described above form the satellite Earth observation archives, supporting meteorology and remote sensing in the strictest sense but also military reconnaissance.

With the growing stockpile of such images comes the need for simple referencing principles. The principles adopted may be based on the themes treated in the images – clouds, water, land – or on the geographical coordinates of the region surveyed, this being the method favoured for indexing data from the major satellite remote-sensing systems: Landsat with the World Reference System and SPOT with the SPOT Reference Grid (*Grille de Référence SPOT*, GRS). Within these overall structures, straightforward criteria are then required for classification to cope with the different acquisition modes and – a greater challenge still – the very many forms of image processing. These criteria are needed to assign images to categories for subsequent archiving but also for educational purposes.

Scalar levels

A simple approach to the classification of satellite images is based on the concept of scale – scale of field and resolution. Strictly speaking, the concept of scale is fully meaningful only when applied to images in the very literal pictorial sense. The photographic film used in the KVR 100 camera on SPIN 2 (SPace INformation) or in the Metric and Large Format Cameras on Shuttle Spacelab missions records images for which there is a geometric relationship between ground distances and the distances on the original film. But in the case of a scanner sweeping the ground and collecting signals which define the luminescence of the instantaneous field of view on the ground, the concept of scale begins to blur, coming to mean the basic ground unit – the area represented by one picture element (pixel), the basic constituent of an image. An image consisting of a patchwork of pixels can be built up to the desired scale in the same way that a true photograph taken by the Metric Camera or MKF 6 can be enlarged or reduced at will.

While the criterion of scale is satisfactory for true photographs, a more helpful concept when referring to images obtained by scanning is that of the ground area – field or spot – corresponding to a pixel or the closely related idea of Ground Sample Distance (GSD), both of which are sometimes regarded by extension as units of resolution.

Working from the concepts of global field of view, instantaneous field of view and resolution, three main scalar levels can be defined (fig. 9.14):

A: planetary images: these cover an expanse of the planet from limb to limb and are acquired from a high altitude – generally from geostationary orbit some 36 000 km above the Earth. Images at this level have kilometre-range resolution. They are used in meteorology and global climate research (radiation budget etc.).

B: Images at regional level: these cover smaller expanses and are acquired from lower altitudes. Ground resolution is in the 100 m or 10 m range. Applications here are cartography and terrestrial resource surveying.

images		ground resolution	altitude (km)	examples of satellites	major uses
limb to limb field	planetary A	10 - 2.5 km	36 000	Meteosat GOES Kosmos 1940	meteorology climatology oceanography
limb to limb field		1 - 2 km	1500 to 800	Meteor NOAA 15	meteorology Earth resources studies
limited field	regional B	120 - 25 m	900 to 700	MOS Landsat 1-5 JERS Almaz ERS	cartography Earth resources studies
limited field		20 - 10 m	900 to 250	SPOT 1-4 Shuttle (metric camera)	cartography Earth resources studies
limited field	local C	4 - 0.3 m	800 to 150	Yantar KH 11 Helios 1 A-B Ikonos 2	military observation arms control verification urban cartography

Figure 9.14. **Classification of satellite images by scalar level.**

C: Images at local level: these cover restricted areas and are acquired by vertical or oblique viewing, often from very low altitudes. Metre and decimetre range resolutions are obtained. Initially used for military reconnaissance purposes, these images are now finding civil applications at metre-level resolution.

Between these main levels there are many intermediate levels. Limb-to-limb images acquired by the AVHRR sensor on NOAA satellites are, for example, situated between levels **A** and **B**. Each pixel represents a ground area in excess of 1 km^2 and utilisation is a mix of meteorology and terrestrial resource surveying (location of local upwellings, study of Sahel vegetation etc.). Similarly, SPOT imagery and photos taken from the Mir space station are positioned somewhere between levels **B** and **C**.

Types of processing

Another system of classification is based on the nature of the image itself and the type of processing applied to it. This scheme rests on a distinction between primary and derived or secondary images. Images in the first category are those which have retained the geometry of acquisition and have undergone no specialised processing. They can be termed 'primary images', although their primary status is difficult to define. The negative of a photographic film remains a primary image even after development. But can an image obtained by visualising an analog recording on film still be regarded as a primary image? When an image is digitised by sampling, is the output a primary or a secondary image?

Primary images acquired by satellite remote sensing are used far less widely than the derived products, in sharp contrast to the situation in aerial photography interpretation. Secondary images, which exist in much larger quantities, can be classed according to various criteria. One of the most straightforward is whether only one or more than one primary image was used to produce the derived image. Images derived from a single image are described as monogenic and those resulting from a combination of images are called polygenic (fig. 9.15).

Monogenic secondary images are derived from a single image. They may be the product of enlargement, reduction, correcting geometric anamorphosis, change in projection or contrast adjustments. Contrast may be accentuated by extending the grey scale in such a way that the darkest grey in the entire image is shown as black and the lightest grey as white. This operation can also be performed in a small mobile window to heighten local contrasts in the image without regard to relative values in the image as a whole. Changes in the grey scale do not necessarily obey a monotonic function and may substitute for initially rising values, values which rise and fall alternately.

Other modifications may be governed by probability. Images may for example be created in which each grey tone occupies an equal surface area or relates to a gaussian distribution.

The original images may be simplified by reducing intensity levels or by sampling information elements and reducing the number of pixels. It is also possible by filtering to

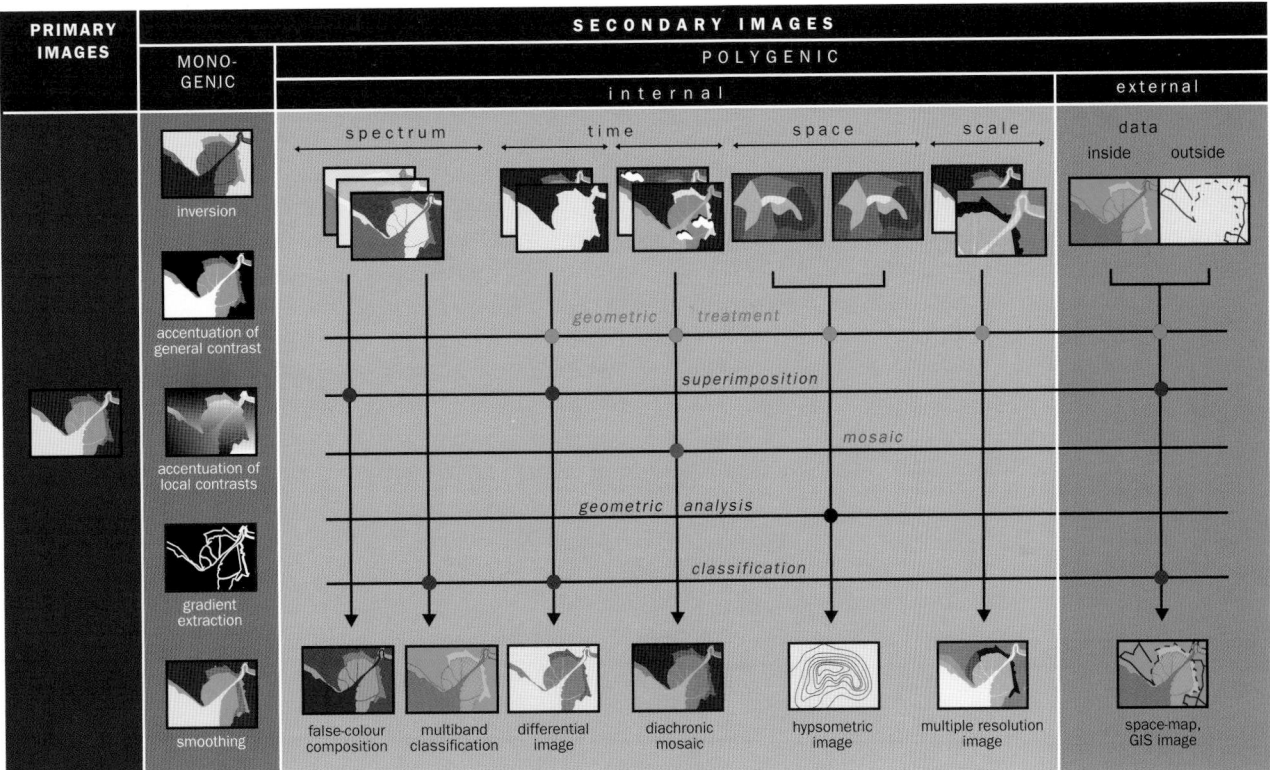

Figure 9.15. **Types of satellite remote-sensing image resulting from the most common forms of processing.**

replace the primary image with an image conveying the differences in value between adjacent pixels (a gradient image) or alternatively with a smoothed image in which each original pixel is replaced by the pixel most frequently encountered in a specified area.

Polygenic secondary images are composite products built up from several primary images in an internal process relying solely on remote-sensing data.

The most common type is the conventional false-colour composite in which each of three primary images acquired simultaneously in three spectral bands by a multiband sensor is translated by one of the sensitive layers of a colour film using the subtractive (cyan, magenta, yellow) colour process or by the colours of the TV screen using additive (green, blue, red) colour mixing. As computer displays have come into more widespread use, subtractive synthesis has given way to additive mixing of the primary colours. This is essentially a spectral technique for deriving images and spectral polygenic processing can form the basis for many other types of processing, such as classification by segmentation or by clustering around mobile points.

Diachronic processing – another form of polygenic treatment – involves combining images acquired at different dates. This may simply involve superimposing a positive image acquired at one date and a negative image acquired at another. Differences between the two images thus appear as areas of lighter or darker shading. Another approach is to use diachronic data to reference pixels so as to characterise families by their evolution. Images of this kind have been produced to define areas of coastal erosion and accretion in the time interval between two given successive satellite passes or again to monitor the evolution of vegetation frontiers or shrub die-back in zones subject to desertification. Derivation of images by diachronic composition has other applications too, one being to fill localised lacunae in particular images. Small isolated clouds may for example mask certain points on the surface of the Earth but there is every chance that the cloud positions will be different on images acquired on subsequent days: by taking the maximum values for thermographic images of the same locations several days running, it is generally easy enough to eliminate the offending clouds – which are colder than the ground. Similarly, it becomes possible to draw up notional maps of small regional units by combining the changing faces of landscapes over a year.

Directional polygenic treatment involves combining images of the same area acquired from different viewpoints. This is the only available means of reconstituting terrain contours from two images. A pair of SPOT images acquired at different angles in the course of two different orbital revolutions can thus be used to produce digital elevation models. Digital models of this kind, in conjunction with the colours deduced from the radiometric readings indicated by a multispectral image, can by computer calculation be used to derive and display a genuinely panoramic view for any given viewpoint. By computer processing it is thus possible to derive a new image corresponding to a viewpoint independent of the actual viewpoints of observation. What is really useful here is the directional aspect – the diachronic treatment which necessarily goes with it is more of a handicap than anything else as there is always the possibility of some change to objects between two image acquisitions. This problem can be overcome to some extent by an arrangement used on some satellites – on JERS 1 channels B3 and B4 for example – in which forward viewing is combined with nadir viewing in the same revolution; this limits to just a few fractions of a second the time-lag between two stereoscopic acquisitions. The POLDER system equipping the ADEOS series works in the same way, combining fractions of wide-angle images acquired from two viewpoints some distance apart.

Scalar polygenic processing involves combining images at varying scales. Combining an Ikonos multispectral image with a pixel area of 4 m × 4 m and a panchromatic image made up of 1 m × 1 m pixels amounts to double polygenic processing – both spectral and scalar – insofar as the combined data relates to different spectral bands and unequal surface areas.

External polygenic processing comes into play where data acquired otherwise than by satellite remote sensing is combined with satellite data. One impressive example of such treatment is the production of maps in which a satellite image is used as background, information of a topographic or toponymic nature being superimposed on the image. Geographical Information Systems (GIS) which choose to incorporate satellite data often generate considerable demand for polygenic images of this type. In some cases, administrative and statistical data are combined with classifications based on multispectral imagery. This particular technique, which calls for a firm grasp of the geometry of essentially heterogeneous data coupled with a sound understanding of the subject matter, is becoming an increasingly widespread feature of the civil and military exploitation of satellite data.

Images can therefore be differentiated by form of derivation – from one, or more than one primary image – and, in the case of polygenic derivation, by the treatment applied to them, which may be spectral, diachronic, directional or scalar, and by whether non-satellite inputs are included in the process. These forms of differentiation determine the nature of the informational content. Images are distinguished too by the types of treatment applied to the source images in order to generate them.

The widening range of data sources and forms of processing is attributable in part to the growing number of satellites – which feed enormous collections of images – and in part to advances in computer-based image processing techniques. The biggest problem now facing users is that of deciding what images to choose and what treatment to apply to them for a given application.

Meteorology

Meteorology offers a range of practical applications which make a major contribution to human safety and the security of air and sea transport. Thanks to satellite data, the paths of tropical cyclones for example can now be predicted far more reliably. On the basis of such predictions, shipping is rerouted and preventive action is taken in the zones through which hurricanes pass. Meteorological information is also of considerable importance for the conduct of military operations and for the planning of photographic reconnaissance missions – hence the existence of dedicated military programmes such as the Defense Meteorological Satellite Program (DMSP).

There is a permanent demand for meteorological data from the media with their requirement for short-term forecasts for the general public. They are of interest too in agriculture as an input to forecasting of frost, rainfall and fog and to research into snowcover and the effects of drought.

Climatology is another area in which meteorological satellite data is of interest. While this discipline is less demanding than meteorology in terms of how quickly data must be supplied, it does require much longer time sequences.

The contribution which observation from space could make to meteorology was recognised at an early stage, prompting the launch of non-geostationary satellites in the first instance and, from 1966 onwards, geostationary satellites (fig. 9.16).

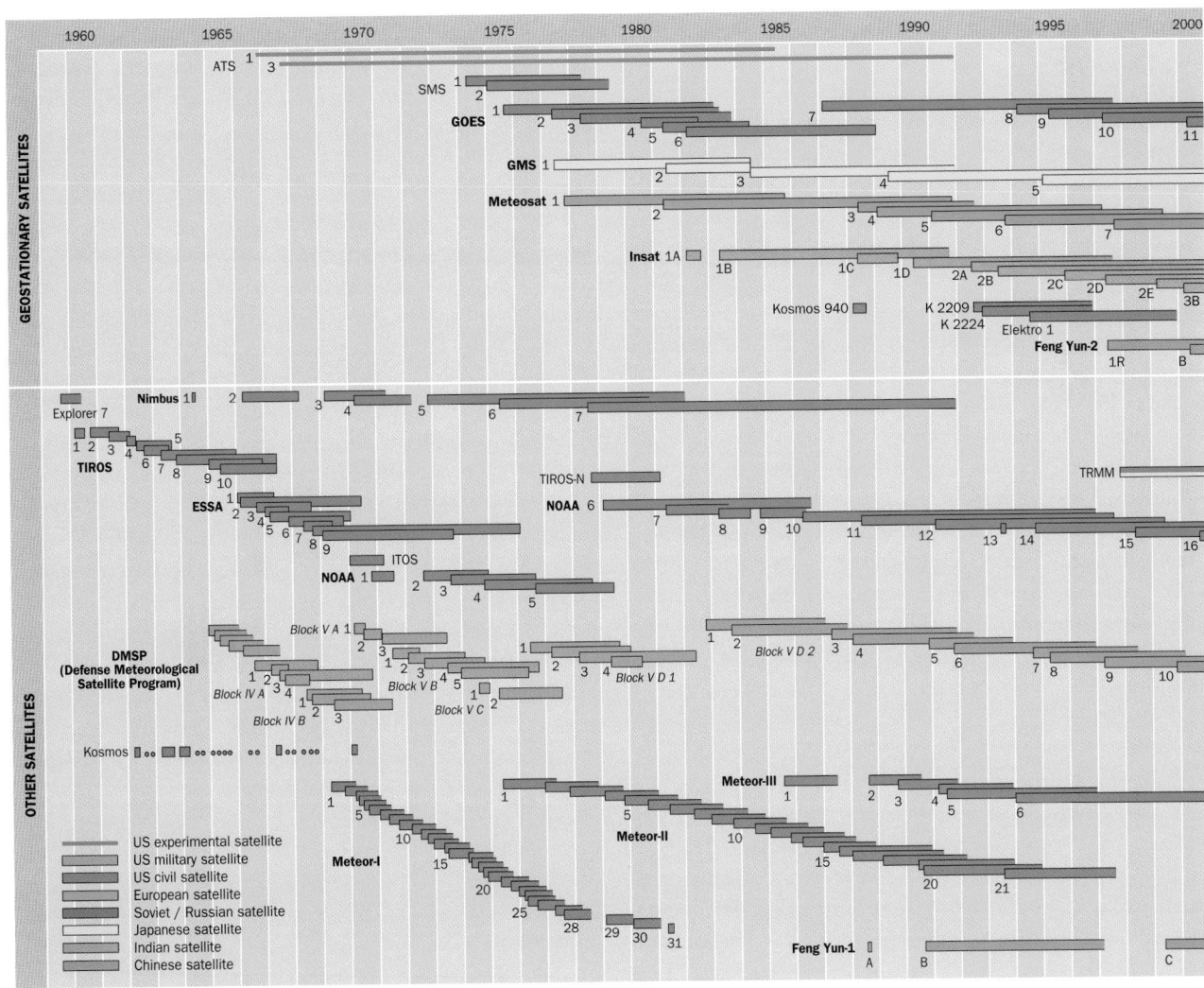

Figure 9.16. **Time chart of meteorological satellites.**

Non-geostationary satellites

Explorer 7, a satellite which measured the overall radiation of the Earth and Sun, was launched as early as 1959, followed as from 1960, by the TIROS series, a family of experimental satellites designed to gather images of the cloud system. This was the first family of satellites to couple measurements in the visible and the thermal infrared. The early satellites in the TIROS series (1–7), with an inclination of around 40°, could not provide full coverage of the Earth but they flew at sufficiently low altitude to support adequate ground resolution, and this despite the relatively poor angular resolution offered by the infrared sensors available at that time.

TIROS 8 saw a switch to Automatic Pictures Transmission (APT) to a large number of image receiving stations and, as from TIROS 9, it was decided to opt for the sun-synchronous orbit, from which the Earth can be observed almost in its entirety with constant solar illumination.

The TIROS satellites paved the way for the Environmental Science Services Administration (ESSA) series, characterised by improved spatial resolution despite higher altitudes of around 1600 km. In parallel to this effort on the civil front, the Americans were also proceeding with a first series of launches under the Defense Meteorological Satellite Programme (DMSP).

With its Nimbus experimental satellites, NASA was able to explore numerous innovations, with particular emphasis on techniques for vertical sounding of the atmosphere and, with the last satellite in the series, on the detection of atmospheric and marine pollution. A number of sensors developed thanks to the Nimbus programme went on to equip the ITOS-NOAA operational series managed by the meteorological administration; the replacement as from NOAA 6 of the VHRR sensor by the AVHRR version saw a sharp improvement in the performance offered by these satellites.

These weather satellites may also carry various kinds of measuring instrument. TIROS-N and the subsequent NOAA satellites were equipped with a TIROS Operational Vertical Sounder (TOVS) system consisting of three instruments:

- the High Resolution Infrared Radiation Sounder (HIRS) for establishing the humidity and temperature profile of the atmosphere up to an altitude at which atmospheric pressure is no more than 10 hectopascals;
- the Microwave Sounding Unit (MSU) for establishing the temperature profile of the atmosphere in the presence of cloud cover;
- the Stratospheric Sounding Unit (SSU) for establishing the temperature profile of the stratosphere.

The Soviet Union, and subsequently Russia, built up substantial space meteorology programmes, though these got underway a little later than their American counterparts. Launch of a number of Kosmos experimental craft – three-axis stabilisation being introduced as from 1966 with Kosmos 122 and inclinations close to 81° being achieved on Kosmos 144, 156 and 184 – was followed by that of three series of Meteor satellites, Meteor I as from 1969, Meteor II from 1975

and Meteor III from 1985. There can be no doubt that the integration of civil and military activity within the Soviet military–industrial complex made for synergies between the various programmes and today, in a difficult economic environment, is helping to maintain a basic capability.

In the United States, with 40 or so dedicated military satellites already launched, a May 1994 presidential directive gave official impetus to efforts to forge closer links between the NOAA and DMSP programmes. The first satellite in the resulting new series is scheduled for delivery in 2005. This should reduce by half the number of satellites that are needed, with an accompanying improvement in performance.

The low orbits occupied by polar satellites ensure high ground resolution and hence a detailed description of cloud formations, landforms and seascapes and of the distribution of surface temperatures. Comparison of data acquired at different dates in both the visible and the thermal infrared bands is also facilitated by the mostly sun-synchronous orbits occupied.

An ability to determine cloud and ground temperatures is of great interest, as is information about temperatures at the surface of the oceans – and here meteorological satellites today make a substantial contribution to oceanography. Data acquired in the visible and the thermal infrared by, for example, NOAA satellites have contributed much to an improved understanding of upwellings and of the horizontal movements of water masses of differing temperatures. Such data is also widely used in the study of landscape evolution. The AVHRR sensors equipping NOAA satellites supply data in either Global Area Coverage (GAC) or more frequently Local Area Coverage (LAC) form. GAC data relates to 4 km × 3.3 km ground patch areas at the nadir point while LAC data describe a smaller basic unit, 1.1 km × 1.1 km at the nadir point.

The Feng Yun 1A, 1B and 1C satellites launched by China in 1988, 1990 and 1999 respectively operate from sun-synchronous orbits at an altitude of 890–900 km and have similar functions to those of the American NOAA satellites.

Orbview 1, launched in April 1995 by Orbimage Corporation, is the world's first commercial meteorological satellite. From its sun-synchronous orbit at an altitude of 740 km, Orbview 1 supplies, thanks to its Optical Transient Detector (OTD), meteorological forecasting data but also cloud-to-cloud lightning imagery which it would be impossible to obtain from the ground. Lightning imaging research has also been performed – with greater frequency though limited to the intertropical zone – by the US–Japanese Tropical Rainfall Measuring Mission (TRMM) satellite launched in 1997 (see chapter 7). TRMM features a number of NASA-supplied instruments – a 5-channel Visible Infra-Red Scanner (VIRS) offering 2 km resolution, the 9-channel TRMM Microwave Imager (TMI) and a Lightning Imaging Sensor (band centred on 0.7774 μm) offering 5 km resolution at the nadir – and the NASDA-supplied Precipitation Radar (PR) instrument. MEGHA-Tropics is a project allowing for most frequent passages above tropical cyclones studied by ESA.

Lastly, the European organisation Eumetsat is pressing ahead with the METeorological OPerational Satellite (METOP) programme, which involves providing sun-synchronous satellites to meet, in cooperation with NOAA, meteorological requirements over the years 2001–2010. The METOP 1 satellite, which will effect its descending node pass at 09:00, is built round a downsized Envisat platform carrying a series of sensors, including an AVHRR 3 (6 spectral bands from 0.68 μm to 12.5 μm), an HIRS 3 (20 spectral bands from 0.69 μm to 14.95 μm) and two Advanced Microwave Sounder Units (AMSU A1 and A2) – operating 13 and 2 channels respectively – which will establish atmospheric humidity and temperature profiles.

Geostationary satellites

Geostationary meteorological satellites came on the scene somewhat later than their non-geostationary meteorological counterparts. Geostationary satellites were first used for meteorological purposes by NASA with its Application Technology Satellites (ATS) designed to test technologies that could benefit from the geostationary environment for various types of application. Although the bulk of the tests were devoted to telecommunications, meteorology was by no means absent from this American programme and indeed it was ATS 1 which in 1966 supplied the first-ever images of the Earth from 36 000 km and ATS 3 which a year later sent back the first colour images from the same orbit. Meteorology from the Clarke orbit made its operational debut with the Synchronous Meteorological Satellites (SMS 1 and 2) launched in 1974 and 1975. The SMS satellites form part of the ring of geostationary meteorological satellites providing complete coverage of planet Earth with the exception of the very high latitudes, which are not observable from this specific equatorial orbit.

This ring coverage is supplied by satellites occupying the following five stations (fig. 9.17):

- 140° east: Geostationary Meteorological Satellite, Japan (GMS 5 since 1995);
- 74° east: Insat, India (Insat 2D since 1997; Insat 2E is on standby since 1999 at 83° east);
- 0°: Meteosat, Europe (Meteosat 7 since 1997);
- 75° west: Geostationary Operational Environmental Satellites, USA (GOES 10 since 1997);
- 135° west: Geostationary Operational Environmental Satellites, USA (GOES 9 since 1996).

In addition to these nominal orbital positions, most satellites are stationed at one time or another in different waiting positions (immediately after launch) or standby positions (at the end of their operational lives or following operating incidents). Thus when GOES 6 (GOES East) failed on 21 January 1989, GOES 7 (GOES West) was moved first, in Spring 1989, from 75° west to 108° west and again, in Autumn the same year, to 98° west. This failure left only four satellites operational in 1989 (fig. 9.18).

These satellites gather data primarily in a visible (fig. 9.19) and a thermal infrared band. In some cases other spectral

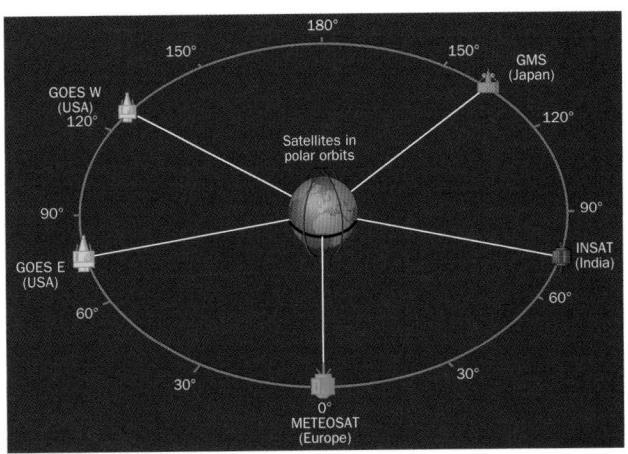

Figure 9.17. **Satellites making up the World Weather Watch.** The long tradition of public service and international cooperation in meteorology entered a new phase with the advent of satellites. Creation of the World Weather Watch is a fine example of this new departure.

The World Weather Watch arose out of a United Nations recommendation made in 1961 to the World Meteorological Organisation (WMO), an agency which had been in existence since 1951. Its brief is to provide operational services supported by a global system for observation of the circulation of cloud masses, by means of geostationary and non-geostationary satellites belonging to various member states of the organisation.

The system is built around a core structure of five geostationary satellites positioned around the Earth in such a way as to provide global coverage, except for the polar caps up to 8° from the poles. The five radii indicate the orbital positions designated by the WMO for the ring of geostationary satellites. They are complemented by a number of satellites in near-polar orbit, which provide coverage of the higher latitudes which cannot be seen from the equatorial plane. They also supply more detailed information about cloud cover and marine/continental temperature distributions.

bands, mainly in the SWIR, are added – GOES 9 for instance gathers data in the following five spectral bands: 0.55 μm–0.75 μm, 3.80 μm–4.00 μm, 6.50 μm–7.00 μm, 10.20 μm–11.20 μm and 11.50 μm–12.50 μm.

Meteosat Second Generation (MSG) satellites are equipped with a number of new sensors, including the Spinning Enhanced Visible/IR Imager (SEVIRI) which will scan the entire visible segment of the globe every quarter of an hour in 12 channels. Another of the sensors, the High Resolution Visible (HRV) will offer a 1 km ground resolution in the 0.50 μm–0.90 μm spectral band. MSG 1 has been launched in August 2002.

Data from geostationary satellites can be received by a large number of stations and indeed satisfactory dissemination and effective utilisation depends on this.

In the case of the Meteosat series for example, data is digitised on board the satellite, transmitted in real time to ground, where it is geometrically corrected, and then retransmitted to Meteosat, which distributes it. Digital processing is performed by Primary Data User Stations (PDUS) and analogue processing by Secondary Data User Stations (SDUS). There are some 2000 secondary stations but a much smaller number of primary stations. SDUS receive Wefax analogue data on a free access basis whereas, for part of the

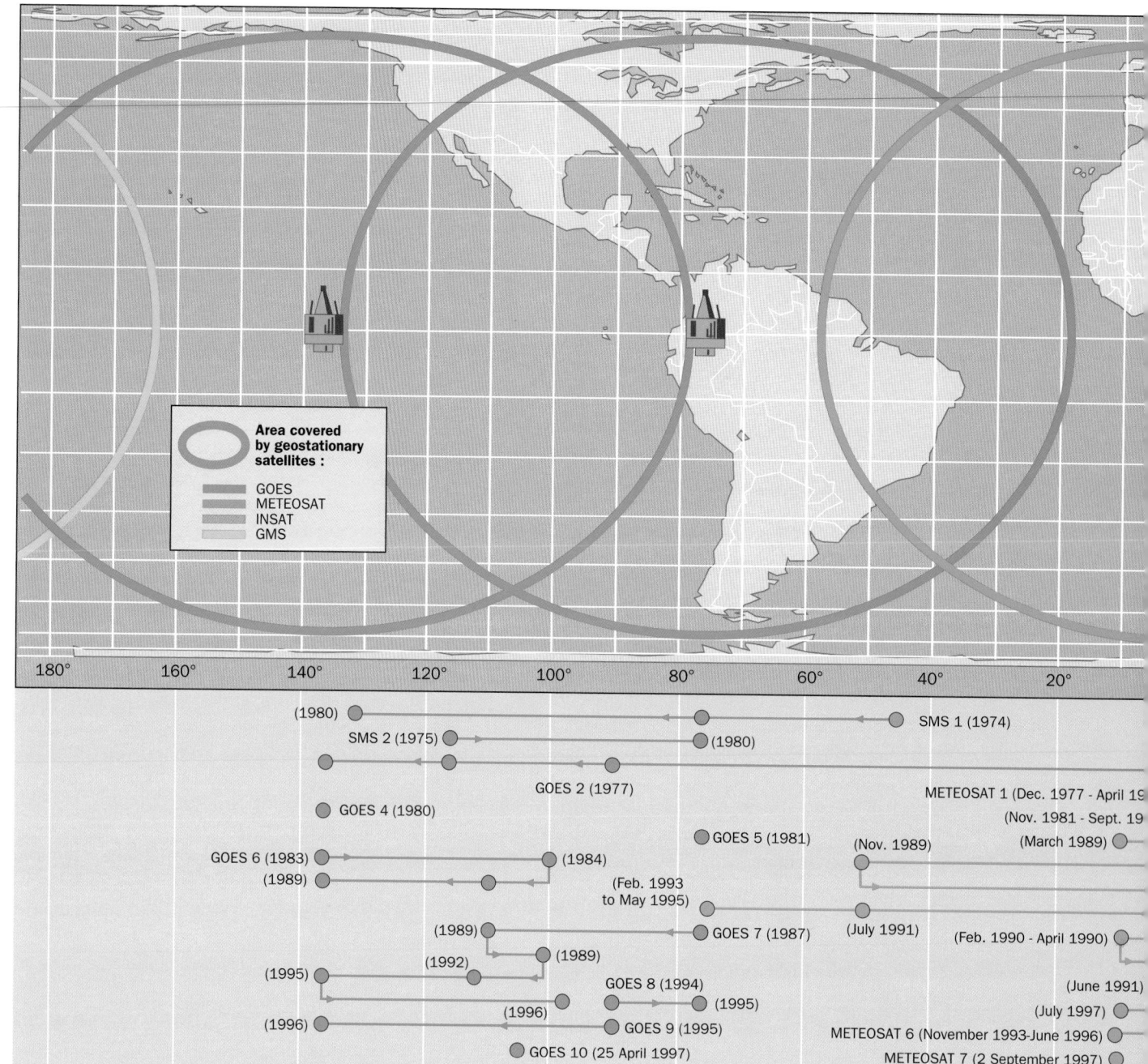

Figure 9.18. **The positions of geostationary meteorology satellites.** The lower part of the figure shows changes in the positions of satellites until their transfer to disposal orbits above the geostationary orbit.

digital data received by PDUS, encryption arrangements and royalty payments have been introduced. Meteosat also distributes data supplied by Data Collection System (DCS) ground platforms.

In view of the success of meteorological satellites, the various experimental programmes have been converted to operational status.

Following the launch in 1977, 1981 and 1988 of the first three Meteosat satellites, Eumetsat – established in 1986 – was given the job of funding the Meteosat Operational Pro-

gramme (MOP) comprising three new operational satellites: Meteosat 4, in service from 1989 to 1995, and Meteosat 5 and 6, launched in 1991 and 1993 respectively. In 1997 Eumetsat launched Meteosat 7 under the Meteosat Transition Programme.

Two further geographical positions – in addition to the five World Weather Watch positions referred to above – are occupied by the Russian Elektro 1 satellite (at 76° east) launched on 31 October 1994 and China's FengYun 2 satellite (at 105° east) launched on 12 June 1997.

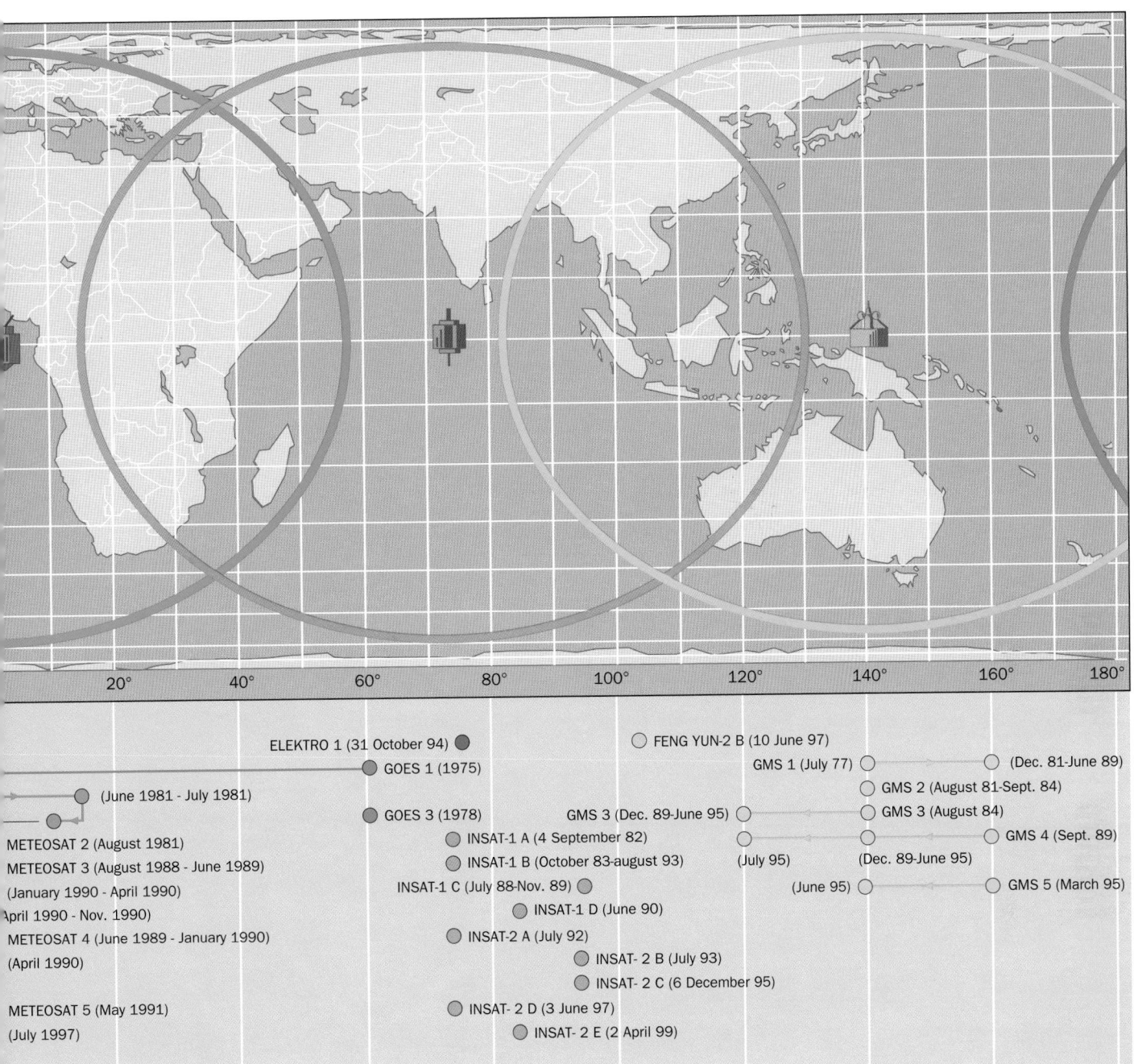

ELEKTRO 1 (31 October 94) ●   ○ FENG YUN-2 B (10 June 97)

○ GOES 1 (1975)   GMS 1 (July 77) ○   ○ (Dec. 81-June 89)

(June 1981 - July 1981) ●   ○ GMS 2 (August 81-Sept. 84)

● GOES 3 (1978)   GMS 3 (Dec. 89-June 95) ○   ○ GMS 3 (August 84)

METEOSAT 2 (August 1981)   ● INSAT-1 A (4 September 82)   ○   ○ GMS 4 (Sept. 89)

METEOSAT 3 (August 1988 - June 1989)   ● INSAT-1 B (October 83-august 93)   (July 95)   (Dec. 89-June 95)

(January 1990 - April 1990)   INSAT-1 C (July 88-Nov. 89) ○

April 1990 - Nov. 1990)   ● INSAT-1 D (June 90)   (June 95) ○   ○ GMS 5 (March 95)

METEOSAT 4 (June 1989 - January 1990)   ○ INSAT-2 A (July 92)

(April 1990)   ○ INSAT- 2 B (July 93)

         ○ INSAT- 2 C (6 December 95)

METEOSAT 5 (May 1991)   ● INSAT- 2 D (3 June 97)

(July 1997)   ○ INSAT- 2 E (2 April 99)

Figure 9.19. **Image produced from data acquired using the 'visible' channel on Meteosat (4 November 1986).** This image, processed by the European Space Agency, took 25 minutes to acquire by a west-to-east sweep; the scanning motion was obtained by rotation of the satellite (see fig. 9.8a), the succession of scanlines being obtained by switching the telescope axis from north to south.

The band of cloud over the Atlantic and West Africa at equatorial latitudes corresponds to rising moist air in the intertropical confluence zone. It extends southwards in Africa and Latin America. The cloud-bands that can be seen at high altitudes are caused by fronts associated with cyclonic conditions. One such front has reached the northerly part of the Western Sahara. The irregular cloud masses off the coast of South Africa are the product of condensation over cold surface water. The areas of clear sky are due to high atmospheric pressure, particularly over Europe. Image courtesy of ESA.

Remote sensing of terrestrial resources

Remote sensing of terrestrial resources is placed, in the field of observation, somewhere between the acquisition of low resolution meteorological data and very high resolution military reconnaissance. Satellites designed for the remote sensing of terrestrial resources saw service in space somewhat later than non-geostationary meteorological satellites and benefited from the experience accumulated with them. But while this technological debt is very clear, there are marked differences in the way the programmes in the two areas are structured. Pursued through international cooperation, satellite meteorology has much in common with a public service, enjoying from the outset very specific users expressing carefully targeted needs. Remote sensing of terrestrial resources on the other hand found itself for some time having to make a tool available to users before there was any real expression of demand, which in any case was often met in part by aerial photography.

And while the vast majority of sensors used for satellite meteorology encompass a field of view stretching from one limb to the other, the overall field of terrestrial resource remote sensors is limited by an instrument look angle, as is also the case for military reconnaissance sensors. Viewing angles are generally narrower for military than for civil applications.

In the very earliest years of the space age, the potential contribution which Earth observation from space could make to geography was beginning to be perceived. When, looking down from low orbit, the astronauts on board the first manned spacecraft – such as Vostok, Voskhod, Mercury and Gemini – saw with their own eyes the Earth unfold before them, they were struck by the implications of this vision for an understanding of the terrestrial landscape. It was common at that time to talk of 'geographical satellites'. A 1956 report by the American Academy of Sciences bore the title 'Spacecraft in Geographic Research' and a study which the same Academy commissioned from the Office of Naval Research and NASA in 1965 was called 'Geography from Space'. And while the qualifier 'geographical' subsequently came to be used less frequently, the fact remains that a whole series of satellites are fundamentally important to geography as a whole, even if the terminological preference is now to refer to resources, the environment or simply 'Earth observation'.

Earth observation was first developed for military purposes (see chapter 12). In a situation in which international air law provided for the sovereignty of states over their airspace, the planetary nature of the movements described by satellites was recognised as a key to the acquisition of information over areas hitherto regarded as inaccessible. The first remote-sensing images did however, in addition to their strategic value, make available new types of information whose relevance to meteorology, geology and cartography soon became apparent to the scientists who had access to this classified data. But to develop a satellite for civil purposes it was necessary, in the United States, to allay the misgivings of those who were worried that the greater accessibility of information would be detrimental to national security and also to persuade potential users to pool their efforts in developing a joint programme. The very different needs of the various categories of user – geologists, geographers, agronomists and environmental experts – created a great many problems when it came to defining the technical characteristics of the sensors required and considerably complicated the arrangements for managing and commercialising the programmes (fig. 9.20).

With the manned space programmes came much greater opportunity for innovative experiments. Using hardware from the successful Apollo Moon missions, the Skylab flights were used to test various new instruments, such as the sensors making up the Earth Resources Experiment Package (EREP). The Salyut and Mir stations and the Space Shuttle also provided platforms for considerable observation activity. The Shuttle for example provided an environment for testing Shuttle Imaging Radars (SIR) and for Spacelab photographic missions involving the Metric Camera and the Large Format Camera (fig. 9.21). The Mir station – which in its basic configuration incorporates a series of high resolution still cameras (Kate 140, Kate 200, KFA 1000, KAP 350 and MKF6 MA) – was able with the arrival of the Priroda module equipped with passive and active remote-sensing instruments to support an operational mission specifically devoted to the environment, oceanography and atmospheric research. The German sensor MOMS 02 was twice used in Spacelab, in April 1993 onboard the Space Shuttle and in 1996 on Priroda (MOMS 02P).

Other experiments were conducted as part of the Heat Capacity Mapping Mission (HCMM), devoted to fairly fine resolution thermal infrared imagery, from a number of Kosmos satellites and above all from Nimbus 7 with its Coastal Zone Colour Scanner (CZCS).

Operational missions in the visible and near infrared, sometimes coupled with the thermal infrared, have been conducted under a number of automatic satellite programmes. The most important of these – those which have succeeded in providing a high level of continuity of image provision and consistent user access to the images acquired – are the American Landsat programme which began as early as 1972, the French SPOT programme which got underway in 1986 and the Japanese MOS and Indian IRS programmes with start dates in 1987 and 1988 respectively. The trend towards involving more and more

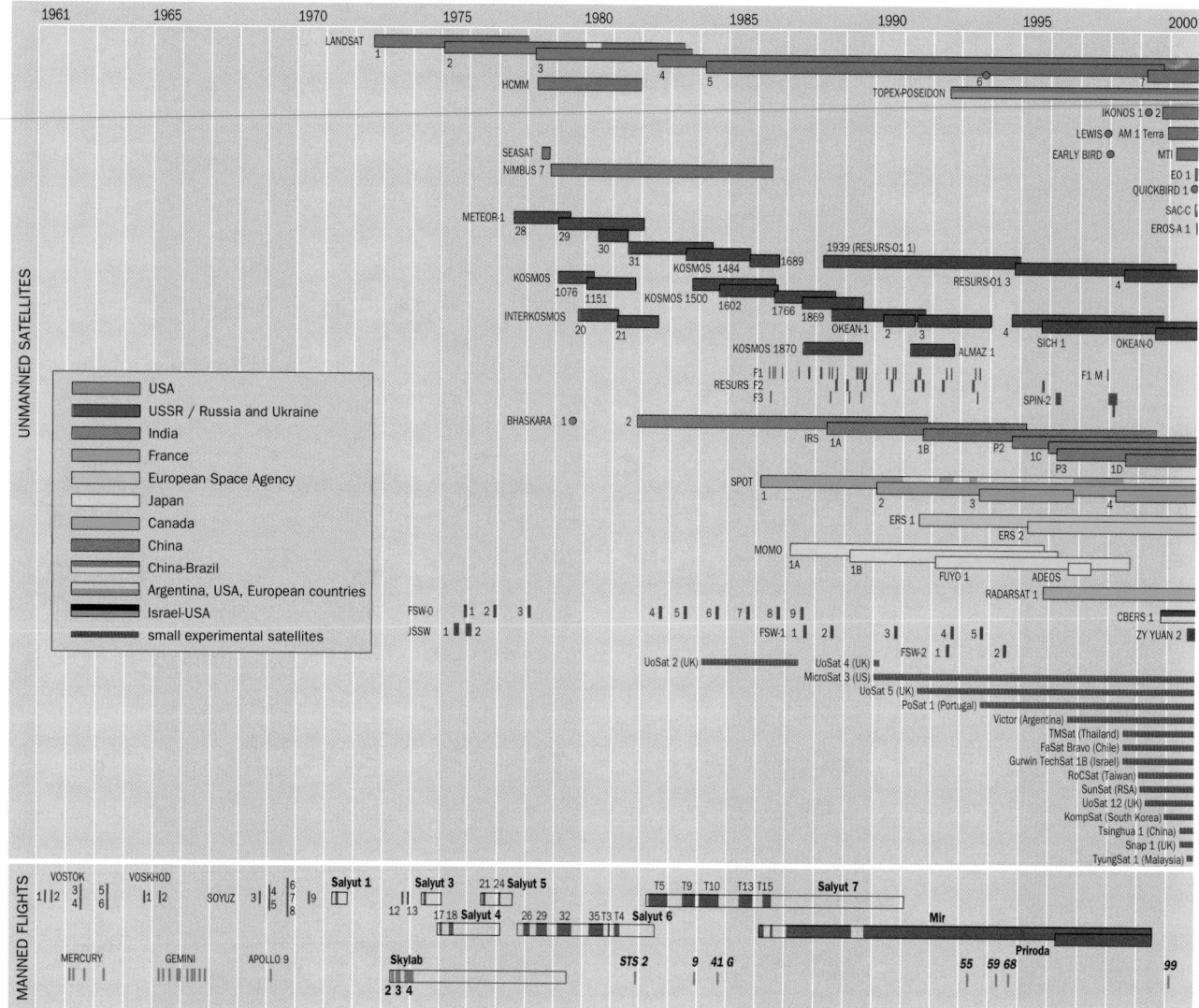

Figure 9.20. Time chart of satellites performing Earth resource remote-sensing missions. A number of Russian military reconnaissance satellites and some weather satellites which also perform terrestrial resource missions have only been shown in the time-charts covering their primary activity. The Chinese FSW satellites shown here also serve military purposes.

countries is continuing with conventional systems such as the Chinese–Brazilian CBERS satellite, launched in 1999, but also thanks to the development of simpler and significantly cheaper systems derived from the Uosat platform developed by the University of Surrey; these less complex systems are intended to allow emerging countries to familiarise themselves, albeit in a limited way, with this type of application.

Progress was rather slower in the radar imaging field. There still seemed little prospect in the 1960s of being able to fly radar imagers onboard satellites in view of the amount of energy required to receive back a measurable signal, following its emission by the onboard sensor and a 'round trip' of more than a thousand miles. Radar as an application seemed to be restricted to aviation, where it was already in very wide use. Subsequent advances in solar array and radar technologies did however make it possible to take this form

of imaging into the satellite domain. Radar imagery from space made its debut with the launch on 26 June 1978 of the American Seasat satellite. The technique used on this and the other satellites in the series was that of Synthetic Aperture Radar, which uses the satellite's motion along its orbit to create a virtual aperture much wider than the real aperture, thereby providing improved resolution (see fig. 9.9c). It is more commonly referred to by its acronym SAR (fig. 9.22).

Setting aside its military applications, the main users of radar imagery for the remote sensing of terrestrial resources are the Russian Almaz programme, the European ERS satellites, the Japanese Earth Remote Sensing Satellite (JERS) programme and the Canadian Radarsat series, although some work has also been done from the Space Shuttle and the Priroda module attached to the Mir space station.

SAR systems are becoming more diverse, with the use of a range of different incidence angles, horizontal or vertical polarisation at reception or emission, and an increasing variety of bands (see fig. 9.13). The specific features required for particular types of use depend on the surface texture of the areas studied, their vegetation cover and water content.

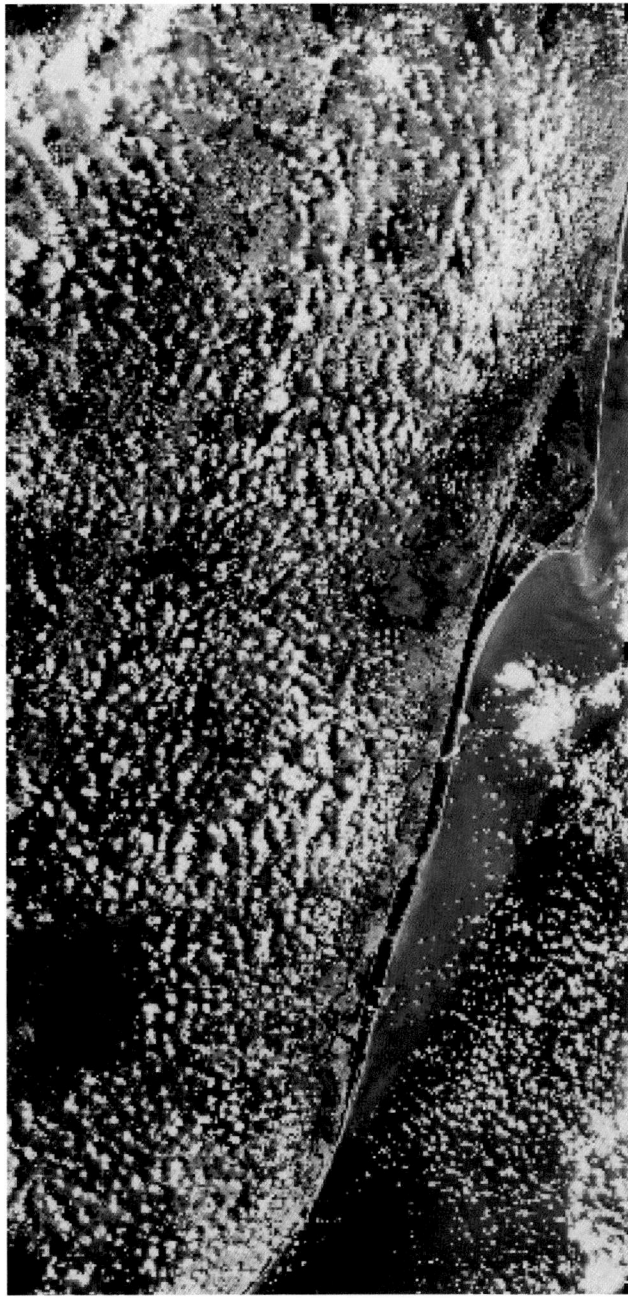

Figure 9.21. **Photograph taken by the Large Format Camera.** This photograph was taken on 8 October 1984, from an altitude of 230 km, on Space Shuttle flight 41G. The image, originally recorded on 23 cm × 46 cm film, covers a ground area 344.67 km in length and 172.33 km in width. It depicts the area round Cape Canaveral in Florida. Another smaller-field image of the same area, supplied by SPOT, is reproduced in part in fig. 5.19. Image courtesy of NASA.

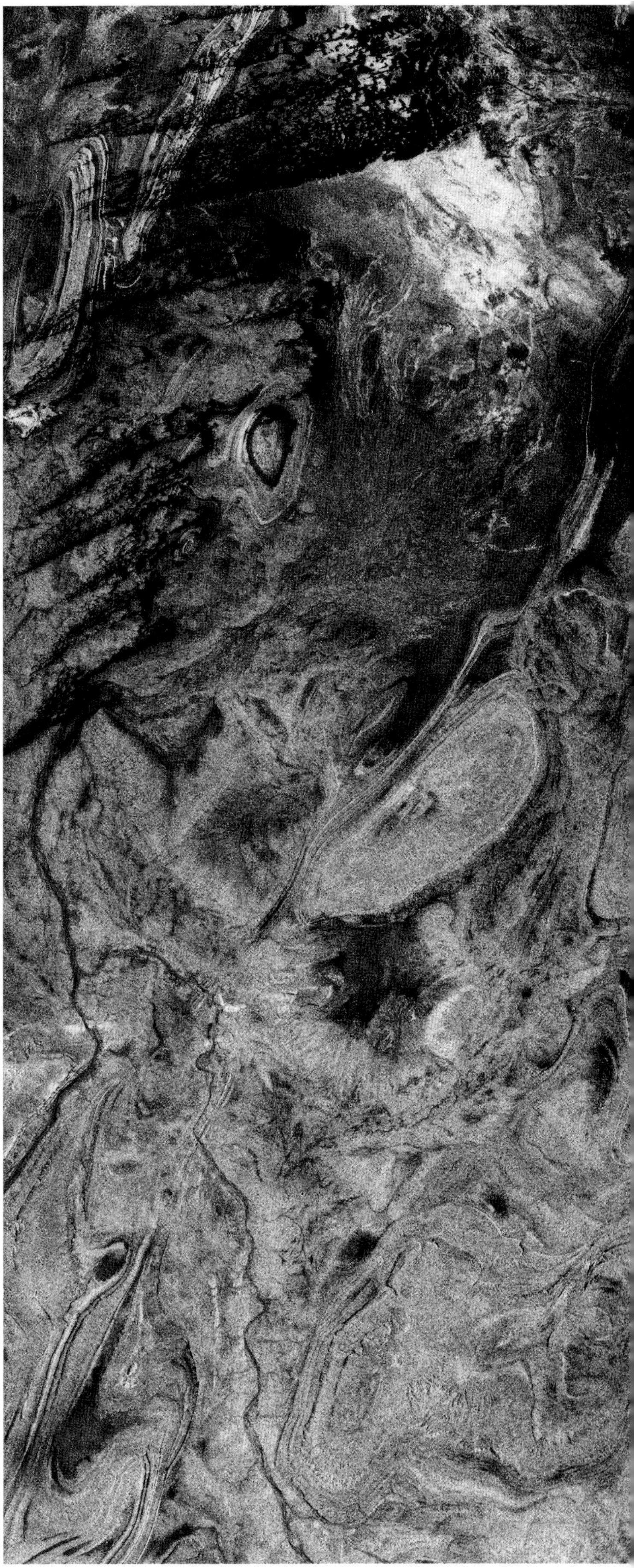

Figure 9.22. **Image acquired by X-SAR Synthetic Aperture Radar.**
This German–Italian experimental radar sensor was flown on Space Shuttle mission STS 59. It uses the X-band. This image, acquired on 12 April 1994, shows an area of the Algerian Sahara in which small relief features associated with outcrops of resistant strata are partly hidden by sand-stripes without backscatter in the X-band. Image by courtesy of DLR/DPAF.

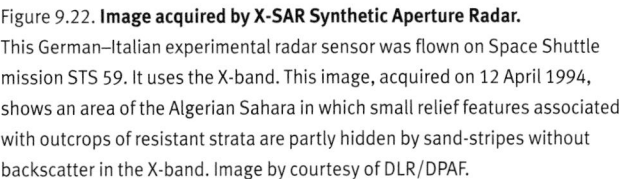

Optical remote-sensing systems

The Landsat programme

The idea of a civil Earth observation programme began to take shape as early as 1962, prompted by the enthusiasm with which photographs of the Earth from manned spacecraft had been greeted by a very wide audience and also by the interest generated among scientists having access to classified data acquired by military reconnaissance satellites. A meteorological satellite programme was seen too as a possible framework for experimentation. And yet it would be another ten years before the first satellite in the Landsat series was lofted. The length of that intervening period suggests how difficult it was, for structural and political reasons, to get from the idea to a system that was up and running.

The period from 1962 to 1966 was taken up at NASA with study of the various technical options for such a programme. The Department of Defense having refused to declassify the technologies embodied in its military satellites and with the bulk of NASA's resources tied up in the manned space programme, it was hardly surprising that thought should first turn to manned spacecraft as a platform for Earth observation experiments. The NASA laboratories were moreover primarily interested in developing new sensor technologies and as this implied relatively long lead times, most of the proposed projects targeted post-Apollo applications. The reasoning at that time was that astronauts would need to test a number of systems before an operational automatic system could be developed. For their part, potential users such as the Geological Survey tended to favour relatively unsophisticated satellites which could be tailored to their requirements and made available in a short space of time. This approach corresponded to NASA's minimalist option for Earth observation based on three types of satellite – large, medium and small. The balance of opinion within NASA was weighted towards the most complex of these structures.

The decision to go ahead with the Landsat programme was in the end taken under pressure from the user agencies, and more particularly the Department of Agriculture and the Department of Interior with of course the Geological Survey. The latter, particularly unhappy with the protracted soul-searching at NASA, even went so far as to petition Congress for funds to launch a satellite which it proposed to develop itself. And although NASA, by reacting immediately, did succeed in retaining responsibility for the project, it found itself committed to developing the satellite to a short timescale. Even so, another year passed before an official decision was taken, in October 1968, on the choice of sensors. The satellite would use existing technologies except in respect of sensors, on-board recorders and the data processing system. The task

of developing the sensors went to the Goddard Space Flight Center which already had experience of meteorological and geophysical satellites but was new to Earth observation – unlike the Johnson Space Center, which continued to concentrate on post-Apollo applications. According to some observers this decision by NASA had one fundamental drawback, which was that the Goddard team would be looking to test its new systems rather than giving priority to meeting the requirements of future users. There were other problems too: the sometimes conflicting demands of the various participants in the project, the need to avoid friction with the Department of Defense and that of complying with national policy directives, all of which meant specifying technical capabilities that would be acceptable to all concerned in what was a highly sensitive area.

A programme for the study of Earth resources from automatic satellites was in the end approved by the US House of Representatives in 1969. In keeping with the political decision which finally emerged, the programme was presented as an investment which would benefit the whole of mankind. And while the first satellite under the programme was initially called ERTS 1 (Earth Resources Technology Satellite), it was later renamed Landsat 1, a designation retained for the rest of the series.

Landsat 1, 2 and 3, launched from the Vandenberg Space Center in 1972, 1975 and 1978 and based on the platform used for the Nimbus meteorological satellites, made up an initial series within this family of satellites (see fig. 9.20). They operated from a circular Sun-synchronous orbit at an altitude of 920 km and effected their descending node pass at 09:30 (fig. 2.27). They had an 18-day orbital cycle (see fig. 3.25).

Landsat 4 and 5, launched in 1982 and 1984, again from Vandenberg, formed a second series. Their orbit, while still circular and Sun-synchronous, was now lower at 690 km. Equator crossing time in the descending pass was 09:37 and the orbital cycle lasted 16 days (see fig. 3.26).

Although some users, familiar with photointerpretation, would have liked the Landsat series to have been equipped with photographic cameras, this option was rejected because of technical constraints and also as a result of opposition from the Pentagon, NASA preferring not to cross swords with it on this issue. The early Landsat craft carried two sensors, the Return Beam Vidicon (RBV) TV camera and the MultiSpectral Scanner (MSS). An RBV camera had earlier seen service in the Ranger lunar observation programme and, at the time the Landsat instrument payload was being elaborated, a more advanced version was being developed, though it had not been assigned any particular mission. The RBV was

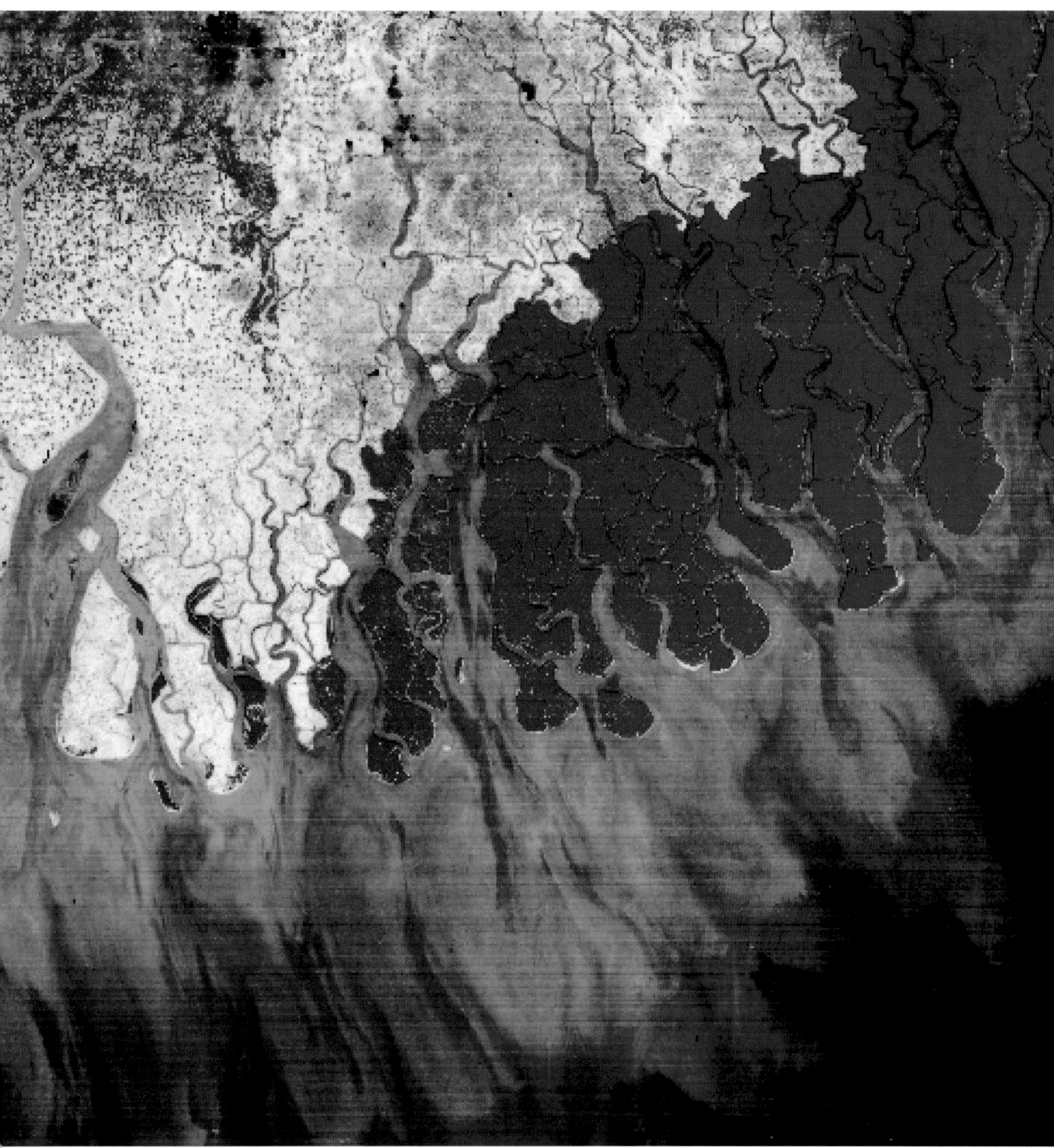

Figure 9.23. **Multispectral image acquired by the Multi Spectral Scanner (MSS) on Landsat 1.** MSS images are made up of pixels each representing a 56 m × 79 m ground area. This image of the Ganges delta is a false-colour composite combining channels 4 (in yellow), 5 (in magenta) and 7 (in cyan). The waters of the Bay of Bengal are roily, which explains the particularly light blue which prevails. The chlorophyll-rich mangrove forests are shown in red while the bare soil and the low-chlorophyll crops characterising the acquisition period (February) appear orange-tawny or yellow. A complete MSS scene is a non-rectangular parallelogram (see fig. 3.2) whose base represents 185 km on the ground. The rectangular extract reproduced here covers a ground area of 172 km. Image: courtesy of NASA.

not moreover one of the instruments used by the military. Use of the sensor attracted support from the Department of the Interior but a fair degree of opposition had to be overcome on the way to its adoption. Landsat 1 and 2 were equipped with one version of the RBV, while the third satellite in that first series carried another model (fig. 9.6 and table 9.1). The results were disappointing however and it was not kept on for the second Landsat series.

The other sensor, MSS, was the product of both civil and military research at the University of Michigan. Scientists at NASA took the view that agricultural studies would benefit particularly from this multispectral scanner whereas Earth science specialists were concerned at the more limited resolution it offered. The Department of Agriculture having put all its weight behind the effort to get the sensor flown, NASA grasped this opportunity to counterbalance the power of the Department of the Interior. In February 1969 the decision to equip the satellites with MSS was taken.

The MSS carried by the second Landsat series scans a 185 km wide swath. The MSS scenes acquired by the Landsat family constitute the first near-exhaustive collection covering the surface of the Earth's exposed land masses at various seasons (fig. 9.23).

The second series was further equipped with a seven-channel radiometer, the thematic mapper (TM), offering improved spatial resolution. TM-acquired scenes are of excellent quality overall and represent a considerable improvement in the precision of remote-sensing scan data.

The two Landsat series differed too in the data transmission modes used. Satellites in the first series either transmitted data in real time direct to ground receiving stations, when they were in their line of sight, or recorded it onboard for subsequent transmission to the ground stations. On-board processing is no longer a feature of satellites in the second series. These rely solely on real-time transmission, either directly or indirectly via a geostationary relay satellite (TDRSS) which retransmits data received to the White Sands station in the United States.

Although the volume of images acquired and their relevance to a wide range of users bear ample witness to the technical success of the Landsat programme, hesitations as to its status have placed a continuing question mark over its future. The intention at the outset was that Landsat data should benefit the international community and charging for images distributed was based almost entirely on the costs of reproduction and forwarding. This policy, which was similar to that adopted for satellite meteorology data, was designed to generate interest among potential users in advance of a commercial system. As early as 1979 the Carter Administration, aware of the NOAA's experience in the management of weather satellites, gave it temporary responsibility for operating the Landsat system as a step towards its eventual privatisation. Despite continuing debate as to whether the Landsat system could be run profitably, Congress adopted in 1984, while Ronald Reagan was in office, a law on the commercialisation of remote sensing data and, in 1985, handed responsibility for the Landsat programme over to a private entity, the Earth Observation Satellite Corporation (EOSAT). Image archiving was left to the Department of Commerce. The US government does however still find itself subsidising the Landsat programme for want of a real market.

Landsat 6, built for EOSAT under NOAA responsibility, was lost on 5 October 1993 during an unsuccessful launch on a Titan II vehicle. The lost satellite had been equipped with the new Enhanced Thematic Mapper (ETM), featuring six bands in the visible, NIR and SWIR ranges offering a spatial resolution of 30 m, one band in the thermal infrared with a resolution of about 100 m and lastly a new panchromatic band with a resolution of 15 m. It was not until 1998 that Landsat 7 was launched. The trend was by this time towards achieving savings by pursuing civil and military objectives in tandem but disagreement between DoD and NASA on the sensors to be flown was initially something of an obstacle. When a decision was taken in 1994, it went in favour of a single sensor, ETM+, prompting a transfer of military funding to NASA which had in the meantime become sole prime contractor for the satellite. Programme organisation and satellite data management are performed jointly by NASA, NOAA and the US Geological Survey, part of the Ministry of the Environment. The EOSAT corporation, itself bought out by Lockheed in 1996, now concentrates on commercialising remote-sensing images.

The ETM+ sensor draws heavily on its forerunner on Landsat 6, though with a number of improvements, including resolution enhancement to 60 m in the thermal infrared.

What the future holds in store after Landsat 7 is only partly clear. There are plans to equip future Landsat craft with an Advanced Land Remote Sensing System (ALRSS) and testing of new instruments can be expected under NASA's New Millennium Programme, basically a research and development programme focusing on remote sensing, communications and power production. The Earth Orbiting 1 (NM EO 1) platform is likely to carry a new-generation Earth Surface Spectroradiometer dubbed Advanced Landsat together with an Integrated Multispectral Atmospheric Sounder (IMAS). These developments are in keeping with the broader role which has now been assigned to Landsat in environmental monitoring. They also allow civil aims to be tied in with high technology projects derived from research pursued under the Strategic Defense Initiative.

The SPOT Programme

The SPOT (*Satellite Pour l'Observation de la Terre*, Earth observation satellite) programme was approved in 1978 by the French government, with Belgium and Sweden as minority partners. A proposal to establish this second satellite system for permanent observation of the Earth had previously been submitted to the European Space Agency but had not gained acceptance.

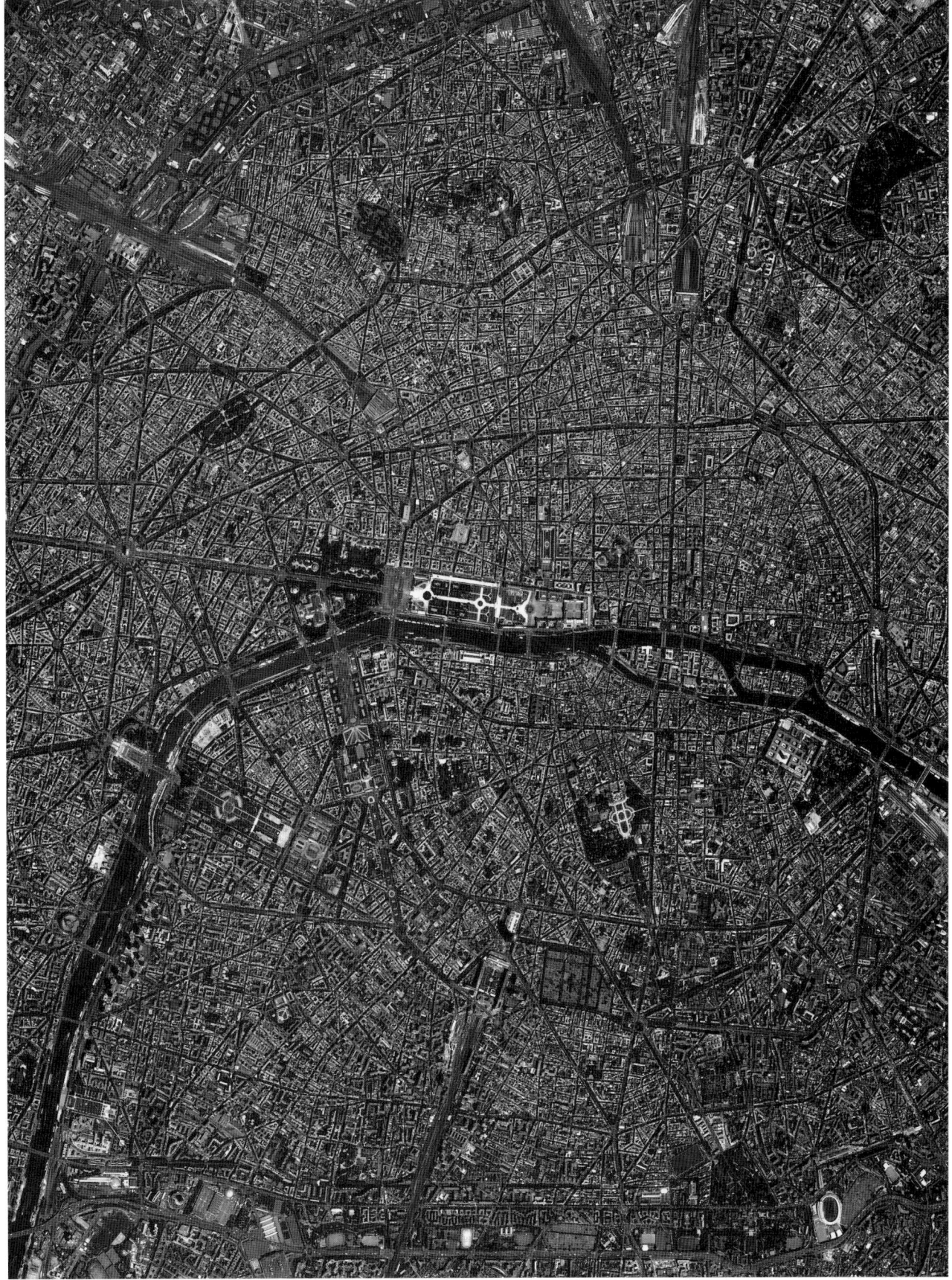

Figure 9.24. **Image of Paris acquired by SPOT 5 (HRG Supermode) on 16 May 2002.** © CNES 2002 Distribution SpotImage.

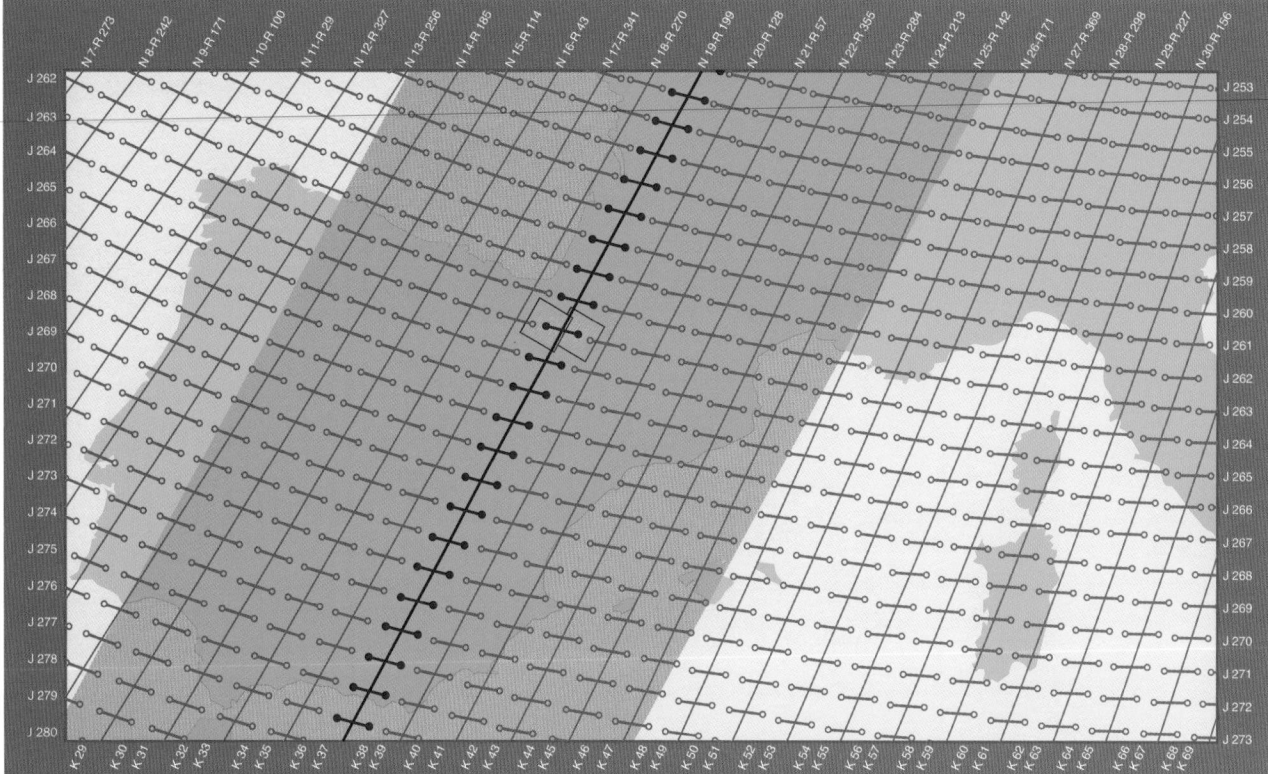

Figure 9.25. **SPOT Grid Reference System (GRS).** The geographical numbering (N series) of the ground tracks applies to the 369 tracks which make up the complete orbital cycle. They are numbered from east to west starting from track N1 which crosses the Equator at 29° 36 east. The chronological numbering (R series) of the tracks applies to the 369 tracks in order of overpass (see fig. 3.26). The observation swaths, relating to twin-HRV vertical viewing, are numbered geographically in the K series. As each ground track is associated with a series of scene pairs, each track in the N series is bordered by two scenes referenced by K values equal to 2N in the east and 2N-1 in the west. The lines in the J series are parallels of latitude. The 950 km wide oblique-viewing corridor straddling ground track N 19 appears in darker green. Scenes obtained by oblique viewing are always centred on a J-series line but their centre may deviate in longitude from the GRS nodal point. The figure shows for J 264 two scenes acquired by vertical viewing with twinned sensors during a pass along N 19. These scenes can be referenced as K 37 and K 38 on J 264.

Figure 9.26. **Panoramic image derived from SPOT stereopairs.** This panoramic view of Reunion Island was obtained from SPOT images acquired between 1986 and 1993. SPOT images © CNES 1986–1993, distributed by SPOT Image, produced by ISTAR.

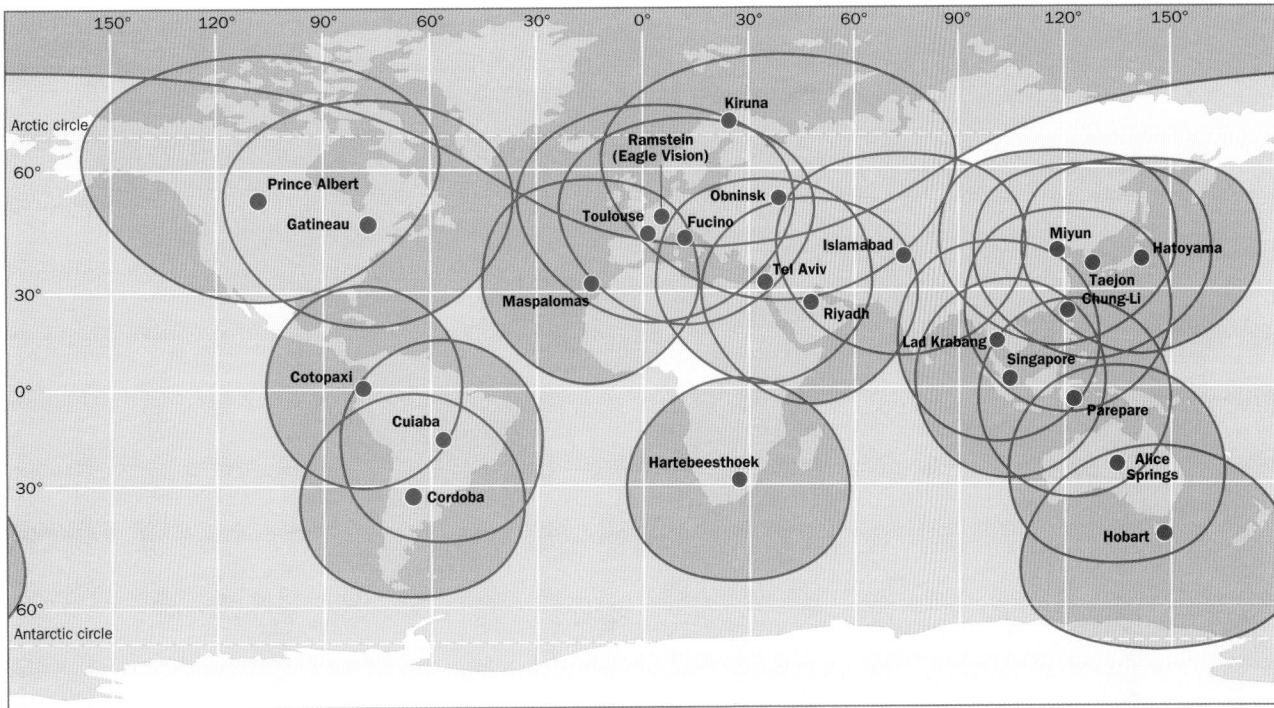

Figure 9.27. **SPOT receiving stations in 2002.** Eagle Vision is a prototype mobile receiving station being tested by Matra. It comprises a 3.6 m antenna and two trailers. This system is suitable for operation with all commercial satellites.

The basic concept was to treat remote sensing as a commercial endeavour rather than as a scientific activity, as had been the case with a number of programmes (Landsat, ERS, MOS) at their inception. The decision to offer better resolution than the Landsat systems was thus driven in part by a desire to enter the international market with a higher performance offering. The same thinking prompted the creation of SpotImage, a private law subsidiary of CNES, a move which had the further advantage of giving CNES responsibility for a national programme clearly distinct from the programmes pursued cooperatively in an ESA framework. The need for public support in the early stages was nevertheless clearly recognised from the outset since the initial funding allocation covered the cost of developing and launching the first two satellites in the series. But it was not until 1990 and the decision on funding of SPOT 3 that the commercial viability of the SPOT programme was officially called into question. SPOT was however already demonstrably an operational programme and this contrasted sharply with the vagaries of American policy on the status of Landsat. The diplomatic benefit to France of possessing the only commercial system in existence was seen as justifying continued investment.

The SPOT programme took a further step forward with the successful launch of SPOT 4 in March 1998. SPOT 5 was launched in May 2002. The SPOT follow-up system known as Pleiades is under active discussion. This involves a constellation of smaller satellites designed to meet national requirements, possibly with a dual approach, and to be developed in synergy with the Italian COSMO-Skymed program.

The decision to proceed simultaneously with development of the civil SPOT satellites and the Hélios defence satellites marked something of a turning point in the programme, again demonstrating a willingness to look for new, untried solutions. The existence of two utilisation environments for space systems is potentially of economic interest since many elements, including the platform, are capable of 'dual use'. But adopting such an approach does at the same time imply a penalty in terms of adaptability. In the case of the SPOT 4/Hélios 1 pair, the advantages outweighed the disadvantages, particularly as the aim on the military side was to acquire an initial capability, which gave the design authority a reasonable degree of freedom. With the military developing its own expertise and hence an ability to elaborate more precise requirements and with the growing emphasis on the flexibility needed to adapt to emerging civil requirements, it seems likely that the joint approach will produce more limited benefits when the programme moves on to SPOT 5/Hélios 2.

SPOT satellites move in a circular, sun-synchronous orbit at an altitude of 830 km. They effect the descending node pass at 10:30 and have a 26-day orbital cycle (see fig. 3.26). SPOT 1, 2 and 3 each carry a High Resolution Visible (HRV) instrument based on linear array sensors (fig. 9.8d). The sensors can operate in two modes. In panchromatic (P) mode the spectral response band is fairly wide (0.51 μm to 0.73 μm) but the ground area represented by individual pixels is small (10 m × 10 m on the ground). The multiband (XS) mode comprises three narrower spectral bands (XS1: 0.50 μm–0.59 μm, XS2: 0.61 μm–0.68 μm and XS3: 0.79 μm–0.89 μm) but the ground area represented by each pixel is four times as large at 20 m × 20 m. Extracts from these images are shown in figures 6.7, 6.9, 6.19, 6.21, 6.30, 6.42 and 6.45. Data acquired in the two modes can also be combined to form multiple resolution images (see fig. 9.15), allying the planimetric accuracy of the XP mode with the fuller spectral signature derived from XS data.

A particularly novel feature of the SPOT system is the ability to perform oblique viewing on either side of the satellite nadir by remotely controlled repositioning of the HRV telescope entrance mirror. Scenes can thus be acquired in a corridor 950 km wide (fig. 9.25). This feature offers a number of advantages, such as the ability to make better use of good local meteorological conditions or again more flexible observation of events at optimum dates: crop development stages, volcanic eruptions, flooding, fires etc.

This capability can again be exploited to acquire stereoscopic image pairs of one and the same scene taken at different angles on successive days. It is also possible to reconstitute landforms using photogrammetry techniques and to obtain digital elevation models giving a three-dimensional vision of the terrain (fig. 9.26 and 9.27).

SPOT 4 is the first of a new generation of SPOT satellites to be built around a new platform MK2 which allows for an expanded payload and extended operational life. The HRVIR instrument operates in four spectral bands: B1 (0.5 μm–0.6 μm), B2 (0.61 μm–0.68 μm), B3 (0.78 μm–0.89 μm) and short-wave infrared (1.58 μm–1.75 μm), offering ground resolution of 20 m. In addition, the M-band, with ground resolution of 10 m, covers the same part of the spectrum as B2 (0.61 μm–0.68 μm).

SPOT 4 is further equipped with a four-channel 'vegetation' instrument, three of the channels being responsive to the same spectral bands as the HRVIR instrument: B2 (0.61 μm–0.69 μm), B3 (0.78 μm–0.89 μm) and short-wave infrared (1.58 μm–1.75 μm), the fourth being responsive to blue: B0 (0.43–0.47 μm). The vegetation instrument offers much cruder resolving power: 1150 m but with a much wider field of view (2200 km), enabling virtually the entire surface of the globe to be covered daily. This system is thus extremely well suited to the task of monitoring the continental biosphere and crops around the planet (see fig. 9.43).

SPOT 5 provides higher ground resolution: 5 m and 2.5 m in panchromatic mode (fig. 9.24) and 10 m in multispectral mode across all three spectral bands in the visible and near infrared. The spectral band in the short-wave infrared band is maintained at a resolution of 20 m due to technical constraints. Its instrumentation is also likely to include an independent High Resolution Stereoscopic (HRS) system with a forward/backward viewing capability and resolution of 10 m.

The philosophy underlying the system is the continuity and improvement of SPOT products. However, it confirms the current trend towards smaller and cheaper satellites providing metre resolution images. In this respect, the new system Pleiades will have a very high resolution capability, allowing it to compete with the new commercial systems and may be a constellation of smaller satellites.

MOS (Momo)

Japan's first remote-sensing programme combines, in a characteristically innovative approach, development of internal capabilities and, like the other major space powers, participation in Earth observation on a global basis. The main features of MOS, and indeed of the JERS programme, reflect Japanese interest in management of planet Earth, from global environment issues to the monitoring of major risks, together with a concern to establish technical credibility at international level. As with its other space programmes, NASDA's first steps in remote sensing took the form of extensive cooperation with the United States, one aspect of which was to reutilise various technologies under US licence, with the gradual introduction of improvements as national expertise grew. The scientific nature of the programme was at the same time emphasised to prevent any disputes arising with the United States under the "Super 301" agreement governing the liberalisation of US–Japanese commercial relations.

The Japanese satellites MOS 1 and MOS 1b (Marine Observation Satellite) – subsequently renamed Momo 1 and 1b, Momo meaning peach blossom – have circular, sun-synchronous orbits at an altitude of 910 km and with a 17-day orbital cycle, 237 orbits being completed in that time.

The satellites carry three systems:

- the Multispectral Electronic Self-Scanning Radiometer (MESSR) which collects data in four spectral bands similar to the Landsat MSS response bands; ground resolution is 50 m (fig. 9.28). The electronic scanning capability is shared with the SPOT HRV instrument. The MESSR comprises two sensors, each observing a strip 100 km wide, with 15 km overlap, giving a total swath width of 100 km + 100 km − 15 km = 185 km.

- the Visible and Thermal Infrared Radiometer (VTIR), which scans a 1500 km wide swath using a rotating mirror mechanism. One spectral band corresponds to the visible portion of the spectrum, the other three to the infrared range: the 6 μm–7 μm band for the study of stratospheric water vapour; the 10.5 μm–11.5 μm and 11.5 μm–12.5 μm bands for thermography of the Earth and its cloud cover.

- the Microwave Scanning Radiometer (MSR), which measures water vapour, snow and ice quantities across a 317 km ground strip.

The IRS programme

India's resolve to develop a national Earth observation capability was first asserted in the late 1970s with the launch in 1979 of Bhaskara 1, followed in 1981 by Bhaskara 2. A far more ambitious programme followed as from 1988 in the shape of the Indian Remote Sensing (IRS) satellite series devoted not only to environmental research but also to land and oceanographic applications.

The originality of the Indian approach resides in the priority assigned to domestic requirements, and in particular more effective land use through improved general cartography and

Figure 9.28. **Mosaic of images acquired by the Momo 1 satellite.** This mosaic, built up from four images acquired by the MESSR sensor on 3 and 20 December 1987, depicts Tokyo Bay and the surrounding area. Image courtesy of NASDA.

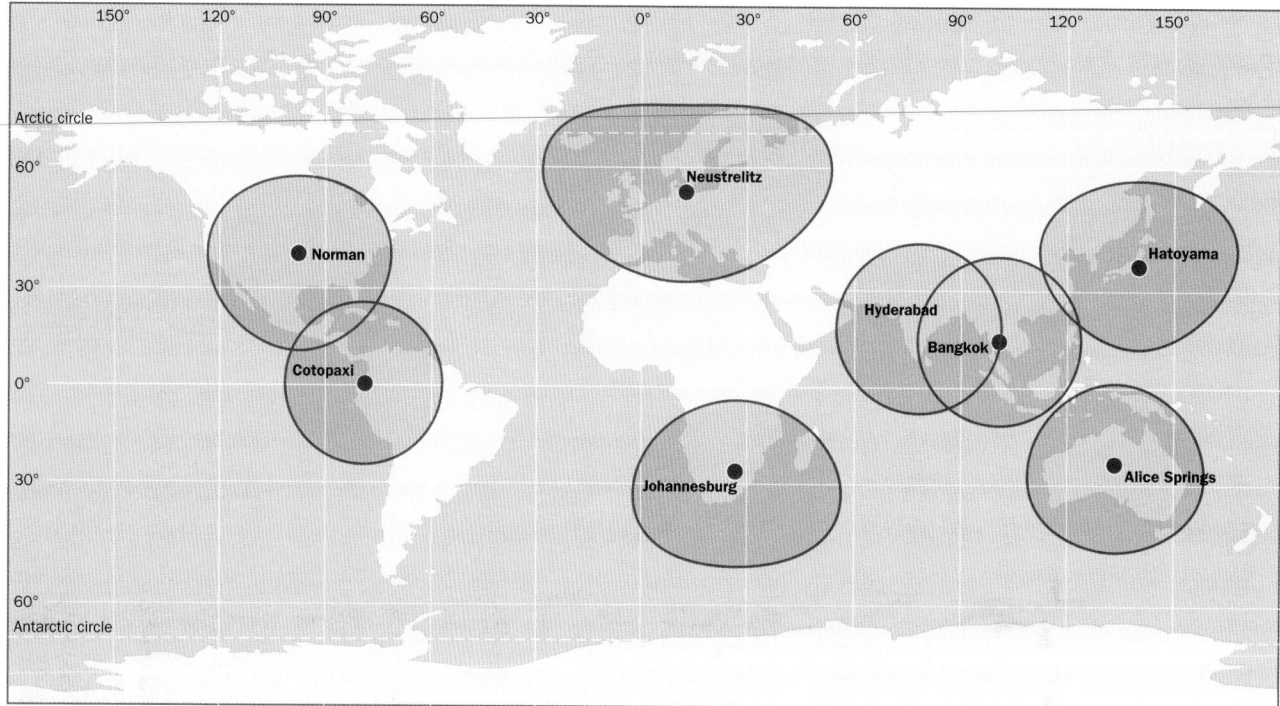

Figure 9.29. **IRS receiving stations in 2002.**

improved monitoring of major risks. Here was a situation in which space capabilities were in harmony with land use management priorities in a developing sub-continent. As has been the case across the entire Indian space programme, the approach chosen was that of encouraging technology transfers and engaging in wide-ranging cooperative projects with the various space powers.

IRS 1A, launched in 1988 aboard a Soviet rocket, occupied a sun-synchronous orbit at an altitude of 900 km. It had a 22-day orbital cycle and executed the descending node pass at 10:25. It was equipped with the LISS 1 and LISS 2 (Linear Imaging Self Scanning) instruments collecting data in four spectral bands. A follow-on satellite having the same characteristics, IRS 1B, was launched in 1991.

The LISS 2 equipped IRS P2 satellite, placed in orbit by an Indian PSLV launcher in 1994, occupied a sun-synchronous orbit at an altitude of 817 km, with a descending node pass at about 10:40 and a 24-day orbital cycle.

IRS 1C, lofted by a Russian launch vehicle in December 1995, took civil remote sensing substantially further down the road towards higher resolving powers, almost ten years on from a similar major contribution by the first SPOT satellite. This it was able to do thanks to the PAN sensor, which offered ground resolution of about 5 m in a panchromatic spectral band (0.50 μm–0.75 μm). Other lower resolution sensors – LISS 3 and the Wide Field Sensor (WiFS) – also formed part of the instrument payload.

IRS data are received by India's Shadnagar station but also by stations around the world (fig. 9.29). They are commercialised in conjunction with Space Imaging-Eosat, already responsible for the commercial distribution of Landsat data.

A second satellite with similar specifications, IRS 1D, was launched on 29 September 1997 (fig. 9.30).

IRS P3, launched in March 1996 on an Indian PSLV vehicle into sun-synchronous orbit at an altitude of 820 km, is pursuing a range of environmental missions, including the study of chlorophyll concentrations in seawater, vegetation indices, cloud cover investigations and geological cartography. The sensors equipping the satellite are designed more for spectral than spatial resolution. Like IRS 1C it features a WiFS, though with an additional sensor in the 1.5 μm–1.7 μm band, and its instrumentation also includes an 18-channel Modular Opto-electronic Scanner (MOS) providing resolutions between 580 m and 2520 m. The next satellite in the series, IRS P4, carrying an Ocean Colour Monitor (OCM) sensor covering 9 spectral bands, was launched on 25 May 1999.

Where optical imagery is concerned, the Indian remote-sensing programme is now emerging as the most ambitious of the government programmes being pursued by the various space powers. No less than three satellites are currently scheduled for launch, one devoted primarily to oceanography. IRS P5 and IRS P6 will pursue complementary missions. Assigned to agricultural applications, IRS P6 will carry a multispectral 'vegetation' sensor with 10 m resolution. IRS P5 will be used for cartographic purposes, with 2.5 m resolution or even better. Launch in around 2003 of Carto-Sat, a satellite with a resolution understood to be as high as 1 m, is also planned. In choosing to operate two separate systems side-by-side, the main consideration for India was no doubt the need to keep within the launch capabilities of its PSLV vehicles, giving it full autonomy in the conduct of its

Figure 9.30. **Image acquired by the PAN sensor on IRS 1D.** The image shows part of the township of Mumbaï in India. Courtesy of NRSA, India.

observation programme. But this approach had the added advantage of greatly increasing the number of information sources and extending coverage.

In addition to meeting domestic demand, India is looking to serve the Asian market, which is predicted to grow rapidly. Attractive prices and finer resolution than its rivals have allowed IRS 1C to make a successful entry in the com-

mercial market, fully vindicating the policy of commercialisation by Space Imaging-Eosat. Nor is it by any means inconceivable that future satellites could also be used for military purposes, a policy development which would not in principle be at odds with the civil nature of the Indian space programme insofar as such military users would be regarded simply as customers.

The Resurs system

The Resurs system is built around two families of satellites, Resurs F and Resurs O, which are seen as offering a complementary range of capabilities for Earth resource surveying purposes. Resurs satellites have to date been limited to passive remote sensing. The programme, which came into being in the heyday of Soviet space power, is now having to contend with drastically reduced funding and consequent uncertainty as to its future prospects. Commercialisation of Resurs products is also proving extremely difficult, due in

part to the unresponsive nature of the commercial structures involved and in part to the increasingly tough competition at international level, leaving little prospect of establishing an adequate alternative source of funding. Manufactured by TsSKB (or OKB Koslov) of Samara, the Resurs F satellites can be traced back to the earlier Vostok craft and are close relatives of the Kosmos reconnaissance satellites (see chapter 12). They occupy low circular orbits at altitudes between 250 km and 350 km and have a short operational life. The altitude at which they fly can if required be lowered to below 200 km to give better resolution. The general characteristics of Resurs satellites are those of third-generation military reconnaissance satellites (see fig. 12.1) but some of their products are used for civil purposes, particularly by the authorities responsible for cartography. This trend has become increasingly marked with the fall-off in funding and

Figure 9.31. **Extract from a KVR 1000 image.** This photograph of the Baghdad region was taken by a KVR camera onboard a Russian Yantar Kometa military reconnaissance satellite. The Baghdad urban area can be seen in considerable detail on the left bank of the Tigris. The extract shown here covers a ground area 18.6 km wide and 16 km long. KVR 1000 image. Processed by Sovinformsputnik. Distributed by SPOT Image.

the growing need to find commercial outlets in order to maintain the system's capabilities.

The first operational satellite in the Resurs F1 series was launched in 1986 and this was followed by many others. Resurs F1 satellites have an average mission duration of only two weeks, compared with a standard one month in orbit for the F2 series, extendable to up to one-and-a-half months. A Resurs F3 version modelled closely on F1 but operating at a lower altitude was used on a few occasions in 1993.

These satellites are equipped with still cameras (Kate 200 and KFA 1000 for Resurs F1, MK 4 for F2) providing spatial resolutions between 15 m and 5 m; the KFA 3000 camera on-board Resurs F3 gave up to 2 m resolution. In conjunction with stellar viewing techniques, the multiband photographs taken by these satellites (fig. 9.31) allow for precise geographical position determination of certain images of the Earth. Exposures are returned to Earth in recoverable capsules. The greatest strength of the photographic equipment carried by Resurs F craft is its very high spatial resolution. The F1M enhanced version scheduled for 1994 did not in the end make its debut until November 1997. Further improved F2M versions were to have come into service beginning 1997 but launch has had to be postponed owing to funding difficulties.

The launch on 17 February 1998 of a Kosmos satellite of the Yantar Kometa class marked a new departure in the funding and commercialisation of this type of system. This was the first of four missions in a joint venture by the Russian Sovinformsputnik organisation, which is responsible for commercialising Russian space products, and the American Aerial Images Inc., a specialist in the development of advanced digital imagery. The aim is to provide 'SPIN 2' imagery covering the world's main capitals and most towns with more than 50 000 inhabitants. Two cameras, a KVR 1000 with 2 m resolution and a TK 350 with 10 m resolution, provide 160 km × 40 km and 300 km × 200 km images respectively. A single TK 350 image thus covers the same area as seven KVR 1000 images and the system also has stereoscopic capability.

The Resurs O satellites use the same platform as the Meteor 3 series. The first operational flight unit, bearing the name Kosmos 1939, was launched in 1988. Resurs O satellites operate from orbital altitudes of between 600 km and 700 km and are fitted with multispectral sensors similar in many ways to those equipping the Landsat series. They are used primarily for coastal waters surveillance, detection of industrial pollution sources and agricultural monitoring. Resurs O satellites have an operational life of 3 to 5 years. Launched in November 1994, Resurs O1 N3 was the first satellite in the series. The first of its two scanners, MSU SK, operates in spectral regions ranging from the visible (0.5 μm–1.1 μm with resolutions of about 170 m) to the thermal infrared (600 m resolution), while the second, MSU E, covers the visible and the near infrared with 45 m resolution.

Resurs O data is commercialised by the Russian consortium Sovzond, whose offering includes current and archived images. The Swedish station at Kiruna – which also receives SPOT images – is equipped to receive some Resurs O data and the European company Eurimage also possesses a data collection of commercial interest.

Resurs O1 N4 was placed in orbit on 10 July 1998 and will be followed in the next few years by Resurs O1 N5. These satellites feature a number of new sensors, some from Belgium, Germany and Italy, but this is still essentially a multimission platform.

The first of the new generation Resurs DK family developed by TsSKB of Samara is scheduled to be launched in 2003. But although the programme was given priority status by the Russian Space Agency in 1996, it remains chronically underfunded, partly as a result of the very small overall space budget and also because what little funds are available have for a long time now been almost entirely eaten up by the international space station programme.

The initial project, a fairly ambitious one, underwent some descoping in 1997 to bring it into line with a shrinking budget but – despite assurances that the technical specifications were now final – funding continued to be a problem. Later in the same year the project had therefore to be simplified yet further, the Director-General of the Russian Space Agency at the same time announcing that the programme had priority status for both civil and military requirements.

Weighing in at 7 tonnes, the satellite will operate from an altitude of between 400 km and 600 km, extending its service life to 3 years. In panchromatic imaging mode it is expected to provide 1 m resolution, this being the minimum currently imposed by the Russian government, with resolutions of 2–3 m in multispectral mode and 6–8 m in the near infrared waveband.

ADEOS

Japan's ADEOS 1 (Advanced Earth Observation Satellite) was launched on 11 August 1996 but contact with the craft was lost on 30 June 1997 with only 20% of the data collection mission completed. This failure has been traced back to structural damage to the solar array paddle.

It was placed in a sun-synchronous orbit at an altitude of 797 km, with a descending node pass at 10:30 and a 41-day orbital cycle. It carried a substantial and varied instrument payload. The Advanced Visible Near Infrared Radiometer (AVNIR) operated in the visible (0.42 μm–0.50 μm, 0.52 μm–0.60 μm, 0.61 μm–0.69 μm) and the near infrared (0.76 μm–0.89 μm) at 16 m resolution and in a panchromatic band (0.52 μm–0.69 μm) at 8 m resolution across an 80 km strip width (fig. 9.32). The Ocean Colour and Temperature Scanner (OCTS) delivered poorer spatial resolution (700 m) and greater (1400 km) strip width but had finer spectral resolving power in the visible (0.402 μm–0.422 μm, 0.433 μm–0.453 μm, 0.479 μm–0.500 μm, 0.511 μm–0.529 μm, 0.555 μm–0.575 μm, 0.660 μm–0.680 μm), the near infrared (0.745 μm–0.785 μm, 0.845 μm–0.885 μm), the short-wave infrared (3.55 μm–

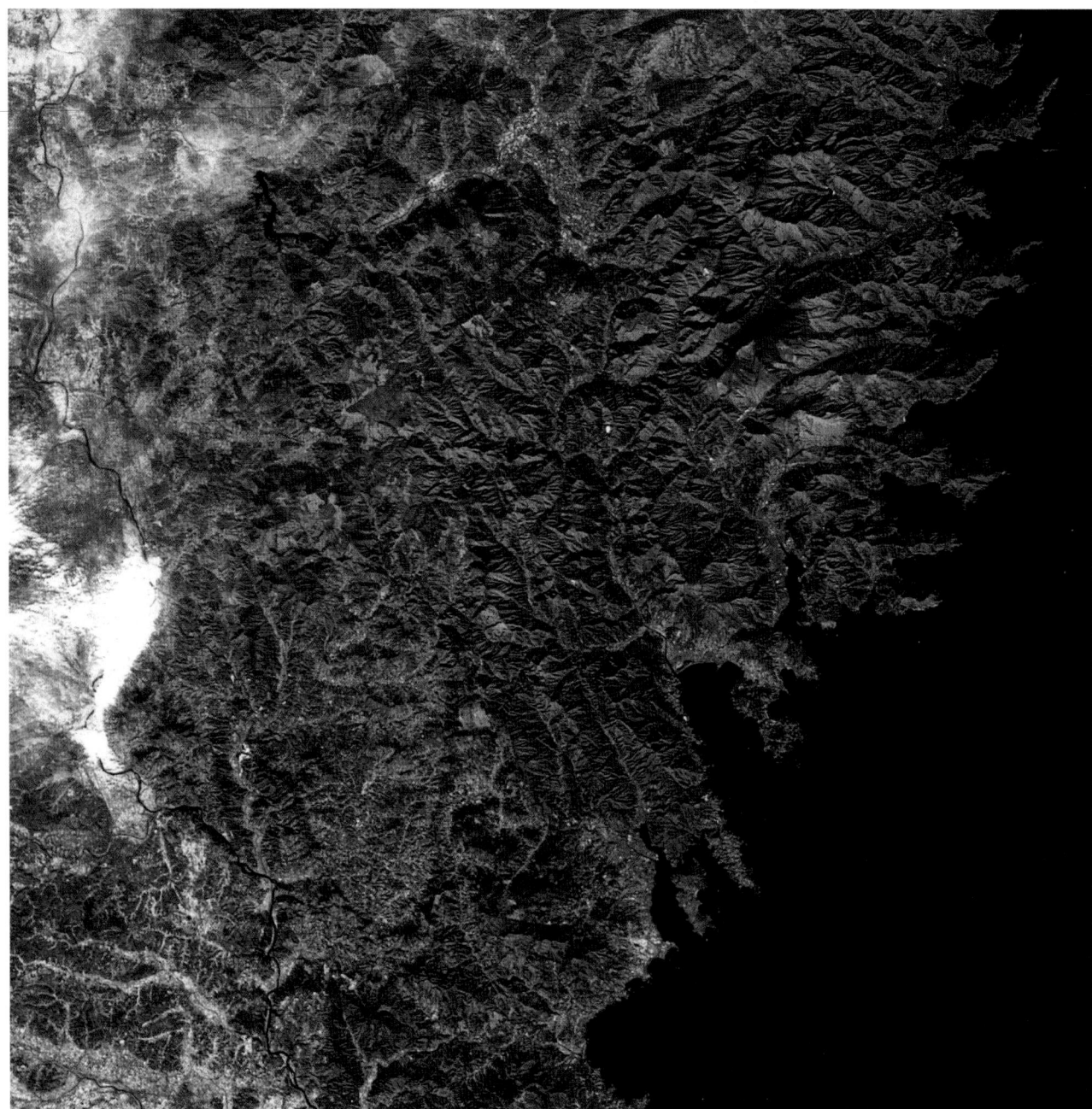

Figure 9.32. Multispectral image acquired by the AVNIR sensor on ADEOS 1.
The image shows the same region (Department of Iwate, Japan) as that appearing in fig. 9.39. © NASDA.

3.88 μm) and the thermal infrared (8.25 μm–8.80 μm, 10.3 μm–11.4 μm, 11.4 μm–12.7 μm). A number of these wavebands are particularly well suited to the study of matter contained in seawater and more particularly chlorophyll distributions. OCTS readings were moreover obtained using a rotating scanner and the resulting oblique viewing angles allowed most specular reflection of solar rays off the ocean surface to be avoided. NASA supplied two instruments: an NSCAT scatterometer derived from the Seasat model but improved for the US Navy as part of the Navy Remote Ocean Sensing System (NROSS) programme (the latter cancelled in 1988); and a Total Ozone Mapping Scatterometer (TOMS). The CNES-supplied Polarisation/Directionality of Earth Reflectance (POLDER) sensor collected data over a 1440 km × 1920 km area, which meant that a specified region could be observed from different angles in successive passes. Measurements were taken in eight spectral bands in the visible and near infrared in natural and polarised light. The data gathered in this way will contribute to a better understanding of the directional properties of solar radiation reflected by the Earth and the atmosphere.

A second satellite in the series, ADEOS 2, is to be launched in 2002 and will operate from a similar orbit to its forerunner. ADEOS 2 will carry sensors developed by NASDA, one of which will be an advanced microwave radiometer (AMSR).

Seastar

The OrbView 2 or Seastar satellite from the firm Orbimage was launched in August 1997. It is equipped with SeaWiFS, an eight-channel sensor covering a spectral range from 402 nm to 885 nm. Data collected is used to determine chlorophyll, humic substance and sedimentation levels in the oceans and inland lakes and the concentration of atmospheric aerosols, the latter information providing a basis for signal correction decisions. OrbView 2 imagery also covers land-based vegetation. Its 2800 km swath is sufficient for full daily coverage of the Earth's surface and colour imagery from the satellite is continuously downlinked in real time and can be acquired by HRPT ground stations.

UoSat

The UoSat series of small satellites are designed and built by Surrey Satellite Technology Ltd. (SSTL), a company set up in 1985 by the University of Surrey to support technology transfer to industry and commercial development of research activity. The UoSat series is also used to test satellite miniaturisation. A number of these mini-satellites are equipped with various remote-sensing instruments, often with low spatial resolution, examples being Kitsat, a joint project with South Korea (1992) and Posat, in association with Portugal (1993). The most powerful satellite in the series is the 325 kg UoSat 12, a cooperative venture with the University of Singapore, launched from Baikonur on 21 April 1999. It carries a four-band multispectral sensor with 32 m resolution and a 10 m resolution panchromatic sensor. The scenes generated cover ground areas of 33 km × 33 km and 10 km × 10 km respectively.

CBERS, ZY 2

The CBERS (China Brazil Earth Resources Satellite) programme is a fine example of creative cooperation by China and Brazil in the development of two satellites in the Landsat category. Approved in 1988, the programme brought China external funding for development, manufacture and launch of the initial satellite. The programme was also for China a fast route to a civil satellite with a 2-year design life – which contrasts favourably with the short observation times available on FSW capsules. For Brazil the agreement was part of a broader policy aimed at introducing Brazilian high-technology products to the Chinese market and was also a means of diversifying cooperative links in satellite imagery, a major priority for the Brazilian space programme since 1978. The programme was to have been operational by 1993 but, owing to delays in Brazilian funding, CBERS 1 did not in fact reach orbit until October 1999. A follow-up satellite is to be launched, with integration likely to be performed in Brazil, and discussions are continuing on the construction of two further units, CBERS 3 and 4.

The CBERS series is unusual for its ability to provide global coverage with its multi-sensor payload combined with a high revisit frequency. The Wide Field Imager (WFI) provides a synoptic view of the Earth every five days with 260 m spatial resolution while the high resolution CCD camera, which has some limited stereoscopic capability, is able to supply images with 20 m resolution in five spectral bands. Phenomena detected in wide view by the WFI can thus be 'zoomed' by the CCD though with a time lag of up to three days. Lastly the Infrared Multispectral Scanner (IRMSS) operates in four spectral bands, imaging a 120 km swath with a resolution of 80 m. The last two sensors provide complete Earth coverage in a 26-day cycle.

The form commercialisation policy will take would still seem to be undecided. Judging by approaches made to Western firms concerning a high resolution sensor, it seems likely however that improved resolution will be a feature of the new satellites.

Independently of future CBERS, China is also pursuing a different programme known as ZY 2. Commercialisation is not a priority in itself but rather expresses the need for the two countries to develop their territories.

Ikonos

Following a series of failures – Early Bird, Lewis and the first Ikonos satellite – and the decision not to proceed with Clark, very high resolution commercial remote sensing got underway in earnest with the successful launch by Space Imaging. of Ikonos 2 on 24 September 1999 (Fig. 9.33).

Ikonos marked an important new departure in two respects: the resolution offered and reliance on private funding. Lockheed Martin, the largest American space enterprise, was developing plans for a commercial remote-sensing system as early as 1991 and in June 1993 it applied for a licence covering images with 1 m resolution. That request was granted the following April by the US Department of Commerce in pursuance of the presidential directive of 23 March 1994. This led in December 1994 to the creation of Space Imaging by Lockheed Martin, Raytheon's E Systems and Mitsubishi. With Lockheed's buy-out of Hughes' shares in Eosat, the company – which with its new affiliate became Space Imaging Eosat – went on to acquire a complete end-to-end capability. Current estimates suggest the overall cost of the Space Imaging operation (2 satellites, 2 launchers, 1 primary and 2 secondary stations, plus imaging hardware and software) may amount to some $500 M, equivalent to about a year's income. The company's strategy would seem to be to fully exploit its technical superiority and financial strength while encouraging regional franchises, such as the one granted Mitsubishi.

Operating from a sun-synchronous circular orbit at an altitude of 681 km, the satellite is supplying images in panchromatic mode (0.45 μm–0.90 μm) with resolution varying from 0.82 m at nadir to 1 m and 1.12 m up to 26° and 45° respectively either side of nadir. It also provides multispectral images with resolutions of 3.2 m to 8 m in a range of bands also used by Landsat: 0.45 μm–0.52 μm, 0.52 μm–0.60 μm, 0.63 μm–0.69 μm, 0.76 μm–0.90 μm. Strip width is limited to 13 km.

Revisit capability using one satellite only is 3.9 days for 1 m resolution in panchromatic mode and 4m in multispec-

Figure 9.33. **First image acquired by the Ikonos-2A satellite.** This metre-range resolution image depicts downtown Washington; the Washington Monument and the Ellipse can be made out on the left and, on the right, the National Museum of American History and the Department of Commerce complex. © Space Imaging.

tral mode; for 0.82 m resolution, the revisit figure is 11 days. When a satellite pair is available, a 3.2 day capability will be achieved, and even a single day in regions at 51° latitude. Finally, the satellite offers very flexible data collection since final programming can be performed up to a few minutes before the satellite pass.

Ikonos data is forwarded by the Norman Station in Oklahoma (USA) to a number of stations around the world: Shadnagar (India), Dubai (United Arab Emirates), Neustrelitz (Germany) and Chung-Li (Taiwan). A limitation was however placed on universal distribution of 1 m resolution images when the US Congress voted in 1996 to outlaw the distribution of images of Israel where these were of higher resolution than those available from international competitors.

EROS

The launch of the Israeli satellite EROS A1 by the Russian launch vehicle Start 1 in December 2000 signalled the arrival of a new player in the market for one metre resolution imagery. The 250 kg satellite follows a circular orbit at altitude 480 km and provides a resolution of 1.8 m in panchromatic mode and 0.8 m using an oversampling technique (fig. 9.34). It constitutes the first element in a constellation comprising at least two EROS A satellites, the second being scheduled for launch in 2002, and six EROS B satellites to be launched over the next five years. Given the agility of the system, a single satellite provides a mean revisiting time of 1.8 days, descending to 1.1 days with two satellites, whilst six satellites will be able to cover every point of the globe at least once a day and eight satellites will allow revisitation more than twice daily.

Although this is a civilian programme, Israel is using experience gained in the field of military reconnaissance. Indeed, one aim is to obtain some financial return on development of the military Offeq satellite. The company responsible for marketing is ImageSat, with headquarters in Cyprus but largely run from Tel Aviv. It is a joint venture between Israeli companies IAI and ELOP, and the US company Core Software Technology. The marketing policy favours sale of images to government bodies within the framework of specifically drawn contracts known as SOPs (Satellite Operating Partnerships). Technical specifications are designed to meet civilian as well as security needs, including urban planning and development of the oil and gas industries. Data from space is intended to complement aerial photography products. This explains the 50 cm resolution planned for EROS B, which may actually be launched in preference to EROS A1 if market requirements should justify it.

EROS data can be acquired via three main channels: acquisition, archiving and distribution (AAD), priority acquisition service (PAS), and satellite operating partnership (SOP). The so-called AAD system is currently based on thirteen relay stations which can downlink, store and forward data. These are set up in South Africa, Argentina, South Korea, Italy, Japan, Russia, Singapore, Sweden, Taiwan, the United States, Norway and Canada. The second option can be exercised by any customer within the bounds of existing SOPs. Priority acquisition service partners have priority in satellite tasking. Partners submit image requests to the ground control station, which places their requests first among those submitted by AAD partners. PAS partners' requests go to the top of the list of images to be acquired by the next available satellite. Conflicts between two geographically adjacent PAS partners are mediated on a first-come, first-served basis, for the most timely response, with a delivery time of 48 hours before the image is made available. The most original type of partnership is the third, whereby the customer can hold exclusive rights to images produced by a given satellite over a zone of radius 2000 kilometres around the ground receiving station. This station with both uplink and downlink capabilities is supplied by ImageSat as part of the SOP program. It allows the customer to control which angles are viewed by the satellite as soon as it flies into the relevant coverage zone. The satellite then downlinks the required images exclusively to the SOP customer. ImageSat insists that there is no possibility of conflicting orders in the same zone since by definition only one country can have an SOP in a given zone. It would only be by increasing the number of satellites that several countries might access the same zone via different satellites.

The Israeli Ministry of Defence is supposed to pay ImageSat International about $15 million for the exclusive rights to all photographs of Israeli territory taken by the EROS 1 civilian satellite. The exclusive receiving rights, signed on 3 January 2001, include not only Israeli territory, but also the area within a 2000 km radius. The region extends to Libya in the west, Sudan in the south, the Gulf of Oman in the east, and CIS countries in the north. The entire territory of Iraq, Iran and Syria is included within the area, as well as almost all of Libya, most of Sudan and the Muslim republics of the former Soviet Union.

The success of the EROS programme thus directly depends on the interest shown by government bodies and the number of SOPs purchased.

Figure 9.34. **Image of Izmir acquired by EROS A1** © ImageSat.

SAR-equipped Earth resource systems

Seasat

The American Seasat 1, launched on 26 June 1978, was the first satellite designed for remote sensing of the Earth's surface using synthetic aperture radar. By the time it ceased functioning on 10 October the same year it had sent back images covering 125 million km² of territory in North America, Western Europe, the North Pacific region, the arctic zones and the North Atlantic. Seasat paved the way for many other programmes.

The Seasat SAR, equipped with a single rightward-pointing radar antenna, offered little flexibility, there being no scope for modifying its wavelength (L-band = 23 cm), incidence angle (20° nadir deviation) or HH polarisation. Despite this shortcoming, the scientific results obtained were of the greatest importance, demonstrating in particular the relevance of SAR imagery to the study of ocean swell directions, the surface manifestations of internal waves, topographic gradients, ground surface roughness and geological structures (see fig. 9.37).

Seasat carried a number of other instruments, including a radar altimeter used to measure wave heights to 10 cm vertical accuracy and a scatterometer for measuring wind speed and direction at ocean surface (fig. 9.35). The instrument payload also featured two passive sensors, a scanning radiometer operating in the visible and the thermal infrared and a Scanning Multichannel Microwave Radiometer (SMMR) similar to the one on Nimbus 7.

Almaz

With the Almaz ('diamond') satellite series, the Soviet Earth observation programme entered new territory. Derived from military stations of the same name – which were subsequently reinstated in the Salyut programme – the Almaz platform was built by OKB Tchelomei, now NPO Machinostroienya.

The first satellite was ready for service in 1981 but its launch was vetoed by the then Defence Minister D. Ustinov who was on particularly bad terms with the engineer-in-chief V.N. Tchelomei. The programme did not resume until 1984, the year in which both men died. A prototype with the designation Kosmos 1870 was launched on 22 July 1987 and underwent a two-year programme of testing. Almaz 1, the first operational unit, was launched from Baikonur on 30 March 1991. It had a theoretical design life of two to three years but in fact fell back in October 1992. This very large spacecraft weighing over 18 tonnes and measuring 15 metres end to end operated from a 300 km orbital altitude at 73° orbital inclination. Orbits could be adjusted to enable the satellite to acquire close-up images, a common procedure with military satellites. With a total propellant load of 1350 kg and routine consumption of only 100 g a day, a substantial reserve was available and this was primarily used for such orbit manoeuvres.

Almaz 1 was equipped with an HH-polarised SAR operating in the X-band and two antennas, one left-pointing, the other right-pointing. The SAR's incidence angle could be adjusted within a 30° to 60° range either side of nadir. Resolution was about 15 m. Almaz data was commercialised by the American Space Commerce Corp., by Almaz Corp. and by other distributors such as Hughes STX. Whether the launch of Almaz 1B, with a more varied payload thought to include two SARs, eventually materialises will depend on whether the US–Russian company Sokol Almaz Radar manages to secure the necessary funding.

The ERS, Envisat programme

The European Space Agency's European Remote Sensing satellite (ERS) programme was given the go-ahead in 1982 by the Agency's then 12 member states and Canada. The decision to opt for radar sensing was largely determined by the meteorological conditions, and in particular the predominance of cloud cover, prevailing in many ESA member states. Germany played a particularly important role in the decision-making process, especially in view of its industrial strength in this area.

ERS 1 occupied a sun-synchronous orbit at an altitude of 785 km. It operated a 35-day orbital cycle until 20 December 1993 but then switched to a 3-day cycle, a higher revisit frequency being more suitable for the study of ice cover during the Arctic winter. The repeat cycle was then extended to 168 days for the geodesic utilisation phase, which continued to the end of the satellite's service life on 10 March 2000. These cycle changes were effected by means of very slight variations in altitude and inclination.

The observation payload included a broad range of advanced technology sensors. The Active Microwave Instrument (AMI) included a Synthetic Aperture Radar (SAR) operating in the C-band (5.3 GHz) with VV polarisation. The SAR could operate in image mode, where it covered strips of the Earth's surface about 100 km across, whose central axis was located some 300 km to the side of the satellite track (see fig. 9.9a). In the second ('wave') operating mode, the 5 km × 5 km field provided discrete sub-images every 200 or 300 km. The two modes were mutually exclusive. The AMI also featured a scatterometer used to determine wind direction and speed at the surface of the sea by means of measurements taken across a 500 km wide swath, with a standoff

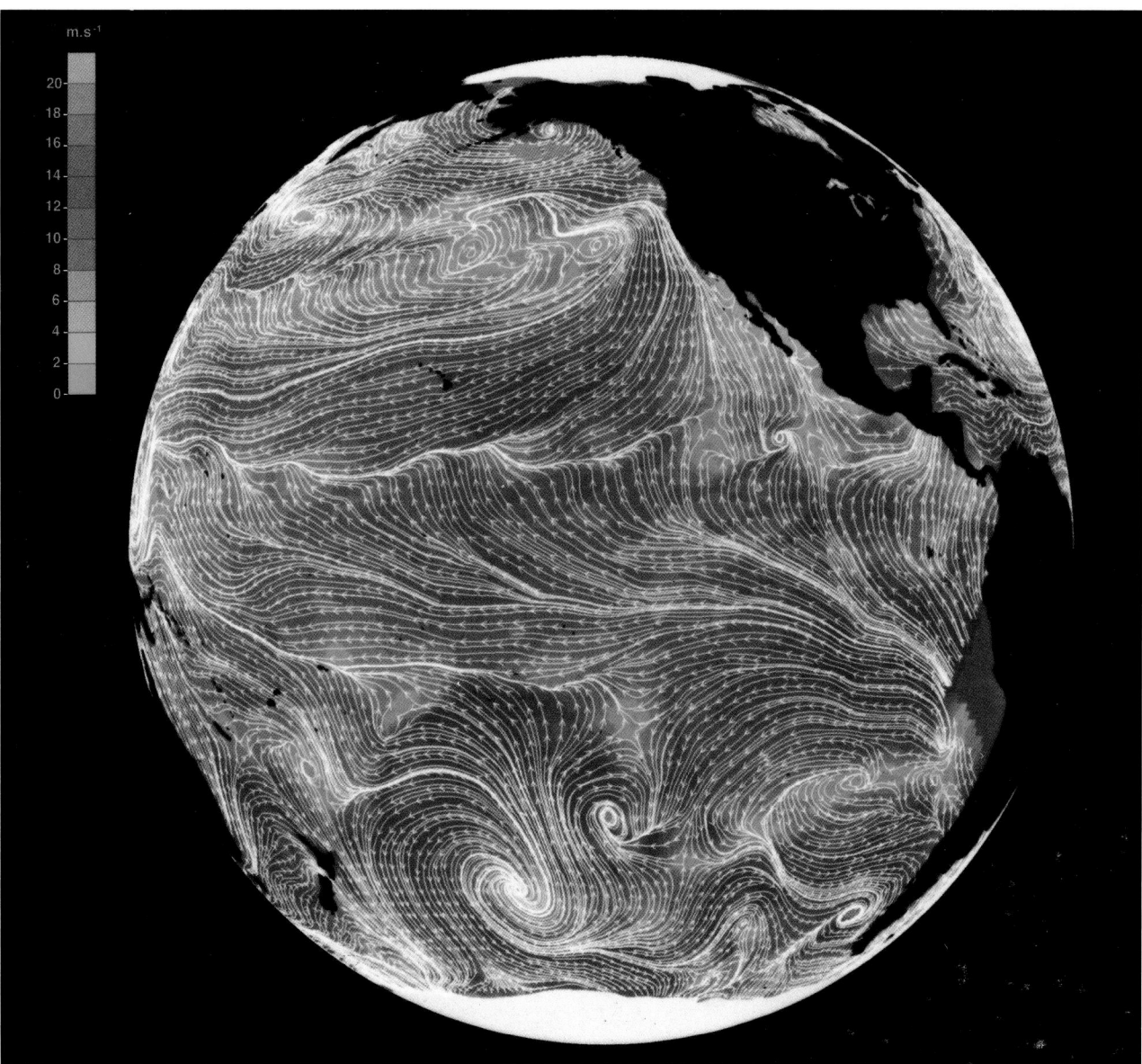

Figure 9.35. **Cartographic representation of winds at the surface of the sea based on Seasat scatterometer readings.** An understanding of wind phenomena at global level is essential to the study of energy exchanges. Work in this area had for a long time to rely on the study of cloud displacement but is now backed up by the mapping of sea-surface winds based on spaceborne scatterometer readings. The earliest studies of this kind were carried out in 1978 using Seasat – this image of wind speed and direction at the surface of the Pacific Ocean was acquired using the Seasat scatterometer. The colours represent wind speed and the arrows wind direction. Image NASA, produced by JPL.

between 250 km and 750 km to the right of the satellite track. Measurements were taken in relation to three axes proceeding from the sub-satellite point. One axis was perpendicular to the satellite track, the other two following the bisectors forward and backward of the angles formed by the ground track and the first axis (see fig. 9.11). Backscatter readings for a given point on the surface of the sea were thus taken in succession (forward beam, perpendicular beam and lastly backward beam) with very short time intervals.

ERS 1 also carries a radar altimeter working in the Ku-band (13.9 GHz) for measurements of vertical distance to 10 cm accuracy. The altimeter can be used to determine the frontier between free-flowing sea and pack ice, the elevation of the polar icecaps or the significant height of ocean swell. Another item in the ERS 1 payload is the Along Track Scanning Radiometer (ATSR), a high-precision passive microwave instrument operating in the infrared. Two final instruments equipping ERS 1 are the Precise Range and Range Rate Equipment (PRARE), which failed to function correctly from the time the satellite was launched, and a laser reflector. Data collected by the SAR far exceed on-board recording capabilities and it has therefore to transmit over 100 million bits of information per second in real time to ground receiving stations.

The ERS satellites are essentially multidisciplinary in scope. Information provided by the infrared radiometer concerning ocean surface temperatures and by the radar altimeter (operating in 'ice' mode) concerning the evolution of polar ice cover is of great relevance to environmental research and

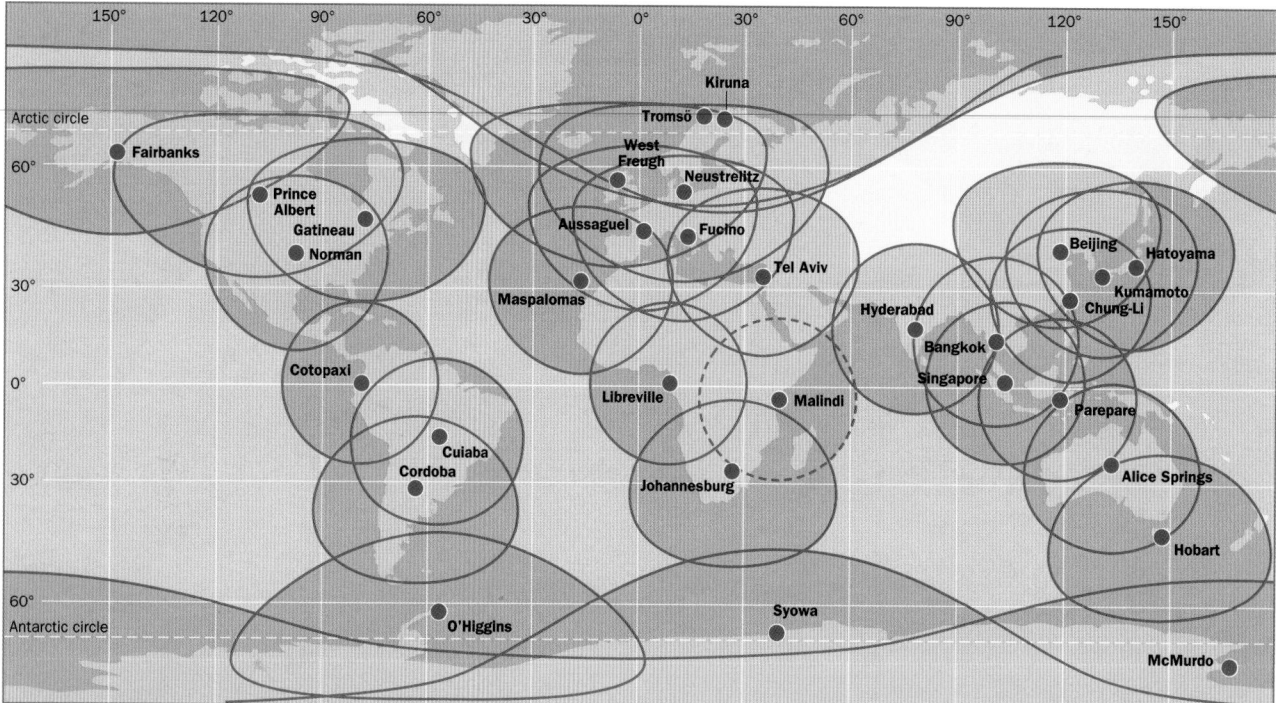

Figure 9.36. **ERS 1 data receiving stations January 2000.**

the study of changes at global level. The prominence given to study of the ice regions is clear from the geographical locations of the receiving stations at Kiruna, Tromsö, Fairbanks, O'Higgins, Syowo and McMurdo (fig. 9.36). In 'ocean' mode, the radar altimeter provides considerable information on marine topography, from which the strength of ocean currents – variations in which are responsible for certain climatic changes – can be derived.

The SAR images supplied by the AMI are probably the most important ERS contribution to environmental knowledge, despite the fact that the AMI can only operate for ten minutes in every orbit and even then only when it is in sight of a receiving station. These images provide immensely useful material for a whole range of disciplines in biology, geology, hydrology and geography (fig. 9.37). The follow-on ERS 2 satellite, launched on 21 April 1995, carries an AMI and radar altimeter similar to those on board ERS-1, an enhanced ATSR, an improved PRARE and a new instrument, the Global Ozone Monitoring Experiment (GOME). The two craft operated in tandem from 16 August 1995 to mid-May 1996. Since 2002, the data (fig. 9.38) are aquired by Envisat 1 (see chapter 7, p. 169).

JERS 1 (Fuyo 1)

The launch by Japan of the MOS satellite in 1987, the JERS radar satellite in 1992 and finally ADEOS 1 in 1996 demonstrates a strong and consistent interest in satellite observation. Development of an 18 m resolution civil radar capability, in the shape of the L-band SAR carried by JERS 1, was a first in this frequency band. The OPS sensor supplies optical data with the same resolution (18 m) and offers stereoscopic capability but these features are much the same as for Landsat.

Wavering by Japan on pricing for JERS data does however reveal some inconsistencies in its national space policy. MITI, which was responsible for the instrument payload, preferred to pitch prices low, whereas NASDA, in charge of the platform, took a more commercial stance along the lines of SPOT. The obligation on Japan, arising out of its special relations with the United States, to refrain from protectionist policies (that is, policies designed to give domestic industry an edge over international competitors) in the commercial arena undoubtedly contributed to the adoption of low prices. Another factor was a concern to be seen as a 'good world citizen' and even to acquire, through generous policies, a measure of regional influence. Finally, the emphasis on maritime rather than land observation reinforced the scientific

Figure 9.37. **Images acquired by the Seasat and ERS 1 SARs.**

a) Seasat: Image of the cove at Aiguillon (in Western France) at high tide acquired by the Seasat SAR (L-band) on 27 August 1978. The areas of weaker backscatter, shown in black, are characteristic of the uniform mud flats at the fringes of the cove. Backscatter is strongest over water. The transient shore line is particularly well delineated where very smooth mud flats are bordered by backscattering water. Image © NASA.

b) ERS 1: Image of the cove at Aiguillon at low tide, acquired by the ERS 1 SAR (C-band) on 8 November 1991. Contrasts in orientation of surface roughness appear clearly on this image. The areas of weaker backscatter, shown in black, are characteristic of the uniform mud flats at the fringes of the cove and of the beaches at the Arçay headland. Stronger backscatter is recorded where ripple marks run perpendicular to the radar beam (the central part of the cove to the north of the Sèvre Niortaise area) or where the bank of a channel is at right angles to the beam direction. Areas of strong backscatter are shown in white on the image. A: Aiguillon cove; P: Arçay headland; S: Sèvre Niortaise.

Image ERS 1, © ESA (R), processed by Géosystèmes.

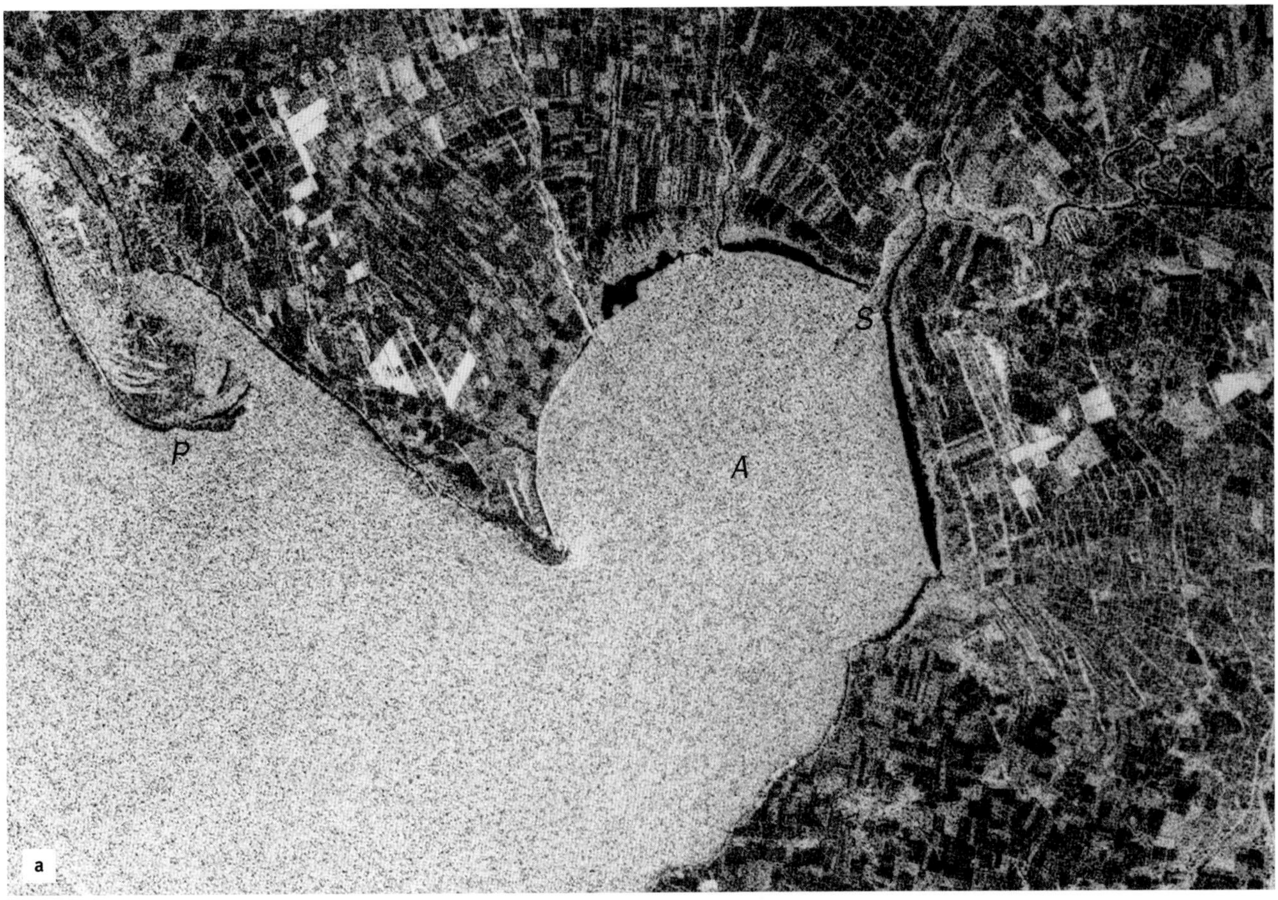
P
A
S
a

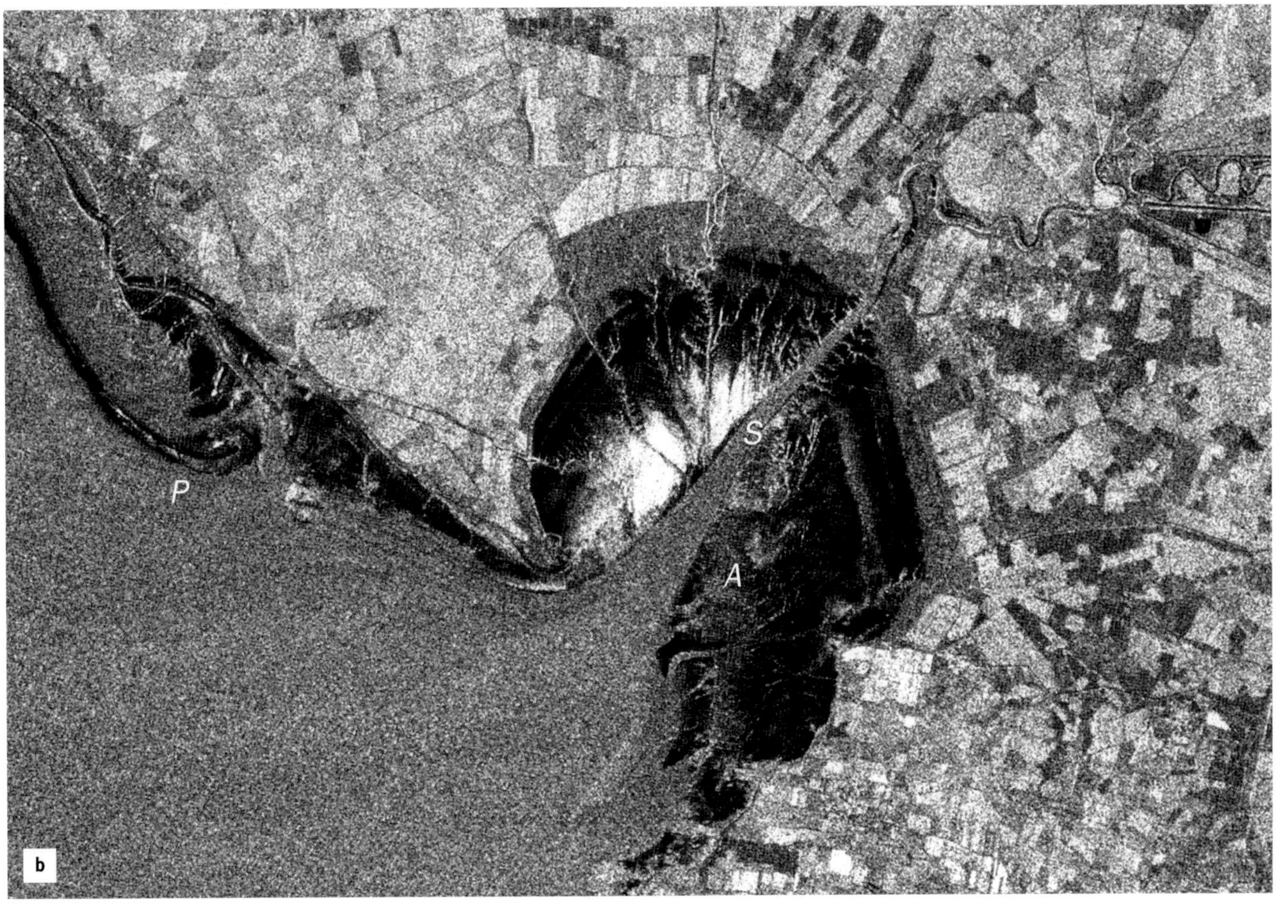
P
A
S
b

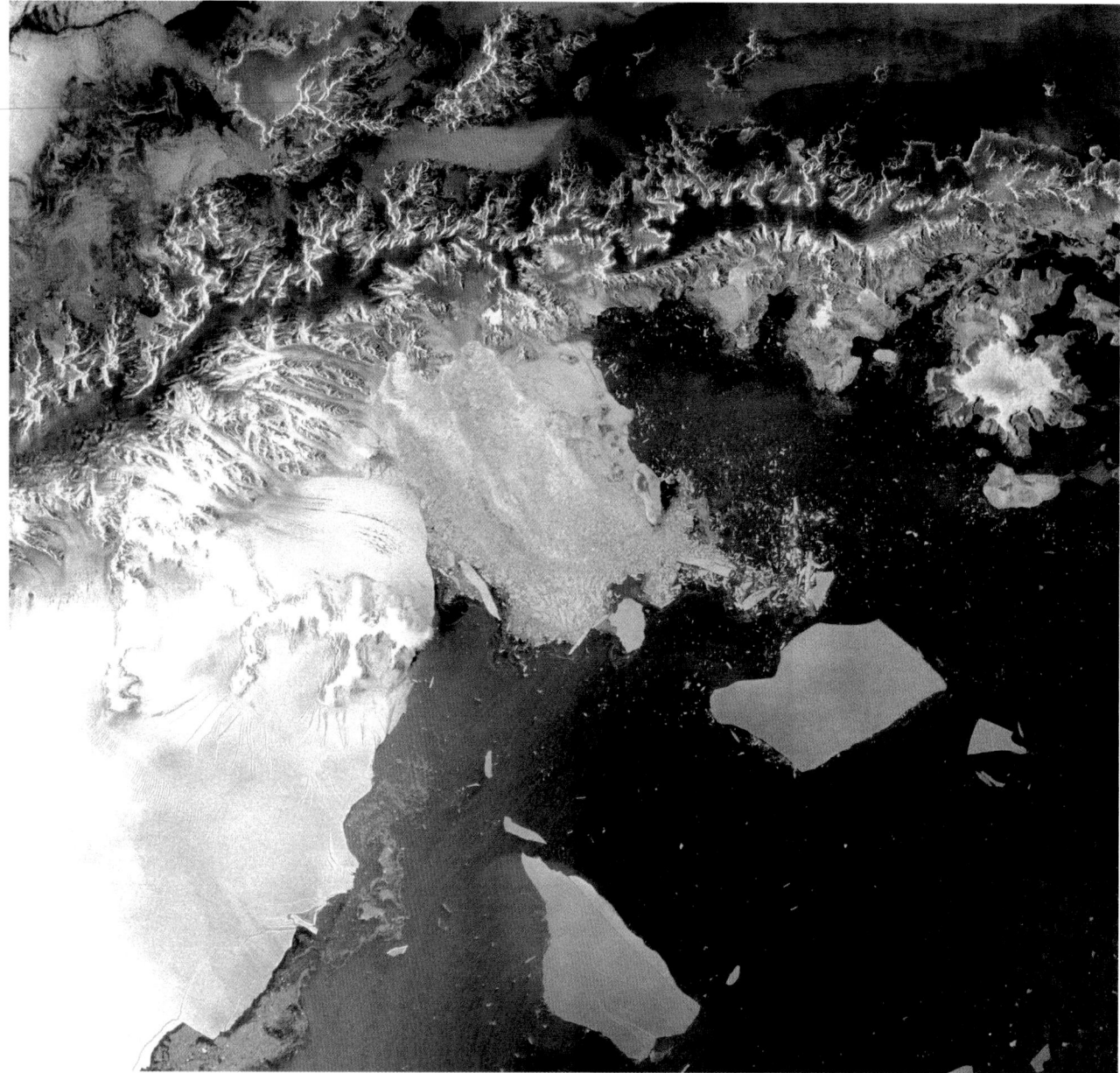

Figure 9.38. **Multispectral image acquired by the Envisat 1 ASAR sensor.**
This image depicts Weddel Sea and Antarctic Peninsula. © ESA, 2002.

and experimental nature of the programme – which in any case never got beyond this first satellite.

JERS 1 (Japanese Earth Resources Satellite), launched in February 1992, operated from a sun-synchronous orbit at an altitude of 568 km. This relatively low altitude implied a need for weekly orbit adjustments in order to keep to a regular 44-day cycle. JERS 1 operations terminated in autumn 1998.

JERS carried an Optical Sensor (OPS) collecting data in the visible and near infrared and a synthetic aperture radar (fig. 9.39). The OPS operates in seven spectral bands: B1 (0.52 μm–0.60 μm), B2 (0.63 μm–0.69 μm), B3 and 4 (0.76 μm–0.86 μm), B5 (1.60 μm–1.71 μm), B6 (2.01 μm–2.12 μm), B7 (2.13 μm–2.25 μm), B8 (2.27 μm–2.40 μm). The B3 sensor is nadir-viewing whereas B4 views

15.3° forward for stereoscopic imaging. Resolution is 18 m in all spectral bands. The scenes obtained cover a 75 km × 75 km area with a resolution of 24.2 m along the track and 18.3 m across the track.

When JERS came to the end of its operational life in autumn 1998, a little over a year after the premature end of the ADEOS 1 mission, Japan's Earth observation capability came temporarily to a halt. And since no high resolution sensor will figure among the instruments carried by ADEOS 2, due to be launched in 2002, the ALOS satellite (10 m SAR, 2.5 m PAN) billed for launch in 2004 becomes an even greater priority, at a time when Japan is giving increasingly serious thought to the prospect of acquiring a high resolution capability for civilian and military reconnaissance.

In this connection, the new political context created by the August 1998 firing of a North-Korean Taepo Dong missile, making regional monitoring a strategic priority for

Figure 9.39. **JERS 1 SAR image.** Acquired on 16 October 1992, the image shows part of the same region as that appearing in fig. 9.32. © NASDA 1992.

Japan, could well shift the programme into higher gear. It is likely however that the programme of surveillance satellites approved in 1999 will evolve more or less in parallel with the civil programmes, insofar as it is prompted by defence objectives which have little in common with the civil space considerations which have shaped the existing space capability. The special relationship between Japan and the United States in security matters – a relationship embodied in a specific defence agreement – may result in suitably adapted American systems being chosen in preference to national products.

Okean and Sich

The Okean O satellites manufactured by the Ukrainian firm NPO Yuzhnoye operate from a circular orbit at an altitude of 650 km. The first operational unit, Okean O 1, was launched in 1988. Designed for ice zone studies and oceanography, these satellites carry a side-looking radar operating in the X-band and a number of scanning radiometers. Their orbit allows them to revisit the same location twice a day. Once Okean data received by Meteor primary stations has been processed, it can be relayed via Ekran geostationary satellites to the ships concerned.

The first of this family of satellites to be built after the break-up of the USSR and Ukrainian accession to independence was Sich 1, launched in 1995. This craft was jointly owned by Russia and the Ukraine but subsequent output is likely to be produced separately for the two countries.

Figure 9.40. **First SAR Image asquired by Radarsat 1.** Courtesy of Canadian Space Agency.

Russia's Okean O satellite, launched on 17 July 1999, is fitted with additional sensors for Earth Observation and atmospheric studies.

Launch of the Ukrainian Sich 2 satellite is currently at the planning stage. It is expected to carry an SAR, unlike the Sich 3 follow-on unit funding for which is in any case still unsure.

Radarsat

The Radarsat programme has a long history and the initial choices were heavily influenced by Canada's physical location and characteristics. Canada's commitment to radar remote sensing is prompted by a number of geographical constraints: a vast and often inaccessible territory, concern to exploit natural resources, heavy cloud cover, priority assigned to ice studies. Radar offered the most effective means of meeting these needs and at the same time an opportunity to develop a special capability in a high-tech niche – a favourite with space sector managers around the world, concerned to find more or less exclusive openings for national products (fig. 9.40).

The objectives set for Radarsat – the first mention of which is in fact to be found in the context of Paxsat, a major treaty verification project dating back to the 1960s – were very ambitious and this no doubt is why the project fell so far behind schedule and went so far over budget. But it is nevertheless the only civil operational spaceborne radar system in existence. Confirmation of the dominant position of

the programme came with the conclusion of an agreement between the Canadian Space Agency and the firm MacDonald Dettwiler for construction of Radarsat-2.

Radarsat-1, launched in November 1995, operates in Sun-synchronous dawn–dusk orbit at an altitude of about 800 km (see fig. 2.29). It carries only one sensor, a synthetic aperture radar. This very advanced instrument operates at C-band wavelength with HH polarisation and provides scope for using a variety of incidence angles, from 20° to 50°, and, for experimental purposes, from 10° to 20° and from 50° to 60°. Swath widths vary too, as does viewing direction. Primarily designed to support study of the arctic regions, Radarsat-1 normally performs oblique viewing to the right – i.e. to the north – of its orbit path. But viewing direction was successfully changed to the left on two occasions to accommodate observation of the Antarctic – and indeed almost total coverage of this region was obtained (fig. 9.11). To do this, the satellite was twice rotated 180° around its vertical axis.

User enthusiasm for Radarsat data and the increased Canadian share in the imaging market has led to the development of the Radarsat-2 programme, although along different lines. In fact, this development officially makes the transition from mainly public to private financing. It has also led to reflection on Radarsat-3 which may correspond to a new approach based on cooperation whilst still moving towards predominantly private investment.

Launch of Radarsat-2, with a claimed 3 m resolution, is scheduled for 2003. This project, which benefits from an innovative funding arrangement whereby a private firm covers development costs while the Canadian government agrees to buy a volume of data fixed in advance, is driven by a combination of competitive and cooperative goals. An initial commitment by the United States to launch the satellite was however reassessed in view of the competitive threat Radarsat may pose to the new LightSAR satellite to be developed by NASA and a private company and of concern being expressed by the Department of Defense about the security implications of the resolution which the new Canadian craft will apparently offer. After some attempts to investigate a link-up with ESA with which it has a cooperation agreement, and having emphasised complementarity with Envisat, the European remote-sensing platform, Radarsat-2 will be launched commercially by the United States. Canada and the US have concluded a specific agreement concerning the distribution of data in order to secure their national interests.

The Canadian Space Agency's decision in February 2001 to carry out a mission feasibility study with MDA for the development of Radarsat-3 corresponds to yet another logic. Radarsat-3 would be launched soon after Radarsat-2 in order to benefit from combined use of the two satellites. The aim here is to enhance Canadian commercialisation capacities in partnership with the value-added sector, whilst at the same time contributing to a global understanding of the environment. Synergy with the European programme GMES, still under definition (see chapter 7), has also been envisaged.

Remote sensing – the way ahead

Over the last ten years the number of remote-sensing systems has grown rapidly and the range of facilities offered has become more diverse. This has been reflected in an expanding user community. At the present time two development trends can be distinguished. A first trend is to exploit the greater resolving power now offered by sensors to survey landforms on a large scale. Civil activity in this area is likely to surge forward with the advent of the very high resolution systems whose products have only recently come onto the market and which are now becoming serious rivals for aerial photography. A second development line is to exploit the global scope of this technology and its integrative potential – allowing vast surface areas to be characterised by single values. Progress here is supported by lengthening time series, a wider range of spectral bands and more sophisticated sensors capable of collecting data on different phenomena simultaneously. This opens up previously undreamed-of possibilities for global modelling.

Widening range of sensors

The earliest remote-sensing missions relied on optical sensors. Their subsequent history has been marked by continuity combined with gradual diversification. The United States and France are moving ahead with their Landsat and SPOT programmes, with a commitment to continuity of service, as expected by customers who have invested in dedicated processing systems and have in some cases even established their own data receiving stations. At the same time the range of suppliers for this type of product has begun to widen: India has opted for development of its own satellites and Japan has acquired an optical – in addition to its radar-imaging – capability. Cooperation has been another line of approach, one example being joint development by Brazil and China of the CBERS satellite. Finally, with small satellites such as Uosat now coming onto the market, the number of countries able to acquire their own satellite sensing capability is set to rise. The need to supply images in standard format is not however inconsistent with evolving sensor designs – which is why Landsat 6 (which failed to reach orbit), Landsat 7 and SPOT 4 and 5 differ from their predecessors while providing data continuity.

Meanwhile a number of civil remote-sensing missions were taking advantage of flight opportunities in the American and Russian manned spaceflight programmes, on board the Shuttle with the successful SIR flights and, until 1999, on the Mir space station with its various specialised modules. The Spektr module for example was equipped with a whole series of instruments for measuring the lower atmosphere and the Earth's surface, while the Priroda module featured a side aperture radar operating in the X- and L-bands together with a number of other very advanced sensors, such as the Ikar II radiometers. But while manned space missions provide scope for innovative experiments and creative observation and while Mir has produced a rich harvest of photographic images, they are ill-suited to systematic, continuous collection of images of the Earth.

The major shift in emphasis seems to have come in the 1990s, with the development of radar sensing and launch of the Almaz, Radarsat, ERS, and JERS satellites – despite the decision not to go ahead with an SAR for EOS. Radar sensing is valued not only for its ability to 'see through' cloud cover but also for the enormous wealth of images which can be obtained at varying incidence angles and with different polarisations. The American LightSAR project is built around an L-band radar offering multipolarisation and a range of resolutions, some down to 1 m to 3 m. Using interferometry techniques, it will be capable of detecting very slight changes on the Earth's surface.

Remote sensing is being pushed forward by new programmes too. Under the European Polar Orbit Earth Observation Mission (POEM), two satellites will be placed in orbit – METOP (see p. 243) and Envisat 1, launched in 2002. For budgetary reasons, the Envisat mission will use the polar platform solely for Earth observation purposes. Envisat will in some ways take over from ERS 1, carrying an Advanced Synthetic Aperture Radar (ASAR) and an advanced Along-Track Scanning Radiometer (ATSR) not unlike those which equipped the earlier craft.

There are also plans to use hyperspectral sensors, designed to break incoming signals down into a large number of spectral bands. The instrument payload on board the Lewis satellite included such a sensor – previously only tested from aircraft – but the mission failure prevented the experiment going ahead. Hyperspectral sensing features in a number of current satellite projects in the USA: NASA is planning to test the Hyperion sensor under the New Millennium programme and the Navy and Air Force are showing a keen interest in the Navy Earth Map Observer (NEMO) and Warfighter 1 projects. Australia too is working on a small satellite which will carry a sensor of this kind, the Australian Resource Information and Environment Satellite (ARIES).

This movement is increasingly accompanied by the development of other non-imaging Earth observation techniques. Radar altimetry, for example, is providing the basis for ever more sophisticated models – but they are useful primarily as a contribution to understanding phenomena at the scale of the planet. With the available images already covering the

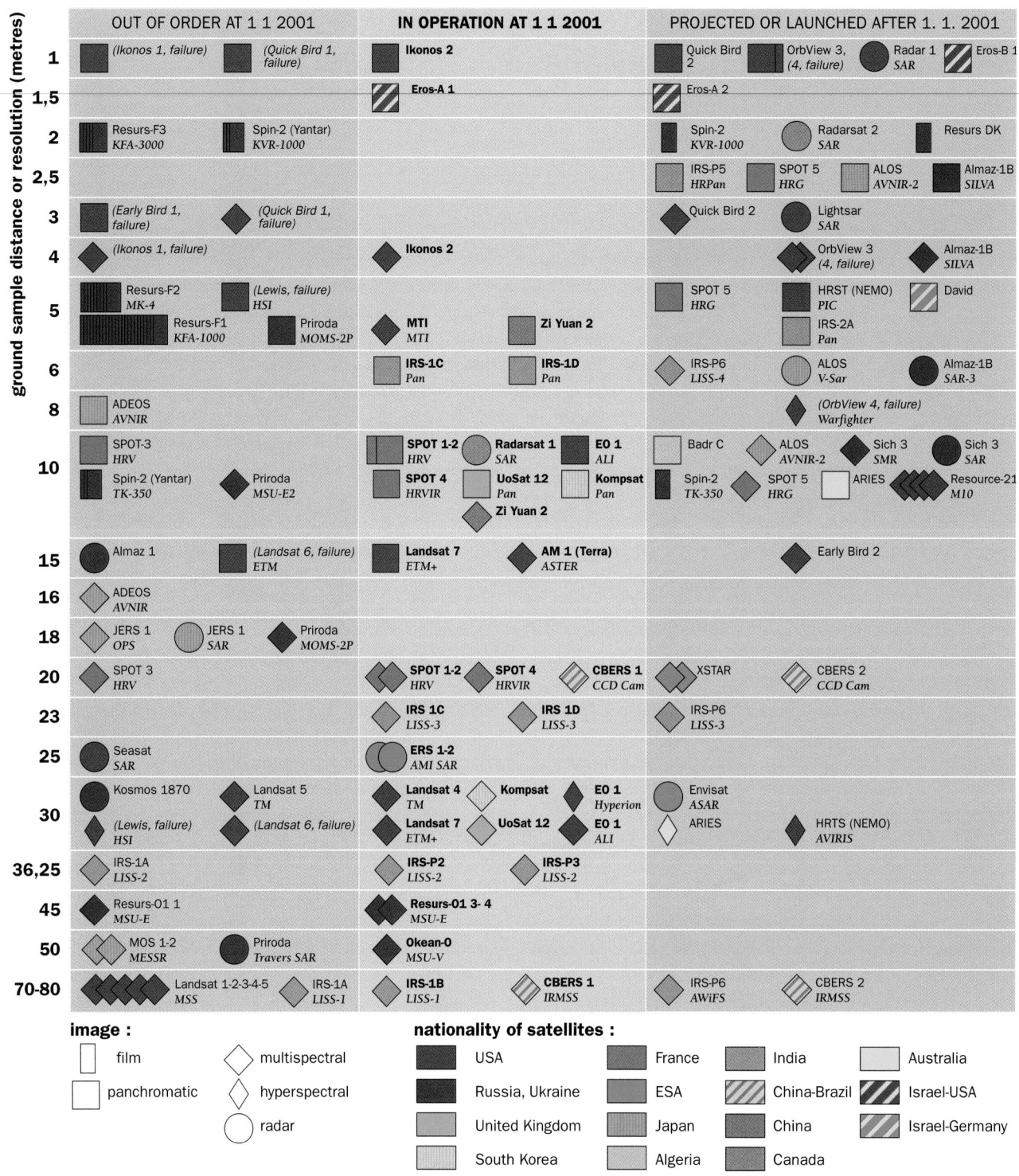

Figure 9.41. **Evolution of high resolution capabilities of sensors in civil Earth observation systems.** Only sensors offering resolutions equal to or less than 80 m are included. The name of the satellite appears on the upper line and that of the sensor, where known, on the lower line. The satellites are all Sun-synchronous, except for Quick Bird, Spin 2, Priroda (Mir module) and Resurs F.

Earth's surface many times over, market saturation would not be far off if demand were limited to cartography, currently the main user of remote-sensing data. But the market for remote-sensing imagery is expanding with the availability of higher resolution images and the emerging requirements for vegetation surveillance at specified dates determined by plant life-cycles or for the monitoring of natural disasters such as floods, fires and volcanic eruptions.

Increasingly high resolving power

With sensors offering increasingly high resolution and correspondingly improved spatial precision, morphological recognition of objects on the ground is now becoming possible. The

10 m resolution delivered by SPOT in panchromatic mode and the 5 m resolution available from India's IRS 1C satellite are symptomatic of this trend and the growing market for this type of image reflects the narrowing gap between satellite remote sensing and conventional photointerpretation. This development in turn facilitates the integration of remote-sensing data in geographical information systems used for management purposes in a great many contexts, such as forests and built-up areas. This general trend is being taken a stage further with the arrival on the scene of a number of civil satellites offering very high resolutions – of the order of 1 to 5 m – such as the commercial satellites from Space Imaging (Ikonos, see p. 263–264), Orbimage (Orbview 3) and Earthwatch (Quick Bird), the latter equipped with a sensor initially developed for the Clarke project (fig. 9.41).

The arrival in the civil domain of resolutions close to those available to states possessing a military space capability marks a clear break with the past, although it is still too early to say just what its impact will be. For while meeting demand from civil users for high resolution data may create a new market, the strategic implications of this type of product are worrying. The special legal regime applying to circumterrestrial space, under which states may not impose prior authorisation to overfly their territory, implies by definition a potential threat to national and perhaps even international security. In granting licences to private firms in pursuance of the Space Commercialization Act of 1994, the US Department of Commerce effectively reopened the debate on freedom of observation from space. It must however be borne in mind that data distribution is still to a large degree under American control. And the fact that the US Congress passed an amendment in 1996 outlawing the collection and distribution of images of Israel and other territories if the proposed resolution was higher than that available from other sources suggests that the principle of universal distribution may take a few knocks. It is worth remembering, finally, that information can sometimes help to defuse a crisis, as can be seen from the history of national military and indeed civil systems.

Integration in global modelling

In a parallel development, Earth observation has, since the late eighties, been increasingly taking on a new dimension based on the 'system' concept, with growing recognition of the need for integrated multidisciplinary study of the mechanisms which influence the global environment of the Earth. The Global Change Programme now underway responds precisely to that need. But while the satellite resource is essential to an undertaking of this kind, it is by no means sufficient given the complexity of the interactions which generate global change and the need for permanent global monitoring. This is why NASA is working on a new mission known as the Earth Science Enterprise, the basic idea of which is to bring the exploratory capacity of deep space systems to bear on planet Earth.

The global environment and its evolution are conditioned by a range of natural factors, including solar radiation and variations in its intensity, galactic energy flows and telluric factors such as terrestrial volcanic activity. In addition to natural variations in the biogeochemical cycle, the ecosystem is subject to increasingly invasive human intervention through industrial activity and a generalised onslaught on landscapes worldwide, marked by deforestation, impoverished biodiversity, changes in soil fertility and the productivity of the oceans.

The range of space resources contributing to an understanding of the environment is enormous but special mention should perhaps be made of NOAA satellite data, which – in conjunction with data from the DMSP's SSM/I microwave imager – form the basis for the work currently being done on three-dimensional, thermodynamic modelling of the atmosphere, as a tool for describing and explaining atmospheric energy transfers. Another source of particularly valuable data, this time on the causes of the 'ozone hole' first detected in 1985, is the UARS satellite launched in 1991. The ozone depletion phenomenon will also be studied by the Total Ozone Mapping Spectrometers (TOMS) carried by the Russian Meteor-35 satellite, having previously equipped Nimbus 7 and Japan's ADEOS 1.

Observation of the oceans is more difficult than that of the atmosphere and has reached a less advanced stage. It relies on thermal mapping of the ocean surface (NOAA and ADEOS 1, see fig. 9.42a) but also on measurements of seawater colour by specialised sensors offering fine spectral resolution, such as the CZCS instrument on the 1978-launched Nimbus 7. The number of such sensors in service has greatly increased since 1996, examples being OCTS on ADEOS, SeaWiFS on Seastar and OCM on Insat P4 (fig. 9.42b). Calibration for satellite altimetry can be effected by means of isolated tidal measurements in coastal zones (Seasat, ERS 1 and 2, TOPEX-Poseidon and Jason).

An understanding of sea ice and its possible melting is a key input to efforts to explain climate change. Progress in this area calls for an evaluation of ocean–atmosphere exchanges – sea ice forming a kind of screen between the two – and of the energy deficit resulting from the high albedo of sea ice.

Observation of continental landmasses is the third field of investigation to which planetary modelling is relevant. And indeed satellite observation is critical in the continental zones, where the impact of human activity is felt with particular intensity. The most widely used indicator for investigation of the biosphere is the Normalized Difference Vegetation Index (NDVI) – by normalising the difference between reflectance values in the near infrared and in the visible, it becomes possible to quantify such phenomena as desertification. This will make for an improved understanding of the various aspects of the carbon cycle and of exchanges between ocean, atmosphere and continental or marine biosphere. The NOAA satellites and the Vegetation instrument onboard SPOT 4 are making a particularly valuable contribution in this area (fig. 9.43).

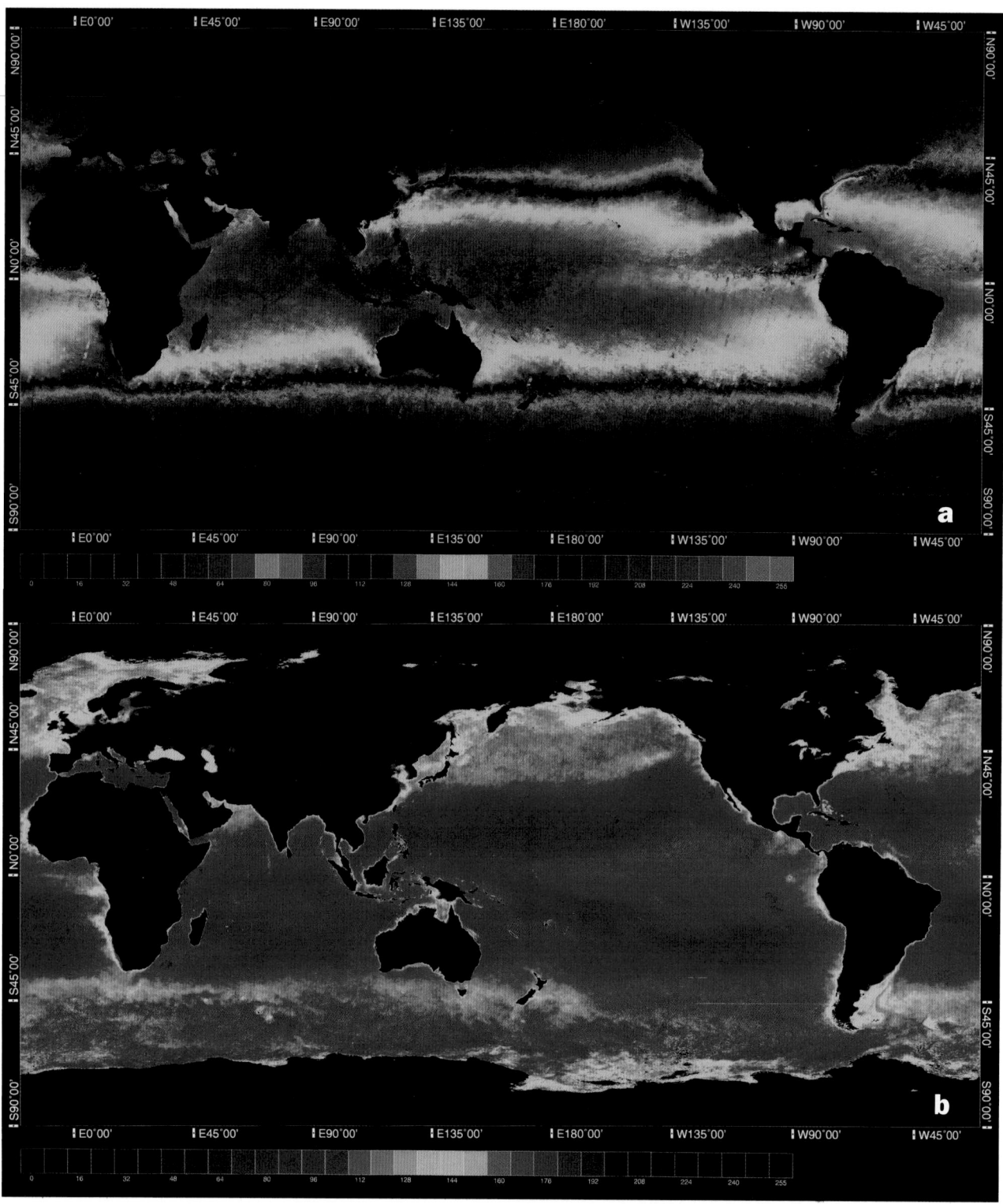

Figure 9.42. **Planispheres, highlighting maritime zones, produced from ADEOS 1 OCTS data for the period January–June 1997.**

a) Sea surface temperatures

b) Chlorophyll-a concentrations.

© NASDA 1997

A number of major international programmes have been launched to coordinate research into global phenomena. The World Climate Research Program (WCRP), which dates back to 1980, deals with the physical aspects of climatic systems while the International Geosphere Biosphere Program (IGBP), given the go-ahead in 1986, addresses the complementary issue of interaction between chemical and biological processes.

Following successive rounds of cuts in the initial budgets and against a background of widespread concern among scientists at the prospect of space activities pushing other science into second place, NASA finally opted for the Earth Science

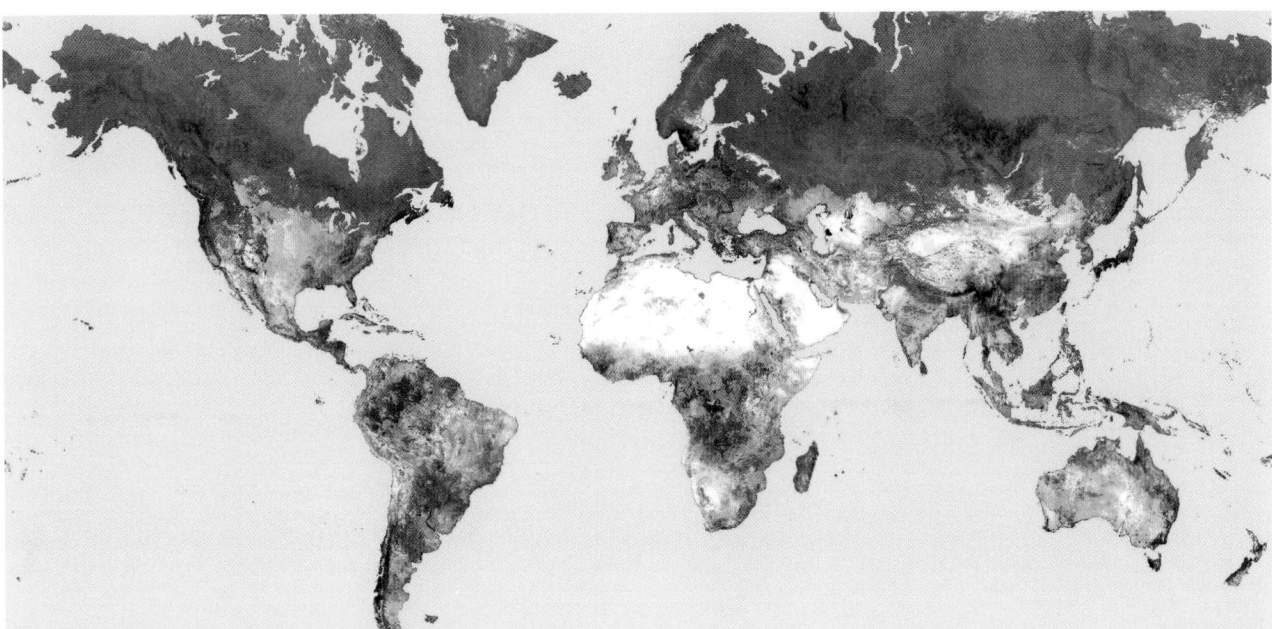

Figure 9.43. **Planisphere, highlighting the continental landmasses, produced from SPOT 4 Vegetation sensor.** False-colour composite produced from data collected from 1–10 March 1999. The three bands used are red (shown as blue), near infrared (shown as red) and medium infrared (shown in green). Snowcover in northern regions is represented by vast expanses in which purple predominates. © CNES 1999.

Enterprise Programme, which embodies a new approach giving greater priority to research and the development of new systems. Past, present and future remote-sensing systems are thus being revisited in a global perspective with the emphasis on environmental research. In this connection the existence of long series of satellite data – the older parts of which are often called heritage archives as opposed to the more recent commercial archives – is of particular importance. Conserving heritage archives implies the need to copy the oldest data, something which has not always been done, particularly by national agencies, for cost reasons. This is certainly to be regretted, bearing in mind the contribution such data can make to the study of changes in the Earth's land-scapes and such phenomena as desert encroachment and urban growth.

The Earth Science Enterprise is, to quote its initiators, 'dedicated to understanding the total Earth System and the effects of humans on the global environment'. The programme seeks primarily to analyse and model the many interactions between land, water, air and life on Earth. In pursuing this aim, NASA has established a wide range of cooperative links – with American partners such as NOAA and USGS, with international bodies like the International Geosphere-Biosphere Programme and the United Nations Environment Programme, or again with many foreign space agencies. The Earth Science Enterprise ties in with ESA's Living Planet Programme. NASA also acts directly through a number of its own programmes, referred to in chapter 7, and uses a number of satellites, such as Landsat 7 and GOES, described in the present chapter.

CHAPTER TEN

Telecommunications

- GEOGRAPHICAL CONSTRAINTS
- FREQUENCIES AND RESERVATIONS
- MISSIONS
- GEOGRAPHY OF SPACE TELECOMMUNICATIONS
- INTERNATIONAL SYSTEMS
- REGIONAL SYSTEMS
- NATIONAL SYSTEMS

Geographical constraints

Space telecommunications stand out amongst other areas of space activity. This can be explained by rapid growth in the sector, combined with the tremendous political and economic issues at stake. In a general context of deregulation, the existence of private investors and the development of commercial concerns in parallel with governmental programmes has introduced new ways of thinking which only go to emphasise this singularity.

By its very nature, a satellite is an ideal means for transmitting information over vast geographical expanses. Its vocation is therefore global, even if there is nothing to prevent its use on a regional or even local basis. Removing the need for expensive infrastructure at ground level, satellites would also appear particularly well suited to the needs of countries with unevenly distributed populations, although they do involve a relatively large basic investment for their construction and launch. A further significant advantage for satellite systems is their immunity to natural disaster, such as flooding, storm and earthquake which so often impair Earth-based networks. This is especially so now that ground control stations are smaller and less complex. From the beginning, satellites have constituted a pioneering element in the setting up of telecommunications networks, although rivalled today by terrestrial techniques like optical fibres when the volume of data reaches a certain level. In terms of television broadcasting, on the other hand, satellites remain the cornerstone, often in conjunction with cable networks.

The development of space-based telecommunications systems across the world is characterised by an ever-increasing number of participating countries, with a clear predominance of American satellites or satellites manufactured under American licence. The complexity of the technology involved and the need for heavy initial investment are the decisive factors that have led to this hegemony. The many governmental programmes for American military and civilian satellites have brought financing for research and development that has only strengthened the lead of US industrial firms. Moreover, large orders and consequent economies of scale resulting from high domestic demand and increasing foreign sales meant that production costs could be cut. Finally, by encouraging competition, deregulation at US initiative has gone even further towards strengthening the position of major private operators.

The change in charter of the first multinational organisations to set up intercontinental commercial telecommunications networks can be viewed in this context, the United States promoting open competition as a way of keeping the cost of services as low as possible. Insofar as technological lead is a determining factor, the appearance of new needs creating new markets such as mobile phone communications, in-company networks and multimedia, has even further reinforced an extremely favourable commercial and industrial context for American business. Whilst the attribution of frequencies and setting up of standards become ever more crucial issues, the dominant position of the United States on the international scene gives it an even further advantage, propitious for multinational industrial collaboration.

In fact, although telecommunications represent a special area of activity in the field of space applications in the sense that there is a genuine commercial aspect, satellites correspond to but a small part of telecommunications as a whole. Indeed, present developments in services are not particularly favourable to the space sector, given the strong competition from optical fibres. Some areas in which space systems were well placed, such as fixed point-to-point services, now offer fewer prospects. This implies a growth in point-to-multipoint links. Amongst new services, the satellite sector will undoubtedly benefit from the Internet boom, but with regard to growth in the mobile phone market, projects for constellations of non-geostationary satellites have not yet proved their profitability. Of course, satellites are not necessarily the tool best suited to fulfilling new needs. Interactive multimedia is an example, where short response delays are a crucial feature. However, the ability to guarantee global services in zones lacking any form of infrastructure is certainly a key asset. In this sense, there is a need today to identify new openings in the market so as to create further opportunities.

Independently of television broadcasting services which they comfortably dominate around the world, satellites must adapt to the new technological and statutory environment of telecommunications. Whereas they were originally confined largely to the developed countries, space systems are today finding new outlets in developing areas of the world and reorientating their transoceanic links towards a predominantly continental coverage.

Links and coverage

Telecommunications satellites provide a way of communicating between two remote locations, be they fixed or in motion, and this without regard for mountain relief or the curvature of the Earth. The only condition is that they must remain within the zone from which the transmitting satellite is visible (the footprint). The idea of using objects placed in space for transmitting data is not a new one. Even in 1945, Arthur C. Clarke had suggested that satellites at rest relative to the Earth's surface (geostationary satellites) could be used in this way. In 1951, the US army were already considering the possibility of passively relaying communications from the surface of the Moon. Hence, from the beginning of the space age, telecommunications featured amongst the first applications to be proposed.

There are two parts in such a system: the ground segment made up of emitting and receiving stations on the surface, and the space segment comprising the satellites themselves and their control stations. Links established between an Earth station and a satellite are referred to as uplinks when directed from Earth to satellite, and downlinks when directed oppositely. The SCORE satellites (Signal Communication by Orbiting Relay Experiment) launched in 1958 and Courier, sent into orbit in 1960, are sometimes considered as the first telecommunications satellites. However, these did not instantaneously relay the messages they received and so did not perfectly fulfill the definition. The former broadcast a previously recorded message and the latter, although it could record messages emitted from the ground, could only reemit them at a later date.

Echo 1 (1960) was the first genuine telecommunications satellite, even if it was a passive satellite, since it merely reflected the energy it received. The significantly reduced signal was a major drawback and this approach was soon abandoned. In contrast, an active satellite amplifies the signals it receives and then reemits them by means of transponders (or repeaters). The satellite Telstar (1962) was the first of this type. These satellites must have an on-board energy supply, generally provided by solar panels. Their payload includes directional antennas with differing degrees of directivity that must be pointed towards Earth to an accuracy of the order of one tenth of a degree. This requires a perfect control of satellite attitude. Once received, the uplinked signal must be transposed to another frequency band, amplified

several billion times and then reemitted. This sequence of operations is carried out by the transponder.

Earth stations are provided with antennas that grow in size as the strength of the signal to be received diminishes. In addition, they must be pointed with great accuracy towards the satellite, both for emission and for reception. This therefore involves satellite tracking procedures. The antennas can rotate about two perpendicular axes which may be orientated to cover azimuth and elevation, or they may lie parallel and perpendicular to the Earth's axis, or both horizontal. In the case of non-geostationary satellites, a great freedom of movement is needed. Even for geostationary satellites, some tracking is required since their positions change due to an unavoidable drift factor, as well as the imperfectly circular shape and slight inclination of their orbits.

The tight constraint imposed by this pointing accuracy is hard to reconcile with violent meteorological phenomena like the wind. At the beginning of the space age, the reflecting dish at Pleumeur-Bodou, in Britanny, France, which measured 29 m high and 54 m long, and weighed 340 t, was protected by a huge radome, 64 m across and itself weighing 27 t. Considerable technological progress has been a decisive factor as far as antennas are concerned, with the emphasis on size, efficiency and cost. The development of small terminals VSAT (Very Small Aperture Terminal), some only 60 cm across, has meant that firms wishing to set up in-company networks can now be equipped with a ground station integrated into a star network. Networking can then be arranged to connect the VSATs directly, without passing through a main station, so that direct links can be made between users via the satellite.

Carrier waves convey signals in frequency bands determined by use and availability. The so-called single access system which originally only allowed one link has been replaced

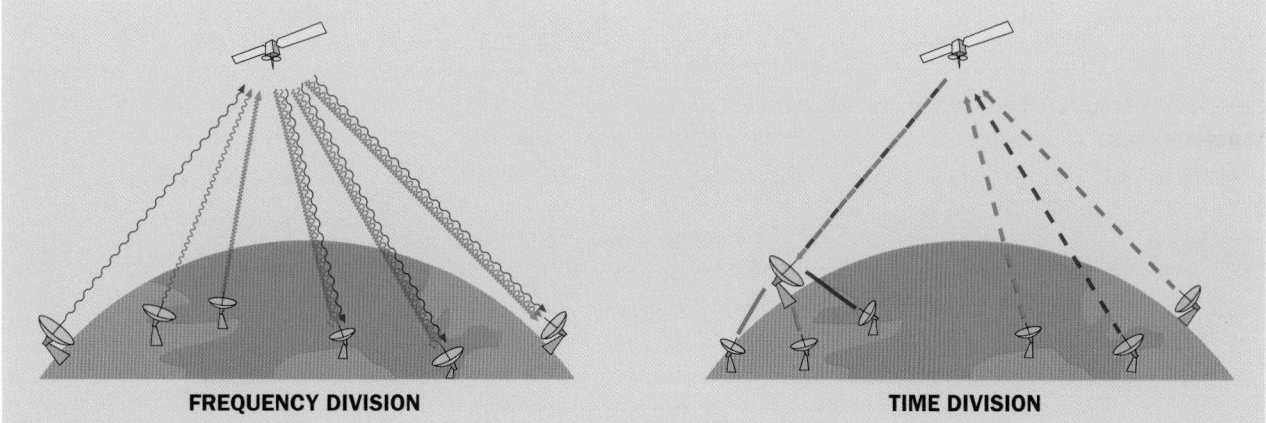

FREQUENCY DIVISION   **TIME DIVISION**

Figure. 10.1. **Types of multiple access.** In Frequency Division Multiple Access (FDMA), the first type of multiple access, each station emits at its own frequencies and the satellite reemits all received messages. These are then sorted by the receiving stations. In Time Division Multiple Access (TDMA), each station can use the same frequency bands, but it emits successive batches of information at predetermined times. Communications are then sorted either at the satellite or the receiving station. More recent modes exist, such as code division multiple access (CDMA), which are designed to be more robust in the face of jamming and interference. Before transmission, the user signal is multiplied by a random but reproducible digital sequence at a much higher frequency (of the order of 1000 times greater than the signal frequency). This increases the signal energy by spreading out its frequency spectrum. The user signal thus occupies several MHz of bandwidth, compared with a few kHz initially. The signal has absolute immunity to interference at levels significantly greater than the initially accepted level. The receiver can reproduce the same digital sequence, and hence decode the signal.

by the multiple access system. In the latter, which exists in various forms (fig. 10.1), satellites simultaneously process several messages that are then reconstituted.

Since radio waves propagate in straight lines, two points to be connected must both lie in direct view. Such a configuration depends on geometrical factors. The spherical portion of the Earth's surface from which the satellite is visible depends on the satellite's altitude (see fig. 9.1). Hence, low-Earth orbiting satellites have a much smaller footprint than geostationary satellites, which cover an area of around 42% of the Earth's surface centred at a point on the equator.

An equal portion of the Earth's surface, that is, 42%, but centred on a point 62° north (see fig. 3.22), is covered by Molniya-type satellites for a period of a few hours at their apogee. At this point they attain comparable altitudes but at a higher latitude. Their geographical location is thus better suited to countries at high latitudes and extending over a wide range of longitudes, such as Russia (fig. 10.2).

In all cases, as the link moves away from the satellite's nadir and closer to the Earth's limb, the distance separating site and satellite increases, whilst the so-called site angle between satellite direction and local horizontal decreases. Both these factors are liable to impair the quality of the link (fig. 10.3).

Other considerations must be added to these cosmographic conditions, this time of a topographic nature. Indeed, the geographic relief surrounding an Earth station can blot out the satellite, and all the more so as the site angle falls. Mountainous areas at high latitudes pose particular problems when selecting sites for communications with geostationary satellites.

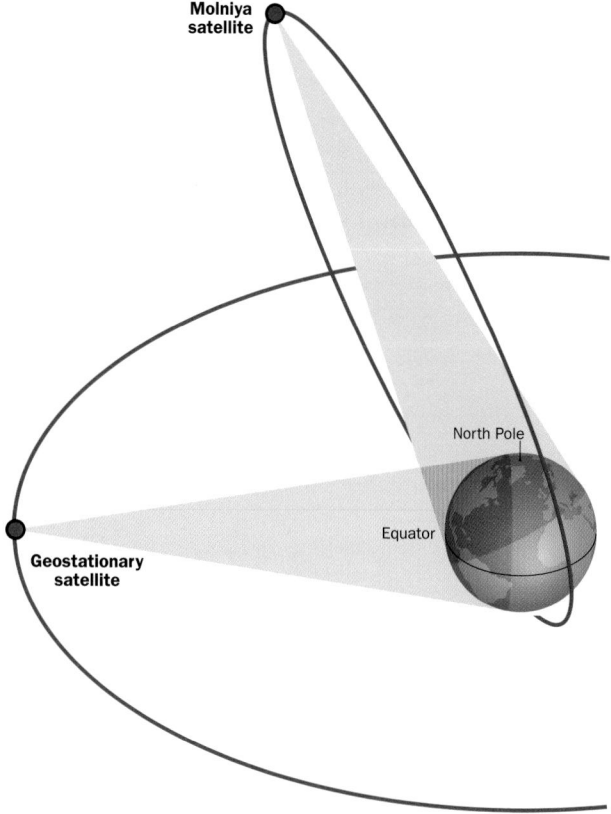

Figure 10.2. **Footprint of a Molniya satellite compared with that of a geostationary satellite.**

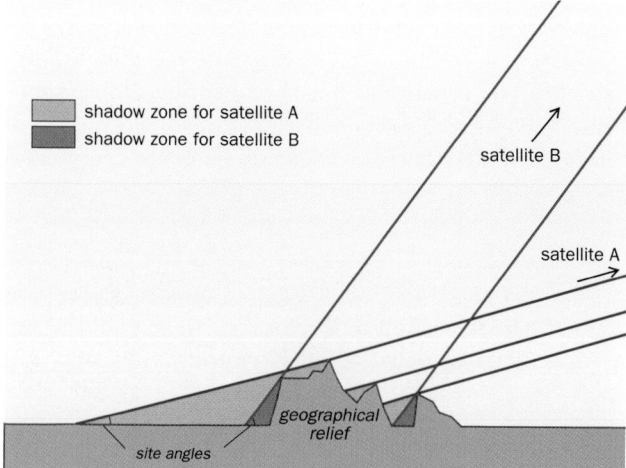

Figure 10.3. **Site angle and the effects of geographical relief.** Geographical relief is particularly restrictive in high latitude regions with mountains and deep valleys.

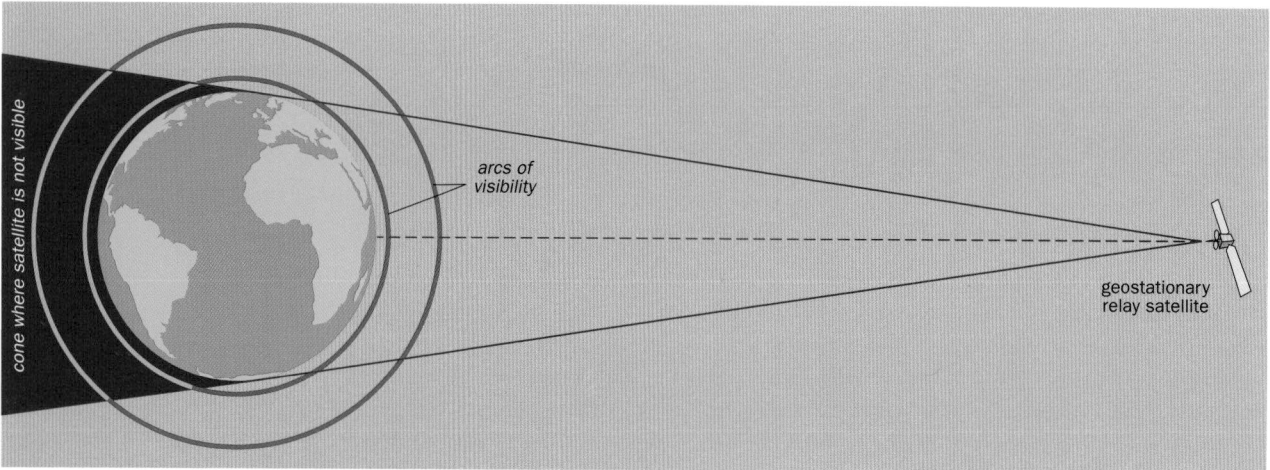

Figure 10. 4. **Intervisibility of a geostationary satellite and a satellite in low circular orbit.** The higher the orbit, the shorter the arc of visibility.

When the link is between a geostationary satellite and a low-orbiting satellite, the arc of the low orbit in which the LEO satellite is visible from the geostationary satellite greatly exceeds the 81° of arc of the visible Earth, and it increases with the altitude of the low orbit (fig. 10.4).

Direct view facilitates several types of link that are used to differing degrees in practice. Links between fixed points on Earth were the first to be exploited and are still the most common. Today, links with mobile phones are taking on increasing importance. Several projects for constellations of non-geostationary satellites are under construction to meet the demand from this new market, even though it would appear to be developing a little more slowly than originally expected.

Another type of link is that between a point on Earth and a low-Earth orbiting satellite operating via a geostationary relay satellite. These are known as Inter-Orbital Links (IOLs). Unfortunately, they are interrupted each time the satellite in low orbit goes behind the Earth and is hidden from the geostationary relay. With four satellites, of which one is a spare, NASA's TDRS system (Tracking Data Relay Satellite) provides permanent contact with the Space Shuttle. A commercial version had been envisaged for the purpose of tracking about twenty satellites in multiple access mode. However, the US Department of Defense (DoD) wished to remove all possibility of electronic jamming and the resulting technical difficulties and greatly increased costs proved discouraging.

Links between two satellites (Inter-Satellite Links or ISLs) can reduce the technical handicap of double hop for geostationary satellites and guarantee very long distance communications without the need for ground-based relays. The latter are not always available in time of crisis when located in foreign territory.

The region from which the satellite is directly visible is divided up into coverage areas by the directivity of its antennas. These are the zones in which signals emitted by the satellite can be efficiently received. The greater their directivity, the more restricted is the zone over which the energy is concentrated and hence, the smaller is the coverage. Although the maximal coverage of a geostationary satellite may include almost one third of the surface of the globe, and so is sometimes called global coverage, the level of the signal produced over this region is rather low and interference may cause problems. Semi-global coverage, also called hemispheric coverage, and regional coverage concern less extensive areas. At the other extreme, narrow or spot coverage requires highly directional emitting antennas (fig. 10.5). However, over limited regions, they may reach a high EIRP (Equivalent Isotropically Radiated Power). One satellite is equipped with several antennas that can simultaneously provide different coverages.

Apart from the distance to the axis of the emitted beam and the length of the path it has to travel, atmospheric effects and certain meteorological phenomena also tend to attenu-

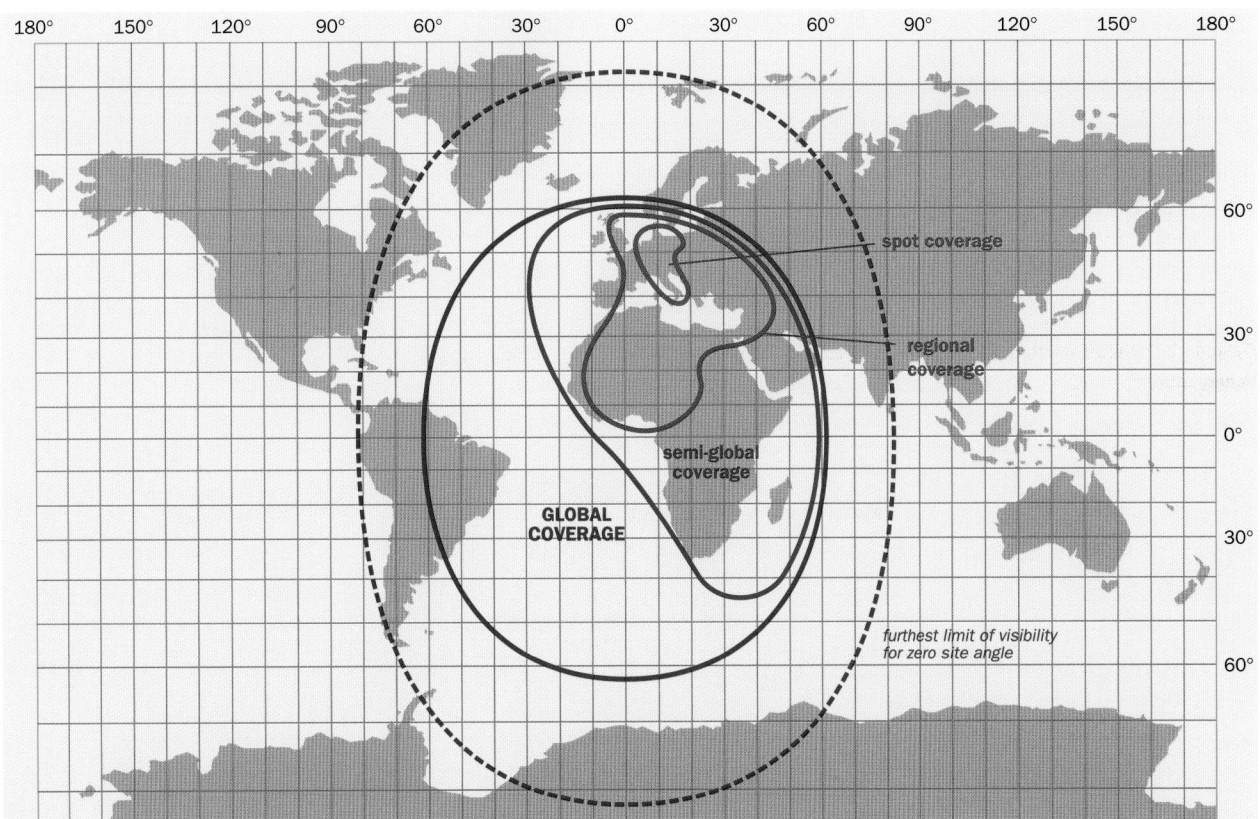

Figure 10.5. **Types of coverage by a geostationary telecommunications satellite located at 0° longitude.** The global coverage limit adopted here corresponds to a site angle of 20° beyond which links are commonly too attenu-ated to be exploited. Restricted forms of coverage – semi-global, regional and spot coverage – have the advantage of concentrating energy over smaller areas and hence providing a stronger signal.

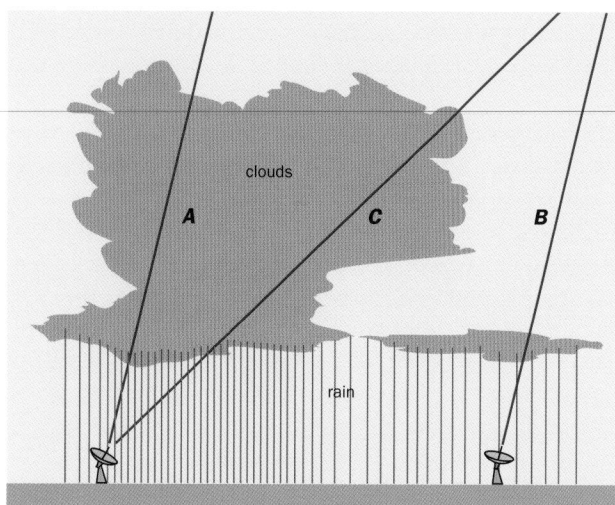

Figure 10.6. **Perturbation by rain.** Signal A is more seriously perturbed than signal B (due to cloud density and height) but less seriously perturbed than signal C (due to path length and angle with respect to rain direction).

ate the signal. High humidity, snow, ice, excessive temperatures and sand storms all lead to a depolarisation and weakening of the signal, with particular effect at high frequencies and for very low site angles. Rain causes the greatest perturbation. The density of rain droplets, their size, precipitation frequency, cloud height and turbulence all interfere with the propagation of waves. The elevation of the satellite can influence such perturbation by changing the length of the trajectory (fig. 10.6). It is generally considered that the site angle must be above 20° in order to guarantee a good quality link, although certain special conditions can considerably modify this value (fig. 10.7).

In addition to the above factors, satellites are affected by fluctuations in solar activity. The periodic return of increased activity (roughly every 11 years), referred to as the solar maximum, perturbs the telecommunications environment. Payloads must therefore be protected against changes in the level of electrostatic charge, upsets and increased radiation dosage.

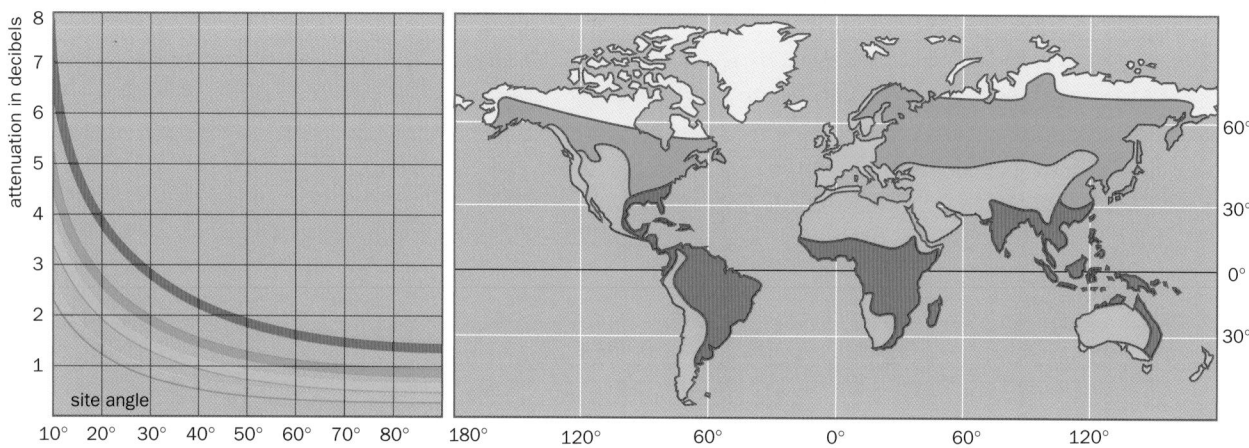

Figure 10.7. **Rain and signal attenuation.** The planisphere shows the geographical distribution of the likely rain attenuation, based on local rain conditions. Isopleths on the graph to the left represent the attenuation in decibels at different site angles for each of the four levels of precipitation (matched by colour). In the least favourable colour zone, site angles should be above 30°, insofar as this is possible, whereas angles of the order of 20° are sufficient for the other zones. It just happens that the least favourable zones are mainly situated in the equatorial regions where site angles for geostationary satellites are generally higher than elsewhere, thus reducing some of the difficulties caused by rainfall. These difficulties can be overcome by technical means mounted on board the satellite or on Earth, and also by judicious choice of frequency bands, avoiding those that are most sensitive to rain (courtesy of the International Telecommunication Union).

Frequencies and reservations

Telecommunications satellites make wide use of two limited natural resources: the geostationary satellite orbit and radio frequency bands. Countries are thus actively concerned about safeguarding their immediate and future use. In fact, the need to regulate telecommunications arose well before the space age. The International Telecommunication Union (ITU) is the oldest intergovernmental organisation, responsible for coordinating the activities of the various users since 1865. It became a specialised institution of the United Nations in 1947, and can claim almost universal membership. It has since played a central role in the field of international telecommunications, including those based in space, with the aim of ensuring 'a rational, equitable, efficient and economical use of the radiofrequency spectrum' whilst at the same time affirming in the opening words of its constitution that it fully recognises the sovereign right of each member state to regulate its own telecommunications.

The basis for the ITU's activities is compromise between member states. The first regulatory step was international allocation of the different frequency bands to services organised into categories. This was adopted by the International Radiotelegraph Conference in Berlin in 1906. Member states retain the authority to assign frequencies on a national level for their radio stations, providing they remain within the technical framework so defined. The obligation to report these national allocations was the second regulatory measure. The 1927 International Radiotelegraph Conference in Washington produced the first complete allocation table of frequency bands that could be used for the various services. Whilst not obligatory, this was to serve as a guide when member states allocated frequencies. In this way, without explicitly stating a right of precedence, the conference implicitly conferred a privileged status on the first stations to inform the ITU that they were using a given frequency band, according to a principle of first come, first served. This flexible process of attribution after the event is still the basis for frequency allocations. The only case of prior planning in space telecommunications was a legislative scheme treating the question of direct broadcasting, adopted in 1977 and amended in 2000.

As a technical organisation taking on the role of an administrative international public service managing telecommunications, the ITU has been faced with a novel political context since the beginning of the 1990s, brought about by deregulation combined with rapid growth in the information society and its economy. A new feature on the technical level is the convergence of technology in telecommunications, computing and audiovisual media, whilst on the industrial level, the arrival of private operators and globalisation of companies has transformed the landscape. At a Plenipotentiary Conference in Geneva in 1992, the ITU therefore carried out a fundamental reform of its own organisation, providing itself with a definitive constitution and an amendable convention, with a view to increasing its own durability. Care was nevertheless taken not to increase the powers of the ITU with respect to its member states.

ITU activities

The ITU has two main areas of activity concerning frequency allocations and orbital positions in the geostationary orbit that can be associated with them. One is their series of World Radio Conferences (WRC). The other is the Radio Regulation Board which implements frequency assignment policy and maintains an international register of frequencies.

The World Radio Conferences have almost legislative power and take place within the framework of international diplomacy. They allocate frequencies to services and not to countries, as the latter retain the fundamental right to assign frequencies on a national level. Meetings generally involve about a thousand delegates, initially representing state-run telecommunications bodies as well as regional and international organisations like the World Meteorological Organisation or Intelsat, these having observer status. The ever-increasing number of private organisations now integrated into national delegations, particularly from deregulated countries, is a good indicator of the impact that commercial interests are having on the WRC. WRCs initiate revisions to the Radio Regulations (RR), which have the weight of an international convention. They contain the full set of technical, operational and regulatory conditions applied to the spectrum and satellite orbits, together with the allocation table specifying which parts of the spectrum can be used by the various services. These allocations are made either on an exclusive or shared basis and can either be put into effect worldwide or limited to a region (fig. 10.8). Conformity with the table published in the RR is an obligation for frequency allocations made by member states if they wish to obtain international protection.

At the Geneva WRC in 1997, the considerable commercial stakes involved in band allocations, and the fact that those bands already exploited were reaching saturation, led member states into difficult negotiations. They also set up different technical solutions, such as opening new, higher frequency bands, and sharing the same frequency bands between geostationary systems and new systems for personal satellite communications using non-geostationary orbits

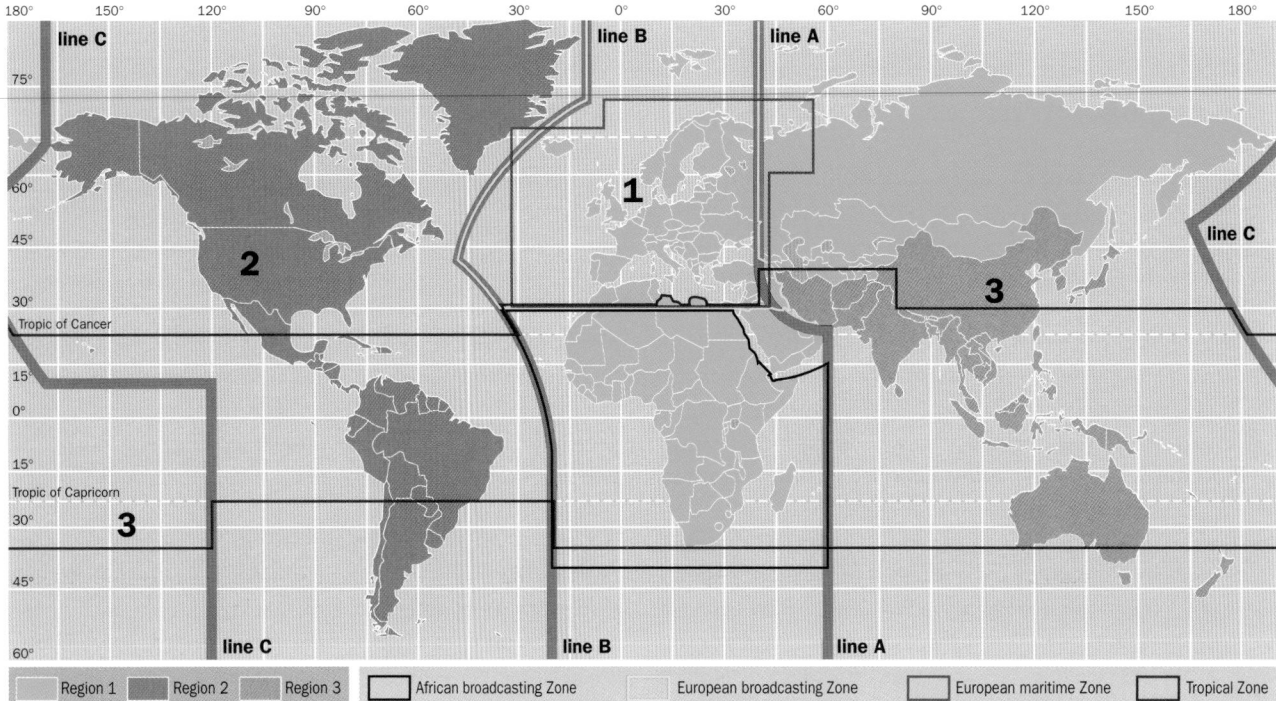

Figure 10.8. **Regions and Zones defined by the International Telecommunication Union.** In order to allocate frequencies, the Radio Regulations divide the world into three ITU Regions by means of the lines A, B and C. Some territories are removed or added according to boundaries also stated in the RR. Moreover, ITU Zones are defined by the RR for specific applications which may bring together countries belonging to different Regions.

(Global Mobile Personal Communications Satellite GMPCS). State-run national telecommunications bodies, the only ones holding the right to vote, play the role of regulator when private operators seek to exploit new services. At the same time, they tend to favour companies falling within their own jurisdiction, a situation which has greatly benefited US operators with their significantly more advanced projects.

The WRC 2000 took place in Istanbul, attended by 2037 delegates from 150 countries, and had four main items on the agenda. It began by confirming the decisions taken in 1997 concerning the rules for new non-geostationary systems (non-GSO systems) to coexist with geostationary systems without undue interference, by sharing the Ku band (10–18 GHz). It also reached an agreement on the granting of frequency bands for third generation international mobile telecommunications (IMT 2000). To wind up, it determined the further frequency assignments required to develop Galileo, a future European satellite radionavigation system.

In parallel, in the context of plans for direct broadcasting, an amendment to the 1977 project was decided. It aimed to widen the user base and at the same time soothe concerns in developing countries, faced with a shortage of frequencies and an increasing number of demands. Each country in Europe and in Africa would now have at its disposal an orbital position associated with ten analogue channels, whilst each country in Asia would have access to an orbital position associated with twelve analogue channels.

At the regulatory stage of coordination and notification affecting allocations made by member states to their space-based stations, the Radio Regulation Board (RRB) is an independent international body which prepares registering procedures. The 1992 reform, applied in 1994, had tended to limit the powers of the RRB relative to its predecessor, the International Frequency Registration Board (IFRB), which had permanent members at its disposal. Notwithstanding, although the RRB's conclusions involve no formal obligation to member states during coordinations and notifications, the RRB carries international weight when it comes to fixing rules for these procedures and, if necessary, helping member countries to find mutually acceptable arrangements. The complex regulatory procedure includes four stages: publication, coordination, notification and assignation. Frequencies are then recorded in the international frequency register. This approach is designed to promote compromise and agreement, rather than simply to attribute operating rights.

However, with increasing requests for fictional satellites, known as paper satellites, the process has proved itself over-complex and inadequate to the task. The paper satellite practice is motivated by a desire to reserve orbital positions and frequencies for future use, even if the states involved have no genuine project for them within the time limits they propose.

Developing countries were the first to apply this method, fearing that all the useful positions on the geostationary orbit would be occupied by the industrialised nations long before they could acquire the necessary technological knowhow.

Requests to reserve frequencies and orbital slots then snowballed as countries quickly recognised the growing economic value of these resources in the new deregulated environment of telecommunications. In 1992, the government of the islands of Tonga thus notified the IFRB of its request for 31 orbital positions, a number quite out of proportion with its national telecommunications needs. It eventually gave in to pressure from the IFRB and accepted six allocations, but immediately began trying to sell off the rights to use these frequencies and orbital positions to other countries, thus raising doubts about what constitutes a sincere interpretation of the ITU texts.

The developed countries, including the United States, Russia and some European countries, also used this approach. For example, they created a situation in which the Ka spectral band devoted to multimedia services was saturated even before satellite construction had actually begun.

Procedures for coordinating fictional satellites have thus tended to spawn in an unproductive way over the past few years. This has delayed approvals for real satellite networks and blocked the board's work with a huge influx of requests. The 1997 WRC, followed by the 1998 Plenipotentiary Conference in Minneapolis, stressed the seriousness of the problem but no genuinely restrictive procedure was acceptable to the member states. Only partial measures were adopted, aimed at limiting deadlines and introducing payments such as a filing fee for processing requests. The most significant was a procedure of administrative due diligence requiring operators to make a regular progress report on their projects and to disclose basic information concerning the network, construction of the spacecraft and the launch vehicle they intended to use. Such a measure is designed to assess the reality of these projects.

Complementing these classic resource management activities, the ITU has also had to evolve in consequence to the increasing presence of the private sector. Representatives from private industry take part in the activities of the committees preparing conferences. But the increasing integration of new private players into the ranks of the longer-standing members became official in 1996 with the creation of a worldwide forum on telecommunications policy. This is a flexible structure with no decision-making powers, responsible for strategic orientations. Flexible regulations, or soft law, devised with the assistance of private operators and which currently govern global satellite constellations, emerged from

this forum. The second forum, in March 1998, was concerned with the trade aspects of telecommunications and consequences of the World Trade Organisation (WTO) agreement on space services.

The next step was to write the role of the private sector as active participants in ITU missions into the ITU constitution and convention. In March 2000, the Reform Advisory Panel (created in 1998) confirmed the unique role of the ITU as the only organisation offering a worldwide forum for all parties: spectrum regulatory authorities, government ministries, users, manufacturers and suppliers of services. Its aim is to support the actions of the ITU as a global organisation harmonising international radiocommunications activities. The RAP recommended the introduction of a genuine partnership between private and public sectors by extending measures already taken to integrate private operators. This reform of the ITU seemed all the more necessary to its advocates in that the organisation looked likely to find itself in competition with other international players.

In this context, the predominant role played by the United States in creating an information society through the Global Information Infrastructure project (GII), proposed in 1994, marked a turning point in the liberalisation of telecommunications and hence also of space-based services. At the G7 conference in Brussels in 1995, the adoption of guidelines aimed at fixing a competitive framework for activities was witness to the high degree of motivation amongst industrialised countries. In 1996, a conference extended to include developing countries stressed the liberal character of the structure which was to frame the new rise in space telecommunications.

For its part, on 15 February 1997, following three years of difficult negotiations, the WTO reached an agreement on basic telecommunications. A genuine world telecommunications market opened with the commitment of 69 governments representing more than 90% of the world's telecommunications networks. This movement was confirmed in 1999 when China agreed to take part in the agreement. In this context, the US Federal Communications Commission (FCC) adopted a new law opening up the domestic telecommunications market to foreign operators of non-US satellites, whilst still requiring them to conform to the broader criteria of national interest and national security.

Telecommunications satellites provide a varied range of services and most systems are multipurpose in this respect (fig. 10.9, and see also fig. 10.14). Traditionally there have always been two markets existing side-by-side that have led to the development of space systems: television broadcasting, and fixed and mobile telecommunications. Although the development of new services like Internet and multimedia carry new possibilities for growth, satellites remain only a marginal feature, given the competition from terrestrial means and from fibre optics in particular. The latter can now claim considerably improved lifetime and capacity, compared with the state of the art in the 1990s. Finally, navigation and positioning activities which lie at the fringe of telecommunications, and for which satellites are particularly well suited, work towards different ends, even if certain new projects envisage a degree of convergence between positioning and telecommunications satellite activities for third generation mobile systems (see chapter 11). In parallel with this, current evolution in the space telecommunications sector, also affected by deregulation, has been marked by the growing number of private operators, who are mainly American but also include increasing links with international finance.

In the beginning, satellites were conceived of as a sideline, used particularly for intercontinental links but often accompanied by ground relays for retransmission over short and medium distances. Gradually, however, they developed their own regular service of high speed and long distance links, both for television broadcasting and point-to-point telecommunications. Alongside these international systems, first national and then regional systems began to appear, as space telecommunications systems became more common and involved a growing number of countries (see fig. 10.10).

At the same time, a great deal of progress was achieved in the area of fibre optics and this strengthened the position of terrestrial cable services. For certain applications, satellites were at a disadvantage owing to transmission delays inherent in their great distance from the surface. This is an even greater handicap when a double hop is required, that is, when the signal travels the return journey from the surface to Clarke's orbit twice. However, continued growth in world demand in the telecommunications sector means that the two approaches can coexist, all the more so in that space systems hold only a tiny part of the market share, of the order of 2%.

Satellites remain the most flexible means for providing occasional links between all points around the globe with a minimum of terrestrial facilities. For certain groups of users, possessing only a minimum of lightweight equipment, this is an appreciable advantage. In contrast, over short distances and for high volume links, the use of satellites can only be justified if the terrestrial network is inadequate.

Satellites have contributed to the downfall of monopoly in the field of telecommunications. Whereas terrestrial systems tended to favour the existence of single systems, thereby facilitating state control, the increasing capacity of satellites, combined with the development of small ground stations known as Very Small Aperture Terminals (VSAT), has allowed private users to bypass traditional communications channels.

The non-localised nature of space activities has its answer in the notion of *teleports*. These group together a variety of means of communication in a single site, including direct access to satellite links. In theory, the location of such a teleport should modify the geographic distribution of economic activities and directly influence national and regional development. Even though the phenomenon has not yet lived up to expectations, it remains true to say that these groupings clustered around telecommunications infrastructures illustrate the strategic position occupied by information. Because of their heavy dependence on telecommunications networks, multinational companies represent a potentially promising market, and a growing number of countries are keen to benefit from the resulting profits. This phenomenon is particularly striking in Asia, where recently industrialised countries such as Malaysia, Taiwan, South Korea and Singapore are liberalising their telecommunications services in order to appear more commercially attractive and speed up the modernisation process. National and regional space systems currently under development in the region are contributing to this trend (see figs. 10.24 and 10.27).

It is sometimes difficult to identify exactly what services are provided, insofar as transponders can be hired and used to supply various needs (see fig. 10.14). Diversification is becoming more and more the order of the day. Naturally, certain satellites would appear to be more specialised, particularly in broadcasting, but on closer examination, this is often just a symptom of the way tasks have been divided up within a global system controlled by the same operator.

The assignment of services to different frequency bands (fig. 10.9) clearly reveals the commercial interests at stake as far as frequency allocations are concerned. This is demonstrated by the preponderance of services in the C and Ku bands. Indeed, fierce competition and a high level of technology combine in such a way that Western manufacturers have preferred to develop Ku band transponders.

Considering the different services available on the market, leaving aside governmental programmes, television represents more than three quarters of the total activity. Following a considerable development of English-speaking television channels on the planetary level, television broadcasting has become more concerned with regional and even local channels. In the case of cultural minorities abroad, this is a way of maintaining contact with the mother culture. Interaction between cable and satellite is such that the latter is now ready to undertake any kind of televisual distribution.

Standard fixed telecommunications, although still present, do not offer particularly attractive prospects. The fibre optics sector now benefits from greater investment than satellite and has been significantly affected by deregulation. Satellite services must therefore diversify by offering point-to-multipoint links. The main advantage of satellite is to be able to develop different types of space application in the form of systems. The

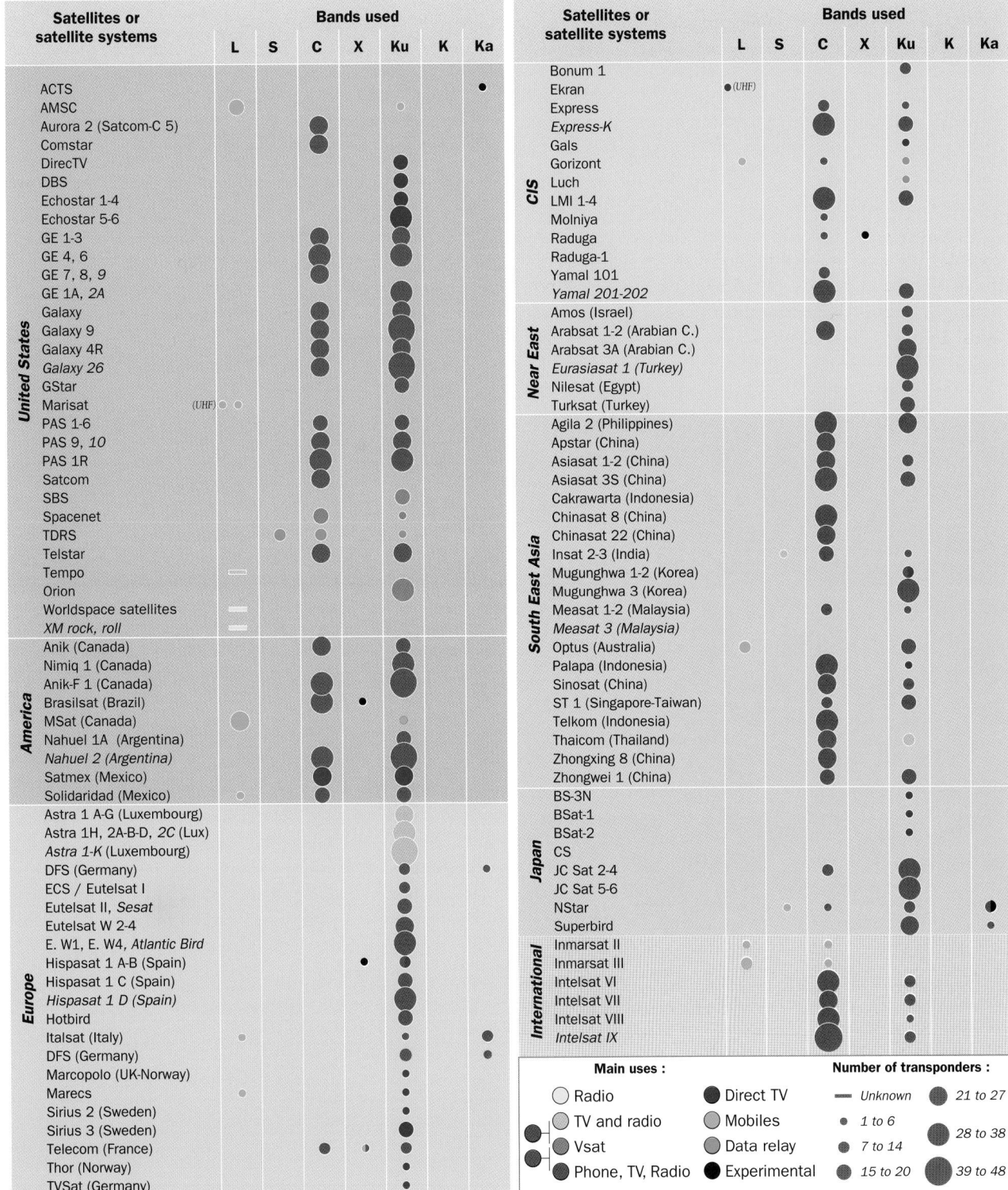

Figure 10.9. **Space services and frequency allocations in 2001.** Satellites of the same family are represented by one sample. New generation satellites are indicated separately. New Skies satellites are given their original Intelsat names. Satellites planned at 1/1/2001 are in italic.

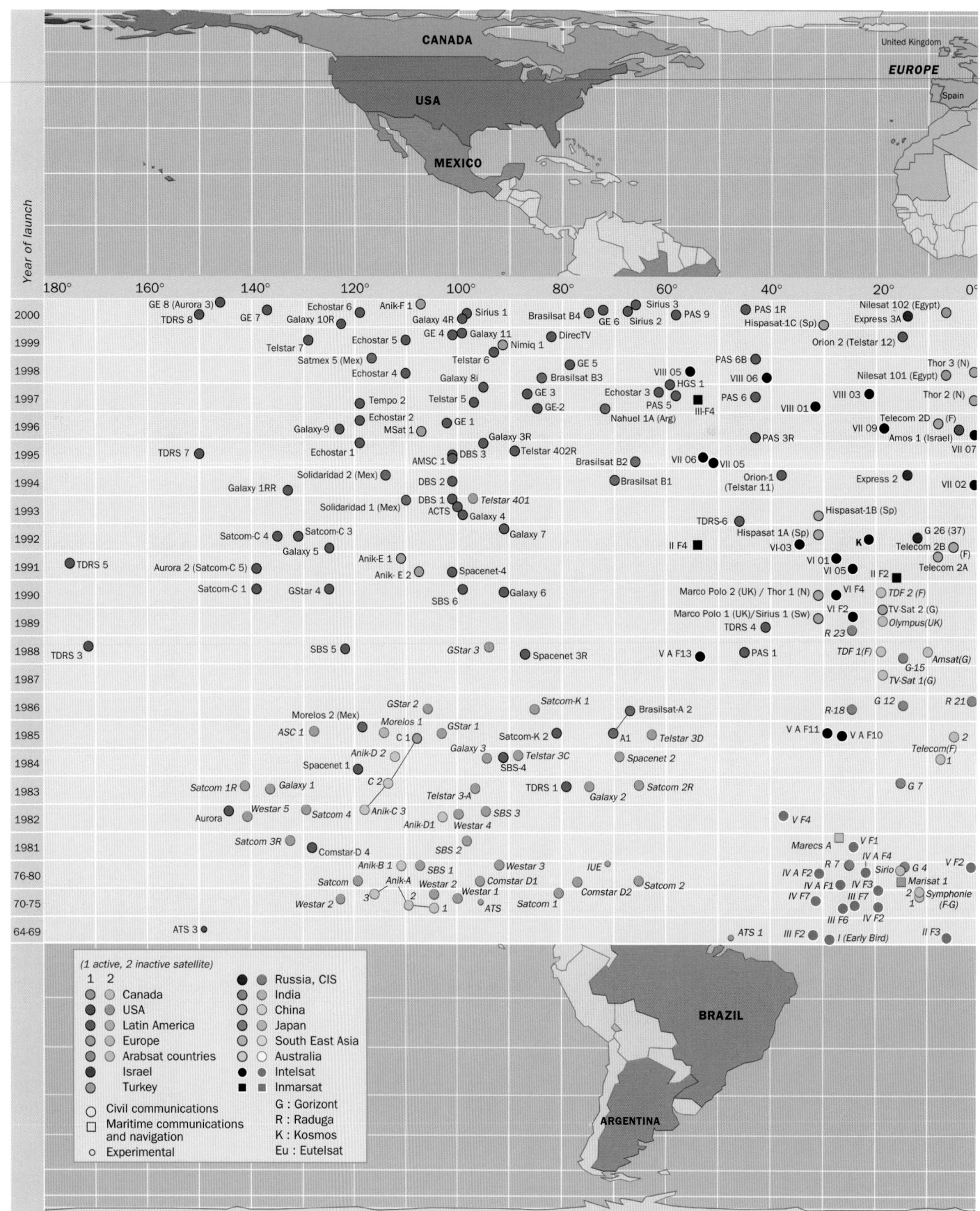

Figure 10.10. **Chronology and initial positions of civil geostationary telecommunications satellites.** Positions on orbits are where satellites were first stationed, although they may since have moved from those positions. Satellites represented by pale colours, with names written in italics, were no

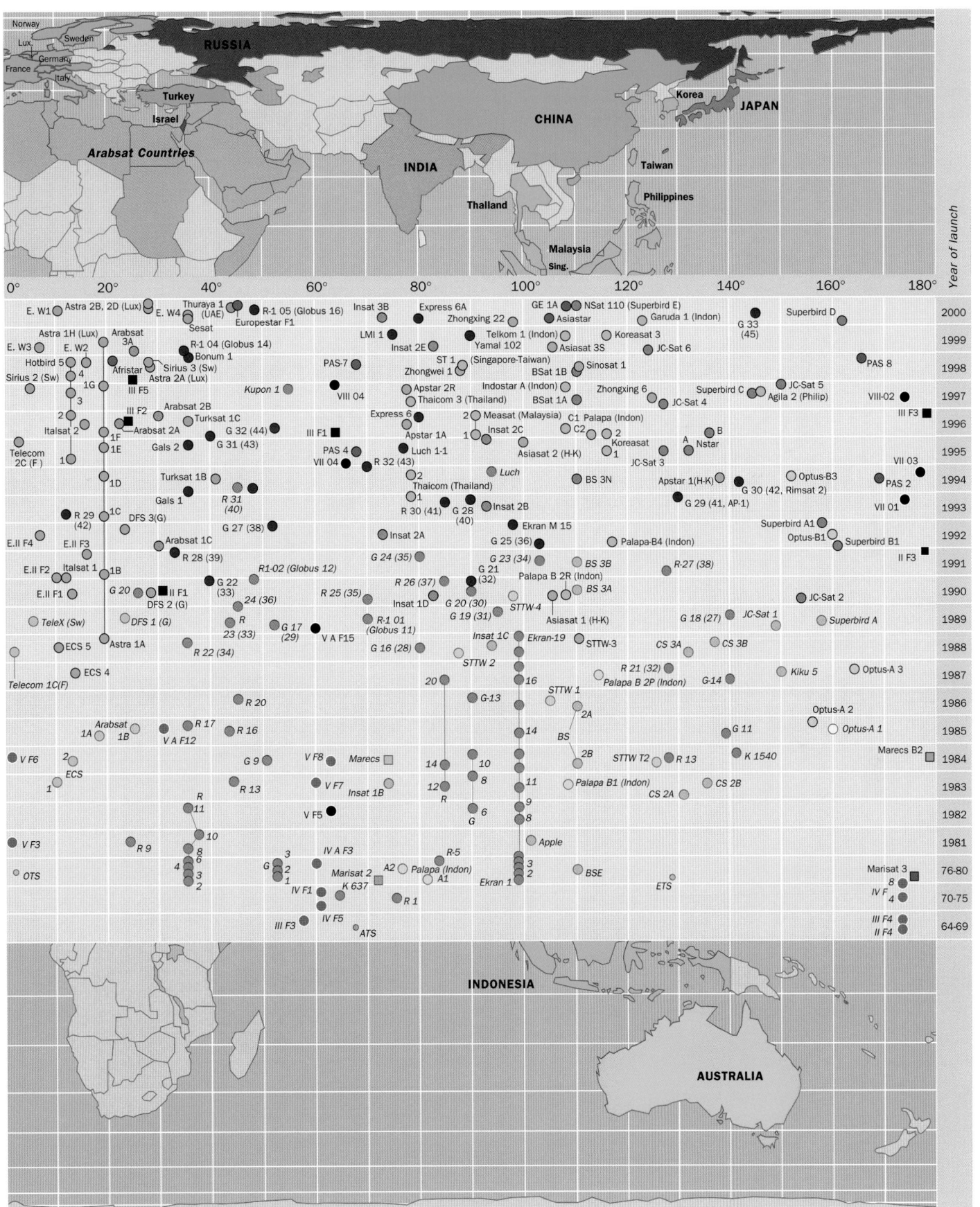

Year of launch

longer in activity at the beginning of the year 2001. Gorizont satellites are
given their traditional numbering.

satellite can thus be envisaged, not just as a simple relay, but as a supplier of information, concerning the weather, positioning, risk management, and so on. Despite this, the global coverage provided by most satellites involved in this type of activity is not continuous and this means that a certain number of technical difficulties remain to be overcome.

Mobile telecommunications certainly offer new opportunities, with an annual growth of 25%. The market share held by satellites in this area, which has already seen drastic deregulation, should be of the same order of magnitude as held across the whole telecommunications market, that is, a few percent. In such conditions, only a small number of systems would appear to be viable, all the more so in that the investment required is enormous. Certain geostationary systems are already capable of providing such services and the constellations which should be able to provide universal coverage find themselves confronted with a lack of real markets in precisely those regions with the poorest terrestrial means of communications, for which satellite is supposed to represent a genuine interest. In addition to this, the deregulatory nature of satellite, which is in theory free of any constraint laid down by local operators, does not help in attracting the latter. The constraint imposed by gateways and the desire to interact with the terrestrial network in programmes like Globalstar (see fig. 10.20) tend to compensate for these handicaps, although rather to the detriment of satellite's specific advantages.

Apart from encouraging prospects associated with the development of Internet, where satellites will certainly find opportunities, the main unknown is undoubtedly the multimedia market. At the present time, interactive applications take place rather on a local scale. Although technically well suited to data distribution, satellite systems must resolve a number of difficulties in order to achieve real interactivity.

Space telecommunications have favoured geostationary systems for over forty years now, their technical characteristics being well suited to the missions involved. With the identification of new services, the idea of low- and medium-Earth orbiting satellite constellations has become more attractive. In this context, mobile phone projects greatly increased in number at the end of the 1990s. But this has not spelt the end for geostationary satellites. Difficulties encountered in financing constellation programmes, as witnessed by the bankruptcy of the Iridium society and also the problems finding financial backing for ICO in 2000, show that it is still hard to assess the importance of this turning point.

Predominance of geostationary satellites

Deploying geostationary satellites is a delicate operation (see chapter 2), but their apparent relative immobility greatly simplifies the ground segment. The size of antennas can be limited as satellite power increases, so that terminals can be developed at lower cost. Another advantage is that lifetimes can be as long as 15 years, which helps to smooth the problem of costs. The launch programme for geostationary satellites shows the rising number of such systems as time goes by and also the wider range of user nationalities, although

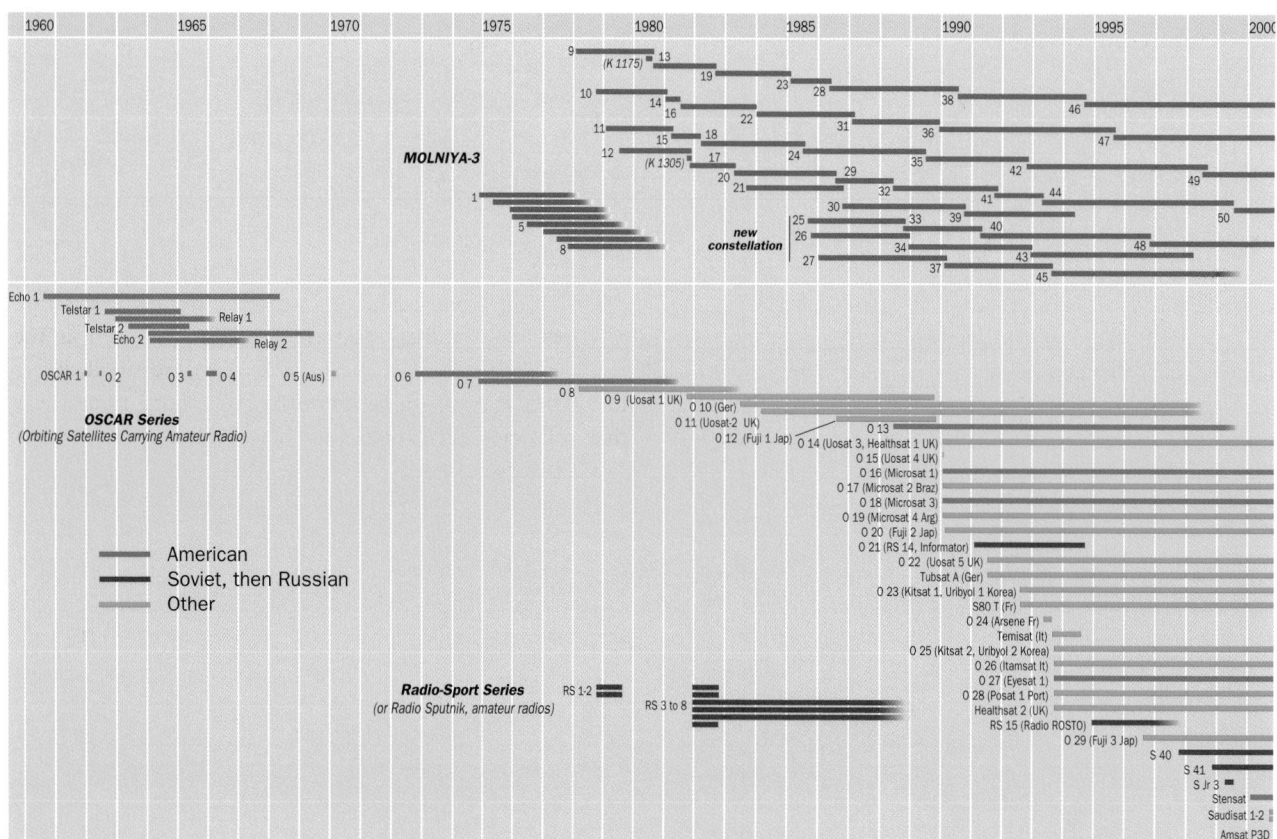

Figure 10.11. **Chronology of civil non-geostationary telecommunications satellites.** Constellations beginning after 1991 are shown in fig. 10.12.

the number of space powers capable of carrying out this type of launch remains extremely limited (fig. 10.10).

Between 1964 and 1975, the United States was the only nation with the technical knowhow to launch and deploy geostationary satellites. From 1966, they have carried out an increasing number of launches for their own programmes or those of their partners, the international company Intelsat, the United Kingdom and Canada. Since this period, the construction and launch of geostationary satellites has represented the main market for space activities. The large number of launches between 1966 and 1969, signalling the strong demand at this time, eventually stabilised at around five launches per year (see fig. 4.4).

The USSR generally used the more economical Molniya orbits, given its shortage of low latitudes. It was not until 1974 that it positioned its first geostationary satellite, and from then until 1978, only launched one or two others per year. The annual number of such launches then reached the same level as the United States, whereupon Russia set about pursuing much the same policy.

The 1980s are marked by the arrival of new players. This was when Japan launched its first geostationary satellite, using a rocket manufactured largely under American licence. Europe for its part became capable of carrying out its own launches in 1981, using the Ariane launch vehicle. It was then able to free itself from the American monopoly as far as commercial space communications were concerned. The last to arrive on the scene was China, placing its first experimental geostationary satellite in orbit in 1984.

Since then, with the exception of India, which developed its own launch vehicle in order to access the GEO orbit, the trend in the 1990s has been towards an increased number of countries and more and more private consortiums merely possessing geostationary satellites, rather than launching satellites themselves. Asia is a case in point; although the trend slowed down somewhat with the financial crisis of 1997–1998, it was soon reaffirmed during 1999–2000. The development of geostationary systems was a decisive factor in the establishment of the space market, both in terms of satellite manufacture and launches; and they maintain a key role in setting up national programmes and distributing traditional services such as television, or more novel services such as access to Internet.

For more than 35 years, GEO satellites have heavily outnumbered other satellite systems, at least for civilian applications, since some Russian and American military systems continue to be developed for purposes far removed from any commercial motivation (see chapter 12). The programme of low-Earth orbiting telecommunications satellites (fig. 10.11) shows that the use of low orbits corresponds to two highly specific cases. The first are small satellites for amateur radio enthusiasts seeking easy low-cost access, and the second are the Molniya satellites.

The existence of low-Earth orbiting satellite programmes providing similar services to those of geostationary satellites is an innovation attributable to Soviet space activities. The Mol-

niya programme, begun in 1965, was an answer to the specific geographical situation of the Soviet Union, and in particular to the need to provide services in high latitude zones. The name of the satellite means 'lightening', corresponding to its apparent path over Russian territory (see fig. 3.22). Indeed, Molniya orbits have perigee at 500 km and apogee at 40 000 km. With a period of 12 hours and reduced displacement speed at apogee, relative to the Earth's rotation, they remain roughly 8 hours over the Soviet Union. The interest of such systems is twofold. Firstly, they can provide services at very high latitudes (see fig. 10.2) whereas geostationary satellite coverage cannot go above 81°. Secondly, it is much easier to place a satellite into Molniya orbit than into geostationary orbit when the launch pad itself is located at a relatively high latitude (see fig. 2.35). However, the lifetime of electronic equipment is reduced by repeated crossings of the Van Allen belt (see fig. 1.2) and more stations must be provided, even if the antennas equipping them are smaller in size. For these reasons, since 1975, several families of geostationary satellites have been installed to complement the Molniya system.

New constellations and new services

During the past few years, technical progress has been made in antenna technology, and also in on-board signal processing capacity and electronics. Combined with the new needs of mobile communications, private users and multimedia, this has led to what some would qualify as a genuine revolution: projects involving constellations of non-geostationary satellites in Low-Earth Orbit (LEO), Medium-Earth Orbit (MEO), and even in High Eccentric Orbit (HEO).

These constellations of satellites with apparent motion across the sky (sometimes referred to as orbiting satellites, as opposed to 'fixed' geostationary satellites, fig. 10.12) illustrate a new commercial philosophy favouring flexible use and small ground-based terminals. Costs must be reduced by mass producing equipment with less demanding technical specifications. A complex set of factors has made this approach possible, including progress in miniaturisation resulting from military research programmes such as the Strategic Defense Initiative. It marks a reverse trend that is likely to continue, although probably in a more modest way than was originally expected in the mid-1990s.

Examining the projects and programmes as of the summer 2000 (fig. 10.13), it is clear that the number of planned constellations is well down compared with the many projects that saw the light of day in the second half of the 1990s. The Iridium programme studied by Motorola as early as 1987 had a decisive influence, stimulating competitive projects with Globalstar in 1989 and Project 21 (which became ICO) in 1991. It then caused a minor crisis in the financial world when bankruptcy struck after just six months in action (see p. 306). Several factors can help to explain this development.

The main theoretical interest of these constellations is to guarantee users flexible links without requiring Earth-based installations at all points of the globe. On the other hand, the

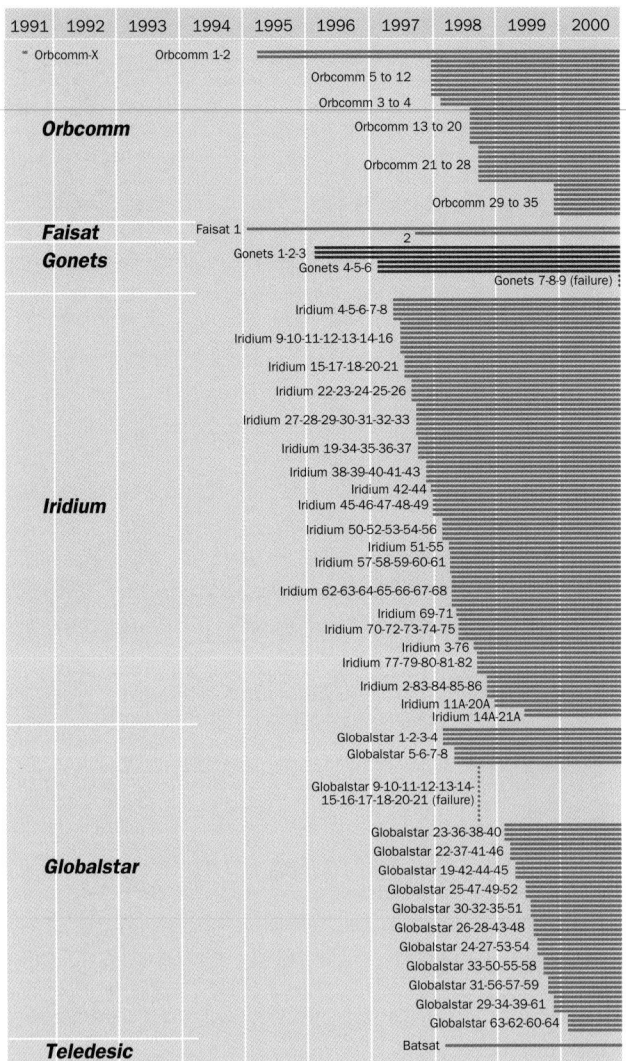

1991 1992 1993 1994 1995 1996 1997 1998 1999 2000

Orbcomm
* Orbcomm-X — Orbcomm 1-2
Orbcomm 5 to 12
Orbcomm 3 to 4
Orbcomm 13 to 20
Orbcomm 21 to 28
Orbcomm 29 to 35

Faisat — Faisat 1 — 2
Gonets — Gonets 1-2-3
Gonets 4-5-6
Gonets 7-8-9 (failure)

Iridium
Iridium 4-5-6-7-8
Iridium 9-10-11-12-13-14-16
Iridium 15-17-18-20-21
Iridium 22-23-24-25-26
Iridium 27-28-29-30-31-32-33
Iridium 19-34-35-36-37
Iridium 38-39-40-41-43
Iridium 42-44
Iridium 45-46-47-48-49
Iridium 50-52-53-54-56
Iridium 51-55
Iridium 57-58-59-60-61
Iridium 62-63-64-65-66-67-68
Iridium 69-71
Iridium 70-72-73-74-75
Iridium 3-76
Iridium 77-79-80-81-82
Iridium 2-83-84-85-86
Iridium 11A-20A
Iridium 14A-21A

Globalstar
Globalstar 1-2-3-4
Globalstar 5-6-7-8
Globalstar 9-10-11-12-13-14-15-16-17-18-20-21 (failure)
Globalstar 23-36-38-40
Globalstar 22-37-41-46
Globalstar 19-42-44-45
Globalstar 25-47-49-52
Globalstar 30-32-35-51
Globalstar 26-28-43-48
Globalstar 24-27-53-54
Globalstar 33-50-55-58
Globalstar 31-56-57-59
Globalstar 29-34-39-61
Globalstar 63-62-60-64

Teledesic — Batsat

Figure 10.12. Chronology of constellations of non-geostationary telecommunications satellites launched between 1991 and 1 January 2001.

profits from such an approach remain limited insofar as the number of potential users remains small, whilst the system itself must be technically complex. Comparing the two programmes so far completed, Iridium and Globalstar, it is clear that the existence of a market is more important than the specific advantages of a space operation. Iridium, designed as an autonomous space service, was equipped with miniaturised switching on board its satellites so that it could operate quite independently of terrestrial networks. In practice, however, terminals proved difficult to set up correctly and relatively expensive. This limited the number of subscribers, particularly as terrestrial cellular telephones have seen a faster and more substantial development than was predicted at the beginning of the 1990s. The latter have been able to reduce charges in consequence, whilst space systems remain relatively costly for private users. In addition, the liberalisation of the telecommunications market related to deregulation has led to roaming agreements between terrestrial mobile phone operators, whereas the self-sufficiency of space systems with regard to the terrestrial network, which may be valued by a limited number of users, in particular, governmental organi-

sations, can only lead to hostility from certain operators. Comparing the Iridium and Globalstar programmes (see pp. 306–307), it would seem that current trends favour space systems that complement terrestrial networks rather than maintaining their independence from them.

The ICO project was originally designed by Inmarsat, an organisation specialised in mobile communications, which has been privatised. Amongst the founding projects, this has been the one the most adversely affected by the Iridium failure. After its bankruptcy in November 1999, ICO became NewICO in May 2000, merging with ICO Teledesic Global.

Although less well known, the Ellipso project also remains plausible insofar as it proposes an original architecture involving satellites in eccentric orbit, thereby ensuring better services in the target regions, where it aims to develop fixed stations with larger but cheaper terminals. At the same time, financial obligations compelled them to enter a coordination with ICO Teledesic Global.

Teledesic or Skybridge correspond to a different type of project. Designed later, they are also intended to provide multimedia services. The thinking behind the two systems once again bears witness to a difference of approach, the space system being autonomous or complementing terrestrial networks.

Teledesic is a joint project from Motorola, Boeing and Matra Marconi Space. It was originally the most ambitious, since it was to comprise 800 satellites with on-board switching and inter-satellite links ensuring continuous and global coverage. The gradual fall in the number of satellites to fewer than 300 in 1998, then fewer than 100 in 2002, is witness to the difficulty of financing this type of project, as well as its still rather experimental nature.

Developed by Alcatel and Loral, the Skybridge system set out to complement rather than compete with terrestrial networks for Internet access. Far less complex, and hence cheaper, it relies on satellite flexibility to fill the gaps in the terrestrial network, for example, providing immediate initial services to new users. The use of both GEO and LEO satellites further strengthens the credibility of the project.

One last feature is striking, namely the preponderance of American companies amongst the operators proposing these systems, whether they aim to provide domestic or international services.

Although Global Mobile Personal Communications Satellite (GMPCS) services and transmission of multimedia data are the main motivation for new constellations, the identification of new services does not only concern non-geostationary satellites, whatever the degree to which programmes may have been implemented. From the standpoint of space occupation, it seems that many projects are based upon geostationary satellites, particularly in Asia. Certain types of service, although sometimes also offered by alternative systems such as mobile phones, would nevertheless appear to favour certain types of orbit.

Hence broadcasting services (radio, television, video) remain very much the domain of projected geostationary satellites whilst messaging applications seem well suited to

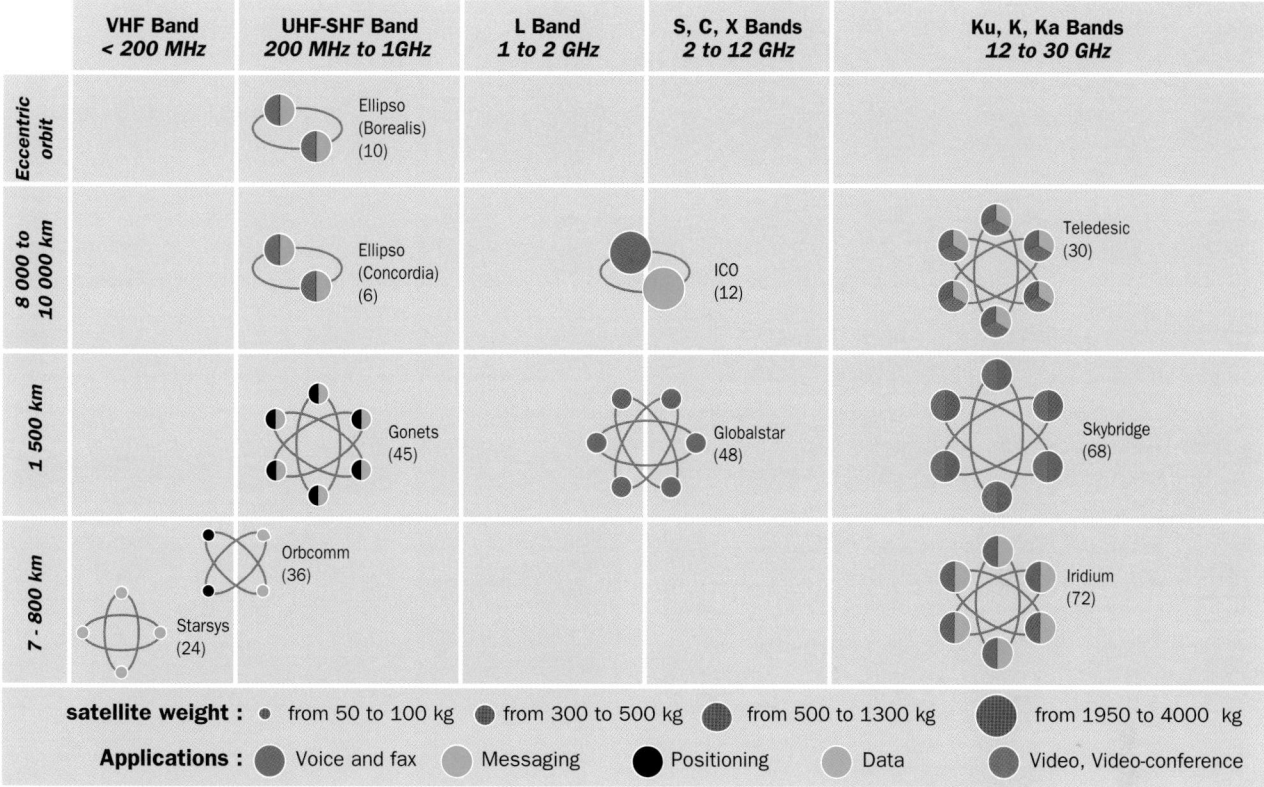

	VHF Band < 200 MHz	UHF-SHF Band 200 MHz to 1GHz	L Band 1 to 2 GHz	S, C, X Bands 2 to 12 GHz	Ku, K, Ka Bands 12 to 30 GHz
Eccentric orbit		Ellipso (Borealis) (10)			
8 000 to 10 000 km		Ellipso (Concordia) (6)	ICO (12)		Teledesic (30)
1 500 km		Gonets (45)		Globalstar (48)	Skybridge (68)
7 - 800 km	Starsys (24)	Orbcomm (36)			Iridium (72)

satellite weight : from 50 to 100 kg · from 300 to 500 kg · from 500 to 1300 kg · from 1950 to 4000 kg

Applications : Voice and fax · Messaging · Positioning · Data · Video, Video-conference

Figure 10.13. **Constellations of non-geostationary telecommunications satellites launched or under development in January 2002.**

low orbits. Continuous reception without the need for switching between two satellites is an advantage of geostationary satellites over the other types of satellite and the investment required favours traditional operators already well placed in this market. On the other hand, the use of low-orbiting satellites seems to be taking the lead in personal communications for which running costs should remain low.

Other cases, such as the provision of multimedia services, are more complex. Technically, the time required for a signal to make its way up to a geostationary satellite and back appears to be a severe disadvantage if the aim is genuine interactivity. Various technical solutions are under investigation and services of this type are available on existing GEO satellites like Cyberstar and GE*Star, whilst projects like Spaceway, Astrolink or West are still in the pipeline. The fact that it is difficult to assess the way technology will develop has led to a proliferation of projects, sometimes proposed by the same operators.

The allocation of frequency bands to services is a response to other considerations (see figs. 10.9 and 10.13). The lower frequencies (VHF and UHF) are mainly reserved for messaging and positioning services. Low transmission rates and the need to favour simple solutions to keep costs down are both decisive factors. In the case of positioning, conformity with the GPS standard is a further argument. The L-bands tend to be used for mobile phones, whilst the S-, C- and X-bands concern broadcasting and telecommunications. Messaging applications, when integrated into a global system of services, are also to be found amongst the users of these bands. Finally, the K-bands are those most often cited for new projects. Apart from their technical advantages, including broad

bandwidth and high transmission rates, they are less used and this means that it is also easier for the regulating bodies to make allocations. Technical difficulties are gradually being overcome, such as signal attenuation due to rain (rain attenuation), and this is helping to reinforce their position.

Traditionally, satellites provide a certain number of advantages. New satellites should be characterised by the implementation of various technological choices still under development. Complete digitalisation, on-board processing capabilities, Asynchronous Transfer Mode (ATM) switching, Application Specific Integrated Circuits (ASIC), intelligent agents, nanotechnologies, optoelectronics, and active antennas are the main lines of research today. An up-to-date needs analysis shows that developments in space telecommunications services must satisfy a range of requirements, including worldwide capacity, very high transmission rates, continuous operation, mobility and flexibility. Whether they involve geostationary or non-geostationary satellites, projected systems all aim to fulfill these criteria in the nearest possible future with a view to taking the lead on new markets. Finally, it is essential to cut costs as far as possible if space telecommunications systems are to be in a position to compete with or connect up with the new terrestrial networks.

To resume, a global approach via the concept of interconnected worldwide telecommunications systems is gaining ground over one which breaks the problem down into different services. To succeed in new markets, satellites must play a complementary role, integrating themselves into a logic of global systems proposed to users.

Geography of space telecommunications

Space systems have developed on different scales, moving from international, through national to regional services, in order of appearance. The use made of the various systems, and indeed combinations of the three systems, reflects a typology of countries around the world (fig. 10.14).

Participation in an international system is by far the most common occurrence. This may be supplemented by the installation of national or regional systems, or both. Countries can be categorised through an analysis of economic factors such as the per capita and total gross national product, and their involvement in world trade, as well as an examination of more directly geographical factors, such as population, surface area, and the existence of population groups living far removed from the main centres of population. The existence of foreign interests is a further criterion with a significant effect on the design and choice of a regional, national or international network.

International systems are the norm since only a very few underprivileged states lying completely outside the current cultural and commercial circuit do not use these services. Consequently, the working principles of international organisations traditionally inspired by a concept of public service and non-discriminatory access, together with an equitable pricing policy, whatever the profitability of the links involved, have been strongly defended by most developing countries. Indeed, the latter often have no alternative, even for national communications. Moreover, the excellent representation of countries and national telecommunications organisations within decision-making authorities comprised an important factor for countries hoping to maintain a maximum level of control. However, with the gradual implementation of the American project for a global information society and the continued spread of deregulation, international organisations have evolved, often subject to privatisation whilst maintaining some obligation to provide a public service (see pages 298–304).

As far as national systems are concerned, it is the combined effect of geographic factors and economic parameters that turns out to be decisive. The proliferation of American national systems and the wide range of Soviet systems, although arising through totally different circumstances, economic in the first case and rather more geographical in the second, have led to the acquisition of unequalled skills that today mark their supremacy.

A further motivation for many countries is the more politically motivated desire to prove their independence by developing their own systems, even though they may be purchased largely ready-made from abroad. In the 1970s, the development of national systems allowed European countries to acquire their own technological knowhow.

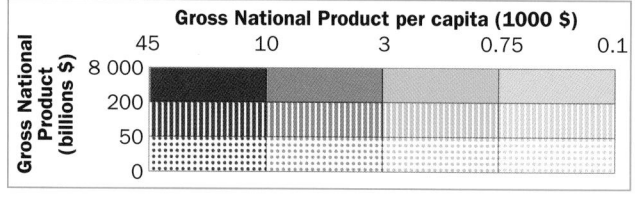

Figure 10.14. **Geography of telecommunications on 1 January 2000.** Satellites are represented by company nationality rather than countries served. Membership of a consortium is indicated.

It also served to implant an industrial base via the commercialisation of satellites. The large number of systems and overall size of the US market has given American manufacturers an undisputed lead, but they are no longer the only suppliers.

Japanese and Indian efforts in this direction reflect the same desire to establish independence and offer diversified systems. The Indian experience, in particular, is of great interest to developing countries facing the same need for distance learning or telemedicine in deprived or inaccessible regions. Along quite other lines, Japanese knowhow in high definition television broadcasting aims rather at acquiring the experience essential in setting up standards, whilst more

recently, private companies like Melco have shown interest in the satellite telecommunications market.

In one form or another, it is the same quest for independence that has inspired many recent programmes. The existence of national telecommunications satellites can often be taken as an indicator of the degree of development of a country and the way it views itself within the group of industrialised nations. Mexico, Korea and Taiwan may serve to illustrate this reasoning. Attempts to involve the country's own manufacturing industry in setting up the system also show that even a partial mastery of space technology carries with it a powerful image.

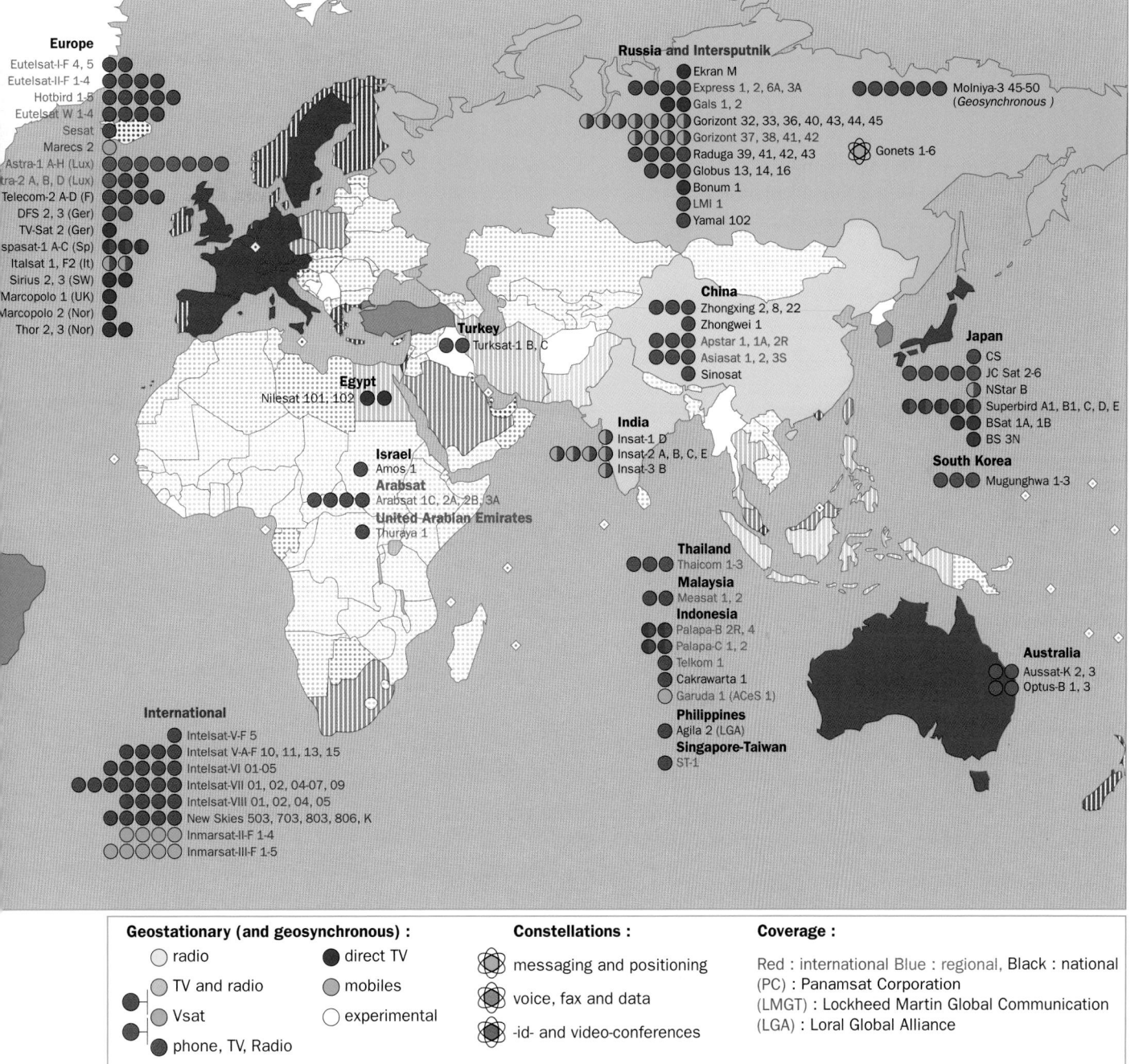

Europe
Eutelsat-I-F 4, 5
Eutelsat-II-F 1-4
Hotbird 1-5
Eutelsat W 1-4
Sesat
Marecs 2
Astra-1 A-H (Lux)
tra-2 A, B, D (Lux)
Telecom-2 A-D (F)
DFS 2, 3 (Ger)
TV-Sat 2 (Ger)
spasat-1 A-C (Sp)
Italsat 1, F2 (It)
Sirius 2, 3 (SW)
Marcopolo 1 (UK)
Marcopolo 2 (Nor)
Thor 2, 3 (Nor)

Russia and Intersputnik
Ekran M
Express 1, 2, 6A, 3A
Gals 1, 2
Gorizont 32, 33, 36, 40, 43, 44, 45
Gorizont 37, 38, 41, 42
Raduga 39, 41, 42, 43
Globus 13, 14, 16
Bonum 1
LMI 1
Yamal 102

Molniya-3 45-50
(*Geosynchronous*)

Gonets 1-6

China
Zhongxing 2, 8, 22
Zhongwei 1
Apstar 1, 1A, 2R
Asiasat 1, 2, 3S
Sinosat

Japan
CS
JC Sat 2-6
NStar B
Superbird A1, B1, C, D, E
BSat 1A, 1B
BS 3N

Turkey
Turksat-1 B, C

Egypt
Nilesat 101, 102

Israel
Amos 1

Arabsat
Arabsat 1C, 2A, 2B, 3A

United Arabian Emirates
Thuraya 1

India
Insat-1 D
Insat-2 A, B, C, E
Insat-3 B

South Korea
Mugunghwa 1-3

Thailand
Thaicom 1-3

Malaysia
Measat 1, 2

Indonesia
Palapa-B 2R, 4
Palapa-C 1, 2
Telkom 1
Cakrawarta 1
Garuda 1 (ACeS 1)

Philippines
Agila 2 (LGA)

Singapore-Taiwan
ST-1

Australia
Aussat-K 2, 3
Optus-B 1, 3

International
Intelsat-V-F 5
Intelsat V-A-F 10, 11, 13, 15
Intelsat-VI 01-05
Intelsat-VII 01, 02, 04-07, 09
Intelsat-VIII 01, 02, 04, 05
New Skies 503, 703, 803, 806, K
Inmarsat-II-F 1-4
Inmarsat-III-F 1-5

Geostationary (and geosynchronous) :
- radio
- direct TV
- TV and radio
- mobiles
- Vsat
- experimental
- phone, TV, Radio

Constellations :
- messaging and positioning
- voice, fax and data
- -id- and video-conferences

Coverage :
Red : international Blue : regional, Black : national
(PC) : Panamsat Corporation
(LMGT) : Lockheed Martin Global Communication
(LGA) : Loral Global Alliance

International systems

The planetary scope of satellites and increased needs with regard to international links were decisive factors behind the establishment of an international space telecommunications organisation, at the initiative of the United States. In his 1961 declaration concerning American satellite communication policy, J.F. Kennedy announced: 'I invite all nations to participate in a satellite communications system, in the interests of world peace and closer brotherhood among peoples of the world.'

In the technical and political context of the day, countries chose to develop a system based upon worldwide cooperation between the specialised administrative bodies controlling telecommunications and holding national monopolies. Hence, even though the United States insisted on introducing trade regulations that favoured the interests of their own national industry, the International Telecommunications Satellite organisation (Intelsat) was conceived as part of the intergovernmental framework. The Intelsat consortium was created as a provisional measure in 1964 at the request of European countries hoping to protect their rights in the face of American domination. In 1973, it was replaced by a genuine international organisation, but with some commercial features. Partners of the United States considered the presence of countries within such an institutional framework as a defence against American influence, with its control of technical knowhow, and as a guarantee that the world system would fulfill its duty as a public service.

In 1971, the Soviet Union created a mirror organisation, called Intersputnik, along classic intergovernmental lines, as a way of combating the monopolistic aspect of Intelsat. In practice, this body involved countries in the Soviet bloc.

In 1979, a new international organisation, the International Maritime Satellite (Inmarsat) was created in the same mould as Intelsat, dealing with mobile satellite communications. This time the Soviet Union was a member and became the second contributor after the United States.

The liberalisation of telecommunications, whose effects began to be felt in the space sector as early as 1984, led to a break with the existing order. The new environment was, and still is, characterised by much more varied relationships between those involved, who now included private companies as well as countries. These companies have become, or are in the process of becoming, owners and operators of space systems. The form of involvement and means of finance are many and varied. At the same time, states are limiting their role to that of regulator. Under the dual effect of deregulation and converging technologies in the areas of communications and the media, the rigid framework of rules applying to services is being relaxed. The regulatory distinction between international and national services is thus disappearing. In such conditions, the organisation of international space telecommunications is emerging under a new guise, with a much more complex combination of commercial and political interests and offering vastly more diversified systems.

At the international level, space systems are all in the process of globalisation and now constitute a rather heterogeneous category. Apart from Intersputnik, which has kept its intergovernmental status despite some measure of evolution, the old space telecommunications organisations have become private companies or are well on the way to doing so. The previously guaranteed obligation to provide a public service is now maintained by a controlling intergovernmental structure. In addition, new private international networks are being set up at the initiative of the three major US satellite manufacturers, Hughes, Lockheed Martin and Loral. These have committed themselves to a process of vertical integration, assuming the role of global systems operators. Other more marginal players have appeared, such as the private company Worldspace which was originally formed with the aim of building a global system based on three geostationary satellites and devoted to a specialised digital radio service. Finally, various companies, mainly American, have taken the lead in developing global constellations to provide new personal communications services via satellite, establishing consortiums to open projects up to wide-ranging international investment.

Transformation of intergovernmental systems

■ Intelsat

At its foundation in 1964, Intelsat brought together 11 countries. In 1987, 138 countries had become members, to be joined by the republics of the ex-USSR in 1992. By this time, Intelsat had reached genuine world proportions, as confirmed by its 300 or so users in 2000. There can be no doubt about the success of the undertaking. However, in answer to the ever more competitive environment, and in particular to the new American policy of liberalising telecommunications, Intelsat was compelled to evolve towards a negotiated privatisation process. The obligation to provide a public service was nevertheless maintained.

In the mind of its patrons, Intelsat would extract profit from the US technological lead by imposing the principle of a single world trade system. The COMmunications SATellite corporation (COMSAT), a private company under governmental control, founded in 1963 to supply national and international needs, represented the US on the Intelsat interim committee. It held 61% of the votes and thus retained a controlling majority.

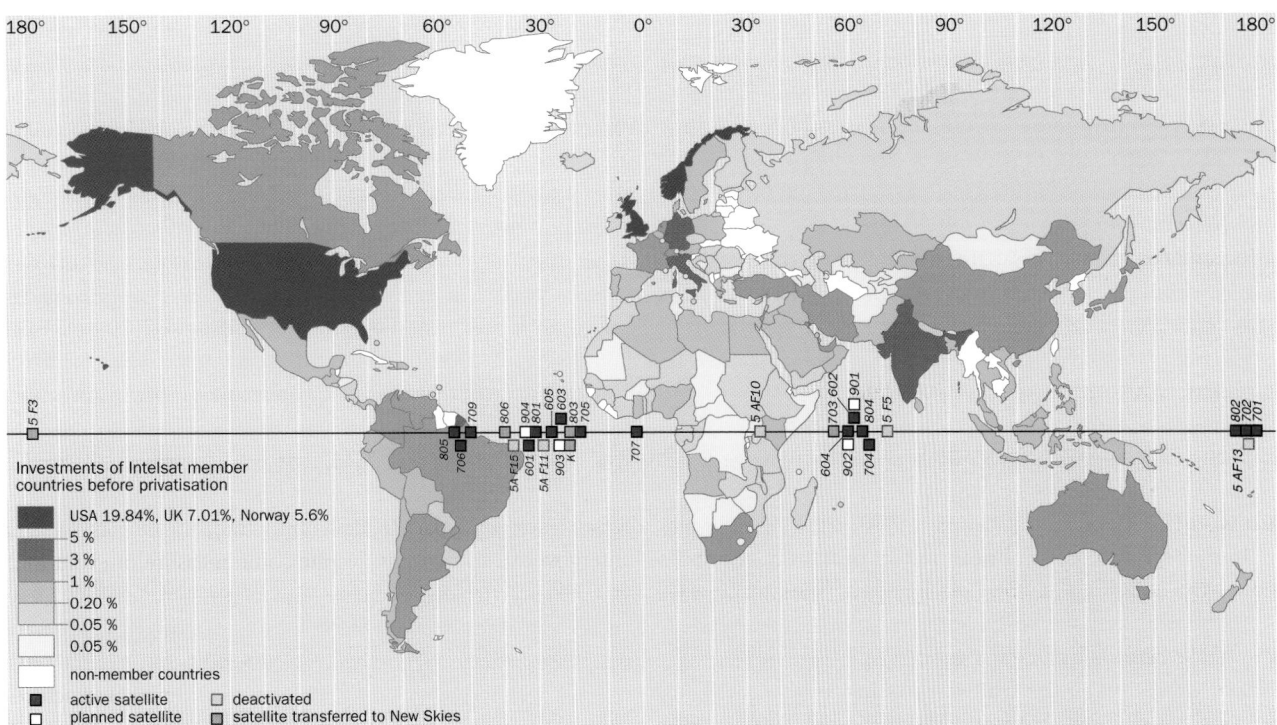

Figure 10.15. **Global communications system provided by Intelsat satellites.** The map shows investments of member countries (public and private) as of 1 January 2000.

The initial project proposed to use non-geostationary satellites, which assumed perfect coordination between the space segment and ground-based antennas. However, the deployment of geostationary satellites had been mastered since the launch of the first Early Bird satellite in 1965, and their use no longer justified American domination, at least on the technical level. Growing reluctance on the part of non-American members finally led in 1973 to the adoption of new articles stipulating that no signatory could henceforth hold more than 40% of the votes, and this marked the end of the original monopolistic ambition.

In its original structure, i.e. until its privatisation in 2001, the non-profit-making commercial cooperative Intelsat provides for public telecommunications services in member countries on the basis of the purchase or lease of transponders. The Meeting of Signatories is made up of representatives from the investors, either governments or national bodies in charge of telecommunications. For its part, the Assembly of Parties is composed of the representatives from all states that signed the 1973 agreement specifying the current Intelsat charter. This is the principal executive organ, in which each state holds one vote from the moment it invests at least 0.05% of the total. Ordinary decisions are passed by a majority vote and more important ones by a two-thirds majority. This is standard procedure in international organisations.

Another principle is applied at meetings of the Board of Governors, whereby votes are weighted according to the level of investment of the country concerned, itself calculated from the percentage use of Intelsat satellites. This governing body is in charge of managing and organising operations carried out by Intelsat. It is made up of about thirty representatives of the bigger users, either representing a single country or a group of countries that have chosen to work together. They exercise a right of direct oversight and appoint the Director General at the head of the executive committee. The principal users thus maintain a strong hold over decision-making, as in a typical commercial organisation.

Considering the investment shares of member countries, which themselves correspond directly to their share in the traffic (fig. 10.15), levels of participation in international trade are clearly apparent, even if the existence of regional systems somewhat blurs the picture. Within the framework of the Intelsat system, the United States are the undisputed leaders with one fifth of the traffic, followed by the United Kingdom and Norway. Italy, Japan, France and Germany form the second group of large users but with a distinctly lower level of traffic.

Third World countries, constituting two-thirds of member nations, are the smallest users by far, since together they only represent one-third of the total traffic. The Intelsat pricing policy is such that costs are the same no matter how much the system is used. This means that regions with the lowest level of trade are not yet penalised, and these users are amongst the most ardent defenders of the present system. Intelsat thus played a key role in African telecommunications in 1999, with more than 50 countries using it for their international telecommunications and 25 of those also depending on the global system for national links.

Worldwide services are one of the fundamental obligations of the organisation. As early as 1966, Intelsat satellites were ready to serve the world over. The spectacular growth in Intelsat traffic has been made possible by a considerable increase in satellite capacity (fig. 10.16). The first Early Bird satellite could handle 240 telephone calls or a single television programme and the size of its receiving antennas (30 m) meant that it was limited to intercontinental links. Intelsat VI satellites can achieve 120 000 circuits. Increased EIRP (Equivalent Isotropically Radiated Power) has led to the possibility of handling greater volumes of information as well as reduced receiving antenna dimensions, which means easier installation of new stations. This capacity could still be greatly increased by installing new on-board systems.

Intelsat benefits from its seniority, its wide range of services and its image as a non-profit-making association. The latest technical advances in its satellites have facilitated a diversification into telephone, television, telex, fax, data transmission, video teleconferencing, and teleprinting. They also help to reduce costs of ground installations which remain at the expense of the country concerned. Intelsat certainly plays a global role with almost 500 Earth receiving stations scattered across the whole planet and a great many users, both private and public. Notwithstanding, the general trend towards deregulation and acceptance by the majority of countries of the main ideas behind a global information

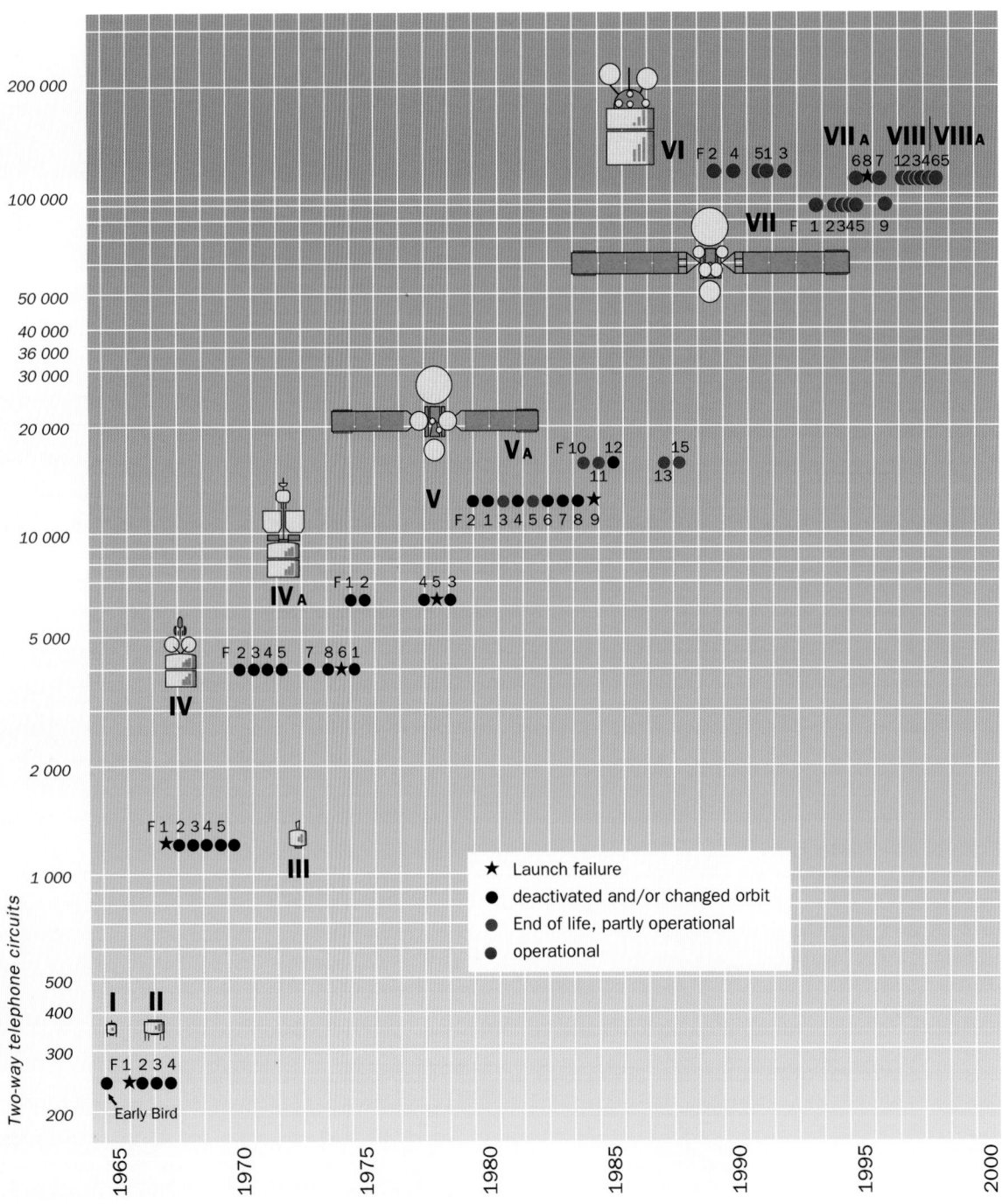

Figure 10.16. **Intelsat satellites: launches and current state of operation.**

infrastructure as proposed by the United States has seriously called the status of the organisation into question. Indeed, the main shareholder, the United States, objects to a certain number of 'privileges' exercised by Intelsat in its capacity as an international organisation. These privileges are now presented as obstacles to free competition. Symptomatic of the reversal in American position, the rule originally imposed upon new systems to ensure that they could cause no harm to the organisation is today vigorously condemned.

Since 1992, a working group created by the Assembly of Parties and named Intelsat 2000 in 1995, has been negotiating to find common ground between those who cling to the obligations of a universal service and those who favour privatisation. In the atmosphere of compromise prevailing at the present time, the desire to continue to guarantee the benefits of a global system, especially to dependent users, has led to recognition that the public service provided by Intelsat should be maintained intact. Three main principles will therefore be respected: a worldwide service, involving coverage of the whole globe on a commercial and non-discriminatory basis, global interconnection allowing all networks to connect together, and uniform pricing, whatever the region provided for.

In March 1998, the Assembly of Parties concluded the first step in the privatisation process when it decided to transfer part of its assets, namely six satellites, to a separate private company called New Skies, capable of developing new multimedia and Internet services. The main concern for the member countries was to ensure independence of the company from Intelsat, in order to guarantee conditions of fair competition between all systems operators. In October 1999, the member countries formally endorsed the transformation of

the current international cooperative into a complex organisation, comprising a commercial company which recovered all the assets but remained under the control of an intergovernmental structure guaranteeing respect for the basic principles of a public service. The Penang working group, made up of shareholders and member governments of Intelsat, was then created to bring restructuring to its conclusion. The Board of Governors took a further step in December 1999, opting for a holding company structure for the future Intelsat, with two exclusively owned subsidiaries. Finally, in November 2000, a historic meeting of the Assembly of Parties unanimously approved the privatisation of Intelsat and the amendments made to its constitutive charter. The transfer was completed in July 2001. The holding company Intelsat Ltd wholly owns two subsidiaries: Intelsat Service Corp, a service company employing the main part of the personnel and in charge of operations, and Intelsat LCC owning orbital slots and licences. A separate and independent intergovernmental office known as the International Telecommunication Satellite Organisation (ITSO) monitors the private company's implementation of its public service agreements.

■ Intersputnik

Intersputnik is an international telecommunications cooperative founded in 1971 on the initiative of the Soviet Union with its head office in Moscow. It has also experienced far-reaching modification, although of a rather different type since it does not concern the structure of the organisation, but rather the creation of a new company called Lockheed Martin Intersputnik (LMI), based in London and destined to exploit the new generation of satellites. Asserting itself as a system with worldwide designs, Intersputnik has witnessed

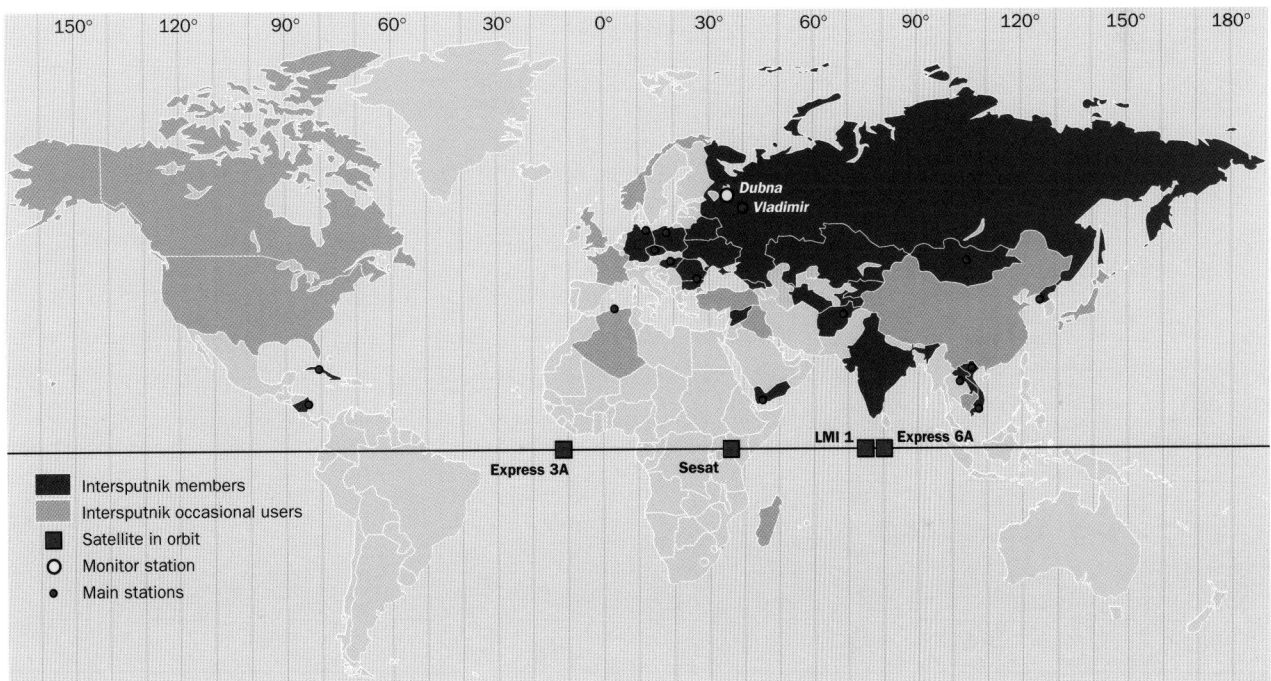

Figure 10.17. **The Intersputnik system in January 2001.**

a growth in traffic of 20% per year since 1991. In 2000, it comprise 23 signatory countries, with Indian membership and more than a hundred users.

Intersputnik was created to express the Soviet Union's refusal of the way Intelsat was run. Indeed, at the time, it broke the Intelsat monopoly and distinguished itself from the latter by allowing its members to join other telecommunications organisations, international or otherwise. Likewise, in contrast to the situation at Intelsat, each member country had one vote in the Intersputnik decision-making process. In fact, apart from the exclusive role played by Comsat (see p. 298), the USSR was at a disadvantage because of its small share in international trade. This meant that it had only a limited influence on the Board of Governors. At a still more fundamental level, the political stakes involved in controlling telecommunications, and especially television, were clearly appreciated by the Soviet government, which therefore sought to obtain some level of autonomy in this area. Even in 1959, N. Khrushchev had written plans for 'cosmic' retransmission of television into his political schedule, and the first official proposal for a universal system was made in 1965.

The Intersputnik organisation originally included the countries already members of the Interkosmos association, created in 1967 to encourage cooperation in the area of space research, together with sister countries from other regions of the world and, since 1991, seven further countries (fig. 10.17). The main decision-making body was the Board in which each member country held one vote. The Directorate, which handled administrative aspects, was chaired by a Director, appointed for four years. It consisted of a small committee that made policy suggestions with regard to equipment and pricing. A two-thirds majority was required if a unanimous result was not obtained. Financing was provided by the various member countries in proportion to the use they made of the system, and this also determined the share of the profits that they should receive, at least in theory. Technical management of the system was guaranteed by 'those countries possessing the necessary infrastructure and technical means', which effectively meant Russia, the only world power outside the United States to have such means at its disposal. The particularly large investment made by the USSR, which then charged services at a level well below the real price, was justified by the 'educational' objectives of the system and the desire to strengthen links with member states. Televisual transmissions between countries represented half the traffic and are still a major component.

For a long time, Intersputnik depended on Molniya 3 satellites to provide links across the Soviet bloc. Today these have been replaced by ten geostationary Radouga, Gorizont, Express and Gals satellites. For the new generation, Western satellites like LMI 1, launched in September 1999, are also used. The reliability and superior performance of these satellites are not the only explanation for their use. The organisa-

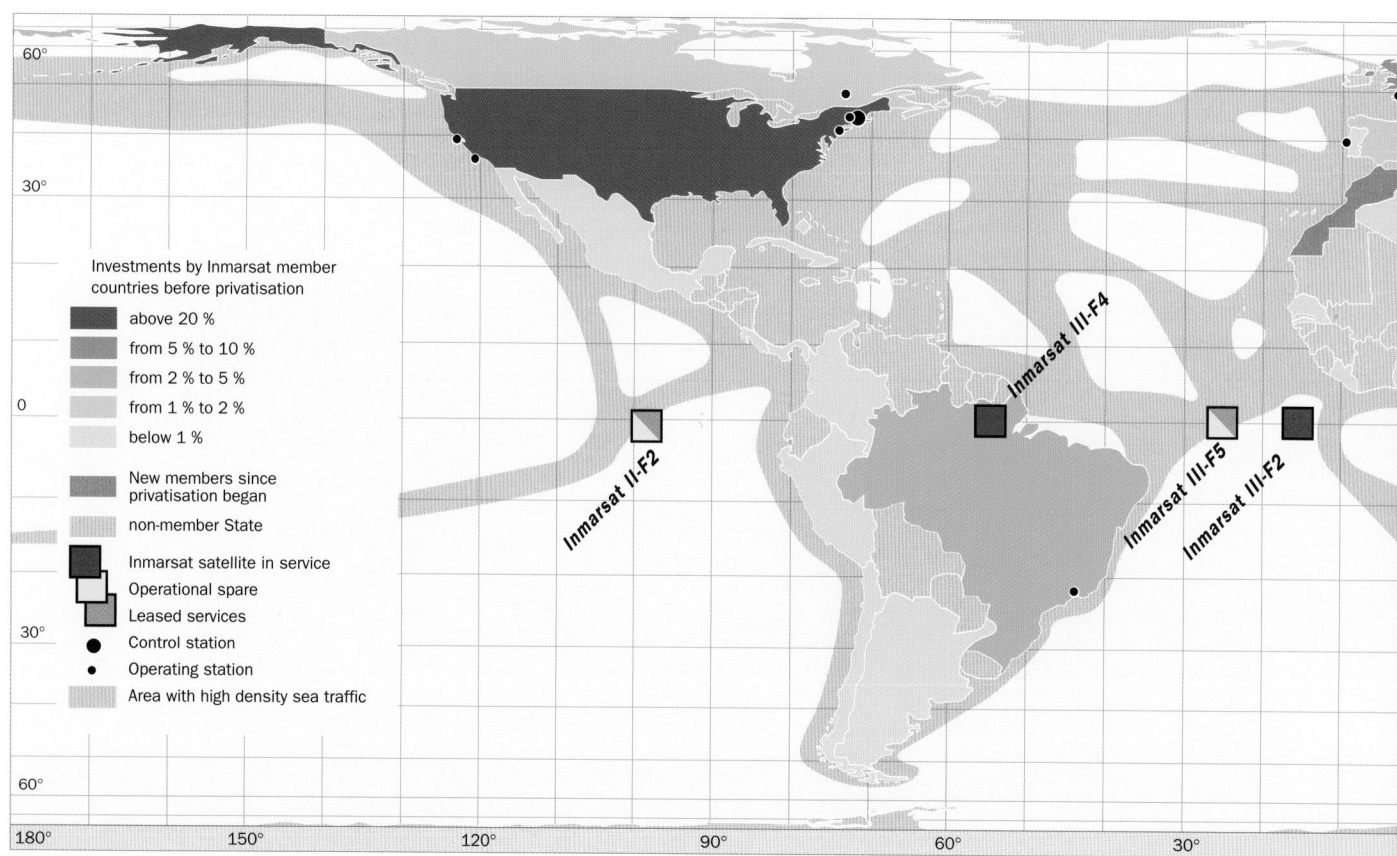

Figure 10.18. **Inmarsat satellites and Earth stations.**

tion has been opened up to foreign investment, allowing a certain revitalisation. In return, it offers the advantage of reserved slots on the geostationary orbit. The current development, striking though it may be, is thus based on purely pragmatic issues. It also marks the end of Russian dominion and reveals the impact of the trend towards deregulation and privatisation of telecommunications in this country.

■ Inmarsat

The Inmarsat organisation (INternational MARitime SATellite) was founded later than the other two international telecommunications organisations, in the East as in the West, since its charter was only ratified in 1979 and it did not begin commercial operations until 1982. Inmarsat is a highly specialised body created on the initiative of a UN agency, the Intergovernmental Maritime Consultative Organisation (IMCO). Its mission is to provide links between maritime mobiles and terrestrial telecommunications networks via geostationary satellites.

A wide range of services is provided including telephone links and telex transmission, but also priority access to the SARSAT–COSPAS search and rescue service (see chapter 11) which has its secretarial offices on Inmarsat's London premises. The organisation's constitution also plans to provide navigation services and several countries, mainly European, are considering the possibility of using the Inmarsat system to develop a sort of civilian GPS, called the Global Navigation Satellite System (GNSS). Two versions, referred to as 1 and 2, in various stages of development are now under discussion.

Inmarsat originally had a similar organisational structure to Intelsat. However, it offered specific services, without being particularly dominated by the United States, and the Soviet Union expressed serious interest for its own navy as early as 1975. The long term policy of the organisation is debated at the Assembly of Parties which bring together all participating members. Each member holds one vote. The Board is the main decision-making body and appoints the Director for a six year period. It is composed of the 18 most important members and four representatives for the other member countries elected by the Assembly. The vote is weighted in accordance with respective shares in the investment. This undoubtedly explains the large initial contribution made by the Soviet Union, which held second place amongst the investors in 1980, behind the United States. The organisation contained 48 members in 1986, 58 in 1990, 64 in 1991, 79 in 1997, and 86 in 2000. This growth rate indicates how successful the undertaking has been and has led to a diversification in its activities. Nevertheless, in the present climate of reorganisation affecting space telecommunications organisations, new communications services with mobiles have inspired the creation of a subsidiary called ICO Global Communications.

The first operational system dates from 1985 and used transponders on six satellites, half of which formed a reserve, leased from the European Space Agency in the case

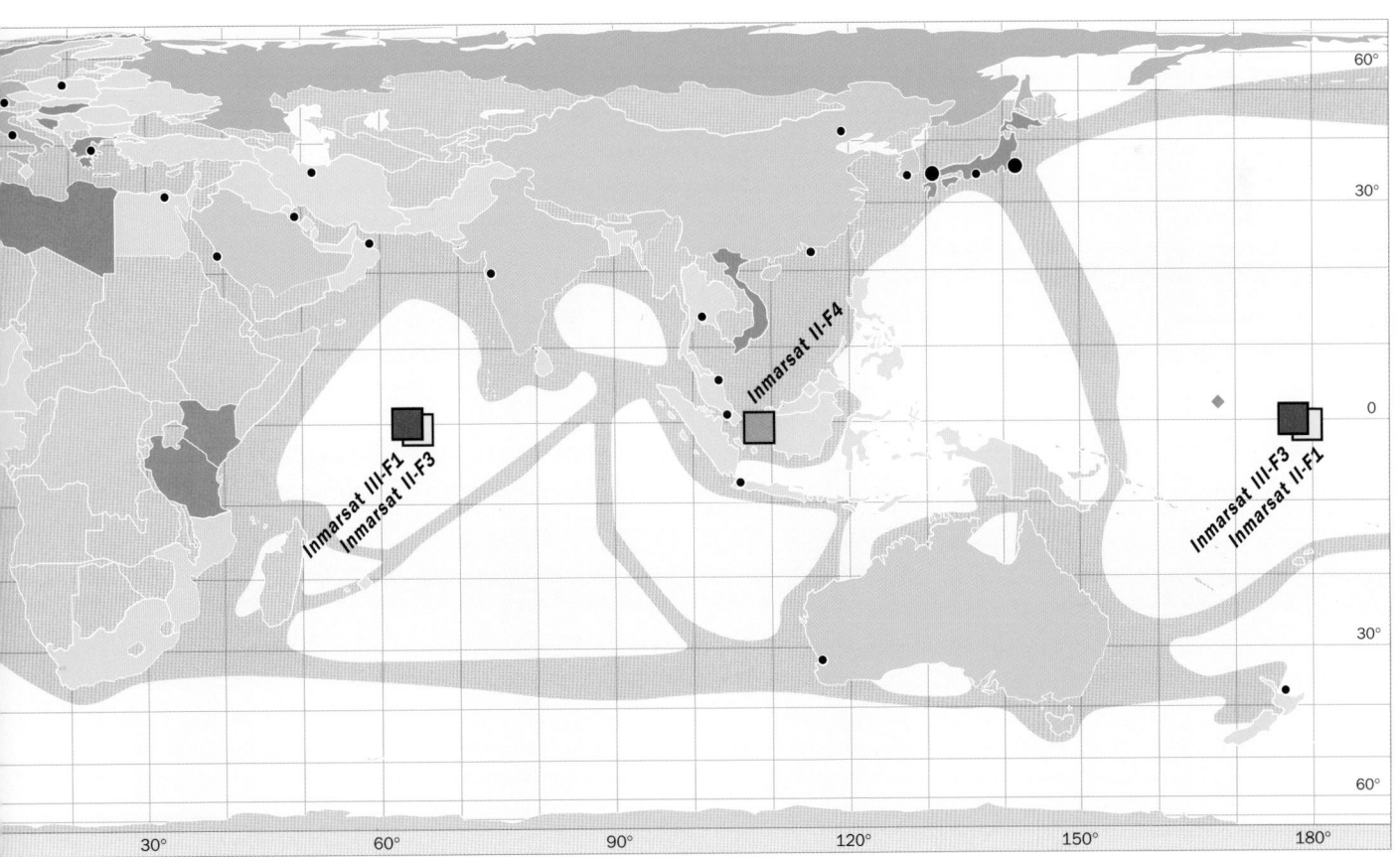

of the Marisat and Marecs satellites, and from Intelsat. Faced with an increasing number of links, the organisation decided in 1983 to obtain its own satellites. Four second generation Inmarsat satellites and three third generation satellites have been launched since that time in addition to leased capacity. The first in the series, Inmarsat II F1, was launched on 30 October 1990 and stationed over the Indian Ocean. In 1991, Inmarsat II F2 was stationed over the Atlantic and Inmarsat II F3 above the Pacific. In 1992, Inmarsat II F4 completed the system, doubling available capacity over the Atlantic, and in 1996, Inmarsat III began with the launch of three satellites, Inmarsat III F1, F2 and F3, followed in 1997 by Inmarsat III F4, and in 1998 by Inmarsat III F5.

The investment share of each member country is a good indicator of its maritime interests and also represents its capacity to provide itself with receivers (fig. 10.18). The latter point may partly explain the fact that Russia, with its growing economic difficulties, gradually fell to eighth position, after France and Germany, with only 4% in 1998, the final year of operation of the intergovernmental organisation. At this time, the United States still held the lead position by a clear margin, supplying slightly less than a quarter of the total investment. It was followed by the United Kingdom and Japan with steadily increasing stakes, the latter having recently displaced Norway from third position. Apart from these countries with a significant maritime tradition, using Inmarsat services for both petrol tankers and fishing boats, increasing interest from other users motivated by airborne and maritime applications has caused a certain readjustment in the hierarchy of member states.

A broadening of Inmarsat's activities to mobiles other than maritime had already been envisaged in 1985, when it was decided at the Assembly that the organisation should acquire the skills needed to accommodate airborne services. The first links of this type were provided in 1989. Eventually, 300 airlines from 170 countries should use the system.

In 1987, when the first frequency allocations for mobile services were undergoing tight negotiations at the ITU, Inmarsat, which could only provide temporary services in exceptional circumstances, considered the possibility of supplying terminals for terrestrial mobiles. The implications of catering for this new type of service were taken more and more seriously at the beginning of the 1990s. Such a decision marks a significant turning point for several reasons. It involved complementing the initial geostationary system with non-geostationary satellites in order to serve high latitude zones, and identifying a legal status that could adequately hold out against large-scale American private enterprise. The latter was in the process of taking over the whole mobile personal communications market through projects like Globalstar, Iridium and others.

Since the American position made direct intervention difficult, it was decided to create a subsidiary. Its role was to set up a constellation dedicated to a global portable telephone system. The decision was taken in May 1994 following long and tedious discussions between the Signatories, complicated by the rather special nature of the undertaking. Several considerations lay behind the final choice. For one thing, the philosophy of international organisations does not generally extend to the idea of such extensive commercial risk. For another, it would have been no more possible to hold the Signatories responsible for such projects than it would have been to expect them to invest in proportion to the use they made of the system, as stipulated in the founding agreement.

For these reasons ICO, formed in January 1995 under British law, features certain unusual characteristics, resulting from a series of compromises. The new high risk services of the future ICO global constellation will be provided by a private-law affiliated company in which Inmarsat cannot hold a majority shareholding. This measure is designed to ensure fair competition with other companies offering the same type of service. Inmarsat has nevertheless retained a certain level of control over its subsidiary, both to protect itself from any untimely infringement upon its own activities, and also to reap some of the profits expected from new activities.

As in the case of Intelsat, the fundamental reconsideration of Inmarsat's organisational status which began in 1994 triggered a negotiated privatisation process. The Assembly of Parties came to an agreement in principle in 1998, and new structures began to take shape in April of the following year. They consist of two entities. The first, Inmarsat Ltd, is a company under British law, two-thirds privately owned, due to be launched on the stock exchange eventually. The other is an intergovernmental organisation called the International Mobile Satellite Organization (IMSO). The latter has a simple structure based on an Assembly of Parties bringing together the 86 current member states, a Director and a Secretariat with a maximum of 11 members. The budget of the organisation is managed by the new private company Inmarsat, and its task is to ensure that the obligations of Inmarsat's public services are respected. Four fundamental principles to be observed by the company are written into the amended articles and also the public service agreement between IMSO and Inmarsat Ltd. These are: continued provision of maritime communications, and in particular, search and rescue activities within the framework of the Global Maritime Distress and Safety System (GMDSS), fully operational since 1999; non-discriminatory services as far as nationality is concerned; a commitment to peaceful purposes; and finally, provision of services to all zones in which satellite communications are needed.

Private international systems

The three large US space manufacturers, Hughes, Lockheed Martin and Loral are preparing for the globalisation of the telecommunications market by launching themselves as operators in the supply of satellite communications services, with the added advantage of profiting from the enormous expected growth in new services, whilst guaranteeing themselves an outlet for their own products. This has involved restructuring the US space telecommunications market through mergers.

■ Hughes Communications Inc

The electronics and defence group Hughes Electronics Corporation was restructured between 1996 and 1997 after merging its defence activities with Raytheon. Through its subsidiary Hughes Space and Communications Company, it thus positioned itself amongst the biggest satellite manufacturers. It proceeded to implement a policy of integrating space applications, forming a further subsidiary, Hughes Network Systems (HNS), dealing with ground installations, and another, Hughes Communications Inc (HCI), handling space service activities. The latter then bought up PanAmsat, owner and operator of the biggest private international satellite system developed independently of Intelsat and devoted essentially to transatlantic traffic. With this operation, HCI became an operator itself, and began to finance new activities at PanAmsat. The new company, PanAmSat Corporation, took over a space system providing both national and international services and bringing together in a single group PAS's international network of 14 satellites, HCI's fleet of 10 national satellites, the Galaxy satellite for DTH direct broadcasting in Latin America and the American DirecTV network, with 5 satellites supplying television programmes in the United States. The latest satellites launched serve to illustrate new zones of expansion, such as China covered by PanAmSat 8, placed in orbit in November 1998. They also highlight privileged satellite applications, such as the distribution service for cable TV provided by Galaxy XR in January 2000 (see fig. 10.14).

Also in January 2000, Boeing Co's takeover of all Hughes' satellite construction activities led to another shake-up in the American space industry. Hughes thereby obtained the finances required to invest in broadband projects and concentrate its resources on provision of new and rapidly expanding communications services. In particular, HCI is involved in the Spaceway project for a global constellation providing fixed point high speed data transmission services in the Ka band. This involves two systems. One is a set of eight satellites in geostationary orbit, Spaceway Exp, and the other a set of orbiting satellites, Spaceway NGSO. For its part, Boeing Co has become the world's biggest satellite manufacturer as a result of this takeover. The satellite industry is now polarised around Boeing and the other large US firm Lockheed Martin.

■ Lockheed Martin Global Telecommunications

The second large private concentration of space activities, with its own global system of communications satellites is the American giant Lockheed Martin (see fig. 10.14). As in the previous case, it has developed its space capabilities through a succession of mergers and buyouts. In 1995, Lockheed merged with Martin Marietta to become the biggest defence manufacturer and a leading contributor to the space industry. Martin Marietta brought with it GE Space, a subsidiary bought from the General Electric group. Lockheed Martin then adjusted its commercial strategy and extended its capabilities as a satellite manufacturer by becoming an operator of satellite communications systems. The new operator Lockheed Martin Global Telecommunications (LMGT) was created in August 1998 with the aim of taking its place in the lucrative market of state-of-the-art satellite services.

In order to bring together the various networks and implement connections between them, LMGT formed a joint venture called GE Americom with the operator GE American Communications, which had itself inherited GE Astro Space. Its role was to run a network comprising not only the 12 national Satcom C and GE satellites, but also satellites covering the South American markets with Nahuelsat and GE 4, and the European and Asian markets. In addition, Lockheed Martin formed a joint venture in 1997 with the international organisation Intersputnik. The new company, LMI, brought together the first elements of a global fleet by leasing satellites from Intersputnik and launching its first satellite in 1999. These covered not only the CIS countries, but also Eastern Europe, Southern Asia and Africa. Finally, Lockheed Martin Global Telecommunications also launched itself into a constellation project called Astrolink, in geostationary orbit, to provide global broadband multimedia and Internet services in Ka band services, creating a new company with the firms TRW and Telespazio.

■ Loral Global Alliance

Loral Space & Communications is the third large US manufacturer to have brought together a group of satellite networks by successive acquisitions. This group was placed under the control of Loral Global Alliance in 1998 (see fig. 10.14). Loral thereby increased its commercial capabilities, offering greater capacity, increased flexibility and wider choice to consumers through a combination of three companies, LoralSkynet formed in 1997 with the purchase of AT&T's Telstar 4 and 5 satellites, CyberStar formed in 1998 by merger with Orion Networks Systems, and finally, Satmex in which it had had a 49% stake since 1997. Loral CyberStar is specialised in the global distribution of Internet access services and high-performance services using Internet protocol. In 1999, to make up for the loss of the Orion 3P satellite, Loral leased the whole telecommunications payload of the Apstar 2R satellite from the APT Satellite Company of Hong Kong, thus obtaining very extensive coverage of Asia, Europe and Africa. Also, in 1998, Loral Space & Communications united with Alcatel in a joint venture Europe*Star Limited to provide an elaborate range of video transmission and telecommunications services to five growing markets, Europe, the Middle East, South Africa, India and South-East Asia. Europe*Star belongs to the Loral Global Alliance. Then, whilst paying particular attention to local and regional needs within its coverage zones, it will also be capable of providing global services. Loral has also invested in low-orbiting constellation projects for satellite mobile communications, especially in Globalstar.

■ **WorldSpace**

WorldSpace is a private venture motivated by a highly specific objective. Its development was made possible by the newly liberalised context of international space telecommunications and represents a novel type of system. It was created in 1990 by N.A. Samara to provide an innovative digital radio service, received directly by private individuals. Designed to answer the needs of regions that remained poorly served in developing countries as well as emergent markets, it uses a technology capable of covering vast zones and providing services for the mass market. Progressive global coverage of three continents is projected using the satellites AfriStar, launched in 1999, for Africa, AsiaStar launched in 2000 for Asia, and AmeriStar to be launched for South America (see fig. 10.14). The private operator Worldspace has a wide range of financial backers, including programme producers and distributors of receiving equipment, but also receives funding from governments, UN sister organisations and private sponsors.

New global personal communications systems

Since 1998, plans by private consortiums to set up new global constellations of low-orbiting satellites have become a reality. Whatever the technology involved – geostationary or non-geostationary satellites, narrow or broad band, global or regional, fixed or mobile services – these systems belong to a category known, to use the terms adopted by the ITU, as GMPCS systems, which represent the latest advances in personal telephone communications.

In almost all cases, American firms are the instigators of such projects. However, the way they have evolved during this first period of operation reveals the complexity of these undertakings and the significant differences in their perspectives. The **Orbcomm** constellation, created by Orbital Sciences Corporation and put into service in December 1998, is a so-called little LEO system. This means that its low-orbiting satellites are light, of the order of 40 kg, and investments reduced in consequence. Its low speed messaging and GPS positioning services were designed to extend and improve existing fixed or mobile services, such as electronic mail communications on a global scale. Results have been satisfactory to date.

From 1987, the American firm Motorola designed and developed the **Iridium** constellation. In order to cope with repeated increases in overall costs (more than 3.5 billion dollars) and handle unexpected technical difficulties, it was forced to form an international consortium with 19 partners. The commercial results of the constellation, inaugurated on 1 November 1998, did not live up to initial forecasts. This was due to the changing context during development of the programme (see pages 293–295). Iridium therefore moved away from its original philosophy of an autonomous system by offering two modes of use, selection being possible from the terminal of the terrestrial network or the spatial network. At the same time, the need to find a market led Motorola to drop the idea of global coverage (fig. 10.19), whilst turning to face direct competition from other systems. By the end of March 1999, Iridium could count upon 10 294 clients, well below its target of 40 000 for the end of 1998. With net losses amounting to around half a billion dollars at the end of the first quarter of 1999, the company was no longer able to pursue normal operations and went into liquidation. It finally ceased activities on 17 March 2000. However, the unique ability of Iridium to provide a worldwide system that is completely independent of terrestrial networks gave the American government a strong incentive to find a financial solution, as happened at the end of the year 2000. The contract between

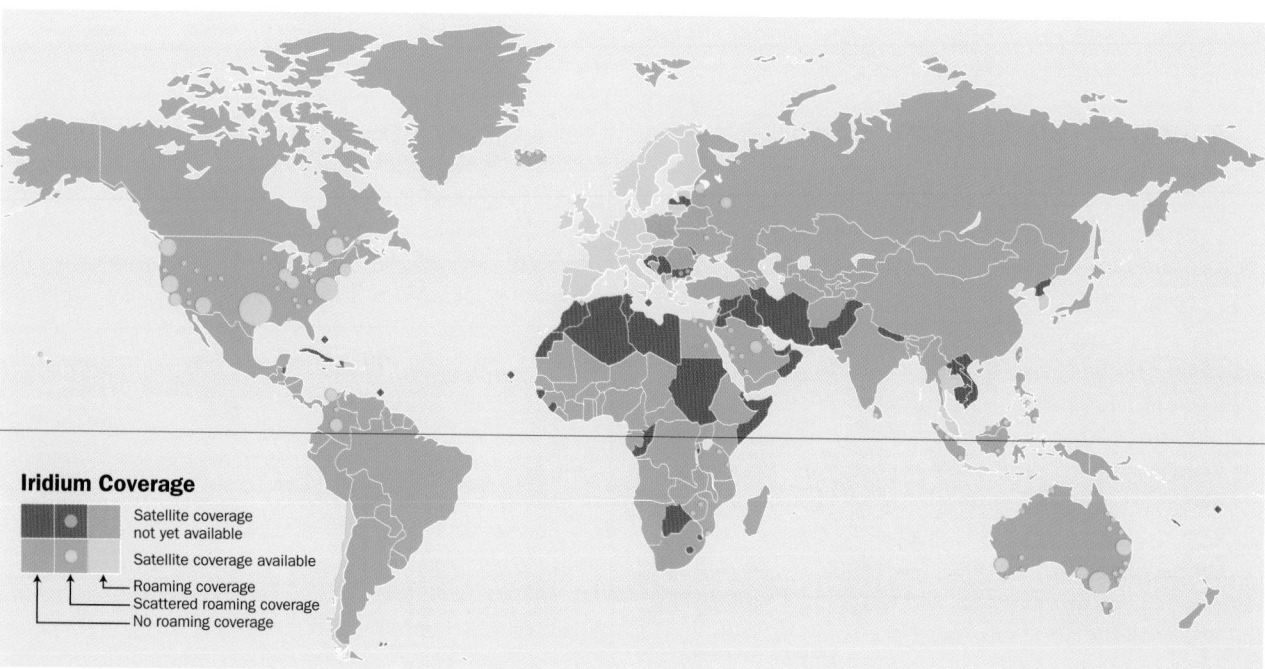

Figure 10.19. **The Iridium system in January 2000. Maritime area is also satellite covered.**

the DoD and the new society, Iridium Limited Liability Corporation, allows the use of satellites by Pentagon and other goverment organisations. With this support, Iridium LLC established services again in 2001 on a commercial basis and launched two new satellites in 2002.

On 8 February 2000, **Globalstar** successfully launched the four reserve satellites that complemented and completed its system of 48 operational satellites in low-Earth orbit. This project dates from 1989, but it was not until 1993 that Loral Space & Communications formed a consortium with the founding operators in the form of a private American company called Globalstar LP. It took a 42% stake for its own part, but European involvement was significant.

The Globalstar programme offers a less ambitious architecture since it was designed at the outset to be run in cooperation with terrestrial networks. This greatly simplifies technical constraints, both in terms of the number of satellites and the complexity of on-board installations. Under such conditions, the term global coverage is not really applicable, since the system aims at providing a space-based complement to the terrestrial network via different gateways (fig. 10.20). Reductions in terrestrial mobile costs immediately become an advantage and the possibility of providing fixed services through telephone boxes in isolated regions seems better suited to the financial capacities of developing countries, even though services for Subsaharan Africa remain notably absent from the present programme. Globalstar has made choices which give it a semblance of commercial viability that was lacking in the case of Iridium, operating satellites as simple transponders to transmit voice and data to terrestrial networks, and serving as wholesaler in its links with national and regional operators. The fastest uptake has been observed in highly developed countries and especially

those in which personal communications technology satisfies a need for contact over extended geographical regions. On the other hand, those regions considered as the most attractive potential markets are yet to begin or are only just beginning to use services.

Globalstar was selling its terminals and services in more than 100 countries in January 2002, and others have given authorisation, so that distribution circuits are currently being set up. The main markets are in mining and exploration, maritime activities, energy, agribusiness, transport, and search and rescue. The government sector also appears to be important.

Broadband systems or satellite constellations for fixed very high speed data transmission links, multimedia applications and interactive services including Internet are still under development. It is not yet clear what their chances of success might be. Two of the most serious projects, **Teledesic** and **ICO Global Communications** merged in May 2000 in response to major financial and technical difficulties faced by both alike. A new company, ICO–Teledesic Global, was created. Its mission is to develop a two-stage programme: the first, early in 2003, will lead to the deployment of those ICO satellites ready for launch, whilst the second will begin in 2004 with the installation of Teledesic's high speed system.

At the World Radio Conference of 2000, the originally European project **Skybridge** received guarantees that it would be allowed to use the frequencies it required. Skybridge LP reached a partnership agreement with Boeing in December 1999. Boeing became a shareholder and supplier of launch services providing an opening on the American market. In April 2000, a similar agreement was signed with the Euro-Russian company Starsem, thereby securing the launch of the 80-satellite constellation ready to begin commercial operations in 2003.

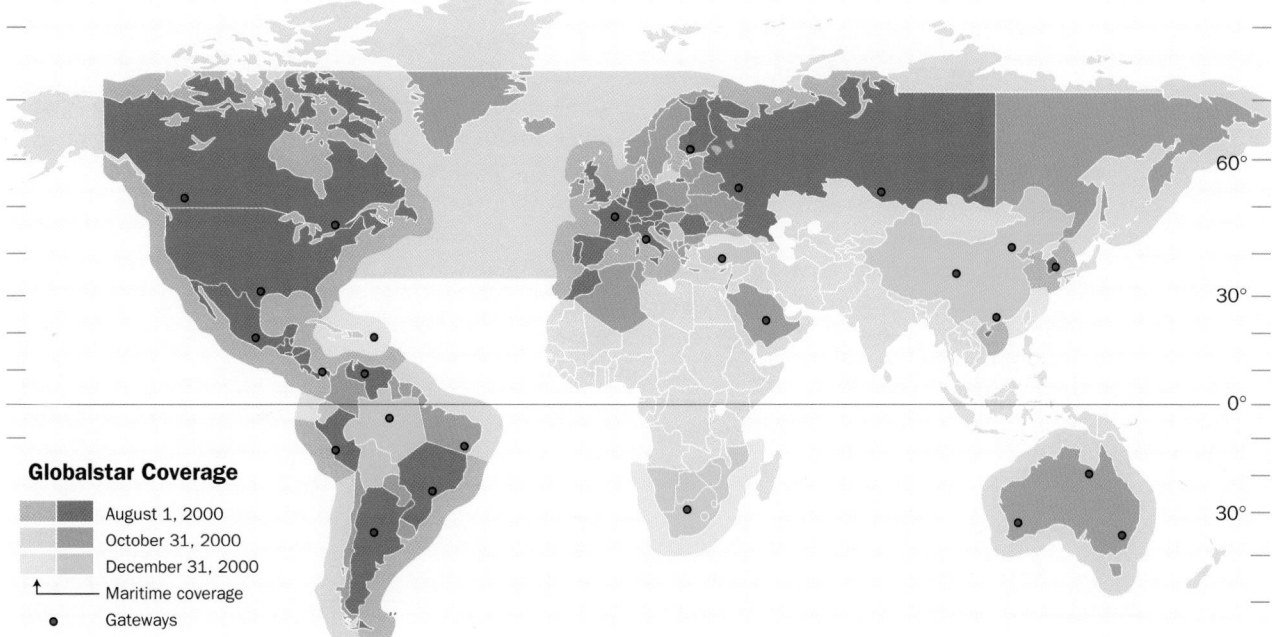

Figure 10.20. **The Globalstar system in 2000.**

Regional systems

Regional systems form a relatively complex category in the sense that there is a tendency for technical developments and commercial operations to become increasingly tied up with those in national and international systems. Composed of intergovernmental organisations like Eutelsat or Arabsat, or depending on privately-owned companies like Astra or Asiasat, regional systems are experiencing similar developments to those affecting international systems in the current deregulatory climate, that is, gradual privatisation of intergovernmental organisations and growing numbers of private investors, particularly in Asia.

Intergovernmental systems

■ Eutelsat

The significance of telecommunications satellites was realised early on in European countries. Since the 1960s, various experimental satellite projects had been envisaged by the European Space Research Organisation (ESRO). Preliminary studies had focused on developing television services but cost-effectiveness seemed uncertain. The main European countries thus committed themselves rather to national programmes, like Italy with the Sirio satellite, or bilateral projects, like France and Germany who set up their common programme, Symphonie.

By the time Europe had organised its space activities, the idea of establishing a regional telecommunications system came back into the foreground. In 1979, the European Space Agency decide to build five satellites called European Communication Satellites (ECS) to be managed by the European organisation Eutelsat (*EUropean TELecommunication SATellite organisation*) created in 1977. The ECS system had a double role to play. The satellites had to satisfy the needs of the post office and telecommunications administration, and also had to broadcast television programmes planned by the European Broadcasting Union (EBU). In order to avoid frequency allocation problems, which would have been inevitable in the heavily used 4–6 GHz band, the satellites were designed to operate at frequencies above 10 GHz. Another advantage of this choice was that European industry could acquire new skills in state-of-the-art technology.

A first satellite, the Orbital Test Satellite (OTS), was used in 1978 to carry out link tests with small Earth stations with the help of a powerful antenna on board the satellite. It also provided the first data transmission services for industry. Despite initial reluctance on the part of the telecommunications administration, who feared with some reason that their services might be bypassed, the ECS took up and improved these services, changing their name to Eutelsat I once Eutelsat had taken over the operation.

Since 1985, when Eutelsat was officially set up with a total of 17 members, the organisation has grown to 44 member countries and its services have considerably diversified. Apart from television services, which represent a significant share of the traffic, the satellites provide fixed regional and international telephone services as well as data transmission. Specific services have been developed, for example, in business communications (Satellite Multiservices System) or links with land mobiles (Euteltracs) since 1991. This incursion into the highly competitive market of mobile telephony demonstrates the adaptive capacity of the organisation, in search of new sources of income and ready to assert itself as a world leader.

Since 1990, Eutelsat's regional vocation has been strengthened by an increasing membership of Eastern European countries and improved services in this region through Eutelsat II F 4, deployed in 1992 (fig. 10.21). This policy has been pursued and with the launch of the new Hot Bird satellites, coverage will be guaranteed from Finland to Morocco, and from Lisbon to Moscow, for 70 cm antennas. The launch on 17 April 2000 of the Siberian European SATellite (SESAT) by the Russian company NPO PM in association with Alcatel shows that Eutelsat's European vision, extending from the Atlantic to the Ural mountains, has not faltered since the fall of the iron curtain.

In its original form, the mission aimed to complement the terrestrial telephone network of European countries without transgressing the rules laid down by Intelsat, since Eutelsat's signatories used the latter for their international communications. In practice, the Eutelsat system made most of its profits from the high demand for television, rather than through telephone services, the former yielding 70% of its takings. On this market, Eutelsat found itself in direct competition with the Astra satellite developed by the Compagnie Luxembourgeoise de Télédiffusion (CLT), and which has a particularly strong commercial base. In addition, in 1999, Eutelsat began to deploy its Atlantic Gate mission, with satellites located at 12.5° east. This will eventually allow the organisation to set up links connecting North and South America to Europe. It will thus meet the great demand from business networks on the transatlantic Internet market, and for television services for the general public. At the same time, it is strengthening its position on the European continent.

With the general shake-up in space telecommunications organisations, quite different questions have become relevant today. Eutelsat must adapt to cope with competition and develop its own internal organisation accordingly. A privatisation process was initiated which, in 1998, led the Assembly of Parties to approve the transformation of Eutelsat into a limited liability private company under French law. In 1999,

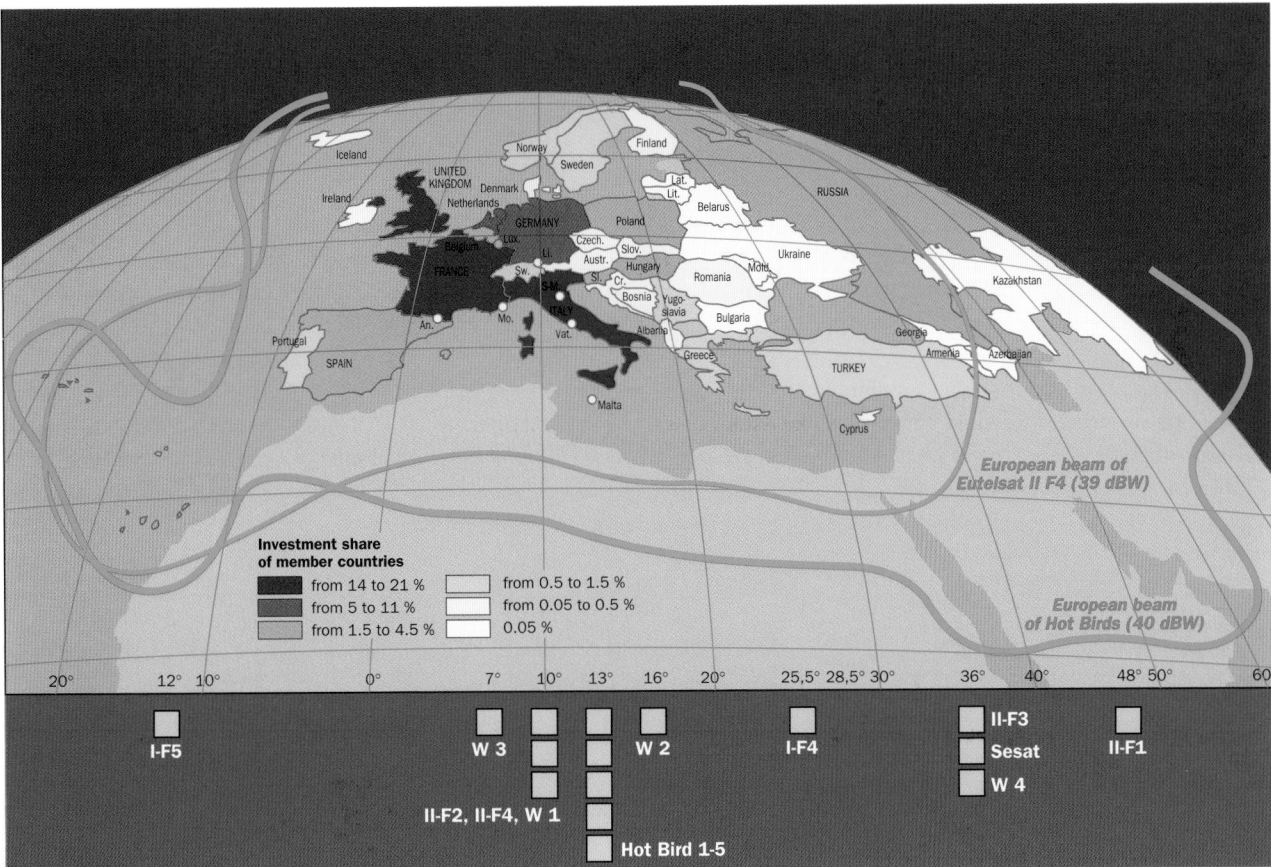

Figure 10.21. **The Eutelsat system in January 2000.**

the date for transferring the assets and activities of Eutelsat was brought forward by general consensus to 2 July 2001. A restricted level of intergovernmental authority has been maintained to ensure that Eutelsat SA continues to respect the basic principles of the European system, namely, the obligation to provide universal services, pan-European coverage, equitable treatment for all concerned and a guarantee of fair competition.

■ Arabsat

The Arab Satellite Communications Organisation (Arabsat) was also created through a political desire to develop regional links, although in a very different context, since it reflects the particularly strong urge felt by certain Arabic countries to encourage cultural unity and strengthen Arabic identity. It was officially created in 1976, following nine years in the pipeline, 30 years after the foundation of the Arab Postal Union, with space television broadcasting as its main activity. Not only does the system show how techniques have evolved, but it also serves to express a new trend developing since the beginning of the 1970s. Although the notion of cultural and historic unity is still strongly felt, the ideal of total reunification is giving way to a quest for a more modest and more effective form of coalition between the Arab nations.

The Arabsat programme sprang from a combination of economic and political factors. In 1976, encouraged by their initial success on the petroleum market and strengthened by the new source of revenue, 22 countries decided to build, launch and operate geostationary telecommunications satellites for their own purposes. At the time, the Arabsat project was a great innovation for, although Canada and the United States possessed satellites other than those in the Intelsat system, the aim was purely to provide for communications needs on their home markets, that is, on a national basis.

On the technical level, petrodollars were reinvested in such a way as to exploit competition on the space market, whilst preserving a diplomatic concern for diversification, with an order for three satellites from an international consortium. The French company Aérospatiale was the project coordinator and launches were entrusted to Ariane and the Space Shuttle Discovery. Since 1983, receiving stations have often been built by Japanese companies.

The time required to deliver such projects meant that the first satellite was launched in February 1985 in a completely different context, which had by then become perfectly favourable from an international standpoint: Ariane had been operating successfully since 1981 and the Intelsat monopoly had been officially broken in 1984.

Meanwhile, however, the position of Arabic countries had considerably deteriorated. On a regional level, the distribution of receiving stations and the related question of investment shares had brought out political differences (fig. 10.22). In particular, Egypt had held a rather unusual position in the Arab world since negotiations had begun with

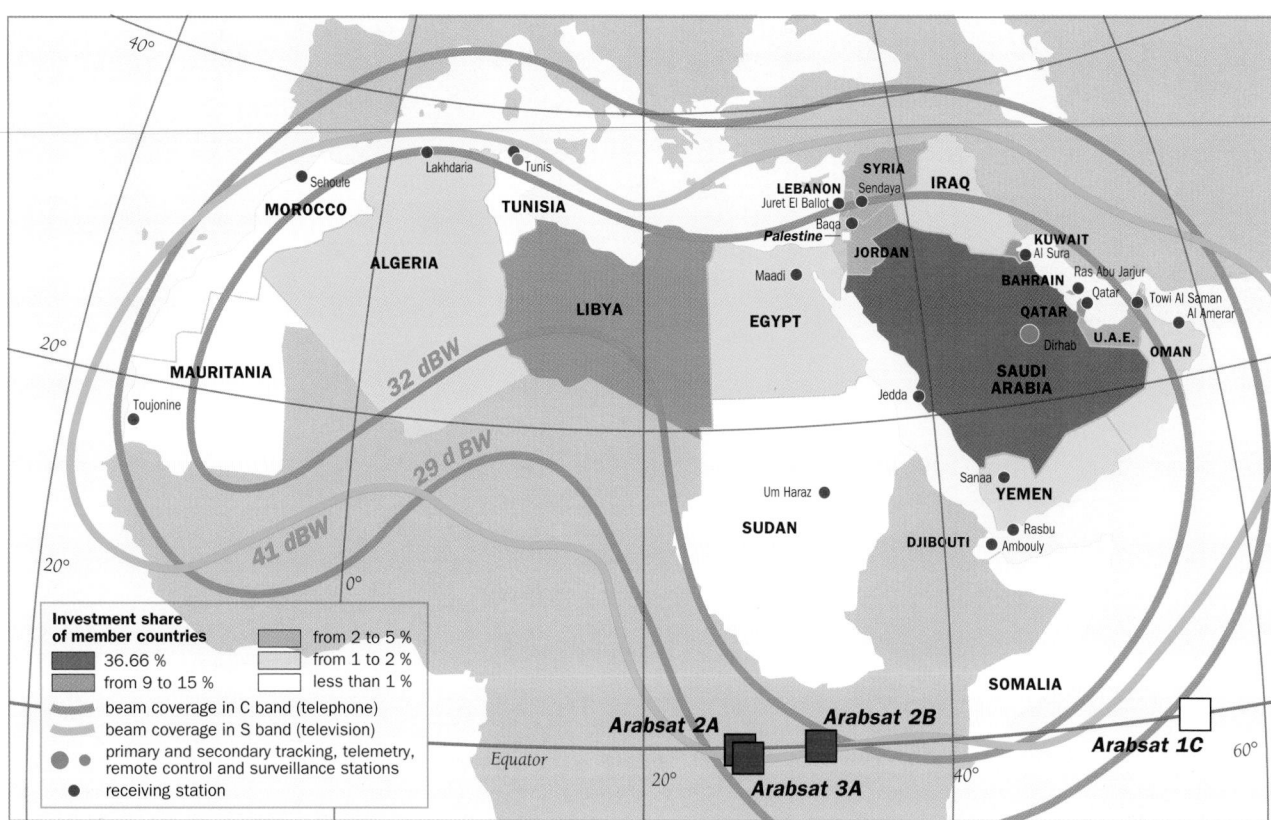

Figure 10.22. **The Arabsat system in January 2001.**

Israel, and when the system was being set up, did not possess its own station. Financial considerations soon led the organisation to adopt a more flexible approach. In addition, there were many delays in installing ground stations, stemming from financial problems as much as from political disagreement. Hence, Libya still has no ground station, despite the fact that it holds a good place amongst the investors.

Apart from relaying telephone calls, the Arabsat satellites were also supposed to transmit a common broadcasting programme that would express and strengthen Arabic identity. However, problems also arose with this last type of service due to disagreement between countries. For a long time, only one weekly and one daily programme of television news were actually broadcast, produced in the Tunisian television and radio studios. They were made up of information transmitted by satellite from the different countries and could be either partially or fully broadcast via the national television network. On the whole, the original ambition for conciliation between Arab nations is far from being achieved, even though recent improvements could be said to belie an accusation of failure. The consolidation of telephone links remains limited and certain countries continue to favour Intelsat satellites.

In fact, since 1991, the philosophy behind the system has evolved significantly. For a certain time, the possibility of direct broadcasting by Arabsat satellites had been neutralised because member governments hoped to maintain total control. However, the emitting power of foreign satellites and

the unhindered proliferation of receiving antennas had shown that such a position could no longer be justified in the face of this kind of competition. Moreover, the development of a wide-coverage national satellite like the Egyptian Nilesat contributed to more flexible political relations between Arabic countries, who could use the system as a way of juxtaposing national services.

After several financially testing years, the Arabsat organisation appears to have found new life by emphasising the commercial and profit-making aspects of its services. This is exemplified by the sale of the first generation satellite 1C to India. The 1996 launch of two second generation Arabsat satellites with part of their capacity open to lease is also witness to this new attitude. Services proposed are precisely like those of other systems, including telephone, data transmission, radio and television broadcasting and business networks. The first third generation satellite was launched in 1999, devoted to direct broadcasting in the Ku band, with wide coverage of the Middle East, North Africa and Europe. This extension of online services to 20 television channels is part of an effort to broaden the user base, at the same time re-establishing services to Iraq and opening up to foreign television companies, which nevertheless remain in the minority. In parallel, the project to set up an Arab information highway highlights a partnership between Arabsat, the Arab Union of radio broadcasters and the Egyptian company Nilesat. Profits made by the organisation in 2000, mainly from television activities, go hand in hand with a more sig-

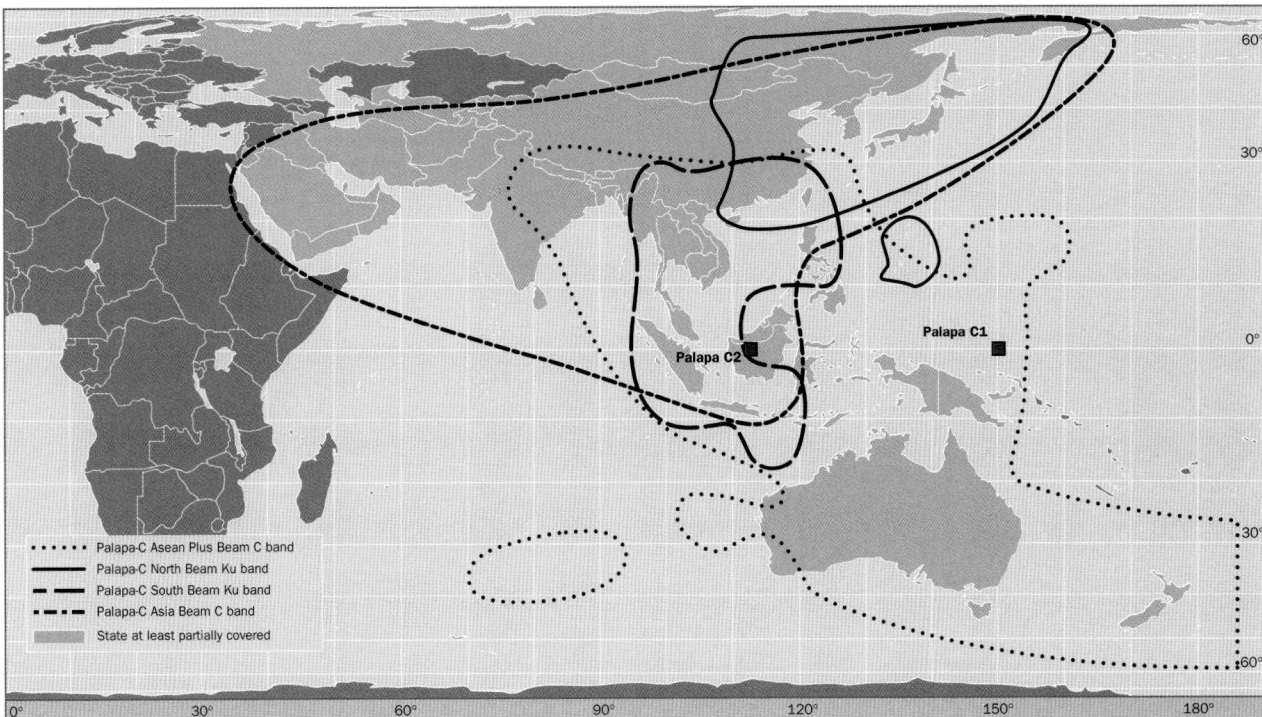

Figure 10.23. **The Palapa system in January 2002.**

nificant interpenetration of Arab and foreign television stations in the satellite bouquets broadcast by the three Arabsat satellites, in combination with wider broadcasting of Arab channels by Western satellites. The announcement in November 2001 of an order for two new-generation satellites to be launched by 2004 attests to a clear determination to uphold their presence on the international market. Apart from asserting a more commercial outlook, this new orientation may also serve to strengthen a cultural presence around the world, as demonstrated by the wide-ranging broadcasts of the Qatar-based channel Al Jazeera during the Afghanistan conflict in the autumn of 2001.

■ Palapa

Historically, Palapa was the second telecommunications system with a regional vocation. Its approach is different again. The programme was originally developed by Indonesia, a country with widely scattered territory and population, factors which made satellite an ideal tool for treating the communication problem. Indeed terrestrial solutions had been far from successful. The first Palapa satellites were launched in 1976, and were initially intended to cater for national needs in television and telephone lines.

Two years later, the Philippines signed an agreement with Indonesia whereby Palapa would provide for a part of their national coverage. Thailand and Malaysia followed suit and in their turn became users of the space segment. Finally, in 1979, Palapa was officially recognised by Intelsat as a regional system, even though certain characteristics of its use and even its motivation differed fundamentally from those of Eutelsat, the first organisation of its kind.

As time went by, Palapa began more and more to resemble a national system capable of providing services to countries in the Association of Southeast Asian Nations ASEAN, (fig. 10.23). The third generation of satellites (Palapa C) was brought into service in 1996 with the aim of extending coverage across the whole of Asia, supplying new Ku band broadcasting services. At the same time, the second generation satellites (Palapa B) were gradually sold off to another consortium which included the original administrator of the Palapa system. The system is now run by a new private operator created in 1993, Satelindo, forming a strategic partnership between both state and privately funded national and international telecommunications companies.

Private regional systems

Since the notion of regional systems does not exist in the United States, Europe was first to offer an example of a private regional system when Astra was set up by the Société Européenne des Satellites (SES), whose main shareholder is the grand duchy of Luxembourg. They lease transponders to broadcasting companies who can either use the satellite alone or in association with a cable network. Today, this commercial approach and the limits of a European market subject to competition from Eutelsat have led the company to acquire interests in international consortiums that are active on new markets.

It is in the developing countries, keen to profit from new information technologies, as exemplified by many Asian countries, that privately run systems are the most common. Indeed, the lack of telecommunications infrastructure in Asia and the presence of natural barriers make this region particularly well suited to satellite services. After a period of economic difficul-

60¡
30¡
0¡
30¡
60¡
30¡
60¡
90¡
120¡
150¡
180¡
Asiasat 1 northern beam
Asiasat 1 southern beam
Asiasat 2 C band
Asiasat 2 Ku band
Asiasat 3 S C band
Asiasat 3 S Ku band East Asia beam
Asiasat 3 S Ku band South Asia beam
Asiasat 3 S Ku band steerable beam
ACeS
Measat 1 C band 29 dBW
Measat 2 C band 28 dBW
Measat 2 Ku band
Palapa-C 2 Asean Plus Beam C band
Palapa-C 2 North Beam Ku band
Palapa-C 2 South Beam Ku band
Palapa-C 2 Asia Beam C band
ST 1 C band 38 dBW
ST 1 Ku (K1) band 50 dBW
ST 1 Ku (K2) band 50 dBW
Thaicom 3 C band
Thaicom 3, 2,1A Ku band Thailand Beam
Thaicom 3 steerable Beam
Thaicom 1 A C band regional beam
Thaicom 2 C band regional beam

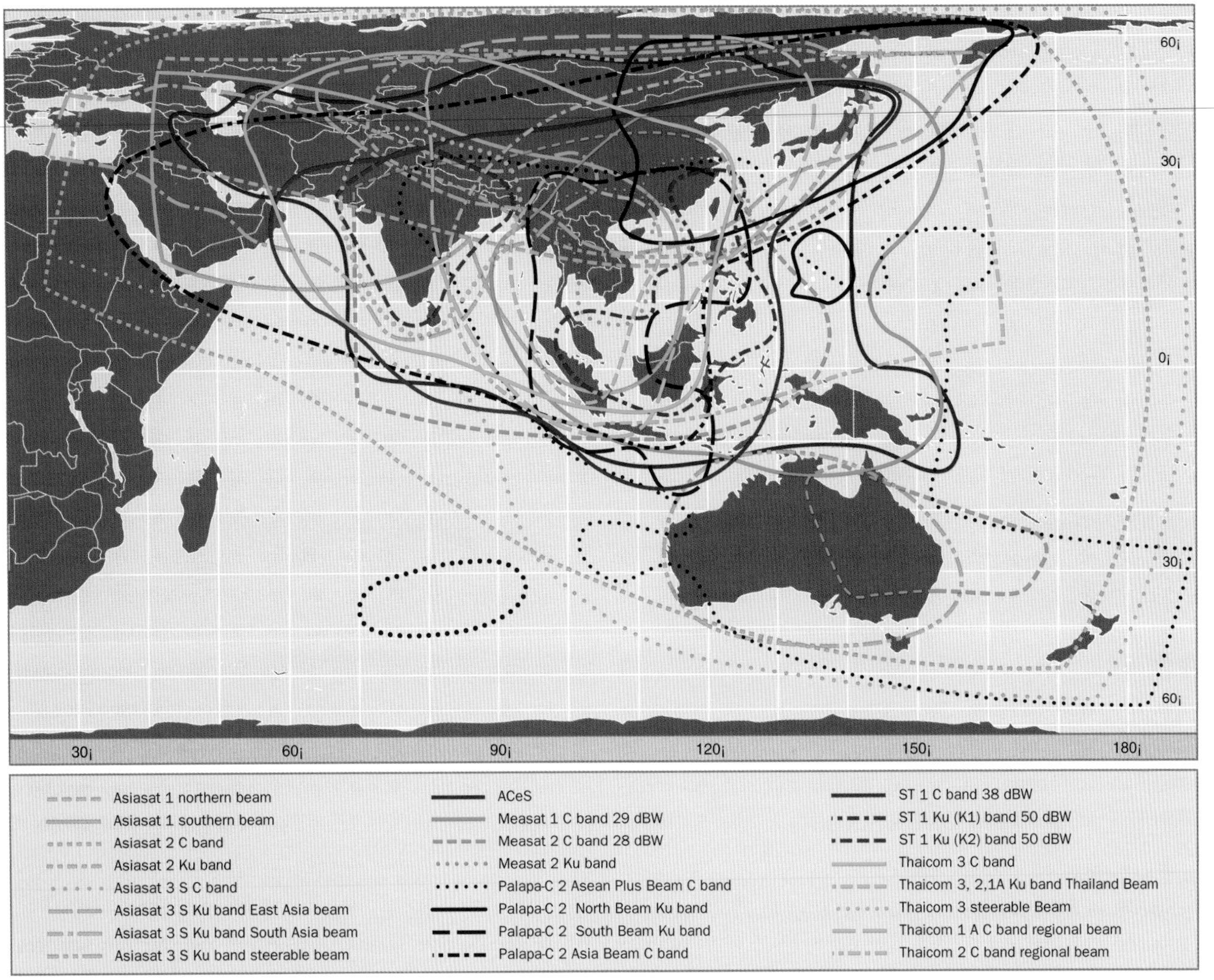

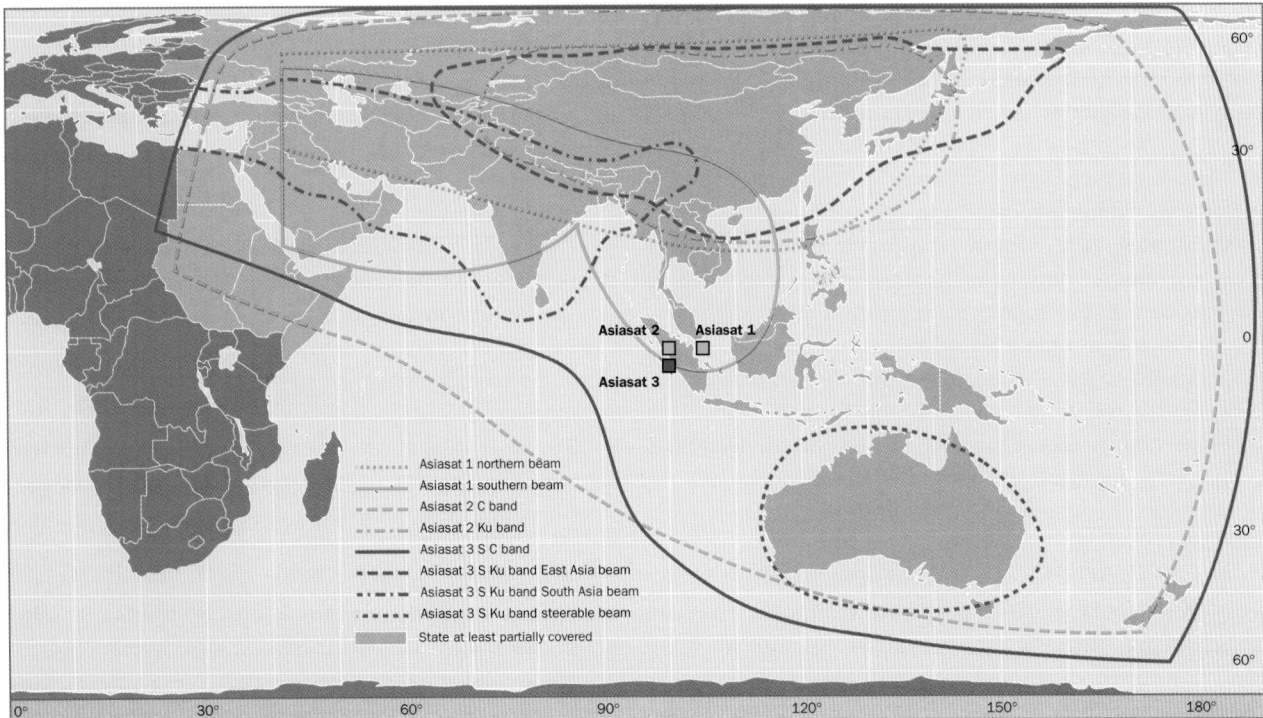

Figure 10.24. **Telecommunications coverage in Asia in 2000.** Some systems are also shown separately in Figures 10.23, 10.25 and 10.26.

60°
30°
0
30°
60°
0°
30°
60°
90°
120°
150°
180°
Asiasat 2
Asiasat 1
Asiasat 3
Asiasat 1 northern beam
Asiasat 1 southern beam
Asiasat 2 C band
Asiasat 2 Ku band
Asiasat 3 S C band
Asiasat 3 S Ku band East Asia beam
Asiasat 3 S Ku band South Asia beam
Asiasat 3 S Ku band steerable beam
State at least partially covered

Figure 10.25. **The Asiasat system in January 2000.**

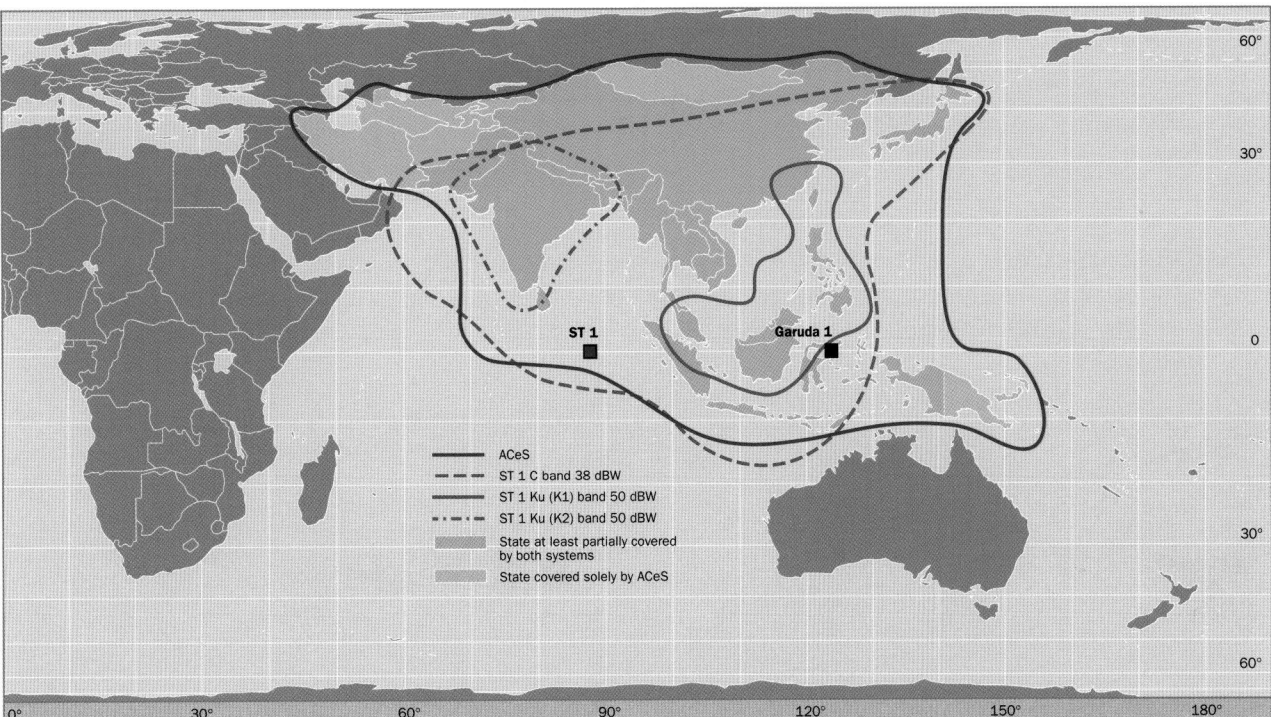

Figure 10.26. **The ACeS and Singapore-Taiwan systems in January 2000.**

Legend:
- ACeS
- ST 1 C band 38 dBW
- ST 1 Ku (K1) band 50 dBW
- ST 1 Ku (K2) band 50 dBW
- State at least partially covered by both systems
- State covered solely by ACeS

ties, space-based projects began to increase in number, generally taking on regional dimensions, in response to the trend towards privatisation and globalisation (fig. 10.24).

■ **SES-ASTRA**

As the United States never developed any regional-like satellite system per se, Europe offered the first example of a genuine regional system based on the Astra system managed by the Société Européenne de Satellites (SES). With Luxemburg as its main shareholder, it is renting transponders that are made available to radio broadcasting companies than can use them on their own or in association in a network. The business-orientated activity as well as the limitations inherent to a strongly competitive European market partly occupied by Eutelsat, have led SES to buy shares in international consortiums involved in new markets. In 2002, SES-Astra, equipped with 12 satellites positioned on 3 orbital slots, is providing now a significant number of very powerful transponders to broadcasters or to multimedia companies. More over, new partnerships with operators in Scandinavian countries, Latin America and Asia, as well as the acquisition of the US operator Americom, led to the creation of SES-Global. The company is now active in 4 continents and is presenting itself as a strategic investment for a significant number of satellite operators.

■ **Asiasat**

In Asia, the first effects of deregulation were felt in Hong Kong in 1990 with the creation of the consortium Asia Satellite Telecommunications Company, or Asiasat. Set up to exploit commercial space telecommunications, the company was formed by the China International Trust and Investment Corporation (CITIC), together with two other shareholders,

Hutchinson Whampoa of Hong Kong and Cable and Wireless. In 1999, the latter two investors brought in the European company SES, keeping a 49.5% stake. This operation marked a new need for space operators to obtain global coverage. Asiasat 1, the former Westar satellite, was taken over and relaunched in 1990, whilst Asiasat 2 was launched in 1995 and Asiasat 3 on 21 March 1999 to replace Asiasat 1. According to its operators, it covers two-thirds of the world population and its latest satellite covers more than 50 countries in Asia (including China and India), the Middle East, and CIS countries, as well as Australia (fig. 10.25). The services provided are highly diversified, ranging from public telephones to private VSAT services, and including high speed Internet and multimedia links.

■ **Measat**

In 1996, the Malaysian private firm Binariang Satellite Systems sent two satellites into orbit to form a new regional network MEASAT, providing broadcasting and telecommunications services in the Far East, India, the Philippines and Australia (fig. 10.24). Two further satellites extend coverage to China, the Middle East and Africa.

ACeS

Asia Cellular Satellite (ACeS) is the first regional satellite system dedicated to mobile personal telecommunications, with coverage exclusively aimed at the Asia–Pacific region (fig. 10.26). It is the joint property of several Asian partners. Launched on 12 February 2000, Garuda 1 provides state-of-the-art mobile telephone services as well as data usable on private terminals at very low costs. Agreements have been reached with more than 43 terrestrial cellular network operators in 26 countries, including China, and a second satellite is planned to serve both as a backup and also to extend ACeS coverage to Central Asia, Eastern Europe and part of North Africa. The system has aimed at a regional approach capable of competing with Iridium and Globalstar's international policy.

■ Thuraya

The regional system Thuraya, named after a star cluster in the constellation of Taurus, was launched by the private firm Thuraya Satellite Communications Company, with a first satellite in May 2000, to provide GSM mobile communications services to almost 40% of the world population. The company is mainly composed of shareholders from the Arabic countries, including those involved in Arabsat, but also the American giant Hughes Space and Communications, who manufactured the satellite. The service is in principle guaranteed for 15 years, given the lifetime of GEO satellites. Such a low cost system should provide services to more than 100 countries, across a zone covering the Indian subcontinent, Central Asia, the Middle East, North and Central Africa, and Europe.

National systems

The very idea of a national system has also changed considerably through the new role played by private entities in competition with state-run organisations. The American case undoubtedly serves as an extreme example, spelling the end of a specific definition for this type of service, but even if the dividing lines are less clear, combination of national systems into international consortiums such as Global Alliance and the extension of coverage beyond national territories represent significant changes.

National systems have developed in response to a range of different concerns (fig. 10.27). They were originally established by developed countries needing to cover a vast

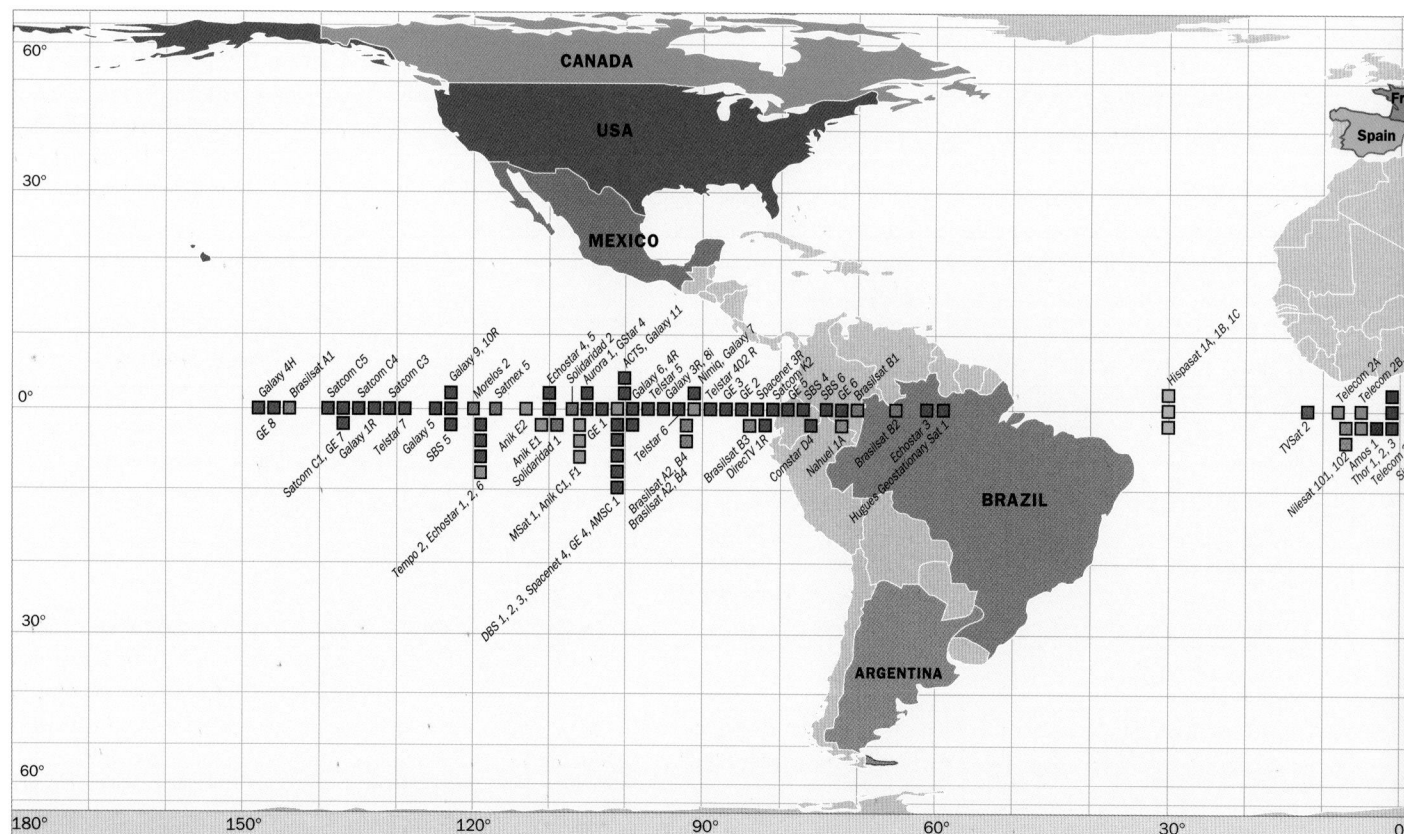

Figure 10.27. **National space telecommunications systems in January 2000.** Satellites are shown in the colour of the country they belong to.

and unequally occupied territory with telephone and television broadcasting services (see fig. 10.14). The first such countries were Canada, the United States and the Soviet Union. From the 1970s, they had complemented their terrestrial networks with space systems. Other national systems were set up later, between 1983 and 1985, by the more advanced amongst the developing countries, such as India, Indonesia, China, Mexico and Brazil, and also Western countries like France, Australia and the United Kingdom, the latter concentrating on military satellites. The trend continued from 1989 to 1992 with Japan, Sweden, the United Kingdom, Germany, Hong Kong, Spain and Italy establishing their own space networks. All these cases represent countries keen to strengthen their trade capacity and develop their television broadcasting capabilities beyond their own borders, when necessary. The latest systems brought into service in Thailand, Turkey, South Korea, Israel, Malaysia, Argentina, and Egypt correspond to the development of increasing national needs, but also to political and commercial concerns. This is the case in Egypt where Nilesat, the first direct broadcasting Arabic satellite, is used to relay national cinema productions.

The announcement of the first African satellite launch for 2002 by the regional organisation Rascom shows that the African continent still lags behind. This is due to the low level of its own needs but also, above all, to a lack of investment. Indeed, national satellites are not expected in the near future.

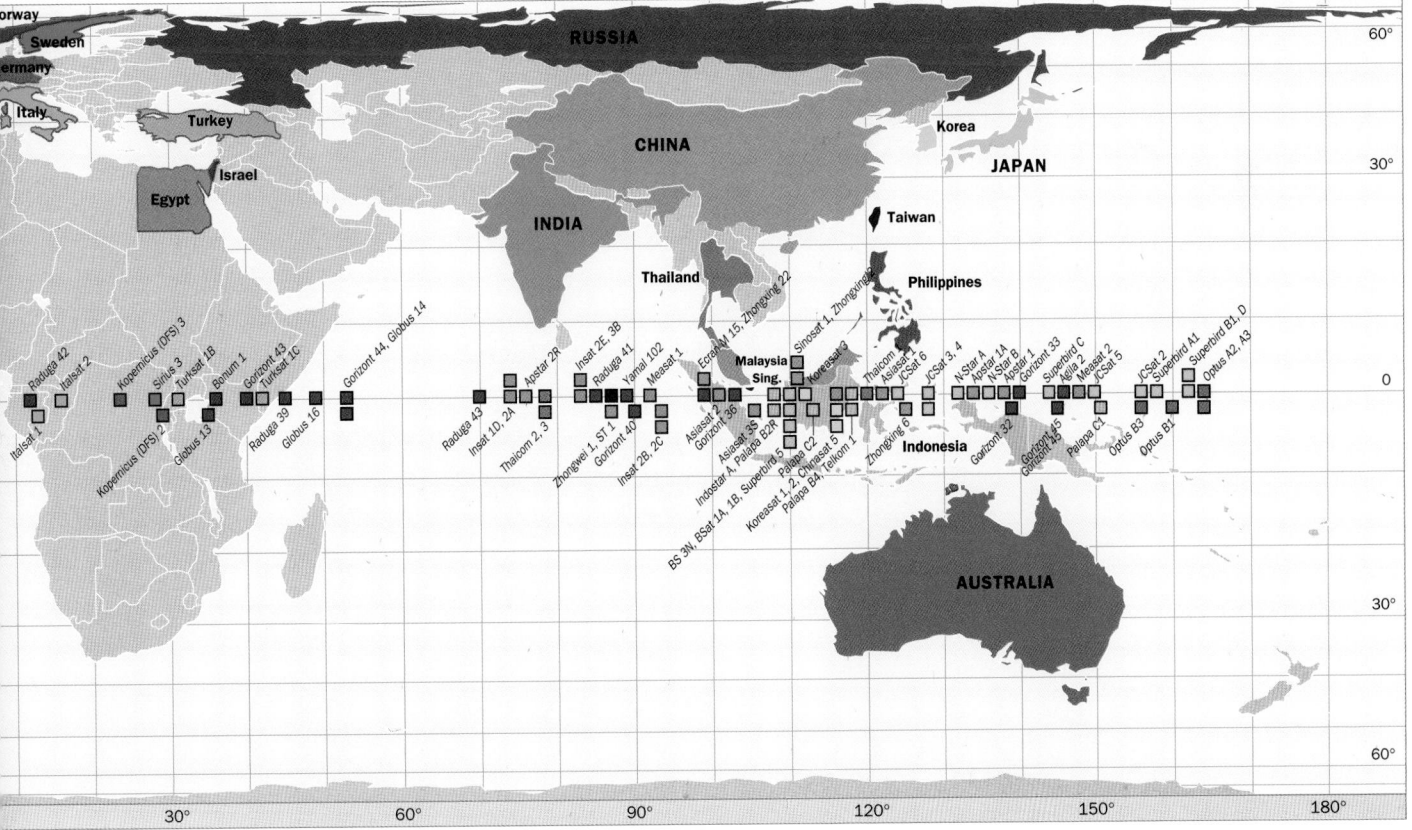

CHAPTER ELEVEN

Positioning and navigation

- **OVERVIEW**

- **SYSTEMS USING DOWNLINK SIGNALS**

- **SYSTEMS USING UPLINK SIGNALS**

Overview

Navigation is the art of determining the direction to be taken by a craft on the basis of an exact knowledge of its position. Traditional use of the Sun, Moon and stars to take bearings is gradually being superseded by services available from artificial satellites. These have greatly improved possibilities for identifying positions within an accurate universal reference system that extends everywhere over land and sea, as well as in the air, regardless of weather conditions.

Only the United States and the Soviet Union have developed worldwide satellite navigation systems, viz., GPS and GLONASS. Although initially set up for military purposes, civilian applications soon became commonplace. Indeed, since the 1990s, these systems have grown into one of the main methods for positioning thanks to their accuracy and availability.

Although navigation was quickly identified by European industrialists as a key space application, the financial and technological difficulties involved in this type of system and the absence of any decisive military need have meant that there has never been an independent European programme. The relative lack of interest can also be explained by the fact that the US system is freely open to all, without charge, even though data available to civilian users was less accurate than that available to the US military up until May 2000. It is only recently that Europe has really become aware of what is viewed as a strategic dependence, given the important applications for aviation and land-based activities, and considered developing its own complementary but interoperable system. Carried at last in March 2002 by the EU and ESA, Galileo should be operational in 2008. In the same way, China's launch of the beginnings of a constellation in 2001 and Japan's continued research in key areas such as the development of atomic clocks, a key element of the system, clearly indicate a universal interest in this type of space activity.

For the established systems, the situation is quite different. Russia has gone in search of international collaboration that will bring financial backing for its system, whilst the United States, the only country to provide international services in this domain, is seeking to maintain its technological lead and a suitable balance between civilian and military needs.

Systems set up for radiopositioning and radionavigation by satellite comprise three segments. The first, on the ground, accurately determines the orbits of satellites (orbitography) and controls their positions, as well as providing users with satellite ephemerides, among other things. The second is made up of the satellites forming the reference network, whilst the third refers to all users, whether they be on land, at sea, in the air or even in space.

Several systems fit into this general scenario.

- Uplink systems: the ground-based transmitter may be simple, but the onboard receiver carried by each satellite is complex and its transmission can be scrambled. Examples are Argos, COSPAS–SARSAT, and DORIS.
- Downlink systems: satellites carry a simple payload, emitting signals timed by an ultrastable clock. Examples are Transit, Kosmos–Tsikada, NAVSTAR–GPS (Global Positioning System) and GLONASS.

The position of the mobile is deduced from variations in the Doppler effect. The satellite supplies its exact position as a function of time and also the frequency of the signals, making it possible to calculate the latitude and longitude of the receiver. Extremely high precision less than 1 m in the best cases can be achieved.

Systems using downlink signals

This is the main type of system used for navigation by satellite, since users can determine their position and velocity at any moment. Such systems were first developed by the United States and the Soviet Union to satisfy military needs and subsequently opened up to the civilian sector (fig. 11.9).

Global navigation satellite systems can be used in several domains such as geodesy, aircraft navigation, and time synchronisation. The number of users is on the increase. The most dependent applications are geographical surveying and mapping, together with the monitoring of agricultural land

and trawler fishing. Satellite-based navigation services are also used in automobile navigation systems (an accuracy of 5 to 20 metres is needed). Combined with radio information about traffic levels, these systems can optimise routes and make good estimates of journey times. Ever more precise, such services are today capable of competing with ground beacons for aircraft take-off and landing and even of monitoring rockets during launch. However, the idea of using a single system assumes a level of reliability that has not yet been attained, in particular with regard to integrity parameters, that is, the system's ability to inform the user sufficiently rapidly when failure occurs. At the same time, navigation and telecommunications applications are tending to merge, thus giving access to new markets such as mobile phone services.

Operating principles

In the case of the Tsikada and Transit navigation services, the first systems in this category to be developed, the satellite motion is known and the user position computed from measurements of successive Doppler or apparent frequency shifts of the signal as the satellite approaches or passes the user. This yields the latitude and the longitude (fig. 11.1). The satellite continuously emits two frequencies. These serve to transmit satellite positions and time references by phase modulation, but also to calculate distance differences by measurement of Doppler frequency shifts. The receiver counts the number of cycles received during a given time interval and the calculation is carried out iteratively by comparing relatively measured distance differences with those that would have been measured to the estimated point. Successive approximations are made and the calculation stops when discrepancies grow smaller than some previously chosen value.

For GPS and GLONASS, the user has to measure a time difference between the instant a satellite broadcasts its signal and the instant the user receives the signal. This time interval multiplied by the speed of light corresponds to a range, called the pseudo-range, from the receiver to the space vehicle. With this information, the user's position is located on a sphere with radius the measured range, centred on the satellite. Navigation messages contain parameters called ephemerides specifying spacecraft orbits. Therefore, a user receiving three signals from three different satellites will be able to determine its position as the intersection of the three spheres (fig. 11.2).

However, setting aside the theoretical aspect, there are two problems. First, the user and satellites have to be synchronised in time. The ground segment synchronises the constellation (e.g., the beginning of the initial GPS time reference was 6 January 1980, which then changed to 22 August 1999). Naturally, the user cannot be exactly synchronised with GPS time, since it costs too much to procure very accurate clocks. Therefore a fourth measured pseudo-range is needed to determine a last state parameter, the user clock bias. The other problem is that additional delays occur in the range measurements, such as atmospheric effects. Navigation messages contain correction parameters that can be used to estimate these delays although not to determine the exact bias. The position computed using a navigation satellite system will not therefore be the true position. The better the additional delays can be estimated, the more accurate the computed position will be.

Transit navigation satellite system

The Navy Navigation Satellite System, also called Transit, was the world's first operational satellite navigation service. This project was developed in the early 1960s to meet the precise navigation requirements of the US Navy's ballistic missile submarine fleet. The first navigation sets were installed in

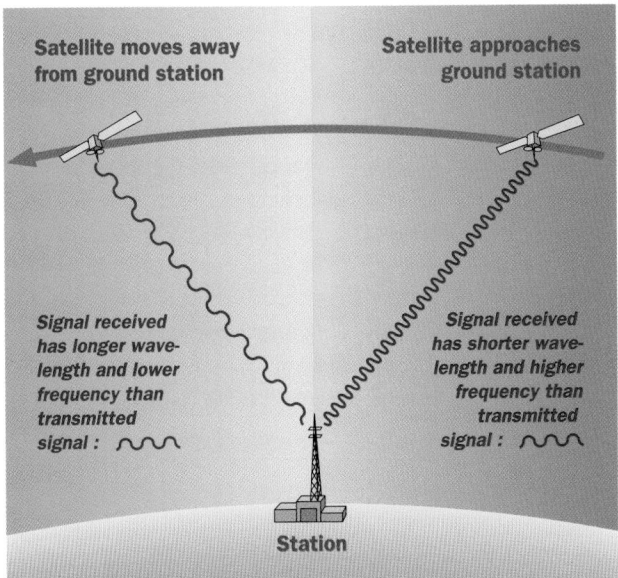

Figure 11.1. **The Doppler effect.** The Doppler effect is used to determine the component of the satellite's velocity along the line of sight. The frequency shift due to the satellite's motion can be measured by counting the number of cycles in a given time interval, for example. This approach works for low-Earth orbiting satellites, but cannot be used for geostationary satellites.

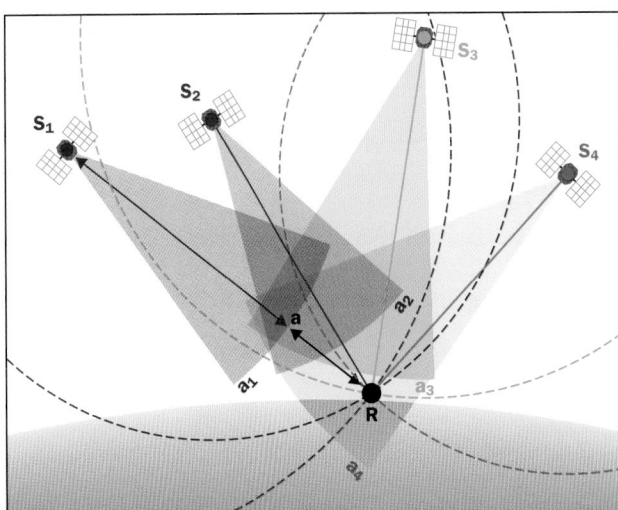

$S_1 \longleftrightarrow a$: Measured range (pseudo-distance)
$a \longleftrightarrow R$: Receiver clock bias
S_1R : Estimated range from point R to satellite S_1

Figure 11.2. **GPS ranging measurements.**

ballistic missile submarines and aircraft carriers in 1962 with accuracy better than 100 m. Many other types of navy and scientific ships used this all-weather navigation service. The system was declared operational in 1964 for the benefit of the US Department of Defense, but was made available to civilian users as early as 1967.

The space segment was composed of six satellites describing 1000 km polar orbits. Of the six satellites, three provided navigation services while three others were on-orbit spares. The last Transit satellite was launched in August 1988. The Transit ground segment had three ground-based monitor stations. These tracked each satellite while it remained in view and provided the tracking information necessary to update satellite orbital parameters every 12 hours. Transit satellite signals were monitored by the Naval Astronautics Group (NAG) at Point Mugu, California, which served as the ground control facility for the satellite constellation.

The signal in space contained satellite ephemeris information and was broadcast continuously on two frequencies (150 and 400 MHz). With one frequency, the user was able to determine its position but with both frequencies together, accuracy was enhanced. To compute position, the receiver measured successive Doppler, or apparent frequency shifts of the signal, as the satellite approached or passed the user.

Predictable positioning accuracy was 500 metres for a single frequency receiver and 25 metres for a dual frequency receiver. Repeatable positioning accuracy was 50 metres for a single frequency and 15 metres for a dual frequency receiver.

Despite its high level of performance, the Transit system was limited in several ways. For one thing, the speed and course of the vessel had to be known. Furthermore, the system could not be used continuously. The satellites orbited at relatively low altitude so that they were visible only for brief periods and moved too rapidly across the sky. Finally, receivers that were themselves moving too quickly, such as aircraft, could not use the system. The Transit programme thus terminated navigation service on 31 December, 1996, when the next-generation Navstar–GPS system was ready to take over this role.

Parus and Tsikada

Various constellations of Tsyklon satellites, followed later by Parus and Tsikada satellites, provided similar services to those of Transit (see fig. 11.9). They were based on Doppler shift VHF transmissions (150 and 400 MHz).

According to Western experts, the first Soviet navigation satellite seems to have been Kosmos 158, launched in 1967, eight years after its American equivalent Transit. Known as Tsyklon (meaning 'cyclone'), this series of navigation satellites was continued up until 1978 and involved two distinct devices.

The second group, known as Parus (meaning 'sail') is a military system. The Parus constellation, first launched in 1974 with Kosmos 700, uses six orbital planes spaced at 30° longitude intervals and thus covering 180°. The remaining 180° of longitude comprises four orbital planes at 45° intervals, occupied by the civilian Tsikada (meaning 'cicada') system, first launched in 1976 with Kosmos 883. (fig. 11.4) Certain satellites in this last family, named Nadezhda ('hope'), contribute to the international search and rescue system COSPAS–SARSAT. Although distinct, the two systems use the same beacons. However, the military space segment is replaced twice as often as the civilian part. Moreover, the civilian constellation can be used by the military if need be, insofar as the civilian satellites are arranged in opposition. The Tsikada system is widely used by the Russian merchant navy and the user's location can be calculated to an accuracy of 100 m.

Unlike Transit, the Parus–Tsikada system is still operational, although satellite replacements have been drastically reduced since 1993, especially in the civilian sector. These systems only supply users with their latitude and longitude, and to accuracies well below those of GPS and GLONASS, which also give the altitude.

NAVSTAR–GPS

The US Department of Defense began developing this system in 1973 in order to provide a continuous and uniform service in three dimensions. It is based on radio ranging to a constellation of satellites called NAVSTAR (NAVigation System with Time And Ranging).

The GPS space segment is composed of 24 satellites with an orbital period of 12h sidereal time and altitude around

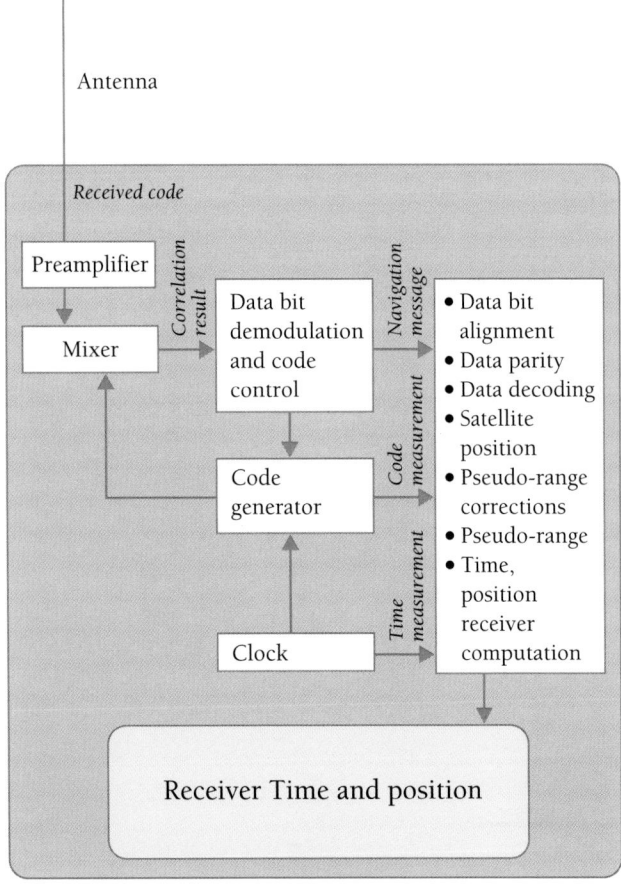

Figure 11.3. **Block diagram for GPS receiver.**

20 000 km. Since Block II, the satellites have been inclined at 55° to the equator (fig. 11.5). This architecture guarantees permanent visibility of at least five satellites from any point on the globe, with a possibility of viewing up to 11 satellites. On the basis of signals from at least four satellites, the receiver can calculate the user's latitude, longitude, altitude and time (fig. 11.3).

The system uses CDMA (Code Division Multiple Access). The key role of the space vehicles is to transmit a precisely timed GPS signal at two L-band frequencies 1.575 42 GHz (L1) and 1.227 6 GHz (L2).

Various satellites have been developed. The first NTS (Navigation Technology Satellites) were used to investigate space technology, carrying oscillators and computers. The Block I NDS (Navigation Development Satellites) were launched by Atlas rockets. Originally designed to last for three years, several survived for over ten years.

The first operational satellites were the Operational Satellites Block II, superseded by the Replacement Operational Satellites Block II-R. The latter have more autonomy and carry protection against natural and manmade radiation. They have autonomous navigation systems and generate their own navigation message data.

American authorities are currently working on the GPS Block II-F, which will be launched between 2003 and 2010. The design lifetime will be 15 years and autonomy will be increased in order to limit the need for ground contacts. Two new civilian signals will be added, significantly enhancing the service and allowing civilians to compute ionospheric delays exactly, whilst simultaneously reducing receiver noise.

The GPS control segment has the following objectives: to maintain each of the satellites in its proper orbit via a limited number of small manoeuvres; to make corrections and adjustments to satellite clocks and payload when necessary; to track GPS satellites and generate and upload navigation data; and to carry out major relocations in the event of satellite failure, so as to minimise impact damage.

The master control facility is located at Schriever Air Force Base (formerly Falcon AFB) in Colorado. The master control and four other monitor stations around the world pick up navigation data from a particular satel-

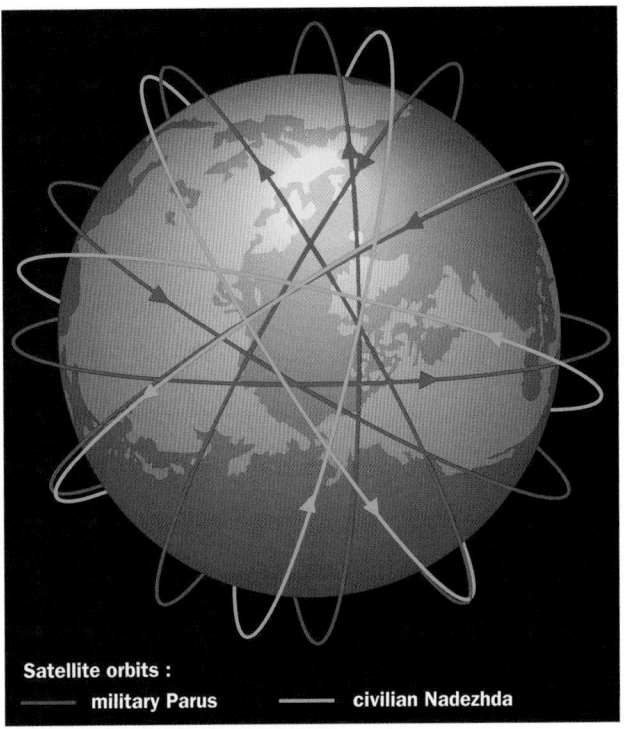

Satellite orbits :
— military Parus   — civilian Nadezhda

Figure 11.4. **The Tsikada–Parus system.**

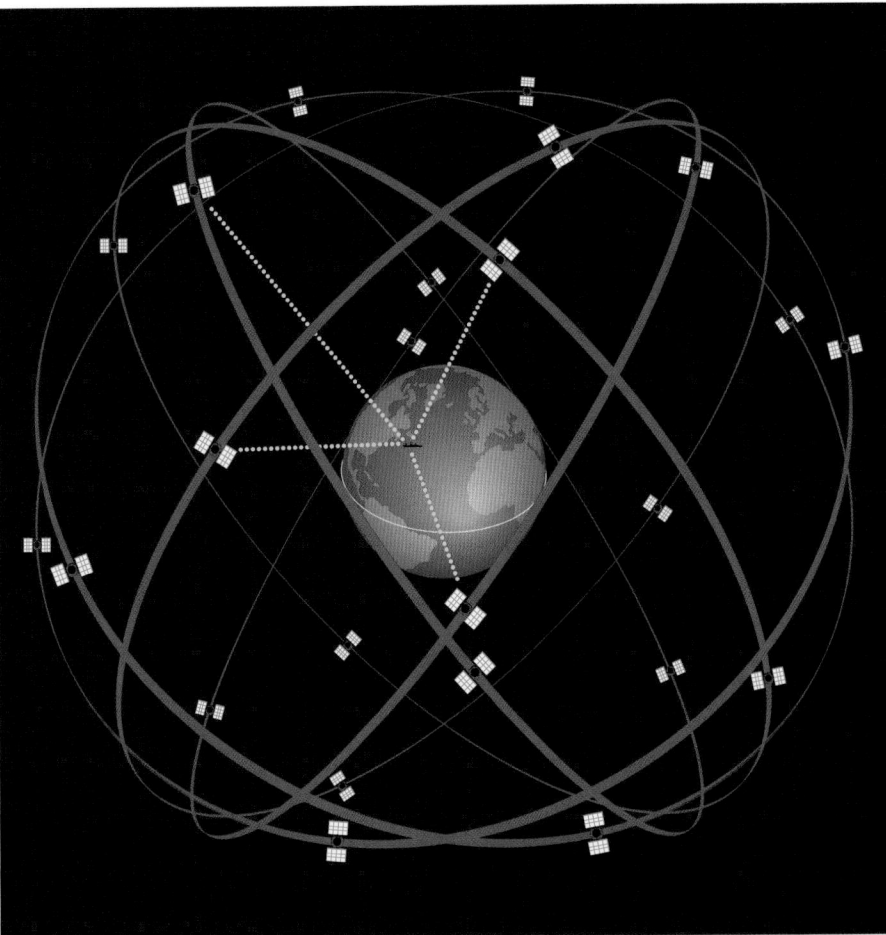

Figure 11.5 **Configuration of the NAVSTAR–GPS constellation.** 24 satellites are distributed over six orbital planes.

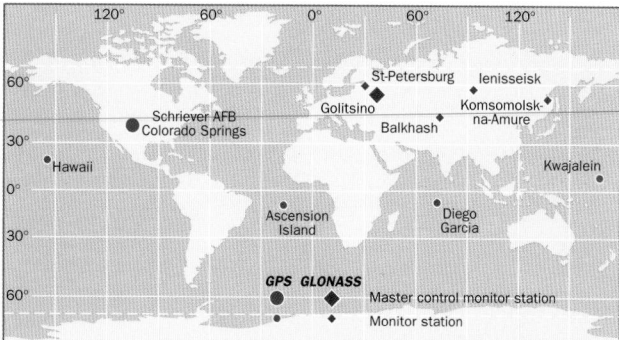

Figure 11.6. **GPS and GLONASS master control and monitor station network.**

Impact in metres			
SPS (civilian) before May 1999	SPS (civilian) after May 1999	PPS (military)	ERROR SOURCE
			SATELLITE
3.6	3.6	1.5	Ephemeris bias: discrepancy between true and calculated positions of the satellite ; Clock bias: error in the time given by the satellite
20	0	0	Selective availability: deliberate error introduced into time given by the satellite
			MEDIUM
7	7	0.01	Ionospheric delay
0.7	0.7	0.7	Tropospheric delay
			RECEIVER
1.5	1.5	0.3	Receiver noise: perturbation of carrier wave
1.2	1.2	0.6	Multipath bias: interference between direct signal and reflected signals
?	?	?	Receiver clock bias: error in reception time
34	14	3.1	UERE : User Equivalent Range Error: mean error remaining when all corrections have been made
100	42?	22	Horizontal accuracy

Table 11.1 **Main delays in GPS ranging measurements.**

- Clock bias: To begin with, the satellites are not exactly synchronised. Correction parameters are contained in navigation messages. Secondly, for civilian users, the GPS system was deliberately downgraded by the US Department of Defense under their Selective Availability policy until May 2000. This meant that a random bias was introduced to render inferior services to civilian users.

- Ephemeris bias: Orbital parameters broadcast in navigation messages are not exact. Spacecraft orbits are influenced by solar wind and other phenomena.

- Atmospheric bias: Broadcast signals move more slowly in the ionosphere and troposphere than in a perfect vacuum. Theories like the Klobuchar model for ionospheric delays and the Hopfield model for tropospheric delays permit civilian users to estimate such delays. By measuring two frequency ranges, military users can precisely determine ionospheric delays.

lite at the same time. With this data, the master control facility can determine the position of the satellite and the clock error (11.6).

GPS signals have navigation data embedded in them. The user receiver can therefore determine both satellite time and position at the time of transmission. Redundant atomic oscillators are carried to ensure excellent stability of GPS satellite clocks. The two signals L1 and L2 are generated synchronously so that the user who receives both signals can directly compute the ionospheric delay and apply appropriate corrections. However, most civilian users only have access to the primary frequency (L1). For this reason, parameters describing the state of the ionosphere are also diffused in navigation messages. The two broadcasts can each have two modulations at the same time (called phase quadrature).

Current implementation has two modulations on the L1 frequency but only one (protected) on L2:

- C/A or Clear Acquisition code. This is a short pseudorandom noise code broadcast at a bit of 1.023 MHz. It is the principal civil ranging signal and it is always uncrypted. The use of this signal is called the Standard Positioning Service (or SPS). It is always available and most of the time degraded. At the time of writing, the C/A code is available only on L1 frequency.

- P or Precise code. This is a very long code that is broadcast at 10.23 MHz. Because of its higher modulation bandwidth, the code ranging signal is more accurate. It reduces the noise in the received signal but not the inaccuracies caused by biases. This service is called the Precise Positioning Service (or PPS) and is used for military purposes.

Table 11.1 shows the impact in metres of all parameters on the User Equivalent Range Error (mean error made in the pseudo-range measurement) and on the horizontal accuracy.

In order to augment GPS services, networks of reference stations have been developed around the world. These stations are receivers with a known fixed position. They compute corrections and broadcast them locally together with information concerning service integrity in order to make the service more accurate. For example, in the United States, NAVCEN operates the Coast Guard Maritime Differential GPS Service, consisting of two control centres and over 50 remote broadcast sites. This service broadcasts correction signals on marine radiobeacon frequencies to improve the accuracy and integrity of GPS-derived positions. Suitably equipped receivers typically provide better than 10 metre accuracy, especially benefiting harbour entrance and approach navigation. The system provides coastal coverage for the continental United States, the Great Lakes, Puerto Rico, portions of Alaska and Hawaii, and portions of the Mississippi river basin (since 1996).

Declared operational in 1990, with seven satellites ready for use, the system provided two-dimensional coverage from 14 to 22h. During the Gulf War, fifteen GPS satellites enabled coalition forces to navigate, manoeuvre and fire with unprecedented accuracy. Indeed, SA was suppressed between August 1990 and July 1991 because of the abrupt increase in the number of users and the increased use of civilian receivers by the allied forces. In September 1991, taking into account the huge potential for civilian applications, the US offered to make the GPS Standard Positioning Service (SPS) available to the international community by 1993 on a continuous worldwide basis with no direct user charges for a minimum of ten years. In doing so, the DoD accepted the responsibility for preserving US national security. In 1994, the US Department of Transport became a stakeholder in the GPS management system in order to represent civilian interests, and GPS was then integrated into the US National Airspace System. The US DoD nevertheless maintained ultimate control over the system.

The rising number of GPS users and a growing awareness of the political and economic stakes have led to worldwide reflection on the role of global navigation satellite systems. In an attempt to maintain their lead in this field, the United States has increasingly taken civilian interests into account, with the government insisting on free services and availability of military accuracy for all users as of May 2000. Needless to say, they maintain jamming capabilities allowing them to discontinue these services in the event of a threat to national security.

GLONASS

The GLONASS programme (GLObal Navigation Satellite System) arose from military considerations, like similar systems in the United States, and is managed by the Russian Space Forces (VKS). The political decision was made in 1976 and the first satellite launched in 1982. Although it was declared operational in 1993, it was not really completed until 1995 and even then only remained so for a brief period. Budgetary problems have meant that the system has often operated in reduced form with seven or eight satellites, or even as few as four in the worst cases (fig. 11.7). However, its importance is in no doubt. This is proven, at least according to official declarations, by the Russian Defence Minister's determination to increase the constellation size to 12 whilst developing a new generation known as GLONASS M with a lifetime of seven years, and scheduling a third generation GLONASS K with a lifetime of ten years. Leaving its national needs aside, the existence of an operational system gives Russia some international credibility, even if lack of finance leads to delays, especially at a time when regional systems are under study and there is a general interest in acquiring technology, frequencies and orbital slots. In the same spirit, the Russian Space Agency (RKA) was made responsible for GLONASS applications and developments destined for civilian users or international cooperation.

Completely deployed, the GLONASS constellation comprises 24 satellites known as Uragan (or 'hurricane') with a lifetime of 3 to 5 years in three orbital planes whose ascendant nodes are 120° apart. Eight satellites are equally spaced in each plane with latitude arguments distributed at 45° intervals. The minimal configuration for full service is 18 satellites. The satellites operate in circular 19 100 km orbits at an inclination of 64.8° and each satellite completes its orbit in approximately 11h15m. Given these orbital characteristics, GLONASS gives better accuracy than GPS at high latitudes, but poorer accuracy in the equatorial regions.

The constellation is run by a ground-based control complex which includes the system control centre (Golitsyno 2, Moscow region) and a network of command and tracking stations located throughout Russia, the five main ones being in Moscow, St Petersburg, Eniseisk, Komsomoslk-na-Amure and Balkhash. The control segment provides monitoring of GLONASS constellation status, corrections to orbital parameters and uploading of navigation data.

The interface between the space segment and user equipment consists of an L-band radio link. Each GLONASS satel-

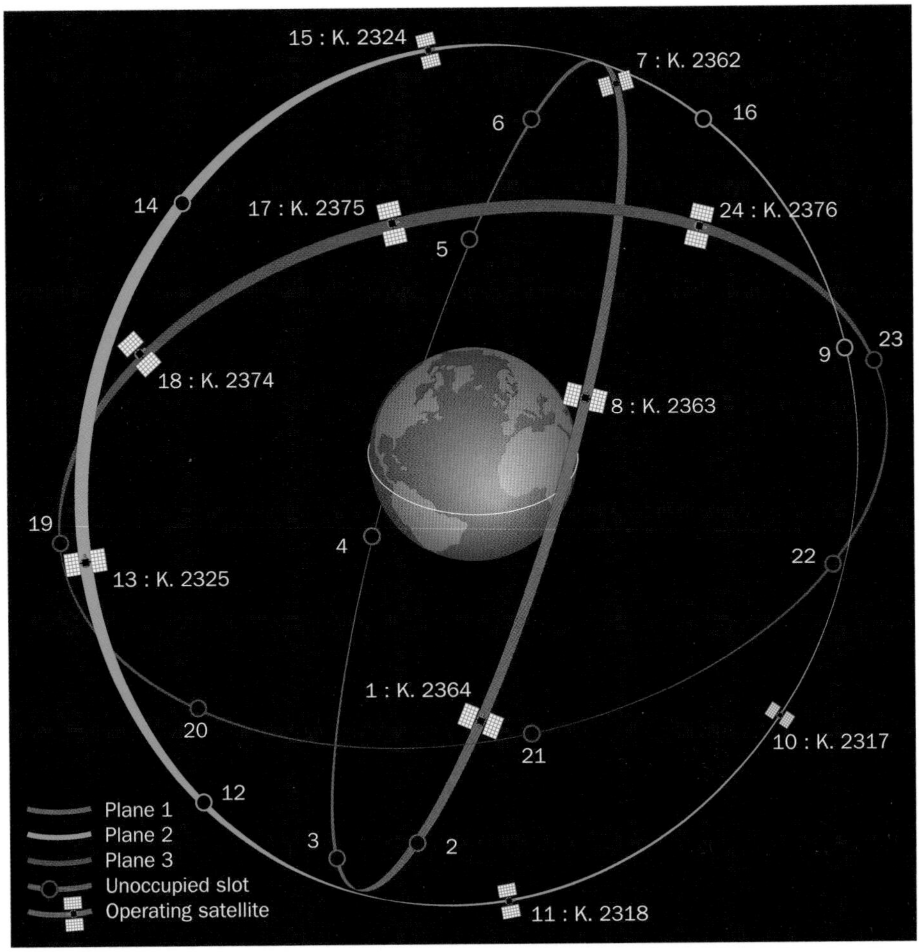

Figure 11.7. **Configuration of the GLONASS constellation.**

lite transmits two types of signal: standard precision (SP) and high precision (HP) in two sub-bands of the L-band (L1 = 1.6 GHz and L2 = 1.2 GHz). The use of Frequency Division Multiple Access (FDMA) means that each satellite can transmit a signal at its own frequency, different from those of the other satellites. However, some satellites do transmit at the same frequency, but these are placed in antipodal slots on orbital planes and are never simultaneously visible to the user. Each satellite has a frequency channel defining the frequency used: 1602 MHz + K*562.5 kHz and 1246 MHz + K*437.5 kHz.

GLONASS receivers automatically receive navigation signals from at least four satellites and measure their pseudoranges and velocities. They simultaneously select and process navigation messages from satellite signals. The computer handling the GLONASS receiver processes all the input data and calculates the three position coordinates, the three components of the velocity vector and the precise time. In contrast to GPS, no frequency is subject to restricted access but the service can be discontinued at anytime.

Since the operating principle is basically the same as for GPS, it is not difficult to develop hybrid receivers for the two systems. However, in order to achieve interoperability, adjustment procedures must be introduced to cater for the fact that each system uses its respective national time scale as reference and different geocentric Cartesian coordinate frames to express the positions of its satellites.

Research into differential GLONASS had already started in the 1970s when the system was first set up. However, the absence of SA and a standard accuracy of around ten metres that was sufficient for most marine and airborne users explains why it was only when differential GPS had become widespread at the beginning of the 1990s that it was seriously reconsidered. The initial trend favoured the development of Local Area Differential Systems (LADS) which provided accuracies better than 1 m over limited zones measuring about 80 km across. The system was aimed at specific groups of users such as air traffic controllers and shipping services, but was subsequently extended to the regional level (RADS) with accuracies in the range of 5–10 m over 3500 km. Insofar as financial constraints made it impossible to extend these services over the whole territory, another approach was conceived for setting up Wide Area Differential Systems (WADS), using the ground-based infrastructure of the spacecraft control complex. In 1994, the concept of differential GLONASS was finally established by the VKS, the RKA and the Transport Ministry of the Russian federation, coordinating the various differential systems and preparing to integrate them into the framework of the United Differential System (UDS) set up by the United States. This provides for a three-level structure of LADS, RADS and WADS. Each level is designed to operate independently of the others and the combination of the three meets a certain pre-established level of performance.

Satellite-Based Augmentation Services

GPS performance levels have opened the way to new applications and led to the idea of creating a Global Navigation Satellite System (GNSS 1) that could replace existing ground systems. This is achieved by improving existing capabilities through a Satellite-Based Augmentation Service (SBAS). Several systems have been set up for the benefit of their international partners: the US WAAS, the European EGNOS, and the Japanese MSAS. GNSS 2 would then be a further stage corresponding to autonomous and complementary systems such as Galileo for the Europeans.

The GPS Augmentation Service, called WAAS (Wide Area Augmentation Service) (fig 11.8), was studied in the United States in 1989–1990 and laid before the Federal Aviation Administration in 1991. This system provides navigation services for all phases of flight through Category I precision approach landings. It uses the GPS space segment and GEO satellites. It is based on differential GPS but the corrections are broadcast via GEO satellites to make the service available all over North America. The first goal of this service is to make the GPS service the principal means of aircraft navigation. Indeed, GPS cannot be used on its own as the only navigation tool because no exact integrity parameters are broadcast. With the WAAS, users have precise information on integrity and also complementary corrections making the service more accurate. The FAA established the U.S. National Satellite Test Bed (NSTB) in 1994 to develop, verify and demonstrate satellite-based navigation for all phases of flight. Since 1995, the NSTB has expanded to include 18 reference stations throughout the continental US, two master stations (one in Virginia and one in California), and four radio frequency uplinks (one in Maryland, two in California, and one in the state of Washington), together with three reference stations in Canada, five in Alaska, two in Hawaii and one in Iceland.

In April 1999, the WAAS provided a horizontal 95% accuracy of 3 metres and vertical 95% accuracy of 5 metres using two Inmarsat satellites, Pacific Ocean Region (POR) and Atlantic Ocean Region (AOR-W), and further tests demonstrated significant progress toward achieving WAAS's system capabilities. However, difficulties involved in quantifying the integrity of the signal and the potential risk of jamming by the army in the case of conflict constitute distinct disadvantages for WAAS users. In addition to these problems, the budgetary overrun has been significant and this has led to renewed interest in the LAAS (Local Area Augmentation System) which only guides aircraft in the approach and landing stage. The economic and industrial stakes are high and many proposals have been made in connection with a GCNSS (Global Communication Navigation Surveillance System) which would not necessarily exclude ground systems.

An SBAS navigation test bed has also been developed by the US in partnership with Chile, Mexico, and many Central and South American nations who are currently developing a GNSS Implementation Plan for the entire Caribbean and South American Region. South Africa, Asia/Pacific and Mid-

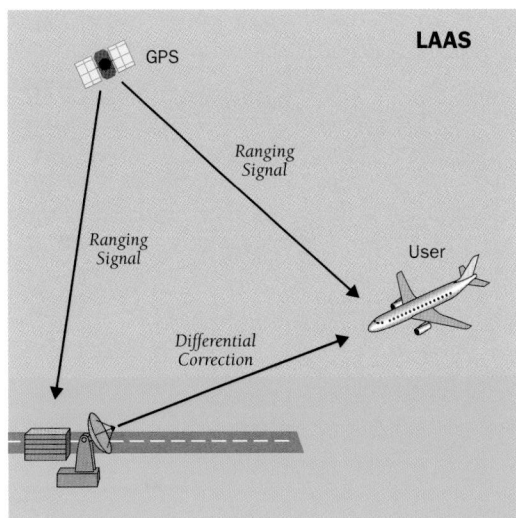
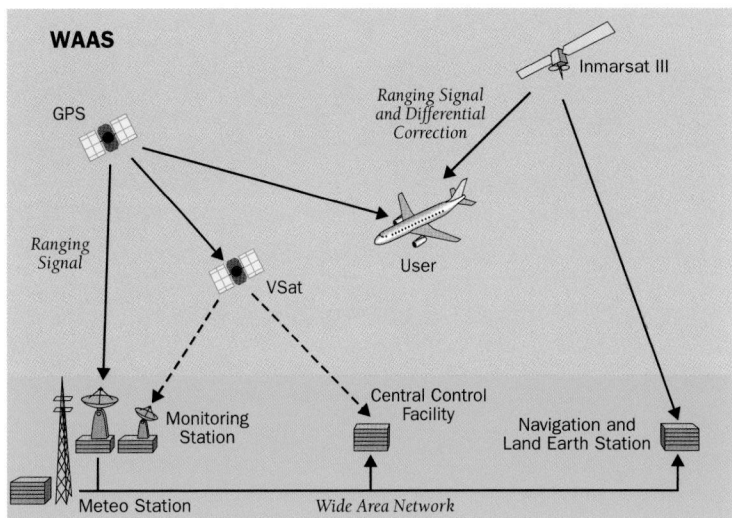

Figure 11.8. **Local Area Augmentation Service (LAAS) and Wide Area Augmentation Service (WAAS).**

dle Eastern nations have also expressed interest in adopting SBAS technology.

For their part, European countries are developing a GPS/GLONASS Augmentation Service. Specifications are the same as for the WAAS, allowing users to switch between the two systems during a flight between Paris and New York, for example. This project, called EGNOS (European Geostationary Navigation Overlay Service) should be operational in 2003. It is being developed by various European industries for a tripartite group composed of Eurocontrol, the European Commission and the European Space Agency. Some preliminary work has been carried out within the framework of the EURIDIS project (run by the French space agency CNES), developing a specific payload on a satellite (Inmarsat III) to broadcast a GPS-like ranging service. With one more satellite in view, users obtain more accurate results. The other advantage of this service is that it is always available for users in the GEO satellite region. This service was running as a prototype in 2002. Operational qualification of the ranging function was carried out in January 1999. EURIDIS provides an additional GPS-like signal twenty-four hours a day with a User Equivalent Range Error better than 12 m.

The GEO transponders are flown aboard three complementary satellites: two Inmarsat III satellites located at longitude 64° east (Indian Ocean Region, IOR) and 15.5° west (Atlantic Ocean Region, AOR), and aboard the European Artemis satellite launched in 2001. The first broadcast tests began in November 1999. The EGNOS Test Bed is composed of eight monitoring stations receiving both GPS and GLONASS signals, a central control facility computing the correction and a Navigation Land Earth Station (NLES) broadcasting information to users via the EURIDIS system on board Inmarsat III. The operational service will involve 34 monitoring stations, four control and supervision centres, and four NLES. The objective is to provide horizontal positioning better than 8 m and vertical positioning better than 10 m over a geographical zone covering western Europe (latitude 40° to 60°, longitude

5° west to 10° east). A redundant broadcast capability is also implemented via the Inmarsat IOR satellites thanks to the Italian Mediterranean Test Bed contribution. Europe, Africa and Asia are the zones covered by the space segment.

A third system, the Japanese Multifunctional Satellite-Based Augmentation System (MSAS), has been selected by the Japanese Civil Aviation Board to cover its Asia/Pacific airspace. The US–Japan Joint Statement of September 1998 regarding cooperation in the use of GPS determines the framework for this programme. The MSAS employs a ranging function to generate GPS-like signals and thus enables aircraft to use the Japanese Multifunctional Transport Satellite (MTSAT 1R) as a 25th GPS satellite. Indeed, the Japanese case is original in the sense that they are developing a specific satellite that will be, in 2003, the first air navigation and telecommunications space system to meet the 1991 requirements of the International Civil Aviation Organization (ICAO). The great similarity between the WAAS and MSAS systems, in which US and European companies play a major technical role, demonstrates a profound concern for interoperability. This should facilitate Japanese participation in the GNSS and confirms the growing Japanese awareness that there are significant economic and strategic advantages in possessing its own knowhow in the area of navigation by satellite.

The European system Galileo

In the mid-1990s, Europe decided to contribute to the elaboration of an independent Global positioning and Navigation Satellite System known as GNSS 2, which marks a second stage of development. Once sufficient knowhow has been acquired to develop an alternative/complementary service to GPS, via the EGNOS system, GNSS 2 will be based on a European satellite constellation under civilian control, named Galileo.

GalileoSat is ESA's specific contribution to the European Galileo Initiative. It covers the definition, development and validation of the space and associated ground segment. It will

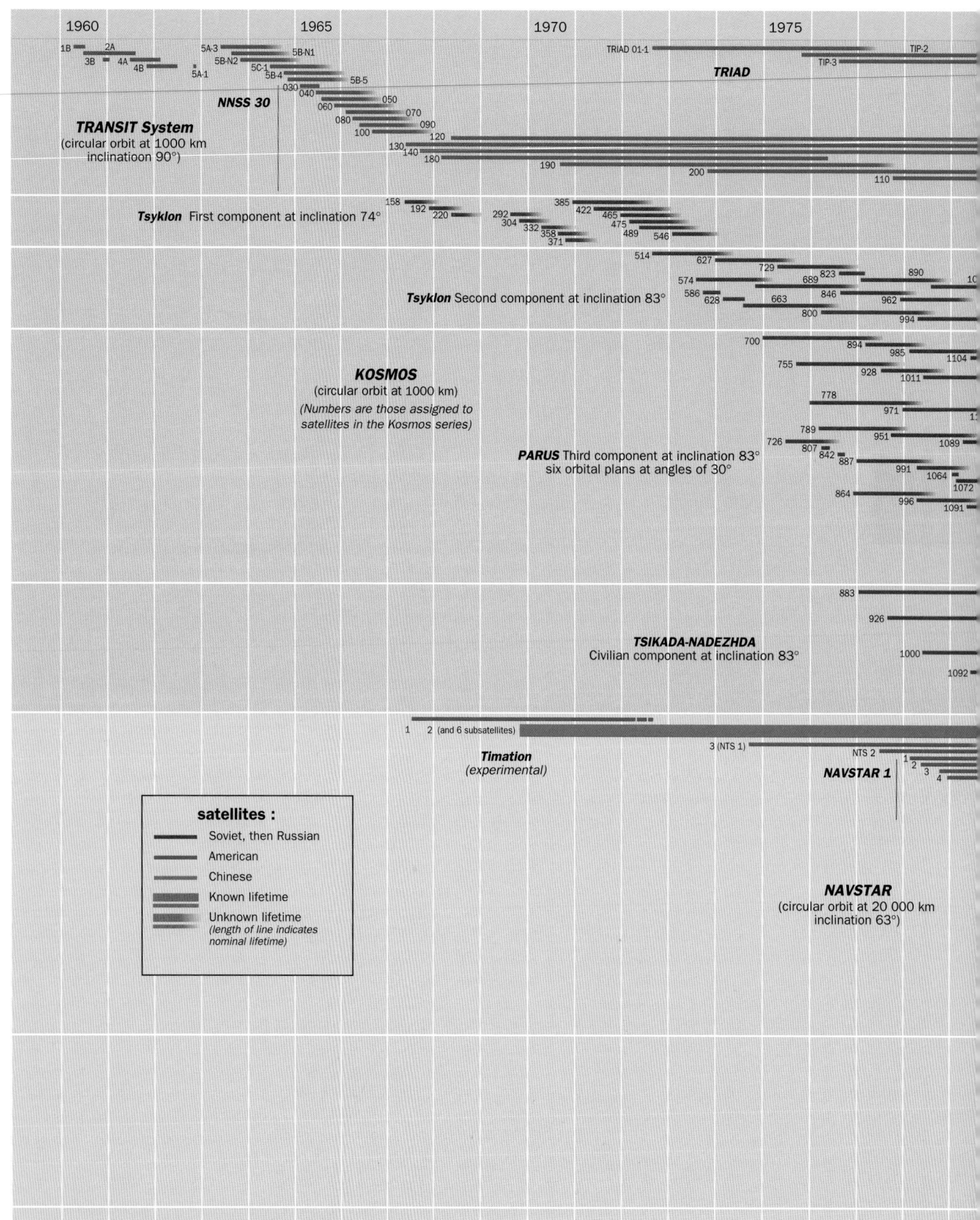

Figure 11.9. **Chronology of navigation and positioning satellites.**

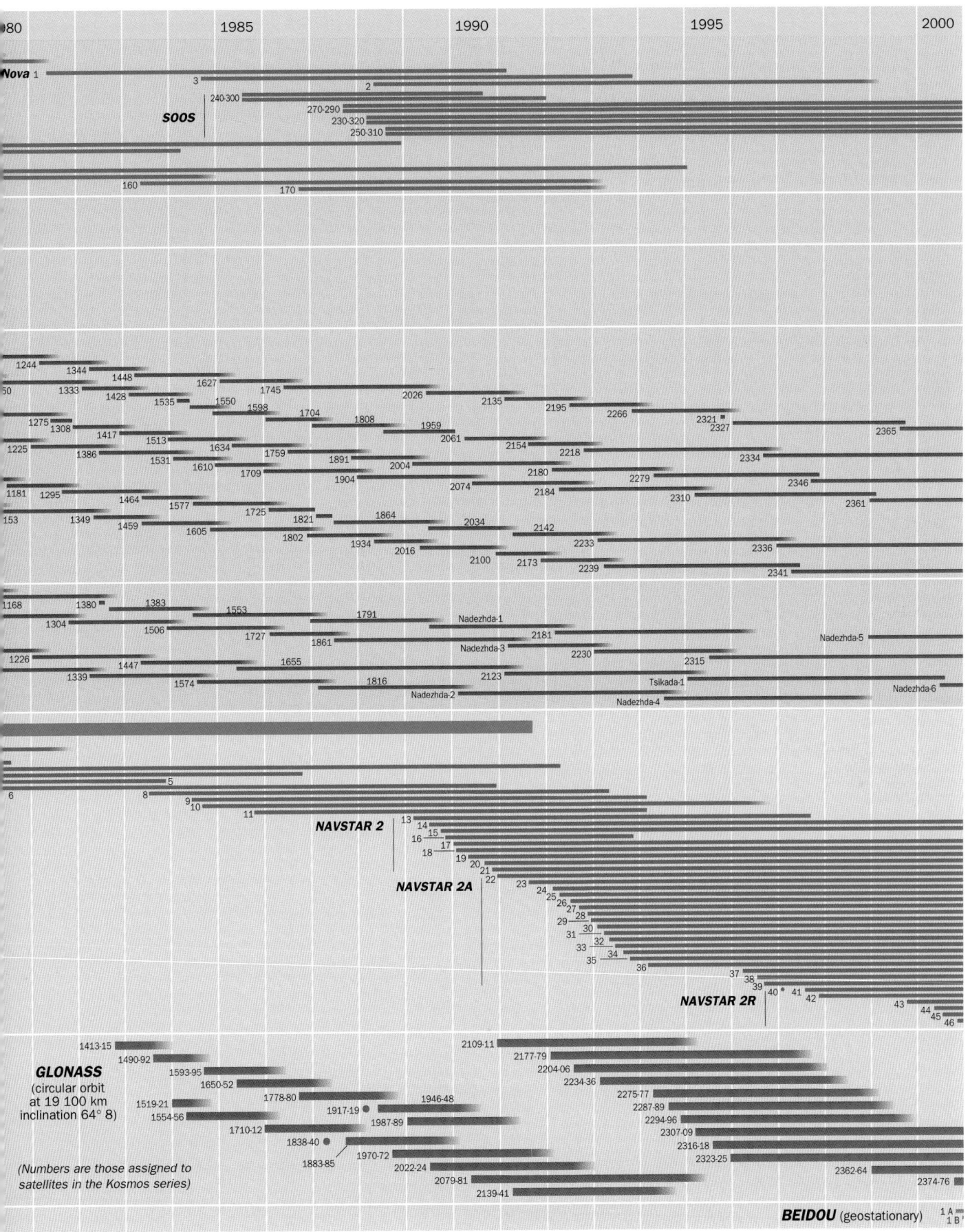

be composed of a space segment (21 or 24 MEO satellites) possibly complemented by geostationary satellites and associated ground infrastructure based on the experience acquired during the EGNOS development. Many aspects remain to be dealt with, such as frequency allocation, type of space vehicles to use, and composition of navigation messages.

Security is also a sensitive issue. The idea of a military frequency raises a certain number of internal European problems, given the limited degree to which the common defence policy has so far been implemented. It also raises difficulties with regard to the United States, if GPS and Galileo are to remain interoperable. In this context, the question of security directly affects Europe's image of strategic independence.

Setting aside the technical and financial difficulties involved in organising the Galileo programme, it symbolises European resolve to maintain strategic independence. The term should be taken in a much broader sense than the purely military one, given the widely ranging commercial applications of navigation systems. Many challenges must be faced, related to the complexity of European construction. Hence, on an organisational level, the communication of the European Commission on Galileo issued on 10 February 1999 developed the various aspects of European strategy affected by extension of the EGNOS programme. By 2003, the latter should provide a service indicating the state of the GPS and GLONASS constellations, as well as considerably improving measurement integrity and realising the autonomous navigation satellite programme, Galileo. A definition phase began in June 1999, when the European Union Transport Council met in Luxembourg, entrusting leadership of the operation to the Directorate General for Transport. When this phase ended in 2000, a new entity had to be set up, creating a common legal framework for the European Union and the European Space Agency. The former were responsible for policy control and implementing the project, whilst the latter represented the management structure and held technical responsibility. By the summer of 2001, it seemed that a solution had been found, bringing together the idea of an ESA optional project and that of a joint undertaking according to Article 171 of the European Treaty. A way of financing the programme also seems to have been determined. The concept of a Private Funding Initiative (PFI), guaranteeing a return on investment to the private sector whereby the public authorities buy services, seems likely to replace that of a Public/Private Partnership. The latter would be difficult to realise in the present case because the basic investments are so large and provision of GPS services is supposed to be free.

The Chinese system

Two Beidou (Northern Dipper) navigation satellites were launched at the end of the year 2000. These are placed in geostationary orbit and, according to the Chinese authorities, they constitute the first independent national navigation system, known as the Beidou Navigation System (BNS). Located at 140° east and 80° east, the two satellites are designed to localise rail and shipping fleets in latitude, longitude and altitude.

Insofar as China already possesses GPS and GLONASS receivers, which Chinese manufacturers claim to be able to produce themselves, the new navigation system seems to be an assertion of national independence. It belongs to the November 2000 development programme in which China set out to become a major player in space. The system is quite similar in principle to Japanese and European regional systems and the idea of setting up such a complementary capability may be the first step towards an autonomous system at some later stage. Although the technological and financial problems involved in building a complete constellation will doubtless lead to delays, cooperation programmes may serve to reduce them at least partially.

Systems using uplink signals

Systems in this category are mainly used to locate beacons and relay collected data to ground stations. They are carried as part of the payload aboard satellites with other missions.

Argos

The Argos system is a positioning and data relay programme set up jointly by the United States and France (NASA/NOAA/CNES) and becoming operational in 1978. Apart from positioning, it also gathers data for surveillance and protection of the environment (fig. 11.10). Fixed or mobile platforms are equipped with sensors collecting meteorological data (temperature, humidity, wind speed, ocean currents, thickness of snow or ice, etc.) and a transmitter that sends out its coded message at regular intervals. This message is received by sun-synchronous weather satellites in the NOAA series. At least two of these, a morning and an evening satellite, are always in service at any given time. Each is visible from a spherical portion of the Earth of diameter 5000 km and each transmitting beacon is overflown 28 times a day in the vicinity of the pole and 6 to 8 times a day near the equator.

Satellites up to NOAA 14 can process four simultaneous messages. The Argos system evolved and satellites NOAA 15, launched in 1998, and NOAA 16 launched in 2001, carried a second generation payload, Argos 2, with three times the receiver bandwidth and capable of processing eight simultaneous signals rather than just four. Argos 2 will also equip the next three satellites of the NOAA series. This new generation was successfully tested in November 1996, in a joint project between the CNES (France) and NASDA (Japan), and it will be carried aboard the Japanese satellite ADEOS 2. A later Argos 2 enhancement will allow two-way communications with beacons. Further developments are projected with a third generation which should equip Eumetsat's METOP satellites from 2003, and very likely NASDA's ADEOS 3. These new satellites will increase capacity by supplying more data and more often.

Argos has applications in many areas, from communication with offshore platforms and shipping to tracking of wild animals, like sea turtles, birds, bears, dolphins and whales, for scientific purposes. The system has become better known through its surveillance role in yacht races, such as the Vendée Globe Challenge single-handed race around the world.

Figure 11.10. **The Argos system.** Information is uplinked to the satellite where it is stored in memory (1). When the satellite is visible from one of the acquisition stations, the latter sends a data transmission command for data collected with the identity of the emitter, and the frequency and time of the measurements (2). This station transmits the data to the National Oceanic and Atmospheric Administration (NOAA) processing centre (3), which then forwards it to the Largo centre (near Washington, US) and Toulouse centre (France) to be processed (4). The emitter position is calculated and results sent to users in the form of computer files (5). Two secondary centres in Australia and Japan are connected to the centre in Toulouse.

With more than 5000 beacons registered and over 200 permanent users in almost 60 countries, the system is currently running under full capacity. It can supply data within 20 minutes via its global processing centres in the United States and France and regional centres connected to regional receiving stations which immediately process messages relayed by satellites (fig 11.11).

COSPAS–SARSAT

The COSPAS–SARSAT programme results from international efforts to improve search and rescue operations around the world by supplying distress and positioning information.

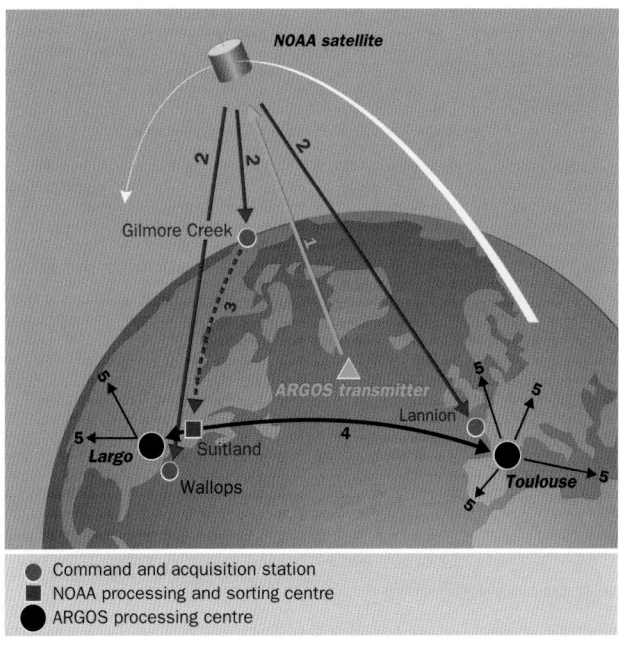

Figure 11.11. **Argos ground stations.**

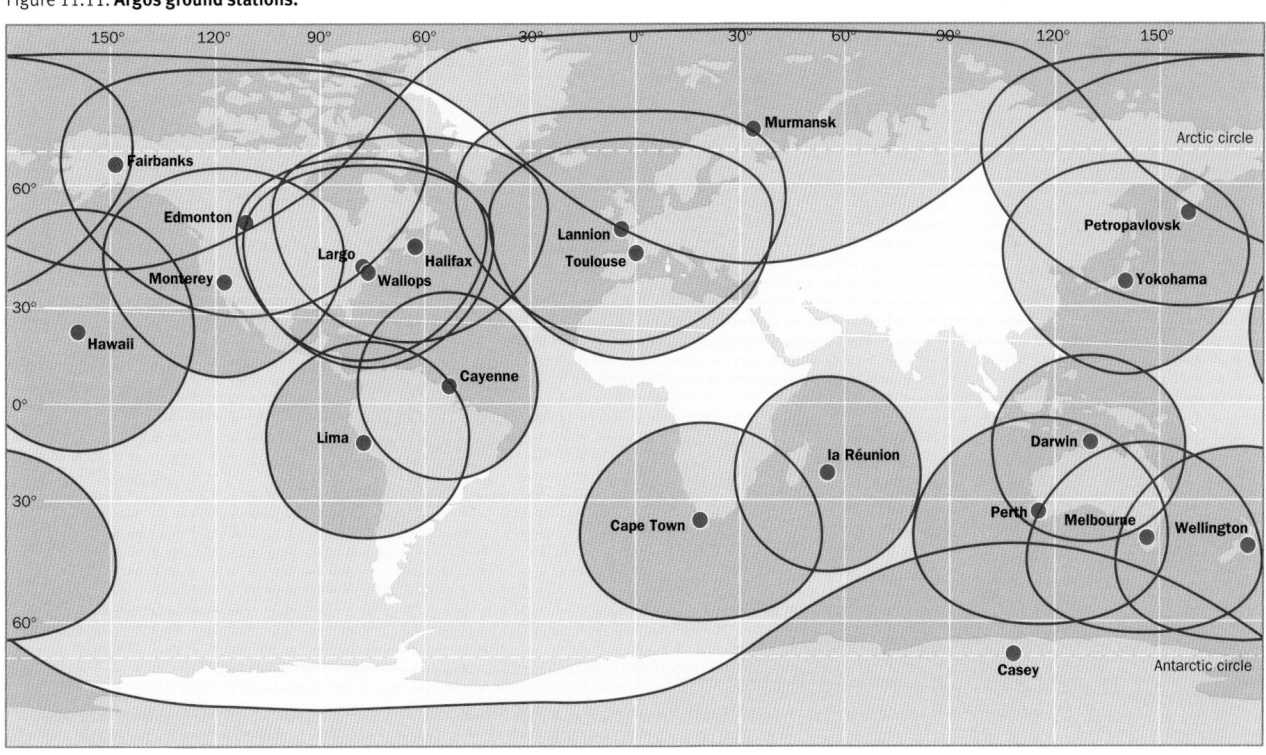

The Russian system COSPAS originally joined up with the SARSAT programme (Search And Rescue Satellite Aided Tracking), jointly run by the United States, France and Canada. The first Soviet COSPAS repeater was placed in orbit aboard the civilian navigation satellite Kosmos 1383 in June 1982, whilst the first Western SARSAT was carried by the American satellite TIROS N in March 1983. Subsequently, a series of American NOAA polar satellites and Russian Nadezhda polar satellites was launched to ensure continuous beacon location. It was only in 1988 that COSPAS–SARSAT was declared operational and a permanent management organisation set up. At the same time, an agreement within the framework of Inmarsat established a dedicated search and rescue satellite system, integrated into the Global Maritime Distress and Safety System (GMDSS).

Since then, 24 other countries have become members, including Norway, the United Kingdom, Finland, Bulgaria, India, Sweden and Italy, committed to providing the Earth-based components of the system. Between 1982 and the end of 2000, the system helped to rescue more than 11 000 people, mainly at sea.

Like Argos, the system uses one-way communications, beacons being activated either inertially upon impact, by immersion, or manually when this is possible. Today it is made up of two complementary systems, LEOSAR and GEOSAR. LEOSAR uses low-Earth orbiting American NOAA satellites and Russian Nadezhda satellites whose relatively

low altitude (850 km) facilitates use of the Doppler effect. The system can localise signals from a geographical region of radius 3000 km around each ground facility, or LEOLUT (LEO Local User Terminal) (fig. 11.12).

GEOSAR uses American geostationary GOES East and West satellites and geostationary Indian Insat satellites with transponders built into them. Insat covers the Indian Ocean, whilst the American GOES satellites cover the Atlantic and Pacific Oceans. These satellites provide permanent surveillance but cannot use the Doppler effect to locate distress beacons since they are fixed relative to the Earth. Positions can be ascertained if they are encoded into the message from the beacon, provided that the beacon is capable of calculating its position via the GPS system, or they can be derived from the LEOSAR system, with possible delays.

At the beginning of the year 2000, 250 000 beacons used the frequency 406 MHz for uplinks to GEOSAR and LEOSAR satellites, and 600 000 beacons exclusively used the frequency 121.5 MHz for uplinks to LEOSAR satellites, a frequency scheduled for withdrawal in 2009.

DORIS

DORIS (Doppler Orbitography and Radiopositioning Integrated by Satellite) was developed by French organisations (the French space agency CNES, the Groupement de Recherches Géodésiques et Spatiales GRGS and the Institut Géographique National). It is designed to make accurate deter-

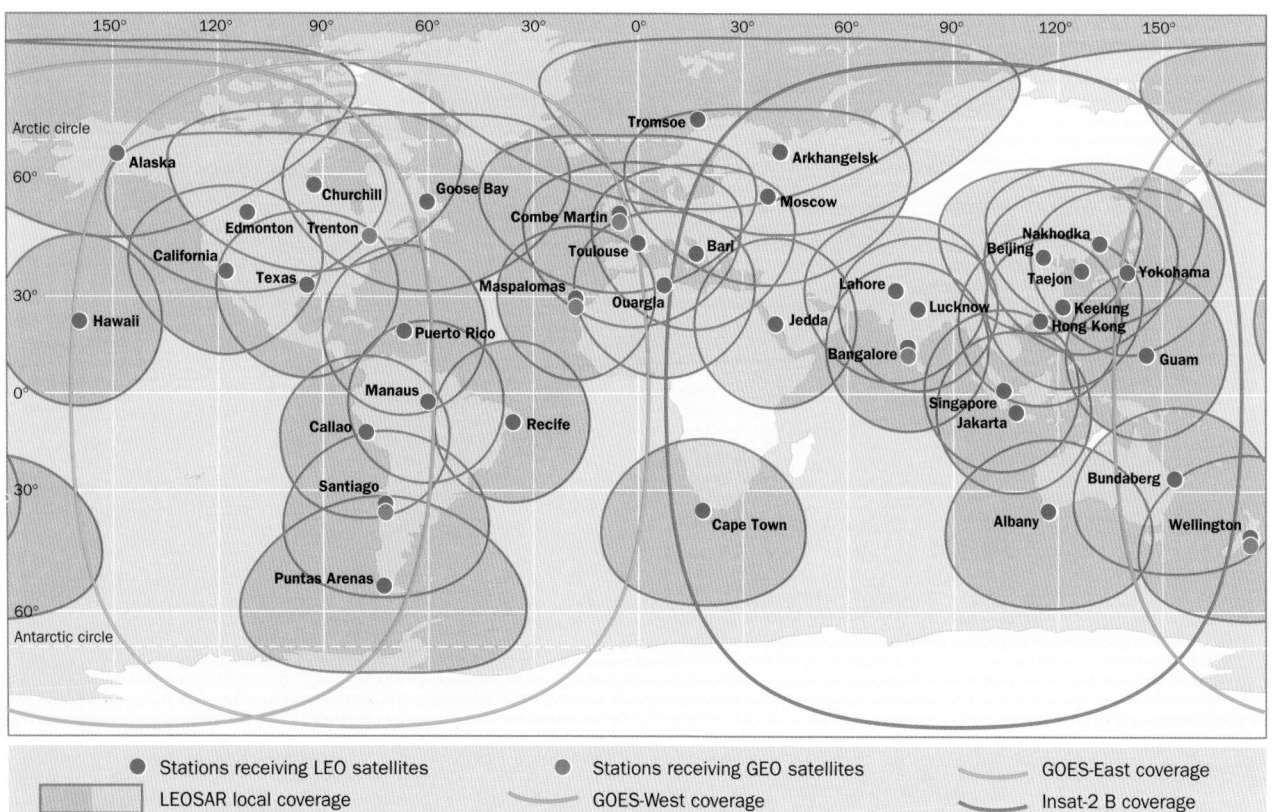

Figure 11.12. **The COSPAS–SARSAT stations.** Regions covered in 2001 by the network of ground stations where distress signals can be immediately relayed by satellite.

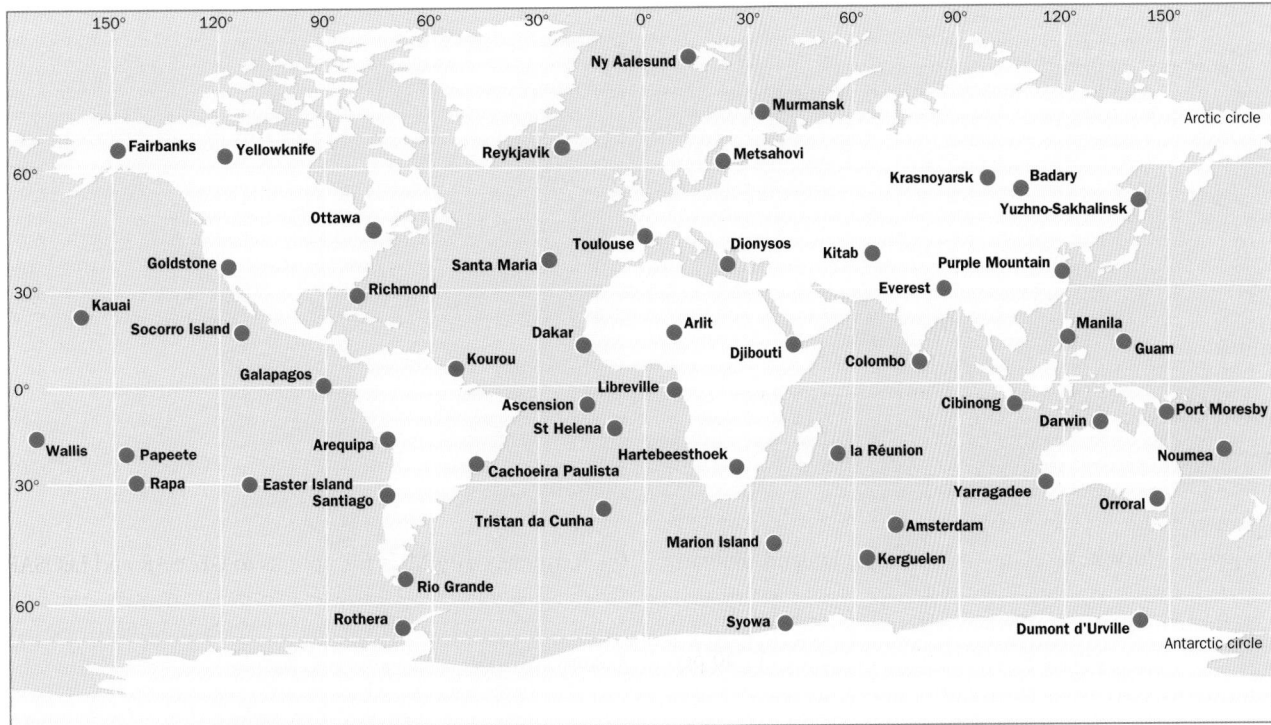

Figure 11.13. **The DORIS permanent beacon network.**

minations of orbits and to localise beacons. Satellite orbits are precisely calculated by measuring the Doppler shift at two frequencies (400 MHz and 2 Ghz) transmitted by ground stations. Two frequencies are used in order to reduce the error introduced when the signal propagates across the ionosphere. The system requires a network of about 50 orbitography beacons, evenly distributed across the surface of the planet. These transmit signals to the satellite whose position is to be determined and can achieve accuracies of the order of 10 cm. The DORIS device has been successively carried by the SPOT 2, SPOT 3, and SPOT 4 satellites, as well as the TOPEX–POSEIDON satellite. In the latter case, it was able to calculate the orbit to within about 2 cm, a quite exceptional radial accuracy. It will be carried aboard both Jason and Envisat 1.

Once the orbit has been accurately determined, it becomes possible to find the position of a beacon, already localised to within a few kilometres, with accuracy better than 10 cm relative to the three axes of a geocentric world reference system, using the approximate location of an associated Argos beacon if need be.

Coupled with onboard software capable of real time orbital calculations, DORIS can be used for autonomous satellite navigation. In this capacity, it even provides an alternative to GPS when establishing relative positions within constellations. This has led to increased interest in the system on a European and international level (fig. 11.13).

CHAPTER TWELVE

Military applications of space

- **STATUS OF MILITARY ACTIVITY**

- **COLLECTION OF INFORMATION**

- **TELECOMMUNICATIONS**

- **SPACE: THE NEW BATTLEFIELD?**

Status of military activity

Military satellites, whose existence was for a long time officially denied, are often referred to by the general public as spy satellites. This term, which only identifies the category of surveillance satellites, does immediately raise a question concerning the legitimacy of military uses for space (fig. 12.1).

The treaty 'on principles governing the activities of states in the exploration and use of outer space, including the Moon and other celestial bodies' remains the basic reference for such questions. Often referred to as the Outer Space Treaty or the 1967 Treaty, this first great treaty recognises 'the common interest of all mankind in the progress of the exploration and use of outer space for peaceful purposes'.

As far as international treaties go, this one was drawn up at very short notice. Furthermore, since there was no possibility of referring to common practice, the 1967 Treaty is characterised by a much more liberal attitude than can be found in corresponding treaties overseeing maritime or airspace law. This approach was strongly upheld by United States representatives on UN committees responsible for setting out the treaty, but initially met with opposition from a great many countries still lacking in any space capabilities. These countries saw themselves irredeemably condemned to second place with respect to the space powers. For its part, the Soviet Union also expressed concern over the American position.

In fact, a question of sovereignty is raised with regard to territories overflown by satellites. The particular technical constraints that affect satellite orbits in space (see Chapters 2 and 3) mean that any non-geostationary satellite must by definition fly over every region of the Earth's surface with latitude less than the inclination of its orbit (see Chapter 3). In addition, the fact that the Soviet Union was unable to prevent satellites from flying over its territory and their tacit acceptance of this state of affairs, as witnessed by their participation in the International Geophysical Year, and even more clearly by the precedent set by sending Sputnik into orbit, rendered further opposition to the principle impossible. The idea of defending national interests therefore had to be weighed against more practical considerations. Those upholding the sovereign rights of states only obtained the guarantee that no weapons of mass destruction would be allowed, and an obligation to use space for peaceful purposes.

The absence of any real consensus and the desire to find a compromise are clearly apparent in the ambiguity of the final text, which remains open to a wide range of interpretations. According to a strict reading, the term 'peaceful' excludes all military use at the outset. This is the position still adopted today by India, for example, which considers that 'peaceful purposes' can by definition only concern civilian applications. According to most other interpretations, however, only offensive military activities are thereby ruled out. The Agreement governing the activities of States on the Moon and other celestial bodies, or Moon Treaty, put forward for signature in 1979, would eventually confirm this looser interpretation, specifying that activities must be of an exclusively peaceful nature, but that the use of military personnel or equipment is not forbidden, provided that it serves only peaceful purposes. For various reasons, however, this agreement was signed by very few countries. Military satellites are therefore allowed to exist insofar as the expression 'peaceful purposes' is understood to be the antonym of 'aggressive purposes'. The ambiguity remains and practice has only strengthened the position of military users. Over the last 30 years, a considerable number of military satellites have been sent into space, and countries other than the United States and Russia now possess military space capabilities, although sometimes limited, or their own programmes.

Nevertheless, even if it is accepted that 'peaceful' can be taken to mean 'non-aggressive', the dividing line between offensive and defensive uses of a space system is also rather poorly defined. Observations may serve to prepare an attack just as they may serve to warn of one, and navigation–positioning systems could be used to improve the accuracy of a missile just as effectively as they could locate a vessel in difficulty.

Among existing military space systems, it appears that many military satellites fulfill very similar roles to their civilian counterparts, i.e., scientific, telecommunications, weather, observation, and navigation satellites. They do include certain rather particular technical specifications, often of a defensive nature, such as satellite and data protection systems, or else they satisfy rather specific operational requirements. Other missions are exclusively military in aim, but may help to stabilise relations between countries. Examples are the early warning systems set up in the context of nuclear dissuasion, or electronic intelligence systems designed for times of crisis. Alongside this first category, other uses have been at least tacitly admitted, characterised by a potentially offensive capability. These include antisatellite and antiballistic missile systems (ASAT and ABM), or more symbolically at the time, the US Strategic Defense Initiative referred to as Star Wars, and today the National Missile Defense (NMD). In the latter cases, only the claim that they serve a dissuasive purpose can be invoked in their support.

Since the beginning of the 1990s, a new dimension has been introduced by satellites carrying both military and civilian capabilities, as described by the terms dual-use or dual-purpose. One example is given by localisation and navigation systems considered separately (Chapter 11). Military satellites are recognised as force multipliers and, through the development of civilian applications, have come within reach of an increasing number of countries. In such conditions, finding itself to be the only significant military space power remaining after the collapse of the USSR, the United States considers that space has become a key element in its national security and that all possible risk must be excluded. The concept of 'space control' has thus become fundamental. The idea that satellites could guarantee peace, originally put forward to justify the existence of military satellites, was transformed. The American position as world peace-keeper was sufficient to sustain the kind of monopoly envisaged. This reversed trend has been confirmed by a proliferation of programmes treating space as a potential battle field. At the same time, the basic principles of unrestricted use, and even free access to space by all states, have been tacitly questioned. In the coming decades, this evolution in the fundamental principles controlling military use of space, encouraged by some, whilst being perceived as a destabilising factor by others, may well embody one of the main changes in space development.

New states, in particular certain European countries, have expressed the intention of acquiring their own means of surveillance in a world context plagued by instability. The relative strengths of the space powers are therefore likely to undergo a gradual transformation, with the United States confirming its current role as world policeman.

Figure 12.1. **Image of a Soviet shipyard taken by an American military reconnaissance satellite.**

Collection of information

Amongst the wide range of space vehicles accomplishing military missions, a great many satellites are concerned with surveillance missions, such as high resolution observation, detection of nuclear explosions, early warning in the event of ballistic missile launch, maritime surveillance and communications monitoring.

Although both the United States and the Soviet Union developed the same type of system for gathering military information, deeper analysis reveals different strategies in line with their differing national capabilities and the role they hope to play on the international stage. Since the beginning of the 1990s, the faltering Russian economy has led to a blatant imbalance between the United States and the rest of the world, despite the appearance of new actors whose modest capacities nevertheless counter the threat of total American hegemony.

Reconnaissance satellites

Earth observation was first developed for military reasons. Satellites offered an unexpected opportunity to acquire strategic information from poorly accessible regions. The American experimental reconnaissance satellites Discoverer and SAMOS were launched as early as 1959, whilst the Soviets remained hostile to the idea of unrestricted observation

from space, since they could more readily obtain information from other countries using quite traditional means. Nikita Khrushchev immediately threatened to shoot the satellites down, together with any U2 spy planes, before choosing to develop his own space-based observational capability in the form of the Kosmos 4 satellite, launched in 1962.

Reconnaissance satellites constitute the largest category of military satellites, from a purely numerical point of view. The main families are the American Key Hole satellites (KH) and the Soviet, then Russian, Cosmos satellites. There are several types of spacecraft, whose characteristics have evolved with technological advances and also with changing mission objectives. The latter range from strategic and tactical observation to checking that disarmament treaties have been respected, and assessing conflict damage.

In 1995, first Israel and then France, in cooperation with Spain and Italy, broke the Russian–American oligopoly that had stood for more than 35 years. In 2000, following a modest incursion with its FSW recoverable capsule programme, China launched its first photo-reconnaissance satellite returning data to ground stations digitally. Zi Yuan 2, which means 'resources', very probably uses the China–Brazil Earth Resources Satellite bus (see chapter 9), known also by the Chinese as ZY 1. Launched at 500 km altitude, the satellite

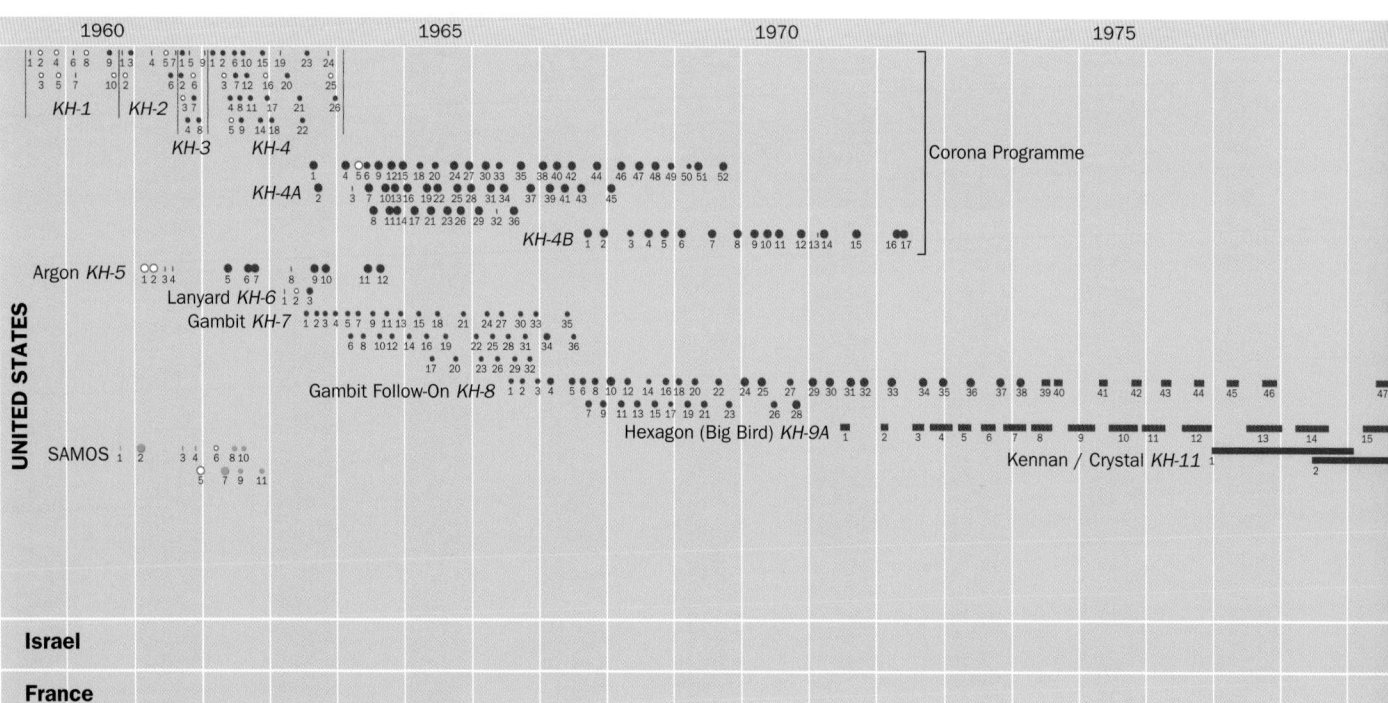

Figure 12.2. **Chronology of reconnaissance satellites.** (United States, Israel and France)

336

very likely has resolution between 5 and 10 m, which is rather inadequate for military purposes.

Soviet reconnaissance satellites always vastly outnumbered their American counterparts. Their different lifetimes, based on a quite different approach to the problem, explain this disparity. In the Soviet Union, mass production of satellites and launch vehicles allowed steady output at lower costs. For their part, the United States chose to develop a limited number of much more sophisticated models, with lifetimes as long as three years. When the Soviet Union collapsed, this situation was gradually turned round, since Russia no longer had the means to sustain continuous spaceborne surveillance.

■ In the literature, American reconnaissance satellites (fig. 12.2) have been classed a posteriori under the heading KH, except for the SAMOS programme run by the US Air Force, about which little is known and which came to an end in 1962 when the CIA's Corona programme proved successful. In February 1995, when President Clinton decided to make 860 000 pictures publicly available from the Corona, Argon and Lanyard programmes, some of the mystery surrounding this first generation was finally dispelled. It was an opportunity to form a more precise idea of the difficulties encountered in these programmes, with regard to satellite orbit acquisition, photographic development or data transmission.

The Corona programme lasted for 13 years and comprised four satellite generations labelled KH 1 to KH 4. The last of these families was subdivided into three series, KH 4, KH 4A and KH 4B. Approved by President Eisenhower in February 1958, the project aimed to acquire information about Soviet territory and Eastern Bloc countries in a period

of considerable international tension. The missile gap was one of the main preoccupations of the day. The first satellites in the Corona series were equipped with panoramic camera systems and exposed film was loaded into recovery vehicles that were later collected by the US Air Force. Image resolution was initially between 10 and 12 m. At first, these satellites operated for no longer than one week, but their lifetime was gradually extended to 19 days with the KH 4B series at the end of the 1960s and the beginning of the 1970s. By then, resolution was down to 2 or 3 m. To a large extent, these satellites and their successors were used to obtain a better knowledge of Soviet military potential, and eventually led up to the SALT and ABM arms control agreements in the 1970s, providing a means of verification for the agreements that could be carried out in no other way. The rather implicit recognition of military space surveillance systems mentioned in the text of the agreements, under the somewhat obscure term 'national technical means', dates from this same period.

The first known launch inaugurating the KH 1 series was in June 1959, with a total of ten launches taking place by September 1960. Only one was successful. These first experimental satellites, judged to be of the utmost strategic importance by the President himself, were concealed behind the Discoverer capsule programme, which generated a great deal of innocent public interest.

The KH 2 and KH 3 series showed significant improvements although many problems were far from being completely solved. There were still difficulties with both orbit acquisition and the photographic mission, as well as subsequent film recovery. At the beginning of 1962, the first

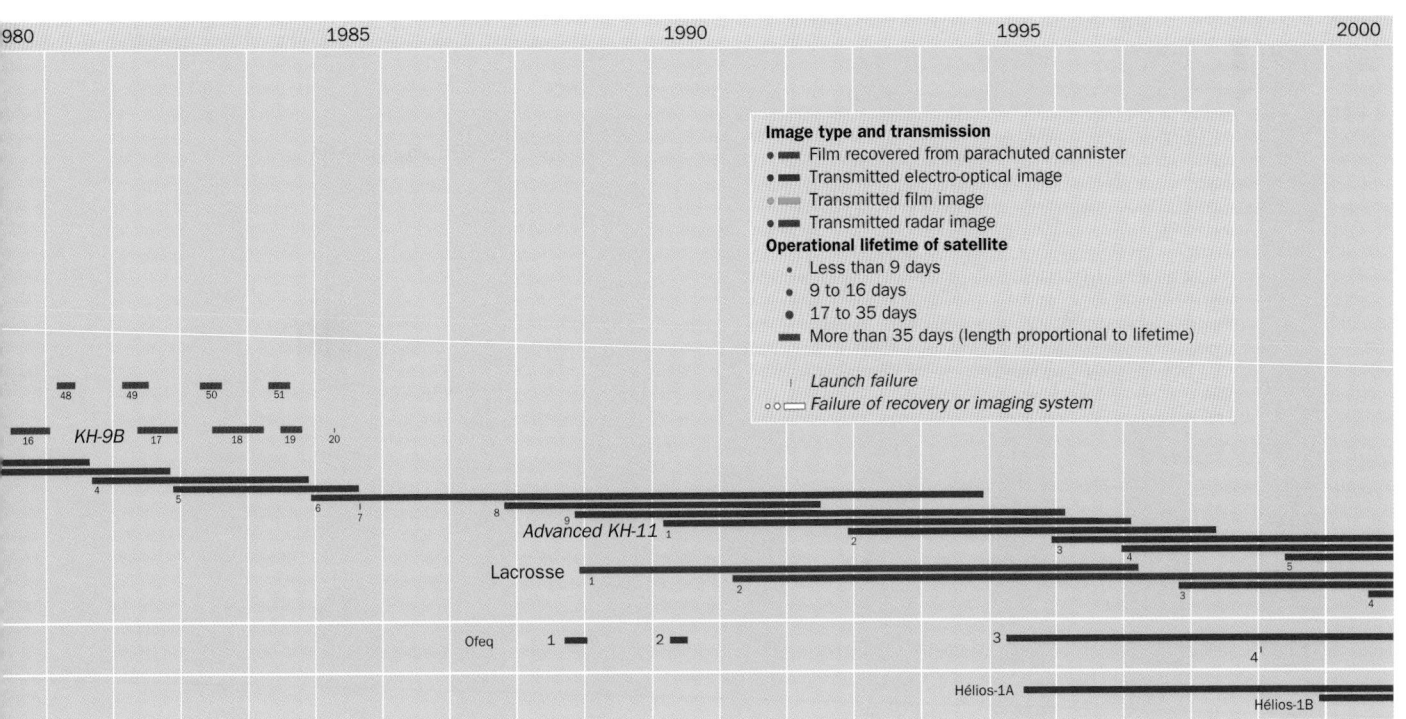

experimental period ended and the KH 4 series, with all its developments, marked a turning point.

The KH 4 generation corresponds to the last period of the Corona/Discoverer programme. The first KH 4 series, with lifetime still below 7 days, consisted of about 20 successful launches (between 18 and 23, according to the experts). Two panoramic camera systems were incorporated from then on, and even though some difficulties remained, the system could benefit from the experience of its operators.

With the KH 4A series, the use of two recovery vehicles made it possible to extend the mission to two weeks. About 50 satellites of this type were launched between 1963 and 1969, some following retrograde orbits in order to benefit from a constant illumination of the Earth's surface (see Sun-synchronous orbit).

Finally, with 16 launches between 1967 and 1972, the KH 4B series reached a lifetime of 19 days, and thereby provided the highest success rate of the whole series.

The KH 5 satellite series, also known by the code name Argon, had a complementary mission. They were designed for large-scale mapmaking. Equipped with a single, low resolution camera, they had a lifetime of around 29 days. Of the 12 satellites launched between May 1962 and August 1964, seven were successful and managed to locate Corona images in their environment.

The KH 6 satellite series, known as Lanyard satellites, was a rather brief experiment carried out in 1963, with three launches and only one success. Equipped with just one frame camera, the satellite was supposed to obtain high resolution stereo images (about 60 cm). The single operational Lanyard had a lifetime of 33 days.

Under the code name Gambit, the KH 7 and KH 8 series were the earliest programmes not yet to have been declassified, so that the information available about these two systems is severely limited.

Many KH 7 satellites were launched in the period 1963–1967, with 36 successes in 38 attempts. These satellites had a short lifetime, about 5 days, and a perigee as low as 150 km, so that they may have supplied images with 50 cm resolution.

An improved version, going by the name KH 8, made intensive use of the system with more than 50 launches between 1966 and 1984. It had a lifetime greater than 30 days, subsequently extended to 50 days, and an equally low perigee (135 km), allowing image resolutions in the 15 cm range. Right up to the last KH 8, these satellites contributed fully to the scheme of optimised use governing space applications from the end of the 1970s and the beginning of the 1980s.

The KH 9 satellites, also known as Big Bird, appeared in 1971 (KH 9A). The last in the series (KH 9B) was launched in 1986 but did not reach its orbit. With these satellites, spaceborne reconnaissance photography became fully operational. The gradually extended lifetime of the satellite was one of its main features.

Whilst the lifetime was 52 days (continuing where the KH 8 left off) for the first six in the series, launched between 1971 and 1973, 90 days was achieved for the two following satellites, then 138 days for all those launched after October 1974. This trend continued, despite one or two setbacks, and the first in the KH 9B series, launched in June 1980, remained in operation for a record 216 days. In such conditions, the launch rate fell. Whereas the series went on for another 15 years, only about 20 satellites were put into orbit. This should be compared with the 53 KH 8 launches over a period only slightly longer, which shows the tremendous progress achieved in this domain.

Operating at a relatively low altitude, ranging between 114 and 271 km, the KH 9 satellites could supply images at a resolution of about 50 cm and it had stereo capabilities when using the KH 10 camera originally intended for the Manned Orbital Laboratory (MOL), planned by the US Department of Defense (DoD).

The KH 11 family, under the code names Kennan and Crystal, marked a further turning point at the end of 1976. The first in the series was equipped with CCDs and transmitted data by radio. These satellites had lifetimes as long as three years from 1981, and even ten years in the rather exceptional case of the sixth satellite in the series, the last successful launch before the Shuttle accident. The KH 11 operated in pairs and observations were carried out alternately in the morning and afternoon. Equipped with mirrors similar to the one used on the Space Telescope, their sensors may have achieved decimeter resolution.

For about ten years, this satellite was the flagship of American space-based reconnaissance. A typical scenario involved launching KH 8 and KH 9 satellites in the spring to complement KH 11 data. The disappearance of the previously used wide-field systems can be explained by the development of the civilian satellites Landsat in 1972 and then Spot in 1986, both providing the same type of service.

At the end of the 1980s, the currently used system was finally established with the launch of a new version of the KH 11 (the Advanced KH 11), sometimes called KH 12, and the appearance of the first radar satellites, going by the code name Lacrosse.

The first recognised example of the new KH system dates from November 1992, and operated from a Sun-synchronous orbit at altitudes lying between 300 and 500 km. Their great manoeuverability allowed a corresponding flexibility of use. However, the limits imposed on optical systems by the problem of visibility could only be overcome by a radar system, able to function in all weathers. Further launches in the series seem to have been carried out in December 1995 and 1996, then in May 1999 and October 2001.

Tested in 1982 and approved in 1983, the Lacrosse system became available from the end of 1988. With an inclination of 57°, corresponding to the constraints of launch by the Space Shuttle, and an altitude in the range 650–700 km, the first radar imager established an almost unlimited observational capacity. A second satellite was launched in March 1991, whilst another in 1997 may have replaced the first, launched nine years previously, a new launch in August 2000 improving the radar capacity.

In 2002, the United States thus benefits from a continuous flow of real time data, mainly in the visible region of the spectrum, but also in the near and thermal infrared, together with radar. This ability to observe at a range of length scales, with resolutions anywhere above ten centimetres for optical systems and a metre for radar, is quite unique in the world today.

A new turning point may be forthcoming in the history of spaceborne reconnaissance programmes, with the latest KH systems planned for the period 2001–2003. The 8X project (also known as the Enhanced Imaging System), which aims to replace each KH in turn by new, more flexible systems of the same performance level, was not unanimously accepted at the US Congress. An increasing number of representatives are demanding deeper reflection concerning this form of intelligence. Traditional positions in the intelligence community may no longer be relevant within the new strategic situation. In the specific context of observation satellites, this questioning has led to a reappraisal of the idea of small satellites in an integrated framework, known as Future Imagery Architecture (FIA). The appearance of new, high resolution commercial satellite programmes, such as Earthwatch, Orbimage or Space Imaging, has created a need for global analysis. In parallel, a new defence agency has been set up, the National Imagery and Mapping Agency (see pp. 83–84), whose task is to coordinate the new arrivals of data with more traditional sources. The use of images should now benefit the whole American community, both military and civilian, with the new agency eventually becoming a highly integrated institution capable of responding at the shortest notice, once official acceptance has been given, to all requests from the US or abroad.

■ Soviet, then Russian, space-based observation provides a rather contrasting picture when compared with the American situation, even though certain general trends may seem to be similar, particularly in the early days (fig. 12.3).

The basic principle of space-based observation originally represented a threat in the eyes of Soviet political leaders, offering the Americans new means of intelligence concerning their strictly closed and jealously guarded regime. The first reconnaissance programmes must thus be viewed as a response to an American initiative they were in no position to prevent.

The information available about Soviet military activities in space is still somewhat fragmentary, despite the opening up of the space sector at the beginning of the 1990s. It remains a sensitive area, just as in the United States, where the existence of the National Reconnaissance Office was only officially recognised by the US government in 1992, 32 years after it was set up. Soviet capabilities in the field of spaceborne reconnaissance were only known to a very small number of adepts. Most of those involved often had only an incomplete knowledge of the systems used. A table of launches can be established by studying the orbits of the various satellites, but it remains difficult to assess their technical characteristics and interpretations often disagree, even over the identification of families.

The first two generations of Kosmos satellites were equipped with Zenit 2 and 4 cameras.

The first Zenit 2 was launched on 24 April 1962, and it was regularly used until 1967. These recoverable satellites orbited at an altitude of 200 to 300 km on average and had a lifetime of about 8 days. They provided images whose resolutions have been evaluated at 2 to 3 m, making them comparable with the first KH systems.

The first Zenit 4 was launched at the end of 1963 and it had a slightly longer lifetime of up to 13 days. They are thought to have been low resolution systems. They remained in use until 1970, complementing the first and then the third generation of satellites.

The third generation of Kosmos satellites appeared in 1968 and proved to be exceptionally long-lasting, since the last in the series was apparently launched as recently as 1994. These satellites had characteristic inclinations of 63°, 73°, 82° for launch from Plesetsk and 65°–70° for launch from Baikonur. They acquired civilian remote-sensing data with resolutions between 5 and 10 m, and greatly resembled the Resurs satellites, considered to be civilian since 1991. In addition, they carried out mapping for military purposes, also allowing the occasional higher resolution image. Several names were used, probably depending on the aims of the mission. Data was transmitted via Vostok capsules, usually de-orbited after 10 to 14 days, although missions sometimes lasted as long as one month. The launch and deployment flexibility of these satellites, carried through intact to the fourth generation, is a characteristic feature of Russian space-based reconnaissance.

The fourth generation, also known as Yantar, or Kometa for those involved in topographic survey, made its debut in 1975. With a system of detachable canisters that could be successively released, the Soyuz capsule could be used for longer periods of between six and eight weeks. The inclinations of these satellites lay in the range 62° to 70°. They were highly manoeuverable in order to accomplish close range, high resolution observations. In time of crisis, some of these satellites could make one transit per day over regions considered sensitive. More than 130 were launched between 1982 and 2001 with a gradual reduction in numbers since 1993, doubtless related to restrictions on the Russian defence budget and Russia's relative decline as an player on the international stage.

The fifth generation was first launched in 1982 and was supposed to be the first capable of transmitting images to Earth in digital form. With a maximum lifetime of eight months, these satellites were designed to cover extensive regions. Their great manoeuverability was often called upon in time of crisis. Twenty-two such satellites were launched between 1982 and 2000. The Yantar 4K-S1 satellites are said to carry out photo-digital surveillance, but they have a significantly shorter lifetime than their theoretical US counterpart KH 11. The Arkon 1 satellite may be the first electro-optical transmission system designed to have a long lifetime, even though it was soon to break down.

The sixth generation Orletz-1 appeared in 1989 and comprised systems rather similar to the fourth generation Kometa satellites, with lifetimes between two and three months. Launched on an occasional basis about once a year up to 1994, these satellites were not subsequently used except once in 1997.

Apparently, a seventh generation (Orletz-2) made its appearance in 1994 with the launch of a high resolution satel-

lite, Kosmos 2290. Its mission may have lasted some eight months, and the second in the series was only launched in 2000.

In the wake of this impressive armada of reconnaissance satellites, at least during certain periods, the fading Russian presence in this type of mission is quite remarkable. Between the spring of 1996 and the summer of 1997, and following two quite untypical launch failures, the only available data was provided by satellites of the Resurs or Priroda type, that had since become 'civilian' (see chapter 9). The launch in June 1997 of an Orletz-1 type satellite shows that Russia has not abandoned this type of activity. This is confirmed by the launch of a new generation Arkon in 1997 and the launches of one Yantar and two Kometa in 1998 then one Yantar in 1999. However, the need to economise scarce launch and

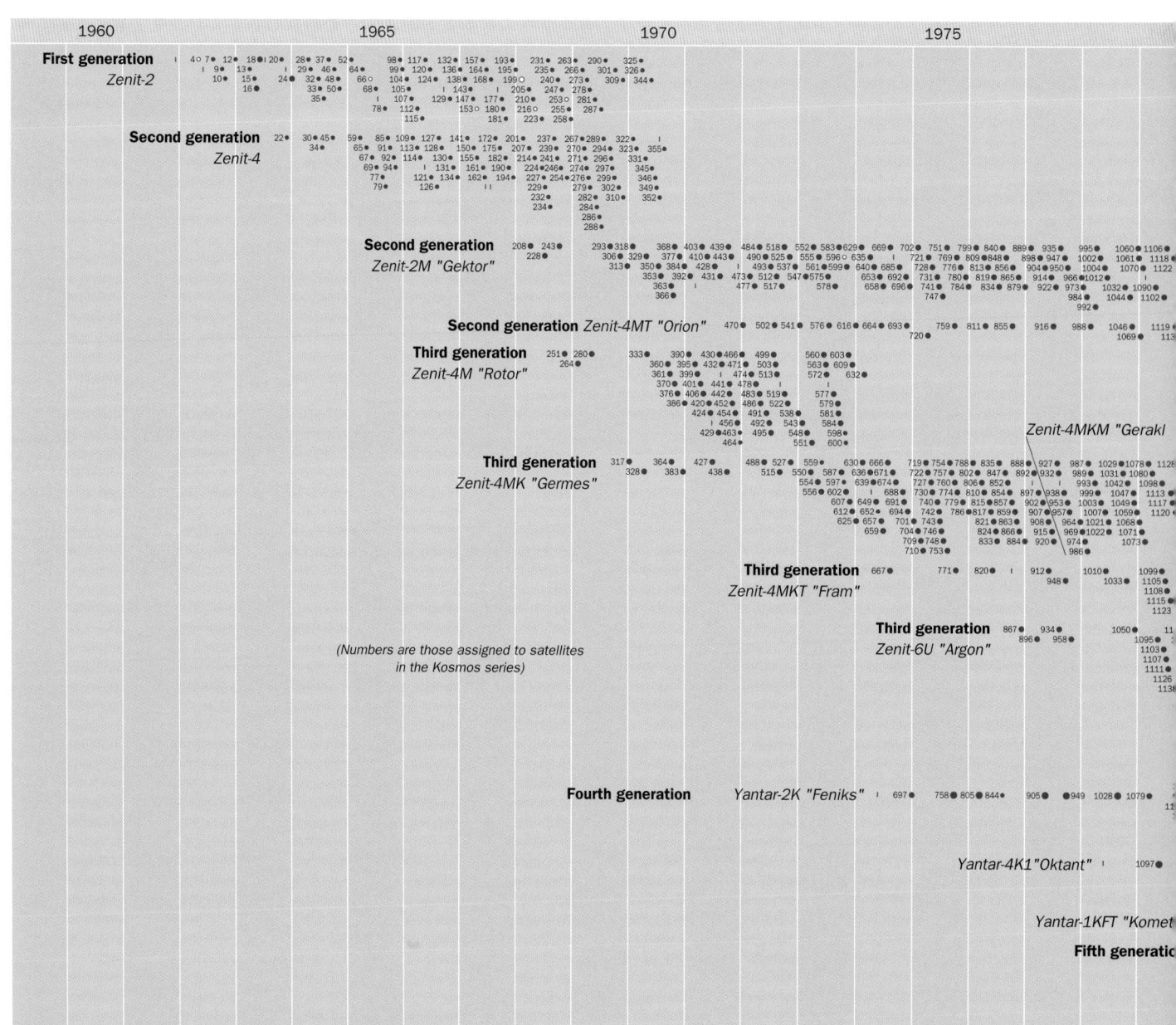

Figure 12.3. **Chronology of reconnaisance satellites.** (Former USSR, then Russia. Kosmos 2331 is sometimes included in the third Kosmos generation.)

satellite resources must certainly weigh heavily. It has also meant a rather irregular involvement in this field with the launches of three satellites in 2000 (one Yantar, two Orletz) and two Yantar in 2001. The Geyzer relay satellite, launched in 2000, is devoted to image transmission and should allow observations to continue for longer periods and prolong the lifetime of the latest generation of satellites.

■ Other nations have played a much lesser role with regard to these activities. The launch in 1995 of the Israeli satellite Offeq 3 and the French satellite Hélios 1 (a joint effort with Spain and Italy) mark the appearance of new and independent sources of information from space. These two systems possess different technical features, determined by the specific national policies they embody (see fig. 12.2).

Hélios had its beginnings in 1986, eight years after a proposal for the SAMRO project (SAtellite Militaire de Reconnaissance Optique), which was eventually abandoned in 1982. The original mission was to be surveillance of the Soviet Bloc. Launched in 1995, in a completely different geostrategic context, the satellite was eventually reorientated towards a world surveillance mission. The Gulf War and European inability to access independent information sources further bolstered politcal interest in the programme. In 1991, the Western European Union (WEU) created the satellite observation centre in Torrejon. Its task was to process civilian and Hélios images in special circumstances and it corresponds to a first stage of European cooperation.

Hélios 1A follows a Sun-synchronous orbit, like SPOT whose platform it uses, but at an altitude of 685 km. It is

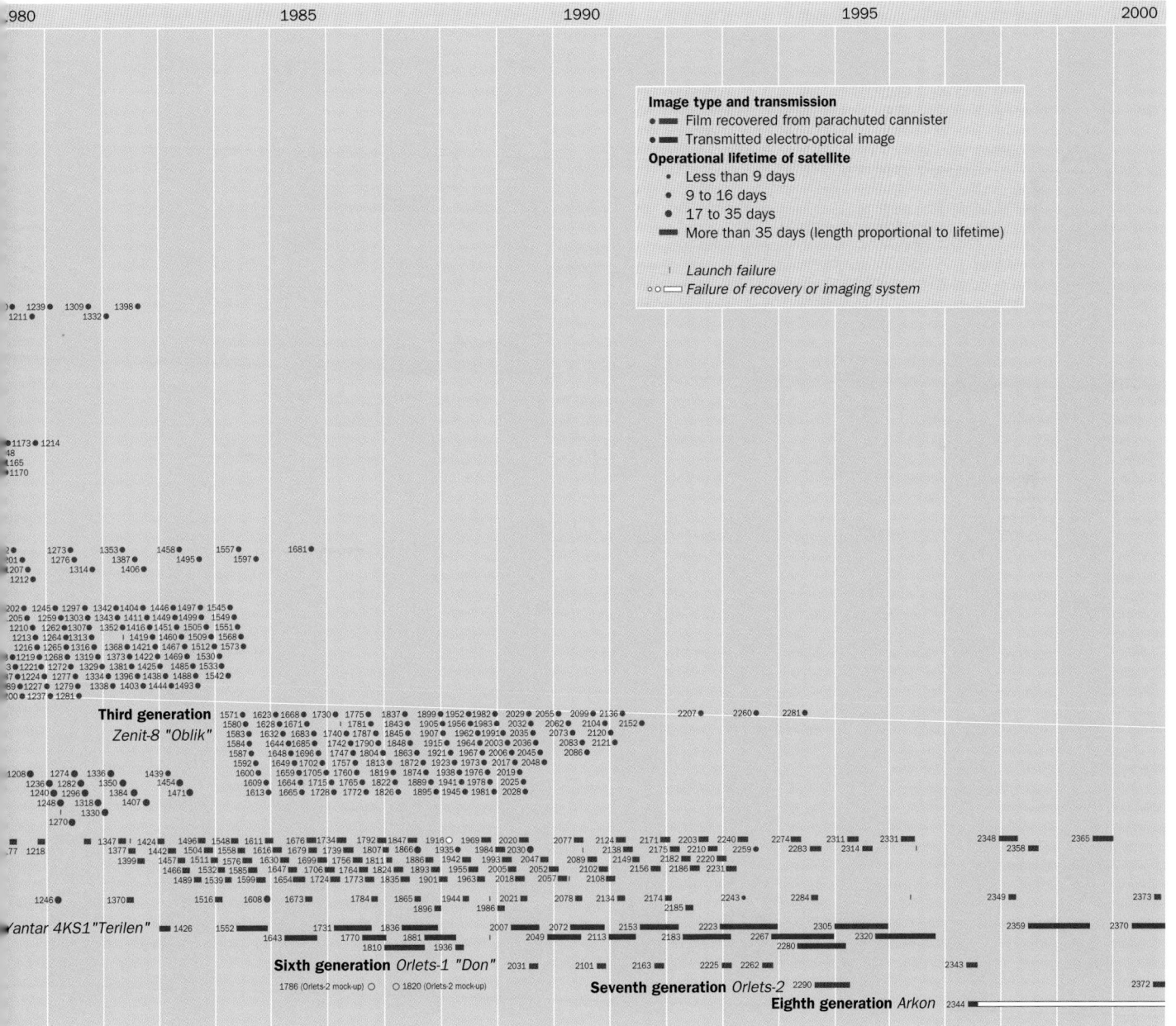

equipped with a metre-resolution sensor and two on-board recording instruments. Its manoeuvering capabilities allow it to observe the same site every other day using wide-field and narrow-field cameras. It can take fifteen images on a daily basis within certain preselected regions (Europe, Africa, Middle East), on behalf of the three countries participating in the programme (France 80%, Italy 13%, Spain 7%). The launch of a second Hélios 1B satellite, in 1999, whilst Hélios 1 was still operational, has led to improved coverage and reduced times for image acquisition. The same site can be imaged every day, weather permitting.

The Hélios programme was to move into the next stage with the joint development of Hélios 2 by France and Horus by Germany. Technical skills were complementary, with Hélios operating in the optical region of the spectrum, and Horus using radar. This combined effort was intended to strengthen Franco-German cooperation and its role in European defence. However, political and budgetary difficulties in Germany meant that the project could not be brought to fruition, despite official commitments made in 1998. The European Security and Defence Initiative (ESDI) provides a suitable framework for rekindling interest in satellite reconnaissance programmes. In 2001, the Torrejon satellite centre has become an element of the European Union's common foreign and security policy. Still the developments are organised rather on a national basis. Hence, France is pursuing the Hélios 2 programme, with participation of Belgium and Spain (2.5% each), due for launch in 2003, whilst other European countries are considering development of their own dual capabilities. The German SAR-Lupe programme still in definition in 2001 includes five radar satellites of 700 kg at an altitude of 500 km. The first satellite is supposed to be launched at the end of 2004 and the system to be operational in 2006. The Italian COSMO-Skymed (COnstellation of small Satellites for Mediterranean basin Observation) includes four dual radar satellites supposed to be used on a complementary basis with the optical satellites of the French Pléiades programme. The formal agreement between France

Figure 12.4. **Chronology of nuclear detection and early warning satellites.**

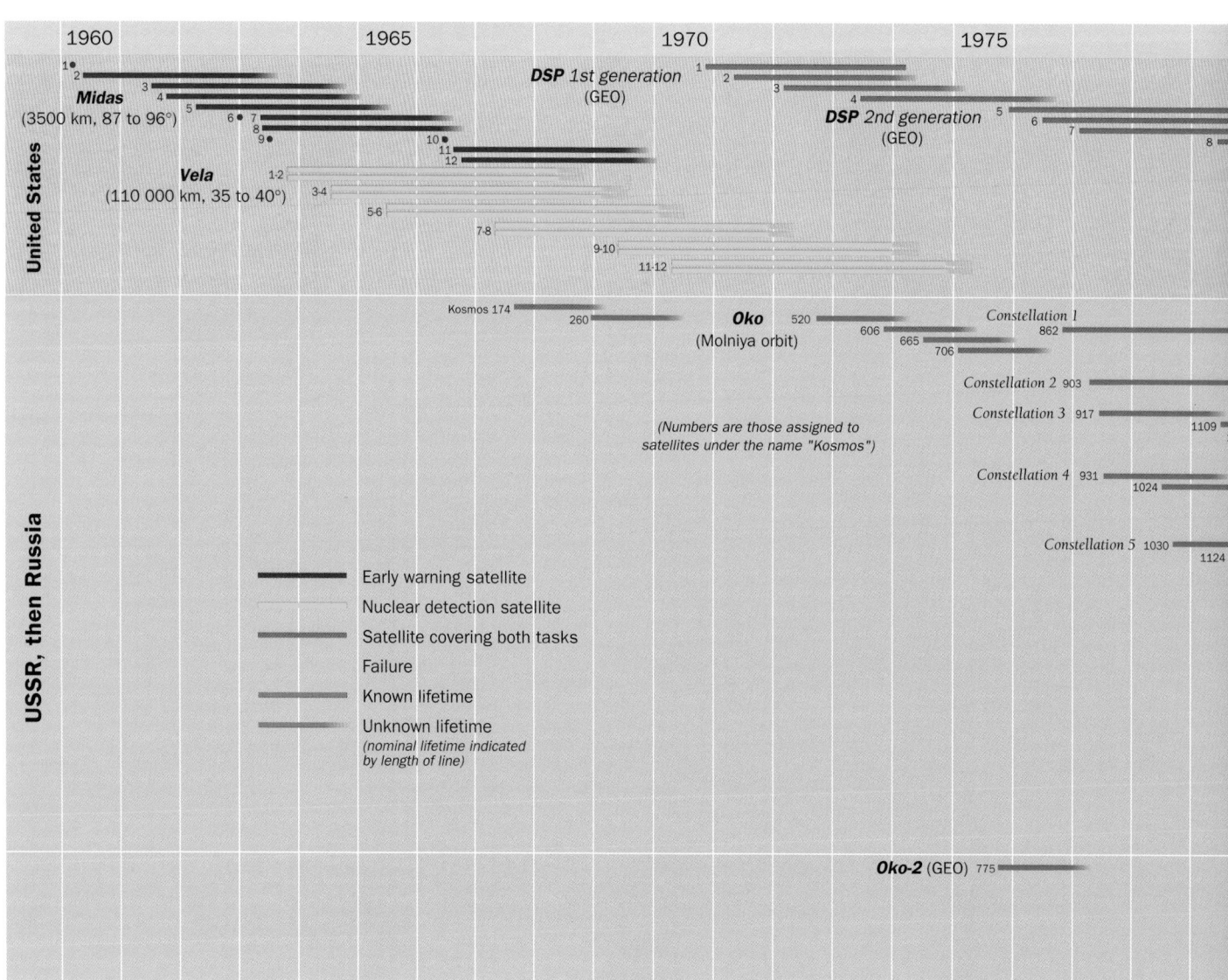

and Italy is open to new participants and Belgium, Sweden and Austria have expressed their interest in 2002.

The Hélios system is fairly close to conventional military reconnaissance systems, although it has developed in a rather novel way. Indeed, it breaks with a hitherto undisputed tradition whereby civilian satellites were invariably subordinate to their military counterparts. Experience acquired through the civilian programme SPOT (platform, CCD sensor technology, on-board recording techniques and digital data transmission systems), combined with a genuinely profit-orientated financing, show that another logic is perfectly feasible, in which military capabilities derive from civilian ones. Other countries in possession of civilian remote-sensing satellites, such as Japan, India and China, may also be tempted to follow the same line, provided it falls in with their political outlook. At the same time, the case of the Israeli satellite Offeq demonstrates that small satellites resulting from relatively simple technological developments can also fulfill certain fairly limited requirements with regard to the pressing concerns of national security.

Following two experimental satellites, the deployment of Offeq 3 in 1995 marked the beginning of a novel programme aimed at space-based military observation of just part of the Earth's surface, motivated by regional rather than worldwide geostrategic considerations. With limited financial resources, they had to face significant physical constraints. The location of the Palmahim base means that satellites can only be launched into retrograde inclinations, if Arab territories are to be avoided during the launch phase. This is also a disadvantage for the launch vehicle, which was developed for the Jericho missile programme and has only a limited launch power (see chapter 6). The Offeq 3 satellite, with a mass of 225 kg at launch, and a 36 kg payload, was placed in an orbit with apogee at 369 km in such a way as to cover a region lying between latitudes 30° north and 30° south. The launch of Offeq 4 in 1998 was to bring marked improvements to the system, still operating in optical mode, with resolutions down to about one metre. However, it failed and the launch of Offeq 5 is expected for 2002 (see fig. 12.2).

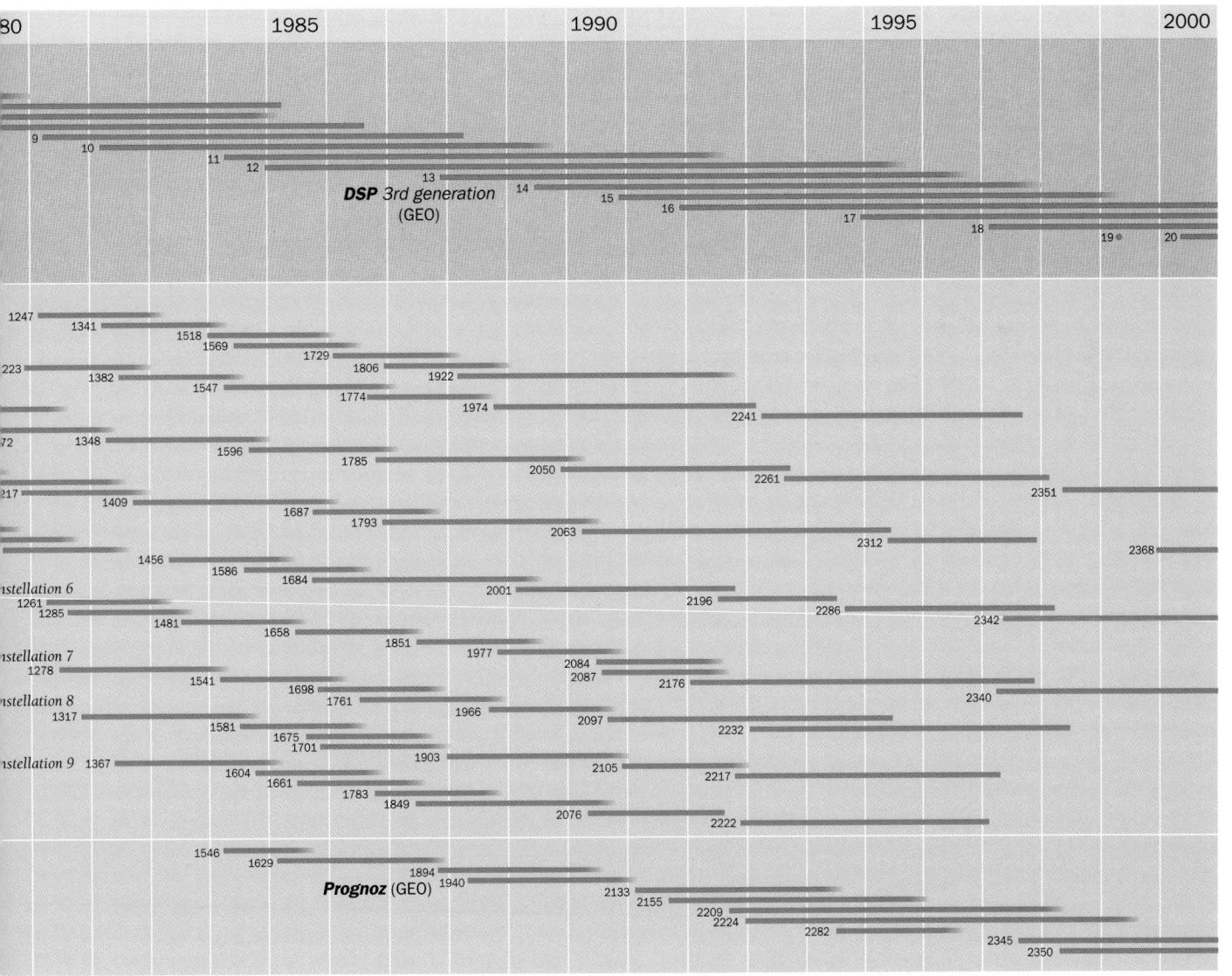

The appearance of new metre-resolution commercial projects (see chapter 9) has introduced a new element into the problem, even though three fundamental features, the degree of control entrusted to users, continuity of data acquisition, and data transmission times, make all the difference with a military system. The Israeli–US programme EROS, launched in December 2000, forms part of this new logic. The situation may be even more complex in the context of an officially dual-use programme, i.e., a programme devoted to both civilian and military objectives, as in the case of the Japanese Information Gathering Satellite (IGS). This option was chosen when North Korea launched its missile Taepo Dong in November 1998. The tendency to use American technology shows how far most countries still depend on the main space powers, but it does not rule out the possibility of subsequent autonomous development. The same trend can be observed in South Korean or Taiwanese projects, also involved in the technological development of small satellites (see chapter 4). The launch in October 2001 of the experimental Indian satellite TES (Technology Experiment Satellite) with one metre resolution indicates a new military interest for officially civilian programmes

Early warning satellites

Whereas observation satellites are today being used by an increasing number of countries, early warning systems belong to much more restricted circles (fig. 12.4). Their mission is closely bound to the question of nuclear dissuasion. With the development of Soviet intercontinental missiles at the end of the 1950s, one of the primary concerns of the United States was to set up a network of early warning satellites. Their task was to provide rapid information about the launch of ballistic missiles. By overcoming the obstacle posed by the Earth's curvature, the best systems were able to double the warning time as compared with Earth-based radar systems, extending it to around thirty minutes. The long series of satellite families is witness to the many technical problems encountered in designing and gradually improving these alert systems.

The MIDAS satellites (MIssile Defense Alarm System), deployed in circular polar orbits, were originally equipped with infrared sensors designed to detect heat emissions produced by missiles during their combustion phase. However, it proved difficult to restrict identifications to the phenomenon in question, with other natural factors likely to interfere, such as the reflection of the Sun's rays from high altitude clouds. In order to reduce the number of false alerts, television cameras were also carried on board. Later, their successors followed circular orbits at an altitude of 36 000 km, but with inclination close to 10°, to provide improved coverage of the high latitudes where Soviet missile launches were based. At the beginning of the 1970s, the IMEWS system (Integrated Missile Early Warning Satellite) marked the beginning of a new phase. These satellites were placed in geostationary orbit and simultaneously provided early warning and detection of nuclear explosions, a task that had previously been carried out by the Vela family within the framework of the 1963 treaty banning nuclear tests in the atmosphere and in space. These were in turn succeeded by the DSP satellites (Defense Support Program), the name being attributed in 1976. DSP combined alert and detection of nuclear explosions (fig. 12.4). In 1991, two of the five geostationary DSP satellites were reserves, clearly illustrating the importance of this mission for the American government. The DSP system has since undergone many improvements. The first concerned data relay with the Talon Shield programme, implemented in 1996. This programme was entrusted to a specially created military unit, and mainly consisted in simplifying the procedures and stages involved in relaying alert data. To this end, data reception, processing and retransmission to units based abroad were centralised at the US Air Force base on Cheyenne Mountain in Colorado. Eventually, specially equipped vehicles will be sent to future operational theatres to receive satellite data directly. Often better known, the second phase of this modernisation programme concerns the satellites themselves. As part of the SBIRS warning programme (Space Based InfraRed System), improvements in satellite performances (see fig. 12.11) now take into account the threat of medium range ballistic missiles. Two features must be improved in consequence. The first is infrared detection of short range missiles in the atmosphere, which is not yet adequate, as witnessed by difficulties encountered in the 1990 Gulf War. The second is the frequency with which data is collected. At the present rate of once every ten seconds for DSP satellites, the trajectories of these missiles, propelled for much shorter periods than intercontinental missiles, cannot be correctly ascertained.

Beyond such improvements, new categories of satellites should see the light of day in the period 2004–2010. Two further early warning satellites are planned for highly elliptical orbits that will provide coverage of the North Pole. These satellites are intended to supersede infrared detectors probably carried on board the Jumpseat intelligence satellites, in a similar orbit (Heritage programme). Perhaps more significant, an as yet undetermined number of low-orbiting satellites should complete the US defence panoply. Their main task would be to provide precise determinations of reentrant enemy missile trajectories for the purposes of exo-atmospheric interception. If detector performances are adequate, other missions could be envisaged for these satellites, such as the detection and characterisation of shots fired on the battle field (technical intelligence). This component would be a fully integrated into the kind of tactical intelligence imagined for the years to come, that is, based on ever closer cross referencing of the complete data set gathered by satellite, including visible and infrared data, signal analysis, remote sensing, intelligence and so on (see fig. 12.11).

Soviet systems took much longer to set up. This was partly due to technical problems in developing an infrared system, but political choices undoubtedly played a part in the delay.

The first genuine early warning system was not established until the second half of the 1970s, and even that was somewhat limited. After deploying the first Oko satellites on Molniya orbits (fig. 12.4), the full constellation comprising nine satellites only became operational in 1987. Since 1989, surveillance by non-geostationary systems has been complemented by the Prognoz geostationary satellites. The existence of such systems was only officially recognised in 1993, even though the corresponding positions on the Clarke orbit had been reserved under this name at the ITU (International Telecommunication Union) since 1981. In 1992, two satellites were launched, followed by one more each year, but this would not appear sufficient to maintain the whole system. Difficulties encountered by certain Prognoz satellites combined with inadequate replenishment of the Oko satellite constellation have continued to limit present Russian capabilities (fig. 12.5). The changing strategic environment, particularly with regard to the United States, explains why this warning programme is no longer a priority. The inability of Russian defence to finance new systems better suited to countering the threat of medium range ballistic missiles is a clear sign of their gradual decline as a military space power. At the same time, US development of the NMD programme has led to a new appreciation of Russian needs. Efforts have been made to maintain a minimal capacity and since 1998 there has even been a certain degree of cooperation with the United States in the guise of the RAMOS research programme (Russian AMerican Observational Satellite).

Electronic intelligence satellites

These satellites, often called ELINT (ELectronic INTelligence), are designed to listen in to communications, whether they be military or otherwise, in order to obtain information about the size, deployment and level of readiness of enemy forces. This mission is known as COMINT, which stands for COMmunication INTelligence. Within the context of another activity known as SIGINT (SIGnal INTelligence), they also serve to locate anti-aircraft and antimissile radars, identifying their frequencies with a view to neutralising them by suitable countermeasures. All these parameters, under the command name of battle intelligence, are continually monitored in peacetime in order to give immediate warning of any unusual activity. In time of crisis, the aim is to obtain a clearer assessment of the real intentions a potential enemy may have, and in time of war, to identify the type of attack under preparation and thereby set up appropriate countermeasures, such as jamming of enemy communications.

ELINT satellites, launched by the United States since 1961 and five years later for the USSR, thus have an essential part to play in information gathering (fig. 12.6). Little is known about these programmes, and what there is generally comes from indirect sources. Soviet systems are mainly identified via orbital parameters. These systems, invariably going by the name of Cosmos, are all officially referred to as scientific satellites, even though no results have ever been published. The appearance of the first prototypes as early as 1967 shows that ELINT satellites, together with photographic reconnais-

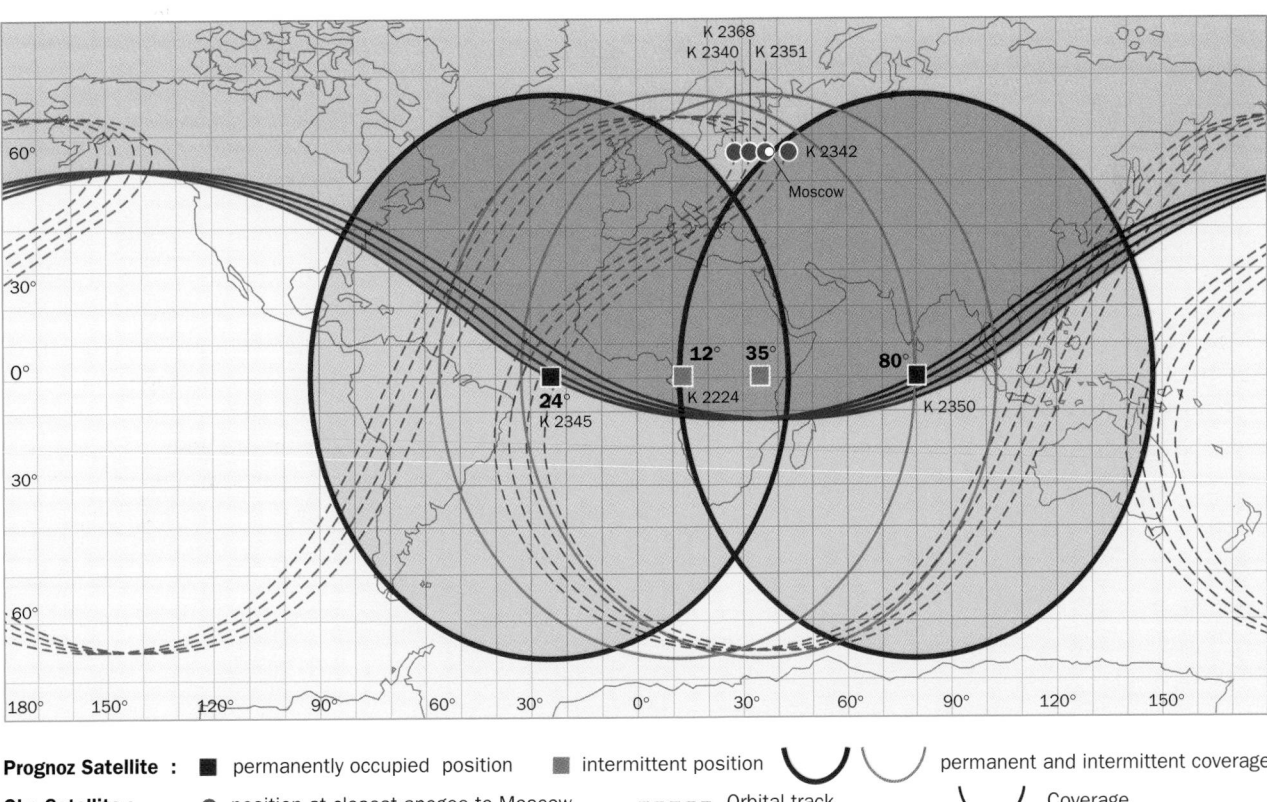

Prognoz Satellite : ■ permanently occupied position ■ intermittent position ⌣⌣ permanent and intermittent coverage

Oko Satellite : ● position at closest apogee to Moscow - - - - Orbital track ⌣ Coverage

Figure 12.5. **Russian early warning satellites. Trajectories and surveillance zones.** In 2000, the four Oko satellites were operating but the Prognoz positions did not appear to be occupied by any active satellites.

sance satellites, were granted political priority. Satellites launched in the 1980s belong to the third and fourth generations, and appear in considerable numbers. This can be explained by the preponderance of non-geostationary satellites, with the task of providing global intelligence almost in real time. Furthermore, although in constant progression,

the lifetime of these satellites remains relatively short. Since the second half of the 1980s two constellations of Tselina satellites (Tselina D and 2) have been exploited in parallel. The Tselina 2 system, apparently designed to cover the whole ELINT mission, is still being deployed. Launches continue, although at an ever reducing rate since 1992 (between one

Figure 12.6. **Chronology of ELINT and ocean surveillance satellites.**

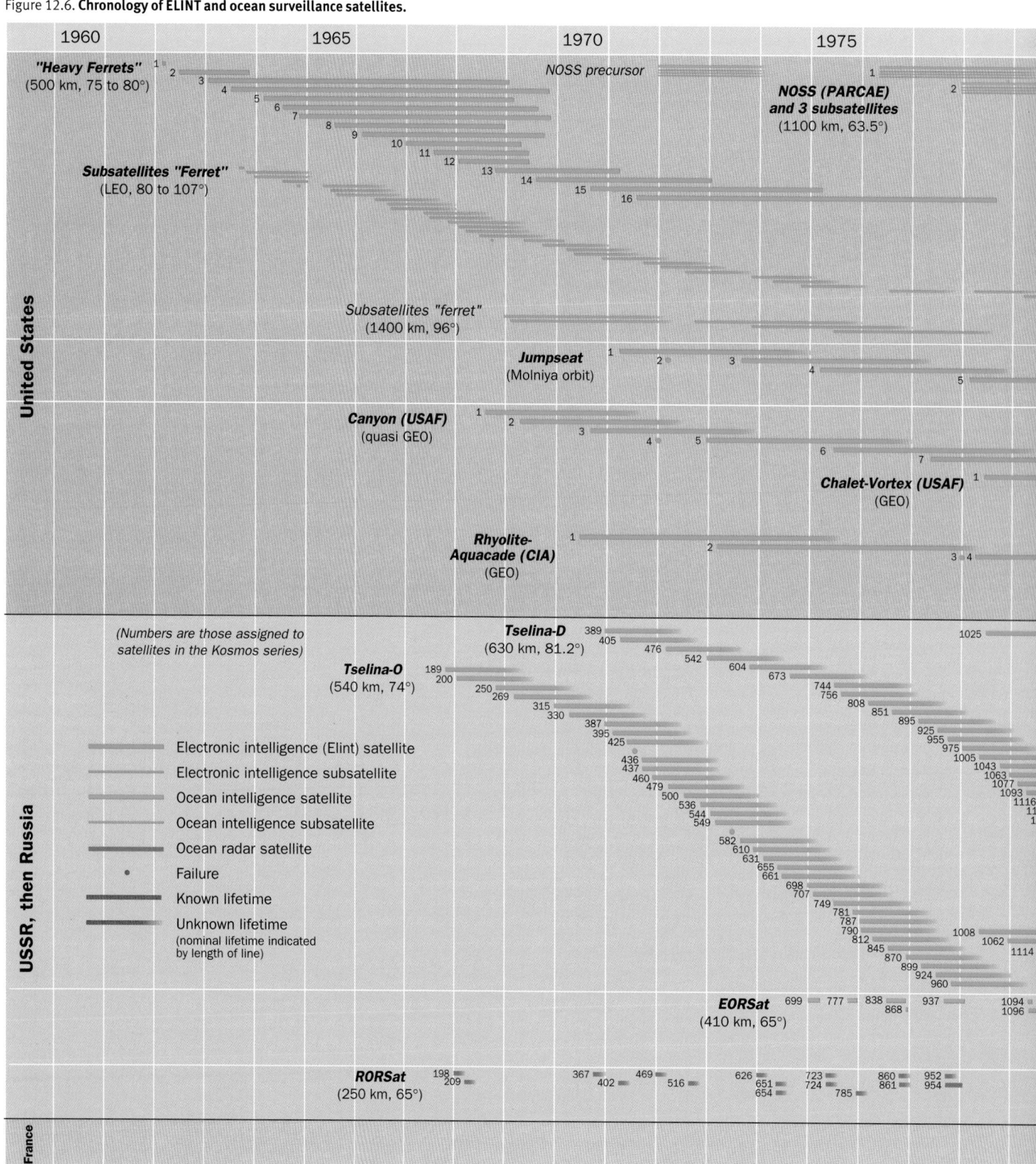

and three per year), and it is quite certain that the system is operating well under its full capacity. However the launches of 2000–2001 may appear as a beginning of recovery.

The first American satellites are known in the literature as Ferrets, a name which gives a good idea of their purpose. They originally served to intercept signals concerning air defence, antiballistic missiles and early warning radars. The first 17 satellites of this type were launched between 1962 and 1971 in low orbits with inclinations varying between 75° and 82°. From this point on, only Ferret subsatellites were sent into orbit, generally at the same time as KH type reconnaissance satellites, and their numbers are decreasing as the years go by. Geostationary satellites belonging to the Chalet and Magnum programmes, deployed since the second

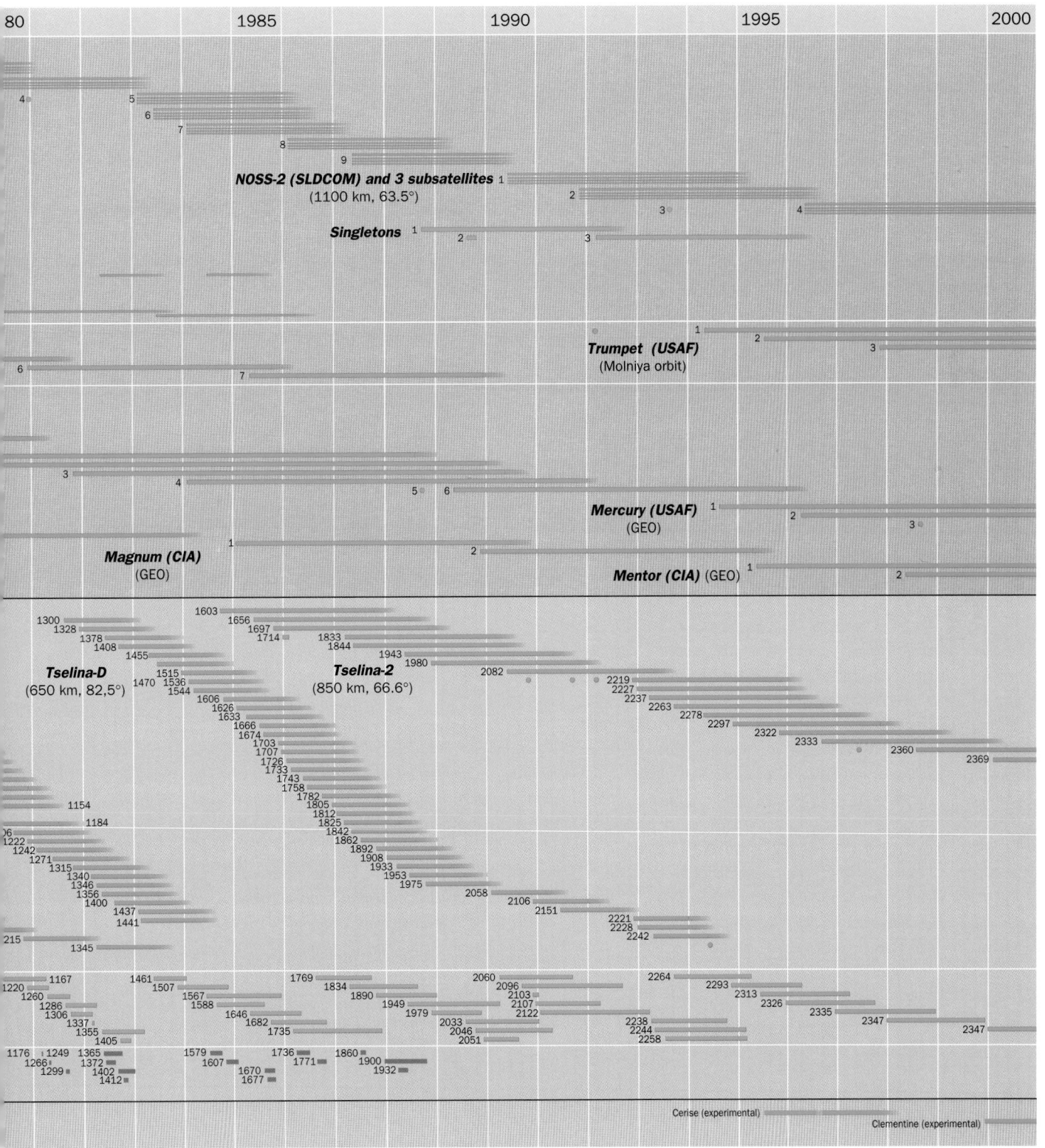

half of the 1970s, today form the main constellation. It is complemented by satellites in highly eccentric orbits, like the Jumpseat series. These are used to listen in to high latitude regions and also the Soviet Molniya telecommunications satellites in the same type of orbit. In the United States, the increasing multiplicity and complexity of electronic sources under investigation may lead to the implementation of a combined architecture for intelligence and signal analysis. This would bring together all the components so far developed in a single set of low-orbiting satellites better suited to intercepting a wide spectrum of data. It seems that one such project, designated IOSA 2 (Integrated Overhead SIGINT Architecture), is currently under study at the National Reconnaissance Office.

Among the space powers beginning to build up military capabilities, France has developed two experimental satellites, Cerise in 1995 and Clémentine in 1999, launched at the same time as the reconnaissance satellites Hélios 1 and 2. Only the second is actually operational. A swarm of Comint microsatellites (ESSAIM programme) is scheduled for 2004, but these will only provide warning signals in times of crisis.

Ocean surveillance satellites

Chronologically speaking, ocean surveillance constitutes the third main sector of spaceborne military activities. The great number of Soviet satellites expresses their particular desire to track naval units, a key component in US and NATO forces. The Rorsat programme thus consists in carrying out radar surveillance from low orbits, at altitudes around 250 km, given the energy requirements of radar systems. This is the name attributed to it by the Western world, an allusion to its mission (Radar Ocean Reconnaissance Satellite), although the Russian code name may well be USA. Intelligence obtained by the Eorsat system (Electronic Ocean Reconnaissance Satellite) – or USP according to the Russian name – from an orbit at 400 km, but with the same inclination of about 65°, was originally designed to complement this radar detection. In fact, technical difficulties in developing the nuclear reactors intended to supply Rorsat's energy needs have led to at least a temporary discontinuation in the programme. There has been a corresponding increase in the strength of the Eorsat programme, which varies between four and six satellites distributed in two different planes.

The US ocean surveillance system has long been made up exclusively of the White Cloud satellites, also called NOSS after the name of the programme they belong to (Naval Ocean Surveillance System). The system usually comprises a main satellite and three subsatellites, all orbiting within a few hundred kilometres of one another. Boats are then located by triangulation, and identified by analysing emitted signals. Five primary and secondary systems are theoretically operational at the same time, from an orbit at altitude 1000 km, and inclined at 63°.

Several complementary satellite projects have been studied, in particular by the US Marines and Air Force, with a view to wider ocean surveillance. At the beginning of the 1990s, in the context of the Space Based Wide Area Surveillance (SB-WASS) programme, a long debate was fought out over whether the sensors should be infrared or radar. The result of these discussions is not precisely known. However, launches carried out during the first part of the 1990s, involving various Singleton or Triplet type satellites, would seem to suggest a compromise.

New openings for global surveillance

The United States is now perfecting and increasing the number of programmes designed for tactical surveillance of theatres of war. These new technical capabilities should give the US military better information concerning their enemies in every situation and in any weather. One particular aim is improved tracking of moving targets on the ground, often difficult to locate and almost impossible to track by traditional spaceborne means. Lessons learned in the Gulf War, and in subsequent conflicts in ex-Yugoslavia, have led American strategists to think about extending the use of satellite-based radar surveillance, making it a key element in all future space-based information gathering systems.

These new needs have been invoked to justify financial backing for new radar satellites, possibly to be used in a constellation. Hence, the main objective of the American programme Discoverer II, run jointly with the US Air Force, DARPA, and NRO, is the tracking and global surveillance of military targets. This programme is a continuation of the Starlite project (Surveillance, TArgeting and Reconnaissance SatelLITE) launched in 1997 by the Pentagon research agency. It would use 24 satellites in low orbits, equipped with radars called Ground Moving Target Indicators (GMTI), which can detect and follow moving targets, and synthetic aperture radars capable of high resolution imaging and high precision numerical data acquisition concerning ground elevation. One of the main tasks of the programme is to be able to revisit any zone between latitudes 65° north and 65° south within just 15 minutes. In February 1999, a first study phase was granted to a team made up of three manufacturers (Lockheed–Martin, TRW and Spectrum Astro) for platforms weighing between 1000 and 1500 kg at costs below 100 million dollars each. The schedule includes a test period in orbit extending from 2003 right through to a fully operational deployment expected for 2007. Even if the manufacturers keep to these deadlines, this does not mean that the project will be carried out in its entirety. Its cost and complexity remained the subject of much controversy in 2000. The lack of consensus on this point may lead to the programme being abandoned in its present form, although there seems to be agreement that radar surveillance is needed.

In parallel with this, the warning and detection system SBIRS has also undergone studies aimed at increasing the tactical potential of infrared detection, in addition to the localisation of ballistic missile launches. Going beyond considerations of new programmes, efforts are also being made to improve the way these systems work together. In the longer term, a genuine architecture for global surveillance may be the result.

Telecommunications

Satellites have become an essential tool for conducting modern military operations and responding to a wide range of crisis situations. By providing links between units in the field and the command centre, or transmitting data from a reconnaissance satellite in almost real time, satellites play the role of force multipliers, overcoming the usual terrestrial constraints, particularly in inaccessible or uncontrolled regions. American telecommunications satellites thus handle around three-quarters of all long distance military links, and cable plays a rather complementary role, increasing security by diversifying the means available. The huge diversification in information sources and the need to integrate the resulting data are a further spur to developing such systems, and it has become impossible to imagine military activity taking place without them.

The increasing role of geostationary satellites in telecommunications activities does not mean that other types of orbit are likely to disappear (fig. 12.7).

Various arguments support the use of such systems. The first is of a geographic nature, since non-geostationary satellites are the only ones capable of covering zones at latitudes above 70°. The most common are the Soviet, then Russian, Molniya satellites, with a very characteristic orbit providing civilian and military links from the Baltic to the Pacific, over a complete range of services extending from telephone to television. The SDS satellites (Satellite Data System), transmitting US communications to and from their Arctic military bases, have the same type of orbit.

Moreover, from a purely economic standpoint, these satellites can be lightweight, launched in clusters or together with a heavy payload, when the launch vehicle places several instruments in orbit. This further reduces the overall cost of such missions. These light satellite systems have always been preferred by amateur radio enthusiasts.

Most non-geostationary satellites in operation are Russian, generally in low circular orbits, but with eccentric geosynchronous and geostationary orbits since the end of the 1980s. Confusion between civilian and military activities, a traditional feature of Soviet, then Russian, space operations, makes it difficult to determine exactly what mission a given constellation is intended to achieve. However, it would appear that the most diverse services, ranging from telephone, through governmental and military communications to television, are guaranteed by Molniya satellites.

Low circular orbits tend to be reserved for the needs of the military and in particular for direct tactical communications linking military command centres on national territory with ships, aeroplanes and ground units. Some satellites seem to be more especially designed to provide long range communications and gather information from low power transmitters, storing them for subsequent relay to ground-based stations, very often almost in real time.

The whole range of telecommunications links is provided by three main types of closely integrated system (fig. 12.8). Most of these low-orbiting satellites, referred to as Strela satellites, are launched at altitude 1500 km and inclination 74°. They are launched in clusters of eight small Kosmos satellites, some of which are still operational. They are complemented by a constellation of slightly heavier Kosmos satellites, launched individually. The first three of these were positioned at altitudes 780–810 km, and inclinations of 74°, moving in three orbital planes at angles of 120° to one another. Finally, since 1985, a new system of satellites has been launched in clusters of six in two orthogonal planes at altitude 1400 km and with an inclination of 83°. It would appear to be destined to supersede the first generation with its clusters of eight satellites. The Raduga and Potok satellites (as they are called when reservations are made at the ITU) complete the panoply of Russian military space telecommunications systems. The launch in 2000 of a new Russian Geyzer relay satellite shows that new systems continue to be implemented, despite delays.

US non-geostationary systems only include military satellites in the IDSCS and SDS series. The first satellites in the Initial Defense Satellite Communication System appeared in 1976, in the plane of the equator but at altitudes slightly below the geostationary orbit. The whole constellation therefore drifts on a daily basis, thus allowing coverage even if one member should fail. The advantage of this system was to provide security at a relatively low cost. It provided long range tactical communications (up to a maximum of 16 000 km) and transmission of high resolution images.

American satellites in the SDS family have highly eccentric orbits and have often been launched at the same time as reconnaissance satellites in the KH series. Indeed, they long provided the downlink for the KH satellites. Initially carried out by two satellites better known by the name of Jumpseat, the last two of which were launched in February 1987 and November 1988, these missions were gradually transferred to NASA's TDRS (Tracking and Data Relay Satellites) in geostationary orbits. This type of system involves considerable savings and small satellite projects of the Lightsat type, which became fashionable again in the 1980s, are close to the original idea behind IDSCS.

As for other military applications, future satellite communications systems should have quite different designs from existing systems. New trends in this domain follow two main lines.

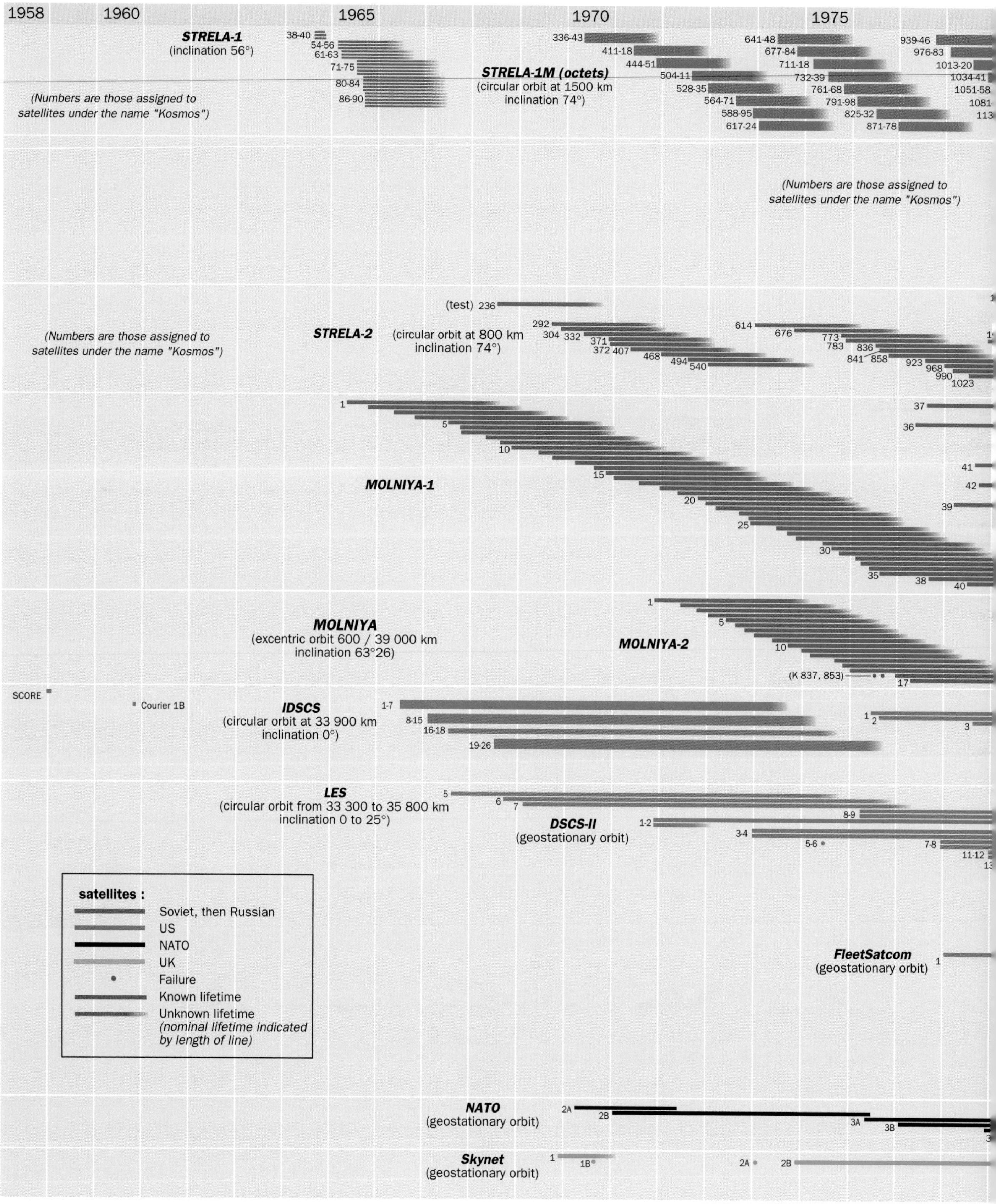

Figure 12.7. **Chronology of military telecommunications satellites.**

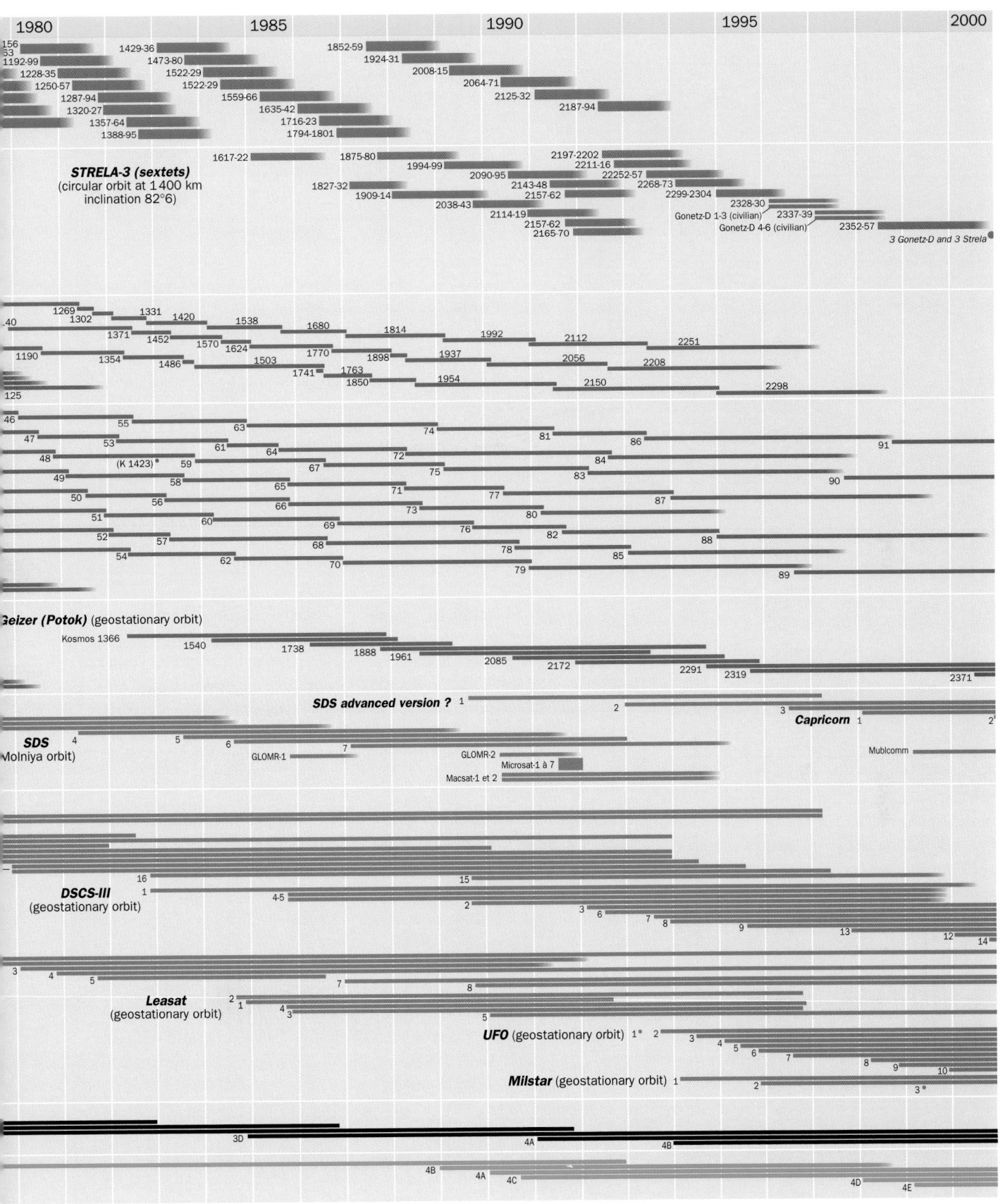

1980 **1985** **1990** **1995** **2000**

STRELA-3 (sextets)
(circular orbit at 1 400 km
inclination 82°6)

3 Gonetz-D and 3 Strela

Geizer (Potok) (geostationary orbit)

Kosmos 1366

SDS advanced version ?

Capricorn

SDS
(Molniya orbit)

GLOMR-1   GLOMR-2   Mubicomm
Microsat-1 à 7
Macsat-1 et 2

DSCS-III
(geostationary orbit)

Leasat
(geostationary orbit)

UFO (geostationary orbit)

Milstar (geostationary orbit)

Circular orbit at 800 km, inclination 74° followed by a single satellite
Circular orbit at 1 400 km, inclination 82,6° followed by a cluster of 6 satellites
Diverging circular orbits at 1 500 km, inclination 74°, followed by a cluster of 8 satellites

Figure 12.8. **Orbits of non-geostationary Kosmos satellites.**

In the first place, the likelihood of major conflict has considerably diminished and, since the Gulf War, sources of threat have been perceived in a new light. This, combined with a need for shortlived actions carried out on the regional scale, has meant that certain military activities can be significantly curtailed. Hence, the main core of protected military communications, once extending over a wide range of different frequencies, now tends to fall within a narrow band of the very high frequency region of the spectrum (EHF, i.e., 20–44 GHz), more appropriate for discreet, small scale operations.

The American programme Milstar, which began in 1982, today represents the state-of-the-art in this frequency band. Two satellites are already operational, to be joined by four others by 2002. The completed system will form the backbone of future American EHF telecommunications. This satellite system will also be equipped to operate in the lowest frequency bands and should provide worldwide coverage, whilst using only narrow beams to facilitate discreet reception via portable devices weighing less than ten kilograms. The US Department of Defense is already experimenting with high speed links (1.544 Mb/s) in the frequency band 20–30 GHz, using portable antennas measuring between 0.25 and 1.25 metres in diameter. This is part of NASA's ACTS programme (Advanced Communication Technology Satellite). The so-called Advanced EHF satellite programme is at present under study as a continuation for the Milstar programme. The gigantic costs incurred by Milstar (around

17 billion dollars in all) testify that it was originally planned with nuclear war in mind. At the same time, a global restructuring of existing means in other frequency bands is currently underway. It is particularly concerned with the 7.2–8.4 GHz range, used until now by the DSCS satellite series (Defense Satellite Communication System). This family has been the subject of constant improvements. Examples are the appearance in 1982 of the third generation DSCS III satellites (also carrying a UHF capability), or the current DSCS development programme which rests upon improvements in the tactical performance of the satellites (extended pass band, low-noise amplifiers, etc., included in the DSCS Service Life Enhancement Programme, known as SLEP). Likewise, in the period 2008–2010, commercial satellites operating in the Ka-band are due to be purchased as part of the Wideband Gapfiller programme (WG) and the Advanced Wideband Gapfiller programme (AWG). These are expected to transmit a large fraction of unprotected military communications. The second main development results from advances made in mass market electronics and the advent of commercial satellite constellations which may eventually call into question the use of specifically military satellites in these frequency bands. The quest for budgetary savings, together with the increased number of common military actions, has incited the United States to offer these new systems as a standard for use during allied military operations according to protocols still under study. This is also an aim of the Global Broadcasting Service (GBS), intended to broadcast large volumes of data to all military users across the board, and possibly right down to the soldier in the field. Such a system remains flexible whilst providing a quite enormous capacity. It should meet the challenge raised by digital data transmission, expected to increase ten-fold by 2002.

The GBS programme can be divided into three main stages. The first involved testing military broadcasting via civilian satellites. Operations of this type were carried out in 1996, for example, in the context of the Bosnia Control and Command Augmentation System by the Defense Information Systems Agency (DISA) and the Advanced Research Project Agency (ARPA) of the US DoD. In the second stage, three of the US Marine's UHF Follow-On satellites (UFO 8, 9, and 10) were launched with GBS payloads working in the K-band. Finally, in the third stage, planned for 2006, the GBS system should be associated with future American military satellites of the AWG type and a corresponding network of ground-based installations.

The GBS programme testifies to a changing awareness on the part of American military leaders during the period 1994–1995. They felt the time had come to end the technological and budgetary excesses of great telecommunications programmes along the lines of Milstar. The new programme has been imposed upon the space telecommunications architecture conceived by the US Air Force as part of a new desire to adopt a customer approach and at the same time make allowance for the improved performance of technologies available on the civilian market. Internal debates are not yet over, however. At the present time, the Global Broadcasting System could only be considered as a complement to existing or planned, exclusively military programmes.

As far as other countries are concerned, civilian satellites are equipped with special transponders for provision of military communications, as in the French case, where the Syracuse 1 and 2 systems are carried by civilian telecommunications satellites. With the same aim in mind, an experimental EHF payload will be tested on the experimental satellite Stentor. The United Kingdom is an exception to this rule, having deployed the Skynet system very early on, with satellites similar to the NATO series. In 1997, attempts were made to organise European cooperation in the area of military space telecommunications, but the scheme failed when Great Britain withdrew from the project in August 1998. Although

several options remain, some motivated by purely European aspirations and some having rather more transatlantic origin, none has yet been given the final go ahead. Given the exceedingly high volume transmission facilities in the general architecture of military command and control systems, especially when joint operations are run with the participation of several countries, it seems likely that the final choice will be made in accordance with the way allied alliance systems evolve, whether it be within the framework of NATO or the gestating European Security and Defence Initiative (ESDI). In the meantime, programmes are pursued on a national or bilateral basis. The first purely military French satellite, Syracuse 3, is due for launch in 2006 and an interim satellite, Gapfiller, is under study. The United Kingdom is developing its national programme with Skynet 5 and Germany is studying a national satellite known as Dmilsatcom.

Space: the new battlefield?

States possessing the kind of system just described, with a military vocation, present them as an essential element in their national security. Telecommunications and navigation satellites (see chapter 11), together with reconnaissance systems, play a crucial supporting role in military actions. Current thinking, especially with regard to the organisation of forces, tends to treat space as an ever more decisive factor in battle management. The American authorities openly admit as much: without spaceborne capabilities, the world's most powerful country would have no advantage whatsoever in any large-scale conventional conflict. Naturally, any sufficiently advanced country with knowhow in the area of space technology (e.g., launch, tracking, detection) may be tempted to attack enemy satellites in the hope of considerably weakening its opponent. Moreover, satellites are vulnerable. They have exactly predictable orbits, apart from the rare case of manoeuvrable satellites, on-board subsystems are fragile, and it is not always possible to arrange for reserve capabilities. Finally, ground links constitute another weak point, especially those transmitting commands up to the satellite. It is no surprise, therefore, that the surveillance (fig. 12.9) and protection of military space-based systems has today become a strategic priority.

In the United States, the term space control is used to invoke the new idea of maintaining an American advantage in space. Two main lines would appear to be favoured. The first is better protection for both civilian and military, strate-

gic space-based systems, thus preserving what is now called information dominance. The second point is the ability to control access to space and its uses, if the need arises. This is the first step towards what might genuinely be referred to as space weapons, which aim to disable enemy means either temporarily or permanently.

This last idea is not a new one. It has reappeared regularly since the inception of space travel. At the beginning of the 1960s, for example, the US Air Force was an enthusiastic proponent of orbiting assault weapons and other spaceborne arms programmes, claiming that space, like the oceans, would soon become the arena of conflict between the great powers. The advent of the atomic age made such developments irrelevant and these ideas were regularly rejected by successive political authorities. The first known studies concerning spaceborne interception systems, or even assault weapons, date from the end of the 1950s. Very little has ever come of them, however, for both technical and political reasons.

On the American side, the first half of the 1960s was marked by a few antisatellite missile tests, but the technical knowhow available was insufficient to produce anything very effective. For example, the fixed bases for the Zeus and Nike-X missiles used at the time meant that they could only target satellites when they passed at low altitude through a field of action determined by the location of these ground bases. Worse still, they proceeded by triggering a nuclear explosion, which did not allow a selective action, so that not only the target but its whole environment was also affected. As a result, tests carried out by the US Army on the Kwajalein atoll in the Marshall Islands were discontinued in 1968. The corresponding US Air Force tests carried out on Johnston Island in Polynesia using Thor missiles were finally brought to an end in 1975. Such systems should be considered as a spaceborne antibomb defense rather than an antisatellite system.

In fact, the possibility of launching a nuclear warhead in such a way that it would follow part of an orbit was never completely abandoned at the beginning of the space age. Soviet tests referred to as Fractional Orbit Bombardment System (FOBS) seem to have been carried out between 1966 and 1971. The technical characteristics of a launch from space would even seem to offer a tactical advantage. It is much easier to attain orbits in the same direction as the Earth's rotation, so that the United States would have been attacked from the Pacific, thus circumventing their most sophisticated means of defense. The shortcomings of the system, lack of operational flexibility, inaccuracy and control problems, proved too great and eventually led to its abandonment. The new ballistic missiles were largely superior.

Since the beginnings of space conquest, antisatellite weapons have always been at the forefront, with a particular desire in the Soviet Union to neutralise reconnaissance satellites. The development of antisatellite satellites was therefore favoured. Known today as Poliot, these were first tested in 1963 and 1964.

Tests carried out between 1968 and 1971 proved Soviet capacity to approach, and if need be, destroy a satellite, by exploding a shrapnel bomb nearby. However, the technology required for such missions can also be tested in a more discreet manner during rendezvous missions, and these are commonplace as part of manned space programmes. Ameri-

can rendezvous tests between Gemini and Agena spacecraft, carried out in 1966, showed that the United States was in possession of precisely the same capability as the Soviets. These tests, carried out by NASA, were supported by the US Department of Defense which, a few years previously, had studied the Satellite Inspector Technique programme. This project, whose first tests with an Agena target did not take place as planned in 1962, proved to be unrealistic, since an inspection satellite would have been required for each Soviet satellite.

During this period, it seems that the United States had no desire to develop an ASAT system that would appear too threatening, since this would have encouraged the Soviets to develop similar capabilities. The Americans no doubt had a greater dependence on spaceborne military systems and those in charge may well have preferred to maintain as far as possible the idea of space as a haven. Moreover, Soviet means of interception only threatened low-orbiting satellites, and for these a certain number of countermeasures were available.

A new tactical approach was adopted by the Soviet Union in 1976, after tests had been discontinued for a period of four years. They developed an intercepting satellite capable of much greater manoeuvrability, since it could travel around different orbits before settling itself on that of the target satellite (fig. 12.10). This ability to change from one orbit to another was aimed at complicating the opponent's task. It became impossible to determine exactly which satellite was at risk, and hence to plan protection for it.

The American response to this type of threat was the Miniature Homing Vehicle (MHV), an antisatellite weapon in two stages, fired from an aeroplane. This was given the go ahead by President Ford in 1977. The main interest of such a

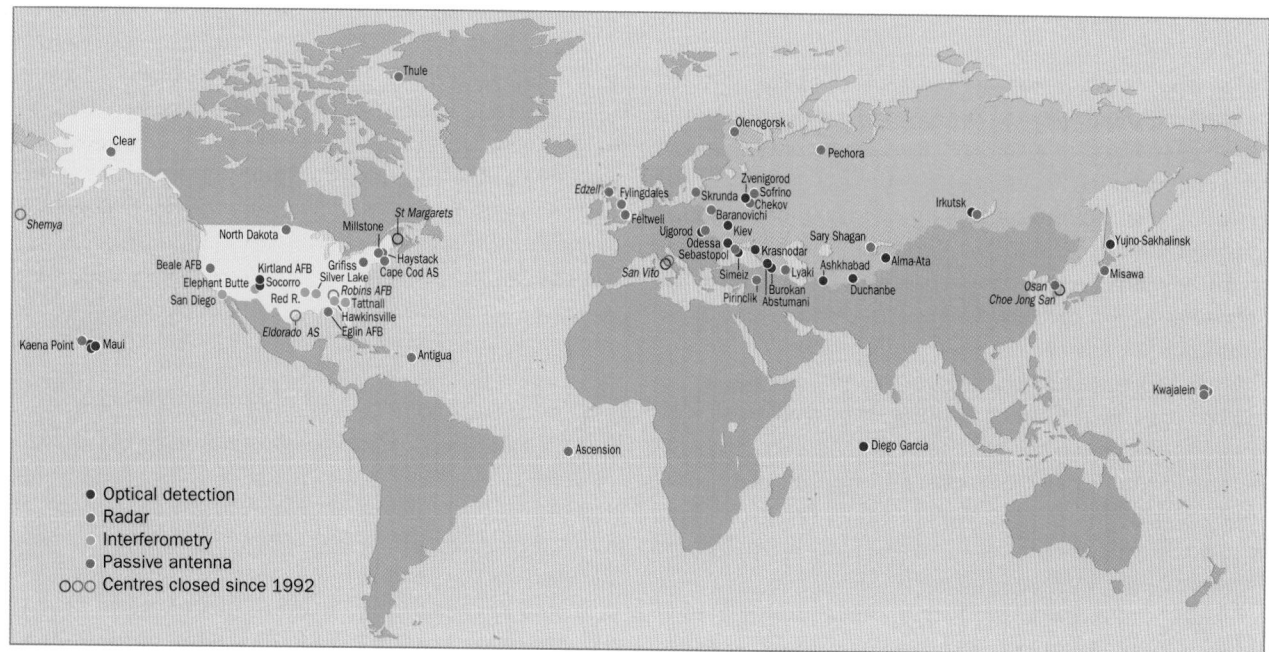

Figure 12.9. **US and Russian surveillance networks for spaceborne objects.**

system is that it is freed of standard satellite deployment constraints. Indeed, airborne launch allows much greater flexibility in the way the system is implemented, and more flexible positioning relative to the target itself, even though altitude remains a limitation.

The appearance of new Soviet surveillance satellites, in particular for ocean surveillance, was deemed to be a significant threat for the security of conventional American forces, and interest in this programme was renewed. Notwithstanding, confirmations by the Carter administration were accompanied by a recommendation that negotiations be taken up with the Soviet authorities. The American antisatellite programme could then serve as a negotiating point, to be traded off against some concession or held as a guarantee in case of a breakdown in the talks.

Spaceborne arms projects again aroused interest at the beginning of the 1980s, first in the second half of the Carter administration, a particularly tense period for Soviet–American relations, and then under the political attack launched by the Strategic Defense Initiative, which was announced in March 1983. Once again, results fell short of the ambitions expressed, although efforts were maintained at a steady level. Apart from these temporally isolated incidents, research in the antisatellite domain was reorganised as a rather limited part of the antimissile programme.

The strategic significance of antiballistic systems is a subject closely linked to the state of international affairs. Antimissile systems are directly related to nuclear dissuasion, whose basic principle is the mutual vulnerability of both camps. The Soviet and American limited defence systems Galosh and Safeguard, respectively, formed part of the follow-up to the 1972 ABM Treaty. This banned any extended antimissile defense programme and forbade, in particular, the use of spaceborne antimissile weapons, as well as any warning or battle space characterisation system conferring ABM capabilities on existing defences. However, the treaty did not ban research into new technologies, and such was pursued in both the United States and the Soviet Union. The difficulties surrounding Soviet–American talks at the end of the 1970s had the effect of preparing the ground for a more radical conception of antimissile protection. Among other things, it led to renewed interest in the Space Shuttle programme under the Carter administration. Right from the beginning, one of the main justifications for the Shuttle was its use as a platform for experimenting with new weapons.

In Helsinki, in 1978, bilateral talks over the question of defining and deploying antisatellite weapons failed after only three sessions. Differences concerning the definition of these weapons were too great, and the Soviets took up further testing in the period 1980 to 1982, at which point they opted for a unilateral moratorium. For their part, the Americans pursued their MHV programme. Experimented with some success in 1985, this system, like the Soviet ASAT satellite system, was nevertheless restricted to low altitudes. On the other hand, the situation today is that a very great number of

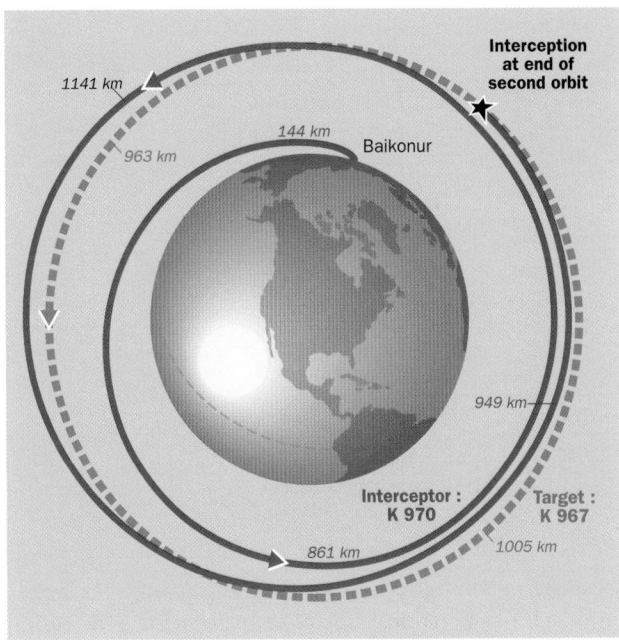

Figure 12.10. **Example interception of a target satellite.**

satellites, particularly reconnaissance and ocean surveillance satellites, and even the Mir space station, remain within its potential field of action.

Whilst the US Congress had become more and more reluctant to support MHV since 1986, the Department of Defense put forward a plan to reorganise the antisatellite programme, advocating an improved version that would use a new launch vehicle, either Earth-based or airborne, capable of reaching geostationary orbits. This project was not well received in political circles. It was expensive and it suffered due to the more widespread enthusiasm expressed for SDi. In 1988, this programme was therefore officially shelved, although five interceptors remained available.

For their part, in 1989, the Soviets set up an armada of satellite killers at their Baikonur base, using the Tsyklon launch vehicle, and thereby proved that traditional means had lost none of their relevance. According to assessments made by the US Department of Defense, the Soviet Union would have had sufficient capacity to destroy about ten American satellites in low orbit, on a time scale varying between two days and one week, depending on how many launches were made per day. Even if Soviet launch capacity remained unequalled, their system suffered from a serious weakness, namely, the vulnerability of their own base during the time required for an attack.

Since 1983 and President Reagan's announcement of the Strategic Defense Initiative, the very idea of space-based weapons has taken on a new dimension, with all the publicity carried out on the Star Wars theme.

The original idea rested upon the construction of a complete protection system for United States territory. This was to be provided by ground-based or airborne systems, as well as those set up in space. No violation of the 1967 Treaty

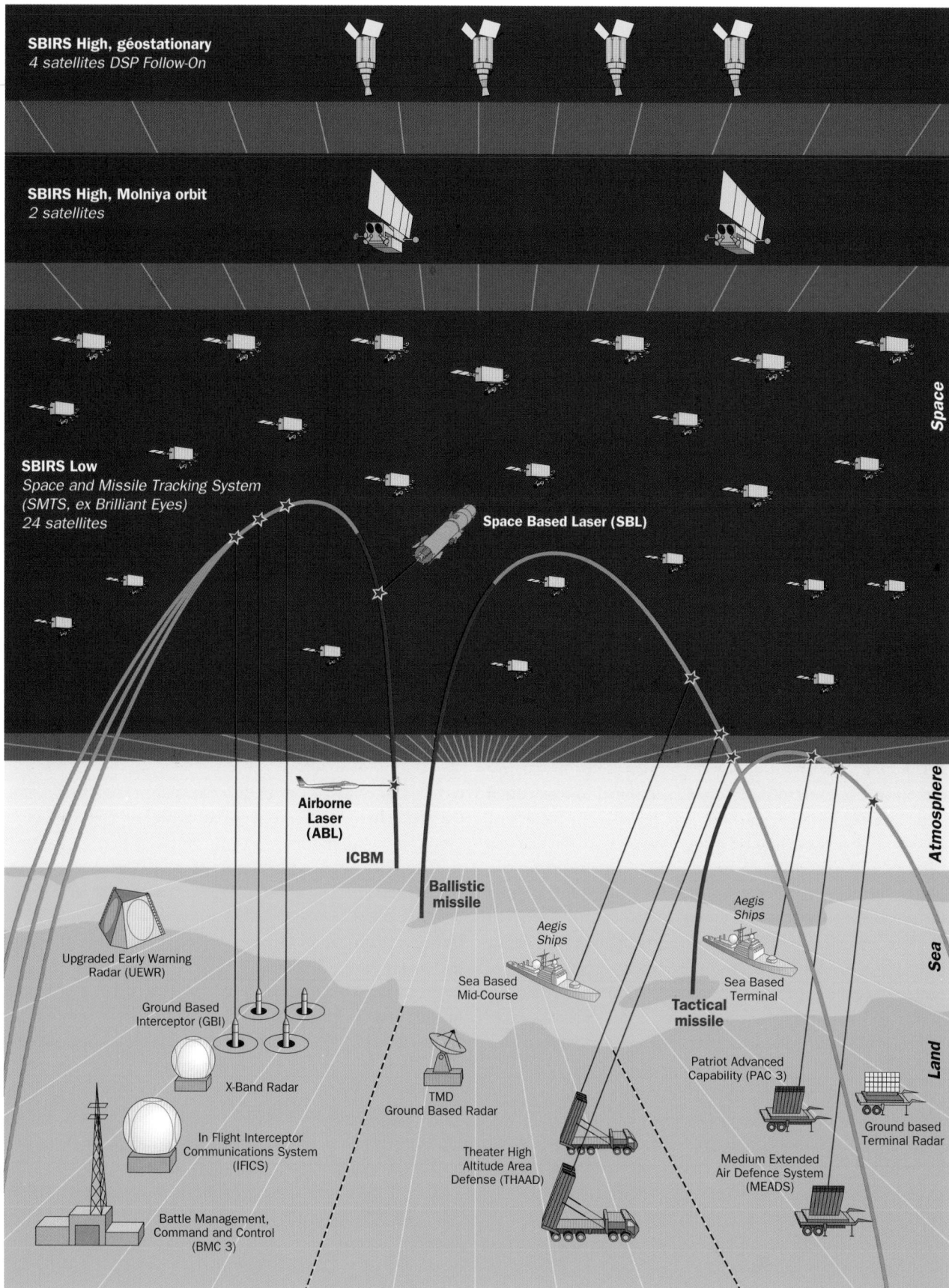

Figure 12.11. **The American National Missile Defense (NMD) programme.**

could really be identified in this respect, since this treaty only forbids the deployment of weapons of mass destruction.

This proliferation of defensive means, with the installation in space of several hundred early warning, interception and communications satellites, was aimed at total interception of all enemy missiles, the response being triggered right from the first phase of their launch and continuing through each of the subsequent phases (fig. 12.11). In the final phase, only a handful of warheads would remain to be picked off by terminal defences.

Behind this grand idea, several problems were raised by the very notion of SDi. To begin with, by attempting to turn American territory into a perfect sanctuary, SDi contradicted what had become one of the foundation stones in the doctrine of nuclear dissuasion. Indeed, it meant that the vulnerability of the opposing forces would no longer be a mutual affair. Given the greatly reduced chances of a return bout of destruction, there would be nothing to prevent the dominant side from entering into a full scale conflict.

This argument was thoroughly developed by the Soviet Union, not without causing some consternation amongst European countries. It nevertheless remained rather theoretical insofar as the feasibility of the American project had soon confined it to the realms of Utopia, at least as far as the experts were concerned.

From this standpoint, SDi did not have the destabilising effect on the Soviet regime that some were to claim. As soon as it became a real possibility to saturate defences, and this at relatively low cost, Soviet national security was no longer an issue. On the other hand, the technological ambitions advertised by such a programme could not fail to confront Russian MIC (Military Industrial Complex) authorities with their own limitations at a time when they were wondering whether or not they should pursue a parity policy, albeit in a case-by-case manner, with respect to the United States.

In a wider sense, even if SDi had little chance of ever becoming operational, the antiballistic systems to be developed in its name would provide undeniable antisatellite potential. Indeed, the number of satellites, predictability of their motion in terms of speed and position, and their large size considerably simplifies the kind of problem faced when intercepting ballistic missiles. If a single country were to possess such a capability, even indirectly, other space users could not but feel threatened.

From this point on, supporters of the SDi programme made every effort to clarify the difference between offensive and defensive space-based weapons. The first could be used in peacetime to destroy an opponent's assets. The second, such as terminal terrestrial defences against ballistic missiles, should only be usable against enemy weaponry. Despite these attempts, the distinction remains rather subtle.

The main theme behind SDi became a much less sensitive issue in 1987, when the deployment plans were presented. Given the insurmountable technical difficulties it involved, the IDS project underwent significant changes at the begin-ning of the 1990s. It amounted at this time to a combination of a great number of small satellites, called Brilliant Eyes and Brilliant Pebbles. The first were equipped with sensors capable of detecting Soviet missiles and computers with similar capacity to Cray 1 type supercomputers. Their aim was to observe and then predict enemy missile trajectories. They were supposed to transmit interception coordinates to the second set of satellites, which would then hurry to the scene. Such a system offered many advantages. Launching them into several orbits could theoretically dissuade potential enemies from large-scale attack and would at least increase chances of survival. The Brilliant Pebbles were mass-produced and hence relatively inexpensive. According to some commentators, they could even replace the costly satellites making up the originally planned Booster Surveillance and Tracking System. Finally, their ability to guide themselves in a quite autonomous way towards the target by means of an on-board computer served to economise the use of communications systems with a ground station and also reduced the time required for response.

The system was far from operational at this time, and did not come into being either within the SDi framework or in the context of the new programme supported by G. Bush, the Global Protection Against Limited Strikes (GPALS) programme, whose aims and means remained poorly determined. Even though President Bush supported the project with more conviction than some analysts have claimed, this programme also proved to be virtually impossible to bring to fruition. The initial deployment phase, intended for 1993, was to establish both ground-based and space-based weapon systems, together with sensors whose main task was to protect missile arsenals on the ground. This was to be a partial defence scheme and its strategic implications were therefore greatly reduced.

The Gulf War, the gradual decline of Russia as a super-power and the redefinition of the ballistic threat to the United States led to a reorientation of the American antiballistic programme, and the SDi was officially brought to an end by President Clinton in 1993.

The American administration is advocating a complete Missile Defense (MD) encompassing both the former National Missile Defense (NMD) and Theater Missile Defense (TMD), to be set up progressively. It would closely combine space-based surveillance and missile tracking with a panoply of ground-based antimissile missiles (fig. 12.11). The space segment is called the Space Based Infrared System (SBIRS) and should replace the satellites of the Defense Support Program over the period 2004–2005. It will comprise two distinct layers, SBIRS-High and SBIRS-Low, the former comprising four geostationary satellites and two satellites in highly eccentric orbits.

The main contribution to the existing system would be an improvement in infrared detection capabilities and a better system for keeping data up to date (data refreshment). The performance of the system would be considerably boosted in comparison with the former Defense Support Program (DSP). The latter is poorly suited to short range missile detection for two reasons: its detectors are not sensitive enough and its scanning frequency is limited by the fact that the satellites spin once every ten seconds. The upper layer of SBIRS could thus begin the task of tracking enemy missiles, whilst the lower layer, consisting of LEO satellites, would then fine-tune trajectory analysis. The second component, called the Space and Missile Tracking System (SMTS) or SBIRS-Low, is in fact based on the Brilliant Eyes concept developed for SDi. These satellites would operate in a constellation and would be able to intercommunicate and hence supply ground-based defence systems with the latest information. The latter would then guide GBI (Ground Based Interception) or THAAD type intercepting missiles (Theater High Altitude Area Defense missiles) towards their target. The last stages of interception would be made possible by the active infrared guidance systems equipping these missiles.

Still under study, the lower layer would include 18 to 24 satellites, depending on the altitude finally chosen. However, there is no assurance that this layer will actually be implemented. During the Clinton presidency only the US Congress seemed to be in favour of an all-inclusive programme, whilst the government itself was pushing for a more restricted system for protecting troops deployed abroad. Furthermore, the existence of space-based systems directly connected to interceptor missiles violates the terms of the 1972 SALT 1 Treaty, and steps in this direction must show some restraint.

The current Bush administration decided to endorse the congressional view. This new antiballistic programme, including TMD, is considerably less ambitious than SDi, and does not involve the same space-based arms proliferation. The only real antisatellite capability remaining to the United States would be the occasional use of interceptor warheads on extra-atmospheric missiles or certain devices still under study, such as the spaceborne or airborne laser. It should still be emphasised that these elements are not expected to be operational before the period 2015–2020, and that they are primarily designed for ballistic interception.

In fact, the basic framework for space-based weapons, founded on the notion of space control since the end of 1996, is undergoing a transformation, although this does not mean that legal difficulties have been properly resolved. Those defending the idea of space-based weapons are now directly invoking the need to adapt space law so that it can protect American satellites. According to this lobby, the strategic interests of the country rest equally upon both military satellites and constellations of low-Earth orbiting civilian satellites planned for the present century. It is thus that, across the Atlantic, there is new talk of implementing a kinetic energy ASAT satellite, KE-Sat, which would become the first American killer satellite designed with this end in mind. Other approaches advocate further work on lasers, whether they be airborne or used directly from space (airborne and spaceborne lasers).

Considering the high costs involved, space weapons programmes must be based on sound arguments if they are to be accepted. Changes occurred since the Cold War. The Russian programme, held back by so many difficulties, is a faint shadow of the former Soviet threat, and few countries declared hostile by the US authorities would seem to be in any position to cause them much serious concern. European military space programmes, French for the main part, are comparatively modest in scope and Asian space powers could hardly be said to represent a genuine threat, at least in the short term.

However, after some self-restraint exerted by the previous Democrat administration on this issue, the Bush government has recently called for an increased effort in the militarization of space. One difficulty remains to be settled. Those advocating space weapons programmes find themselves in direct competition with other space programmes in the context of a global military budget where consolidation is the order of the day.

Chapter Thirteen

Living in space

The idea of human travel in space and even of settling in other worlds is by no means new. It can be traced back up the centuries as far as the Ancient World and has at different times in history inspired the most fanciful tales of exploration. Even before it was popularised and developed by recent science-fiction writers, the concept of a human presence in space appeared fundamentally natural to the pioneers of space history – for example the great theorist Konstantin Tsiolkovski or, one step nearer the technical realities, the engineer Robert Goddard. Speaking even before the space age got underway, Tsiolkovski said the Earth was the cradle of humanity – and it could not stay in its cradle indefinitely. Goddard, though still engaged in his very earliest experiments, was already convinced that the Moon would be an ideal base from which humans could explore other planets.

Thanks to the development of sufficiently powerful launchers, in situ discoveries by automatic satellites about the real conditions prevailing in the space environment and journeys by animals into space lasting a few orbital revolutions, Yuri Gagarin was able in his 1961 flight to take the long human quest to drive back existing frontiers a decisive stage further. In a gradual process, leaving the Earth to live for a few hours, then a few months and at last more than a year in an alien environment, the space pioneers opened a new chapter in the history of exploration. And while forays into space were still confined to the innermost reaches of circum-terrestrial space, the ecumene – in the traditional sense of the inhabited domain – was no longer confined to planet Earth.

Almost forty years on from the first human spaceflight, and even if Europe, China and Japan are officially committed to going into space on their own account in the next few years, Russia and the United States are still the only countries with the infrastructure needed for people to travel in space and stay alive there. And if other countries have been represented, however briefly, in space, they have owed this to the access they have been offered to Russian and American training programmes and related facilities in space (fig. 13.1).

The Soviet Union was the first country to offer flight opportunities to foreign nationals – from communist bloc countries as from 1978, from a Western nation, France, in 1982, and from a non-aligned state, India, in 1984. This policy was prompted by a concern to obtain a return, in terms of international prestige, on a major asset which no other country possessed, namely a comprehensive programme of spaceflight and space occupancy. The nationalities of the guest cosmonauts reflected the closeness of political, economic and scientific ties with the countries concerned and the changing face of international diplomacy.

As from 1983 and the flight of the European laboratory Spacelab on board the Space Shuttle, the United States had the resources to pursue a similar policy, driven by identical considerations. Between then and the disbanding of the Soviet Union in 1991 – which at the same time brought to an end the head-on rivalry between the two superpowers – seven non-Americans took the Shuttle route to space. That figure fell far short of the 15 Soviet invitations and this despite the fact that the Shuttle could accommodate up to eight crew members and flew something like twice as frequently as the corresponding Soviet vehicles. Part of the explanation for this imbalance lies in the Shuttle's dual role as a crew-carrying transport system and launch vehicle, some of whose missions – on the military side in particular – did not lend themselves to guest flights.

In the period from 1961 to end-1990, 240 men and women got to see the Earth from space. This relatively small number is understandable enough considering the heavy training investment, the special importance attached to experience and the need to seek a return on this human capital. The frequency of flight invitations also reflects the relative importance of countries on the international scene. The nations of Europe are thus heavily represented among the eighteen countries from around the world to have received such invitations. The countries of Africa and South America on the other hand are still very much on the sidelines.

The Soviet Union's demise marked a turning-point in the history of the human presence in space, which ceased to function as a vehicle for rivalry between the two blocs. Hard cash now became a means of securing a ticket to the Mir space station, a development symbolically prefigured by the visit of a Japanese journalist in 1990. Above all, the revised International Space Station concept of a joint US–Russian venture supported by the partners in the initial project (Europe, Japan, Canada) gave formal status to use of the Mir station as a training post for missions of varying duration. The official closure of the station and its scheduled deorbiting in March 2000 was postponed until March 2001. The introduction of the station as a 'tourist' destination was considered for some time. However this would have been only for a relatively short period considering the age of Mir (15 years' service by February 2001).

It is clear therefore that the role of permanent 'home from home' in space for humanity now shifts to the International Space Station.

In programmes directed at an ongoing human presence in space, a fairly uniform approach could reasonably have been expected, the shared goal being to develop a station and

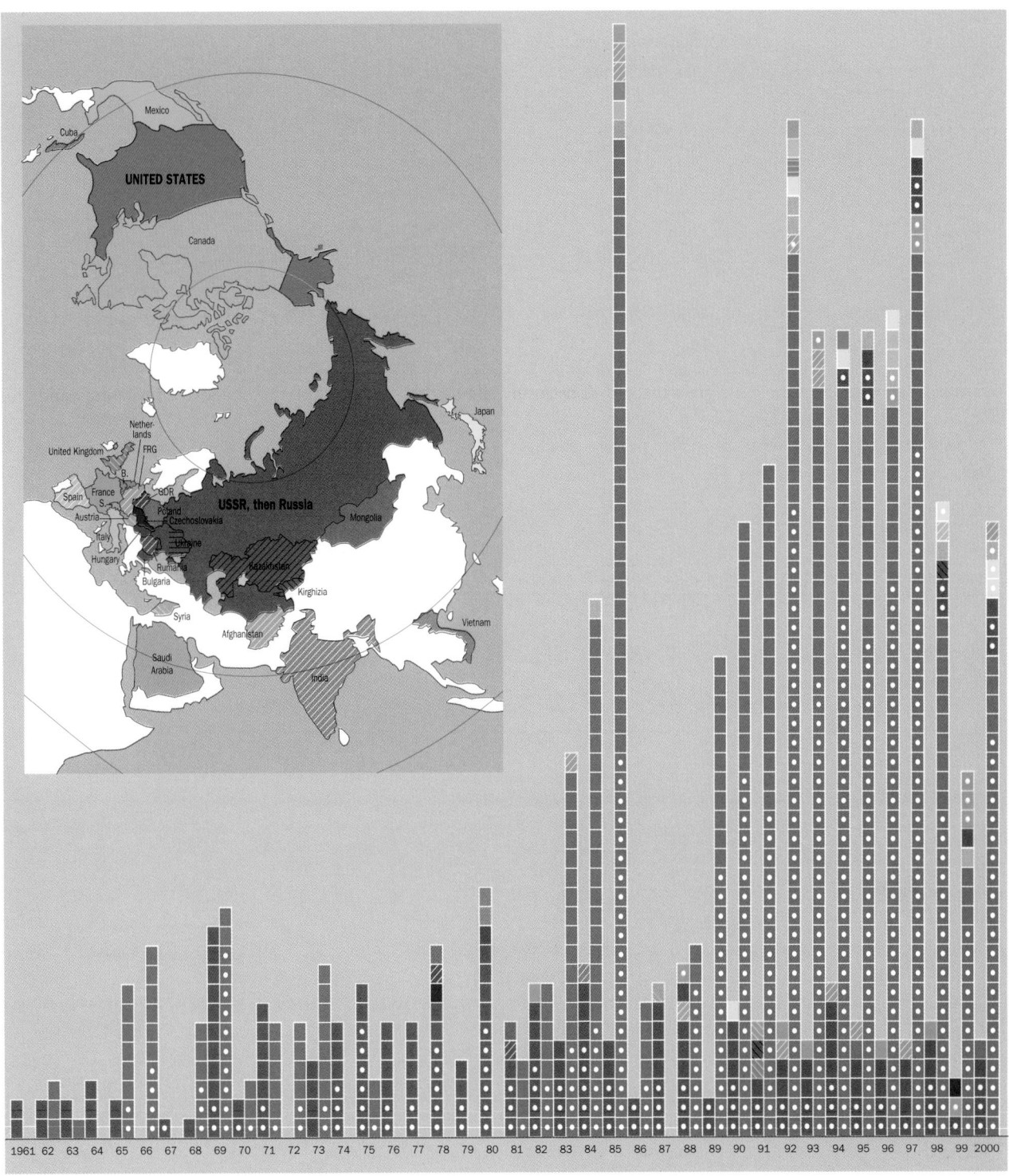

Figure 13.1. **Numbers of astronauts in space per year and their nationalities.** The white dots indicate that this is not the astronaut's first stay in space. Astronauts from reunified Germany are shown in the colour assigned to the FRG.

acquire the systems required to service it. And yet it can be seen from the history of the space endeavour that the United States and the Soviet Union have addressed the problem differently, essentially for domestic political reasons. The Russians have preferred to focus on the orbital infrastructure, while in America innovation has centred on transportation. And yet in the late 1980s a more balanced situation seemed to be taking shape, with the flight in 1988 of Buran, the Russian space shuttle, and approval of Freedom, the American space station programme (fig. 13.2). But the broad pic-

ture, from about the middle of the 1970s, was one of near-Earth space occupied on an ongoing basis by a handful of Soviet and later Russian long-stay specialists while Americans began in increasing numbers to voyage in space on board the Shuttle, but on much shorter missions lasting about ten days. A few guests benefited from the occasional flight opportunities made available by both space powers.

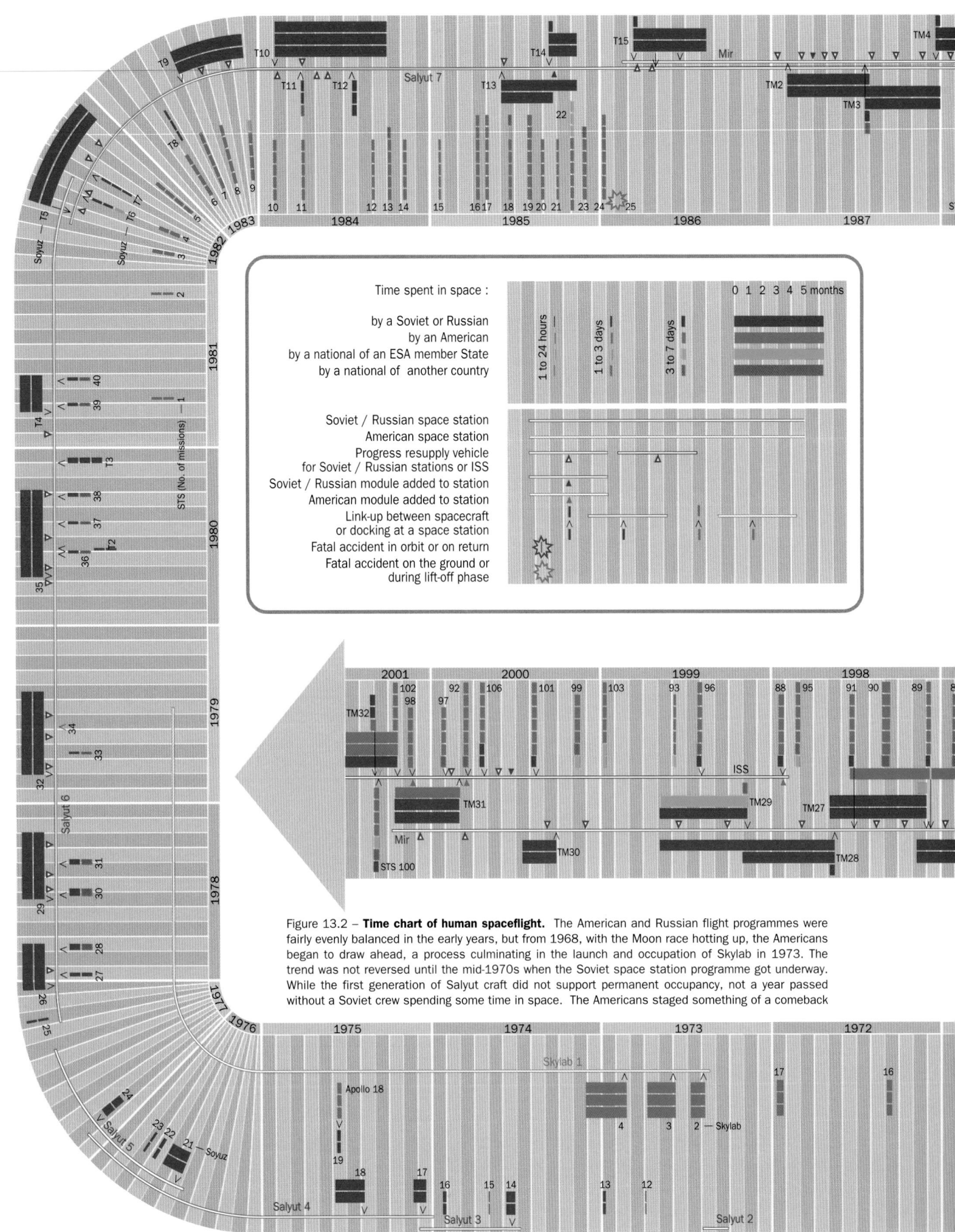

Time spent in space :

by a Soviet or Russian
by an American
by a national of an ESA member State
by a national of another country

1 to 24 hours
1 to 3 days
3 to 7 days
0 1 2 3 4 5 months

Soviet / Russian space station
American space station
Progress resupply vehicle
for Soviet / Russian stations or ISS
Soviet / Russian module added to station
American module added to station
Link-up between spacecraft
or docking at a space station
Fatal accident in orbit or on return
Fatal accident on the ground or
during lift-off phase

Figure 13.2 – **Time chart of human spaceflight.** The American and Russian flight programmes were fairly evenly balanced in the early years, but from 1968, with the Moon race hotting up, the Americans began to draw ahead, a process culminating in the launch and occupation of Skylab in 1973. The trend was not reversed until the mid-1970s when the Soviet space station programme got underway. While the first generation of Salyut craft did not support permanent occupancy, not a year passed without a Soviet crew spending some time in space. The Americans staged something of a comeback

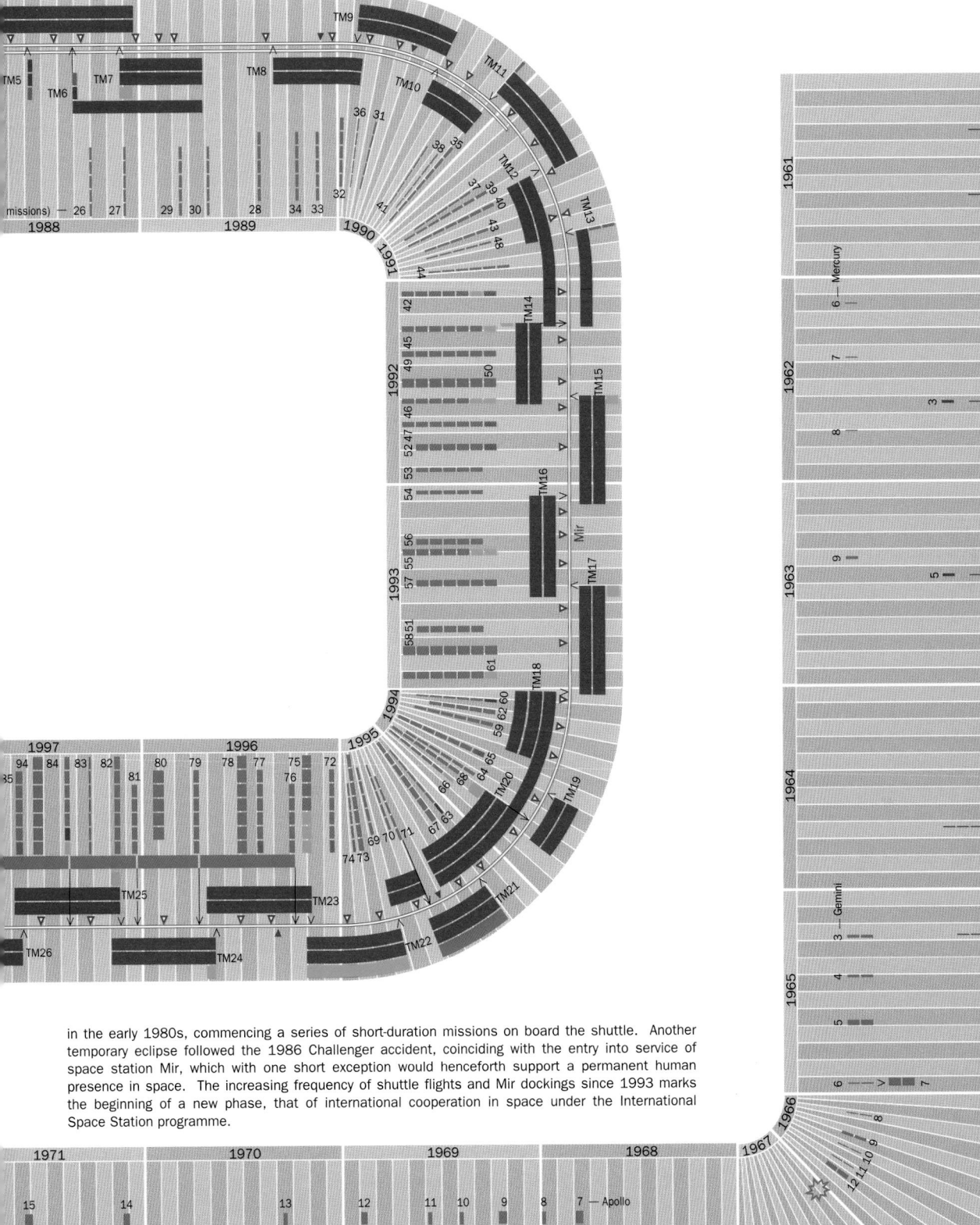

in the early 1980s, commencing a series of short-duration missions on board the shuttle. Another temporary eclipse followed the 1986 Challenger accident, coinciding with the entry into service of space station Mir, which with one short exception would henceforth support a permanent human presence in space. The increasing frequency of shuttle flights and Mir dockings since 1993 marks the beginning of a new phase, that of international cooperation in space under the International Space Station programme.

With the collapse of the Soviet Union in 1991 came a period of radical change. US–Russian cooperation developed rapidly with special emphasis on human presence in space. There was a growing realisation that the resources available in the two countries were complementary. Shortage of national funding forced a halt to the Buran space shuttle programme while the United States took only a little longer to announce their intention to discontinue the initial Freedom programme in favour of an international programme. Initially called Alpha, this programme was seen as an opportunity to make full use of some of the technologies already mastered by Russia – and indeed the name R-Alpha was used at one point. The cooperative arrangements which finally emerged in the shape of the ISS programme were very much dependent on American financial support for Russia's human spaceflight activity. This in turn provided substantial scope for long-duration missions by Americans to the Mir space station. The record for the longest stay in space by an American had reached six months by 1996. Prompted again by the need to attract new funds, the Russians also offered European astronauts mission opportunities lasting several weeks on board Mir, the longest such stay – 188 days – being undertaken by a Frenchman in 1999. Nor is it uncommon for missions to be extended, giving guest astronauts more time to familiarise themselves with the space environment and allowing the hosts to limit the number of Soyuz launches required for station servicing purposes.

First steps in space

Looking at the graph summarising the history of human presence in space, a striking feature of the very early years is the extent to which the unfolding Vostok and Mercury programmes mirrored one another. A concern to maximise public relations impact – for domestic and international political reasons – was soon firmly established as one of the main forces driving crewed space missions and the Russians were quick to exploit their lead on the launcher front to loft heavy payloads and above all to demonstrate their ability to maintain a consistently high launch rate.

The flight of Yuri Gagarin in particular earned the USSR enormous international prestige. This encouraged its leaders to look increasingly for world firsts, exploiting their media potential to the full whenever occasion presented: first man in space, first launch of two spacecraft one day apart coming within 5 km of each other, first woman to leave the Earth, first spacewalk, first triple flight. In fact, these successes were more a matter of making maximum use of the existing resources than a consistent record of technological progress. The Voskhod spacecraft for example were if anything less advanced than the Gemini vehicles with their more sophisticated flight control system (fig. 13.3).

Capitalising on space exploits was a feature of domestic policy too. The cosmonaut revered by all the peoples of the Union came to acquire an almost mythical dimension, becoming the first true Soviet hero, a cult figure transcending national boundaries. This is a recurring theme in official addresses of the period, with their eschewal of Russian nationhood in favour of an emerging Soviet identity.

Rivalry between the United States and the USSR dictated programme priorities and the relative emphasis on space activity from an early stage. Then towards the middle of the 1960s came the first clear signs of a significant shift in policy on human exploration. When Nikita Khrushchev – who was largely responsible for the high media profile given to the human space programme – held a telephone conversation with the crew of Voskhod-1 on 12 October 1964, a conversation broadcast around the world, he did not know he was engaging in his last public act as First Secretary of the Soviet Communist Party. On their return to Earth, the cosmonauts were welcomed by Leonid Brezhnev who, as the secretary in charge of the Defence Industries in the Central Committee, was already associated with the achievements of the space programme. In his congratulatory address, the new First Secretary emphasised that space exploration would no longer be undertaken for reasons of prestige but in pursuit of scientific and humanitarian goals. The exploitation of human spaceflight for ideological purposes was already shifting to another continent.

For the American people the early string of Soviet successes were so many blows to national pride. So in electing John F. Kennedy to the presidency they were voting to restore America's battered image and, by mobilising the country's scientific and industrial potential, to reestablish world leadership. The occupation of space proved a natural theme for the new President, with his appeal to the pioneering spirit so firmly rooted in the national culture, a spirit which would now be revived in the conquest of a new frontier in space. In

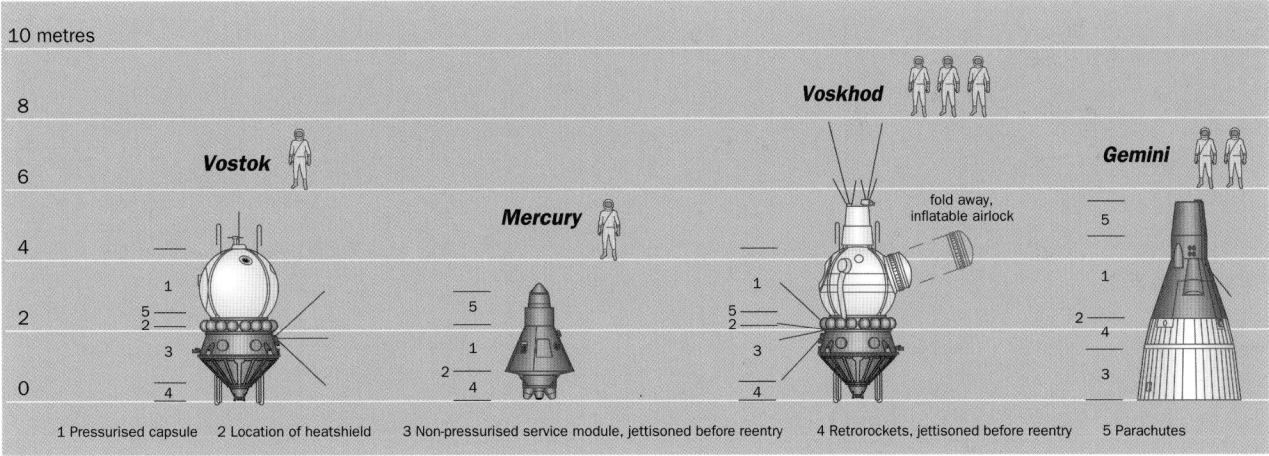

Figure 13.3. **The first crewed spacecraft.** The symbols indicate maximum crew size. The earliest such craft could transport one person only in view of the capsule diameter, which was limited to about two metres. The capsule was the only spacecraft element to return to Earth, all other hardware being jettisoned prior to atmospheric reentry in order to lighten the load and uncover the heat shield. A manoeuvring capability was not introduced until the Gemini programme, the Voskhod series being equipped with a manual landing system only.

declaring its resolve to conquer our only natural satellite – the Moon – the United States in turn committed itself to seeking a systematic ideological return on American forays into space.

The American lunar programme went from strength to strength in the years 1965–66, at a time when the Soviet programme was in temporary eclipse owing to technical problems with development of the new Soyuz vehicle. The two programmes continued, all the same, to share many characteristics until as late as 1969. The end of the lunar programme marked a first break in the history of the human occupation of space and it was not until the mid-1970s that the first of a series of Soviet and later Russian space stations entered service in Earth orbit. Circling the Earth on a permanent basis, they were occupied by cosmonauts for ever longer periods of time. A few years more would pass before Americans began, as from 1981, to ply the space highway, doing so in ever increasing numbers, though only for missions of up to ten days. But without the permanent backing which a station provides, the technical contingencies to which space shuttles are subject can have dramatic results, as became starkly apparent in 1986 when the Challenger accident grounded American crewed missions for a full two years. Since the beginning of the 1990s the context has changed yet again, characterised now by international cooperation under US leadership.

The race to the Moon

The Moon race played a decisive role in the development of human spaceflight. The public relations benefits to be derived from projects associated with the Earth's natural satellite were apparent from the earliest exploratory missions by automatic spacecraft. The first photographs of the dark side of the Moon taken by Luna-3, for example, or again the first successful spacecraft landings (see pp. 193–197) exercised a fascination which it is sometimes difficult to recapture today. On a visit to New York at that time, First Secretary Khrushchev even presented President Eisenhower with a medal identical to the one placed on the Moon by Luna-2.

The decision to land a human being on the Moon did however represent a startling change of gear. When America vowed, in a solemn announcement by President John F. Kennedy, to go to the Moon within a decade – by the end of the 1960s – it was quite clearly throwing down the gauntlet. And while the Kremlin did not officially take up the challenge, the game rules by which they were playing left it with no choice but to pursue the same objective. In the course of time the very scale of the undertaking did however uncover various structural weaknesses in the Soviet space programme. These had to do with the way decisions were taken but also the organisation of production.

Winning the race meant developing a series of new technologies ranging from a heavy launcher to new crewed spacecraft capable of staying in orbit for almost three weeks and possessing fairly sophisticated manoeuvring capabilities (fig. 13.4). Long before Kennedy's May 1961 decision, prior even to the creation of NASA in 1958, the team headed by Wernher von Braun had already investigated the prospects for human exploration of the Moon and Mars. Their initial findings were taken over by the Office of Future Projects at the Marshall Space Flight Center when the team became part of NASA. On the Soviet side, Sergey Korolev and his team had drawn up in 1958 a plan for human exploration of the planets including, in an initial six-year period, automatic flights to Mars and Venus as a prelude to a human voyage of discovery. This would lead ultimately to a Moon landing and construction of a base in Earth orbit.

The political authorities having opted for a lunar landing scenario and set a timeframe of 1967–68 for achievement of that goal, the question of how to get there now became crucial. Various possibilities were considered and the first two to be selected for further investigation – in both the United States and the USSR – were direct launch from the Earth of a

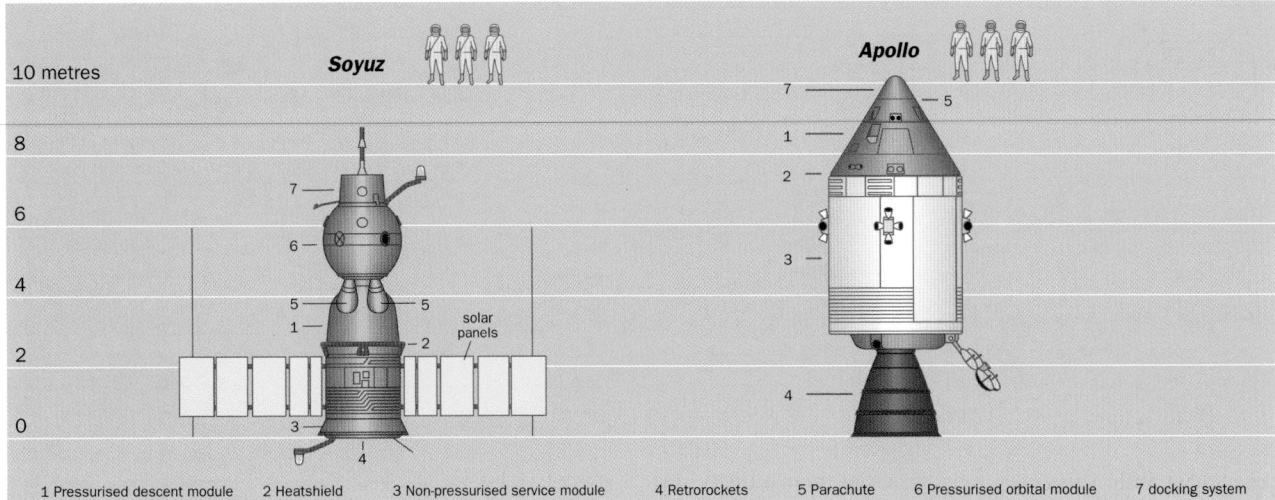

| 1 Pressurised descent module | 2 Heatshield | 3 Non-pressurised service module | 4 Retrorockets | 5 Parachute | 6 Pressurised orbital module | 7 docking system |

Figure 13.4. Lunar spacecraft. Although originally designed for this purpose, Soyuz was not used in the human conquest of the Moon. The main innovation on the Apollo and Soyuz vessels was the docking system; designed for the lunar descent module, this system also made it possible to service the Skylab and Salyut space stations. The spacecraft were equipped with orbital manoeuvring engines for attitude changes, supporting in particular complete reversal of the vehicle during braking phases, and also featured a more powerful engine for changing orbit. As with the previous vessels, only a pressurised module returned to Earth.

spacecraft bound for the surface of the Moon and rendezvous and assembly in low-Earth orbit of the various components of a space train. The proposed direct approach assumed development of a heavy launcher with a lift capability of some 100 tonnes, allowing the USA to catch up with and even distance the Soviets with their R-7 vehicle (see chapter 6). This was regarded as achievable. What appeared more problematic was to devise a method for placing on the Moon itself a spacecraft carrying the propellant required for the return leg and further weighed down by a thermal shield. One option was to build a staging post in Earth orbit. Such a rendezvous infrastructure could be assembled using a smaller launcher – with a lift of about 20 tonnes – and would subsequently be available for missions to other planets.

An alternative concept turned on rendezvous in lunar orbit, an idea which reemerged in the United States some time in 1961. This had the advantage over the direct approach of facilitating the task of landing on the Moon, since a large proportion of the space train would remain in orbit. This concept gradually gained ground within NASA (see fig. 8.3) though it continued to attract a hostile response from many scientists. They argued, not unconvincingly perhaps, that a launcher capable of carrying more than 100 tonnes would be of no interest for subsequent space activities.

On the Soviet side, the broad thrust of technical thinking seems to have been much the same and here too the chief engineers were divided in their support for the various options. Another source of disagreement was Chief Designer Korolev's initial commitment to developing a cryogenic propulsion system for the N-1 vehicle capable of delivering the same power as the Saturn heavy-lift launcher. Valentin Glushko, head of propulsion design, was however sceptical of the use of the liquid hydrogen/oxygen combination as propellant.

Where the two countries differed most was in their organisational arrangements. In the United States, getting an astronaut to the Moon was identified as the leading priority in the space programme. By the end of 1962, the programme architecture had been established, the work of the industrial firms and design bureaus being closely coordinated by NASA. The picture was very different on the Soviet side. There personal power relations between design chiefs and the political authorities fundamentally structured the space programme. And initially at least the challenge coming from America was underestimated. Engineers and scientists alike were slow to realise how far the American programme had come and failed to appreciate fully the technical qualities of their opposite numbers in the USA. This left them – and apparently also the political leaders – for a long time with a misguided sense of optimism. Funding for the Soviet lunar programme was much lower too and in political circles the Moon endeavour was not viewed as an essential priority. The MOM – set up to bring a measure of coordination to space activities – did not come into being until 1965 (see pp. 73–74) and even then many of the capabilities demanded by the programme depended on other ministries. Finally, the vacuum left by Sergei Korolev on his death in 1966 and the ensuing struggle to succeed him, further encouraged personal rivalries. According to an account by Gregory Mishin, Korolev's successor and luckless head of the lunar programme, the final decision on the Soviet side also went in favour of rendezvous in lunar orbit but this did not happen until early 1967 and in the meantime too many variants had been developed in parallel.

The first space stations

The Soviet response to the success of the Apollo missions was to declare that they had never had any intention of sending a crew to the Moon, preferring to concentrate on automatic exploration under the Zond and Luna programmes. It should be added that this exploratory activity was itself far more limited in scope than that undertaken by the Apollo crews (see fig. 8.4). It would seem in fact that sometime in 1967 the Soviet authorities, aware by then that their human lunar programme was slipping badly behind, began to look seriously at a new option. This was for a cosmonaut to circle the Moon in a crewed version of Zond, itself directly derived from the Soyuz capsules. This relatively unambitious mission would have done much to offset the expected impact of the Apollo programme. But the last launch window prior to the departure of Apollo 7 in December 1968 came and went unused. This option no longer had any interest and work on it was taken no further.

The outstandingly successful Apollo series marked a major turning point in the American and Soviet programmes for the human occupation of space. Following the brilliant demonstration by the United States, through the race to the Moon, of its technical and financial capabilities, political interest in space waned considerably. The Skylab space station was primarily an opportunity to reuse hardware from the Apollo programme and, even then, the sharp drop in funding ruled out any possibility of carrying out the full range of missions initially planned.

Meanwhile in the USSR, the development of space stations designed to support a permanent human presence in space remained the avowed objective. The official line continued to be that no lunar programme had ever been announced. And indeed the development of outposts in Earth orbit can be seen as a logical step in any overall plan of space conquest, in contrast to the less immediately necessary goal of placing astronauts on the Moon. The Soviet space station programme represents a continuum, as can be seen from the launch between 1971 and 1986 of a succession of space stations – the seven Salyut stations followed by Mir – with varying degrees of success. Each new station bore witness to a growing command of the systems required for human beings to live in space. On the American side, on the other hand, conversion of the lunar programme into a space station programme was not taken very far. The Skylab venture was left without any follow-up and the Americans chose instead to explore an entirely new approach with the development of a space shuttle. This new type of transportation system would revolutionise traditional thinking about access to space.

On the domestic and international stage, human space exploration had by now lost much of its potency as a political statement. The Apollo–Soyuz link-up in 1975 was seen as a symbol of détente – the thaw in US–Soviet relations perceptible in that period. At a time when the Soviet presence was gaining ground, the Americans did not seem unduly perturbed by their inability through to 1981 to send crews into space (see fig. 13.2). It seems clear all the same that a concern not to vacate the 'last frontier' entirely – even if only temporarily – helped keep the shuttle programme on track, even if the arguments actually advanced were different. Similarly, the decision taken by the USSR in 1976 to give priority to the Buran shuttle to the detriment of other programmes (Moon base, multi-purpose on-orbit infrastructure, Mars expedition) was ultimately prompted by the American example. And finally the space station Freedom project announced by President Reagan in 1984 and the Space Exploration Initiative launched in 1989 by his successor President Bush showed clearly that the desire to demonstrate comprehensive superiority was still very much alive in the United States. In Europe, the initial decision in 1985 in favour of a station module, Columbus, intended ultimately to become autonomous, was followed by signs of mounting uncertainty. This wavering pointed, as the decade drew to an end, to widespread misgivings as to the real value of a human space programme. In this context, with the Cold War now a thing of the past, thought turned to a new form of international cooperation, led by the USA with the Russians on board.

Skylab

In both countries the first-generation space station programme was essentially a vast drive to recycle the lunar programme. Past achievements would form the basis for establishing a permanent human presence in space. The main aim of the Apollo programme, to restore America's image, had been met by 1969. The first steps of an American on the Moon, relayed around the world in a carefully prepared media operation, had generated enormous enthusiasm at a time when the USA was sinking ever more deeply into the Vietnam War. The Space Task Group, charged by President Nixon with considering the future course of US space activity in the post-Apollo period, came up with a conventional strategy of exploration, with a range of technical options. And while timescales and the extent of reliance on existing hardware varied, all the alternatives shared a common goal, that of a flight to Mars. In each case a station in Earth obit and another in lunar orbit were identified as intermediate stages. The existence of a space shuttle/orbital station complex was thus an essential first step in this concept of gradual expansion towards Mars.

The political authorities took a very different approach. For them, the time had now come to make savings. The final Apollo science missions were cancelled and the Post-Apollo Application Programme was redefined as a cost-effective reutilisation of existing hardware – Saturn launchers and Apollo spacecraft. It was in this context that a military programme to establish an inhabited station, the Manned Orbital Laboratory (MOL), was terminated in favour of Skylab, a less ambitious but nonetheless impressive 75-tonne structure which would take the USA to another first (fig. 13.5).

Launched in 1973, the station's official assignment was research into the impact of extended stays in space on human behaviour and physiology. Experience had to be acquired in this area for future programmes. Three successive crews lived and worked on board the station and in the course of their stays the various problems associated with life in micrograv-

ity conditions were probed. These included changes in the circulatory system, decalcification, adaptation difficulties, loss of appetite and sleep disorders. A series of successful – if very demanding – Extra Vehicular Activity (EVA) assignments showed that spaceworkers could address a variety of tasks outside the station and were in this respect irreplaceable. In addition to the many astronomical observations carried out from it, the station also accommodated almost 200 experiments in such areas as space environment research, Earth observation, animal behaviour (experiments involving flies, mice and spiders), plant and crystal growth. Though the results were judged to be very interesting by the scientific community, the station only ever received the three crews manifested at the outset. Following an 84-day record stay, Skylab was abandoned in 1974, having first been lifted to a higher orbit in the hope that by 1978 the Space Transportation System programme would have advanced sufficiently for a shuttle to be sent to boost its altitude.

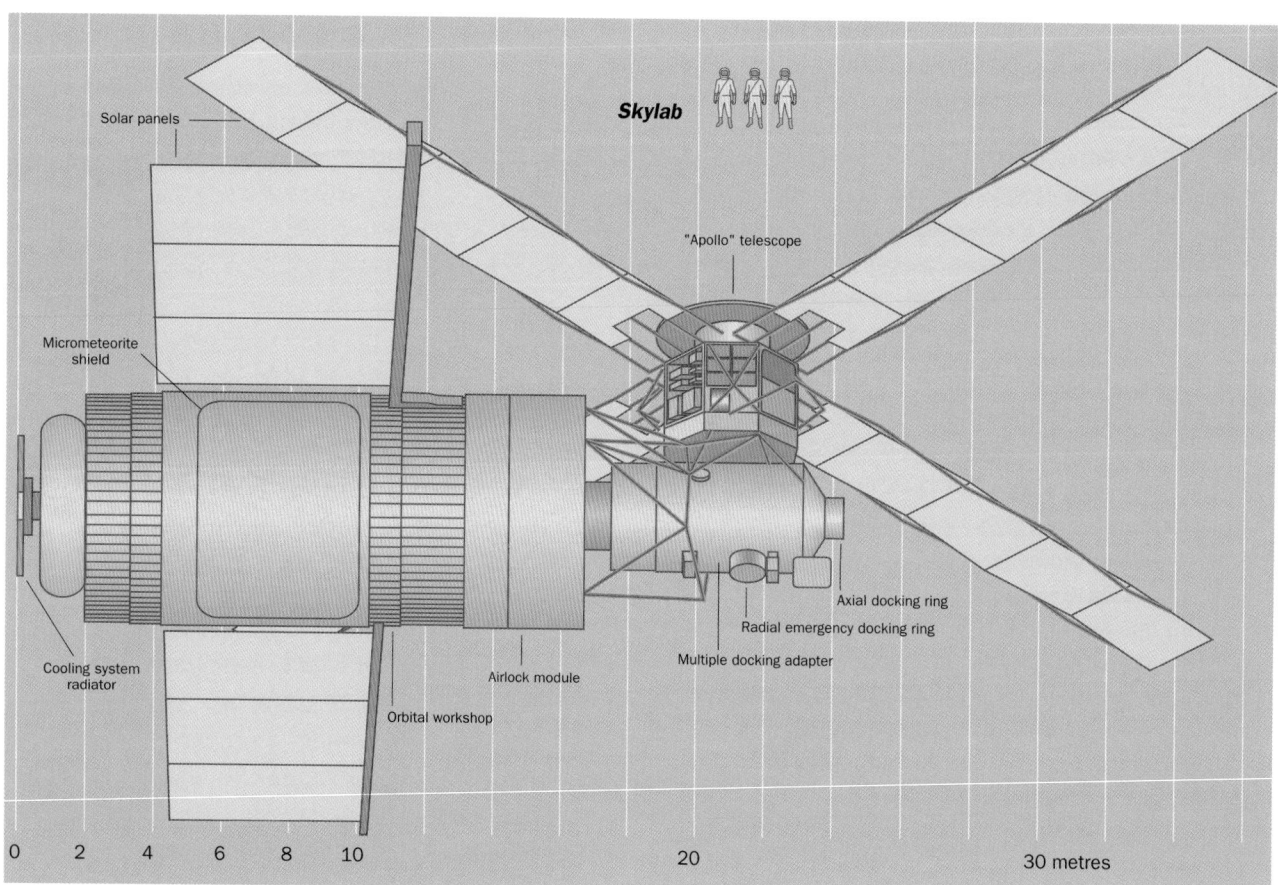

Skylab

Solar panels

"Apollo" telescope

Micrometeorite shield

Axial docking ring

Radial emergency docking ring

Multiple docking adapter

Cooling system radiator

Airlock module

Orbital workshop

0   2   4   6   8   10   20   30 metres

Figure 13.5. **Skylab, the first American space station.** Based on the third stage of the Saturn-V launcher, Skylab comprised some 321 m³ of pressurised accommodation, making it the largest station ever carried aloft up to that time. The left-hand solar panel and the meteoroid shield – still shown in the illustration - were torn off in air turbulence at lift-off. The 270 m³ orbital workshop replaced the hydrogen tank, while the oxygen tank was reused for waste collection. Skylab was launched without passengers but carrying the water, air (70% oxygen, 30% nitrogen), food supplies and instrumentation required for the intended three missions. Apart from the emergency docking port there was only one docking station and this was designed neither for in-flight resupply nor for replacement of the three-cosmonaut crews – which came and left in the Apollo module. The damage incurred during launch did not in the end detract from the success of the three missions, thanks to repairs performed by the cosmonauts. The most important mission results concerned human ability to live in the space environment, solar observation using the eight instruments associated with the Apollo telescope, remote-sensing with the Earth Resources Experiment Package (EREP), including the first radar altimetry test, and a substantial number of human and animal biology studies. Following the third mission, Skylab was abandoned and crashed to Earth in 1979 – none of the rescue projects developed having been carried through.

The Salyut stations

Meanwhile the Soviet space community had to contend with a whole range of problems, technical, financial and organisational. The space station programme was built up from the various elements that were available, namely the multi-purpose Soyuz spacecraft in service since 1967, the planned crewed orbital station Almaz and the Proton launcher. When the decision was taken in 1974 to call a halt to the final N-1 launcher tests – with success close at hand according to the development team – the only realistic launching option was the Proton rocket and it could only loft relatively small stations, perhaps on a modular basis. But the media impact of human spaceflight in low-Earth orbit had in any case waned considerably and ambitions had been lowered. Space stations would henceforth be designed for a specific range of tasks rather than as a staging post for journeys beyond. The programme ran into some early problems too and the first success, a two-astronaut mission on Salyut 3, did not come until June 1974. By that time a third US crew had already completed its stay on Skylab. This greatly reduced the impact of the Soviet achievement.

The successful launch and occupation as from 1977 of Salyut 6 and 7, the purely civil, second-generation stations (fig. 13.6) marked a second stage, with the acquisition of unique expertise, especially in docking techniques and life in a weightless environment. The enthusiasm of the general public and of the players themselves was however seriously dented by this time, a crucial weakness which would make itself felt as the human spaceflight programme unfolded. And while the medium-term aspiration was still to pave the way for crewed missions to other planets, the more immediate objective was now simply to achieve longer and longer stays in orbit. By 1984, the record had risen to eight months. Development activity centred on improving the existing water and air recycling systems, while on-station attention focused primarily on medical and biological experiments, though some work continued to be done on observation of the Earth and the space environment.

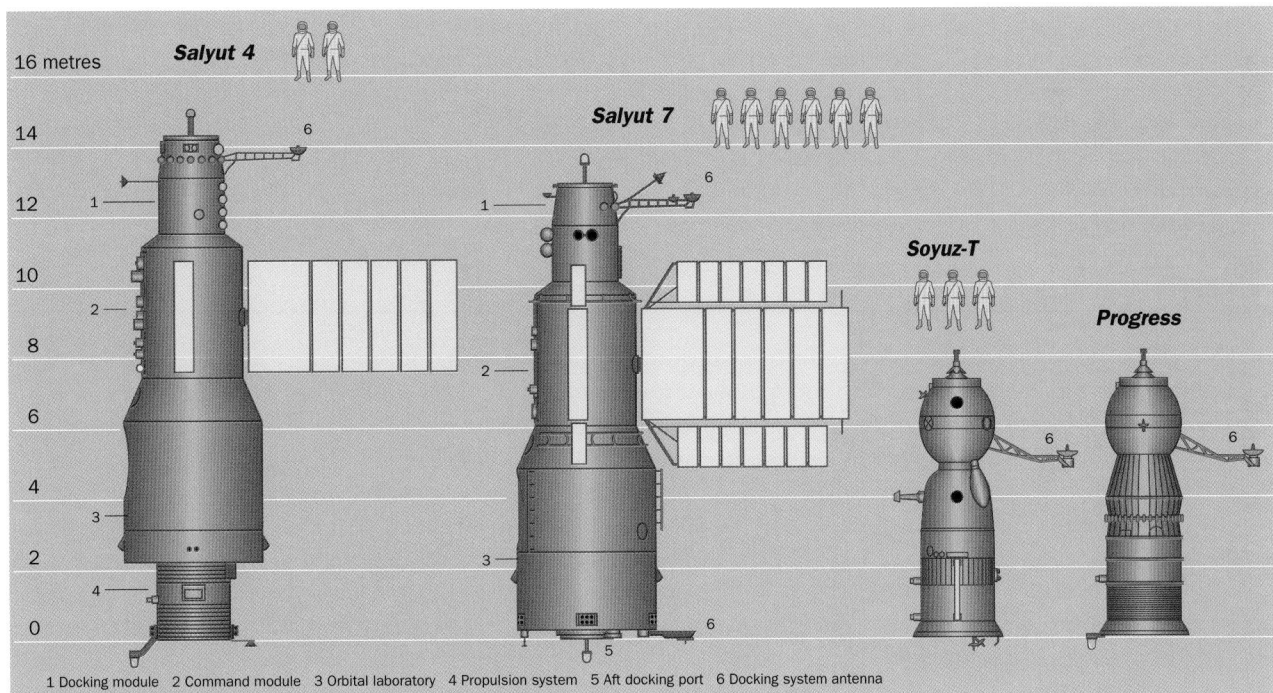

1 Docking module   2 Command module   3 Orbital laboratory   4 Propulsion system   5 Aft docking port   6 Docking system antenna

Figure 13.6. **The Soviet Salyut stations and their servicing vehicles.** The Salyut stations had a useful volume one third that of Skylab. The first generation comprised both military (Salyut-3 and 5) and civil (Salyut-4) stations. They had only one docking port, precluding in-flight resupply and crew replacement, and their lifespan was limited by the altitude at which they flew (around 230 km) and the small volume of on-board propellant. Salyut-6 and 7, the civil stations making up the second generation, featured two docking ports to receive the new Soyuz spacecraft with their crew of three and the Progress automatic cargocraft. The resupply vehicles carried water, food supplies, air regenerators and scientific material. They also transported the propellant required to maintain a correct orbit or, docked to the station, provided any reboost that might be required. As they were not equipped with a heat-shield, they disintegrated on reentry into the atmosphere, together with waste unloaded from the station. As the programme unfolded, increasingly long missions became possible, culminating in the 236-day Salyut-7 mission in 1984. The crew could be visited by Soyuz spacecraft during their stay and could be relieved fully or in part. Long-stay missions were particularly suitable for the medical research required for interplanetary missions and for materials science experiments performed in microgravity.

The Space Shuttle, Mir and Buran

The US Space Shuttle

The Space Transportation System (STS), better known as the Space Shuttle, represented a new approach to human space activity. The priority now shifted to developing a recoverable system designed to carry people into space and back again but also transport civil and military payloads. These might be for release into orbit or form an integrated payload, as with Spacelab. The Shuttle's capabilities would even extend to space system maintenance, including recovery of satellites stowed in its hold for return to Earth (fig. 13.7).

Very ambitious technology and a multiple mission dimension were among the essential features of this programme. Faced with the problem of preserving the potential it had built up in the course of the Apollo programme, NASA needed to secure responsibility for another substantial programme, one which would attract the necessary funding and provide an adequate workload. The modest objectives assigned to the Skylab programme had brought home to planners at NASA the diminished political commitment to human space activity. It was no longer possible in these circumstances to make out a high-profile case for space exploration but NASA, while insisting that it was now setting its sights much lower, was determined all the same to maintain the wherewithal to resume its pioneering efforts, sooner or later. And yet the presidential authorities proved unable for reasons of prestige to withdraw completely from the space arena. With presidential elections looming, simply cancelling the contracts which guaranteed employment levels in the major aerospace companies, above all in California, was not a serious option.

So in 1972 the Space Shuttle programme was finally accepted by President Richard Nixon. In securing that decision however, NASA had had to commit itself to exacting targets, to attract military interest in the project and guarantee that the shuttle would ultimately produce substantial savings. These commitments would in some cases prove contradictory and indeed came perilously close to sinking the programme in the early days of the Carter presidency. And if the programme survived, this had very little to do with the issue of living in space. It went ahead primarily thanks to an understanding thrashed out between President Carter and Congress at the time of the disarmament discussions in the late 1970s. SALT-2, the Strategic Arms Limitation Treaty being negotiated by the two superpowers, brought with it entirely new verification requirements and these, combined with renewed controversy about Soviet anti-satellite tests, conferred on the Shuttle a new role in the debate on US defence strategy. Initially imposed as the sole transportation

option for the future KH 11 range of heavy military surveillance satellites, the Shuttle also attracted interest from a very early stage as an ideal tool for testing possible future space weapons. This set the scene for President Reagan's 1983 address on the Strategic Defense Initiative (see pp. 355–357). It was against this background that the political decision was taken to go for the 'Shuttle only' option. That decision, despite considerable opposition even in military circles, effectively put conventional launch systems out of business. When the Challenger accident came in 1986, leaving America powerless to get its satellites – and especially those needed for intelligence-gathering – to orbit, conventional launchers returned to grace with the US military, though the Shuttle continued to be used to transport the very largest reconnaissance craft. Challenger led also to a complete reappraisal of the STS programme, resulting in a statement by President Bush in 1992 that no further Shuttles would be built. Protests from the NASA Administrator R. Truly, himself a former astronaut, were of no avail and he had no option but to resign.

These conflicting demands have undoubtedly prevented the Space Transportation System developing its full potential. But they have also prompted the United States to imagine and enact new forms of human activity in space. With the Shuttle, in-flight human intervention to deal with problems prior to satellite release is now possible, opening the way for satellite inspection, repair, resupply and even recovery operations. It should be remembered too that while the Shuttle is restricted to short-duration stays in space, this 100-tonne vehicle is comparable in volume terms with Mir in full modular configuration. True, cutting the number of flights and giving priority to military missions meant deferring or even cancelling entire science programmes. However Spacelab missions with Europeans on-board offered an opportunity for Europe to acquire valuable experience.

For NASA, the Shuttle programme was always intimately tied up with the space station programme. President Reagan's official announcement of the space station Freedom programme in 1984 was seen by many as a return to space endeavour on the grand scale and his call for international cooperation under the American banner revived a long tradition of US leadership. The international context had changed and human spaceflight programmes could no longer be justified solely on prestige grounds but programmes of this kind still carried considerable political weight. In giving the go-ahead for a big international station, President Reagan appealed to a new generation of pioneers. They would take up the challenge of the 'high frontier'.

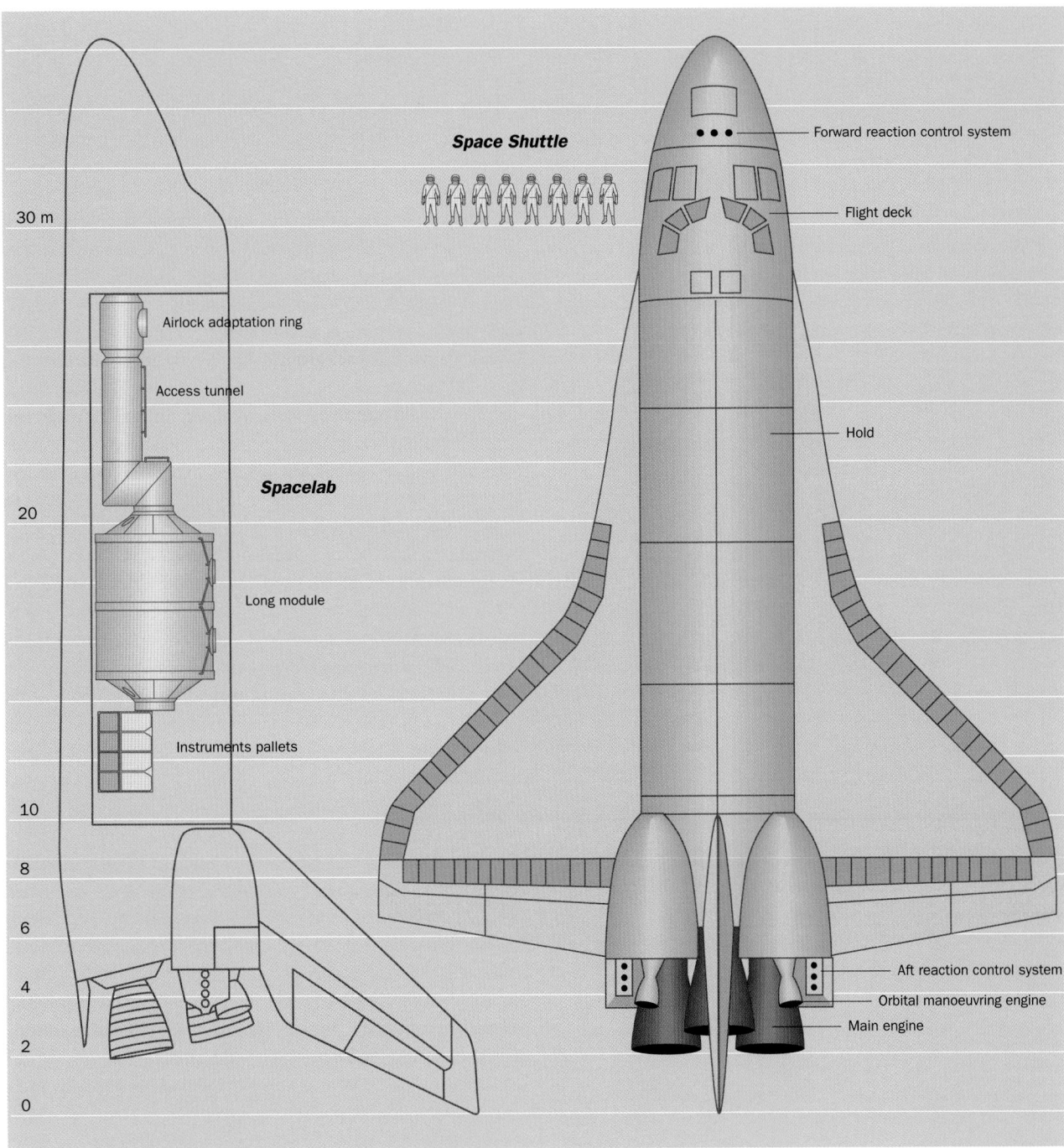

Figure 13.7. **The US Space Shuttle and the European laboratory Spacelab.**
In addition to its role as a heavy launcher, the Shuttle can also serve as a temporary space station, accommodating special hardware units – Spacelab for example – for experiments in microgravity conditions.

Spacelab, developed by the European Space Agency with a majority contribution (55%) by Germany, represented the bulk of Europe's contribution to the shuttle programme. A cooperation agreement on Spacelab was concluded with NASA in 1973. The system was of modular design with configurations built up from a pressurised module measuring 2.70 metres across at mid-point and varying numbers of instrument pallets. A connecting tunnel provided 'shirt-sleeve' access to the module from the orbiter cabin mid-deck; access to the pallets on the other hand was available only to astronauts wearing space-suits, their purpose being to provide a mounting for experiments requiring direct exposure to the harsh space environment. Lastly, an instrument pointing sub-system allowed instruments of different sizes and weights (up to almost 7000 kg) to be pointed at stars, the Sun, the Earth or any other target with arc second precision. The energy supplied by the Shuttle's hydrogen–oxygen fuel cells was limited – 7 kW with 12 kW peaks – and this placed constraints on the types of mission that could be performed and their duration.

Spacelab offered a range of configurations – a long module consisting of two central segments and two pallets, a short module in combination with three pallets, five pallets but no habitable module. In the latter configuration, a pressurised, temperature-controlled housing called the 'igloo' provided connections for data collection, communications, power supply etc. Vertically attached to the forward end of the first pallet, the igloo was 2.4 m in height and 1.1 m in diameter.

Spacelab 1 and 2, the second of these purchased by NASA, flew 25 times between 1983 and 1998. Activities pursued in the course of those flights included experiments in biology, medicine and materials processing, astronomical observations, upper atmosphere research and Earth observation.

This was a clear break with the policies earlier pursued by President Carter – never a champion of large-scale human endeavours in space – and no opportunity was lost to bring this point home. At the same time space exploration itself was coming back into favour, under pressure from the president's advisors. The various committees set up to review this issue took much the same line as NASA itself had done in the late 1960s – and yet this time the space administration's reaction was if anything hostile. Many top-level managers in NASA took the view that too close a link between the space station and exploration programmes could well lead to cuts

in the Freedom budget – on the grounds that exploration was a longer-term affair and funding could therefore be spread out.

Four years later, President Bush dropped an apparent political bombshell. In a July 1989 address he made it plain that human space exploration was officially back on the US agenda and the corresponding technical capability would be developed. This new departure was dubbed the Space Exploration Initiative (SEI). The presidential address on this theme was however very much tied to a particular event – the celebrations to mark the 20th anniversary of the Apollo Moon-landing – and no specific deadlines were set for the new project.

In a foreword to the '90 Day Study on Human Exploration of the Moon and Mars' commissioned by the National Space Council, NASA put forward various ways in which the presidential strategy could be approached. For the Agency, the first step had to be the establishment of the space station Freedom in preparation for a return to the Moon. Only then should the journey to Mars be undertaken. The linkage between the station and the rest of the SEI and the case for visiting the Moon 'on the way' did however prove to be two of the most hotly disputed issues and further adjustments to the station plans (see fig. 13.13) – which had already undergone considerable descoping – could not be ruled out.

Clearly NASA's technological reasoning remained faithful to the concepts it had developed in the past, a continuity of thought which seems to have proved politically effective since the presidential proposal endorsed a strategy of gradually pushing back the limits of human activity in the solar system. On the downside, the proposed reliance on sometimes unduly conventional technologies revealed the limitations of a project which was perhaps too heavily influenced by thinking from the 1970s. NASA was particularly vulnerable on this point insofar as one of the main arguments for space activity was precisely its capacity to stimulate technological innovation, firing the imagination of scientists and engineers alike. This in turn was supposed to help rebuild America's lead in the high-tech industries.

The early 1980s were years of decisive development. Entry into service of the Space Shuttle heralded America's return to space and in the period immediately prior to the Challenger accident, the number of flights reached a remarkable level. The emphasis was however on relatively short missions, usually lasting about a week. This contrasted sharply with the Soviet effort, which had for a number of years been focusing on long-duration stays in space. Real experience of living in orbit was thus confined largely to the USSR.

Mir and Buran

The launch of space station Mir in 1986 ushered in a third generation of orbital stations and, according to statements at the time, marked the beginning of a new phase in the settlement of space. For this station, unlike its predecessors, featured a number of docking ports for add-on modules supplying services to the station or offering specialised

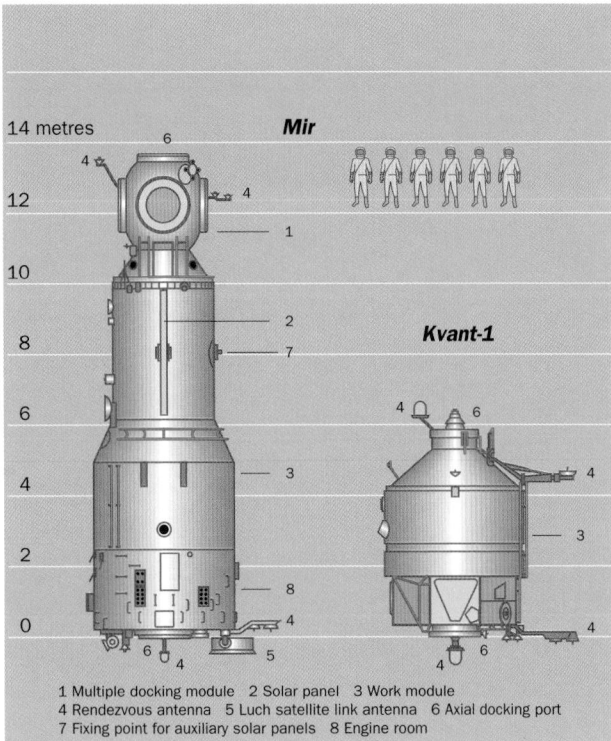

1 Multiple docking module 2 Solar panel 3 Work module
4 Rendezvous antenna 5 Luch satellite link antenna 6 Axial docking port
7 Fixing point for auxiliary solar panels 8 Engine room

Figure 13.8. **Space station Mir and the Kvant-1 astrophysics module.** Mir represents a new generation of station based on a modular design. The aft docking port is much the same as on the later Salyut stations but at the forward end a multiple docking module, comprising one axial and four lateral ports, is now provided. Modules are delivered to the forward axial port and are then transferred to a lateral port using a mechanical arm. The final configuration consists of the following five modules (see also fig. 13.9): Kvant-1 (astrophysics) docked at the aft collar, and, at the four lateral rings on the docking module: Kvant-2 (extension), Kristall (materials production in microgravity), Spektr (atmospheric research) and Priroda (Earth observation). This leaves the two longitudinal docking ports at the forward end of Mir and at the aft end of Kvant-1 available to receive the new Soyuz-TM and Progress-M vehicles, both little changed from their predecessors.

With Mir, the Soviets acquired the ability to occupy space on a permanent basis. Apart from a four-month interruption in 1989, crews of two cosmonauts replaced each other regularly from 1987 to 1997, a new record for length of stay in space of 437 days and 18 hours being set by Valeri Poliakov in 1995. The third passenger on Soyuz is often a foreign paying guest. The pressurised volume of all the modules combined stands at 380 m², just 60 m² more than Skylab. Available onboard power is 34 kW, compared with the 10 kW available to Skylab prior to loss of one of the solar panels.

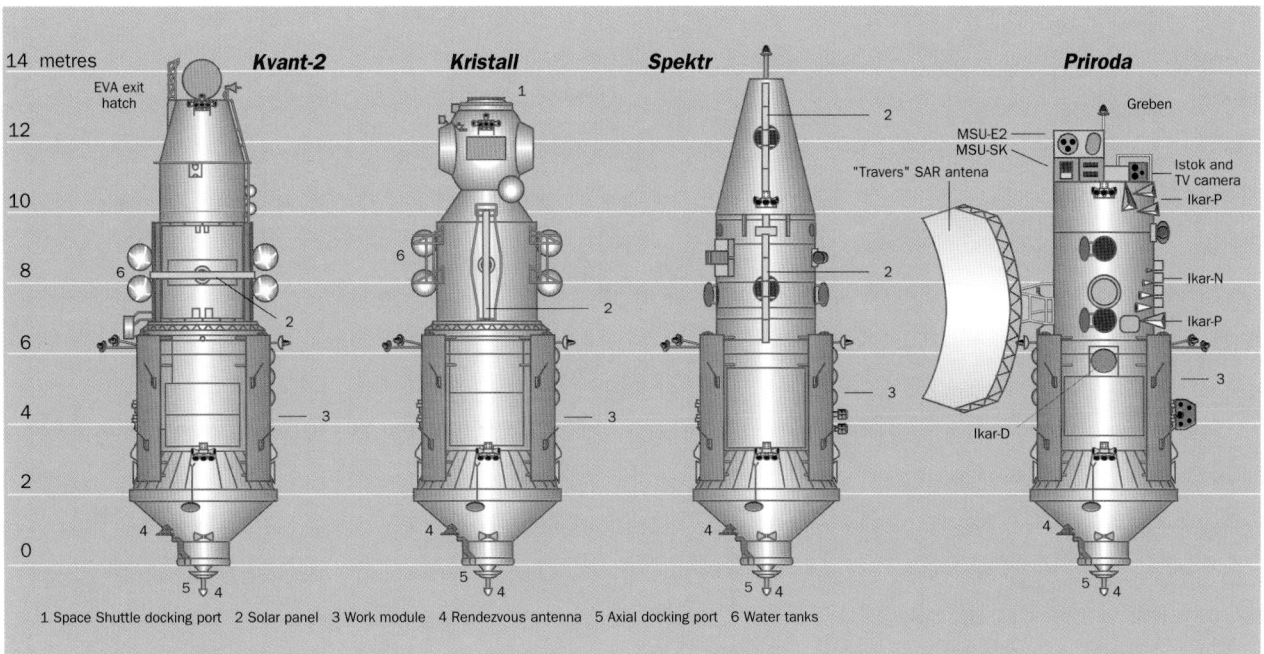

Figure 13.9. Mir's four radial modules. US shuttles dock at the port on the Kristall module – which was to be have been used by Buran – equipped with a docking adaptor not shown in the figure. The main instruments making up the Priroda payload are: the various Ikar radiometers, the MSU-SK and E2 imaging scanners, the Istok 1 spectroradiometer offering video mode operation, the Greben ocean altimeter and Ozon-Mir, an ozone and aerosol detection instrument. These have been joined by the SAR and by the German MOMS-02P imager.

accommodation for astrophysics studies, Earth observation, production of materials in microgravity and biomedical experiments (figs. 13.8 and 13.9). The concept of a station assembled from building blocks each weighing in at some 20 tonnes made for greater operational flexibility. There was also a new emphasis on international cooperation involving, from the earliest days of Mir, many countries in Eastern and Western Europe.

While pressing ahead with its station programme, the Soviet Union decided also to diversify its space transport capability. The new heavy launcher Energya had its first test flight in 1987, followed in November 1988 by the first and only flight of the Buran shuttle (fig. 13.10), in automatic mode. The Energya–Buran programme, given the go-ahead in 1976, was prompted by a concern to acquire a system equivalent to the STS, which Soviet analysts saw as a high-technology programme offering significant military applications. As in the United States, the space community had in fact already developed various spaceplane concepts, with a greater or lesser emphasis on military aspects. They had concluded that if the aim was to establish settlements in space, there was little alternative to a reusable spacecraft. The Soviet Union had also learnt much from Spiral, the two-stage aerospace vehicle programme, and from the piloted orbital flight experiments derived from that programme in the 1970s. This left it well placed to successfully develop a shuttle and get back to parity with the United States. The shuttle specifications were finalised in March 1977 and in December 1981 the programme was classed a national priority with a target

of 1985 for the maiden flight. Though Buran was three years late in getting off the ground, its unpiloted first flight went without a hitch. But by that time the economic environment had changed and the programme could no longer be reconciled with Mikhail Gorbachev's new political priorities.

The space goals that now took shape were far less ambitious and there was no longer any talk of launching Mir-2. When M. Gorbachev, speaking in May 1989, presented the Mars-94 programme as a Soviet economic priority, he was concerned no doubt to match earlier declarations by George Bush but was also keenly aware that this project was well suited to international cooperation, with the access to funding this implied. At the same time, the human spaceflight programme, being one of the least sensitive areas of space activity, lent itself particularly well to the transfer of industrial capacity from the military–industrial complex to the civil sector. As events unfolded, space did however become increasingly remote from immediate political concerns and the firms working on human spaceflight – an area which was attracting particular criticism in a strained political and economic context – had to draw on their own resources (see Chapter 5).

It was clear by 1992 that the money needed to replace the Mir-1 core module at the end of its design life in 1995 would not in fact be forthcoming. The overriding objective now became to save whatever could be saved of the capabilities built up over the years. The policy of inviting 'paying guests' to form international crews was typical of a new drive to make space pay and the charge of $10 million per

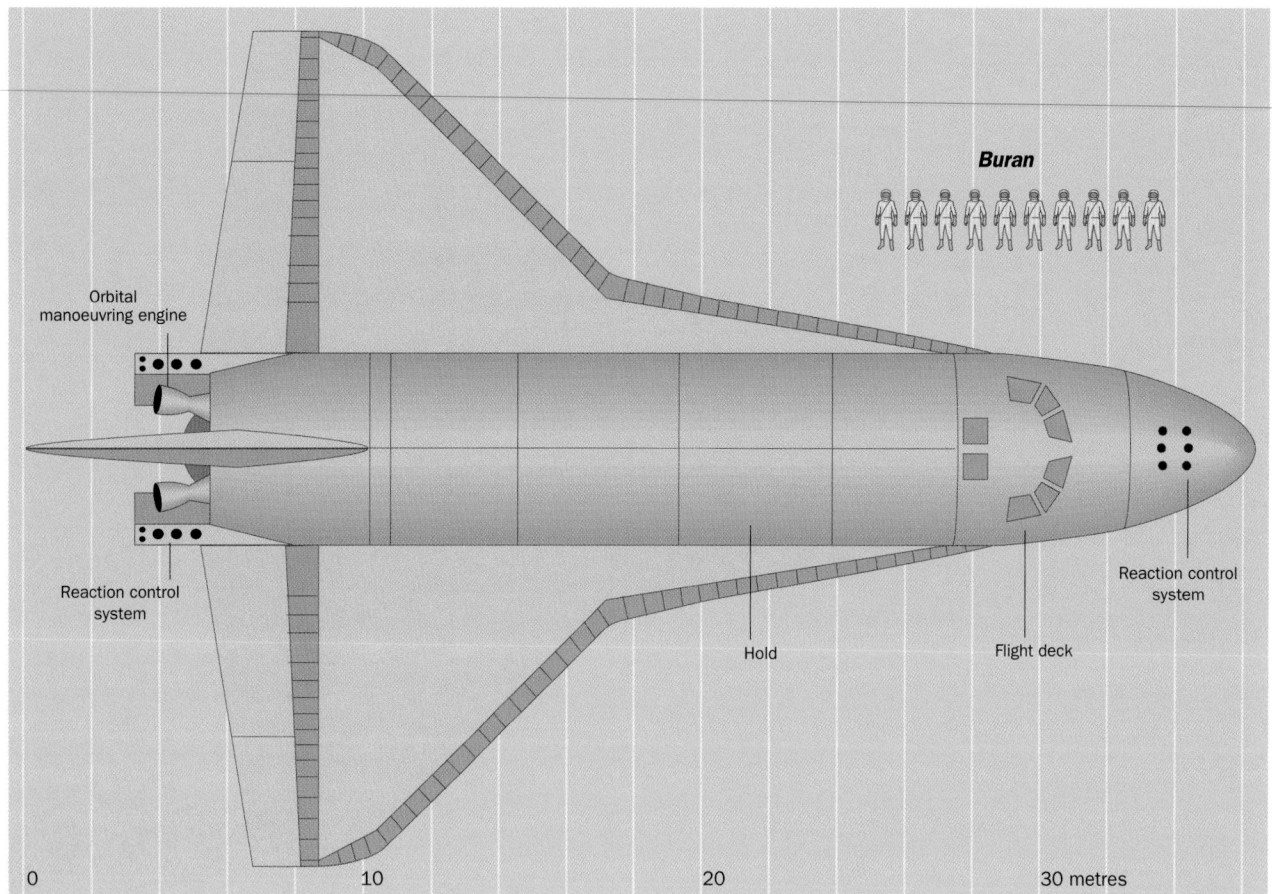

Buran

Orbital manoeuvring engine

Reaction control system

Hold

Flight deck

Reaction control system

0   10   20   30 metres

Figure 13.10. **The Soviet space shuttle Buran.** The Soviet and American shuttles are similar not only in design and dimensions but also in terms of the military involvement in their development. The main difference lies in the fact that Buran does not have its own propulsion system but is launched into orbit by Energia. Designed to carry up to eight cosmonauts, Buran flew only once, in automatic mode.

head helped offset the cost of sending new crews to the station. If at the end of the 1980s, plans to establish a human presence in space were still on the agenda in the USA, despite some descoping, the situation in Russia had by that time become particularly critical.

As a new decade started, there was no doubt that human spaceflight had lost much of its impact. Some symbolic value did however remain, as seemed clear from the insistence with which the Russians, Americans and, in Europe, the French continued to defend the terms 'cosmonaut', 'astronaut' and 'spationaute', with their powerful national connotations. Space missions by fellow nationals still got coverage in the national press and attracted continuing public interest. Since the summer of 1990 the Mir station had no difficulty in attracting guests prepared to pay for the round journey, whether lay visitors looking to acquire prestige – the trip by a Japanese journalist is a case in point – or space professionals wishing to train for their countries' future programmes. The same phenomenon is even to be found in countries of the former Soviet Union. In seeking a place for one of their nationals on a Mir mission, the new independent republics seemed to be saying that more than 30 years on from the first human spaceflight and despite, or perhaps because of, a difficult economic context, a place on a space mission is still a subject of national pride.

The value placed on symbols was again apparent in Russia's reluctance to proceed as planned with the destruction of Mir at the end of the first phase of cooperation with its Western partners. The March 2000 deadline for official closure and deorbiting of the station came and went and, after a number of failed attempts to find sponsors, a private company, MirCorp, announced a first flight to the station in the framework of the Citizen Explorer programme supposed to bring in $20 million. But even disregarding the funding difficulties with which Russia is having to contend as it struggles also to meet its ISS commitments, the age of the station – in service since 1986 – precluded anything other than a relatively short-term future for this programme. Planned and controlled deorbit finally took place on 23 March 2001 and the remains of Mir fell into the South Pacific.

The International Space Station

The Russian space station programme was from the outset a purely national venture. It was shaped by the conditions prevailing in the wake of the Moon race and its successful outcome for the US. Meanwhile in the United States the space station programme – which was supposed in theory to grow out of the shuttle programme – was in practice weighed down by the technical difficulties and cost overruns which were plaguing the shuttle development. Against this background, the station programme can to some extent be seen as a necessary stage in the long-term advance into space but has also to be viewed in relation to Soviet achievements.

From Freedom to the International Space Station

On 25 January 1984, President Reagan invited America's friends and allies to come in on development of the US space station. This proposition was prompted by a number of concerns. From a national viewpoint, Freedom was perceived as NASA's new major programme but it lacked clearly defined objectives and this had fuelled controversy as to its scientific and technical value. And failure to gain recognition as a political priority exposed the programme to increasing budgetary pressure from Congress. Internationalising the programme meant redefining the pursuit of leadership in a new cooperative framework. Cooperation provided scope, in principle at least, for reducing the cost to the USA of developing and operating the station while for NASA the involvement of foreign partners increased the programme's chances of survival. All the proposals put forward were however consistently shaped by a commitment to maintaining sovereignty over subsequent utilisation of the station and exclusive command of the advanced technologies essential to the project.

The Europeans registered their interest as early as 1985 and went on to confirm this when ESA's ruling Council met at ministerial level in The Hague on 10 November 1987. The fundamental ambivalence which has always been a mark of the partnership was however already apparent from the Inter-Governmental Agreement and the three Memorandums of Understanding adopted in 1988. While each party maintained control of its own contribution, the United States led on operations and played a pivotal role in the administration of the agreements, of which it was the depository. This was the atmosphere in which the Europeans, again at the 1987 conference in The Hague, set their sights on a major programme to achieve autonomous access to space. The European Space Agency proposed to develop the Columbus orbital laboratory in two versions, one docked to the US station and the other – the Man-Tended Free Flyer – capable of operating independently. And above all there was the plan to build the 23-tonne Hermes spaceplane. This would be launched atop Ariane 5 and would be capable of flying autonomously on 10-day missions with crews of three and transporting 2.5 tonnes of payload to low-Earth orbit. The European partners' determination to acquire the fundamental technologies required for human missions to space soon began to falter however in the face of mounting technological and budgetary difficulties. This led to abandonment of the Hermes project at the ESA ministerial conference in Granada in 1992.

With the countries of Europe pulling back, repeated questioning on the US side of the basic station design and programme timetable provoked a series of crises. Again in the United States, growing hostility from Congress, faced at one and the same time with cost escalation and an increasingly apparent lack of genuine consensus among the partners, placed the future of the Freedom programme in serious jeopardy. Moreover, the final demand for the station to be descoped – made by President Clinton in 1992 – coming after so many other revisions of the original plans, deprived Freedom of its main grounds for existence: as a staging post for subsequent missions of exploration.

Against this background, the new international situation and the closer links between the United States and Russia would soon lead to a comprehensive reassessment of human activity in space and a growing belief in the value of cooperation. The fact that the human space programmes had on a number of occasions in the past provided a propitious environment for international cooperation did much to help that belief take root. As far back as 1975, the Apollo–Soyuz Test Program (ASTP) flight, in which the emissaries of the two major space powers met as equals in orbit, was widely perceived as a symbol of the thaw in international relations. And with the policy of in-space hospitality pursued in subsequent years, cooperation between nations in this area rapidly became commonplace. In a parallel development towards the end of the 1980s, both the United States and Russia formally invited international participation in their future Mars exploration programmes. This confirmed the increasingly open nature of those programmes, even in the case of the so-called 'precursor' missions reliant on automatic satellites. And yet the framework for collaborative ventures of this kind was somewhat ambiguous.

The new American proposals for cooperation were driven largely by non-space considerations, the need to stabilise Russia through external aid being viewed as an international security imperative. In this connection, reinforcing a civilian space sector with heavy reliance on US financial support offered a particularly attractive means of limiting Russia's military–industrial potential, which might otherwise reemerge as a threat at some point in the future. The United States could in this way hope to benefit from past investments in human spaceflight and the broad and varied experience acquired on the way.

Russia's human space programme was by now at low ebb. National support had dried up completely and there was little option but to accept the proposals from the USA. For their part, the Russian authorities could see some political mileage in the negotiations, which seemed to prove that in the area of human spaceflight at least a US–Russian duopoly continued to exist. With attempts by Russia to build a range of collaborative links foundering on the desire of potential partners – Europe for instance – to develop equivalent capabilities, cooperation with the United States offered an unexpectedly attractive solution. Other players in the space community were admittedly dismayed at the prospect of a significant proportion of a limited national budget being invested in an area offering little or no industrial or commercial outlook, but their objections were of no avail in the absence of any viable alternative.

The proposal announced at the US–Russian space summit of 2 September 1993 to build a space station in worldwide cooperation was thus an eloquent demonstration of how heavily international political considerations weigh in space policy decisions.

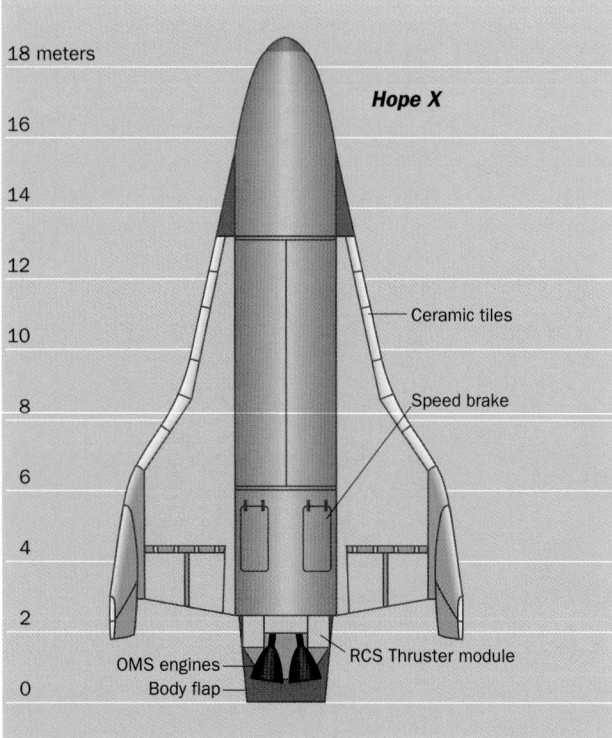

Figure 13.11. **HOPE-X.** The experimental phase in work on a space transport vehicle. The vehicle is designed to dock at the station and land on a 3 km runway fully automatically. The project schedule is heavily dependent on the status of the H-II launcher project.

The International Space Station (ISS)

The new-generation space station originally planned by Russia was to have taken the shape of a huge modular complex – Cosmograd. This was to provide a base from which to explore

Figure 13.12. **ISS artist's view.** Courtesy NASA

first Mars and then the other planets and would perhaps even support genuine industrial activity in space. It was to have been composed of a 120-tonne core module launched by Energya and various specialised modules orbited by Zenith and Proton launchers and serviced by Buran. The new international station, a less ambitious affair altogether, was based in large part on the supply by the United States and Russia of elements which were already being developed under the

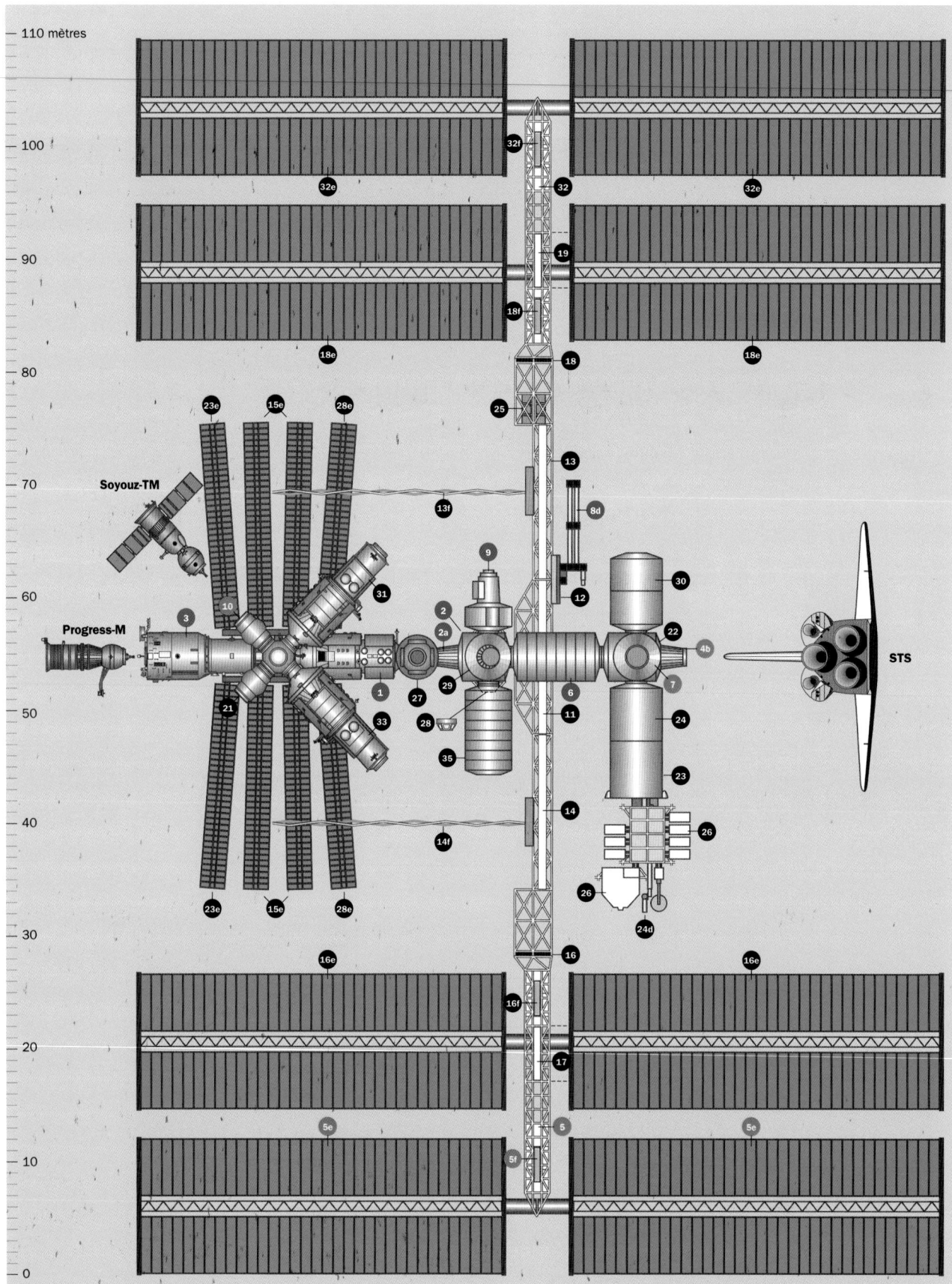

Figure 13.13. **ISS Assembly sequences.** Numbers are in the same order as the combination process of the station. Elements that reached their definitive position January the 1st 2002 are marked in blue. Elements that have been launched but which have not reached their definitive position yet are marked in orange, and are placed at their final position. Elements engineered in partner countries and which are given to the US in return for launchings are signalled as American ones.

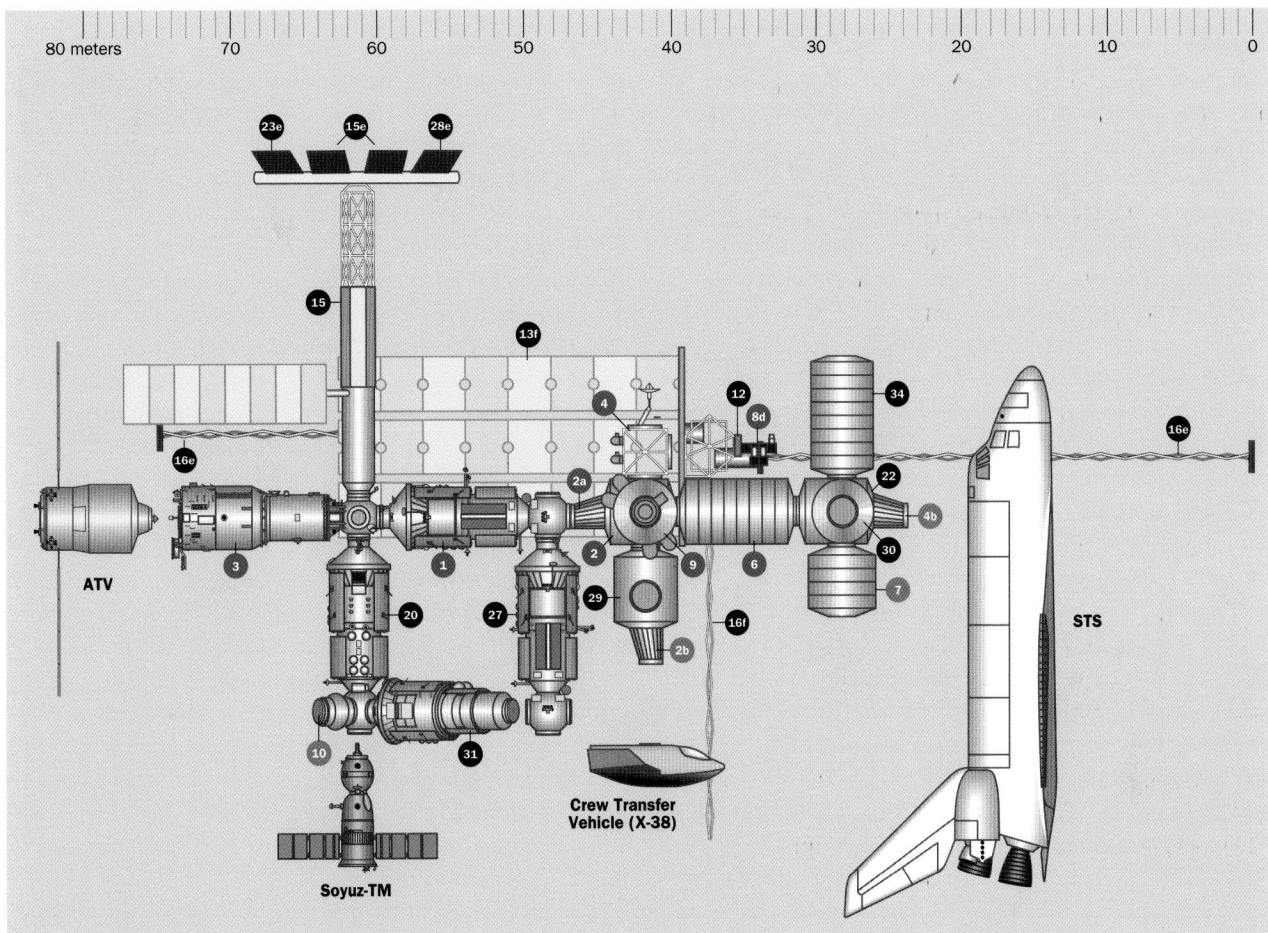

Sequence	Nationality	Elements	Sequence	Nationality	Elements
1	Russia	Functional Cargo Block / Funktsionalny Gruzovoi Blok (FGB) : Zaria, *launched 20 11 1998*	15 e	Russia	SPP Solar arrays
2	USA	node-1 : Unity, *launched 04 12 1998*	16	USA	Integrated Truss Structure (ITS) : P3-P4
2 a	USA	Pressurized Mating Adapter (PMA)-1 *launched 04 12 1998*	16 e	USA	Photovoltaic Module P4
2 b	USA	Pressurized Mating Adapter(PMA)-2 *temporarily installed on 2, launched 04 12 1998*	16 f	USA	Thermal control system (TCS) radiator
3	Russia	Service module : Zvezda, *launched 12 07 2000*	17	USA	Integrated Truss Structure (ITS) : P5
4	USA	Integrated Truss Structure (ITS) : Z1, *launched 11 10 2000*	18	USA	Integrated Truss Structure (ITS) : S3-S4
4 b	USA	Pressurized Mating Adapter (PMA)-3 *temporarily installed on 2, launched 11 10 2000*	18 e	USA	Photovoltaic Module S4
5	USA	Integrated Truss Structure (ITS) : P6 *temporarily installed on 4, launched 30 11 2000*	18 f	USA	Thermal control system (TCS) radiator
5 e	USA	Photovoltaic Module P6 *temporarily installed on 4, launched 30 11 2000*	19	USA	Integrated Truss Structure (ITS) : S5
5 f	USA	Thermal control system (TCS) radiator *temporarily installed on 4, launched 30 11 2000*	20	Russia	Universal Docking Module (UDM)
6	USA	Laboratory module : Destiny, *launched 07 02 2001*	21	Russia	Docking Compartment 2 (DC-2)
7	Italy	Mini Pressurized Logistic Module (MPLM) : Leonardo, *launched 08 03 2001*	22	USA	Node 2
8 d	Canada	Space Station Remote Manipulator System, *launched 19 04 2001*	23	Japan	Experimental Logistic Module (ELM), part of Japanese Experiment Module (JEM)
9	USA	Airlock: Quest, *launched 12 07 2001*	23 e	Russia	SPP Solar arrays
10	Russia	Docking Compartment 1 (DC-1) : PIRS, *launched 14 09 2001 temporarily installed on 3*	24	Japan	Japanese Experiment Module (JEM) : Kibo
			24 d	Japan	Remote Manipulator System : RMS part of JEM
11	USA	Integrated Truss Structure (ITS) : S0	25	Brazil	Express pallet
12	Canada	Mobile Base System (MBS) *to complete the Canadian Mobile Servicing System*	26	Japan	Japanese Experiment Module Exposed Facilities (JEM EF)
13	USA	Integrated Truss Structure (ITS) : S1	27	Russia	Docking and Stowage Module (DSM)
13 f	USA	Thermal Control System (TCS) radiator *temporarily stowed*	28	USA	Cupola
14	USA	Integrated Truss Structure (ITS) : P1	28 e	Russia	SPP Solar arrays
14 f	USA	Thermal Control System (TCS) radiator *temporarily stowed*	29	USA	Node 3
15	Russia	Science Power Plateform (SPP)	30	ESA	Attached Pressurized Module (APM) : Columbus (COF)
			31	Russia	Research Module 1
			32	USA	Integrated Truss Structure (ITS) : S6
			32 e	USA	Photovoltaic Module S6
			32 f	USA	Thermal control system (TCS) radiator
			33	Russia	Research Module 2
			34	USA	Centrifuge Accomodations Module (CAM)
			35	USA	Habitation Module

Freedom programme or were covered by the intended modernisation of Mir. Reconfiguration of the programme was now driven primarily by Russia, with its unrivalled expertise in this area, to the detriment of the Europeans. The station's orbit would now be inclined at 51.6°, rather than the 28.8° inclination originally planned for Freedom, to accommodate the launch constraints on Russian vehicles. Energya and Buran would not however be called upon and launch services would in the first instance be supplied by the US Shuttle and Russian conventional launchers only.

The main partners in the original Freedom project had themselves to review their involvement in the new International Space Station (ISS) programme. Japan, a long-standing partner whose commitment had never wavered, maintained its plans to build the Japanese Experiment Module Kibo (meaning 'Hope' in Japanese) . Kibo, designed to allow four astronauts to conduct long-duration experiments, would itself be made up of two modules – a pressurised module and an exposed facility – plus logistics modules attached to each and a remote-controlled manipulator. In return for launch of Kibo onboard the shuttle, NASDA would deliver a centrifuge for use in the American laboratory. Development work also continued on the HTV (H-2 Transfer Vehicle), which would contribute to automatic servicing of the station and would be captured by the manipulator arm. The HOPE-X (H-II Orbiting Plane-Experimental) vehicle is the next step in the Japanese reusable space vehicle programme (fig. 13.11). Early 2004 is the target for a test flight.

The Europeans for their part decided at the 1995 ESA Council meeting in Toulouse, with further clarification at a further such meeting in Brussels in May 1999, to go through with development work on the Columbus Orbital Facility (COF) and the Automated Transfer Vehicle and fund definition studies for a Crew Transfer Vehicle. Italy is building the MPLMs (Mini Pressurised Logistic Module) Leonardo, Rafaelo, Donatello which will travel back and forth between the earth and the Station in the Shuttle bay. Finally, Canada would be making an essential contribution to station assembly in the form of the robotic arm, which had already seen service on-board the shuttle.

As in the early days of human spaceflight, the political environment continues therefore to shape space policy at the most fundamental level. The decision to discontinue the Freedom programme, after more than $10 million had been spent on studies, in favour of a special cooperative relationship with Russia was prompted by a period of revived political interest, in a changed international environment. But with the various partners at different points on the learning curve and against the background of the difficulties encountered with Freedom, some redefinition of the concept of international cooperation was felt necessary. The new approach assumed partnership between states enjoying equal status and making balanced contributions to the project. The United States and Russia would be equal-ranking as the partners responsible for the elements 'serving as the foundation

for the International Space Station'. They were followed in order of importance by the European partner and Japan, which would supply elements 'significantly enhancing the space station's capabilities'. Finally the Canadian contribution was 'an essential part of the space station'.

In practice however, the American approach, shaped by persistent congressional criticism and obvious security considerations, has been to secure control of the key station elements. In addition to the corresponding Russian craft already in existence, a Crew Return Vehicle is being developed jointly with the Europeans for emergency rescue operations. Orders have been placed for four units with delivery scheduled in 2003. Similarly, NASA has developed its own Interim Control Module in response to the various delays to supply of Zvezda, the Russian service module. But the US partner has also registered delays and achieving autonomy in relation to Russian technologies may take longer than expected. It seems likely for example that the propulsion module, which was to have been launched in 2003 and supply an orbit reboost capability in place of the Progress supply ships, will not be available until 2005 at the earliest. The uncertainties of cooperation, with Russia in particular, are tending therefore to exert an upward pressure on costs. But this has to be weighed against the fundamental benefit of the enhanced security provided by the powerful web of commitments binding the various participants.

The space station programme was organised in three phases, in the first of which the special status of the United States and Russia – the only partners with a proven human spaceflight capability – was particularly apparent.

The first phase ran from June 1994 to June 1998 and consisted of nine missions in which the US shuttle visited and docked at Mir to acquire the experience needed for future station assembly and exploitation. The second phase got underway on 20 November 1998 with the launch of the Zarya (Dawn) module on a Proton launcher and the assembly, on 7 December 1998, of the American Unity connecting node by the US shuttle. At this point the international station, though still in embryonic form, was already a reality (fig. 13.12). Following two additional shuttle flights, in December 1999 and May 2000, made necessary by delays to the Russian module and the impact of more intense solar activity than had been predicted, the Zvezda (Star) service module was launched in July 2000, ensuring the station remained on orbit.

According to August 2000 planning data, station assembly proper will get underway in the 2004–05 timeframe, beginning with the supporting structure for the solar panels and the heat exchangers. It will then be time to launch the Japanese and European modules. Europe's Columbus Orbital Facility will leave Earth on an Ariane 5 launcher, in 2004 on current plans. In the final flight phase, an Automated Transfer Vehicle (ATV) will carry it to a station node. Developed by the European Space Agency, the ATV will first fly in 2004 when it will dock at the Russian segment. Once fully assembled, and this will take more than 30 shuttle and 40 Russian launcher missions, the station will house a crew of six on a permanent basis and will provide service for a planned 10 years (fig.13.13). Weighing in at over 400 tonnes in its final configuration, the station will offer 1200m³ of habitable space and enjoy a 110 kW power supply.

Human spaceflight, with its focus on projects having a powerful media impact and relying on a particularly broad range of complex programmes, has often been perceived as competing with less spectacular endeavours in the fields of scientific research or automatic space exploration. This has left it open to fairly regular criticism. The human space programmes must respond by maintaining public interest, thereby retaining political support and access to the substantial funding on which they depend – taken together, the shuttle and station budgets, for example, have for a number of years now used up almost half of total NASA funding. At the same time however, the emphasis on political considerations itself introduces uncertainty and can lead to dead-end situations. Stop–go development is thus a typical feature of human space exploration programmes, in the United States in particular but also to a lesser extent in Europe. The absence of any real goal is a further source of weakness and construction of a station for its own sake cannot exert the same attraction as a station programme explicitly billed as a staging post for subsequent exploration.

Human presence in space – a generation on from the short-lived forays to the Moon – is today confined to a belt encircling the Earth between 57° north and south at an altitude of less than 500 km. And yet the dream of human occupation of space is now a reality, however incomplete that reality may be. It is clear too from the expanding participation in the International Space Station programme – Brazil for example joined in 1997 – but also from the existence of the emerging European, Japanese and Chinese programmes that this stage in development is an essential component of space power.

Future Chinese programme

The specifics of China's geopolitical situation initially prevented it from joining human space missions on a cooperative basis although since 1978 China has made no secret of its resolve to send human beings into space. By 1976 China had already, through the FSW (Fanhui Shi Weixing) programme of retrievable capsules used for photoreconnaissance missions and microgravity experiments, acquired the necessary atmospheric reentry technology. And while the Chinese human spaceflight programme was officially put on hold in 1980 for cost reasons, there were various allusions from the mid-1980s onwards to the construction of a station to rival the Americans and the Soviets – though it would apparently be more like a Gemini capsule capable of accommodating two- to four-member crews.

The State Council announced in 1982 that China would by the end of the decade launch a crewed vehicle, demonstrating its status as a major space power. Project 921, as it was called, consisted of a habitable capsule (921-1) – with a maiden flight scheduled for October 1999 – and an orbital laboratory (921-2). The project assumed modifications to the CZ-2E launcher, redesignated CZ-2F.

The initial project evolved as opportunities arose for international cooperation. Cooperative links were established as from 1994 with Russia, which was anxious to get an economic return on its space investment. In addition to supplying Soyuz capsules, docking systems and other hardware, Russia provided training in 1996 for two Chinese astronauts accompanied by a team of experts. The 921-1 capsule was consequently modified to integrate the Soyuz capsule and the other Russian hardware elements.

A key milestone was reached on 20 November 1999 with the return to Earth, after 14 orbits and more than 21 hours in space, of the habitable capsule dubbed Shenzhou or 'divine craft' by President Jian Zemin – a name evocative also of the 'divine country', China. Like Soyuz, Shenzhou comprises three distinct sections, a service module, a retrievable capsule and an orbital compartment. Its slightly larger dimensions and a number of original features (a more cylindrical than spherical shape, a large lateral hatch and window and a forward docking collar) suggest this may be a transition towards a Chinese design closer to the initial 1992 concept (fig. 13.14).

Whatever the date of the first crewed flight, which could be anywhere between 2002 and 2005 depending on the degree of autonomy sought for future craft in relation to the Russian model, the unwavering Chinese commitment to a human presence in space amply demonstrates that the forces which drove the human spaceflight programmes in the very earliest days are still very much at work some 30 years later.

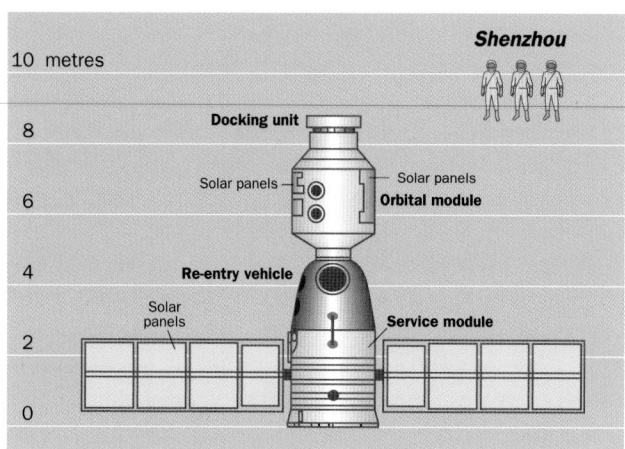

10 metres

Shenzhou

8

Docking unit

Solar panels — Solar panels
Orbital module

6

4

Re-entry vehicle

Solar
panels

Service module

2

0

Figure 13.14. **The Chinese Shenzhou spacecraft.** Heavily influenced by Soyuz, Shenzhou nevertheless includes a larger orbital module capable of remaining in orbit on the departure of the reentry vehicle. The dimensions shown are approximate.

Sending human beings into space remains a potent expression of national status. This is borne out too by the choice of launch date, the 50th anniversary of the proclamation of the People's Republic of China on 1 October 1999. Similar thinking presumably informs suggestions that the future Chinese craft may dock at the International Space Station or that an independent Chinese station may be developed. Some experts do not rule out the possibility that China may also proceed with a military space station programme along the lines of the US Manned Orbiting Laboratory (MOL) programme of the 1970s, which was finally abandoned. Finally, use of the term 'taikonaut', as proposed by Chinese and Malaysian internet users and enthusiastically relayed by Western media, is a clear indication – even if Chinese specialists tend to favour the term 'yuhangyuan' (space navigator) – that for public opinion human presence in space is still a powerful national symbol.

Bibliography

BOOKS

BAKER J.C., O'CONNELL K.M. and WILLIAMSON R.A. (editors) 2001 *Commercial Observation Satellites: At the Leading Edge of Global Transparency*, Rand and ASPRS, Washington DC and Bethesda MD, 643 p.

BILSTEIN R.E. 1996 *Stages to Saturn. A Technological History of the Apollo/Saturn Launch Vehicles*, The NASA History Series, NASA, Washington, 513 p.

Bilt C., Peyrelevade J., Späth L. 2000 *Toward a Space Agency for the European Union*, Report to the ESA Director General, ESA, Paris, 58 p.

BLAMONT J. 1987 *Vénus dévoilée*, Odile Jacob, Paris, 370 p.

BONN F. and ROCHON G. 1992 *Précis de Télédétection*, Volume I: *Principes et Méthodes*, Presses de l'Université Laval/AUPELF, 486 p.

BULKELEY R. 1991 *The Sputnik Crisis and Early United States Space Policy*, Indiana University Press, Bloomington and Indianapolis, 286 p.

BULKELEY R. and SPINARDI G. 1986 *Space Weapons, Deterrence or Delusion?* Polity Press, Cambridge, 378 p.

BURROWS W.E. 1986 *Deep Black, Space Espionage and National Security*, Berkeley Books, New York, 406 p.

CAPDEROU M. *Satellites: Orbite et Échantillonnage*, Springer-Verlag, Paris, 2002.

CAZENAVE A. and FEIGL K. 1994 *Formes et Mouvements de la Terre, satellites et géodésie*, Belin, Paris, 160 p.

COLLINS J.M. 1989 *Military Space Forces. The next 50 years*, Pergamon Brassey's, Mc Lean, 242 p.

COLTON T.J. and GUSTAFSON T. [editors] 1990 *Soldiers and the Soviet State*, Princeton University Press, Princeton, New Jersey, 370 p.

COLWELL R.N. (editor) 1983 *Manual of Remote Sensing*, American Society of Photogrammetry, Falls Church, Virginia, 2440 p. 2 volumes.

COURTEIX S. [coordinator] 1997 *Le Cadre Institutionnel des Activités spatiales des États*, Éditions A. Pedone, Paris, 384 p.

CROVISIER J. and ENCRENAZ T. 2000 *Comet Science: The Study of Remnants from the birth of the Solar System*, Cambridge University Press, Cambridge, 173 p.

DURCH W.J. 1984 *National Interests and the Military Use of Space*, Ballinger Publishing Co, Cambridge, Massachusetts, 286 p.

ENCRENAZ T. and BIBRING J.-P. 1987 *Le Système Solaire*, InterEditions et Éditions du CNRS, Paris, 392 p.

FRANKLIN H.B. 1988 *War Stars – The Superweapon and the American Imagination*, Oxford University Press, Oxford, New York, 256 p.

GARCIN T. 2001 *Les Enjeux Stratégiques de l'Espace*, Bruylant, Bruxelles, LGDJ, Paris, 164 p.

GLOUCHKO V.P. 1985 *Kosmonavtika Enciklopedija* [Encyclopedia of Astronautics], Sovetskaja enciklopedija, Moscow, 528 p.

HARLAND D. 1997 *The Mir Station: A Precursor to Space Colonization*, Wiley, Chichester, 439 p.

HARVEY B. 1998 *The Chinese Space Programme: from Conception to Future Capabilities*, Praxis Publishing, Chichester, 182 p.

HARVEY B. 2000 *The Japanese and Indian Space Programmes: Two Roads into Space*, Springer Praxis Publishing, Chichester, 210 p.

HOUSTON A. and RYCROFT M. (editors) 1999 *Keys to Space: an Interdisciplinary Approach to Space Studies*, McGraw-Hill, Boston.

HU W.-R. 1997 *Space Science in China*, Gordon and Breach, Amsterdam, 1997, 439 p.

JASANI B. (editor) 1982 *Outer Space – A New Dimension of the Arms Race*, SIPRI, Taylor and Francis Ltd, London, 424 p.

JASANI B. (editor) 1984 *Space Weapons – The Arms Control Dilemma*, Taylor and Francis Ltd, London, 256 p.

JASANI B. (editor) 1987 *Space Weapons and International Security*, Oxford University Press, New York and London, 382 p.

JOHNSON D.J., PACE S., GABBARD C.B. 1998 *Space, Emerging Options for National Power*, Rand, 90 p.

JOHNSON N.L. 1987 *Soviet Military Strategy in Space*, Jane's Publishing Company, London, 287 p.

JOHNSON N.L. 1987 *Soviet Space Programs 1980–1985*, Science and Technology Series, vol. 66. An American Astronautical Society Publication, 288 p.

JOHNSON N.L. 1991 *The Soviet Year in Space* 1990, Teledyne Brown Engineering, Colorado Springs, 172 p.

KAHN Ph. [coordinator] 1992 *L'exploitation Commerciale de l'Espace, Droit Positif, Droit Prospectif*, Litec, Paris, 500 p.

KREPON M. (editor) 1990 *Commercial Observation Satellites and International Security*, Mac Millan Press, London, 230 p.

KRIEGE J., RUSSO A. 1994 *Europe in Space* 1960–1973, ESA Publications Division, Noordwijk, 142 p.

LAFFERANDERIE G. and CROWTHER D. 1997 *Outlook on Space Law over the Next 30 Years – Essays Published for the 30th Anniversary of the Outer Space Treaty*, Kluwer Law International, 473 p.

LARDIER C. 1992 *L'astronautique Soviétique*, Armand Colin, Paris, 322 p.

LEBEAU A. 1986 *L'espace en Héritage*, Odile Jacob, Paris, 445 p.

LEBEAU A. 1998 *L'espace, les Enjeux et les Mythes*, Hachette littératures, Paris, 312 p.

LI K.L. 1997 *World Wide Space Law Bibliography*, Update, De Daro Publishing, Montreal, [loose leaf].

LOGSDON J.M. 1970 *The Decision to go to the Moon: Project Apollo and National Interest*, Cambridge, MIT Press, 187 p.

LOGSDON J.M. 1995, 1996, 1998, 1999 *Exploring the Unknown*. Vols. 1,2,3,4, The NASA history series, NASA, Washington D.C., 795 p., 636 p., 608 p., 684 p.

LOUCHET A. 1988 *La Planète Mars, Description Géographique*, Masson, Paris, 138 p.

MCCURDY H.E. 1997 *Space and the American Imagination*, Smithsonian Institution Press, Washington and London, 294 p.

MACK P.E. 1990 *Viewing the Earth. The Social Construction of the Landsat Satellite System*, MIT Press, Cambridge, Massachusetts, 270 p.

MADDERS K. 1997 *A New Force at the Frontier: Europe's Development in the Space Field in the Light of its Main Actors, Policies, Law and Activities from its Beginnings up to the Present*, Cambridge University Press, Cambridge

MALAVIALLE A.-M., PASCO X. and SOURBÈS-VERGER I. 1999 *Espace et Puissance*, Ellipses, Paris, 205 p.

McDOUGALL W. A. 1985 *The Heavens and the Earth: A Political History of the Space Age*, Basic Books, New York, 556 p.

MCLEAN A. 1992 *Western European Military Space Policy*, Dartmouth, Aldershot, 186 p.

MASSEY H. 1986 *History of British Space Science*, Cambridge University Press, Cambridge, 514 p.

MICHINE V.P. 1993 *Pourquoi nous ne sommes pas allés sur la Lune*, Cepadues 98 p.

MURRAY B. 1989 *Journey into Space*, W.W. Norton and Company, New York and London, 382 p.

NEWKIRK D. 1990 *Almanac of Soviet Manned Space Flight*, Gulf Publishing Company, 391 p.

NIE Rong-Zen 1984 *Nie Rong-Zen Huiyilu* [*Memoires of Nie Rong-Zen*], [Publisher of the Chinese People's Liberation Army], Peking, 873 p.

PACE S. 1990 *US Access to Space, Launch Vehicle Choices for 1990–2010*, The Rand Corporation, Santa Monica, CA, 227 p.

PASCO X. 1997 *La Politique Spatiale des États-Unis, 1958–1995: Technologies, Intérêt National et Débat Public*, L'Harmattan, Paris, 300 p.

PAYNE K.B. 1986 *Strategic Defence: "Star Wars" in Perspective*, Hamilton Press, London, 250 p.

PEEBLES C. 1987 *Guardians: Strategic Reconnaissance Satellite*, Praesidio Press, Novatto, 418 p.

REVOL H., senator, 2001 *L'espace: Une Ambition Politique et Stratégique pour l'Europe*, Office Parlementaire d'Évaluation des Choix Scientifiques et Technologiques, Rapport no. 293, 478 p.

RICHELSON J.T. 1990 *America's Secret Eyes in Space: The US Keyhole Spy Satellite Program*, Harper and Row, New York, 376 p.

RICHELSON J.T. 2001 *America's Space Sentinels: DSP Satellites and National Security*, University Press of Kansas, 350 p.

RUFNER K.C. (editor) 1995 *Corona: America's First Satellite Program, Center for the Study of Intelligence*, Central Intelligence Agency, Washington D.C. 360 p.

RUSSIAN SPACE AGENCY & EUROPEAN SPACE AGENCY 1996 *Directory of Russian Space Industry 1996–1997*, third edition, Sevig Press, 240 p.

RYNIN N. A. 1971 *Interplanetary flight and Communication*, Vol. 1, no. 2. Spacecraft in Science Fiction, Leningrad, 1926, translated from Russian, Israel Program for Scientific Translation, Jerusalem, 180 p.

Sidiqi A.A. 2000 *Challenge to Apollo: The Soviet Union and the Space Race, 1945–1974*, NASA, Washington D.C., 1011 p.

SIPRI 1978 *Outer Space, Battlefield of the Future*, Taylor and Francis Ltd, London, 202 p.

SMITH R.W. 1989 *The Space Telescope. A study of NASA*, science, technology and politics, Cambridge University Press, Cambridge, 478 p.

SNYDER J.P. 1987 *Map Projections – A Working Manual*, Professional paper 1395, US Geological Survey, US Government Printing Office, Washington, 384 p.

STARES P.B. 1985 *The Militarization of Space: US Policy 1945–1984*, Cornell University Press, Ithaca, 334 p.

STARES P.B. 1987 *Space and National Security*, The Brookings Institution, Washington, 219 p.

UNIDIR 1987 *Disarmament: Problems related to Outer Space*, United Nations, New York, 190 p.

VAN FENEMA H.P. 1999 *The International Trade in Launch Services*, Leiden University, Leiden, 473 p.

VON DER DUNK F.G. 1998 *Private Enterprise and Public Interest in the European "Spacescape"*, Leiden University, Leiden, 390 p.

WILHELMS E. 1987 *The Geological History of the Moon*, US Geological Survey, Professional Paper 1938, US Government Printing Office, Washington, 302 p. + 24 plates.

WILSON A. (editor) 1996 *Jane's Space Directory, 1996–97*, Jane's Information Group Ltd, Coulsdon, 524 p.

YU Q.Y. 1999 *The Implementation of China's Science and Technology Policy*, Westport, Conn., 233 p.

ZHANG Jun *et al.* 1984 *Dangdai Zhongguo Hangtian* [*Space policy in Contemporary China*], [Publisher of Chinese social sciences], Peking, 598 p.

Periodicals and collections

American Astronautical Society Publication.

Annales de droit aérien et spatial.

Aerospace America.

Aerospace Daily.

Air and Space Europe.

Air and Space Law.

Air et Cosmos.

Aviation Week and Space Technology.

Ciel et Espace.

CNES Magazine.

Cospar Information Bulletin.

Earth Sensing Report.

Earth Space Review.

ESA Bulletin.

ESA Journal.

ESA Earth observation.

Flight International.

Jane's Intelligence Review.

Journal canadien de télédétection.

Journal of the Astronautical Sciences.

Journal of Space Law.

Journal of the British Interplanetary Society.

Lettre d'Eurisy (La).

NASA history series.

Nouvelles de l'U.I.T.

Spot Magazine.

Photo-interprétation, Images aériennes et spatiales.

Pour la Science.

Proceedings of the Colloquium on the Law of Outer Space.

Recherche (La).

Revue aérospatiale.

Russian Space Bulletin.

Satellite News.

Science et Avenir.

Space and Missile Defense Report.

Space Calendar.

Space News.

Space Policy.

Via Satellite.

Zeitschrift für Luft- und Weltraum-recht, German Journal of Air and Space Law.

Internet sites

National bodies

Argentina
CONAE-Comisión Nacional de Actividades Espaciales
http://www.conae.gov.ar/
INVAP
http://invap.bariloche.com.ar/space/index-e.html

Australia
CSIRO- Commonwealth Scientific and Industrial Research Organisation
http://www.csiro.au/
Links to Australian space resources
http://www.gbnet.net/orgs/seds/links/links-australia.html

Austria
Austrian Aerospace
http://www.space.at/
ASA-Austrian Space Agency
http://www.asaspace.at/
Austrian Remote Sensing Data Centre
http://ofd.ac.at/ofd/hmofd_er.html

Bangladesh
Bangladesh Space Research and Remote Sensing Organization
http://www.sparrso.org/main.htm

Belgium
Belgian Federal Office for Scientific Technical and Cultural Affairs
http://www.belspo.be/
Belgian Space Agency
http://www.belspo.be/

Brazil
INPE-Instituto Nacional de Pesquisas Espaciais
http://www.inpe.br/

Bulgaria
Bulgarian Aerospace Agency
http://members.tripod.com/basa_resac/webs/Home.htm

Canada
Canadian Aeronautics and Space Institute
http://www.casi.ca
Canadian Space Agency
http://www.space.gc.ca/
Canada Centre for Remote Sensing
http://www.ccrs.nrcan.gc.ca/ccrs/homepg.pl?f
MacDonald Dettwiler
http://www.mda.ca/index.shtml
Telesat Canada
http://www.telesat.ca/intro.htm

China
China Academy of Launch Technology
http://www.calt.com/english.htm

China National Space Administration
http://www.cnsa.gov.cn/
COSTIND State Commission of Science, Technology, and Industry for
    National Defense
http://www.costind.gov.cn
Great Wall Company
http://www.cgwic.com.cn/
Chinese space information
http://www.spacechina.com/EnglishPage.htm
Sinosat satellite communication company
http://www.sinosatcom.com/indexe.htm

Costa Rica
Fundacion para la Ciencia y la Educacion Espacial
http://www.crc.co.cr/ciencia/fuces/fucese.html

Denmark
DSRI-Danish Space Research Institute
http://www.dsri.dk/
Politique spatiale
http://www.fsk.dk/fsk/publ/1997/space/

Egypt
National Authority for Remote Sensing and Space Sciences
http://www.narss.org
Nilesat
http://www.nilesat.com.eg

Europe
Alcatel
http://www.alcatel.com/
Arianespace
http://www.arianespace.com/index1.htm
Astrium
http://www.astrium-space.com/
EADS
http://www.eads.net/
Earth Observation Data Policy and Europe (EOPOLE)
http://www.geog.ucl.ac.uk/eopole
Eurospace
http://eurospace.org
Eurimage
http://www.eurimage.com
European Geophysical Society (EGS)
http://www.copernicus.org/EGS/EGS.html
ESA-European Space Agency
http://www.esrin.esa.it/
Eumetsat, European Organisation for the Exploitation of Meteorological
    Satellites
http://www.eumetsat.de/en
Galileo
www.galileo-pgm.org/
Global Monitoring for Environment and Security
http://gmes.jrc.it
Joint Research Center: Space coordination group

http://www.europa.eu.int/comm/jrc/space
European Space Policy
http://europa.eu.int/comm/space/index_en.html
SES-Astra
http://www.astra.lu/
European Union site devoted to the information society
http://www.ispo.cec.be/
Space Applications Institute of the JRC
http://www.sai.jrc.it/
Thales
http://www.thalesgroup.com/

Finland
Laboratory of Space Technology
http://avasun.hut.fi/
National Land Survey of Finland
http://www.novosat.com/

France
CNES-Centre National d'Études Spatiales
http://www.cnes.fr/
SPOT Image
http://www.spotimage.fr/

Germany
DLR-Deutsche Forschungsanstalt für Luft- und Raumfahrt
http://www.dlr.de/
University of Cologne
http://www.uni-koeln.de/jur-fak/instluft/index-e.html

Greece
Institute for Space Applications & Remote Sensing
http://www.space.noa.gr/mainframeset.htm

Hungary
Hungarian Space Organization
http://sas2.elte.hu/eng4.html

India
Space agency
http://www.isro.org/
Agrani Satellite Services Limited, ASSL
http://www.agrani.com
Antrix Corp.
http://www.isro.org/commercial.htm
National Remote Sensing Agency
http://www.stph.net/nrsa/

Indonesia
Indonesian Aerospace
http://www.indonesian-aerospace.com/
Space agency
http://www.lapan.go.id/
Indonesian Center for Air and Space Law
http://www2.elga.net.id/~webleged/icasl.htm

Israel
ELOP
http://www.el-op.co.il/new/spce-pays.html
Israel Space Agency
http://www.iami.org.il/isa/isahome.html
IAI-Israel Aircraft Industries
http://www.iai.co.il/dows/dows/serve/level/English/1.1.html

Imagesat
http://www.westindianspace.com/satellites.html
http://www.imagesatintl.com

Italy
ASI-Agenzia Spaziale Italiana
http://www.asi.it/
Italian Aerospace Research Centre
http://www.cira.it/
Alenia
http://www.aleniaspazio.com/

Japan
ISAS Institute of Space and Astronautical Science
http://www.isas.ac.jp/index-eS.html
Information on national sites
http://www.eoc.nasda.go.jp/links/links_e.html
Ishikawajima-Harima Heavy Ind. Co. Ltd.
http://www.ihi.co.jp/index-e.html
JASPEC Japan Marine Science and Technology Center
http://www.jamstec.go.jp/
Mitsubishi Heavy Industries Ltd
http://www.mhi.co.jp/indexe.html
National Aerospace Laboratory of Japan
http://www.nal.go.jp/Welcome-e.html
NASDA-National Space Development Agency
http://www.nasda.go.jp/
NEC
http://www.nec.com
Toshiba
http://www.toshiba.co.jp/worldwide/
Toyota
http://global.toyota.com/

Jordan
Jordanian Astronomical Society
http://www.jas.org.jo/index.html#islam

Libya
Libyan Center for Remote Sensing and Space Science
http://www.lcrsss.org/

Malaysia
MACRES-Malaysian Centre for Remote Sensing
http://www.macres.gov.my/
Malaysian Communications and Multimedia Commission
http://www.cmc.gov.my
Malaysia East Asia Satellite
http://www.measat.com.my
Space Science Studies Division
http://www.baksa.gov.my/

Morocco
Centre Royal de Télédétection Spatiale
http://www.crts.gov.ma

Netherlands
DESC-Dutch Experiment Support Center
http://www.desc.med.vu.nl
International Institute of Air and Space Law, Leiden University
http://ruljis.leidenuniv.nl/group/jflr/www/
NLR-National Aerospace Laboratory
http://www.nlr.nl/

NIVR-Netherlands Agency for Aerospace Programs
http://www.nivr.nl/index.htm
SRON-Space Research Organization Netherlands
http://www.sron.nl/

Nigeria
National space research agency
http://www.nasrda.gov.ng/

Norway
Norwegian Space Center
http://www.spacecentre.no/

Pakistan
SUPARCO-Space and Upper Atmosphere Research Commission
http://www.suparco.gov.pk/

Peru
Comision Nacional de Investigacion y Desarollo Aeroespacial (CONIDA)
http://www.conida.gob.pe/

Poland
Space Research Centre
http://www.cbk.waw.pl/

Romania
Romanian space agency
http://www.rosa.ro/english/space_program/ebspacesci.htm

Russia
Space agency
http://www.rosaviakosmos.ru/english/eindex.htm
Energia
http://www.energia.ru/english/index.html
IKI-Institute of Space Research
http://arc.iki.rssi.ru/Welcome.html
Glonass
http://mx.iki.rssi.ru/SFCSIC/english.html
Khrunichev State Space Center
http://www.intertec.co.at/itc2/partners/KHRUNICHEV/Default.htm
Russian aerospace guide
http://home.attbi.com/~rusaerog/space_centers.html
Russia/FSU Space-related on-line resources
http://www.friends-partners.org/~jgreen/rusres.html

Saudi Arabia
Saudi Center for Remote Sensing
http://scrscbs.kacst.edu.sa/
Space Research Institute
http://www.kasct.edu.sa/en/institutes/sri/index.asp

Singapore
Centre for Remote Imaging, Sensing and Processing
http://www.crisp.nus.edu.sg/
Singapore Telecom Satellite System
http://www.mlesat.com/st_1

South Korea
Korea Aerospace Industries
http://www.koreaaero.com/index_main_eng.html/
Korean Aerospace Institute
http://www.kari.re.kr/index.html
SaTReC-Satellite Technology Research Center
http://satrec.kaist.ac.kr/english/SaTReC.html

Spain
Hispasat
http://www.hispasat.es/
INTA-Instituto Nacional de Técnica Aeroespacial
http://www.inta.es/
INSA-Ingenieria Y Servicios Aerospaciales
http://www.insa.es/
CDTI - Centro para el Desarrollo Tecnológico Industrial
http://www.cdti.es/webCDTI/esp/index.html

Sweden
Saab Ericsson Space
http://www.space.se/
SSC-Swedish Space Corporation
http://www.ssc.se/

Switzerland
Swiss space agency
http://www.sso.admin.ch/

Taiwan
Center for Space and Remote Sensing Research
http://www.csrsr.ncu.edu.tw/english.ver/index.html
National Space Program Office
http://www.nspo.gov.tw/e50/home/

Thailand
Thailand Remote Sensing Center
http://www.nrct.go.th/remoteSensing/intro.html
Thai-Paht Satellite Research Center
http://www.mut.ac.th/~wwwtmsat/

Tunisia
Centre National de Télédétection
http://www.cnt.nat.tn

United Arab Emirates
Etisalat
http://www.etisalat.co.ae/

United Kingdom
British National Space Centre
http://www.bnsc.gov.uk/
Lloyd's satellite constellations
http://www.ee.surrey.ac.uk/Personal/L.Wood/constellations/overview.html
NRSC-National Remote Sensing Centre
http://www.nrsc.co.uk/
Surrey Space Centre
http://www.ee.surrey.ac.uk/SSC/
UK space industry web site
http://www.ukspace.com

United States of America
American Institute of Aeronautics and Astronautics
http://www.aiaa.org/
American Satellite Industry Association
http://www.sia.org/
Astrolink
http://www.astrolink.com/
Boeing
http://www.boeing.com/flash.html
Congressional Research Service
http://cnie.org/nle/crsreports/Science/

Earthwatch
http://www.digitalglobe.com/
Federal Communication Commission
http://www.fcc.gov/ib/
Global Positioning System Overview (GPS)
http://www.colorado.edu/geography/gcraft/notes/gps/gps_f.html
Globalstar : low orbit global constellation
http://www.globalstar.com/
Hughes Electronics Corp.
http://www.hughes.com/home/default.xml
Hughes Communications, Inc.
http://www.hughescommunications.com/
Landsat Program
http://geo.arc.nasa.gov/sge/landsat/landsat.html
Lockeed Martin
http://www.lockheedmartin.com/
Loral
http://www.loral.com/
Loral Skynet
http://www.loralskynet.com/
NASA-National Aeronautics and Space Administration
http://www.nasa.gov/
NASA's Earth Observing System
http://eospso.gsfc.nasa.gov/
NASA-human spaceflight
http://www.spaceflight.nasa.gov/station
NASA-Jet Propulsion Laboratory
http://www.jpl.nasa.gov/
NIMA-National Imagery and Mapping Agency
http://www.nima.mil/
NOAA-National Oceanographic and Atmospheric Administration
http://www.noaa.gov/
NOAA Satellite Active Archive
http://www.saa.noaa.gov/
National Reconnaissance Office
http://www.nro.gov/
National Security Space Architect
http://www.acq.osd.mil/nssa/
Naval Earth Map Observer (NEMO) remote sensing program
http://nemo.nrl.navy.mil/public/index.html
NORAD-North American Aerospace Defense Command
http://www.peterson.af.mil/norad/index.htm
Northrop Grumman
http://www.northgrum.com/
ORBIMAGE-Orbital Imaging Corporation
http://www.orbimage.com/
Orbital
http://www.orbital.com/
Raytheon
http://www.raytheon.com/
Skybridge
http://www.skybridgesatellite.com/
Teledesic
http://www.teledesic.com/
TERRA - The EOS Flagship
http://eos-am.gsfc.nasa.gov/
TRW
http://www.trw.com/
US Army Space Command
http://www.armyspace.army.mil/index.asp
U.S. Naval Space Command
http://www.navspace.navy.mil/

US Space Command
http://www.spacecom.af.mil/usspace/index.htm
USGS Global Land Information System
http://edcwww.cr.usgs.gov/webglis

Regional and international organisations

Arabsat
http://www.arabsat.com/
Asia Cellular Satellite (Aces) International (Garuda)
http:///www.aces.co.id
Asia Pacific Space Center
http:///www.apsc2orbit.com
Asia Satellite Telecommunications Company Limited
http:///www.asiasat.com
CEO Centre for Earth Observation
http://www.ceo.org/
COSPAR-Committee on space research
http://www.cosparhq.org
Eutelsat
http://www.eutelsat.org
Global Aerospace Database Industry
http://x-cd.com/
IAA-International Academy of Astronautics
http://www.iaanet.org
IAF-International Astronautical Federation
http://www.iafastro.com
IAU-International Astronomical Union
http://www.iau.org/
ICO
http://www.ico.com/
IGBP-International Geosphere–Biosphere Programme
http://www.igbp.kva.se
INMARSAT
http://www.inmarsat.org/
INTELSAT
http://www.intelsat.int
Intersputnik: International Organization of Space Communications
http://www.intersputnik.com/
ISU- International Space University
http://www.isunet.edu/
IUGG-International Union of Geodesy and Geophysics
http://www.iugg.org/
IUGS-International Union of Geological Sciences
http://www.iugs.org/
OOSA-Office for Outer Space Affairs (ONU)
http://www.oosa.unvienna.org/OOSA/oosa.html
Palapa
http://www.satelindo.co.id/
Panamsat Corporation
http://www.panamsat.com/
Regional African Satellite Communication Organisation
http://www.rascom.org
SES-Global (Astra, GE Americom)
http://www.ses-global.com
Sea Launch
http://www.sea-launch.com/
Shin Satellite Public Company Limited (Thaicom)
http:///www.thaicom.net
Starsem
http://www.starsem.com
Thuraya
http://www.thuraya.com/

UIT-Union Internationale des Télécommunications
http://www.itu.ch/
Worldspace- Global geostationary system
http://www.worldspace.com/

General information

Ad Astra
http://www.nss.org/adastra/
Andrews Technical Service
http://www.spaceandtech.com/
Archimedes Institute: Space Law and Policy Library
http://www.permanent.com/archimedes/LawLibrary.html
Asia Pacific Aerospace Consultant
http://www.apac.com.au/
Asian Technology Information Program (ATIP)
http://www.atip.or.jp/
Astronomy and science links
http://www.sciencepresse.qc.ca/cyber-express/520d.html
Astronomy links
http://www.bdl.fr/webastro.html
Astrosurf
http://www.astrosurf.com/spacenews/entree.html
CHAART - Current and Future Sensor Systems
http://geo.arc.nasa.gov/sge/health/sensor/cfsensor.html
Encyclopedia astronautica
http://www.astronautix.com/
Etats-Unis Espace, French space agency (CNES) in Washington
http://www.france-science.org/usa-espace
Federation of American Scientists (Space Policy Project)
http://www.fas.org/spp
Fondation pour la Recherche Stratégique
http://www.frstrategie.org/
Galileo Magazine
http://www.galileosworld.com/galileosworld/
Global Aerospace Database
http://www.x-cd.com/xcdnew.htm
Global Security
http://www.globalsecurity.org
Go Taikonauts : an unofficial Chinese website
http://www.geocities.com/CapeCanaveral/Launchpad/1921/
General information
http://www.floridatoday.com
Legal and general information
http://www.spacebizstation.com/
Information concerning Asia
http://www.satnewsasia.com/asiamain.htm
Information concerning the Arab world
http://www.arabicnews.com
Information concerning space transportation
http://www.orbireport.com/
Institute of Air and Space Law, McGill University
http://www.iasl.mcgill.ca/spacelaw.htm
Institute for cooperation in space
http://www.peaceinspace.com
Jonathan's Space Home Page

http://hea-www.harvard.edu/QEDT/jcm/space/space.html
Johnson Space Center
http://www.jsc.nasa.gov
Journalist's guide to remote sensing resources on the Internet
http://www.american.edu/academic.depts/soc/radiowave/recommen.htm
Laboratoire Communication et Politique / French research centre on space policy
http://www.lcp.cnrs.fr
Launch sites
http://www.orbireport.com/Linx/Sites.html
Lists of telecommunications satellites launched
http://www.tbs-satellite.com/tse/online/mis_telecom_geo.html
http://www.lyngsat.com/launches.shtml
News about satellites
http://www.latrobe.edu.au/www/crcss/news.html
Orbitography
http://www.hal-pc.org/~sattrack/keplinks.htm
http://celestrak.com/NORAD/elements/tle-new.txt
Remote sensing and environmental treaties
http://sedac.ciesin.columbia.edu/rs-treaties/pubs.html
Remote sensing links
http://www.geo.tudelft.nl/frs/rs2/rs.html
Remote sensing satellite images and data sets
http://www.itc.nl/~bakker/satellite.htmlSpace industry news from around the world
http://www.spacenewsfeed.co.uk
Satellite communications
http://satellite.about.com/industry/satellite/mbody.htm
Satellite imagery available through the Internet
http://umbc7.umbc.edu/~tbenja1/freedata.html
Satellite news in brief
http://www.sat-nd.com/
Space Daily
http://www.spacedaily.com/
Spaceflight Now
http://www.spaceflightnow.com/
Space News
http://www.space.com/
Space Policy Institute, Georges Washington University
http://www.gwu.edu/~spi/
Space Today
http://www.spacetoday.com
Space Resources
http://www.globalsecurity.org/space/index.html
The Satellite Encyclopedia
http://www.tbs-satellite.com/tse/
Universe Today
http://www.universetoday.com

The publisher has used its best endeavours to ensure that the URLs for external websites referred to in this book are correct and active at the time of going to press. However, the publisher has no responsibility for the websites and can make no guarantee that a site will remain live or that the content is or will remain appropriate.

Index

Bold face entries refer to figures and figure captions.

HANDBOOK OF
APPLIED MATHEMATICS

HANDBOOK OF
APPLIED MATHEMATICS

FOURTH EDITION

Edited by
EDWARD E. GRAZDA, B.E.E.
Editor, Electronic Design

MORRIS BRENNER, B.E.E., M.Ad.E., Assoc. Editor
Licensed Professional Engineer
Burndy Engineering Co., Inc., Norwalk, Conn.

WILLIAM R. MINRATH, A.B., Ch.E.
Vice President, D. Van Nostrand Co., Inc.

BASED ON THE ORIGINAL WORK BY
MARTIN E. JANSSON
HERBERT D. HARPER   PETER L. AGNEW

 VAN NOSTRAND REINHOLD COMPANY
New York   Cincinnati   Toronto   London   Melbourne

VAN NOSTRAND REINHOLD COMPANY REGIONAL OFFICES:
New York   Cincinnati   Chicago   Millbrae   Dallas

VAN NOSTRAND REINHOLD COMPANY INTERNATIONAL OFFICES:
London   Toronto   Melbourne

Published by VAN NOSTRAND REINHOLD COMPANY
450 West 33rd Street, New York, N.Y. 10001

Published simultaneously in Canada by VAN NOSTRAND REINHOLD LTD.

15 14 13 12 11 10 9 8 7 6 5 4

Preface to Fourth Edition

This book has been prepared to demonstrate how readily mathematics lends itself to the solution of practical problems. While it does not illustrate every type of problem, it seeks to develop logical reasoning which, if properly cultivated, will enable the reader to analyze his own problems and arrive at their solution by the most direct method. It is also a reference book with a wealth of specific information on many subjects. Whether the book is used for reference or as a text for self-instruction, the reader is urged to read the Introduction with some care for it contains the key to the handling of mathematical problems.

Every effort has been made to retain the working usefulness of the original book in this fourth edition. At the same time a vast amount of new material has been added to meet the present day needs of workers in many fields. The new developments include plate glass in home construction, new methods of building-block construction, new insulating materials, complete information on transistors, television in black-and-white and color, and stereophonic music systems.

Most of the other chapters have been extensively revised and brought up to date. New tables and detailed estimating data have been added on materials such as plastic tile, plywood, rubber-base paints, glass brick, copper tubing, modern heating systems, and electrical wiring. Several topics in the section on Business Mathematics including insurance, foreign exchange, small loans, stock and bond transactions, and loans were revised in accordance with present day rates and practices. A feature

which should greatly assist the user of this handbook is a separate
index of all the tables.

Many manufacturers, technical societies, trade associations,
and the University of Illinois Small Homes Council have provided a great deal of valuable reference material. Special mention is made of Mr. William M. Milazzo of the Triangle Conduit
and Cable Co., the American Plywood Assn., the Flat Glass
Jobbers Assn., Mr. J. F. McCullough of the Union Carbide Corp.,
the Cast Iron Soil Pipe Inst., the Libbey-Owens-Ford Glass Co.,
and Mr. Fred Apostolos of the Herbert L. Jamison Co.

W. R. M.

Princeton, New Jersey
September, 1966

Contents

I

Introduction

HOW TO USE THIS BOOK

Mathematics underlies virtually all phases of the complex age we live in. From the intricate calculations involved in designing an atomic power plant to the everyday additions and subtractions performed on the job, in a small business, or in figuring out a home budget, mathematics plays a vital role. A working knowledge of this important science can be a valuable aid in job advancement, increasing business profits, and in simplifying "do-it-yourself" tasks around the home.

The purpose of this book is to promote a better understanding of this rewarding subject and to show how it is actually applied in solving practical problems in many fields. The first six chapters are devoted to a review of the operations of arithmetic, algebra, geometry, trigonometry, and differential calculus. Detailed descriptions of the principles of these branches of mathematics have been left to more specialized books, many of which are cited as references. These chapters lay the foundation for a sound understanding of the remainder of the book. The reader who has not been a constant user of mathematics will do well to read these chapters carefully, since they contain the key to the applications that follow.

The remainder of the book is divided into sections each covering some special trade or art, such as carpentry, electricity, machine-shop work, etc. Each of these sections has been made to cover the subject with a logical development from the elementary to the more complex. If then, for example, an electrical worker wishes to study the applications of mathematics to his entire

1

field of work, he can do no better than to start at the beginning of the electrical section and follow it through to its conclusion. On the other hand, the man who is interested in the solution of a specific problem will find the index the best guide to the section dealing with this or a similar problem. Liberal use of the index is recommended because some subjects are covered in widely separated parts; for example, roofing occurs in both the carpentry and sheet metal sections.

Any man who operates an automobile or a machine of any kind, knows that it is to his advantage to "know what it can do" when operating under various conditions. Similarly, this book will increase in reference value to the man who knows what information it contains and where it is located. No man can carry a great amount of statistical information in his head, and, more important than knowing facts, is knowing where to find facts quickly.

The first step in solving a problem by any method is to picture the problem in its entirety and determine in what terms the final result is desired. This is particularly true in mathematics. The final result must be kept constantly in mind or energy may be needlessly expended in arriving at unnecessary partial results.

Having determined what is wanted in the way of a solution, the next step is to examine the data from which the problem is to be solved. Perhaps this is not sufficiently complete. Then it must be supplemented by information contained in this book or obtained from some other source, or the solution must proceed based on assumptions.

Simple problems may be solved most conveniently by setting them up as one expression. As an illustration, consider the problem of finding the distance which a train will travel in $2\frac{1}{4}$ hours when running at an average speed of 36 miles per hour. If we let S represent the speed, t the time, and D the distance, then

$$D = S \times t$$

The problem is now completely set up and all that is required to find the answer is to substitute the correct values and perform the indicated operations. "Correct values" implies proper units. We may then substitute as follows:

$$D = S \times t = 36 \times 2\tfrac{1}{4} = 81 \text{ miles.}$$

This example illustrates another very important principle in the use of mathematics. That is, if a problem is set up as above, how will one know in what units the answer will result? This is simple, because the unit designations may be cancelled, raised to powers or have their roots extracted in a manner similar to the operations performed on numbers. Thus, if we are finding the length of a surface whose area is 136 square inches and whose breadth is 8 inches, we may write, Length $= \dfrac{136}{8} = 17$. To find the units of the answer we may set up the units as we did the numerical problem. Thus we have, Length $= \dfrac{\text{in.}^2}{\text{in.}}$ or $\dfrac{\text{in.} \times \text{in.}}{\text{in.}}$. Cancelling "in." in both the numerator and the denominator, the answer is in inches.

In these considerations the word "per" has the same significance as the bar or line of a decimal. In fact, "miles per hour" may be written, "miles/hour." Then, in the previous illustration, when $D = S \times t$, we may write $D = \dfrac{\text{miles}}{\text{hour}} \times \text{hour}$. The hours cancel and the answer is in miles.

More involved problems and particularly those requiring the addition of many parts, can best be solved by attacking them step by step. Thus, in estimating the quantities of material required for a building construction job, it is necessary to compute the separate quantities required for the various parts of the building and then find the sum of these quantities for the final result.

Once a problem has been set up and the steps and operations determined, the processes of multiplication and division and the finding of powers, roots and reciprocals of numbers may proceed by any one of several methods. They may be performed by arithmetic, by algebra, by the use of tables, by logarithms, by the logarithmic slide rule, or by a computing machine.

This book does not attempt to dictate which method should be used, but generally shows the problem set up for arithmetical solution with the understanding that the reader will select the method which he can handle most readily and which is most suitable for his particular problem. Arithmetical solution is the longest process, except for simple calculations, and the practical man will do well to acquaint himself with other shorter methods and the types of problems to which they are most applicable.

Logarithms may be used most effectively when a problem calls for the multiplication or division or the handling of roots and powers of several factors. Thus, the operations indicated by

$$\frac{(25.136)^2 \times 728 \times 1728 \times 0.005679}{33,485 \times 36}$$ may be performed logarith-

mically with greater ease than by any other method. Logarithms are also particularly adapted to the extraction of roots. Thus,

solving $\sqrt[5]{\dfrac{838.75}{0.658}}$ is a very simple matter with this method. If,

however, the addition and subtraction of a number of terms is interposed in an expression also involving multiplication and division, the use of logarithms may *not* be a time-saver. For example, the operations indicated by

$$\frac{0.125 \times 367 + 36.25 \times 450.3 + 0.825 \times 380}{750 \times 45.38}$$

are a border case, since the finding of the anti-logarithms to perform the additions may consume more than the time saved by performing the multiplications by logarithms. The use of logarithms is recommended whenever the multiplication or division of trigonometric functions is involved. Thus, in solving

$$252.67 \times \cos 67° 36',$$

the logarithm of the cosine of 67° 36' may be found from the tables with no more effort than would be required in finding the cosine itself.

The ordinary slide rule is a convenient instrument for multiplying and dividing when accuracy to greater than three significant figures is not a matter of great concern. Calculations may be made very rapidly with a slide rule and it is of great value in making rough estimates and checking results. It is not to be assumed from these remarks that a slide rule is a crude instrument and inherently inaccurate. This is not true, but it is rather a case of the inability of the human eye to evaluate the relative lengths of short distances with greater accuracy which limits its usefulness.

Computing machines have come into considerable favor in many offices and where one is available it will pay a man to learn how to perform the various operations on it. One of their particular merits is that addition and subtraction may be performed on them as well as multiplication, division, raising to powers and extracting square root. When computing machines are used to compute quantities which must be checked by another person within very narrow limits of discrepancy, as when computing certain land measurements, it is necessary to record all of the figures which appear on the machine, no matter if this results in nine or ten decimal places; and also to decide on a convention as to whether the nearest even or the nearest odd number should be recorded when the last figure of an eliminated decimal is 5.

As with the use of many other tools, the application of a liberal amount of common sense is necessary with the use of mathematics. Thus, it would be foolish to compute to the nearest cubic foot the quantity of sand required for a job, when sand is sold by five-ton truckloads. Also it would be wasted effort to measure farm land worth $50.00 an acre with the same care as would be used in measuring city property worth thousands of dollars an acre.

The illustrations of the last paragraph indicate that there is an economic reason for the use of mathematics. Such is the case. Correct application of mathematics leads to accuracy, accuracy results in less waste and fewer rejections, hence a higher return for work done.

There is another field of applied mathematics which is not governed by economic considerations. That is the calculation of the strength, proportions, or security of machinery or structures on which the safety of life and property depend. Here not only must the most care be exercised, but computations must be checked by responsible persons and should then be preserved in legible form. In the event of disaster, a court of inquiry to fix responsibility will ask, "Was accepted practice followed; and was due diligence exercised in arriving at results?" Accurate and well-preserved computations may be a big aid in establishing affirmative answers to these questions.

One of the points in dealing with figures at which common sense comes into greatest play is in the evaluation of the true accuracy of figures. "Figures do not lie," is a common expression but not always a true one.

Let us illustrate with an example. Suppose a piece of lumber is measured with a carpenter's rule and is found to be $3\frac{3}{8}$ inches wide and $1\frac{5}{8}$ inches thick. Changing these figures to decimals, as is common in performing computations, they become 3.375 inches and 1.625 inches, respectively. Now, if we want to obtain the cross-sectional area of this piece of lumber, we multiply the breadth of the board by the thickness and obtain $3.375 \times 1.625 = 5.484375$ square inches. Many of the figures of this decimal have no significance and the retention of the right number of figures requires the exercise of judgment and a knowledge of the accuracy with which the measurements were made and the purpose for which the figures are to be used. In the illustration we were told that the wood was measured with a carpenter's rule and presumably only to the nearest $\frac{1}{16}$ inch. Then, it will be entirely accurate to state that the cross-sectional area is $5\frac{1}{2}$ square inches.

To be precise the preceding problem would be written $3.375 \times 1.625 = 5.484375$, the number of square inches. However, in practical problems correct mathematical notation is usually disregarded for the sake of brevity. Throughout this work the answers are given in units which, theoretically, would not result from operations with abstract numbers.

The units of a dimension often indicate the degree of accuracy. Thus, if we are told without further qualification that a man is 5 feet 8 inches or 68 inches tall we know that he is between $67\frac{1}{2}$ inches and $68\frac{1}{2}$ inches tall. In other words, his height has been measured or estimated to the nearest one-half inch and we have no right to assume a more exact measurement. However, if we are told that he is $68\frac{3}{16}$ inches tall we know that his height has been measured to the nearest $\frac{1}{32}$ inch and that the actual height is between $68\frac{5}{32}$ and $68\frac{7}{32}$ inches.

When dimensions are stated in decimals, the decimal is an index of its accuracy. Thus, if we are told that a bolt is 0.318 inch in diameter we can feel reasonably sure that the measurement is correct to the nearest half of a thousandth of an inch or to 0.0005 inch. This would indicate that the measurement had been made with a micrometer caliper. However, if the diameter is given as 0.325 inch, the 5 in the last place raises a question as to whether the measurement was actually made to thousandths or to half-hundredths or to quarter-tenths. As a matter of fact, vernier calipers would give such a measurement since they are usually graduated to spaces 0.025 inch long.

It is equally important that the final results of a problem be expressed in rational practical units. Thus, quantities of lumber should be given in board feet or thousand board feet, cement in barrels, sand and gravel in cubic yards, etc. This does not imply that it is not perfectly proper to deal with fractional quantities during the course of the solution of a problem. This is particularly true when arriving at *unit quantities*. For instance, we may state that the quantities of materials required for one cubic yard of concrete are: 0.61 bbl. cement, 7.32 cu. ft. sand, 10.93 cu. ft. stone. These are unit quantities, but after being multiplied by the number of cubic yards of concrete to be made, the quantity of cement should be given to the next nearest whole barrel and the quantities of aggregates to the next nearest whole cubic yards.

These brief remarks indicate that clear logical thinking is a necessary adjunct to the use of mathematics. It may be added that nothing stimulates such mental procedure more than does mathematics and hence its use will result in many indirect benefits.

II

Arithmetic

Definitions.—Arithmetic is the science and application of numbers. Numbers are said to be *concrete* when they apply to *things*, *objects*, or *quantities* (examples, 12 bolts, 8 bricks, and 25 watts) and *abstract* when they do not so apply (examples, 12, 8, and 25).

The four *fundamental operations* of mathematics are *addition*, *subtraction*, *multiplication* and *division*; all necessary in performing *calculations*. A *proposition* is a statement set forth either with or without *demonstration*. It may be (1) an *axiom*, or self-evident truth, without demonstration; (2) a *theorem*, or truth by demonstration; (3) a *problem*, or question for solution; (4) an *hypothesis*, or tentative or preliminary proposition.

Signs and Symbols Used in Arithmetic.—Mathematical operations are largely indicated by signs and symbols. Thus $+$ placed between two numbers means that they are to be added and $\times$ between two numbers means that they are to be multiplied by each other.

The common mathematical symbols of arithmetic together with illustrations of their use are as follows:

$=$ Equals, sign of equality, is equal to, as 100 cents $=$ 1 dollar

$+$ Plus, sign of addition, as $3 + 4 = 7$; positive, as $+ \frac{1}{2} = + 0.5$

$-$ Minus, sign of subtraction, as $4 - 1 = 3$; negative, as $-\frac{1}{2} = - 0.5$; contraction, as $\frac{1}{6} = 0.17 -$

$\pm$ Plus or minus, as $\sqrt{4} = \pm 2$

× Times, multiplication sign, multiplied by, as $3 \times 2 = 6$

÷ Divided by, division sign, as $8 \div 2 = 4$; also $\frac{8}{2} = 4$; and $8/2 = 4$

∴ Therefore, hence, as if $2 + 2 = 4 \therefore 4 - 2 = 2$

∵ Because

. Decimal point

: Is to, sign of division, in ratio as $3 : 6$

:: Formerly used in proportion for the equality sign as $2 : 3 ::$ $4 : 6$ (read " 2 is to 3 as 4 is to 6 "), which means $2 : 3 = 4 : 6$ or $\frac{2}{3} = \frac{4}{6}$

> Is greater than, as $4 > 3$; reads " 4 is greater than 3 "

< Is less than, as $3 < 4$; reads " 3 is less than 4 "

≅ Congruent sign, coincides with

∞ Infinity, as $\frac{3}{0} = \infty$

| Bar $\left.\begin{array}{l} \\ \\ \\ \\ \\ \end{array}\right\}$ These symbols denote that quantities covered or enclosed must be "taken together"

— Vinculum

( ) Parentheses

[ ] Brackets

{ } Braces

$$3 \times 4 + 3 = 3 \times 7 = 21$$
$$3 \times (4 + 3) = 3 \times 7 = 21$$
$$2[3 \times (4 + 3)] = 2[3 \times 7] = 42$$
$$4\{2[5(7 + 3) + 8] + 6$$
$$= 4\{2[50 + 8] + 6\}$$
$$= 4\{116 + 6\} = 488$$

$\sqrt{}$ Radical sign or square root, as $\sqrt{9} = 3$

$\sqrt[3]{}$ Cube root

$\sqrt[n]{}$ nth root

a^2 A squared or second power of a, as $a \times a$

a^3 A cubed or third power of a, as $a \times a \times a$

x^n nth power of a

$\dfrac{1}{n}$ Reciprocal value of n

π Pi = 3.1416 (more accurately 3.14159265359) = $\dfrac{\text{circumference}}{\text{diameter}}$

Notation and Numeration.—Notation is a system of representing numbers by symbols while numeration is a system of naming or reading numbers.

There are two methods of notation in use, (1) the Roman and (2) the Arabic. The Roman has little use, the Arabic being the notation commonly used.

Roman notation is a method of notation by letters,

I	V	X	L	C	D	M
1	5	10	50	100	500	1,000

Repeating a letter repeats its value, i.e., I = 1, II = 2, III = 3.

Placing a letter of less value before one of greater value diminishes the value of the greater by the lesser, i.e.,

$$IX = 9, \qquad XC = 90$$

Placing the lesser after the greater increases the value of the greater by that of the lesser, i.e.,

$$VIII = 8, \qquad XIV = 14, \qquad LXX = 70$$

Placing a vinculum or horizontal line over a letter increases its value one thousand times, i.e.:

$$\overline{V} = 5000, \qquad \overline{X} = 10,000, \qquad \overline{M} = 1,000,000$$

Arabic method of notation uses ten characters or figures, i.e.:

1	2	3	4	5	6	7	8	9	0
one	two	three	four	five	six	seven	eight	nine	zero

Numeration.—In the Arabic method of reading numbers, the value of numbers increases from left to right in a ten-fold ratio. The successive figures from right to left or from left to right are called orders of units, the value of any order being ten times the value of one of the order next to its right, and one-tenth the value of one of the order next to its left.

Hundreds of Billions	Tens of Billions	Billions	Hundreds of Millions	Tens of Millions	Millions	Hundreds of Thousands	Tens of Thousands	Thousands	Hundreds	Tens	Units
4	9	5	8	6	7	5	0	1	3	2	

To read an integral number expressed in figures, begin at the right and separate the figures by commas into periods of three figures each. Then begin at the left and read each period as if it stood alone, adding the name of each period except the name of the period of units.

ILLUSTRATION: Read the number 49,586,750,132.

Forty-nine billion, five hundred eighty-six million, seven hundred fifty thousand, one hundred thirty-two. Note: The names beyond billions are in order: trillions, quadrillions, quintillions, sextillions, etc.

Addition.—Addition is the process of finding the sum of two or more numbers. To add several numbers, place the numbers in a vertical column with units under units, tens under tens, hundreds under hundreds, etc. Then add the figures in the right-hand column (column of units) and place the sum under this column. If there be more than one figure in this sum write down only the right-hand one and " carry " the others to the next column to the left. Repeat until each column has been added.

```
 438
1273
  46
 391
2148  Ans.
121   carried
```

The accuracy of the addition may be checked by writing the sums of the columns as shown below and adding

$$
\begin{array}{lr}
\text{Sum of column (1)} & 18 \\
\text{Sum of column (2)} & 23 \\
\text{Sum of column (3)} & 9 \\
\text{Sum of column (4)} & 1 \\
\hline
\text{Sum} = & 2148
\end{array}
$$

Subtraction.—Subtraction is the process of finding the difference of two numbers by taking one number from another. Example: $15 - 7 = 8$. The *minuend* is the number from which the other is to be taken (15 is the minuend in the example). The *subtrahend* is the number which is to be taken from the minuend (7 is the subtrahend in the example). The *remainder* is the number which remains after the subtrahend has been taken from the minuend (8 is the remainder in the example).

In order to subtract two figures write the subtrahend under the minuend so that the units of one are under the units of the other, tens under tens, etc. Take the figure in the subtrahend from the corresponding figure in the minuend and write the remainder directly underneath as follows:

$$
\begin{array}{lr}
\text{Minuend:} & 56387 \\
\text{Subtrahend:} & -12265 \\
\hline
\text{Remainder:} & 44122
\end{array}
$$

If, however, the figure in the subtrahend is larger than the figure directly above it, it is necessary to borrow one unit from the next figure to the left. This is illustrated in the following operation:

$$
\begin{array}{lrcccc}
 & & 4 & & 8 & \\
\text{Minuend} & 4 & \not{5}\,^{1}3 & \not{9}\,^{1}6 \\
\text{Subtrahend} & -2 & 4 & 7 & 5 & 8 \\
\hline
\text{Remainder} & 2 & 0 & 6 & 3 & 8
\end{array}
$$

A subtraction may be checked by adding the subtrahend to the remainder. This sum should always equal the minuend. The following example illustrates this operation:

Minuend	6356
Subtrahend	−1728
Remainder	4628
Subtrahend	+1728
Minuend	6356 (Check)

Multiplication.—Multiplication is the process of taking or increasing one number a certain number of times and the result is called the *product*. The number which is multiplied or taken a certain number of times is called the *multiplicand* and the number by which it is multiplied is the *multiplier*. The multiplicand and the multiplier are known as *factors*.

In performing multiplication, the multiplier is written below the multiplicand, the units of one under the units of the other, tens under tens, etc. Each figure of the multiplicand, beginning at the right, is multiplied by each figure of the multiplier and the right-hand figure of each partial product is placed in turn directly under the figure used as a multiplier. Partial products are placed on different lines. The sum of the partial products will equal the required product.

Illustration:

1653	multiplicand
247	multiplier
11571	
6612	
3306	
408291	product

Division.—Division is the process of finding how many times one number is contained in another. The number to be divided is called the *dividend* and the number by which it is divided, the *divisor*. The result of the operation, or the number of times the divisor is contained in the dividend, is known as the *quotient*.

When the divisor contains but one figure, the method commonly used is known as *short division*. In performing this, place the divisor to the left of the dividend, separated by a line, and draw a line under the dividend. Divide the first or the first two figures of the dividend, as is necessary, by the divisor and place the quotient under the line. If the divisor does not go a whole number of times, the remainder is prefixed to the next figure in the dividend and the process is repeated.

Illustration: divide 21372 by 6

Solution: 6)21372
                  ‾‾‾‾‾‾‾
                  3562 quotient

When the divisor contains two or more figures, the method used is known as *long division*. This is performed as follows: Place the divisor at the left of the dividend, separated by a line, and place the quotient either above or to the right of the dividend. Divide the first group of figures which gives a number larger than the divisor by the divisor, place the first figure of the quotient above the dividend, multiply this figure by the divisor and place this product below the figures divided into and subtract. The remainder prefixed to the next figure brought down from the dividend forms the new trial dividend. Repeat until all figures of the dividend are brought down.

Illustration: divide 2841020 by 364

$$
\begin{array}{r}
7805 \quad \text{quotient} \\
364\overline{)2841020} \\
2548 \\
\overline{2930} \\
2912 \\
\overline{1820} \\
1820 \\
\overline{}
\end{array}
$$

It is very common in both short and long division that the

divisor will not go into the last trial dividend a whole number of times. It is then necessary to express the remainder as a fraction.

EXAMPLE: Divide 327 by 18

$$\begin{array}{r} 18\tfrac{1}{6} \text{ quotient} \\ 18\overline{)327} \\ 18 \\ \hline 147 \\ 144 \\ \hline \end{array}$$

SOLUTION:

$$\text{Remainder} = \frac{3}{18} = \frac{1}{6}$$

Fractions.—A fraction is a part of any object or unit. It consists of three essential elements, a number called a *denominator* which denotes the number of equal parts into which the object or unit is divided, a horizontal line above the denominator, called the *fraction line*, and a number above the line known as the *numerator* which denotes how many of the equal parts are to be taken. Thus in the fraction $\frac{3}{4}$, 3 is the numerator and 4 the denominator. This type of fraction is usually called a common fraction. To read a common fraction, read the numerator and then the denominator.

ILLUSTRATION: $\frac{1}{2}$, $\frac{3}{4}$, $\frac{7}{11}$ are read, one-half, three-fourths or three-quarters, seven-elevenths.

A *proper fraction* is one whose numerator is less than the denominator, as $\frac{2}{3}$.

An *improper fraction* is one whose numerator is greater than the denominator, as $\frac{5}{2}$.

A *mixed number* consists of a whole number and a fraction written together, as $2\frac{1}{2}$.

Reduction of Fractions.—A fraction may be reduced to its lowest form (without changing its value) by dividing both the numerator and the denominator by their *greatest common divisor*

(G.C.D.). Thus the G.C.D. of $\frac{12}{30}$ is 6 and if the numerator and denominator are both divided by this number, the fraction becomes, $\dfrac{12}{30} = \dfrac{12 \div 6}{30 \div 6} = \dfrac{2}{5}.$

A mixed number may be reduced to an improper fraction by multiplying the whole number by the denominator and adding the numerator to form a new numerator. Thus,

$$4\tfrac{1}{2} = \frac{(2 \times 4) + 1}{2} = \frac{9}{2}.$$

To change an improper fraction to a mixed number, divide the numerator by the denominator. The quotient is the whole number and the remainder is the new numerator. Thus, $\frac{177}{32} = 5\frac{17}{32}.$

Addition and Subtraction of Fractions.—To add fractions, the *least common multiple* (L.C.M.) of all denominators must first be determined to find a common denominator. The L.C.M. is found by multiplying the product of the prime factors (numbers divisible only by themselves and one) of the largest denominator by the product of the prime factors which occur in the other denominators but not in the largest. Thus the prime factors of the denominators of the fractions $\dfrac{3}{4}, \dfrac{5}{6},$ and $\dfrac{7}{12},$ are $\dfrac{3}{2 \times 2}, \dfrac{5}{3 \times 2},$ and $\dfrac{7}{2 \times 2 \times 3}.$ In this case $2 \times 2 \times 3$ contains all of the factors in the required number of times, so 12 is the least common denominator of these fractions.

The next step is to expand both terms of each fraction proportionately so that their denominators will be equal. Thus, $\dfrac{3 \times 3}{4 \times 3} = \dfrac{9}{12}, \dfrac{5 \times 2}{6 \times 2} = \dfrac{10}{12},$ and $\dfrac{7 \times 1}{12 \times 1} = \dfrac{7}{12}.$ Then all of the expanded numerators may be placed over the common denominator and the numerators added, thus,

$$\frac{9 + 10 + 7}{12} = \frac{26}{12} = 2\frac{2}{12} = 2\frac{1}{6}.$$

When mixed numbers are added they may be changed to

improper fractions and then the same procedure as above followed. Thus,

$$2\frac{1}{2} + 3\frac{3}{4} = \frac{5}{2} + \frac{15}{4} = \frac{5 \times 2}{2 \times 2} + \frac{15}{4} = \frac{10}{4} + \frac{15}{4} = \frac{10 + 15}{4} = \frac{25}{4} = 6\frac{1}{4}.$$

Fractions are subtracted by reducing to the smallest common denominator as for addition and then finding the difference of the new numerators. Thus, $\frac{15}{16} - \frac{3}{8} = \frac{15}{16} - \frac{6}{16} = \frac{9}{16}$; and $6\frac{1}{4} - 3\frac{7}{16} = \frac{25}{4} - \frac{55}{16} = \frac{100}{16} - \frac{55}{16} = \frac{45}{16} = 2\frac{13}{16}.$

Multiplication and Division of Fractions.—To multiply a fraction by a whole number, multiply the numerator by the whole number. The product will be the new numerator over the old denominator. Thus, $5 \times \frac{3}{4} = \frac{5 \times 3}{4} = \frac{15}{4} = 3\frac{3}{4}.$

To divide a fraction by a whole number, multiply the denominator of the fraction by the whole number. The quotient will be the old numerator over the new denominator. Thus,

$$\frac{1}{2} \div 5 = \frac{1}{2 \times 5} = \frac{1}{10}; \frac{7}{8} \div 3 = \frac{7}{8 \times 3} = \frac{7}{24}.$$

To multiply one fraction by another fraction, place the product of the numerators over the product of the denominators and reduce to required form. Thus,

$$\frac{5}{8} \times \frac{2}{3} = \frac{5 \times 2}{8 \times 3} = \frac{10}{24} = \frac{5}{12}; \frac{3}{16} \times \frac{1}{2} = \frac{3 \times 1}{16 \times 2} = \frac{3}{32}.$$

When mixed numbers are to be multiplied by fractions, it is advisable to change the mixed numbers to improper fractions before using the above procedure. Thus,

$$5\frac{3}{8} \times \frac{3}{4} = \frac{40 + 3}{8} \times \frac{3}{4} = \frac{43 \times 3}{8 \times 4} = \frac{129}{32} = 4\frac{1}{32}.$$

To divide a whole number or a fraction by a fraction, invert the divisor and multiply. Thus, $\frac{1}{8} \div \frac{1}{2} = \frac{1}{8} \times \frac{2}{1} = \frac{2}{8} = \frac{1}{4}$; $\frac{5}{9} \div \frac{2}{3} = \frac{5}{9} \times \frac{3}{2} = \frac{15}{18} = \frac{5}{6}$; $12\frac{1}{4} \div \frac{1}{6} = \frac{49}{4} \div \frac{1}{6} = \frac{49}{4} \times \frac{6}{1} = \frac{294}{4} = 73\frac{1}{2}.$

Cancellation.—In practical operations where the multiplication of various kinds of numbers, including fractions is expressed, the process may often be shortened by cancellation. This consists of taking out common factors above and below the fraction line before multiplying. As an example, take the expression $\dfrac{10 \times 4 \times 12}{25 \times 3 \times 8}$. This would require several operations to simplify without cancellation. However, it will be noted that 5 is a common factor in the 10 and the 25, and that 4 can be factored out of the 4 and the 8 and the 3 can be factored out of the 12. The operation is performed by striking out the numbers and writing above or below the remaining portion as follows:

$$\dfrac{\overset{2}{\cancel{10}} \times \overset{4}{\cancel{4}} \times \cancel{12}}{\underset{5}{\cancel{25}} \times \cancel{3} \times \underset{2}{\cancel{8}}} = \dfrac{4}{5}.$$

Cancellation can be made as long as factors remain which will cancel each other, but there is, however, a medium point where cancellation may sometimes cease for simplicity of operation.

Decimal Fractions.—A fraction which has for its denominator the number 10, 100, 1000, etc., may be expressed by writing only one number and using a period or a decimal point to indicate whether the fraction is tenths, hundredths, etc. Thus, .1 is $\frac{1}{10}$, .01 is $\frac{1}{100}$, .17 is $\frac{17}{100}$, .125 is $\frac{125}{1000}$, etc. These are called decimal fractions or simply decimals. When written alone, 0 is usually placed to the left of the decimal point, 0.125.

To read a decimal expressed in figures, read the decimal as if a whole number, and add the fractional name of the lowest place. For example, 6.18 is read 6 and 18 hundredths; 6.0018, 6 and 18 ten-thousandths.

Changing Common Fractions into Decimals.—Since the fraction line indicates division it is easy to see that a fraction can be reduced to a decimal simply by performing the indicated operation and dividing the numerator by the denominator and writing the quotient in decimal form. Thus, $\dfrac{3}{8} = \dfrac{8)3.000}{0.375} = 0.375.$

In this example the quotient came out exactly and the decimal is the exact equivalent of the fraction. Some decimals will not come out exactly and the division should then be carried out only as far as the nature of the work requires. Decimals are seldom carried out to more than five places. When the value of a decimal correct to the nearest tenth, hundredth, thousandth, etc., is required, 1 is added to the last required figure if the next figure is

TABLE 1

DECIMAL EQUIVALENTS OF COMMON FRACTIONS

Fraction				Decimal	Fraction				Decimal
			$\frac{1}{64}$	0.015625				$\frac{33}{64}$	0.515625
		$\frac{1}{32}$		.03125			$\frac{17}{32}$		.53125
			$\frac{3}{64}$	.046875				$\frac{35}{64}$	.546875
	$\frac{1}{16}$			.0625		$\frac{9}{16}$			.5625
			$\frac{5}{64}$	.078125				$\frac{37}{64}$	.578125
		$\frac{3}{32}$		.09375			$\frac{19}{32}$		.59375
			$\frac{7}{64}$	.109375				$\frac{39}{64}$	.609375
$\frac{1}{8}$				.125	$\frac{5}{8}$				.625
			$\frac{9}{64}$	.140625				$\frac{41}{64}$	.640625
		$\frac{5}{32}$		.15625			$\frac{21}{32}$		.65625
			$\frac{11}{64}$	.171875				$\frac{43}{64}$	.671875
	$\frac{3}{16}$			.1875		$\frac{11}{16}$			.6875
			$\frac{13}{64}$	.203125				$\frac{45}{64}$	.703125
		$\frac{7}{32}$		.21875			$\frac{23}{32}$		.71875
			$\frac{15}{64}$	.234375				$\frac{47}{64}$	.734375
$\frac{1}{4}$				.25	$\frac{3}{4}$				.75
			$\frac{17}{64}$	.265625				$\frac{49}{64}$	.765625
		$\frac{9}{32}$		.28125			$\frac{25}{32}$		.78125
			$\frac{19}{64}$	.296875				$\frac{51}{64}$	.796875
	$\frac{5}{16}$			.3125		$\frac{13}{16}$			.8125
			$\frac{21}{64}$	.328125				$\frac{53}{64}$	.828125
		$\frac{11}{32}$		.34375			$\frac{27}{32}$		.84375
			$\frac{23}{64}$	.359375				$\frac{55}{64}$	.859375
$\frac{3}{8}$				.375	$\frac{7}{8}$				.875
			$\frac{25}{64}$	.390625				$\frac{57}{64}$	.890625
		$\frac{13}{32}$		.40625			$\frac{29}{32}$		.90625
			$\frac{27}{64}$	.421875				$\frac{59}{64}$	.921875
	$\frac{7}{16}$			.4375		$\frac{15}{16}$			.9375
			$\frac{29}{64}$	.453125				$\frac{61}{64}$	.953125
		$\frac{15}{32}$		.46875			$\frac{31}{32}$		.96875
			$\frac{31}{64}$	.484375				$\frac{63}{64}$	.984375
$\frac{1}{2}$				.5	1				1.

five or more.   Thus, 0.375 correct to the nearest tenth is 0.4;
correct to the nearest hundredth is 0.38.

Addition and Subtraction of Decimals.—In the addition and
subtraction of decimals, the numbers are written one above the
other in such a manner that the decimal points are always directly
in a vertical column.   The operations are then performed in the
ordinary manner, care being taken that the decimal point in the
sum or the remainder is also directly in line with those above.

Example of addition: 2.0625
        315.25
         0.0375
        —————
        317.3500

Zeros to the right of the last significant figure in a decimal may
be stricken out when they have no significance without changing
the value of the number.

Example of subtraction: 24.325
         5.7036
       —————
        18.6214

Multiplication of Decimals.—In multiplication of decimals,
the points are not required to fall under each other and the fractions are placed so that the right-hand figures of the multiplier
and multiplicand are in the same column as when dealing with
whole numbers.   The multiplication is then performed as with
whole numbers and the product has as many decimal places as the
multiplicand and the multiplier combined.   That is, if the multiplicand has three figures to the right of the decimal point and the
multiplier has two figures to the right of the decimal point, then the
product will have $3 + 2 = 5$ figures to the right of the decimal point.

Examples: 8.475   1.26
      2.25   0.0012
     —————   —————
     42375   252
     16950   126
     16950   —————————
    —————   0.001512 product
   19.06875 product

Decimals, or any other number, may be multiplied by 10 by simply moving the decimal point one place to the right; by 100 by moving the decimal point two places to the right, etc. Examples: $10 \times 46.75 = 467.5$; $1000 \times 0.0627 = 62.7$.

Division of Decimals.—To divide decimals, multiply or divide the divisor and the dividend by some power of 10 (10, 100, 1000, etc.) so as to make the divisor a whole number. Mark the new decimal point in the dividend by a caret ($\wedge$) and proceed as with the division of whole numbers, placing the decimal point of the quotient above or below the caret depending on whether long or short division is used. The quotient will then have as many decimal places as the new dividend.

Examples: Divide 43.28 by 400.

$$400)_{\wedge}43.28$$
$$\overline{\quad 0.1082 \quad} \text{ (Ans.)}$$

Divide 1728.5 by 1.356 to the nearest thousandth.

$$\frac{1274.705 \quad \text{(Ans.)}}{1_x356)1728_x500_\wedge000_\wedge}$$
$$\begin{array}{r} 1356 \\ \hline 3725 \\ 2712 \\ \hline 10130 \\ 9492 \\ \hline 6380 \\ 5424 \\ \hline 9560 \\ 9492 \\ \hline 6800 \\ 6780 \\ \hline \end{array}$$

Divide 43.28 by 0.004.

$$0.004)43.280_{\wedge}$$
$$\overline{\quad 10820. \quad} \text{ (Ans.)}$$

Changing Decimals to Common Fractions.—*Exact Decimals,* that is, a decimal whose denominator is contained in the numerator without a remainder. For the numerator of the fraction, use the significant figures of the decimal, the denominator being 1 with as many ciphers as there are decimal places in the decimal; reduce to lowest terms.

Examples: $0.75 = \frac{75}{100} = \frac{3}{4}$; $0.375 = \frac{375}{1000} = \frac{3}{8}$.

Table 1 will be found convenient for finding the equivalent fraction to many decimals.

Repeating Decimals.—A common fraction can be expressed exactly by a decimal if the denominator contains no other factors than 2 or 5; otherwise it cannot. For example, when the fraction $\frac{3}{11}$ is expressed as a decimal the quotient obtained by dividing 3 by 11 is 0.27272727, etc., however far it is carried.

A decimal that contains a constantly recurring figure or series of figures is called a repeating decimal. In the case given above, 0.27272727, etc. is a repeating decimal, the series of figures constantly recurring being 27. In writing a repeating decimal dots are usually placed over the first and last figures of the repetend, i.e., the figure or series of figures that constantly recurs. Thus, $0.272727\ldots$ would be written $0.\dot{2}\dot{7}$ and $0.333\ldots0.\dot{3}$.

To Reduce a Repeating Decimal to a Common Fraction.— Treat the *non-repeating* and the *first repeating* groups as a whole number; subtract from this the non-repeating group treated as a whole number; the difference will be the numerator of the fraction. The denominator will be composed of as many 9's as there are repeating figures in the group, followed by as many 0's as there are non-repeating figures. Reduce to lowest terms.

Example: Reduce $0.\dot{3}$ to a fraction

$$-\ 0$$
numer. $\dfrac{3}{9} = \dfrac{1}{3}$ (Ans.)
denom.

Example: Reduce $0.\dot{2}\dot{7}$ to a fraction

$$-\ 00$$
numer. $\dfrac{27}{99} = \dfrac{3}{11}$ (Ans.)
denom.

Example: Reduce 0.79054054

$$\begin{array}{r} -\quad 79 \\ \hline \end{array}$$

numer. $\dfrac{78975}{99900} = \dfrac{117}{148}$ (Ans.)
denom.

Compound or Denominate Numbers.—A quantity expressed in units of two or more denominations is called a compound quantity or a compound denominate number. Thus, $4\frac{1}{2}$ feet is a simple quantity; but its equivalent 4 feet 6 inches is a compound quantity.

The process of changing the denomination in which a quantity is expressed, without changing the value of the quantity, is called reduction.

Reduction Descending.—To reduce a compound number to a lower denomination, multiply the number by as many units of the lower denomination as makes one of the higher.

Examples: Reduce $4\frac{1}{2}$ feet to inches: $4\frac{1}{2} \times 12 = 54$ inches.

Reduce $3\frac{1}{4}$ pecks to quarts: $3\frac{1}{4} \times 8 = 26$ quarts.

When the given number is expressed in more than one denomination, proceed in steps from the highest denomination to the next lower, and so on to the lowest, adding in the units of each denomination as the operation proceeds.

Example: Reduce 10 gallons, 1 quart, 1 pint, to pints.

$10 \times 4 = 40, +1 = 41, 41 \times 2 = 82, +1 = 83$ pints. (Ans.)

Reduction Ascending.—To express a number of a lower denomination in terms of a higher, divide the number by the number of units of the lower denomination contained in one of the next higher; the quotient is in the higher denomination, and the remainder, if any, is in the lower.

Example: Reduce 227 pints to higher units.

$$227 \div 2 = 113 \text{ qts.,} \quad +1 \text{ pt.,} \quad 113 \div 4 = 28 \text{ gal.} + 1 \text{ qt.}$$
$$28 \text{ gal. 1 qt. 1 pt. (Ans.)}$$

To express the results in decimals of the higher denomination, divide the given number by the number of units of the given denomination contained in one of the required denomination, carrying the result to as many places as required.

Example: Reduce 1 inch to feet.   Give result in ten-thousandths. $1 \div 12 = 0.0833$ ft. (Ans.)

Addition of Compound Quantities :—

Example: Add 12 feet $4\frac{1}{4}$ inches, 6 feet $8\frac{5}{8}$ inches, and 15 feet $3\frac{1}{2}$ inches.

ft.	in.	8
12	$4\frac{1}{4}$	2
6	$8\frac{5}{8}$	5
15	$3\frac{1}{2}$	4
33	15	$\dfrac{11}{8} = 1\frac{3}{8}$

$$+ \ 1\frac{3}{8}$$
$$16\frac{3}{8} \qquad = 1 \text{ ft. } 4\frac{3}{8} \text{ in.}$$

$$33 \text{ ft.} + 1 \text{ ft. } 4\frac{3}{8} \text{ in.} = 34 \text{ ft. } 4\frac{3}{8} \text{ in. (Ans.)}$$

Subtraction of Compound Quantities.—

Example: Subtract 4 yds. 1 ft. 3 in. from 6 yd. 7 ft. 1 in.

yd.	ft.	in.
	6	
6	7	12 + 1 (1 ft. or 12 inches is borrowed
4	1	3 from 7 ft.)
2	5	10

Therefore, the required difference is 2 yds. 5 ft. 10 in.   (Ans.)

Multiplication of Compound Quantities.—

Example: Multiply 3 ft. $4\frac{5}{16}$ in. by 8.

$$3' \qquad 4\frac{5}{16}'' \qquad \frac{5}{16} \times 8 = \frac{40}{16} = 2\frac{1}{2}''$$
$$\underline{\times\, 8}$$
$$24' \quad 32$$
$$\underline{+\ 2\frac{1}{2}}$$
$$34\frac{1}{2}'' = 2'\ 10\frac{1}{2}''$$
$$24' +\quad 2'\ 10\frac{1}{2}'' = 26'\ 10\frac{1}{2}''$$

Therefore, the product is 26 ft. $10\frac{1}{2}$ in. (Ans.)

Division of Compound Quantities.—

Example: Divide 122 bu. 2 pk. 7 qt. 1 pt. by 5.

	bu.	pk.	qt.	pt.
5	122	2	7	1
	24	2	1	1

Therefore, the quotient is 24 bu. 2 pk. 1 qt. 1 pt. (Ans.)

Example: Divide 12 ft. 4 in. by 5 to the nearest $\frac{1}{16}$ in.

$$12'\ 4'' \times 12 = 148''$$

$$148'' \div 5 = \frac{148}{5} = 29\frac{3''}{5} = 2'\ 5\frac{3''}{5}$$

$$\frac{3}{5} \times 16 = \frac{48}{5} \quad \text{or} \quad \frac{9\frac{3}{5}}{16} = \frac{10}{16} = \frac{5}{8}$$

Therefore, the quotient is 2 ft. $5\frac{5}{8}$ in. (Ans.)

Powers.—When a number is multiplied by itself once it is said to be *squared* and the product is called the square of the number. Thus, in $3 \times 3 = 9$, the 9 is the square of 3. The same number has been used twice as a factor. The operation of squaring a number is usually indicated by a small number called an *exponent*, thus $3^2 = 9$. A number multiplied by itself once is said to have been raised to the second power.

Similarly, a number may be multiplied by itself twice. It is then used three times as a factor and is said to have been *cubed* or raised to the third power and the operation is indicated thus $3^3 = 27$, which means $3 \times 3 \times 3 = 27$.

A number may be raised to any power, the power being indicated by the proper exponent. Thus, 4^6 is four to the sixth power or $4 \times 4 \times 4 \times 4 \times 4 \times 4$, 3^{10} is three to the tenth, etc.

Roots.—A number may be divided into several equal factors. Thus, 36 is the product of 6×6. Each of the equal factors of a number is called a *root* of the number. If a number is divided into two equal factors, the root is said to be the *square root*; if three equal factors, the *cube-root*; if four equal factors, the fourth root, etc.

A root is indicated by the symbol $\sqrt{}$ called the *radical sign* and the degree of the root is indicated by a small number called the *root index* thus $\sqrt[3]{}$. When the radical sign has no index number, the square root is meant, which could also be indicated by writing $\sqrt[2]{}$. Thus, $\sqrt{25} = 5$ or $\sqrt[2]{25} = 5$. In all other cases an index number must be used, as $\sqrt[3]{27} = 3$ and $\sqrt[4]{16} = 2$.

The values of roots may be determined by arithmetical computation, by the use of logarithms, or by reference to tables containing values already computed. Square roots and cube roots are those most commonly needed and for most practical purposes, the average man will find that the tables of these values fill his needs. Such a table will be found on pages 23 to 36 and the values may be read directly. The computation of square root is described in the next paragraph, and the finding of roots by logarithms (the most convenient method for higher roots) is dealt with on page 62.

Square Root.—The square root of a number is extracted as follows:

Point off the number into periods of two figures each, beginning with the units; if there are decimals, begin at the decimal point, separating the whole number to the left and the decimal to the right into such periods, supplying as many ciphers in groups of two as may be desired in the decimal.

Find the greatest number whose square is less than the first left-hand period and place this to the right of the given number as the first figure of the root. Subtract its square from the first left-hand period and to the remainder annex the second period for a dividend.

Place before this as a partial divisor, double the root figure just found. Find how many times the dividend, exclusive of its right-hand figure, contains the divisor, and place the quotient as the second figure of the root, and also at the right of the partial divisor.

Multiply the divisor thus completed, by the second root figure and subtract the product from the dividend. To this remainder annex the next period for a new dividend, and double the two root figures for a new partial divisor. Proceed as before until all the periods have been brought down.

Example: Extract the square root of 5386.3928 to 3 decimal places.

$$
\begin{array}{r}
53'86.39'28'00\ (73.392 \quad \text{(Ans.)} \\
49 \\
\end{array}
$$

$$
\begin{array}{rl}
143 &)\ \ 486 \\
& \ \ \ 429 \\
1463 &)\ \ 5739 \\
& \ \ \ 4389 \\
14669 &)\ \ 135028 \\
& \ \ \ 132021 \\
146782 &)\ \ \ \ 300700 \\
& \ \ \ \ \ 293564 \\
& \ \ \ \ \ \overline{\ \ 7136} \\
\end{array}
$$

Extracting Square Root of a Fraction.—The square root of a fraction is the square root of its numerator over the square root of its denominator. Thus, $\sqrt{\dfrac{9}{16}} = \dfrac{\sqrt{9}}{\sqrt{16}} = \dfrac{3}{4}$. When neither the numerator nor the denominator is a perfect square a convenient short cut is to multiply both by a common number to convert one or the other to a perfect square. Thus,

$$\sqrt{\frac{2}{3}} = \sqrt{\frac{2 \times 3}{3 \times 3}} = \sqrt{\frac{6}{9}} = \frac{\sqrt{6}}{\sqrt{9}} = \frac{\sqrt{6}}{3} = \frac{2.449}{3}.$$

Since the square root of a fraction often results in decimals, it is

TABLE 2

SQUARES, CUBES, SQUARE ROOTS, CUBE ROOTS, OF NUMBERS 1 TO 1600.

No.	Square	Cube.	Sq. Rt.	Cu. Rt.	No.	Square	Cube.	Sq. Rt.	Cu. Rt.
0	0	0	0.0000000	0.0000000	65	42 25	274 625	8.0622577	4.0207256
1	1	1	1.0000000	1.0000000	6	43 56	287 496	.1240384	.0412401
2	4	8	.4142136	.2599210	7	44 89	300 763	.1853528	.0615480
3	9	27	.7320508	.4422496	8	46 24	314 432	.2462113	.0816551
4	16	64	2.0000000	5874011	9	47 61	328 509	.3066239	.1015661
5	25	125	2.2360680	1.7099759	70	49 00	343 000	8.3666003	4.1212853
6	36	216	.4494897	.8171206	1	50 41	357 911	.4261498	.1408178
7	49	343	.6457513	.9129312	2	51 84	373 248	.4852814	.1601676
8	64	512	.8284271	2.0000000	3	53 29	389 017	.5440037	.1793392
9	81	729	3.0000000	.0800837	4	54 76	405 224	.6023253	.1983364
10	1 00	1 000	3.1622777	2.1544347	75	56 25	421 875	8.6602540	4.2171633
11	1 21	1 331	.3166248	.2239801	6	57 76	438 976	.7177979	.2358236
12	1 44	1 728	.4641016	.2894286	7	59 29	456 533	.7749644	.2543210
13	1 69	2 197	.6055513	.3513347	8	60 84	474 552	.8317609	.2726586
14	1 96	2 744	.7416574	.4101422	9	62 41	493 039	.8881944	2908404
15	2 25	3 375	3.8729833	2.4662121	80	64 00	512 000	8.9442719	4.3088695
16	2 56	4 096	4.0000000	.5198421	1	65 61	531 441	9.0000000	.3267487
17	2 89	4 913	.1231056	.5712816	2	67 24	551 368	.0553851	.3444815
18	3 24	5 832	.2426407	.6207414	3	68 89	571 787	.1104336	.3620707
19	3 61	6 859	.3588989	.6684016	4	70 56	592 704	.1651514	.3795191
20	4 00	8 000	4.4721360	2.7144177	85	72 25	614 125	9.2195445	4.3968296
1	4 41	9 261	.5825757	.7589243	6	73 96	636 056	.2736185	.4140049
2	4 84	10 648	.6904158	.8020393	7	75 69	658 503	.3273791	.4310476
3	5 29	12 167	7958315	.8438670	8	77 44	681 472	.3808315	.4479602
4	5 76	13 824	.8989795	.8844991	9	79 21	704 969	.4339811	.4647451
25	6 25	15 625	5.0000000	2.9240177	90	81 00	729 000	9.4868330	4.4814047
6	6 76	17 576	.0990195	.9624960	1	82 81	753 571	.5393920	.4979414
7	7 29	19 683	.1961524	3.0000000	2	84 64	778 688	.5916630	.5143574
8	7 84	21 952	.2915026	.0365889	3	86 49	804 357	.6436508	.5306549
9	8 41	24 389	.3851648	.0723168	4	88 36	830 584	.6953597	.5468359
30	9 00	27 000	5.4772256	3.1072325	95	90 25	857 375	9.7467943	4.5629026
1	9 61	29 791	.5677644	.1413806	6	92 16	884 736	.7979590	.5788570
2	10 24	32 768	.6568542	.1748021	7	94 09	912 673	.8488578	.5947009
3	10 89	35 937	.7445626	.2075343	8	96 04	941 192	.8994949	.6104363
4	11 56	39 304	.8309519	.2396118	9	98 01	970 299	.9498744	.6260659
35	12 25	42 875	5.9160798	3.2710663	100	1 00 00	1 000 000	10.0000000	4.6415888
6	12 96	46 656	6.0000000	.3019272	1	1 02 01	1 030 301	.0498756	.6570095
7	13 69	50 653	.0827625	.3322218	2	1 04 04	1 061 208	.0995049	.6723287
8	14 44	54 872	.1644140	.3619754	3	1 06 09	1 092 927	.1488916	.6875482
9	15 21	59 319	2449980	.3912114	4	1 08 16	1 124 864	.1980390	.7026694
40	16 00	64 000	6.3245553	3.4199519	105	1 10 25	1 157 625	10.2469508	4.7176940
1	16 81	68 921	.4031242	.4482172	6	1 12 36	1 191 016	.2956301	.7326235
2	17 64	74 088	.4807407	.4760266	7	1 14 49	1 225 043	.3440804	.7474594
3	18 49	79 507	.5574385	.5033981	8	1 16 64	1 259 712	.3923048	.7622032
4	19 36	85 184	.6332483	.5303483	9	1 18 81	1 295 029	.4403065	.7768562
45	20 25	91 125	6.7082039	3.5568933	110	1 21 00	1 331 000	10.4880885	4.7914199
6	21 16	97 336	.7823300	.5830479	11	1 23 21	1 367 631	.5356538	.8058955
7	22 09	103 823	.8556546	.6088261	12	1 25 44	1 404 928	.5830052	.8202845
8	23 04	110 592	.9282032	.6342411	13	1 27 69	1 442 897	.6301458	.8345881
9	24 01	117 649	7.0000000	.6593057	14	1 29 96	1 481 544	.6770783	.8488076
50	25 00	125 000	7.0710678	3.6840314	115	1 32 25	1 520 875	10.7238053	4.8629442
1	26 01	132 651	.1414284	.7084298	16	1 34 56	1 560 896	.7703296	.8769990
2	27 04	140 608	.2111026	.7325111	17	1 36 89	1 601 613	8166538	.8909732
3	28 09	148 877	.2801099	.7562858	18	1 39 24	1 643 032	.8627805	.9048681
4	29 16	157 464	.3484692	.7797621	19	1 41 61	1 685 159	.9087121	.9186847
55	30 25	166 375	7.4161985	3.8029525	120	1 44 00	1 728 000	10.9544512	4.9324242
6	31 36	175 616	.4833148	.8258624	1	1 46 41	1 771 561	11.0000000	9460874
7	32 49	185 193	.5498344	.8485011	2	1 48 84	1 815 848	.0453610	.9596757
8	33 64	195 112	.6157731	.8708766	3	1 51 29	1 860 867	.0905365	.9731898
9	34 81	205 379	.6811457	.8929965	4	1 53 76	1 906 624	.1355287	.9866310
60	36 00	216 000	7.7459667	3.9148676	125	1 56 25	1 953 125	11.1803399	5.0000000
1	37 21	226 981	.8102497	.9364972	6	1 58 76	2 000 376	.2249722	.0132979
2	38 44	238 328	.8740079	.9578915	7	1 61 29	2 048 383	.2694277	.0265527
3	39 69	250 047	9372539	.9790571	8	1 63 84	2 097 152	3137085	.0396842
4	40 96	262 144	8.0000000	4.0000000	9	1 66 41	2 146 689	.3578167	.0527743
65	42 25	274 625	8.0622577	4.0207256	130	1 69 00	2 197 000	11.4017543	5.0657970

28

2. —Squares, Cubes, Square Roots, Cube Roots, of Numbers
1 to 1600—Continued.

No.	Square	Cube.	Sq. Rt.	Cu. Rt.	No.	Square	Cube.	Sq. Rt.	Cu. Rt.
130	1 69 00	2 197 000	11.4017543	5.0657979	195	3 80 25	7 414 875	13.9642400	5.7988900
1	1 71 61	2 248 091	.4455231	.0787531	6	3 84 16	7 529 536	14.0000000	.8087857
2	1 74 24	2 299 968	.4891253	.0916434	7	3 88 09	7 645 373	.0356688	.8186479
3	1 76 89	2 352 637	.5325626	.1044687	8	3 92 04	7 762 392	.0712473	.8284767
4	1 79 56	2 406 104	.5758369	.1172299	9	3 96 01	7 880 599	.1067360	.8382725
135	1 82 25	2 460 375	11.618C500	5.1299278	200	4 00 00	8 000 000	14.1421356	5.8480355
6	1 84 96	2 515 456	.6619038	.1425632	1	4 04 01	8 120 601	.1774469	.8577660
7	1 87 69	2 571 353	.7046999	.1551367	2	4 08 04	8 242 408	.2126704	.8674643
8	1 90 44	2 628 072	.7473401	.1676493	3	4 12 09	8 365 427	.2478068	.8771307
9	1 93 21	2 685 619	.7898261	.1801015	4	4 16 16	8 489 664	.2828569	.8867653
140	1 96 00	2 744 000	11.8321596	5.1924941	205	4 20 25	8 615 125	14.3178211	5.8963685
1	1 98 81	2 803 221	.8743422	.2048279	6	4 24 36	8 741 816	.3527001	.9059406
2	2 01 64	2 863 288	.9163753	.2171034	7	4 28 49	8 869 743	.3874946	.9154817
3	2 04 49	2 924 207	.9582607	.2293215	8	4 32 64	8 998 912	.4222051	.9249921
4	2 07 36	2 985 984	12.0000000	.2414828	9	4 36 81	9 129 329	.4568323	.9344721
145	2 10 25	3 048 625	12.0415946	5.2535879	210	4 41 00	9 261 000	14.4913767	5.9439220
6	2 13 16	3 112 136	.0830460	.2656374	11	4 45 21	9 393 931	.5258390	.9533418
7	2 16 09	3 176 523	.1243557	.2776321	12	4 49 44	9 528 128	.5602198	.9627320
8	2 19 04	3 241 792	.1655251	.2895725	13	4 53 69	9 663 597	.5945195	.9720926
9	2 22 01	3 307 949	.2065556	.3014592	14	4 57 96	9 800 344	.6287388	.9814240
150	2 25 00	3 375 000	12.2474487	5.3132928	215	4 62 25	9 938 375	14.6628783	5.9907264
1	2 28 01	3 442 951	.2882057	.3250740	16	4 66 56	10 077 696	.6969385	6.0000000
2	2 31 04	3 511 808	.3288280	.3368033	17	4 70 89	10 218 313	.7309199	.0092450
3	2 34 09	3 581 577	.3693169	.3484812	18	4 75 24	10 360 232	.7648231	.0184617
4	2 37 16	3 652 264	.4096736	.3601084	19	4 79 61	10 503 459	.7986486	.0276502
155	2 40 25	3 723 875	12.4498996	5.5716854	220	4 84 00	10 648 000	14.8323970	6.0368107
6	2 43 36	3 796 416	.4899960	.3832126	1	4 88 41	10 793 861	.8660687	.0459435
7	2 46 49	3 869 893	.5299641	.3946097	2	4 92 84	10 941 048	.8996644	.0550489
8	2 49 64	3 944 312	.5698051	.4061202	3	4 97 29	11 089 567	.9331845	.0641270
9	2 52 81	4 019 679	.6095202	.4175015	4	5 01 76	11 239 424	.9666295	.0731779
160	2 56 00	4 096 000	12.6491106	5.4288352	225	5 06 25	11 390 625	15.0000000	6.0822020
1	2 59 21	4 173 281	.6885775	.4401218	6	5 10 76	11 543 176	.0332964	.0911994
2	2 62 44	4 251 528	.7279221	.4513618	7	5 15 29	11 697 083	.0665192	.1001702
3	2 65 69	4 330 747	.7671453	.4625556	8	5 19 84	11 852 352	.0996689	.1091147
4	2 68 96	4 410 944	.8062485	.4737037	9	5 24 41	12 008 989	.1327460	.1180332
165	2 72 25	4 492 125	12.8452326	5.4848066	230	5 29 00	12 167 000	15.1657509	6.1269257
6	2 75 56	4 574 296	.8840987	.4958647	1	5 33 61	12 326 391	.1986842	.1357924
7	2 78 89	4 657 463	.9228480	.5068784	2	5 38 24	12 487 168	.2315462	.1446337
8	2 82 24	4 741 632	.9614814	.5178484	3	5 42 89	12 649 337	.2643375	.1534495
9	2 85 61	4 826 809	13.0000000	.5287748	4	5 47 56	12 812 904	.2970585	.1622401
170	2 89 00	4 913 000	13.0384048	5.5396583	235	5 52 25	12 977 875	15.3297097	6.1710058
1	2 92 41	5 000 211	.0766968	.5504991	6	5 56 96	13 144 256	.3622915	.1797466
2	2 95 84	5 088 448	.1148770	.5612978	7	5 61 69	13 312 053	.3948043	.1884628
3	2 99 29	5 177 717	.1529464	.5720546	8	5 66 44	13 481 272	.4272486	.1971544
4	3 02 76	5 268 024	.1909060	.5827702	9	5 71 21	13 651 919	.4596248	.2058218
175	3 06 25	5 359 375	13.2287566	5.5934447	240	5 76 00	13 824 000	15.4919334	6.2144650
6	3 09 76	5 451 776	.2664992	.6040787	1	5 80 81	13 997 521	.5241747	.2230843
7	3 13 29	5 545 233	.3041347	.6146724	2	5 85 64	14 172 488	.5563492	.2316797
8	3 16 84	5 639 752	.3416641	.6252263	3	5 90 49	14 348 907	.5884573	.2402515
9	3 20 41	5 735 339	.3790882	.6357408	4	5 95 36	14 526 784	.6204994	.2487998
180	3 24 00	5 832 000	13.4164079	5.6462162	245	6 00 25	14 706 125	15.6524758	6.2573248
1	3 27 61	5 929 741	.4536240	.6566528	6	6 05 16	14 886 936	.6843871	.2658266
2	3 31 24	6 028 568	.4907376	.6670511	7	6 10 09	15 069 223	.7162336	.2743054
3	3 34 89	6 128 487	.5277493	.6774114	8	6 15 04	15 252 992	.7480157	.2827613
4	3 38 56	6 229 504	.5646600	.6877340	9	6 20 01	15 438 249	.7797338	.2911946
185	3 42 25	6 331 625	13.6014705	5.6980192	250	6 25 00	15 625 000	15.8113883	6.2996053
6	3 45 96	6 434 856	.6381817	.7082675	1	6 30 01	15 813 251	.8429795	.3079935
7	3 49 69	6 539 203	.6747943	.7184791	2	6 35 04	16 003 008	.8745079	.3163596
8	3 53 44	6 644 672	.7113092	.7286543	3	6 40 09	16 194 277	.9059737	.3247035
9	3 57 21	6 751 269	.7477271	.7387936	4	6 45 16	16 387 064	.9373775	.3330256
190	3 61 00	6 859 000	13.7840488	5.7488971	255	6 50 25	16 581 375	15.9687194	6.3413257
1	3 64 81	6 967 871	.8202750	.7589652	6	6 55 36	16 777 216	16.0000000	.3496042
2	3 68 64	7 077 888	.8564065	.7689982	7	6 60 49	16 974 593	.0312195	.3578611
3	3 72 49	7 189 057	.8924440	.7789966	8	6 65 64	17 173 512	.0623784	.3660968
4	3 76 36	7 301 384	.928.888.3	.7889604	9	6 70 81	17 373 979	.0934769	.3743111
195	3 80 25	7 414 875	13.9642400	5.7988900	260	6 76 00	17 576 000	16.1245155	6.3825043

2.—Squares, Cubes, Square Roots, Cube Roots, of Numbers
1 to 1600—Continued.

No.	Square	Cube.	Sq. Rt.	Cu. Rt.	No.	Square	Cube.	Sq. Rt.	Cu. Rt.
260	6 76 00	17 576 000	16.1245155	6.3825043	325	10 56 25	34 328 125	18.0277564	6.8753443
1	6 81 21	17 779 581	.1554944	.3906765	6	10 62 76	34 645 976	.0554701	.8823888
2	6 86 44	17 984 728	.1864141	.3988279	7	10 69 29	34 965 783	.0831413	.8894188
3	6 91 69	18 191 447	.2172747	.4069585	8	10 75 84	35 287 552	.1107703	.8964345
4	6 96 96	18 399 744	.2480768	.4150687	9	10 82 41	35 611 289	.1383571	.9034359
265	7 02 25	18 609 625	16.2788206	6.4231583	330	10 89 00	35 937 000	18.1659021	6.9104232
6	7 07 56	18 821 096	.3095064	.4312276	1	10 95 61	36 264 691	.1934054	.9173964
7	7 12 89	19 034 163	.3401346	.4392767	2	11 02 24	36 594 368	.2208672	.9243556
8	7 18 24	19 248 832	.3707055	.4473057	3	11 08 89	36 926 037	.2482876	.9313008 ·
9	7 23 61	19 465 109	.4012195	.4553148	4	11 15 56	37 259 704	.2756669	.9382321
270	7 29 00	19 683 000	16.4316767	6.4633041	335	11 22 25	37 595 375	18.3030052	6.9451496
1	7 34 41	19 902 511	.4620776	.4712736	6	11 28 96	37 933 056	.3303028	.9520533
2	7 39 84	20 123 648	.4924225	.4792236	7	11 35 69	38 272 753	.3575598	.9589434
3	7 45 29	20 346 417	.5227116	.4871541	8	11 42 44	38 614 472	.3847763	.9658198
4	7 50 76	20 570 824	.5529454	.4950653	9	11 49 21	38 958 219	.4119526	.9726825
275	7 56 25	20 796 875	16.5831240	6.5029572	340	11 56 00	39 304 000	18 4390889	6.9795321
6	7 61 76	21 024 576	.6132477	.5108300	1	11 62 81	39 651 821	.4661853	.9863681
7	7 67 29	21 253 933	.6433170	.5186839	2	11 69 64	40 001 688	.4932420	.9931906
8	7 72 84	21 484 952	.6733320	.5265189	3	11 76 49	40 353 607	.5202592	7.0000000
9	7 78 41	21 717 639	.7032931	.5343351	4	11 83 36	40 707 584	.5472370	.0067962
280	7 84 00	21 952 000	16.7332005	6.5421326	345	11 90 25	41 063 625	18.5741756	7.0135791
1	7 89 61	22 188 041	.7630546	.5499116	6	11 97 16	41 421 736	.6010752	.0203490
2	7 95 24	22 425 768	.7928556	.5576722	7	12 04 09	41 781 923	.6279360	.0271058
3	8 00 89	22 665 187	.8226038	.5654144	8	12 11 04	42 144 192	.6547581	.0338497
4	8 06 56	22 906 304	.8522995	.5731385	9	12 18 01	42 508 549	.6815417	.0405806
285	8 12 25	23 149 125	16.8819430	6.5808443	350	12 25 00	42 875 000	18.7082869	7.0472987
6	8 17 96	23 393 656	.9115345	.5885323	1	12 32 01	43 243 551	.7349940	.0540041
7	8 23 69	23 639 903	.9410743	.5962023	2	12 39 04	43 614 208	.7616630	.0606967
8	8 29 44	23 887 872	.9705627	.6038545	3	12 46 09	43 986 977	.7882942	.0673767
9	8 35 21	24 137 569	17.0000000	.6114890	4	12 53 16	44 361 864	.8148877	.0740440
290	8 41 00	24 389 000	17.0293864	6.6191060	355	12 60 25	44 738 875	18.8414437	7.0806988
1	8 46 81	24 642 171	.0587221	.6267054	6	12 67 36	45 118 016	.8679623	.0873411
2	8 52 64	24 897 088	.0880075	.6342874	7	12 74 49	45 499 293	.8944436	.0939709
3	8 58 49	25 153 757	.1172428	.6418522	8	12 81 64	45 882 712	.9208879	.1005885
4	8 64 36	25 412 184	.1464282	.6493998	9	12 88 81	46 268 279	.9472953	.1071937
295	8 70 25	25 672 375	17.1755640	6.6569302	360	12 96 00	46 656 000	18.9736660	7.1137866
6	8 76 16	25 934 336	2046505	.6644437	1	13 03 21	47 045 881	19.0000000	.1203674
7	8 82 09	26 198 073	.2336879	.6719403	2	13 10 44	47 437 928	.0262976	.1269360
8	8 88 04	26 463 592	.2626765	.6794200	3	13 17 69	47 832 147	.0525589	.1334925
9	8 94 01	26 730 899	.2916165	.6868831	4	13 24 96	48 228 544	.0787840	.1400370
300	9 00 00	27 000 000	17.3205081	6.6943295	365	13 32 25	48 627 125	19.1049732	7.1465695
1	9 06 01	27 270 901	.3493516	.7017593	6	13 39 56	49 027 896	.1311265	.1530901
2	9 12 04	27 543 608	.3781472	.7091729	7	13 46 89	49 430 863	.1572441	.1595988
3	9 18 09	27 818 127	.4068952	.7165700	8	13 54 24	49 836 032	.1833261	.1660957
4	9 24 16	28 094 464	.4355958	.7239508	9	13 61 61	50 243 409	.2093727	.1725809
305	9 30 25	28 372 625	17.4642492	6.7313155	370	13 69 00	50 653 000	19.2353841	7.1790544
6	9 36 36	28 652 616	.4928557	.7386641	1	13 76 41	51 064 811	.2613603	.1855162
7	9 42 49	28 934 443	.5214155	.7459967	2	13 83 84	51 478 848	.2873015	.1919663
8	9 48 64	29 218 112	.5499288	.7533134	3	13 91 29	51 895 117	.3132079	.1984050
9	9 54 81	29 503 629	.5783958	.7606143	4	13 98 76	52 313 624	.3390796	.2048322
310	9 61 00	29 791 000	17.6068169	6.7678995	375	14 06 25	52 734 375	19.3649167	7.2112479
11	9 67 21	30 080 231	.6351921	.7751690	6	14 13 76	53 157 376	.3907194	.2176522
12	9 73 44	30 371 328	.6635217	.7824229	7	14 21 29	53 582 633	.4164878	.2240450
13	9 79 69	30 664 297	.6918060	.7896613	8	14 28 84	54 010 152	.4422221	.2304268
14	9 85 96	30 959 144	.7200451	.7968844	9	14 36 41	54 439 939	.4679223	.2367972
315	9 92 25	31 255 875	17.7482393	6.8040921	380	14 44 00	54 872 000	19.4935887	7.2431565
16	9 98 56	31 554 496	.7763888	.8112847	1	14 51 61	55 306 341	.5192213	.2495045
17	10 04 89	31 855 013	.8044938	.8184620	2	14 59 24	55 742 968	5448203	.2558415
18	10 11 24	32 157 432	.8325545	.8256242	3	14 66 89	56 181 887	.5703858	.2621675
19	10 17 61	32 461 759	.8605711	.8327714	4	14 74 56	56 623 104	.5959179	.2684824
320	10 24 00	32 768 000	17.8885438	6.8399037	385	14 82 25	57 066 625	19 6214169	7.2747864
1	10 30 41	33 076 161	.9164729	.8470213	6	14 89 96	57 512 456	.6468827	.2810794
2	10 36 84	33 386 248	.9443584	.8541240	7	14 97 69	57 960 603	.6723156	.2873617
3	10 43 29	33 698 267	.9722008	.8612120	8	15 05 44	58 411 072	.6977156	.2936330
4	10 49 76	34 012 224	18.0000000	.8682855	9	15 13 21	58 863 869	.7230829	.2998936
325	10 56 25	34 328 125	18.0277564	6.8753443	390	15 21 00	59 319 000	19.7484177	7.3061436

2. —Squares, Cubes, Square Roots, Cube Roots, of Numbers
1 to 1600—Continued.

No.	Square	Cube.	Sq. Rt.	Cu. Rt.	No.	Square	Cube	Sq. Rt.	Cu. Rt.
390	15 21 00	59 319 000	19.7484177	7.3061436	455	20 70 25	94 196 375	21.3307290	7.6913717
1	15 28 81	59 776 471	.7737199	.3123828	6	20 79 36	94 818 816	.3541565	.6970023
2	15 36 64	60 236 288	.7989899	.3186114	7	20 88 49	95 443 993	.3775583	.7026246
3	15 44 49	60 698 457	.8242276	.3248295	8	20 97 64	96 071 912	.4009346	.7082388
4	15 52 36	61 162 984	.8494332	.3310369	9	21 06 81	96 702 579	.4242853	.7138448
395	15 60 25	61 629 875	19.8746069	7.3372339	460	21 16 00	97 336 000	21.4476106	7.7194426
6	15 68 16	82 099 136	.8997487	.3434205	1	21 25 21	97 972 181	.4709106	.7250325
7	15 76 09	62 570 773	.9248588	.3495966	2	21 34 44	98 611 128	.4941853	.7306141
8	15 84 04	63 044 792	.9499373	.3557624	3	21 43 69	99 252 847	.5174348	.7361877
9	15 92 01	63 521 199	.9749844	.3619178	4	21 52 96	99 897 344	.5406592	.7417532
400	16 00 00	64 000 000	20.0000000	7.3680630	465	21 62 25	100 544 625	21.5638587	7.7473109
1	16 08 01	64 481 201	.0249844	.3741079	6	21 71 56	101 194 696	.5870331	.7528606
2	16 16 04	64 964 808	.0499377	.3803227	7	21 80 89	101 847 563	.6101828	.7584023
3	16 24 09	65 450 827	.0748599	.3864373	8	21 90 24	102 503 232	.6333077	.7639361
4	16 32 16	65 939 264	.0997512	.3925418	9	21 99 61	103 161 709	.6564078	.7694620
405	16 40 25	66 430 125	20.1246118	7.3986363	470	22 09 00	103 823 000	21.6794834	7.7749801
6	16 48 36	66 923 416	.1494417	.4047206	1	22 18 41	104 487 111	.7025344	.7804904
7	16 56 49	67 419 143	.1742410	.4107950	2	22 27 84	105 154 048	.7255610	.7859928
8	16 64 64	67 917 312	.1990099	.4168595	3	22 37 29	105 823 817	.7485632	.7914875
9	16 72 81	68 417 929	.2237484	.4229142	4	22 46 76	106 496 424	.7715411	.7969745
410	16 81 00	68 921 000	20.2484567	7.4289589	475	22 56 25	107 171 875	21.7944947	7.8024538
11	16 89 21	69 426 531	.2731349	.4349938	6	22 65 76	107 850 176	.8174242	.8079254
12	16 97 44	69 934 528	.2977831	.4410189	7	22 75 29	108 531 333	.8403297	.8133892
13	17 05 69	70 444 997	.3224014	.4470342	8	22 84 84	109 215 352	.8632111	.8188456
14	17 13 96	70 957 944	.3469899	.4530399	9	22 94 41	109 902 239	.8860686	.8242942
415	17 22 25	71 473 375	20.3715488	7.4590359	480	23 04 00	110 592 000	21.9089023	7.8297353
16	17 30 56	71 991 296	.3960781	.4650223	1	23 13 61	111 284 641	.9317122	.8351688
17	17 38 89	92 511 713	.4205779	.4709991	2	23 23 24	111 980 168	.9544984	.8405949
18	17 47 24	73 034 632	.4450483	.4769664	3	23 32 89	112 678 587	.9772610	.8460134
19	17 55 61	73 560 059	.4694895	.4829242	4	23 42 56	113 379 904	22.0000000	.8514244
420	17 64 00	74 088 000	20.4939015	7.4888724	485	23 52 25	114 084 125	22.0227155	7.8568281
1	17 72 41	74 618 461	.5182845	.4948113	6	23 61 96	114 791 256	.0454077	.8622242
2	17 80 84	75 151 448	.5426386	.5007406	7	23 71 69	115 501 303	.0680765	.8676130
3	17 89 29	75 686 967	.5669638	.5066607	8	23 81 44	116 214 272	.0907220	.8729944
4	17 97 76	76 225 024	.5912603	.5125715	9	23 91 21	116 930 169	.1133444	.8783684
425	18 06 25	76 765 625	20.6155281	7.5184730	490	24 01 00	117 649 000	22.1359436	7.8837352
6	18 14 76	77 308 776	.6397674	.5243652	1	24 10 81	118 370 771	.1585198	.8890946
7	18 23 29	77 854 483	.6639783	.5302482	2	24 20 64	119 095 488	.1810730	.8944468
8	18 31 84	78 402 752	.6881609	.5361221	3	24 30 49	119 823 157	.2036033	.8997917
9	18 40 41	78 953 589	.7123152	.5419867	4	24 40 36	120 553 784	.2261108	.9051294
430	18 49 00	79 507 000	20.7364414	7.5478423	495	24 50 25	121 287 375	22.2485955	7.9104599
1	18 57 61	80 062 991	.7605395	.5536608	6	24 60 16	122 023 936	.2710575	.9157832
2	18 66 24	80 621 568	.7846097	.5595263	7	24 70 09	122 763 473	.2934968	.9210994
3	18 74 89	81 182 737	.8086520	.5653548	8	24 80 04	123 505 992	.3159136	.9264085
4	18 83 56	81 746 504	.8326667	.5711743	9	24 90 01	124 251 499	.3383079	.9317104
435	18 92 25	82 312 875	20.8566536	7.5769849	500	25 00 00	125 000 000	22.3606798	7.9370053
6	19 00 96	82 881 856	.8806130	.5827865	1	25 10 01	125 751 501	.3830293	.9422931
7	19 09 69	83 453 453	.9045450	.5885793	2	25 20 04	126 506 008	.4053565	.9475739
8	19 18 44	84 027 672	.9284495	.5943633	3	25 30 09	127 263 527	.4276615	.9528477
9	19 27 21	84 604 519	.9523268	.6001385	4	25 40 16	128 024 064	.4499443	.9581144
440	19 36 00	35 184 000	20.9761770	7.6059049	505	25 50 25	128 787 625	22.4722051	7.9633743
1	19 44 81	85 766 121	21.0000000	.6116626	6	25 60 36	129 554 216	.4944438	.9686271
2	19 53 64	86 350 888	.0237960	.6174116	7	25 70 49	130 323 843	.5166605	.9738731
3	19 62 49	86 938 307	.0475652	.6231519	8	25 80 64	131 096 512	.5388553	.9791122
4	19 71 36	87 528 384	.0713075	.6288837	9	25 90 81	131 872 229	.5610283	.9843444
445	19 80 25	88 121 125	21.0950231	7.6346067	510	26 01 00	132 651 000	22.5831796	7.7895697
6	19 89 16	88 716 536	.1187121	.6403213	11	26 11 21	133 432 831	.6053091	.9947883
7	19 98 09	89 314 623	.1423745	.6460272	12	26 21 44	134 217 728	.6274170	8.0000000
8	20 07 04	89 915 392	.1660105	.6517247	13	26 31 69	135 005 697	.6495033	.0052049
9	20 16 01	90 518 849	.1896201	.6574138	14	26 41 96	135 796 744	.6715681	.0104032
450	20 25 00	91 125 000	21.2132034	7.6630943	515	26 52 25	136 590 875	22.6936114	8.0155946
1	20 34 01	91 723 851	.2367606	.6687665	16	26 62 56	137 388 096	.7156334	.0207794
2	20 43 04	92 345 408	.2602916	.6744303	17	26 72 89	138 188 413	.7376340	.0259574
3	20 52 09	92 959 677	.2837967	.6800857	18	26 83 24	138 991 832	.7596134	.0311287
4	20 61 16	93 576 664	.3072758	.6857328	19	26 93 61	139 798 359	.7815715	.0362935
455	20 70 25	94 196 375	21.3307290	.6913717	520	27 04 00	140 608 000	22.8035085	8.0414515

2.—SQUARES, CUBES, SQUARE ROOTS, CUBE ROOTS, OF NUMBERS
1 TO 1600—Continued.

No.	Square	Cube.	Sq. Rt.	Cu. Rt.	No.	Square	Cube.	Sq. Rt.	Cu. Rt.
520	27 04 00	140 608 000	22.8035085	8.0414515	585	34 22 25	200 201 625	24.1867732	8.3634466
1	27 14 41	141 420 761	.8254244	.0466030	6	34 33 96	201 230 056	.2074369	.3682095
2	27 24 84	142 236 648	.8473193	.0517479	7	34 45 69	202 262 003	2280829	.3729668
3	27 35 29	143 055 667	.8691933	.0568862	8	34 57 44	203 297 472	.2487113	.3777188
4	27 45 76	143 877 824	.8910463	.0620180	9	34 69 21	204 336 469	.2693222	.3824653
525	27 56 25	144 703 125	22.9128785	8.0671432	590	34 81 00	205 379 000	24.2899156	8.3872065
6	27 66 76	145 531 576	.9346899	.0722620	1	34 92 81	206 425 071	.3104916	.3919423
7	27 77 29	146 363 183	.9564806	.0773743	2	35 04 64	207 474 688	.3310501	.3966729
8	27 87 84	147 197 952	.9782506	.0824800	3	35 16 49	208 527 857	.3515913	.4013981
9	27 98 41	148 035 889	23.0000000	.0875794	4	35 28 36	209 584 584	.3721152	.4061180
530	28 09 00	148 877 000	23.0217289	8.0926723	595	35 40 25	210 644 875	24.3926218	8.4108326
1	28 19 61	149 721 291	.0434372	0977589	6	35 52 16	211 708 736	.4131112	.4155419
2	28 30 24	150 568 768	.0651252	.1028390	7	35 64 09	212 776 173	.4335834	.4202460
3	28 40 89	151 419 437	.0867928	.1079128	8	35 76 04	213 847 192	.4540385	.4249448
4	28 51 56	152 273 304	.1084400	.1129803	9	35 88 01	214 921 799	.4744765	.4296383
535	28 62 25	153 130 375	23.1300670	8 1180414	600	36 00 00	216 000 000	24.4948974	8.4343267
6	28 72 96	153 990 656	.1516738	.1230962	1	36 12 01	217 081 801	.5153013	.4390098
7	28 83 69	154 854 153	.1732605	.1281447	2	36 24 04	218 167 208	.5356883	.4436877
8	28 94 44	155 720 872	.1948270	.1331870	3	36 36 09	219 256 227	.5560583	.4483605
9	29 05 21	156 590 819	.2163735	.1382230	4	36 48 16	220 348 864	.5764115	.4530281
540	29 16 00	157 464 000	23.2379001	8.1432529	605	36 60 25	221 445 125	24.5967478	8.4576906
1	29 26 81	158 340 421	.2594067	.1482765	6	36 72 36	222 545 016	.6170673	.4623479
2	29 37 64	159 220 088	.2808935	.1532939	7	36 84 49	223 648 543	.6373700	.4670000
3	29 48 49	160 103 007	.3023604	.1583051	8	36 96 64	224 755 712	.6576560	.4716471
4	29 59 36	160 989 184	.3238076	.1633102	9	37 08 81	225 866 529	.6779254	.4762892
545	29 70 25	161 878 625	23.3452351	8.1683092	610	37 21 00	226 981 000	24.6981781	8.4809261
6	29 81 16	162 771 336	.3666429	.1733020	11	37 33 21	228 099 131	.7184142	.4855579
7	29 92 09	163 667 323	.3880311	.1782888	12	37 45 44	229 220 928	.7386338	.4901848
8	30 03 04	164 566 592	.4093998	.1832695	13	37 57 69	230 346 397	.7588368	.4948065
9	30 14 01	165 469 149	.4307490	.1882441	14	37 69 96	231 475 544	.7790234	.4994233
550	30 25 00	166 375 000	23.4520788	8.1932127	615	37 82 25	232 608 375	24.7991935	8.5040350
1	30 36 01	167 284 151	.4733892	.1981753	16	37 94 56	233 744 896	.8193473	.5086417
2	30 47 04	168 196 608	.4946802	.2031319	17	38 06 89	234 885 113	.8394847	.5132435
3	30 58 09	169 112 377	.5159520	.2080825	18	38 19 24	236 029 032	.8596058	.5178403
4	30 69 16	170 031 464	.5372046	.2130271	19	38 31 61	237 176 659	.8797106	.5224321
555	30 80 25	170 953 875	23.5584380	8.2179657	620	38 44 00	238 328 000	24.8997992	8.5270189
6	30 91 36	171 879 616	.5796522	.2228985	1	38 56 41	239 483 061	.9198716	.5316009
7	31 02 49	172 808 693	.6008474	.2278254	2	38 68 84	240 641 848	.9399278	.5361780
8	31 13 64	173 741 112	.6220236	.2327463	3	38 81 29	241 804 367	.9599679	5407501
9	31 24 81	174 676 879	.6431808	.2376614	4	38 93 76	242 970 624	.9799920	.5453173
560	31 36 00	175 616 000	23.6643191	8.2425706	625	39 06 25	244 140 625	25.0000000	8.5498797
1	31 47 21	176 558 481	.6854386	.2474740	6	39 18 76	245 314 376	.0199920	.5544372
2	31 58 44	177 504 328	.7065392	.2523715	7	39 31 29	246 491 883	.0399681	.5589899
3	31 69 69	178 453 547	.7276210	.2572633	8	39 43 84	247 673 152	.0599282	.5635377
4	31 80 96	179 406 144	.7486842	.2621492	9	39 56 41	248 858 189	.0798724	.5680807
565	31 92 25	180 362 125	23.7697286	8.2670294	630	39 69 00	250 047 000	25.0998008	8.5726189
6	32 03 56	181 321 496	.7907545	.2719039	1	39 81 61	251 239 591	.1197134	.5771523
7	32 14 89	182 284 263	.8117618	.2767726	2	39 94 24	252 435 968	.1396102	.5816809
8	32 26 24	183 250 432	.8327506	.2816355	3	40 06 89	253 636 137	.1594913	.5862047
9	32 37 61	184 220 009	.8537209	.2864928	4	40 19 56	254 840 104	.1793566	.5907238
570	32 49 00	185 193 000	23.8746728	8.2913444	635	40 32 25	256 047 875	25.1992063	8.5952380
1	32 60 41	186 169 411	.8956063	.2961903	6	40 44 96	257 259 456	.2190404	.5997476
2	32 71 84	187 149 248	.9165215	.3010304	7	40 57 69	258 474 853	.2388589	.6042525
3	32 83 29	188 132 517	.9374184	.3058651	8	40 70 44	259 694 072	.2586619	.6087526
4	32 94 76	189 119 224	.9582971	.3106941	9	40 83 21	260 917 119	.2784493	.6132480
575	33 06 25	190 109 375	23.9791576	8.3155175	640	40 96 00	262 144 000	25.2982213	8.6177388
6	33 17 76	191 102 976	24.0000000	.3203353	1	41 08 81	263 374 721	.3179778	.6222248
7	33 29 29	192 100 033	.0208243	.3251475	2	41 21 64	264 609 288	.3377189	.6267063
8	33 40 84	193 100 552	.0416306	.3299542	3	41 34 49	265 847 707	.3574447	.6311830
9	33 52 41	194 104 539	.0624188	.3347553	4	41 47 36	267 089 984	.3771551	.6356551
580	33 64 00	195 112 000	24.0831891	8.3395509	645	41 60 25	268 336 125	25.3968502	8.6401226
1	33 75 61	196 122 941	.1039416	.3443410	6	41 73 16	269 586 136	.4165301	.6445855
2	33 87 24	197 137 368	.1246762	.3491256	7	41 86 09	270 840 023	.4361947	.6490437
3	33 98 89	198 155 287	.1453929	.3539047	8	41 99 04	272 097 792	.4558441	.6534974
4	34 10 56	199 176 704	.1660919	.3586784	9	42 12 01	273 359 449	.4754784	.6579465
585	34 22 25	200 201 625	24.1867732	8.3634466	650	42 25 00	274 625 000	25.4950976	8.66235:.

2.—SQUARES, CUBES, SQUARE ROOTS, CUBE ROOTS, OF NUMBERS
1 TO 1600—Continued.

No.	Square	Cube.	Sq. Rt.	Cu. Rt.	No.	Square	Cube.	Sq. Rt.	Cu. Rt.
650	42 25 00	274 625 000	25.4950976	8.6623911	715	51 12 25	365 525 875	26.7394839	8.9420140
1	42 38 01	275 894 451	.5147016	.6668310	16	51 26 56	367 061 696	.7581763	.9461809
2	42 51 04	277 167 808	.5342907	.6712665	17	51 40 89	368 601 813	.7768557	.9503438
3	42 64 09	278 445 077	.5538647	.6756974	18	51 55 24	370 146 232	.7955220	.9545029
4	42 77 16	279 726 264	.5734237	.6801237	19	51 69 61	371 694 959	.8141754	.9586581
655	42 90 25	281 011 375	25.5929678	8.6845456	720	51 84 00	373 248 000	26.8328157	8.9628095
6	43 03 36	282 300 416	.6124969	.6889630	1	51 98 41	374 805 361	.8514432	.9669570
7	43 16 49	283 593 393	.6320112	.6933759	2	52 12 84	376 367 048	.8700577	.9711007
8	43 29 64	284 890 312	.6515107	.6977843	3	52 27 29	377 933 067	.8886593	.9752406
9	43 42 81	286 191 179	.6709953	.7021882	4	52 41 76	379 503 424	.9072481	.9793766
660	43 56 00	287 496 000	25.6904652	8.7065877	725	52 56 25	381 078 125	26.9258240	8.9835089
1	43 69 21	288 804 781	.7099203	.7109827	6	52 70 76	382 657 176	.9443872	.9876373
2	43 82 44	290 117 528	.7293607	.7153734	7	52 85 29	384 240 583	.9629375	.9917620
3	43 95 69	291 434 247	.7487864	.7197596	8	52 99 84	385 828 352	.9814751	.9958829
4	44 08 96	292 754 944	.7681975	.7241414	9	53 14 41	387 420 489	27.0000000	9.0000000
665	44 22 25	294 079 625	25.7875939	8.7285137	730	53 29 00	389 017 000	27.0185122	9.0041134
6	44 35 56	295 408 296	.8069758	.7328918	1	53 43 61	390 617 891	.0370117	.0082229
7	44 48 89	296 740 963	.8263431	.7372604	2	53 58 24	392 223 168	.0554985	.0123288
8	44 62 24	298 077 632	.8456960	.7416246	3	53 72 89	393 832 837	.0739727	.0164309
9	44 75 61	299 418 309	.8650343	.7459846	4	53 87 56	395 446 904	.0924344	.0205293
670	44 89 00	300 763 000	25.8843582	8.7503401	735	54 02 25	397 065 375	27.1108834	9.0246229
1	45 02 41	302 111 711	.9036677	.7546913	6	54 16 96	398 688 256	.1293199	.0287149
2	45 15 84	303 464 448	.9229628	.7590383	7	54 31 69	400 315 553	.1477439	.0328021
3	45 29 29	304 821 217	.9422435	.7633809	8	54 46 44	401 947 272	.1661554	.0368857
4	45 42 76	306 182 024	.9615100	.7677192	9	54 61 21	403 583 419	.1845544	.0409655
675	45 56 25	307 546 875	25.9807621	8.7720532	740	54 76 00	405 224 000	27.2029410	9.0450447
6	45 69 76	308 915 776	26.0000000	.7763830	1	54 90 81	406 869 021	.2213152	.0491142
7	45 83 29	310 288 733	.0192237	.7807084	2	55 05 64	408 518 488	.2396769	.0531831
8	45 96 84	311 665 752	.0384331	.7850296	3	55 20 49	410 172 407	.2580263	.0572482
9	46 10 41	313 046 839	.0576284	.7893466	4	55 35 36	411 830 784	.2763634	.0613098
680	46 24 00	314 432 000	26.0768096	8.7936593	745	55 50 25	413 493 625	27.2946881	9.0653677
1	46 37 61	315 821 241	.0959767	.7979679	6	55 65 16	415 160 936	.3130006	.0694220
2	46 51 24	317 214 568	.1151297	.8022721	7	55 80 09	416 832 723	.3313007	.0734726
3	46 64 89	318 611 987	.1342687	.8065722	8	55 95 04	418 508 992	.3495887	.0775197
4	46 78 56	320 013 504	.1533937	.8108681	9	56 10 01	420 189 749	.3678644	.0815631
685	46 92 25	321 419 125	26.1725047	8.8151598	750	56 25 00	421 875 000	27.3861279	9.0856030
6	47 05 96	322 828 856	.1916017	.8194474	1	56 40 01	423 564 751	.4043792	.0896392
7	47 19 69	324 242 703	.2106848	.8237307	2	56 55 04	425 259 008	.4226184	.0936719
8	47 33 44	325 660 672	.2297541	.8280099	3	56 70 09	426 957 777	.4408455	.0977010
9	47 47 21	327 082 769	.2488095	.8322850	4	56 85 16	428 661 064	.4590604	.1017265
690	47 61 00	328 509 000	26.2678511	8.8365559	755	57 00 25	430 368 875	27.4772633	9.1057485
1	47 74 81	329 939 371	.2868789	.8408227	6	57 15 36	432 081 216	.4954542	.1097669
2	47 88 64	331 373 888	.3058929	.8450854	7	57 30 49	433 798 093	.5136330	.1137818
3	48 02 49	332 812 557	.3248932	.8493440	8	57 45 64	435 519 512	.5317998	.1177931
4	48 16 36	334 255 384	.3438797	.8535985	9	57 60 81	437 245 479	.5499546	.1218010
695	48 30 25	335 702 375	26.3628527	8.8578486	760	57 76 00	438 976 000	27.5680975	9.1258053
6	48 44 16	337 153 536	.3818119	.8620952	1	57 91 21	440 711 081	.5862284	.1298061
7	48 58 09	338 608 873	.4007576	.8663375	2	58 06 44	442 450 728	.6043475	.1338034
8	48 72 04	340 068 392	.4196896	.8705757	3	58 21 69	444 194 947	.6224546	.1377971
9	48 86 01	341 532 099	.4386081	.8748099	4	58 36 96	445 943 744	.6405499	.1417874
700	49 00 00	343 000 000	26.4575131	8.8790400	765	58 52 25	447 697 125	27.6586334	9.1457742
1	49 14 01	344 472 101	.4764046	.8832661	6	58 67 56	449 455 096	.6767050	.1497576
2	49 28 04	345 948 408	.4952826	.8874882	7	58 82 89	451 217 663	.6947648	.1537375
3	49 42 09	347 428 927	.5141472	.8917063	8	58 98 24	452 984 832	.7128129	.1577139
4	49 56 16	348 913 664	.5329983	.8959204	9	59 13 61	454 756 609	.7308492	.1616869
705	49 70 25	350 402 625	26.5518361	8.9001304	770	59 29 00	456 533 000	27.7488739	9.1656565
6	49 84 36	351 895 816	.5706605	.9043366	1	59 44 41	458 314 011	.7668868	.1696225
7	49 98 49	353 393 243	.5894716	.9085387	2	59 59 84	460 099 648	.7848880	.1735852
8	50 12 64	354 894 912	.6082694	.9127369	3	59 75 29	461 889 917	.8028775	.1775445
9	50 26 81	356 400 829	.6270539	.9169311	4	59 90 76	463 684 824	.8208555	.1815003
710	50 41 00	357 911 000	26.6458252	8.9211214	775	60 06 25	465 484 375	27.8388218	9.1854527
11	50 55 21	359 425 431	.6645833	.9253078	6	60 21 76	467 288 576	.8567766	.1894018
12	50 69 44	360 944 128	.6833281	.9294902	7	60 37 29	469 097 433	.8747197	.1933474
13	50 83 69	362 467 097	.7020598	.9336687	8	60 52 84	470 910 952	.8926514	.1972897
14	50 97 96	363 994 344	.7207784	.9378433	9	60 68 41	472 729 139	.9105715	.2012286
715	51 12 25	365 525 875	26.7394839	8.9420140	780	60 84 00	474 552 000	27.9284801	9.2051641

2.—Squares, Cubes, Square Roots, Cube Roots, of Numbers
1 to 1600—Continued.

No.	Square	Cube	Sq. Rt.	Cu. Rt.	No.	Square	Cube	Sq. Rt.	Cu. Rt.
780	60 84 00	474 552 000	27.9284801	9.2051641	845	71 40 25	603 351 125	29.0688837	9.4540719
1	60 99 61	476 379 541	.9463772	.2090962	6	71 57 16	605 495 736	.0860791	.4577999
2	61 15 24	478 211 768	.9642629	.2130250	7	71 74 09	607 645 423	.1032644	.4615249
3	61 30 89	480 048 687	.9821372	.2169505	8	71 91 04	609 800 192	.1204396	.4652470
4	61 46 56	481 890 304	28.0000000	.2208726	9	72 08 01	611 960 049	.1376046	.4689661
785	61 62 25	483 736 625	28.0178515	9.2247914	850	72 25 00	614 125 000	29.1547595	9.4726824
6	61 77 96	485 587 656	.0356915	.2287068	1	72 42 01	616 295 051	.1719043	.4763957
7	61 93 69	487 443 403	.0535203	.2326189	2	72 59 04	618 470 208	.1890390	.4801061
8	62 09 44	489 303 872	.0713377	.2365277	3	72 76 09	620 650 477	.2061637	.4838136
9	62 25 21	491 169 069	.0891438	.2404333	4	72 93 16	622 835 864	.2232784	.4875182
790	62 41 00	493 039 000	28.1069386	9.2443355	855	73 10 25	625 026 375	29.2403830	9.4912200
1	62 56 81	494 913 671	.1247222	.2482344	6	73 27 36	627 222 016	.2574777	.4949188
2	62 72 64	496 793 088	.1424946	.2521300	7	73 44 49	629 422 793	.2745623	.4986147
3	62 88 49	498 677 257	.1602557	.2560224	8	73 61 64	631 628 712	.2916370	.5023078
4	63 04 36	500 566 184	.1780056	.2599114	9	73 78 81	633 839 779	.3087018	.5059980
795	63 20 25	502 459 875	28.1957444	9.2637973	860	73 96 00	636 056 000	29.3257566	9.5096854
6	63 36 16	504 358 336	.2134720	.2676798	1	74 13 21	638 277 381	.3428015	.5133699
7	63 52 09	506 261 573	.2311884	.2715592	2	74 30 44	640 503 928	.3598365	.5170515
8	63 68 04	508 169 592	.2488938	.2754352	3	74 47 69	642 735 647	.3768616	.5207303
9	63 84 01	510 082 399	.2665881	.2793081	4	74 64 96	644 972 544	.3938769	.5244063
800	64 00 00	512 000 000	28.2842712	9.2831777	865	74 82 25	647 214 625	29.4108823	9.5280794
1	64 16 01	513 922 401	.3019434	.2870440	6	74 99 56	649 461 896	.4278779	.5317497
2	64 32 04	515 849 608	.3196045	.2909072	7	75 16 89	651 714 363	.4448637	.5354172
3	64 48 09	517 781 627	.3372546	.2947671	8	75 34 24	653 972 032	.4618397	.5390818
4	64 64 16	519 718 464	.3548938	.2986239	9	75 51 61	656 234 909	.4788059	.5427437
805	64 80 25	521 660 125	28.3725219	9.3024775	870	75 69 00	658 503 000	29.4957624	9.5464027
6	64 96 36	523 606 616	.3901391	.3063278	1	75 86 41	660 776 311	.5127091	.5500589
7	65 12 49	525 557 943	.4077454	.3101750	2	76 03 84	663 054 848	.5296461	.5537123
8	65 28 64	527 514 112	.4253408	.3140190	3	76 21 29	665 338 617	.5465734	.5573630
9	65 44 81	529 475 129	.4429253	.3178599	4	76 38 76	667 627 624	.5634910	.5610108
810	65 61 00	531 441 000	28.4604989	9.3216975	875	76 56 25	669 921 875	29.5803989	9.5646559
11	65 77 21	533 411 731	.4780617	.3255320	6	76 73 76	672 221 376	.5972972	.5682982
12	65 93 44	535 387 328	.4956137	.3293634	7	76 91 29	674 526 133	.6141858	.5719377
13	66 09 69	537 367 797	.5131549	.3331916	8	77 08 84	676 836 152	.6310648	.5755745
14	66 25 96	539 353 144	.5306852	.3370167	9	77 26 41	679 151 439	.6479342	.5792085
815	66 42 25	541 343 375	28.5482048	9.3408386	880	77 44 00	681 472 000	29.6647939	9.5828397
16	66 58 56	543 338 496	.5657137	.3446575	1	77 61 61	683 797 841	.6816442	.5864682
17	66 74 89	545 338 513	.5832119	.3484731	2	77 79 24	686 128 968	.6984848	.5900939
18	66 91 24	547 343 432	.6006993	.3522857	3	77 96 89	688 465 387	.7153159	.5937169
19	67 07 61	549 353 259	.6181760	.3560952	4	78 14 56	690 807 104	.7321375	.5973373
820	67 24 00	551 368 000	28.6356421	9.3599016	885	78 32 25	693 154 125	29.7489496	9.6009548
1	67 40 41	553 387 661	.6530976	.3637049	6	78 49 96	695 506 456	.7657521	.6045696
2	67 56 84	555 412 248	.6705424	.3675051	7	78 67 69	697 864 103	.7825452	.6081817
3	67 73 29	557 441 767	.6879766	.3713022	8	78 85 44	700 227 072	.7993289	.6117911
4	67 89 76	559 476 224	.7054002	.3750963	9	79 03 21	702 595 369	.8161030	.6153977
825	68 06 25	561 515 625	28.7228132	9.3788873	890	79 21 00	704 969 000	29.8328678	9.6190017
6	68 22 76	563 559 976	.7402157	.3826752	1	79 38 81	707 347 971	.8496231	.6226030
7	68 39 29	565 609 283	.7576077	.3864600	2	79 56 64	709 732 288	.8663690	.6262016
8	68 55 84	567 663 552	.7749891	.3902419	3	79 74 49	712 121 957	.8831056	.6297975
9	68 72 41	569 722 789	.7923601	.3940206	4	79 92 36	714 516 984	.8998328	.6333907
830	68 89 00	571 787 000	28.8097206	9.3977964	895	80 10 25	716 917 375	29.9165506	9.6369812
1	69 05 61	573 856 191	.8270706	.4015691	6	80 28 16	719 323 136	.9332591	.6405690
2	69 22 24	575 930 368	.8444102	.4053387	7	80 46 09	721 734 273	.9499583	.6441542
3	69 38 89	578 009 537	.8617394	.4091054	8	80 64 04	724 150 792	.9666481	.6477367
4	69 55 56	580 093 704	.8790582	.4128690	9	80 82 01	726 572 699	.9833287	.6513166
835	69 72 25	582 182 875	28.8963666	9 4166297	900	81 00 00	729 000 000	30.0000000	9.6548938
6	69 88 96	584 277 056	.9136646	.4203873	1	81 18 01	731 432 701	.0166620	.6584684
7	70 05 69	586 376 253	.9309523	.4241420	2	81 36 04	733 870 808	.0333148	.6620403
8	70 22 44	588 480 472	.9482297	.4278936	3	81 54 09	736 314 327	.0499584	.6656096
9	70 39 21	590 589 719	.9654967	.4316423	4	81 72 16	738 763 264	.0665928	.6691762
840	70 56 00	592 704 000	28.9827535	9.4353880	905	81 90 25	741 217 625	30.0832179	9.6727403
1	70 72 81	594 823 321	29.0000000	.4391307	6	82 08 36	743 677 416	.0998339	.6763017
2	70 89 64	596 947 688	.0172363	.4428704	7	82 26 49	746 142 643	.1164407	.6798604
3	71 06 49	599 077 107	.0344623	.4466072	8	82 44 64	748 613 312	.1330383	.6834166
4	71 23 36	601 211 584	.0516781	.4503410	9	82 62 81	751 089 429	.1496269	.6869701
845	71 40 25	603 351 125	29.0688837	9.4540719	910	82 81 00	753 571 000	30.1662063	9.6905211

2. —Squares, Cubes, Square Roots, Cube Roots, of Numbers
1 to 1600—Continued.

No.	Square	Cube	Sq. Rt.	Cu. Rt.	No.	Square	Cube	Sq. Rt.	Cu. Rt.
910	82 81 00	753 571 000	30.1662063	9.6905211	975	95 06 25	926 859 375	31.2249900	9.9159624
11	82 99 21	756 058 031	.1827765	.6940694	6	95 25 76	929 714 176	.2409987	.9193513
12	83 17 44	758 550 528	.1993377	.6976151	7	95 45 29	932 574 833	.2569992	.9227379
13	83 35 69	761 048 497	.2158899	.7011583	8	95 64 84	935 441 352	.2729915	.9261222
14	83 53 96	763 551 944	.2324329	.7046989	9	95 84 41	938 313 739	.2889757	.9295042
915	83 72 25	766 060 875	30.2489669	9.7082369	980	96 04 00	941 192 000	31.3049517	9.9328839
16	83 90 56	768 575 296	.2654919	.7117723	1	96 23 61	944 076 141	.3209195	.9362613
17	84 08 89	771 095 213	.2820079	.7153051	2	96 43 24	946 966 168	.3368792	.9396363
18	84 27 24	773 620 632	.2985148	.7188354	3	96 62 89	949 862 087	.3528308	.9430092
19	84 45 61	776 151 559	.3150128	.7223631	4	96 82 56	952 763 904	.3687743	.9463797
920	84 64 00	778 688 000	30.3315018	9.7258883	985	97 02 25	955 671 625	31.3847097	9.9497479
1	84 82 41	781 229 961	.3479818	.7294109	6	97 21 96	958 585 256	.4006369	.9531138
2	85 00 84	783 777 448	.3644529	.7329309	7	97 41 69	961 504 803	.4165561	.9564775
3	85 19 29	786 330 467	.3809151	.7364484	8	97 61 44	964 430 272	.4324673	.9598389
4	85 37 76	788 889 024	.3973683	.7399634	9	97 81 21	967 361 669	.4483704	.9631981
925	85 56 25	791 453 125	30.4138127	9.7434758	990	98 01 00	970 299 000	31.4642654	9.9665549
6	85 74 76	794 022 776	.4302481	.7469857	1	98 20 81	973 242 271	.4801525	.9699005
7	85 93 29	796 597 983	.4466747	.7504930	2	98 40 64	976 191 488	.4960315	.9732619
8	86 11 84	799 178 752	.4630924	.7539979	3	98 60 49	979 146 657	.5119025	.9766120
9	86 30 41	801 765 089	.4795013	.7575002	4	98 80 36	982 107 784	.5277655	.9799599
930	86 49 00	804 357 000	30.4959014	9.7610001	995	99 00 25	985 074 875	31.5436206	9.9833055
1	86 67 61	806 954 491	.5122926	.7644974	6	99 20 16	988 047 936	.5594677	.9866488
2	86 86 24	809 557 568	.5286750	.7679922	7	99 40 09	991 026 973	.5753068	.9899900
3	87 04 89	812 166 237	.5450487	.7714845	8	99 60 04	994 011 992	.5911330	.9933289
4	87 23 56	814 780 504	.5614136	.7749743	9	99 80 01	997 002 999	.6069613	.9966656
935	87 42 25	817 400 375	30.5777697	9.7784616	1000	1 00 00 00	1 000 000 000	31.6227766	10.0000000
6	87 60 96	820 025 856	.5941171	.7819466	1	1 00 20 01	1 003 003 001	.6385840	.0033322
7	87 79 69	822 656 953	.6104557	.7854288	2	1 00 40 04	1 006 012 008	.6543836	.0066622
8	87 98 44	825 293 672	.6267857	.7889087	3	1 00 60 09	1 009 027 027	.6701752	.0099899
9	88 17 21	827 936 019	.6431069	.7923861	4	1 00 80 16	1 012 048 064	.6859590	.0133155
940	88 36 00	830 584 000	30.6594194	9.7958611	1005	1 01 00 25	1 015 075 125	31.7017349	10.0166389
1	88 54 81	833 237 621	.6757233	.7993336	6	1 01 20 36	1 018 108 216	.7175030	.0199601
2	88 73 64	835 896 888	.6920185	.8028036	7	1 01 40 49	1 021 147 343	.7332633	.0232791
3	88 92 49	838 561 807	.7083051	.8062711	8	1 01 60 64	1 024 192 512	.7490157	.0265958
4	89 11 36	841 232 384	.7245830	.8097362	9	1 01 80 81	1 027 243 729	.7647604	.0299104
945	89 30 25	843 908 625	30.7408523	9.8131989	1010	1 02 01 00	1 030 301 000	31.7804972	10.0332228
6	89 49 16	846 590 536	.7571130	.8166591	11	1 02 21 21	1 033 364 331	.7962262	.0365330
7	89 68 09	849 278 123	.7733651	.8201169	12	1 02 41 44	1 036 433 728	.8119474	.0398410
8	89 87 04	851 971 392	.7896086	.8235723	13	1 02 61 69	1 039 509 197	.8276609	.0431469
9	90 06 01	854 670 349	.8058436	.8270252	14	1 02 81 96	1 042 590 744	.8433666	.0464506
950	90 25 00	857 375 000	30.8220700	9.8304757	1015	1 03 02 25	1 045 678 375	31.8590646	10.0497521
1	90 44 01	860 085 351	.8382879	.8339238	16	1 03 22 56	1 048 772 096	.8747549	.0530514
2	90 63 04	862 801 408	.8544972	.8373695	17	1 03 42 89	1 051 871 913	.8904374	.0563485
3	90 82 09	865 523 177	.8706981	.8408127	18	1 03 63 24	1 054 977 832	.9061123	.0596435
4	91 01 16	868 250 664	.8868904	.8442536	19	1 03 83 61	1 058 089 859	.9217794	.0629364
955	91 20 25	870 983 875	30.9030743	9.8476920	1020	1 04 04 00	1 061 208 000	31.9374388	10.0662271
6	91 39 36	873 722 816	.9192497	.8511250	21	1 04 24 41	1 064 332 261	.9530906	.0695156
7	91 58 49	876 467 493	.9354166	.8545617	22	1 04 44 84	1 067 462 648	.9687347	.0728020
8	91 77 64	879 217 912	.9515751	.8579929	23	1 04 65 29	1 070 599 167	.9843712	.0760863
9	91 96 81	881 974 079	.9677251	.8614218	24	1 04 85 76	1 073 741 824	32.0000000	.0793684
960	92 16 00	884 736 000	30.9838668	9.8648483	1025	1 05 06 25	1 076 890 625	32.0156212	10.0826484
1	92 35 21	887 503 681	31.0000000	.8682724	26	1 05 26 76	1 080 045 576	.0312348	.0859262
2	92 54 44	890 277 128	.0161248	.8716941	27	1 05 47 29	1 083 206 683	.0468407	.0892019
3	92 73 69	893 056 347	.0322413	.8751135	28	1 05 67 84	1 086 373 952	.0624391	.0924755
4	92 92 96	895 841 344	.0483494	.8785305	29	1 05 88 41	1 089 547 389	.0780298	.0957469
965	93 12 25	898 632 125	31.0644491	9.8819451	1030	1 06 09 00	1 092 727 000	32.0936131	10.0990163
6	93 31 56	901 428 696	.0805405	.8853574	1	1 06 29 61	1 095 912 791	.1091887	.1022835
7	93 50 89	904 231 063	.0966236	.8887673	2	1 06 50 24	1 099 104 768	.1247568	.1055487
8	93 70 24	907 039 232	.1126984	.8921749	3	1 06 70 89	1 102 302 937	.1403247	.1088117
9	93 89 61	909 853 209	.1287648	.8955801	4	1 06 91 56	1 105 507 304	.1558704	.1120726
970	94 09 00	912 673 000	31.1448230	9.8989830	1035	1 07 12 25	1 108 717 875	32.1714159	10.1153314
1	94 28 41	915 498 611	.1608729	.9023835	36	1 07 32 96	1 111 934 656	.1869539	.1185882
2	94 47 84	918 330 048	.1769145	.9057817	37	1 07 53 69	1 115 157 653	.2024844	.1218428
3	94 67 29	921 167 317	.1929477	.9091776	38	1 07 74 44	1 118 386 872	.2180074	.1250953
4	94 86 76	924 010 424	.2089731	.9125712	39	1 07 95 21	1 121 622 319	.2335229	.1283457
975	95 06 25	926 859 375	31.2249900	9.9159624	1040	1 08 16 00	1 124 864 000	32.2490310	10.1315941

2.—Squares, Cubes, Square Roots, Cube Roots, of Numbers
1 to 1600—Continued.

No.	Square	Cube.	Sq. Rt.	Cu. Rt.	No.	Square	Cube.	Sq. Rt.	Cu. Rt.
1040	1 08 16 00	1 124 864 000	32.2490310	10.1315941	1105	1 22 10 25	1 349 232 625	33.2415403	10.3384181
41	1 08 36 81	1 128 111 921	.2645316	.1348403	6	1 22 32 36	1 352 899 016	.2565783	.3415358
42	1 08 57 64	1 131 366 088	.2800248	.1380845	7	1 22 54 49	1 356 572 043	.2716095	.3446517
43	1 08 78 49	1 134 626 507	.2955105	.1413266	8	1 22 76 64	1 360 251 712	.2866339	.3477657
44	1 08 99 36	1 137 893 184	.3109888	.1445667	9	1 22 98 81	1 363 938 029	.3016516	.3508778
1045	1 09 20 25	1 141 166 125	32.3264598	10.1478047	1110	1 23 21 00	1 367 631 000	33.3166625	10.3539880
46	1 09 41 16	1 144 445 336	.3419233	.1510406	11	1 23 43 21	1 371 330 631	.3316666	.3570964
47	1 09 62 09	1 147 730 823	.3573794	.1542744	12	1 23 65 44	1 375 036 928	.3466640	.3602029
48	1 09 83 04	1 151 022 592	.3728281	.1575062	13	1 23 87 69	1 378 749 897	.3616546	.3633076
49	1 10 04 01	1 154 320 649	.3882695	.1607359	14	1 24 09 96	1 382 469 544	.3766385	.3664103
1050	1 10 25 00	1 157 625 000	32.4037035	10.1639636	1115	1 24 32 25	1 386 195 875	33.3916157	10.3695113
51	1 10 46 01	1 160 935 651	.4191301	.1671893	16	1 24 54 56	1 389 928 896	.4065862	.3726103
52	1 10 67 04	1 164 252 608	.4345495	.1704129	17	1 24 76 89	1 393 668 613	.4215499	.3757076
53	1 10 88 09	1 167 575 877	.4499615	.1736344	18	1 24 99 24	1 397 415 032	.4365070	.3788030
54	1 11 09 16	1 170 905 464	.4653662	.1768539	19	1 25 21 61	1 401 168 159	.4514573	.3818965
1055	1 11 30 25	1 174 241 375	32.4807635	10.1800714	1120	1 25 44 00	1 404 928 000	33.4664011	10.3849882
56	1 11 51 36	1 177 583 616	.4961536	.1832868	21	1 25 66 41	1 408 694 561	.4813381	.3880781
57	1 11 72 49	1 180 932 193	.5115364	.1865002	22	1 25 88 84	1 412 467 848	.4962684	.3911661
58	1 11 93 64	1 184 287 112	.5269119	.1897116	23	1 26 11 29	1 416 247 867	.5111921	.3942523
59	1 12 14 81	1 187 648 379	.5422802	.1929209	24	1 26 33 76	1 420 034 624	.5261092	.3973366
1060	1 12 36 00	1 191 016 000	32.5576412	10.1961283	1125	1 26 56 25	1 423 828 125	33.5410196	10.4004192
61	1 12 57 21	1 194 389 981	.5729949	.1993336	26	1 26 78 76	1 427 628 376	.5559234	.4034999
62	1 12 78 44	1 197 770 328	.5883415	.2025369	27	1 27 01 29	1 431 435 383	.5708206	.4065787
63	1 12 99 69	1 201 157 047	.6036807	.2057382	28	1 27 23 84	1 435 249 152	.5857112	.4096557
64	1 13 20 96	1 204 550 144	.6190129	.2089375	29	1 27 46 41	1 439 069 689	.6005952	.4127310
1065	1 13 42 25	1 207 949 625	32.6343377	10.2121347	1130	1 27 69 00	1 442 897 000	33.6154726	10.4158044
66	1 13 63 56	1 211 355 496	.6496554	.2153300	31	1 27 91 61	1 446 731 091	.6303434	.4188760
67	1 13 84 89	1 214 767 763	.6649659	.2185233	32	1 28 14 24	1 450 571 968	.6452077	.4219458
68	1 14 06 24	1 218 186 432	.6802693	.2217146	33	1 28 36 89	1 454 419 637	.6600653	.4250138
69	1 14 27 61	1 221 611 509	.6955654	.2249039	34	1 28 59 56	1 458 274 104	.6749165	.4280800
1070	1 14 49 00	1 225 043 000	32.7108544	10.2280912	1135	1 28 82 25	1 462 135 375	33.6897610	10.4311443
71	1 14 70 41	1 228 480 911	.7261363	.2312766	36	1 29 04 96	1 466 003 456	.7045991	.4342069
72	1 14 91 84	1 231 925 248	.7414111	.2344599	37	1 29 27 69	1 469 878 353	.7194306	.4372677
73	1 15 13 29	1 235 376 017	.7566787	.2376413	38	1 29 50 44	1 473 760 072	.7342556	.4403267
74	1 15 34 76	1 238 833 224	.7719392	.2408207	39	1 29 73 21	1 477 648 619	.7490741	.4433839
1075	1 15 56 25	1 242 296 875	32.7871926	10.2439981	1140	1 29 96 00	1 481 544 000	33.7638860	10.4464393
76	1 15 77 76	1 245 766 976	.8024389	.2471735	41	1 30 18 81	1 485 446 221	.7786915	.4494929
77	1 15 99 29	1 249 243 533	.8176782	.2503470	42	1 30 41 64	1 489 355 288	.7934905	.4525448
78	1 16 20 84	1 252 726 552	.8329103	.2535186	43	1 30 64 49	1 493 271 207	.8082830	.4555948
79	1 16 42 41	1 256 216 039	.8481354	.2566881	44	1 30 87 36	1 497 193 984	.8230691	.4586431
1080	1 16 64 00	1 259 712 000	32.8633535	10.2598557	1145	1 31 10 25	1 501 123 625	33.8375486	10.4616896
81	1 16 85 61	1 263 214 441	.8785644	.2630213	46	1 31 33 16	1 505 060 136	.8526218	.4647343
82	1 17 07 24	1 266 723 368	.8937684	.2661850	47	1 31 56 09	1 509 003 523	.8673884	.4677773
83	1 17 28 89	1 270 238 787	.9089653	.2693467	48	1 31 79 04	1 512 953 792	.8821487	.4708185
84	1 17 50 56	1 273 760 704	.9241553	.2725065	49	1 32 02 01	1 516 910 949	.8969025	.4738579
1085	1 17 72 25	1 277 289 125	32.9393382	10.2756644	1150	1 32 25 00	1 520 875 000	33.9116499	10.4768955
86	1 17 93 96	1 280 824 056	.9545141	.2788203	51	1 32 48 01	1 524 845 951	.9263909	.4799314
87	1 18 15 69	1 284 365 503	.9696830	.2819743	52	1 32 71 04	1 528 823 808	.9411255	.4829656
88	1 18 37 44	1 287 913 472	.9848450	.2851264	53	1 32 94 09	1 532 808 577	.9558537	.4859980
89	1 18 59 21	1 291 467 969	33.0000000	.2882765	54	1 33 17 16	1 536 800 264	.9705755	.4890286
1090	1 18 81 00	1 295 029 000	33.0151480	10.2914247	1155	1 33 40 25	1 540 798 875	33.9852910	10.4920575
91	1 19 02 81	1 298 596 571	.0302891	.2945709	56	1 33 63 36	1 544 804 416	34.0000000	.4950847
92	1 19 24 64	1 302 170 688	.0454233	.2977153	57	1 33 86 49	1 548 816 893	.0147027	.4981101
93	1 19 46 49	1 305 751 357	.0605505	.3008577	58	1 34 09 64	1 552 836 312	.0293990	.5011337
94	1 19 68 36	1 309 338 584	.0756708	.3039982	59	1 34 32 81	1 556 862 679	.0440890	.5041556
1095	1 19 90 25	1 312 932 375	33.0907842	10.3071368	1160	1 34 56 00	1 560 896 000	34.0587727	10.5071757
96	1 20 12 16	1 316 532 736	.1058907	.3102735	61	1 34 79 21	1 564 936 281	.0734501	.5101942
97	1 20 34 09	1 320 139 673	.1209903	.3134083	62	1 35 02 44	1 568 983 528	.0881211	.5132109
98	1 20 56 04	1 323 753 192	.1360830	.3165411	63	1 35 25 69	1 573 037 747	.1027858	.5162259
99	1 20 78 01	1 327 373 299	.1511689	.3196721	64	1 35 48 96	1 577 098 944	.1174442	.5192391
1100	1 21 00 00	1 331 000 000	33.1662479	10.3228012	1165	1 35 72 25	1 581 167 125	34.1320963	10.5222506
1	1 21 22 01	1 334 633 301	.1813200	.3259284	66	1 35 95 56	1 585 242 296	.1467422	.5252604
2	1 21 44 04	1 338 273 208	.1963853	.3290537	67	1 36 18 89	1 589 324 463	.1613817	.5282685
3	1 21 66 09	1 341 919 727	.2114438	.3321770	68	1 36 42 24	1 593 413 632	.1760150	.5312749
4	1 21 88 16	1 345 572 864	.2264955	.3352985	69	1 36 65 61	1 597 509 809	.1906420	.5342795
1105	1 22 10 25	1 349 232 625	33.2415403	10.3384181	1170	1 36 89 00	1 601 613 000	34.2052627	10.5372825

2.—Squares, Cubes, Square Roots, Cube Roots, of Numbers
1 to 1600—Continued.

No.	Square	Cube	Sq. Rt.	Cu. Rt.	No.	Square	Cube	Sq. Rt.	Cu. Rt.
1170	1 36 89 00	1 601 613 000	34.2052627	10.5372825	1235	1 52 52 25	1 883 652 875	35.1425668	10.7289112
71	1 37 12 41	1 605 723 211	.2198773	.5402837	36	1 52 76 96	1 888 232 256	.1567917	.7318062
72	1 37 35 84	1 609 840 448	.2344855	.5432832	37	1 53 01 69	1 892 819 053	.1710108	.7346997
73	1 37 59 29	1 613 964 717	.2490875	.5462810	38	1 53 26 44	1 897 413 272	.1852242	.7375916
74	1 37 82 76	1 618 096 024	.2636834	.5492771	39	1 53 51 21	1 902 014 919	.1994318	.7404819
1175	1 38 06 25	1 622 234 375	34.2782730	10.5522715	1240	1 53 76 00	1 906 624 000	35.2136337	10.7433707
76	1 38 29 76	1 626 379 776	.2928564	.5552642	41	1 54 00 81	1 911 240 521	.2278299	.7462579
77	1 38 53 29	1 630 532 233	.3074336	.5582552	42	1 54 25 64	1 915 864 488	.2420204	.7491436
78	1 38 76 84	1 634 691 752	.3220046	.5612445	43	1 54 50 49	1 920 495 907	.2562051	.7520277
79	1 39 00 41	1 638 858 339	.3365694	.5642322	44	1 54 75 36	1 925 134 784	.2703842	.7549103
1180	1 39 24 00	1 643 032 000	34.3511281	10.5672181	1245	1 55 00 25	1 929 781 125	35.2845575	10.7577913
81	1 39 47 61	1 647 212 741	.3656805	.5702024	46	1 55 25 16	1 934 434 936	.2987252	.7606708
82	1 39 71 24	1 651 400 568	.3802268	.5731849	47	1 55 50 09	1 939 096 223	.3128872	.7635488
83	1 39 94 89	1 655 595 487	.3947670	.5761658	48	1 55 75 04	1 943 764 992	.3270435	.7664252
84	1 40 18 56	1 659 797 504	.4093011	.5791449	49	1 56 00 01	1 948 441 249	.3411941	.7693001
1185	1 40 42 25	1 664 006 625	34.4238289	10.5821225	1250	1 56 25 00	1 953 125 000	35.3553391	10.7721735
86	1 40 65 96	1 668 222 856	.4383507	.5850983	51	1 56 50 01	1 957 816 251	.3694784	.7750453
87	1 40 89 69	1 672 446 203	.4528663	.5880725	52	1 56 75 04	1 962 515 008	.3836120	.7779156
88	1 41 13 44	1 676 676 672	.4673759	.5910450	53	1 57 00 09	1 967 221 277	.3977400	.7807843
89	1 41 37 21	1 680 914 269	.4818793	.5940158	54	1 57 25 16	1 971 935 064	.4118624	.7836516
1190	1 41 61 00	1 685 159 000	34.4963766	10.5969850	1255	1 57 50 25	1 976 656 375	35.4259792	10.7865173
91	1 41 84 81	1 689 410 871	.5108678	.5999525	56	1 57 75 36	1 981 385 216	.4400903	.7893815
92	1 42 08 64	1 693 669 888	.5253530	.6029184	57	1 58 00 49	1 986 121 593	.4541958	.7922441
93	1 42 32 49	1 697 936 057	.5398321	.6058826	58	1 58 25 64	1 990 865 512	.4682957	.7951053
94	1 42 56 36	1 702 209 384	.5543051	.6088451	59	1 58 50 81	1 995 616 979	.4823900	.7979649
1195	1 42 80 25	1 706 489 875	34.5687720	10.6118060	1260	1 58 76 00	2 000 376 000	35.4964787	10.8008230
96	1 43 04 16	1 710 777 536	.5832329	.6147652	61	1 59 01 21	2 005 142 581	.5105618	.8036797
97	1 43 28 09	1 715 072 373	.5976879	.6177228	62	1 59 26 44	2 009 916 728	.5246393	.8065348
98	1 43 52 04	1 719 374 392	.6121366	.6206788	63	1 59 51 69	2 014 698 447	.5387113	.8093884
99	1 43 76 01	1 723 683 599	.6265794	.6236331	64	1 59 76 96	2 019 487 744	.5527777	.8122404
1200	1 44 00 00	1 728 000 000	34.6410162	10.6265857	1265	1 60 02 25	2 024 284 625	35.5668385	10.8150909
1	1 44 24 01	1 732 323 601	.6554469	.6295367	66	1 60 27 56	2 029 089 096	.5808937	.8179400
2	1 44 48 04	1 736 654 408	.6698716	.6324860	67	1 60 52 89	2 033 901 163	.5949434	.8207876
3	1 44 72 09	1 740 992 427	.6842904	.6354338	68	1 60 78 24	2 038 720 832	.6089876	.8236336
4	1 44 96 16	1 745 337 664	.6987031	.6383799	69	1 61 03 61	2 043 548 109	.6230262	.8264782
1205	1 45 20 25	1 749 690 125	34.7131099	10.6413244	1270	1 61 29 00	2 048 383 000	35.6370593	10.8293213
6	1 45 44 36	1 754 049 816	.7275107	.6442672	71	1 61 54 41	2 053 225 511	.6510869	.8321629
7	1 45 68 49	1 758 416 743	.7419055	.6472085	72	1 61 79 84	2 058 075 648	.6651090	.8350030
8	1 45 92 64	1 762 790 912	.7562944	.6501480	73	1 62 05 29	2 062 933 417	.6791255	.8378416
9	1 46 16 81	1 767 172 329	.7706773	.6530860	74	1 62 30 76	2 067 798 824	.6931366	.8406788
1210	1 46 41 00	1 771 561 000	34.7850543	10.6560223	1275	1 62 56 25	2 072 671 875	35.7071421	10.8435144
11	1 46 65 21	1 775 956 931	.7994253	.6589570	76	1 62 81 76	2 077 552 576	.7211422	.8463485
12	1 46 89 44	1 780 360 128	.8137904	.6618902	77	1 63 07 29	2 082 440 933	.7351367	.8491812
13	1 47 13 69	1 784 770 597	.8281495	.6648217	78	1 63 32 84	2 087 336 952	.7491258	.8520125
14	1 47 37 96	1 789 188 344	.8425028	.6677516	79	1 63 58 41	2 092 240 639	.7631095	.8548422
1215	1 47 62 25	1 793 613 375	34.8568501	10.6706799	1280	1 63 84 00	2 097 152 000	35.7770876	10.8576704
16	1 47 86 56	1 798 045 696	.8711915	.6736066	81	1 64 09 61	2 102 071 041	.7910603	.8604972
17	1 48 10 89	1 802 485 313	.8855271	.6765317	82	1 64 35 24	2 106 997 768	.8050276	.8633225
18	1 48 35 24	1 806 932 232	.8998567	.6794552	83	1 64 60 89	2 111 932 187	.8189894	.8661464
19	1 48 59 61	1 811 386 459	.9141805	.6823771	84	1 64 86 56	2 116 874 304	.8329457	.8689687
1220	1 48 84 00	1 815 848 000	34.9284984	10.6852973	1285	1 65 12 25	2 121 824 125	35.8468966	10.8717897
21	1 49 08 41	1 820 316 861	.9428104	.6882160	86	1 65 37 96	2 126 781 656	.8608421	.8746091
22	1 49 32 84	1 824 793 048	.9571166	.6911331	87	1 65 63 69	2 131 746 903	.8747822	.8774271
23	1 49 57 29	1 829 276 567	.9714169	.6940486	88	1 65 89 44	2 136 719 872	.8887169	.8802436
24	1 49 81 76	1 833 767 424	.9857114	.6969625	89	1 66 15 21	2 141 700 569	.9026461	.8830587
1225	1 50 06 25	1 838 265 625	35.0000000	10.6998748	1290	1 66 41 00	2 146 689 000	35.9165699	10.8858723
26	1 50 30 76	1 842 771 176	.0142828	.7027855	91	1 66 66 81	2 151 685 171	.9304884	.8886845
27	1 50 55 29	1 847 284 083	.0285598	.7056947	92	1 66 92 64	2 156 689 088	.9444015	.8914952
28	1 50 79 84	1 851 804 352	.0428309	.7086023	93	1 67 18 49	2 161 700 757	.9583092	.8943044
29	1 51 04 41	1 856 331 989	.0570963	.7115083	94	1 67 44 36	2 166 720 184	.9722115	.8971123
1230	1 51 29 00	1 860 867 000	35.0713558	10.7144127	1295	1 67 70 25	2 171 747 375	35.9861084	10.8999186
31	1 51 53 61	1 865 409 391	.0856096	.7173155	96	1 67 96 16	2 176 782 336	36.0000000	.9027235
32	1 51 78 24	1 869 959 168	.0998575	.7202168	97	1 68 22 09	2 181 825 073	.0138862	.9055269
33	1 52 02 89	1 874 516 337	.1140997	.7231165	98	1 68 48 04	2 186 875 592	.0277671	.9083290
34	1 52 27 56	1 879 080 904	.1283361	.7260146	99	1 68 74 01	2 191 933 899	.0416426	.9111296
1235	1 52 52 25	1 883 652 875	35.1425668	10.7289112	1300	1 69 00 00	2 197 000 000	36.0555128	10.9139287

2.—Squares, Cubes, Square Roots, Cube Roots, of Numbers
1 to 1600—Continued.

No.	Square	Cube	Sq. Rt.	Cu. Rt.	No.	Square	Cube	Sq. Rt.	Cu. Rt.
1300	1 69 00 00	2 197 000 000	36.0555128	10.9139287	1365	1 86 32 25	2 543 302 125	36.9459064	11.0928775
1	1 69 26 01	2 202 073 901	.0693776	.9167265	66	1 86 59 56	2 548 895 896	.9594372	.0955857
2	1 69 52 04	2 207 155 608	.0832371	.9195228	67	1 86 86 89	2 554 497 863	.9729631	.0982926
3	1 69 78 09	2 212 245 127	.0970913	.9223177	68	1 87 14 24	2 560 108 032	.9864840	.1009982
4	1 70 04 16	2 217 342 464	.1109402	.9251111	69	1 87 41 61	2 565 726 409	37.0000000	.1037025
1305	1 70 30 25	2 222 447 625	36.1247837	10.9279031	1370	1 87 69 00	2 571 353 000	37.0135110	11.1064054
6	1 70 56 36	2 227 560 616	.1386220	.9306937	71	1 87 96 41	2 576 987 811	.0270172	.1091070
7	1 70 82 49	2 232 681 443	.1524550	.9334829	72	1 88 23 84	2 582 630 848	.0405184	.1118073
8	1 71 08 64	2 237 810 112	.1662826	.9362706	73	1 88 51 29	2 588 282 117	.0540146	.1145064
9	1 71 34 81	2 242 946 629	.1801050	.9390569	74	1 88 78 76	2 593 941 624	.0675060	.1172041
1310	1 71 61 00	2 248 091 000	36.1939221	10.9418418	1375	1 89 06 25	2 599 609 375	37.0809924	11.1199004
11	1 71 87 21	2 253 243 231	.2077340	.9446253	76	1 89 33 76	2 605 285 376	.0944740	.1225955
12	1 72 13 44	2 258 403 328	.2215406	.9474074	77	1 89 61 29	2 610 969 633	.1079506	.1252893
13	1 72 39 69	2 263 571 297	.2353419	.9501880	78	1 89 88 84	2 616 662 152	.1214224	.1279817
14	1 72 65 96	2 268 747 144	.2491379	.9529673	79	1 90 16 41	2 622 362 939	.1348893	.1306729
1315	1 72 92 25	2 273 930 875	36.2629287	10.9557451	1380	1 90 44 00	2 628 072 000	37.1483512	11.1333628
16	1 73 18 56	2 279 122 496	.2767143	.9585215	81	1 90 71 61	2 633 789 341	.1618084	.1360514
17	1 73 44 89	2 284 322 013	.2904946	.9612965	82	1 90 99 24	2 639 514 968	.1752606	.1387386
18	1 73 71 24	2 289 529 432	.3042697	.9640701	83	1 91 26 89	2 645 248 887	.1887079	.1414246
19	1 73 97 61	2 294 744 759	.3180396	.9668423	84	1 91 54 56	2 650 991 104	.2021505	.1441093
1320	1 74 24 00	2 299 968 000	36.3318042	10.9696131	1385	1 91 82 25	2 656 741 625	37.2155881	11.1467926
21	1 74 50 41	2 305 199 161	.3455637	.9723825	86	1 92 09 96	2 662 500 456	.2290209	.1494747
22	1 74 76 84	2 310 438 248	.3593179	.9751505	87	1 92 37 69	2 668 267 603	.2424489	.1521555
23	1 75 03 29	2 315 685 267	.3730670	.9779171	88	1 92 65 44	2 674 043 072	.2558720	.1548350
24	1 75 29 76	2 320 940 224	.3868108	.9806823	89	1 92 93 21	2 679 826 869	.2692903	.1575133
1325	1 75 56 25	2 326 203 125	36.4005494	10.9834462	1390	1 93 21 00	2 685 619 000	37.2827037	11.1601903
26	1 75 82 76	2 331 473 976	.4142829	.9862087	91	1 93 48 81	2 691 419 471	.2961124	.1628659
27	1 76 09 29	2 336 752 783	.4280112	.9889696	92	1 93 76 64	2 697 228 288	.3095162	.1655403
28	1 76 35 84	2 342 039 552	.4417343	.9917293	93	1 94 04 49	2 703 045 457	.3229152	.1682134
29	1 76 62 41	2 347 334 289	.4554523	.9944876	94	1 94 32 36	2 708 870 984	.3363094	.1708852
1330	1 76 89 00	2 352 637 000	36.4691650	10.9972445	1395	1 94 60 25	2 714 704 875	37.3496988	11.1735558
31	1 77 15 61	2 357 947 691	.4828727	11.0000000	96	1 94 88 16	2 720 547 136	.3630834	.1762250
32	1 77 42 24	2 363 266 368	.4965752	.0027541	97	1 95 16 09	2 726 397 773	.3764632	.1788930
33	1 77 68 89	2 368 593 037	.5102725	.0055069	98	1 95 44 04	2 732 256 792	.3898382	.1815598
34	1 77 95 56	2 373 927 704	.5239647	.0082583	99	1 95 72 01	2 738 124 199	.4032084	.1842252
1335	1 78 22 25	2 379 270 375	36.5376518	11.0110082	1400	1 96 00 00	2 744 000 000	37.4165738	11.1868894
36	1 78 48 96	2 384 621 056	.5513338	.0137569	1	1 96 28 01	2 749 884 201	.4299345	.1895523
37	1 78 75 69	2 389 979 753	.5650106	.0165041	2	1 96 56 04	2 755 776 808	.4432904	.1922139
38	1 79 02 44	2 395 346 472	.5786823	.0192500	3	1 96 84 09	2 761 677 827	.4566416	.1948743
39	1 79 29 21	2 400 721 219	.5923489	.0219945	4	1 97 12 16	2 767 587 264	.4699880	.1975334
1340	1 79 56 00	2 406 104 000	36.6060104	11.0247377	1405	1 97 40 25	2 773 505 125	37.4833296	11.2001913
41	1 79 82 81	2 411 494 821	.6196668	.0274795	6	1 97 68 36	2 779 431 416	.4966665	.2028479
42	1 80 09 64	2 416 893 688	.6333181	.0302199	7	1 97 96 49	2 785 366 143	.5099987	.2055032
43	1 80 36 49	2 422 300 607	.6469644	.0329590	8	1 98 24 64	2 791 309 312	.5233261	.2081573
44	1 80 63 36	2 427 715 584	.6606056	.0356967	9	1 98 52 81	2 797 260 929	.5366487	.2108101
1345	1 80 90 25	2 433 138 625	36.6742416	11.0384330	1410	1 98 81 00	2 803 221 000	37.5499667	11.2134617
46	1 81 17 16	2 438 569 736	.6878726	.0411680	11	1 99 09 21	2 809 189 531	.5632799	.2161120
47	1 81 44 09	2 444 008 923	.7014986	.0439017	12	1 99 37 44	2 815 166 528	.5765885	.2187611
48	1 81 71 04	2 449 456 192	.7151195	.0466339	13	1 99 65 69	2 821 151 997	.5898922	.2214089
49	1 81 98 01	2 454 911 549	.7287353	.0493649	14	1 99 93 96	2 827 145 944	.6031913	.2240554
1350	1 82 25 00	2 460 375 000	36.7423461	11.0520945	1415	2 00 22 25	2 833 148 375	37.6164857	11.2267007
51	1 82 52 01	2 465 846 551	.7559519	.0548227	16	2 00 50 56	2 839 159 296	.6297754	.2293448
52	1 82 79 04	2 471 326 208	.7695526	.0575497	17	2 00 78 89	2 845 178 713	.6430604	.2319876
53	1 83 06 09	2 476 813 977	.7831483	.0602752	18	2 01 07 24	2 851 206 632	.6563407	.2346292
54	1 83 33 16	2 482 309 864	.7967390	.0629995	19	2 01 35 61	2 857 243 059	.6696164	.2372696
1355	1 83 60 25	2 487 813 875	36.8103246	11.0657222	1420	2 01 64 00	2 863 288 000	37.6828874	11.2399087
56	1 83 87 36	2 493 326 016	.8239053	.0684437	21	2 01 92 41	2 869 341 461	.6961536	.2425465
57	1 84 14 49	2 498 846 293	.8374809	.0711639	22	2 02 20 84	2 875 403 448	.7094153	.2451831
58	1 84 41 64	2 504 374 712	.8510515	.0738828	23	2 02 49 29	2 881 473 967	.7226722	.2478185
59	1 84 68 81	2 509 911 279	.8646172	.0766003	24	2 02 77 76	2 887 553 024	.7359245	.2504527
1360	1 84 96 00	2 515 456 000	36.8781778	11.0793165	1425	2 03 06 25	2 893 640 625	37.7491722	11.2530856
61	1 85 23 21	2 521 008 881	.8917335	.0820314	26	2 03 34 76	2 899 736 776	.7624152	.2557173
62	1 85 50 44	2 526 569 928	.9052842	.0847449	27	2 03 63 29	2 905 841 483	.7756535	.2583478
63	1 85 77 69	2 532 139 147	.9188299	.0874571	28	2 03 91 84	2 911 954 752	.7888873	.2609770
64	1 86 04 96	2 537 716 544	.9323706	.0901679	29	2 04 20 41	2 918 076 589	.8021163	.2636050
1365	1 86 32 25	2 543 302 125	36.9459064	11.0928775	1430	2 04 49 00	2 924 207 000	37.8153408	11.2662318

2. —Squares, Cubes, Square Roots, Cube Roots, of Numbers
1 to 1600—Continued.

No.	Square	Cube.	Sq. Rt.	Cu. Rt.	No.	Square	Cube.	Sq. Rt.	Cu. Rt.
1430	2 04 49 00	2 924 207 000	37.8153408	11.2662318	1495	2 23 50 25	3 341 362 375	38.6652299	11.4344092
31	2 04 77 61	2 930 345 991	.8285606	.2688573	96	2 23 80 16	3 348 071 936	.6781593	.4369581
32	2 05 06 24	2 936 493 568	.8417759	.2714816	97	2 24 10 09	3 354 790 473	.6910843	.4395059
33	2 05 34 89	2 942 649 737	.8549864	.2741047	98	2 24 40 04	3 361 517 992	.7040050	.4420525
34	2 05 63 56	2 948 814 504	.8681924	.2767266	99	2 24 70 01	3 368 254 499	.7169214	.4445980
1435	2 05 92 25	2 954 987 875	37.8813938	11.2793472	1500	2 25 00 00	3 375 000 000	38.7298335	11.4471424
36	2 06 20 96	2 961 169 856	.8945906	.2819666	1	2 25 30 01	3 381 754 501	.7427412	.4496857
37	2 06 49 69	2 967 360 453	.9077828	.2845849	2	2 25 60 04	3 388 518 008	.7556447	.4522278
38	2 06 78 44	2 973 559 672	.9209704	.2872019	3	2 25 90 09	3 395 290 527	.7685439	.4547688
39	2 07 07 21	2 979 767 519	.9341535	.2898177	4	2 26 20 16	3 402 072 064	.7814389	.4573087
1440	2 07 36 00	2 985 984 000	37.9473319	11.2924323	1505	2 26 50 25	3 408 862 625	38.7943294	11.4598474
41	2 07 64 81	2 992 209 121	.9605058	.2950457	6	2 26 80 36	3 415 662 216	.8072158	.4623850
42	2 07 93 64	2 998 442 888	.9736751	.2976579	7	2 27 10 49	3 422 470 843	.8200978	.4649215
43	2 08 22 49	3 004 685 307	.9868398	.3002688	8	2 27 40 64	3 429 288 512	.8329757	.4674568
44	2 08 51 36	3 010 936 384	38.0000000	.3028786	9	2 27 70 81	3 436 115 229	.8458491	.4699911
1445	2 08 80 25	3 017 196 125	38.0131556	11.3054871	1510	2 28 01 00	3 442 951 000	38.8587184	11.4725242
46	2 09 09 16	3 023 464 536	.0263067	.3080945	11	2 28 31 21	3 449 795 831	.8715834	.4750562
47	2 09 38 09	3 029 741 623	.0394532	.3107006	12	2 28 61 44	3 456 649 728	.8844442	.4775871
48	2 09 67 04	3 036 027 392	.0525952	.3133056	13	2 28 91 69	3 463 512 697	.8973006	.4801169
49	2 09 96 01	3 042 321 849	.0657326	.3159094	14	2 29 21 96	3 470 384 744	.9101529	.4826455
1450	2 10 25 00	3 048 625 000	38.0788655	11.3185119	1515	2 29 52 25	3 477 265 875	38.9230009	11.4851731
51	2 10 54 01	3 054 936 851	.0919939	.3211132	16	2 29 82 56	3 484 156 096	.9358447	.4876995
52	2 10 83 04	3 061 257 408	.1051178	.3237134	17	2 30 12 89	3 491 055 413	.9486841	.4902249
53	2 11 12 09	3 067 586 677	.1182371	.3263124	18	2 30 43 24	3 497 963 832	.9615194	.4927491
54	2 11 41 16	3 073 924 664	.1313519	.3289102	19	2 30 73 61	3 504 881 359	.9743505	.4952722
1455	2 11 70 25	3 080 271 375	38.1444622	11.3315067	1520	2 31 04 00	3 511 808 000	38.9871774	11.4977942
56	2 11 99 36	3 086 626 816	.1575681	.3341022	21	2 31 34 41	3 518 743 761	39.0000000	.5003151
57	2 12 28 49	3 092 990 993	.1706693	.3366964	22	2 31 64 84	3 525 688 648	.0128184	.5028348
58	2 12 57 64	3 099 363 912	.1837662	.3392894	23	2 31 95 29	3 532 642 667	.0256326	.5053535
59	2 12 86 81	3 105 745 579	.1968585	.3418813	24	2 32 25 76	3 539 605 824	.0384426	.5078711
1460	2 13 16 00	3 112 136 000	38.2099463	11.3444719	1525	2 32 56 25	3 546 578 125	39.0512483	11.5103876
61	2 13 45 21	3 118 535 181	.2230297	.3470614	26	2 32 86 76	3 553 559 576	.0640499	.5129030
62	2 13 74 44	3 124 943 128	.2361085	.3496497	27	2 33 17 29	3 560 550 183	.0768473	.5154173
63	2 14 03 69	3 131 359 847	.2491829	.3522368	28	2 33 47 84	3 567 549 952	.0896406	.5179305
64	2 14 32 96	3 137 785 344	.2622529	.3548227	29	2 33 78 41	3 574 558 889	.1024296	.5204425
1465	2 14 62 25	3 144 219 625	38.2753184	11.3574075	1530	2 34 09 00	3 581 577 000	39.1152144	11.5229535
66	2 14 91 56	3 150 662 696	.2883794	.3599911	31	2 34 39 61	3 588 604 291	.1279951	.5254634
67	2 15 20 89	3 157 114 563	.3014360	.3625735	32	2 34 70 24	3 595 640 768	.1407716	.5279722
68	2 15 50 24	3 163 575 232	.3144881	.3651547	33	2 35 00 89	3 602 686 437	.1535439	.5304799
69	2 15 79 61	3 170 044 709	.3275358	.3677347	34	2 35 31 56	3 609 741 304	.1663120	.5329865
1470	2 16 09 00	3 176 523 000	38.3405790	11.3703136	1535	2 35 62 25	3 616 805 375	39.1790760	11.5354920
71	2 16 38 41	3 183 010 111	.3536178	.3728914	36	2 35 92 96	3 623 878 656	.1918359	.5379965
72	2 16 67 84	3 189 506 048	.3666522	.3754679	37	2 36 23 69	3 630,961 153	.2045915	.5404998
73	2 16 97 29	3 196 010 817	.3796821	.3780433	38	2 36 54 44	3 638 052 872	.2173431	.5430021
74	2 17 26 76	3 202 524 424	.3927076	.3806175	39	2 36 85 21	3 645 153 819	.2300905	.5455033
1475	2 17 56 25	3 209 046 875	38.4057287	11.3831906	1540	2 37 16 00	3 652 264 000	39.2428337	11.5480034
76	2 17 85 76	3 215 578 176	.4187454	.3857625	41	2 37 46 81	3 659 383 421	.2555728	.5505025
77	2 18 15 29	3 222 118 333	.4317577	.3883332	42	2 37 77 64	3 666 512 088	.2683078	.5530004
78	2 18 44 84	3 228 667 352	.4447656	.3909028	43	2 38 08 49	3 673 650 007	.2810387	.5554973
79	2 18 74 41	3 235 225 239	.4577691	.3934712	44	2 38 39 36	3 680 797 184	.2937654	.5579931
1480	2 19 04 00	3 241 792 000	38.4707681	11.3960384	1545	2 38 70 25	3 687 953 625	39.3064880	11.5604878
81	2 19 33 61	3 248 367 641	.4837627	.3986045	46	2 39 01 16	3 695 119 336	.3192065	.5629815
82	2 19 63 24	3 254 952 168	.4967530	.4011695	47	2 39 32 09	3 702 294 323	.3319208	.5654740
83	2 19 92 89	3 261 545 587	.5097390	.4037332	48	2 39 63 04	3 709 478 592	.3446311	.5679655
84	2 20 22 56	3 268 147 904	.5227206	.4062959	49	2 39 94 01	3 716 672 149	.3573373	.5704559
1485	2 20 52 25	3 274 759 125	38.5356977	11.4088574	1550	2 40 25 00	3 723 875 000	39.3700394	11.5729453
86	2 20 81 96	3 281 379 256	.5486705	.4114177	51	2 40 56 01	3 731 087 151	.3827373	.5754336
87	2 21 11 69	3 288 008 303	.5616389	.4139769	52	2 40 87 04	3 738 308 608	.3954312	.5779208
88	2 21 41 44	3 294 646 272	.5746030	.4165349	53	2 41 18 09	3 745 539 377	.4081210	.5804069
89	2 21 71 21	3 301 293 169	.5875627	.4190918	54	2 41 49 16	3 752 779 464	.4208067	.5828919
1490	2 22 01 00	3 307 949 000	38.6005181	11.4216476	1555	2 41 80 25	3 760 028 875	39.4334883	11.5853759
91	2 22 30 81	3 314 613 771	.6134691	.4242022	56	2 42 11 36	3 767 287 616	.4461658	.5878588
92	2 22 60 64	3 321 287 488	.6264158	.4267556	57	2 42 42 49	3 774 555 693	.4588393	.5903407
93	2 22 90 49	3 327 970 157	.6393582	.4293079	58	2 42 73 64	3 781 833 112	.4715087	.5928215
94	2 23 20 36	3 334 651 784	.6522962	.4318591	59	2 43 04 81	3 789 119 879	.4841740	.5953013
1495	2 23 50 25	3 341 362 375	38.6652299	11.4344092	1560	2 43 36 00	3 796 416 000	39.4968353	11.5977799

2.—SQUARES, CUBES, SQUARE ROOTS, CUBE ROOTS, OF NUMBERS
1 TO 1600—Concluded.

No.	Square	Cube.	Sq. Rt	Cu. Rt	No.	Square	Cube.	Sq. Rt.	Cu. Rt.
1560	2 43 36 00	3 796 416 000	39.4968353	11.5977799	1580	2 49 64 00	3 944 312 000	39.7492138	11.6471329
61	2 43 67 21	3 803 721 481	.5094925	.6002576	81	2 49 95 61	3 951 805 941	.7617907	.6495895
62	2 43 98 44	3 811 036 328	.5221457	.6027342	82	2 50 27 24	3 959 309 368	.7743636	.6520452
63	2 44 29 69	3 818 360 547	.5347948	.6052097	83	2 50 58 89	3 966 822 287	.7869325	.6544998
64	2 44 60 96	3 825 694 144	.5474399	.6076841	84	2 50 90 56	3 974 344 704	.7994975	.6569534
1565	2 44 92 25	3 833 037 125	39.5600809	11.6101575	1585	2 51 22 25	3 981 876 625	39.8120585	11.6594059
66	2 45 23 56	3 840 389 496	.5727179	.6126299	86	2 51 53 96	3 989 418 056	.8246155	.6618574
67	2 45 54 89	3 847 751 263	.5853508	.6151012	87	2 51 85 69	3 996 969 003	.8371686	.6643079
68	2 45 86 24	3 855 122 432	.5979797	.6175715	88	2 52 17 44	4 004 529 472	.8497177	.6667574
69	2 46 17 61	3 862 503 009	.6106046	.6200407	89	2 52 49 21	4 012 099 469	.8622628	.6692058
1570	2 46 49 00	3 869 893 000	39.6232255	11.6225088	1590	2 52 81 00	4 019 679 000	39.8748040	11.6716532
71	2 46 80 41	3 877 292 411	.6358424	.6249759	91	2 53 12 81	4 027 268 071	.8873413	.6740996
72	2 47 11 84	3 884 701 248	.6484552	.6274420	92	2 53 44 64	4 034 866 688	.8998747	.6765449
73	2 47 43 29	3 892 119 517	.6610640	.6299070	93	2 53 76 49	4 042 474 857	.9124041	.6789892
74	2 47 74 76	3 899 547 224	.6736688	.6323710	94	2 54 08 36	4 050 092 584	.9249295	.6814325
1575	2 48 06 25	3 906 984 375	39.6862696	11.6348339	1595	2 54 40 25	4 057 719 875	39.9374511	11.6838748
76	2 48 37 76	3 914 430 976	.6988665	.6372957	96	2 54 72 16	4 065 356 736	.9499687	.6863161
77	2 48 69 29	3 921 887 033	.7114593	.6397566	97	2 55 04 09	4 073 003 173	.9624824	.6887563
78	2 49 00 84	3 929 352 552	.7240481	.6422164	98	2 55 36 04	4 080 659 192	.9749922	.6911955
79	2 49 32 41	3 936 827 539	.7366329	.6446751	99	2 55 68 01	4 088 324 799	.9874980	.6936337
1580	2 49 64 00	3 944 312 000	39.7492138	11.6471329	1600	2 56 00 00	4 096 000 000	40.0000000	11.6960709

2a.—SQUARES OF NUMBERS 1600 TO 1810.

No.	Square.	No.	Square.	No.	Square.	No.	Square.	No.	Square.	No.	Square.
1600	2560000	1635	2673225	1670	2788900	1705	2907025	1740	3027600	1775	3150625
01	2563201	36	2676496	71	2792241	06	2910436	41	3031081	76	3154176
02	2566404	37	2679769	72	2795584	07	2913849	42	3034564	77	3157729
03	2569609	38	2683044	73	2798929	08	2917264	43	3038049	78	3161284
04	2572816	39	2686321	74	2802276	09	2920681	44	3041536	79	3164841
1605	2576025	1640	2689600	1675	2805625	1710	2924100	1745	3045025	1780	3168400
06	2579236	41	2692881	76	2808976	11	2927521	46	3048516	81	3171961
07	2582449	42	2696164	77	2812329	12	2930944	47	3052009	82	3175524
08	2585664	43	2699449	78	2815684	13	2934369	48	3055504	83	3179089
09	2588881	44	2702736	79	2819041	14	2937796	49	3059001	84	3182656
1610	2592100	1645	2706025	1680	2822400	1715	2941225	1750	3062500	1785	3186225
11	2595321	46	2709316	81	2825761	16	2944656	51	3066001	86	3189796
12	2598544	47	2712609	82	2829124	17	2948089	52	3069504	87	3193369
13	2601769	48	2715904	83	2832489	18	2951524	53	3073009	88	3196944
14	2604996	49	2719201	84	2835856	19	2954961	54	3076516	89	3200521
1615	2608225	1650	2722500	1685	2839225	1720	2958400	1755	3080025	1790	3204100
16	2611456	51	2725801	86	2842596	21	2961841	56	3083536	91	3207681
17	2614689	52	2729104	87	2845969	22	2965284	57	3087049	92	3211264
18	2617924	53	2732409	88	2849344	23	2968729	58	3090564	93	3214849
19	2621161	54	2735716	89	2852721	24	2972176	59	3094081	94	3218436
1620	2624400	1655	2739025	1690	2856100	1725	2975625	1760	3097600	1795	3222025
21	2627641	56	2742336	91	2859481	26	2979076	61	3101121	96	3225616
22	2630884	57	2745649	92	2862864	27	2982529	62	3104644	97	3229209
23	2634129	58	2748964	93	2866249	28	2985984	63	3108169	98	3232804
24	2637376	59	2752281	94	2869636	29	2989441	64	3111696	99	3236401
1625	2640625	1660	2755600	1695	2873025	1730	2992900	1765	3115225	1800	3240000
26	2643876	61	2758921	96	2876416	31	2996361	66	3118756	01	3243601
27	2647129	62	2762244	97	2879809	32	2999824	67	3122289	02	3247204
28	2650384	63	2765569	98	2883204	33	3003289	68	3125824	03	3250809
29	2653641	64	2768896	99	2886601	34	3006756	69	3129361	04	3254416
1630	2656900	1665	2772225	1700	2890000	1735	3010225	1770	3132900	1805	3258025
31	2660161	66	2775556	01	2893401	36	3013696	71	3136441	06	3261636
32	2663424	67	2778889	02	2896804	37	3017169	72	3139984	07	3265249
33	2666689	68	2782224	03	2900209	38	3020644	73	3143529	08	3268864
34	2669956	69	2785561	04	2903616	39	3024121	74	3147076	09	3272481
1635	2673225	1670	2788900	1705	2907025	1740	3027600	1775	3150625	1810	3276100

2b.—Square Roots and Cube Roots of Numbers 1600 to 1860.

No.	Sq. Rt.	Cu. Rt.	No.	Sq. Rt.	Cu. Rt.	No.	Sq. Rt.	Cu. Rt.	No.	Sq. Rt.	Cu. Rt.
1600	40.0000	11.6961	1665	40.8044	11.8524	1730	41.5933	12.0046	1795	42.3674	12.1531
1	.0125	.6985	66	.8167	.8547	31	.6053	.0069	96	.3792	.1554
2	.0250	.7009	67	.8289	.8571	32	.6173	.0093	97	.3910	.1576
3	.0375	.7034	68	.8412	.8595	33	.6293	.0116	98	.4028	.1599
4	.0500	.7058	69	.8534	.8618	34	.6413	.0139	99	.4146	.1622
1605	40.0625	11.7082	1670	40.8656	11.8642	1735	41.6533	12.0162	1800	42.4264	12.1644
6	.0749	.7107	71	.8779	.8666	36	.6653	.0185	1	.4382	.1667
7	.0874	.7131	72	.8901	.8689	37	.6773	.0208	.2	.4500	.1689
8	.0999	.7155	73	.9023	.8713	38	.6893	.0231	3	.4617	.1712
9	.1123	.7180	74	.9145	.8737	39	.7013	.0254	4	.4735	.1734
1610	40.1248	11.7204	1675	40.9268	11.8760	1740	41.7133	12.0277	1805	42.4853	12.1757
11	.1373	.7228	76	.9390	.8784	41	.7253	.0300	6	.4971	.1779
12	.1497	.7252	77	.9512	.8808	42	.7373	.0323	7	.5088	.1802
13	.1622	.7277	78	.9634	.8831	43	.7493	.0346	8	.5206	.1824
14	.1746	.7301	79	.9756	.8855	44	.7612	.0369	9	.5323	.1846
1615	40.1871	11.7325	1680	40.9878	11.8878	1745	41.7732	12.0392	1810	42.5441	12.1869
16	.1995	.7350	81	41.0000	.8902	46	.7852	.0415	11	.5558	.1891
17	.2119	.7373	82	.0122	.8926	47	.7971	.0438	12	.5676	.1914
18	.2244	.7398	83	.0244	.8949	48	.8091	.0461	13	.5793	.1936
19	.2368	.7422	84	.0366	.8973	49	.8210	.0484	14	.5911	.1959
1620	40.2492	11.7446	1685	41.0488	11.8996	1750	41.8330	12.0507	1815	42.6028	12.1981
21	.2616	.7470	86	.0609	.9020	51	.8450	.0530	16	.6146	.2003
22	.2741	.7494	87	.0731	.9043	52	.8569	.0553	17	.6263	.2026
23	.2865	.7518	88	.0853	.9067	53	.8688	.0576	18	.6380	.2048
24	.2989	.7543	89	.0974	.9090	54	.8808	.0599	19	.6497	.2071
1625	40.3113	11.7567	1690	41.1096	11.9114	1755	41.8927	12.0622	1820	42.6615	12.2093
26	.3237	.7591	91	.1218	.9137	56	.9047	.0645	21	.6732	.2115
27	.3361	.7615	92	.1339	.9161	57	.9166	.0668	22	.6849	.2138
28	.3485	.7639	93	.1461	.9184	58	.9285	.0690	23	.6966	.2160
29	.3609	.7663	94	.1582	.9208	59	.9404	.0713	24	.7083	.2182
1630	40.3733	11.7687	1695	41.1704	11.9231	1760	41.9524	12.0736	1825	42.7200	12.2205
31	.3856	.7711	96	.1825	.9255	61	.9643	.0759	26	.7317	.2227
32	.3980	.7735	97	.1947	.9278	62	.9762	.0782	27	.7434	.2249
33	.4104	.7759	98	.2068	.9301	63	.9881	.0805	28	.7551	.2272
34	.4228	.7783	99	.2189	.9325	64	42.0000	.0828	29	.7668	.2294
1635	40.4351	11.7807	1700	41.2311	11.9348	1765	42.0119	12.0850	1830	42.7785	12.2316
36	.4475	.7831	1	.2432	.9372	66	.0238	.0873	31	.7902	.2338
37	.4599	.7855	2	.2553	.9395	67	.0357	.0896	32	.8019	.2361
38	.4722	.7879	3	.2674	9418	68	.0476	.0919	33	.8135	.2383
39	.4846	.7903	4	.2795	.9442	69	.0595	.0942	34	.8252	.2405
1640	40.4969	11.7927	1705	41.2916	11.9465	1770	42.0714	12.0964	1835	42.8369	12.2427
41	.5093	.7951	6	.3038	.9489	71	.0833	.0987	36	.8486	.2450
42	.5216	.7975	7	.3159	.9512	72	.0951	.1010	37	.8602	.2472
43	.5339	.7999	8	.3280	.9535	73	.1070	.1033	38	.8719	.2494
44	.5463	.8023	9	.3401	.9559	74	.1189	.1056	39	.8836	.2516
1645	40.5586	11.8047	1710	41.3521	11.9582	1775	42.1307	12.1078	1840	42.8952	12.2539
46	.5709	.8071	11	.3642	.9605	76	.1426	.1101	41	.9069	.2561
47	.5832	.8095	12	.3763	.9628	77	.1545	.1124	42	.9185	.2583
48	.5956	.8119	13	.3884	.9652	78	.1663	.1146	43	.9302	.2605
49	.6079	.8143	14	.4005	.9675	79	.1782	.1169	44	.9418	.2627
1650	40.6202	11.8167	1715	41.4126	11.9698	1780	42.1900	12.1192	1845	42.9535	12.2649
51	.6325	.8190	16	.4246	.9722	81	.2019	.1215	46	.9651	.2672
52	.6448	.8214	17	.4367	.9745	82	.2137	.1237	47	.9767	.2694
53	.6571	.8238	18	.4488	.9768	83	.2256	.1260	48	.9884	.2716
54	.6694	.8262	19	.4608	.9791	84	.2374	.1283	49	43.0000	.2738
1655	40.6817	11.8286	1720	41.4729	11.9815	1785	42.2493	12.1305	1850	43.0116	12.2760
56	.6940	.8310	21	.4849	.9838	86	.2611	.1328	51	.0232	.2782
57	.7063	.8333	22	.4970	.9861	87	.2729	.1350	52	.0349	.2804
58	.7185	.8357	23	.5090	.9884	88	.2847	.1373	53	.0465	.2826
59	.7308	.8381	24	.5211	.9907	89	.2966	.1396	54	.0581	.2849
1660	40.7431	11.8405	1725	41.5331	11.9931	1790	42.3084	12.1418	1855	43.0697	12.2871
61	.7554	.8429	26	.5452	.9954	91	.3202	.1441	56	.0813	.2893
62	.7676	.8452	27	.5572	.9977	92	.3320	.1464	57	.0929	.2915
63	.7799	.8476	28	.5692	12.0000	93	.3438	.1486	58	.1045	.2937
64	.7922	.8500	29	.5812	.0023	94	.3556	.1509	59	.1161	.2959
1665	40.8044	11.8524	1730	41.5933	12.0046	1795	42.3674	12.1531	1860	43.1277	12.2981

2c.—Squares of Mixed Numbers from $\frac{1}{64}$ to 12, by 64ths

I. Squares of Mixed Numbers from $\frac{1}{64}$ to 6.

	0	1	2	3	4	5
1/64	0.00024	1.03149	4.06274	9.09399	16.12524	25.15649
1/32	0.00098	1.06348	4.12598	9.18848	16.25098	25.31348
3/64	0.00220	1.09595	4.18970	9.28345	16.37720	25.47095
1/16	0.00391	1.12891	4.25391	9.37891	16.50391	25.62891
5/64	0.00610	1.16235	4.31860	9.47485	16.63110	25.78735
3/32	0.00879	1.19629	4.38379	9.57129	16.75879	25.94629
7/64	0.01196	1.23071	4.44946	9.66821	16.88696	26.10571
1/8	0.01562	1.26562	4.51562	9.76562	17.01562	26.26562
9/64	0.01978	1.30103	4.58228	9.86353	17.14478	26.42603
5/32	0.02441	1.33691	4.64941	9.96191	17.27441	26.58691
11/64	0.02954	1.37329	4.71704	10.06079	17.40454	26.74829
3/16	0.03516	1.41016	4.78516	10.16016	17.53516	26.91016
13/64	0.04126	1.44751	4.85376	10.26001	17.66626	27.07251
7/32	0.04785	1.48535	4.92285	10.36035	17.79785	27.23535
15/64	0.05493	1.52368	4.99243	10.46118	17.92993	27.39868
1/4	0.06250	1.56250	5.06250	10.56250	18.06250	27.56250
17/64	0.07056	1.60181	5.13306	10.66431	18.19556	27.72681
9/32	0.07910	1.64160	5.20410	10.76660	18.32910	27.89160
19/64	0.08813	1.68188	5.27563	10.86938	18.46313	28.05688
5/16	0.09766	1.72266	5.34766	10.97266	18.59766	28.22266
21/64	0.10767	1.76392	5.42017	11.07642	18.73267	28.38892
11/32	0.11816	1.80566	5.49316	11.18066	18.86816	28.55566
23/64	0.12915	1.84790	5.56663	11.28540	19.00415	28.72290
3/8	0.14062	1.89062	5.64062	11.39062	19.14062	28.89062
25/64	0.15259	1.93384	5.71509	11.49634	19.27759	29.05884
13/32	0.16504	1.97754	5.79004	11.60254	19.41504	29.22754
27/64	0.17798	2.02173	5.86548	11.70923	19.55298	29.39673
7/16	0.19141	2.06641	5.94141	11.81641	19.69141	29.56641
29/64	0.20532	2.11157	6.01782	11.92407	19.83032	29.73657
15/32	0.21973	2.15723	6.09473	12.03223	19.96973	29.90723
31/64	0.23462	2.20337	6.17212	12.14087	20.10962	30.07837
1/2	0.25000	2.25000	6.25000	12.25000	20.25000	30.25000
33/64	0.26587	2.29712	6.32837	12.35962	20.39087	30.42212
17/32	0.28223	2.34473	6.40723	12.46973	20.53223	30.59473
35/64	0.29907	2.39282	6.48657	12.58032	20.67407	30.76782
9/16	0.31641	2.44141	6.56641	12.69141	20.81641	30.94141
37/64	0.33423	2.49048	6.64673	12.80298	20.95923	31.11548
19/32	0.35254	2.54004	6.72754	12.91504	21.10254	31.29004
39/64	0.37134	2.59009	6.80884	13.02759	21.24634	31.46509
5/8	0.39062	2.64062	6.89062	13.14062	21.39062	31.64062
41/64	0.41040	2.69165	6.97290	13.25415	21.53540	31.81665
21/32	0.43066	2.74316	7.05566	13.36816	21.68066	31.99316

2c.—Squares of Mixed Numbers from $\frac{1}{64}$ to 6—*Continued*:

	o	**1**	**2**	**3**	**4**	**5**
4⅜₆₄	0.45142	2.79517	7.13892	13.48267	21.82642	32.17017
1¹⁄₁₆	0.47266	2.84766	7.22266	13.59766	21.97266	32.34766
4⁵⁄₆₄	0.49438	2.90063	7.30688	13.71313	22.11938	32.52563
2³⁄₃₂	0.51660	2.95410	7.39160	13.82910	22.26660	32.70410
4⁷⁄₆₄	0.53931	3.00806	7.47681	13.94556	22.41431	32.88306
¾	0.56250	3.06250	7.56250	14.06250	22.56250	33.06250
4⁹⁄₆₄	0.58618	3.11743	7.64868	14.17993	22.71118	33.24243
2⁵⁄₃₂	0.61035	3.17285	7.73535	14.29785	22.86035	33.42285
5¹⁄₆₄	0.63501	3.22876	7.82251	14.41626	23.01001	33.60376
1³⁄₁₆	0.66016	3.28516	7.91016	14.53516	23.16016	33.78516
5³⁄₆₄	0.68579	3.34204	7.99829	14.65454	23.31079	33.96704
2⁷⁄₃₂	0.71191	3.39941	8.08691	14.77441	23.46191	34.14941
5⁵⁄₆₄	0.73853	3.45728	8.17603	14.89478	23.61363	34.33228
⅞	0.76562	3.51562	8.26562	15.01562	23.76562	34.51562
5⁷⁄₆₄	0.79321	3.57446	8.35571	15.13696	23.91821	34.69946
2⁹⁄₃₂	0.82129	3.63379	8.44629	15.25879	24.07129	34.88379
5⁹⁄₆₄	0.84985	3.69360	8.53735	15.38110	24.22485	35.06860
1⁵⁄₁₆	0.87891	3.75391	8.62891	15.50391	24.37891	35.25391
6¹⁄₆₄	0.90845	3.81470	8.72095	15.62720	24.53345	35.43970
3¹⁄₃₂	0.93848	3.87598	8.81348	15.75098	24.68848	35.62598
6³⁄₆₄	0.96899	3.93774	8.90649	15.87524	24.84399	35.81274

2d.—II. Squares of Mixed Numbers from $6\frac{1}{64}$ to 12

	6	**7**	**8**	**9**	**10**	**11**
¹⁄₆₄	36.18774	49.21899	64.25024	81.28149	100.31274	121.34399
¹⁄₃₂	36.37598	49.43848	64.50098	81.56348	100.62598	121.68848
³⁄₆₄	36.56470	49.65845	64.75220	81.84595	100.93970	122.03345
¹⁄₁₆	36.75391	49.87891	65.00391	82.12891	101.25391	122.37891
⁵⁄₆₄	36.94360	50.09985	65.25610	82.41235	101.56860	122.72485
³⁄₃₂	37.13379	50.32129	65.50879	82.69629	101.88379	123.07129
⁷⁄₆₄	37.32446	50.54321	65.76196	82.98071	102.19946	123.41821
⅛	37.51562	50.76562	66.01562	83.26562	102.51562	123.76562
⁹⁄₆₄	37.70728	50.98853	66.26978	83.55103	102.83228	124.11353
⁵⁄₃₂	37.89941	51.21191	66.52441	83.83691	103.14941	124.46191
11⁄₆₄	38.09204	51.43579	66.77954	84.12329	103.46704	124.81079
³⁄₁₆	38.28516	51.66016	67.03516	84.41016	103.78516	125.16016
13⁄₆₄	38.47876	51.88501	67.29126	84.69751	104.10376	125.51001
⁷⁄₃₂	38.67285	52.11035	67.54785	84.98535	104.42285	125.86035
15⁄₆₄	38.86743	52.33618	67.80493	85.27368	104.74243	126.21110
¼	39.06250	52.56250	68.06250	85.56250	105.06250	126.56250

2d.—Squares of Mixed Numbers from 6$\frac{1}{64}$ to 12—*Continued*

	6	7	8	9	10	11
17/64	39.25806	52.78931	68.32056	85.85181	105.38306	126.91431
9/32	39.45410	53.01660	68.57910	86.14160	105.70410	127.26660
19/64	39.65063	53.24438	68.83813	86.43188	106.02563	127.61938
5/16	39.84766	53.47266	69.09766	86.72266	106.34766	127.97266
21/64	40.04517	53.70142	69.35767	87.01392	106.67017	128.32642
11/32	40.24316	53.93066	69.61816	87.30566	106.99316	128.68066
23/64	40.44165	54.16040	69.87915	87.59790	107.31665	129.03540
3/8	40.64062	54.39062	70.14062	87.89062	107.64062	129.39062
25/64	40.84009	54.62134	70.40259	88.18384	107.96509	129.74634
13/32	41.04004	54.85254	70.66504	88.47754	108.29004	130.10254
27/64	41.24048	55.08423	70.92798	88.77173	108.61548	130.45923
7/16	41.44141	55.31641	71.19141	89.06641	108.94141	130.81641
29/64	41.64282	55.54907	71.45532	89.36157	109.26782	131.17407
15/32	41.84473	55.78223	71.71973	89.65723	109.59473	131.53223
31/64	42.04712	56.01587	71.98462	89.95337	109.92212	131.89087
1/2	42.25000	56.25000	72.25000	90.25000	110.25000	132.25000
33/64	42.45337	56.48462	72.51587	90.54712	110.57837	132.60962
17/32	42.65723	56.71973	72.78223	90.84473	110.90723	132.96973
35/64	42.86157	56.95532	73.04907	91.14282	111.23657	133.33032
9/16	43.06641	57.19141	73.31641	91.44141	111.56641	133.69141
37/64	43.27173	57.42798	73.58423	91.74048	111.89673	134.05298
19/32	43.47754	57.66504	73.85254	92.04004	112.22754	134.41504
39/64	43.68384	57.90259	74.12134	92.34009	112.55884	134.77759
5/8	43.89062	58.14062	74.39062	92.64062	112.89062	135.14062
41/64	44.09790	58.37915	74.66040	92.94165	113.22290	135.50415
21/32	44.30566	58.61816	74.93066	93.24316	113.55566	135.86816
43/64	44.51392	58.85767	75.20142	93.54517	113.88892	136.23267
11/16	44.72266	59.09766	75.47266	93.84766	114.22266	136.59766
45/64	44.93188	59.33813	75.74438	94.15063	114.55688	136.96313
23/32	45.14160	59.57910	76.01660	94.45410	114.89160	137.32910
47/64	45.35181	59.82056	76.28931	94.75806	115.22681	137.69556
3/4	45.56250	60.06250	76.56250	95.06250	115.56250	138.06250
49/64	45.77368	60.30493	76.83618	95.36743	115.89868	138.42993
25/32	45.98535	60.54785	77.11035	95.67285	116.23535	138.79785
51/64	46.19751	60.79126	77.38501	95.97876	116.57251	139.16626
13/16	46.41016	61.03516	77.66016	96.28516	116.91016	139.53516
53/64	46.62329	61.27954	77.93579	96.59204	117.24829	139.90454
27/32	46.83691	61.52441	78.21191	96.89941	117.58691	140.27441
55/64	47.05103	61.76978	78.48853	97.20728	117.92603	140.64478
7/8	47.26562	62.01562	78.76562	97.51562	118.26562	141.01562
57/64	47.48071	62.26196	79.04321	97.82446	118.60571	141.38696
29/32	47.69629	62.50879	79.32129	98.13379	118.94629	141.75879
59/64	47.91235	62.75610	79.59985	98.44360	119.28735	142.13110
15/16	48.12891	63.00391	79.87891	98.75391	119.62891	142.50391
61/64	48.34595	63.25220	80.15845	99.06470	119.97095	142.87720
31/32	48.56348	63.50098	80.43848	99.37598	120.31348	143.25098
63/64	48.78149	63.75024	80.71899	99.68774	120.65649	143.62524

frequently most convenient to change the fraction to a decimal first and then extract the square root.

Ratio and Proportion.—The *ratio* of two numbers is the relation which the value of the first bears to the value of the second and this relation is indicated by the sign (:). Thus, 3 : 4 is the ratio of 3 to 4. Ratio is equivalent to the fraction obtained by dividing the first number by the second. Thus, $\frac{3}{4}$ also expresses the ratio of 3 to 4.

An expression consisting of two equal ratios is called a *proportion*. It is written, 3 : 4 = 9 : 12, and read, " 3 is to 4 as 9 is to 12." The first and last, or the " end," numbers are called the extremes and the second and third, or the middle, numbers are called the *means*. Since a ratio may also be expressed as a fraction, then a proportion may also be set upon, $\frac{1}{8} = \frac{9}{12}$.

Illustration: If the diameter of a gear is 13.53 inches and the circumference is 42.5 inches, find the ratio of the diameter to the circumference.

$$\frac{13.53}{42.5} = 0.3183$$

Therefore, the ratio of the diameter to the circumference is 0.3183. The above value is the same as that obtained by dividing 1 by 3.1416; that is, in any circle the ratio of the diameter to the circumference is $1 \div \pi$. Thus, it is evident that ratio is always the quotient obtained by dividing the first number by the second.

Proportion is one of the most useful tools in mathematical calculation. It is the *key* to many of its operations. Indeed, practically all mathematical problems may be expressed in proportion.

Rules of Proportion.—Proportion derives its great usefulness from the fundamental rule which states that *the product of the means equals the product of the extremes*. Thus, in the proportion 3 : 4 = 9 : 12, according to the rule, 4×9 (the product of the means) $= 3 \times 12$ (the product of the extremes) $= 36$. Then, when three terms of a proportion are known, the fourth can be found. For example, if it takes twenty days to build five lathes, how long will it take to build fifteen lathes at the same rate?

$$x : 20 = 15 : 5$$

whence
$$x = \frac{20 \times \overset{3}{\cancel{15}}}{\cancel{5}} = 60 \text{ days (Ans.)}$$

Where one extreme and both means are known, to find the other extreme, divide the product of the means by the known extreme.

Where both extremes and one mean are known, to find the other mean, divide the product of the extremes by the known mean.

For the purpose of illustrating these rules, replace the figures in a proportion by the letters A, B, C, D, and write $A : B = C : D$; then

$$A \times D = B \times C, \frac{A}{B} = \frac{C}{D}, \quad A = \frac{B \times C}{D},$$

$$D = \frac{B \times C}{A}, \quad B = \frac{A \times D}{C}, \quad C = \frac{A \times D}{B}.$$

Triangles may be used advantageously in illustrating ratio and proportion. Thus, let us say, if a train travels 260 miles in

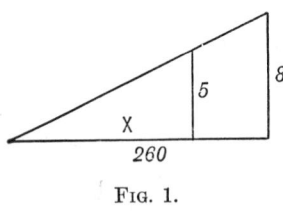

Fig. 1.

8 hours, how far will it travel in 5 hours? Draw a triangle letting the base represent the distance (260 mi.) and a leg the time 8 hours. Then draw another leg parallel to the first and of a length in proportion to the first as 5 is to 8. Then the distance x represents the distance which the train will travel in 5 hours because from similar triangles, and

$$x : 260 = 5 : 8$$

whence
$$x = \frac{5 \times 260}{8} = 162.5 \text{ miles (Ans.)}$$

Inverse Proportion.—In the preceding problems the ratio of the elements of one figure was equal to the ratio of the corresponding elements of the other figure, that is, directly proportional. When the ratio is equal to the inverse of that ratio the elements are said to be inversely proportional.

The speed of pulleys connected by belts are inversely proportional to their diameters, i.e., the smaller pulley rotates faster than the larger pulley.

ILLUSTRATION: A 24-inch pulley fixed to a line shaft which makes 400 revolutions per minute (R.P.M.) is belted to a 6-inch pulley. Find the number of R.P.M. of the smaller pulley.

R.P.M. of Driven Pulley		R.P.M. of Driving Pulley	Diameter of Driving Pulley		Diameter of Driven Pulley
x	:	400	=	24	:	6

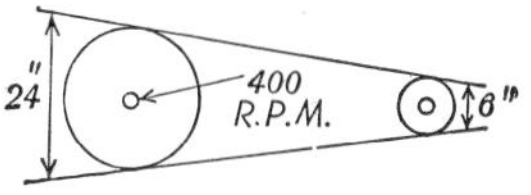

FIG. 2.

$$\text{whence } x = \frac{400 \times \overset{4}{24}}{\underset{}{6}} = 1600 \text{ R.P.M. (Ans.)}$$

Likewise, the speeds of gears running together are inversely proportional to their number of teeth.

ILLUSTRATION: A driving gear with 48 teeth meshes with a driven gear with 16 teeth. If the driving gear makes 100 R.P.M. find the number of R.P.M. made by the driven gear.

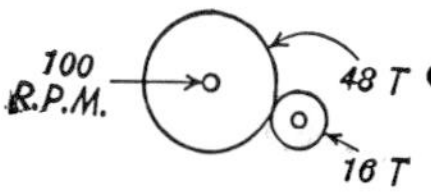

FIG. 3.

R.P.M. of Driven Gear		R.P.M. of Driving Gear	No. of Teeth of Driving Gear		No. of Teeth on Driven Gear
x	:	100	=	48	:	16

whence $\quad x = \dfrac{100 \times \overset{3}{\cancel{48}}}{\cancel{16}} = 300$ R.P.M. (Ans.)

Pulley Train.—A pulley train is a series of pulleys connected by belting, the power coming from one of the pulleys.

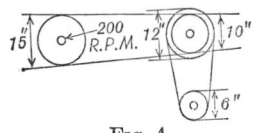

Fig. 4.

Illustration: In the sketch at the left, find the R.P.M. of the 6-inch pulley.

R.P.M. of Last Driven Pulley	:	R.P.M. of First Driving Pulley	=	Product of Diameters of All Driving Pulleys	:	Product of Diameters of All Driven Pulleys
x	:	200	=	(15×12)	:	(10×6)

whence $\quad x = \dfrac{\overset{20}{\cancel{200}} \times 15 \times \overset{2}{\cancel{12}}}{\cancel{10} \times \cancel{6}} = 600$ R.P.M. (Ans.)

Gear Train.—A gear train is a series of gears running together.

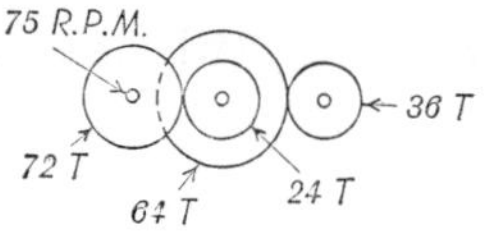

Illustration: In the sketch at the right find the R.P.M. of the 36 T. gear.

75 R.P.M.

36 T

72 T

6 + T

24 T

Fig. 5.

R.P.M. of Last Driven Gear	:	R.P.M. of First Driving Gear	=	Product of Number of Teeth of Driving Gears	:	Product of Number of Teeth of Driven Gears
x	:	75	=	(72×64)	:	(24×36)

whence $\quad x = \dfrac{\overset{25}{\cancel{75}} \times \overset{2}{\cancel{72}} \times \overset{8}{\cancel{64}}}{\underset{3}{\cancel{24}} \times \cancel{36}} = 400$ R.P.M. (Ans.)

Inverse proportion can be used to solve other types of problems. For instance in manufacturing plants the time per week is in an inverse proportion to the number of men employed; the shorter the time, the more men.

ILLUSTRATION: A factory employing 300 men completes a given number of vacuum cleaners weekly, the number of working hours being 40 per week. How many men would be required for the same production if the working hours were reduced to 30 per week?

$$x : 300 = 40 : 30$$

whence $$x = \frac{\overset{10}{\cancel{300}} \times 40}{\cancel{30}} = 400$$

therefore, 400 men would be needed for the same production.

Compound Proportion.—A compound proportion is a proportion which has one of its ratios a compound ratio, that is, a ratio expressed by a fraction that is the product of fractions representing given ratios. Thus, the ratios 3 : 4 and 5 : 7 are represented by the fractions $\frac{3}{4}$ and $\frac{5}{7}$; and the ratio 15 : 28 which is represented by $\frac{15}{28}$, the product of $\frac{3}{4}$ and $\frac{5}{7}$, is said to be compounded of the ratios 3 : 4 and 5 : 7.

Problems in compound proportion are solved by the cause and effect method which is based on the following principle. Like causes produce like effects; and the ratio between any two causes equals the ratio between the effects produced.

ILLUSTRATION: If a mechanic who machines 70 pieces in an 8-hour day is paid $1.47 per hour, find how much a man ought to be paid who machines 80 similar pieces in a 7-hour day if paid in the same proportion.

Make up a table with four columns headed " First Cause," " First Effect," " Second Cause," " Second Effect," and place under each the respective factors given in the problem. In the example above, the table would be as follows:

First Cause	First Effect	Second Cause	Second Effect
1 man 8 hours 147 cents	70 pieces	1 man 7 hours x cents	80 pieces

whence $(1 \times 8 \times 147) : 70 = (1 \times 7 \times x) : 80$

and $x = \dfrac{1 \times 8 \times \overset{\overset{3}{21}}{147} \times 80}{70 \times 1 \times 7} = 192$ cents $= \$1.92$

Therefore, the second operator should receive \$1.92 an hour.

Reciprocals.—The use of reciprocals facilitates computations in long division particularly when many different dividends are to be divided by the same divisor.

ILLUSTRATION: $7246 \div 1572$.

From the table on page 58 find the reciprocal of 1572. *Ans.:* 0.000636132.

TABLE 3

3.—RECIPROCALS, 1 TO 200

No.	Reciprocal	No.	Reciprocal	No.	Reciprocal	No.	Reciprocal
1	1.0000000	51	0.0196078	101	0.0099010	151	0.0066225
2	0.5000000	52	0.0192308	102	0.0098039	152	0.0065789
3	0.3333333	53	0.0188679	103	0.0097087	153	0.0065359
4	0.2500000	54	0.0185185	104	0.0096154	154	0.0064935
5	0.2000000	55	0.0181818	105	0.0095238	155	0.0064516
6	0.1666667	56	0.0178571	106	0.0094340	156	0.0064103
7	0.1428571	57	0.0175439	107	0.0093458	157	0.0063694
8	0.1250000	58	0.0172414	108	0.0092593	158	0.0063291
9	0.1111111	59	0.0169492	109	0.0091743	159	0.0062893
10	0.1000000	60	0.0166667	110	0.0090909	160	0.0062500
11	0.0909091	61	0.0163934	111	0.0090090	161	0.0062112
12	0.0833333	62	0.0161290	112	0.0089286	162	0.0061728
13	0.0769231	63	0.0158730	113	0.0088496	163	0.0061350
14	0.0714286	64	0.0156250	114	0.0087719	164	0.0060976
15	0.0666667	65	0.0153846	115	0.0086957	165	0.0060606
16	0.0625000	66	0.0151515	116	0.0086207	166	0.0060241
17	0.0588235	67	0.0149254	117	0.0085470	167	0.0059880
18	0.0555556	68	0.0147059	118	0.0084746	168	0.0059524
19	0.0526316	69	0.0144928	119	0.0084034	169	0.0059172
20	0.0500000	70	0.0142857	120	0.0083333	170	0.0058823
21	0.0476190	71	0.0140845	121	0.0082645	171	0.0058480
22	0.0454545	72	0.0138889	122	0.0081967	172	0.0058140
23	0.0434783	73	0.0136986	123	0.0081301	173	0.0057803
24	0.0416667	74	0.0135135	124	0.0080645	174	0.0057471
25	0.0400000	75	0.0133333	125	0.0080000	175	0.0057143
26	0.0384615	76	0.0131579	126	0.0079365	176	0.0056818
27	0.0370370	77	0.0129870	127	0.0078740	177	0.0056497
28	0.0357143	78	0.0128205	128	0.0078125	178	0.0056180
29	0.0344828	79	0.0126582	129	0.0077519	179	0.0055866
30	0.0333333	80	0.0125000	130	0.0076923	180	0.0055556
31	0.0322581	81	0.0123457	131	0.0076336	181	0.0055249
32	0.0312500	82	0.0121951	132	0.0075758	182	0.0054945
33	0.0303030	83	0.0120482	133	0.0075188	183	0.0054645
34	0.0294118	84	0.0119048	134	0.0074627	184	0.0054348
35	0.0285714	85	0.0117647	135	0.0074074	185	0.0054054
36	0.0277778	86	0.0116279	136	0.0073529	186	0.0053763
37	0.0270270	87	0.0114943	137	0.0072993	187	0.0053476
38	0.0263158	88	0.0113636	138	0.0072464	188	0.0053191
39	0.0256410	89	0.0112360	139	0.0071942	189	0.0052910
40	0.0250000	90	0.0111111	140	0.0071429	190	0.0052632
41	0.0243902	91	0.0109890	141	0.0070922	191	0.0052356
42	0.0238095	92	0.0108696	142	0.0070423	192	0.0052083
43	0.0232558	93	0.0107527	143	0.0069930	193	0.0051813
44	0.0227273	94	0.0106383	144	0.0069444	194	0.0051546
45	0.0222222	95	0.0105263	145	0.0068966	195	0.0051282
46	0.0217391	96	0.0104167	146	0.0068493	196	0.0051020
47	0.0212766	97	0.0103093	147	0.0068027	197	0.0050761
48	0.0208333	98	0.0102041	148	0.0067568	198	0.0050505
49	0.0204082	99	0.0101010	149	0.0067114	199	0.0050251
50	0.0200000	100	0.0100000	150	0.0066667	200	0.0050000

3.—Reciprocals, 201 to 400

No.	Reciprocal	No.	Reciprocal	No.	Reciprocal	No.	Reciprocal
201	0.0049751	251	0.0039841	301	0.0033223	351	0.0028490
202	0.0049505	252	0.0039683	302	0.0033113	352	0.0028409
203	0.0049261	253	0.0039526	303	0.0033003	353	0.0028329
204	0.0049020	254	0.0039370	304	0.0032895	354	0.0028249
205	0.0048780	255	0.0039216	305	0.0032787	355	0.0028169
206	0.0048544	256	0.0039063	306	0.0032680	356	0.0028090
207	0.0048309	257	0.0038911	307	0.0032573	357	0.0028011
208	0.0048077	258	0.0038760	308	0.0032468	358	0.0027933
209	0.0047847	259	0.0038610	309	0.0032362	359	0.0027855
210	0.0047619	260	0.0038462	310	0.0032258	360	0.0027778
211	0.0047393	261	0.0038314	311	0.0032154	361	0.0027701
212	0.0047170	262	0.0038168	312	0.0032051	362	0.0027624
213	0.0046948	263	0.0038023	313	0.0031949	363	0.0027548
214	0.0046729	264	0.0037879	314	0.0031847	364	0.0027473
215	0.0046512	265	0.0037736	315	0.0031746	365	0.0027397
216	0.0046296	266	0.0037594	316	0.0031646	366	0.0027322
217	0.0046083	267	0.0037453	317	0.0031546	367	0.0027248
218	0.0045872	268	0.0037313	318	0.0031447	368	0.0027174
219	0.0045662	269	0.0037175	319	0.0031348	369	0.0027100
220	0.0045455	270	0.0037037	320	0.0031250	370	0.0027027
221	0.0045249	271	0.0036900	321	0.0031153	371	0.0026954
222	0.0045045	272	0.0036765	322	0.0031056	372	0.0026882
223	0.0044843	273	0.0036630	323	0.0030960	373	0.0026810
224	0.0044643	274	0.0036496	324	0.0030864	374	0.0026738
225	0.0044444	275	0.0036364	325	0.0030769	375	0.0026667
226	0.0044248	276	0.0036232	326	0.0030675	376	0.0026596
227	0.0044053	277	0.0036101	327	0.0030581	377	0.0026525
228	0.0043860	278	0.0035971	328	0.0030488	378	0.0026455
229	0.0043668	279	0.0035842	329	0.0030395	379	0.0026385
230	0.0043478	280	0.0035714	330	0.0030303	380	0.0026316
231	0.0043290	281	0.0035587	331	0.0030211	381	0.0026247
232	0.0043103	282	0.0035461	332	0.0030120	382	0.0026178
233	0.0042918	283	0.0035336	333	0.0030030	383	0.0026110
234	0.0042735	284	0.0035211	334	0.0029940	384	0.0026042
235	0.0042553	285	0.0035088	335	0.0029851	385	0.0025974
236	0.0042373	286	0.0034965	336	0.0029762	386	0.0025907
237	0.0042194	287	0.0034843	337	0.0029674	387	0.0025840
238	0.0042017	288	0.0034722	338	0.0029586	388	0.0025773
239	0.0041841	289	0.0034602	339	0.0029499	389	0.0025707
240	0.0041667	290	0.0034483	340	0.0029412	390	0.0025641
241	0.0041494	291	0.0034364	341	0.0029326	391	0.0025575
242	0.0041322	292	0.0034247	342	0.0029240	392	0.0025510
243	0.0041152	293	0.0034130	343	0.0029155	393	0.0025445
244	0.0040984	294	0.0034014	344	0.0029070	394	0.0025381
245	0.0040816	295	0.0033898	345	0.0028986	395	0.0025316
246	0.0040650	296	0.0033784	346	0.0028902	396	0.0025253
247	0.0040486	297	0.0033670	347	0.0028818	397	0.0025189
248	0.0040323	298	0.0033557	348	0.0028736	398	0.0025126
249	0.0040161	299	0.0033445	349	0.0028653	399	0.0025063
250	0.0040000	300	0.0033333	350	0.0028571	400	0.0025000

3.—Reciprocals, 401 to 600

No.	Reciprocal	No.	Reciprocal	No.	Reciprocal	No.	Reciprocal
401	0.0024938	451	0.0022173	501	0.0019960	551	0.0018149
402	0.0024876	452	0.0022124	502	0.0019920	552	0.0018116
403	0.0024814	453	0.0022075	503	0.0019881	553	0.0018083
404	0.0024752	454	0.0022026	504	0.0019841	554	0.0018051
405	0.0024691	455	0.0021978	505	0.0019802	555	0.0018018
406	0.0024631	456	0.0021930	506	0.0019763	556	0.0017986
407	0.0024570	457	0.0021882	507	0.0019724	557	0.0017953
408	0.0024510	458	0.0021834	508	0.0019685	558	0.0017921
409	0.0024450	459	0.0021786	509	0.0019646	559	0.0017889
410	0.0024390	460	0.0021739	510	0.0019608	560	0.0017857
411	0.0024331	461	0.0021692	511	0.0019569	561	0.0017825
412	0.0024272	462	0.0021645	512	0.0019531	562	0.0017794
413	0.0024213	463	0.0021598	513	0.0019493	563	0.0017762
414	0.0024155	464	0.0021552	514	0.0019455	564	0.0017731
415	0.0024096	465	0.0021505	515	0.0019417	565	0.0017699
416	0.0024038	466	0.0021459	516	0.0019380	566	0.0017668
417	0.0023981	467	0.0021413	517	0.0019342	567	0.0017637
418	0.0023923	468	0.0021368	518	0.0019305	568	0.0017606
419	0.0023866	469	0.0021322	519	0.0019268	569	0.0017575
420	0.0023810	470	0.0021277	520	0.0019231	570	0.0017544
421	0.0023753	471	0.0021231	521	0.0019194	571	0.0017513
422	0.0023697	472	0.0021186	522	0.0019157	572	0.0017483
423	0.0023641	473	0.0021142	523	0.0019120	573	0.0017452
424	0.0023585	474	0.0021097	524	0.0019084	574	0.0017422
425	0.0023529	475	0.0021053	525	0.0019048	575	0.0017391
426	0.0023474	476	0.0021008	526	0.0019011	576	0.0017361
427	0.0023419	477	0.0020964	527	0.0018975	577	0.0017331
428	0.0023364	478	0.0020921	528	0.0018939	578	0.0017301
429	0.0023310	479	0.0020877	529	0.0018904	579	0.0017271
430	0.0023256	480	0.0020833	530	0.0018868	580	0.0017241
431	0.0023202	481	0.0020790	531	0.0018832	581	0.0017212
432	0.0023148	482	0.0020747	532	0.0018797	582	0.0017182
433	0.0023095	483	0.0020704	533	0.0018762	583	0.0017153
434	0.0023041	484	0.0020661	534	0.0018727	584	0.0017123
435	0.0022989	485	0.0020619	535	0.0018692	585	0.0017094
436	0.0022936	486	0.0020576	536	0.0018657	586	0.0017065
437	0.0022883	487	0.0020534	537	0.0018622	587	0.0017036
438	0.0022831	488	0.0020492	538	0.0018587	588	0.0017007
439	0.0022779	489	0.0020450	539	0.0018553	589	0.0016978
440	0.0022727	490	0.0020408	540	0.0018519	590	0.0016949
441	0.0022676	491	0.0020367	541	0.0018484	591	0.0016920
442	0.0022624	492	0.0020325	542	0.0018450	592	0.0016892
443	0.0022573	493	0.0020284	543	0.0018416	593	0.0016863
444	0.0022523	494	0.0020243	544	0.0018382	594	0.0016835
445	0.0022472	495	0.0020202	545	0.0018349	595	0.0016807
446	0.0022422	496	0.0020161	546	0.0018315	596	0.0016779
447	0.0022371	497	0.0020121	547	0.0018282	597	0.0016750
448	0.0022321	498	0.0020080	548	0.0018248	598	0.0016722
449	0.0022272	499	0.0020040	549	0.0018215	599	0.0016694
450	0.0022222	500	0.0020000	550	0.0018182	600	0.0016667

3.—Reciprocals, 601 to 800

No.	Reciprocal	No.	Reciprocal	No.	Reciprocal	No.	Reciprocal
601	0.0016639	651	0.0015361	701	0.0014265	751	0.0013316
602	0.0016611	652	0.0015337	702	0.0014245	752	0.0013298
603	0.0016584	653	0.0015314	703	0.0014225	753	0.0013280
604	0.0016556	654	0.0015291	704	0.0014205	754	0.0013263
605	0.0016529	655	0.0015267	705	0.0014184	755	0.0013245
606	0.0016502	656	0.0015244	706	0.0014164	756	0.0013228
607	0.0016474	657	0.0015221	707	0.0014144	757	0.0013210
608	0.0016447	658	0.0015198	708	0.0014124	758	0.0013193
609	0.0016420	659	0.0015175	709	0.0014104	759	0.0013175
610	0.0016393	660	0.0015152	710	0.0014085	760	0.0013158
611	0.0016367	661	0.0015129	711	9.0014065	761	0.0013141
612	0.0016340	662	0.0015106	712	0.0014045	762	0.0013123
613	0.0016313	663	0.0015083	713	0.0014025	763	0.0013106
614	0.0016287	664	0.0015060	714	0.0014006	764	0.0013089
615	0.0016260	665	0.0015038	715	0.0013986	765	0.0013072
616	0.0016234	666	0.0015015	716	0.0013966	766	0.0013055
617	0.0016207	667	0.0014993	717	0.0013947	767	0.0013038
618	0.0016181	668	0.0014970	718	0.0013928	768	0.0013021
619	0.0016155	669	0.0014948	719	0.0013908	769	0.0013004
620	0.0016129	670	0.0014925	720	0.0013889	770	0.0012987
621	0.0016103	671	0.0014903	721	0.0013870	771	0.0012970
622	0.0016077	672	0.0014881	722	0.0013850	772	0.0012953
623	0.0016051	673	0.0014859	723	0.0013831	773	0.0012937
624	0.0016026	674	0.0014837	724	0.0013812	774	0.0012920
625	0.0016000	675	0.0014815	725	0.0013793	775	0.0012903
626	0.0015974	676	0.0014793	726	0.0013774	776	0.0012887
627	0.0015949	677	0.0014771	727	0.0013755	777	0.0012870
628	0.0015924	678	0.0014749	728	0.0013736	778	0.0012853
629	0.0015898	679	0.0014728	729	0.0013717	779	0.0012837
630	0.0015873	680	0.0014706	730	0.0013699	780	0.0012821
631	0.0015848	681	0.0014684	731	0.0013680	781	0.0012804
632	0.0015823	682	0.0014663	732	0.0013661	782	0.0012788
633	0.0015798	683	0.0014641	733	0.0013643	783	0.0012771
634	0.0015773	684	0.0014620	734	0.0013624	784	0.0012755
635	0.0015748	685	0.0014599	735	0.0013605	785	0.0012739
636	0.0015723	686	0.0014577	736	0.0013587	786	0.0012723
637	0.0015699	687	0.0014556	737	0.0013569	787	0.0012706
638	0.0015674	688	0.0014535	738	0.0013550	788	0.0012690
639	0.0015649	689	0.0014514	739	0.0013532	789	0.0012674
640	0.0015625	690	0.0014493	740	0.0013514	790	0.0012658
641	0.0015601	691	0.0014472	741	0.0013495	791	0.0012642
642	0.0015576	692	0.0014451	742	0.0013477	792	0.0012626
643	0.0015552	693	0.0014430	743	0.0013459	793	0.0012610
644	0.0015528	694	0.0014409	744	0.0013441	794	0.0012594
645	0.0015504	695	0.0014388	745	0.0013423	795	0.0012579
646	0.0015480	696	0.0014368	746	0.0013405	796	0.0012563
647	0.0015456	697	0.0014347	747	0.0013387	797	0.0012547
648	0.0015432	698	0.0014327	748	0.0013369	798	0.0012531
649	0.0015408	699	0.0014306	749	0.0013351	799	0.0012516
650	0.0015385	700	0.0014286	750	0.0013333	800	0.0012500

Then arrange a small table of its multiples up to nine times and use this as a multiplication table.

$$0.000636132 \times 1 = 0.000636132$$
$$0.000636132 \times 2 = 0.001272264$$
$$0.000636132 \times 3 = 0.001908396$$
$$0.000636132 \times 4 = 0.002544528$$
$$0.000636132 \times 5 = 0.003180660$$
$$0.000636132 \times 6 = 0.003816792$$
$$0.000636132 \times 7 = 0.004452924$$
$$0.000636132 \times 8 = 0.005088956$$
$$0.000636132 \times 9 = 0.005696188$$

Dividend 7246

Take from above table 6.......... .003816792
4.......... 0.02544528
2.......... 00.1272264
7.......... 004.452924

4.609412472

Correct quotient by direct division to hundred thousandths 4.60941.

Percentage.—*Percent* means *hundredths* and rate percent means any given number of hundredths. Thus, 5 per cent, or 5%, means .05 or $\frac{5}{100}$, in which 5 is the rate. It may also be expressed in true ratio, 5 : 100, meaning 5 *parts* of the 100, both terms being of the same denomination. The percents commonly used may be written in fractional form as follows:

$6\frac{1}{4}\% = \frac{1}{16}$	$12\frac{1}{2}\% = \frac{1}{8}$	$25\ \% = \frac{1}{4}$	$62\frac{1}{2}\% = \frac{5}{8}$
$6\frac{2}{3}\% = \frac{1}{15}$	$14\frac{2}{7}\% = \frac{1}{7}$	$33\frac{1}{3}\% = \frac{1}{3}$	$66\frac{2}{3}\% = \frac{2}{3}$
$8\frac{1}{3}\% = \frac{1}{12}$	$16\frac{2}{3}\% = \frac{1}{6}$	$50\ \% = \frac{1}{2}$	$83\frac{1}{3}\% = \frac{5}{6}$
$10\ \% = \frac{1}{10}$	$20\ \% = \frac{1}{5}$	$37\frac{1}{2}\% = \frac{3}{8}$	$100\ \% = 1$

Percentage covers the operations of finding the part of a given number at a given rate percent, as 4 percent of 650, $650 \times .04 = 26$; of finding what percent one number is of another; as, what percent of 560 is 32?

$$32 \div 560 = .057 = 5.7 \text{ percent};$$

of ascertaining a number when an amount is given, which is a given percent of that number; as, 112 is 24 percent of what number?

$$112 \div .24 = 467.$$

Logarithms of Numbers.—This section will not attempt to describe in detail the principles upon which logarithms are founded but will confine itself to a brief exposition of the *use* of logarithms.

The *logarithm* of any given number is the exponent of the power to which another fixed number, called the *base*, must be raised in order to produce the given number. A system of logarithms may be founded on any base. Two systems are in use, namely, *common logarithms* and *Naperian* or *natural logarithms*. Common logarithms are on the base 10. In other words, the logarithm of a number indicates the power to which 10 must be raised to produce the given number. In this system

$10^0 = 1$	$\log \quad 1 = 0$
$10^1 = 10$	$\log \quad 10 = 1$
$10^2 = 100$	$\log \quad 100 = 2$
$10^3 = 1000$, etc.	$\log 1000 = 3$, etc.

This system is in general use for all practical purposes. When logarithms are mentioned without further qualification, common logarithms are meant.

Natural or Naperian logarithms are founded on a base $e = 2.7182818+$. It is used in pure mathematical discussion and in steam and electrical engineering.

Common Logarithms.—The logarithm of a number is composed of the *characteristic*, or integral portion to the left of the decimal point, and the *mantissa* or decimal fraction. The mantissa is all that appears in any table of logarithms and the degree of accuracy is dependent upon the number of decimal places used in the mantissa. Table 4, following, to five decimal places will be found compact and convenient, where the result to five significant figures is sufficiently accurate. Where greater accuracy is required, *Vega's* tables to seven decimal places are recommended.

In the logarithm of any number, the mantissa is independent of the position of the decimal point, while on the contrary the characteristic is dependent only on the position of the first significant figure of the number with relation to the decimal point. Thus in the following examples:

(a) log 3456.2 $\quad= 3.53859$

(b) log 345.62 $\quad= 2.53859$

(c) log 34.562 $\quad= 1.53589$

(d) log 3.4562 $\quad= 0.53859$

(e) log .34562 $\quad= \overline{1}.53859 = 9.53859 - 10$

(f) log .034562 $= \overline{2}.53859 = 8.53859 - 10$

The use of the positive characteristic is generally preferred, omitting the (−10) in ordinary cases.

it will be seen that the characteristic is equal, *algebraically*, to the number of places minus one, which the first significant figure of the number occupies to the *left* of the decimal point. In (a) the characteristic is 3; in (b), 2; in (d) 0; in (e), −1; and in (f), −2. Some mathematicians prefer the use of the negative characteristic, but most of them employ the "positive," by algebraically adding 10 to the integer and placing −10 to the right of the mantissa or omitting the latter (−10) altogether. For example, log .040217 = 8.60441, the −10 being understood and the value of the characteristic being, of course, −2. In the case of finding the root of (or dividing) a pure decimal, however, the −10 must be employed.

To Find the Logarithm of a Number.—Example: Find the log of 357.46. Solution: The characteristic is 3 − 1 = 2. The mantissa for the first four figures, 3574, is read directly from Table 4 and is .55315. To this, however, must be added $\frac{6}{10}$ (the next figure of the number is 6) of the difference between .55315 and the log of 3575, or .55328. This difference is 13 and in the proportional parts (P.P.) column under 13 and opposite 6 will be found the value 8, which, added to .55315 in the last place, gives .55323. Hence, the log of 357.46 is 2.55323 (Ans.).

To Find the Anti-logarithm (number corresponding to a logarithm).—Example: What is the number whose logarithm is 1.73821? Solution: This is the reverse of finding the logarithm of a number. Neglecting, for the present, the characteristic, the next lower mantissa to .73821 is .73815 and the number corresponding is 5472. The difference between .73815 and the next higher mantissa in Table 4, .73823, is 8, and the proportional difference $\dfrac{.73821 - .73815}{.73823 - .73815} = \dfrac{6}{8}$ calls for .8 to be added to the fourth figure, i.e., 8 to the fifth place of the number, disregarding the decimal point, is 54728. The characteristic, 1, calls for two places to the left of the decimal point, hence the antilog of 1.73821 is 54.728 (Ans.).

Multiplication with Logarithms.—*To multiply two or more numbers, add the logarithms of the numbers and the sum is the logarithm of the product.*

Example: Multiply 25.316 by 42.18

Solution: log 25.316 = 1.40339
    log 42.18 = 1.62511
       Sum = 3.02850

Product = antilog 3.02851 = 1067.9 (Ans.).

Division with Logarithms.—*To divide one number by another, subtract the logarithm of the divisor from the logarithm of the dividend; the difference is the logarithm of the quotient.*

Example: Divide 458.62 by 86.25

Solution: log 458.62 = 2.66145
    log 86.25 = 1.93576
    Difference = 0.72569

Quotient = antilog 0.72569 = 5.3173 (Ans.).

Raising to Powers with Logarithms.—*To raise a number to a certain power, multiply the logarithm of the number by the exponent of the power; the product is the logarithm of the number raised to the required power.*

Example: What is the value of 4.53^5?

Solution: log 4.53 = 0.65610
            Exponent of power =   5
                                  ─────────────
                              Product = 3.28050

Number raised to the 5th power = antilog 3.28050 = 1907.65 (Ans.).

To Extract the Root of a Number.—*To extract the root of a number, divide the logarithm of the number by the index of the root; the quotient is the logarithm of the root.*

Example: What is $\sqrt[5]{356.07}$?

Solution: log 356.07 = 2.55153

$$5\overline{)2.55153}$$
$$.51031$$

Root = antilog .51031 = 3.2382 (Ans.).

Example: What is $\sqrt{.2516}$?

Solution: log .2516 = 9.40071 − 10

$$2\overline{)9.40071 - 10}$$
$$4.70035 - \quad 5 = \bar{1}.70035$$

Root = antilog $\bar{1}$.70035 = .50159 (Ans.).

TABLE 4

LOGARITHMS

No.	L. O	1	2	3	4	5	6	.7	8	9	P. P.	No.	Log.	Dif.
100	00 000	043	087	130	173	217	260	303	346	389	44 43 42	1.00	.00000	995
1	432	475	518	561	604	647	689	732	775	817	1 4 4 4	1.01	.00995	985
2	860	903	945	988	030	072	115	157	199	242	2 9 9 8	1.02	.01980	976
3	01 284	326	368	410	452	494	536	578	620	662	3 13 13 13	1.03	.02956	966
4	703	745	787	828	870	912	953	995	036	078	4 18 17 17	1.04	.03922	957
105	02 119	160	202	243	284	325	366	407	449	490	5 22 22 21	1.05	.04879	948
6	531	572	612	653	694	735	776	816	857	898	6 26 26 25	1.06	.05827	939
7	938	979	019	060	100	141	181	222	262	302	7 31 30 29	1.07	.06766	930
8	03 342	383	423	463	503	543	583	623	663	703	8 35 34 34	1.08	.07696	922
9	743	782	822	862	902	941	981	021	060	100	9 40 39 38	1.09	.08618	913
110	04 139	179	218	258	297	336	376	415	454	493	41 40 39	1.10	.09531	905
1	532	571	610	650	689	727	766	805	844	883	1 4 4 4	1.11	.10436	897
2	922	961	999	038	077	115	154	192	231	269	2 8 8 8	1.12	.11333	889
3	05 308	346	385	423	461	500	538	576	614	652	3 12 12 12	1.13	.12222	881
4	690	729	767	805	843	881	918	956	994	032	4 16 16 16	1.14	.13103	873
115	06 070	108	145	183	221	258	296	333	371	408	5 21 20 20	1.15	.13976	866
6	446	483	521	558	595	633	670	707	744	781	6 25 24 23	1.16	.14842	858
7	819	856	893	930	967	004	041	078	115	151	7 29 28 27	1.17	.15700	851
8	07 188	225	262	298	335	372	408	445	482	518	8 33 32 31	1.18	.16551	844
9	555	591	628	664	700	737	773	809	846	882	9 37 36 35	1.19	.17395	837
120	918	954	990	027	063	099	135	171	207	243	38 37 36	1.20	.18232	830
1	08 279	314	350	386	422	458	493	529	565	600	1 4 4 4	1.21	.19062	823
2	636	672	707	743	778	814	849	884	920	955	2 8 7 7	1.22	.19885	816
3	991	026	061	096	132	167	202	237	272	307	3 11 11 11	1.23	.20701	810
4	09 342	377	412	447	482	517	552	587	621	656	4 15 15 14	1.24	.21511	803
125	691	726	760	795	830	864	899	934	968	003	5 19 19 18	1.25	.22314	797
6	10 037	072	106	140	175	209	243	278	312	346	6 23 22 22	1.26	.23111	791
7	380	415	449	483	517	551	585	619	653	687	7 27 26 25	1.27	.23902	784
8	721	755	789	823	857	890	924	958	992	025	8 30 30 29	1.28	.24686	778
9	11 059	093	126	160	193	227	261	294	327	361	9 34 33 32	1.29	.25464	772
130	394	428	461	494	528	561	594	628	661	694	35 34 33	1.30	.26236	767
1	727	760	793	826	860	893	926	959	992	024	1 4 3 3	1.31	.27003	760
2	12 057	090	123	156	189	222	254	287	320	352	2 7 7 7	1.32	.27763	755
3	385	418	450	483	516	548	581	613	646	678	3 11 10 10	1.33	.28518	749
4	710	743	775	808	840	872	905	937	969	001	4 14 14 13	1.34	.29267	743
135	13 033	066	098	130	162	194	226	258	290	322	5 18 17 17	1.35	.30010	738
6	354	386	418	450	481	513	545	577	609	640	6 21 20 20	1.36	.30748	733
7	672	704	735	767	799	830	862	893	925	956	7 25 24 23	1.37	.31481	727
8	988	019	051	082	114	145	176	208	239	270	8 28 27 26	1.38	.32208	722
9	14 301	333	364	395	426	457	489	520	551	582	9 32 31 30	1.39	.32930	717
140	613	644	675	706	737	768	799	829	860	891	32 31 30	1.40	.33647	712
1	922	953	983	014	045	076	106	137	168	198	1 3 3 3	1.41	.34359	707
2	15 229	259	290	320	351	381	412	442	473	503	2 6 6 6	1.42	.35066	701
3	534	564	594	625	655	685	715	746	776	806	3 10 9 9	1.43	.35767	697
4	836	866	897	927	957	987	017	047	077	107	4 13 12 12	1.44	.36464	692
145	16 137	167	197	227	256	286	316	346	376	406	5 16 16 15	1.45	.37156	688
6	435	465	495	524	554	584	613	643	673	702	6 19 19 18	1.46	.37844	682
7	732	761	791	820	850	879	909	938	967	997	7 22 22 21	1.47	.38526	678
8	17 026	056	085	114	143	173	202	231	260	289	8 26 25 24	1.48	.39204	674
9	319	348	377	406	435	464	493	522	551	580	9 29 28 27	1.49	.39878	669
150	609	638	667	696	725	754	782	811	840	869		1.50	.40547	

4.—LOGARITHMS—*Continued*

No.	Common Logarithms of Numbers										P. P.	Naperian		Dif
	L. O	1	2	3	4	5	6	7	8	9		No.	Log.	
150	17 609	638	667	696	725	754	782	811	840	869	29 28	1.50	.40547	664
1	898	926	955	984	013	041	070	099	127	156	1 3 3	1.51	.41211	660
2	18 184	213	241	270	298	327	355	384	412	441	2 6 6	1.52	.41871	656
3	469	498	526	554	583	611	639	667	696	724	3 9 8	1.53	.42527	651
4	752	780	808	837	865	893	921	949	977	005	4 12 11	1.54	.43178	647
155	19 033	061	089	117	145	173	201	229	257	285	5 15 14	1.55	.43825	644
6	312	340	368	396	424	451	479	507	535	562	6 17 17	1.56	.44469	639
7	590	618	645	673	700	728	756	783	811	838	7 20 20	1.57	.45108	634
8	866	893	921	948	976	003	030	058	085	112	8 23 22	1.58	.45742	631
9	20 140	167	194	222	249	276	303	330	358	385	9 26 25	1.59	.46373	627
160	412	439	466	493	520	548	575	602	629	656	27 26	1.60	.47000	623
1	683	710	737	763	790	817	844	871	898	925	1 3 3	1.61	.47623	620
2	952	978	005	032	059	085	112	139	165	192	2 5 5	1.62	.48243	615
3	21 219	245	272	299	325	352	378	405	431	458	3 8 8	1.63	.48858	612
4	484	511	537	564	590	617	643	669	696	722	4 11 10	1.64	.49470	608
165	748	775	801	827	854	880	906	932	958	985	5 14 13	1.65	.50078	604
6	22 011	037	063	089	115	141	167	194	220	246	6 16 16	1.66	.50682	600
7	272	298	324	350	376	401	427	453	479	505	7 19 18	1.67	.51282	597
8	531	557	583	608	634	660	686	712	737	763	8 22 21	1.68	.51879	594
9	789	814	840	866	891	917	943	968	994	019	9 24 23	1.69	.52473	590
170	23 045	070	096	121	147	172	198	223	249	274	25 24	1.70	.53063	586
1	300	325	350	376	401	426	452	477	502	528	1 3 2	1.71	.53649	583
2	553	578	603	629	654	679	704	729	754	779	2 5 5	1.72	.54232	580
3	805	830	855	880	905	930	955	980	005	030	3 8 7	1.73	.54812	577
4	24 055	080	105	130	155	180	204	229	254	279	4 10 10	1.74	.55389	573
175	304	329	353	378	403	428	452	477	502	527	5 13 12	1.75	.55962	569
6	551	576	601	625	650	674	699	724	748	773	6 15 14	1.76	.56531	567
7	797	822	846	871	895	920	944	969	993	018	7 18 17	1.77	.57098	563
8	25 042	066	091	115	139	164	188	212	237	261	8 20 19	1.78	.57661	561
9	285	310	334	358	382	406	431	455	479	503	9 23 22	1.79	.58222	557
180	527	551	575	600	624	648	672	696	720	744	24 23	1.80	.58779	554
1	768	792	816	840	864	888	912	935	959	983	1 2 2	1.81	.59333	551
2	26 007	031	055	079	102	126	150	174	198	221	2 5 5	1.82	.59884	548
3	245	269	293	316	340	364	387	411	435	458	3 7 7	1.83	.60432	545
4	482	505	529	553	576	600	623	647	670	694	4 10 9	1.84	.60977	542
185	717	741	764	788	811	834	858	881	905	928	5 12 12	1.85	.61519	539
6	951	975	998	021	045	068	091	114	138	161	6 14 14	1.86	.62058	536
7	27 184	207	231	254	277	300	323	346	370	393	7 17 16	1.87	.62594	533
8	416	439	462	485	508	531	554	577	600	623	8 19 18	1.88	.63127	531
9	646	669	692	715	738	761	784	807	830	852	9 22 21	1.89	.63658	527
190	875	898	921	944	967	989	012	035	058	081	22 21	1.90	.64185	525
1	28 103	126	149	171	194	217	240	262	285	307	1 2 2	1.91	.64710	523
2	330	353	375	398	421	443	466	488	511	533	2 4 4	1.92	.65233	519
3	556	578	601	623	646	668	691	713	735	758	3 7 6	1.93	.65752	517
4	780	803	825	847	870	892	914	937	959	981	4 9 8	1.94	.66269	514
.195	29 003	026	048	070	092	115	137	159	181	203	5 11 11	1.95	.66783	511
6	226	248	270	292	314	336	358	380	403	425	6 13 13	1.96	.67294	509
7	447	469	491	513	535	557	579	601	623	645	7 15 15	1.97	.67803	507
8	667	688	710	732	754	776	798	820	842	863	8 18 17	1.98	.68310	503
9	885	907	929	951	973	994	016	038	060	081	9 20 19	1.99	.68813	502
200	30 103	125	146	168	190	211	233	255	276	298		2.00	.69315	

4.—LOGARITHMS—*Continued*

No.	L. O	1	2	3	4	5	6	7	8	9	P. P.	No.	Log.	Dif.
200	30 103	125	146	168	190	211	233	255	276	298	22	2.00	.69315	498
1	320	341	363	384	406	428	449	471	492	514	1 2	2.01	.69813	497
2	535	557	578	600	621	643	664	685	707	728	2 4	2.02	.70310	494
3	750	771	792	814	835	856	878	899	920	942	3 7	2.03	.70804	491
4	963	984	006	027	048	069	091	112	133	154	4 9	2.04	.71295	489
205	31 175	197	218	239	260	281	302	323	345	366	5 11	2.05	.71784	487
6	387	408	429	450	471	492	513	534	555	576	6 13	2.06	.72271	484
7	597	618	639	660	681	702	723	744	765	785	7 15	2.07	.72755	482
8	806	827	848	869	890	911	931	952	973	994	8 18	2.08	.73237	479
9	32 015	035	056	077	098	118	139	160	181	201	9 20	2.09	.73716	478
210	222	243	263	284	305	325	346	366	387	408	21	2.10	.74194	475
1	428	449	469	490	510	531	552	572	593	613	1 2	2.11	.74669	473
2	634	654	675	695	715	736	756	777	797	818	2 4	2.12	.75142	470
3	838	858	879	899	919	940	960	980	001	021	3 6	2.13	.75612	469
4	33 041	062	082	102	122	143	163	183	203	224	4 8	2.14	.76081	466
215	244	264	284	304	325	345	365	385	405	425	5 11	2.15	.76547	464
6	445	465	486	506	526	546	566	586	606	626	6 13	2.16	.77011	462
7	646	666	686	706	726	746	766	786	806	826	7 15	2.17	.77473	459
8	846	866	885	905	925	945	965	985	005	025	8 17	2.18	.77932	458
9	34 044	064	084	104	124	143	163	183	203	223	9 19	2.19	.78390	456
220	242	262	282	301	321	341	361	380	400	420	20	2.20	.78846	453
1	439	459	479	498	518	537	557	577	596	616	1 2	2.21	.79299	452
2	635	655	674	694	713	733	753	772	792	811	2 4	2.22	.79751	449
3	830	850	869	889	908	928	947	967	986	005	3 6	2.23	.80200	448
4	35 025	044	064	083	102	122	141	160	180	199	4 8	2.24	.80648	445
225	218	238	257	276	295	315	334	353	372	392	5 10	2.25	.81093	443
6	411	430	449	468	488	507	526	545	564	583	6 12	2.26	.81536	442
7	603	622	641	660	679	698	717	736	755	774	7 14	2.27	.81978	440
8	793	813	832	851	870	889	908	927	946	965	8 16	2.28	.82418	437
9	984	003	021	040	059	078	097	116	135	154	9 18	2.29	.82855	436
230	36 173	192	211	229	248	267	286	305	324	342	19	2.30	.83291	434
1	361	380	399	418	435	455	474	493	511	530	1 2	2.31	.83725	432
2	549	568	586	605	624	642	661	680	698	717	2 4	2.32	.84157	430
3	736	754	773	791	810	829	847	866	884	903	3 6	2.33	.84587	428
4	922	940	959	977	996	014	033	051	070	088	4 8	2.34	.85015	427
235	37 107	125	144	162	181	199	218	236	254	273	5 10	2.35	.85442	424
6	291	310	328	346	365	383	401	420	438	457	6 11	2.36	.85866	423
7	475	493	511	530	548	566	585	603	621	639	7 13	2.37	.86289	421
8	658	676	694	712	731	749	767	785	803	822	8 15	2.38	.86710	419
9	840	858	876	894	912	931	949	967	985	003	9 17	2.39	.87129	418
240	38 021	039	057	075	093	112	130	148	166	184	18	2.40	.87547	416
1	202	220	238	256	274	292	310	328	346	364	1 2	2.41	.87963	414
2	382	399	417	435	453	471	489	507	525	543	2 4	2.42	.88377	412
3	561	578	596	614	632	650	668	686	703	721	3 5	2.43	.88789	411
4	739	757	775	792	810	828	846	863	881	899	4 7	2.44	.89200	409
245	917	934	952	970	987	005	023	041	058	076	5 9	2.45	.89609	407
6	39 094	111	129	146	164	182	199	217	235	252	6 11	2.46	.90016	406
7	270	287	305	322	340	358	375	393	410	428	7 13	2.47	.90422	404
8	445	463	480	498	515	533	550	568	585	602	8 14	2.48	.90826	402
9	620	637	655	672	690	707	724	742	759	777	9 16	2.49	.91228	401
250	794	811	829	846	863	881	898	915	933	950		2.50	.91629	

4.—Logarithms—*Continued*

No.	L. O	1	2	3	4	5	6	7	8	9	P.P.	No.	Log.	Dif.
						Common Logarithms of Numbers.							Naperian.	
250	39 794	811	829	846	863	881	898	915	933	950	18	2.50	.91629	399
1	967	985	002	019	037	054	071	088	106	123	1 2	2.51	.92028	398
2	40 140	157	175	192	209	226	243	261	278	295	2 4	2.52	.92426	396
3	312	329	346	364	381	398	415	432	449	466	3 5	2.53	.92822	394
4	483	500	518	535	552	569	586	603	620	637	4 7	2.54	.93216	393
255	654	671	688	705	722	739	756	773	790	807	5 9	2.55	.93609	392
6	824	841	858	875	892	909	926	943	960	976	6 11	2.56	.94001	390
7	993	010	027	044	061	078	095	111	128	145	7 13	2.57	.94391	388
8	41 162	179	196	212	229	246	263	280	296	313	8 14	2.58	.94779	387
9	330	347	363	380	397	414	430	447	464	481	9 16	2.59	.95166	385
260	497	514	531	547	564	581	597	614	631	647	17	2.60	.95551	384
1	664	681	697	714	731	747	764	780	797	814	1 2	2.61	.95935	382
2	830	847	863	880	896	913	929	946	963	979	2 3	2.62	.96317	381
3	996	012	029	045	062	078	095	111	127	144	3 5	2.63	.96698	380
4	42 160	177	193	210	226	243	259	275	292	308	4 7	2.64	.97078	378
265	325	341	357	374	390	406	423	439	455	472	5 9	2.65	.97456	377
6	488	504	521	537	553	570	586	602	619	635	6 10	2.66	.97833	375
7	651	667	684	700	716	732	749	765	781	797	7 12	2.67	.98208	374
8	813	830	846	862	878	894	911	927	943	959	8 14	2.68	.98582	372
9	975	991	008	024	040	056	072	088	104	120	9 15	2.69	.98954	371
270	43 136	152	169	185	201	217	233	249	265	281	16	2.70	.99325	370
1	297	313	329	345	361	377	393	409	425	441	1 2	2.71	.99695	368
2	457	473	489	505	521	537	553	569	584	600	2 3	2.72	1.00063	367
3	616	632	648	664	680	696	712	727	743	759	3 5	2.73	1.00430	366
4	775	791	807	823	838	854	870	886	902	917	4 6	2.74	1.00796	364
275	933	949	965	981	996	012	028	044	059	075	5 8	2.75	1.01160	363
6	44 091	107	122	138	154	170	185	201	217	232	6 10	2.76	1.01523	362
7	248	264	279	295	311	326	342	358	373	389	7 11	2.77	1.01885	360
8	404	420	436	451	467	483	498	514	529	545	8 13	2.78	1.02245	359
9	560	576	592	607	623	638	654	669	685	700	9 14	2.79	1.02604	358
280	716	731	747	762	778	793	809	824	840	855	15	2.80	1.02962	356
1	871	886	902	917	932	948	963	979	994	010	1 2	2.81	1.03318	356
2	45 025	040	056	071	086	102	117	133	148	163	2 3	2.82	1.03674	354
3	179	194	209	225	240	255	271	286	301	317	3 5	2.83	1.04028	352
4	332	347	362	378	393	408	423	439	454	469	4 6	2.84	1.04380	352
285	484	500	515	530	545	561	576	591	606	621	5 8	2.85	1.04732	350
6	637	652	667	682	697	712	728	743	758	773	6 9	2.86	1.05082	349
7	788	803	818	834	849	864	879	894	909	924	7 11	2.87	1.05431	348
8	939	954	969	984	000	015	030	045	060	075	8 12	2.88	1.05779	347
9	46 090	105	120	135	150	165	180	195	210	225	9 14	2.89	1.06126	345
290	240	255	270	285	300	315	330	345	359	374	14	2.90	1.06471	344
1	389	404	419	434	449	464	479	494	509	523	1 1	2.91	1.06815	343
2	538	553	568	583	598	613	627	642	657	672	2 3	2.92	1.07158	342
3	687	702	716	731	746	761	776	790	805	820	3 4	2.93	1.07500	341
4	835	850	864	879	894	909	923	938	953	967	4 6	2.94	1.07841	340
295	982	997	012	026	041	056	070	085	100	114	5 7	2.95	1.08181	338
6	47 129	144	159	173	188	202	217	232	246	261	6 8	2.96	1.08519	337
7	276	290	305	319	334	349	363	378	392	407	7 10	2.97	1.08856	336
8	422	436	451	465	480	494	509	524	538	553	8 11	2.98	1.09192	335
9	567	582	596	611	625	640	654	669	683	698	9 13	2.99	1.09527	334
300	712	727	741	756	770	784	799	813	828	842		3.00	1.09861	

4.—LOGARITHMS—Continued

No.	L. O	1	2	3	4	5	6	7	8	9	P. P.		No.	Log.	Dif.
														Common Logarithms of Numbers. / Naperian.	
300	47 712	727	741	756	770	784	799	813	828	842		15	3.00	1.09861	333
1	857	871	885	900	914	929	943	958	972	986	1	2	3.01	1.10194	332
2	48 001	015	029	044	058	073	087	101	116	130	2	3	3.02	1.10526	330
3	144	159	173	187	202	216	230	244	259	273	3	5	3.03	1.10856	330
4	287	302	316	330	344	359	373	387	401	416	4	6	3.04	1.11186	328
305	430	444	458	473	487	501	515	530	544	558	5	8	3.05	1 11514	327
6	572	586	601	615	629	643	657	671	686	700	6	9	3.06	1.11841	327
7	714	728	742	756	770	785	799	813	827	841	7	11	3.07	1.12168	325
8	855	869	883	897	911	926	940	954	968	982	8	12	3.08	1.12493	324
9	996	010	024	038	052	066	080	094	108	122	9	14	3 09	1.12817	323
310	49 136	150	164	178	192	206	220	234	248	262		14	3.10	1.13140	322
1	276	290	304	318	332	346	360	374	388	402	1	1	3.11	1.13462	321
2	415	429	443	457	471	485	499	513	527	541	2	3	3.12	1.13783	320
3	554	568	582	596	610	624	638	651	665	679	3	4	3.13	1.14103	319
4	693	707	721	734	748	762	776	790	803	817	4	6	3.14	1.14422	318
315	831	845	859	872	886	900	914	927	941	955	5	7	3.15	1.14740	317
6	969	982	996	010	024	037	051	065	079	092	6	8	3.16	1.15057	316
7	50 106	120	133	147	161	174	188	202	215	229	7	10	3.17	1.15373	315
8	243	256	270	284	297	311	325	338	352	365	8	11	3.18	1.15688	314
9	379	393	406	420	433	447	461	474	488	501	9	13	3.19	1.16002	313
320	515	529	542	556	569	583	596	610	623	637		13	3.20	1.16315	312
1	651	664	678	691	705	718	732	745	759	772	1	1	3.21	1.16627	311
2	786	799	813	826	840	853	866	880	893	907	2	3	3.22	1.16938	310
3	920	934	947	961	974	987	001	014	028	041	3	4	3.23	1.17248	309
4	51 055	068	081	095	108	121	135	148	162	175	4	5	3.24	1.17557	308
325	188	202	215	228	242	255	268	282	295	308	5	7	3.25	1.17865	308
6	322	335	348	362	375	388	402	415	428	441	6	8	3.26	1.18173	306
7	455	468	481	495	508	521	534	548	561	574	7	9	3.27	1.18479	305
8	587	601	614	627	640	654	667	680	693	706	8	10	3.28	1.18784	305
9	720	733	746	759	772	786	799	812	825	838	9	12	3.29	1.19089	303
330	851	865	878	891	904	917	930	943	957	970		13	3.30	1.19392	303
1	983	996	009	022	035	048	061	075	088	101	1	1	3.31	1.19695	301
2	52 114	127	140	153	166	179	192	205	218	231	2	3	3.32	1.19996	301
3	244	257	270	284	297	310	323	336	349	362	3	4	3.33	1.20297	300
4	375	388	401	414	427	440	453	466	479	492	4	5	3.34	1.20597	299
335	504	517	530	543	556	569	582	595	608	621	5	7	3.35	1.20896	299
6	634	647	660	673	686	699	711	724	737	750	6	8	3.36	1.21194	297
7	763	776	789	802	815	827	840	853	866	879	7	9	3.37	1.21491	297
8	892	905	917	930	943	956	969	982	994	007	8	10	3.38	1.21788	295
9	53 020	033	046	058	071	084	097	110	122	135	9	12	3.39	1.22083	295
340	148	161	173	186	199	212	224	237	250	263		12	3.40	1.22378	293
1	275	288	301	314	326	339	352	364	377	390	1	1	3.41	1.22671	293
2	403	415	428	441	453	466	479	491	504	517	2	2	3.42	1.22964	292
3	529	542	555	567	580	593	605	618	631	643	3	4	3.43	1.23256	291
4	656	668	681	694	706	719	732	744	757	769	4	5	3.44	1.23547	290
345	782	794	807	820	832	845	857	870	882	895	5	6	3.45	1.23837	290
6	908	920	933	945	958	970	983	995	008	020	6	7	3.46	1.24127	288
7	54 033	045	058	070	083	095	108	120	133	145	7	8	3.47	1.24415	288
8	158	170	183	195	208	220	233	245	258	270	8	10	3.48	1.24703	287
9	283	295	307	320	332	345	357	370	382	394	9	11	3.49	1.24990	286
350	407	419	432	444	456	469	481	494	506	518			3.50	1.25276	

4.—LOGARITHMS—Continued

Common Logarithms of Numbers.

No.	L. O	1	2	3	4	5	6	7	8	9	P. P.	
350	54 407	419	432	444	456	469	481	494	506	518	13	
1	531	543	555	568	580	593	605	617	630	642	1	1
2	654	667	679	691	704	716	728	741	753	765	2	3
3	777	790	802	814	827	839	851	864	876	888	3	4
4	900	913	925	937	949	962	974	986	998	011	4	5
355	55 023	035	047	060	072	084	096	108	121	133	5	7
6	145	157	169	182	194	206	218	230	242	255	6	8
7	267	279	291	303	315	328	340	352	364	376	7	9
8	388	400	413	425	437	449	461	473	485	497	8	10
9	509	522	534	546	558	570	582	594	606	618	9	12
360	630	642	654	666	678	691	703	715	727	739		
1	751	763	775	787	799	811	823	835	847	859	1	
2	871	883	895	907	919	931	943	955	967	979	2	
3	991	003	015	027	038	050	062	074	086	098	3	
4	56 110	122	134	146	158	170	182	194	205	217	4	
365	229	241	253	265	277	289	301	312	324	336	5	
6	348	360	372	384	396	407	419	431	443	455	6	
7	467	478	490	502	514	526	538	549	561	573	7	
8	585	597	608	620	632	644	656	667	679	691	8	
9	703	714	726	738	750	761	773	785	797	808	9	
370	820	832	844	855	867	879	891	902	914	926	12	
1	937	949	961	972	984	996	008	019	031	043	1	1
2	57 054	066	078	089	101	113	124	136	148	159	2	2
3	171	183	194	206	217	229	241	252	264	276	3	4
4	287	299	310	322	334	345	357	368	380	392	4	5
375	403	415	426	438	449	461	473	484	496	507	5	6
6	519	530	542	553	565	576	588	600	611	623	6	7
7	634	646	657	669	680	692	703	715	726	738	7	8
8	749	761	772	784	795	807	818	830	841	852	8	10
9	864	875	887	898	910	921	933	944	955	967	9	11
380	978	990	001	013	024	035	047	058	070	081		
1	58 092	104	115	127	138	149	161	172	184	195	1	
2	206	218	229	240	252	263	274	286	297	309	2	
3	320	331	343	354	365	377	388	399	410	422	3	
4	433	444	456	467	478	490	501	512	524	535	4	
385	546	557	569	580	591	602	614	625	636	647	5	
6	659	670	681	692	704	715	726	737	749	760	6	
7	771	782	794	805	816	827	838	850	861	872	7	
8	883	894	906	917	928	939	950	961	973	984	8	
9	995	006	017	028	040	051	062	073	084	095	9	
390	59 106	118	129	140	151	162	173	184	195	207	11	
1	218	229	240	251	262	273	284	295	306	318	1	1
2	329	340	351	362	373	384	395	406	417	428	2	2
3	439	450	461	472	483	494	506	517	528	539	3	3
4	550	561	572	583	594	605	616	627	638	649	4	4
395	660	671	682	693	704	715	726	737	748	759	5	6
6	770	780	791	802	813	824	835	846	857	868	6	7
7	879	890	901	912	923	934	945	956	966	977	7	8
8	988	999	010	021	032	043	054	065	076	086	8	9
9	60 097	108	119	130	141	152	163	173	184	195	9	10
400	206	217	228	239	249	260	271	282	293	304		

Naperian.

No.	Log.	Dif.
3.50	1.25276	286
3.51	1.25562	284
3.52	1.25846	284
3.53	1.26130	283
3.54	1.26413	283
		282
3.55	1.26695	281
3.56	1.26976	281
3.57	1.27257	279
3.58	1.27536	279
3.59	1.27815	279
		278
3.60	1.28093	278
3.61	1.28371	276
3.62	1.28647	276
3.63	1.28923	275
3.64	1.29198	275
		275
3.65	1.29473	273
3.66	1.29746	273
3.67	1.30019	272
3.68	1.30291	272
3.69	1.30563	270
		270
3.70	1.30833	270
3.71	1.31103	269
3.72	1.31372	269
3.73	1.31641	268
3.74	1.31909	268
		267
3.75	1.32176	266
3.76	1.32442	265
3.77	1.32707	265
3.78	1.32972	265
3.79	1.33237	263
		263
3.80	1.33500	263
3.81	1.33763	262
3.82	1.34025	261
3.83	1.34286	261
3.84	1.34547	260
		260
3.85	1.34807	260
3.86	1.35067	258
3.87	1.35325	259
3.88	1.35584	257
3.89	1.35841	257
		257
3.90	1.36098	256
3.91	1.36354	255
3.92	1.36609	255
3.93	1.36864	254
3.94	1.37118	254
		254
3.95	1.37372	252
3.96	1.37624	253
3.97	1.37877	251
3.98	1.38128	251
3.99	1.38379	251
		250
4.00	1.38629	

4.—LOGARITHMS—*Continued*

No.	L. O	1	2	3	4	5	6	7	8	9	P. P.	No.	Log.	Dif.
400	60 206	217	228	239	249	260	271	282	293	304	11	4.00	1.38629	250
1	314	325	336	347	358	369	379	390	401	412	1 1	4.01	1.38879	249
2	423	433	444	455	466	477	487	498	509	520	2 2	4.02	1.39128	249
3	531	541	552	563	574	584	595	606	617	627	3 3	4.03	1.39377	247
4	638	649	660	670	681	692	703	713	724	735	4 4	4.04	1.39624	248
405	746	756	767	778	788	799	810	821	831	842	5 6	4.05	1.39872	246
6	853	863	874	885	895	906	917	927	938	949	6 7	4.06	1.40118	246
7	959	970	981	991	002	013	023	034	045	055	7 8	4.07	1.40364	246
8	61 066	077	087	098	109	119	130	140	151	162	8 9	4.08	1.40610	244
9	172	183	194	204	215	225	236	247	257	268	9 10	4.09	1.40854	245
410	278	289	300	310	321	331	342	352	363	374		4.10	1.41099	243
1	384	395	405	416	426	437	448	458	469	479	1	4.11	1.41342	243
2	490	500	511	521	532	542	553	563	574	584	2	4.12	1.41585	243
3	595	606	616	627	637	648	658	669	679	690	3	4.13	1.41828	242
4	700	711	721	731	742	752	763	773	784	794	4	4.14	1.42070	241
415	805	815	826	836	847	857	868	878	888	899	5	4.15	1.42311	241
6	909	920	930	941	951	962	972	982	993	003	6	4.16	1.42552	240
7	62 014	024	034	045	055	066	076	086	097	107	7	4.17	1.42792	239
8	118	128	138	149	159	170	180	190	201	211	8	4.18	1.43031	239
9	221	232	242	252	263	273	284	294	304	315	9	4.19	1.43270	238
420	325	335	346	356	366	377	387	397	408	418	10	4.20	1.43508	238
1	428	439	449	459	469	480	490	500	511	521	1	4.21	1.43746	238
2	531	542	552	562	572	583	593	603	613	624	2	4.22	1.43984	236
3	634	644	655	665	675	685	696	706	716	726	3	4.23	1.44220	236
4	737	747	757	767	778	788	798	808	818	829	4	4.24	1.44456	236
425	839	849	859	870	880	890	900	910	921	931	5	4.25	1.44692	235
6	941	951	961	972	982	992	002	012	022	033	6	4.26	1.44927	235
7	63 043	053	063	073	083	094	104	114	124	134	7	4.27	1.45161	234
8	144	155	165	175	185	195	205	215	225	236	8	4.28	1.45395	234
9	246	256	266	276	286	296	306	317	327	337	9	4.29	1.45629	234
430	347	357	367	377	387	397	407	417	428	438		4.30	1.45861	233
1	448	458	468	478	488	498	508	518	528	538	1	4.31	1.46094	232
2	548	558	568	579	589	599	609	619	629	639	2	4.32	1.46326	231
3	649	659	669	679	689	699	709	719	729	739	3	4.33	1.46557	230
4	749	759	769	779	789	799	809	819	829	839	4	4.34	1.46787	231
435	849	859	869	879	889	899	909	919	929	939	5	4.35	1.47018	229
6	949	959	969	979	988	998	008	018	028	038	6	4.36	1.47247	229
7	64 048	058	068	078	088	098	108	118	128	137	7	4.37	1.47476	229
8	147	157	167	177	187	197	207	217	227	237	8	4.38	1.47705	228
9	246	256	266	276	286	296	306	316	326	335	9	4.39	1.47933	227
440	345	355	365	375	385	395	404	414	424	434	9	4.40	1.48160	227
1	444	454	464	473	483	493	503	513	523	532	1	4.41	1.48387	227
2	542	552	562	572	582	591	601	611	621	631	2	4.42	1.48614	226
3	640	650	660	670	680	689	699	709	719	729	3	4.43	1.48840	225
4	738	748	758	768	777	787	797	807	816	826	4	4.44	1.49065	225
445	836	846	856	865	875	885	895	904	914	924	4.5	4.45	1.49290	225
6	933	943	953	963	972	982	992	002	011	021	5.4	4.46	1.49515	224
7	65 031	040	050	060	070	079	089	099	108	118	6	4.47	1.49739	223
8	128	137	147	157	167	176	186	196	205	215	7	4.48	1.49962	223
9	225	234	244	254	263	273	283	292	302	312	8	4.49	1.50185	223
450	321	331	341	350	360	369	379	389	398	408		4.50	1.50408	

Common Logarithms of Numbers.   Naperian.

4.—LOGARITHMS—Continued

No.	L. O	1	2	3	4	5	6	7	8	9	P.P.	No.	Log.	Dif.
450	65 321	331	341	350	360	369	379	389	398	408	10	4.50	1.50408	222
1	418	427	437	447	456	466	475	485	495	504	1	4.51	1.50630	221
2	514	523	533	543	552	562	571	581	591	600	2	4.52	1.50851	221
3	610	619	629	639	648	658	667	677	686	696	3	4.53	1.51072	221
4	706	715	725	734	744	753	763	772	782	792	4	4.54	1.51293	220
455	801	811	820	830	839	849	858	868	877	887	5	4.55	1.51513	219
6	896	906	916	925	935	944	954	963	973	982	6	4.56	1.51732	219
7	992	001	011	020	030	039	049	058	068	077	7	4.57	1.51951	219
8	66 087	096	106	115	124	134	143	153	162	172	8	4.58	1.52170	218
9	181	191	200	210	219	229	238	247	257	266	9	4.59	1.52388	218
460	276	285	295	304	314	323	332	342	351	361		4.60	1.52606	217
1	370	380	389	398	408	417	427	436	445	455	1	4.61	1.52823	216
2	464	474	483	492	502	511	521	530	539	549	2	4.62	1.53039	217
3	558	567	577	586	596	605	614	624	633	642	3	4.63	1.53256	215
4	652	661	671	680	689	699	708	717	727	736	4	4.64	1.53471	216
465	745	755	764	773	783	792	801	811	820	829	5	4.65	1.53687	215
6	839	848	857	867	876	885	894	904	913	922	6	4.66	1.53902	214
7	932	941	950	960	969	978	987	997	006	015	7	4.67	1.54116	214
8	67 025	034	043	052	062	071	080	089	099	108	8	4.68	1.54330	213
9	117	127	136	145	154	164	173	182	191	201	9	4.69	1.54543	213
470	210	219	228	237	247	256	265	274	284	293	9	4.70	1.54756	213
1	302	311	321	330	339	348	357	367	376	385	1	4.71	1.54969	212
2	394	403	413	422	431	440	449	459	468	477	2	4.72	1.55181	212
3	486	495	504	514	523	532	541	550	560	569	3	4.73	1.55393	211
4	578	587	596	605	614	624	633	642	651	660	4	4.74	1.55604	210
475	669	679	688	697	706	715	724	733	742	752	4.5	4.75	1.55814	211
6	761	770	779	788	797	806	815	825	834	843	5.4	4.76	1.56025	210
7	852	861	870	879	888	897	906	916	925	934	6	4.77	1.56235	209
8	943	952	961	970	979	988	997	006	015	024	7	4.78	1.56444	209
9	68 034	043	052	061	070	079	088	097	106	115	8	4.79	1.56653	209
480	124	133	142	151	160	169	178	187	196	205		4.80	1.56862	208
1	215	224	233	242	251	260	269	278	287	296	1	4.81	1.57070	207
2	305	314	323	332	341	350	359	368	377	386	2	4.82	1.57277	208
3	395	404	413	422	431	440	449	458	467	476	3	4.83	1.57485	206
4	485	494	502	511	520	529	538	547	556	565	4	4.84	1.57691	207
485	574	583	592	601	610	619	628	637	646	655	5	4.85	1.57898	206
6	664	673	681	690	699	708	717	726	735	744	6	4.86	1.58104	205
7	753	762	771	780	789	797	806	815	824	833	7	4.87	1.58309	206
8	842	851	860	869	878	886	895	904	913	922	8	4.88	1.58515	204
9	931	940	949	958	966	975	984	993	002	011	9	4.89	1.58719	205
490	69 020	028	037	046	055	064	073	082	090	099	8	4.90	1.58924	203
1	108	117	126	135	144	152	161	170	179	188	1	4.91	1.59127	204
2	197	205	214	223	232	241	249	258	267	276	2 1.6	4.92	1.59331	203
3	285	294	302	311	320	329	338	346	355	364	3 2.4	4.93	1.59534	203
4	373	381	390	399	408	417	425	434	443	452	4 3	4.94	1.59737	202
495	461	469	478	487	496	504	513	522	531	539	5 4	4.95	1.59939	202
6	548	557	566	574	583	592	601	609	618	627	6 5	4.96	1.60141	201
7	636	644	653	662	671	679	688	697	705	714	7 5.6	4.97	1.60342	201
8	723	732	740	749	758	767	775	784	793	801	8 6.4	4.98	1.60543	201
9	810	819	827	836	845	854	862	871	880	888	9 7	4.99	1.60744	200
500	897	906	914	923	932	940	949	958	966	975		5.00	1.60944	

4.—LOGARITHMS—*Continued*

Common Logarithms of Numbers.

No.	L. O	1	2	3	4	5	6	7	8	9	P. P.
500	69 897	906	914	923	932	940	949	958	966	975	9
1	984	992	001	010	018	027	036	044	053	062	1 1
2	70 070	079	088	096	105	114	122	131	140	148	2 2
3	157	165	174	183	191	200	209	217	226	234	3 3
4	243	252	260	269	278	286	295	303	312	321	4 4
505	329	338	346	355	364	372	381	389	398	406	5 4.5
6	415	424	432	441	449	458	467	475	484	492	6 5.4
7	501	509	518	526	535	544	552	561	569	578	7 6
8	586	595	603	612	621	629	638	646	655	663	8 7
9	672	680	689	697	706	714	723	731	740	749	9 8
510	757	766	774	783	791	800	808	817	825	834	
1	842	851	859	868	876	885	893	902	910	919	1
2	927	935	944	952	961	969	978	986	995	003	2
3	71 012	020	029	037	046	054	063	071	079	088	3
4	096	105	113	122	130	139	147	155	164	172	4
515	181	189	198	206	214	223	231	240	248	257	5
6	265	273	282	290	299	307	315	324	332	341	6
7	349	357	366	374	383	391	399	408	416	425	7
8	433	441	450	458	466	475	483	492	500	508	8
9	517	525	533	542	550	559	567	575	584	592	9
520	600	609	617	625	634	642	650	659	667	675	8
1	684	692	700	709	717	725	734	742	750	759	1 1
2	767	775	784	792	800	809	817	825	834	842	2 1.6
3	850	858	867	875	883	892	900	908	917	925	3 2.4
4	933	941	950	958	966	975	983	991	999	008	4 3
525	72 016	024	032	041	049	057	066	074	082	090	5 4
6	099	107	115	123	132	140	148	156	165	173	6 5
7	181	189	198	206	214	222	230	239	247	255	7 5.6
8	263	272	280	288	296	304	313	321	329	337	8 6.4
9	346	354	362	370	378	387	395	403	411	419	9 7
530	428	436	444	452	460	469	477	485	493	501	1
1	509	518	526	534	542	550	558	567	575	583	
2	591	599	607	616	624	632	640	648	656	665	2
3	673	681	689	697	705	713	722	730	738	746	3
4	754	762	770	779	787	795	803	811	819	827	4
535	835	843	852	860	868	876	884	892	900	908	5
6	916	925	933	941	949	957	965	973	981	989	6
7	997	006	014	022	030	038	046	054	062	070	7
8	73 078	086	094	102	111	119	127	135	143	151	8
9	159	167	175	183	191	199	207	215	223	231	9
540	239	247	255	263	272	280	288	296	304	312	7
1	320	328	336	344	352	360	368	376	384	392	1 0.7
2	400	408	416	424	432	440	448	456	464	472	2 1.4
3	480	488	496	504	512	520	528	536	544	552	3 2
4	560	568	576	584	592	600	608	616	624	632	4 3
545	640	648	656	664	672	679	687	695	703	711	5 3.5
6	719	727	735	743	751	759	767	775	783	791	6 4.2
7	799	807	815	823	830	838	846	854	862	870	7 5
8	878	886	894	902	910	918	926	933	941	949	8 5.6
9	957	965	973	981	989	997	005	013	020	028	9 6.3
550	74 036	044	052	060	068	076	084	092	099	107	

Naperian.

No.	Log.	Dif.
5.00	1.60944	200
5.01	1.61144	199
5.02	1.61343	199
5.03	1.61542	199
5.04	1.61741	199
		198
5.05	1.61939	198
5.06	1.62137	197
5.07	1.62334	197
5.08	1.62531	197
5.09	1.62728	197
		196
5.10	1.62924	196
5.11	1.63120	195
5.12	1.63315	196
5.13	1.63511	194
5.14	1.63705	
		195
5.15	1.63900	194
5.16	1.64094	193
5.17	1.64287	194
5.18	1.64481	192
5.19	1.64673	
		193
5.20	1.64866	192
5.21	1.65058	192
5.22	1.65250	191
5.23	1.65441	191
5.24	1.65632	
		191
5.25	1.65823	190
5.26	1.66013	190
5.27	1.66203	190
5.28	1.66393	189
5.29	1.66582	
		189
5.30	1.66771	188
5.31	1.66959	188
5.32	1.67147	188
5.33	1.67335	188
5.34	1.67523	
		187
5.35	1.67710	186
5.36	1.67896	187
5.37	1.68083	186
5.38	1.68269	186
5.39	1.68455	
		185
5.40	1.68640	185
5.41	1.68825	185
5.42	1.69010	184
5.43	1.69194	184
5.44	1.69378	
		184
5.45	1.69562	183
5.46	1.69745	183
5.47	1.69928	183
5.48	1.70111	182
5.49	1.70293	
		182
5.50	1.70475	

4.—LOGARITHMS—*Continued*

No.	L. O	1	2	3	.4	5	6	7	8	.9	P. P.		No.	Log.	Dif.
550	74 036	044	052	060	068	076	084	092	099	107			5.50	1.70475	
1	115	123	131	139	147	155	162	170	178	186	1		5.51	1.70656	181
2	194	202	210	218	225	233	241	249	257	265	2		5.52	1.70838	182
3	273	280	288	296	304	312	320	327	335	343	3		5.53	1.71019	181
4	351	359	367	374	382	390	398	406	414	421	4		5.54	1.71199	180
															181
555	429	437	445	453	461	468	476	484	492	500	5		5.55	1.71380	
6	507	515	523	531	539	547	554	562	570	578	6		5.56	1.71560	180
7	586	593	601	609	617	624	632	640	648	656	7		5.57	1.71740	180
8	663	671	679	687	695	702	710	718	726	733	8		5.58	1.71919	179
9	741	749	757	764	772	780	788	796	803	811	9		5.59	1.72098	179
															179
560	819	827	834	842	850	858	865	873	881	889	8		5.60	1.72277	
1	896	904	912	920	927	935	943	950	958	966	1 .8		5.61	1.72455	178
2	974	981	989	997	005	012	020	028	035	043	2 1.6		5.62	1.72633	178
3	75 051	059	066	074	082	089	097	105	113	120	3 2.4		5.63	1.72811	178
4	128	136	143	151	159	166	174	182	189	197	4 3		5.64	1.72988	177
															178
565	205	213	220	228	236	243	251	259	266	274	5 4		5.65	1.73166	
6	282	289	297	305	312	320	328	335	343	351	6 5		5.66	1.73342	176
7	358	366	374	381	389	397	404	412	420	427	7 5.6		5.67	1.73519	177
8	435	442	450	458	465	473	481	488	496	504	8 6.4		5.68	1.73695	176
9	511	519	526	534	542	549	557	565	572	580	9 7		5.69	1.73871	176
															176
570	587	595	603	610	618	626	633	641	648	656			5.70	1.74047	
1	664	671	679	686	694	702	709	717	724	732	1		5.71	1.74222	175
2	740	747	755	762	770	778	785	793	800	808	2		5.72	1.74397	175
3	815	823	831	838	846	853	861	868	876	884	3		5.73	1.74572	175
4	891	899	906	914	921	929	937	944	952	959	4		5.74	1.74746	174
															174
575	967	974	982	989	997	005	012	020	027	035	5		5.75	1.74920	
6	76 042	050	057	065	072	080	087	095	103	110	6		5.76	1.75094	174
7	118	125	133	140	148	155	163	170	178	185	7		5.77	1.75267	173
8	193	200	208	215	223	230	238	245	253	260	8		5.78	1.75440	173
9	268	275	283	290	298	305	313	320	328	335	9		5.79	1.75613	173
															173
580	343	350	358	365	373	380	388	395	403	410	7		5.80	1.75786	
1	418	425	433	440	448	455	462	470	477	485	1 0.7		5.81	1.75958	172
2	492	500	507	515	522	530	537	545	552	559	2 1.4		5.82	1.76130	172
3	567	574	582	589	597	604	612	619	626	634	3 2		5.83	1.76302	172
4	641	649	656	664	671	678	686	693	701	708	4 3		5.84	1.76473	171
															171
585	716	723	730	738	745	753	760	768	775	782	5 3.5		5.85	1.76644	
6	790	797	805	812	819	827	834	842	849	856	6 4.2		5.86	1.76815	171
7	864	871	879	886	893	901	908	916	923	930	7 5		5.87	1.76985	170
8	938	945	953	960	967	975	982	989	997	004	8 5.6		5.88	1.77156	171
9	77 012	019	026	034	041	048	056	063	070	078	9 6.3		5.89	1.77326	170
															169
590	085	093	100	107	115	122	129	137	144	151			5.90	1.77495	
1	159	166	173	181	188	195	203	210	217	225	1		5.91	1.77665	170
2	232	240	247	254	262	269	276	283	291	298	2		5.92	1.77834	169
3	305	312	320	327	335	342	349	357	364	371	3		5.93	1.78002	168
4	379	386	393	401	408	415	422	430	437	444	4		5.94	1.78171	169
															168
595	452	459	466	474	481	488	495	503	510	517	5		5.95	1.78339	
6	525	532	539	546	554	561	568	576	583	590	6		5.96	1.78507	168
7	597	605	612	619	627	634	641	648	656	663	7		5.97	1.78675	168
8	670	677	685	692	699	706	714	721	728	735	8		5.98	1.78842	167
9	743	750	757	764	772	779	786	793	801	808	9		5.99	1.79009	167
															167
600	815	822	830	837	844	851	859	866	873	880			6.00	1.79176	

4.—Logarithms—*Continued*

No.	L. O	1	2	3	4	5	6	7	8	9	P. P.	No.	Log.	Dif.
					Common Logarithms of Numbers.								Naperian.	
600	77 815	822	830	837	844	851	859	866	873	880	8	6.00	1.79176	166
1	887	895	902	909	916	924	931	938	945	952	1 1	6.01	1.79342	167
2	960	967	974	981	988	996	003	010	017	025	2 1.6	6.02	1.79509	166
3	78 032	039	046	053	061	068	075	082	089	097	3 2.4	6.03	1.79675	166
4	104	111	118	125	132	140	147	154	161	168	4 3	6.04	1.79840	165
														166
605	176	183	190	197	204	211	219	226	233	240	5 4	6.05	1.80006	165
6	247	254	262	269	276	283	290	297	305	312	6 5	6.06	1.80171	165
7	319	326	333	340	347	355	362	369	376	383	7 5.6	6.07	1.80336	165
8	390	398	405	412	419	426	433	440	447	455	8 6.4	6.08	1.80500	164
9	462	469	476	483	490	497	504	512	519	526	9 7	6.09	1.80665	165
														164
610	533	540	547	554	561	569	576	583	590	597		6.10	1.80829	164
1	604	611	618	625	633	640	647	654	661	668	1	6.11	1.80993	163
2	675	682	689	696	704	711	718	725	732	739	2	6.12	1.81156	163
3	746	753	760	767	774	781	789	796	802	810	3	6.13	1.81319	163
4	817	824	831	838	845	852	859	866	873	880	4	6.14	1.81482	163
														163
615	888	895	902	909	916	923	930	937	944	951	5	6.15	1.81645	163
6	858	965	972	979	986	993	000	007	014	021	6	6.16	1.81808	162
7	79 029	036	043	050	057	064	071	078	085	092	7	6.17	1.81970	162
8	099	106	113	120	127	134	141	148	155	162	8	6.18	1.82132	162
9	169	176	183	190	197	204	211	218	225	232	9	6.19	1.82294	162
														161
620	239	246	253	260	267	274	281	288	295	302	7	6.20	1.82455	161
1	309	316	323	330	337	344	351	358	365	372	0.7	6.21	1.82616	161
2	379	386	393	400	407	414	421	428	435	442	1.4	6.22	1.82777	161
3	449	456	463	470	477	484	491	498	505	511	2	6.23	1.82938	160
4	518	525	532	539	546	553	560	567	574	581	3	6.24	1.83098	160
														160
625	588	595	602	609	616	623	630	637	644	650	5 3.5	6.25	1.83258	160
6	657	664	671	678	685	692	699	706	713	720	6 4.2	6.26	1.83418	160
7	727	734	741	748	754	761	768	775	782	789	7 5	6.27	1.83578	160
8	796	803	810	817	824	831	837	844	851	858	8 5.6	6.28	1.83737	159
9	865	872	879	886	893	900	906	913	920	927	9 6.3	6.29	1.83896	159
														159
630	934	941	948	955	962	969	975	982	989	996		6.30	1.84055	159
1	80 003	010	017	024	030	037	044	051	058	065	1	6.31	1.84214	158
2	072	079	085	092	099	106	113	120	127	134	2	6.32	1.84372	158
3	140	147	154	161	168	175	182	188	195	202	3	6.33	1.84530	158
4	209	216	223	229	236	243	250	257	264	271	4	6.34	1.84688	158
														157
635	277	284	291	298	305	312	318	325	332	339	5	6.35	1.84845	158
6	346	353	359	366	373	380	387	393	400	407	6	6.36	1.85003	157
7	414	421	428	434	441	448	455	462	468	475	7	6.37	1.85160	157
8	482	489	496	502	509	516	523	530	536	543	8	6.38	1.85317	156
9	500	557	564	570	577	584	591	598	604	611	9	6.39	1.85473	157
														156
640	618	625	632	638	645	652	659	665	672	679	6	6.40	1.85630	156
1	686	693	699	706	713	720	726	733	740	747	1 0.6	6.41	1.85786	156
2	754	760	767	774	781	787	794	801	808	814	2 1.2	6.42	1.85942	155
3	821	828	835	841	848	855	862	868	875	882	3 1.8	6.43	1.86097	156
4	889	895	902	909	916	922	929	936	943	949	4 2.4	6.44	1.86253	155
														155
645	956	963	969	976	983	990	996	003	010	017	5 3	6.45	1.86408	155
6	81 023	030	037	043	050	057	064	070	077	084	6 3.6	6.46	1.86563	155
7	090	097	104	111	117	124	131	137	144	151	7 4.2	6.47	1.86718	154
8	158	164	171	178	184	191	198	024	211	218	8 4.8	6.48	1.86872	154
9	224	231	238	245	251	258	265	271	278	285	9 5.4	6.49	1.87026	154
														154
650	291	298	305	311	318	325	331	338	345	351		6.50	1.87180	

4.—Logarithms—*Continued*

No.	L. O	1	2	3	4	5	6	7	8	9	P. P.		No.	Log.	Dif.
650	81 291	298	305	311	318	325	331	338	345	351			6.50	1.87180	154
1	358	365	371	378	385	391	398	405	411	418	1		6.51	1.87334	153
2	425	431	438	445	451	458	465	4/1	478	485	2		6.52	1.87487	154
3	491	498	505	511	518	525	531	538	544	551	3		6.53	1.87641	153
4	558	564	571	578	584	591	598	604	611	617	4		6.54	1.87794	153
655	624	631	637	644	651	657	664	671	677	684	5		6.55	1.87947	152
6	690	697	704	710	717	723	730	737	743	750	6		6.56	1.88099	152
7	757	763	770	776	783	790	796	803	809	816	7		6.57	1.88251	152
8	823	829	836	842	849	856	862	869	875	882	8		6.58	1.88403	152
9	889	895	902	908	915	921	928	935	941	948	9		6.59	1.88555	152
660	954	961	968	974	981	987	994	000	007	014	7		6.60	1.88707	151
1	82 020	027	033	040	046	053	060	066	073	079	1 0.7		6.61	1.88858	152
2	086	092	099	105	112	119	125	132	138	145	2 1.4		6.62	1.89010	150
3	151	158	164	171	178	184	191	197	204	210	3 2		6.63	1.89160	151
4	217	223	230	236	243	249	256	263	269	276	4 3		6.64	1.89311	151
665	282	289	295	302	308	315	321	328	334	341	5 3.5		6.65	1.89462	150
6	347	354	360	367	373	380	387	393	400	406	6 4.2		6.66	1.89612	150
7	413	419	426	432	439	445	452	458	465	471	7 5		6.67	1.89762	150
8	478	484	491	497	504	510	517	523	530	536	8 5.6		6.68	1.89912	150
9	543	549	556	562	569	575	582	588	595	601	9 6.3		6.69	1.90061	149
670	607	614	620	627	633	640	646	653	659	666			6.70	1.90211	150
1	672	679	685	692	698	705	711	718	724	730	1		6.71	1.90360	149
2	737	743	750	756	763	769	776	782	789	795	2		6.72	1.90509	149
3	802	808	814	821	827	834	840	847	853	860	3		6.73	1.90658	149
4	866	872	879	885	892	898	905	911	918	924	4		6.74	1.90806	148
675	930	937	943	950	956	963	969	975	982	988	5		6.75	1.90954	148
6	995	001	008	014	020	027	033	040	046	052	6		6.76	1.91102	148
7	83 059	065	072	078	085	091	097	104	110	117	7		6.77	1.91250	148
8	123	129	136	142	149	155	161	168	174	181	8		6.78	1.91398	147
9	187	193	200	206	213	219	225	232	238	245	9		6.79	1.91545	147
680	251	257	264	270	276	283	289	296	302	308	6		6.80	1.91692	147
1	315	321	327	334	340	347	353	359	366	372	1 0.6		6.81	1.91839	147
2	378	385	391	398	404	410	417	423	429	436	2 1.2		6.82	1.91986	146
3	442	448	455	461	467	474	480	487	493	499	3 1.8		6.83	1.92132	147
4	506	512	518	525	531	537	544	550	556	563	4 2.4		6.84	1.92279	146
685	569	575	582	588	594	601	607	613	620	626	5 3		6.85	1.92425	146
6	632	639	645	651	658	664	670	677	683	689	6 3.6		6.86	1.92571	145
7	696	702	708	715	721	727	734	740	746	753	7 4.2		6.87	1.92716	146
8	759	765	771	778	784	790	797	803	809	816	8 4.8		6.88	1.92862	145
9	822	828	835	841	847	853	860	866	872	879	9 5.4		6.89	1.93007	145
690	885	891	897	904	910	916	923	929	935	942			6.90	1.93152	145
1	948	954	960	967	973	979	985	992	998	004	1		6.91	1.93297	145
2	84 011	017	023	029	036	042	048	055	061	067	2		6.92	1.93442	144
3	073	080	086	092	098	105	111	117	123	130	3		6.93	1.93586	144
4	136	142	148	155	161	167	173	180	186	192	4		6.94	1.93730	144
695	198	205	211	217	223	230	236	242	248	255	5		6.95	1.93874	144
6	261	267	273	280	286	292	298	305	311	317	6		6.96	1.94018	144
7	323	330	336	342	348	354	361	367	373	379	7		6.97	1.94162	143
8	386	392	398	404	410	417	423	429	435	442	8		6.98	1.94305	143
9	448	454	460	466	473	479	485	491	497	504	9		6.99	1.94448	143
700	510	516	522	528	535	541	547	553	559	566	.		7.00	1.94591	

4.—Logarithms—*Continued*

No.	L. O	1	2	3	4	5	6	7	8	9	P. P.		No.	Log.	Dif.
700	84 510	516	522	528	535	541	547	553	559	566		**7**	7.00	1.94591	143
1	572	578	584	590	597	603	609	615	621	628	1	0.7	7.01	1.94734	142
2	634	640	646	652	658	665	671	677	683	689	2	1.4	7.02	1.94876	143
3	696	702	708	714	720	726	733	739	745	751	3	2	7.03	1.95019	142
4	757	763	770	776	782	788	794	800	807	813	4	3	7.04	1.95161	142
705	819	825	831	837	844	850	856	862	868	874	5	3.5	7.05	1.95303	141
6	880	887	893	899	905	911	917	924	930	936	6	4.2	7.06	1.95444	142
7	942	948	954	960	967	973	979	985	991	997	7	5	7.07	1.95586	141
8	85 003	009	016	022	028	034	040	046	052	058	8	5.6	7.08	1.95727	142
9	065	071	077	083	089	095	101	107	114	120	9	6.3	7.09	1.95869	140
710	126	132	138	144	150	156	163	169	175	181	1		7.10	1.96009	141
1	187	193	199	205	211	217	224	230	236	242	1		7.11	1.96150	141
2	248	254	260	266	272	278	285	291	297	303	2		7.12	1.96291	140
3	309	315	321	327	333	339	345	352	358	364	3		7.13	1.96431	140
4	370	376	382	388	394	400	406	412	418	425	4		7.14	1.96571	140
715	431	437	443	449	455	461	467	473	479	485	5		7.15	1.96711	140
6	491	497	503	509	516	522	528	534	540	546	6		7.16	1.96851	140
7	552	558	564	570	576	582	588	594	600	606	7		7.17	1.96991	139
8	612	618	625	631	637	643	649	655	661	667	8		7.18	1.97130	139
9	673	679	685	691	697	703	709	715	721	727	9		7.19	1.97269	139
720	733	739	745	751	757	763	769	775	781	788		**6**	7.20	1.97408	139
1	794	800	806	812	818	824	830	836	842	848	1	0.6	7.21	1.97547	138
2	854	860	866	872	878	884	890	896	902	908	2	1.2	7.22	1.97685	139
3	914	920	926	932	938	944	950	956	962	968	3	1.8	7.23	1.97824	138
4	974	980	986	992	998	004	010	016	022	028	4	2.4	7.24	1.97962	138
725	86 034	040	046	052	058	064	070	076	082	088	5	3	7.25	1.98100	138
6	094	100	106	112	118	124	130	136	141	147	6	3.6	7.26	1.98238	138
7	153	159	165	171	177	183	189	195	201	207	7	4.2	7.27	1.98376	137
8	213	219	225	231	237	243	249	255	261	267	8	4.8	7.28	1.98513	137
9	273	279	285	291	297	303	308	314	320	326	9	5.4	7.29	1.98650	137
730	332	338	344	350	356	362	368	374	380	386			7.30	1.98787	137
1	392	398	404	410	415	421	427	433	439	445	1		7.31	1.98924	137
2	451	457	463	469	475	481	487	493	499	504	2		7.32	1.99061	137
3	510	516	522	528	534	540	546	552	558	564	3		7.33	1.99198	136
4	570	576	581	587	593	599	605	611	617	623	4		7.34	1.99334	136
735	629	635	641	646	652	658	664	670	676	682	5		7.35	1.99470	136
6	688	694	700	705	711	717	723	729	735	741	6		7.36	1.99606	136
7	747	753	759	764	770	776	782	788	794	800	7		7.37	1.99742	135
8	806	812	817	823	829	835	841	847	853	859	8		7.38	1.99877	136
9	864	870	876	882	888	894	900	906	911	917	9		7.39	2.00013	135
740	923	929	935	941	947	953	958	964	970	976		**5**	7.40	2.00148	135
1	982	988	994	999	005	011	017	023	029	035	1	0.5	7.41	2.00283	135
2	87 040	046	052	058	064	070	075	081	087	093	2	1	7.42	2.00418	135
3	099	105	111	116	122	128	134	140	146	151	3	1.5	7.43	2.00553	134
4	157	163	169	175	181	186	192	198	204	210	4	2	7.44	2.00687	134
745	216	221	227	233	239	245	251	256	262	268	5	2.5	7.45	2.00821	135
6	274	280	286	291	297	303	309	315	320	326	6	3	7.46	2.00956	133
7	332	338	344	349	355	361	367	373	379	384	7	3.5	7.47	2.01089	134
8	390	396	402	408	413	419	425	431	437	442	8	4	7.48	2.01223	134
9	448	454	460	466	471	477	483	489	495	500	9	4.5	7.49	2.01357	133
750	506	512	518	523	529	535	541	547	552	558			7.50	2.01490	

Common Logarithms of Numbers. — Naperian.

4.—Logarithms—*Continued*

No.	L.	O	1	2	3	4	5	6	7	8	9	P. P.		No.	Log.	Dif.
750	87	506	512	518	523	529	535	541	547	552	558	1		7.50	2.01490	134
1		564	570	576	581	587	593	599	604	610	616	1		7.51	2.01624	133
2		622	628	633	639	645	651	656	662	668	674	2		7.52	2.01757	133
3		679	685	691	697	703	708	714	720	726	731	3		7.53	2.01890	133
4		737	743	749	754	760	766	772	777	783	789	4		7.54	2.02022	132
																133
755		795	800	806	812	818	823	829	835	841	846	5		7.55	2.02155	132
6		852	858	864	869	875	881	887	892	898	904	6		7.56	2.02287	132
7		910	915	921	927	933	938	944	950	955	961	7		7.57	2.02419	132
8		967	973	978	984	990	996	001̄	007̄	013̄	018̄	8		7.58	2.02551	132
9	88	024	030	036	041	047	053	058	064	070	076	9		7.59	2.02683	132
760		081	087	093	098	104	110	116	121	127	133		6	7.60	2.02815	131
1		138	144	150	156	161	167	173	178	184	190	1	0.6	7.61	2.02946	132
2		195	201	207	213	218	224	230	235	241	247	2	1.2	7.62	2.03078	131
3		252	258	264	270	275	281	287	292	298	304	3	1.8	7.63	2.03209	131
4		309	315	321	326	332	338	343	349	355	360	4	2.4	7.64	2.03340	131
																131
765		366	372	377	383	389	395	400	406	412	417	5	3	7.65	2.03471	130
6		423	429	434	440	446	451	457	463	468	474	6	3.6	7.66	2.03601	131
7		480	485	491	497	502	508	513	519	525	530	7	4.2	7.67	2.03732	130
8		536	542	547	553	559	564	570	576	581	587	8	4.8	7.68	2.03862	130
9		593	598	604	610	615	621	627	632	638	643	9	5.4	7.69	2.03992	130
																130
770		649	655	660	666	672	677	683	689	694	700			7.70	2.04122	130
1		705	711	717	722	728	734	739	745	750	756	1		7.71	2.04253	130
2		762	767	773	779	784	790	795	801	807	812	2		7.72	2.04381	130
3		818	824	829	835	840	846	852	857	863	868	3		7.73	2.04511	129
4		874	880	885	891	897	902	908	913	919	925	4		7.74	2.04640	129
																129
775		930	936	941	947	953	958	964	969	975	981	5		7.75	2.04769	129
6		986	992	997	003̄	009̄	014̄	020̄	025̄	031̄	037̄	6		7.76	2.04898	129
7	89	042	048	053	059	064	070	076	081	087	092	7		7.77	2.05027	129
8		098	104	109	115	120	126	131	137	143	148	8		7.78	2.05156	128
9		154	159	165	170	176	182	187	193	198	204	9		7.79	2.05284	128
																128
780		209	215	221	226	232	237	243	248	254	260		5	7.80	2.05412	128
1		265	271	276	282	287	293	298	304	310	315	1	0.5	7.81	2.05540	128
2		321	326	332	337	343	348	354	360	365	371	2	1	7.82	2.05668	128
3		376	382	387	393	398	404	409	415	421	426	3	1.5	7.83	2.05796	128
4		432	437	443	448	454	459	465	470	476	481	4	2	7.84	2.05924	127
																127
785		487	492	498	504	509	515	520	526	531	537	5	2.5	7.85	2.06051	128
6		542	548	553	559	564	570	575	581	586	592	6	3	7.86	2.06179	127
7		597	603	609	614	620	625	631	636	642	647	7	3.5	7.87	2.06306	127
8		653	658	664	669	675	680	686	691	697	702	8	4	7.88	2.06433	127
9		708	713	719	724	730	735	741	746	752	757	9	4.5	7.89	2.06560	126
																126
790		763	768	774	779	785	790	796	801	807	812			7.90	2.06686	127
1		818	823	829	834	840	845	851	856	862	867	1		7.91	2.06813	126
2		873	878	883	889	894	900	905	911	916	922	2		7.92	2.06939	126
3		927	933	938	944	949	955	960	966	971	977	3		7.93	2.07065	126
4		982	988	993	998	004̄	009̄	015̄	020̄	026̄	031̄	4		7.94	2.07191	126
																126
795	90	037	042	048	053	059	064	069	075	080	086	5		7.95	2.07317	126
6		091	097	102	108	113	119	124	129	135	140	6		7.96	2.07443	125
7		146	151	157	162	168	173	179	184	189	195	7		7.97	2.07568	126
8		200	206	211	217	222	227	233	238	244	249	8		7.98	2.07694	125
9		255	260	266	271	276	282	287	293	298	304	9		7.99	2.07819	125
																125
800		309	314	320	325	331	336	342	347	352	358			8.00	2.07944	

4.—LOGARITHMS—*Continued*

No.	L. O	1	2	3	4	5	6	7	8	9	P. P.		No.	Log.	Dif.
800	90 309	314	320	325	331	336	342	347	352	358	1		8.00	2.07944	125
1	363	369	374	380	385	390	396	401	407	412			8.01	2.08069	125
2	417	423	428	434	439	445	450	455	461	466	2		8.02	2.08194	124
3	472	477	482	488	493	499	504	509	515	520	3		8.03	2.08318	125
4	526	531	536	542	547	553	558	563	569	574	4		8.04	2.08443	124
805	580	585	590	596	601	607	612	617	623	628	5		8.05	2.08567	124
6	634	639	644	650	655	660	666	671	677	682	6		8.06	2.08691	124
7	687	693	698	703	709	714	720	725	730	736	7		8.07	2.08815	124
8	741	747	752	757	763	768	773	779	784	789	8		8.08	2.08939	124
9	795	800	806	811	816	822	827	832	838	843	9		8.09	2.09063	123
810	849	854	859	865	870	875	881	886	891	897		6	8.10	2.09186	124
1	902	907	913	918	924	929	934	940	945	950	1	0.6	8.11	2.09310	123
2	956	961	966	972	977	982	983	993	998	004	2	1.2	8.12	2.09433	123
3	91 009	015	020	025	030	036	041	046	052	057	3	1.8	8.13	2.09556	123
4	062	068	073	078	084	089	094	100	105	110	4	2.4	8.14	2.09679	123
815	116	121	126	132	137	142	148	153	158	164	5	3	8.15	2.09802	122
6	169	174	180	185	190	196	201	206	212	217	6	3.6	8.16	2.09924	123
7	222	228	233	238	243	249	254	259	265	270	7	4.2	8.17	2.10047	122
8	275	281	286	291	297	302	307	312	318	323	8	4.8	8.18	2.10169	122
9	328	334	339	344	350	355	360	365	371	376	9	5.4	8.19	2.10291	122
820	381	387	392	397	403	408	413	418	424	429			8.20	2.10413	122
1	434	440	445	450	455	461	466	471	477	482	1		8.21	2.10535	122
2	487	492	498	503	508	514	519	524	529	535	2		8.22	2.10657	123
3	540	545	551	556	561	566	572	577	582	587	3		8.23	2.10779	121
4	593	598	603	609	614	619	624	630	635	640	4		8.24	2.10900	121
825	645	651	656	661	666	672	677	682	687	693	5		8.25	2.11021	121
6	698	703	709	714	719	724	730	735	740	745	6		8.26	2.11142	121
7	751	756	761	766	772	777	782	787	793	798	7		8.27	2.11263	121
8	803	808	814	819	824	829	834	840	845	850	8		8.28	2.11384	121
9	855	861	866	871	876	882	887	892	897	903	9		8.29	2.11505	121
830	908	913	918	924	929	934	939	944	950	955		5	8.30	2.11626	120
1	960	965	971	976	981	986	991	997	002	007	1	0.5	8.31	2.11746	120
2	92 012	018	023	028	033	038	044	049	054	059	2	1	8.32	2.11866	120
3	065	070	075	080	085	091	096	101	106	111	3	1.5	8.33	2.11986	120
4	117	122	127	132	137	143	148	153	158	163	4	2	8.34	2.12106	120
835	169	174	179	184	189	195	200	205	210	215	5	2.5	8.35	2.12226	120
6	221	226	231	236	241	247	252	257	262	267	6	3	8.36	2.12346	119
7	273	278	283	288	293	298	304	309	314	319	7	3.5	8.37	2.12465	119
8	324	330	335	340	345	350	355	361	366	371	8	4	8.38	2.12585	119
9	376	381	387	392	397	402	407	412	418	423	9	4.5	8.39	2.12704	119
840	428	433	438	443	449	454	459	464	469	474			8.40	2.12823	119
1	480	485	490	495	500	505	511	516	521	526	1		8.41	2.12942	119
2	531	536	542	547	552	557	562	567	572	578	2		8.42	2.13061	119
3	583	588	593	598	603	609	614	619	624	629	3		8.43	2.13180	118
4	634	639	645	650	655	660	665	670	675	681	4		8.44	2.13298	119
845	686	691	696	701	706	711	716	722	727	732	5		8.45	2.13417	118
6	737	742	747	752	758	763	768	773	778	783	6		8.46	2.13535	118
7	788	793	799	804	809	814	819	824	829	834	7		8.47	2.13653	118
8	840	845	850	855	860	865	870	875	881	886	8		8.48	2.13771	118
9	891	896	901	906	911	916	921	927	932	937	9		8.49	2.13889	118
850	942	947	952	957	962	967	973	978	983	988			8.50	2.14007	118

4.—Logarithms—*Continued*

Common Logarithms of Numbers.

No.	L. O	1	2	3	4	5	6	7	8	9	P. P.
850	92 942	947	952	957	962	967	973	978	983	988	6
1	993	998	003	008	013	018	024	029	034	039	1 0.6
2	93 044	049	054	059	064	069	075	080	085	090	2 1.2
3	095	100	105	110	115	120	125	131	136	141	3 1.8
4	146	151	156	161	166	171	176	181	186	192	4 2.4
855	197	202	207	212	217	222	227	232	237	242	5 3
6	247	252	258	263	268	273	278	283	288	293	6 3.6
7	298	303	308	313	318	323	328	334	339	344	7 4.2
8	349	354	359	364	369	374	379	384	389	394	8 4.8
9	399	404	409	414	420	425	430	435	440	445	9 5.4
860	450	455	460	465	470	475	480	485	490	495	1
1	500	505	510	515	520	526	531	536	541	546	
2	551	556	561	566	571	576	581	586	591	596	2
3	601	606	611	616	621	626	631	636	641	646	3
4	651	656	661	666	671	676	682	687	692	697	4
865	702	707	712	717	722	727	732	737	742	747	5
6	752	757	762	767	772	777	782	787	792	797	6
7	802	807	812	817	822	827	832	837	842	847	7
8	852	857	862	867	872	877	882	887	892	897	8
9	902	907	912	917	922	927	932	937	942	947	9
870	952	957	962	967	972	977	982	987	992	997	5
1	94 002	007	012	017	022	027	032	037	042	047	1 0.5
2	052	057	062	067	072	077	082	086	091	096	1
3	101	106	111	116	121	126	131	136	141	146	1.5
4	151	156	161	166	171	176	181	186	191	196	2
875	201	206	211	216	221	226	231	236	240	245	2.5
6	250	255	260	265	270	275	280	285	290	295	3 5
7	300	305	310	315	320	325	330	335	340	345	3 5
8	349	354	359	364	369	374	379	384	389	394	4
9	399	404	409	414	419	424	429	433	438	443	4 5
880	448	453	458	463	468	473	478	483	488	493	
1	498	503	507	512	517	522	527	532	537	542	1
2	547	552	557	562	567	571	576	581	586	591	2
3	596	601	606	611	616	621	626	630	635	640	3
4	645	650	655	660	665	670	675	680	685	689	4
885	694	699	704	709	714	719	724	729	734	738	5
6	743	748	753	758	763	768	773	778	783	787	6
7	792	797	802	807	812	817	822	827	832	836	7
8	841	846	851	856	861	866	871	876	880	885	8
9	890	895	900	905	910	915	919	924	929	934	9
890	939	944	949	954	959	963	968	973	978	983	4
1	988	993	998	002	007	012	017	022	027	032	0.4
2	95 036	041	046	051	056	061	066	071	075	080	2 0.8
3	085	090	095	100	105	109	114	119	124	129	3 1.2
4	134	139	143	148	153	158	163	168	173	177	4 1.6
895	182	187	192	197	202	207	211	216	221	226	2
6	231	236	240	245	250	255	260	265	270	274	6 2.4
7	279	284	289	294	299	303	308	313	318	323	7 2.8
8	328	332	337	342	347	352	357	361	366	371	8 3.2
9	376	381	386	390	395	400	405	410	415	419	9 3.6
900	424	429	434	439	444	448	453	458	463	468	

Naperian.

No.	Log.	Dif.
8.50	2.14007	117
8.51	2.14124	118
8.52	2.14242	117
8.53	2.14359	117
8.54	2.14476	117
8.55	2.14593	117
8.56	2.14710	117
8.57	2.14827	116
8.58	2.14943	117
8.59	2.15060	116
8.60	2.15176	116
8.61	2.15292	117
8.62	2.15409	115
8.63	2.15524	116
8.64	2.15640	116
8.65	2.15756	115
8.66	2.15871	116
8.67	2.15987	115
8.68	2.16102	115
8.69	2.16217	115
8.70	2.16332	115
8.71	2.16447	115
8.72	2.16562	115
8.73	2.16677	114
8.74	2.16791	114
8.75	2.16905	115
8.76	2.17020	114
8.77	2.17134	114
8.78	2.17248	113
8.79	2.17361	114
8.80	2.17475	114
8.81	2.17589	113
8.82	2.17702	114
8.83	2.17816	113
8.84	2.17929	113
8.85	2.18042	113
8.86	2.18155	112
8.87	2.18267	113
8.88	2.18380	113
8.89	2.18493	112
8.90	2.18605	112
8.91	2.18717	113
8.92	2.18830	112
8.93	2.18942	112
8.94	2.19054	111
8.95	2.19165	112
8.96	2.19277	112
8.97	2.19389	111
8.98	2.19500	111
8.99	2.19611	111
9.00	2.19722	

4.—Logarithms—*Continued*

No.					Common Logarithms of Numbers.						P. P.	Naperian.		
	L. O	1	2	3	4	5	6	7	8	9		No.	Log.	Dif.
900	95 424	429	434	439	444	448	453	458	463	468		9.00	2.19722	112
1	472	477	482	487	492	497	501	506	511	516	1	9.01	2.19834	110
2	521	525	530	535	540	545	550	554	559	564	2	9.02	2.19944	111
3	569	574	578	583	588	593	598	602	607	612	3	9.03	2.20055	111
4	617	622	626	631	636	641	646	650	655	660	4	9.04	2.20166	110
905	665	670	674	679	684	689	694	698	703	708	5	9.05	2.20276	111
6	713	718	722	727	732	737	742	746	751	756	6	9.06	2.20387	110
7	761	766	770	775	780	785	789	794	799	804	7	9.07	2.20497	110
8	809	813	818	823	828	832	837	842	847	852	8	9.08	2.20607	110
9	856	861	866	871	875	880	885	890	895	899	9	9.09	2.20717	110
910	904	909	914	918	923	928	933	938	942	947	5	9.10	2.20827	110
1	952	957	961	966	971	976	980	985	990	995	1 0.5	9.11	2.20937	110
2	999	004	009	014	019	023	028	033	038	042	2 1	9.12	2.21047	110
3	96 047	052	057	061	066	071	076	080	085	090	3 1.5	9.13	2.21157	109
4	095	099	104	109	114	118	123	128	133	137	4 2	9.14	2.21266	109
915	142	147	152	156	161	166	171	175	180	185	5 2.5	9.15	2.21375	110
6	190	194	199	204	209	213	218	223	227	232	6 3	9.16	2.21485	109
7	237	242	246	251	256	261	265	270	275	280	7 3.5	9.17	2.21594	109
8	284	289	294	298	303	308	313	317	322	327	8 4	9.18	2.21703	109
9	332	336	341	346	350	355	360	365	369	374	9 4.5	9.19	2.21812	109
920	379	384	388	393	398	402	407	412	417	421		9.20	2.21920	108
1	426	431	435	440	445	450	454	459	464	468	1	9.21	2.22029	109
2	473	478	483	487	492	497	501	506	511	515	2	9.22	2.22138	109
3	520	525	530	534	539	544	548	553	558	562	3	9.23	2.22246	108
4	567	572	577	581	586	591	595	600	605	609	4	9.24	2.22354	108
925	614	619	624	628	633	638	642	647	652	656	5	9.25	2.22462	108
6	661	666	670	675	680	685	689	694	699	703	6	9.26	2.22570	108
7	708	713	717	722	727	731	736	741	745	750	7	9.27	2.22678	108
8	755	759	764	769	774	778	783	788	792	797	8	9.28	2.22786	108
9	802	806	811	816	820	825	830	834	839	844	9	9.29	2.22894	107
930	848	853	858	862	867	872	876	881	886	890	4	9.30	2.23001	108
1	895	900	904	909	914	918	923	928	932	937	1 0.4	9.31	2.23109	107
2	942	946	951	956	960	965	970	974	979	984	2 0.8	9.32	2.23216	108
3	988	993	997	002	007	011	016	021	025	030	3 1.2	9.33	2.23324	107
4	97 035	039	044	049	053	058	063	067	072	077	4 1.6	9.34	2.23431	107
935	081	086	090	095	100	104	109	114	118	123	5 2	9.35	2.23538	107
6	128	132	137	142	146	151	155	160	165	169	6 2.4	9.36	2.23645	106
7	174	179	183	188	192	197	202	206	211	216	7 2.8	9.37	2.23751	107
8	220	225	230	234	239	243	248	253	257	262	8 3.2	9.38	2.23858	107
9	267	271	276	280	285	290	294	299	304	308	9 3.6	9.39	2.23965	106
940	313	317	322	327	331	336	340	345	350	354		9.40	2.24071	106
1	359	364	368	373	377	382	387	391	396	400	1	9.41	2.24177	107
2	405	410	414	419	424	428	433	437	442	447	2	9.42	2.24284	106
3	451	456	460	465	470	474	479	483	488	493	3	9.43	2.24390	106
4	497	502	506	511	516	520	525	529	534	539	4	9.44	2.24496	105
945	543	548	552	557	562	566	571	575	580	585	5	9.45	2.24601	106
6	589	594	598	603	607	612	617	621	626	630	6	9.46	2.24707	106
7	635	640	644	649	653	658	663	667	672	676	7	9.47	2.24813	105
8	681	685	690	695	699	704	708	713	717	722	8	9.48	2.24918	106
9	727	731	736	740	745	749	754	759	763	768	9	9.49	2.25024	105
950	772	777	782	786	791	795	800	804	809	813		9.50	2.25129	

4.—LOGARITHMS—*Concluded*

No.	L. O	1	2	3	4	5	6	7	8	9	P. P.		No.	Log.	Dif.
950	97 772	777	782	786	791	795	800	804	809	813			9.50	2.25129	105
1	818	823	827	832	836	841	845	850	855	859	1		9.51	2.25234	105
2	864	868	873	877	882	886	891	896	900	905	2		9.52	2.25339	105
3	909	914	918	923	928	932	937	941	946	950	3		9.53	2.25444	105
4	955	959	964	968	973	978	982	987	991	996	4		9.54	2.25549	105
955	98 000	005	009	014	019	023	028	032	037	041	5		9.55	2.25654	105
6	046	050	055	059	064	068	073	078	082	087	6		9.56	2.25759	105
7	091	096	100	105	109	114	118	123	127	132	7		9.57	2.25863	104
8	137	141	146	150	155	159	164	168	173	177	8		9.58	2.25968	105
9	182	186	191	195	200	204	209	214	218	223	9		9.59	2.26072	104
960	227	232	236	241	245	250	254	259	263	268	5		9.60	2.26176	104
1	272	277	281	286	290	295	299	304	308	313	1 0.5		9.61	2.26280	104
2	318	322	327	331	336	340	345	349	354	358	2 1		9.62	2.26384	104
3	363	367	372	376	381	385	390	394	399	403	3 1.5		9.63	2.26488	104
4	408	412	417	421	426	430	435	439	444	448	4 2		9.64	2.26592	104
965	453	457	462	466	471	475	480	484	489	493	5 2.5		9.65	2.26696	103
6	498	502	507	511	516	520	525	529	534	538	6 3		9.66	2.26799	104
7	543	547	552	556	561	565	570	574	579	583	7 3.5		9.67	2.26903	103
8	588	592	597	601	605	610	614	619	623	628	8 4		9.68	2.27006	103
9	632	637	641	646	650	655	659	664	668	673	9 4.5		9.69	2.27109	104
970	677	682	686	691	695	700	704	709	713	717			9.70	2.27213	103
1	722	726	731	735	740	744	749	753	758	762	1		9.71	2.27316	103
2	767	771	776	780	784	789	793	798	802	807	2		9.72	2.27419	102
3	811	816	820	825	829	834	838	843	847	851	3		9.73	2.27521	103
4	856	860	865	869	874	878	883	887	892	896	4		9.74	2.27624	103
975	900	905	909	914	918	923	927	932	936	941	5		9.75	2.27727	102
6	945	949	954	958	963	967	972	976	981	985	6		9.76	2.27829	103
7	989	994	998	003	007	012	016	021	025	029	7		9.77	2.27932	102
8	99 034	038	043	047	052	056	061	065	069	074	8		9.78	2.28034	102
9	078	083	087	092	096	100	105	109	114	118	9		9.79	2.28136	102
980	123	127	131	136	140	145	149	154	158	162	4		9.80	2.28238	102
1	167	171	176	180	185	189	193	198	202	207	1 0.4		9.81	2.28340	102
2	211	216	220	224	229	233	238	242	247	251	2 0.8		9.82	2.28442	102
3	255	260	264	269	273	277	282	286	291	295	3 1.2		9.83	2.28544	102
4	300	304	308	313	317	322	326	330	335	339	4 1.6		9.84	2.28646	101
985	344	348	352	357	361	366	370	374	379	383	5 2		9.85	2.28747	102
6	388	392	396	401	405	410	414	419	423	427	6 2.4		9.86	2.28849	101
7	432	436	441	445	449	454	458	463	467	471	7 2.8		9.87	2.28950	101
8	476	480	484	489	493	498	502	506	511	515	8 3.2		9.88	2.29051	101
9	520	524	528	533	537	542	546	550	555	559	9 3.6		9.89	2.29152	101
990	564	568	572	577	581	585	590	594	599	603			9.90	2.29253	101
1	607	612	616	621	625	629	634	638	642	647	1		9.91	2.29354	101
2	651	656	660	664	669	673	677	682	686	691	2		9.92	2.29455	101
3	695	699	704	708	712	717	721	726	730	734	3		9.93	2.29556	101
4	739	743	747	752	756	760	765	769	774	778	4		9.94	2.29657	100
995	782	787	791	795	800	804	808	813	817	822	5		9.95	2.29757	101
6	826	830	835	839	843	848	852	856	861	865	6		9.96	2.29858	100
7	870	874	878	883	887	891	896	900	904	909	7		9.97	2.29958	100
8	913	917	922	926	930	935	939	944	948	952	8		9.98	2.30058	100
9	957	961	965	970	974	978	983	987	991	996	9		9.99	2.30158	100
1000	000 000	043	087	130	174	217	260	304	347	391			10.00	2.302585	

Common Logarithms of Numbers. — Naperian.

SLIDE RULE

The Slide Rule.—The fact that we are able to accomplish multiplication with logarithms by the *addition* of two numbers suggests the possibility of a simple mechanical instrument to carry this out. Such an instrument is the *slide rule.*

Let us consider first an ordinary ruler. If we place our finger on the two-inch mark and then *add* three inches to these two inches, our finger moves to the five-inch mark. Then, if we have a ruler which, instead of being graduated in inches, is graduated by the *logarithms* of numbers, it is evident that if we add to a distance representing the logarithm of 2 a distance representing the logarithm of 3, the sum of these logarithms is the logarithm of 6, and by the addition of these logarithms we have performed *multiplication.* This is the principle of the slide rule and the operation just described is illustrated in Fig. 6.

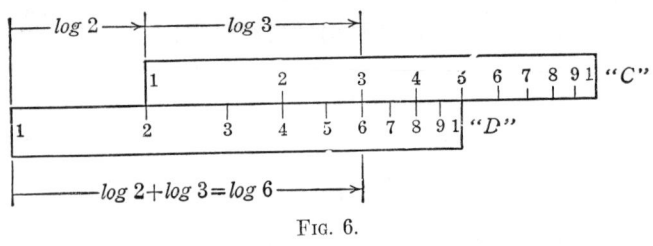

Fɪɢ. 6.

The addition of one distance to another is readily accomplished by having the slide rule made in two parts, one of which is called the *stock* and remains stationary, while the other is the *slide* which is free to move parallel with the stock. (See Fig. 7.)

Both the stock and the slide have graduations, but while the distances between graduations are proportioned according to logarithms, the *numbers* on the scales represent antilogs. Thus the answers to the operations are read directly without the need of conversion.

There are many forms of slide rules. With regard to shape, they may be straight, circular or cylindrical. With regard to size they may be short or long. Then there are many types of special graduations. The most common slide rule is the 10-inch straight rule of the Mannheim type illustrated in Fig. 8. It will be noted that this has four scales, two on the stock and two on the slide. These we shall refer to as the A, B, C, and D scales reading downward, and they are usually so stamped on the rules.

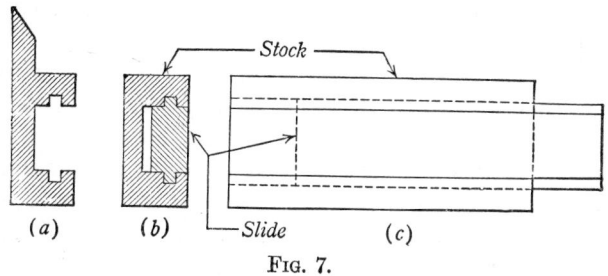

(a)   *(b)*   ⎣—*Slide*   *(c)*

Fɪɢ. 7.

Straight slide rules other than the Mannheim usually have scales corresponding to these and differ only in that they have in addition a number of other scales. At the extreme end of each scale is the figure "1." At the left of the scale, this point is called the *left index*; at the right is the *right index*. A glass with a hair-line stretching across the scales is attached to the rule in such a

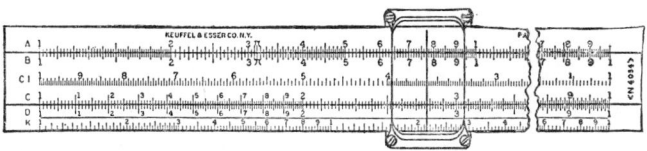

Fɪɢ. 8.—Mannheim Slide Rule.

manner that it may be moved along the scale to any position. This is called the indicator or runner and is a great aid in setting and reading values.

It will be noted that on the two lower scales on the slide rule in Fig. 8 the numbers begin at the left with 1, 1, 2, 3, 4, etc. These numbers represent 10, 11, 12, 13, 14, respectively, or 1.0, 1.1, 1.2, 1.3, 1.4, or 100, 110, 120, 130, 140. The extra "1" before the small numbers is omitted to save space. The space between each of these numbers is divided into ten spaces for the next significant figure. To the right of this first series of figures which terminate with "9" or "19" by the above representation, the numbers continue 2, 3, 4, 5, etc. If the preceding "9" is taken to represent "19" (not "1.9" or "190") then these numbers represent 20, 30, 40, 50, etc. The spaces between these numbers are divided into ten spaces. These are again subdivided, some into fifths, others into halves.

The operation of a slide rule cannot be mastered without a rule at hand, and even then, considerable practice is required to develop speed and accuracy. The following examples assume that the reader has a slide rule before him with the conventional A, B, C, and D scales found on the Mannheim and Polyphase rules.

C	*Set 1*	*R to 12*
D	*To 28*	*Read 336*

(a)

C	*Set 1*	*R to Multiplier*
D	*To Multiplicand*	*Read Product*

(b)

Fig. 9.

How to Multiply with the Slide Rule.—Let us assume that we wish to multiply 28 by 12. Move the slide to the right and set 1 (the index) of the C scale to 28 on the D scale. Then move the runner to 12 on the C scale and read the product on the D scale at this point. It is 336. These operations can be set up in the form of a diagram as shown in Fig. 9a. From this we can derive a general form for *all* multiplication as shown in Fig. 9b. Expressed in words we may say multiplication is carried out as follows:

(1) To *Multiplicand* on D set C index.
(2) To *Multiplier* on C set runner indicator.
(3) At indicator on D read *Product*.

Let us take another example. Multiply 52 by 25. Proceeding as in the previous example we find that by moving the slide to the right, the multiplier (25) falls beyond the end of the D scale. It is necessary then in this and all similar cases that the slide be moved to the *left* and the right index of the C scale set on the multiplicand. The answer on the D scale is then 1300.

C	Set 1	R to 25
D	to 52	Read *1300*
		——(Ans.)

Fig. 10.

By using the runner, R, it is possible to perform continued multiplication without having to read the intermediate products.

ILLUSTRATION: Multiply 12 × 8 × 18.

C	Set 1	R to 8	1 to R	Under 18
D	to 12			Read 1728
				——(Ans.)

Fig. 11.

With a ten-inch slide rule most numbers can be read directly to only two significant figures and the third figure must be estimated. Thus, if we multiply 854 by 537 we find in setting the index to the multiplicand that there is no line which represents 854. There is one which represents 850 and another 855 and it is necessary to set the index as closely as possible to a point which is estimated by eye to be $\frac{4}{5}$ of the distance from the smaller to the larger of these numbers. Similarly, the position to place the runner to represent 537 must be estimated by eye for the last figure. The product reads 459,000 (the last significant figure being estimated). With practice and careful operation the last place can be determined with remarkable accuracy.

| C | Set | Div-
isor | Under 1
Right or Left |
| D | Over | Divi-
dend | Read
Quotient |

| C | 25 | Under 1 |
| D | 1300 | Read 52 (*Ans.*) |

Fig. 12.

Division with the Slide Rule.—Division is, of course, the reverse of multiplication and it is to be expected that it is carried out on the slide rule by performing the operations for multiplication in the reverse order. This is the case. For example, let us divide 1300 by 25.

To dividend (1300) on the D scale set divisor (25) on the C scale. Under 1 on the C scale read the quotient (52).

Another example, divide 1648 by 536. To 1648 on the D scale, set 536 on the C scale. Under the index on the C scale read the quotient. This appears to be 3.075.

Calculations involving continued multiplication and division can be performed on the slide rule without having to read the intermediate results.

ILLUSTRATION: Find the value of

$$\frac{150 \times 72 \times 10}{8 \times 6}$$

There are two methods:

C	8	R to 72	6 to R	Under 10
D	150			Read 2250 (Ans.)

C	Set 1	R to 72	1 to R	R to 10	8 to R	R to 1	6 to R	Under 1
D	150							Read 2250 (Ans.)

FIG. 13.

Locating Decimal Point in Slide Rule Multiplication and Division.—The preceding examples have illustrated the manipulation of the slide rule in arriving at products and quotients of numbers without any mention of how the decimal point is located in the result. We shall state briefly the rules governing this and then illustrate by a few examples. First, a definition is necessary. In the rules we shall use the word *characteristic* of a number, which is not to be confused with the characteristic of a logarithm. The characteristic of a number is the number of digits before the decimal point, the characteristic of a decimal fraction is the number of ciphers immediately after the decimal point and is negative.

RULE I: *When the slide projects to the right in multiplication the characteristic of the product is one less than the sum of the characteristics of the factors.*

Thus, in the first example of multiplication we found the product of 28 and 12, the slide projecting to the right. The characteristic of each of these numbers is 2, the sum is 4, one less than the sum is 3, which is the characteristic of the product. Thus the product has three figures to the left of the decimal point (336).

In another example, 23×0.415, the characteristics are 2 and 0, respectively, and the sum less 1 is $2 + 0 - 1 = 1$. Thus the product (9.55) has one digit to the left of the decimal point.

In still another example, 0.0328×0.0024, the characteristics are -1 and -2, respectively. The sum less 1 is $-1 - 2 - 1 = -4$. Then the product is 0.0000787 with four ciphers following the decimal point.

These examples illustrate the cases which are apt to occur.

RULE II. *When the slide projects to the left in multiplication the characteristic of the product equals the sum of the characteristics of the factors.*

This rule requires no illustration in view of the foregoing.

RULE III: *When the slide projects to the right in division the characteristic of the quotient equals the characteristic of the dividend minus that of the divisor, plus 1.*

As an illustration, divide 6850 by 37.2. The characteristic of the dividend is 4 and that of the divisor 2. Then, according to the rule, $4 - 2 + 1 = 3$ and the quotient has three digits to the left of the decimal point—in this case 184.1.

As an illustration involving decimals, take the division of 47 by 0.024. The characteristic of the dividend is 2 and that of the divisor -1. Then $2 - (-1) + 1 = +4$ as the characteristic of the quotient. The quotient is then 1957.

The division of one decimal by another is illustrated by the following $0.0074 \div 0.026$. The characteristic of the dividend is -2 and that of the divisor -1. Then, $-2 - (-1) + 1 = 0$ and thus there are no digits to the left of the decimal point and no ciphers to the right, the quotient being 0.2847.

RULE IV: *When the slide projects to the left in division the characteristic of the quotient equals the characteristic of the dividend minus that of the divisor.*

The four rules may be combined into the following chart for ready reference:

Characteristic of result	Slide LEFT	Slide RIGHT
Multiplication	Sum of Characteristics of 2 Factors	Sum -1
Division	Characteristic of Dividend $-$ that of divisor	Difference $+1$

Squares, Cubes, and Roots.—The square of a number can, of course, be computed with a slide rule by multiplying the number by itself with the C and D scales. Likewise, the cube may be determined by multiplying the square so found by the original number. However, by the use of the D and A scales the square of any number on the D scale can be found by simply moving the runner to that number and reading the square on the A scale at the cross-line on the runner. Thus 4 on the A scale is directly opposite 2 on the D scale, 9 opposite 3, etc.

The following examples indicate how the slide rule can be used for evaluating such expressions as $x^2 y$ and $\sqrt{\dfrac{a}{b}}$.

ILLUSTRATION: Find the value of $6^2 \times 5$.

A		Read 180 (Ans.)
B	Set 1 (right)	Over 5
C		
D	Over 6	

ILLUSTRATION: Find the value of $\sqrt{\frac{3}{4}}$.

A	Under 3	
B	Set 4	Under 1 (right)
C		
D		Read 0.866 (Ans.)

The cube of a number is found by setting the runner on the number on the D scale, then setting either the left or right index of the B scale on the cross-line of the runner and reading the cube on the A scale opposite the original number on the B scale.   Thus, the process consists of finding the square of the number and then performing a multiplication on the A and B scales of the square with the original number.

It will be noted that A and B scales have indexes not only at the left and right ends but also one in the middle. If the left index is taken as 1, the middle index is 10 and the right index is 100, or if the left index is taken as 100, the middle index is 1000 and the right index is 10,000. The left index may never be taken as 10 or 1000 because the square roots of these numbers are 3.16 and 31.6, respectively, and this occurs on the D scale only at the middle.

Thus it becomes apparent that whenever the square root of a number is to be found with the A and D scales it is very important to decide whether this number should be selected on the left- or the right-hand portions of the A scale. This is determined by first pointing off the digits of the number whose square root is to be found into groups of two's, beginning at the decimal point, and moving to the right. For example, 25,346 pointed off is 2,53,46. The square root will have as many digits to the left of the decimal point as there are groups.   Decimal fractions are pointed off to the right from the decimal point thusly: 0.02758 becomes .02,75,8. The last group may have either one or two digits. Then if we call the left half of the A scale A1 and the right half A2, we may write the

> RULE: *If the last group contains one figure, use A1 for finding the square root, if it contains two use A2. In either case the characteristic of the square root read on D equals the number of groups in the given number.*

As an illustration, take the first number cited above, 2,53,46. This has one figure in the last group, so it is located on the left-hand (A1) scale and the runner placed on it. The square read on the D scale appears to be 159.1 and since there are three groups of figures in the original number, the root has three digits to the left of the decimal point. For decimal fractions we have the following.

> RULE: *After pairing off the digits to the right of the decimal point, at first disregard the groups immediately following the point which contain only ciphers. If in the first group containing other figures the first figure is a cipher, use A1. If the first figure is not a cipher, use A2. In the root there is one cipher immediately after the decimal point for each group consisting wholly of ciphers in the given number.*

As an example find the square root of the number 0.000625. Pointing off, .00,06,25. The first group containing significant figures (06) has first a cipher so the A1 scale is used. The square root is 0.025. Since the first group in the original number has only ciphers, the first digit of the root is a cipher.

By reversing the above rules we obtain a rule for locating the decimal point when computing squares.

> RULE: *If the square is on A1 the characteristic of the square is 1 less than 2 times that of the number, if on A2 it is twice that of the number. This applies to both positive and negative characteristics.*

References—The reader who is interested in a more comprehensive treatment of arithmetic is referred to the book ARITHMETIC FOR THE PRACTICAL MAN, by Mr. J. E. Thompson, published by the D. Van Nostrand Company. The same author has also written an excellent book entitled A MANUAL OF THE SLIDE RULE, which is also published by the D. Van Nostrand Company.

III

Algebra

Algebraic Symbols.—Algebra is the shorthand of mathematics. Letters and symbols take the place of cumbersome numbers, and many of the ordinary operations of arithmetic take a simpler and more compact form. In addition, algebra can be used to advantage in some problems where arithmetical solution would be extremely involved. In arithmetic the Greek letter π is used to designate the number $3.14159+$ and multiplication, division, etc., is performed with it. Similarly, the letters, a, b, c, . . . can be used to represent certain quantities. The first letters of the alphabet are usually used to represent known quantities and the last letters, x, y, z, to represent unknown quantities.

The number of times that a single algebraic quantity is to be taken is indicated by a number before the letter. This number is called the *coefficient*. Thus, in $3b$, the 3 is the coefficient and the expression equivalent to $b + b + b$.

Signs of Algebra.—Whereas in arithmetic it is common to deal only with positive numbers, both positive and negative numbers are used in algebra and it thus becomes necessary to employ symbols to indicate the sign of the quantity. Thus, $+a$, $+b$, etc., denotes that the quantity is positive and $-a$, $-b$, etc., denotes that the quantity is negative. When no sign precedes a number or quantity it is understood to be positive. Powers and roots are indicated as in arithmetic.

Parentheses.—When a number of quantities are enclosed in parentheses with a positive sign before, the parentheses may be removed without altering the expression. Thus, $+(a + b)$ becomes $+a + b$. However, if the sign before is negative, the sign of each quantity must be changed when the parentheses are removed.   Thus, $-(a + b)$ becomes $-a - b$, and $-(a - b)$ becomes $-a + b$.

Addition of Algebraic Quantities.—A number of like algebraic terms of like sign may be added by arranging in a column and adding together the coefficients, the sum having the same sign as the parts. Thus,

$$
\begin{array}{ccc}
+\ 7b & & -12c \\
+\ 3b & \text{and} & -\ 4c \\
+\ 5b & & -\ 6c \\
\hline
+15b & & -22c
\end{array}
$$

If some of the quantities are unlike in sign, proceed as before, but regard the negative coefficients as being subtracted from the positive. Thus,

$$
\begin{array}{ccc}
+9a & & +\ 5b \\
+3a & \text{and} & +\ 4b \\
-4a & & -12b \\
\hline
+8a & & -\ 3b
\end{array}
$$

When compound quantities (that is, quantities containing more than one term, as $2a - 4b$) are to be added, like terms must be placed in the same column and then added as above. For example, if $5a + 14b + 10c$, $2b - 6c$, $3a - 9c + 3x$, and $-12b - 11c - x$ are to be added, the procedure is as follows:

$$
\begin{array}{l}
5a + 14b + 10c \\
\quad\ \ 2b -\ \ 6c \\
3a \qquad\ \ -\ \ 9c + 3x \\
\quad\ -\ 12b - 11c -\ \ x \\
\hline
8a +\ \ 4b - 16c + 2x \quad \text{(Ans.)}
\end{array}
$$

Subtraction of Algebraic Quantities.—To subtract algebraic quantities, change the sign of the number to be subtracted and then combine the two numbers as in addition.

Example: Subtract $6x$ from $15x$.

$$\begin{array}{r} 15x \\ 6x \\ \hline 9x \end{array}$$

changing the sign of $6x$ makes it $-6x$. Adding $15x$ and $-6x$ gives $9x$.

Example: Subtract $6x$ from $-15x$.

$$\begin{array}{r} -15x \\ 6x \\ \hline -21x \end{array}$$

changing the sign of $6x$ makes it $-6x$. Adding $-15x$ and $-6x$ gives $-21x$.

Example: From $7x - 3y$ take $5x + 12y$.

$$\begin{array}{r} 7x - 3y \\ 5x + 12y \\ \hline 2x - 15y \end{array}$$

write like terms under each other and proceed with each pair of like terms as explained above.

Multiplication of Simple Quantities.—The parts of an algebraic expression separated by plus and minus signs are called *terms*. An expression consisting of one term is known as *monomial*, one of two terms, a *binomial*, one of three terms, a *trinomial*, and one of many terms a *polynomial*.

If two quantities to be multiplied have like signs, the sign of the product is plus; if they have unlike signs, that of the product is minus. Thus, $+a$ multiplied by $+b$ is $+ab$ (the multiplication sign ($\times$) is usually omitted between letters of a term in algebra), $-a$ multiplied by $-b$ is $+ab$, but $-a$ multiplied by $+b$ is $-ab$.

When multiplying monomial expressions, multiply the coefficients together and prefix the product by the proper sign as outlined above. Examples:

Multiply $-a$ $-b$.   Product equals $+ab$.
Multiply $+4b$ by $-c$.   Product equals $-4bc$.
Multiply $+6b$ by $+3c$.   Product equals $+18bc$.
Multiply $-4ax$ by $+5ab$.   Product equals $-20aabx = -20a^2bx$

Multiplication of Compound Quantities.—To multiply one polynomial by another, it is necessary to multiply each term of the multiplicand by all of the terms of the multiplier one after the other as by the former rule. The products are then collected into one sum for the required product.

Example: Multiply $3x - 2y$ by $x + 4y$.

Solution:
$$
\begin{array}{l}
3x \ - \ \ 2y \\
\ \ x \ + \ \ 4y \\
\hline
3x^2 - \ 2xy \\
\ \ \ \ \ + 12xy - 8y^2 \\
\hline
3x^2 + 10xy - 8y^2 \ \text{(Ans.)}
\end{array}
$$

Example: Multiply $x - y + z$ by $x + y - z$.

Solution:
$$
\begin{array}{l}
x \ - \ \ y + z \\
x \ + \ \ y - z \\
\hline
x^2 - xy + xz \\
\ \ \ \ + xy \ \ \ \ \ \ \ - y^2 + \ yz \\
\ \ \ \ \ \ \ \ \ - xz \ \ \ \ \ \ \ + yz - z^2 \\
\hline
x^2 \ \ \ \ \ \ \ \ \ \ \ \ \ \ \ \ - y^2 + 2yz - z^2 \ \text{(Ans.)}
\end{array}
$$

Division of Monomials.—One monomial is divided by another by simply writing the dividend over the divisor as a fraction and cancelling out common factors as in arithmetic. Thus,

$$
12ax \div 6a = \frac{\overset{2}{\cancel{12}}ax}{\cancel{6}a} = 2x, \quad \text{and} \quad \frac{\overset{3}{\cancel{9}6}}{\underset{x}{\cancel{36}x}} = \frac{3}{x}
$$

Since $x^2 = x \times x$ and $y^3 = y \times y \times y$, powers may be factored and the common factors cancelled. Then,

$$
\frac{x^2 y^4}{xy^2} \text{ may be written } \frac{\cancel{x} \times x \times \cancel{y} \times \cancel{y} \times y \times y}{\cancel{x} \times \cancel{y} \times \cancel{y}} = xy^2
$$

It is evident from this example that the same result can be arrived at by subtracting the exponent of the smaller number from the exponent of the larger. Thus,

$$\frac{x^2y^4}{xy^2} = x^{(2-1)}y^{(4-2)} = xy^2$$

This is the method actually used in dividing monomials higher than the first power.

Examples: $\quad \dfrac{4a^2b^5}{a^3b^2x^2} = \dfrac{4b^3}{ax^2}; \quad \dfrac{a^2}{a^4} = \dfrac{1}{a^2}; \quad \dfrac{3ab^2x}{ab^2x} = 3$

Division of Polynomials.—A polynomial may be divided by a monomial by dividing each term of the polynomial by the monomial. Thus, $2a^2x^3 + 3ax^2 + 5x$ divided by ax may be written $\dfrac{2a^2x^3}{ax} + \dfrac{3ax^2}{ax} + \dfrac{5x}{ax}$. Cancelling out like terms, the quotient becomes $2ax^2 + 3x + \dfrac{5}{a}$.

To divide a polynomial by a polynomial, arrange both the dividend and divisor according to the ascending or descending powers of some letter and keep this arrangement throughout the operation. Divide the first term of the dividend by the first term of the divisor, and write the result as the first term of the quotient.

Multiply all the terms of the divisor by the first term of the quotient and subtract the product from the dividend. If there is a remainder, consider it as a new dividend and proceed as before.

Example: Divide $2x^3 + 4x^2y - xy - 2y^2$ by $x + 2y$

Solution: These expressions are already arranged according to descending powers of x. Then,

$$x + 2y) \quad \begin{array}{lllll} 2x^3 & + 4x^2y & - xy & - 2y^2 & \underline{}\ 2x^2 - y \text{ (Ans.)} \\ \underline{-+2x^3 -+ 4x^2y} & & & \\ & & - xy & - 2y^2 \\ & & \underline{+- xy +- 2y^2} \end{array}$$

Multiply $x + 2y$ by $2x^2$ and obtain $+2x^3 + 4x^2y$, which is to be subtracted from $2x^3 + 4x^2y$ in the dividend. Changing the signs of $2x^3 + 4x^2y$ so that this term becomes $-2x^3 - 4x^2y$ proceed as in addition. Then multiply $x + 2y$ by $-y$ and obtain $-xy - 2y^2$ which is to be subtracted from $-xy - 2y^2$. Changing the signs so that this term becomes $+xy + 2y^2$ proceed as in addition.

If the division is not exact and there is a remainder after the last operation has been performed, write the divisor beneath it to form a fraction which is the last term of the quotient.

Example: Divide $4x^2y - 3xy + 6y^2$ by $x^2 - y$.

Solution:

$$x^2 - y \,)\overline{\,4x^2y - 3xy + 6y^2\,}^{\,4y}$$
$$\underline{4x^2y \qquad\quad - 4y^2}$$
$$-3xy + 10y^2$$
$$= 4y + \frac{-3xy + 10y^2}{x^2 - y} \quad \text{(Ans.)}$$

Factoring.—When a number is the product of two other numbers, the component parts are known as *factors*. Thus, in the expression $3a^2$, 3, a, and a, are the factors. Separating a number into its factors is called *factoring*.

Factoring is useful in solving equations, as will be discussed later, and also in simplifying complicated expressions. The operation of removing a monomial factor consists of scrutinizing each term of an expression with a view to determining common factors and then dividing each term by the common factor and placing it before the parentheses which contain the several quotients.

Example: Factor $12a^3x^2 + 33a^2x^2 - 18ax^3 + 9ax$.

Solution: Inspection reveals that a factor common to each term is $3ax$. Then, dividing each term by $3ax$, the expression becomes, $3ax(4a^2x + 11ax - 6x^2 + 3)$.

It is often the case that no single factor can be found common to all the terms of an expression. Then the terms must be examined and compared with a view to grouping them and removing factors common to the group. Thus, in the expression $3x^2 + 9bx + 24xy + 4ax + 12ab + 32ay$, there is no factor common to all terms, but a further examination shows that the first three terms have the common factor $3x$ and the last three terms the common factor $4a$. Removing these factors from the respective terms, the expression becomes,

$$3x(x + 3b + 8y) + 4a(x + 3b + 8y)$$

which may then be consolidated to,

$$(3x + 4a)(x + 3b + 8y)$$

Certain trinomials which are the product of two binomials lend themselves to ready recognition and factoring. Examples of such trinomials are, $(x + 5)(x + 2) = x^2 + 7x + 10$; $(x - 3)(x + 6) = x^2 + 3x - 18$; $(x + y)(x + y) = x^2 + 2xy + y^2$; and $(x - y)(x - y) = x^2 - 2xy + y^2$.

The first of these trinomials, $x^2 + 7x + 10$, could be written $x^2 + 5x + 2x + 10$ and the first two and the last two groups factored as, $x(x + 5) + 2(x + 5) = (x + 2)(x + 5)$. Further examination of this example leads to the observation that the coefficient of the middle term of the trinomial is the sum of the last terms $(2 + 5 = 7)$ of the factors, and the last term of the trinomial is the product of these last terms $(2 \times 5 = 10)$. This is the key to the factoring of factorable expressions of this type. Thus:

$$x^2 + 2x - 8 = (x + 4)(x - 2)$$
$$x^2 + x - 20 = (x + 5)(x - 4)$$
$$x^2 + 3xy + 2y^2 = (x + y)(x + 2y)$$

A ready recognition of a few other special forms is also valuable. These are,

$$x^2 + 2xy + y^2 = (x + y)(x + y) = (x + y)^2$$
$$x^2 - 2xy + y^2 = (x - y)(x - y) = (x - y)^2$$
$$x^2 - y^2 = (x + y)(x - y)$$

Powers and Exponents.—When a quantity is multiplied by itself several times, the resulting product is called a *power* and the quantity itself is called the *root*. Thus, in $ax \times ax \times ax \times ax = a^4x^4$, ax is the root and a^4x^4 is the power. A small number called the *exponent* is used to indicate how many times a number has been multiplied by itself.

The sign of the product of two positive numbers is plus $(+a \times + a = + a^2)$ and the sign of the product of two negative numbers is also plus $(-a \times - a = + a^2)$, but the product of a positive and a negative number is minus $(+a \times - a = - a^2)$. If, then, we raise a negative number to an odd power, for example to the third, as in $-a \times - a \times - a$ it is evident that the first product of $-a \times - a$ results in a positive number and then when this is multiplied again by $-a$ the product becomes negative. Hence, we derive the rule that the sign of an even power of a negative number is positive and the sign of an odd power of a negative number is negative. Examples: $(-a)^2 = + a^2$; $(-a)^3 = - a^3$; $(-a)^4 = + a^4$; $(-a)^5 = - a^5$, etc. The sign of any power of a positive number is, of course, plus.

The product of two or more powers of any quantity is the quantity with an exponent equal to the sum of the exponents of the powers. Examples: $x^2 \times x^3 = x^5$; $x^2y \times xy = x^3y^2$; $4xy \times (-3xz) = - 12x^2yz$.

In a similar manner, the quotient of two powers is the difference of their exponents. Thus, $x^5 \div x^3 = x^{5-3} = x^2$, and $6x^4 \div 2x^3 = \dfrac{6x^4}{2x^3} = 3x$. Then it is apparent that if the exponent of the divisor is greater than the exponent of the dividend, the exponent of the quotient becomes a negative number. Thus, $x^2 \div x^3 = x^{2-3} = x^{-1}$, or $\dfrac{x^2}{x^3} = \dfrac{1}{x} = x^{-1}$. In other words, if a power appears in the denominator with a positive exponent it may be shifted to the numerator by changing the sign of the exponent, as $\dfrac{2ab}{x^3} = 2abx^{-3}$. The law holds equally true for the reverse operation

If we divide one power by an equal power we have this interesting situation $x^3 \div x^3 = x^{3-3} = x^0$. But $\dfrac{x^3}{x^3} = 1$. Then $x^0 = 1$ and the general rule may be stated, that any quantity raised to the zero power is equal to 1.

When a quantity with an exponent is raised to a power, the exponent of the resulting quantity is the product of the exponent of the original quantity and the exponent of the power to which it was raised. This can be well understood from the following illustrations:

$$(x^2)^3 = x^2 \times x^2 \times x^2 = x^6; \quad (y^5)^2 = y^5 \times y^5 = y^{10}.$$

The square of the sum of two quantities is the sum of their squares plus twice their product. Thus,

$$(x + y)^2 = x^2 + y^2 + 2xy; \quad (3x + 4y)^2 = 9x^2 + 16y^2 + 24xy.$$

The square of the difference of two quantities is the sum of their squares minus twice their product. Thus,

$$(x - y)^2 = x^2 + y^2 - 2xy; \quad (2x - 5y)^2 = 4x^2 + 25y^2 - 20xy.$$

The square of a trinomial is equal to the sum of the squares of each term plus twice the product of each term by each of the other terms. Examples:

$$(x + y + z)^2 = x^2 + y^2 + z^2 + 2xy + 2xz + 2yz$$
$$(x - y - z)^2 = x^2 + y^2 + z^2 - 2xy - 2xz + 2yz$$

Roots.—The opposite operation to finding the power of an expression is called finding or extracting a root. The symbol used is the radical sign the same as in arithmetic, $\sqrt{}$, with a small number called the root index, $\sqrt[3]{}$, to indicate the number of times the root is contained as a factor in the power. When no index number is shown in the hook of the radical sign, the square root is intended.

The root of a product is equal to the product of the roots of the factors. Thus, $\sqrt{144} = \sqrt{9 \times 16} = \sqrt{9} \times \sqrt{16} = 3 \times 4 = 12$, $\sqrt{xy} = \sqrt{x} \times \sqrt{y}$, and $\sqrt{a^2b} = \sqrt{a^2} \times \sqrt{b} = a\sqrt{b}$. However, the root of the *sum* of several terms is *not* the sum of the roots of the individual terms. Thus, $\sqrt{x+y}$ is not $\sqrt{x} + \sqrt{y}$. A polynomial expression under a radical sign must be treated as a whole unless it can be simplified.

In the preceding section it was shown that when a quantity with an exponent is raised to a power the exponent of the resulting quantity is the product of the exponent of the original quantity and the exponent to which it was raised, as $(a^3)^6 = a^{18}$. Then, if we give a quantity a fractional exponent, for example $\frac{1}{2}$, and square the quantity we get this interesting result: $(x^{\frac{1}{2}})^2 = x^{\frac{2}{2}} = x$. But $(\sqrt{x})^2$ also equals x: Then $\sqrt{x} = x^{\frac{1}{2}}$ and the exponent $\frac{1}{2}$ is another way of indicating square root. Similarly, it can be shown that $x^{\frac{1}{3}} = \sqrt[3]{x}$, $x^{\frac{1}{4}} = \sqrt[4]{x}$, etc.

If we multiply, for example, $x^{\frac{1}{3}}$ by $x^{\frac{1}{3}}$ we obtain $x^{\frac{1}{3}} \times x^{\frac{1}{3}} = (x^{\frac{1}{3}})^2 = x^{\frac{2}{3}}$. Expressed in words this is, " the cube root of the square of x " and can be written $\sqrt[3]{x^2}$. Other fractional exponents can be similarly expressed, as $a^{\frac{3}{2}} = \sqrt{a^3}$, $b^{\frac{3}{4}} = \sqrt[4]{b^3}$.

In the preceding section it was shown that while the square of a positive number is positive, the square of a negative number is also positive. Then, if we are confronted with a positive power, as 25, it is impossible to tell whether its square root is positive or negative. Therefore, when the square root of a number has been found, it is necessary to precede it by a plus or minus sign. Thus, $\sqrt{25} = \pm 5$, and $\sqrt{x^2} = \pm x$. It was also found that the odd power of a negative number was negative. Then the odd root of a negative number is negative, as $\sqrt[3]{-8} = -2$, $\sqrt[5]{-243} = -3$. The odd root of a positive number is always positive, but the even root of a positive number may be either negative or positive.

The even root of a negative number cannot be determined and is said to be an *imaginary* number. Thus, the square root of -25 does not exist. Such expressions do, however, sometimes occur and then for the sake of simplicity may be treated as follows: $\sqrt{-25} = \sqrt{25 \times (-1)} = \sqrt{25} \times \sqrt{-1} = 5\sqrt{-1} = 5i$. The letter i is a symbol used to designate $\sqrt{-1}$.

Simple Equations.—If one algebraic expression is equal in value to another, the two, if written with an equality sign between them, constitute an algebraic *equation*, as $a + b = c + d$.

Both sides of an equation may be changed equally by addition, subtraction, multiplication, or division without disturbing the equality. To illustrate, if

$$a + b = c + d$$
$$\text{then} \quad a + b + x = c + d + x,$$
$$a + b - x = c + d - x,$$
$$x(a + b) = x(c + d)$$
$$\text{and} \quad \frac{a + b}{x} = \frac{c + d}{x}$$

Thus, if we have the equation, $x + 3y = 10$, and want to know the value of x, it is only necessary to subtract $3y$ from both sides of the equation. Then

$$x + 3y - 3y = 10 - 3y$$
$$x = 10 - 3y$$

From this it is apparent that any term of an equation may be changed from one side to the other provided its sign is moved. This is called transposition.

Solution of Simple Equations.—When the value of an unknown symbol in an equation is determined, the equation is said to be solved. Equations containing only one unknown quantity may be solved as follows: Transpose all the terms containing the unknown quantity to the left side of the equation, and all the other terms to the right side. Combine like terms, and divide both sides of the equation by the coefficient of the unknown quantity.

ILLUSTRATIONS:

$$9x - 18 = 12 - 6 + 3x$$
$$9x - 3x = 12 - 6 + 18 \quad \text{(transposing)}$$
$$6x = 24 \quad \quad \text{(collecting terms)}$$
$$\frac{6x}{6} = \frac{24}{6} = \quad \quad \text{(dividing by coefficient)}$$
$$x = 4$$

$$3y + 4 = 8y + 36$$

$$3y - 8y = 36 - 4 \qquad \text{(transposing)}$$

$$-5y = 32 \qquad \text{(collecting terms)}$$

$$-\frac{5y}{5} = 32 \qquad \text{(dividing by coefficient)}$$

$$y = -6\frac{2}{5} \qquad \text{(changing signs of both sides)}$$

$$3\tfrac{1}{2}z - 14 = 8 + 3z$$

$$3\tfrac{1}{2}z - 3z = 8 + 14$$

$$\tfrac{1}{2}z = 22$$

$$\frac{\tfrac{1}{2}z}{\tfrac{1}{2}} = \frac{22}{\tfrac{1}{2}}$$

$$z = \frac{22}{\tfrac{1}{2}} = 22 \times \frac{2}{1} = 44$$

Solution of Simultaneous Simple Equations.—If an equation contains two unknown quantities, an indefinite number of pairs of values for them may be found, which will satisfy the equation. For example, in the equation, $x + y = 12$, when x is 4, y is 8; when x is 9, y is 3; when x is 16, y is -4; etc. However, if a second equation containing the same unknowns is given, a single pair of values may be found which will satisfy both equations. Equations solved for common values of their unknowns are called simultaneous equations.

The process of solving two simultaneous equations of two unknowns is to eliminate temporarily one of the unknowns by combining the two equations into one equation containing the other unknown only. One method of doing this is *elimination by addition or subtraction*. This proceeds as follows: Multiply the equations by such a number as will make the coefficients of one of the unknown quantities equal in both. Add or subtract the two equations according to whether the unknown quantities of equal coefficients have unlike or like signs. Solve the resulting equation of the remaining unknown in the regular manner and substitute the value found in one of the original equations to determine the value of the second unknown.

ILLUSTRATION: Find the values of x and y in the simultaneous equations

$$3x - 2y = 30$$
$$4x + 4y = 20$$

Multiply 1st by 4 $\qquad 12x - 8y = 120$

Multiply 2nd by 3 $\qquad 12x + 12y = 60$

$$-20y = 60$$
$$y = -3$$

Substituting value of y in first equation

$$3x + 6 = 30$$
$$3x = 24$$
$$x = 8$$

Substituting the values found, $x = 8$, $y = -3$ in the other original equation to check results,

$$4 \times 8 + 4(-3) = 20$$
$$32 - 12 = 20$$
$$20 = 20$$

Another method is *elimination by comparison*. From each equation obtain the value of one of the unknown quantities in terms of the other. Form an equation from these equal values of the same unknown quantity and reduce and solve in the regular manner and substitute the value found in one of the original equations to determine the value of the second unknown.

ILLUSTRATION: Find the values of x and y in the simultaneous equations

$$2x + 3y = 7 \quad (1)$$
$$4x - 5y = 3 \quad (2)$$

From (1) $\qquad x = \dfrac{7 - 3y}{2}$

From (2) $\qquad x = \dfrac{3 + 5y}{4}$

Equating these, $\dfrac{7 - 3y}{2} = \dfrac{3 + 5y}{4}$

Multiplying by 4, $14 - 6y = 3 + 5y$

$$11y = 11$$
$$y = 1$$

Substituting in one of the original equations:

$$2x + (3 \times 1) = 7$$
$$2x = 4$$
$$x = 2$$

The answer is, $x = 2$, $y = 1$, and may be checked by substituting these values in the two original equations.

A third method is *elimination by substitution*.   From one of the original equations obtain the value of one of the unknown quantities in terms of the other.   Substitute this value of this unknown quantity for it in the other equation and reduce the resulting equations.

ILLUSTRATION: Find the values of x and y in the simultaneous equations

$$4x - 6y = 28 \quad (1)$$
$$2x - 8y = 24 \quad (2)$$

From (1) $x = \dfrac{28 + 6y}{4}$

Substituting this value in (2)

$$2 \times \dfrac{28 + 6y}{4} - 8y = 24$$
$$14 + 3y - 8y = 24$$
$$-5y = 10$$
$$y = -2$$

Substituting this value in (1)

$$4x + 12 = 28$$
$$4x = 16$$
$$x = 4$$

The answer is, $x = 4$, $y = -2$.

The solution of equations containing three unknowns requires three simultaneous equations. Essentially the same methods may be applied as for the solution of two simultaneous equations. One of the unknown quantities must be eliminated between two pairs of the equations, then a second between the two resulting equations.

Quadratic Equations.—Equations containing the square or the second power of the unknown quantity but no higher power are called *quadratic equations*. A *pure quadratic* contains only the square; an *affected* or *complete quadratic* contains both the square and the first power. The equation $25x^2 + 18 = 3x^2 - 8$ is a pure quadratic; $50x^2 - 5x = 125$ is a complete or affected quadratic.

Solution of Pure Quadratic Equations.—To solve a pure quadratic collect the unknown quantities on the left side and the known quantities on the right side; divide by the coefficient of the unknown quantity and extract the square root of each side of the resulting equation. Examples:

Solve $\quad 6x^2 - 2x^2 = \quad 64$

$$4x^2 = \quad 64 \quad \text{(Combining terms)}$$
$$x^2 = \quad 16 \quad \text{(Dividing by coefficient)}$$
$$x = \pm\ 4 \quad \text{(Extracting square root)}$$

Solve $\quad 5x^2 - 55 = 0$

$$5x^2 = 55$$
$$x^2 = 11$$
$$x = \pm\ \sqrt{11}$$

The root which is indicated, but can only be found approximately, is called a *surd*.

Solve $\quad 8x^2 + 64 = 0$

$$8x^2 = -\ 64$$
$$x^2 = -\ 8$$
$$x = \sqrt{-8}$$

The square root of a negative number cannot be found **even** approximately and the root which is indicated is called *imaginary*.

Solution of Affected or Complete Quadratics.—Several methods of solution are applicable to complete quadratics. We shall consider first equations which may be solved by *factoring*. All of the terms are first transposed to the left-hand side leaving zero on the right and we obtain an equation of this type.

$$x^2 + 8x + 15 = 0$$

By the process previously described, the middle term may be separated into the sum of two terms. We then have

$$x^2 + 3x + 5x + 15 = 0$$

then grouping $(x^2 + 3x) + (5x + 15) = 0$

and factoring $x(x + 3) + 5(x + 3) = 0$

$$(x + 5)(x + 3) = 0$$

Any number multiplied by zero is equal to zero. Then in order for the product of these two factors to equal zero, either $(x + 5)$ or $(x + 3)$ or both must equal zero.

If $x + 5 = 0$, then $x = -5$

If $x + 3 = 0$, then $x = -3$

If we substitute $x = -5$ into the original equation we obtain

$$(-5)^2 + 8(-5) + 15 = 0$$

$$25 - 40 + 15 \quad\quad = 0$$

Similarly, if we substitute $x = -3$,

$$(-3)^2 + 8(-3) + 15 = 0$$

$$9 - 24 + 15 \quad\quad = 0$$

Thus, there are *two* solutions to the equation since either $x = -5$ or $x = -3$ satisfy it.

All complete quadratics may be solved by the method of *completing the square*. First transpose all of the terms containing the unknown to the left-hand side of the equation and the known quantities to the right-hand side. Arrange the unknown quantities in the order of their exponents and change signs, if necessary, so that the term containing the square will be positive. Divide all terms by the coefficient of the square of the unknown quantity. Complete the square by adding to both sides of the equation the square of half the coefficient of the first power of the unknown. The left-hand side will then be a perfect square. Extract the square root of both sides of the equation and solve the resulting simple equation. Examples:

Solve $2x^2 + 4x - 70 = 0$.

$$2x^2 + 4x = 70 \qquad \text{(Transposition)}$$
$$x^2 + 2x = 35 \qquad \text{(Dividing by coefficient of } x^2\text{)}$$
$$x^2 + 2x + 1 = 35 + 1 \qquad \text{(Adding square of } \tfrac{1}{2} \text{ coefficient of } x\text{)}$$
$$(x + 1)^2 = 36$$
$$x + 1 = \pm 6 \qquad \text{(Extracting square root)}$$
$$x = -1 \pm 6$$
$$\left. \begin{array}{l} x = -7 \\ \text{or } x = +5 \end{array} \right\} \quad \text{(Ans.)}$$

Here again we find that the equation has two solutions. Both solutions may be correct. Moreover, in some practical problems one answer may be correct and the other inconsistent with the conditions of the problem.

Example: A park which is in the form of a right triangle has one side twenty-five feet longer than the other. If the area ($\frac{1}{2}$[base $\times$ height]) is 625 square feet, find the length of the sides.

Let $\quad x = $ shorter side

$x + 25 = $ longer side

$$\frac{x(x + 25)}{2} = 625$$
$$x^2 + 25x = 1250$$
$$x^2 + 25x - 1250 = 0$$
$$(x + 50)(x - 25) = 0$$

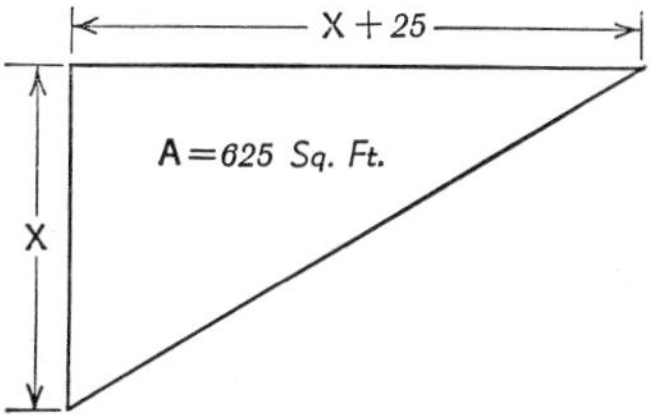

Fig. 1.

$$x = -50 \text{ ft.}, \; x + 25 = -25 \text{ ft.}$$
$$x = 25 \text{ ft.}, \; x + 25 = 50 \text{ ft.}$$

The -50 and -25 do not satisfy the conditions of the problem and therefore should be neglected.

A third method of solution is by the use of the *quadratic formula*. The terms of a complete quadratic equation when collected on one side of the equality sign constitute a trinomial consisting of one term with the unknown to the second power, one term with the unknown to the first power, and the third term of known quantities. This may be written in the general form

$$ax^2 + bx + c = 0$$

The coefficients a and b and the term c may be numerical or literal numbers, positive or negative, monomials or polynomials. The roots of this equation by the quadratic formula are

$$x = \frac{-b + \sqrt{b^2 - 4ac}}{2a}$$

$$x = \frac{-b - \sqrt{b^2 - 4ac}}{2a}$$

Examples:

Solve $2x^2 + 3x + 1 = 0$.

$$x = \frac{-3 + \sqrt{(3)^2 - 4 \times 2 \times 1}}{2 \times 2}$$

$$= \frac{-3 + \sqrt{9 - 8}}{4} = \frac{-3 + 1}{4} = -\frac{1}{2}$$

$$x = \frac{-3 - \sqrt{(3)^2 - 4 \times 2 \times 1}}{2 \times 2}$$

$$= \frac{-3 - \sqrt{9 - 8}}{4} = \frac{-3 - 1}{4} = -1$$

The roots of the equation are, $x = -\frac{1}{2}$, $x = -1$, both real and rational numbers.

Solve $3x^2 + 5x - 4 = 0$.

$$x = \frac{-5 + \sqrt{25 + 48}}{6} = \frac{-5 + \sqrt{73}}{6}$$

$$= \frac{-5 + 8.544+}{6} = \frac{3.544}{6} = .590+$$

$$x = \frac{-5 - \sqrt{25 + 48}}{6} = \frac{-5 - \sqrt{73}}{6}$$

$$= \frac{-5 - 8.544+}{6} = \frac{-13.544}{6} = -2.257+$$

In this example the roots are real, but since $(b^2 - 4ac)$ is not a perfect square, they are not rational, that is, they terminate in never-ending decimals.

Solve $-4x^2 + 4x - 8 = 0$.

$$-x^2 + x - 2 = 0$$
$$x^2 - x + 2 = 0$$
$$x = \frac{1 + \sqrt{1 - 8}}{2} = \frac{1 + \sqrt{-7}}{2}$$
$$x = \frac{1 - \sqrt{1 - 8}}{2} = \frac{1 - \sqrt{-7}}{2}$$

In this example $(b^2 - 4ac)$ is less than zero (negative) and since the square root of a negative number is an imaginary, the roots of the equation are imaginary.

Reference.—ALGEBRA FOR THE PRACTICAL MAN, by Mr. J. E. Thompson (D. Van Nostrand Company), covers the subjects dealt with above, as well as many others, with a simplicity particularly suited for home study.

IV

Geometry

Geometry is the science which treats of the properties of lines, angles, surfaces, and solids. It is based on a number of theorems and constructions for which formal proofs have been developed.

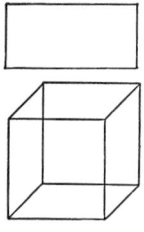

FIG. 1.

These proofs are of little concern to the practical man. Hence, this section will present the most important definitions and conclusions without proofs, and then pass on to mensuration or the measurement of lines, areas and volumes, which is of great practical value to everyone, and then to geometrical construction which is very useful to the man in the shop and at the drafting table.

Definitions.—A *point* indicates position but has no magnitude, nor dimensions; neither length, breadth, nor thickness.

A *line* has length but no breadth or thickness. It may be

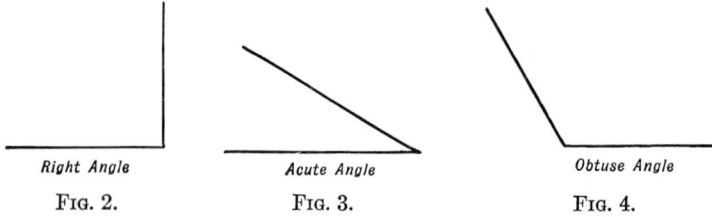

Right Angle	Acute Angle	Obtuse Angle
FIG. 2.	FIG. 3.	FIG. 4.

straight, curved, or mixed. A straight line is the shortest distance between two points. A curve continually changes its direction between its extreme points. When a line is mentioned simply, it means a straight line.

A *surface* has length and breadth but no thickness. It may be either plane or curved.

A *solid* or body is a figure of three *dimensions*, namely, length, breadth, and depth or thickness.

An *angle* is formed by the intersection of two lines. The point of intersection is called the vertex.

A *right* angle is formed when one of the lines is perpendicular to or makes an angle of 90 degrees with the other line. An *acute* angle is less than a right angle. An *obtuse* angle is greater than a right angle. Acute and obtuse angles are also said to be *oblique*.

A *plane* is that with which a straight line may every way coincide, or, if the line touches the plane at two points, it will touch it at every point.

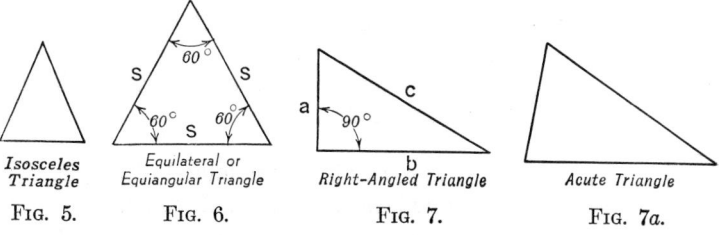

Isosceles Triangle	Equilateral or Equiangular Triangle	Right-Angled Triangle	Acute Triangle
Fig. 5.	Fig. 6.	Fig. 7.	Fig. 7a.

Plane figures are bounded either by straight lines or curves. Plane figures that are bounded by straight lines have names according to their number of sides or of their angles, for they have as many sides as angles, the least number being three.

A plane figure bounded by three sides is called a *triangle*.

An *equilateral triangle* has three sides "S" equal. Its three angles are also equal and each has a value of 60 degrees.

An *isosceles triangle* has two equal sides, called its legs. The angles between each leg of the isosceles triangle and the third side are called the base angles and are equal.

A *scalene triangle* has no sides equal.

A *right-angled triangle* has two sides perpendicular to each other making the angle between them a right angle or 90 degrees. The side opposite the right angle is called the *hypotenuse*, the other two sides are called the *legs*. The square of the length of the hypotenuse is equal to the sum of the squares of the lengths of the legs, or in Fig. 7, $c^2 = a^2 + b^2$.

All triangles other than right-angled triangles are *oblique-angled* and are *obtuse-angled* if they have one obtuse angle and *acute-angled* if all three angles are acute.

A figure of four sides and angles is called a *quadrangle* or *quadrilateral*.

A *parallelogram* is a quadrilateral which has both of its pairs of opposite sides parallel, and it takes the following particular names: rectangle, square, rhomboid, and rhombus.

FIG. 9.   A *rectangle* is a parallelogram, having right angles.

FIG. 10.   A *square* is an equilateral rectangle, having its length and breadth equal.

FIG. 11.   A *rhomboid* is an oblique-angled parallelogram.

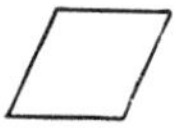

FIG. 12.   A *rhombus* is an equilateral rhomboid, having all its sides equal but its angles oblique.

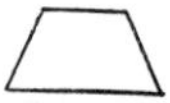

FIG. 13.   A *trapezoid* is a quadrilateral which has only one pair of opposite sides parallel.

FIG. 14.   A *trapezium* is a quadrilateral which has no opposite sides parallel.

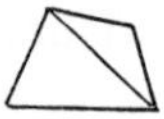

FIG. 15.   A *diagonal* is a line joining any two opposite angles of a quadrilateral.

Plane figures having more than four sides are, in general, called *polygons* and they receive their names according to their number of sides or angles. Thus, a *pentagon* is a polygon of five sides; a *hexagon* of six sides; a *heptagon*, seven; an *octagon*, eight; a *nonagon*, nine; a *decagon*, ten, etc. A *regular polygon* has all its sides equal and all its angles equal.

A *circle* is a plane figure bounded by a curved line called the *circumference* or periphery which is everywhere equidistant from a certain point within called its *center* (point c in Fig. 16).

The *radius* of a circle is a line drawn from the center to the circumference (cf in Fig. 16).

The *diameter* of a circle is a line drawn through the center and terminating at the circumference on both sides (ecd in Fig. 16). It is equal to twice the radius.

An *arc* of a circle is any part of the circumference (as ab or bd in Fig. 16).

A *chord* is a straight line joining the extremities of an arc (ab in Fig. 16).

A *segment* is any part of a circle bounded by an arc and its chord (as shaded area between a and b, Fig. 16).

A *sector* is any part of a circle bounded by an arc and two radii drawn to its extremities (as shaded area between cd, cf, and fd, Fig. 16).

A *semicircle* is half the circle, or a segment cut off by a diameter. The half circumference is sometimes called the semicircumference.

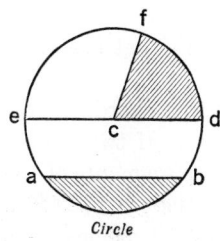

Circle

FIG. 16.

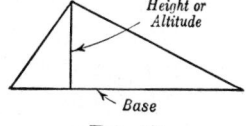

Height or Altitude

Base

FIG. 17.

The *height* or *altitude* of a figure is a perpendicular let fall from an angle or its vertex, to the opposite side, called the *base*.

Geometrical Propositions.—A great many of the practical

problems in this book are based upon the following geometrical propositions:

If a triangle is equilateral, it is equiangular, and vice versa.

If a straight line from the vertex of an isosceles triangle bisects the base it bisects the vertical angle and is perpendicular to the base.

The sum of the three angles in a triangle always equals 180 degrees.

If two triangles are mutually equiangular, they are similar and their corresponding sides are proportional.

In every triangle, that angle is greater which is opposite a longer side. In every triangle, that side is greater which is opposite a greater angle.

In every triangle, the sum of the lengths of two sides is always greater than the length of the third side.

In a right triangle the square on the hypotenuse is equal to the sum of the squares on the other two sides.

The areas of triangles having equal base and equal height are equal.

If a triangle is inscribed in a semicircle, it is right-angled.

In a quadrilateral, the sum of the interior angles equals four right angles or 360 degrees.

In a parallelogram, the opposite sides are equal; the opposite angles are equal; it is bisected by its diagonal and its diagonals bisect each other.

The areas of two parallelograms which have equal base and height are equal.

If the diameter of a circle is at right angles to a chord, then it bisects or divides the chord into two equal parts. If two chords intersect each other in a circle, the rectangle of the segments of the one equals the rectangle of the segments of the other.

If an angle is formed by a tangent of any chord, it is measured by one-half of the arc intercepted by the chord; that is, it is equal to half the angle at the center subtended by the chord.

If two circles are tangent to each other, then the straight line which passes through the centers of the two circles must also pass through the point of tangency.

The length of circular arcs of the same circle are proportional to the corresponding angles at the center.

The circumference of two circles are proportional to their radii.

The areas of two circles are proportional to the squares of their radii.

Mensuration.—This subject deals with the finding of lengths, areas, and volumes, of lines, surfaces, and solids, respectively. We need a few more definitions of solids before proceeding.

A *prism* is a solid of which the sides are parallelograms and the ends equal, similar, and parallel plane figures. The figure of the ends gives the name to the prism; if the ends are triangular, the prism is triangular, etc. If the sides and ends of a prism be all equal squares, the prism is called a *cube*; and if the base or ends be parallelograms, the prism is called a *parallelepiped*. The *cylinder* is a round prism having circular ends. A *right prism* has its axis perpendicular to the base.

The *pyramid* has any plane figure for its base, and its sides triangles of which all the vertices meet in a point at the top called the *vertex* of the pyramid. A *right pyramid* has its axis perpendicular to the base.

A *cone* is a solid figure having a circle for its base and terminated in a vertex.

A *sphere* or globe is a solid bounded by one continued curved surface, every point of which is equally distant from a point within the sphere called the *center*.

The *axis* of a solid is a straight line drawn through the solid, from the middle of one end to the middle of the opposite.

The *height* of a solid is a line drawn from the vertex perpendicular to the base or the plane on which the base rests.

The *segment* of a solid is a part cut off by a plane, parallel to the base; and the *frustum* is the part remaining after the segment is cut off.

Properties of the Circle.—The *circumference* of a circle is divided into 360 equal parts, called *degrees*; each degree into 60 *minutes*, each minute into 60 *seconds*. Hence a semicircle contains 180 degrees, and a quarter of a circle, or a *quadrant*, 90 degrees.

The ratio of the length of the circumference of a circle to its diameter is a constant and has the value, 3.14159265+. For nearly all practical computations, this number is shortened to 3.1416. This ratio is called *pi* and is represented by the Greek letter π. If we let D represent the diameter of a circle and r the radius, then we may write

or,
$$\text{circumference} = \pi \times D = 3.1416D$$
$$\text{circumference} = \pi \times 2r = 2 \times 3.1416r$$

ILLUSTRATION: What is the circumference of a circle whose radius is 6 inches?

$$\text{circumference} = \pi \times 2r = 2 \times 6 \times 3.1416 = 37.7 \text{ in. (Ans.)}$$

The *area* of a circle is equal to $\frac{1}{4}\pi D^2$ or πr^2.

ILLUSTRATION: What is the area of a circle whose diameter is 5 inches?

$$\text{area} = \frac{1}{4}\pi D^2 = \frac{1}{4} \times 3.1416 \times 25 = 19.6 \text{ sq.in.}$$

ILLUSTRATION: What is the area of a circle whose radius is $\frac{1}{8}$ inch?

$$\text{area} = \pi \times r^2 = 3.1416 \times \tfrac{1}{64} = 0.049 \text{ sq.in. (Ans.)}$$

To find the *area of a sector* when (I) the length of the arc is known, and (II) when the angle of the sector is known:

CASE I. Multiply the length of the arc by $\frac{1}{2}$ the radius. Then, when A = area, l = length of arc, and r = radius,

$$A = \frac{rl}{2}.$$

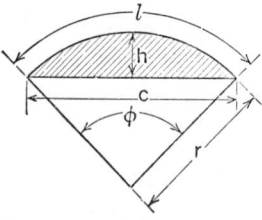

FIG. 18.

ILLUSTRATION: The length of arc of a sector is 40 feet on a circle whose diameter is 300 feet. What is the area of the sector?

$$A = \frac{rl}{2} = \frac{150 \times 40}{2} = 3000 \text{ sq.ft. (Ans.)}$$

CASE II. The area of a sector of a circle is to the area of the whole circle as the number of degrees in the arc of the sector is to 360 degrees. Then if ϕ = angle of sector, and area of circle = πr^2,

$$\frac{A}{\pi r^2} = \frac{\phi}{360}, \quad A = \frac{\phi}{360}\pi r^2$$

ILLUSTRATION: What is the area of a 60-degree sector of a circle whose diameter is 12 inches?

$$A = \frac{\phi}{360}\pi r^2 = \frac{6\cancel{0}}{3\cancel{6}\cancel{0}} \times 3.1416 \times \cancel{6} \times \cancel{6} =$$

$$6 \times 3.1416 = 18.85 \text{ sq. in. (Ans.)}$$

The *area of a segment* of a circle in terms of its height, h, length of arc, l, length of chord, c, and radius of circle, r, is

$$A = \tfrac{1}{2}[r(l - c) + hc]$$

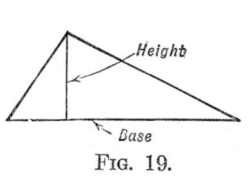

FIG. 19.

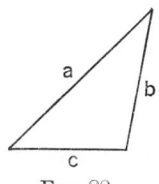

FIG. 20.

Properties of Triangles.—The *area of any triangle* is one-half the product of the base and the height

$$\text{Area} = \tfrac{1}{2}(\text{base} \times \text{height})$$

ILLUSTRATION: What is the area of a triangular lot whose base is 40 feet and whose height is 48 feet?

$$A = \tfrac{1}{2}(b \times h) = \tfrac{1}{2}(40 \times 48) = 960 \text{ sq.ft. (Ans.)}$$

The *area of a right triangle* is one-half of the product of the two legs.

The *area of any triangle* whose three sides are known can be found by subtracting from one-half the sum of the three sides each side severally, then extracting the square root of the product of the three remainders and the half-sum of the sides. Thus when

$$s = \tfrac{1}{2}(a + b + c)$$
$$\text{Area} = \sqrt{s(s - a)(s - b)(s - c)}$$

ILLUSTRATION: What is the area of a triangle whose sides are 5, 7, and 8 inches long?

$$s = \frac{a + b + c}{2} = \frac{5 + 7 + 8}{2} = 10$$

$$A = \sqrt{10(10 - 5)(10 - 7)(10 - 8)}$$
$$= \sqrt{10 \times 5 \times 3 \times 2} = \sqrt{300} = 17.32 \text{ sq.in. (Ans.)}$$

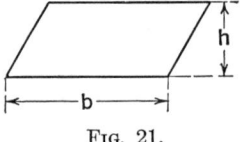

FIG. 21.

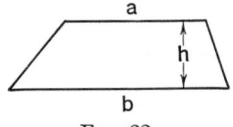

FIG. 22.

Properties of Quadrilaterals.—The *area of any parallelogram* is the product of the altitude and the base. $A = b \times h$.

ILLUSTRATION: What is the area of a rhomboid whose base is 8 inches and whose height is $3\tfrac{1}{2}$ inches?

$$A = b \times h = 8 \times 3\tfrac{1}{2} = 28 \text{ sq.in. (Ans.)}$$

ILLUSTRATION: What is the area of a square whose side is $4\tfrac{1}{4}$ inches?

$$A = b \times h = 4\tfrac{1}{4} \times 4\tfrac{1}{4} = 18.0625 \text{ sq.in. (Ans.)}$$

The *area of a trapezoid* is the product of one-half the sum of the two parallel sides and the height. $A = \tfrac{1}{2}(a + b) \times h.$

The *area of a trapezium* can only be found by drawing the trapezium to scale and then drawing a diagonal the length of which is measured by the same scale and then solving for the separate areas of the two resulting triangles by

$$A = \sqrt{S(S - a)(S - b)(S - c)}$$

Areas of Regular Polygons.—The areas of regular polygons may readily be calculated with the use of Table 1. The area is equal to the product of the square of the length of one side and the corresponding factor in the third column of the table.

TABLE 1

No. of Sides	Name of Polygon	Factor (F)
3	Triangle	0.4330127
4	Tetragon	1.0000000
5	Pentagon	1.7204774
6	Hexagon	2.5980762
7	Heptagon	3.6339124
8	Octagon	4.8284271
9	Nonagon	6.1818242
10	Decagon	7.6942088
11	Undecagon	9.3656405
12	Dodecagon	11.1961524

ILLUSTRATION: What is the area of a regular octagon the length of whose side is 6 inches?

$$A = s^2 \times F = 6 \times 6 \times 4.828 = 173.81 \text{ sq. in. (Ans.)}$$

Properties of Prisms and Cylinders.—The *volume of any prism* or *cylinder* is the product of the area of the base and the altitude.

The *volume of a circular cylinder* is then, $V = \pi r^2 h$, when h is the altitude and r the radius of the base.

ILLUSTRATION: What is the volume of an oil drum 20 inches in diameter and 30 inches high?

$$V = \pi r^2 h = \pi 10^2 \times 30 = 3000 \times 3.1418 = 9,425 \text{ cu. in. (Ans.)}$$

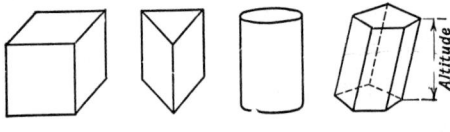

FIG. 23.

ILLUSTRATION: What is the volume of a prism whose height is 12 inches and whose base is a right triangle with legs 5 inches and 8 inches long?

Area of base = $\frac{1}{2} \times 5 \times 8 = 20$ sq. in.   Volume = $A \times h = 20 \times 12 = 240$ cu. in. (Ans.)

The *surface area* of a right prism or cylinder is the product of the height and the perimeter of a base plus the area of the two bases.   The surface area of a cylinder is then, $A = 2\pi rh + 2\pi r^2 = 2\pi r(h + r)$ or $\pi Dh + \frac{1}{2}\pi D^2 = \pi D\left(h + \frac{D}{2}\right)$.

ILLUSTRATION: What is the surface area of pole 12 inches in diameter and 9 feet long?

$$A = \pi D\left(h + \frac{D}{2}\right) = \pi \times 1 \times (9 + \tfrac{1}{2})$$
$$= 9.5 \times 3.1416 = 29.8 \text{ sq. ft. (Ans.)}$$

ILLUSTRATION: What is the surface area of a hexagonal bar 1 inch on the side and 8 inches long?

Area of end $= S^2 \times F = 1^2 \times 2.598 = 2.6$ sq. in.   Area of 2 ends $= 5.2$ sq in.

Perimeter $= 6 \times 1 = 6$ in. Area of sides $= 6 \times 8 = 48$ sq in. Total area $= 48 + 5.2 = 53.2$ sq. in. (Ans.)

Properties of the Sphere.—The volume of a sphere is $\frac{4}{3}\pi r^3$ or $\frac{1}{6}\pi D^3$.

ILLUSTRATION: What are the cubical contents of a spherical balloon 50 feet in diameter?

$$V = \frac{1}{6}\pi D^3 = \frac{125,000}{6} \times 3.1416$$
$$= 65,450 \text{ cu.ft. (Ans.)}$$

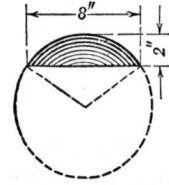

Segment of a Sphere

FIG. 24.

The surface of a sphere is πD^2 or $4\pi r^2$.

ILLUSTRATION: What is the area of a spherical water tank 22 feet in diameter?

$$A = \pi D^2 = 3.1416 \times 22 \times 22 = 1521 \text{ sq.ft. (Ans.)}$$

The volume of a segment of a sphere is three times the square of the radius of the base plus the square of the height, this sum multiplied by the height and by 0.5236. If r is the radius of the base and h is the height, then volume $= 0.5236h(3r^2 + h^2)$.

ILLUSTRATION: What is the volume of the segment shown in Fig. 24?

Here, $r = 4$ in., $h = 2$ in.   Then,

$$V = 0.5236h(3r^2 + h^2) = 0.5236 \times 2(3 \times 16 + 4)$$
$$= 54.45 \text{ cu.in. (Ans.)}$$

Properties of Pyramids and Frustums of Pyramids.—The volume of any pyramid is one-third the product of the area of the base and the altitude. $V = \frac{1}{3}Ah$.

ILLUSTRATION: What is the volume of a pyramid whose base is a square, 8 feet on a side, and whose altitude is 4 feet?

$$V = \frac{1}{3} \times Ah = \frac{1}{3} \times 8 \times 8 \times 4 = 85.33 \text{ cu. ft. (Ans.)}$$

The slanted surface of a regular pyramid is one-half the product of the perimeter of the base and the slant height of a side (not the slant height of an edge).

The total surface area of a pyramid is the sum of the slanted surface and the area of the base.

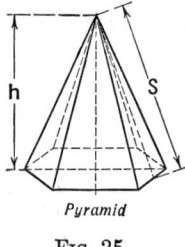

Pyramid

FIG. 25.

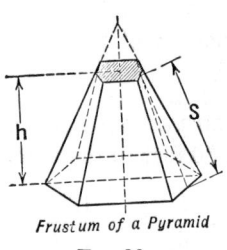

Frustum of a Pyramid

FIG. 26.

The volume of a frustum of a pyramid when a is the area of the small end, A the area of the large end, and h the perpendicular distance between the ends is, $V = \dfrac{h}{3}(a + A + \sqrt{Aa})$.

The area of the slanted surface of a frustum of a pyramid is the

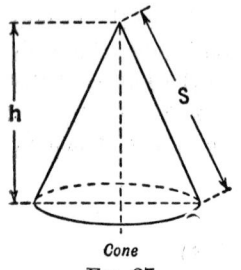

Cone

FIG. 27.

sum of the perimeter of the small end and the perimeter of the large end multiplied by the slant height and divided by two.

Properties of Cones and Frustums of Cones.—The volume of a cone is one-third the product of the area of the base and the altitude. Then, $V = \frac{1}{3}\pi r^2 h$ or $\frac{1}{12}\pi D^2 h$.

ILLUSTRATION: What is the volume of a conical pile of coal 30 feet in diameter and 14 feet high?

$V = \frac{1}{12}\pi D^2 h = \frac{1}{12} \times 3.1416 \times 30^2 \times 14 = 3299$ cu.ft. (Ans.)

The area of the curved surface of a cone is one-half the product of the circumference and the slant height. If S = slant height, then, $A = \frac{1}{2}\pi D S$.

The volume of a frustum of a cone when R is the radius of the large end, r the radius of the small end, and h the perpendicular distance between the ends is, $V = (R^2 + r^2 + Rr)\pi\dfrac{h}{3}$.

The area of the curved surface of the frustum of a cone when R, r, and h have the same significance as above, is,

$$\text{Curved area} = (R + r)\pi\sqrt{(R - r)^2 + h^2}$$

Conic Sections.—A cone has already been defined as a solid figure having a circle for its base and terminated in a vertex. Conic sections are the figures made by a plane cutting a cone. Depending on the different positions of the cutting plane, there arise five different figures or sections, namely, a triangle, a circle,

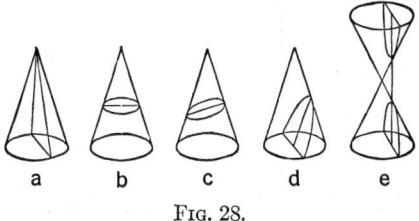

a   b   c   d   e

Fig. 28.

an ellipse, an hyperbola, and a parabola, only the last three of which are usually called *conic sections*.

If the plane passes through the vertex and any part of the base, the section will be a *triangle* as in Fig. 28a. When the plane cuts the cone parallel to the base, the section will be a *circle* as in Fig. 28b. When the cutting plane makes an angle with the base of less inclination than the side of the cone, as in Fig. 28c, the section will be an *ellipse*. When the cutting plane and the side of the cone make equal angles with the base, the section will be a *parabola* as in Fig. 28d. The section is a *hyperbola* when the cutting plane makes a greater angle with the base than the side of the cone makes, Fig. 28e. If the sides of the cone be continued through the vertex, forming an opposite equal cone, and the plane also continued to cut the opposite cone, this latter section will be the opposite hyperbola to the former.

Conic sections have considerable practical usefulness. Reinforced concrete arch bridges are often elliptical, parabolic, or even hyperbolic in section. Where curves with large diameters are needed such as for the cross-section of a pavement, the camber of a bridge, or the upper chord of a truss bridge, a parabolic curve is usually used instead of a circular curve because it is more readily computed and laid out. If a source of rays is placed at a certain point called a focus within a parabolic surface, these rays will be reflected in parallel lines. This principle is made use of in heat and light reflectors.

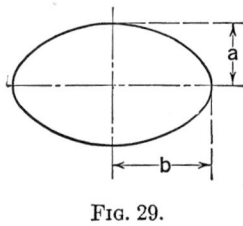

FIG. 29.

The subject of conic sections belongs to the study of analytical geometry which cannot be covered in this book.

Circumference and Area of an Ellipse.—The approximate circumference of an ellipse may be found by the following equation when a is half the smallest diameter and b half the largest diameter:

$$\text{Circumference} = \pi \sqrt{2(a^2 + b^2)}$$

The area of an ellipse is given by

$$\text{Area} = \pi \times a \times b$$

Geometrical Drawing.—Euclidean geometry is based on constructions using as the only tools a pencil, a pair of compasses, and a straight-edge or ruler. These constructions are simple and very useful. For instance, a building foreman may be confronted with the problem of laying out a line perpendicular to another line and of lengths too great for the effective use of the carpenter's square. Then, knowing the principles of geometrical construction and using a string for compasses, a sight-line between two nails, or a board, for a straight-edge, and a pencil, he can erect the perpendicular just as readily as it can be drawn on paper.

The following are the more important constructions:

To divide a straight line into a given number of equal parts. (See Fig. 30.)

Given line *a b*, which is to be divided into a given number of equal parts. Draw the line *b c*, of indefinite length, and point off from *b* the required number of equal parts, as *h, g, f, e, d, c′*; join *c′* and *a*, and draw the other lines parallel to *c′ a*.

To erect a perpendicular at a given point on a straight line. (See Fig. 31.)

Given line *a b* and the point *x*. The required perpendicular is *x y*.

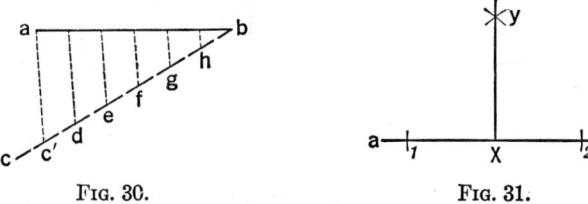

| Fig. 30. | Fig. 31. |

SOLUTION:

With *x* as center and any radius, as *x* 1, cut the line *a b* at 1 and 2. With 1 and 2 as centers and with a radius somewhat greater than 1 to *x*, describe arcs intersecting each other at *y*. Draw *x y*. This will be the required perpendicular.

From a given point without a straight line to draw a perpendicular to the line. (See Fig. 32.)

Given line *a b* and the point *c*. The required perpendicular is *x*.

SOLUTION:

With the point *c* as center and any radius as *c* 1, strike the arc 1 to 2. With 1 and 2 as centers and any suitable radius, describe arcs intersecting each other at *n*, lay the straight-edge through points *n* and *c* and draw the perpendicular *x*.

To erect a perpendicular at the extremity of a straight line. (See Fig. 33.)

Given line $a\,b$.   The required perpendicular is x.

SOLUTION:

From any point, as c, with radius as $a\,c$, draw the circle. From point of intersection, n, through center, c, draw the diameter $n\,p$. From the point a, through the point of intersection at p, draw the perpendicular x.

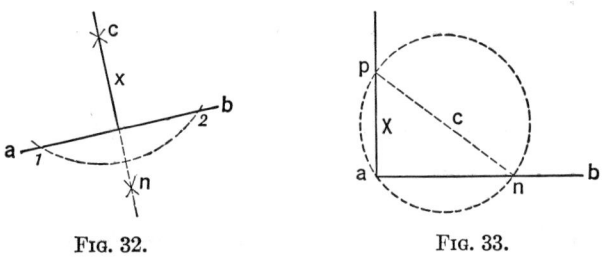

FIG. 32.   FIG. 33.

The correctness of this construction is founded on the principle that inside a half circle no other angle but an angle of 90° can simultaneously touch three points in the circumference when two of these points are in the point of intersection with the diameter and the circumference and the third one anywhere on the circumference of the half circle.   The pattern maker is making

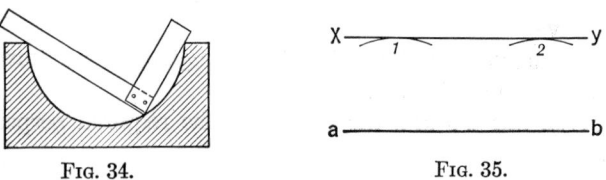

FIG. 34.   FIG. 35.

practical use of this geometrical principle, when he by a common carpenter's square is trying the correctness of a semi-circular core box, as shown in Fig. 34.

Draw a line parallel to a given line.   (See Fig. 35.)

Given line $a\,b$.   The required line $x\,y$.

SOLUTION:

Describe with the compass from the line $a\ b$, the arcs 1 and 2; draw line $x\ y$, touching these arcs.

To divide a given angle into two equal angles. (Fig. 36).

The given angle, $a\ b\ c$, is divided by the line $b\ d$.

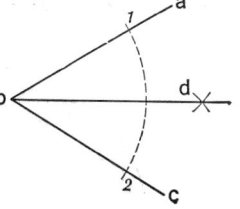

SOLUTION:

With b as center and any radius, as $b\ 1$, describe the arc 1 to 2. With 1 and

FIG. 36.

2 as centers and any suitable radius, describe arcs cutting each other at d. Draw line $b\ d$, which will divide the angle into two equal parts.

To draw an angle equal to a given angle. (Fig. 37).

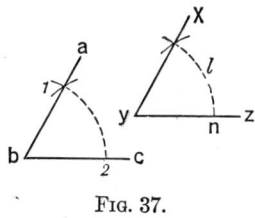

FIG. 37.

Given angle $a\ b\ c$. Construct angle $x\ y\ z$.

With b as center and any radius, as $b\ 1$, describe the arc 1 to 2, using y as center and without altering the compass describe the arc 1, intersecting $y\ z$. Measuring the distance from 2 to 1 on the given angle, transfer this measure to the arc 1, through the point of intersection. Draw the line $y\ x$, and this angle will be equal to the first angle.

NOTE.—Angles are usually measured by a tool called a protractor, looking somewhat like Fig. 38 or 39, usually made from metal, and supplied by dealers in drafting instruments. A protractor may also be constructed on paper and used for measuring angles, but it should then always be made on as large a scale as convenient.

To draw a protractor with a division of 5°. (See Fig. 39.)

Construct an angle of exactly 90°, divide the arc into nine

equal parts, then each part is 10°; divide each part into two equal parts and each is 5°.

Prove that the sum of the three angles in a triangle consists of 180°. (See Fig. 40.)

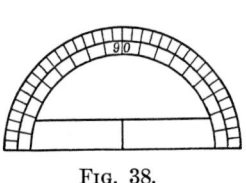

Fig. 38.

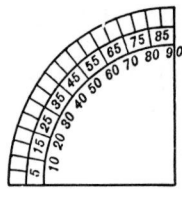

Fig. 39.

SOLUTION:

In the triangle $a\ b\ c$, extend the base line to i. Draw the line $o\ p$, parallel to the side $a\ b$, thereby the angle g will be equal to the angle d, and the angle h must be equal to the angle c. The angle f is one angle in the triangle and $f + g + h = 180°$.

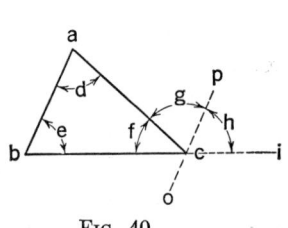

Fig. 40.

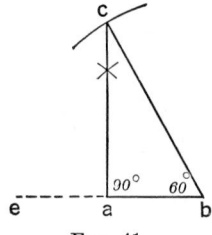

Fig. 41.

To draw on a given base line a triangle having angles 90°, 30°, and 60°. (See Fig. 41.)

Given line $a\ b$, required triangle is a, c, b.

SOLUTION:

Extend the line $a\ b$ to twice its length, to the point e. With e and b as centers strike arcs intersecting each other and erect the perpendicular $a\ c$. With b as center and a radius be draw an arc intersecting ac at c. Connect b and c. This will complete the triangle.

To draw a square inside a given circle.   (See Fig. 42.)

SOLUTION:

Draw the line *a b* through the center of the circle. From points of intersection at *a* and *b*, describe with any suitable radius arcs intersecting at *n* and *m*. Draw through the points the line *c d*. Connect the points of intersection on the circle, and the required square is constructed.

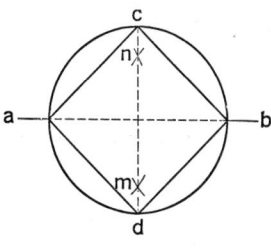

FIG. 42.

To draw a square outside a given circle.   (See Fig. 43.)

SOLUTION:

Draw lines *a b* and *c d*, and from points of intersection at *b* and *c*, describe half circles; their points of intersection determine the sides of the square.

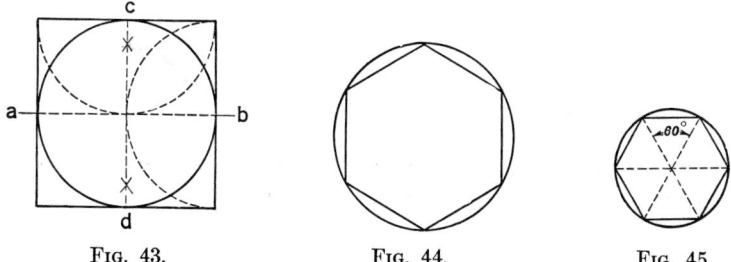

FIG. 43.   FIG. 44.   FIG. 45.

To draw a hexagon within a given circle.   (See Fig. 44.)

Apply the radius as a chord successively about the circle; the resulting figure will be a hexagon.

To inscribe in a circle a regular polygon of any given number of sides.

SOLUTION:

Divide 360 by the number of sides, and the quotient is the number of degrees, minutes, and seconds contained in the center angle of a triangle, of which one side will make one of the sides in the polygon. For instance, draw a hexagon by this method. (See Fig. 45.)

$$\frac{360}{6} = 60°$$

To find the center in a given circle.   (See Fig. 46.)

SOLUTION:

Draw anywhere on the circumference of the circle two chords at approximately right angles to each other; bisect these by the

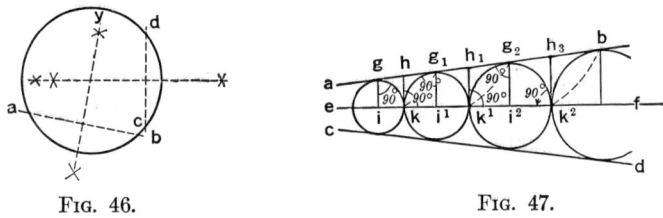

FIG. 46.   FIG. 47.

perpendiculars x and y, and their point of intersection is the center of the circle.

To draw any number of circles between two inclined lines touching each other and the lines.   (See Fig. 47.)

SOLUTION:

Bisect the inclination of the given lines $a\,b$, $c\,d$ by the line $e\,f$. From a point i in this line draw the perpendicular $i\,g$ to the line $a\,b$ and at i describe the circle $g\,e$ touching the lines and cutting the center line at k.   From k draw $k\,h$ perpendicular to the center line and cutting $a\,b$ at h and from h describe an arc $k\,g'$ cutting $a\,b$ at $g'\,l''$ parallel to $g\,i$ the center of the next circle to be described with radius $k\,i'$ and so on for the next.

To draw a circle through three given points.   (See Fig. 48.)
The given points are a, b, and c.

SOLUTION:

From a and b as centers with suitable radius, describe arcs intersecting at $e\,e$. Draw a line through these points. From b and c as centers, describe arcs intersecting at $d\,d$; draw a line through these points. The point where these two lines intersect is the center of the circle.

To draw two tangents to a circle from a given point without same circle. (See Fig. 49.)

Given point a, and the circle with the center n. The required tangents are $a\,d$ and $a\,b$.

SOLUTION:

Bisect line $n\,a$. With c as center and radius $a\,c$, describe the arc $b\,d$ through the center of the circle. The points of intersection

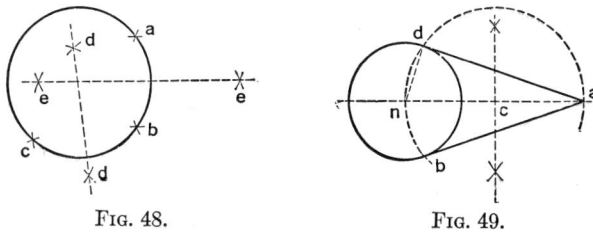

FIG. 48. FIG. 49.

at b and d are the points where the required tangents $a\,b$ and $a\,d$ will touch the circle.

To draw a tangent to a given point in a given circle. (See Fig. 50.)

Given circle and the point h, $x\,y$ is required.

SOLUTION:

The radius is drawn to the point h and a line constructed perpendicular to it at the point h. This perpendicular, touching the circle at h, is called a *tangent*.

To draw a circle of a certain size that will touch the periphery of two given circles. (See Fig. 51.)

Given the diameter of circles a, b, and c.   Locate the center for circle c, when centers for a and b are given.

SOLUTION:

From center of a, describe an arc with a radius equal to the sum of radii of a and c.   From b as center, describe another arc using a radius equal to the sum of the radii of b and c.   The point of intersection of those two arcs is the center of the circle c.

NOTE.—This construction is useful when locating the center for an intermediate gear.   For instance, if a and b are the pitch circles of two gears, c would be the pitch circle located in correct position to connect a and b.

To draw an ellipse, the longest and shortest diameter being given.   The diameters a b and c d are given.   The required ellipse is constructed thus (see Fig. 52):

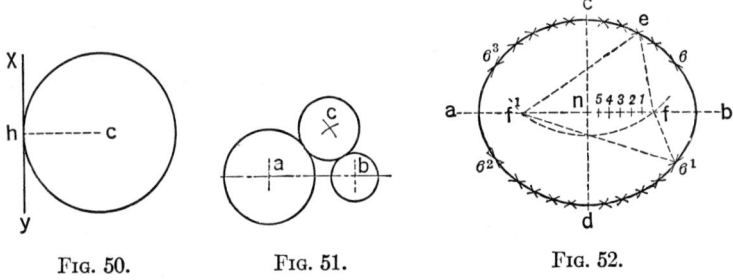

FIG. 50.   FIG. 51.   FIG. 52.

From c as center with a radius a n, describe an arc $f^1 f$.   The points where this arc intersect a b are foci.   The distance $f n$ is divided into any number of parts, as 1, 2, 3, 4, 5.   With radius 1 to b, and the focus f as center, describe arcs 6 and 6^1 with the same radius and with f^1 as center describe arcs 6^2 and 6^3.   With radius 1 to a and f^1 as center, describe arcs intersecting at 6 and 6^1; with the same radius and with f as center, describe arcs intersecting at 6^2 and 6^3.   Continue this operation for points 2, 3, etc., and when all the points for the circumference are in this way marked out, draw the ellipse by using a scroll.   It is a property

with ellipses that the sum of any two lines drawn from the foci
to any point in the circumference is equal to the largest diameter.
For instance:

$$f^1 e + f e, \; = a\,b, \quad \text{or} \quad f\,6^1 + f^1\,6^1, \; = a\,b.$$

Cycloids.—Suppose that a round disc, c, rolls on a straight line,
a, b, and that a lead pencil is fastened at the point r; it will then
describe a curved line, a, l, r, n, b. This line is called a *cycloid.*
(See Fig. 53.)

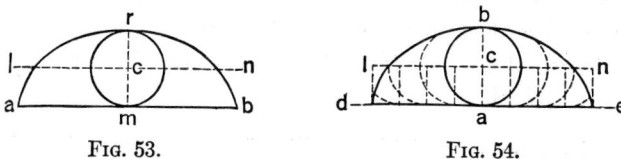

FIG. 53.　　　　　FIG. 54.

This supposed disc is usually called the *generating circle.*
The line $a\,b$ is the base line of the cycloid and is equal in length
to π times $m\,r$, or practically 3.1416 times the diameter of the
generating circle. The length of the curved line a, l, r, n, b is four
times $r\,m$ (four times as long as the diameter of the generating
circle).

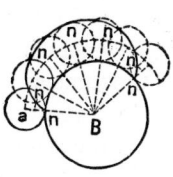

FIG. 55.

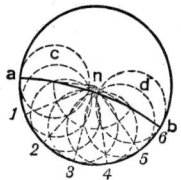

FIG. 56.

A circle rolling on a straight line generates a cycloid. (See
Figs. 53 and 54.)

A circle rolling upon another circle is generating an *epicycloid.*
(See Fig. 55.)

A circle rolling within another circle generates a *hypocycloid.*
(See Fig. 56.)

To draw a cycloid, the generating circle being given.

SOLUTION:

Divide the diameter of the rolling circle in 7 equal parts. Set off 11 of these parts on each side of a on the line $d\,e$. This will give a base line practically equal to the circumference. Divide the base line from the point a into any number of equal parts; erect the perpendiculars; with center-line as centers and a radius equal to the radius of the generating circle describe the arcs. On the first arc from d or e set off one part of the base line. On the second arc set off two parts of the base line; on the third arc, three parts, etc. This will give the points through which to draw the cycloid.

To draw an epicycloid (see Fig. 55), the generating circle a and the fundamental circle B being given.

SOLUTION:

Concentric with the circle B, describe an arc through the center of the generating circle. Divide the circumference of the generating circle into any number of equal parts and set this off on the circumference of the circle B. Through those points draw radial lines extending until they intersect the arc passing through the center of the generating circle. These points of intersection give the centers for the different positions of the generating circle, and for the rest, the construction is essentially the same as the cycloids. In Fig. 55 the generating circle is shown in seven different positions, and the point n, in the circumference of the generating circle, may be followed from the position at the extreme left for one full rotation to the position where it again touches the circle B.

To draw a hypocycloid. (See Fig. 56.)

The hypocycloid is the line generated by a point in a circle rolling within another larger circle, and is constructed thus (see Fig. 56):

Divide the circumference of the generating circle into any number of equal parts. Set off these on the circumference of the fundamental circle. From each point of division draw radial lines, 1, 2, 3, 4, 5, 6. From n as center describe an arc through the center of the generating circle, as the arc $c\,d$. The point of intersection between this arc and the radial lines are centers for the different positions of the generating circle. The distance from 1 to a on the fundamental circle is set off from 1 on the generating circle in its first new position; the distance 2 to a on the fundamental circle is set off from 2 on the generating circle in its second position, etc. For the rest, the construction is substantially the same as Figs. 54 and 55.

NOTE.—If the diameter of the generating circle is equal to the radius of the fundamental circle, the hypocycloid will be a straight line, which is the diameter of the fundamental circle.

Involute.—An involute is a curved line which may be assumed to be generated in the following manner: Suppose a string be placed

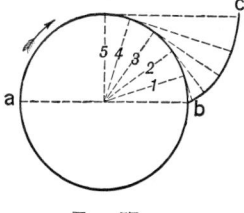

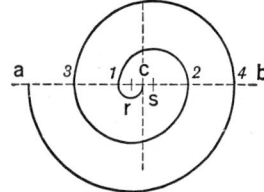

FIG. 57.   FIG. 58.

around a cylinder from a to b, in the direction of the arrow (see Fig. 57), and having a pencil attached at b; keep the string tight and move the pencil toward c, and the involute, $b\,c$, is generated.

To draw an involute.

SOLUTION:

From the point b (see Fig. 57) set off any number of radial lines at equal distances, as 1, 2, 3, 4, 5. From points of intersection draw the tangents (perpendicular to the radial lines). Set off on the first tangent the length of the arc 1 to b; on the second tangent the arc 2 to b, etc.   This will give the points through which to draw the involute.

To draw a spiral from a given point, c.

SOLUTION:

Draw the line $a\,b$ through the point c.   Set off the centers r and S, one-fourth as far from c as the distance is to be between two lines in the spiral.   Using r as center, describe the arc from c to 1; and using S as center, describe the arc from 1 to 2; using r as center, describe the arc from 2 to 3, etc.

V

Trigonometry

Trigonometry is that branch of geometry which deals with angles and with the solution of triangles by means of trigonometric functions.

Angles.—The opening between two straight intersecting lines is an *angle*. An angle may be designated in any one of several ways. Thus, in Fig. 1 we may speak of the angle *B*, the angle *ABC*, or the angle *a*, and refer in each instance to the same angle.

Angles are measured in *degrees*. One degree is $\frac{1}{360}$ of a whole angle, or angle describing a full circle. Then a 90-degree angle is one-quarter of a whole angle. It is called a right angle and the legs are perpendicular to each other. An angle of 180 degrees is equal to the sum of two right angles and is therefore a straight line. It is sometimes called a straight angle.

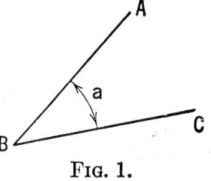

Fig. 1.

Trigonometric Functions.—If we have a right triangle whose acute angles are each 45 degrees and whose legs are each 1 unit long we know from geometry that the length of the hypotenuse is equal to the square root of the sum of the squares of the two sides. Then, in this case, the hypotenuse is equal to $\sqrt{2}$ units. Then, if we have *any* equilateral right triangle, the ratio of the length of legs to the length of the hypotenuse is $1 : \sqrt{2}$. This ratio may then be used to find the hypotenuse if the leg is given, and vice versa. Thus, if the hypotenuse of a 45-degree-angled

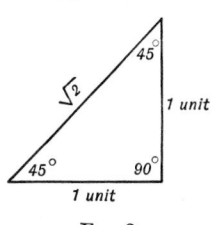

Fig. 2.

right triangle is 9 inches, the leg is $9 \times \dfrac{1}{\sqrt{2}}$ or 6.4 inches. Simi-

larly, if the leg is given as 8 inches, the hypotenuse is $8 \times \dfrac{\sqrt{2}}{1}$ or 11.3 inches.

For a 45-degree-angled right triangle, the ratio of a side to the hypotenuse is *always* $\dfrac{1}{\sqrt{2}} = \dfrac{1}{1.414} = 0.707$, and the ratio of the hypotenuse to a side is *always* $\dfrac{\sqrt{2}}{1} = \dfrac{1.414}{1} = 1.414$.

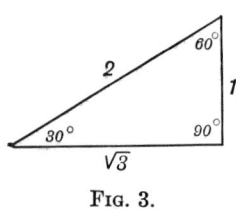

FIG. 3.

Let us now consider a right triangle whose angles are 30, 60, and 90 degrees. If the short side is 1 unit long, the hypotenuse is 2 units and the long leg $\sqrt{3}$ or 1.732 units long. Then, if we are given any 30-60-90 degree triangle and the length of one side, we can readily solve for the other sides. For example, if the hypot-

enuse is 12 inches, the short side is $12 \times \frac{1}{2}$ or 6 inches, and the

long leg is $12 \times \dfrac{1.732}{2} = 10.4$ inches.

We have shown how the ratios of one side of a right triangle to another may be used in solving triangles. These ratios are called *trigonometric functions*. Not only are there definite ratios between the sides of right triangles with angles of 30 degrees, 45 degrees, and 60 degrees, as we have shown, but definite ratios exist for right triangles of *any* angle.

There are six fundamental trigometric functions known as (with abbreviations) *sine* (sin), *cosine* (cos), *tangent* (tan), *cotangent* (cot), *secant* (sec), and *cosecant* (csc).

The sine of an acute angle of a right triangle is the opposite side divided by the hypotenuse, or, in fractional form, opposite side over hypotenuse.

The cosine is the adjacent side over the hypotenuse.

The tangent is the opposite side over the adjacent side.

The cotangent is the adjacent side over the opposite side, or one over the tangent.

The secant is the hypotenuse over the adjacent side, or one over the cosine.

The cosecant is the hypotenuse over the opposite side, or one over the sine.

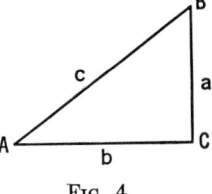

FIG. 4.

In Fig. 4 let a, b, and c represent the lengths of the sides of any right triangle, ABC. Then,

$$\sin A = \frac{a}{c} \qquad\qquad \cot A = \frac{b}{a}$$

$$\cos A = \frac{b}{c} \qquad\qquad \sec A = \frac{c}{b}$$

$$\tan A = \frac{a}{b} \qquad\qquad \csc A = \frac{c}{a}$$

Relations of Functions.—We notice that the cotangent, secant, and cosecant are reciprocals respectively of the tangent, cosine, and sine. Other relations between functions of one angle or of several angles, such as the functions of the sum of two angles, half an angle, twice an angle, etc., are very important and we give a few of them here:

Functions of one angle (A)

$$\sin^2 A + \cos^2 A = 1$$
$$\sec^2 A - \tan^2 A = 1$$
$$\csc^2 A - \cot^2 A = 1$$

Functions of the sum of two angles $(A + B)$

$$\sin (A + B) = \sin A \cos B + \cos A \sin B$$
$$\cos (A + B) = \cos A \cos B - \sin A \sin B$$
$$\tan (A + B) = \frac{\tan A + \tan B}{1 - \tan A \tan B}$$
$$\cot (A + B) = \frac{\cot A \cot B - 1}{\cot B + \cot A}$$

Functions of the difference of two angles $(A - B)$

$$\sin (A - B) = \sin A \cos B - \cos A \sin B$$
$$\cos (A - B) = \cos A \cos B + \sin A \sin B$$
$$\tan (A - B) = \frac{\tan A - \tan B}{1 + \tan A \tan B}$$
$$\cot (A - B) = \frac{\cot A \cot B + 1}{\cot B - \cot A}$$

Functions of one-half an angle $(\tfrac{1}{2}A)$

$$\sin \tfrac{1}{2}A = \frac{\sin A}{2 \cos \tfrac{1}{2}A} = \pm \sqrt{\frac{1 - \cos A}{2}}$$

$$\cos \tfrac{1}{2}A = \frac{\sin A}{2 \sin \tfrac{1}{2}A} = \pm \sqrt{\frac{1 + \cos A}{2}}$$

$$\tan \tfrac{1}{2}A = \frac{1 - \cos A}{\sin A} = \pm \sqrt{\frac{1 - \cos A}{1 + \cos A}}$$

$$\cot \tfrac{1}{2}A = \pm \sqrt{\frac{1 + \cos A}{1 - \cos A}}$$

Functions of twice an angle $(2A)$

$$\sin 2A = 2 \sin A \cos A = \frac{2 \tan A}{1 + \tan^2 A}$$
$$\cos 2A = \cos^2 A - \sin^2 A = 1 - 2 \sin^2 A$$
$$= 2 \cos^2 A - 1 = \frac{1 - \tan^2 A}{1 + \tan^2 A}$$
$$\tan 2A = \frac{2 \tan A}{1 - \tan^2 A} = \frac{\sin 3A - \sin A}{\cos 3A + \cos A}$$
$$\cot 2A = \frac{\cot^2 A - 1}{2 \cot A}$$

Functions of three times an angle $(3A)$

$$\sin 3A = 3 \sin A - 4 \sin^3 A$$
$$\cos 3A = 4 \cos^3 A - 3 \cos A$$

$$\tan 3A = \frac{3 \tan A - \tan^3 A}{1 - 3 \tan^2 A}$$

$$\cot 3A = \frac{\cot^3 A - 3 \cot A}{3 \cot^2 - 1}$$

Tables of Natural and Logarithmic Trigonometric Functions.— Tables for practical use need consist only of the values for sines, cosines, and tangents since the other functions can readily be obtained from these.

The *natural* functions are the actual values of the trigonometric functions themselves. The logarithms of these values are called the *logarithmetic* functions.

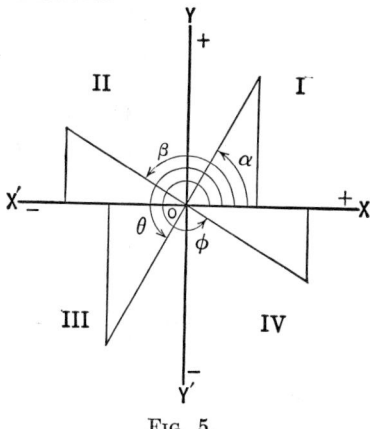

FIG. 5.

Table 2 contains the natural sines, tangents, cotangents and cosines. The functions from 0 degrees to 45 degrees are read *down* the page and the functions from 45 degrees to 90 degrees are read *up* the page.

The solution of problems with trigonometric functions often involves logarithmetic computations. A table giving directly the *logarithms* of the sines, cosines and tangents is a great convenience in such cases; these logarithmic functions are given in Table 3. The use of these tables will be illustrated later in the solution of triangles.

If a circle be imagined as divided into four quadrants and these numbered I, II, III, and IV as shown in Fig. 5, then an angle, such as α, which is less than 90 degrees is said to lie in the first quadrant. An angle between 90 degrees and 180 degrees, such as β, is said to lie in the second quadrant; an angle between 180 degrees and 270 degrees, in the third quadrant; and an angle between 270 degrees and 360 degrees in the fourth quadrant. The function of any angle may be reduced to the function of an angle not greater than 90 degrees by the use of Table 1 paying careful attention to signs.

TABLE 1

	1st Quadrant	2nd Quadrant
sin	$\sin \alpha = \cos (90°-\alpha)$	$\begin{cases} \sin \beta = \sin (180°- \beta) \\ \sin \beta = \cos (\beta -90°) \end{cases}$
cos	$\cos \alpha = \sin (90°-\alpha)$	$\begin{cases} \cos \beta = -\cos (180°- \beta) \\ \cos \beta = -\sin (\beta -90°) \end{cases}$
tan	$\tan \alpha = \cot (90°-\alpha)$	$\begin{cases} \tan \beta = -\tan (180°- \beta) \\ \tan \beta = -\cot (\beta -90°) \end{cases}$
cot	$\cot \alpha = \tan (90°-\alpha)$	$\begin{cases} \cot \beta = -\cot (180°- \beta) \\ \cot \beta = -\tan (\beta -90°) \end{cases}$
	3rd Quadrant	4th Quadrant
sin	$\begin{cases} \sin \theta = -\sin (\theta -180°) \\ \sin \theta = -\cos (270°- \theta) \end{cases}$	$\begin{cases} \sin \phi = -\sin (360°- \phi) \\ \sin \phi = -\cos (\phi -270°) \end{cases}$
cos	$\begin{cases} \cos \theta = -\cos (\theta -180°) \\ \cos \theta = -\sin (270°- \theta) \end{cases}$	$\begin{cases} \cos \phi = \cos (360°- \phi) \\ \cos \phi = \sin (\phi -270°) \end{cases}$
tan	$\begin{cases} \tan \theta = \tan (\theta -180°) \\ \tan \theta = \cot (270°- \theta) \end{cases}$	$\begin{cases} \tan \phi = -\tan (360°- \phi) \\ \tan \phi = -\cot (\phi -270°) \end{cases}$
cot	$\begin{cases} \cot \theta = \cot (\theta -180°) \\ \cot \theta = \tan (270°- \theta) \end{cases}$	$\begin{cases} \cot \phi = -\cot (360°- \phi) \\ \cot \phi = -\tan (\phi -270°) \end{cases}$

TABLE 2

TABLE OF NATURAL TRIGONOMETRIC FUNCTIONS

2.—Natural Sines, Tangents, Cotangents, Cosines.

(Versed sine = 1 − cosine; coversed sine = 1 − sine.)

0°   **1°**

′	Sine.	Tang.	Cotang.	Cosine.	′′	′	Sine.	Tang.	Cotang.	Cosine.	
0	.0000000	.000000	Infinite	1.000000	60	0	.0174524	.017455	57.28996	.9998477	60
1	.0002909	.000291	3437.746	1.000000	59	1	.0177432	.017746	56.35059	.9998426	59
2	.0005818	.000582	1718.873	.9999998	58	2	.0180341	.018037	55.44151	.9998374	58
3	.0008727	.000872	1145.915	.9999996	57	3	.0183249	.018328	54.56130	.9998321	57
4	.0011636	.001163	859.4363	.9999993	56	4	.0186158	.018619	53.70858	.9998267	56
5	.0014544	.001454	687.5488	.9999989	55	5	.0189066	.018910	52.88211	.9998213	55
6	.0017453	.001745	572.9572	.9999985	54	6	.0191974	.019201	52.08067	.9998157	54
7	.0020362	.002036	491.1060	.9999979	53	7	.0194883	.019492	51.30315	.9998101	53
8	.0023271	.002327	429.7175	.9999973	52	8	.0197791	.019783	50.54850	.9998044	52
9	.0026180	.002618	381.9709	.9999966	51	9	.0200699	.020074	49.81572	.9997986	51
10	.0029089	.002908	343.7737	.9999958	50	10	.0203608	.020365	49.10388	.9997927	50
11	.0031998	.003199	312.5213	.9999949	49	11	.0206516	.020656	48.41208	.9997867	49
12	.0034907	.003490	286.4777	.9999939	48	12	.0209424	.020947	47.73950	.9997807	48
13	.0037815	.003781	264.4408	.9999928	47	13	.0212332	.021238	47.08534	.9997745	47
14	.0040724	.004072	245.5519	.9999917	46	14	.0215241	.021529	46.44886	.9997683	46
15	.0043633	.004363	229.1816	.9999905	45	15	.0218149	.021820	45.82935	.9997620	45
16	.0046542	.004654	214.8576	.9999892	44	16	.0221057	.022111	45.22614	.9997555	44
17	.0049451	.004945	202..2187	.9999878	43	17	.0223965	.022402	44.63859	.9997492	43
18	.0052360	.005236	190.9841	.9999863	42	18	.0226873	.022693	44.06611	.9997426	42
19	.0055268	.005527	180.9322	.9999847	41	19	.0229781	.022984	43.50812	.9997360	41
20	.0058177	.005817	171.8854	.9999831	40	20	.0232690	.023275	42.96407	.9997292	40
21	.0061086	.006108	163.7001	.9999813	39	21	.0235598	.023566	42.43444	.9997224	39
22	.0063995	.006399	156.2590	.9999795	38	22	.0238506	.023857	41.91579	.9997156	38
23	.0066904	.006690	149.4650	.9999776	37	23	.0241414	.024148	41.41058	.9997086	37
24	.0069813	.006981	143.2371	.9999756	36	24	.0244322	.024439	40.91741	.9997015	36
25	.0072721	.007272	137.5075	.9999736	35	25	.0247230	.024730	40.43583	.9996943	35
26	.0075630	.007563	132.2185	.9999714	34	26	.0250138	.025021	39.96546	.9996871	34
27	.0078539	.007854	127.3213	.9999692	33	27	.0253046	.025312	39.50589	.9996798	33
28	.0081448	.008145	122.7739	.9999668	32	28	.0255954	.025603	39.05677	.9996724	32
29	.0084357	.008436	118.5401	.9999644	31	29	.0258862	.025894	38.61773	.9996649	31
30	.0087265	.008726	114.5886	.9999619	30	30	.0261769	.026185	38.18845	.9996573	30
31	.0090174	.009017	110.8920	.9999593	29	31	.0264677	.026477	37.76861	.9996497	29
32	.0093083	.009308	107.4264	.9999567	28	32	.0267585	.026768	37.35789	.9996419	28
33	.0095992	.009599	104.1709	.9999539	27	33	.0270493	.027059	36.95600	.9996341	27
34	.0098900	.009890	101.1069	.9999511	26	34	.0273401	.027350	36.56265	.9996262	26
35	.0101809	.010181	98.21794	.9999482	25	35	.0276309	.027641	36.17759	.9996182	25
36	.0104718	.010472	95.48947	.9999452	24	36	.0279216	.027932	35.80055	.9996101	24
37	.0107627	.010763	92.90848	.9999421	23	37	.0282124	.028223	35.43128	.9996020	23
38	.0110535	.011054	90.46333	.9999389	22	38	.0285032	.028514	35.06954	.9995937	22
39	.0113444	.011345	88.14357	.9999357	21	39	.0287940	.028805	34.71511	.9995854	21
40	.0116353	.011636	85.93979	.9999323	20	40	.0290847	.029097	34.36777	.9995770	20
41	.0119261	.011927	83.84350	.9999289	19	41	.0293755	.029388	34.02730	.9995684	19
42	.0122170	.012217	81.84704	.9999254	18	42	.0296662	.029679	33.69350	.9995599	18
43	.0125079	.012508	79.94343	.9999218	17	43	.0299570	.029970	33.36619	.9995512	17
44	.0127987	.012799	78.12634	.9999181	16	44	.0302478	.030261	33.04517	.9995424	16
45	.0130896	.013090	76.39000	.9999143	15	45	.0305385	.030552	32.73026	.9995336	15
46	.0133805	.013381	74.72916	.9999105	14	46	.0308293	.030843	32.42129	.9995247	14
47	.0136713	.013672	73.13899	.9999065	13	47	.0311200	.031135	32.11809	.9995157	13
48	.0139622	.013963	71.61507	.9999025	12	48	.0314108	.031426	31.82051	.9995066	12
49	.0142530	.014254	70.15333	.9998984	11	49	.0317015	.031717	31.52839	.9994974	11
50	.0145439	.014545	68.75008	.9998942	10	50	.0319922	.032008	31.24157	.9994881	10
51	.0148348	.014836	67.40185	.9998900	9	51	.0322830	.032299	30.95992	.9994788	9
52	.0151256	.015127	66.10547	.9998856	8	52	.0325737	.032591	30.68330	.9994693	8
53	.0154165	.015418	64.85800	.9998812	7	53	.0328644	.032882	30.41158	.9994598	7
54	.0157073	.015709	63.65674	.9998766	6	54	.0331552	.033173	30.14461	.9994502	6
55	.0159982	.016000	62.49915	.9998720	5	55	.0334459	.033464	29.88229	.9994405	5
56	.0162890	.016291	61.38290	.9998673	4	56	.0337366	.033755	29.62449	.9994307	4
57	.0165799	.016582	60.30582	.9998625	3	57	.0340274	.034047	29.37110	.9994209	3
58	.0168707	.016873	59.26587	.9998577	2	58	.0343181	.034338	29.12200	.9994110	2
59	.0171616	.017164	58.26117	.9998527	1	59	.0346088	.034629	28.87708	.9994009	1
60	.0174524	.017455	57.28996	.9998477	0	60	.0348995	.034920	28.63625	.9993908	0
	Cosine.	Cotang	Tang.	Sine.	′		Cosine.	Cotang	Tang.	Sine.	′

89°   **88°**

Note.—Secant = 1 ÷ cosine; cosecant = 1 ÷ sine.

2.—Natural Sines, Tangents, Cotangents, Cosines.—(Continued).

(Versed sine = 1 − cosine; coversed sine = 1 − sine.)

2°   3°

′	Sine.	Tang.	Cotang.	Cosine.	′	′	Sine.	Tang.	Cotang.	Cosine.	′
0	.0348995	.034920	28.63625	.9993908	60	0	.0523360	.052407	19.08113	.9986295	60
1	.0351902	.035212	28.39939	.9993806	59	1	.0526264	.052699	18.97552	.9986143	59
2	.0354809	.035503	28.16642	.9993704	58	2	.0529169	.052991	18.87106	.9985989	58
3	.0357716	.035794	27.93723	.9993600	57	3	.0532074	.053282	18.76775	.9985835	57
4	.0360623	.036085	27.71174	.9993495	56	4	.0534979	.053574	18.66556	.9985680	56
5	.0363530	.036377	27.48985	.9993390	55	5	.0537883	.053866	18.56447	.9985524	55
6	.0366437	.036668	27.27148	.9993284	54	6	.0540788	.054158	18.46447	.9985367	54
7	.0369344	.036959	27.05655	.9993177	53	7	.0543693	.054449	18.36553	.9985209	53
8	.0372251	.037250	26.84498	.9993069	52	8	.0546597	.054741	18.26765	.9985050	52
9	.0375158	.037542	26.63669	.9992960	51	9	.0549502	.055033	18.17080	.9984891	51
10	.0378065	.037833	26.43160	.9992851	50	10	.0552406	.055325	18.07497	.9984731	50
11	.0380971	.038124	26.22963	.9992740	49	11	.0555311	.055616	17.98015	.9984570	49
12	.0383878	.038416	26.03073	.9992629	48	12	.0558215	.055908	17.88631	.9984408	48
13	.0386785	.038707	25.83482	.9992517	47	13	.0561119	.056200	17.79344	.9984245	47
14	.0389692	.038998	25.64183	.9992404	46	14	.0564024	.056492	17.70152	.9984081	46
15	.0392598	.039290	25.45170	.9992290	45	15	.0566928	.056784	17.61055	.9983917	45
16	.0395505	.039581	25.26436	.9992176	44	16	.0569832	.057075	17.52051	.9983751	44
17	.0398411	.039872	25.07975	.9992060	43	17	.0572736	.057367	17.43138	.9983585	43
18	.0401318	.040164	24.89782	.9991944	42	18	.0575640	.057659	17.34315	.9983418	42
19	.0404224	.040455	24.71851	.9991827	41	19	.0578544	.057951	17.25580	.9983250	41
20	.0407131	.040746	24.54175	.9991709	40	20	.0581448	.058243	17.16933	.9983082	40
21	.0410037	.041038	24.36750	.9991590	39	21	.0584352	.058535	17.08372	.9982912	39
22	.0412944	.041329	24.19571	.9991470	38	22	.0587256	.058827	16.99895	.9982742	38
23	.0415850	.041621	24.02632	.9991350	37	23	.0590160	.059119	16.91502	.9982571	37
24	.0418757	.041912	23.85927	.9991228	36	24	.0593064	.059410	16.83191	.9982398	36
25	.0421663	.042203	23.69453	.9991106	35	25	.0595967	.059702	16.74961	.9982225	35
26	.0424569	.042495	23.53205	.9990983	34	26	.0598871	.059994	16.66811	.9982051	34
27	.0427475	.042786	23.37177	.9990859	33	27	.0601775	.060286	16.58739	.9981877	33
28	.0430382	.043078	23.21366	.9990734	32	28	.0604678	.060578	16.50745	.9981701	32
29	.0433288	.043369	23.05767	.9990609	31	29	.0607582	.060870	16.42827	.9981525	31
30	.0436194	.043660	22.90376	.9990482	30	30	.0610485	.061162	16.34985	.9981348	30
31	.0439100	.043952	22.75189	.9990355	29	31	.0613389	.061454	16.27217	.9981170	29
32	.0442006	.044243	22.60201	.9990227	28	32	.0616292	.061746	16.19522	.9980991	28
33	.0444912	.044535	22.45409	.9990098	27	33	.0619196	.062038	16.11899	.9980811	27
34	.0447818	.044826	22.30809	.9989968	26	34	.0622099	.062330	16.04348	.9980631	26
35	.0450724	.045118	22.16398	.9989837	25	35	.0625002	.062622	15.96866	.9980450	25
36	.0453630	.045409	22.02171	.9989706	24	36	.0627905	.062914	15.89454	.9980267	24
37	.0456536	.045701	21.88125	.9989573	23	37	.0630808	.063206	15.82110	.9980084	23
38	.0459442	.045992	21.74256	.9989440	22	38	.0633711	.063498	15.74833	.9979900	22
39	.0462347	.046284	21.60563	.9989306	21	39	.0636614	.063790	15.67623	.9979716	21
40	.0465253	.046575	21.47040	.9989171	20	40	.0639517	.064082	15.60478	.9979530	20
41	.0468159	.046867	21.33685	.9989035	19	41	.0642420	.064375	15.53398	.9979343	19
42	.0471065	.047158	21.20494	.9988899	18	42	.0645323	.064667	15.46381	.9979156	18
43	.0473970	.047450	21.07466	.9988761	17	43	.0648226	.064959	15.39427	.9978968	17
44	.0476876	.047741	20.94596	.9988623	16	44	.0651129	.065251	15.32535	.9978779	16
45	.0479781	.048033	20.81882	.9988484	15	45	.0654031	.065543	15.25705	.9978589	15
46	.0482687	.048325	20.69322	.9988344	14	46	.0656934	.065835	15.18934	.9978399	14
47	.0485592	.048616	20.56911	.9988203	13	47	.0659836	.066127	15.12224	.9978207	13
48	.0488498	.048908	20.44648	.9988061	12	48	.0662739	.066419	15.05572	.9978015	12
49	.0491403	.049199	20.32530	.9987919	11	49	.0665641	.066712	14.98978	.9977821	11
50	.0494308	.049491	20.20555	.9987775	10	50	.0668544	.067004	14.92441	.9977627	10
51	.0497214	.049782	20.08719	.9987631	9	51	.0671446	.067296	14.85961	.9977433	9
52	.0500119	.050074	19.97021	.9987486	8	52	.0674349	.067588	14.79537	.9977237	8
53	.0503024	.050366	19.85459	.9987340	7	53	.0677251	.067880	14.73167	.9977040	7
54	.0505929	.050657	19.74029	.9987194	6	54	.0680153	.068173	14.66852	.9976843	6
55	.0508835	.050949	19.62729	.9987046	5	55	.0683055	.068465	14.60591	.9976645	5
56	.0511740	.051241	19.51558	.9986898	4	56	.0685957	.068757	14.54383	.9976445	4
57	.0514645	.051532	19.40513	.9986748	3	57	.0688859	.069049	14.48227	.9976245	3
58	.0517550	.051824	19.29592	.9986598	2	58	.0691761	.069342	14.42123	.9976044	2
59	.0520455	.052116	19.18793	.9986447	1	59	.0694663	.069634	14.36069	.9975843	1
60	.0523360	.052407	19.08113	.9986295	0	60	.0697565	.069926	14.30066	.9975641	0

| | Cosine. | Cotang | Tang. | Sine. | ′ | ′ | | Cosine. | Cotang | Tang. | Sine. | ′ |

87°   86°

Note.—Secant = 1 ÷ cosine.   Cosecant = 1 ÷ sine.

2. —Natural Sines, Tangents, Cotangents, Cosines.—(Continued).

(Versed sine = 1 — cosine; coversed sine = 1 — sine.)

4°   5°

′	Sine.	Tang.	Cotang.	Cosine.		′	Sine.	Tang.	Cotang.	Cosine.	
0	.0697565	.069926	14.30066	.9975641	60	0	.0871557	.087488	11.43005	.9961947	60
1	.0700467	.070219	14.24113	.9975437	59	1	.0874455	.087781	11.39188	.9961693	59
2	.0703368	.070511	14.18209	.9975233	58	2	.0877353	.088074	11.35397	.9961438	58
3	.0706270	.070803	14.12353	.9975028	57	3	.0880251	.088368	11.31630	.9961183	57
4	.0709171	.071096	14.06545	.9974822	56	4	.0883148	.088661	11.27888	.9960926	56
5	.0712073	.071388	14.00785	.9974615	55	5	.0886046	.088954	11.24171	.9960669	55
6	.0714974	.071680	13.95071	.9974408	54	6	.0888943	.089247	11.20478	.9960411	54
7	.0717876	.071973	13.89404	.9974199	53	7	.0891840	.089540	11.16808	.9960152	53
8	.0720777	.072265	13.83782	.9973990	52	8	.0894738	.089834	11.13163	.9959892	52
9	.0723678	.072558	13.78206	.9973780	51	9	.0897635	.090127	11.09541	.9959631	51
10	.0726580	.072850	13.72673	.9973569	50	10	.0900532	.090420	11.05943	.9959370	50
11	.0729481	.073143	13.67185	.9973357	49	11	.0903429	.090713	11.02367	.9959107	49
12	.0732382	.073435	13.61740	.9973145	48	12	.0906326	.091007	10.98815	.9958844	48
13	.0735283	.073727	13.56339	.9972931	47	13	.0909223	.091300	10.95285	.9958580	47
14	.0738184	.074020	13.50979	.9972717	46	14	.0912119	.091593	10.91777	.9958315	46
15	.0741085	.074312	13.45662	.9972502	45	15	.0915016	.091887	10.88292	.9958049	45
16	.0743986	.074605	13.40386	.9972286	44	16	.0917913	.092180	10.84828	.9957783	44
17	.0746887	.074897	13.35151	.9972069	43	17	.0920809	.092473	10.81387	.9957515	43
18	.0749787	.075190	13.29957	.9971851	44	18	.0923706	.092767	10.77967	.9957247	42
19	.0752688	.075482	13.24803	.9971633	41	19	.0926602	.093060	10.74568	.9956978	41
20	.0755589	.075775	13.19688	.9971413	40	20	.0929499	.093354	10.71191	.9956708	40
21	.0758489	.076068	13.14612	.9971193	39	21	.0932395	.093647	10.67834	.9956437	39
22	.0761390	.076360	13.09575	.9970972	38	22	.0935291	.093940	10.64499	.9956165	38
23	.0764290	.076653	13.04576	.9970750	37	23	.0938187	.094234	10.61184	.9955892	37
24	.0767190	.076945	12.99616	.9970528	36	24	.0941083	.094527	10.57889	.9955620	36
25	.0770091	.077238	12.94692	.9970304	35	25	.0943979	.094821	10.54615	.9955345	35
26	.0772991	.077531	12.89805	.9970080	34	26	.0946875	.095114	10.51360	.9955070	34
27	.0775891	.077823	12.84955	.9969854	33	27	.0949771	.095408	10.48126	.9954794	33
28	.0778791	.078116	12.80141	.9969628	32	28	.0952666	.095701	10.44911	.9954517	32
29	.0781691	.078409	12.75363	.9969401	31	29	.0955562	.095995	10.41715	.9954240	31
30	.0784591	.078701	12.70620	.9969173	30	30	.0958458	.096289	10.38539	.9953962	30
31	.0787491	.078994	12.65912	.9968945	29	31	.0961353	.096582	10.35382	.9953683	29
32	.0790391	.079287	12.61239	.9968715	28	32	.0964248	.096876	10.32244	.9953403	28
33	.0793290	.079579	12.56599	.9968485	27	33	.0967144	.097169	10.29125	.9953122	27
34	.0796190	.079872	12.51994	.9968254	26	34	.0970039	.097463	10.26024	.9952840	26
35	.0799090	.080165	12.47422	.9968022	25	35	.0972934	.097757	10.22942	.9952557	25
36	.0801989	.080458	12.42883	.9967789	24	36	.0975829	.098050	10.19878	.9952274	24
37	.0804889	.080750	12.38376	.9967555	23	37	.0978724	.098344	10.16833	.9951990	23
38	.0807788	.081043	12.33902	.9967321	22	38	.0981619	.098638	10.13805	.9951705	22
39	.0810687	.081336	12.29460	.9967085	21	39	.0984514	.098932	10.10795	.9951419	21
40	.0813587	.081629	12.25050	.9966849	20	40	.0987408	.099225	10.07803	.9951132	20
41	.0816486	.081922	12.20671	.9966612	19	41	.0990303	.099519	10.04828	.9950844	19
42	.0819385	.082215	12.16323	.9966374	18	42	.0993197	.099813	10.01871	.9950556	18
43	.0822284	.082507	12.12006	.9966135	17	43	.0996092	.100107	9.989305	.9950266	17
44	.0825183	.082800	12.07719	.9965895	16	44	.0998986	.100400	9.960072	.9949976	16
45	.0828082	.083093	12.03462	.9965655	15	45	.1001881	.100694	9.931008	.9949685	15
46	.0830981	.083386	11.99234	.9965414	14	46	.1004775	.100988	9.902112	.9949393	14
47	.0833880	.083679	11.95037	.9965172	13	47	.1007669	.101282	9.873382	.9949101	13
48	.0836778	.083972	11.90868	.9964929	12	48	.1010563	.101576	9.844816	.9948807	12
49	.0839677	.084265	11.86728	.9964685	11	49	.1013457	.101870	9.816414	.9948513	11
50	.0842576	.084558	11.82616	.9964440	10	50	.1016351	.102164	9.788173	.9948217	10
51	.0845474	.084851	11.78533	.9964195	9	51	.1019245	.102458	9.760092	.9947921	9
52	.0848373	.085144	11.74477	.9963948	8	52	.1022138	.102752	9.732171	.9947625	8
53	.0851271	.085437	11.70450	.9963701	7	53	.1025032	.103046	9.704407	.9947327	7
54	.0854169	.085730	11.66449	.9963453	6	54	.1027925	.103339	9.676800	.9947028	6
55	.0857067	.086023	11.62476	.9963204	5	55	.1030819	.103634	9.649347	.9946729	5
56	.0859966	.086316	11.58529	.9962954	4	56	.1033712	.103928	9.622048	.9946428	4
57	.0862864	.086609	11.54609	.9962704	3	57	.1036605	.104222	9.594902	.9946127	3
58	.0865762	.086902	11.50715	.9962452	2	58	.1039499	.104516	9.567906	.9945825	2
59	.0868660	.087195	11.46847	.9962200	1	59	.1042392	.104810	9.541061	.9945523	1
60	.0871557	.087488	11.43005	.9961947	0	60	.1045285	.105104	9.514364	.9945219	0
	Cosine.	Cotang	Tang.	Sine.	′		Cosine.	Cotang	Tang.	Sine.	′

85°   84°

Note.—Secant = 1 ÷ cosine.   Cosecant = 1 ÷ sine.

2. –Natural Sines, Tangents, Cotangents, Cosines.—(Continued).

(Versed sine = 1−cosine; coversed sine = 1−sine.)

6°   7°

'	Sine.	Tang.	Cotang.	Cosine.		'	Sine.	Tang.	Cotang.	Cosine.	
0	.1045285	.105104	9.514364	.9945219	60	0	.1218693	.122784	8.144346	.9925462	60
1	.1048178	.105398	9.487814	.9944914	59	1	.1221581	.123079	8.124807	.9925107	59
2	.1051070	.105692	9.461411	.9944609	58	2	.1224468	.123375	8.105359	.9924751	58
3	.1053963	.105986	9.435153	.9944303	57	3	.1227355	.123670	8.086004	.9924394	57
4	.1056856	.106280	9.409038	.9943996	56	4	.1230241	.123965	8.066739	.9924037	56
5	.1059748	.106575	9.383066	.9943688	55	5	.1233128	.124261	8.047564	.9923679	55
6	.1062641	.106869	9.357235	.9943379	54	6	.1236015	.124556	8.028479	.9923319	54
7	.1065533	.107163	9.331545	.9943070	53	7	.1238901	.124852	8.009483	.9922959	53
8	.1068425	.107457	9.305993	.9942760	52	8	.1241788	.125147	7.990575	.9922599	52
9	.1071318	.107751	9.280580	.9942448	51	9	.1244674	.125442	7.971755	.9922237	51
10	.1074210	.108046	9.255303	.9942136	50	10	.1247560	.125738	7.953022	.9921874	50
11	.1077102	.108340	9.230162	.9941823	49	11	.1250446	.126033	7.934375	.9921511	49
12	.1079994	.108634	9.205156	.9941510	48	12	.1253332	.126329	7.915815	.9921147	48
13	.1082885	.108929	9.180283	.9941195	47	13	.1256218	.126624	7.897339	.9920782	47
14	.1085777	.109223	9.155543	.9940880	46	14	.1259104	.126920	7.878948	.9920416	46
15	.1088669	.109517	9.130934	.9940563	45	15	.1261990	.127216	7.860642	.9920049	45
16	.1091560	.109812	9.106456	.9940246	44	16	.1264875	.127511	7.842419	.9919682	44
17	.1094452	.110106	9.082107	.9939928	43	17	.1267761	.127807	7.824279	.9919314	43
18	.1097343	.110401	9.057886	.9939610	42	18	.1270646	.128103	7.806221	.9918944	42
19	.1100234	.110695	9.033793	.9939290	41	19	.1273531	.128398	7.788245	.9918574	41
20	.1103126	.110989	9.009826	.9938969	40	20	.1276416	.128694	7.770350	.9918204	40
21	.1106017	.111284	8.985984	.9938648	39	21	.1279302	.128990	7.752536	.9917832	39
22	.1108908	.111578	8.962266	.9938326	38	22	.1282186	.129285	7.734802	.9917459	38
23	.1111799	.111873	8.938672	.9938003	37	23	.1285071	.129581	7.717148	.9917086	37
24	.1114689	.112168	8.915200	.9937679	36	24	.1287956	.129877	7.699573	.9916712	36
25	.1117580	.112462	8.891850	.9937355	35	25	.1290841	.130173	7.682076	.9916337	35
26	.1120471	.112757	8.868620	.9937029	34	26	.1293725	.130469	7.664658	.9915961	34
27	.1123361	.113051	8.845510	.9936703	33	27	.1296609	.130764	7.647317	.9915584	33
28	.1126252	.113346	8.822518	.9936375	32	28	.1299494	.131060	7.630053	.9915206	32
29	.1129142	.113641	8.799644	.9936047	31	29	.1302378	.131356	7.612865	.9914828	31
30	.1132032	.113935	8.776887	.9935719	30	30	.1305262	.131652	7.595754	.9914449	30
31	.1134922	.114230	8.754246	.9935389	29	31	.1308146	.131948	7.578717	.9914069	29
32	.1137812	.114525	8.731719	.9935058	28	32	.1311030	.132244	7.561756	.9913688	28
33	.1140702	.114819	8.709307	.9934727	27	33	.1313913	.132540	7.544869	.9913306	27
34	.1143592	.115114	8.687008	.9934395	26	34	.1316797	.132836	7.528057	.9912923	26
35	.1146482	.115409	8.664822	.9934062	25	35	.1319681	.133132	7.511317	.9912540	25
36	.1149372	.115703	8.642747	.9933728	24	36	.1322564	.133428	7.494651	.9912155	24
37	.1152261	.115998	8.620783	.9933393	23	37	.1325447	.133724	7.478057	.9911770	23
38	.1155151	.116293	8.598929	.9933057	22	38	.1328330	.134020	7.461535	.9911384	22
39	.1158040	.116588	8.577183	.9932721	21	39	.1331213	.134316	7.445085	.9910997	21
40	.1160929	.116883	8.555546	.9932384	20	40	.1334096	.134612	7.428706	.9910610	20
41	.1163818	.117178	8.534017	.9932045	19	41	.1336979	.134909	7.412397	.9910221	19
42	.1166707	.117473	8.512594	.9931706	18	42	.1339862	.135205	7.396159	.9909832	18
43	.1169596	.117767	8.491277	.9931367	17	43	.1342744	.135501	7.379990	.9909442	17
44	.1172485	.118062	8.470065	.9931026	16	44	.1345627	.135797	7.363891	.9909051	16
45	.1175374	.118357	8.448957	.9930685	15	45	.1348509	.136094	7.347861	.9908659	15
46	.1178263	.118652	8.427953	.9930342	14	46	.1351392	.136390	7.331898	.9908266	14
47	.1181151	.118947	8.407051	.9929999	13	47	.1354274	.136686	7.316004	.9907873	13
48	.1184040	.119242	8.386251	.9929655	12	48	.1357156	.136983	7.300178	.9907478	12
49	.1186928	.119537	8.365553	.9929310	11	49	.1360038	.137279	7.284418	.9907083	11
50	.1189816	.119832	8.344955	.9928965	10	50	.1362919	.137575	7.268725	.9906687	10
51	.1192704	.120127	8.324457	.9928618	9	51	.1365801	.137872	7.253098	.9906290	9
52	.1195593	.120423	8.304058	.9928271	8	52	.1368683	.138168	7.237537	.9905893	8
53	.1198481	.120718	8.283757	.9927922	7	53	.1371564	.138465	7.222042	.9905494	7
54	.1201368	.121013	8.263554	.9927573	6	54	.1374445	.138761	7.206611	.9905095	6
55	.1204256	.121308	8.243448	.9927224	5	55	.1377327	.139058	7.191245	.9904694	5
56	.1207144	.121603	8.223438	.9926873	4	56	.1380208	.139354	7.175943	.9904293	4
57	.1210031	.121898	8.203523	.9926521	3	57	.1383089	.139651	7.160705	.9903891	3
58	.1212919	.122194	8.183704	.9926169	2	58	.1385970	.139947	7.145530	.9903489	2
59	.1215806	.122489	8.163978	.9925816	1	59	.1388850	.140244	7.130419	.9903085	1
60	.1218693	.122784	8.144346	.9925462	0	60	.1391731	.140540	7.115369	.9902681	0
	Cosine.	Cotang	Tang.	Sine.	'		Cosine.	Cotang	Tang.	Sine.	'

83°   82°

Note.—Secant = 1÷cosine.   Cosecant = 1÷sine.

2.–Natural Sines, Tangents, Cotangents, Cosines.—(Continued.)

(Versed sine = 1 − cosine; coversed sine = 1 − sine.)

8°   9°

'	Sine.	Tang.	Cotang.	Cosine.		'	Sine.	Tang.	Cotang.	Cosine.	
0	.1391731	.140540	7.115369	.9902681	60	0	.1564345	.158384	6.313751	.9876883	60
1	.1394612	.140837	7.100382	.9902275	59	1	.1567218	.158682	6.301886	.9876428	59
2	.1397492	.141134	7.085457	.9901869	58	2	.1570091	.158980	6.290065	.9875972	58
3	.1400372	.141430	7.070593	.9901462	57	3	.1572963	.159279	6.278286	.9875514	57
4	.1403252	.141727	7.055790	.9901055	56	4	.1575836	.159577	6.266551	.9875057	56
5	.1406132	.142024	7.041048	.9900646	55	5	.1578708	.159875	6.254858	.9874598	55
6	.1409012	.142321	7.026366	.9900237	54	6	.1581581	.160174	6.243208	.9874138	54
7	.1411892	.142617	7.011744	.9899826	53	7	.1584453	.160472	6.231600	.9873678	53
8	.1414772	.142914	6.997180	.9899415	52	8	.1587325	.160770	6.220034	.9873216	52
9	.1417651	.143211	6.982678	.9899003	51	9	.1590197	.161069	6.208510	.9872754	51
10	.1420531	.143508	6.968233	.9898590	50	10	.1593069	.161367	6.197027	.9872291	50
11	.1423410	.143805	6.953847	.9898177	49	11	.1595940	.161666	6.185586	.9871827	49
12	.1426289	.144102	6.939519	.9897762	48	12	.1598812	.161964	6.174186	.9871363	48
13	.1429168	.144399	6.925248	.9897347	47	13	.1601683	.162263	6.162827	.9870897	47
14	.1432047	.144696	6.911035	.9896931	46	14	.1604555	.162561	6.151508	.9870431	46
15	.1434926	.144993	6.896879	.9896514	45	15	.1607426	.162860	6.140230	.9869964	45
16	.1437805	.145290	6.882780	.9896096	44	16	.1610297	.163159	6.128992	.9869496	44
17	.1440684	.145587	6.868737	.9895677	43	17	.1613167	.163457	6.117794	.9869027	43
18	.1443562	.145884	6.854750	.9895258	42	18	.1616038	.163756	6.106636	.9868557	42
19	.1446440	.146181	6.840819	.9894838	41	19	.1618909	.164055	6.095517	.9868087	41
20	.1449319	.146478	6.826943	.9894416	40	20	.1621779	.164353	6.084438	.9867615	40
21	.1452197	.146775	6.813122	.9893994	39	21	.1624650	.164652	6.073397	.9867143	39
22	.1455075	.147072	6.799356	.9893572	38	22	.1627520	.164951	6.062396	.9866670	38
23	.1457953	.147369	6.785644	.9893148	37	2₃	.1630390	.165250	6.051434	.9866196	37
24	.1460830	.147667	6.771986	.9892723	36	24	.1633260	.165548	6.040510	.9865722	36
25	.1463708	.147964	6.758382	.9892298	35	25	.1636129	.165847	6.029624	.9865246	35
26	.1466585	.148261	6.744831	.9891872	34	26	.1638999	.166146	6.018777	.9864770	34
27	.1469463	.148559	6.731334	.9891445	33	27	.1641868	.166445	6.007967	.9864293	33
28	.1472340	.148856	6.717889	.9891017	32	28	.1644738	.166744	5.997195	.9863815	32
29	.1475217	.149153	6.704496	.9890588	31	29	.1647607	.167043	5.986461	.9863336	31
30	.1478094	.149451	6.691156	.9890159	30	30	.1650476	.167342	5.975764	.9862856	30
31	.1480971	.149748	6.677867	.9889728	29	31	.1653345	.167641	5.965104	.9862375	29
32	.1483848	.150045	6.664630	.9889297	28	32	.1656214	.167940	5.954481	.9861894	28
33	.1486724	.150343	6.651444	.9888865	27	33	.1659082	.168239	5.943895	.9861412	27
34	.1489601	.150640	6.638310	.9888432	26	34	.1661951	.168539	5.933345	.9860929	26
35	.1492477	.150938	6.625225	.9887998	25	35	.1664819	.168838	5.922832	.9860445	25
36	.1495353	.151235	6.612191	.9887564	24	36	.1667687	.169137	5.912355	.9859960	24
37	.1498230	.151533	6.599208	.9887128	23	37	.1670556	.169436	5.901913	.9859475	23
38	.1501106	.151830	6.586273	.9886692	22	38	.1673423	.169735	5.891508	.9858988	22
39	.1503981	.152128	6.573389	.9886255	21	39	.1676291	.170035	5.881138	.9858501	21
40	.1506857	.152426	6.560553	.9885817	20	40	.1679159	.170334	5.870804	.9858013	20
41	.1509733	.152723	6.547767	.9885378	19	41	.1682026	.170633	5.860505	.9857524	19
42	.1512608	.153021	6.535029	.9884939	18	42	.1684894	.170933	5.850241	.9857035	18
43	.1515484	.153319	6.522339	.9884498	17	43	.1687761	.171232	5.840011	.9856544	17
44	.1518359	.153617	6.509698	.9884057	16	44	.0690628	.171532	5.829817	.9856053	16
45	.1521234	.153914	6.497104	.9883615	15	45	.1693495	.171831	5.819657	.9855561	15
46	.1524109	.154212	6.484558	.9883172	14	46	.1696362	.172130	5.809531	.9855068	14
47	.1526984	.154510	6.472059	.9882728	13	47	.1699228	.172430	5.799440	.9854574	13
48	.1529858	.154808	6.459607	.9882284	12	48	.1702095	.172730	5.789382	.9854079	12
49	.1532733	.155106	6.447201	.9881838	11	49	.1704961	.173029	5.779358	.9853583	11
50	.1535607	.155404	6.434842	.9881392	10	50	.1707828	.173329	5.769368	.9853087	10
51	.1538482	.155701	6.422530	.9880945	9	51	.1710694	.173628	5.759412	.9852590	9
52	.1541356	.155999	6.410263	.9880497	8	52	.1713560	.173928	5.749488	.9852092	8
53	.1544230	.156297	6.398042	.9880048	7	53	.1716425	.174228	5.739598	.9851593	7
54	.1547104	.156595	6.385866	.9879599	6	54	.1719291	.174527	5.729741	.9851093	6
55	.1549978	.156893	6.373735	.9879148	5	55	.1722156	.174827	5.719917	.9850593	5
56	.1552851	.157191	6.361650	.9878697	4	56	.1725022	.175127	5.710125	.9850091	4
57	.1555725	.157490	6.349609	.9878245	3	57	.1727887	.175427	5.700366	.9849589	3
58	.1558598	.157788	6.337612	.9877792	2	58	.1730752	.175727	5.690639	.9849086	2
59	.1561472	.158086	6.325660	.9877338	1	59	.1733617	.176027	5.680944	.9848582	1
60	.1564345	.158384	6.313751	.9876883	0	60	.1736482	.176327	5.671281	.9848078	0
	Cosine.	Cotang	Tang.	Sine.	'		Cosine.	Cotang	Tang.	Sine.	'

81°   80°

Note.—Secant = 1 ÷ cosine.   Cosecant = 1 ÷ sine.

2.—Natural Sines, Tangents, Cotangents, Cosines.—(Continued.)

(Versed sine = 1 − cosine; coversed sine = 1 − sine).

10° 11°

	Sine.	Tang.	Cotang.	Cosine.				Sine.	Tang.	Cotang.	Cosine.	
0	.1736482	.176327	5.671281	.9848078	60		0	.1908090	.194380	5.144554	.9816272	60
1	.1739346	.176626	5.661650	.9847572	59		1	.1910945	.194682	5.136576	.9815716	59
2	.1742211	.176926	5.652051	.9847066	58		2	.1913801	.194984	5.128622	.9815160	58
3	.1745075	.177226	5.642483	.9846558	57		3	.1916656	.195286	5.120692	.9814603	57
4	.1747939	.177527	5.632947	.9846050	56		4	.1919510	.195588	5.112785	.9814045	56
5	.1750803	.177827	5.623442	.9845542	55		5	.1922365	.195890	5.104902	.9813486	55
6	.1753667	.178127	5.613968	.9845032	54		6	.1925220	.196192	5.097042	.9812927	54
7	.1756531	.178427	5.604524	.9844521	53		7	.1928074	.196494	5.089206	.9812366	53
8	.1759395	.178727	5.595112	.9844010	52		8	.1930928	.196796	5.081392	.9811805	52
9	.1762258	.179027	5.585730	.9843498	51		9	.1933782	.197098	5.073602	.9811243	51
10	.1765121	.179327	5.576378	.9842985	50		10	.1936636	.197400	5.065835	.9810680	50
11	.1767984	.179628	5.567057	.9842471	49		11	.1939490	.197703	5.058090	.9810116	49
12	.1770847	.179928	5.557766	.9841956	48		12	.1942344	.198005	5.050369	.9809552	48
13	.1773710	.180228	5.548505	.9841441	47		13	.1945197	.198307	5.042670	.9008986	47
14	.1776573	.180529	5.539274	.9840924	46		14	.1948050	.198610	5.034993	.9808420	46
15	.1779435	.180829	5.530072	.9840407	45		15	.1950903	.198912	5.027339	.9807853	45
16	.1782298	.181129	5.520900	.9839889	44		16	.1953756	.199214	5.019707	.9807285	44
17	.1785160	.181430	5.511757	.9839370	43		17	.1956609	.199517	5.012098	.9806716	43
18	.1788022	.181730	5.502644	.9838850	42		18	.1959461	.199819	5.004511	.9806147	42
19	.1790884	.182031	5.493560	.9838330	41		19	.1962314	.200122	4.996945	.9805576	41
20	.1793746	.182331	5.484505	.9837808	40		20	.1965166	.200424	4.989402	.9805005	40
21	.1796607	.182632	5.475478	.9837286	39		21	.1968018	.200727	4.981881	.9804433	39
22	.1799469	.182933	5.466481	.9836763	38		22	.1970870	.201030	4.974381	.9803860	38
23	.1802330	.183233	5.457512	.9836239	37		23	.1973722	.201332	4.966903	.9803286	37
24	.1805191	.183534	5.448571	.9835715	36		24	.1976573	.201635	4.959447	.9802712	36
25	.1808052	.183835	5.439659	.9835189	35		25	.1979425	.201938	4.952012	.9802136	35
26	.1810913	.184135	5.430775	.9834663	34		26	.1982276	.202240	4.944599	.9801560	34
27	.1813774	.184436	5.421918	.9834136	33		27	.1985127	.202543	4.937206	.9800983	33
28	.1816635	.184737	5.413090	.9833608	32		28	.1987978	.202846	4.929835	.9800405	32
29	.1819495	.185038	5.404290	.9833079	31		29	.1990829	.203149	4.922485	.9799827	31
30	.1822355	.185339	5.395517	.9832549	30		30	.1993679	.203452	4.915157	.9799247	30
31	.1825215	.185639	5.386771	.9832019	29		31	.1996530	.203755	4.907849	.9798667	29
32	.1828075	.185940	5.378053	.9831487	28		32	.1999380	.204058	4.900562	.9798086	28
33	.1830935	.186241	5.369363	.9830955	27		33	.2002230	.204361	4.893295	.9797504	27
34	.1833795	.186542	5.360699	.9830422	26		34	.2005080	.204664	4.886049	.9796921	26
35	.1836654	.186843	5.352062	.9829888	25		35	.2007930	.204967	4.878824	.9796337	25
36	.1839514	.187144	5.343452	.9829353	24		36	.2010779	.205270	4.871620	.9795752	24
37	.1842373	.187446	5.334869	.9828818	23		37	.2013629	.205573	4.864435	.9795167	23
38	.1845232	.187747	5.326313	.9828282	22		38	.2016478	.205876	4.857271	.9794581	22
39	.1848091	.188048	5.317783	.9827744	21		39	.2019327	.206180	4.850128	.9793994	21
40	.1850949	.188349	5.309279	.9827206	20		40	.2022176	.206483	4.843004	.9793406	20
41	.1853808	.188650	5.300801	.9826668	19		41	.2025024	.206786	4.835901	.9792818	19
42	.1856666	.188952	5.292350	.9826128	18		42	.2027873	.207090	4.828817	.9792228	18
43	.1859524	.189253	5.283925	.9825587	17		43	.2030721	.207393	4.821753	.9791638	17
44	.1862382	.189554	5.275525	.9825046	16		44	.2033569	.207696	4.814709	.9791047	16
45	.1865240	.189855	5.267151	.9824504	15		45	.2036418	.208000	4.807685	.9790455	15
46	.1868098	.190157	5.258803	.9823961	14		46	.2039265	.208303	4.800680	.9789862	14
47	.1870956	.190458	5.250480	.9823417	13		47	.2042113	.208607	4.793695	.9789268	13
48	.1873813	.190760	5.242183	.9822873	12		48	.2044961	.208910	4.736730	.9788674	12
49	.1876670	.191061	5.233911	.9822327	11		49	.2047808	.209214	4.779783	.9788079	11
50	.1879528	.191363	5.225664	.9821781	10		50	.2050655	.209518	4.772856	.9787483	10
51	.1882385	.191664	5.217442	.9821234	9		51	.2053502	.209821	4.765949	.9786886	9
52	.1885241	.191966	5.209245	.9820686	8		52	.2056349	.210125	4.759060	.9786288	8
53	.1880898	.192268	5.201073	.9820137	7		53	.2059195	.210429	4.752190	.9785689	7
54	.1890954	.192569	5.192926	.9819587	6		54	.2062042	.210733	4.745340	.9785090	6
55	.1893811	.192871	5.184803	.9819037	5		55	.2064888	.211036	4.738508	.9784490	5
56	.1896667	.193173	5.176705	.9818485	4		56	.2067734	.211340	4.731695	.9783889	4
57	.1899523	.193474	5.168631	.9817933	3		57	.2070580	.211644	4.724901	.9783287	3
58	.1902379	.193776	5.160581	.9817380	2		58	.2073426	.211948	4.718125	.9782684	2
59	.1905234	.194078	5.152555	.9816826	1		59	.2076272	.212252	4.711368	.9782080	1
60	.1908090	.194380	5.144554	.9816272	0		60	.2079117	.212556	4.704630	.9781476	0

| | Cosine. | Cotang | Tang. | Sine. | ′ | | | Cosine. | Cotang | Tang. | Sine. | |

79° 78°

Note.—Secant = 1 ÷ cosine. Cosecant = 1 ÷ sine.

2.—Natural Sines, Tangents, Cotangents, Cosines.—(Continued.)

(Versed sine = 1 − cosine; coversed sine = 1 − sine.)

12°   13°

′	Sine.	Tang.	Cotang.	Cosine.		′	Sine.	Tang.	Cotang.	Cosine.	
0	.2079117	.212556	4.704630	.9781476	60	0	.2249511	.230868	4.331475	.9743701	60
1	.2081962	.212860	4.697910	.9780871	59	1	.2252345	.231174	4.325734	.9743046	59
2	.2084807	.213164	4.691208	.9780265	58	2	.2255179	.231481	4.320007	.9742390	58
3	.2087652	.213468	4.684524	.9779658	57	3	.2258013	.231787	4.314295	.9741734	57
4	.2090497	.213773	4.677859	.9779050	56	4	.2260846	.232094	4.308597	.9741077	56
5	.2093341	.214077	4.671212	.9778441	55	5	.2263680	.232400	4.302913	.9740419	55
6	.2096186	.214381	4.664583	.9777832	54	6	.2266513	.232707	4.297244	.9739760	54
7	.2099030	.214685	4.657972	.9777222	53	7	.2269346	.233014	4.291588	.9739100	53
8	.2101874	.214990	4.651378	.9776611	52	8	.2272179	.233321	4.285947	.9738439	52
9	.2104718	.215294	4.644803	.9775999	51	9	.2275012	.233627	4.280319	.9737778	51
10	.2107561	.215598	4.638245	.9775386	50	10	.2277844	.233934	4.274706	.9737116	50
11	.2110405	.215903	4.631705	.9774773	49	11	.2280677	.234241	4.269107	.9736453	49
12	.2113248	.216207	4.625183	.9774159	48	12	.2283509	.234547	4.263521	.9735789	48
13	.2116091	.216512	4.618678	.9773544	47	13	.2286341	.234854	4.257950	.9735124	47
14	.2118934	.216816	4.612190	.9772928	46	14	.2289172	.235161	4.252392	.9734458	46
15	.2121777	.217121	4.605720	.9772311	45	15	.2292004	.235468	4.246848	.9733792	45
16	.2124619	.217425	4.599268	.9771693	44	16	.2294835	.235775	4.241317	.9733125	44
17	.2127462	.217730	4.592832	.9771075	43	17	.2297666	.236082	4.235800	.9732457	43
18	.2130304	.218035	4.586414	.9770456	42	18	.2300497	.236390	4.230297	.9731789	42
19	.2133146	.218340	4.580012	.9769836	41	19	.2303328	.236697	4.224808	.9731119	41
20	.2135988	.218644	4.573628	.9769215	40	20	.2306159	.237004	4.219331	.9730449	40
21	.2138829	.218949	4.567261	.9768593	39	21	.2308989	.237311	4.213869	.9729777	39
22	.2141671	.219254	4.560911	.9767970	38	22	.2311819	.237618	4.208419	.9729105	38
23	.2144512	.219559	4.554577	.9767347	37	23	.2314649	.237926	4.202983	.9728432	37
24	.2147353	.219864	4.548260	.9766723	36	24	.2317479	.238233	4.197560	.9727759	36
25	.2150194	.220169	4.541960	.9766098	35	25	.2320309	.238541	4.192151	.9727084	35
26	.2153035	.220474	4.535677	.9765472	34	26	.2323138	.238848	4.186754	.9726409	34
27	.2155876	.220779	4.529410	.9764845	33	27	.2325967	.239156	4.181371	.9725733	33
28	.2158716	.221084	4.523160	.9764217	32	28	.2328796	.239463	4.176001	.9725056	32
29	.2161556	.221389	4.516926	.9763589	31	29	.2331625	.239771	4.170644	.9724378	31
30	.2164396	.221694	4.510708	.9762960	30	30	.2334454	.240078	4.165299	.9723699	30
31	.2167236	.221999	4.504507	.9762330	29	31	.2337282	.240386	4.159968	.9723020	29
32	.2170076	.222305	4.498322	.9761699	28	32	.2340110	.240694	4.154650	.9722339	28
33	.2172915	.222610	4.492153	.9761067	27	33	.2342938	.241001	4.149344	.9721658	27
34	.2175754	.222915	4.486000	.9760435	26	34	.2345766	.241309	4.144051	.9720976	26
35	.2178593	.223221	4.479863	.9759802	25	35	.2348594	.241617	4.138771	.9720294	25
36	.2181432	.223526	4.473742	.9759168	24	36	.2351421	.241925	4.133504	.9719610	24
37	.2184271	.223831	4.467637	.9758533	23	37	.2354248	.242233	4.128249	.9718926	23
38	.2187110	.224137	4.461548	.9757897	22	38	.2357075	.242541	4.123007	.9718240	22
39	.2189948	.224442	4.455475	.9757260	21	39	.2359902	.242849	4.117778	.9717554	21
40	.2192786	.224748	4.449418	.9756623	20	40	.2362729	.243157	4.112561	.9716867	20
41	.2195624	.225054	4.443376	.9755985	19	41	.2365555	.243465	4.107356	.9716180	19
42	.2198462	.225359	4.437350	.9755345	18	42	.2368381	.243773	4.102164	.9715491	18
43	.2201300	.225665	4.431339	.9754706	17	43	.2371207	.244081	4.096985	.9714802	17
44	.2204137	.225971	4.425343	.9754065	16	44	.2374033	.244390	4.091817	.9714112	16
45	.2206974	.226276	4.419364	.9753423	15	45	.2376859	.244698	4.086662	.9713421	15
46	.2209811	.226582	4.413399	.9752781	14	46	.2379684	.245006	4.081519	.9712729	14
47	.2212648	.226888	4.407450	.9752138	13	47	.2382510	.245315	4.076389	.9712036	13
48	.2215485	.227194	4.401516	.9751494	12	48	.2385335	.245623	4.071270	.9711343	12
49	.2218321	.227500	4.395597	.9750849	11	49	.2388159	.245932	4.066164	.9710649	11
50	.2221158	.227806	4.389694	.9750203	10	50	.2390984	.246240	4.061070	.9709953	10
51	.2223994	.228112	4.383805	.9749556	9	51	.2393808	.246549	4.055987	.9709258	9
52	.2226830	.228418	4.377931	.9748909	8	52	.2396633	.246857	4.050917	.9708561	8
53	.2229666	.228724	4.372073	.9748261	7	53	.2399457	.247166	4.045859	.9707863	7
54	.2232501	.229030	4.366229	.9747612	6	54	.2402280	.247475	4.040812	.9707165	6
55	.2235337	.229336	4.360400	.9746962	5	55	.2405104	.247783	4.035777	.9706466	5
56	.2238172	.229642	4.354586	.9746311	4	56	.2407927	.248092	4.030755	.9705766	4
57	.2241007	.229949	4.348786	.9745660	3	57	.2410751	.248401	4.025744	.9705065	3
58	.2243842	.230255	4.343001	.9745008	2	58	.2413574	.248710	4.020744	.9704363	2
59	.2246676	.230561	4.337231	.9744355	1	59	.2416396	.249019	4.015757	.9703660	1
60	.2249511	.230868	4.331475	.9743701	0	60	.2419219	.249328	4.010780	.9702957	0
	Cosine.	Cotang	Tang.	Sine.			Cosine.	Cotang	Tang.	Sine.	′

77°   76°

Note.—Secant = 1 ÷ cosine.   Cosecant = 1 ÷ sine.

2.—**Natural Sines, Tangents, Cotangents, Cosines.**—(Continued.)

(Versed sine = 1 − cosine; coversed sine = 1 − sine.)

14°   15°

'	Sine.	Tang.	Cotang.	Cosine.		'	Sine.	Tang.	Cotang.	Cosine.	
0	.2419219	.249328	4.010780	.9702957	60	0	.2588190	.267949	3.732050	.9659258	60
1	.2422041	.249637	4.005816	.9702253	59	1	.2591000	.268261	3.727713	.9658505	59
2	.2424863	.249946	4.000863	.9701548	58	2	.2593810	.268572	3.723384	.9657751	58
3	.2427685	.250255	3.995922	.9700842	57	3	.2596619	.268884	3.719065	.9656996	57
4	.2430507	.250564	3.990992	.9700135	56	4	.2599428	.269196	3.714756	.9656240	56
5	.2433329	.250873	3.986073	.9699428	55	5	.2602237	.269508	3.710455	.9655484	55
6	.2436150	.251182	3.981166	.9698720	54	6	.2605045	.269820	3.706164	.9654726	54
7	.2438971	.251491	3.976271	.9698011	53	7	.2607853	.270132	3.701883	.9653968	53
8	.2441792	.251801	3.971386	.9697301	52	8	.2610662	.270444	3.697610	.9653209	52
9	.2444613	.252110	3.966513	.9696591	51	9	.2613469	.270757	3.693346	.9652449	51
10	.2447433	.252420	3.961651	.9695879	50	10	.2616277	.271069	3.689092	.9651689	50
11	.2450254	.252729	3.956801	.9695167	49	11	.2619085	.271381	3.684847	.9650927	49
12	.2453074	.253038	3.951961	.9694453	48	12	.2621892	.271694	3.680611	.9650165	48
13	.2455894	.253348	3.947133	.9693740	47	13	.2624699	.272006	3.676384	.9649402	47
14	.2458713	.253658	3.942315	.9693025	46	14	.2627506	.272318	3.672166	.9648638	46
15	.2461533	.253967	3.937509	.9692309	45	15	.2630312	.272631	3.667957	.9647873	45
16	.2464352	.254277	3.932714	.9691593	44	16	.2633118	.272943	3.663757	.9647108	44
17	.2467171	.254587	3.927929	.9690875	43	17	.2635925	.273256	3.659566	.9646341	43
18	.2469990	.254896	3.923156	.9690157	42	18	.2638730	.273569	3.655384	.9645574	42
19	.2472809	.255206	3.918393	.9689438	41	19	.2641536	.273881	3.651211	.9644806	41
20	.2475627	.255516	3.913642	.9688719	40	20	.2644342	.274194	3.647046	.9644037	40
21	.2478445	.255826	3.908901	.9687998	39	21	.2647147	.274507	3.642891	.9643268	39
22	.2481263	.256136	3.904171	.9687277	38	22	.2649952	.274820	3.638744	.9642497	38
23	.2484081	.256446	3.899451	.9686555	37	23	.2652757	.275133	3.634606	.9641726	37
24	.2486899	.256756	3.894742	.9685832	36	24	.2655561	.275445	3.630477	.9640954	36
25	.2489716	.257066	3.890044	.9685108	35	25	.2658366	.275758	3.626356	.9640181	35
26	.2492533	.257376	3.885357	.9684383	34	26	.2661170	.276071	3.622244	.9639407	34
27	.2495350	.257686	3.880680	.9683658	33	27	.2663973	.276385	3.618141	.9638633	33
28	.2498167	.257997	3.876014	.9682931	32	28	.2666777	.276698	3.614046	.9637858	32
29	.2500984	.258307	3.871358	.9682204	31	29	.2669581	.277011	3.609960	.9637081	31
30	.2503800	.258617	3.866713	.9681476	30	30	.2672384	.277324	3.605883	.9636305	30
31	.2506616	.258928	3.862078	.9680748	29	31	.2677187	.277637	3.601814	.9635527	29
32	.2509432	.259238	3.857453	.9680018	28	32	.2677989	.277951	3.597754	.9634748	28
33	.2512248	.259548	3.852839	.9679288	27	33	.2680792	.278264	3.593702	.9633969	27
34	.2515063	.259859	3.848235	.9678557	26	34	.2683594	.278578	3.589659	.9633189	26
35	.2517879	.260169	3.843642	.9677825	25	35	.2686396	.278891	3.585624	.9632408	25
36	.2520694	.260480	3.839059	.9677092	24	36	.2689198	.279205	3.581597	.9631626	24
37	.2523508	.260791	3.834486	.9676358	23	37	.2692000	.279518	3.577579	.9630843	23
38	.2526323	.261101	3.829923	.9675624	22	38	.2694801	.279832	3.573569	.9630060	22
39	.2529137	.261412	3.825370	.9674888	21	39	.2697602	.280145	3.569568	.9629275	21
40	.2531952	.261723	3.820828	.9674152	20	40	.2700403	.280459	3.565574	.9628490	20
41	.2534766	.262034	3.816295	.9673415	19	41	.2703204	.280773	3.561590	.9627704	19
42	.2537579	.262345	3.811773	.9672678	18	42	.2706004	.281087	3.557613	.9626917	18
43	.2540393	.262656	3.807260	.9671939	17	43	.2708805	.281401	3.553644	.9626130	17
44	.2543206	.262967	3.802758	.9671200	16	44	.2711605	.281715	3.549684	.9625342	16
45	.2546019	.263278	3.798266	.9670459	15	45	.2714404	.282029	3.545732	.9624552	15
46	.2548832	.263589	3.793783	.9669718	14	46	.2717204	.282343	3.541788	.9623762	14
47	.2551645	.263900	3.789310	.9668977	13	47	.2720003	.282657	3.537852	.9622972	13
48	.2554458	.264211	3.784848	.9668234	12	48	.2722802	.282971	3.533925	.9622180	12
49	.2557270	.264522	3.780395	.9667490	11	49	.2725601	.283285	3.530005	.9621387	11
50	.2560082	.264833	3.775951	.9666746	10	50	.2728400	.283599	3.526093	.9620594	10
51	.2562894	.265145	3.771518	.9666001	9	51	.2731198	.283914	3.522190	.9619800	9
52	.2565705	.265456	3.767094	.9665255	8	52	.2733997	.284228	3.518294	.9619005	8
53	.2568517	.265768	3.762680	.9664508	7	53	.2736794	.284543	3.514407	.9618210	7
54	.2571328	.266079	3.758276	.9663761	6	54	.2739592	.284857	3.510527	.9617413	6
55	.2574139	.266390	3.753881	.9663012	5	55	.2742390	.285172	3.506655	.9616616	5
56	.2576950	.266702	3.749496	.9662263	4	56	.2745187	.285486	3.502791	.9615818	4
57	.2579760	.267014	3.745120	.9661513	3	57	.2747984	.285801	3.498935	.9615019	3
58	.2582570	.267325	3.740754	.9660762	2	58	.2750781	.286115	3.495087	.9614219	2
59	.2585381	.267637	3.736398	.9660011	1	59	.2753577	.286430	3.491247	.9613418	1
60	.2588190	.267949	3.732050	.9659258	0	60	.2756374	.286745	3.487414	.9612617	0
	Cosine.	Cotang	Tang.	Sine.			Cosine.	Cotang	Tang.	Sine.	'

75°   74°

Note.—Secant = 1 ÷ cosine.   Cosecant = 1 ÷ sine.

2.-Natural Sines. TANGENTS, COTANGENTS, COSINES.—(Continued.)

(Versed sine = 1 − cosine; coversed sine = 1 − sine.)

16° 17°

	Sine.	Tang.	Cotang.	Cosine.				Sine.	Tang.	Cotang.	Cosine.	
0	.2756374	.286745	3.487414	.9612617	60		0	.2923717	.305730	3.270852	.9563048	60
1	.2759170	.287060	3.483589	.9611815	59		1	.2926499	.306048	3.267452	.9562197	59
2	.2761965	.287375	3.479772	.9611012	58		2	.2929280	.306367	3.264059	.9561345	58
3	.2764761	.287690	3.475963	.9610208	57		3	.2932061	.306685	3.260672	.9560492	57
4	.2767556	.288005	3.472161	.9609403	56		4	.2934842	.307003	3.257292	.9559639	56
5	.2770352	.288320	3.468367	.9608598	55		5	.2937623	.307321	3.253918	.9558785	55
6	.2773147	.288635	3.464581	.9607792	54		6	.2940403	.307640	3.250550	.9557930	54
7	.2775941	.288950	3.460802	.9606984	53		7	.2943183	.307958	3.247189	.9557074	53
8	.2778736	.289265	3.457031	.9606177	52		8	.2945963	.308277	3.243834	.9556218	52
9	.2781530	.289580	3.453267	.9605368	51		9	.2948743	.308595	3.240486	.9555361	51
10	.2784324	.289896	3.449512	.9604558	50		10	.2951522	.308914	3.237143	.9554502	50
11	.2787118	.290211	3.445763	.9603748	49		11	.2954302	.309233	3.233807	.9553643	49
12	.2789911	.290526	3.442022	.9602937	48		12	.2957081	.309551	3.230478	.9552784	48
13	.2792704	.290842	3.438289	.9602125	47		13	.2959859	.309870	3.227154	.9551923	47
14	.2795497	.291157	3.434563	.9601312	46		14	.2962638	.310189	3.223837	.9551062	46
15	.2798290	.291473	3.430844	.9600499	45		15	.2965416	.310508	3.220526	.9550199	45
16	.2801083	.291789	3.427132	.9599684	44		16	.2968194	.310827	3.217221	.9549336	44
17	.2803875	.292104	3.423429	.9598869	43		17	.2970971	.311146	3.213922	.9548473	43
18	.2806667	.292420	3.419735	.9598053	42		18	.2973749	.311465	3.210630	.9547608	42
19	.2809459	.292736	3.416044	.9597236	41		19	.2976526	.311784	3.207344	.9546743	41
20	.2812251	.293052	3.412362	.9596418	40		20	.2979303	.312103	3.204063	.9545876	40
21	.2815042	.293368	3.408688	.9595600	39		21	.2982079	.312422	3.200789	.9545009	39
22	.2817833	.293683	3.405021	.9594781	38		22	.2984856	.312742	3.197521	.9544141	38
23	.2820624	.293999	3.401361	.9593961	37		23	.2987632	.313061	3.194259	.9543273	37
24	.2823415	.294316	3.397708	.9593140	36		24	.2990408	.313381	3.191003	.9542403	36
25	.2826205	.294632	3.394063	.9592318	35		25	.2993184	.313700	3.187754	.9541533	35
26	.2828995	.294948	3.390424	.9591496	34		26	.2995959	.314020	3.184510	.9540662	34
27	.2831785	.295264	3.386792	.9590672	33		27	.2998734	.314339	3.181272	.9539790	33
28	.2834575	.295580	3.383169	.9589848	32		28	.3001509	.314659	3.178040	.9538917	32
29	.2837364	.295897	3.379553	.9589023	31		29	.3004284	.314979	3.174814	.9538044	31
30	.2840153	.296213	3.375943	.9588197	30		30	.3007058	.315298	3.171594	.9537170	30
31	.2842942	.296529	3.372340	.9587371	29		31	.3009832	.315618	3.168380	.9536294	29
32	.2845731	.296846	3.368745	.9586543	28		32	.3012606	.315938	3.165172	.9535418	28
33	.2848520	.297163	3.365156	.9585715	27		33	.3015380	.316258	3.161970	.9534542	27
34	.2851308	.297479	3.361575	.9584886	26		34	.3018153	.316578	3.158774	.9533664	26
35	.2854096	.297796	3.358000	.9584056	25		35	.3020926	.316898	3.155584	.9532786	25
36	.2856884	.298112	3.354433	.9583226	24		36	.3023699	.317218	3.152399	.9531907	24
37	.2859671	.298429	3.350872	.9582394	23		37	.3026471	.317538	3.149220	.9531027	23
38	.2862458	.298746	3.347319	.9581562	22		38	.3029244	.317859	3.146047	.9530146	22
39	.2865246	.299063	3.343772	.9580729	21		39	.3032016	.318179	3.142880	.9529264	21
40	.2868032	.299380	3.340232	.9579895	20		40	.3034788	.318499	3.139719	.9528382	20
41	.2870819	.299697	3.336699	.9579060	19		41	.3037559	.318820	3.136563	.9527499	19
42	.2873605	.300014	3.333173	.9578225	18		42	.3040331	.319140	3.133414	.9526615	18
43	.2876391	.300331	3.329654	.9577389	17		43	.3043102	.319461	3.130270	.9525730	17
44	.2879177	.300648	3.326141	.9576552	16		44	.3045872	.319781	3.127131	.9524844	16
45	.2881963	.300965	3.322636	.9575714	15		45	.3048643	.320102	3.123999	.9523958	15
46	.2884748	.301283	3.319137	.9574875	14		46	.3051413	.320423	3.120872	.9523071	14
47	.2887533	.301600	3.315645	.9574035	13		47	.3054183	.320744	3.117750	.9522183	13
48	.2890318	.301917	3.312159	.9573195	12		48	.3056953	.321064	3.114635	.9521294	12
49	.2893103	.302235	3.308681	.9572354	11		49	.3059723	.321385	3.111525	.9520404	11
50	.2895887	.302552	3.305209	.9571512	10		50	.3062492	.321706	3.108421	.9519514	10
51	.2898671	.302870	3.301743	.9570669	9		51	.3065261	.322027	3.105322	.9518623	9
52	.2901455	.303187	3.298285	.9569825	8		52	.3068030	.322348	3.102229	.9517731	8
53	.2904239	.303505	3.294833	.9568981	7		53	.3070798	.322670	3.099141	.9516838	7
54	.2907022	.303823	3.291387	.9568136	6		54	.3073566	.322991	3.096059	.9515944	6
55	.2909805	.304141	3.287948	.9567290	5		55	.3076334	.323312	3.092983	.9515050	5
56	.2912588	.304458	3.284516	.9566443	4		56	.3079102	.323633	3.089912	.9514154	4
57	.2915371	.304776	3.281090	.9565595	3		57	.3081869	.323955	3.086846	.9513258	3
58	.2918153	.305094	3.277671	.9564747	2		58	.3084636	.324276	3.083786	.9512361	2
59	.2920935	.305412	3.274258	.9563898	1		59	.3087403	.324598	3.080732	.9511464	1
60	.2923717	.305730	3.270852	.9563048	0		60	.3090170	.324919	3.077683	.9510565	0
	Cosine.	Cotang	Tang.	Sine.				Cosine.	Cotang	Tang.	Sine.	

73° 72°

Note.—Secant = 1.÷cosine.   Cosecant = 1÷sine.

2. Natural Sines, Tangents, Cotangents, Cosines.—(Continued.)

(Versed sine = 1 − cosine; coversed sine = 1 − sine.)

18°   19°

'	Sine.	Tang.	Cotang.	Cosine.	'	Sine.	Tang.	Cotang.	Cosine.	
0	.3090170	.324919	3.077683	.9510565	0	.3255682	.344327	2.904210	.9455186	60
1	.3092936	.325241	3.074640	.9509666	1	.3258432	.344653	2.901468	.9454238	59
2	.3095702	.325563	3.071602	.9508766	2	.3261182	.344978	2.898731	.9453290	58
3	.3098468	.325884	3.068569	.9507865	3	.3263932	.345304	2.895998	.9452341	57
4	.3101234	.326206	3.065542	.9506963	4	.3266681	.345629	2.893270	.9451391	56
5	.3103999	.326528	3.062520	.9506061	5	.3269430	.345955	2.890546	.9450441	55
6	.3106764	.326850	3.059503	.9505157	6	.3272179	.346281	2.887827	.9449489	54
7	.3109529	.327172	3.056492	.9504253	7	.3274928	.346606	2.885113	.9448537	53
8	.3112294	.327494	3.053487	.9503348	8	.3277676	.346932	2.882403	.9447584	52
9	.3115058	.327816	3.050486	.9502443	9	.3280424	.347258	2.879697	.9446630	51
10	.3117822	.328138	3.047491	.9501536	10	.3283172	.347584	2.876997	.9445675	50
11	.3120586	.328461	3.044501	.9500629	11	.3285919	.347910	2.874300	.9444720	49
12	.3123349	.328783	3.041517	.9499721	12	.3288666	.348236	2.871608	.9443764	48
13	.3126112	.329105	3.038538	.9498812	13	.3291413	.348563	2.868921	.9442807	47
14	.3128875	.329428	3.035564	.9497902	14	.3294160	.348889	2.866238	.9441849	46
15	.3131638	.329750	3.032595	.9496991	15	.3296906	.349215	2.863560	.9440890	45
16	.3134400	.330073	3.029632	.9496080	16	.3299653	.349542	2.860886	.9439931	44
17	.3137163	.330395	3.026673	.9495168	17	.3302398	.349868	2.858216	.9438971	43
18	.3139925	.330718	3.023720	.9494255	18	.3305144	.350195	2.855551	.9438010	42
19	.3142686	.331041	3.020772	.9493341	19	.3307889	.350521	2.852891	.9437048	41
20	.3145448	.331363	3.017830	.9492426	20	.3310634	.350848	2.850234	.9436085	40
21	.3148209	.331686	3.014892	.9491511	21	.3313379	.351175	2.847583	.9435122	39
22	.3150969	.332009	3.011960	.9490595	22	.3316123	.351501	2.844935	.9434157	38
23	.3153730	.332332	3.009033	.9489678	23	.3318867	.351828	2.842292	.9433192	37
24	.3156490	.332655	3.006110	.9488760	24	.3321611	.352155	2.839653	.9432227	36
25	.3159250	.332978	3.003193	.9487842	25	.3324355	.352482	2.837019	.9431260	35
26	.3162010	.333302	3.000282	.9486922	26	.3327098	.352809	2.834389	.9430293	34
27	.3164770	.333625	2.997375	.9486002	27	.3329841	.353136	2.831763	.9429324	33
28	.3167529	.333948	2.994473	.9485081	28	.3332584	.353464	2.829142	.9428355	32
29	.3170288	.334271	2.991576	.9484159	29	.3335326	.353791	2.826525	.9427386	31
30	.3173047	.334595	2.988685	.9483237	30	.3338069	.354118	2.823912	.9426415	30
31	.3175805	.334918	2.985798	.9482313	31	.3340810	.354446	2.821304	.9425444	29
32	.3178563	.335242	2.982916	.9481389	32	.3343552	.354773	2.818700	.9424471	28
33	.3181321	.335566	2.980040	.9480464	33	.3346293	.355101	2.816100	.9423498	27
34	.3184079	.335889	2.977168	.9479538	34	.3349034	.355428	2.813504	.9422525	26
35	.3126836	.336213	2.974301	.9478612	35	.3351775	.355756	2.810913	.9421550	25
36	.3189593	.336537	2.971439	.9477684	36	.3354516	.356084	2.808326	.9420575	24
37	.3192350	.336861	2.968583	.9476756	37	.3357256	.356411	2.805743	.9419598	23
38	.3195106	.337185	2.965731	.9475827	38	.3359996	.356739	2.803164	.9418621	22
39	.3197863	.337509	2.962884	.9474897	39	.3362735	.357067	2.800590	.9417644	21
40	.3200619	.337833	2.960042	.9473966	40	.3365475	.357395	2.798019	.9416665	20
41	.3203374	.338157	2.957205	.9473035	41	.3368214	.357723	2.795453	.9415686	19
42	.3206130	.338481	2.954372	.9472103	42	.3370953	.358051	2.792891	.9414705	18
43	.3208885	.338805	2.951545	.9471170	43	.3373691	.358380	2.790333	.9413724	17
44	.3211640	.339129	2.948722	.9470236	44	.3376429	.358708	2.787780	.9412743	16
45	.3214395	.339454	2.945905	.9469301	45	.3379167	.359036	2.785230	.9411760	15
46	.3217149	.339778	2.943092	.9468366	46	.3381905	.359365	2.782685	.9410777	14
47	.3219903	.340103	2.940284	.9467430	47	.3384642	.359693	2.780144	.9409793	13
48	.3222657	.340427	2.937480	.9466493	48	.3387379	.360022	2.777606	.9408808	12
49	.3225411	.340752	2.934682	.9465555	49	.3390116	.360350	2.775073	.9407822	11
50	.3228164	.341077	2.931888	.9464616	50	.3392852	.360679	2.772544	.9406835	10
51	.3230917	.341401	2.929099	.9463677	51	.3395589	.361008	2.770019	.9405848	9
52	.3233670	.341726	2.926315	.9462736	52	.3398325	.361337	2.767499	.9404860	8
53	.3236422	.342051	2.923535	.9461795	53	.3401060	.361666	2.764982	.9403871	7
54	.3239174	.342376	2.920761	.9460854	54	.3403796	.361994	2.762469	.9402881	6
55	.3241926	.342701	2.917990	.9459911	55	.3406531	.362324	2.759960	.9401891	5
56	.3244678	.343026	2.915225	.9458968	56	.3409265	.362653	2.757456	.9400899	4
57	.3247429	.343351	2.912464	.9458023	57	.3412000	.362982	2.754955	.9399907	3
58	.3250180	.343677	2.909708	.9457078	58	.3414734	.363311	2.752458	.9398914	2
59	.3252931	.344002	2.906957	.9456132	59	.3417468	.363640	2.749966	.9397921	1
60	.3255682	.344327	2.904210	.9455186	60	.3420201	.363970	2.747477	.9396926	0

	Cosine.	Cotang	Tang.	Sine.	'		Cosine.	Cotang	Tang.	Sine.	'

71°   70°

Note.—Secant = 1 ÷ cosine.   Cosecant = 1 ÷ sine.

2.—Natural Sines, TANGENTS, COTANGENTS, COSINES.—(Continued.)

(Versed sine = 1 − cosine; coversed sine = 1 − sine.)

20°

′	Sine.	Tang.	Cotang.	Cosine.	
0	.3420201	.363970	2.747477	.9396926	60
1	.3422935	.364299	2.744992	.9395931	59
2	.3425668	.364629	2.742512	.9394935	58
3	.3428400	.364958	2.740035	.9393938	57
4	.3431133	.365288	2.737562	.9392940	56
5	.3433865	.365618	2.735093	.9391942	55
6	.3436597	.365948	2.732628	.9390943	54
7	.3439329	.366277	2.730167	.9389943	53
8	.3442060	.366607	2.727710	.9388942	52
9	.3444791	.366937	2.725256	.9387940	51
10	.3447521	.367268	2.722807	.9386938	50
11	.3450252	.367598	2.720362	.9385934	49
12	.3452982	.367928	2.717920	.9384930	48
13	.3455712	.368258	2.715482	.9383925	47
14	.3458441	.368589	2.713048	.9382920	46
15	.3461171	.368919	2.710618	.9381913	45
16	.3463900	.369250	2.708192	.9380906	44
17	.3466628	.369580	2.705789	.9379898	43
18	.3469357	.369911	2.703351	.9378889	42
19	.3472085	.370242	2.700936	.9377880	41
20	.3474812	.370572	2.698525	.9376869	40
21	.3477540	.370903	2.696118	.9375858	39
22	.3480267	.371234	2.693714	.9374846	38
23	.3482994	.371565	2.691314	.9373833	37
24	.3485720	.371896	2.688919	.9372820	36
25	.3488447	.372227	2.686526	.9371806	35
26	.3491173	.372559	2.684138	.9370790	34
27	.3493898	.372890	2.681753	.9369774	33
28	.3496624	.373221	2.679372	.9368758	32
29	.3499349	.373553	2.676995	.9367740	31
30	.3502074	.373884	2.674621	.9366722	30
31	.3504798	.374216	2.672251	.9365703	29
32	.3507523	.374547	2.669885	.9364683	28
33	.3510246	.374879	2.667522	.9363662	27
34	.3512970	.375211	2.665163	.9362641	26
35	.3515693	.375543	2.662808	.9361618	25
36	.3518416	.375875	2.660456	.9360595	24
37	.3521139	.376207	2.658108	.9359571	23
38	.3523862	.376539	2.655764	.9358547	22
39	.3526584	.376871	2.653423	.9357521	21
40	.3529306	.377203	2.651086	.9356495	20
41	.3532027	.377536	2.648753	.9355468	19
42	.3534748	.377868	2.646423	.9354440	18
43	.3537469	.378201	2.644096	.9353412	17
44	.3540190	.378533	2.641774	.9352382	16
45	.3542910	.378866	2.639454	.9351352	15
46	.3545630	.379198	2.637139	.9350321	14
47	.3548350	.379531	2.634827	.9349289	13
48	.3551070	.379864	2.632518	.9348257	12
49	.3553789	.380197	2.630213	.9347223	11
50	.3556508	.380530	2.627912	.9346189	10
51	.3559226	.380863	2.625614	.9345154	9
52	.3561944	.381196	2.623319	.9344119	8
53	.3564662	.381529	2.621028	.9343082	7
54	.3567380	.381862	2.618741	.9342045	6
55	.3570097	.382196	2.616457	.9341007	5
56	.3572814	.382529	2.614176	.9339968	4
57	.3575531	.382863	2.611899	.9338928	3
58	.3578248	.383196	2.609625	.9337888	2
59	.3580964	.383530	2.607355	.9336846	1
60	.3583679	.383864	2.605089	.9335804	0

	Cosine.	Cotang	Tang.	Sine.	

69°

21°

′	Sine.	Tang.	Cotang.	Cosine.	
0	.3583679	.383864	2.605089	.9335804	60
1	.3586395	.384197	2.602825	.9334761	59
2	.3589110	.384531	2.600565	.9333718	58
3	.3591825	.384865	2.598309	.9332673	57
4	.3594540	.385199	2.596056	.9331628	56
5	.3597254	.385533	2.593806	.9330582	55
6	.3599968	.385867	2.591560	.9329535	54
7	.3602682	.386202	2.589317	.9328488	53
8	.3605395	.386536	2.587078	.9327439	52
9	.3608108	.386870	2.584842	.9326390	51
10	.3610821	.387205	2.582609	.9325340	50
11	.3613534	.387539	2.580380	.9324290	49
12	.3616246	.387874	2.578153	.9323238	48
13	.3618958	.388209	2.575931	.9322186	47
14	.3621669	.388543	2.573711	.9321133	46
15	.3624380	.388878	2.571495	.9320079	45
16	.3627091	.389213	2.569283	.9319024	44
17	.3629802	.389548	2.567073	.9317969	43
18	.3632512	.389883	2.564867	.9316912	42
19	.3635222	.390218	2.562664	.9315855	41
20	.3637932	.390554	2.560464	.9314797	40
21	.3640641	.390889	2.558268	.9313739	39
22	.3643351	.391224	2.556075	.9312679	38
23	.3646059	.391560	2.553885	.9311619	37
24	.3648768	.391895	2.551699	.9310558	36
25	.3651476	.392231	2.549516	.9309496	35
26	.3654184	.392567	2.547335	.9308434	34
27	.3656891	.392902	2.545159	.9307370	33
28	.3659599	.393238	2.542985	.9306306	32
29	.3662306	.393574	2.540815	.9305241	31
30	.3665012	.393910	2.538647	.9304176	30
31	.3667719	.394246	2.536483	.9303109	29
32	.3670425	.394582	2.534323	.9302042	28
33	.3673130	.394918	2.532165	.9300974	27
34	.3675836	.395255	2.530011	.9299905	26
35	.3678541	.395591	2.527859	.9298835	25
36	.3681246	.395928	2.525711	.9297765	24
37	.3683950	.396264	2.523566	.9296694	23
38	.3686654	.396601	2.521424	.9295622	22
39	.3689358	.396937	2.519286	.9294549	21
40	.3692061	.397274	2.517150	.9293475	20
41	.3694765	.397611	2.515018	.9292401	19
42	.3697468	.397868	2.512889	.9291326	18
43	.3700170	.398285	2.510762	.9290250	17
44	.3702872	.398622	2.508639	.9289173	16
45	.3705574	.398959	2.506519	.9288096	15
46	.3708276	.399296	2.504403	.9287017	14
47	.3710977	.399634	2.502289	.9285938	13
48	.3713678	.399971	2.500178	.9284858	12
49	.3716379	.400308	2.498070	.9283778	11
50	.3719079	.400646	2.495966	.9282696	10
51	.3721780	.400984	2.493864	.9281614	9
52	.3724479	.401321	2.491766	.9280531	8
53	.3727179	.401659	2.489670	.9279447	7
54	.3729878	.401997	2.487578	.9278363	6
55	.3732577	.402335	2.485488	.9277277	5
56	.3735275	.402673	2.483402	.9276191	4
57	.3737973	.403011	2.481319	.9275104	3
58	.3740671	.403349	2.479238	.9274016	2
59	.3743369	.403687	2.477161	.9272928	1
60	.3746066	.404026	2.475086	.9271839	0

	Cosine.	Cotang	Tang.	Sine.	

68°

Note.—Secant = 1 ÷ cosine.   Cosecant = 1 ÷ sine.

2.—**Natural Sines**, Tangents, Cotangents, Cosines.—(Continued.)

(Versed sine = 1 − cosine; coversed sine = 1 − sine.)

22°   23°

′	Sine.	Tang.	Cotang.	Cosine.		′	Sine.	Tang.	Cotang.	Cosine.	
0	.3746066	.404026	2.475086	.9271839	60	0	.3907311	.424474	2.355852	.9205049	60
1	.3748763	.404364	2.473015	.9270748	59	1	.3909889	.424818	2.353948	.9203912	59
2	.3751459	.404703	2.470947	.9269658	58	2	.3912666	.425161	2.352046	.9202774	58
3	.3754156	.405041	2.468881	.9268566	57	3	.3915343	.425505	2.350148	.9201635	57
4	.3756852	.405380	2.466819	.9267474	56	4	.3918019	.425848	2.348251	.9200496	56
5	.3759547	.405719	2.464759	.9266380	55	5	.3920695	.426192	2.346358	.9199356	55
6	.3762243	.406057	2.462703	.9265286	54	6	.3923371	.426536	2.344467	.9198215	54
7	.3764938	.406396	2.460649	.9264192	53	7	.3926047	.426880	2.342578	.9197073	53
8	.3767632	.406735	2.458598	.9263096	52	8	.3928722	.427223	2.340692	.9195931	52
9	.3770327	.407074	2.456551	.9262000	51	9	.3931397	.427568	2.338809	.9194788	51
10	.3773021	.407413	2.454506	.9260902	50	10	.3934071	.427912	2.336928	.9193644	50
11	.3775714	.407753	2.452464	.9259805	49	11	.3936745	.428256	2.335050	.9192499	49
12	.3778408	.408092	2.450425	.9258706	48	12	.3939419	.428600	2.333174	.9191353	48
13	.3781101	.408431	2.448389	.9257606	47	13	.3942093	.428944	2.331301	.9190207	47
14	.3783794	.408771	2.446355	.9256506	46	14	.3944766	.429289	2.329431	.9189060	46
15	.3786486	.409110	2.444325	.9255405	45	15	.3947439	.429633	2.327563	.9187912	45
16	.3789178	.409450	2.442298	.9254303	44	16	.3950111	.429978	2.325697	.9186763	44
17	.3791870	.409790	2.440273	.9253201	43	17	.3952783	.430323	2.323834	.9185614	43
18	.3794562	.410129	2.438251	.9252097	42	18	.3955455	.430668	2.321974	.9184464	42
19	.3797253	.410469	2.436233	.9250993	41	19	.3958127	.431012	2.320116	.9183313	41
20	.3799944	.410809	2.434217	.9249888	40	20	.3960798	.431357	2.318260	.9182161	40
21	.3802634	.411149	2.432204	.9248782	39	21	.3963468	.431703	2.316407	.9181009	39
22	.3805324	.411489	2.430193	.9247676	38	22	.3966139	.432048	2.314557	.9179855	38
23	.3808014	.411830	2.428186	.9246568	37	23	.3968809	.432393	2.312709	.9178701	37
24	.3810704	.412170	2.426181	.9245460	36	24	.3971479	.432738	2.310863	.9177546	36
25	.3813393	.412510	2.424180	.9244351	35	25	.3974148	.433084	2.309020	.9176391	35
26	.3816082	.412851	2.422181	.9243242	34	26	.3976818	.433429	2.307180	.9175234	34
27	.3818770	.413191	2.420185	.9242131	33	27	.3979486	.433775	2.305342	.9174077	33
28	.3821459	.413532	2.418191	.9241020	32	28	.3982155	.434120	2.303506	.9172919	32
29	.3824147	.413872	2.416201	.9239908	31	29	.3984823	.434466	2.301673	.9171760	31
30	.3826834	.414213	2.414213	.9238795	30	30	.3987491	.434812	2.299842	.9170601	30
31	.3829522	.414554	2.412228	.9237682	29	31	.3990158	.435158	2.298014	.9169440	29
32	.3832209	.414895	2.410246	.9236567	28	32	.3992825	.435504	2.296188	.9168279	28
33	.3834895	.415236	2.408267	.9235452	27	33	.3995492	.435850	2.294365	.9167118	27
34	.3837582	.415577	2.406290	.9234336	26	34	.3998158	.436196	2.292544	.9165955	26
35	.3840268	.415918	2.404316	.9233220	25	35	.4000825	.436542	2.290725	.9164791	25
36	.3842953	.416259	2.402345	.9232102	24	36	.4003490	.436889	2.288909	.9163627	24
37	.3845639	.416601	2.400377	.9230984	23	37	.4006156	.437235	2.287095	.9162462	23
38	.3848524	.416942	2.398411	.9229865	22	38	.4008821	.437582	2.285284	.9161297	22
39	.3851008	.417284	2.396449	.9228745	21	39	.4011486	.437928	2.283475	.9160130	21
40	.3853693	.417625	2.394488	.9227624	20	40	.4014150	.438275	2.281669	.9158963	20
41	.3856377	.417967	2.392531	.9226503	19	41	.4016814	.438622	2.279865	.9157795	19
42	.3859060	.418309	2.390576	.9225381	18	42	.4019478	.438969	2.278063	.9156626	18
43	.3861744	.418650	2.388625	.9224258	17	43	.4022141	.439316	2.276264	.9155456	17
44	.3864427	.418992	2.386675	.9223134	16	44	.4024804	.439663	2.274467	.9154286	16
45	.3867110	.419334	2.384729	.9222010	15	45	.4027467	.440010	2.272672	.9153115	15
46	.3869792	.419676	2.382785	.9220884	14	46	.4030129	.440357	2.270880	.9151943	14
47	.3872474	.420019	2.380844	.9219758	13	47	.4032791	.440705	2.269090	.9150770	13
48	.3875156	.420361	2.378906	.9218632	12	48	.4035453	.441052	2.267303	.9149597	12
49	.3877837	.420703	2.376970	.9217504	11	49	.4038114	.441400	2.265518	.9148422	11
50	.3880518	.421046	2.375037	.9216375	10	50	.4040775	.441747	2.263735	.9147247	10
51	.3883199	.421388	2.373106	.9215246	9	51	.4043436	.442095	2.261955	.9146072	9
52	.3885880	.421731	2.371179	.9214116	8	52	.4046096	.442443	2.260177	.9144895	8
53	.3888560	.422073	2.369254	.9212986	7	53	.4048756	.442791	2.258401	.9143718	7
54	.3891240	.422416	2.367331	.9211854	6	54	.4051416	.443139	2.256628	.9142540	6
55	.3893919	.422759	2.365411	.9210722	5	55	.4054075	.443487	2.254857	.9141361	5
56	.3896598	.423102	2.363494	.9209589	4	56	.4056734	.443835	2.253088	.9140181	4
57	.3899277	.423445	2.361580	.9208455	3	57	.4059393	.444183	2.251322	.9139001	3
58	.3901955	.423788	2.359668	.9207320	2	58	.4062051	.444531	2.249558	.9137819	2
59	.3904633	.424131	2.357759	.9206185	1	59	.4064709	.444880	2.247796	.9136637	1
60	.3907311	.424474	2.355852	.9205049	0	60	.4067366	.445228	2.246036	.9135455	0
	Cosine.	Cotang	Tang.	Sine.	′		Cosine.	Cotang	Tang.	Sine.	′

67°   66°

Note.—Secant = 1 ÷ cosine.   Cosecant = 1 ÷ sine.

2.—Natural Sines, TANGENTS, COTANGENTS, COSINES.—(Continued.)

(Versed sine = 1 − cosine; coversed sine = 1 − sine.)

24°　　　　　　25°

'	Sine.	Tang.	Cotang.	Cosine.		'	Sine.	Tang.	Cotang.	Cosine.	
0	.4067366	.445226	2.246036	.9135455	60	0	.4226183	.466307	2.144506	.9063078	60
1	.4070024	.445577	2.244279	.9134271	59	1	.4228819	.466661	2.142379	.9061848	59
2	.4072681	.445926	2.242524	.9133087	58	2	.4231455	.467016	2.141253	.9060618	58
3	.4075337	.446274	2.240772	.9131902	57	3	.4234090	.467370	2.139630	.9059386	57
4	.4077993	.446623	2.239021	.9130716	56	4	.4236725	.467725	2.138008	.9058154	56
5	.4080649	.446972	2.237273	.9129529	55	5	.4239360	.468079	2.136389	.9056922	55
6	.4083305	.447321	2.235523	.9128342	54	6	.4241994	.468434	2.134771	.9055668	54
7	.4085960	.447670	2.233784	.9127154	53	7	.4244628	.468789	2.133155	.9054454	53
8	.4088615	.448020	2.232043	.9125965	52	8	.4247262	.469143	2.131542	.9053219	52
9	.4091269	.448369	2.230304	.9124775	51	9	.4249895	.469498	2.129930	.9051983	51
10	.4093923	.448718	2.228567	.9123584	50	10	.4252528	.469853	2.128321	.9050746	50
11	.4096577	.449068	2.226833	.9122393	49	11	.4255161	.470209	2.126713	.9049509	49
12	.4099230	.449417	2.225100	.9121201	48	12	.4257793	.470564	2.125108	.9048271	48
13	.4101883	.449767	2.223370	.9120008	47	13	.4260425	.470919	2.123504	.9047032	47
14	.4104536	.450117	2.221643	.9118815	46	14	.4263056	.471275	2.121903	.9045792	46
15	.4107189	.450467	2.219917	.9117620	45	15	.4265687	.471630	2.120303	.9044551	45
16	.4109841	.450817	2.218192	.9116425	44	16	.4268318	.471986	2.118705	.9043310	44
17	.4112492	.451167	2.216473	.9115229	43	17	.4270949	.472342	2.117110	.9042068	43
18	.4115144	.451517	2.214754	.9114033	42	18	.4273579	.472697	2.115516	.9040825	42
19	.4117795	.451867	2.213037	.9112835	41	19	.4276208	.473053	2.113924	.9039582	41
20	.4120445	.452217	2.211323	.9111637	40	20	.4278838	.473409	2.112334	.9038310	40
21	.4123096	.452568	2.209611	.9110438	39	21	.4281467	.473765	2.110747	.9037093	39
22	.4125745	.452918	2.207901	.9109238	38	22	.4284095	.474122	2.109161	.9035847	38
23	.4128395	.453269	2.206193	.9108038	37	23	.4286723	.474478	2.107577	.9034600	37
24	.4131044	.453620	2.204487	.9106837	36	24	.4289351	.474834	2.105995	.9033353	36
25	.4133693	.453970	2.202784	.9105635	35	25	.4291979	.475191	2.104415	.9032105	35
26	.4136342	.454321	2.201083	.9104432	34	26	.4294606	.475548	2.102836	.9030856	34
27	.4138990	.454672	2.199384	.9103228	33	27	.4297233	.475904	2.101260	.9029606	33
28	.4141638	.455023	2.197687	.9102024	32	28	.4299859	.476261	2.099686	.9028356	32
29	.4144285	.455375	2.195992	.9100819	31	29	.4302485	.476618	2.098114	.9027105	31
30	.4146932	.455726	2.194299	.9099613	30	30	.4305111	.476975	2.096543	.9025853	30
31	.4149579	.456077	2.192609	.9098406	29	31	.4307736	.477332	2.094975	.9024600	29
32	.4152226	.456428	2.190921	.9097199	28	32	.4310361	.477689	2.093408	.9023347	28
33	.4154872	.456780	2.189234	.9095990	27	33	.4312986	.478047	2.091843	.9022092	27
34	.4157517	.457132	2.187551	.9094781	26	34	.4315610	.478404	2.090280	.9020838	26
35	.4160163	.457483	2.185869	.9093572	25	35	.4318234	.478762	2.088720	.9019582	25
36	.4162808	.457835	2.184189	.9092361	24	36	.4320857	.479119	2.087161	.9018325	24
37	.4165453	.458187	2.182511	.9091150	23	37	.4323481	.479477	2.085603	.9017068	23
38	.4168097	.458539	2.180836	.9089938	22	38	.4326103	.479835	2.084048	.9015810	22
39	.4170741	.458891	2.179163	.9088725	21	39	.4328726	.480193	2.082495	.9014551	21
40	.4173385	.459243	2.177492	.9087511	20	40	.4331348	.480551	2.080943	.9013292	20
41	.4176028	.459596	2.175822	.9086297	19	41	.4333970	.480909	2.079394	.9012032	19
42	.4178671	.459948	2.174155	.9085082	18	42	.4336591	.481267	2.077846	.9010770	18
43	.4181313	.460301	2.172491	.9083866	17	43	.4339212	.481625	2.076300	.9009508	17
44	.4183956	.460653	2.170828	.9082649	16	44	.4341832	.481984	2.074756	.9008246	16
45	.4186597	.461006	2.169167	.9081432	15	45	.4344453	.482342	2.073214	.9006982	15
46	.4189239	.461359	2.167509	.9080214	14	46	.4347072	.482701	2.071674	.9005718	14
47	.4191880	.461711	2.165852	.9078995	13	47	.4349692	.483060	2.070135	.9004453	13
48	.4194521	.462064	2.164198	.9077775	12	48	.4352311	.483418	2.068599	.9003188	12
49	.4197161	.462417	2.162546	.9076554	11	49	.4354930	.483777	2.067064	.9001921	11
50	.4199801	.462771	2.160895	.9075333	10	50	.4357548	.484136	2.065531	.9000654	10
51	.4202441	.463124	2.159247	.9074111	9	51	.4360166	.484495	2.064000	.8999386	9
52	.4205080	.463477	2.157601	.9072888	8	52	.4362784	.484855	2.062471	.8998117	8
53	.4207719	.463831	2.155957	.9071665	7	53	.4365401	.485214	2.060944	.8996848	7
54	.4210358	.464184	2.154315	.9070440	6	54	.4368018	.485573	2.059418	.8995578	6
55	.4212996	.464538	2.152675	.9069215	5	55	.4370634	.485933	2.057895	.8994307	5
56	.4215634	.464891	2.151037	.9067989	4	56	.4373251	.486293	2.056373	.8993035	4
57	.4218272	.465245	2.149402	.9066762	3	57	.4375866	.486652	2.054853	.8991763	3
58	.4220909	.465599	2.147768	.9065535	2	58	.4378482	.487012	2.053334	.8990489	2
59	.4223546	.465953	2.146136	.9064307	1	59	.4381097	.487372	2.051818	.8989215	1
60	.4226183	.466307	2.144506	.9063078	0	60	.4383711	.487732	2.050303	.8987940	0
	Cosine.	Cotang	Tang.	Sine.	'		Cosine.	Cotang	Tang.	Sine.	'

65°　　　　　　64°

Note.—Secant = 1 ÷ cosine.　　Cosecant = 1 ÷ sine.

2.—Natural Sines, Tangents, Cotangents, Cosines.—(Continued.)

(Versed sine = 1 − cosine; coversed sine = 1 − sine.)

26°   27°

'	Sine.	Tang.	Cotang.	Cosine.		'	Sine.	Tang.	Cotang.	Cosine.	
0	.4383711	.487732	2.050303	.8987940	60	0	.4539905	.509525	1.962610	.8910065	60
1	.4386326	.488092	2.048791	.8986665	59	1	.4542497	.509891	1.961200	.8908744	59
2	.4388940	.488453	2.047280	.8985389	58	2	.4545088	.510258	1.959791	.8907423	58
3	.4391553	.488813	2.045770	.8984112	57	3	.4547679	.510625	1.958383	.8906100	57
4	.4394166	.489173	2.044263	.8982834	56	4	.4550269	.510991	1.956978	.8904777	56
5	.4396779	.489534	2.042757	.8981555	55	5	.4552859	.511358	1.955573	.8903453	55
6	.4399392	.489894	2.041254	.8980276	54	6	.4555449	.511725	1.954171	.8902128	54
7	.4402004	.490255	2.039751	.8978996	53	7	.4558038	.512093	1.952770	.8900803	53
8	.4404615	.490616	2.038251	.8977715	52	8	.4560627	.512460	1.951371	.8899476	52
9	.4407227	.490977	2.036753	.8976433	51	9	.4563216	.512827	1.949973	.8898149	51
10	.4409838	.491338	2.035256	.8975151	50	10	.4565804	.513195	1.948577	.8896822	50
11	.4412448	.491699	2.033761	.8973868	49	11	.4568392	.513562	1.947112	.8895493	49
12	.4415059	.492061	2.032268	.8972584	48	12	.4570979	.513930	1.945789	.8894164	48
13	.4417668	.492422	2.030776	.8971299	47	13	.4573566	.514298	1.944398	.8892834	47
14	.4420278	.492783	2.029287	.8970014	46	14	.4576153	.514665	1.943008	.8891503	46
15	.4422887	.493145	2.027799	.8968727	45	15	.4578739	.515033	1.941620	.8890171	45
16	.4425496	.493507	2.026313	.8967440	44	16	.4581325	.515401	1.940233	.8888839	44
17	.4428104	.493868	2.024828	.8966153	43	17	.4583910	.515770	1.938848	.8887506	43
18	.4430712	.494230	2.023346	.8964864	42	18	.4586496	.516138	1.937464	.8886172	42
19	.4433319	.494592	2.021865	.8963575	41	19	.4589080	.516506	1.936082	.8884838	41
20	.4435927	.494954	2.020386	.8962285	40	20	.4591665	.516875	1.934702	.8883503	40
21	.4438534	.495317	2.018908	.8960994	39	21	.4594248	.517244	1.933323	.8882166	39
22	.4441140	.495679	2.017433	.8959703	38	22	.4596832	.517612	1.931945	.8880830	38
23	.4443746	.496041	2.015959	.8958411	37	23	.4599415	.517981	1.930569	.8879492	37
24	.4446352	.496404	2.014486	.8957118	36	24	.4601998	.518350	1.929195	.8878154	36
25	.4448957	.496766	2.013016	.8955824	35	25	.4604581	.518719	1.927822	.8876815	35
26	.4451562	.497129	2.011547	.8954529	34	26	.4607162	.519089	1.926451	.8875475	34
27	.4454167	.497492	2.010080	.8953234	33	27	.4609744	.519458	1.925081	.8874134	33
28	.4456771	.497855	2.008615	.8951938	32	28	.4612325	.519827	1.923713	.8872793	32
29	.4459375	.498218	2.007151	.8950641	31	29	.4614906	.520197	1.922347	.8871451	31
30	.4461978	.498581	2.005689	.8949344	30	30	.4617486	.520567	1.920982	.8870108	30
31	.4464581	.498944	2.004229	.8948045	29	31	.4620066	.520936	1.919618	.8868765	29
32	.4467184	.499308	2.002771	.8946746	28	32	.4622646	.521306	1.918256	.8867420	28
33	.4469786	.499671	2.001314	.8945446	27	33	.4625225	.521676	1.916896	.8866075	27
34	.4472388	.500035	1.999859	.8944146	26	34	.4627804	.522046	1.915537	.8864730	26
35	.4474990	.500398	1.998406	.8942844	25	35	.4630352	.522417	1.914179	.8863383	25
36	.4477591	.500762	1.996953	.8941542	24	36	.4632960	.522787	1.912823	.8862036	24
37	.4480192	.501126	1.995503	.8940240	23	37	.4635538	.523157	1.911469	.8860688	23
38	.4482792	.501490	1.994055	.8938936	22	38	.4638115	.523528	1.910116	.8859339	22
39	.4485392	.501854	1.992608	.8937632	21	39	.4640692	.523899	1.908764	.8857989	21
40	.4487992	.502218	1.991163	.8936326	20	40	.4643269	.524269	1.907414	.8856639	20
41	.4490591	.502583	1.989720	.8935021	19	41	.4645845	.524640	1.906066	.8855288	19
42	.4493190	.502947	1.988278	.8933714	18	42	.4648420	.525011	1.904719	.8853936	18
43	.4495789	.503312	1.986838	.8932406	17	43	.4650996	.525382	1.903373	.8852584	17
44	.4498387	.503676	1.985399	.8931098	16	44	.4653571	.525754	1.902029	.8851230	16
45	.4500984	.504041	1.983963	.8929789	15	45	.4656145	.526125	1.900687	.8849876	15
46	.4503582	.504406	1.982528	.8928480	14	46	.4658719	.526496	1.899346	.8848522	14
47	.4506179	.504771	1.981095	.8927169	13	47	.4661293	.526868	1.898006	.8847166	13
48	.4508775	.505136	1.979663	.8925858	12	48	.4663866	.527240	1.896668	.8845810	12
49	.4511372	.505501	1.978233	.8924546	11	49	.4666439	.527612	1.895332	.8844453	11
50	.4513967	.505866	1.976805	.8923234	10	50	.4669012	.527983	1.893997	.8843095	10
51	.4516563	.506232	1.975378	.8921920	9	51	.4671584	.528356	1.892663	.8841736	9
52	.4519158	.506597	1.973953	.8920606	8	52	.4674156	.528728	1.891331	.8840377	8
53	.4521753	.506963	1.972529	.8919291	7	53	.4676727	.529100	1.890000	.8839017	7
54	.4524347	.507329	1.971107	.8917975	6	54	.4679298	.529472	1.888671	.8837656	6
55	.4526941	.507694	1.969687	.8916659	5	55	.4681869	.529845	1.887343	.8836295	5
56	.4529535	.508060	1.968268	.8915342	4	56	.4684439	.530217	1.886017	.8834933	4
57	.4532128	.508426	1.966851	.8914024	3	57	.4687009	.530590	1.884692	.8833569	3
58	.4534721	.508792	1.965436	.8912705	2	58	.4689578	.530963	1.883369	.8832206	2
59	.4537313	.509159	1.964022	.8911385	1	59	.4692147	.531336	1.882047	.8830841	1
60	.4539905	.509525	1.962610	.8910065	0	60	.4694716	.531709	1.880726	.8829476	0
	Cosine.	Cotang	Tang.	Sine.	'		Cosine.	Cotang	Tang.	Sine.	'

63°   62°

Note.—Secant = 1 ÷ cosine.   Cosecant = 1 ÷ sine.

2.—**Natural Sines**, Tangents, Cotangents, Cosines.—(**Continued**.)

(Versed sine = 1 − cosine; coversed sine = 1 − sine.)

28°   29°

'	Sine.	Tang.	Cotang.	Cosine.		'	Sine.	Tang.	Cotang.	Cosine.	
0	.4694716	.531709	1.880726	.8829476	60	0	.4848096	.554309	1.804047	.8746197	60
1	.4697284	.532082	1.879407	.8828110	59	1	.4850640	.554689	1.802810	.8744786	59
2	.4699852	.532455	1.878089	.8826743	58	2	.4853184	.555069	1.801575	.8743375	58
3	.4702419	.532829	1.876773	.8825376	57	3	.4855727	.555450	1.800340	.8741963	57
4	.4704986	.533202	1.875458	.8824007	56	4	.4858270	.555831	1.799107	.8740550	56
5	.4707553	.533576	1.874145	.8822638	55	5	.4860812	.556211	1.797875	.8739137	55
6	.4710119	.533950	1.872833	.8821269	54	6	.4863354	.556592	1.796645	.8737722	54
7	.4712685	.534324	1.871523	.8819898	53	7	.4865895	.556973	1.795416	.8736307	53
8	.4715250	.534698	1.870214	.8818527	52	8	.4868436	.557355	1.794188	.8734891	52
9	.4717815	.535072	1.868906	.8817155	51	9	.4870977	.557736	1.792961	.8733475	51
10	.4720380	.535446	1.867600	.8815782	50	10	.4873517	.558117	1.791736	.8732058	50
11	.4722944	.535820	1.866295	.8814409	49	11	.4876057	.558499	1.790512	.8730640	49
12	.4725508	.536195	1.864992	.8813035	48	12	.4878597	.558881	1.789289	.8729221	48
13	.4728071	.536569	1.863690	.8811660	47	13	.4881136	.559262	1.788067	.8727801	47
14	.4730634	.536944	1.862389	.8810284	46	14	.4883674	.559644	1.786847	.8726381	46
15	.4733197	.537319	1.861090	.8808907	45	15	.4886212	.560026	1.785628	.8724960	45
16	.4735759	.537694	1.859792	.8807530	44	16	.4888750	.560409	1.784410	.8723538	44
17	.4738321	.538069	1.858496	.8806152	43	17	.4891288	.560791	1.783194	.8722116	43
18	.4740882	.538444	1.857201	.8804774	42	18	.4893825	.561173	1.781979	.8720693	42
19	.4743443	.538819	1.855908	.8803394	41	19	.4896361	.561556	1.780765	.8719269	41
20	.4746004	.539195	1.854615	.8802014	40	20	.4898897	.561939	1.779552	.8717844	40
21	.4748564	.539570	1.853325	.8800633	39	21	.4901433	.562321	1.778340	.8716419	39
22	.4751124	.539946	1.852035	.8799251	38	22	.4903968	.562704	1.777130	.8714993	38
23	.4753683	.540322	1.850747	.8797869	37	23	.4906503	.563087	1.775921	.8713566	37
24	.4756242	.540698	1.849461	.8796486	36	24	.4909038	.563471	1.774714	.8712138	36
25	.4758801	.541074	1.848176	.8795102	35	25	.4911572	.563854	1.773507	.8710710	35
26	.4761359	.541450	1.846892	.8793717	34	26	.4914105	.564237	1.772302	.8709281	34
27	.4763917	.541826	1.845609	.8792332	33	27	.4916638	.564621	1.771098	.8707851	33
28	.4766474	.542202	1.844328	.8790946	32	28	.4919171	.565005	1.769895	.8706420	32
29	.4769031	.542579	1.843049	.8789559	31	29	.4921704	.565388	1.768694	.8704989	31
30	.4771588	.542955	1.841770	.8788171	30	30	.4924236	.565772	1.767494	.8703557	30
31	.4774144	.543332	1.840494	.8786783	29	31	.4926767	.566156	1.766295	.8702124	29
32	.4776700	.543709	1.839218	.8785394	28	32	.4929298	.566541	1.765097	.8700691	28
33	.4779255	.544086	1.837944	.8784004	27	33	.4931829	.566925	1.763901	.8699256	27
34	.4781810	.544463	1.836671	.8782613	26	34	.4934359	.567309	1.762705	.8697821	26
35	.4784364	.544840	1.835399	.8781222	25	35	.4936889	.567694	1.761511	.8696385	25
36	.4786919	.545217	1.834129	.8779830	24	36	.4939419	.568079	1.760318	.8694949	24
37	.4789472	.545595	1.832861	.8778437	23	37	.4941948	.568463	1.759126	.8693512	23
38	.4792026	.545972	1.831593	.8777043	22	38	.4944476	.568848	1.757936	.8692074	22
39	.4794579	.546350	1.830327	.8775649	21	39	.4947005	.569233	1.756747	.8690636	21
40	.4797131	.546728	1.829062	.8774254	20	40	.4949532	.569619	1.755559	.8689196	20
41	.4799683	.547106	1.827799	.8772858	19	41	.4952060	.570004	1.754372	.8687756	19
42	.4802235	.547484	1.826537	.8771462	18	42	.4954587	.570389	1.753186	.8686315	18
43	.4804786	.547862	1.825276	.8770064	17	43	.4957113	.570775	1.752002	.8684874	17
44	.4807337	.548240	1.824017	.8768666	16	44	.4959639	.571161	1.750819	.8683431	16
45	.4809888	.548618	1.822759	.8767268	15	45	.4962165	.571547	1.749637	.8681988	15
46	.4812438	.548997	1.821502	.8765868	14	46	.4964690	.571933	1.748456	.8680544	14
47	.4814987	.549375	1.820247	.8764468	13	47	.4967215	.572319	1.747276	.8679100	13
48	.4817537	.549754	1.818993	.8763067	12	48	.4969740	.572705	1.746098	.8677655	12
49	.4820086	.550133	1.817740	.8761665	11	49	.4972264	.573091	1.744921	.8676209	11
50	.4822634	.550512	1.816489	.8760263	10	50	.4974787	.573478	1.743745	.8674762	10
51	.4825182	.550891	1.815239	.8758859	9	51	.4977310	.573864	1.742570	.8673314	9
52	.4827730	.551270	1.813990	.8757455	8	52	.4979833	.574251	1.741396	.8671866	8
53	.4830277	.551650	1.812743	.8756051	7	53	.4982355	.574638	1.740224	.8670417	7
54	.4832824	.552029	1.811496	.8754645	6	54	.4984877	.575025	1.739053	.8668967	6
55	.4835370	.552409	1.810252	.8753239	5	55	.4987399	.575412	1.737883	.8667517	5
56	.4837916	.552789	1.809008	.8751832	4	56	.4989920	.575799	1.736714	.8666066	4
57	.4840462	.553168	1.807766	.8750425	3	57	.4992441	.576187	1.735546	.8664614	3
58	.4843007	.553548	1.806525	.8749016	2	58	.4994961	.576574	1.734380	.8663161	2
59	.4845552	.553928	1.805286	.8747607	1	59	.4997481	.576962	1.733214	.8661708	1
60	.4848096	.554309	1.804047	.8746197	0	60	.5000000	.577350	1.732050	.8660254	0
	Cosine.	Cotang	Tang.	Sine.			Cosine.	Cotang	Tang.	Sine.	'

61°   60°

Note.—Secant = 1 ÷ cosine.   Cosecant = 1 ÷ sine.

2.—Natural Sines, TANGENTS, COTANGENTS, COSINES.—(Continued.)

(Versed sine = 1 − cosine; coversed sine = 1 − sine.)

30°   31°

′	Sine.	Tang.	Cotang.	Cosine.			′	Sine.	Tang.	Cotang.	Cosine.	
0	.5000000	.577350	1.732050	.8660254	60	0	.5150381	.600860	1.664279	.8571673	60
1	.5002519	.577738	1.730887	.8658799	59	1	.5152874	.601256	1.663183	.8570174	59
2	.5005037	.578126	1.729726	.8657344	58	2	.5155367	.601652	1.662088	.8568675	58
3	.5007556	.578514	1.728565	.8655887	57	3	.5157859	.602049	1.660994	.8567175	57
4	.5010073	.578902	1.727406	.8654430	56	4	.5160351	.602445	1.659901	.8565674	56
5	.5012591	.579291	1.726247	.8652973	55	5	.5162842	.602841	1.658809	.8564173	55
6	.5015107	.579679	1.725090	.8651514	54	6	.5165333	.603238	1.657718	.8562671	54
7	.5017624	.580068	1.723934	.8650055	53	7	.5167824	.603635	1.656629	.8561168	53
8	.5020140	.580457	1.722779	.8648595	52	8	.5170314	.604032	1.655540	.8559664	52
9	.5022655	.580846	1.721626	.8647134	51	9	.5172804	.604429	1.654452	.8558160	51
10	.5025170	.581235	1.720473	.8645673	50	10	.5175293	.604826	1.653366	.8556655	50
11	.5027685	.581624	1.719322	.8644211	49	11	.5177782	.605224	1.652280	.8555149	49
12	.5030199	.582013	1.718172	.8642748	48	12	.5180270	.605621	1.651196	.8553643	48
13	.5032713	.582403	1.717023	.8641284	47	13	.5182758	.606019	1.650112	.8552135	47
14	.5035227	.582793	1.715875	.8639820	46	14	.5185246	.606417	1.649030	.8550627	46
15	.5037740	.583182	1.714728	.8638355	45	15	.5187733	.606814	1.647949	.8549119	45
16	.5040252	.583572	1.713582	.8636889	44	16	.5190219	.607213	1.646868	.8547609	44
17	.5042765	.583962	1.712438	.8635423	43	17	.5192705	.607611	1.645789	.8546099	43
18	.5045276	.584352	1.711294	.8633956	42	18	.5195191	.608009	1.644711	.8544588	42
19	.5047788	.584743	1.710152	.8632488	41	19	.5197676	.608408	1.643633	.8543077	41
20	.5050298	.585133	1.709011	.8631019	40	20	.5200161	.608806	1.642557	.8541564	40
21	.5052809	.585524	1.707871	.8629549	39	21	.5202646	.609205	1.641482	.8540051	39
22	.5055319	.585914	1.706732	.8628079	38	22	.5205130	.609604	1.640408	.8538538	38
23	.5057828	.586305	1.705595	.8626608	37	23	.5207613	.610003	1.639335	.8537023	37
24	.5060338	.586696	1.704458	.8625137	36	24	.5210096	.610402	1.638263	.8535508	36
25	.5062846	.587087	1.703323	.8623664	35	25	.5212579	.610801	1.637191	.8533992	35
26	.5065355	.587473	1.702189	.8622191	34	26	.5215061	.611201	1.636121	.8532475	34
27	.5067863	.587870	1.701055	.8620717	33	27	.5217543	.611601	1.635052	.8530958	33
28	.5070370	.588261	1.699923	.8619243	32	28	.5220024	.612000	1.633984	.8529440	32
29	.5072877	.588653	1.698792	.8617768	31	29	.5222505	.612400	1.632917	.8527921	31
30	.5075384	.589045	1.697663	.8616292	30	30	.5224986	.612800	1.631851	.8526402	30
31	.5077890	.589436	1.696534	.8614815	29	31	.5227466	.613201	1.630786	.8524881	29
32	.5080396	.589828	1.695406	.8613327	28	32	.5229945	.613601	1.629722	.8523360	28
33	.5082901	.590221	1.694280	.8611859	27	33	.5232424	.614001	1.628659	.8521839	27
34	.5085406	.590613	1.693155	.8610380	26	34	.5234903	.614402	1.627597	.8520316	26
35	.5087910	.591005	1.692030	.8608901	25	35	.5237381	.614803	1.626536	.8518793	25
36	.5090414	.591398	1.690907	.8607420	24	36	.5239859	.615204	1.625476	.8517269	24
37	.5092918	.591791	1.689785	.8605939	23	37	.5242336	.615605	1.624417	.8515745	23
38	.5095421	.592183	1.688664	.8604457	22	38	.5244813	.616006	1.623359	.8514219	22
39	.5097924	.592576	1.687544	.8602975	21	39	.5247290	.616407	1.622302	.8512693	21
40	.5100426	.592969	1.686426	.8601491	20	40	.5249766	.616809	1.621246	.8511167	20
41	.5102928	.593363	1.685308	.8600007	19	41	.5252241	.617210	1.620192	.8509639	19
42	.5105429	.593756	1.684191	.8598523	18	42	.5254717	.617612	1.619138	.8508111	18
43	.5107930	.594150	1.683076	.8597037	17	43	.5257191	.618014	1.618085	.8506582	17
44	.5110431	.594543	1.681962	.8595551	16	44	.5259665	.618416	1.617033	.8505053	16
45	.5112931	.594937	1.680848	.8594064	15	45	.5262139	.618818	1.615982	.8503522	15
46	.5115431	.595331	1.679736	.8592576	14	46	.5264613	.619221	1.614932	.8501991	14
47	.5117930	.595725	1.678625	.8591088	13	47	.5267085	.619623	1.613882	.8500459	13
48	.5120429	.596119	1.677515	.8589599	12	48	.5269558	.620026	1.612834	.8498927	12
49	.5122927	.596514	1.676406	.8588109	11	49	.5272030	.620429	1.611787	.8497394	11
50	.5125425	.596908	1.675298	.8586619	10	50	.5274502	.620832	1.610741	.8495860	10
51	.5127923	.597303	1.674192	.8585127	9	51	.5276973	.621235	1.609696	.8494325	9
52	.5130420	.597697	1.673086	.8583635	8	52	.5279443	.621638	1.608652	.8492790	8
53	.5132916	.598092	1.671981	.8582143	7	53	.5281914	.622041	1.607609	.8491254	7
54	.5135413	.598487	1.670878	.8580649	6	54	.5284383	.622445	1.606567	.8489717	6
55	.5137908	.598882	1.669775	.8579155	5	55	.5286853	.622848	1.605526	.8488179	5
56	.5140404	.599278	1.668674	.8577660	4	56	.5289322	.623252	1.604485	.8486641	4
57	.5142899	.599673	1.667574	.8576164	3	57	.5291790	.623656	1.603446	.8485102	3
58	.5145393	.600069	1.666474	.8574668	2	58	.5294258	.624060	1.602408	.8483562	2
59	.5147887	.600464	1.665376	.8573171	1	59	.5296726	.624465	1.601370	.8482022	1
60	.5150381	.600860	1.664279	.8571673	0	60	.5299193	.624869	1.600334	.8480481	0

| | Cosine. | Cotang | Tang. | Sine. | ′ | | | Cosine. | Cotang | Tang. | Sine. | ′ |

59°   58°

Note.—Secant = 1 ÷ cosine.   Cosecant = 1 ÷ sine.

2.—Natural Sines, Tangents, Cotangents, Cosines.—(Continued.)

(Versed sine = 1 − cosine; coversed sine = 1 − sine.)

32° 33°

'	Sine.	Tang.	Cotang.	Cosine.		'	Sine.	Tang.	Cotang.	Cosine.	
0	.5299193	.624869	1.600334	.8480481	60	0	.5446390	.649407	1.539865	.8386706	60
1	.5301659	.625273	1.599299	.8478939	59	1	.5448830	.649821	1.538884	.8385121	59
2	.5304125	.625678	1.598264	.8477397	58	2	.5451269	.650235	1.537905	.8383536	58
3	.5306591	.626083	1.597231	.8475853	57	3	.5453707	.650649	1.536927	.8381950	57
4	.5309057	.626488	1.596198	.8474309	56	4	.5456145	.651063	1.535949	.8380363	56
5	.5311521	.626893	1.595167	.8472765	55	5	.5458583	.651477	1.534972	.8378775	55
6	.5313986	.627298	1.594136	.8471219	54	6	.5461020	.651891	1.533996	.8377187	54
7	.5316450	.627704	1.593107	.8469673	53	7	.5463456	.652306	1.533021	.8375598	53
8	.5318913	.628109	1.592078	.8468126	52	8	.5465892	.652721	1.532047	.8374009	52
9	.5321376	.628515	1.591050	.8466579	51	9	.5468328	.653136	1.531074	.8372418	51
10	.5323839	.628921	1.590023	.8465030	50	10	.5470763	.653551	1.530102	.8370827	50
11	.5326301	.629327	1.588997	.8463481	49	11	.5473198	.653996	1.529130	.8369236	49
12	.5328763	.629733	1.587973	.8461932	48	12	.5475632	.654381	1.528160	.8367643	48
13	.5331224	.630139	1.586949	.8460381	47	13	.5478066	.654797	1.527190	.8366050	47
14	.5333685	.630546	1.585926	.8458830	46	14	.5480499	.655212	1.526221	.8364456	46
15	.5336145	.630953	1.584904	.8457278	45	15	.5482932	.655628	1.525253	.8362862	45
16	.5338605	.631359	1.583883	.8455726	44	16	.5485365	.656044	1.524286	.8361266	44
17	.5341065	.631766	1.582862	.8454172	43	17	.5487797	.656460	1.523320	.8359670	43
18	.5343523	.632173	1.581843	.8452618	42	18	.5490228	.656877	1.522354	.8358074	42
19	.5345982	.632581	1.580825	.8451064	41	19	.5492659	.657293	1.521389	.8356476	41
20	.5348440	.632988	1.579807	.8449508	40	20	.5495090	.657710	1.520426	.8354878	40
21	.5350898	.633395	1.578791	.8447952	39	21	.5497520	.658127	1.519463	.8353279	39
22	.5353355	.633803	1.577776	.8446395	38	22	.5499950	.658544	1.518501	.8351680	38
23	.5355812	.634211	1.576761	.8444838	37	23	.5502379	.658961	1.517540	.8350080	37
24	.5358268	.634619	1.575747	.8443279	36	24	.5504807	.659378	1.516579	.8348479	36
25	.5360724	.635027	1.574735	.8441720	35	25	.5507236	.659796	1.515620	.8346877	35
26	.5363179	.635435	1.573723	.8440161	34	26	.5509663	.660213	1.514661	.8345275	34
27	.5365634	.635844	1.572712	.8438600	33	27	.5512091	.660631	1.513703	.8343672	33
28	.5368089	.636252	1.571702	.8437039	32	28	.5514518	.661049	1.512746	.8342068	32
29	.5370543	.636661	1.570693	.8435477	31	29	.5516944	.661467	1.511790	.8340463	31
30	.5372996	.637070	1.569685	.8433914	30	30	.5519370	.661885	1.510835	.8338858	30
31	.5375449	.637479	1.568678	.8432351	29	31	.5521795	.662304	1.509880	.8337252	29
32	.5377902	.637888	1.567672	.8430787	28	32	.5524220	.662722	1.508927	.8335646	28
33	.5380354	.638297	1.566666	.8429222	27	33	.5526645	.663141	1.507974	.8334038	27
34	.5382806	.638707	1.565662	.8427657	26	34	.5529069	.663560	1.507022	.8332430	26
35	.5385257	.639116	1.564659	.8426091	25	35	.5531492	.663979	1.506071	.8330822	25
36	.5387708	.639526	1.563656	.8424524	24	36	.5533915	.664398	1.505121	.8329212	24
37	.5390158	.639936	1.562654	.8422956	23	37	.5536338	.664817	1.504171	.8327602	23
38	.5392608	.640346	1.561654	.8421388	22	38	.5538760	.665237	1.503222	.8325991	22
39	.5395058	.640756	1.560654	.8419819	21	39	.5541182	.665657	1.502275	.8324380	21
40	.5397507	.641167	1.559655	.8418249	20	40	.5543603	.666076	1.501328	.8322768	20
41	.5399955	.641577	1.558657	.8416679	19	41	.5546024	.666496	1.500382	.8321155	19
42	.5402403	.641988	1.557660	.8415108	18	42	.5548444	.666917	1.499436	.8319541	18
43	.5404851	.642399	1.556663	.8413536	17	43	.5550864	.667337	1.498492	.8317927	17
44	.5407298	.642810	1.555668	.8411963	16	44	.5553283	.667758	1.497548	.8316312	16
45	.5409745	.643221	1.554674	.8410390	15	45	.5555702	.668178	1.496605	.8314696	15
46	.5412191	.643632	1.553680	.8408816	14	46	.5558121	.668599	1.495663	.8313080	14
47	.5414637	.644044	1.552688	.8407241	13	47	.5560539	.669020	1.494722	.8311463	13
48	.5417082	.644456	1.551696	.8405666	12	48	.5562956	.669441	1.493782	.8309845	12
49	.5419527	.644867	1.550705	.8404090	11	49	.5565373	.669863	1.492842	.8308226	11
50	.5421971	.645279	1.549715	.8402513	10	50	.5567790	.670284	1.491903	.8306607	10
51	.5424415	.645691	1.548726	.8400936	9	51	.5570206	.670706	1.490965	.8304987	9
52	.5426859	.646104	1.547738	.8399357	8	52	.5572621	.671128	1.490028	.8303366	8
53	.5429302	.646516	1.546751	.8397778	7	53	.5575036	.671550	1.489092	.8301745	7
54	.5431744	.646929	1.545764	.8396199	6	54	.5577451	.671972	1.488157	.8300123	6
55	.5434187	.647341	1.544779	.8394618	5	55	.5579865	.672394	1.487222	.8298500	5
56	.5436628	.647754	1.543794	.8393037	4	56	.5582279	.672816	1.486288	.8296877	4
57	.5439069	.648167	1.542810	.8391455	3	57	.5584692	.673239	1.485355	.8295252	3
58	.5441510	.648580	1.541828	.8389873	2	58	.5587105	.673662	1.484423	.8293628	2
59	.5443951	.648994	1.540846	.8388290	1	59	.5589517	.674085	1.483491	.8292002	1
60	.5446390	.649407	1.539865	.8386706	0	60	.5591929	.674508	1.482561	.8290376	0
	Cosine.	Cotang	Tang.	Sine.	'		Cosine.	Cotang	Tang.	Sine.	'

57° 56°

Note.—Secant = 1 ÷ cosine. Cosecant = 1 ÷ sine.

2.–Natural Sines, Tangents, Cotangents, Cosines.—(Continued.)

(Versed sine = 1 − cosine; coversed sine = 1 − sine.)

34°   **35°**

′	Sine.	Tang.	Cotang.	Cosine.		′	Sine.	Tang.	Cotang.	Cosine.	
0	.5591929	.674508	1.482561	.8290376	60	0	.5735764	.700207	1.428148	.8191520	60
1	.5594340	.674931	1.481631	.8288749	59	1	.5738147	.700641	1.427264	.8189852	59
2	.5596751	.675355	1.480702	.8287121	58	2	.5740529	.701074	1.426381	.8188182	58
3	.5599162	.675779	1.479773	.8285493	57	3	.5742911	.701508	1.425498	.8186512	57
4	.5601572	.676202	1.478846	.8283864	56	4	.5745292	.701943	1.424617	.8184841	56
5	.5603981	.676626	1.477919	.8282234	55	5	.5747672	.702377	1.423736	.8183169	55
6	.5606390	.677050	1.476993	.8280603	54	6	.5750053	.702811	1.422856	.8181497	54
7	.5608798	.677475	1.476068	.8278972	53	7	.5752432	.703246	1.421976	.8179824	53
8	.5611206	.677899	1.475144	.8277340	52	8	.5754811	.703681	1.421097	.8178151	52
9	.5613614	.678324	1.474221	.8275708	51	9	.5757190	.704116	1.420220	.8176476	51
10	.5616021	.678749	1.473298	.8274074	50	10	.5759568	.704551	1.419342	.8174801	50
11	.5618428	.679174	1.472376	.8272440	49	11	.5761946	.704986	1.418466	.8173125	49
12	.5620834	.679599	1.471455	.8270806	48	12	.5764323	.705422	1.417590	.8171449	48
13	.5623239	.680024	1.470535	.8269170	47	13	.5766700	.705858	1.416715	.8169772	47
14	.5625645	.680450	1.469615	.8267534	46	14	.5769076	.706294	1.415840	.8168094	46
15	.5628049	.680875	1.468696	.8265897	45	15	.5771452	.706730	1.414967	.8166416	45
16	.5630453	.681301	1.467778	.8264260	44	16	.5773827	.707166	1 414094	.8164736	44
17	.5632857	.681727	1.466861	.8262622	43	17	.5776202	.707602	1.413222	.8163056	43
18	.5635260	.682153	1.465945	.8260983	42	18	.5778576	.708039	1 412350	.8161376	42
19	.5637663	.682580	1.465029	.8259343	41	19	.5780950	.708476	1.411479	.8159695	41
20	.5640066	.683006	1.464114	.8257703	40	20	.5783323	.708913	1.410609	.8158013	40
21	.5642467	.683433	1.463200	.8256062	39	21	.5785696	.709350	1.409740	.8156330	39
22	.5644869	.683860	1.462287	.8254420	38	22	.5788069	.709787	1.408871	.8154647	38
23	.5647270	.684287	1.461374	.8252778	37	23	.5790440	.710225	1.408003	.8152963	37
24	.5649670	.684714	1.460463	.8251135	36	24	.5792812	.710663	1.407136	.8151278	36
25	.5652070	.685141	1.459552	.8249491	35	25	.5795183	.711100	1.406270	.8149593	35
26	.5654469	.685569	1.458642	.8247847	34	26	.5797553	.711539	1 405404	.8147906	34
27	.5656868	.685996	1.457732	.8246202	33	27	.5799923	.711977	1.404539	.8146220	33
28	.5659267	.686424	1.456824	.8244556	32	28	.5802292	.712415	1.403674	.8144532	32
29	.5661665	.686852	1.455916	.8242909	31	29	.5804661	.712854	1.402811	.8142844	31
30	.5664062	.687281	1.455009	.8241262	30	30	.5807030	.713293	1.401948	.8141155	30
31	.5666459	.687709	1.454102	.8239614	29	31	.5809397	.713732	1.401086	.8139466	29
32	.5668856	.688137	1.453197	.8237965	28	32	.5811765	.714171	1.400224	.8137775	28
33	.5671252	.688566	1.452292	.8236316	27	33	.5814132	.714610	1.399363	.8136084	27
34	.5673648	.688995	1.451388	.8234666	26	34	.5816498	.715050	1.398503	.8134393	26
35	.5676043	.689424	1.450485	.8233015	25	35	.5818864	.715490	1.397644	.8132701	25
36	.5678437	.689853	1.449582	.8231364	24	36	.5821230	.715929	1.396785	.8131008	24
37	.5680832	.690283	1.448680	.8229712	23	37	.5823595	.716369	1.395927	.8129314	23
38	.5683225	.690712	1.447779	.8228059	22	38	.5825959	.716810	1.395069	.8127620	22
39	.5685619	.691142	1.446879	.8226405	21	39	.5828323	.717250	1.394213	.8125925	21
40	.5688011	.691572	1.445980	.8224751	20	40	.5830687	.717691	1.393357	.8124229	20
41	.5690403	.692002	1.445081	.8223096	19	41	.5833050	.718131	1.392501	.8122532	19
42	.5692795	.692432	1.444183	.8221440	18	42	.5835412	.718572	1.391647	.8120835	18
43	.5695187	.692863	1.443286	.8219784	17	43	.5837774	.719014	1.390793	.8119137	17
44	.5697577	.693293	1.442389	.8218127	16	44	.5840136	.719455	1.389940	.8117439	16
45	.5699968	.693724	1.441494	.8216469	15	45	.5842497	.719897	1.389087	.8115740	15
46	.5702357	.694155	1.440599	.8214811	14	46	.5844857	.720338	1.388235	.8114040	14
47	.5704747	.694586	1.439704	.8213152	13	47	.5847217	.720780	1.387384	.8112339	13
48	.5707136	.695018	1.438811	.8211492	12	48	.5849577	.721222	1.386534	.8110638	12
49	.5709524	.695449	1.437918	.8209832	11	49	.5851936	.721665	1.385684	.8108936	11
50	.5711912	.695881	1.437026	.8208170	10	50	.5854294	.722107	1.384835	.8107234	10
51	.5714299	.696313	1.436135	.8206509	9	51	.5856652	.722550	1.383986	.8105530	9
52	.5716686	.696745	1.435245	.8204846	8	52	.5859010	.722993	1.383139	.8103826	8
53	.5719073	.697177	1.434355	.8203183	7	53	.5861367	.723436	1.382292	.8102122	7
54	.5721459	.697609	1.433466	.8201519	6	54	.5863724	.723879	1.381445	.8100416	6
55	.5723844	.698042	1.432578	.8199854	5	55	.5866080	.724322	1.380600	.8098710	5
56	.5726229	.698474	1.431690	.8198189	4	56	.5868435	.724766	1.379755	.8097004	4
57	.5728614	.698907	1.430803	.8196523	3	57	.5870790	.725210	1.378910	.8095296	3
58	.5730998	.699340	1.429917	.8194856	2	58	.5873145	.725654	1.378067	.8093588	2
59	.5733381	.699774	1.429032	.8193189	1	59	.5875499	.726098	1.377224	.8091879	1
60	.5735764	.700207	1.428148	.8191520	0	60	.5877853	.726542	1.376381	.8090170	0
	Cosine.	Cotang	Tang.	Sine.	′		Cosine.	Cotang	Tang.	Sine.	′

55°   **54°**

Note.—Secant = 1 ÷ cosine.   Cosecant = 1 ÷ sine.

2.–Natural Sines, Tangents, Cotangents, Cosines.—(Continued.)

(Versed sine = 1 − cosine; coversed sine = 1 − sine.)

36°   37°

′	Sine.	Tang.	Cotang.	Cosine.		′	Sine.	Tang.	Cotang.	Cosine.	
0	.5877853	.726542	1.376381	.8090170	60	0	.6018150	.753554	1.327044	.7986355	60
1	.5880206	.726987	1.375540	.8088460	59	1	.6020473	.754010	1.326242	.7984604	59
2	.5882558	.727431	1.374699	.8086749	58	2	.6022795	.754466	1.325439	.7982853	58
3	.5884910	.727876	1.373859	.8035037	57	3	.6025117	.754923	1.324638	.7981100	57
4	.5887262	.728321	1.373019	.8083325	56	4	.6027439	.755379	1.323837	.7979347	56
5	.5889613	.728767	1.372180	.8081612	55	5	.6029760	.755836	1.323036	.7977594	55
6	.5891964	.729212	1.371342	.8079899	54	6	.6032080	.756294	1.322237	.7975839	54
7	.5894314	.729658	1.370504	.8078185	53	7	.6034400	.756751	1.321437	.7974084	53
8	.5896663	.730104	1.369667	.8076470	52	8	.6036719	.757209	1.320639	.7972329	52
9	.5899012	.730550	1.368831	.8074754	51	9	.6039038	.757666	1.319841	.7970572	51
10	.5901361	.730996	1.367995	.8073038	50	10	.6041356	.758124	1.319044	.7968811	50
11	.5903709	.731442	1.367161	.8071321	49	11	.6043674	.758582	1.318247	.7967058	49
12	.5906057	.731889	1.366326	.8069602	48	12	.6045991	.759041	1.317451	.7965299	48
13	.5908404	.732336	1.365493	.8067885	47	13	.6048308	.759499	1.316655	.7963540	47
14	.5910750	.732783	1.364660	.8066166	46	14	.6050624	.759958	1.315861	.7961780	46
15	.5913096	.733230	1.363827	.8064446	45	15	.6052940	.760417	1.315066	.7960020	45
16	.5915442	.733677	1.362996	.8062726	44	16	.6055255	.760876	1.314273	.7958259	44
17	.5917787	.734125	1.362165	.8061005	43	17	.6057570	.761336	1.313480	.7956497	43
18	.5920132	.734573	1.361335	.8059283	42	18	.6059884	.761795	1.312687	.7954735	42
19	.5922476	.735021	1.360505	.8057560	41	19	.6062198	.762255	1.311895	.7952972	41
20	.5924819	.735469	1.359676	.8055837	40	20	.6064511	.762715	1.311104	.7951208	40
21	.5927163	.735917	1.358848	.8054113	39	21	.6066824	.763175	1.310314	.7949444	39
22	.5929505	.736366	1.358020	.8052389	38	22	.6069136	.763636	1.309523	.7947678	38
23	.5931847	.736814	1.357193	.8050664	37	23	.6071447	.764096	1.308734	.7945913	37
24	.5934189	.737263	1.356367	.8048938	36	24	.6073758	.764557	1.307945	.7944146	36
25	.5936530	.737712	1.355541	.8047211	35	25	.6076069	.765018	1.307157	.7942379	35
26	.5938871	.738162	1.354716	.8045484	34	26	.6078379	.765480	1.306369	.7940611	34
27	.5941211	.738611	1.353891	.8043756	33	27	.6080689	.765941	1.305582	.7938843	33
28	.5943550	.739061	1.353068	.8042028	32	28	.6082998	.766403	1.304796	.7937074	32
29	.5945889	.739511	1.352244	.8040299	31	29	.6085306	.766864	1.304010	.7935304	31
30	.5948228	.739961	1.351422	.8038569	30	30	.6087614	.767327	1.303225	.7933533	30
31	.5950566	.740411	1.350600	.8036838	29	31	.6089922	.767789	1.302440	.7931762	29
32	.5952904	.740861	1.349779	.8035107	28	32	.6092229	.768251	1.301656	.7929990	28
33	.5955241	.741312	1.348958	.8033375	27	33	.6094535	.768714	1.300873	.7928218	27
34	.5957577	.741763	1.348139	.8031642	26	34	.6096841	.769177	1.300090	.7926445	26
35	.5959913	.742214	1.347319	.8029909	25	35	.6099147	.769640	1.299308	.7924671	25
36	.5962249	.742665	1.346501	.8028175	24	36	.6101452	.770103	1.298526	.7922896	24
37	.5964584	.743117	1.345683	.8026440	23	37	.6103756	.770567	1.297745	.7921121	23
38	.5966918	.743568	1.344865	.8024705	22	38	.6106060	.771030	1.296964	.7919345	22
39	.5969252	.744020	1.344049	.8022969	21	39	.6108363	.771494	1.296185	.7917569	21
40	.5971586	.744472	1.343233	.8021232	20	40	.6110666	.771958	1.295405	.7915792	20
41	.5973919	.744924	1.342417	.8019495	19	41	.6112969	.772423	1.294627	.7914014	19
42	.5976251	.745377	1.341602	.8017756	18	42	.6115270	.772887	1.293848	.7912235	18
43	.5978583	.745829	1.340788	.8016018	17	43	.6117572	.773352	1.293071	.7910456	17
44	.5980915	.746282	1.339975	.8014278	16	44	.6119873	.773817	1.292294	.7908676	16
45	.5983246	.746735	1.339162	.8012538	15	45	.6122173	.774282	1.291517	.7906896	15
46	.5985577	.747188	1.338350	.8010797	14	46	.6124473	.774748	1.290742	.7905115	14
47	.5987906	.747642	1.337538	.8009056	13	47	.6126772	.775213	1.289966	.7903333	13
48	.5990236	.748095	1.336727	.8007314	12	48	.6129071	.775679	1.289192	.7901550	12
49	.5992565	.748549	1.335917	.8005571	11	49	.6131369	.776145	1.288418	.7899767	11
50	.5994893	.749003	1.335107	.8003827	10	50	.6133666	.776611	1.287644	.7897983	10
51	.5997221	.749457	1.334298	.8002083	9	51	.6135964	.777078	1.286871	.7896198	9
52	.5999549	.749911	1.333490	.8000338	8	52	.6138260	.777544	1.286099	.7894413	8
53	.6001876	.750366	1.332682	.7998593	7	53	.6140556	.778011	1.285327	.7892627	7
54	.6004202	.750821	1.331875	.7996847	6	54	.6142852	.778478	1.284556	.7890841	6
55	.6006528	.751276	1.331068	.7995100	5	55	.6145147	.778946	1.283786	.7889054	5
56	.6008854	.751731	1.330262	.7993352	4	56	.6147442	.779413	1.283016	.7887266	4
57	.6011179	.752186	1.329457	.7991604	3	57	.6149736	.779881	1.282246	.7885477	3
58	.6013503	.752642	1.328652	.7989855	2	58	.6152029	.780349	1.281477	.7883688	2
59	.6015827	.753098	1.327848	.7988105	1	59	.6154322	.780817	1.280709	.7881898	1
60	.6018150	.753554	1.327044	.7986355	0	60	.6156615	.781285	1.279941	.7880108	0

| Cosine. | Cotang | Tang. | Sine. | ′ | | Cosine. | Cotang | Tang. | Sine. | ′ |

53°   52°

Note.—Secant = 1 ÷ cosine.   Cosecant = 1 ÷ sine.

2.—Natural Sines, Tangents, Cotangents, Cosines.—(Continued.)

(Versed sine = 1 − cosine; coversed sine = 1 − sine.)

38°   39°

′	Sine.	Tang.	Cotang.	Cosine.	′	Sine.	Tang.	Cotang.	Cosine.	
0	.6156615	.781285	1.279941	.7880108	0	.6293204	.809784	1.234897	.7771460	60
1	.6158907	.781754	1.279174	.7878316	1	.6295464	.810265	1.234162	.7769629	59
2	.6161198	.782222	1.278407	.7876524	2	.6297724	.810747	1.233429	.7767797	58
3	.6163489	.782691	1.277641	.7874732	3	.6299983	.811230	1.232696	.7765965	57
4	.6165780	.783161	1.276876	.7872939	4	.6302242	.811712	1.231963	.7764132	56
5	.6168069	.783630	1.276111	.7871145	5	.6304500	.812195	1.231231	.7762298	55
6	.6170359	.784100	1.275347	.7869350	6	.6306758	.812678	1.230499	.7760464	54
7	.6172648	.784570	1.274583	.7867555	7	.6309015	.813161	1.229768	.7758629	53
8	.6174936	.785040	1.273820	.7865759	8	.6311272	.813644	1.229038	.7756794	52
9	.6177224	.785510	1.273057	.7863963	9	.6313528	.814128	1.228308	.7754957	51
10	6179511	.785980	1.272295	.7862165	10	.6315784	.814611	1.227578	.7753121	50
11	.6181798	.786451	1.271534	.7860367	11	.6318039	.815095	1.226849	.7751283	49
12	.6184084	.786922	1.270773	.7858569	12	.6320293	.815580	1.226121	.7749445	48
13	.6186370	.787393	1.270013	.7856770	13	.6322547	.816064	1.225393	.7747606	47
14	.6188655	.787864	1.269253	.7854970	14	.6324800	.816549	1.224665	.7745767	46
15	.6190939	.788336	1.268494	.7853169	15	.6327053	.817034	1.223938	.7743926	45
16	.6193224	.788808	1.267735	.7851368	16	.6329306	.817519	1.223212	.7742086	44
17	.6195507	.789280	1.266977	.7849566	17	.6331557	.818004	1.222486	.7740244	43
18	.6197790	.789752	1.266219	.7847764	18	.6333809	.818490	1.221761	.7738402	42
19	.6200073	.790224	1.265462	.7845961	19	.6336059	.818976	1.221036	.7736559	41
20	.6202355	.790697	1.264706	.7844157	20	.6338310	.819462	1.220312	.7734716	40
21	.6204636	.791170	1.263950	.7842352	21	.6340559	.819948	1.219588	.7732872	39
22	.6206917	.791643	1.263195	.7840547	22	.6342808	.820435	1.218865	.7731027	38
23	.6209198	.792116	1.262440	.7838741	23	.6345057	.820922	1.218142	.7729182	37
24	.6211478	.792590	1.261686	.7836935	24	.6347305	.821409	1.217419	.7727336	36
25	.6213757	.793064	1.260932	.7835127	25	.6349553	.821896	1.216698	.7725489	35
26	.6216036	.793537	1.260179	.7833320	26	.6351800	.822384	1.215976	.7723642	34
27	.6218314	.794012	1.259426	.7831511	27	.6354046	.822871	1.215256	.7721794	33
28	.6220592	.794486	1.258674	.7829702	28	.6356292	.823359	1.214535	.7719945	32
29	.6222870	.794961	1.257923	.7827892	29	.6358537	.823847	1.213816	.7718096	31
30	.6225146	.795435	1.257172	.7826082	30	.6360782	.824336	1.213097	.7716246	30
31	.6227423	.795911	1.256421	.7824270	31	.6363026	824825	1.212378	.7714395	29
32	.6229698	.796386	1.255672	.7822459	32	.6365270	.825314	1.211660	.7712544	28
33	.6231974	.796861	1.254922	.7820646	33	.6367513	.825803	1.210942	.7710692	27
34	.6234248	.797337	1.254174	.7818833	34	.6369756	.826292	1.210225	.7708840	26
35	.6236522	.797813	1.253426	.7817019	35	.6371998	.826782	1.209508	.7706986	25
36	.6238796	.798289	1.252678	.7815205	36	.6374240	.827271	1.208792	.7705132	24
37	.6241069	.798765	1.251931	.7813390	37	.6376481	.827762	1.208076	.7703278	23
38	.6243342	.799242	1.251184	.7811574	38	.6378721	.828252	1.207361	.7701423	22
39	.6245614	.799719	1.250438	.7809757	39	.6380961	.828742	1 206646	.7699567	21
40	.6247885	.800196	1.249693	.7807940	40	.6383201	.829233	1.205932	.7697710	20
41	.6250156	.800673	1.248948	.7806123	41	.6385440	.829724	1.205219	.7695853	19
42	.6252427	.801151	1.248204	.7804304	42	.6387678	.830216	1.204505	.7693996	18
43	.6254696	.801628	1.247460	.7802485	43	.6389916	.830707	1.203793	.7692137	17
44	.6256966	.802106	1.246716	.7800665	44	.6392153	.831199	1.203081	.7690278	16
45	.6259235	.802584	1.245974	.7798845	45	.6394390	.831691	1.202369	.7688418	15
46	.6261503	.803063	1.245232	.7797024	46	.6396626	.832183	1.201658	.7686558	14
47	.6263771	.803541	1.244490	.7795202	47	.6398862	.832675	1.200947	.7684697	13
48	.6266038	.804020	1.243749	.7793380	48	.6401097	.833168	1.200237	.7682835	12
49	.6268305	.804499	1.243008	.7791557	49	.6403332	.833661	1 199527	.7680973	11
50	.6270571	.804979	1.242268	.7789733	50	.6405566	.834154	1 198818	.7679110	10
51	.6272837	.805458	1.241529	.7787909	51	.6407799	.834648	1 198109	.7677246	9
52	.6275102	.805938	1.240790	.7786084	52	.6410032	.835141	1 197401	.7675382	8
53	.6277366	.806418	1.240051	.7784258	53	.6412264	.835635	1.196693	.7673517	7
54	.6279631	.806898	1 239313	.7782431	54	.6414496	.836129	1 195986	.7671652	6
55	.6281894	.807378	1.238576	.7780604	55	.6416728	.836624	1.195279	.7669785	5
56	.6284157	.807859	1 237839	.7778777	56	.6418958	.837118	1.194573	.7667918	4
57	.6286420	.808340	1.237103	.7776949	57	.6421189	.837613	1 193867	.7666051	3
58	.6288682	.808821	1.236367	.7775120	58	.6423418	.838108	1.193162	.7664183	2
59	.6290943	.809302	1.235631	.7773290	59	.6425647	.838604	1 192457	.7662314	1
60	.6293204	809784	1.234897	.7771460	60	.6427876	.839099	1.191753	.7660444	0
	Cosine.	Cotang	Tang.	Sine.	′	Cosine.	Cotang	Tang.	Sine.	′

51°   50°

Note.—Secant = 1 ÷ cosine.   Cosecant = 1 ÷ sine

2.—Natural Sines, Tangents, Cotangents, Cosines.—(Continued.)

(Versed sine = 1 − cosine; coversed sine = 1 − sine.)

40°   41°

'	Sine.	Tang.	Cotang.	Cosine.		'	Sine.	Tang.	Cotang.	Cosine.	
0	.6427876	.839099	1.191753	.7660444	60	0	.6560590	.869286	1.150368	.7547096	60
1	.6430104	.839595	1.191049	.7658574	59	1	.6562785	.869797	1.149692	.7545187	59
2	.6432332	.840091	1.190346	.7656704	58	2	.6564980	.870308	1.149017	.7543278	58
3	.6434559	.840587	1.189643	.7654832	57	3	.6567174	.870820	1.148342	.7541368	57
4	.6436785	.841084	1.188941	.7652960	56	4	.6569367	.871331	1.147668	.7539457	56
5	.6439011	.841581	1.188239	.7651087	55	5	.6571560	.871843	1.146994	.7537546	55
6	.6441236	.842078	1.187538	.7649214	54	6	.6573752	.872355	1.146321	.7535634	54
7	.6443461	.842575	1.186837	.7647340	53	7	.6575944	.872868	1.145648	.7533721	53
8	.6445685	.843073	1.186136	.7645465	52	8	.6578135	.873380	1.144976	.7531808	52
9	.6447909	.843570	1.185437	.7643590	51	9	.6580326	.873893	1.144304	.7529894	51
10	.6450132	.844068	1.184737	.7641714	50	10	.6582516	.874406	1.143632	.7527980	50
11	.6452355	.844567	1.184038	.7639838	49	11	.6584706	.874920	1.142961	.7526065	49
12	.6454577	.845065	1.183340	.7637960	48	12	.6586895	.875433	1.142290	.7524149	48
13	.6456798	.845564	1.182642	.7636082	47	13	.6589083	.875947	1.141620	.7522233	47
14	.6459019	.846063	1.181944	.7634204	46	14	.6591271	.876462	1.140950	.7520316	46
15	.6461240	.846562	1.181247	.7632325	45	15	.6593458	.876976	1.140281	.7518398	45
16	.6463460	.847062	1.180551	.7630445	44	16	.6595645	.877491	1.139612	.7516480	44
17	.6465679	.847561	1.179855	.7628564	43	17	.6597831	.878006	1.138944	.7514561	43
18	.6467898	.848061	1.179159	.7626683	42	18	.6600017	.878521	1.138276	.7512641	42
19	.6470116	.848562	1.178464	.7624802	41	19	.6602202	.879037	1.137608	.7510721	41
20	.6472334	.849062	1.177769	.7622919	40	20	.6604386	.879552	1.136941	.7508800	40
21	.6474551	.849563	1.177075	.7621036	39	21	.6606570	.880068	1.136274	.7506879	39
22	.6476767	.850064	1.176382	.7619152	38	22	.6608754	.880585	1.135608	.7504957	38
23	.6478984	.850565	1.175688	.7617268	37	23	.6610936	.881101	1.134942	.7503034	37
24	.6481199	.851066	1.174996	.7615383	36	24	.6613119	.881618	1.134277	.7501111	36
25	.6483414	.851568	1.174303	.7613497	35	25	.6615300	.882135	1.133612	.7499187	35
26	.6485628	.852070	1.173612	.7611611	34	26	.6617482	.882653	1.132947	.7497262	34
27	.6487842	.852572	1.172920	.7609724	33	27	.6619662	.883170	1.132283	.7495337	33
28	.6490056	.853075	1.172229	.7607837	32	28	.6621842	.883688	1.131620	.7493411	32
29	.6492268	.853577	1.171539	.7605949	31	29	.6624022	.884206	1.130957	.7491484	31
30	.6494480	.854080	1.170849	.7604060	30	30	.6626200	.884725	1.130294	.7489557	30
31	.6496692	.854583	1.170160	.7602170	29	31	.6628379	.885244	1.129632	.7487629	29
32	.6498903	.855087	1.169471	.7600280	28	32	.6630557	.885763	1.128970	.7485701	28
33	.6501114	.855591	1.168782	.7598389	27	33	.6632734	.886282	1.128308	.7483772	27
34	.6503324	.856095	1.168094	.7596498	26	34	.6634910	.886801	1.127647	.7481842	26
35	.6505533	.856599	1.167407	.7594606	25	35	.6637087	.887321	1.126987	.7479912	25
36	.6507742	.857103	1.166720	.7592713	24	36	.6639262	.887841	1.126327	.7477981	24
37	.6509951	.857608	1.166033	.7590820	23	37	.6641437	.888361	1.125667	.7476049	23
38	.6512158	.858113	1.165347	.7588926	22	38	.6643612	.888882	1.125008	.7474117	22
39	.6514366	.858618	1.164661	.7587031	21	39	.6645785	.889403	1.124349	.7472184	21
40	.6516572	.859123	1.163976	.7585136	20	40	.6647959	.889924	1.123690	.7470251	20
41	.6518778	.859629	1.163291	.7583240	19	41	.6650131	.890445	1.123032	.7468317	19
42	.6520984	.860135	1.162607	.7581343	18	42	.6652304	.890967	1.122375	.7466382	18
43	.6523189	.860641	1.161923	.7579446	17	43	.6654475	.891489	1.121718	.7464446	17
44	.6525394	.861148	1.161240	.7577548	16	44	.6656646	.892011	1.121061	.7462510	16
45	.6527598	.861655	1.160557	.7575650	15	45	.6658817	.892534	1.120405	.7460574	15
46	.6529801	.862162	1.159874	.7573751	14	46	.6660987	.893056	1.119749	.7458636	14
47	.6532004	.862669	1.159192	.7571851	13	47	.6663156	.893579	1.119094	.7456699	13
48	.6534206	.863176	1.158511	.7569951	12	48	.6665325	.894103	1.118439	.7454760	12
49	.6536408	.863684	1.157830	.7568050	11	49	.6667493	.894626	1.117784	.7452821	11
50	.6538609	.864192	1.157149	.7566148	10	50	.6669661	.895150	1.117130	.7450881	10
51	.6540810	.864700	1.156469	.7564246	9	51	.6671828	.895674	1.116476	.7448941	9
52	.6543010	.865209	1.155789	.7562343	8	52	.6673994	.896199	1.115823	.7446999	8
53	.6545209	.865718	1.155110	.7560439	7	53	.6676160	.896723	1.115170	.7445058	7
54	.6547408	.866227	1.154431	.7558535	6	54	.6678326	.897248	1.114518	.7443115	6
55	.6549607	.866736	1.153753	.7556630	5	55	.6680490	.897773	1.113866	.7441173	5
56	.6551804	.867246	1.153075	.7554724	4	56	.6682655	.898299	1.113214	.7439229	4
57	.6554002	.867755	1.152397	.7552818	3	57	.6684818	.898825	1.112563	.7437285	3
58	.6556198	.868265	1.151721	.7550911	2	58	.6686981	.899351	1.111912	.7435340	2
59	.6558395	.868775	1.151044	.7549004	1	59	.6689144	.899877	1.111262	.7433394	1
60	.6560590	.869286	1.150368	.7547096	0	60	.6691305	.900404	1.110612	.7431448	0
	Cosine.	Cotang	Tang.	Sine.	'		Cosine.	Cotang	Tang.	Sine.	'

49°   48°

Note.—Secant = 1 ÷ cosine.   Cosecant = 1 ÷ sine.

2.—Natural Sines, Tangents, Cotangents, Cosines.—(Continued.)

(Versed sine = 1 − cosine; coversed sine = 1 − sine.)

42°   43°

'	Sine.	Tang.	Cotang.	Cosine.	'	Sine.	Tang.	Cotang.	Cosine.	
0	.6691306	.900404	1.110612	.7431448	0	.6819984	.932515	1.072368	.7313537	60
1	.6693468	.900930	1.109963	.7429502	1	.6822111	.933059	1.071743	.7311553	59
2	.6695628	.901458	1.109314	.7427554	2	.6824237	.933603	1.071118	.7309568	58
3	.6697789	.901985	1.108665	.7425606	3	.6826363	.934147	1.070494	.7307583	57
4	.6699948	.902513	1.108017	.7423658	4	.6828489	.934692	1.069870	.7305597	56
5	.6702108	.903041	1.107369	.7421708	5	.6830613	.935238	1.069246	.7303610	55
6	.6704266	.903569	1.106721	.7419758	6	.6832738	.935783	1.068623	.7301623	54
7	.6706424	.904097	1.106075	.7417808	7	.6834861	.936329	1.068000	.7299635	53
8	.6708582	.904626	1.105428	.7415857	8	.6836984	.936875	1.067377	.7297646	52
9	.6710739	.905155	1.104782	.7413905	9	.6839107	.937421	1.066755	.7295657	51
10	.6712895	.905685	1.104136	.7411953	10	.6841229	.937968	1.066134	.7293668	50
11	.6715051	.906214	1.103491	.7410000	11	.6843350	.938515	1.065512	.7291677	49
12	.6717206	.906744	1.102846	.7408046	12	.6845471	.939062	1.064891	.7289686	48
13	.6719361	.907274	1.102201	.7406092	13	.6847591	.939610	1.064271	.7287695	47
14	.6721515	.907805	1.101557	.7404137	14	.6849711	.940157	1.063651	.7285703	46
15	.6723668	.908336	1.100914	.7402181	15	.6851830	.940706	1.063031	.7283710	45
16	.6725821	.908867	1.100270	.7400225	16	.6853948	.941254	1.062411	.7281716	44
17	.6727973	.909398	1.099628	.7398268	17	.6856066	.941803	1.061792	.7279722	43
18	.6730125	.909930	1.098985	.7396311	18	.6858184	.942352	1.061174	.7277728	42
19	.6732276	.910461	1.098343	.7394353	19	.6860300	.942901	1.060556	.7275732	41
20	.6734427	.910994	1.097702	.7392394	20	.6862416	.943451	1.059938	.7273736	40
21	.6736577	.911526	1.097060	.7390435	21	.6864532	.944001	1.059320	.7271740	39
22	.6738727	.912059	1.096420	.7388475	22	.6866647	.944551	1.058703	.7269743	38
23	.6740876	.912592	1.095779	.7386515	23	.6868761	.945102	1.058086	.7267745	37
24	.6743024	.913125	1.095139	.7384553	24	.6870875	.945653	1.057470	.7265747	36
25	.6745172	.913659	1.094500	.7382592	25	.6872988	.946204	1.056854	.7263748	35
26	.6747319	.914192	1.093861	.7380629	26	.6875101	.946755	1.056238	.7261748	34
27	.6749466	.914727	1.093222	.7378666	27	.6877213	.947307	1.055623	.7259748	33
28	.6751612	.915261	1.092584	.7376703	28	.6879325	.947859	1.055008	.7257747	32
29	.6753757	.915796	1.091946	.7374738	29	.6881435	.948411	1.054394	.7255746	31
30	.6755902	.916331	1.091308	.7372773	30	.6883546	.948964	1.053780	.7253744	30
31	.6758046	.916866	1.090671	.7370808	31	.6885655	.949517	1.053166	.7251741	29
32	.6760190	.917402	1.090034	.7368842	32	.6887765	.950070	1.052553	.7249738	28
33	.6762333	.917937	1.089398	.7366875	33	.6889873	.950624	1.051940	.7247734	27
34	.6764476	.918474	1.088762	.7364908	34	.6891981	.951178	1.051327	.7245729	26
35	.6766618	.919010	1.088126	.7362940	35	.6894089	.951732	1.050715	.7243724	25
36	.6768760	.919547	1.087491	.7360971	36	.6896195	.952287	1.050103	.7241719	24
37	.6770901	.920084	1.086857	.7359002	37	.6898302	.952842	1.049492	.7239712	23
38	.6773041	.920621	1.086222	.7357032	38	.6900407	.953397	1.048880	.7237705	22
39	.6775181	.921159	1.085588	.7355061	39	.6902512	.953952	1.048270	.7235698	21
40	.6777320	.921696	1.084955	.7353090	40	.6904617	.954508	1.047659	.7233690	20
41	.6779459	.922235	1.084322	.7351118	41	.6906721	.955064	1.047049	.7231681	19
42	.6781597	.922773	1.083689	.7349146	42	.6908824	.955620	1.046440	.7229671	18
43	.6783734	.923312	1.083057	.7347173	43	.6910927	.956177	1.045831	.7227661	17
44	.6785871	.923851	1.082425	.7345199	44	.6913029	.956734	1.045222	.7225651	16
45	.6788007	.924391	1.081793	.7343225	45	.6915131	.957291	1.044613	.7223640	15
46	.6790143	.924930	1.081162	.7341250	46	.6917232	.957849	1.044005	.7221628	14
47	.6792278	.925470	1.080532	.7339275	47	.6919332	.958407	1.043397	.7219615	13
48	.6794413	.926010	1.079901	.7337299	48	.6921432	.958965	1.042790	.7217602	12
49	.6796547	.926550	1.079271	.7335322	49	.6923531	.959524	1.042183	.7215589	11
50	.6798681	.927091	1.078642	.7333345	50	.6925629	.960082	1.041576	.7213574	10
51	.6800813	.927632	1.078013	.7331367	51	.6927728	.960642	1.040970	.7211559	9
52	.6802946	.928173	1.077384	.7329388	52	.6929825	.961201	1.040364	.7209544	8
53	.6805078	.928715	1.076756	.7327409	53	.6931922	.961761	1.039758	.7207528	7
54	.6807209	.929257	1.076128	.7325429	54	.6934018	.962321	1.039153	.7205511	6
55	.6809339	.929799	1.075500	.7323449	55	.6936114	.962881	1.038548	.7203494	5
56	.6811469	.930342	1.074873	.7321467	56	.6938209	.963442	1.037944	.7201476	4
57	.6813599	.930884	1.074246	.7319486	57	.6940304	.964003	1.037340	.7199457	3
58	.6815728	.931428	1.073620	.7317503	58	.6942398	.964565	1.036736	.7197438	2
59	.6817856	.931971	1.072994	.7315521	59	.6944491	.965126	1.036133	.7195418	1
60	.6819984	.932515	1.072368	.7313537	60	.6946584	.965688	1.035530	.7193398	0
	Cosine.	Cotang	Tang.	Sine.	'	Cosine.	Cotang	Tang.	Sine.	'

47°   46°

Note.—Secant = 1 ÷ cosine.   Cosecant = 1 ÷ sine.

2.—Natural Sines, Tangents, Cotangents, Cosines.—(Concluded.)

(Versed sine = 1 − cosine; coversed sine = 1 − sine.)

44°   **44°**

′	Sine.	Tang.	Cotang.	Cosine.	′	Sine.	Tang.	Cotang.	Cosine.	
0	.6946584	.965688	1.035530	.7193398	30	.7009093	.982697	1.017607	.7132504	30
1	.6948676	.966251	1.034927	.7191377	31	.7011167	.983269	1.017015	.7130465	29
2	.6950767	.966813	1.034325	.7189355	32	.7013241	.983841	1.016423	.7128426	28
3	.6952858	.967376	1.033723	.7187333	33	.7015314	.984414	1.015832	.7126385	27
4	.6954949	.967939	1.033122	.7185310	34	.7017387	.984987	1.015241	.7124344	26
5	.6957039	.968503	1.032520	.7183287	35	.7019459	.985560	1.014651	.7122303	25
6	.6959128	.969067	1.031919	.7181263	36	.7021531	.986133	1.014061	.7120260	24
7	.6961217	.969631	1.031319	.7179238	37	.7023601	.986707	1.013471	.7118218	23
8	.6963305	.970196	1.030719	.7177213	38	.7025672	.987282	1.012881	.7116174	22
9	.6965392	.970761	1.030119	.7175187	39	.7027741	.987856	1.012292	.7114130	21
10	.6967479	.971326	1.029520	.7173161	40	.7029811	.988431	1.011703	.7112086	20
11	.6969565	.971891	1.028921	.7171134	41	.7031879	.989006	1.011115	.7110041	19
12	.6971651	.972457	1.028322	.7169106	42	.7033947	.989582	1.010527	.7107995	18
13	.6973736	.973023	1.027724	.7167078	43	.7036014	.990158	1.009939	.7105948	17
14	.6975821	.973590	1.027126	.7165049	44	.7038081	.990734	1.009352	.7103901	16
15	.6977905	.974156	1.026528	.7163019	45	.7040147	.991311	1.008764	.7101854	15
16	.6979988	.974724	1.025931	.7160989	46	.7042213	.991888	1.008178	.7099806	14
17	.6982071	.975291	1.025334	.7158959	47	.7044278	.992465	1.007591	.7097757	13
18	.6984153	.975859	1.024738	.7156927	48	.7046342	.993042	1.007005	.7095707	12
19	.6986234	.976427	1.024141	.7154895	49	.7048406	.993620	1.006420	.7093657	11
20	.6988315	.976995	1.023546	.7152863	50	.7050469	.994199	1.005834	.7091607	10
21	.6990396	.977564	1.022950	.7150830	51	.7052532	.994777	1.005249	.7089556	9
22	.6992476	.978133	1.022355	.7148796	52	.7054594	.995356	1.004665	.7087504	8
23	.6994555	.978702	1.021760	.7146762	53	.7056655	.995935	1.004080	.7085451	7
24	.6996633	.979272	1.021166	.7144727	54	.7058716	.996515	1.003496	.7083398	6
25	.6998711	.979842	1.020572	.7142691	55	.7060776	.997095	1.002913	.7081345	5
26	.7000789	.980412	1.019978	.7140655	56	.7062835	.997675	1.002329	.7079291	4
27	.7002866	.980983	1.019385	.7138618	57	.7064894	.998256	1.001746	.7077236	3
28	.7004942	.981554	1.018792	.7136581	58	.7066953	.998837	1.001164	.7075180	2
29	.7007018	.982125	1.018199	.7134543	59	.7069011	.999418	1.000581	.7073124	1
30	.7009093	.982697	1.017607	.7132504	60	.7071068	1.00000	1.000000	.7071068	0

| | Cosine. | Cotang | Tang. | Sine. | ′ | | Cosine. | Cotang | Tang. | Sine. | ′ |

45°   **45°**

Note.—Secant = 1 ÷ cosine.   Cosecant = 1 ÷ sine.

TABLE 3

TABLE OF LOGARITHMIC SINES

3.—Logarithmic Sines, TANGENTS, COTANGENTS, COSINES. (SECANTS, COSECANTS.)*

0°						1°					
'	Sine.	Tang.	Cotang.	Cosine.		'	Sine.	Tang.	Cotang.	Cosine.	
0	Inf. Neg.	Inf. Neg.	Infinite.	10.00000	60	0	8.24186	8.24192	11.75808	9.99993	60
1	6.46373	6.46373	13.53627	.00000	59	1	.24903	.24910	.75090	.99993	59
2	.76476	.76476	.23524	.00000	59	2	.25609	.25616	.74384	.99993	58
3	6.94085	6.94085	13.05915	.00000	57	3	.26304	.26312	73688	.99993	57
4	7.06579	7.06579	12.93421	.00000	56	4	.26988	.26996	.73004	.99992	56
5	7.16270	7.16270	12.83730	10.00000	55	5	8.27661	8.27669	11.72331	9.99992	55
6	.24188	24188	.75812	.00000	54	6	.28324	.28332	.71668	.99992	54
7	.30882	30882	69118	.00000	53	7	.28977	.28986	.71014	.99992	53
8	.36682	.36682	.63318	.00000	52	8	.29621	.29629	.70371	.99992	52
9	.41797	.41797	.58203	.00000	51	9	.30255	.30263	.69737	.99991	51
10	7.46373	7.46373	12.53627	10.00000	50	10	8.30879	8.30888	11.69112	9.99991	50
11	.50512	.50512	.49488	.00000	49	11	.31495	.31505	.68495	.99991	49
12	54291	54291	.45709	00000	48	12	.32103	.32112	.67888	.99990	48
13	.57767	.57767	.42233	.00000	47	13	.32702	.32711	.67289	.99990	47
14	.60985	.60986	.39014	.00000	46	14	.33292	.33302	.66698	.99990	46
15	7.63982	7.63982	12.36018	10.00000	45	15	8.33875	8.33886	11.66114	9.99990	45
16	.66784	.66785	.33215	.99999	44	16	.34450	.34461	.65539	.99989	44
17	.69417	.69418	.30582	9.99999	43	17	.35018	.35029	.64971	.99989	43
18	.71900	.71900	.28100	.99999	42	18	.35578	.35590	.64410	.99989	42
19	.74248	.74248	.25752	.99999	41	19	.36131	.36143	.63857	.99989	41
20	7.76475	7.76476	12.23524	9.99999	40	20	8.36678	8.36689	11.63311	9.99988	40
21	.78594	.78595	.21405	.99999	39	21	.37217	.37229	.62771	.99988	39
22	.80615	.80615	.19385	.99999	38	22	.37750	.37762	.62238	.99988	38
23	.82546	.82546	.17454	.99999	37	23	.38276	.38289	.61711	.99987	37
24	.84393	.84394	.15606	.99999	36	24	.38796	.38809	.61191	.99987	36
25	7.86166	7.86167	12.13833	9.99999	35	25	8.39310	8.39323	11.60677	9.99987	35
26	.87870	.87871	.12129	.99999	34	26	.39818	.39832	.60168	.99986	34
27	.89509	.89510	.10490	.99999	33	27	.40320	.40334	.59666	.99986	33
28	.91088	.91089	.08911	.99999	32	28	.40816	.40830	.59170	.99986	32
29	.92612	.92613	.07387	.99998	31	29	.41307	.41321	.58679	.99985	31
30	7.94086	7.94086	12.05914	9.99998	30	30	8.41792	8.41807	11.58193	9.99985	30
31	.95508	.95510	.04490	.99998	29	31	.42272	.42287	.57713	.99985	29
32	.96887	.96888	.03111	.99998	28	32	.42746	.42762	.57238	.99984	28
33	.98223	.98225	.01775	.99998	27	33	.43216	.43232	.56768	.99984	27
34	7.99520	7.99522	12.00478	.99998	26	34	.43680	.43696	.56304	.99984	26
35	8.00779	8.00781	11.99219	9.99998	25	35	8.44139	8.44156	11.55844	9.99983	25
36	.02002	.02004	.97996	.99998	24	36	.44594	.44611	.55389	.99983	24
37	.03192	.03194	.96806	.99997	23	37	.45044	.45061	.54939	.99983	23
38	.04350	.04353	.95647	.99997	22	38	.45489	.45507	.54493	.99982	22
39	.05478	.05481	.94519	.99997	21	39	.45930	.45948	.54052	.99982	21
40	8.06578	8.06581	11.93419	9.99997	20	40	8.46366	8.46385	11.53615	9.99982	20
41	.07650	.07653	.92347	9.99997	19	41	.46799	.46817	.53183	.99981	19
42	.08696	.08700	.91300	.99997	18	42	.47226	.47245	.52755	.99981	18
43	.09718	.09722	.90278	.99997	17	43	.47650	.47669	.52331	.99981	17
44	.10717	.10720	.89280	.99997	16	44	.48069	.48089	.51911	.99980	16
45	8.11693	8.11696	11.88304	9.99996	15	45	8.48485	8.48505	11.51495	9.99980	15
46	.12647	.12651	.87349	.99996	14	46	.48896	.48917	.51083	.99979	14
47	.13581	.13585	.86415	.99996	13	47	.49304	.49325	.50675	.99979	13
48	.14495	.14500	.85500	.99996	12	48	.49708	.49729	.50271	.99979	12
49	.15391	.15395	.84605	.99996	11	49	.50108	.50130	.49870	.99978	11
50	8.16268	8.16273	11.83727	9.99995	10	50	8.50504	8.50527	11.49473	9.99978	10
51	.17128	.17133	.82867	.99995	9	51	.50897	.50920	.49080	.99977	9
52	.17971	.17976	82024	.99995	8	52	.51287	.51310	.48690	.99977	8
53	.18798	.18804	.81196	.99995	7	53	.51673	.51696	.48304	.99977	7
54	.19610	.19616	80384	.99995	6	54	.52055	.52079	.47921	.99976	6
55	8.20407	8.20413	11.79587	9.99994	5	55	8.52434	8.52459	11.47541	9.99976	5
56	.21189	.21195	.78805	.99994	4	56	.52810	.52835	.47165	.99975	4
57	.21958	.21964	.78036	.99994	3	57	.53183	.53208	.46792	.99975	3
58	.22713	.22720	.77280	.99994	2	58	.53552	.53578	.46422	.99974	2
59	.23456	.23462	76538	.99994	1	59	.53919	.53945	.46055	.99974	1
60	8.24186	8.24192	11.75808	9.99993	0	60	8.54282	8.54308	11.45692	9.99974	0
	Cosine.	Cotang	Tang.	Sine.	'		Cosine.	Cotang.	Tang.	Sine.	'
		89°						88°			

*Log secant = colog cosine = 1 − log cosine; log cosecant = colog sine = 1 − log sine.
Ex.—Log sec 0°- 30' = 10.00002. Ex.—Log cosec 0°- 30' = 12.05916.

3.—Logarithmic Sines, Tangents, Cotangents, Cosines.
(Secants, Cosecants.)*—(Cont'd.)

2° · **3°**

'	Sine.	Tang.	Cotang.	Cosine.		'	Sine.	Tang.	Cotang.	Cosine.	
0	8.54282	8.54308	11.45692	9.99974	60	0	8.71880	8.71940	11.28060	9.99940	60
1	.54642	.54669	.45331	.99973	59	1	.72120	.72181	.27819	.99940	59
2	.54999	.55027	.44973	.99973	58	2	.72359	.72420	.27580	.99939	58
3	.55354	.55382	.44618	.99972	57	3	.72597	.72659	.27341	.99938	57
4	.55705	.55734	.44266	.99972	56	4	.72834	.72896	.27104	.99938	56
5	8.56054	8.56083	11.43917	9.99971	55	5	8.73069	8.73132	11.26868	9.99937	55
6	.56400	.56429	.43571	.99971	54	6	.73303	.73366	.26634	.99936	54
7	.56743	.56773	.43227	.99970	53	7	.73535	.73600	.26400	.99936	53
8	.57084	.57114	.42886	.99970	52	8	.73767	.73832	.26168	.99935	52
9	.57421	.57452	.42548	.99969	51	9	.73997	.74063	.25937	.99934	51
10	8.57757	8.57788	11.42212	9.99969	50	10	8.74226	8.74292	11.25708	9.99934	50
11	.58089	.58121	.41879	.99968	49	11	.74454	.74521	.25479	.99933	49
12	.58419	.58451	.41549	.99968	48	12	.74680	.74748	.25252	.99932	48
13	.58747	.58779	.41221	.99967	47	13	.74906	.74974	.25026	.99932	47
14	.59072	.59105	.40895	.99967	46	14	.75130	.75199	.24801	.99931	46
15	8.59395	8.59428	11.40572	9.99967	45	15	8.75353	8.75423	11.24577	9.99930	45
16	.59715	.59749	.40251	.99966	44	16	.75575	.75645	.24355	.99929	44
17	.60033	.60068	.39932	.99966	43	17	.75795	.75867	.24133	.99929	43
18	.60349	.60384	.39616	.99965	42	18	.76015	.76087	.23913	.99928	42
19	.60662	.60698	.39302	.99964	41	19	.76234	.76306	.23694	.99927	41
20	8.60973	8.61009	11.38991	9.99964	40	20	8.76451	8.76525	11.23475	9.99926	40
21	.61282	.61319	.38681	.99963	39	21	.76667	.76742	.23258	.99926	39
22	.61589	.61626	.38374	.99963	38	22	.76883	.76958	.23042	.99925	38
23	.61894	.61931	.38069	.99962	37	23	.77097	.77173	.22827	.99924	37
24	.62196	.62234	.37766	.99962	36	24	.77310	.77387	.22613	.99923	36
25	8.62497	8.62535	11.37465	9.99961	35	25	8.77522	8.77600	11.22400	9.99923	35
26	.62795	.62834	.37166	.99961	34	26	.77733	.77811	.22189	.99922	34
27	.63091	.63131	.36869	.99960	33	27	.77943	.78022	.21978	.99921	33
28	.63385	.63426	.36574	.99960	32	28	.78152	.78232	.21768	.99920	32
29	.63678	.63718	.36282	.99959	31	29	.78360	.78441	.21559	.99920	31
30	8.63968	8.64009	11.35991	9.99959	30	30	8.78568	8.78649	11.21351	9.99919	30
31	.64256	.64298	.35702	.99958	29	31	.78774	.78855	.21145	.99918	29
32	.64543	.64585	.35415	.99958	28	32	.78979	.79061	.20939	.99917	28
33	.64827	.64870	.35130	.99957	27	33	.79183	.79266	.20734	.99917	27
34	.65110	.65154	.34846	.99956	26	34	.79386	.79470	.20530	.99916	26
35	8.65391	8.65435	11.34565	9.99956	25	35	8.79588	8.79673	11.20327	9.99915	25
36	.65670	.65715	.34285	.99955	24	36	.79789	.79875	.20125	.99914	24
37	.65947	.65993	.34007	.99955	23	37	.79990	.80076	.19924	.99913	23
38	.66223	.66269	.33731	.99954	22	38	.80189	.80277	.19723	.99913	22
39	.66497	.66543	.33457	.99954	21	39	.80388	.80476	.19524	.99912	21
40	8.66769	8.66816	11.33184	9.99953	20	40	8.80585	8.80674	11.19326	9.99911	20
41	.67039	.67087	.32913	.99952	19	41	.80782	.80872	.19128	.99910	19
42	.67308	.67356	.32644	.99952	18	42	.80978	.81068	.18932	.99909	18
43	.67575	.67624	.32376	.99951	17	43	.81173	.81264	.18736	.99909	17
44	.67841	.67890	.32110	.99951	16	44	.81367	.81459	.18541	.99908	16
45	8.68104	8.68154	11.31846	9.99950	15	45	8.81560	8.81653	11.18347	9.99907	15
46	.68367	.68417	.31583	.99949	14	46	.81752	.81846	.18154	.99906	14
47	.68627	.68678	.31322	.99949	13	47	.81944	.82038	.17962	.99905	13
48	.68886	.68938	.31062	.99948	12	48	.82134	.82230	.17770	.99904	12
49	.69144	.69196	.30804	.99948	11	49	.82324	.82420	.17580	.99904	11
50	8.69400	8.69453	11.30547	9.99947	10	50	8.82513	8.82610	11.17390	9.99903	10
51	.69654	.69708	.30292	.99946	9	51	.82701	.82799	17201	.99902	9
52	.69907	.69962	.30038	.99946	8	52	.82888	.82987	.17013	.99901	8
53	70159	.70214	.29786	.99945	7	53	.83075	.83175	.16825	.99900	7
54	.70409	.70465	.29535	.99944	6	54	.83261	.83361	.16639	99899	6
55	8.70658	8.70714	11.29286	9.99944	5	55	8.83446	8.83547	11.16453	9.99898	5
56	.70905	.70962	.29038	.99943	4	56	.83630	.83732	.16268	.99898	4
57	.71151	.71208	.28792	.99942	3	57	.83813	.83916	.16084	.99897	3
58	.71395	.71453	.28547	.99942	2	58	.83996	.84100	15900	.99896	2
59	.71638	.71697	.28303	.99941	1	59	.84177	.84282	.15718	.99895	1
60	8.71880	8.71940	11.28060	9.99940	0	60	8.84358	8.84464	11.15536	9.99894	0
	Cosine.	Cotang.	Tang.	Sine.	'		Cosine.	Cotang.	Tang.	Sine.	

87° · **86°**

*Log secant = colog cosine = 1 − log cosine; log cosecant = colog sine = 1 − log sine.

Ex.—Log sec 2°-30' = 10.00041   Ex.—Log cosec 2°-30' = 11.36032.

3. —Logarithmic Sines, Tangents, Cotangents, Cosines.—(Cont'd.)
(Secants, Cosecants.)*

4°   **5°**

'	Sine.	Tang.	Cotang.	Cosine.	‖	'	Sine.	Tang.	Cotang.	Cosine.	
0	8.84358	8.84464	11.15536	9.99894	60	0	8.94030	8.94195	11.05805	9.99834	60
1	.84539	.84646	.15354	.99893	59	1	.94174	.94340	.05660	.99833	59
2	.84718	.84826	.15174	.99892	58	2	.94317	.94485	.05515	.99832	58
3	.84897	.85006	.14994	.99891	57	3	.94461	.94630	.05370	.99831	57
4	.85075	.85185	.14815	.99891	56	4	.94603	.94773	.05227	.99830	56
5	8.85252	8.85363	11.14637	9.99890	55	5	8.94746	8.94917	11.05083	9.99829	55
6	.85429	.85540	.14460	.99889	54	6	.94887	.95060	.04940	.99828	54
7	.85605	.85717	.14283	.99888	53	7	.95029	.95202	.04798	.99827	53
8	.85780	.85893	.14107	.99887	52	8	.95170	.95344	.04656	.99825	52
9	.85955	.86069	.13931	.99886	51	9	.95310	.95486	.04514	.99824	51
10	8.86128	8.86243	11.13757	9.99885	50	10	8.95450	8.95627	11.04373	9.99823	50
11	.86301	.86417	.13583	.99884	49	11	.95589	.95767	.04233	.99822	49
12	.86474	.86591	.13409	.99883	48	12	.95728	.95908	.04092	.99821	48
13	.86645	.86763	.13237	.99882	47	13	.95867	.96047	.03953	.99820	47
14	.86816	.86935	.13065	.99881	46	14	.96005	.96187	.03813	.99819	46
15	8.86987	8.87106	11.12894	9.99880	45	15	8.96143	8.96325	11.03675	9.99817	45
16	.87156	.87277	.12723	.99879	44	16	.96280	.96464	.03536	.99816	44
17	.87325	.87447	.12553	.99879	43	17	.96417	.96602	.03398	.99815	43
18	.87494	.87616	.12384	.99878	42	18	.96553	.96739	.03261	.99814	42
19	.87661	.87785	.12215	.99877	41	19	.96689	.96877	.03123	.99813	41
20	8.87829	8.87953	11.12047	9.99876	40	20	8.96825	8.97013	11.02987	9.99812	40
21	.87995	.88120	.11880	.99875	39	21	.96960	.97150	.02850	.99810	39
22	.88161	.88287	.11713	.99874	38	22	.97095	.97285	.02715	.99809	38
23	.88326	.88453	.11547	.99873	37	23	.97229	.97421	.02579	.99808	37
24	.88490	.88618	.11382	.99872	36	24	.97363	.97556	.02444	.99807	36
25	8.88654	8.88783	11.11217	9.99871	35	25	8.97496	8.97691	11.02309	9.99806	35
26	.88817	.88948	.11052	.99870	34	26	.97629	.97825	.02175	.99804	34
27	.88980	.89111	.10889	.99869	33	27	.97762	.97959	.02041	.99803	33
28	.89142	.89274	.10726	.99868	32	28	.97894	.98092	.01908	.99802	32
29	.89304	.89437	.10563	.99867	31	29	.98026	.98225	.01775	.99801	31
30	8.89464	8.89598	11.10402	9.99866	30	30	8.98157	8.98358	11.01642	9.99800	30
31	.89625	.89760	.10240	.99865	29	31	.98288	.98490	.01510	.99798	29
32	.89784	.89920	.10080	.99864	28	32	.98419	.98622	.01378	.99797	28
33	.89943	.90080	.09920	.99863	27	33	.98549	.98753	.01247	.99796	27
34	.90102	.90240	.09760	.99862	26	34	.98679	.98884	.01116	.99795	26
35	8.90260	8.90399	11.09601	9.99861	25	35	8.98808	8.99015	11.00985	9.99793	25
36	.90417	.90557	.09443	.99860	24	36	.98937	.99145	.00855	.99792	24
37	.90574	.90715	.09285	.99859	23	37	.99066	.99275	.00725	.99791	23
38	.90730	.90872	.09128	.99858	22	38	.99194	.99405	.00595	.99790	22
39	.90885	.91029	.08971	.99857	21	39	.99322	.99534	.00466	.99788	21
40	8.91040	8.91185	11.08815	9.99856	20	40	8.99450	8.99662	11.00338	9.99787	20
41	.91195	.91340	.08660	.99855	19	41	.99577	.99791	.00209	.99786	19
42	.91349	.91495	.08505	.99854	18	42	.99704	8.99919	11.00081	.99785	18
43	.91502	.91650	.08350	.99853	17	43	.99830	9.00046	10.99954	.99783	17
44	.91655	.91803	.08197	.99852	16	44	8.99956	9.00174	.99826	.99782	16
45	8.91807	8.91957	11.08043	9.99851	15	45	9.00082	9.00301	10.99699	9.99781	15
46	.91959	.92110	.07890	.99850	14	46	.00207	.00427	.99573	.99780	14
47	.92110	.92262	.07738	.99848	13	47	.00332	.00553	.99447	.99778	13
48	.92261	.92414	.07586	.99847	12	48	.00456	.00679	.99321	.99777	12
49	.92411	.92565	.07435	.99846	11	49	.00581	.00805	.99195	.99776	11
50	8.92561	8.92716	11.07284	9.99845	10	50	9.00704	9.00930	10.99070	9.99775	10
51	.92710	.92866	.07134	.99844	9	51	.00828	.01055	.98945	.99773	9
52	.92859	.93016	.06984	.99843	8	52	.00951	.01179	.98821	.99772	8
53	.93007	.93165	.06835	.99842	7	53	.01074	.01303	.986J7	.99771	7
54	.93154	.93313	.06687	.99841	6	54	.01196	.01427	.98573	.99769	6
55	8.93301	8.93462	11.06538	9.99840	5	55	9.01318	9.01550	10.98450	9.99768	5
56	.93448	.93609	.06391	.99839	4	56	.01440	.01673	.98327	.99767	4
57	.93594	.93756	.06244	.99838	3	57	.01561	.01796	.98204	.99765	3
58	.93740	.93903	.06097	.99837	2	58	.01682	.01918	.98082	.99764	2
59	.93885	.94049	.05951	.99836	1	59	.01803	.02040	.97960	.99763	1
60	8.94030	8.94195	11.05805	9.99834	0	60	9.01923	9.02162	10.97838	9.99761	0

| | Cosine. | Cotang. | Tang. | Sine. | ' | ‖ | | Cosine. | Cotang. | Tang. | Sine. | ' |

85°   **84°**

*Log secant = colog cosine = 1 − log cosine; log cosecant = colog sine = 1 − log sine.
Ex.—Log sec 4°- 30′ = 10.00134.   Ex.—Log cosec 4°- 30′ = 11.10536.

3.—Logarithmic Sines, TANGENTS, COTANGENTS, COSINES.—(Cont'd.)
(SECANTS, COSECANTS.)*

6°   **7°**

	Sine.	Tang.	Cotang.	Cosine.		′	Sine.	Tang.	Cotang.	Cosine.	
0	9.01923	9.02162	10.97838	9.99761	60	0	9.08589	9.08914	10.91086	9.99675	60
1	.02043	.02283	.97717	.99760	59	1	.08692	.09019	.90981	.99674	59
2	.02163	.02404	.97596	.99759	58	2	.08795	.09123	.90877	.99672	58
3	.02283	.02525	.97475	.99757	57	3	.08897	.09227	.90773	.99670	57
4	.02402	.02645	.97355	.99756	56	4	.08999	.09330	.90670	.99669	56
5	9.02520	9.02766	10.97234	9.99755	55	5	9.09101	9.09434	10.90566	9.99667	55
6	.02639	.02885	.97115	.99753	54	6	.09202	.09537	.90463	.99666	54
7	.02757	.03005	.96995	.99752	53	7	.09304	.09640	.90360	.99664	53
8	.02874	.03124	.96876	.99751	52	8	.09405	.09742	.90258	.99663	52
9	.02992	.03242	.96758	.99749	51	9	.09506	.09845	.90155	.99661	51
10	9.03109	9.03361	10.96639	9.99748	50	10	9.09606	9.09947	10.90053	9.99659	50
11	.03226	.03479	.96521	.99747	49	11	.09707	.10049	.89951	.99658	49
12	.03342	.03597	.96403	.99745	48	12	.09807	.10150	.89850	.99656	48
13	.03458	.03714	.96286	.99744	47	13	.09907	.10252	.89748	.99655	47
14	.03574	.03832	.96168	.99742	46	14	.10006	.10353	.89647	.99653	46
15	9.03690	9.03948	10.96052	9.99741	45	15	9.10106	9.10454	10.89546	9.99651	45
16	.03805	.04065	.95935	.99740	44	16	.10205	.10555	.89445	.99650	44
17	.03920	.04181	.95819	.99738	43	17	.10304	.10656	.89344	.99648	43
18	.04034	.04297	.95703	.99737	42	18	.10402	.10756	.89244	.99647	42
19	.04149	.04413	.95587	.99736	41	19	.10501	.10856	.89144	.99645	41
20	9.04262	9.04528	10.95472	9.99734	40	20	9.10599	9.10956	10.89044	9.99643	40
21	.04376	.04643	.95357	.99733	39	21	.10697	.11056	.88944	.99642	39
22	.04490	.04758	.95242	.99731	38	22	.10795	.11155	.88845	.99640	38
23	.04603	.04873	.95127	.99730	37	23	.10893	.11254	.88746	.99638	37
24	.04715	.04987	.95013	.99728	36	24	.10990	.11353	.88647	.99637	36
25	9.04828	9.05101	10.94899	9.99727	35	25	9.11087	9.11452	10.88548	9.99635	35
26	.04940	.05214	.94786	.99726	34	26	.11184	.11551	.88449	.99633	34
27	.05052	.05328	.94672	.99724	33	27	.11281	.11649	.88351	.99632	33
28	.05164	.05441	.94559	.99723	32	28	.11377	.11747	.88253	.99630	32
29	.05275	.05553	.94447	.99721	31	29	.11474	.11845	.88155	.99629	31
30	9.05386	9.05666	10.94334	9.99720	30	30	9.11570	9.11943	10.88057	9.99627	30
31	.05497	.05778	.94222	.99718	29	31	.11666	.12040	.87960	.99625	29
32	.05607	.05890	.94110	.99717	28	32	.11761	.12138	.87862	.99624	28
33	.05717	.06002	.93998	.99716	27	33	.11857	.12235	.87765	.99622	27
34	.05827	.06113	.93887	.99714	26	34	.11952	.12332	.87668	.99620	26
35	9.05937	9.06224	10.93776	9.99713	25	35	9.12047	9.12428	10.87572	9.99618	25
36	.06046	.06335	.93665	.99711	24	36	.12142	.12525	.87475	.99617	24
37	.06155	.06445	.93555	.99710	23	37	.12236	.12621	.87379	.99615	23
38	.06264	.06556	.93444	.99708	22	38	.12331	.12717	.87283	.99613	22
39	.06372	.06666	.93334	.99707	21	39	.12425	.12813	.87187	.99612	21
40	9.06481	9.06775	10.93225	9.99705	20	40	9.12519	9.12909	10.87091	9.99610	20
41	.06589	.06885	.93115	.99704	19	41	.12612	.13004	.86996	.99608	19
42	.06696	.06994	.93006	.99702	18	42	.12706	.13099	.86901	.99607	18
43	.06804	.07103	.92897	.99701	17	43	.12799	.13194	.86806	.99605	17
44	.06911	.07211	.92789	.99699	16	44	.12892	.13289	.86711	.99603	16
45	9.07018	9.07320	10.92680	9.99698	15	45	9.12985	9.13384	10.86616	9.99601	15
46	.07124	.07428	.92572	.99696	14	46	.13078	.13478	.86522	.99600	14
47	.07231	.07536	.92464	.99695	13	47	.13171	.13573	.86427	.99598	13
48	.07337	.07643	.92357	.99693	12	48	.13263	.13667	.86333	.99596	12
49	.07442	.07751	.92249	.99692	11	49	.13355	.13761	.86239	.99595	11
50	9.07548	9.07858	10.92142	9.99690	10	50	9.13447	9.13854	10.86146	9.99593	10
51	.07653	.07964	.92036	.99689	9	51	.13539	.13948	.86052	.99591	9
52	.07758	.08071	.91929	.99687	8	52	.13630	.14041	.85959	.99589	8
53	.07863	.08177	.91823	.99686	7	53	.13722	.14134	.85866	.99588	7
54	.07968	.08283	.91717	.99684	6	54	.13813	.14227	.85773	.99586	6
55	9.08072	9.08389	10.91611	9.99683	5	55	9.13904	9.14320	10.85680	9.99584	5
56	.08176	.08495	.91505	.99681	4	56	.13994	.14412	.85588	.99582	4
57	.08280	.08600	.91400	.99680	3	57	.14085	.14504	.85496	.99581	3
58	.08383	.08705	.91295	.99678	2	58	.14175	.14597	.85403	.99579	2
59	.08486	.08810	.91190	.99677	1	59	.14266	.14688	.85312	.99577	1
60	9.08589	9.08914	10.91086	9.99675	0	60	9.14356	9.14780	10.85220	9.99575	0

	Cosine.	Cotang.	Tang.	Sine.	′		Cosine.	Cotang.	Tang.	Sine.	′

83°   **82°**

*Log secant = colog cosine = 1 − log cosine; log cosecant = colog sine = 1 − log sine.

Ex.—Log sec 6°- 30′ = 10.00280.   Ex.—Log cosec 6°- 30′ = 10.94614.

3. —**Logarithmic Sines,** Tangents, Cotangents, Cosines.—(Cont'd.)
(Secants, Cosecants.)*

8°   9°

'	Sine.	Tang.	Cotang.	Cosine.		'	Sine.	Tang.	Cotang.	Cosine.	
0	9.14356	9.14780	10.85220	9.99575	60	0	9.19433	9.19971	10.80029	9.99462	60
1	.14445	.14872	.85128	.99574	59	1	.19513	.20053	.79947	.99460	59
2	.14535	.14963	.85037	.99572	58	2	.19592	.20134	.79866	.99458	58
3	.14624	.15054	.84946	.99570	57	3	.19672	.20216	.79784	.99456	57
4	.14714	.15145	.84855	.99568	56	4	.19751	.20297	.79703	.99454	56
5	9.14803	9.15236	10.84764	9.99566	55	5	9.19830	9.20378	10.79622	9.99452	55
6	.14891	.15327	.84673	.99565	54	6	.19909	.20459	.79541	.99450	54
7	.14980	.15417	.84583	.99563	53	7	.19988	.20540	.79460	.99448	53
8	.15069	.15508	.84492	.99561	52	8	.20067	.20621	.79379	.99446	52
9	.15157	.15598	.84402	.99559	51	9	.20145	.20701	.79299	.99444	51
10	9.15245	9.15688	10.84312	9.99557	50	10	9.20223	9.20782	10.79218	9.99442	50
11	.15333	.15777	.84223	.99556	49	11	.20302	.20862	.79138	.99440	49
12	.15421	.15867	.84133	.99554	48	12	.20380	.20942	.79058	.99438	48
13	.15508	.15956	.84044	.99552	47	13	.20458	.21022	.78978	.99436	47
14	.15596	.16046	.83954	.99550	46	14	.20535	.21102	.78898	.99434	46
15	9.15683	9.16135	10.83865	9.99548	45	15	9.20613	9.21182	10.78818	9.99432	45
16	.15770	.16224	.83776	.99546	44	16	.20691	.21261	.78739	.99429	44
17	.15857	.16312	.83688	.99545	43	17	.20768	.21341	.78659	.99427	43
18	.15944	.16401	.83599	.99543	42	18	.20845	.21420	.78580	.99425	42
19	.16030	.16489	.83511	.99541	41	19	.20922	.21499	.78501	.99423	41
20	9.16116	9.16577	10.83423	9.99539	40	20	9.20999	9.21578	10.78422	9.99421	40
21	.16203	.16665	.83335	.99537	39	21	.21076	.21657	.78343	.99419	39
22	.16289	.16753	.83247	.99535	38	22	.21153	.21736	.78264	.99417	38
23	.16374	.16841	.83159	.99533	37	23	.21229	.21814	.78186	.99415	37
24	.16460	.16928	.83072	.99532	36	24	.21306	.21893	.78107	.99413	36
25	9.16545	9.17016	10.82984	9.99530	35	25	9.21382	9.21971	10.78029	9.99411	35
26	.16631	.17103	.82897	.99528	34	26	.21458	.22049	.77951	.99409	34
27	.16716	.17190	.82810	.99526	33	27	.21534	.22127	.77873	.99407	33
28	.16801	.17277	.82723	.99524	32	28	.21610	.22205	.77795	.99404	32
29	.16886	.17363	.82637	.99522	31	29	.21685	.22283	.77717	.99402	31
30	9.16970	9.17450	10.82550	9.99520	30	30	9.21761	9.22361	10.77639	9.99400	30
31	.17055	.17536	.82464	.99518	29	31	.21836	.22438	.77562	.99398	29
32	.17139	.17622	.82378	.99517	28	32	.21912	.22516	.77484	.99396	28
33	.17223	.17708	.82292	.99515	27	33	.21987	.22593	.77407	.99394	27
34	.17307	.17794	.82206	.99513	26	34	.22062	.22670	.77330	.99392	26
35	9.17391	9.17880	10.82120	9.99511	25	35	9.22137	9.22747	10.77253	9.99390	25
36	.17474	.17965	.82035	.99509	24	36	.22211	.22824	.77176	.99388	24
37	.17558	.18051	.81949	.99507	23	37	.22286	.22901	.77099	.99385	23
38	.17641	.18136	.81864	.99505	22	38	.22361	.22977	.77023	.99383	22
39	.17724	.18221	.81779	.99503	21	39	.22435	.23054	.76946	.99381	21
40	9.17807	9.18306	10.81694	9.99501	20	40	9.22509	9.23130	10.76870	9.99379	20
41	.17890	.18391	.81609	.99499	19	41	.22583	.23206	.76794	.99377	19
42	.17973	.18475	.81525	.99497	18	42	.22657	.23283	.76717	.99375	18
43	.18055	.18560	.81440	.99495	17	43	.22731	.23359	.76641	.99372	17
44	.18137	.18644	.81356	.99494	16	44	.22805	.23435	.76565	.99370	16
45	9.18220	9.18728	10.81272	9.99492	15	45	9.22878	9.23510	10.76490	9.99368	15
46	.18302	.18812	.81188	.99490	14	46	.22952	.23586	.76414	.99366	14
47	.18383	.18896	.81104	.99488	13	47	.23025	.23661	.76339	.99364	13
48	.18465	.18979	.81021	.99486	12	48	.23098	.23737	.76263	.99362	12
49	.18547	.19063	.80937	.99484	11	49	.23171	.23812	.76188	.99359	11
50	9.18628	9.19146	10.80854	9.99482	10	50	9.23244	9.23887	10.76113	9.99357	10
51	.18709	.19229	.80771	.99480	9	51	.23317	.23962	.76038	.99355	9
52	.18790	.19312	.80688	.99478	8	52	.23390	.24037	.75963	.99353	8
53	.18871	.19395	.80605	.99476	7	53	.23462	.24112	.75888	.99351	7
54	.18952	.19478	.80522	.99474	6	54	.23535	.24186	.75814	.99348	6
55	9.19033	9.19561	10.80439	9.99472	5	55	9.23607	9.24261	10.75739	9.99346	5
56	.19113	.19643	.80357	.99470	4	56	.23679	.24335	.75665	.99344	4
57	.19193	.19725	.80275	.99468	3	57	.23752	.24410	.75590	.99342	3
58	.19273	.19807	.80193	.99466	2	58	.23823	.24484	.75516	.99340	2
59	.19353	.19889	.80111	.99464	1	59	.23895	.24558	.75442	.99337	1
60	9.19433	9.19971	10.80029	9.99462	0	60	9.23967	9.24632	10.75368	9.99335	0

	Cosine.	Cotang.	Tang.	Sine.			Cosine.	Cotang.	Tang.	Sine.	

81°   80°

*Log secant = colog cosine = 1 − log cosine; log cosecant = colog sine = 1 − log sine.
Ex.—Log sec 8°- 30′ = 10.00480.   Ex.—Log cosec 8°- 30′ = 10.83030.

3.—Logarithmic Sines, Tangents, Cotangents, Cosines.—(Cont'd.)
(Secants, Cosecants.)*

10°   **11°**

'	Sine.	Tang.	Cotang.	Cosine.	'	Sine.	Tang.	Cotang.	Cosine.	
0	9.23967	9.24632	10.75368	9.99335	0	9.28060	9.28865	10.71135	9.99195	60
1	.24039	.24706	.75294	.99333	1	.28125	.28933	.71067	.99192	59
2	.24110	.24779	.75221	.99331	2	.28190	.29000	.71000	.99190	58
3	.24181	.24853	.75147	.99328	3	.28254	.29067	.70933	.99187	57
4	.24253	.24926	.75074	.99326	4	.28319	.29134	.70866	.99185	56
5	9.24324	9.25000	10.75000	9.99324	5	9.28384	9.29201	10.70799	9.99182	55
6	.24395	.25073	.74927	.99322	6	.28448	.29268	.70732	.99180	54
7	.24466	.25146	.74854	.99319	7	.28512	.29335	.70665	.99177	53
8	.24536	.25219	.74781	.99317	8	.28577	.29402	.70598	.99175	52
9	.24607	.25292	.74708	.99315	9	.28641	.29468	.70532	.99172	51
10	9.24677	9.25365	10.74635	9.99313	10	9.28705	9.29535	10.70465	9.99170	50
11	.24748	.25437	.74563	.99310	11	.28769	.29601	.70399	.99167	49
12	.24818	.25510	.74490	.99308	12	.28833	.29668	.70332	.99165	48
13	.24888	.25582	.74418	.99306	13	.28896	.29734	.70266	.99162	47
14	.24958	.25655	.74345	.99304	14	.28960	.29800	.70200	.99160	46
15	9.25028	9.25727	10.74273	9.99301	15	9.29024	9.29866	10.70134	9.99157	45
16	.25098	.25799	.74201	.99299	16	.29087	.29932	.70068	.99155	44
17	.25168	.25871	.74129	.99297	17	.29150	.29998	.70002	.99152	43
18	.25237	.25943	.74057	.99294	18	.29214	.30064	.69936	.99150	42
19	.25307	.26015	.73985	.99292	19	.29277	.30130	.69870	.99147	41
20	9.25376	9.26086	10.73914	9.99290	20	9.29340	9.30195	10.69805	9.99145	40
21	.25445	.26158	.73842	.99288	21	.29403	.30261	.69739	.99142	39
22	.25514	.26229	.73771	.99285	22	.29466	.30326	.69674	.99140	38
23	.25583	.26301	.73699	.99283	23	.29529	.30391	.69609	.99137	37
24	.25652	.26372	.73628	.99281	24	.29591	.30457	.69543	.99135	36
25	9.25721	9.26443	10.73557	9.99278	25	9.29654	9.30522	10.69478	9.99132	35
26	.25790	.26514	.73486	.99276	26	.29716	.30587	.69413	.99130	34
27	.25858	.26585	.73415	.99274	27	.29779	.30652	.69348	.99127	33
28	.25927	.26655	.73345	.99271	28	.29841	.30717	.69283	.99124	32
29	.25995	.26726	.73274	.99269	29	.29903	.30782	.69218	.99122	31
30	9.26063	9.26797	10.73203	9.99267	30	9.29966	9.30846	10.69154	9.99119	30
31	.26131	.26867	.73133	.99264	31	.30028	.30911	.69089	.99117	29
32	.26199	.26937	.73063	.99262	32	.30090	.30975	.69025	.99114	28
33	.26267	.27008	.72992	.99260	33	.30151	.31040	.68960	.99112	27
34	.26335	.27078	.72922	.99257	34	.30212	.31104	.68896	.99109	26
35	9.26403	9.27148	10.72852	9.99255	35	9.30275	9.31168	10.68832	9.99106	25
36	.26470	.27218	.72782	.99252	36	.30336	.31233	.68767	.99104	24
37	.26538	.27288	.72712	.99250	37	.30398	.31297	.68703	.99101	23
38	.26605	.27357	.72643	.99248	38	.30459	.31361	.68639	.99099	22
39	.26672	.27427	.72573	.99245	39	.30521	.31425	.68575	.99096	21
40	9.26739	9.27496	10.72504	9.99243	40	9.30582	9.31489	10.68511	9.99093	20
41	.26806	.27566	.72434	.99241	41	.30643	.31552	.68448	.99091	19
42	.26873	.27635	.72365	.99238	42	.30704	.31616	.68384	.99088	18
43	.26940	.27704	.72296	.99236	43	.30765	.31679	.68321	.99086	17
44	.27007	.27773	.72227	.99233	44	.30826	.31743	.68257	.99083	16
45	9.27073	9.27842	10.72158	9.99231	45	9.30887	9.31806	10.68194	9.99080	15
46	.27140	.27911	.72089	.99229	46	.30947	.31870	.68130	.99078	14
47	.27206	.27980	.72020	.99226	47	.31008	.31933	.68067	.99075	13
48	.27273	.28049	.71951	.99224	48	.31068	.31996	.68004	.99072	12
49	.27339	.28117	.71883	.99221	49	.31129	.32059	.67941	.99070	11
50	9.27405	9.28186	10.71814	9.99219	50	9.31189	9.32122	10.67878	9.99067	10
51	.27471	.28254	.71746	.99217	51	.31250	.32185	.67815	.99064	9
52	.27537	.28323	.71677	.99214	52	.31310	.32248	.67752	.99062	8
53	.27602	.28391	.71609	.99212	53	.31370	.32311	.67689	.99059	7
54	.27668	.28459	.71541	.99209	54	.31430	.32373	.67627	.99056	6
55	9.27734	9.28527	10.71473	9.99207	55	9.31490	9.32436	10.67564	9.99054	5
56	.27799	.28595	.71405	.99204	56	.31549	.32498	.67502	.99051	4
57	.27864	.28662	.71338	.99202	57	.31609	.32561	.67439	.99048	3
58	.27930	.28730	.71270	.99200	58	.31669	.32623	.67377	.99046	2
59	.27995	.28798	.71202	.99197	59	.31728	.32685	.67315	.99043	1
60	9.28060	9.28865	10.71135	9.99195	60	9.31788	9.32747	10.67253	9.99040	0

	Cosine.	Cotang.	Tang.	Sine.	'		Cosine.	Cotang.	Tang.	Sine.	'

79°   **78°**

*Log secant = colog cosine = 1 − log cosine; log cosecant = colog sine = 1 − log sine.

Ex.—Log sec 10°- 30' = 10.00733.   Ex.—Log cosec 10°- 30' = 10.73937.

3.—Logarithmic Sines, Tangents, Cotangents, Cosines.—(Cont'd.)
(Secants, Cosecants.)*

12°						13°					
′	Sine.	Tang.	Cotang.	Cosine.	′	′	Sine.	Tang.	Cotang.	Cosine.	′
0	9.31788	9.32747	10.67253	9.99040	60	0	9.35209	9.36336	10.63664	9.98872	60
1	.31847	.32810	.67190	.99038	59	1	.35263	.36394	.63606	.98869	59
2	.31907	.32872	.67128	.99035	58	2	.35318	.36452	.63548	.98867	58
3	.31966	.32933	.67067	.99032	57	3	.35373	.36509	.63491	.98864	57
4	.32025	.32995	.67005	.99030	56	4	.35427	.36566	.63434	.98861	56
5	9.32084	9.33057	10.66943	9.99027	55	5	9.35481	9.36624	10.63376	9.98858	55
6	.32143	.33119	.66881	.99024	54	6	.35536	.36681	.63319	.98855	54
7	.32202	.33180	.66820	.99022	53	7	.35590	.36738	.63262	.98852	53
8	.32261	.33242	.66758	.99019	52	8	.35644	.36759	.63205	.98849	52
9	.32319	.33303	.66697	.99016	51	9	.35698	.36852	.63148	.98846	51
10	9.32378	9.33365	10.66635	9.99013	50	10	9.35752	9.36909	10.63091	9.98843	50
11	.32437	.33426	.66574	.99011	49	11	.35806	.36966	.63034	.98840	49
12	.32495	.33487	.66513	.99008	48	12	.35860	.37023	.62977	.98837	48
13	.32553	.33548	.66452	.99005	47	13	.35914	.37080	.62920	.98834	47
14	.32612	.33609	.66391	.99002	46	14	.35968	.37137	.62863	.98831	46
15	9.32670	9.33670	10.66330	9.99000	45	15	9.36022	9.37193	10.62807	9.98828	45
16	.32728	.33731	.66269	.98997	44	16	.36075	.37250	.62750	.98825	44
17	.32786	.33792	.66208	.98994	43	17	.36129	.37306	.62694	.98822	43
18	.32844	.33853	.66147	.98991	42	18	.36182	.37363	.62637	.98819	42
19	.32902	.33913	.66087	.98989	41	19	.36236	.37419	.62581	.98816	41
20	9.32960	9.33974	10.66026	9.98986	40	20	9.36289	9.37476	10.62524	9.98813	40
21	.33018	.34034	.65966	.98983	39	21	.36342	.37532	.62468	.98810	39
22	.33075	.34095	.65905	.98980	38	22	.36395	.37588	.62412	.98807	38
23	.33133	.34155	.65845	.98978	37	23	.36449	.37644	.62356	.98804	37
24	.33190	.34215	.65785	.98975	36	24	.36502	.37700	.62300	.98801	36
25	9.33248	9.34276	10.65724	9.98972	35	25	9.36555	9.37756	10.62244	9.98798	35
26	.33305	.34336	.65664	.98969	34	26	.36608	.37812	.62188	.98795	34
27	.33362	.34396	.65604	.98967	33	27	.36660	.37868	.62132	.98792	33
28	.33420	.34456	.65544	.98964	32	28	.36713	.37924	.62076	.98789	32
29	.33477	.34516	.65484	.98961	31	29	.36766	.37980	.62020	.98786	31
30	9.33534	9.34576	10.65424	9.98958	30	30	9.36819	9.38035	10.61965	9.98783	30
31	.33591	.34635	.65365	.98955	29	31	.36871	.38091	.61909	.98780	29
32	.33647	.34695	.65305	.98953	28	32	.36924	.38147	.61853	.98777	28
33	.33704	.34755	.65245	.98950	27	33	.36976	.38202	.61798	.98774	27
34	.33761	.34814	.65186	.98947	26	34	.37028	.38257	.61743	.98771	26
35	9.33818	9.34874	10.65126	9.98944	25	35	9.37081	9.38313	10.61687	9.98768	25
36	.33874	.34933	.65067	.98941	24	36	.37133	.38368	.61632	.98765	24
37	.33931	.34992	.65008	.98938	23	37	.37185	.38423	.61577	.98762	23
38	.33987	.35051	.64949	.98936	22	38	.37237	.38479	.61521	.98759	22
39	.34043	.35111	.64889	.98933	21	39	.37289	.38534	.61466	.98756	21
40	9.34100	9.35170	10.64830	9.98930	20	40	9.37341	9.38589	10.61411	9.98753	20
41	.34156	.35229	.64771	.98927	19	41	.37393	.38644	.61356	.98750	19
42	.34212	.35288	.64712	.98924	18	42	.37445	.38699	.61301	.98746	18
43	.34268	.35347	.64653	.98921	17	43	.37497	.38754	.61246	.98743	17
44	.34324	.35405	.64595	.98919	16	44	.37549	.38808	.61192	.98740	16
45	9.34380	9.35464	10.64536	9.98916	15	45	9.37600	9.38863	10.61137	9.98737	15
46	.34436	.35523	.64477	.98913	14	46	.37652	.38918	.61082	.98734	14
47	.34491	.35581	.64419	.98910	13	47	.37703	.38972	.61028	.98731	13
48	.34547	.35640	.64360	.98907	12	48	.37755	.39027	.60973	.98728	12
49	.34602	.35698	.64302	.98904	11	49	.37806	.39082	.60918	.98725	11
50	9.34658	9.35757	10.64243	9.98901	10	50	9.37858	9.39136	10.60864	9.98722	10
51	.34713	.35815	.64185	.98898	9	51	.37909	.39190	.60810	.98719	9
52	.34769	.35873	.64127	.98896	8	52	.37960	.39245	.60755	.98517	8
53	.34824	.35931	.64069	.98893	7	53	.38011	.39299	.60701	.98712	7
54	.34879	.35989	.64011	.98890	6	54	.38062	.39353	.60647	.98709	6
55	9.34934	9.36047	10.63953	9.98887	5	55	9.38113	9.39407	10.60593	9.98706	5
56	.34989	.36105	.63895	.98884	4	56	.38164	.39461	.60539	.98703	4
57	.35044	.36163	.63837	.98881	3	57	.38215	.39515	.60485	.98700	3
58	.35099	.36221	.63779	.98878	2	58	.38266	.39569	.60431	.98697	2
59	.35154	.36279	.63721	.98875	1	59	.38317	.39623	.60377	.98694	1
60	9.35209	9.36336	10.63664	9.98872	0	60	9.38368	9.39677	10.60323	9.98690	0
	Cosine.	Cotang.	Tang.	Sine.	′		Cosine.	Cotang.	Tang.	Sine.	′

77°   76°

*Log secant = colog cosine = 1 – log cosine; log cosecant = colog sine = 1 – log sine.
  Ex.—Log sec 12°– 30′ = 10.01042. *Ex.*—Log cosec 12°– 30′ = 10.66466.

3. —Logarithmic Sines, TANGENTS, COTANGENTS, COSINES.—(Cont'd.)
(SECANTS, COSECANTS.)*

14°

	Sine.	Tang.	Cotang.	Cosine.	
0	9.38368	9.39677	10.60323	9.98690	60
1	.38418	.39731	.60269	.98687	59
2	.38469	.39785	.60215	.98684	58
3	.38519	.39838	.60162	.98681	57
4	.38570	.39892	.60108	.98678	56
5	9.38620	9.39945	10.60055	9.98675	55
6	.38670	.39999	.60001	.98671	54
7	.38721	.40052	.59948	.98668	53
8	.38771	.40106	.59894	.98665	52
9	.38821	.40159	.59841	.98662	51
10	9.38871	9.40212	10.59788	9.98659	50
11	.38921	.40266	.59734	.98656	49
12	.38971	.40319	.59681	.98652	48
13	.39021	.40372	.59628	.98649	47
14	.39071	.40425	.59575	.98646	46
15	9.39121	9.40478	10.59522	9.98643	45
16	.39170	.40531	.59469	.98640	44
17	.39220	.40584	.59416	.98636	43
18	.39270	.40636	.59364	.98633	42
19	.39319	.40689	.59311	.98630	41
20	9.39369	9.40742	10.59258	9.98627	40
21	.39418	.40795	.59205	.98623	39
22	.39467	.40847	.59153	.98620	38
23	.39517	.40900	.59100	.98617	37
24	.39566	.40952	.59048	.98614	36
25	9.39615	9.41005	10.58995	9.98610	35
26	.39664	.41057	.58943	.98607	34
27	.39713	.41109	.58891	.98604	33
28	.39762	.41161	.58839	.98601	32
29	.39811	.41214	.58786	.98597	31
30	9.39860	9.41266	10.58734	9.98594	30
31	.39909	.41318	.58682	.98591	29
32	.39958	.41370	.58630	.98588	28
33	.40006	.41422	.58578	.98584	27
34	.40055	.41474	.58526	.98581	26
35	9.40103	9.41526	10.58474	9.98578	25
36	.40152	.41578	.58422	.98574	24
37	.40200	.41629	.58371	.98571	23
38	.40249	.41681	.58319	.98568	22
39	.40297	.41733	.58267	.98565	21
40	9.40346	9.41784	10.58216	9.98561	20
41	.40394	.41836	.58164	.98558	19
42	.40442	.41887	.58113	.98555	18
43	.40490	.41939	.58061	.98551	17
44	.40538	.41990	.58010	.98548	16
45	9.40586	9.42041	10.57959	9.98545	15
46	.40634	.42093	.57907	.98541	14
47	.40682	.42144	.57856	.98538	13
48	.40730	.42195	.57805	.98535	12
49	.40778	.42246	.57754	.98531	11
50	9.40825	9.42297	10.57703	9.98528	10
51	.40873	.42348	.57652	.98525	9
52	.40921	.42399	.57601	.98521	8
53	.40968	.42450	.57550	.98518	7
54	.41016	.42501	.57499	.98515	6
55	9.41063	9.42552	10.57448	9.98511	5
56	.41111	.42603	.57397	.98508	4
57	.41158	.42653	.57347	.98505	3
58	.41205	.42704	.57296	.98501	2
59	.41252	.42755	.57245	.98498	1
60	9.41300	9.42805	10.57195	9.98494	0

	Cosine.	Cotang.	Tang.	Sine.	'

75°

15°

H		Sine.	Tang.	Cotang.	Cosine.	
60	0	9.41300	9.42805	10.57195	9.98494	60
59	1	.41347	.42856	.57144	.98491	59
58	2	.41394	.42906	.57094	.98488	58
57	3	.41441	.42957	.57043	.98484	57
56	4	.41488	.43007	.56993	.98481	56
55	5	9.41535	9.43057	10.56943	9.98477	55
54	6	.41582	.43108	.56892	.98474	54
53	7	.41628	.43158	.56842	.98471	53
52	8	.41675	.43208	.56792	.98467	52
51	9	.41722	.43258	.56742	.98464	51
50	10	9.41768	9.43308	10.56692	9.98460	50
49	11	.41815	.43358	.56642	.98457	49
48	12	.41861	.43408	.56592	.98453	48
47	13	.41908	.43458	.56542	.98450	47
46	14	.41954	.43508	.56492	.98447	46
45	15	9.42001	9.43558	10.56442	9.98443	45
44	16	.42047	.43607	.56393	.98440	44
43	17	.42093	.43657	.56343	.98436	43
42	18	.42140	.43707	.56293	.98433	42
41	19	.42186	.43756	.56244	.98429	41
40	20	9.42232	9.43806	10.56194	9.98426	40
39	21	.42278	.43855	.56145	.98422	39
38	22	.42324	.43905	.56095	.98419	38
37	23	.42370	.43954	.56046	.98415	37
36	24	.42416	.44004	.55996	.98412	36
35	25	9.42461	9.44053	10.55947	9.98409	35
34	26	.42507	.44102	.55898	.98405	34
33	27	.42553	.44151	.55849	.98402	33
32	28	.42599	.44201	.55799	.98398	32
31	29	.42644	.44250	.55750	.98395	31
30	30	9.42690	9.44299	10.55701	9.98391	30
29	31	.42735	.44348	.55652	.98388	29
28	32	.42781	.44397	.55603	.98384	28
27	33	.42826	.44446	.55554	.98381	27
26	34	.42872	.44495	.55505	.98377	26
25	35	9.42917	9.44544	10.55456	9.98373	25
24	36	.42962	.44592	.55408	.98370	24
23	37	.43008	.44641	.55359	.98366	23
22	38	.43053	.44690	.55310	.98363	22
21	39	.43098	.44738	.55262	.98359	21
20	40	9.43143	9.44787	10.55213	9.98356	20
19	41	.43188	.44836	.55164	.98352	19
18	42	.43233	.44884	.55116	.98349	18
17	43	.43278	.44933	.55067	.98345	17
16	44	.43323	.44981	.55019	.98342	16
15	45	9.43367	9.45029	10.54971	9.98338	15
14	46	.43412	.45078	.54922	.98334	14
13	47	.43457	.45126	.54874	.98331	13
12	48	.43502	.45174	.54826	.98327	12
11	49	.43546	.45222	.54778	.98324	11
10	50	9.43591	9.45271	10.54729	9.98320	10
9	51	.43635	.45319	.54681	.98317	9
8	52	.43680	.45367	.54633	.98313	8
7	53	.43724	.45415	.54585	.98309	7
6	54	.43769	.45463	.54537	.98306	6
5	55	9.43813	9.45511	10.54489	9.98302	5
4	56	.43857	.45559	.54441	.98299	4
3	57	.43901	.45606	.54394	.98295	3
2	58	.43946	.45654	.54346	.98291	2
1	59	.43990	.45702	.54298	.98288	1
0	60	9.44034	9.45750	10.54250	9.98284	0

	Cosine.	Cotang.	Tang.	Sine.	'

74°

*Log secant = colog cosine = 1 — log cosine, log cosecant = colog sine =
1 — log sine.
Ex.—Log sec 14°- 30′ = 10.01406.   *Ex.*—Log cosec 14°- 30′ = 10.60140.

3.—Logarithmic Sines, Tangents, Cotangents, Cosines.—(Cont'd.)
(Secants, Cosecants.)*

16°					17°				
′ | Sine. | Tang. | Cotang. | Cosine. |					∥ ′ | Sine. | Tang. | Cotang. | Cosine. |				

′	Sine.	Tang.	Cotang.	Cosine.		′	Sine.	Tang.	Cotang.	Cosine.	
0	9.44034	9.45750	10.54250	9.98284	60	0	9.46594	9.48534	10.51466	9.98060	60
1	.44078	.45797	.54203	.98281	59	1	.46635	.48579	.51421	.98056	59
2	.44122	.45845	.54155	.98277	58	2	.46676	.48624	.51376	.98052	58
3	.44166	.45892	.54108	.98273	57	3	.46717	.48669	.51331	.98048	57
4	.44210	.45940	54060	.98270	56	4	.46758	.48714	.51286	.98044	56
5	9.44253	9.45987	10.54013	9.98266	55	5	9.46800	9.48759	10.51241	.98040	55
6	.44297	.46035	.53965	.98262	54	6	.46841	.48804	.51196	.98036	54
7	.44341	.46082	.53918	.98259	53	7	.46882	.48849	.51151	.98032	53
8	.44385	.46130	.53870	.98255	52	8	.46923	.48894	.51106	.98029	52
9	.44428	.46177	.53823	.98251	51	9	.46964	.48939	.51061	.98025	51
10	9.44472	9.46224	10.53776	9.98248	50	10	9.47005	9.48984	10.51016	9.98021	50
11	.44516	.46271	.53729	.98244	49	11	.47045	.49029	.50971	.98017	49
12	.44559	.46319	.53681	.98240	48	12	.47086	.49073	.50927	.98013	48
13	.44602	.46366	.53634	.98237	47	13	.47127	.49118	.50882	.98009	47
14	.44646	.46413	.53587	.98233	46	14	.47168	.49163	.50837	.98005	46
15	9.44689	9.46460	10.53540	9.98229	45	15	9.47209	9.49207	10.50793	9.98001	45
16	.44733	.46507	.53493	.98226	44	16	.47249	.49252	.50748	.97997	44
17	.44776	.46554	.53446	.98222	43	17	.47290	.49296	.50704	.97993	43
18	.44819	.46601	.53399	.98218	42	18	.47330	.49341	.50659	.97989	42
19	.44862	.46648	.53352	.98215	41	19	.47371	.49385	.50615	.97986	41
20	9.44905	9.46694	10.53306	9.98211	40	20	9.47411	9.49430	10.50570	9.97982	40
21	.44948	.46741	.53259	.98207	39	21	.47452	.49474	.50526	.97978	39
22	.44992	.46788	.53212	.98204	38	22	.47492	.49519	.50481	.97974	38
23	.45035	.46835	.53165	.98200	37	23	.47533	.49563	.50437	.97970	37
24	.45077	.46881	.53119	.98196	36	24	.47573	.49607	.50393	.97966	36
25	9.45120	9.46928	10.53072	9.98192	35	25	9.47613	9.49652	10.50348	9.97962	35
26	.45163	.46975	.53025	.98189	34	26	.47654	.49696	.50304	.97958	34
27	.45206	.47021	.52979	.98185	33	27	.47694	.49740	.50260	.97954	33
28	.45249	.47068	.52932	.98181	32	28	.47734	.49784	.50216	.97950	32
29	.45292	.47114	.52886	.98177	31	29	.47774	.49828	.50172	.97946	31
30	9.45334	9.47160	10.52840	9.98174	30	30	9.47814	9.49872	10.50128	9.97942	30
31	.45377	.47207	.52793	.98170	29	31	.47854	.49916	.50084	.97938	29
32	.45419	.47253	.52747	.98166	28	32	.47894	.49960	.50040	.97934	28
33	.45462	.47299	.52701	.98162	27	33	.47934	.50004	.49996	.97930	27
34	.45504	.47346	.52654	.98159	26	34	.47974	.50048	.49952	.97926	26
35	9.45547	9.47392	10.52608	9.98155	25	35	9.48014	9.50092	10.49908	9.97922	25
36	.45589	.47438	.52562	.98151	24	36	.48054	.50136	.49864	.97918	24
37	.45632	.47484	.52516	.98147	23	37	.48094	.50180	.49820	.97914	23
38	.45674	.47530	.52470	.98144	22	38	.48133	.50223	.49777	.97910	22
39	.45716	.47576	.52424	.98140	21	39	.48173	.50267	.49733	.97906	21
40	9.45758	9.47622	10.52378	9.98136	20	40	9.48213	9.50311	10.49689	9.97902	20
41	.45801	.47668	.52332	.98132	19	41	.48252	.50355	.49645	.97898	19
42	.45843	.47714	.52286	.98129	18	42	.48292	.50398	.49602	.97894	18
43	.45885	.47760	.52240	.98125	17	43	.48332	.50442	.49558	.97890	17
44	.45927	.47806	.52194	.98121	16	44	.48371	.50485	.49515	.97886	16
45	9.45969	9.47852	10.52148	9.98117	15	45	9.48411	9.50529	10.49471	9.97882	15
46	.46011	.47897	.52103	.98113	14	46	.48450	.50572	.49428	.97878	14
47	.46053	.47943	.52057	.98110	13	47	.48490	.50616	.49384	.97874	13
48	.46095	.47989	.52011	.98106	12	48	.48529	.50659	.49341	.97870	12
49	.46136	.48035	.51965	.98102	11	49	.48568	.50703	.49297	.97866	11
50	9.46178	9.48080	10.51920	9.98098	10	50	9.48607	9.50746	10.49254	9.97861	10
51	.46220	.48126	.51874	.98094	9	51	.48647	.50789	.49211	.97857	9
52	.46262	.48171	.51829	.98090	8	52	.48686	.50833	.49167	.97853	8
53	.46303	.48217	.51783	.98087	7	53	.48725	.50876	.49124	.97849	7
54	.46345	.48262	.51738	.98083	6	54	.48764	.50919	.49081	.97845	6
55	9.46386	9.48307	10.51693	9.98079	5	55	9.48803	9.50962	10.49038	9.97841	5
56	.46428	.48353	.51647	.98075	4	56	.48842	.51005	.48995	.97837	4
57	.46469	.48398	.51602	.98071	3	57	.48881	.51048	.48952	.97833	3
58	.46511	.48443	.51557	.98067	2	58	.48920	.51092	.48908	.97829	2
59	.46552	.48489	.51511	.98063	1	59	.48959	.51135	.48865	.97825	1
60	9.46594	9.48534	10.51466	9.98060	0	60	9.48998	9.51178	10.48822	9.97821	0

	Cosine.	Cotang.	Tang.	Sine.	′		Cosine.	Cotang.	Tang.	Sine.	′

73°						72°					

*Log secant = colog cosine = 1 − log cosine; log cosecant = colog sine = 1 − log sine.

Ex.—Log sec 16°- 30′ = 10.01826.   *Ex.*—Log cosec 16°- 30′ = 10.54666.

3.—Logarithmic Sines, Tangents, Cotangents, Cosines.—(Cont'd.)
(Secants, Cosecants.)*

18°   **19°**

'	Sine.	Tang.	Cotang.	Cosine.		'	Sine.	Tang.	Cotang.	Cosine.	
0	9.48998	9.51178	10.48822	9.97821	60	0	9.51264	9.53697	10.46303	9.97567	60
1	.49037	.51221	.48779	.97817	59	1	.51301	.53738	.46262	.97563	59
2	.49076	.51264	.48736	.97812	58	2	.51338	.53779	.46221	.97558	58
3	.49115	.51306	.48694	.97808	57	3	.51374	.53820	.46180	.97554	57
4	.49153	.51349	.48651	.97804	56	4	.51411	.53861	.46139	.97550	56
5	9.49192	9.51392	10.48608	9.97800	55	5	9.51447	9.53902	10.46098	9.97545	55
6	.49231	.51435	.48565	.97796	54	6	.51484	.53943	.46057	.97541	54
7	.49269	.51478	.48522	.97792	53	7	.51520	.53984	.46016	.97536	53
8	.49308	.51520	.48480	.97788	52	8	.51557	.54025	.45975	.97532	52
9	.49347	.51563	.48437	.97784	51	9	.51593	.54065	.45935	.97528	51
10	9.49385	9.51606	10.48394	9.97779	50	10	9.51629	9.54106	10.45894	9.97523	50
11	.49424	.51648	.48352	.97775	49	11	.51666	.54147	.45853	.97519	49
12	.49462	.51691	.48309	.97771	48	12	.51702	.54187	.45813	.97515	48
13	.49500	.51734	.48266	.97767	47	13	.51738	.54228	.45772	.97510	47
14	.49539	.51776	.48224	.97763	46	14	.51774	.54269	.45731	.97506	46
15	9.49577	9.51819	10.48181	9.97759	45	15	9.51811	9.54309	10.45691	9.97501	45
16	.49615	.51861	.48139	.97754	44	16	.51847	.54350	.45650	.97497	44
17	.49654	.51903	.48097	.97750	43	17	.51883	.54390	.45610	.97492	43
18	.49692	.51946	.48054	.97746	42	18	.51919	.54431	.45569	.97488	42
19	.49730	.51988	.48012	.97742	41	19	.51955	.54471	.45529	.97484	41
20	9.49768	9.52031	10.47969	9.97738	40	20	9.51991	9.54512	10.45488	9.97479	40
21	.49806	.52073	.47927	.97734	39	21	.52027	.54552	.45448	.97475	39
22	.49844	.52115	.47885	.97729	38	22	.52063	.54593	.45407	.97470	38
23	.49882	.52157	.47843	.97725	37	23	.52099	.54633	.45367	.97466	37
24	.49920	.52200	.47800	.97721	36	24	.52135	.54673	.45327	.97461	36
25	9.49958	9.52242	10.47758	9.97717	35	25	9.52171	9.54714	10.45286	9.97457	35
26	.49996	.52284	.47716	.97713	34	26	.52207	.54754	.45246	.97453	34
27	.50034	.52326	.47674	.97708	33	27	.52242	.54794	.45206	.97448	33
28	.50072	.52368	.47632	.97704	32	28	.52278	.54835	.45165	.97444	32
29	.50110	.52410	.47590	.97700	31	29	.52314	.54875	.45125	.97439	31
30	9.50148	9.52452	10.47548	9.97696	30	30	9.52350	9.54915	10.45085	9.97435	30
31	.50185	.52494	.47506	.97691	29	31	.52385	.54955	.45045	.97430	29
32	.50223	.52536	.47464	.97687	28	32	.52421	.54995	.45005	.97426	28
33	.50261	.52578	.47422	.97683	27	33	.52456	.55035	.44965	.97421	27
34	.50298	.52620	.47380	.97679	26	34	.52492	.55075	.44925	.97417	26
35	9.50336	9.52661	10.47339	9.97674	25	35	9.52527	9.55115	10.44885	9.97412	25
36	.50374	.52703	.47297	.97670	24	36	.52563	.55155	.44845	.97408	24
37	.50411	.52745	.47255	.97666	23	37	.52598	.55195	.44805	.97403	23
38	.50449	.52787	.47213	.97662	22	38	.52634	.55235	.44765	.97399	22
39	.50486	.52829	.47171	.97657	21	39	.52669	.55275	.44725	.97394	21
40	9.50523	9.52870	10.47130	9.97653	20	40	9.52705	9.55315	10.44685	9.97390	20
41	.50561	.52912	.47088	.97649	19	41	.52740	.55355	.44645	.97385	19
42	.50598	.52953	.47047	.97645	18	42	.52775	.55395	.44605	.97381	18
43	.50635	.52995	.47005	.97640	17	43	.52811	.55434	.44566	.97376	17
44	.50673	.53037	.46963	.97636	16	44	.52846	.55474	.44526	.97372	16
45	9.50710	9.53078	10.46922	9.97632	15	45	9.52881	9.55514	10.44486	9.97367	15
46	.50747	.53120	.46880	.97628	14	46	.52916	.55554	.44446	.97363	14
47	.50784	.53161	.46839	.97623	13	47	.52951	.55593	.44407	.97358	13
48	.50821	.53202	.46798	.97619	12	48	.52986	.55633	.44367	.97353	12
49	.50858	.53244	.46756	.97615	11	49	.53021	.55673	.44327	.97349	11
50	9.50896	9.53285	10.46715	9.97610	10	50	9.53056	9.55712	10.44288	9.97344	10
51	.50933	.53327	.46673	.97606	9	51	.53092	.55752	.44248	.97340	9
52	.50970	.53368	.46632	.97602	8	52	.53126	.55791	.44209	.97335	8
53	.51007	.53409	.46591	.97597	7	53	.53161	.55831	.44169	.97331	7
54	.51043	.53450	.46550	.97593	6	54	.53196	.55870	.44130	.97326	6
55	9.51080	9.53492	10.46508	9.97589	5	55	9.53231	9.55910	10.44090	9.97322	5
56	.51117	.53533	.46467	.97584	4	56	.53266	.55949	.44051	.97317	4
57	.51154	.53574	.46426	.97580	3	57	.53301	.55989	.44011	.97312	3
58	.51191	.53615	.46385	.97576	2	58	.53336	.56028	.43972	.97308	2
59	.51227	.53656	.46344	.97571	1	59	.53370	.56067	.43933	.97303	1
60	9.51264	9.53697	10.46303	9.97567	0	60	9.53405	9.56107	10.43893	9.97299	0

	Cosine.	Cotang.	Tang.	Sine.	'			Cosine.	Cotang.	Tang.	Sine.	'
					71°							**70°**

*Log secant = colog cosine = 1 − log cosine; log cosecant = colog sine = 1 − log sine.

Ex.—Log sec 18°- 30' = 10.02304.   *Ex.*—Log cosec 18°- 30' = 10.49852.

3.—Logarithmic Sines, Tangents, Cotangents, Cosines.—(Cont'd.)
(Secants, Cosecants.)*

20°   **21°**

'	Sine.	Tang.	Cotang.	Cosine.		'	Sine.	Tang.	Cotang.	Cosine.	
0	9.53405	9.56107	10.43893	9.97299	60	0	9.55433	9.58418	10.41582	9.97015	60
1	.53440	.56146	.43854	.97294	59	1	.55466	.58455	.41545	.97010	59
2	.53475	.56185	.43815	.97289	58	2	.55499	.58493	.41507	.97005	58
3	.53509	.56224	.43776	.97285	57	3	.55532	.58531	.41469	.97001	57
4	.53544	.56264	43736	.97280	56	4	.55564	.58569	.41431	.96996	56
5	9.53578	9.56303	10.43697	9.97276	55	5	9.55597	9.58606	10.41394	9.96991	55
6	.53613	.56342	.43658	.97271	54	6	.55630	.58644	.41356	.96986	54
7	.53647	.56381	.43619	.97266	53	7	.55663	.58681	.41319	.96981	53
8	.53682	.56420	.43580	.97262	52	8	.55695	.58719	.41281	.96976	52
9	.53716	.56459	.43541	.97257	51	9	.55728	.58757	.41243	.96971	51
10	9.53751	9.56498	10.43502	9.97252	50	10	9.55761	9.58794	10.41206	9.96966	50
11	.53785	.56537	.43463	.97248	49	11	.55793	.58832	.41168	.96962	49
12	.53819	.56576	.43424	.97243	48	12	.55826	.58869	.41131	.96957	48
13	.53854	.56615	.43385	.97238	47	13	.55858	.58907	.41093	.96952	47
14	.53888	.56654	.43346	.97234	46	14	.55891	.58944	.41056	.96947	46
15	9.53922	9.56693	10.43307	9.97229	45	15	9.55923	9.58981	10.41019	9.96942	45
16	.53957	.56732	.43268	.97224	44	16	.55956	.59019	.40981	.96937	44
17	.53991	.56771	.43229	.97220	43	17	.55988	.59056	.40944	.96932	43
18	.54025	.56810	.43190	.97215	42	18	.56021	.59094	.40906	.96927	42
19	.54059	.56849	.43151	.97210	41	19	.56053	.59131	.40869	.96922	41
20	9.54093	9.56887	10.43113	9.97206	40	20	9.56085	9.59168	10.40832	9.96917	40
21	.54127	.56926	.43074	.97201	39	21	.56118	.59205	.40795	.96912	39
22	.54161	.56965	.43035	.97196	38	22	.56150	.59243	.40757	.96907	38
23	.54195	.57004	.42996	.97192	37	23	.56182	.59280	.40720	.96903	37
24	.54229	.57042	.42958	.97187	36	24	.56215	.59317	.40683	.96898	36
25	9.54263	9.57081	10.42919	9.97182	35	25	9.56247	9.59354	10.40646	9.96893	35
26	.54297	.57120	.42880	.97178	34	26	.56279	.59391	.40609	.96888	34
27	.54331	.57158	.42842	.97173	33	27	.56311	.59429	.40571	.96883	33
28	.54365	.57197	.42803	.97168	32	28	.56343	.59466	.40534	.96878	32
29	.54399	.57235	.42765	.97163	31	29	.56375	.59503	.40497	.96873	31
30	9.54433	9.57274	10.42726	9.97159	30	30	9.56408	9.59540	10.40460	9.96868	30
31	.54466	.57312	.42688	.97154	29	31	.56440	.59577	.40423	.96863	29
32	.54500	.57351	.42649	.97149	28	32	.56472	.59614	.40386	.96858	28
33	.54534	.57389	.42611	.97145	27	33	.56504	.59651	.40349	.96853	27
34	.54567	.57428	.42572	.97140	26	34	.56536	.59688	.40312	.96848	26
35	9.54601	9.57466	10.42534	9.97135	25	35	9.56568	9.59725	10.40275	9.96843	25
36	.54635	.57504	.42496	.97130	24	36	.56599	.59762	.40238	.96838	24
37	.54668	.57543	.42457	.97126	23	37	.56631	.59799	.40201	.96833	23
38	.54702	.57581	.42419	.97121	22	38	.56663	.59835	.40165	.96828	22
39	.54735	.57619	.42381	.97116	21	39	.56695	.59872	.40128	.96823	21
40	9.54769	9.57658	10.42342	9.97111	20	40	9.56727	9.59909	10.40091	9.96818	20
41	.54802	.57696	.42304	.97107	19	41	.56759	.59946	.40054	.96813	19
42	.54836	.57734	.42266	.97102	18	42	.56790	.59983	.40017	.96808	18
43	.54869	.57772	.42228	.97097	17	43	.56822	.60019	.39981	.96803	17
44	.54903	.57810	.42190	.97092	16	44	.56854	.60056	.39944	.96798	16
45	9.54936	9.57849	10.42151	9.97087	15	45	9.56886	9.60093	10.39907	9.96793	15
46	.54969	.57887	.42113	.97083	14	46	.56917	.60130	.39870	.96788	14
47	.55003	.57925	.42075	.97078	13	47	.56949	.60166	.39834	.96783	13
48	.55036	.57963	.42037	.97073	12	48	.56980	.60203	.39797	.96778	12
49	.55069	.58001	.41999	.97068	11	49	.57012	.60240	.39760	.96772	11
50	9.55102	9.58039	10.41961	9.97063	10	50	9.57044	9.60276	10.39724	9.96767	10
51	.55136	.58077	.41923	.97059	9	51	.57075	.60313	.39687	.96762	9
52	.55169	.58115	.41885	.97054	8	52	.57107	.60349	.39651	.96757	8
53	.55202	.58153	.41847	.97049	7	53	.57138	.60386	.39614	.96752	7
54	.55235	.58191	.41809	.97044	6	54	.57169	.60422	.39578	.96747	6
55	9.55268	9.58229	10.41771	9.97039	5	55	9.57201	9.60459	10.39541	9.96742	5
56	.55301	.58267	.41733	.97035	4	56	.57232	.60495	.39505	.96737	4
57	.55334	.58304	.41696	.97030	3	57	.57264	.60532	.39468	.96732	3
58	.55367	.58342	.41658	.97025	2	58	.57295	.60568	.39432	.96727	2
59	.55400	.58380	.41620	.97020	1	59	.57326	.60605	.39395	.96722	1
60	9.55433	9.58418	10.41582	9.97015	0	60	9.57358	9.60641	10.39359	9.96717	0

	Cosine.	Cotang.	Tang.	Sine.	'			Cosine.	Cotang.	Tang.	Sine.	'

69°   **68°**

*Log secant = colog cosine = 1 − log cosine; log cosecant = colog sine = 1 − log sine.

Ex.—Log sec 20°- 30′ = 10.02841.   *Ex.*—Log cosec 20°- 30′ = 10.45567.

3.—Logarithmic Sines, Tangents, Cotangents, Cosines.—(Cont'd.)
(Secants, Cosecants.)*

22°   23°

′	Sine.	Tang.	Cotang.	Cosine.	′	′	Sine.	Tang.	Cotang.	Cosine.	′
0	9.57358	9.60641	10.39359	9.96717	60	0	9.59188	9.62785	10.37215	9.96403	60
1	.57389	.60677	.39323	.96711	59	1	.59218	.62820	.37180	.96397	59
2	.57420	.60714	.39286	.96706	58	2	.59247	.62855	.37145	.96392	58
3	.57451	.60750	.39250	.96701	57	3	.59277	.62890	.37110	.96387	57
4	.57482	.60786	.39214	.96696	56	4	.59307	.62926	.37074	.96381	56
5	9.57514	9.60823	10.39177	9.96691	55	5	9.59336	9.62961	10.37039	9.96376	55
6	.57545	.60859	.39141	.96686	54	6	.59366	.62996	.37004	.96370	54
7	.57576	.60895	.39105	.96681	53	7	.59396	.63031	.36969	.96365	53
8	.57607	.60931	.39069	.96676	52	8	.59425	.63066	.36934	.96360	52
9	.57638	.60967	.39033	.96670	51	9	.59455	.63101	.36899	.96354	51
10	9.57669	9.61004	10.38996	9.96665	50	10	9.59484	9.63135	10.36865	9.96349	50
11	.57700	.61040	.38960	.96660	49	11	.59514	.63170	.36830	.96343	49
12	.57731	.61076	.38924	.96655	48	12	.59543	.63205	.36795	.96338	48
13	.57762	.61112	.38888	.96650	47	13	.59573	.63240	.36760	.96333	47
14	.57793	.61148	.38852	.96645	46	14	.59602	.63275	.36725	.96327	46
15	9.57824	9.61184	10.38816	9.96640	45	15	9.59632	9.63310	10.36690	9.96322	45
16	.57855	.61220	.38780	.96634	44	16	.59661	.63345	.36655	.96316	44
17	.57885	.61256	.38744	.96629	43	17	.59690	.63379	.36621	.96311	43
18	.57916	.61291	.38708	.96624	42	18	.59720	.63414	.36586	.96305	42
19	.57947	.61328	.38672	.96619	41	19	.59749	.63449	.36551	.96300	41
20	9.57978	9.61364	10.38636	9.96614	40	20	9.59778	9.63484	10.36516	9.96294	40
21	.58008	.61400	.38600	.96608	39	21	.59808	.63519	.36481	.96289	39
22	.58039	.61436	.38564	.96603	38	22	.59837	.63553	.36447	.96284	38
23	.58070	.61472	.38528	.96598	37	23	.59866	.63588	.36412	.96278	37
24	.58101	.61508	.38492	.96593	36	24	.59895	.63623	.36377	.96273	36
25	9.58131	9.61544	10.38456	9.96588	35	25	9.59924	9.63657	10.36343	9.96267	35
26	.58162	.61579	.38421	.96582	34	26	.59954	63692	.36308	.96262	34
27	.58192	.61615	.38385	.96577	33	27	.59983	.63726	.36274	.96256	33
28	.58223	.61651	.38349	.96572	32	28	.60012	.63761	.36239	.96251	32
29	.58253	.61687	.38313	.96567	31	29	.60041	.63796	.36204	.96245	31
30	9.58284	9.61722	10.38278	9.96562	30	30	9.60070	9.63830	10.36170	9.96240	30
31	.58314	.61758	.38242	.96556	29	31	.60099	.63865	.36135	.96234	29
32	.58345	.61794	.38206	.96551	28	32	.60128	.63899	.36101	.96229	28
33	.58375	.61830	.38170	.96546	27	33	.60157	.63934	.36066	.96223	27
34	.58406	.61865	.38135	.96541	26	34	.60186	.63968	.36032	.96218	26
35	9.58436	9.61901	10.38099	9.96535	25	35	9.60215	9.64003	10.35997	9.96212	25
36	.58467	.61936	.38064	.96530	24	36	.60244	.64037	.35963	.96207	24
37	.58497	.61972	.38028	.96525	23	37	.60273	.64072	.35928	.96201	23
38	.58527	.62008	.37992	.96520	22	38	.60302	.64106	.35894	.96196	22
39	.58557	.62043	.37957	.96514	21	39	.60331	.64140	.35860	.96190	21
40	9.58588	9.62079	10.37921	9.96509	20	40	9.60359	9 64175	10.35825	9.96185	20
41	.58618	.62114	.37886	.96504	19	41	.60388	.64209	.35791	.96179	19
42	.58648	.62150	.37850	.96498	18	42	.60417	.64243	.35757	.96174	18
43	.58678	.62185	.37815	.96493	17	43	.60446	.64278	.35722	.96168	17
44	.58709	.62221	.37779	.96488	16	44	.60474	.64312	.35688	.96162	16
45	9.58739	9.62256	10.37744	9.96483	15	45	9.60503	9.64346	10.35654	9.96157	15
46	.58769	.62292	.37708	.96477	14	46	.60532	.64381	.35619	.96151	14
47	.58799	.62327	.37673	.96472	13	47	.60561	.64415	.35585	.96146	13
48	.58829	.62362	.37638	.96467	12	48	.60589	.64449	.35551	.96140	12
49	.58859	.62398	.37602	.96461	11	49	.60618	.64483	.35517	.96135	11
50	9.58889	9.62433	10.37567	9.96456	10	50	9.60646	9.64517	10.35483	9.96129	10
51	.58919	.62468	.37532	.96451	9	51	.60675	.64552	.35448	.96123	9
52	.58949	.62504	.37496	.96445	8	52	.60704	.64586	.35414	.96118	8
53	.58979	.62539	.37461	.96440	7	53	.60732	.64620	.35380	.96112	7
54	.59009	.62574	.37426	.96435	6	54	.60761	.64654	.35346	.96107	6
55	9.59039	9.62609	10.37391	9.96429	5	55	9.60789	9.64688	10.35312	9.96101	5
56	.59069	.62645	.37355	.96424	4	56	.60818	.64722	.35278	.96095	4
57	.59098	.62680	.37320	.96419	3	57	.60846	.64756	.35244	.96090	3
58	.59128	.62715	.37285	.96413	2	58	.60875	.64790	.35210	.96084	2
59	.59158	.62750	.37250	.96408	1	59	.60903	.64824	.35176	.96079	1
60	9.59188	9.62785	10.37215	9.96403	0	60	9.60931	9.64858	10.35142	9.96073	0
	Cosine.	Cotang.	Tang.	Sine.	′		Cosine.	Cotang.	Tang.	Sine.	′

67°   66°

*Log secant = colog cosine = 1 − log cosine; log cosecant = colog sine = 1 − log sine.

Ex.—Log sec 22°- 30′ = 10.03438.   Ex.—Log cosec 22°- 30′ = 10.41716.

3.—Logarithmic Sines, Tangents, Cotangents, Cosines.—(Cont'd.)
(Secants, Cosecants.)*

24° 25°

'	Sine.	Tang.	Cotang.	Cosine.		'	Sine.	Tang.	Cotang.	Cosine.	
0	9.60931	9.64858	10.35142	9.96073	60	0	9.62595	9.66867	10.33133	9.95728	60
1	.60960	.64892	.35108	.96067	59	1	.62622	.66900	.33100	.95722	59
2	.60988	.64926	.35074	.96062	58	2	.62649	.66933	.33067	.95716	58
3	.61016	.64960	.35040	.96056	57	3	.62676	.66966	.33034	.95710	57
4	.61045	.64994	.35006	.96050	56	4	.62703	.66999	.33001	.95704	56
5	9.61073	9.65028	10.34972	9.96045	55	5	9.62730	9.67032	10.32968	9.95698	55
6	.61101	.65062	.34938	.96039	54	6	.62757	.67065	.32935	.95692	54
7	.61129	.65096	.34904	.96034	53	7	.62784	.67098	.32902	.95686	53
8	.61158	.65130	.34870	.96028	52	8	.62811	.67131	.32869	.95680	52
9	.61186	.65164	.34836	.96022	51	9	.62838	.67163	.32837	.95674	51
10	9.61214	9.65197	10.34803	9.96017	50	10	9.62865	9.67196	10.32804	9.95668	50
11	.61242	.65231	.34769	.96011	49	11	.62892	-.67229	.32771	.95663	49
12	.61270	.65265	.34735	.96005	48	12	.62918	.67262	.32738	.95657	48
13	.61298	.65299	.34701	.96000	47	13	.62945	.67295	.32705	.95651	47
14	.61326	.65333	.34667	.95994	46	14	.62972	.67327	.32673	.95645	46
15	9.61354	9.65366	10.34634	9.95988	45	15	9.62999	9.67360	10.32640	9.95639	45
16	.61382	.65400	.34600	.95982	44	16	.63026	.67393	.32607	.95633	44
17	.61411	.65434	.34566	.95977	43	17	.63052	.67426	.32574	.95627	43
18	.61438	.65467	.34533	.95971	42	18	.63079	.67458	.32542	.95621	42
19	.61466	.65501	.34499	.95965	41	19	.63106	.67491	.32509	.95615	41
20	9.61494	9.65535	10.34465	9.95960	40	20	9.63133	9.67524	10.32476	9.95609	40
21	.61522	.65568	.34432	.95954	39	21	.63159	.67556	.32444	.95603	39
22	.61550	.65602	.34398	.95948	38	22	.63186	.67589	.32411	.95597	38
23	.61578	.65636	.34364	.95942	37	23	.63213	.67622	.32378	.95591	57
24	.61606	.65669	.34331	.95937	36	24	.63239	.67654	.32346	.95585	36
25	9.61634	9.65703	10.34297	9.95931	35	25	9.63266	9.67687	10.32313	9.95579	35
26	.61662	.65736	.34264	.95925	34	26	.63292	.67719	.32281	.95573	34
27	.61689	.65770	.34230	.95920	33	27	.63319	.67752	.32248	.95567	33
28	.61717	.65803	.34197	.95914	32	28	.63345	.67785	.32215	.95561	32
29	.61745	.65837	.34163	.95908	31	29	.63372	.67817	.32183	.95555	31
30	9.61773	9.65870	10.34130	9.95902	30	30	9.63398	9.67850	10.32150	9.95549	30
31	.61800	.65904	.34096	.95897	29	31	.63425	.67882	.32118	.95543	29
32	.61828	.65937	.34063	.95891	28	32	.63451	.67915	.32085	.95537	28
33	.61856	.65971	.34029	.93885	27	33	.63478	.67947	.32053	.95531	27
34	.61883	.66004	.33996	.95879	26	34	.63504	.67980	.32020	.95525	26
35	9.61911	9.66038	10.33962	9.95873	25	35	9.63531	9.68012	10.31988	9.95519	25
36	.61939	.66071	.33929	.95868	24	36	.63557	.68044	.31956	.95513	24
37	.61966	.66104	.33896	.95862	23	37	.63583	.68077	.31923	.95507	23
38	.61994	.66138	.33862	.95856	22	38	.63610	.68109	.31891	.95500	22
39	.62021	.66171	.33829	.95850	21	39	.63636	.68142	.31858	.95494	21
40	9.62049	9.66204	10.33796	9.95844	20	40	9.63662	9.68174	10.31826	9.95488	20
41	.62076	.66238	.33762	.95839	19	41	.63689	.68206	.31794	.95482	19
42	.62104	.66271	.33729	.95833	18	42	.63715	.68239	.31761	.95476	18
43	.62131	.66304	.33696	.95827	17	43	.63741	.68271	.31729	.95470	17
44	.62159	.66337	.33663	.95821	16	44	.63767	.68303	.31697	.95464	16
45	9.62186	9.66371	10.33629	9.95815	15	45	9.63794	9.68336	10.31664	9.95458	15
46	.62214	.66404	.33596	.95810	14	46	.63820	.68368	.31632	.95452	14
47	.62241	.66437	.33563	.95804	13	47	.63846	.68400	.31600	.95446	13
48	.62268	.66470	.33530	.95798	12	48	.63872	.68432	.31568	.95440	12
49	.62296	.66503	.33497	.95792	11	49	.63898	.68465	.31535	.95434	11
50	9.62323	9.66537	10.33463	9.95786	10	50	9.63924	9.68497	10.31503	9.95427	10
51	.62350	.66570	.33430	.95780	9	51	.63950	.68529	.31471	.95421	9
52	.62377	.66603	.33397	.95775	8	52	.63976	.68561	.31439	.95415	8
53	.62405	.66636	.33364	.95769	7	53	.64002	.68593	.31407	.95409	7
54	.62432	.66669	.33331	.95763	6	54	.64028	.68626	.31374	.95403	6
55	9.62459	9.66702	10.33298	9.95757	5	55	9.64054	9.68658	10.31342	9.95397	5
56	.62486	.66735	.33265	.95751	4	56	.64080	.68690	.31310	.95391	4
57	.62513	.66768	.33232	.95745	3	57	.64106	.68722	.31278	.95384	3
58	.62541	.66801	.33199	.95739	2	58	.64132	.68754	.31246	.95378	2
59	.62568	.66834	.33166	.95733	1	59	.64158	.68786	.31214	.95372	1
60	9.62595	9.66867	10.33133	9.95728	0	60	9.64184	9.68818	10.31182	9.95366	0
	Cosine.	Cotang.	Tang.	Sine.	'		Cosine.	Cotang.	Tang.	Sine.	'

65° 64°

*Log secant = colog cosine = 1 − log cosine; log cosecant = colog sine = 1 − log sine.

Ex.—Log sec 24°- 30′ = 10.04098.   *Ex.*—Log cosec 24°- 30′ = 10.38227.

3.—Logarithmic Sines, TANGENTS, COTANGENTS, COSINES.—(Cont'd.)
(SECANTS, COSECANTS.)*

26° 27°

'	Sine.	Tang.	Cotang.	Cosine.	|| '	Sine.	Tang.	Cotang.	Cosine.	
0	9.64184	9.68818	10.31182	9.95366	60	0	9.65705	9.70717	10.29283	9.94988	60
1	.64210	.68850	.31150	.95360	59	1	.65729	.70748	.29252	.94982	59
2	.64236	.68882	.31118	.95354	58	2	.65754	.70779	.29221	.94975	58
3	.64262	.68914	.31086	.95348	57	3	.65779	.70810	.29190	.94969	57
4	.64288	.68946	.31054	.95341	56	4	.65804	.70841	.29159	.94962	56
5	9.64313	9.68978	10.31022	9.95335	55	5	9.65828	9.70873	10.29127	9.94956	55
6	.64339	.69010	.30990	.95329	54	6	.65853	.70904	.29096	.94949	54
7	.64365	.69042	.30958	.95323	53	7	.65878	.70935	.29065	.94943	53
8	.64391	.69074	.30926	.95317	52	8	.65902	.70966	.29034	.94936	52
9	.64417	.69106	.30894	.95310	51	9	.65927	.70997	.29003	.94930	51
10	9.64442	9.69138	10.30862	9.95304	50	10	9.65952	9.71028	10.28972	9.94923	50
11	.64468	.69170	.30830	.95298	49	11	.65976	.71059	.28941	.94917	49
12	.64494	.69202	.30798	.95292	48	12	.66001	.71090	.28910	.94911	48
13	.64519	.69234	.30766	.95286	47	13	.66025	.71121	.28879	.94904	47
14	.64545	.69266	.30734	.95279	46	14	.66050	.71153	.28847	.94898	46
15	9.64571	9.69298	10.30702	9.95273	45	15	9.66075	9.71184	10.28816	9.94891	45
16	.64596	.69329	.30671	.95267	44	16	.66099	.71215	.28785	.94885	44
17	.64622	.69361	.30639	.95261	43	17	.66124	.71246	.28754	.94878	43
18	.64647	.69393	.30607	.95254	42	18	.66148	.71277	.28723	.94871	42
19	.64673	.69425	.30575	.95248	41	19	.66173	.71308	.28692	.94865	41
20	9.64698	9.69457	10.30543	9.95242	40	20	9.66197	9.71339	10.28661	9.94858	40
21	.64724	.69489	.30511	.95236	39	21	.66221	.71370	.28630	.94852	39
22	.62749	.69520	.30480	.95229	38	22	.66246	.71401	.28599	.94845	38
23	.64775	.69552	.30448	.95223	37	23	.66270	.71431	.28569	.94839	37
24	.64800	.69584	.30416	.95217	36	24	.66295	.71462	.28538	.94832	36
25	9.64826	9.69615	10.30385	9.95211	35	25	9.66319	9.71493	10.28507	9.94826	35
26	.64851	.69647	.30353	.95204	34	26	.66343	.71524	.28476	.94819	34
27	.64877	.69679	.30321	.95198	33	27	.66368	.71555	.28445	.94813	33
28	.64902	.69710	.30290	.95192	32	28	.66392	.71586	.28414	.94806	32
29	.64927	.69742	.30258	.95185	31	29	.66416	.71617	.28383	.94799	31
30	9.64953	9.69774	10.30226	9.95179	30	30	9.66441	9.71648	10.28352	9.94793	30
31	.64978	.69805	.30195	.95173	29	31	.66465	.71679	.28321	.94786	29
32	.65003	.69837	.30163	.95167	28	32	.66489	.71709	.28291	.94780	28
33	.65029	.69868	.30132	.95160	27	33	.66513	.71740	.28260	.94773	27
34	.65054	.69900	.30100	.95154	26	34	.66537	.71771	.28229	.94767	26
35	9.65079	9.69932	10.30068	9.95148	25	35	9.66562	9.71802	10.28198	9.94760	25
36	.65104	.69963	.30037	.95141	24	36	.66586	.71833	.28167	.94753	24
37	.65130	.69995	.30005	.95135	23	37	.66610	.71863	.28137	.94747	23
38	.65155	.70026	.29974	.95129	22	38	.66634	.71894	.28106	.94740	22
39	.65180	.70058	.29942	.95122	21	39	.66658	.71925	.28075	.94734	21
40	9.65205	9.70089	10.29911	9.95116	20	40	9.66682	9.71955	10.28045	9.94727	20
41	.65230	.70121	.29879	.95110	19	41	.66706	.71986	.28014	.94720	19
42	.65255	.70152	.29848	.95103	18	42	.66731	.72017	.27983	.94714	18
43	.65281	.70184	.29816	.95097	17	43	.66755	.72048	.27952	.94707	17
44	.65306	.70215	.29785	.95090	16	44	.66779	.72078	.27922	.94700	16
45	9.65331	9.70247	10.29753	9.95084	15	45	9.66803	9.72109	10.27891	9.94694	15
46	.65356	.70278	.29722	.95078	14	46	.66827	.72140	.27860	.94687	14
47	.65381	.70309	.29691	.95071	13	47	.66851	.72170	.27830	.94680	13
48	.65406	.70341	.29659	.95065	12	48	.66875	.72201	.27799	.94674	12
49	.65431	.70372	.29628	.95059	11	49	.66899	.72231	.27769	.94667	11
50	9.65456	9.70404	10.29596	9.95052	10	50	9.66922	9.72262	10.27738	9.94660	10
51	.65481	.70435	.29565	.95046	9	51	.66946	.72293	.27707	.94654	9
52	.65506	.70466	.29534	.95039	8	52	.66970	.72323	.27677	.94647	8
53	.65531	.70498	.29502	.95033	7	53	.66994	.72354	.27646	.94640	7
54	.65556	.70529	.29471	.95027	6	54	.67018	.72384	.27616	.94634	6
55	9.65580	9.70560	10.29440	9.95020	5	55	9.67042	9.72415	10.27585	9.94627	5
56	.65605	.70592	.29408	.95014	4	56	.67066	.72445	.27555	.94620	4
57	.65630	.70623	.29377	.95007	3	57	.67090	.72476	.27524	.94614	3
58	.65655	.70654	.29346	.95001	2	58	.67113	.72506	.27494	.94607	2
59	.65680	.70685	.29315	.94995	1	59	.67137	.72537	.27463	.94600	1
60	9.65705	9.70717	10.29283	9.94988	0	60	9.67161	9.72567	10.27433	9.94593	0

| | Cosine. | Cotang. | Tang. | Sine. | ' || | Cosine. | Cotang. | Tang. | Sine. | ' |

63° 62°

*Log secant = colog cosine = 1 − log cosine; log cosecant = colog sine = 1 − log sine.
Ex.—Log sec 26°- 30' = 10.04821. Ex.—Log cosec 26°- 30' = 10.35047.

3.—**Logarithmic Sines, Tangents, Cotangents, Cosines.**—(Cont'd.)
(Secants, Cosecants.)*

28°   29°

'	Sine.	Tang.	Cotang.	Cosine.	||	'	Sine.	Tang.	Cotang.	Cosine.	
0	9.67161	9.72567	10.27433	9.94593	60	0	9.68557	9.74375	10.25625	9.94182	60
1	.67185	.72598	.27402	.94587	59	1	.68580	.74405	.25595	.94175	59
2	.67208	.72628	.27372	.94580	58	2	.68603	.74435	.25565	.94168	58
3	.672з2	.72659	.27341	.94573	57	3	.63625	.74465	.25535	.94161	57
4	.67256	.72689	.27311	.94567	56	4	.68648	.74494	.25506	.94154	56
5	9.67280	9.72720	10.27280	9.94560	55	5	9.68671	9.74524	10.25476	9.94147	55
6	.67303	.72750	.27250	.94553	54	6	.68694	74554	.25446	.94140	54
7	.67327	.72780	.27220	.94546	53	7	.68716	.74583	.25417	.94133	53
8	.67350	.72811	.27189	.94540	52	8	.68739	.74613	.25387	.94126	52
9	.67374	.72841	.27159	.94533	51	9	.68762	.74643	.25357	.94119	51
10	9.67398	9.72872	10.27128	9.94526	50	10	9.68784	9.74673	10.25327	9.94112	50
11	.67421	.72902	.27098	.94519	49	11	.68807	.74702	.25298	.94105	49
12	.67445	.72932	.27068	.94513	48	12	.68829	.74732	.25268	.94098	48
13	.67468	.72963	.27037	.94506	47	13	.68852	.74762	.25238	94090	47
14	.67492	.72993	.27007	.94499	46	14	.68875	.74791	.25209	.94083	46
15	9.67515	9.73023	10.26977	9.94492	45	15	9.68897	9.74821	10.25179	9.94076	45
16	.67539	.73054	.26946	.94485	44	16	.68920	.74851	.25149	.94069	44
17	.67562	.73084	.26916	.94479	43	17	.68942	.74880	.25120	.94062	43
18	.67586	.73114	.26886	.94472	42	18	.68965	.74910	.25090	.94055	42
19	.67609	.73144	.26856	.94465	41	19	.68987	.74939	.25061	.94048	41
20	9.67633	9.73175	10.26825	9.94458	40	20	9.69010	9.74969	10.25031	9.94041	40
21	.67656	.73205	.26795	.94451	39	21	.69032	.74998	.25002	.94034	39
22	.67680	.73235	.26765	.94445	38	22	.69055	.75028	.24972	.94027	38
23	.67703	.73265	.26735	.94438	37	23	.69077	.75058	.24942	.94020	37
24	.67726	.73295	.26705	.94431	36	24	.69100	.75087	.24913	.94012	36
25	9.67750	9.73326	10.26674	9.94424	35	25	9.69122	9.75117	10.24883	9.94005	35
26	.67773	.73356	.26644	.94417	34	26	.69144	.75146	.24854	.93998	34
27	.67796	.73386	.26614	.94410	33	27	.69167	.75176	.24824	.93991	33
28	.67820	.73416	.26584	.94404	32	28	.69189	.75205	.24795	.93984	32
29	.67843	.73446	.26554	.94397	31	29	.69212	.75235	.24765	.93977	31
30	9.67866	9.73476	10.26524	9.94390	30	30	9.69234	9.75264	10.24736	9.93970	30
31	.67890	.73507	.26493	.94383	29	31	.69256	.75294	.24706	.93963	29
32	.67913	.73537	.26463	.94376	28	32	.69279	75323	.24677	.93955	28
33	.67936	.73567	.26433	.94369	27	33	.69301	.75353	.24647	.93948	27
34	.67959	.73597	.26403	.94362	26	34	.69323	.75382	.24618	.93941	26
35	9.67982	9.73627	10.26373	9.94355	25	35	9.69345	9.75411	10.24589	9.93934	25
36	.68006	.73657	.26343	.94349	24	36	.69368	.75441	.24559	.93927	24
37	.68029	.73687	.26313	.94342	23	37	.69390	.75470	.24530	.93920	23
38	.68052	.73717	.26283	.94335	22	38	.69412	.75500	.24500	.93912	22
39	.68075	.73747	.26253	.94328	21	39	.69434	.75529	.24471	.93905	21
40	9.68098	9.73777	10.26223	9.94321	20	40	9.69456	9.75558	10.24442	9.93898	20
41	.68121	.73807	.26193	.94314	19	41	.69479	.75588	.24412	.93891	19
42	.68144	.73837	.26163	.94307	18	42	.69501	.75617	.24383	.93884	18
43	.68167	.73867	.26133	.94300	17	43	.69523	.75647	.24353	.93876	17
44	.68190	.73897	.26103	.94293	16	44	.69545	.75676	.24324	.93869	16
45	9.68213	9.73927	10.26073	9.94286	15	45	9.69567	9.75705	10.24295	9.93862	15
46	.68237	.73957	.26043	.94279	14	46	.69589	.75735	.24265	.93855	14
47	.68260	.73987	.26013	.94273	13	47	.69611	.75764	.24236	.93847	13
48	.68283	.74017	.25983	.94266	12	48	.69633	.75793	.24207	.93840	12
49	.68305	.74047	.25953	.94259	11	49	.69655	.75822	.24178	.93833	11
50	9.68328	9.74077	10.25923	9.94252	10	50	9.69677	9.75852	10.24148	9.93826	10
51	.68351	.74107	.25893	.94245	9	51	.69699	.75881	.24119	.93819	9
52	.68374	.74137	.25863	.94238	8	52	.69721	.75910	.24090	.93811	8
53	.68397	.74166	.25834	.94231	7	53	.69743	.75939	.24061	.93804	7
54	.68420	.74196	.25804	.94224	6	54	.69765	.75969	.24031	.93797	6
55	9.68443	9.74226	10.25774	9.94217	5	55	9.69787	9.75998	10.24002	9.93789	5
56	.68466	.74256	.25744	.94210	4	56	.69809	.76027	.23973	.93782	4
57	.68489	.74286	.25714	.94203	3	57	.69831	.76056	.23944	.93775	3
58	.68512	.74316	.25684	.94196	2	58	.69853	.76086	.23914	.93768	2
59	.68534	.74345	.25655	.94189	1	59	.69875	.76115	.23885	.93760	1
60	9.68557	9.74375	10.25625	9.94182	0	60	9.69897	9.76144	10.23856	9.93753	0

| | Cosine. | Cotang. | Tang. | Sine. | ' || | | Cosine. | Cotang. | Tang. | Sine. | ' |

61°   60°

*Log secant=colog cosine=1—log cosine; log cosecant=colog sine=1—log sine.
Ex.—Log sec 28°-30' = 10.05610.   Ex.—Log cosec 28°-30' = 10.32184

3.—Logarithmic Sines, Tangents, Cotangents, Cosines.—(Cont'd.)
(Secants, Cosecants.)*

30°   31°

'	Sine.	Tang.	Cotang.	Cosine.	'	Sine.	Tang.	Cotang.	Cosine.	
0	9.69897	9.76144	10.23856	9.93753	0	9.71184	9.77877	10.22123	9.93307	60
1	.69919	.76173	.23827	.93746	1	.71205	.77906	.22094	.93299	59
2	.69941	.76202	.23798	.93738	2	.71226	.77935	.22065	.93291	58
3	.69963	.76231	.23769	.93731	3	.71247	.77963	.22037	.93284	57
4	.69984	.76261	.23739	.93724	4	.71268	.77992	.22008	.93276	56
5	9.70006	9.76290	10.23710	9.93717	5	9.71289	9.78020	10.21980	9.93269	55
6	.70028	.76319	.23681	.93709	6	.71310	.78049	.21951	.93261	54
7	.70050	.76348	.23652	.93702	7	.71331	.78077	.21923	.93253	53
8	.70072	.76377	.23623	.93695	8	.71352	.78106	.21894	.93246	52
9	.70093	.76406	.23594	.93687	9	.71373	.78135	.21865	.93238	51
10	9.70115	9.76435	10.23565	9.93680	10	9.71393	9.78163	10.21837	9.93230	50
11	.70137	.76464	.23536	.93673	11	.71414	.78192	.21808	.93223	49
12	.70159	.76493	.23507	.93665	12	.71435	.78220	.21780	.93215	48
13	.70180	.76522	.23478	.93658	13	.71456	.78249	.21751	.93207	47
14	.70202	.76551	.23449	.93650	14	.71477	.78277	.21723	.93200	46
15	9.70224	9.76580	10.23420	9.93643	15	9.71498	9.78306	10.21694	9.93192	45
16	.70245	.76609	.23391	.93636	16	.71519	.78334	.21666	.93184	44
17	.70267	.76639	.23361	.93628	17	.71539	.78363	.21637	.93177	43
18	.70288	.76668	.23332	.93621	18	.71560	.78391	.21609	.93169	42
19	.70310	.76697	.23303	.93614	19	.71581	.78419	.21581	.93161	41
20	9.70332	9.76725	10.23275	9.93606	20	9.71602	9.78448	10.21552	9.93154	40
21	.70353	.76754	.23246	.93599	21	.71622	.78476	.21524	.93146	39
22	.70375	.76783	.23217	.93591	22	.71643	.78505	.21495	.93138	38
23	.70396	.76812	.23188	.93584	23	.71664	.78533	.21467	.93131	37
24	.70418	.76841	.23159	.93577	24	.71685	.78562	.21438	.93123	36
25	9.70439	9.76870	10.23130	9.93569	25	9.71705	9.78590	10.21410	9.93115	35
26	.70461	.76899	.23101	.93562	26	.71726	.78618	.21382	.93107	34
27	.70482	.76928	.23072	.93554	27	.71747	.78647	.21353	.93100	33
28	.70504	.76957	.23043	.93547	28	.71767	.78675	.21325	.93092	32
29	.70525	.76986	.23014	.93539	29	.71788	.78704	.21296	.93084	31
30	9.70547	9.77015	10.22985	9.93532	30	9.71809	9.78732	10.21268	9.93077	30
31	.70568	.77044	.22956	.93525	31	.71829	.78760	.21240	.93069	29
32	.70590	.77073	.22927	.93517	32	.71850	.78789	.21211	.93061	28
33	.70611	.77101	.22899	.93510	33	.71870	.78817	.21183	.93053	27
34	.70633	.77130	.22870	.93502	34	.71891	.78845	.21155	.93046	26
35	9.70654	9.77159	10.22841	9.93495	35	9.71911	9.78874	10.21126	9.93038	25
36	.70675	.77188	.22812	.93487	36	.71932	.78902	.21098	.93030	24
37	.70697	.77217	.22783	.93480	37	.71952	.78930	.21070	.93022	23
38	.70718	.77246	.22754	.93472	38	.71973	.78959	.21041	.93014	22
39	.70739	.77274	.22726	.93465	39	.71994	.78987	.21013	.93007	21
40	9.70761	9.77303	10.22697	9.93457	40	9.72014	9.79015	10.20985	9.92999	20
41	.70782	.77332	.22668	.93450	41	.72034	.79043	.20957	.92991	19
42	.70803	.77361	.22639	.93442	42	.72055	.79072	.20928	.92983	18
43	.70824	.77390	.22610	.93435	43	.72075	.79100	.20900	.92976	17
44	.70846	.77418	.22582	.93427	44	.72096	.79128	.20872	.92968	16
45	9.70867	9.77447	10.22553	9.93420	45	9.72116	9.79156	10.20844	9.92960	15
46	.70888	.77476	.22524	.93412	46	.72137	.79185	.20815	.92952	14
47	.70909	.77505	.22495	.93405	47	.72157	.79213	.20787	.92944	13
48	.70931	.77533	.22467	.93397	48	.72177	.79241	.20759	.92936	12
49	.70952	.77562	.22438	.93390	49	.72198	.79269	.20731	.92929	11
50	9.70973	9.77591	10.22409	9.93382	50	9.72218	9.79297	10.20703	9.92921	10
51	.70994	.77619	.22381	.93375	51	.72238	.79326	.20674	.92913	9
52	.71015	.77648	.22352	.93367	52	.72259	.79354	.20646	.92905	8
53	.71036	.77677	.22323	.93360	53	.72279	.79382	.20618	.92897	7
54	.71058	.77706	.22294	.93352	54	.72299	.79410	.20590	.92889	6
55	9.71079	9.77734	10.22266	9.93344	55	9.72320	9.79438	10.20562	9.92881	5
56	.71100	.77763	.22237	.93337	56	.72340	.79466	.20534	.92874	4
57	.71121	.77791	.22209	.93329	57	.72360	.79495	.20505	.92866	3
58	.71142	.77820	.22180	.93322	58	.72381	.79523	.20477	.92858	2
59	.71163	.77849	.22151	.93314	59	.72401	.79551	.20449	.92850	1
60	9.71184	9.77877	10.22123	9.93307	60	9.72421	9.79579	10.20421	9.92842	0

| | Cosine. | Cotang. | Tang. | Sine. | ' | | Cosine. | Cotang. | Tang. | Sine. | ' |

59°   58°

*Log secant = colog cosine = 1 − log cosine; log cosecant = colog sine = 1 − log sine.

Ex.—Log sec 30°- 30′ = 10.06468.   *Ex.*—Log cosec 30°- 30′ = 10.29453.

3. —Logarithmic Sines, TANGENTS, COTANGENTS, COSINES.—(Cont'd.)
(SECANTS, COSECANTS.)*

32°   **33°**

'	Sine.	Tang.	Cotang.	Cosine.		'	Sine.	Tang.	Cotang.	Cosine.	
0	9.72421	9.79579	10.20421	9.92842	60	0	9.73611	9.81252	10.18748	9.92359	60
1	.72441	.79607	.20393	.92834	59	1	.73630	.81279	.18721	.92351	59
2	.72461	.79635	.20365	.92826	58	2	.73650	.81307	.18693	.92343	58
3	.72482	.79663	.20337	.92818	57	3	.73669	.81335	.18665	.92335	57
4	.72502	.79691	.20309	.92810	56	4	.73689	.81362	.18638	.92326	56
5	9.72522	9.79719	10.20281	9.92803	55	5	9.73708	9.81390	10.18610	9.92318	55
6	.72542	.79747	.20253	.92795	54	6	.73727	.81418	.18582	.92310	54
7	.72562	.79776	.20224	.92787	53	7	.73747	.81445	.18555	.92302	53
8	.72582	.79804	.20196	.92779	52	8	.73766	.81473	.18527	.92293	52
9	.72602	.79832	.20168	.92771	51	9	.73785	.81500	.18500	.92285	51
10	9.72622	9.79860	10.20140	9.92763	50	10	9.73805	9.81528	10.18472	9.92277	50
11	.72643	.79888	.20112	.92755	49	11	.73824	.81556	.18444	.92269	49
12	.72663	.79916	.20084	.92747	48	12	.73843	.81583	.18417	.92260	48
13	.72683	.79944	.20056	.92739	47	13	.73863	.81611	.18389	.92252	47
14	.72703	.79972	.20028	.92731	46	14	.73882	.81638	.18362	.92244	46
15	9.72723	9.80000	10.20000	9.92723	45	15	9.73901	9.81666	10.18334	9.92235	45
16	.72743	.80028	.19972	.92715	44	16	.73921	.81693	.18307	.92227	44
17	.72763	.80056	.19944	.92707	43	17	.73940	.81721	.18279	.92219	43
18	.72783	.80084	.19916	.92699	42	18	.73959	.81748	.18252	.92211	42
19	.72803	.80112	.19888	.92691	41	19	.73978	.81776	.18224	.92202	41
20	9.72823	9.80140	10.19860	9.92683	40	20	9.73997	9.81803	10.18197	9.92194	40
21	.72843	.80168	.19832	.92675	39	21	.74017	.81831	.18169	.92186	39
22	.72863	.80195	.19805	.92667	38	22	.74036	.81858	.18142	.92177	38
23	.72883	.80223	.19777	.92659	37	23	.74055	.81886	.18114	.92169	37
24	.72903	.80251	.19749	.92651	36	24	.74074	.81913	.18087	.92161	36
25	9.72922	9.80279	10.19721	9.92643	35	25	9.74093	9.81941	10.18059	9.92152	35
26	.72942	.80307	.19693	.92635	34	26	.74113	.81968	.18032	.92144	34
27	.72962	.80335	.19665	.92627	33	27	.74132	.81996	.18004	.92136	33
28	.72982	.80363	.19637	.92619	32	28	.74151	.82023	.17977	.92127	32
29	.73002	.80391	.19609	.92611	31	29	.74170	.82051	.17949	.92119	31
30	9.73022	9.80419	10.19581	9.92603	30	30	9.74189	9.82078	10.17922	9.92111	30
31	.73041	.80447	.19553	.92595	29	31	.74208	.82106	.17894	.92102	29
32	.73061	.80474	.19526	.92587	28	32	.74227	.82133	.17867	.92094	28
33	.73081	.80502	.19498	.92579	27	33	.74246	.82161	.17839	.92086	27
34	.73101	.80530	.19470	.92571	26	34	.74265	.82188	.17812	.92077	26
35	9.73121	9.80558	10.19442	9.92563	25	35	9.74284	9.82215	10.17785	9.92069	25
36	.73140	.80586	.19414	.92555	24	36	.74303	.82243	.17757	.92060	24
37	.73160	.80614	.19386	.92546	23	37	.74322	.82270	.17730	.92052	23
38	.73180	.80642	.19358	.92538	22	38	.74341	.82298	.17702	.92044	22
39	.73200	.80669	.19331	.92530	21	39	.74360	.82325	.17675	.92035	21
40	9.73219	9.80697	10.19303	9.92522	20	40	9.74379	9.82352	10.17648	9.92027	20
41	.73239	.80725	.19275	.92514	19	41	.74398	.82380	.17620	.92019	19
42	.73259	.80753	.19247	.92506	18	42	.74417	.82407	.17593	.92010	18
43	.73278	.80781	.19219	.92498	17	43	.74436	.82435	.17565	.92002	17
44	.73298	.80808	.19192	.92490	16	44	.74455	.82462	.17538	.91993	16
45	9.73318	9.80836	10.19164	9.92482	15	45	9.74474	9.82489	10.17511	9.91985	15
46	.73337	.80864	.19136	.92473	14	46	.74493	.82517	.17483	.91976	14
47	.73357	.80892	.19108	.92465	13	47	.74512	.82544	.17456	.91968	13
48	.73377	.80919	.19081	.92457	12	48	.74531	.82571	.17429	.91959	12
49	.73396	.80947	.19053	.92449	11	49	.74549	.82599	.17401	.91951	11
50	9.73416	9.80975	10.19025	9.92441	10	50	9.74568	9.82626	10.17374	9.91942	10
51	.73435	.81003	.18997	.92433	9	51	.74587	.82653	.17347	.91934	9
52	.73455	.81030	.18970	.92425	8	52	.74606	.82681	.17319	.91925	8
53	.73474	.81058	.18942	.92416	7	53	.74625	.82708	.17292	.91917	7
54	.73494	.81086	.18914	.92408	6	54	.74644	.82735	.17265	.91908	6
55	9.73513	9.81113	10.18887	9.92400	5	55	9.74662	9.82762	10.17238	9.91900	5
56	.73533	.81141	.18859	.92392	4	56	.74681	.82790	.17210	.91891	4
57	.73552	.81169	.18831	.92384	3	57	.74700	.82817	.17183	.91883	3
58	.73572	.81196	.18804	.92376	2	58	.74719	.82844	.17156	.91874	2
59	.73591	.81224	.18776	.92367	1	59	.74737	.82871	.17129	.91866	1
60	9.73611	9.81252	10.18748	9.92359	0	60	9.74756	9.82899	10.17101	9.91857	0

	Cosine.	Cotang.	Tang.	Sine.	'			Cosine.	Cotang.	Tang.	Sine.	

57°   **56°**

*Log secant=colog cosine=1—log cosine; log cosecant=colog sine=1—log sine.
Ex.—Log sec 32°-30' = 10.07397.   Ex.—Log cosec 32°-30' = 10.26978.

3. —Logarithmic Sines, Tangents, Cotangents, Cosines.—(Cont'd.)
(Secants, Cosecants.)*

34°   **35°**

'	Sine.	Tang.	Cotang.	Cosine.		'	Sine.	Tang.	Cotang.	Cosine.	
0	9.74756	9.82899	10.17101	9.91857	60	0	9.75859	9.84523	10.15477	9.91336	60
1	.74775	.82926	.17074	.91849	59	1	.75877	.84550	.15450	.91328	59
2	.74794	.82953	.17047	.91840	58	2	.75895	.84576	.15424	.91319	58
3	.74812	.82980	.17020	.91832	57	3	.75913	.84603	.15397	.91310	57
4	.74831	.83008	.16992	.91823	56	4	.75931	.84630	.15370	.91301	56
5	9.74850	9.83035	10.16965	9.91815	55	5	9.75949	9.84657	10.15343	9.91292	55
6	.74868	.83062	.16938	.91806	54	6	.75967	.84684	.15316	.91283	54
7	.74887	.83089	.16911	.91798	53	7	.75985	.84711	.15289	.91274	53
8	.74906	.83117	.16883	.91789	52	8	.76003	.84738	.15262	.91266	52
9	.74924	.83144	.16856	.91781	51	9	.76021	.84764	.15236	.91257	51
10	9.74943	9.83171	10.16829	9.91772	50	10	9.76039	9.84791	10.15209	9.91248	50
11	.74961	.83198	.16802	.91763	49	11	.76057	.84818	.15182	.91239	49
12	.74980	.83225	.16775	.91755	48	12	.76075	.84845	.15155	.91230	48
13	.74999	.83252	.16748	.91746	47	13	.76093	.84872	.15128	.91221	47
14	.75017	.83280	.16720	.91738	46	14	.76111	.84899	.15101	.91212	46
15	9.75036	9.83307	10.16693	9.91729	45	15	9.76129	9.84925	10.15075	9.91203	45
16	.75054	.83334	.16666	.91720	44	16	.76146	.84952	.15048	.91194	44
17	.75073	.83361	.16639	.91712	43	17	.76164	.84979	.15021	.91185	43
18	.75091	.83388	.16612	.91703	42	18	.76182	.85006	.14994	.91176	42
19	.75110	.83415	.16585	.91695	41	19	.76200	.85033	.14967	.91167	41
20	9.75128	9.83442	10.16558	9.91686	40	20	9.76218	9.85059	10.14941	9.91158	40
21	.75147	.83470	.16530	.91677	39	21	.76236	.85086	.14914	.91149	39
22	.75165	.83497	.16503	.91669	38	22	.76253	.85113	.14887	.91141	38
23	.75184	.83524	.16476	.91660	37	23	.76271	.85140	.14860	.91132	37
24	.75202	.83551	.16449	.91651	36	24	.76289	.85166	.14834	.91123	36
25	9.75221	9.83578	10.16422	9.91643	35	25	9.76307	9.85193	10.14807	9.91114	35
26	.75239	.83605	.16395	.91634	34	26	.76324	.85220	.14780	.91105	34
27	.75258	.83632	.16368	.91625	33	27	.76342	.85247	.14753	.91096	33
28	.75276	.83659	.16341	.91617	32	28	.76360	.85273	.14727	.91087	32
29	.75294	.83686	.16314	.91608	31	29	.76378	.85300	.14700	.91078	31
30	9.75313	9.83713	10.16287	9.91599	30	30	9.76395	9.85327	10.14673	9.91069	30
31	.75331	.83740	.16260	.91591	29	31	.76413	.85354	.14646	.91060	29
32	.75350	.83768	.16232	.91582	28	32	.76431	.85380	.14620	.91051	28
33	.75368	.83795	.16205	.91573	27	33	.76448	.85407	.14593	.91042	27
34	.75386	.83822	.16178	.91565	26	34	.76466	.85434	.14566	.91033	26
35	9.75405	9.83849	10.16151	9.91556	25	35	9.76484	9.85460	10.14540	9.91023	25
36	.75423	.83876	.16124	.91547	24	36	.76501	.85487	.14513	.91014	24
37	.75441	.83903	.16097	.91538	23	37	.76519	.85514	.14486	.91005	23
38	.75459	.83930	.16070	.91530	22	38	.76537	.85540	.14460	.90996	22
39	.75478	.83957	.16043	.91521	21	39	.76554	.85567	.14433	.90987	21
40	9.75496	9.83984	10.16016	9.91512	20	40	9.76572	9.85594	10.14406	9.90978	20
41	.75514	.84011	.15989	.91504	19	41	.76590	.85620	.14380	.90969	19
42	.75533	.84038	.15962	.91495	18	42	.76607	.85647	.14353	.90960	18
43	.75551	.84065	.15935	.91486	17	43	.76625	.85674	.14326	.90951	17
44	.75569	.84092	.15908	.91477	16	44	.76642	.85700	.14300	.90942	16
45	9.75587	9.84119	10.15881	9.91469	15	45	9.76660	9.85727	10.14273	9.90933	15
46	.75605	.84146	.15854	.91460	14	46	.76677	.85754	.14246	.90924	14
47	.75624	.84173	.15827	.91451	13	47	.76695	.85780	.14220	.90915	13
48	.75642	.84200	.15800	.91442	12	48	.76712	.85807	.14193	.90906	12
49	.75660	.84227	.15773	.91433	11	49	.76730	.85834	.14166	.90896	11
50	9.75678	9.84254	10.15746	9.91425	10	50	9.76747	9.85860	10.14140	9.90887	10
51	.75696	.84280	.15720	.91416	9	51	.76765	.85887	.14113	.90878	9
52	.75714	.84307	.15693	.91407	8	52	.76782	.85913	.14087	.90869	8
53	.75733	.84334	.15666	.91398	7	53	.76800	.85940	.14060	.90860	7
54	.75751	.84361	.15639	.91389	6	54	.76817	.85967	.14033	.90851	6
55	9.75769	9.84388	10.15612	9.91381	5	55	9.76835	9.85993	10.14007	9.90842	5
56	.75787	.84415	.15585	.91372	4	56	.76852	.86020	.13980	.90832	4
57	.75805	.84442	.15558	.91363	3	57	.76870	.86046	.13954	.90823	3
58	.75823	.84469	.15531	.91354	2	58	.76887	.86073	.13927	.90814	2
59	.75841	.84496	.15504	.91345	1	59	.76904	.86100	.13900	.90805	1
60	9.75859	9.84523	10.15477	9.91336	0	60	9.76922	9.86126	10.13874	9.90796	0

| | Cosine. | Cotang. | Tang. | Sine. | ' | | Cosine. | Cotang. | Tang. | Sine. | ' |

55°   **54°**

*Log secant = colog cosine = 1 − log cosine; log cosecant = colog sine = 1 − log sine.

Ex.—Log sec 34°- 30' = 10.08401.   Ex.—Log cosec 34°- 30' = 10.24687.

3.—Logarithmic Sines, Tangents, Cotangents, Cosines—(Cont'd.)
(Secants, Cosecants.)*

36°

'	Sine.	Tang.	Cotang.	Cosine.	
0	9.76922	9.86126	10.13874	9.90796	60
1	.76939	.86153	.13847	.90787	59
2	.76957	.86179	.13821	.90777	58
3	.76974	.86206	.13794	.90768	57
4	.76991	.86232	.13768	.90759	56
5	9.77009	9.86259	10.13741	9.90750	55
6	.77026	.86285	.13715	.90741	54
7	.77043	.86312	.13688	.90731	53
8	.77061	.86338	.13662	.90722	52
9	.77078	.86365	.13635	.90713	51
10	9.77095	9.86392	10.13608	9.90704	50
11	.77112	.86418	.13582	.90694	49
12	.77130	.86445	.13555	.90685	48
13	.77147	.86471	.13529	.90676	47
14	.77164	.86498	.13502	.90667	46
15	9.77181	9.86524	10.13476	9.90657	45
16	.77199	.86551	.13449	.90648	44
17	.77216	.86577	.13423	.90639	43
18	.77233	.86603	.13397	.90630	42
19	.77250	.86630	.13370	.90620	41
20	9.77268	9.86656	10.13344	9.90611	40
21	.77285	.86683	.13317	.90602	39
22	.77302	.86709	.13291	.90592	38
23	.77319	.86736	.13264	.90583	37
24	.77336	.86762	.13238	.90574	36
25	9.77353	9.86789	10.13211	9.90565	35
26	.77370	.86815	.13185	.90555	34
27	.77387	.86842	.13158	.90546	33
28	.77405	.86868	.13132	.90537	32
29	.77422	.86894	.13106	.90527	31
30	9.77439	9.86921	10.13079	9.90518	30
31	.77456	.86947	.13053	.90509	29
32	.77473	.86974	.13026	.90499	28
33	.77490	.87000	.13000	.90490	27
34	.77507	.87027	.12973	.90480	26
35	9.77524	9.87053	10.12947	9.90471	25
36	.77541	.87079	.12921	.90462	24
37	.77558	.87106	.12894	.90452	23
38	.77575	.87132	.12868	.90443	22
39	.77592	.87158	.12842	.90434	21
40	9.77609	9.87185	10.12815	9.90424	20
41	.77626	.87211	.12789	.90415	19
42	.77643	.87238	.12762	.90405	18
43	.77660	.87264	.12736	.90396	17
44	.77677	.87290	.12710	.90386	16
45	9.77694	9.87317	10.12683	9.90377	15
46	.77711	.87343	.12657	.90368	14
47	.77728	.87369	.12631	.90358	13
48	.77744	.87396	.12604	.90349	12
49	.77761	.87422	.12578	.90339	11
50	9.77778	9.87448	10.12552	9.90330	10
51	.77795	.87475	.12525	.90320	9
52	.77812	.87501	.12499	.90311	8
53	.77829	.87527	.12473	.90301	7
54	.77846	.87554	.12446	.90292	6
55	9.77862	9.87580	10.12420	9.90282	5
56	.77879	.87606	.12394	.90273	4
57	.77896	.87633	.12367	.90263	3
58	.77913	.87659	.12341	.90254	2
59	.77930	.87685	.12315	.90244	1
60	9.77946	9.87711	10.12289	9.90235	0

	Cosine.	Cotang.	Tang.	Sine.	'

53°

37°

'	Sine.	Tang.	Cotang.	Cosine.	
0	9.77946	9.87711	10.12289	9.90235	60
1	.77963	.87738	.12262	.90225	59
2	.77980	87764	.12236	.90216	58
3	.77997	.87790	.12210	.90206	57
4	.78013	.87817	.12183	.90197	56
5	9.78030	9.87843	10.12157	9.90187	55
6	.78047	.87869	.12131	.90178	54
7	.78063	.87895	.12105	.90168	53
8	.78080	.87922	.12078	.90159	52
9	.78097	.87948	.12052	.90149	51
10	9.78113	9.87974	10.12026	9.90139	50
11	.78130	.88000	.12000	.90130	49
12	.78147	.88027	.11973	.90120	48
13	.78163	.88053	.11947	.90111	47
14	.78180	.88079	.11921	.90101	46
15	9.78197	9.88105	10.11895	9.90091	45
16	.78213	.88131	.11869	.90082	44
17	.78230	.88158	.11842	.90072	43
18	.78246	.88184	.11816	.90063	42
19	.78263	.88210	.11790	.90053	41
20	9.78280	9.88236	10.11764	9.90043	40
21	.78296	.88262	.11738	.90034	39
22	.78313	.88289	.11711	.90024	38
23	.78329	.88315	.11685	.90014	37
24	.78346	.88341	.11659	.90005	36
25	9.78362	9.88367	10.11633	9.89995	35
26	.78379	.88393	.11607	.89985	34
27	.78395	.88420	.11580	.89976	33
28	.78412	.88446	.11554	.89966	32
29	.78428	.88472	.11528	.89956	31
30	9.78445	9.88498	10.11502	9.89947	30
31	.78461	.88524	.11476	.89937	29
32	.78478	.88550	.11450	.89927	28
33	.78494	.88577	.11423	.89918	27
34	.78510	.88603	.11397	.89908	26
35	9.78527	9.88629	10.11371	9.89898	25
36	.78543	.88655	.11345	.89888	24
37	.78560	.88681	.11319	.89879	23
38	.78576	.88707	.11293	.89869	22
39	.78592	.88733	.11267	.89859	21
40	9.78609	9.88759	10.11241	9.89849	20
41	.78625	.88786	.11214	.89840	19
42	.78642	.88812	.11188	.89830	18
43	.78658	.88838	.11162	.89820	17
44	.78674	.88864	.11136	.89810	16
45	9.78690	9.88890	10.11110	9.89801	15
46	.78707	.88916	.11084	.89791	14
47	.78723	.88942	.11058	.89781	13
48	.78739	.88968	.11032	.89771	12
49	.78756	.88994	.11006	.89761	11
50	9.78772	9.89020	10.10980	9.89752	10
51	.78788	.89046	.10954	.89742	9
52	.78805	.89073	.10927	.89732	8
53	.78821	.89099	.10901	.89722	7
54	.78837	.89125	.10875	.89712	6
55	9.78853	9.89151	10.10849	9.89702	5
56	.78869	.89177	.10823	.89693	4
57	.78886	.89203	.10797	.89683	3
58	.78902	.89229	.10771	.89673	2
59	.78918	.89255	.10745	.89663	1
60	9.78934	9.89281	10.10719	9.89653	0

	Cosine.	Cotang.	Tang.	Sine.	'

52°

*Log secant = colog cosine = 1 − log cosine; log cosecant = colog sine = 1 − log sine.

Ex.—Log sec 36°- 30′ = 10.09482.   Ex.—Log cosec 36°- 30′ = 10.22561

3.—Logarithmic Sines, TANGENTS, COTANGENTS, COSINES.—(Cont'd.)
(SECANTS, COSECANTS.)*

38°   **39°**

'	Sine.	Tang.	Cotang.	Cosine.		'	Sine.	Tang.	Cotang.	Cosine.	
0	9.78934	9.89281	10.10719	9.89653	60	0	9.79887	9.90837	10.09163	9.89050	60
1	.78950	.89307	.10693	.89643	59	1	.79903	.90863	.09137	.89040	59
2	.78967	.89333	.10667	.89633	58	2	.79918	.90889	.09111	.89030	58
3	.78983	.89359	.10641	.89624	57	3	.79934	.90914	.09086	.89020	57
4	.78999	.89385	.10615	.89614	56	4	.79950	.90940	.09060	.89009	56
5	9.79015	9.89411	10.10589	9.89604	55	5	9.79965	9.90966	10.09034	9.88999	55
6	.79031	.89437	.10563	.89594	54	6	.79981	.90992	.09008	.88989	54
7	.79047	.89463	.10537	.89584	53	7	.79996	.91018	.08982	.88978	53
8	.79063	.89489	.10511	.89574	52	8	.80012	.91043	.08957	.88968	52
9	.79079	.89515	.10485	.89564	51	9	.80027	.91069	.08931	.88958	51
10	9.79095	9.89541	10.10459	9.89554	50	10	9.80043	9.91095	10.08905	9.88948	50
11	.79111	.89567	.10433	.89544	49	11	.80058	.91121	.08879	.88937	49
12	.79128	.89593	.10407	.89534	48	12	.80074	.91147	.08853	.88927	48
13	.79144	.89619	.10381	.89524	47	13	.80089	.91172	.08828	.88917	47
14	.79160	.89645	.10355	.89514	46	14	.80105	.91198	.08802	.88906	46
15	9.79176	9.89671	10.10329	9.89504	45	15	9.80120	9.91224	10.08776	9.88896	45
16	.79192	.89697	.10303	.89495	44	16	.80136	.91250	.08750	.88886	44
17	.79208	.89723	.10277	.89485	43	17	.80151	.91276	.08724	.88875	43
18	.79224	.89749	.10251	.89475	42	18	.80166	.91301	.08699	.88865	42
19	.79240	.89775	.10225	.89465	41	19	.80182	.91327	.08673	.88855	41
20	9.79256	9.89801	10.10199	9.89455	40	20	9.80197	9.91353	10.08647	9.88844	40
21	.79272	.89827	.10173	.89445	39	21	.80213	.91379	.08621	.88834	39
22	.79288	.89853	.10147	.89435	38	22	.80228	.91404	.08596	.88824	38
23	.79304	.89879	.10121	.89425	37	23	.80244	.91430	.08570	.88813	37
24	.79319	.89905	.10095	.89415	36	24	.80259	.91456	.08544	.88803	36
25	9.79335	9.89931	10.10069	9.89405	35	25	9.80274	9.91482	10.08518	9.88793	35
26	.79351	.89957	.10043	.89395	34	26	.80290	.91507	.08493	.88782	34
27	.79367	.89983	.10017	.89385	33	27	.80305	.91533	.08467	.88772	33
28	.79383	.90009	.09991	.89375	32	28	.80320	.91559	.08441	.88761	32
29	.79399	.90035	.09965	.89364	31	29	.80335	.91585	.08415	.88751	31
30	9.79415	9.90061	10.09939	9.89354	30	30	9.80351	9.91610	10.08390	9.88741	30
31	.79431	.90086	.09914	.89344	29	31	.80366	.91636	.08364	.88730	29
32	.79447	.90112	.09888	.89334	28	32	.80382	.91662	.08338	.88720	28
33	.79463	.90138	.09862	.89324	27	33	.80397	.91688	.08312	.88709	27
34	.79478	.90164	.09836	.89314	26	34	.80412	.91713	.08287	.88699	26
35	9.79494	9.90190	10.09810	9.89304	25	35	9.80428	9.91739	10.08261	9.88688	25
36	.79510	.90216	.09784	.89294	24	36	.80443	.91765	.08235	.88678	24
37	.79526	.90242	.09758	.89284	23	37	.80458	.91791	.08209	.88668	23
38	.79542	.90268	.09732	.89274	22	38	.80473	.91816	.08184	.88657	22
39	.79558	.90294	.09706	.89264	21	39	.80489	.91842	.08158	.88647	21
40	9.79573	9.90320	10.09680	9.89254	20	40	9.80504	9.91868	10.08132	9.88636	20
41	.79589	.90346	.09654	.89244	19	41	.80519	.91893	.08107	.88626	19
42	.79605	.90371	.09629	.89233	18	42	.80534	.91919	.08081	.88615	18
43	.79621	.90397	.09603	.89223	17	43	.80550	.91945	.08055	.88605	17
44	.79636	.90423	.09577	.89213	16	44	.80565	.91971	.08029	.88594	16
45	9.79652	9.90449	10.09551	9.89203	15	45	9.80580	9.91996	10.08004	9.88584	15
46	.79668	.90475	.09525	.89193	14	46	.80595	.92022	.07978	.88573	14
47	.79684	.90501	.09499	.89183	13	47	.80610	.92048	.07952	.88563	13
48	.79699	.90527	.09473	.89173	12	48	.80625	.92073	.07927	.88552	12
49	.79715	.90553	.09447	.89162	11	49	.80641	.92099	.07901	.88542	11
50	9.79731	9.90578	10.09422	9.89152	10	50	9.80656	9.92125	10.07875	9.88531	10
51	.79746	.90604	.09396	.89142	9	51	.80671	.92150	.07850	.88521	9
52	.79762	.90630	.09370	.89132	8	52	.80686	.92176	.07824	.88510	8
53	.79778	.90656	.09344	.89122	7	53	.80701	.92202	.07798	.88499	7
54	.79793	.90682	.09318	.89112	6	54	.80716	.92227	.07773	.88489	6
55	9.79809	9.90708	10.09292	9.89101	5	55	9.80731	9.92253	10.07747	9.88478	5
56	.79825	.90734	.09266	.89091	4	56	.80746	.92279	.07721	.88468	4
57	.79840	.90759	.09241	.89081	3	57	.80762	.92304	.07696	.88457	3
58	.79856	.90785	.09215	.89071	2	58	.80777	.92330	.07670	.88447	2
59	.79872	.90811	.09189	.89060	1	59	.80792	.92356	.07644	.88436	1
60	9.79887	9.90837	10.09163	9.89050	0	60	9.80807	9.92381	10.07619	9.88425	0
	Cosine.	Cotang.	Tang.	Sine.	'		Cosine.	Cotang.	Tang.	Sine.	'

51°   **50°**

*Log secant = colog cosine = 1 − log cosine; log cosecant = colog sine = 1 − log sine.

Ex.—Log sec 38°- 30′ = 10 10646.   *Ex.*—Log cosec 38°- 30′ = 10.20585.

3. —Logarithmic Sines, Tangents, Cotangents, Cosines.—(Cont'd.)
(Secants, Cosecants.)*

40° **41°**

′	Sine.	Tang.	Cotang.	Cosine.	′	′	Sine.	Tang.	Cotang.	Cosine.	′
0	9.80807	9.92381	10.07619	9.88425	60	0	9.81694	9.93916	10.06084	9.87778	60
1	.80822	.92407	.07593	.88415	59	1	.81709	.93942	.06058	.87767	59
2	.80837	.92433	.07567	.88404	58	2	.81723	.93967	.06033	.87756	58
3	.80852	.92458	.07542	.88394	57	3	.81738	.93993	.06007	.87745	57
4	.80867	.92484	.07516	.88383	56	4	.81752	.94018	.05982	.87734	56
5	9.80882	9.92510	10.07490	9.88372	55	5	9.81767	9.94044	10.05956	9.87723	55
6	.80897	.92535	.07465	.88362	54	6	.81781	.94069	.05931	.87712	54
7	.80912	.92561	.07439	.88351	53	7	.81796	.94095	.05905	.87701	53
8	.80927	.92587	.07413	.88340	52	8	.81810	.94120	.05880	.87690	52
9	.80942	.92612	.07388	.88330	51	9	.81825	.94146	.05854	.87679	51
10	9.80957	9.92638	10.07362	9.88319	50	10	9.81839.	9.94171	10.05829	9.87668	50
11	.80972	.92663	.07337	.88308	49	11	.81854	.94197	.05803	.87657	49
12	.80987	.92689	.07311	.88298	48	12	.81868	.94222	.05778	.87646	48
13	.81002	.92715	.07285	.88287	47	13	.81882	.94248	.05752	.87635	47
14	.81017	.92740	.07260	.88276	46	14	.81897	.94273	.05727	.87624	46
15	9.81032	9.92766	10.07234	9.88266	45	15	9.81911	9.94299	10.05701	9.87613	45
16	.81047	.92792	.07208	.88255	44	16	.81926	.94324	.05676	.87601	44
17	.81061	.92817	.07183	.88244	43	17	.81940	.94350	.05650	.87590	43
18	.81076	.92843	.07157	.88234	42	18	.81955	.94375	.05625	.87579	42
19	.81091	.92868	.07132	.88223	41	19	.81969	.94401	.05599	.87568	41
20	9.81106	9.92894	10.07106	9.88212	40	20	9.81983	9.94426	10.05574	9.87557	40
21	.81121	.92920	.07080	.88201	39	21	.81998	.94452	.05548	.87546	39
22	.81136	.92945	.07055	.88191	38	22	.82012	.94477	.05523	.87535	38
23	.81151	.92971	.07029	.88180	37	23	.82026	.94503	.05497	.87524	37
24	.81166	.92996	.07004	.88169	36	24	.82041	.94528	.05472	.87513	36
25	9.81180	9.93022	10.06978	9.88158	35	25	9.82055	9.94554	10.05446	9.87501	35
26	.81195	.93048	.06952	.88148	34	26	.82069	.94579	.05421	.87490	34
27	.81210	.93073	.06927	.88137	33	27	.82084	.94604	.05396	.87479	33
28	.81225	.93099	.06901	.88126	32	28	.82098	.94630	.05370	.87468	32
29	.81240	.93124	.06876	.88115	31	29	.82112	.94655	.05345	.87457	31
30	9.81254	9.93150	10.06850	9.88105	30	30	9.82126	9.94681	10.05319	9.87446	30
31	.81269	.93175	.06825	.88094	29	31	.82141	.94706	.05294	.87434	29
32	.81284	.93201	.06799	.88083	28	32	.82155	.94732	.05268	.87423	28
33	.81299	.93227	.06773	.88072	27	33	.82169	.94757	.05243	.87412	27
34	.81314	.93252	.06748	.88061	26	34	.82184	.94783	.05217	.87401	26
35	9.81328	9.93278	10.06722	9.88051	25	35	9.82198	9.94808	10.05192	9.87390	25
36	.81343	.93303	.06697	.88040	24	36	.82212	.94834	.05166	.87378	24
37	.81358	.93329	.06671	.88029	23	37	.82226	.94859	.05141	.87367	23
38	.81372	.93354	.06646	.88018	22	38	.82240	.94884	.05116	.87356	22
39	.81387	.93380	.06620	.88007	21	39	.82255	.94910	.05090	.87345	21
40	9.81402	9.93406	10.06594	9.87996	20	40	9.82269	9.94935	10.05065	9.87334	20
41	.81417	.93431	.06569	.87985	19	41	.82283	.94961	.05039	.87322	19
42	.81431	.93457	.06543	.87975	18	42	.82297	.94986	.05014	.87311	18
43	.81446	.93482	.06518	.87964	17	43	.82311	.95012	.04988	.87300	17
44	.81461	.93508	.06492	.87953	16	44	.82326	.95037	.04963	.87288	16
45	9.81475	9.93533	10.06467	9.87942	15	45	9.82340	9.95062	10.04938	9.87277	15
46	.81490	.93559	.06441	.87931	14	46	.82354	.95088	.04912	.87266	14
47	.81505	.93584	.06416	.87920	13	47	.82368	.95113	.04887	.87255	13
48	.81519	.93610	.06390	.87909	12	48	.82382	.95139	.04861	.87243	12
49	.81534	.93636	.06364	.87898	11	49	.82396	.95164	.04836	.87232	11
50	9.81549	9.93661	10.06339	9.87887	10	50	9.82410	9.95190	10.04810	9.87221	10
51	.81563	.93687	.06313	.87877	9	51	.82424	.95215	.04785	.87209	9
52	.81578	.93712	.06288	.87866	8	52	.82439	.95240	.04760	.87198	8
53	.81592	.93738	.06262	.87855	7	53	.82453	95266	.04734	.87187	7
54	.81607	.93763	.06237	.87844	6	54	.82467	.95291	.04709	.87175	6
55	9 81622	9.93789	10 06211	9.87833	5	55	9.82481	9.95317	10.04683	9.87164	5
56	.81636	.93814	.06186	.87822	4	56	.82495	.95342	.04658	.87153	4
57	.81651	.93840	.06160	.87811	3	57	.82509	.95368	.04632	.87141	3
58	81665	.93865	.06135	.87800	2	58	.82523	.95393	.04607	.87130	2
59	.81680	.93891	.06109	.87789	1	59	.82537	.95418	.04582	.87119	1
60	9.81694	9.93916	10.06084	9.87778	0	60	9.82551	9.95444	10.04556	9.87107	0

	Cosine.	Cotang.	Tang.	Sine.	′		Cosine.	Cotang.	Tang.	Sine.	′

49° **48°**

*Log secant = colog cosine = 1 − log cosine; log cosecant = colog sine = 1 − log sine.

Ex.—Log sec 40°- 30′ = 10.11895.   *Ex.*—Log cosec 40°- 30′ = 10.18746.

3. –Logarithmic Sines, TANGENTS, COTANGENTS, COSINES.—(Cont'd.)
(SECANTS, COSECANTS.)*

42°　　　　　　　　　　　　43°

	Sine.	Tang.	Cotang.	Cosine.				Sine.	Tang.	Cotang.	Cosine.	
0	9.82551	9.95444	10.04556	9.87107	60		0	9.83378	9.96966	10.03034	9.86413	60
1	.82565	.95469	.04531	.87096	59		1	.83392	.96991	.03009	.86401	59
2	.82579	.95495	.04505	.87085	58		2	.83405	.97016	.02984	.86389	58
3	.82593	.95520	.04480	.87073	57		3	.83419	.97042	.02958	.86377	57
4	.82607	.95545	.04455	.87062	56		4	.83432	.97067	.02933	.86366	56
5	9.82621	9.95571	10.04429	9.87050	55		5	9.83446	9.97092	10.02908	9.86354	55
6	.82635	.95596	.04404	.87039	54		6	.83459	.97118	.02882	.86342	54
7	.82649	.95622	.04378	.87038	53		7	.83473	.97143	.02857	.86330	53
8	.82663	.95647	.04353	.87016	52		8	.83486	.97168	.02832	.86318	52
9	.82677	.95672	.04328	.87005	51		9	.83500	.97193	.02807	.86306	51
10	9.82691	9.95698	10.04302	9.86993	50		10	9.83513	9.97219	10.02781	9.86295	50
11	.82705	.95723	.04277	.86982	49		11	.83527	.97244	.02756	.86283	49
12	.82719	.95748	.04252	.86970	48		12	.83540	.97269	.02731	.86271	48
13	.82733	.95774	.04226	.86959	47		13	.83554	.97295	.02705	.86259	47
14	.82747	.95799	.04201	.86947	46		14	.83567	.97320	.02680	.86247	46
15	9.82761	9.95825	10.04175	9.86936	45		15	9.83581	9.97345	10.02655	9.86235	45
16	.82775	.95850	.04150	.86924	44		16	.83594	.97371	.02629	.86223	44
17	.82788	.95875	.04125	.86913	43		17	.83608	.97396	.02604	.86211	43
18	.82802	.95901	.04099	.86902	42		18	.83621	.97421	.02579	.86200	42
19	.82816	.95926	.04074	.86890	41		19	.83634	.97447	.02553	.86188	41
20	9.82830	9.95952	10.04048	9.86879	40		20	9.83648	9.97472	10.02528	9.86176	40
21	.82844	.95977	.04023	.86867	39		21	.83661	.97497	.02503	.86164	39
22	.82858	.96002	.03998	.86855	38		22	.83674	.97523	.02477	.86152	38
23	.82872	.96028	.03972	.86844	37		23	.83688	.97548	.02452	.86140	37
24	.82885	.96053	.03947	.86832	36		24	.83701	.97573	.02427	.86128	36
25	9.82899	9.96078	10.03922	9.86821	35		25	9.83715	9.97598	10.02402	9.86116	35
26	.82913	.96104	.03896	.86809	34		26	.83728	.97624	.02376	.86104	34
27	.82927	.96129	.03871	.86793	33		27	.83741	.97649	.02351	.86092	33
28	.82941	.96155	.03845	.86786	32		28	.83755	.97674	.02326	.86080	32
29	.82955	.96180	.03820	.86775	31		29	.83768	.97700	.02300	.86068	31
30	9.82968	9.96205	10.03795	9.86763	30		30	9.83781	9.97725	10.02275	9.86056	30
31	.82982	.96231	.03769	.86752	29		31	.83795	.97750	.02250	.86044	29
32	.82996	.96256	.03744	.86740	28		32	.83808	.97776	.02224	.86032	28
33	.83010	.96281	.03719	.86728	27		33	.83821	.97801	.02199	.86020	27
34	.83023	.96307	.03693	.86717	26		34	.83834	.97826	.02174	.86008	26
35	9.83037	9.96332	10.03668	9.86705	25		35	9.83848	9.97851	10.02149	9.85996	25
36	.83051	.96357	.03643	.86694	24		36	.83861	.97877	.02123	.85984	24
37	.83065	.96383	.03617	.86682	23		37	.83874	.97902	.02098	.85972	23
38	.83078	.96408	.03592	.86670	22		38	.83887	.97927	.02073	.85960	22
39	.83092	.96433	.03567	.86659	21		39	.83901	.97953	.02047	.85948	21
40	9.83106	9.96459	10.03541	9.86647	20		40	9.83914	9.97978	10.02022	9.85936	20
41	.83120	.96484	.03516	.86635	19		41	.83927	.98003	.01997	.85924	19
42	.83133	.96510	.03490	.86624	18		42	.83940	.98029	.01971	.85912	18
43	.83147	.96535	.03465	.86612	17		43	.83954	.98054	.01946	.85900	17
44	.83161	.96560	.03440	.86600	16		44	.83967	.98079	.01921	.85888	16
45	9.83174	9.96586	10.03414	9.86589	15		45	9.83980	9.98104	10.01896	9.85876	15
46	.83188	.96611	.03389	.86577	14		46	.83993	.98130	.01870	.85864	14
47	.83202	.96636	.03364	.86565	13		47	.84006	.98155	.01845	.85851	13
48	.83215	.96662	.03338	.86554	12		48	.84020	.98180	.01820	.85839	12
49	.83229	.96687	.03313	.86542	11		49	.84033	.98206	.01794	.85827	11
50	9.83242	9.96712	10.03288	9.86530	10		50	9.84046	9.98231	10.01769	9.85815	10
51	.83256	.96738	.03262	.86518	9		51	.84059	.98256	.01744	.85803	9
52	.83270	.96763	.03237	.86507	8		52	.84072	.98281	.01719	.85791	8
53	.83283	.96788	.03212	.86495	7		53	.84085	.98307	.01693	.85779	7
54	.83297	.96814	.03186	.86483	6		54	.84098	.98332	.01668	.85766	6
55	9.83310	9.96839	10.03161	9.86472	5		55	9.84111	9.98357	10.01643	9.85754	5
56	.83324	.96864	.03136	.86460	4		56	.84125	.98383	.01617	.85742	4
57	.83338	.96890	.03110	.86448	3		57	.84138	.98408	.01592	.85730	3
58	.83351	.96915	.03085	.86436	2		58	.84151	.98433	.01567	.85718	2
59	.83365	.96940	.03060	.86425	1		59	.84164	.98458	.01542	.85706	1
60	9.83378	9.96966	10.03034	9.86413	0		60	9.84177	9.98484	10.01516	9.85693	0

| | Cosine. | Cotang. | Tang. | Sine. | ' | | | Cosine. | Cotang. | Tang. | Sine. | |

47°　　　　　　　　　　　　46°

*Log secant = colog cosine = 1 − log cosine; log cosecant = colog sine = 1 − log sine.
Ex.—Log sec 42°- 30' = 10.13237.　Ex.—Log cosec 42°- 30' = 10.17032.

3.—Logarithmic Sines, Tangents, Cotangents, Cosines.—(Concl'd.)
(Secants, Cosecants.)

44°   44°

'	Sine.	Tang.	Cotang.	Cosine.		'		Sine.	Tang.	Cotang.	Cosine.	
0	9.84177	9.98484	10.01516	9.85693	60	30	9.84566	9.99242	10.00758	9.85324	30
1	.84190	.98509	.01491	.85681	59	31	.84579	.99267	.00733	.85312	29
2	.84203	.98534	.01466	.85669	58	32	.84592	.99293	.00707	.85299	28
3	.84216	.98560	.01440	.85657	57	33	.84605	.99318	.00682	.85287	27
4	.84229	.98585	.01415	.85645	56	34	.84618	.99343	.00657	.85274	26
5	9.84242	9.98610	10.01390	9.85632	55	35	9.84630	9.99368	10.00632	9.85262	25
6	.84255	.98635	.01365	.85620	54	36	.84643	.99394	.00606	.85250	24
7	.84269	.98661	.01339	.85608	53	37	.84656	.99419	.00581	.85237	23
8	.84282	.98686	.01314	.85596	52	38	.84669	.99444	.00556	.85225	22
9	.84295	.98711	.01289	.85583	51	39	.84682	.99469	.00531	.85212	21
10	9.84308	9.98737	10.01263	9.85571	50	40	9.84694	9.99495	10.00505	9.85200	20
11	.84321	.98762	.01238	.85559	49	41	.84707	.99520	.00480	.85187	19
12	.84334	.98787	.01213	.85547	48	42	.84720	.99545	.00455	.85175	18
13	.84347	.98812	.01188	.85534	47	43	.84733	.99570	.00430	.85162	17
14	.84360	.98838	.01162	.85522	46	44	.84745	.99596	.00404	.85150	16
15	9.84373	9.98863	10.01137	9.85510	45	45	9.84758	9.99621	10.00379	9.85137	15
16	.84385	.98888	.01112	.85497	44	46	.84771	.99646	.00354	.85125	14
17	.84398	.98913	.01087	.85485	43	47	.84784	.99672	.00328	.85112	13
18	.84411	.98939	.01061	.85473	42	48	.84796	.99697	.00303	.85100	12
19	.84424	.98964	.01036	.85460	41	49	.84809	.99722	.00278	.85087	11
20	9.84437	9.98989	10.01011	9.85448	40	50	9.84822	9.99747	10.00253	9.85074	10
21	.84450	.99015	.00985	.85436	39	51	.84835	.99773	.00227	.85062	9
22	.84463	.99040	.00960	.85423	38	52	.84847	.99798	.00202	.85049	8
23	.84476	.99065	.00935	.85411	37	53	.84860	.99823	.00177	.85037	7
24	.84489	.99090	.00910	.85399	36	54	.84873	.99848	.00152	.85024	6
25	9.84502	9.99116	10.00884	9.85386	35	55	9.84885	9.99874	10.00126	9.85012	5
26	.84515	.99141	.00859	.85374	34	56	.84898	.99899	.00101	.84999	4
27	.84528	.99166	.00834	.85361	33	57	.84911	.99924	.00076	.84986	3
28	.84540	.99191	.00809	.85349	32	58	.84923	.99949	.00051	.84974	2
29	.84553	.99217	.00783	.85337	31	59	.84936	.99975	.00025	.84961	1
30	9.84566	9.99242	10.00758	9.85324	30	60	.84949	10.00000	10.00000	9.84949	0
	Cosine.	Cotang.	Tang.	Sine.	'		Cosine.	Cotang.	Tang.	Sine.	'

45°   45°

3a.—Table for Finding the Logarithmic Sines and Tangents of Small Angles.
[Values of S and T in Formulas Below.*]

A.	A(sec.).	S.	A.	A(sec.).	T.	A.	A(sec.).	T.
0°00'00"	0000"	4.68557	0°00'00"	000"	4.68557	1°25'40"	5140"	4.68566
0°40'10"	2410"	57	0°03'00"	180"	57	1°25'50"	5150"	67
0°40'20"	2420"	56	0°03'20"	200"	58	1°30'20"	5420"	67
0°57'00"	3420"	56	0°28'40"	1720"	58	1°30'30"	5430"	68
0°57'10"	3430"	55	0°28'50"	1730"	59	1°34'40"	5680"	68
1°09'50"	4190"	55	0°40'20"	2420"	59	1°34'50"	5690"	69
1°10'00"	4200"	54	0°40'30"	2430"	60	1°39'00"	5940"	69
1°20'40"	4840"	54	0°49'30"	2970"	60	1°39'10"	5950"	70
1°20'50"	4850"	53	0°49'40"	2980"	61	1°43'00"	6180"	70
1°30'10"	5410"	53	0°57'10"	3430"	61	1°43'10"	6190"	71
1°30'20"	5420"	52	0°57'20"	3440"	62	1°46'50"	6410"	71
1°38'50"	5930"	52	1°03'50"	3830"	62	1°47'00"	6420"	72
1°39'00"	5940"	51	1°04'00"	3840"	63	1°50'40"	6640"	72
1°46'50"	6410"	51	1°10'00"	4200"	63	1°50'50"	6650"	73
1°47'00"	6420"	50	1°10'10"	4210"	64	1°54'10"	6850"	73
1°54'10"	6850"	50	1°15'30"	4530"	64	1°54'20"	6860"	74
1°54'20"	6860"	49	1°15'40"	4540"	65	1°57'40"	7060"	74
2°01'00"	7260"	49	1°20'50"	4850"	65	1°57'50"	7070"	75
2°01'10"	7270"	48	1°21'00"	4860"	66	2°01'10"	7270"	75

* Log sin A = log A (seconds) + S.   Log tan A = log A (seconds) + T.

The Solution of Right-Angled Triangles.—Let triangle $A\,B\,C$ of Fig. 4 (p. 141) represent any right-angled triangle and a, b, and c, the lengths of its sides. Then, with any two sides, or any one side and one acute angle known, the missing information can be obtained by the following formulas:

TABLE 4

SOLUTION OF RIGHT-ANGLED TRIANGLES

Sides and Angles Known	Formulas for Sides and Angles to be Found		
Sides c and a	$b = \sqrt{c^2 - a^2}$	$\sin A = \dfrac{a}{c}$	$B = 90° - A$
Sides c and b	$a = \sqrt{c^2 - b^2}$	$\sin B = \dfrac{b}{c}$	$A = 90° - B$
Sides a and b	$c = \sqrt{a^2 + b^2}$	$\tan A = \dfrac{a}{b}$	$B = 90° - A$
Side c Ang. A	$c = c \times \sin A$	$b = c \times \cos A$	$B = 90° - A$
Side c Ang. B	$a = c \times \cos B$	$b = c \times \sin B$	$A = 90° - B$
Side a Ang. A	$c = \dfrac{a}{\sin A}$	$b = a \times \cot A$	$B = 90° - A$
Side a Ang. B	$c = \dfrac{a}{\cos B}$	$b = a \times \tan B$	$A = 90° - B$
Side b Ang. A	$c = \dfrac{b}{\cos A}$	$a = b \times \tan A$	$B = 90° - A$
Side b Ang. B	$c = \dfrac{b}{\sin B}$	$a = b \times \cot B$	$A = 90° - B$

ILLUSTRATION: A gabled roof has a pitch of 45 degrees. What is the length of the rafters if the span is 20 feet?

In this case A and B
  = 45 degrees, C = 20 feet.
Length of rafter
  = $a = c \times \sin A$
  $a = 20 \times .707 = 14.14$ feet (Ans.)

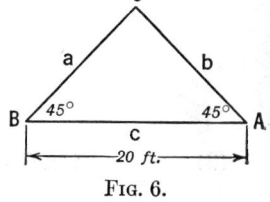

FIG. 6.

ILLUSTRATION: Figure 7 shows a method used to measure the distance CB across a river. A surveying party sets points A and C, then with a transit at C, a right angle is turned and point B set. The distance b is measured and also the angle A. What is the distance across the river (a) if b is 487.32 feet and $\angle A$ is 35°17′?

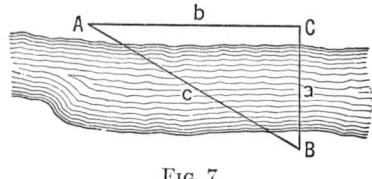

FIG. 7.

$$a = b \times \tan A$$
$$\log 487.32 = 2.68782$$
$$\log \tan 35°17' = \underline{9.84979}$$
$$\log a = 2.53761$$
$$a = 344.85 \text{ feet (Ans.)}$$

Solution of Any Plane Triangle.—Not only right triangles but any plane triangle may be solved by trigonometric formulas if two sides and an angle, or two angles and a side, or three sides are given. Four cases will be considered.

CASE I. Given any two sides b and c and their included angle A. Use any one of the following sets of formulas:

(1)
$$\tfrac{1}{2}(B + C) = 90° - \tfrac{1}{2}A$$

$$\tan \tfrac{1}{2}(B - C) = \frac{b - c}{b + c} \tan \tfrac{1}{2}(B + C)$$

$$B = \tfrac{1}{2}(B + C) + \tfrac{1}{2}(B - C)$$
$$C = \tfrac{1}{2}(B + C) - \tfrac{1}{2}(B - C)$$

$$a = \frac{b \sin A}{\sin B}$$

(2)
$$\tan C = \frac{c \sin A}{b - c \cos A}$$

$$B = 180° - (A + C)$$

$$a = \frac{c \sin A}{\sin C}$$

(3)

$$a = \sqrt{b^2 + c^2 - 2bc \cos A}$$

$$\sin B = \frac{b \sin A}{a}$$

$$C = 180° - (A + B)$$

Case II.   Given any two angles A and B and any side c.

$$C = 180° - (A + B)$$

$$a = \frac{c \sin A}{\sin C}$$

$$b = \frac{c \sin B}{\sin C}$$

Case III.   Given the three sides a, b, and c.   Use either of the following sets of formulas:

(1)

$$\cos A = \frac{b^2 + c^2 - a^2}{2bc}$$

$$\cos B = \frac{a^2 + c^2 - b^2}{2ac}$$

$$C = 180° - (A + B)$$

(2)   Let

$$s = \tfrac{1}{2}(a + b + c)$$

$$r = \sqrt{\frac{(s - a)(s - b)(s - c)}{s}}$$

$$\tan \tfrac{1}{2}A = \frac{r}{s - a}$$

$$\tan \tfrac{1}{2}B = \frac{r}{s - b}$$

$$\tan \tfrac{1}{2}C = \frac{r}{s - c}$$

(3)   Following also comment for case III, let c be longest side, and $a > b$.   Then (see Fig. 8) or similarly for any triangle:

$$g = \tfrac{1}{2}\left[\frac{(a + b)(a - b)}{c} + c\right]$$

$$s = c - g$$

$$\cos A = \frac{s}{b}$$

$$\cos B = \frac{g}{a}$$

$$C = 180° - (A + B)$$

Case IV.   Given any two sides a and b and an angle A opposite either one of these.

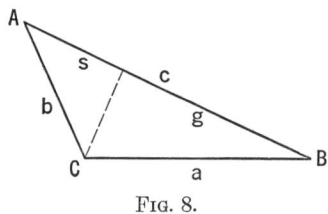

Fig. 8.

$$\sin C = \frac{c \sin A}{a}$$

$$B = 180° - (A + C)$$

$$b = \frac{a \sin B}{\sin A}$$

Note.   There may be two values for the angle C.   If, however, one solution is such that $A + C > 180°$, use only the other value.

Reference.—An excellent treatise on trigonometry is also contained in the set of mathematics books by J. E. Thompson.   This is entitled Trigonometry for the Practical Man and is published by the D. Van Nostrand Company.

VI

Differential Calculus

In arithmetic we deal with numbers that represent fixed quantities or *constants*. In algebra, the numbers may be constants or they may vary (variables), but in any given problem, the numbers remain constant throughout the consideration of that particular problem.

Other problems often arise in which the quantities involved, or the symbol expressing these quantities, is continually changing and therefore cannot be solved by arithmetic or algebra. For example, if a weight is dropped and allowed to fall freely, its speed steadily increases. If thrown directly upward, it moves slower and slower as it rises until it stops. Then it begins to fall and moves more and more rapidly until it hits the ground. Many other such examples could be cited.

The branch of mathematics which is very helpful in understanding these phenomena is known as the calculus. Calculus is concerned with the study of rates of change of quantities. In this chapter, some of the basic ideas underlying calculus are explained and a few practical applications are cited.

Differentiation.—If a straight line of length x is divided into an enormous number of parts, then the length of each of these infinitesimal parts is represented by the symbol dx. This expression is known as the *differential* of x. Thus the differential of a line x is dx. (Note that dx does not signify d times x, but is a single symbol in itself.)

Now suppose a square is formed whose sides are x units long. The area (x times x) will be x^2. If the length of each side is increased by an infinitesimally small amount dx, then the length of each side becomes $x + dx$, and the new area is $(x + dx)(x + dx)$ or $x^2 + 2xdx + (dx)^2$. The first term is the original

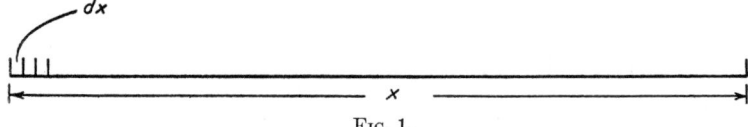

Fig. 1.

area, and the second and third terms represent the additional area formed by increasing the two sides of the square by the amount dx.

For practical purposes we can neglect $(dx)^2$ because its magnitude is negligible (for example, if dx equals 0.001 inch, then $(dx)^2$ equals 0.000001 inch and since dx actually is very much smaller than 0.001, when squared, it becomes negligible). The remaining terms represent the original area (x^2) plus the increase, or change, in the area $(2xdx)$.

The expression $2xdx$ is known as the *differential* of x^2. Visually (see Fig. 2), $2xdx$ is the sum of each area strip, neglecting

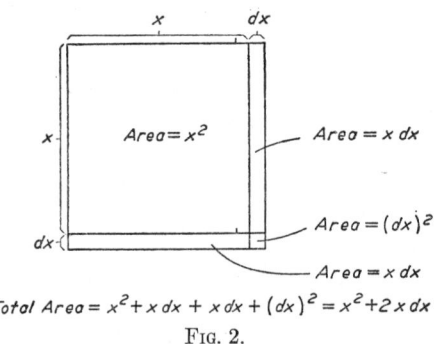

Fig. 2.

the corner $(dx)^2$, formed by changing each side of the square by an amount dx. Similarly the differential of x^3 (representing, for example, the volume of a cube whose sides are each x units long) can be shown to be $3x^2dx$. Thus, we find a pattern forming:

the differential of x is dx
the differential of x^2 is $2\,xdx$
the differential of x^3 is $3x^2dx$
the differential of x^4 is $4x^3dx$
etc.

We now can devise a formula which will tell us the differential of x raised to any power n. It can be written:

the differential of x^n equals $nx^{n-1}dx$

This formula can be applied to the area problem shown in Fig. 2. Given a square with sides x and area A, what happens to the area if each side is increased by an amount dx?

$$A = x^2$$

Taking the differential of both sides of the above equation we get

$$\frac{d(A)}{dx} = \frac{d(x^2)}{dx} = 2xdx$$

The expression d/dx means the differential of a variable with respect to x. For example, dy/dx would be the differential of y with respect to x, or the change in y caused by a change in x. This is referred to as the derivative of y with respect to x.

Differentiation of Constants.—A constant quantity, having no rate of change, cannot be differentiated; therefore its differential is zero. If the constant is a coefficient of a variable, as 6 in the expression $6x$, then the constant which multiplies the variable remains unchanged in the differentiating process. Thus:

the differential of 6 = zero
the differential of $6x = 6dx$
the differential of $6x^2 = 6(2xdx) = 12xdx$

Differential of Sum or Difference.—To differentiate an algebraic expression consisting of several terms with positive or negative signs before them such as:

$$x^2 - 2x + 6 + 3x^4$$

each term must be differentiated separately. This is so because each term is separate and distinct from the other terms and therefore its differential or rate of growth will be distinct or separate from the other terms. Thus·

the differential of $x^2 - 2x + 6 + 3x^4 = 2xdx - 2dx + 0 + 12x^3dx$

Differentiation of a Product.—To illustrate how the product of two variables is differentiated, consider a rectangle whose sides are x and y, such as the one shown in Fig. 3.

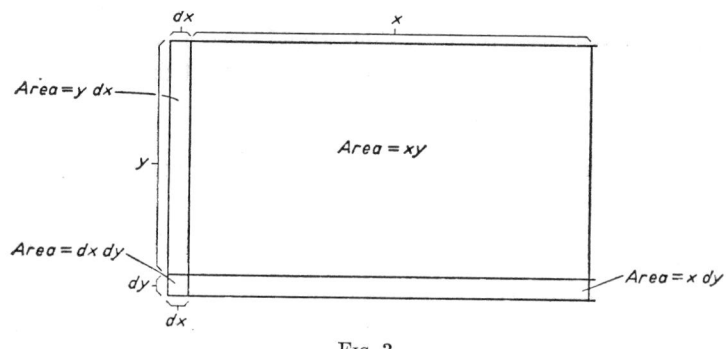

Fig. 3.

Its area is equal to the product xy. If the sides are increased by the amounts dx and dy respectively, then:

New area $= xy$ (old area) $+ ydx + xdy + dydx$

The last term $dydx$ is a very small quantity and therefore can be safely neglected. Therefore, the differential of the original area xy is $xdy + ydx$. This can be generalized for every case:

The differential of the product of two variables is equal to the first multiplied by the differential of the second, plus the second multiplied by the differential of the first. Thus, the differential of $x^2y = x^2dy + y(2xdx) = x^2dy + 2yxdx$

This law also holds for any number of variables. For example, the differential of $xyz = xydz + xzdy + yzdx$.

Differential of a Fraction.—To differentiate a fraction, say x/y, the expression is changed to a product by writing it in the form xy^{-1}, using the negative exponent to eliminate the fraction. Then the product is differentiated in the manner just described. Thus:

$$\text{differential } xy^{-1} = -xy^{-2}dy + y^{-1}dx$$

$$= -xdy/y^2 + dx/y$$

reducing to a common denominator:

$$= \frac{(ydx - xdy)}{y^2}$$

This can be expressed in the following way:

The differential of a fraction is equal to the denominator times the differential of the numerator, minus the numerator times the differential of the denominator, all divided by the square of the denominator.

By further mathematical analysis the differentials for other expressions can be found. The following table summarizes those we have considered and also includes several others:

Expression	Differential	Derivative
$x + y + z + \cdots$	$dx + dy + dz + \cdots$	
$y = C$	$dy = 0$	$dy/dx = 0$
$y = x + C$	$dy = dx + 0$	$dy/dx = 1$
$y = \pm x$	$dy = \pm dx$	$dy/dx = \pm 1$
$y = Cx$	$dy = Cdx$	$dy/dx = C$
$y = x^n$	$dy = nx^{n-1}dx$	$dy/dx = nx^{n-1}$
$y = x^{1/n}$	$dy = (1/n)x^{(1/n)-1}dx$	$dy/dx = (1/n)x^{(1/n)-1}$
$z = xy$	$dz = xdy + ydx$	
$z = x/y$	$dz = \dfrac{ydx - xdy}{y^2}$	

It should be noted that differentials can also be found for trigonometric functions and logarithmic functions.

Maximum and Minimum Problems.—One of the most important and practical applications of differential calculus is the solution of problems involving maximum and minimum values. Suppose we had a hollow box into which we placed an uninflated balloon. If we inflate the balloon, its volume V changes at every instant with respect to the box volume B. The rate of change of the volume occupied by the balloon with respect to the volume of the box is dV/dB and it has some significant value so long as the balloon continues to be inflated. When the balloon completely fills the box, reaching its maximum possible size, the volume V no longer changes and at this point dV/dB is zero, and V is now at its maximum value.

Then, if we let the air out of the balloon, V again changes (now decreasing with respect to B) and dV/dB again has a significant value. When all the air has escaped from the balloon, dV/dB again becomes zero, for once again V is no longer changing with respect to B, now having reached its minimum value.

This example illustrates one of the most important rules in calculus: When the derivative of an expression is zero, the value of the expression is at a maximum or a minimum. In the example just cited, when the balloon filled the box, dV/dB was zero and the volume of the balloon was at its maximum. Similarly, when the balloon was deflated, dV/dB again became zero and its volume was now at its minimum value.

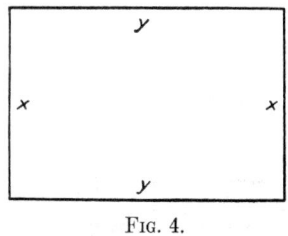

FIG. 4.

Maximum and Minimum Examples.—The following two practical problems further illustrate the application of maximum and minimum rules.

EXAMPLE 1. What are the dimensions of the largest possible rectangle which can be enclosed by a fence 600 ft long (Fig. 4)?

Perimeter of rectangle $= 2x + 2y = 600$ ft; area of rectangle $= xy = A$.

$$2x = 600 - 2y \qquad\qquad A = xy = (300 - y)y$$
$$x = 300 - y \qquad\qquad A = -y^2 + 300y$$

To find maximum area, differentiate the equation:

$$A = -y^2 + 300y$$

$$\frac{dA}{dy} = -2y + 300$$

Now set the derivative equal to zero to find the value of y which gives a maximum A (the minimum value in this case is zero).

$$-2y + 300 = 0$$

$$2y = 300$$

$$y = 150$$

Substituting in the perimeter equation,

$$x = 300 - y$$

we obtain

$$x = 300 - 150 = 150$$

Therefore, to enclose the maximum area, the rectangle measures 150 ft $\times$ 150 ft, or 22,500 sq ft. Note that, for a given perimeter, the rectangle which will enclose the maximum area turns out to be a square.

EXAMPLE 2. Transmitting electricity from the powerhouse to the consumer involves the cost of the heat loss and the cost of the copper wire used to carry the electrical energy. The heat loss is inversely proportional to the cross-section of the wire. The cost of the copper wire is directly proportional to the cross-section of the wire. In a certain electrical system, the cost C turns out to be

$$C = \frac{2}{A} + \frac{2A}{3}$$

where C is the cost and A is the cross-section of the wire. Find the area A which makes the cost C a minimum.

To find minimum C differentiate the expression:

$$C = \frac{2}{A} + \frac{2A}{3}.$$

$$\frac{dC}{dA} = \frac{-2}{A^2} + \frac{2}{3}$$

Now set the derivative equal to zero and solve for A.

$$\frac{-2}{A^2} + \frac{2}{3} = 0$$

$$\frac{A^2}{2} = \frac{3}{2}$$

$$A^2 = 3$$

$$A = \sqrt{3} = 1.732 \text{ sq. inches}$$

In this chapter we have presented some of the basic ideas underlying the calculus. Understandably we have only scratched the surface of this fascinating and useful branch of mathematics. For a further treatment of the subject (including *integration* which is the exact opposite of differentiation) the reader is referred to *The Calculus for the Practical Man* by J. E. Thompson, D. Van Nostrand and Co., New York, N. Y.

VII

Mechanics

Mechanics.—Mechanics is a science which treats of the action of forces and their effect upon bodies. A force is defined as a push or pull which tends to change the velocity or direction of a body's motion. The units by which a force is measured are pounds or tons. Distance measured in linear units and time expressed in seconds, minutes, etc., are two other elementary quantities in mechanics from which numerous compound quantities are derived.

Work is the product of force by distance. The units for measuring work are derived from the units of force and distance. In the British system, the unit of work is the foot-pound.

Power is the time rate of doing work. In mechanics it is the product of force by distance divided by time. Power is commonly expressed as inch-pounds per minute, foot-pounds per minute or second, etc. Horsepower, H.P., is the unit of power adopted for engineering work. One horsepower = 33,000 foot-pounds per minute = 550 foot-pounds per second.

Velocity is the time rate of motion. It is distance divided by time, and is expressed in feet per minute, miles per hour, etc.

Stress and Strain.—An external force applied to a body, so as to pull it apart, is resisted by an internal force, or resistance, and the action of these forces cause a displacement of the molecules, or deformation. The external forces are called stresses while the alteration produced by the stresses is called by the term strain. For example, a load on a steel column tends to compress or crush the column. At the same time, the column reacts against the tendency of the load to crush it and exerts a force opposite to the load. The external force or the tendency of the outside load to change the shape of the column is called stress. The internal force or the resistance of the column to the tendency of the outside load to change its shape is called strain.

There are five kinds of stresses:

1. Tensile stress, or pull, is a force which tends to elongate a piece of material.

2. Compressive stress, or push, is a force which tends to shorten a piece of material.

3. Shearing stress is a force which tends to force one part of a piece of material to slide over an adjacent part.

4. Torsional stress, a form of shearing stress, is a force which tends to twist a piece of material.

5. Transverse stress, a combination of tension and compression, is a force which tends to bend a piece of material.

All stresses to which a material is subjected cause a deformation in it. If the stress is not too great, however, the material will return to its original shape and dimensions when the external stress is removed. The property which enables a material to return to its original shape and dimensions is called its elasticity.

The elastic limit is the unit stress beyond which the material will not return to its original shape when the load is removed.

There is a law, called Hooke's Law, which expresses the relation between the amount of stress applied to a body and the amount of strain it produces.

Hooke's Law. The amount of change in the shape of an elastic body is proportional to the force applied, provided that the elastic limit is not exceeded. In other words the strain is directly proportional to the stress.

For different stresses the rule becomes:

Tensile stress, the stretch is proportional to the force applied.

Torsional stress, the twist is proportional to the stress causing it.

Transverse stress, the deflections are proportioned to the loads causing them.

ILLUSTRATION. If a weight of one pound is hung on a spring it lengthens the spring 1.5 inch; what weight would lengthen it 0.75 inch?

$$x : 1 = 0.75 : 1.5$$

$$x = \frac{1 \times 0.75}{1.5} = 0.5$$

Therefore, $\frac{1}{2}$ pound weight would lengthen the spring $\frac{3}{4}$ inch.

Modulus of Elasticity.—The modulus of elasticity is a term expressing the relation between the amount of extension or compression of a material and the load producing that extension or compression. It is defined as the load per unit of section divided by the extension per unit of length.

The following table gives the moduli of elasticity for various materials.

Brass, cast.......	9,170,000	Tin, cast........	4,600,000
Copper..........	15,000,000	Iron, cast........	12,000,000
Lead............	1,000,000	Steel............	28,000,000

The following rule may be used to find the modulus of elasticity, commonly designated by E.

Divide the stress per square inch by the elongation in one inch caused by this stress. Expressed as a formula:

$$E = \frac{P}{e}$$

where E = modulus of elasticity in pounds

P = stress

e = elongation in inches

ILLUSTRATION: If the elongation of 0.02 inch is produced in a bar 10 inches long by a load of 48,000 pounds per square inch of cross section of the bar, find the modulus of elasticity.

$$E = \frac{P}{e}$$
$$= \frac{48,000 \times 10}{0.02}$$
$$= 24,000,000$$

Therefore, the modulus of elasticity is 24,000,000 pounds.

Graphical Representation of Forces.—Forces may be represented geometrically by straight lines, proportional to the forces. The three characteristics which, when known, determine a force are (1) direction, (2) place of application, and (3) magnitude. These three are defined as follows:

1. The direction of a force is the direction in which it tends to move the body upon which it acts.

2. The place of application is usually assumed to be a point such as the center of gravity.

3. The magnitude is measured in pounds.

Composition of Forces.—The operation of finding a single force whose effect is the same as that of two or more given forces is called the composition of forces. This single force is called the resultant of the given forces. The separate forces which can be so combined are called the components.

Resolution of Forces.—The operation of finding two or more components of a given force is called the resolution of forces.

Straight lines, drawn to a convenient scale, may be used to represent the forces and arrowheads the direction of the force, the length of the line being its magnitude. The point of application may be any point on the line, although usually it is more convenient to assume the point to be at one end.

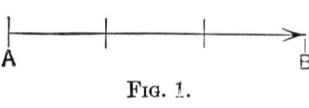

In the sketch at the left a force is supposed to act along $A\ B$ in a direction from left to right.

Fig. 1.

ILLUSTRATION: In the above sketch if A is assumed to be the point of application, the force is exerted as a pull; but if point B is assumed to be the point of application, it would indicate that the force is exerted as a push. If the line is 3 units long and if each unit represents 5 pounds, the line $A B$ represents a force of fifteen pounds applied at A.

Composition and Resolution of Forces.—The following rules may be used in the composition and resolution of forces:

1. The resultant of two forces acting in the same direction, is equal to the sum of the forces.

ILLUSTRATION: Two forces $A B$ equal to two pounds and $A C$ equal to four pounds are both applied at point A. Find the resultant $A D$.

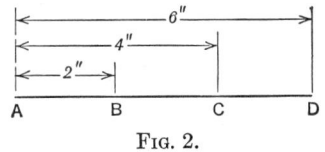

FIG. 2.

$A D$ = Sum of the forces

$= 2 + 4 = 6$

Therefore, the resultant equals 6 pounds.

2. If two forces act in opposite directions, then their resultant is equal to their difference, and the direction of the resultant is the same as the direction of the greater of the two forces.

ILLUSTRATION: Two forces one $A B$ equal to 3 pounds and one $A C$ equal to 5 pounds are both applied at A. Find the resultant.

$A D$ = Difference of two forces

$= 5 - 3 = 2$

Therefore, the resultant is 2 pounds and acts in the direction of $A C$.

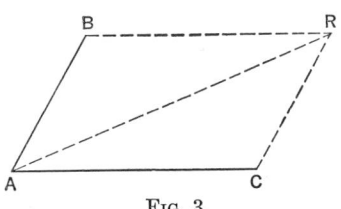

FIG. 3.

Parallelogram of Forces.—If two forces acting on a point are represented in magnitude and direction by the adjacent sides of a parallelogram $A B$ and $A C$ in the sketch on the left, the resultant will be represented in magnitude and direction by the diagonal $A R$ drawn from the intersection of the two component forces.

ILLUSTRATION: If in the figure at the left, two forces, one
A C of 4 pounds acting in the direction of the arrow, and, one

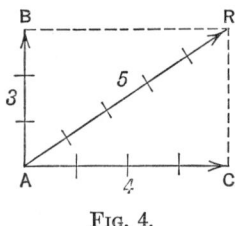

A B of 3 pounds acting in the direction
of the arrow, are both applied at A, find
the resultant A R.

Use the geometrical proposition relative to the right triangle i.e., the square
on the hypotenuse is equal to the sum
of the squares on the other two sides.
Expressed as a formula:

FIG. 4.

$$\overline{A\ R}^2 = \overline{A\ C}^2 + \overline{A\ B}^2$$

$$\overline{A\ R} = \sqrt{\overline{A\ C}^2 + \overline{A\ B}^2}$$

$$= \sqrt{(4 \times 4) + (3 \times 3)}$$

$$= \sqrt{25} = 5$$

Therefore, the resultant is equal to 5 pounds.

Factor of Safety.—A factor of safety is defined as the ratio
in which the load that is just sufficient to overcome instantly the
strength of a piece of material is greater than the greatest safe
ordinary working load. The character of the loading determines
in a large degree the margin that should be left for safety. The
following table gives the factor of safety for some metals which
have been determined by an analytical method:

Cast-iron and other castings.......... 4

Wrought iron or mild steel........... 3

Oil-tempered or nickel steel.......... $2\frac{1}{4}$

Hardened steel...................... 3

Bronze or brass, rolled or forged....... 3

TABLE 1
Average Ultimate Strength of Common Metals; Pounds per Square Inch

Material	Tension	Compression	Shear	Modulus of Elasticity
Cast iron............	15,000	80,000	18,000	12,000,000
Wrought iron.........	48,000	46,000	40,000	27,000,000
Steel castings.........	70,000	70,000	60,000	30,000,000
Steel structural.......	60,000	60,000	50,000	29,000,000
Cast brass..........	24,000	30,000	36,000	9,000,000

TABLE 2
General Factors of Safety

Material	Steady load	Load varying from zero to maximum in one direction	Load varying from zero to maximum in both directions	Suddenly varying loads
Cast iron............	6	10	15	20
Wrought iron.........	4	6	8	12
Steel...............	5	6	8	12

Symbols and Formulas for the Strength of Materials.—The following symbols are commonly used in the formulas:

A = area of cross section of material in square inches;

E = modulus of elasticity;

I = moment of inertia of section about an axis passing through the center of gravity;

I_p = polar moment of inertia of section;

M_b = maximum bending moment in inch-pounds;

M_t = moment of force tending to twist (torsional moment) in inch-pounds;

P = total stress in pounds;

Y = distance from center of gravity to most remote fiber;
S = permissible working stress in pounds per square inch;
Z = section modulus for bending (moment of resistance);
Z_p = section modulus for torsion;
e = elongation or shortening in inches;
l = length in inches.

These formulas may be used to calculate strength of materials:

For tension and compression: $P = A \times S$; $e = \dfrac{Pl}{AE}$

For shear: $P = A \times S$

For torsion: $M_t = \dfrac{SI_p}{Y} = SZ_p$

For bending: $M_b = \dfrac{SI}{Y} = SZ$

Combined bending and torsion:

$$\text{Combined moment} = \sqrt{M_b{}^2 + M_t{}^2} = SZ$$

The following group of illustrative problems shows how these formulas are employed in practice. The calculations are simple and straightforward, and the reader should have no difficulty in following the steps involved.

ILLUSTRATIVE PROBLEMS

(1) Find the diameter of a wrought iron bar which is to support (in tension) a load of 32,000 pounds if the load is gradually applied and after reaching its maximum value gradually removed.

$$48,000 \div 6 = 8,000$$
$$P = A \times S$$
$$A = \frac{P}{S} = \frac{32,000}{8,000} = 4 \text{ sq. in.}$$

Diameter = $1.128\sqrt{4} = 2.256$ inches

Divide 48,000 obtained from the ultimate strength table on page 209 by 6, the factor of safety obtained from table on same page. $48,000 \div 6 = 8,000$ pounds per square inch. Then dividing 32,000 by 8,000 obtain the answer of 4 square inches. The diameter of a circle of this area is $2\frac{1}{4}$ inches approx.

(2) In the above problem what would be the total elongation of the bar under full load if the bar were 6 feet long?

Multiply 32,000, the load in the above problem, by 6, the length of the bar, and then by 12, the number of inches in one foot. Divide this product by the product of the area 4 and the modulus of elasticity, 27,000,000, obtained from the table on page 209. The quotient, 0.021 inch, is the total elongation.

$$e = \frac{P}{AE}$$

$$= \frac{32,000 \times 6 \times 12}{4 \times 27,000,000}$$

$$= 0.021 \text{ in.}$$

(3) A square bar 3 feet long firmly fixed at one end, is supporting a load of 4,000 pounds at the outer free end. If the bar is to be made of structural steel and the load is steady, find the size of bar required for safe loading.

$$4,000 \times 36 = 144,000$$

$$60,000 \div 5 = 12,000$$

$$M_b = \frac{SI}{Y}$$

$$144,000 = \frac{12,000 \times s^4}{12 \times \frac{1}{2}s} = \frac{12,000s^3}{6}$$

$$s^3 = 72, \qquad s = 4.16$$

$M_b =$ load $\times$ lever arm in inches, in this case $4,000 \times 36$ or 144,000 inch-pounds.

S = safe stress = 60,000 obtained from the table divided by 5, the factor of safety, for a safe load for steel, or 12,000;

$I = s^4 \div 12$ for a square, if s = side of square;

$Y = s \div 2$ in this case.

Substituting these values in the equation find size of bar to be 4.16 inches.

(4) A square bar made of structural steel is subjected to a steady torsional moment of 80,000 inch-pounds. Find the size of bar required for safe loading.

$$M_t = 80,000$$

Ultimate strength in shear for torsion = $60,000 \times \frac{4}{5}$ = 48,000.

$$S = 48,000 \div 5 \text{ (factor of safety for steel)} = 9,600$$

$$Z_p = \tfrac{2}{9} s^3 \text{ for a square, if } s = \text{side of square}$$

Substituting the above values in the equation for M and evaluating find $s^3 = 32.5$ and $s = 3.17$ in.

$$M_t = S Z_p$$
$$80,000 = 9,600 \times \tfrac{2}{9} s^3$$
$$s^3 = 32.5, \quad s = 3.17$$

(5) If the bar in the two previous problems is subjected to combined bending and torsion find the size of square bar required to withstand the combined moment safely.

$$M_b = 144,000 \text{ inch-pounds in (3)}$$

$$M_t = 80,000 \text{ inch-pounds in (4)}$$

$$\text{Combined Moment} = \sqrt{144,000^2 + 80,000^2}$$

$$= 165,000 \text{ approx.}$$

$$165,000 = SZ$$

thus
$$165,000 = 12,000 \times \frac{s^3}{6}$$

and
$$s^3 = 82.5, \quad s = 4.35 \text{ in.}$$

Thus 12,000 obtained from (3) is the safe load for steel. $\dfrac{s^3}{6}$ is the formula for section modulus. Substituting these values in the equation obtain 4.35 in., the size of the required bar.

Simple Machines.—A machine is a device by which useful work is done in such a way that the operator gains in effort, speed or convenience. A machine is a simple one when it contains but one moving part. The six fundamental *simple machines* are the *lever*, the *pulley*, the *screw*, the *inclined plane*, the *wedge*, and the *wheel and axle*. Practically all of these machines are used in the machine shop in some form or other. It is, therefore, desirable to know their properties.

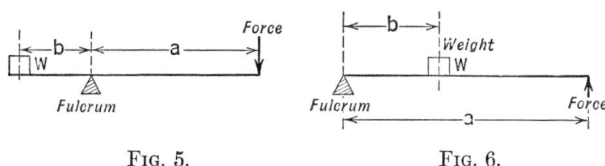

Fig. 5. Fig. 6.

Levers.—A lever is an inflexible rod capable of motion about a fixed point, called a *fulcrum*. The rod may be straight, curved or bent at any angle.

There are three kinds or classes of levers which differ in the respective locations of the applied force, the moved weight and the fulcrum.

In the *lever of the first class*, the fulcrum lies between the points at which the force and the load act (Fig. 5). An example of this type of lever is a claw hammer pulling a nail.

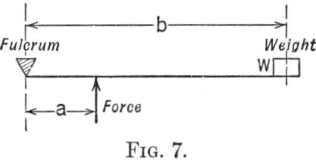

Fig. 7.

In the *lever of the second class*, the load acts at a point between the fulcrum and the force (Fig. 6). An example of this type of lever is a wheelbarrow in which the wheel axle is the fulcrum.

In the *lever of the third class* the force acts between the fulcrum and the load.

Levers are usually used to gain power at the expense of time or motion. Thus, in a first class lever, if the distance from the fulcrum to the force is five times the distance from the fulcrum to the weight, it will give five times the power, but the force will have to move a distance five times greater than the weight.

Levers of the third class involve a mechanical disadvantage as the power must always be greater than the weight. However, there is a gain in motion.

Law of the Lever.—The force multiplied by its distance to the fulcrum is equal to the weight multiplied by its distance to the fulcrum.

The law for bent levers is the same as for straight levers but the

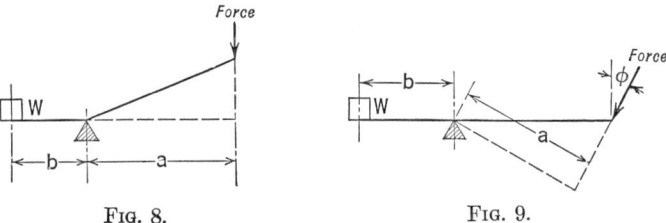

FIG. 8.   FIG. 9.

length of arms is computed on lines from the fulcrum at right angles to the direction in which the power and weight act. (See Figs. 8 and 9).

Letting P = power or force;

   a = power arm or distance from the fulcrum to the point where power is applied;

  W = weight or resistance;

   b = weight arm or distance from the fulcrum to the point where the weight or resistance is applied;

the law may then be stated as follows:

$$P \times a = W \times b$$

From this the following relations may be obtained by transposition.

$$P = \frac{W \times b}{a}, \qquad W = \frac{P \times a}{b},$$

$$a = \frac{W \times b}{P}, \qquad b = \frac{P \times a}{W}.$$

ILLUSTRATION: What force in pounds is applied at the brake shoe shown in Fig. 10 if a force of 50 pounds is exerted on the pedal?

In this case, $P = 50$ lb, $a = 14$ in., and $b = 5$ in.

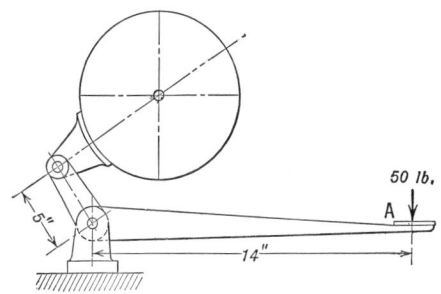

FIG. 10.

Then $W = \dfrac{P \times a}{b} = \dfrac{50 \times 14}{5} = 140$ lb   (Ans.)

This is an example of a lever of the first class.

ILLUSTRATION: What force will be exerted by the rod "A" in Fig. 11 if a force of 40 pounds is exerted at the handle of the lever?

Here, $P = 40$ lb, $a = 6 + 20 = 26$ in., $b = 6$ in.

Then, $W = \dfrac{P \times a}{b} = \dfrac{40 \times 26}{6} = 173$ lb   (Ans.)

This is a lever of the second class.

ILLUSTRATION: Figure 12 shows an air brake layout. If the piston in the air cylinder is 10 inches in diameter and the air pressure is 100 pounds per square inch, what is the force on the brake shoe?

Area of piston $= \pi r^2 = \pi \times 5 \times 5 = 25\pi$ sq. in.

Force on brake rod $= P = 100 \times 25\pi = 2500\pi$ lb.

$$a = 12 \text{ in.} \quad b = 12 + 8 = 20 \text{ in.}$$

Then $W = \dfrac{P \times a}{b} = \dfrac{2500\pi \times 12}{20} = 1500\pi = 4712$ lb.   (Ans.)

This is the lever of the third class.

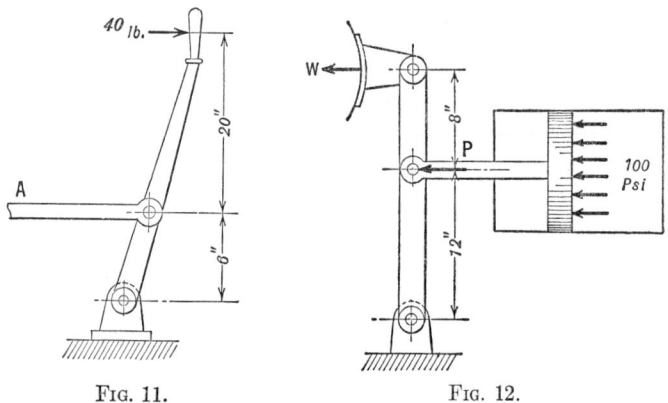

FIG. 11.   FIG. 12.

Wheel and Axle.—This is simply an application of the lever of the first order so that the power and resistance may act through greater distances; the radius of the wheel is the lever arm of the power and that of the axle at the bearing, the lever arm of the resistance. The hoist on a derrick, the capstan on a ship, and the dumbwaiter hoist are common examples of this type of machine.

In considering the wheel and axle, the same formulas are used, the radius of the wheel, R, and the radius of the axle, r, being used for power arm and weight arm. Then

$$P : W = r : R$$

and $$P = \frac{W \times r}{R}$$

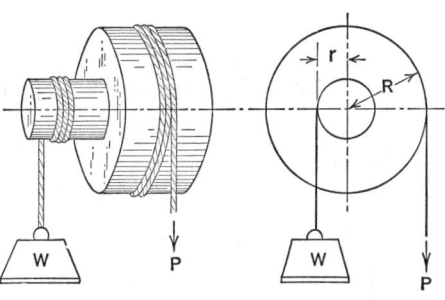

ILLUSTRATION: If the radius of a drum on which is wound the lifting rope of a windlass is 2 inches, find the power that must be exerted at the periphery

FIG. 13.

of a wheel 20 inches in diameter when mounted on the same shaft as the drum and transmitting power to it if 1800 pounds is to be lifted.

$$P = \frac{1800 \times 2}{10} = 360 \text{ lb.} \quad \text{(Ans.)}$$

Pulleys.—A pulley is a wheel mounted to revolve on an axle and has a grooved rim in which a cord, band or chain is passed to transmit the force applied in another direction. A pulley block is a device for holding one or more pulleys as a unit.

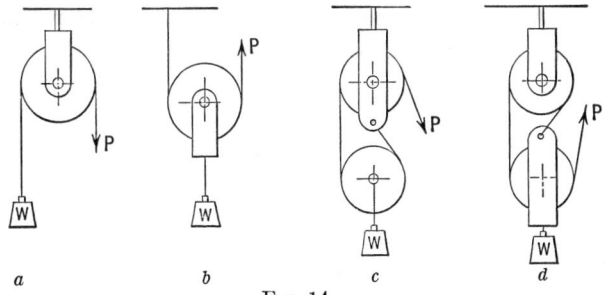

FIG. 14.

Pulleys are either fixed or movable, depending on whether they are held in a fixed position or move with the load. Fig. 14a shows a fixed pulley and Fig. 14b a movable pulley. In the case of the former the only mechanical advantage is the change in the direction of the applied force.

Figure 14c and d shows two combinations of fixed and movable pulleys.   In each of these arrangements, the weight will move through half the distance through which the pulling force acts.

Rule for Pulleys.—The force (P) multiplied by the number of moving strands equals the weight that can be raised.   Stated as a formula this is,

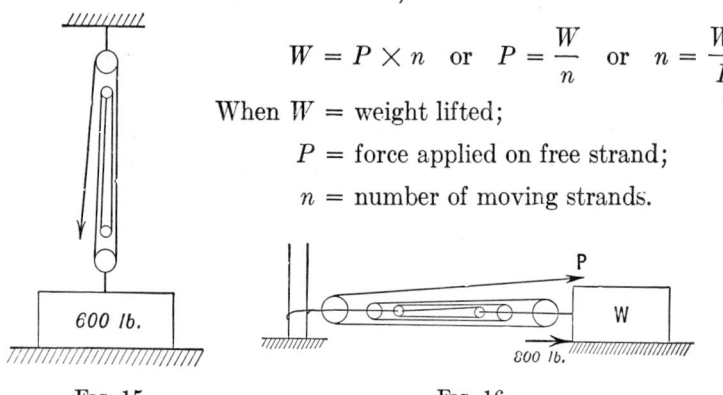

$$W = P \times n \quad \text{or} \quad P = \frac{W}{n} \quad \text{or} \quad n = \frac{W}{P}$$

When W = weight lifted;

P = force applied on free strand;

n = number of moving strands.

| FIG. 15. | FIG. 16. |

ILLUSTRATION: How many moving strands will be required to lift a weight of 600 pounds with a force of 150 pounds?

$$n = \frac{W}{P} = \frac{600}{150} = 4 \text{ moving strands.} \quad \text{(Ans.)}$$

ILLUSTRATION: A weight offers a resistance of 800 pounds to being pulled along a floor.   What force will be required to pull it if a block and tackle with six moving strands is attached?   (Fig. 16).

$$P = \frac{W}{n} = \frac{800}{6} = 133 \text{ lb.} \quad \text{(Ans.)}$$

Differential Pulley.—Figure 17 shows a differential pulley which has great general usefulness. In this device an endless chain sprocketed to the pulley wheels replaces the rope. The two pulleys at the top are slightly different diameters and are attached so that they rotate as a unit.

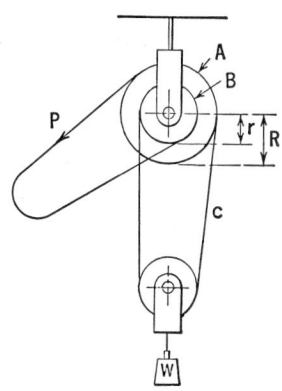

Fig. 17.

When the chain is drawn over the larger pulley it passes around the lower pulley and up over the small wheel from which it is unwound, causing the loop in which the movable pulley rests to shorten by an amount equal to the difference in circumference of the two upper wheels, when they have made one revolution. The weight is moved by an amount equal to one-half this difference.

This may be condensed into a formula as follows:

$$P = \frac{W(R - r)}{2R} \quad \text{or} \quad W = \frac{2PR}{R - r}$$

ILLUSTRATION: A weight of 800 pounds is to be lifted by a differential pulley whose upper wheels are 16 inches and 15 inches in diameter, respectively. What pull or force will be required?

$$P = \frac{W(R - r)}{2R} = \frac{800(\frac{16}{2} - \frac{15}{2})}{2 \times \frac{16}{2}} = \frac{800}{32} = 25 \text{ lb.} \quad \text{(Ans.)}$$

What is the ratio of load to power in this illustration?

$$\text{Ratio} = \frac{\text{Load}}{\text{Force}} = \frac{800}{25} = \frac{32}{1} \quad \text{(Ans.)}$$

Inclined Planes.—An inclined plane is a flat surface sloping or inclined from the horizontal. A body moving up an inclined plane is opposed both by gravity and friction, while one moving down an inclined plane is assisted by gravity and opposed by only friction.

When the force which is being applied is exerted in a direction parallel to the inclined surface as in Fig. 18, it is evident that the power must move through the distance equal to the length of the incline in order to raise the weight through the distance H. The gain in power will then be equal to the length of the incline divided by the height, or

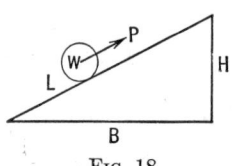

Fig. 18.

$$\frac{P}{W} = \frac{H}{L}$$

from which

$$P = \frac{W \times H}{L}, \quad \text{and} \quad W = \frac{P \times L}{H}.$$

ILLUSTRATION: A roll of paper weighing 500 pounds is to be rolled up onto a 3-foot loading platform by the use of an incline 12 feet long. What force will be required if it acts parallel to the incline? (Fig. 18.)

$$P = \frac{W \times H}{L} = \frac{500 \times 3}{12} = 125 \text{ lb.} \quad \text{(Ans.)}$$

If a force acts along a line parallel to the base as in Fig. 19 then

$$\frac{P}{W} = \frac{H}{B} \quad \text{and} \quad P = \frac{W \times H}{B}, \quad \text{and} \quad W = \frac{P \times B}{H}$$

ILLUSTRATION: What force will be required in the above problem if the force moving the roll of paper acts horizontally? (Fig. 19).

If $L = 12$ and $H = 3$, then by the law of right triangles, B is the square root of the differences of the squares of the hypotenuse and the opposite side, or

$$B = \sqrt{12^2 - 3^2} = \sqrt{144 - 9} = \sqrt{135} = 11.62 \text{ ft.}$$

Then, $P = \dfrac{W \times H}{B} = \dfrac{500 \times 3}{11.62} = 129.1 \text{ lb.} \quad \text{(Ans.)}$

If a force acts at any angle to the plane as X in Fig. 20 and the angle of the incline makes Y degrees with the horizontal, then

$$\frac{P}{W} = \frac{\sin Y}{\cos X}$$

From which

$$P = \frac{W \times \sin Y}{\cos X}, \quad W = \frac{P \times \cos X}{\sin Y} \quad \text{and} \quad \cos X = \frac{W \times \sin Y}{P}$$

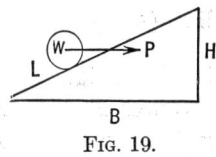

Fig. 19.

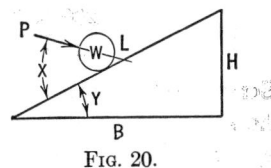

Fig. 20.

ILLUSTRATION: A boiler drum weighing one ton is to be rolled up a 10-degree incline. What force will be required (ignoring friction) if X is 20 degrees?

$$\sin Y = \sin 10° = 0.1736$$
$$\cos X = \cos 20° = 0.9397$$

Then,

$$P = \frac{W \times \sin Y}{\cos X} = \frac{2000 \times 0.1736}{0.9397} = \frac{347.2}{0.9397} = 369.5 \text{ lb.} \quad \text{(Ans.)}$$

Wedges.—A wedge is a pair of inclined planes united at their bases. The power is usually applied by a blow of a heavy body or by pressure. Wedges are used for splitting logs and stones and raising heavy weights short distances. Due to excessive friction, they are not very efficient.

Ignoring friction, the relations of weight and force may be expressed,

$$\frac{P}{W} = \frac{T}{L}, \quad \text{or} \quad P = \frac{W \times T}{L}$$

when, P = power applied;
$\quad W$ = weight or resistance;
$\quad T$ = thickness of wedge at base;
$\quad L$ = length of wedge.

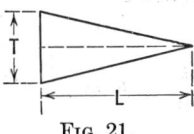

Fig. 21.

This may be expressed in the following forms.

$$W = \frac{P \times L}{T}, \quad L = \frac{W \times T}{P}, \quad \text{and} \quad T = \frac{L \times P}{W}$$

ILLUSTRATION: What force will be required to drive a wedge 4 inches long and $\frac{3}{8}$ inch thick to raise a 200-pound casting?

$$P = \frac{W \times T}{L} = \frac{200 \times \frac{3}{8}}{4}$$

$$P = \frac{200 \times 3}{4 \times 8} = \frac{600}{32} = 19 \text{ lb.} \quad \text{(Ans.)}$$

Screws.—A screw is a modified form of inclined plane. The lead of the screw, or the distance the thread advances in going around once, corresponds to the height of the incline, and the distance around the screw measured on the thread is the length of the incline.

When a force is applied to raise a weight or overcome resistance by means of a screw or nut, either the screw or nut may be fixed, the other being movable. The force is generally applied at the end of a wrench or lever arm or at the circumference of a wheel. The ratio of the power to weight is independent of the diameter of the screw. In actual work, a considerable proportion of the power transmitted is lost through friction.

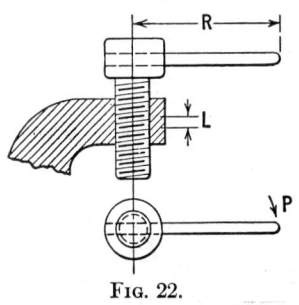

FIG. 22.

Ignoring friction, the force multiplied by the circumference of the circle through which the force arm moves, equals the weight or resulting force multiplied by the lead of the screw. This may be expressed as an equation:

$$\frac{P}{W} = \frac{L}{2\pi R}$$

When P = power applied;

L = lead of screw; in single threads the lead is equal to the pitch, in double threads the lead is twice the pitch, etc.

R = length of bar, wrench, or radius of hand wheel used to operate screw;

W = resulting force or weight moved.

Note. All lengths must be expressed in the same unit and all forces in one unit.

The equation may also be expressed in the following forms:

$$P = \frac{W \times L}{2\pi R}, \quad \text{and} \quad W = \frac{P \times 2\pi R}{L}$$

Illustration: What is the pressure produced in a milling machine vise if the screw has six single threads per inch, the handle a length of 10 inches, if a pressure of 50 pounds is applied and the loss through friction is 40 per cent?

If 40 per cent of the power is lost in friction only 50 − (50 × 0.40) = 30 pounds of pressure remains for useful work.

Since there are six single threads per inch, the lead (L) is $\frac{1}{6}$ inch.

Then, $$W = \frac{P \times 2\pi R}{L} = \frac{30 \times 2 \times 10}{\frac{1}{6}} \times 3.14$$

$W = 300 \times 6 \times 2 \times 3.14 = 3,600 \times 3.14 = 11,310$ lb.   (Ans.)

Mechanical Advantage.—The mechanical advantage of a perfect machine is the number obtained by dividing the resistance by the effort. Expressed as a formula:

$$\text{Mechanical advantage} = \frac{\text{resistance}}{\text{effort}}$$

Many machine problems may be solved by using the principle of mechanical advantage. If a machine has a mechanical advantage of 5, an effort of 20 pounds will lift 5 times as much weight, or 100 pounds.

In the lever the mechanical advantage is found by dividing the length of effort arm by the length of resistance arm; in the wheel and axle by dividing the radius of the wheel by the radius of the axle. The mechanical advantage of a fixed pulley is 1, of a single movable pulley is 2.

Mechanical Efficiency.—The efficiency of a machine is a fraction expressing the ratio of the useful work to the whole work performed which is equal to the energy expended.

$$\text{Efficiency of a machine} = \frac{\text{useful output}}{\text{input}}$$

Efficiency in machines is always expressed by a percent. Thus, if 100 units of work are put into a machine and only 95 units are gotten out, the efficiency of the machine is $\frac{95}{100}$ or 95%.

Friction is the chief cause of the loss of efficiency in most machines.

Pages 228–235 contain tables which give moments of inertia and other properties of different cross sections (of such outlines) frequently met with in structural steel shapes and in cast iron designs.

The Moment of Inertia.—The moment of inertia of any cross-section may be defined as the sum of the products obtained by multiplying each of the elementary areas of which the section is composed by the square of the distance of the center of gravity of the elementary area to the neutral axis of the section. The moment of inertia varies, in the same body, according to the position of the axis. It is the least possible when the axis passes through the center of gravity.

The Section-Modulus or Section-Factor.—The strength of sections to resist strains either as girders or as columns, depends not only on the area but also on the form of the section, and the property which forms the basis of the constants used in the formulas for the strength of girders and columns to express the effect of form, is the moment of inertia about its neutral axis. The modulus of resistance of any section to transverse bending is its moment of inertia divided by the normal distance of the extreme fiber from the neutral axis.

Radius of Gyration.—The effect of the form of the cross-

TABLE 3

COEFFICIENTS OF DEFLECTION OF STEEL BEAMS FOR UNIFORMLY
DISTRIBUTED LOADS

Span in feet	Fiber Stress, pounds per square inch		Span in feet	Fiber Stress, pounds per square inch	
	16,000	12,500		16,000	12,500
1	0.017	0.013	21	7.299	5.703
2	0.066	0.052	22	8.011	6.259
3	0.149	0.166	23	8.756	6.841
4	0.265	0.207	24	9.534	7.448
5	0.414	0.323	25	10.345	8.082
6	0.596	0.466	26	11.189	8.741
7	0.811	0.634	27	12.066	9.427
8	1.059	0.828	28	12.977	10.138
9	1.341	1.047	29	13.920	10.875
10	1.655	1.293	30	14.897	11.638
11	2.003	1.565	31	15.906	12.427
12	2.383	1.862	32	16.949	13.241
13	2.797	2.185	33	18.025	14.082
14	3.244	2.534	34	19.134	14.948
15	3.724	2.909	35	20.276	15.841
16	4.237	3.310	36	21.451	16.759
17	4.783	3.737	37	22.659	17.703
18	5.363	4.190	38	23.901	18.672
19	5.975	4.668	39	25.175	19.668
20	6.621	5.172	40	26.483	20.690

section of a column on its strength is determined by a quantity called the radius of gyration, which is the normal distance from the neutral axis to the center of gyration. The center of gyration is defined as the point where the entire area might be concentrated and have the same moment of inertia as the actual distributed area.

The following notation is used:

A = the area of section in square inches;

d = the depth of cross-section in inches;

I = the moment of inertia in inches;

r = the radius of gyration in inches;

S = the section modulus in inches;

X_{1z} = the distance of the center of gravity of section from extreme fiber in inches.

Deflection of Steel Beams.—To find the deflection in inches of a section symmetrical about the neutral axis, such as the section of an I beam, channel, zee, etc., divide the coefficient in the table corresponding to the given span and fiber-stress by the depth of the section in inches.

ILLUSTRATION: Find the deflection in a 10-inch 25-pound beam of a 10-foot span, under its maximum distributed load of 13 tons, the fiber-stress being taken at 12,500 pounds per square inch.

The table of coefficients, page 225, gives the deflection of a 10-foot span as 1.293 for a fiber stress of 12,500. Therefore, $1.293 \div 10 = 0.1293$ the deflection at the middle.

Another method of calculating the deflection of steel beams is to use Table 3A on page 227. When given the length and depth of a beam, girder, or truss, and the design unit stress; then the corresponding factor from Table 3A, multiplied by the length in feet, will give the center deflection in inches.

For unit stress values not shown in Table 3A, multiply the factor for 10,000 psi by the ratio of the design unit stress to 10,000. For example, if the design unit stress is 13,000 psi, mul-

TABLE 3A*

COEFFICIENTS OF DEFLECTION OF STEEL BEAMS

Ratio of Depth/Span	Maximum Fibre Stress in Lbs. per Sq. In.					
	10,000	12,000	14,000	16,000	18,000	20,000
1/4	.0034	.0041	.0048	.0054	.0061	.0068
1/5	.0043	.0051	.0060	.0068	.0077	.0085
1/6	.0051	.0061	.0072	.0082	.0092	.0102
1/7	.0060	.0072	.0084	.0096	.0107	.0119
1/8	.0068	.0082	.0095	.0109	.0123	.0136
1/9	.0077	.0092	.0107	.0123	.0138	.0153
1/10	.0085	.0102	.0119	.0136	.0153	.0170
1/11	.0094	.0112	.0131	.0150	.0169	.0187
1/12	.0102	.0122	.0143	.0164	.0184	.0204

* Courtesy American Institute of Steel Construction.

tiply the factors listed in the column headed "10,000" by 13,000/10,000 or 1.3.

Table 3A assumes uniformly distributed loading. For a single load at center, multiply these factors by 0.8; for two equal loads at the third points, by 1.02. The factors shown are strictly correct for beams of constant section; close for cover-plated beams and girders; and reasonably approximate for trusses.

The following pages give tables of the moments of inertia and other properties of different cross-section of such outlines as are most frequently met with in structural steel shapes, together with the formulas used to determine bending moments and deflections for steel beams. From these the total safe load may be determined and the proper size beam may be selected from tables in the *Steel Construction Manual* of the American Institute of Steel Construction, New York, N. Y.

TABLE 4

PROPERTIES OF VARIOUS SECTIONS

Sections.	Area of Section. A	Distance from Neutral Axis to Extremities of Section. x and x_1
	a^2	$x_1 = \dfrac{a}{2}$
	a^2	$x_1 = a$
	$a^2 - a_1^2$	$x_1 = \dfrac{a}{2}$
	a^2	$x_1 = \dfrac{a}{\sqrt{2}} = .707a$
	bd	$x_1 = \dfrac{d}{2}$
	bd	$x_1 = d$
	$bd - b_1 d_1$	$x_1 = \dfrac{d}{2}$
	bd	$x_1 = \dfrac{bd}{\sqrt{b^2 + d^2}}$

TABLE 4.—Properties of Various Sections—*Continued*

Moment of Inertia. I	Section Modulus. $S = \dfrac{I}{x_1}.$	Radius of Gyration. $r = \sqrt{\dfrac{I}{A}}.$
$\dfrac{a^4}{12}$	$\dfrac{a^3}{6}$	$\dfrac{a}{\sqrt{12}} = .289a$
$\dfrac{a^4}{3}$	$\dfrac{a^3}{3}$	$\dfrac{a}{\sqrt{3}} = .577a$
$\dfrac{a^4 - a_1^4}{12}$	$\dfrac{a^4 - a_1^4}{6a}$	$\sqrt{\dfrac{a^2 + a_1^2}{12}}$
$\dfrac{a^4}{12}$	$\dfrac{a^3}{6\sqrt{2}} = .118a^3$	$\dfrac{a}{\sqrt{13}} = .289a$
$\dfrac{bd^3}{12}$	$\dfrac{bd^2}{6}$	$\dfrac{d}{\sqrt{12}} = .289d$
$\dfrac{bd^3}{3}$	$\dfrac{bd^2}{3}$	$\dfrac{d}{\sqrt{3}} = .577d$
$\dfrac{bd^3 - b_1 d_1^3}{12}$	$\dfrac{bd^3 - b_1 d_1^3}{6d}$	$\sqrt{\dfrac{bd^3 - b_1 d_1^3}{12(bd - b_1 d_1)}}$
$\dfrac{b^3 d^3}{6(b^2 + d^2)}$	$\dfrac{b^2 d^2}{6\sqrt{b^2 + d^2}}$	$\dfrac{bd}{\sqrt{6(b^2 + d^2)}}$

TABLE 4.—Properties of Various Sections—*Continued*

Sections.	Area of Section. A	Distance from Neutral Axis to Extremities of Section. x and x_1
	bd	$x_1 = \dfrac{d\cos\alpha + b\sin\alpha}{2}$
	$\dfrac{bd}{2}$	$x = \dfrac{d}{3}$ $x_1 = \dfrac{2d}{3}$
	$\dfrac{bd}{2}$	$x_1 = d$
	$\dfrac{\pi d^2}{4} = .785d^2$	$x_1 = \dfrac{d}{2}$
	$\dfrac{\pi(d^2 - d_1^2)}{4} = .785(d^2 - d_1^2)$	$x_1 = \dfrac{d}{2}$
	$\dfrac{\pi d^2}{8} = .393d^2$	$x = \dfrac{2d}{3\pi} = .212d$ $x_1 = \dfrac{(3\pi - 4)d}{6\pi} = .288d$
	$\dfrac{b + b_1}{2} \cdot d$	$x = \dfrac{b + 2b_1}{b + b_1} \cdot \dfrac{d}{3}$ $x_1 = \dfrac{b_1 + 2b}{b + b_1} \cdot \dfrac{d}{3}$

TABLE 4.—Properties of Various Sections—*Continued*

Moment of Inertia. I	Section Modulus. $S = \dfrac{I}{x_1}$.	Radius of Gyration. $r = \sqrt{\dfrac{I}{A}}$
$\dfrac{bd}{12}\,(d^2\cos^2 a + b^2 \sin^2 a)$	$\dfrac{db}{6}\left(\dfrac{d^2\cos^2 a + b^2\sin^2 a}{d\cos a + b\,\sin a}\right)$	$\sqrt{\dfrac{d^2\cos^2 a + b^2\sin^2 a}{12}}$
$\dfrac{bd^3}{36}$	$\dfrac{bd^2}{24}$	$\dfrac{d}{\sqrt{18}} = .236d$
$\dfrac{bd^3}{12}$	$\dfrac{bd^2}{12}$	$\dfrac{d}{\sqrt{6}} = .408d$
$\dfrac{\pi d^4}{64} = .049d^4$	$\dfrac{\pi d^3}{32} = .098d^3$	$\dfrac{d}{4}$
$\dfrac{\pi(d^4 - d_1^4)}{64} = .049\,(d^4 - d_1^4)$	$\dfrac{\pi}{32}\dfrac{(d^4 - d_1^4)}{d} = .098\dfrac{(d^4 - d_1^4)}{d}$	$\dfrac{\sqrt{d^2 + d_1^2}}{4}$
$\dfrac{9\pi^2 - 64}{1152\pi}\cdot d^4 = .007d^4$	$\dfrac{9\pi^2 - 64}{192\,(3\pi - 4)}\cdot d^3 = .024d^3$	$\dfrac{\sqrt{9\pi^2 - 64}}{12\pi}\cdot d = .132d$
$\dfrac{b^2 + 4bb_1 + b_1^2}{36\,(b + b_1)}\cdot d^3$	$\dfrac{b^2 + 4bb_1 + b_1^2}{12\,(b_1 + 2b)}\cdot d^2$	$\dfrac{d}{6(b + b_1)}\sqrt{2(b^2 + 4bb_1 + b_1^2)}$

TABLE 4.—Properties of Various Sections—*Continued*

Sections.	Area of Section. A	Distance from Neutral Axis to Extremities of Section. x and x_1
	$\frac{3}{2}$ d² tan. 30° = .866d²	$x_1 = \frac{d}{2}$
	$\frac{3}{2}$ d² tan. 30° = .866d²	$x_1 = \frac{d}{2 \cos 30^0} = .577d$
	2d² tan. 22½° = .828d²	$x_1 = \frac{d}{2}$
	$\frac{\pi bd}{4}$ = .785 bd	$x_1 = \frac{d}{2}$
	td + 2b' (s + n')	$x_1 = \frac{d}{2}$
	td + 2b' (s + n')	$x_1 = \frac{b}{2}$
	td + b' (s + n')	$x_1 = \frac{d}{2}$
	td + b' (s + n')	$x = [b^2s + \frac{ht^2}{2} + \frac{g}{3}(b-t)^2$ $(b + 2t)] \div A$ $x_1 = b - x$

TABLE 4.—Properties of Various Sections—*Continued*

Moment of Inertia. I	Section Modulus. $S = \dfrac{I}{x_1}$	Radius of Gyration. $r = \sqrt{\dfrac{I}{A}}$
$\dfrac{A}{12}\left[\dfrac{d^2(1+2\cos^2 30°)}{4\cos^2 30°}\right]$ $= .06d^4$	$\dfrac{A}{6}\left[\dfrac{d(1+2\cos^2 30°)}{4\cos^2 30°}\right] = .12d^3$	$\dfrac{d}{4\cos 30°}\sqrt{\dfrac{1+2\cos^2 30°}{3}}$ $= .261d$
$\dfrac{A}{12}\left[\dfrac{d^2(1+2\cos^2 30°)}{4\cos^2 30°}\right]$ $= .06d^4$	$\dfrac{A}{6}\left[\dfrac{d(1+2\cos^2 30°)}{4\cos 30°}\right]$ $= .104d^3$	$\dfrac{d}{4\cos 30°}\sqrt{\dfrac{1+2\cos^2 30°}{3}}$ $= .261d$
$\dfrac{A}{12}\left[\dfrac{d^2(1+2\cos^2 22\frac{1}{2}°)}{4\cos^2 22\frac{1}{2}°}\right]$ $= .055d^4$	$\dfrac{A}{6}\left[\dfrac{d(1+2\cos^2 22\frac{1}{2}°)}{4\cos 22\frac{1}{2}°}\right]$ $= .109d^3$	$\dfrac{d}{4\cos 22\frac{1}{2}°}\sqrt{\dfrac{1+2\cos^2 22\frac{1}{2}°}{3}}$ $= .257d$
$\dfrac{\pi bd^3}{64} = .049bd^3$	$\dfrac{\pi bd^2}{32} = .098bd^2$	$\dfrac{d}{4}$
$\dfrac{1}{12}\left[bd^3 - \dfrac{1}{4g}(h^4 - l^4)\right]$	$\dfrac{2I}{d}$	$r = \sqrt{\dfrac{I}{A}}$
$\dfrac{1}{12}\left[b^3(d-h)+lt^3 + \dfrac{g}{4}(b^4 - t^4)\right]$	$\dfrac{2I}{b}$	$r = \sqrt{\dfrac{I}{A}}$
$\dfrac{1}{12}\left[bd^3 - \dfrac{1}{8g}(h^4 - l^4)\right]$	$\dfrac{2I}{d}$	$r = \sqrt{\dfrac{I}{A}}$
$\dfrac{1}{3}\left[2sb^3 + lt^3 + \dfrac{g}{2}(b^4 - t^4)\right]$ $- Ax^2$	$\dfrac{I}{b-x}$	$r = \sqrt{\dfrac{I}{A}}$

TABLE 4.—Properties of Various Sections—*Continued*

Sections.	Area of Section. A	Distance from Neutral Axis to Extremities of Section. x and x_1
	$bd - h(b - t)$	$x, x_1 = \dfrac{d}{2}$
	$bd - h(b - t)$	$x_1 = \dfrac{b}{2}$
	$bd - h(b - t)$	$x_1 = \dfrac{d}{2}$
	$bd - h(b - t)$	$x = \dfrac{2b^2s + ht^2}{2A}$ $x_1 = b - x$
	$td + s(b - t)$	$x_1 = \dfrac{d}{2}$
	$bs + ht$	$x = \dfrac{d^2t + s^2(b - t)}{2A}$ $x_1 = d - x$
	$bs + ht + b_1s$	$x = \dfrac{td^2 + s^2(b - t) + s(b_1 - t)(2d - s)}{2A}$ $x_1 = d - x$
	$bs + \dfrac{h(t + t_1)}{2}$	$x =$ $\dfrac{3bs^2 + 3th(d + s) + h(t_1 - t)(h + 3s)}{6A}$ $x_1 = d - x$

TABLE 4.—Properties of Various Sections—*Concluded*

Moment of Inertia. I	Section Modulus. $S = \dfrac{I}{x_1}$	Radius of Gyration. $r = \sqrt{\dfrac{I}{A}}$
$\dfrac{bd^3 - h^3(b-t)}{12}$	$\dfrac{bd^3 - h^3(b-t)}{6d}$	$\sqrt{\dfrac{bd^3 - h^3(b-t)}{12\,[bd - h(b-t)]}}$
$\dfrac{2sb^3 + ht^3}{12}$	$\dfrac{2sb^3 + ht^3}{6b}$	$\sqrt{\dfrac{2sb^3 + ht^3}{12\,[bd - h(b-t)]}}$
$\dfrac{bd^3 - h^3(b-t)}{12}$	$\dfrac{bd^3 - h^3(b-t)}{6d}$	$\sqrt{\dfrac{bd^3 - h^3(b-t)}{12\,[bd - h(b-t)]}}$
$\dfrac{2sb^3 + ht^3}{3} - Ax^2$	$\dfrac{I}{b-x}$	$\sqrt{\dfrac{I}{A}}$
$\dfrac{td^3 + s^3(b-t)}{12}$	$\dfrac{td^3 + s^3(b-t)}{6d}$	$\sqrt{\dfrac{td^3 + s^3(b-t)}{12\,[td + s(b-t)]}}$
$\dfrac{tx_1^3 + bx^3 - (b-t)(x-s)^3}{3}$	$\dfrac{I}{d-x}$	$\sqrt{\dfrac{tx_1^3 + bx^3 - (b-t)(x-s)^3}{3\,(bs + ht)}}$
$\dfrac{bx^3 + b_1x_1^3 - (b-t)(x-s)^3}{3} - \dfrac{(b_1-t)(x_1-s)^3}{3}$	$\dfrac{I}{d-x}$	$\left[\dfrac{bx^3 + b_1x_1^3 - (b-t)(x-s)^3}{3\,(bs + ht + b_1s)} - \dfrac{(b_1-t)(x_1-s)^3}{3\,(bs + ht + b_1s)}\right]^{\frac{1}{2}}$
$\dfrac{4bs^3 + h^3(3t + t_1)}{12} - A(x-s)^2$	$\dfrac{I}{d-x}$	$\sqrt{\dfrac{I}{A}}$

TABLE 5

BENDING MOMENTS AND DEFLECTIONS FOR BEAMS OF UNIFORM SECTION

W = Total Load, in lbs., uniformly distributed, including the weight of beam.

W_1 = Total Superimposed or Live Load, in lbs., uniformly distributed.

W_2 = Total Weight of Beam or Dead Load, in lbs., uniformly distributed.

P, P_1, P_2, P_3 = Loads, in lbs., concentrated at any points.

M = Total Bending Moment, in inch-lbs.

M_{w1}, M_p = Bending Moments, in inch-lbs., due to Weights W_1 and P respectively.

I = Moment of Inertia, in inches⁴.

l = Length of Span, in inches.

E = Modulus of Elasticity, in lbs. per square inch = 29 000 000 for steel.

W_s = Total Safe Load, in lbs., uniformly distributed, including weight of beam = Total Safe Load of Tables.

The ordinates in diagrams give the bending moments for corresponding points on beam. For superimposed load only, make W_2 in formulæ equal to zero.

---

(1) Beam Supported at both ends and Uniformly Loaded.

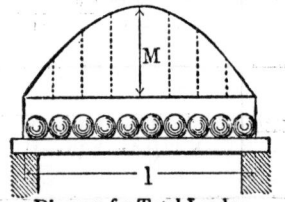

Diagram for Total Load :—
Draw parabola having $M = \dfrac{Wl}{8}$

Safe Superimposed Load, in lbs., uniformly distributed, $W'_s = W_s - W_2$.

Maximum Bending Moment at middle of beam $= M = \dfrac{Wl}{8} = \dfrac{(W_1 + W_2) l}{8}$.

Maximum Shear at points of support $= \dfrac{W}{2} = \dfrac{W_1 + W_2}{2}$.

Maximum Deflection $= \dfrac{5}{384} \dfrac{Wl^3}{EI} = \dfrac{5}{384} \dfrac{(W_1 + W_2) l^3}{EI}$.

---

(2) Beam Supported at both ends with Load Concentrated at the Middle.

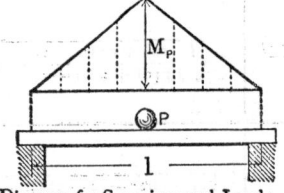

Diagram for Superimposed Load :—
Draw triangle having $M_p = \dfrac{Pl}{4}$

Diagram for Dead Load similar to Case (1)

Safe Superimposed Load, in lbs., concentrated, $P_s = \dfrac{W_s - W_2}{2}$.

Maximum Bending Moment at middle of beam $= M = \dfrac{Pl}{4} + \dfrac{W_2 l}{8}$.

Maximum Shear at points of support $= \dfrac{P + W_2}{2}$.

Max. Deflection $= \dfrac{Pl^3}{48EI} + \dfrac{5}{384} \dfrac{W_2 l^3}{EI}$.

---

(3) Beam fixed at one end, Unsupported at the other and Uniformly Loaded.

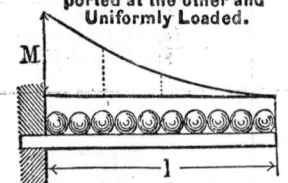

Diagram for Total Load :—
Draw Parabola having $M = \dfrac{Wl}{2}$

Safe Superimposed Load, in lbs., uniformly distributed, $W'_s = \dfrac{W_s}{4} - W_2$.

Maximum Bending Moment at point of support $= \dfrac{Wl}{2} = \dfrac{(W_1 + W_2) l}{2}$.

Maximum Shear at point of support $= W = W_1 + W_2$.

Max. Deflection $= \dfrac{Wl^3}{8EI} = \dfrac{(W_1 + W_2) l^3}{8EI}$.

TABLE 5.—BENDING MOMENTS AND DEFLECTIONS FOR BEAMS OF
UNIFORM SECTION.—*Continued*

W = Total Load, in lbs., uniformly distributed, including the weight of beam.	M = Total Bending Moment, in inch lbs.
W_1 = Total Superimposed or Live Load, in lbs., uniformly distributed.	M_{w1}, M_p = Bending Moments, in inch-lbs., due to Weights W_1 and P respectively.
W_2 = Total Weight of Beam or Dead Load, in lbs., uniformly distributed.	I = Moment of Inertia, in inches⁴. l = Length of Span, in inches. E = Modulus of Elasticity, in lbs. per square inch = 29 000 000 for steel.
P, P_1, P_2, P_3 = Loads, in lbs., concentrated at any points.	W_s = Total Safe Load, in lbs., uniformly distributed, including weight of beam = Total Safe Load of Tables.

The ordinates in diagrams give the bending moments for corresponding points on beam. For superimposed load only, make W_2 in formulæ equal to zero.

(4) Beam fixed at one end, and Unsupported at the other, with Load Concentrated at the free end.

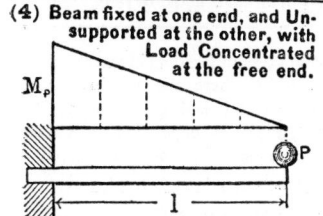

M_p

Diagram for Superimposed Load:—
Draw triangle having M_p = Pl.
Diagram for Dead Load similar to Case(3).

Safe Superimposed Load, in lbs., concentrated, $P_s = \dfrac{W_s - 4W_2}{8}$.

Maximum Bending Moment at point of support $= Pl + \dfrac{W_2 \, l}{2}$.

Maximum Shear at point of support $= P + W_2$.

Maximum Deflection $= \dfrac{Pl^3}{3EI} + \dfrac{W_2 \, l^3}{8EI}$.

(5) Beam Supported at both ends with Load Concentrated at any point.

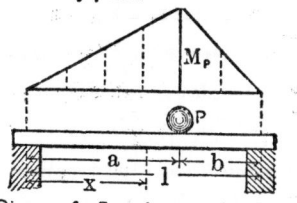

M_p

Diagram for Superimposed Load:—

Draw triangle having $M_p = \dfrac{Pab}{l}$.

Diagram for Dead Load similar to Case (1)

Safe Superimposed Load, in lbs., concentrated, $P_s = \dfrac{W_s \, l^2 - 4a \, W_2 \, (l - a)}{8ab}$.

Maximum Bending Moment under load
$= \dfrac{a \, (2 \, Pb + W_2 l - W_2 a)}{2l}$.

Max. Shear at Sup. near a $= \dfrac{Pb}{l} + \dfrac{W_2}{2}$.

Max. Shear at Sup. near b $= \dfrac{Pa}{l} + \dfrac{W_2}{2}$.

Deflection at distance x from left
support $= \dfrac{1}{3lEI} \left[\dfrac{2al - a^2}{3} \right]^{\frac{3}{2}}$

$\left[Pb + \dfrac{W_2}{8} \left(2l - \sqrt{\dfrac{2al - a^2}{3}} - \dfrac{3l^3}{2al - a^2} \right) \right]$

$x = \sqrt{\dfrac{2al - a^2}{3}}$ = Distance, from left support, of point of maximum deflection for superimposed load.

(6) Beam Supported at both ends with two Symmetrical Loads.

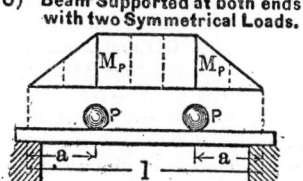

M_p M_p

Diagram for Superimposed Load:—
Draw trapezoid having M_p = Pa.
Diagram for Dead Load similar to Case(1)

Safe Superimposed Load, in lbs., concentrated, each, $P_s = \dfrac{W_s \, l - W_2 l}{8a}$.

Maximum Bending Moment at center of beam $= Pa + \dfrac{W_2 l}{8}$.

Maximum Shear at points of support $= \dfrac{2P + W_2}{2}$.

Maximum Deflection $=$
$\dfrac{Pa}{24EI} \left(3l^2 - 4a^2 \right) + \dfrac{5}{384} \dfrac{W_2 \, l^3}{EI}$.

TABLE 5.—Bending Moments and Deflections for Beams of
Uniform Section.—*Continued*

W = Total Load, in lbs., uniformly distributed, including the weight of beam.	M = Total Bending Moment, in inch-lbs.
W_1 = Total Superimposed or Live Load, in lbs., uniformly distributed.	M_{w1}, M_p = Bending Moments, in inch-lbs., due to Weights W_1 and P respectively.
W_2 = Total Weight of Beam or Dead Load, in lbs., uniformly distributed.	I = Moment of Inertia, in inches⁴.
P, P_1, P_2, P_3 = Loads, in lbs., concentrated at any points.	l = Length of Span, in inches.
	E = Modulus of Elasticity, in lbs., per square inch = 29 000 000 for steel.
	W_s = Total Safe Load, in lbs., uniformly distributed, including the weight of beam = Total Safe Load of Tables.

The ordinates in diagrams give the bending moments for corresponding points on beam. For superimposed load only, make W_2 in formulæ equal to zero.

---

(7) Beam Supported at both ends with Loads Concentrated at various Points.

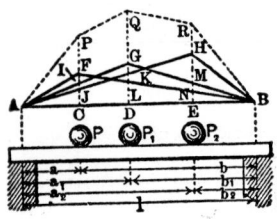

The total bending moment at any point produced by all the weights is equal to the sum of the moments at that point produced by each of the weights separately.

Diagram for Dead Load similar to Case (1)

The Maximum Bending Moment occurs at the point where the vertical shear equals zero and will be at one of the loads P, P_1, or P_2 depending upon their amounts and spacing if W_2 is neglected.

Let R = Reaction at Left Support.

Bending Moment at P =
$$M_p = Ra - \frac{W_2\, a^2}{2l}.$$

Bending Moment at P_1 =
$$M_{p1} = Ra_1 - \left[\frac{W_2\, a_1^2}{2l} + P\,(a_1 - a)\right].$$

Bending Moment at $P_2 = M_{p2} = Ra_2 -$
$$\left[\frac{W_2\, a_2^2}{2l} + P_1\,(a_2 - a_1) + P\,(a_2 - a)\right].$$

Shear or Reaction at Left Support =
$$\frac{P_2\, b_2 + P_1\, b_1 + Pb}{l} + \frac{W_2}{2}.$$

Shear or Reaction at Right Support =
$$\frac{P_2\, a_2 + P_1\, a_1 + Pa}{l} + \frac{W_2}{2}.$$

Diagram for Superimposed Load:—Draw as in Case (5) the Ordinates FC, GD and HE representing the bending moments due to loads P, P_1 and P_2 respectively. Produce FC to P, making PC = FC + IC + JC; GD to Q, making QD = GD + KD + LD; and HE to R, making RE = HE + ME + NE. Join the points A, P, Q, R and B, then the ordinates between A B and polygon A P Q R B will represent the bending moments for corresponding points on beam.

TABLE 5.—BENDING MOMENTS AND DEFLECTIONS FOR BEAMS OF
UNIFORM SECTION.—*Concluded*

W = Total Load, in lbs., uniformly distributed, including the weight of beam.

W_1 = Total Superimposed or Live Load, in lbs., uniformly distributed.

W_2 = Total Weight of Beam or Dead Load, in lbs., uniformly distributed.

P, P_1, P_2, P_3 = Loads, in lbs., concentrated at any points.

M = Total Bending Moment, in inch-lbs.

M_{w1}, M_p = Bending Moments, in inch-lbs., due to Weights W_1 and P respectively.

I = Moment of Inertia, in inches^4.

l = Length of Span, in inches.

E = Modulus of Elasticity, in lbs., per square inch = 29 000 000 for steel.

W_s = Total Safe Load, in lbs., uniformly distributed, including the weight of beam = Total Safe Load of Tables.

The ordinates in diagrams give the bending moments for corresponding points on beam. For superimposed load only, make W_2 in formulæ equal to zero.

(8) Beam Fixed at both ends and Uniformly Loaded.

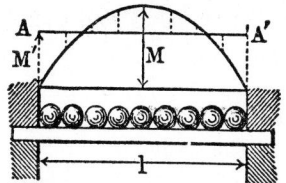

Diagram for Total Load:—Draw parabola having $M = \dfrac{Wl}{8}$. Also A A' parallel to base and at a distance $M' = \dfrac{Wl}{12}$. The Vertical distances between the parabola and line A A' are the moments for corresponding points on beam.

Safe Superimposed Load, in lbs., uniformly distributed, $W'_s = \frac{2}{3} W_s - W_2$.

Distance of points of contra-flexure from supports = $.2113l$.

Maximum Bending Moment at points of support = $\dfrac{Wl}{12} = \dfrac{(W_1 + W_2) l}{12}$.

Bending Moment at middle of beam = $\dfrac{Wl}{24} = \dfrac{(W_1 + W_2) l}{24}$.

Maximum Shear at points of support = $\dfrac{W_1 + W_2}{2}$.

Maximum Deflection $= \dfrac{Wl^3}{384EI} = \dfrac{(W_1 + W_2) l^3}{384EI}$

(9) Beam Fixed at both ends with Load Concentrated at the Middle.

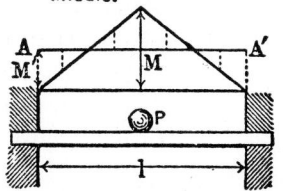

Diagram for Superimposed Load:—Draw triangle having $M = \dfrac{Pl}{4}$. Also A A' parallel to base and at a distance $M' = \dfrac{Pl}{8}$. The Vertical distances between the triangle and line A A' are the moments for corresponding points on beam.

Diagram for Dead Load similar to Case (8)

Safe Superimposed Load, in lbs., concentrated, $P_s = W_s - \frac{2}{3} W_2$.

Distance of points of contra-flexure from supports = $\frac{1}{4}l$.

Maximum Bending Moment at points of support = $\dfrac{Pl}{8} + \dfrac{W_2 l}{12}$.

Bending Moment at middle of beam = $\dfrac{Pl}{8} + \dfrac{W_2 l}{24}$.

Maximum Shear at points of support = $\dfrac{P + W_2}{2}$.

Maximum Deflection = $\dfrac{Pl^3}{192EI} + \dfrac{W_2 l^3}{384EI}$.

VIII

Weights and Measures

In the United States and other English-speaking countries a customary system of weights and measures has been in use for many hundreds of years. Its basis is the yard as a measure of length and the pound as a measure of weight. From these units many other units are derived by the familiar tables shown in the following pages.

The troy pound, which is accepted as standard in the United States, contains 5760 grains, and is the same as the Imperial troy pound of Great Britain. The avoirdupois pound (commercial) of the United States contains 7000 grains, and agrees with the British avoirdupois pound within 0.001 grain.

The U. S. yard differs from the British yard by less than 2 parts per million. However, certain derived measures, such as the liquid and dry measures differ widely. Therefore there is widespread use in both countries of the metric system, which was made a permissive system for use in the U. S. by an Act of Congress of 1866. That Act also established equivalents between the two systems. After some conferences between the U. S., Great Britain, Canada, and Australia, an agreement was reached defining the inch as 25.4 millimeters and the pound as 453.6 grams in all these countries. These equivalents were then established in the U. S. by proclamation, but not by law.

The name units of length and mass in the metric system are the meter and the gram; all larger and smaller units are multiples of ten, their names being formed by using prefixes. Thus deka- is 10 times, hecto- is 100 times, kilo- is 1000 times, deci- is $\frac{1}{10}$, centi- is $\frac{1}{100}$, milli- is $\frac{1}{1000}$, etc. Examples are kilogram and millimeter.

The legal equivalent of the meter as established by Act of Congress is 39.37 inches = 3.28083 feet = 1.093611 yards.

Long Measure—Measures of Length

12 inches (in.)	= 1 foot (ft.)
3 feet	= 1 yard (yd.)
1760 yards, or 5280 feet	= 1 mile (mi.)

Additional measures of length occasionally used are:

1000 mils = 1 inch; 3 inches = 1 palm; 4 inches = 1 hand
9 inches = 1 span; $2\frac{1}{2}$ feet = 1 military space
$5\frac{1}{2}$ yards or $16\frac{1}{2}$ feet = 1 rod; 2 yards = 1 fathom;
a cable length = 120 fathoms = 720 feet;
1 inch = 0.0001157 cable length = 0.013889 fathom = 0.111111 span.

Old Land or Surveyors' Measure*

7.92 inches = 1 link (l.)
100 links, or 66 feet, or 4 rods = 1 chain (ch.)
10 chains or 220 yards = 1 furlong
8 furlongs or 80 chains = 1 mile (mi.)

Nautical Measure

6080.26 feet or 1.15156 statute miles = 1 nautical mile or knot †
3 nautical miles = 1 league
60 nautical miles, or 69.169 statute miles = 1 degree at the equator
360 degrees = circumference of the earth at the equator

* Sometimes called Gunter's Chain.
† The value varies according to different measures of the earth's diameter.

Square Measure—Measures of Surface*

144 square inches (sq. in.) = 1 square foot (sq. ft.)

9 square feet   = 1 square yard (sq. yd.)

$30\frac{1}{4}$ square yards
or
$272\frac{1}{4}$ square feet   } = 1 square rod (sq. rd.)

160 square rods
or
43,560 square feet   } = 1 acre (A.)

640 acres   = 1 square mile (sq. mi.)

Surveyors' Measure

16 square rods   = 1 square chain (sq. ch.)

10 square chains = 1 acre (A.)

640 acres   = 1 square mile (sq. mi.)

1 square mile   = 1 section (sec.)

36 sections   = 1 township (tp.)

Measures used for Diameters and Areas of Electric Wires

Circular inch: a circular inch is the area of a circle 1 inch in diameter.

1 circular inch = 0.7854 square inch

1 square inch   = 1.2732 circular inches

1 circular inch = 1,000,000 circular mils = 1000 MCM

Circular mil: a circular mil is the area of a circle one mil, or 0.001 inch in diameter. In larger cable sizes, 1000 circular mils usually is written 1 MCM.

* Square measures are used in computing area or surfaces, as land, lumber painting, etc.

Solid or Cubic Measure—Measures of Volume*

1728 cubic inches (cu. in.) = 1 cubic foot (cu. ft.)
27 cubic feet   = 1 cubic yard (cu. yd.)

The following measures are also used for wood and masonry.

1 cord of wood   = a pile, 4 × 4 × 8 feet = 128 cubic feet
1 perch of masonry = $16\frac{1}{2}$ × $1\frac{1}{2}$ × 1 foot = $24\frac{3}{4}$ cubic feet

Shipping Measure

Register Ton—For register tonnage or for measuring entire internal capacity of a ship or vessel:

100 cubic feet = 1 register ton

Shipping Ton—For the measurement of cargo.

40 cubic feet = 1 United States shipping ton = 32.143 U. S. bushels

42 cubic feet = 1 British shipping ton = 32.719 imperial bushels.

Carpenter's Rule—To find the weight a vessel will carry multiply the length of keel by the breadth at main beam by the depth of the hold in feet and divide by 95 (the cubic feet allowed for a ton). The result will be the tonnage.

Dry Measure—United States†

2 pints (pt.) = 1 quart (qt.)
8 quarts   = 1 peck (pk.)
4 pecks   = 1 bushel (bu.)

* This table is used in measuring bodies having three dimensions; length, breadth, and height or depth.

† This measure is used in measuring grain, fruit and other articles not liquid. The standard U. S. bushel is the Winchester bushel, which is, in cylinder form $18\frac{1}{2}$ inches in diameter and 8 inches deep and contains 2150.42 cubic inches. A struck bushel contains 2150.42 cubic inches = 1.2445 cubic feet; 1 cubic foot = 0.80356 struck bushel.

The British Imperial bushel = 8 imperial gallons or 2,219.360 cubic inches = 1.2837 cubic feet. The British quarter = 8 imperial bushels.

Liquid Measure

4 gills (gi.)   = 1 pint (pt.)
2 pints   = 1 quart (qt.)
4 quarts   = 1 gallon (gal.) $\begin{cases} \text{U. S. 231 cubic inches} \\ \text{British 277.274 cubic inches} \end{cases}$
1 cubic foot = 7.48 U. S. gallons

Old Liquid Measure

$31\frac{1}{2}$ gallons   = 1 barrel (bbl.)
42 gallons   = 1 tierce
2 barrels or 63 gallons   = 1 hogshead (hhd.)
84 gallons or 2 tierces   = 1 puncheon
2 hogsheads or 4 barrels or 126 gallons = 1 pipe or butt
2 pipes or 3 puncheons   = 1 tun

Apothecaries' Fluid Measure

60 minims = 1 fluid drachm;   8 drachms = 1 fluid ounce

1 U. S. fluid ounce = 8 drachms = 1.805 cubic inch = $\frac{1}{128}$ U. S. gallon. The fluid ounce in Great Britian is 1.732 cubic inches.

Measures of Weight

Avoirdupois or Commercial Weight

16 drachms or 437.5 grains = 1 ounce (oz.)
16 ounces or 7000 grains   = 1 pound (lb.)
2000 pounds   = 1 net or short ton
2240 pounds   = 1 gross or long ton
2204.6 pounds   = 1 metric ton

Measures of weight occasionally used in collecting duties on foreign goods at U. S. custom houses and in freighting coal are:

1 hundredweight = 4 quarters = 112 pounds (1 gross or long ton = 20 hundredweight); 1 quarter = 28 pounds; 1 stone = 14 pounds; 1 quintal = 100 pounds.

Troy Weight *

24 grains	= 1 pennyweight (pwt.)
20 pennyweights	= 1 ounce (oz.)
12 ounces or 5760 grains	= 1 pound (lb.)

A carat of the jewelers, for precious stones = 3.2 grains in the United States. The International carat = 3.168 grains or 200 milligrams. In avoirdupois, apothecaries' and troy weights, the grain is the same, 1 pound troy = 0.82286 pound avoirdupois.

Apothecaries' Weight †

20 grains (gr.)	= 1 scruple ($\ominus$)
3 scruples	= 1 drachm ($\mathfrak{z}$)
8 drachms	= 1 ounce ($\mathfrak{z}$)
12 ounces	= 1 pound troy (lb.)

Measures of Time

one millionth of a second	= 1 microsecond (μsec.)
one thousandth of a second	= 1 millisecond (msec.)
1/3600 hour	= 1 second (sec.)
60 seconds (sec.)	= 1 minute (min.)
60 minutes	= 1 hour (hr.)
24 hours	= 1 day (da.)
7 days	= 1 week (wk.)
365 days	= 1 solar year (yr.)
366 days	= 1 leap-year (every four years)
100 years	= 1 century

By the Gregorian calender every year whose number is divisible by 4 is a leap year except that the centesimal years (each 100 years: 1800, 1900, 2000, etc.) are leap-years only when the number of the year is divisible by 400.

* Used for weighing gold, silver, jewels, etc.

† This table is used in compounding medicines and prescriptions.

A solar day is measured by the rotation of the earth upon its axis, with respect to the sun. In astronomical calculations and in nautical time the day begins at noon.

In civil calculations the day commences at midnight, and is divided into two parts of 12 hours each. A mean lunar month, or lunation of the moon, is 29 days, 12 hours, 44 minutes, 2 seconds, and 5.24 thirds (29.53 days).

In one hour a point on the earth's surface describes $\frac{1}{24}$ of $360° = 15°$, in one minute $\frac{1}{60}$ of $15° = 15'$, and in one second $\frac{1}{60}$ of $15' = 15''$.

In military calculations, the day begins at midnight and the hours are numbered around the clock from 0000 to 2400. Thus 8:46 A.M. is written as 0846 hours, 4:36 P.M. is written as 1636 hours and 10:32 P.M. is written as 2232 hours. The United States Army adopted this 24-hour clock system July 1, 1942. This method of time notation is easily used by noting that all times prior to noon are written directly (8:46 A.M. = 0846 hours), and all times past noon are written by adding 12 to the hour involved (4:36 P.M. = 12 + 436 = 1636 hours). No colon is used in this notation.

Circular and Angular Measures *

60 seconds ($''$)	= 1 minute ($'$)
60 minutes	= 1 degree ($°$)
90 degrees	= 1 quadrant
360 degrees	= 1 circumference

A second is usually sub-divided into tenths and hundreths. A minute of the earth's circumference is a geographical mile.

---

* This table is used for measuring angles and arcs, and for determining latitude and longitude.

TABLE 1

Rectangular Tanks
Capacity in U. S. Gallons Per Foot of Depth

Widths, Feet	Length of Tank—in Feet						
	2	2½	3	3½	4	4½	5
2	29.92	37.40	44.88	52.36	59.84	67.32	74.81
2½	—	46.75	56.10	65.45	74.81	84.16	93.51
3	—	—	67.32	78.55	89.77	101.0	112.2
3½	—	—	—	91.64	104.7	117.8	130.9
4	—	—	—	—	119.7	134.6	149.6
4½	—	—	—	—	—	151.5	168.3
5	—	—	—	—	—	—	187.0

	5½	6	6½	7	7½	8	8½
2	82.29	89.77	97.25	104.7	112.2	119.7	127.2
2½	102.9	112.2	121.6	130.9	140.3	149.6	159.0
3	123.4	134.6	145.9	157.1	168.3	179.5	190.8
3½	144.0	157.1	170.2	183.3	196.4	209.5	222.5
4	164.6	179.5	194.5	209.5	224.4	239.4	254.3
4½	185.1	202.0	218.8	235.6	252.5	269.3	286.1
5	205.7	224.4	243.1	261.8	280.5	299.2	317.9
5½	226.3	246.9	267.4	288.0	308.6	329.1	349.7
6	—	269.3	291.7	314.2	336.6	359.1	381.5
6½	—	—	316.1	340.4	364.7	389.0	413.3
7	—	—	—	366.5	392.7	418.9	445.1
7½	—	—	—	—	420.8	448.8	476.9
8	—	—	—	—	—	478.8	508.7
8½	—	—	—	—	—	—	540.5

	9	9½	10	10½	11	11½	12
2	134.6	142.1	149.6	157.1	164.6	172.1	179.5
2½	168.3	177.7	187.0	196.4	205.7	215.1	224.4
3	202.0	213.2	224.4	235.6	246.9	258.1	269.3
3½	235.6	248.7	261.8	274.9	288.0	301.1	314.2
4	269.3	284.3	299.2	314.2	329.1	344.1	359.1
4½	303.0	319.3	336.6	353.5	370.3	387.1	403.9
5	336.6	355.3	374.0	392.7	411.4	430.1	448.8
5½	370.3	390.9	411.4	432.0	452.6	473.1	493.7
6	403.9	426.4	448.8	471.3	493.7	516.2	538.6
6½	437.6	461.9	486.2	510.5	534.9	559.2	583.5
7	471.3	497.5	523.6	549.8	576.0	602.2	628.4
7½	504.9	533.0	561.0	589.1	617.1	645.2	673.2
8	538.6	568.5	598.4	628.4	658.3	688.2	718.1
8½	572.3	604.1	635.8	667.6	699.4	731.2	763.0
9	605.9	639.6	673.2	706.9	740.6	774.2	807.9
9½	—	675.1	710.6	746.2	781.7	817.2	852.3
10	—	—	748.1	785.5	822.9	860.3	897.7
10½	—	—	—	824.7	864.0	903.3	942.5
11	—	—	—	—	905.1	946.3	987.4
11½	—	—	—	—	—	989.3	1032
12	—	—	—	—	—	—	1077

U. S. Gallon of water weighs 8.34523 Pounds Avoirdupois at 4° C.

TABLE 2

CIRCULAR TANKS

Capacity in U. S. Gallons Per Foot of Depth

Diam., Ft. In.		Gallons	Diam., Ft. In.		Gallons	Diam., Ft. In.		Gallons
1		5.875	3	6	71.97	5	11	205.7
1	1	6.895	3	7	75.44	6		211.5
1	2	7.997	3	8	78.99	6	3	229.5
1	3	9.180	3	9	82.62	6	6	248.2
1	4	10.44	3	10	86.33	6	9	267.7
1	5	11.79	3	11	90.13	7		287.9
1	6	13.22	4		94.00	7	3	308.8
1	7	14.73	4	1	97.96	7	6	330.5
1	8	16.32	4	2	102.0	7	9	352.9
1	9	17.99	4	3	106.1	8		376.0
1	10	19.75	4	4	110.3	8	3	399.9
1	11	21.58	4	5	114.6	8	6	424.5
2		23.50	4	6	119.0	8	9	449.8
2	1	25.50	4	7	123.4	9		475.9
2	2	27.58	4	8	127.9	9	3	502.7
2	3	29.74	4	9	132.6	9	6	530.2
2	4	31.99	4	10	137.3	9	9	558.5
2	5	34.31	4	11	142.0	10		587.5
2	6	36.72	5		146.9	10	3	617.3
2	7	39.21	5	1	151.8	10	6	647.7
2	8	41.78	5	2	156.8	10	9	679.0
2	9	44.43	5	3	161.9	11		710.9
2	10	47.16	5	4	167.1	11	3	743.6
2	11	49.98	5	5	172.4	11	6	777.0
3		52.88	5	6	177.7	11	9	811.1
3	1	55.86	5	7	183.2	12		846.0
3	2	58.92	5	8	188.7	12	3	881.6
3	3	62.06	5	9	194.2	12	6	918.0
3	4	65.28	5	10	199.9	12	9	955.1
3	5	68.58						

U. S. Gallon of water weighs 8.34523 Pounds Avoirdupois at 4° C.

Water Conversion Factors

U. S. gallons	× 8.33	= pounds
U. S. gallons	× 0.13368	= cubic feet
U. S. gallons	× 231	= cubic inches
U. S. gallons	× 0.83	= English gallons
U. S. gallons	× 3.78	= liters
English gallons (Imperial)	× 10	= pounds
English gallons (Imperial)	× 0.16	= cubic feet
English gallons (Imperial)	× 277.274	= cubic inches
English gallons (Imperial)	× 1.2	= U. S. gallons
English gallons (Imperial)	× 4.537	= liters
Cubic inches of water (39.1°) ×	0.036024	= pounds
Cubic inches of water (39.1°) ×	0.004329	= U. S. gallons
Cubic inches of water (39.1°) ×	0.003607	= English gallons
Cubic inches of water (39.1°) ×	0.576384	= ounces
Cubic feet (of water) (39.1°) ×	62.425	= pounds
Cubic feet (of water) (39.1°) ×	7.48	= U. S. gallons
Cubic feet (of water) (39.1°) ×	6.232	= English gallons
Cubic feet (of water) (39.1°) ×	0.028	= tons
Pounds of water	× 27.72	= cubic inches
Pounds of water	× 0.01602	= cubic feet
Pounds of water	× 0.12	= U. S. gallons
Pounds of water	× 0.10	= English gallons

Miscellaneous Tables

Numbers

12 units = 1 dozen
12 dozen = 1 gross
12 gross = 1 great gross
20 units = 1 score

Paper

$$
\begin{array}{rl}
24 \text{ sheets} & = 1 \text{ quire} \\
20 \text{ quires} & = 1 \text{ ream} \\
2 \text{ reams} & = 1 \text{ bundle} \\
5 \text{ bundles} & = 1 \text{ bale}
\end{array}
$$

Books

A book of sheets folded in:

2 leaves is a folio

4 leaves is a quarto

8 leaves is an octavo

12 leaves is a duodecimo

16 leaves is a 16mo.

The Metric System

The metric system is a system of weights and measures based upon a unit called a meter and expressed in the decimal scale. The meter was intended to be one ten millionth of the distance from the equator to either pole, but more careful measurements show that this distance is 10,001,887 meters. The value of the meter, as authorized by the United States Government, is 39.37 inches.

The names of derived metric denominations are formed by prefixing to the name of the primary unit of measure:

$$
\begin{array}{ll}
\text{Micro, a millionth} & = \frac{1}{1,000,000} \\
\text{Milli, a thousandth} & = \frac{1}{1000} \\
\text{Centi, a hundredth} & = \frac{1}{100} \\
\text{Deci, a tenth} & = \frac{1}{10} \\
\text{Deca, ten} & = 10 \\
\text{Hecto, one hundred} & = 100 \\
\text{Kilo, one thousand} & = 1000 \\
\text{Myria, ten thousand} & = 10,000 \\
\text{Mega, one million} & = 1,000,000
\end{array}
$$

The principal units of the metric system are:

> The meter for lengths
> The square meter for surfaces
> The cubic meter for large volumes
> The liter for small volumes
> The gram for weights

Measures of Length

10 millimeters (mm.)	= 1 centimeter (cm.)
10 centimeters	= 1 decimeter (dm.)
10 decimeters	= 1 meter (m.)
10 meters	= 1 decameter (Dm.)
10 decameters	= 1 hectometer (Hm.)
10 hectometers	= 1 kilometer (Km.)
10 kilometers	= 1 myriameter

A meter is used in ordinary measurements; the centimeter or millimeter in calculating very small distances; and the kilometer for long distances.

Square Measures—Measures of Surface

100 square millimeters (mm.^{2})	= 1 square centimeter (cm.^{2})
100 square centimeters	= 1 square decimeter (dm.^{2})
100 square decimeters	= 1 square meter (m.^{2})
100 centiares, or square meters	= 1 are (a.)
100 ares	= 1 hectare (ha.)

The square meter is used for ordinary surfaces; the are, a square, each of whose sides is 10 meters, is the unit of land measure.

Cubic Measure—Measures of Volume

1000 cubic millimeters (mm.^{3})	= 1 cubic centimeter (cm.^{3} or cc.)
1000 cubic centimeters	= 1 cubic decimeter (dm.^{3})
1000 cubic decimeters	= 1 cubic meter (m.^{3})

The term stere is used to designate the cubic meter in measuring wood and timber.   A tenth of a stere is a decistere, and ten steres are a decastere.

Liquid and Dry Measures—Measures of Capacity

$$
\begin{aligned}
10 \text{ milliliters (ml.)} &= 1 \text{ centiliter (cl.)} \\
10 \text{ centiliters} &= 1 \text{ deciliter (dl.)} \\
10 \text{ deciliters} &= 1 \text{ liter (l.)} \\
10 \text{ liters} &= 1 \text{ decaliter (Dl.)} \\
10 \text{ decaliters} &= 1 \text{ hectoliter (Hl.)} \\
10 \text{ hectoliters} &= 1 \text{ kiloliter (Kl.)}
\end{aligned}
$$

The liter, which is a cube each of whose edges is $\frac{1}{10}$ of a meter in length, is the principal unit of measures of capacity.   The hectoliter is the unit that is used in measuring large quantities of grain, fruits, roots, and liquids.

Measures of Weight

$$
\begin{aligned}
10 \text{ milligrams (mg.)} &= 1 \text{ centigram (cg.)} \\
10 \text{ centigrams} &= 1 \text{ decigram (dg.)} \\
10 \text{ decigrams} &= 1 \text{ gram (g.)} \\
10 \text{ grams} &= 1 \text{ decagram (Dg.)} \\
10 \text{ decagrams} &= 1 \text{ hectogram (Hg.)} \\
10 \text{ hectograms} &= 1 \text{ kilogram (Kg.)} \\
1000 \text{ kilograms} &= 1 \text{ (metric) ton (T.)}
\end{aligned}
$$

The gram, which is the primary unit of weights, is the weight of one cubic centimeter of pure distilled water at a temperature of 39.2° F., the kilogram is the weight of 1 liter of water; the ton is the weight of 1 cubic meter of water.   The gram is used in weighing gold, jewels, and small quantities of things.   The kilogram, commonly called kilo for brevity, is used by grocers; the ton is used for weighing heavy articles.

Heat and Power Equivalents

1 Horsepower =
- 746 watts
- 0.746 kilowatt
- 33,000 foot pounds per minute
- 550 foot pounds per second
- 2546.5 heat units per hour
- 42.4 heat units per minute
- 0.707 heat unit per second
- 0.175 pound carbon oxidized per hour
- 2.64 pounds of water evaporated per hour from and at 212° F.

1 Heat unit (British thermal unit) =
- 778 foot pounds
- 1,055 watt second
- 0.000293 kilowatt hour
- 0.000393 horsepower hour
- 0.001036 pound water evaporated from or at 212° F.
- 107.6 kilogram meters

Heat unit per square foot per minute =
- 0.122 watt per square inch
- 0.0176 kilowatt per square foot
- 0.0236 horsepower per square foot

1 Horsepower-hour =
- 0.746 kilowatt hour
- 1,980,000 foot pounds
- 2546.5 heat units
- 2.64 pounds water evaporated from and at 212° F.
- 17.0 pounds water raised from 62° F. to 212° F.

1 Pound of water evaporated from and at 212° F =
- 0.283 kilowatt hour
- 0.379 horsepower hour
- 965.2 heat units
- 1,019,000 joules
- 751,300 foot pounds

Measures of Pressure

1 Pound per square inch =

- 144 pounds per square foot
- 0.068 atmosphere
- 2.042 inches of mercury at 62° F.
- 27.7 inches of water at 62° F.
- 2.31 feet of water at 62° F.

1 Atmosphere =

- 30 inches of mercury at 62° F.
- 14.7 pounds per square inch
- 2116.3 pounds per square foot
- 33.95 feet of water at 62° F.

1 Foot of water at 62° F. =

- 62.355 pounds per square foot
- 0.433 pound per square inch

1 Inch of mercury at 62° F.

- 1.132 foot of water
- 13.58 inches of water
- 0.491 pound per square inch

METRIC AND ENGLISH CONVERSION TABLE

Measures of Length

1 millimeter	= 0.03937 inch
1 centimeter	= 0.3937 inch

1 meter =
- 39.37 inches
- 3.2808 feet
- 1.0936 yards

1 kilometer	= 0.6214 mile

1 inch =
- 25.4 millimeters
- 2.54 centimeters

1 foot =
- 304.8 millimeters
- 0.3048 meter

1 yard	= 0.9144 meter
1 mile	= 1.609 kilometer

Square Measure—Measures of Surface

1 square millimeter	=	0.00155 square inch
1 square centimeter	=	0.155 square inch

1 square meter $= \begin{cases} 10.764 \text{ square feet} \\ 1.196 \text{ square yard} \end{cases}$

1 are $= \begin{cases} 0.0247 \text{ acre} \\ 1076.4 \text{ square feet} \end{cases}$

1 hectare $= \begin{cases} 2.471 \text{ acres} \\ 107,640 \text{ square feet} \end{cases}$

1 square kilometer $= \begin{cases} 0.3861 \text{ square mile} \\ 247.1 \text{ acres} \end{cases}$

1 square inch $= \begin{cases} 6.452 \text{ square centimeters} \\ 645.2 \text{ square millimeters} \end{cases}$

1 square foot $= \begin{cases} 0.0929 \text{ square meter} \\ 9.290 \text{ square centimeters} \end{cases}$

1 square yard $= \quad 0.836$ square meter

1 acre $= \begin{cases} 0.4047 \text{ hectare} \\ 40.47 \text{ ares} \end{cases}$

1 square mile $= \quad 2.5899$ square kilometers

Cubic Measure—Measures of Volume and Capacity

1 cubic centimeter $= \quad 0.061$ cubic inch

1 cubic decimeter $= \begin{cases} 61.023 \text{ cubic inches} \\ 0.0353 \text{ cubic foot} \end{cases}$

1 cubic meter $= \begin{cases} 35.314 \text{ cubic feet} \\ 1.308 \text{ cubic yards} \\ 264.2 \text{ U. S. gallons} \end{cases}$

$$
1 \text{ liter} = \begin{cases}
1 \text{ cubic decimeter} \\
61.023 \text{ cubic inches} \\
0.0353 \text{ cubic foot} \\
1.0567 \text{ U. S. quarts} \\
0.2642 \text{ U. S. gallons} \\
2.202 \text{ lbs. of water at } 62° \text{ F.}
\end{cases}
$$

1 cubic inch   =   16.383 cubic centimeters

$$
1 \text{ cubic foot} = \begin{cases}
0.02832 \text{ cubic meter} \\
28.317 \text{ cubic decimeters} \\
28.317 \text{ liters}
\end{cases}
$$

1 cubic yard	=	0.7645 cubic meter
1 gallon U. S.	=	3.785 liters
1 gallon British	=	4.543 liters

Measures of Weight

$$
1 \text{ gram} = \begin{cases}
0.03216 \text{ ounce troy} \\
0.03527 \text{ ounce avoirdupois} \\
15.432 \text{ grains}
\end{cases}
$$

$$
1 \text{ kilogram} = \begin{cases}
2.2046 \text{ pounds avoirdupois} \\
35.274 \text{ ounces avoirdupois}
\end{cases}
$$

$$
1 \text{ metric ton} = \begin{cases}
0.9842 \text{ ton of 2,240 pounds} \\
19.68 \text{ hundredweight} \\
2204.6 \text{ pounds} \\
1.1023 \text{ tons of 2,000 pounds}
\end{cases}
$$

1 grain	=	0.0648 gram
1 ounce troy	=	31.103 grams
1 ounce avoirdupois	=	28.35 grams

$$
1 \text{ pound} = \begin{cases}
0.4536 \text{ kilogram} \\
453.6 \text{ grams}
\end{cases}
$$

$$
1 \text{ ton of 2240 pounds} = \begin{cases}
1.016 \text{ metric tons} \\
1016 \text{ kilograms}
\end{cases}
$$

TABLE 3

INCHES AND EQUIVALENTS IN MILLIMETERS

Inches	MM	Inches	MM	Inches	MM
1/64	.397	45/64	17.859	26	660.4
1/32	.794	23/32	18.256	27	685.8
3/64	1.191	47/64	18.653	28	711.2
1/16	1.588	3/4	19.050	29	637.6
5/64	1.984	49/64	19.447	30	762.0
3/32	2.381	25/32	19.844	31	787.4
7/64	2.778	51/64	20.241	32	812.8
1/8	3.175	13/16	20.638	33	838.2
9/64	3.572	53/64	21.034	34	863.6
5/32	3.969	27/32	21.431	35	889.0
11/64	4.366	55/64	21.828	36	914.4
3/16	4.763	7/8	22.225	37	939.8
13/64	5.159	57/64	22.622	38	965.2
7/32	5.556	29/32	23.019	39	990.6
15/64	5.953	59/64	23.416	40	1016.0
1/4	6.350	15/16	23.813	41	1041.4
17/64	6.747	61/64	24.209	42	1066.8
9/32	7.144	31/32	24.606	43	1092.2
19/64	7.540	63/64	25.003	44	1117.6
5/16	7.938	1	25.400	45	1143.0
21/64	8.334	2	50.8	46	1168.4
11/32	8.731	3	76.2	47	1193.8
23/64	9.128	4	101.6	48	1219.2
3/8	9.525	5	127.0	49	1244.6
25/64	9.922	6	152.4	50	1270.0
13/32	10.319	7	177.8	51	1295.4
27/64	10.716	8	203.2	52	1320.8
7/16	11.113	9	228.6	53	1346.2
29/64	11.509	10	254.0	54	1371.6
15/32	11.906	11	279.4	55	1397.0
31/64	12.303	12	304.8	56	1422.4
1/2	12.700	13	330.2	57	1447.8
33/64	13.097	14	355.6	58	1473.2
17/32	13.494	15	381.0	59	1498.6
35/64	13.891	16	406.4	60	1524.0
9/16	14.288	17	431.8	61	1549.4
37/64	14.684	18	457.2	62	1574.8
19/32	15.081	19	482.6	63	1600.2
39/64	15.478	20	508.0	64	1625.6
5/8	15.875	21	533.4	65	1651.0
41/64	16.272	22	558.8	66	1676.4
21/32	16.669	23	584.2	67	1701.8
43/64	17.066	24	609.6	68	1727.2
11/16	17.463	25	635.0	69	1752.6

3.—Inches and Equivalents in Millimeters—*Continued*

Inches	MM	Inches	MM	Inches	MM
70	1778.0	114	2895.6	158	4013.2
71	1803.4	115	2921.0	159	4038.6
72	1828.8	116	2946.4	160	4064.0
73	1854.2	117	2971.8	161	4089.4
74	1879.6	118	2997.2	162	4114.8
75	1905.0	119	3022.6	163	4140.2
76	1930.4	120	3048.0	164	4165.6
77	1955.8	121	3073.4	165	4191.0
78	1981.2	122	3098.8	166	4216.4
79	2006.6	123	3124.2	167	4241.8
80	2032.0	124	3149.6	168	4267.2
31	2057.4	125	3175.0	169	4292.6
32	2082.8	126	3200.4	170	4318.0
83	2108.2	127	3225.8	171	4343.4
84	2133.6	128	3251.2	172	4368.8
85	2159.0	129	3276.6	173	4394.2
86	2184.4	130	3302.0	174	4419.6
87	2209.8	131	3327.4	175	4445.0
88	2235.2	132	3352.8	176	4470.4
89	2260.6	133	3378.2	177	4495.8
90	2286.0	134	3403.6	178	4521.2
91	2311.4	135	3429.0	179	4546.6
92	2336.8	136	3454.4	180	4572.0
93	2362.2	137	3479.8	181	4597.4
94	2387.6	138	3505.2	182	4622.8
95	2413.0	139	3530.6	183	4648.2
96	2438.4	140	3556.0	184	4673.6
97	2463.8	141	3581.4	185	4699.0
98	2489.2	142	3606.8	186	4724.4
99	2514.6	143	3632.2	187	4749.8
100	2540.0	144	3657.6	188	4775.2
101	2565.4	145	3683.0	189	4800.6
102	2590.8	146	3708.4	190	4826.0
103	2616.2	147	3733.8	191	4851.4
104	2641.6	148	3759.2	192	4876.8
105	2667.0	149	3784.6	193	4902.2
106	2692.4	150	3810.0	194	4927.6
107	2717.8	151	3835.4	195	4953.0
108	2743.2	152	3860.8	196	4978.4
109	2768.6	153	3886.2	197	5003.8
110	2794.0	154	3911.6	198	5029.2
111	2819.4	155	3937.0	199	5054.6
112	2844.8	156	3962.4	200	5080.0
113	2870.2	157	3987.8		

TABLE 4

MILLIMETERS AND EQUIVALENTS IN INCHES

MM	Inches	MM	Inches	MM	Inches
1/100	.0004	45/100	.0177	89/100	.0350
2/100	.0008	46/100	.0181	90/100	.0354
3/100	.0012	47/100	.0185	91/100	.0358
4/100	.0016	48/100	.0189	92/100	.0362
5/100	.0020	49/100	.0193	93/100	.0366
6/100	.0024	50/100	.0197	94/100	.0370
7/100	.0028	51/100	.0201	95/100	.0374
8/100	.0031	52/100	.0205	96/100	.0378
9/100	.0035	53/100	.0209	97/100	.0382
10/100	.0039	54/100	.0213	98/100	.0386
11/100	.0043	55/100	.0217	99/100	.0390
12/100	.0047	56/100	.0221	1	.0394
13/100	.0051	57/100	.0224	2	.0787
14/100	.0055	58/100	.0228	3	.1181
15/100	.0059	59/100	.0232	4	.1575
16/100	.0063	60/100	.0236	5	.1969
17/100	.0067	61/100	.0240	6	.2362
18/100	.0071	62/100	.0244	7	.2756
19/100	.0075	63/100	.0248	8	.3150
20/100	.0079	64/100	.0252	9	.3543
21/100	.0083	65/100	.0256	10	.3937
22/100	.0087	66/100	.0260	11	.4331
23/100	.0091	67/100	.0264	12	.4724
24/100	.0094	68/100	.0268	13	.5118
25/100	.0098	69/100	.0272	14	.5512
26/100	.0102	70/100	.0276	15	.5906
27/100	.0106	71/100	.0280	16	.6299
28/100	.0110	72/100	.0284	17	.6693
29/100	.0114	73/100	.0287	18	.7087
30/100	.0118	74/100	.0291	19	.7480
31/100	.0122	75/100	.0295	20	.7874
32/100	.0126	76/100	.0299	21	.8268
33/100	.0130	77/100	.0303	22	.8661
34/100	.0134	78/100	.0307	23	.9055
35/100	.0138	79/100	.0311	24	.9449
36/100	.0142	80/100	.0315	25	.9843
37/100	.0146	81/100	.0319	26	1.0236
38/100	.0150	82/100	.0323	27	1.0630
39/100	.0154	83/100	.0327	28	1.1024
40/100	.0158	84/100	.0331	29	1.1417
41/100	.0161	85/100	.0335	30	1.1811
42/100	.0165	86/100	.0339	31	1.2205
43/100	.0169	87/100	.0343	32	1.2598
44/100	.0173	88/100	.0347	33	1.2992

4.—Millimeters and Equivalents in Inches—*Continued*

MM	Inches	MM	Inches	MM	Inches
34	1.3386	78	3.0709	122	4.8031
35	1.3780	79	3.1102	123	4.8425
36	1.4173	80	3.1496	124	4.8819
37	1.4567	81	3.1890	125	4.9213
38	1.4961	82	3.2283	126	4.9606
39	1.5354	83	3.2677	127	5.0000
40	1.5748	84	3.3071	128	5.0394
41	1.6142	85	3.3465	129	5.0787
42	1.6535	86	3.3858	130	5.1181
43	1.6929	87	3.4252	131	5.1575
44	1.7323	88	3.4646	132	5.1968
45	1.7717	89	3.5039	133	5.2362
46	1.8110	90	3.5433	134	5.2756
47	1.8504	91	3.5827	135	5.3150
48	1.8898	92	3.6220	136	5.3543
49	1.9291	93	3.6614	137	5.3937
50	1.9685	94	3.7008	138	5.4331
51	2.0079	95	3.7402	139	5.4724
52	2.0472	96	3.7795	140	5.5118
53	2.0866	97	3.8189	141	5.5512
54	2.1260	98	3.8583	142	5.5905
55	2.1654	99	3.8976	143	5.6299
56	2.2047	100	3.9370	144	5.6693
57	2.2441	101	3.9764	145	5.7087
58	2.2835	102	4.0157	146	5.7480
59	2.3228	103	4.0551	147	5.7874
60	2.3622	104	4.0945	148	5.8268
61	2.4016	105	4.1339	149	5.8661
62	2.4409	106	4.1732	150	5.9055
63	2.4803	107	4.2126	151	5.9449
64	2.5197	108	4.2520	152	5.9842
65	2.5591	109	4.2913	153	6.0236
66	2.5984	110	4.3307	154	6.0630
67	2.6378	111	4.3701	155	6.1024
68	2.6772	112	4.4094	156	6.1417
69	2.7165	113	4.4488	157	6.1811
70	2.7559	114	4.4882	158	6.2205
71	2.7953	115	4.5276	159	6.2598
72	2.8346	116	4.5669	160	6.2992
73	2.8740	117	4.6063	161	6.3386
74	2.9134	118	4.6457	162	6.3779
75	2.9528	119	4.6850	163	6.4173
76	2.9921	120	4.7244	164	6.4567
77	3.0315	121	4.7638	165	6.4961

4.—Millimeters and Equivalents in Inches—*Concluded*

MM	Inches	MM	Inches	MM	Inches
166	6.5354	211	8.3071	256	10.079
167	6.5748	212	8.3464	257	10.118
168	6.6142	213	8.3858	258	10.157
169	6.6535	214	8.4252	259	10.197
170	6.6929	215	8.4646	260	10.236
171	6.7323	216	8.5039	261	10.276
172	6.7716	217	8.5433	262	10.315
173	6.8110	218	8.5827	263	10.354
174	6.8504	219	8.6220	264	10.394
175	6.8898	220	8.6614	265	10.433
176	6.9291	221	8.7008	266	10.472
177	6.9685	222	8.7401	267	10.512
178	7.0079	223	8.7795	268	10.551
179	7.0472	224	8.8189	269	10.591
180	7.0866	225	8.8583	270	10.630
181	7.1260	226	8.8976	271	10.669
182	7.1653	227	8.9370	272	10.709
183	7.2047	228	8.9764	273	10.748
184	7.2441	229	9.0157	274	10.787
185	7.2835	230	9.0551	275	10.827
186	7.3228	231	9.0945	276	10.866
187	7.3622	232	9.1338	277	10.905
188	7.4016	233	9.1732	278	10.945
189	7.4409	234	9.2126	279	10.984
190	7.4803	235	9.2520	280	11.024
191	7.5197	236	9.2913	281	11.063
192	7.5590	237	9.3307	282	11.102
193	7.5984	238	9.3701	283	11.142
194	7.6378	239	9.4094	284	11.181
195	7.6772	240	9.4488	285	11.220
196	7.7165	241	9.4882	286	11.260
197	7.7559	242	9.5275	287	11.299
198	7.7953	243	9.5669	288	11.339
199	7.8346	244	9.6063	289	11.378
200	7.8740	245	9.6457	290	11.417
201	7.9134	246	9.6850	291	11.457
202	7.9527	247	9.7244	292	11.496
203	7.9921	248	9.7638	293	11.535
204	8.0315	249	9.8031	294	11.575
205	8.0709	250	9.8425	295	11.614
206	8.1102	251	9.8819	296	11.654
207	8.1496	252	9.9212	297	11.693
208	8.1890	253	9.9606	298	11.732
209	8.2283	254	10.000	299	11.772
210	8.2677	255	10.039		

Useful Factors, English Measures

Inches............	×	0.08333	= feet
"	×	0.02778	= yards
"	×	0.00001578	= miles
Square inches......	×	0.00695	= square feet
" "	×	0.0007716	= square yards
Cubic inches.......	×	0.00058	= cubic feet
" "	×	0.0000214	= cubic yards
" "	×	0.004329	= U. S. gallons
Feet.............	×	0.3334	= yards
"	×	0.00019	= miles
Square feet........	×	144.0	= square inches
" "	×	0.1112	= square yards
Cubic feet.........	×	1,728	= cubic inches
" "	×	0.03704	= cubic yards
" "	×	7.48	= U. S. gallons
Yards.............	×	36	= inches
"	×	3	= feet
"	×	0.0005681	= miles
Square yards.......	×	1,296	= square inches
" "	×	9	= square feet
Cubic yards.......	×	46,656	= cubic inches
" "	×	27	= cubic feet
Miles.............	×	63,360	= inches
"	×	5,280	= feet
"	×	1,760	= yards
Avoirdupois ounces.	×	0.0625	= pounds
" " .	×	0.00003125	= tons
" pounds.	×	16	= ounces
" " .	×	.001	= hundredweight
" " .	×	.0005	= tons
" " :	×	27.681	= cubic inches of water at 39.2° F
" tons...	×	32,000	= ounces
" " ...	×	2,000	= pounds
Watts............	×	0.00134	= horse power
Horse power.......	×	746	= watts

Weight of round iron per foot = square of diameter in quarter inches ÷ 6.
Weight of flat iron per foot = width × thickness× 19⅔.
Weight of flat plates per square foot = 5 pounds for each ⅛ inch thickness.

Millimeters × 0.03937	= inches
Millimeters ÷ 25.4	= inches
Centimeters × 0.3937	= inches
Centimeters ÷ 2.54	= inches
Meters × 39.37	= inches
Meters × 3.281	= feet
Meters × 1.094	= yards
Kilometers × 0.621	= miles
Kilometers ÷ 1.6093	= miles
Kilometers × 3280.7	= feet
Square millimeters × 0.0155	= square inches
Square millimeters ÷ 645.1	= square inches
Square centimeters × 0.155	= square inches
Square centimeters ÷ 6.451	= square inches
Square meters × 10.764	= square feet
Square kilometers × 247.1	= acres
Hectares × 2.471	= acres
Cubic centimeters ÷ 16.385	= cubic inches
Cubic centimeters ÷ 3.69	= fluid drachms, U. S. Pharmacopœia
Cubic centimeters ÷ 29.57	= fluid ounce U. S. Pharmacopœia
Cubic meters × 35.315	= cubic feet
Cubic meters × 1.038	= cubic yards
Cubic meters × 264.2	= gallons, United States
Liters × 61.022	= cubic inches
Liters × 33.84	= fluid ounces
Liters × 0.2642	= gallons, United States

Liters ÷ 3.78	= gallons, United States
Liters ÷ 28.316	= cubic feet
Hectoliters × 3.531	= cubic feet
Hectoliters × 2.84	= bushels, United States
Hectoliters × 0.131	= cubic yards
Hectoliters × 26.42	= gallons, United States
Grams × 15.432	= grains
Grams (water) ÷ 29.57	= fluid ounces
Grams ÷ 28.35	= ounces, avoirdupois
Kilograms × 2.2046	= pounds
Kilograms × 35.3	= ounces, avoirdupois
Kilograms ÷ 1102.3	= tons, 2000 pounds

Specific Gravity

The relative heaviness of substances is of much practical importance to the industrial world.   In the metal industry research workers are constantly seeking for relatively light materials that possess great strength.

Weight measures the earth's pull upon body, and depends upon the body's mass.   But substances which are equal in volume vary in heaviness.   Thus, it is evident that the pull of gravity is stronger on some substances than on others.   As the weight of a body is the measure of the pull between all bodies and the earth, or gravity, the specific gravity of a substance is found by comparing the weight of a certain volume of that substance with the weight of an equal volume of another substance taken as a standard.

The specific gravity of a substance is its weight as compared with the weight of an equal bulk of pure water.

RULE.—To calculate the specific gravity of a substance, find the weight of the body in air and divide by the difference of the weight of the body in air and the weight of the body submerged in water.

Expressed as a formula:

$$\text{Specific gravity} = \frac{W}{W - w}$$

where W = weight of body in air
w = weight of body submerged in water

ILLUSTRATION: Find the specific gravity of a lump of coal that weighs 150 grams in air and 60 grams immersed in water.

$$\text{Specific gravity} = \frac{W}{W - w}$$

$$= \frac{150}{150 - 60}$$

$$= \frac{150}{90} = 1.66$$

Specific gravity determinations are usually referred to the standard of the weight of water at 62° F., 62.355 pounds per cubic feet. The formula becomes:

$$\text{Specific gravity} = \frac{\text{weight of solid}}{\text{weight of equal volume of water}}$$

ILLUSTRATION: Find the specific gravity of a cube of steel 1 foot on a side and weighing 489.6 pounds per cubic foot.

$$\text{Specific gravity} = \frac{\text{weight of solid}}{\text{weight of equal volume of water}}$$

$$= \frac{489.6}{62.355} = 7.85$$

The following tables give the specific gravities and weights of various substances.

TABLE 5

SPECIFIC GRAVITIES AND WEIGHTS OF VARIOUS SUBSTANCES

The Basis for Specific Gravities is Pure Water at 62 Degrees Fah., Barometer 30 Inches. Weight of One Cubic Foot, 62.355 Pounds.	Average Specific Gravity. Water = 1.	Average Weight of One Cubic Foot. Pounds.
Air, atmospheric at 60 degrees F., under pressure of one atmosphere, or 14.7 pounds per square inch, weighs $\frac{1}{815}$th as much as water	.00123	.0765
Aluminum..............................	2.6	162
Anthracite, 1.3 to 1.84; of Penna., 1.3 to 1.7.	1.5	93.5
" broken, of any size, loose	....	52 to 56
" " moderately shaken	....	56 to 60
" " heaped bushel, loose, 77 to 83 pounds.........	....	....
" " a ton loose occupies 40 to 43 cubic feet	....	....
Antimony, cast..........................	6.70	418
" native......................	6.67	416
Ash, perfectly dry	.752	47
" American White, dry	.61	38
Ashes of soft coal, solidly packed		40 to 45
Asphaltum, 1 to 1.8	1.4	87.3
Brass (copper and zinc), cast, 7.8 to 8.4.....	8.1	504
" rolled	8.4	524
Brick, best pressed	....	150
" common and hard...................	....	125
" soft inferior	....	100
Brickwork, pressed brick, fine joints.......	....	140
" medium quality.................	....	125
" coarse, inferior, soft...........	....	100
" at 125 pounds per cubic foot, 1 cubic yard equals 1.507 tons, and 17.92 cubic feet equal 1 ton....	....	....
Bronze, copper 8, tin 1 (gun metal)........	8.5	529
Cement, hydraulic. American, Rosendale, ground and loose..............	....	56
" hydraulic. American, Rosendale, U. S. struck bush., 70 pounds ...	....	....
" hydraulic. American, Rosendale, Louisville bushel, 62 pounds	....	....
" hydraulic. American, Cumberland, ground, loose	....	65
" hydraulic. American, Cumberland, ground, thoroughly shaken......	....	85
" hydraulic. English Portland (U.S. struck bushel, 100 to 128)	....	81 to 102

TABLE 5.—Specific Gravities and Weights of Various Substances—
Continued

The Basis for Specific Gravities is Pure Water at 62 Degrees Fah., Barometer 30 Inches. Weight of One Cubic Foot, 62.355 Pounds.	Average Specific Gravity. Water = 1.	Average Weight of One Cubic Foot. Pounds.
Cement, hydraulic. English Portland, a barrel, 400 to 430 pounds	....	....
" hydraulic. American Portland, loose	....	88
" hydraulic. American Portland, thoroughly shaken...............	....	110
Charcoal of pines and oaks	....	15 to 30
Chalk	2.5	156
Cherry, perfectly dry....................	.672	42
Clay, potters', dry, 1.8 to 2.1	1.9	119
" dry in lump, loose.................	....	63
Coal, bituminous, solid, 1.2 to 1.5..........	1.35	84
" bituminous, solid, Cambria Co., Pa., 1.27–1.34	....	79 to 84
" bituminous, broken, of any size, loose..	....	47 to 52
" bituminous, moderately shaken.......	....	51 to 56
" bituminous, a heaped bushel, loose, 70 to 78...................	....	....
" bituminous, 1 ton occupies 43 to 48 cubic feet	....	....
Coke, loose, good quality...	....	23 to 32
" loose, a heaped bushel, 35 to 42......	....	....
" 1 ton occupies 80 to 97 cubic feet	....	....
Corundum, pure, 3.8 to 4	3.9	....
Copper, cast, 8.6 to 8.8	8.7	542
" rolled, 8.8 to 9..................	8.9	555
Cork, dry	.24	15
Earth, common loam, perfectly dry, loose...	....	72 to 80
" " " perfectly dry, shaken..	....	82 to 92
" " " perfectly dry, rammed.	....	90 to 100
" " " slightly moist, loose ..	....	70 to 76
" " " more moist, loose	....	66 to 68
" " " more moist, shaken...	....	75 to 90
" " " more moist, packed...	....	90 to 100
" " " as soft flowing mud...	....	104 to 112
" " " as soft flowing mud well pressed......	....	110 to 120
Elm, perfectly dry	.56	35
Flint	2.6	162
Glass, 2.5 to 3.45	2.98	186
" common window	2.52	157
Gneiss, common, 2.62 to 2.76	2.69	168

TABLE 5.—Specific Gravities and Weights of Various Substances—
Continued

The Basis for Specific Gravities is Pure Water at 62 Degrees Fah., Barometer 30 Inches. Weight of One Cubic Foot, 62.355 Pounds.	Average Specific Gravity. Water = 1.	Average Weight of One Cubic Foot. Pounds.
Gneiss, in loose piles	....	96
Gold, cast, pure or 24 karat...............	19.258	1204
" pure, hammered	19.5	1217
Granite, 2.56 to 2.88......................	2.72	170
Greenstone, trap, 2.8 to 3.2...............	3.00	187
Gypsum, plaster of Paris; 2.24 to 2.30	2.27	141.6
Hickory, perfectly dry	.85	53
Ice, .917 to .922	.92	57.4
Iron, cast, 6.9 to 7.4	7.15	446
" grey foundry, cold.................	7.21	450
" " molten	6.94	433
" wrought	7.69	480
Lead, commercial	11.38	709.6
Lignumvitæ (dry)......................	.65–1.33	41 to 83
Limestone and marble	2.6	164.4
Lime, quick	1.5	95
" quick, ground, well shaken, per struck bushel 80 pounds.................	....	64
" quick, ground, thoroughly shaken, per struck bushel 93¾ pounds	....	75
Locust, dry	.71	44
Mahogany, Spanish, dry	.85	53
" Honduras, dry	.56	35
Maple, dry	.79	49
Marble (see Limestone).		
Masonry of granite or limestone, well-dressed	....	165
" of granite, well-scabbled mortar rubble, about ⅕ of mass will be mortar	....	154
" of granite, well-scabbled dry rubble	....	138
" of granite, roughly scabbled mortar rubble, about ¼ to ⅓ of mass will be mortar	....	150
" of granite, scabbled dry rubble....	....	125
" of sandstone, ⅛ less than granite..	....	....
Masonry of brickwork		
Mercury, at 32 degrees Fah	13.62	849
Mica, 2.75 to 3.1......................	2.93	183
Mortar, hardened, 1.4 to 1.9...............	1.65	103
Mud, dry, close	....	80 to 110
" wet, moderately pressed.............	....	110 to 130
" " fluid......................	....	104 to 120

TABLE 5.—Specific Gravities and Weights of Various Substances—
Concluded

The Basis for Specific Gravities is Pure Water at 62 Degrees Fah., Barometer 30 Inches. Weight of One Cubic Foot, 62.355 Pounds.	Average Specific Gravity. Water = 1.	Average Weight of One Cubic Foot. Pounds.
Oak, live, perfectly dry, .88–1.02 (see note below)	.95	59.3
" Red, Black, perfectly dry	....	32 to 45
Petroleum	.878	54.8
Pitch	1.15	71.7
Poplar, dry (see note below)	.47	29
Platinum	21.5	1342
Quartz	2.65	165
Rosin	1.10	68.6
Salt, coarse, (per struck bushel, Syracuse, N. Y., 56 pounds)	....	45
Sand, of pure quartz, perfectly dry and loose	....	90 to 106
" " " voids full of water	....	118 to 129
" " " very large and small grains, dry	....	117
Sandstone, 2.1 to 2.73, 131 to 171	2.41	151
" quarried and piled, 1 measure solid makes 1¾ (about) piled.	....	86
Snow, fresh fallen	....	5 to 12
" moistened, compacted by rain	....	15 to 50
Sycamore, perfectly dry (see note below)	.59	37
Shales, red or black, 2.4 to 2.8	2.6	162
Silver	10.5	655
Slate, 2.7 to 2.9	2.8	175
Soapstone, 2.65 to 2.8	2.73	170
Steel	7.85	490
Sulphur	2.00	125
Tallow	.94	58.6
Tar	1	62.355
Tin, cast, 7.2 to 7.5	7.35	459
Walnut, Black, perfectly dry (see note below)	.61	38
Water, pure rain, distilled, at 32 degrees F., Bar. 30 inches.	....	62.417
" " " at 62 degrees F., Bar. 30 inches.	1	62.355
" " " at 212 degrees F., Bar. 30 inches	....	59.7
" sea, 1.026 to 1.030	1.028	64.08
Zinc or spelter, 6.8 to 7.2	7.00	437.5

Note.—Green timbers usually weigh from one-fifth to nearly one-half more than dry; ordinary building timbers, tolerably seasoned, one-sixth more.

When the specific gravity of a substance is known the weight per cubic foot of the substance can be found by multiplying the specific gravity by 62.355; the weight of one cubic inch by multiplying the specific gravity by 0.0361 the weight of one cubic inch of pure water at 62° F.

ILLUSTRATION: From the table, page 268, the specific gravity of cast iron is given as 7.2. Find the weight of 6 cubic inches of cast iron.

$$7.2 \times 0.0361 \times 6 = 1.5586 \text{ pounds}$$

If the weight per cubic foot of a substance is known, the specific gravity can be calculated by multiplying this weight by 0.01604.

ILLUSTRATION: Find the specific gravity of a cubic foot of cast tin that weighs 455 pounds.

$$455 \times 0.01604 = 7.29$$

Specific Gravity of Liquids. The specific gravity of liquids is the number which indicates how much a certain volume of the liquid weighs compared with an equal volume of water.

TABLE 6

Specific Gravity of Liquids

Liquid	Sp. Gr.	Liquid	Sp. Gr.	Liquid	Sp. Gr.
Acetic acid...........	1.06	Fluoric acid....	1.50	Petroleum oil...	0.82
Alcohol, commerical...	0.83	Gasoline......	0.70	Phosphoric acid.	1.78
Alcohol, pure.........	0.79	Kerosene......	0.80	Rape oil.......	0.92
Ammonia.............	0.89	Linseed oil....	0.94	Sulphuric acid...	1.84
Benzine..............	0.69	Mineral oil....	0.92	Tar............	1.00
Bromine.............	2.97	Muriatic acid..	1.20	Turpentine oil...	0.87
Carbolic acid.........	0.96	Naphtha......	0.76	Vinegar........	1.08
Carbon disulphide.....	1.26	Nitric acid.....	1.22	Water..........	1.00
Cotton-seed oil........	0.93	Olive oil.......	0.92	Water, sea......	1.03
Ether, sulphuric.......	0.72	Palm oil.......	0.97	Whale oil.......	0.92

There are three methods of determining the specific gravity of liquids:

(1) Hydrometer method, in which the specific gravity of the liquid tested is read as the scale division marking the liquid level on the stem of the hydrometer.

(2) Bottle method, in which the specific gravity

$$= \frac{\text{weight of liquid in a full bottle}}{\text{weight of water in a full bottle}}$$

(3) Displacement method in which specific gravity

$$= \frac{\text{weight of liquid displaced by a body}}{\text{weight of equal volume of water displaced by the body}}$$

Specific Gravity of Gases.—The specific gravity of gases is the number which indicates their weight in comparison with that of an equal volume of air. The specific gravity of air is 1, and the comparison is made at 32° F.

TABLE 7
Specific Gravity of Gases at 32 degrees F.

Gas	Sp. Gr.	Gas	Sp. Gr.	Gas	Sp. Gr.
Air................	1.000	Ether vapor.....	2.586	Marsh gas......	0.555
Acetylene..........	0.920	Ethylene........	0.967	Nitrogen........	0.971
Alcohol vapor.......	1.601	Hydrofluoric acid.	2.370	Nitric oxide.....	1.039
Ammonia.....	0.592	Hydrochloric acid.	1.261	Nitrous oxide...	1.527
Carbon dioxide......	1.520	Hydrogen........	0.069	Oxygen.........	1.106
Carbon monoxide....	0.967	Illuminating gas..	0.400	Sulphur dioxide.	2.250
Chlorine............	2.423	Mercury vapor...	6.940	Water vapor....	0.623

1 cubic foot of air at 32 degrees F. and atmospheric pressure weighs 0.0807 pound.

Weights of Materials

The weight of any object may be found by calculating its volume in cubic inches or cubic feet and multiplying this volume by the unit of weight, that is, the weight per cubic foot or cubic inch of the material of which the object is made.

Weight of Square Bars:

ILLUSTRATION: (1) Find the weight of a wrought iron bar 1 foot long and 1 inch square if one cubic inch weighs 0.2778 pound.

$$0.2778 \times 12 = 3.33$$

Therefore, a wrought iron bar 1 inch square and one foot long weighs 3.33 pounds.

(2) Find the weight of a steel bar 1 foot long and 2 inches square if one cubic inch weighs 0.2835 pound.

$$0.2835 \times (2 \times 2) \times 12 = 13.63$$

Therefore, a steel bar 2 inches square, the cross section area 4 sq. in., and 1 foot long weighs 13.63 pounds.

Weight of Sheet Metal.—The weight of one square foot of sheet iron equals $40 \times$ thickness in thousandths of an inch. A square sheet of iron plate 1 inch thick and measuring 1 foot on each side contains:

$$12 \times 12 \times 1 = 144 \text{ cubic inches}$$
$$144 \times 0.2778 \text{ (the weight of 1 cubic inch of iron)} = 40$$

Therefore, the weight of a sheet of iron plate 1 inch thick and 1 foot on each side weighs 40 pounds.

ILLUSTRATION: What is the weight of 1 sq. ft. of sheet iron, No. 20 gage, i.e., 0.032 inch thick.

$$40 \times 0.032 = 1.28$$

Therefore the weight is 1.28 pounds.

ILLUSTRATION: Find the weight of a sheet of steel 6 feet 8 inches long, 2 feet 6 inches wide and No. 2 gage, i.e., 0.2576 inch thick.

$$6 \text{ feet 8 inches} = 80 \text{ inches}, \quad 2 \text{ feet 6 inches} = 30 \text{ inches}$$
$$80 \times 30 \times 0.2576 \times 0.2835 = 165.26$$

Therefore, the weight of the bar is 165.26 pounds.

Weight of Round Bars.—The weight of round bars are found by a similar method used in square bars, the only difference being that the area of the end of the bar is the area of a circle whose diameter is given.

ILLUSTRATION: Find the weight of a steel bar 1 inch in diameter and 1 foot long.

$$0.2835 \times (1^2 \times 0.7854) \times 12 = 2.67$$

Therefore a round steel bar 1 inch in diameter and 1 foot long weighs 2.67 pounds.

Table 6 may also be used to calculate the weights of round, square and hexagon steel bars.

ILLUSTRATION: Find the weight of a steel bar 1 inch in diameter and 1 foot long.

From the table, weight per inch of a 1 inch round bar is 0.2227 lb. Therefore, $12 \times 0.2227 = 2.67$ **pounds.**

TABLE 8

Weights and Areas of Round, Square and Hexagon Steel

Weight of one cubic inch = 0.2836 lb
Weight of one cubic foot = 490 lb

Thickness or Diameter	Area = Diam.² × 0.7854			Area = Side² × 1		Area = Diam.² × 0.866	
	Round			Square		Hexagon	
	Weight Per Inch	Area Square Inches	Circum-ference Inches	Weight Per Inch	Area Square Inches	Weight Per Inch	Area Square Inches
$\frac{1}{32}$	0.0002	0.0008	0.0981	0.0003	0.0010	0.0002	0.0008
$\frac{1}{16}$	.0009	.0031	.1963	.0011	.0039	.0010	.0034
$\frac{3}{32}$	.0020	.0069	.2995	.0025	.0088	.0022	.0076
$\frac{1}{8}$	.0035	.0123	.3927	.0044	.0156	.0038	.0135
$\frac{5}{32}$	.0054	.0192	.4908	.0069	.0244	.0060	.0211
$\frac{3}{16}$	.0078	.0276	.5890	.0101	.0352	.0086	.0304
$\frac{7}{32}$	.0107	.0376	.6872	.0136	.0479	.0118	.0414
$\frac{1}{4}$	.0139	.0491	.7854	.0177	.0625	.0154	.0540
$\frac{9}{32}$	.0176	.0621	.8835	.0224	.0791	.0194	.0686
$\frac{5}{16}$	.0218	.0767	.9817	.0277	.0977	.0240	.0846
$\frac{11}{32}$	.0263	.0928	1.0799	.0335	.1182	.0290	.1023
$\frac{3}{8}$	.0313	.1104	1.1781	.0405	.1406	.0345	.1218
$\frac{13}{32}$	.0368	.1296	1.2762	.0466	.1651	.0405	.1428
$\frac{7}{16}$	.0426	.1503	1.3744	.0543	.1914	.0470	.1658
$\frac{15}{32}$	.0489	.1726	1.4726	.0623	.2197	.0540	.1903
$\frac{1}{2}$	.0557	.1963	1.5708	.0709	.2500	.0614	.2161
$\frac{17}{32}$	.0629	.2217	1.6689	.0800	.2822	.0693	.2444
$\frac{9}{16}$	.0705	.2485	1.7671	.0897	.3164	.0777	.2743
$\frac{19}{32}$	.0785	.2769	1.8653	.1036	.3526	.0866	.3053
$\frac{5}{8}$	.0870	.3068	1.9635	.1108	.3906	.0959	.3383
$\frac{21}{32}$	.0959	.3382	2.0616	.1221	.4307	.1058	.3730
$\frac{11}{16}$	.1053	.3712	2.1598	.1340	.4727	.1161	.4093
$\frac{23}{32}$	.1151	.4057	2.2580	.1465	.5166	.1270	.4474
$\frac{3}{4}$	.1253	.4418	2.3562	.1622	.5625	.1382	.4871
$\frac{25}{32}$	.1359	.4794	2.4543	.1732	.6103	.1499	.5286
$\frac{13}{16}$	.1470	.5185	2.5525	.1872	.6602	.1620	.5712
$\frac{27}{32}$	.1586	.5591	2.6507	.2019	.7119	.1749	.6165
$\frac{7}{8}$	.1705	.6013	2.7489	.2171	.7656	.1880	.6631

TABLE 8—(*Continued*)

Thickness or Diameter	Area = Diam.² × 0.7854 Round			Area = Side² × 1 Square		Area = Diam.² × 0.866 Hexagon	
	Weight Per Inch	Area Square Inches	Circumference Inches	Weight Per Inch	Area Square Inches	Weight Per Inch	Area Square Inches
29/32	0.1829	0.6450	2.8470	0.2329	0.8213	0.2015	0.7112
15/16	.1958	.6903	2.9452	.2492	.8789	.2159	.7612
31/32	.2090	.7371	3.0434	.2661	.9384	.2305	.8127
1	.2227	.7854	3.1416	.2836	1.0000	.2456	.8643
1 1/16	.2515	.8866	3.3379	.3201	1.1289	.2773	.9776
1 1/8	.2819	.9940	3.5343	.3589	1.2656	.3109	1.0973
1 3/16	.3141	1.1075	3.7306	.4142	1.4102	.3464	1.2212
1 1/4	.3480	1.2272	3.9270	.4431	1.5625	.3838	1.3531
1 5/16	.3837	1.3530	4.1233	.4885	1.7227	.4231	1.4919
1 3/8	.4211	1.4849	4.3197	.5362	1.8906	.4643	1.6373
1 7/16	.4603	1.6230	4.5160	.5860	2.0664	.5076	1.7898
1 1/2	.5012	1.7671	4.7124	.6487	2.2500	.5526	1.9485
1 9/16	.5438	1.9175	4.9087	.6930	2.4414	.5996	2.1143
1 5/8	.5882	2.0739	5.1051	.7489	2.6406	.6480	2.2847
1 11/16	.6343	2.2365	5.3014	.8076	2.8477	.6994	2.4662
1 3/4	.6821	2.4053	5.4978	.8685	3.0625	.7521	2.6522
1 13/16	.7317	2.5802	5.6941	.9316	3.2852	.8069	2.8450
1 7/8	.7831	2.7612	5.8905	.9970	3.5156	.8635	3.0446
1 15/16	.8361	2.9483	6.0868	1.0646	3.7539	.9220	3.2509
2	.8910	3.1416	6.2832	1.1342	4.0000	.9825	3.4573
2 1/16	.9475	3.3410	6.4795	1.2064	4.2539	1.0448	3.6840
2 1/8	1.0058	3.5466	6.6759	1.2806	4.5156	1.1091	3.9106
2 3/16	1.0658	3.7583	6.8722	1.3570	4.7852	1.1753	4.1440
2 1/4	1.1276	3.9761	7.0686	1.4357	5.0625	1.2434	4.3892
2 5/16	1.1911	4.2000	7.2649	1.5165	5.3477	1.3135	4.6312
2 3/8	1.2564	4.4301	7.4613	1.6569	5.6406	1.3854	4.8849
2 7/16	1.3234	4.6664	7.6575	1.6849	5.9414	1.4593	5.1454
2 1/2	1.3921	4.9087	7.8540	1.7724	6.2500	1.5351	5.4126
2 5/8	1.5348	5.4119	8.2467	1.9541	6.8906	1.6924	5.9674
2 3/4	1.6845	5.9396	8.6394	2.1446	7.5625	1.8574	6.5493
2 7/8	1.8411	6.4918	9.0321	2.3441	8.2656	2.0304	7.1590
3	2.0046	7.0686	9.4248	2.5548	9.0000	2.2105	7.7941

TABLE 8—(*Concluded*)

Thickness or Diameter	Area = Diam.² × 0.7854			Area = Side² × 1		Area = Diam.² × 0.866	
	Round			Square		Hexagon	
	Weight Per Inch	Area Square Inches	Circumference Inches	Weight Per Inch	Area Square Inches	Weight Per Inch	Area Square Inches
3⅛	2.1752	7.6699	9.8175	2.7719	9.7656	2.3986	8.4573
3¼	2.3527	8.2958	10.2102	2.9954	10.5625	2.5918	9.1387
3⅜	2.5371	8.9462	10.6029	3.2303	11.3906	2.7977	9.8646
3½	2.7286	9.6211	10.9956	3.4740	12.2500	3.0083	10.6089
3⅝	2.9269	10.3206	11.3883	3.7265	13.1407	3.2275	11.3798
3¾	3.1323	11.0447	11.7810	3.9880	14.0625	3.4539	12.1785
3⅞	3.3446	11.7932	12.1737	4.2582	15.0156	3.6880	13.0035
4	3.5638	12.5664	12.5664	4.5374	16.0000	3.9298	13.8292
4⅛	3.7900	13.3640	12.9591	4.8254	17.0156	4.1792	14.7359
4¼	4.0232	14.1863	13.3518	5.1223	18.0625	4.4364	15.6424
4⅜	4.2634	15.0332	13.7445	5.4280	19.1406	4.7011	16.5761
4½	4.5105	15.9043	14.1372	5.7426	20.2500	4.9736	17.5569
4⅝	4.7345	16.8002	14.5299	6.0662	21.3906	5.2538	18.5249
4¾	5.0255	17.7205	14.9226	6.6276	22.5625	5.5416	19.5397
4⅞	5.2935	18.6655	15.3153	6.7397	23.7656	5.8371	20.5816
5	5.5685	19.6350	15.7080	7.0897	25.0000	6.1403	21.6503
5⅛	5.8504	20.6290	16.1007	7.4496	26.2656	6.4511	22.7456
5¼	6.1392	21.6475	16.4934	7.8164	27.5624	6.7697	23.8696
5⅜	6.4351	22.6905	16.8861	8.1930	28.8906	7.0959	25.0198
5½	6.7379	23.7583	17.2788	8.5786	30.2500	7.4298	26.1971
5⅝	7.0476	24.8505	17.6715	8.9729	31.6406	7.7713	27.4013
5¾	7.3643	25.9672	18.0642	9.3762	33.0625	8.1214	28.6361
5⅞	7.6880	27.1085	18.4569	9.7883	34.5156	8.4774	29.8913
6	8.0186	28.2743	18.8496	10.2192	36.0000	8.8420	31.1765
6¼	8.7007	30.6796	19.6350	11.0877	39.0625	9.5943	33.8291
6½	9.4107	33.1831	20.4204	11.9817	42.2500	10.3673	36.5547
6¾	10.1485	35.7847	21.2058	12.9211	45.5625	11.1908	39.4584
7	10.9142	38.4845	21.9912	13.8960	49.0000	12.0351	42.4354
7½	12.5291	44.1786	23.5620	15.9520	56.2500	13.8158	48.7142
8	14.2553	50.2655	25.1328	18.1497	64.0000	15.7192	55.3169

Multiply above weights by 0.993 for wrought iron, 0.918 for cast iron, 1.0331 for cast brass, 1.1209 for copper, 1.1748 for phos. bronze, and 0.3265 for aluminum.

IX

Excavation and Foundations

Excavation.—Excavation of earth and rock involves three or four general operations on the excavated material; viz., (*a*) loosening, (*b*) loading, (*c*) hauling, and (*d*) dumping. Rock, hardpan, and frozen ground may be loosened most economically with explosives, although pneumatic spades may be used on the latter two where explosives are not permitted.

In soft ground, loosening and loading become one operation. On small work, picks and shovels are used to break up the ground

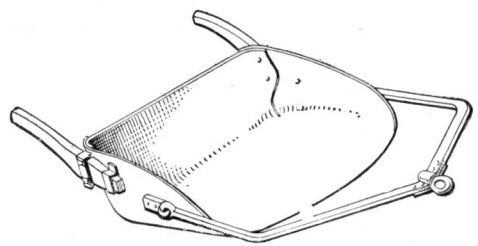

Fig. 1.—Western Slip or Drag Scraper.

and load it into dump wagons. Drag scrapers such as shown in Fig. 1 are also widely used on small building excavation, particularly when the dirt may be disposed of close at hand. In the case of excavations for larger buildings, diesel or gasoline shovels are generally used. These dump into trucks which have access to the hole by ramps or elevators.

Three special types of excavations will be considered in the following paragraphs; namely, foundation, right-of-way cut, and borrow pit excavations.

Laying Out a Foundation.—The first step preparatory to excavating for the foundation of a small building is to set stakes into the ground on the lines of the excavation and some distance back from the corners as shown in Fig. 2.   These stakes should be set by an engineer or surveyor.   When lines are stretched between the stakes, the diagonals between the corners are equal when the excavation is rectangular.   When the corners are supposed to be square, the angle may be checked by laying off a distance of 6 feet from the corner along one line and 8 feet from the corner on the other.   The distance between these two points should measure 10 feet.

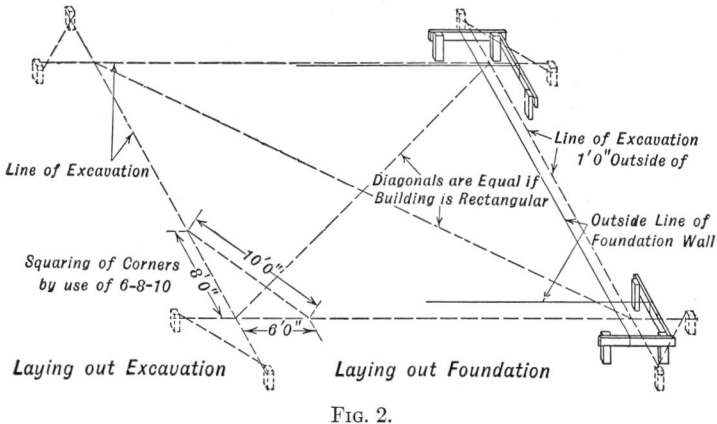

Fig. 2.

Excavation lines should be set 1 foot outside of the foundation lines to allow sufficient working space.

Estimating Quantity of Excavated Material—Material removed from an excavation is measured by cubic yards " in place."   That is, it is measured as solid ground and not as the loose material which is hauled away and dumped.   The reason for this is that the latter occupies a volume about 25 percent greater than its original volume.   The problem of measuring the amount of material excavated becomes then a case of determining the volume of the resulting hole.   When the ground is level and the figure regular, the computation is quite simple. It can best be illustrated by a few examples.

ILLUSTRATION: Figure 3 shows the plan of a building whose outside dimensions are 20 feet and 32 feet. What is the volume of excavation if the depth is uniformly eight feet and the lines of excavation are one foot outside the building lines?

The dimensions of the hole are 8 ft. × 22 ft. × 34 ft.
Volume = 8 × 22 × 34 = 5984 cu. ft.

Changing to cu. yd., volume = $\dfrac{5984}{27}$ = 222 cu. yd. (Ans.)

ILLUSTRATION: Figure 4 gives the dimensions of the plan of a T-shaped building. What is the volume of excavation if the

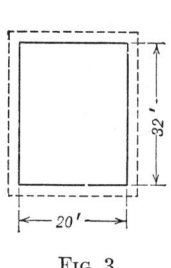

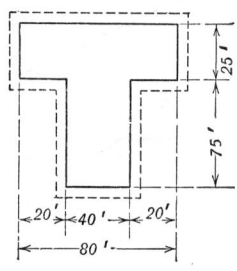

FIG. 3.   FIG. 4.

excavating line is one foot outside the building line and the depth is nine feet?

The area of the excavation can be computed most readily by mentally dividing its plan into two rectangles, one 82 feet by 27 feet and the other 75 feet by 42 feet. The areas of these are

$$82 \times 27 = 2214 \text{ sq. ft.}$$
$$75 \times 42 = 3150 \text{ sq. ft.}$$

Total   5364 sq. ft.

The volume in cubic feet is then the total area times the depth of 9 feet. This is changed to cubic yards by dividing by 27. If these operations are set up together, the computation may be completed mentally:

$$\frac{5364 \times \overset{3}{\cancel{9}}}{\underset{3}{\cancel{27}}} = 1788 \text{ cu. yd. (Ans.)}$$

ILLUSTRATION: A building of the dimensions shown in Fig. 5 is to be built on a triangular lot. If the excavation is eight feet deep and one foot outside the building line, what volume of earth will have to be removed?

The problem gives us the three sides of an oblique-angled triangle, but we do not know any of the dimensions of the larger triangle represented by the excavation line. Determining these dimensions would be a tedious operation not warranted by this problem. A practical solution is to solve for the area of the triangle represented by the building line and add to this the area of a strip one foot wide and slightly longer than the perimeter of the triangle.

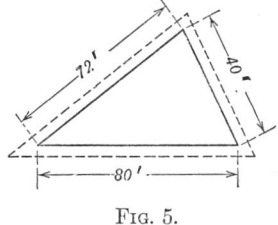

FIG. 5.

From geometry we know that the area of any triangle, whose three sides are represented by a, b, and c, is

$$\sqrt{S(S - a)(S - b)(S - c)}$$

when $S = \frac{1}{2}(a + b + c)$. Using this, we proceed to find the area of the inner triangle.

$$S = \tfrac{1}{2}(a + b + c) = \tfrac{1}{2}(72 + 40 + 80) = 96$$

$$\text{Area} = \sqrt{96(96 - 72)(96 - 40)(96 - 80)}$$

$$= \sqrt{96 \times 24 \times 56 \times 16} = 1437 \text{ sq. ft.}$$

In computing the area of the one-foot strip around this triangle, let us arbitrarily add 3 feet to the sum of the lengths of the sides of the foundation wall. Then the area is

area of strip $= 1 \times (72 + 40 + 80 + 3) = 195$ sq. ft.

Adding this to the area of the triangle we obtain, $1437 + 195 = 1632$ sq ft. The volume of the excavation is then this area times the depth, 8 feet, and divided by 27 to change to cubic yards, or

$$\text{Volume} = \frac{1632 \times 8}{27} = 483 \text{ cu. yd. (Ans.)}$$

Average End Area Method of Estimating Earthwork.—The preceding paragraphs have considered only excavations regular in shape and with vertical sides such as are common in foundation work for buildings. Vertical faces of earth will, however, only remain standing a short time and when a permanent depression

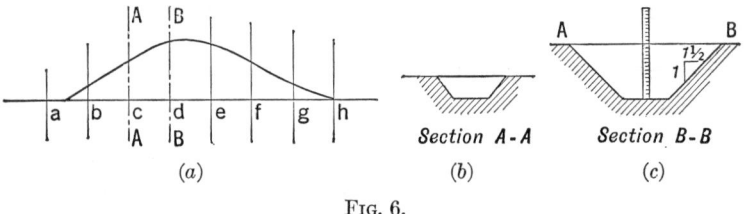

(a)   (b)   (c)

Section A-A   Section B-B

Fig. 6.

in earth is desired without retaining walls, the sides of the excavation must be sloped.

The slope which a loose material will naturally assume and at which it will remain stable, is called the *angle of repose*, referred to the horizontal. Sand has an angle of repose of about 34 degrees, a mixture of sand, gravel, and clay, an angle of about 45 degrees, while sound rock will stand vertical or at an angle of 90 degrees.

An irregular excavation or a uniform excavation through irregular ground is usually measured by dividing the total volume into small prisms and arriving at the sum of the volumes of these prisms. For example, let Fig. 6 (a) represent the profile of a·hill through which a driveway is to be cut, and Fig. 6 (b) a cross-

section at A–A while (c) is a cross-section at B–B.   The volume
to be excavated between sections A–A and B–B is a six-sided
prism whose shape is approximately as shown in Fig. 7.   In the
average-end-area method of computing this volume, the area of
$ABCD$ is averaged with the area of $EFGH$, resulting in the area
of the mid-section $IJKL$.   This is multiplied by the distance
between the cross-sections (CG or DE) to obtain the volume.   If
then, in Fig. 6 (a) we average the areas of the sections, a and b,
b and c, c and d, d and e, e and f, f and g, g and h, and multiply
each average by the distance between its respective end areas,
we will obtain the volume of the entire excavation.

It is to be noted that the result is only approximately correct
and that the error increases as the difference in areas of the end
sections increases.   However, the
method represents accepted practice
in engineering work.

Right-of-Way Excavations.—
The method outlined in the preceding paragraphs is equally applicable
to cuts for driveways, highways,
railways, canals, etc., which we

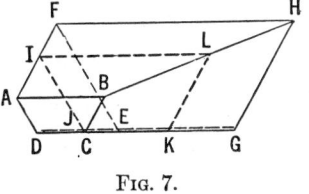

Fig. 7.

shall call right-of-way excavations for want of a more descriptive
term.   These excavations, or " cuts," as they are called, have
the common property of being generally uniform in shape of
cross-section, the only major variation being the depth of the
cut.   This being true, it has been possible to develop tables so
that the volumes can be estimated with a minimum of computation and without the use of surveying instruments.

Whether the tables or direct computation are used, a longitudinal line is first laid out along the centerline of the work and
stakes or markers are set at horizontal intervals of 100 feet along
this line.   These points are called *stations*.   If the ground is very
irregular, the intervals may be only 50 feet, and for rock excavation the interval is often only 25 feet.

The use of the tables requires a knowledge of the width of the
base of a roadway, the slope of the sides, and the depth of the cut

TABLE 1

LEVEL SECTIONS (EARTHWORK); HEIGHT, 0–60 FT.
BASE OF ROADWAY, 16 FT., SIDE SLOPE, 1 TO 1

Note.—The last two columns enable us to use any other base than 14 ft.:
Ex.—Given height, 20.3 ft.; roadway 14 ft. Then we have, 2729.2—
(148.15+ 2.22) = 2578.8 cu. yds.

[Cu. Yds. per 100-Ft. Station.]

Left margin note: These quantities in this table may all be multiplied by the same factor; thus, for base of 12 ft. and slopes ¾ to 1, we have, 3264 cu. yds. — Note that Base, Slope, and Cu. Yds. in this table may all be multiplied by the same factor; using factor of ¾ for height of 27.2 ft.

Right margin note: Add for Tenths of Feet in Height. — Use with preceding column only.

Ht. Ft.	.0	.1	.2	.3	.4	.5	.6	.7	.8	.9	Width of 2 Ft. Cu.Yds.
0	......	6.0	12.0	18.1	24.3	30.6	36.9	43.3	49 8	56 3	
1	63.0	69.7	76.4	83.3	90.2	97.2	104.3	111.4	118.7	126 0	7.41
2	133.3	140.8	148.3	155.9	163.6	171.3	179.1	187.0	195.0	203.0	14.81
3	211.1	219.3	227.6	235.9	244.3	252.8	261.3	270.0	278.7	287.4	22.22
4	296.3	305.2	314.2	323.3	332.4	341.7	351.0	360.3	369.8	379.3	29.63
5	388.9	398.6	408.3	418.1	428.0	438.0	448.0	458.1	468.3	478.6	37.04
6	488.9	499.3	509.8	520.3	531.0	541.7	552.4	563.3	574.2	585.2	44.44
7	596.3	607.4	618.7	630.0	641.3	652.8	664.3	675.9	687.6	699.3	51.85
8	711.1	723.0	735.0	747.0	759.1	771.3	783.6	795.9	808.3	820.8	59.26
9	833.3	846.0	858.7	871.4	884.3	897.2	910.2	923.3	936.4	949.7	66.67
10	963.0	976.3	989.8	1003.3	1016.9	1030.6	1044.3	1058.1	1072.0	1086.0	74.07
11	1100.0	1114.1	1128.3	1142.6	1156.9	1171.3	1185.8	1200.3	1215.0	1229.7	81.48
12	1244.4	1259.3	1274.2	1289.2	1304.3	1319.4	1334.7	1350.0	1365.3	1380.8	88.89
13	1396.3	1411.9	1427.6	1443.3	1459.1	1475.0	1491.0	1507.0	1523.1	1539.3	96.30
14	1555.6	1571.9	1588.3	1604.8	1621.3	1638.0	1654.7	1671.4	1688.3	1705.2	103.70
15	1722.2	1739.3	1756.4	1773.7	1791.0	1808.3	1825.8	1843.3	1860.9	1878.6	111.11
16	1896.3	1914.1	1932.0	1950.0	1968.0	1986.1	2004.3	2022.6	2040.9	2059.3	118.52
17	2077.8	2096.3	2115.0	2133.7	2152.4	2171.3	2190.2	2209.2	2228.3	2247.4	125.93
18	2266.7	2286.0	2305.3	2324.8	2344.3	2363.9	2383.6	2403.3	2423.1	2443.0	133.33
19	2463.0	2483.0	2503.1	2523.3	2543.6	2563.9	2584.3	2604.8	2625.3	2646.0	140.74
20	2666.7	2687.4	2708.3	2729.2	2750.2	2771.3	2792.4	2813.7	2835.0	2856.3	148.15
21	2877.8	2899.3	2920.9	2942.6	2964.3	2986.1	3008.0	3030.0	3052.0	3074.1	156.56
22	3096.3	3118.6	3140.9	3163.3	3185.8	3208.3	3231.0	3253.7	3276.4	3299.3	162.96
23	3322.2	3345.2	3368.2	3391.4	3414.7	3438.0	3461.3	3484.8	3508.3	3531.9	170.37
24	3555.6	3579.3	3603.1	3627.0	3651.0	3675.0	3699.1	3723.3	3747.6	3771.9	177.78
25	3796.3	3820.8	3845.3	3870.0	3894.7	3919.4	3944.3	3969.2	3994.2	4019.3	185.19
26	4044.4	4069.7	4095.0	4120.3	4145.8	4171.3	4196.9	4222.6	4248.3	4274.1	192.59
27	4300.0	4326.0	4352.0	4378.1	4404.3	4430.6	4456.9	4483.3	4509.8	4536.3	200.00
28	4563.0	4589.7	4616.4	4643.3	4670.2	4697.2	4724.3	4751.4	4778.7	4806.0	207.41
29	4833.3	4860.8	4888.3	4915.9	4943.6	4971.3	4999.1	5027.0	5055.0	5083.0	214.81
30	5111.1	5139.3	5167.6	5195.9	5224.3	5252.8	5281.3	5310.0	5338.7	5367.4	222.22
31	5396.3	5425.2	5454.2	5483.3	5512.4	5541.7	5571.0	5600.3	5629.8	5659.3	229.63
32	5688.9	5718.6	5748.3	5778.1	5808.0	5838.0	5868.0	5898.1	5928.3	5958.6	237.04
33	5988.9	6019.3	6049.8	6080.3	6111.0	6141.7	6172.4	6203.3	6234.2	6265.2	244.44
34	6296.3	6327.4	6358.7	6390.0	6421.3	6452.8	6484.3	6515.9	6547.6	6579.3	251.85
35	6611.1	6643.0	6675.0	6707.0	6738.1	6771.3	6803.6	6835.9	6868.3	6900.8	259.26
36	6933.3	6966.0	6998.7	7031.4	7064.3	7097.2	7130.2	7163.3	7196.4	7229.7	266.67
37	7263.0	7296.3	7329.8	7363.3	7396.9	7430.6	7464.3	7498.1	7532.0	7566.0	274.07
38	7600.0	7634.1	7668.3	7702.6	7736.9	7771.3	7805.8	7840.3	7875.0	7909.7	281.48
39	7944.4	7979.3	8014.2	8049.2	8084.3	8119.4	8154.7	8190.0	8225.3	8260 8	288.89
40	8296.3	8331.9	8367.6	8403.3	8439.1	8475.0	8511.0	8547.0	8583.1	8619.3	296.30
41	8655.6	8691.9	8728.3	8764.8	8801.3	8838.0	8874.7	8911.4	8948.3	8985.2	303.70
42	9022.2	9059.3	9096.4	9133.7	9171.0	9208.3	9245.8	9283.3	9320.9	9358.6	311.11
43	9396.3	9434.1	9472.0	9510.0	9548.0	9586.1	9624.3	9662.6	9700.9	9739.3	318.52
44	9777.8	9816.3	9855.0	9893.7	9932.4	9971.3	10010	10049	10088	10127	325.93
45	10167	10206	10245	10285	10324	10364	10404	10443	10483	10523	333.33
46	10563	10603	10643	10683	10724	10764	10804	10845	10885	10926	340.74
47	10967	11007	11048	11089	11130	11171	11212	11254	11295	11336	348.15
48	11378	11419	11461	11503	11544	11586	11628	11670	11712	11754	355.56
49	11796	11839	11881	11923	11966	12008	12051	12094	12136	12179	362.96
50	12222	12265	12308	12351	12395	12438	12481	12525	12568	12612	370.37
51	12656	12699	12743	12787	12831	12875	12919	12963	13008	13052	377.78
52	13096	13141	13185	13230	13275	13319	13364	13409	13454	13499	385.19
53	13544	13590	13635	13680	13726	13771	13817	13863	13908	13954	392.59
54	14000	14046	14092	14138	14184	14231	14277	14323	14370	14416	400.00
55	14463	14510	14556	14603	14650	14697	14744	14791	14839	14886	407.41
56	14933	14981	15028	15076	15124	15171	15219	15267	15315	15363	414.81
57	15411	15459	15508	15556	15604	15653	15701	15750	15799	15847	422.22
58	15896	15945	15994	16043	16092	16142	16191	16240	16290	16339	429.63
59	16389	16439	16488	16538	16588	16638	16688	16738	16788	16839	437.04
60	16889	16939	16990	17040	17091	17142	17192	17243	17294	17345	444.44

P. P.
7.41

1	.74
2	1.48
3	2.22
4	2.96
5	3.70
6	4.44
7	5.19
8	5.93
9	6.67

TABLE 2

Level Sections (Earthwork); Height, 0–60 Ft.
Base of Roadway, 16 Ft., Side Slopes, 1½ to 1

Note.—The last two columns enable us to use any other base than 16 ft.:
Ex.—Given height, 39.7 ft.; roadway 14 ft. Then we have, 11109 —
(288.89 + 5.19) = 10815 cu. yds.

[Cu. Yds. per 100-Ft. Station.]

Note that Base, Slope, and Cu. Yds. in this table may all be multiplied by the same factor; thus, using factor of 1½ for height of 39.1 ft., we have, 16215 cu. yds. for base of 24 ft. and slopes 2¼ to 1.

Add for Tenths of Feet in Height.

Ht. Ft.	.0	.1	.2	.3	.4	.5	.6	.7	.8	.9	Width of 2 Ft. Cu.Yds
0	......	6.0	12.1	18.3	24.6	31.0	37.6	44.2	51.0	57.8	......
1	64.8	71.9	79.1	86.4	93.9	101.4	109.0	116.8	124.7	132.6	7.41
2	140.7	148.9	157.3	165.7	174.2	182.9	191.6	200.5	209.5	218.6	14.81
3	227.8	237.1	246.5	256.1	265.7	275.5	285.3	295.3	305.4	315.6	22.22
4	325.9	336.4	346.9	357.5	368.3	379.2	390.1	401.2	412.4	423.8	29.63
5	435.2	446.7	458.4	470.1	482.0	494.0	506.1	518.3	530.6	543.0	37.04
6	555.6	568.2	581.0	593.8	606.8	619.9	633.1	646.4	659.9	673.4	44.44
7	687.0	700.8	714.7	728.6	742.7	756.9	771.3	785.7	800.2	814.9	51.85
8	829.6	844.5	859.5	874.6	889.8	905.1	920.5	936.1	951.7	967.5	59.26
9	983.3	999.3	1015.4	1031.6	1047.9	1064.4	1080.9	1097.5	1114.3	1131.2	66.67
10	1148.1	1165.2	1182.4	1199.8	1217.2	1234.7	1252.4	1270.1	1288.0	1306.0	74.07
11	1324.1	1342.3	1360.6	1379.0	1397.6	1416.2	1435.0	1453.8	1472.8	1491.9	81.48
12	1511.1	1530.4	1549.9	1569.4	1589.0	1608.8	1628.7	1648.6	1668.7	1688.9	88.89
13	1709.3	1729.7	1750.2	1770.9	1791.6	1812.5	1833.5	1854.6	1875.8	1897.1	96.30
14	1918.5	1940.1	1961.7	1983.5	2005.3	2027.3	2049.4	2071.6	2093.9	2116.4	103.70
15	2138.9	2161.5	2184.3	2207.2	2230.1	2253.2	2276.4	2299.8	2323.2	2346.7	111.11
16	2370.4	2394.1	2418.0	2442.0	2466.1	2490.3	2514.6	2539.0	2563.6	2588.2	118.52
17	2613.0	2637.8	2662.8	2687.9	2713.1	2738.4	2763.9	2789.4	2815.0	2840.8	125.93
18	2866.7	2892.6	2918.7	2944.9	2971.3	2997.7	3024.2	3050.9	3077.6	3104.5	133.33
19	3131.5	3158.6	3185.8	3213.1	3240.5	3268.0	3295.7	3323.5	3351.3	3379.3	140.74
20	3407.4	3435.6	3463.9	3492.4	3520.9	3549.5	3578.3	3607.2	3636.1	3665.2	148.15
21	3694.4	3723.8	3753.2	3782.7	3812.4	3842.1	3872.0	3902.0	3932.1	3962.3	156.56
22	3992.6	4023.0	4053.6	4084.2	4115.0	4145.8	4176.8	4207.9	4239.1	4270.4	162.96
23	4301.9	4333.4	4365.0	4396.8	4428.7	4460.6	4492.7	4524.9	4557.3	4589.7	170.37
24	4622.2	4654.9	4687.6	4720.5	4753.5	4786.6	4819.8	4853.1	4886.5	4920.1	177.78
25	4953.7	4987.5	5021.3	5055.3	5089.4	5123.6	5157.9	5192.4	5226.9	5261.5	185.19
26	5296.3	5331.2	5366.1	5401.2	5436.4	5471.8	5507.2	5542.7	5578.4	5614.1	192.59
27	5650.0	5686.0	5722.1	5758.3	5794.6	5831.0	5867.6	5904.2	5941.0	5977.8	200.00
28	6014.8	6051.9	6089.1	6126.4	6163.9	6201.4	6239.0	6276.8	6314.7	6352.6	207.41
29	6390.7	6428.9	6467.3	6505.7	6544.2	6582.9	6621.6	6660.5	6699.5	6738.6	214.81
30	6777.8	6817.1	6856.6	6896.1	6935.7	6975.5	7015.3	7055.3	7095.4	7135.6	222.22
31	7175.9	7216.4	7256.9	7297.5	7338.3	7379.2	7420.1	7461.2	7502.4	7543.8	229.63
32	7585.2	7626.7	7668.4	7710.1	7752.0	7794.0	7836.1	7878.3	7920.6	7963.0	237.04
33	8005.6	8048.2	8091.0	8133.8	8176.8	8219.9	8263.1	8306.4	8349.9	8393.4	244.44
34	8437.0	8480.8	8524.7	8568.6	8612.7	8656.9	8701.3	8745.7	8790.2	8834.9	251.85
35	8879.6	8924.5	8969.5	9014.6	9059.8	9105.1	9150.5	9196.1	9241.7	9287.5	259.26
36	9333.3	9379.3	9425.4	9471.6	9517.9	9564.4	9610.9	9657.5	9704.3	9751.2	266.67
37	9798.1	9845.2	9892.4	9939.8	9987.2	10035	10082	10130	10178	10226	274.07
38	10274	10322	10371	10419	10468	10516	10565	10614	10663	10712	281.48
39	10761	10810	10860	10909	10959	11009	11059	11109	11159	11209	288.89
40	11259	11310	11360	11411	11462	11513	11563	11615	11666	11717	296.30
41	11769	11820	11872	11923	11975	12027	12079	12132	12184	12236	303.70
42	12289	12341	12394	12447	12500	12553	12606	12660	12713	12767	311.11
43	12820	12874	12928	12982	13036	13090	13145	13199	13254	13308	318.52
44	13363	13418	13473	13528	13583	13638	13694	13749	13805	13861	325.93
45	13917	13973	14029	14085	14141	14198	14254	14311	14368	14425	333.33
46	14481	14539	14596	14653	14711	14768	14826	14883	14941	14999	340.74
47	15057	15116	15174	15232	15291	15350	15408	15467	15526	15585	348.15
48	15644	15704	15763	15823	15882	15942	16002	16062	16122	16182	355.56
49	16243	16303	16364	16424	16485	16546	16607	16668	16729	16790	362.96
50	16852	16913	16975	17037	17099	17161	17223	17285	17347	17410	370.37
51	17472	17535	17598	17661	17723	17787	17850	17913	17977	18040	377.78
52	18104	18167	18231	18295	18359	18424	18488	18552	18617	18682	385.19
53	18746	18811	18876	18941	19006	19072	19137	19203	19268	19334	392.59
54	19400	19466	19532	19598	19665	19731	19798	19864	19931	19998	400.00
55	20065	20132	20199	20266	20334	20401	20469	20537	20605	20673	407.41
56	20741	20809	20877	20946	21014	21083	21152	21221	21289	21359	414.81
57	21428	21497	21567	21636	21706	21775	21845	21915	21985	22056	422.22
58	22126	22196	22267	22338	22408	22479	22550	22621	22692	22764	429.63
59	22835	22907	22978	23050	23122	23194	23266	23338	23411	23483	437.04
60	23556	23628	23701	23774	23847	23920	23993	24066	24140	24213	444.44

Use with preceding column only.

P. P.
7.41

1	.74
2	1.48
3	2.22
4	2.96
5	3.70
6	4.44
7	5.19
8	5.93
9	6.67

TABLE 3

LEVEL SECTIONS (EARTHWORK); HEIGHT, 0–60 FT.

BASE OF ROADWAY, 28 FT., SIDE SLOPES, 1 TO 1

Note.—The last two columns enable us to use any other base than 28 ft.: Ex.—Given height, 57.5 ft.; roadway 26 ft. Then we have, 18208−(422.22+3.70)=17782 cu. yds.

[Cu. Yds. per 100-Ft. Station.]

Note that Base, Slope, and Cu. Yds. in this table may all be multiplied by the same factor; thus, using factor of ½ for height of 45.1 ft., we have, 6105 cu. yds. for base of 14 ft. and slopes ½ to 1.

Ht. Ft.	.0	.1	.2	.3	.4	.5	.6	7	.8	.9	Width of 2 Ft. Cu.Yds
0	......	10.4	20.9	31.4	42.1	52.8	63.6	74.4	85.3	96.3	......
1	107.4	118.6	129.8	141.1	152.4	163.9	175.4	187.0	198.7	210.4	7.41
2	222.2	234.1	246.1	258.1	270.2	282.4	294.7	307.0	319.4	331.9	14.81
3	344.4	357.1	369.8	382.6	395.4	408.3	421.3	434.4	447.6	460.8	22.22
4	474.1	487.4	500.9	514.4	528.0	541.7	555.4	569.2	583.1	597.1	29.63
5	611.1	625.2	639.4	653.7	668.0	682.4	696.9	711.4	726.1	740.8	37.04
6	755.6	770.4	785.4	800.4	815.5	830.6	845.8	861.1	876.5	891.9	44.44
7	907.5	923.0	938.7	954.5	970.3	986.1	1002.1	1018.1	1034.2	1050.4	51.85
8	1066.7	1083.0	1099.4	1115.9	1132.4	1149.1	1165.8	1182.6	1199.4	1216.3	59.26
9	1233.3	1250.4	1267.6	1284.8	1302.1	1319.4	1336.9	1354.4	1372.0	1389.7	66.67
10	1407.4	1425.2	1443.1	1461.1	1479.1	1497.2	1515.4	1533.7	1552.0	1570.4	74.07
11	1588.9	1607.4	1626.1	1644.8	1663.6	1682.4	1701.3	1720.3	1739.4	1758.6	81.48
12	1777.8	1797.1	1816.4	1835.9	1855.4	1875.0	1894.7	1914.4	1934.2	1954.1	88.89
13	1974.1	1994.1	2014.2	2034.4	2054.7	2075.0	2095.4	2115.9	2136.4	2157.1	96.30
14	2177.8	2198.6	2219.4	2240.3	2261.3	2282.4	2303.6	2324.8	2346.1	2367.4	103.70
15	2388.9	2410.4	2432.0	2453.7	2475.4	2497.2	2519.1	2541.1	2563.1	2585.2	111.11
16	2607.4	2629.7	2652.0	2674.4	2696.9	2719.4	2742.1	2764.8	2787.6	2810.4	118.52
17	2833.3	2856.3	2879.4	2902.6	2925.8	2949.1	2972.4	2995.9	3019.4	3043.0	125.93
18	3066.7	3090.4	3114.2	3138.1	3162.1	3186.1	3210.2	3234.4	3258.7	3283.0	133.33
19	3307.4	3331.9	3356.4	3381.1	3405.8	3430.6	3455.4	3480.3	3505.3	3530.4	140.74
20	3555.6	3580.8	3606.1	3631.4	3656.9	3682.4	3708.0	3733.7	3759.4	3785.2	148.15
21	3811.1	3837.1	3863.1	3889.2	3915.4	3941.7	3968.0	3994.4	4020.9	4047.4	155.56
22	4074.1	4100.8	4127.6	4154.4	4181.3	4208.3	4235.4	4262.6	4289.8	4317.1	162.96
23	4344.4	4371.9	4399.4	4427.0	4454.7	4482.4	4510.2	4538.1	4566.1	4594.1	170.37
24	4622.2	4650.4	4678.7	4707.0	4735.4	4763.9	4792.4	4821.1	4849.8	4878.6	177.78
25	4907.4	4936.3	4965.3	4994.4	5023.6	5052.8	5082.1	5111.4	5140.9	5170.4	185.19
26	5200.0	5229.7	5259.4	5289.2	5319.1	5349.1	5379.1	5409.2	5439.4	5469.7	192.59
27	5500.0	5530.4	5560.9	5591.4	5622.1	5652.8	5683.6	5714.4	5745.3	5776.3	200.00
28	5807.4	5838.6	5869.8	5901.1	5932.4	5963.9	5995.4	6027.0	6058.7	6090.4	207.41
29	6122.1	6154.1	6186.1	6218.1	6250.2	6282.4	6314.7	6347.0	6379.4	6411.9	214.81
30	6444.4	6477.1	6509.8	6542.6	6575.4	6608.3	6641.3	6674.4	6707.6	6740.8	222.22
31	6774.1	6807.4	6840.9	6874.4	6908.0	6941.7	6975.4	7009.2	7043.1	7077.1	229.63
32	7111.1	7145.2	7179.4	7213.7	7248.0	7282.4	7316.9	7351.4	7386.1	7420.8	237.04
33	7455.6	7490.4	7525.3	7560.3	7595.4	7630.6	7665.8	7701.1	7736.4	7771.9	244.44
34	7807.4	7843.0	7878.7	7914.4	7950.2	7986.1	8022.1	8058.1	8094.2	8130.4	251.85
35	8166.7	8203.0	8239.4	8275.9	8312.4	8349.1	8385.8	8422.6	8459.4	8496.3	259.26
36	8533.3	8570.4	8607.6	8644.8	8682.1	8719.4	8756.9	8794.4	8832.0	8869.7	266.67
37	8907.4	8945.2	8983.1	9021.1	9059.1	9097.2	9135.4	9173.7	9212.0	9250.4	274.07
38	9288.9	9327.4	9366.1	9404.8	9443.6	9482.4	9521.3	9560.3	9599.4	9638.6	281.48
39	9677.8	9717.1	9756.4	9795.9	9835.4	9875.0	9914.7	9954.4	9994.2	10034	288.89
40	10074	10114	10154	10194	10235	10275	10315	10356	10396	10437	296.30
41	10478	10519	10559	10600	10641	10682	10724	10765	10806	10847	303.70
42	10889	10930	10972	11014	11055	11097	11139	11181	11223	11265	311.11
43	11307	11350	11392	11434	11477	11519	11562	11605	11648	11690	318.52
44	11733	11776	11819	11863	11906	11949	11992	12036	12079	12123	325.93
45	12167	12210	12254	12298	12342	12386	12430	12474	12519	12563	333.33
46	12607	12652	12696	12741	12786	12831	12875	12920	12965	13010	340.74
47	13056	13101	13146	13191	13237	13282	13328	13374	13419	13465	348.15
48	13511	13557	13603	13649	13695	13742	13788	13834	13881	13927	355.56
49	13974	14021	14068	14114	14161	14208	14255	14303	14350	14397	362.96
50	14444	14492	14539	14587	14635	14682	14730	14778	14826	14874	370.37
51	14922	14970	15019	15067	15115	15164	15212	15261	15310	15359	377.78
52	15407	15456	15505	15554	15604	15653	15702	15751	15801	15850	385.19
53	15900	15950	15999	16049	16099	16149	16199	16249	16299	16350	392.59
54	16400	16450	16501	16551	16602	16653	16704	16754	16805	16856	400.00
55	16907	16959	17010	17061	17112	17164	17215	17267	17319	17370	407.41
56	17422	17474	17526	17578	17630	17682	17735	17787	17839	17892	414.81
57	17944	17997	18050	18103	18155	18208	18261	18314	18368	18421	422.22
58	18474	18527	18581	18634	18688	18742	18795	18849	18903	18957	429.63
59	19011	19065	19119	19174	19228	19282	19337	19391	19446	19501	437.04
60	19556	19610	19665	19720	19775	19831	19886	19941	19996	20052	444.44

Add for Tenths of Feet in Height.

P. P. 7.41

1	.74
2	1.48
3	2.22
4	2.96
5	3.70
6	4.44
7	5.19
8	5.93
9	6.67

Use with preceding column only.

TABLE 4

LEVEL SECTIONS (EARTHWORK); HEIGHT, 0–60 FT.
BASE OF ROADWAY, 28 FT., SIDE SLOPES, 1½ TO 1

Note.—The last two columns enable us to use any other base than 28 ft.:
Ex.—Given height, 33.6 ft.; roadway 30 ft. Then we have, 9756.4+
(244.44+4.44)=10005.3 cu. yds.

[Cu. Yds. per 100-Ft. Station.]

so that Base, Slope, and Cu. Yds. in this table may all be multiplied by the same factor; thus, using factor of ½ for height of 51.4 ft. we have, 10004 cu. yds. for base of 14 ft. and slopes ¾ to 1.

Add for Tenths of Feet in Height.

Ht. Ft.	.0	.1	.2	.3	.4	.5	.6	.7	.8	.9	Width of 2 Ft. Cu.Yds
0	......	10.4	21.0	31.6	42.4	53.2	64.2	75.3	86.5	97.9	......
1	109.3	120.8	132.5	144.3	156.1	168.1	180.2	192.4	204.8	217.2	7.41
2	229.6	242.3	255.0	267.9	280.9	294.0	307.2	320.5	334.0	347.5	14.81
3	361.2	374.9	388.8	402.8	416.9	431.1	445.4	459.9	474.4	489.1	22.22
4	503.7	518.6	533.6	548.6	563.9	579.3	594.7	610.2	625.8	641.6	29.63
5	657.5	673.4	689.5	705.7	722.1	738.5	755.0	771.7	788.4	805.3	37.04
6	822.2	839.3	856.5	873.8	891.2	908.8	926.4	944.2	962.0	980.0	44.44
7	998.1	1016.4	1034.7	1053.1	1071.6	1090.3	1109.0	1127.9	1146.9	1166.0	51.85
8	1185.2	1204.5	1223.9	1243.5	1263.1	1282.9	1302.7	1322.7	1342.8	1363.0	59.26
9	1383.3	1403.8	1424.3	1444.9	1465.7	1486.6	1507.6	1528.6	1549.8	1571.2	66.67
10	1592.6	1614.1	1635.8	1657.5	1679.4	1701.4	1723.5	1745.7	1768.0	1790.4	74.07
11	1813.0	1835.6	1858.4	1881.2	1904.2	1927.3	1950.5	1973.8	1997.3	2020.8	81.48
12	2044.4	2068.2	2092.1	2116.1	2140.1	2164.3	2188.7	2213.1	2237.6	2262.3	88.89
13	2287.0	2311.9	2336.9	2362.0	2387.2	2412.5	2437.9	2463.5	2489.1	2514.9	96.30
14	2540.7	2566.7	2592.8	2619.0	2645.3	2671.8	2698.3	2724.9	2751.7	2778.6	103.70
15	2805.6	2832.6	2859.9	2887.2	2914.6	2942.1	2969.8	2997.5	3025.4	3053.4	111.11
16	3081.5	3109.7	3138.0	3166.4	3195.0	3223.6	3252.4	3281.2	3310.2	3339.3	118.52
17	3368.5	3397.8	3427.3	3456.8	3486.4	3516.2	3546.1	3576.1	3606.1	3636.4	125.93
18	3666.5	3697.1	3727.6	3758.3	3789.0	3819.9	3850.9	3882.0	3913.2	3944.5	133.33
19	3975.9	4007.5	4039.1	4070.9	4102.7	4134.7	4166.8	4199.0	4231.3	4263.8	140.74
20	4296.3	4328.9	4361.5	4394.4	4427.6	4460.6	4493.9	4527.2	4560.6	4594.1	148.15
21	4627.8	4661.5	4695.4	4729.4	4763.5	4797.7	4832.0	4866.4	4901.0	4935.6	155.56
22	4970.4	5005.2	5040.2	5075.3	5110.5	5145.8	5181.3	5216.8	5252.4	5288.2	162.96
23	5324.1	5360.1	5396.1	5432.4	5468.7	5505.1	5541.6	5578.3	5615.0	5651.9	170.37
24	5688.9	5726.0	5763.2	5800.5	5837.9	5875.5	5913.1	5950.9	5988.7	6026.7	177.78
25	6064.8	6103.0	6141.3	6179.8	6218.3	6256.9	6295.7	6334.6	6373.6	6412.6	185.19
26	6451.9	6491.2	6530.6	6570.1	6609.8	6649.5	6689.4	6729.4	6769.5	6809.7	192.59
27	6850.0	6890.4	6931.0	6971.6	7012.4	7053.2	7094.2	7135.3	7176.5	7217.8	200.00
28	7259.3	7300.8	7342.4	7384.2	7426.1	7463.1	7510.1	7552.4	7594.7	7637.1	207.41
29	7679.6	7722.3	7765.0	7807.9	7850.9	7894.0	7937.2	7980.5	8023.9	8067.5	214.81
30	8111.1	8154.9	8198.7	8242.7	8286.8	8331.0	8375.3	8419.8	8464.3	8508.9	222.22
31	8553.7	8598.6	8643.6	8688.6	8733.9	8779.2	8824.6	8870.1	8915.8	8961.5	229.63
32	9007.4	9053.4	9099.5	9145.7	9192.0	9238.4	9285.0	9331.6	9378.4	9425.2	237.04
33	9472.2	9519.3	9566.5	9613.8	9661.3	9708.8	9756.4	9804.2	9852.1	9900.1	244.44
34	9948.1	9996.4	10045	10093	10142	10190	10239	10288	10337	10386	251.85
35	10435	10484	10534	10583	10633	10683	10732	10782	10832	10882	259.26
36	10933	10983	11034	11084	11135	11186	11237	11288	11339	11391	266.67
37	11443	11494	11546	11598	11649	11701	11753	11806	11858	11910	274.07
38	11963	12016	12068	12121	12174	12227	12281	12334	12387	12441	281.48
39	12494	12548	12602	12656	12710	12764	12819	12873	12928	12982	288.89
40	13037	13092	13147	13202	13257	13312	13368	13423	13479	13535	296.30
41	13591	13647	13703	13759	13815	13872	13928	13985	14042	14099	303.70
42	14156	14213	14270	14327	14385	14442	14500	14558	14615	14673	311.11
43	14731	14790	14848	14906	14965	15024	15082	15141	15200	15259	318.52
44	15318	15378	15437	15497	15556	15616	15676	15736	15796	15856	325.93
45	15917	15977	16038	16098	16159	16220	16281	16342	16403	16465	333.33
46	16526	16587	16649	16711	16773	16835	16897	16959	17021	17084	340.74
47	17146	17209	17272	17335	17398	17461	17524	17587	17651	17714	348.15
48	17778	17842	17905	17969	18033	18098	18162	18226	18291	18356	355.56
49	18420	18485	18550	18615	18680	18746	18811	18877	18942	19008	362.96
50	19074	19140	19206	19272	19339	19405	19472	19538	19605	19672	370.37
51	19739	19806	19873	19940	20008	20075	20143	20211	20279	20347	377.78
52	20415	20483	20551	20620	20688	20757	20826	20894	20963	21032	385.19
53	21102	21171	21241	21310	21380	21450	21519	21589	21659	21730	392.59
54	21800	21870	21941	22012	22082	22153	22224	22295	22366	22438	400.00
55	22509	22581	22652	22724	22796	22868	22940	23012	23085	23157	407.41
56	23230	23302	23375	23448	23521	23594	23667	23741	23814	23888	414.81
57	23961	24035	24109	24183	24257	24331	24405	24480	24554	24629	422.22
58	24704	24779	24854	24929	25004	25079	25155	25230	25306	25381	429.63
59	25457	25533	25609	25686	25762	25838	25915	25992	26068	26145	437.04
60	26222	26299	26376	26454	26531	26609	26686	26764	26842	26920	444.44

P. P.
7.41

1	.74
2	1.48
3	2.22
4	2.96
5	3.70
6	4.44
7	5.19
8	5.93
9	6.67

Use with preceding column only.

at the centerline of each station.   The latter may be obtained by scaling the depth on a profile drawing or sighting on a graduated rod as from A to B in Fig. 6 (c).   Side slopes of an earth excavation are usually about 45 degrees and the slope is given on the drawings as $1\frac{1}{2}$ to 1, 1 to 1, etc., which means " $1\frac{1}{2}$ foot horizontal to 1 foot vertical," " 1 foot horizontal to 1 foot vertical," etc. See Fig. 6 (c).

Tables 1 and 2 give the cubic yards of excavation per 100-foot stations for a 16-foot roadway and side slopes of 1 to 1 and $1\frac{1}{2}$ to 1, respectively.   Tables 3 and 4 give the corresponding data on roadways 28 feet wide.   These tables are also applicable to the determination of the volumes of fills, since the inverted cross-section of a typical cut is the cross-section of a typical fill.

ILLUSTRATION: How many cubic yards of excavation are involved in the cut shown in Fig. 6 (a) if the roadway is 16 feet wide, the side slopes $1\frac{1}{2}$ to 1 and the centerline depth in feet at the various stations 100 feet apart as follows: a, 0.0; b, 4.7; c, 10.4; d, 15.3; e, 14.7; f, 12.1; g, 6.2; h, 1.2?

Table 2 applies to the conditions of this problem.   Taking the values from this table for the depths (or heights) corresponding to each station, we obtain the following total:

Station	Height Feet	Cubic Yards per 100-Ft Station
a	0.0	0.0
b	4.7	401.2
c	10.4	1217.2
d	15.3	2207.2
e	14.7	2071.6
f	12.1	1530.4
g	6.2	581.0
h	1.2	79.1
		8087.7 cu. yd. (Ans.)

This would be given in an estimate as 8100 cu. yd.

Borrow Pit Excavation.—When a fill of earth is to be made, the material is taken from what is called a " borrow pit." It is often necessary to measure the amount of material which has been removed from such a pit, and since its shape is generally irregular, tables cannot be used and the average-end-area method is often applied.

As in the case of the measurement of right-of-way excavations, the determination of the volume of a borrow pit requires the use of a base line and a determination of the profiles of the ground before and after excavation at right angles to and at regular intervals along the base line.   Figure 8 shows the plan of a borrow pit with the base line and stations.   In practice, the setting of the base lines and the measurement of the profiles is the work of a

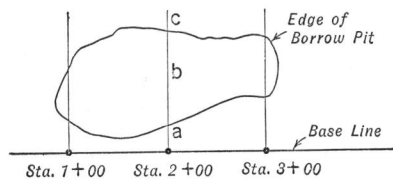

FIG. 8.

surveying party and this phase is beyond the scope of this book. We only propose to show how the volume of excavation is computed after the surveying notes have been made and plotted. For the sake of simplicity only a minimum number of cross-sections will be used and each of these as elementary as possible.

In Fig. 9 let abc represent a profile of the original ground surface of a borrow pit at a point such as at Sta. 2 + 00 in Fig. 8, and let $adec$ represent the final ground surface.   The two form a cross-section.   It will be noted that by referring the points in this figure to a reference line such as fh and dropping perpendiculars, a number of trapezoids are formed.   From geometry we know that the area of trapezoid $abgf$ is

$$\frac{(af + \overline{bg})}{2} \times \overline{fg} \quad \text{or} \quad \tfrac{1}{2}(\overline{af} \times \overline{fg} + \overline{bg} \times \overline{fg})$$

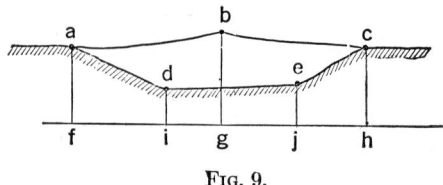

Fig. 9.

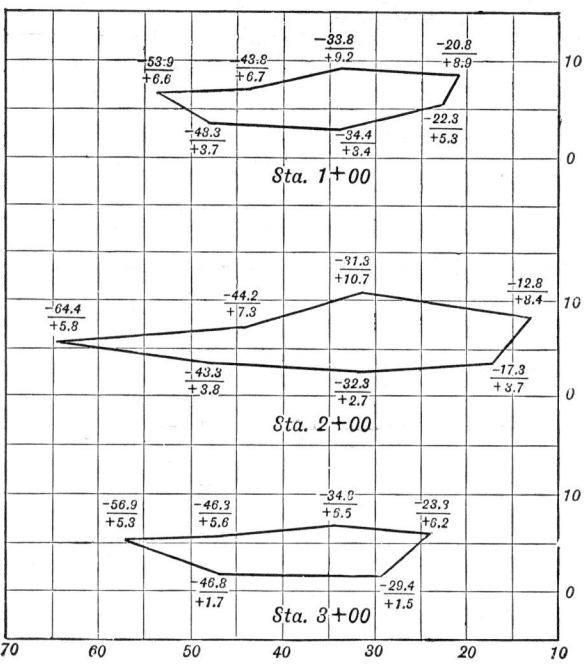

Fig. 10.

We can similarly find the area of each of the other trapezoids in the figure by multiplying half the sum of the two sides by the base. Then, if we subtract the sum of the areas *adif*, *deji*, and *echj*, from the sum of the areas of *abgf* and *bchg*, it is obvious that the remainder will be the area sought, *abced*. This is the principle on which is based the method of computation described in the next paragraph.

Figure 10 represents three cross-sections such as might be obtained from an excavation such as is shown in plan in Fig. 8. Each " break " in the ground level is represented by a point on the plotted cross-section. At each point the figure above the line is the distance from the base line (distances to the left are marked as − and those to the right as +) while the figure below the line is the elevation above or below an arbitrarily selected grade (marked + if above and − if below). In this case the base line is represented by the right-hand margin. Beginning at any point, proceed clockwise around the figure multiplying each elevation by the distance for the point next in advance minus the distance for the preceding point, with due observance of algebraic signs. The algebraic sum of these products divided by 2 is the area.

ILLUSTRATION (Station 1 + 00, Fig. 10)

$$
\begin{array}{lrr}
 & + & - \\
+6.6[-43.8-(-48.3)] = & 29.7 & \\
+6.7[-33.8-(-53.9)] = & 134.7 & \\
+9.2[-20.8-(-43.8)] = & 211.6 & \\
+8.9[-22.3-(-33.8)] = & 102.4 & \\
+5.3[-34.4-(-20.8)] = & & 72.1 \\
+3.4[-48.3-(-22.3)] = & & 88.4 \\
+3.7[-53.9-(-34.4)] = & & 72.2 \\
\hline
 & +\ 478.4 & -\ 232.7 \\
 & -\ 232.7 & \\
 & 2)\overline{245.7} & \\
 & 122.8 \text{ sq. ft.} &
\end{array}
$$

By carrying out a similar computation for the areas of the cross-sections at stations 2 + 00 and 3 + 00 we find that these are 220.6 sq. ft. and 112.4 sq. ft. respectively.   Then, by the average-end-area method, the volume is the product of the average of the areas of two adjacent cross-sections and the distance between them.   In this case the distance between cross-sections is 100 feet. In actual practice these computations involve many cross-sections and they can be handled most conveniently in tabular form as follows:

Station	Area, Square Feet	Average Area, Square Feet	Distance, Feet	Volume, Cubic Feet
1+00	122.8			
		171.7	100	17,170
2+00	220.6			
		166.5	100	16,650
3+00	112.4			27)33,820
				1,253 cu. yd.
				(Ans.)

Planimeter Measurements.—If cross-sections are plotted on coordinate paper to a scale of 1 in. = 10 ft, then 1 sq.in. on the paper represents 100 sq.ft.   An instrument known as a planimeter,

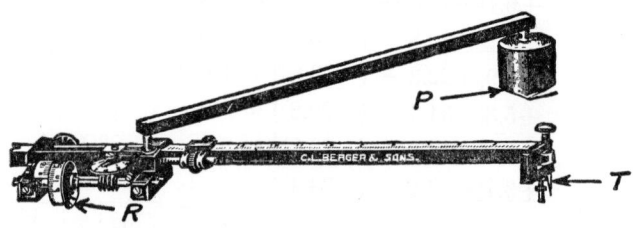

Fig. 11.

of which one form is shown in Fig. 11, is a convenient and fairly accurate device which may be used for measuring directly the areas plotted on paper.   It consists of a point P which is held

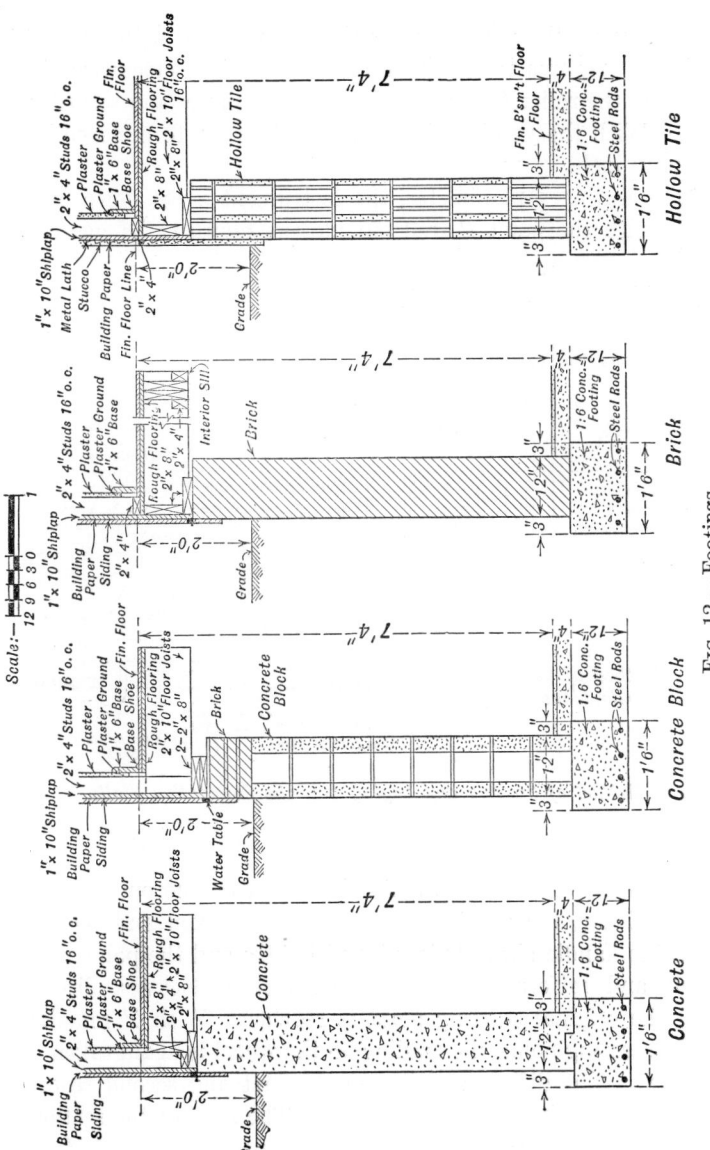

FIG. 12.—Footings.

stationary on the paper by a weight, a point T with which the outline of an area to be measured is traced in a clockwise direction, and the roller R which slides over the paper and records the area. With the vernier, the area may be read to hundredths of a square inch.

In operating a planimeter, the instrument is set down on the paper with the point P well outside the figure to be measured.

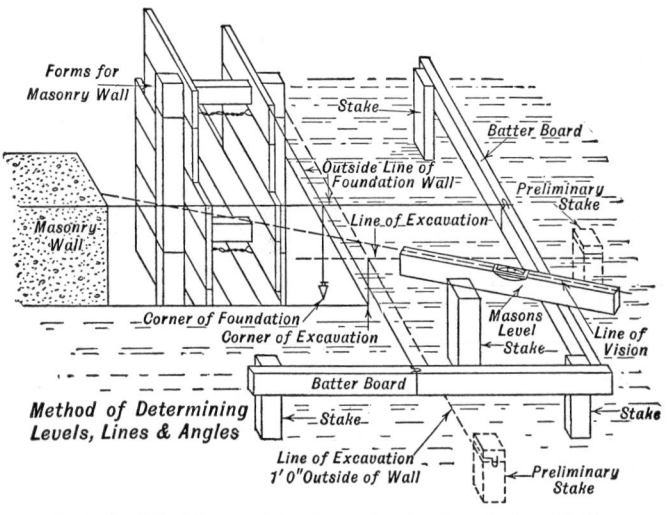

Method of Staking and Laying out the Foundation Walls

Fɪɢ. 13.

Then point T is moved to a starting point on the outline of the figure. Next, the reading on the roller is taken and recorded. Then the handle above T is gripped lightly and the point moved slowly and uniformly along the outline in a clockwise direction until the original starting point is reached. With T remaining on the starting point, the reading on the wheel is again taken and recorded. The difference between the first and second readings is the area outlined in square inches. The area of the section is then converted to square feet by multiplying by the scale factor.

Dwelling Foundations.—The type of foundation selected for a structure depends on the weight of the structure and the allowable bearing capacity of the soil.   In the case of ordinary dwellings, however, the weight of the building distributed over the foundation wall results in a unit pressure on the soil so low that no special treatment is necessary for adequate support.   Of more concern in this case is the matter of even settlement to prevent cracks in walls and plaster.   It is therefore a rather general practice to build a footing somewhat wider than the foundation wall as shown in Fig. 12.   The use of steel rods at the bottom makes the footing act as a beam and results in better distribution of pressures.

The foundation is staked out on the ground after the excavation has been completed.   This consists of setting batter boards as shown in Fig. 13 and stretching lines between them when needed to align the walls.   Here again a right angle formed by the lines can be checked by the method discussed on page 280.   If a grade stake has been set by an instrument man and properly preserved, elevations may be checked by the use of a mason's level as shown in Fig. 13.

The footing, if concrete, may be poured directly into the excavation made for it without form work.   It then serves the useful function of being a solid and level support for concrete forms which may be erected on it as shown in Fig. 14.

Heavy Foundations.—Foundations which must carry heavy loads require careful attention to the bearing capacity of the soil. Safe bearing capacities in tons per square foot are:

Material	Allowable Bearing Tons per Sq. Ft.
Quicksand and alluvial soil	$\frac{1}{2}$
Soft clay	1
Moderately dry clay, fine sand	2
Firm and dry loam or clay	3
Compact coarse sand or stiff gravel	4
Coarse gravel	6
Gravel and sand, well-cemented	8
Good hardpan or hard shale	10
Very hard native bedrock	20
Rock under caissons	25

The unit soil pressure exerted by a foundation is computed by dividing the total weight by the bearing area.

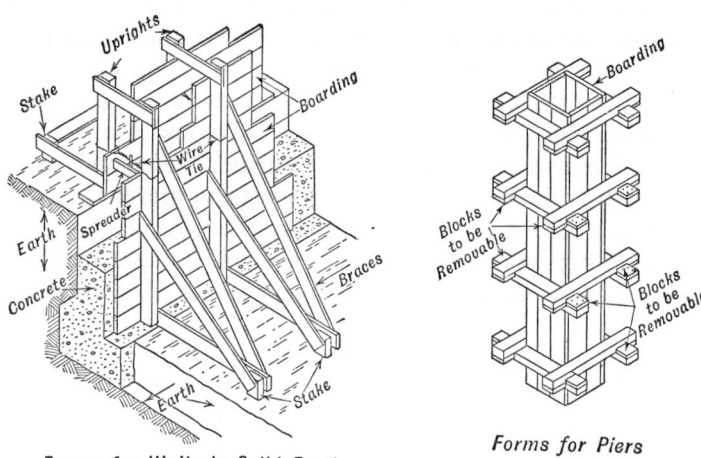

Forms for Walls in Solid Earth

Forms for Piers

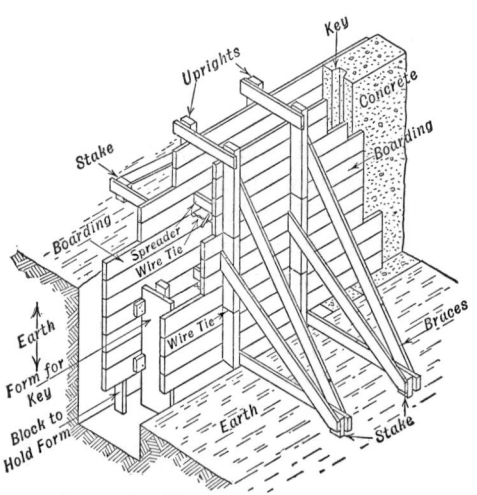

Forms for Wall in Soft Earth and
Method of Keying Wall for a Halt in Concrete

FIG. 14.

ILLUSTRATION: A building weighing 100 tons rests on a footing 18 inches wide and of a total length of 84 feet. What is the unit soil pressure?

Bearing area = 1.5 × 84 = 126 sq. ft.

Unit soil pressure = $\dfrac{100}{126}$ = 0.8 ton per sq. ft. (Ans.)

ILLUSTRATION: A spread footing 8 feet square carries a column load of 55 tons. What is the unit soil pressure?

Bearing area = 8 × 8 = 64 sq. ft.

Unit soil pressure = $\dfrac{55}{64}$ = 0.86 ton per sq. ft. (Ans.)

ILLUSTRATION: What area of bearing is required to support a load of 72 tons on coarse gravel?

The safe bearing on coarse gravel is 6 tons per square foot. Then, required bearing area is

$$\frac{72}{6} = 12 \text{ sq. ft. (Ans.)}$$

ILLUSTRATION: What area of bearing is required for a load of 536,000 pounds on firm clay?

$$\text{Changing to tons, } \frac{536,000}{2000} = 268 \text{ tons}$$

Safe bearing on firm clay is 3 tons per square foot. Then, bearing area required is,

$$\frac{268}{3} = 89\tfrac{1}{3} \text{ sq. ft. (Ans.)}$$

Weights of Structures.—It is obvious from these illustrations that the determination of the weights of structures is a necessary preliminary to the determination of bearing pressures. Average unit weights of building materials in pounds per cubic foot which may be used for this purpose, are shown in Table 5:

TABLE 5

Average Weights of Various Materials

Material	Weight, lb. per cu. ft.	Material	Weight, lb. per cu. ft.
Masonry Construction		**Timber, U. S. Seasoned**	
Ashlar, granite..............	165	(Moisture, 15–20 per cent)	
Ashlar, limestone, marble....	160	Ash....................	45
Ashlar, sandstone, bluestone.	140	Beech.................	46
Dry rubble, granite..........	130	Birch.................	32
Dry rubble, limestone, marble.	125	Butternut..............	26
Dry rubble, sandstone, blue-		Cedar.................	39
stone...................	110	Cherry................	44
Mortar rubble, granite.......	155	Chestnut..............	41
Mortar rubble, limestone,		Cypress................	30
marble..................	150	Elm...................	40
Mortar rubble, sandstone,		Fir, Douglas Spruce......	34
bluestone................	130	Hemlock...............	29
Brick masonry, common brick	120	Hickory...............	48
Brick masonry, pressed brick.	140	Mahogany.............	44
Concrete masonry:		Maple, hard............	42
cinder concrete..........	115	Maple, white...........	33
slag concrete.............	130	Oak, red, black.........	42
stone or gravel concrete,		Oak, white.............	46
plain.................	145	Pine, Oregon...........	32
stone or gravel concrete,		Pine, red..............	30
reinforced.............	150	Pine, white.............	28
		Pine, yellow, long leaf....	44
Stone, Quarried, Piled		Pine, yellow, short leaf...	38
Basalt, granite, gneiss........	96	Poplar.................	27
Limestone, marble, quartz....	95	Redwood..............	28
Sandstone.................	82	Spruce................	27

A common building material omitted from this list is steel, which has a unit weight of 490 lb per cu ft. The reason for the omission is that its weight in structures is not estimated on a volume basis. The weight of each member is determined separately from the tables of weights per linear foot of structural steel which are found in such handbooks as the AISC Steel Construction Manual.

In estimating the weights of buildings and other structures for the purpose of determining bearing pressures, the quantity of each material must often be computed separately. This divided by the unit weight in the above table gives the total weight of that material. Particular attention must be paid to the *distribution* of weights so that the computed bearing on a pedestal footing, for example, will represent only the load transmitted to that particular footing.

The weight of walls, floors, roofs, partitions, and all permanent construction is called *dead load*. All other loads are variable loads or *live loads*. Live loads per square foot of floor should be figured with the following *minimum* values:

Type of Structure	Live Load, Pounds per Square Foot
Dwelling, apartment, hotel	40
Hospitals, operating rooms	60
Office building, first floor	100
" " , all floors above the first	80
School, classrooms	40
Assembly halls, fixed seats	60
Ordinary stores, light manufacturing, light storage	125
Dining rooms, public	100
Theaters, orchestra and balconies	60

When computing bearing pressures, both the dead loads and the live loads must be taken into consideration in the following relation: For warehouses and factories, full dead and full live loads; for stores, light factories, churches, school houses, and places of public amusement or assembly, full dead and 75 percent of live loads; for office buildings, hotels, dwellings, apartment houses, full dead load and 60 percent of live load.

Pile Foundations.—When a ground surface does not have sufficient bearing power to support a structure but is underlaid by a stratum of satisfactory material, it is common practice to support the structure on piles. Wooden piles should be sound straight timbers, not less than six inches in diameter at the point or less than twelve inches at the butt. Piles are driven point downward by the successive blows of a hammer which is either permitted to fall a considerable distance or is actuated by steam pressure through short strokes.

When piles are driven to rock or hardpan, that is, driven to " refusal," the safe sustaining power is that of the pile as a column, provided that the maximum load on any pile should not exceed 20 tons. When a pile is not driven to refusal, its bearing capacity may be determined by the following formulas, known as the ENGINEERING NEWS formulas:

The safe load in pounds when a drop hammer is used is,

$$p = \frac{2wh}{s+1}$$

when w = weight of hammer in pounds
h = fall of the hammer in feet
s = penetration in inches under the last blow
p = safe load in pounds

ILLUSTRATION: What is the safe bearing power of a pile which penetrates $\frac{3}{4}$ inch under the last blow of a 3000-lb hammer dropping 20 feet?

$$p = \frac{2wh}{s+1} = \frac{2 \times 3000 \times 20}{\frac{3}{4}+1}$$

$$p = \frac{120\,000}{1.75} = 68,000 \text{ lb} = 34.3 \text{ tons}$$

This exceeds the maximum allowable of 20 tons, so the pile cannot be counted on to carry more than 20 tons. (Ans.)

When a steam hammer is used, the safe load P per pile may be computed from the following formula:

$$P = \frac{2h(w + am - b)}{s + k}$$

in which w = weight of striking part of hammer, in lb.

h = stroke, or height of fall, in ft.

a = effective area of piston, in sq. in.

m = mean effective pressure of the steam on the downward stroke, in lb. per sq. in.

b = total back pressure, in lb.

k = a constant, sometimes taken at from 0.1 to 0.3. The Navy Department uses $k = 0.3W/w$, in which W is the weight of the pile, in lb., and w is the same as above.

X

Concrete

Definitions.—The word " concrete " now generally refers to masonry material which is made from Portland cement, water, sand, and stone or gravel. *Portland Cement* is the product formed by pulverizing the clinker produced by heating to incipient fusion a properly proportioned mixture of siliceous, argillaceous, and calcareous material. *Water* for concrete must be fresh water and free from injurious salts or organic material. *Sand,* for concrete, also referred to as *fine aggregate,* must be graded, clean, and free from clay, loam, or organic impurities. The *coarse aggregate* is gravel or crushed rock. This should be made up of strong and unlaminated particles no larger in size than half the thickness of the thinnest section of concrete to be poured.

The *proportion* of other material to concrete is usually given by volume. Cement is measured by the bag, which contains 1 cu. ft., and the aggregates in cubic feet. It is standard practice to designate a concrete mix by three numbers, such as 1 : 3.75 : 5. The first number always refers to the number of bags of cement; the second number is the number of cubic feet of fine aggregate, usually sand, and the third number is the number of cubic feet of coarse aggregate, such as pebbles, crushed stone, steam cinders, etc.

Strength of Concrete.—In many concrete structures, the concrete is called upon to carry a certain amount of load. In other structures such as dwelling foundations, strength is a secondary factor, but perviousness or water-tightness is an important consideration. Whether density or strength is desired, both are arrived at simultaneously by proper proportioning of the ingredients of the mix.

Field and laboratory experiments have determined that with concrete of a given plastic consistency (not so stiff as to be harsh and not soft enough to permit the aggregates to separate) the strength will depend upon the ratio of water to cement in the mixture; the smaller this ratio, the stronger the resulting concrete.

In heavy construction work such as hydroelectric dams, bridge foundations, and large buildings, the water-cement ratio is calculated from engineering data and carefully maintained throughout the job. Climatic conditions, exposure to sea water, and other environmental factors are all taken into account in determining the correct water-cement ratio.

Research and practical experience under actual job conditions have produced concrete mixes suitable for a number of ordinary types of jobs. Several of these mixes are shown in Table 1 which includes the recommended quantities of water for different classes of work as well as the proportions of cement to sand and coarse aggregate that have proved successful for each type of construction.

A trial batch should be mixed and tested for workability before the final proportions of a particular mix are fixed. It may be found that the exact proportion of water given in Table 1 results in a mixture that is too stiff, too wet, or lacking in smoothness and workability. This situation may be remedied easily by changing the amount of the aggregates slightly, but *not the water*. If the mix is too wet, add sand and pebbles slowly until the desired stiffness is obtained. If the mix is too stiff, cut down the proportion of the sand and coarse aggregate *slightly* in another trial batch.

In the selection of trial proportions of the aggregates, it should be remembered that increasing the proportion of coarse aggregate up to a certain point reduces the cement factor. Beyond this point the saving in cement is very slight, while the deficiency in mortar increases the labor cost of placing and finishing. An excess

TABLE 1

QUANTITIES OF CEMENT AND AGGREGATES FOR VARIOUS JOBS

Kind of Concrete Work	Mix by Vol. Materials Cubic Feet			Workability of the Mix	Gals. Water per Bag When Mixing	One Bag Batch Makes This Much Concrete — Cu. Ft.	Materials for One Cubic Yard of Concrete			
	Cement Bags	Sand Cu. Ft.	Stone Gravel Cu. Ft.				Cement Bags	Sand Cu. Ft.	Stone Gravel Cu. Ft.	Gals. Water When Mixing
Footings, Heavy Foundations	1	3.75	5	Stiff	6.4	6.2	4.3	16.3	21.7	27.6
Watertight Construction, Walls	1	2.5	3.5	Med.	5	4.5	6.0	15.0	21.0	30
Driveways Floors Walks } One Course	1	2.5	3	Stiff	4.4	4.1	6.5	16.3	19.5	28.7
Driveways Floors Walks } Two Courses	1	Top 2	0	Stiff	3.6	2.14	12.6	25.2	0	45.3
	1	Bottom 2.5	4	Stiff	4.9	4.8	5.7	14.2	22.8	27.8
Pavements	1	2.2	3.5	Stiff	4.3	4.2	6.4	14.1	22.4	27.5
Watertight Construction for Tanks, Wells, Cast units like posts, slabs	1	2	3	Med.	4.1	3.8	7.1	14.2	21.3	29.3
				Wet	4.9	3.9	6.9	13.8	20.7	33.7
Heavy Duty Floor, Barns, Shops, etc.	1	1.25	2	Stiff	3.4	2.8	9.8	12.3	19.6	33.9
Mortar for Brick, Concrete Blocks	1	6	1 Sack 50 lb. Hydrated Lime	Med.	12.5	5.5	4.9	29.4	5	61.2

of coarse material will produce mixtures that are undersanded and harsh. (See Fig. 1.)

Estimating Quantities. Table 1 may be used for estimating the quantities of cement and aggregates needed for a job, and hence the cost of materials, based on local prices. For example,

A concrete mixture in which there is not sufficient cement-sand mortar to fill spaces between pebbles. Such a mixture will be hard to work and will result in rough, honeycombed surfaces.

A concrete mixture which contains correct amount of cement-sand mortar. With light troweling all spaces between pebbles are filled with mortar. Note appearance on edges of pile. This is a good workable mixture and will give maximum yield of concrete with a given amount of cement.

A concrete mixture in which there is an excess of cement-sand mortar. While such a mixture is plastic and workable and will produce smooth surfaces, the yield of concrete will be low. Such concrete is also likely to be porous.

Fig. 1

a footing 27' long, 2' wide, and 1' thick requires 54 cu. ft. or 2 cu. yds. of concrete. A 1 : 3.75 : 5 mix is recommended for this type of work by Table 1. To make 1 cu. yd. of this mix it is necessary to use 4.3 bags of cement, 16.3 cu. ft. of sharp sand, and 21.7 cu. ft. of gravel or crushed stone, and 27.6 gal. of clean water. Although the total volume of the materials when dry is much greater than 1 cu. yd., the addition of water reduces the net volume by causing the sand and cement to fill in the voids in the coarse aggregate. To obtain the 2 cu. yds. of concrete required for the footing it is necessary to double the quantities just given for one cubic yard. No allowance for waste is included in Table 1 and this factor cannot be ignored. It is customary to allow about 5% to 10% for material waste, on the job, so the following quantities should be purchased for the footing: cement, 9 bags; sand, 1⅓ cu. yds.; gravel or crushed stone, 1¾ cu. yds.

Mixing concrete or mortar involves several simple operations and requires some forethought. Excellent concrete can be mixed by hand, although machine mixing is preferred because better quality concrete results.

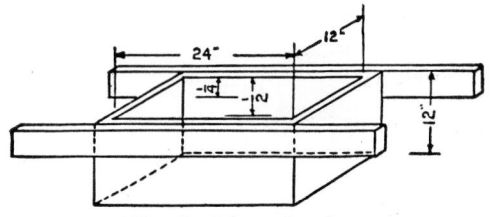

FIG. 2.—Measuring box.

Measuring Box. A simple and convenient measure for proportioning aggregates is a bottomless measuring box, shown in Fig. 2. This measure is simply a strong box with no bottom, made to exact *inside* measurements. The handles help in lifting or moving the box. As shown it has a capacity of 2 cu. ft., with marks on the inside of the box for 1 cu. ft. and for ½ cu. ft.

Selection of Consistency.—Consistency relates to the fluidity of the concrete. Figure 3 illustrates three samples of concrete with consistencies varying from "stiff" to "wet." It is important that the consistency be suited to the work to be done. Stiff

Fig. 3.—Showing Stiff, Medium, and Wet Mixtures of Concrete.

The stiff consistency is recommended for footings, walls, and pavements. The medium mix is suitable for tank walls, floors, slabs, beams, etc. The wet mix is suitable for thin walls and columns.

concrete is best suited for foundations, pavements, and massive walls. When poured into forms, it may require considerable spading to obtain smooth faces. Concrete of medium consistency is recommended for beams, slabs, and walls, where a smooth face and good bond with reinforcing steel is essential. A wet consistency should only be used in thin walls or columns or where the reinforcing is so heavy that spading is difficult or impossible.

Mixing the Concrete.—Except for small quantities mixed by hand, practically all concrete is mixed in mixing drums of the batch type. Portable mixers are usually provided with skips for charging the drum and one of the problems is that of measuring the aggregates to be dumped into the skip. In highway work a batching plant is often used and the properly proportioned aggregates are dumped into the skip from batch boxes. In charging from open stock piles, the wheelbarrow serves the double purpose of conveying the material and measuring it at the same time. The ordinary wheelbarrow used in construction work holds two cubic feet when struck off and three cubic feet when heaped. If there is any doubt as to its capacity, a wheelbarrow can readily be calibrated by filling it with sand from a box with 12 inches for each internal dimension.

On construction projects with central mixing plants the aggregates are measured in hoppers which must be calibrated by computing the internal volumes if they are of the fixed type, or adjusted according to the manufacturer's rating if they are of the patented automatic type.

A mixer should not be charged with material in excess of its rated capacity. Not only does overloading prevent the aggregates from mixing properly, but rich mortar is apt to be lost by slopping out. There is the additional disadvantage that an overloaded mixer may stall its driving engine and the mixer can then usually be started again only after most of its load has been laboriously shoveled out.

Placing Concrete.—Concrete should be deposited into a form at several points so that no appreciable flow results within the form. Such flow causes segregation of the materials and results in porous concrete. The concrete should not be allowed to fall more than a few feet into a form. It should be spaded the minimum amount necessary to make it flow around reinforcing bars and into corners of the form. If water accumulates in a form it should be drained off carefully so that cement will not be lost. A small hole drilled into the form may amply serve this purpose.

Concreting operations should not be undertaken in freezing weather unless provision is made for heating the aggregates and for protecting the newly placed concrete from frost for seventy-two hours.

Concrete will develop greater strength and wearing qualities if it is " cured " by being kept moist for a week or ten days after pouring. This may effectively be accomplished by covering with burlap and sprinkling or, in the case of pavements, by building earth dikes and permitting pools of water to stand on the concrete.

Practical Project—Concrete Stoop.—In order to show how the previous information is used in practice, a practical project will be considered—that of constructing a concrete stoop and steps such as the one shown in Fig. 4. It is a stoop consisting of three steps and is intended for installation at the rear or side entrance of a small house.*

The platform, or upper step, of this stoop is 3' square. Each step is 3' wide, and has an 8" riser, and a 12" tread. All the dimensions of this project are shown in the accompanying illustra-

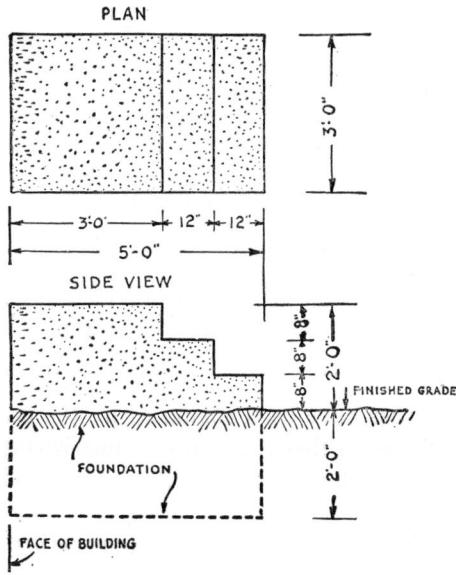

Fig. 4.—Dimensions of Stoop and Steps.

tion (Fig. 4). Note that the foundation extends 2' below grade (ground level).

The materials needed to build this project consist of the lumber for the forms, shown in Fig. 5, and the materials for making the concrete. The quantities of these materials are listed in Table 2.

* This project is taken from *Masonry* by K. H. Bailey, D. Van Nostrand Co., Inc., New York.

TABLE 2

BILL OF MATERIAL

Number Re- quired	Part	Size	Kind of Material
2	Sides of Top Step	$3' 1'' \times 8'' \times 1''$	Wood
2	Sides of Middle Step	$4' 1'' \times 8'' \times 1''$	"
2	Sides of Bottom Step	$5' 1'' \times 8'' \times 1''$	"
3	Front of Steps (risers)	$3' \times 8'' \times 1''$	"
1	Brace	$3' \times 4'' \times 2''$	"
6	Stakes for Holding Form	$1' 6'' \times 2'' \times 4''$	"
12	Cleats	Various	"
9		Bags	Cement
36		Cubic Feet	Sand
45		" "	Gravel & Crushed Stone
56		Gallons	Water

These quantities have been computed by the use of Table 1, page 292, which shows that a 1 : 3.75 : 5 mix is suitable for this kind of work. A batch amounting to 2 cu. yds. of mixed concrete will be obtained.

The Excavation. The first step in constructing this stoop is the excavation. It should be made at the point where the steps are to be installed. It should be 2' deep and its dimensions are $5' \times 3'$. When digging, take care to cut the sides of the hole so that no dirt or stones fall into the space, because these earth sides are used to hold the concrete, so that no form is necessary below grade. The foundation of the building forms one end of the hole, and must be well cleaned of all dirt so that the concrete can bond with it. Tamp the earth in the bottom of the hole, unless it is already firm, to produce a sound base for the concrete.

The Form. When the excavation is complete, work can begin on the form. Naturally, it must be made to the exact dimensions given, with all measurements made on the inside. It is usually convenient to build the form in a shop or workroom and place it on the foundation afterwards. This is easy to do, if it is nailed together securely, as shown in Fig. 5, before it is moved.

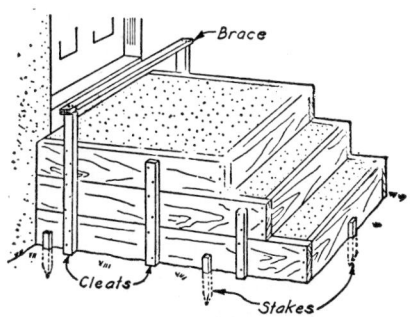

Fig. 5.—The Form.

In building the form, nail the three side pieces together first, using the cleats for that purpose, as shown in Fig. 5. Then make the other side in exactly the same way. Finally, the two sides are nailed to the front pieces, the risers, which must be carefully cut to size. The joints should be made as close as possible to prevent leaks. The last step in making the form is to nail on the brace that appears in Fig. 5. This is the piece of $2'' \times 4''$ exactly 3' long, which acts as a spreader to keep the sides of the form at the proper distance apart until some concrete is poured. Be sure to remove it before the concrete is high enough to touch it.

Pouring the Foundation. The first step in concrete work is to pour the foundation, the part below ground level, and let it set for at least a day before doing the work above grade. Then the form is set in place and the job is finished. To mix the concrete for the foundation, take 60% of the total materials given, that is, $5\frac{1}{2}$ bags of cement, $21\frac{1}{2}$ cu. ft. of sand, 27 cu. ft. of gravel or crushed stone, and $33\frac{1}{2}$ gals. of water. Mix these materials, and pour the mixture into the excavation, following the methods explained earlier in this section. Remember that thorough mixing is most important. More troubles result from too little mixing than from too much, especially with stiff concrete such as is used for this job.

After this foundation has set for at least one day, set up the form on it. The form must be placed so that it is level and square with the house before it is fastened and braced firmly into place. It is secured by means of the 2″ × 4″ stakes illustrated, and also by 3 or 4 diagonal braces (not illustrated) running from the top edge of the sides to stakes in the ground outside. Put these wherever the form needs support to prevent moving, or spreading under weight of the concrete. Chips of brick or slate should be used to wedge up the form by inserting them between the bottom edges of the side pieces and foundation. As a final check on the form after it has been placed in position, but before the stakes and braces are secured, measure the diagonals across the square that will form the stoop. If both are the same length the sides are square, parallel, and ready to be fastened into position as rigidly as possible.

Platform and Steps. When the form has been properly placed, the next step is to spread a thick cement paste over the concrete foundation to bond it with the new concrete for the steps. To make this, mix ½ bag of cement with 3 gals. of water. Then spread the paste evenly over the top of the foundation. This is done immediately before pouring the final batch of concrete, which is prepared by mixing the remainder of the materials, 3½ bags of cement, 14½ cu. ft. of sand, 18 cu. ft. of gravel or crushed stone, and 22½ gals. of water. This pouring must be done very carefully to avoid hitting or jarring the form out of position. After the operation is complete, the form should be filled to the top of the steps and platform. If water collects on these flat surfaces it should be removed with a sponge or brush, so that the concrete is left moist, but never puddled.

As soon as the concrete has set a little, that is, after a few hours, the platform and steps may be finished with a wood float.

This produces an even, yet gritty surface that is not slippery in wet weather. A small trowel or dull knife blade should be run along the inside edge of the form between the boards and concrete to make a smooth and slightly rounded edge on all exposed surfaces.

Leave the form in position for a week. Then remove it, and cure the concrete by covering it with straw, or burlap bags, and keep damp for a week or 10 days. After this curing, the stoop is ready for use. If desired, an additional finish may be applied by stripping the forms after 24 hours, and trowelling a thick cement paste onto the damp concrete with a wood float. After this finish has had time to harden, the work is cured in the usual manner.

Transit-Mixed Concrete.—The work of installing concrete can be materially reduced by ordering ready mixed concrete to be delivered by truck. This service is available even for relatively small jobs, since most of the suppliers deliver quantities as small as 1 cubic yard. However, the truck cannot "stand by," so that the job must be ready to receive the entire quantity ordered at one pouring. Therefore, if the site is not on a road or driveway, the best plan is to place planking (which can be rented in some localities) so that the truck can drive to the site. Of course, arrangements might be made to have enough men and wheelbarrows on hand to transfer the concrete from the truck to the site, but this plan is seldom desirable for smaller jobs.

In any case, the grade and quantity of the concrete must be carefully determined in advance. The suppliers of transit-mixed concrete do not usually classify their mixes according to the proportions of cement, fine aggregate and coarse aggregate, as explained earlier in this chapter, but according to bearing strength in pounds per square foot. Grades usually offered are 1000, 1500, 2000, 2500, 3000, 3500, 4000, 4500 and 5000 pound strengths. These figures are not, however, actual test results, but include safety factors. Therefore, the supplier should be consulted in choosing the grade for a particular job. One should note, however, that these grades increase in density as they increase in strength—for that reason the higher grades are more resistant to surface abrasion. Thus for footings for residential structures, i.e., homes, garages, etc., the 2000 pound grade is usually considered adequate, while for garage and cellar floors, walks, etc., the 3000 pound grade is usually the minimum specified, even though the recommended practice of using reinforcement to prevent temperature and tension cracking is followed. (Concrete is much weaker in tension than in compression.)

In ordering concrete for transit-delivery, the cubical contents of the volume to be filled are calculated, and an additional percentage added to compensate for any spillage that may occur. The percentage depends upon the method of placement and the bounding surfaces of the volume to be filled. If the surfaces bounding the volume consist entirely of wooden forms, and if the arrangements for pouring do not indicate that much loss is likely, the excess figure can be held to 10%. On the other hand, if one bounding surface is the coarse aggregate foundation for a cellar or garage floor, the increase should be 20 or 25%.

Reinforcement for concrete is of steel, which has the high tensile strength necessary to balance the weak tensile strength of concrete. The two important types of reinforcement are rods, used, for example, in beams and columns, and mesh, used in floors and walks; in roads rods are often used as well as mesh, to carry the tension between sections.

The correct placement of reinforcing is based upon the principle, established by stress analysis, that the greatest tensile stresses occur close to the outside surfaces. The application to the placement of reinforcing bars in columns and beams, and to reinforcing mesh is floors and roads, is shown in Fig. 6.

The use of additives in concrete is often found to improve special properties. The incorporation of asphalt and pitch emulsions, while producing a dark color, gives improved resistance to abrasion and thermal cracking. Waterproofing can also be effected by additives, in addition to the membrane method shown in the road sketched in Fig. 6. A representative formulation for a waterproofing additive is calcium chloride 25%. sodium silicate

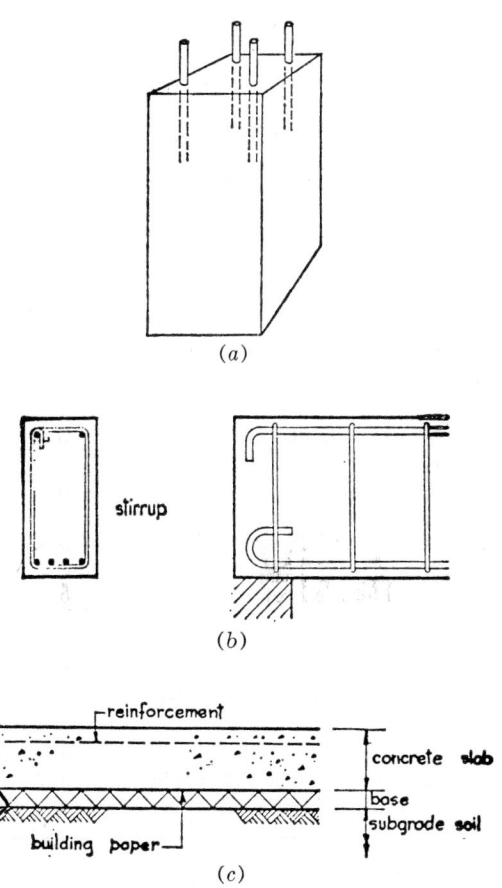

Fig. 6.—Steel Reinforcing for Concrete as Placed in (a) Columns; (b) Beams; and (c) Roads and Walks.

3%, ground silica 3%, and water 69%. One quart of this mixture is added to the concrete per bag of cement used. A 2% solution of ammonium stearate is also used for this purpose. Additives are also used for coloring concrete, according to the color desired as shown in Table 3.

TABLE 3

FORMULAS FOR COLORING CONCRETE

| | | Pounds per Bag of Cement | |
Desired Color	Pigment	Light Shade	Medium Shade
Black, Blue-black and Grays	Germantown Lampblack	½	1
	Carbon Black	½	1
	Black Oxide of Manganese	1	2
Blue	Ultramarine Blue	5	9
Brownish Red to Dull Brick Red	Red Oxide of Iron	5	9
Bright Red to Vermilion	Mineral Turkey Red	5	9
Brown to Reddish Brown	Metallic Brown (oxide)	5	9
Buff, Colonial Tint	Yellow Ochre with not less than 15% Yellow Oxide of Iron	5	9
Cream	Small Quantity of Yellow Oxide of Iron		
Green	Chromium Oxide	5	9
	Greenish-blue Ultramarine	6	
Pink	Red Oxide of Iron	4	8

XI

Brickwork

Uses of Brickwork.—Brickwork is well adapted to many kinds of masonry construction and is used extensively in building walls, tunnel linings, small arches, culverts, street paving, sewers, etc. The convenience in handling and laying brick, in forming arches and " rounding " corners makes it particularly useful in these classes of construction. In fire-resisting qualities it is superior to most natural building stone and in general durability it has a high rating.

Bond.—Bricks laid longitudinally in a wall are called *stretchers*. Bricks laid across the wall are called *headers*. (See Fig. 1.)

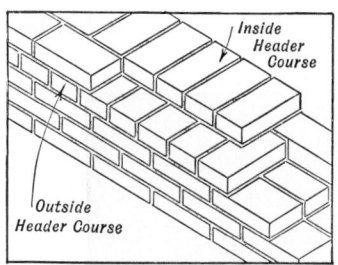

Inside Header Course

Outside Header Course

Fig. 1.—Method of lapping inside and outside header courses in 12″ solid brick walls, common bond.

The method of arrangement of the stretchers and headers in the same or adjacent courses of a wall is spoken of as the *bond*. Cost and appearance are the chief considerations in the selection of style of bond.

Common Bond.—This consists of one course of headers to every four to six courses of stretchers. Local building codes usually specify how many stretcher courses are permitted to each header course, but placing a header course at every sixth course is a safe rule. The header course may be either plain or " Flemish."

English Bond.—This bond is composed of alternate courses of headers and stretchers, the headers centering on the stretchers or the joints between them. This is considered the strongest bond.

315

Flemish Bond.—Each course is made up of alternate headers and stretchers. This bond gives a strong construction and a pleasing appearance.

Types of Joints.—There are four types of joints commonly used in brickwork: (1) shoved joints, (2) grouted joints, (3) open joints, and (4) dry joints.

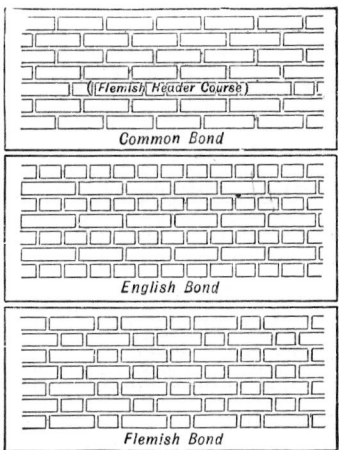

Common Bond

English Bond

Flemish Bond

Fig. 2.

Shoved Joints.—The brick is laid on a bed of mortar a little thicker than the finished joint will be. It is then pressed downward and sideways, the soft mortar rising and filling the vertical joints. This joint produces strong and watertight masonry. (See Fig. 3.)

Grouted Joint.—The brick is laid on a level bed of mortar and the vertical joints are filled with mortar to which water has been added till it is of a " soupy " consistency.

Fig. 3.—Method of Forming Shoved Joint.

Shoved and grouted joints constitute the two types of *filled joints.* This type construction is used in fire, party and division walls, also in chimneys and piers or walls designed for heavy loads.

Open Joints.—This type of joint is often permitted above ground in dwelling construction. It is used principally with common bond and consists of laying the stretcher course on flat beds of mortar and leaving the middle vertical joint unfilled. Each header course has, however, filled joints.

Dry Joints.—This consists of laying a course of bricks directly on top of the lower course with no mortar in between. It is some-

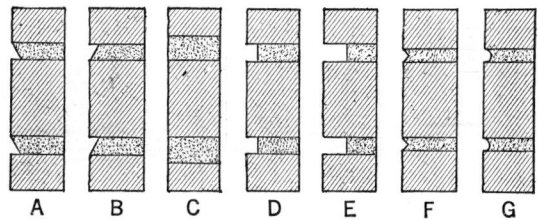

FIG. 4.—Common Types of Joints. (a) Struck joint, (b) weathered joint, (c) flush or plain cut joint, (d) raked joint, (e) stripped joint, (f) "V" joint, (g) concave joint.

times used on the interior face of every sixth course in cheap construction work, but its use is not recommended.

Types of Exposed Joints.—Exposed joints are finished in a variety of ways for aesthetic effect or structural strength. The various types of exposed joints are illustrated in Fig. 4. Some can be made with the trowel and others require special implements.

Fireplaces.—A fireplace that will burn fuel properly and radiate maximum heat can be designed based on proportions gained from experience. The width of the opening should not be too great for moderate size rooms. For a living room with 300 sq. ft. floor space, the width should be 30 to 36 in. The height of the opening should be from 30 to 34 in., regardless of the width.

The combustion chamber must be properly proportioned for

Dimensions for various
fireplace openings.

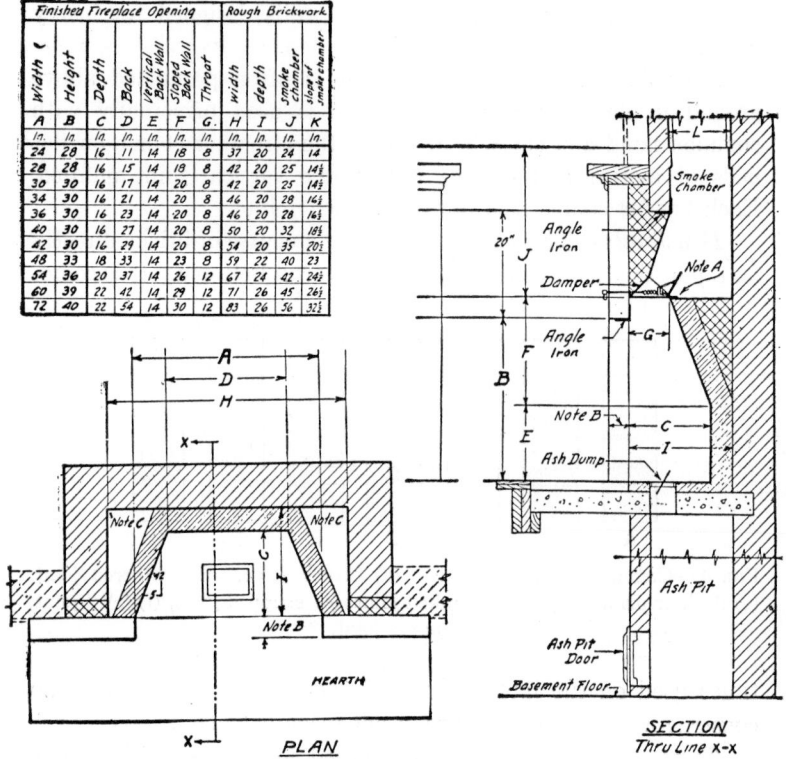

Finished Fireplace Opening							Rough Brickwork			
Width ℄	Height	Depth	Back	Vertical Back Wall	Sloped Back Wall	Throat	width	depth	smoke chamber	slope of smoke chamber
A	B	C	D	E	F	G.	H	I	J	K
In.	In.	In.	In.	In.	In.	In.	In.	In.	In.	In.
24	28	16	11	14	18	8	37	20	24	14
28	28	16	15	14	18	8	42	20	25	14½
30	30	16	17	14	20	8	42	20	25	14½
34	30	16	21	14	20	8	46	20	28	16½
36	30	16	23	14	20	8	46	20	28	16½
40	30	16	27	14	20	8	50	20	32	18½
42	30	16	29	14	20	8	54	20	35	20½
48	33	18	33	14	23	8	59	22	40	23
54	36	20	37	14	26	12	67	24	42	24½
60	39	22	42	14	29	12	71	26	45	26½
72	40	22	54	14	30	12	83	26	56	32½

Fig. 5.—Construction Sketch of Fireplace.

Note A. The back flange of the damper must be protected from intense heat, by being fully supported and shielded by the back wall masonry.

Note B. The drawing shows the brick fireplace front as 4″ thick. No dimensions are given for this particular part of the fireplace because of the variety of materials that may be used to face the opening, such as marble, stone, tile, etc., all of which have different thicknesses.

Note C. These hollow spaces should be filled with masonry to form a solid backing. If it is necessary to include one or more furnace flues in either space beside the fireplace, the overall dimensions of the chimney must be increased to allow at least 8″ of solid brickwork on all sides of each flue.

the sake of both proper draft and heat radiation. The upper part on all sides should slope in gently to the size of the throat. This slope should preferably be no greater than about 30 degrees from the vertical or to a ratio of approximately 3 inches horizontal to 5 inches vertical. The slope should start from a point a little less than halfway up from the hearth to the throat. Not only should the sides slope toward the center, but they should be splayed toward the back. The amount of splay which gives the maximum radiation is about 5 inches for each 12 inches of depth.

A smoke shelf is an essential part of a fireplace. This performs the function of deflecting upward the downward air currents in the chimney. A damper placed at the throat is a further aid to keep smoke from being blown back into the room. A combination metal throat and damper is on the market and is a valuable aid in fireplace construction.

Flue Sizes for Fireplaces.—It is desirable to have a relatively high velocity of the gases through the throat of a fireplace. This

TABLE 1

Fireplace Dimensions			Flue Sizes			
			Rectangular		Circular	
Width of opening, inches	Approximate height, inches	Depth of opening, inches	Outside dimensions, inches	Effective area, square inches	Diameter, inches	Effective area, square inches
24	28	17—20	$8\frac{1}{2}\times8\frac{1}{2}$	41	10	78
28	28	17—20	$8\frac{1}{2}\times13$	70	10	78
30	30	17—21	$8\frac{1}{2}\times13$	70	12	113
34	30	17—21	$8\frac{1}{2}\times13$	70	12	113
36	30	21	$8\frac{1}{2}\times18$	97	12	113
40	30	21—24	$8\frac{1}{2}\times18$	97	15	177
42	30	21—25	$8\frac{1}{2}\times18$	97	15	177
48	32	21—26	$13\ \ \times13$	100	15	177

requires a flue of proper proportions. For a chimney 30 feet or more in height, the flue area should be about one-twelfth the area of the fireplace opening; where the chimney is 20 feet high or less, one-tenth the area is more desirable. In a flue of rectangular cross-section the gases travel up only in the center so that the *effective area* is considerably less than the cross-sectional area. Table 1 shows the sizes of flues required for various sizes of fireplaces with proper reductions made to obtain effective area.

Chimney Construction.—A chimney should be built up from a footing in the basement. Not more than two flues should occupy the same chimney space. Where three or more flues are necessary, a 4-inch partition called a *withe* should be incorporated. Every fireplace, stove, or furnace should have a separate flue.

A chimney should extend at least three feet above the highest point at which it comes into contact with the roof and at least two feet higher than any ridge within ten feet of the chimney.

The thickness of the chimney wall may be 4 inches when a flue lining of fire clay is used, but must be at least 8 inches thick with joints carefully pointed if no flue lining is used.

Mortar for Brickwork.—The mortar used in brickwork may vary considerably depending upon the class of work. For instance, the National Board of Fire Underwriters' Chimney Ordinance requires that chimney mortar be composed of two bags of Portland cement and one bag hydrated lime mixed dry and added to three times its volume of clean sharp sand, and mortar where a structure is exposed to a considerable amount of stress should be composed of one part Portland cement and two parts sand, while for most ordinary brickwork, one part fresh, well-slaked lime to $2\frac{1}{2}$ to 3 parts sand will answer. Between these limits there are various mixtures of Portland cement, natural cement, common lime, and sand. Table 2 gives a few of these mixtures.

Estimating Brickwork.—The standard size of a common brick is 8 in. $\times$ $2\frac{1}{4}$ in. $\times$ $3\frac{3}{4}$ in. and the most common thickness of joint is $\frac{1}{2}$ inch. It is readily seen that the number of bricks in a wall depends on the thickness of the wall, the thickness of the

TABLE 2

Class	Portland Cement	Natural Cement	Common Lime	Clean, Sharp Sand
A.	1	. . . .	. . . .	2
A₁.	1	. . . .	. . . .	2½
A₂.	1	. . . .	. . . .	3
B.	. . . .	1	. . . .	2
B₁.	. . . .	1	. . . .	2½
B₂.	. . . .	1	. . . .	3
C.	. . . .	. . . .	1	2
C₁.	. . . .	. . . .	1	2½
C₂.	. . . .	. . . .	1	3
D.	1	1	. . . .	4
D₁.	1	1	. . . .	5
D₂.	1	1	. . . .	6
			Lime Paste	
E, etc.	1	. . . .	1	2
F, etc.	. . . .	1	1	2

Class A is used in superior building construction, for railroad masonry in general, tunnel lining and sewers; class E for building work of the highest class; class C for common brickwork as in buildings.

joints and the type of the bond. Estimating the quantities for brickwork is greatly facilitated by the use of tables which are here reproduced by courtesy of the Common Brick Manufacturers Association. Table 4 gives the number of courses in brick walls of various heights. Table 3 gives the quantities of brick and mortar in footings, piers, and chimneys of various proportions. Table 5 gives the weights of solid brick walls of unit areas. Table 6 gives the number of bricks in solid walls of various thicknesses. Table 7 gives the quantities of material needed for the mortar.

Use of the Tables—Estimating Problem.—Figure 6 shows a simple brick dwelling with 12-inch walls resting on footings and reaching to the first floor level. From the first floor to the roof the walls are 8 inches thick. The problem is to determine the approximate quantity of brick and mortar needed. Regard the remote sides of the house as having door and window openings equivalent to those shown.

Since the footing is symmetrical about the center line of each wall, the actual length of the footing will be the sum of the distances from center to center of each wall. Since the walls at the base are 12 inches thick, we can obtain the distances from center

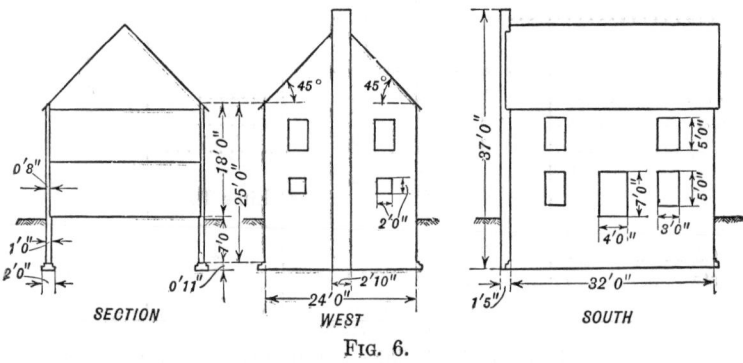

SECTION   WEST   SOUTH

Fig. 6.

to center of each wall by subtracting 6 inches or $\frac{1}{2}$ foot from each end. Then the length of the footing for the south side of the house is $32 - \frac{1}{2} - \frac{1}{2} = 31$ feet. Similarly, the length of the west footing is 23 feet. The total length of footings all around the house is then $31 + 31 + 23 + 23 = 108$ feet. Referring to Table 3 we note that the quantities for footings are given in terms of 100 feet. In this example we have $\frac{108}{100} = 1.08$ hundreds of feet. The quantities given in the table for footings for a 12-inch wall are 2812 bricks and 48 cubic feet of mortar. We multiply these two figures by 1.08 and obtain

$2812 \times 1.08 = 3037$ bricks for the footings
$48 \times 1.08 = $ 51.8 cubic feet of mortar for the footings.

TABLE 3

QUANTITIES OF BRICK AND MORTAR IN FOOTINGS, PIERS, AND CHIMNEYS*

Construction	Number of Brick	Mortar Cu. Ft.
8" Wall	2272	39
12" Wall	2812	48
16" Wall	4592	78

PIERS — Quantities for 10 Ft. Height

Construction	Number of Brick	Mortar Cu. Ft.
8" x 12" Solid	124	2¼
12" x 12" Solid	185	3¼
12" x 16" Solid	247	4½
10¾" x 10¾" Hollow Brick Laid on Edge	113	1

CHIMNEYS — Quantities for 10 Ft. Height

Construction	Number of Brick	Mortar Cu. Ft.
8" x 8" Flue	259	4½
12" x 12" Flue	345	6
12" x 12" and 8" x 8" Flues	539	8½
8" x 8" Flue	173	3
12" x 12" Flue	238	4
12" x 12" and 8" x 12" Flues	367	6½

* Quantities are for **100-foot** lengths of footing.

TABLE 4.

HEIGHT OF SOLID AND IDEAL BRICKWORK BY COURSES

Based on Standard Brick 2¼″×3¾″×8″

Height from Bottom of Mortar Joint to Bottom of Mortar Joint

Number of Courses	⅜″ Joints		½″ Joints		⅝″ Joints		¾″ Joints	Number of Courses
	Brick flat	Brick on edge	Brick flat	Brick on edge	Brick flat	Brick on edge	Brick flat	
1	2⅝″	4⅛″	2¾″	4¼″	2⅞″	4⅜″	3″	1
2	5¼″	8¼″	5½″	8½″	5¾″	8¾″	6″	2
3	7⅞″	1′— 0⅜″	8¼″	1′— 0¾″	8⅝″	1′— 1⅛″	9″	3
4	10½″	1′— 4½″	11″	1′— 5″	11½″	1′— 5½″	1′—0″	4
5	1′— 1⅛″	1′— 8⅝″	1′— 1¾″	1′— 9¼″	1′— 2⅜″	1′— 9⅞″	1′—3″	5
6	1′— 3¾″	2′— 0¾″	1′— 4½″	2′— 1½″	1′— 5¼″	2′— 2¼″	1′—6″	6
7	1′— 6⅜″	2′— 4⅞″	1′— 7¼″	2′— 5¾″	1′— 8⅛″	2′— 6⅝″	1′—9″	7
8	1′— 9″	2′— 9″	1′—10″	2′—10″	1′—11″	2′—11″	2′—0″	8
9	1′—11⅝″	3′— 1⅛″	2′— 0¾″	3′— 2¼″	2′— 1⅞″	3′— 3⅜″	2′—3″	9
10	2′— 2¼″	3′— 5¼″	2′— 3½″	3′— 6½″	2′— 4¾″	3′— 7¾″	2′—6″	10
11	2′— 4⅞″	3′— 9⅜″	2′— 6¼″	3′—10¾″	2′— 7⅝″	4′— 0⅛″	2′—9″	11
12	2′— 7½″	4′— 1½″	2′— 9″	4′— 3″	2′—10½″	4′— 4½″	3′—0″	12
13	2′—10⅛″	4′— 5⅝″	2′—11¾″	4′— 7¼″	3′— 1⅜″	4′— 8⅞″	3′—3″	13
14	3′— 0¾″	4′— 9¾″	3′— 2½″	4′—11½″	3′— 4¼″	5′— 1¼″	3′—6″	14
15	3′— 3⅜″	5′— 1⅞″	3′— 5¼″	5′— 3¾″	3′— 7⅛″	5′— 5⅝″	3′—9″	15
16	3′— 6″	5′— 6″	3′— 8″	5′— 8″	3′—10″	5′—10″	4′—0″	16
17	3′— 8⅝″	5′—10⅛″	3′—10¾″	6′— 0¼″	4′— 0⅞″	6′— 2⅜″	4′—3″	17
18	3′—11¼″	6′— 2¼″	4′— 1½″	6′— 4½″	4′— 3¾″	6′— 6¾″	4′—6″	18
19	4′— 1⅞″	6′— 6⅜″	4′— 4¼″	6′— 8¾″	4′— 6⅝″	6′—11⅛″	4′—9″	19
20	4′— 4½″	6′—10½″	4′— 7″	7′— 1″	4′— 9½″	7′— 3½″	5′—0″	20
21	4′— 7⅛″	7′— 2⅝″	4′— 9¾″	7′— 5¼″	5′— 0⅜″	7′— 7⅞″	5′—3″	21
22	4′— 9¾″	7′— 6¾″	5′— 0½″	7′— 9½″	5′— 3¼″	8′— 0¼″	5′—6″	22
23	5′— 0⅜″	7′—10⅞″	5′— 3¼″	8′— 1¾″	5′— 6⅛″	8′— 4⅝″	5′—9″	23

24	6'—0"	8'—9"	5'—9"	8'—6"	5'—6"	8'—3"	5'—3"
25	6'—3"	9'—1 3/8"	5'—11 7/8"	8'—10 1/4"	5'—8 3/4"	8'—7 1/8"	5'—5 5/8"
26	6'—6"	9'—5 3/4"	6'—2 3/4"	9'—2 1/2"	5'—11 1/2"	8'—11 1/4"	5'—8 1/4"
27	6'—9"	9'—10 1/8"	6'—5 5/8"	9'—6 3/4"	6'—2 1/4"	9'—3 3/8"	5'—10 7/8"
28	7'—0"	10'—2 1/2"	6'—8 1/2"	9'—11"	6'—5"	9'—7 1/2"	6'—1 1/2"
29	7'—3"	10'—6 7/8"	6'—11 3/8"	10'—3 1/4"	6'—7 3/4"	9'—11 5/8"	6'—4 1/8"
30	7'—6"	10'—11 1/4"	7'—2 1/4"	10'—7 1/2"	6'—10 1/2"	10'—3 3/4"	6'—6 3/4"
31	7'—9"	11'—3 5/8"	7'—5 1/8"	10'—11 3/4"	7'—1 1/4"	10'—7 7/8"	6'—9 3/8"
32	8'—0"	11'—8"	7'—8"	11'—4"	7'—4"	11'—0"	7'—0"
33	8'—3"	12'—0 3/8"	7'—10 7/8"	11'—8 1/4"	7'—6 3/4"	11'—4 1/8"	7'—2 5/8"
34	8'—6"	12'—4 3/4"	8'—1 3/4"	12'—0 1/2"	7'—9 1/2"	11'—8 1/4"	7'—5 1/4"
35	8'—9"	12'—9 1/8"	8'—4 5/8"	12'—4 3/4"	8'—0 1/4"	12'—0 3/8"	7'—7 7/8"
36	9'—0"	13'—1 1/2"	8'—7 1/2"	12'—9"	8'—3"	12'—4 1/2"	7'—10 1/2"
37	9'—3"	13'—5 7/8"	8'—10 3/8"	13'—1 1/4"	8'—5 3/4"	12'—8 5/8"	8'—1 1/8"
38	9'—6"	13'—10 1/4"	9'—1 1/4"	13'—5 1/2"	8'—8 1/2"	13'—0 3/4"	8'—3 3/4"
39	9'—9"	14'—2 5/8"	9'—4 1/8"	13'—9 3/4"	8'—11 1/4"	13'—4 7/8"	8'—6 3/8"
40	10'—0"	14'—7"	9'—7"	14'—2"	9'—2"	13'—9"	8'—9"
41	10'—3"	14'—11 3/8"	9'—9 7/8"	14'—6 1/4"	9'—4 3/4"	14'—1 1/8"	8'—11 5/8"
42	10'—6"	15'—3 3/4"	10'—0 3/4"	14'—10 1/2"	9'—7 1/2"	14'—5 1/4"	9'—2 1/4"
43	10'—9"	15'—8 1/8"	10'—3 5/8"	15'—2 3/4"	9'—10 1/4"	14'—9 3/8"	9'—4 7/8"
44	11'—0"	16'—0 1/2"	10'—6 1/2"	15'—7"	10'—1"	15'—1 1/2"	9'—7 1/2"
45	11'—3"	16'—4 7/8"	10'—9 3/8"	15'—11 1/4"	10'—3 3/4"	15'—5 5/8"	9'—10 1/8"
46	11'—6"	16'—9 1/4"	11'—0 1/4"	16'—3 1/2"	10'—6 1/2"	15'—9 3/4"	10'—0 3/4"
47	11'—9"	17'—1 5/8"	11'—3 1/8"	16'—7 3/4"	10'—9 1/4"	16'—1 7/8"	10'—3 3/8"
48	12'—0"	17'—6"	11'—6"	17'—0"	11'—0"	16'—6"	10'—6"
49	12'—3"	17'—10 3/8"	11'—8 7/8"	17'—4 1/4"	11'—2 3/4"	16'—10 1/8"	10'—8 5/8"
50	12'—6"	18'—2 3/4"	11'—11 3/4"	17'—8 1/2"	11'—5 1/2"	17'—2 1/4"	10'—11 1/4"
60	15'—0"	21'—10 1/2"	14'—4 1/2"	21'—3"	13'—9"	20'—7 1/2"	13'—1 1/2"
70	17'—6"	25'—6 1/4"	16'—9 1/4"	24'—9 1/2"	16'—0 1/2"	24'—0 3/4"	15'—3 3/4"
80	20'—0"	29'—2"	19'—2"	28'—4"	18'—4"	27'—6"	17'—6"
90	22'—6"	32'—9 3/4"	21'—6 3/4"	31'—10 1/2"	20'—7 1/2"	30'—11 1/4"	19'—8 1/4"
100	25'—0"	36'—5 1/2"	23'—11 1/2"	35'—5"	22'—11"	34'—4 1/2"	21'—10 1/2"

The lengths of the basement walls may also be regarded as running from center to center of each wall. The total length of the basement wall is, then, also 108 feet. The height of the basement wall is 7 feet. The area of this wall is then $108 \times 7 = 756$ square feet. Referring to Table 6 we find that for a 12-inch wall the quantities are,

for 700 sq. ft. $= 12,937$ bricks, 220 cu. ft. mortar
for 50 sq. ft. $=$ 925 bricks, 16 cu. ft. mortar
for   6 sq. ft. $= 6 \times 1 =$ 111 bricks,   2 cu. ft. mortar

or a total for the basement wall of 13,973 bricks and 238 cubic feet mortar.

Turning our attention now to 8-inch walls, we subtract only 4 inches or one-third foot from each end to obtain the lengths of the sides from center to center of walls. The south wall is then $32 - \frac{1}{3} - \frac{1}{3} = 31\frac{1}{3}$ feet long. The height of this wall is 18 feet. The area is $31\frac{1}{3} \times 18 = 564$ square feet. From this figure must be subtracted the openings. There are four windows 3 feet by 5 feet and a door 4 feet by 7 feet. The total area of the window space is

$$3 \times 5 \times 4 = 60 \text{ sq. ft.}$$
$$\text{Area of door} = \quad 4 \times 7 = 28 \text{ sq. ft.}$$
$$\text{Total area } \overline{88} \text{ sq. ft.}$$

The net area of the south wall is then $564 - 88 = 476$ square feet. The north wall has a like area.

The west and east 8-inch walls may be considered as consisting of a rectangle $23\frac{1}{3}$ feet by 18 feet in size and a 45-degree right triangle whose hypotenuse is $23\frac{1}{3}$ feet long. The area of the rectangle is $23\frac{1}{3} \times 18 = 420$ square feet. The areas of the window openings to be subtracted are

$$3 \times 5 \times 2 = 30 \text{ sq. ft.}$$
$$\text{and} \quad\quad 2 \times 2 \times 2 = \underline{8} \text{ sq. ft.}$$
$$\text{Total } \overline{38} \text{ sq. ft.}$$

TABLE 5

AVERAGE WEIGHT OF SOLID BRICK WALLS

Brick Assumed to Weigh $4\frac{1}{2}$ lb. each. $\frac{1}{2}''$ Joints Filled with Mortar

Weight in Pounds per Square Foot of Wall Area

4" Wall	8" Wall	12" Wall
36.782	78.808	115.414

Thus from Table 5 a 125.5 square foot wall, 8 inches deep, weighs 125.5 × 78.808 or 9890.4 pounds.

TABLE 6

BRICKS AND MORTAR FOR SOLID WALLS IN ALL BONDS

Half-Inch Joints—All Joints Filled with Mortar

Square Feet Area of Wall	4-Inch Wall		8-Inch Wall		12-Inch Wall		16-Inch Wall		Square Feet Area of Wall
	Number of bricks	Cubic feet of mortar	Number of bricks	Cubic feet of mortar	Number of bricks	Cubic feet of mortar	Number of bricks	Cubic feet of mortar	
1	6.160	0.075	12.320	0.195	18.481	0.314	24.641	0.433	1
10	62	1	124	2	185	$3\frac{1}{2}$	247	$4\frac{1}{2}$	10
20	124	2	247	4	370	$6\frac{1}{2}$	493	9	20
30	185	$2\frac{1}{2}$	370	6	555	$9\frac{1}{2}$	740	13	30
40	247	$3\frac{1}{2}$	493	8	740	13	986	$17\frac{1}{2}$	40
50	309	4	617	10	925	16	1,233	22	50
60	370	5	740	12	1,109	19	1,479	26	60
70	432	$5\frac{1}{2}$	863	14	1,294	22	1,725	31	70
80	493	$6\frac{1}{2}$	986	16	1,479	25	1,972	35	80
90	555	7	1,109	18	1,664	28	2,218	39	90
100	617	8	1,233	20	1,849	32	2,465	44	100
200	1,233	15	2,465	39	3,697	63	4,929	87	200
300	1,849	23	3,697	59	5,545	94	7,393	130	300
400	2,465	30	4,929	78	7,393	126	9,857	173	400
500	3,081	38	6,161	98	9,241	157	12,321	217	500
600	3,697	46	7,393	117	11,089	189	14,786	260	600
700	4,313	53	8,625	137	12,937	220	17,250	303	700
800	4,929	61	9,857	156	14,786	251	19,714	347	800
900	5,545	68	11,089	175	16,634	283	22,178	390	900
1,000	6,161	76	12,321	195	18,482	314	24,642	433	1,000
2,000	12,321	151	24,642	390	36,963	628	49,284	866	2,000
3,000	18,482	227	36,963	584	55,444	942	73,926	1299	3,000
4,000	24,642	302	49,284	779	73,926	1255	98,567	1732	4,000
5,000	30,803	377	61,605	973	92,407	1569	123,209	2165	5,000
6,000	36,963	453	73,926	1168	110,888	1883	147,851	2599	6,000
7,000	43,124	528	86,247	1363	129,370	2197	172,493	3032	7,000
8,000	49,284	604	98,567	1557	147,851	2511	197,124	3465	8,000
9,000	55,444	679	110,888	1752	166,332	2825	221,776	3898	9,000
10,000	61,605	755	123,209	1947	184,813	3139	246,418	4331	10,000

TABLE 7

QUANTITIES OF MATERIALS REQUIRED IN MORTAR

| Cubic Feet of Mortar | Lime Mortar | | | | | | Cubic Feet Mortar |
| | 1 : 2½ | | | 1 : 3 | | | |
	180 lb barrels lump lime	or	50 lb sacks hydrated lime	Cubic yards sand	180 lb barrels lump lime	or	50 lb sacks hydrated lime	Cubic yard sand	
1	0.057	or	0.350	0.037	0.1	or	0.3	0.037	1
2	.1	or	.7	.1	.1	or	.6	.1	2
3	.2	or	1.1	.1	.1	or	.9	.1	3
4	.2	or	1.4	.1	.2	or	1.2	.1	4
5	.3	or	1.8	.2	.2	or	1.5	.2	5
6	.3	or	2.1	.2	.3	or	1.8	.2	6
7	.4	or	2.5	.3	.3	or	2.0	.3	7
8	.5	or	2.8	.3	.4	or	2.3	.3	8
9	.5	or	3.2	.3	.4	or	2.6	.3	9
10	.6	or	3.5	.4	.5	or	2.9	.4	10
11	.7	or	3.9	.4	.5	or	3.2	.4	11
12	.7	or	4.2	.4	.6	or	3.5	.4	12
13	.7	or	4.6	.5	.6	or	3.8	.5	13
14	.8	or	4.9	.5	.7	or	4.1	.5	14
15	.9	or	5.3	.6	.7	or	4.4	.6	15
16	.9	or	5.6	.6	.8	or	4.7	.6	16
17	1.0	or	6.0	.6	.8	or	5.0	.6	17
18	1.0	or	6.3	.7	.9	or	5.3	.7	18
19	1.1	or	6.7	.7	.9	or	5.5	.7	19
20	1.1	or	7.0	.7	.9	or	5.8	.7	20
27	1.5	or	9.5	1.0	1.3	or	7.8	1.0	27
30	1.7	or	10.5	1.1	1.4	or	8.7	1.1	30
40	2.3	or	14.0	1.5	1.9	or	11.7	1.5	40
50	2.8	or	17.5	1.9	2.4	or	14.6	1.9	50
60	3.4	or	21.0	2.2	2.8	or	17.5	2.2	60
70	3.9	or	24.5	2.6	3.3	or	20.4	2.6	70
80	4.6	or	28.0	3.0	3.8	or	23.3	3.0	80
90	5.1	or	31.5	3.3	4.3	or	26.3	3.3	90
100	6	or	35	4	5	or	29	4	100
200	11	or	70	7	9	or	58	7	200
300	17	or	105	11	14	or	88	11	300
400	23	or	140	15	19	or	117	15	400
500	29	or	175	19	24	or	146	19	500
600	34	or	210	22	28	or	175	22	600
700	40	or	245	26	33	or	204	26	700
800	46	or	280	30	38	or	233	30	800
900	51	or	315	33	43	or	263	33	900
1000	57	or	350	37	47	or	292	37	1000

NOTES.—Quantities of lime are based on the use of good quality lime. Lime quantities are approximate and will vary with the grade of lime and the size of particles composing the sand. In the cement mortars, ⅒ of the cement by weight is replaced by dry hydrated lime or its equivalent in lump lime paste.

TABLE 7—*Continued*

Cubic Feet Mortar	Cement-Lime Mortar				Cement Mortar				Cubic Feet Mortar
	1 : 1 : 6				1 : 2				
	94 lbs net sacks cement	180 lb barrels lump lime	or	50 lb sacks hydrated lime	Cubic yards sand	94 lb net sacks cement	180 lb barrels lump lime	or	50 lb sacks hydrated lime	Cubic yards sand	
1	0.18	0.023	or	0.145	0.037	0.4	0.1	or	0.1	0.1	1
2	.4	.1	or	.3	.1	.9	.1	or	.2	.1	2
3	.5	.1	or	.4	.1	1.3	.1	or	.3	.1	3
4	.7	.1	or	.6	.1	1.8	.1	or	.4	.1	4
5	.9	.1	or	.7	.2	2.2	.1	or	.5	.2	5
6	1.1	.1	or	.9	.2	2.7	.1	or	.6	.2	6
7	1.3	.2	or	1.0	.3	3.1	.1	or	.6	.2	7
8	1.4	.2	or	1.2	.3	3.5	.1	or	.7	.3	8
9	1.6	.2	or	1.3	.3	3.9	.1	or	.8	.3	9
10	1.8	.2	or	1.5	.4	4.4	.2	or	.9	.3	10
11	2.0	.3	or	1.6	.4	4.9	.2	or	1.0	.4	11
12	2.2	.3	or	1.7	.4	5.3	.2	or	1.1	.4	12
13	2.3	.3	or	1.9	.5	5.7	.2	or	1.2	.4	13
14	2.5	.3	or	2.0	.5	6.2	.2	or	1.3	.5	14
15	2.7	.4	or	2.2	.6	6.6	.2	or	1.4	.5	15
16	2.9	.4	or	2.3	.6	7.1	.3	or	1.5	.5	16
17	3.1	.4	or	2.5	.6	7.5	.3	or	1.6	.6	17
18	3.2	.4	or	2.6	.7	8.0	.3	or	1.7	.6	18
19	3.4	.5	or	2.8	.7	8.4	.3	or	1.8	.6	19
20	3.6	.5	or	2.9	.7	8.9	.3	or	1.8	.7	20
27	4.1	.64	or	3.94	1.0	12.0	.4	or	2.5	.9	27
30	5.4	.7	or	4.4	1.1	13.3	.5	or	2.8	1.0	30
40	7.2	.9	or	5.8	1.5	17.7	.7	or	3.7	1.3	40
50	9.0	1.2	or	7.3	1.9	22.1	.8	or	4.6	1.7	50
60	10.8	1.4	or	8.8	2.2	26.6	1.0	or	5.5	2.1	60
70	12.6	1.7	or	10.2	2.6	31.0	1.1	or	6.5	2.4	70
80	14.4	1.9	or	11.7	3.0	35.4	1.3	or	7.4	2.8	80
90	16.2	2.1	or	13.1	3.3	39.8	1.5	or	8.3	3.1	90
100	18	2	or	15	4	44	2	or	9	3	100
200	36	5	or	29	7	88	3	or	18	7	200
300	54	7	or	44	11	132	5	or	28	10	300
400	72	10	or	58	15	177	7	or	37	14	400
500	90	12	or	73	19	221	8	or	46	17	500
600	108	14	or	88	22	265	10	or	55	21	600
700	126	17	or	102	26	310	11	or	65	24	700
800	144	19	or	117	30	354	13	or	74	28	800
900	162	21	or	131	33	398	15	or	83	31	900
1000	180	24	or	146	37	442	16	or	92	34	1000

NOTES.—Quantities of lime are based on the use of good quality lime. Lime quantities are approximate and will vary with the grade of lime and the size of particles composing the sand. In the cement mortars, $\frac{1}{10}$ of the cement by weight is replaced by dry hydrated lime or its equivalent in lump lime paste.

TABLE 7—*Concluded*

	Cement Mortar									
	1 : 3					1 : 4				
Cubic Feet Mortar	94 lb net sacks cement	180 lb barrels lump lime	or	50 lb sacks hydrated lime	Cubic yards sand	94 lb net sacks cement	180 lb barrels lump lime	or	50 lb sacks hydrated lime	Cubic yards sand	Cubic Feet Mortar
1	0.3	0.1	or	0.1	0.1	0.3	0.1	or	0.1	0.1	1
2	.7	.1	or	.1	.1	.5	.1	or	.1	.1	2
3	1.0	.1	or	.2	.1	.8	.1	or	.2	.1	3
4	1.3	.1	or	.3	.2	1.1	.1	or	.2	.2	4
5	1.7	.1	or	.3	.2	1.3	.1	or	.3	.2	5
6	2.0	.1	or	.4	.2	1.6	.1	or	.3	.2	6
7	2.3	.1	or	.5	.3	1.8	.1	or	.4	.3	7
8	2.6	.1	or	.6	.3	2.1	.1	or	.4	.3	8
9	3.0	.1	or	.6	.3	2.4	.1	or	.5	.4	9
10	3.3	.1	or	.7	.4	2.6	.1	or	.6	.4	10
11	3.6	.1	or	.8	.4	2.9	.1	or	.6	.5	11
12	4.0	.1	or	.8	.5	3.2	.1	or	.7	.5	12
13	4.3	.2	or	.9	.5	3.4	.1	or	.7	.5	13
14	4.6	.2	or	.9	.5	3.7	.1	or	.8	.6	14
15	5.0	.2	or	1.0	.6	4.0	.2	or	.8	.6	15
16	5.3	.2	or	1.1	.6	4.2	.2	or	.9	.7	16
17	5.6	.2	or	1.2	.7	4.6	.2	or	.9	.7	17
18	6.0	.2	or	1.2	.7	4.8	.2	or	1.0	.7	18
19	6.3	.2	or	1.3	.7	5.0	.2	or	1.0	.8	19
20	6.6	.2	or	1.4	.8	5.3	.2	or	1.1	.8	20
27	8.9	.3	or	1.9	1.1	7.1	.3	or	1.5	1.1	27
30	9.9	.4	or	2.1	1.2	7.9	.3	or	1.7	1.2	30
40	13.2	.5	or	2.8	1.6	10.6	.4	or	2.2	1.6	40
50	16.5	.6	or	3.5	1.9	13.2	.5	or	2.8	2.1	50
60	19.8	.7	or	4.1	2.3	15.8	.6	or	3.3	2.5	60
70	23.1	.9	or	4.8	2.7	18.5	.7	or	3.9	2.9	70
80	26.4	1.0	or	5.5	3.1	21.1	.8	or	4.4	3.3	80
90	29.8	1.1	or	6.2	3.5	23.8	.9	or	5.0	3.7	90
100	33	1.2	or	7	4	26	1	or	6	4	100
200	66	2	or	14	8	53	2	or	11	8	200
300	99	4	or	21	12	79	3	or	17	12	300
400	132	5	or	28	16	106	4	or	22	16	400
500	165	6	or	35	19	132	5	or	28	21	500
600	198	7	or	41	23	158	6	or	33	25	600
700	231	9	or	48	27	184	7	or	39	29	700
800	265	10	or	55	31	211	8	or	44	33	800
900	298	11	or	62	35	238	9	or	50	37	900
1000	331	12	or	69	39	264	10	or	55	41	1000

Notes.—Quantities of lime are based on the use of good quality lime. Lime quantities are approximate and will vary with the grade of lime and the size of particles composing the sand. In the cement mortars, 1/10 of the cement by weight is replaced by dry hydrated lime or its equivalent in lump lime paste.

This makes the net area of the rectangle 420 − 38 = 382 square feet.

In a 45-degree right triangle the sides bear a relation to the hypotenuse of $1 : \sqrt{2}$. In the case of the wall we know the hypotenuse to be $23\frac{1}{3}$ or 23.33 feet. Letting L represent the sloping side, we may set up the proportion,

$$\frac{L}{23.33} = \frac{1}{\sqrt{2}}$$

Transposing,

$$L = \frac{23.33}{\sqrt{2}} = \frac{23.33}{1.414} = 16.50 \text{ ft}$$

The area of the triangle is

$$\frac{16.5 \times 16.5}{2} = 136 \text{ sq ft}$$

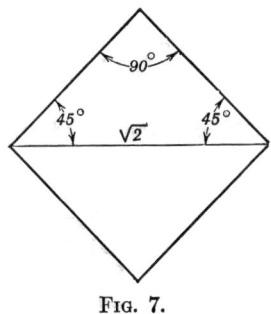

Fig. 7.

This makes the area of each end wall 136 + 382 = 518 sq. ft.

We are now ready to add the areas of all the 8-inch walls:

$$
\begin{array}{lll}
\text{South wall} = & 476 \text{ sq. ft.} \\
\text{North wall} = & 476 \text{ sq. ft.} \\
\text{East wall} = & 518 \text{ sq. ft.} \\
\text{West wall} = & 518 \text{ sq. ft.} \\
\hline
\text{Total area} & 1988 \text{ sq. ft.}
\end{array}
$$

Turning again to Table 6, but this time to the column for the 8-inch wall, we obtain the following quantities:

$$
\begin{array}{lll}
\text{for 1000 sq. ft.} = & 12{,}321 \text{ bricks} & 195 \quad \text{cu. ft. mortar} \\
900 \text{ sq. ft.} = & 11{,}089 \text{ bricks} & 175 \quad \text{cu. ft. mortar} \\
80 \text{ sq. ft.} = & 986 \text{ bricks} & 16 \quad \text{cu. ft. mortar} \\
8 \text{ sq. ft.} = & 99 \text{ bricks} & 1.6 \text{ cu. ft. mortar}
\end{array}
$$

The totals for 8-inch wall are 24,495 bricks; 387.6 cu. ft. mortar.

This leaves only the chimney to be estimated. It will be noted that Table 4 gives the quantities for chimneys by 10-foot heights. Since the chimney in this problem is 37 feet high it is 3.7 10-foot lengths. The unit quantities given in the table for a chimney 1 foot 5 inches by 2 feet 10 inches in cross-section are 367 bricks and $6\frac{1}{2}$ cubic feet mortar. Multiplying these quantities by 3.7 we obtain,

$$367 \times 3.7 = 1358 \text{ bricks for chimney}$$
$$6.5 \times 3.7 = 24.0 \text{ cu. ft. of mortar for chimney}$$

It remains only to make a recapitulation of all the quantities.

	Bricks	Mortar
Footings............	3,037	51.8
12-in walls..........	13,973	238.0
8-in walls...........	24,495	387.6
Chimney............	1,358	24.0
Totals...	42,863	701.4

With the quantity of mortar known, the amount of cement, lime, and sand required can readily be found from Table 7.

Concrete Blocks.—Portland cement concrete blocks are a comparatively cheap and easily handled building material. Because of the ease with which they may be manufactured, the blocks are available everywhere. Practically every building code in the country permits the use of Portland cement blocks, although some codes restrict their use to small buildings. A standard block is usually 8″ wide, 8″ high, and 16″ long. Half blocks, corners, and other shapes may be made to meet any need. The aggregate may be sand or steam cinders. The cinder block has several advantages such as lighter weight, porous surface for plaster, and excellent insulating qualities.

The standard concrete block produces a wall 8″ thick, in 8″ courses when laid in a single thickness. For 100 sq. ft. of wall area, one hundred and ten 8″ × 8″ × 16″ blocks are required. Concrete building tile are smaller and lighter than concrete blocks.

They are usually 12″ long and 8″ wide, but their height varies. The 3½″ and 5″ heights are the most common. Two hundred and twenty 5″, or three hundred, 3½″ tiles are required for 100 sq. ft. of wall area. Table 8 gives a complete tabulation of materials including mortar for joints, for estimating quantities of material per hundred square feet of wall area. It should be remembered that the table does not include wall opening such as doors and windows. These should be deducted before estimating the net wall area.

Reinforcement.—The increasing use of concrete block construction has extended not only to installations of wider and higher walls, but also to more severe external conditions (e.g., in hurricane and earthquake areas), and more troublesome structural conditions (e.g., settlement of footings, distortions due to expansion of floors and roofs). This situation has led to greater

TABLE 8

DATA ON VARIOUS CONCRETE MASONRY WALLS PER 100 SQ. FT. OF WALL AREA

Concrete Block

Description		Wall Thickness	Weight per Unit (lb.)	Number of Units (per 100 sq. ft. of wall area)	Mortar* (cu. ft.)	Weight, lb. (per 100 sq. ft. of wall area)
8″× 8″×16″		8″	50	110	3.25	5850
8″× 8″×12″		8″	38	146	3.50	6000
8″×12″×16″		12″	85	110	3.25	9700
8″× 3″×16″		3″	20	110	2.75	2600
9″× 3″×18″	Hollow	3″	26	87	2.50	2500
12″× 3″×12″	Parti-	3″	23	100	2.50	2550
8″× 3″×12″	tion	3″	15	146	3.50	2550
8″× 4″×16″	Block	4″	28	110	3.25	3450
9″× 4″×18″		4″	35	87	3.25	3350
12″× 4″×12″		4″	31	100	3.25	3450
8″× 4″×12″		4″	21	146	4.00	3500
8″× 6″×16″		6″	42	110	3.25	5000

* These figures are based on ⅜-inch mortar joints, 25% wastage included. Weight of mortar assumed at 103 pounds per cubic foot. For ½-inch mortar joints, use one-fourth more mortar than specified in the table.

TABLE 8—*Continued*

Concrete Tile

Description	Wall Thickness	Weight per Unit (lb.)	Number of Units (per 100 sq. ft. of wall area)	Mortar* (cu. ft.)	Weight, lb. (per 100 sq. ft. of wall area)
5″×8″×12″	8″	19.9	220	5.00	4900
5″×4″×12″	4″	9.9	220	5.00	2700
5″×6″×12″	6″	14.9	220	4.00	3800
3½″×8″×12″	8″	16.5	300	6.00	5550
3½″×4″×12″	4″	8.5	300	5.00	3050
3½″×6″×12″	6″	12.5	300	5.50	4300

Light Weight Concrete Block

Description	Wall Thickness	Weight per Unit (lb.)	Number of Units (per 100 sq. ft. of wall area)	Mortar* (cu. ft.)	Weight, lb. (per 100 sq. ft. of wall area)
8″× 8″×16″	8″	27–32	110	3.25	3300–3850
8″× 8″×12″	8″	21–24	146	3.50	3400–3850
8″×12″×16″	12″	46–54	110	3.25	5400–6300
8″× 3″×16″	3″	11–13	110	2.75	1500–1700
9″× 3″×18″	3″	14–17	87	2.50	1450–1700
12″× 3″×12″	3″	12–15	100	2.50	1450–1750
8″× 3″×12″	3″	8–10	146	3.50	1500–1800
8″× 4″×16″	4″	15–18	110	3.25	2000–2350
9″× 4″×18″	4″	19–22	87	3.25	2000–2250
12″× 4″×12″	4″	17–20	100	3.25	2050–2350
8″× 4″×12″	4″	11–13	146	4.00	2000–2300
8″× 6″×16″	6″	23–27	110	3.25	2900–3300

Light Weight Concrete Tile

Description	Wall Thickness	Weight per Unit (lb.)	Number of Units (per 100 sq. ft. of wall area)	Mortar* (cu. ft.)	Weight, lb. (per 100 sq. ft. of wall area)
5″×8″×12″	8″	10.8–12.7	220	5.00	2900–3300
5″×4″×12″	4″	5.4– 6.3	220	5.00	1700–1950
5″×6″×12″	6″	8.1– 9.5	220	4.00	2200–2500
3½″×8″×12″	8″	8.9–10.6	300	6.00	3300–3800
3½″×4″×12″	4″	4.6– 5.4	300	5.50	2000–2200
3½″×6″×12″	6″	6.8– 8.0	300	5.50	2600–3000

Concrete Brick

Description	Wall Thickness	Weight per Unit (lb.)	Number of Units (per 100 sq. ft. of wall area)	Mortar* (cu. ft.)	Weight, lb. (per 100 sq. ft. of wall area)
2¼″×3¾″×8″	8″	5	1300	18.10	8350

* These figures are based on ⅜-inch mortar joints, 25% wastage included. Weight of mortar assumed at 103 pounds per cubic foot. For ½-inch mortar joints, use one-fourth more mortar than specified in the table

use of reinforcing in concrete block construction. While the extent to which such reinforcing is necessary has no general answer, it is considered especially desirable in cases where plaster is to be applied directly to concrete block.

Therefore there is greater use of both rods and shaped sections for reinforcement. Fig. 8 shows an installation of vertical stabilization in a Vibrapac® pilaster block of the Besser Company, and Fig. 9 shows their unit designed to control the volume

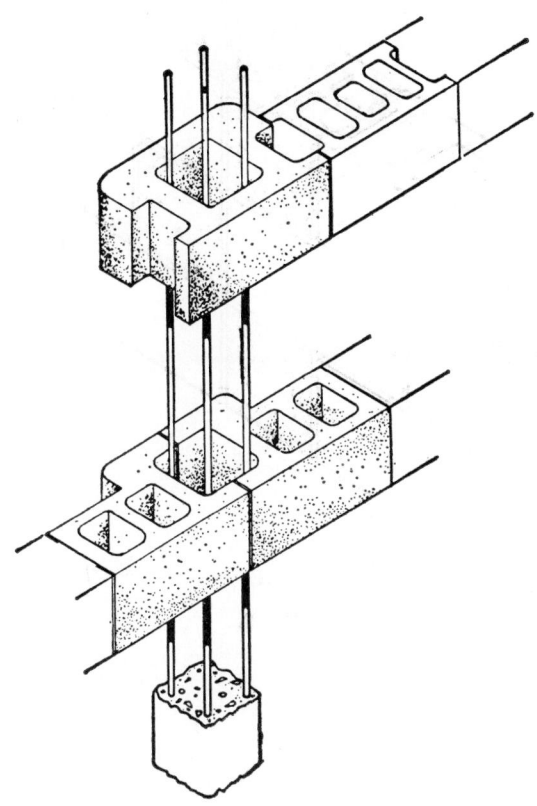

Fig. 8.

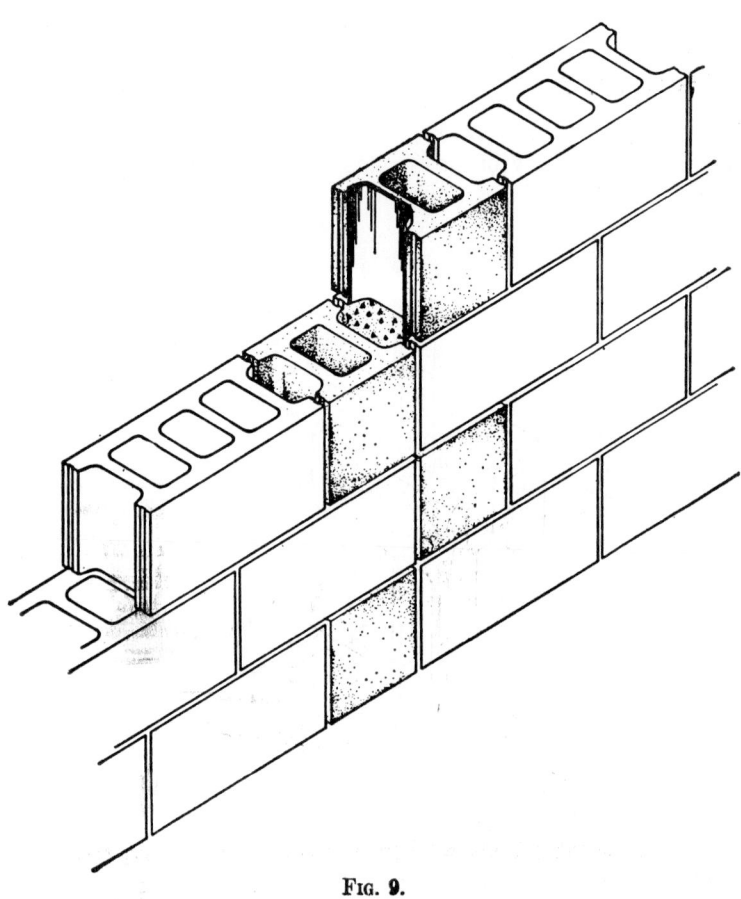

FIG. 9.

changes of a wall—this block being placed at specific intervals to relieve the volume changes in the wall. Figure 10 shows their block designed to permit construction of a 8″ to 16″ deep continuous reinforced concrete bond around entire building. This arrangement can be combined with vertical reinforced concrete column to attain masonry walls of great strength.

Figure 11 shows various installations of the Dur-O-Wal® reinforcement of the Dur-O-Wal Company. Fig. 11(a) shows single side rods 16″ center to center with cross rods; Fig. 11(b) shows a similar installation with double side rods; while Fig. 11(c) shows an 8″ wall with a pilaster and control joint; and 11(d) shows a corner. Figure 12 shows the design of a unit of Dur-O-Wal reinforcing.

Painting Block Walls.—An attractive waterproof finish for exterior walls consists of two coats of Portland cement base paint, applied as follows:

Uniformly dampen wall and apply paint, mixed to creamy consistency. Scrub into surface with a fiber brush with 1½″ to 2″ bristles. Dampen painted surfaces the morning following application with a fog spray. Apply second coat as above to the dampened wall not less than two days after application of first coat. Dampen second coat with fog spray the morning after application.

Interior Walls.—While plaster is often applied directly to cement block walls, the opinion is quite general that, unless the wall is heavily reinforced, cracking is likely to occur. This situation is not nearly so likely if more porous blocks of cinders or slag are used, such as the Waylite® blocks, made by the Waylite Company. These blocks, having larger air spaces, are more effective insulators. Even for them, however, a metal lath installation is recommended if the wall is to be plastered. This is feasible for these blocks, because nails can be driven directly into them. Of course a wallboard interior surface provides the necessary insulating dead air space for all these types of building blocks.

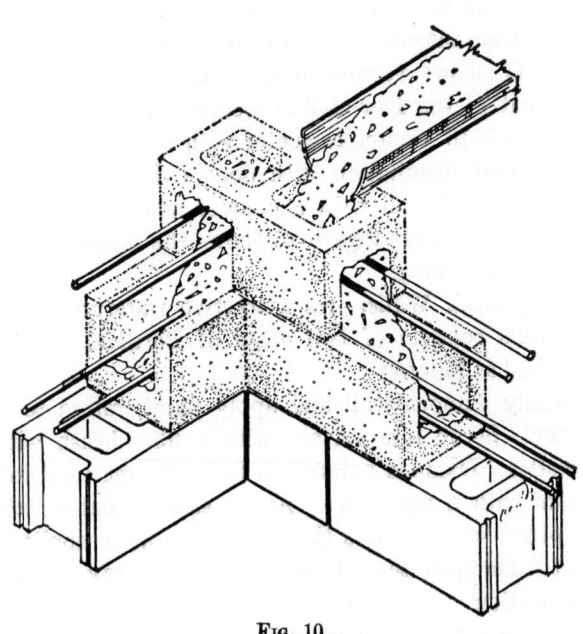

Fig. 10.

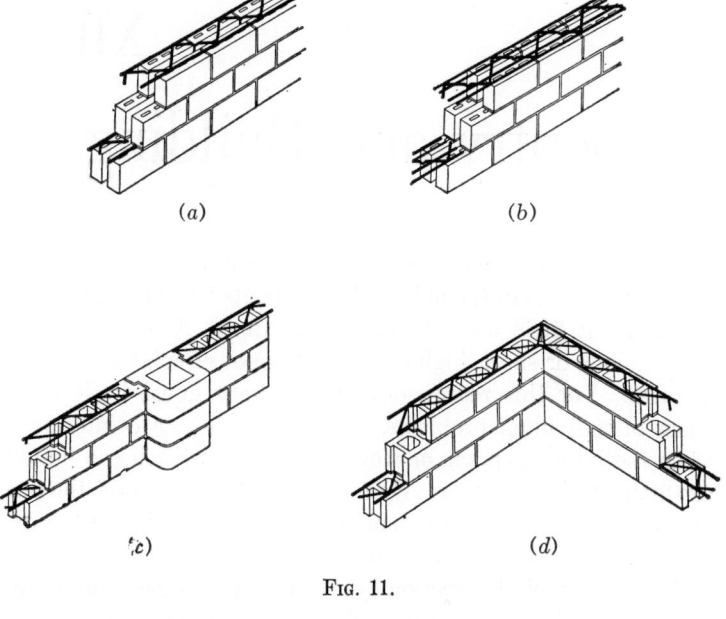

(a)

(b)

(c)

(d)

Fig. 11.

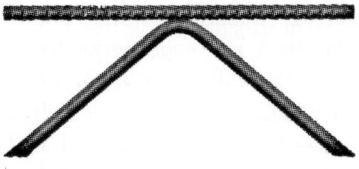

Fig. 12.

XII

Carpentry and Building

Carpentry finds its greatest expression in house building, and, although it is one of the oldest of the arts, its basic principles have changed but little with the passing of time. A man may build a house long or short, high or low, square or circular as his fancy dictates. For this reason practically every house built has its series of individual problems. Not all of these are fully solved when the building plans or working drawings reach the building foreman or the estimator. He must be able to interpret the plans in their proper light and independently find the missing information by computation or estimation.

The object of this section is to show how these computations are made and how estimates of material are arrived at. After a general discussion of board measure, it takes up house framing and surface covering, including walls, floors and roofs, all of which may be classified as *rough carpentry*. Interior trim and millwork, which may be called *finish carpentry*, does not present many problems in which mathematics is helpful.

Board Measure.—Lumber is measured in terms of *board feet*, abbreviated *fbm* (feet board measure). A board 12 inches wide, 12 inches long, and one inch thick contains one board foot of lumber. Similarly, a board 6 inches wide, 24 inches long, and 1 inch thick is also one board foot. The rule for determining the number of board feet in a piece of lumber may then be stated: *Multiply the length in feet by the width in inches and the thickness in inches ($\frac{1}{2}$ inch or over) and divide the product by twelve.* Stated as a formula, this is,

TABLE 1

BOARD MEASURE

Size	Length in feet							
	12	14	16	18	20	22	24	26
	Square feet							
1× 8	8	9⅓	10⅔	12	13⅓	14⅔	16	17⅓
1×10	10	11⅔	13⅔_	15	16⅔	18⅓	20	21⅔
1×12	12	14	16	18	20	22	24	26
1×14	14	16⅓	18⅔	21	23⅓	25⅔	28	30⅓
1×16	16	18⅔	21⅓	24	26⅔	29⅓	32	34⅔
2× 3	6	7	8	9	10	11	12	13
2× 4	8	9⅓	10⅔	12	13⅓	14⅔	16	17⅓
2× 6	12	14	16	18	20	22	24	26
2× 8	16	18⅔	21⅓	24	26⅔	29⅓	32	34⅔
2×10	20	23⅓	26⅔	30	33⅓	36⅔	40	43⅓
2×12	24	28	32	36	40	44	48	52
2×14	28	32⅔	37⅓	42	46⅔	51⅓	56	60⅔
2×16	32	37⅓	42⅔	48	53⅓	58⅔	64	69⅓
3× 4	12	14	16	18	20	22	24	26
3× 6	18	21	24	27	30	33	36	39
3× 8	24	28	32	36	40	44	48	52
3×10	30	35	40	45	50	55	60	65
3×12	36	42	48	54	60	66	72	78
3×14	42	49	56	63	70	77	84	91
3×16	48	56	64	72	80	88	96	104
4× 4	16	18⅔	21⅓	24	26⅔	29⅓	32	34⅔
4× 6	24	28	32	36	40	44	48	52
4× 8	32	37⅓	42⅔	48	53⅓	58⅔	64	69⅓
4×10	40	46⅔	53⅓	60	66⅔	73⅓	80	86⅔
4×12	48	56	64	72	80	88	96	104
4×14	56	65⅓	74⅔	84	93⅓	102⅔	112	121⅓
4×16	64	74⅔	85⅓	96	106⅔	117⅓	128	138⅔
6× 6	36	42	48	54	60	66	72	78
6× 8	48	56	64	72	80	88	96	104
6×10	60	70	80	90	100	110	120	130
6×12	72	84	96	108	120	132	144	156
6×14	84	98	112	126	140	154	168	182
6×16	96	112	128	144	160	176	192	208
8× 8	64	74⅔	85⅓	96	106⅔	117⅓	128	138⅔
8×10	80	93⅓	106⅔	120	133⅓	146⅔	160	173⅓
8×12	96	112	128	144	160	176	192	208
8×14	112	130⅔	149⅓	168	186⅔	205⅓	224	242⅔
8×16	128	149⅓	170⅔	192	213⅓	234⅔	256	277⅓
10×10	100	116⅔	133⅓	150	166⅔	183⅓	200	216⅔
10×12	120	140	160	180	200	220	240	260
10×14	140	163⅓	186⅔	210	233⅓	256⅔	280	303⅓
10×16	160	186⅔	213⅓	240	266⅔	293⅓	320	346⅔
12×12	144	168	192	216	240	264	288	312
12×14	168	196	224	252	280	308	336	364
12×16	192	224	256	288	320	352	384	416
14×14	196	228⅔	261⅓	294	326⅔	359⅓	392	424⅔
⁻4×16	224	261⅓	298⅔	336	373⅓	410⅔	448	485⅓
16×16	256	298⅔	341⅓	384	426⅔	469⅓	512	554⅔

TABLE 1

BOARD MEASURE — *(Continued)*

Size	Length in feet						
	28	30	32	34	36	38	40
	Square feet						
1× 8	$18\frac{2}{3}$	20	$21\frac{1}{3}$	$22\frac{2}{3}$	24	$25\frac{1}{3}$	$26\frac{2}{3}$
1×10	$23\frac{1}{3}$	25	$26\frac{2}{3}$	$28\frac{1}{3}$	30	$31\frac{2}{3}$	$33\frac{1}{3}$
1×12	28	30	32	34	36	38	40
1×14	$32\frac{2}{3}$	35	$37\frac{1}{3}$	$39\frac{2}{3}$	42	$44\frac{1}{3}$	$46\frac{2}{3}$
1×16	$37\frac{1}{3}$	40	$42\frac{2}{3}$	$45\frac{1}{3}$	48	$50\frac{2}{3}$	$53\frac{1}{3}$
2× 3	14	15	16	17	18	19	20
2× 4	$18\frac{2}{3}$	20	$21\frac{1}{3}$	$22\frac{2}{3}$	24	$25\frac{1}{3}$	$26\frac{2}{3}$
2× 6	28	30	32	34	36	38	40
2× 8	$37\frac{1}{3}$	40	$42\frac{2}{3}$	$45\frac{1}{3}$	48	$50\frac{2}{3}$	$53\frac{1}{3}$
2×10	$46\frac{2}{3}$	50	$53\frac{1}{3}$	$56\frac{2}{3}$	60	$63\frac{1}{3}$	$66\frac{2}{3}$
2×12	56	60	64	68	72	76	80
2×14	$65\frac{1}{3}$	70	$72\frac{2}{3}$	$79\frac{1}{3}$	84	$88\frac{2}{3}$	$93\frac{1}{3}$
2×16	$74\frac{2}{3}$	80	$85\frac{1}{3}$	$90\frac{2}{3}$	96	$101\frac{1}{3}$	$106\frac{2}{3}$
3× 4	28	30	32	34	36	38	40
3× 6	42	45	48	51	54	57	60
3× 8	56	60	64	68	72	76	80
3×10	70	75	80	85	90	95	100
3×12	84	90	96	102	108	114	120
3×14	98	105	112	119	126	133	140
3×16	112	120	128	136	144	152	160
4× 4	$37\frac{1}{3}$	40	$42\frac{2}{3}$	$45\frac{1}{3}$	48	$50\frac{2}{3}$	$53\frac{1}{3}$
4× 6	56	60	64	68	72	76	80
4× 8	$74\frac{2}{3}$	80	$85\frac{1}{3}$	$90\frac{2}{3}$	96	$101\frac{1}{3}$	$106\frac{2}{3}$
4×10	$93\frac{1}{3}$	100	$106\frac{2}{3}$	$113\frac{1}{3}$	120	$126\frac{2}{3}$	$133\frac{1}{3}$
4×12	112	120	128	136	144	152	160
4×14	$130\frac{2}{3}$	140	$149\frac{1}{3}$	$158\frac{2}{3}$	168	$177\frac{1}{3}$	$186\frac{2}{3}$
4×16	$149\frac{1}{3}$	160	$170\frac{2}{3}$	$181\frac{1}{3}$	192	$202\frac{2}{3}$	$213\frac{1}{3}$
6× 6	84	90	96	102	108	114	120
6× 8	112	120	128	136	144	152	160
6×10	140	150	160	170	180	190	200
6×12	168	180	192	204	216	228	240
6×14	196	210	224	238	252	266	280
6×16	224	240	256	272	288	304	320
8× 8	$149\frac{1}{3}$	160	$170\frac{2}{3}$	$181\frac{1}{3}$	192	$202\frac{2}{3}$	$213\frac{1}{3}$
8×10	$186\frac{2}{3}$	200	$213\frac{1}{3}$	$226\frac{2}{3}$	240	$253\frac{1}{3}$	$266\frac{2}{3}$
8×12	224	240	256	272	288	304	320
8×14	$261\frac{1}{3}$	280	$298\frac{2}{3}$	$317\frac{1}{3}$	336	$354\frac{2}{3}$	$373\frac{1}{3}$
8×16	$298\frac{2}{3}$	320	$341\frac{1}{3}$	$362\frac{2}{3}$	384	$405\frac{1}{3}$	$426\frac{2}{3}$
10×10	$233\frac{1}{3}$	250	$266\frac{2}{3}$	$283\frac{1}{3}$	300	$316\frac{2}{3}$	$333\frac{1}{3}$
10×12	280	300	320	340	360	380	400
10×14	$326\frac{2}{3}$	350	$373\frac{1}{3}$	$396\frac{2}{3}$	410	$443\frac{1}{3}$	$466\frac{2}{3}$
10×16	$373\frac{1}{3}$	400	$426\frac{2}{3}$	$453\frac{1}{3}$	480	$506\frac{2}{3}$	$533\frac{1}{3}$
12×12	336	360	384	408	432	456	480
12×14	392	420	448	476	504	532	560
12×16	448	480	512	544	576	608	640
14×14	$457\frac{1}{3}$	490	$522\frac{2}{3}$	$555\frac{1}{3}$	588	$620\frac{2}{3}$	$653\frac{1}{3}$
14×16	$522\frac{1}{3}$	560	$597\frac{1}{3}$	$634\frac{2}{3}$	672	$709\frac{1}{3}$	$746\frac{2}{3}$
16×16	$597\frac{1}{3}$	640	$682\frac{2}{3}$	$725\frac{1}{3}$	768	$810\frac{2}{3}$	$853\frac{1}{3}$

NOTE. —By simply multiplying or dividing the above amounts, the number of feet contained in other dimensions can be obtained.

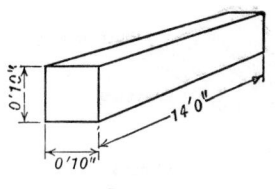

$$\text{fbm} = \frac{L \times w \times t}{12}$$

where L = length in feet;

w = width in inches;

t = thickness in inches.

FIG. 1.

ILLUSTRATION: What is the board measure of a timber 10 inches by 10 inches and 14 feet long?

$$\text{fbm} = \frac{L \times w \times t}{12} = \frac{14 \times 10 \times 10}{12} = 116.7 \text{ fbm} \quad \text{(Ans.)}$$

Lumber is measured on the basis of "rough stock." When lumber is "dressed" or planed, $\frac{1}{8}$ inch is taken off each side if the lumber is $1\frac{1}{2}$ inches or greater in thickness, and $\frac{1}{16}$ inch if the thickness is less than $1\frac{1}{2}$ inches. The purchaser pays, however, on the basis of its measurement before planing.

Thicknesses less than one inch are regarded as one inch in measuring lumber.

In measuring width of boards, fractions of an inch, one-half or greater are regarded as a whole inch, while fractions less than one-half inch are ignored. For example, a board $4\frac{1}{2}$ inches or $4\frac{3}{4}$ inches wide would be called 5 inches, while a board $4\frac{3}{8}$ inches wide would be measured as but 4 inches.

Building lumber is sold in standard lengths which are multiples of two feet from 10 to 24 feet, that is 10, 12, 14, etc. feet.

Lumber dealt with in large quantities is measured and sold by the thousand board feet (M fbm). Board feet are changed to thousand board feet by simply shifting the decimal point three places to the left. Thus, 28,500 fbm = 28.5 M fbm.

ILLUSTRATION: How many thousand board feet are there in 1200 pieces 2 inches × 4 inches and 18 feet long?

$$\text{fbm (one piece)} = \frac{L \times w \times t}{12} = \frac{18 \times 4 \times 2}{12} = 12 \text{ fbm}$$

$$1200 \times 12 = 14,400 \text{ fbm} = 14.4 \text{ M fbm} \quad \text{(Ans.)}$$

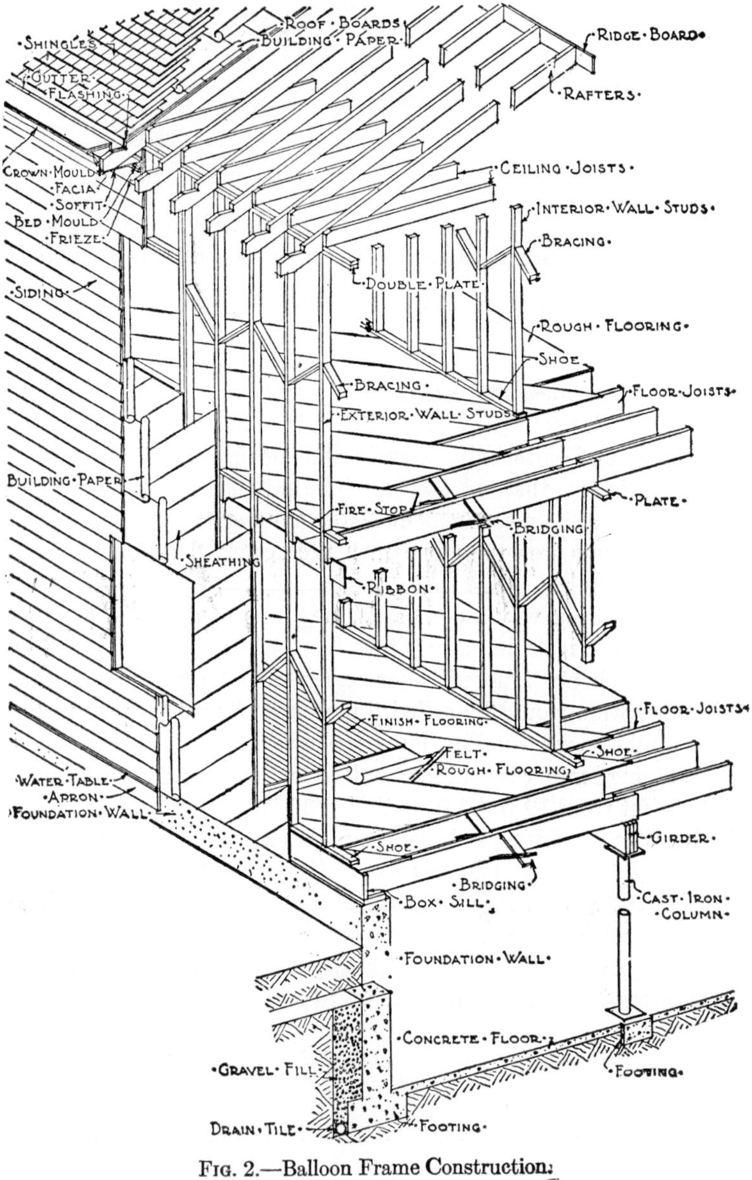

·SHINGLES·
ROOF·BOARDS·
·BUILDING·PAPER·
·RIDGE·BOARD·
·GUTTER·
·FLASHING·
·RAFTERS·
·CROWN·MOULD·
·FACIA·
·SOFFIT·
·BED·MOULD·
·FRIEZE·
·CEILING·JOISTS·
·INTERIOR·WALL·STUDS·
·BRACING·
·SIDING·
·DOUBLE·PLATE·
·ROUGH·FLOORING·
·SHOE·
·FLOOR·JOISTS·
·BRACING·
·EXTERIOR·WALL·STUDS·
·BUILDING·PAPER·
·FIRE·STOP·
·PLATE·
·BRIDGING·
·SHEATHING·
·RIBBON·
·FINISH·FLOORING·
·FLOOR·JOISTS·
·FELT·
·SHOE·
·ROUGH·FLOORING·
·WATER·TABLE·
·APRON·
·FOUNDATION·WALL·
·SHOE·
·GIRDER·
·BRIDGING·
·BOX·SILL·
·CAST·IRON·
·COLUMN·
·FOUNDATION·WALL·
·CONCRETE·FLOOR·
·GRAVEL·FILL·
·FOOTING·
·DRAIN·TILE·
·FOOTING·

Fig. 2.—Balloon Frame Construction.

House Framing.—The details of frame dwelling construction have been so well standardized by building codes and convention that it is entirely feasible to make fairly accurate estimates of the quantities of material required from the general dimensions of the structure. In preparing orders for material for a building, it is well to bear in mind that the use of standard sizes is most economical and that a further saving is often effected by them in the elimination of unnecessary sawing and handling. When listing lumber, it is common practice to give the number of pieces first, then the width and thickness in inches and the length in feet or feet and inches. Thus, 24 pieces 2 × 4 in. by 16 ft 0 in.

This section will concern itself with a few typical details representing accepted standard practice. Figure 2 shows a corner of what is known as "balloon frame construction" and illustrates the terminology used in house framing and the general location of the various members. The following paragraphs will proceed to deal with the details separately.

Sills.—The first carpentry on a frame building usually begins after the completion of the foundation and consists of laying the sill. The sill may be either a solid timber, as a 4 in. by 6 in., or 4 in. by 8 in., or may be built up as from two 2 in. by 6 in., or 3 in. × 6 in. planks. The sill should be placed about an inch from the outer edge of the foundation, and should be bedded in mortar to secure even bearing and be securely anchored to the masonry. Joints at the corners are made by halving the sills as shown for both types in Fig. 3.

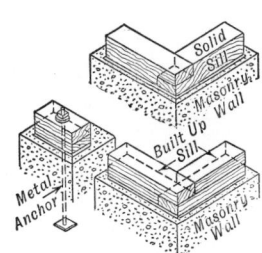

Halving of Sills at Corner

Fig. 3.

The length of sill required is, for practical purposes, the sum of the lengths of the outside walls, or the girth, plus an allowance of six inches in each length for splices. This will, of course, result in one-foot splices.

ILLUSTRATION: If 4-inch by 6-inch timbers are to be used for the sills in the building shown in plan in Fig. 4, how many board

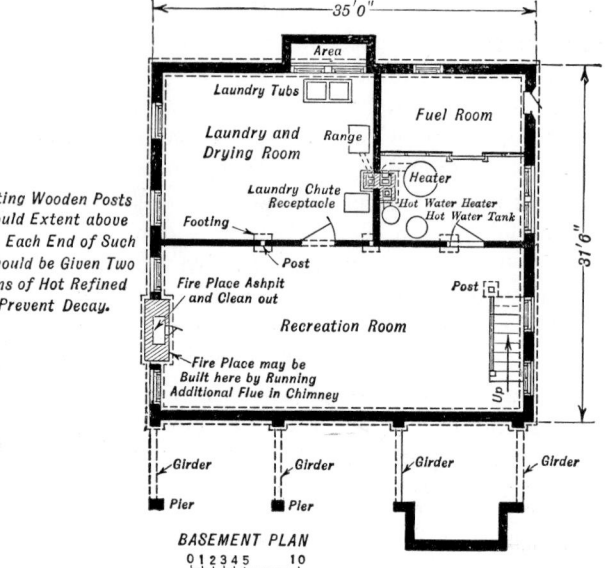

NOTE:—
Footings Supporting Wooden Posts and Columns Should Extent above the Finish Floor. Each End of Such Post or Column Should be Given Two Brush Applications of Hot Refined Creosote to Prevent Decay.

BASEMENT PLAN

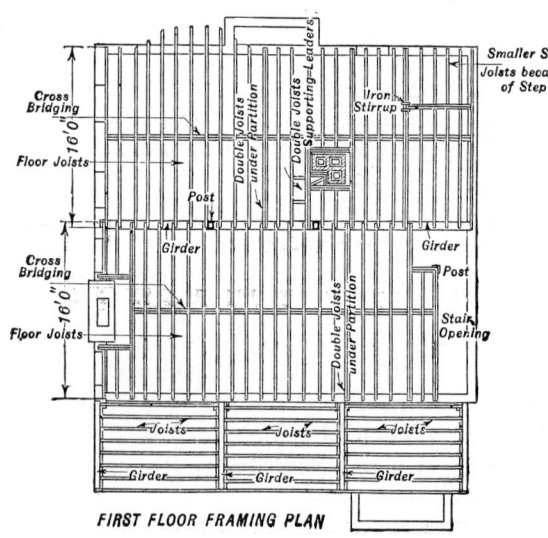

FIRST FLOOR FRAMING PLAN

Fig. 4.

feet will be required and what lengths of pieces can be used advantageously?

The girth of the building is, in round figures, $35 + 35 + 32 + 32 = 134$ feet. One joint at about the middle of each wall will obviously be needed. This will add one-half foot to each of eight timbers, making the total length $134 + 4 = 138$ feet. This allows nothing for waste and assumes that commercial lengths will fit.

Turning our attention to specific lengths of timbers needed for the house, we note that for the front and back, two timbers each $\frac{35}{2} + \frac{1}{2} = 18$ feet long will fit each of these walls without waste, or a total of 4 18-foot timbers. On the sides, $\frac{32}{2} + \frac{1}{2} = 16\frac{1}{2}$ feet, but the next larger commercial length is 18 feet. However, one 18-foot piece and one 16-foot piece will take care of each side nicely with a total waste of only about 2 feet. The bill of material for the sill would then read:

$$\left. \begin{array}{l} \text{6 pieces 4 by 6 in. 18 ft long} \\ \text{2 pieces 4 by 6 in. 14 ft long} \end{array} \right\} \quad \text{(Ans.)}$$

The original estimate of 138 linear feet must now be revised by the addition of 2 feet to make a total of 140 feet. Converting this to board feet we obtain,

$$\text{fbm} = \frac{L \times w \times t}{12} = \frac{140 \times 6 \times 4}{12} = 280 \text{ fbm}$$

Floor Joists.—Floor joists form the support for the floor, as their name implies, and, in the case of those for the first floor, rest on edge on the sills. The joists may be anywhere from 2 in. by 6 in. to 3 in. by 14 in. in cross-section depending on the load, the span, and the extreme bending stress allowed by the building code for the kind and grade of lumber used. These factors also determine the spacing, which may be 12 in., 16 in., 20 in., or 24 in. center to center. Sixteen-inch spacing is the most common in dwelling construction because it conveniently connects up with the favored spacing of studding.

In the case of narrow buildings, joists span the entire width
and rest on the side sills. In larger buildings where the span
would be too great the joists have one end resting on wall sills
and the other on girders supported by columns as shown in Fig. 4.

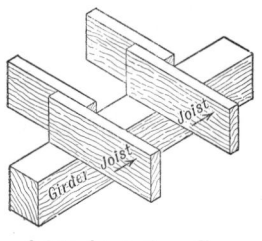

Joists Lapped on Top
of Girder

Fig. 5.

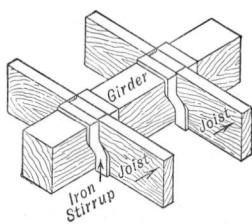

Joists Hung on Girder
with Iron Stirrups

Fig. 6.

When sufficient basement headroom is available, they can be
made to lap over the girder as shown in Fig. 5. This makes for
a minimum amount of sawing since it is not material how far
the end of the joist extends beyond the bearing on the girder.

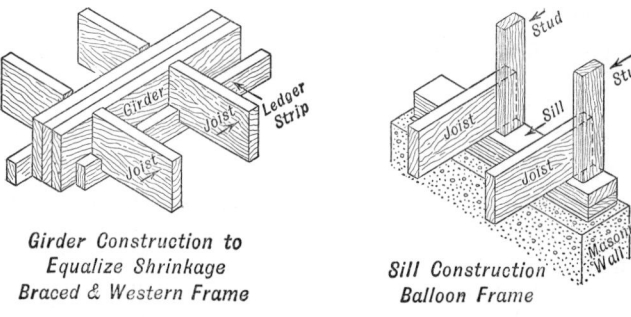

Girder Construction to
Equalize Shrinkage
Braced & Western Frame

Fig. 7.

Sill Construction
Balloon Frame

Fig. 8.

Figures 6 and 7 show other girder connections which require less
headroom.

At the wall bearing end, joists may either rest directly on the
sills as shown in Fig. 8, or may be dapped a small amount as

shown in Fig. 9 to bring their top surfaces to an absolutely level plane.

Under partitions and around floor openings, heavier members than the regular joists are required. These are called *trimmer beams*, but the required reinforcement is often accomplished by using double joists as shown in Fig. 4. The members around

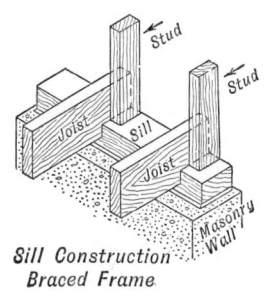

Sill Construction
Braced Frame

Fig. 9.

openings which are placed transverse to the direction of the joists are called *headers*, and these, too, are often made up of double joist timbers.

When a joist spacing of twelve inches is used, the number of joists required will be equal to the length of the opening in feet plus one, plus one for each point at which the joists are doubled.

ILLUSTRATION: A building with a floor space 17 feet wide and 60 feet long is to have joists spaced 12 inches center to center spanning the width. How many joists will be required for a floor if there are no floor openings but eight partitions to be supported?

Joists required = length in feet + 1 + number of partitions
Joists required = 60 + 1 + 8 = 69   (Ans.)

The number of joists required when the spacing is 16 inches may be estimated by multiplying the distance of the opening across the joists in feet by $\frac{3}{4}$, adding 1 and adding further 1 for each doubling of joists.

ILLUSTRATION: A floor 20 feet wide by 32 feet long is to have joists spaced 16 inches center to center transverse to the length of the house. How many joists will be required, if there are six points at which they must be doubled up?

Joists required = length in feet $\times \frac{3}{4}$ + 1 + no. of doublings
Joists required = $32 \times \frac{3}{4}$ + 1 + 6 = 24 + 7 = 31   (Ans.)

Another method of estimating the number of floor joists is, of course, to count them from the plans. Thus in Fig. 4 it is an easy matter to determine that the equivalent of some 60 long joists will be required, with a slight addition for headers, and 24 short joists for the porch floor.

Floor joists are given lateral support by *cross bridging* consisting of $1\frac{1}{2}$ in. by 3 in. pieces nailed as shown in Fig. 10 in rows not more than 8 feet apart or from the supporting wall.

Studding.—The vertical members of the walls and partitions of a frame dwelling are called *studs*. These usually consist of 2 in. by 4 in. pieces of lumber spaced 16 inches center to center. In the outside walls they may be continuous from the sill to the

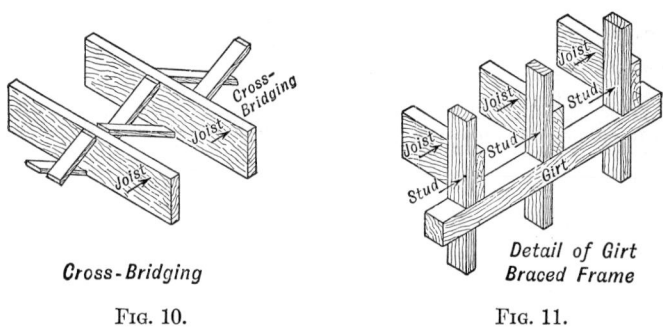

Cross-Bridging

Fig. 10.

Detail of Girt
Braced Frame

Fig. 11.

roof plate as shown in Fig. 2, or they may terminate at the ceiling level and be capped by a plate or girt as shown in Fig. 11. Studding is doubled around openings and at corners although the construction at corners shown in Figs. 2 and 12 gives more convenient nailing surfaces for the lath. Studding is braced at the midpoint between floor and ceiling either by straight diagonal bridging or by herringbone bracing as shown in Fig. 2.

Studding spaced 16 inches center to center may be estimated by multiplying the lineal lengths of the walls and partitions by $\frac{3}{4}$ and adding one for each corner and opening. However, the more common and sufficiently accurate practice is to estimate one stud per lineal foot of walls and partitions, the surplus being sufficient for doubling at corners and openings.

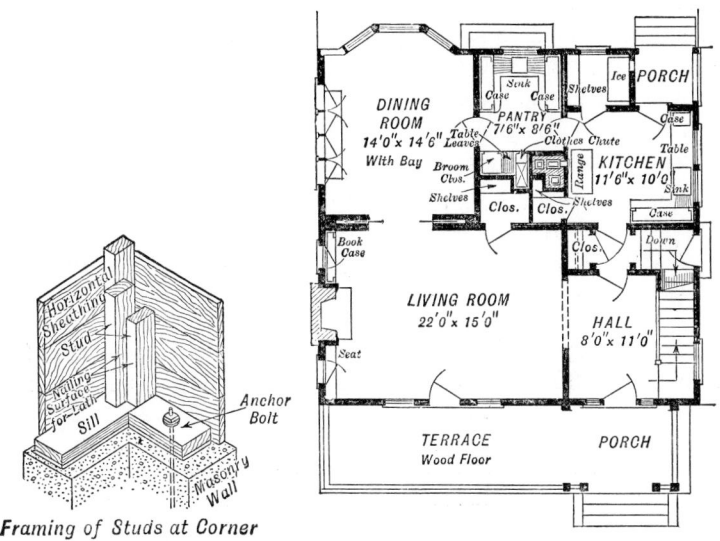

Framing of Studs at Corner
Fig. 12.   Fig. 13.

ILLUSTRATION: The floor plan shown in Fig. 13 is the first floor plan of the same building as shown in Fig. 4. Estimate the approximate number of studs needed for the walls and partitions of this floor.

Length of outside walls = 35 + 35 + 32 + 32 = 134 feet

Center transverse partition................. =   35 feet

Living room-hall partition.................. =   15 feet

Hall-stair partition = 8 + 6............... =   14 feet

Dining room-pantry partition.............. =   14 feet

Pantry-kitchen partition................... =   14 feet

Pantry-closet partition.................... =   8 feet

Total length of walls and partitions........   234 feet

Then, if one stud is allowed for each lineal foot of wall and partition, the number required in this case will be 234.   (Ans.)

Framing for Wall Openings.—Openings in walls and partitions are framed as shown in Fig. 14.   The architect's plans or working drawings usually indicate the sizes of doors and windows by the size of the finished opening, and sometimes in the case of the latter, by the glass size.   Then, when framing an opening, an allowance must be made for doors of 5 inches in width and 3 inches in height and for windows 6 inches in width and 4 inches in height over the finished opening size.   If glass size is shown, an additional 4 inches for bottom rail and 2 inches for stiles, check rail and top rail must be added.

ILLUSTRATION: Working drawings show door openings 2 feet 6 inches by 6 feet 6 inches.   What size opening should be made in framing the partitions?

Width of door........	2 ft.	6 in.
Add...........		5 in.
Width of opening.....	2 ft. 11 in.	(Ans.)
Height of door.......	6 ft.	6 in.
Add...........		3 in.
Height of opening.....	6 ft.	9 in. (Ans.)

ILLUSTRATION: A working drawing shows a window opening 2 feet 4 inches by 4 feet 10 inches.   What size opening should be provided in framing the wall?

Width of window....	2 ft.	4 in.
Add...........		6 in.
Width of opening....	2 ft. 10 in.	
Height of window....	4 ft.	10 in.
Add...........		4 in.
Height of opening....	5 ft.	2 in. (Ans.)

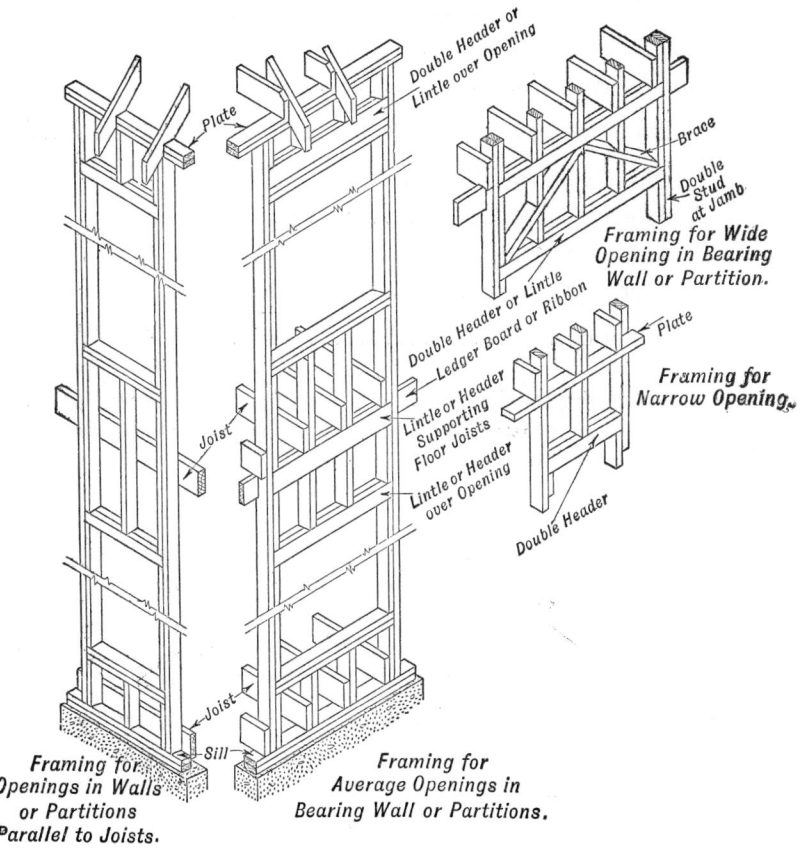

METHODS OF FRAMING AROUND OPENINGS IN WALLS AND PARTITIONS

FIG. 14.

ILLUSTRATION: Architect's drawings show a two-light window with glass sizes 24 inches by 20 inches.   What size opening should be provided in framing the wall?

Glass width......................	24 in.
Add for stiles = 2 + 2............	4 in.
Add for trim.....................	6 in.
Width of opening.................	34 in. 2 ft. 10 in.(Ans.)

Height of glass = 20 + 20........	40 in.
Add for bottom rail..............	4 in.
Add for check & top rails = 2 + 2..	4 in.
Add for trim.....................	4 in.
Height of opening...............	52 in. 4 ft. 4 in. (Ans.)

Roof Framing.—The elements of a roof and the terms pertaining to them are illustrated in Fig. 15.   The *span* is the distance

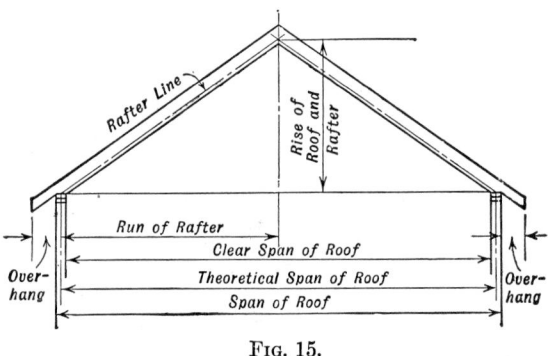

FIG. 15.

between the outer edges of the side walls supporting a roof.   The *rise* is the vertical distance between the ridge and the plates supporting the roof.   The *run* is the horizontal distance between the ridge and the outside edge of the plate supporting the roof.

The *pitch* of a roof is the slope of the rafters expressed as a ratio of the rise to the span. Thus, to find the pitch of a roof when the rise and span are given, merely substitute the known values in this equation,

$$\text{Pitch} = \frac{\text{rise}}{\text{span}}$$

ILLUSTRATION: What is the pitch of a roof whose rise is 6 feet and span 18 feet?

$$\text{Pitch} = \frac{\text{rise}}{\text{span}} = \frac{6}{18} = \frac{1}{3} \quad \text{(Ans.)}$$

To find the rise when the pitch and span are known, use the equation,

$$\text{Rise} = \text{pitch} \times \text{span}$$

ILLUSTRATION: What is the rise of a roof whose pitch is $\frac{2}{3}$ and span 24 feet?

$$\text{Rise} = \text{pitch} \times \text{span} = \frac{2}{3} \times 24 = 16$$

With these relationships in mind it is a simple matter to compute the length of the rafters by extracting the square root of the sum of the squares of the rise and the run, since these form a right triangle. Thus,

$$\text{Rafter length} = \sqrt{(\text{rise})^2 + (\text{run})^2}$$

The overhang for the eaves, if any, must then be added to this figure.

Another convenient method of determining the length of rafters is to let the inches on a steel square represent the rise and run in feet. Thus, in Fig. 16 the run of 20 feet is represented by 20 inches on the square and the rise of 10 feet is represented by 10 inches. Then the length of the diagonal in inches may be measured with a rule and this represents the length of the rafter in feet.

Flat Roof.—A flat roof or lean-to has but one pitch and is used widely on sheds, porches, dormers, etc. The slope is often just sufficient for drainage and the length of the rafters may be computed by either of the above methods.

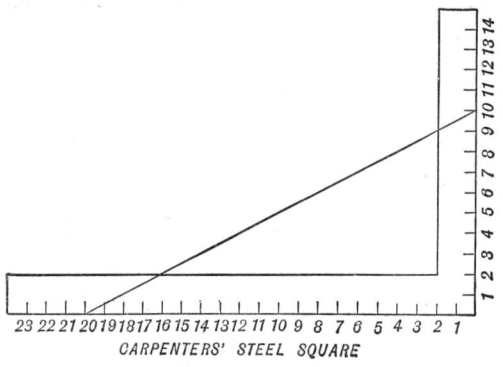

CARPENTERS' STEEL SQUARE

FIG. 16.

ILLUSTRATION: The roof shown in Fig. 17 has a rise of 18 inches and a run of 15 feet. How long must the rafters be if the overhang front and back is 8 inches?

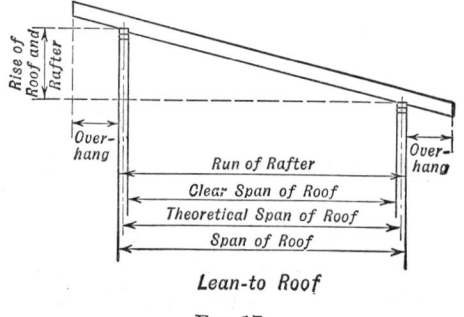

Lean-to Roof

FIG. 17.

Length of rafters = $\sqrt{(\text{rise})^2 + (\text{run})^2}$

Length of rafters =

$$\sqrt{(1\tfrac{1}{2})^2 + (15)^2} = \sqrt{2.25 + 225} = \sqrt{227.25} = 15.075 \text{ ft}$$

Converting the decimal to inches and fractions of an inch by multiplying by 12 and referring to Table 1, page 19, we obtain a length of 15 ft. $0\frac{7}{8}$ in. To this must be added 16 inches for the overhangs.

$$\begin{array}{r} 15 \text{ ft. } 0\frac{7}{8} \text{ in.} \\ 16 \text{ in.} \\ \hline \end{array}$$

Total length of rafters....... 16 ft. $4\frac{7}{8}$ in.

This problem illustrates that when the pitch is small, the length of the rafters will very nearly equal the run, so that in sheds and unimportant structures, where the exact amount of overhang is not of great concern, the overhang added to the run may be used for the length of the rafters. However, the calculation illustrated is important in the case of roofs of greater pitch and in dwelling construction.

Gable Roofs.—A gable roof has two sloping surfaces which meet at the ridge. Figure 18 shows an end view of such a roof.

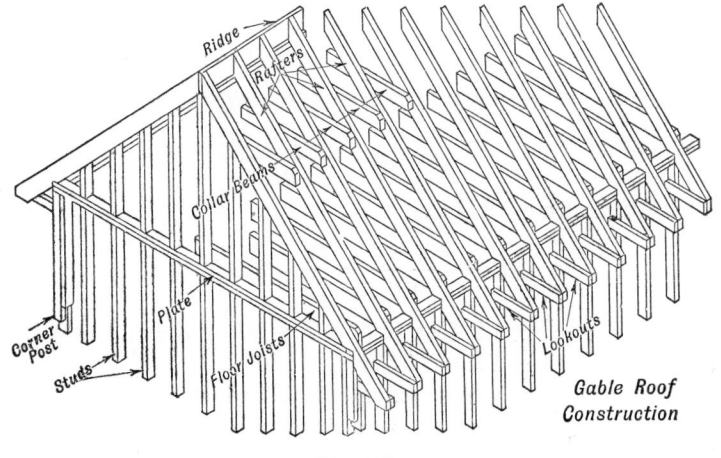

FIG. 18.

The length of the rafters is computed as for a flat roof except, of course, that an overhang occurs on only one end.

ILLUSTRATION: What is the length of the rafters of a roof which has a rise of 10 feet, a run of 12 feet and an overhang of 1 foot?

$$\text{Length of rafter} = \sqrt{(\text{rise})^2 + (\text{run})^2} = \sqrt{(10)^2 + (12)^2}$$
$$\sqrt{100 + 144} = \sqrt{244} = 15.62 \text{ ft.}$$

Changing the decimal 0.62 to inches by multiplying by 12, we obtain a length of 15 feet $7\frac{1}{2}$ inches to which must be added the overhang, making a total of 16 ft. $7\frac{1}{2}$ in.   (Ans.)

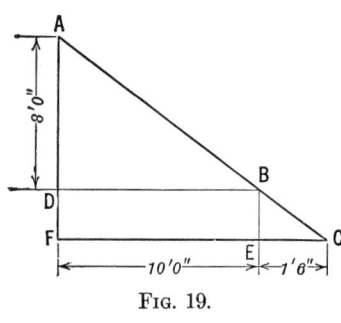

Fig. 19.

ILLUSTRATION: A roof has a rise of 8 feet and a run of 10 feet and eaves projecting a horizontal distance of 1 foot 6 inches. What is the length of the rafters?

From geometry we know that ABD and BCE (Fig. 19) are similar triangles and that therefore the sides of one are proportional to the sides of the other.   Then, $BE : AD = EC : DB$ and

$$BE \times DB = AD \times EC$$
$$BE = \frac{AD \times EC}{DB}$$

Substituting known values,

$$BE = \frac{8 \times 1.5}{10} = \frac{12}{10} = 1.2 \text{ ft.}$$

Then DF is also 1.2 feet and $AD + DF = 8 + 1.2 = 9.2$ ft.; $FE + EC = 10 + 1.5 = 11.5$ ft.

We have then a new triangle, ACF, which can be solved in the regular manner for the side AC which represents the entire length of the rafter including overhang.

$$AC = \sqrt{(9.2)^2 + (11.5)^2} = \sqrt{84.64 + 132.25}$$
$$AC = \sqrt{216.89} = 14.727 = 14 \text{ ft. } 8\frac{3}{4} \text{ in.}   (Ans.)$$

Frequently two gable roofs will meet at right angles as shown in Fig. 20.

This construction calls for a *valley rafter* at the intersection of the roof surfaces. The valley rafter may be represented by the hypotenuse of a right triangle one of whose legs is the length of

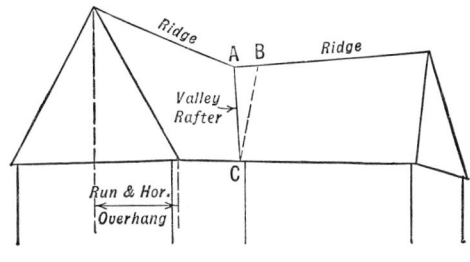

FIG. 20.

the common rafter *BC*, and the other leg the distance *AB* from the intersection of the ridges to a point in a plane with the extremities of the rafters of the gable perpendicular to *AB*. The length of the valley rafter is then

$$AC = \sqrt{(AB)^2 + (BC)^2}$$

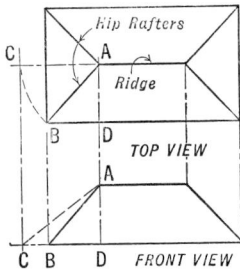

FIG. 21.—Hipped Roof.

It will be noted that when two gables intersect at exactly right angles, the distance *AB* is equal to the run plus horizontal overhang of the intersecting gable.

Hip Roofs.—A hip roof has surfaces sloping toward all four walls as shown in Fig. 21. The only new problem which this involves is the calculation of the length of the hip rafters.

If a roof drawing is made to scale the length of the hip rafter can be found by scribing radius *AB* in Top View to point *C* on ridge center line. By dropping a vertical line to the line of plate in the Front View the actual length of hip rafter can be measured along *AC*.

The length of the hip rafter can be computed when the length of the common rafter AD and the distance BD are known.   Then,

$$\text{Length of hip rafter} = \sqrt{(AD)^2 + (BD)^2}$$

When the pitches of the intersecting roof surfaces are equal, as they usually are, the run of the hip rafter (See Fig. 22) is the

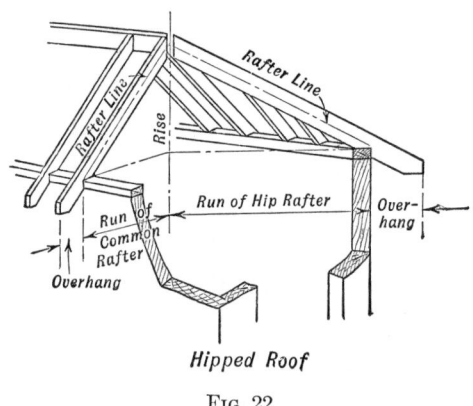

Hipped Roof

Fig. 22.

hypotenuse of the isosceles right triangle whose legs are the run of the common rafters and the distance BD along the plate.   Then,

$$\text{Run of hip rafters} \quad = \text{run of common rafters} \times \sqrt{2},$$
and,
$$\text{Length of hip rafters} = \sqrt{(\text{rise})^2 + 2(\text{run of common rafters})}$$

ILLUSTRATION: A hip roof of equal pitch all around has a rise of 10 feet and a run of 14 feet.   What is the length of the hip rafters?

$$
\begin{aligned}
\text{Length of hip rafters} &= \sqrt{(\text{rise})^2 + 2(\text{run})^2} \\
&= \sqrt{(10)^2 + 2(14)^2} = \sqrt{100 + 2 \times 196} \\
&= \sqrt{492} = 22.181 \text{ ft.} = 22 \text{ ft. } 2\tfrac{3}{16} \text{ in. (Ans.)}
\end{aligned}
$$

The length of the hip rafter can also be found without computation by scaling the distance *AB* on a plan or top view drawing, laying this distance off to a scale of 1 in. = 1 ft on one leg of a carpenters' square, as in Fig. 16, and laying the rise off on the other leg. Then the diagonal distance between these points is the scale length of the hip rafter.

Stair Construction.—The proportioning and construction of stairs present several nice problems of calculation. The elements of a stairway are shown in Fig. 23 and the details of framing in Fig. 24.

The ideal angle for a stairway is between 30 degrees and 35 degrees with the horizontal, although both steeper and flatter stairways are sometimes necessary. However, regardless of the angle of stair, a certain relationship between the rise and the run of each step must prevail. That is, the sum of the rise and the run shall not be less than 17 inches nor more than 18 inches. (It is to be noted that the run does not include the nosing.) Then, if a step has a rise of 7 inches, its run will be between 10 and 11 inches.

When the distance between two floors or the rise of the stair is known, and the approximate amount of the rise of each step has been determined, then the number of steps required may be found by dividing the rise of the stair by the rise of the step. If the quotient is not an even number, divide the rise of the stair by the nearest whole number of the quotient to obtain the exact rise of the step.

ILLUSTRATION: The distance between two floors is 12 feet 4 inches. How many steps will be required if the rise is to be about $7\frac{1}{4}$ inches?

$$12 \text{ ft. } 4 \text{ in.} = 148 \text{ in.} \qquad 148 \div 7.25 = 20.4$$

Then, since the quotient is not a whole number, divide the rise of the stair by 20. $148 \div 20 = 7.4$ or approximately $7\frac{13}{32}$.

The result shows that 20 steps each with a rise of $7\frac{13}{32}$ inches are required. (Ans.)

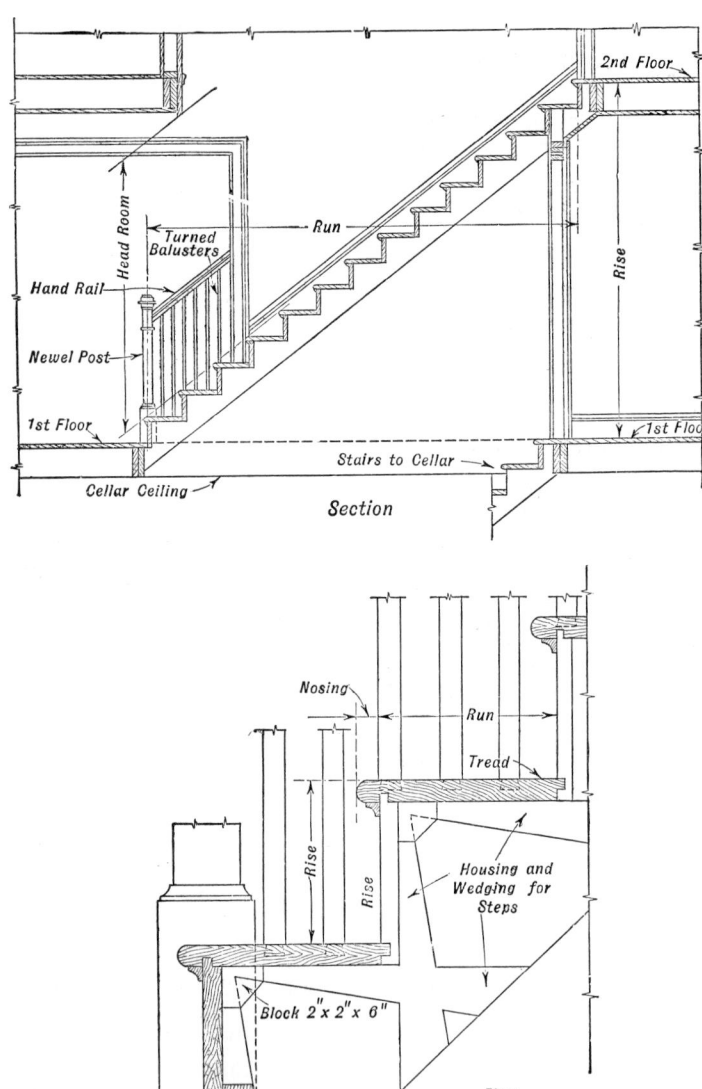

FIG. 23.—Stair Details.

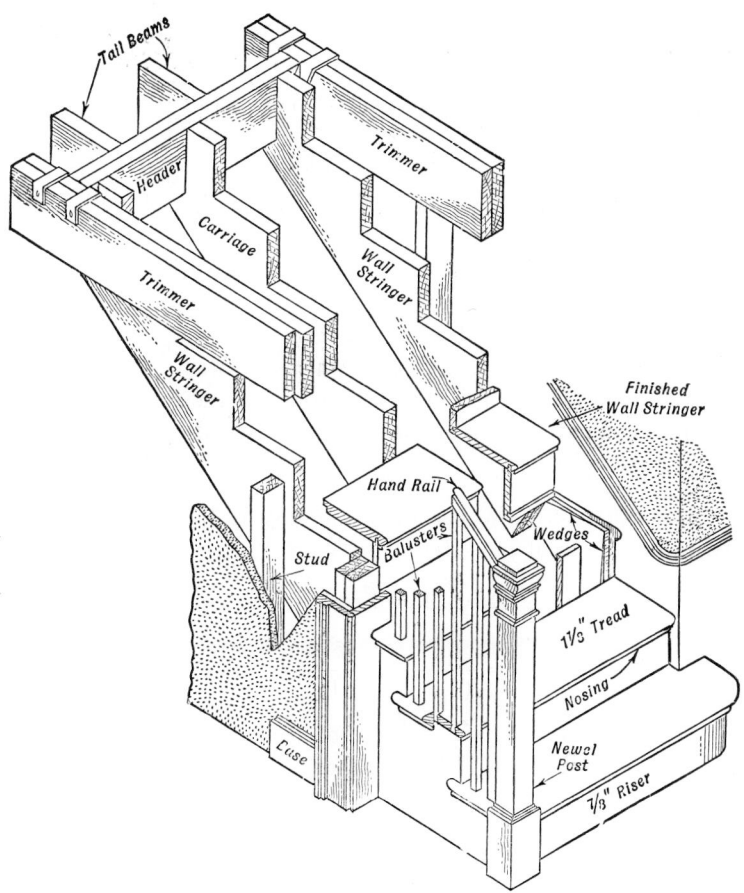

Front Elevation Frame of the Stairs

FIG. 24.

ILLUSTRATION: How many steps will be required between two floors with a difference in elevation of 9 feet 7 inches, if the rise is to be about 7 inches?

$$9 \text{ ft. } 7 \text{ in. } = 115 \text{ in.} \qquad \tfrac{115}{7} = 16.4$$

$$\tfrac{115}{16} = 7\tfrac{3}{16} \text{ in.}$$

The result shows that 16 steps are required, each with a rise of $7\frac{3}{16}$ inches. (Ans.)

The computations in the preceding illustrations instead of actually arriving at the number of steps, arrived at the number of risers. The top landing is not regarded as a step, and thus there is one less tread than riser in a stairway. Reference to Fig. 23 makes this clear. Then the width of the run of each step is equal to the total run of the stairway divided by one less than the number of risers.

ILLUSTRATION: The run of a stairway is 13 feet 1½ inches. What is the run of each step if there are 16 risers?

$$13 \text{ ft. } 1\tfrac{1}{2} \text{ in. } = 157.5 \text{ in.}$$

$$\text{Run of step} = \frac{157.5}{16-1} = \frac{157.5}{15} = 10\tfrac{1}{2} \text{ in.} \quad \text{(Ans.)}$$

ILLUSTRATION: What is the run of each step if a stairway has 20 risers, a total rise of 12 feet 6 inches, and a slope of 35 degrees?

Since the length of the run of the stairway is lacking, it must be found by trigonometry. It is evident from the triangle in Fig. 25 that

$$\frac{\text{run}}{\text{rise}} = \text{cotangent } 35°$$

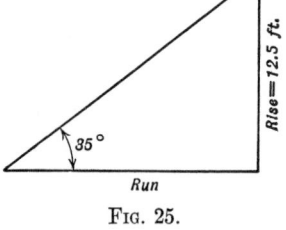

FIG. 25.

Plywood, Wallboard, etc.—In recent years the use of laminated and specially treated large-surface covering materials, such as plywood and wallboard, has become increasingly popular. They are readily available and lend themselves to quick, easy installation on both interiors and exteriors.

Calculating the amount of material required is a relatively simple and straightforward procedure. Plywood can be furnished in panels $2\frac{1}{2}$ to 4 feet wide, by 5 to 12 feet long, in thicknesses from $\frac{3}{16}$ inch to $1\frac{1}{8}$ inches. To determine the number and size of panels required, calculate the square feet of area to be covered and divide it by the number of square feet in the panel to be used. The particular panel size selected is usually determined by the dimensions of the wall being sheathed.

ILLUSTRATION: A solid wall 8 feet high by 18 feet long with studs 16 inches on center to be sheathed with $\frac{5}{16}$-inch thick plywood. Find the number and size of panels required.

$$\text{Wall area} = 8 \times 18 = 144 \text{ sq ft}$$

Since the wall is 8 feet high, use panels 8 feet long. Since the studs are on 16-inch centers, use panels 4 feet wide.

Divide the wall area by the area of the panel selected (4×8, or 32 sq ft) to find the number of panels required.

$$\frac{144}{32} = 4\frac{1}{2} \text{ or 5 panels required}$$

ANSWER: 5 panels measuring 4×8 feet required.

Plywood is manufactured in two basic types (exterior and interior), and in several grades. The *type* refers to the type of bond between plies. Exterior-type panels are completely waterproof, and interior types are moisture resistant but not waterproof. The *grade* refers to the appearance of the wood veneer in the outer plies.

Standards for plywood have been set up by the American Plywood Association, a nonprofit organization of plywood manufacturers. Table 2 lists the different types, grades, and sizes of plywood commercially available. The trade names of the various grades are also defined below.

TABLE 2—Data on Douglas Fir Plywood

EXT-DFPA·A-A. Use where the appearance of both sides is important. Fences, built-ins, signs, boats, cabinets, commercial refrigerators, shipping containers, tote boxes and ducts.

EXT-DFPA·A-B. For uses similar to EXT·A-A panels, but where the appearance of one side is less important.

EXT-DFPA·A-C. Use where the appearance of only one side is important. Siding, soffits, fences, structural uses. Box car and truck lining and farm buildings.

EXT-DFPA·PLYFORM. Concrete form grade, with high re-use factor. Red painted edges, mill-oiled unless otherwise specified.

EXT-DFPA·B-C. An outdoor utility panel. For farm service and work buildings. Box car and truck linings, containers and agricultural equipment.

EXT-DFPA·C-C, PLUGGED, or EXT-DFPA·UNDER-LAYMENT. Use as a base for resilient flooring, mosaic tile, carpeting, where unusual moisture conditions exist, such as bathrooms, utility rooms, kitchens. Use for conventional or single layer subfloor-underlayment. Other uses include pallets, pallet bins and farm service buildings.

EXT-DFPA·C-C. Unsanded grade with waterproof bond. Use for backing, rough construction, farm service buildings, crating, pallets and pallet bins.

INT-DFPA·A-A. For interior applications where both sides will be viewed. Built-ins, cabinets, furniture, and partitions.

INT-DFPA·A-B. For uses similar to INT·A-A panels, but where the appearance of one side is less important and two smooth surfaces are necessary.

INT-DFPA·A-D. Interior use where appearance of only one side is important. Paneling, built-ins, shelving, partitions and flow racks.

INT-DFPA·PLYFORM. Re-usable form plywood. Glue is moisture-resistant, not waterproof. Green painted edges, mill-oiled unless otherwise specified.

INT-DFPA·B-B. Interior utility panel used where two smooth sides are required.

INT-DFPA·B-D. Interior utility panel used where one smooth side is required. Good for backing, sides of built-ins. Industry: separator boards and bins.

INT-DFPA·UNDERLAYMENT. Base for resilient flooring and carpeting. A backing material where moisture is not a problem. Used for conventional or single layer subfloor-underlayment.

INT-DFPA·PLYSCORD. Unsanded structural grade panel for sheathing, subflooring, limited exposure crates, containers, pallets and dunnage.

INT-DFPA·PLYSCORD WITH EXTERIOR GLUE. Same as PlyScord, but with waterproof glue. Not a substitute for Exterior-type plywood.

TEXTURE 1-11, EXT-DFPA. Exterior type, sanded or unsanded, shiplapped edges with parallel grooves $\frac{1}{4}''$ deep, $\frac{3}{8}''$ wide. Grooves $2''$ or $4''$ o.c. Available in $8'$ and $10'$ lengths and MD Overlay. Use for siding, gable ends, fences, interior paneling.

EXT-DFPA·303, SPECIALTY SIDING. Exterior siding or fencing panel with special surface treatment such as V-groove, channel groove, striated, brushed, Lauan One-Eleven, rough sawn and many other decorative surface patterns. Available in several wood face veneers including fir, red cedar, Philippine mahogany (lauan) and MD Overlay.

INT-DFPA·BRUSHED, INT-DFPA·EMBOSSED, INT-DFPA·STRIATED. Textured paneling, for accent walls, built-ins, counter facings, and displays. An effect of depth and dimension can be accomplished with the use of brushed, embossed or striated faced panels. For outdoor applications, use only Exterior-type.

EXT-DFPA·MD, OVERLAY. Exterior-type Medium Density Overlaid plywood with a smooth, fused, resin-fiber overlay. Especially good surface for painting. Exterior siding, soffits, kitchen cabinets, signs.

EXT-DFPA·HD, OVERLAY. Exterior-type High Density Overlaid plywood with hard, translucent, fused resin-fiber overlay. Slight grain pattern. Abrasion resistant. Painting not ordinarily required. Good for concrete forms, acid tanks, cabinets.

EXT-DFPA·PLYRON, INT-DFPA·PLYRON. Hardboard face on one or both sides. For concrete forms, counter tops, shelving, cabinet doors and built-ins.

INT-DFPA·N-N. Natural finish cabinet quality. Both sides select all heartwood veneer. Special jointed core construction. For furniture having a natural finish, cabinet doors, built-ins, etc. (5)

INT-DFPA·N-A. Same as N-N except one side is A faced for economy. Special jointed core construction. For furniture having a natural finish, cabinet doors, built-ins, etc. (5).

INT-DFPA·N-D. One side special select, all heartwood veneer. For wall paneling. (5)

MARINE·EXT-DFPA. Waterproof panel with limitations on core gaps and with special solid jointed core construction. Made especially for marine use. (2) Available also in overlay grades. Especially good for boat hulls.

DFPA·2·4·1, DFPA·2·4·1, WITH EXT. GLUE, DFPA·2·4·1 T & G, DFPA·2·4·1 T & G, WITH EXT. GLUE. Combination subfloor and underlayment. Base for resilient finish flooring, carpeting and wood strip flooring. Available in square edges; or with tongue and grooved sides or sides and ends. Use 2·4·1 with Exterior Glue in baths, kitchens or utility areas where moisture may be a problem (or when construction delays are anticipated). (1)

The thicknesses available of the first three types are $\frac{1}{4}''$, $\frac{3}{8}''$, $\frac{1}{2}''$, $\frac{5}{8}''$, $\frac{3}{4}''$ and $1''$. The next 11 types are available in all these thicknesses except, $1''$. The other types are special veneers and other items whose available thicknesses must be checked with the supplier.

Standard size plywood panels are 4×8 feet. Other sizes are available upon order. King size panels 12', 14', 16', 20', and longer are also available. Panels $\frac{3}{8}''$ and thinner have a minimum of 3 plies; $\frac{5}{8}''$ to $\frac{3}{4}''$ have 5-ply minimum; thicker panels have 7-ply minimum.

Surface Covering

Up to this point, only structural members of buildings have been considered and the main concerns of these are strength and conformity with building regulations. The measurement of these members has generally been by the piece. Surface covering, on the other hand, while it is purchased by the board foot by nominal dimensions, covers areas only in proportion to its actual dimensions. Surface measure is made in square feet, or, for the sake of smaller figures, in *squares,* one square being a surface 10 feet by 10 feet or 100 square feet.

Certain factors pertaining to surface covering with common boards and strips are common to sheathing, rough flooring, and roof boarding. Thus, in any of these uses, a seven-inch board will cover a space less than seven inches wide and a ten-inch board will cover a space less than ten inches wide. When the area to be covered has been calculated, the following percentages must be added to make up for the scant widths:

Width of Board, Inches	Percentage to be Added	Width of Board, Inches	Percentage to be Added
3	14.39	8	6.66
4	10.34	9	5.88
5	8.11	10	5.26
6	6.66	11	4.76
7	5.66	12	4.35

This table does not provide for waste resulting from short ends. An additional 5 percent should be added for waste when sheathing is placed horizontally or when rough flooring is laid parallel to the walls. Ten percent should be added for waste when these coverings are laid diagonally.

Sheathing.—Sheathing may be nailed to the studding of a frame building either diagonally as shown in Fig. 2, or horizontally as shown in Fig. 12. It may be either matched or unmatched lumber ⅞ inch thick and planed on at least one side.

In estimating the amount of lumber needed for sheathing the procedure is to calculate the net wall surfaces and add the proper percentage for waste and scant widths. The area of the triangular surface under the end of a gable roof is, by geometry, one-half the product of the rise and the span.

ILLUSTRATION: The bungalow shown in Fig. 26 is to be sheathed diagonally with 1-inch by 6-inch common boards. How many board feet of lumber will be required? (Assume door and window openings on far sides equal in area to those on the near sides.)

Area of side wall = 23 ft. 10 in. × 10 ft. 2 in. − openings
$$= 23.83 \times 10.17 - (3.17 \times 4.92 + 3.17 \times 5.42)$$
$$= 242.35 - 32.78 = 180.58 \text{ sq. ft.}$$

Area of end wall = 18 ft. 0 in. × 10 ft. 2 in.
$$+ \tfrac{1}{2}(18 \text{ ft. } 0 \text{ in.} \times 6 \text{ ft. } 4 \text{ in.})$$
$$- 2(5 \text{ ft. } 5 \text{ in.} \times 3 \text{ ft. } 2 \text{ in.})$$
$$- 3 \text{ ft. } 4 \text{ in.} \times 8 \text{ ft. } 0 \text{ in.}$$
$$- 3 \text{ ft. } 4 \text{ in.} \times 1 \text{ ft. } 10 \text{ in.}$$
$$= 18 \times 10.17 + \tfrac{1}{2}(18 \times 6.33) - 2(5.42 \times 3.17)$$
$$- 3.33 \times 8 - 3.33 \times 1.83$$
$$= 183.06 + 57 - 34.4 - 26.6 - 6.1$$
$$= 240.06 - 67.09 = 172.97 \text{ sq. ft.}$$

Total surface = 2 sides @ 180.58 sq. ft. = 361.16 sq. ft.
$$= 2 \text{ ends @ } 172.97 \text{ sq. ft.} = 345.94 \text{ sq. ft.}$$

Total = 707.10 sq. ft.

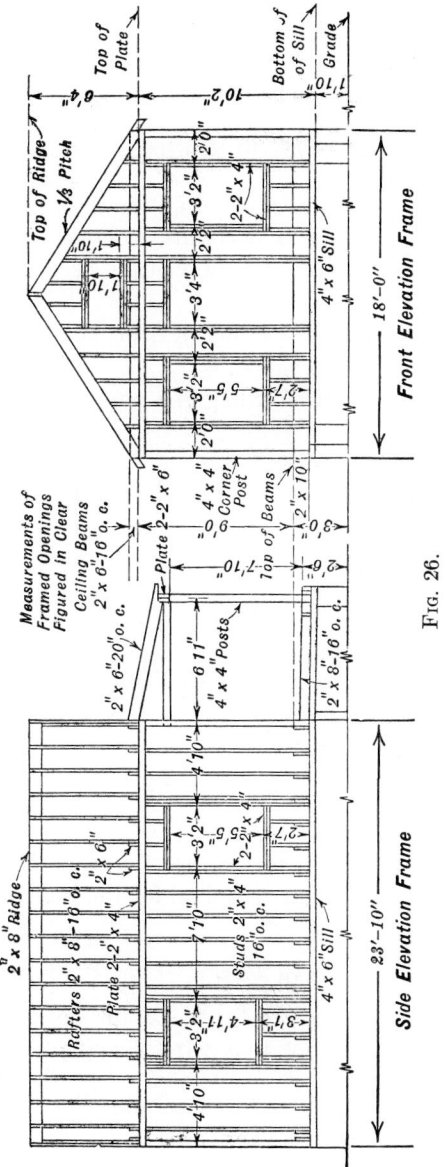

FIG. 26.

This area must then be increased by 10 percent for waste and by 6.66 percent (according to the above table) for scant widths or a total of 16.66 percent. The lumber needed will then be

$$707.10 + 707.10 \times 0.1666 = 825 \text{ fbm} \quad \text{(Ans.)}$$

Siding.—Exterior walls of wood may be either siding or shingles. Siding is laid in horizontal courses outside of a layer of building paper which has previously been attached to the sheathing.

Figure 27 shows cross-sections of the common bevel siding and several patterns of drop siding.

The usual size of bevel siding is a nominal width of 6 inches, a thickness of $\frac{1}{2}$ inch at the bottom edge and $\frac{1}{4}$ inch at the top edge. It is lapped on the wall as shown in Fig. 2. When laid with $4\frac{1}{2}$ inches exposed to the weather, 33 percent must be added to the area of the wall to obtain the area of siding required. With 4 inches exposed to the weather, 50 percent must be added. In both cases an additional 5 percent should be added for waste.

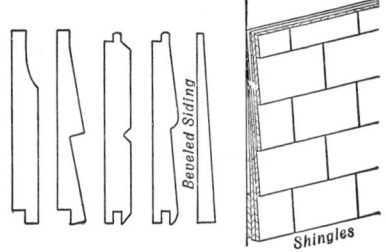

Fig. 27.—Siding.

ILLUSTRATION: How many board feet of bevel siding laid 4 inches to the weather are required for the bungalow in the previous illustration?

Net wall area = 707.10 sq. ft.

Add 50% for lap and 5% for waste; total of 55%

Lumber required = $707.10 + 707.10 \times 0.55 = 1096$ fbm (Ans.)

For drop siding with a $5\frac{3}{16}$ face add 16.3 percent for scant width and 5 percent for waste.

ILLUSTRATION: How many board feet of $5\frac{3}{16}$-inch drop siding would be required for the bungalow of the preceding exercises?

Net wall area = 707.10 square feet

Lumber required = $707.10 + 707.10 \times 0.213 = 858$ fbm. (Ans.)

Flooring.—Rough flooring should be laid diagonally on the floor joists. The lumber required is estimated in exactly the same manner as the sheathing.

ILLUSTRATION: How many board feet of lumber are required for a floor 26 feet by 28 feet if 7-inch common lumber is used and laid diagonally?

$$\text{Area} = 26 \times 28 = 728 \text{ sq. ft.}$$

Add for scant width..............	5.66%
Add for waste..................	10.00%
Total.....................	15.66%

Lumber required = $728 + (728 \times 0.1566) = 842$ fbm   (Ans.)

A finish flooring of hard maple, beech, birch or oak provides a substantial wearing surface. It is laid directly on top of the rough flooring at right angles to the direction of the floor joists, but never parallel to the rough flooring. It is nailed at intervals of 12 or 16 inches with 8-penny steel-cut flooring nails driven at an angle of 45 degrees and starting just above the tongue.

Hardwood flooring comes in thicknesses of $\frac{3}{8}$ in., $\frac{1}{2}$ in., $\frac{5}{8}$ in. and $\frac{25}{32}$ in. and in face widths of $1\frac{1}{2}$ in., 2 in., $2\frac{1}{4}$ in. and $3\frac{1}{4}$ in. The scant width loss due to the tongue and groove is considerable and the following percentages must be added when estimating the flooring required:

Face Width, Inches	Allowance, Percent
$1\frac{1}{2}$	50
2	37.5
$2\frac{1}{4}$	33.3
$3\frac{1}{4}$	24

An additional 3 to 5 percent must be added for waste in cutting and fitting.

ILLUSTRATION: How many board feet of flooring are required to lay 1252 square feet of $\frac{25}{32}$-in. by $2\frac{1}{4}$-in. flooring and allowing 5 percent for waste?

$$\begin{aligned}
\text{Scant width loss} &= 33.3\% \\
\text{Waste loss} &= \underline{5.0\%} \\
\text{Total loss} &\quad 38.3\% = .383
\end{aligned}$$

Flooring required $= 1252 + 1252 \times 0.383 = 1732$ fbm (Ans.)

Roofing.—The area of a gable roof is the sum of the two sloping surfaces. The area of one of these surfaces is equal to the product of the length of the roof and the slope length or the rafter length.

ILLUSTRATION: What is the area of a gable roof whose length is 35 feet and whose rafters are 18 feet long?

Area of $\frac{1}{2}$ of roof $= 35 \times 18 = 630$ sq. ft.

Area of whole roof $= 630 \times 2 = 1260$ sq. ft. $= 12.6$ squares (Ans.)

A hip roof has the same area as a gable roof of the same pitch, overhang and plate dimensions. Therefore, the area of a hip roof is equal to twice the product of the length of rafters on the long side and the length of the eaves on the long side.

A dormer having the same roof pitch as the main roof adds only the amount of the overhang to the area which would obtain if the dormer did not exist.

Roof rafters are covered with boarding as a support for the roof covering material. This boarding is usually tight sheathing as in Fig. 28 for slate or composition roofing.

Roof sheathing is estimated in the same manner as side-wall sheathing, the allowances for scant widths given at the head of this section being used, and 5 percent allowed for waste.

ILLUSTRATION: How many board feet of sheathing are required to cover a hip roof 35 feet long and a rafter length of 17 feet, if 1-inch by 6-inch boards are used?

Area of roof = $2 \times 35 \times 17 = 1190$ sq. ft.

Add for scant widths............. 6.66%

Add for waste.................. 5.00%

Total..................... 11.66%

Lumber required for sheathing

$$= 1190 + 1190 \times 0.1166 = 1329 \text{ fbm} \quad \text{(Ans.)}$$

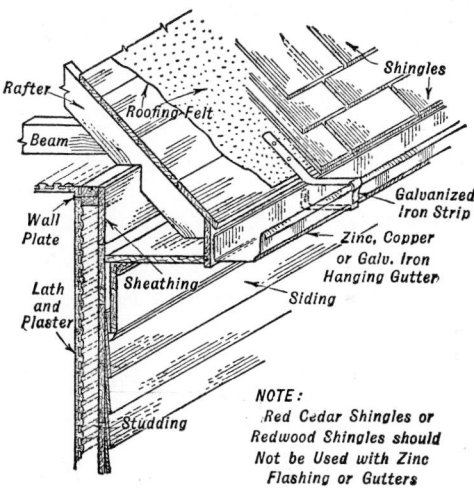

FIG. 28.

There is little unanimity on the question as to whether or not solid sheathing should be used under wood shingles. The alternative construction is the use of 1 in. by 4 in. shingle lath spaced an inch apart, as shown in Fig. 29.

Since the actual width of a 4-inch board is $3\frac{5}{8}$ inches, if 1 inch is left open, only $\dfrac{3\frac{5}{8}}{1 + 3\frac{5}{8}} = \dfrac{3.625}{4.625} = 0.784 = 78.4$ percent of the roof area will be covered. When computing the lumber required for covering a roof with 1-in. by 4-in. shingle lath spaced 1 inch apart, only 78.4 percent of the actual area is considered. The usual factors for scant widths and waste still apply, however.

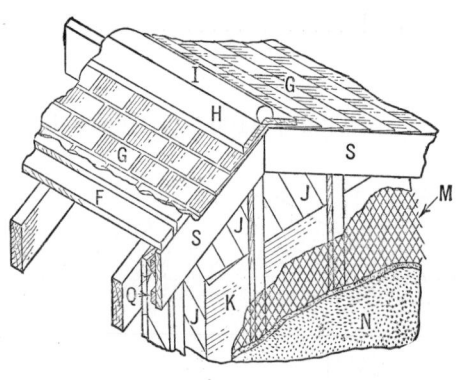

Fig. 29.

ILLUSTRATION: How many board feet of lumber are required to cover a roof of 1450 square feet with 1-inch by 4-inch shingle lath spaced 1 inch apart?

First the area must be reduced to 78.4% of its actual area.

$$1450 \times 0.784 = 1136.8 \text{ sq. ft.}$$

Allowance for scant widths..... = 10.34%

Allowance for waste.......... = 5.00%

Total...................   15.34%

Lumber required = $1136.8 + 1136.8 \times 0.1534 = 1311$ fbm   (Ans.)

Shingles.—Cedar or cypress shingles form a roof covering of great durability. Shingles are sold in bundles which contain the equivalent of 250 shingles 4 inches wide. Actually they are of random widths. They come in lengths of 16, 18, and 24 inches and in butt thicknesses of from $\frac{5}{16}$ inch to $\frac{1}{2}$ inch. Shingles are listed in this fashion:

24-in. Royals, 4/2 in.

16-in. Perfects, 5/2 in.

The first figure gives the length of the shingle; (4/2 in.) means that 4 shingles measure 2 inches at the butts, and (5/2 in.) means that 5 shingles measure 2 inches at the butts.

The amount of roof surface which a bundle of shingles will cover depends on the amount exposed to the weather. Sixteen-inch roof shingles are laid 4 in., $4\frac{1}{2}$ in., and 5 in. to the weather. Twenty-four-inch shingles are usually used for siding and laid $7\frac{1}{2}$ in. or even 10 in. to the weather. The number of bundles of shingles required for each square of roof area including an allowance of 10 percent for waste is, for various exposures, as follows:

Exposure, Inches	Bundles per Square
4	4.0
$4\frac{1}{2}$	3.6
5	3.2
6	2.7
$7\frac{1}{2}$	2.1
10	1.6

ILLUSTRATION: How many bundles of shingles are required to cover a roof of 2240 square feet when $4\frac{1}{2}$ inches are exposed to the weather?

2240 sq. ft. = 22.4 squares

22.4 × 3.6 = 81 bundles   (Ans.)

Nails Required.—The quantity of nails required for the various operations in the construction of a house may be obtained from Table 3.

ILLUSTRATION: What kind and how many pounds of nails are required for nailing 2400 fbm of 1-inch by 6-inches sheathing on 16 inches center to center studding?

The table shows 8d common to be the proper size and 32 pounds per 1000 fbm as the unit quantity.   Then,

$$2.4 \times 32 = 77 \text{ lb for 2400 fbm} \quad \text{(Ans.)}$$

ILLUSTRATION: What kind and how many pounds of nails are required for nailing 1700 fbm of 1-inch by 8-inches drop siding nailed on 12-inch centers?

The table gives 8d casing as the proper size and 23 lb per 1000 fbm as the unit quantity.   Then,

$$1.7 \times 23 = 39 \text{ lb for 1700 fbm} \quad \text{(Ans.)}$$

Interior Trim.—This work includes door jambs and trim, window frames, sash and trim, baseboards and mouldings.   Frames and sash are seldom made up on the job these days, and dealers supply even door and window trim already cut and bundled. Baseboards, mouldings, etc., should be estimated to the nearest 100 feet in excess of the actual length wanted.

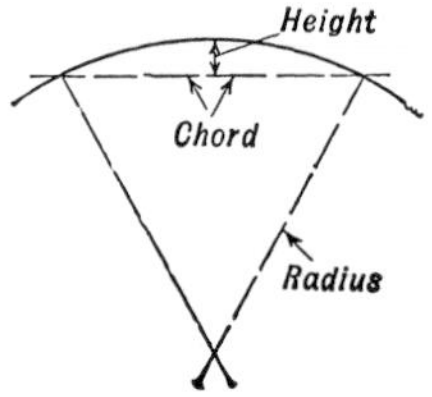

Fig. 30.

Determining Radius.—In making a bend as for a moulding or baseboard, of a known chord and height, the radius must be known so that a line can be struck to which to work.

To determine radius, add the square of half the chord to the square of the height and divide by twice the height.   Thus, in Fig. 30, if the chord is 8 feet and the height 1 foot,

$$\text{Radius} = \frac{4^2 + 1^2}{2 \times 1} = 8.5 \text{ feet} \quad \text{(Ans.)}$$

A slight bend can be made in a board if soaked in hot water 30 minutes.   Sharp bends can be made after wood has been cooked or steamed for at least 6 hours.

TABLE 3

Wire Nails—Kinds and Quantities Required *

Length, in inches	Am. Steel & Wire Co.'s Steel Wire Gauge	No. to lbs. Approx.	Nailings	Sizes and Kinds of Material	Trade Names	Pounds per 1000 feet B. M. on center as follows:				
						12"	16"	20' (Pounds)	36"	48'
2½	10¼	106	2	1 x 4	8d common	60	48	37	23	20
2½	10¼	106	2	1 x 6	8d common	40	32	25	16	13
2½	10¼	106	2	1 x 8	8d common	31	27	20	12	10
2½	10¼	106	2	1 x 10	8d common	25	20	16	10	8
2½	10¼	106	3	1 x 12	8d common	31	24	20	12	10
4	6	31	2	2 x 4	20d common	105	80	65	60	33
4	6	31	2	2 x 6	20d common	70	54	43	27	22
4	6	31	2	2 x 8	20d common	53	40	53	21	17
4	6	31	3	2 x 10	20d common	60	50	40	25	20
4	6	31	3	2 x 12	20d common	52	41	33	21	17
6	2	11	2	3 x 4	60d common	197	150	122	76	61
6	2	11	2	3 x 6	60d common	131	97	82	52	42
6	2	11	2	3 x 8	60d common	100	76	61	38	34
6	2	11	3	3 x 10	60d common	178	137	110	70	55
6	2	11	3	3 x 12	60d common	145	115	92	58	46
2½	12½	189	2	Base, per 100 ft. lin.	8d finish	...	1	...	...	...
2½	10¼	106	2	Byrket lath	8d common	...	48	...	...	...

Sizes and Kinds of Material notes:
I. Used square edge, as platforms, floors, sheathing, or shiplap.
II. When used D. & M., blind nailed, only ⅔ quantity named required.

* Courtesy American Steel and Wire Company.

Wire Nails—Kind and Quantities Required—Cont.

Length, in inches	Am. Steel & Wire Co.'s Steel Wire Gauge	Approx. No. to lbs.	Nailings	Sizes and Kinds of Material	Trade Names	Pounds per 1000 feet B. M. on center as follows:				
						12"	16"	20"	36"	48"
								Pounds		
2½	12½	189	1	Ceiling, ¾ x 4	8d finish	18	14			
2	13	309	1	Ceiling, ½ and ⅝	6d finish	11	8			
2½	12½	189	2	Finish, ⅞	8d finish	25	12			
3	11½	121	2	Finish, 1⅛	10d finish	12	10			
2½	10	99	1	Flooring, 1 x 3	8d floor brads	42	32			
2½	10	99	1	Flooring, 1 x 4	8d floor brads	32	26			
2½	10	99	1	Flooring, 1 x 6	8d floor brads	22	18			
4	6	31	}	Framing, 2x4 to 2x16 requires 3 or more sizes and vary greatly	20d common	20	16	14		
3½	8	49			16d common	10	10	8		
3	9	69			10d common	8	6	5		
6	2	11		Framing, 3x4 to 3x14	60d common	30	25	20		
2½	11½	145	2	Siding, drop, 1 x 4	8d casing	45	35			
2½	11½	145	2	Siding, drop, 1 x 6	8d casing	30	25			
2½	11½	145	2	Siding, drop, 1 x 8	8d casing	23	18			
2	13	309	1	Siding, bevel, ½ x 4	6d finish	23	18			
2	13	309	1	Siding, bevel, ½ x 6	6d finish	15	13			
2	13	309	1	Siding, bevel, ½ x 8	6d finish	12	10			
				Casing, per opening	6d and 8d casing	About ½ pound per side.				

Wire Nails—Kinds and Quantities Required—Cont.

1¼	14	568	12" o.c.	Flooring, ⅜ x 2....	3d brads.........	About 10 pounds per 1000 square feet.
1⅛	15	778	16" o.c.	Lath, 48"..........	3d fine............	6 pounds per 1000 pieces.
⅞	12	469	2" o.c.	Ready roofing......	Barbed roofing.....	¾ of a pound to the square.
⅞	12	469	1" o.c.	Ready roofing......	Barbed roofing.....	1½ pounds to the square.
⅞	12	180	2" o.c.	Ready roofing (⅝ heads)......	American felt roofing	1½ pounds to the square.
⅞	12	180	1" o.c.	Ready roofing (⅝ heads)......	American felt roofing	3 pounds to the square.
1¼	13	429	....	Shingles†.........	3d shingle........	4½ pounds; about 2 nails to each 4 inches.
1½	12	274	....	Shingles..........	4d shingle........	7½ pounds; about 2 nails to each 4 inches.
⅞	12	180	4"	Shingles..........	American felt roofing	12 lbs., 4 nails to shingle.
⅞	12	469	4"	Shingles..........	Barbed roofing.....	4½ lbs., 4 nails to shingle.
1	16	1150	2"	Wall board, around entire edge....	2d Barbed Berry, flat head	5 pounds, per 1,000 square feet.
1	15½	1010	3" o.c.	Wall board, intermediate nailings..	2d casing or floor brad	2½ lbs., per 1.000 square feet.

†Wood shingles vary in width; asphalt are usually 8 inches wide. Regardless of width 1000 shingles are the equivalent of 1000 pieces 4 inches wide.

XIII

Lathing and Plastering

Laths form the supporting structure for plaster on walls and ceilings when the plaster cannot be applied directly to a firm base to which it will bind. Laths may be of either wood or metal and are nailed either to furring strips or to the studding of walls and partitions and to the under side of floor joists to form ceilings.

Wood Laths.—Wood laths are strips $1\frac{1}{2}$ in. wide, $\frac{1}{4}$ in. or $\frac{3}{8}$ in. thick, and 48 in. long sawed from pine, spruce, or hemlock. This length permits the lath to cover, without cutting, three spans between studs when these are placed on 16-inch centers. Laths for lime plaster are spaced $\frac{1}{4}$ in. or $\frac{3}{8}$ in. and closer for gypsum plaster. A bundle of 100 laths

spaced $\frac{1}{4}$ in., will cover 6.48 sq.yd.: equal to 1543 laths per 100 sq.yd.
spaced $\frac{3}{8}$ in., will cover 6.94 sq.yd.: equal to 1441 laths per 100 sq.yd.

About 10 pounds of fine lath nails are required per 100 square yards of lathing.

ILLUSTRATION: How many bundles of laths will be required for lathing the walls and ceiling of a room 12 feet × 18 feet, ceiling 9 feet high, if the areas of the windows and doorways total 12 square yards and the spacing of the lath is $\frac{1}{4}$ inch? Allow 5% for waste.

$$\text{Area of ceiling} \quad = 4 \times 6 \quad\quad = 24 \text{ sq. yd.}$$
$$\text{Area of side walls} = 3 \times 6 \times 2 = 36 \text{ sq. yd.}$$
$$\text{Area of end walls} = 3 \times 4 \times 2 = 24 \text{ sq. yd.}$$

$$\text{Total} \quad 84 \text{ sq. yd.}$$

$$\begin{array}{lr}
\text{Total carried forward} & 84 \quad \text{sq. yd.} \\
\text{Area of openings} & 12 \\
\hline
& 72 \quad \text{sq. yd.} \\
5\% \text{ for waste} & 3.6 \\
\hline
& 75.6 \text{ sq. yd.}
\end{array}$$

If one bundle covers 6.48 sq. yd., then the number of bundles required is $\dfrac{75.6}{6.48} = 11.7$ and the next larger whole number is, of course, 12 bundles. (Ans.)

ILLUSTRATION: A room to be lathed has two window openings 2 ft. 10 in. by 5 ft. 2 in. and two door openings 3 ft. 0 in. by 7 ft. 0 in. What quantity of nails and how many bundles of lath will be required if the size of the room is 13 ft by 12 ft 6 in. and the height of the ceiling is 9 ft 6 in. and the spacing is $\frac{3}{8}$ inch? Allow 5% for waste.

In this problem it is more convenient to change the inches to tenths of a foot and compute the total area in square feet and reduce to square yards by dividing by 9.

$$\begin{array}{lll}
\text{Area of two windows} & = 2 \times 2.83 \times 5.17 & = 29.3 \text{ sq. ft.} \\
\text{Area of two doors} & = 2 \times 3 \times 7 & = 42.0 \text{ sq. ft.} \\
\hline
& \text{Total} & 71.3 \text{ sq. ft.}
\end{array}$$

$$\begin{array}{lll}
\text{Area of ceiling} & = 13 \times 12.5 & = 162.5 \text{ sq. ft.} \\
\text{Area of end walls} & = 2 \times 9.5 \times 12.5 & = 237.5 \text{ sq. ft.} \\
\text{Area of side walls} & = 2 \times 9.5 \times 13.0 & = 247.0 \text{ sq. ft.} \\
\hline
& \text{Total} & 647.0 \text{ sq. ft.} \\
& \text{Area of openings} & 71.3 \text{ sq. ft.} \\
\hline
& & 575.7 \text{ sq. ft.} \\
& 5\% \text{ for waste} & 28.8 \text{ sq. ft.} \\
\hline
& & 604.5 \text{ sq. ft.}
\end{array}$$

Changing to square yards,

$$\text{area} = \frac{604.5}{9} = 67.17 \text{ sq. yd.}$$

If one bundle at $\frac{3}{8}$ in. spacing covers 6.94 sq. yd., then the number of bundles required will be

$$\frac{67.17}{6.94} = 9.7 \text{ or } 10 \text{ whole bundles} \quad \text{(Ans.)}$$

If 10 pounds of nails are required for 100 sq. yd., this room will require $10 \times \dfrac{67}{100} = 6.7$ pounds of nails. (Ans.)

Metal Lath.—Metal lath is manufactured in two general forms, as a wire mesh and as expanded metal (Figs. 1 and 2). Both forms are protected from corrosion by being painted, japanned or galvanized. Metal lath is not only a base for plaster but also serves as reinforcing. It is universally used in fireproof construction and is particularly adapted for thin partition walls and suspended ceilings.

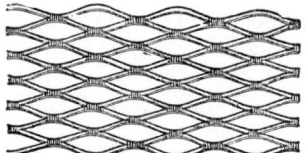

Fig. 1.—Expanded Metal Lath.

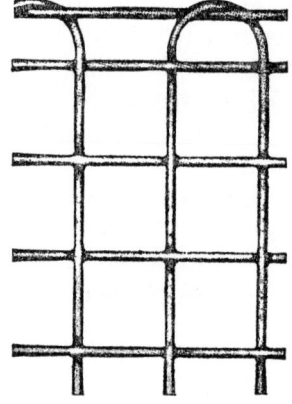

Fig. 2.—Wire Mesh Lath.*

Both wire lath and expanded metal lath are attached to steel furring with No. 18 gage annealed galvanized wire lacing and to wooden furring, studding, or floor joists with No. 13 gage galvanized wire staples spaced about six inches apart. The following are average quantities of lacing and staples required per 100 square yards of metal lath:*

* Courtesy Wickwire Spencer Steel Company.

Spacing of Furring, Inches, Center to Center	No. 18 Galvanized Wire Lacing, Pounds	1¼-In. No. 13 Galvanized Wire Staples, Pounds
12	6	9½
14	5	8
16	4½	7

Wire Lath.—Wire lath is woven from No. 18 to No. 21 Washburn & Moen gage wire with 2 and 2½ meshes per lineal inch in each direction. Some forms have V-shaped metal stiffeners attached at intervals of 8 inches to provide the fabric with greater rigidity. The lath usually comes in rolls 150 feet long and 36 inches wide. Thus one roll will cover 50 square yards.

With 12-inch spacing of furring, a No. 19 gage plain wire lath is recommended, while the No. 18 gage is more suitable when the spacing of furring is 14 or 16 inches. If lath with V-stiffeners is used, a No. 20 gage wire is sufficient.

ILLUSTRATION: An auditorium 50 feet by 100 feet with a 20-foot ceiling is to be lathed with wire lath on metal furring, 12 inches on centers. How many square yards of lath, how many rolls, and how many pounds of lacing will be required if the total area of doors and windows is 50 square yards?

$$\text{Area of ceiling} \quad = 50 \times 100 \quad = \quad 5{,}000 \text{ sq. ft.}$$
$$\text{Area of end walls} = 2 \times 20 \times 50 \quad = \quad 2{,}000 \text{ sq. ft.}$$
$$\text{Area of side walls} = 2 \times 20 \times 100 = \quad 4{,}000 \text{ sq. ft.}$$

$$\text{Total area} \quad 11{,}000 \text{ sq. ft.}$$

Reducing to square yards,

$$\frac{11{,}000}{9} = 1222 \text{ sq. yd.}$$

less openings   50 sq. yd.

Net area = 1172 sq. yd. of lath required (**Ans.**)

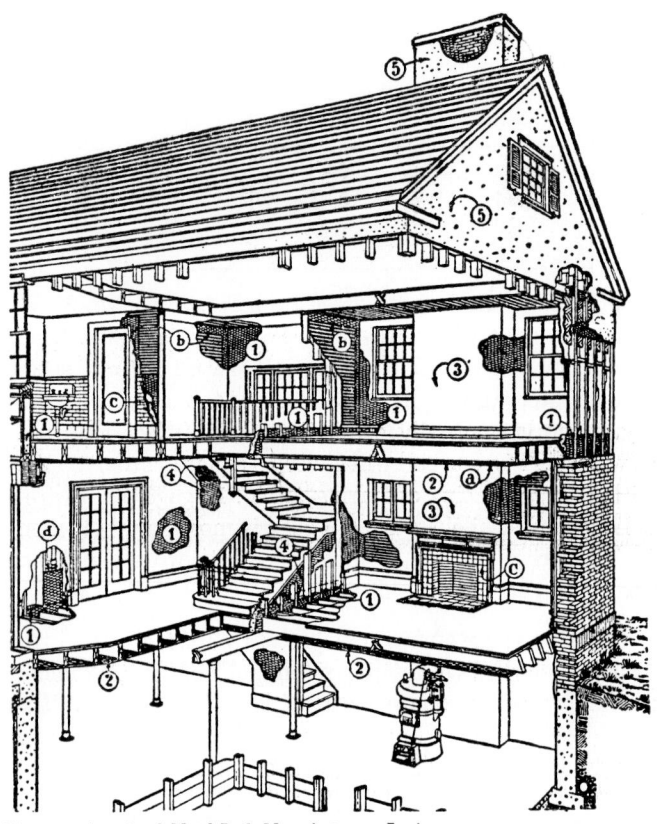

(Courtesy Associated Metal Lath Manufacturers, Inc.)

Fig. 3.—Most Advantageous Positions for Metal Lath for Fire Stops and Crack Prevention.

For Fire Stops—

    (1) On all stud bearing partitions and walls and fire stops between studs. (Fire stops to be metal lath basket-shaped to fit between studs, coated with plaster or cement and filled with incombustible materials.)

       (2) On ceilings under inhabited floors, especially over heating plants and coal bins.

       (3) At chimney breasts, around flues and back of kitchen ranges.

       (4) For stair-wells and under stairs.

       (5) As a base and reinforcement for exterior stucco.

For Crack Prevention—

       (a) On ceilings of prominent rooms.

       (b) Lap 4 in. on either side of wall and partition angles, and around door bucks.

       (c) Back of wainscots and tile mantels.

       (d) Across plumbing pipes and heat ducts.

       (e) Proper construction of exterior stud walls for successful stucco.

If each roll contains 50 square yards

$$\frac{1172}{50} = 23.4 \text{ or } 24 \text{ whole rolls required (Ans.)}$$

Wire lacing required at 6 pounds per 100 square yards is,

$$6 \times 11.72 = 70\tfrac{1}{4} \text{ pounds (Ans.)}$$

ILLUSTRATION: A ceiling is to be lathed on joists spaced 16 inches center to center. What size of plain or reinforced wire lath should be used?

No. 18 gage plain or No. 20 gage reinforced (Ans.)

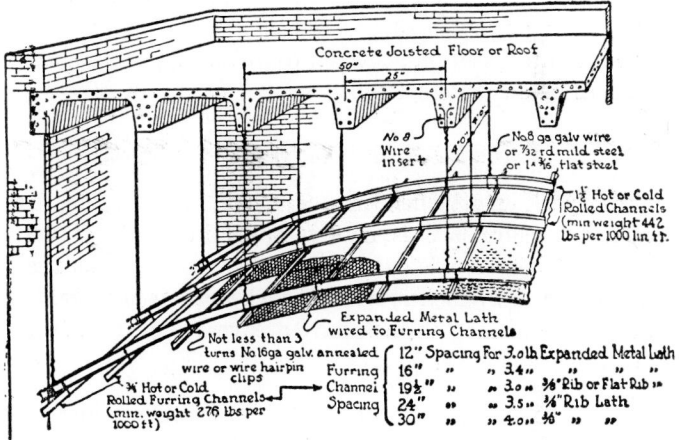

(*Courtesy Associated Metal Lath Manufacturers, Inc.*)

FIG. 4.—Metal Lath Used for Suspended Ceiling.

Expanded Metal Lath.—Expanded metal lath is made by punching and stamping sheet metal and then pulling it so that the punched slits open up as holes which hold the plaster. Ribs are quite frequently stamped into the metal to obtain greater rigidity.

The uses of expanded metal lath are illustrated in Fig. 3. It will be noted that not only is it used to support plaster by itself, but also in corners in combination with wood lath to prevent cracks. Fig. 4 shows the application to suspended ceiling.

Generally, the weight of the expanded metal per unit area is about one-half or less of the unit weight of the original sheet. The following are the minimum weights per square yard recommended for various uses:

Expanded Metal Lath for Interior Work

For vertical position attached to metal studs spaced not to exceed 12 in. on centers, 2.2 lb.

For vertical position attached to wood or metal studs not to exceed 16 in. on centers, 2.5 lb.

For horizontal position attached to metal supports spaced not to exceed 16 in. on centers, 3.4 lb.

For horizontal position attached to metal supports spaced not to exceed 12 in. on centers, 3.0 lb.

Expanded Metal Lath for Exterior Work

For any position attached to wood, metal, masonry, etc., 3.4 lb.

Expanded metal lath is manufactured in sheets of various dimensions, a common length being 8 feet, and widths ranging from 15 inches to 27 inches, with 24 inches as an average. It is sold in bundles of sheets which have a coverage of from 10 to 25 square yards per bundle.

ILLUSTRATION: A room 30 feet by 70 feet with a ceiling 18 feet high is to be lathed with expanded metal lath on metal furring on 12-inch centers. What total weight of lath will be required if the area of doors and windows is 34 square yards and a skylight 18 square yards?

$$\text{Area of ceiling} = 30 \times 70 = 2100 \text{ sq. ft.}$$

$$\frac{2100}{9} = 233 \text{ sq. yd.}$$

Subtracting skylight area,

$$233 - 18 = 215 \text{ sq. yd. (net area)}$$

The weight of lath required for a horizontal position on metal supports spaced 12 inches on centers is 3.0 pounds per square yard. Then the weight of lath required for ceiling is,

$$3.0 \times 215 = 645 \text{ lb.}$$

Area of end walls $= 2 \times 18 \times 30 = 1080$ sq. ft.

Area of side walls $= 2 \times 18 \times 70 = 2520$ sq. ft.

Total area 3600 sq. ft.

Reducing to square yards,

$$\text{Area} = \frac{3600}{9} = 400 \text{ sq. yd.}$$

Net wall area $= 400 - 34 = 366$ sq. yd.

The weight of lath which may be used on this vertical surface is 2.2 pounds per square yard. Then the total weight required for the walls is,

$$2.2 \times 366 = 805 \text{ lb.}$$

The sum of the weights required for the ceiling and walls is

$$645 + 805 = 1450 \text{ lb. total weight of lath (Ans.)}$$

Plastering.—Plastering usually consists of three coats (Fig. 5), viz., (1) the rough or "scratch" coat which is applied directly to the wood or metal lath; (2) the "brown" coat which is floated onto the scratch coat, which has been scratched with a comb in order to roughen it so the brown coat will adhere better and (3) the finishing or "skim" coat which is applied to the brown coat after it has been finely scratched or roughened. When plaster is applied to a masonry wall, the scratch coat is often omitted, the brown coat being applied directly to the masonry. Plaster prepared in sheets and commonly known as plaster board or gypsum lath shipped ready for nailing is often substituted for the scratch coat and sometimes for both the scratch coat and the brown coat.

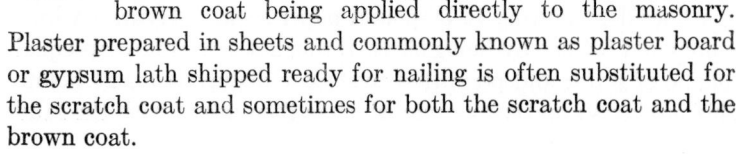
Fig. 5

Scratch Coat.—The scratch coat is applied with sufficient force to insure good key to the lath, and is composed of a mixture of slaked lime, clear river or pit sand free from salt and long cattle or goat hair (wood fiber, jute or asbestos is sometimes used instead of hair on cheap work). These are mixed in the proportions of one part lime paste to two parts sand, with $1\frac{1}{2}$ bushels of hair to each barrel of unslaked lime. Unslaked lime (quicklime) comes in lumps and is sold in barrels containing from 200 to 260 pounds. A barrel of Rockland, Me., lime weighs 220 pounds net, contains about $3\frac{1}{2}$ cubic feet and will make about 2.6 barrels or 9 cubic feet of paste. A barrel of 200 pounds will make about 8 cubic feet of paste. Approximately 9 cubic feet of lime paste, 18 cubic feet of sand, and 4 bushels of hair will cover about 40 square yards about $\frac{3}{8}$ inch thick on wooden laths and about 30 square yards on metal laths.

ILLUSTRATION: What quantities of materials will be required for the scratch coat in a building having 520 square yards of wood-lathed walls?

If one 220-pound barrel of lime, 18 cubic feet of sand, and 4 bushels of hair will cover 40 square yards, then $\frac{520}{40} = 13$ times these quantities will give the total amounts required.

$13 \times 1 \quad = 13 \quad$ 220-pound barrels of quicklime   (Ans.)

$\dfrac{13 \times 18}{27} = 8.7$ cubic yards of sand   (Ans.)

$13 \times 4 = 52$ bushels of hair   (Ans.)

Quicklime must be slaked and aged before using. To obviate the delays incident to these operations, a *hydrated lime* may be used which has been slaked by the manufacturer and is marketed as a flocculent powder in 50-pound paper sacks. Hydrated lime is prepared for use by being sifted through a screen into an equal volume of water and permitted to soak undisturbed for 24 hours. This produces a putty or paste which is then mixed with the sand and hair.

The proportions of materials for the scratch coat using hydrated lime are: 1 sack (50 lb.) hydrated lime; 200 pounds of dry plastering sand; $\frac{1}{2}$ pound of hair or fiber. This will produce about 2.3 cu. ft. or 0.085 cu. yd. of plaster and will cover about $4\frac{1}{2}$ square yards on wood lath with a thickness of about $\frac{3}{8}$ inch, or $3\frac{1}{3}$ square yards on metal lath. The weight of a cubic foot of sand is about 100 pounds.

ILLUSTRATION: What quantities of hydrated lime, sand and hair are required to apply a scratch coat on wood lath to 243 square yards of surface?

Since the quantities given in the statement of the proportions of materials produce a coverage of $4\frac{1}{2}$ square yards on wood lath, the factor obtained by dividing 243 by $4\frac{1}{2}$ when multiplied by these figures will give the total quantities required.

$$\frac{243}{4.5} = 54$$

Then,

$$54 \times 1 = 54 \text{ sacks of hydrated lime (Ans.)}$$

$$54 \times 200 = 10,800 \text{ lb. sand}$$

$$\frac{10,800}{100} = 108 \text{ cu. ft.} = \frac{108}{27} = 4 \text{ cu. yd. sand (Ans.)}$$

$$54 \times 0.5 = 27 \text{ lb. hair (Ans.)}$$

Brown Coat.—The brown coat is usually leaner in lime and has a smaller percentage of hair than the scratch coat. It is applied after the scratch coat has dried and is generally $\frac{1}{4}$ inch to $\frac{3}{8}$ inch thick. Considerable care is exercised in its application so that the surface produced will be straight and true and within about $\frac{1}{8}$ inch of the final finished surface or grounds.

When hydrated lime is used for the brown coat, the recommended proportions are: 1 sack (50 lb.) hydrated lime; 250 pounds of dry plastering sand and $\frac{1}{4}$ pound of hair. This will produce about 2.7 cubic feet or 0.1 cubic yard and will cover about 10 square yards to a thickness of $\frac{3}{8}$ inch.

ILLUSTRATION: What quantities of material are required to cover 340 square yards of wall space with a brown coat of plaster $\frac{3}{8}$ inch thick?

$$\frac{340}{10} = 34$$

Then,

$34 \times 1 = 34$ sacks of hydrated lime (Ans.)

$34 \times 250 = 8500$ lb. sand

$$\frac{8500}{100 \times 27} = 3.15 \text{ cu. yd. sand (Ans.)}$$

$34 \times \frac{1}{4} = 8\frac{1}{2}$ lb. hair (Ans.)

Finish Coat.—The skim coat or finish coat is usually $\frac{1}{8}$ inch thick and contains no hair. It may be made with one part of slaked lime to two parts of clear white sand or marble dust. However, a harder finish may be obtained by using any of the patent plasters on the market. These are composed principally of plaster of Paris or gypsum. Hydrated lime is mixed with these to retard the time of set. The materials and proportions used depend on the type of finish desired.

White Smooth Finish.—This finish may be obtained by mixing 4 sacks (200 lb.) hydrated lime with 50 pounds of plaster of Paris. The resulting putty will cover about 45 square yards to a thickness of $\frac{1}{8}$ inch.

Sand Finish.—A mixture of $2\frac{1}{2}$ cubic feet each of lime, plaster of Paris, and white sand or marble dust will skim-coat about 100 square yards from $\frac{1}{16}$ in. to $\frac{1}{8}$ in. thick.

A coarser sand finish may be produced by mixing 2 sacks (100 lb.) of hydrated lime with 3 cubic feet (300 lb.) of plastering sand. This will cover about 65 square yards of surface.

Textured Finish.—A textured finish is made by first applying a sand finish coat and then a second heavier coat, and the texture desired worked in with tools or hands. This second or texture coat may be proportioned as follows: 3 sacks (150 lb.) of hydrated lime to 50 pounds of plaster of Paris.

ILLUSTRATION: What quantities of materials will be required for a white smooth finish coat of plaster on 355 square yards of surface?

Using the above proportions which yield a coverage of 45 square yards, we obtain,

$$\frac{355}{45} = 7.9 = \text{factor for multiplying ingredients in the mix.}$$

Then,

$7.9 \times 4 = 31.6 = 32$ whole bags of hydrated lime (Ans.)

$7.9 \times 50 = 395$ lb. plaster of Paris (Ans.)

Plaster of Paris is often sold in 100-pound bags. Four bags would be required in this case.

ILLUSTRATION: What quantity of materials would be required to make a finishing plaster composed of equal parts of lime, plaster of Paris, and sand to cover 1150 square yards of surface?

A mixture given above with ingredients in this proportion covers 100 square yards when $2\frac{1}{2}$ cu. ft. sand, $2\frac{1}{2}$ cu. ft. plaster of Paris, and $2\frac{1}{2}$ cu. ft. lime are mixed together.

Then,

$$\frac{1150}{100} = 11.5$$

and

$11.5 \times 2.5 = 28.75$ cu. ft. lime (Ans.)

$11.5 \times 2.5 = 28.75$ cu. ft. plaster of Paris (Ans.)

$$\frac{11.5 \times 2.5}{27} = 1.06 \text{ cu. yd. sand (Ans.)}$$

Thickness of Plaster.—The minimum total thickness of plaster on wood or metal lath should be $\frac{7}{8}$ inch from the face of the lath to the grounds divided as follows:

Scratch coat, average, $\frac{3}{8}$ inch

Brown coat, average, $\frac{3}{8}$ inch

Finish coat, average, $\frac{1}{8}$–$\frac{3}{8}$ inch according to finish

On brick, stone, hollow tile, concrete blocks or poured concrete, the minimum total thickness from the normal masonry line to the grounds should be $\frac{3}{4}$ inch for two-coat work divided as follows:

Brown coat, average, $\frac{3}{8}$ inch

Finish coat, average, $\frac{3}{8}$ inch

Stucco.—Plaster made with Portland cement is used in interior work only as a base coat to support bathroom, kitchen, or ornamental tile. In exterior work, however, such plaster, called *stucco*, is widely used in finishing buildings.

Stucco should always be supported on painted or galvanized metal lath on a wooden structure. It may be applied directly to masonry structures.

The first (scratch) and second (brown) coats each $\frac{3}{8}$ inch thick are usually composed of one part of Portland cement to three parts clean well-graded sand. Eight pounds of hydrated lime per sack of cement are often added to aid the plasticity of the mix. One sack of cement mixed with three cubic feet of sand and eight pounds of hydrated lime will cover about 11 square yards $\frac{3}{8}$ inch thick.

The same proportions or somewhat richer may be used for the finish coat, which may be from $\frac{1}{8}$ inch to $\frac{1}{4}$ inch thick depending on the finish. Smooth troweled, sand floated, rough trowel floated, rough cast, and pebble dash are some of the finishes effected.

Illustration: What quantities of materials are required for a three-coat stucco job, the finish coat being $\frac{1}{8}$ inch, smooth troweled and the total area of the houses to be stuccoed, 1400 square yards?

Since a scratch coat of one sack of cement, 3 cubic feet of sand and 8 pounds of hydrated lime will cover 11 square yards $\frac{3}{8}$ inch thick, then

$$\frac{1400}{11} = 127$$

and

$$127 \times 1 = 127 \text{ sacks of cement}$$

$$\frac{127 \times 3}{27} = 14.1 \text{ cu. yd. sand}$$

$$127 \times 8 = 1015 \text{ lb. hydrated lime}$$

The second coat will duplicate these quantities and the third coat will be one-third of these quantities.   Then the total materials required are:

	Cement, Sacks	Sand, Cubic Yards	Hydrated Lime, Pounds
First coat...............	127	14.1	1015
Second coat..............	127	14.1	1015
Third coat...............	43	4.7	338
Totals..............	297	32.9	2368

Reducing these quantities to purchasable units, figuring 4 sacks of cement per barrel and 50 pounds of hydrated lime per bag, we have

$$\text{Cement,} \ \frac{297}{4} = 75 \text{ barrels (Ans.)}$$

Sand, 33 cubic yards (Ans.)

$$\text{Lime,} \ \frac{2368}{50} = 48 \text{ bags (Ans.)}$$

XIV

Painting, Paperhanging, Glazing

Paint and Its Uses.—Paint is a liquid medium having in suspension solid particles of various kinds known as pigments, and so formulated that it may be spread evenly and easily over a surface by brushes, rollers or sprays. After the liquid film on the surface dries, it should leave behind a coating that is uniform in appearance, and that will resist deterioration by its environment for a long period. A special requirement of exterior paints is that as they do undergo weathering over the years, they should do so by chalking (i.e., breaking down to form small particles) rather than by flaking or cracking. Paint is used to protect exterior wooden or metallic surfaces from decay or corrosion, and to enhance the appearance of all surfaces.

Composition of Paints.—The four major functional components of paints are the vehicle, the pigment or mixture of pigments, the drier and the thinner. The most general classification of paints is based upon differences in the vehicle. On this basis the major types are paints having oil-based vehicles, solvent-based vehicles and emulsion-based vehicles.

Oil Paints.—This class of paints includes the linseed oil paints which have been in use for hundreds of years, and owe their effectiveness to the ease with which linseed oil "dries," i.e., is oxidized by the air to form a hard adherent film. Other vegetable oils have been discovered which are equally or more effective as drying oils than linseed oil and many of them are used in present-day paints. They include soya, castor, safflower, coconut and tall oil, the latter being a product of the pine tree. The most important development of recent years, however, has not been the use of the new oils, but the production of oil vehicles containing in solution other substances to improve the durability and other properties of the paint film. Prominent in this group of additives are the synthetic resins, and paints based upon one of the most widely used types of resins, the alkyd resins, are found classified according to the percentage of resin used, as the long oil type, the medium oil type and the short oil type. These terms can be understood by considering that the more oil contained in the vehicle, the less its resin content, so that the long oil type is lowest in resin and highest in oil, while the short oil type being lowest in oil has the highest amount of resin. The long oil type is used where slower drying time and more even penetration is desired, while the short oil type is best for producing a hard finish on a metallic surface where rapid drying is not objectionable and may be an advantage, as for application by spraying.

Other resins besides the alkyd types are used in oil type paints, including especially the melamine and the urea resins. Moreover, these other resins are often used in combination with the alkyd resins to produce very hard finishes.

Solvent Based Paints.—In designating this group of paints by the word solvent, one must remember that the oils used in an oil paint are themselves solvents. The term solvent is used here to distinguish the other organic solvents from the oils. It should also be noted that it is a regular practice to add various organic solvents such as benzene, and toluene to oil based paints. What is meant here, however, are the products that contain solvents with very little oil. These are primarily the lacquer paints based on nitrocellulose. Such paints are used mostly for painting furniture and fixtures, and not for walls and similar surfaces, either interior or exterior. Nitrocellulose based paints vary widely in viscosity, to provide for differences in film thickness and methods of application. In addition to the various viscosity grades of nitrocellulose and differences in the solvents used, which are chiefly organic esters such as ethyl acetate, butyl acetate, amyl acetate and amyl butyrate, they include plasticizers as well. There are a great variety of these plasticizers and they range from simple vegetable oils such as castor oil, to high boiling solvents such as dibutyl or dioctyl phthalate and tricresyl phosphate. Many of them are themselves resins, such as the ester gums.

Emulsion Paints.—Emulsion paints are also called water based paints because they consist of a suspension in water of all the components of the paint. This suspension is so finely divided that it does not separate under any ordinary conditions. However, emulsion paints should never be exposed to freezing. The emulsion is often called by the word, "latex," though it is not usually a natural rubber but a synthetic resin with similar properties, such as the styrene-butadiene polymer. This resin is particularly useful where resistance to staining is important. However, it does tend to "yellow" and is therefore often replaced in formulas for emulsion paints by such other resins as polyvinyl acetate and acrylic resins. Rubber solutions and chlorinated rubber solutions are also used widely in emulsion paints.

Pigments.—With all of the types of vehicles discussed the types of pigments used are much the same. Their most important property is their covering power, and therefore the pigments of high covering power are called hiding pigments, while those of lower covering power are known as extender pigments. Pigments in the first class are the various types of titanium dioxide as well as zinc sulfide, zinc oxide and white lead, to mention only the white pigments. In the extender class would be such minerals as talc, clay, silica and gypsum. It should not be concluded that the extender pigments are merely used to cheapen the product; on the contrary, they may well make an important contribution to the hiding power, since this is not simply an additive property. They also may contribute greatly to other important properties of the finished coating, such as its mechanical strength and electrical resistance.

Driers.—Driers are substances used to speed up the reaction of a drying oil with the air, so that the film hardens more rapidly. It is important, however, to avoid adding too much drier, which would cause the paint film to crack or wrinkle, or to dry rapidly on the outside so that blisters would form from within.

Thinners.—The general purpose of thinners is to facilitate the application of the paint by reducing its viscosity. The aromatic hydrocarbons, such as benzene, toluene and xylene are excellent thinners for lacquer type paints For oil type paints turpentine is still the standard thinner. Emulsion type paints, especially those based on chlorinated rubber, tolerate very little addition of thinners and no such addition should be made other than in the manufacturing plant to this type of paint.

Varnishes.—In the past, varnishes were produced entirely by the high temperature processing of natural resins. However, at the present time the term is used for any finish which does not contain suspended pigments and which are based upon synthetic resins, oils and solvents. The phenolic resins have been used for years in high oil-content varnishes, because without high oil the phenolic resins give a finish that is too hard and brittle. They also tend to "yellow" and should only be used on very dark woods. More recently the melamine and epoxy resins have come into use in general purpose varnishes. Like oil paints varnishes are available in short, medium and long oil types which are described as "10 gallon varnish" "20 gallon varnish," etc., the numbers being the number of gallons of oil used per 100 pounds of synthetic resin. Of course, natural resins are still used in varnishes to some extent, especially to modify the properties of the synthetics.

Calcimine.—This material is a solution of chalk with glue as a binder and water as the thinner. It is an inexpensive material for interior walls and ceilings and cannot be cleaned by washing. It must be removed before redecorating. Formerly quite popular, it is being supplanted by more modern coatings which require no special surface preparation.

Shellac.—This versatile and widely used material consists of a natural resinous substance known as "lac" dissolved in alcohol which acts as the thinner. Sealing and finishing floors, sizing furniture or trim before painting or varnishing, and touching up "hot spots" on plaster are common applications of shellac.

Aluminum Paint.—This is a ready-mixed paint that has aluminum as its pigment. Three types of aluminum paint are commonly used: aluminum house paint for weather-exposed wood surfaces, metal and masonry aluminum paint for hard surfaces such as machinery, metal roofs, etc., and aluminum enamel for inside applications where high gloss or high heat-reflecting surfaces are desired. For increased roof surface protection and added heat reflection, special aluminum-asphalt paints are available.

Whitewash.—Common whitewash for sheds and barns can be made either by slaking one-half bushel (38 pounds) of common lime and straining, or by mixing one sack (50 pounds) of hydrated lime with water and adding a solution of 15 pounds of common salt in $7\frac{1}{2}$ gallons of water and subsequently thinning with water as desired. If a disinfectant or insecticidal whitewash is desired, one or two quarts of crude carbolic acid should be added.

Spreading Rates.—The area over which a certain quantity of paint will spread depends on the nature and consistency of the paint and the porosity and roughness of the surface to which it is applied. Only approximate figures for average conditions can be given. Table 1, furnished through the courtesy of the National Paint, Varnish, and Lacquer Association, Inc., Washington, D. C., lists the average covering power of various types of paints.

Estimating Paint Requirements.—The quantity of paint required for a job may be estimated by dividing the area to be covered by the spreading rate of the particular paint for the kind of surface to be covered and the number of coats to be applied, as given in Table 1.

TABLE 1

Spreading Rates of Various Paints *

Surface and Product	Average Coverage in Square Feet per Gallon		
	1st (or Primer) Coat	2nd Coat	3rd Coat
Frame Siding			
Exterior House Paint	468	540	630
Aluminum Paint	550	600	
Trim (Exterior)			
Exterior Trim Paint	850	900	972
Porch Floors and Steps			
Porch and Deck Paint	378	540	576
Asbestos Wall Shingles			
Exterior House Paint	180	400	
Shingle Siding			
Exterior House Paint	342	423	
Shingle Stain	150	225	
Shingle Roofs			
Exterior Oil Paint	150	250	
Shingle Stain	120	200	
Brick (Exterior)			
Exterior Oil Paint	200	400	
Cement Water Paint	100	150	
Exterior Emulsion	215		
Cement and Cinder Block			
Cement Water Paint	100	140	
Exterior Oil Paint	180	240	
Medium Texture Stucco			
Exterior Oil Paint	153	360	360
Cement Water Paint	99	135	
Aluminum Paint	300	400	
Cement Floors and Steps (Exterior)			
Porch and Deck Paint	450	600	600
Color Stain and Finish	510	480	

TABLE 1   (*Continued*)

SPREADING RATES OF VARIOUS PAINTS *

Surface and Product	Average Coverage in Square Feet per Gallon		
	1st (or Primer) Coat	2nd Coat	3rd Coat
Doors and Windows (Interior)			
Enamel	603	405	504
Picture Molding, Chair Rails and Other Trim (Coverage per gal. in linear ft.)	1200	810	810
Floors, Hardwood (Interior)			
Oil Paint	540	450	
Shellac	540	675	765
Varnish	540	540	540
Linoleum			
Varnish	540	558	
Walls, Smooth-Finish Plaster			
Flat Oil Paint	630 Primer	540	630
Gloss or Semi-Gloss Oil Paint	630 Primer	540	540
Calcimine	720 Size	240	
Emulsion Paint (Latex Base Paints, Kemtone, etc.)	540	700	
Casein Water Paint	540	700	
Aluminum Paint	450	600	

* Courtesy National Paint, Varnish, and Lacquer Association, Inc., Washington, D. C.

ILLUSTRATION: How many gallons of flat finish oil paint are required for two coats on the plaster walls of one room 14 feet by

22 feet and two rooms 13 feet by 15 feet if the ceilings are 9 feet high? Assume door and window openings to total 200 square feet.

Large room areas

$$
\begin{aligned}
\text{End walls} &= 2 \times 14 \times 9 = 252 \text{ sq. ft.} \\
\text{Side walls} &= 2 \times 22 \times 9 = 396 \text{ sq. ft.} \\
\text{Ceiling} &= 14 \times 22 \quad\quad = 308 \text{ sq. ft.}
\end{aligned}
$$

Two smaller room areas

$$
\begin{aligned}
\text{End walls} &= 4 \times 13 \times 9 = 468 \text{ sq. ft.} \\
\text{Side walls} &= 4 \times 15 \times 9 = 540 \text{ sq. ft.} \\
\text{Ceiling} &= 2 \times 13 \times 15 = 390 \text{ sq. ft.}
\end{aligned}
$$

$$
\begin{aligned}
\text{Total area} &\ldots\ldots\ldots\ldots \quad 2354 \text{ sq. ft.} \\
\text{Area of openings} &\ldots\ldots\ldots \quad 200 \text{ sq. ft.} \\[4pt]
&\quad\quad\quad\quad\quad\quad 2154 \text{ sq. ft.}
\end{aligned}
$$

Spreading rate per gallon for the first (or primer) coat is 630 sq. ft. from Table 1.

Paint required for first coat $= \frac{2154}{630} = 3.4$ gal. (approx.).

Spreading rate per gallon for the second coat is 540 sq. ft. from Table 1.

Paint required for second coat $= \frac{2154}{540} = 4.0$ gal. (approx.).

Total paint required for the two coats $= 3.4 + 4.0$

$$= 7.4 = 8 \text{ gal.}$$

ILLUSTRATION: How much varnish is needed for two coats on a hardwood floor 60 feet by 40 feet?

$$\text{Area} = 60 \times 40 = 2400 \text{ sq. ft.}$$

Spreading rate per gallon for the first coat is 540 sq. ft. from Table 1.

Varnish required for the first coat $= \frac{2400}{540} = 4.5$ gal.

Since the spreading rate is the same for the second coat, the same amount of varnish will be required for the second coat. Therefore the total amount of varnish required is $4.5 \times 2 = 9$ gal.

TABLE 2 *

SPREADING RATES OF ROOF PAINTS

Type of Roof	Square Feet Covered by One Gallon	
	Asbestos or Fibered Roof Coating	Roof or Metal Paint (Nonfibered)
Composition Roofing (Felt or Paper)	50–75	75–100
Concrete	50–75	50–75
Metal	75–100	100–250
Slag or Gravel	25–50	20–35

* Circular 736, National Paint, Varnish, and Lacquer Association, Inc. Washington, D. C.

ILLUSTRATION: How much aluminum paint is required for two coats on a wooden silo 12 feet in diameter and 30 feet high if it has a conical roof with a rise of 4 feet and an overhang of one foot?

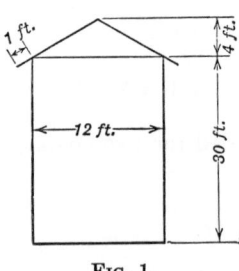

FIG. 1.

Computation of the roof area as a cone whose area is one-half of the product of the slant height and the circumference of the base, would be a refinement not warranted by the problem. It is sufficiently accurate to regard the roof as a disc 14 feet in diameter. Then,

$$\text{Area of roof} = \tfrac{1}{4}\pi D^2 = \frac{14 \times 14\pi}{4} = 49\pi = \ 154 \text{ sq. ft.}$$

$$\text{Area of cylinder} = \pi Dh = 12 \times 30 \times \pi = \underline{1130 \text{ sq. ft.}}$$

$$\text{Total} \quad \overline{1284 \text{ sq. ft.}}$$

Spreading rate first coat = 550 sq. ft. per gallon

$$\text{Paint required first coat} = \frac{1284}{550} = 2.3 \text{ gallons}$$

Spreading rate second coat = 600 sq. ft. per gallon

$$\text{Paint required second coat} = \frac{1284}{600} = 2.1 \text{ gallons}$$

Total pa'nt required = 2.3 + 2.1 = 4.4 = 5 gallons (Ans.)

ILLUSTRATION: A smooth hemispherical dome 32 feet in diameter is to be given one coat of asbestos roof paint. How many gallons of paint will be required?

The area of a sphere is πD^2, then the area of a hemisphere is $\dfrac{\pi D^2}{2}$ and,

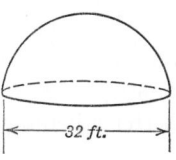

FIG. 2.

$$\text{Area of dome} = \frac{\pi D^2}{2} = \frac{\pi \times 32 \times 32}{2} = \pi \times 512 = 1610$$

Spreading rate = 100 sq. ft. per gallon (Table 2)

$$\text{Paint required} = \frac{1610}{100} = 16 \text{ gallons (Ans.)}$$

For a two-coat repainting job on the exterior of a nouse of moderate size and in good condition, it is fairly safe to estimate that as many gallons of paint as there are rooms in the house will be required. Half again as many gallons will be required for a three-coat job.

PAPER HANGING

Papers for the walls of rooms are printed with distemper color and with oil colors; the cheaper papers are made by machine, the more expensive are hand blocked.

Wall paper is usually made in rolls 18 inches wide and, single rolls, 8 yards long, double rolls, 16 yards long. A roll of border is the same length as a roll of wall paper.

Calculating the Number of Rolls.—There are several methods of figuring the numbers of rolls required to paper a room. Moreover, the methods of measurement vary in different localities. Some measure all the walls as solid, without deductions for the ordinary openings; others deduct one-half of a single roll for each ordinary door or window. Some do not deduct for openings less than 20 square feet in order to compensate for cutting and fitting; others add 15 percent to the area to allow for waste.

There is always waste in matching which must be allowed for; and the height of the room has a great deal to do with the number of strips that can be cut from a roll. Often a double roll cuts to better advantage than a single roll.

ILLUSTRATION: Find the number of rolls of paper for a room 9 feet in height, 15 feet long and 12 feet wide, if the room has one door and three windows each $3\frac{1}{2}$ feet wide.

First Method:

$$\text{Perimeter of room} = 2 \times (12 \text{ ft.} + 15 \text{ ft.}) = 54 \text{ ft.}$$
$$\text{Width of door and windows} = 4 \times 3\tfrac{1}{2} \text{ ft.} = 14 \text{ ft.}$$

Perimeter less door and windows $= 40$ ft.

Allowing one double roll or two single rolls for every seven feet,

$$40 \div 7 = 5\tfrac{5}{7}$$

Therefore, 6 double rolls will be required.

Second Method:

Perimeter of room = 54 ft.
Wall surface = 54 ft. × 9 ft.   = 486 sq. ft.
Allowing 20 sq. ft. per opening = 4 × 20   = 80 sq. ft.
Area of single roll = 24 × 1½ = 30 square feet = 406 sq. ft.

$$406 \div 30 = 13\tfrac{1}{2}$$

Therefore 14 single rolls will be required.

Third Method:

Perimeter of room in yards = 2 × (4 + 5) = 18   yards
Subtract width of doors and windows, ap-
    proximately   =   4½ yards
                                                                       ———
                                                                   13½ yards

Because a roll is ½ yard wide, the number of strips = 13½ × 2 = 27

Because the room is 9 feet high, each strip will be 9 feet or 3 yards long.

27 × 3 = 81 yards required

81 ÷ 16 (the number of yards in a double roll) = $5\tfrac{5}{16}$

Therefore 6 double rolls will be required.

Since the distance around the room is 54 feet or 18 yards, and a 2-strip roll of border contains 16 yards, 18 ÷ 16 or $1\tfrac{1}{8}$ rolls of border are required.

The amount of wall paper needed for the ceiling is found by finding the area, 12 feet × 15 feet = 180 square feet.

Dividing the area of the ceiling in feet by the area of 1 roll in feet, 24 × 1½ = 36.   Then, 180 ÷ 36 = 5 rolls.

Allowing 1 roll for trimming and matching, 6 single rolls would be required.

TABLE 3

Dimensions of Room in Feet	Height of Ceiling in Feet	Number of Doors	Number of Windows	Rolls of Paper	Yards of Border
7 × 9	9	1	1	7	11
7 × 9	10	1	1	8	11
8 × 10	9	1	1	8	12
8 × 10	10	1	1	9	12
9 × 11	9	1	1	10	14
9 × 11	10	1	1	11	14
10 × 12	9	1	1	10	15
10 × 12	10	1	1	11	15
11 × 12	9	2	2	9	16
11 × 12	10	2	2	10	16
12 × 13	9	2	2	10	17
12 × 13	10	2	2	11	17
12 × 15 or 13 × 14	9	2	2	11	18
12 × 15 or 13 × 14	10	2	2	13	18
13 × 15	9	2	2	11	19
13 × 15	10	2	2	13	19
14 × 16	9	2	2	12	20
14 × 16	10	2	2	14	20
14 × 18	9	2	2	13	22
14 × 18	10	2	2	15	22
15 × 16	10	2	2	15	21
15 × 17	12	2	2	19	22

* 18″ rolls; papering of ceilings not included.

OTHER SURFACE COVERINGS

In addition to painting and paperhanging, walls, ceilings, and floors can be covered by a wide variety of other materials. These include wood, plastic, metal, and special compositions which come in many sizes. A number of the commercially available types are listed in Tables 4 and 5.

TABLE 4

FLOOR COVERINGS

Tiles	Tile Sizes in Inches								
	4 × 4	6 × 6	6 × 12	9 × 9	9 × 18	12 × 12	12 × 24	18 × 24	18 × 36
Asphalt			x	x		x	x	x	
Cork		x	x	x		x			
Linoleum				x					
Plastic		x	x	x		x	x	x	
Rubber	x	x	x	x	x	x			x

CONTINUOUS COVERINGS

Linoleum and Vinyl floor coverings are available in continuous rolls in several standard widths, some of which are:

Linoleum: 2', 3', 6', 9', and 12'
Vinyl: 6' and 9'

TABLE 5

CEILING AND WALL COVERINGS

Acoustic Materials		Tile Sizes in Inches			
Type	Thickness	6 × 12	12 × 12	12 × 24	24 × 24
"Cushiontone" *	½", ¾", 1"	x	x	x	x
"Travertone" *	1¹⁄₁₆", 1³⁄₁₆"	x	x		
"Arrestone" *	2½"		x	x	
"Corkoustic" *	1¼"	x	x		
"Celotex" †					
Perf. Asbestos Bd.	³⁄₁₆", ⅛"		x	x	x
Perf. Acoustic Tile	⅝"		x		
Perf. Acoustic Tile	1"		x	x	

Paneling		Panel Length in Feet						
Type	Width	5	6	7	8	9	10	12
Wall Plank	8", 10", 12", 16"				x		x	x
½" Building Board	4'		x	x	x	x	x	x
1" Building Board	4'				x		x	x
³⁄₁₆"–¾" Plywood	2½', 3', 3½', 4'	x	x	x	x	x	x	x
¹⁄₈₅" "Flexwood"	18", 24"				x		x	x

WALL TILE

Aluminum—4¼" × 4¼", 5" × 5", 5" × 10", 10" × 10"
Plastic—4¼" × 4¼", 4¼" × 8½", 8½" × 8½"
Plastic Asbestos—4" × 12", 6" × 6", 6" × 12", 9" × 18", 12" × 12", 18" × 24"
"Cork Wall"—6" × 12", 9" × 9", 12" × 24", 24" × 48"
Steel, Aluminum, Stainless Steel—4¼" × 4¼", 6" × 6", 8½" × 8½"

* Armstrong Cork Company.
† Johns Manville.

Generally speaking, the calculations involved in estimating the amount of material needed to cover a given surface are simple and straightforward. Areas of surfaces are calculated in a manner similar to that described for painting and paperhanging.

Allowances are made for openings, and the net area to be covered is divided by the area of the tile, panel, or strip being used.

For example, if it is desired to cover a $10' \times 15'$ floor with $9'' \times 9''$ tiles the problem is solved as follows:

$$\frac{\text{Area of floor in square feet}}{\text{Area of tile in square feet}} = \frac{10 \times 15}{\frac{9}{12} \times \frac{9}{12}} = \frac{150}{0.75 \times 0.75} = \frac{150}{0.5625}$$

$$= 266.67 \quad = 267 \text{ tiles required.}$$

To allow for wastage and trim it is wise to add approximately 5% to the calculated figure, which would bring the total number of tiles required to about 280.

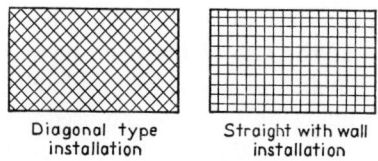

Diagonal type   Straight with wall
installation   installation

FIG. 3.—Diagonal and Straight Type Tile Arrangements.

Diagonal layouts such as that shown in Fig. 3 are often desirable from the standpoint of appearance and are suggested where floor boards are wider than $2''$. In such layouts, approximately $4''$ should be added to the length and width of the room when calculating the room area to allow for the additional trim needed.

WINDOW GLASS AND GLAZING

Common window glass is technically known as sheet glass or cylinder glass. It is usually set with putty and fastened with triangular pieces of zinc called glazier's points, driven into the wood over the glass and covered with putty.

Besides common window glass there are other kinds used in building construction, such as plate glass, wire glass, ornamental and colored glass, skylight glass, etc.

The best quality of window glass is specified as AA, the second as A, and the third as B. It is graded as double-thick or single-thick, and each thickness is further divided into three qualities, first, second, or third. This grading is based upon the color and brilliancy, and the presence or absence of flaws in the material. Single-thick window glass is approximately $\frac{1}{16}$ inch in thickness, double-thick being approximately $\frac{1}{8}$ inch.

Stock Sizes of Window Glass.—The regular stock sizes vary by inches from 6 inches to 16 inches in width. Above that they vary by even inches up to 60 inches in width and 70 inches in length for double thickness, and up to 30 inches by 50 inches for single thickness.

Cost Calculations.—Window glass is sold by the box containing about 50 square feet of glass. The price per square foot increases rapidly as the size of the pane increases.

To find the number of boxes of window glass of a given required size the following rule may be used:

Divide the product of 50×144 by the product obtained by multiplying the length and width of each pane.

ILLUSTRATION: Find the number of boxes of glass required to furnish glass for 15 windows consisting of 4 panes of glass, each 13 inches $\times$ 28 inches.

$$50 \times 144 = 7200$$

$$13 \times 28 = 364$$

$$7200 \div 364 = 20 \text{ (approximately)}$$

Therefore, 1 box of glass will contain 20 panes.

$$15 \times 4 = 60 \text{ panes required.}$$

$$60 \div 20 = 3$$

Thus, 3 boxes of glass are needed.

Glass Blocks.—Many modern architectural designs call for the use of glass blocks. This material provides a modern appearance

TABLE 6

Sizes and Number of Panes in a Box of Window Glass

Size in Inches	Panes in Box	Size in Inches	Panes in Box	Size in Inches	Panes in Box	Size in Inches	Panes in Box
6 × 8	150	12 × 19	32	16 × 20	23	24 × 44	7
7 × 9	115	12 × 20	30	16 × 22	20	24 × 50	6
8 × 10	90	12 × 21	29	16 × 24	19	24 × 56	5
8 × 11	82	12 × 22	27	16 × 30	15	26 × 36	8
8 × 12	75	12 × 23	26	16 × 36	12	26 × 40	7
9 × 10	80	12 × 24	25	16 × 40	11	26 × 48	6
9 × 11	72	13 × 14	40	18 × 20	20	26 × 54	5
9 × 12	67	13 × 15	37	18 × 22	18	28 × 34	8
9 × 13	62	13 × 16	35	18 × 24	17	28 × 40	6
9 × 14	57	13 × 17	33	18 × 26	15	28 × 46	6
9 × 15	53	13 × 18	31	18 × 34	12	28 × 50	5
9 × 16	50	13 × 19	29	18 × 36	11	30 × 40	6
10 × 10	72	13 × 20	28	18 × 40	10	30 × 44	5
10 × 12	60	13 × 21	26	18 × 44	9	30 × 48	4
10 × 13	55	13 × 22	25	20 × 22	16	30 × 54	4
10 × 14	52	13 × 24	23	20 × 24	15	32 × 42	5
10 × 15	48	14 × 15	34	20 × 25	14	32 × 44	5
10 × 16	45	14 × 16	32	20 × 26	14	32 × 46	5
10 × 17	42	14 × 18	29	20 × 28	13	32 × 48	5
10 × 18	40	14 × 19	27	20 × 30	12	32 × 50	4
11 × 11	59	14 × 20	26	20 × 34	11	32 × 54	4
11 × 12	55	14 × 22	23	20 × 36	10	32 × 56	4
11 × 13	50	14 × 24	22	20 × 40	9	34 × 60	4
11 × 14	47	14 × 28	18	20 × 44	8	34 × 40	5
11 × 15	44	14 × 32	16	20 × 50	7	34 × 44	5
11 × 16	41	14 × 36	14	22 × 24	14	34 × 46	5
11 × 17	39	14 × 40	13	22 × 26	13	34 × 50	4
11 × 18	36	15 × 16	30	22 × 28	12	34 × 52	4
12 × 12	50	15 × 18	27	22 × 36	9	34 × 56	4
12 × 13	46	15 × 20	24	22 × 40	8	36 × 44	5
12 × 14	43	15 × 22	22	22 × 50	7	36 × 50	4
12 × 15	40	15 × 24	20	24 × 28	11	36 × 56	4
12 × 16	38	15 × 30	16	24 × 30	10	36 × 60	3
12 × 17	35	15 × 32	15	24 × 32	9	36 × 64	3
12 × 18	33	16 × 18	25	24 × 36	8	40 × 60	3

and admits diffused daylight over a wide area, at the same time affording complete privacy.

Glass blocks are hollow "all-glass" units which have fused seals. The interior of the blocks is relatively free of water vapor, and the dry dead air space acts as an effective heat insulator. A single cavity glass block has an insulating value greater than that of an 8-inch brick wall, and more than twice that of ordinary windows. Double cavity blocks which have a fibrous glass screen insert that divides the dead air space into two pockets also are available. These provide even better thermal insulation.

A special resilient plastic coating on all mortar edges forms a permanent bond between glass and mortar, insuring a high degree of wind resistance and weather-tightness. The glass block edge construction forms a "key-lock" mortar joint which allows a full bed of mortar and a visible joint of only about $\frac{1}{4}$ inch.

Glass blocks are made in various patterns and in three sizes: $5\frac{3}{4}'' \times 5\frac{3}{4}''$, $7\frac{3}{4}'' \times 7\frac{3}{4}''$, and $11\frac{3}{4}'' \times 11\frac{3}{4}''$ (generally referred to as 6'', 8'', and 12''). All units are $3\frac{7}{8}''$ thick. Special shapes are available for turning corners and for building curved panels. Typical estimating data for glass blocks are shown in Table 7.

TABLE 7

Glass Block Estimating Data

(For 100 sq. ft. of panel laid with $\frac{1}{4}''$ visible mortar joints)

Size of block	6''	8''	12''
Number of blocks	400	225	100
Weight of panel	2000 lbs.	1800 lbs.	1900 lbs.
Volume of mortar	4.3 cu. ft.	3.2 cu. ft.	2.2 cu. ft.

†**Plate Glass and Thermopane**®.—The trend toward the construction of buildings with larger portions of their external structures composed of glass, including picture windows with wide and high panes, has resulted in the use in construction of various types of plate glass, as well as double-glass with insulating air spaces between the two panes of glass. In this way one obtains the greater strength necessary to withstand the greater wind and thermal stresses on the greater areas of glass, and also the lower transmission of heat through the window, that is necessary to reduce heat losses in winter and heat gains in summer.

The calculation of the thickness of plate glass required to withstand wind loadings can be computed from tables which show the relation between the maximum area of glass in square feet and the wind loading in pounds per square foot. This loading is established by local building codes and ordnances, and must be found out for the local area from the building authorities. Assume that in a given locality the maximum permissible load is 30 lbs. per square foot, one would use the graphs of Fig. 4 to find the maximum area that could be covered with a given thickness of glass. Thus for a 9.7′ × 10′ window, there would be an area of 97 sq. ft. Find on the bottom of Fig. 4 the vertical coordinate for 30 lbs. per sq. ft. loading, and run up that line until it intersects the horizontal coordinate for 97 sq. ft. This point is above the graph for ⅜″ plate, but below that for ½″ plate, so the latter is specified.

Reference to Fig. 4 also discloses the words "Design Factor 2.5." This means that the graphs in this figure have been constructed for the average pressure divided by 2.5, which is thus the factor of safety. When a design factor other than 2.5 is chosen, the graphs in Fig. 4 may be used providing the design load is adjusted as follows:

$$\begin{bmatrix} \text{design load for} \\ \text{use with graphs} \end{bmatrix} = \begin{bmatrix} \dfrac{\text{actual design load}}{2.5} \end{bmatrix} \times \begin{bmatrix} \text{chosen design} \\ \text{factor} \end{bmatrix}$$

† Acknowledgement is made to the Libbey-Owens-Ford Glass Co. for the information on which this section is based.

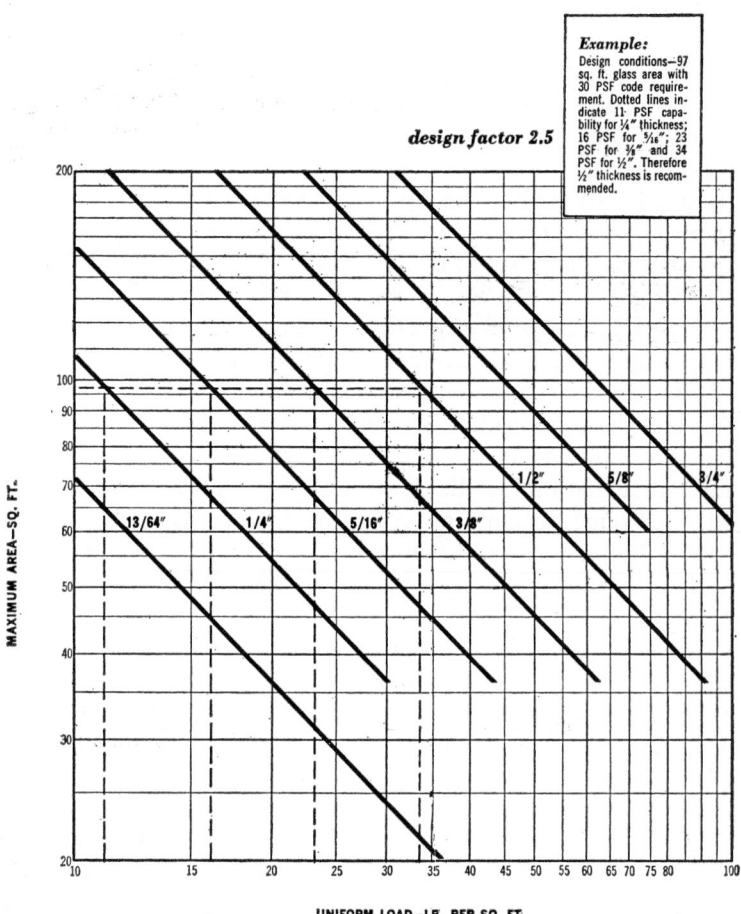

design factor 2.5

Example:
Design conditions—97 sq. ft. glass area with 30 PSF code requirement. Dotted lines indicate 11 PSF capability for ¼" thickness; 16 PSF for ⁵/₁₆"; 23 PSF for ⅜" and 34 PSF for ½". Therefore ½" thickness is recommended.

MAXIMUM AREA—SQ. FT.

UNIFORM LOAD—LB. PER SQ. FT.

FIG. 4.—Thickness of Glass Required for Various Areas. (Graphs apply to square and rectangular lights of glass when the length is not more than 5 times the width.)

For example: Assume a building code requires the choice of glass areas and thicknesses to be based on a design load of 25 psf and the architect or engineer decides the appropriate design factor is 2.0. The adjusted design load for use with the graphs would be determined as follows:

$$\text{design load for use with graphs} = \left[\frac{25}{2.5} \right] \times [2.0]$$

$$= 20 \text{ lbs. per sq. ft.}$$

The types of glass for which these graphs were prepared are Libbey-Owens-Ford Polished Plate Glass for the $1\frac{3}{64}''$ and $\frac{1}{4}''$ thicknesses, and L-O-F Heavy-Duty-Plate Glass® for the greater thicknesses. The data can be used for other types of glass by multiplying the strength found by suitable factors, as follows:

Type of Glass	Multiplying Factor
Tuf-flex®	4.0
Thermopane®	1.5
Vitrolux®	2.0
Wired	0.5
Laminated	0.6
Rough Plate	1.0
Sandblasted	0.4

The complete data on the varieties of L-O-F polished plate glass for residential glazing are given in Table 8.

Laminated safety glass is made in two lights of polished plate or sheet glass bonded by a plastic, which is a polyvinyl butyral resin selected for strength, durability and adhesion. It is suitable for use not only in automobiles, but also in bathroom partitions and doors because of the greater thermal stresses and mechanical shocks to which they may be subjected. Table 9 gives selection data on laminated safety glass.

Thermopane® is manufactured in two styles, differing in the method of sealing the two panels together. In the GlasSeal® edge the two panes are fused together, and therefore the sizes of the lights are more restricted than are those of the Bonder-

TABLE 8

Plate Glass	Thickness	Quality	Thickness Tolerance	Maximum Size — Standard	Maximum Size — Special	Approx. Weight Lbs. per Sq. Ft.	Luminous Illuminant C (Average Daylight) Transmitt.	Avg. Solar Radiation Transmittance — Ultra-violet Radiation	Avg. Solar Radiation Transmittance — Total Solar Radiation
Parallel-O-Plate	1/4″	Silvering	±1/32″	up to 25 sq. ft.					
	1/4″	Mir. Glazing	±1/32″	up to 75 sq. ft.		3.27	89.1	67.8	79.9
Regular Plate	1/4″	Glazing	±1/32″	124″ × 170″	124″ × 252″	1.64	90.6	75.9	86.1
Parallel-O-Grey	3/8″	Glazing	±1/32″	72″ × 74″	74″ × 120″	2.66	50.0	44.0	52.2
	13/64″	Glazing	±1/32″	84″ × 120″	120″ × 192″	3.27	44.2	39.0	46.6
Parallel-O-Bronze	1/4″	Glazing	±1/32″	96″ × 138″	120″ × 192″	2.66	54.4	32.5	47.0
	13/64″	Glazing	±1/32″	84″ × 120″	96″ × 138″	3.27	48.5	27.5	41.0
Heat Absorbing	1/4″	Glazing	±1/32″	96″ × 138″		3.27	74.7	44.9	46.3
Regular Plate	1/4″	Glazing	±1/32″	84″ × 120″	72″ × 120″	3.27	89.1	67.8	79.9
Grey	1/4″	Glazing	±1/32″	±1/16″	72″ × 120″	3.27	44.2	39.0	46.6
Bronze	1/4″	Glazing	±1/32″	±1/16″	72″ × 120″	3.27	48.5	27.5	41.0
Heat Absorbing	1/4″	Glazing	±1/32″	±1/16″	72″ × 120″	3.27	74.7	44.9	46.3

TABLE 9

Types	Nominal Thickness	Thickness Range	Max. Size	Net Wt. Lbs. per Sq. Ft.
Thin Safety Sheet	$\frac{9}{64}''$	.120–.160	7 sq. ft.	1.92
S. S. Safety Sheet	$\frac{7}{32}''$	.185–.250	15 sq. ft.	2.49
Combination Safety Sheet (S.S. & D.S.)	$\frac{15}{64}''$	.200–.260	15 sq. ft.	2.91
D. S. Safety Sheet	$\frac{1}{4}''$	.220–.280	15 sq. ft.	3.32
E-Z-Eye Safety Sheet	$\frac{15}{64}''$	.200–.260	15 sq. ft.	2.91
Safety Plate (Clear and E-Z-Eye)	$\frac{1}{4}''$	.220–.280	$72'' \times 138''$	3.16
Heavy Safety Plate	$\frac{3}{8}''$	$\pm\frac{1}{32}''$	$72'' \times 138''$	4.88
	$\frac{1}{2}''$	$\pm\frac{1}{16}''$	$72'' \times 138''$	6.54
	$\frac{5}{8}''$	$\pm\frac{1}{16}''$	$72'' \times 138''$	8.13
	$\frac{3}{4}''$	$\pm\frac{1}{16}''$	$72'' \times 138''$	9.81
	$\frac{7}{8}''$	$\pm\frac{1}{16}''$	$72'' \times 138''$	11.45
	$1''$	$\pm\frac{1}{16}''$	$72'' \times 138''$	13.08

metic® style, which have a glass-to-metal edge. These sizes are shown in Table 10. (Note that SSA means "single-strength A-quality" and DSA means "double-strength A-quality.")

In computing the heat savings in winter, the factor used is represented by U_t which is the heat transfer coefficient for Thermopane®. Under winter conditions, it is 0.69 B.T.U. per hour per sq. ft. per °F. temperature difference for GlasSeal® Thermopane (with $\frac{3}{16}''$ air space), 0.58 for Bondermetic® Thermopane (with $\frac{1}{2}''$ air space), both as against a single glass pane value of 1.13 in the same units. Under summer conditions, U_t is 0.64 B.T.U. per hour per sq. ft. per °F. temperature difference for the GlasSeal® Thermopane, 0.56 for the Bondermetic® Thermopane, and 1.06 for a single glass pane.

Using these the applicable figures above, the reduction in heating load due to the use of Thermopane can be calculated by the following equation:

$$\text{Reduction in Btu per hr.} = (U_s - U_t) \times (t_i - t_0)$$

TABLE 10

Bondermetic

2 panes of ¼″ polished Parallel-O-Plate ½″ air space. 6.5 lbs. per sq. ft.

Width	Height	Width	Height
33″	× 76¾″	64½″	× 50″
35½″	× 36″	64½″	× 58″
35½″	× 48⅛″	64½″	× 66″
35½″	× 60⅜″	66⅝″	× 47¾″
42″	× 66″	66⅝″	× 60⅛″
42″	× 72″	68¾″	× 36″
44½″	× 36″	68¾″	× 48⅛″
44½″	× 46″	68¾″	× 60⅜″
44½″	× 48⅛″	68¾″	× 72¾″
44½″	× 60⅜″	70⅛″	× 52½″
44½″	× 72¾″	70⅛″	× 56½″
45″	× 76¾″	72″	× 48″
45⅜″	× 52″	72″	× 60″
46⅛″	× 52½″	72½″	× 46″
47¾″	× 50⅜″	72½″	× 50″
48″	× 48″	72½″	× 58″
48″	× 60″	72½″	× 66″
48½″	× 42″	75″	× 36″
48½″	× 46″	75″	× 48⅛″
48½″	× 50″	75″	× 60⅜″
48½″	× 58″	80½″	× 50″
50⅜″	× 47¾″	80½″	× 58″
50⅜″	× 60⅛″	84″	× 66″
55¼″	× 36″	84″	× 72″
55¼″	× 48⅛″	93″	× 36″
55¼″	× 60⅜″	93″	× 48⅛″
56½″	× 42″	93″	× 60⅜″
56½″	× 46⅛″	93″	× 72¾″
56½″	× 50″	96″	× 66″
56½″	× 58⅛″	96″	× 72″
56½″	× 66″	96½″	× 50″
57″	× 76¾″	96½″	× 58″
58⅛″	× 52½″	116½″	× 58″
64½″	× 46″		

2 panes of DSA window glass ¼″ air space. 3.25 per sq. ft.

21¾″	× 62¾″	42½″	× 22½″
25¾″	× 62¾″	45½″	× 25½″

2 panes of 3/16″ "A" heavy sheet glass with ½″ air space. 5 lbs. per sq. ft.

35½″	× 60⅜″	56½″	× 46⅛″
48½″	× 42″	56½″	× 50″
48½″	× 50″	56½″	× 42″

Glasseal

SSA window glass with 3/16″ air space (thickness .375″ ± .050″) 2.4 lbs. per sq. ft.

Width	Height	Width	Height
16″	× 20″	28″	× 16″
16″	× 22″	28″	× 18″
16″	× 24″	28″	× 20″
16″	× 32″	28″	× 22″
16″	× 36″	28″	× 24″
16″	× 48″	32″	× 16″
16″	× 60″	32″	× 20″
16 9/16″	× 24⅝″	32″	× 22″
16 9/16″	× 30 13/16″	32″	× 24″
16 9/16″	× 36 13/16″	36″	× 14″
16 9/16″	× 49″	36″	× 16″
16 9/16″	× 61 3/16″	36″	× 18″
19″	× 15″	36″	× 20″
19½″	× 53″	36″	× 22″
20″	× 16″	36″	× 24″
20″	× 20″	36″	× 30″
20″	× 22″	36⅝″	× 14¼″
20″	× 24″	36⅝″	× 18¼″
20″	× 32″	36⅝″	× 22¼″
20″	× 36″	36⅝″	× 30¼″
20″	× 48″	39″	× 14″
20″	× 60″	39″	× 18
21 1/16″	× 24⅝″	39″	× 22″
21 1/16″	× 30 13/16″	39″	× 30″
21 1/16″	× 36 13/16″	39⅝″	× 14¼″
21 1/16″	× 49″	39⅝″	× 18¼″
21 1/16″	× 61 3/16″	39⅝″	× 22¼″
22″	× 18″	39⅝″	× 30¼″
22″	× 55 9/16″	40″	× 16″
24″	× 16″	40″	× 20″
24″	× 20″	40″	× 24″
24″	× 22″	42½″	× 22½″
24″	× 24″	44″	× 14″
24″	× 32″	44″	× 16″
24″	× 36″	44″	× 18″
24″	× 48″	44″	× 22″
24″	× 60″	44″	× 30″
24¼″	× 15¼″	44⅝″	× 14¼″
27¼″	× 14¼″	44⅝″	× 18¼″
27¼″	× 18¼″	44⅝″	× 22¼″
27¼″	× 22¼″	44⅝″	× 30¼″
27¼″	× 30¼″	45½″	× 25½″
28″	× 14″		

where U_s = overall coefficient of heat transmission for single glass,

Btu per hr. per sq. ft. per deg. F.

U_t = overall coefficient of heat transmission for Thermopane®,
Btu per hr.—sq. ft. per deg. F.

A = Area of glass, sq. ft.

t_i = indoor design temperature, °F.

t_0 = winter outdoor design temperature, °F.

Installation.—The use of plate glass, Thermopane® and other heavy types of glass requires special attention to its installation. While putty is still one of the important glazing compounds, it is essential that the wooden groove be clean and dry and then coated with a good priming paint before the glass is placed to prevent absorption into the wood of the oil from the putty. Window frames that are not to be painted should be primed with varnish. Other than putty, the principal glazing compounds, which are often superior for heavy glass installations, are the mastics. There are two groups: the elastic glazing compounds which are formulated to remain plastic over long periods of time, and the non-skinning compounds which have a resin base of polybutene. This type, however, is not to be used in combination glazing with curing type sealants. These sealants are available in three major types: (1) the chemical curing type, which is based on silicones or urethane resins and which cures by a chemical reaction to form a firm resilient seal; (2) the solvent release type, which dries by evaporation of its solvent, leaving behind the resin content, which may be of the acrylic, Neoprene® or vinyl type in a soft and pliable state; and (3) the two-part polymer base sealants, which cure more rapidly than the other types. These types are available both for use by pouring and in "sealing guns." For heavy work the preformed sealants which are made of natural or synthetic rubber and from various plastics are particularly useful. They are available in preformed tapes, beads, ribbons or mastics, and in both resilient and non-resilient types.

XV

Plumbing

Introduction.—Plumbing is defined as the art of installing in buildings the pipes, fixtures, and other apparatus for bringing in the water supply and removing liquid and water-carried wastes. It has developed within the span of a generation from a task which could be handled by a handy man or a lead wiper to a trade which requires a sound fundamental knowledge of hydraulics, mechanics, and building.

In new building construction the plumber locates the pipes and fixtures as shown on the plans, but he must be on the alert to insure that the installations do not violate the local plumbing code or the principles of good practice. When installing plumbing in an old building, even greater responsibility rests on him. Here it may be his lot to determine all of the pipe sizes and locations. In either case his work may even include bringing water from an independent source of supply and disposing of the wastes by an independent system.

As we have suggested, a complete plumbing installation consists of two mutually independent systems, one the water supply, the other the disposal of wastes, often called the *sanitary plumbing*. These two systems will be considered separately.

Figure 1 shows a typical water supply and drainage layout for a two-story house with a basement. This layout conforms to the Uniform Plumbing Code for Housing. The plumbing layout for a bathroom in a one-story house without a basement is shown in Fig. 2.

Water Systems.—A water system is made up of a source of supply (city main, well and pump, etc.), a distribution and drain-

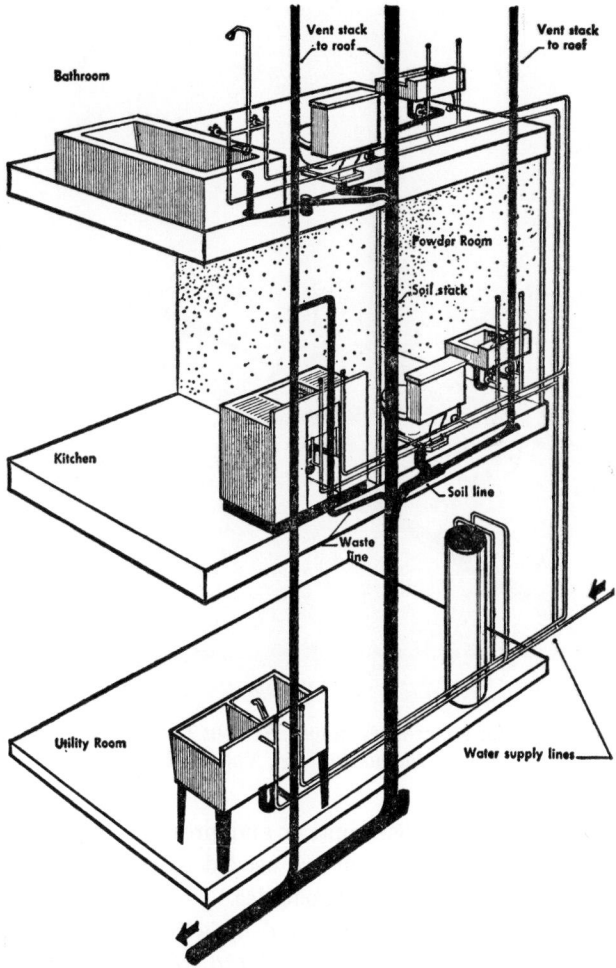

Fig. 1.—A Typical Water Supply and Drainage Layout for a Two-story House with a Basement. This layout conforms with the Uniform Plumbing Code for Housing. (From *University of Illinois Bulletin,* Vol. 48, No. 15.)

age system (Figs. 1 and 2), and various types of fixtures (closet, bathtub, shower, sink, etc.). The size of all these elements de-

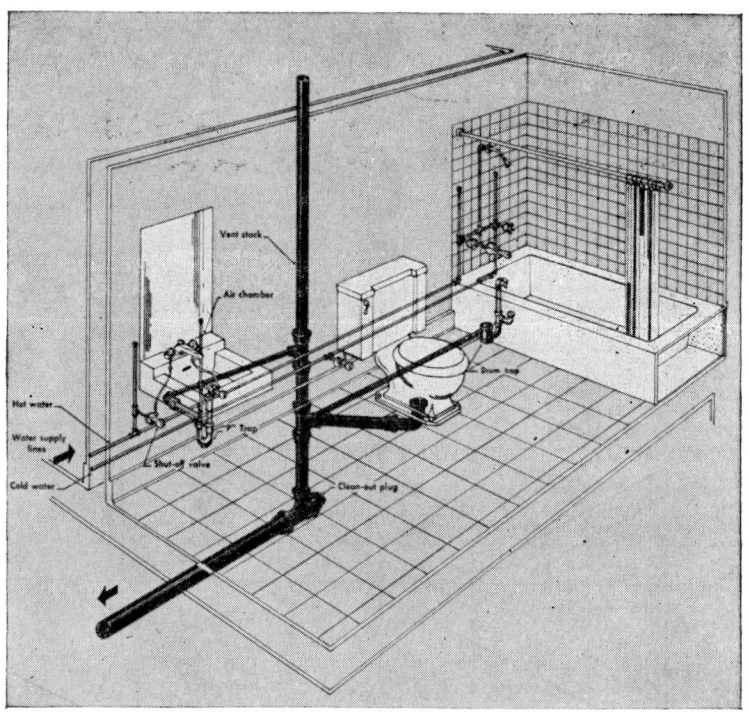

FIG. 2.—Layout for a One-story House Without Basement. (From *University of Illinois Bulletin*, Vol. 48, No. 15.)

pends on the water consumption and pressure required. Estimates on water consumption are needed to calculate pipe sizes, tanks, reservoirs, and pump capacities.

The quantity of water required by a family can readily be computed. Each person uses from 20 to 40 gallons per day; this includes requirements for cooking and laundry. The amounts required for some specific uses are shown in Table 1.

In some plumbing installations such as sprinkler systems for theaters, office buildings, and warehouses, it is required to have a storage unit capable of holding a large quantity of water.

TABLE 1

WATER CONSUMPTION (COMBINED HOT AND COLD WATER) IN U. S. GALLONS PER FIXTURE *

	Type	Size of Inlet	Pressure at Outlet in Psi 10	20	30	40	50
Water closets							
Tank types	All types		18				
Flush valve types	Wash down		30				
	Siphon jet			30			
	Blow-out			40			
Urinals							
Tank types	All types		5				
Flush valve types	Stall	¾	5				
	Wash-down		7				
	Siphon jet			25			
	Blow-out			30			
Showers	2¼″ head	½		4			
	4 "	"		5½			
	5 "	"		6½			
	6 "	"		7¼			
	8 "	¾		9			
	8″, tubular	"		16			
Lavatory	18 x 20		5				
	21 x 24		9				
Kitchen sink			7				
Bathtub			15				
Laundry tray			9				
Hoses, lawn	Solid stream	¾	6				
	with spray	¾	4				

* *Plumbing Practice and Design,* Vol. II, S. Plum, John Wiley & Sons, Inc., p. 110.

Cisterns and reservoirs are sometimes used for the storage of rainwater in rural areas. The water is collected from the roofs of buildings by a system of gutters or downspouts which lead the water to the cistern. Table 2 gives the capacities of plain cylindrical cisterns and tanks and may be used to calculate the size of such containers.

TABLE 2

CAPACITIES OF PLAIN CYLINDRICAL CISTERNS AND TANKS

Depth of Cistern or Tank in Feet	Diameter of Cistern or Tank in Feet								
	4	5	6	7	8	9	10	11	12
	Capacity of Cistern or Tank in Gallons								
4	376	588	846	1152	1504	1904	2350	2844	3384
5	470	735	1058	1439	1880	2380	2938	3555	4230
6	564	881	1269	1727	2256	2855	3525	4265	5076
7	658	1028	1481	2015	2632	3331	4113	4976	5922
8	752	1175	1692	2303	3008	3807	4700	5687	6768
9	846	1322	1904	2591	3384	4283	5288	6398	7614
10	940	1469	2115	2879	3760	4759	5875	7109	8460
11	1034	1616	2327	3167	4132	5235	6463	7820	9306
12	1128	1763	2537	3455	4512	5711	7050	8531	10152

ILLUSTRATION: In a certain sprinkler system it is required that 7,000 gallons of water be stored on the roof of the building. What size tank is required?

Referring to Table 2, the size of the tank with the next larger capacity is selected. This tank is 11 ft. in diameter and 10 ft. deep.

Hydropneumatic Water Systems.—In rural and suburban areas where there is no central source of water (reservoirs), hydropneumatic water systems are used to supply the required water.

The water is usually obtained from a deep or shallow well and is raised to the surface and distributed by a motor- or engine-driven pump. A tank, originally filled with air, is used with the pump (see Fig. 3). When pumped partly full with water, the air compresses and the pressure thus created is used to force the water through the pipes. The effective capacity of the hydropneumatic tank is increased if an initial air pressure is used.

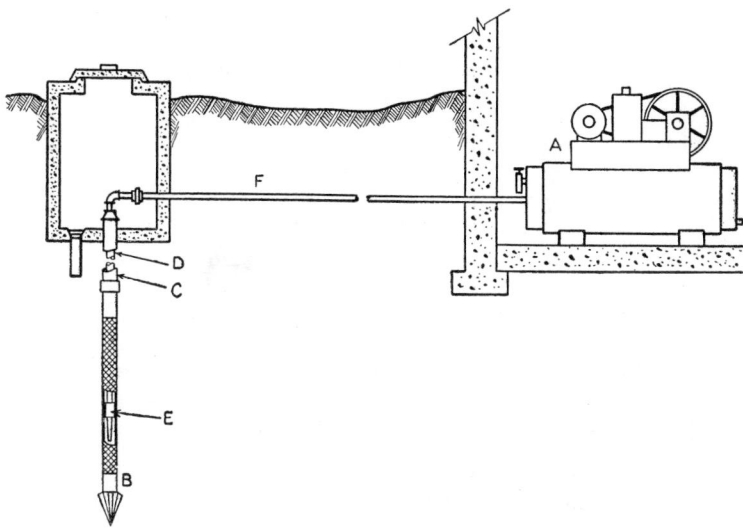

Fig. 3.—A Typical Hydropneumatic Water System. A—motordriven pump mounted directly on a pressure tank; B—well point (driven into ground); C—drive pipe or well casing; D—suction pipe inside well casing; E—foot valve which acts like a check valve to prevent loss of head in casing; F— horizontal suction pipe from well to pump. In this type of installation, the vertical distance from the pump to the water level must not exceed 22 ft.

Table 3 gives the capacity of a tank with and without initial air pressure. Then, if a pump is to be run every other day to supply the requirements of a family using 120 gallons per day with a working range of pressures from 10 pounds to 50 pounds gage pressure, the tank capacity may be figured as follows: At

TABLE 3

WATER CAPACITY OF A HYDROPNEUMATIC TANK WITH AND WITHOUT AN
INITIAL AIR PRESSURE *

Gage Pressure, Pounds per Square Inch	Water in Tank When No Initial Air Pressure Is Provided, Percent	Water in Tank With 10 Lb Initial Air Pressure, Percent
50	76.9	61.5
45	75.0	58.3
40	72.7	54.5
35	70.0	50.0
30	66.7	44.4
25	62.5	37.5
20	57.1	28.6
15	50.0	16.7
10	40.0	....
5	25.0	....

* From Circular 303, University of Illinois, College of Agriculture.

50 pounds pressure the tank will hold 61.5 percent of its capacity
and at 10 pounds it will be empty. The total quantity of water
needed between pumpings is $120 \times 2 = 240$ gallons. Then the
total capacity of tank required is $240 \div 61.5$ percent which is
390 gallons. From Table 4, the dimensions of the tank next larger
in capacity are 30 inches in diameter by 12 feet long. When a
pump on a hydropneumatic system is operated by an electric
motor, provision is usually made for starting the motor automatically when the pressure drops to a certain point.

Some idea of the delivery capacity, suction lift, and size of
motors required to drive this type of pumping unit can be gained
from Table 5. This table lists the performance of Montgomery
Ward's shallow well jet pump (a compact, centrifugal type pump
specially designed for water systems).

TABLE 4

STANDARD SIZES OF TANKS FOR HYDROPNEUMATIC WATER SYSTEMS

Diameter, Inches	Length, Feet	Size, Gallons	Diameter, Inches	Length, Feet	Size, Gallons
24	6	140	42	8	575
24	8	190	42	10	720
24	10	235	42	12	865
30	6	220	42	14	1000
30	8	295	48	10	940
30	10	365	48	12	1128
30	12	440	48	14	1300
36	6	315	48	16	1500
36	8	420	48	18	1700
36	10	525	48	20	1880
36	12	630	48	24	2260
36	14	735			

TABLE 5

PERFORMANCE OF SHALLOW WELL JET PUMPS

Suction Lift	Discharge Pressure					
	20 lbs.		30 lbs.		40 lbs.	
	¼-HP	½-HP	¼-HP	½-HP	¼-HP	½-HP
	Delivery Capacity in Gallons per Hour					
5 ft.	500	930	460	880	260	530
10 ft.	430	820	400	785	230	440
15 ft.	360	700	340	690	180	360
20 ft.	300	590	280	585	150	290

Municipal Water Supply Connections.—City plumbing installations usually connect with a water main in the street as a source of supply. This connection is made in some cities by the water company or the water department, and in others, by the plumber.

There are three types of connections: taps, wet-connections and three-way branches. With the use of special tools, a hole may be bored in a water main and a tap inserted without interrupting the service. Taps are of brass and are made in the following sizes: ⅝ in., ¾ in., 1 in., 1½ in., and 2 in. The tap connects with the service pipe by a lead gooseneck as shown in Fig. 4 so bent that settlement of either pipe will not loosen the

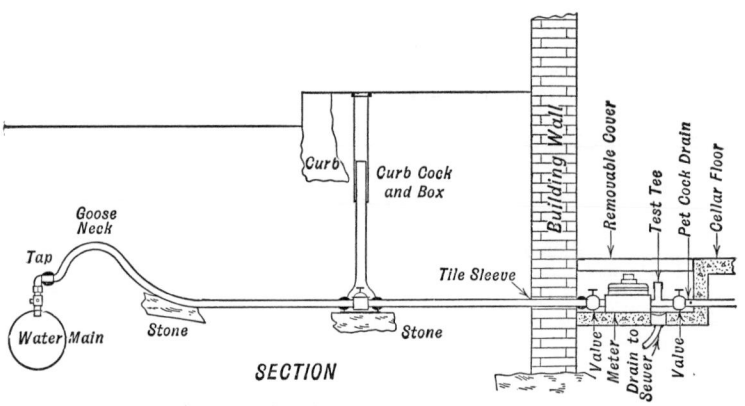

Fig. 4.—House Service Connection.

connection or break the pipe. Taps are commonly used for buildings requiring less than 200 gallons per minute. For larger connections such as for apartment buildings, factories or office buildings, several taps may be used leading to a common service pipe. Wet-connections and three-way branches are also only used for these larger demands.

Distributing Systems.—After delivery to the building, water is distributed to the fixtures by branch pipes and risers as shown in Fig. 5. In dwellings, small apartment buildings and low struc-

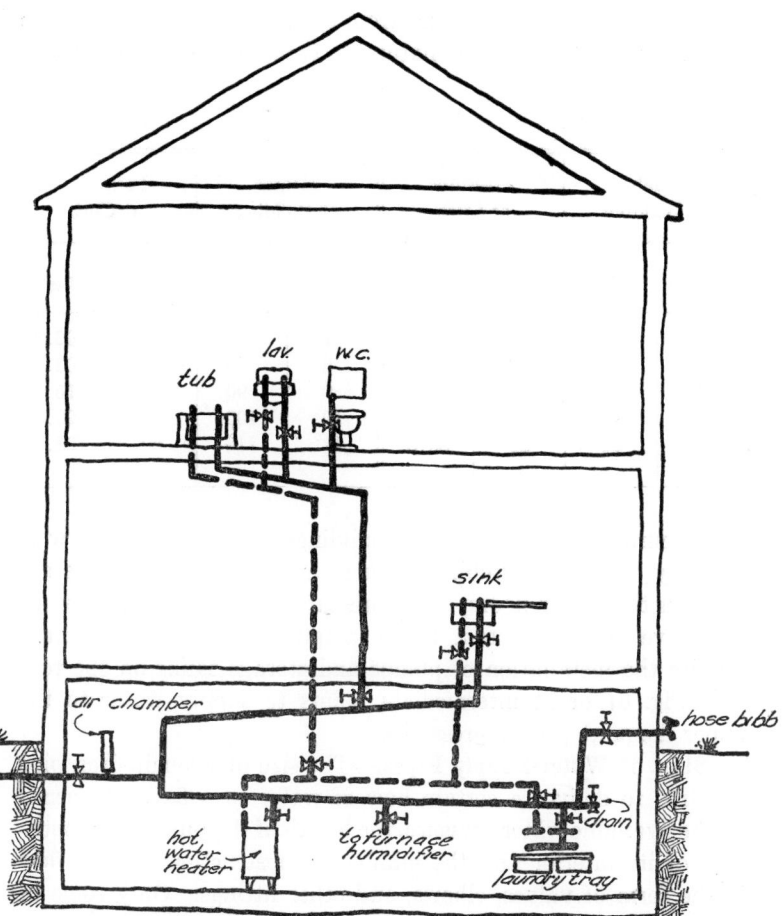

Fig. 5.—Sketch of Typical Hot and Cold Water Distribution System for a Private Dwelling. This house has a full bathroom A (tub, sink and water closet) on the second floor; a half bathroom B (sink and water closet only), and a kitchen with sink C on the first floor. The basement equipment includes a water heater D, laundry tub E, and two hose connections F and G (for lawn sprinkling, car washing, etc.). A jet pump with pressure tank H supplies water from an outside well. Where city water is available, this pump would not be needed.

tures of any kind, the pressure from the city mains is relied upon
to deliver the water to the fixtures. Table 6 shows the pressure

TABLE 6

Height of Building		Pressures at Curb		Height of Building		Pressures at Curb	
Stories	Feet	Pounds	Feet	Stories	Feet	Pounds	Feet
2	20	15	34.5	7	70	40	92.0
3	30	20	46.0	8	80	45	103.5
4	40	25	57.5	9	90	50	115.0
5	50	30	69.0	10	100	55	126.5
6	60	35	80.5	11	110	60	138.0

at the curb necessary to supply buildings of various heights with
properly designed plumbing. Where automatic flush valves are
used on the upper floors, these pressures should be increased at
least five pounds or roof tanks installed. Very tall buildings
require the installation of pumps and storage tanks on the roof
(see Fig. 6) or at intermediate points to supply water to the
upper floors at proper pressures.

Sizes of Water-Supply Pipes.—The size of wrought-iron pipe
required to deliver a certain flow of water to a fixture depends
on the available water pressure, the length of the pipe, the smoothness of its interior and the number of obstructions to the flow
in the form of valves, elbows, and other fittings. Only some of
these factors can be determined with any degree of accuracy and
the deficiency must be supplied by experience and the exercise of
good judgment.

Information as to the pressure of water available at a building
site can usually be obtained from the city water department's
office. Water pressure is measured in pounds per square inch and

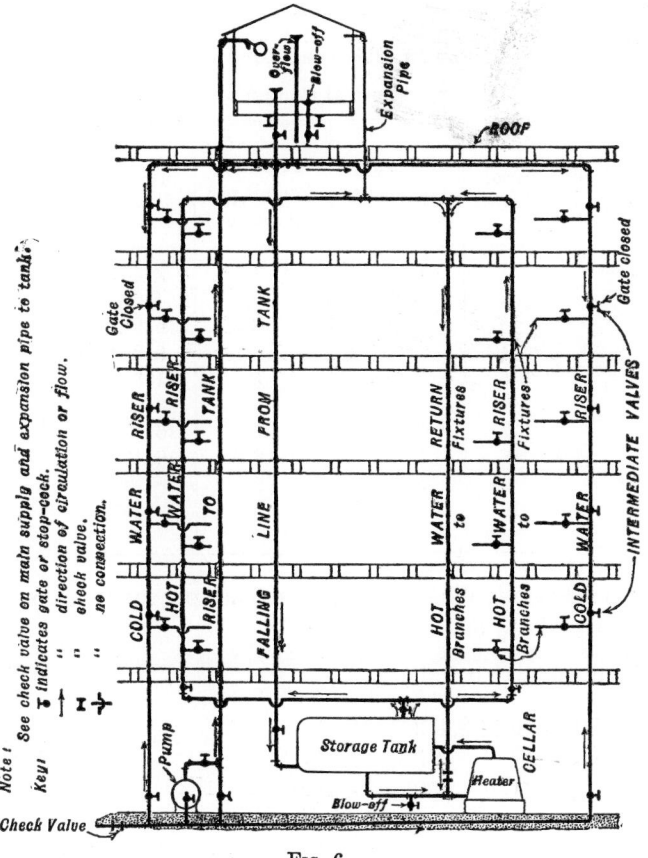

FIG. 6.

in feet of head. Table 7 may be used to convert values in one unit to those in the other.

TABLE 7

HEAD AND PRESSURE EQUIVALENTS

(Water Assumed at 62.5 Lb per Cu Ft)

Head Feet	Pressure, Pounds per Square Inch	Pressure, Pounds per Square Inch	Head Feet
1	0.434	1	2.304
2	0.868	2	4.608
3	1.302	3	6.912
4	1.736	4	9.216
5	2.170	5	11.520
6	2.604	6	13.824
7	3.038	7	16.128
8	3.472	8	18.432
9	3.906	9	20.736
10	4.340	10	23.040

ILLUSTRATION: What pressure in pounds per square inch corresponds to a head of 85.3 feet?

From table (left half)

$$80 = 10 \times 3.472 = 34.72$$
$$5 = 2.170$$
$$.3 = .1 \times 1.302 = .1302$$

Head of 85.3 ft.   37.02   lb. per sq. in.   (Ans.)

ILLUSTRATION: How many feet of head is the equivalent of a water pressure of 45 pounds per square inch?

From table (right half)

$$40 = 10 \times 9.216 = 92.16$$
$$5 = 11.52$$

Pressure = 103.68 ft. of head   (Ans.)

Table 8 gives the recommended rates of supply to plumbing fixtures, Table 9 the recommended sizes of water-supply pipes to fixtures, and Table 10 the sizes of branch water-supply pipes to

TABLE 8

RECOMMENDED RATES OF SUPPLY TO PLUMBING FIXTURES

(Gallons per Minute)

Fixture	H. E. Babbitt *	A. Buenger †			W. S. Timmis ‡	Copper and Brass §	Plum **
		Fair	Good	Excellent			
Bath tub............	10	3	4	6	15	10	15
Wash basin.........	2	2	3	4	4	5	7
Manicure table......	....	1	1½	2	....	....	....
Slop sink...........	5	3	4	6	15	....	10
Pantry sink.........	1	2	4	6	....	....	....
Kitchen sink........	5	....	....	....	15	10	9
Shower bath........	6	4	6	8	8	5	9
Bidet..............	1	....	....	....	....	....	....
Drinking fountain...	1	....	....	....	....	....	....
Laundry tray.......	5	4	6	8	....	10	9
Urinal.............	4	....	....	....	6	....	7
Hot-water heater....	5	....	....	....	....	....	....
Water-closet........	5	....	....	....	8	5	18
Water-closet flush valve............	50	....	....	....	30	30	40
Garden hose........	12	....	....	....	....	10	....

* *Plumbing*, McGraw-Hill, Second Edition.

† *Jour. Am. Soc. Heat. Vent. Engrs.*, Vol. 26, p. 701.

‡ *Ibid.*, Vol. 28, p. 397.

§ *Practical Brass Plumbing*, Copper and Brass Research Association.

** *Plumbing Practice and Design*, Vol. II, John Wiley & Sons, Inc., p. 110.

fixtures. There should be neither an increase nor a decrease in the size of a branch pipe between the fixture it serves and the riser or branch from which it obtains its water.

TABLE 9

RECOMMENDED SIZES OF WATER-SUPPLY PIPES TO FIXTURES *

(Standard Wrought Pipe)

Sizes based on pressure drop of 30 lb per 100 ft.
Hot-water faucets to be disregarded when estimating sizes of risers and mains.

Fixture	Number of Fixtures								
	1	2	4	8	12	16	24	32	40
Water closet:									
Tank:									
Gpm	8	16	24	48	60	80	96	128	150
Pipe size, inches	½	¾	1	1¼	1½	1½	2	2	2
Flush valve:									
Gpm	30	50	80	120	140	160	200	250	300
Pipe size, inches	1	1¼	1½	2	2	2	2½	2½	2½
Urinal:									
Tank:									
Gpm	6	12	20	32	42	56	72	90	120
Pipe size, inches	½	¾	1	1¼	1¼	1¼	1½	2	2
Flush valve:									
Gpm	25	37	45	75	85	100	125	150	175
Pipe size, inches	1	1¼	1¼	1½	1½	2	2	2	2
Wash basin: †									
Gpm	4	8	12	24	30	40	48	64	75
Pipe size, inches	½	½	¾	1	1	1¼	1¼	1½	1½
Bath tub:									
Gpm	15	30	40	80	96	112	144	192	240
Pipe size, inches	¾	1	1¼	1½	2	2	2	2½	2½
Shower bath:									
Gpm	8	16	32	64	96	128	192	256	320
Pipe size, inches	½	¾	1¼	1½	2	2	2½	2½	3
Sinks,† slop, kitchen:									
Gpm	15	25	40	64	84	96	120	150	200
Pipe size, inches	¾	1	1¼	1½	1½	2	2	2	2½

* W. S. Timmis, *Jour. Am. Soc. Heat. Vent. Engrs.*, Vol. 28, p. 397.
† Each faucet.

The determination of the proper size of a riser or branch to serve a number of fixtures involves a consideration of the probability of simultaneous use of these fixtures. Thus, in the case of the one-family house shown in Fig. 5, it is conceivable that the bath tub, the water-closet, the sink, and the garden hose might be used simultaneously. What should then be the size of the riser beyond the water pump? Referring to Table 10, next to the last

TABLE 10

SIZES OF BRANCH WATER-SUPPLY PIPES TO FIXTURES *

Description of the fixture	U. S. Department of Commerce recommendation, minimum size	Recommendation by W. S. L. Cleverdon †		Recommendation for sizes for different rates of pressure loss		
		Pressure, 5–15 lb.	Pressure, over 15 lb.	$H = 0.5L$ ‡	$H = L$ ‡	$H = 5L$ ‡
	In.	In.	In.	In.	In.	In.
Water closet................	⅜	½	⅜	½	⅜	⅜
Urinal.....................	..	½	⅜	⅜	⅜	⅜
Bathtub 4 ft. long...........	½	¾	½ to ⅝	1	¾	½
Bathtub 7 ft. long...........	..	....	....	1¼	1	¾
Wash basin.................	⅜	½	⅜	⅜	⅜	⅜
Laundry tray...............	½	¾ to 1	½ to ¾	½	½	⅜
Kitchen sink, small..........	½	⅝ to ¾	½ to ⅝	¾	½	⅜
Kitchen sink, hotel..........	..	....	....	¾	½	⅜
Slop sink..................	..	¾	½ to ⅝	¾	½	⅜
Hot-water heater...........	½	¾ to 1	⅝ to ¾	¾	½	½
Bidet......................	..	⅝ to ¾	½	⅜	⅜	⅜
Shower....................	..	....	....	¾	½	½
Garden hose...............	..	....	....	¾	¾	½
Drinking fountain...........	..	....	....	⅜	⅜	⅜
Water-closet flushometer.....	..	1¼ to 1½	1	1½	1¼	1
Pantry sink................	..	½	⅜	⅜	⅜	⅜
Urinal flush valve...........	..	⅝ to ¾	½			
Foot bath..................	..	⅝ to ¾	½	½	⅜	⅜

* *Plumbing*, H. E. Babbitt, Second Edition, 1950, McGraw-Hill, p. 55.
† *Plumbers Trade Journal*, Vol. 72, p. 867.
‡ H = head loss per unit length; L = length.

column (except for the garden hose), we find that the individual pipe sizes for the fixtures named would be ¾ in., ½ in., ¾ in., and

½ in., respectively. The next problem is then to determine what size pipe will carry as much water as these four combined. From Table 10 we note that one ¾-inch pipe is equivalent in capacity to 2.8 ½-inch pipes. Using this table we can reduce each branch pipe to terms of equivalent ½-inch pipes. Thus,

$$1—\tfrac{3}{4}\text{-inch pipe} = 2.8—\tfrac{1}{2}\text{-inch pipes}$$
$$1—\tfrac{1}{2}\text{-inch pipe} = 1.0—\tfrac{1}{2}\text{-inch pipe}$$
$$1—\tfrac{3}{4}\text{-inch pipe} = 2.8—\tfrac{1}{2}\text{-inch pipes}$$
$$1—\tfrac{1}{2}\text{-inch pipe} = 1.0—\tfrac{1}{2}\text{-inch pipe}$$

Sum equivalent to 7.6 ½-inch pipes

Referring this sum to Table 11 we find that the corresponding single pipe would be between 1 inch and 1¼ inches. Experience would probably dictate that the 1-inch size would be ample.

TABLE 11

EQUIVALENT PIPE SIZES

(The number of ½-in. pipes which will discharge as a single pipe of another size for the same pressure loss.)

Size of Pipe, Inches	½	⅝	¾	1	1¼	1½	2	2½	3	4	5	6
Number of ½-in. pipes with same capacity....	1	1.7	2.8	5.7	10.0	15.6	32.0	55.8	88.3	181	316	498

Let us consider another example, that of a riser leading to a theater washroom and serving 4 water-closets with flush valves, 8 urinals with flush valves, and 2 wash basins. During the intermission of a performance, these facilities would be so heavily taxed that simultaneous usage would be the safest assumption in computing the size of the riser. Referring this time to Table 9, we note that the sizes of pipes to serve these groups of fixtures are

$1\frac{1}{2}$ in., $1\frac{1}{2}$ in., and $\frac{1}{2}$ in., respectively. Referring next to Table 11, the equivalent number of $\frac{1}{2}$-inch pipes are,

$1\frac{1}{2}$-in. $=$ 15.6 $\frac{1}{2}$-inch pipes
$1\frac{1}{2}$-in. $=$ 15.6 $\frac{1}{2}$-inch pipes
$\frac{1}{2}$-in. $=$ 1.0 $\frac{1}{2}$-inch pipe
—————
Total   32.2 $\frac{1}{2}$-inch pipes (or one 2-inch pipe)

Sizes of Copper and Brass Pipe.—Iron pipe carrying soft or corrosive water will rust even if galvanized. The rust forms a spongy mass which seriously impedes the flow of water. The calculations for the wrought-iron pipe sizes above, took a certain amount of this reduction into account. Therefore, if copper, brass, or lead pipe is used a smaller pipe size than that arrived at by these calculations may be used. Table 12 gives the equivalent non-ferrous pipe and tubing sizes which may be used.

TABLE 12

Non-Ferrous Pipes of Capacities Equal to Wrought-Iron or Steel Pipes, Inches

Copper Tubing, Any Water	Copper, Brass or Lead Pipe, Any Water	Wrought-Iron or Steel Pipe		
		Hard Water	Soft Water	Corrosive Water or Softened Water
. . . .	$\frac{3}{8}$	$\frac{3}{8}$	$\frac{1}{2}$	$\frac{5}{8}$
$\frac{3}{8}$	$\frac{1}{2}$	$\frac{1}{2}$	$\frac{5}{8}$	$\frac{3}{4}$
. . . .	$\frac{5}{8}$	$\frac{5}{8}$	$\frac{3}{4}$	1
$\frac{1}{2}$	$\frac{3}{4}$	$\frac{3}{4}$	1	$1\frac{1}{4}$
$\frac{3}{4}$	1	1	$1\frac{1}{4}$	$1\frac{1}{2}$

Piping Installation.—Every plumbing installation should follow a building plan or a sketch showing the pipe sizes and the locations and general arrangement of fixtures. Figures 7 and 8

Plumbing symbols			
Symbol	Plan	Initials	Item
———————	O	D	Drainage line
— — — —	O	VS	Vent line
— — —	◎		Tile pipe
—•——•——•—	O	C.W.	Cold water line
—— —— ——	O	H.W.	Hot water line
— —— — ——	O	H.W.R.	Hot water return
✕—✕—✕	◉	G	Gas pipe
•• —•• —•• —••	O	D.W.	Ice water supply
••• —••• —•••	O	D.R.	Ice water return
✓—✓—✓—✓	O	F.L.	Fire line
❯—❯—❯—❯	⊕	I.W.	Indirect waste
— ı —— ı —	⊕	I.S.	Industrial sewer
-—\—\—	⊘	AW	Acid waste
— •◉• — •◉• —	Ⓐ	A	Air line
—◦◦◦◦—◦◦◦◦—	Ⓥ	V	Vacuum line
←—←—←—←	ⓡ	R	Refrigerator waste
—◌—◌—			Gate valves
—◌—◌—◌—			Check valves
—◦C.O. ⌐C.O.		C.O.	Cleanout
◻ F.D.		F.D.	Floor drain
◉ R.D.		R.D.	Roof drain
◉ R.E.F.		R.E.F	Refrigerator drain
☒		S.D.	Shower drain
◉ G.T.		G.T.	Grease trap
⊢ S.C.		S.C.	Sill cock
⊢ G		G	Gas outlet
⊢ VAC.		VAC.	Vacuum outlet
—⊖—		M	Meter
▣			Hydrant
⊂⊃ H.R.		H.R.	Hose rack
⊏▭ H.R.		H.R.	Hose rack, built in
⋕ L		L	Leader
⊕		H.W.T.	Hot water tank
⊛		W.H.	Water heater
⊜		W.M.	Washing machine
⊛		R.B.	Range boiler

Fɪɢ. 7.—Standard Plumbing Symbols. (From *Plumbing*, H. E. Babbitt, Second Edition, 1950, McGraw-Hill, p. 13.)

illustrate the standard conventions by which piping and fixtures are shown on building drawings. In new buildings, the pipes that are to be concealed should go in after the framing is erected.

Piping		Pipe Fittings and Valves (cont'd)	Screwed
Plumbing		140 *Single Sweep Tee*	
100 *Soil,Waste,or Leader* .... (*above Grade*)		141 *Double Sweep Tee*	
101 *Soil,Waste, or Leader* .... (*below Grade*)		142 *Reducing Elbow*	
102 *Vent*		143 *Tee*	
103 *Cold Water*		144 *Tee – Outlet Up*	
104 *Hot Water*		145 *Tee – Outlet Down*	
105 *Hot Water Return*		146 *Side Outlet Tee* ...... Outlet Up	
106 *Fire Line*	—F—F—	147 *Side Outlet Tee* ...... Outlet Down	
107 *Gas*	—G—G—	148 *Cross*	
108 *Acid Waste*	Acid	149 *Reducer*	
109 *Drinking Water Flow*		150 *Eccentric Reducer*	
110 *Drinking Water Return* ..		151 *Lateral*	
111 *Vacuum Cleaning*	—V—V—	152 *Gate Valve*	
112 *Compressed Air*	—A—	153 *Globe Valve*	
Sprinklers		154 *Angle Globe Valve*	
120 *Main Supplies*	—S—	155 *Angle Gate Valve*	
121 *Branch and Head*		156 *Check Valve*	
122 *Drain*	—S---S—	157 *Angle Check Valve*	
Pneumatic Tubes		158 *Stop Cock*	
123 *Tube Runs*		159 *Safety Valve*	
Pipe Fittings and Valves	Screwed	160 *Quick Opening Valve*	
130 *Joint*		161 *Float Operating Valve* ...	
131 *Elbow – 90 deg.*		162 *Motor Operated Gate* ... Valve	
132 *Elbow – 45 deg.*		163 *Expansion Joint Flange* ..	
133 *Elbow – Turned Up*		164 *Reducing Flange*	
134 *Elbow – Turned Down*		165 *Union*	
135 *Elbow – Long Radius*		166 *Bushing*	
136 *Side Outlet* .... *Elbow – Outlet Down*			
137 *Side Outlet* .... *Elbow – Outlet Up*			
138 *Base Elbow*			
139 *Double Branch Elbow*			

Fig. 8.—Graphical Symbols for Use on Drawings.

If the drawings do not show the " roughing in " dimensions of the fixtures, these should be obtained. Piping should be so located that there will be no danger of water freezing it in a building normally heated. " Horizontal " water pipes should be pitched

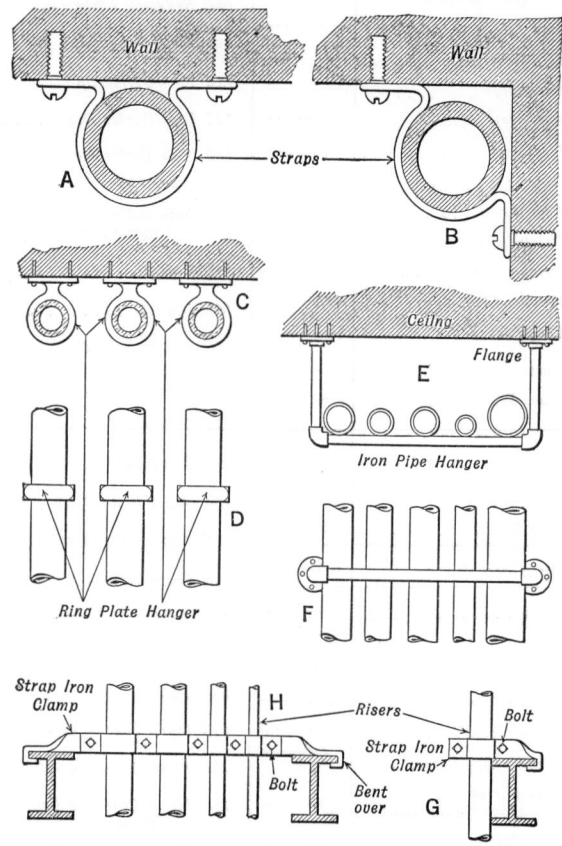

FIG. 9.—Types of Pipe Hangers.

$\frac{1}{10}$ inch per foot towards the supply pipe so that the entire system may be drained by a stop-and-waste valve just inside the cellar wall. Soil and waste pipes should be sloped at least ¼ inch per foot toward the sewer. No sags or pockets in which water will freeze should be permitted.

Pipe Supports.—A sure way of insuring proper slope on alignment of pipes is adequate support. For horizontal pipes ¾-inch

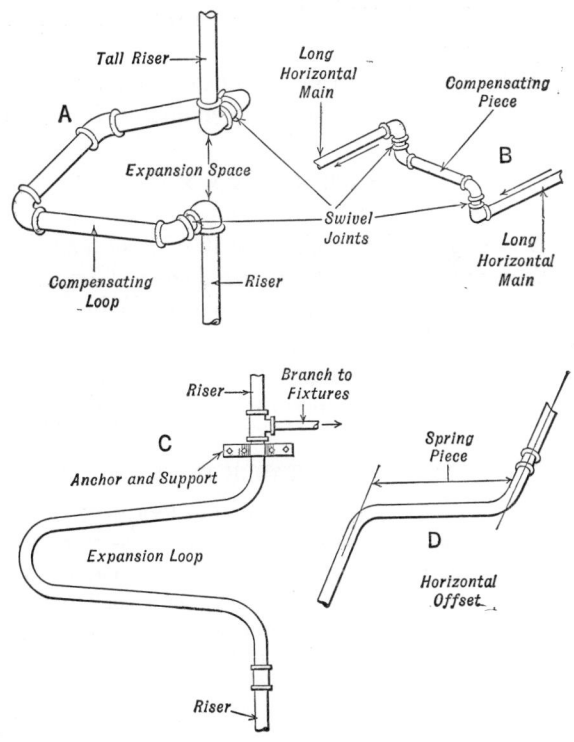

Copper and Brass Research Ass'n.

Fig. 10.—Loops for Expansion.

and larger, pipe hangers should be about 10 feet apart; for ½-inch and ⅜-inch pipe not more than 6 or 8 feet apart. Figure 9 shows several different types of pipe hangers.

Expansion of Pipe.—Pipes expand with an increase in temperature, and when a pipe is long or the change in temperature apt to be great, definite provision must be made to care for this expansion. Pipes passing through concrete or plastered walls should be given freedom of movement by passing them through a sleeve. Either expansion joints, loops, or swing joints (Fig. 10) must be used between fixed supports when the movement is apt to be great.

The change in length or the linear expansion in inches for a pipe 100 feet long can readily be computed from the formula

$$E = 100 \times 12 \times k \times (t_1 - t_2)$$

when E = expansion in inches per 100 feet of length;

k = the rate of increase per degree of temperature, called the coefficient of expansion (this varies with the material, see Table 13);

t_1 = the highest temperature the pipe will reach;

t_2 = the lowest temperature the pipe will reach.

TABLE 13

COEFFICIENTS OF LINEAR EXPANSION

Material	Coefficient of Expansion per degree Fahrenheit	Change in length per 100 ft per 100 degree change in temperature, in.
Wrought iron	.0000067	$1\frac{3}{16}$
Copper	.0000093	$1\frac{1}{8}$
Brass	.0000104	$1\frac{1}{4}$
Cast iron	.0000059	$1\frac{1}{16}$
Steel	.0000067	$1\frac{3}{16}$
Lead	.0000159	$1\frac{15}{16}$

ILLUSTRATION: In Fig. 11 what will be the change in length of the riser between the first and the fifth floors if its temperature changes from 32° F to 212° F?

$E = 100 \times 12 \times k \times (t_1 - t_2)$
$E = 100 \times 12 \times 0.0000104 \times (212 - 32)$
$E = 2.25$ inches per 100 ft.

Then the expansion for 50 ft. is $\dfrac{2.25}{2} = 1\frac{1}{8}$ in.

(Ans.)

ILLUSTRATION: What would be the change of length of a wrought-iron riser in place of the brass in the preceding illustration?

$E = 100 \times 12 \times k \times (t_1 - t_2)$
$E = 100 \times 12 \times 0.0000067 \times (212 - 32)$
$E = 1.447$ inches per 100 ft

Then the expansion per 50 ft is

$$\frac{1.447}{2} = \tfrac{23}{32} \text{ in. (Ans.)}$$

Wrought Pipe.—Pipes for water supply and waste disposal are made of wrought iron, wrought steel, cast iron, copper, brass, and lead. Wrought-iron pipe is, whether galvanized or black, most commonly used for the water supply plumbing of buildings. Wrought steel is less commonly used be-

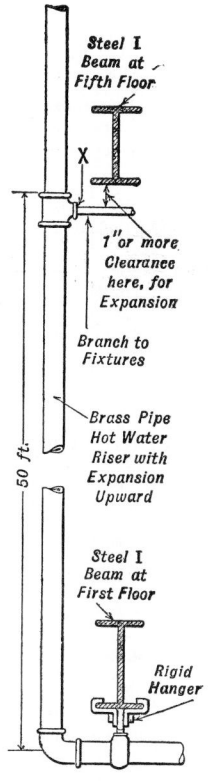

Copper and Brass Research Ass'n.

FIG. 11.—Clearance for Expansion.

cause it has a greater tendency to rust. However, where very high pressures are encountered its use may be the more desirable. Wrought pipe is specified by nominal inside diameters up to twelve inches; above twelve inches the pipe is known as O.D., or outside diameter pipe and is specified accordingly with the desired thickness of walls. " Standard " pipe is used for pressures up to 125 pounds per square inch. " Extra strong " and " double extra

TABLE 14

STANDARD WROUGHT PIPE

Size, In.	Diameters, Inches		Thickness, Inches	Weight per Foot, Pounds		Threads per Inch	Length of Thread, Inches (Distance E in Fig. 17, p. 441)		Taper per Foot, Inch	Hydrostatic Test, Pounds
	External	Internal		Plain ends	Threads and couplings					
⅛	.405	.265	.070	.244	.245	27	⅜	.375	¾	750
¼	.540	.360	.090	.424	.425	18	9⁄16	.569	¾	750
⅜	.675	.489	.093	.567	.568	18	9⁄16	.574	¾	750
½	.840	.618	.111	.850	.852	14	¾	.748	¾	750
¾	1.050	.820	.115	1.130	1.134	14	¾	.760	¾	750
1	1.315	1.043	.136	1.678	1.684	11½	15⁄16	.944	¾	750
1¼	1.660	1.374	.143	2.272	2.281	11½	1	.968	¾	750
1½	1.900	1.604	.148	2.717	2.731	11½	1	.984	¾	750
2	2.375	2.059	.158	3.652	3.678	11½	1	1.017	¾	1000
2½	2.875	2.459	.208	5.793	5.819	8	1½	1.512	¾	1000
3	3.500	3.058	.221	7.575	7.616	8	1 9⁄16	1.575	¾	1000
3½	4.000	3.538	.231	9.109	9.202	8	1⅝	1.625	¾	1000
4	4.500	4.016	.242	10.790	10.889	8	1 11⁄16	1.675	¾	1000
4½	5.000	4.496	.252	12.538	12.642	8	1¾	1.725	¾	1000
5	5.563	5.037	.263	14.617	14.810	8	1¾	1.781	¾	1000
6	6.625	6.053	.286	18.974	19.185	8	1⅞	1.887	¾	1000
7	7.625	7.011	.307	23.544	23.769	8	2	1.987	¾	1000
*8	8.625	8.059	.283	24.696	25.000	8	2 1⁄16	2.087	¾	800
8	8.625	7.967	.329	28.554	28.809	8	2 1⁄16	2.087	¾	1000
9	9.625	8.927	.349	33.907	34.188	8	2 3⁄16	2.187	¾	900
*10	10.750	10.182	.284	31.201	32.000	8	2 5⁄16	2.300	¾	600
*10	10.750	10.124	.313	34.240	35.000	8	2 5⁄16	2.300	¾	800
10	10.750	10.006	.372	40.483	41.132	8	2 5⁄16	2.300	¾	900
11	11.750	10.986	.382	45.557	46.247	8	2⅜	2.400	¾	800
*12	12.750	12.078	.336	43.773	45.000	8	2½	2.500	¾	600
12	12.750	11.986	.382	49.562	50.706	8	2½	2.500	¾	800
14	14.000	13.250	.375	53.510	55.712	8	2⅝	2.625	¾	700
15	15.000	14.250	.375	57.437	59.859	8	2¾	2.725	¾	700
16	16.000	15.250	.375	61.364	63.927	8	2 13⁄16	2.825	¾	600
17	17.000	16.250	.375	65.292	69.436	8	2 15⁄16	2.925	¼	550
18	18.000	17.250	.375	69.219	73.681	8	3	3.025	¾	550
20	20.000	19.250	.375	77.073	82.078	8	3¼	3.225	¾	550

* Unless specified the lighter weight will not be furnished.

strong " pipes are used for higher pressures. The extra thickness is gained by making the bore smaller, the nominal diameter remaining the same. Figure 12 shows a comparison of the cross-sections of ¾-inch pipe of the three different weights. Table 14 gives the dimensions of standard wrought pipe. This pipe is sold

Standard   *Extra Heavy*   *Double Extra Heavy*

Fig. 12.—Full Size Sections of ¾-in. Pipe.

in random lengths averaging 20 feet, threaded unless otherwise ordered and with one coupling on each length. Extra strong and double extra strong are also sold in random lengths but generally with plain ends. Table 15 gives the dimensions of these sizes.

TABLE 15

Extra Strong and Double Extra Strong Pipe

Nominal Size, In.	External Diameter, Inches	Internal Diameter		Nominal Size, In.	External Diameter, Inches	Internal Diameter	
		Extra Strong	Double Extra Strong			Extra Strong	Double Extra Strong
⅛	0.405	0.215	....	1½	1.900	1.500	1.100
¼	0.540	0.302	....	2	2.375	1.939	1.503
⅜	0.675	0.423	....	2½	2.875	2.323	1.771
½	0.840	0.546	0.252	3	3.500	2.900	2.300
¾	1.050	0.742	0.434	3½	4.00	3.364	2.728
1	1.315	0.957	0.599	4	4.50	3.826	3.152
1¼	1.660	1.278	0.896				

Cast-Iron Pipe.—Pipe of cast iron is universally used in municipal water distribution systems. In the case of a large building or factory, the service pipe may be of this material. Cast-iron pipe of lighter weight is also generally used for the drainage plumbing of buildings. Sizes of the water pipe range from 3 inches to 84 inches nominal inside diameter and the standard length of bell-and-spigot sections is 12 feet. The dimensions of this pipe and fittings have been standardized. Some of these dimensions are shown in Tables 16 and 17.

PIPE OF OTHER MATERIALS

Copper and Brass Pipe.—Copper and brass pipe are excellent plumbing materials, but rather expensive. Therefore they are not generally used except where corrosive water conditions or local Code requirements make their use necessary. Copper pipe is made of 99.9% pure copper in standard and extra-heavy wrought pipe sizes as given in Tables 14 and 15. It is sometimes used for the water service (the pipe joining the city main or well with the house distribution system), because of its great durability. Its resistance to corrosion permits the use of a smaller pipe size than would be needed if wrought iron or steel pipe were used, for a given capacity as indicated in Table 12.

Brass is an alloy of copper and zinc. It has all the advantages of iron pipe plus the fact that it does not rust. Commonly available brasses are made in various grades from "yellow brass," containing about 67% copper, to "red brass" containing about 85% copper.

Brass pipe is replacing galvanized iron pipe to a large degree because of its greater durability. A system using brass pipe will generally last the life of the building in which it is installed. The smooth interior surface of brass pipe permits it to carry water with less friction. This means that a smaller brass pipe size can be used to carry a given volume of water than that which would be required if steel pipe were used.

TABLE 16

STANDARD DIMENSIONS OF BELLS, SOCKETS, SPIGOT BEADS, AND OUTSIDE
DIAMETERS OF PIT CAST PIPE

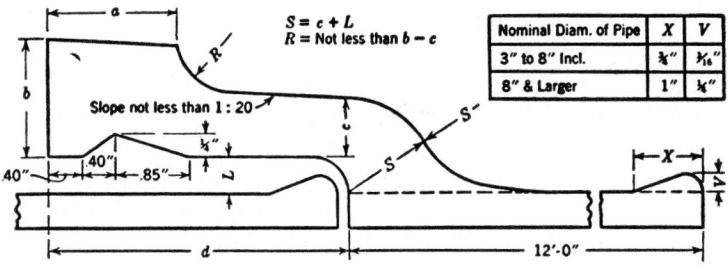

$S = c + L$
$R = $ Not less than $b - c$

Nominal Diam. of Pipe	X	V
3″ to 8″ Incl.	⅜″	⅜₆″
8″ & Larger	1″	¼″

Slope not less than 1 : 20

Nominal Diam	Thickness of Pipe		Outside Diam of Pipe	Dimensions of Bells					
	From	To		Diam of Socket	Thickness of Joint L	Depth of Socket d	a	b	c
3	0.37	0.45	3.80	4.60	0.40	3.50	1.25	1.30	0.65
	0.46	0.53	3.96	4.76	0.40	3.50	1.25	1.30	0.65
4	0.40	0.45	4.80	5.60	0.40	3.50	1.50	1.30	0.65
	0.46	0.55	5.00	5.80	0.40	3.50	1.50	1.30	0.65
6	0.43	0.50	6.90	7.70	0.40	3.50	1.50	1.40	0.70
	0.51	0.60	7.10	7.90	0.40	3.50	1.50	1.40	0.70
	0.61	0.66	7.22	8.02	0.40	4.00	1.50	1.75	0.75
	0.67	0.74	7.38	8.18	0.40	4.00	1.50	1.85	0.85
8	0.46	0.57	9.05	9.85	0.40	4.00	1.50	1.50	0.75
	0.58	0.70	9.30	10.10	0.40	4.00	1.50	1.50	0.75
	0.71	0.76	9.42	10.22	0.40	4.00	1.50	1.85	0.85
	0.77	0.85	9.60	10.40	0.40	4.00	1.50	1.95	0.95
10	0.50	0.60	11.10	11.90	0.40	4.00	1.50	1.50	0.75
	0.61	0.75	11.40	12.20	0.40	4.00	1.50	1.60	0.80
	0.76	0.85	11.60	12.40	0.40	4.50	1.75	1.95	0.95
	0.86	0.97	11.84	12.64	0.40	4.50	1.75	2.05	1.05
12	0.54	0.65	13.20	14.00	0.40	4.00	1.50	1.60	0.80
	0.66	0 80	13.50	14.30	0.40	4.00	1.50	1.70	0.85
	0.81	0 94	13.78	14.58	0.40	4.50	1.75	2.05	1.05
	0.95	1.09	14.08	14.88	0.40	4.50	1.75	2.20	1.20

All dimensions given in inches.

From American Standard A21.2. Complete tables are available from
American Water Works Association or American Standards Association.

A comparison of pipe sizes required for different water conditions is shown in Table 12. Brass pipe is manufactured in 12-ft. lengths in standard pipe sizes.

Lead Pipe.—The use of lead pipe is decreasing. Factors which have contributed to its unpopularity include the high cost of the material, the skill required to install lead pipe, and the possible dangers from lead poisoning. Some of its desirable features are flexibility, durability, vibration resistance, and its great resistance to corrosion. The size of lead pipe is designated by the inside diameter. Service and supply pipes in three grades range in size from $\frac{1}{2}''$ ID through $2''$ ID. Waste pipes range in size from $1\frac{1}{4}''$ ID through $6''$ ID. Lead pipe is smooth, lasting, and pliable, but requires skilled workmanship to achieve smooth soldered or wiped joints.

Vitrified-Clay Pipe.—This type of pipe is generally used for draining sewage, industrial wastes, and storm water. It is manufactured with nominal internal diameters ranging from $4''$ to $36''$ ID. The *Handbook of Vitrified Clay Sewer Pipe and Kindred Clay Products* of the Clay Sewer Pipe Association, Inc., contains detailed dimensions and weights of fittings for this type of pipe.

Plastic Pipe.—Plastic pipe offers a number of valuable features. They include freedom from corrosion; less likelihood of cracking when water freezes within the pipe; lower weight; and greater ease of handling and assembly. A number of kinds are available, including polyvinyl chloride (PVC), unplasticized polyvinyl chloride (UPVC), polyethylene (PE), polypropylene (PP), chlorinated polyether (trade name Penton®), acrylonitrile-butadiene-styrene (ABS), cellulose-acetate-butyrate (CAB), polycarbonate, polyacetal, polyvinyl chloride (trade name High-Temp Geon®) and polyvinylidene fluoride (trade name Kynar®).

These types of piping differ somewhat in their strength, corrosion resistance and other properties, so that for industrial applications, especially in the chemical industry, data on the resistance to specific conditions of temperature, pressure and chemical action of the various types of plastic pipe, as well as plastic-lined pipe, should be obtained from a general manufacturer, such as Tube Turns Plastics, Inc., 30th and Magazine Sts.,

Louisville, Ky., 40211. For residential use, all the kinds mentioned here have suitable properties, so that selection is largely a question of availability and cost. It is to be noted that for conditions in which the pipe is exposed to mechanical shock, the PVC and UPVC types are available in both normal impact and high impact types.

A number of methods are available for joining plastic pipe and fittings. Like metallic pipe, plastic pipe may be threaded and joined directly to fittings; in the larger industrial sizes flanged connections are used. The thermoplastic types, which include polyethylene, polypropylene and chlorinated polyether, can be joined (for most fittings, after screwing the threaded ends together by hand) by thermal bonding, a process in which they are fused together by a special heating unit called the THERMO-SEAL® tool, which may be obtained from Cabot Piping Systems, Louisville, Ky., 40201. The same company provides solvent cements for joining plastic piping and fittings by cementing them together, although these cements can also be obtained from other suppliers of plastic piping.

Plastic piping of the PE and PP types, which are among those which may be connected by thermal sealing, is available in the sizes shown in Table 16A, in the end dimensions shown in Table 16B, and in the fitting types shown in Fig. 13-14. The fittings for cemented plastic pipe are similar, but are not threaded.

TABLE 16A

POLYETHYLENE (PE) AND POLYPROPYLENE (PP) PIPE
IN STANDARD 10′ AND 20′ LENGTHS

Pipe Size	Outside Diam.	Wall Thickness	Approx. Wt./Ft.
1½	1.900	0.145	.32
2	2.375	0.154	.43
3	3.500	0.216	.90
4	4.500	0.237	1.26
6	6.625	0.280	2.24

TABLE 16B

END DIMENSIONS OF THERMO-SEAL® FITTINGS

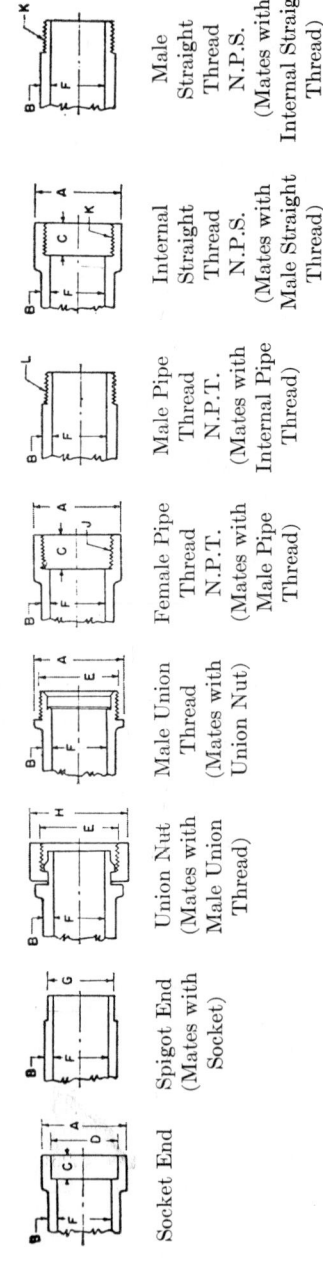

| | | Socket End | Spigot End (Mates with Socket) | Union Nut (Mates with Male Union Thread) | Male Union Thread (Mates with Union Nut) | Female Pipe Thread N.P.T. (Mates with Male Pipe Thread) | Male Pipe Thread N.P.T. (Mates with Internal Pipe Thread) | Internal Straight Thread N.P.S. (Mates with Male Straight Thread) | Male Straight Thread N.P.S. (Mates with Internal Straight Thread) |

Nom. Pipe Size	Outside Diam.			Inside Diam.		Wall Thickness B	Depth of Socket C	Threads			
	A	G	H	D	F			E	J	K	L
1½	2½	1.900	2¹³⁄₁₆	1.902	1.610	.218	¹¹⁄₁₆	2⅜–12N2	1½–11½NPT	1½–11½NPS	1½–11½NPT
2	3	2.375	3¼	2.380	2.067	.231	¾	2¾–12N2	2 –11½NPT	2 –11½NPS	2 –11½NPT
3	4¼	3.500	4¹¹⁄₁₆	3.495	3.068	.324	1.200	4 – 8N2	3 – 8 NPT	3 – 8 NPS	3 – 8 NPT
4	5²¹⁄₆₄	4.500	5⅞	4.495	4.093	.375	1.300	5⅛– 8N2	4 – 8 NPT	4 – 8 NPS	4 – 8 NPT

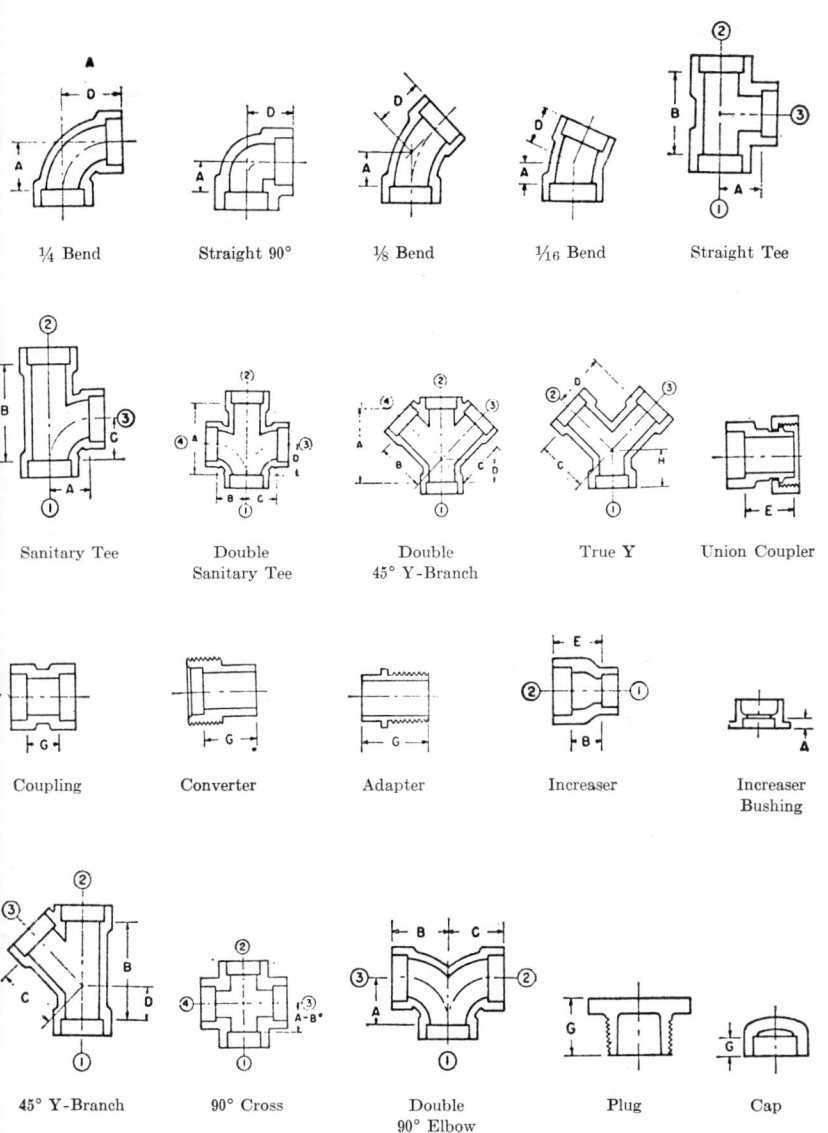

¼ Bend   Straight 90°   ⅛ Bend   ¹⁄₁₆ Bend   Straight Tee

Sanitary Tee   Double Sanitary Tee   Double 45° Y-Branch   True Y   Union Coupler

Coupling   Converter   Adapter   Increaser   Increaser Bushing

45° Y-Branch   90° Cross   Double 90° Elbow   Plug   Cap

Fig. 13-14. Types of Plastic Pipe Thermo-Seal® Fittings.

Other Materials.—A number of other materials are used to manufacture special types of pipe. Cement pipe, concrete pipe, asbestos-cement pipe ("Transite"), and bituminized-fiber pipe, as well as a wide variety of specially coated pipes such as rubber- or glass-lined pipe are commercially available. All of these types have special properties which fit them for special applications.

Pipe Joints.—Wrought-iron, wrought-steel, brass, and copper pipes are joined to each other by threaded or flanged couplings, unions, or fittings, or by welding. These pipes are usually cut and threaded on the job to suit the requirements. Bell-and-spigot cast-iron pipes are joined by first ramming a strand of oakum or jute into the bottom of the joint space and then filling the remainder of the space with poured lead or calked lead wool. (See Table 17.)

The cutting, threading, calking, and joining of pipes requires a certain number of specialized plumbing tools.

TABLE 17

STANDARD THICKNESSES OF CAST-IRON PIT CAST PIPE

Size Inches	Class 50 50 Lb Pressure 115 Ft Head Thickness Inches	Class 100 100 Lb Pressure 231 Ft Head Thickness Inches	Class 150 150 Lb Pressure 346 Ft Head Thickness Inches	Class 200 200 Lb Pressure 462 Ft Head Thickness Inches	Class 250 250 Lb Pressure 577 Ft Head Thickness Inches	Class 300 300 Lb Pressure 693 Ft Head Thickness Inches	Class 350 350 Lb Pressure 808 Ft Head Thickness Inches
3	0.37	0.37	0.37	0.37	0.37	0.37	0.37
4	0.40	0.40	0.40	0.40	0.40	0.40	0.40
6	0.43	0.43	0.43	0.43	0.43	0.46	0.50
8	0.46	0.46	0.46	0.46	0.50	0.54	0.58
10	0.50	0.50	0.54	0.58	0.63	0.68	0.73
12	0.54	0.54	0.58	0.63	0.68	0.73	0.79

Pipe Threads.—Pipes and fittings are threaded to the American Standard Pipe Thread shown in cross-section in Fig. 16.

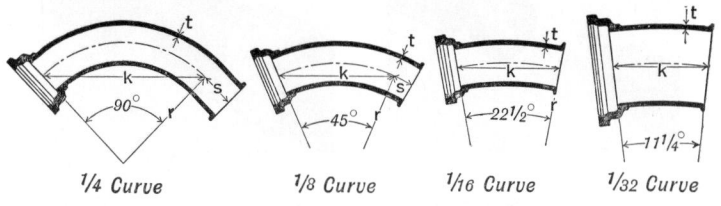

¼ Curve   ⅛ Curve   1/16 Curve   1/32 Curve

Fig. 15.—Standard Bell-and-spigot Curves.

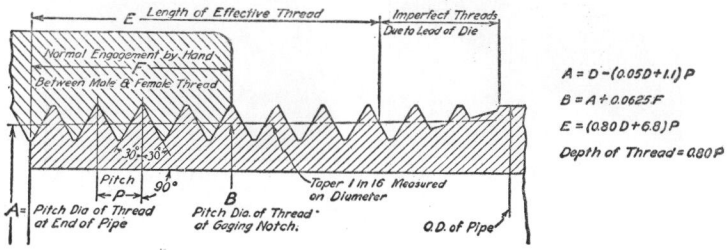

$$A = D - (0.05D + 1.1)P$$
$$B = A + 0.0625F$$
$$E = (0.80D + 6.8)P$$
$$\text{Depth of Thread} = 0.80P$$

Fig. 16.—American Standard Pipe Thread (ASA B2.1).

Female threads are cut on the fittings by the manufacturer so that the plumber's only practical concern is the cutting of male threads. on pipes. This he does by the use of a die selected for the proper pipe size and held in a stock. When the full length of the die is run onto the pipe so that the pipe end is flush with the face of the die, the correct length of thread is cut. The number of threads per inch and the effective length E (Fig. 16) for pipes of various sizes is given in columns 7, 8, and 9 in Table 14. The latter figures will be used in connection with piping measurements.

Hubless Cast-Iron Soil Pipe.—In addition to the bell-and-spigot types of cast-iron pipe joints which are joined by the older lead or combination calking methods, a method superior in many respects has been developed by the Cast Iron Pipe Institute. Known by their trade name of No-Hub® pipe, it provides

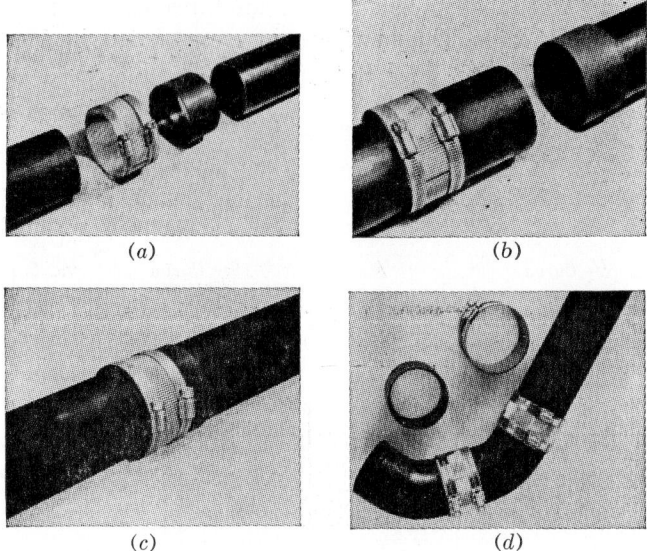

(a)

(b)

(c)

(d)

Fig. 17.

a mechanical joint composed of a Neoprene sleeve gasket and stainless steel shield-clamp rather than the lead and oakum just described. The method is shown in Fig. 17 in four illustrations. Fig. 17(a) shows the parts of the joint: the two pipe-ends, the shield-clamp and the Neoprene sleeve gasket. Fig. 17(b) shows the sleeve-gasket placed on the end of the right-hand pipe-end and the shield-clamp on the left-hand pipe. The joint is then completed by butting the pipe ends against an integrally-molded shoulder inside the sleeve-gasket, sliding the shield-clamp into position, and tightening it to make the completed joint of Fig. 17(c). Fittings are jointed and fastened in the same way (Fig. 17(d)).

Pipe Fittings.—As already shown for cast iron and plastic pipe, the couplings tees, elbows, crosses, etc., used for joining pipes, making branches, turns, etc., are called fittings. Small pipes are usually "made up" with screwed or threaded fittings. The sizes of fittings are identified by the nominal pipe size. In specifications for reducing tees and crosses, the size of the

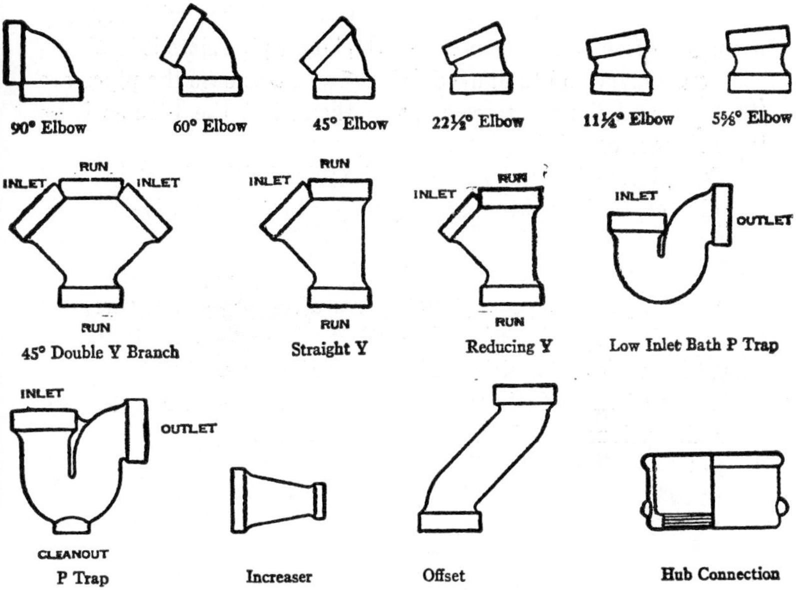

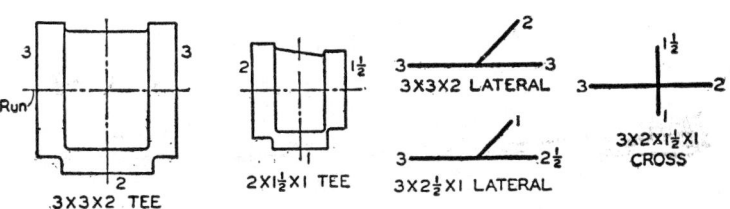

90° Elbow
60° Elbow
45° Elbow
22½° Elbow
11¼° Elbow
5⅝° Elbow
RUN
INLET
INLET
RUN
45° Double Y Branch
INLET
RUN
RUN
Straight Y
INLET
RUN
RUN
Reducing Y
INLET
OUTLET
Low Inlet Bath P Trap
INLET
OUTLET
CLEANOUT
P Trap
Increaser
Offset
Hub Connection

FIG. 18.—Cast-iron Drainage Fittings. Screwed for wrought pipe.

largest run opening is given first, followed by the size of the opening at the other end of the run. Where the fitting is a tee,

3
3
Run
2
3X3X2 TEE
2
1½
1
2X1½X1 TEE
2
3
3
3X3X2 LATERAL
1
3
2½
3X2½X1 LATERAL
1½
3
2
1
3X2X1½X1
CROSS

FIG. 19.—Specifications of Fittings.

the size of the outlet is given next. Where the fitting is a cross, the largest opening is the third dimension followed by the opposite opening. Fig. 19 illustrates these conventions.

The assembling of pipes larger than 4 inches with screwed fittings is often cumbersome. For the larger pipe sizes it is, therefore, customary to use flanged fittings. In this case the pipes are threaded and flanges screwed onto them, but the fittings have their flanges cast into place.

Two pieces of straight pipe may be joined either by a threaded coupling or, when making up the last joint or where it is desirable to have a joint which may readily be dissembled, by a union, either screwed or flanged.

A short piece of pipe (usually less than 12 inches) threaded at the ends is called a *nipple*. A *close nipple* is about twice the length E in Fig. 17 and is threaded all the way.

Many of these fittings are illustrated in Figs. 20 to 27 and their dimensions are given in Tables 18 to 20.

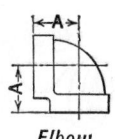

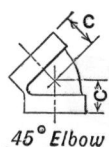

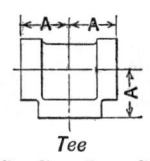

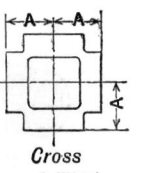

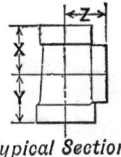

A
C
Elbow
45° Elbow
Tee
Cross
X
Y
Z
Typical Section

FIG. 20.—125-lb. Cast-Iron Screwed Fittings.

TABLE 18

DIMENSIONS OF CAST-IRON SCREWED FITTINGS *

(See Fig. 20)

All dimensions given in inches.

Nonimal pipe size, in.	Standard (125 lb)		Extra Heavy (250 lb)		Nominal pipe size, in.	Standard (125 lb)		Extra Heavy (250 lb)	
	A	C	A	C		A	C	A	C
¼	13/16	¾	15/16	13/16	3½	3 7/16	2 3/8	3 ¾	2 5/8
3/8	15/16	13/16	1 1/16	7/8	4	3 13/16	2 5/8	4 1/8	2 13/16
½	1 1/8	7/8	1¼	1	5	4½	3 1/16	4 7/8	3 3/16
¾	1 5/16	1	1 7/16	1 1/8	6	5 1/8	3 7/16	5 5/8	3½
1	1½	1 1/8	1 5/8	1 5/16	8	6 9/16	4¼	7	4 5/16
1¼	1 ¾	1 5/16	1 15/16	1½	10	8 1/16	5 3/16	8 5/8	5 3/16
1½	1 15/16	1 7/16	2 1/8	1 11/16	12	9½	6	10	6
2	2¼	1 11/16	2½	2	14 O.D.	10 3/8	.....	11	
2½	2 11/16	1 15/16	2 15/16	2¼	16 O.D.	11 13/16	.....	12½	
3	3 1/16	2 3/16	3 3/8	2½					

* U. S. Govt. master specifications.

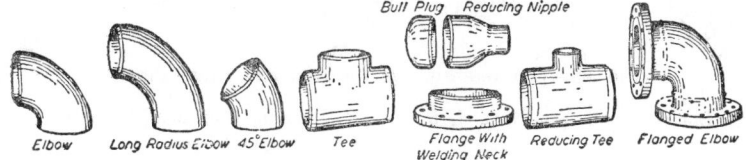

Bull Plug
Reducing Nipple
Elbow
Long Radius Elbow
45° Elbow
Tee
Flange With Welding Neck
Reducing Tee
Flanged Elbow

FIG. 21.—Welded Pipe Fittings.

90°
45°
45°
OPEN VEE
S
2½ t min.
¼ t min.
OPEN SQUARE
2 t min.
¼ t min.
S
2 t min.
¼ t min.
S
2½ t min.
CLOSED VEE
No Spacing

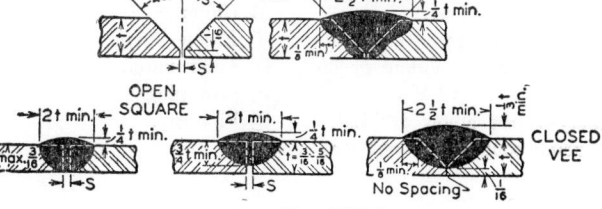

FIG. 22.—Pipe Welds.

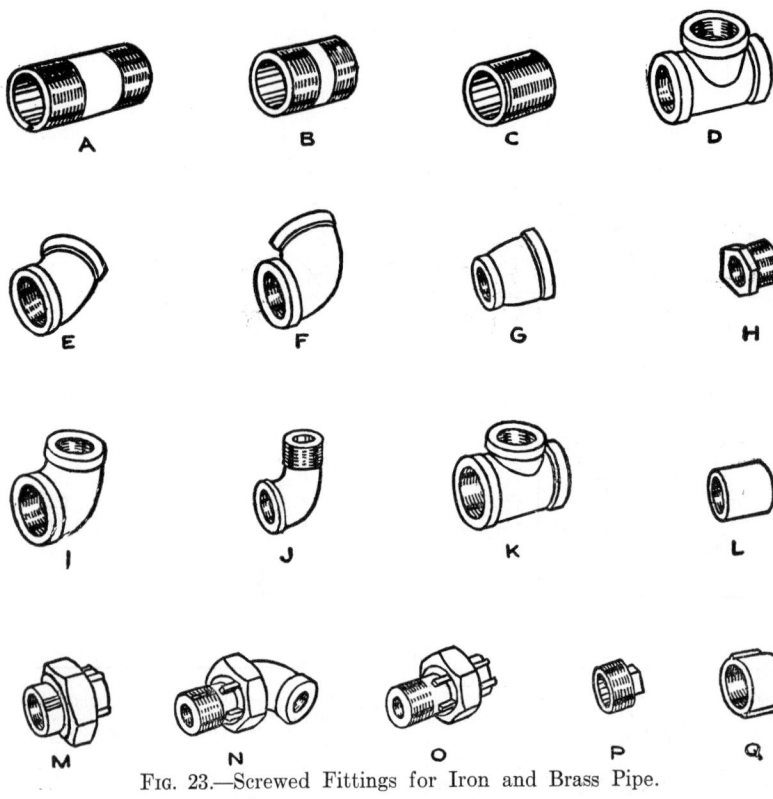

Fig. 23.—Screwed Fittings for Iron and Brass Pipe.

A. Long nipple, usually 3, 4, 5 or 6″ long.

B. Short nipple (length varies with diameter of pipe).

C. Close nipple (length varies with diameter of pipe).

D. Tee or "T" fitting.

E. Forty-five degree elbow or 45° "L."

F. Ninety-degree elbow or "L."

G. Reducing coupling or reducer.

H. Bushing (hex head bushing).

I. Reducing elbow.

J. Street elbow.

K. Reducing tee.

L. Coupling.

M. Union.

N. Elbow union.

O. Tank union.

P. Pipe plug.

Q. Pipe cap.

Reducing tees: Give size A, then B, and last C.

Either size of B or C may be reduced or both.

Example: $1 \times \frac{3}{4} \times \frac{3}{4}$ "T"

Hex head bushings: Give size A and B.

Example: $1\frac{1}{4} \times \frac{3}{4}$ bushing.

Reducing couplings: Give size A and B.

Example: $1 \times \frac{3}{4}$ coupling.

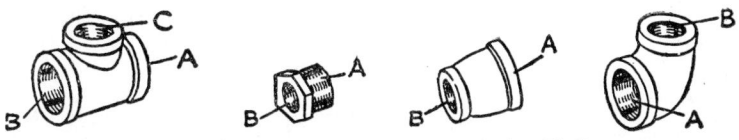

Fig. 24.—Designating the Size of Reducing Fittings.

Reducing fittings usually reduce only one size, from a given size to the next size smaller. However, bushings which reduce two or more sizes are quite common.

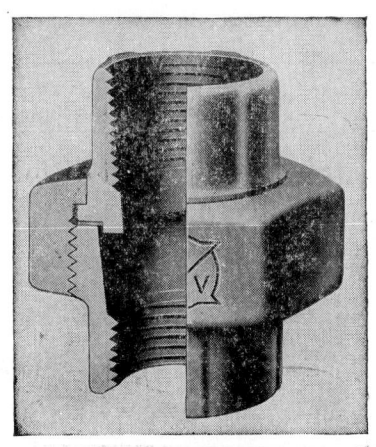

Fig. 25.—Common Gasket Type Union. *Courtesy The Kennedy Valve Mfg. Co.*

Fig. 26.—Union with Ground Brass Seat. *Courtesy E. M. Dart Mfg. Co.*

Unions are made in two general types: (a) the old style gasket type, which requires a fiber gasket at the joint of the two members and (b) the brass-seated union, which requires no gasket, as a watertight joint is obtained by the brass inserts at the seat of the fitting. The latter has replaced the former in most plumbing systems, and is used universally in systems made up of brass pipe and fittings.

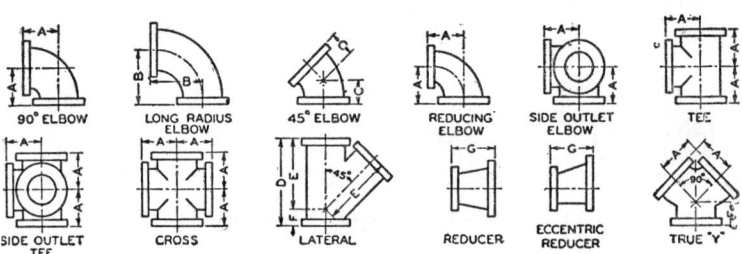

FIG. 27.—Flanged Fittings.

TABLE 19

DIMENSIONS OF AMERICAN STANDARD 125 LB. CAST IRON FLANGED FITTINGS

Nominal Pipe Size	A	B	C	D	E	F	G
1	$3^{1}/_{2}$	5	$1^{3}/_{4}$	$7^{1}/_{2}$	$5^{3}/_{4}$	$1^{3}/_{4}$	. . . .
$1^{1}/_{4}$	$3^{3}/_{4}$	$5^{1}/_{2}$	2	8	$6^{1}/_{4}$	$1^{3}/_{4}$	. . . .
$1^{1}/_{2}$	4	6	$2^{1}/_{4}$	9	7	2	. . . .
2	$4^{1}/_{2}$	$6^{1}/_{2}$	$2^{1}/_{2}$	$10^{1}/_{2}$	8	$2^{1}/_{2}$	5
$2^{1}/_{2}$	5	7	3	12	$9^{1}/_{2}$	$2^{1}/_{2}$	$5^{1}/_{2}$
3	$5^{1}/_{2}$	$7^{3}/_{4}$	3	13	10	3	6
$3^{1}/_{2}$	6	$8^{1}/_{2}$	$3^{1}/_{2}$	$14^{1}/_{2}$	$11^{1}/_{2}$	3	$6^{1}/_{2}$
4	$6^{1}/_{2}$	9	4	15	12	3	7
5	$7^{1}/_{2}$	$10^{1}/_{4}$	$4^{1}/_{2}$	17	$13^{1}/_{2}$	$3^{1}/_{2}$	8
6	8	$11^{1}/_{2}$	5	18	$14^{1}/_{2}$	$3^{1}/_{2}$	9
8	9	14	$5^{1}/_{2}$	22	$17^{1}/_{2}$	$4^{1}/_{2}$	11
10	11	$16^{1}/_{2}$	$6^{1}/_{2}$	$25^{1}/_{2}$	$20^{1}/_{2}$	5	12
12	12	19	$7^{1}/_{2}$	30	$24^{1}/_{2}$	$5^{1}/_{2}$	14
14 O.D.	14	$21^{1}/_{2}$	$7^{1}/_{2}$	33	27	6	16
16 O.D.	15	24	8	$36^{1}/_{2}$	30	$6^{1}/_{2}$	18
18 O.D.	$16^{1}/_{2}$	$26^{1}/_{2}$	$8^{1}/_{2}$	39	32	7	19

TABLE 20

Dimensions of 125-Lb Cast-Iron Reducing Tees (See Fig. 20)

Nominal Pipe Sizes	Center to End			Nominal Pipe Sizes	Center to End		
	X	Y	Z		X	Y	Z
1/2 × 1/2 × 3/4	1-1/4	1-1/4	1-3/16	1-1/2 × 3/4 × 1-1/2	1-15/16	1-3/4	1-15/16
1/2 × 1/2 × 3/8	1-1/16	1-1/16	1	1-1/2 × 3/4 × 1-1/4	1-13/16	1-5/8	1-7/8
3/4 × 3/4 × 1	1-7/16	1-7/16	1-3/8	1-1/2 × 1/2 × 1-1/2	1-15/16	1-11/16	1-15/16
3/4 × 3/4 × 1/2	1-3/16	1-1/4	1-1/4	2 × 2 × 3	2-7/8	2-7/8	2-1/2
3/4 × 3/4 × 3/8	1-1/8	1-1/8	1-1/8	2 × 2 × 2-1/2	2-5/8	2-5/8	2-3/8
3/4 × 1/2 × 3/4	1-5/16	1-1/4	1-5/16	2 × 2 × 1-1/2	2	2	2-3/16
3/4 × 1/2 × 1/2	1-3/16	1-1/8	1-1/4	2 × 2 × 1-1/4	1-7/8	1-7/8	2-1/8
1 × 1 × 1-1/2	1-13/16	1-13/16	1-5/8	2 × 2 × 1	1-3/4	1-3/4	2
1 × 1 × 1-1/4	2-1/8	2-1/8	1-7/8	2 × 2 × 3/4	1-5/8	1-5/8	2
1 × 1 × 3/4	1-3/8	1-3/8	1-7/16	2 × 2 × 1/2	1-1/2	1-1/2	1-7/8
1 × 1 × 1/2	1-1/4	1-1/4	1-3/8	2 × 1-1/2 × 2-1/2	2-5/8	2-1/2	2-3/8
1 × 1 × 3/8	1-3/16	1-3/16	1-1/4	2 × 1-1/2 × 2	2-1/4	2-3/16	2-1/4
1 × 3/4 × 1	1-1/2	1-7/16	1-1/2	2 × 1-1/2 × 1-1/2	2	1-15/16	2-3/16
1 × 3/4 × 3/4	1-3/8	1-5/16	1-7/16	2 × 1-1/2 × 1-1/4	1-7/8	1-13/16	2-1/8
1 × 3/4 × 1/2	1-1/4	1-3/16	1-3/8	2 × 1-1/2 × 1	1-3/4	1-5/8	2
1 × 1/2 × 1	1-1/2	1-3/8	1-1/2	2 × 1-1/2 × 3/4	1-5/8	1-1/2	2
1 × 1/2 × 3/4	1-3/8	1-1/4	1-7/16	2 × 1-1/2 × 1/2	1-1/2	1-7/16	1-7/8
1 × 3/8 × 1	1-1/2	1-1/4	1-1/2	2 × 1-1/4 × 2	2-1/4	2-1/8	2-1/4
1-1/4 × 1-1/4 × 2	1-11/16	1-11/16	1-9/16	2 × 1-1/4 × 1-1/2	2	1-7/8	2-3/16
1-1/4 × 1-1/4 × 1-1/2	1-7/8	1-7/8	1-13/16	2 × 1-1/4 × 1-1/4	1-7/8	1-3/4	2-1/8
1-1/4 × 1-1/4 × 1	1-9/16	1-9/16	1-11/16	2 × 1-1/4 × 1	1-3/4	1-9/16	2
1-1/4 × 1-1/4 × 3/4	1-7/16	1-7/16	1-5/8	2 × 1 × 2	2-1/4	2	2-1/4
1-1/4 × 1-1/4 × 1/2	1-5/16	1-5/16	1-1/2	2 × 1 × 1-1/4	2	1-13/16	2-1/8
1-1/4 × 1 × 1-1/2	1-7/8	1-13/16	1-13/16	2 × 3/4 × 2	2-1/4	2	2-1/4
1-1/4 × 1 × 1-1/4	1-3/4	1-11/16	1-3/4	2-1/2 × 2-1/2 × 4	3-1/2	3-1/2	3-1/16
1-1/4 × 1 × 1	1-9/16	1-1/2	1-11/16	2-1/2 × 2-1/2 × 3	3	3	2-13/16
1-1/4 × 1 × 3/4	1-7/16	1-3/8	1-5/8	2-1/2 × 2-1/2 × 2	2-3/8	2-3/8	2-5/8
1-1/4 × 1 × 1/2	1-5/16	1-1/4	1-1/2	2-1/2 × 2-1/2 × 1-1/2	2-3/16	2-3/16	2-1/2
1-1/4 × 3/4 × 1-1/4	1-3/4	1-5/8	1-3/4	2-1/2 × 2-1/2 × 1-1/4	2-1/16	2-1/16	2-7/16
1-1/4 × 3/4 × 1	1-9/16	1-7/16	1-11/16	2-1/2 × 2-1/2 × 1	1-7/8	1-7/8	2-3/8
1-1/4 × 1/2 × 1-1/4	1-3/4	1-1/2	1-3/4	2-1/2 × 2-1/2 × 3/4	1-3/4	1-3/4	2-5/16
1-1/2 × 1-1/2 × 2	2-3/16	2-3/16	2	2-1/2 × 2 × 2-1/2	2-11/16	2-5/8	2-11/16
1-1/2 × 1-1/2 × 1-1/4	1-13/16	1-13/16	1-7/8	2-1/2 × 2 × 2	2-3/8	2-1/4	2-5/8
1-1/2 × 1-1/2 × 1	1-5/8	1-5/8	1-13/16	2-1/2 × 2 × 1-1/2	2-3/16	2	2-1/2
1-1/2 × 1-1/2 × 3/4	1-1/2	1-1/2	1-3/4	2-1/2 × 2 × 1-1/4	2-1/16	1-7/8	2-7/16
1-1/2 × 1-1/2 × 1/2	1-7/16	1-7/16	1-11/16	2-1/2 × 2 × 1	1-7/8	1-3/4	2-3/8
1-1/2 × 1-1/4 × 1-1/2	1-15/16	1-7/8	1-15/16	2-1/2 × 2 × 3/4	1-3/4	1-5/8	2-9/16
1-1/2 × 1-1/4 × 1-1/4	1-13/16	1-3/4	1-7/8	2-1/2 × 2 × 1/2	1-5/8	1-1/2	2-1/4
1-1/2 × 1-1/4 × 1	1-5/8	1-9/16	1-13/16	2-1/2 × 1-1/2 × 2-1/2	2-11/16	2-1/2	2-11/16
1-1/2 × 1-1/4 × 3/4	1-1/2	1-7/16	1-3/4	2-1/2 × 1-1/2 × 2	2-3/8	1-15/16	2-5/8
1-1/2 × 1-1/4 × 1/2	1-7/16	1-5/16	1-11/16	2-1/2 × 1-1/2 × 1-1/2	2-3/16	1-15/16	2-1/2
1-1/2 × 1 × 2	2-3/16	2	2	3 × 3 × 4	3-5/8	3-5/8	3-5/16
1-1/2 × 1 × 1-1/2	1-15/16	1-13/16	1-15/16	3 × 3 × 3-1/2	3-5/16	3-5/16	3-3/16
1-1/2 × 1 × 1-1/4	1-13/16	1-11/16	1-7/8	3 × 3 × 2-1/2	2-13/16	2-13/16	3
1-1/2 × 1 × 1	1-5/8	1-1/2	1-13/16				

All dimensions given in inches.

TABLE 20—*Continued*

Nominal Pipe Sizes	Center to End X	Y	Z
3 × 3 × 2	2 1/2	2 1/2	2 7/8
3 × 3 × 1 1/2	2 5/16	2 5/16	2 13/16
3 × 3 × 1 1/4	2 3/16	2 3/16	2 3/4
3 × 3 × 1	2	2	2 5/8
3 × 3 × 3/4	1 7/8	1 7/8	2 5/8
3 × 2 1/2 × 3	3 1/16	3	3 1/16
3 × 2 1/2 × 2 1/2	2 13/16	2 11/16	3
3 × 2 1/2 × 2	2 1/2	2 3/8	2 7/8
3 × 2 1/2 × 1 1/2	2 5/16	2 3/16	2 13/16
3 × 2 1/2 × 1 1/4	2 3/16	2 1/16	2 3/4
3 × 2 1/2 × 1	2	1 7/8	2 11/16
3 × 2 × 3	3 1/16	3	3 1/16
3 × 2 × 2 1/2	2 13/16	2 5/8	3
3 × 2 × 2	2 1/2	2 1/4	2 7/8
3 × 2 × 1 1/2	2 5/16	2	2 13/16
3 × 1 1/2 × 3	3 1/16	2 13/16	3 1/16
3 × 1 × 3	3 1/16	2 11/16	3 1/16
3 1/2 × 3 1/2 × 3	3 9/16	3 3/16	3 9/16
3 1/2 × 3 1/2 × 2 1/2	2 15/16	2 15/16	3 1/4
3 1/2 × 3 1/2 × 2	2 5/8	2 5/8	3 1/8
3 1/2 × 3 1/2 × 1 1/2	2 3/8	2 3/8	3 1/16
3 1/2 × 3 1/2 × 1 1/4	2 1/4	2 1/4	3
3 1/2 × 3 × 3	3 3/16	3 1/16	3 9/16
3 1/2 × 3 × 2 1/2	2 15/16	2 13/16	3 1/4
3 1/2 × 3 × 2	2 5/8	2 1/2	3 1/8
3 1/2 × 3 × 1 1/2	2 3/8	3 9/16	3 1/16
3 1/2 × 2 × 3 1/2	3 7/16	3 1/8	3 7/16
3 1/2 × 1 1/2 × 3 1/2	3 7/16	3 1/16	3 7/16
4 × 4 × 6	4 15/16	4 15/16	4 1/8
4 × 4 × 5	4 7/16	4 7/16	4
4 × 4 × 3 1/2	3 9/16	3 9/16	3 11/16
4 × 4 × 3	3 5/16	3 5/16	3 5/8
4 × 4 × 2 1/2	3 1/16	3 1/16	3 1/2
4 × 4 × 2	2 3/4	2 3/4	3 7/16
4 × 4 × 1 1/2	2 1/2	2 1/2	3 9/16
4 × 4 × 1 1/4	2 3/8	2 3/8	3 1/4
4 × 4 × 1	2 1/4	2 1/4	3 3/16
4 × 3 1/2 × 3	3 5/16	3 3/16	3 5/8
4 × 3 1/2 × 2 1/2	3 1/16	2 15/16	3 1/2
4 × 3 1/2 × 2	2 3/4	2 5/8	3 7/16
4 × 3 × 4	3 13/16	3 5/8	3 13/16
4 × 3 × 3	3 5/16	3 1/16	3 3/8
4 × 3 × 2	2 3/4	2 1/2	3 7/16
4 × 2 1/2 × 4	3 13/16	3 1/2	3 13/16
4 × 2 1/2 × 2 1/2	3 1/16	2 11/16	3 1/2

Nominal Pipe Sizes	Center to End X	Y	Z
4 × 2 × 4	3 13/16	3 7/16	3 13/16
4 × 1 1/2 × 4	3 13/16	3 9/16	3 13/16
4 × 1 1/4 × 4	3 13/16	3 1/4	3 13/16
5 × 5 × 6	5	5	4 5/8
5 × 5 × 4	4	4	4 7/16
5 × 5 × 3 1/2	3 3/4	3 3/4	4 9/16
5 × 5 × 3	3 1/2	3 1/2	4 1/4
5 × 5 × 2 1/2	3 1/4	3 1/4	4 1/8
5 × 5 × 2	2 15/16	2 15/16	4
5 × 5 × 1 1/2	2 3/4	2 3/4	3 15/16
5 × 4 × 5	4 1/2	4 7/16	4 1/2
5 × 4 × 4	4	3 13/16	4 7/16
5 × 4 × 3	3 1/2	3 5/16	4 1/4
5 × 4 × 2	2 15/16	2 3/4	4
5 × 3 × 5	4 1/2	4 1/4	4 1/2
5 × 3 × 4	4	3 5/8	4 7/16
5 × 3 × 3	3 1/2	3 1/16	4 1/4
5 × 2 × 5	4 1/2	4	4 1/2
6 × 6 × 8	6 3/8	6 3/8	5 9/16
6 × 6 × 5	4 7/16	4 7/16	5
6 × 6 × 4	4 1/8	4 1/8	4 15/16
6 × 6 × 3	3 5/8	3 5/8	4 3/4
6 × 6 × 2 1/2	3 3/8	3 3/8	4 11/16
6 × 6 × 2	3 1/16	3 1/16	4 9/16
6 × 5 × 5	4 5/8	4 1/2	5
6 × 5 × 4	4 1/8	4	4 15/16
6 × 4 × 6	5 1/8	4 15/16	5 1/8
6 × 4 × 4	4 1/8	3 13/16	4 15/16
6 × 3 × 6	5 1/8	4 3/4	5 1/8
6 × 2 × 6	5 1/8	4 9/16	5 1/8
8 × 8 × 6	5 9/16	5 9/16	6 3/8
8 × 8 × 5	5	5	6 1/4
8 × 8 × 4	4 1/2	4 1/2	6 3/16
8 × 8 × 3	4	4	6 1/8
8 × 8 × 2 1/2	3 11/16	3 11/16	6
8 × 8 × 2	3 7/16	3 7/16	5 13/16
8 × 6 × 8	6 9/16	6 3/8	6 9/16
8 × 6 × 6	5 9/16	5 1/8	6 3/8
10 × 10 × 8	7	7	7 7/8
10 × 10 × 6	6	6	7 11/16
10 × 10 × 4	4 15/16	4 15/16	7 1/2
12 × 12 × 8	7 7/16	7 7/16	9 1/4
12 × 12 × 6	6 7/16	6 7/16	8 1/2

All dimensions given in inches.

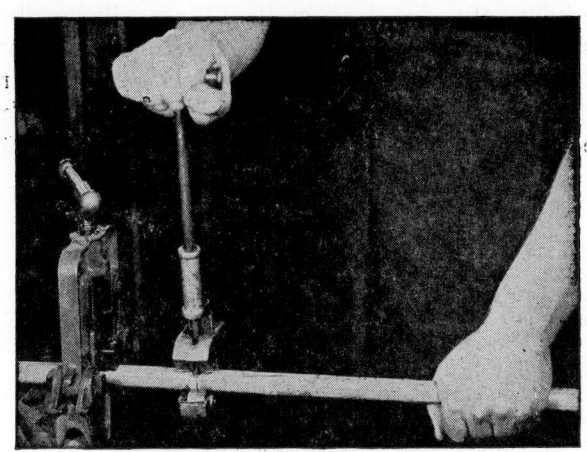

Cutting pipe with a pipe cutter.

TABLE 21

NORMAL ENGAGEMENT BETWEEN MALE AND FEMALE THREADS TO MAKE TIGHT JOINTS †

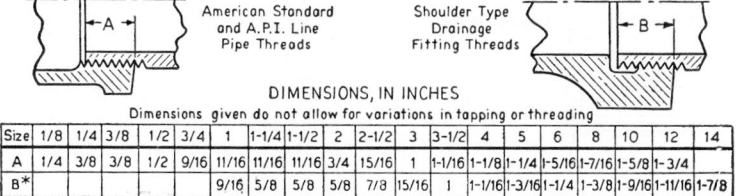

American Standard and A.P.I. Line Pipe Threads

Shoulder Type Drainage Fitting Threads

DIMENSIONS, IN INCHES

Dimensions given do not allow for variations in tapping or threading

Size	1/8	1/4	3/8	1/2	3/4	1	1-1/4	1-1/2	2	2-1/2	3	3-1/2	4	5	6	8	10	12	14
A	1/4	3/8	3/8	1/2	9/16	11/16	11/16	11/16	3/4	15/16		1-1/16	1-1/8	1-1/4	1-5/16	1-7/16	1-5/8	1-3/4	
B*						9/16	5/8	5/8	5/8	7/8	15/16	1	1-1/16	1-3/16	1-1/4	1-3/8	1-9/16	1-11/16	1-7/8

*Using American Standard Taper Male Thread with Crane Shoulder Type Drainage Fittings. The external thread, however, should not be threaded small to gage and not more than one turn large.

† Crane Co., Chicago, Illinois.

Measuring Pipes.—When making a piping installation it is important that the pipes be cut to the proper lengths to insure obtaining proper slopes, locations of the fittings, and to eliminate the possibility of undue strain on the fittings. Piping drawings, as shown in Fig. 28, give the sizes of pipes and fittings and the distances from center to center of fittings and pipes. Determining the lengths of pipes to be cut from these center-line dimensions is done by applying the dimensions of the pipe fittings and the

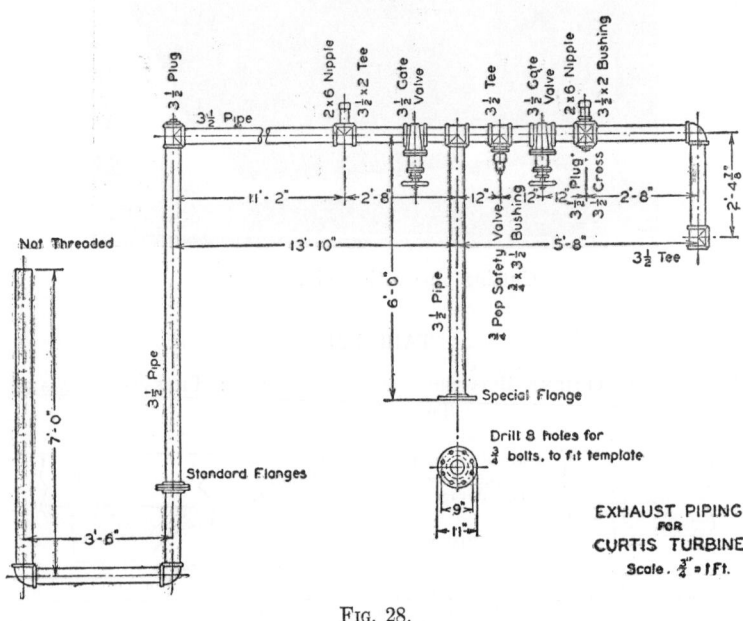

Fig. 28.

length of thread as given in the preceding tables. This is illustrated in Fig. 29 where D is the center-to-center distance between pipes, A_1 and A_2 the dimensions of screwed fittings as given in Tables 18 to 20, E the length of thread as given in Table 14, and L the desired length of pipe. Then,

$$L = D - (A_1 + A_2 - 2E)$$

ILLUSTRATION: If the pipe in Fig. 29 is $2\frac{1}{2}$-inch, the fittings

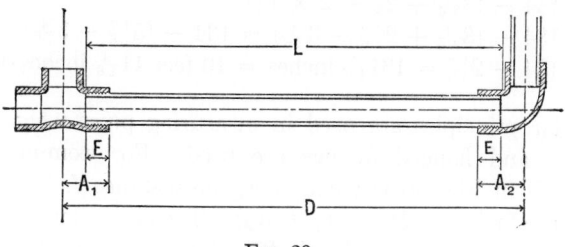

FIG. 29.

125-pound cast iron, and the distance D is 8 feet 6 inches, what length L should the pipe be cut?

$D = 8 \times 12 + 6 = 102$ inches
$A_1 = 2\frac{11}{16}$ inches. From Table 18
$A_2 = 2\frac{11}{16}$ inches. From Table 18
$E = 1\frac{1}{2}$ inches. From Table 14

Then,

$L = D - (A_1 + A_2 - 2E)$
$L = 102 - (2\frac{11}{16} + 2\frac{11}{16} - 2 \times 1\frac{1}{2})$
$L = 102 - (5\frac{3}{8} - 3)$
$L = 102 - 2\frac{3}{8} = 99\frac{5}{8}$ inches $= 8$ feet $3\frac{5}{8}$ inches (Ans.)

ILLUSTRATION: What is the actual length of the pipe in Fig. 28 situated between the two tees whose center-to-center distance is 11 feet 2 inches? The pipe is standard wrought and the fittings are 125-pound cast iron.

In this problem

$D = 11 \times 12 + 2 = 134$ inches
$A_1 = 3\frac{7}{16}$ inches. From Table 18
$A_2 = 2\frac{5}{8}$ inches. From Table 18
$E = 1\frac{5}{8}$ inches. From Table 14

Then,

$$L = D - (A_1 + A_2 - 2E)$$
$$L = 134 - (3\tfrac{7}{16} + 2\tfrac{5}{8} - 2 \times 1\tfrac{5}{8})$$
$$L = 134 - (3\tfrac{7}{16} + 2\tfrac{10}{16} - 3\tfrac{4}{16}) = 134 - (5\tfrac{17}{16} - 3\tfrac{4}{16})$$
$$L = 134 - 2\tfrac{13}{16} = 131\tfrac{3}{16} \text{ inches} = 10 \text{ feet } 11\tfrac{3}{16} \text{ inches (Ans.)}$$

Similar principles are used in measuring pipes when flanged couplings and flanged fittings are used. For example, if the distance D in Fig. 30 is fixed, then the distance L between the faces of the fittings is $D - (A_1 + A_2)$. The dimensions A_1 and A_2 may both be found from Table 27. Then, if the pipe B is cut $\frac{1}{4}$ inch shorter than the distance L, that quarter inch may be distributed as follows: $\frac{1}{16}$ inch clearance between each end of the pipe and the face of its screwed flange, and $\frac{1}{16}$ inch space for each of two gaskets. There is no substitute for experience and judgment in making the proper allowances for clearances.

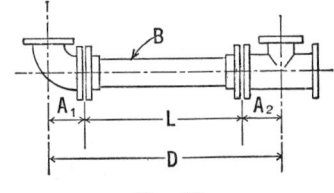

Fig. 30.

ILLUSTRATION: If the pipe shown in Fig. 30 has a nominal diameter of 6 inches and the fitting to the left is a long radius elbow, what is the length L if the center-to-center distance D is 7 feet 9 inches?

$$D = 7 \times 12 + 9 = 93 \text{ inches}$$
$A_1 = 11\frac{1}{2}$ inches.   "B" for 6-in. pipe in Table 27
$A_2 = 8$ inches.   "A" for 6-in. pipe in Table 27

Then,

$$L = D - (A_1 + A_2)$$
$$L = 93 - (11\tfrac{1}{2} + 8)$$
$$L = 93 - 19\tfrac{1}{2} = 73\tfrac{1}{2} \text{ inches} = 6 \text{ feet } 1\tfrac{1}{2} \text{ inches (Ans.)}$$

The principles are again applied to the measurement of bell-and-spigot type cast-iron water pipe.

Illustration: What is the length L of the pipe in Fig. 31 if the distance D is 9 feet 3 inches and the pipe and fittings are standard cast-iron water pipe of a nominal diameter of 6 inches?

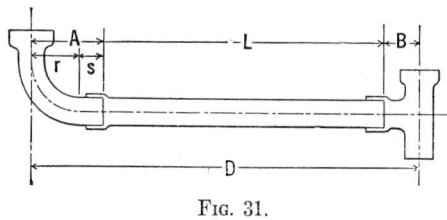

Fig. 31.

From the figure, $L = D - (A + B)$. A is made up of the two dimensions r and s. From standard references, these turn out to be 16 inches and 24 inches, respectively; and B is found to be 12 inches. Then we may write

$$L = D - (A + B)$$
$$L = 9 \times 12 - (16 + 24 + 12)$$
$$L = 108 - 52 = 56 \text{ inches} = 4 \text{ feet } 8 \text{ inches (Ans.)}$$

Measuring Diagonal Pipe.—If two offset pipes are to be con-connected by a diagonal pipe and the angle of the fittings and one of the dimensions A, B, or C (Fig. 32) are known, the other dimensions may readily be found. Without going into the principles of trigonometry back of it we offer Table 22 as a short-cut to these calculations. The table applies equally well to offsets from Y-connections. The numbers in the table are calculated from the trigonometry of the triangle A, B, C.

Knowing the angle of the fitting and the length of either A or B, the other dimensions can be found by multiplying the known length by the proper figure in the table.

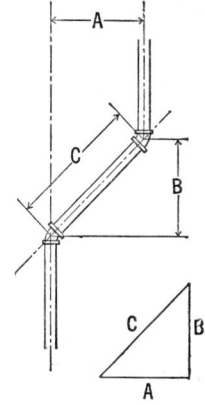

Fig. 32.

ILLUSTRATION: If the fittings in Fig. 32 are $22\frac{1}{2}$-degree elbows and the offset A is 2 feet 6 inches, what are the lengths of B and C? A is then $2 \times 12 + 6 = 30$ inches.

TABLE 22

Angle of Fittings (See Fig. 32)		Length of B when $A = 1$	Length of A when $B = 1$	Length of C when $A = 1$	Length of C when $B = 1$	Length of A when $C = 1$	Length of B when $C = 1$
$\frac{1}{64}$ curve	$5\frac{5}{8}°$	10.1531	0.0985	10.2033	1.0048	0.098	0.9952
$\frac{1}{32}$ curve	$11\frac{1}{4}°$	5.0273	0.1989	5.1258	1.0196	0.1951	0.9809
$\frac{1}{16}$ curve	$22\frac{1}{2}°$	2.4142	0.4142	2.6131	1.0828	0.3826	0.9239
$\frac{1}{12}$ curve	$30°$	1.7320	0.5773	2.0000	1.1547	0.5000	0.866
$\frac{1}{8}$ curve	$45°$	1.0000	1.0000	1.4142	1.4142	0.7071	0.7071
$\frac{1}{6}$ curve	$60°$	0.5773	1.7320	1.1547	2.0000	0.866	0.5000
$\frac{3}{16}$ curve	$67\frac{1}{2}°$	0.4142	2.4142	1.0824	2.6131	0.9239	0.3826

In column 3 of Table 22 opposite $22\frac{1}{2}$ degrees we find the factor 2.4142. Then

$$B = 30 \times 2.4142 = 76.43 \text{ in.} = 6 \text{ ft. } 0\frac{7}{16} \text{ in.}$$

(Use Table 1, page 19, for conversion.)   (Ans.)

Similarly, from column 5,

$$C = 30 \times 2.6131 = 78.393 \text{ in.} = 6 \text{ ft. } 6\frac{3}{8} \text{ in.}   \text{(Ans.)}$$

Having found the length of C, the actual length of the pipe may be found by the method of the preceding paragraphs.

ILLUSTRATION: If the fittings in Fig. 32 are 60-degree elbows and B is 10 inches, what are the lengths A and C?

From column 4 of Table 22,

$$A = 10 \times 1.7320 = 17.320 \text{ in.} = 17\frac{5}{16} \text{ in.}   \text{(Ans.)}$$

Then, from column 6 of Table 22,

$$C = 10 \times 2 = 20 \text{ in.}   \text{(Ans.)}$$

Copper Tubing.—A plumbing material that has come into widespread use recently is copper tubing. Although copper tubes with flanged fittings have been used in automotive, gasoline, and oil lines for many years, and copper water tubing has been used successfully in Europe and other parts of the world, its use in plumbing systems in the United States has been comparatively recent.

One of the main reasons for the popularity of copper tubing is its ease of installation. It is soft enough to be bent easily around obstructions, and it can be run between studding and flooring beams without the use of fittings. Another advantage is that there is no weakening of the tubing at the joint as in threaded pipe, where the original thickness of the pipe is reduced. Copper tubes with sweated or flanged fittings can be much thinner than threaded pipe without being weak at the joints.

Copper tubing is also rigid enough to be strung up in long lengths without undue sagging, if hard temper tubing is employed. It can withstand the rough usage that pipes usually get when being installed. Under normal conditions, it is resistant to corrosion. Hard-drawn copper tubes, like iron pipe, can be damaged by freezing water. Soft or annealed copper tubes, however, resist the bursting pressure caused by freezing water.

Because of its flexibility and ductility, soft or annealed copper tubing is especially suitable for underground water service. In such installations, there is always a certain amount of settlement of the ground, and the flexible tubing readily accommodates itself to any ground movements. If used under normal conditions, and if installed with ordinary skill and care, copper tubes can last the life of the building in which they are installed.

There are two widely used types of copper tubing, and these are classified by wall thicknesses. Type L (light) is furnished in either hard or soft temper for general plumbing and heating applications. Type K (heavy) is available in either hard or soft temper and is used for underground as well as for general plumbing and heating purposes. The hard tubes are usually installed with solder fittings, while the soft tubes can be installed with solder or flared fittings.

Type K and Type L hard temper tubes are furnished in straight lengths of 20 ft. The soft temper tubes are available in 20-ft. lengths, and for sizes up to 1¼ in. inclusive, in coils of 60 ft. A thinner copper tube known as Type M also is available,

TABLE 23

SIZES AND WEIGHTS OF COPPER WATER TUBE

Standard Water Tube Size	Actual Outside Diameter	Nominal Wall Thickness			Theoretical Weight		
		Type K	Type L	Type M	Type K	Type L	Type M
Inches	Inches	Inches	Inches	Inches	Lb./Ft.	Lb./Ft.	Lb./Ft.
⅜	0.500	0.049	0.035	.....	0.269	0.198	
½	.625	.049	.040	.....	.344	.285	.....
⅝	.750	.049	.042	.....	.418	.362	.....
¾	.875	.065	.045	.....	.641	.455	.....
1	1.125	.065	.050	.....	.839	.655	.....
1¼	1.375	.065	.055	.....	1.04	.884	.....
1½	1.625	.072	.060	.....	1.36	.114	.....
2	2.125	.083	.070	.....	2.06	1.75	.....
2½	2.625	.095	.080	0.065	2.93	2.48	2.03
3	3.125	.109	.090	.072	4.00	3.33	2.68
3½	3.625	.120	.100	.083	5.12	4.29	3.58
4	4.125	.134	.110	.095	6.51	5.38	4.66
5	5.125	.160	.125	.109	9.67	7.61	6.66
6	6.125	.192.	.140	.122	13.9	10.2	8.92
8	8.125	.271	.200	.170	25.9	19.3	16.5
10	10.125	.338	.250	.212	40.3	30.1	25.6
12	12.125	.405	.280	.254	57.8	40.4	36.7

but only in sizes above 2 in. Table 23 shows the sizes and weights of copper water tubing for these three types.

TABLE 24

SIZES OF COPPER TUBE WATER SUPPLY BY SHORT BRANCHES TO PLUMBING FIXTURES *

Fixture	Pressures		
	High Over 60 lbs.	Medium 30 to 60 lbs	Low Under 30 lbs.
	Inch	Inch	Inch
To Baths	½	¾	¾
Lavatories	⅜	½	½
Tank Closets	⅜	⅜	½
Valve Closets	1	1	1¼
Pantry Sinks	½	½	½
Kitchen Sinks	½	½	¾
Slop Sinks	½	¾	¾
Showers	½	½	¾
Urinals (Flush Tank)	½	¾	¾
Urinals (Flush Valve)	¾	¾	¾
Drinking Fountains	⅜	⅜	½

* *Copper Tube,* T. N. Thomson, Copper & Brass Research Assn., 1949, p. 24.

TABLE 25

RELATIVE SIZES OF BRANCHES AND WATER MAINS *

Size of Main	Number & Size of Branches Main Will Supply—Running Full
½″	Two—⅜″
¾″	Two—½″
1″	Two—¾″
1¼″	Two—1″ *or* One—1″ and Two—¾″
1½″	Two—1¼″ *or* One—1¼″ and Two ¾″
2″	Two—1½″ *or* One—1½″ and Two—1¼″
2½″	Two—1½″ and Two—1¼″ *or* One—2″ and Two—1¼″
3″	One—2½″ and One—2″ *or* Two—2″ and Two—1½″
3½″	Two—2½″ *or* One—3″ and One—2″ *or* Four—2″
4″	One—3½″ and One—2½″ *or* Two—3″ *or* Three—2½″ and One—2″ *or* Six—2″

* *Copper Tube,* T. N. Thomson, Copper & Brass Research Assn., 1949, p. 24.

The sizes of tubes commonly used as branches to fixtures vary with the pressures at the fixtures and are shown in Table 24. Table 25 is used to determine the size of a main that will service a given number and size of branches.

Copper Tube Fittings.—A wide variety of fittings for copper tubing are available. They come in practically every form in which screwed fittings are supplied. There are also available

Fig. 33.—Soldered Type Fitting for Copper Tube.

special adapters for joining copper tube to threaded pipe of iron pipe size. A typical soldered type fitting is shown in Fig. 33 and a flanged type fitting appears in Fig. 34.

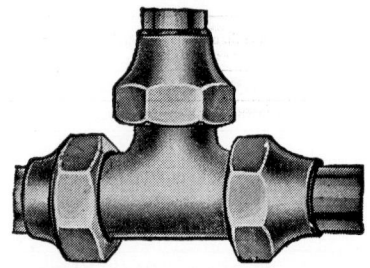

Fig. 34.—Flanged Type Fitting for Copper Tube.

Hot Water Supply.—There are two main ways of providing hot water for small dwellings or apartments. In one system, water is heated directly in a separate unit by gas, coal, oil, or electricity, and then is stored in a hot water storage tank. In

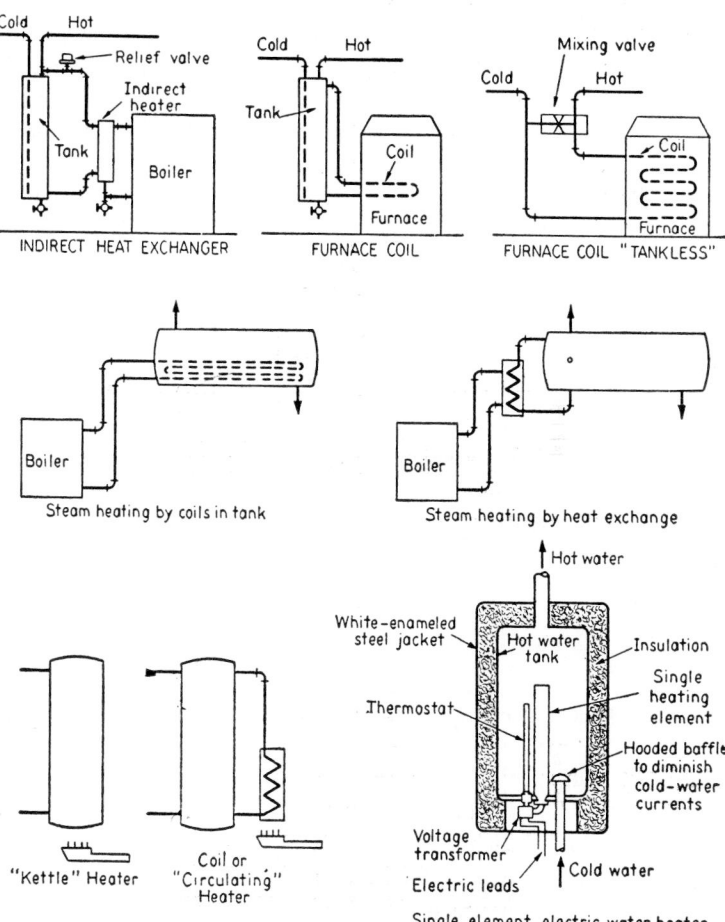

FIG. 35.—Various Methods of Obtaining Hot Water. At the top are shown three ways in which the boiler or furnace is used as the source of heat. In the center, steam is used as the source. At the bottom are shown three types of separate hot water heaters.

the second method, the heating system boiler is used to heat the water indirectly by contact heaters consisting of copper coils surrounded by the boiler water. This type of installation eliminates the need for a separate water heater, and in certain types of installations also eliminates the storage tank. Fig. 35 shows some typical methods for obtaining hot water.

Small tanks which stand vertically are called range boilers and have been standardized by the Division of Simplified Practice of the Department of Commerce to provide one shell tapping 6 inches from the top and one 6 inches up from the bottom (measurements to be made from the edge of the shell plate), and two tappings in the top and one in the bottom. All tappings are

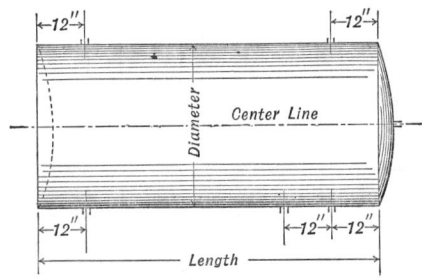

Fig. 36.—Standard Location of Openings for Hot Water Storage Tanks.

one inch. On special order, tanks with four side openings may be obtained. These are placed in line 6, 18, and 26 inches from the bottom and 6 inches from the top.

Range boilers are made in two classes, "standard" for 85 pounds working pressure, and "extra heavy" for 150 pounds working pressure. Dimensions are given in Table 26.

Storage tanks of larger capacity are mounted horizontally and have been standardized as to tappings as shown in Fig. 36. The standard dimensions of these tanks are also given in Table 26. They are made for a "standard" working pressure of 65 pounds and "extra heavy" of 100 pounds.

The hot-water-supply system may consist of one branch pipe leading to each fixture as shown in Fig. 5. In this case an interval of time elapses after opening a faucet before the hot water arrives and a certain amount of water is wasted. This can be obviated by providing a loop, as shown in Fig. 6, through which the hot water constantly circulates. However, in this system a great

TABLE 26

DIMENSIONS OF RANGE BOILERS AND HOT WATER STORAGE TANKS

Range Boilers			Storage tanks		
Diameter, inches	Length, inches	Capacity, gallons	Diameter, inches	Length, feet	Capacity, gallons
12	36	18	20	5	82
12	48	24	24	5	118
12	60	30	24	6	141
14	48	32	30	6	220
14	60	40	30	8	294
16	48	42	36	6	318
16	60	52	36	8	423
18	60	66	42	7	504
20	60	82	42	8	576
22	60	100	42	10	720
24	60	120	42	14	1008
24	72	144	48	10	940
24	96	192	48	16	1504
			48	20	1880

Diameters refer to inside measurements; lengths are mean lengths of sheets, not over-all length of tank.

amount of heat is lost and fuel wasted which overcomes the advantages.

The circulation of water in a loop or between the heater and the storage tank depends on the fact that water is slightly less in weight when hot than when cold. Therefore, the hot water tends to rise and the cold to sink, thus providing the circulation. However, a free circulation requires that the pipes be pitched properly, and humps or air pockets must be avoided.

In selecting a water heater it must be borne in mind that the capacity of the heater depends on the grate area and the efficiency with which the heat may be absorbed from the fuel by the water.

A certain type of heating unit known as an instantaneous or tankless heater is sometimes employed to furnish hot water. These units have large heat absorbing surfaces and large heaters which permit very rapid heating of water to meet peak demands without the need for a storage tank. They are always automatic in operation and depend upon either the flow of water or a thermostatic control to maintain desired water temperature.

A mixing valve may be placed in the hot water output line to maintain constant water temperature during the heating season when the boiler temperature may be higher than is required for domestic hot water. This valve mixes cold water with the hot water to give the desired temperature.

TABLE 27 *

Cost Comparison for Water Heaters on Yearly Basis

	COAL	OIL	GAS Natural	GAS Manufactured	GAS L.P. (Bottled)	ELECTRICITY
Fuel Used Per Year	3600 pounds	270 gallons	32,450 cubic feet	61,800 cubic feet	1475 pounds	6658 kwh
Cost @	$ 9/ton –$16.20 $12/ton – 21.60 $15/ton – 27.00 $18/ton – 32.40 $21/ton – 37.80	10¢/gal. – $27.00 11¢/gal. – 29.70 12¢/gal. – 32.40 13¢/gal. – 35.10 14¢/gal. – 37.80	5¢/100 cu. ft. –$16.22 6¢/100 cu. ft. – 19.46 7¢/100 cu. ft. – 22.70 8¢/100 cu. ft. – 25.94 9¢/100 cu. ft. – 29.18	7¢/100 cu. ft. –$43.26 8¢/100 cu. ft. – 49.44 9¢/100 cu. ft. – 55.62 10¢/100 cu. ft. – 61.80 11¢/100 cu. ft. – 67.98	5¢/lb. –$ 73.75 6¢/lb. – 88.50 7¢/lb. – 103.25 8¢/lb. – 118.00 9¢/lb. – 132.75	1½¢/kwh –$ 99.87 2¢ /kwh – 133.16 2½¢/kwh – 166.45 3¢ /kwh – 199.74 3½¢/kwh – 233.03
Efficiency Assumed	50%	60%	70%	70%	70%	100%
Fuel Value	12,500 Btu/lb.	140,000 Btu/gal.	1000Btu/cu. ft.†	525 Btu/cu. ft.	22,000 Btu/lb.	3412 Btu/kwh

University of Illinois Small Homes Council Circular G5.0.
Computations based on 50 gal. per day, 100° F. temperature rise, and 50% stand-by and piping loss. Efficiency of burners indicated above by fuel.

† A therm is equal to 100,000 Btu.

The cost of obtaining hot water varies widely depending upon geographical location, fuel cost, and type of heating system employed. Table 27 shows the annual cost for operating water heaters using a variety of fuels (heaters not connected to the house system). It can be seen that the most economical type of heater depends upon the most economical fuel for a given part of the country. No one fuel is the "best" for the job.

Although electric water heaters have the greatest thermal efficiency, this advantage is offset by the greater cost of electricity. In certain areas this is compensated for by obtaining favorable electric rates for water heating. Another method of reducing the cost of operating electric water heaters is to have them controlled to receive current only during certain periods of the day or night when the demand for electric power from the public utility is normally at a minimum.

A large, well-insulated storage tank is needed with this arrangement to provide hot water between periods when electric power is not available.

Drainage Plumbing.—Drainage plumbing, using the term in a broad sense, consists of the waste pipes which carry the used water not containing human excrement from such fixtures as sinks, wash basins, etc., soil pipes which carry the wastes from water-closets and urinals, vent pipes which admit air to the system, and traps which prevent the foul air in the pipes from entering the house. A vertical drainage pipe is known as a "stack." Fig. 37 illustrates the elements of the drainage plumbing for a one-family house.

The physics of drainage plumbing is rather complicated, but recommendations based on experimental work done largely by the Bureau of Standards are easy to understand.

Fixture Units.—In order to compare the discharges of various fixtures for determining the sizes of traps and pipes the so-called *fixture unit* has been devised. This unit is equivalent to a discharge of about 7.5 gallons per minute. The rate of discharge for various fixtures in terms of fixture units is given in Table 28.

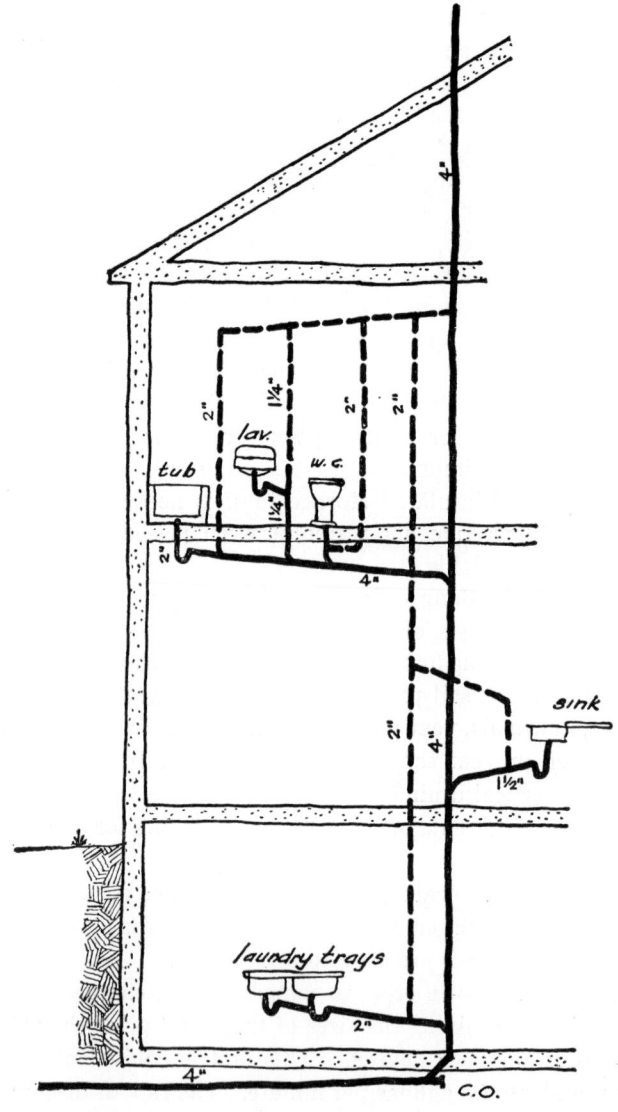

FIG. 37.—Sketch of Typical Drainage Plumbing for a One-family House.

TABLE 28

RATE OF DISCHARGE OF PLUMBING FIXTURES IN FIXTURE UNITS

Fixture	Units	Fixture	Units
One lavatory or wash basin..	1	One urinal................	3
One kitchen sink...........	1½	One floor drain............	3
One bathtub...............	2	One shower bath...........	3
One laundry tray...........	3	One slop sink.............	4
One combination fixture.....	3	One water-closet...........	6

One bathroom group consisting of one water-closet, one lavatory, and one bathtub and overhead shower; or one water-closet, one lavatory, and one shower compartment is regarded as having a combined discharge of eight fixture units. One hundred eighty square feet of roof or drained area in horizontal projection counts as one fixture unit.

Capacities of Vertical and Horizontal Drains.—The capacity of vertical soil stacks depending on the type of inlet fittings has been determined experimentally as given in Table 29. It is evi-

TABLE 29

CAPACITY OF SOIL STACKS IN FIXTURE UNITS

Diameter, inches	Single or double sanitary T fittings	Single or double Y, combination Y, and one-eighth bend fittings
2	6	12
3	13.5	27
4	24	48

dent from this table that a three-inch soil stack is adequate for any ordinary dwelling. It also emphasizes the effect of type of inlet fixture on the capacity of the stack. These figures presume, of course, that the outlet at the bottom is clear and of sufficient capacity to carry off the discharge without backing it up into the soil stack.

The capacity of horizontal drains depends on the slope as well as the size of pipes. Slopes flatter than one-quarter inch fall per foot are not recommended. Table 30 gives capacities of horizontal drains.

TABLE 30

CAPACITIES OF HORIZONTAL DRAINS IN FIXTURE UNITS

Diameter of drain, inches	Slope, $\frac{1}{8}$-in. fall per foot	Slope, $\frac{1}{4}$-in. fall per foot	Slope, $\frac{1}{2}$-in. fall per foot	Diameter of drain, inches	Slope, $\frac{1}{8}$-in. fall per foot	Slope, $\frac{1}{4}$-in. fall per foot	Slope, $\frac{1}{2}$-in. fall per foot
3	15	18	21	8	990	1,392	2,220
4	84	96	114	10	1,800	2,520	3,900
5	162	216	264	12	3,084	4,320	6,912
6	300	450	600				

Traps.—Good practice and most plumbing codes provide that each fixture must have an individual trap except that laundry trays may have a common trap. In general, these traps must provide a seal of at least one inch under all operating conditions. The minimum trap diameters and drain sizes for various fixtures are given in Table 31. Class 1 applies to private installations, residences, apartments, etc.; class 2 applies to semipublic installations, office buildings, factories, dormitories, etc.; and class 3 applies to public installations, schools, railroad stations, public comfort stations, etc.

Vent Pipes.—The main purpose of vents in plumbing systems is to release the suction which results when water flows through the drainage pipes and thus prevents the water seals in the traps from being siphoned out. Common practice is to make the vent a continuation of the soil stack as shown in Fig. 37. Most building codes require that any fixture more than five feet from the

TABLE 31

Minimum Trap Diameters, Minimum Drain Sizes, and Fixture Unit Values

Fixture and class of installation	Minimum nominal trap diameter, inches	Minimum nominal diameter, inches, individual drain	Fixture units
1 lavatory or washbasin, class 1........	$1\frac{1}{4}$	$1\frac{1}{4}$	1
1 lavatory or washbasin, class 2 or 3....	$1\frac{1}{4}$	$1\frac{1}{4}$	2
1 water-closet, class 1.................	3	3	3
1 water-closet, class 2.................	3	3	5
1 water-closet, class 3.................	3	3	6
1 bathtub, class 1....................	$1\frac{1}{2}$	$1\frac{1}{2}$	3
1 bathtub, class 2 or 3...............	2	2	4
1 shower stall, shower head only, class 1.	$1\frac{1}{2}$	$1\frac{1}{2}$	2
1 shower stall, multiple spray, class 1...	2	2	4
1 shower stall, head only, class 2 or 3...	2	2	3
1 shower stall, multiple spray, class 2 or 3.	3	3	6
1 urinal, lip, or each 2 feet of trough or gutter............................	$1\frac{1}{2}$	$1\frac{1}{2}$	2
1 urinal, stall or wall hung with tank or flush-valve supply.................	2	2	4
1 urinal, pedestal or blow out..........	3	3	5

soil stack must have a separate vent. Figure 38 illustrates good and bad practice in such venting. The line xy is the hydraulic gradient when the bowl is full and $x'y$ the gradient when the bowl is almost empty. The vent connection should come above the line xy to prevent back-flow into the vent pipe. Figure 39 illustrates types of plumbing installations including venting recommended by the U. S. Department of Commerce.

The size of vent required depends on its length and the load on the soil stack. Experiments have shown that for a three-inch soil stack carrying a capacity load of 40 fixture units a 2-inch vent 80 feet long or a $1\frac{1}{2}$-inch vent 20 feet long is satisfactory.

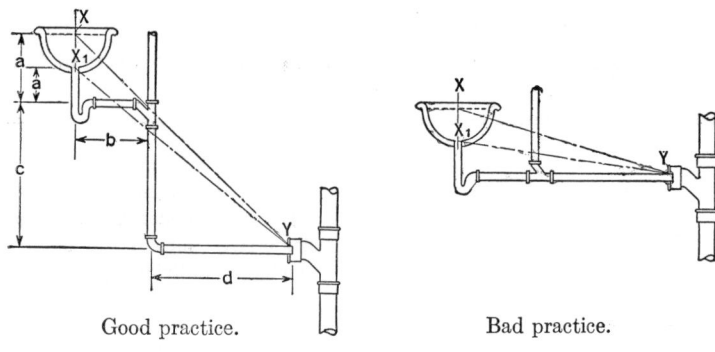

Good practice. Bad practice.

Fig. 38.

The use of a vent stack less than $1\frac{1}{4}$ inches in diameter is not recommended. Table 32 gives size and length of main vents for variously loaded soil stacks.

TABLE 32

Size and Length of Main Vents

Diameter of Soil or Waste Stack (Inches)	Number of Fixture Units on Soil or Waste Stack	Maximum Permissible Developed Length of Vent								
		$1\frac{1}{4}$-in. vent	$1\frac{1}{2}$-in. vent	2-in. vent	$2\frac{1}{2}$-in. vent	3-in. vent	4-in. vent	5-in. vent	6-in. vent	8-in. vent
		Feet	Feet	Feet	Feet	Feet	Feet	Feet	Feet	Feet
$1\frac{1}{4}$	2	75								
$1\frac{1}{2}$	8	70	150							
2	24	28	70	300						
3	40	..	20	80	260	650				
3	80	..	18	75	240	600				
4	310	..	..	30	95	240	1000			
4	620	..	..	22	70	180	750			
5	750	..	..	..	28	70	320	1000		
5	1500	..	..	..	20	50	240	750		
6	1440	..	..	..	..	20	95	240	1000	
6	2880	..	..	..	..	18	70	180	750	
8	3100	..	..	..	..	..	30	80	350	1100
8	6200	..	..	..	..	..	25	60	250	800

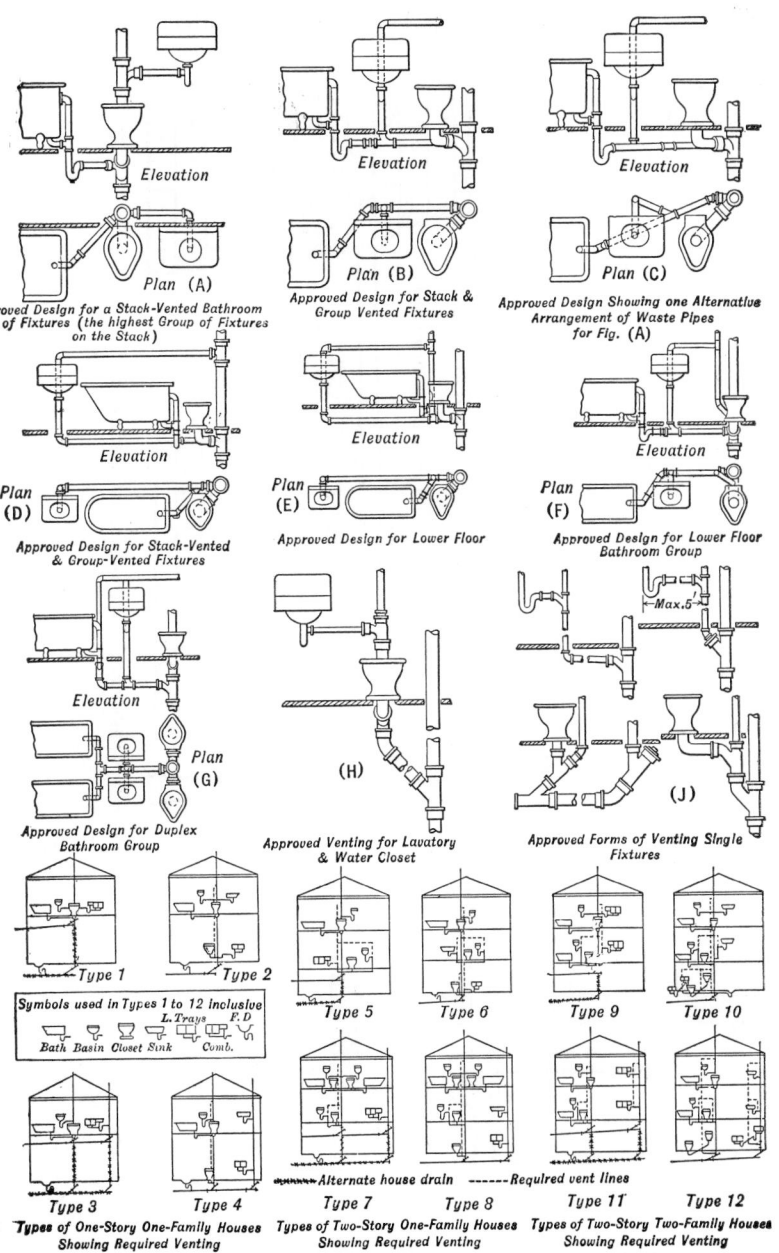

Elevation

Plan (A)

Approved Design for a Stack-Vented Bathroom Group of Fixtures (the highest Group of Fixtures on the Stack)

Elevation

Plan (B)

Approved Design for Stack & Group Vented Fixtures

Elevation

Plan (C)

Approved Design Showing one Alternative Arrangement of Waste Pipes for Fig. (A)

Elevation

Plan (D)

Approved Design for Stack-Vented & Group-Vented Fixtures

Elevation

Plan (E)

Approved Design for Lower Floor

Elevation

Plan (F)

Approved Design for Lower Floor Bathroom Group

Elevation

Plan (G)

Approved Design for Duplex Bathroom Group

(H)

Approved Venting for Lavatory & Water Closet

Max.5'

(J)

Approved Forms of Venting Single Fixtures

Type 1   Type 2

Symbols used in Types 1 to 12 inclusive
L. Trays   F. D
Bath Basin Closet Sink   Comb.

Type 3   Type 4

Types of One-Story One-Family Houses Showing Required Venting

Type 5   Type 6

Type 7   Type 8

Alternate house drain   ------ Required vent lines

Types of Two-Story One-Family Houses Showing Required Venting

Type 9   Type 10

Type 11   Type 12

Types of Two-Story Two-Family Houses Showing Required Venting

FIG. 39.—Plumbing Installations Recommended by "Hoover Report."

485

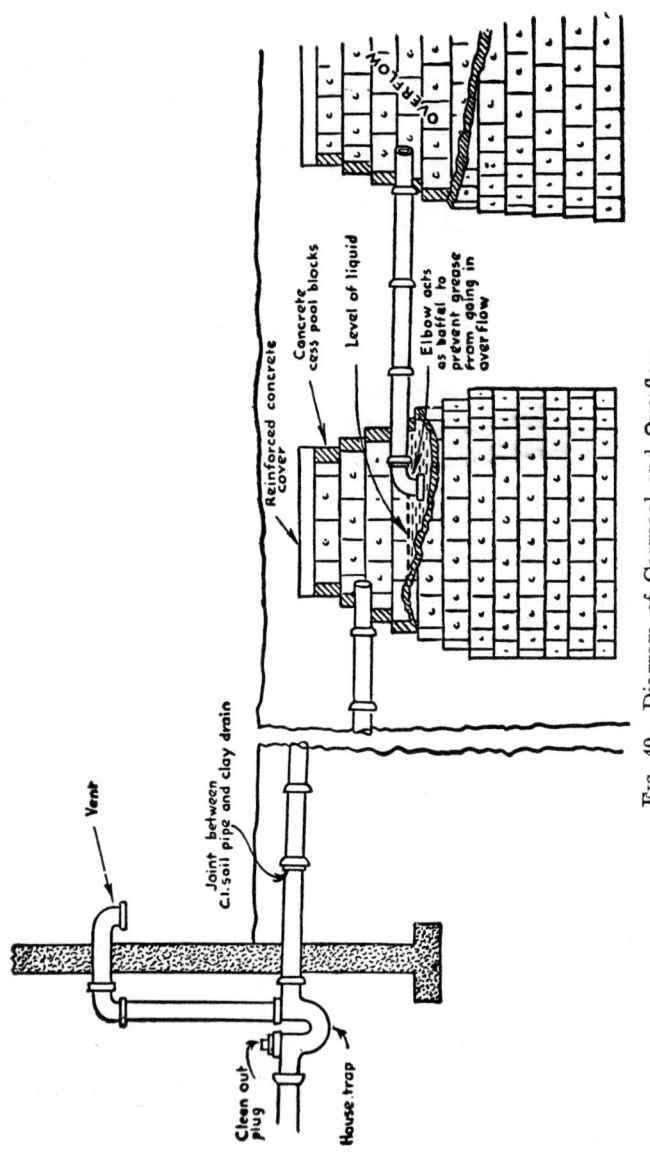

Fig. 40.—Diagram of Cesspool and Overflow.

Essential Parts

1. HOUSE SEWER — Pipe line which carries household wastes from the house to the septic tank.

2. SEPTIC TANK — A water-tight container in which sewage is disintegrated by bacterial action. (The enlarged view of the single-chamber tank shows how sewage decomposes.)

3. OUTLET SEWER LINE — Tile line which carries the liquid wastes from the septic tank to the disposal area.

4. DISTRIBUTION BOX (Needed only for multiple disposal lines) — A small box, with outlets, which discharges an equal amount of the liquid wastes simultaneously into several disposal lines.

5. DISPOSAL LINE — Drain-tile lines which allow the liquid wastes to seep into the soil through open joints of tile.

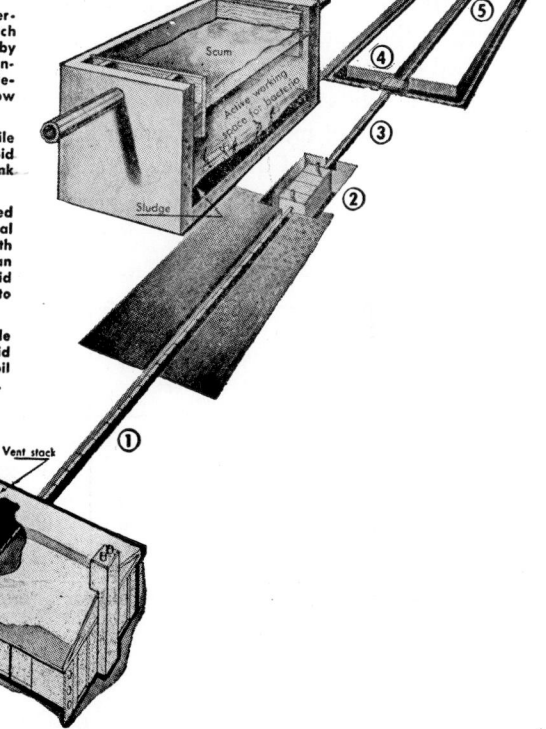

Fig. 41.—A Typical Septic Tank System. (*Univ. of Illinois Small Homes Council Circular* G5.5.)

Cesspools.—When access to public sewer systems is not available, some means must be provided to dispose of wastes. One common method is the use of a "leeching" cesspool, shown in Fig. 40. This is a pit in the ground, about 15 or 20 ft. from the house, made of stone, wood, brick, or concrete blocks, whose function is to drain liquid wastes through its walls into the surrounding ground. It slowly fills with solid matter that must eventually be removed.

Mortar is not used in the construction of the cesspool, since openings must be left for drainage purposes. A jacket of sand around the cesspool to filter the wastes is desirable.

Septic Tanks.—A more satisfactory method for disposing of sewage is the septic tank, shown in Fig. 41. This is a specially designed water-tight tank in which natural bacterial action changes the solids to liquid which then are easily drained. Septic tanks are made of metal, brick, concrete, or masonry. Table 33 lists the capacity and dimensions of single chamber rectangular septic tanks.

TABLE 33‡

CAPACITY AND DIMENSIONS FOR RECTANGULAR SEPTIC TANKS

No. of people	Sewage flow per person per day	Capacity* for 72-hour retention		Effective cross-section		Length*
	gals.	gals.	cu. ft.	width	depth†	
7 (or less)	26 —	540	72	3' x	4'	6' 0"
9	23 +	630	84	3' x	4'	7' 0"
12	20	720	96	3' x	4'	8' 0"

‡ *University of Illinois Small Homes Council Circular* G5.5.

* Capacity and length apply to single-chamber tank and first chamber of double-chamber tank. Capacity and length of second chamber are half that of the first.

† Depth is from the bottom of tank to the outlet tile.

Plumbing Fixtures.—A discussion of plumbing fixtures and other appurtenances is beyond the scope of this book. Such information, together with roughing-in dimensions, is available from fixture manufacturers, as well as in books on plumbing and pipe-fitting.

XVI

Heating

Heat and Temperature.—The problem of the man devising or constructing a heating system is to transfer the heat energy of a fuel into the air of the building as efficiently as possible. A knowledge of some of the basic properties of heat is extremely helpful in understanding the principles which govern the operation of house heating plants, and particularly in diagnosing the trouble when these fail to function properly. A physicist would define heat as "molecular energy" but we are not as much concerned with its definition as we are with the effects it produces.

Intensity of heat is measured in terms of degrees Fahrenheit, the freezing point of water being 32° F. and the boiling point 212° F.

Quantity of heat is measured by the British thermal unit (Btu) and one Btu is that quantity of heat which will raise the temperature of one pound of water one degree Fahrenheit.

Quantity of heat must not be confused with intensity. For example, a cupful of water at 150° F. will contain a *smaller* quantity of heat than a pailful of water at 70° F.

Effects of Heat on Fluids.—When air is heated it expands. When water above 39.2° F. is heated it also expands. Both of these substances are fluids. When fluids expand they become less dense, that is, they weigh less per cubic foot, and if they are free to move, the lighter fluid will rise to the top and the denser fluid will flow to the bottom to take its place. This principle is employed in hot air heating plants and hot water heating plants. In either case the lighter heated fluid rises and loses its heat in the rooms of a house and upon cooling becomes more dense and

489

descends to the heating plant. In such a system a pound of water going into a radiator at a temperature of 180° F. and coming out of it at a temperature of 90° F. has given up 90 Btu of heat.

If the temperature of a pound of water in a steam boiler under atmospheric pressure is 150° F. and 62 Btu of heat are added to it, the water will increase in temperature to about 212° F. Adding a small additional amount of heat to this water will neither increase the temperature of the water nor convert the whole pound of it to steam. As much as 970 Btu must be added to this pound of water at 212° F. to change it to steam at 212° F. This additional heat is called the *latent heat of vaporization*. In heating houses by steam, most of the heat in the rooms is derived from this latent heat of vaporization which is given up by the steam in the radiators in changing back to water.

Heat Transfer.—In a heating plant, for example, a hot water system, the heat from the fuel is transferred to the casing of the boiler, then to the water in the boiler, then to the water in the radiator, to the casing of the radiator, to the room which is being heated, then finally through the walls and windows to the outdoors where it is dissipated. There are three ways by which heat is transferred, by radiation, by conduction, and by convection.

Radiation.—Heat travels in direct rays from a source much the same as light does. This is best illustrated by the heat which comes from the sun or from a fire in an open fireplace. In either case if the direct rays are cut off by an object, a heat shadow is formed and the same intensity of the heat is not felt in the shadow.

Conduction.—If the end of an iron rod is heated, the heat will be transferred from one iron particle to the next until the heat has traveled the whole length of the rod. This is called conduction. Some materials conduct heat more readily than others. Copper is a particularly good conductor. Materials which are poor conductors, such as asbestos, and mineral wool, are used for insulation.

Convection.—Heat transfer by convection depends on the circulation of a fluid, the warmed particles of the fluid mingling with the unwarmed particles. Thus the circulation of warm air from a hot air furnace is an example of heat transfer by convection. So is also the circulation of water in a hot water heating system.

Estimating Heating Requirements.—When a public utility company builds an electric power plant, a gas plant, or a water supply system it must first estimate the probable *demand* which the consumer will place upon the service. The design of a heating plant is approached from much the same angle. First the heat *demand* of the building must be determined and then the radiators, pipes, boilers, etc., must be selected to satisfy this demand completely yet economically. The heat demand of a building depends on the following factors: *

1. Outside temperature.
2. Rain or snow.
3. Sunshine or cloudiness. } *Outside Conditions (The Weather)*
4. Wind velocity.
5. Heat transmission of exposed parts of buildings.
6. Infiltration of air through cracks, crevices, and open doors and windows.
7. Heat capacity of materials. } *Building Construction*
8. Rate of absorption of solar radiation of exposed material.
9. Inside temperatures.
10. Stratification.
11. Type of heating system.
12. Ventilation requirements. } *Inside Conditions*
13. Period and nature of occupancy.
14. Temperature regulation.

It will be noticed that many of these factors are variable and this leads to a great many combinations of circumstances. Values for many of these factors have been established by the American Society of Heating and Ventilating Engineers, the Heating and Piping Contractors National Association and university research bureaus so that the heat required by a room or a house in terms of Btu per hour can be set up in practically a single equation.

* From the Guide of A.S.H. & V.E.

Needless to say this equation is long and its solution tedious. The Heating and Piping Contractors National Association has therefore compiled a Standard Radiation Estimating Table which shortens the work materially. By the use of this table the heat requirements of a room may be translated directly into square feet of steam radiation (see below) without going into the intermediate step of estimating the number of Btu's required. Before describing this method of estimating, we shall discuss some of the factors entering into the estimate and define what is meant by radiation.

Temperature, Wind and Exposure.—The amount of heat lost from a room depends partly on the difference between the inside and the outside temperatures. The average outside temperature during the heating season varies, of course, with the locality. Experience has shown that periods of intense cold are generally of short duration so that the factor which is used as the base temperature in the calculations is several degrees higher.

Desirable inside temperatures have been fairly well standardized. These are listed in Table 1.

TABLE 1

INSIDE TEMPERATURES

Type of room or building	Temperature, degrees F.
Warm air baths	120
Steam baths	110
Hospital operating rooms	85
Bath rooms	80
Paint shops	80
Hospitals	72 to 75
Public buildings	68 to 72
Residences	70
Schools	70
Factories	65
Stores	65
Gymnasia	55 to 60
Machine shops	60 to 65
Foundries, boiler shops, etc.	50 to 60

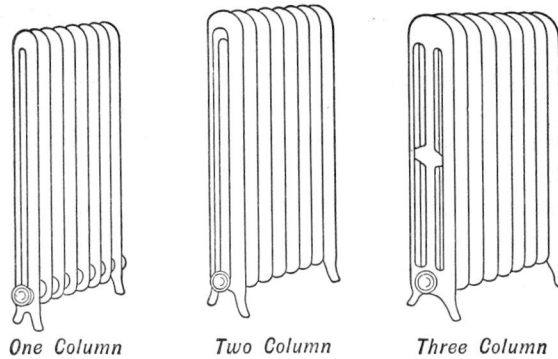

One Column   Two Column   Three Column

FIG. 1.—Typical Old-style Cast-iron Radiators No Longer Manufactured.
(*Courtesy American Oil Burner Association.*)

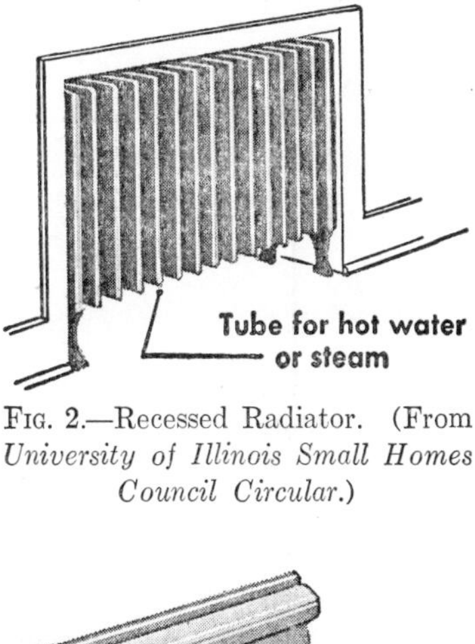

Tube for hot water or steam

Fins

Core

Cabinet

FIG. 2.—Recessed Radiator. (From *University of Illinois Small Homes Council Circular.*)

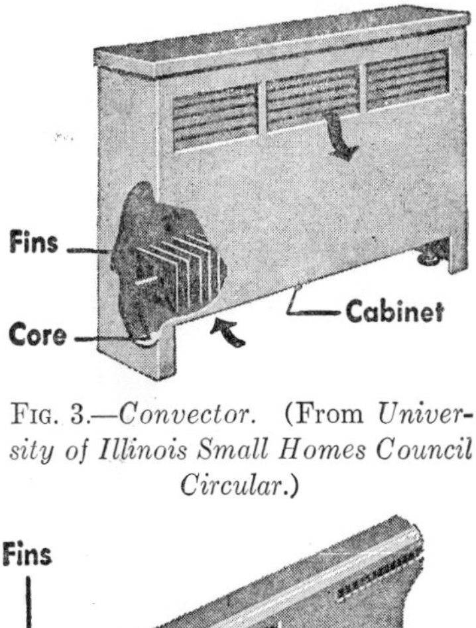

FIG. 3.—Convector. (From *University of Illinois Small Homes Council Circular.*)

Face

Channel for hot water or steam

Fins

Face

Channel for hot water or steam

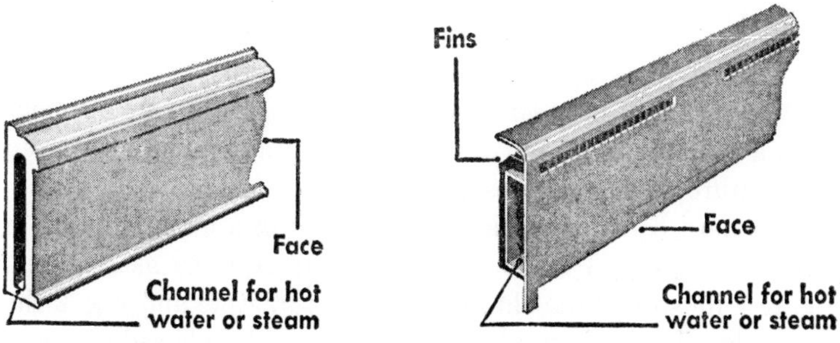

FIG. 4.—Hollow-type Baseboard Units. (From *University of Illinois Small Homes Council Circular.*)

Wind increases the loss of heat by transmission through walls and increases the infiltration through cracks. Most localities are subjected to prevailing winds of certain intensity during the winter months. The factors for base temperature and exposure to prevailing winds have been combined in a single tabulation in Table 16.

Radiation.—Radiators for steam and hot water systems are rated in square feet by the amount of heat they are capable of giving off. One square foot of radiation is equal to an emission of 240 Btu per hour when a radiator is filled with steam under standard conditions. However, radiators seldom operate under standard conditions and manufacturers' ratings sometimes vary, so that in actual practice an emission of only 225 Btu per square foot of rated area is counted. The tables which follow are made up on this basis.

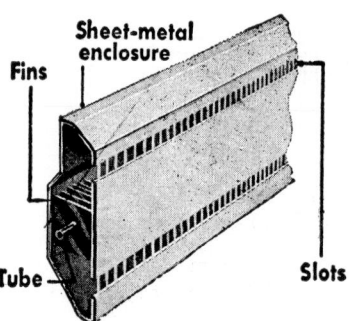

FIG. 5.—Finned-Tube Baseboard Unit. (From *University of Illinois Small Homes Council Circular.*)

A square foot of steam radiation gives off 150 Btu per hour when a radiator is used in a hot water heating system at a mean temperature of 170 degrees.

The most common radiators are of cast iron manufactured in several heights and made up of as many sections as required. Figure 1 illustrates the old style of such radiators and Fig. 2 the new style. Table 2 gives the rating of the old style radiator and the rating of the new style may be obtained from Table 3.

TABLE 2

Approximate Rating of Old-Style Radiators, Square Feet per Section

No. of Columns	Height in inches										
	13	15	16	18	20	22	23	26	32	38	45
1	.....	.....	.....	.....	1.5	.....	1.7	2.0	2.5	3.0	.....
2	.....	1.5	.....	.....	2.0	2.3	2.3	2.7	3.3	4.0	5.0
3	.....	.....	.....	2.3	.....	3.0	.....	3.8	4.5	5.0	6.0
4	.....	.....	.....	3.0	.....	4.0	.....	5.0	6.5	8.0	10.0
5	.....	.....	.....	4.7	.....	.....	.....	7.0	.....	10.0	.....
6	3.0	.....	3.8	4.5	5.0	.....	.....	.....	.....	.....	.....

ILLUSTRATION: A computation shows that a certain room will require 45 square feet of radiation and it is desired to use 3-column old-style radiators 32 inches high. How many sections will be required?

From Table 2 the radiating surface of one section of 3-column, 32-inch old-style radiator is 4.5 square feet. Then,

$$\frac{45}{4.5} = 10 \text{ sections required (Ans.)}$$

ILLUSTRATION: Another room requires 18 square feet of radiation and it is desired to use a small-tube cast-iron radiator 25 inches high. How many sections and tubes per section will be required?

The rating of a section depends upon the number of tubes per section used. From Table 3, the rating of a 25-inch high section is 1.6 square feet for 3 tubes per section, 2.0 square feet for 4 tubes per section, 2.4 square feet for 5 tubes per section, and 3.0 square

TABLE 3

SMALL-TUBE CAST-IRON RADIATORS *

NUMBER OF TUBES PER SECTION	CATALOG RATING PER SECTION[a]		SECTION DIMENSIONS				
			A Height[c]	*B* Width		*C* Spacing[b]	*D* Leg Height[c]
				Minimum	Maximum		
	Sq Ft	Btu/hr	In.	In.	In.	In.	In.
3[d]	1.6	384	25	3¼	3½	1¾	2½
4[d]	1.6	384	19	4 7/16	4 13/16	1¾	2½
	1.8	432	22	4 7/16	4 13/16	1¾	2½
	2.0	480	25	4 7/16	4 13/16	1¾	2½
5[d]	2.1	504	22	5 5/8	6 5/16	1¾	2½
	2.4	576	25	5 5/8	6 5/16	1¾	2½
6[d]	2.3	552	19	6 13/16	8	1¾	2½
	3.0	720	25	6 13/16	8	1¾	2½
	3.7	888	32	6 13/16	8	1¾	2½

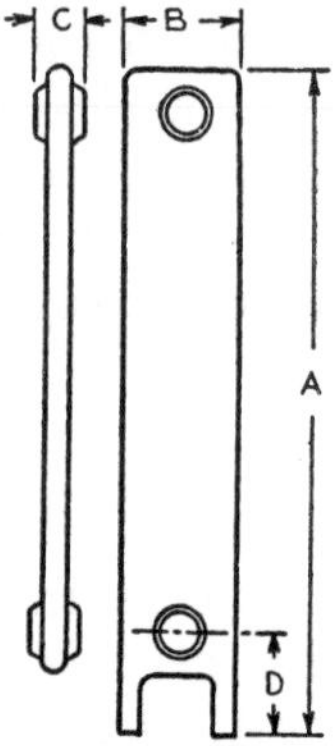

* *A.S.H.V.E.*

[a] These ratings are based on steam at 215° F. and air at 70° F. They apply only to installed radiators exposed in a normal manner; not to radiators installed behind enclosures, grilles, or under shelves.

[b] Length equals number of sections times 1¾ in.

[c] Overall height and leg height, as produced by some manufacturers, are one inch (1 in.) greater than shown in Columns A and D. Radiators may be furnished without legs. Where greater than standard leg heights are required this dimension shall be 4½ in.

[d] Or equal.

feet for 6 tubes per section. In this problem 18 square feet of radiation are required. There are several possibilities:

(1) $18/1.6 = 1\frac{1}{8}$ (or 2) sections of 3 tubes per section.
(2) $18/2 = 9$ sections of 4 tubes per section.
(3) $18/2.4 = 7\frac{1}{2}$ (or 8) sections of 5 tubes per section.
(4) $18/3 = 6$ sections of 6 tubes per section.

In these solutions, (1) or (3) gives a larger radiation than is necessary. The answers arrived at in (2) or (4)—9 sections of 4 tubes per section or 6 sections of 6 tubes per section—can be used. The final selection will depend upon the space available and the cost of the unit.

Another style of radiator is a so-called "wall radiator" which is hung on the walls or ceiling to conserve space. These are usually two-column affairs and both have several coils cast as one unit or are made up in units from separate sections.

Table 4 gives the radiation areas of such units.

TABLE 4

RATINGS OF CAST-IRON WALL RADIATOR UNITS, SQUARE FEET

Height, inches	Length or width, inches	Thickness, inches	Heating surface, square feet
$13\frac{1}{4}$	$16\frac{1}{2}$	3	$6\frac{1}{2}$
$13\frac{1}{4}$	22	3	8
$13\frac{1}{4}$	29	3	11
22	$13\frac{1}{4}$	3	8
29	$13\frac{1}{4}$	3	11

Heating coils are sometimes also made up from standard pipe or a pipe riser may be used to heat a small room. Table 5 gives the heating surface of standard pipe.

ILLUSTRATION: A bathroom requires five square feet of radiation. If the headroom available is 8 feet, how large a pipe riser will be required to provide the necessary radiation?

Since 5 square feet of radiation are required from 8 feet of pipe, then $\frac{5}{8} = 0.625$ square foot is required per foot of pipe. Referring this per-foot figure to Table 5 to obtain the diameter of pipe, we find that the 2-inch pipe fills the need very closely. (Ans.)

TABLE 5

HEATING SURFACE OF STANDARD PIPE, SQUARE FEET

Length of pipe, feet	Nominal diameter of pipe, inches									
	$\frac{3}{4}$	1	$1\frac{1}{4}$	$1\frac{1}{2}$	2	$2\frac{1}{2}$	3	4	5	6
1	.275	.346	.434	.494	.622	.753	.916	1.175	1.455	1.739

Estimating Radiation.—Estimating radiation requirements is simple with the aid of the tables. Let us take as an example the room shown in Fig. 6. This represents the dining room of a house

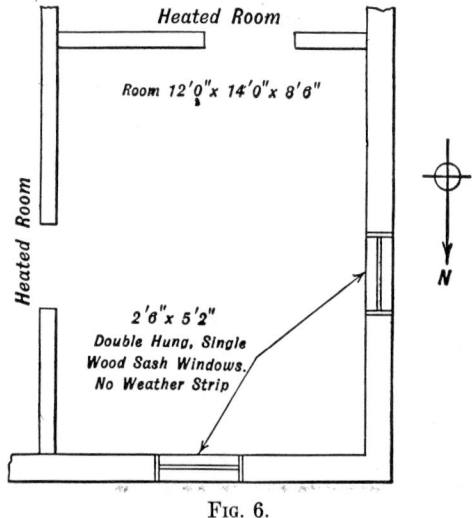

Heated Room

Room 12'0"x 14'0"x 8'6"

Heated Room

2'6"x 5'2"
Double Hung, Single
Wood Sash Windows.
No Weather Strip

N

FIG. 6.

in Philadelphia. The outside walls of frame construction, 1-inch sheathing, and brick veneer. Inside walls are plastered on wood lath on studding. The floor above has heated rooms. The problem is now to find how many square feet of steam radiation will

TABLE 6

Summarizing the whole estimate we have:	Area or lin. ft.	Sq. ft. radiation	
North wall, $12 \times 8.5 = 102$ sq. ft. $- 12.9$ sq. ft.			
(window)...............................	89.1 sq. ft.	8	sq. ft.
1 window, 2 ft. 6 in. by 5 ft. 2 in.............	12.9 sq. ft.	4	sq. ft.
Infiltration, cracks..........................	17.8 ft.	5	sq. ft.
Total for north wall without exposure factor		17	sq. ft.
Exposure factor........................		0.94	sq. ft.
Total................................		15.98	sq. ft.
West wall, $14 \times 8.5 = 119$ sq. ft. $- 12.9$ sq. ft.			
(window)...............................	106.1 sq. ft.	9	sq. ft.
1 window, 2 ft. 6 in. by 5 ft. 2 in.............	12.9 sq. ft.	4	sq. ft.
Infiltration, cracks..........................	17.8 lin. ft.	5	sq. ft.
Total for west wall without exposure factor		18	sq. ft.
Exposure factor........................		0.94	sq. ft.
Total................................		16.92	sq. ft.
The total for the room is			
North wall...		15.98	sq. ft.
West wall..		16.92	sq. ft.
Total...		32.90	sq. ft.

be required to heat this room. This may then be translated into other terms as desired.

The area of the north wall is $12 \times 8.5 = 102$ square feet. The area of the window ($2'6'' \times 5'2''$) is 12.9 square feet. The net wall area is $102 - 12.9 = 89.1$ square feet.

Then we look through Tables 7 to 18 to find the one which has the figures for this type of construction. In Table 7, Wall No. 27, we find "Brick Veneer, 1-inch Wood Sheathing." Following this line across to Column C, which represents plaster on wood lath on studding, we find a figure 0.27 which is called a

TABLE 7

COEFFICIENTS OF TRANSMISSION (U) OF FRAME WALLS *

EXTERIOR FINISH	INTERIOR FINISH	TYPE OF SHEATHING: GYPSUM (½ IN. THICK) A	PLYWOOD (5/16 IN. THICK) B	WOOD[f] (25/32 IN. THICK) BLDG. PAPER C	INSULATING BOARD (25/32 IN. THICK) D	WALL NUMBER
WOOD SIDING (Clapboard) 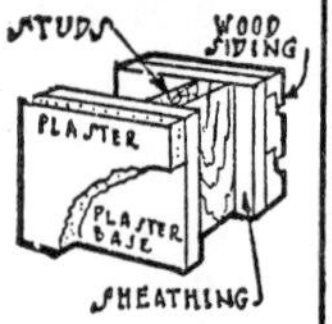	Metal Lath and Plaster[b]	0.33	0.32	0.26	0.20	1
	Gypsum Board (⅜ in.) Decorated	0.32	0.32	0.26	0.20	2
	Wood Lath and Plaster	0.31	0.31	0.25	0.19	3
	Gypsum Lath (⅜ in.) Plastered[c]	0.31	0.30	0.25	0.19	4
	Plywood (⅜ in.) Plain or Decorated	0.30	0.30	0.24	0.19	5
	Insulating Board (½ in.) Plain or Decorated	0.23	0.23	0.19	0.16	6
	Insulating Board Lath (½ in.) Plastered[c]	0.22	0.22	0.19	0.15	7
	Insulating Board Lath (1 in.) Plastered[c]	0.17	0.17	0.15	0.12	8
WOOD[d] SHINGLES 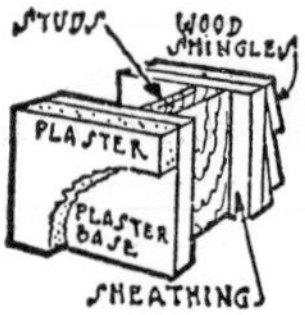	Metal Lath and Plaster[b]	0.25	0.25	0.26	0.17	9
	Gypsum Board (⅜ in.) Decorated	0.25	0.25	0.26	0.17	10
	Wood Lath and Plaster	0.24	0.24	0.25	0.16	11
	Gypsum Lath (⅜ in.) Plastered[c]	0.24	0.24	0.25	0.16	12
	Plywood (⅜ in.) Plain or Decorated	0.24	0.24	0.24	0.16	13
	Insulating Board (½ in.) Plain or Decorated	0.19	0.19	0.19	0.14	14
	Insulating Board Lath (½ in.) Plastered[c]	0.19	0.18	0.19	0.13	15
	Insulating Board Lath (1 in.) Plastered[c]	0.14	0.14	0.15	0.11	16
STUCCO 	Metal Lath and Plaster[b]	0.43	0.42	0.32	0.23	17
	Gypsum Board (⅜ in.) Decorated	0.42	0.41	0.31	0.23	18
	Wood Lath and Plaster	0.40	0.39	0.30	0.22	19
	Gypsum Lath (⅜ in.) Plastered[c]	0.39	0.39	0.30	0.22	20
	Plywood (⅜ in.) Plain or Decorated	0.39	0.38	0.29	0.22	21
	Insulating Board (½ in.) Plain or Decorated	0.27	0.27	0.22	0.18	22
	Insulating Board Lath (½ in.) Plastered[c]	0.26	0.26	0.22	0.17	23
	Insulating Board Lath (1 in.) Plastered[c]	0.19	0.19	0.16	0.14	24
BRICK VENEER[e] 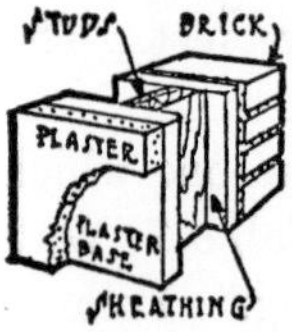	Metal Lath and Plaster[b]	0.37	0.36	0.28	0.21	25
	Gypsum Board (⅜ in.) Decorated	0.36	0.36	0.28	0.21	26
	Wood Lath and Plaster	0.35	0.34	0.27	0.20	27
	Gypsum Lath (⅜ in.) Plastered[c]	0.34	0.34	0.27	0.20	28
	Plywood (⅜ in.) Plain or Decorated	0.34	0.33	0.27	0.20	29
	Insulating Board (½ in.) Plain or Decorated	0.25	0.25	0.21	0.17	30
	Insulating Board Lath (½ in.) Plastered[c]	0.24	0.24	0.20	0.16	31
	Insulating Board Lath (1 in.) Plastered[c]	0.18	0.18	0.15	0.13	32

* Table footnotes on p. 504.

TABLE 8

COEFFICIENTS OF TRANSMISSION (U) OF FRAME WALLS WITH INSULATION
BETWEEN FRAMING *, a

COEFFICIENT WITH *NO* INSULATION BETWEEN FRAMING	COEFFICIENT WITH INSULATION BETWEEN FRAMING				NUMBER
	MINERAL WOOL OR VEGETABLE FIBERS IN BLANKET OR BAT FORM b (Thickness below)			3⅛ IN. MINERAL WOOL BETWEEN FRAMING c	
	1 IN.	2 IN.	3 IN.		
	A	B	C	D	
0.11	0.078	0.063	0.054	0.051	33
0.13	0.088	0.070	0.058	0.055	35
0.15	0.097	·0.075	0.062	0.059	37
0.17	0.10	0.080	0.066	0.062	39
0.19	0.11	0.084	0.069	0.065	41
0.21	0.12	0.088	0.072	0.067	43
0.23	0.12	0.091	0.074	0.069	45
0.25	0.13	0.094	0.076	0.071	47
0.27	0.14	0.097	0.078	0.073	49
0.29	0.14	0.10	0.080	0.075	51
0.31	0.14	0.10	0.081	0.076	53
0.33	0.15	0.10	0.083	0.077	55
0.35	0.15	0.11	0.084	0.078	57
0.37	0.16	0.11	0.085	0.080	59
0.39	0.16	0.11	0.086	0.081	61
0.41	0.16	0.11	0.087	0.082	63
0.43	0.17	0.11	0.088	0.082	65

These coefficients are expressed in Btu per (hour)(square feet)(Fahrenheit degree difference in temperature between the air on the two sides), and are based on an outside wind velocity of 15 mph.

* *A.S.H.V.E.*

a This table may be used for determining the coefficients of transmisson of frame constructions with the types and thicknesses of insulation indicated in Columns A to D inclusive between framing. Columns A, B and C may be used for walls, ceilings or roofs with only one air space between framing but are not applicable to ceilings with no flooring above. (See Table 13.) Column D is applicable to walls only. Example: Find the coefficient of transmission of a frame wall consisting of wood siding, 25/32 in. insulating board sheathing studs, gypsum lath and plaster, with 2 in. blanket insulation between studs. According to Table 7, a wall of this construction with no insulation between studs has a coefficient of 0.19 (Wall No. 4D). Referring to Column B above it will be found that a wall of this value with 2 in. blanket insulation between the studs has a coefficient of 0.084.

Coefficients corrected for 2×4 framing, 16 in. on centers—15 percent of surface area.

b Based on one air space between framing.

c No air space.

TABLE 9

COEFFICIENTS OF TRANSMISSION (*U*) OF MASONRY WALLS *

TYPE OF MASONRY	THICKNESS OF MASONRY INCHES	INTERIOR FINISH (PLUS INSULATION WHERE INDICATED)									WALL NUMBER
		Plain Walls—No Interior Finish	Plaster (½ in.) on Walls	Metal Lath and Plaster/ Furred[A]	Gypsum Board (⅜ in.) Decorated—Furred[A]	Gypsum Lath (⅜ in.) Plastered—Furred[A]	Insulating Board (½ in.) Plain or Decorated—Furred[A]	Insulating Board Lath (½ in.) Plastered—Furred[A]	Insulating Board (1 in.) Plastered—Furred[A]	Gypsum Lath[b] Plastered Plus 1 In. Blanket Insulation—Furred[A]	
		A	B	C	D	E	F	G	H	I	
Solid[a] Brick	8	0.50	0.46	0.32	0.31	0.30	0.22	0.22	0.16	0.14	67
	12	0.36	0.34	0.25	0.25	0.24	0.19	0.19	0.14	0.13	68
	16	0.28	0.27	0.21	0.21	0.20	0.17	0.16	0.13	0.12	69
Hollow[b] Tile (Stucco Exterior Finish)	8	0.40	0.37	0.27	0.27	0.26	0.20	0.20	0.15	0.13	70
	10	0.39	0.37	0.27	0.27	0.26	0.20	0.19	0.15	0.13	71
	12	0.30	0.28	0.22	0.22	0.21	0.17	0.17	0.13	0.12	72
	16	0.24	0.24	0.19	0.19	0.18	0.15	0.15	0.12	0.11	73
Stone[c]	8	0.70	0.64	0.39	0.38	0.36	0.26	0.25	0.18	0.16	74
	12	0.57	0.53	0.35	0.34	0.33	0.24	0.23	0.17	0.15	75
	16	0.49	0.45	0.31	0.31	0.29	0.22	0.22	0.16	0.14	76
	24	0.37	0.35	0.26	0.26	0.25	0.19	0.19	0.15	0.13	77
Poured Concrete[d]	6	0.79	0.71	0.42	0.41	0.39	0.27	0.26	0.19	0.16	78
	8	0.70	0.64	0.39	0.38	0.36	0.26	0.25	0.18	0.16	79
	10	0.63	0.58	0.37	0.36	0.34	0.25	0.24	0.18	0.15	80
	12	0.57	0.53	0.35	0.34	0.33	0.24	0.23	0.17	0.15	81
Hollow Concrete Blocks	Gravel Aggregate										
	8	0.56	0.52	0.34	0.34	0.32	0.24	0.23	0.17	0.15	82
	12	0.49	0.46	0.32	0.31	0.30	0.22	0.22	0.16	0.14	83
	Cinder Aggregate										
	8	0.41	0.39	0.28	0.28	0.27	0.21	0.20	0.15	0.13	84
	12	0.38	0.36	0.26	0.26	0.25	0.20	0.19	0.15	0.13	85
	Light Weight Aggregate[e]										
	8	0.36	0.34	0.26	0.25	0.24	0.19	0.19	0.15	0.13	86
	12	0.34	0.33	0.25	0.24	0.24	0.19	0.18	0.14	0.13	87

* Table footnotes on p. 504.

TABLE 10

COEFFICIENTS OF TRANSMISSION (U) OF BRICK AND STONE VENEER MASONRY WALLS *

TYPICAL CONSTRUCTION	FACING	BACKING	INTERIOR FINISH (PLUS INSULATION WHERE INDICATED)									WALL NUMBER
			Plain Walls—no Interior Finish	Plaster (½ in.) on Walls	Metal Lath and Plaster—Furred	Gypsum Board (⅜ in.), Decorated—Furred	Gypsum Lath (⅜ in.) Plastered—Furred	Insulating Board (½ in.) Plain or Decorated—Furred	Insulating Board Lath (½ in.) Plastered—Furred	Insulating Board Lath (1 in.) Plastered—Furred	Gypsum Lath Plastered Plus 1 in. Blanket Insulation—Furred	
			A	B	C	D	E	F	G	H	I	
		6 in. Hollow Tile	0.35	0.34	0.25	0.25	0.24	0.19	0.18	0.14	0.13	88
		8 in. Hollow Tile	0.34	0.32	0.25	0.24	0.23	0.19	0.18	0.14	0.13	89
	4 in. Brick Veneer	6 in. Concrete	0.59	0.54	0.35	0.35	0.33	0.24	0.23	0.17	0.15	90
		8 in. Concrete	0.54	0.50	0.33	0.33	0.31	0.23	0.23	0.17	0.15	91
		8 in. Concrete Blocks (Gravel Aggregate)	0.44	0.41	0.29	0.29	0.28	0.21	0.21	0.16	0.14	92
		8 in. Concrete Blocks (Cinder Aggregate)	0.34	0.33	0.25	0.24	0.24	0.19	0.18	0.14	0.13	93
		8 in. Concrete Blocks (Light Weight Aggregate)	0.31	0.29	0.23	0.23	0.22	0.18	0.17	0.14	0.12	94
		6 in. Hollow Tile	0.37	0.35	0.26	0.26	0.25	0.19	0.19	0.15	0.13	95
		8 in. Hollow Tile	0.36	0.34	0.25	0.25	0.24	0.19	0.19	0.14	0.13	96
	4 in. Cut Stone Veneer	6 in. Concrete	0.63	0.58	0.37	0.36	0.34	0.25	0.24	0.18	0.15	97
		8 in. Concrete	0.57	0.53	0.35	0.34	0.33	0.24	0.23	0.17	0.15	98
		8 in. Concrete Blocks (Gravel Aggregate)	0.47	0.44	0.30	0.30	0.29	0.22	0.21	0.16	0.14	99
		8 in. Concrete Blocks (Cinder Aggregate)	0.36	0.34	0.25	0.25	0.24	0.19	0.19	0.15	0.13	100
		8 in. Concrete Blocks (Light Weight Aggregate)	0.32	0.30	0.23	0.23	0.22	0.18	0.17	0.14	0.12	101

* Table footnotes on p. 504.

These coefficients are expressed in Btu per (hour)(square feet)(Fahrenheit degree difference in temperature between the air on the two sides), and are based on an outside wind velocity of 15 mph.

No insulation between studs [a] (see Table 8).

[a] Coefficients not weighted; effect of studding neglected.

[b] Plaster assumed ¾ in. thick.

[c] Plaster assumed ½ in. thick.

[d] Furring strips (1 in. nominal thickness) between wood shingles and all sheathings except wood.

[e] Small air space and mortar between building paper and brick veneer neglected.

[f] Nominal thickness, 1 in.

TABLE FOOTNOTES FOR TABLE 9, P. 502

Coefficients are expressed in Btu per (hour)(square foot)(Fahrenheit degree difference in temperature between the air on the two sides), and are based on an outside wind velocity of 15 mph.

[a] Based on 4 in. hard brick and remainder common brick.

[b] The 8 in. and 10 in. tile figures are based on two cells in the direction of heat flow. The 12 in. tile is based on three cells in the direction of heat flow. The 16 in. tile consists of one 10 in. and one 6 in. tile, each having two cells in the direction of heat flow.

[c] Limestone or sandstone.

[d] These figures may be used with sufficient accuracy for concrete walls with stucco exterior finish.

[e] Expanded slag, burned clay or pumice.

[f] Thickness of plaster assumed ¾ in.

[g] Thickness of plaster assumed ½ in.

[h] Based on 2 in. furring strips; one air space.

TABLE FOOTNOTES FOR TABLE 10, P. 503

Coefficients are expressed in Btu per (hour)(square foot)(Fahrenheit degree difference in temperature between the air on the two sides), and are based on an outside wind velocity of 15 mph.

[a] Calculation based on ½ in. cement mortar between backing and facing, except in the case of the concrete backing which is assumed to be poured in place.

[b] The hollow tile figures are based on two air cells in the direction of heat flow.

[c] Hollow concrete blocks.

[d] Expanded slag, burned clay or pumice.

[e] Thickness of plaster assumed ¾ in.

[f] Thickness of plaster assumed ½ in.

[g] Based on 2 in. furring strips; one air space.

TABLE 11

COEFFICIENTS OF TRANSMISSION (U) OF FRAME PARTITIONS OF INTERIOR WALLS *, a

INTERIOR FINISH	SINGLE PARTITION (Finish on one side only of studs)	DOUBLE PARTITION (Finish on both sides of studs)		PARTITION NUMBER
		No INSULATION BETWEEN STUDS	1 IN. BLANKET d BETWEEN STUDS. ONE AIR SPACE.	
	A	B	C	
Metal Lath and Plaster b	0.69	0.39	0.16	1
Gypsum Board (⅜ in.) Decorated	0.67	0.37	0.16	2
Wood Lath and Plaster	0.62	0.34	0.15	3
Gypsum Lath (⅜ in.) Plastered c	0.61	0.34	0.15	4
Plywood (⅜ in.) Plain or Decorated	0.59	0.33	0.15	5
Insulating Board (½ in.) Plain or Decorated	0.36	0.19	0.11	6
Insulating Board Lath (½ in.) Plastered c	0.35	0.13	0.11	7
Insulating Board Lath (1 in.) Plastered c	0.23	0.12	0.082	8

Coefficients are expressed in Btu per (hour)(square foot)(Fahrenheit degree difference in temperature between the air on the two sides), and are based on still air (no wind) conditions on both sides.

* A.S.H.V.E.

a Coefficients not weighted; effect of studding neglected.

b Plaster assumed ¾ in. thick.

c Plaster assumed ½ in. thick.

d For partitions with other insulations between studs refer to Table 8, using values in Column B of above table, in left-hand column of Table 8. Example: What is the coefficient of transmission (U) of a partition consisting of gypsum lath and plaster on both sides of studs with 2 in. blanket between studs? Solution: According to above table, this partition with no insulation between studs (No. 4B) has a coefficient of 0.34. Referring to Table 8, it will be found that a wall having a coefficient of 0.34 with no insulation between studs, will have a coefficient of 0.10 with 2 in. of blanket insulation between studs (No. 56B).

TABLE 12

COEFFICIENTS OF TRANSMISSION (U) OF MASONRY PARTITIONS *

TYPE OF PARTITION		THICKNESS OF MASONRY (INCHES)	TYPE OF FINISH			PARTITION NUMBER
			No FINISH (Plain walls)	PLASTER ONE SIDE	PLASTER BOTH SIDES a	
			A	B	C	
HOLLOW CLAY TILE		3 4	0.50 0.45	0.47 0.42	0.43 0.40	9 10
HOLLOW GYPSUM TILE		3 4	0.35 0.29	0.33 0.28	0.32 0.27	11 12
HOLLOW CONCRETE TILE OR BLOCKS	Cinder Aggregate	3 4	0.50 0.45	0.47 0.42	0.43 0.40	13 14
	Light Weight Aggregate b	3 4	0.41 0.35	0.39 0.34	0.37 0.32	15 16
COMMON BRICK		4	0.50	0.46	0.43	17

Coefficients are expressed in Btu per (hour)(square foot)(Fahrenheit degree difference between the air on the two sides) and are based on still air (no wind) conditions on both sides.

* *A.S.H.V.E.*

a 2 in. solid plaster partition, U = 0.53.

b Expanded slag, burned clay or pumice.

TYPE OF CEILING	INSULATION BETWEEN, OR ON TOP OF, JOISTS (No Flooring Above)												WITH FLOORING[g] (On Top of Ceiling Joists)		NUMBER
	None	Insulating Board on Top of Joists		Blanket or Bat Insulation Between Joists			Vermiculite Insulation Between Joists			Mineral Wool Insulation Between Joists			Single Wood Floor	Double Wood Floor	
		½ In.	1 In.	1 In.	2 In.	3 In.	2 In.	3 In.	4 In.	2 In.	3 In.	4 In.			
	A	B	C	D	E	F	G	H	I	J	K	L	M	N	
No Ceiling		0.37	0.24										0.45	0.34	1
Metal Lath and Plaster[d]	0.69	0.26	0.19	0.19	0.12	0.093	0.18	0.14	0.11	0.12	0.093	0.077	0.30	0.25	2
Gypsum Board (⅜ in.) Plain or Decorated	0.67	0.26	0.18	0.19	0.12	0.092	0.18	0.13	0.10	0.12	0.092	0.077	0.30	0.24	3
Wood Lath and Plaster[d]	0.62	0.25	0.18	0.19	0.12	0.091	0.17	0.13	0.10	0.12	0.091	0.076	0.28	0.24	4
Gypsum Lath (⅜ in.) Plastered[e]	0.61	0.25	0.18	0.19	0.12	0.091	0.17	0.13	0.10	0.12	0.091	0.076	0.38	0.24	5
Plywood (⅜ in.) Plain or Decorated	0.59	0.24	0.18	0.19	0.12	0.091	0.17	0.13	0.10	0.12	0.091	0.076	0.28	0.23	6
Insulating Board (⅜ in.) Plain or Decorated	0.36	0.19	0.15	0.16	0.10	0.082	0.14	0.12	0.097	0.10	0.082	0.069	0.22[a]	0.19[a]	7
Insulating Board Lath (½ in.) Plastered[e]	0.35	0.19	0.15	0.15	0.10	0.081	0.14	0.11	0.096	0.11	0.081	0.068	0.21	0.18	8
Insulating Board Lath (1 in.) Plastered[e]	0.23	0.15	0.12	0.12	0.089	0.072	0.12	0.097	0.084	0.089	0.072	0.061	0.16	0.14	9

Coefficients are expressed in Btu per (hour)(square foot)(Fahrenheit degree difference between the air on the two sides) and are based on still air (no wind) conditions on both sides.

* A.S.H.V.E.

a Coefficients corrected for framing on basis of 15 percent area, 2 in. × 4 in. (nominal) framing, 16 in. on centers.

b 25/32 in. yellow pine or fir.

c 25/32 in. pine or fir sub-flooring plus 13/16 in. hardwood flooring.

d Plaster assumed ¾ in. thick.

e Plaster assumed ½ in. thick.

f Based on insulation in contact with ceiling, and consequently no air space between.

g For coefficients for constructions in Columns M and N (except No. 1) with insulation between joists, refer to Table 8. Example: The coefficient for No. 3-N of Table 12 is 0.24. With 2 in. blanket insulation between joists, the coefficient will be 0.093. (See Table 8.) (Column D of Table 8 applicable only for 3⅝ in. joists.)

h For 25/32 in. insulating board sheathing applied to the under side of the joists, the coefficient for single wood floor (Column M) is 0.18 and for double wood floor (Column N) is 0.16. For coefficients with insulation between joists, see Table 8.

TABLE 14

CoEFFICIENTS OF TRANSMISSION (U) OF CONCRETE CONSTRUCTION FLOORS
AND CEILINGS *

TYPE OF CEILING	THICKNESS OF CONCRETE (INCHES)	TYPE OF FLOORING					NUMBER
		No Flooring (Concrete Bare)	Tile[a] or Terrazzo Flooring on Concrete	⅛ In. Asphalt Tile[b] Directly on Concrete	Parquett[c] Flooring in Mastic on Concrete	Double Wood Floor on Sleepers[d]	
		A	B	C	D	E	
No Ceiling..........................	3	0.68	0.65	0.66	0.45	0.25	1
	6	0.59	0.56	0.58	0.41	0.23	2
	10	0.50	0.48	0.49	0.36	0.22	3
½ in. Plaster Applied to Underside of Concrete.........................	3	0.62	0.59	0.60	0.43	0.24	4
	6	0.54	0.52	0.53	0.39	0.22	5
	10	0.46	0.44	0.45	0.34	0.21	6
Metal Lath and Plaster[e]—Suspended or Furred.........................	3	0.38	0.37	0.37	0.30	0.19	7
	6	0.35	0.34	0.35	0.28	0.18	8
	10	0.32	0.31	0.32	0.26	0.17	9
Gypsum Board (⅜ in.) and Plaster[f]—Suspended or Furred..............	3	0.36	0.35	0.35	0.28	0.19	10
	6	0.33	0.32	0.33	0.27	0.18	11
	10	0.30	0.29	0.30	0.24	0.17	12
Insulating Board Lath (½ in.) and Plaster[f]—Suspended or Furred	3	0.25	0.24	0.25	0.21	0.15	13
	6	0.23	0.23	0.23	0.20	0.15	14
	10	0.22	0.21	0.22	0.19	0.14	15

Coefficients are expressed in Btu per (hour)(square foot)(Fahrenheit degree difference in temperature between the air on the two sides), and are based on still air (no wind) conditions on both sides.

* *A.S.H.V.E.*

[a] Thickness of tile assumed to be 1 in.

[b] Conductivity of asphalt tile assumed to be 3.1.

[c] Thickness of wood assumed to be 1³⁄₁₆ in.; thickness of mastic, ⅛ in. (k = 4.5). Column D may also be used for concrete covered with carpet.

[d] Based on 2⁵⁄₃₂ in. yellow pine or fir sub-flooring and 1³⁄₁₆ in. hardwood finish flooring with an air space between sub-floor and concrete.

[e] Thickness of plaster assumed to be ¾ in.

[f] Thickness of plaster assumed to be ½ in.

coefficient. Then turning to Table 19 and looking along the top line of the center section for a column headed by 0.27 we find 0.26 and then 0.28. Either column may be used with sufficient accuracy. For our purpose, let us look down the column headed

TABLE 15

COEFFICIENTS OF TRANSMISSION (U) OF CONCRETE BASEMENT FLOORS ON GROUND WITH VARIOUS TYPES OF FINISH FLOORING *

$U = 0.10^a$ Btu per (hr) (sq ft) (Fahrenheit degree temperature difference between the ground and the air over the floor).

---

* *A.S.H.V.E.*

a Since few data are available, a coefficient of 0.10 is frequently used for all types of basement concrete floors on the ground, with or without insulation. For basement wall below grade, use the same average coefficient (0.10). A lower ground temperature should, however, be used for walls than floors. For further data see A.S.H.V.E. RESEARCH REPORT No. 1213—Heat Loss Through Basement Walls and Floors, by F. C. Houghten, S. I. Taimuty, Carl Gutberlet and C. J. Brown (A.S.H.V.E. TRANSACTIONS, Vol. 48, 1942, p. 369).

by 0.28 to find the figure coming closest to 89.1, the net wall area. We find the figure 92.0 and following this line to the extreme left we arrive at the figure 8 which represents the square feet of radiation required for the heat lost by transmission through the wall.

Next considering the loss of heat through the window we look down the second column from the left in the same table for the figure closest to 12.9, the area of the window. This is 11.7, and following this line to the left we find that 4 square feet of radiation will be required to care for the heat lost by transmission through the window.

This double hung window has 17.8 lineal feet of cracks (the sum of the lengths of two vertical and three horizontal cracks). Referring to the small table on infiltration on page 481 we find, opposite "double hung wood sash" the figure 50 which represents cubic feet per hour per lineal foot of crack. Then referring to Table 19 under infiltration, finding the column headed by 50 and looking for the figure in this column closest to 17.8 we find 17.9. Following this to the left we find that 5 square feet of radiation are required for the infiltration.

TABLE 16

Coefficients of Transmission (U) of Flat Roofs Covered with Built-up Roofing. No Ceiling—Under Side of Roof Exposed *

(See Table 17 for Flat Roofs with Ceilings)

Type of Roof Deck	Thick-ness of Roof Deck (Inches)	No In-sula-tion	Insulation on Top of Deck (Covered with Built-Up Roofing)							Num-ber
			Insulating Board (Thickness Below)				Corkboard (Thickness Below)			
			½ In.	1 In.	1½ In.	2 In.	1 In.	1½ In.	2 In.	
		A	B	C	D	E	F	G	H	
Flat Metal Roof Deck[a] 		0.94	0.39	0.24	0.18	0.14	0.23	0.17	0.13	
Precast Cement Tile 	1½ in.	0.84	0.3	0.24	0.17	0.14	0.22	0.16	0.13	2
Concrete 	2 in. 4 in. 6 in.	0.82 0.72 0.65	0.36 0.34 0.33	0.24 0.23 0.22	0.17 0.17 0.16	0.14 0.13 0.13	0.22 0.21 0.21	0.16 0.16 0.15	0.13 0.12 0.12	3 4 5
Gypsum Fiber Concrete[b] on ½ in. Gypsum Board 	2½ in. 3½ in.	0.38 0.31	0.24 0.21	0.18 0.16	0.14 0.13	0.12 0.11	0.17 0.15	0.13 0.12	0.11 0.10	6 7
Wood[c] 	1 in. 1½ in. 2 in. 3 in.	0.49 0.37 0.32 0.23	0.28 0.24 0.22 0.17	0.20 0.17 0.16 0.14	0.15 0.14 0.13 0.11	0.12 0.11 0.11 0.096	0.19 0.17 0.16 0.13	0.14 0.13 0.12 0.11	0.12 0.11 0.10 0.091	8 9 10 11

* Table footnotes on p. 512.

TABLE 17

Coefficients of Transmission (U) of Flat Roofs Covered with Built-up
Roofing. With Lath and Plaster Ceilings [a] *

(See Table 16 for Flat Roofs with No Ceilings)

Type of Roof Deck	Thickness of Roof Deck (Inches)	No Insulation	Insulation on Top of Deck (Covered with Built-up Roofing)							Number
			Insulating Board (Thickness Below)				Corkboard (Thickness Below)			
			½ In.	1 In.	1½ In.	2 In.	1 In.	1½ In.	2 In.	
		A	B	C	D	E	F	G	H	
Flat Metal Roof Deck		0.46	0.27	0.19	0.15	0.12	0.18	0.14	0.11	12
Precast Cement Tile	1½ in.	0.43	0.26	0.19	0.15	0.12	0.18	0.14	0.11	13
Concrete	2 in. 4 in. 6 in.	0.42 0.40 0.37	0.26 0.25 0.24	0.19 0.18 0.18	0.14 0.14 0.14	0.12 0.12 0.11	0.18 0.17 0.17	0.14 0.13 0.13	0.11 0.11 0.11	14 15 16
Gypsum Fiber Concrete[b] on ½ in. Gypsum Board	2½ in. 3½ in.	0.27 0.23	0.19 0.17	0.15 0.14	0.12 0.11	0.10 0.097	0.14 0.13	0.12 0.11	0.097 0.091	17 18
Wood[c]	1 in. 1½ in. 2 in. 3 in.	0.31 0.26 0.24 0.18	0.21 0.19 0.17 0.14	0.16 0.15 0.14 0.12	0.13 0.12 0.11 0.10	0.11 0.10 0.097 0.087	0.15 0.14 0.13 0.11	0.12 0.11 0.11 0.095	0.10 0.095 0.092 0.082	19 20 21 22

* Table footnotes on p. 512.

Coefficients are expressed in Btu per (hour)(square foot)(Fahrenheit degree difference in temperature between the air on two sides), and are based on an outside wind velocity of 15 mph.

* *A.S.H.V.E.*

a Coefficient of transmission of bare corrugated iron (no roofing) is 1.50 Btu per (hr.)(sq. ft. of projected area)(F. deg. difference in temperature) based on an outside wind velocity of 15 mph.

b 87½ percent gypsum, 12½ percent wood fiber. Thickness indicated includes ½ in. gypsum board.

c Nominal thicknesses specified—actual thicknesses used in calculations.

Coefficients are expressed in Btu per (hour)(square foot)(Fahrenheit degree difference in temperature between the air on two sides), and are based on an outside wind velocity of 15 mph.

* *A.S.H.V.E.*

a Calculations based on metal lath and plaster ceilings, but coefficients may be used with sufficient accuracy for gypsum lath or wood lath and plaster ceilings. It is assumed that there is an air space between the under side of the roof deck and the upper side of the ceiling.

b 87½ percent gypsum, 12½ percent wood fiber. Thickness indicated includes ½ in. gypsum board.

c Nominal thicknesses specified—actual thicknesses used in calculations.

We have now found three separate radiation figures 8, 4, and 5 which total 17 square feet. This must now be multiplied by a factor for exposure and temperature. This is found in Table 20 in the "N" column, since this is a north wall, and opposite Philadelphia. The factor is 0.94. Then $17 \times 0.94 = 15.98$ square feet.

The radiation for the west wall is estimated in a similar manner. It happens that in this particular case the temperature and exposure factor is the same as for the north wall.

The radiation required for each room in the house may be estimated in a similar manner.

TABLE 18

COEFFICIENTS OF TRANSMISSION (U) OF PITCHED ROOFS *

Roofing / Roof Sheathing / ceiling

TYPE OF CEILING (APPLIED DIRECTLY TO ROOF RAFTERS)	WOOD SHINGLES (ON 1 x 4 WOOD STRIPS SPACED 2 IN. APART)				ASPHALT SHINGLES OR ROLL ROOFING (ON SOLID WOOD SHEATHING)				SLATE OR TILE (ON SOLID WOOD SHEATHING)				NUMBER
	None	Blanket or Bat (Thickness Below)			None	Blanket or Bat (Thickness Below)			None	Blanket or Bat (Thickness Below)			
		1 In.	2 In.	3 In.		1 In.	2 In.	3 In.		1 In.	2 In.	3 In.	
	A	B	C	D	E	F	G	H	I	J	K	L	
No Ceiling Applied to Rafters	0.48^f	0.15	0.10	0.081^a	0.52^f	0.15	0.11	0.084^a	0.55^f	0.16	0.11	0.085^a	1
Metal Lath and Plaster^d	0.31	0.14	0.10	0.081	0.33	0.15	0.10	0.083	0.34	0.15	0.10	0.083	2
Gypsum Board (3/8 in.) Decorated	0.30	0.14	0.10	0.080	0.32	0.15	0.10	0.082	0.33	0.15	0.10	0.083	3
Wood Lath and Plaster^e	0.29	0.14	0.10	0.080	0.31	0.14	0.10	0.081	0.32	0.15	0.10	0.082	4
Gypsum Lath (3/8 in.) Plastered^e	0.29	0.14	0.10	0.079	0.31	0.14	0.10	0.081	0.32	0.15	0.10	0.082	5
Plywood (3/8 in.) Plain or Decorated	0.29	0.14	0.099	0.079	0.30	0.14	0.10	0.081	0.31	0.13	0.10	0.081	6
Insulating Board (1/4 in.) Plain or Decorated	0.22	0.12	0.090	0.072	0.23	0.12	0.091	0.074	0.24	0.12	0.092	0.074	7
Insulating Board Lath (1/2 in.) Plastered^e	0.22	0.12	0.088	0.072	0.22	0.12	0.090	0.073	0.23	0.12	0.091	0.074	8
Insulating Board Lath (1 in.) Plastered^e	0.16	0.10	0.078	0.064	0.17	0.10	0.079	0.065	0.17	0.10	0.080	0.066	9

Coefficients are expressed in Btu per (hour)(square foot)(Fahrenheit degree difference in temperature between the air on the two sides), and are based on an outside wind velocity of 15 mph.

* A.S.H.V.E.

^{a} Coefficients corrected for framing on basis of 15 percent area, 2 in. × 4 in. (nominal), 16 in. on centers.

^{b} Figures in Columns I, J, K and L may be used with sufficient accuracy for rigid asbestos shingles on wood sheathing. Layer of slater's felt neglected.

^{c} Sheathing and wood strips assumed 25/32 in. thick.

^{d} Plaster assumed 3/4 in. thick.

^{e} Plaster assumed 1/2 in. thick.

^{f} No air space included in 1-A, 1-E or 1-I; all other coefficients based on one air space,

513

Special Cases.—If the east wall of the room in Fig. 6 had been a solid partition of wood lath and plaster on each side of the studding and the room on the other side *unheated*, additional radiation would be necessary for the loss of heat through this wall. This radiation is estimated by the use of Table 11 which, for the particular conditions of this problem, gives in column "B" the coefficient 0.34. This coefficient is now referred to the *right-hand* portion of Table 19. We find there a column headed by 0.35, which is sufficiently close. Looking down this column till we reach a number equal to the area of the wall (14 × 8.5 = 119 sq. ft.) we note that this lies midway between 110 and 128 and that the radiation (last column) is, therefore, 6½ square feet for the loss through this wall.

Then there is the case when the space below or the space above a room is unheated. Let us take the case of the room shown in Fig. 6 if this room has an unheated attic above it and a ceiling of plaster on insulating board lath with no insulation between the joists and no flooring in the attic. This case is covered in

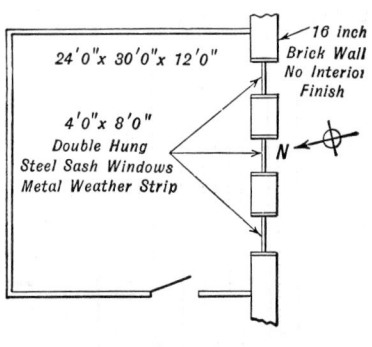

24'0"x 30'0"x 12'0"

16 inch
Brick Wall
No Interior
Finish

4'0"x 8'0"
Double Hung
Steel Sash Windows
Metal Weather Strip

N

FIG. 7.

Table 13 and we find in line 8, column "A" that the coefficient is 0.35. Then again consulting the *right-hand* portion of Table 19 we find a column headed by this figure. Looking down it for a figure representing the area of the ceiling (12 × 14 = 168) we

TABLE 19.—Heating and Piping Contractors National

Showing Radiation Required

3 Col. 38" Steam Rad.	Glass		Infiltration																				
	Window or Door	Skylight	Rate per Lineal Foot																				
			25	50	100	200																	
	1.1	1.3	0.45	0.9	1.8	3.6	0.08	0.10	0.12	0.14	0.16	0.18	0.20	0.22	0.24	0.26	0.28	0.30	0.35	0.40	0.45	0.50	0.55
225	77.0	91.0	31.5	63.0	126	252	5.60	7.00	8.40	9.80	11.2	12.6	14.0	15.4	16.8	18.2	19.6	21.0	24.5	28.0	31.5	35.0	38.5
1	2.92	2.47	7.14	3.57	1.79	0.89	40.2	32.1	26.8	23.0	20.1	17.9	16.1	14.6	13.4	12.4	11.5	10.7	9.18	8.04	7.14	6.43	5.84
2	5.84	4.94	14.3	7.14	3.58	1.78	80.4	64.2	53.6	46.0	40.2	35.8	32.2	29.2	26.8	24.8	23.0	21.4	18.4	16.1	14.3	12.9	11.7
3	8.76	7.41	21.4	10.7	5.37	2.67	121	96.3	80.4	69.0	60.3	53.7	48.3	43.8	40.2	37.2	34.5	32.1	27.5	24.1	21.4	19.3	17.5
4	11.7	9.88	28.6	14.2	7.16	3.56	161	128	107	92.0	80.4	71.6	64.4	58.4	53.6	49.6	46.0	42.8	36.7	32.2	28.6	25.7	23.3
5	14.6	12.4	35.7	17.9	8.95	4.45	201	160	134	115	101	89.5	80.5	73.0	67.0	62.0	57.5	53.5	45.9	40.2	35.7	32.1	29.2
6	17.5	14.8	42.8	21.4	10.7	5.34	241	193	161	138	121	107	96.6	87.6	80.4	74.4	69.0	64.2	55.1	48.2	42.8	38.6	35.0
7	20.4	17.3	49.9	25.0	12.5	6.23	281	225	188	161	141	125	113	102	93.8	86.8	80.5	74.9	64.3	56.3	50.0	45.0	40.9
8	23.4	19.8	57.1	28.6	14.3	7.12	322	257	214	184	161	143	129	117	107	99.2	92.0	85.6	73.4	64.3	57.1	51.4	46.7
9	26.3	22.2	64.3	32.1	16.1	8.01	362	289	241	207	181	161	145	131	121	112	103	96.3	82.6	72.4	64.3	57.9	52.6
10	29.2	24.7	71.4	35.7	17.9	8.90	402	321	268	230	201	179	161	146	134	124	115	107	91.8	80.4	71.4	64.3	58.4
11	32.1	27.2	78.5	39.3	19.7	9.79	442	353	295	253	221	197	177	161	147	136	126	118	101	88.4	78.5	70.7	64.2
12	35.0	29.6	85.7	42.8	21.5	10.7	482	385	322	276	241	215	193	175	161	149	138	128	110	96.5	85.7	77.2	70.1
13	38.0	32.1	92.8	46.4	23.2	11.6	523	417	348	299	261	233	209	190	174	161	149	139	119	105	92.8	83.6	75.9
14	40.9	34.6	100	50.0	25.1	12.5	563	449	375	322	281	251	225	204	188	174	161	150	129	113	100	90.0	81.8
15	43.8	37.1	107	53.6	26.9	13.4	603	481	402	345	301	268	241	219	201	186	172	160	138	121	107	96.4	87.6
16	46.7	39.5	114	57.1	28.6	14.2	643	514	429	368	322	286	258	234	214	198	184	171	147	129	114	103	93.4
17	49.6	42.0	121	60.7	30.4	15.1	683	546	456	391	342	304	274	248	228	211	195	182	156	137	121	109	99.3
18	52.6	44.5	129	64.3	32.2	16.0	724	578	482	414	362	322	290	263	241	223	207	193	165	145	129	116	105
19	55.5	46.9	136	67.9	34.0	16.9	764	610	509	437	382	340	306	277	255	236	218	203	174	153	136	122	111
20	58.4	49.4	143	71.4	35.8	17.8	804	642	536	460	402	358	322	292	268	248	230	214	184	161	143	129	117
21	61.3	51.9	150	75.0	37.6	18.7	844	674	563	483	422	376	338	307	281	260	241	225	193	169	150	135	123
22	64.2	54.3	157	78.5	39.4	19.6	884	706	590	506	442	394	354	321	295	273	253	235	202	177	157	141	128
23	67.2	56.8	164	82.1	41.2	20.5	925	738	616	529	462	412	370	336	308	285	264	246	211	185	164	148	134
24	70.8	59.3	171	85.7	43.0	21.4	965	770	643	552	482	430	386	350	322	298	276	257	220	193	171	154	140
25	73.0	61.8	179	89.3	44.8	22.3	1005	802	670	575	502	447	402	365	335	310	287	267	229	201	178	161	146

ASSOCIATION STANDARD RADIATION ESTIMATING TABLE

for Quantities Indicated

0.60	0.65	0.70	0.75	0.80	0.85	0.90	1.00	0.10	0.15	0.20	0.25	0.30	0.35	0.40	0.50	0.60	0.70	0.80	0.90	U
42.0	45.5	49.0	52.5	56.0	59.5	63.0	70.0	3.50	5.25	7.00	8.75	10.5	12.3	14.0	17.5	21.0	24.5	28.0	31.5	$U(T_1\text{-}T_0)$
5.36	4.95	4.59	4.29	4.02	3.78	3.57	3.21	64.3	42.9	32.1	25.7	21.4	18.3	16.1	12.9	10.7	9.18	8.04	7.14	1
10.7	9.90	9.18	8.58	8.04	7.56	7.14	6.42	129	85.8	64.2	51.4	42.8	36.6	32.2	25.8	21.4	18.4	16.1	14.3	2
16.1	14.8	13.8	12.9	12.1	11.3	10.7	9.63	193	129	96.3	77.1	64.2	54.9	48.3	38.7	32.1	27.5	24.1	21.4	3
21.4	19.8	18.4	17.2	16.1	15.1	14.3	12.8	257	172	128	103	85.6	73.2	64.4	51.6	42.8	36.7	32.2	28.6	4
26.8	24.7	22.9	21.4	20.1	18.9	17.8	16.0	321	214	160	128	107	91.5	80.5	64.5	53.5	45.9	40.2	35.7	5
32.2	29.7	27.5	25.7	24.1	22.7	21.4	19.3	386	257	193	154	128	110	96.6	77.4	64.2	55.1	48.2	42.8	6
37.5	34.6	32.1	30.0	28.1	26.5	25.0	22.5	450	300	225	180	150	128	113	90.3	74.9	64.3	56.3	50.0	7
42.9	39.6	36.7	34.3	32.2	30.2	28.6	25.7	514	343	257	205	171	146	129	103	85.6	73.4	64.3	57.1	8
48.2	44.5	41.3	38.6	36.2	34.0	32.1	28.9	579	386	289	231	193	165	145	116	96.3	82.6	72.4	64.3	9
53.6	49.5	45.9	42.9	40.2	37.8	35.7	32.1	643	429	321	257	214	183	161	129	107	91.8	80.4	71.4	10
59.0	54.4	50.5	47.2	44.2	41.6	39.3	35.3	707	472	353	282	235	201	177	142	118	101	88.4	78.5	11
64.3	59.4	55.1	51.5	48.2	45.4	42.8	38.5	772	515	385	308	257	220	193	155	128	110	96.5	85.7	12
69.7	64.3	59.7	55.8	52.3	49.1	46.4	41.7	836	558	417	334	278	238	209	168	139	119	105	92.8	13
75.0	69.3	64.3	60.1	56.3	52.9	50.0	44.9	900	601	449	359	300	256	225	181	150	129	113	100	14
80.4	74.2	68.8	64.3	60.3	56.7	53.6	48.1	964	643	481	385	321	274	241	193	160	138	121	107	15
85.8	79.2	73.4	68.6	64.3	60.5	57.1	51.4	1029	686	514	411	342	293	258	206	171	147	129	114	16
91.1	84.1	78.0	72.9	68.3	64.3	60.7	54.6	1093	729	546	436	364	311	274	219	182	156	137	121	17
96.5	89.1	82.6	77.2	72.4	68.0	64.3	57.8	1157	772	578	462	385	329	290	232	193	165	145	129	18
102	94.0	87.2	81.5	76.4	71.8	67.8	61.0	1222	815	610	488	407	348	306	245	203	174	153	136	19
107	99.0	91.8	85.8	80.4	75.6	71.4	64.2	1286	858	642	514	428	366	322	258	214	184	161	143	20
113	104	96.4	90.1	84.4	79.4	75.0	67.4	1350	901	674	539	449	384	338	271	225	193	169	150	21
118	109	101	94.4	88.4	83.2	78.5	70.6	1415	944	706	565	471	403	354	283	235	202	177	157	22
123	114	106	98.7	92.5	86.9	82.1	73.8	1479	987	738	591	492	421	370	296	246	211	185	164	23
129	119	110	103	96.5	90.7	85.7	77.0	1543	1030	770	616	514	439	386	309	257	220	193	171	24
134	124	115	107	100	94.5	89.2	80.2	1607	1072	802	642	535	457	402	322	267	229	201	178	25

←— These Items Figured on Basis of $U\left(\dfrac{T_1 - T_0}{2}\right)$ —→

TABLE 20

Combined Temperature and Exposure Factors

City	Base Temp.	Temp. Factor	N	NE	E	SE	S	SW	W	NW
Albany, N. Y............	+ 5°	0.93	1.02	1.02	0.97	0.93	0.93	0.93	1.02	1.02
Baltimore, Md..........	+30°	0.57	0.80	0.80	0.74	0.57	0.74	0.74	0.80	0.80
Birmingham, Ala.......	+30°	0.57	0.66	0.66	0.57	0.57	0.57	0.60	0.66	0.66
Boston, Mass...........	+15°	0.79	1.02	0.86	0.79	0.79	0.79	1.02	1.02	1.02
Buffalo, N. Y..........	0°	1.00	1.00	1.00	1.00	1.00	1.25	1.40	1.40	1.40
Chicago, Ill............	+10°	0.86	1.07	0.86	0.86	0.86	0.99	1.16	1.16	1.16
Cincinnati, Ohio........	+15°	0.79	0.86	0.79	0.79	0.79	1.06	1.06	1.06	0.94
Cleveland, Ohio........	+ 5°	0.93	1.07	1.00	1.00	0.93	1.00	1.07	1.07	1.07
Columbus, Ohio........	+15°	0.79	0.94	0.90	0.79	0.94	1.07	1.07	1.07	0.94
Denver, Colo.*........	+20°	0.80	1.04	1.04	0.96	1.00	1.00	1.00	0.80	1.04
Detroit, Mich..........	0°	1.00	1.10	1.00	1.00	1.00	1.10	1.10	1.10	1.10
Eastport, Me..........	+10°	0.86	1.24	1.03	1.03	0.86	0.86	1.24	1.24	1.24
Grand Rapids, Mich....	+15°	0.79	0.87	0.79	0.79	0.79	0.84	0.87	0.87	0.87
Green Bay, Wis.........	− 5°	1.07	1.07	1.07	1.07	1.07	1.12	1.18	1.18	1.18
Greensboro, N. C.......	+35°	0.50	0.60	0.60	0.60	0.50	0.50	0.60	0.60	0.60
Houston, Texas........	+40°	0.43	0.81	0.56	0.51	0.43	0.43	0.43	0.81	0.81
Indianapolis, Ind.......	+15°	0.79	1.03	0.84	0.84	0.79	0.90	0.99	1.03	1.03
Ithaca, N. Y...........	+15°	0.79	0.87	0.79	0.84	0.84	0.84	0.84	0.87	0.87
Kansas City, Mo.......	+15°	0.79	1.14	1.06	0.79	0.79	0.86	0.86	1.14	1.14
Los Angeles, Cal.......	+50°	0.29	0.43	0.43	0.43	0.29	0.29	0.29	0.43	0.43
Louisville, Ky..........	+20°	0.71	0.93	0.93	0.71	0.75	0.75	1.04	1.04	1.04
Madison, Wis..........	+ 5°	0.93	1.16	1.07	1.02	0.93	1.02	1.16	1.16	1.16
Memphis, Tenn........	+30°	0.57	0.80	0.68	0.63	0.57	0.74	0.74	0.80	0.80
Milwaukee, Wis........	+10°	0.86	1.07	0.86	0.86	0.86	0.99	1.16	1.16	1.16
New Orleans, La........	+45°	0.36	0.54	0.50	0.45	0.36	0.36	0.36	0.54	0.54
New York, N. Y........	+10°	0.86	1.29	1.07	0.86	0.86	0.86	1.14	1.29	1.29
Norfolk, Va............	+30°	0.57	0.86	0.74	0.68	0.57	0.57	0.68	0.86	0.86
Philadelphia, Pa........	+15°	0.79	0.94	0.86	0.86	0.79	0.79	0.79	0.94	0.94
Pittsburgh, Pa.........	+15°	0.79	1.02	0.79	0.79	0.79	1.02	1.06	1.06	1.06
Portland, Ore..........	+25°	0.64	0.64	0.64	0.64	0.64	0.64	0.64	0.64	0.64
Providence, R. I........	+15°	0.79	1.18	0.98	0.79	0.79	0.86	0.98	1.18	1.18
Richmond, Va..........	+30°	0.57	0.77	0.71	0.71	0.57	0.57	0.74	0.77	0.77
Rochester, N. Y........	+10°	0.86	0.90	0.86	0.86	0.86	1.07	1.11	1.11	1.11
St. Louis, Mo..........	+20°	0.71	0.93	0.86	0.71	0.86	0.86	0.86	0.93	0.93
St. Paul, Minn.........	− 5°	1.07	1.28	1.07	1.07	1.07	1.07	1.18	1.28	1.28
Sacramento, Cal........	+45°	0.35	0.45	0.42	0.42	0.42	0.42	0.35	0.45	0.45
Salt Lake City, Utah...	+25°	0.64	0.71	0.64	0.71	0.71	0.71	0.64	0.71	0.71
San Antonio, Texas.....	+45°	0.36	0.61	0.61	0.50	0.36	0.36	0.36	0.61	0.61
San Diego, Cal.........	+55°	0.20	0.20	0.23	0.23	0.23	0.23	0.23	0.27	0.27
San Francisco, Cal......	+45°	0.36	0.43	0.43	0.43	0.36	0.36	0.36	0.36	0.41
Seattle, Wash..........	+25°	0.64	0.64	0.64	0.64	0.80	0.80	0.80	0.64	0.64
Syracuse, N. Y.........	0°	1.00	1.10	1.00	1.00	1.00	1.05	1.10	1.10	1.10
Washington, D. C.......	+20°	0.71	0.86	0.71	0.71	0.71	0.71	0.71	0.86	0.86
Wichita, Kans..........	+10°	0.86	1.03	1.03	0.94	0.86	0.86	0.86	0.86	1.03

* Denver base temperature and exposure factors based on actual Weather Bureau records, but due to rapid changes and high altitude both temperature factors and combined temperature and exposure factors have been corrected to care for these conditions.

(Copyright, Heating, Piping and Air Conditioning Contractors' National Association)

find 165 and to the right in this line is the corresponding radiation of 9 sq. ft. This must be added to the total radiation already computed.

The tables which have been used are based on an inside temperature of 70 degrees. If a different inside temperature is desired, the radiation is computed by the tables in the regular manner and then multiplied by the proper factor from Table 21. Let us illustrate by an example. Figure 7 represents a room in Grand Rapids, Michigan, used as a gymnasium in which a temperature of 60 degrees is desired. The problem is to find the radiation required.

Solution: (proceed as in the previous problem).

	Area or lin. ft.	Sq. ft. radiations
South wall 30 × 12 = 360 sq. ft. − 96 sq. ft. (3 windows at 32 sq. ft.)...................	264 sq. ft.	23 sq. ft.
3 windows 4 ft. × 8 ft......................	96 sq. ft.	33 sq. ft.
Infiltration...............................	84 lin. ft.	24 sq. ft.

Total without either exposure or temperature factors...... 80 sq. ft.

Exposure and temperature factor............... 0.84

Total.................... 67.2 sq. ft.

INFILTRATION TABLE

Stationary Wood Sash........	25	Rolled Section Steel Windows....	100*
Double Hung Wood Sash......	50	French Doors..................	100
Double Hung Steel Sash.......	100	Outside Doors, Residences.......	100
Casement Windows, Wood....	100	Same with Storm Doors.........	50
Casement Windows, Steel.....	50	Same with Inner Vestibule Doors.	50
		Outside Doors, Store, etc.	200

Metal Weather Strip Deducts 50 per cent

* Per foot of crack of Ventilating Sash.

Up to this point the procedure has been the same as in the previous estimate, that is, the radiation has been estimated for a room to be kept at a temperature of 70 degrees. This radiation is now multiplied by a factor found in Table 21. We look first

TABLE 21

Conversion Factors

Base Temperature	Room Temperature								
	80	75	70	65	60	55	50	45	40
− 5°	1.219	1.104	1	0.093	0.811	0.725	0.646	0.572	0.498
0°	1.228	1.111	1	0.896	0.801	0.712	0.628	0.549	0.472
+ 5°	1.239	1.119	1	0.892	0.791	0.698	0.608	0.525	0.447
+10°	1.253	1.123	1	0.886	0.780	0.680	0.586	0.498	0.415
+15°	1.269	1.13	1	0.878	0.765	0.659	0.569	0.465	0.375
+20°	1.289	1.14	1	0.870	0.748	0.634	0.528	0.427	0.332
+25°	1.312	1.151	1	0.859	0.728	0.604	0.489	0.380	0.277
+30°	1.343	1.166	1	0.845	0.702	0.566	0.44	0.312	0.207
+35°	1.380	1.183	1	0.829	0.669	0.519	.....	.....	.....
+40°	1.433	1.21	1	0.806	0.627	0.453	.....	.....	.....
+45°	1.504	1.243	1	0.773	0.561	0.363	.....	.....	.....

(Copyright, Heating, Piping and Air Conditioning Contractors' National Association)

FORMULA

$$\text{Factor} = \frac{Tr - Tb}{70 - Tb} \times \frac{Ts - 70}{Ts - Tr}$$

Tr = Room Temperature
Tb = Base Temperature
Ts = 215 deg.

To calculate amount of radiation required for other room temperatures than 70 deg., compute the amount for 70 deg. and multiply by the factor shown corresponding to room temperature desired and *proper base temperature.*

for the proper column (60 degrees) and then follow down till we reach the line of the proper base temperature (+15°, see Table 20). The factor is 0.765.

Then, $67.2 \times 0.765 = 51.41$ sq. ft. of radiation required. (Ans.)

Similarly, if the amount of radiation is wanted when the room is to be heated by wall coils, by indirect steam radiation, by vapor radiation, by hot water radiation, etc., the factors for conversion given in Table 22 are used.

TABLE 22

RADIATOR TRANSMISSION FACTORS

For Room Temperature of......................   70 Deg. F.
And Steam Pressure of.........................   1 Lb. Gage
Direct Steam Radiation (Standard 3 Col. 38 in.
  High)......................................   225 Btu per square foot.

Multiply by the following factors for the equivalent of 3 Col. 38 in. radiation of the following types.

Wall Coil................... 0.75
Double Wall Coil........... 0.90
Ceiling Coil................ 1.00
Wall Radiator.............. 0.82
Double Wall Radiators....... 1.00
Wall Radiator (Ceiling)...... 1.00

	Increase Surface
Indirect Steam Radiation..............	50 percent
Direct Indirect Steam Radiation........	25 percent

Vapor Radiation: Open return line vapor systems, on which thermostatic traps are not used, require 10 percent to 20 percent additional surface in each radiator to act as a condenser and prevent the flow of steam into the return main.

Hot Water Radiation: In figuring hot water radiators, assume mean temperature of the water in the radiators to be 170 deg. Under this condition the amount of hot water radiating surface may be determined by adding 50 percent to the amount of steam radiating surface figured.

Approximate Method of Estimating Radiation.—A method of estimating radiation known as the "2–20–200 Method," formerly widely used, is not accurate but is presented here because it may be used for quick rough estimates. It calls for one square foot of radiating surface for each 2 square feet of glass surface, one for each 20 square feet of net outside walls and one for each 200 cubic feet of room contents. We may express it in this fashion.

$$\text{Steam radiation, sq. ft.} = \frac{G}{2} + \frac{W}{20} + \frac{C}{200}$$

where G = glass area in sq. ft.

W = net exposed wall area in sq. ft.

C = cubical contents of room in cu. ft.

Applying this to the problem of Fig. 6 solved on page 483 we have

$$\text{Steam radiation} = \frac{26}{2} + \frac{195}{20} + \frac{1428}{200} = 30 \text{ sq. ft.} \quad \text{(Ans.)}$$

In designing a heating system the equivalent steam radiation for each room of a house must, of course, be estimated. These figures are used not only to determine the sizes of the radiators required but also the sizes of pipes and boilers.

Selecting Size of Boiler.—The boiler of a heating system must provide capacity for:

1. The radiators which heat the rooms.

2. The heat lost in the pipes.

3. The heat consumed in water heaters and other appliances.

4. Reserve capacity needed for starting up a cold system, for intermittent firing, and careless operation.

Since the capacities of commercial heating boilers are rated in terms of square feet of equivalent direct radiation it is convenient to reduce all of the factors to these terms. The radiation for a house is, of course, the sum of the radiation required by each room.

The loss of heat from the pipes varies with the installation and the degree and kind of pipe covering, if any. However, a flat allowance of 25 percent for steam systems and 35 percent for hot water systems, of the total radiation for the house is considered good practice for general installations.

The allowance of equivalent radiation for water-heating appliances is made on the following basis: * for water boilers with coil

* From *The Ideal Fitter,* American Radiator Co.

in firebox, 2½ sq. ft. equivalent direct water radiation per gallon of storage tank capacity; for externally attached water heaters below water-line of steam boilers, 1½ sq. ft. of equivalent direct steam radiation per gallon of tank capacity; and for externally attached water heaters below the water line of steam boilers *without* storage tank, 4 square feet for each gallon of water heated per hour.

The reserve capacity needed for small boilers is from 50 to 65 percent of the total capacity needed for other purposes.

ILLUSTRATION: A building has an estimated direct radiation requirement for steam heating of 440 square feet and an externally attached water heater connected to a 120-gallon storage tank.

What rated boiler capacity will it require if a reserve capacity of 60 percent is deemed ample?

440 sq. ft. steam radiation............ = 440 sq. ft. edr.*
Piping loss (440 × 0.25)............. = 110 sq. ft. edr.
Water heater (120 × 1½)............. = 180 sq. ft. edr.
                                         ———
    Total......................... = 730 sq. ft. edr.
Capacity for warming up (730 × 0.60). = 438 sq. ft. edr.
    Required capacity of boiler...... = 1168 sq. ft. edr. (Ans.)

Reputable boiler manufacturers have accurate performance records for all of their boilers and are prepared to guarantee the rated capacities.

Warm Air Heating Systems.—The design of warm air heating systems has been more or less standardized and published in the form of a Code by the National Warm Air Heating Association. It is possible here to illustrate only the essential steps. These involve the determination of the following items: †

---

* Equivalent direct radiation.

† From the *Guide,* American Society of Heating and Ventilating Engineers.

1. The heat loss in Btu per hour from each room in the building.

2. Area and diameter in inches of warm-air pipes in basement (known as leaders).

3. Area and dimensions in inches of vertical pipes (known as wall stacks).

4. Free and gross area and dimensions in inches of warm-air registers.

5. Area and dimensions of recirculating or outside air ducts in inches.

6. Free and gross area and dimensions in inches of recirculating registers.

7. Size of furnace necessary to supply the warm air required to overcome the heat loss from the building. This *size* should include square inches of leader pipe area which furnace must supply. It is also desirable to call for a minimum bottom fire-pot diameter in inches, which is the nominal grate diameter.

8. Area and dimensions in inches of chimney and smoke pipe.

Heat Loss in Btu per Hour.—The heat loss in Btu per hour can be arrived at conveniently and with sufficient accuracy by using the tables for estimating steam radiation and multiplying by 225, the equivalent Btu emission per hour on which the tables are based. Thus, in the problem of estimating the heat requirements for the room in Fig. 6, we found that the direct steam radiation required was 32.9 square feet. Multiplying this by 225 we find the heat loss to be about 7400 Btu per hour.

Size of Leader Pipes.—When a warm air system is designed to give an air temperature of 175 degrees Fahrenheit at the registers, and H represents the heat in Btu per hour to be supplied to a room, then the approximate area of the leaders should be:

> For the first floor, $0.009H$ sq. in.
> For the second floor, $0.006H$ sq. in.
> For the third floor, $0.005H$ sq. in.

ILLUSTRATION: A first-floor room requires 7400 Btu per hour. What size of leader pipe will it require?

0.009 × 7400 = 66.6 sq. in. = area of leader

$$2\sqrt{\frac{66.6}{\pi}} = 9\tfrac{1}{4} \text{ inches (approx.) = diameter of leader}$$

A 9-inch leader would be used in this case. They are installed only to the nearest inch and no leaders smaller than 8 inches in diameter should be used.

Stacks and Registers.—The sizes of wall stacks and registers do not lend themselves to mathematical determination. However, accepted practice is to make the area of stacks greater than 70 percent of the area of the leaders to which they are connected. Registers should have a net area not less than the area of the leader which connects with it.

SUPPLY REGISTERS

RETURN-AIR INTAKES

FORCED WARM-AIR REGISTER

This type can be used for ceiling, sidewall, and baseboard installations.

BASEBOARD GRAVITY REGISTER

FLOOR REGISTER

RETURN-AIR INTAKE
(for gravity system)

RETURN-AIR INTAKE
(for forced system)

Fig. 8.—Different Types of Registers and Intakes Used in Warm Air Systems. (From *University of Illinois Small Homes Council Circular* G3.1.)

Gravity Warm Air.—In a gravity warm air heating system the warm air rises upward through the house from a centrally located furnace. The cooler air in the rooms flows downward through the return air intakes back to the furnace. This type of system is economical to install and lends itself to low-cost homes. It is simple to operate, has no electric motors or controls, and responds rapidly to changes in outdoor temperatures. A house with a compact floor plan will heat well with gravity warm air since the leader pipes and return air ducts will be very short.

This type of system is not suitable for heating basementless homes or for heating basement rooms since the furnace must be below the level of the rooms to be heated in order to function

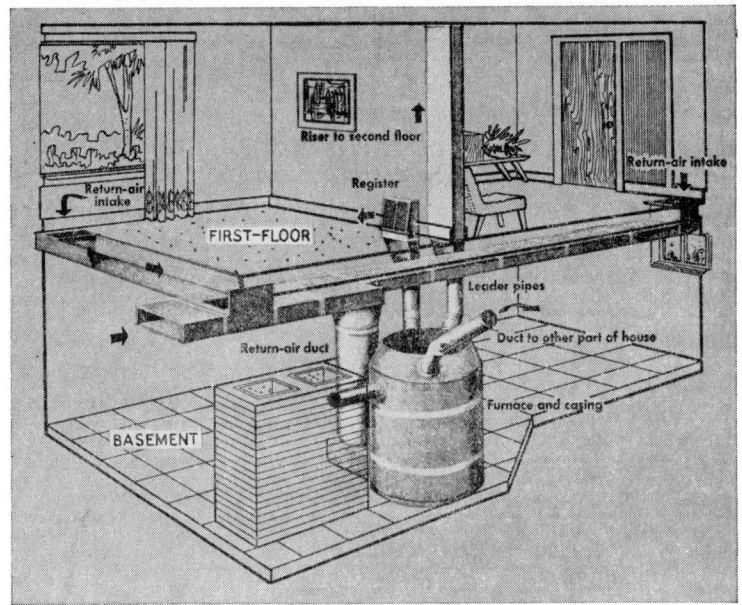

Univ. of Illinois Circular

FIG. 9.—Gravity Warm-air Heating System.

properly. By adding a blower it can be converted to a forced-air system. When making such a conversion, filters are usually added and some minor changes made in the duct system.

Forced Warm-Air System.—Similar in operation to the gravity warm-air system, the difference lies in that the air is circulated by a blower. The cooler air in the rooms is sucked down through the return air intakes and ducts to the furnace where it is heated and then recirculated to the house under slight pressure. The system responds quickly to outside temperature changes and can be used to heat basementless houses, large structures, and basement rooms since air circulation is maintained by the blower.

This type of installation costs more than the gravity system, but requires less space for the ducts and furnace. Also the furnace

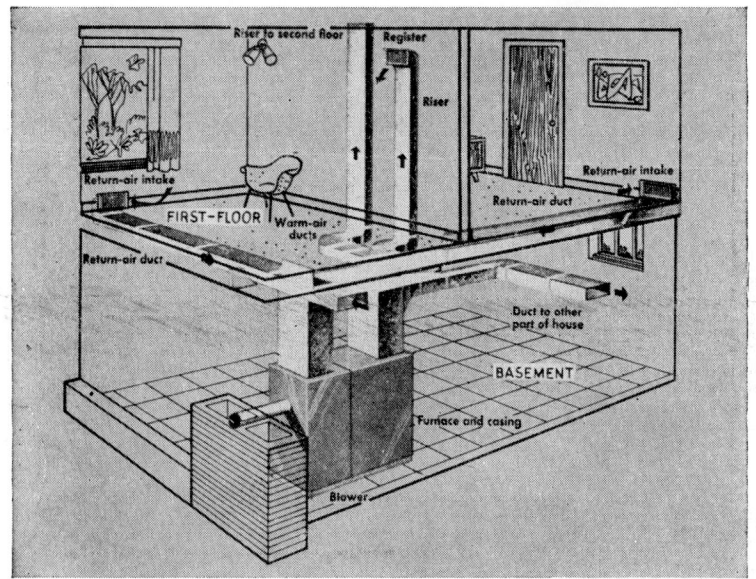

Univ. of Illinois Circular

Fig. 10.—A Forced Warm-air Heating System.

does not have to be centrally located to be assured of proper heat distribution. A humidifier and filters can be easily used, whereas in a gravity system filters would restrict the flow of air.

Perimeter Heating System.—Perimeter heating lends itself best to basementless houses built on a concrete slab. The duct system, embedded in the concrete, encircles the slab. Warm air in the ducts enters the rooms through floor registers while the cool air returns to the furnace through intakes at high locations on the inside walls. As with the forced warm-air system a blower is required to maintain proper circulation of the air.

A perimeter heating system is economical to install and needs very little floor space if a furnace specifically designed for it is used. It eliminates cold floors, can be used with a humidifier and filters, and lends itself to standard thermostatic controls. A well-constructed concrete slab laid on proper fill is essential

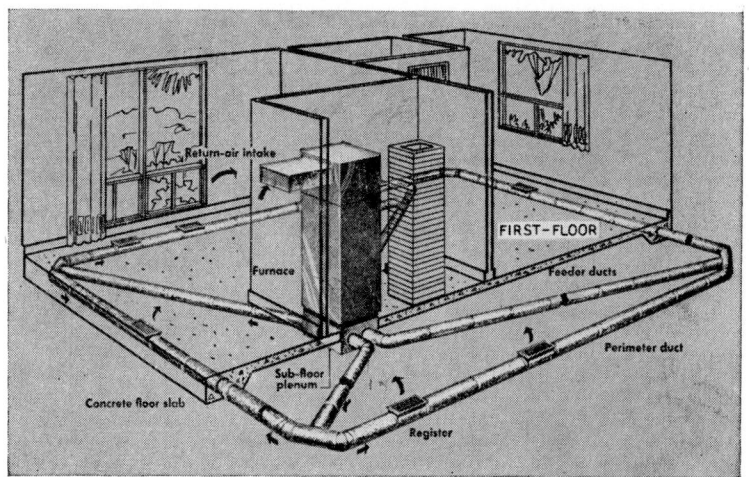

Univ. of Illinois Circular

Fig. 11.—Warm-air Perimeter Installation.

for proper operation. Several duct arrangements are possible. The ducts can be one of the standard precast forms such as concrete pipe or vitrified tile, or sheet metal.

Warm-Air Panel System.—A hollow sheet metal panel is placed in the ceiling, floor, or wall, through which warm air is circulated from the furnace by a blower and duct system. Heat is transmitted from the panel to the room by radiation and convection currents. No registers are required, but they are sometimes used.

Since a panel is part of the house, it can only be installed in a new building. The system can be used with or without a basement, but has definite temperature limitations. Floors cannot exceed 85° F., while walls and ceilings should not go higher than approximately 120° F. Proper insulation in back of the panel is essential for economical operation of this type of system. While well suited for standard thermostatic control, panels are subject to overheating or underheating where rapid temperature changes may occur.

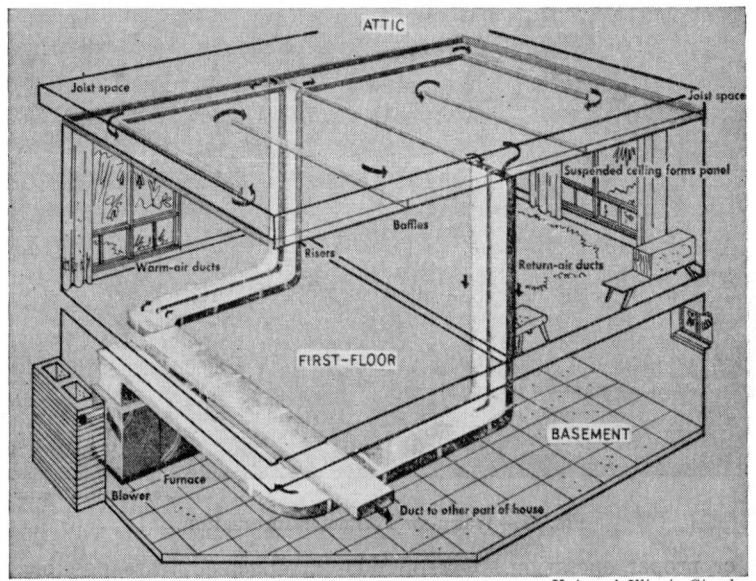

Univ. of Illinois Circular

FIG. 12.—Ceiling Type Warm-air Panel Installation.

One-Pipe Steam System.—Steam generated in the boiler rises to the room heating units where it condenses, forms water, and returns to the boiler. When it condenses it gives up its "heat of vaporization" to the radiator, which in turn heats the room. In a one-pipe system the same pipe which carries the steam to the room also returns the water (condensed steam) to the boiler. The pipes must therefore be larger than those used in other types of systems.

This type of system is simple and economical to install. No auxiliary motors, pumps, or blowers are required other than those needed for operation of an automatic fuel burner (if one is used). Domestic hot water can be supplied year-round if an automatic burner is used. Radiator temperatures cannot be controlled and they must be completely on, or off, for proper operation. Steam

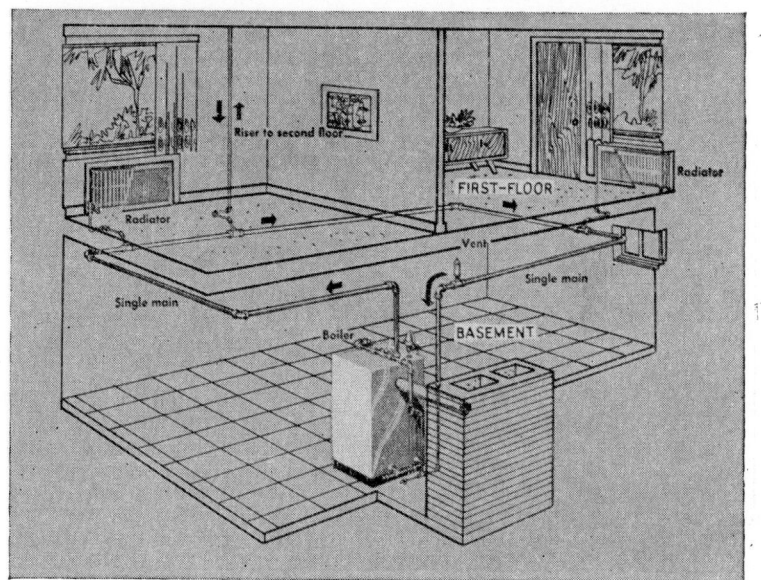

Univ. of Illinois Circular

Fig. 13.—A One-pipe Steam Heating System.

heat is not recommended for heating basementless homes, or basement rooms, because the boiler must be below the level of the radiators.

Gravity Hot-Water System.—The entire system is full of water. When the water in the boiler is heated, it rises to the room heating units. The cooler water in the rooms flows downward back to the boiler. Circulation is maintained by *gravity* or natural convection.

In an *open* system the expansion tank is placed above the highest radiator and the water is exposed to the air, or "open." In a *closed* system, such as that shown in Fig. 14, the expansion tank is located near the boiler. The heated water compresses the air in the tank, raising the pressure which in turn raises the boiling point of the water. This permits a higher temperature system and smaller heating units.

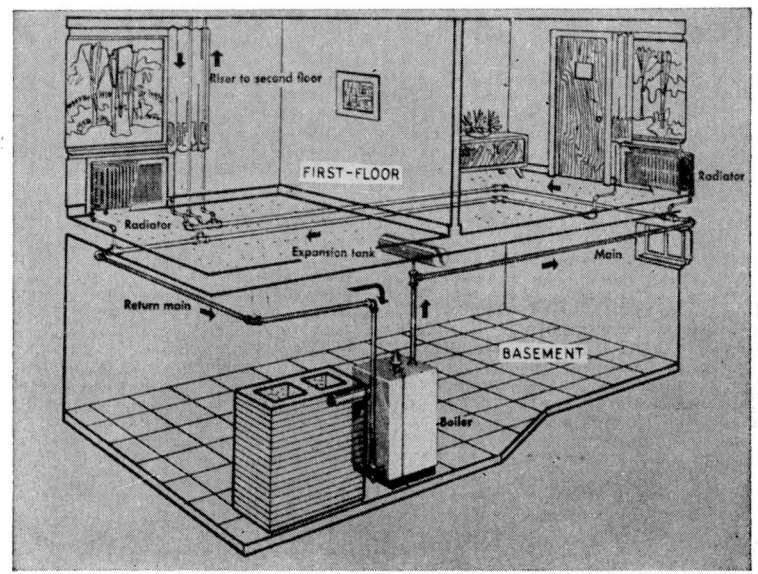

Univ. of Illinois Circular

FIG. 14.—A Gravity Hot-water Heating System (Closed Type).

It is economical to install, requiring no auxiliary motors other than those used for an automatic burner. Large supply and return mains are required for good circulation. It has a slow response to temperature changes and is not recommended for basementless houses or heating of basement rooms.

Forced Hot-Water Heating System.—As with the gravity system all pipes and heating units are full of water. The hot water from the boiler is forced through the system by a circulating pump. The one-pipe system shown in Fig. 15 has one main which supplies hot water to the heating units, and also returns the cold water to the boiler. In a two-pipe system one main supplies hot water, while the other returns the cold water back to the boiler.

This type of system responds rapidly to temperature changes and can be used for domestic hot water year-round if an automatic fuel burner is used. Since circulation is maintained by a pump,

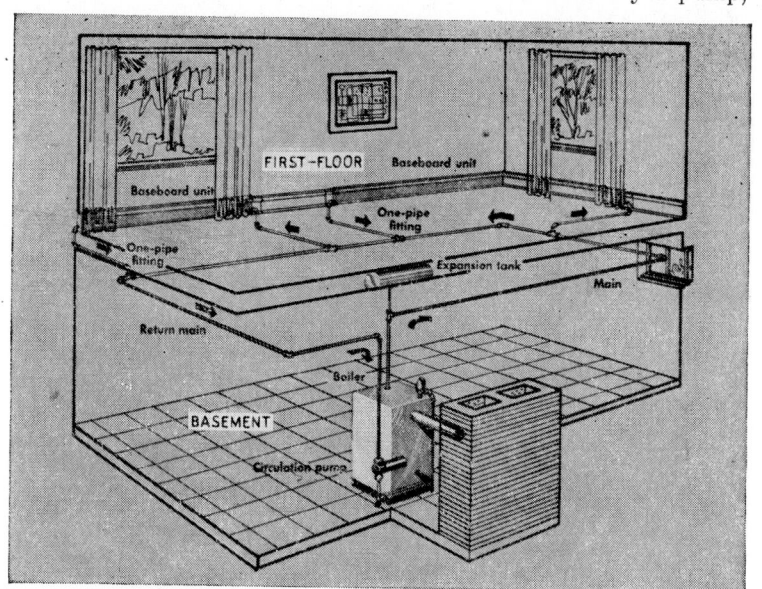

Univ. of Illinois Circular

FIG. 15.—A Forced Hot-water Heating System (One-pipe Type).

it lends itself to heating basementless houses and basement rooms. Smaller pipes can also be used because the pump is capable of maintaining circulation against relatively high friction heads. It is more expensive to install than the gravity system because of the pump and need for special fittings with the one-pipe system. The smaller pipes will compensate for this, however.

Hot-Water Panel System.—With this type of system hot water from the boiler is circulated by a pump through pipe coils (usually tubing) which are buried in the floor, ceiling, or wall. No radiators, convectors, or baseboard units are required because the tubing acts as the heating unit, transmitting heat to the room by direct radiation and indirectly by convection.

Hot-water panels can be used in houses with or without basements, but they must be built into the house. As with warm air panels temperatures cannot exceed 85° F. on floors and 120° F.

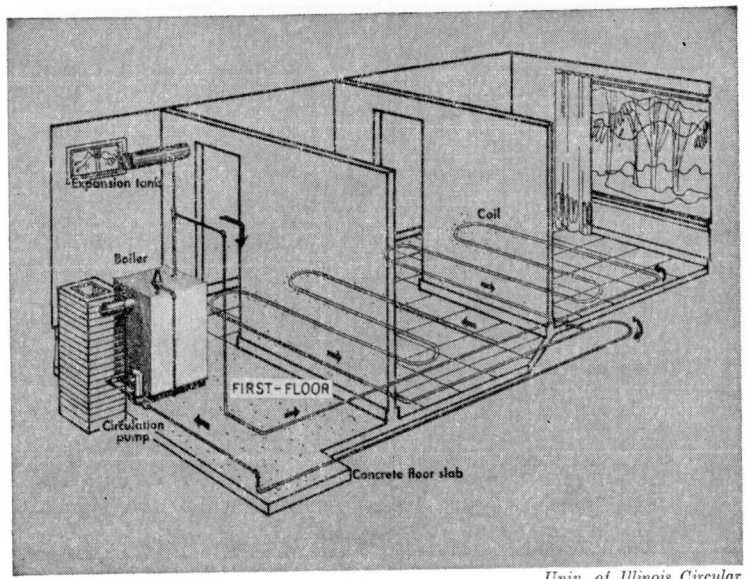

Univ. of Illinois Circular

Fig. 16.—Floor Installation of a Hot-water Panel System.

for walls or ceilings. For economical operation proper insulation in back of the panel is essential to keep heat losses down. While well suited to standard thermostatic control, the panels may overheat or underheat. This is particularly true where rapid temperature changes may occur, such as in houses with large glass areas or in certain regions of the country.

Insulation.—While several references are made in this Chapter to the effect of insulation upon heat transfer coefficients of walls, roofs, etc., and thus upon heating requirements, the subject of insulation itself has been deferred to this section. Here it can be discussed as a unit.

There are a number of materials from which insulation is made, metallurgical products like rock wool, which is made from slag; glass fiber products, such as Fiberglas®; vegetable products such as kapok; mineral products such as asbestos; and a number of blown plastic products such as polystyrene. All these materials possess the common property of having small air spaces which have low heat transmission rates. The differences between them in heat transmission are usually not the decisive consideration in choosing a type of insulation. Preference is given to other considerations, for example, vegetable materials may be eaten by rodents or vermin, and are objectionable from this point of view. Some materials absorb moisture more readily than others—from this point of view the glass and slag products are superior. In general, however, the decisive consideration in choosing an insulation is its physical form, which may be that of (1) particles (that is, granules or flocs); (2) blankets of various widths which are made by placing the particles or loosely woven fibers of insulation within containing materials, such as paper or metal foil; (3) batts, which are similar to blankets but prepared in standard lengths and widths, so they are sealed on all edges; and (4) slab insulation, which is prepared from the various insulating materials mentioned by combining them with a binder to form the slab and yet to retain their insulating properties.

The choice of one of these four forms of insulation depends somewhat upon the type of installation. The loose materials, i.e., fibrous or granular, have the advantage that they can be blown by use of a blower and hose into spaces within walls. In new house construction, however, one has a considerable choice of the type of insulation to be used. In the following tables of Fiberglas® building insulation, the various types are classified by their R values. The symbol "R" is a measure of the resistance to heat transmission, and therefore the higher the R value the better the performance. These values are specified for various types of insulation for various Fiberglas® products as shown in Table 4.

The types of insulation cited in the table are described in Table 5.

Vapor Barriers.—To enhance the effectiveness of insulation the use of vapor barriers is strongly recommended. A vapor barrier near the inside of all exposed walls, floors and ceilings can keep moisture vapor from penetrating to a point where it can condense. Two factors are vital in making vapor barriers effective:

> (1) Vapor permeance must be less than 1 perm (a measure of the ability of materials to stop moisture).
>
> (2) Vapor barrier must be kept warm enough with insulation on cold side to be sure vapor cannot condense on it.

There are a number of types of vapor barriers, including the following:

Insulation Vapor Barriers. Insulation with a kraft-asphalt vapor barrier provides good condensation protection if carefully applied.

Polyethylene Film. Use of unfaced Friction-Fit insulation with large sheets of film applied inside offers installation economies and a separate vapor barrier.

Foil-Backed Gypsum Board. Aluminum foil adhered to gypsum board in the factory provides a continuous noncombustible vapor barrier on the warm side of the wall construction and is installed in the standard manner.

Paint and Other Films. Vapor resistant paints on interior walls and ceilings offer some condensation protection as do vinyl wall coverings and other vapor-impervious interiors.

TABLE 4

	Air Conditioned Electrically Heated Quality Home Standard	Heated Only Moderate Comfort and Economy Standard	Minimum Comfort Standard
Recommended for	All electrically heated dwellings (including resistance heating, and heat pump systems) and all air conditioned dwellings. Also in Alaska (and Canadian provinces) and wherever fuel costs economically justify it. Also for homes in a price and quality bracket where the finest of products and standards prevail.	All homes heated but not air conditioned in moderate climate and rate areas. These standards help assure good comfort levels in the North and Central areas in winter and in the South in the summer. There is marked improvement in comfort and substantial reductions in heating costs compared to minimum standard at right.	Minimum housing in mild climates where low cost fuels and minimum income buyers make improved insulation unprofitable or unsalable to builder. Since low-income buyers need the fuel savings better insulation can bring, and more buyers can qualify when housing expense is cut, this standard is rarely desirable to the builder.
Roof—Ceilings	Fiberglas 6 inch Batts (R = 19), or when air conditioned but conventionally heated, Fiberglas Thick Double Foil Batts (R = 19). R = 24 also available for severe climate and high rate areas.	Fiberglas Thick Standard Batts or Roll Blankets (R = 13), or Fiberglas Medium Thick Double Foil (R = 16).	Fiberglas Medium Standard Batts or Roll Blankets (R = 9).

TABLE 4 (Cont.)

	Air Conditioned Electrically Heated Quality Home Standard	Heated Only Moderate Comfort and Economy Standard	Minimum Comfort Standard
WALLS frame	Fiberglas Thick Standard Batts or Roll Blankets (R=11), or Fiberglas Medium Foil Faced Batts or Roll Blankets (R=11s). R=13 also available.	Fiberglas Medium Standard Batts or Roll Blankets (R=8), or Fiberglas Economy Foil Faced Batts or Roll Blankets (R=8s).	Fiberglas Economy Batts or Roll Blankets (R=7).
masonry	Fiberglas Masonry Wall Batts (R=3) between nominal 1" furring strips. In cavity walls, Fiberglas Prescored Perimeter Insulation; 1⅝" thick (R=7) in 2" or wider cavity.	Fiberglas Masonry Wall Batts (R=3) between nominal 1" furring strips. In cavity walls, Fiberglas Prescored Perimeter Insulation: 1" thick (R=4) in 2" or wider cavity.	Fiberglas Masonry Wall Batts (R=3), between nominal 1" furring strips.
sill detail	Fiberglas Sill Sealer under wood sills on foundation.	Fiberglas Sill Sealer under wood sills on foundation.	Fiberglas Sill Sealer (To save caulking expense only) under wood sills on foundation.

TABLE 4 (Cont.)

	Air Conditioned Electrically Heated Quality Home Standard	Heated Only Moderate Comfort and Economy Standard	Minimum Comfort Standard
FLOORS over basements	With heated basement, no insulation. Over unheated basement areas, Fiberglas Reverse Flange Batts, Thick (R=13), installed between floor joists.	With heated basement, no insulation. Over unheated basement areas, Fiberglas Reverse Flange Batts, Medium (R=9), installed between floor joists.	With heated basement, no insulation. Over unheated basement areas Fiberglas Economy Batts or Roll Blankets (R=7), installed between floor joists.
vented crawl spaces	If used as heating plenum, Fiberglas Prescored Perimeter Insulation, 1⅝" thick (R=7), installed around inside of crawl space wall to 12" below outside grade.	If used as heating plenum, Fiberglas Prescored Perimeter Insulation 1" thick (R=4), installed around inside of crawl space to 12 inches below outside grade.	If used as heating plenum, Fiberglas Perimeter Insulation, ¾" thick (R=3), installed around inside of crawl space walls to 12 inches below outside grade.
	If vented, Fiberglas Reverse Flange Batts, Thick (R=13) installed between floor joists.	If vented, Fiberglas Reverse Flange Batts, Medium (R=9) installed between floor joists.	If vented, Fiberglas Economy Batts or Roll Blankets (R=7) installed between floor joists.
slab-on-grade	Fiberglas Prescored Perimeter Insulation, 1⅝" Thick (R=7), installed between slab edge and foundation and under slab for a total width of 24 inches.	Fiberglas Prescored Perimeter Insulation, 1" thick (R=4), installed between slab edge and foundation and under slab for a total width of 24 inches.	None in the South, unless to separate exposed sun-lit slab from house slab. In Central and North States, Fiberglas Perimeter Insulation ¾" thick (R=3) installed between slab edge and foundation and under slab for a total width of 24 inches.

TABLE 5

Application	Product	Description	Performance Data
Ceilings Floors Walls	Paper Blanket Insulation	Kraft faced product for nailing to studs or joists or unrolling in open joist spaces from above. Strong asphalted kraft paper provides a vapor barrier and folded nailing flanges for attachment. Packaged: rolls, batts in rolls and tubes.	Installed Thermal Resistance Product ↔ Thick R=13 R=11 15", 19", 23" Medium R=9 R=8 15", 19", 23" Economy R=7 R=7 15", 19", 23"
Floors	Paper Blanket Insulation with nailing flanges	Nailing flange on breather-paper face of insulation opposite vapor barrier for installation from the cold side between floor joists in crawl spaces or over garages or other unheated spaces. Can be used in wall constructions but requires a separate vapor barrier on warm side.	Installed Thermal Resistance Product ↔ Widths Thick R=13 15", 23" Medium R=9 15", 23"
Ceilings Floors Walls	Foil-Faced Insulation	An insulation product with a genuine aluminum foil vapor barrier asphalted to one side. Full insulating value is obtained when installed with an air space of at least ¾" facing the foil. Medium and Standard for walls, and thicknesses for floors and ceiling. Packaged in rolls. Super Thick packaged batts in tubes only.	Installed Thermal Resistance Product ↓ ↓ ↔ Widths Super Thick R=23s R=25s ... 15", 23" Extra R=14s R=16s R=11 15", 23" Full R=13s R=15s R=10 15", 23" Medium R=10s R=13s R=11s 15", 23" Standard R=9s R=11s R=9s 15", 23"

TABLE 5 (Cont.)

Application	Product	Description	Performance Data
Ceilings	Double-Foil Insulation	Utilizing the heat reflective value of genuine aluminum foil, this product has a solid foil sheet asphalted to one side, providing a vapor barrier, and a vapor porous foil breather paper on the other side. Specifically designed to give maximum thermal performance in minimum space. Packaged: rolls and batts in tubes.	**Installed Thermal Resistance** Product · · Widths Thick · R=15s · R=19s · 15″, 19″, 23″ Medium · R=11s · R=16s · 15″, 23″
Ceilings Floors	Six-inch Batts	For ceilings or floors between joists especially for air-conditioned homes or other applications where high thermal efficiency is desired. Has an asphalted kraft paper vapor barrier with nailing flanges on one side only. Special heavier density product called R=24 provides top thermal effectiveness. Packaged: batts in tubes.	**Installed Thermal Resistance** Product · · Widths 6″ · R=19 · 15″, 23″ R=24 · R=24 · 15″, 23″

XVII

Ventilating and
Air Conditioning

Ventilation is defined * in part as the process of supplying air to, or removing air from, any space by natural or mechanical means. The word itself implies quantity, but air must be of the proper quality also. The American Society of Heating and Ventilating Engineers Code of Minimum Requirements for Comfort Air Conditioning defines air conditioning ". . . as the process by which simultaneously the temperature, moisture content, movement and quality of the air in enclosed spaces intended for human occupancy may be maintained within required limits."

Fans.—A fan is the most economical method of mechanically removing air from, or supplying air to, a room or building. Fans are also used to "move air around," thus providing a desired local cooling effect. Fans are available in a wide variety of sizes and shapes, depending on where they are to be installed and how they are to be used. They are rated by the amount of air they can move in a given time: "cfm" or cubic feet per minute. Table 1 gives ventilation standards which may be used as the basis of minimum fan selection. Night air cooling by an attic fan is a popular way of cooling a house at relatively little cost. The size of the fan to be used depends on the floor area of the house and how many air changes are desired. Table 2 gives the approximate capacities of attic fans for houses of different floor areas.

* *A.S.H.V.E. Guide.*

TABLE 1

Fan Capacities Recommended *

Amount of Air Fan Should Deliver †

Floor Area of House, sq. ft.	For Regions of Cool Nights, cu. ft./min.		For Regions of Warm Nights, cu. ft./min.
800	3000	to	6500
1000	4000	to	8000
1200	5000	to	9500
1400	5500	to	11000
1600	6500	to	13000
1800	7000	to	14500

* *University of Illinois Small Homes Council Circular G6.0.*

† Capacities are for so-called "free air delivery." If obstructions or resistances to flow of air exist (such as louvers, screens, ducts), then the actual deliveries will be less than the "free air delivery." In general, low-speed fans provide quieter operation than small, high-speed units.

Air Conditioning.—Air conditioning systems cool and dehumidify warm, humid air by passing it over a cooling coil. All air contains moisture (water vapor). When the air is warm it can hold more moisture than when it is cold. In an air conditioning unit warm air, from the building or outside, is cooled by passing it over a cold evaporator coil through which refrigerant constantly circulates (the same as in an ordinary refrigerator). Since cool air cannot hold as much moisture as warm air the excess water vapor condenses out on the evaporator coil. The cooled and dehumidified air is then sent into the room or building by a fan.

The size unit needed for a particular installation will depend upon several factors such as the prevailing temperatures in the region, the area of the room or building to be cooled, the number of windows exposed to the sun or in the shade, the height of the rooms, thickness of the walls, number of people who will use the rooms, how the rooms will be used (apartment, office, shop, etc.), whether or not the rooms are directly under the roof, how the

TABLE 2

VENTILATION STANDARDS [a]

| Application | Smoking | Cfm per Person [b] | | Cfm per Sq. [b] Ft. of Floor |
		Recommended	Minimum [c]	Minimum [c]
Apartment, Average	Some	20	10	....
Deluxe	Some	30	25	0.33
Banking Space	Occasional	10	7½	....
Barber Shops	Considerable	15	10	....
Beauty Parlors	Occasional	10	7½	....
Brokers' Board Rooms	Very Heavy	50	20	....
Cocktail Bars	Heavy	40	25	....
Corridors (Supply or Exhaust)		..	..	0.25
Department Stores	None	7½	5	0.05
Directors' Rooms	Extreme	50	30	....
Drug Stores [e]	Considerable	10	7½	....
Factories [d,f]	None	10	7½	0.10
Five and Ten Cent Stores	None	7½	5	....
Funeral Parlors	None	10	7½	....
Garages [d]		..	..	1.0
Hospitals, Operating Rooms [f,g]	None	..	..	2.0
Private Rooms	None	30	25	0.33
Wards	None	20	10	....
Hotel Rooms	Heavy	30	25	0.33
Kitchens, Restaurant		..	..	4.0
Residence		..	..	2.0
Laboratories [e]	Some	20	15	....
Meeting Rooms	Very Heavy	50	30	1.25
Offices, General	Some	15	10	....
Private	None	25	15	0.25
Private	Considerable	30	25	0.25
Restaurant, Cafeteria [e]	Considerable	12	10	....
Dining Room [e]	Considerable	15	12	....
School Rooms [d]	None	..	..	....
Shop, Retail	None	10	7½	....
Theater [d]	None	7½	5	....
Theater	Some	15	10	....
Toilets [d] (Exhaust)		..	..	2.0

[a] *A.S.H.V.E. Guide.* Taken from present-day practice or large air conditioning companies. [b] This is contaminant-free air and may be either outdoor air or recirculated air which has been appropriately purified. [c] When minimum is used, take the larger of the two. [d] See local codes which may govern. [e] May be governed by exhaust. [f] May be governed by special sources of contamination or local codes. [g] All outside air recommended to overcome explosion hazard of anesthetics. See *National Board of Fire Underwriters'* pamphlet No. 56.

building is insulated, etc. All these factors contribute to what is known as the *cooling load* which the air conditioner must be able to handle. Cooling load is expressed in *tons* (of refrigeration) or in Btu per hour. A ton of refrigeration is the amount of cooling equal to the melting of a ton of ice per day. It is also equal to 200 Btu per minute, or 12,000 Btu per hour.

Calculation of Cooling Load.—The air-conditioning unit must be large enough to absorb all the heat and moisture generated in a room or building. To determine the proper size unit it is therefore necessary to calculate the maximum amount of heat it is anticipated must be removed. A detailed procedure for doing this can be found in the *A.S.H.V.E. Guide*.

A simpler method, particularly suitable for room air-conditioners, is suggested by the Air-Conditioning and Refrigeration Institute. Table 3 shows a cooling load estimate form which, by proper use of the factors, will make it possible to calculate the cooling load of a room. For example, it is desired to calculate the cooling load and select a room air-conditioner for the bedroom shown in Fig. 1. It will be occupied by three people.

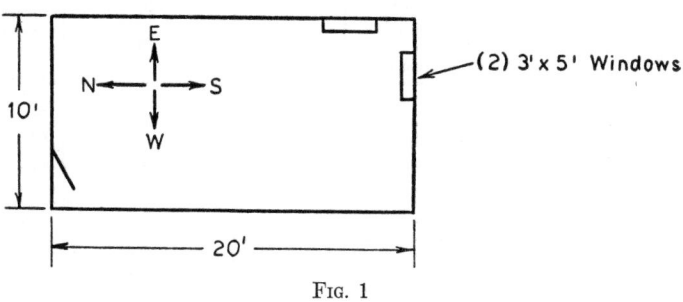

Fig. 1

The appropriate places in Table 3 are filled out as shown in the table, in the quantity column. They are then multiplied by the indicated factors and cooling units result in the last column. These values are totaled for the cooling load of the room in Btu per hour. For this bedroom the cooling load is 8,375 Btu per hour. To determine what size unit is required divide this by 12,000 Btu per hour to get tons of refrigeration. Thus, 8,375/12,000 = 0.698. The closest standard unit is 0.75 or ¾ ton.

TABLE 3

COOLING LOAD ESTIMATE FORM FOR ROOM AIR-CONDITIONERS

Recommended Practice for Members of Room Air-Conditioner Section of Air-Conditioning and Refrigeration Institute. This estimate is suitable for comfort air-conditioning jobs not requiring specific conditions of temperature and humidity.

Customer..Buyer..
Address..Space to be used for..
Estimate by................................Date................Approval................................Date................

ITEM	Quantity	Factor		Cooling Units
		Inside Shades	Outside Awnings	
1. WINDOWS EXPOSED TO SUN, FACING*				
a. East, Southeast, or South	_15_ sq ft	45	25	
b. Southwest	_15_ sq ft	65	40	_975_
c. West	.........sq ft	100	60	....
d. Northwest	.........sq ft	35	25	....
*Use only the exposure with the largest load.				
2. WINDOWS FACING NORTH OR IN SHADE (Include all windows not included in Item 1.) .	_0_ sq ft	14		....
3. WALLS (based on lineal feet of wall)				
a. Light construction, exposed to sun* . .	_10_ ft	90		_900_
b. Heavy construction, exposed to sun* . .	.........ft	50		....
c. Shaded walls or partitions (Include all walls not included in 3a or 3b.) . . .	_30_ ft	30		_900_
*Use for only that exposure used in Item 1.				
4. ROOF OR CEILING (Use one only.)				
a. Roof, uninsulated	.........sq ft	16		....
b. Roof, with one inch or more insulation .	_200_ sq ft	7		_1400_
c. Ceiling, with occupied space above . .	.........sq ft	3		....
d. Ceiling, with attic space above . . .	.........sq ft	10		....
5. FLOOR (Neglect floor directly on ground or over unheated basement.)	_200_ sq ft	3		_600_
6. PEOPLE AND VENTILATION—Number of people	_3_	900		_2700_
7. LIGHTS AND ELECTRICAL EQUIPMENT IN USE	_300_ watts	3		_900_
8. DOORS AND ARCHES CONTINUOUSLY OPEN TO UNCONDITIONED SPACE (lineal feet of width)	_0_ ft	300		....
9. TOTAL LOAD in cooling units to be used for selection of room air conditioner (s) . .	xxxx	xxxx		_8375_

Air Conditioning Systems.—Several types of systems are available, depending on the cooling load and the type of installation desired.   Central air conditioning is available in many different sizes and shapes for summer use only or for year-round use.   They usually require water-cooled condensers and the use of a water-conserving device known as a cooling tower.   Fig. 2 shows how a central air conditioner operates to keep the entire house cool.

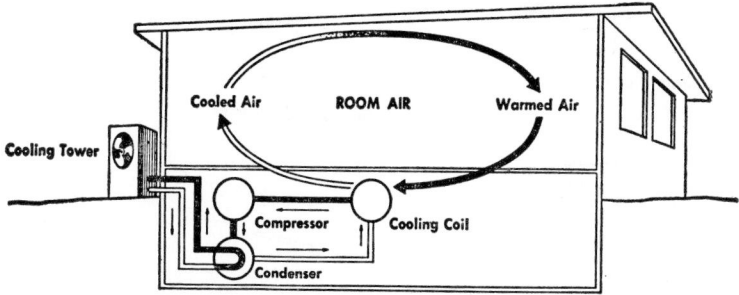

Fig. 2.—Passage of Heat and Cooled Air in a Central Air-conditioning System.   In an air-conditioned house, the warm air from the rooms is passed over a cooling coil and is then returned to the rooms as cooled air.   The heat removed from the air is eventually discharged from the house by means of a condenser.   With water-cooled condensers, a cooling tower makes re-use of water possible.   (From *Univ. of Illinois Circular.*)

If a warm air system, with ducts, exists it may be possible to install air conditioner into this system.   It will be necessary to check to be certain that the ducts have ample capacity for cooling.   If the ducts do not have sufficient capacity it will be necessary to add additional ducts and supply outlets.   The most popular types of outlets are diffusers, and registers with fixed or adjustable vanes.   Fig. 3 shows different types of outlets and how they are installed.

Room air conditioners of the window and cabinet type are used where it is desired to cool one room, or a small section of a house.   The window units fit into a window opening and are available in sizes from ⅓ to 1 ton of cooling capacity.   The room air is circulated through the unit where it is cooled, dehumidified,

and filtered. Since the condenser is air cooled no water pipes
are necessary. Similarly since the moisture condensed from the
room air is evaporated directly to the outdoors no drainage system
is required. Cabinet units stand on the floor and require a wall

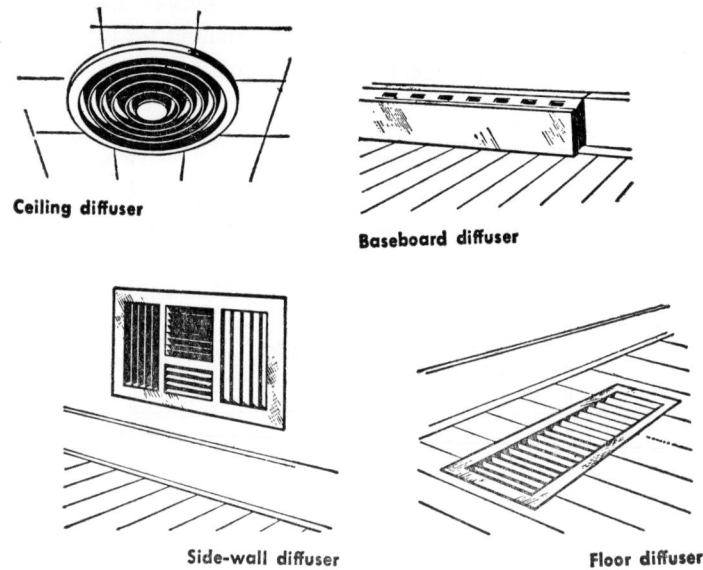

Ceiling diffuser

Baseboard diffuser

Side-wall diffuser

Floor diffuser

Fig. 3.—Different Types of Air-conditioning Outlets. (From *Univ. of Illinois Circular.*)

opening through which the air to cool the condenser is brought
in and discharged. Units of these types can usually be plugged
into existing house wiring if it is of adequate capacity. If the
wiring is inadequate it will be necessary to run a special branch
circuit for the air conditioner.

Heat pumps are a new type of year-round air conditioning
system which remove heat from the house during the summer
and transfer it to the outside ground, well, air, river, or lake.
During the winter heat is withdrawn from the ground (or other
source) and pumped into the house. Since the heat pump is
electrically operated it is economical at the present time only in
areas where the cost of electricity is low.

XVIII

Machine Shop Work

Measuring Instruments.—A knowledge of measuring instruments is one of the first things needed in machine shop work because the foundation of present-day machinery manufacture is based upon measuring instruments. All types of calipers, outside, inside, hermaphrodite, thread, vernier, micrometer, etc., and all types of gages, caliper, collar and plug, limit, internal and external threads, etc., are required to make possible the modern production system of interchangeable parts.

Methods of Measuring

Measuring is an art which must be mastered in order to produce good machine work. It can be mastered only by patient and careful practice. Proficiency in measurement will save many a job from being spoiled or rejected.

Calipers.—A steel scale and outside calipers are commonly used to measure diameters on lathe work. First, the calipers are accurately set to the required measurement as shown in Fig. **1.** Then they are tried on the work, care being taken that they are held in a plane perpendicular to the longitudinal axis of the work. When the work is of the desired size the calipers should slide snugly across the cylinder without forcing. If a number of pieces of work of the same diameter are to be measured, it is often desirable to set the calipers on a standard test cylinder. The " feel " of the calipers on the test cylinder should be carefully gaged and when the work is measured the same " feel " should be obtained when the work is of the required size.

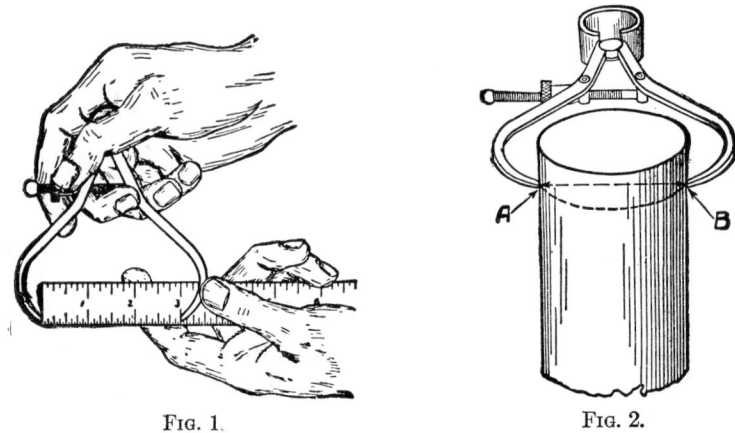

FIG. 1.   FIG. 2.

Inside calipers (Fig. 3) are used for measuring the diameters of holes. They, too, are set on steel scales and require a sensitive feel for accuracy. When a shaft is being turned to fit a certain

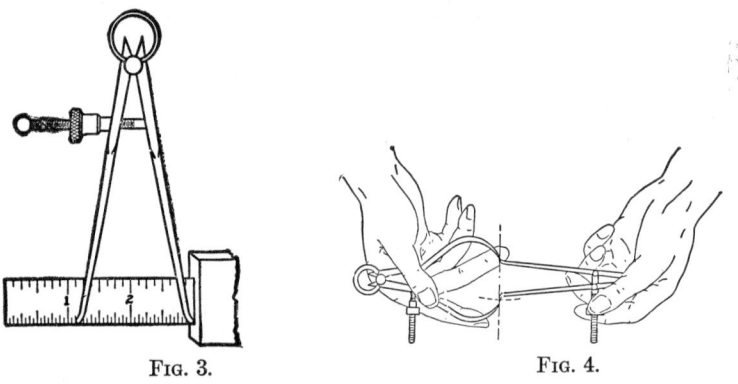

FIG. 3.   FIG. 4.

hole, the measurement of the hole may be transferred to a pair of outside calipers as shown in Fig. 4. Micrometers may also be set from the inside calipers in a similar manner.

Micrometer Calipers.—When greater accuracy is required than can be obtained by using a caliper and scale, a micrometer caliper such as shown in Fig. 5 is used. This is a precision instrument which requires the most careful treatment. The micrometer screw has 40 threads to the inch. If the sleeve or thimble is turned one complete revolution, the spindle will advance $\frac{1}{40}$ inch, which is equivalent to 0.025 inch. The sleeve has 25 graduations so that if it is turned one graduation or

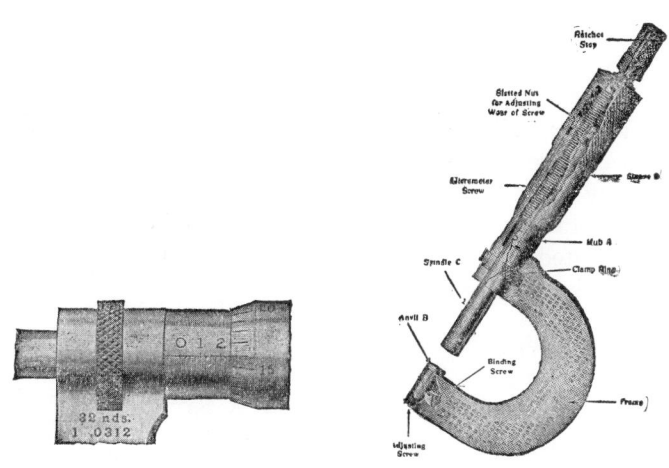

Fig. 5.—Micrometer.

$\frac{1}{25}$ of a revolution the spindle will move $\frac{1}{25}$th of $\frac{1}{40}$th of an inch or 0.001 inch. Four complete turns of the sleeve will move the spindle 4×0.025 inch or 0.100 inch. The hub or barrel of the micrometer has a number at each 0.100 inch division and four unnumbered spaces between, each representing 0.025 inch.

To read a micrometer, add the readings on the hub and the sleeve.

ILLUSTRATION: What is the reading of the micrometer shown in Fig. 6?

Numbered graduations........	0.200 inch
Unnumbered graduations......	0.050 inch
Sleeve graduations...........	0.008 inch
Total...................	0.258 inch (Ans.)

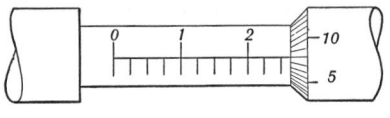

FIG. 6.

Vernier Caliper.—A caliper based on the principle that the eye can more readily judge the coincidence of two lines than it can visually interpolate between graduations on a scale is known as a vernier caliper. A simple form is shown in Fig. 7. A more complicated adaptation is used to measure gear teeth and is usually known as a gear-tooth micrometer.

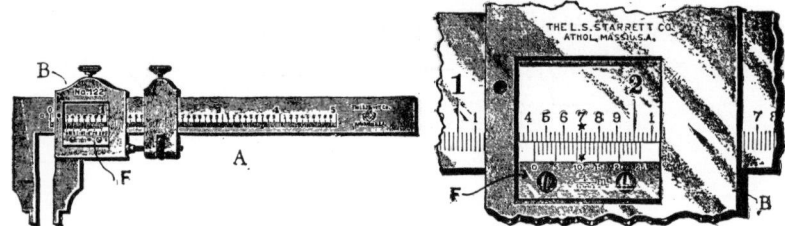

FIG. 7.—Vernier Caliper.

A vernier has a fixed scale and a sliding scale so graduated that the sum total length of its divisions is exactly equal to the sum of the lengths of one fewer divisions on the fixed scale. We will illustrate by a specific example.

On the fixed or (A) scale of Fig. 8 let the major numbered divisions represent inches. Then the minor numbered divisions are tenths of inches and the unnumbered divisions are quarters of tenths, $\frac{1}{40}$th or 0.025 inch each point. It will be noted that the movable scale has twenty-five equal divisions which aggregate a

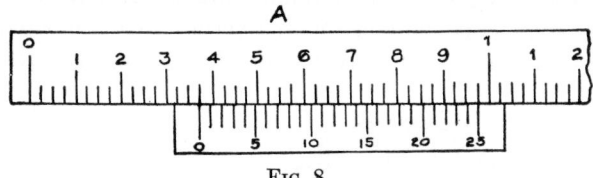

Fig. 8.

length exactly equal to twenty-four divisions on the fixed scale. The difference in length between one division on the fixed scale and one division on the sliding scale is $\frac{1}{25}$ of $\frac{1}{40}$ or 0.001 inch. Then if the sliding scale in Fig. 8 moves slightly to the right so that its second line will coincide with (4) it will have moved 0.001 inch.

To read a vernier, note the number of divisions and calibrated parts of divisions up to the zero or index of the sliding scale.

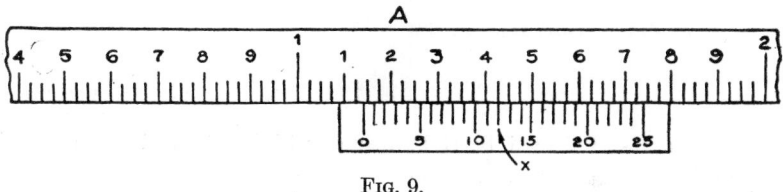

Fig. 9.

The fraction of the space on which the zero rests may be read directly on the sliding scale at the point of coincidence of any two lines.

ILLUSTRATION: What is the reading of the vernier shown in Fig. 9?

Whole inches.....................	1.000
Tenths.........................	0.100
Thousandths...................	0.025
Vernier........................	0.012
Total......................	1.137 in. (Ans.)

Vernier Bevel Protractor

A protractor is used for dividing circles into any number of equal parts or degrees and determining angles.   A bevel protractor, which is commonly combined with a vernier, is a graduated semi-circular protractor with a pivoted arm for measuring angles.

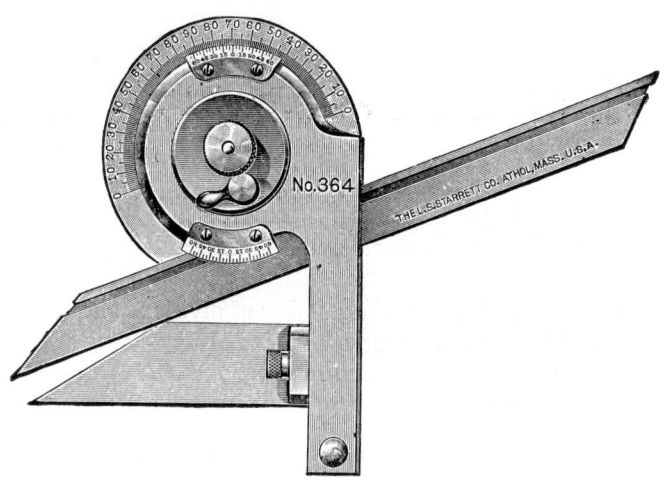

Fig. 10—Universal Bevel Protractors with Vernier and Acute
Angle Attachment.
(Courtesy of The L. S. Starrett Company).

The disc of a vernier bevel protractor is graduated in degrees from 0 to 90° each way.   The vernier plate is graduated so that 12 divisions on the vernier occupy the same space as 23 degrees on the disc.   The difference between the width of one of the 12 spaces on the vernier and two of the 23 spaces on the disc is therefore $\frac{1}{12}$ of a degree.

Each space on the vernier is $\frac{1}{12}$ of a degree, or five minutes shorter than two spaces on the disc.   If a line on the vernier coincides with a line on the disc and the protractor is rotated until the next line on the vernier coincides with the next line but one on the disc, the vernier has been moved through an arc of $\frac{1}{12}$ of a degree, or 5 minutes.

To read the protractor, note on the disc the number of whole degrees between 0 on the disc and 0 on the vernier. Then count in the same direction the number of spaces from 0 on the vernier to a line that coincides with a line on the disc. Multiply this number by 5 and the product will be the number of minutes to be added to the number of whole degrees.

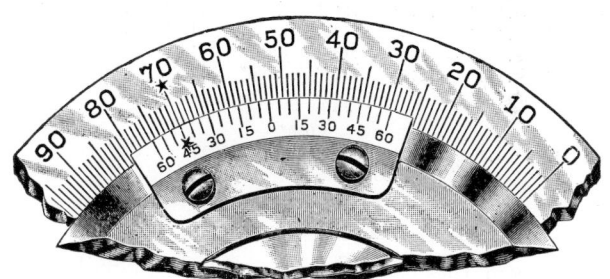

Fig. 11.—How to Read Universal Bevel Protractor with Vernier.

(Courtesy of The L. S. Starrett Company)

ILLUSTRATION: What is the reading of the vernier bevel protractor in Fig. 11.

Whole degrees...................	52°	
Minutes, 9 × 5.................		45′
		52° 45′

The starred 45 line, the 9th from zero, is the one that coincides with a line on the disc.

Gage Blocks.—Johansson Gage Blocks are rectangular pieces of tool steel, hardened, ground, stabilized, and finished to an accuracy of a few millionths of an inch. They are used to check micrometers, snap gages, sine bar settings, and any other place where extremely precise measurements are required.

A full set of eighty-one blocks, which have surfaces flat and parallel within .000008 of an inch, is made up of four series:

FIRST SERIES

.1001″	.1002″	.1003″	.1004″	.1005″	.1006″	.1007″	.1008″	.1009″

SECOND SERIES

.101″	.102″	.103″	.104″	.105″	.106″	.107″	.108″	.109″	.110″
.111″	.112″	.113″	.114″	.115″	.116″	.117″	.118″	.119″	.120″
.121″	.122″	.123″	.124″	.125″	.126″	.127″	.128″	.129″	.130″
.131″	.132″	.133″	.134″	.135″	.136″	.137″	.138″	.139″	.140″
.141″	.142″	.143″	.144″	.145″	.146″	.147″	.148″	.149″	

THIRD SERIES

.050″

.100″	.200″	.300″	.400″	.500″	.600″	.700″	.800″	.900″
.150″	.250″	.350″	.450″	.550″	.650″	.750″	.850″	.950″

FOURTH SERIES

1.000″	2.000″	3.000″	4.000″

Blocks from these series are combined to build up to the dimension which is being checked. It is always desirable to build the combination with the fewest number of blocks. To do this begin with the right-hand figure of the specified size and continue working from right to left. For example, it is desired to build a combination of blocks to check a 1.2721-inch dimension. The following block combinations will do the job:

		.1008	.1006	.1007
	.1009	.1003	.1005	.1004
.1001	.1002	.139	.138	.141
.149	.147	.132	.133	.130
.123	.124	.100	.500	.600
.900	.800	.700	.300	.200
1.2721	1.2721	1.2721	1.2721	1.2721

These blocks are made so accurately that it is necessary to wipe each of the blocks together when assembling them.

Metal Cutting

Cutting Tools.—Machine work practice has developed a number of different tools for cutting metal, all of which have, however, several points of similarity. (See Fig. 12.) They consist in general of a shank by which they are held in the cutting machine

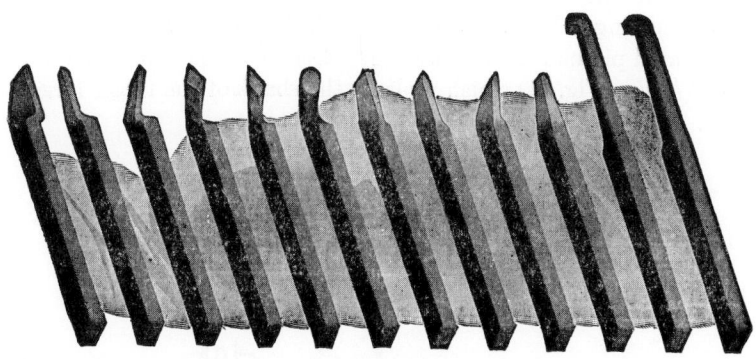

Fig. 12.

and a cutting edge which engages the metal being cut, and shears off the shaving. Figure 13 shows the shape of cutting edge of a standard forged lathe tool. The shape to which a tool is ground depends on the machine in which it is to be used, the type of cut

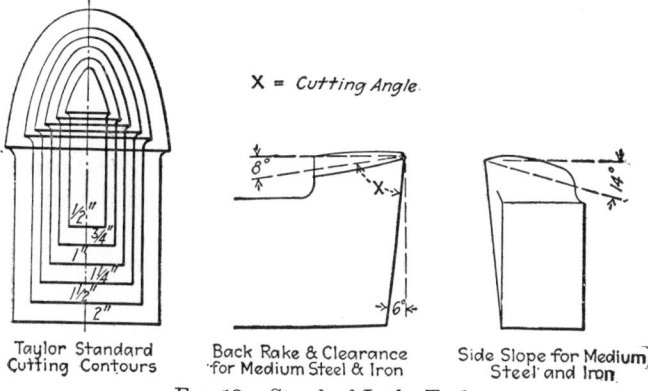

X = *Cutting Angle*

Taylor Standard
Cutting Contours

Back Rake & Clearance
for Medium Steel & Iron

Side Slope for Medium
Steel and Iron

Fig. 13.—Standard Lathe Tool.

it is to make and the hardness of the metal which is to be cut. The tool illustrated in Fig. 13 is a round-nose roughing tool. The angle marked 8° is called the *back rake* or *front top rake*; the angle marked 6° is known as the *clearance* or *front rake*; and the angle marked 14° the *side slope* or top *side rake*. Forged lathe tools such as we have discussed are used mainly for large work involving heavy cutting. For more delicate work the cutting edge is ground on a small piece of metal known as a *tool bit* which is inserted into a *tool holder* (Fig. 14) which replaces the shank of the larger forged tool.

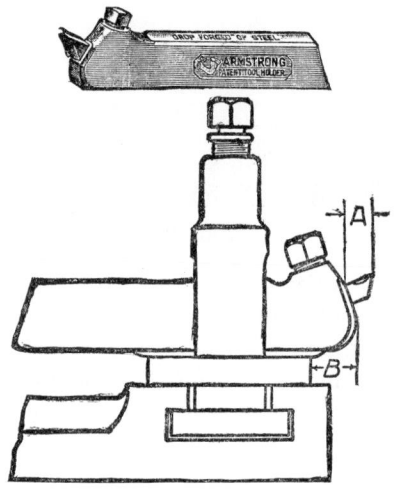

FIG. 14.

Tool steels or high carbon steels contain from 0.60 percent to 1.50 percent carbon, the hardness increasing with the amount of carbon.

High speed steels contain several other ingredients such as tungsten, chromium, manganese, silicon, molybdenum, vanadium and nickel. Tungsten and chromium in particular give the steel the property of retaining its cutting ability under very high speeds or heat.

Cutting Speed.—Cutting speed is the velocity with which a cutting tool engages the work and is always given in feet per minute (f.p.m.). The term feet per minute has somewhat different meanings for different machines. In turning work on a lathe it means the number of linear feet, measured on the surface of the work, which passes the edge of a cutting tool in one minute.

On a shaper it means the rate in f.p.m. at which the tool passes the work, while on a planer it means the rate in f.p.m. at which the work passes the tool.

On a milling machine it means the surface speed of the cutter, i.e., the speed of a point on the rim of the cutter.

The following formula may be used to calculate cutting speeds of lathes and milling machines.

$$C = \frac{\pi R D}{12}$$

where C = cutting speed

R = revolutions per minute

D = diameter of work, or diameter of cutter in inches

π = $3\frac{1}{7}$ or 3.1416

ILLUSTRATION: What is the cutting speed if a piece of work $\frac{1}{2}$ inch in diameter is turning at 458 revolutions per minute?

$$C = \frac{\pi D R}{12}$$

$$= \frac{22 \times 1 \times 458}{7 \times 2 \times 12} = 60 \text{ feet per minute} \quad \text{(Ans.)}$$

If a certain cutting speed is wanted, the proper revolutions per minute may be found by the following transposition of the preceding formula:

$$R = \frac{12C}{\pi D}$$

ILLUSTRATION: A cutting speed of 80 feet per minute is desired on a piece of work whose average diameter is 2 inches. What speed of the machine will be required?

TABLE 1

TABLE OF SPEEDS

Diam. In.	Cutting Speeds in Feet per Minute								
	20	30	40	50	60	70	80	90	100
	Revolutions per Minute								
1/4	306	458	611	764	916	1070	1222	1376	1528
3/8	204	306	407	509	612	712	814	916	1019
1/2	153	229	306	382	458	534	612	688	764
5/8	122	183	244	306	366	428	488	550	611
3/4	102	153	204	255	306	356	408	458	509
7/8	87	131	175	218	262	306	350	392	437
1	76	115	153	191	230	268	306	344	382
1 1/8	68	102	136	170	204	238	272	306	340
1 1/4	61	92	122	153	184	214	244	274	306
1 3/8	56	83	111	139	167	194	222	250	278
1 1/2	51	76	102	127	152	178	204	228	255
1 5/8	47	71	94	118	141	165	188	212	235
1 3/4	44	65	87	109	130	152	174	196	218
1 7/8	41	61	82	102	122	143	163	183	204
2	38	57	76	95	114	134	152	172	191
2 1/8	36	54	72	90	108	126	144	162	180
2 1/4	34	51	68	85	102	119	136	153	170
2 3/8	32	48	64	80	97	112	129	145	161
2 1/2	31	46	61	76	92	106	122	134	153
2 5/8	29	44	58	73	88	102	117	130	146
2 3/4	28	42	56	70	83	97	111	125	139
2 7/8	27	40	53	67	80	93	106	119	133
3	25	38	51	64	76	90	102	114	127

Diam. In.	Cutting Speeds in Feet per Minute							
	110	120	130	140	150	160	170	180
	Revolutions per Minute							
1/4	1681	1833	1986	2139	2292	2462	2615	2780
3/8	1120	1222	1324	1426	1528	1632	1735	1836
1/2	840	917	993	1070	1146	1221	1298	1374
5/8	672	733	794	856	917	976	1036	1098
3/4	560	611	662	713	764	816	867	918
7/8	480	524	568	611	655	699	742	786
1	420	458	497	535	573	611	649	687
1 1/8	373	407	441	475	509	542	576	610
1 1/4	336	367	397	428	458	489	520	551
1 3/8	306	333	361	389	417	444	472	500
1 1/2	280	306	331	357	382	407	433	458
1 5/8	259	282	306	329	353	377	400	423
1 3/4	240	262	284	306	327	349	371	393
1 7/8	224	244	265	285	306	326	346	366
2	210	229	248	267	287	306	324	344
2 1/8	198	216	234	252	270	288	306	323
2 1/4	187	204	221	238	255	272	289	306
2 3/8	177	193	210	225	241	257	273	290
2 1/2	168	183	199	214	229	244	260	275
2 5/8	160	175	189	204	218	233	248	262
2 3/4	153	167	181	194	208	222	236	250
2 7/8	146	159	173	186	199	213	226	239
3	140	153	166	178	191	204	216	229

$$R = \frac{12C}{\pi D}$$

$$= \frac{12 \times 80 \times 7}{2 \times 22} = \frac{1680}{11} = 153 \text{ r.p.m.} \quad \text{(Ans.)}$$

ILLUSTRATION: Find the cutting speed of a side facing milling cutter 6 inches in diameter running at 30 revolutions per minute.

$$C = \frac{\pi R D}{12}$$

$$= \frac{22 \times 30 \times 6}{7 \times 12} = 47 \text{ feet per minute} \quad \text{(Ans.)}$$

Cutting speeds may conveniently be found directly by reference to Table 1.

Proper Cutting Speed.—It can readily be seen that if the cutting speed in machine work is too slow, the parts produced per day will be fewer and the costs will mount. In competitive manufacturing it is, then, necessary to run the cutting operations at the *maximum safe cutting speed*. What this speed is cannot be definitely stated. In general the maximum safe cutting speed may be defined as a speed slightly lower than that at which the tool or the work may be injured by excessive heat and the cutting edge dulled too rapidly.

Cutting speeds depend on the following conditions:

1. Kind of steel used, whether tool steel or high-speed steel.
2. Shape of tool, whether narrow or broadnosed.
3. Lip angle of tool or inclined angle of nose.
4. Position of tool in the tool post.
5. Sharpness of tool.
6. Depth of cut and amount of feed.
7. Material to be cut, whether soft, medium or hard, or whether brass, cast iron, or steel.
8. Cooling medium, whether used or not, the amount of cooling and lubricating effect produced.
9. Heat treatment of steel.

10. Elasticity of work or tool, which causes chattering.

11. Rigidity with which work is held.

12. Condition of machine to be used.

The proper cutting speed of a lathe with modern high speed tools, can be found by using the following empirical formula:

$$V = \frac{H \times S}{(\sqrt[3]{D + Y})\,(\sqrt[2]{F - Z})}$$

when V = cutting speed in feet per minute

D = depth of cut, taking $\frac{1}{64}$ inch as a unit

F = feed, taking $\frac{1}{64}$ inch per revolution as a unit

H = constant for hardness of material to be cut:

> Hard cast iron or steel, 0.6
> Medium cast iron or steel, 1.0
> Soft cast iron or steel, 2.0

S = constant for size of tool:

> 232 for $\frac{3}{4}$ in. sq. tool on cast iron
> 215 for $\frac{1}{2}$ in. sq. tool on cast iron
> 325 for $\frac{3}{4}$ in. sq. tool on steel
> 288 for $\frac{1}{2}$ in. sq. tool on steel

Y = constant:

> 3 for $\frac{3}{4}$ in. sq. tool on cast iron
> 8 for $\frac{1}{2}$ in. sq. tool on cast iron
> -2 for $\frac{3}{4}$ in. sq. tool on steel
> 0 for $\frac{1}{2}$ in. sq. tool on steel

Z = constant:

> 0 for $\frac{3}{4}$ in. sq. tool on cast iron
> 0.3 for $\frac{1}{2}$ in. sq. tool on cast iron
> 0.3 for $\frac{3}{4}$ in. sq. tool on steel
> 0.5 for $\frac{1}{2}$ in. sq. tool on steel

With carbon tool steel, the cutting speed is one-half of the above amount.

TABLE 2

CHART SHOWING APPROXIMATE CUTTING SPEEDS IN FEET PER MINUTE FOR
VARIOUS MACHINES AND MATERIALS

Material	Machine	High Speed Steel Tools	Tool Steel Tools
		Speed in Feet per Minute	Speed in Feet per Minute
Tool steel............	Drill press	50–60	20–30
	Lathe	50–70	25–35
	Miller	50–60	20–30
	Shaper	40–50	20–25
	Gear cutter	....	....
	Planer	40–50	20–50
	Screw machine	60–70	25–35
Cast iron..............	Drill press	100–170	40–80
	Lathe	75–175	40–80
	Miller	100–150	60–80
	Shaper	80–100	50–60
	Gear cutter	60–80	30–50
	Planer	70–90	40–50
	Screw machine	100–150	50–70
Machine steel..........	Drill press	100–120	50–60
	Lathe	100–150	50–70
	Miller	100–125	50–70
	Shaper	60–80	50–60
	Gear cutter	60–80	30–40
	Planer	50–70	40–50
	Screw machine	100–150	50–70
Brass, bronze..........	Drill press	200–300	100–150
	Lathe	150–300	70–150
	Miller	150–250	80–125
	Shaper	100–120	60–80
	Gear cutter	100–125	50–60
	Planer	90–100	60–70
	Screw machine	200–300	100–150
Aluminum..............	Drill press	200–300	100–150
	Lathe	200–300	100–150
	Miller	200–350	100–175
	Shaper	125–200	80–125
	Gear cutter	150–200	70–100
	Planer	150–200	75–100
	Screw machine	200–300	100–150

The above speeds should be increased or decreased according to the nature
of the work, tool, lubricant, machine, etc.

ILLUSTRATION: What is the proper cutting speed of a $\frac{1}{2}$ in. square high speed tool in a lathe when the depth of cut is $\frac{1}{32}$ inch and the feed per revolution is $\frac{1}{64}$ inch upon a piece of medium steel?

$$H = 1.0 \qquad\quad D = 2 \qquad\quad F = 1$$
$$S = 288 \qquad\quad Y = 0 \qquad\quad Z = 0.5$$

$$V = \frac{H \times S}{(\sqrt[3]{D + Y})(\sqrt[2]{F - Z})} = \frac{1 \times 288}{(\sqrt[3]{2 + 0})(\sqrt[2]{1 - 0.5})}$$

$$V = \frac{288}{(\sqrt[3]{2})(\sqrt[2]{0.5})} = \frac{288}{1.26 \times 0.71}$$

$$V = \frac{288}{0.895} = 322 \text{ ft. per minute} \quad (\textbf{Ans.})$$

Table 2 shows approximate cutting speed for various machines and materials.

Estimating Time of Making Cut.—To find the time in minutes required to take one complete cut over a part to be turned, the following formula may be used.

$$T = \frac{L}{R \times F}$$

when
T = time in minutes
L = total length of cut in inches
R = revolutions per minute
F = feed per revolution of the machine

Feed may be expressed in terms of the distance which the cutting tool advances along the work for each revolution or stroke of the machine; for example, a feed of 0.020 inch.

This is the form to be used in the above formula. Feed may also be expressed in terms of number of revolutions per inch of side motion of the cutting tool; for instance, a feed of 100 means that the cutting tool moves one inch for each 100 revolutions of the machine or the motion per revolution is 0.010 inch.

ILLUSTRATION: What will be the time required to make a cut 8 inches long if the speed of the machine is 60 r.p.m. and the feed is 0.008 inch?

$$T = \frac{L}{R \times F}$$

$$= \frac{8}{60 \times 0.008} = 17 \text{ minutes} \quad \text{(Ans.)}$$

Power Required for Cutting.—The power required to remove a given amount of metal depends on the shape and sharpness of the cutting tool, hardness of the work, depth and feed of cut, lubrication of cutting point, and also upon the kind and condition of machine.

The average horsepower required to drive the machine can be determined by the product of the amount of chips (W) multiplied by two constants (Y, Z). The quantity (Y) varies with the kind of material to be cut and (Z) with the kind of machine to be used.

Horsepower required $= YZW$

When W = weight of metal removed in pounds per hour.

Y = constant, 1.0 for cast iron
1.3 for mild steel
2.0 for tool steel
0.7 for bronze

Z = constant, 0.035 for lathe
0.030 for shaper
0.025 for miller
0.030 for drill

ILLUSTRATION: What power will be required to run a lathe at 80 r.p.m. to turn a piece of cast iron 6 inches in diameter with a $\frac{1}{32}$ inch feed and $\frac{3}{32}$ inch depth of cut?

The first problem will be to find the amount of metal removed per hour. This is represented by a ribbon $\frac{1}{32}$ inch wide, $\frac{3}{32}$ inch thick and a length represented by the cutting speed in inches per hour.

Cutting speed $= \pi \times D \times$ r.p.m. $\times 60 = \pi \times 6 \times 80 \times 60$ $= 28,800\pi$ inches per hour. This represents the length of the ribbon. The volume of the ribbon is then,

$$28,800\pi \times \tfrac{1}{32} \times \tfrac{3}{32} = \tfrac{3}{1024} \times 28,800\pi = 84.37\pi \text{ cubic inches.}$$

The weight of one cubic foot of cast iron is 450 pounds. The weight of one cubic inch is $\tfrac{450}{1728} = 0.26$ pound.

Weight (W) removed per hour is $84.37 \times 0.26 \times \pi = 68.91$ pounds per hour. $Y = 1.0$. $Z = 0.035$.
Then,

$$\text{horsepower} = YZW = 1 \times 0.035 \times 68.91 = 2.41 \text{ hp. (Ans.)}$$

TABLE 3

APPROXIMATE HORSEPOWER ELECTRIC MOTOR REQUIRED TO DRIVE VARIOUS
TYPES OF MACHINES

Drill Presses

Sensitive drill up to ½ in.	¼ to	¾ hp
12 in. to 20 in.	1	hp
24 in. to 28 in.	2	hp
30 in. to 32 in.	3	hp

Shapers

10 in. to 14 in. stroke	1	to 2	hp
16 in. to 18 in. stroke	2	to 3	hp
20 in. to 24 in. stroke	3	to 5	hp
30 in. stroke	5	to 7½	hp

Lathes

6 in. to 10 in.		1	hp
12 in. to 14 in.	1	to 2	hp
16 in. to 20 in.	2	to 3	hp
22 in. to 27 in.	3	to 5	hp
30 in. to 36 in.	7½	to 10	hp

Planers

22 in.		3	hp
24 in. to 27 in.	3	to 5	hp
30 in.	5	to 7½	hp
36 in.	10	to 15	hp
42 in.	15	to 20	hp

Universal Milling Machines

No. 1	1	to 2	hp
" 1½	2	to 3	hp
" 2	3	to 5	hp
" 3	5	to 7½	hp
" 4	7½	to 10	hp

Gear Cutters

36 in. × 9 in.	2	to 3	hp
48 in. × 10 in.	3	to 5	hp
60 in. × 12 in.	5	to 7½	hp
72 in. × 14 in.	7½	to 10	hp

Grinders

8 in. to 10 in. wheel	5	hp
12 in. to 14 in. wheel	7½	hp
16 in. to 20 in. wheel	10	hp

Taper Calculations

A piece of work is said to taper when there is a gradual and uniform increase or decrease in its diameter or thickness. Examples are, a wedge which has two plane surfaces separating at a uniform rate from the edge to the base, and a cone or lathe center (Fig. 15) whose diameter increases at a uniform rate from the apex to the base.

Fig. 15.—Wedge and Lathe Center.

Wedge-shaped pieces are used in machine design for keys to attach wheels to shafts and as tapered gibs for adjusting sliding bearings.

Conical tapers, in addition to their use on lathe centers, find a wide use on shanks of twist drills, reamers, etc. (Fig. 16.)

Fig. 16.

Amount of Taper.—The amount of taper is expressed as a certain number of inches or parts of an inch per foot and indicates a variation in diameter or thickness of that amount in twelve inches of length. For example, if a truncated cone twelve inches long is 4 inches in diameter at the small end and 5 inches in diameter at the large end, the taper is $\dfrac{5 - 4}{1} = 1$ inch. If another cone has end diameters of 4 inches and 5 inches, respectively, but is only six inches long, the taper is $\dfrac{5 - 4}{\frac{1}{2}} = 2$ inches. Tapers are also expressed in terms of degrees of the angle which one side makes with the center line axis of the work.

Standard Tapers.—Lathe centers, drilling machine spindles, tapered-shank milling cutters, and many other machine shop tools have tapers. In order to provide a degree of interchangeability of parts, machine and tool manufacturers have standardized on a few tapers which we will define.

TABLE 4

MORSE TAPERS

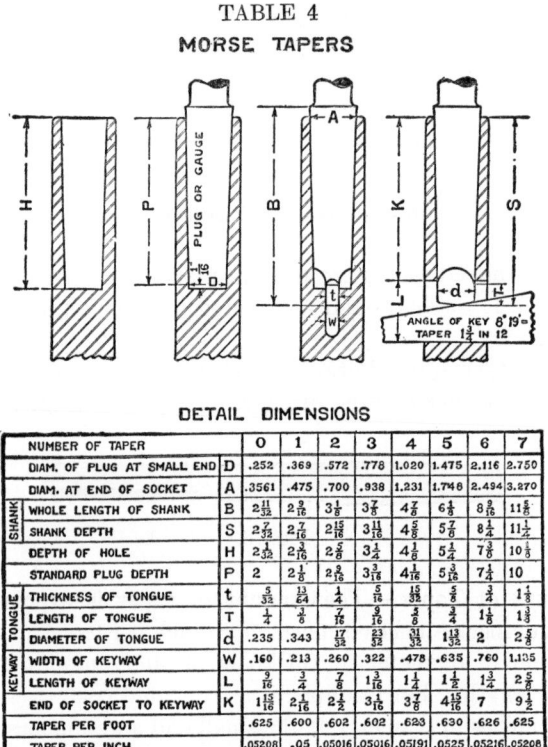

DETAIL DIMENSIONS

NUMBER OF TAPER		0	1	2	3	4	5	6	7
DIAM. OF PLUG AT SMALL END	D	.252	.369	.572	.778	1.020	1.475	2.116	2.750
DIAM. AT END OF SOCKET	A	.3561	.475	.700	.938	1.231	1.748	2.494	3.270
SHANK — WHOLE LENGTH OF SHANK	B	$2\frac{11}{32}$	$2\frac{9}{16}$	$3\frac{1}{8}$	$3\frac{7}{8}$	$4\frac{7}{8}$	$6\frac{1}{8}$	$8\frac{9}{16}$	$11\frac{5}{8}$
SHANK DEPTH	S	$2\frac{7}{32}$	$2\frac{7}{16}$	$2\frac{15}{16}$	$3\frac{11}{16}$	$4\frac{5}{8}$	$5\frac{7}{8}$	$8\frac{1}{4}$	$11\frac{1}{4}$
DEPTH OF HOLE	H	$2\frac{5}{32}$	$2\frac{5}{16}$	$2\frac{7}{8}$	$3\frac{1}{4}$	$4\frac{1}{8}$	$5\frac{1}{4}$	$7\frac{3}{8}$	$10\frac{5}{8}$
STANDARD PLUG DEPTH	P	2	$2\frac{1}{8}$	$2\frac{9}{16}$	$3\frac{3}{16}$	$4\frac{1}{16}$	$5\frac{3}{16}$	$7\frac{1}{4}$	10
TONGUE — THICKNESS OF TONGUE	t	$\frac{5}{32}$	$\frac{13}{64}$	$\frac{1}{4}$	$\frac{5}{16}$	$\frac{15}{32}$	$\frac{5}{8}$	$\frac{3}{4}$	$1\frac{1}{8}$
LENGTH OF TONGUE	T	$\frac{1}{4}$	$\frac{3}{8}$	$\frac{7}{16}$	$\frac{9}{16}$	$\frac{5}{8}$	$\frac{3}{4}$	$1\frac{1}{8}$	$1\frac{3}{8}$
DIAMETER OF TONGUE	d	.235	.343	$\frac{17}{32}$	$\frac{23}{32}$	$\frac{31}{32}$	$1\frac{13}{32}$	2	$2\frac{5}{8}$
KEYWAY — WIDTH OF KEYWAY	W	.160	.213	.260	.322	.478	.635	.760	1.135
LENGTH OF KEYWAY	L	$\frac{9}{16}$	$\frac{3}{4}$	$\frac{7}{8}$	$1\frac{3}{16}$	$1\frac{1}{4}$	$1\frac{1}{2}$	$1\frac{3}{4}$	$2\frac{5}{8}$
END OF SOCKET TO KEYWAY	K	$1\frac{15}{16}$	$2\frac{1}{16}$	$2\frac{1}{2}$	$3\frac{1}{16}$	$3\frac{7}{8}$	$4\frac{15}{16}$	7	$9\frac{1}{2}$
TAPER PER FOOT		.625	.600	.602	.602	.623	.630	.626	.625
TAPER PER INCH		.05208	.05	.05016	.05016	.05191	.0525	.05216	.05208
NUMBER OF KEY		0	1	2	3	4	5	6	7

SOUTH BEND LATHE WORKS

FIG. 17.—Morse Tapers.

The *Morse* standard has a taper of approximately $\frac{5}{8}$ inch per foot. This taper is further defined as No. 1, No. 2, etc., depending on the diameter at the small end. Figure 17 and Table 4 give the chief characteristics of this taper.

Brown & Sharpe is another standard, with a taper of $\frac{1}{2}$ inch per foot. This is also specified by numbers as follows:

No. of taper.......... 4   5   7   9
Diameter at small end.. 0.35 in.   0.45 in.   0.60 in.   0.90 in.

Three other tapers are: *Jarno,* 0.6 inch per foot; *Sellers and Pipe* taper, $\frac{3}{4}$ inch per foot; and *Pratt & Whitney* pins, $\frac{1}{4}$ inch per foot.

Formulas for Calculating Tapers

$$\text{T.P.I.} = \frac{\text{T.P.F.}}{12} \qquad\qquad \text{T.P.F.} = \frac{12(D - d)}{l}$$

$$\text{T.P.L.} = \frac{l \times \text{T.P.F.}}{12} \qquad\qquad l = \frac{12(D - d)}{\text{T.P.F.}}$$

$$D = d + \frac{(l \times \text{T.P.F.})}{12} \qquad\qquad d = D - \frac{(l \times \text{T.P.F.})}{12}$$

when   T.P.I. = taper per inch D = larger diameter
T.P.F. = taper per foot d = smaller diameter
T.P.L. = taper in any length l = length of taper

ILLUSTRATION: In a taper bushing, $D = 2$ inches, $d = 1\frac{1}{2}$ inches, and $1 = 3$ inches. Find the taper per foot.

$$\text{T.P.F.} = \frac{12(D - d)}{l}$$

$$= \frac{12(2 - 1\frac{1}{2})}{3}$$

$$= \frac{12 \times 1}{2 \times 3} = 2$$

Therefore, the taper per foot is 2 inches.

ILLUSTRATION: If the taper of the shank of an end mill is 0.625 inch per foot and $D = \frac{3}{4}$ inch and $d = \frac{1}{2}$ inch, find the length of the taper.

$$l = \frac{12(D - d)}{\text{T.P.F.}}$$

$$= \frac{12(\frac{3}{4} - \frac{1}{2})}{\text{T.P.F.}}$$

$$= \frac{12 \times 1}{4 \times .625} = 4.8.$$

Therefore, the length of the taper is 4.8 inches.

Taper Turning in Lathe.—There are three ways of turning tapers on a lathe, (1) by offsetting the tailstock, (2) by using a taper attachment, and (3) by using the compound rest.

Offsetting the Tailstock.—The tailstock or dead center is moved out of alignment with the line center by means of screws on the base of the tailstock.

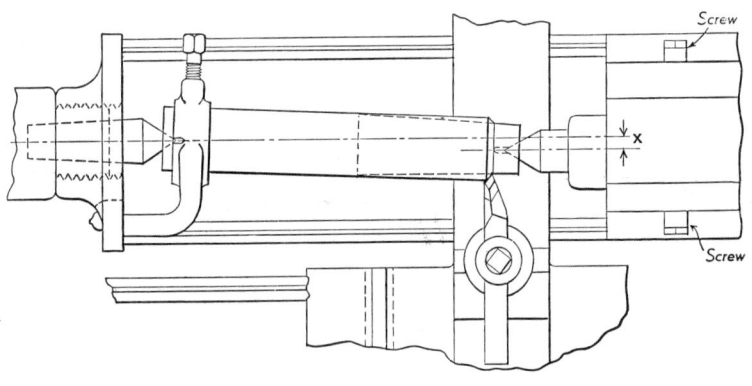

FIG. 18.

Formulas for Calculating the Amount of Offset

(a) When the taper runs the entire length of the bar.

$$O = \frac{D - d}{2}$$

where $O = $ offset

ILLUSTRATION: Find the offset if a bar is to be turned taper to diameters of $1\frac{1}{2}$ inches and $\frac{7}{8}$ inch respectively.

$$O = \frac{D - d}{2}$$

$$= \frac{1\frac{1}{2} - \frac{7}{8}}{2}$$

$$= \frac{5}{8 \times 2} = \frac{5}{16}$$

Therefore the offset is $\frac{5}{16}$ inches

(b) When the taper runs only part of the length of a bar.

$$O = \frac{(D - d)L}{2l}$$

where L is the total length of the bar in inches or the total distance between the centers of the lathe.

ILLUSTRATION: Find the offset if a taper 3 inches long with diameter of 2 inches and $1\frac{1}{2}$ inches respectively is to be turned on a bar 12 inches long.

$$O = \frac{(D - d)L}{2l}$$

$$= \frac{(2 - 1\frac{1}{2})12}{2 \times 3}$$

$$= \frac{1 \times 12}{2 \times 2 \times 3} = 1$$

Therefore the tailstock offset is 1 inch.

(c) **When a bar is tapered to a given taper per foot.**

$$O = \frac{\text{T.P.F.} \times L}{24}$$

ILLUSTRATION: Find the offset if a T.P.F. of $\frac{1}{2}$ inch is to be turned on a bar 6 inches long.

$$O = \frac{\text{T.P.F.} \times L}{24}$$

$$= \frac{1 \times 6}{2 \times 24} = \frac{1}{8}$$

Therefore, the tailstock offset is $\frac{1}{8}$ inch.

NOTE: The above formulas are only exact between the ends of the centers. As Fig. 18 shows, the centers penetrate a short distance into the stock, thus the formulas give only a close approximation.

(d) By using a taper attachment. This device permits the tool to feed transversely at the same time that it feeds longitudinally, thus turning a taper. The guide bar is swiveled on a central pin an amount proportional to the taper, without considering the length of the stock to be turned. There are graduations at either end of the plate upon which the guide swivels indicate the amount of taper. Thus, in setting a taper attachment, only the taper per foot must be obtained.

(e) By using the compound rest. This part of a lathe permits the cutting tool to be set at any desired angle, thus making possible the turning of very steep tapers. The slide of the compound rest is set at the complementary angle to the angle which the taper makes with the center line of the lathe.

Taper Angle.—A steep taper is usually referred to as an angle. Angles up to 10° are commonly designated as tapers, while a larger angle is stated either as the included angle or as the angle with the center line.

Fig. 19.

In the above sketch ABC is the included angle and the angle b is the angle with the center line.

The following formulas may be used to calculate b, the angle with the center line.

when the T.P.F. is known

$$\tan b = \frac{\text{T.P.F.}}{24}$$

when the diameters and length of the taper are known

$$\tan b = \frac{D - d}{2l}$$

ILLUSTRATION: If the taper per foot is $\frac{1}{2}$ inch, find the angle with the center line and the included angle.

$$\tan b = \frac{\text{T.P.F.}}{24}$$

$$= \frac{\frac{1}{2}}{24}$$

$$= \frac{1}{48} = 0.02083$$

From the table of tangents, $0.02083 = 1° \ 11' \ 35''$.

Therefore, b, the angle with the center line is $1° \ 11' \ 35''$ and the included angle is $2 \times b$ or $2° \ 23' \ 10''$.

ILLUSTRATION: If $D = 1\frac{1}{8}$ inch, $d = \frac{1}{2}$ inch, and $1 = 1\frac{1}{4}$ inch, find the angle with the center line and the included angle.

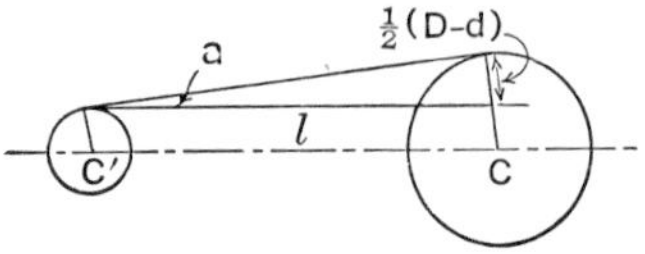

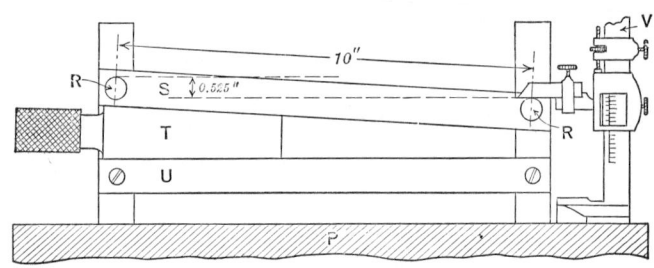

$$\tan b = \frac{D - d}{2l}$$

$$= \frac{1\frac{1}{8} - \frac{1}{2}}{2 \times 1\frac{1}{4}}$$

FIG. 20

$$= \frac{5 \times 1 \times 4}{8 \times 2 \times 5} = \frac{1}{4} = 0.25000$$

From the table of tangents $0.2500 = 14°\ 2'$.

Therefore, b the angle with the center line is $14°\ 2'$ and the included angle is $2 \times b$ or $28°\ 4'$.

The table on page 533 shows tapers per foot and corresponding angles.

Measuring Tapers with a Sine Bar.—An instrument known as a *sine bar* is often used to measure the angle of a taper.

FIG. 21

P, scraped surface plate; *R, R*, plugs; *S,* hardened-steel sine bar; *T*, taper plug gage; *U*, straight edge; *V*, vernier height gage

The taper to be measured is placed on the straight edge *U*, which is parallel to the surface plate *P*, and the sine bar *S*, which has two plugs *R, R* set $10''$ apart, is clamped along the taper. Then *r*, the difference in height in inches between the plugs, is found by means of the height gage *V*. Letting *A* be the included angle, we have the following formulas:

$$\sin A = \frac{r}{10} \qquad r = 10 \sin A$$

For example, in the above figure $r = 0.525''$, and we have

$$\sin A = \frac{r}{10} = \frac{0.525}{10} = 0.0525; \text{ whence } A = 3° 1'.$$

Therefore the included angle of the taper plug gage is $3° 1'$.

TABLE 5

TAPERS AND ANGLES

Taper per Foot	Included			With Center Line			Taper	Taper per Inch from Center Line
	Deg.	Min.	Sec.	Deg.	Min.	Sec.		
⅛	0	35	48	0	17	54	0.010416	0.005203
3⁄16	0	53	44	0	26	52	.015625	.007812
¼	1	11	36	0	35	48	.020833	.010416
5⁄16	1	29	30	0	44	45	.026042	.013021
⅜	1	47	24	0	53	42	.031250	.015625
7⁄16	2	5	18	1	2	39	.036458	.018229
½	2	23	10	1	11	35	.041667	.020833
9⁄16	2	41	4	1	20	32	.046875	.023438
⅝	2	59	42	1	29	51	.052084	.026042
11⁄16	3	16	54	1	38	27	.057292	.028646
¾	3	34	44	1	47	22	.062500	.031250
13⁄16	3	52	38	1	56	19	.067708	.033854
⅞	4	10	32	2	5	16	.072917	.036456
15⁄16	4	28	24	2	14	12	.078125	.039063
1	4	46	18	2	23	9	.083330	.041667
1¼	5	57	48	2	58	54	.104666	.052084
1½	7	9	10	3	34	35	.125000	.062500
1¾	8	20	26	4	10	13	.145833	.072917
2	9	31	36	4	45	48	.666666	.083332
2½	11	53	36	5	56	48	.208333	.104166
3	14	15	0	7	7	30	.250000	.125000
3½	16	35	40	8	17	50	.291666	.145833
4	18	55	28	9	27	44	.333333	.166666
4½	21	14	2	10	37	1	.375000	.187500
5	23	32	12	11	46	6	.416666	.208333
6	28	4	2	14	2	1	.500000	.250000

Testing Tapers.—To test a taper for a given angle, the difference in height r of the plugs is found from the second formula, and bar S is set to this distance by means of the height gage. The taper is then tested between bars S and U.

For example, what should be the difference in height of the plugs for testing a taper which is to have an included angle of 26° 30′?

We have
$$r = 10 \sin A = 10 \times 0.4462 = 4.462.$$

Hence the difference in height of the plugs should be 4.462″. This result can be found in the table on page 576 under the column headed constant 26 degrees and opposite 30 in the column headed minutes.

Measuring Tapers with Discs. The angle of a taper may also be measured by means of two discs of unequal diameters.

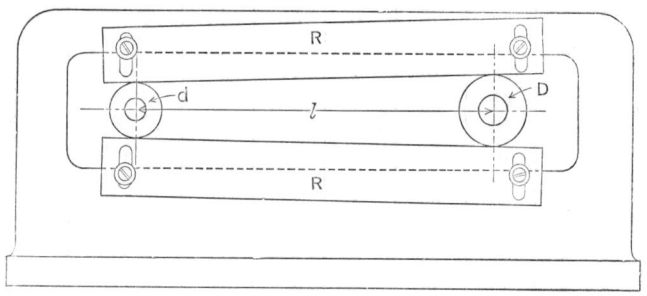

Fig. 22.—Measuring Tapers with Discs.
R, R, hardened-steel edges; D, d, discs of different diameters; l, distance between centers of discs.

The discs are placed as shown above, and the straight edges R, R, which are made of hardened steel and carefully ground, are adjusted so that the tangent lines form the taper.

Taking a as the angle with the center axis, D as the larger diameter, d as the smaller diameter, and l as the distance between the centers, as shown in the figure below, we have

$$\sin a = \frac{\frac{1}{2}(D - d)}{l};$$

whence $$\sin a = \frac{D - d}{2l}.$$

Angle a can then be found from a table of sines, and from it we can find $2a$, the included angle of the taper.

Furthermore, from the formula for $\sin a$ we have

$$l = \frac{D - d}{2 \sin a},$$

so that, given D, d, and the angle with the axis, we can find l

Screw Threads

Screw threads are familiar to every mechanic. They are used on bolts to hold pieces of machinery together, on testing machines to transmit power and in the micrometer caliper for measuring purposes. The threads on a bolt are known as " outside threads " and those in a nut as " inside threads."

The principal parts of a thread, established by the National Screw Thread Commission are shown in Fig. 25. The pro-

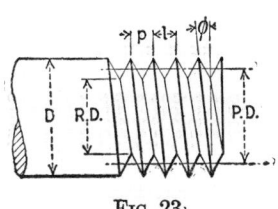

Fig. 23.

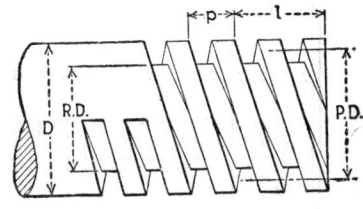

Fig. 24.—Double Square Thread.

truding edge is known as the *crest*. The base of the groove is called the *root*. The *depth of thread*, i.e., the perpendicular dis-

Table 6 is calculated for degrees and minutes with sines based on a radius of 10. In the preceding problem, instead of looking up the sine of A and multiplying by 10, the table gives the result without computation.

TABLE 6.—SINE BAR TABLE

Min.	Constant, 0 Deg.	Constant, 1 Deg.	Constant, 2 Deg.	Constant, 3 Deg.	Constant, 4 Deg.	Constant, 5 Deg.	Constant, 6 Deg.	Constant, 7 Deg.	Constant, 8 Deg.
0	.0000	.1745	.3490	.5234	.6976	.8716	1.0453	1.2187	1.3917
1	.0029	.1774	.3519	.5263	.7005	.8745	1.0482	1.2216	1.3946
2	.0058	.1803	.3548	.5292	.7034	.8774	1.0511	1.2245	1.3975
3	.0087	.1832	.3577	.5321	.7063	.8803	1.0540	1.2274	1.4004
4	.0116	.1862	.3606	.5350	.7092	.8831	1.0569	1.2302	1.4033
5	.0145	.1891	.3635	.5379	.7121	.8860	1.0597	1.2331	1.4061
6	.0175	.1920	.3664	.5408	.7150	.8889	1.0626	1.2360	1.4090
7	.0204	.1949	.3693	.5437	.7179	.8918	1.0655	1.2389	1.4119
8	.0233	.1978	.3723	.5466	.7208	.8947	1.0684	1.2418	1.4148
9	.0262	.2007	.3752	.5495	.7237	.8976	1.0713	1.2447	1.4177
10	.0291	.2036	.3781	.5524	.7266	.9005	1.0742	1.2476	1.4205
11	.0320	.2065	.3810	.5553	.7295	.9034	1.0771	1.2504	1.4234
12	.0349	.2094	.3839	.5582	.7324	.9063	1.0800	1.2533	1.4263
13	.0378	.2123	.3868	.5611	.7353	.9092	1.0829	1.2562	1.4292
14	.0407	.2132	.3897	.5640	.7382	.9121	1.0858	1.2591	1.4320
15	.0436	.2181	.3926	.5669	.7411	.9150	1.0887	1.2620	1.4349
16	.0465	.2211	.3955	.5698	.7440	.9179	1.0916	1.2649	1.4378
17	.0495	.2240	.3984	.5727	.7469	.9208	1.0945	1.2678	1.4407
18	.0524	.2269	.4013	.5756	.7498	.9237	1.0973	1.2706	1.4436
19	.0553	.2298	.4042	.5785	.7527	.9266	1.1002	1.2735	1.4464
20	.0582	.2327	.4071	.5814	.7556	.9295	1.1031	1.2764	1.4493
21	.0611	.2356	.4100	.5844	.7585	.9324	1.1060	1.2793	1.4522
22	.0640	.2385	.4129	.5873	.7614	.9353	1.1089	1.2822	1.4551
23	.0669	.2414	.4159	.5902	.7643	.9382	1.1118	1.2851	1.4580
24	.0698	.2443	.4188	.5931	.7672	.9411	1.1147	1.2880	1.4608
25	.0727	.2472	.4217	.5960	.7701	.9440	1.1176	1.2908	1.4637
26	.0756	.2501	.4246	.5980	.7730	.9469	1.1205	1.2937	1.4666
27	.0785	.2530	.4275	.6018	.7739	.9498	1.1234	1.2966	1.4695
28	.0814	.2560	.4304	.6047	.7788	.9527	1.1269	1.2995	1.4723
29	.0844	.2589	.4333	.6076	.7817	.9556	1.1291	1.3024	1.4752
30	.0873	.2618	.4362	.6105	.7846	.9585	1.1320	1.3053	1.4781

	.09	.26	.43	.61	.78	.96	1.13	1.30	1.48
31	.0902	.2647	.4391	.6134	.7875	.9614	1.1349	1.3081	1.4810
32	.0931	.2676	.4420	.6163	.7904	.9642	1.1378	1.3110	1.4838
33	.0960	.2705	.4449	.6192	.7933	.9671	1.1407	1.3139	1.4867
34	.0989	.2734	.4478	.6221	.7962	.9700	1.1436	1.3168	1.4896
35	.1018	.2763	.4507	.6250	.7993	.9729	1.1465	1.3197	1.4925
36	.1047	.2792	.4536	.6279	.8020	.9758	1.1494	1.3226	1.4954
37	.1076	.2821	.4565	.6308	.8049	.9787	1.1523	1.3254	1.4982
38	.1105	.2850	.4594	.6337	.8078	.9816	1.1552	1.3283	1.5011
39	.1134	.2879	.4623	.6360	.8107	.9845	1.1580	1.3312	1.5040
40	.1164	.2908	.4653	.6395	.8136	.9874	1.1609	1.3341	1.5069
41	.1193	.2938	.4682	.6424	.8165	.9903	1.1638	1.3370	1.5097
42	.1222	.2967	.4711	.6453	.8194	.9932	1.1667	1.3399	1.5126
43	.1251	.2996	.4740	.6482	.8223	.9961	1.1696	1.3427	1.5155
44	.1280	.3025	.4769	.6511	.8252	.9990	1.1725	1.3456	1.5184
45	.1309	.3054	.4798	.6540	.8281	1.0019	1.1754	1.3485	1.5212
46	.1338	.3083	.4827	.6569	.8310	1.0048	1.1783	1.3514	1.5241
47	.1367	.3112	.4856	.6598	.8339	1.0077	1.1812	1.3543	1.5270
48	.1396	.3141	.4885	.6627	.8368	1.0106	1.1840	1.3572	1.5299
49	.1425	.3170	.4914	.6656	.8397	1.0135	1.1869	1.3600	1.5327
50	.1454	.3199	.4943	.6685	.8426	1.0164	1.1898	1.3629	1.5356
51	.1483	.3228	.4972	.6714	.8455	1.0192	1.1927	1.3658	1.5385
52	.1513	.3257	.5001	.6743	.8484	1.0221	1.1956	1.3687	1.5414
53	.1542	.3286	.5030	.6773	.8513	1.0250	1.1985	1.3716	1.5442
54	.1571	.3316	.5059	.6802	.8542	1.0279	1.2014	1.3744	1.5471
55	.1600	.3345	.5088	.6831	.8571	1.0308	1.2043	1.3773	1.5500
56	.1629	.3374	.5117	.6860	.8600	1.0337	1.2071	1.3802	1.5529
57	.1658	.3403	.5146	.6889	.8629	1.0366	1.2100	1.3831	1.5557
58	.1687	.3432	.5175	.6918	.8658	1.0395	1.2129	1.3860	1.5586
59	.1716	.3461	.5205	.6947	.8687	1.0424	1.2158	1.3889	1.5615
60	.1745	.3490	.5234	.6976	.8716	1.0453	1.2187	1.3917	1.5643

Table 6.—Sine Bar Table—*Continued*

Min.	Constant, 9 Deg.	Constant, 10 Deg.	Constant, 11 Deg.	Constant, 12 Deg.	Constant, 13 Deg.	Constant, 14 Deg.	Constant, 15 Deg.	Constant, 16 Deg.	Constant, 17 Deg.
0	1.5643	1.7365	1.9081	2.0791	2.2495	2.4192	2.5882	2.7564	2.9237
1	1.5672	1.7393	1.9109	2.0820	2.2523	2.4220	2.5910	2.7592	2.9265
2	1.5701	1.7422	1.9138	2.0848	2.2552	2.4249	2.5938	2.7620	2.9293
3	1.5730	1.7451	1.9167	2.0877	2.2580	2.4277	2.5966	2.7648	2.9321
4	1.5758	1.7479	1.9195	2.0905	2.2608	2.4305	2.5994	2.7676	2.9348
5	1.5787	1.7508	1.9224	2.0933	2.2637	2.4333	2.6022	2.7704	2.9376
6	1.5816	1.7537	1.9252	2.0962	2.2665	2.4362	2.6050	2.7731	2.9404
7	1.5845	1.7565	1.9281	2.0990	2.2693	2.4390	2.6079	2.7759	2.9432
8	1.5873	1.7594	1.9309	2.1019	2.2722	2.4418	2.6107	2.7787	2.9460
9	1.5902	1.7623	1.9338	2.1047	2.2750	2.4446	2.6135	2.7815	2.9487
10	1.5931	1.7651	1.9366	2.1076	2.2778	2.4474	2.6163	2.7843	2.9515
11	1.5959	1.7680	1.9395	2.1104	2.2807	2.4503	2.6191	2.7871	2.9543
12	1.5988	1.7708	1.9423	2.1132	2.2835	2.4531	2.6219	2.7899	2.9571
13	1.6017	1.7737	1.9452	2.1161	2.2863	2.4559	2.6247	2.7927	2.9599
14	1.6046	1.7766	1.9481	2.1189	2.2892	2.4587	2.6275	2.7955	2.9626
15	1.6074	1.7794	1.9509	2.1218	2.2920	2.4615	2.6303	2.7983	2.9654
16	1.6100	1.7823	1.9538	2.1246	2.2948	2.4644	2.6331	2.8011	2.9682
17	1.6132	1.7852	1.9566	2.1275	2.2977	2.4672	2.6359	2.8039	2.9710
18	1.6160	1.7880	1.9595	2.1303	2.3005	2.4700	2.6387	2.8067	2.9737
19	1.6189	1.7909	1.9623	2.1331	2.3033	2.4728	2.6415	2.8095	2.9765
20	1.6218	1.7937	1.9652	2.1360	2.3062	2.4756	2.6443	2.8123	2.9793
21	1.6246	1.7966	1.9680	2.1388	2.3090	2.4784	2.6471	2.8150	2.9821
22	1.6275	1.7995	1.9709	2.1417	2.3118	2.4813	2.6500	2.8178	2.9849
23	1.6304	1.8023	1.9737	2.1445	2.3146	2.4841	2.6528	2.8206	2.9876
24	1.6333	1.8052	1.9760	2.1474	2.3175	2.4869	2.6556	2.8234	2.9904
25	1.6361	1.8081	1.9794	2.1502	2.3203	2.4897	2.6584	2.8262	2.9932
26	1.6390	1.8109	1.9823	2.1530	2.3231	2.4925	2.6612	2.8290	2.9960
27	1.6419	1.8138	1.9831	2.1559	2.3260	2.4954	2.6640	2.8318	2.9987
28	1.6447	1.8166	1.9880	2.1587	2.3288	2.4982	2.6668	2.8346	3.0015
29	1.6476	1.8195	1.9908	2.1616	2.3316	2.5010	2.6696	2.8374	3.0043
30	1.6505	1.8224	1.9937	2.1644	2.3345	2.5038	2.6724	2.8402	3.0071

N									
31	1.6533	1.8252	1.9965	2.1672	2.3373	2.5066	2.6752	2.8429	3.0098
32	1.6562	1.8281	1.9994	2.1701	2.3401	2.5094	2.6780	2.8457	3.0126
33	1.6591	1.8309	2.0022	2.1729	2.3429	2.5122	2.6808	2.8485	3.0154
34	1.6620	1.8338	2.0051	2.1758	2.3458	2.5151	2.6836	2.8513	3.0182
35	1.6648	1.8367	2.0079	2.1786	2.3486	2.5179	2.6864	2.8541	3.0209
36	1.6677	1.8395	2.0108	2.1814	2.3514	2.5207	2.6892	2.8569	3.0237
37	1.6706	1.8424	2.0136	2.1843	2.3542	2.5235	2.6920	2.8597	3.0265
38	1.6734	1.8452	2.0165	2.1871	2.3571	2.5263	2.6948	2.8625	3.0292
39	1.6763	1.8481	2.0193	2.1899	2.3599	2.5291	2.6976	2.8652	3.0320
40	1.6792	1.8509	2.0222	2.1928	2.3627	2.5320	2.7004	2.8680	3.0348
41	1.6820	1.8538	2.0250	2.1956	2.3656	2.5348	2.7032	2.8708	3.0376
42	1.6849	1.8567	2.0279	2.1985	2.3684	2.5376	2.7060	2.8736	3.0403
43	1.6878	1.8595	2.0307	2.2013	2.3712	2.5404	2.7088	2.8764	3.0431
44	1.6906	1.8624	2.0336	2.2041	2.3740	2.5432	2.7116	2.8792	3.0459
45	1.6935	1.8652	2.0364	2.2070	2.3769	2.5460	2.7144	2.8820	3.0486
46	1.6964	1.8681	2.0393	2.2098	2.3797	2.5488	2.7172	2.8847	3.0514
47	1.6992	1.8710	2.0421	2.2126	2.3825	2.5516	2.7200	2.8875	3.0542
48	1.7021	1.8738	2.0450	2.2155	2.3853	2.5545	2.7228	2.8903	3.0570
49	1.7050	1.8767	2.0478	2.2183	2.3882	2.5573	2.7256	2.8931	3.0597
50	1.7078	1.8795	2.0507	2.2212	2.3910	2.5601	2.7284	2.8959	3.0625
51	1.7107	1.8824	2.0535	2.2240	2.3938	2.5629	2.7312	2.8987	3.0653
52	1.7136	1.8852	2.0563	2.2268	2.3966	2.5657	2.7340	2.9015	3.0680
53	1.7164	1.8881	2.0592	2.2297	2.3995	2.5685	2.7368	2.9042	3.0708
54	1.7193	1.8910	2.0620	2.2325	2.4023	2.5713	2.7396	2.9070	3.0736
55	1.7222	1.8938	2.0649	2.2353	2.4051	2.5741	2.7424	2.9098	3.0763
56	1.7250	1.8967	2.0677	2.2382	2.4079	2.5769	2.7452	2.9126	3.0791
57	1.7279	1.8995	2.0706	2.2410	2.4108	2.5798	2.7480	2.9154	3.0819
58	1.7307	1.9024	2.0734	2.2438	2.4136	2.5826	2.7508	2.9182	3.0846
59	1.7336	1.9052	2.0763	2.2467	2.4164	2.5854	2.7536	2.9209	3.0874
60	1.7365	1.9081	2.0791	2.2495	2.4192	2.5882	2.7564	2.9237	3.0902

TABLE 6.—SINE BAR TABLE—*Continued*

Min.	Constant, 18 Deg.	Constant, 19 Deg.	Constant, 20 Deg.	Constant, 21 Deg.	Constant, 22 Deg.	Constant, 23 Deg.	Constant, 24 Deg.	Constant, 25 Deg.	Constant, 26 Deg.
0	3.0902	3.2557	3.4202	3.5837	3.7461	3.9073	4.0674	4.2262	4.3837
1	3.0929	3.2584	3.4229	3.5864	3.7488	3.9100	4.0700	4.2288	4.3863
2	3.0957	3.2612	3.4257	3.5891	3.7515	3.9127	4.0727	4.2315	4.3889
3	3.0986	3.2639	3.4284	3.5918	3.7542	3.9153	4.0753	4.2341	4.3916
4	3.1012	3.2667	3.4311	3.5945	3.7569	3.9180	4.0780	4.2367	4.3942
5	3.1040	3.2694	3.4339	3.5973	3.7595	3.9207	4.0806	4.2394	4.3968
6	3.1068	3.2722	3.4366	3.6000	3.7622	3.9234	4.0833	4.2420	4.3994
7	3.1095	3.2749	3.4393	3.6027	3.7649	3.9260	4.0860	4.2446	4.4020
8	3.1123	3.2777	3.4421	3.6054	3.7676	3.9287	4.0886	4.2473	4.4046
9	3.1151	3.2804	3.4448	3.6081	3.7703	3.9314	4.0913	4.2499	4.4072
10	3.1178	3.2832	3.4475	3.6108	3.7730	3.9341	4.0939	4.2525	4.4098
11	3.1206	3.2859	3.4503	3.6135	3.7757	3.9367	4.0966	4.2552	4.4124
12	3.1233	3.2887	3.4530	3.6162	3.7784	3.9394	4.0992	4.2578	4.4151
13	3.1261	3.2914	3.4557	3.6190	3.7811	3.9421	4.1019	4.2604	4.4177
14	3.1289	3.2942	3.4584	3.6217	3.7838	3.9448	4.1045	4.2631	4.4203
15	3.1316	3.2969	3.4612	3.6244	3.7865	3.9474	4.1072	4.2657	4.4229
16	3.1344	3.2997	3.4639	3.6271	3.7892	3.9501	4.1098	4.2683	4.4255
17	3.1372	3.3024	3.4666	3.6298	3.7919	3.9528	4.1125	4.2709	4.4281
18	3.1399	3.3051	3.4694	3.6325	3.7946	3.9555	4.1151	4.2736	4.4307
19	3.1427	3.3079	3.4721	3.6352	3.7973	3.9581	4.1178	4.2762	4.4333
20	3.1454	3.3106	3.4748	3.6379	3.7999	3.9608	4.1204	4.2788	4.4359
21	3.1482	3.3134	3.4775	3.6406	3.8026	3.9635	4.1231	4.2815	4.4385
22	3.1510	3.3161	3.4803	3.6434	3.8053	3.9661	4.1257	4.2841	4.4411
23	3.1537	3.3189	3.4830	3.6461	3.8080	3.9688	4.1284	4.2867	4.4437
24	3.1565	3.3216	3.4857	3.6488	3.8107	3.9715	4.1310	4.2894	4.4464
25	3.1593	3.3244	3.4884	3.6515	3.8134	3.9741	4.1337	4.2920	4.4490
26	3.1620	3.3271	3.4912	3.6542	3.8161	3.9768	4.1363	4.2946	4.4516
27	3.1648	3.3298	3.4939	3.6569	3.8188	3.9795	4.1390	4.2972	4.4542
28	3.1675	3.3326	3.4966	3.6596	3.8215	3.9822	4.1416	4.2999	4.4568
29	3.1703	3.3353	3.4993	3.6623	3.8241	3.9848	4.1443	4.3025	4.4594
30	3.1730	3.3381	3.5021	3.6650	3.8268	3.9875	4.1469	4.3051	4.4620

31	3.1758	3.3408	3.5048	3.6677	3.8295	3.9902	4.1496	4.3077	4.4646
32	3.1786	3.3436	3.5075	3.6704	3.8322	3.9928	4.1522	4.3104	4.4672
33	3.1813	3.3463	3.5102	3.6731	3.8349	3.9955	4.1549	4.3130	4.4698
34	3.1841	3.3490	3.5130	3.6758	3.8376	3.9982	4.1575	4.3156	4.4724
35	3.1868	3.3518	3.5157	3.6785	3.8403	4.0008	4.1602	4.3182	4.4750
36	3.1896	3.3545	3.5184	3.6812	3.8430	4.0035	4.1628	4.3209	4.4776
37	3.1923	3.3573	3.5211	3.6839	3.8456	4.0062	4.1655	4.3235	4.4802
38	3.1951	3.3600	3.5239	3.6867	3.8483	4.0088	4.1681	4.3261	4.4828
39	3.1979	3.3627	3.5266	3.6894	3.8510	4.0115	4.1707	4.3287	4.4854
40	3.2006	3.3655	3.5293	3.6921	3.8537	4.0141	4.1734	4.3313	4.4880
41	3.2034	3.3682	3.5320	3.6948	3.8564	4.0168	4.1760	4.3340	4.4906
42	3.2061	3.3710	3.5347	3.6975	3.8591	4.0195	4.1787	4.3366	4.4932
43	3.2089	3.3737	3.5375	3.7002	3.8617	4.0221	4.1813	4.3392	4.4958
44	3.2116	3.3764	3.5402	3.7029	3.8644	4.0248	4.1840	4.3418	4.4984
45	3.2144	3.3792	3.5429	3.7056	3.8671	4.0275	4.1866	4.3445	4.5010
46	3.2171	3.3819	3.5456	3.7083	3.8698	4.0301	4.1892	4.3471	4.5036
47	3.2199	3.3846	3.5484	3.7110	3.8725	4.0328	4.1919	4.3497	4.5062
48	3.2227	3.3874	3.5511	3.7137	3.8752	4.0355	4.1945	4.3523	4.5088
49	3.2254	3.3901	3.5538	3.7164	3.8778	4.0381	4.1972	4.3549	4.5114
50	3.2282	3.3929	3.5565	3.7191	3.8805	4.0408	4.1998	4.3575	4.5140
51	3.2309	3.3956	3.5592	3.7218	3.8832	4.0434	4.2024	4.3602	4.5166
52	3.2337	3.3983	3.5619	3.7245	3.8859	4.0461	4.2051	4.3628	4.5192
53	3.2364	3.4011	3.5647	3.7272	3.8886	4.0488	4.2077	4.3654	4.5218
54	3.2392	3.4038	3.5674	3.7299	3.8912	4.0514	4.2104	4.3680	4.5243
55	3.2419	3.4065	3.5701	3.7326	3.8939	4.0541	4.2130	4.3706	4.5269
56	3.2447	3.4093	3.5728	3.7353	3.8966	4.0567	4.2156	4.3733	4.5295
57	3.2474	3.4120	3.5755	3.7380	3.8993	4.0594	4.2183	4.3759	4.5321
58	3.2502	3.4147	3.5782	3.7407	3.9020	4.0621	4.2209	4.3785	4.5347
59	3.2529	3.4175	3.5810	3.7434	3.9046	4.0647	4.2235	4.3811	4.5373
60	3.2557	3.4202	3.5837	3.7461	3.9073	4.0674	4.2262	4.3837	4.5399

TABLE 6.—SINE BAR TABLE—*Continued*

Min.	Constant, 27 Deg.	Constant, 28 Deg.	Constant, 29 Deg.	Constant, 30 Deg.	Constant, 31 Deg.	Constant, 32 Deg.	Constant, 33 Deg.	Constant, 34 Deg.	Constant, 35 Deg.
0	4.5399	4.6947	4.8481	5.0000	5.1504	5.2992	5.4464	5.5919	5.7358
1	4.5425	4.6973	4.8506	5.0025	5.1529	5.3017	5.4488	5.5943	5.7381
2	4.5451	4.6999	4.8532	5.0050	5.1554	5.3041	5.4513	5.5968	5.7405
3	4.5477	4.7024	4.8557	5.0076	5.1579	5.3066	5.4537	5.5992	5.7429
4	4.5503	4.7050	4.8583	5.0101	5.1604	5.3091	5.4561	5.6016	5.7453
5	4.5529	4.7076	4.8608	5.0126	5.1628	5.3115	5.4586	5.6040	5.7477
6	4.5554	4.7101	4.8634	5.0151	5.1653	5.3140	5.4610	5.6064	5.7501
7	4.5580	4.7127	4.8659	5.0176	5.1678	5.3164	5.4635	5.6088	5.7524
8	4.5606	4.7153	4.8684	5.0201	5.1703	5.3189	5.4659	5.6112	5.7548
9	4.5632	4.7178	4.8710	5.0227	5.1728	5.3214	5.4683	5.6136	5.7572
10	4.5658	4.7204	4.8735	5.0252	5.1753	5.3238	5.4708	5.6160	5.7596
11	4.5684	4.7229	4.8761	5.0277	5.1778	5.3263	5.4732	5.6184	5.7619
12	4.5710	4.7255	4.8786	5.0302	5.1803	5.3288	5.4756	5.6208	5.7643
13	4.5736	4.7281	4.8811	5.0327	5.1828	5.3312	5.4781	5.6232	5.7667
14	4.5762	4.7306	4.8837	5.0352	5.1852	5.3337	5.4805	5.6256	5.7691
15	4.5787	4.7332	4.8862	5.0377	5.1877	5.3361	5.4829	5.6280	5.7715
16	4.5813	4.7358	4.8888	5.0403	5.1902	5.3386	5.4854	5.6305	5.7738
17	4.5839	4.7383	4.8913	5.0428	5.1927	5.3411	5.4878	5.6329	5.7762
18	4.5865	4.7409	4.8938	5.0453	5.1952	5.3435	5.4902	5.6353	5.7786
19	4.5891	4.7434	4.8964	5.0478	5.1977	5.3460	5.4927	5.6377	5.7810
20	4.5917	4.7460	4.8989	5.0503	5.2002	5.3484	5.4951	5.6401	5.7833
21	4.5942	4.7486	4.9014	5.0528	5.2026	5.3509	5.4975	5.6425	5.7857
22	4.5968	4.7511	4.9040	5.0553	5.2051	5.3534	5.4999	5.6449	5.7881
23	4.5994	4.7537	4.9065	5.0578	5.2076	5.3558	5.5024	5.6473	5.7904
24	4.6020	4.7562	4.9090	5.0603	5.2101	5.3583	5.5048	5.6497	5.7928
25	4.6046	4.7588	4.9116	5.0628	5.2126	5.3607	5.5072	5.6521	5.7952
26	4.6072	4.7614	4.9141	5.0654	5.2151	5.3632	5.5097	5.6545	5.7976
27	4.6097	4.7639	4.9166	5.0679	5.2175	5.3656	5.5121	5.6569	5.7999
28	4.6123	4.7665	4.9192	5.0704	5.2200	5.3681	5.5145	5.6593	5.8023
29	4.6149	4.7690	4.9217	5.0729	5.2225	5.3705	5.5169	5.6617	5.8047
30	4.6175	4.7716	4.9242	5.0754	5.2250	5.3730	5.5194	5.6641	5.8070

31	4.6201	4.7741	4.9268	5.0779	5.2275	5.3754	5.5218	5.6665	5.8094
32	4.6226	4.7767	4.9293	5.0804	5.2299	5.3779	5.5242	5.6689	5.8118
33	4.6252	4.7793	4.9318	5.0829	5.2324	5.3804	5.5266	5.6713	5.8141
34	4.6278	4.7818	4.9344	5.0854	5.2349	5.3828	5.5291	5.6736	5.8165
35	4.6304	4.7844	4.9369	5.0879	5.2374	5.3853	5.5315	5.6760	5.8189
36	4.6330	4.7869	4.9394	5.0904	5.2399	5.3877	5.5339	5.6784	5.8212
37	4.6355	4.7895	4.9419	5.0929	5.2423	5.3902	5.5363	5.6808	5.8236
38	4.6381	4.7920	4.9445	5.0954	5.2448	5.3926	5.5388	5.6832	5.8260
39	4.6407	4.7946	4.9470	5.0979	5.2473	5.3951	5.5412	5.6856	5.8283
40	4.6433	4.7971	4.9495	5.1004	5.2498	5.3975	5.5436	5.6880	5.8307
41	4.6458	4.7997	4.9521	5.1029	5.2522	5.4000	5.5460	5.6904	5.8330
42	4.6484	4.8022	4.9546	5.1054	5.2547	5.4024	5.5484	5.6928	5.8354
43	4.6510	4.8048	4.9571	5.1079	5.2572	5.4049	5.5509	5.6952	5.8378
44	4.6536	4.8073	4.9596	5.1104	5.2597	5.4073	5.5533	5.6976	5.8401
45	4.6561	4.8099	4.9622	5.1129	5.2621	5.4097	5.5557	5.7000	5.8426
46	4.6587	4.8124	4.9647	5.1154	5.2646	5.4122	5.5581	5.7024	5.8449
47	4.6613	4.8150	4.9672	5.1179	5.2671	5.4146	5.5605	5.7047	5.8472
48	4.6639	4.8175	4.9697	5.1204	5.2696	5.4171	5.5630	5.7071	5.8496
49	4.6664	4.8201	4.9723	5.1229	5.2720	5.4195	5.5654	5.7095	5.8519
50	4.6690	4.8226	4.9748	5.1254	5.2745	5.4220	5.5678	5.7119	5.8543
51	4.6716	4.8252	4.9773	5.1279	5.2770	5.4244	5.5702	5.7143	5.8567
52	4.6742	4.8277	4.9798	5.1304	5.2794	5.4269	5.5726	5.7167	5.8590
53	4.6767	4.8303	4.9824	5.1329	5.2819	5.4293	5.5750	5.7191	5.8614
54	4.6793	4.8328	4.9849	5.1354	5.2844	5.4317	5.5775	5.7215	5.8637
55	4.6819	4.8354	4.9874	5.1379	5.2869	5.4342	5.5799	5.7238	5.8661
56	4.6844	4.8379	4.9899	5.1404	5.2893	5.4366	5.5823	5.7262	5.8684
57	4.6870	4.8405	4.9924	5.1429	5.2918	5.4391	5.5847	5.7286	5.8708
58	4.6896	4.8430	4.9950	5.1454	5.2943	5.4415	5.5871	5.7310	5.8731
59	4.6921	4.8456	4.9975	5.1479	5.2967	5.4440	5.5895	5.7334	5.8755
60	4.6947	4.8481	5.0000	5.1504	5.2992	5.4464	5.5919	5.7358	5.8779

TABLE 6.—SINE BAR TABLE—Continued

Min.	Constant, 36 Deg.	Constant, 37 Deg.	Constant, 38 Deg.	Constant, 39 Deg.	Constant, 40 Deg.	Constant, 41 Deg.	Constant, 42 Deg.	Constant, 43 Deg.	Constant, 44 Deg.
0	5.8779	6.0182	6.1566	6.2932	6.4279	6.5606	6.6913	6.8200	6.9466
1	5.8802	6.0205	6.1589	6.2955	6.4301	6.5628	6.6935	6.8221	6.9487
2	5.8826	6.0228	6.1612	6.2977	6.4323	6.5650	6.6956	6.8242	6.9508
3	5.8849	6.0251	6.1635	6.3000	6.4346	6.5672	6.6978	6.8264	6.9529
4	5.8873	6.0274	6.1658	6.3022	6.4368	6.5694	6.6999	6.8285	6.9549
5	5.8896	6.0298	6.1681	6.3045	6.4390	6.5716	6.7021	6.8306	6.9570
6	5.8920	6.0321	6.1704	6.3068	6.4412	6.5738	6.7043	6.8327	6.9591
7	5.8943	6.0344	6.1726	6.3090	6.4435	6.5759	6.7064	6.8349	6.9612
8	5.8967	6.0367	6.1749	6.3113	6.4457	6.5781	6.7086	6.8370	6.9633
9	5.8990	6.0390	6.1772	6.3136	6.4479	6.5803	6.7107	6.8391	6.9654
10	5.9014	6.0414	6.1795	6.3158	6.4501	6.5825	6.7129	6.8412	6.9675
11	5.9037	6.0437	6.1818	6.3180	6.4524	6.5847	6.7151	6.8434	6.9690
12	5.9061	6.0460	6.1841	6.3203	6.4546	6.5869	6.7172	6.8455	6.9717
13	5.9084	6.0483	6.1864	6.3225	6.4568	6.5891	6.7194	6.8476	6.9737
14	5.9108	6.0506	6.1887	6.3248	6.4590	6.5913	6.7215	6.8497	6.9758
15	5.9131	6.0529	6.1909	6.3271	6.4612	6.5935	6.7237	6.8518	6.9779
16	5.9154	6.0553	6.1932	6.3293	6.4635	6.5956	6.7258	6.8539	6.9800
17	5.9178	6.0576	6.1965	6.3316	6.4657	6.5978	6.7280	6.8561	6.9821
18	5.9201	6.0599	6.1978	6.3338	6.4679	6.6000	6.7301	6.8582	6.9842
19	5.9225	6.0622	6.2001	6.3361	6.4701	6.6022	6.7323	6.8603	6.9862
20	5.9248	6.0645	6.2024	6.3383	6.4723	6.6044	6.7344	6.8624	6.9885
21	5.9272	6.0668	6.2046	6.3400	6.4746	6.6066	6.7366	6.8645	6.9904
22	5.9295	6.0691	6.2069	6.3428	6.4768	6.6088	6.7387	6.8666	6.9925
23	5.9318	6.0714	6.2092	6.3451	6.4790	6.6109	6.7409	6.8688	6.9946
24	5.9342	6.0738	6.2115	6.3473	6.4812	6.6131	6.7430	6.8709	6.9966
25	5.9365	6.0761	6.2138	6.3496	6.4834	6.6153	6.7452	6.8730	6.9987
26	5.9389	6.0784	6.2160	6.3518	6.4856	6.6175	6.7473	6.8751	7.0008
27	5.9412	6.0807	6.2183	6.3540	6.4878	6.6197	6.7495	6.8772	7.0029
28	5.9436	6.0830	6.2206	6.3563	6.4901	6.6218	6.7516	6.8793	7.0049
29	5.9459	6.0853	6.2229	6.3585	6.4923	6.6240	6.7538	6.8814	7.0070
30	5.9482	6.0876	6.2251	6.3608	6.4945	6.6262	6.7559	6.8835	7.0091

7.0112	6.8857	6.7580	6.6284	6.4967	6.3630	6.2274	6.0899	5.9506	31
7.0132	6.8878	6.7602	6.6306	6.4989	6.3653	6.2297	6.0922	5.9529	32
7.0153	6.8899	6.7623	6.6327	6.5011	6.3675	6.2320	6.0945	5.9552	33
7.0174	6.8920	6.7645	6.6349	6.5033	6.3698	6.2342	6.0968	5.9576	34
7.0195	6.8941	6.7666	6.6371	6.5055	6.3720	6.2365	6.0991	5.9599	35
7.0215	6.8962	6.7688	6.6393	6.5077	6.3742	6.2388	6.1015	5.9622	36
7.0236	6.8983	6.7709	6.6414	6.5100	6.3765	6.2411	6.1038	5.9646	37
7.0257	6.9004	6.7730	6.6436	6.5122	6.3787	6.2433	6.1061	5.9669	38
7.0277	6.9025	6.7752	6.6458	6.5144	6.3810	6.2456	6.1084	5.9693	39
7.0298	6.9046	6.7773	6.6480	6.5166	6.3832	6.2479	6.1107	5.9716	40
7.0319	6.9067	6.7795	6.6501	6.5188	6.3854	6.2502	6.1130	5.9739	41
7.0339	6.9088	6.7816	6.6523	6.5210	6.3877	6.2524	6.1153	5.9763	42
7.0360	6.9109	6.7837	6.6545	6.5232	6.3899	6.2547	6.1176	5.9786	43
7.0381	6.9130	6.7859	6.6566	6.5254	6.3922	6.2570	6.1199	5.9809	44
7.0401	6.9151	6.7880	6.6588	6.5276	6.3944	6.2592	6.1222	5.9832	45
7.0422	6.9172	6.7901	6.6610	6.5298	6.3966	6.2615	6.1245	5.9856	46
7.0443	6.9193	6.7923	6.6632	6.5320	6.3989	6.2638	6.1268	5.9879	47
7.0463	6.9214	6.7944	6.6653	6.5342	6.4011	6.2660	6.1291	5.9902	48
7.0484	6.9235	6.7965	6.6675	6.5364	6.4033	6.2683	6.1314	5.9926	49
7.0505	6.9256	6.7987	6.6697	6.5386	6.4056	6.2706	6.1337	5.9949	50
7.0525	6.9277	6.8008	6.6718	6.5408	6.4078	6.2728	6.1360	5.9972	51
7.0546	6.9298	6.8029	6.6740	6.5430	6.4100	6.2751	6.1383	5.9995	52
7.0567	6.9319	6.8051	6.6762	6.5452	6.4123	6.2774	6.1406	6.0019	53
7.0587	6.9340	6.8072	6.6783	6.5474	6.4145	6.2796	6.1429	6.0042	54
7.0608	6.9361	6.8093	6.6805	6.5496	6.4167	6.2819	6.1451	6.0065	55
7.0628	6.9382	6.8115	6.6827	6.5518	6.4190	6.2842	6.1474	6.0089	56
7.0649	6.9403	6.8136	6.6848	6.5540	6.4212	6.2864	6.1497	6.0112	57
7.0670	6.9424	6.8157	6.6870	6.5562	6.4234	6.2887	6.1520	6.0135	58
7.0690	6.9445	6.8179	6.6891	6.5584	6.4256	6.2909	6.1543	6.0158	59
7.0711	6.9466	6.8200	6.6913	6.5606	6.4279	6.2932	6.1566	6.0182	60

tance from the crest to the bottom of the groove, is represented
by H in Fig. 25. Twice the depth is called the double depth.
The diameter measured over the crests is the *outside diameter*
or *major diameter* (indicated by D in Fig. 23). The diameter
measured at the root is the *root diameter* or *minor diameter*. The
pitch diameter (indicated by PD in Fig. 23) is the diameter measured between the mid-points between the crest and the root of

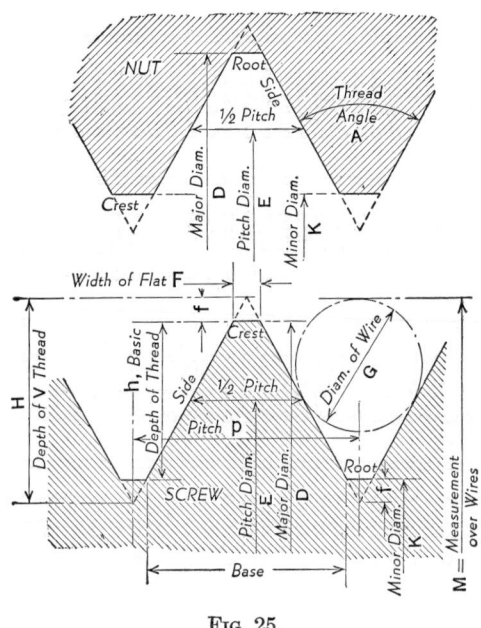

Fig. 25.

the thread. It is equal to $D-d$. The *pitch*, P, is the longitudinal
distance between any point on one thread and the corresponding
point on the adjacent thread. *Lead, l*, is the distance which a
screw advances when turned one complete revolution. In a sin-
gle-thread screw it is equal to the pitch; in a double-thread screw
it is twice the pitch, etc. (See Figs. 23 and 24.)

Symbols

For use in formulas for expressing relations of screw threads, and for use on drawings and for similar purposes, the following symbols should be used:

Major diameter........................... D
Corresponding radius...................... d
Pitch diameter............................ E
Corresponding radius...................... e
Minor diameter............................ K
Corresponding radius...................... k
Angle of thread........................... A
One half angle of thread.................. a
Number of turns per inch.................. N
Number of threads per inch................ n

Lead..................................... $L = \dfrac{1}{N}$

Pitch or thread interval.................. $p = \dfrac{1}{n}$

Helix angle............................... s

Tangent of helix angle.................... $S = \dfrac{L}{3.14159 \times E}$

Width of basic flat at top, crest, or root...... F
Depth of basic truncation................. f
Depth of sharp V thread................... H
Depth of American National form of thread... h
Length of engagement...................... Q
Included angle of taper................... Y
One half included angle of taper.......... y

There are different forms of screw threads—Sharp V, American National, Whitworth, Square, Acme, American National Pipe, etc. The methods of calculating the elements of these threads are shown in the following pages.

Sharp V Thread.—This thread is shown in Figs. 23 and 26. The sides of the thread form an angle of 60 degrees with each other and are theoretically sharp at the top and bottom.

FIG. 26.

The pitch and depth of the thread are found by the following formulas:

$$\text{Pitch} = \frac{1}{\text{No. of threads per inch}}$$

$$\text{Depth} = \frac{0.866}{\text{No. of threads per inch}} \text{ or } 0.8660 \times \text{pitch}$$

ILLUSTRATION: What is the depth of a V-thread of $\frac{1}{8}$-inch pitch?

$$\text{Depth} = 0.866 \text{ pitch} = 0.866 \times \tfrac{1}{8} = 0.108.$$

Therefore the depth is 0.108 inch.

ILLUSTRATION: What is the root diameter of a $\frac{3}{4}$ inch $\times$ 10-V thread? ($\frac{3}{4}$ inch $\times$ 10 means 1 inch in diameter and 10 threads to the inch). From the figures 17 and 18 it is evident that the root diameter is equal to the outside or major diameter minus the double depth of the thread.

$$\text{depth} = 0.866 \times \tfrac{1}{10} = 0.0866, \ 0.0866 \times 2 = 0.173$$

$$\text{R.D.} = 0.750 - 0.173 = 0.577 \text{ in. (Ans.)}$$

Unified and American Thread.—After World War II the need was felt for a screw thread system which would permit interchangeability of parts with other nations. Representatives of the United Kingdom, Canada, and the United States set up what is known as a Unified Thread Series to obtain this interchangeability.

The thread system agreed upon permits some minor variations in British and American practice but still permits the desired interchangeability. The Unified Thread Series can be identified in Tables 7 and 8 by the bold face type. The sides of the thread

TABLE 7a

UNIFIED AND AMERICAN STANDARD COARSE-THREAD SERIES—BASIC DIMENSIONS

Sizes	Basic Major Diameter,	Thds. per Inch,	Basic Pitch Diameter,	Minor Diameter Ext. Thds.	Minor Diameter Int. Thds.	Lead Angle at Basic Pitch Diameter,		Section at Minor Diameter	Tensile Stress Area
	Inches		Inches	Inches	Inches	Deg	Min	Sq In.	Sq In.
1 (.073)	0.0730	64	0.0629	0.0538	0.0561	4	31	0.0022	0.0026
2 (.086)	0.0860	56	0.0744	0.0641	0.0667	4	22	0.0031	0.0036
3 (.099)	0.0990	48	0.0855	0.0734	0.0764	4	26	0.0041	0.0048
4 (.112)	0.1120	40	0.0958	0.0813	0.0849	4	45	0.0050	0.0060
5 (.125)	0.1250	40	0.1088	0.0943	0.0979	4	11	0.0067	0.0079
6 (.138)	0.1380	32	0.1177	0.0997	0.1042	4	50	0.0075	0.0090
8 (.164)	0.1640	32	0.1437	0.1257	0.1302	3	58	0.0120	0.0139
10 (.190)	0.1900	24	0.1629	0.1389	0.1449	4	39	0.0145	0.0174
12 (.216)	0.2160	24	0.1889	0.1649	0.1709	4	1	0.0206	0.0240
1/4	0.2500	20	0.2175	0.1887	0.1959	4	11	0.0269	0.0317
5/16	0.3125	18	0.2764	0.2443	0.2524	3	40	0.0454	0.0522
3/8	0.3750	16	0.3344	0.2983	0.3073	3	24	0.0678	0.0773
7/16	0.4375	14	0.3911	0.3499	0.3602	3	20	0.0933	0.1060
1/2	0.5000	13	0.4500	0.4056	0.4167	3	7	0.1257	0.1416
9/16	0.5625	12	0.5084	0.4603	0.4723	2	59	0.1620	0.1816
5/8	0.6250	11	0.5660	0.5135	0.5266	2	56	0.2018	0.2256
3/4	0.7500	10	0.6850	0.6273	0.6417	2	40	0.3020	0.3340
7/8	0.8750	9	0.8028	0.7387	0.7547	2	31	0.4193	0.4612
1	1.0000	8	0.9188	0.8466	0.8647	2	29	0.5510	0.6051
1 1/8	1.1250	7	1.0322	0.9497	0.9704	2	31	0.6931	0.7627
1 1/4	1.2500	7	1.1572	1.0747	1.0954	2	15	0.8898	0.9684
1 3/8	1.3750	6	1.2667	1.1705	1.1946	2	24	1.0541	1.1538
1 1/2	1.5000	6	1.3917	1.2955	1.3196	2	11	1.2938	1.4041
1·3/4	1.7500	5	1.6201	1.5046	1.5335	2	15	1.7441	1.8983
2	2.0000	4 1/2	1.8557	1.7274	1.7594	2	11	2.3001	2.4971
2 1/4	2.2500	4 1/2	2.1057	1.9774	2.0094	1	55	3.0212	3.2464
2 1/2	2.5000	4	2.3376	2.1933	2.2294	1	57	3.7161	3.9976
2 3/4	2.7500	4	2.5876	2.4433	2.4794	1	46	4.6194	4.9326
3	3.0000	4	2.8376	2.6933	2.7294	1	36	5.6209	5.9659
3 1/4	3.2500	4	3.0876	2.9433	2.9794	1	29	6.7205	7.0992
3 1/2	3.5000	4	3.3376	3.1933	3.2294	1	22	7.9183	8.3268
3 3/4.	3.7500	4	3.5876	3.4433	3.4794	1	16	9.2143	9.6546
4	4.0000	4	3.8376	3.6933	3.7294	1	11	10.6084	11.0805

Courtesy American Society of Mechanical Engineers.

In British practice, the term "Effective Diameter" is used instead of "Pitch Diameter."

The area designated as "Stress Area" is based upon a diameter that is a mean of the pitch and minor diameters to allow for the strengthening effect of the threads.

Figures in bold type indicate Unified threads.

TABLE 7b

UNIFIED AND AMERICAN STANDARD FINE-THREAD SERIES—BASIC DIMENSIONS

Sizes	Basic Major Diameter,	Thds. per Inch,	Basic Pitch Diameter,	Minor Diameter Ext. Thds.	Minor Diameter Int. Thds.	Lead Angle at Basic Pitch Diameter,		Section at Minor Diameter	Tensile Stress Area
	Inches		Inches	Inches	Inches	Deg	Min	Sq In.	Sq In.
0 (.060)	0.0600	80	0.0519	0.0447	0.0465	4	23	0.0015	0.0018
1 (.073)	0.0730	72.	0.0640	0.0560	0.0580	3	57	0.0024	0.0027
2 (.086)	0.0860	64	0.0759	0.0668	0.0691	3	45	0.0034	0.0039
3 (.099)	0.0990	56	0.0874	0.0771	0.0797	3	43	0.0045	0.0052
4 (.112)	0.1120	48	0.0985	0.0864	0.0894	3	51	0.0057	0.0065
5 (.125)	0.1250	44	0.1102	0.0971	0.1004	3	45	0.0072	0.0082
6 (.138)	0.1380	40	0.1218	0.1073	0.1109	3	44	0.0087	0.0101
8 (.164)	0.1640	36	0.1460	0.1299	0.1339	3	28	0.0128	0.0146
10 (.190)	0.1900	32	0.1697	0.1517	0.1562	3	21	0.0175	0.0199
12 (.216)	0.2160	28	0.1928	0.1722	0.1773	3	22	0.0226	0.0257
1/4	0.2500	28	0.2268	0.2062	0.2113	2	52	0.0326	0.0362
5/16	0.3125	24	0.2854	0.2614	0.2674	2	40	0.0524	0.0579
3/8	0.3750	24	0.3479	0.3239	0.3299	2	11	0.0809	0.0876
7/16	0.4375	20	0.4050	0.3762	0.3834	2	15	0.1090	0.1185
1/2	0.5000	20	0.4675	0.4387	0.4459	1	57	0.1486	0.1597
9/16	0.5625	18	0.5264	0.4943	0.5024	1	55	0.1888	0.2026
5/8	0.6250	18	0.5889	0.5568	0.5649	1	43	0.2400	0.2555
3/4	0.7500	16	0.7094	0.6733	0.6823	1	36	0.3513	0.3724
7/8	0.8750	14	0.8286	0.7874	0.7977	1	34	0.4805	0.5088
1	1.0000	14	0.9536	0.9124	0.9227	1	22	0.6464	0.6791
1	1.0000	12	0.9459	0.8978	0.9098	1	36	0.6245	0.6624
1 1/8	1.1250	12	1.0709	1.0228	1.0348	1	25	0.8118	0.8549
1 1/4	1.2500	12	1.1959	1.1478	1.1598	1	16	1.0237	1.0721
1 3/8	1.3750	12	1.3209	1.2728	1.2848	1	9	1.2602	1.3137
1 1/2	1.5000	12	1.4459	1.3978	1.4098	1	3	1.5212	1.5799

TABLE 8

Selected Combinations of Special Diameter and Pitch—Class 2B*

(UN, N, NEF, and NS Threads)

Size	Threads per Inch	Thread Symbol	L_e 9 X Pitch	Minor Diameter Min	Minor Diameter Max	Minor Diameter Tolerance	Pitch Diameter Min	Pitch Diameter Max	Pitch Diameter Tolerance	Major Diameter Min
10 (.190)	28	NS-2B	0.3214	0.1513	0.1604	0.0091	0.1668	0.1711	0.0043	0.1900
10 (.190)	36	NS-2B	0.2500	0.1599	0.1669	0.0070	0.1720	0.1759	0.0039	0.1900
10 (.190)	40	NS-2B	0.2250	0.1629	0.1691	0.0062	0.1738	0.1775	0.0037	0.1900
10 (.190)	48	NS-2B	0.1875	0.1674	0.1725	0.0051	0.1765	0.1799	0.0034	0.1900
10 (.190)	56	NS-2B	0.1607	0.1707	0.1749	0.0042	0.1784	0.1816	0.0032	0.1900
12 (.216)	32	NEF-2B	0.2812	0.1822	0.1895	0.0073	0.1957	0.1998	0.0041	0.2160
12 (.216)	36	NS-2B	0.2500	0.1859	0.1923	0.0064	0.1980	0.2019	0.0039	0.2160
12 (.216)	40	NS-2B	0.2250	0.1889	0.1946	0.0057	0.1998	0.2035	0.0037	0.2160
12 (.216)	48	NS-2B	0.1875	0.1934	0.1981	0.0047	0.2025	0.2059	0.0034	0.2160
12 (.216)	56	NS-2B	0.1607	0.1967	0.2006	0.0039	0.2044	0.2076	0.0032	0.2160
1/4	24	NS-2B	0.3750	0.2049	0.2139	0.0090	0.2229	0.2277	0.0048	0.2500
1/4	32	NEF-2B	0.2812	0.2162	0.2229	0.0067	0.2297	0.2339	0.0042	0.2500
1/4	36	UN-2B	0.2500	0.2199	0.2258	0.0059	0.2320	0.2360	0.0040	0.2500
1/4	40	NS-2B	0.2250	0.2229	0.2282	0.0053	0.2338	0.2376	0.0038	0.2500
1/4	48	NS-2B	0.1875	0.2274	0.2317	0.0043	0.2365	0.2401	0.0036	0.2500
1/4	56	NS-2B	0.1607	0.2307	0.2343	0.0036	0.2384	0.2417	0.0033	0.2500
5/16	20	NS-2B	0.4500	0.2584	0.2680	0.0096	0.2800	0.2852	0.0052	0.3125
5/16	28	NS-2B	0.3214	0.2738	0.2807	0.0069	0.2893	0.2937	0.0044	0.3125
5/16	32	NEF-2B	0.2812	0.2787	0.2847	0.0060	0.2922	0.2964	0.0042	0.3125
5/16	36	UN-2B	0.2500	0.2824	0.2877	0.0053	0.2945	0.2985	0.0040	0.3125
5/16	40	NS-2B	0.2250	0.2854	0.2902	0.0048	0.2963	0.3001	0.0038	0.3125
5/16	48	NS-2B	0.1875	0.2899	0.2940	0.0041	0.2990	0.3026	0.0036	0.3125
3/8	18	NS-2B	0.5000	0.3149	0.3246	0.0097	0.3389	0.3445	0.0056	0.3750
3/8	20	NS-2B	0.4500	0.3209	0.3297	0.0088	0.3425	0.3479	0.0054	0.3750
3/8	28	NS-2B	0.3214	0.3363	0.3426	0.0063	0.3518	0.3564	0.0046	0.3750
3/8	32	NEF-2B	0.2812	0.3412	0.3469	0.0057	0.3547	0.3591	0.0044	0.3750
3/8	36	UN-2B	0.2500	0.3449	0.3501	0.0052	0.3570	0.3612	0.0042	0.3750
3/8	40	NS-2B	0.2250	0.3479	0.3527	0.0048	0.3588	0.3628	0.0040	0.3750
7/16	16	NS-2B	0.5625	0.3698	0.3800	0.0102	0.3969	0.4028	0.0059	0.4375
7/16	18	NS-2B	0.5000	0.3774	0.3865	0.0091	0.4014	0.4070	0.0056	0.4375
7/16	24	NS-2B	0.3750	0.3924	0.3994	0.0070	0.4104	0.4153	0.0049	0.4375
7/16	28	UNEF-2B	0.3214	0.3988	0.4051	0.0063	0.4143	0.4189	0.0046	0.4375
7/16	32	NS-2B	0.2812	0.4037	0.4094	0.0057	0.4172	0.4216	0.0044	0.4375
1/2	12	N-2B	0.7500	0.4098	0.4223	0.0125	0.4459	0.4529	0.0070	0.5000
1/2	14	NS-2B	0.6429	0.4227	0.4336	0.0109	0.4536	0.4601	0.0065	0.5000
1/2	16	NS-2B	0.5625	0.4323	0.4419	0.0096	0.4594	0.4655	0.0061	0.5000
1/2	18	NS-2B	0.5000	0.4399	0.4485	0.0086	0.4639	0.4697	0.0058	0.5000
1/2	24	NS-2B	0.3750	0.4549	0.4619	0.0070	0.4729	0.4780	0.0051	0.5000
1/2	28	UNEF-2B	0.3214	0.4613	0.4676	0.0063	0.4768	0.4816	0.0048	0.5000
1/2	32	NS-2B	0.2812	0.4662	0.4719	0.0057	0.4797	0.4842	0.0045	0.5000
9/16	14	NS-2B	0.6429	0.4852	0.4956	0.0104	0.5161	0.5226	0.0065	0.5625
9/16	16	NS-2B	0.5625	0.4948	0.5040	0.0092	0.5219	0.5280	0.0061	0.5625
9/16	20	NS-2B	0.4500	0.5084	0.5162	0.0078	0.5300	0.5355	0.0055	0.5625
9/16	24	NEF-2B	0.3750	0.5174	0.5244	0.0070	0.5354	0.5405	0.0051	0.5625
9/16	28	UN-2B	0.3214	0.5238	0.5301	0.0063	0.5393	0.5441	0.0048	0.5625
9/16	32	NS-2B	0.2812	0.5287	0.5344	0.0057	0.5422	0.5467	0.0045	0.5625
5/8	12	N-2B	0.7500	0.5348	0.5463	0.0115	0.5709	0.5780	0.0071	0.6250
5/8	14	NS-2B	0.6429	0.5477	0.5577	0.0100	0.5786	0.5852	0.0066	0.6250
5/8	16	NS-2B	0.5625	0.5573	0.5662	0.0089	0.5844	0.5906	0.0062	0.6250
5/8	20	NS-2B	0.4500	0.5709	0.5787	0.0078	0.5925	0.5981	0.0056	0.6250
5/8	24	NEF-2B	0.3750	0.5799	0.5869	0.0070	0.5979	0.6031	0.0052	0.6250
5/8	28	UN-2B	0.3214	0.5863	0.5926	0.0063	0.6018	0.6067	0.0049	0.6250
5/8	32	NS-2B	0.2812	0.5912	0.5969	0.0057	0.6047	0.6093	0.0046	0.6250
11/16	12	N-2B	0.7500	0.5973	0.6085	0.0112	0.6334	0.6405	0.0071	0.6875
11/16	24	NEF-2B	0.3750	0.6424	0.6494	0.0070	0.6604	0.6656	0.0052	0.6875

* American Society of Mechanical Engineers. Internal thread limits of size applicable to lengths of engagement of from 5 to 15 times the pitch, inclusive. For 8-, 12- and 16-thread series sizes not shown, use dimensions

TABLE 8

Selected Combinations of Special Diameter and Pitch—Class 2B *—
Continued

(UN, N, NEF, and NS Threads)

Size	Threads per Inch	Thread Symbol	L_e 9 X Pitch	Minor Diameter Min	Minor Diameter Max	Tolerance	Pitch Diameter Min	Pitch Diameter Max	Tolerance	Major Diameter Min
3/4	12	N-2B	0.7500	0.6598	0.6707	0.0109	0.6959	0.7031	0.0072	0.7500
3/4	14	NS-2B	0.6429	0.6727	0.6822	0.0095	0.7036	0.7103	0.0067	0.7500
3/4	18	NS-2B	0.5000	0.6899	0.6980	0.0081	0.7139	0.7199	0.0060	0.7500
3/4	20	UNEF-2B	0.4500	0.6959	0.7037	0.0078	0.7175	0.7232	0.0057	0.7500
3/4	24	NS-2B	0.3750	0.7049	0.7119	0.0070	0.7229	0.7282	0.0053	0.7500
3/4	28	UN-2B	0.3214	0.7113	0.7176	0.0063	0.7268	0.7318	0.0050	0.7500
3/4	32	NS-2B	0.2812	0.7162	0.7219	0.0057	0.7297	0.7344	0.0047	0.7500
13/16	12	N-2B	0.7500	0.7223	0.7329	0.0106	0.7584	0.7656	0.0072	0.8125
13/16	16	UN-2B	0.5625	0.7448	0.7533	0.0085	0.7719	0.7782	0.0063	0.8125
13/16	20	UNEF-2B	0.4500	0.7584	0.7662	0.0078	0.7800	0.7857	0.0057	0.8125
7/8	10	NS-2B	0.9000	0.7667	0.7789	0.0122	0.8100	0.8178	0.0078	0.8750
7/8	12	N-2B	0.7500	0.7848	0.7952	0.0104	0.8209	0.8281	0.0072	0.8750
7/8	16	UN-2B	0.5625	0.8073	0.8158	0.0085	0.8344	0.8407	0.0063	0.8750
7/8	18	NS-2B	0.5000	0.8149	0.8230	0.0081	0.8389	0.8449	0.0060	0.8750
7/8	20	UNEF-2B	0.4500	0.8209	0.8287	0.0078	0.8425	0.8482	0.0057	0.8750
7/8	24	NS-2B	0.3750	0.8299	0.8369	0.0070	0.8479	0.8532	0.0053	0.8750
7/8	28	UN-2B	0.3214	0.8363	0.8426	0.0063	0.8518	0.8568	0.0050	0.8750
7/8	32	NS-2B	0.2812	0.8412	0.8469	0.0057	0.8547	0.8594	0.0047	0.8750
15/16	12	UN-2B	0.7500	0.8473	0.8575	0.0102	0.8834	0.8908	0.0074	0.9375
15/16	16	UN-2B	0.5625	0.8698	0.8783	0.0085	0.8969	0.9034	0.0065	0.9375
15/16	20	UNEF-2B	0.4500	0.8834	0.8912	0.0078	0.9050	0.9109	0.0059	0.9375
1	10	NS-2B	0.9000	0.8917	0.9037	0.0120	0.9350	0.9430	0.0080	1.0000
1	16	UN-2B	0.5625	0.9323	0.9408	0.0085	0.9594	0.9659	0.0065	1.0000
1	18	NS-2B	0.5000	0.9399	0.9480	0.0081	0.9639	0.9701	0.0062	1.0000
1	20	UNEF-2B	0.4500	0.9459	0.9537	0.0078	0.9675	0.9734	0.0059	1.0000
1	24	NS-2B	0.3750	0.9549	0.9619	0.0070	0.9729	0.9784	0.0055	1.0000
1	28	UN-2B	0.3214	0.9613	0.9676	0.0063	0.9768	0.9820	0.0052	1.0000
1	32	NS-2B	0.2812	0.9662	0.9719	0.0057	0.9797	0.9846	0.0049	1.0000
1 1/16	12	UN-2B	0.7500	0.9723	0.9823	0.0100	1.0084	1.0158	0.0074	1.0625
1 1/16	16	UN-2B	0.5625	0.9948	1.0033	0.0085	1.0219	1.0284	0.0065	1.0625
1 1/16	18	NEF-2B	0.5000	1.0024	1.0105	0.0081	1.0264	1.0326	0.0062	1.0625
1 1/8	10	NS-2B	0.9000	1.0167	1.0287	0.0120	1.0600	1.0680	0.0080	1.1250
1 1/8	14	NS-2B	0.6429	1.0477	1.0565	0.0088	1.0786	1.0855	0.0069	1.1250
1 1/8	16	UN-2B	0.5625	1.0573	1.0658	0.0085	1.0844	1.0910	0.0066	1.1250
1 1/8	18	NEF-2B	0.5000	1.0649	1.0730	0.0081	1.0889	1.0951	0.0062	1.1250
1 1/8	20	UN-2B	0.4500	1.0709	1.0787	0.0078	1.0925	1.0984	0.0059	1.1250
1 1/8	24	NS-2B	0.3750	1.0799	1.0869	0.0070	1.0979	1.1034	0.0055	1.1250
1 1/8	28	UN-2B	0.3214	1.0863	1.0926	0.0063	1.1018	1.1070	0.0052	1.1250
1 3/16	12	UN-2B	0.7500	1.0973	1.1073	0.0100	1.1334	1.1409	0.0075	1.1875
1 3/16	16	UN-2B	0.5625	1.1198	1.1283	0.0085	1.1469	1.1535	0.0066	1.1875
1 3/16	18	NEF-2B	0.5000	1.1274	1.1355	0.0081	1.1514	1.1577	0.0063	1.1875
1 1/4	10	NS-2B	0.9000	1.1417	1.1537	0.0120	1.1850	1.1932	0.0082	1.2500
1 1/4	14	NS-2B	0.6429	1.1727	1.1815	0.0088	1.2036	1.2106	0.0070	1.2500
1 1/4	16	UN-2B	0.5625	1.1823	1.1908	0.0085	1.2094	1.2160	0.0066	1.2500
1 1/4	18	NEF-2B	0.5000	1.1899	1.1980	0.0081	1.2139	1.2202	0.0063	1.2500
1 1/4	20	UN-2B	0.4500	1.1959	1.2037	0.0078	1.2175	1.2236	0.0061	1.2500
1 1/4	24	NS-2B	0.3750	1.2049	1.2119	0.0070	1.2229	1.2285	0.0056	1.2500
1 5/16	12	UN-2B	0.7500	1.2223	1.2323	0.0100	1.2584	1.2659	0.0075	1.3125
1 5/16	16	UN-2B	0.5625	1.2448	1.2533	0.0085	1.2719	1.2785	0.0066	1.3125
1 5/16	18	NEF-2B	0.5000	1.2524	1.2605	0.0081	1.2764	1.2827	0.0063	1.3125

and symbols shown in tables for UNC, NC, UNF, and NF thread series. Bold type indicates Unified threads-UN.

In British practice, the term "Effective Diameter" is used instead of "Pitch Diameter."

TABLE 8

Selected Combinations of Special Diameter and Pitch—Class 2B— *Continued*

(UN, N, NEF, and NS Threads)

Size	Threads per Inch	Thread Symbol	L_e 9 X Pitch	Minor Diameter Min	Minor Diameter Max	Tolerance	Pitch Diameter Min	Pitch Diameter Max	Tolerance	Major Diameter Min
1 3/8	10	NS-2B	0.9000	1.2667	1.2787	0.0120	1.3100	1.3182	0.0082	1.3750
1 3/8	14	NS-2B	0.6429	1.2977	1.3065	0.0088	1.3286	1.3356	0.0070	1.3750
1 3/8	16	UN-2B	0.5625	1.3073	1.3158	0.0085	1.3344	1.3410	0.0066	1.3750
1 3/8	18	NEF-2B	0.5000	1.3149	1.3230	0.0081	1.3389	1.3452	0.0063	1.3750
1 7/16	12	UN-2B	0.7500	1.3473	1.3573	0.0100	1.3834	1.3910	0.0076	1.4375
1 7/16	16	UN-2B	0.5625	1.3698	1.3783	0.0085	1.3969	1.4037	0.0068	1.4375
1 7/16	18	NEF-2B	0.5000	1.3774	1.3855	0.0081	1.4014	1.4079	0.0065	1.4375
1 1/2	10	NS-2B	0.9000	1.3917	1.4037	0.0120	1.4350	1.4433	0.0083	1.5000
1 1/2	14	NS-2B	0.6429	1.4227	1.4315	0.0088	1.4536	1.4608	0.0072	1.5000
1 1/2	16	UN-2B	0.5625	1.4323	1.4408	0.0085	1.4594	1.4662	0.0068	1.5000
1 1/2	18	NEF-2B	0.5000	1.4399	1.4480	0.0081	1.4639	1.4704	0.0065	1.5000
1 1/2	20	UN-2B	0.4500	1.4459	1.4537	0.0078	1.4675	1.4737	0.0062	1.5000
1 1/2	24	NS-2B	0.3750	1.4549	1.4619	0.0070	1.4729	1.4787	0.0058	1.5000
1 9/16	16	N-2B	0.5625	1.4948	1.5033	0.0085	1.5219	1.5287	0.0068	1.5625
1 9/16	18	NEF-2B	0.5000	1.5024	1.5105	0.0081	1.5264	1.5329	0.0065	1.5625
1 5/8	6	NS-2B	1.5000	1.4446	1.4646	0.0200	1.5167	1.5272	0.0105	1.6250
1 5/8	7	NS-2B	1.2857	1.4704	1.4875	0.0171	1.5322	1.5420	0.0098	1.6250
1 5/8	10	NS-2B	0.9000	1.5167	1.5287	0.0120	1.5600	1.5683	0.0083	1.6250
1 5/8	12	N-2B	0.7500	1.5348	1.5448	0.0100	1.5709	1.5785	0.0076	1.6250
1 5/8	14	NS-2B	0.6429	1.5477	1.5565	0.0088	1.5786	1.5858	0.0072	1.6250
1 5/8	16	N-2B	0.5625	1.5573	1.5658	0.0085	1.5844	1.5912	0.0068	1.6250
1 5/8	18	NEF-2B	0.5000	1.5649	1.5730	0.0081	1.5889	1.5954	0.0065	1.6250
1 5/8	20	NS-2B	0.4500	1.5709	1.5787	0.0078	1.5925	1.5987	0.0062	1.6250
1 5/8	24	NS-2B	0.3750	1.5799	1.5869	0.0070	1.5979	1.6037	0.0058	1.6250
1 11/16	16	N-2B	0.5625	1.6198	1.6283	0.0085	1.6469	1.6538	0.0069	1.6875
1 11/16	18	NEF-2B	0.5000	1.6274	1.6355	0.0081	1.6514	1.6580	0.0066	1.6875
1 3/4	6	NS-2B	1.5000	1.5696	1.5896	0.0200	1.6417	1.6523	0.0106	1.7500
1 3/4	7	NS-2B	1.2857	1.5954	1.6125	0.0171	1.6572	1.6671	0.0099	1.7500
1 3/4	8	UN-2B	1.1250	1.6147	1.6297	0.0150	1.6688	1.6781	0.0093	1.7500
1 3/4	10	NS-2B	0.9000	1.6417	1.6537	0.0120	1.6850	1.6934	0.0084	1.7500
1 3/4	12	UN-2B	0.7500	1.6598	1.6698	0.0100	1.6959	1.7037	0.0078	1.7500
1 3/4	14	NS-2B	0.6429	1.6727	1.6815	0.0088	1.7036	1.7109	0.0073	1.7500
1 3/4	16	UNEF-2B	0.5625	1.6823	1.6908	0.0085	1.7094	1.7163	0.0069	1.7500
1 3/4	18	NS-2B	0.5000	1.6899	1.6980	0.0081	1.7139	1.7205	0.0066	1.7500
1 3/4	20	UN-2B	0.4500	1.6959	1.7037	0.0078	1.7175	1.7238	0.0063	1.7500
1 13/16	16	N-2B	0.5625	1.7448	1.7533	0.0085	1.7719	1.7788	0.0069	1.8125
1 7/8	6	NS-2B	1.5000	1.6946	1.7146	0.0200	1.7667	1.7773	0.0106	1.8750
1 7/8	7	NS-2B	1.2857	1.7204	1.7375	0.0171	1.7822	1.7921	0.0099	1.8750
1 7/8	8	NS-2B	1.1250	1.7397	1.7547	0.0150	1.7938	1.8031	0.0093	1.8750
1 7/8	10	NS-2B	0.9000	1.7667	1.7787	0.0120	1.8100	1.8184	0.0084	1.8750
1 7/8	12	N-2B	0.7500	1.7848	1.7948	0.0100	1.8209	1.8287	0.0078	1.8750
1 7/8	14	NS-2B	0.6429	1.7977	1.8065	0.0088	1.8286	1.8359	0.0073	1.8750
1 7/8	16	N-2B	0.5625	1.8073	1.8158	0.0085	1.8344	1.8413	0.0069	1.8750
1 7/8	18	NS-2B	0.5000	1.8149	1.8230	0.0081	1.8389	1.8455	0.0066	1.8750
1 7/8	20	NS-2B	0.4500	1.8209	1.8287	0.0078	1.8425	1.8488	0.0063	1.8750
1 15/16	16	N-2B	0.5625	1.8698	1.8783	0.0085	1.8969	1.9039	0.0070	1.9375

TABLE 8

Selected Combinations of Special Diameter and Pitch—Class 2B—
Continued

(UN, N, NEF, and NS Threads)

Designation			L_e	Internal Thread Limits of Size						Major Diameter
				Minor Diameter			Pitch Diameter			
Size	Threads per Inch	Thread Symbol	9 X Pitch	Limits		Tolerance	Limits		Tolerance	
				Min	Max		Min	Max		Min
2	6	NS-2B	1.5000	1.8196	1.8396	0.0200	1.8917	1.9025	0.0108	2.0000
2	7	NS-2B	1.2857	1.8454	1.8625	0.0171	1.9072	1.9172	0.0100	2.0000
2	8	UN-2B	1.1250	1.8647	1.8797	0.0150	1.9188	1.9282	0.0094	2.0000
2	10	NS-2B	0.9000	1.8917	1.9037	0.0120	1.9350	1.9435	0.0085	2.0000
2	12	UN-2B	0.7500	1.9098	1.9198	0.0100	1.9459	1.9538	0.0079	2.0000
2	14	NS-2B	0.6429	1.9227	1.9315	0.0088	1.9536	1.9610	0.0074	2.0000
2	16	UNEF-2B	0.5625	1.9323	1.9408	0.0085	1.9594	1.9664	0.0070	2.0000
2	18	NS-2B	0.5000	1.9399	1.9480	0.0081	1.9639	1.9706	0.0067	2.0000
2	20	UN-2B	0.4500	1.9459	1.9537	0.0078	1.9675	1.9739	0.0064	2.0000
2 1/16	16	N-2B	0.5625	1.9948	2.0033	0.0085	2.0219	2.0289	0.0070	2.0625
2 1/8	12	N-2B	0.7500	2.0348	2.0448	0.0100	2.0709	2.0788	0.0079	2.1250
2 1/8	16	N-2B	0.5625	2.0573	2.0658	0.0085	2.0844	2.0914	0.0070	2.1250
2 3/16	16	N-2B	0.5625	2.1198	2.1283	0.0085	2.1469	2.1539	0.0070	2.1875
2 1/4	6	NS-2B	1.5000	2.0696	2.0896	0.0200	2.1417	2.1525	0.0108	2.2500
2 1/4	7	NS-2B	1.2857	2.0954	2.1125	0.0171	2.1572	2.1672	0.0100	2.2500
2 1/4	8	UN-2B	1.1250	2.1147	2.1297	0.0150	2.1688	2.1782	0.0094	2.2500
2 1/4	10	NS-2B	0.9000	2.1417	2.1537	0.0120	2.1850	2.1935	0.0085	2.2500
2 1/4	12	UN-2B	0.7500	2.1598	2.1698	0.0100	2.1959	2.2038	0.0079	2.2500
2 1/4	14	NS-2B	0.6429	2.1727	2.1815	0.0088	2.2036	2.2110	0.0074	2.2500
2 1/4	16	UN-2B	0.5625	2.1823	2.1908	0.0085	2.2094	2.2164	0.0070	2.2500
2 1/4	18	NS-2B	0.5000	2.1899	2.1980	0.0081	2.2139	2.2206	0.0067	2.2500
2 1/4	20	UN-2B	0.4500	2.1959	2.2037	0.0078	2.2175	2.2239	0.0064	2.2500
2 5/16	16	N-2B	0.5625	2.2448	2.2533	0.0085	2.2719	2.2791	0.0072	2.3125
2 3/8	12	N-2B	0.7500	2.2848	2.2948	0.0100	2.3209	2.3290	0.0081	2.3750
2 3/8	16	N-2B	0.5625	2.3073	2.3158	0.0085	2.3344	2.3416	0.0072	2.3750
2 7/16	16	N-2B	0.5625	2.3698	2.3783	0.0085	2.3969	2.4041	0.0072	2.4375
2 1/2	6	NS-2B	1.5000	2.3196	2.3396	0.0200	2.3917	2.4026	0.0109	2.5000
2 1/2	7	NS-2B	1.2857	2.3454	2.3625	0.0171	2.4072	2.4174	0.0102	2.5000
2 1/2	8	UN-2B	1.1250	2.3647	2.3797	0.0150	2.4188	2.4284	0.0096	2.5000
2 1/2	10	NS-2B	0.9000	2.3917	2.4037	0.0120	2.4350	2.4437	0.0087	2.5000
2 1/2	12	UN-2B	0.7500	2.4098	2.4198	0.0100	2.4459	2.4540	0.0081	2.5000
2 1/2	14	NS-2B	0.6429	2.4227	2.4315	0.0088	2.4536	2.4612	0.0076	2.5000
2 1/2	16	UN-2B	0.5625	2.4323	2.4408	0.0085	2.4594	2.4666	0.0072	2.5000
2 1/2	18	NS-2B	0.5000	2.4399	2.4480	0.0081	2.4639	2.4708	0.0069	2.5000
2 1/2	20	UN-2B	0.4500	2.4459	2.4537	0.0078	2.4675	2.4741	0.0066	2.5000
2 5/8	12	N-2B	0.7500	2.5348	2.5448	0.0100	2.5709	2.5790	0.0081	2.6250
2 5/8	16	N-2B	0.5625	2.5573	2.5658	0.0085	2.5844	2.5916	0.0072	2.6250
2 3/4	6	NS-2B	1.5000	2.5696	2.5896	0.0200	2.6417	2.6526	0.0109	2.7500
2 3/4	7	NS-2B	1.2857	2.5954	2.6125	0.0171	2.6572	2.6674	0.0102	2.7500
2 3/4	8	UN-2B	1.1250	2.6147	2.6297	0.0150	2.6688	2.6784	0.0096	2.7500
2 3/4	10	NS-2B	0.9000	2.6417	2.6537	0.0120	2.6850	2.6937	0.0087	2.7500
2 3/4	12	UN-2B	0.7500	2.6598	2.6698	0.0100	2.6959	2.7040	0.0081	2.7500
2 3/4	14	NS-2B	0.6429	2.6727	2.6815	0.0088	2.7036	2.7112	0.0076	2.7500
2 3/4	16	UN-2B	0.5625	2.6823	2.6908	0.0085	2.7094	2.7166	0.0072	2.7500
2 3/4	18	NS-2B	0.5000	2.6899	2.6980	0.0081	2.7139	2.7208	0.0069	2.7500
2 7/8	12	N-2B	0.7500	2.7848	2.7948	0.0100	2.8209	2.8291	0.0082	2.8750
2 7/8	16	N-2B	0.5625	2.8073	2.8158	0.0085	2.8344	2.8417	0.0073	2.8750

TABLE 8

Selected Combinations of Special Diameter and Pitch—Class 2B—
Continued

(UN, N, NEF, and NS Threads)

Size	Threads per Inch	Thread Symbol	L_e 9 X Pitch	Minor Diameter Min	Minor Diameter Max	Minor Tolerance	Pitch Diameter Min	Pitch Diameter Max	Pitch Tolerance	Major Diameter Min
3	6	NS-2B	1.5000	2.8196	2.8396	0.0200	2.8917	2.9028	0.0111	3.0000
3	7	NS-2B	1.2857	2.8454	2.8625	0.0171	2.9072	2.9176	0.0104	3.0000
3	8	UN-2B	1.1250	2.8647	2.8797	0.0150	2.9188	2.9286	0.0098	3.0000
3	10	NS-2B	0.9000	2.8917	2.9037	0.0120	2.9350	2.9439	0.0089	3.0000
3	12	UN-2B	0.7500	2.9098	2.9198	0.0100	2.9459	2.9541	0.0082	3.0000
3	14	NS-2B	0.6429	2.9227	2.9315	0.0088	2.9536	2.9613	0.0077	3.0000
3	16	UN-2B	0.5625	2.9323	2.9408	0.0085	2.9594	2.9667	0.0073	3.0000
3	18	NS-2B	0.5000	2.9399	2.9480	0.0081	2.9639	2.9709	0.0070	3.0000
3 1/8	12	N-2B	0.7500	3.0348	3.0448	0.0100	3.0709	3.0791	0.0082	3.1250
3 1/8	16	N-2B	0.5625	3.0573	3.0658	0.0085	3.0844	3.0917	0.0073	3.1250
3 1/4	6	NS-2B	1.5000	3.0696	3.0896	0.0200	3.1417	3.1528	0.0111	3.2500
3 1/4	7	NS-2B	1.2857	3.0954	3.1125	0.0171	3.1572	3.1676	0.0104	3.2500
3 1/4	8	UN-2B	1.1250	3.1147	3.1297	0.0150	3.1688	3.1786	0.0098	3.2500
3 1/4	10	NS-2B	0.9000	3.1417	3.1537	0.0120	3.1850	3.1939	0.0089	3.2500
3 1/4	12	UN-2B	0.7500	3.1598	3.1698	0.0100	3.1959	3.2041	0.0082	3.2500
3 1/4	14	NS-2B	0.6429	3.1727	3.1815	0.0088	3.2036	3.2113	0.0077	3.2500
3 1/4	16	UN-2B	0.5625	3.1823	3.1908	0.0085	3.2094	3.2167	0.0073	3.2500
3 1/4	18	NS-2B	0.5000	3.1899	3.1980	0.0081	3.2139	3.2209	0.0070	3.2500
3 3/8	12	N-2B	0.7500	3.2848	3.2948	0.0100	3.3209	3.3293	0.0084	3.3750
3 3/8	16	N-2B	0.5625	3.3073	3.3158	0.0085	3.3344	3.3419	0.0075	3.3750
3 1/2	6	NS-2B	1.5000	3.3196	3.3396	0.0200	3.3917	3.4030	0.0113	3.5000
3 1/2	7	NS-2B	1.2857	3.3454	3.3625	0.0171	3.4072	3.4177	0.0105	3.5000
3 1/2	8	UN-2B	1.1250	3.3647	3.3797	0.0150	3.4188	3.4287	0.0099	3.5000
3 1/2	10	NS-2B	0.9000	3.3917	3.4037	0.0120	3.4350	3.4440	0.0090	3.5000
3 1/2	12	UN-2B	0.7500	3.4098	3.4198	0.0100	3.4459	3.4543	0.0084	3.5000
3 1/2	14	NS-2B	0.6429	3.4227	3.4315	0.0088	3.4536	3.4615	0.0079	3.5000
3 1/2	16	UN-2B	0.5625	3.4323	3.4408	0.0085	3.4594	3.4669	0.0075	3.5000
3 1/2	18	NS-2B	0.5000	3.4399	3.4480	0.0081	3.4639	3.4711	0.0072	3.5000
3 5/8	12	N-2B	0.7500	3.5348	3.5448	0.0100	3.5709	3.5793	0.0084	3.6250
3 5/8	16	N-2B	0.5625	3.5573	3.5658	0.0085	3.5844	3.5919	0.0075	3.6250
3 3/4	6	NS-2B	1.5000	3.5696	3.5896	0.0200	3.6417	3.6530	0.0113	3.7500
3 3/4	7	NS-2B	1.2857	3.5954	3.6125	0.0171	3.6572	3.6677	0.0105	3.7500
3 3/4	8	UN-2B	1.1250	3.6147	3.6297	0.0150	3.6688	3.6787	0.0099	3.7500
3 3/4	10	NS-2B	0.9000	3.6417	3.6537	0.0120	3.6850	3.6940	0.0090	3.7500
3 3/4	12	UN-2B	0.7500	3.6598	3.6698	0.0100	3.6959	3.7043	0.0084	3.7500
3 3/4	14	NS-2B	0.6429	3.6727	3.6815	0.0088	3.7036	3.7115	0.0079	3.7500
3 3/4	16	UN-2B	0.5625	3.6823	3.6908	0.0085	3.7094	3.7169	0.0075	3.7500
3 3/4	18	NS-2B	0.5000	3.6899	3.6980	0.0081	3.7139	3.7211	0.0072	3.7500
3 7/8	12	N-2B	0.7500	3.7848	3.7948	0.0100	3.8209	3.8294	0.0085	3.8750
3 7/8	16	N-2B	0.5625	3.8073	3.8158	0.0085	3.8344	3.8420	0.0076	3.8750
4	6	NS-2B	1.5000	3.8196	3.8396	0.0200	3.8917	3.9031	0.0114	4.0000
4	7	NS-2B	1.2857	3.8454	3.8625	0.0171	3.9072	3.9178	0.0106	4.0000
4	8	UN-2B	1.1250	3.8647	3.8797	0.0150	3.9188	3.9288	0.0100	4.0000
4	10	NS-2B	0.9000	3.8917	3.9037	0.0120	3.9350	3.9441	0.0091	4.0000
4	12	UN-2B	0.7500	3.9098	3.9198	0.0100	3.9459	3.9544	0.0085	4.0000
4	14	NS-2B	0.6429	3.9227	3.9315	0.0088	3.9536	3.9616	0.0080	4.0000
4	16	UN-2B	0.5625	3.9323	3.9408	0.0085	3.9594	3.9670	0.0076	4.0000
4 1/4	4	UN-2B	2.2500	3.9794	4.0094	0.0300	4.0876	4.1014	0.0138	4.2500
4 1/4	6	NS-2B	1.5000	4.0696	4.0896	0.0200	4.1417	4.1531	0.0114	4.2500
4 1/4	7	NS-2B	1.2857	4.0954	4.1125	0.0171	4.1572	4.1678	0.0106	4.2500
4 1/4	8	UN-2B	1.1250	4.1147	4.1297	0.0150	4.1683	4.1788	0.0100	4.2500
4 1/4	10	NS-2B	0.9000	4.1417	4.1537	0.0120	4.1850	4.1941	0.0091	4.2500
4 1/4	12	UN-2B	0.7500	4.1598	4.1698	0.0100	4.1959	4.2044	0.0085	4.2500
4 1/4	14	NS-2B	0.6429	4.1727	4.1815	0.0088	4.2036	4.2116	0.0080	4.2500
4 1/4	16	UN-2B	0.5625	4.1823	4.1908	0.0085	4.2094	4.2170	0.0076	4.2500

TABLE 8

Selected Combinations of Special Diameter and Pitch—Class 2B—
Continued

(UN, N, NEF, and NS Threads)

Size	Threads per Inch	Thread Symbol	L_e 9 X Pitch	Minor Diameter Limits Min	Minor Diameter Limits Max	Minor Diameter Tolerance	Pitch Diameter Limits Min	Pitch Diameter Limits Max	Pitch Diameter Tolerance	Major Diameter Min
4 1/2	4	UN-2B	2.2500	4.2294	4.2594	0.0300	4.3376	4.3514	0.0138	4.5000
4 1/2	6	NS-2B	1.5000	4.3196	4.3396	0.0200	4.3917	4.4031	0.0114	4.5000
4 1/2	7	NS-2B	1.2857	4.3454	4.3625	0.0171	4.4072	4.4178	0.0106	4.5000
4 1/2	8	UN-2B	1.1250	4.3647	4.3797	0.0150	4.4188	4.4288	0.0100	4.5000
4 1/2	10	NS-2B	0.9000	4.3917	4.4037	0.0120	4.4350	4.4441	0.0091	4.5000
4 1/2	12	UN-2B	0.7500	4.4098	4.4198	0.0100	4.4459	4.4544	0.0085	4.5000
4 1/2	14	NS-2B	0.6429	4.4227	4.4315	0.0088	4.4536	4.4616	0.0080	4.5000
4 1/2	16	UN-2B	0.5625	4.4323	4.4408	0.0085	4.4594	4.4670	0.0076	4.5000
4 3/4	4	UN-2B	2.2500	4.4794	4.5094	0.0300	4.5876	4.6016	0.0140	4.7500
4 3/4	6	NS-2B	1.5000	4.5696	4.5896	0.0200	4.6417	4.6533	0.0116	4.7500
4 3/4	7	NS-2B	1.2857	4.5954	4.6125	0.0171	4.6572	4.6681	0.0109	4.7500
4 3/4	8	UN-2B	1.1250	4.6147	4.6297	0.0150	4.6688	4.6791	0.0103	4.7500
4 3/4	10	NS-2B	0.9000	4.6417	4.6537	0.0120	4.6850	4.6944	0.0094	4.7500
4 3/4	12	UN-2B	0.7500	4.6598	4.6698	0.0100	4.6959	4.7046	0.0087	4.7500
4 3/4	14	NS-2B	0.6429	4.6727	4.6815	0.0088	4.7036	4.7119	0.0083	4.7500
4 3/4	16	NS-2B	0.5625	4.6823	4.6908	0.0085	4.7094	4.7173	0.0079	4.7500
5	4	UN-2B	2.2500	4.7294	4.7594	0.0300	4.8376	4.8516	0.0140	5.0000
5	6	NS-2B	1.5000	4.8196	4.8396	0.0200	4.8917	4.9033	0.0116	5.0000
5	7	NS-2B	1.2857	4.8454	4.8625	0.0171	4.9072	4.9181	0.0109	5.0000
5	8	UN-2B	1.1250	4.8647	4.8797	0.0150	4.9188	4.9291	0.0103	5.0000
5	10	NS-2B	0.9000	4.8917	4.9037	0.0120	4.9350	4.9444	0.0094	5.0000
5	12	UN-2B	0.7500	4.9098	4.9198	0.0100	4.9459	4.9546	0.0087	5.0000
5	14	NS-2B	0.6429	4.9227	4.9315	0.0088	4.9536	4.9619	0.0083	5.0000
5	16	UN-2B	0.5625	4.9323	4.9408	0.0085	4.9594	4.9673	0.0079	5.0000
5 1/4	4	UN-2B	2.2500	4.9794	5.0094	0.0300	5.0876	5.1016	0.0140	5.2500
5 1/4	6	NS-2B	1.5000	5.0696	5.0896	0.0200	5.1417	5.1533	0.0116	5.2500
5 1/4	7	NS-2B	1.2857	5.0954	5.1125	0.0171	5.1572	5.1681	0.0109	5.2500
5 1/4	8	UN-2B	1.1250	5.1147	5.1297	0.0150	5.1688	5.1791	0.0103	5.2500
5 1/4	10	NS-2B	0.9000	5.1417	5.1537	0.0120	5.1850	5.1944	0.0094	5.2500
5 1/4	12	UN-2B	0.7500	5.1598	5.1698	0.0100	5.1959	5.2046	0.0087	5.2500
5 1/4	14	NS-2B	0.6429	5.1727	5.1815	0.0088	5.2036	5.2119	0.0083	5.2500
5 1/4	16	UN-2B	0.5625	5.1823	5.1908	0.0085	5.2094	5.2173	0.0079	5.2500
5 1/2	4	UN-2B	2.2500	5.2294	5.2594	0.0300	5.3376	5.3516	0.0140	5.5000
5 1/2	6	NS-2B	1.5000	5.3196	5.3396	0.0200	5.3917	5.4033	0.0116	5.5000
5 1/2	7	NS-2B	1.2857	5.3454	5.3625	0.0171	5.4072	5.4181	0.0109	5.5000
5 1/2	8	UN-2B	1.1250	5.3647	5.3797	0.0150	5.4188	5.4291	0.0103	5.5000
5 1/2	10	NS-2B	0.9000	5.3917	5.4037	0.0120	5.4350	5.4444	0.0094	5.5000
5 1/2	12	UN-2B	0.7500	5.4098	5.4198	0.0100	5.4459	5.4546	0.0087	5.5000
5 1/2	14	NS-2B	0.6429	5.4227	5.4315	0.0088	5.4536	5.4619	0.0083	5.5000
5 1/2	16	UN-2B	0.5625	5.4323	5.4408	0.0085	5.4594	5.4673	0.0079	5.5000
5 3/4	4	UN-2B	2.2500	5.4794	5.5094	0.0300	5.5876	5.6018	0.0142	5.7500
5 3/4	6	NS-2B	1.5000	5.5696	5.5896	0.0200	5.6417	5.6535	0.0118	5.7500
5 3/4	7	NS-2B	1.2857	5.5954	5.6125	0.0171	5.6572	5.6683	0.0111	5.7500
5 3/4	8	UN-2B	1.1250	5.6147	5.6297	0.0150	5.6688	5.6793	0.0105	5.7500
5 3/4	10	NS-2B	0.9000	5.6417	5.6537	0.0120	5.6850	5.6946	0.0096	5.7500
5 3/4	12	UN-2B	0.7500	5.6598	5.6698	0.0100	5.6959	5.7049	0.0090	5.7500
5 3/4	14	NS-2B	0.6429	5.6727	5.6815	0.0088	5.7036	5.7121	0.0085	5.7500
5 3/4	16	NS-2B	0.5625	5.6823	5.6908	0.0085	5.7094	5.7175	0.0081	5.7500
6	4	UN-2B	2.2500	5.7294	5.7594	0.0300	5.8376	5.8518	0.0142	6.0000
6	6	NS-2B	1.5000	5.8196	5.8396	0.0200	5.8917	5.9035	0.0118	6.0000
6	7	NS-2B	1.2857	5.8454	5.8625	0.0171	5.9072	5.9183	0.0111	6.0000
6	8	UN-2B	1.1250	5.8647	5.8797	0.0150	5.9188	5.9293	0.0105	6.0000
6	10	UN-2B	0.9000	5.8917	5.9037	0.0120	5.9350	5.9446	0.0096	6.0000
6	12	UN-2B	0.7500	5.9098	5.9198	0.0100	5.9459	5.9549	0.0090	6.0000
6	14	NS-2B	0.6429	5.9227	5.9315	0.0088	5.9536	5.9621	0.0085	6.0000
6	16	UN-2B	0.5625	5.9323	5.9408	0.0085	5.9594	5.9675	0.0081	6.0000

form an angle of 60 degrees with each other as shown in Figs. 27 and 28. The root contour may be either flat or rounded, the exact

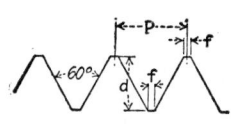

FIG. 27.

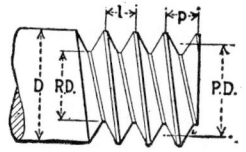

FIG. 28.

shape of the rounding not being specified since in practice it will vary with tool wear.

General formulas for the basic dimensions of the Unified and American Standard Screw Threads are:

$$\text{Pitch} = \frac{1}{\text{No. of threads per inch}}$$

Depth external thread = 0.61343 × pitch

Depth internal thread = 0.54127 × pitch

Flat at crest, external thread = 0.125 × pitch

Flat at crest, internal thread = 0.25 × pitch

Flat at root, internal thread = 0.125 × pitch

ILLUSTRATION: Find the depth, pitch, and the flat of a $\frac{1}{4}$ inch × 20 Unified Screw Thread.

$$\text{Pitch} = \frac{1}{\text{No. of threads per inch}} = \frac{1}{20}$$

$$\text{Depth external thread} = 0.61343 \times \text{pitch}$$

$$= 0.61343 \times \tfrac{1}{20}$$

$$= 0.03077$$

$$\text{Depth internal thread} = 0.54127 \times \tfrac{1}{20}$$

$$= 0.02706$$

$$\text{Flat at crest, external thread} = 0.125 \times \tfrac{1}{20}$$

$$= 0.00625$$

Flat at crest, internal thread = $0.25 \times \frac{1}{20} = 0.0125$

Flat at root, internal thread = $0.125 \times \frac{1}{20} = 0.00625$

ILLUSTRATION: Find the tap drill size for a 1 inch $\times$ 14 Unified Screw Thread.

Note: The tap drill size is equal to the root diameter when a full thread is desired. Standard tables usually give the tap drill size as 75% of a full thread.

Depth external thread = $0.61343 \times$ pitch = $0.61343 \times \frac{1}{14}$

$$= 0.0438$$

Double depth = $0.0438 \times 2 = 0.0876$

Root diameter = $1.000 - 0.0876 = 0.9124.$

Therefore the depth is 0.0438 inch and tap drill size, root diameter, is 0.9124 inch.

Whitworth Standard Thread.—This is the British Standard thread. As shown in Fig. 29, the roots and the crests are rounded and the sides form an angle of 55° with each other. If the thread were carried to a sharp point top and bottom, the rounded part would take $\frac{1}{6}$ at the top and $\frac{1}{6}$ at the bottom. Thus, $\frac{2}{3}$ is left for the depth of the thread. In such a thread the pitch, the depth and the radius are found by the following formulas.

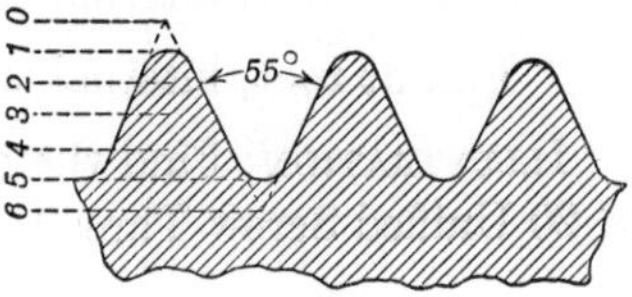

FIG. 29.

$$\text{Pitch} = \frac{1}{\text{No. threads per inch}}$$

$$\text{Depth} = \frac{0.6403}{\text{No. threads per inch}} \quad \text{or} \quad 0.6403 \times \text{pitch}$$

$$\text{Radius} = \frac{0.1373}{\text{No. threads per inch}} \quad \text{or} \quad 0.1373 \times \text{pitch}$$

ILLUSTRATION: Find the pitch, depth, and radius for a $\frac{11}{16}$ inch × 11 Whitworth Standard Screw Thread.

$$\text{Pitch} = \frac{1}{\text{No. threads per inch}} = \frac{1}{11}$$

Depth = $0.6403 \times$ pitch = $0.6403 \times \frac{1}{11}$ = 0.0582 in. (Ans.)

Radius = $0.1373 \times$ pitch = $0.1373 \times \frac{1}{11}$ = 0.0125 in. (Ans.)

Square Thread.—The sides of the square thread are parallel and the depth of the thread is equal to the width of the space

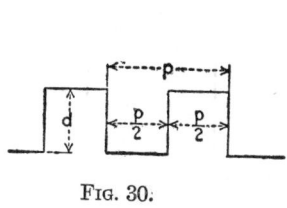

FIG. 30.

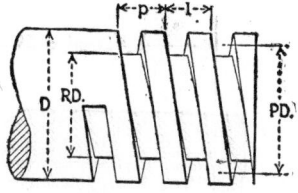

FIG. 31.—Single Square Thread:

between the teeth. (See Figs. 30, 31.) This space is theoretically equal to one-half of the pitch. It is necessary in practice to make the space in the nut a trifle wider than the thread so as to have a running fit between the screw and the nut.

Acme Thread.—The Acme Thread (see Fig. 32) has to a large extent replaced the square thread because of greater ease in cutting and of the greater strength secured. The angle between the sides is 29°, $14\frac{1}{2}$° on each side of the vertical.

The following formulas are used in calculating measurements of Acme Screw Threads and tap threads.

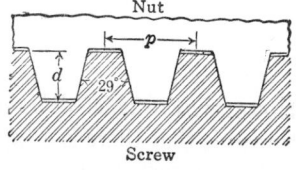

FIG. 32.—Acme Thread.

For Screws:

$d = \frac{1}{2}p + 0.0100$

$f = 0.3707p$

$c = 0.3707p - 0.0052$ in.

For Taps:

$d = \frac{1}{2}p + 0.0200$ in.

$f = 0.3707p - 0.0052$ in.

$c = 0.3707p - 0.0052$ in.

when d = depth of thread, f = width of flat at top of thread, and c = width of flat at root of thread.

Diameter of tap = diameter of screw + 0.200 inch

Diameter at root of thread (tap and screw) = diameter of screw − (p + 0.020 inch)

American Standard Taper Pipe Thread.—This was formerly known as the Briggs standard pipe thread. These threads are similar to the American National Thread, the sides making an angle of 60 degrees, but the root and crest are slightly rounded.

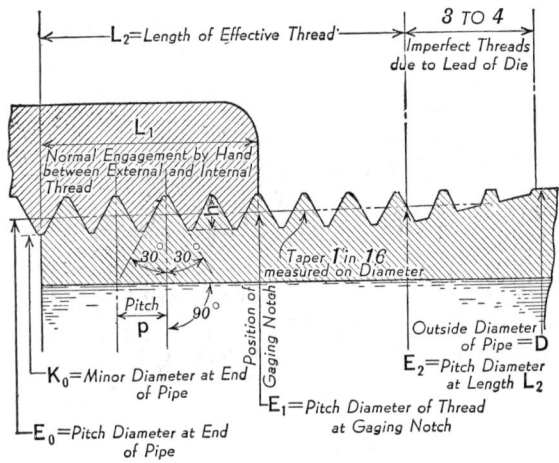

Fɪɢ. 33.

However, the chief difference between pipe thread and ordinary thread is that there is a taper on the diameter equal to $\frac{1}{16}$ inch per inch or $\frac{3}{4}$ inch per foot. The thread depth equals 0.8 × pitch of thread. The number of threads per inch for various pipe sizes is given in Table 9 with the other elements.

Pipe threads are employed on standard iron, steel, and brass pipe where the joint is subject to internal pressure from the liquid or gas which the pipe is carrying. These joints are usually made up with a special joint compound and "plumber's cotton."

TABLE 9.

Nominal size of pipe in inches	Number of threads per inch, n	Pitch, p	Depth of thread, h	Outside diameter of pipe, D	Length of normal engagement by hand, L_1	Length of effective thread, L_2	Increase in diameter per thread, $\frac{0.0625}{n}$	Pitch diameters				Basic minor diameter at small end of pipe, K_0
								At end of pipe, or at length L_1 from end of coupling, $E_0 = D - \frac{0.05D + 1.1}{n}$		At length L_1 on pipe, or at end of coupling, $E_1 = E_0 + \frac{L_1}{16}$		
								Basic	Maximum	Basic	Minimum	
1	2	3	4	5	6	7	8	9	10	11	12	13
		Inch	Inch	Inches	Inches	Inches	Inch	Inches	Inches	Inches	Inches	Inches
⅛	27	0.03704	0.02963	0.405	0.180	0.26385	0.00231	0.36351	0.37823	0.37476	0.37129	0.33388
¼	18	.05556	.04444	.540	.200	.40178	.00347	.47739	.49510	.48989	.48468	.43294
⅜	18	.05556	.04444	.675	.240	.40778	.00347	.61201	.63222	.62701	.62181	.56757
½	14	.07143	.05714	.840	.320	.53371	.00446	.75843	.78513	.77843	.77173	.70129
¾	14	.07143	.05714	1.050	.339	.54571	.00446	.96768	.99556	.98887	.98217	.91054
1	11½	.08696	.06957	1.315	.400	.68278	.00543	1.21363	1.24678	1.23863	1.23048	1.14407
1¼	11½	.08696	.06957	1.660	.420	.70678	.00543	1.55713	1.59153	1.58338	1.57523	1.48757
1½	11½	.08696	.06957	1.900	.420	.72348	.00543	1.79609	1.83049	1.82234	1.81418	1.72652
2	11½	.08696	.06957	2.375	.436	.75652	.00543	2.26902	2.30442	2.29627	2.28812	2.19946
2½	8	.12500	.10000	2.875	.682	1.13750	.00781	2.71953	2.77388	2.76216	2.75044	2.61953
3	8	.12500	.10000	3.500	.766	1.20000	.00781	3.34062	3.40022	3.38850	3.37678	3.24063
3½	8	.12500	.10000	4.000	.821	1.25000	.00781	3.83710	3.90053	3.88881	3.87709	3.73750
4	8	.12500	.10000	4.500	.844	1.30000	.00781	4.33438	4.39584	4.38712	4.37541	4.23438
4½	8	.12500	.10000	5.000	.875	1.35000	.00781	4.83125	4.89766	4.88594	4.87422	4.73125
5	8	.12500	.10000	5.563	.937	1.40630	.00781	5.39073	5.46101	5.44929	5.43757	5.29073
6	8	.12500	.10000	6.625	.958	1.51250	.00781	6.44609	6.51769	6.50597	6.49425	6.34609
7	8	.12500	.10000	7.625	1.000	1.61250	.00781	7.43964	7.51406	7.50234	7.49062	7.33984
8	8	.12500	.10000	8.625	1.063	1.71250	.00781	8.43359	8.51175	8.50003	8.48831	8.33359
9	8	.12500	.10000	9.625	1.130	1.81250	.00781	9.42734	9.50969	9.49797	9.48625	9.32734
10	8	.12500	.10000	10.750	1.210	1.92500	.00781	10.54531	10.63266	10.62094	10.60922	10.44531
11	8	0.12500	0.10000	11.750	1.285	2.02500	0.00781	11.53906	11.63109	11.61938	11.60766	11.43906
12	8	.12500	.10000	12.750	1.360	2.12500	.00781	12.53281	12.62953	12.61781	12.60609	12.43281
14 O.D.	8	.12500	.10000	14.000	1.562	2.25000	.00781	13.77500	13.88434	13.87262	13.86091	13.67500
15 O.D.	8	.12500	.10000	15.000	1.687	2.35000	.00781	14.76875	14.88591	14.87419	14.86247	14.66875
16 O.D.	8	.12500	.10000	16.000	1.812	2.45000	.00781	15.76250	15.88747	15.87575	15.86403	15.66250
17 O.D.	8	.12500	.10000	17.000	1.900	2.55000	.00781	16.76625	16.88672	16.87500	16.86328	16.65625
18 O.D.	8	.12500	.10000	18.000	2.000	2.65000	.00781	17.75000	17.88672	17.87500	17.86328	17.65000
20 O.D.	8	.12500	.10000	20.000	2.125	2.85000	.00781	19.73750	19.88203	19.87031	19.85855	19.63750
22 O.D.	8	.12500	.10000	22.000	2.250	3.05000	.00781	21.72500	21.87734	21.86562	21.85391	21.62500
24 O.D.	8	.12500	.10000	24.000	2.375	3.25000	.00781	23.71250	23.87266	23.86094	23.84922	23.61250
26 O.D.	8	.12500	.10000	26.000	2.500	3.45000	.00781	25.70000	25.86797	25.85625	25.84453	25.60000
28 O.D.	8	.12500	.10000	28.000	2.625	3.65000	.00781	27.68750	27.86328	27.85156	27.83984	27.58750
30 O.D.	8	.12500	.10000	30.000	2.750	3.85000	.00781	29.67500	29.85859	29.84688	29.83516	29.57500

† Given as information for use in selecting tap drills.

Measuring Screw Threads.—There are several methods of measuring screw threads, depending on what instruments are available. The number of threads per inch may be determined by means of a steel scale as shown in Fig. 34 or by a pitch gage as shown in Fig. 35.

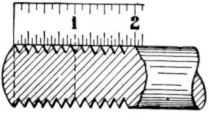

FIG. 34.

FIG. 35.

Pitch diameter is one of the most important measurements of a screw. This may be read directly from a special thread micrometer caliper. However, if such an instrument is not available, an accurate measurement may be obtained with an ordinary micrometer by the three-wire method. Three wires of equal diameter

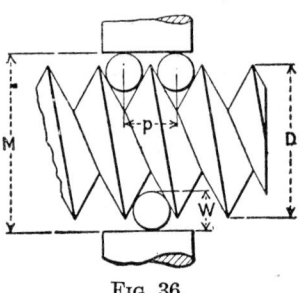

Fig. 36.

are arranged as shown in Fig. 36, one wire being placed in the angle of thread on one side of the screw and the other two on the opposite side, then measuring over the whole with a micrometer.

When W = diameter of wire,

M = micrometer reading,

pitch diameter of the American National thread is:

$$\text{P.D.} = M - 3W + \frac{0.8660}{N}$$

Other equations derived from substitution of relations pertaining to this thread are:

$$D = M - 3W + 1.5155p$$
$$M = D - 1.5155p + 3W$$

ILLUSTRATION: What will be the correct micrometer reading of a $\frac{1}{2}$ in. $\times$ 12 (American National) thread if the three-wire system is used and the diameter of the wires is 0.070 in.?

$W = 0.070$; $p = \frac{1}{12}$; $D = \frac{1}{2}$

$$M = D - 1.5155p + 3W = \frac{1}{2} - \frac{1.5155}{12} + 3 \times 0.070$$

$$M = 0.5 - 0.1263 + 0.210 = 0.5837 \text{ in.} \quad \text{(Ans.)}$$

ILLUSTRATION: What is the pitch diameter of the threads in the above illustration?

$$\text{P.D.} = M - 3W + \frac{0.8660}{N} = 0.5837 - 3 \times 0.070 + \frac{0.8660}{12}$$

$$\text{P.D.} = 0.5837 - 0.210 + 0.0723 = 0.4460 \text{ in.} \quad \text{(Ans.)}$$

Similar equations have been developed for use when measuring the Sharp V thread by the three-wire system. They are:

$$\text{P.D.} = M - 3W + \frac{0.8660}{N} \quad \text{(as before)}$$

But, $$D = M - 3W + 1.7320p,$$

and $$M = D - 1.7320p + 3W$$

In the three-wire system of measurement, any wire which will project above the crest of the thread and which has a diameter less than the pitch may be used. However, the best results will be obtained when the wire is of such size that it is tangent to the sides of the thread at the mid-points between the root and the crests. A wire which meets this qualification is known as the *best wire*. It can readily be demonstrated that the best-wire diameter is equal to two-thirds of the depth of a V thread. Since the depth of the V thread equals $\frac{0.866}{N}$, the best-thread diameter W is $\frac{2}{3} \times \frac{0.866}{N} = \frac{0.57735}{N}$. This formula also holds true for the proper size of wire for measuring American National threads.

ILLUSTRATION: What is the best-wire size for measuring a $\frac{1}{2}$ in. $\times$ 13 American National thread bolt by the three-wire method?

$$W = \frac{0.57735}{N} = \frac{0.57735}{13} = 0.04441 \text{ inch diameter} \quad \textbf{(Ans.)}$$

Screw Thread Angle.—The *angle of the helix* is designated by ϕ in Fig. 42. This angle varies with the pitch diameter and the lead of the screw.

$$\text{Tangent of helix angle} = \frac{\text{lead}}{\text{P.D.} \times \pi}$$

ILLUSTRATION: What is the helix angle of a $\frac{5}{8}$ in. $\times$ 11 American National thread?

$$\text{lead} = l = \tfrac{1}{11}$$

$$\text{P.D.} = D - \frac{0.6495}{N} = \frac{5}{8} - \frac{0.6495}{11}$$

$$\tan \phi = \frac{\frac{1}{11}}{\pi\left(\frac{5}{8} - \frac{0.6495}{11}\right)} = \frac{1}{\pi \times 11 \times (0.625 - 0.059)}$$

$$\tan \phi = \frac{7 \times 1}{22 \times 11 \times 0.566} = 0.051125$$

$$\phi = 2°\ 56'\ \text{(Ans.)}$$

Taps and Tap Drills.—Internal threads less than three-quarter inch in diameter may be cut by the use of taps, shown in Fig. 37,

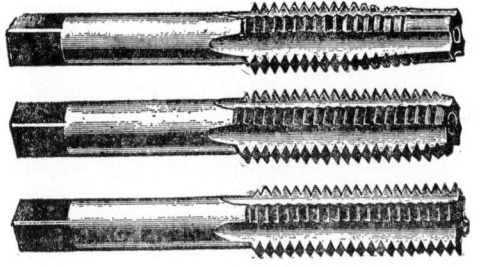
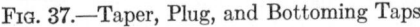
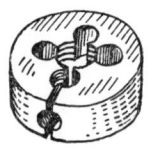

FIG. 37.—Taper, Plug, and Bottoming Taps.   FIG. 38.

and the corresponding external threads may be cut with a die. (Fig. 38.)

When drilling a hole preparatory to tapping, the theoretical size of the drill is the root diameter of the screw which is to fit the

TABLE 10

Sizes of Twist Drills with Decimal Equivalents

Size	Decimal Equivalent	Size	Decimal Equivalent	Size	Decimal Equivalent	Size	Decimal Equivalent
$\frac{1}{2}''$	0.5000''	$\frac{1}{4}''$	0.2500''	# 26	0.1470''	# 56	0.0465''
$\frac{31}{64}''$	.4844	E	.2500	# 27	.1440	# 57	.0430
$\frac{15}{32}''$	.4688	D	.2460	$\frac{9}{64}''$	.1406	# 58	.0420
$\frac{29}{64}''$	.4531	C	.2420	# 28	.1405	# 59	.0410
$\frac{7}{16}''$	.4375	B	.2380	# 29	.1360	# 60	.0400
$\frac{27}{64}''$	.4219	$\frac{15}{64}''$	.2344	# 30	.1285	# 60½	.0390
Z	.4130	A	.2340	$\frac{1}{8}''$	.1250	# 61	.0380
$\frac{13}{32}''$	.4063	# 1	.2280	# 31	.1200	# 62	.0370
Y	.4040	# 2	.2210	# 32	.1160	# 63	.0360
X	.3970	$\frac{7}{32}''$	.2188	# 33	.1130	# 64	.0350
$\frac{25}{64}''$	.3906	# 3	.2130	# 34	.1110	# 65	.0330
W	.3860	# 4	.2090	# 35	.1100	# 66	.0320
V	.3770	# 5	.2055	$\frac{7}{64}''$	.1094	$\frac{1}{32}''$	.0313
$\frac{3}{8}''$	.3750	# 6	.2040	# 36	.1065	# 67	.0310
U	.3680	$\frac{13}{64}''$	.2031	# 37	.1040	# 68	.0300
$\frac{23}{64}''$	.3594	# 7	.2010	# 38	.1015	# 68½	.0295
T	.3580	# 8	.1990	# 39	.0995	# 69	.0290
S	.3480	# 9	.1960	# 40	.0980	# 69½	.0280
$\frac{11}{32}''$	.3438	# 10	.1935	# 41	.0960	# 70	.0270
R	.3390	# 11	.1910	$\frac{3}{32}''$	.0938	# 71	.0260
Q	.3320	# 12	.1890	# 42	.0935	# 71½	.0250
$\frac{21}{64}''$	.3281	$\frac{3}{16}''$	.1875	# 43	.0890	# 72	.0240
P	.3230	# 13	.1850	# 44	.0860	# 73	.0230
O	.3160	# 14	.1820	# 45	.0820	# 73½	.0225
$\frac{5}{16}''$	.3125	# 15	.1800	# 46	.0810	# 74	.0220
N	.3020	# 16	.1770	# 47	.0785	# 74½	.0210
$\frac{19}{64}''$	.2969	# 17	.1730	$\frac{5}{64}''$	.0781	# 75	.0200
M	.2950	$\frac{11}{64}''$	.1719	# 48	.0760	# 76	.0180
L	.2900	# 18	.1695	# 49	.0730	# 77	.0160
$\frac{9}{32}''$	.2813	# 19	.1660	# 50	.0700	$\frac{1}{64}''$	.0156
K	.2810	# 20	.1610	# 51	.0670	# 78	.0150
J	.2770	# 21	.1590	# 52	.0635	# 78½	.0145
I	.2720	# 22	.1570	$\frac{1}{16}''$	.0625	# 79	.0140
H	.2660	$\frac{5}{32}''$	.1563	# 53	.0595	# 79½	.0135
$\frac{17}{64}''$	.2656	# 23	.1540	# 54	.0550	# 80	.0130
G	.2610	# 24	.1520	# 55	.0520	...	....
F	.2570	# 25	.1495	$\frac{3}{64}''$	.0469	...	....

tapped hole. In actual practice the drill is a little larger to prevent excessive strain on the tap and facilitate production. Table 10 gives sizes of twist drills with decimal equivalents; Table 11 the proper tap drill sizes of American National Threads.

TABLE 11

TAP DRILL SIZES FOR THREADS OF AMERICAN STANDARD FORM *

Thread Diameter	Threads to 1″	Diameter of Commercial Drill	Thread Diameter	Threads to 1″	Diameter of Commercial Drill
¼″	20	0.2010″	1″	8	0.8750″
⁵⁄₁₆	18	.2570	1⅛	7	0.9844
⅜	16	.3125	1¼	7	1.1094
⁷⁄₁₆	14	.3680	1⅜	6	1.2187
½	13	.4210	1½	6	1.3437
⁹⁄₁₆	12	.4844	1⅝	5½	1.4531
⅝	11	.5312	1¾	5	1.5625
¾	10	.6562	1⅞	5	1.6875
⅞	9	.7656	2	4½	1.7812

* These tap drill diameters permit approximately 75% of a full thread.

Cutting Threads on a Lathe.—If a thread-cutting tool is brought up to a piece of work previously turned and the feed is thrown in, the threads cut will correspond to the threads on the lead screw. If the lead screw has six threads per inch and makes six revolutions, the carriage will travel one inch. The threading tool will travel the same distance along the piece to be threaded. If the spindle and the lead screw are geared one to one, the spindle will make the same number of revolutions as the lead screw. If the gear on the stud is one-half that on the lead screw, the spindle makes twice as many revolutions as the feed screw, the spindle revolving twelve times while the tool moves one inch. Therefore twelve threads will be cut.

The rate of the feed may be changed by inserting different

gears to transmit the motion from the stud to the lead screw. These gears, of which a number are provided for each machine are called " change gears " and are arranged as shown in Fig. 39. When a single idler gear connects the gears, the spindle and the lead screw, the arrangement is called *simple gearing*. To find the gear ratio between the stud and lead screw in simple gearing, the following formula is used:

$$\frac{\text{threads per in. of lead screw}}{\text{threads per in. to be cut}} = \frac{\text{teeth in gear on stud}}{\text{teeth in gear on lead screw}}$$

ILLUSTRATION: What gear ratio will be required to cut 16 threads per inch on a lathe which has a lead screw with 6 threads per inch?

$$\frac{\text{threads per in. of lead screw}}{\text{threads per in. to be cut}} = \frac{6}{16} = \frac{3}{8} \text{ (Ans.)}$$

Having obtained the ratio of the gears, it is necessary to multiply the numerator and denominator by some number so that the result will represent gears in stock. From the above illustration,

$$\frac{3}{8} \times \frac{8}{8} = \frac{24}{64} = \frac{\text{teeth in gear on stud}}{\text{teeth in gear on lead screw}}$$

If gears with 24 and 64 teeth are not available, some other number must be tried. Gears with 30 and 80 teeth, respectively, would serve equally well as seen below.

$$\frac{3}{8} \times \frac{10}{10} = \frac{30}{80} = \frac{\text{teeth in gear on stud}}{\text{teeth in gear on lead screw}}$$

Sometimes it is not possible to obtain the correct ratio with two gears, particularly when a very small or a very large number of threads per inch are to be cut. Then it is necessary to insert two additional gears keyed to the same shaft, either replacing the idler as shown in Fig. 39 or in addition to the idler.   This is called *compound gearing*.

For compound gearing, the same formula as given for simple gearing may be used except that both the numerator and the denominator are divided into two factors.

ILLUSTRATION: What change gears are required to cut a screw thread with 30 threads per inch on a lathe with a lead screw of 5 threads per inch?

$$\frac{5}{30} = \frac{1}{6} = \frac{1 \times 1}{3 \times 2}$$

These factors are then multiplied separately by numbers which will give suitable gear teeth numbers as follows:

$$\frac{1}{3} \times \frac{30}{30} = \frac{30}{90} \quad \text{and} \quad \frac{1}{2} \times \frac{25}{25} = \frac{25}{50}$$

The numbers 30 and 25 represent the teeth on driving gears; 90 and 50 the teeth on driven gears.

The number of threads per inch on the lead screw varies with the make of machine, hence the necessity of having different sets of change gears.   The following are the standard gears supplied with a Reed lathe having a 5-pitch screw:

25–30–35–40–40–45–50–55–60–65–69–70–75–80–90

The following are the standard gears supplied with the Pratt & Whitney lathe having a 6-pitch screw:

30–40–50–60–65–70–75–80–90–95–100–105–110–115–120

ILLUSTRATION: What change gears can be used to cut 13 threads to the inch with a lathe that has a lead screw with four threads to the inch, using a stud gear of 20 teeth?   From the proportion on page 560.

$$x : 20 = 13 : 4$$
$$x = \frac{20 \times 13}{4} = \textbf{65}$$

Therefore, a 65 T gear is used on lead screw.

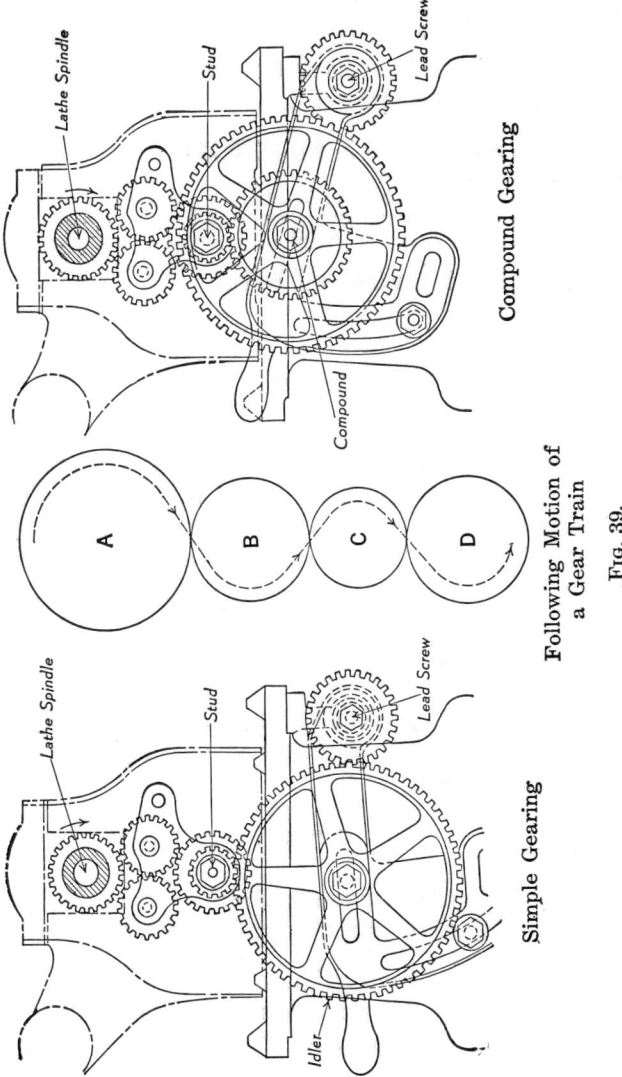

Compound Gearing

Following Motion of
a Gear Train

Fig. 39.

Simple Gearing

ILLUSTRATION: Using a 110-tooth gear on the lead screw and a 75 on the stud, with compound driven and driver gears of 50 and 80 teeth, respectively, how many threads per inch will be cut if the lead screw has 6 threads per inch?

$$x : 6 = (110 \times 50) : (75 \times 80)$$

$$x = \frac{6 \times 110 \times 50}{75 \times 80} = 5.5$$

Therefore 5.5 threads per inch will be cut.

Milling

Simple Indexing.—In machine shop milling it is often necessary to machine a piece of work on several faces with considerable accuracy. This is usually accomplished by attaching the work to a dividing or index head so that it may be rotated into any position. (See Fig. 40.) On all standard dividing heads it requires 40 turns of the index crank to revolve the dividing head spindle once.

If a piece of work is to be cut at any number of points equidistant apart on its periphery, then to find the number of turns of the index crank for these divisions, divide the number of turns required for one revolution of the dividing head (40) by the number of divisions wanted.

$$R = \frac{40}{N}$$

when N = number of divisions required;
R = number of turns of the crank for given division.

ILLUSTRATION: A 57-toothed gear is to be turned on a milling machine. How many turns of the crank will be required to turn the work from one tooth to the next?

$$R = \frac{40}{N} = \frac{40}{57} \text{ revolution (Ans.)}$$

FIG. 40. —Above, Simple Indexing; Below, Differential Indexing.

The last illustration brings up the question of how the crank is to be stopped accurately at fractional revolutions, in this case $\frac{40}{17}$ of a revolution. This is accomplished by the perforated index plate shown in Fig. 40. This plate has small holes evenly spaced along concentric circles. There are generally three interchangeable plates with each dividing head. The following list gives the number of holes per circle on the three plates used on a standard machine.

Plate	Number of Holes in the Various Circles
1.........	15 — 16 — 17 — 18 — 19 — 20
2.........	21 — 23 — 27 — 29 — 31 — 33
3.........	37 — 39 — 41 — 43 — 47 — 49

Some dividing heads have only one plate. In this case the plate has holes on each side as follows:

one side 24–25–28–30–34–37–38–39–41–42–43

and on the other side

46–47–49–51–53–54–57–58–59–62–66

The crank is provided with an index pin which engages the desired hole and holds the crank stationary.

ILLUSTRATION: What is the simple indexing for 330 divisions?

$R = \dfrac{40}{N} = \dfrac{40}{330} = \dfrac{4}{33}$ revolution, or 4 spaces on the circle with 33 holes or 8 spaces on the circle with 66 holes. (Ans.)

In order to obtain a number of divisions that cannot be obtained with ordinary index plates a process of differential indexing is used. By this process the index plate is revolved by suitable gears which connect it to the dividing head spindle, the stop pin holding the index plate being disengaged altogether. (See Fig. 40.)

The rotary or differential motion of the index plate takes place when the crank is turned, which turns the plate either forward or backward as may be required. The result is that the actual movement of the crank, in indexing, is either more or less than the movement in relation to the index plate.

The differential method cannot be used in connection with spiral milling, because the dividing head spindle is geared to the lead screw of the milling machine.

The amount of rotation of the index plate may be regulated by the difference in velocity ratios of the change gears.

ILLUSTRATION: Find the indexing required for 81 divisions. By simple indexing the index crank would be rotated through $\frac{40}{81}$ of a turn for each division, but as there is no plate with 81 divisions, the spacing is impossible: therefore, another fraction is selected whose value is near $\frac{40}{81}$, for example, $\frac{40}{81}$ or $\frac{10}{21}$, then a 21-hole circle can be used.

In simple indexing for 80 divisions the movement of the index crank is $\frac{40}{80}$ or $\frac{1}{2}$ of a turn for each cut.

If the crank is given $\frac{1}{2}$ of a turn eighty-one times, it makes $40\frac{1}{2}$ turns or $\frac{1}{2}$ of a turn more than the 40 turns required for one revolution of the work. Hence the index plate must move backward $\frac{1}{2}$ of a revolution while the work revolves once

$$\frac{40}{80} = \frac{1}{2}, \quad 81 \times \frac{1}{2} = 40\frac{1}{2}$$

$$41\frac{1}{2} - 40 = 1\frac{1}{2}$$

Hence the ratio of the gears is $1 : 2$.

$$\frac{1}{2} \times \frac{24}{24} = \frac{24}{48}$$

A 24 T gear (driving) is placed on the special differential indexing center in the spindle of the dividing head; and the 48 T gear (driven) is placed on the worm shaft which turns the index plate. (See gear on spindle and gear on worm in Fig. 40.)

TABLE 12

Leads, Change Gears and Angles for Cutting Spirals

Lead of Spiral, Inches	Gear on Worm	1st Intermediate Gear	2d Intermediate Gear	Gear on Screw	Diameter of Work, Inches									
					⅛	¼	⅜	½	⅝	¾	⅞	1	1¼	1½
0.67	24	86	24	100	30¼	...	...	...	Approximate Angles for Milling Machine Table					
0.78	24	86	28	100	26	44½	...	...						
0.89	24	86	32	100	23½	41	...	...						
1.12	24	86	40	100	19	34½	...	...						
1.34	24	86	48	100	16	30¼	41½	...	...	...	...	...	...	...
1.46	24	64	28	72	14¾	28	38½	...	...	...	...	...	...	...
1.56	24	86	56	100	13¾	26½	37	...	...	...	...	...	...	...
1.67	24	64	32	72	12¾	25	34¾	43¼	...	...	...	...	...	...
1.94	32	64	28	72	11¼	21¾	31	39	45	...	...	...	...	...
2.08	24	64	40	72	10¼	20½	29½	37	43¼	...	...	...	...	...
2.22	32	56	28	72	9¾	19¼	27½	35	41¼	...	...	...	...	...
2.50	24	64	48	72	8¾	17	25	32	38	43¼	...	...	...	...
2.78	40	56	28	72	8	15½	23	29½	35¼	40½	44¾	...	...	...
2.92	24	64	56	72	7½	15	21¾	28¼	34	39	43¼	...	...	...
3.24	40	48	28	72	6¾	13¼	19¾	25¾	31¼	36	40½	44¼	...	...
3.70	40	48	32	72	6	11¾	17½	23	28	32½	36½	40½	...	...
3.89	56	48	24	72	5½	11¼	16¾	22	26¾	31¼	35¼	39	...	...
4.17	40	72	48	64	5¼	10½	15¾	20½	25¼	29½	33½	37	43¼	...
4.46	48	40	32	86	4¾	9¾	14¾	19¼	23¾	27¾	31½	35	41½	...
4.86	40	64	56	72	4½	9	13½	17¾	22	25¾	29½	33	39	44¼
5.33	48	40	32	72	4	8¼	12¼	16½	20¼	23¾	27¼	30½	36½	41½
5.44	56	40	28	72	4	8	12	16	20	23½	26¾	30	36	41
6.12	56	40	28	64	3½	7¼	11	14½	17¾	21	24¼	27	33	37¾
6.22	56	40	32	72	3½	7	10¾	14¼	17½	20¾	23¾	26¾	32½	37¼
6.48	56	48	40	72	3¼	6¾	10¼	13½	16¾	20	23	25¾	31½	36¼
6.67	64	48	28	56	3¼	6½	10	13¼	16½	19½	22½	25¼	30¾	35¼
7.29	56	48	40	64	3	6¼	9¼	12¼	15	18	20½	23½	28½	33
7.41	64	48	40	72	3	6	9	12	14¾	17¾	20¼	22¾	28¼	32½
7.62	64	48	32	56	2¾	5¾	8¾	11½	14½	17¼	19¾	22¼	27½	32
8.33	48	32	40	72	2½	5¼	8	10½	13¼	15¾	18¼	20½	25½	29½
8.95	86	48	28	56	2½	5	7½	10	12½	14¾	17	19¼	24	28
9.33	56	40	48	72	2¼	4¾	7¼	9½	11¾	14	16¼	18½	23	27
9.52	64	48	40	56	2¼	4½	7	9¼	11½	13¾	16	18¼	22½	26½
10.29	72	40	32	56	2	4¼	6½	8¾	10¾	12¾	15	17¼	21	24¾
10.37	64	48	56	72	2	4¼	6½	8½	10½	12¾	14¾	17	20¾	24½
10.50	48	40	56	64	2	4¼	6¼	8½	10½	12½	14½	16¾	20½	24¼
10.67	64	40	48	72	2	4	6¼	8¼	10¼	12¼	14¼	16½	20¼	24
10.94	56	32	40	64	2	4	6	8¼	10¼	12	14	16¼	20	23½
11.11	64	32	40	72	2	4	6	8	10	11¾	13¾	16	19¾	23
11.66	56	32	48	72	1¾	3¾	5¾	7½	9½	11¼	13¼	15¼	18¾	22
12.00	72	40	32	48	1¾	3¾	5½	7¼	9¼	11	12¾	15	18¼	21½
13.12	56	32	48	64	1½	3½	5¼	6¾	8½	10¼	11¾	13½	16¾	20
13.33	56	28	48	72	1½	3¼	5	6½	8¼	10	11½	13¼	16½	19½
13.71	64	40	48	56	1½	3¼	4¾	6½	8	9¾	11¼	13	16	19
15.24	64	28	48	72	1½	3	4½	5¾	7¼	8¾	10¼	11¾	14½	17¼
15.56	64	32	56	72	1¼	2¾	4¼	5¾	7¼	8¾	10	11½	14¼	17
15.75	56	64	72	40	1¼	2¾	4¼	5½	7	8½	9¾	11¼	14	16¾
16.87	72	32	48	64	1¼	2½	4	5¼	6¾	7¾	9¼	10½	13¼	15¾
17.14	64	32	48	56	1¼	2½	4	5¼	6½	7¾	9	10¼	13	15½
18.75	72	32	40	48	1	2¼	3½	4¾	6	7¼	8¼	9½	12	14¼
19.29	72	32	48	56	1	2¼	3½	4½	5¾	7	8	9¼	11½	13¾
19.59	64	28	48	56	1	2¼	3¼	4½	5¾	6¾	8	9¼	11½	13½
19.69	72	32	56	64	1	2¼	3¼	4½	5¾	6¾	8	9	11½	13½
21.43	72	24	40	56	1	2	3¼	4¼	5¼	6¼	7½	8½	10½	12½
22.50	72	28	56	64	1	2	3	4	5	6	7	8	10	12
23.33	64	32	56	48	1	2	3	4	5	5¾	6¾	7¾	9¾	11½
26.25	72	24	56	64	1	1¾	2¾	3½	4¼	5	6	7	8½	10¼
26.67	64	28	56	48	¾	1¾	2¾	3½	4¼	5	6	6¾	8½	10
28.00	64	32	56	40	¾	1¾	2½	3¼	4	4¾	5¾	6½	8	9½
30.86	72	28	48	40	¾	1½	2¼	3	3¾	4½	5	5¾	7¼	8¾

As the motion of the index plate must be in the direction opposite to the movement of the index crank, idler gears must be used. These do not affect the ratio.

The following gears are generally available for differential indexing: 24–24–28–32–40–44–48–56–64–72–86–100.

Angular Indexing.—Sometimes a milling job calls for making cuts at intervals of a certain number of degrees around the periphery of a piece of work. With a standard index head, where 40 turns of the index crank are required for one revolution of the work one turn of the crank equals $\frac{1}{40}$ of 360 degrees or 9 degrees.

Thus, if one complete turn of the crank equals 9 degrees, 2 holes in the 18 circle or 3 holes in the 27 circle must equal 1 degree, or 1 hole in the 18 circle will equal $\frac{1}{2}$ degree or 30 minutes, and 1 hole in the 27 circle will equal $\frac{1}{3}$ of a degree or 20 minutes.

ILLUSTRATION: What is the angular indexing for 19 degrees?

If 1 turn equals 9 degrees, 2 turns equal 18 degrees. Add 2 holes on 18 circle or 3 holes on 27 circle.

Indexing for 19 degrees is then, 2 turns + 2 holes on 18 circle or 2 turns + 3 holes on 27 circle. (Ans.)

ILLUSTRATION: What is the angular indexing for 7 degrees 40 minutes?

40 minutes $= \frac{2}{3}$ degree.

then $7\frac{2}{3} \div 9 = \frac{23}{27}$.

Therefore, the indexing for 7 deg. 40 min. is 23 holes on 27 circle. (Ans.)

Table 13 gives the plain and differential indexing of the numbers up to 370.

Spiral Milling.—Cutting a helical milling cutter as shown in Fig. 41, or a twist drill, is called *spiral milling* and can be attained by the use of an index head so geared to the longitudinal feed screw of the milling machine as to impart a rotary motion to the work as it is fed along under the cutter by the action of a train of gears.

The *lead of a helix* or *spiral* is the distance, measured along the axis of the work, which the spiral makes in one full turn around the work.

TABLE 13

Number of Divisions	Index Circle	No. of Turns of Index	Gear on Worm	No. 1 Hole		Gear on Spindle	Idlers	
				First Gear on Stud	Second Gear on Stud		No. 1 Hole	No. 2 Hole
2	Any	20						
3	39	$13\frac{13}{39}$						
4	Any	10						
5	Any	8						
6	39	$6\frac{26}{39}$						
7	49	$5\frac{35}{49}$						
8	Any	5						
9	27	$4\frac{12}{27}$						
10	Any	4						
11	33	$3\frac{21}{33}$						
12	39	$3\frac{13}{39}$						
13	39	$3\frac{3}{39}$						
14	49	$2\frac{42}{49}$						
15	39	$2\frac{26}{39}$						
16	20	$2\frac{10}{20}$						
17	17	$2\frac{6}{17}$						
18	27	$2\frac{6}{27}$						
19	19	$2\frac{2}{19}$						
20	Any	2						
21	21	$1\frac{19}{21}$						
22	33	$1\frac{27}{33}$						
23	23	$1\frac{17}{23}$						
24	39	$1\frac{26}{39}$						
25	20	$1\frac{12}{20}$						
26	39	$1\frac{21}{39}$						
27	27	$1\frac{13}{27}$						
28	49	$1\frac{21}{49}$						
29	29	$1\frac{11}{29}$						
30	39	$1\frac{13}{39}$						
31	31	$1\frac{9}{31}$						
32	20	$1\frac{5}{20}$						
33	33	$1\frac{7}{33}$						
34	17	$1\frac{3}{17}$						
35	49	$1\frac{7}{49}$						
36	27	$1\frac{3}{27}$						
37	37	$1\frac{3}{37}$						
38	19	$1\frac{1}{19}$						
39	39	$1\frac{1}{39}$						
40	Any	1						
41	41	$\frac{40}{41}$						
42	21	$\frac{20}{21}$						

Table 13.—Continued

| Number of Divisions | Index Circle | No. of Turns of Index | Gear on Worm | No. 1 Hole | | Gear on Spindle | Idlers | |
				First Gear on Stud	Second Gear on Stud		No. 1 Hole	No. 2 Hole
125	39	$\frac{13}{39}$	24			40	24	44
126	39	$\frac{13}{39}$	24			48	24	44
127	39	$\frac{13}{39}$	24			56	24	44
128	16	$\frac{5}{16}$						
129	39	$\frac{13}{39}$	24			72	24	44
130	39	$\frac{12}{39}$						
131	20	$\frac{6}{20}$	40			28	44	
132	33	$\frac{10}{33}$						
133	21	$\frac{6}{21}$	24			48	44	
134	21	$\frac{6}{21}$	28			48	44	
135	27	$\frac{8}{27}$						
136	17	$\frac{5}{17}$						
137	21	$\frac{6}{21}$	28			24	56	
138	21	$\frac{6}{21}$	56			32	44	
139	21	$\frac{6}{21}$	56	32	48	24		
140	49	$\frac{14}{49}$						
141	18	$\frac{5}{18}$	48			40	44	
142	21	$\frac{6}{21}$	56			32	24	44
143	21	$\frac{6}{21}$	28			24	24	44
144	18	$\frac{5}{18}$						
145	29	$\frac{8}{29}$						
146	21	$\frac{6}{21}$	28			48	24	44
147	21	$\frac{6}{21}$	24			48	24	44
148	37	$\frac{10}{37}$						
149	21	$\frac{6}{21}$	28			72	24	44
150	15	$\frac{4}{15}$						
151	20	$\frac{5}{20}$	32			72	44	
152	19	$\frac{5}{19}$						
153	20	$\frac{5}{20}$	32			56	44	
154	20	$\frac{5}{20}$	32			48	44	
155	31	$\frac{8}{31}$						
156	39	$\frac{10}{39}$						
157	20	$\frac{5}{20}$	32			24	56	
158	20	$\frac{5}{20}$	48			24	44	
159	20	$\frac{5}{20}$	64	32	56	28		
160	20	$\frac{5}{20}$						
161	20	$\frac{5}{20}$	64	32	56	28		24
162	20	$\frac{5}{20}$	48			24	24	44
163	20	$\frac{5}{20}$	32			24	24	44
164	41	$\frac{10}{41}$						
165	33	$\frac{8}{33}$						

TABLE 13.—*Continued*

Number of Divisions	Index Circle	No. of Turns of Index	Gear on Worm	No. 1 Hole		Gear on Spindle	Idlers	
				First Gear on Stud	Second Gear on Stud		No. 1 Hole	No. 2 Hole
166	20	$\frac{5}{20}$	32			48	24	44
167	20	$\frac{5}{20}$	32			56	24	44
168	21	$\frac{5}{21}$						
169	20	$\frac{5}{20}$	32			72	24	44
170	17	$\frac{4}{17}$						
171	21	$\frac{5}{21}$	56			40	24	44
172	43	$\frac{10}{43}$						
173	18	$\frac{4}{18}$	72	56	32	64		
174	18	$\frac{4}{18}$	24			32	56	
175	18	$\frac{4}{18}$	72	40	32	64		
176	18	$\frac{4}{18}$	72	24	24	64		
177	18	$\frac{4}{18}$	72			48	24	
178	18	$\frac{4}{18}$	72			32	44	
179	18	$\frac{4}{18}$	72	24	48	32		
180	18	$\frac{4}{18}$						
181	18	$\frac{4}{18}$	72	24	48	32		24
182	18	$\frac{4}{18}$	72			32	24	44
183	18	$\frac{4}{18}$	48			32	24	44
184	23	$\frac{5}{23}$						
185	37	$\frac{8}{37}$						
186	18	$\frac{4}{18}$	48			64	24	44
187	18	$\frac{4}{18}$	72	48	24	56		24
188	47	$\frac{10}{47}$						
189	18	$\frac{4}{18}$	32			64	24	44
190	19	$\frac{4}{19}$						
191	20	$\frac{4}{20}$	40			72	24	
192	20	$\frac{4}{20}$	40			64	44	
193	20	$\frac{4}{20}$	40			56	44	
194	20	$\frac{4}{20}$	40			48	44	
195	39	$\frac{8}{39}$						
196	49	$\frac{4}{49}$						
197	20	$\frac{4}{20}$	40			24	56	
198	20	$\frac{4}{20}$	56	28	40	32		
199	20	$\frac{4}{20}$	100	40	64	32		
200	20	$\frac{4}{20}$						
201	20	$\frac{4}{20}$	72	24	40	24		24
202	20	$\frac{4}{20}$	72	24	40	48		24
203	20	$\frac{4}{20}$	40			24	24	44
204	20	$\frac{4}{20}$	40			32	24	44
205	41	$\frac{8}{41}$						
206	20	$\frac{4}{20}$	40			48	24	44

TABLE 13.—*Continued*

Number of Divisions	Index Circle	No. of Turns of Index	Gear on Worm	No. 1 Hole		Gear on Spindle	Idlers	
				First Gear on Stud	Second Gear on Stud		No. 1 Hole	No. 2 Hole
207	20	$\frac{4}{20}$	40			56	24	44
208	20	$\frac{4}{20}$	40			64	24	44
209	20	$\frac{4}{20}$	40			72	24	44
210	21	$\frac{4}{21}$						
211	16	$\frac{8}{16}$	64			28	44	
212	43	$\frac{8}{43}$	86	24	24	48		
213	27	$\frac{5}{27}$	72			40	44	
214	20	$\frac{4}{20}$	40	56	32	64		24
215	43	$\frac{8}{43}$						
216	27	$\frac{5}{27}$						
217	21	$\frac{4}{21}$	48			64	24	44
218	16	$\frac{8}{16}$	64			56	24	44
219	21	$\frac{4}{21}$	28			48	24	44
220	33	$\frac{6}{33}$						
221	17	$\frac{9}{17}$	24			24	56	
222	18	$\frac{8}{18}$	24			72	44	
223	43	$\frac{8}{43}$	86	48	24	64		24
224	18	$1\frac{8}{18}$	24			64	44	
225	27	$\frac{2}{27}$	24			40	24	44
226	18	$\frac{9}{18}$	24			56	44	
227	49	$\frac{4}{49}$	56	64	28	72		
228	18	$1\frac{8}{18}$	24			48	44	
229	18	$1\frac{8}{18}$	24			44	48	
230	23	$2\frac{4}{23}$						
231	18	$1\frac{3}{18}$	32			48	44	
232	29	$2\frac{5}{29}$						
233	18	$1\frac{3}{18}$	48			56	44	
234	18	$1\frac{3}{18}$	24			24	56	
235	47	$4\frac{8}{47}$						
236	18	$1\frac{3}{18}$	48			32	44	
237	18	$1\frac{3}{18}$	48			24	44	
238	18	$1\frac{3}{18}$	72			24	44	
239	18	$1\frac{3}{18}$	72	24	64	32		
240	18	$1\frac{8}{18}$						
241	18	$1\frac{3}{18}$	72	24	64	32		24
242	18	$1\frac{3}{18}$	72			24	24	44
243	18	$1\frac{3}{18}$	64			32	24	44
244	18	$1\frac{3}{18}$	48			32	24	44
245	49	$4\frac{8}{49}$						
246	18	$1\frac{3}{18}$	24			24	24	44
247	18	$1\frac{3}{18}$	48			56	24	44

TABLE 13.—*Continued*

Number of Divisions	Index Circle	No. of Turns of Index	Gear on Worm	First Gear on Stud	Second Gear on Stud	Gear on Spindle	No. 1 Hole (Idler)	No. 2 Hole (Idler)
248	31	$3\frac{5}{31}$						
249	18	$1\frac{3}{18}$	32			48	24	44
250	18	$1\frac{3}{18}$	24			40	24	44
251	18	$1\frac{3}{18}$	48	44	32	64		24
252	18	$1\frac{3}{18}$	24			48	24	44
253	33	$3\frac{5}{33}$	24			40	56	
254	18	$1\frac{3}{18}$	24			56	24	44
255	18	$1\frac{3}{18}$	48	40	24	72		24
256	18	$1\frac{3}{18}$	24			64	24	44
257	49	$4\frac{9}{49}$	56	48	28	64		24
258	43	$4\frac{7}{43}$	32			64	24	44
259	21	$2\frac{1}{21}$	24			72	44	
260	39	$3\frac{6}{39}$						
261	29	$2\frac{9}{29}$	48	64	24	72		
262	20	$2\frac{3}{20}$	40			28	44	
263	49	$4\frac{9}{49}$	56	64	28	72		24
264	33	$3\frac{5}{33}$						
265	21	$2\frac{8}{21}$	56	40	24	72		
266	21	$2\frac{8}{21}$	32			64	44	
267	27	$2\frac{4}{27}$	72			32	44	
268	21	$2\frac{8}{21}$	28			48	44	
269	20	$2\frac{8}{20}$	64	32	40	28		24
270	27	$2\frac{4}{27}$						
271	21	$2\frac{8}{21}$	56			72	24	
272	21	$2\frac{8}{21}$	56			64	24	
273	21	$2\frac{8}{21}$	24			24	56	
274	21	$2\frac{8}{21}$	56			48	44	
275	21	$2\frac{8}{21}$	56			40	44	
276	21	$2\frac{8}{21}$	56			32	44	
277	21	$2\frac{3}{21}$	56			24	44	
278	21	$2\frac{3}{21}$	56	32	48	24		
279	27	$2\frac{4}{27}$	24			32	24	44
280	49	$4\frac{7}{49}$						
281	21	$2\frac{3}{21}$	72	24	56	24		24
282	43	$4\frac{6}{43}$	86	24	24	56		
283	21	$2\frac{3}{21}$	56			24	24	44
284	21	$2\frac{8}{21}$	56			32	24	44
285	21	$2\frac{8}{21}$	56			40	24	44
286	21	$2\frac{3}{21}$	56			48	24	44
287	21	$2\frac{8}{21}$	24			24	24	44
288	21	$2\frac{8}{21}$	28			32	24	44

TABLE 13.—*Continued*

Number of Divisions	Index Circle	No. of Turns of Index	Gear on Worm	No. 1 Hole		Gear on Spindle	Idlers	
				First Gear on Stud	Second Gear on Stud		No. 1 Hole	No. 2 Hole
289	21	$\frac{8}{21}$	56			72	24	44
290	29	$\frac{4}{29}$						
291	15	$1\frac{2}{15}$	40			48	44	
292	21	$\frac{8}{21}$	28			48	24	44
293	15	$\frac{2}{15}$	48	32	40	56		
294	21	$\frac{8}{21}$	24			48	24	44
295	15	$1\frac{1}{15}$	48			32	44	
296	37	$\frac{5}{37}$						
297	33	$\frac{4}{33}$	28	48	24	56		
298	21	$\frac{9}{21}$	28			72	24	44
299	23	$\frac{3}{23}$	24			24	56	
300	15	$\frac{2}{15}$						
301	43	$\frac{6}{43}$	24			48	24	44
302	16	$\frac{1}{16}$	32			72	24	
303	15	$\frac{2}{15}$	72	24	40	48		24
304	16	$\frac{2}{16}$	24			48	44	
305	15	$\frac{2}{15}$	48			32	24	44
306	15	$\frac{2}{15}$	40			32	24	44
307	15	$\frac{2}{15}$	72	48	40	56		24
308	16	$\frac{2}{16}$	32			48	44	
309	15	$\frac{1}{15}$	40			48	24	44
310	31	$\frac{4}{31}$						
311	16	$\frac{2}{16}$	64	24	24	72		
312	39	$\frac{5}{39}$						
313	16	$\frac{2}{16}$	32			28	56	
314	16	$\frac{2}{16}$	32			24	56	
315	16	$\frac{2}{16}$	64			40	24	
316	16	$\frac{1}{16}$	64			32	44	
317	16	$\frac{1}{16}$	64			24	44	
318	16	$\frac{2}{16}$	56	28	48	24		
319	29	$\frac{4}{29}$	48	64	24	72		24
320	16	$\frac{2}{16}$						
321	16	$\frac{2}{16}$	72	24	64	24		24
322	23	$\frac{3}{23}$	32			64	24	44
323	16	$\frac{1}{16}$	64			24	24	44
324	16	$\frac{2}{16}$	64			32	24	44
325	16	$\frac{2}{16}$	64			40	24	44
326	16	$\frac{1}{16}$	32			24	24	44
327	16	$\frac{1}{16}$	32			28	24	44
328	41	$\frac{5}{41}$						
329	16	$\frac{2}{16}$	64	24	24	72		24

TABLE 13.—*Continued*

Number of Divisions	Index Circle	No. of Turns of Index	Gear on Worm	No. 1 Hole First Gear on Stud	No. 1 Hole Second Gear on Stud	Gear on Spindle	Idlers No. 1 Hole	Idlers No. 2 Hole
330	33	$\frac{4}{33}$						
331	16	$1\frac{2}{16}$	64	44	24	48		24
332	16	$1\frac{2}{16}$	32			48	24	44
333	18	$1\frac{8}{18}$	24			72	44	
334	16	$1\frac{2}{16}$	32			56	24	44
335	33	$4\frac{4}{33}$	72	48	44	40		24
336	16	$1\frac{2}{16}$	32			64	24	44
337	43	$4\frac{5}{43}$	86	40	32	56		
338	16	$1\frac{2}{16}$	32			72	24	44
339	18	$1\frac{2}{18}$	24			56	44	
340	17	$1\frac{7}{17}$						
341	43	$4\frac{5}{43}$	86	24	32	40		
342	18	$1\frac{2}{18}$	32			64	44	
343	15	$1\frac{5}{15}$	40	64	24	86		24
344	43	$4\frac{5}{43}$						
345	18	$1\frac{2}{18}$	24			40	56	
346	18	$1\frac{2}{18}$	72	56	32	64		
347	43	$4\frac{5}{43}$	86	24	32	40		24
348	18	$1\frac{2}{18}$	24			32	56	
349	18	$1\frac{2}{18}$	72	44	24	48		
350	18	$1\frac{2}{18}$	72	40	32	64		
351	18	$1\frac{2}{18}$	24			24	56	
352	18	$1\frac{2}{18}$	72	24	24	64		
353	18	$1\frac{2}{18}$	72			56	24	
354	18	$1\frac{2}{18}$	72			48	24	
355	18	$1\frac{2}{18}$	72			40	24	
356	18	$1\frac{2}{18}$	72			32	24	
357	18	$1\frac{2}{18}$	72			24	44	
358	18	$1\frac{2}{18}$	72	32	48	24		
359	43	$4\frac{5}{43}$	86	48	32	100		24
360	18	$1\frac{2}{18}$						
361	19	$1\frac{9}{19}$	32			64	44	
362	18	$1\frac{2}{18}$	72	28	56	32		24
363	18	$1\frac{2}{18}$	72			24	24	44
364	18	$1\frac{2}{18}$	72			32	24	44
365	20	$2\frac{0}{20}$	32	48	24	56		
366	18	$1\frac{2}{18}$	48			32	24	44
367	18	$1\frac{2}{18}$	72			56	24	24
368	18	$1\frac{2}{18}$	72	24	24	64		24
369	41	$4\frac{4}{41}$	32	56	28	64		
370	37	$\frac{4}{37}$						

Table 13.—*Continued*

Number of Divisions	Index Circle	No. of Turns of Index	Gear on Worm	No. 1 Hole First Gear on Stud	No. 1 Hole Second Gear on Stud	Gear on Spindle	Idlers No. 1 Hole	Idlers No. 2 Hole
371	21	$\frac{2}{21}$	32	56	24	64		
372	18	$\frac{2}{18}$	48			64	24	44
373	20	$\frac{2}{20}$	40	48	32	72		
374	18	$\frac{2}{18}$	72	64	32	56		24
375	18	$\frac{2}{18}$	24			40	24	44
376	47	$\frac{5}{47}$						
377	29	$\frac{3}{29}$	24			24	56	
378	18	$\frac{2}{18}$	32			64	24	44
379	20	$\frac{2}{20}$	48	56	40	72		
380	19	$\frac{2}{19}$						
381	18	$\frac{2}{18}$	24			56	24	44
382	20	$\frac{2}{20}$	40			72	24	
383	20	$\frac{2}{20}$	40			68 [1]	44	
384	20	$\frac{2}{20}$	40			64	44	
385	20	$\frac{2}{20}$	32			48	44	
386	20	$\frac{2}{20}$	40			56	44	
387	43	$\frac{4}{43}$	32	56	28	64		
388	20	$\frac{2}{20}$	40			48	44	
389	20	$\frac{2}{20}$	40			44	56	
390	39	$\frac{4}{39}$						
391	20	$\frac{2}{20}$	48	24	40	72		
392	49	$\frac{5}{49}$						
393	20	$\frac{2}{20}$	40			28	44	
394	20	$\frac{2}{20}$	40			24	56	
395	20	$\frac{2}{20}$	64			32	44	
396	20	$\frac{2}{20}$	56	28	40	32		
397	20	$\frac{2}{20}$	64	24	40	32		
398	20	$\frac{2}{20}$	100	40	64	32		
399	21	$\frac{2}{21}$	32			64	44	
400	20	$\frac{2}{20}$						
401	21	$\frac{2}{21}$	56	32	24	76 [1]		
402	21	$\frac{2}{21}$	28			48	44	
403	20	$\frac{2}{20}$	64	24	40	32		24
404	20	$\frac{2}{20}$	72	24	40	48		24
405	20	$\frac{2}{20}$	64			32	24	44
406	20	$\frac{2}{20}$	40			24	24	44
407	20	$\frac{2}{20}$	40			28	24	44
408	20	$\frac{2}{20}$	40			32	24	44
409	20	$\frac{2}{20}$	40	24	32	48		24
410	41	$\frac{4}{41}$						

NOTE. Special gears in this and following tables are 46, 47, 52, 58, 68, 70, 76, 84.　[1] Special gear.

TABLE 13.—*Continued*

Number of Divisions	Index Circle	No. of Turns of Index	Gear on Worm	No. 1 Hole First Gear on Stud	No. 1 Hole Second Gear on Stud	Gear on Spindle	Idlers No. 1 Hole	Idlers No. 2 Hole
411	21	$\frac{2}{21}$	28			24	56	
412	20	$\frac{2}{20}$	40			48	24	44
413	21	$\frac{2}{21}$	48			32	44	
414	21	$\frac{2}{21}$	56			32	44	
415	20	$\frac{2}{20}$	32			48	24	44
416	20	$\frac{2}{20}$	40			64	24	44
417	21	$\frac{2}{21}$	56	32	48	24		
418	20	$\frac{2}{20}$	40			72	24	44
419	33	$\frac{3}{33}$	44	28	24	72		
420	21	$\frac{2}{21}$						
421	20	$\frac{2}{20}$	48	56	40	72		24
422	20	$\frac{2}{20}$	40	44	32	64		24
423	21	$\frac{2}{21}$	72	24	56	48		24
424	43	$\frac{4}{43}$	86	24	24	48		
425	21	$\frac{2}{21}$	72	48	56	40		24
426	21	$\frac{2}{21}$	56			32	24	44
427	20	$\frac{2}{20}$	40	48	32	72		24
428	20	$\frac{2}{20}$	40	56	32	64		24
429	21	$\frac{2}{21}$	28			24	24	44
430	43	$\frac{4}{43}$						
431	21	$\frac{2}{21}$	72	44	28	48		24
432	20	$\frac{2}{20}$	40	56	28	64		24
433	20	$\frac{2}{20}$	40	44	24	72		24
434	21	$\frac{2}{21}$	48			64	24	44
435	21	$\frac{2}{21}$	28			40	24	44
436	20	$\frac{2}{20}$	40	48	24	72		24
437	23	$\frac{2}{23}$	32			64	44	
438	21	$\frac{2}{21}$	28			48	24	44
439	43	$\frac{4}{43}$	86	24	24	72		24
440	33	$\frac{3}{33}$						
441	21	$\frac{2}{21}$	32			64	24	44
442	20	$\frac{2}{20}$	40	56	24	72		24
443	20	$\frac{2}{20}$	40	48	24	86		24
444	21	$\frac{2}{21}$	56	48	24	64		24
445	33	$\frac{3}{33}$	64	32	44	40		24
446	33	$\frac{3}{33}$	44			24	24	48
447	21	$\frac{2}{21}$	28			72	24	44
448	20	$\frac{2}{20}$	40	64	24	72		24
449	33	$\frac{3}{33}$	64	32	44	72		24
450	33	$\frac{3}{33}$	44			40	24	32

Table 13.—*Continued*

Number of Divisions	Index Circle	No. of Turns of Index	Gear on Worm	No. 1 Hole First Gear on Stud	No. 1 Hole Second Gear on Stud	Gear on Spindle	Idlers No. 1 Hole	Idlers No. 2 Hole
451	33	$\frac{8}{33}$	24			24	24	44
452	33	$\frac{8}{33}$	44			48	24	40
453	33	$\frac{8}{33}$	44			52 [1]	24	40
454	49	$\frac{4}{49}$	56	64	28	72		
455	49	$\frac{4}{49}$	28	40	32	64		
456	21	$\frac{2}{21}$	56	64	24	72		24
457	33	$\frac{3}{33}$	44			68 [1]	24	40
458	33	$\frac{3}{33}$	44			72	24	24
459	27	$\frac{2}{27}$	24	48	24	72		
460	23	$\frac{2}{23}$						
461	33	$\frac{8}{33}$	44	28	24	72		24
462	33	$\frac{8}{33}$	32			64	24	44
463	21	$\frac{2}{21}$	56	64	24	86		24
464	33	$\frac{3}{33}$	44	48	28	56		24
465	33	$\frac{3}{33}$	44	24	24	100		24
466	49	$\frac{4}{49}$	56	48	28	64		
467	33	$\frac{8}{33}$	44	48	32	72		24
468	39	$\frac{8}{39}$	28	48	24	56		
469	49	$\frac{4}{49}$	28			48	44	
470	47	$\frac{4}{47}$						
471	49	$\frac{4}{49}$	56	32	28	76 [1]		
472	49	$\frac{4}{49}$	56	32	28	72		
473	33	$\frac{3}{33}$	48	64	32	72		24
474	49	$\frac{4}{49}$	56	32	28	64		
475	49	$\frac{4}{49}$	56	40	28	48		
476	49	$\frac{4}{49}$	56			64	24	
477	27	$\frac{2}{27}$	24	48	24	56		
478	49	$\frac{4}{49}$	56	24	28	64		
479	49	$\frac{4}{49}$	56	32	28	44		
480	49	$\frac{4}{49}$	56	32	28	40		
481	37	$\frac{8}{37}$	24			24	56	
482	33	$\frac{3}{33}$	44	56	24	72		24
483	49	$\frac{4}{49}$	56			32	44	
484	49	$\frac{4}{49}$	56	24	28	32		
485	23	$\frac{2}{23}$	46 [1]	24	24	100		24
486	27	$\frac{2}{27}$	32	56	28	64		
487	39	$\frac{8}{39}$	24	72	52 [1]	44		
488	33	$\frac{3}{33}$	44	64	24	72		24
489	23	$\frac{2}{23}$	46 [1]	58 [1]	32	64		24
490	49	$\frac{4}{49}$						

[1] Special gear.

Table 13.—*Continued*

| Number of Divisions | Index Circle | No. of Turns of Index | Gear on Worm | No. 1 Hole | | Gear on Spindle | Idlers | |
				First Gear on Stud	Second Gear on Stud		No. 1 Hole	No. 2 Hole
491	33	$\frac{3}{33}$	44	68 [1]	24	72		24
492	41	$\frac{4}{41}$	28	48	24	56		
493	29	$\frac{2}{29}$	32	64	24	72		
494	39	$\frac{3}{39}$	32			64	44	
495	27	$\frac{2}{27}$	32	40	24	64		
496	49	$\frac{4}{49}$	56	24	28	32		24
497	49	$\frac{4}{49}$	56			32	24	44
498	27	$\frac{2}{27}$	48	56	24	64		
499	49	$\frac{4}{49}$	56	24	28	48		24
500	49	$\frac{4}{49}$	56	32	28	40		24
501	49	$\frac{4}{49}$	56	32	28	44		24
502	49	$\frac{4}{49}$	56	32	28	48		24
503	23	$\frac{2}{23}$	46 [1]	64	32	86		24
504	49	$\frac{4}{49}$	56			64	24	24
505	49	$\frac{4}{49}$	56	40	28	48		24
506	49	$\frac{4}{49}$	56	32	28	64		24
507	39	$\frac{8}{39}$	24			24	56	
508	49	$\frac{4}{49}$	56	32	28	72		24
509	49	$\frac{4}{49}$	56	32	28,	76 [1]		24
510	49	$\frac{4}{49}$	56	40	28	64		24
511	49	$\frac{4}{49}$	28			48	24	44
512	49	$\frac{4}{49}$	56	44	28	64		24
513	27	$\frac{2}{27}$	32			64	44	
514	49	$\frac{4}{49}$	56	48	28	64		24
515	27	$\frac{2}{27}$	72	32	24	100		
516	43	$\frac{8}{43}$	32	56	28	64		
517	49	$\frac{4}{49}$	56	48	28	72		24
518	49	$\frac{4}{49}$	28			64	24	44
519	27	$\frac{2}{27}$	72	56	32	64		
520	39	$\frac{3}{39}$						
521	27	$\frac{2}{27}$	72	76 [1]	48	64		
522	29	$\frac{2}{29}$	48	64	24	72		
523	27	$\frac{2}{27}$	72	68 [1]	48	64		
524	27	$\frac{2}{27}$	72	32	24	64		
525	27	$\frac{2}{27}$	72	40	32	64		
526	49	$\frac{4}{49}$	56	64	28	72		24
527	31	$\frac{3}{31}$	32	64	24	72		
528	27	$\frac{2}{27}$	72	24	24	64		
529	27	$\frac{2}{27}$	72	44	48	64		
530	15	$\frac{1}{15}$	24	56	32	64		

[1] Special gear.

TABLE 13.—*Continued*

Number of Divisions	Index Circle	No. of Turns of Index	Gear on Worm	No. 1 Hole		Gear on Spindle	Idlers	
				First Gear on Stud	Second Gear on Stud		No. 1 Hole	No. 2 Hole
531	27	$\frac{2}{27}$	72			48	24	
532	27	$\frac{2}{27}$	72	32	48	64		
533	27	$\frac{2}{27}$	72	32	48	56		
534	27	$\frac{2}{27}$	72			32	44	
535	27	$\frac{2}{27}$	72	32	48	40		
536	39	$\frac{8}{39}$	52 ¹			64	24	44
537	27	$\frac{2}{27}$	72	28	56	32		
538	29	$\frac{4}{29}$	58 ¹	56	24	72		
539	49	$\frac{4}{49}$	28	48	24	56		24
540	27	$\frac{2}{27}$						
541	39	$\frac{8}{39}$	52 ¹	56	32	48		24
542	39	$\frac{8}{39}$	52 ¹	44	32	64		24
543	27	$\frac{2}{27}$	72	24	48	32		24
544	15	$\frac{1}{15}$	40	56	24	64		
545	15	$\frac{1}{15}$	32	44	24	64		
546	39	$\frac{8}{39}$	32			64	24	44
547	27	$\frac{2}{27}$	72	32	48	56		24
548	27	$\frac{2}{27}$	72	32	48	64		24
549	27	$\frac{2}{27}$	72			48	24	24
550	15	$\frac{1}{15}$	32	40	24	64		
551	29	$\frac{2}{29}$	32			64	44	
552	27	$\frac{2}{27}$	72	24	24	64		24
553	49	$\frac{4}{49}$	28	48	24	72		24
554	27	$\frac{2}{27}$	72	56	48	64		24
555	15	$\frac{1}{15}$	24			72	44	
556	15	$\frac{1}{15}$	24	44	40	64		
557	15	$\frac{1}{15}$	40	32	24	86		
558	27	$\frac{2}{27}$	48			64	24	44
559	39	$\frac{3}{39}$	24			72	24	44
560	43	$\frac{8}{43}$	86	40	32	64		
561	27	$\frac{2}{27}$	72	56	32	64		24
562	27	$\frac{2}{27}$	72	44	24	64		24
563	29	$\frac{2}{29}$	58 ¹			68 ¹	44	
564	43	$\frac{8}{43}$	86	24	24	56		
565	15	$\frac{1}{15}$	24			56	44	
566	43	$\frac{3}{43}$	86	24	24	44		
567	15	$\frac{1}{15}$	32	44	40	64		
568	15	$\frac{1}{15}$	40	32	24	64		
569	29	$\frac{2}{29}$	58 ¹			44	24	
570	15	$\frac{1}{15}$	32			64	44	

¹ Special gear.

TABLE 13.—*Continued*

				No. 1 Hole			Idlers	
Number of Divisions	Index Circle	No. of Turns of Index	Gear on Worm	First Gear on Stud	Second Gear on Stud	Gear on Spindle	No. 1 Hole	No. 2 Hole
571	43	$4\frac{8}{43}$	86	28	64	32		
572	15	$1\frac{1}{5}$	40	28	24	64		
573	15	$1\frac{1}{5}$	40			72	24	
574	41	$4\frac{1}{41}$	32			64	24	44
575	15	$1\frac{1}{5}$	24			40	44	
576	15	$1\frac{1}{5}$	40			64	24	
577	43	$4\frac{8}{43}$	86	32	64	44		24
578	15	$1\frac{1}{5}$	48	44	40	64		
579	15	$2\frac{1}{9}$	40			56	44	
580	29	$1\frac{1}{5}$						
581	15	$1\frac{1}{5}$	48	32	40	76[1]		
582	15	$2\frac{2}{7}$	40			48	44	
583	27	$1\frac{1}{5}$	72	64	24	86		24
584	15	$1\frac{1}{5}$	48	32	40	64		
585	15	$1\frac{1}{5}$	24			24	56	
586	15	$2\frac{2}{9}$	72	48	40	56		
587	29	$1\frac{1}{5}$	58[1]			28	24	44
588	15	$1\frac{1}{5}$	40			32	44	
589	15	$1\frac{1}{5}$	72	44	40	48		
590	15	$1\frac{1}{5}$	48			32	44	
591	15	$1\frac{1}{5}$	40			24	44	
592	16	$1\frac{1}{6}$	24			72	44	
593	15	$1\frac{1}{5}$	72	28	40	48		
594	33	$3\frac{2}{33}$	32	56	28	64		
595	15	$1\frac{1}{5}$	72			24	44	
596	15	$1\frac{1}{5}$	72	24	40	32		
597	33	$2\frac{2}{33}$	44	56	24	72		
598	16	$1\frac{1}{6}$	64	56	24	72		
599	43	$4\frac{3}{43}$	86	44	24	84		24
600	15	$1\frac{1}{5}$						
601	29	$2\frac{2}{9}$	58[1]	56	48	72		24
602	43	$4\frac{3}{43}$	32			64	24	44
603	15	$1\frac{1}{5}$	72	24	40	24		24
604	16	$1\frac{1}{6}$	32			72	24	
605	15	$1\frac{1}{5}$	72			24	24	44
606	15	$1\frac{1}{5}$	72	24	40	48		24
607	15	$1\frac{1}{5}$	72	28	40	48		24
608	16	$1\frac{1}{6}$	32			64	44	
609	15	$1\frac{1}{5}$	40			24	24	44
610	15	$1\frac{1}{5}$	48			32	24	44

[1] Special gear.

TABLE 13.—*Continued*

Number of Divisions	Index Circle	No. of Turns of Index	Gear on Worm	No. 1 Hole First Gear on Stud	No. 1 Hole Second Gear on Stud	Gear on Spindle	No. 2 Hole No. 1 Hole	No. 2 Hole No. 2 Hole
611	15	$\frac{1}{15}$	72	44	40	48		24
612	15	$\frac{1}{15}$	40			32	24	44
613	16	$\frac{1}{16}$	64	48	32	72		
614	15	$\frac{1}{15}$	72	48	40	56		24
615	15	$\frac{1}{15}$	24			24	24	44
616	16	$\frac{1}{16}$	32			48	44	
617	33	$\frac{3}{33}$	44	32	24	86		
618	15	$\frac{1}{15}$	40			48	24	44
619	16	$\frac{1}{16}$	48	28	32	72		
620	31	$\frac{3}{31}$						
621	15	$\frac{1}{15}$	40			56	24	44
622	16	$\frac{1}{16}$	64	24	24	72		
623	16	$\frac{1}{16}$	64	24	24	68 [1]		
624	16	$\frac{1}{16}$	24			24	56	
625	15	$\frac{1}{15}$	24			40	24	44
626	16	$\frac{1}{16}$	32			28	56	
627	15	$\frac{1}{15}$	40			72	24	44
628	16	$\frac{1}{16}$	32			24	56	
629	16	$\frac{1}{16}$	64			44	24	
630	16	$\frac{1}{16}$	64			40	24	
631	16	$\frac{1}{16}$	64	28	56	72		
632	16	$\frac{1}{16}$	64			32	44	
633	16	$\frac{1}{16}$	64			28	44	
634	16	$\frac{1}{16}$	64			24	44	
635	15	$\frac{1}{15}$	24			56	24	44
636	16	$\frac{1}{16}$	56	28	48	24		
637	49	$\frac{4}{49}$	24			24	56	
638	29	$\frac{2}{29}$	48	64	24	72		24
639	33	$\frac{3}{33}$	44	28	32	64		
640	16	$\frac{1}{16}$						
641	33	$\frac{3}{33}$	44	32	48	76 [1]		
642	16	$\frac{1}{16}$	72	24	64	24		24
643	16	$\frac{1}{16}$	64	28	56	24		24
644	49	$\frac{4}{49}$	56			32	44	
645	15	$\frac{1}{15}$	24			72	24	44
646	16	$\frac{1}{16}$	64			24	24	44
647	16	$\frac{1}{16}$	64			28	24	44
648	16	$\frac{1}{16}$	64			32	24	44
649	33	$\frac{3}{33}$	72			48	24	
650	16	$\frac{1}{16}$	64			40	24	44

[1] Special gear.

TABLE 13.—*Continued*

Number of Divisions	Index Circle	No. of Turns of Index	Gear on Worm	No. 1 Hole First Gear on Stud	No. 1 Hole Second Gear on Stud	Gear on Spindle	No. 2 Hole No. 1 Hole	No. 2 Hole No. 2 Hole
651	16	$\frac{1}{16}$	64			44	24	24
652	16	$\frac{1}{16}$	32			24	24	44
653	33	$\frac{2}{33}$	72	28	44	48		
654	16	$\frac{1}{16}$	64			56	24	44
655	16	$\frac{1}{16}$	64	40	32	48		24
656	16	$\frac{1}{16}$	24			24	24	44
657	18	$\frac{1}{18}$	32	48	24	56		
658	16	$\frac{1}{16}$	64	24	24	72		24
659	16	$\frac{1}{16}$	64	24	24	76 [1]		24
660	33	$\frac{2}{33}$						
661	16	$\frac{1}{16}$	64	56	48	72		24
662	16	$\frac{1}{16}$	64	44	24	48		24
663	17	$\frac{1}{17}$	24			24	56	
664	16	$\frac{1}{16}$	32			48	24	44
665	49	$\frac{4}{49}$	56			40	24	44
666	18	$\frac{1}{18}$	24			72	44	
667	16	$\frac{1}{16}$	64	48	32	72		24
668	16	$\frac{1}{16}$	32			56	24	44
669	33	$\frac{2}{33}$	44			24	24	24
670	33	$\frac{2}{33}$	72	48	44	40		24
671	33	$\frac{2}{33}$	72			48	24	24
672	18	$\frac{1}{18}$	24			64	44	
673	16	$\frac{1}{16}$	48	44	32	72		24
674	33	$\frac{2}{33}$	72	56	44	48		24
675	33	$\frac{2}{33}$	44			40	24	24
676	16	$\frac{1}{16}$	32			72	24	44
677	18	$\frac{1}{18}$	48	32	24	86		
678	18	$\frac{1}{18}$	24			56	44	
679	49	$\frac{8}{49}$	28			44	24	40
680	17	$\frac{1}{17}$						
681	33	$\frac{2}{33}$	44			56	24	24
682	33	$\frac{2}{33}$	48			64	24	24
683	16	$\frac{1}{16}$	32			86	24	44
684	18	$\frac{1}{18}$	32			64	44	
685	18	$\frac{1}{18}$	24	56	48	40		
686	15	$\frac{1}{15}$	40	64	24	86		24
687	18	$\frac{1}{18}$	24			44	48	
688	16	$\frac{1}{16}$	24			72	24	44
689	39	$\frac{2}{39}$	24	48	24	56		
690	18	$\frac{1}{18}$	24			40	56	

[1] Special gear.

TABLE 13.—*Continued*

Number of Divisions	Index Circle	No. of Turns of Index	Gear on Worm	No. 1 Hole		Gear on Spindle	Idlers	
				First Gear on Stud	Second Gear on Stud		No. 1 Hole	No. 2 Hole
691	18	$\frac{1}{18}$	48	32	24	58 [1]		
692	18	$\frac{1}{18}$	72	56	32	64		
693	18	$\frac{1}{18}$	32			48	44	
694	17	$\frac{1}{17}$	68 [1]			56	24	44
695	18	$\frac{1}{18}$	72	24	24	100		
696	18	$\frac{1}{18}$	24			32	56	
697	17	$\frac{1}{17}$	24			24	24	44
698	18	$\frac{1}{18}$	72	44	24	48		
699	18	$\frac{1}{18}$	48			56	44	
700	18	$\frac{1}{18}$	72	40	32	64		
701	17	$\frac{1}{17}$	68 [1]	48	32	56		24
702	18	$\frac{1}{18}$	24			24	56	
703	19	$\frac{1}{19}$	24			72	44	
704	18	$\frac{1}{18}$	72	24	24	64		
705	18	$\frac{1}{18}$	48			40	44	
706	18	$\frac{1}{18}$	72			56	24	
707	18	$\frac{1}{18}$	72			52 [1]	24	
708	18	$\frac{1}{18}$	72			48	24	
709	18	$\frac{1}{18}$	72			44	24	
710	18	$\frac{1}{18}$	72			40	24	
711	18	$\frac{1}{18}$	64			32	44	
712	18	$\frac{1}{18}$	72			32	24	
713	18	$\frac{1}{18}$	72			28	44	
714	18	$\frac{1}{18}$	72			24	44	
715	18	$\frac{1}{18}$	72	32	64	40		
716	18	$\frac{1}{18}$	72	28	56	32		
717	18	$\frac{1}{18}$	72	24	64	32		
718	33	$\frac{3}{33}$	44	58 [1]	24	64		24
719	17	$\frac{1}{17}$	68 [1]	52 [1]	24	72		24
720	18	$\frac{1}{18}$						
721	21	$\frac{2}{21}$	24	64	32	68 [1]		
722	19	$\frac{1}{19}$	32			64	44	
723	18	$\frac{1}{18}$	72	24	64	32		24
724	18	$\frac{1}{18}$	72	28	56	32		24
725	18	$\frac{1}{18}$	72	24	48	40		24
726	18	$\frac{1}{18}$	72			24	24	44
727	18	$\frac{1}{18}$	72			28	24	44
728	18	$\frac{1}{18}$	72			32	24	44
729	18	$\frac{1}{18}$	64			32	24	44
730	20	$\frac{1}{20}$	32	48	24	56		

[1] Special gear.

By the *lead of the milling machine* is meant the distance the table will travel while the index head spindle makes one complete revolution when the gear ratio between the feed screw and the worm gear stud is 1 to 1.

Fig. 41.—Spiral Milling. A—Gear on worm (driven); B—First gear on stud (driver). C—Second gear on stud (driven). D—Gear on screw (driver).

The lead of the milling machine equals the revolutions of the feed screw required for one revolution of the index head spindle with equal gears, times the lead of the feed screw.

$$\frac{\text{Lead of spiral}}{\text{Lead of machine}} = \frac{\text{product of driven gears}}{\text{product of driving gears}}$$

In finding the change gears to be used in a compound train, place the lead to be cut in the numerator, and the lead of milling

machine in the denominator, then resolve the fraction into its
factors and multiply each pair of factors by the same number
until suitable numbers of teeth in change gears are obtained.

The following change gears are available on most milling
machines: 24–24–28–32–40–44–48–56- 64–72–86–100.

ILLUSTRATION: What change gears are required for a spiral
index head to cut a 36-inch lead with a 10-inch lead milling
machine?

$$\frac{36}{10} = \frac{4 \times 9}{2 \times 5} = \frac{4}{2} \times \frac{16}{16} = \frac{64}{32}$$

$$\frac{9}{5} \times \frac{8}{8} = \frac{72}{40}$$

The 64 and 72 are driven gears and 32 and 40 are driving
gears. Then place the 72 T gear on worm, the 40 T gear on
screw, the 32 T first gear on stud and 64 T second gear on stud
(See Fig. 41.)

ILLUSTRATION: What lead or spiral can be cut with the
following gears if the lead on the machine is 10 inches;
gear on worm, 40; first gear on stud, 24; second gear on stud, 24:
gear on screw, 32?

$$\text{Driven gears} = 40 \times 24 = 960$$

$$\text{Driving gears} = 24 \times 32 = 768$$

Then,

$$\frac{\text{Lead of spiral}}{10} = \frac{960}{768}$$

$$\text{Lead of spiral} = \frac{10 \times 960}{768} = \frac{10 \times 5}{4} = 12.5 \text{ in. (Ans.)}$$

The Angle of Helix.—This is the angle which the spiral makes
with the axis of the work. The swiveled milling machine table

must be set to this angle when cutting a helix. This **angle** (π) may be found by the following formula:

$$\text{tangent of helix angle} = \frac{\pi \times \text{diameter of work}}{\text{lead of helix}}$$

ILLUSTRATION: A helix with a 24-inch lead is to be cut on a piece of work 3 inches in diameter. What is the angle of helix?

$$\tan \phi = \frac{\pi \times \text{diameter of work}}{\text{lead of helix}} = \frac{3.1416 \times 3}{24}$$

$$\tan \phi = \frac{3.1416}{8} = 0.3927$$

$$\phi = 21° \, 26' \text{ (Ans.)}$$

NOTE: Because the scale by which angle ϕ is set is usually graduated only to fourths of a degree, the table would be set $21\frac{1}{2}°$.

The angle of the helix may be found graphically as follows: Draw a base line equivalent to the lead and a vertical line equal to the circumference. If the two lines are then connected by a hypotenuse the helix angle (ϕ) which the hypotenuse makes with the base may be measured with a protractor.

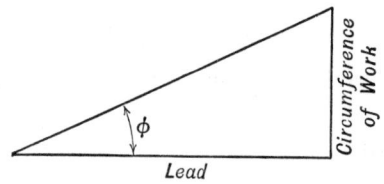

If the drawing is made on paper, the triangle may be cut out and wrapped around the work with the side representing the circumference encircling the work. The hypotenuse will trace out the helix on the work.

TABLE 14

Pitches and Approximate Angles for Cutting Spirals on the Universal Milling Machine

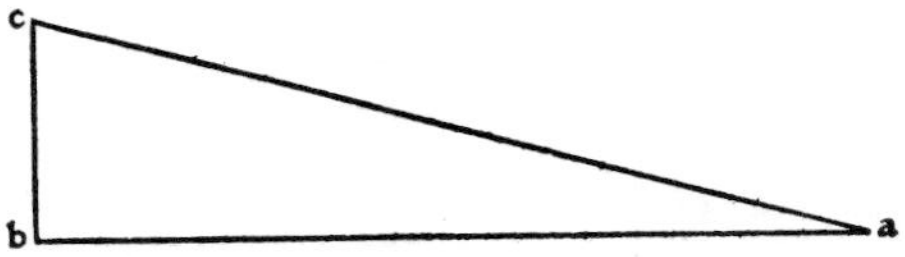

To find the angle for cutters of a larger diameter than given in the table, make a drawing as shown in the diagram; the angle b being a right angle. Let $b\,c$ equal the circumference. Let $a\,b$ equal the pitch. Connect $c\,a$ by a line, and measure the angle a with a protractor; or divide the circumference by the lead and the quotient will be the tangent of the angle. Find the angle in a table of tangents.

Diameter of Mill, Cutter, or Drill to be Cut

Inches

Values Given Under Diameters are Angles in Degrees

Gear on Worm	First Gear on Stud	Second Gear on Stud	Gear on Screw	Pitch in Inches to one Turn	1/8	1/4	3/8	1/2	5/8	3/4	7/8	1	1 1/4	1 1/2	1 3/4	2	2 1/4	2 1/2	2 3/4	3	3 1/4	3 1/2	3 3/4	4
24	64	24	72	1.25	17 1/2	32 1/4																		
24	64	28	72	1.46	14 3/4	28	38 1/2																	
24	64	30	72	1.56	14 1/4	26 3/4	37																	
24	64	32	72	1.67	12 3/4	25	34 3/4	43 1/4																

$$\text{The lead in inches in one turn} = \frac{10 \times \text{Gear on Worm} \times \text{2nd Gear on Stud}}{\text{Gear on Screw} \times \text{1st Gear on Stud}}$$

1.94	72	28	64	32
2.08	72	40	64	24
2.22	72	28	56	32
2.50	72	48	64	24
2.78	72	28	56	40
2.92	72	56	64	24
3.24	72	28	48	40
3.70	72	32	48	00
3.89	64	24	72	56
4.17	72	48	48	40
4.53	72	28	64	56
4.86	72	56	40	48
5.33	72	32	40	56
5.44	64	28	40	56
6.12	72	28	48	56
6.22	76	32	48	56
6.48	56	40	48	64
6.67	64	28	48	56
7.29	72	40	48	64
7.41	56	32	32	48
7.62	72	40	30	72
8.33	64	24	48	56
9.00	72	48	40	64
9.33	56	40	48	72
9.52	72	32	48	64
10.29	56	56	40	48
10.37	72	56		
10.50	64	56		

TABLE 14

PITCHES AND APPROXIMATE ANGLES FOR CUTTING SPIRALS ON THE UNIVERSAL MILLING MACHINE—*Concluded*

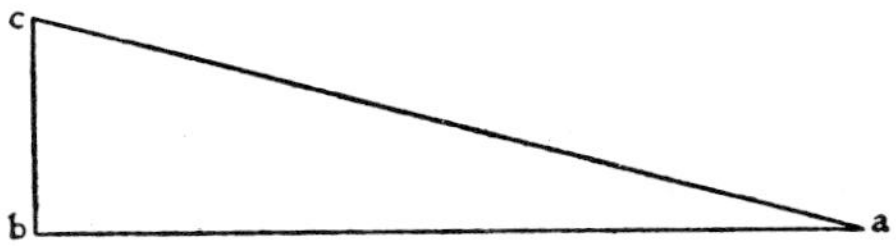

To find the angle for cutters of a larger diameter than given in the table, make a drawing as shown in the diagram; the angle *b* being a right angle. Let *b c* equal the circumference. Let *a b* equal the pitch. Connect *c a* by a line, and measure the angle *a* with a protractor; or divide the circumference by the lead and the quotient will be the tangent of the angle. Find the angle in a table of tangents.

Diameter of Mill, Cutter, or Drill to be Cut

Inches

Values Given Under Diameters are Angles in Degrees

Gear on Worm	First Gear on Stud	Second Gear on Stud	Gear on Screw	Pitch in Inches to one Turn	1/8	1/4	3/8	1/2	5/8	3/4	7/8	1	1 1/4	1 1/2	1 3/4	2	2 1/4	2 1/2	2 3/4	3	3 1/4	3 1/2	3 3/4	4
64	40	48	72	10.67	2	4	6 1/4	8 1/4	10 1/4	12 1/4	14 1/4	16 1/2	20 1/4	24	27 1/4	30 1/2	33 1/2	36 1/2	39	41 1/2	43 3/4			
56	32	40	64	10.94	2	4	6	8 1/4	10 1/4	12	14	16 1/4	20	23 1/2	26 3/4	30	33	35 3/4	38 1/4	40 3/4	43			
64	32	40	72	11.11	2	4	6	8	10	11 3/4	13 3/4	16	19 3/4	23	26 1/2	29 1/2	32 1/2	35 1/4	38	40 1/4	42 1/2	44 3/4		
56	32	48	72	11.66	1 3/4	3 3/4	5 3/4	7 1/2	9 1/2	11 1/4	13 1/4	15 1/4	18 3/4	22	25 1/4	28 1/2	31 1/4	34	36 1/2	39	41 1/4	43 1/2		

Gears

Types of Gears.—There is a great variety of gears with regard to shapes, sizes, and uses. They may, however, be classified under four general groups: spur gears, bevel gears, worm gearing, and spiral or helical gears.

Spur Gears are the most commonly used gears and are used to transmit positive rotary motion between parallel shafts. They are cylindrical in shape and the teeth are cut parallel with the axis.

FIG. 42. Spur Gear.   FIG. 43.—Bevel Gears.

Bevel Gears are used to transmit positive rotary motion to shafts at an angle to each other, and in the same plane.

The teeth of a bevel gear are made on a frustum of a cone whose apex is the same point as the intersection of the axes of the shafts.

Bevel gears usually connect shafts running at right angles. When the angle of the shafts is 90 degrees and the velocity ratio is 1 to 1, then both gears are of the same size and are called *miter gears*. If the velocity ratio between two gears is other than 1 to 1, the smaller gear is called the *pinion*.

Worm Gearing is used to transmit power between two shafts at 90 degrees to each other, but not in the same plane, and is generally used when it is desired to obtain smoothness of action and great speed reduction from one shaft to another.

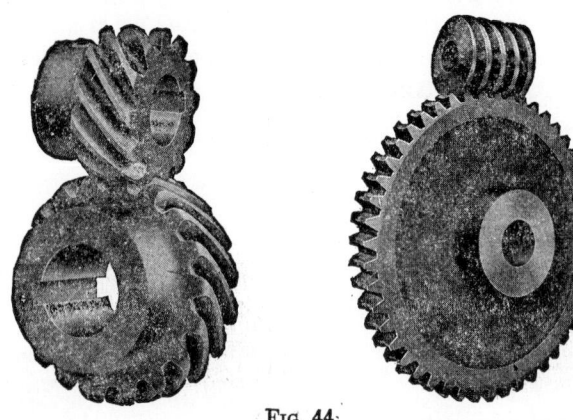

Fig. 44.

The greatest objection to worm-gear drive is the excessive friction between the teeth, making them very inefficient and subject to heating.

A *worm* is a screw so cut as to mesh properly with the teeth of a worm wheel, the included angle of the sides being 29 degrees.

The *worm wheel* is similar to a spiral spur gear. It usually has a concave face and the tooth spaces are concave and at an angle other than 90 degrees to the side of the gear.

Spiral or Helical Gears are used to drive shafts parallel to each other and in the same plane, or shafts at angles to each other but not in the same plane.

Herringbone Gears conform to two spiral gears fastened to each other, one right hand and the other left hand, thus equalizing the side thrust. They are used to transmit power between two parallel shafts. Herringbone gears are very quiet in action because some parts of their teeth are always in full action.

Efficiencies of Gears.—In relative efficiency, the different styles of gearing rank as follows, from the most efficient to the least efficient: spur, herringbone, bevel, spiral or helical, and worm.

Gearing Definitions.—The *center distance* of a pair of gears is the shortest distance between the centers of the shafts on which they are mounted.

The *pitch circles* of a pair of gears have the same diameters as a pair of friction rolls which would fill the same center distance and revolve at the same velocity ratio.

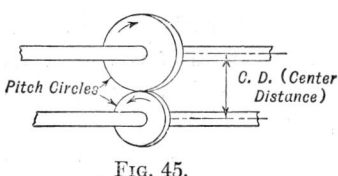

Fig. 45.

The *pitch diameter* of a gear is the diameter of its pitch circle.

The *diametral pitch* is the number of teeth a gear has per inch of pitch diameter. To find the diametral pitch, divide the number of teeth by the pitch diameter. The pitch diameter may in turn be found by dividing the number of teeth by the diametral pitch.

The *circular pitch* is the distance from the center of one tooth to the center of the next, measured along the pitch line. To find the circular pitch, divide the pitch circle by the number of teeth, or divide π by the diametral pitch.

The *addendum* is the height of the tooth above the pitch line.

The *dedendum* is the depth below the pitch line to which the tooth of the mating gear extends.

The *size of gear tooth* is designated by its pitch; thus, a 10-pitch tooth has an addendum of $\frac{1}{10}$ inch and a dedendum of $\frac{1}{10}$ inch.

Note: The term "pitch" when used alone always refers to the diametral pitch.

The *tooth thickness* is measured along the pitch line and is one-half the circular pitch.

The *working depth* is the depth in the tooth space to which the tooth of the mating gear extends, and is equal to the addendum plus the dedendum.

The *clearance* is the distance from the point of the tooth to the bottom of the space in the mating gear.

The *whole depth* is the distance from the top of the tooth to the bottom of the same tooth and consists of the addendum, dedendum, and clearance.

The *outside diameter* is found by adding twice the addendum to the pitch diameter.

The *root diameter* is the diameter at the bottom of the tooth space.

The *face* of the gear tooth is that part of the gear tooth outline which extends above the pitch line.

The *flank* is that part of the gear tooth outline below the pitch line.

The *fillet* is the rounded corner where the flank of the tooth runs to the bottom of the tooth space.

The *base circle* is the circle from which the involute curve is generated. It is drawn tangent to the pressure line. Its position will vary according to the pressure angle used. The two common pressure angles are $14\frac{1}{2}$ degrees and 20 degrees. The former is the more common, while the latter is used on the so-called "stub-tooth." For a $14\frac{1}{2}$-degree pressure angle tooth gear, the base circle will lie inside the pitch circle a distance equal to $\frac{1}{60}$ of the pitch diameter.

Tooth Curves.—The shape of gear teeth is usually either involute or cycloidal.

The *involute curve* is the more desirable because it will allow a certain amount of variation in the center distance, and is for this reason used almost universally.

The way actually to draw this curve on paper with drawing instruments is explained on page 137.

Cycloid Gear Teeth will not be described in detail at this point since this principle is used mostly in large cast gears of one-inch circular pitch or more. These gears must always meet on the pitch line in both gears and racks. This means that there can be no variation in the pitch diameter.

Cycloidal teeth are constructed by making the outline of the face a part of an epicycloid and the flanks a part of a hypocycloid.

With these definitions in mind, we may proceed to a study of the characteristics of individual gear types.

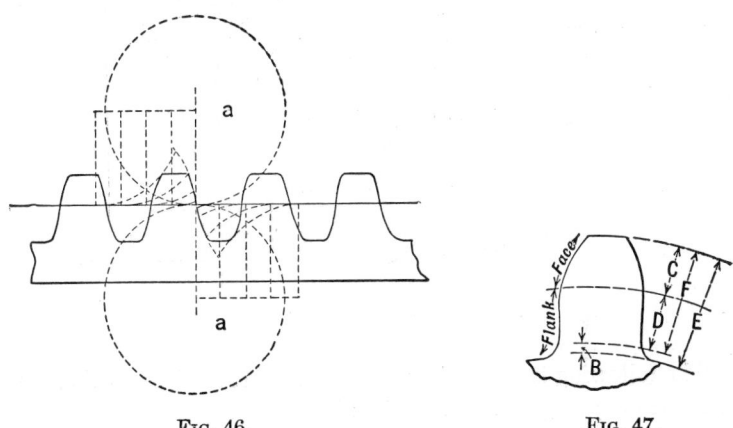

FIG. 46.   FIG. 47.

Characteristics of the Spur Gear.—The preceding definitions as applied to the spur gear and as illustrated in Figs. 47 and 48 are:

A = Circular pitch or distance from center of one tooth to the next, measured on the pitch line.

B = Clearance.

C = Addendum—the height of a tooth above the pitch circle.

D = Dedendum—bottom of tooth between pitch diameter and clearance.

E = Whole depth—addendum, dedendum, and clearance.

F = Working depth—addendum and dedendum.

G = Thickness of tooth—width of tooth from outside to outside on pitch line.

H = Outside diameter.

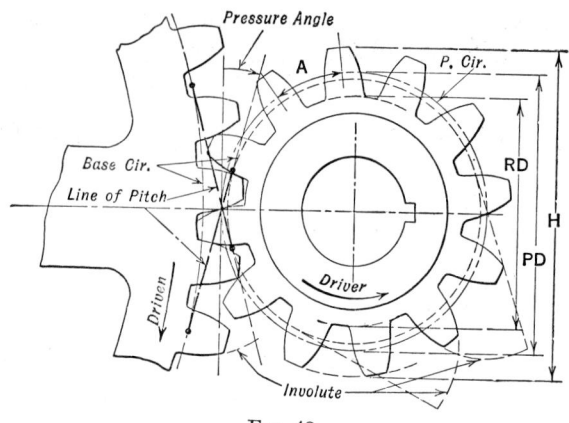

Fig. 48.

The following is a list of symbols and abbreviations used in the formulas of spur gear relationships:

P = Diametral pitch, or pitch.

O.D. = Outside diameter.

N = Number of teeth.

Np = Number of teeth in pinion.

Ng = Number of teeth in gear.

N.R. = Number of teeth in rack.

L = Length of rack.

P.D. = Pitch diameter.

C.D. = Center distance.

C.P. = Circular pitch.

Wh.D. = Whole depth.

Wg.D. = Working depth.

Add. = Addendum.

Ded. = Dedendum.

C = Clearance.

Th. = Thickness of tooth.

R.D. = Root diameter.

The following are formulas for dimensions of spur gears.

$P = \pi \div$ C.P. or N $\div$ P.D.

O.D. $= (N + 2) \div P$ or $(N + 2) \times$ C.P. $\div \pi$ or P.D. $+$ (2 $\times$ Add.)

C.P. $= \pi \div P$ or P.D. $\times \pi \div N$

P.D. $= N \div P$ or $N \times$ C.P. $\div \pi$ or O.D. $- (2 \times$ Add.)

C.D. $= (Ng + Np) \div 2P$ or $(Ng + Np) \times$ C.P. $\div 6.2832$

Clear. $= 0.157 \div P$ or C.P. $\div 20$

Add. $= 1 \div P$ or C.P. $\div \pi$ or C.P. $\times 0.318$

Ded. $= 1 \div P$ or C.P. $\div \pi$ or C.P. $\times 0.318$

Wh.D. $= 2.157 \div P$ or $0.6866 \times$ C.P.

Th. $= 1.5708 \div P$ or C.P. $\div 2$

$N = P \times$ P.D. or $\pi \times$ P.D. $\div$ C.P.

$L = \pi \times$ N.R. $\div P$ or N. $\times$ C.P.

R.D. $=$ O.D. $- 2$ Wh.D. or P.D. $- 2(\text{Ded.} + C)$

ILLUSTRATION: How many teeth are there in a gear of 4 pitch 8-in. pitch diameter?

$P = 4 =$ no. teeth per in. of pitch diameter.

P.D. $= 8$ in.

Then, $N = P \times$ P.D. $= 4 \times 8 = 32$ teeth (Ans.)

ILLUSTRATION: What are the addendum, dedendum and clearance of a 4-pitch gear?

$$\text{Addendum} = \frac{1}{P} = \frac{1}{4} = 0.25 \text{ in. (Ans.)}$$

$$\text{Dedendum} = \frac{1}{P} = \frac{1}{4} = 0.25 \text{ in. (Ans.)}$$

$$\text{Clearance} = \frac{0.157}{P} = \frac{0.157}{4} = 0.0392 \text{ in. (Ans.)}$$

ILLUSTRATION: What is the approximate outside diameter of a gear whose circular pitch is 0.500 in. and which has 60 teeth

$$\text{C.P.} = 0.500,$$

and

$$P = \frac{\pi}{0.500} = 6.2832$$

Then O.D. $= \dfrac{N + 2}{p} = \dfrac{62}{6.2832} = 10$ inches (approx.) (Ans.)

ILLUSTRATION: What is the center distance of two gears of 40 and 60 teeth, 10 pitch?

$$\text{Center distance} = \frac{Np + Ng}{2P} = \frac{40 + 60}{2 \times 10} = 5 \text{ in. (Ans.)}$$

ILLUSTRATION: Given approximate center distance of two gears of $5\frac{1}{8}$ in., ratio 15 to 26, 8 pitch; find pitch diameter, outside diameter and number of teeth in each gear.

NOTE: The subscripts g for " gear " and p for " pinion " are added to indicate the symbol applies to the gear or to the pinion. Thus, P.D.$_g$ is the pitch diameter of the gear.

$$\text{P.D.}_g = 2V_p \times \frac{\text{C.D.}}{V_g + V_p} \qquad\qquad \text{P.D.}_p = 2V_g \times \frac{\text{C.D.}}{V_g + V_p}$$

$$= 2 \times 26 \times \frac{5.125}{15 + 26} \qquad\qquad = 2 \times 15 \times \frac{5.125}{15 + 26}$$

$$= 52 \times 0.125 = 6.5 \text{ in.} \qquad\qquad - 30 \times 0.125 = 3.75 \text{ in.}$$

$$N = 8 \times 6.5 = 52 \text{ teeth} \qquad\qquad N = 8 \times 3.75 = 30 \text{ teeth}$$

$$\text{O.D.} = \frac{52 + 2}{8} = \frac{54}{8} = 6.75 \text{ in.} \qquad \text{O.D.} = \frac{30 + 2}{8} = \frac{32}{8} = 4 \text{ in.}$$

Cutting Spur Teeth.—Smooth-running involute gear teeth may be cut on a milling machine by the use of standard gear cutters. A separate set is required for each pitch and there are eight cutters to each set. These cutters are adapted to cut gears ranging from 12-tooth to a rack. The following table can be used to select the proper number of cutter when the number of teeth to be cut is known:

No. of cutter	No. of teeth	No. of cutter	No. of teeth
1	135 to rack	5	21 to 25
2	55 to 134	6	17 to 20
3	35 to 54	7	14 to 16
4	26 to 34	8	12 to 13

ILLUSTRATION: What number of cutter should be used to cut (a) an 18-tooth gear; (b) a 48-tooth gear?

(a) No. 6. (Ans.)

(b) No. 3. (Ans.)

The depth to which the slot between the teeth is cut depends upon the diametral pitch. All gears of one pitch have the same depth of slot. Table 15 gives the depths to which the spaces should be cut in gears of various pitch.

TABLE 15

DEPTH OF SPACES IN GEARS

Diametral pitch	Depth to be cut in gear, inches	Thickness of tooth on pitch line, in.	Diametral pitch	Depth to be cut in gear, inches	Thickness of tooth on pitch line, in.
2	1.078	0.785	12	0.180	0.131
2½	0.863	0.628	14	0.154	0.112
3	0.719	0.523	16	0.135	0.098
3½	0.616	0.448	18	0.120	0.087
4	0.539	0.393	20	0.108	0.079
5	0.431	0.314	22	0.098	0.071
6	0.359	0.262	24	0.090	0.065
7	0.307	0.224	26	0.083	0.060
8	0.270	0.196	28	0.077	0.056
9	0.240	0.175	30	0.072	0.052
10	0.216	0.157	32	0.067	0.049
11	0.196	0.143			

Characteristics of Bevel Gears.—When the pitch of two bevel gears is the same, they will mesh properly regardless of the number of teeth, providing they have twelve or more teeth. A gear with less than twelve teeth must be specially cut to avoid interference of teeth while rolling.

The pitch, outside diameter, and pitch diameter of a bevel gear are always calculated on the large end of the tooth.

Figure 49 shows a cross section of a bevel gear and pinion.

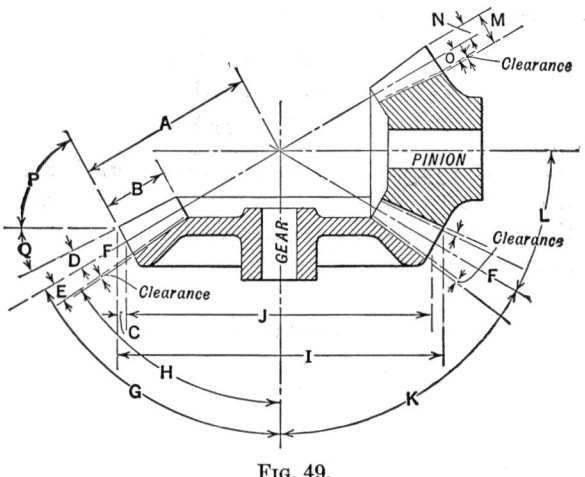

FIG. 49.

The following is a list of symbols and abbreviations of the bevel gear parts and a key to these parts in the figure:

$$\begin{aligned}
&\text{P.C.R.} = \text{Pitch cone radius} \ldots \ldots = A \\
&\text{W. of F.} = \text{Width of face} \ldots \ldots \ldots = B \\
&\text{Ang. add.} = \text{Angular addendum} \ldots \ldots = C \\
&\text{Add. ang.} = \text{Addendum angle} \ldots \ldots = D \\
&\text{Ded. ang.} = \text{Dedendum angle} \ldots \ldots = E \\
&\text{P. line} = \text{Pitch line} \ldots \ldots \ldots = F \\
&\text{P.C. ang.} = \text{Pitch cone angle} \ldots \ldots = G \\
&\text{Cut. ang.} = \text{Cutting angle} \ldots \ldots = H \\
&\text{O.D.} = \text{Outside diameter} \ldots \ldots = I \\
&\text{P.D.} = \text{Pitch diameter} \ldots \ldots = J
\end{aligned}$$

P.C. ang. $G.$ = Pitch cone angle of gear.. = K
P.C. ang. P. = Pitch cone angle of pinion = L
Wh.D. = Whole depth............ = M
Add. = Addendum............. = N
Ded. = Dedendum............. = O
E. ang. = Edge angle............ = P
F. ang. = Face angle............. = Q
Ng = Number of teeth in gear
Np = Number of teeth in pinion
N = Number of teeth
P = Diametral pitch, or pitch
T = Thickness of tooth
N' = Number of teeth for which to select cutter

The principal bevel gear formulas are:

Tangent of P.C. ang. of pinion = $Np \div Ng$
Tangent of P.C. ang. of gear = $Ng \div Np$
Pitch diameter.............. = $N \div P$
Addendum................. = $1 \div P$ or C.P. $\times 0.318$ or
                                     C.P. $\div \pi$
Dedendum................. = $1 \div P$ or C.P. $\times 0.318$ or
                                     C.P. $\div \pi$
Whole depth of tooth........ = $2.157 \div P$ or C.P. $\times 0.687$
Pitch cone radius............ = P.D. $\div (2 \times$ sin P.C. ang.$)$
Thickness of tooth........... = $1.571 \div P$ or C.P. 2
Small addendum............ = (P.C.R. $- B) \div$ P.C.R. $\times$ Add.
Small thickness of tooth...... = (P.C.R. $- B) \div$ P.C.R. $\times$
                                     thickness
Tangent ang. of addendum... = Add. $\div$ P.C.R.
Tangent ang. of dedendum... = Ded. $\div$ P.C.R.
Face angle.................. = 90 deg. $-$ (P.C. ang. $+$ Add.
                                     ang.)
Cutting angle............... = P.C. ang. $-$ Ded. ang.
Angular addendum.......... = Cos. of P.C. ang. $\times$ Add.
Outside diameter............ = Ang. add. $\times 2 +$ P.D.
No. of teeth for which to select
  cutter................... $= \dfrac{N}{\text{Cos. of P.C. Ang.}}$

ILLUSTRATION: What is the pitch cone radius, addendum angle, and outside diameter of a bevel gear whose pitch diameter is 4 inches, pitch cone angle 60 degrees, and which is 10 pitch?

Summarizing the known factors:

$$P.D. = 4 \text{ in.}$$
$$P.C. \text{ ang.} = 60°$$
$$P = 10$$

Then, pitch cone radius (P.C.R.) = P.D. $\div$ (2 $\times$ sin P.C. ang.)

$$P.C.R. = \frac{4}{2 \times \sin 60°} = \frac{2}{0.866} = 2.309 \text{ in. (Ans.)}$$

$$\text{Addendum} = \frac{1}{P} = \frac{1}{10} = 0.10 \text{ in.}$$

Tangent addendum angle = Add. $\div$ P.C.R. $= \dfrac{0.100}{2.309} = 0.04331$

Addendum angle = 2° 29′ (Ans.)

Angular addendum = cos P.C. ang. $\times$ Add.

Ang. add. = cos 60° $\times$ 0.10 = 0.50 $\times$ 0.10 = 0.050 in.

Outside diameter = Ang. add. $\times$ 2 + P.D.

O.D. = 0.05 $\times$ 2 + 4 = 4.10 in. (Ans.)

ILLUSTRATION: What is the whole depth of tooth at the small end of a bevel gear with 30 teeth, 6 pitch and a pitch cone angle of 54 degrees and a width of face of 1 inch?

Since all of the dimensions of a gear tooth (except width of face) gradually decrease until they are zero at the intersection of the centerline axes, we can best solve this problem by finding the whole depth at the large end of the tooth and multiplying this by $\dfrac{P.C.R. - B}{P.C.R.}$. (See Fig. 49.)

$$\text{Whole depth at large end} = \frac{2.157}{P} = \frac{2.157}{6} = 0.3595 \text{ in}$$

$$\text{Pitch diameter} = \frac{N}{P} = \frac{30}{6} = 5 \text{ in.}$$

$$\text{Pitch cone radius} = \frac{\text{P.D.}}{2 \times \sin \text{P.C. ang.}} = \frac{5}{2 \times \sin 54°}$$

$$\text{P.C.R.} = \frac{5}{2 \times 0.809} = 3.09$$

$$\text{P.C.R.} - B = 3.09 - 1 = 2.09$$

Then, whole depth at small end $= 0.3595 \times \dfrac{2.09}{3.09} = 0.2432$ in.

(Ans.)

ILLUSTRATION: A pair of 2-pitch bevel gears with shafts at 90 degrees have a velocity ratio of $2\frac{1}{2}$ to 1 and the pinion has 24 teeth. What is the face angle, pitch cone angle and cutting angle of the larger gear?

$$Np = 24 \text{ teeth}$$

then, $\qquad Ng = 2\frac{1}{2} \times 24 = 60 \text{ teeth}$

Tangent pitch cone angle of gear $= \dfrac{Ng}{Np} = \dfrac{60}{24} = 2.50$

Pitch cone angle $\quad = 68° 12'$ (Ans.)

Pitch diameter $\quad = \dfrac{N}{P} = \dfrac{60}{2} = 30$ in.

Pitch cone radius $= \dfrac{\text{P.D.}}{2 \times \sin \text{P.C. ang.}} = \dfrac{30}{2 \times \sin 68° 12'}$

$$\text{P.C.R.} = \frac{30}{2 \times 0.9285} = 16.1551$$

Addendum $= \dfrac{1}{P} = \dfrac{1}{2} = 0.50$

Tan Angle of Addendum $= \dfrac{\text{Add.}}{\text{P.C.R.}} = \dfrac{0.50}{16.1551} = 0.03095$

Angle of Addendum = 1° 46′

Face angle = 90° − (P.C. ang. + Add. ang.) = 90°
       − (68° 12′ + 1° 46′)

Face angle = 90° − 69° 58′ = 20° 2′ (Ans.)

Add. angle = Ded. angle

Then, Cutting angle = P.C. ang. − Add. ang. = 68° 12′ − 1° 46′
= 66° 26′ (Ans.)

Characteristics of Worms and Worm Wheels.—The worm wheel or gear is similar to a spiral spur gear. It usually has a concave face and the tooth spaces are concave and at an angle other than 90 degrees to the side of the gear.

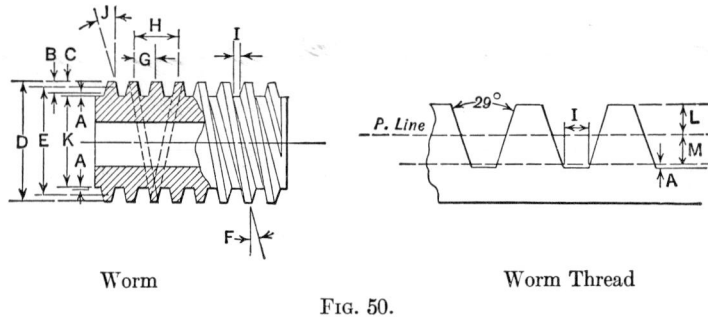

Worm   Worm Thread

Fig. 50.

The *linear pitch* is the distance from the center of one tooth to the center of the next, measured on the pitch circle. The ratio between the linear pitch and the diameter of the worm is arbitrary. It may be four times the circular pitch of a worm gear for single thread; five times the circular pitch of the worm gear for double thread; six times the circular pitch of the worm gear for triple thread.

The *lead* sometimes differs from the pitch and it is the distance a tooth on the worm would advance in one revolution, or the distance the worm wheel advances in one complete turn of a worm.

Parts of the worm with reference to Fig. 50 are:

A = Clearance
B = Working depth of tooth
C = Whole depth of tooth
D = Outside diameter of worm
E = Pitch diameter of worm
F = Angle of helix
G = Linear pitch
H = Lead
I = Thickness of end of tool at bottom of space
J = Half angle of tooth
K = Root diameter of worm
L = Addendum
M = Dedendum

Worm relations are:

Lead = linear pitch × no. of separate threads on the worm
Linear pitch = lead ÷ no. of separate threads on the worm
Addendum = linear pitch × 0.3183
Whole depth of thread = linear pitch × 0.6866

Width of threading tool at end or width of bottom of space = linear pitch × 0.31

O.D. = P.D. + (2 × Add.)
P.D. = O.D. − (2 × Add.)
P.D. = 2 × center distance − P.D. of gear
Root diameter = O.D. − 2 × whole depth of tooth

Cotangent of angle of worm tooth or gashing angle of wheel = P.D. × π ÷ lead.

ILLUSTRATION: What is the root diameter of a worm whose outside diameter is $1\frac{1}{4}$ inches and whose linear pitch is 0.25 inch?

Whole depth of tooth = $P \times 0.6866 = 0.25 \times 0.6866 = 0.17165$ in.

Root diameter = O.D. − 2 × whole depth

Root diameter = $1.25 - 2 \times 0.17165 = 0.9067$ in.   (Ans.)

ILLUSTRATION: What is the width of a thread tool at its cutting edge for a worm whose linear pitch is 0.215 inch?

Width of thread tool = $P \times 0.31 = 0.215 \times 0.31 = 0.06665$ in.
(Ans.)

ILLUSTRATION: What is the angle of worm tooth or gashing angle of wheel, if the outside diameter of the worm is $1\frac{3}{4}$ inches, the linear pitch is 0.60 inch and the screw is double thread?

Addendum = $P \times 0.3183 = 0.60 \times 0.3183 = 0.1910$ in.

P.D. = O.D. − (2 × Add.) = $1.75 - 2 \times 0.1910 = 1.368$ in.

Lead = pitch × no. of separate threads

Lead = $0.60 \times 2 = 1.20$ inches

Cotangent of angle of worm = $\dfrac{\text{P.D.} \times \pi}{\text{lead}} = \dfrac{1.368 \times \pi}{1.20}$

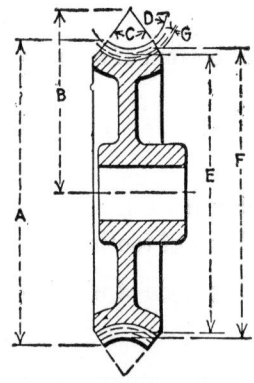

FIG. 51.—Worm Wheel.

Cotangent   = 3.5814

Angle of worm = 15° 36′   (Ans.)

The following list indicates the meaning of the symbols used as dimensions in Fig. 51.

A = O.D. of worm wheel

B = Center distance of worm and worm wheel

C = Angle of face

D = Throat radius

E = Pitch diameter

F = Throat diameter

G = Clearance

Worm wheel formulas are:

P.D. = (no. of teeth in wheel × linear pitch of worm) ÷ π

Throat diameter = P.D. of worm wheel + 2 × Add.

Radius of throat = $\frac{1}{2}$ of O.D. of worm − (2 × Add. of worm)

Center distance = (P.D. of worm + P.D. of wheel) ÷ 2

O.D. = (throat radius − throat radius × cosine $\frac{1}{2}$ face angle) × 2 + throat diameter of wheel.

Addendum of worm wheel = addendum of worm.

ILLUSTRATION: What is the pitch diameter of worm wheel with 48 teeth and a linear pitch of 0.350 inch?

$$P.D. = (no.\ teeth \times linear\ pitch) \div \pi$$

$$P.D. = \frac{48 \times 0.350}{3.1416} = 5.3475\ in.\quad (Ans.)$$

ILLUSTRATION: What is the radius of curvature of worm wheel throat if the pitch of the worm is 0.150 inch and the outside diameter is 1 inch?

Addendum of worm = linear pitch × 0.3183

Addendum = 0.150 × 0.3183 = 0.04775 in.

Then, radius of throat = $\frac{1}{2}$ of O.D. of worm − (2 × Add. of worm)

radius of throat = $\frac{1}{2}$ × 1 − 2 × 0.04775

radius of throat = 0.5000 − 0.0955 = 0.4045 in.   (Ans.)

ILLUSTRATION: What is the outside diameter of a worm wheel whose face angle is 70 degrees, throat radius 0.500 inch, number of teeth 32 and linear pitch of worm 0.200 inch?

Addendum = linear pitch × 0.3183 = 0.200 × 0.3183
= 0.07366 in.

Pitch diameter of gear = (no. of teeth in wheel × linear pitch of worm) ÷ π

$$\text{Pitch diameter} = \frac{32 \times 0.200}{\pi} = \frac{6.4}{3.1416} = 2.0372$$

Throat diameter = P.D. of worm wheel + 2 × Add.

Throat diameter = 2.0372 + 2 × 0.07366

Throat diameter = 2.0372 + 0.1473 = 2.1845 in.

Outside diameter, diameter to sharp corners, = (throat radius − throat radius × cosine of $\frac{1}{2}$ face angle) × 2 + throat diameter of wheel.

Outside Diameter = (0.5 − 0.5 × cos 35°) × 2 + 2.1845

Outside Diameter = (0.5 − 0.5 × 0.8192) × 2 + 2.1845

Outside Diameter = (0.5000 − 0.4096) × 2 + 2.1845

Outside Diameter = 0.0904 × 2 + 2.1845 = 2.3653 in. (Ans.)

Planing and Shaping

Dovetail.—One of the problems of planing and shaping which lends itself to mathematical solution is the measurement of dovetail slides. The dimensions of these are usually given as shown in Fig. 52, but it is difficult to make these measurements on the work with any great accuracy because the edges are not uniformly

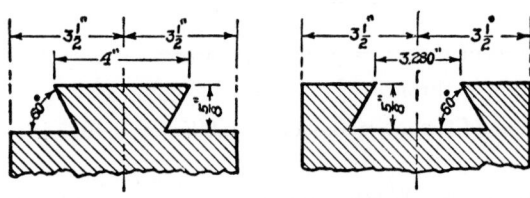

Fig. 52.

sharp. The method used is to measure between rods of equal diameter in the case of the slot, as shown in Fig. 53, and over rods on its counterpart.

To obtain X and Y (Fig. 53) which are used in the practical measuring of dovetail slides, the following formulas may be used.

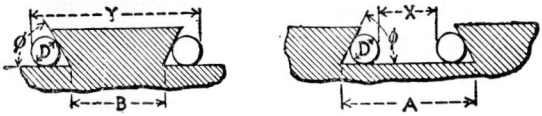

FIG. 53.

$$X = A - [D(1 + \cot \tfrac{1}{2}\phi)]$$
$$Y = D(1 + \cot \tfrac{1}{2}\phi) + B$$

The best size of plug or rod to use is one whose diameter is two-thirds the depth of the slot.

ILLUSTRATION: What is the overall length in measuring a male dovetail, if the following data are given on the blue print: angle 66°, width at bottom 2.956 inches, if plugs $\tfrac{3}{4}$ inch in diameter are used?

$$Y = D(1 + \cot \tfrac{1}{2}\phi) + B = 0.75(1 + \cot 33°) + 2.956$$
$$Y = 0.75(1 + 1.5399) + 2.956 = 4.861 \text{ in. (Ans.)}$$

ILLUSTRATION: What is the distance between $\tfrac{5}{8}$ inch plugs placed in a female dovetail which is cut to a 2.125 inch width at the bottom and has an included angle of 50 degrees?

$$X = A - [D(1 + \cot \tfrac{1}{2}\phi)] = 2.125 - [0.625(1 + \cot 25°)]$$
$$X = 2.125 - [0.625(1 + 2.1445)] = 0.160 \text{ inch (Ans.)}$$

Grinding

Finishing by Grinding.—Machine work is often turned or planed oversize by an amount of from 0.002 to 0.010 inch and the excess removed by grinding. In the grinding operation, cuts of 0.001 or less can easily be made and the result is a finish of greater smoothness and accuracy than can readily be obtained with a cutting tool. Wheels of emery or silicon carbide are most commonly used in finishing metal work.

Speed of Grinding Wheel.—Grinding wheels do good work at surface speeds of 5000 to 6000 feet per minute. The surface speed depends on the speed of revolution and the diameter of the wheel.

The following formulas may be used to find the surface speed in feet per minute of a wheel.

$$S = \frac{\pi R D}{12}$$

S = Surface speed

R = Revolutions per minute

D = Diameter in inches

π = $3\frac{1}{7}$ or 3.14

ILLUSTRATION: What is the surface speed of a 9-inch grinding wheel revolving at a speed of 2500 revolutions per minute?

$$S = \frac{\pi R D}{12}$$

$$= \frac{22 \times 2500 \times 9}{7 \times 12} = 5893 \text{ feet per minute.}$$

If a certain surface speed of a given wheel is desired, to find the number of revolutions of the wheel spindle,

$$R = \frac{12S}{\pi D}$$

ILLUSTRATION: A surface speed of 5500 feet per minute is desired from an 18 inch grinding wheel. How many revolutions per minute should it turn?

$$R = \frac{12S}{\pi D}$$

$$= \frac{5500 \times 12}{3.14 \times 18} = 1168 \text{ revolutions per minute} \quad \text{(Ans.)}$$

Table 16 gives the necessary revolutions per minute for obtaining certain surface speeds from wheels of various diameters.

TABLE 16

GRINDING WHEEL SPEEDS

Diameter of wheel, inches	Revolutions per minute for surface speed of 4000 feet	Revolutions per minute for surface speed of 5000 feet	Revolutions per minute for surface speed of 6000 feet	Revolutions per minute for surface speed of 9000 feet
1	15,279	19,099	22,918	34,377
2	7,639	9,549	11,459	17,188
3	5,093	6,366	7,639	11,459
4	3,820	4,775	5,730	8,595
5	3,056	3,820	4,584	6,876
6	2,546	3,183	3,820	5,729
7	2,183	2,728	3,274	4,911
8	1,910	2,387	2,865	4,297
10	1,528	1,910	2,292	3,438
12	1,273	1,592	1,910	2,864
14	1,091	1,364	1,637	2,455
16	955	1,194	1,432	2,149
18	849	1,061	1,273	1,910
20	764	955	1,146	1,719
22	694	868	1,042	1,562
24	637	796	955	1,433
30	509	637	764	1,146
36	424	531	637	954

The revolutions per minute at which wheels are run is dependent on conditions and style of machine and the work to be ground.

Fits

Types of Fits.—In the mating of two parts of a machine, the perfection of the mating is called the *fit*. Sometimes the pieces are assembled so that there may be motion between them, as, for instance, a shaft in a bearing or an engine crosshead in its frame. In other cases two parts may be assembled so that they can act only in unison, as a flywheel on a shaft or a tire on a locomotive wheel.

Fits may be classified broadly as *running fits*, *wringing fits*, *pressed fits* and *shrinking fits*.

Running Fit.—To make a running fit, like a bearing, an allowance may be made of about two thousandths of an inch for a shaft one inch in diameter, and one thousandth more for each inch the shaft is increased in diameter.

If D = diameter of the hole in inches and d = diameter of the shaft, then, $d = D - [(D - 1) \times 0.001 + 0.002]$

ILLUSTRATION: A shaft is to run in a self-aligning and self-oiling bearing 6 inches in diameter. What should be the diameter of the shaft?

$$d = D - [(D - 1) \times 0.001 + 0.002] = 6 - (5 \times 0.001 + 0.002)$$

$$d = 5.9975 \text{ inches} \quad (\text{Ans.})$$

Wringing Fit.—In a fit of this type, the shaft is made the same size as the hole into which it is to fit.

Pressed Fit.—The force required to press a shaft into a hole made for a press fit will depend not only on the allowance made on the fit, but also on the kind of material, the length of the fit, the finish, etc. Press fits are frequently made so that a pressure of from 5 to 10 tons per inch diameter is required to force the shaft into its hole.

When the length of the fit is from two to three times its diameter, and the finish is good and smooth, an allowance of three-quarters to one and one-quarter of a thousandth of an inch may do well for pressing a one-inch shaft of machinery steel into a hole in cast iron or machinery steel, and as the shaft increases in size, the allowance may be increased about half of one-thousandth for each inch the shaft is increased in diameter. There is no hard and fast rule for making these allowances; judgment and experience alone will dictate what modifications to make.

Setting up the above rule in equation form with average values when D = diameter of hole in inches and d = diameter of shaft, we get

$$d = D + [(D - 1) \times 0.0005 + 0.001]$$

ILLUSTRATION: A shaft is to be turned for a press fit into a 3-inch hole. To what diameter should it be turned?

$$d = D + [(D - 1) \times 0.0005 + 0.001] = 3 + [2 \times 0.0005 + 0.001]$$
$$d = 3 + 0.002 = 3.002 \text{ in.} \quad \text{(Ans.)}$$

Shrinking Fit.—The allowance to be made for a shrinking fit will vary more or less according to the nature of work and the judgment of the designer.

When shrinking a collar on a shaft or similar work, an allowance of 0.002 inch to 0.003 inch will do for a shaft of one inch diameter, and as the shaft increases in diameter add 0.0005 inch to the allowance for each inch the diameter is increased.

$$d = D + [(D - 1) \times 0.0005 + 0.0025]$$

ILLUSTRATION: A shaft is to have a collar shrunk onto it with a 7-inch hole. What should be the diameter of the shaft?

$$d = D + [(D - 1) \times 0.0005 + 0.0025]$$
$$= 7 + [(7 - 1) \times 0.0005 + 0.0025]$$
$$d = 7 + 0.0030 + 0.0025 = 7.0055 \text{ in.} \quad \text{(Ans.)}$$

References.—ENGINEERING TOOLS AND PROCESSES by H. C. Hesse; BLUEPRINT READING: UNDERSTANDING SHOP PRACTICES by F. Nicholson and F. Jones; THE MACHINISTS' AND DRAFTSMEN'S HANDBOOK and MACHINE SHOP: THEORY AND PRACTICE, both by A. Wagener and H. R. Arthur; all published by D. Van Nostrand Company, contain much valuable information on the subject of Machine Shop Work.

XIX

Sheet Metal Work

Sheet metal work makes abundant use of geometry in that flat sheets must be made into the common geometrical shapes of cones, cylinders, etc. The plans or drawings usually give the dimensions of the finished shape, and the problem is one of laying out a design on the flat metal so that when it is cut and bent it will result in the desired shape.

When the surface of a solid is thus opened or flattened out it is said to be developed. The following figures will indicate the meaning of the term development as applied to the surfaces of different solids. Moreover, a knowledge of volume is necessary in sheet metal work, for example; a tinsmith is required to make some cylinder shaped cans to hold one gallon each and to be 6 inches high. What radius should be used in laying out the base? Practically all formulas of surface and cubic measure apply to problems in sheet metal work. Some are given on pages 115–125.

A cube is shown in Fig. 1 together with its development.

The following formulas may be used:

Volume, $V = S^3$

Side, $S = \sqrt[3]{V}$

Lateral Surface, $L =$ area of two ends $+$ areas of all side faces.

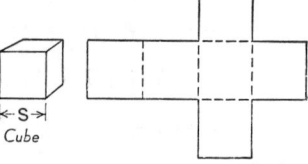

Fig. 1.

ILLUSTRATIONS: 1. Find the volume of a cube whose side is 9.5 inches.

$$V = S^3 = 9.5^3 = 9.5 \times 9.5 \times 9.5 = 857.38 \text{ cu. in.}$$

2. If the volume of a cube is 231 cubic inches, find the length of the side.

$$S = \sqrt[3]{V} = \sqrt[3]{231}$$

from the table on page 29 find 6.136 in.

3. If in Fig. 1 each side is 1 foot, find the lateral surface.

L = area of two ends + areas of all side faces.
= 2 sq. ft. + 4 sq. ft. = 6 sq. ft.

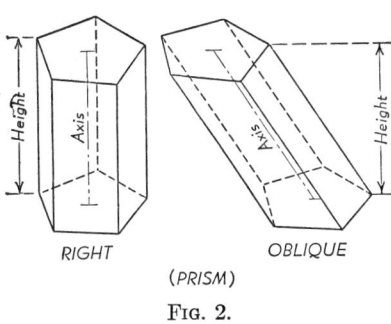

RIGHT   OBLIQUE

(PRISM)

FIG. 2.

A pentagonal prism is shown on the left, Fig. 2, and its development in Fig. 3. Prisms are named triangular, square, pentagonal, etc., in accordance with the shape of the base.

The following formulas may be used which apply to all prisms:

Volume, $V = A$, area of end surface $\times h$, height

Lateral Surface, L = area of two ends = areas of all side faces

The area of a pentagon can be found by multiplying the length of the side by 1.7204. See table, page 651, for constants to determine the area of polygons.

ILLUSTRATIONS: 1. If a pentagonal prism measures 1.5 feet on a side 6 feet in height, find the volume.

Area of end surface = 1.7204 $\times$ 1.5 = 2.58 sq. ft.
$V = Ah$
= 2.58 $\times$ 6 = 15.48 cu. ft.

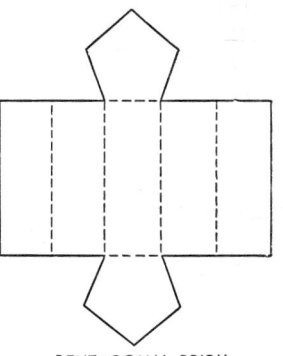

PENTAGONAL PRISM
(RIGHT)
FIG. 3.

2. Find the lateral surface in the above illustration.

L = area of two ends + area of all side faces

from preceding problem, area of one end surface = 2.58 square feet, therefore,

$2 \times 2.58 = 5.16$ square feet = area of two ends

In a right prism the area of the side faces = perimeter of base $\times$ height.

Area of one side face = $1.5 \times 6 = 9.0$ sq. ft.
Area of 5 faces   = $9 \times 5 = 45$ sq. ft.

or, perimeter of base, $(1.5 \times 5) = 7.5 \times$ height $(6) = 45$ sq. ft.

$L = 5.16 + 45 = 50.16$ sq. ft.

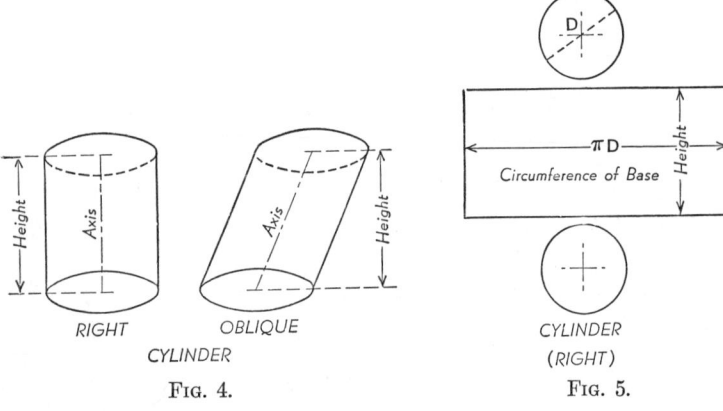

RIGHT   OBLIQUE   CYLINDER
     CYLINDER   (RIGHT)
       Fig. 4.   Fig. 5.

A cylinder is shown in Fig. 4 and its development in Fig. 5. The following formulas may be used:

Volume, $V = 3.1416r^2h = 0.7854d^2h$.

Lateral or cylindrical surface = perimeter of base $\times$ height $= 3.1416dh$.

Total area, A, lateral or cylindrical surface and end surfaces $= 6.2832r(r + h)$.

ILLUSTRATIONS: 1. Find the volume of a cylinder whose diameter is $2\frac{1}{2}$ inches and height is 20 inches.

$V = 0.7854d^2h = 0.7854\,(2.5)^2 \times 20$
$$= 0.7854 \times 6.25 \times 20 = 98.18 \text{ cu. in.}$$

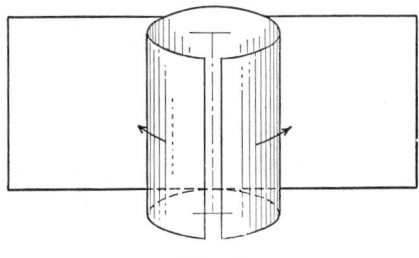

FIG. 6.

2. Find the lateral surface in the above illustration.

$L = 3.1416dh = 3.1416 \times 2.5 \times 20 = 157.08$ sq. in.

3. In illustration 1 find the total lateral area.

$A = 6.2832r\,(r + h) = 6.2832 \times 1.25\,(1.25 + 20) = 166.9$ sq. in.

When the volume of a prism and the area of the base is known the height may be found by the following formula:

$$\text{Height} = \frac{\text{volume}}{\text{area of base}}$$

ILLUSTRATION: Find the height of a cylinder 2 feet 2 inches in diameter to contain 6500 cubic inches.

$$\text{Height} = \frac{6500}{531} = 12.24 \text{ in.}$$

Aid.   Area of base $= \pi r^2 = 3.14 \times 13 \times 13 = 531$ sq. in.

A square pyramid is shown in Fig. 7 and its development in Fig. 8.   Pyramids are named triangular, square, pentagonal, etc. in accordance with the shape of the base.   The following formulas may be used.

Volume, $V = \frac{1}{3}h \times$ area of base

Lateral surface, $L =$ area of the base $+$ areas of all the triangular faces, or, $\frac{1}{2} \times$ perimeter of base $\times$ slant height.

NOTE.   In a right pyramid all triangular faces are isosceles.

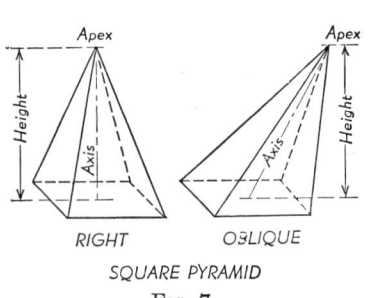

RIGHT   OBLIQUE
SQUARE PYRAMID
FIG. 7.

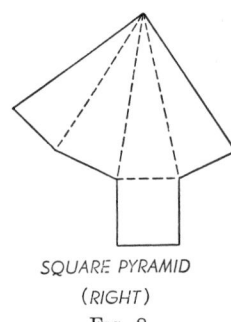

SQUARE PYRAMID
(RIGHT)
FIG. 8.

ILLUSTRATIONS: 1. A pyramid whose base is 2 feet square has a height of 6 feet.   Find the volume.

Area of base $= 2 \times 2 = 4$ sq. ft.

$V = \frac{1}{3}h \times$ area of base $= \frac{1}{3} \times 6 \times 4 = 8$ cu. ft.

2. In illustration 1 find the lateral surface.

Area of base, from above, $= 4$ square feet

Fig. 9 indicates that the lateral surface is made of four isosceles triangles similar to ADE.   The base of each triangle is a side of the base of the pyramid.   The lateral surface of the pyramid is obtained by multiplying the area of one of the triangles by 4.

Area of a triangle $= \frac{1}{2}$ base $\times$ height

If the pyramid is 6 feet high and the base is 2 feet square, then the base of the triangle $ADE = 2$ feet but the height line of the triangle is the line AB, called the slant height.

Figure 10 shows the pyramid with one quarter removed so that the actual height AC and the slant height AB can be seen. From this figure it is evident that triangle ABC is a right triangle with the slant height AB as the hypotenuse. The height of altitude, AC, of this right triangle is 6 feet, the height of the pyramid. The base, BC, of the triangle is half the distance across the square or $\frac{1}{2} \times 2 = 1$. The hypotenuse

$$AB = \sqrt{AC^2 + BC^2} = \sqrt{6^2 + 1^2} = \sqrt{37} = 6.08 \text{ feet}$$

the slant height, which is the height or altitude of the triangle ADE.

Area of triangle $ADE = \frac{1}{2} \times$ base $\times$ height
$$= \frac{1}{2} \times 2 \times 6.08 = 6.08 \text{ sq. ft.}$$

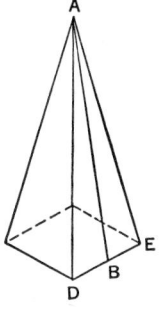

Fig. 9.

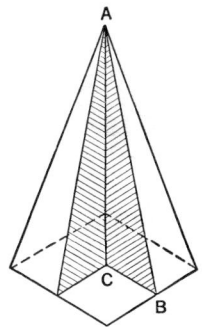

Fig. 10.

The lateral surface of the square pyramid is four times the area of one of the sides, therefore $4 \times 6.08 = 24.32$ square feet.

By the other formula,

Lateral surface $= \frac{1}{2} \times$ perimeter of base $\times$ slant height
$$= \frac{1}{2} \times (4 \times 2) \times 6.08 = 24.32 \text{ sq. ft.}$$

A cone is shown in Fig. 11 and its development in Fig. 12. The volume of a cone, like that of a pyramid, is one-third the volume of a cylinder of the same size, thus the formulas are similar to those used in pyramids.

Volume, $V = \frac{1}{3}h \times$ area of base

Lateral surface $= \frac{1}{2} \times$ perimeter of base $\times$ slant height

ILLUSTRATIONS: 1. Find the volume and the lateral surface of a cone the base of which is a circle 6 feet in diameter and whose height is 4 feet.

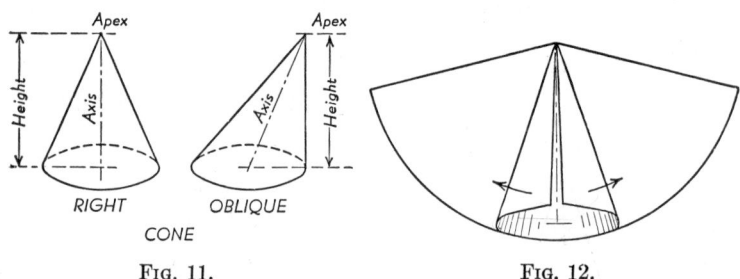

RIGHT   OBLIQUE

CONE

FIG. 11.   FIG. 12.

Area of base $= 0.7854d^2$

$= 0.7854 \times 6 \times 6 = 28.27$

$V = \frac{1}{3} \times 4 \times 28.27 = 37.7$ cu. ft.

Perimeter of base $= 3.1416 \times 6 = 18.8496$

Lateral surface $=$ perimeter of base $\times$ slant height

$= \frac{1}{2} \times 18.8496 \times \sqrt{3^2 + 4^2}$

$= 9.4248 \times \sqrt{25}$

$= 9.4248 \times 5$

$= 47.124$ sq. ft.

2. Making a conical ventilator top which will be 24 inches in diameter and 6 inches high. What shape of metal should be cut, allowing $1\frac{1}{2}$ inches for a lap joint?

First it is necessary to determine two other dimensions of the cone, the slant height and the circumference of the base.

The slant height $= \sqrt{6^2 + 12^2} = \sqrt{36 + 144}$
$$= \sqrt{180} = 13.4164 = 13\tfrac{7}{16} \text{ in.}$$

The circumference $= \pi \times 24 = 3.1416 \times 24 = 75.3984 = 75\tfrac{3}{8}$ in.

Then, using the slant height as a radius, draw a circle on the metal to be cut. The length $75\tfrac{3}{8}$ inches plus the $1\tfrac{1}{2}$ inches for lap may be measured off on the circumference of this circle and the metal cut.

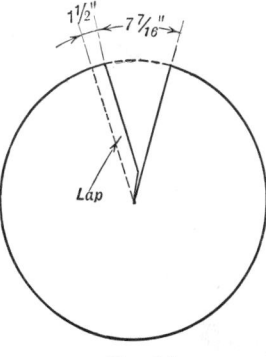

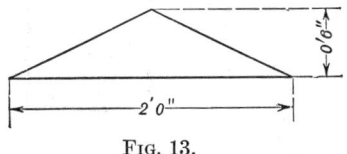

Fig. 13.

Fig. 14.

However, if the length of the circumference of the circle just drawn is computed, and the $75\tfrac{3}{8}$ plus $1\tfrac{1}{2}$ inches subtracted from this length, the difference provides a shorter measurement along the circumference. Thus

Circumference of flat circle

$$= 13.4164 \times 2 \times \pi = 84.294 = 84\tfrac{5}{16} \text{ in.}$$

Then, $84\tfrac{5}{16} - (75\tfrac{3}{8} + 1\tfrac{1}{2}) = 7\tfrac{7}{16}$ inches as the distance to be measured along the circumference.

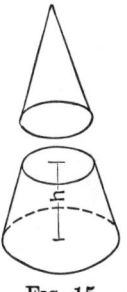

The part of a regular pyramid or of a cone which is left after its top has been cut off by a plane parallel to its base is called the frustum of the pyramid or cone. In practical work the frustum of a pyramid or cone has more applications than the pyramid or cone. The height is the shortest distance between the bases which are the base of the pyramid or cone and

Fig. 15.

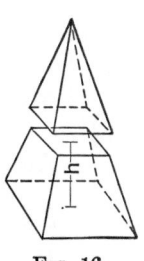

Fig. 16.

the section made by the cutting plane.   The lateral faces of a frustum of a regular pyramid are trapezoids.

The following formulas may be used:

$$\text{Volume, } V = \frac{h}{3} \ (B + b + \sqrt{B \times b})$$

where B = area of large base and b = area of small base.

Lateral surface = average perimeter of bases × slant height

ILLUSTRATIONS: 1. Find the volume of the frustum of a cone 5 inches high, the upper base being 4 inches and the lower base, 8 inches in diameter.

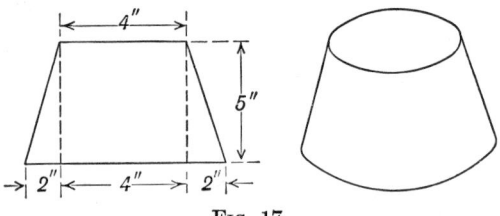

FIG. 17.

Area of upper base, b = 3.14 × 2 × 2 = 12.56
Area of lower base, B = 3.14 × 4 × 4 = 50.24

$$V = \frac{h}{3} \ (B + b + \sqrt{B \times b})$$

$$= \frac{5}{3} \ (50.24 + 12.56 + \sqrt{50.24 \times 12.55})$$

$$= \frac{5}{3} \times 87.92$$

$$= 146.6 \text{ cu. in.}$$

In illustration 1, find the lateral surface.

Perimeter of upper base, b = 3.14 × 8 = 25.12 inches
Perimeter of lower base, B = 3.14 × 4 = 12.56 inches
                                              37.68 inches

Average perimeter = $\dfrac{37.68}{2}$ = 18.84 in.

Slant height   = $\sqrt{5^2 + 2^2}$ = $\sqrt{29}$ = 5.38 in.
Lateral surface   = 18.84 × 5.38 = 101.36 sq. in.

Frequently a sheet metal pattern maker is required to design a container of a certain capacity and is required to calculate the height.

The volume formula is transposed to read:

$$h = \frac{3 \times \text{volume}}{B + b + \sqrt{B \times b}}$$

ILLUSTRATION: 2. A container shaped like the frustum of a cone is to contain 1 cubic foot. If the upper base is 12 inches in diameter and the lower base 16 inches in diameter find the height.

Area of $B = 3.14 \times 6 \times 6 = 113.04$ sq. in.

Area of $b = 3.14 \times 8 \times 8 = 200.96$ sq. in.

then,

$$h = \frac{3 \times \text{volume (in cubic inches)}}{B + b + \sqrt{B \times b}}$$

$$= \frac{3 \times 17.28}{200.96 + 113.10 + \sqrt{200.96 \times 113.04}}$$

$$= \frac{5184}{314.00 + \sqrt{22,739.886}} = \frac{5184}{464.07} = 11.19 \text{ in.}$$

Figure 18 is the frustum of a hexagonal pyramid and its development is shown in Fig. 19. The formulas used are the same as those used for the frustum of a cone, i.e.;

Volume, $V = \dfrac{h}{3} (B + b + \sqrt{B \times b})$

Lateral surface = Average perimeter of bases × slant height.

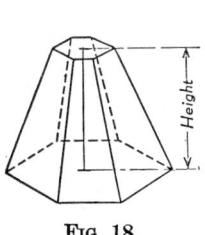

ILLUSTRATION: Find the volume of the frustum of a hexagonal pyramid 9 inches high, the side of the upper base being 2 inches and the side of the lower base 4 inches.

FIG. 18.

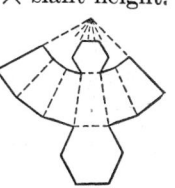

TRUNCATED HEXAGONAL PYRAMID

FIG. 19.

Area of upper base, $b = 2.5980 \times 2 \times 2 = 10.392$

Area of lower base, $B = 2.5980 \times 4 \times 4 = 41.568$

$$V = \frac{9}{3}(41.568 + 10.392 + \sqrt{10.392 \times 41.568})$$

$$= \frac{9}{3}(51.96 + \sqrt{431.9747})$$

$$= \frac{9}{3}(51.96 + 20.78)$$

$$= 218.22 \text{ cu. in.}$$

The following table may be used to lay out regular polygons and to calculate their area. Notice that 2.5980 used for finding the area of the bases in preceding problem is a constant taken from this table.

TABLE 1

ELEMENTS OF REGULAR POLYGONS

Number of sides	Name of figure	Diameter of circle that will just enclose when side is 1	Diameter of circle that will just go inside when side is 1	Length of side where diameter of enclosure circle equals 1	Length of side where inside circle equals 1	Angle formed by lines drawn from center to corners	Angle formed by outer sides of figures	To find area of figure multiply side by itself and by the number in this column
3	Triangle....	1.1546	0.5774	0.8660	1.7320	120°	60°	0.4330
4	Square.....	1.4142	1.000	0.7071	1.0000	90	90	1.0000
5	Pentagon...	1.7012	1.3764	0.5878	0.7265	72	108	1.7204
6	Hexagon...	2.0000	1.7320	0.5000	0.5774	60	120	2.5980
7	Heptagon...	2.3048	2.0766	0.4338	0.4815	51°–26′	128°–34′	3.6339
8	Octagon....	2.6132	2.4142	0.3827	0.4142	45°	135°	4.8284
9	Nonagon...	2.9238	2.7474	0.3420	0.3639	40	140	6.1818
10	Decagon....	3.2360	3.0776	0.3090	0.3247	36	144	7.6942
11	Undecagon..	3.5494	3.4056	0.2817	0.2936	32°–43′	147°–17′	9.3656
12	Dodecagon..	3.8638	3.7320	0.2858	0.2679	30°	150°	11.1961

A sphere is a solid in which all points on the surface are at the same distance from an internal point called the center. The volume and lateral surface may be found by the following formulas:

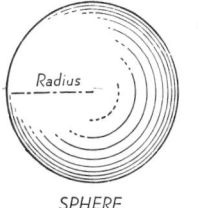

SPHERE

FIG. 20.

$$\text{Volume, } V = \frac{4\pi r^3}{3} = 4.1888r^3 \quad \text{or,}$$

$$\frac{\pi d^3}{6} = 0.5236d^3$$

$$\text{Lateral surface, } L = 4\pi r^2 = 3.1416d^2$$

ILLUSTRATION: Find the volume and lateral surface of a sphere $6\frac{1}{2}$ inches in diameter.

$$V = 0.5236d^3 = 0.5236 \times 6.5 \times 6.5 \times 6.5 = 143.79 \text{ cu. in.}$$
$$L = 3.1416d^2 = 3.1416 \times 6.5 \times 6.5 = 132.73 \text{ sq. in.}$$

There are problems in sheet metal work that occur in daily practice in which the rules of mensuration must be used before the pattern draftsman can make the development. Among these problems are transition pieces for heating, ventilating, blower and exhaust work together with the sizes and areas of outlets.

ILLUSTRATIONS:

1. Find the radius a tinsmith should use in laying out a circular hole for a pipe, the cross-section of which is 166 square inches. The formula $r = \sqrt{0.32A}$, which is derived from $A = \pi r^2$, can be used.

$$\text{radius} = \sqrt{0.32 \times \text{area}}$$
$$= \sqrt{0.32 \times 166}$$
$$= \sqrt{53.12} = 7\frac{7}{32} \text{ in.}$$

2. A tinsmith is required to make some cylindric cans to hold 1 gallon (231 cu. in.) each and to be 8 inches high. What radius should be used in laying out the base?

Allowance for seams are neglected.

$$\text{radius} = \sqrt{\frac{231}{8} \times 0.32}$$

$$= \sqrt{9.24} = 3\frac{1}{32} \text{ in.}$$

3. Find the height of a flaring measure required to hold 3 gallons and whose top diameter is 7 inches, the bottom diameter $11\frac{1}{2}$ inches and the diameter in the center is $9\frac{1}{4}$ inches.

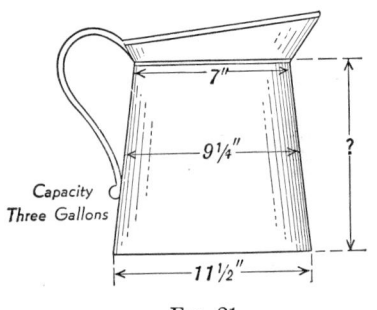

FIG. 21.

RULE.—Divide the capacity in cubic inches by the sum of the areas of the top and bottom diameters plus 4 times the area of the center section.   Then multiply the quotient by 6.

Capacity $= 3 \times 231 = 693$ cu. in.
Area of top $= 7 \times 7 \times 0.7854 = 38.485$ sq. in.
Area of bottom $= 11.5 \times 11.5 \times 0.7854 = 103.87$ sq. in.
Area of middle section $= \dfrac{7 + 11.5}{2} = 9.25,\ 9.25 \times 9.25 \times 0.7854$

$$= 67.20 \text{ sq. in.}$$

$$67.20 \times 4 = 268.80$$
$$38.485 + 103.87 + 268.80 = 411.155$$

$693 \div 411.55 = 1.684,\ 1.684 \times 6 = 10.11$ in.   or $10\frac{1}{8}$ inches, the required height of the measure.

The table of areas and circumferences of circles page 683 is convenient for finding the area of circular-shaped vessels without computation. In the preceding case to find the area of the top, look under the column headed diameter and after 7 to the right read 38.4846 in the area column.

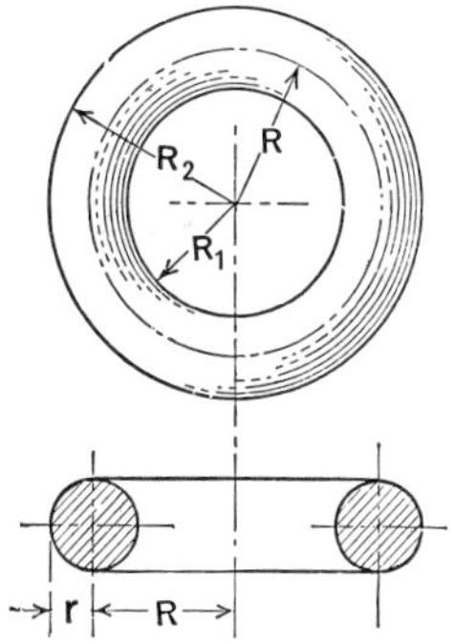

4. Find the volume and lateral surface of a cylindric ring whose outside diameter is 12 inches and whose inside diameter is 8 inches.

$$\text{Volume} = \pi r^2 \times (2\pi R)$$

where,

$$R = \frac{R_1 + R_2}{2} = \frac{4 + 6}{2} = 5$$

$$r = \frac{R_2 - R_1}{2} = \frac{6 - 4}{2} = 1$$

Fig. 22.

Then,

$$V = (3.14 \times 1 \times 1) \times (2 \times 3.14 \times 5)$$
$$= 3.14 \times 31.4 = 98.6 \text{ cu. in.}$$

Lateral Surface, $L = (2\pi r) \times (2\pi R)$
$$= (2 \times 3.14 \times 1) \times (2 \times 3.14 \times 5)$$
$$= 6.28 \times 31.4 = 197.2 \text{ sq. in.}$$

5. In the offset boot shown in Fig. 23, find the length of the rectangular pipe in order that its dimension will equal the area of the 10-inch round pipe if the width of the rectangular pipe is 4 inches.

Area of round pipe $= \pi r^2 = 3.14 \times 5 \times 5 = 78.5$ sq. in.

Then, $78.5 \div 4 = 19.625$ or $19\frac{5}{8}$ in.

Therefore, the size of the rectangular riser of equal area to the 10-inch round pipe is 4 in. $\times$ $19\frac{5}{8}$ in.

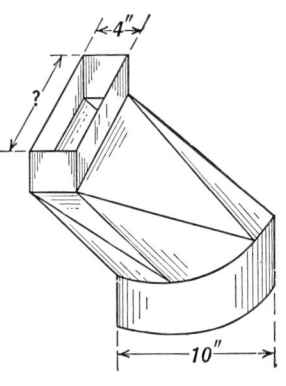

Fig. 23.

6. Find the volume and lateral surface of a ring with a 6-inch diameter and a square cross-section of 1 inch.

$$\text{Volume} = \pi H(R_2{}^2 - R_1{}^2)$$
$$= 3.14 \times 1(3 \times 3 - 2 \times 2)$$
$$= 3.14 \times 5 = 15.7 \text{ cu. in.}$$
$$\text{Lateral surface} = 2(B + H)(2\pi R)$$
$$= 2(1 + 1)(2 \times 3.14 \times 2.5)$$
$$= 4 \times 15.7 = 62.8 \text{ sq. in.}$$

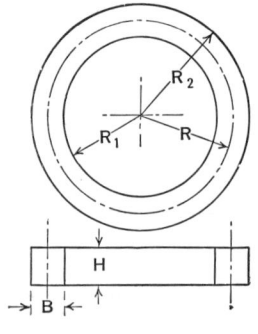

Fig. 24.

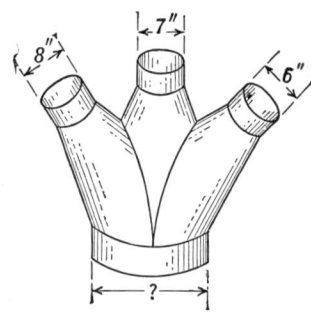

Fig. 25.

7. Find the diameter of a main pipe whose capacity will equal the combined capacity of three branches whose diameters are 6 inches, 7 inches, and 8 inches respectively.

Areas of circles vary as the squares of their diameters, therefore,

$$\text{Diameter of main pipe} = \sqrt{8^2 + 7^2 + 6^2}$$
$$= \sqrt{149} = 12.2 \text{ or } 12\tfrac{1}{4} \text{ in.}$$

The square root can be found in table of squares, square roots, etc. on page 29.

Another method which makes use of the tables of areas of circles on page 683 is:

Area of 6″ pipe = 28.2744 square inches
Area of 7″ pipe = 38.4846 square inches
Area of 8″ pipe = 50.2656 square inches

Combined areas = 117.0246 square inches

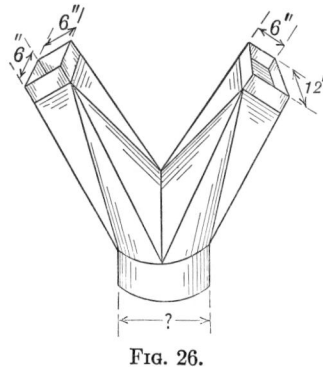

Taking the nearest number in the table to 117.0246, i.e., 117.859, which is the diameter of a pipe $12\frac{1}{4}$ inches.

8. Find the diameter of a round main to equal the area of one square and one rectangular branch of a two-branched prong, Fig. 26, when one branch is 6 inches square and the other measures 6 inches × 12 inches.

Fig. 26.

Area of square branch = 6 × 6 = 36 square inches
Area of rectangular branch = 6 × 12 = 72 square inches

Combined area = 108 square inches

The nearest calculation from the table on page 683 is 108.43, the diameter of a $11\frac{3}{4}$-inch circle.

9. Find the missing dimension of a rectangular pipe of area equal to that of two round branches, 8 inches and 12 inches in diameter when one side of the rectangular pipe measures 10 inches. See Fig. 27.

From the table,

The area of the 8-inch pipe = 50.265 square inches
The area of the 12-inch pipe = 113.098 square inches

The combined area = 163.363 square inches

The missing dimension will be, 163.363 ÷ 10 or 16.33 = $16\frac{5}{16}$ in.

10. Find the size of a square pipe having an area equal to that of two round branches whose diameters are $13\frac{5}{8}$ inches and 16 inches respectively. See Fig. 28.

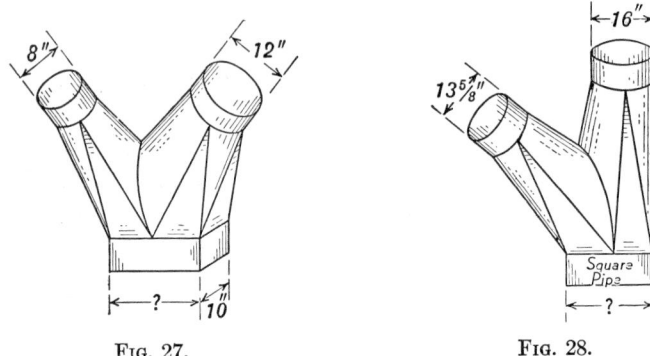

FIG. 27.   FIG. 28.

The area of the $13\frac{5}{8}$-inch pipe = 145.80 square inches
The area of the 16-inch pipe   = 201.06 square inches

The combined area = 346.86 square inches

The $\sqrt{346.86}$ = 18.6 or $18\frac{5}{8}$ inches, thus making the required size of the square main pipe, $18\frac{5}{8}$ in. × $18\frac{5}{8}$ in.

11. Find the increased sizes of the ducts. **A, B, and C,** shown in the ventilating system, in order to take care of the 8-inch, 10-inch and 15-inch branches respectively.

From the tables of areas of circles, page 683.

Area of 6-inch pipe = 28.2743 square inches
Area of 8-inch pipe = 50.2655 square inches

Combined area = 78.5398 square inches

this is the area of a 10-inch circle. Thus pipe A should have a diameter of 10 inches.

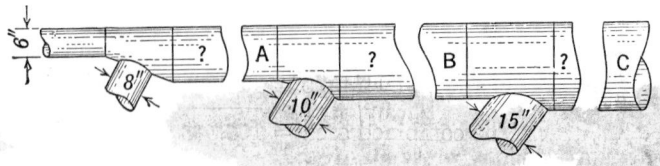

FIG. 29.

Area of B = area of pipe A + area of 10-inch branch. Because both are 10 inches in diameter, $78.539 \times 2 = 157.078$ square inches, the combined areas. From the table, page 683, the nearest number to 157.078 is 159.48, the area of a $14\frac{1}{4}$-inch circle. Thus, pipe B should have a diameter of $14\frac{1}{4}$ inches.

Area of C = area of pipe B + area of 15-inch branch
$$= 159.48 + 176.71$$
$$= 336.19 \text{ sq. in.}$$

From the table, page 683, find 336.19 equal to $20\frac{11}{16}$. Thus pipe C should have a diameter of $20\frac{11}{16}$ inches.

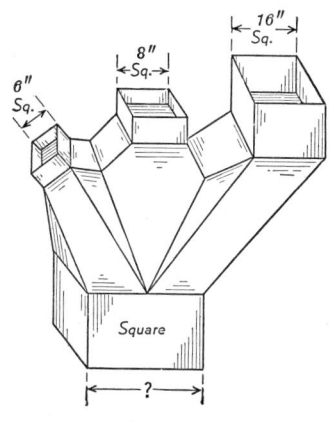

Fɪɢ. 30.

Find the size of a square main in a three-branched fitting whose outlets are, 16 inches $\times$ 16 inches, 8 inches $\times$ 8 inches and 6 inches $\times$ 6 inches respectively. From the tables, page 29,

Area of 6-inch $\times$ 6-inch outlet = 36 square inches
Area of 8-inch $\times$ 8-inch outlet = 64 square inches
Area of 16-inch $\times$ 16-inch outlet = 256 square inches

Combined area = 356 square inches

Side of square main = $\sqrt{356}$ = 18.86.

In practical work the size of the square would be taken as $18\frac{7}{8}$ inches $\times$ $18\frac{7}{8}$ inches.

13. Find the dimensions of a rectangular vertical flue having an equal area to the combined areas of five horizontal vent ducts shown in Fig. 31.

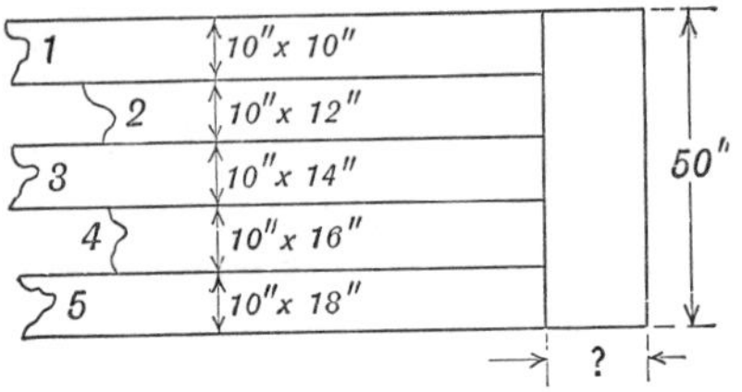

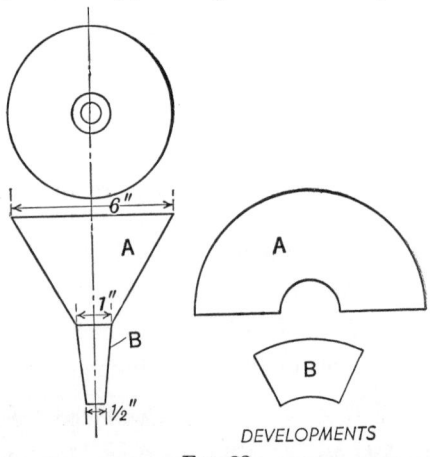

FIG. 31.

Area of first duct $= 10 \times 10 = 100$ square inches
Area of second duct $= 10 \times 12 = 120$ square inches
Area of third duct $= 10 \times 14 = 140$ square inches
Area of fourth duct $= 10 \times 16 = 160$ square inches
Area of fifth duct $= 10 \times 18 = \underline{180}$ square inches

Combined area $= \overline{700}$ square inches

As all ducts are set in 10-inch way, the space taken up is $5 \times 10 = 50$ inches.

Therefore, $700 \div 50 = 14$. Thus the flue with the required area will be 14 inches $\times$ 50 inches.

14. Find the amount of tin required for the funnel shown in the sketch if the slant height of the upper piece is 4.5 inches, and the slant height of the lower piece is 3.5 inches. Allow $\frac{1}{2}$ inch on the length and width of each piece for locks.

DEVELOPMENTS

FIG. 32.

The formula from page 651.

Lateral surface = average perimeter of bases × slant height

Perimeter of upper large base = 3.14 × 6 = 18.84 inches

Perimeter of upper small base = 3.14 × 1 = 3.14 inches

$$\overline{21.98\ \text{inches}}$$

Average perimeter = $\dfrac{21.98}{2}$ = 10.99 inches

Perimeter of lower large base = 3.14 × 1 = 3.14 inches

Perimeter of lower small base = 3.14 × 0.5 = 1.57 inches

Average perimeter = $\dfrac{4.71}{2}$ = 2.35 inches

Upper slant height $4\frac{1}{2}$ inches + $\frac{1}{2}$ inch = 5 inches

Lower slant height $3\frac{1}{2}$ inches + $\frac{1}{2}$ inch = 4 inches

10.99 + 0.5 = 11.49 × 5 = 57.45 square inches

 2.35 + 0.5 = 2.85 × 4 = 11.40 square inches

$$\overline{68.85\ \text{square inches}}$$

Therefore 68.85 square inches of tin are required.

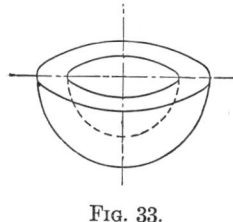

15. Find the volume of the hemispherical bowl shown in Fig. 33, when the outside diameter is 12 inches and the inside diameter is 8 inches.

Aid. Treat the solid as the differences of two hemispheres.

FIG. 33.

From page 654. Volume of a sphere = $0.5236d^3$

whence V = 0.5236 × 12 × 12 × 12 = 904.78 cubic inches

V = 0.5236 × 8 × 8 × 8   = 268.08 cubic inches

904.72 ÷ 2 = 452.36 cubic inches in outside hemisphere

268.08 ÷ 2 = 134.04 cubic inches in inside hemisphere

$$\overline{318.32\ \text{cubic inches, the volume of the bowl.}}$$

TABLE 2.—Capacity of Tanks in United States Gallons

Decimal Equivalents of Fractional Parts of a Gallon

0.03125 of a gallon	=	1	gill		0.53125 of a gallon	=	17	gills
0.06250 " " "	=	½	pint		0.56250 " " "	=	4½	pints
0.09375 " " "	=	3	gills		0.62500 " " "	=	5	pints
0.12500 " " "	=	1	pint		0.59375 " " "	=	19	gills
0.15625 " " "	=	5	gills		0.65625 " " "	=	21	gills
0.18750 " " "	=	1½	pints		0.68750 " " "	=	5½	pints
0.21875 " " "	=	7	gills		0.71875 " " "	=	23	gills
0.2500 " " "	=	1	quart		0.75000 " " "	=	3	quarts
0.28125 " " "	=	9	gills		0.78125 " " "	=	25	gills
0.31250 " " "	=	2½	pints		0.81250 " " "	=	6½	pints
0.34375 " " "	=	11	gil s		0.84375 " " "	=	27	gills
0.37500 " " "	=	3	pin.s		0.87500 " " "	=	7	pints
0.40625 " " "	=	13	gills		0.90625 " " "	=	29	gills
0.43750 " " "	=	3½	pi nts		0.93750 " " "	=	7½	pints
0.46875 " " "	=	15	gills		0.968750 " " "	=	31	gills
0.50000 " " "	=	½	gallon		1.00000 " " "	=	1	gallon

Tin Roofing.—Pure block tin is not used for common building purposes; but thin plates of sheet iron covered with it on both sides constitute the *tinned plates*, or, as they are called, the *tin,* used for covering roofs, rain pipes and many domestic utensils.   For roofs it is laid on boards.

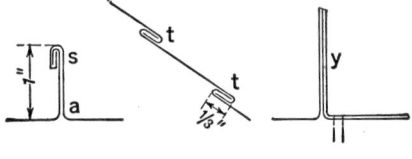

Fig. 34.

The sheets of tin are united as shown in Fig. 34.   First, several sheets are joined together in the shop, end for end, as at *tt,* by being first bent over, then hammered flat, and then soldered. These are then formed into a roll to be carried to the roof, a roll being long enough to reach from the peak to the eaves.   Different rolls being spread up and down the roof are then united along

TYPES OF JOINTS & EDGES
FOR SHEET METAL

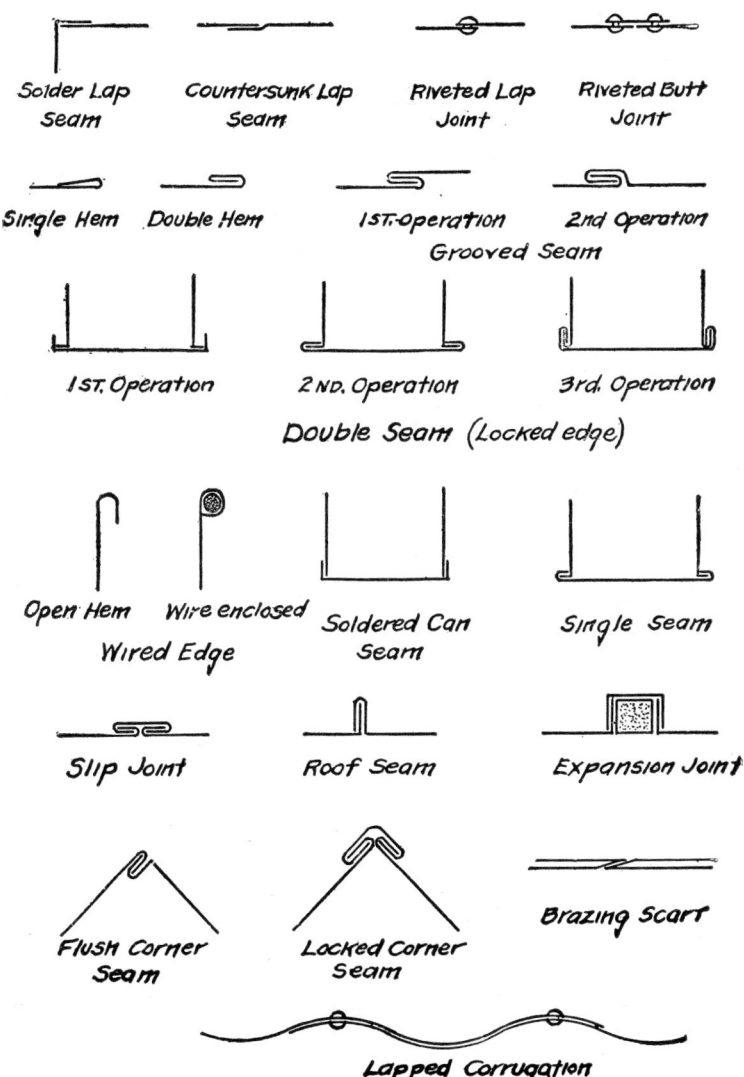

FIG. 35.—Types of Joints and Edges for Sheet Metal.

their sides by simply being bent as at a and s, by a tool for that purpose. The roofers call the bending at s a *double groove*, or *double lock*; and the more simple ones at t, a *single groove*, or lock.

To hold the tin securely to the sheeting boards, pieces of the tin 3 or 4 inches long by 2 inches wide, called cleats, are nailed to the boards at about every 18 inches along the joints of the rolls that are to be united, and are bent over with the double groove s. This will be understood from y, where the middle piece is the cleat.

Flat-Seam Tin Roofing.—When a sheet of tin 14 × 20 inches with $\frac{1}{2}$-inch edges is edged or folded, it will measure 13 inches × 19 inches or 247 square inches of area. However, when this sheet is joined to other sheets on the roof, its covering capacity is only $12\frac{1}{2}$ inches × $18\frac{1}{2}$ inches or 231.25 square inches. A box of 112 sheets, 14 inches × 20 inches, laid this way will cover, approximately, 180 square feet.

ILLUSTRATION: Find the number of sheets of tin 14 inches × 20 inches required for one square (100 square feet) using flat seams with $\frac{1}{2}$-inch edge.

100 × 144 (the number of square inches in a square foot) = 14,400
$$14,400 \div 231.25 = 63$$

Standing-Seam Tin Roofing.—When standing-seams edged $1\frac{1}{4}$ inches and $1\frac{1}{2}$ inches are used $2\frac{3}{4}$ inches is taken off the width; and the flat cross-seams edged $\frac{3}{8}$ inches take $1\frac{1}{8}$ inches off the length of the sheet. The covering capacity of each 14 × 20-inch sheet is, therefore, $11\frac{1}{4}$ inches × $18\frac{7}{8}$ inches or 212.34 square inches. A box of 112 sheets 14 inches × 20 inches laid in this will cover 165 square feet.

ILLUSTRATION: Find the number of sheets of tin required for one square when using standing seams.

$$14,400 \div 212.34 = 68$$

NOTE: The weight of sheet metal is calculated on page 272 in the chapter on weights and measures.

Corrugated sheets of iron and steel are used not only for roofing but also for siding of sheds, mills and other structures. These sheets are carried in stock in 4-foot, 5-foot, 6-foot, 8-foot, 9-foot and 10-foot lengths, the 8-foot length being the most commonly used. The usual width of sheets is 24 inches between the centers of the outer corrugations, so that the covering width is 24 inches when one corrugation is used for the side lap. Ordinary corrugated sheets should have a lap $1\frac{1}{2}$ or 2 corrugations side lap for roofing in order to secure water-tight side seams. For covering roofs, either 3-inch, $2\frac{1}{2}$-inch or 2-inch corrugations should be used, the 2-inch corrugation being the most common size. No. 28 gage corrugated iron is generally used for applying to wooden buildings. When laid on a roof, corrugated sheets should have a lap on the lower end from 3 to 6 inches, according to the pitch of the roof.

TABLE 3

NUMBER OF SQUARE FEET OF CORRUGATED SHEETS TO COVER
100 SQUARE FEET OF ROOF

End Laps..............	1 inch	2 inches	3 inches	4 inches	5 inches
	Sq. Ft.	Sq. Ft.	Sq. Ft.	Sq. Ft.	Sq. Ft.
Side lap, 1 corrugation...	110	111	112	113	114
Side lap, $1\frac{1}{2}$ corrugations...	116	117	117	119	120
Side lap, 2 corrugations...	123	124	125	126	127

CORRUGATED IRON ROOFING

B. W. Gauge	Weight per square (100 square feet). Plain	Galvanized
Number	Pounds	
28	97	
26	105	Weights from 5 to 15 per
24	128	cent heavier than plain,
22	150	according to the number
20	185	B. W. G.
18	270	
16	340	

Allow one-third the net width for lapping and for corrugations. From $2\frac{1}{2}$ to $3\frac{1}{2}$ pounds for rivets will be required per square.

The best plates, both for tinning and for ternes, are made of charcoal iron, which, being tough, bears bending better. Coke is used for making cheaper plates, but is inferior as regards bending.

Much use is made of what is called leaded tin, or ternes, for roofing. It is simply sheet iron coated with an alloy of lead and tin, lead being less expensive. In one standard brand the alloy is 32% tin, 68% lead.

TABLE 4

GALVANIZED SHEET IRON

Am. Galv. Iron Assn. B. W. G.

No.	Ounces avoir. per square foot	Square feet per 2240 pounds	No.	Ounces avoir. per square foot	Square feet per 2240 pounds	No.	Ounces avoir. per square foot	Square feet per 2240 pounds
29	12	2987	24	17	2108	19	33	1084
28	13	2757	23	19	1886	18	38	943
27	14	2560	22	21	1706	17	43	833
26	15	2389	21	24	1493	16	48	746
25	16	2240	20	28	1280	14	60	597

TABLE 5.—TIN REQUIRED FOR FLAT SEAMS

No. of square feet..	100	110	120	130	140	150	160	170	180	190	200
Sheets required.....	63	69	75	81	88	94	100	106	112	119	125
No. of square feet..	210	220	230	240	250	260	270	280	290	300	310
Sheets required.....	131	137	144	150	156	162	169	175	181	187	193
No. of square feet..	320	330	340	350	360	370	380	390	400	410	420
Sheets required.....	200	206	212	218	224	231	237	243	249	256	262
No. of square feet..	430	440	450	460	470	480	490	500	510	520	530
Sheets required.....	268	274	281	287	293	299	305	312	318	324	330
No. of square feet..	540	550	560	570	580	590	600	610	620	630	640
Sheets required.....	337	343	349	355	362	368	374	380	386	393	396
No. of square feet..	650	660	670	680	690	700	710	720	730	740	750
Sheets required.....	405	411	418	424	430	436	442	448	455	461	467
No. of square feet..	760	770	780	790	800	810	820	830	840	850	860
Sheets required.....	474	480	486	492	499	505	511	517	523	530	536
No. of square feet..	870	880	890	900	910	920	930	940	950	960	970
Sheets required.....	542	548	554	561	567	573	579	586	592	598	604
No. of square feet..	980	990	1000	....	....	....	....	....	....	....	....
Sheets required.....	610	617	625	....	....	....	....	....	....	....	....

A box of 112 sheets 14 by 20 in laid in this way will cover 180 sq ft.

TABLE 5.—TIN REQUIRED FOR FLAT SEAMS—*Continued*

No. of square feet..	100	110	120	130	140	150	160	170	180	190	200
Sheets required.....	30	33	36	39	42	45	47	50	53	56	59
No. of square feet..	210	220	230	240	250	260	270	280	290	300	310
Sheets required.....	62	65	68	71	74	77	80	83	86	89	92
No. of square feet..	320	330	340	350	360	370	380	390	400	410	420
Sheets required.....	94	97	100	103	106	109	112	115	118	121	124
No. of square feet..	430	440	450	460	470	480	490	500	510	520	530
Sheets required.....	127	130	133	136	139	141	144	147	150	153	156
No. of square feet..	540	550	560	570	580	590	600	610	620	630	640
Sheets required.....	159	162	165	168	171	174	177	180	183	186	188
No. of square feet..	650	660	670	680	690	700	710	720	730	740	750
Sheets required.....	191	194	197	200	203	206	209	212	215	218	221
No. of square feet..	760	770	780	790	800	810	820	830	840	850	860
Sheets required.....	224	227	230	233	235	238	241	244	247	250	253
No. of square feet..	870	880	890	900	910	920	930	940	950	960	970
Sheets required...	256	259	262	265	268	271	274	277	280	282	285
No. of square feet..	980	990	1000	....	....	....	....	....	....	....	....
Sheets required.....	288	291	294	....	....	....	....	....	....	....	....

A box of 112 sheets 28 by 20 in laid in this way will cover 381 sq ft.

TABLE 6.—TIN REQUIRED FOR STANDING SEAMS

No. of square feet..	100	110	120	130	140	150	160	170	180	190	200
Sheets required.....	68	75	82	89	95	102	109	116	123	129	136
No. of square feet..	210	220	230	240	250	260	270	280	290	300	310
Sheets required.....	143	150	156	163	170	177	184	190	197	204	211
No. of square feet..	320	330	340	350	360	370	380	390	400	410	420
Sheets required.....	218	224	231	238	245	251	258	265	271	279	285
No. of square feet..	430	440	450	460	470	480	490	500	510	520	530
Sheets required.....	292	299	306	312	319	326	333	340	346	353	360
No. of square feet..	540	550	560	570	580	590	600	610	620	630	640
Sheets required.....	367	374	379	387	393	401	407	414	421	428	435
No. of square feet..	650	660	670	680	690	700	710	720	730	740	750
Sheets required.....	441	447	455	462	468	475	482	489	495	501	509
No. of square feet..	760	770	780	790	800	810	820	830	840	850	860
Sheets required.....	515	523	529	536	543	550	557	563	570	577	584
No. of square feet..	870	880	890	900	910	920	930	940	950	960	970
Sheets required.....	590	597	604	611	618	623	630	637	644	651	658
No. of square feet..	980	990	1000	....	....	....	....	....	....	....	....
Sheets required.....	665	672	679	....	....	....	....	....	....	....	....

A box of 112 sheets 14 by 20 in laid in this way will cover 165 sq ft.

TABLE 6.—Tin Required for Standing Seams—*Continued*

No. of square feet..	100	110	120	130	140	150	160	170	180	190	200
Sheets required.....	32	35	38	41	44	47	50	53	56	59	62
No. of square feet..	210	220	230	240	250	260	270	280	290	300	310
Sheets required.....	65	68	71	74	77	80	84	87	90	94	97
No. of square feet..	320	330	340	350	360	370	380	390	400	410	420
Sheets required.....	100	103	106	109	112	115	118	121	125	128	131
No. of square feet..	430	440	450	460	470	480	490	500	510	520	530
Sheets required.....	134	137	141	144	147	150	153	156	159	162	165
No. of square feet...	540	550	560	570	580	590	600	610	620	630	640
Sheets required.....	168	171	174	177	180	184	187	190	193	196	199
No. of square feet..	650	660	670	680	690	700	710	720	730	740	750
Sheets required.....	202	205	208	211	214	218	221	224	227	230	233
No. of square feet..	760	770	780	790	800	810	820	830	840	850	860
Sheets required.....	236	239	242	245	249	252	255	258	261	265	268
No. of square feet..	870	880	890	900	910	920	930	940	950	960	970
Sheets required.....	271	274	277	280	283	286	289	292	296	299	302
No. of square feet...	980	990	....	....	....	....	....	....	....	....	....
Sheets required.....	305	308	....	....	....	.. .	....	....	....	....	....

A box of 112 sheets 28 by 20 in laid in this way will cover 360 sq ft.

In giving orders, it is important to specify whether charcoal plates or coke ones are required; also whether *tinned* plates, or *ternes*.

Tinned and leaded sheets of Bessemer and other cheap steel are now much used. They are sold at about the price of charcoal tin and terne plates.

If the tin is laid with a flat-seam or flat lock, the roof should have an incline of $\frac{1}{2}$ inch or more to a foot. If laid with a standing seam, there should be an incline of not less than 2 inches to a foot.

This is put up in rolls 14, 20, and 28 inches wide for the convenience of roofers. Each roll contains 108 square feet. The following table shows the number of sheets required per lineal foot for 20- and 28-inch widths.

Roof Flashings.—Flashings are pieces of tin, lead or copper, let into the joints of a wall so as to lap over gutters and in places where leaks are likely to occur such as around chimneys, dormers, skylights and in valleys. In shingle work, the valley flashings are usually 14 inches wide, while the length depends upon the length

TABLE 7.—Sizes and Weight of Sheet Tin

Mark	Number of sheets in box	Dimension		Weight of box, pounds
		Length, inches	Breadth, inches	
1C................	225	13¾	10	112
11C...............	225	13¼	9¾	105
111C..............	225	12¾	9½	98
1X................	225	13¾	10	140
1XX...............	225	13¾	10	161
1XXX.............	225	13¾	10	182
1XXXX..........	225	13¾	10	203
DC...............	100	16¾	12½	105
DX...............	100	16¾	12½	126
DXX.............	100	16¾	12½	147
DXXX...........	100	16¾	12½	168
DXXXX..........	100	16¾	12½	189
5DC..............	200	15	11	168
5DX..............	200	15	11	189
5DXX............	200	15	11	210
5DXXX..........	200	15	11	231
1CW..............	225	13¾	10	112

A box containing 225 sheets, 13¾ by 10, contains 214.84 square feet; but allowing for seams it will cover only 150 square feet of roof.

of the valley. The sides of dormers, chimneys and all intersections are flashed with tin cut so as to turn up 3½ inches on the vertical and 3 inches on the roof. Flashings are measured by the number of square feet.

There are also in use for roofing, certain compound metals which resist tarnish better than either lead, tin, or zinc but which are so fusible as to be liable to be melted by large burning cinders falling on the roof from a neighboring conflagration.

A roof covered with tin or other metal should, if possible, slope not much *less* than five degrees, or about an inch to a foot;

TABLE 8.—Tin in Rolls or Gutter Strips

Feet	Widths 20	Widths 28	Feet	Widths 20	Widths 28	Feet	Widths 20	Widths 28	Hundred feet	Widths 20	Widths 28
1	1	1	35	16	23	69	31	44	2	89	128
2	1	2	36	16	23	70	32	45	3	134	192
3	2	2	37	17	24	71	32	45	4	178	256
4	2	3	38	17	24	72	32	46	5	223	320
5	3	4	39	18	25	73	33	47	6	267	384
6	3	4	40	18	26	74	33	47	7	312	444
7	4	5	41	19	27	75	34	48	8	356	512
8	4	5	42	19	27	76	34	48	9	401	576
9	4	6	43	20	28	77	35	49	10	445	640
10	5	7	44	20	28	78	35	50	11	495	704
11	5	7	45	20	29	79	36	50	12	540	768
12	6	8	46	21	29	80	36	51	13	585	832
13	6	9	47	21	30	81	36	52	14	630	896
14	7	9	48	22	31	82	37	52	15	675	960
15	7	10	49	22	31	83	37	53	16	720	1 024
16	8	11	50	23	32	84	38	54	17	765	1 088
17	8	11	51	23	33	85	38	54	18	810	1 152
18	8	12	52	24	33	86	39	55	19	855	1 216
19	9	12	53	24	34	87	39	55	20	900	1 280
20	9	13	54	24	34	88	40	56	21	945	1 344
21	10	14	55	25	35	89	40	57	22	990	1 408
22	10	14	56	25	36	90	40	57	23	1 035	1 472
23	11	15	57	26	36	91	41	58	24	1 080	1 536
24	11	16	58	26	37	92	41	59	25	1 135	1 600
25	12	16	59	27	38	93	42	59	26	1 170	1 664
26	12	17	60	27	38	94	42	60	27	1 215	1,738
27	12	18	61	28	39	95	43	61	28	1 260	1 792
28	13	18	62	28	40	96	43	62	29	1 305	1 856
29	13	19	63	28	40	97	44	62	30	1 350	1 920
30	14	19	64	29	41	98	44	63	31	1 395	1 984
31	14	20	65	29	41	99	44	64	32	1 440	2 048
32	15	21	66	30	42	100	45	64	33	1 485	2 112
33	15	21	67	30	43	.....	.....	....	34	1 530	2 176
34	16	22	68	31	43	.....	.....	....	35	1 575	2 240

and at the eaves there should be a sudden fall into the rain-gutter, to prevent rain from backing up so as to overtop the double-groove joint s, and thus cause leaks. When coal is used for fuel, tin roofs should receive two coats of paint when first put up, and a coat every 2 or 3 years after. Where wood only is used, this is not necessary; and a tin roof with a good pitch will last 20 or 30 years.

Two good workmen can put on, and paint outside, from 250 to 300 square feet of tin roof, per day of 8 hours.

Tinned iron plates are sold by the box. These boxes, unlike glass, have *not* equal areas of contents. They may be designated or ordered either by their names or sizes. Many makers, however, have their private brands in addition; and some of these have a much higher reputation than others.

TABLE 9.—Weights of Sheet Steel and Iron
United States Standard Gage

Number of Gage	App. Thickness	Weight per Sq. Foot		No. of Gage	App. Thickness	Weight per Sq. Foot	
		Steel	Iron			Steel	Iron
0000000	.5	20.320	20.00	17	.05625	2.286	2.25
000000	.46875	19.050	18.75	18	.05	2.032	2.
00000	.4375	17.780	17.50	19	.04375	1.778	1.75
0000	.40625	16.510	16.25	20	.0375	1.524	1.50
000	.375	15.240	15.00	21	.03437	1.397	1.375
00	.34375	13.970	13.75	22	.03125	1.270	1.25
0	.3125	12.700	12.50	23	.02812	1.143	1.125
1	.28125	11.430	11.25	24	.025	1.016	1.
2	.26562	10.795	10.625	25	.02187	.903	.875
3	.25	10.160	10.00	26	.01875	.762	.75
4	.23437	9.525	9.375	27	.01718	.698	.687
5	.21875	8.890	8.75	28	.01562	.635	.623
6	.20312	8.255	8.125	29	.01406	.571	.562
7	.1875	7.620	7.5	30	.0125	.508	.5
8	.17187	6.985	6.875	31	.01093	.440	.437
9	.15625	6.350	6.25	32	.01015	.413	.406
10	.14062	5.715	5.625	33	.00937	.381	.375
11	.125	5.080	5.00	34	.00859	.349	.343
12	.10937	4.445	4.375	35	.00781	.317	.312
13	.09375	3.810	3.75	36	.00703	.285	.281
14	.07812	3.175	3.125	37	.00664	.271	.265
15	.0703	2.857	2.812	38	.00625	.254	.25
16	.0625	2.540	2.50				

Weight of 1 cubic foot is assumed to be 487.7 lbs. for steel plates and 480 lbs. for iron plates.

Aluminum Roofing.—Aluminum roofing and siding have become very popular, especially for farm and industrial buildings. As a roofing and siding material, aluminum offers a number of important advantages. It resists normal outdoor exposure, it does not need painting, it does not stain or streak painted buildings, and for decorative purposes it can be painted after about three months' exposure to weather. The lighter weight of aluminum roofing also makes possible economies in the supporting framework of a building.

In the summer, aluminum reflects the sun's rays and helps to keep the building interior cooler. In the winter it helps to retain heat inside buildings. Another useful feature of aluminum is that it will not adversely affect the purity of rain water for cistern collection. In installing aluminum roofing or siding, aluminum nails with neoprene washers are generally used.

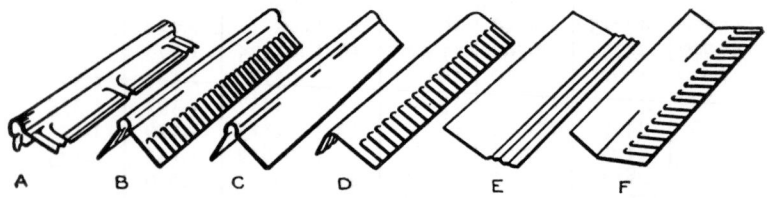

A   B   C   D   E   F

Fɪɢ. 36.—Typical Accessories for Finishing Roofing Jobs.

For roofing and siding applications, aluminum sheet is corrugated or crimped to add rigidity and strength. It comes in sheets of various sizes and thicknesses. Typical types available are 1¼″ corrugated, 0.019″ thick (equivalent to 26 gage steel); 2½″ corrugated, 0.024″ thick (equivalent to 24 gage steel); and a V-crimp style, 0.019″ thick (equivalent to 26 gage steel). All of these types are available in sheets 26 inches wide by 6, 7, 8, 10, and 12 feet long. A wide variety of accessories such as ridge rolls to cover the top of a ridge on finished roofing jobs, side wall and end wall flashings, and narrow rolls of aluminum for valleys and flashings are available to simplify roofing finishing. Some of these are shown in Fig. 36.

TABLE 10

Gauge No.	American or Brown & Sharpe's	Birmingham or Stubs	Wash.&Moen	Imperial S. W. G.	London or Old English	United States Standard	Gauge No.
0000000			.490	.500		.500	0000000
000000	.5800		.460	.464		.46875	000000
00000	.5165		.430	.432		.4375	00000
0000	.4600	.454	.3938	.400	.454	.40625	0000
000	.4096	.425	.3625	.372	.425	.375	000
00	.3648	.380	.3310	.348	.38	.34375	00
0	.3249	.340	.3065	.324	.34	.3125	0
1	.2893	.300	.2830	.300	.3	.28125	1
2	.2576	.284	.2625	.276	.284	.265625	2
3	.2294	.259	.2437	.252	.259	.25	3
4	.2043	.238	.2253	.232	.238	.234375	4
5	.1819	.220	.2070	.212	.22	.21875	5
6	.1620	.203	.1920	.192	.203	.203125	6
7	.1443	.180	.1770	.176	.18	.1875	7
8	.1285	.165	.1620	.160	.165	.171875	8
9	.1144	.148	.1483	.144	.148	.15625	9
10	.1019	.134	.1350	.128	.134	.140625	10
11	.09074	.120	.1205	.116	.12	.125	11
12	.08081	.109	.1055	.104	.109	.109375	12
13	.07196	.095	.0915	.092	.095	.09375	13
14	.06408	.083	.0800	.080	.083	.078125	14
15	.05707	.072	.0720	.072	.072	.0703125	15
16	.05082	.065	.0625	.064	.065	.0625	16
17	.04526	.058	.0540	.056	.058	.05625	17
18	.04030	.049	.0475	.048	.049	.05	18
19	.03589	.042	.0410	.040	.040	.04375	19
20	.03196	.035	.0348	.036	.035	.0375	20
21	.02846	.032	.03175	.032	.0315	.034375	21
22	.02535	.028	.0286	.028	.0295	.03125	22
23	.02257	.025	.0258	.024	.027	.028125	23
24	.02010	.022	.0230	.022	.025	.025	24
25	.01790	.020	.0204	.020	.023	.021875	25
26	.01594	.018	.0181	.018	.0205	.01875	26
27	.01420	.016	.0173	.0164	.0187	.0171875	27
28	.01264	.014	.0162	.0148	.0165	.015625	28
29	.01126	.013	.0150	.0136	.0155	.0140625	29
30	.01003	.012	.0140	.0124	.01372	.0125	30
31	.008928	.010	.0132	.0116	.0122	.0109375	31
32	.007950	.009	.0128	.0108	.0112	.01015625	32
33	.007080	.008	.0118	.0100	.0102	.009375	33
34	.006305	.007	.0104	.0092	.0095	.00859375	34
35	.005615	.005	.0095	.0084	.009	.0078125	35
36	.005000	.004	.0090	.0076	.0075	.00703125	36
37	.004453		.0085	.0068	.0065	.006640625	37
38	.003965		.008	.0060	.0057	.00625	38
39	.003531		.0075	.0052	.005		39
40	.003145		.007	.0048	.0045		40
41	.002800			.0044			41
42	.002494			.004			42
43	.002221			.0036			43
44	.001978			.0032			44
45	.001761			.0028			45
46	.001568			.0024			46
47	.001397			.002			47
48	.001244			.0016			48
49	.001018			.0012			49
50	.0009863			.001			50

TABLE 11

WEIGHTS OF STEEL, WROUGHT IRON, BRASS AND COPPER PLATES

BIRMINGHAM OR STUBS' GAGE

No. of Gage	Thickness in Inches	Steel	Iron	Brass	Copper
		WEIGHT IN LBS. PER SQUARE FOOT			
0000	.454	18.52	18.16	19.431	20.556
000	.425	17.34	17.00	18.190	19.253
00	.380	15.30	15.20	16.264	17.214
0	.340	13.87	13.60	14.552	15.402
1	.300	12.24	12.00	12.840	13.590
2	.284	11.59	11.36	12.155	12.865
3	.259	10.57	10.36	11.085	11.733
4	.238	9.71	9.52	10.186	10.781
5	.220	8.98	8.80	9.416	9.966
6	.203	8.28	8.12	8.689	9.196
7	.180	7.34	7.20	7.704	8.154
8	.165	6.73	6.60	7.062	7.475
9	.148	6.04	5.92	6.334	6.704
10	.134	5.47	5.36	5.735	6.070
11	.120	4.90	4.80	5.137	5.436
12	.109	4.45	4.36	4.667	4.933
13	.095	3.88	3.80	4.066	4.303
14	.083	3.39	3.32	3.552	3.769
15	.072	2.94	2.88	3.081	3.262
16	.065	2.65	2.60	2.782	2.945
17	.058	2.37	2.32	2.482	2.627
18	.049	2.00	1.96	2.097	2.220
19	.042	1.71	1.68	1.797	1.902
20	.035	1.43	1.40	1.498	1.585
21	.032	1.31	1.28	1.369	1.450
22	.028	1.14	1.12	1.198	1.270
23	.025	1.02	1.00	1.070	1.132
24	.022	.898	.88	.941	.997
25	.020	.816	.80	.856	.906
26	.018	.734	.72	.770	.815
27	.016	.653	.64	.685	.725
28	.014	.571	.56	.599	.634
29	.013	.530	.52	.556	.589
30	.012	.490	.48	.514	.544
31	.010	.408	.40	.428	.453
32	.009	.367	.36	.385	.408
33	.008	.326	.32	.342	.362
34	.007	.286	.28	.2996	.317
35	.005	.204	.20	.214	.227
36	.004	.163	.16	.171	.181

TABLE 12.

WEIGHTS OF STEEL, WROUGHT IRON, BRASS AND COPPER PLATES

AMERICAN OR BROWN & SHARPE GAGE

No. of Gage	Thickness in Inches	WEIGHT IN LBS. PER SQUARE FOOT			
		Steel	Iron	Brass	Copper
0000	.46	18.77	18.40	19.688	20.838
000	.4096	16.71	16.38	17.533	18.557
00	.3648	14.88	14.59	15.613	16.525
0	.3249	13.26	13.00	13.904	14.716
1	.2893	11.80	11.57	12.382	13.105
2	.2576	10.51	10.30	11.027	11.670
3	.2294	9.39	9.18	9.819	10.392
4	.2043	8.34	8.17	8.745	9.255
5	.1819	7.42	7.28	7.788	8.242
6	.1620	6.61	6.48	6.935	7.340
7	.1443	5.89	5.77	6.175	6.536
8	.1285	5.24	5.14	5.499	5.821
9	.1144	4.67	4.58	4.898	5.183
10	.1019	4.16	4.08	4.361	4.616
11	.0908	3.70	3.63	3.884	4.110
12	.0808	3.30	3.23	3.458	3.660
13	.0720	2.94	2.88	3.080	3.260
14	.0641	2.62	2.56	2.743	2.903
15	.0571	2.33	2.28	2.442	2.585
16	.0508	2.07	2.03	2.175	2.302
17	.0453	1.85	1.81	1.937	2.050
18	.0403	1.64	1.61	1.725	1.825
19	.0359	1.46	1.44	1.536	1.626
20	.0320	1.31	1.28	1.367	1.448
21	.0285	1.16	1.14	1.218	1.289
22	.0253	1.03	1.01	1.085	1.148
23	.0226	.922	.904	.966	1.023
24	.0201	.820	.804	.860	.910
25	.0179	.730	.716	.766	.811
26	.0159	.649	.636	.682	.722
27	.0142	.579	.568	.608	.643
28	.0126	.514	.504	.541	.573
29	.0113	.461	.452	.482	.510
30	.0100	.408	.400	.429	.454
31	.0089	.363	.356	.382	.404
32	.0080	.326	.320	.340	.360
33	.0071	.290	.284	.303	.321
34	.0063	.257	.252	.269	.286
35	.0056	.228	.224	.240	.254
36	.0050	.190	.188	.214	.226
37	.0045	.169	.167	.191	.202
38	.0040	.151	.149	.170	.180
39	.0035	.134	.132	.151	.160
40	.0031	.119	.118	.135	.142

TABLE 13.

Diam. of Circle, D	Side of Square, S	Area of Circle or Square	Diam. of Circle, D	Side of Square, S	Area of Circle or Square	Diam. of Circle, D	Side of Square, S	Area of Circle or Square
½	0.44	0.196	20½	18.17	330.06	40½	35.89	1288.25
1	0.89	0.785	21	18.61	346.36	41	36.34	1320.25
1½	1.33	1.767	21½	19.05	363.05	41½	36.78	1352.65
2	1.77	3.142	22	19.50	380.13	42	37.22	1385.44
2½	2.22	4.909	22½	19.94	397.61	42½	37.66	1418.63
3	2.66	7.069	23	20.38	415.48	43	38.11	1452.20
3½	3.10	9.621	23½	20.83	433.74	43½	38.55	1486.17
4	3.54	12.566	24	21.27	452.39	44	38.99	1520.53
4½	3.99	15.904	24½	21.71	471.44	44½	39.44	1555.28
5	4.43	19.635	25	22.16	490.87	45	39.88	1590.43
5½	4.87	23.758	25½	22.60	510.71	45½	40.32	1625.97
6	5.32	28.274	26	23.04	530.93	46	40.77	1661.90
6½	5.76	33.183	26½	23.49	551.55	46½	41.21	1698.23
7	6.20	38.485	27	23.93	572.56	47	41.65	1734.94
7½	6.65	44.179	27½	24.37	593.96	47½	42.10	1772.05
8	7.09	50.265	28	24.81	615.75	48	42.54	1809.56
8½	7.53	56.745	28½	25.26	637.94	48½	42.98	1847.45
9	7.98	63.617	29	25.70	660.52	49	43.43	1885.74
9½	8.42	70.882	29½	26.14	683.49	49½	43.87	1924.42
10	8.86	78.540	30	26.59	706.86	50	44.31	1963.50
10½	9.31	86.590	30½	27.03	730.62	50½	44.75	2002.96
11	9.75	95.033	31	27.47	754.77	51	45.20	2042.82
11½	10.19	103.87	31½	27.92	779.31	51½	45.64	2083.07
12	10.64	113.10	32	28.36	804.25	52	46.08	2123.72
12½	11.08	122.72	32½	28.80	829.58	52½	46.53	2164.75
13	11.52	132.73	33	29.25	855.30	53	46.97	2206.18
13½	11.96	143.14	33½	29.69	881.41	53½	47.41	2248.01
14	12.41	153.94	34	30.13	907.93	54	47.86	2290.22
14½	12.85	165.13	34½	30.57	934.82	54½	48.30	2332.83
15	13.29	176.71	35	31.02	962.11	55	48.74	2375.83
15½	13.74	188.69	35½	31.46	989.80	55½	49.19	2419.22
16	14.18	201.06	36	31.90	1017.88	56	49.63	2463.01
16½	14.62	213.82	36½	32.35	1046.35	56½	50.07	2507.19
17	15.07	226.98	37	32.79	1075.21	57	50.51	2551.76
17½	15.51	240.53	37½	33.23	1104.47	57½	50.96	2596.72
18	15.95	254.47	38	33.68	1134.11	58	51.40	2642.08
18½	16.40	268.80	38½	34.12	1164.16	58½	51.84	2687.83
19	16.84	283.53	39	34.56	1194.59	59	52.29	2733.97
19½	17.28	298.65	39½	35.01	1225.42	59½	52.73	2780.51
20	17.72	314.16	40	35.45	1256.64	60	53.17	2827.43

TABLE 14.

Gauge Numbers and Millimeter Equivalents

Gauge No.	American or Brown & Sharpe's		Birmingham or Stubs	
	Inches	Millimeters	Inches	Millimeters
000000	.5800	14.732		
00000	.5165	13.119		
0000	.4600	11.684	.454	11.532
000	.4096	10.404	.425	10.795
00	.3648	9.266	.380	9.652
0	.3249	8.252	.340	8.636
1	.2893	7.348	.300	7.620
2	.2576	6.543	.284	7.214
3	.2294	5.827	.259	6.579
4	.2043	5.189	.238	6.045
5	.1819	4.620	.220	5.588
6	.1620	4.115	.203	5.156
7	.1443	3.665	.180	4.572
8	.1285	3.264	.165	4.191
9	.1144	2.906	.148	3.759
10	.1019	2.588	.134	3.404
11	.09074	2.305	.120	3.048
12	.08081	2.053	.109	2.769
13	.07196	1.828	.095	2.413
14	.06408	1.628	.083	2.108
15	.05707	1.450	.072	1.829
16	.05082	1.291	.065	1.651
17	.04526	1.150	.058	1.473
18	.04030	1.024	.049	1.245
19	.03589	.912	.042	1.067
20	.03196	.812	.035	.889
21	.02846	.723	.032	.813
22	.02535	.644	.028	.711
23	.02257	.573	.025	.635
24	.02010	.511	.022	.559
25	.01790	.455	.020	.508
26	.01594	.405	.018	.457
27	.01420	.361	.016	.406
28	.01264	.321	.014	.356
29	.01126	.286	.013	.330
30	.01003	.255	.012	.305
31	.008928	.227	.010	.254
32	.007950	.202	.009	.229
33	.007080	.180	.008	.203
34	.006305	.160	.007	.178
35	.005615	.143	.005	.127
36	.005000	.127	.004	.102
37	.004453	.113		
38	.003965	.101		
39	.003531	.090		
40	.003145	.080		
41	.002800	.071		
42	.002494	.063		
43	.002221	.056		
44	.001978	.050		

TABLE 15.

1 gallon = 231 cu in.　1 cu ft = 7.4805 gal

Diameter in inches *	For 1 ft in length Cu ft, also area in sq ft	U. S. gal 231 cu in	Diameter in inches	For 1 ft in length Cu ft, also area in sq ft	U. S. gal, 231 cu in	Diameter in inches	For 1 ft in length Cu ft, also area in sq ft	U. S. gal, 231 cu in
¼	0.0003	0.0025	6¾	0.2485	1.859	19	1.969	14.73
5⁄16	0.0005	0.0040	7	0.2673	1.999	19½	2.074	15.51
3⁄8	0.0008	0.0057	7¼	0.2867	2.145	20	2.182	16.32
7⁄16	0.0010	0.0078	7½	0.3068	2.295	20½	2.292	17.15
½	0.0014	0.0102	7¾	0.3276	2.450	21	2.405	17.99
9⁄16	0.0017	0.0129	8	0.3491	2.611	21½	2.521	18.86
5⁄8	0.0021	0.0159	8¼	0.3712	2.777	22	2.640	19.75
11⁄16	0.0026	0.0193	8½	0.3941	2.948	22½	2.761	20.66
¾	0.0031	0.0230	8¾	0.4176	3.125	23	2.885	21.58
13⁄16	0.0036	0.0269	9	0.4418	3.305	23½	3.012	22.53
⅞	0.0042	0.0312	9¼	0.4667	3.491	24	3.142	23.50
15⁄16	0.0048	0.0359	9½	0.4022	3.682	25	3.409	25.50
1	0.0055	0.0408	9¾	0.5185	3.879	26	3.687	27.58
1¼	0.0085	0.0638	10	0.5454	4.080	27	3.976	29.74
1½	0.0123	0.0918	10¼	0.5730	4.286	28	4.276	31.99
1¾	0.0167	0.1249	10½	0.6013	4.498	29	4.587	34.31
2	0.0218	0.1632	10¾	0.6303	4.715	30	4.909	36.72
2¼	0.0276	0.2066	11	0.6600	4.937	31	5.241	39.21
2½	0.0341	0.2550	11¼	0.6903	5.164	32	5.585	41.78
2¾	0.0412	0.3085	11½	0.7213	5.396	33	5.940	44.43
3	0.0491	0.3672	11¾	0.7530	5.633	34	6.305	47.16
3¼	0.0576	0.4309	12	0.7854	5.875	35	6.681	49.98
3½	0.0668	0.4998	12½	0.8522	6.375	36	7.069	52.88
3¾	0.0767	0.5738	13	0.9218	6.895	37	7.467	55.86
4	0.0873	0.6528	13½	0.9940	7.436	38	7.876	58.92
4¼	0.0985	0.7369	14	1.0690	7.997	39	8.296	62.06
4½	0.1134	0.8263	14½	1.1470	8.578	40	8.727	65.28
4¾	0.1231	0.9206	15	1.2270	9.180	41	9.168	68.58
5	0.1364	1.0200	15½	1.3100	9.801	42	9.621	71.97
5¼	0.1503	1.1250	16	1.3960	10.440	43	10.085	75.44
5½	0.1650	1.2340	16½	1.4850	11.110	44	10.559	78.99
5¾	0.1803	1.3490	17	1.5760	11.790	45	11.045	82.62
6	0.1963	1.4690	17½	1.6700	12.490	46	11.541	86.33
6¼	0.2131	1.5940	18	1.7680	13.220	47	12.048	90.13
6½	0.2304	1.7240	18½	1.8670	13.960	48	12.566	94.00

* Actual.

TABLE 16.

Number of U. S. Gallons in Rectangular Tanks
For One Foot in Depth
1 cu ft = 7.4805 gal

Width, ft	Length of tank, ft										
	2	2.5	3	3.5	4	4.5	5	5.5	6	6.5	7
2	29.92	37.40	44.88	52.36	59.84	67.32	74.81	82.29	89.77	97.25	104.73
2.5	....	46.75	56.10	65.45	74.80	84.16	93.51	102.80	112.21	121.56	130.91
3	....	....	67.32	78.54	89.77	100.99	112.21	123.43	134.65	145.87	157.09
3.5	....	....	....	91.64	104.73	117.82	130.91	144.00	157.09	170.18	183.27
4	....	....	....	....	119.69	134.65	149.61	164.57	179.53	194.49	209.45
4.5	....	....	....	....	....	151.48	168.31	185.14	201.97	218.80	235.63
5	....	....	....	....	....	....	187.01	205.71	224.41	243.11	261.82
5.5	....	....	....	....	....	....	....	226.28	246.86	267.43	288.00
6	....	....	....	....	....	....	....	....	269.30	291.74	314.18
6.5	....	....	....	....	....	....	....	....	....	316.05	340.36
7	....	....	....	....	....	....	....	....	....	....	366.54

Width, ft	Length of tank, ft									
	7.5	8	8.5	9	9.5	10	10.5	11	11.5	12
2	112.21	119.69	127.17	134.65	142.13	149.61	157.09	164.57	172.05	179.53
2.5	140.26	149.61	158.96	168.31	177.66	187.01	196.36	205.71	215.06	224.41
3	168.31	179.53	190.75	202.97	213.19	224.41	235.63	246.86	258.07	269.30
3.5	196.36	209.45	222.54	235.63	248.73	261.82	274.90	288.00	301.09	314.18
4	224.41	239.37	254.34	269.30	284.26	299.22	314.18	329.14	344.10	359.06
4.5	252.47	269.30	286.13	302.96	319.79	336.62	353.45	370.28	387.11	403.94
5	280.52	299.22	317.92	336.62	355.32	374.03	392.72	411.43	430.13	448.83
5.5	308.57	329.14	349.71	370.28	390.85	411.43	432.00	452.57	473.14	493.71
6	336.62	359.06	381.50	403.94	426.39	448.83	471.27	493.71	516.15	538.59
6.5	364.67	388.98	413.30	437.60	461.92	486.23	510.54	534.85	550.16	583.47
7	392.72	418.91	445.09	471.27	497.45	523.64	549.81	575.00	602.18	628.36
7.5	420.78	448.83	476.88	504.93	532.98	561.04	589.08	617.14	645.19	673.24
8	....	478.75	508.67	538.59	568.51	598.42	628.36	658.28	688.20	718.12
8.5	....	....	540.46	572.25	604.05	635.84	667.63	690.42	731.21	703.00
9	....	....	....	605.92	639.58	673.25	706.90	740.56	774.23	807.89
9.5	....	....	....	....	675.11	710.65	746.17	781.71	817.24	852.77
10	....	....	....	....	....	748.05	785.45	822.86	860.26	897.66
10.5	....	....	....	....	....	....	824.73	864.00	903.26	942.56
11	....	....	....	....	....	....	....	905.14	946.27	987.43
11.5	....	....	....	....	....	....	....	....	989.29	1032.3
12	....	....	....	....	....	....	....	....	....	1077.2

To find weight of water in pounds at 62° F., multiply the number of gallons by 8¼.

References.—GENERAL METAL WORK by Alfred B. Grayshon, published by the D. Van Nostrand Company, contains additional material on sheet metal work.

TABLE 17

CIRCUMFERENCE AND AREAS OF CIRCLES

Diameter.	Circumference.	Area.	Diameter.	Circumference.	Area.
$\frac{1}{64}$	0.0491	0.00019	$\frac{43}{64}$	2.1108	0.35454
$\frac{1}{32}$	0.0982	0.00077	$1\frac{1}{16}$	2.1598	0.37122
$\frac{3}{64}$	0.1473	0.00173	$\frac{45}{64}$	2.2089	0.38829
$\frac{1}{16}$	0.1964	0.00307	$\frac{23}{32}$	2.2580	0.40574
$\frac{5}{64}$	0.2454	0.00479	$\frac{47}{64}$	2.3071	0.42357
$\frac{3}{32}$	0.2945	0.00690	$\frac{3}{4}$	2.3562	0.44179
$\frac{7}{64}$	0.3436	0.00940	$\frac{49}{64}$	2.4053	0.46039
$\frac{1}{8}$	0.3927	0.01227	$\frac{25}{32}$	2.4544	0.47937
$\frac{9}{64}$	0.4418	0.01553	$\frac{51}{64}$	2.5035	0.49874
$\frac{5}{32}$	0.4909	0.01918	$1\frac{3}{16}$	2.5525	0.51849
$\frac{11}{64}$	0.5400	0.02320	$\frac{53}{64}$	2.6016	0.53862
$\frac{3}{16}$	0.5890	0.02761	$\frac{27}{32}$	2.6507	0.55914
$\frac{13}{64}$	0.6381	0.03241	$\frac{55}{64}$	2.6998	0.58004
$\frac{7}{32}$	0.6872	0.03758	$\frac{7}{8}$	2.7489	0.60132
$\frac{15}{64}$	0.7363	0.04314	$\frac{57}{64}$	2.7980	0.62299
$\frac{1}{4}$	0.7854	0.04909	$\frac{29}{32}$	2.8471	0.64504
$\frac{17}{64}$	0.8345	0.05542	$\frac{59}{64}$	2.8962	0.66747
$\frac{9}{32}$	0.8836	0.06213	$1\frac{5}{16}$	2.9452	0.69029
$\frac{19}{64}$	0.9327	0.06922	$\frac{61}{64}$	2.9943	0.71349
$\frac{5}{16}$	0.9818	0.07670	$\frac{31}{32}$	3.0434	0.73708
$\frac{21}{64}$	1.0308	0.08456	$\frac{63}{64}$	3.0925	0.76105
$\frac{11}{32}$	1.0799	0.09281	1	3.1416	0.78540
$\frac{23}{64}$	1.1290	0.10144	$1\frac{1}{64}$	3.1907	0.81013
$\frac{3}{8}$	1.1781	0.11045	$1\frac{1}{32}$	3.2398	0.83525
$\frac{25}{64}$	1.2272	0.11984	$1\frac{3}{64}$	3.2889	0.86075
$\frac{13}{32}$	1.2763	0.12962	$1\frac{1}{16}$	3.3379	0.88664
$\frac{27}{64}$	1.3254	0.13979	$1\frac{5}{64}$	3.3870	0.91291
$\frac{7}{16}$	1.3744	0.15033	$1\frac{3}{32}$	3.4361	0.93956
$\frac{29}{64}$	1.4235	0.16126	$1\frac{7}{64}$	3.4852	0.96660
$\frac{15}{32}$	1.4726	0.17258	$1\frac{1}{8}$	3.5343	0.99402
$\frac{31}{64}$	1.5217	0.18427	$1\frac{9}{64}$	3.5834	1.02182
$\frac{1}{2}$	1.5708	0.19635	$1\frac{5}{32}$	3.6325	1.05001
$\frac{33}{64}$	1.6199	0.20881	$1\frac{11}{64}$	3.6816	1.07858
$\frac{17}{32}$	1.6690	0.22166	$1\frac{3}{16}$	3.7306	1.10753
$\frac{35}{64}$	1.7181	0.23489	$1\frac{13}{64}$	3.7797	1.13687
$\frac{9}{16}$	1.7671	0.24850	$1\frac{7}{32}$	3.8288	1.16659
$\frac{37}{64}$	1.8162	0.26250	$1\frac{15}{64}$	3.8779	1.19670
$\frac{19}{32}$	1.8653	0.27688	$1\frac{1}{4}$	3.9270	1.22718
$\frac{39}{64}$	1.9144	0.29165	$1\frac{17}{64}$	3.9761	1.25806
$\frac{5}{8}$	1.9635	0.30680	$1\frac{9}{32}$	4.0252	1.28931
$\frac{41}{64}$	2.0126	0.32233	$1\frac{19}{64}$	4.0743	1.32095
$\frac{21}{32}$	2.0617	0.33824	$1\frac{5}{16}$	4.1233	1.35297

TABLE 17—*Continued*

Diameter.	Circumference.	Area.	Diameter.	Circumference.	Area.
$1\frac{21}{64}$	4.1724	1.38538	$2\frac{1}{8}$	6.6759	3.5466
$1\frac{11}{32}$	4.2215	1.41817	$2\frac{3}{16}$	6.8722	3.7584
$1\frac{23}{64}$	4.2706	1.45134	$2\frac{1}{4}$	7.0686	3.9761
$1\frac{3}{8}$	4.3197	1.48489	$2\frac{5}{16}$	7.2649	4.2
$1\frac{25}{64}$	4.3688	1.51883	$2\frac{3}{8}$	7.4613	4.4301
$1\frac{13}{32}$	4.4179	1.55316	$2\frac{7}{16}$	7.6576	4.6664
$1\frac{27}{64}$	4.4670	1.58786	$2\frac{1}{2}$	7.8540	4.9087
$1\frac{7}{16}$	4.5160	1.62295	$2\frac{9}{16}$	8.0503	5.1573
$1\frac{29}{64}$	4.5651	1.65843	$2\frac{5}{8}$	8.2467	5.4119
$1\frac{15}{32}$	4.6142	1.69428	$2\frac{11}{16}$	8.4430	5.6727
$1\frac{31}{64}$	4.6633	1.73052	$2\frac{3}{4}$	8.6394	5.9396
$1\frac{1}{2}$	4.7124	1.76715	$2\frac{13}{16}$	8.8357	6.2126
$1\frac{33}{64}$	4.7615	1.80415	$2\frac{7}{8}$	9.0321	6.4918
$1\frac{17}{32}$	4.8106	1.84154	$2\frac{15}{16}$	9.2284	6.7772
$1\frac{35}{64}$	4.8597	1.87932	3	9.4248	7.0686
$1\frac{9}{16}$	4.9087	1.91748	$3\frac{1}{16}$	9.6211	7.3662
$1\frac{37}{64}$	4.9578	1.95602	$3\frac{1}{8}$	9.8175	7.6699
$1\frac{19}{32}$	5.0069	1.99494	$3\frac{3}{16}$	10.0138	7.9798
$1\frac{39}{64}$	5.0560	2.03425	$3\frac{1}{4}$	10.2102	8.2958
$1\frac{5}{8}$	5.1051	2.07394	$3\frac{5}{16}$	10.4066	8.6179
$1\frac{41}{64}$	5.1542	2.11402	$3\frac{3}{8}$	10.6029	8.9462
$1\frac{21}{32}$	5.2033	2.15448	$3\frac{7}{16}$	10.7992	9.2807
$1\frac{43}{64}$	5.2524	2.19532	$3\frac{1}{2}$	10.9956	9.6211
$1\frac{11}{16}$	5.3014	2.23654	$3\frac{9}{16}$	11.1919	9.9678
$1\frac{45}{64}$	5.3505	2.27815	$3\frac{5}{8}$	11.3883	10.3206
$1\frac{23}{32}$	5.3996	2.32015	$3\frac{11}{16}$	11.5846	10.6796
$1\frac{47}{64}$	5.4487	2.36252	$3\frac{3}{4}$	11.7810	11.0447
$1\frac{3}{4}$	5.4978	2.40528	$3\frac{13}{16}$	11.9773	11.4160
$1\frac{49}{64}$	5.5469	2.44843	$3\frac{7}{8}$	12.1737	11.7933
$1\frac{25}{32}$	5.5960	2.49195	$3\frac{15}{16}$	12.3701	12.1768
$1\frac{51}{64}$	5.6450	2.53586	4	12.5664	12.5664
$1\frac{13}{16}$	5.6941	2.58016	$4\frac{1}{16}$	12.7628	12.9622
$1\frac{53}{64}$	5.7432	2.62483	$4\frac{1}{8}$	12.9591	13.3641
$1\frac{27}{32}$	5.7923	2.66989	$4\frac{3}{16}$	13.1554	13.7721
$1\frac{55}{64}$	5.8414	2.71534	$4\frac{1}{4}$	13.3518	14.1863
$1\frac{7}{8}$	5.8905	2.76117	$4\frac{5}{16}$	13.5481	14.6066
$1\frac{57}{64}$	5.9396	2.80738	$4\frac{3}{8}$	13.7445	15.0330
$1\frac{29}{32}$	5.9887	2.85397	$4\frac{7}{16}$	13.9408	15.4656
$1\frac{59}{64}$	6.0377	2.90095	$4\frac{1}{2}$	14.1372	15.9043
$1\frac{15}{16}$	6.0868	2.94831	$4\frac{9}{16}$	14.3335	16.3492
$1\frac{61}{64}$	6.1359	2.99606	$4\frac{5}{8}$	14.5299	16.8002
$1\frac{31}{32}$	6.1850	3.04418	$4\frac{11}{16}$	14.7262	17.2573
$1\frac{63}{64}$	6.2341	3.0927	$4\frac{3}{4}$	14.9226	17.7206
2	6.2832	3.1416	$4\frac{13}{16}$	15.1189	18.19
$2\frac{1}{16}$	6.4795	3.3410	$4\frac{7}{8}$	15.3153	18.6655

TABLE 17—*Continued*

Diameter.	Circumfer- ence.	Area.	Diameter.	Circumfer- ence.	Area.
4 15/16	15.5116	19.1472	10 1/2	32.9868	86.5908
5	15.7080	19.6350	10 5/8	33.3795	88.6643
5 1/8	16.1007	20.6290	10 3/4	33.7722	90.7625
5 1/4	16.4934	21.6476	10 7/8	34.1649	92.8858
5 3/8	16.8861	22.6907	11	34.5576	95.0334
5 1/2	17.2788	23.7583	11 1/8	34.9503	97.2055
5 5/8	17.6715	24.8505	11 1/4	35.343	99.4019
5 3/4	18.0642	25.9673	11 3/8	35.7357	101.6234
5 7/8	18.4569	27.1084	11 1/2	36.1284	103.8691
6	18.8496	28.2744	11 5/8	36.5211	106.1394
6 1/8	19.2423	29.4648	11 3/4	36.9138	108.4338
6 1/4	19.635	30.6797	11 7/8	37.3065	110.7537
6 3/8	20.0277	31.9191	12	37.6992	113.098
6 1/2	20.4204	33.1831	12 1/4	38.4846	117.859
6 5/8	20.8131	34.4717	12 1/2	39.2700	122.719
6 3/4	21.2058	35.7848	12 3/4	40.0554	127.677
6 7/8	21.5985	37.1224	13	40.8408	132.733
7	21.9912	38.4846	13 1/4	41.6262	137.887
7 1/8	22.3839	39.8713	13 1/2	42.4116	143.139
7 1/4	22.7766	41.2826	13 3/4	43.1970	148.490
7 3/8	23.1693	42.7184	14	43.9824	153.938
7 1/2	23.5620	44.1787	14 1/4	44.7678	159.485
7 5/8	23.9547	45.6636	14 1/2	45.5532	165.130
7 3/4	24.3474	47.1731	14 3/4	46.3386	170.874
7 7/8	24.7401	48.7071	15	47.1240	176.715
8	25.1328	50.2656	15 1/4	47 9094	182.655
8 1/8	25.5255	51.8487	15 1/2	48.6948	188.692
8 1/4	25.9182	53.4561	15 3/4	49.4802	194.828
8 3/8	26.3109	55.0884	16	50.2656	201.062
8 1/2	26.7036	56.7451	16 1/4	51.051	207.395
8 5/8	27.0963	58.4264	16 1/2	51.8364	213.825
8 3/4	27.489	60.1319	16 3/4	52.6218	220.354
8 7/8	27.8817	61.8625	17	53.4072	226.981
9	28.2744	63.6174	17 1/4	54.1926	233.706
9 1/8	28.6671	65.3968	17 1/2	54.9780	240.529
9 1/4	29.0598	67.2008	17 3/4	55.7634	247.450
9 3/8	29.4525	69.0293	18	56.5488	254.470
9 1/2	29.8452	70.8823	18 1/4	57.3342	261.587
9 5/8	30.2379	72.7599	18 1/2	58.1196	268.803
9 3/4	30.6306	74.6619	18 3/4	58.905	276.117
9 7/8	31.0233	76.5888	19	59.6904	283.529
10	31.4166	78.5400	19 1/4	60.4758	291.040
10 1/8	31.8087	80.5158	19 1/2	61.2612	298.648
10 1/4	32.2014	82.5158	19 3/4	62.0466	306.355
10 3/8	32.5941	84.5409	20	62.8320	314.16

TABLE 17—*Continued*

Diameter.	Circumference.	Area.	Diameter.	Circumference.	Area.
21	65.9736	346.361	66	207.34	3421.19
22	69.1152	380.134	67	210.49	3525.65
23	72.2568	415.477	68	213.63	3631.68
24	75.3984	452.39	69	216.77	3739.28
25	78.540	490.87	70	219.91	3848.45
26	81,681	530.93	71	223.05	3959.19
27	84.823	572.56	72	226.19	4071.50
28	87.965	615.75	73	229.34	4185.39
29	91.106	660.52	74	232.48	4300.84
30	94.248	706.86	75	235.62	4417.86
31	97.389	754.77	76	238.76	4536.46
32	100.53	804.25	77	241.90	4656.63
33	103.67	855.30	78	245.04	4778.36
34	106.81	907.92	79	248.19	4901.67
35	109.96	962.11	80	251.33	5026.55
36	113.10	1017.88	81	254.47	5153.00
37	116.24	1075.21	82	257.61	5281.02
38	119.38	1134.11	83	260.75	5410.61
39	122.52	1194.59	84	263.89	5541.77
40	125.66	1256.64	85	267.04	5674.50
41	128.81	1320.25	86	270.18	5808.80
42	131.95	1385.44	87	273.32	5944.68
43	135.09	1452.20	88	276.46	6082.12
44	138.23	1520.53	89	279.60	6221.14
45	141.37	1590.43	90	282.74	6361.73
46	144.51	1661.90	91	285.88	6503.88
47	147.65	1734.94	92	289.03	6647.61
48	150.80	1809.56	93	292.17	6792.91
49	153.94	1885.74	94	295.31	6939.78
50	157.08	1963.50	95	298.45	7088.22
51	160.22	2042.82	96	301.59	7238.23
52	163.36	2123.72	97	304.73	7389.81
53	166.50	2206.18	98	307.88	7542.96
54	169.65	2290.22	99	311.02	7697.69
55	172.79	2375.83	100	314.16	7853.98
56	175.93	2463.01	101	317.30	8011.85
57	179.07	2551.76	102	320.44	8171.28
58	182.21	2642.08	103	323.58	8332.29
59	185.35	2733.97	104	326.73	8494.87
60	188.50	2827.43	105	329.87	8659.01
61	191.64	2922.47	106	333.01	8824.73
62	194.78	3019.07	107	336.15	8992.02
63	197.92	3117.25	108	339.29	9160.88
64	201.06	3216.99	109	342.43	9331.32
65	204.20	3318.31	110	345.58	9503.32

XX

Electricity

Electricity has more useful and universal application than any other natural phenomenon, and the end of the range of its applications is not yet in sight. Its increasing importance need not be emphasized here, but it is significant to note that many even recent developments have been the result of new study of the fundamental principles of the subject. It is also significant that the applications of electricity which enter into the daily life of the average person range from the simple heating elements and the dry cell which operates the door bell, to the more intricate motors and vacuum tubes.

Hence, it is important for the practical man to understand the fundamental principles of the subject in order to appreciate the rules which have been laid down for the applications. This section devotes a substantial amount of space to these fundamentals and with each step shows how they are applied and how the calculations pertaining to them are made.

The Nature of Electricity.—Electricity is, as we have suggested, a phenomenon of nature. It exists all about us like the air we breathe, but why it exists or what it actually is, we are unable to say. We do know, however, something of what it can do and how it acts under certain conditions, and that is the more important concern in adapting it to the uses of mankind. Since electricity already exists, it is obvious that we cannot create it. We can, however, create a *flow* of electricity as we create a flow of water through a pipe by means of a pump. This flow or current of electricity is created by mechanical, thermal, or chemical means. The energy of these agents is transformed into electrical energy capable of doing work. The work of the man dealing with electricity may be epitomized as the proper control of electrical energy while performing useful service.

Units.—To cause a current of electricity to flow, there must be a pressure. This is known as electromotive force and is measured in *volts*; it is therefore often referred to as *voltage*. The current flow is measured in *amperes* and the resistance to such a flow is measured in *ohms*. These are the fundamental units, and they are defined as follows:

A *volt* is a unit of electrical pressure or potential difference (pd) or the electromotive force (emf) required to cause a current of one ampere to flow through a resistance of one ohm.

An *ampere* is a unit of current strength, or the quantity of flow, or the quantity of current which will flow through a resistance of one ohm under an electromotive force of one volt.

An *ohm* is a unit of resistance, or the resistance of a conductor through which a current of one ampere will pass under an electromotive force of one volt.

Thus the three units depend on one another. One of the three must therefore be stated independently, and this one is the *ohm*. The ohm is usually defined as the resistance of a certain conductor of a particular material, size and form.

Ohm's Law.—This is a statement of the relation between volts, amperes and ohms. It may be expressed as

$$I = \frac{E}{R}, \quad \text{or} \quad E = IR, \quad \text{or} \quad R = \frac{E}{I}$$

where I = current in amperes

E = electromotive force in volts

R = resistance in ohms

These are the standard algebraic letter symbols and are **not** to be confused with the abbreviations for these quantities.

ILLUSTRATION: How many amperes of current are flowing in a circuit with a resistance of 25 ohms when the pressure is 110 volts?

$$I = \frac{E}{R}, \quad I = \frac{110}{25} = 4.4 \text{ amperes} \quad \text{(Ans.)}$$

ILLUSTRATION: A circuit has a resistance of 200 ohms. What is the applied voltage when 0.03 ampere of current is flowing?

$$E = IR, \ E = 200 \times 0.03 = 6 \text{ volts} \quad \text{(Ans.)}$$

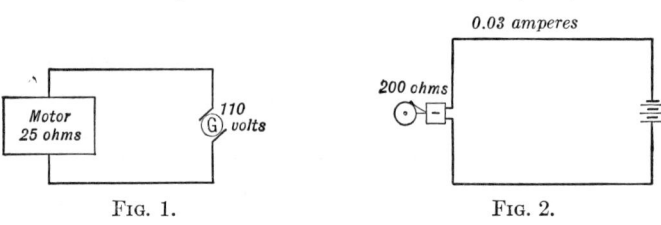

FIG. 1.   FIG. 2.

Electric Circuits

Series Connections.—Two or more pieces of electrical apparatus in a circuit one after the other are said to be in series.

The resistance of a series combination in a circuit is the sum of the resistances of the separate parts.

Current through a series combination is the same as the current through each part.

Voltage across a series combination is the sum of the voltages across separate parts.

ILLUSTRATION: It is desired to use 110-volt lights in a street car which operates on a current of 550 volts. How many lights must be placed in series?

By the last rule above, the voltage across a series combination is the sum of the voltages across separate parts.

Number of lights (separate parts) = $\frac{550}{110}$ = 5 lights   (**Ans.**)

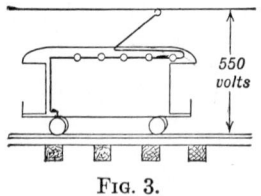

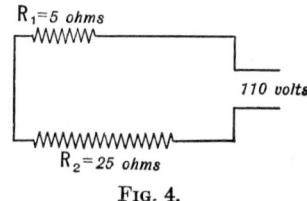

FIG. 3.   FIG. 4.

ILLUSTRATION: A 110-volt circuit has a resistance R_1 of 5 ohms and a resistance R_2 of 25 ohms. What current flows through it?

$$\text{Total resistance} = R_1 + R_2$$

$$I = \frac{110}{R_1 + R_2}, \quad I = \frac{110}{5 + 25}, \quad I = \frac{110}{30} = 3.67 \text{ amperes} \quad (\text{Ans.})$$

ILLUSTRATION: A current of 4 amperes flows through the circuit in Fig. 5. What is the voltage at the terminals?

Volts across 8-ohm resistance $= 4 \times 8 = 32$ volts
Volts across 30-ohm resistance $= 4 \times 30 = 120$ volts
Volts across 15-ohm resistance $= 4 \times 15 = 60$ volts

Volts across circuit at terminals $ = 212$ volts (Ans.)

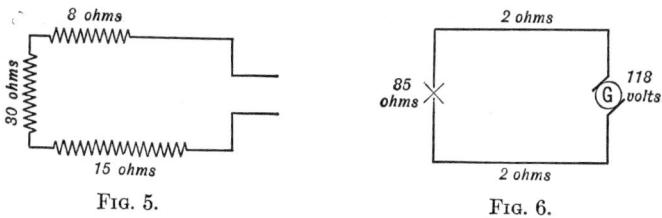

FIG. 5.   FIG. 6.

ILLUSTRATION: In Fig. 6 a generator is producing an electric current with a pressure of 188 volts. An arc light on the circuit has a resistance of 85 ohms and the resistance of each wire leading to it is 2 ohms. What is the voltage at the lamp terminals?

Total current

$$= I = \frac{E}{R_1 + R_2 + R_3} = \frac{118}{2 + 85 + 2} = \frac{118}{89} = 1.3 \text{ amperes}$$

Volts lost through $R_1 = 1.3 \times 2 = 2.6$ volts
Volts lost through $R_3 = 1.3 \times 2 = 2.6$ volts

Total potential drop through wires $= 5.2$ volts

Voltage at lamp terminals then $= 118 - 5.2 = 112.8$ volts (Ans.)

Parallel Connections.—Two or more pieces of electrical apparatus in a circuit so connected that the current is divided between them are said to be in parallel.

The resistance of a parallel circuit is the reciprocal of the sum of the reciprocals for the various resistances. This joint resistance is less than any branch resistance.

Because mathematical difficulties arise when finding joint resistance by using reciprocals, another method is to find the product divided by the sum of two resistances at a time. This is explained later.

The current through a parallel circuit is the sum of the currents through the separate branches.

Voltage across a parallel circuit is the same as the voltage across each branch.

Illustration: What is the joint resistance of the circuit shown in Fig. 7?

By the rule, the joint resistance is the reciprocal of the sum of the reciprocals of the parts, or

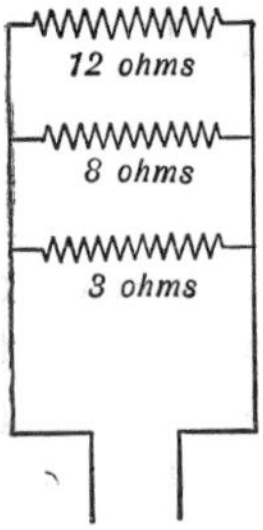

12 ohms

8 ohms

3 ohms

Fig. 7.

$$\frac{1}{R} = \frac{1}{R_1} + \frac{1}{R_2} + \frac{1}{R_3}$$

Then, $\frac{1}{R} = \frac{1}{12} + \frac{1}{8} + \frac{1}{3}$, and, reducing to common denominator,

$$\frac{1}{R} = \frac{2}{24} + \frac{3}{24} + \frac{8}{24} = \frac{13}{24}$$

R is then the reciprocal of $\frac{13}{24}$ or $\frac{24}{13}$ which reduces to 1.85 ohms   (Ans.)

The problem can be set up in one equation as follows:

$$\text{Joint resistance} = \frac{1}{\dfrac{2+3+8}{24}} = \frac{1}{\dfrac{13}{24}} = \frac{24}{13} = 1.85 \text{ ohms}\quad\text{(Ans.)}$$

Joint resistance of the same circuit found by the product over the sum method:

Joint resistance of 12 ohms and 8 ohms $= \dfrac{12 \times 8}{12 + 8} = \dfrac{96}{20} = 4.8$ ohms

Joint resistance of 4.8 ohms and 3 ohms

$$= \dfrac{4.8 \times 3}{4.8 + 3} = \dfrac{14.4}{7.8} = 1.85 \text{ ohms} \quad \text{(Ans.)}$$

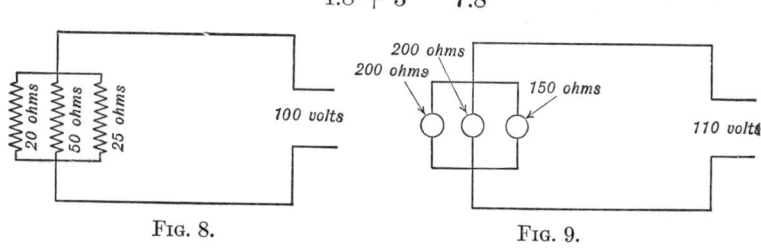

FIG. 8.   FIG. 9.

ILLUSTRATION: A current is flowing in the circuit shown in Fig. 8 under a pressure of 100 volts. How great is this current?

Current flowing through 20-ohm resistance.................... $= \frac{100}{20} = 5$ amperes

Current flowing through 50-ohm resistance..................... $= \frac{100}{50} = 2$ amperes

Current flowing through 25-ohm resistance.................... $= \frac{100}{25} = 4$ amperes

Current through the combination.. $= $ 11 amperes   (Ans.)

ILLUSTRATION: What is the total or joint resistance of the above circuit?

According to the rule, $\dfrac{1}{R} = \dfrac{1}{R_1} + \dfrac{1}{R_2} + \dfrac{1}{R_3}.$ Then,

$$\dfrac{1}{R} = \dfrac{1}{20} + \dfrac{1}{50} + \dfrac{1}{25} = \dfrac{5 + 2 + 4}{100} = \dfrac{11}{100}$$

and, joint resistance $R = \frac{100}{11} = 9.1$ ohms   (Ans.)

ILLUSTRATION: The circuit in Fig. 9 has two lamps with resistances of 200 ohms each and one lamp with a resistance of 150 ohms. What is the joint resistance of the combination and total current if the pressure is 110 volts?

Current flowing through each 200-

ohm lamp.................. $= \dfrac{110}{200} = 0.55$ ampere

Current flowing through 150-ohm

lamp........................ $= \dfrac{110}{150} = 0.73$ ampere

Current through the combination.. $= 0.55 + 0.55 + 0.73 = 1.83$ amperes

Resistance of the circuit.......... $= \dfrac{110}{1.83} = 60.1$ ohms (Ans.)

Series-Parallel Connections.—In many actual installations electrical apparatus instead of being in a simple parallel or series connection, is in a combination of these.

In a series-parallel circuit each part must be considered separately when computing the current, voltage and resistance of the entire circuit.

ILLUSTRATION: In the circuit shown in Fig. 10 a string of 110-volt lamps is connected to a 220-volt circuit. Two lamps are

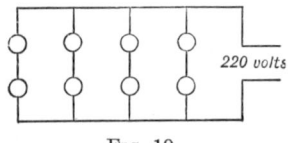

FIG. 10.

placed in series and each set of two is in parallel in the circuit. What is the total current if the resistance of each lamp is 200 ohms?

The resistance of each set of two lamps in series $= 200 + 200 = 400$ ohms.

The current flowing through each set of two-series lamps is,

$\frac{220}{400} = 0.55$ ampere

Since there are four identical sets in parallel the total current of the circuit is,

$4 \times 0.55 = 2.2$ amperes (Ans.)

ILLUSTRATION: In the circuit shown in **Fig. 11, what** is the total resistance and the total current?

Resistance of the parallel combination =

$$\frac{1}{\frac{1}{16} + \frac{1}{24}} = \frac{1}{\frac{3}{48} + \frac{2}{48}} = \frac{1}{\frac{5}{48}} = \frac{48}{5} = 9.6 \text{ ohms}$$

Total resistance of the circuit.. = 9.6 + 30 + 10 = 49.6 ohms

(Ans.)

Total current............... $= \dfrac{42}{49.2} = 0.86$ ampere   (Ans.)

Line Drop.—Wire used in electirc circuits offers a certain resistance to the passage of the electric current and results in loss of pressure or voltage which must often be taken into account. The amount of the resistance varies with the material and the temperature of the con- ductor. This will be treated more specifically in a later section. Now

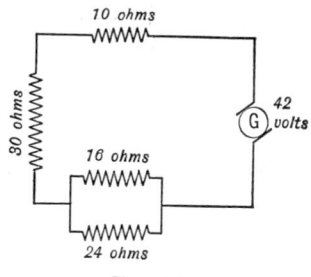

FIG. 11.

we are concerned only with the computation of typical " line drops."

The loss of voltage is equal to the product of the current flowing in a conductor and the resistance of the conductor between any two points.

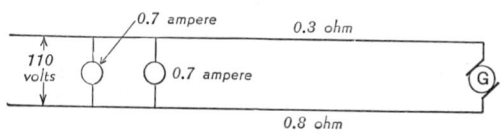

FIG. 12.

ILLUSTRATION: Two lamps shown in Fig. 12 each require 0.7 ampere of current. It is desired that they operate at 110 volts. What voltage must the generator produce if the resistance of each conductor is 0.8 ohm?

Amperes in the circuit $= 2 \times 0.7 \ldots\ldots = 1.4$ amperes
Volts lost in upper wire $= 1.4 \times 0.8 \ldots\ldots = \quad 1.12$ volts
Volts lost in lower wire $= 1.4 \times 0.8 \ldots\ldots = \quad 1.12$ volts
Volts required across lamps$\ldots\ldots\ldots\ldots = 110.00$ volts
Voltage which must be produced at the
  generator$\ldots\ldots\ldots\ldots\ldots\ldots\ldots \quad 112.24$ volts (Ans.)

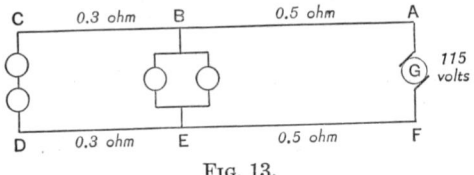

FIG. 13.

ILLUSTRATION: In the circuit shown in Fig. 13, each lamp takes 0.5 ampere of current. What is the voltage drop between A and B, B and C, D and E, and E and F?

Current through $BE = 2 \times 0.5 = 1.0$ ampere
Current through CD $\quad\quad\quad = 0.5$ ampere
Current through AB and EF $= 1.5$ amperes

Line drop A to $B = 0.5 \times 1.5 = 0.75$ volt (Ans.)
Line drop E to $F = 0.5 \times 1.5 = 0.75$ volt (Ans.)
Line drop B to $C = 0.3 \times 0.5 = 0.15$ volt (Ans.)
Line drop D to $E = 0.3 \times 0.5 = 0.15$ volt (Ans.)

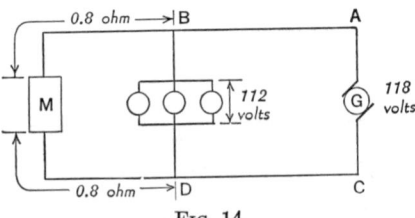

FIG. 14.

ILLUSTRATION: In the circuit shown in Fig. 14 find the resistance of the line between the generator and the lamps if each lamp takes 0.6 ampere of current and the motor 3.2 amperes.

Loss of voltage between generator and lamps
$$= 118 - 112 = 6 \text{ volts}$$
Loss of voltage in each wire $= \frac{6}{2} = 3$ volts

Total amperes in line $AB = 1.8 + 3.2 = 5$ amperes

Resistance of line $AB = \dfrac{E}{I} = \dfrac{3}{5} = 0.60$ ohm

Resistance of both lines AB and $DC = 0.60 \times 2 = 1.2$ ohm

Electric Insulators and Conductors.—Resistance has been defined as opposition to the flow of electricity. Some materials or substances do not permit the passage of electricity through them at all, and are called *insulators*. Such materials as glass, porcelain, dry wood, paper, wax, rubber, most gases, many liquids and minerals, plastics, etc., are insulators. These materials cover, separate, or support parts of electrical apparatus and circuits to prevent the escape or undesired flow of electricity.

Substances which allow the passage of electricity are called *conductors*. Most metals and some liquids, gases and minerals are conductors. Conductors which require high electric pressure (voltage) to send an electric current through them are said to be poor conductors, or to have high resistance. Otherwise, they are said to be good conductors, or to have low resistance.

Most metals are good conductors, the best (those of lowest resistance) among the common metals being silver, copper, gold, aluminum, tungsten, zinc, and brass, in the order named. Of these copper is the most plentiful and consequently the most used.

The resistance of a conductor varies with its temperature, increasing as the conductor is heated and decreasing as it is cooled. The resistance of a particular conductor of any one material depends also upon the size and shape of the conductor.

For most purposes conductors in the form of wires or rods are used. The calculations of resistances and sizes of wires form an important subject in themselves and are dealt with in the following section.

Electrical Measuring Instruments

Current and Voltage.—A *voltmeter* is an instrument which shows the electromotive force impressed upon its terminals.

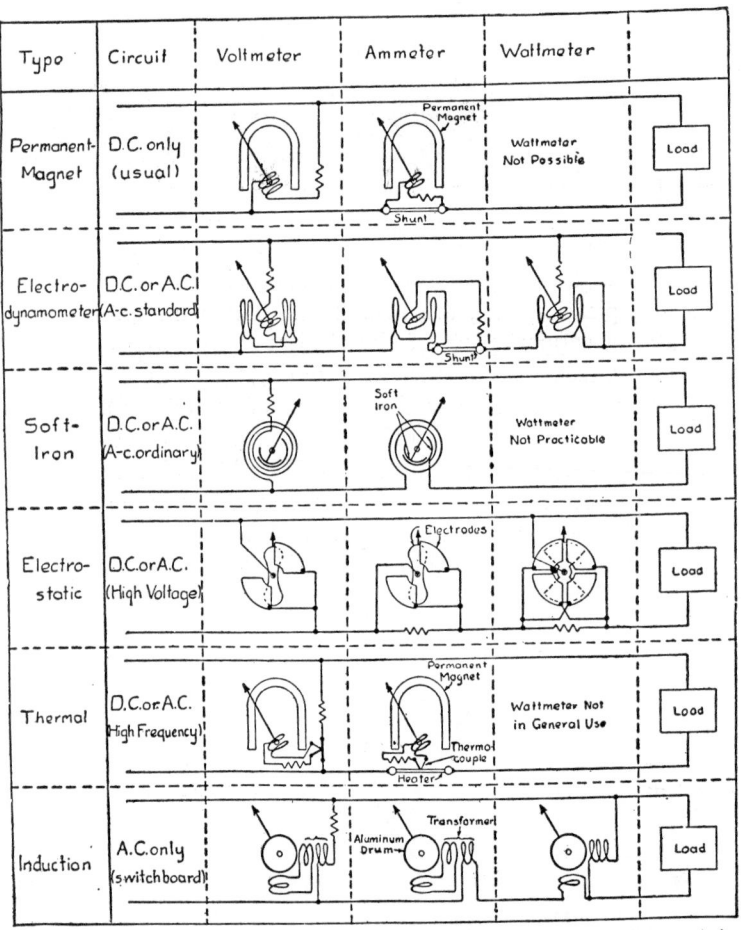

From "Electrical Engineering," by L. A. Hazeltine. Reproduced by special permission of The Macmillan Company, publishers.

Fig. 15.—Types of Voltmeter, Ammeter and Wattmeter.

An *ammeter* is an instrument which shows the strength of the current flowing through it.

The different types of voltmeters and ammeters are illustrated in Fig. 15. The similarity in construction and principle of operation of the two meters will be noted in the case of one type which is based on the D'Arsonval galvanometer. It consists fundamentally of a coil mounted on a pivot placed between the poles of a horseshoe magnet. When a current flows through the coil, the coil has a magnetic field and attempts

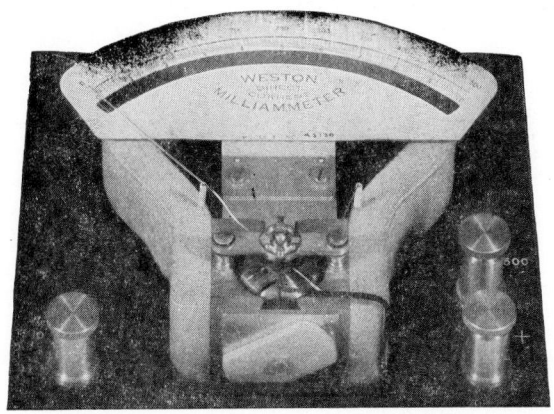

Fig. 16.—A Direct Current Milliammeter.

to turn so that its lines of force are in line with those of the horseshoe magnet. A pointer attached to the coil moves along a graduated scale.

In an *ammeter* the angle through which the coil turns is proportional to the strength of the current. Doubling the strength of the current makes the coil turn through twice as great an angle. An ammeter may be constructed and calibrated to measure thousandths of an ampere. Such an instrument is known as a *milliammeter*. The prefix " milli " always means one-thousandth.

To convert amperes into milliamperes, multiply by 1000.

ILLUSTRATION: How many milliamperes are 0.055 ampere?

$$0.055 \times 1000 = 55 \text{ milliamperes} \quad (Ans.)$$

An ammeter must always be connected on one side of a circuit in series with one conductor, so that all of the current passes through it. *Never connect an ammeter across the line.*

A voltmeter differs from an ammeter chiefly in that it has a high resistance in series with the coil and that its scale is, of course, calibrated for volts. A voltmeter measures the voltage or fall of potential between the two points to which it is connected. It

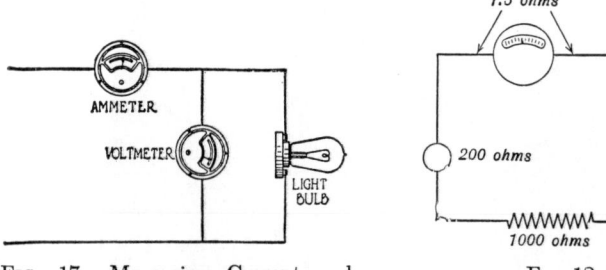

FIG. 17.—Measuring Current and Voltage of an Electric Light Bulb.

FIG. 18.

should therefore be connected *across the line,* not in the line. A voltmeter may be calibrated to measure millivolts and it is then known as a millivoltmeter.

To convert volts into millivolts, multiply by 1000.

ILLUSTRATION: How many millivolts are 0.28 volt?

$$0.28 \times 1000 = 280 \text{ millivolts} \quad (Ans.)$$

ILLUSTRATION: In the circuit shown in Fig. 18 the battery generates current at a pressure of 4.224 volts. What is the current flowing through the circuit and what is the potential drop through the meter in millivolts?

Total resistance of circuit $= 1000 + 200 + 1.5 = 1201.5$ ohms.

$$I = \frac{E}{R} \qquad I = \frac{4.224}{1201.5} = 0.0035 \text{ ampere} \quad \text{(Ans.)}$$

Potential drop through meter $= IR = 0.0035 \times 1.5 = 0.00525$

$$0.00525 \times 1000 = 5.25 \text{ millivolts} \quad \text{(Ans.)}$$

Resistance.—The simplest method of measuring resistance is by the use of an ammeter and a voltmeter. It is necessary only to obtain the voltage and the current and then apply Ohm's Law,
$R = \dfrac{E}{I}$.

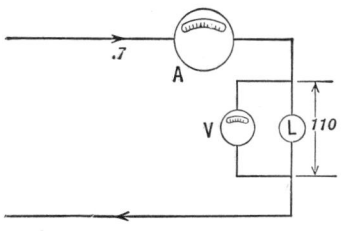

Fig. 19.

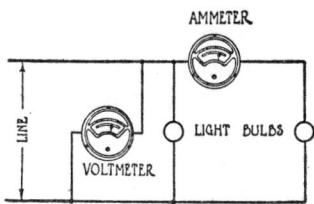

Fig. 20.—Two Light Bulbs in Parallel. Measuring Resistance of One.

ILLUSTRATION: In the partial circuit shown in Fig. 19 the potential measures 110 volts and the current 0.7 ampere. What is the resistance of the light?

$$R = \frac{E}{I} \qquad R = \frac{110}{0.7} = 157 \text{ ohms} \quad \text{(Ans.)}$$

ILLUSTRATION: If the potential measures 110 volts and the current is 0.5 ampere, what are the resistances of the two lights in Fig. 20?

Inspection will reveal that the ammeter measures only the current used by the right-hand bulb. The resistance of this bulb is, therefore,

$$R = \frac{E}{I} = \frac{110}{0.5} = 220 \text{ ohms} \quad \text{(Ans.)}$$

The resistance of the left-hand bulb if the same size as the other, will also take 0.5 ampere and would have the same resistance.

The *Wheatstone bridge* is the most accurate instrument for measuring resistances. It consists essentially of a device for providing two parallel paths for an electric current to pass through as shown in Fig. 21. The unknown resistance (R_1) is inserted into one of the paths and the two paths are then so balanced with known resistances R_2, R_3, and R_4 that an equal amount of current passes through each path. This state is detected by closing the key to the galvanometer on line bc. The galvanometer needle will remain stationary when the circuits are balanced.

However, R_2, R_3, and R_4 varied until the galvanometer needle is at zero. Then the voltage drop from a to b is the same as from a to c, and from b to d is the same as from c to d.

Expressed mathematically:

Let I = amperes in the line from the battery

I_x = amperes in R_1 and R_2 branch

I_y = amperes in R_3 and R_4 branch

Then for balance, or when the needle is at zero

$$R_1 I_x = R_3 I_y \quad (1)$$

and $$R_2 I_x = R_4 I_y \quad (2)$$

Dividing equation 1 by equation 2

$$\frac{R_1 I_x}{R_2 I_x} = \frac{R_3 I_y}{R_4 I_y}$$

whence

$$\frac{R_1}{R_2} = \frac{R_3}{R_4}$$

and

$$R_1 = \frac{R_2 R_3}{R_4}$$

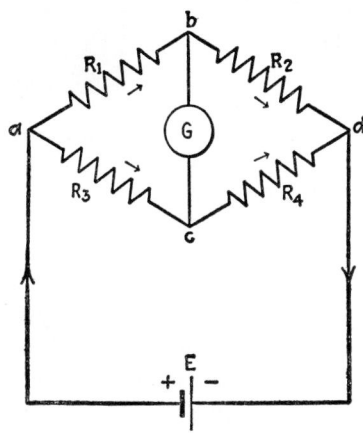

Fig. 21.—Wheatstone Bridge.

ILLUSTRATION: What is the value of R_1 if $R_2 = 1000$ ohms, $R_3 = 783$ ohms, and $R_4 = 100$ ohms?

$$R_1 = \frac{R_2 R_3}{R_4}$$

$$R_1 = \frac{1000 \times 783}{100} = 7830 \text{ ohms} \quad (\text{Ans.})$$

The Wheatstone bridge is made up in a variety of forms for commercial testing, one of which is shown in Fig. 22.

FIG. 22.—Dial Type Wheatstone Bridge or Testing Set.

FIG. 23.—Weston Ohmmeter.

An *ohmmeter* is an instrument for the direct measurement of electrical resistances. It is based on the principles of the Wheatstone bridge.

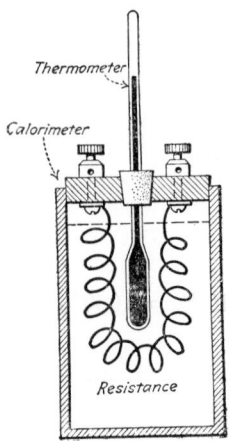

FIG. 24.—Calorimeter.

Electrical Heat.—Heat is generated by the passage of electricity through a conductor, and the amount of heat so generated can be measured by a calorimeter such as that shown in Fig. 24. In this device a coil of wire of known resistance is immersed in a known weight of water, and the resulting rise in temperature of the water is measured. It has been determined experimentally from such an apparatus that one ampere flowing through one ohm for one second always develops 0.24 *calorie* of heat—a calorie being the amount of heat required to raise the temperature of one gram of water one degree centigrade.

Electrical Work and Power

Force.—Force may be defined as that which changes or tends to change the state of motion of a body. We immediately recognize a number of applications of the common conceptions of force which fall within this definition. A pitcher exerts a force on a ball and causes it to fly through the air. The catcher exerts a force which arrests the motion of the ball. The batter transmits a force through the bat which causes a change in direction in the flight of the ball. The batter may lean on his bat and exert a considerable force on it without producing any motion.

Force is manifested in several different forms. There is the *force of gravity* resulting from the attraction between the earth and other bodies; *muscular force,* such as exerted by a horse pulling a wagon; and *mechanical force,* such as that exerted by a locomotive pulling a train. When dynamite breaks up rock in a quarry the action is due to *chemical force;* a force which tends to produce a flow of electricity is *electromotive force;* and the attraction between a magnet and a piece of iron is *magneto-mechanical force. The ordinary unit of force is the pound.*

Mass and Weight.—The *mass* of a body is the quantity of matter in it. The *weight* of a body is the force exerted by gravity on the mass of the body. A ball of iron may weigh five pounds at sea level, but if it is taken to an altitude of seven miles in a balloon, the gravitational force on it will be much decreased and hence it will weigh less. However, the mass will in both cases be the same. *The unit of weight is the pound.*

Work.—Work is done when force overcomes a resistance and moves a body on which it acts. *Work is force acting through space.* The amount of work done is measured by the product of the magnitude of the force and the distance through which it acts:

work = force × distance

work = pounds × feet = foot-pounds.

ILLUSTRATION: It takes a force of 125 pounds to pull a cable through a conduit. What work is done if a section between two manholes 200 feet apart is pulled through?

Work = $F \times D$ = 125 $\times$ 200 = 25,000 foot-pounds   (Ans.)

ILLUSTRATION: A man weighing 150 pounds climbs to the top of a 22-foot telegraph pole. What work is done?

Work = $F \times D$ = 150 $\times$ 22 = 3,300 foot-pounds.   (Ans.)

It must be noted that work is not always done when a force acts. For instance, if a man exerts all of his force to lift a 500-pound weight, no work is done if the resistance is not overcome and no motion results.

We have seen that force manifests itself in several different forms. There is a corresponding variety of work. A steam engine does work when it operates a derrick; dynamite does work when it throws stone through a distance; a gasoline engine does work when it propels an automobile; chemical action in a battery sets up a force which causes an electrical current to flow. However, whether the work is done mechanically, chemically, thermally, or electrically, it can be expressed in foot-pounds. Work done takes no account of time. A man may lift a 25-pound weight two feet in a minute or in an hour. The work done in each case is 50 foot-pounds.

Power.—*Power is the rate of doing work.* It is not to be confused with the total work done.

$$\text{Power} = \frac{\text{work}}{\text{time}}$$

or $\qquad \dfrac{\text{foot-pounds}}{\text{time}}$ = foot-pounds per unit of time.

Any convenient unit of time may be used. In mechanical work it may be foot-pounds per minute or foot-pounds per second. An arbitrarily selected unit of power is the *horsepower*, (H. P.).

One horsepower = 33,000 foot-pounds per minute

or $\qquad \dfrac{33,000}{60}$ = 550 foot-pounds per second.

ILLUSTRATION: A mine hoist raises a cage weighing 800 pounds at the rate of 300 feet per minute. What horsepower is it expending?

$$\text{H.P.} = \frac{800 \times 300}{33,000} = \frac{240,000}{33,000} = 7.3 \quad \text{(Ans.)}$$

ILLUSTRATION: A tractor exerts a pull of 500 pounds in pulling a motor on skids at a rate of five miles per hour. What horsepower is it exerting?

Since there are 5280 feet in a mile and 60 minutes in an hour,

$$5 \text{ mph} = \frac{5 \times 5280}{60} = 440 \text{ feet per minute.}$$

Power $= 500 \times 440 = 220,000$ foot-pounds per minute.

$$\text{Horsepower} = \frac{220,000}{33,000} = 6.7 \quad \text{(Ans.)}$$

Difference between Energy, Force, Work, and Power.—It is important to have a clear understanding of the differences of the meanings of these terms. *Energy* is the capacity to do work. *Force* is one of the factors of work and has to be exerted through a distance to do work. *Work* is done when energy is expended and is reckoned as the product of the magnitude of a force and the distance through which it acts in overcoming a resistance. *Power* is the rate of doing work.

Electrical Work.—Work is, as we have seen, force acting through space, or energy expended in overcoming resistance. However, force may exist without work being done, as for example, when a man pushes against a table but does not move it.

An electrical force exists between the two terminals of a battery tending to send an electrical current through the air. However, the resistance of the air is too great and no current flows, hence no work is done. The same is true when a generator is running on an open circuit. When a wire is connected across the terminals, the force is able to overcome the resistance of the wire, a current flows, and electrical work is done. In this case, the work takes the form of generation of heat. If an electric lamp is connected in the circuit, the work manifests itself as heat and light. If a motor is connected in the circuit, the work is that done by the motor in turning its shaft and pulley or gear under load.

The unit of electrical work is the amount of work performed by a current of one ampere flowing for one second under a pressure of one volt, and is called a joule.

Electrical work = volts × amperes × seconds

One joule has been found by experiment to be equivalent to 0.7375 foot-pound of mechanical work.

Electrical work is a subordinate factor in applied electricity to electrical *power*, which we shall now consider.

Electrical Power.—Power is, as we have seen, the rate at which work is done, and is independent of the total amount of work accomplished. A few paragraphs back we saw that

$$\text{Power} = \frac{\text{work}}{\text{time}}$$

Then, as we may expect,

$$\text{Electrical power} = \frac{\text{electrical work}}{\text{time}}$$

The unit of electrical power is the *watt*. It is equivalent to one joule of electrical work per second. Then,

$$\text{Watts} = \frac{\text{joules}}{\text{seconds}} = \frac{\text{volts} \times \text{amperes} \times \text{seconds}}{\text{seconds}}$$

The " seconds " cancel out and the equation becomes,

$$\text{Watts} = \text{volts} \times \text{amperes}.$$

One watt, therefore, equals one volt multiplied by one ampere.

To find the rate in watts at which energy is expended in a circuit:

Multiply the current in amperes by the pressure in volts causing it to flow.

In general, electrical power = voltage × current.

When P = watts expended

I = current in amperes

E = pressure in volts

Then $P = E \times I$.

ILLUSTRATION: A 110-volt circuit has fifty incandescent lamps connected in parallel, each with a resistance of 220 ohms and two electric toasters each with resistances of 18.33 ohms. How many watts of power are consumed by the circuit?

By Ohm's Law $I = \dfrac{E}{R}$. Then $I = \frac{110}{220} = \frac{1}{2}$ ampere current for each lamp.

$50 \times \frac{1}{2} = 25$ amperes of current for fifty lamps.

$$I = \frac{110}{18.33} = 6 \text{ amperes for each toaster.}$$

Then

$$P = E \times I = 110 \times (25 + 6 + 6) = 4070 \text{ watts} \quad \text{(Ans.)}$$

ILLUSTRATION: If it requires 12 amperes of current to operate the heaters on a street car from a 550-volt circuit, what power will be required?

$$P = E \times I. \quad P = 550 \times 12 = 6,600 \text{ watts.} \quad \text{(Ans.)}$$

To find the current when the power and the voltage are known: *Divide the watts expended by the voltage causing the current to flow.*

$$\text{Current} = \frac{\text{watts}}{\text{volts}} \qquad I = \frac{P}{E}$$

ILLUSTRATION: What current does a 75-watt lamp require when operating on a 110-volt circuit?

$$I = \frac{P}{E} = \frac{75}{110} = 0.68 \text{ ampere} \quad \text{(Ans.)}$$

ILLUSTRATION: What current does a 550-watt electric flatiron require when connected to a 115-volt circuit?

$$I = \frac{P}{E} = \tfrac{550}{115} = 4.78 \text{ amperes} \quad \text{(Ans.)}$$

To find the voltage when the power and current are known: *Divide the watts expended by the current flowing.*

$$\text{Volts} = \frac{\text{watts}}{\text{amperes}}. \qquad E = \frac{P}{I}$$

ILLUSTRATION: A 1200-watt motor requires a current of 10 amperes. What voltage is necessary to operate it?

$$E = \frac{P}{I} = \tfrac{1200}{10} = 120 \text{ volts} \quad \text{(Ans.)}$$

Electrical Horsepower.—The relationship between mechanical work and the expenditure of electrical energy has been determined by calorimeter experiments. From the results thus obtained, the following relationship has been established.

$$1 \text{ watt} = 0.7375 \text{ foot-pound per second}$$

or

$$1 \text{ foot-pound per second} = \frac{1}{0.7375} = 1.356 \text{ watts}$$

Since 550 foot-pounds per second are equivalent to 1 mechanical horsepower, an equivalent rate of electrical power would be:

$$\frac{550}{0.7375} = 746 \text{ watts} = 1 \text{ electrical horsepower.}$$

The electrical horsepower is a convenient unit since the watt is very small.

To find the electrical horsepower maintained in any circuit or part of a circuit:

Multiply the volts causing the current to flow by the current expressed in amperes and divide this product by 746.

$$\text{H.P.} = \frac{\text{watts}}{746} = \frac{\text{volts} \times \text{amperes}}{746} = \frac{E \times I}{746}$$

ILLUSTRATION: A motor on a 220-volt circuit requires 28 amperes of current. What horsepower is it using?

$$\text{H.P.} = \frac{E \times I}{746} = \frac{220 \times 28}{746} = \frac{6160}{746} = 8.26 \quad \text{(Ans.)}$$

ILLUSTRATION: A generator maintains a pressure of 110 volts across an electric light circuit and the ammeter indicates 75 amperes. What horsepower is being generated by the generator?

$$\text{H.P.} = \frac{E \times I}{746} = \frac{110 \times 75}{746} = \frac{8250}{746} = 11.06 \quad \text{(Ans.)}$$

The Kilowatt.—The *kilowatt* is a still larger unit of power. *One kilowatt equals 1000 watts.* The following relations are immediately obvious:

$$\text{Kilowatts (kw.)} = \frac{\text{watts}}{1000} = \frac{E \times I}{1000}$$

$$\text{Watts} = \text{kw.} \times 1000$$

$$1 \text{ h.p.} = 0.746 \text{ kw.}$$

$$1 \text{ kw.} = \frac{1}{0.746} = 1.34 \text{ h.p.}$$

ILLUSTRATION: What is the capacity in kilowatts of a generator carrying a load of 400 amperes at a pressure of 220 volts?

$$\text{Kw.} = \frac{E \times I}{1000} = \frac{220 \times 400}{1000} = 88 \quad \text{(Ans.)}$$

ILLUSTRATION: At full load how many amperes can be delivered by a 60-kilowatt generator at a pressure of 110 volts?

$$\text{Watts} = \text{kw.} \times 1000 = 60 \times 1000 = 60,000 \text{ watts} = P$$

$$I = \frac{P}{E} = \frac{60,000}{110} = 545 \text{ amperes} \quad \text{(Ans.)}$$

The Watt-hour and Kilowatt-hour.—Electrical energy or power when sold to a consumer is usually measured in terms of *watt-hours* or *kilowatt-hours* since the joule is too small a unit for practical use in this connection. A watt-hour is one watt expended for one hour. It is equivalent to 3600 watt-seconds (or joules) and also to 60 watt-minutes.

$$\text{Watt-hours} = \text{watts} \times \text{hours}$$

A *kilowatt-hour* is a larger unit of electrical work and equal to 1000 watts maintained for one hour, or 500 watts maintained for two-hours, etc.

$$\text{Kilowatt-hours} = \text{kw.} \times \text{hours}$$

An electrical *horsepower-hour* is one electrical horsepower maintained for one hour or 746 watts maintained for one hour.

$$\text{Horsepower-hours} = \text{h.p.} \times \text{hour}$$

The dials of a consumer's meter, by which the electrical energy used for light and power is measured, generally record kilowatt-hours. In Fig. 25 the dial face of a watt-hour meter has four

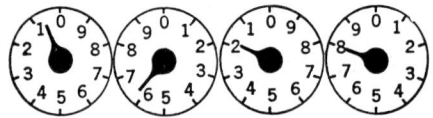

FIG. 25.—Dial of Watt-hour Meter.

circles. In the preceding figure each denotes the amount of energy in kilowatt-hours measured by the movement of the pointer over one division of the corresponding scale. One complete revolution of the pointer of any scale moves the pointer of the next scale immediately to its left over one division. It will be noted that some of the pointers turn in a clockwise direction and that the others turn counter-clockwise. In reading a pointer on a circle, it is necessary to look at the pointer immediately to its right to determine whether or not it has reached the point on which it appears to rest. For example, in Fig. 25 it is almost impossible to tell by looking at the second pointer from the right whether it has passed the "2" or failed to reach it. However, by looking at the circle to its right, it is apparent that it has not yet reached the "2."

ILLUSTRATION: How many kilowatt-hours does the meter in Fig. 25 read? 0618 kilowatt-hours. (Ans.)

Electrical Power Calculations.—A number of formulas are here presented which have a great variety of practical applications in the calculation of electrical power. These rules and formulas have been derived either by transposition of the formulas presented on the preceding pages or by combining them with the formulas expressing Ohm's Law. They are applicable equally well to the whole or a part of a circuit. *Caution must be exercised in the use of these formulas to use the volts lost in only the particular part of the circuit considered, and also the resistance of, and the current through, this part only.*

Given current and pressure, to find the watts expended:

The watts lost or expended in any circuit equal the product of the current and the voltage causing it to flow.

$P = E \times I$

when P = watts expended

I = current in amperes

E = pressure in volts

The use of this formula is illustrated on page 707.

Given current and resistance, to find the energy expended in watts:

The watts lost or expended in any circuit are equal to the current

*squared multiplied by the resistance. This is often called the
" I-square R loss."*

$$P = I^2 \times R$$

ILLUSTRATION: The resistance of the field magnets of a dynamo
is 430 ohms and the magnetizing current is 3 amperes. What
power is used in magnetizing the field?

$$P = I^2R = 3 \times 3 \times 430 = 3870 \text{ watts} \quad \text{(Ans.)}$$

ILLUSTRATION: An electric light has a resistance of 121 ohms
and uses 0.909 ampere. How many watts does it use?

$$P = I^2R = 0.909 \times 0.909 \times 121 = 100 \text{ watts} \quad \text{(Ans.)}$$

Given resistance and pressure, to find the watts expended:
*The watts lost or expended in any circuit are equal to the square
of the voltage divided by the resistance.*

$$P = \frac{E^2}{R}$$

ILLUSTRATION: An electromagnet with a resistance of 40 ohms
is operated on the current from a 6-volt storage battery. What
power is expended and what current does the magnet require?

$$P = \frac{E^2}{R} = \frac{6 \times 6}{40} = 0.90 \text{ watt} \quad \text{(Ans.)}$$

$$I = \frac{E}{R} = \frac{6}{40} = 0.15 \text{ ampere} \quad \text{(Ans.)}$$

ILLUSTRATION: The resistance of a solenoid is 60 ohms.
What power will be expended in it if a current passes through it
under a pressure of 110 volts?

$$P = \frac{E^2}{R} = \frac{110 \times 110}{60} = 201.67 \text{ watts} \quad \text{(Ans.)}$$

Given watts expended and current, to find the resistance:

The resistance is equal to watts expended divided by the square of the current.

$$R = \frac{P}{I^2}$$

ILLUSTRATION: An electric flatiron uses 660 watts of power and draws 6 amperes current. What is its resistance?

$$R = \frac{P}{I^2} = \frac{660}{6 \times 6} = 18.33 \text{ ohms} \quad \text{(Ans.)}$$

ILLUSTRATION: A 75-watt incandescent lamp requires 0.682 ampere. What is its resistance?

$$R = \frac{P}{I^2} = \frac{75}{0.682 \times 0.682} = \frac{75}{0.465} = 161.3 \text{ ohms} \quad \text{(Ans.)}$$

Given watts expended and resistance, to find the current:

The current equals the square root of the quotient of the watts divided by the resistance.

$$I = \sqrt{\frac{P}{R}}$$

ILLUSTRATION: The resistance of a 55-watt lamp is 220 ohms. What current will it require?

$$I = \sqrt{\frac{P}{R}} = \sqrt{\frac{55}{220}} = \sqrt{\frac{1}{4}} = \tfrac{1}{2} \text{ ampere} \quad \text{(Ans.)}$$

ILLUSTRATION: A printshop glue pot has a resistance of 88 ohms and draws 8,800 watts of power. What current does it require?

$$I = \sqrt{\frac{P}{R}} = \sqrt{\frac{8,800}{88}} = \sqrt{100} = 10 \text{ amperes} \quad \text{(Ans.)}$$

Given watts expended and pressure, to find the resistance:

The resistance equals the square of the voltage divided by the watts expended.

$$R = \frac{E^2}{P}$$

ILLUSTRATION: What is the resistance of a 55-watt, 110-volt incandescent lamp?

$$R = \frac{E^2}{P} = \frac{110 \times 110}{55} = 220 \text{ ohms} \quad \text{(Ans.)}$$

ILLUSTRATION: An electric furnace uses 7.2 kilowatts of power when operating at a pressure of 80 volts. What is its resistance?

$$R = \frac{E^2}{P} = \frac{80 \times 80}{7,200} = 0.889 \text{ ohm} \quad \text{(Ans.)}$$

All of the above formulas are applicable to problems where the power is given or wanted in electrical horsepower, by remembering that 1 horsepower = 746 watts = 0.746 kw, and 1 kw = 1.34 hp.

ILLUSTRATION: A calcium carbide electric furnace uses 3,500 amperes of current at 110 volts. How much horsepower is expended?

$$P = 3,500 \times 110 = 385,000 \text{ watts}$$

Since   1 hp = 746 watts,

$$\text{hp} = \frac{385,000}{746} = 516 \quad \text{(Ans.)}$$

ELECTRIC CELLS AND BATTERIES

Dry Cells.—The simplest method of producing electric current is by chemical means. A simple primary cell may be made by placing two dissimilar metals into an acid or alkaline solution. If the two metals are then connected by a piece of wire, a chemical reaction will result between the metals and the liquid solution and an electric current will flow through the wire.

Primary Cells—Carbon-Zinc.—The primary cells on the market, however, are all various types of dry cells. The standard carbon-zinc Leclanché type dry cell is now, and is likely to remain, the most widely used system because of low cost and reliable performance. Its electrochemical system uses a zinc anode, a manganese-dioxide cathode, and an electrolyte of ammonium chloride and zinc chloride dissolved in water. Powdered carbon is used in the depolarizing mix, usually in the form of acetylene black, to improve conductivity of the mix and to retain moisture.

The standard carbon-zinc dry battery is considered a primary type, i.e., not efficiently reversible. The basic cell is made in many shapes and sizes but two general categories exist:

1. Round Cells—available as unit cells or in assembled batteries.

2. Flat Cells—available in multi-cell batteries only.

The difference between the round and flat cells is mostly physical. The chemical ingredients are the same in both cases—carbon, depolarizing mix separator, electrolyte and zinc. The round-type cell is shown in Fig. 26, and the flat-type in Fig. 26A.

In flat cells, carbon is coated on a zinc plate to form a duplex electrode—a combination of the zinc of one cell and the carbon of the adjacent one. The "Mini-Max" cell (Figure 26A) contains no expansion chambers or carbon rod as does the round cell. This increases the amount of depolarizing mix available per unit cell volume and therefore the energy content. In addition the flat cell, because of its rectangular form, reduces waste space in assembled batteries. The energy to volume ratio of a battery utilizing round cells is inherently poor because of the voids occurring between cells. These two factors account for an energy to volume improvement of nearly 100% for "Mini-Max" cells compared to round cell assemblies.

The nominal voltage of a carbon-zinc cell is 1.5 v. Present battery types are available in voltages ranging from 1.5 to 510 v. Cells and batteries may, of course, be connected in series to obtain higher voltages, in parallel to achieve greater service capacity, or in series-parallel to obtain higher voltage and greater service capacity.

Primary Cells—Alkaline Type.—The electrochemical system of alkaline cells is comprised of a zinc anode of large surface area, a manganese-dioxide cathode of high density, and a potassium-hydroxide electrolyte. This alkaline primary differs from

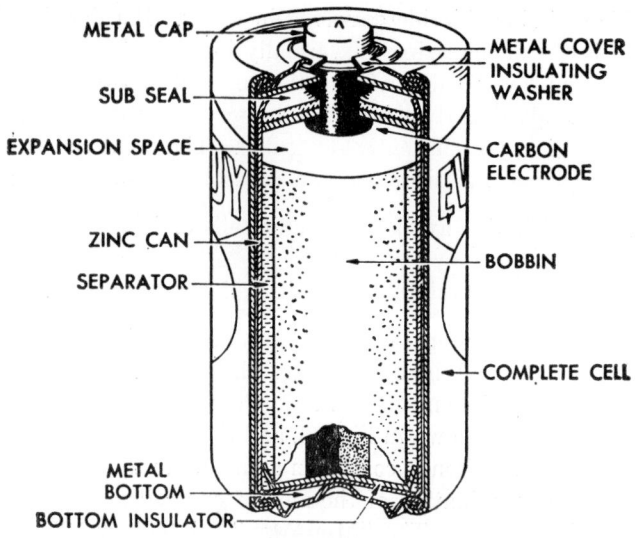

Fig. 26.—Standard Round Carbon-Zinc Cell.

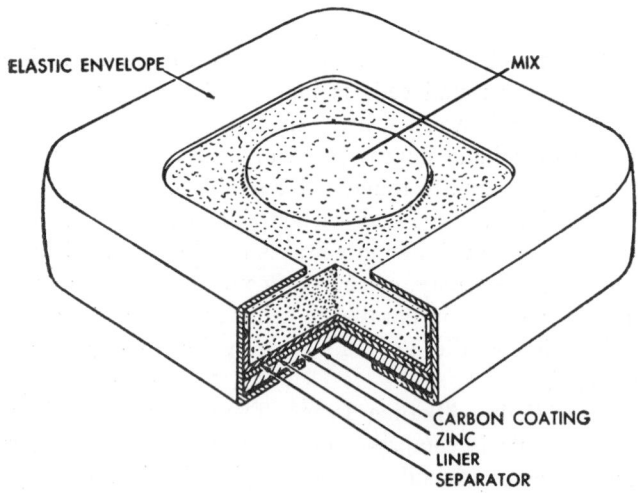

Fig. 26A.—"Mini-Max"® Flat Carbon-Zinc Cell.

the conventional Leclanché cell primarily in the highly alkaline electrolyte that is used. The cell is a high rate source of electrical energy. Its outstanding advantages are derived from the unique assembly of components and construction methods. (See Figure 27.)

Two principal features are a manganese-dioxide cathode of high density in conjunction with a steel can which serves as a cathode current collector, and a zinc anode of extra high surface area in contact with the electrolyte. These features, coupled with the use of a potassium-hydroxide electrolyte of high conductivity give these cells their very low internal resistance and impedance and high service capacity. Nominal voltage of an alkaline-manganese dioxide primary cell is 1.5 v in standard N, AA, C, ½D, D, and G cell sizes. Batteries are available with voltages up to 6 v and in a number of different service capacities. The alkaline-manganese primary cell is for applications requiring more power or longer life than can be obtained from carbon-zinc batteries. Alkaline cells contain 50 to 100 per cent more total energy than a carbon-zinc cell of the same size.

Primary Cells—Mercury Type.—The mercury battery consists essentially of a depolarizing mercuric-oxide cathode, an anode of pure amalgamated zinc, and a concentrated aqueous electrolyte of potassium hydroxide saturated with zincate. The nominal voltage is 1.35 v for a mercury cell with a depolarizer of 100 percent mercuric oxide and 1.4 v for a cell with a mixture of mercuric oxide and manganese dioxide.

The fundamental components of the mercury cell are a pressed mercuric-oxide cathode (in sleeve or pellet form) and pressed cylinders, or pellets, of powdered zinc with steel enclosures. These provide precise mechanical assemblies having maximum dimensional stability and marked improvements in performance over dry batteries of Leclanché (carbon-zinc) type. Cells are currently produced in two different designs using either flat or cylindrical types of pressed powder electrodes. Electrochemically both cells are the same, differing only in case design and internal electrode arrangements.

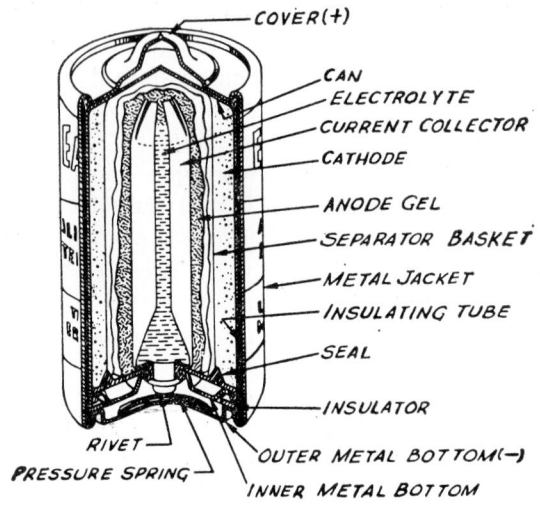

Fig. 27.—Primary Alkaline Cell.

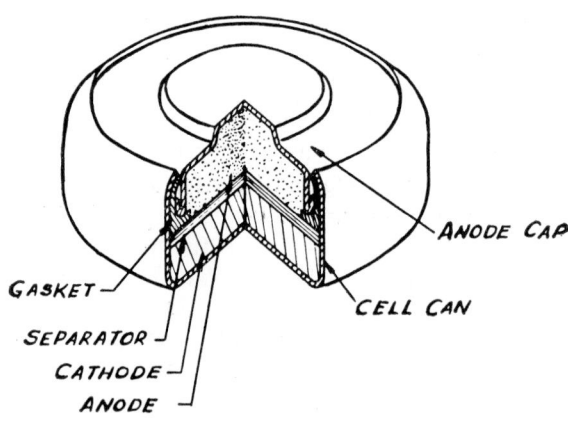

Fig. 27A.—Silver Oxide Cell.

1. Depolarizing cathodes of mercuric oxide, to which a small percentage of graphite is added, are shaped and either consolidated to the cell case (for flat electrode types), or pressed into the cases of the cylindrical types.

2. Anodes are formed of amalgamated zinc powder of high purity, pressed into either flat or cylindrical shapes.

3. A permeable barrier of specially selected material prevents migration of any solid particles in the cell, thereby contributing effectively to long shelf and service life.

4. Insulating and sealing gaskets are molded of polyethylene or Neoprene, depending on the application for which the cell or battery will be used.

5. Inner cell tops are plated with materials which provide an internal surface to which zinc will form a zinc amalgam bond.

6. Cell cases and outer tops of nickel-plated steel are used to resist corrosion, and to provide greatest passivity to internal cell materials.

7. An outer, nickel-plated, steel jacket is generally used for single cell "A" battery types. This outer jacket is a necessary component for the "self-venting construction" which provides a means of releasing excessive gas in the cell. Venting occurs if operating abnormalities, such as reverse currents or short circuits, produce excessive gas in the cell.

Primary Cells—Silver Type.—The silver oxide battery consists of a depolarizing silver oxide cathode, a zinc anode of high surface area and a highly alkaline electrolyte. The electrolyte is potassium hydroxide in hearing aid batteries. This is used to obtain maximum power density at hearing aid current drains. Sodium hydroxide is used in watch batteries for long term reliability. Mixtures of silver oxide and manganese dioxide may be tailored to provide a flat discharge curve or increased service hours.

Silver oxide batteries are well suited, for example, for use in hearing aids, instruments, photo electric exposure devices, electric watches and as reference voltage sources.

A cutaway of a silver oxide cell is shown in Figure 27A.

The milliampere-hour capacity is the same or a little greater than that of mercury batteries, but the silver oxide battery operates at 1.5 v while mercury batteries operate at about 1.3 v. Thus, silver oxide batteries offer 15 to 20 percent more power than mercury batteries of the same size. Present types are available in 1.5 v cells in capacities of 60 to 165 na-hr. A 9 volt battery rated at 165 milliampere-hours is also obtainable. Maximum current output ranges up to 10 ma.

Connecting Cells in Groups.—Primary cells are often connected in groups to produce greater voltages or currents. There are three methods of connection—series, parallel and series parallel.

In series connection the positive terminal of one cell is connected to the negative of the next succeeding cell and the line is connected to the remaining terminals. When the cells are so connected, the voltage of the battery is the sum of the voltages of all the cells. The current, or amperage, of the battery is equal to the amperage of one cell.

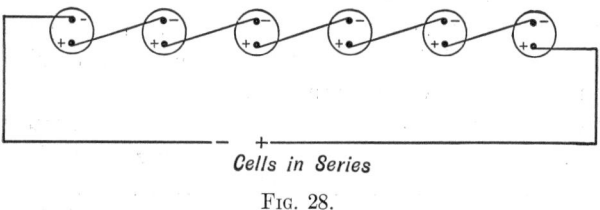

Cells in Series

FIG. 28.

ILLUSTRATION: What current and what voltage may be obtained from the battery shown in Fig. 28 if the voltage of each cell is 1½ volts and the maximum amperage 25 amperes?

$$\text{Voltage} = 1\frac{1}{2} + 1\frac{1}{2} + 1\frac{1}{2} + 1\frac{1}{2} + 1\frac{1}{2} + 1\frac{1}{2}$$

or

$$6 \times 1\frac{1}{2} = 9 \text{ volts} \quad \text{(Ans.)}$$

$$\text{Amperage} = 25 \text{ amperes} \quad \text{(Ans.)}$$

In *parallel connection* the positive terminals of all cells are connected to one line and the negative terminals to the other line. When cells are connected in parallel, every cell should be of the same voltage. The voltage of the battery is the same as the voltage of one cell. The amperage is the sum of the amperages of each cell.

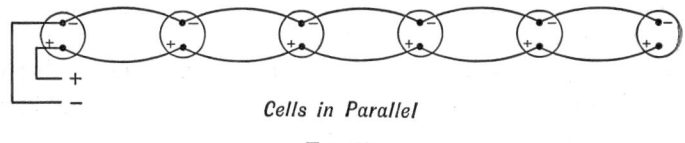

Cells in Parallel

Fig. 29.

ILLUSTRATION: What is the voltage and amperage of the battery shown in Fig. 29 if the voltage of each cell is 1½ volts and the amperage of each cell 25 amperes?

Voltage of battery = 1½ volts (Ans.)

Amperage of battery = 6 × 25 = 150 amperes (Ans.)

A series-parallel connection is made up of sets of cells connected in series with each set connected in parallel in the circuit as shown in Fig. 30. Each set must have the same voltage or an equal number of similar cells.

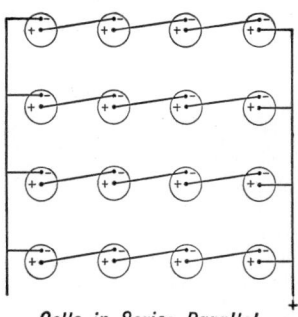

Cells in Series Parallel

Fig. 30.

The voltage of a series-parallel battery is equal to the voltage of each set connected in series, and the amperage is equal to the sum of the amperes delivered by each set.

ILLUSTRATION: In the battery shown in Fig. 30 the voltage of each cell is $1\frac{1}{2}$ volts and the amperage is 25 amperes. What is the voltage and amperage of the battery?

Voltage of each series set $= 4 \times 1\frac{1}{2} = 6$ volts. Therefore, the voltage of the battery is 6 volts. (Ans.)

Amperage of each series set $= 25$ amperes. The number of sets $= 4$. Therefore, the amperage of the battery $= 4 \times 25 = 100$ amperes. (Ans.)

Secondary Cells—Lead Type.—Secondary cells are distinguished from primary cells in that the latter are designed to be recharged, while the former are not. The rechargeable feature is a great advantage in cells operating as parts of prime movers and other moving machines, because it is often feasible to install a generator to be driven by the machinery, as in the automobile, and thus to keep the battery charged (if the load on it is not too great). The type of storage battery used on automobiles consists usually of three lead-type secondary cells connected to each other in a case, as shown in Fig. 31. The lead secondary cell consists of two plates in an electrolyte of diluted sulfuric acid.

When a lead storage cell has a full charge, the plate which forms the positive terminal is lead peroxide and the plate which forms the negative terminal is lead. The lead peroxide is made

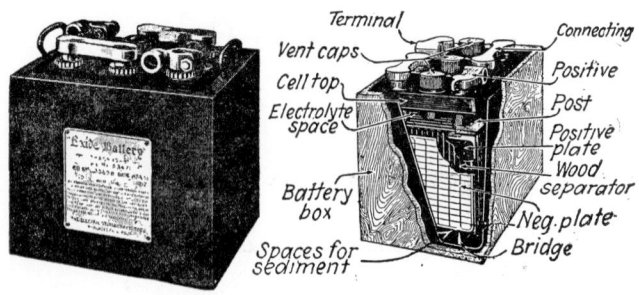

FIG. 31.

into a paste by mixing it with sulphuric acid and the lead in the negative plate is spongy to facilitate the chemical action. Neither of these substances is hard enough to be made into plates so they are pressed into the gridwork of cast lead plates as shown in Fig. 32.

When current is being drawn from a storage cell, the sulphuric acid (H_2SO_4) breaks up into its component parts of ions of H^+_2

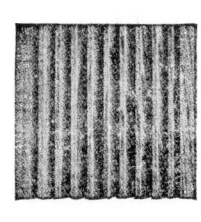

Fig. 32.—Plates of a Storage Battery. A Separator Is Shown in the Center.

and SO^-_4. The SO_4 goes to the lead plate giving it a negative charge and combining with the lead to form lead sulphate. The hydrogen goes to the lead peroxide plate giving it a positive charge and combining with the oxygen of the hydrogen peroxide to form water (H_2O).

It is obvious that as the discharging process continues, the electrolyte increases its proportion of water and decreases its proportion of sulphuric acid. It is possible to take advantage of this to measure the strength of a cell because sulphuric acid is heavier than water.

Specific Gravity.—Sulphuric acid is 1.835 times as heavy as water. This ratio is called the specific gravity of sulphuric acid. The *specific gravity* of any substance is the ratio of its weight to the weight of an equal volume of water at 39.1 degrees Fahrenheit.

Any floating object displaces its own weight of the fluid in which it floats. This principle is employed in the instrument known as the *hydrometer* (Fig. 33). It consists of a sealed glass tube weighted at one end and having a graduated scale in the other end. When this tube is floated in water it will sink to a certain depth. When it is floated in a liquid heavier than water, such as sulphuric acid, it will displace its own weight without sinking so deeply into the liquid. The scale, which is read at the level of the liquid on the tube, can be calibrated to give the specific gravity directly.

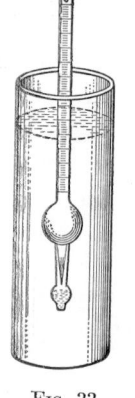

The electrolyte of a storage cell is a mixture of water and sulphuric acid; so its specific gravity can be expected to be something between 1.000 and 1.835. In a fully charged cell for automobile ignition the specific gravity of the electrolyte should not exceed 1.300; in a cell for radio work, the reading should not exceed 1.280. Liquids generally expand with an increase in temperature and are therefore less dense. For practical purposes 70° F. has been set as the standard temperature for the comparison of specific gravities of storage battery electrolytes.

Fig. 33.—
Hydrometer.

When the temperature of an electrolyte is greater than 70° F add one point to the fourth figure of the measured specific gravity for each 3° above 70° to obtain the actual specific gravity.

Illustration: The temperature of an electrolyte is 94° F. and the hydrometer reading is 1.280. What is the correct specific gravity?

$$94° - 70° = 24°. \quad {}^{24}\!/_{3} = 8$$

Therefore, the actual specific gravity is

$$1.280 + 0.008 = 1.288 \quad \text{(Ans.)}$$

Similarly, when the electrolyte is *colder* than 70°, *subtract* one point from the fourth place of the measured specific gravity for each 3° below 70° to obtain the actual specific gravity.

Illustration: The temperature of an electrolyte is 40° and the hydrometer reading is 1.270. What is the actual specific gravity?

$$70° - 40° - 30°. \quad {}^{30}\!/_{3} = 10$$

The actual specific gravity is

$$1.270 - 0.010 = 1.260 \quad \text{(Ans.)}$$

The readings of the hydrometer show the condition of the battery in accordance with the following table:

READING	CONDITION
1.280–1.300	Full charge
1.250	¼ Discharged
1.215	½ Discharged
1.180	¾ Discharged
1.150	Discharged

Other Types of Secondary Cells—The Edison Cell.—The storage cell devised by Thomas A. Edison employs nickel oxide for the positive electrode and finely divided iron for the negative electrode. These materials are packed into pockets carried by steel grids. The electrolyte is a solution of potassium hydroxide. As the cell discharges, the nickel oxide becomes reduced and the iron becomes oxidized, but the electrolyte remains unchanged. The chemical reactions are complex and not fully understood.

This cell is lighter than the lead storage cell, less subject to mechanical derangement, and it not injured by freezing. Moreover, cells of this type can be discharged and left in that condition without injury. They are therefore particularly suited for application in vehicles such as delivery trucks, fork lift trucks, subways, etc. The voltage output is lower than that of the lead storage cell and drops somewhat during discharge, having an average value of about 1.2 volts.

The Alkaline Secondary Cell.—While the Edison Cell employs an alkaline electrolyte the term *alkaline cell* is now applied to the newer Eveready® alkaline-manganese dioxide cell. The alkaline-manganese dioxide batteries, already discussed for their primary cell types, are also designed as secondary (storage) batteries, and are maintenance free, hermetically sealed, and will operate in any position. These batteries have been specifically designed for electronic and electrical applications where low initial cost and a low operating cost are of paramount importance.

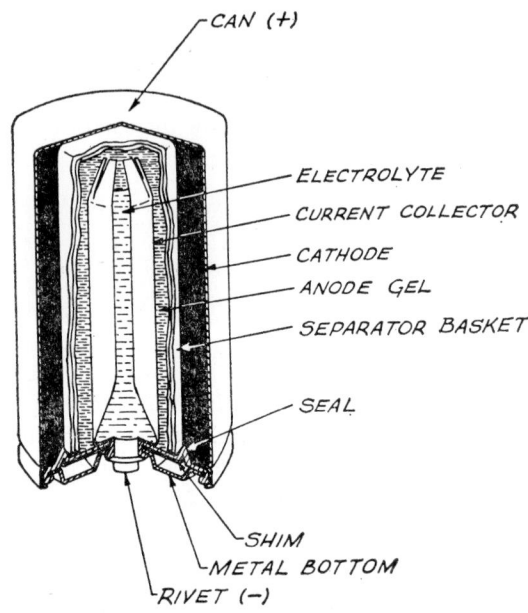

CAN (+)

ELECTROLYTE

CURRENT COLLECTOR

CATHODE

ANODE GEL

SEPARATOR BASKET

SEAL

SHIM

METAL BOTTOM

RIVET (−)

Fig. 33A.—Alkaline Secondary Cell.

Like the primary alkaline cells, these secondary (storage) cells also have zinc and manganese dioxide electrodes, and a potassium hydroxide electrolyte, as shown in Fig. 33A.

Nickel-Cadmium Secondary Cells.—In the uncharged condition the positive electrode of a nickel-cadmium cell is nickelous hydroxide, the negative cadmium hydroxide. In the charged condition the positive electrode is nickelic hydroxide, the negative metallic cadmium. The electrolyte is potassium hydroxide. The average operating voltage of the cell under normal discharge conditions is about 1.2 volts. The over-all chemical reaction of the nickel-cadmium system can be considered as:

$$Cd + 2NiOOH + 2KOH \rightleftarrows Cd(OH)_2 + 2NiO + 2KOH$$

Charged   Discharged

During the latter part of a recommended charge cycle and during overcharge, nickel-cadmium batteries generate gas. Oxygen is generated at the positive (nickel) electrode after it becomes fully charged and hydrogen is formed at the negative (cadmium) electrode when it reaches full charge.

A conventional vented type nickel-cadmium battery will liberate oxygen and hydrogen plus entrained electrolyte fumes through a valve. In order to hermetically seal a nickel-cadmium cell it is necessary to develop means of using up this gas inside the cell. This is acomplished as follows:

1. The battery is constructed with excess ampere-hour capacity in the cadmium electrode.

2. Starting with both electrodes in the fully discharged state, charging the battery causes the positive (nickel) electrode to reach full charge first and it starts oxygen generation. Since the negative (cadmium) electrode has not yet reached full charge it cannot cause hydrogen to be generated.

3. The cell is designed so that the oxygen formed can reach the surface of the metallic cadmium electrode where it reacts, forming electrochemical equivalents of cadmium oxide.

4. Thus in overcharge the cadmium electrode is oxidized at a rate just sufficient to offset input energy, keeping the cell in equilibrium at full charge.

This process can continue for long periods. The level of oxygen pressure thus established in the cell is determined by the charge rate used.

"Eveready"® nickel-cadmium cells are available in three basic configurations—button, cyclindrical and rectangular. The range of capacities for each type is as follows:

Button: (See Figs. 33B and 33C.) 20–300 milliampere-hours
Cylindrical: 450–8000 milliampere-hours
Rectangular: 1.6–23 ampere-hours

When required, two to ten button cells may be assembled into a higher voltage series stack by special factor welding techniques. (Under special conditions more than ten cells may be stacked.)

Rating of Storage Batteries.—Storage batteries are rated in ampere-hours. A current of one ampere flowing for one hour is an ampere-hour. Batteries are rated on the basis of the current which they can deliver continuously for a period of 8 hours. In other words, a 120-ampere-hour battery will deliver 15 amperes of current for eight hours. It will not, however, deliver 120 amperes for one hour or 60 amperes for two hours. The ampere-

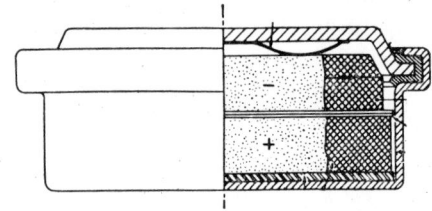

Fig. 33B.—Standard Nickel-Cadmium Button Cell.

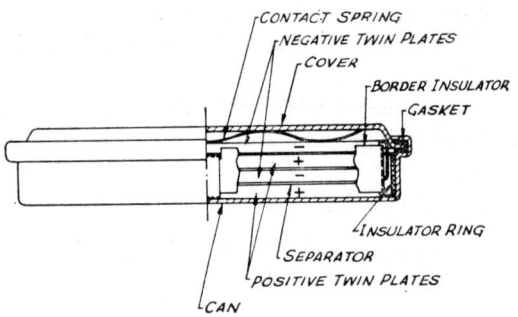

Fig. 33C.—High-Rate Nickel-Cadmium Button Cell.

hour life of a battery is governed by the rate at which it is discharged. If it is permitted to discharge at a very low rate its total ampere-hours of life will probably exceed its rated capacity. If, however, a heavy demand for current is placed upon it, such as when operating an automobile-engine starting motor, its life will be very short if the period of the demand is for any considerable length of time.

Starting batteries for automobiles have rated capacities from 80 to 160 ampere-hours.

ILLUSTRATION: A battery delivers 5 amperes of current for 22 hours. What is its capacity?

Capacity = 5 × 22 = 110 ampere-hours   (Ans.)

Storage Battery Voltage.—The voltage of a lead storage cell when fully charged is about 2.2 volts. During discharge it will give current at a nearly constant pressure of 2 volts. The difference is lost in internal resistance.

ILLUSTRATION: What voltage will a cell give when discharging 15 amperes of current if its internal resistance is 0.013 ohm and its open circuit electromotive force 2.195 volts?

Volts required to send 15 amperes through a resistance of 0.013 ohm = 15 × 0.013 = 0.195.

Terminal voltage = 2.195 − 0.195 = 2.000 volts.

Storage cells are connected in series to give batteries which will give higher voltage than a single cell. Three separate cells are commonly made with only one jar, which is provided with partitions that divide it into three separate compartments. The three-cell six-volt battery is the most common in automotive use.

Charging Storage Batteries.—A storage battery may be charged by connecting the positive wire of a direct-current 110-volt circuit to the positive terminal of the battery and the negative wire to the negative terminal, provided that suitable resistances are placed in the circuit to reduce the voltage and control the charging rate. A 6-ampere rate is satisfactory for small batteries and a 10-ampere rate for 100-ampere-hour batteries.

ILLUSTRATION: A 3-cell battery is to be charged at a 10-ampere rate with a charging voltage of 2.5 volts per cell. What resistance will be required if the battery is being charged from a 110-volt line?

Total charging voltage of battery = 3 × 2.5 = 7.5 volts.

Voltage through external resistances = 110 − 7.5 = 102.5 volts. Then, by Ohm's law

$$R = \frac{102.5}{10} = 10.25 \text{ ohms resistance required} \quad (\textbf{Ans.})$$

Current from Dry Cells.—*The current which a cell will deliver to a circuit is equal to the voltage divided by the sum of the internal and external resistances.*

ILLUSTRATION: In the circuit shown in Fig. 34a the internal resistance of the cell is 0.1 ohm and the resistance of the bell is 25 ohms. What current flows in the circuit?

$$I = \frac{1.5}{0.1 + 25} = \frac{1.5}{25.1} = 0.060 \text{ ampere} \quad \text{(Ans.)}$$

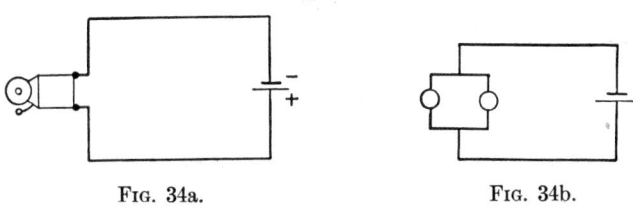

FIG. 34a.   FIG. 34b.

ILLUSTRATION: In the circuit in Fig. 34b the internal resistance of a 1.6-volt cell is 0.085 ohm. What current flows in the circuit if the resistance of each light is 5 ohms?

The joint resistance of the lamps by the law of the reciprocal of the sum of the reciprocals of the individual resistances is,

Reciprocal of 5 ohms $= \frac{1}{5}$

Sum of reciprocals $= \frac{1}{5} + \frac{1}{5} = \frac{2}{5}$

Joint resistance is reciprocal of sum $= \frac{5}{2} = 2.5$ ohms

Current in circuit

$$= \frac{1.6}{2.5 + 0.085} = \frac{1.6}{2.585} = 0.619 \text{ ampere} \quad \text{(Ans.)}$$

Voltage from Primary Cells.—The electrical pressure or voltage produced by a primary cell must overcome both the internal resistance and the external resistance. The voltage *rating*, or that usually referred to, is known as the " open circuit voltage " and is that obtained across the terminals when the cell is not delivering current. When delivering current the terminal voltage, that available for external resistance, is equal to the open circuit voltage minus the voltage across the internal resistance.

ILLUSTRATION: The cell shown in the circuit in Fig. 35 has an open circuit voltage of 1.52 volts and an internal resistance of 0.09 ohm. How many volts are used up by the internal resistance and what is the voltage across the resistance?

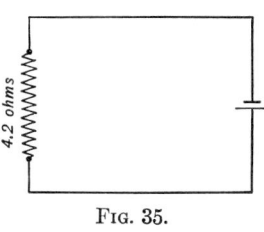

FIG. 35.

By Ohm's Law, total current in the circuit is

$$I = \frac{E}{r + R}. \qquad I = \frac{1.52}{0.09 + 4.2} = \frac{1.52}{4.29} = 0.354 \text{ ampere}$$

Voltage required to force 0.354 ampere, the internal resistance

$$E = Ir. \qquad E = 0.354 \times 0.09 = 0.032 \text{ volt}$$

Therefore, voltage across external resistance =

$$1.52 - 0.032 = 1.498 \text{ volts} \quad \text{(Ans.)}$$

Cells Arranged in Series.—When a number of cells are connected in series and to an external circuit, the current flowing through the external circuit will also pass through each cell. Since each cell has a certain resistance, a portion of the total electromotive force will be used up in overcoming the internal resistance.

To find the total internal resistance of a number of similar cells connected in series:

Multiply the resistance of one cell by the number of cells in the series.

The total internal resistance = $r \times ns$

When r = internal resistance

ns = number of cells in series

ILLUSTRATION: Eight dry cells (Fig. 36) each with an internal resistance of 0.095 ohm are connected in series. What is the total internal resistance?

Total resistance $= r \times ns = 0.095 \times 8 = 0.76$ ohm   (Ans.)

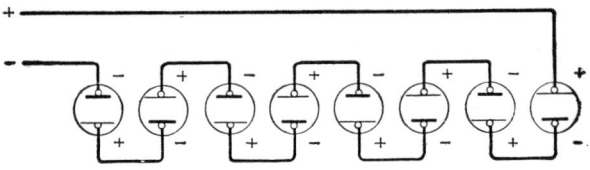

FIG. 36.

Current from Cells in Series.—To find the current that will be maintained in an external circuit by a number of cells in series:

Multiply the electromotive force of one cell by the number connected in series. Find the total internal resistance as above. Then by Ohm's Law the current is equal to the total electromotive force divided by the total resistance.

Expressed as an equation, this rule is,

$$I = \frac{E \times ns}{(r \times ns) + R}$$

When $E =$ electromotive force of one cell

$r =$ internal resistance of one cell

$ns =$ number of cells in series

$R =$ external resistance.

ILLUSTRATION: If the cells shown in Fig. 36 each have an electromotive force of 1.45 volts and an internal resistance of 0.095 ohm, what is the total current if a 15-ohm resistance is connected in the circuit?

$$I = \frac{E \times ns}{(r \times ns) + R} = \frac{1.45 \times 8}{(0.095 \times 8) + 15} = \frac{11.60}{15.76} = 0.74 \text{ ampere}$$

(Ans.)

Cells in Parallel.—When a number of similar cells are connected in parallel, as in Fig. 37, and to an external circuit, the total current does not have to overcome the total internal resistance of all the cells as is the case when they are connected in series, but is divided evenly among the cells in parallel. With a path for the current through each cell, the total resistance is much less than that for one cell.

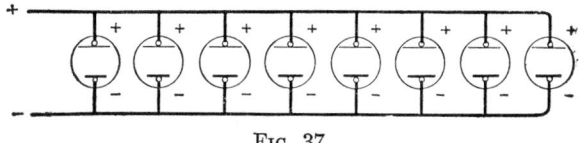

Fig. 37.

To find the internal resistance of a number of cells in parallel: *Divide the resistance of one cell by the number connected in parallel.*

$$\text{Total resistance} = \frac{r}{nq}$$

When r = internal resistance of one cell

nq = number of cells in parallel

ILLUSTRATION: What is the total internal resistance of the cells shown in Fig. 37 if the resistance of each cell is 0.2 ohm?

$$\text{Total resistance} = \frac{r}{nq} = \frac{0.2}{8} = 0.025 \text{ ohm} \quad (\text{Ans.})$$

Current from Cells in Parallel.—To find the current that will be maintained in an external circuit by a number of cells connected in parallel:

Divide the electromotive force of one cell by the sum of the external and internal resistances of a circuit.

This rule set up as an equation becomes,

$$I = \frac{E}{\dfrac{r}{nq} + R}$$

When E = total electromotive force of one cell

r = internal resistance of one cell

R = total external resistance

nq = number of cells in parallel

The quantity $\dfrac{r}{nq}$ in this equation will be recognized as the total internal resistance of the cells in parallel.

ILLUSTRATION: The cells shown in Fig. 37, each having an internal resistance of 0.2 ohm and an electromotive force of 1.5 volts, are connected to an external circuit with a total resistance of 12 ohms.   What is the total current in the external circuit?

By the above rule,

$$I = \frac{E}{\dfrac{r}{nq} + R} = \frac{1.5}{\dfrac{0.2}{8} + 12} = \frac{1.5}{12.025} = 0.125 \text{ ampere}   \text{(Ans.)}$$

Advantage of Cells in Parallel Connection.—Cells are connected in parallel when it is desired to obtain the maximum current through an external circuit of low resistance.   When cells are connected in parallel their zinc or negative plates are all connected to each other and their carbon or positive elements are also connected to each other.   The result is that the group of cells is the equivalent of one large cell, the positive and negative plates of which are equal in area to the sum of the areas of the respective plates in the separate cells.   This grouping is, therefore, capable of giving a large quantity of electrical current.   When the external resistance is small the strength of the current will be great; when the resistance is large, it will be small.

ILLUSTRATION: If a dry cell with an internal resistance of 0.3 ohm and an electromotive force of 1.5 volts is connected to an external circuit with a total resistance of 0.1 ohm, what will be the resultant flow of current?

$$\text{Current} = \frac{E}{r+R} = \frac{1.5}{0.3+0.1} = 3.75 \text{ amperes} \quad \text{(Ans.)}$$

ILLUSTRATION: If eight dry cells with similar characteristics are substituted for the single cell in the above illustration, what is then the current in the external circuit?

$$\text{Current} = \frac{E}{\dfrac{r}{nq}+R} = \frac{1.5}{\dfrac{0.3}{8}+0.1} = \frac{1.5}{0.1375} = 10.91 \text{ amperes} \quad \text{(Ans.)}$$

Advantage of Cells in Series Connection.—A series grouping of cells is employed when the external resistance is the principal one to be overcome and a maximum current strength in the circuit is desired. The advantage of this type of connection is shown by the following examples.

ILLUSTRATION: A dry cell with an electromotive force of 1.5 volts and an internal resistance of 0.3 ohm is connected to an external circuit with a total resistance of 100 ohms. What current will flow through the circuit?

$$I = \frac{E}{r+R} = \frac{1.5}{0.3+100} = \frac{1.5}{100.3} = 0.014955 \text{ ampere} \quad \text{(Ans.)}$$

ILLUSTRATION: What current will flow in the external circuit if ten cells with similar characteristics are connected in parallel in the circuit of the above illustration instead of the single cell?

$$I = \frac{E}{\dfrac{r}{nq}+R} = \frac{1.5}{\dfrac{0.3}{10}+100} = \frac{1.5}{100.03} = 0.014994 \text{ ampere} \quad \text{(Ans.)}$$

It will be seen, therefore, that there is little to be gained in the amount of current produced by substituting ten cells in parallel for the single cell. However, let us substitute ten cells in series in the next example.

ILLUSTRATION: Substitute ten cells of like characteristics in series connection for the cells in the above example. What will then be the current in the external circuit?

$$I = \frac{E \times ns}{(r \times ns) + R} = \frac{1.5 \times 10}{(0.3 \times 10) + 100} = \frac{15}{103} = 0.14563 \text{ ampere}$$

(Ans.)

It is evident from these illustrations that nearly ten times the current from cells in parallel connection passes through the circuit when the same cells are connected in series.

Cells Grouped in Parallel-Series.—It is sometimes desirable to group cells in a combination of series and parallel to give either the

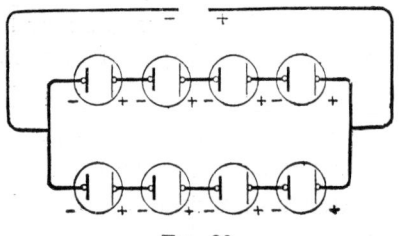

FIG. 38.

maximum current through an external resistance or to increase the capacity of the cells for maintaining a current in a circuit for a long period of time. Figure 38 shows such a connection consisting of two parallel sets of four cells in series. This is sometimes called a *multiple-series* combination.

If 6 volts are required to light a small lamp, four dry cells of 1.5 volts each connected in series will produce (neglecting internal resistance) the required 6 volts and will operate the lamp for a period of possibly 4 hours. If, however, eight cells are connected in parallel series as in Fig. 38 the total electromotive force will still be 6 volts, but the lamp will now be illuminated for a period of 8 hours.

To find the internal resistance of any multiple-series combination of cells:

Multiply the resistance of one cell by the number of cells in one group and divide the product by the number of groups in parallel or multiple.

$$\text{The total resistance} = \frac{r \times ns}{nq}$$

when, r = resistance of one cell

ns = number of cells in series in one group

nq = number of groups in parallel

ILLUSTRATION: What is the internal resistance of the combination of eight cells shown in Fig. 38 if the resistance of each cell is 0.2 ohm?

$$\text{Total resistance} = \frac{r \times ns}{nq} = \frac{0.2 \times 4}{2} = 0.4 \text{ ohm} \quad \text{(Ans.)}$$

Current Strength from Cells in Parallel-Series Combinations.— To find the current that will be maintained in an external circuit by any parallel-series combination of cells:

Divide the total electromotive force of one series group by the sum of the combined internal and external resistances.

Expressed as an equation, this rule becomes

$$I = \frac{E \times ns}{\dfrac{r \times ns}{nq} + R}$$

when, I = current in the external circuit

E = electromotive force of one cell

ns = number of cells in series in one group

nq = number of groups in parallel

r = internal resistance of one cell

R = external resistance

ILLUSTRATION: Fifteen cells are so connected that five cells are in series and three sets of five are in parallel. What current will flow through an external circuit connected to these cells if the total external resistance is 8 ohms and the electromotive force of each cell is 1.5 volts and the internal resistance 0.1 ohm?

$$I = \frac{E \times ns}{\dfrac{r \times ns}{nq} + R} = \frac{1.5 \times 5}{\dfrac{0.1 \times 5}{3} + 8} = \frac{7.5}{0.167 + 8} = 0.918 \text{ ampere}$$
(Ans.)

Groups of cells in parallel may also be connected in series as shown in Fig. 39. This is called a series-parallel or a series-multiple connection.

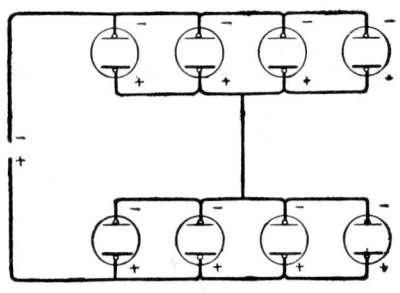

FIG. 39.

To find the current that will be maintained in an external circuit from any series-parallel combination of cells, several progressive steps are necessary as follows:

Find the internal resistance of one parallel group and consider the result as data for one " equivalent " cell (group). Calculate the total electromotive force and resistance for the parallel groups in series and determine the current by Ohm's Law.

ILLUSTRATION: The series-parallel combination shown in Fig. 39 is connected to a circuit with an external resistance of 2 ohms. The cells are four in parallel, two groups in series. Each has an electromotive force of 1.5 volts and an internal resistance of 0.1 ohms. What current will flow through the circuit?

The electromotive force of 1 group = 1.5 volts

The electromotive force of 2 groups in series = $1.5 \times 2 = 3.0$ volts

The internal resistance of 1 group = $\dfrac{r}{nq} = \dfrac{0.1}{4} = 0.025$ ohm

The internal resistance of 2 groups in series

$$= r \times ns = 0.025 \times 2 = 0.05 \text{ ohm.}$$

By Ohm's Law

$$I = \frac{E}{r + R} = \frac{3}{0.05 + 2} = \frac{3}{2.05} = 1.463 \text{ amperes} \quad \text{(Ans.)}$$

Secondary Cells.—Primary cells (dry cells) produce electric currents as a result of chemical action. Secondary cells do not in themselves produce current but have the property of " storing " electric current with which they may be charged and will later give up the current which has been accumulated. Such cells are called *storage cells* or *accumulators*. When these cells are connected in groups of two or more, the group is called a *storage battery*.

Storage batteries are widely used for stand-by emergency service in power substations and in telephone and telegraph work. However, to a great majority of people the storage battery connotes an ignition unit of the automobile or internal combustion engine. For this reason a full treatment of this subject will be found on page 708 of this book.

ELECTROMAGNETS

Magnetization of Iron and Steel by an Electric Current.— When a number of turns of insulated wire are wound around a soft iron bar and a current is sent through the wire, the bar will attract iron filings. This property is called *magnetism*. The wire wound around the iron core is called an *electromagnet*. An electromagnet differs from a permanent iron magnet in that it has magnetic properties only when a current flows through the wire. If a piece of cardboard is fitted around the longitudinal axis of an, electromagnet and iron filings sprinkled around generously (Fig. 40), the filings will not only be attracted by both ends of the magnet, called the *poles*, but will also arrange themselves in a regular order at some distance from the coil. The lines which these filings form represent the *lines of force* of the *magnetic field* about the

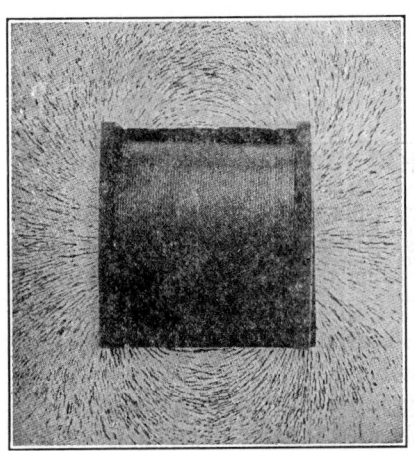

FIG. 40.—Lines of Force Around a Coil.

magnet. If a piece of iron is laid in the magnetic field the lines of force will converge to it at both ends. It is apparent, therefore, that the lines of force find it easier to pass through the piece of iron than through the air or through the filings. *The capability of any substance for conducting magnetic lines of force is termed its permeability.*

Solenoids. — A coil of wire wound on an insulating spool is called a *solenoid*. The winding is always in the same direction, layer upon layer, similar to the thread on a spool. If a solenoid is suspended by a thread from the midpoint of its longitudinal axis it will swing into a position with one end pointing north and the other end south. The pole to the north, or the north-seeking-pole is called the N-pole and the opposite end the S-pole. It is a phenomenon of magnetism that unlike poles of magnets attract each other and like poles repel each other.

A solenoid with an iron core is, as we have seen, an *electro-magnet*. If the core is fitted loosely so that it may be pulled out it will be subjected to a strong "sucking" action when a current is passed through the coil (Fig. 42). This principle is extensively employed to operate the feeding mechanism of electric arc lights, to close switches at a distance for remote control purposes, and in automatic circuit breakers.

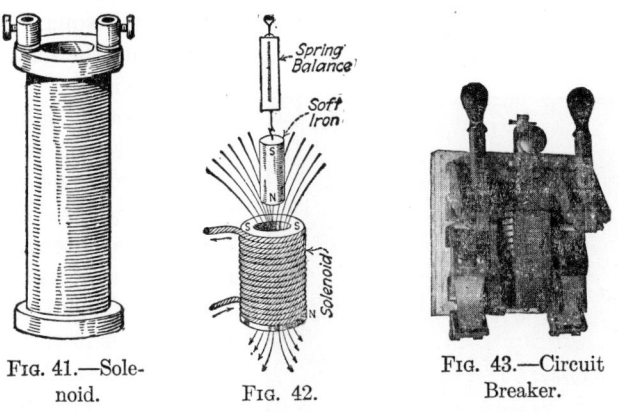

FIG. 41.—Sole-
noid.   FIG. 42.   FIG. 43.—Circuit
                                              Breaker.

A circuit breaker is used to protect electrical circuits against abnormal conditions arising therein. The most common form is the overload type which opens the circuit when the current becomes excessive. Circuit breakers are also used for opening the circuit if the voltage falls below a certain value or if the polarity of a direct-current circuit is reversed.

Applications of the Electromagnet. — If, instead of winding a coil around a straight bar, a bar in the form of a horseshoe is used, bringing the N-pole and S-pole close together, a much stronger

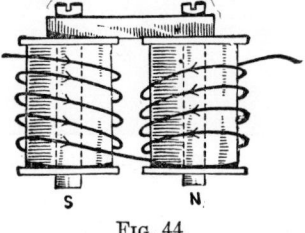

FIG. 44.

magnet will result. In actual practice, the wire is not wound onto the bar but is wound onto spools which are slipped over the bar. The bar need not be in one piece and is commonly made up of three pieces as shown in Fig. 44. These are called the pole pieces and the yoke. The electromagnet finds many applications in this form in electric bells, buzzers, telegraph sounders and relays, etc. Electromagnets of very powerful attractions are built for industrial use in handling iron and steel. These consist of a steel casting having a groove turned to receive the exciting coil. The lifting power of an electromagnet is proportional to the square of the product of the amperes flowing in the magnet and the number of turns of wire.

FIG. 45.—Lifting Magnet.

The most important use of electromagnets is in generators and motors, where they are used to create the intense magnetic fields necessary for the development of electrical power in the case of the generator and the rotation of the armature in the case of a motor.

Magnetomotive Force. —Magnetism or *magnetic flux* (total number of lines of force) depends upon the number of turns of wire in the coil of an electromagnet as well as upon the current strength; the current and number of turns being jointly responsible for the force that drives the magnetic flux around the magnetic circuit, just as an electromotive force drives an electric current around an electric circuit. The magnetizing force set up by a current flowing through a solenoid or any coil of wire is called the *magnetomotive force* (abbreviated mmf). It is directly proportional to the current and to the number of turns on a solenoid. The magnetomotive force is, therefore, proportional to the product of the number of turns and the current strength. That is, one ampere flowing through ten coils or turns will produce the same magnetomotive force as ten amperes flowing through one turn. The magnetomotive force may be expressed in a unit called the *ampere-turn*. The relationship may be expressed by the formula,

$$\text{mmf in ampere-turns} = I \times T$$

when I = current in amperes

T = number of turns on the coil.

ILLUSTRATION: What is the magnetomotive force of a coil with 50 turns through which a current of 3 amperes is passing?

$$\text{mmf} = I \times T = 3 \times 50 = 150 \text{ ampere-turns (Ans.)}$$

It is evident from this relationship that a magnet with a certain magnetomotive force can be made with heavy wire of low resistance and few turns or with smaller wire of high resistance and many turns. Electric bell, telephone, and telegraph instruments are usually made of fine wire since they are usually located some distance from the battery so that the current may be very small. When it is desired to operate a small magnet on a 110-volt circuit it is wound with fine wire so that its resistance will be high and the current consumed small.

Field Intensity.—In magnetic calculations, the magnetomotive force per unit length of the magnetic circuit is called the *intensity of the magnetic field*. This field intensity is the magnetomotive force divided by the length (l) of the magnetic path and is represented by the letter $\mathscr{H}$. It has been determined experimentally that one ampere-turn will produce 1.257 lines of force through an air-path one centimeter in length and one square centimeter in cross-sectional area.

Therefore, the field intensity is,

$$\mathscr{H} = \frac{\text{mmf}}{l} = \frac{1.257 \times I \times T}{l}$$

where l is the length of the path in centimeters and T the number of turns.

If the length (l) of the magnetic path of a solenoid is known, the mmf necessary to produce a desired field intensity ($\mathscr{H}$), is obtained by multiplying $\mathscr{H} \times l$.

ILLUSTRATION: The coil shown in Fig. 46 has a core which forms a complete ring so that there are no free poles. Each line of force has then a complete path inside the core so that the

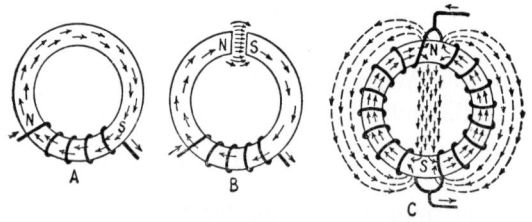

FIG. 46.—Magnetic Polarity of an Iron Ring.

length of the magnetic circuit can easily be measured. If the coil has 30 turns and the current is 15 amperes then

$$\text{mmf} = I \times T = 15 \times 30 = 450 \text{ ampere-turns.}$$

If the mean length of the magnetic circuit is 18 centimeters, then the magnetomotive force per centimeter length is

$$\mathscr{H} = \frac{1.257 \times I \times T}{l} = \frac{1.257 \times 15 \times 30}{18} = 31.4$$

This means that a uniform magnetic field is produced in the solenoid of 31.4 lines per square centimeter.

The difference between the two formulas which have been given should be kept distinctly in mind. The quantity $\mathscr{H}$ represents the force magnetizing a unit length of the core of a solenoid, or the strength of field in lines of force per square centimeter within a coil with an air coil. The quantity mmf represents the force (magnetic pressure) that tends to drive the lines of force throughout the entire path of any kind of material.

If l is given in inches, then $\mathscr{H}$ becomes

$$\mathscr{H} = \frac{.495 \times I \times T}{l}, \text{ in which } l, \text{ is in inches.}$$

Law of the Magnetic Circuit.—Just as electric pressure (emf) is the force that moves electricity through an electric circuit, so magnetic pressure (mmf) drives lines of force through a magnetic circuit. All magnetic substances offer more or less resistance to the passage through them of magnetic lines of force. This magnetic "resistance" is called *reluctance* and its symbol is $\mathscr{R}$. The *total number* of lines of force set up in a magnetic substance is termed the *magnetic flux*. Magnetic flux, or total number of lines of force, is treated as a *magnetic* current flowing in a magnetic circuit.

The calculation of the magnetic flux, which will be represented by N, is similar to the calculation of current in an electric circuit by Ohm's Law. In the latter case, the strength of the electric current equals the electromotive force divided by the resistance, or, $I = \dfrac{E}{R}$. Similarly, in a magnetic circuit

$$\text{magnetic flux} = \frac{\text{magnetomotive force}}{\text{reluctance}}$$

or,

$$N = \frac{\text{mmf}}{\mathscr{R}}$$

Magnetic Density, Permeability and Reluctance.—It is sometimes necessary to specify the flux density in any part of a magnetic circuit, that is, the number of lines passing through a unit area measured at right angles to their direction, whether that part of the circuit is air or some other substance. This number is termed the *magnetic density* or *magnetic induction* of a substance and is denoted by the letter $\mathscr{B}$. If the total flux N is known, and the area A through which it is uniformly distributed, is also known, then the flux density is

$$\mathscr{B} = \frac{N}{A}$$

If A is expressed in square inches, then the flux density will be in number of lines per square inch.

The magnetic density produced in *air* by a solenoid depends upon the magnetic field alone. The magnetic density or induction $\mathscr{B}$ produced in a *magnetic* substance when placed in a solenoid depends also upon the *permeability* of the substance.

The permeability of a magnetic substance is the ratio of the magnetic density $\mathscr{B}$ in the substance to the intensity of the magnetic field $\mathscr{H}$ acting upon the substance; that is a ratio of the number of lines of force per unit area, set up in the material, to the number that would be set up in air under the same conditions. The symbol for permeability is the Greek letter μ (pronounced mu), and its value for any magnetic substance is expressed in the equation

$$\mu = \frac{\mathscr{B}}{\mathscr{H}}$$

If the value of μ and $\mathscr{H}$ are known, the magnetic density is

$$\mathscr{B} = \mu \times \mathscr{H}$$

The permeability of air or nonmagnetic substances is unity or 1; since through air the flux density $\mathscr{B} = \mathscr{H}$, or $\dfrac{\mathscr{B}}{\mathscr{H}} = 1$.

Soft iron under a field intensity of $\mathscr{H} = 10$ (this corresponds to 20.3 ampere-turns per inch) has a flux density $\mathscr{B} = 14,000$ lines per square centimeter. Consequently, the permeability is

$$\mu = \frac{\mathscr{B}}{\mathscr{H}} = \frac{14,000}{10} = 1400$$

In magnetic materials, the value of the permeability does not remain the same for all flux densities. It varies as shown in **Table 1** below:

TABLE 1

FLUX DENSITY AND PERMEABILITY

Flux Density		Permeability		
Lines per square inch	Lines per square centimeter	Annealed sheet steel	Cast steel	Cast iron
20,000	3,100	2600	1400	280
30,000	4,650	2900	1500	230
40,000	6,200	3100	1400	160
50,000	7,750	3200	1350	110
60,000	9,300	3100	1250	80
70,000	10,850	2400	1100	65
80,000	12,400	1800	750	50
90,000	14,000	1400	500	
100,000	15,500	750	280	
110,000	17,400	320	145	
120,000	18,600	160	70	
130,000	20,150	75		

The reluctance of a magnetic circuit depends upon three quantities: the *length* of the circuit, the cross-sectional *area* of the circuit, and the *permeability* of the material of the circuit. The reluctance *increases* as the length of the magnetic circuit increases, and *decreases* as the cross-sectional area is increased and the permeability increases. That is, the reluctance is directly proportional to the length of the magnetic circuit, is inversely proportional to the cross-sectional area and varies as the material of the circuit. This may be expressed by the following formula:

$$\mathscr{R} = \frac{l}{A \times \mu}$$

when $\mathscr{R}$ represents the reluctance, l the length of the magnetic circuit in inches, A the sectional area of the circuit in square inches, and μ the permeability of the material constituting the circuit.

Attractive Force of an Electromagnet.—The magnetism of an electromagnet increases as the current through it is increased, up to a saturation point, but is not directly proportional to the current; that is, if when one ampere is passed through a certain magnet, a force of 56 pounds is required to detach its keeper, then when two amperes are passed through it, not twice the force, or 112 pounds is required, but usually much less.

The lifting or adhesive power of an electromagnet is called its *tractive force*. The tractive force is proportional to the square of the density of lines of force per square inch, and the area of surface contact. To determine the tractive force or " pull " in pounds of an electromagnet, let

$\mathscr{B}$ = flux density or lines of force per square inch.

A = area of contact in square inches.

Then, the pull in pounds is

$$P = \frac{\mathscr{B}^2 \times A}{72,134,000}$$

ILLUSTRATION: What is the tractive force of a magnet if the density of the lines of force per square inch is 96,750 and the area of contact is one square inch?

$$P = \frac{\mathscr{B}^2 \times A}{72,134,000} = \frac{(96,750)^2}{72,134,000} \times A$$

$$P = \frac{9,360,562,500}{72,134,000} = 129.7 \text{ lb.} \text{ (Ans.)}$$

Table 2 gives the traction of electromagnets for various degrees of magnetizations.

GENERATORS AND MOTORS

Dynamo.—A dynamo is a machine which converts either mechanical energy into electrical energy or electrical energy into mechanical energy. A dynamo which converts mechanical energy into electrical energy is called a *generator*. A dynamo which converts electrical energy into mechanical energy in the form of rotation is called a *motor*.

TABLE 2

Magnetization and Traction of Electromagnets

$\mathscr{B}$ Lines per Sq. Cm.	$\mathscr{B}''$ Lines per Sq. Inch	Dynes per Sq. Cm.	Grammes per Sq. Cm.	Kilogs per Sq. Cm.	Pounds per Sq. Inch.
1,000	6,450	39,790	40.56	.04056	.577
2,000	12,900	159,200	162.3	.1623	2.308
3,000	19,350	358,100	365.1	.3651	5.190
4,000	25,800	636,600	648.9	.6489	9.228
5,000	32,250	994,700	1,014	1.014	14.39
6,000	38,700	1,432,000	1,460	1.460	20.75
7,000	45,150	1,950,000	1,987	1.987	28.26
8,000	51,600	2,547,000	2,596	2.596	36.95
9,000	58,050	3,223,000	3,286	3.286	46.72
10,000	64,500	3,979,000	4,056	4.056	57.68
11,000	70,950	4,815,000	4,907	4.907	69.77
12,000	77,400	5,730,000	5,841	5.841	83.07
13,000	83,850	6,725,000	6,855	6.855	97.47
14,000	90,300	7,800,000	7,550	7.550	113.1
15,000	96,750	8,953,000	9,124	9.124	129.7
16,000	103,200	10,170,000	10,390	10.390	147.7
17,000	109,650	11,500,000	11,720	11.720	166.6
18,000	116,100	12,890,000	13,140	13.140	186.8
19,000	122,550	14,360,000	14,630	14.630	208.1
20,000	129,000	15,920,000	16,230	16.230	230.8

The electric generator operates on the principle of *electromagnetic induction*. This principle is illustrated in Fig. 47. A loop of wire revolving between the two poles of a magnet cuts the magnetic lines of force. This sets up an electromotive force in the loop and causes a current to flow around it. In the position *ABCD* (Fig. 47) the loop cuts no lines of force and therefore no current is induced. However, during the first quarter of the revo-

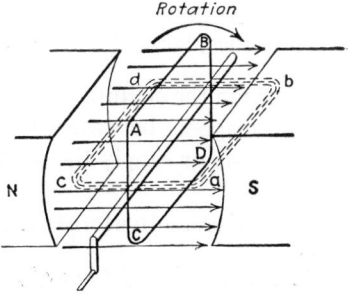

Fig. 47.—Direction and Magnitude of the Induced emf in a Generator.

lution from this point the lines of force are cut at a gradually increasing rate till the loop is in the position *abcd* when the rate of change, and also the electromotive force is a maximum. During the next quarter revolution the cutting of the lines of force gradually decreases until at the end of a half revolution the electromotive force is again zero. During the course of this half revolution the current flows in only one direction, from *a* to *c*, to *d*, to *b*, but the strength has constantly changed from zero to a maximum and back to zero again. During the second half revolution, the same variations in electromotive force occur but the induced current is in the opposite direction. The current is, therefore, reversed twice in every revolution, or an *alternating current flows around the loop*.

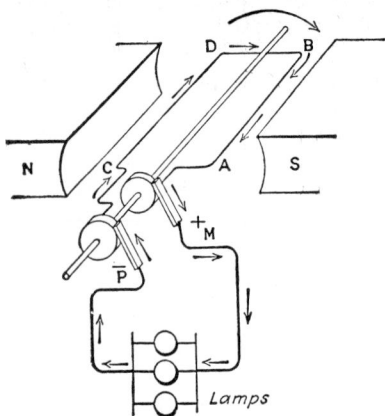

FIG. 48.—Simple Alternating-current Generator. At the instant depicted in the revolution, brush *M* is positive.

Simple Alternating Current Generators.—In order to use in an external circuit the current generated in the revolving loop it is necessary to employ a connecting device consisting of two *collector rings* and brushes insulated from the shaft and from each other. Figures 48 and 49 show the elements of an alternating current generator and the two positions of the loop illustrate the reversal of current in the circuit.

The magnets between which the loop revolves are called the *field magnets* or simply the *field*. The revolving loop is called the *armature*.

The electromotive force produced by a generator depends upon:

1. *The number of lines of force cut by the armature wires.*

2. *The number and length of the cutting wires,*

3. *The speed at which the armature revolves and the lines of force are cut.*

It is apparent, therefore, that if instead of the single loop shown in Fig. 48 an iron core with many turns of wire is substituted, the lines of force between the field magnets will be in-

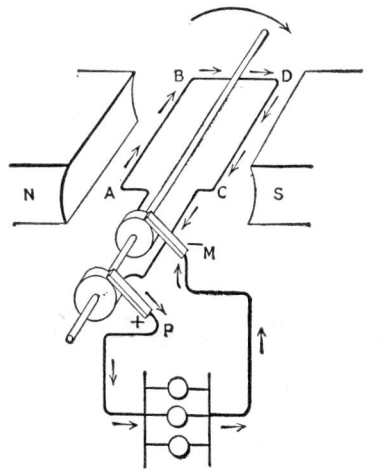

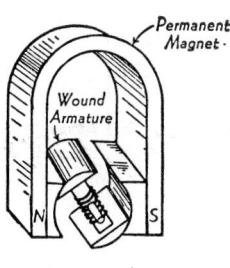

Fig. 49.—Simple Alternating-current Generator. Direction of current in coil at one-half revolution from the positive in Fig. 48; brush *M* is now negative.

Fig. 50.—Principle of Armature-type Magneto.

creased and the number of wires cutting these lines will be increased. The result is that the electromotive force is greatly increased. The *magneto generator* (Fig. 50) is constructed on this principle. It consists of a coil of wire revolving between permanent magnets.

Simple Direct Current Generators.—In order to obtain current flowing in only one direction from a generator, it is necessary to intercept the current from the revolving loop in such a manner that the electromotive force generated by each half revolution is transmitted to separate branches of the external circuit. This is accomplished by substituting one split ring for the two collector rings as shown in Fig. 51. This split ring is called the *commutator*. Brushes rest on the ring at diametrically opposite points, one having a *positive* polarity and the other *negative*.

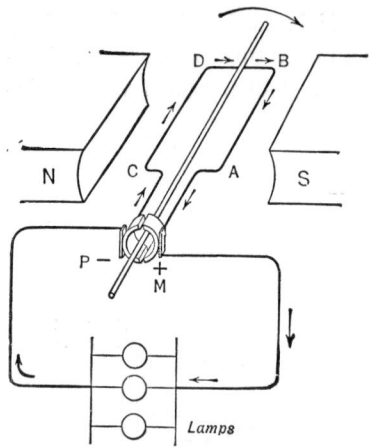

FIG. 51.—Simple Direct-current Generator. At the instant depicted in the revolution, brush *M* is positive.

Principle of the Motor.— If an electric current is passed through a coil or a loop it will create a magnetic field with an N-pole on one side and an S-pole on the other. If this loop is then placed between the poles of a magnet as in Fig. 52, it will tend to turn until its lines of force are in line with the lines of force of the field magnets. When it reaches this point the rotation stops. In order to obtain continuous rotation it is necessary to reverse the current in the loop at the instant that the turning effect ceases. These reversals are automatically performed by the commutator when the brushes are correctly set and adjusted.

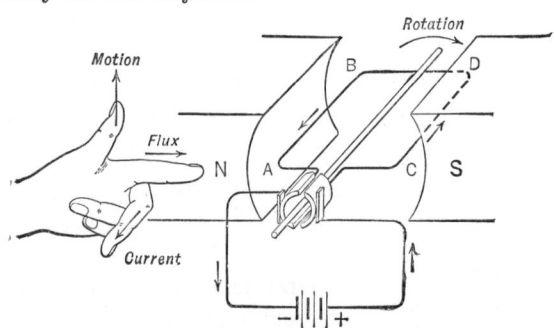

FIG. 52.—Single Loop Armature Driven as a Motor.

The direction of rotation of a motor can be found by the left-hand rule as illustrated in Fig. 52. When the polarity of the field magnets and the direction of the current through the armature have been determined, place the left hand so that the fingers correspond with the polarity and direction of current in the single armature coil motor, and it is found that the loop will rotate in the direction of the hands of a clock. The direction of rotation of a motor can be changed by reversing the current either through the armature or through the fields, but not through both.

Classification of Dynamos According to Their Field Excitation. —Practical dynamos are different in several respects from the elemental forms which have been discussed in the preceding paragraphs. Instead of the revolving loop, the armature consists of a number of coils; instead of a split ring, the commutator consists of a number of segments or sections; and instead of permanent magnets, the field consists of electromagnets. The field magnets may be magnetized by current from a separate generator or by the machine itself and the generator would be styled a *separately-excited* or a *self-excited* generator, respectively. Generators may be classified according to methods used to excite the field magnets as follows:

(a) *Magneto Machines* (Fig. 50).—The field magnets are permanent magnets of horseshoe form and the armature is designed for either direct or alternating current. Such machines supply limited power and are used chiefly in gasoline engine work, telephone signalling, testing of circuits, and firing electric blasting detonators.

(b) *Series Machines* (Fig. 53) (*Constant Current*).—The field magnets are connected in series with the armature and wound with a few turns of heavy wire having a low resistance, so as to present little opposition to the main current flowing through them. Series generators are used only for series arc street-lighting circuits and in the Thury system of high-voltage direct-current power transmission.

In a *constant-current* circuit supplied by a series generator, the current is maintained constant through the external circuit while the electromotive force varies with each change in the resistance of the circuit. The series constant-current generator is now little used.

(c) *Shunt Machines* (Fig. 54) (*Constant Potential.*)—The field magnets are connected in parallel or shunt with the armature and are wound with many turns of small wire; they have a high resistance, compared with the armature, since only a small portion of the current need flow through them.

(d) *Separately-excited Machines* (Figs. 55 and 56) (*Constant*

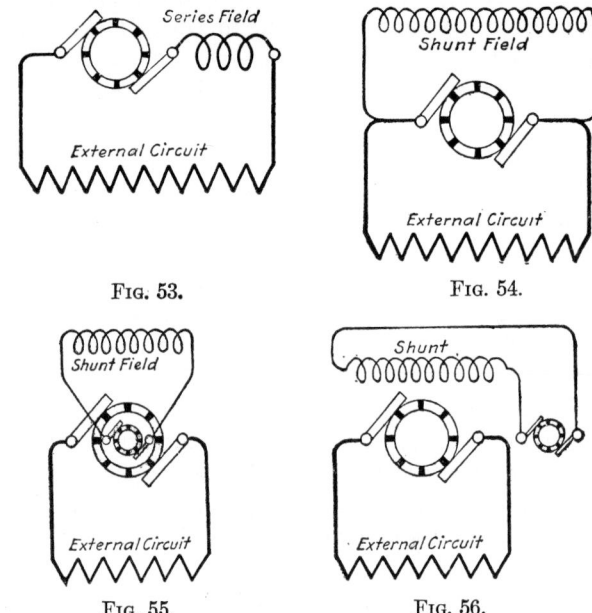

FIG. 53.   FIG. 54.

FIG. 55.   FIG. 56.

FIGS. 53–56.—Classification of Generators according to the Method of Exciting the Field Magnets.

Potential).—Current for the field magnets is supplied from a separate generator. In Fig. 55 this generator forms a part of the main machine by having a separate armature on the same shaft, while in Fig. 56 the field is supplied by a distinct machine called an *exciter*.

(e) *Compound Short-shunt Machines* (Fig. 57) (*Constant Potential*).—The field cores contain two independent spools. One is wound with a few turns of heavy wire, forming the *series coil*, and connected in series with the main circuit; the other with a great many turns of smaller wire, forming the *shunt coil*, and connected in shunt with the armature.

(*f*) *Compound Long-shunt Machines* (Fig. 58) (*Constant Poten-tial*).—The same as (*e*) except that the shunt field bridges not

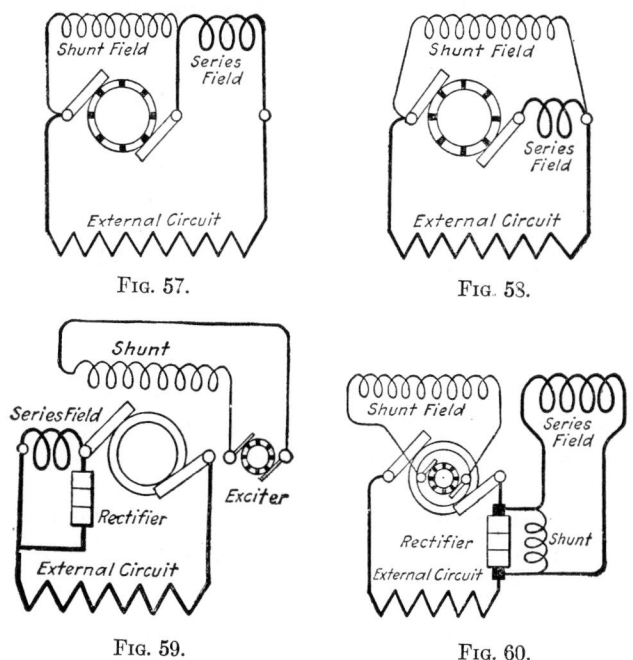

FIG. 57.

FIG. 58.

FIG. 59.

FIG. 60.

FIGS. 57–60.—Classification of Generators according to the Method of Exciting the Field Magnets.

only the armature but also the series field; hence it is called a *long shunt*.

(*g*) *Separately-excited Alternating-current Generators* (Figs. 55 and 56).—The field magnets are excited by direct current from a separate exciter. Alternating current generators, or alterna-tors, always require an exciter, since the alternating current cannot be employed to excite the fields. The exciter may be either a separate generator or an independent direct-current winding upon the alternator shaft.

(*h*) *Compound Separately-excited Alternating-current Generators* (Fig. 59).—Two independent field windings correspond to the series and shunt coils of Fig. 57. The shunt coil is supplied from an exciter, while the main current, commuted, flows through the series field coils. This method is employed in the *composite-wound* alternators, a portion of the main alternating current is commuted by a device called a *rectifier*, located on the armature shaft. Its function is to change that portion of the alternating current intended for the series coils into a direct current for producing the magnetization. Figure 60 shows a self-contained composite wound alternator.

Generators may be further divided into the following three classes according to their mechanical arrangement:

1. *A stationary field magnet and a revolving armature,*

2. *A stationary armature and a revolving field magnet,*

3. *A stationary armature and a stationary field magnet, between which is revolved a toothed iron core.*

Induced Voltage of a Generator.—It has been pointed out that the voltage or the electromotive force produced by a generator depends upon the following three conditions:

1. The number of lines of force cut by the armature wires,

2. The number and length of the cutting wires,

3. The speed at which the armature revolves.

It has been determined experimentally that an electromotive force of one volt is generated when one turn of wire cuts 100,000,000 (usually written 10^8) lines of force in one second. The induced voltage is then the product of the number of lines of force, or flux, and the number of times these are cut by the wire in one second, divided by 10^8.

ILLUSTRATION: How many volts are generated in a wire which cuts 4,000,000 lines of force 1,200 times a minute?

The rate of cutting is $\dfrac{1,200}{60} = 20$ times a second.

Then,

$$\text{Induced voltage} = \frac{4,000,000 \times 20}{100,000,000} = 0.8 \text{ volt} \quad (\textbf{Ans.})$$

From these relationships it is possible to develop the following formula for the volts developed in the armature of a generator when the number of poles is the same as the number of paths through the armature:

$$E = \frac{CNR}{10^8}$$

when

E = generated electromotive force in volts

C = the number of active armature conductors

N = the flux per pole

R = the speed of the armature in revolutions per second.

ILLUSTRATION: What voltage is generated by a dynamo having 175 active conductors on its armature if the flux per pole is 4,000,000 lines and the speed of rotation 1500 rpm?

In this case, $C = 175$, $N = 4,000,000$, and $R = \dfrac{1500}{60}$

Then,

$$E = \frac{CNR}{10^8}$$

$$E = \frac{175 \times 4,000,000 \times 1500}{100,000,000 \times 60}$$

$$E = 175 \text{ volts} \quad (\textbf{Ans.})$$

ILLUSTRATION: An armature generates 220 volts of electro-motive force when rotating at a speed of 1200 rpm. What is the flux per pole if there are 250 active armature conductors?

In this case, $E = 220$, $C = 250$, $R = \dfrac{1200}{60} = 20$

The formula $E = \dfrac{CNR}{10^8}$ may be transposed to

$$N = \frac{E \times 10^8}{CR}$$

Substituting known values,

$$N = \frac{220 \times 100,000,000}{250 \times 20}$$

$$N = 220 \times 20,000$$

$$N = 4,400,000 \quad \text{(Ans.)}$$

Action of a Shunt Generator.—Since a part of the current generated by a shunt generator is used to energize the field magnets, the voltage in the external circuit is something less than the induced electromagnetic force. If the potential of the external circuit measures 112 volts, the induced electromotive force will be $112 + I \times r$, where I equals the current through the fields and r equals the armature resistance.

A field rheostat is used to adjust the voltage in the external circuit. If this is set with the main circuit open so that the voltage will be, for example, 112 volts and the switch is closed so that more current flows from the armature, the voltmeter will at once indicate a lower potential of about 108 volts. If the speed is the same as before, the loss is due to two causes: first, there is an increased drop in the armature due to the additional current flowing through it, which lowers the potential difference at the brushes; second, the potential difference at the brushes being lowered, less current flows around the field so that there are not quite as many lines of force as before.

A statement of the voltage of a generator at no load and when carrying full load is spoken of as its *voltage regulation*. The *percentage regulation* is the ratio of the change in voltage between no-load and full-load to the voltage at full-load.

% voltage regulation =

$$\frac{(\text{no-load voltage}) - (\text{full-load voltage}) \times 100}{\text{full-load voltage}}$$

ILLUSTRATION: The voltage of a shunt generator when operating at no load is 112 and when operating at full load is 108. What is its voltage regulation?

Percent regulation =

$$\frac{112 - 108}{108} = \frac{4}{108} = 0.037 = 3.7 \text{ percent} \quad (\textbf{Ans.})$$

Shunt generators are adapted only to installations where the load is fairly constant, when they require very little attention after the proper adjustment of the field rheostat has been made.

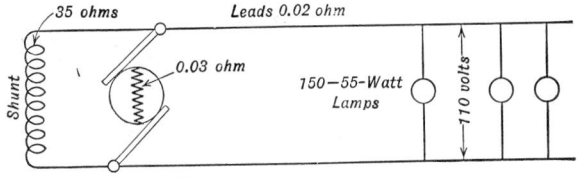

FIG. 61.

SHUNT GENERATOR PROBLEM: A shunt generator, Fig. 61, maintains 110 volts across 150 incandescent lamps joined in parallel, requiring 55 watts and 110 volts each. The lamps are located a distance from the generator and the resistance of the leads is 0.02 ohm. Resistance of the armature is 0.03 ohm and of the field coils is 35 ohms.

1. What is the potential difference at the brushes?

$$I = \frac{P}{E} = \frac{55}{110} \times 150 = 75 \text{ amperes for lamps}$$

$E = I \times R = 75 \times 0.02 = 1.5$ volt drop in leads

$110 + 1.5 = 111.5$ volts potential difference at brushes (Ans.)

2. What is the total electromotive force generated?

$$I = \frac{E}{R} = \frac{111.5}{35} = 3.19 \text{ amperes through the fields}$$

$75 + 3.19 = 78.19$ amperes through armature

$E = I \times R = 78.19 \times 0.03 = 2.35$ volts drop in armature

$111.5 + 2.35 = 113.85$ volts total emf (Ans.)

3. What are the watts lost in the armature?

$P = E \times I = 2.35 \times 78.19 =$
183.7 watts lost in armature (Ans.)

4. What are the watts lost in the field?

$P = E \times I = 111.5 \times 3.19 =$
355.7 watts lost in the field (Ans.)

5. What watts are lost in the leads?

$P = E \times I = 1.5 \times 75 = 112.5$ watts lost in leads (Ans.)

6. What power is supplied to lamps?

$P = E \times I = 110 \times 75 = 8,250$ watts supplied to lamps   (Ans.)

Compound Machines.—The compound-wound generator possesses the characteristics of both the series and the shunt dynamos. It is designed to give automatically a better regulation of voltage on constant-potential circuits than is possible with a shunt machine. The shunt field is the same as in the shunt generator and independent series field spools are added, through which the main current flows. When current flows in the external circuit, the voltage at the brushes is not lowered, as in the shunt generator, since the series winding strengthens the field by the current flowing through it and thus raises the voltage in proportion to the increased current. By a proper selection of the number of turns in the series coils, the voltage is thus kept automatically constant for wide fluctuations in load. If a greater number of turns is used in the series coil than required for constant terminal voltage at all loads, the voltage will rise as the load is increased and thus make up for the loss on the transmission lines, so that a constant voltage will be maintained at some point distant from the generator. The machine is then said to be *over-compounded*.

Compound-wound direct-current generators are used extensively in electric lighting and power stations and in electric railway power stations where the load is very fluctuating.

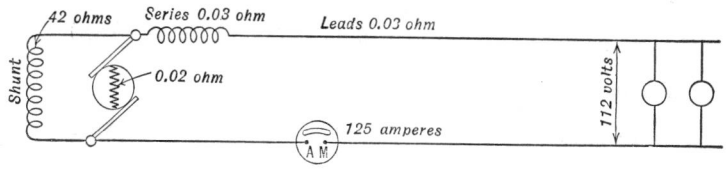

Fig. 62.

COMPOUND GENERATOR PROBLEM: A compound generator, Fig. 62, supplies 125 amperes at 112 volts to a group of lamps located a distance from the generator. The resistances are: Leads, 0.03 ohm; armature, 0.02 ohm; series coil, 0.03 ohm; and shunt coil, 42 ohms.

1. What is the potential difference at the brushes?

$$E = I \times R = 125 \times 0.03 = 3.75 \quad \text{volts} \quad \text{drop} \quad \text{in leads.}$$

$112 + 3.75 = 115.75$ volts potential difference at terminals

$$E = I \times R = 125 \times 0.03 = 3.75 \quad \text{volts} \quad \text{drop} \quad \text{in series field.}$$

$115.75 + 3.75 = 119.50$ volts pd at brushes (Ans.)

2. What is the total electromotive force generated?

$$I = \frac{E}{R} = \frac{119.50}{42} = 2.8 \text{ amperes through shunt field}$$

$125 + 2.8 = 127.8$ amperes total current through armature

$$E = I \times R = 127.8 \times 0.02 = 2.556 \text{ volts drop in armature}$$

Total emf = 112 volts (lamps) + 3.75 volts (leads) + 3.75 volts (series coil) + 2.556 volts (armature) =

122.06 volts (Ans.)

3. What are the watts lost in the leads?

$P = I^2 \times R = 125 \times 125 \times 0.03 = 468.75$ watts (Ans.)

4. What are the watts lost in the series coil?

$P = I^2 \times R = 125 \times 125 \times 0.03 = 468.75$ watts (Ans.)

5. What are the watts lost in the shunt coil?

$P = I^2 \times R = 2.8 \times 2.8 \times 42 = 329.28$ watts (Ans.)

6. What are the watts lost in the armature?

$P = I^2 \times R = 127.8 \times 127.8 \times 0.02 = 326.66$ watts (Ans.)

7. What is the power supplied to the external circuit?

$P = E \times I = 115.75 \times 125 = 14{,}468.75$ watts (Ans.)

Losses in a Dynamo.—The losses of power in a dynamo fall into two general classes:

(1) *Mechanical Losses,*
(2) *Electrical Losses.*

(1) The mechanical losses include the friction between the armature shaft and its bearings, windage, and the friction of the brushes on the commutator. These friction losses are practically constant for all speeds.

(2) The electrical losses include the I^2R losses in the armature and fields and at the brush contacts, the losses due to eddy currents and hysteresis. The losses in the field rheostat when it is in series with the field magnets of a generator should be included, even in separately-excited machines.

All the losses may then be summed up as due to:

(1) Mechanical friction
(2) Electrical friction (resistance)
(3) Magnetic friction (hysteresis)

Efficiency of a Generator.—The efficiency of a generator is the ratio of the power output to the power input. When specific load conditions are not referred to it is always understood that the efficiency is expressed as of full or rated load. Instead of attempting to determine the mechanical power input of a generator, it is sometimes more convenient to obtain an equivalent figure indirectly by adding the value of the losses to the output. We may then state,

$$\text{efficiency} = \frac{\text{output}}{\text{input}} = \frac{\text{output}}{\text{output \& losses}}$$

$$\text{efficiency} = \frac{P}{P + p}$$

when P = output of generator in watts

p = total losses of generator in watts

ILLUSTRATION: If it requires 57 kw to drive a 50-kw generator, what is its efficiency?

Here $P + p = 57$

Then, $\quad \text{eff} = \dfrac{P}{P + p} = \dfrac{50}{57} = 0.88 = 88$ percent (Ans.)

Two efficiencies are recognized with electrical machinery, *conventional efficiency* and *directly-measured efficiency*. Unless other-

wise specified, conventional efficiency is the one employed. Conventional efficiency of machinery is the ratio of the output to the
sum of the output and the losses; or of the input minus the
losses, to the input. In either case conventional values are assigned
to one or more of the losses. This is necessary because it is practically impossible to measure some of the losses in electrical
machinery.

The efficiency of a generator varies with the size of the machine
and the load it is supplying. For example, a 5-kw dynamo may
have as low an efficiency as 80 percent; a well-designed 40-kw
machine, 90 percent, and a 500-kw generator, 94 percent. Again,
a certain 200-kw generator may have an efficiency of 93 percent
at full load, 92 percent at three-quarter load, 90 percent at half
load, and 84 percent at one-quarter load.

Direct Current Motors.—The principle of the operation of a
motor is described on page 744 and much of the descriptive matter
in the preceding paragraphs on generators applies equally well
to motors. Motors may be classified as (a) *series* wound, (b)
shunt wound, and (c) *compound wound*.

Counter Electromotive Force of a Motor.—The wires of a
motor armature, rotating in its own magnetic field, cut the lines
of force just as if the armature were being driven as in a generator.
Hence, there is an induced electromotive force in the wires.
This induced pressure is in a direction opposite to that of the
current applied to the armature. It is called the *counter electromotive force* and is always in such direction as to oppose the current applied at the terminals. A motor with no load will run at
such a speed that the counter electromotive force is nearly equal
to the applied pressure.

The counter electromotive force of a motor running at any
speed will be the same as when it is run as a generator at this
speed, provided the field strength is the same in both cases.
Hence, to find the counter emf of a motor at any speed, run it as
a generator at this speed and measure the induced emf by a voltmeter.

The counter emf in a motor can never equal the applied emf,

but is less by an amount equal to the drop in the motor armature. To find the current flowing through the armature of a motor:

Subtract the counter emf from the applied emf and divide this result by the armature resistance. This, Ohm's Law applied to a motor, may be expressed:

$$I = \frac{E - \mathscr{E}}{r} = \frac{\text{voltage drop in armature}}{r}$$

when E = emf applied at motor brushes

$\mathscr{E}$ = counter emf developed by motor

I = current through motor armature

r = internal resistance of motor armature

ILLUSTRATION: A motor is connected to a 110-volt circuit. Its counter emf is 105 volts at a particular speed. What current is being supplied to the motor if the resistance of the armature is 1 ohm?

$$I = \frac{E - \mathscr{E}}{r} = \frac{110 - 105}{1} = 5 \text{ amperes (Ans.)}$$

The speed which any motor attains is such that the counter emf developed and the drop in the armature are exactly equal to the applied emf. This may be expressed by a transposition of the preceding formula.

Counter emf + $(I \times r)$ = applied emf,

or $\mathscr{E} + (I \times r) = E.$

The voltage drop in the armature of a motor is a small percentage of the applied pressure, perhaps 2 percent of the terminal pressure in a 500-kw motor and about 5 percent in a 1-kw motor, so that the counter emf is not much different from the applied emf. Since the power driving a motor equals the applied pressure times the current, most of which is usefully expended in mechanical output, the counter emf is an essential and valuable feature of a motor rather than a detriment.

To find the counter electromotive force of a motor:

Multiply the resistance of the armature by the current flowing through it and subtract this product from the emf applied to the motor brushes. This may be expressed as follows by again transposing the preceding formulas:

$$\mathscr{E} = E - (I \times r)$$

ILLUSTRATION: The armature resistance of a shunt-wound motor is 0.7 ohm; and at a certain load 10 amperes flow through it; the voltage at the motor brushes is 112 volts. What is the counter emf?

$$\mathscr{E} = E - (I \times r) = 112 - (10 \times 0.7) = 105 \text{ volts} \quad \text{(Ans.)}$$

When a motor is just starting, it is obvious that it has no counter emf. Then, if it were directly connected to the supply

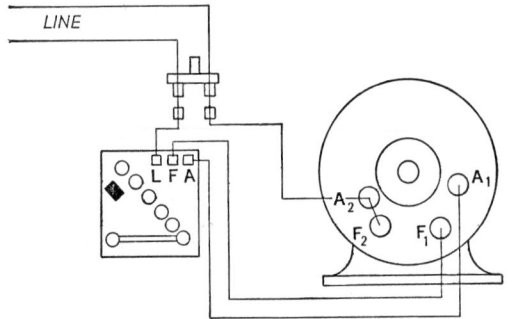

A$_1$ Armature Terminal   A$_2$–F$_3$–F$_1$– Field Terminal
L–Line Terminal,   F–Field Terminal   A–Armature Terminal

FIG. 62-A.

mains, a tremendous amount of current would flow through the armature since its resistance is very low. This might result in considerable damage to the windings before a sufficient counter emf has been built up to check the flow. The problem is solved by using a rheostat called a *starting box* to limit the current or lower the voltage until the motor attains its proper running speed. Such starting boxes are always used in the armature circuits of large shunt motors.

Mechanical Power of a Motor.—To find the mechanical power developed by a motor:

Multiply the counter emf by the current through the armature.

$$P = \mathscr{E} \times I$$

The mechanical power developed includes that dissipated as mechanical friction losses and the power which is expended in eddy currents and hysteresis.

ILLUSTRATION: A small 110-volt motor whose armature resistance is 0.5 ohm runs at a speed to develop a counter emf of 105 volts.

1. What power is developed by this motor?

$$I = \frac{E - \mathscr{E}}{r} = \frac{110 - 105}{0.5} = 10 \text{ amperes}$$

then $P = \mathscr{E} \times I = 105 \times 10 = 1050$ watts (Ans.)

2. What power is supplied to this motor?

$$P = E \times I = 110 \times 10 = 1100 \text{ watts} \text{(Ans.)}$$

Large motors are tested for output by coupling them to generators and measuring the power which is developed by the latter.

Output and Efficiency of Motors.—The capacity of motors to perform useful work is rated according to the amount of power they will maintain at full load at their pulleys, within the limit of permissible heating. The efficiency of a motor, as in the case of the generator, is the ratio of output to input. The energy furnished to a motor is readily measured and from this must be subtracted the losses in the motor to obtain the available energy. These losses are, (1) the I^2R losses in the armature and fields, and the stray power loss, which includes friction, eddy currents and hysteresis.

$$\text{Efficiency} = \frac{\text{output}}{\text{input}} = \frac{\text{input} - \text{losses}}{\text{input}}$$

ILLUSTRATION: A 6-H.P. 110-volt shunt-wound motor has an armature resistance of 0.2 ohm and a field resistance of 40 ohms. The counter emf for a certain speed under load is 100 volts and the stray power loss is 300 watts.

(1) What is the efficiency?

$$\text{Armature current} = I = \frac{E - \mathscr{E}}{r} = \frac{110 - 100}{0.2} = 50 \text{ amperes}$$

$$\text{Field current} = I = \frac{E}{R} = \frac{110}{40} = 2.75 \text{ amperes}$$

Voltage drop in armature $= 110 - 100 = 10$ volts

Power loss in armature $= P = E \times I = 10 \times 50 = 500$ watts

Power loss in field $= P = E \times I = 110 \times 2.75 = 302.5$ watts

Stray power loss $= 300$ watts

Total loss $= 500 + 302.5 + 300 = 1102.5$ watts

Power input in armature $= P = E \times I = 110 \times 50 =$
$$5500 \text{ watts}$$

Power input in field $= 302.5$ watts

Total power input $= 5500 + 302.5 = 5802.5$ watts

$$\text{Efficiency} = \frac{\text{input} - \text{losses}}{\text{input}} = \frac{5802.5 - 1102.5}{5802.5}$$

$$= \frac{4700}{5802.5} = 0.81 = 81\% \quad \text{(Ans.)}$$

(2) What is the power output?

$$\text{Motor output} = \frac{4700}{1000} = 4.7 \text{ kw. or } \frac{4700}{746} = 6.3 \text{ H.P. (Ans.)}$$

Current Required by Motor.—When the output, efficiency and voltage are known, the current required by the motor can be determined by the following rule:

If the output of the motor is expressed in kilowatts (kw.), *mul-*

tiply the kw. rating by 1000 *and divide by the voltage of a motor and its efficiency.* Expressing this as a formula,

$$I = \frac{\text{kw.} \times 1000}{E \times \%M}$$

when E = voltage required by the motor,

kw. = kilowatt rating of the motor,

$\%M$ = efficiency of the motor expressed as a decimal.

ILLUSTRATION: What current is required by a 30-kw., 220-volt motor whose efficiency is 85%?

$$I = \frac{\text{kw.} \times 1000}{E \times \%M} = \frac{30 \times 1000}{220 \times 0.85} = 160 \text{ amperes} \quad (\text{Ans.})$$

When the rating is given in horsepower (H.P.), *multiply the H.P. by* 746 *and divide this product by the voltage of the motor and by its efficiency.* This becomes,

$$I = \frac{\text{H.P.} \times 746}{E \times \%M},$$

when H.P. = horsepower of the motor and the other factors are as above.

ILLUSTRATION: What current will be required by a 2-H.P. 110-volt motor whose efficiency is 90%?

$$I = \frac{\text{H.P.} \times 746}{E \times \%M} = \frac{2 \times 746}{110 \times 0.90} = 15 \text{ amperes} \quad (\text{Ans.})$$

ALTERNATING CURRENTS

Advantages of Alternating Current.—An *alternating current* of electricity is a current which changes its direction of flow at regular intervals of time, usually much shorter than one second.

Alternating current has several advantages over direct current principally in transmission and distribution and for this reason, nearly all of the current generated today is alternating current.

The following problem illustrates the economy which can be effected in the transmission of power by the use of high voltages obtainable only with alternating current.

ILLUSTRATION: 50,000 watts (50 kw.) of power are to be transmitted with a line drop of 2 percent. If the weight of copper required when the energy is delivered at 100 volts is assumed to be 1000 pounds, then the amounts of copper necessary for other voltages are as follows:

Line Voltage, E	Line Current, I amperes	Line Drop, e, volts	Power Loss, Ie, watts	Line Resistance, $R = \dfrac{e}{I}$, ohms	Copper, Pounds
100	500	2	1000	0.004	1000
200	250	4	1000	0.016	250
500	100	10	1000	0.100	40
1000	50	20	1000	0.400	10

These figures show that the *weight of copper wire required for conducting a certain amount of energy with the same percentage loss on the line is inversely proportional to the square of the transmitting voltage.*

It can also be observed that for the transmission of the same amount of power, the increase in line voltage, E, is accompanied by a proportionate decrease in line current. For the same power loss of 1000 watts the reduction of the current from 500 amperes to 250 amperes effects a saving in wire size. This is shown in the resistance column. For 500 amperes a line of 0.004 ohm resistance is used and for 250 amperes a line four times this resistance or 0.016 ohm is used. This indicates that wire of only one-quarter the weight is used to transmit 250 amperes as compared with 500 amperes. The figures in the last column show this fact.

It is not feasible to build direct-current generators to deliver current at higher than 5000 volts, the limitation being in insulation and commutation. Therefore, in order to obtain the economies of high-voltage power transmission, it is necessary to use alternating current. Alternators can be designed for as much as 20,000 volts because the stationary armature can be more readily insulated. Another factor in this consideration is that the voltage of direct current can be changed only by the coupling of two machines in a motor-generator set. On the other hand, transformers can be used to change alternating current efficiently over a wide range.

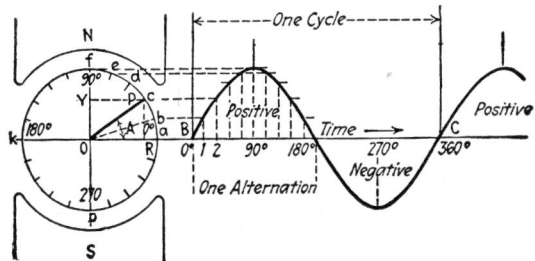

FIG. 63.—Plotting a Sine Curve.

Cycles and Frequency of Alternating Current.—In the discussion of elemental generators it was seen that the electromotive force produced in each coil of an armature rises from zero to a maximum, then declines gradually to zero again, reverses in direction, gradually attaining a maximum in the reversed direction, and then returning to zero. If the value of the electromotive force of one revolution is plotted as the ordinate and time as abscissa, the resulting curve will be as shown in Fig. 63. This is called a *sinusoid* or *sine curve*.

When the alternating current or emf has passed from zero to its maximum value in one direction, to zero, then to its maximum value in the other direction, and back to zero, the complete set of values passed through in that time is called a *cycle*. This cycle of changes takes place in a certain length of time called a *period*. The number of complete cycles in one second is called the *frequency* of the voltage or current. Frequency is, then, *cycles per second* and is sometimes spoken of merely as *cycles*. That is, if an alternator performs the cycle of events depicted in Fig. 63 from *B* to *C* sixty times a second, it is said to have a *frequency* of 60 *cycles*. This would mean 120 changes in direction or *alternations* per second. Frequencies of 25 and 60 cycles are standard in the United States.

To find the frequency in cycles of the voltage or current from any alternating current generator:

Multiply the number of pairs of poles by the speed of the armature in revolutions per second. This may be expressed as

$$f = P \times \frac{N}{60} = \frac{p}{2} \times \frac{N}{60} = \frac{p \times N}{120}$$

when f = frequency (cycles per second)

  P = number of *pairs* of poles

  N = speed in revolutions per *minute*

  p = number of *poles*.

ILLUSTRATION: What is the frequency of the current furnished by an alternator having 24 poles and running at a speed of 300 revolutions per minute?

$$f = P \times \frac{N}{60} = \frac{24}{2} \times \frac{300}{60} = 12 \times 5 = 60 \text{ cycles. (Ans.)}$$

With both the current and emf of alternating current constantly fluctuating, instantaneous values of these qualities are not of great practical concern. Meters used to measure alternating current voltage, amperage and wattage, measure only the average or *effective* values. Alternating currents are expressed in terms of the value of the direct current which would produce the same heating effect and this is called the *effective* value.

Phase and Polyphase.—When the current and the voltage of an alternating current both reach a maximum at the same time they are said to be *in phase*. (Fig. 64a.) If they do not reach a maximum at the same time they are said to be *out of phase*. Figure 64b, c, d, shows three cases of the current being out of phase; in b it is said to *lag* behind, in c it is said to *lead* the voltage, and in d the curves are in *opposite* phase. This lag or lead may be

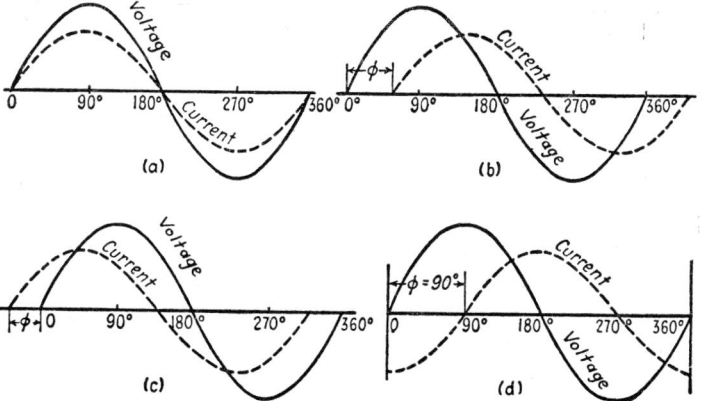

Fig. 64.—Current and Voltage Relations in Alternating-current Circuits. (a) Current in phase with voltage, (b) current lags behind impressed voltage, (c) current leads the voltage, (d) current lags 90 degrees.

expressed as an angle and is usually represented by ϕ, and is called *angular displacement* or *difference in phase*. The angle ϕ is then called the *phase angle*.

"Phase" is also used to express the displacement of two or more different emf's or currents of equal frequency but lacking coincidence in time of rise and fall. An alternator which generates a single voltage is called a *single phase* alternator; a machine which generates two or more separate emf's is called a *polyphase* generator.

Three-phase generators are very widely used. In this case three single-phase currents 120 degrees apart, as shown in Fig. 65, are generated. Theoretically three sets of two wires are required for the conduction of the current, but since the algebraic sum of the currents in the three circuits (if balanced) is at every instant equal to zero, the three return wires, one on each circuit may be dispensed with, leaving but three wires.

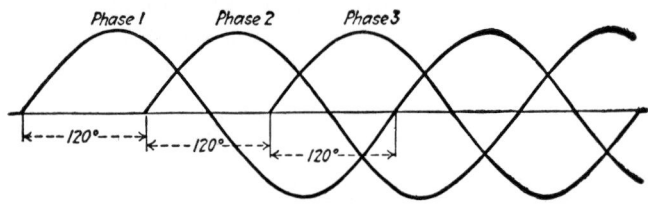

Fig. 65.—Sine emf Curves of a Three-phase Alternator.

Power Factor.—In the study of direct current we saw that the power expended in a circuit was the product of the applied emf and the current or $E \times I$. In the alternating current circuits met with in practice there exists not only resistance, but other influencing forces which are called *inductance* and *capacitance*. (These are defined and discussed later.) The latter two cause the current to be out of phase with the impressed emf. As a result of this, the actual power is reduced.

If we let P = power, E = effective voltage, and I = effective current, then $E \times I$ is called the *apparent power* and is expressed in volt-amperes or kilovolt-amperes (kva). However, if the current I has a lag of ϕ degrees behind the emf, the *actual power* expended in the circuit is

$$P = E \times I \times \cos \phi$$

The factor $\cos \phi$ is called the power factor of the circuit and is usually expressed in percent. Transposing the above equation,

$$\text{Power factor} = \cos \phi = \frac{P}{E \times I},$$

from which we may define power factor as the ratio of the actual power to the apparent power.

ILLUSTRATION: What current will a 220-volt alternator produce in a circuit which has a power factor of 85 percent and takes 1 kilowatt?

In this problem $P = 1000$, $E = 220$, and $\cos \phi = 0.85$. Then,

$$P = E \times I \times \cos \phi$$
$$1000 = 220 \times I \times 0.85$$
$$I = \frac{1000}{220 \times 0.85} = 5.35 \text{ amperes} \quad \text{(Ans.)}$$

Inductance.—We have already referred to the fact that the flow of alternating current depends not only on the resistance but also on *inductance*.

When a current flows through a wire it sets up a magnetic field about the wire. If the current is broken the change in the magnetic field is capable of inducing an emf in a nearby wire. This property is called *inductance*: its symbol is L and the unit is the *henry*. In a wire carrying an alternating current the current is broken many times a second. Not only does this tend to induce current in nearby wires, but in the current-carrying wire itself. This is called self induction and, moreover, the induced emf is opposite in direction to the current emf. The resulting opposition to the flow of the current may be considered as an apparent additional resistance and is called *inductive reactance* to distinguish it from the resistance of the conductor.

The value of inductive reactance is expressed in ohms and it depends on the factors given in the following formula:

$$X_L = 2\pi \times f \times L$$

Where
$X_L =$ inductive reactance in ohms
$f =$ frequency (cycles per second)
$L =$ inductance (henrys)
$\pi = 3.1416$

This formula is also useful in the transposed form

$$L = \frac{X_L}{2\pi \times f}$$

ILLUSTRATION: What would be the inductive reactance of a coil of wire having an inductance of 0.03 henry when connected to an emf of 60 cycles?

$$X_L = 2\pi \times f \times L = 2\pi \times 60 \times 0.03 = 11.32 \text{ ohms} \quad \text{(Ans.)}$$

ILLUSTRATION: What is the inductance of a coil which has an inductive reactance of 2.5 ohms when connected to an emf of 25 cycles?

$$L = \frac{X_L}{2\pi \times f} = \frac{2.5}{2\pi \times 25} = \frac{0.1}{2\pi} = 0.0159 \text{ henry} \quad \text{(Ans.)}$$

Resistance.—Resistance in an alternating-current circuit has exactly the same effect as it has in a direct current circuit. This property of an electric circuit always occasions a loss which appears as heat. If an alternating current of I amperes (effective value) flows through a resistance of R ohms, the loss will be I^2R watts.

Components of Impressed emf. —The emf of a circuit must be sufficiently large to overcome the resistance and to overcome the inductive reactance. It may be regarded as having two components, one devoted to each of these functions, as shown in Fig. 66. The relationship between these components is given in the following definitions:

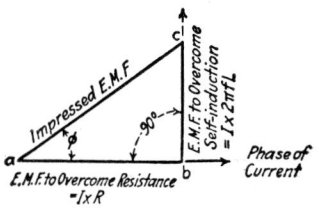

FIG. 66.—Components of emf Impressed on Inductive Circuit.

Resistance is that quantity which, when multiplied by the current, gives that component of the impressed emf which is in phase with the current.

Reactance is that quantity which, when multiplied by the current, gives that component of the impressed emf which is at right angles to the current. Then, when

$$ab = E_r = RI = \text{resistance drop}$$

$$bc = E_L = 2\pi fLI = \text{reactance drop}$$

$$ac = E = \text{impressed emf}$$

According to the "hypotenuse square" rule of right triangles, therefore

$$E = \sqrt{E_r{}^2 + E_L{}^2}$$

or

$$E = \sqrt{(I \times R)^2 + (I \times 2\pi fL)^2}$$

In the following circuit the various elements are represented:

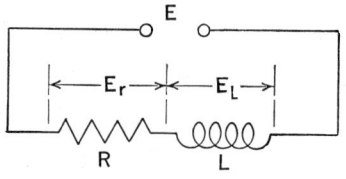

Fig. 67-A.

These equations show that the voltage drop due to resistance and that due to reactance cannot be added arithmetically, but must be added geometrically at right angles to each other to obtain the total voltage on the circuit.

Impedance.—The combined effect of resistance and reactance is called impedance to distinguish it from its two components which may also be represented graphically at right angles as in Fig. 67. Impedance has the symbol Z and is expressed in ohms.

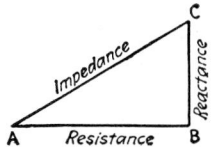

Fig. 67.—Graphical Representation of Impedance

Then,

$$Z = \sqrt{R^2 + X_L{}^2}$$

Other variations of the above formula are also useful. Thus: Given the impedance and resistance, to find the reactance, use:

$$X_L = \sqrt{Z^2 - R^2}$$

Given the impedance and reactance, to find resistance, use:

$$R = \sqrt{Z^2 - X_L{}^2}$$

Capacitance.—Most circuits have the faculty of storing an electrical charge and a momentary flow of current takes place after the circuit is opened. This property is called *capacitance* and is utilized in condensers. It has been found that the current increases directly with the increase in capacitance and also with the increase of frequency. Therefore the apparent resistance due to the condenser, called *capacitive reactance*, decreases with, that is, is inversely proportional to these quantities and hence directly opposite in effect to inductive reactance. Then, if C is the capacitance in farads, and f the frequency, the capacitive reactance will be

$$X = \frac{1}{2\pi \times f \times C}$$

ILLUSTRATION: What is the capacitive reactance of a 40-microfarad condenser to an alternating current of 60 cycles? (1 microfarad = one-millionth part of a farad)

40 microfarads = 0.000040 farad.

$$X_c = \frac{1}{2\pi \times f \times c} = \frac{1}{2 \times 3.1416 \times 60 \times 0.000040} = \frac{1}{0.0151}$$
$$= 66.3 \text{ ohms} \quad \text{(Ans.)}$$

Circuits Having Inductance, Capacitance and Resistance.— When a circuit contains both inductance and capacitance, the net reactance, X, is equal to the arithmetical difference between the inductive reactance, X_L, and the capacitive reactance, X_c, or $X = X_L - X_c$.

Therefore the impedance of a circuit containing inductance, capacitance and resistance is equal to the square root of the quantity

[resistance² + (inductive reactance − capacitive reactance)²], or

$$Z = \sqrt{R^2 + X^2} = \sqrt{R^2 + (X_L - X_c)^2}$$

ILLUSTRATION: What would be the combined impedance of a circuit, having a coil of 3 ohms resistance and of 0.01 henry inductance in series with a condenser of 60-microfarad capacity to an alternating current of 60 cycles?

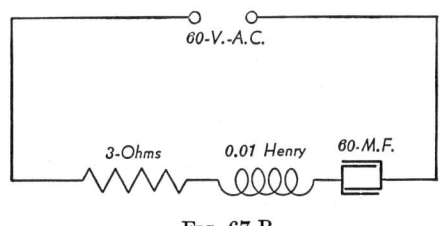

FIG. 67-B

$$X_L = 2\pi \times f \times L = 2 \times 3.1416 \times 60 \times 0.01 = 3.77 \text{ ohms}$$
$$X_c = \frac{1}{2\pi \times f \times C} = \frac{1}{2 \times 3.1416 \times 60 \times 0.000060} = 44.2 \text{ ohms}$$
$$Z = \sqrt{R^2 + (X_L - X_c)^2} = \sqrt{3^2 + (3.77 - 44.2)^2}$$
$$= \sqrt{3^2 + (-40.43)^2}$$
$$Z = \sqrt{9 + 1635.36} = 40.55 \text{ ohms} \quad \text{(Ans.)}$$

Ohm's Law for Alternating-Current Circuits.—In the early pages of this section Ohm's Law applying to direct currents was stated in the three forms,

$$I = \frac{E}{R}, \quad E = I \times R, \quad \text{and} \quad R = \frac{E}{I}$$

We have seen that instead of simple resistance we have in the case of alternating current a number of influences which when grouped together are called impedance and designated by the letter Z. Then Ohm's Law for alternating currents may be expressed:

$$I = \frac{E}{Z}, \quad E = I \times Z, \quad \text{and} \quad Z = \frac{E}{I}$$

When $\quad E$ = emf or the pressure applied to any circuit

$\qquad Z$ = impedance of the circuit expressed in ohms

$\qquad I$ = current strength in that circuit

ILLUSTRATION: (a) What current will flow through a coil with a resistance of 10 ohms and a reactance of 18 ohms when connected to a 60-cycle 110-volt circuit?   (b) What current would flow if this coil were connected across a 110-volt direct-current circuit?

$$Z = \sqrt{R^2 + X_L{}^2} = \sqrt{10^2 + 18^2} = \sqrt{100 + 324} = 20.6 \text{ ohms}$$

$$(a) \ I = \frac{E}{Z} = \frac{110}{20.6} = 5.3 \text{ amperes} \quad \text{(Ans.)}$$

$$(b) \ I = \frac{E}{R} = \frac{110}{10} = 11.0 \text{ amperes} \quad \text{(Ans.)}$$

Impedance may be measured by the volt ammeter method in the same way as resistance is measured in a direct-current circuit, using, of course, an alternating-current voltmeter and ammeter, the impedance being calculated from $Z = E \div I$.

Transformers.—It has already been pointed out that one of the advantages of alternating current is that its voltage may be transformed at will to higher or to lower potentials. This is accomplished by a device called a *transformer* which consists of two windings insulated from each other, but so situated that the magnetic flux developed by one of the windings threads through the other. By running an alternating current through the first winding, there is a constant change in the magnetic flux which induces a current in the second. The two windings are called the *primary* and the *secondary,* the primary being the winding which receives the energy from the supply circuit and the secondary that which receives the energy by induction from the primary.

Figure 68, illustrating three types of power transformers, shows the relation of the two windings to each other and to the core built up from annealed punchings of thin sheet steel. Small transformers such as are placed on poles in power distribution circuits are contained in a cast iron or sheet steel case which is then filled with oil. The oil serves the double purpose of adding further insulation to the windings and of transmitting the heat to the case, where it is dissipated by radiation and air circulation. Such transformers are called self-cooled. Larger transformers such as are used in substations may have the oil cooled by circulating water or air or may be cooled by a blast of air circulated through the windings.

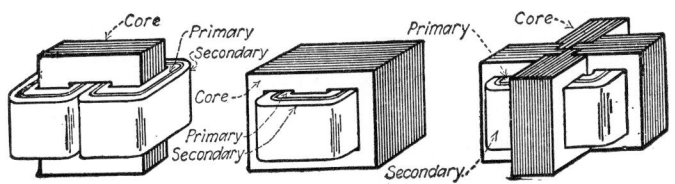

FIG. 68.—Types of Transformers. Left—core type; center—shell type; right—combined core and shell type.

The signal transformers, which are used in radio, television and other communication circuits are available in a considerably larger number of types than the power transformers. These types range from simple air-wound coils, which may consist of only a hollow tube of paperboard, plastic or ceramic on which one coil is wound upon or beside the other, with paper or other insulation between the coils. On the other hand, many types of signal transformers require cores of high permeability material, such as a stack of steel laminations, a mass iron powder or, less commonly in this type of transformer, a solid mass of iron.

The reason for these many different designs is because of the many purposes for which transformers are used in communication circuits. Thus transformers operating at radio frequencies (several hundred kilocycles per second or greater) do not require cores, since the inductive reactance is proportional to the frequency. For the same reason transformers operating at audio frequencies (100 to 15,000 cycles per second) do require cores to transmit the magnetic flux from one winding (the primary) to the other (the secondary). Among the other special requirements are those of the audio output transformer, which links the output stage of a receiver to the input of a speaker, and must therefore operate at a high impedance ratio (roughly 5000 ohms on the receiver side to 5 ohms in the speaker circuit.) Still another type is the "transformer coupler" which links the output of one amplifier stage to the input of the next. As a final example there is the transformer which supplies two or more secondary windings from a single primary winding. A transformer of this type is often used to supply the filament current for a number of tubes. While several secondary windings can be placed about a single primary, another common method is the E type, in which the inner bar of the E carries the primary winding, and the outer bars the two secondaries, with cores between each secondary and the primary.

The transformation of the current from one voltage to another is accomplished by having more turns on one winding than the other. Thus, if the primary winding has 250 turns and the secondary has 1000 turns, then the voltage available at the secondary terminals will be $1000 \div 250 = 4$ times as great as the voltage impressed upon the primary. If we let n_2 represent the number of turns on the high-voltage winding and n_1 the number of turns on the low-voltage winding, then the ratio $n_2 \div n_1 = r$ is called the *ratio of transformation*, and

$$r = \frac{n_2}{n_1} = \frac{E_2}{E_1}$$

when E_2 and E_1 are the respective voltages of the two windings. When a transformer is used to deliver a current at a voltage higher than that it receives, it is called a *step-up* transformer, and when it delivers a current at a lower potential it is called a *step-down* transformer.

Transformers are very efficient in their operation, often rating over 98 percent, so that for many practical calculations the losses may be ignored and the power output regarded as equal to the power input. Then, since power equals volts times amperes we may write

$$P = E_1 \times I_1 = E_2 \times I_2$$

where I_1 and I_2 are the currents in the low and high voltage windings, respectively. From this we may derive the following ratios:

$$\frac{E_1}{E_2} = \frac{I_2}{I_1}$$

which states in effect that the ratio of the voltage is the inverse ratio of the currents in the two windings.

ILLUSTRATION: The primary voltage of a 15-kw. transformer used to supply electricity to a 220-volt circuit is 2200 volts. What is the ratio of this transformer and what are the full-load currents in the two windings, neglecting losses?

This is, of course, a step-down transformer and $E_1 = 220$ volts, $E_2 = 2200$ volts, $P = 15,000$ watts. Then

$$r = \frac{E_2}{E_1} = \frac{2200}{220} = 10 \quad \text{(Ans.)}$$

$$P = E_1 \times I_1$$

then $\qquad I_1 = \frac{P}{E_1} = \frac{15,000}{220} = 68 \text{ amperes} \quad \text{(Ans.)}$

and $\qquad I_2 = \frac{P}{E_2} = \frac{15,000}{2200} = 6.8 \text{ amperes} \quad \text{(Ans.)}$

ILLUSTRATION: What are the full-load currents in the two windings of a 30-kw. transformer used to supply electricity to a 110-volt circuit if the primary voltage is 3300 volts?

$$I_1 = \frac{P}{E_1} = \frac{30,000}{110} = 273 \text{ amperes} \quad \text{(Ans.)}$$

$$I_2 = \frac{P}{E_2} = \frac{30,000}{3300} = 9.1 \text{ amperes} \quad \text{(Ans.)}$$

Transformer Design Calculations.—While the calculations in the precise design of a transformer are quite lengthy, involving consideration of temperature effects and several other variables, the major steps are outlined below for those who may wish to wind a small transformer.

1. Compute the voltage and current requirements of the secondary circuit from the components in it.

2. Choose a core size which is suitable for the volt-ampere rating just calculated. This choice requires comparison with other small transformers, since transformers for radio and television receivers are made as small as possible without overheating (200° F. is maximum for paper and phenolic resin insulation).

3. Determine the flux density in the core. This depends upon the material of the core (70,000 lines per sq. in. for silicon steel).

4. Calculate the number of turns required in the primary to produce this flux.

5. From the volt-ampere rating of the secondary as calculated in (1), calculate the volt-ampere rating of the primary by multiplying by an efficiency factor (85–95%) for iron-core small transformers. From this figure and the primary circuit, compute the primary current when full load is on the secondary.

6. Calculate the no-load current on the primary to avoid overloading.

7. Choose a wire size from Table 4 to take the full-load current.

8. By the equations given in this section, calculate the number of turns and wire sizes for the secondary.

Alternators.—The principles of the alternating current generator have already been discussed and the three principal types classified and described. Revolving field alternators are used practically to the exclusion of all other types in power generating stations. Their field magnets wound in slots revolve inside a stationary armature similarly wound. This results in a well-balanced machine of low resistance which can be successfully operated in connection with high-speed turbines. The revolving field magnets are energized by direct current which reaches them through slip rings. This current is often of a much lower potential than that received from the stationary armature.

Alternating-current generators are usually rated in kilovolt-amperes (kva) instead of kilowatts, since it is impossible for the manufacturer to know in advance the amount of inductance and capacitance of the circuits to which the alternator is required to furnish power.

Engine and hydraulic turbine-driven alternators are in the slow-speed class, and are characterized by large diameter, short length, and many poles. The steam turbine-driven alternator is a high-speed machine having a length larger than its diameter. Standard speeds of turbine-driven alternators range from 1200 to 3600 rpm, with 1800 rpm very common practice. To supply the d-c for the field, a source of d-c at 110–250 volts is necessary.

Conversion.—While practically all electric power is generated as alternating current, some functions are best served by direct current and it is convenient to have some means of changing the alternating current to direct current. This is called *conversion*.

Urban subways usually operate on direct current and the power supplied is most frequently converted to direct current by a machine called a *rotary converter*. This is essentially an alternator and a direct-current generator combined in one machine. Its revolving armature receives alternating current through slip rings and by tapping the armature coils at proper points and connecting them with a segmented commutator, direct current may be taken off by means of brushes.

When only a small amount of direct current is required from an alternating-current source, a device known as a rectifier may be used. This permits the current to pass in only one direction. The four common types in use are, the mercury-arc rectifier, the vibrating rectifier, the tungar rectifier and an electrolytic rectifier. These find use in electroplating, storage-battery charging and radio work.

Alternating Current Motors.—A detailed description of alternating current motors is beyond the scope of this work because the mathematical problems connected with these machines are the concern chiefly of the designer and engineer. However, for the sake of completeness we will list the important types.

1. The *polyphase induction motor* of the squirrel-cage armature type is the most widely used alternating current motor in industrial service. It consists of a wound stationary part called the *stator*, which corresponds to the field magnets of a direct-current motor, and a rotating member called the *rotor*, which corresponds to the armature. Polyphase alternating currents flowing through the stator set up a rotating magnetic field which induces a current in copper bars parallel to the axis of the rotor and the reaction of the magnetic flux of these rotor conductors against the rotating field produces rotation of the rotor. Some motors of this type have a wound rotor to inject resistance into the rotor winding and obtain a higher starting torque.

2. The *single-phase induction motor* differs from the polyphase motor chiefly in that provision must be made for starting the motor and bringing it up to a speed corresponding to the frequency in the stator windings. This is done by one of three methods. (1) the split-phase methods in which an auxiliary stator winding is provided for starting purposes only, (2) an auxiliary winding may be connected to the single-phase line through an external inductance to split the phase, and (3) by providing a wound rotor and a commutator for starting as a repulsion motor.

3. *Single-phase commutator motors* may be divided into three sub-types: plain repulsion, single-phase series, and repulsion induction motors. Of these the second is the simplest form and in general design is practically the same as the direct-current series

motor. It may be operated on either direct current or alternating current and for this reason it is widely used for operating household appliances and small tools.

4. The *synchronous motor* is constructed in practically the same manner as a corresponding alternator, and any alternator may be run as a synchronous motor. However, some auxiliary means must be provided for bringing this type of motor up to synchronous speed before it is connected to the alternating current. This is usually accomplished by attaching to the rotor an auxiliary cage winding similar to the rotor winding of a squirrel-cage induction motor.

WIRE CALCULATIONS

Mil-foot.—In calculating the resistance of wire, the standard unit used is a wire $\frac{1}{1000}$ inch in diameter and one foot long. Such a piece of wire is called a *mil-foot*. The word "mil," however used, means one-thousandth. The cross-sectional area of a wire whose diameter is one mil is one *circular mil*. Since areas of like-shaped surfaces vary as the squares of their dimensions, it follows that the cross-sectional area of a circle whose diameter is 2 mils, is 4 circular mils (See Fig. 69); one whose diameter is 3 mils has an area of 9 circular mils, etc. From this we may devise the rule that:

When d represents the diameter of a wire in mils, d^2 is its cross-sectional area in circular mils.

The resistance of one mil-foot of copper wire is 10.79 ohms at 75° Fahrenheit. The resistance of ten feet will be 107.9 ohms. One foot of copper wire $\frac{2}{1000}$ inch in diameter will have one-fourth the resistance, or 10.79 divided by 4 = 2.70 ohms. *Resistance of a conductor varies directly as the length, inversely as the cross-sectional area, with the material of the conductor, and with its temperature.*

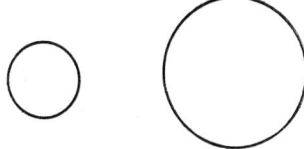

FIG. 69.—The Diameter of the Larger Circle is Twice as Great as that of the Smaller, but the Area is Four Times as Great.

Calculating Resistance of Wires.—Given the length and area of any wire, to find its resistance:

The resistance of any wire at a given temperature is equal to its length in feet multiplied by the resistance of a mil-foot (K) and this product divided by its area in circular mils.

$$R = \frac{K \times L}{d^2}$$

When R = resistance in ohms

K = resistance of one mil-foot in ohms

L = length of wire in feet

d = diameter in mils

d^2 = area in circular mils

ILLUSTRATION: What is the resistance of 500 feet of copper wire having a cross-sectional area of 4107 circular mils?

$$K \text{ for copper} = 10.79$$
$$d^2 = 4107$$

Then $R = \dfrac{K \times L}{d^2} = \dfrac{10.79 \times 500}{4107} = 1.31$ ohms   (Ans.)

ILLUSTRATION: Find the resistance of a copper wire 10.03 mils in diameter and 85 feet long.

$$K = 10.79$$
$$d = 10.03$$
$$d^2 = 100.5$$

Then $R = \dfrac{K \times L}{d^2} = \dfrac{10.79 \times 85}{100.5} = 9.13$ ohms   (Ans.)

The value K is constant for the same wire, but different for each metal. We have seen that it is 10.79 ohms for copper at 75° Fahrenheit. The value of K for other metals is given in Table 3. The variation of resistance with temperature is roughly proportional to the absolute temperature. The following table is based on a temperature of 68° Fahrenheit.

TABLE 3

RESISTANCE OF A MIL-FOOT OF METALS (VALUES OF K)

Silver, 9.84	Zinc, 36.69	German Silver, 128.29
Copper, 10.79	Platinum, 59.02	Platinoid, 188.93
Aluminum, 17.21	Iron, 63.35	Mercury, 586.24

ILLUSTRATION: Substitute iron wire for copper wire in the preceding illustration. It then calls for the resistance of an iron wire 10.03 mils in diameter and 85 feet long.

$$K = 63.35$$
$$d = 10.03$$
$$d^2 = 100.5$$

Then $R = \dfrac{K \times L}{d^2} = \dfrac{63.35 \times 85}{100.5} = 53.57$ ohms (Ans.)

From this it is seen that the resistance of iron is about six times that of copper.

Wire Gage.—This is the term used in describing the size of wire. There are a number of wire gages which have been developed by different manufacturers. The American standard for electrical purposes is the B. & S. gage (Brown & Sharpe Manufacturing Company).

Wires larger than No. 0000 B. & S. are seldom made solid but are built up of a number of small wires. The group of wires is called a " strand "; the term " wire " being reserved for the individual wires of the strand. Strands are usually built up of wires of such a size that the cross-section of the metal in the strand is the same as the cross-section of a solid wire having the same gage number. The sizes of wire larger than No. 0000 are given only in circular mils.

Wire Calculations.—Given the resistance and area of a wire, to find the length.

The length of any wire is equal to its resistance multiplied by its circular mil area, and this product divided by the resistance of a mil-foot (K).

$$L = \frac{R \times d^2}{K}$$

TABLE 4

WIRE TABLE, STANDARD ANNEALED COPPER AT A TEMPERATURE
OF 25° CENTIGRADE (77° FAHRENHEIT)

American Wire Gage (Brown & Sharpe)

Gage No.	Diam. in Mils d	AREA Cir. Mils d²	WEIGHT Lbs. per 1000 ft.	Lbs. per ohm	LENGTH Feet per lb.	Feet per ohm	RESISTANCE Ohms per 1000 ft.	Ohms per lb.
0000	460.0	211660.	640.5	12810.	1.561	20010.	0.04998	0.00007805
000	409.6	167800.	507.9	8057.	1.968	15870.	.06303	.0001217
00	364.8	133100.	402.8	5067.	2.482	12580.	.07947	.0001935
0	324.9	105500.	319.5	3187.	3.130	9979.	.1002	.0003138
1	289.3	83690.	253.3	2004.	3.947	7913.	.1264	.0004990
2	257.6	66370.	200.9	1260.	4.977	6276.	.1594	.0007934
3	229.4	52640.	159.3	792.7	6.276	4977.	.2009	.001262
4	204.3	41740.	126.4	498.6	7.914	3947.	.2534	.002008
5	181.9	33100.	100.2	313.5	9.980	3130.	.3195	.003189
6	162.0	26250.	79.46	197.2	12.58	2482.	.4029	.005071
7	144.3	20820.	63.02	124.0	15.87	1968.	.5080	.008064
8	128.5	16510.	49.98	77.99	20.01	1561.	.6406	.01282
9	114.4	13090.	39.63	49.05	25.23	1233.	.8078	.02039
10	101.9	10380.	31.43	30.85	31.82	981.8	1.019	.03242
11	90.74	8234.	24.92	19.40	40.12	778.5	1.284	.05155
12	80.81	6530.	19.77	12.20	50.59	617.4	1.620	.08196
13	71.96	5178.	15.68	7.673	63.80	489.6	2.042	.1303
14	64.08	4107.	12.43	4.826	80.44	388.3	2.576	.2072
15	57.07	3257.	9.858	3.035	101.4	307.9	3.248	.3295
16	50.82	2583.	7.818	1.909	127.9	244.2	4.095	.5239
17	45.26	2048.	6.200	1.200	161.3	193.7	5.164	.8330
18	40.30	1624.	4.917	0.7549	203.4	153.6	6.512	1.325
19	35.89	1288.	3.899	.4748	256.5	121.8	8.210	2.106
20	31.96	1022.	3.092	.2986	323.4	96.59	10.35	3.349
21	28.46	810.1	2.452	.1878	407.8	76.60	13.06	5.325
22	25.35	642.4	1.945	.1181	514.2	60.74	16.46	8.467
23	22.57	509.5	1.542	.07427	648.4	48.17	20.76	13.46
24	20.10	404.0	1.223	.04671	817.7	38.20	26.18	21.41
25	17.90	320.4	0.9699	.02938	1031.	30.30	33.01	34.04
26	15.94	254.1	.7692	.01847	1300.	24.02	41.62	54.13
27	14.20	201.5	.6100	.01162	1639.	19.05	52.43	86.07
28	12.64	159.8	.4837	.007307	2067.	15.11	66.18	136.8
29	11.26	126.7	.3836	.004595	2607.	11.98	83.46	217.6
30	10.03	100.5	.3042	.002900	3287.	9.503	105.2	346.0
31	8.928	79.70	.2413	.001818	4145.	7.536	132.7	550.2
32	7.950	63.21	.1913	.001143	5227.	5.976	167.3	874.8
33	7.080	50.13	.1517	.0007189	6591.	4.739	211.0	1391.
34	6.305	39.75	.1203	.0004521	8310.	3.759	266.1	2212.
35	5.615	31.52	.09542	.0002843	10480.	2.981	335.5	3517.
36	5.000	25.00	.07568	.0001788	13210.	2.364	423.0	5592.
37	4.453	19.83	.06001	.0001125	16660.	1.874	533.5	8892.
38	3.965	15.72	.04759	.00007074	21010.	1.487	672.7	14140.
39	3.531	12.47	.03774	.00004448	26500.	1.179	848.2	22480.
40	3.145	9.888	.02993	.00002798	33410.	0.9349	1070.	35740.

ILLUSTRATION: What is the length of a German silver wire wound on a spool if its resistance is 30 ohms and the size of the wire is No. 20 B. & S.?

$K = 128.29$ for German silver (See Table 3)

No. 20 B. & S. = 1022 circular mils (See Table 4)

Then $L = \dfrac{R \times d^2}{K} = \dfrac{30 \times 1022}{128.29} = 239$ feet (Ans.)

Given the length and resistance of a wire, to find the area:

The area in circular mils of any wire is equal to its length multiplied by the resistance of a mil-foot (K) and this product divided by its resistance.

$$d^2 = \frac{L \times K}{R}$$

ILLUSTRATION: A reel of 800 feet of copper wire has a resistance of 5 ohms at 75° F. What is its circular mil area?

$K = 10.79$ for copper at 75° F.

Then $d^2 = \dfrac{L \times K}{R} = \dfrac{800 \times 10.79}{5} = 1,726$ circular mils (Ans.)

ILLUSTRATION: A mile of aluminum wire on a power line has a resistance of 1.086 ohms. What is its circular mil area?

1 mile = 5280 feet

$K = 17.21$ for aluminum (See Table 3)

Then $d^2 = \dfrac{L \times K}{R} = \dfrac{5280 \times 17.21}{1.086} = 83,673$ circular mils (Ans.)

This is evidently a No. 1 wire whose area is 83,690 circular mils. When the area in circular mils is known, the square root of this number is the diameter in mils, or thousandths of an inch.

Given the area of a wire, to find its weight:

The weight per mile (5280 feet) of any bare copper wire in pounds is equal to the area in circular mils divided by the constant 62.5.

$$\text{Pounds per mile} = \frac{d^2}{62.5}$$

ILLUSTRATION: Copper telegraph wire 14-gage B. & S. is furnished in coils containing 1.20 miles. What is the weight of such a coil?

$d^2 = 4107$ circular mils for 14-gage wire (See Table 4)

Then, weight of 1 mile $= \dfrac{d^2}{62.5} = \dfrac{4107}{62.5} = 66$ pounds

weight of coil $= 66 \times 1.2 = 79.2$ pounds   (Ans.)

Copper weighs about 555 pounds per cubic foot and iron about 480 pounds. Therefore, the weight of a length of iron wire would be $\frac{480}{555} = \frac{32}{37}$ times that of a corresponding length of copper wire.

ILLUSTRATION: If the wire in the preceding illustration were iron instead of copper, what would be its weight?

$79.2 \times \frac{32}{37} = 68.5$ pounds   (Ans.)

Finding Size of Wire Required.—The formula given on page 781 for the determination of the area of a wire needed, $d^2 = \dfrac{L \times K}{R}$, may be transformed for more practical application by expressing the resistance (R) in terms of current and voltage drop. From Ohm's Law we have $R = \dfrac{E}{I}$

whence, $d^2 = \dfrac{L \times K}{\dfrac{E}{I}} = \dfrac{L \times K \times I}{E}$

ILLUSTRATION: A power line 800 feet long is run to a motor requiring 25 amperes. The voltage drop in the line must not exceed 20 volts. What size wire will be required?

$d^2 = \dfrac{L \times K \times I}{E} = \dfrac{800 \times 2 \times 10.79 \times 25}{20} = 21{,}580$ circular mils

Referring to Table 4, the wire size next larger than this area is gage 6, and this should, therefore, be the wire used.

Wire Calculations Using Tables.—Where many wire size calculations are required, use of the foregoing formulas requires much tedious work. Tables 5, 6, and 7 have been devised to simplify and reduce this work by giving the proper copper wire size and circuit length for a given load.

These tables are based on a maximum voltage drop of 1%, and practical conductor operating temperatures of 50° C. and 60° C. This means that the wire size and wire length arrived at by using these tables will not cause a voltage drop greater than 1% at the load. The operating temperature of the circuit also will not exceed 60° C. on lighting circuits and 50° C. on other circuits.

Table 5 covers lighting circuits on 3-wire and 4-wire, balanced 115-volt lighting loads. Branch circuit sizes are shown. Table 6 covers single-phase, a-c, 3-wire, 115/230-volt, 60-cycle, 100% power factor systems. Feeder sizes generally used for lighting loads are listed. Table 7 covers 3-phase, a-c, 230-volt, 60-cycle, delta, 85% power factor systems, usually power loads such as motors.

It is necessary to know the load requirement in amperes to use the tables. The ampere load is located in the first column of the proper system voltage table. On this line going horizontally to the right, the required circuit length is found, and at the top of the latter column the proper wire size is found.

ILLUSTRATION: A 115-volt, 3-wire lighting circuit, 35 feet long, has to carry a balanced lighting load of 15 amperes. What wire size should be used so that the voltage drop will not exceed 1% (1.15 volts)?

Using Table 5, 15 amperes is found in the first column. Reading across horizontally, it can be seen that 35 feet does not appear in the table. The next higher value, 40 feet, is therefore selected. The top of this column indicates that No. 12 wire is the proper size to use.

Installation of Interior Wiring.—All interior wiring must be installed in such a manner that it will be protected from mechanical injury and be safe as regards fire hazard or danger to life.

TABLE 5

AMPERES (A), WATTS (W), WITH CONDUIT CONDUCTOR (C), FILLS (F)

MAXIMUM OVERCURRENT CIRCUIT PROTECTION	INTERMITTENT LOADS			CONTINUOUS LOADS		
	100% F 2-3 C	80% F 4-6 C	70% F 7-9 C	100% F 2-3 C	80% F 4-6 C	70% F 7-9 C
15 A	15 A 1725 W	12 A 1380 W	10.5 A 1207 W	12 A 1380 W	9.6 A 1104 W	8.4 Amps. 966 Watts
20 A	20 A 2300 W	16 A 1840 W	14 A 1610 W	16 A 1840 W	12.8 A 1472 W	11.2 A 1288 W

LOADS AND LENGTHS IN FEET FOR 1% DROP ON 3 AND 4 WIRE 115 V. CIRCUITS

AMPERE LOAD	#10 WIRE	#12 WIRE	#14 WIRE
1	946	596	374
2	474	298	188
3	316	198	124
4	236	148	94
5	190	120	76
6	158	100	62
7	136	86	54
8	118	74	46
9	106	66	42
10	94	60	38
11	86	54	34
12	78	50	32
13	72	46	28
14	68	42	26
15	64	40	24
16	60	38	
17	56	36	
18	52	34	
19	50	32	
20	48	30	
21	46		
22	44		
23	42		
24	40		
25	38		
26	36		
27	36		
28	34		
29	32		
30	32		

On 4-wire, 3-phase "Y" power and light service with 2-ph, 3-wire circuits use ⅔ table circuit lengths for 2-ph circuits when tapped off 4-wire system.

For 2-wire, 115-volt circuits, use one-half table circuit lengths.

Example: 2-Wire, 115-volt, No. 14 wire circuit

Length = $\dfrac{1.15 \text{ (Volt drop)} \times 4107 \text{(CM of } \#14W)}{15 \text{ (Amps.)} \times 2 \times 13 \text{ (R of wire @ 60°C)}}$

= 12.1 feet

TABLE 6

Table of average circuit lengths for single-phase, A.C., 3-wire (230-115-230 v.), 60 cycle, 100% power factor system. Feeders with 1% (1.15 volts) drop for balanced 3-wire loads. For 2-wire, 230-volt single-phase loads, voltage drop is 2.3 volts. Circuit lengths in feet.

Copper resistance "R"—12.5 ohms per "CM" ft. at 50°C (122°F.)
Reactance and impedance losses calculated for each wire.
Conductors closely grouped in metallic conduit.

AMPERE LOAD	WIRE SIZE—CIRCULAR MILS					WIRE SIZE—B & S or A.W.G.									
	500	400	350	300	250	4/0	3/0	2/0	1/0	1	2	3	4	6	8
40	1106	898	788	669	558	475	378	299	239	188	150	119	94	59	38
50	885	719	630	535	447	380	303	240	191	150	120	91	75	47	30
60	737	599	525	446	372	317	252	200	159	125	100	79	62	39	
70	632	513	450	382	319	271	216	171	136	107	86	68	53	34	
80	553	449	394	334	279	238	189	150	119	94	75	59	47		
90	491	399	350	297	248	211	168	133	106	83	67	53	42		
100	442	359	315	267	223	190	151	120	95	75	60	47			
110	402	327	286	243	203	173	138	109	87	68	55				
120	369	299	263	223	186	158	126	100	79	63					
130	340	276	242	206	172	146	116	92	73	58					
140	316	257	225	191	159	136	108	86	68						
150	295	240	210	178	149	127	101	80	64						
160	276	225	197	167	140	119	95	75	60						
170	260	211	185	157	131	112	89	70							
180	246	200	175	148	124	106	84	66							
190	233	189	166	140	117	100	80								
200	221	180	157	134	112	95	76								
210	211	171	150	127	106	90									
220	201	163	143	122	101	86									
230	192	156	137	116	97	83									
240	184	150	131	111	93										
250	177	144	126	107	89										
260	170	138	121	103	80										
270	164	133	117	99											
280	158	128	112	96											
290	152	124	109	92											
300	147	120	105												
310	143	116	102												
320	138	112													
330	134	109													
340	130	106													
350	126														
360	123														
370	119														
380	116														

AMPERES PER KW. at 100% to 70% P.F.

POWER FACTOR	1-PHASE	3-PHASE
1.00	4.35	2.51
.95	4.58	2.65
.90	4.83	2.80
.85	5.12	2.90
.80	5.44	3.14
.75	5.80	3.35
.70	6.21	3.59

APPROX. COPPER "R"—OHMS PER "CM" FT.

Temp. °C	Temp. °F	"R"—per CM FT.
20	68	10.4 ohms
25	77	10.8 ohms
30	86	11.0 ohms
40	104	11.5 ohms
50	122	12.5 ohms
60	140	13.0 ohms
75	167	13.6 ohms

FORMULAE FOR AMP. PER KW. AT 100% P.F.

1-PHASE $A = \dfrac{1 \ (KW) \times 1000 \ (W)}{230 \ (V) \times 1.00 \ (P.F.)}$

3-PHASE $A = \dfrac{1 \ (KW) \times 1000 \ (W)}{230 \ (V) \times 1.00 \ (P.F.) \times 1.732}$

Conductivity of Copper

Soft Annealed—all sizes 98.3%
Medium Hard—.32 to .04 Dia. 96.7%
Hard Drawn—.32 to .04 Dia. 96.3%

FORMULAE FOR CALCULATING DISTANCE:

Copper "R"—12.5 ohms/CM ft. at 50°C.
A.C. Distance Formulae give only approximate lengths.

D.C. Distance $= \dfrac{\text{Volts drop} \times \text{wire size (circular mils)}}{\text{Ampere load} \times 2 \times R \ (\text{Resistance per CM ft.} = 12.5)}$

1-PHASE Distance $= \dfrac{\text{Volts drop} \times \text{wire size (circular mils)}}{\text{Amps. (at given PF)} \times 2 \times R \ (12.5 + \text{Reactance} + \text{Impedance})}$

For 3-Phase distance on 3-Phase Delta Circuits - use 1-Phase amperes x .866 $\left(\dfrac{\sqrt{3}}{2}\right)$

3-Phase Distance $=$ 1.15 x 1-Phase distance, approximately

TABLE 7

AMPERE LOAD	WIRE SIZE—CIRCULAR MILS					WIRE SIZE—B & S or A.W.G.									
	500	400	350	300	250	4/0	3/0	2/0	1/0	1	2	3	4	6	8
40	710	625	584	530	475	429	364	303	253	208	173	139	113	75	49
50	568	500	467	424	380	343	291	242	203	167	139	111	90	60	39
60	473	417	389	353	317	286	243	202	169	139	115	93	75	50	
70	406	357	333	303	271	245	208	173	145	119	99	79	64	43	
80	355	312	292	265	238	214	182	151	127	104	87	69	56		
90	316	278	259	235	211	191	162	134	113	93	77	62	45		
100	284	250	233	212	190	172	146	121	101	83	69	55			
110	258	227	212	193	173	156	132	110	92	76	63				
120	237	208	195	177	158	143	121	101	84	69	58				
130	218	192	180	163	146	132	112	93	78	64					
140	203	179	167	151	136	123	104	86	72						
150	189	168	156	141	127	114	97	81	67						
160	177	156	146	132	119	107	91	76							
170	167	147	137	125	112	101	86	71							
180	158	139	130	118	106	95	81	67							
190	149	132	123	112	100	90	77								
200	142	125	117	106	95	86	73								
210	135	119	111	101	90	82									
220	129	114	106	96	86	78									
230	123	109	101	92	83	75									
240	118	104	97	88	79										
250	114	100	93	85	76										
260	109	96	90	81	73										
270	105	93	86	78											
280	101	89	83	76											
290	98	86	80	73											
300	95	83	78												
310	92	81	75												
320	89	78													
330	86	76													
340	83	73													
350	81														
360	79														
370	77														
380	75														

Approx. Resistance—Ohms per "CM" Ft. at Copper Temperatures of 50°C to 75°C.

WIRE SIZE	R @ 50°C	R @ 60°C	R @ 75°C
No. 14 to 6	11.6	12.1	12.7
No. 4 to 1/0	12.0	12.5	13.1
2/0 to 4/0	12.4	12.9	13.5
300 MCM	12.7	13.2	13.8
400 MCM	13.0	13.5	14.2
500 MCM	13.4	13.9	14.6
AVERAGE "R"	12.5	13.0	13.6

NOTES:

For 208-volt, 4-wire, "Y" feeders — multiply table circuit lengths by 90%

For 230 volt, 1-phase feeders—multiply table circuit lengths by 85%

For 460-volt, 3- or 4-wire feeders — multiply table circuit lengths by 2

For 2 percent voltage dop — double 1% table circuit lengths

For 3 percent voltage drop—triple 1% table circuit lengths

For aluminum wire which is same size as copper, use 70% of table circuit lengths; or use circuit length of copper wire which is two sizes smaller than aluminum wire under consideration. (Aluminum wire—ohms per CM ft. @ 60°C is 17)

Example:
A 4/0 aluminum wire with 40-ampere load has a circuit length of 303 ft. same as a 2/0 copper wire in table with a voltage drop of one percent.

816

Only approved materials may be used and the work must conform to the local building codes or fire ordinances and to the rules of the National Board of Fire Underwriters as set forth in its "National Electrical Code." This code is in effect throughout the United States and Canada, and gives definite rules for the installation of all kinds of wiring. It also specifies carefully the kind of material, such as wire, conduit, fuses, etc., that may be installed. Copies of the code may be obtained by applying to the National Fire Protection Association, 60 Batterymarsh St., Boston, Massachusetts, 02110.

Installation of wiring for light or power service, at voltages not exceeding 600 volts, may be done by any of the following plans, all of which are approved by the code, but the use of some of them is restricted in special places.

Open or Exposed Wiring.—Wires are supported on porcelain knobs or cleats; the knobs or cleats should separate the wires about $2\frac{1}{2}$ inches and should be $\frac{1}{2}$ inch from the surface along which they run.

Molding Work.—Wires are run in a wood or metal molding. The metal molding consists of a sheet steel trough or backing and a steel cover which is snapped on the backing after the wires are in place. Wood molding consists of a backing with grooves for the wires and a capping which is nailed to the backing after the wires are in place; this molding is made for two and three wires. Molding work is particularly adapted to the wiring of buildings after their completion and has the advantage of cheapness, simplicity and accessibility.

Rigid Conduit.—Wires are run in unlined conduits which are free from scale on the inside and are coated with enamel on the inside and outside; the outside is sometimes galvanized when used where the pipe is exposed to the weather. Conduits must be continuous from outlet to outlet, at which places metal junction boxes made for the purpose are located; the conduit must properly enter and be secured to all fittings, and the system must be mechanically strong. Conduit affords the best protection to the wires from mechanical injury and may be used for all classes of service. It is chiefly used in buildings of fireproof construction where wires are concealed; it is also frequently used for circuits run exposed in power houses and industrial establishments. Conduit systems must be grounded, that is, connected to the earth, by connecting the conduit to a water pipe (on the street side of the meter); grounding is necessary so that in the case of a breakdown of the wire insulation, the conduit will not be charged to a dangerous potential. Table 8 applying to complete conduit systems shows the size of conduit required for several wires.

Flexible Conduit.—Wires are installed in a flexible conduit that is made of steel strips wound spirally to form a tube; the edges of the strip interlock in such a manner that the tube can be concealed work where rigid conduit could not be used. It is not water-tight and therefore is not as suitable as the rigid conduit where exposed to moisture.

Armored Cable.—A flexible armor similar to the above flexible conduit is placed directly upon the wire. The wire is rubber insulated and covered with a braid the same as the wire used in metal conduit systems. This armored cable is made with either single, double, or triple conductors and is used for the same classes of service as the flexible conduit; in fact, it is used more frequently than the flexible conduit since it is cheaper and easier to install.

Plastic Insulation.—A number of types of wiring insulated with thermoplastic materials are available, some of which are classed as UF (Underground Feeder) in the 1965 National Electric Code. For example, there is a small diameter building wire insulated with polyvinylchloride; designated by the National Electric Code letters TW, made by the Triangle Conduit and Cable Co., which is approved for both wet and dry locations for circuits not exceeding 600 volts and temperatures not exceeding 75° F. The Code permits the use of this type of wire for underground applications in concrete slabs or other masonry in direct contact with earth, in wet locations, and where condensation and moisture accumulation within the raceway may occur. This wire

TABLE 8*

MAXIMUM NUMBER OF CONDUCTORS
IN TRADE SIZES OF CONDUIT OR TUBING

New Work or Rewiring—Types RF-2, RFH-2, R, RH, RW, RHH,
RHW, RH-RW†
New Work—FEP, FEPB, RUH, RUW, T, TF, THHN, THW, THWN, TW

Size AWG or MCM	Maximum Number of Conductors in Conduit or Tubing (Based upon % conductor fill, Table 3, Chap. 9, for new work)											
	½ Inch	¾ Inch	1 Inch	1¼ Inch	1½ Inch	2 Inch	2½ Inch	3 Inch	3½ Inch	4 Inch	5 Inch	6 Inch
18	7	12	20	35	49	80	115	176				
16	6	10	17	30	41	68	98	150				
14	4	6	10	18	25	41	58	90	121	155		
12	3	5	8	15	21	34	50	76	103	132	208	
10	1	4	7	13	17	29	41	64	86	110	173	
8	1	3	4	7	10	17	25	38	52	67	105	152
6	1	1	3	4	6	10	15	23	32	41	64	93
4	1	1	1	3‡	5	8	12	18	24	31	49	72
3		1	1	3	4	7	10	16	21	28	44	63
2		1	1	3	3	6	9	14	19	24	38	55
1		1	1	1	3	4	7	10	14	18	29	42
0			1	1	2	4	6	9	12	16	25	37
00			1	1	1	3	5	8	11	14	22	32
000			1	1	1	3	4	7	9	12	19	27
0000				1	1	2	3	6	8	10	16	23
250				1	1	1	3	5	6	8	13	19
300				1	1	1	3	4	5	7	11	16
350				1	1	1	1	3	5	6	10	15
400					1	1	1	3	4	6	9	13
500					1	1	1	3	4	5	8	11
600						1	1	1	3	4	6	9
700						1	1	1	3	3	6	8
750						1	1	1	3	3	5	8
800						1	1	1	2	3	5	7
900						1	1	1	1	3	4	7
1000						1	1	1	1	3	4	6
1250							1	1	1	1	3	5
1500								1	1	1	3	4
1750								1	1	1	2	4
2000								1	1	1	1	3

*1965 National Electrical Code.

†For symbols, see Table 10.

‡Where an existing service run of conduit or electrical metallic tubing does not exceed 50 ft. in length and does not contain more than the equivalent of two quarter bends from end to end, two No. 4 insulated and one No. 4 bare conductors may be installed in 1-inch conduit or tubing.

TABLE 9

Not More than Three Conductors in Raceway or Cable or
Direct Burial (Based on Room Temperature of 30° C. 86° F.

Size	Temperature Rating of Conductor					
AWG MCM	60° C (140° F)	75° C (167° F)	85°–90° C (185° F)	110° C (230° F)	125° C (257° F)	200° C (392° F)
14	15	15	25†	30	30	30
12	20	20	30†	35	40	40
10	30	30	40†	45	50	55
8	40	45	50	60	65	70
6	55	65	70	80	85	95
4	70	85	90	105	115	120
3	80	100	105	120	130	145
2	95	115	120	135	145	165
1	110	130	140	160	170	190
0	125	150	155	190	200	225
00	145	175	185	215	230	250
000	165	200	210	245	265	285
0000	195	230	235	275	310	340
250	215	255	270	315	335	...
300	240	285	300	345	380	...
350	260	310	325	390	420	...
400	280	335	360	420	450	...
500	320	380	405	470	500	...
600	355	420	455	525	545	...
700	385	460	490	560	600	...
750	400	475	500	580	620	...
800	410	490	515	600	640	...
900	435	520	555	...	...	...
1000	455	545	585	680	730	...
1250	495	590	645	...	...	...
1500	520	625	700	785	...	...
1750	545	650	735	...	...	...
2000	560	665	775	840	...	...
Correction Factors, Room Temps. Over 30° C. 86° F.						
C.　F.						
40　104	.82	.88	.90	.94	.95	...
45　113	.71	.82	.85	.90	.92	...
50　122	.58	.75	.80	.87	.89	...
55　131	.41	.67	.74	.83	.86	...
60　140	...	.58	.67	.79	.83	.91
70　158	...	.35	.52	.71	.76	.87
75　167	...	...	.43	.66	.72	.86
80　176	...	...	.30	.61	.69	.84
90　194	...	...	...	.50	.61	.80
100　212	...	...	...	...	.51	.77
120　248	...	...	...	...	...	.69
140　284	...	...	...	...	...	.59

†National Electric Code, 1965. Note that "ampacity" means capacity in
amperes. Note also that the types of cable to which the various temperature
ratings apply are given in Table 10.

TABLE 9A

Trade Name	Type Letter	Max. Operating Temp.	Application Provisions
Rubber-Covered Fixture Wire	†RF-1	60°C 140°F	Fixture wiring. Limited to 300 V.
Solid or 7-Strand	†RF-2	60°C 140°F	Fixture wiring, and as permitted in Section 310-8.
Rubber-Covered Fixture Wire	†FF-1	60°C 140°F	Fixture wiring. Limited to 300 V.
Flexible Stranding	†FF-2	60°C 140°F	Fixture wiring, and as permitted in Section 310-8.
Heat-Resistant Rubber-Covered Fixture Wire	†RFH-1	75°C 167°F	Fixture wiring. Limited to 300 V.
Solid or 7-Strand	†RFH-2	75°C 167°F	Fixture wiring, and as permitted in Section 310-8.
Heat-Resistant Rubber-Covered Fixture Wire	†FFH-1	75°C 167°F	Fixture wiring. Limited to 300 V.
Flexible Stranding	†FFH-2	75°C 167°F	Fixture wiring, and as permitted in Section 310-8.
Thermoplastic-Covered Fixture Wire—Solid or Stranded	†TF	60°C 140°F	Fixture wiring, and as permitted in Section 310-8, and for circuits as permitted in Article 725.
Thermoplastic-Covered Fixture Wire—Flexible Stranding	†TFF	60°C 140°F	Fixture wiring, and as permitted in Section 310-8, and for circuits as permitted in Article 725.
Cotton-Covered, Heat-Resistant, Fixture Wire	*CF	90°C 194°F	Fixture wiring. Limited to 300 V.
Asbestos-Covered Heat-Resistant, Fixture Wire	*AF	150°C 302°F	Fixture wiring. Limited to 300 V. and Indoor Dry Location.

Trade Name	Type Letter	Max. Operating Temp.	Application Provisions
Silicone Rubber Insulated Fixture Wire	*SF-1	200°C 392°F	Fixture wiring. Limited to 300 V.
Solid or 7 Strand	*SF-2	200°C 392°F	Fixture wiring and as permitted in Section 310-8.
Silicone Rubber Insulated Fixture Wire	*SFF-1	150°C 302°F	Fixture wiring. Limited to 300 V.
Flexible Stranding	*SFF-2	150°C 302°F	Fixture wiring and as permitted in Section 310-8.
Code Rubber	R	60°C 140°F	Dry locations.
Heat-Resistant Rubber	RH	75°C 167°F	Dry locations.
Heat Resistant Rubber	RHH	90°C 194°F	Dry locations.
Moisture- Resistant Rubber	RW	60°C 140°F	Dry and wet locations. For over 2000 volts, insulation shall be ozone-resistant.
Moisture and Heat Resistant Rubber	RH-RW	60°C 140°F	Dry and wet locations. For over 2000 volts, insulation shall be ozone-resistant.
		75°C 167°F	Dry locations. For over 2000 volts, insulation shall be ozone-resistant.
Moisture and Heat Resistant Rubber	RHW	75°C 167°F	Dry and wet locations. For over 2000 volts, insulation shall be ozone-resistant.
Latex Rubber	RU	60°C 140°F	Dry locations.
Heat Resistant Latex Rubber	RUH	75°C	Dry locations.

Trade Name	Type Letter	Max. Operating Temp.	Application Provisions
Moisture Resistant Latex Rubber	RUW	60°C 140°F	Dry and wet locations.
Thermoplastic	T	60°C 140°F	Dry locations.
Moisture-Resistant Thermoplastic	TW	60°C 140°F	Dry and wet locations.
Heat-Resistant Thermoplastic	THHN	90°C 194°F	Dry locations.
Moisture and Heat-Resistant Thermoplastic	THW	75°C 167°F	Dry and wet locations.
Moisture and Heat-Resistant Thermoplastic	THWN	75°C 167°F	Dry and wet locations.
Thermoplastic and Asbestos	TA	90°C 194°F	Switchboard wiring only.
Thermoplastic and Fibrous Outer Braid	TBS	90°C 194°F	Switchboard wiring only.
Synthetic Heat-Resistant	SIS	90°C 194°F	Switchboard wiring only.
Mineral Insulation (Metal Sheathed)	MI	85°C 185°F 250°C 482°F	Dry and wet locations with Type O termination fittings. For special application.
Silicone- Asbestos	SA	90°C 194°F 125°C 257°F	Dry locations. For special application.

CONDUCTOR APPLICATION*

Trade Name	Type Letter	Max. Operating Temp.	Application Provisions
Fluorinated Ethylene Propylene	FEP or FEPB	90°C 194°F 200°C 392°F	Dry locations. Dry locations—special applications.
Varnished Cambric	V	85°C 185°F	Dry locations only. Smaller than No. 6 by special permission.
Asbestos and Varnished Cambric	AVA	110°C 230°F	Dry locations only.
Asbestos and Varnished Cambric	AVL	110°C 230°F	Dry and wet locations.
Asbestos and Varnished Cambric	AVB	90°C 194°F	Dry locations only.
Asbestos	A	200°C 392°F	Dry locations only. In raceways, only for leads to or within apparatus. Limited to 300 V.
Asbestos	AA	200°C 392°F	Dry locations only. Open wiring. In raceways, only for leads to or within apparatus. Limited to 300 V.
Asbestos	AI	125°C 257°F	Dry locations only. In raceways, only for leads to or within apparatus. Limited to 300 V.
Asbestos	AIA	125°C 257°F	Dry locations only. Open wiring. In raceways, only for leads to or within apparatus.
Paper		85°C 185°F	For underground service conductors, or by special permission.

*1965 National Electric Code.

†Fixture wires are not intended for installation as branch circuit conductors nor for the connection of portable or stationary appliances.

is available in American Wire Gauge sizes ranging from 8 to 14 (single strand), as well as in seven-strand, 19-strand, 37-strand and 61-strand types as needed for the high current ratings required in industry or feeder lines.

Pertinent Code provisions applying to this type of wire are:

(a) Underground feeder and branch circuit cable may be used underground, including direct burial in the earth, as feeder or branch circuit cable when provided with overcurrent protection of the rated ampacity.

(b) Where single conductor cables are installed, all cables of the feeder circuit, sub-feeder circuit, or branch circuit, including the neutral conductor, if any, shall be run together in the same trench or raceway.

(c) A minimum depth of 18 inches shall be maintained for conductors and cables buried directly in the earth, when supplementary protection from physical injury such as a covering board, concrete pad, raceway, etc., is not provided.

Demand Calculations for Feeder or Service Wires.—Sizes of feeder wires to supply both light and power loads are determined on a basis of the type of building they are to serve and the floor areas. For example, the minimum watts per unit area and demand factors for single-family dwellings are:

Three watts per square foot, plus 1500 watts for appliances.

For area of 3000 or less square feet, demand 100 percent; for the next 117,000 square feet, 35 percent.

No demand shall be applied in connection with appliance loads.

Calculations for Single Family Dwelling.—The requirements of the National Electrical Code (which is distributed by the National Fire Protection Association, 60 Batterymarsh St., Boston, Mass., 02110) govern the entire operation of planning and installation of wiring systems. They give specifications for wire sizes and circuit loading in terms of the power requirements of the lights and appliances to be installed. As a first step in computing the power requirements of appliances, the following list of average requirements of representative appliances shown in Table 9B should be used:

TABLE 9B

Appliance	Average Wattage	Appliance	Average Wattage
Blanket..................	150	Immersion heater.........	300
Bread mixer..............	200	Iron, household...........	1000
Clocks...................	3	Iron, travelers'...........	330
Cigar lighter.............	100	Ironer....................	1320
Coffee maker.............	550	Kitchen mixer with grinder.	200
Coffee percolator..........	450	Mechanical exerciser......	500
Curling iron..............	20	Phonograph..............	40
Chafing dish..............	600	Piano player.............	125
Cream whipper...........	75	Range....................	8000
Dish washer..............	100	Refrigerator.............	300
Egg boiler................	250	Radio....................	100
Fan, 8-inch..............	30	Roaster..................	1320
Fan, 10-inch.............	35	Sewing machine..........	75
Fan, 12-inch.............	50	Soldering iron............	200
Frying pan...............	600	Sun lamp................	450
Griddle..................	450	Teakettle................	400
Grill....................	600	Teapot...................	400
Hair drier................	50	Toaster..................	450
Heater (radiant)..........	1000	Vacuum cleaner..........	160
Heating pad..............	50	Vibrator.................	50
Hot plate................	660	Washing Machine.........	175
Humidifier...............	500	Water heater.............	2000
Ice-cream freezer..........	300	Waffle iron..............	660

To illustrate the method of calculation, it will be applied to a small family house, consisting of two floors and a basement, of which the plan views are shown in Figs. 70, 71 and 72.

Beginning with the problem of general lighting, we can apply the following Code regulation:

General Illumination. . . . not less than one branch circuit be installed for each 500 square feet of floor area. . . .

Since the floor space referred to here means floor space intended for occupancy, the first step is to determine the number of square feet contained in the first and second floors of the house. The first floor, without the dinette, measures $31' \times 15'\ 6''$, outside dimensions. Thus the first floor, without the dinette,

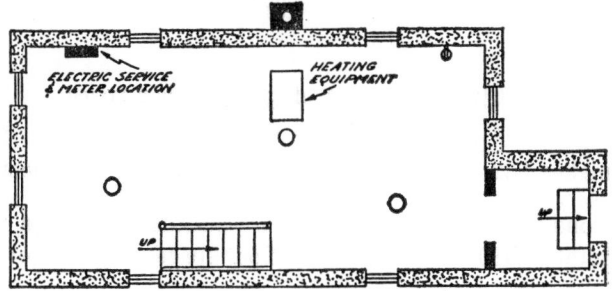

Fig. 70.—Floor Plan—Basement.

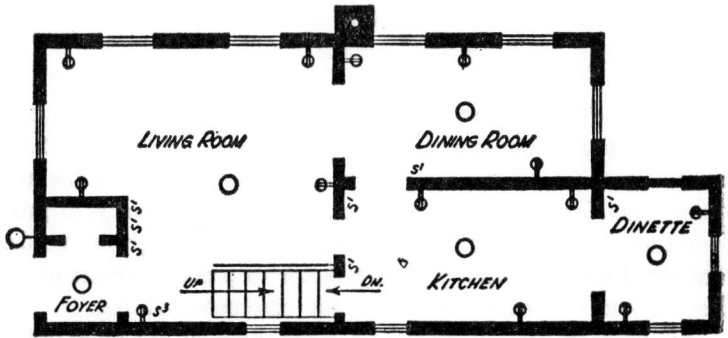

Fig. 71.—Floor Plan—First Floor.

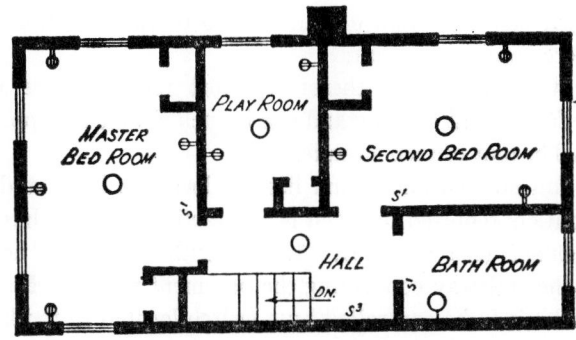

Fig. 72.—Floor Plan—Second Floor.

contains 480½ square feet. The dinette measures 6' 6" × 8', or
52 square feet.

480½ square feet of general floor space
plus   52   square feet of dinette floor space
makes 532½ square feet, total floor space, first floor

The second floor measures 31' × 15' 6", outside dimensions, making a total of 480½ square feet of floor space. Therefore:

480½ square feet of floor space, second floor
plus   532½ square feet of floor space, first floor
makes 1013   square feet of floor space, for both
occupied floors

Now, the Code suggests one branch circuit for each 500 square
feet of occupied floor space. So: 1013 divided by 500 equals 2.02
circuits for general lighting. Since we have exceeded the requirements for two branch circuits, it will be necessary to provide three branch circuits to take care of general lighting needs.
The third circuit is thus a safety factor to prevent circuit overloading and will also provide for the basement lighting.

So far, then, our calculations call for three branch circuits for
general lighting. In addition, however, we must make provision
for a small-appliance circuit, as required by the Code. Our
calculated small appliance load is 12,000 watts, so we plan a
20 ampere appliance circuit.

Our total provision, therefore, must include three branch
circuits for general lighting and one branch circuit for appliances,
or a total of four branch circuits.

We can now go ahead and list the location, kind, and number of outlets required for each floor, starting with the basement.

Basement (see Fig. 70). 1. One outlet (ceiling) near the
foot of the stairway connecting the basement with the first floor.
This outlet is to be controlled by a switch located in the kitchen,
near the entrance to the basement stairway.

2. One outlet (ceiling) over the heating equipment, so placed that it will illuminate the gauges. This outlet will have key-socket control.

3. One outlet (ceiling) in the open space near the rear entrance to the basement; this will be key-socket controlled.

4. One receptacle outlet (side-wall) for an electric washing machine, located in the corner of the basement. This outlet is to be of the 3-pole type.

First Floor (see Fig. 71). 1. One outside outlet (side-wall) at the main entrance door to the house, to be controlled by a switch located in the living room, near the entrance door.

2. One outlet (ceiling) in the foyer, to be controlled by a switch located in the living room, near the entrance door.

3. One outlet (ceiling) in the living room, to be controlled by a switch located near the living-room main entrance door.

4. One switch located at the foot of the stairs leading to the second floor to control the landing hall outlet on the second floor.

5. One outlet (ceiling) in the dining room, to be controlled by a switch located near the door between the dining room and the kitchen, on the dining-room side.

6. One switch to be located in the kitchen, near the entrance door leading to the basement, to control the basement light at the foot of the basement stairs.

7. One outlet (ceiling) in the kitchen, to be controlled by a switch located near the door between the kitchen and the living room, on the kitchen side.

8. One outlet (ceiling) in the dinette, to be controlled by a switch located near the opening between the kitchen and the dinette, on the dinette side.

Receptacles for First Floor. Living room: 16′ × 14′ 6″; total (gross) distance around room is 61′. Thus, 61′ divided by 12 equals 5.0, or 5 receptacles.

Dining room: 13′ 6″ × 7′; total (gross) distance around room is 41′. Thus, 41′ divided by 12 equals 3.4, or 3 receptacles.

Kitchen: 13′ 6″ × 7′; total (gross) distance around room is 41′. Thus, 41′ divided by 12 equals 3.4, or 3 receptacles.

Dinette: 6′ × 7′; total (gross) distance around room is 26′. Thus, 26′ divided by 12 equals 2.1, or 2 receptacles.

Second Floor (see Fig. 72). 1. One outlet (ceiling) in the master bedroom, to be controlled by a switch located in the bedroom, near the entrance door.

2. One outlet (ceiling) in the playroom, to be pull-chain controlled at the fixture.

3. One outlet (ceiling) in the second bedroom, to be controlled by a switch located on the bedroom side of the entrance door.

4. One outlet (wall-bracket) in the bathroom, to be located over the medicine chest and to be controlled by a switch located on the bathroom side of the entrance door.

5. One outlet (ceiling) in the hall, to be controlled by a switch located near the stair landing, and the same fixture to be also controlled by the switch on the first floor, located near the foot of the stairs.

Receptacles for Second Floor—Master bedroom: 9′ 6″ × 14′ 6″; total (gross) distance around room is 48′. Thus, 48′ divided by 12 equals 4.0, or 4 receptacles.

Playroom: 6′ 6″ × 8′ 6″; total (gross) distance around room is 30′. Thus, 30′ divided by 12 equals 2.5, or 2 receptacles.

Second bedroom: 13′ × 8′ 6″; total (gross) distance around room is 43′. Thus, 43′ divided by 12 equals 3.5, or 3 receptacles.

The total of the various kinds of electrical outlets needed for the installation is:

Ceiling outlets and wall-brackets 14
Receptacle outlets 23

We are providing four branch circuits for fourteen ceiling and wall-bracket outlets and twenty-three receptacle outlets. Remember, the Code requires that receptacle outlets in the dining room, kitchen, and dinette, and also the washing machine outlet in the basement, be placed on a special circuit. Thus nine of the twenty-three receptacle outlets must be placed on an independent circuit. That leaves fourteen ceiling and wall-bracket outlets and fourteen receptacle outlets to be divided among three branch circuits.

Proportioning the Load. The next step, then, is to proportion the load—to divide the fourteen lighting outlets and the fourteen convenience receptacle outlets among three branch circuits so that no single branch circuit will be overloaded. There are more ways than one of proportioning the load. However, it is necessary to bear in mind one special point: It is advisable, where possible, to place a ceiling outlet on one circuit, and the receptacle outlets in the same room on another circuit, so that a fuse blow-out in either circuit will not cut off the room entirely from electric service. This method may involve the use of a little more armored cable, but it will prove to be worth the extra cost.

A wiring diagram showing every outlet, floor by floor, will be helpful, because by means of such wiring diagrams we can plan the distribution of the load, circuit by circuit. Let us then examine Figure 73.

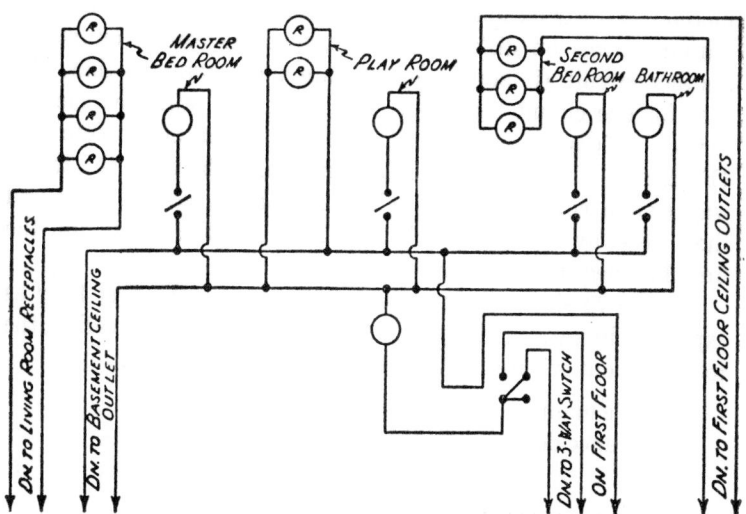

Fig. 73.—Wiring Diagram for Second Floor.

According to the specification, the following outlets are to be installed on the second floor:

Master bedroom ..... ceiling outlet
switch to control ceiling outlet
4 base-receptacles

Playroom .... ceiling outlet, pull-chain control
2 base-receptacles

Second bedroom ..... ceiling outlet
switch to control ceiling outlet
3 base-receptacles

Bathroom .... wall-bracket
switch to control wall-bracket

Hall .... ceiling outlet
3-way switch to control ceiling outlet

Note that the four master bedroom receptacles are connected to the circuit which picks up the living room receptacles (Figure 73), the playroom receptacles are connected to the second floor ceiling outlet circuit, and the second bedroom receptacle outlets are connected to the first floor ceiling outlet circuit. By this method of distribution—placing the convenience outlets in the various rooms on different circuits, and the ceiling outlets on another circuit (if possible)—we gain the advantage that, if a fuse for one circuit blows out, each room will still have electric service, unless, of course, the main fuse blows out. Note also (Figure 73) that a 3-wire cable must be run between the three-way switch controlling the hall outlet on the second floor and the three-way switch controlling the same outlet located on the first floor near the stairway.

Figure 74 shows the circuit distribution for the first floor as expressed in a wiring diagram. According to the specification, the following outlets are to be installed on the first floor:

Outside main
entrance door ... wall-bracket
switch to control wall-bracket

Foyer .... ceiling outlet
switch to control ceiling outlet

Living room ...... ceiling outlet
                   switch to control ceiling outlet
                   5 base-receptacles
                   3-way switch for second-floor hall outlet
Dining room ..... ceiling outlet
                   switch to control ceiling outlet
                   3 base-receptacles (appliance circuit)
Kitchen .... ceiling outlet
                   switch to control ceiling outlet
                   3 receptacles (appliance circuit)
                   switch to control basement outlet
Dinette .... ceiling outlet
                   switch to control ceiling outlet
                   2 base-receptacles (appliance circuit)

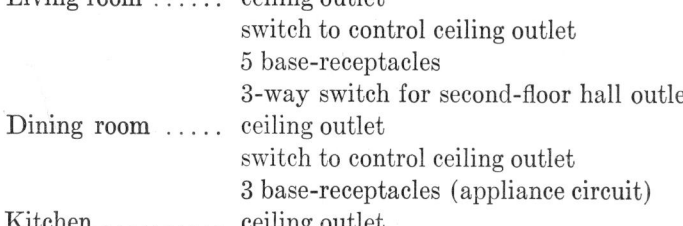

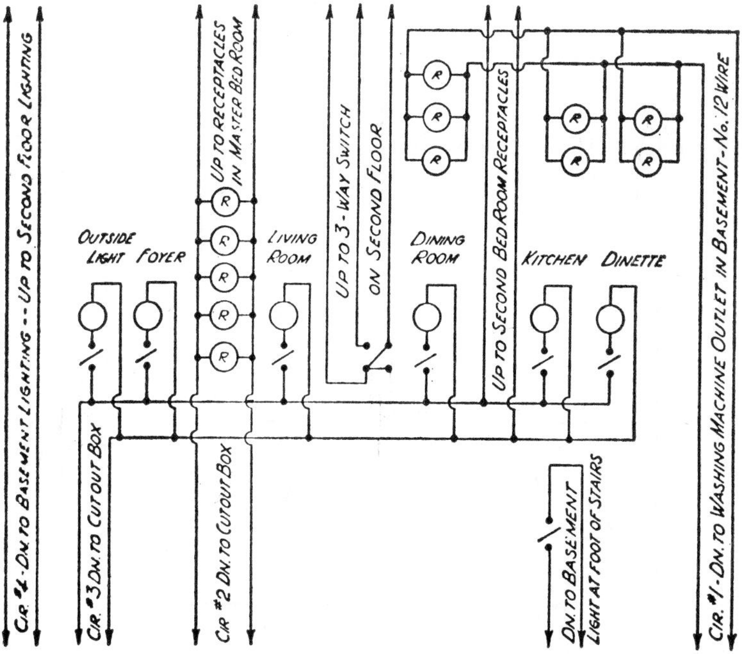

Fig. 74.—Wiring Diagram for First Floor.

Note, in Figure 74, that each circuit has been numbered to assist in clarifying the wiring diagram. Circuit 1 has been chosen as the appliance circuit, and the receptacles in dining room, kitchen, and dinette have therefore been placed on Circuit, which is continued down to the washing machine outlet in the basement (see Fig. 75) and then to the branch circuit cutout box at the service installation. This circuit, which services receptacles in dining room, kitchen, dinette, and laundry only, must be installed with not less than No. 12 wire, to comply with the National Electrical Code.

Circuit 2 runs directly from the branch cutout box in the basement service installation to the living-room base-receptacles, where it is tide-in with the four base-receptacles in the master bedroom on the second floor. Thus Circuit 2 includes nine convenience receptacles in the living room and the master bedroom.

Circuit 3 includes all the ceiling lighting on the first floor and

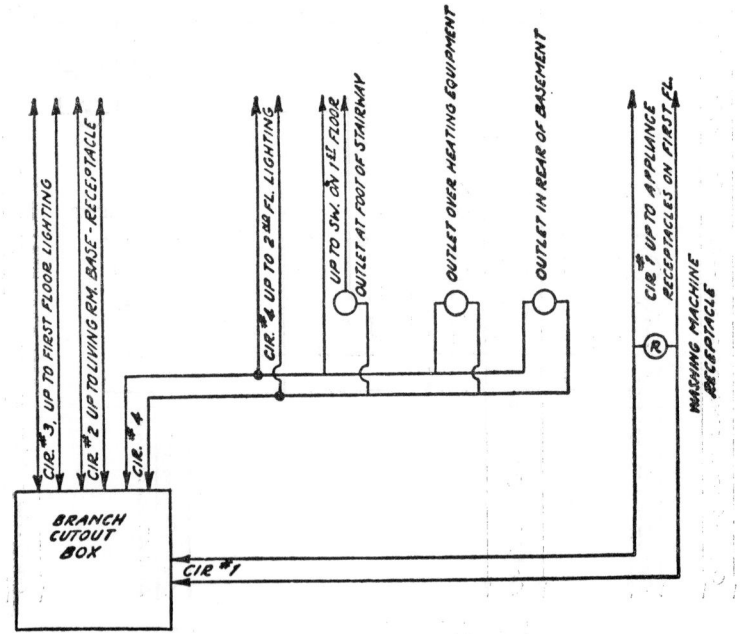

Fig. 75.—Wiring Diagram for Basement.

the three convenience receptacles in the second-floor second bedroom. This includes a total of nine outlets: the outside wall-bracket, foyer ceiling outlet, living room ceiling outlet, dining room ceiling outlet, kitchen ceiling outlet, dinette ceiling outlet, and the three convenience receptacles in the second bedroom on the second floor. It must be remembered and understood, of course, that, since switches consume no current, they are naturally not included in these calculations for circuit loading.

In distributing the general lighting load on the first and second floors among three branch circuits, it is necessary to keep in mind the three basement outlets, which must also be cared for in one or another of these three branch circuits. Living-room, dining-room, and kitchen ceiling outlets generally use higher-wattage lamps than do those in bedrooms and bathrooms. We therefore consider it advisable to make one circuit of all the ceiling outlets on the first floor (that is, Circuit 3), plus the three convenience receptacles in the second bedroom, and place all the basement outlets and the ceiling outlets on the second floor, plus the two playroom convenience receptacles, on Circuit 4.

To make the wiring diagram complete for the first floor, we have included the wiring for the switch in the kitchen which controls the basement outlet at the foot of the basement stairs and have also included that for the three-way switch in the living room near the foot of the stairway leading to the second floor, that is, the switch that controls the hall outlet on the second floor. This completes, then, the picture of the total wiring needs for the first floor.

Figure 75 shows the circuit distribution for the basement. According to the specification, a total of three ceiling outlets and one appliance receptacle outlet is to be installed, as follows:

Basement .... ceiling outlet near the foot of the stairs leading from the first floor, controlled by a switch on the first floor

ceiling outlet over heating equipment, key-socket controlled

ceiling outlet in open cellar space, key-socket controlled

appliance receptacle outlet service equipment

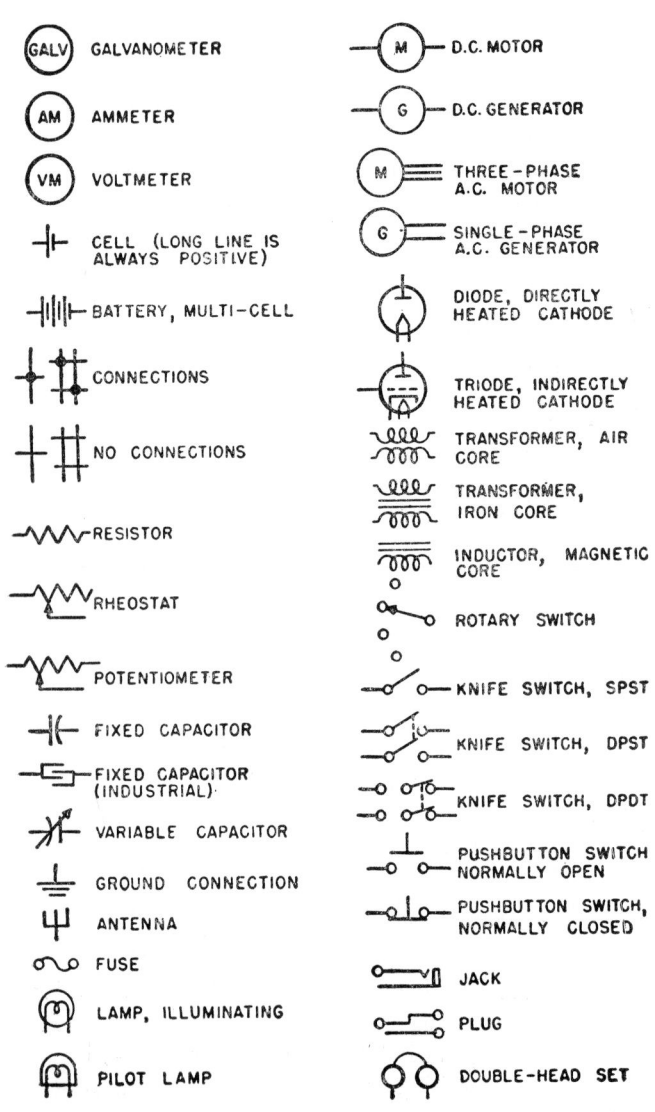

FIG. 70.—Electrical Symbols.

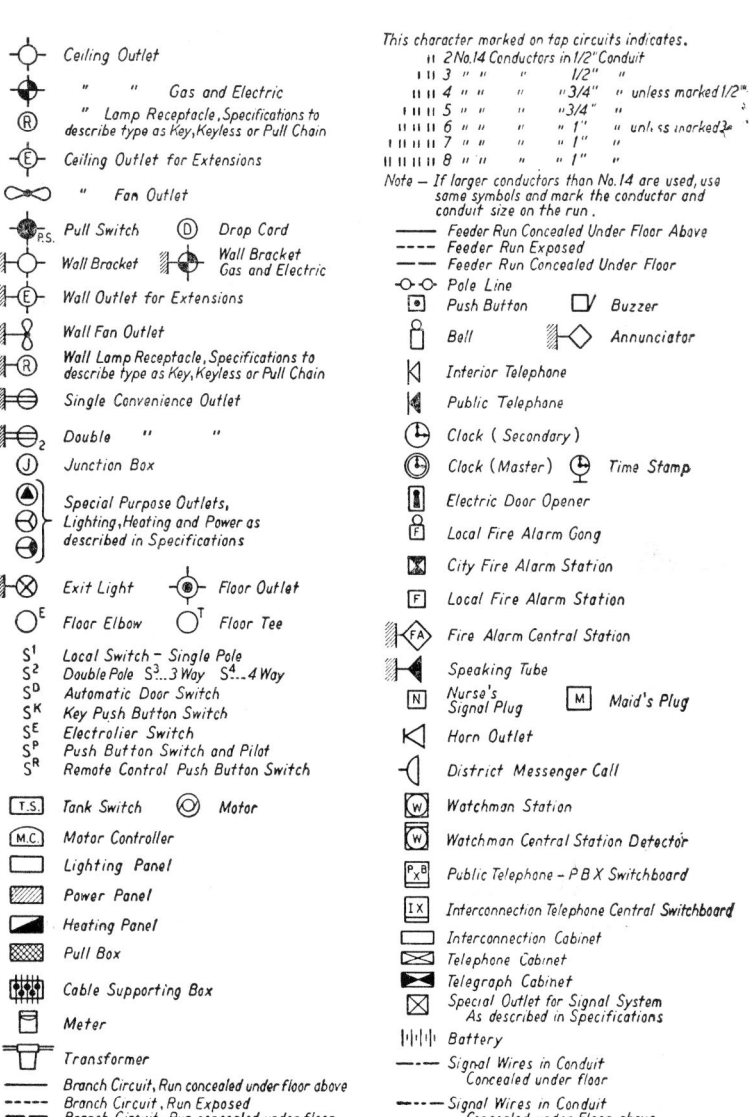

FIG. 71.—Electrical Symbols for Architectural Plans.

ILLUSTRATION: What minimum size of feeder is required for a single-family dwelling having a floor area of 3800 square feet exclusive of unoccupied cellars, unfinished attics, and open porches?

Area in sq. ft., 3800 × 3 watt per sq. ft. = 11,400 watts
Allowance for appliances   = 1500 watts
                                                   ———————
Computed load   = 12,900 watts

Demand selected for this occupancy, first 3000 square feet = demand 100 percent; excess over 3000 square feet = demand 35 percent.   Then

3000 sq. ft. at 3 watts per sq. ft. × 1   = 9000 watts
800 sq. ft. at 3 watts per sq. ft. × 0.35 =   840 watts
Allowance for appliances   = 1500 watts
                                             ———————
Load after applying demand   = 11,340 watts

For 115-volt, 2-wire system:

11,340 watts ÷ 115 volts = 98.8 amperes

Size of Type R conductors (Table 9) = No. 1 for each wire (Ans.)

For 230-volt, 2-wire system:

11,340 watts ÷ 230 volts = 49.4 amperes

Size of Type R conductors (Table 9) = No. 6 for each wire (Ans.)

For 115–230-volt, 3-wire system:

11,340 watts ÷ 2 × 115 volts = 49.4 amperes

Size of Type R conductors (Table 9) = No. 6 for each wire (Ans.)

For 120–208 volt, 4-wire, 3-phase system:

11,340 ÷ 3 × 120 volts = 31.5 amperes

Size of Type R conductors (Table 9) = No. 8 for each wire (Ans.)

The specifications for buildings other than single-family dwellings are contained in the "National Electrical Code" and the computations are carried out in the same manner as above.

Grounding.—In installing electric wiring and appliances, the matter of grounding is so important that the pertinent provisions of the 1965 National Electrical Code should always be followed. Therefore selected provisions of the Code are reproduced below.

250-32. Service Conductor Enclosures. Service raceways, service cable sheaths or armoring, when of metal, shall be grounded.

250-33. Other Conductor Enclosures. Metal enclosures for conductors shall be grounded, except they need not be grounded in runs of less than 25 feet which are free from probable contact with ground, grounded metal, metal lath or conductive thermal insulation and which, where within reach from grounded surfaces, are guarded against contact by persons.

250-42. Fixed Equipment-General. Under any of the following conditions, exposed, noncurrent-carrying metal parts of fixed equipment, which are liable to become energized, shall be grounded:

(a) Where equipment is supplied by means of metal-clad wiring;

(b) Where equipment is located in a wet location and is not isolated;

(c) Where equipment is located within reach of a person who can make contact with any grounded surface or object;

(d) Where equipment is located within reach of a person standing on the ground;

(e) Where equipment is in a hazardous location;

(f) Where equipment is in electrical contact with metal or metal lath;

(g) Where equipment operates with any terminal at more than 150 volts to ground, except as follows:

(1) Enclosures for switches or circuit breakers where accessible to qualified persons only;

(2) Metal frames of electrically heated devices, exempted by special permission, in which case the frames shall be permanently and effectively insulated from ground;

(3) Transformers mounted on wooden poles at a height of more than 8 feet from the ground.

250-43. Fixed Equipment—Specific. Exposed, noncurrent-carrying metal parts of the following kinds of equipment, regardless of voltage, shall be grounded:

(a) Frames of motors;

(b) Controller cases for motors, except lined covers of snap switches;

(c) Electric equipment of elevators and cranes;

(d) Electric equipment in garages, theatres and motion picture studios, except pendant lampholders on circuits of not more than 150 volts to ground;

(e) Motion-picture projection equipment;

(f) Electric signs and associated equipment, unless these are inaccessible to unauthorized persons and are also insulated from ground and from other conductive objects;

250-45. Portable Equipment. Under any of the following conditions, exposed noncurrent-carrying metal parts of portable equipment, which are liable to become energized, shall be grounded:

(a) In hazardous locations;

(b) When operated at more than 150 volts to ground, except:
(1) Motors, where guarded;
(2) Metal frames of electrically heated appliances;

(c) In residential occupancies, (1) clothes-washing, clothes-drying, and dish-washing machines, sump pumps, and (2) portable, hand held, motor operated tools and appliances of the following types: drills, hedge clippers, lawn mowers, wet scrubbers, sanders and saws.

Exception: Such tools and appliances protected by an approved system of double insulation, or its equivalent, need not be grounded. Where such an approved system is employed the equipment shall be distinctively marked.

Portable tools or appliances not provided with special insulating or grounding protection are not intended to be used in damp, wet or conductive locations.

(d) In other than residential occupancies, (1) portable appliances used in damp or wet locations, or by persons standing on the ground or on metal floors or working inside of metal tanks or boilers, and (2) portable tools which are likely to be used in wet and conductive locations shall be grounded except that they need not be grounded where supplied through an insulating transformer with ungrounded secondary of not over 50 volts.

It is recommended that the frames of all portable motors which operate at more than 50 volts to ground be grounded.

Methods of Grounding. 250-51. Effective Grounding. The path to ground from circuits, equipment, and conductor enclosures shall (1) be permanent and continuous and (2) shall have ample carrying capacity to conduct safely any currents liable to be imposed on it, and (3) shall have impedance sufficiently low to limit the potential above ground and to facilitate the operation of the overcurrent devices in the circuit.

250-52. Location of System Ground Connection. The grounding conductor may be connected to the grounded conductor of the wiring system at any convenient point on the premises on the supply side of the service disconnecting means.

It is recommended that high capacity services have the grounding conductor connected to the grounded conductor of the system within the service entrance equipment enclosure.

250-53. Common Use of Grounding Conductor. The grounding conductor of a wiring system shall also be used for grounding equipment, conduit and other metal raceways or enclosures for conductors, including service conduit or cable sheath and service equipment.

250-56. Short Sections of Raceway. Isolated sections of metal raceway or cable armor, where required to be grounded, shall preferably be grounded by connecting to other grounded raceway or armor, but may be grounded in accordance with Section 250-57.

250-57. Fixed Equipment.

(a) Metal boxes, cabinets and fittings, or noncurrent-carrying metal parts of other fixed equipment, where metallically connected to grounded cable armor or metal raceway, are considered to be grounded by such connection.

(b) Where not so connected they may be grounded in one of the following ways:

(1) By a grounding conductor run with circuit conductors; this conductor may be uninsulated, but where it is provided with an individual covering, the covering shall be finished a continuous green color or a continuous green color with a yellow stripe.

(2) By a separate grounding conductor installed the same as a grounding conductor for conduit and the like;

250-58. Equipment on Structural Metal.

(a) Electric equipment secured to and in contact with the grounded structural metal frame of a building, shall be deemed to be grounded.

(b) Metal car frames supported by metal hoisting cables attached to or running over sheaves or drums of elevator machines shall be deemed to be grounded where the machine is grounded in accordance with this Code.

250-59. Portable Equipment. Noncurrent-carrying metal parts of portable equipment may be grounded in any one of the following ways:

(a) By means of the metal enclosure of the conductors feeding such equipment, provided an approved grounding-type attachment plug is used, one fixed contacting member being for the purpose of grounding the metal enclosure, and provided, further, that the metal enclosure of the conductors is attached to the attachment plug and to the equipment by connectors approved for the purpose;

Exception: The grounding contacting member of grounding type attachment plugs on the power supply cord of hand-held tools or hand-held appliances may be of the movable self-restoring type.

Attachment plug caps are not intended to be used as terminations for metal-clad cable or flexible metal conduit.

(b) By means of a grounding conductor run with the power supply conductors in a cable assembly or flexible cord that is properly terminated in an approved grounding-type attachment plug having a fixed grounding contacting member. The grounding conductor in a cable assembly may be uninsulated; but where an individual covering is provided for such conductors it shall be finished a continuous green color or a continuous green color with a yellow stripe.

Exception: The grounding contacting member of grounding type attachment plugs on the power supply cord of hand-held tools or hand-held appliances may be of the movable self-restoring type.

(c) A separate flexible wire or strap, insulated or bare, protected as well as practicable against physical damage may be used only by special permission except where a part of an approved portable equipment.

250-60. Frames of Electric Ranges and Electric Clothes Dryers. Frames of electric ranges and electric clothes dryers shall be grounded by any of the means provided for in Sections 250-57 and 250-59 or where served by 120-240 volt, three-wire branch circuits, they may be grounded by connection to the grounded circuit conductors, provided the grounded circuit conductors are not smaller than 10 AWG. The frames of wall-mounted ovens and counter-mounted cooking units shall be grounded and may be grounded in the same manner as electric ranges.

Wire Sizes for Branch Circuits.—That portion of the supply conductors which extends from the street or duct or transformers to the service switch of the building supplied is called the *service circuit*. That portion of the wiring system which extends beyond the final automatic overload protective device (fuse box) is called the *branch circuit*.

The sizes of wire required for lighting circuits or combination lighting and power circuits for dwellings and apartments connected to separate meters may be computed as above. However, most local codes and good practice require a minimum size of No. 14 wire for these circuits while No. 18 flexible wire is permitted in fixtures and drop cords.

These minimum sizes are usually the governing factors for ordinary requirements. Where, however, special heating or power units are to be used, the sizes of wire must be computed or obtained from a table. If a circuit is to be run for a motor, and the voltage and the current which the motor will use are known, the size of the wire required may be found in Table 10.

ILLUSTRATION: What minimum size of Type R insulated copper wire would be required for a motor with a full-load current rating of 40 amperes?

Running down column 1 of Table 10 until 40 is reached, size of Type R wire is found in column 2 to be No. 6. (Ans.)

If the wiring is being done for a motor whose power requirements are not known, but whose horsepower is known, the current required may be found from Table 11, 12, or 13.

ILLUSTRATION: What size of Type RH wire would be required for a 50-horsepower, 220-volt, 3-phase, induction-type, alternating-current motor?

Current required (from Table 13) = 125 amperes

Size of wire (from Table 10)   = No. 00 (Ans.)

TABLE 10

<small>CONDUCTOR SIZES AND OVERCURRENT PROTECTION FOR MOTORS *</small>

Col. No. 1	2	3	5	6	7	8	9	10
					Maximum Allowable Rating of Branch Circuit Fuses			
Full load current rating of motor amperes	Minimum size conductor in raceways. For conductors in air or for other insulations see tables 1 and 2 AWG and MCM. Type R Type T	Type RH	For Running Protection of Motors****. Maximum rating of non-adjustable protective devices. Am-peres	Maximum setting of adjustable protective device. Am-peres	WITH CODE LETTERS Single-phase and squirrel cage and synchronous. Full voltage, resistor and reactor starting, Code letters F to V inc. WITHOUT CODE LETTERS Same as above.	WITH CODE LETTERS Single-phase, squirrel cage and synchronous. Full voltage, resistor or reactor start, Code letters B–E inc. Auto-transformer start, Code letters F to V inc. WITHOUT CODE LETTERS Squirrel cage and synchronous, auto-transformer start, high reactance squirrel cage. Both not more than 30 amperes	WITH CODE LETTERS Squirrel cage and synchronous auto-transformer start, Code letters B–E inc. WITHOUT CODE LETTERS Squirrel cage and synchronous auto-transformer start, high reactance squirrel cage.*** Both more than 30 amperes	WITH CODE LETTERS All motors. Code letter A. WITHOUT CODE LETTERS DC and wound-rotor motors.
1**	14	14	2*	1.25*	15	15	15	15
2**	14	14	3*	2.50*	15	15	15	15
3**	14	14	4*	3.75*	15	15	15	15
4**	14	14	6*	5.0 *	15	15	15	15
5**	14	14	8*	6.25*	15	15	15	15
6**	14	14	8*	7.50*	20	15	15	15
7	14	14	10*	8.75*	25	20	15	15
8	14	14	10*	10.0 *	25	20	20	15
9	14	14	12*	11.25*	30	25	20	15
10	14	14	15*	12.50*	30	25	20	15
11	14	14	15*	13.75*	35	30	25	20
12	14	14	15	15.00	40	30	25	20
13	12	12	20	16.25	40	35	30	20
14	12	12	20	17.50	45	35	30	25
15	12	12	20	18.75	45	40	30	25
16	12	12	20	20.00	50	40	35	25
17	10	10	25	21.25	60	45	35	30
18	10	10	25	22.50	60	45	40	30
19	10	10	25	23.75	60	50	40	30
20	10	10	25	25.0	60	50	40	30
22	10	10	30	27.50	70	60	45	35
24	10	10	30	30.00	80	60	50	40
26	8	10	35	32.50	80	70	60	40
28	8	10		35.00	90	70	60	45

* 1953 *National Electrical Code.* For further information on the use of this table, see the latest edition of the Code.

TABLE 10

Conductor Sizes and Overcurrent Protection for Motors *—*Continued*

Col. No. 1	2	3	5	6	7	8	9	10
					Maximum Allowable Rating of Branch Circuit Fuses			
Full load current rating of motor amperes	Minimum size conductor in raceways. For conductors in air or for other insulations see tables 1 and 2 AWG and MCM		For Running Protection of Motors****		With Code Letters Single-phase and squirrel cage and synchronous. Full voltage, resistor and reactor starting, Code letters F to V inc. Without Code Letters Same as above.	With Code Letters Single-phase, squirrel cage and synchronous. Full voltage, resistor or reactor start, Code letters B–E inc. Auto-transformer start, Code letters F to V inc. Without Code Letters Squirrel cage and synchronous, auto-transformer start, high reactance squirrel cage.*** Both not more than 30 amperes	With Code Letters Squirrel cage and synchronous Auto-transformer start, Code letters B–E inc. Without Code Letters Squirrel cage and synchronous auto-transformer start, high reactance squirrel cage.*** Both more than 30 amperes.	With Code Letters All motors. Code letter A. Without Code Letters DC and wound-rotor motors.
	Type R Type T	Type RH	Maximum rating of non-adjustable protective devices Amperes	Maximum setting of adjustable protective device Amperes				
30	8	8	40	37.50	90	70	60	45
32	8	8	40	40.00	100	80	70	50
34	6	8	45	42.50	110	90	70	60
36	6	8	45	45.00	110	90	80	60
38	6	6	50	47.50	125	100	80	60
40	6	6	50	50.00	125	100	80	60
42	6	6	50	52.50	125	110	90	70
44	6	6	60	55.0	125	110	90	70
46	4	6	60	57.50	150	125	100	70
48	4	6	60	60.0	150	125	100	80
50	4	6	60	62.50	150	125	100	80
52	4	6	70	65.0	175	15	110	80
54	4	4	70	67.50	175	150	110	90
56	4	4	70	70.00	175	150	120	90
58	3	4	70	72.50	175	150	120	90
60	3	4	80	75.00	200	150	120	90
62	3	4	80	77.50	200	175	125	100
64	3	4	80	80.00	200	175	150	100
66	2	4	80	82.50	200	175	150	100
68	2	4	90	85.00	225	175	150	110
70	2	3	90	87.50	225	175	150	110
72	2	3	90	90.00	225	200	150	110
74	2	3	90	92.50	225	200	150	125
76	2	3	100	95.00	250	200	175	125
78	1	3	100	97.50	250	200	175	125
80	1	3	100	100.00	250	200	175	125
82	1	2	110	102.50	250	225	175	125
84	1	2	110	105.00	250	225	175	150
86	1	2	110	107.50	300	225	175	150
88	1	2	110	110.00	300	225	200	150
90	0	2	110	112.50	300	225	200	150
92	0	2	125	115.00	300	250	200	150
94	0	1	125	117.50	300	250	200	150
96	0	1	125	120.00	300	250	200	150
98	0	1	125	122.50	300	250	200	150
100	0	1	125	125.00	300	250	200	150
105	00	1	150	131.5	350	300	225	175

* 1953 *National Electrical Code.* For further information on the use of this table, see the latest edition of the Code.

TABLE 10

CONDUCTOR SIZES AND OVERCURRENT PROTECTION FOR MOTORS—*Continued*

Col. No. 1	2	3	5	6	7	8	9	10
					Maximum Allowable Rating of Branch Circuit Fuses			
Full load current rating of motor amperes	Minimum size conductor in raceways For conductors in air or for other insulations see tables 1 and 2 AWG and MCM		For Running Protection of Motors****		WITH CODE LETTERS Single-phase and squirrel cage and synchronous. Full voltage, resistor and reactor starting, Code letters F to V inc. WITHOUT CODE LETTERS Same as above.	WITH CODE LETTERS Single-phase, squirrel cage and synchronous. Full voltage, resistor or reactor start, Code letters B-E inc. Auto-transformer start, Code letters F to V inc. WITHOUT CODE LETTERS Squirrel cage and synchronous, auto-transformer start, high reactance squirrel cage.*** Both not more than 30 amperes	WITH CODE LETTERS Squirrel cage and synchronous Auto-transformer start, Code letters B-E inc. WITHOUT CODE LETTERS Squirrel cage and synchronous auto-transformer start, high reactance squirrel cage.*** Both more than 30 amperes	WITH CODE LETTERS All motors. Code letter A. WITHOUT CODE LETTERS DC and wound-rotor motors.
	Type R Type T	Type RH	Maximum rating of non-adjustable protective devices	Maximum setting of adjustable protective device				
	Type R Type T		Amperes	Amperes				
110	00	0	150	137.5	350	300	225	175
115	00	0	150	144.0	350	300	250	175
120	000	0	150	150.0	400	300	250	200
125	000	00	175	156.5	400	350	250	200
130	000	00	175	162.5	400	350	300	200
135	0000	00	175	169.0	450	350	300	225
140	0000	00	175	175.0	450	350	300	225
145	0000	000	200	181.5	450	400	300	225
150	0000	000	200	187.5	450	400	300	225
155	0000	000	200	194.0	500	400	350	250
160	250	000	200	200.0	500	400	350	250
165	250	0000	225	206.	500	450	350	250
170	250	0000	225	213.	500	450	350	300
175	300	0000	225	219.	600	450	350	300
180	300	0000	225	225.	600	450	400	300
185	300	0000	250	231.	600	500	400	300
190	300	250	250	238.	600	500	400	300
195	350	250	250	244.	600	500	400	300
200	350	250	250	250.	600	500	400	300
210	400	300	250	263.	...	600	450	350
220	400	300	300	275.	...	600	450	350
230	500	300	300	288.	...	600	500	350
240	500	350	300	300.	...	600	500	400
250	500	350	300	313.	...	...	500	400
260	600	400	350	325.	...	...	600	400
270	600	400	350	338.	...	...	600	450
280	600	500	350	350.	...	...	600	450
290	700	500	350	363.	...	...	600	450
300	700	500	400	375.	...	...	600	450
320	750	600	400	400.	...	...	...	500
340	900	600	450	425.	...	...	...	600
360	1000	700	450	450.	...	...	...	600
380	1250	750	500	475.	...	...	...	600
400	1500	900	500	500.	...	...	...	600
420	1750	1000	600	525.	...	...	...	...
440	2000	1250	600	550.	...	...	...	...
460	....	1250	600	575.	...	...	...	...
480	....	1500	600	600.	...	...	...	...
500	....	1500	...	625.	...	...	...	...

TABLE 11

FULL-LOAD CURRENT * DIRECT-CURRENT MOTORS †

HP	115V	230V	550V
½	4.6	2.3	
¾	6.6	3.3	1.4
1	8.6	4.3	1.8
1½	12.6	6.3	2.6
2	16.4	8.2	3.4
3	24.	12.	5.0
5	40.	20.	8.3
7½	58.	29.	12.0
10	76.	38.	16.0
15	112.	56.	23.0
20	148.	74.	31.
25	184.	92.	38.
30	220.	110.	46.
40	292.	146.	61.
50	360.	180.	75.
60	430.	215.	90.
75	536.	268.	111.
100		355.	148.
125		443.	184.
150		534.	220.
200		712.	295.

* These values for full-load current are average for all speeds.
† 1953 *National Electrical Code.*

TABLE 12

FULL-LOAD CURRENT * SINGLE-PHASE A.C. MOTORS †

HP	115V	230V	440V
1/6	3.2	1.6	
¼	4.6	2.3	
½	7.4	3.7	
¾	10.2	5.1	
1	13.	6.5	
1½	18.4	9.2	
2	24.	12.	
3	34.	17.	
5	56.	28.	
7½	80.	40.	21.
10	100.	50.	26.

* These values of full-load current are for motors running at speeds usual for belted motors and motors with normal torque characteristics. Motors built for especially low speeds or high torques may require more running current, in which case the nameplate current rating should be used.

† 1953 *National Electrical Code.*

For full-load currents of 208- and 200-volt motors, increase corresponding 230-volt motor full-load current by 10 and 15 percent, respectively.

TABLE 13*

Full-load Current ‡

HP	Induction Type Squirrel-Cage and Wound Rotor Amperes					Synchronous Type †Unity Power Factor Amperes			
	110V	220V	440V	550V	2300V	220V	440V	550V	2300V
½	4	2	1	.8					
¾	5.6	2.8	1.4	1.1					
1	7	3.5	1.8	1.4					
1½	10	5	2.5	2.0					
2	13	6.5	3.3	2.6					
3		9	4.5	4					
5		15	7.5	6					
7½		22	11	9					
10		27	14	11					
15		40	20	16					
20		52	26	21					
25		64	32	26	7	54	27	22	5.4
30		78	39	31	8.5	65	33	26	6.5
40		104	52	41	10.5	86	43	35	8
50		125	63	50	13	108	54	44	10
60		150	75	60	16	128	64	51	12
75		185	93	74	19	161	81	65	15
100		246	123	98	25	211	106	85	20
125		310	155	124	31	264	132	106	25
150		360	180	144	37		158	127	30
200		480	240	192	48		210	168	40

* 1953 *National Electrical Code.*

For full-load currents of 208- and 200-volt motors, increase the corresponding 200-volt motor full-load current by 6 and 10 percent, respectively.

‡ These values of full-load current are for motors running at speeds usual for belted motors and motors with normal torque characteristics. Motors built for especially low speeds or high torques may require more running current, in which case the nameplate current rating should be used.

† For 90 and 80 percent P.F. the above figures should be multiplied by 1.1 and 1.25 respectively.

TABLE 14

Summary of Basic Electrical Calculations

DC CIRCUIT CHARACTERISTICS

Ohm's Law:

$$E = IR \qquad I = \frac{E}{R} \qquad R = \frac{E}{I}$$

E = voltage impressed on circuit (volts)
I = current flowing in circuit (amperes)
R = circuit resistance (ohms)

Resistances in series:

$$R_t = R_1 + R_2 + R_3 + \cdots \text{ etc.}$$

R_t = total resistance (ohms)

R_1, R_2, etc. = individual resistances (ohms)

Resistances in parallel:

$$R_t = \frac{1}{\dfrac{1}{R_1} + \dfrac{1}{R_2} + \dfrac{1}{R_3} + \cdots \text{ etc.}}$$

Formulas for the conversion of electrical and mechanical power:

$$\text{HP} = \frac{\text{watts}}{746} = \text{watts} \times .00134$$

$$= \frac{\text{kilowatts}}{.746} = \text{kilowatts} \times 1.34$$

Kilowatts = watts ÷ 1000 = HP × .746
Watts = HP × 746

In direct current circuits, electrical power is equal to the product of the voltage and current:

$$P = EI = I^2R = \frac{E^2}{R}$$

P = power (watts)
E = voltage (volts)
I = current (amperes)
R = resistance (ohms)

TABLE 14.—SUMMARY OF BASIC ELECTRICAL CALCULATIONS—*Continued*

Solving the basic formula for I, E, and R gives

$$I = \frac{P}{E} = \sqrt{\frac{P}{R}} \qquad E = \frac{P}{I} = \sqrt{RP} \qquad R = \frac{E^2}{P} = \frac{P}{I^2}$$

ENERGY

Energy is the capacity for doing work. Electrical energy is expressed in kilowatt-hours (kwhr), one kilowatt-hour representing the energy expended by a power source of 1 kw over a period of 1 hour.

EFFICIENCY

Efficiency of a machine, motor or other device is the ratio of the energy output (useful energy delivered by the machine) to the energy input (energy delivered to the machine), usually expressed as a percentage:

$$\text{Efficiency} = \frac{\text{output}}{\text{input}} \times 100\%$$

or

$$\text{Output} = \frac{\text{input} \times \text{efficiency}}{100\%}$$

TORQUE

Torque may be described as the measure of the tendency of a body to rotate. It is expressed in pound-feet or pounds of force acting at a certain radius:

Torque (pound-feet) = force tending to produce rotation (pounds)
    × distance from center of rotation to point at which force is applied (feet)

Relations between torque and horsepower:

$$\text{Torque} = \frac{33,000 \times \text{HP}}{6.28 \times \text{rpm}}$$

$$\text{HP} = \frac{6.28 \times \text{rpm} \times \text{torque}}{33,000}$$

rpm = speed of rotating part (rev. per minute)

TABLE 14.—Summary of Basic Electrical Calculations—*Continued*

AC CIRCUIT CHARACTERISTICS

The instantaneous values of an alternating current or voltage vary from zero to a maximum value each half cycle. In the practical formulae which follow, the "effective value" of current and voltage is used, defined as follows:

Effective value = 0.707 × maximum instantaneous value

Impedance:

Impedance is the total opposition to the flow of alternating current. It is a function of resistance, capacitive reactance and inductive reactance. The following formulae relate these circuit properties:

$$X_L = 2\pi f L \qquad X_C = \frac{1}{2\pi f C} \qquad Z = \sqrt{R^2 + (X_L - X_C)^2}$$

X_L = inductive reactance (ohms)
X_C = capacitive reactance (ohms)
Z = impedance (ohms)
f = frequency (cycles per second)
C = capacitance (farads)
L = inductance (henrys)
R = resistance (ohms)
π = 3.14

In circuits where one or more of the properties L, C or R is absent, the impedance formula is simplified as follows:

Resistance only:	Inductance only:	Capacitance only:
$Z = R$	$Z = X_L$	$Z = X_C$
Resistance and Inductance only:	Resistance and Capacitance only:	Inductance and Capacitance only:
$Z = \sqrt{R^2 + X_L{}^2}$	$Z = \sqrt{R^2 + X_C{}^2}$	$Z = \sqrt{(X_L - X_C)^2}$

Ohm's Law for AC circuits:

$$E = IZ \qquad I = \frac{E}{Z} \qquad Z = \frac{E}{I}$$

Capacitances in parallel:

$$C_t = C_1 + C_2 + C_3 + \cdots \text{ etc.}$$

C_t = total capacitance (farads)
C_1, C_2, etc. = individual capacitances (farads)

TABLE 14.—Summary of Basic Electrical Calculations—*Continued*

Capacitances in series:

$$C_t = \frac{1}{\dfrac{1}{C_1} + \dfrac{1}{C_2} + \dfrac{1}{C_3} + \cdots \text{ etc.}}$$

Inductances in series and parallel:

The resulting circuit inductance of several inductances in series or parallel is determined exactly as the sum of resistances in series and parallel as described above.

PHASE ANGLE

An alternating current through an inductance lags the voltage across the inductance in time by an angle computed as follows:

$$\text{Tangent of angle of lag} = \frac{X_L}{R}$$

An alternating current through a capacitance leads the voltage across the capacitance in time by an angle computed as follows:

$$\text{Tangent of angle of lead} = \frac{X_C}{R}$$

The resultant angle by which a current leads or lags the voltage in an entire circuit is called the phase angle and is computed as follows:

$$\text{Cosine of phase angle} = \frac{R}{Z}$$

POWER FACTOR

Power factor of a circuit or system is the ratio of actual power (watts) to apparent power (volt-amperes), and is equal to the cosine of the phase angle of the circuit:

$$PF = \frac{\text{actual power}}{\text{apparent power}} = \frac{\text{watts}}{\text{volts} \times \text{amperes}} = \frac{KW}{KVA} = \frac{R}{Z}$$

KW = kilowatts
KVA = kilovolt-amperes = volt-amperes × 1000
PF = power factor (expressed as decimal)

TABLE 14.—SUMMARY OF BASIC ELECTRICAL CALCULATIONS—*Continued*

SINGLE PHASE CIRCUITS

$$KVA = \frac{EI}{1000} = \frac{KW}{PF} \qquad KW = KVA \times PF$$

$$I = \frac{P}{E \times PF} \qquad E = \frac{P}{I \times PF} \qquad PF = \frac{P}{E \times I}$$

$P = E \times I \times PF$
$P = $ power (watts)

TWO-PHASE CIRCUITS

$$KVA = \frac{2EI}{1000} = \frac{KW}{PF} \qquad KW = KVA \times PF$$

$$I = \frac{P}{2E \times PF} \qquad E = \frac{P}{2I \times PF} \qquad PF = \frac{P}{E \times I}$$

$P = 2E \times I \times PF$
$E = $ phase voltage (volts)

THREE-PHASE CIRCUITS, BALANCED STAR OR WYE

$$I_N = O \qquad I = I_p \qquad E = \sqrt{3}\,E_p = 1.73E_p$$

$$E_p = \frac{E}{\sqrt{3}} = \frac{E}{1.73} = 0.577E$$

$I_N = $ current in neutral (amperes)
$I = $ line current per phase (amperes)
$I_p = $ current in each phase winding (amperes)
$E = $ voltage, phase to phase (volts)
$E_p = $ voltage, phase to neutral (volts)

THREE-PHASE CIRCUITS, BALANCED DELTA

$$I = \sqrt{3}\,I_p = 1.73I_p \qquad I_p = \frac{I}{\sqrt{3}} = 0.577I \qquad E = E_p$$

TABLE 14.—Summary of Basic Electrical Calculations—*Continued*

POWER: BALANCED 3-WIRE, 3-PHASE CIRCUIT, DELTA OR WYE

For unity power factor (PF = 1.0):

$$P = \sqrt{3}\,EI = 1.73EI$$

$$I = \frac{P}{\sqrt{3}\,E} = \frac{0.577P}{E} \qquad E = \frac{P}{\sqrt{3}\,I} = \frac{0.577P}{I}$$

P = total power (watts)

For any load:

$$P = 1.73EI \times PF \qquad VA = 1.73EI$$

$$E = \frac{P}{PF \times 1.73 \times I} = \frac{0.577 \times P}{PF \times I}$$

$$I = \frac{P}{PF \times 1.73 \times E} = \frac{0.577 \times P}{PF \times E}$$

$$PF = \frac{P}{1.73 \times I \times E} = \frac{0.577 \times P}{I \times E}$$

VA = apparent power (volt-amperes)
P = actual power (watts)
E = line voltage (volts)
I = line current (amperes)

POWER LOSS: ANY AC OR DC CIRCUIT

$$P = I^2R \qquad I = \sqrt{\frac{P}{R}} \qquad R = \frac{P}{I^2}$$

P = power heat loss in circuit (watts)
I = effective current in conductor (amperes)
R = conductor resistance (ohms)

XXI

Electronics

Many kinds of mathematics find application in the field of electronics. They range from simple arithmetic and algebra to highly complex calculations which are beyond the scope of this book. In this chapter we will cover some of the simpler appli-

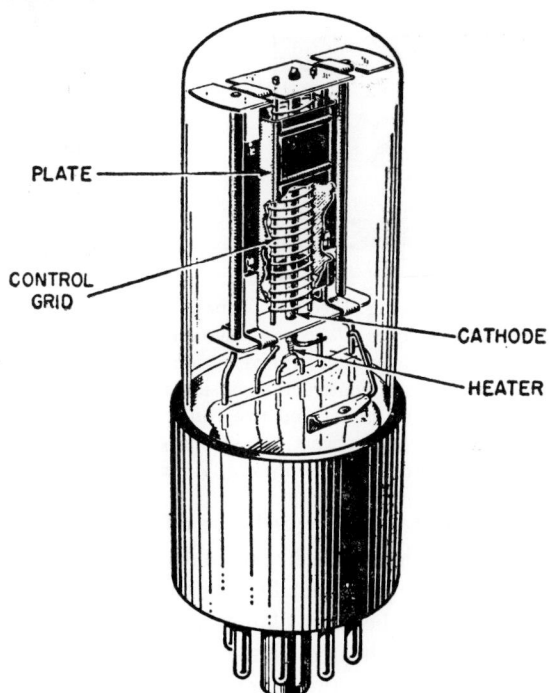

Fig. 1.—Drawing Showing the Construction of a Typical Electron Tube.

857

Labels on figure: PLATE, CONTROL GRID, CATHODE, HEATER

cations of mathematics in electronics. One of these is a special technique which is very useful for rapid solution of many electronic tube performance calculations. This technique is known as graphical calculation. Before discussing this technique, we need to know a few basic facts about electron tubes in general.

Electron Tubes.—An electron tube is a device in which electrons flow from some electron-emitting surface (*cathode*) to a collector surface (*anode* or *plate*), through a vacuum or gas in a closed glass or metal container. (See Figs. 1 and 2.)

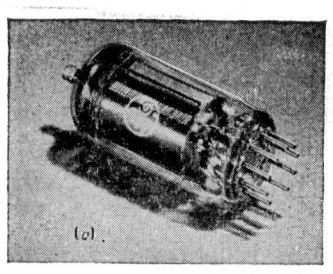

FIG. 2.—Typical Receiving Electron Tubes. (*a*) is a miniature type (*Courtesy General Electric Company*) designed for high frequency applications such as in television receivers. (*b*) is a typical low power receiving tube type whose construction is similar to that shown in Fig. 1 except that these are two plates and two control grids. It is known as a "twin triode" and is actually two tubes in one glass envelope. (*Courtesy Radio Corporation of America*)

Certain materials when heated give off or emit electrons. In electron tubes, the heat which causes this electron emission is produced by an electric current flowing through a wire called the *filament*. In some tubes the filament and the cathode are the same element, and the heated filament is therefore the source of electrons. In other tubes the filament heats a separate cathode which is made of a much better electron-emitting material. This latter type of tube has what is known as an *indirectly heated cathode* and is probably the most common type employed in electronic circuits.

The electron flow can be controlled by varying the voltage applied to the plate. Making the plate more positive with respect to the cathode causes more of the negative electrons being emitted from the cathode to be attracted to the plate, resulting in an increased current flow. When the plate is made negative with respect to the cathode, the negative electrons are repelled and current flow ceases.

Greatly increased control of the electron flow can be achieved by means of a third element known as the *grid* which is placed between the cathode and the plate. The grid is a mesh or coil of fine wires through which electrons can easily pass. By applying different voltages to the grid, the flow of electrons can be increased, decreased, or stopped. Making the grid positive with respect to the cathode increases the flow of electrons to the plate. When the grid is negative with respect to the cathode, the flow of electrons to the plate is reduced. A point can be reached when the grid is sufficiently negative to stop the flow of electrons to the plate completely. Thus the grid acts as a "valve" to control the flow of electrons or current between the cathode and the plate.

The simplest type of electron tube consists of two elements, a cathode and a plate, and is called a *diode*. Its ability to permit current flow only when the plate is positive makes it useful as a *rectifier* to change alternating current into pulsating direct current.

When a grid is added, the tube is known as a *triode*. This type of tube has a very wide variety of uses, the most important of which is amplification of small signals.

Characteristic Curves.—The effect of the grid voltage on the plate current flow for a given tube can be shown by a set of *characteristic curves* as in Fig. 4. In order to obtain these curves, the triode is connected as shown in Fig. 3. One battery is used to furnish plate voltage for the tube, and the other for the grid.

Filament connections are not shown in order to simplify the circuit.

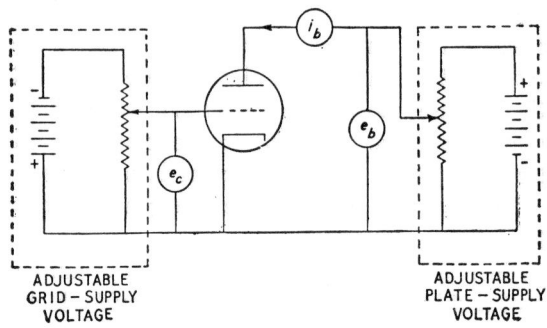

Fig. 3.—Circuit Used to Obtain Triode Characteristic Curves. i_b = plate current; e_c = grid voltage; e_b = plate voltage.

The procedure is as follows: First the plate voltage is set at a certain value, say, 250 volts. Then the grid voltage is varied and the corresponding plate current readings are plotted on a graph. The plate voltage is then set at another value, say, 200 volts. The grid voltage again is varied, and the resulting plate current curves are plotted. Repeating this process for several values of plate voltage results in a group or *family* of characteristic curves. Note that, for any given plate voltage, there is a value of negative grid voltage that will stop the plate current flow. This is referred to as the *cutoff voltage*.

The grid then acts like a valve, controlling the flow of plate current. Varying the grid voltage has a much greater effect upon plate current flow than is obtained when the plate voltage is varied. This means that plate current changes caused by relatively large changes in plate voltage can be produced by relatively small changes in grid voltage.

Because a small voltage on the grid has the same effect as a large voltage on the plate, the triode can be used as an *amplifier* which magnifies or amplifies a small signal on the grid to a large signal on the plate. The amplified output does not come from the tube itself, but from the power source (in this case a battery), connected between the cathode and the plate of the tube. By action of the grid, the tube controls the power from this source and converts it into the form required to do a particular task.

A *load resistor* or *load impedance* must be connected in the plate or output circuit of the tube in order to make use of the controlled power. With this load impedance in the circuit the action is something like this: as the grid voltage is varied by a small amount, the plate current varies, and the resulting varying plate current flowing through the plate load impedance produces large varying voltage drops. These varying voltage drops are magnified or amplified versions of the varying grid voltages. In a typical case a change of one volt in the grid might cause a variation in the voltage across the load impedance of perhaps 50 volts. This would be the same as saying that the input voltage (change in grid voltage) was magnified or *amplified* 50 times in the output circuit.

The performance of tubes depends upon their design, specifically upon the geometry of the different elements. This includes separation of the elements, the shape and size of the elements, the number of turns and the spacing of the grid wires, and several other physical factors. All these factors govern the maximum voltages that can be applied, the cutoff grid voltage, the maximum plate current, etc. They are expressed by a group of numbers known as the *tube constants*.

Tube Constants.—Tube constants differ from tube characteristics. The latter are graphical representations of tube behavior under a given set of conditions. The former are individual numerical ratings resulting from the geometry of the tube. Tubes with similar constants demonstrate similar relationships, but specific values of grid voltage, plate voltage, and plate current necessary to make the tube perform properly may be different for the various tubes. The three most important tube constants are *amplification factor*, *a-c plate resistance*, and *transconductance*. These constants can be measured directly, or the technique of graphical calculation can be used to compute them from the characteristic curves for a given tube. Characteristic curves for tubes are readily available in tube manuals, and by graphical calculation it is a simple matter to find the constants for a given set of operating conditions.

Amplification Factor.—The *amplification factor* or *amplification constant* of a tube is the ratio between a small change in plate voltage and a small change in grid voltage which results in the same change in plate current. It is a measure of the effectiveness of the control grid voltage in controlling the plate current as compared with the plate voltage. In formula form,

$$\text{Amplification Factor} = \frac{\text{small change in plate voltage}}{\text{small change in grid voltage}}$$

The amplification factor is usually designated by the Greek letter μ (mu), pronounced "mew." If a tube has a μ of 100, then the grid voltage change required to produce a certain change in plate current is 100 times less than the plate voltage change required to cause the same change in plate current. For example, suppose that it takes a plate voltage change of 10 volts to produce a 1-milliampere change in plate current, and it takes a grid voltage change of 0.1 volt to produce the same 1-milliampere change in plate current. Then

$$\mu = \frac{\text{change in plate voltage}}{\text{change in grid voltage}} = \frac{10}{0.1} = 100$$

It is important to remember that the changes measured have to be small and that it is the changes in the grid voltage and in plate voltage that are considered in these calculations and not the individual values of plate and grid voltage.

The amplification factor of triodes ranges from about 3 to 100. Low-μ triodes have amplification factors of less than 7 or 8; medium-μ triodes have amplification factors of about 8 to 30; and the amplification factor for high-μ triodes is 30 or more.

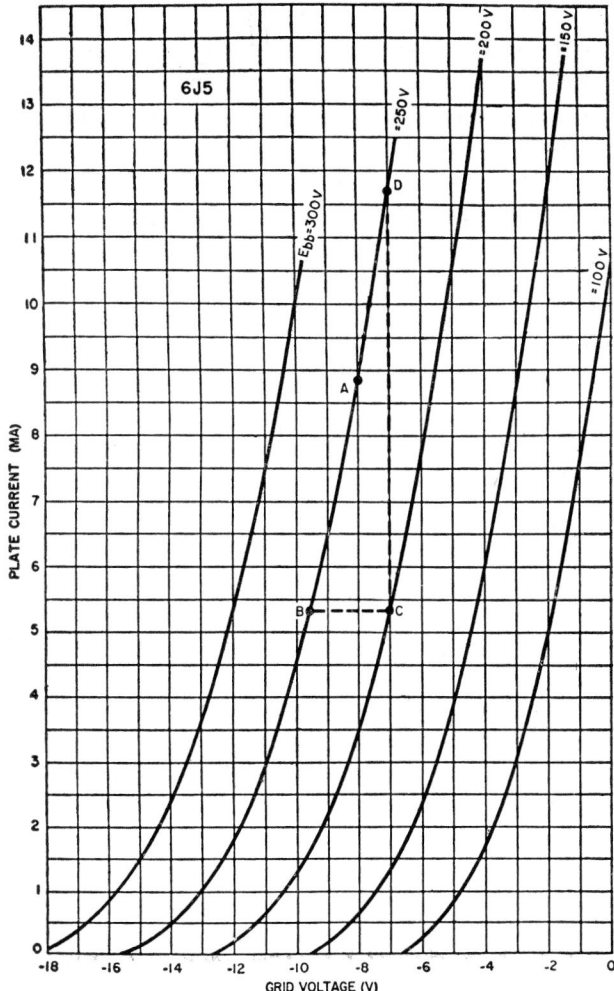

FIG. 4.—Grid Characteristics or Grid Family of Curves of a type 6J5 Triode. These curves can be used to determine the amplification factor of the table by graphical calculations.

Graphical Calculation of μ.—The amplification factor of a tube can be found directly from the characteristic curves for the tube.

Using the grid family of curves, suppose it is required to find the μ of the Type 6J5 tube with 250 volts on its plate and —8 volts on its grid. This corresponds to point A in Fig. 4.

From some convenient point on the straight portion of the $E_{bb} = 250$-volt curve, say point B, a horizontal line parallel to the x axis is drawn until it intersects the next adjacent curve (the 200-volt curve). This is point C. From this intersection a vertical line parallel to the y axis is drawn until it intersects the 250-volt curve again, at point D. The length of the horizontal line represents the change in grid voltage, the length of the vertical line represents the change in plate current, and the change in plate voltage is the difference between the respective voltages of the two voltage curves intersected.

Examining the values represented by points B, C, and D, we see that the change in plate current was from point C (5.35 ma) to point D (11.7 ma), or 6.35 ma. Holding the plate voltage constant at 250 volts, this change in plate current could be produced by varying the grid voltage from B (—9.6 volts) to D (—7 volts), or a total grid voltage change of 2.6 volts. Holding the grid voltage constant at —7 volts, the same change in plate current could be produced by varying the plate voltage from D (250 volts) to C (200 volts), or a total plate voltage change of 50 volts. Using the formula previously given for μ, we obtain

$$\mu = \frac{\text{change in plate voltage}}{\text{change in grid voltage}} = \frac{50 \text{ volts}}{2.6 \text{ volts}} = 19.20$$

The amplification factor or μ of the tube is therefore 19.2. This means that the grid voltage is 19.2 times more effective than the plate voltage in producing a change in plate current. The μ of this tube is specified as 20 in the manufacturer's literature, which means that the graphical calculation produced a result about 4% from the listed value. This is close enough for practical purposes, especially since the μ will vary slightly on the nonlinear portions of the curves anyway.

Plate Resistance.—The *plate resistance* (r_p), another important vacuum tube constant, is a measure of the internal resistance of the tube to the flow of electrons from the cathode to the plate. This constant is expressed in two ways. One of these is the *d-c resistance*, which is the opposition to current flow when steady values of voltage are applied to the tube elements. It is determined by applying Ohm's Law:

$$r_p(\text{ohms}) = \frac{\text{steady-state plate voltage in volts}}{\text{steady-state plate current in amperes}}$$

at any point on the plate current characteristic curve. For example, on the plate characteristic curves or the plate family of curves in Fig. 5, when the grid voltage is —8 volts, we note that

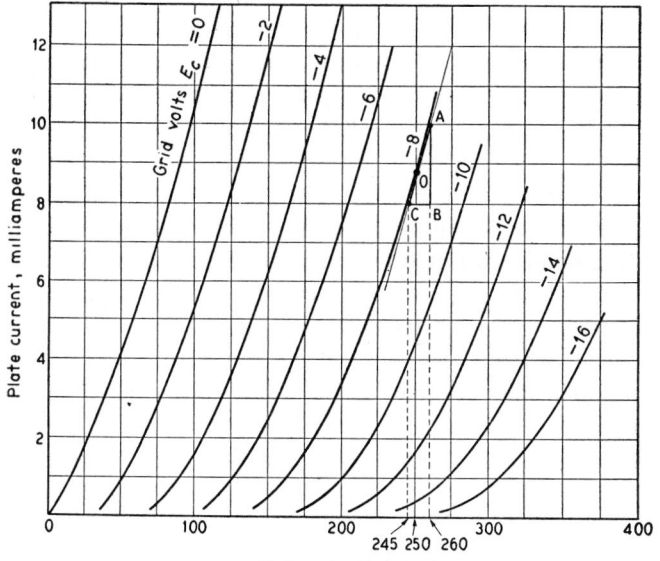

Fig. 5.

for a plate voltage of 250 volts the plate current is 8.9 ma. Then, using Ohm's Law, we obtain

$$r_p = \frac{250 \text{ volts}}{0.0089 \text{ amp.}} = 28,100 \text{ ohms}$$

The *a-c plate resistance* is the ratio of a small change in plate current at a given grid voltage. In formula form it is:

$$r_p(\text{ohms}) = \frac{\text{small change in plate voltage (volts)}}{\text{small change in plate current (amperes)}}$$

Graphical Calculation of r_p.—Like the amplification factor, the plate resistance can also be found from the tube's characteristic curves. In this instance the plate characteristics are used.

The operating point is selected, a tangent to the curve is drawn at this point, and a small right triangle is constructed using the tangent line as the hypotenuse. The ratio of the base of this triangle, measured in volts, to the altitude, measured in amperes, gives the plate resistance at this particular operating point. Thus the plate resistance is really the slope of the plate characteristic curve at a particular operating point.

Referring to Fig. 5, we see that the triangle ABC is constructed at the operating point O which represents a grid voltage of -8 volts and a plate voltage of 250 volts. Note that the tangent line runs practically on top of the curve. This will occur when the operating point is selected on the straight portion of the characteristic curve. In such cases the curve itself can be used as the hypotenuse of the right triangle.

Point A on the triangle corresponds to a plate current of about 10 ma and a plate voltage of 260 volts. At point C the plate current is 8 ma and the plate voltage is 245 volts. Taking these values and using the formula for plate resistance, we obtain

$$r_p = \frac{\text{change in plate voltage}}{\text{change in plate current}} = \frac{260 - 245}{0.010 - 0.008}$$

$$= \frac{15}{0.002} = 7,500 \text{ ohms}$$

The a-c plate resistance will vary, depending upon the operating point selected. A few sample calculations at different operating points will show that the higher the plate voltage, the lower will be the plate resistance; and the more positive the grid voltage (operating at a smaller negative grid voltage), the lower will be the plate resistance.

Transconductance.—The third important tube constant is called *transconductance* or the *mutual conductance*. It expresses the change in plate current caused by a change in grid voltage with the plate voltage held constant. In formula form it is:

$$\text{Transconductance (mhos)} = \frac{\text{change in plate current (amperes)}}{\text{change in grid voltage (volts)}}$$

The symbol commonly used for transconductance is g_m. The unit of transconductance is the *mho* (ohm spelled backward), which represents a condition where a grid voltage change of 1 volt produces a plate current change of 1 amp. This value is too large for use with vacuum tubes and therefore the *micromho*, or the millionth part of a mho, is usually employed. It corresponds to a 1-microampere change in plate current for a 1-volt change in grid voltage. Thus, an electron tube which operates so that a 1-volt change in grid voltage produces a 1-ma change in plate current (1000 microamperes) has a rating of 1000 μmhos, or we say that it has a transconductance of 1000 (micromhos being understood).

Graphical Calculation of g_m.—The transconductance also can be found by graphical means. The plate family of curves is used, and the procedure is quite simple.

On the grid voltage curve, a point is selected in such a way that a line projected upward or downward vertically from this point will intersect an adjacent grid voltage curve. This is the initial point A in Fig. 6, which represents a plate voltage of 235 volts, a grid voltage of —8 volts, and a plate current of 7.1 ma. Going up vertically from this point the next curve is intersected at point B, which represents the same plate voltage, a grid voltage of —6 volts, and a plate current of 12 ma. Using the formula for g_m, we obtain

$$g_m = \frac{\text{change in plate current}}{\text{change in grid voltage}} = \frac{0.012 - 0.0071}{8-6} = \frac{0.0049}{2}$$

$$= 0.00245 \text{ mho, or } 2450 \, \mu\text{mho}$$

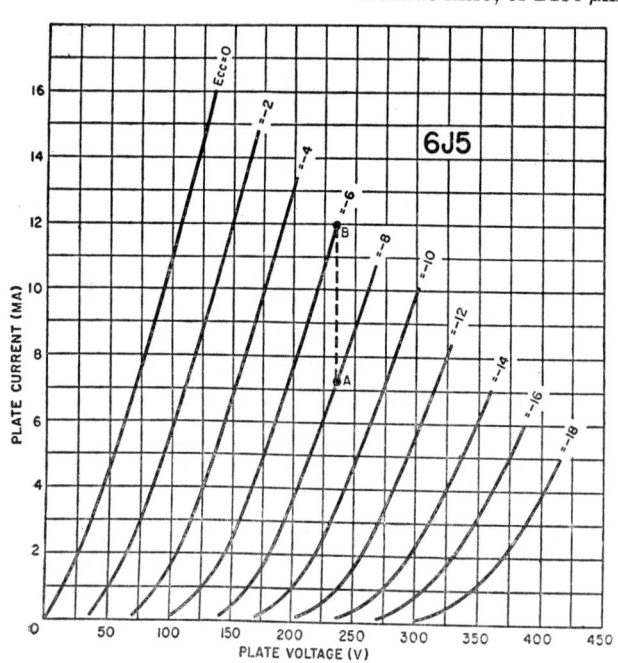

FIG. 6.—Graphically Calculating Transconductance Using the Plate Characteristics.

In terms of plate current change per 1-volt change on the grid, this figure means that a change of 2.45 ma will be obtained for a change of 1 volt on the grid.

Relation of μ, r_p, and g_m.—At a given operating point, the three tube constants: μ, r_p, and g_m are related to one another by the following formulas:

$$\mu = g_m/r_p, \quad r_p = \mu/g_m, \quad g_m = \mu/r_p$$

Therefore if two of the constants are known, the third can be readily found.

The foregoing examples are only a few ways in which graphical techniques are used to simplify and speed up calculations in designing electronic devices. Some of the other applications require a more comprehensive knowledge of electron tubes and their associated circuits and therefore are beyond the scope of this book. The reader is referred to the standard textbooks for additional reading on this subject.

Transistors.—The transistor is an electronic device for amplification and/or control which consists of a semiconducting material to which contact is made by two or more electrodes which are usually metal points or metal surfaces soldered to the semiconductor. The transistor is capable of performing many of the functions filled by the electron tube. Although its operating characteristics depend on temperature, the transistor has many advantages relative to the electron tube since it requires no heater (or filament) current, is small and light weight, can be made mechanically rigid and long lasting, and operates at low voltages with comparatively high efficiencies.

Semiconductors are materials which have electrical conduction properties much poorer than good conductors but, on the other hand, much better than materials classed as insulators. The most common semiconductors in use at present for transistors are germanium and silicon. In a perfect crystal state an atom of germanium (or silicon) has four electrons bounded to electrons of neighboring atoms in which the two electrons in each bond are shared equally with the two atoms at its ends. This bond structure is three dimensional, but the action is indicated in a two dimensional form in Fig. 6. Absorption of energy either as a result of light or heat will cause rupture of a covalent bond releasing an electron and leaving a defect, of "hole," in the otherwise periodic structure. For germanium an absorbed energy of 0.72 electron volts will release an electron from the bond while 1.12 electron volts is required for the same effect in silicon. Once released, the electron can move about the crystal with considerable freedom. Furthermore, it is also possible for an electron in bond adjacent to the bond which has just lost an electron to jump

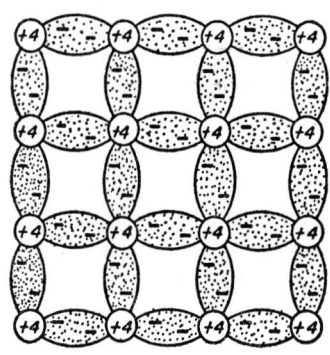

Fig. 6.

into the hole, leaving an electron deficit in the bond which was vacated. By this mechanism it is possible for a hole to move through the crystal at the same time free electrons move.

The hole concept represents a means of describing an incomplete assemblage of electrons and with this fact understood, holes may be considered to have a positive charge equal in magnitude to that possessed by an electron as well as an appropriate mass. These two properties enable calculation of the hole movement to be expected in the presence of electric and magnetic fields. The conductivity (σ mho-cm.$^{-1}$) of a semiconductor can be expressed in terms of the density of holes (p per cubic centimeter) and of electrons (n per cubic centimeter) and the electron and hole mobilities μ_n and μ_p-cm./sec. per volt/cm.), the latter being the drift velocity in a unit electric field. The result is

$$\sigma = q(n\mu_n + p\mu_p)$$

where q is the electron charge. It is clear that an increase in either hole or electron concentration will change the conductivity of the semiconductor although the two densities have different numerical effects due to the difference between hole and electron mobilities ($\mu_p = 1700$, $\mu_n = 3600$ cm./sec. per volts/cm. for germanium; $\mu_p = 400$, $\mu_n = 1200$ cm./sec. per volts/cm. for silicon).

In a pure sample of a semiconductor the holes and electrons are produced in pairs so that the two densities are equal (n_i). The conductivity that results for such an intrinsic semiconductor is called the intrinsic conductivity (σ_i). It is expressible as

$$\sigma_i = qn_i(\mu_n + \mu_p)$$

Transistors depend for their operation on either an excess of free electrons or a deficit of electrons (excess of holes) in atomic bonds. A semiconductor in which the first condition obtains is spoken of as an n type material (excess of negative carriers) whereas one in which the second situation exists is referred to as a p type semiconductor (excess of positive carriers). An excess of electrons or holes is produced by adding minute amounts of impurities to an intrinsic semiconductor. Some atoms (such as arsenic and antimony) have five bonding electrons, and if small amounts of these atoms are added to an otherwise pure sample of germanium or silicon (which, as stated above, have four bonding electrons), four of the five electrons of the impurity atom are shared with neighboring germanium atoms. The fifth valence electron is free to move throughout the crystal. An excess of electrons is thus created by the addition of the impurities; n type germanium now exists.

There are also atoms, such as aluminum and gallium which have only three bonding electrons. When such atoms are added in small amount to pure germanium (or silicon), only three germanium atoms can be bonded to the impurity atom by sharing of electrons. Since the impurity atom occupies a position in the crystal structure normally filled by a germanium atom (which has four bonds), the addition of the impurity will result in the deficiency of one electron in a germanium bond in the proximity of the impurity. A hole has thus been created and p type germanium has resulted. It should be noted that if both five-electron and three-electron atoms are present, the material will exhibit the properties of n or p type material depending on whether the electron density exceeds the hole density or vice versa.

A rod of single crystal germanium with alternating n and p type regions can be obtained by adding suitable impurities to the molten germanium as the rod is withdrawn from the crucible used for melting in the process of forming a crystal. If the two types of impurities are added in succession, it is possible to obtain n-p-n and p-n-p configurations along the rod. By this method, there will be a region created where n type conductivity predominates followed by one where p type prevails in the same germanium crystal. A p-n junction will be produced where the two regions of different conductivity come together. In contrast with these so-called grown junctions, it is also possible to obtain a pair of p-n junctions for transistor action by starting with a thin wafer of germanium of a prescribed conductivity type and fusing impurity regions of the opposite type in localized areas on opposite sides of the wafer, leaving a thin section of the original material in the center of the wafer cross section. Junctions prepared in this manner are called fused alloy junctions.

Before considering the transistor itself which consists of a combination of two p-n junctions (arranged as p-n-p or n-p-n), it is of interest to examine conditions which exist in a single p-n junction. (The three-bonding electron atoms like aluminum are called acceptor atoms, since they are "electron-deficient," while the five-bonding-electron atoms like arsenic are called donor atoms.) Fig. 7 shows a p-n junction including the acceptor and

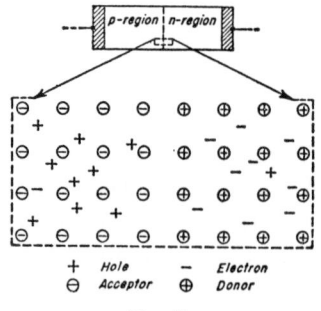

FIG. 7.

donor atoms in the n and p regions, respectively, as well as the holes and electrons which are free to move. It is noted that even in the n type region there are some holes, although a relatively small number, and a corresponding situation exists in the p type region. These carriers, which are of opposite sign to the intended or majority carriers, come into existence due to the rupture of atomic bonds as the result of the thermal energy of the atoms and are called minority carriers. If no external potential difference is applied across the electrodes shown in the figure, an equilibrium condition will be established wherein no electrons or holes exist in the immediate vicinity of the junction. Because of the arrangement of the fixed, oppositely charged donor and acceptor atoms at the boundary of the n and p type materials, an electric dipole is formed yielding an appropriate electric field and corresponding difference of potential across the boundary between the two regions.

If a potential difference of suitable magnitude is applied to the junction such that the electrode connected to the p region is positive with respect to the other electrode (producing a forward biased junction), the majority carriers in each region will be forced to flow across the junction, the net current flow being the sum of the current carried by the holes plus that carried by the electrons. The respective densities of holes and electrons (p and n) are determined by the impurity concentrations in the p and n type regions. By suitable control of the relative amounts of impurities used, the current that flows across the junction in the forward biased condition can be arranged to be composed principally of holes, principally of electrons, or some suitable combination of both. In any event, the flow of conventional current is into the p region and out of the n region. If the polarity of the external potential difference is reversed, the only current that will flow into the terminals of the device will be that contributed by the few minority carriers that are free in the material, which will be a small one indeed compared to the current obtained in the forward biased condition. This current rapidly reaches a limiting value called the saturation current as the reverse bias voltage across the device is increased. Because the current that flows is practically independent of the applied voltage beyond a certain minimum value, the impedance of the reversed bias junction is extremely high. By contrast, the impedance in the forward biased direction is very low.

A transistor of the *n-p-n* type is shown diagrammatically in Fig. 8a. A small *p* type region is contained between *n* type re-

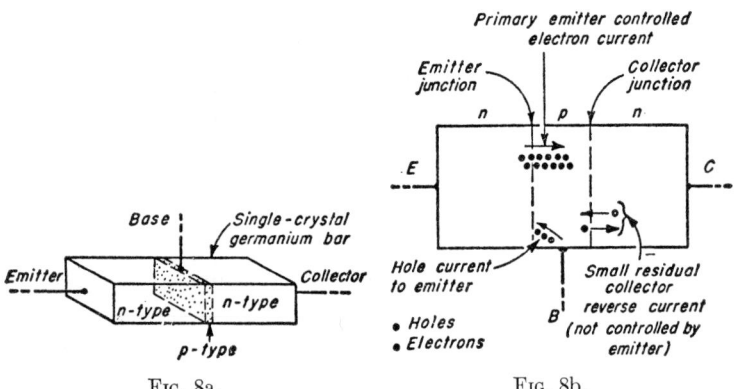

FIG. 8a.   FIG. 8b.

gions and connections are made to all three regions. The terminals are labelled emitter, base, and collector for reasons that will become apparent shortly. In normal operation as an amplifier, the emitter to base junction is forward biased and the collector to base junction is reverse bias. The emitter to base impedance is low whereas the collector to base impedance is high for reasons considered above. Fig. 8b indicates the current flow existing across the junctions. A small saturation current composed of holes and electrons flows across the collector junction; this current is not controlled by the emitter. On the other hand, the electrons coming into the base from the emitter due to the forward biased emitter-base junction represent minority carriers for the collector-base junction and for a thin enough base section these carriers pass through the base layer, cross the collector junction, and contribute to the collector current. A hole current from the base region also flows across the emitter junction to add to the electron current in producing the total emitter current. The essential action is the emitting of carriers from the emitter region and the collection of practically all of these carriers by the collector.

By proper design of the impurity concentrations and base layer width, the ratio of the emitter electron current to the emitter-base hole current can be made very large (values in the order of 100 are attainable). If the input of the amplifier is taken to be between base and emitter with the output taken between collector and emitter, then the input current is the small base-emitter hole current and the output current is the emitter-collector electron current. A significant current gain is thus achieved. If, on the other hand, the input current is applied between emitter and base and the output circuit is connected between collector and base, then no current gain results since the current entering the emitter terminal (sum of hole and electron currents) is only slightly less than the current leaving the collector (electron current from emitter plus collector saturation current). This latter amplifier connection does permit a voltage gain, however, because the high output impedance permits use of a large load resistor through which a current essentially equal to the input current can flow. In contrast with vacuum tubes, transistors control output currents as a result of changes in input current rather than input voltage.

Another type of transistor is shown in Fig. 9. This device is known as a point contact transistor. Rectifying action occurs at both emitter and collector contacts with the germanium surface. A p-n barrier exists at both contacts as shown by the dotted lines as a result of contact between the bulk n type germanium and p type inserts. The base connection is made through a large area contact at the bottom of the material. If the emitter is forward biased and the collector reverse biased with respect to the base, holes are injected from the p region at the emitter into the n type region and are swept to the collector electrode under the influence of the electric field between collector and base. In addition to the high impedance of the collector-circuit which permits voltage gain with the device, there is also a current gain effected. As a result of the current gain, the input resistance of a point contact transistor amplifier can become negative permitting relaxation oscillations under certain conditions.

Characteristic curves are plotted for transistors as they are for electron tubes, with the added property that transistor de-

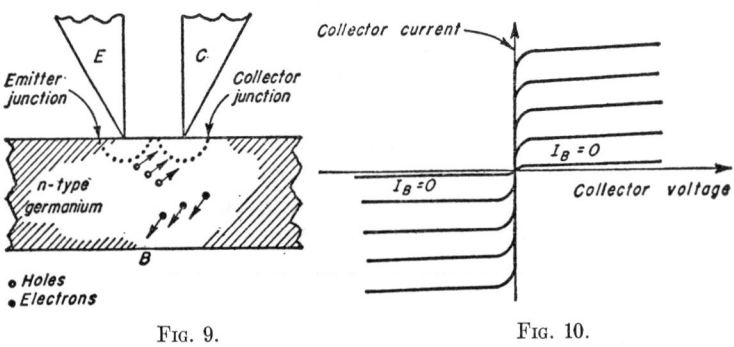

Fig. 9.   Fig. 10.

vices are available in opposite conductivity types, for example, *n-p-n* and *p-n-p* transistors. If two of these devices are made to have identical characteristics except for their conductivity types, it is found that they are symmetrical counterparts of one another in that one has positive currents and voltages whereas the other has identical negative characteristics. One set of collector voltage-current characteristics appears in the first quadrant while the other has odd function symmetry being located in the third quadrant. Fig. 10. shows typical collector characteristics. This property is called complementary symmetry. It is of advantage in arranging for biasing currents and in providing certain circuit functions.

Diodes.—Any two-electrode device, having an anode and a cathode, which has marked unidirectional characteristics is a diode. Many types of diodes are known. Some are electron tubes, so connected as to function as two-electrode devices. There are also many other types. For example:

A *crystal diode* is a diode consisting of a semiconducting material such as a germanium or silicon, as one electrode, and a fine wire "whisker" resting on the semiconductor as the other electrode. Because of its low capacitance, the device finds considerable application as a rectifier or detector of microwave frequencies.

A *semiconductor diode* is a two-electrode semiconductor device having an asymmetrical voltage-current characteristic. A double-base diode is a semiconductor diode in which a potential gradient is produced across the base region by the application of a voltage between two electrodes at either end of the base. The correct polarity and magnitude of this voltage causes the diode to exhibit a controllable, negative resistance between one of the base electrodes and the anode.

A *junction diode* is a semiconductor diode whose nonsymmetrical volt-ampere characteristics are manifested as the result of the junction found between n-type and p-type semiconductor materials. This junction may be either diffused, grown or alloyed.

The Zener diode. A Zener diode is a special type of junction diode in which the reverse breakdown occurs more sharply than in an ordinary diode. In the breakdown region the voltage across the diode is almost independent of current over a wide range of current; a typical value for the incremental resistance, from the slope of the characteristic, is a few ohms. In small low-power diodes the minimum current I_{min} is about 1 milliampere, and the maximum current I_{max} is about 1 ampere. When biased within this range, the Zener diode acts essentially as a battery of voltage V_Z with a low internal resistance.

A *tunnel diode* has a p-n junction with very high densities of both acceptor and donor impurities. A consequence of the high impurity density is to produce a very narrow transition region. This arises because the ionized impurity atoms are close together, and so produce a high space charge density; hence the equilibrium difference in potential is built up in a very short distance.

The resulting current characteristic has a sharp declining portion, which enables the tunnel diode to act as a high speed switch.

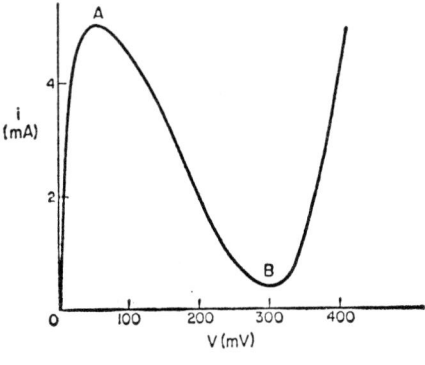

Fig. 11.

Properties of Circuits.—Before treating the circuits used in radio and television equipment, some topics pertaining to electric circuits generally require somewhat more extended treatment than they were given in the preceding chapter. This discussion does not repeat the treatment given there of such fundamentals as resistance, capacitance, inductance, Ohm's Law and similar elementary topics.

Circuit Elements in Multiple.—*Multiple Resistances.*

1. When connected in series, as shown in Figs. 12 and 13, the total resistance of two or more resistors is higher than any of the units.

Expressed as a formula:

$$R_{\text{total}} = R_1 + R_2 + R_3$$

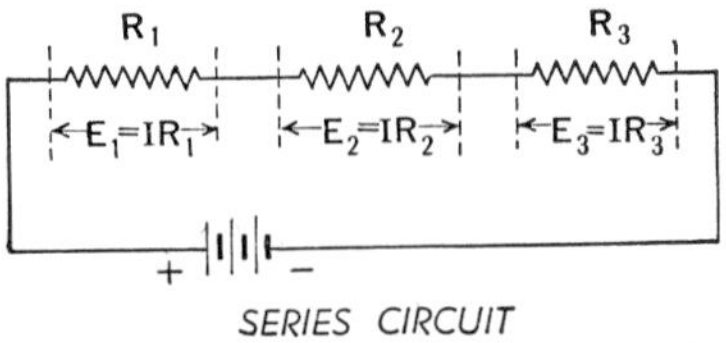

SERIES CIRCUIT

Fig. 12.

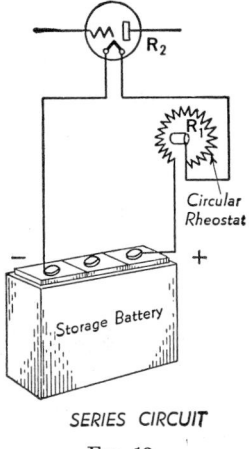

SERIES CIRCUIT

FIG. 13.

2. When connected in parallel as in Figs. 14 and 15, the total resistance of two or more resistors is decreased. Expressed as a formula:

$$R_{\text{total}} = \cfrac{1}{\cfrac{1}{R_1} + \cfrac{1}{R_2} + \cfrac{1}{R_3}}$$

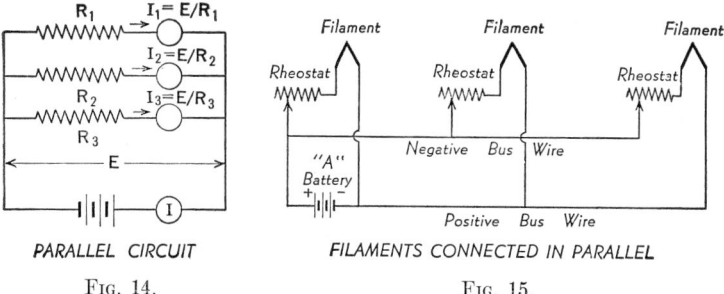

PARALLEL CIRCUIT

FIG. 14.

FILAMENTS CONNECTED IN PARALLEL

FIG. 15.

3. When connected in series parallel as shown in Fig. 16, the total resistance is shown by the following formula:

$$R_{\text{total}} = \cfrac{1}{\cfrac{1}{R_1 + R_2} + \cfrac{1}{R_3 + R_4} + \cfrac{1}{R_5 + R_6} + \cfrac{1}{R_7 + R_8 + R_9}}$$

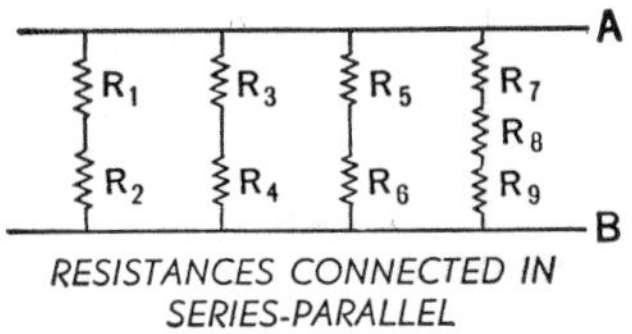

RESISTANCES CONNECTED IN
SERIES-PARALLEL

FIG. 16.

Nearly all radio circuits are combinations of series and parallel circuits. The problems that arise in these complicated circuits can be solved by the use of Ohm's Law. Figure 17 shows a combined circuit.

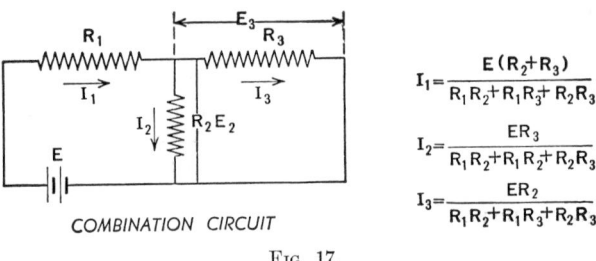

COMBINATION CIRCUIT

$$I_1 = \frac{E(R_2 + R_3)}{R_1 R_2 + R_1 R_3 + R_2 R_3}$$

$$I_2 = \frac{ER_3}{R_1 R_2 + R_1 R_2 + R_2 R_3}$$

$$I_3 = \frac{ER_2}{R_1 R_2 + R_1 R_3 + R_2 R_3}$$

FIG. 17.

Multiple Inductances. When inductances are connected in series, their individual values are added together to find the total inductance. Expressed as a formula:

$$L = L_1 + L_2 + 2M$$

where L = total inductance

M = mutual inductance (usually measured on a Wheatstone bridge)

When inductances are connected in parallel the total inductance may be calculated from the following formula:

$$L = \frac{L_1 \times L_2 - M^2}{L_1 + L_2 - 2M}$$

To find an inductance to match a known capacity the following formula may be used:

$$L = \frac{1}{(2\pi)^2 \times f^2 \times C \times 10^8}$$

where $(2\pi)^2 = 39.47$
f = frequency
C = capacity in microfarads
$10^8 = 100,000,000$

The lumped inductance of coils for transmitting and receiving may be calculated from the following formula:

$$L = \frac{0.2A^2N^2}{3A + 9B + 10C}$$

where L = inductance in microhenries
A = mean diameter of coil in inches
B = length of winding in inches
C = radial depth of winding in inches
N = number of turns of wire

The quantity, C, may be neglected if the coil is a single-layer solenoid.

ILLUSTRATION: Find the inductance of a coil with 35 turns of No. 30 D.S.C. wire on a receiving coil form that has a diameter of 1.5 inches. (From the copper wire table it can be seen that 35 turns of No. 30 D.S.C. wire will give a length of one-half inch.)

$$L = \frac{0.2A^2N^2}{3A + 9B}$$

$$= \frac{0.2 \times (1.5)^2 \times (35)^2}{(3 \times 1.5) \times (9 \times 0.5)}$$

$$= \frac{5512.5}{9} = 61.25 \text{ microhenries.} \quad \text{(Ans.)}$$

TABLE 1

COPPER WIRE TABLE

Gauge No. B. & S.	Diam. in Mils[1]	Circular Mil Area	Turns per Linear Inch[2]				Turns per Square Inch[2]			Feet per Lb.		Ohms per 1000 ft. 250 C.	Current-Carrying Capacity at 1500 C.M. per Amp.[3]	Diam. in mm.	Nearest British S.W.G. No.
			Enamel	S.S.C.	D.S.C. or S.C.C.	D.C.C.	S.C.C.	Enamel S.C.C.	D.C.C.	Bare	D.C.C.				
1	289.3	82690								3.947		.1264	55.7	7.348	1
2	257.6	66370								4.977		.1593	44.1	6.544	3
3	229.4	52640								6.276		.2009	35.0	5.827	4
4	204.3	41740								7.914		.2533	27.7	5.189	5
5	181.9	33100								9.980		.3195	22.0	4.621	7
6	162.0	26250								12.58		.4028	17.5	4.115	8
7	144.3	20820								15.87		.5080	13.8	3.665	9
8	128.5	16510	7.6		7.4	7.1				20.01	19.6	.6405	11.0	3.264	10
9	114.4	13090	8.6		8.2	7.8				25.23	24.6	.8077	8.7	2.906	11
10	101.9	10380	9.5		9.3	8.9				31.82	30.9	1.018	6.9	2.588	12
11	90.74	8234	10.7		10.3	9.8				40.12	38.8	1.284	5.5	2.305	13
12	80.81	6530	12.0		11.5	10.9				50.59	48.9	1.619	4.4	2.053	14
13	71.96	5178	13.5		12.8	12.0				63.80	61.5	2.042	3.5	1.828	15
14	64.08	4107	15.0		14.2	13.3	87.5			80.44	77.3	2.575	2.7	1.628	16
15	57.07	3257	16.8	18.0	15.8	14.7	110	84.8	80.0	101.4	97.3	3.247	2.2	1.450	17
16	50.82	2583	18.9	18.9	17.9	16.4	136	105	97.5	127.9	121	4.094	1.7	1.291	18
17	45.26	2048	21.2	21.6	19.0	18.1	170	131	121	161.3	150	5.163	1.3	1.150	18
18	40.30	1624	23.6	24.4	21.2	19.8	211	162	150	203.4	188	6.510	1.1	1.024	19
19	35.89	1288	26.4	27.2	23.4	21.8	262	198	183	256.5	237	8.210	.86	.9116	20
20	31.96	1022	29.4	29.4	25.8	23.8	321	250	223	323.4	298	10.35	.68	.8118	21
21	28.46	810.1	33.1	32.7	27.0	26.0	395	396	329	407.8	370	13.05	.54	.7230	22
22	25.35	642.4	37.0	36.5	31.3	30.1	493	454	399	514.2	461	16.46	.43	.6438	23
23	22.57	509.5	41.3	40.5	34.1	31.6	592	533	479	648.4	584	20.76	.34	.5733	24
24	20.10	404.0	46.3	45.3	37.6	35.6	775	725	625	817.7	745	26.17	.27	.5106	25
25	17.90	320.4	51.7	50.4	41.5	37.6	940	895	754	1031	903	33.00	.21	.4547	26
26	15.94	254.1	58.0	55.6	45.6	41.8	1150	1070	910	1300	1118	41.62	.17	.4049	27
27	14.20	201.5	64.9	61.5	50.2	45.0	1400	1300	1080	1639	1422	52.48	.13	.3606	29
28	12.64	159.8	72.7	68.8	55.0	48.5	1700	1570	1260	2067	1759	66.44	.11	.3211	30
29	11.26	126.7	81.6	76.3	61.2	51.8	2060	1910	1510	2607	2207	83.44	.084	.2859	31
30	10.03	100.5	90.5	83.3	65.4	55.5	2500	2300	1750	3287	2534	105.2	.067	.2546	33
31	8.928	79.70	101.	92.0	71.5	59.2	3030	2780	2020	4145	2768	132.7	.053	.2268	34
32	7.950	63.21	113.	101.	77.5	62.6	3700	3390	2310	5227	3137	167.3	.042	.2019	36
33	7.080	50.13	127.	110.	83.6	66.3	4340	4060	2700	6591	4697	211.0	.033	.1798	37
34	6.305	39.75	143.	120.	90.3	70.0	5040	4660	3020	8310	6168	266.0	.026	.1601	38
35	5.615	31.52	158.	132.	99.0	73.5	5920	5280		10480	6737	335.0	.021	.1426	38-39
36	5.000	25.00	175.	143.	104.	77.0	7060	6250		13210	7877	423.0	.017	.1270	39-40
37	4.453	19.83	198.	154.	111.	80.3	8120	7360		16660	9309	533.4	.013	.1131	41
38	3.965	15.72	224.	166.	118.	83.6	9600	8310		21010	10666	672.6	.010	.1007	42
39	3.531	12.47	248.	181.	133.	86.6	10900	8700		26500	14222	848.1	.008	.0897	43
40	3.145	9.88	282.	194.	140.	89.7	12200	10700		33410		1069	.006	.0799	44

[1] A mil is 1/1000 (one thousandth) of an inch.

[2] The figures given are approximate only, since the thickness of the insulation varies with different manufacturers.

[3] The current-carrying capacity at 1000 C.M. per ampere is equal to the circular-mil area (Column 3) divided by 1000.

Inductive Reactance. The combined effect of frequency and inductance in coils is termed inductive reactance. To find inductive reactance the following formula may be used:

where $X_L = 2\pi f L$

$X_L =$ the inductive reactance in ohms

$\pi = 3.1416$

$f =$ frequency in cycles per second

$L =$ inductance in henries

ILLUSTRATION: A coil has an inductance of 1000 microhenries. If the frequency of the current is 500 kilocycles per second find the inductive reactance.

$$X_L = 2\pi f L$$
$$= 2 \times 3.1416 \times (500 \times 1000) \times \frac{1000}{1,000,000}$$
$$= 3141.6 \text{ ohms} \quad (\text{Ans.})$$

Multiple Condensers.—Capacitances may be connected in series or in parallel like resistances or inductances. When they are connected in parallel the resultant capacity is the sum of the individual capacities. Expressed as a formula:

$$C_{\text{parallel}} = C_1 + C_2 + C_3$$

The sketch, Fig. 18, illustrates the method of connecting condensers in parallel.

The sketches indicate condensers connected in series (Fig. 19 and 20).

When condensers are connected in series the capacity may be found by the following formula:

$$C_{\text{series}} = \frac{1}{\dfrac{1}{C_1} + \dfrac{1}{C_2} + \dfrac{1}{C_3}}$$

If but two condensers are considered the formula becomes

$$C_{\text{series}} = \frac{C_1 \times C_2}{C_1 + C_2}$$

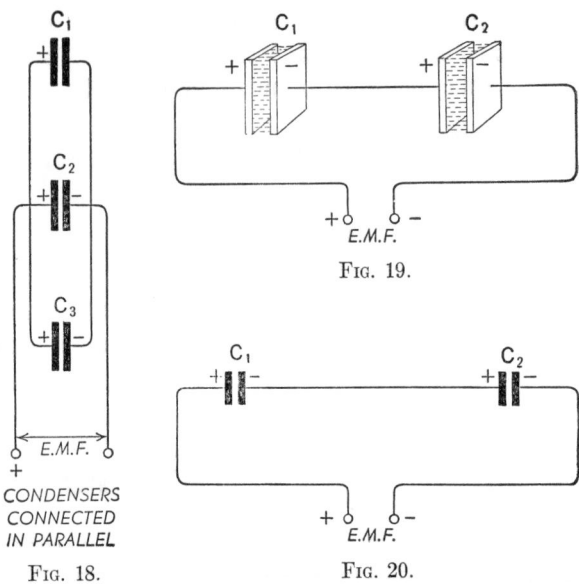

Fig. 18.   Fig. 19.   Fig. 20.

ILLUSTRATION: Find the resultant capacity when condensers of 0.002 microfarad and 0.0015 microfarad capacity are connected in parallel.

$$C = 0.002 + 0.0015 = 0.0035 \ \mu\text{fd.}\quad \text{(Ans.)}$$

ILLUSTRATION: If the condensers in the preceding problem are connected in series find the total capacity.

$$C = \frac{0.002 \times 0.0015}{0.002 + 0.0015} = \frac{0.000003}{0.0035} = 0.00086 \ \mu\text{fd.}\quad \text{(Ans.)}$$

ILLUSTRATION: A radio circuit requires a 0.00035 microfarad condenser but only a 0.0005 microfarad variable condenser is available. What fixed condenser may be used to reduce the maximum capacity in the circuit to the required value? How shall the condenser be connected?

Because the total capacity is to be reduced the fixed condenser must be connected in series with the variable condenser. Then,

$$C = \cfrac{1}{\cfrac{1}{C_1} + \cfrac{1}{C_2}}$$

Removing the reciprocal from the right side and placing it at the left side;

$$\frac{1}{C} = \frac{1}{C_1} + \frac{1}{C_2} \quad \text{and} \quad \frac{1}{C_2} = \frac{1}{C} - \frac{1}{C_1}$$

$$\frac{1}{C_2} = \frac{1}{0.00035} - \frac{1}{0.0005} = \frac{1 - 0.7}{0.00035} = \frac{0.3}{0.00035}$$

$$C = \frac{0.00035}{0.3} = 0.001166 \quad \text{or} \quad 1166 \ \mu\mu\text{fd.} \quad \text{(Ans.)}$$

Capacitive Reactance. Condensers have a reactance that is inversely proportional to the condenser size and to the frequency of the applied voltage.

$$X_c = \frac{1}{2\pi f C}$$

where X_c = capacitive reactance in ohms

$\pi = 3.1416$

f = frequency in cycles per second

C = condenser capacitance in farads

However, the capacitance, in most practical cases, is given in microfarads (μfd.). Then the formula becomes:

$$X_c = \frac{1,000,000}{2\pi f C_{\mu\text{fd.}}}$$

ILLUSTRATION: A 3-plate fixed air condenser with a capacity of 0.0001 microfarad is used in an antenna circuit that is operated on a frequency of 3750 kilocycles. Find the capacitive reactance.

$$X_c = \frac{1,000,000}{2\pi f C_{\mu\text{fd.}}}$$

$$X_c = \frac{1,000,000}{2 \times 3.1416 \times (3,750 \times 1,000) \times 0.0001}$$

$$= \frac{100}{0.1356} = 725 \text{ ohms} \quad \text{(Ans.)}$$

Impedance.—The parts of a radio circuit, such as coils and condensers, are never pure reactances. Some resistance, however small, is always present. The reactance and resistance combined together are called impedance, which is influenced almost entirely by the frequency of the alternating voltages impressed upon the circuit.

To find impedance the following formula may be used:

$$Z = \sqrt{X^2 + R^2}$$

where Z = impedance in ohms

X = total reactance

R = resistance

ILLUSTRATION: What is the impedance in a circuit of 4 ohms resistance and 3 ohms reactance?

$$Z = \sqrt{x^2 + R^2}$$
$$= \sqrt{3^2 + 4^2}$$
$$= \sqrt{9 + 16} = \sqrt{25} = 5 \text{ ohms} \quad \text{(Ans.)}$$

Ohm's Law for Alternating Currents.—If inductances did not have any resistance it could be assumed that the current would be equal to the voltage divided by the reactance. However, this is not the case and Ohm's Law for alternating current becomes:

$$I = \frac{E}{Z}, \quad Z = \frac{E}{I}, \quad E = IZ$$

ILLUSTRATION: A 60 cycle alternating current of 5 amperes flows through a coil whose inductance in 4000 microhenries and whose resistance is 2 ohms. Find the voltage across the coil.

$$I = \frac{E}{Z}, \quad Z = \sqrt{R^2 + x^2}$$

$$X = 2\pi f L$$
$$= 6.28 \times 60 \times 0.004 = 1.507 \text{ ohms}$$
$$Z = \sqrt{4^2 + 1.507^2} = \sqrt{18.25} = 4.27 \text{ ohms}$$
$$E = IZ = 5 \times 4.27 = 21.35 \text{ volts} \quad \text{(Ans.)}$$

Resonance.—A condition of resonance is obtained when a capacity reactance and an inductive reactance of equal magnitude are connected either in series or parallel. The most important circuits in radio are those in which either series or parallel resonance occurs.

The resonant frequency may be determined by the following formula which is frequently called the fundamental equation in radio:

$$f = \frac{1}{2\pi \sqrt{LC}}$$

where f = frequency in cycles per second
$\quad 2\pi = 6.2832$
$\quad L$ = inductance in henries
$\quad C$ = capacitance in farads

Because it is more convenient to use smaller units such as microhenries and microfarads, the formula becomes:

$$f = \frac{1,000,000}{2\pi \sqrt{LC}}$$

where L = microhenries and C = microfarads.

ILLUSTRATION: To what frequency will a circuit tune that has an inductance of 9 microhenries and a capacity of 0.0002 microfarad?

$$f = \frac{1,000,000}{2\pi \sqrt{LC}}$$

$$f = \frac{1,000,000}{6.28 \sqrt{0.0002 \times 9}}$$

$$= \frac{1,000,000}{0.2663} = 3,755,000 \text{ cycles} \quad \text{(Ans.)}$$

$$\text{kilocycles} = \frac{3,755,000}{1000} = 3,755 \quad \text{(Ans.)}$$

Wavelength.—The wavemeter which depends on the principles of resonance is an important instrument in radio measurements. The wavelength is equal to the speed at which electric waves travel (186,000 miles a second, approx.) divided by the frequency in cycles. Expressed as a formula:

$$\lambda = \frac{V}{f}$$

where λ = (pronounced lambda) = wavelength in meters
  f = frequency in cycles
  V = the velocity of propagation of electro-magnetic waves
    —approx. 300,000,000 meters per second

ILLUSTRATION: What frequency in cycles and in kilocycles corresponds to a wavelength of 200 meters?

$$f = \frac{V}{\lambda}$$

$$= \frac{300,000,000}{200} = 1,500,000 \text{ cycles} \quad \text{(Ans.)}$$

$$\text{kilocycles} = \frac{1,500,000}{1000} = 1500 \quad \text{(Ans.)}$$

ILLUSTRATION: What wavelength corresponds to 1500 kilocycles?

$$\lambda = \frac{V}{f}$$

$$= \frac{300,000,000}{1,500,000} = 200 \text{ meters} \quad \text{(Ans.)}$$

The resonance of a tuned circuit is expressed in terms of wavelength as follows:

$$\lambda = 1885 \sqrt{LC}$$

where λ = wavelength in meters
  L = inductance in microhenries
  C = capacitance in microfarads

ILLUSTRATION: A radio circuit has an inductance of 900 micro-henries and a capacity of 0.001 microfarad. Find the wavelength for which this circuit will be resonant.

$$\lambda = 1885 \sqrt{LC}$$
$$= 1885 \sqrt{900 \times 0.001}$$
$$= 1885 \times 0.95 = 1791 \text{ meters.} \quad \text{(Ans.)}$$

Electron Tube Circuits.—Electron tube circuits that is con-nections of resistors, inductors, capacitors, sources of power, and electron tubes assume a wide variety of forms. One convenient method of classifying electron tube circuits employs the ampli-tude of input signal as a basis. Reference to certain fundamental properties of electron tubes is desirable before describing the cir-cuit classifications. For definiteness, the triode vacuum tube will be used as an example. If the grid to cathode voltage of such a tube is not allowed to become negative, the quantities asso-ciated with the device which are of principal interest in describ-ing its behavior are the plate to cathode voltage drop, the plate current, and the grid to cathode voltage drop. The relations among these three variables are usually shown by a graph of two of the three quantities with the third assigned convenient (constant) values. The intersections of the plate characteristics and the load line (the locus of all combinations of the plate volt-age and current permitted by the constraints imposed by the elements external to the tube) establish a relation between grid voltage and plate current known as the transfer characteristic. A typical transfer characteristic is shown in Figure 21.

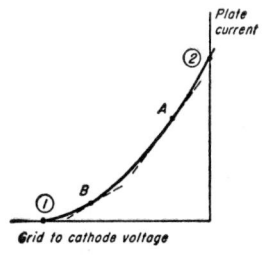

FIG. 21.

Point (1) is commonly referred to as "cut off" for the tube since the plate current is reduced to zero. Point (2) is identified as the "zero bias" point. The difference in grid to cathode voltage between zero bias and cut off is often referred to as the "grid base" of the tube.

This characteristic tells at a glance the plate current to be expected for a given grid to cathode voltage. In practice a grid bias is chosen so that in conjunction with the assigned plate to cathode voltage an appropriate operating point is established about which changes in grid voltage and plate current will take place when an output signal is applied between grid and cathode. The points A and B in Figure 21 are typical operating points. The choice of operating point is an important consideration in the operation of a tube as an amplifier. It is seen from Figure 21 that although the transfer characteristic is not linear, there are certain regions where the slope of the curve does not change appreciably over a fairly large increment of grid voltage. Depending on the location of the operating point, it is possible to choose an amplitude of input signal so that the change in plate current for this amplitude and all smaller values can be considered to be linearly related to the input signal within a prescribed small amount of error. Operation with such a combination of bias and input signal amplitude is called linear operation of the tube. It is clear from the figure that the range of input signal amplitude permitted to achieve a prescribed degree of linearity between grid voltage and plate current depends on the choice of operating point. Point A permits a larger value of input signal for a prescribed degree of Linearity, for example, than does point B.

Vacuum tube circuits can be classified as to whether the tubes operate in a linear fashion corresponding to small signal inputs or whether the input signals are large enough to cause excursions over such a large portion of the transfer characteristic that the tangent to curve at the operating point is not a good approximation to the tube behavior over the entire amplitude range of the input signal. There are thus two principal categories of vacuum tube operation, namely, linear or small signal behavior and non-linear or large signal action. Large signal operation is usually analyzed by the use of graphical constructions utilizing plate characteristics, whereas small signal operation is treated by equivalent circuits by means of which changes in plate current and plate voltage from the quiescent values may be computed.

A trigger or flip-flop circuit is a typical vacuum tube circuit. It has two conditions of stable equilibrium and which can be switched from one stable state to the other by some external influence such as the application of an input pulse. Figure 22

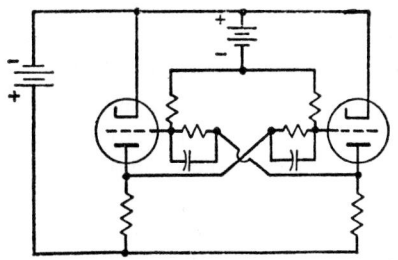

Fig. 22.—Eccles-Jordan circuit.

shows one of the most common flip-flop circuits, the Eccles-Jordan circuit. Two triodes are used in a symmetrical circuit so arranged that one tube is at zero bias and the other is cut off. A positive trigger pulse of sufficient amplitude applied to the grid of the tube cut off, or a negative pulse applied between grid and cathode of the conducting tube, causes a positive feedback action which results in a rapid change of state in each tube. The conducting tube is forced to the cut off state, while at the same time the tube previously cut off is forced into conduction near zero bias. This change of states is effected by the connection from the plate of one tube to the grid of the other. If a tube conducts plate current, the potential at the plate is reduced below the battery voltage by the voltage drop in the resistor connected from the plate to the battery. The grid voltage of the

other tube is thus forced by voltage divider action to be below
the potential previously attained when plate current was cut off
in the first tube. By proper design of the voltage divider and the
negative battery supply, the grid to cathode voltage can be made
to swing from a slightly positive value (approximately zero bias)
to a negative value sufficient to cut the tube off as a result of the
variation of the plate voltage of the other tube as the latter
changes from the cut off value to that at the full conduction
condition.

Another pulse type circuit, which utilizes the transient voltage
produced in a resistor-capacitor circuit, is the sweep circuit employed for deflecting the electron beam in the cathode ray tube
of a television set. Figure 23(a) shows a simple circuit for generating a sweep voltage. The important voltage waveforms are
illustrated in Figure 23(b). Prior to time t_1, the triode conducts
at zero bias, and the voltage across capacitor C is only a small
fraction of the battery voltage because of the voltage drop in the
large resistor R. At time t_1 the grid to cathode voltage is made
negative enough to cut off the flow of plate current. This condition is maintained until time t_2. During the interval of time

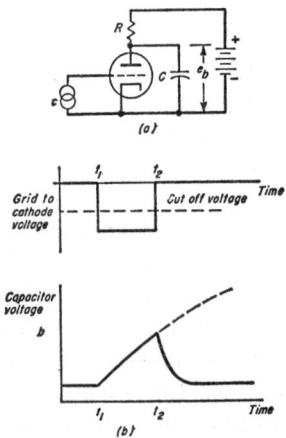

FIG. 23.—(a) Typical sweep generator circuit; (b) Input and output voltage
waveforms for sweep circuits.

that the tube is cut off, the capacitor C is charged from the battery through the resistor R. The voltage across the capacitor increases along the exponential curve associated with the transient in a resistor-capacitor circuit. Ultimately e_b would become equal to the battery voltage. The exponential rise is not permitted to continue after time t_2, however, for when the tube is returned to zero bias at that time, the capacitor discharges through the tube and its voltage rapidly assumes the value which it had at the start of the charging process. Many refinements, including feedback, may be added to the circuit of Figure 23(a) to increase the degree of linearity of the voltage change between t_1 and t_2.

Transistor Circuits.—Transistors may be combined with sources of power and passive elements (resistors, inductors, and capacitors) to form transistor circuits of many forms which are used for the generation, amplification, shaping, and control of electrical signals. Many transistor circuits are similar in form to vacuum tube arrangements designed to perform corresponding functions, but, on the other hand, the unusual properties of transistors also lead to circuit arrangements which have no vacuum tube counterparts. Significant differences between vacuum tubes and transistors make the latter far more attractive in applications where the principal objective is the amplification of low level signals where comparatively little power is associated with the signal itself, the combining and processing of signals as required in electrical computation, and the switching of electrical signals to various paths in accordance with appropriate command signals. The fact that extremely low power is required for a transistor to effect these functions combined with its property of much smaller size and weight than a vacuum tube make possible the fabrication of electronic circuits to perform very complicated operations in a small space with low power consumption. The availability of n-p-n and p-n-p transistors which have the principal charge carriers of opposite sign leads to advantages in cascading transistor amplifiers or data processing circuits on a d-c basis and to simplifications in attaining push-pull operation with its concomitant benefits.

One fundamental difference between vacuum tubes and transistors is the manner in which control of the output current is effected. Vacuum tubes are voltage operated devices, i.e., the plate current is controlled by the voltage impressed between the grid and cathode. The collector current, the usual output current in a transistor, on the other hand, is controlled by the flow of current between base and emitter. The potential difference across the emitter-base junction is a non-linear function of the junction current. As a consequence, to avoid introducing amplitude distortion, the input signal current must be furnished from a high impedance source. This factor alone often results in differences between vacuum tube circuits and the corresponding transistor versions.

As with vacuum tube circuits, when a transistor is employed in conjunction with other circuit elements, a suitable combination of operating parameters must be chosen to establish a quiescent operating point about which the input signal will cause variations in the various electrode currents. The operating point is established by providing appropriate sources of potential in series with resistors in order to obtain the desired steady emitter, base, and collector currents referred to as bias currents.

Operation of transistors may take place in a linear or non-linear fashion depending on the amplitude of the input current. When the variations in the input current are small (defined in the same sense as "small signal" operation with electron tube circuits), equivalent circuits may be used for the calculation of the changes in currents about the operating point. Fig. $24(a)$ illustrates common symbols for the p-n-p and n-p-n transistors. Emitter, base, and collector terminals are indicated by e, b, and c, respectively. Fig. 24 (b) shows a p-n-p unit connected in a common base amplifier circuit, and Figs. $24(c)$ and $24(d)$ indicate (in the dashed outline) one form of equivalent circuit that may be used to compute incremental changes in transistor currents. The representation is spoken of as the T equivalent circuit, and the parameters are defined in terms of the steady emitter-base and collector-base potential differences and currents $(V_e, V_c, I_e, \text{ and } I_c)$ shown on the following page.

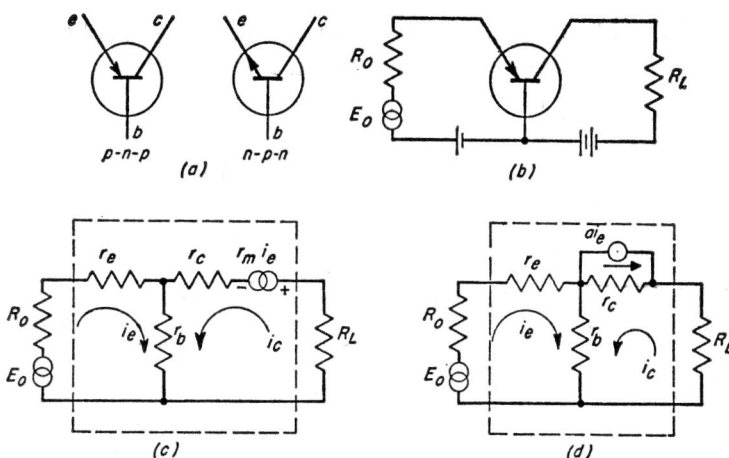

Fig. 24.—(a) Symbols for transistor; (b) Common base amplifier circuit; (c) and (d) Two transistor T equivalent circuit.

Base resistance, $r_b = \dfrac{\partial V_e}{\partial I_c}\bigg|\,I_e\text{ constant}$

Emitter resistance, $r_e = \dfrac{\partial V_e}{\partial I_e}\bigg|\,I_e\text{ constant} \;-\; \dfrac{\partial V_e}{\partial I_e}\bigg|\,I_c\text{ constant}$

Collector resistance, $r_c = \dfrac{\partial V_c}{\partial I_c}\bigg|\,I_c\text{ constant} \;-\; \dfrac{\partial V_e}{\partial I_c}\bigg|\,I_c\text{ constant}$

Mutual resistance, $r_m = \dfrac{\partial V_c}{\partial I_e}\bigg|\,I_c\text{ constant} \;-\; \dfrac{\partial V_e}{\partial I_e}\bigg|\,I_e\text{ constant}$

$$a = \frac{r_m}{r_e}$$

The partial derivatives * appearing in the above expressions may be determined from suitable graphs of the parameters V_e, V_c, I_e,

* The partial derivative is obtained when the process of differentiation (Chapter VI) is applied to expressions in more than one variable, by treating the excess variable or variables (written after the vertical rule in the expression) as constants.

and I_c. The collector-emitter short circuit current amplification factor α_{ce} or simply α, is defined as

$$\alpha = \alpha_{ce} = \left.\frac{\partial I_c}{\partial I_e}\right|\; V_{eb}\; \text{constant}$$

This quantity may also be expressed in terms of r_m, r_b, and r_c as

$$\alpha = \alpha_{ce} = \frac{r_m + r_b}{r_c + r_b}$$

For a junction transistor, r_m and r_c are very much larger than r_b, so that

$$\alpha = \alpha_{ce} \approx \frac{r_m}{r_c} = a$$

For this reason a and α are often used interchangeably in the equivalent circuits of Figs. 1(c) and (d). Another quantity of interest is the collector-base short circuit current amplification factor α_{cb} defined as

$$\alpha_{cb} = \left.\frac{\partial I_c}{\partial I_b}\right|\; V_c\; \text{constant}$$

For a junction transistor the following approximate relation is valid

$$\alpha_{cb} \approx \frac{a}{1 - a} \approx \frac{\alpha}{1 - \alpha}$$

The quantity α_{cb} appears in the quivalent T circuit for the common emitter connection. Fig. 25 indicates the equivalence for

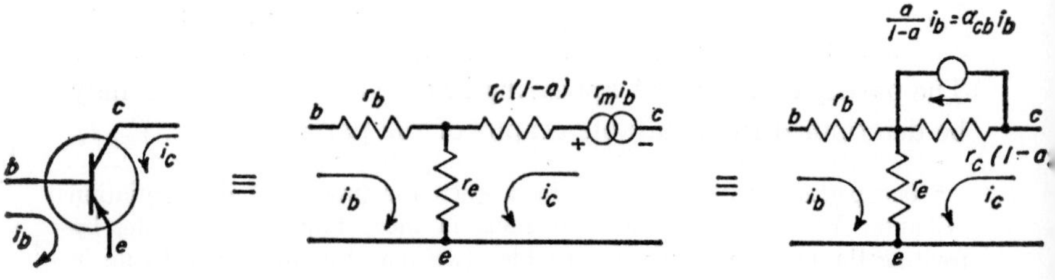

FIG. 25.

this arrangement. It is to be noted that the values of the various parameters in the equivalent circuits vary with the electrode currents as well as with frequency. The circuits presented are valid only at low frequencies. There are effects which occur at high frequencies which require assigning a frequency dependence to α as well as the introduction of capacitors across various elements. Typical low frequency values of r_c and r_m are in megohms; or r_b, several hundred to a thousand ohms; or r_e, less than one hundred ohms; and of α and α_{cb} 0.98 to 0.99 and 50 to 100, respectively.

There are several forms of equivalent circuits that are used to represent transistor small signal operation in addition to the T network form just described. Another representation that finds frequent use is one employing the so-called hybrid parameters. The parameters are "hybrid" in the sense that they do not all have the same dimensions. Fig. 26 indicates the equivalent

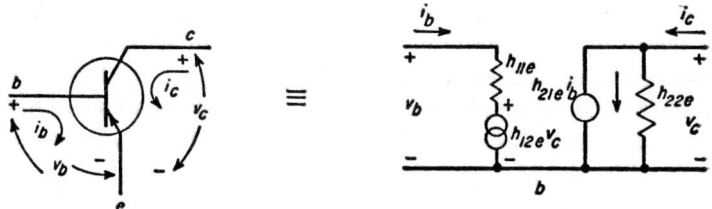

Fig. 26.

circuit for the common emitter connected transistor using these parameters. The base-emitter circuit is represented by an impedance h_{11e} in series with a voltage generator. The collector-emitter circuit is represented by an admittance h_{12e} in parallel with a current generator. The parameters are defined in terms of static currents and voltages as follows:

h_{11e} = input impedance (short circuit across output terminals)

$$= \frac{\partial V_{be}}{\partial I_b} \bigg|\ V_{ce} \text{ constant}$$

h_{22e} = output admittance (no connection across input terminals)

$$= \frac{\partial I_e}{\partial V_{ce}} \bigg|\ I_b \text{ constant}$$

h_{12e} = reverse open circuit voltage amplification factor

$$= \frac{\partial V_{be}}{\partial V_{ce}} \bigg|\ I_b \text{ constant}$$

h_{21e} = forward short circuit current amplification factor

$$= \frac{\partial I_c}{\partial I_e} \bigg|\ V_{ce} \text{ constant} = \alpha_{cb}$$

These parameters may be defined for any form of transistor connection. The subscript e denotes common emitter connection; a corresponding labelling is used to distinguish the parameters for other connections.

The A-M Radio Receiver.—A type of radio receiver that continues to be used extensively is the superheterodyne receiver. It is shown in the diagram of Fig. 27 in a simplified circuit which contains fewer stages than would be provided in an actual set, in order to show clearly the major functional part-circuits. Also, transistors have been used in place of electron tubes, in accordance with present-day design.

This circuit differs from others in that it converts all incoming radio-frequency signals to a common carrier frequency. This is accomplished in the first detector, mixer or converter as it is variously called. The signal from the antenna is fed by a tuned coupled circuit to the mixer unit (or in more elaborate sets a tuned radio-frequency stage may be inserted between the antenna and the mixer). In the mixer stage the incoming signal is heterodyned with a locally generated signal so a beat frequency signal, called the intermediate frequency, is produced. This new frequency signal is radio frequency, ranging from around 450 kc. to several megacycles depending upon the purpose for which the receiver is designed. The intermediate frequency has exactly the same modulation as the original signal. In many

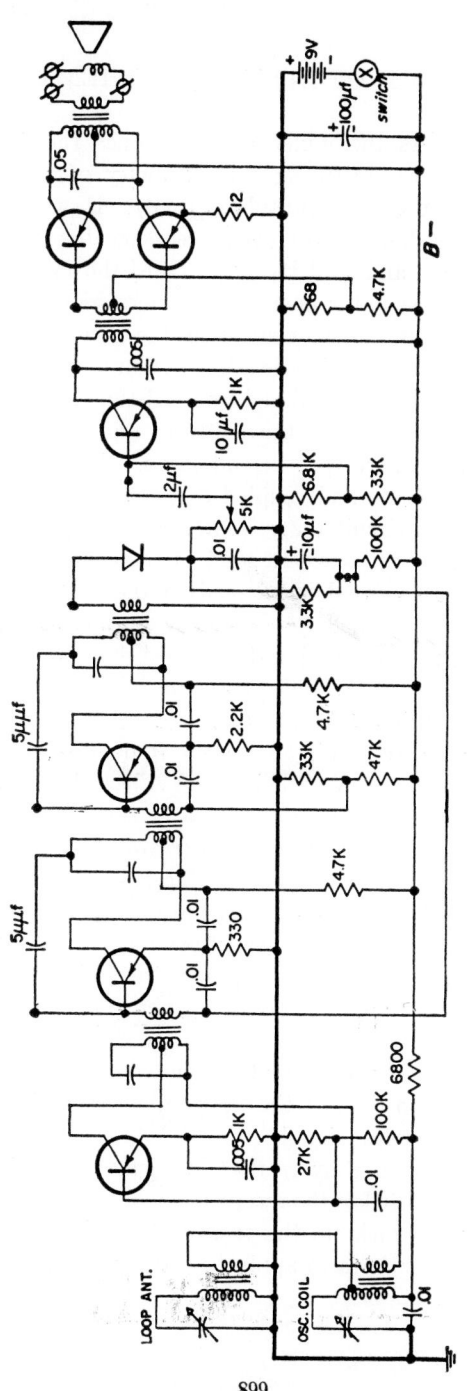

Fig. 27.—Transistor Superheterodyne Circuit.

broadcast receivers the mixer unit combines the functions of mixer and oscillator by using a multiplicity of transistors (in vacuum tube sets, a pentagrid tube is used). However, at higher frequencies it is desirable or even necessary to use a separate unit for oscillator and feed its output into the mixer. Regardless of how the oscillator operates, its frequency is always adjusted by the main tuning control of the receiver so the beat frequency output of the mixer is a fixed value. This intermediate frequency signal is then amplified by fixed-tuned radio-frequency amplifiers and then fed to the detector (commonly called the second detector) where it is demodulated. The audio signal is then further amplified and coupled to the speaker.

The simplified-circuit diagram indicates the various circuits and their relative positions. In some of the cheaper superheterodynes the antenna signal is coupled to the first transistor detector, then the intermediate frequency output of this coupled without further amplification to a grid bias or regenerative detector and hence to the final power unit. Various refinements are often added to the higher quality sets. Among these are automatic volume control, automatic frequency control, noise suppression, tone control, fidelity and selectivity controls, etc. The superheterodyne gives much greater selectivity by its system of frequency changing and also permits the use of circuits having a more uniform response to the sidebands.

Television.—Television is the transmission of scenes, either still or motion, by electrical means, commonly by radio, for instantaneous viewing without permanent recording. For a practical system certain fundamental components or functions are necessary:

1. Camera device to pick up the scene.

2. Tranducer to convert the light impulses of the scene to a corresponding electrical signal.

3. Transmitter to convert the electrical signals into proper form to be transmitted to the receiver.

4. Receiver to pick up the transmitted signals and convert them to the proper form to apply to a transducer.

5. Transducer to convert the electrical signals back into light in a reproduction of the original scene.

The first three of these topics are outside the scope of this book, which deals with signal receivers. Therefore this discussion deals with the equipment required for reception of television signals, and their transformation into sound and pictures on the screen of a cathode-ray tube.

The received scene must be reconstructed from the electrical pulses reaching the receiver. If we assume that the original scanner broke the picture down into ten lines and each line had ten elements side by side (this is determined by the width of the hole, being the width of the picture divided by the hole width), then we have 10 times 10 or 100 elements in the scene and our reproduced picture must be built of 100 blocks. It can readily be seen that this would give a very coarse mosaic effect to the picture since each element is fixed in intensity of light. More lines would give more elements and finer detail in the receiver scene. A close comparison may be drawn between this and the printed pictures of newspapers and magazines. The relatively coarse newspaper pictures are lacking in detail while the usual magazine half-tones give very good detail. The difference between these pictures is the number of dots or elements (clearly visible if the printed picture is viewed through a magnifying glass) of which they are composed.

Besides the number of lines per scene, the rate at which the scenes are repeated is also important. The television pictures must be repeated at a rate high enough to give the illusion of smooth motion. While the motion picture rate of 24 frames per sec. would be satisfactory for this rate, the value should be harmonically related to the power line frequency in order to minimize certain interference effects. The standard power supply frequency of 60 cycles dictates the use of 30 frames per sec. for television. Furthermore, to improve the quality of the reproduced picture, the scanning is not done for adjacent lines in order, but the picture is scanned over alternate lines first and then scanned again over those missed the first time. This double scanning, known as interlaced scanning, is done in the thirtieth-of-a-second period of one frame.

A block diagram of a complete television system is shown in Fig. 28. The original scene is focused on the camera tube by a light lens system. The camera tube converts this light picture into the sequence of electrical elements necessary for transmission.

The very minute electrical signals coming from the camera tube are amplified by wide band video amplifiers, the wide band being necessitated by the great range of frequencies produced by the modern multi-line systems. This wide band amplifier feeds a monitor circuit which reproduces the televised scene on a picture tube so the operator can check the camera circuit operation continuously. It also furnished the input for the modulator which modulates the picture signals on the radio-frequency carrier in a manner very similar to that of the audio modulation of conventional broadcasting. The modulated radio frequency is then supplied to the antenna and radiated into space. At the same time the microphone picks up the sound associated with the scene. This signal is amplified and used to frequency modulate the sound carrier of the television transmitter. The sound-modulated carrier is radiated simultaneously with the picture carrier. Both are then picked up by the receiving antenna, amplified and fed through the first detector and intermediate frequency amplifiers of a superheterodyne receiver. The two types of signal are then separated and each is fed to its proper detector or demodulator. The sound signals, now at audio frequency, are further amplified and drive the loudspeaker. The picture signal circuits are much more complex. In order to reproduce the scene at the receiver, it is necessary for the receiver transducer, whatever its nature, to follow exactly the operation of the camera tube. Thus when the electron beam in the camera scans the scene at a given rate, each scanning line starting at a definite time after the preceding one, the scanning beam in the picture tube in the receiver must retrace the scene in exactly the same order, each line starting at the same time interval after the

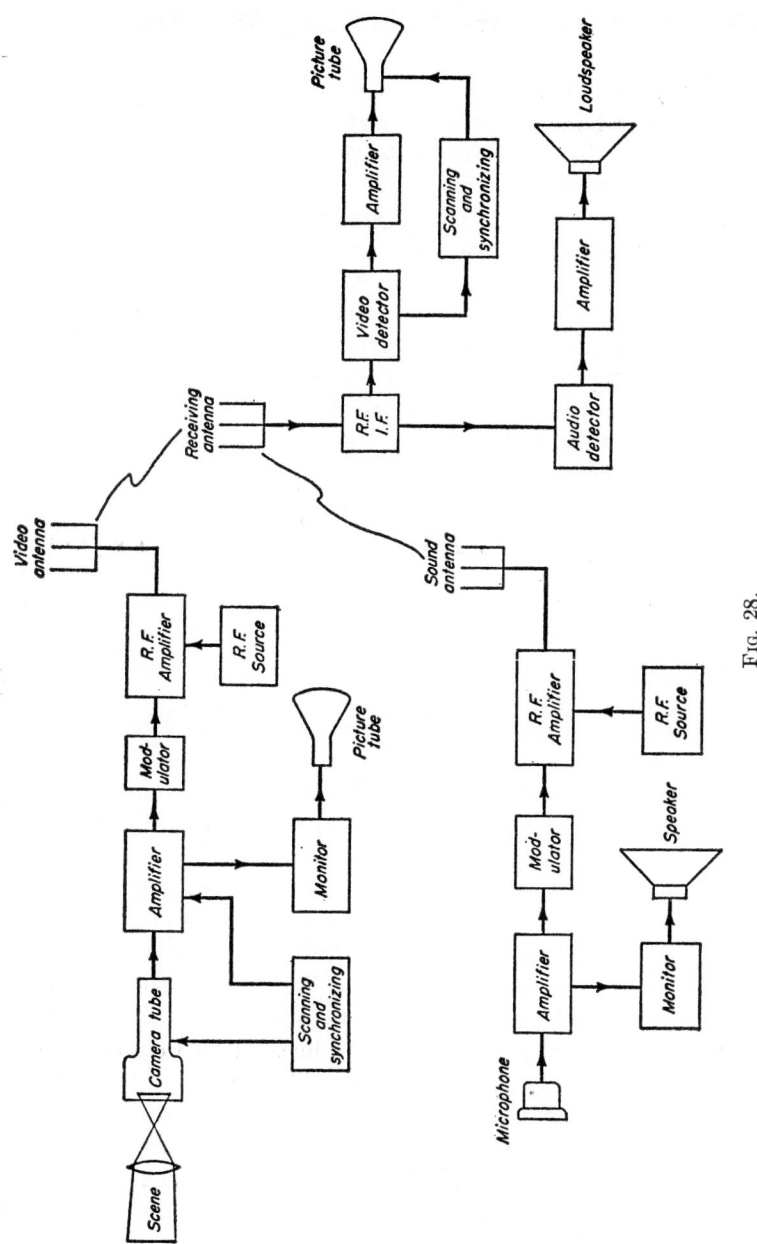

Fig. 28.

903

preceding one. Otherwise the various lines might get badly skewed or out of synchronism and a badly distorted picture would result. To insure accurate synchronization between the transmitting end and the receiving end, synchronizing pulses are transmitted at the end of each scanning line. In addition, synchronizing pulses to govern the return of the scanning to the top of the scene are also transmitted. These various pulses are impressed on the signal in the transmitter during the short time interval while the scanning is returning from the end of one line to the beginning of the next. In the receiver these synchronizing pulses must be separated from the detected signal and routed into the proper channels. They are then used to synchronize the sweep oscillators which provide the scanning at the receiver picture tube. The video signals without the synchronizing pulses are amplified in the proper channels of the circuit and also fed to the picture tube. The electronic picture tube is the cathode ray tube. The electron beam issuing from the gun is modulated in intensity by the picture or video signal so the intensity of the fluorescent spot produced by it on the tube screen is a reproduction of the intensity of the corresponding part of the original scene. The synchronizing and scanning circuit produces a sweep signal which is applied to the picture tube by plates or coils just as in the oscilloscope discussed in the section on cathode ray tubes. This sweep action carries the electron beam relatively slowly across the screen, then blanks it and returns it rapidly, moves it down and repeats, the time for each operation being the same as for the corresponding operation in the camera tube, the two operations being linked together by the synchronizing pulses. After completing the scanning of alternate lines of the picture, the beam is deflected back to the top of the picture and repeats this operation, now filling in the alternate lines which were skipped on the first scanning, again in exact synchronism with the same process in the camera tube. It can be seen, then, that since the intensity of the original scene and the position of the spot corresponds with the position of the original scanning position at the original scene, the reproduced effect on the screen of the picture tube is the scene which was picked up by the camera.

As has been discussed in detail above, television provides for the reproduction of pictures upon the face of a cathode ray tube by control of the brightness of the white light emanating from the tube face. To reproduce scenes in color, that is, to provide color television, not only is transmitting and receiving apparatus of additional complexity needed, but a device capable of providing a colored light output at the scanning location is also required. To provide the basis for the subsequent explanation of a color television system, some attention must be given first to color representation.

Human vision can be expressed in three separate color sensations, each of which may be stimulated in various degrees. Although these sensations always appear to act together, it is believed that if they could be stimulated separately they would be found to be the sensations produced by red, blue, and green. The respective amount of these three primaries needed to match a color of particular wavelength are called the tristimulus coefficients. Since the total of the three quantities is 100%, if any two are known, the other is known also. Thus the three quantities used to specify the amounts of the primaries can be reduced to two, allowing a two-dimensional plot to represent any color. Letting X, Y, and Z be the amount of the primaries needed to match a given color, the trichromatic coefficients x, y, and z are defined as

$$x = \frac{X}{X + Y + Z}$$

$$y = \frac{Y}{X + Y + Z}$$

$$z = \frac{Z}{X + Y + Z}$$

but z is equal to $100\% - (x + y)$. The two-dimensional plot formed by the use of two trichromatic coefficients to represent colors is known as a chromaticity diagram.

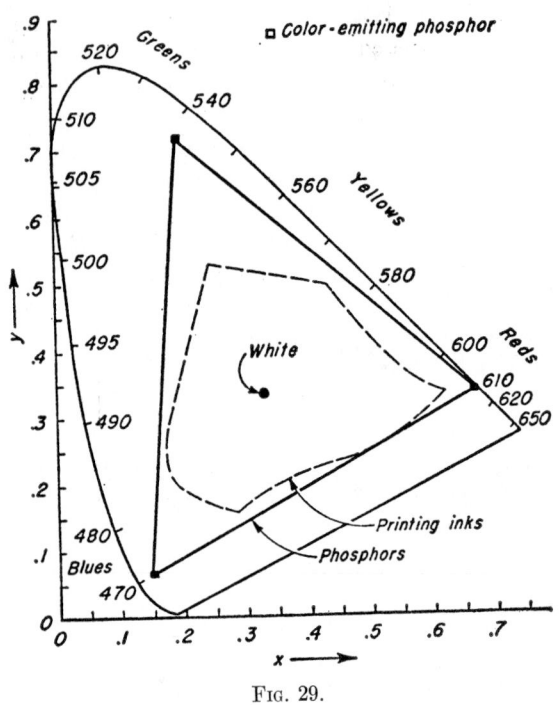

Fig. 29.

The diagram using x and y coordinates is shown in Fig. 29. The horseshoe shaped curve shown contains pure spectral colors; the numbers on the diagram are spectral wavelengths in millimicrons. All colors observable with the human eye are contained within the boundary formed by the horseshoe shaped figure. A color known as equal energy white is one containing equal intensities of the three CIE primaries. (CIE is the French abbreviation for the International Commission on Illumination.) It is shown with the coordinates (0.333, 0.333). The principle of using three arbitrary primary colors to represent a given color can be interpreted in terms of the chromaticity diagram. If the three physical colors are represented by points on the diagram, then the triangle formed by connecting the points contains all possible colors that can be formed from the primaries. In Fig. 4, points are shown for red, blue, and green cathode ray tube phosphors taken as primary color sources. The area of the triangle formed by the use of these points as vertices, compared with the region included by the boundary of the pure spectral colors, indicates the proportion of visible colors that can be produced by the light from three cathode ray tube phospors. Any color within the triangle indicated in the figure can be reproduced by the phosphors. The possible range of color reproduction provided by modern printing inks is shown in the diagram for comparison.

It follows from the foregoing discussion that televising a colored scene required furnishing to the receiving location not only information concerning the color or chromaticity of each picture element in the scene but also data concerning the brightness or luminance of each televised element. One method of generating electrical signals which may be used to reproduce a color picture is shown in Fig. 30. The scene to be televised is viewed through a lens system by three image orthicon tubes, each of which is preceded by a color filter, to separate the required color. Each camera tube is supplied with the same scanning and synchronizing information as is depicted for the camera tube in the black and white television system shown in Fig. 28. The outputs from the three camera tubes provide at each instant

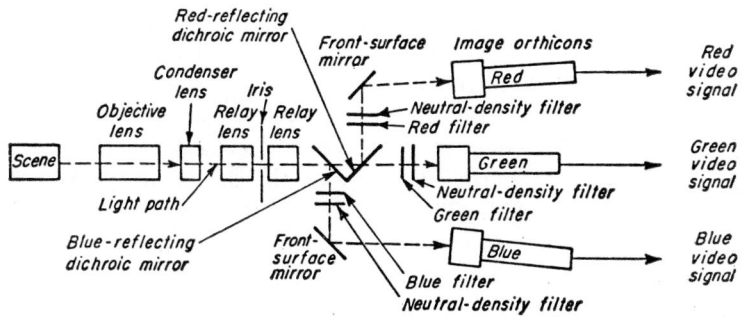

Fig. 30.

of time, just as in the black and white system, signals which are proportional to the light intensity reaching the camera tube from the scene being televised. Because of the color filters that are inserted in front of the camera tubes, the signal from the red channel is proportional to the red content of the various picture elements in the scene being scanned, a corresponding situation exists for the blue and green channels. The three color signals thus produced contain the information from which the scene being televised can be synthesized from sources of red, blue, and green light which are capable of being varied in intensity in accordance with the individual camera signals.

FM Stereo.—The principle of all radio transmission of speech, music, and television pictures is based upon the production of a uniform electromagnetic wave (the carrier) by the sending station and the imposition upon that wave of another wave representing the information to be transmitted, a process called modulation. There are two common methods by which this is done, as shown in Fig. 31. Consider that (a) in that figure represents the carrier wave. Then the signal wave can be added to it by changing the amplitude of its oscillations, as shown in part (b) of the figure. This is Amplitude Modulation (AM). The other method is to vary the frequency of its oscillations, as shown in part (c). At one time all home radio receivers operated on AM.

The number of stations in the frequencies used for AM receivers became so great, however, that their allotted bandwidths were as little as 5000 or 7000 cycles. This meant that the audio frequencies they could transmit were curtailed, especially in the higher frequencies. However, the frequency bands allotted to FM were in the region of megacycles (millions of cycles per second) rather than the hundreds or thousands of cycles per second allotted to AM. Therefore each FM station was allotted a band wide enough to transmit the full range of audible sound. Moreover, FM provides higher signal-to-noise ratios, another characteristic which is advantageous for high fidelity transmission. In stereophonic systems, in which a binaural effect is pro-

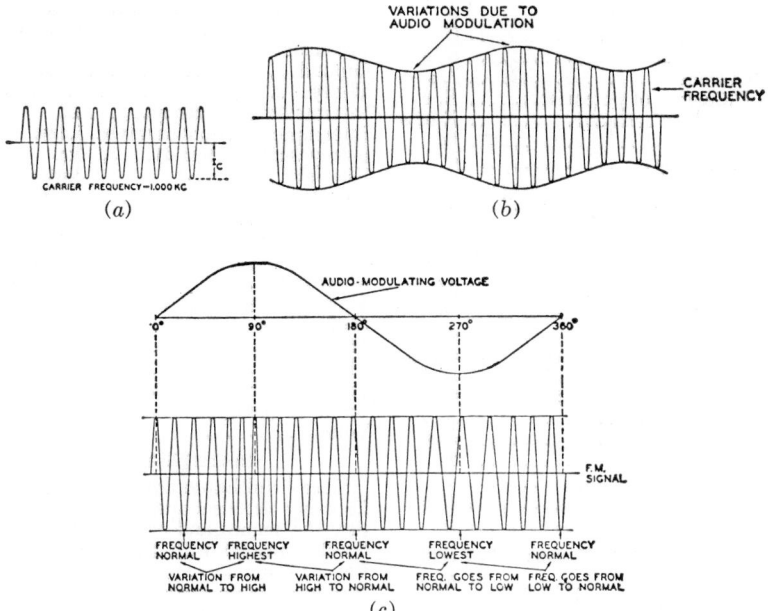

VARIATIONS DUE TO
AUDIO MODULATION

CARRIER
FREQUENCY

CARRIER FREQUENCY—1,000 KC

(a)

(b)

AUDIO-MODULATING VOLTAGE

0°   90°   180°   270°   360°

F.M.
SIGNAL

| FREQUENCY NORMAL | FREQUENCY HIGHEST | FREQUENCY NORMAL | FREQUENCY LOWEST | FREQUENCY NORMAL |

VARIATION FROM NORMAL TO HIGH   VARIATION FROM HIGH TO NORMAL   FREQ. GOES FROM NORMAL TO LOW   FREQ. GOES FROM LOW TO NORMAL

(c)

Fig. 31.

duced by transmitting signals from two microphones so spaced so that the home receiver can reproduce from two speakers effects that give the sound depth sensation of two human ears, fidelity of the sound received to that in the studio is especially important. (It is to be noted that because of the much higher frequencies that are broadcast, FM signals are not appreciably reflected by the atmosphere under normal conditions, so that, unlike AM, effective reception is normally restricted to points that are essentially in line-of-sight of the sending station.)

Because of the much higher signal-to-noise ratio desired, FM receivers (or separate tuners and amplifiers) have many more stages than AM receivers. Moreover, FM stereo equipment requires an entire additional section, called the multiplex section, to separate the inputs to the two speakers.

Standard Letter Symbols for Electrical Quantities

Admittance	Y, y
Angular velocity $(2\pi f)$	ω
Capacitance	C
Conductance	G, g
Current	I, i
Difference of potential	E, e
Dielectric constant	K or ϵ
Energy	W
Frequency	f
Impedance	Z, z
Inductance	L
Magnetic intensity	H
Magnetic flux	Φ
Magnetic flux density	B
Mutual inductance	M
Number of conductors or turns	N
Permeability	μ
Phase displacement	θ or ϕ
Power	P, p
Quantity of electricity	Q, q
Reactance	X, x
Resistance	R, r
Susceptance	b
Speed of rotation	n
Voltage	E, e
Work	W

Letter Symbols for Vacuum Tube Notation

Grid potential	E_g, e_g
Grid current	I_g, i_g
Grid conductance	g_g
Grid resistance	r_g
Grid bias voltage	E_c
Plate potential	E_p, e_p
Plate current	I_p, i_p
Plate conductance	g_p
Plate resistance	r_p
Plate supply voltage	E_b
Emission current	I_s
Mutual conductance	g_m
Amplification factor	1
Filament terminal voltage	E_f
Filament current	I_f
Filament supply voltage	E_a
Grid-plate capacity	C_{gp}
Grid-filament capacity	C_{gf}
Plate-filament capacity	C_{pf}
Grid capacity $(C_{gp} + C_{gf})$	C_g
Plate capacity $(C_{gp} + C_{pf})$	C_p
Filament capacity $(C_{gf} + C_{pf})$	C_f

NOTE.—Small letters refer to instantaneous values.

Abbreviations Commonly Used in Radio

Alternating current	a.c.
Antenna	ant.
Audio frequency	a.f.
Continuous waves	c.w.
Cycles per second	$\smallsmile$
Decibel	db.
Direct current	d.c.
Electromotive force	e.m.f.
Frequency	f.
Ground	gnd.
Henry	h.
Intermediate frequency	i.f.
Interrupted continuous waves	i.c.w.
Kilocycles (per second)	kc.
Kilowatt	kw.
Megohm	$M\Omega$
Microfarad	μfd.
Microhenry	μh.
Micromicrofarad	$\mu\mu$fd.
Microvolt	μv.
Microvolt per meter	μv/m
Milliampere	ma.
Milliwatt	mw.
Ohm	Ω
Power factor	p.f.
Radio frequency	r.f.
Volt	v.

Greek Alphabet

Since Greek letters are used to stand for many electrical and radio quantities, the names and symbols of the Greek alphabet with the equivalent English characters are given.

Greek Letter		Greek Name	English Equivalent
A	α	Alpha	a
B	β	Beta	b
Γ	γ	Gamma	g
Δ	δ	Delta	d
E	ϵ	Epsilon	e
Z	ζ	Zeta	z
H	η	Eta	$\bar{e}$
Θ	ϑ	Theta	th
I	ι	Iota	i
K	κ	Kappa	k
Λ	λ	Lambda	l
M	μ	Mu	m
N	ν	Nu	n
Ξ	ξ	Xi	x
O	o	Omicron	$\breve{o}$
Π	π	Pi	p
P	ρ	Rho	r
Σ	σ	Sigma	s
T	τ	Tau	t
Υ	υ	Upsilon	u
Φ	ϕ	Phi	ph
X	χ	Chi	ch
Ψ	ψ	Psi	ps
Ω	ω	Omega	$\bar{o}$

XXII

Print Shop

The printer's mathematical problems consist mainly of estimating the amount of composition which will be required to set up certain copy, the printing space which this will consume, and estimating quantities of printing and paper. The computations he makes are entirely conventional, but the units he uses are peculiar to his trade.

Type Measure.—Previous to the year 1886 type sizes were indicated by names, *brevier*, *bourgeois*, *pica*, etc. Some of these names have continued in use although the *point* system is now used exclusively. The following is a list of the standard sizes of type with the old name designations:

TABLE 1

Old Name	Point Size	Old Name	Point Size
Excelsior	3	2-line Brevier or Columbian	16
Brilliant	3½	3-line Nonpareil or Great Primer	18
Semi-Brevier	4	2-line Long Primer or Paragon	20
Diamond	4½	2-line Small Pica	22
Pearl	5	2-line Pica	24
Agate	5½	2-line English	28
Nonpareil	6	5-line Nonpareil	30
Minion	7	3-line Pica or Double Great	
Brevier	8	Primer	36
Bourgeois	9	7-line Nonpareil	42
Long Primer	10	4-line Pica or Canon	48
Small Pica	11	9-line Nonpareil	54
Pica	12	5-line Pica	60
English	14	6-line Pica	72

A few of these types sizes are illustrated in Fig. 1.

A *point*, as used in printing, is 0.0138 inch or approximately $\frac{1}{72}$ inch. Thus, an 8-point type has a body (Fig. 2) $\frac{8}{72}$ inch and nine lines of this type measure one inch vertically on a page; a 12-point type has a body $\frac{12}{72}$ inch and equals six lines to the inch.

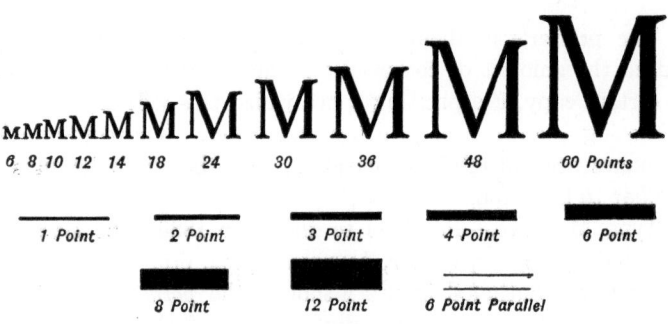

$$\underset{6\ 8\ 10\ 12\ 14}{\text{MMMMM}}\ \underset{18}{M}\ \underset{24}{M}\ \underset{30}{M}\ \underset{36}{M}\ \underset{48}{M}\ \underset{60\ Points}{M}$$

| 1 Point | 2 Point | 3 Point | 4 Point | 6 Point |

| 8 Point | 12 Point | 6 Point Parallel |

Fig. 1.

The number of lines and fractions of lines (in points) to the inch for the more common sizes of type are given below:

TABLE 2

	Lines	Points	
6 point equals	12		to the inch
7 point equals	10	2	to the inch
8 point equals	9		to the inch
9 point equals	8		to the inch
10 point equals	7	2	to the inch
11 point equals	6	6	to the inch
12 point equals	6		to the inch

The size of type as we have defined it is really the size of the block to which the letters are attached. The sizes of the letters themselves depend on the *face* or the cut and shape of the type *family* in question. One 12-point face may have a small body with long descenders while another may have a full heavy body.

Thus:

This line is twelve-point Cloister.

This line is twelve-point Bodoni.

It will be noted that while these two faces are of the same point size, one requires more lateral space for a word than the other. This brings us up to the consideration of the *pica* which is a unit used to measure the length of a line of type and also the size of a type page. A *pica* is 12 points or $\frac{1}{6}$ inch. Thus, a line of type four inches long is said to be 24 picas long. Similarly, a type page 7 inches by $4\frac{1}{2}$ inches is 42 by 27 picas.

ILLUSTRATION: A type page is 54 picas long. How many lines of 12-point type will it accommodate?

54 × 12 (no. of points in a pica) ÷ 12 (no. points of type) =

54 lines   (Ans.)

ILLUSTRATION: How many lines of 10-point type will go on a page 42 picas long?

$42 \times 12 \div 10 = 504 \div 10 = 50\frac{4}{10}$ or 50 lines   (Ans.)

ILLUSTRATION: A type page is to be 6 inches high. How many lines of 8-point type will it hold?

From Table 2, 8-point type gives 9 lines per inch.

Then, $9 \times 6 = 54$ lines   (Ans.)

Another solution is to change the 6 inches to the equivalent number of points and divide by 8. This may be done in one operation as follows:

$$\frac{6 \times \overset{9}{\cancel{72}}}{\cancel{8}} = 6 \times 9 = 54 \quad \text{(Ans.)}$$

Space Measure.—A low type without printing surface is set between words, following the period of a sentence, and between letters of a word when justifying a short line. A *quad* (or *em quad*) is a space type which is just as wide as it is thick. Thus in a 10-point font an em quad is 10 points or four-fifths pica wide. In a 12-point font an em is 12 points or one pica wide. A space half an em in width is called an *en*. Types 5 *to em*, 4 *to em*, and 3 *to em* are also used in spacing.

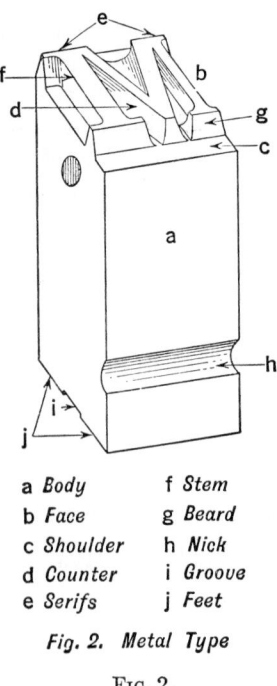

a	*Body*	f	*Stem*
b	*Face*	g	*Beard*
c	*Shoulder*	h	*Nick*
d	*Counter*	i	*Groove*
e	*Serifs*	j	*Feet*

Fig. 2. Metal Type

Fig. 2.

The space between lines of type is often increased for the sake of legibility by the insertion of *lead* rules of any desired thickness. They are furnished in 1-point, 2-point, 3-point, etc., thicknesses as shown in Fig. 1. Type so spaced is said to be leaded. When type is set on a Linotype it may be leaded by having a shoulder cast directly on the slug. Thus, a 10-point type may be cast on a 12-point slug, the effect being the same as inserting a 2-point lead into 10-point type set "solid". Type so set is referred to as "10 on 12", "8 on 10", or whatever the combination of type size to slug size may be. If type is designated as "leaded", without further qualification, a 2-point lead is meant. When computing the number of lines of leaded type to a page, the leading must, of course, be taken into account.

ILLUSTRATION: How many lines of type 9-point on 11 will go onto a page 36 picas long?

$$36 \text{ picas} \times 12 = 432 \text{ points}$$
$$\tfrac{432}{11} = 39\tfrac{3}{11} \text{ or } 39 \text{ lines.} \quad \text{(Ans.)}$$

Estimating Type Space—*Line Method.*—We noted in the discussion of type sizes that considerable difference exists between the lateral space required by different type faces of the same point size. If any considerable amount of type is to be set, this difference may aggregate many pages. The most accurate method of estimating is based on a character count of the copy and the use of a space factor for the type face selected.

Standard office typewriters made in the U. S. (but not usually foreign machines, or electric machines with proportional spacing) produce either 10 or 12 characters to the inch (pica or elite type). These machines write six lines to the inch "single spaced" or three lines to the inch "double spaced." Each punctuation mark takes the same space as a letter. The number of characters on a page of typed manuscript may then be found very readily by measuring the length of an average line in inches and multiplying by ten or twelve, as the case may be, then measuring the length of the type page in inches and multiplying by six if it is single spaced, three if it is double spaced. The product of these products is the number of characters on the page.

ILLUSTRATION: A manuscript page typed single spaced on a "pica" typewriter has an average line length of six inches and a typed page length of eight and one-half inches. How many characters are there to the line and how many characters per page?

$$6 \quad \times 10 = 60 \text{ characters per line.} \quad (Ans.)$$
$$8\frac{1}{2} \times 6 = 51 \text{ lines per page}$$
$$60 \quad \times 51 = 3060 \text{ characters per page} \quad (Ans.)$$

ILLUSTRATION: A manuscript page is typed double spaced on an "elite" typewriter to a depth of nine inches with an average line length of six and one-half inches. How many characters are there to the line and how many characters to the page?

$$6\frac{1}{2} \times 12 = 78 \text{ characters per line} \quad (Ans.)$$
$$9 \quad \times 3 = 27 \text{ lines per page}$$
$$78 \quad \times 27 = 2106 \text{ characters per page} \quad (Ans.)$$

TABLE 3

Average Number of Characters to One Pica

(Each letter, space, and punctuation point is counted as a character)

Type Face	Type Size, Points						
	6	7	8	9	10	11	12
Antique No. 1	3.35	......	2.75	......	2.4	......	2.1
Benedictine and Benedictine Book	3.9	3.5	3.1	2.8	2.5	2.35	2.2
Benedictine Bold	......	......	2.85	......	2.35	......	2.0
Bodoni	3.9	3.4	3.0	......	2.55	......	2.4
Bodoni Book	3.95	3.6	3.2	2.9	2.75	......	2.5
Bodoni Bold	3.6	......	2.8	......	2.4	......	2.2
Caslon	......	3.5	3.2	2.9	2.75	2.4	2.2
Caslon Old Face	4.05	......	3.45	3.1	3.0	2.75	2.4
Caslon No. 3	3.7	......	3.1	......	2.45	......	2.2
Century Expanded	3.45	3.1	2.8	2.6	2.4	2.3	2.1
Century Bold	......	......	2.9	......	2.35	......	2.1
Cheltenham	......	......	3.45	3.15	2.9	2.7	2.55
Cheltenham Wide	3.1	......	2.9	......	2.5	......	2.2
Cheltenham Condensed	......	......	3.5	......	2.85	......	2.6
Cheltenham Bold	3.25	......	2.8	......	2.3	2.2	2.1
Cloister	4.0	......	3.45	......	3.1	2.95	2.85
Cloister Wide and Cloister Bold	3.6	......	3.0	......	2.7	......	2.5
De Vinne	3.5	......	3.0	2.85	2.6	2.3	2.1
Elzevir No. 3	3.75	......	3.0	2.8	2.65	2.4	2.25
Franklin	3.7	3.55	3.2	2.9	2.75	2.6	2.35
Garamond	4.2	......	3.45	3.1	2.9	2.6	2.4
Garamond Bold	3.7	......	3.1	......	2.55	......	2.1
Granjon	......	......	3.45	......	2.9	2.75	2.5
Ionic No. 5	3.16	2.9	2.63	2.45			
Narciss	......	......	......	......	2.4	......	2.0
Number 16	3.15	2.85	2.6	2.4	2.3	......	2.0
Old Style No. 1	3.55	3.25	3.0	2.8	2.65	2.55	2.3
Old Style No. 7	3.8	3.45	3.2	3.0	2.75	2.55	2.35
Original Old Style	3.7	......	3.05	......	2.75	......	2.45
Scotch	3.35	......	3.0	......	2.7	2.55	2.25

In these computations, the short lines at the ends of the paragraphs are regarded as full lines, for if the type is set 20 to 30 picas wide approximately the same number of similar short lines will occur in the type.

Table 3 shows the number of characters in a line for a width of one pica of a number of common Linotype faces. Each letter, space, and punctuation point is counted as a character. Having selected the type face and the width of line, the amount of space which this will take can be computed as shown in the following illustration.

ILLUSTRATION: A manuscript of 75 typewritten pages averaging 2500 characters per page is to be set in Caslon 11-point on 13, 26 picas wide and 40 picas depth of page. How many type pages will this make?

From Table 3 we note that Caslon 11-point type averages 2.4 characters per pica. Then the number of characters per line is,

$$2.4 \times 26 = 63.4 \text{ characters per line of type}$$

The number of lines of type on a 13-point slug which a 40-pica page will hold is,

$$\frac{12 \times 40}{13} = 37 \text{ lines.}$$

Then the number of characters per page of type is the product of the number of characters per line and the number of lines,

$$63.4 \times 37 = 2346 \text{ characters of type per page.}$$

The manuscript of 75 pages averaging 2500 characters per page consists of,

$$75 \times 2500 = 187,500 \text{ characters}$$

The number of type pages is then simply,

$$\frac{187,500}{2346} = 80 \text{ pages} \quad \text{(Ans.)}$$

Estimating Type Space—*Area Method.*—A less accurate method of estimating type space consists of estimating the number of words in a manuscript and using a figure for the number of words per square inch of a certain type size. It takes no account of the variations of the different types of one size and we present it here only because it is still used in a number of shops.

The number of words of a manuscript may be estimated by counting a few lines on a representative page. To obtain the average length of a line in words, count ten lines and move the decimal point one place to the left. Thus, if ten lines of a manuscript contain 115 words, the average line has 11.5 words. The number of lines per page is determined as above, and the number of words per page is then the product of the average number of words per line and the number of lines.

ILLUSTRATION: Ten lines of a manuscript consists of 92 words. How many words are there on the page if there are thirty lines?

$$92 \div 10 = 9.2 \text{ average words per line}$$

$$9.2 \times 30 = 276 \text{ words per page}   \text{(Ans.)}$$

Knowing the number of words of a piece of copy, Table 4 may then be used to determine the space which this will cover.

ILLUSTRATION: A manuscript of 40,000 words is to be set in 10-point on 12 type. How much space will this cover in square inches and how many pages will it cover if the type page is 6 inches by 4 inches?

From Table 4, 10-point leaded type averages 16 words per square inch. Then, the copy will cover

$$\frac{40,000}{16} = 2,500 \text{ square inches}   \text{(Ans.)}$$

A page 6 inches by 4 inches is 24 square inches. Then the number of type pages will be

$$\frac{2,500}{24} = 104\tfrac{1}{6} \text{ pages}   \text{(Ans.)}$$

TABLE 4

WORDS AND EMS TO THE SQUARE INCH

(Approximate number of words of average length for type of average width)

Size	Old Name	Leaded (2-point)	Solid	No. of Ems
4½ point	Diamond......................	74	98	256
5 "	Pearl.........................	46	66	208
5½ "	Agate.........................	40	60	172
6 "	Nonpareil.....................	32	44	144
7 "	Minion........................	26	34	106
8 "	Brevier.......................	22	30	81
9 "	Bourgeois.....................	19	24	64
10 "	Long Primer..................	16	20	52
11 "	Small Pica....................	14	17	43
12 "	Pica..........................	11	14	36
14 "	English,......................	9	11	26
18 "	Great Primer.................	7	8	16
22 "	Double Small Pica............	4	5	11

Book Paper.—The paper generally used for books, magazines, circulars, catalogues, etc., is designated as *book paper*. This is a broad classification which includes a variety of finishes, colors, and weights, both coated and uncoated. The *substance weight* of paper is the weight in pounds of one *ream* (500 sheets) 25 in. by 38 in. in size. Thus, if a 60-pound paper is specified for a job, it means paper of such weight that 500 sheets of it 25 in. by 38 in. weigh 60 pounds.

The price of book paper used to be quoted by the pound in 1000 sheet lots. This has gone into discard by paper dealers, most of whom now quote prices per pound in ream lots. A higher price is demanded for quantities less than 500 sheets.

The weight of paper required for a job may, of course, be computed by arithmetic if the substance weight and sheet size are known. However, this is needless work since most jobs involve the use of standard sizes, and weights may be obtained from Table 5 for 1000 sheets and from Table 7 for the ream.

TABLE 5

WEIGHT IN POUNDS OF 1000 SHEETS OF BOOK PAPER

Sheet size, inches	Substance weight, pounds										
	30	35	40	45	50	60	70	80	90*	100	120*
22×32	44	52	59	67	74	89	104	119	133	148	178
24×36	54	64	72	82	90	110	128	146	164	182	218
25×38	60	70	80	90	100	120	140	160	180	200	240
26×29	48	56	64	72	80	96	112	126	142	158	190
26×40	66	76	88	98	110	132	154	176	198	218	262
28×42	74	86	100	112	124	148	174	198	222	248	298
28×44	78	90	104	118	130	156	182	208	234	260	312
29×52	96	112	128	144	160	192	224	252	284	316	380
30½×41	78	92	106	118	132	158	184	210	236	264	316
32×44	88	104	118	134	148	178	208	238	266	296	356
33×46	96	112	128	144	160	192	224	256	288	320	384
34×44	94	110	126	142	158	188	220	252	284	314	378
35×45	100	116	132	150	166	198	232	266	298	332	398
36×48	108	128	144	164	180	220	256	292	328	364	436
38×50	120	140	160	180	200	240	280	320	360	400	480
41×61	156	184	212	236	264	316	368	420	472	528	632
42×56	148	172	200	224	248	296	348	396	444	496	596
44×56	156	180	208	232	260	312	364	416	468	520	624
44×64	176	208	236	268	296	356	416	476	532	582	712

*Applies only to coated papers.

Sometimes a paper is required which is not one of the standard sizes included in Tables 5 or 7. In that case the weight may be found by determining the weight per square inch per 1000 sheets of paper of the same substance weight and multiplying this by the *area* of the sheet size in question. The product will be the weight of 1000 sheets of that paper. For example, the area of a standard sized sheet is 25 in. $\times$ 38 in. = 950 sq. in. The weight of 1000 sheets of 60-pound paper is, from Table 5, 120 pounds. Then the weight of 1000 sheets per square inch of this paper is 120 ÷ 950 = 0.12632 pound. Table 6 is a list of the unit weights per 1000 sheets of several substance weights.

TABLE 6

WEIGHT PER SQUARE INCH OF 1000 SHEETS OF BOOK PAPER, POUNDS

Substance weight, pounds	Unit weight, pound
50	0.10526
60	0.12632
70	0.14737
80	0.16842
100	0.21053

ILLUSTRATION: 4000 sheets of 70-pound substance weight paper 32 inches by 48 inches are needed for a job. What is the weight of this paper?

This sheet size does not appear in either Tables 5 or 7. Then Table 6 may be used.

Area of sheet $\qquad = 32 \times 48 = 1536$ sq. in.

Weight of 1000 sheets $= 1536 \times .14737 = 226.4$ pounds

Weight of 4000 sheets $= 4 \times 226.4 = 906$ whole pounds (Ans.)

ILLUSTRATION: Five reams of 70-pound paper 28 inches by 44 inches are needed for a job. What is the actual weight of this paper?

From Table 7, one ream of 70-pound paper, 28 inches by 44 inches weighs 91 pounds. Five reams then weigh.

$$5 \times 91 = 455 \text{ pounds}\quad\text{(Ans.)}$$

Cover Papers.—Cover papers are designated by substance weights which refer to the weight of one ream (500 sheets) of a

TABLE 7

WEIGHT IN POUNDS OF ONE REAM OF BOOK PAPER

Size, inches	Substance weight, pounds												
	25	28	30	35	40	45	50	60	70	80	90	100	120
22×32	....	....	22	26	30	34	37	45	52	60	67	74	89
24×36	23	26	27	32	36	41	45	55	64	73	82	91	109
25×38	25	28	30	35	40	45	50	60	70	80	90	100	120
26×29	20	22	24	28	32	36	40	48	56	63	71	79	95
26×40	27	31	33	38	44	49	55	66	77	88	99	109	131
28×42	31	35	37	43	50	56	62	74	87	99	111	124	149
28×44	33	36	39	45	52	58	65	78	91	104	117	130	156
29×52	40	45	48	56	64	72	80	96	112	126	142	158	190
30½×41	33	37	39	46	53	59	66	79	92	105	118	132	158
32×44	37	42	44	52	59	67	74	89	104	119	133	148	178
33×46	40	45	48	56	64	72	80	96	112	128	144	160	192
34×44	39	44	47	55	63	71	79	94	110	126	142	157	189
35×45	42	47	50	58	66	75	83	99	116	133	149	166	199
36×48	46	51	54	64	72	82	90	110	128	146	164	182	218
38×50	50	56	60	70	80	90	100	120	140	160	180	200	240
41×61	66	74	78	92	106	118	132	158	184	210	236	264	316
42×56	62	70	74	86	100	112	124	148	174	198	222	248	298
44×56	66	73	78	90	104	116	130	156	182	208	234	260	312
44×64	74	83	88	104	118	134	148	178	208	238	266	296	356

TABLE 8

WEIGHT OF 1000 SHEETS OF COVER PAPER, POUNDS

Sheet size, inches	Substance weight, pounds						
	25	35	40	50	65	80	130
20×26	50	70	80	100	130	160	260
23×35	78	109	124	155	201	248	402
26×40	100	140	160	200	260	320	520

sheet size 20 in. by 26 in. They, too, are now usually quoted at so much per pound per 1000 sheets. Table 8 gives the weights per 1000 sheets for three standard sizes and Table 9 the corresponding weights per ream.

TABLE 9

WEIGHT OF ONE REAM OF COVER PAPER, POUNDS

Size, inches	Substance weight, pounds						
	25	35	40	50	65	80	90
20×26	25	35	40	50	65	80	90
23×35	39	54½	62	77½	100½	124	139½
26×40	50	70	80	100	130	160	180

Writing Papers.—Bond, writing and ledger papers are referred to a substance weight per ream of a sheet size 17 in. by 22 in. Table 10 gives the weight per 1000 sheets of the more common sheet sizes of writing papers. Many more sizes than those listed in this table are, of course, manufactured and the weights of odd sizes may be determined by obtaining the unit weights per square inch per 1000 sheets as was illustrated with the book papers.

TABLE 10

WEIGHT OF 1000 SHEETS OF WRITING PAPER, POUNDS

Sheet size, inches	Substance weight, pounds						
	13	16	20	24	28	32	36
17×22	26	32	40	48	56	64	72
17×28	33	41	51	61	71	81	92
19×24	32	39	49	59	68	78	88
22×34	52	64	80	96	112	128	144
24×38	64	78	98	118	137	156	176
28×34	66	82	102	122	143	162	184
34×44	104	128	160	192	224	256	288

Table 11 gives the unit weights per 1000 sheets for the more common substance weights.

TABLE 11

WEIGHT PER SQUARE INCH OF 1000 SHEETS OF WRITING PAPER, POUNDS

Substance weight, pounds	Unit weight, pound
16	0.08556
20	0.10695
24	0.12834
28	0.14973
32	0.17112

TABLE 12

SIZES OF PAPER AND COVER PAPER ACCOMMODATING DIFFERENT PAGE SIZES WITH MINIMUM WASTE

Page size, inches	Sheet size, inches	Number of pages	Cover paper size, inches	Number of covers
$3\frac{1}{4} \times 5\frac{1}{8}$	28×44	8, 16, 32	23×35	18
$3\frac{3}{4} \times 5\frac{1}{8}$	32×44	8, 16, 32	23×35	16
3×6	25×38	24	20×26	12
$3\frac{7}{8} \times 5\frac{3}{8}$	33×46	8, 16, 32	23×35	16
$3\frac{3}{4} \times 7$	32×44	24	23×35	12
$4\frac{1}{2} \times 5\frac{7}{8}$	25×38	8, 16, 32	20×26	8
$3\frac{7}{8} \times 7\frac{1}{4}$	33×46	24	23×35	12
$4 \times 9\frac{1}{8}$	25×38	24	20×26	6
$5 \times 6\frac{5}{8}$	28×42	8, 16, 32	23×35	9
$4\frac{7}{8} \times 7\frac{1}{4}$	$30\frac{1}{2} \times 41$	8, 16, 32	20×26	6
$5\frac{1}{4} \times 6\frac{5}{8}$	28×44	8, 16, 32	23×35	9
$4\frac{1}{2} \times 8$	25×38	24	20×26	6
$5\frac{1}{4} \times 7\frac{1}{2}$	32×44	8, 16, 32	23×35	8
$5\frac{1}{2} \times 7\frac{7}{8}$	33×46	8, 16, 32	23×35	8
$4\frac{3}{4} \times 8\frac{7}{8}$	$30\frac{1}{2} \times 41$	24	20×26	4
$5\frac{1}{4} \times 10\frac{1}{4}$	32×44	24	23×35	9
$6 \times 9\frac{1}{8}$	25×38	8, 16, 32	20×26	4
$6\frac{1}{4} \times 9\frac{1}{4}$	26×29	24	20×26	4
$7\frac{1}{2} \times 10\frac{5}{8}$	32×44	8, 16, 32	23×35	4
$8 \times 11\frac{1}{8}$	33×46	8, 16, 32	23×35	4
$9\frac{1}{4} \times 12\frac{1}{8}$	25×38	8, 16	20×26	2

Selecting Paper.—When a job involves printing more than one page at a time, and particularly if a considerable number of impressions are to be made, the selection of paper of suitable size is of utmost importance. This is not only a question of reducing waste of paper, but also of reducing presswork. Thus, an 8-page booklet may be run through a press printing all eight pages at one time. Then the sheet is reversed and run through again with the pages so arranged that when it is cut in half and folded, two complete booklets result. If the page size can be so fitted to the sheet size that no waste beyond the necessary trim occurs, the greatest economy is effected.

Table 12 has been compiled to aid the printer in selecting a size of paper and cover stock which will accommodate 8, 16, 24, and 32 pages of various sizes with a sufficient allowance for folding and trim. The method of determining the number of pieces or pages which may be obtained from a sheet is to find what multiples or near-multiples the dimensions of the piece are of the dimensions of the sheet. The product of these multiples is the number of pieces which may be obtained. Thus, if we have a sheet 32 in. by 44 in. and wish to find how many pieces $7\frac{1}{2}$ in. by $10\frac{1}{2}$ in. we may obtain from it, we write the dimensions as follows:

$$\frac{\cancel{32} \times \cancel{44}}{\cancel{7\tfrac{1}{2}} \times \cancel{10\tfrac{1}{2}}}$$
$$4 \times 4 = 16 \text{ pieces.}$$

Cancelling out the dimensions of the smaller into those of the larger to find the number of *whole* times they are contained therein, we obtained 4 as the multiple in each case in this example. The product of these, 16, is the number of pieces which may be obtained.

ILLUSTRATION: How many pieces 5 inches by $6\frac{1}{2}$ inches may be obtained from a sheet 28 inches by 42 inches in size?

$$\frac{\cancel{28} \times \cancel{42}}{\cancel{6\tfrac{1}{2}} \times \cancel{5}}$$
$$4 \times 8 = 32 \text{ pieces.} \quad \textbf{(Ans.)}$$

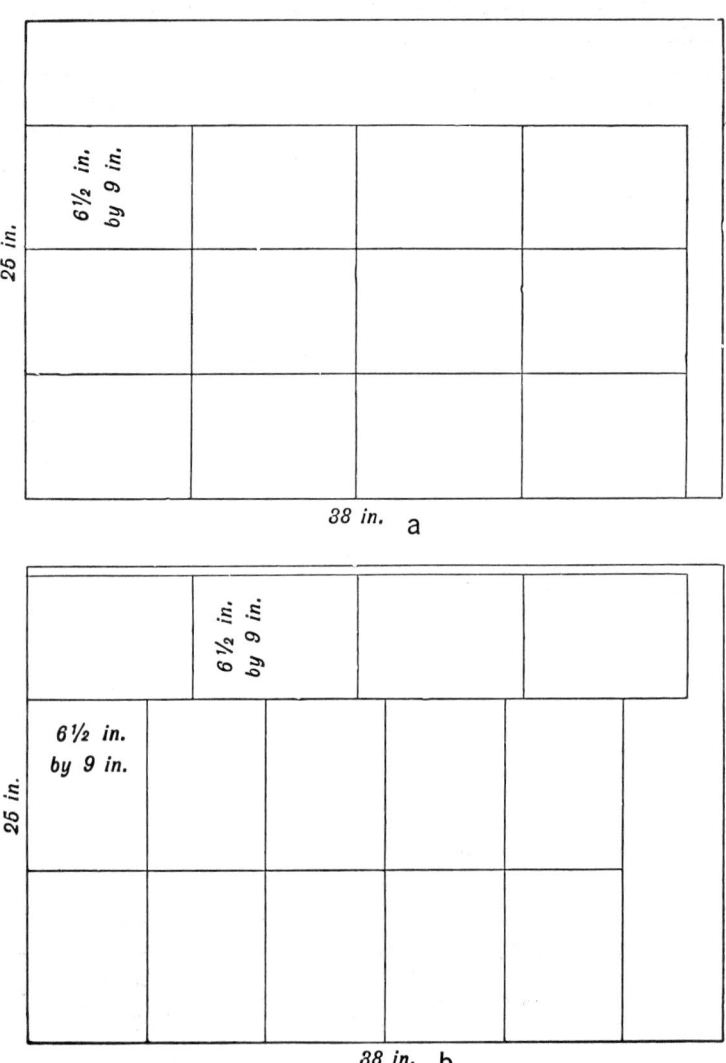

FIG. 3.

The above discussion has concerned itself only with lay-outs which permit straight cuts across the paper in either direction. Sometimes, however, it is necessary to use up a quantity of paper on hand for a certain job which, if it were cut straight across, would entail considerable waste. For example, it is desired to get as many pieces $6\frac{1}{2}$ inches by 9 inches as possible from a sheet 25 inches by 38 inches. By the ordinary computation,

$$\frac{25 \ \times 38}{3 \ \times \ 4} = 12 \text{ pieces},$$
$$6\frac{1}{2} \times \ 9$$

the yield is found to be only 12 pieces as shown in Fig. 3a. However, if we transpose the dimensions of the piece and again cancel, we have,

$$\frac{25 \times 38}{2 \times \ 5} = 10 \text{ pieces.}$$
$$9 \times \ 6\frac{1}{2}$$

There then remains a waste piece 7 inches by 38 inches which is large enough for use. This yields,

$$\frac{7 \ \times 38}{1 \ \times \ 4} = 4 \text{ pieces.}$$
$$6\frac{1}{2} \times \ 9$$

Thus, by cutting the sheet as shown in Fig. 3b, 14 pieces may be obtained.

Paper Allowance for Spoilage.—In each printing and binding operation a certain amount of paper is spoiled for further use. This must be taken into account when ordering stock. As the number of impressions increases the percentage of spoilage decreases. The following are safe values to use in estimating:

Number of copies	Percent spoilage		
	One color	Each add. color	Binding
100 to 250	10	5	5
250 to 500	6	4	4
500 to 1,000	5	$2\frac{1}{2}$	$2\frac{1}{2}$
1,000 to 5,000	$4\frac{1}{2}$	$2\frac{1}{2}$	2
5,000 to 10,000	$3\frac{1}{2}$	$2\frac{1}{2}$	2
Over 10,000	2	2	2

Estimating Quantity of Paper.—ILLUSTRATION: A job calls for 12,000 copies of a 64-page magazine trimmed flush to 6 inches by 9 inches; body stock to be 60-pound machine finished paper; cover stock 80-pound; and one color throughout. How much paper will be required for the job?

Referring to Table 12 we note that a sheet size of 25 in. by 38 in. will accommodate a 6 in. by 9 in. page size economically and conveniently. As we have seen by previous computations, it will take 16 pages on each side or a total of 32 pages. Two such sheets will then be needed for each copy of the magazine. With 12,000 magazines wanted, the sheets needed will be 12,000 × 2 = 24,000. This does not allow for waste. From the foregoing table we note that $3\frac{1}{2}$ percent for printing and 2 percent for binding must be added for waste. Then the total sheets required is,

$$24,000 + 24,000 \times (0.035 + 0.02) = 25,320 \text{ sheets}$$

Referring now to Table 5 we find that this paper weighs 120 pounds per 1000 sheets. Then,

$$120 \times 25.32 = 3038 \text{ pounds paper required.} \quad \text{(Ans.)}$$

From Table 12 we also note that 20 in. by 26 in. cover stock will make 4 covers for a trim size of 6 in. by 9 in. Then, for 12,000 copies, 12,000 ÷ 4 = 3000 sheets will be needed. Again adding a total of $5\frac{1}{2}$ percent for waste for printing and binding we find that it will be prudent to provide

$$3000 + 3000 \times 0.055 = 3165 \text{ sheets}$$

Referring to Table 8 we see that this cover stock weighs 160 pounds per 1000 sheets. Then the weight required will be,

$$160 \times 3.165 = 507 \text{ pounds.} \quad \text{(Ans.)}$$

Illustrations.—When pictures are to be reproduced in letter-press or type printing, a plate called a *halftone* is made by a photo-chemical process which maintains the highlights and shadows of the picture by the use of variable size dots known as a *screen*. These screens, which are placed in the camera between the plate and the object to be photographed, vary in the number of lines to the inch. The standard screens for general commercial work range from 110 lines to the inch to 150 lines to the inch.

A reproduction which shows no highlights or shadows is known as a *line plate* or *line cut* and prints only black solid lines. The purpose of a line cut is to obtain a perfect reproduction of a drawing where lines only are used to suggest the tones that actually exist. The purpose of a halftone is to present a reproduction or drawing in which there are a quantity of tones.

The photoengraver prints the photograph of the original drawing on a smooth metal surface which has been sensitized in a manner similar to photographic paper. The portions which are not to print are eaten away by chemicals, leaving the printing surface composed of a large number of black lines, some large and some small. Tones are produced by the patterns of white and black formed by these lines and the associated white space.

Since photoengraving is a photographic reproduction process, it requires either a drawing, painting, or another photograph as the original material. Reductions in photoengraving are prefer-able because they reduce any imperfections in the same proportion.

Enlargements are undesirable because they increase imperfections of copy. Drawings usually are drawn twice the size of the reproduction, both dimensions being 100 percent larger. Thus a line cut $3'' \times 3''$ would require a drawing $6'' \times 6''$. When the job is made, a one-half reduction in each dimension would be ordered. The 50 percent reduction includes both dimensions.

Offset Printing.—The foregoing discussion of photoengravings and electrotypes refers, of course, to letterpress printing. There are now widely available a number of printing processes in which the plates for printing are made by photographic processes, so that the artwork can be photographed directly, without the necessity for making electrotypes. These processes are known collectively as offset methods, and four well-known varieties of them are Albumin Offset, Deep Etch Offset, Web Offset and Sheet Fed Gravure.

Albumin Offset, so named because albumin plates are used, is best suited to small quantity runs. The reason is that while these plates give good reproduction they do not wear well enough for long runs, and they cannot be stored. In this case, the printer stores the negatives, so that new prints can be made for them for future runs.

Deep Etch Offset uses deep etch plates which have a longer life expectancy than albumin plates, and which may be stored. This method is generally used for runs of intermediate size, approximately 15,000 to 40,000.

Web Offset operates on a web press and is used for runs of large size, most of which are on the order of 100,000 or greater, although the method can be extended to quantities as small as 40,000. It is especially useful where printing is done in more than one color. Unlike the other offset methods, the paper for Web Offset is used in rolls.

Sheet Fed Gravure is used for top quality reproduction of books in limited editions. It is the most costly process and is only used for printing books which are to command a premium price.

Photoengraving.—The dimensions of a desired photoengraving may be found, when both dimensions of the drawing or photograph and one dimension of the plate are known, by using a direct proportion as follows:

$$P_h : P_w = D_h : D_w \quad \text{or} \quad P_h/P_w = D_h/D_w$$

where P_h = height of the plate; P_w = width of the plate; D_h = height of the drawing or photograph; D_w = width of the drawing or photograph.

ILLUSTRATION: A photograph 12″ wide × 6″ high is to be used for a halftone 6″ wide. Find the height of the plate.

$$P_h/P_w = D_h/D_w$$

$$P_h/6 = \tfrac{6}{12}$$

$$P_h = \frac{6 \times 6}{12} = 3$$

Therefore the halftone will be 6″ wide × 3″ high.

When the process is used for line drawings (that is, drawings without gradation of color) the result is called zinc etching. Where such gradation exists, as in photographs, the work is usually done on copper by re-photographing the original photograph or other artwork through a wire screen, which breaks up the light rays so that the metal plate is sensitized in a pattern of dots which vary in size with the darkness of the area in the photograph. Screens vary from 65 to 150 wires ("lines") per inch.

Halftone and line plate photoengravings are priced on the printing face measurement or area of the plate or proof, exclusive of the bevel or margin of the plate as in the case of electrotypes.

The Standard Scale for Photo-Engravers (adopted February 1, 1940) evaluates plate areas from 1 × 1 to 17 × 22 with basic unit values for each ¼″ area. These basic unit values for halftone plates extend from 100 to 1348 and for line plates or zinc from 73 to 809. Some basic unit values are shown in the table on the next page.

Inches	Halftones	Line plates, zinc	Inches	Halftones	Line plates, zinc
3 × 4	128	97	10 × 11	456	270
5 × 6	189	132	11 × 12	528	313
6 × 7	229	151	12 × 13	607	359
7 × 8	275	173	13 × 14	691	409
9 × 10	388	231	14 × 15	784	463

Rule.—To find the selling value of halftones and line plates multiply the basic unit values by the individual shop selling rate.

Illustration: A line plate proof measures $12'' \times 13''$. Find its selling price at \$0.040 per unit.

> From the table a $12'' \times 13''$ area is 359 units
> $359 \times \$0.040 = \$14.36$
> Therefore the selling price is \$14.36

Electrotypes. When a job is to be printed in multiples of the same plate a number of duplicates are made, then locked up in the form and printed on the press. The electrotype is one kind of such duplication. Whenever necessary, a type-set job may be set up several times and duplicated in this manner. Forms set in linotype and monotype have been known to print as many as 25,000 impressions of satisfactory quality. When runs are greater than the accepted standard of impressions from type, electrotypes may be used for duplicates.

Electrotypes are made in several styles, i.e., wax mold, lead mold, nickel types, wax plates, etc. They are sold by the square inch, the area being measured from the printing surface of the plate. When the electrotype surface area is measured from a proof, $\frac{1}{4}''$ is added to provide for a bevel on patent base plates or for a space to drive nails through the electrotype to fasten it to the wooden block.

The Standard Electrotype Scale (adopted 1941) gives complete information regarding the kinds and methods of pricing electrotypes. This scale is built on a ratio unit value of production which is provided for each square inch area of the various kinds of electrotype plates.

Unit values of production extend from 29 for one square inch to 940 for 216 square inches for lead mold steel face electrotypes. Some basic unit values of production are as follows:

Square inches	Line copper	Halftone copper	Halftone steel	Lead mold
1	29	36	44	58
4	36	45	54	72
12	52	65	78	104
24	76	95	114	152
30	89	111	134	178
40	109	136	164	218
44	117	146	176	234
49	128	160	192	256
56	142	178	213	284
64	158	198	237	316

ILLUSTRATION: A copper face electrotype of a type form measures $6\frac{3}{4}'' \times 7\frac{3}{4}''$. If the electrotype has a basic unit value of 142 production units, find its cost at \$0.029 per unit.

$$6\frac{3}{4}'' + \frac{1}{4}'' = 7''$$

$$7\frac{3}{4}'' + \frac{1}{4}'' = 8''$$

$$7 \times 8 = 56$$

From the table, 56 sq. in. is 142 units

$142 \times \$0.029 = \$4.12$

Therefore the cost is \$4.12.

BUSINESS MATHEMATICS

BY

PETER L. AGNEW, A.M., Ed.M.

New York University School of Education
New York, N. Y.

XXIII

Business Mathematics

Business has been defined as "the commercial activity of a community." Naturally, mathematics plays a very important role in the diverse transactions that are executed in this commercial activity.

Invoice.—Perhaps the most common of all business transactions is the buying and selling of commodities, which transaction is generally represented by an invoice which is an itemized list of goods sold by one party to another. The invoice ordinarily carries the following information:

1. Date
2. Name and address of person or firm selling the goods
3. Name and address of person or firm buying the goods
4. Order numbers of both the buyer and the seller
5. Terms and manner of shipment
6. Terms of payment
7. Items, or list of the goods sold, including (*a*) quantity, (*b*) name or brief description of goods sold, (*c*) unit price, (*d*) extension representing the total cost of each article, (*e*) total.

(1) New York, N. Y., Jan. 5, 1955

(2) THE AMERICAN CANDY COMPANY

125 Broadway

NEW YORK CITY

(3) Sold to Fred R. Sterlings   (4) Your order No. 6792
       100 Main Street   Our order No. D873
       Stamford, Connecticut

(5) Delivery: Our Truck   (6) Terms 2/10, n/30

	a	*b*	*c*	*d*
(7)	10	1# Boxes Peppermints	.55	$5.50
	25	2# " Cherries	.84	21.00
				$26.50 *e*

Calculations.—The mathematical phase of the invoice primarily has to do with lower part, the quantity, the unit price, the extension, and the total. In preparing the invoice, the various types of goods covered by the invoice are listed separately with the quantity and unit price of each. The quantity is multiplied by the unit price in order to get what is known as the extension. The extensions are then added to get the total.

ILLUSTRATION: A shoe manufacturer sold a customer 36 pairs of women's pumps at $2.95 per pair, and 36 pairs of women's oxfords at $2.75 per pair. What is the amount of each extension and what is the total?

36 pr.   Women's Pumps @ $2.95 per pair = 36 × 2.95 = $106.20
36 pr.   Women's Oxfords @ $2.75 per pair = 36 × 2.75 =   99.00

                106.20 plus 99.00 $205.20   (Ans.)

Unit Price.—It should be noted that unit prices are sometimes quoted in terms of price per dozen or price per cwt. (hundredweight) or in some other common quantity, while the goods are listed in terms of so many units or so many pounds. Calculations are then slightly more complicated. When price is quoted as so much per special quantity, the price is usually multiplied by the total number of units or pounds and the resulting answer is divided by the number of units in the special quantity, i.e., divided by 100 if the price is quoted per cwt., divided by 2,000 if the price is quoted per ton, etc.

ILLUSTRATION: An invoice lists 6789 lbs. of goods at $6.75 per cwt. What is the amount of the extension?

$6.75 × 6789 = $45,825.75
$45,825.75 ÷ 100 = $458.26   (Ans.)

When the unit price is by the dozen, the quantity is frequently

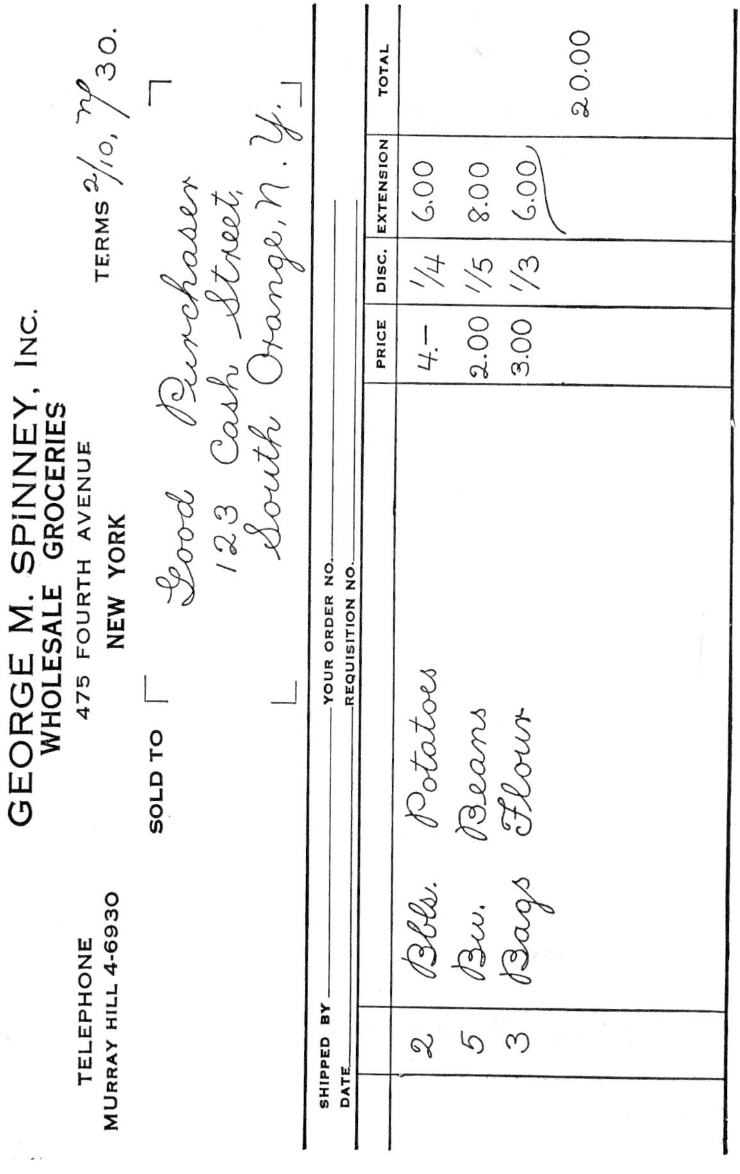

GEORGE M. SPINNEY, INC.
WHOLESALE GROCERIES

475 FOURTH AVENUE
NEW YORK

TELEPHONE
MURRAY HILL 4-6930

TERMS 2/10, n/30.

SOLD TO

Good Purchaser
123 Cash Street,
South Orange, N. Y.

SHIPPED BY _____ YOUR ORDER NO. _____
DATE _____ REQUISITION NO. _____

			PRICE	DISC.	EXTENSION	TOTAL
2	Bbls.	Potatoes	4.—	1/4	6.00	
5	Bu.	Beans	2.00	1/5	8.00	
3	Bags	Flour	3.00	1/3	6.00	
						20.00

Fig. 1.—Invoice

expressed in dozens and fractions thereof.   In such cases, the extension is made in the usual way.

ILLUSTRATION: What is the cost of 12½ doz. black nylon hose @ $8.65 per doz.?

$$\$8.65 \times 12.5 = \$108.13 \quad \text{(Ans.)}$$

Aliquot Parts.—Any number that is contained in another number an equal number of times is called an aliquot part of that number.   Aliquot parts are used extensively in making business calculations, particularly in connection with extensions on invoices as well as discounts and interest calculations.   The aliquot parts of a number are the fractional parts of that number.   The commonly used aliquot parts of the dollar are 50¢ (1/2), 25¢ (1/4), 20¢ (1/5), 16 2/3¢ (1/6), 12½¢ (1/8), 10¢ (1/10), 8 1/3¢ (1/12), 6¼¢ (1/16), 5¢ (1/20), 2¢ (1/50).

ILLUSTRATION: How would the following items of an invoice be calculated using aliquot parts?

$$
\begin{array}{lll}
428 \text{ lbs.} & @\ 25\text{¢} & = \$107.00 \\
192 \text{ lbs.} & @\ 37\tfrac{1}{2}\text{¢} & = \quad 72.00 \\
280 \text{ lbs.} & @\ 70\text{¢} & = \quad 196.00 \\
\hline
\text{Total} & & \$375.00
\end{array}
$$

These extensions should be calculated as follows:

$$
\begin{array}{ll}
25 = \tfrac{1}{4} & \tfrac{1}{4} \text{ of } 428 = \$107.00 \\
37\tfrac{1}{2} = \tfrac{3}{8} & \tfrac{3}{8} \text{ of } 192 = \quad 72.00 \\
70 = \tfrac{7}{10} & \tfrac{7}{10} \text{ of } 280 = \quad 196.00
\end{array}
$$

Invoice and Bill.—These terms are used more or less synonymously; however, the term bill is more frequently applied to a bill for services such as a telephone bill or a lawyer's bill, while the term invoice is almost invariably applied to an itemized listing of goods sold.

Discounts.—Closely associated with invoices are discounts. There are two kinds of commercial discounts: cash discount and trade discount. (There is another type of discount known as "Bank Discount" which is really a form of interest and, therefore, is discussed under the general heading of interest.)

NEW YORK TELEPHONE COMPANY

Addresses: See back of Stub
Telephone: Dial 811 or Call "Business Office"

SEPT. 1, 19

BU 7   C J WOJTAK

4375   1233 FLATBUSH AVE
BROOKLYN 26 N Y

LOCAL SERVICE for One Month in Advance	Message Units Included	75	6 73 *
ADDITIONAL MESSAGE UNITS to Date of Bill	See back of Bill		*
TOLL CALLS AND TELEGRAMS. List enclosed			
OTHER CHARGES OR CREDITS. Explanation enclosed			
BALANCE FROM LAST BILL. Please disregard this amount if paid			

TOTAL   6 73

★ Includes 10% U. S. tax and 3% N. Y. City tax.
See back of bill to determine amount of tax.

FIG. 2.

Cash Discount is a percent of a bill that may be deducted if the bill is paid within a certain specified time. The rate of this discount is stated in the terms of the invoice which includes the rate and the number of days within which the discount may be deducted. Some rather common terms are: 2/10, n/30 (meaning that two percent of the total of the bill may be deducted if it is paid within 10 days. If not paid within 10 days, no discount will be allowed and the full amount of the bill must be paid within 30 days), 5/30, n/60 (meaning that five percent of the total of this bill may be deducted if it is paid within thirty days. If not paid within 30 days, no discount will be allow and the full amount of the bill must be paid within 60 days.)

ILLUSTRATION: The total of an invoice is $897.50 and the terms of payment are 5/30, n/60. How much discount may be deducted and how much must be paid if the invoice is paid within 30 days?

$$
\begin{array}{lr}
\text{Total of invoice} \dots\dots\dots\dots & \$897.50 \\
\text{Less } 5\% \text{ discount} \dots\dots\dots\dots & 44.88 \\
\hline
\text{Net amount to be paid} \dots\dots & \$852.62 \quad \text{(Ans.)}
\end{array}
$$

Applying the principle of aliquot parts mentioned previously, this discount should be calculated as follows:

$$5\% = \tfrac{1}{20} \qquad\qquad \tfrac{1}{20} \text{ of } \$897.50 = \$44.88$$

Trade Discount is a discount granted to a purchaser and is deducted at the time the bill is made out. It is used largely in connection with catalogue and list prices in order that these prices may be brought in line with true market values. Trade discount is also used at times in connection with purchases in large quantities being offered as a special inducement to attract large orders.

These discounts are sometimes in the form of a single rate of discount and sometimes in the form of a series of discounts, each one of the series being deductable from the net amount remaining after the preceding discount has been deducted.

ILLUSTRATION: If an order were placed for 100 hats, the quotation on which was $1.75 less 10%, how would the invoice read?

$$
\begin{array}{llr}
\text{100 Hats} & \$1.75 & \$175.00 \\
& \text{Less } 10\% \text{ discount} & 17.50 \\
& & \hline \\
& \text{Net Amount} & \$157.50 \quad \text{(Ans.)}
\end{array}
$$

Chain Discounts.—Frequently the discount quotation is in the form of a series in which case the quotation might be $1.75 less 25, 10, and 5%. The basic principle involved in chain discounts is that each succeeding discount is based on what is left after the preceding discount is deducted.

ILLUSTRATION: If the quotation on 100 hats was $1.75 less 25, 10, and 5%, what would be the net amount of the invoice? This item could be calculated as follows:

100 Hats	$1.75	$175.00
	Less 25%	43.75
		131.25
	Less 10%	13.13
		118.12
	Less 5%	5.90
		$112.22 (Ans.)

Chain Discount Tables.—Rather than use this long arithmetic process, most business organizations use decimal equivalents for chain discounts. A table of the most common equivalents is shown in Table I. By consulting this table, you will find the decimal equivalent of almost any combination.

ILLUSTRATION: How would the invoice for 100 hats at $1.75 less 25, 10, and 5% be calculated when a table of decimal equivalents is used? By consulting the table of decimal equivalents, one will find that the decimal equivalent of the series 25, 10, and 5%, as listed on the table is 0.64125.

$$100 \text{ less } 0.64125 = 0.35875$$
$$100 \text{ Hats @ } \$1.75 \qquad\qquad = \$175.00$$
$$\text{Less } 0.35875 \ (175 \times 0.35875) = \underline{\quad 62.78}$$
$$\text{Net Amount} \qquad \$112.22$$

Calculating Decimal Equivalents.—When a table of decimal equivalents is not available or when the decimal equivalent of a particular series of chain discounts does not appear on an available table, it may be necessary to calculate the equivalent.

TABLE I.

TABLE OF NET DECIMAL EQUIVALENTS OF CHAIN DISCOUNTS

Multiplying the gross amount by the net decimal equivalent for a chain discount gives the net amount of the invoice. To obtain the discount only, subtract the decimal equivalent given below from 100 and multiply the gross amount by the remainder.

The net equivalent of a chain discount is the same regardless of the sequence of the separate discounts. Example: 60–10–5% is the same as 10–5–60%.

Rate %	5	7½	10	12½	15	16⅔	20	25	30	33⅓	35	37½
	.95	.925	.90	.875	.85	.83333	.80	.75	.70	.66667	.65	.625
2½	.92625	.90188	.8775	.85313	.82875	.8125	.78	.73125	.6825	.65	.63375	.60938
5	.9025	.87875	.855	.83125	.8075	.79166	.76	.7125	.665	.63333	.6175	.59375
5 2½	.87994	.85678	.83363	.81047	.78731	.77187	.741	.69469	.64838	.6175	.60206	.57891
5 5	.85738	.83481	81225	.78969	.76713	.75208	.722	.67688	.63175	.60167	.58663	.56406
5 5 2½	.83594	.81394	.79194	.76995	74795	73328	.70395	.65995	.61596	.58663	.57196	.54996
7½	.87875	.85563	.8325	.80938	.78625	.77083	.74	.69375	.6475	.61667	.60125	.57813
7½ 2½	.85678	.83423	.81169	.78914	.76659	.75156	.7215	.67641	.63131	.60125	.58622	.56367
7½ 5	.83481	.81284	.79088	.76891	.74694	.73229	.703	.65906	.61513	.58583	.57119	.54922
10	.855	.8325	.81	.7875	.765	.75	.72	.675	.63	.6	.585	.5625
10 2½	.83363	.81169	.78975	.76781	.74588	.73125	.702	.65813	.61425	.585	.57038	.54844
10 5	.81225	.79088	.7695	.74813	.72675	.7125	.684	.64125	.5985	.57	.55575	.53438
10 5 2½	.79194	.7711	.75026	.72942	.70858	.69469	.6669	.62522	.58354	.55575	.54186	.52102
10 7½	.79088	.77006	.74925	.72844	.70763	.69375	.666	.62438	.58275	.555	.54113	.52031
10 10	.7695	.74925	.729	.70875	.6885	.675	.648	.6075	.567	.54	.5265	.50625
10 10 5	.73103	.71179	.69255	.67331	.65408	.64125	.6156	.57713	.53865	.513	.50018	.48094
10 10 5 2½	.71275	.69399	.67524	.65648	.63772	.62522	.60021	.5627	.52518	.50018	.48767	.46891

Rate %	40	50	60	62½	65	66⅔	70	75	80	85	87½	90
	.60	.50	.40	.375	.35	.33333	.30	.25	.20	.15	.125	.10
2½	.585	.4875	.39	.36563	.34125	.325	.2925	.24375	.195	.14625	.12188	.0975
5	.57	.475	.38	.35625	.3325	.31667	.285	.2375	.19	.1425	.11875	.095
5 2½	.55575	.46313	.3705	.34734	.32419	.30875	.27788	.23156	.18525	.13894	.11578	.09263
5 5	.5415	.45125	.361	.33844	.31588	.30083	.27075	.22563	.1805	.13538	.11281	.09025
5 5 2½	.52796	.43997	.35198	.32998	.30798	.29331	.26398	.21998	.17599	.13199	.10999	.08799
7½	.555	.4625	.37	.34688	.32375	.30833	.2775	.23125	.185	.13875	.11563	.0925
7½ 2½	.54113	.45094	.36075	.3382	.31566	.30063	.27056	.22547	.18038	.13528	.11273	.09019
7½ 5	.52725	.43938	.3515	.32953	.30756	.29292	.26363	.21969	.17575	.13181	.10984	.08788
10	.54	.45	.36	.3375	.315	.3	.27	.225	.18	.135	.1125	.09
10 2½	.5265	.43875	.351	.32906	.30713	.2925	.26325	.21938	.1755	.13163	.10969	.08775
10 5	.513	.4275	.342	.32063	.29925	.285	.2565	.21375	.171	.12825	.10688	.0855
10 5 2½	.50018	.41631	.33345	.31261	.29177	.27788	.25009	.20841	.16673	.12504	.1042	.08336
10 7½	.4995	.41625	333	.31219	.29138	.2775	.24975	.20813	.1665	.12488	.10406	.08325
10 10	.486	.405	.324	.30375	.2835	.27	.243	.2025	.162	.1215	.10125	.081
10 10 5	.4617	.38475	.3078	.28856	.26933	.2565	.23085	.19238	.1539	.11543	.09619	.07695
10 10 5 2½	.45016	.37513	.30011	.28135	.26259	.25009	.22508	.18757	.15005	.11254	.09378	.07503

From: Instruction Manual, "Burroughs Typewriter Billing Machine," published by Burroughs Adding Machine Company, Detroit, Michigan.

The decimal equivalent of any combination may be calculated by using 100% as the original base, and basing each successive discount on the percent left after the preceding discount has been deducted and finally deducting the final rate from the original 100%.

ILLUSTRATION: What is the decimal equivalent of discount series 25, 10, 5%, and 1% calculated by the above described method?

	100%
Less	25
	75
Less 10%	7.5
	67.5
Less 5%	3.375
	64.125
Less 1%	.64125
	63.48375

100% less 63.48375 = 36.51625% (Ans.)

INTEREST

Interest is money paid for the use of money. The sum upon which the interest is charged, the base amount owed, is called the *principal*. The amount of interest per dollar of principal charged to a borrower or paid to a lender, depends upon (1) the percentage rate per interest period, (2) the length of the period, (3) the time the interest is charged, and (4) the time or times during the loan period when the principal and interest are repaid, in a single payment or in installments.

ILLUSTRATION: A man borrows $1000 for one year, interest to be charged at the rate of 6%. How much interest will be due at the end of the year? What will be the total amount to be paid?

Principal......................... $1000.00

Interest @ 6% (1000 × 0.06).......   60.00

Total Amount............... $1060.00

Bankers' Time.—Most interest calculations are not quite that simple because funds are not usually used for a year; rather are they usually used for a period of days or months, and the interest must be calculated for that length of time. In order to simplify somewhat this calculation, most business organizations, including banks, have adopted the policy of treating the year as if it included 360 days, 12 months of 30 days each. This is usually called bankers' time.

Using bankers' time, one may calculate the interest by multiplying the principal by the number of days that the money was used over 360; by the rate of interest expressed in the form of a fraction. Because of the possibilities for cancelling, this is known as the *cancellation method* of calculating interest.

ILLUSTRATION: $2000 is borrowed for 10 days with interest at the rate of 6% per annum. How much interest must be paid? What amount (principal plus interest) must be paid at the end of 10 days:

$$\$2000 \times \tfrac{10}{360} \times \tfrac{6}{100} = \tfrac{10}{3} = \$3.33 \text{ Interest}$$
$$\$2000 \text{ plus } \$3.33 = \$2003.33 \quad (\text{Amount})$$

60-Day Method.—As suggested previously, most loans are made for a relatively short time. Because of this, business has evolved a simple technique centered around 60 days for calculating interest for short terms. $1000 at interest for one year at the rate of 6% per annum would yield $60.00. For 60 days, (one-sixth of a year $\tfrac{60}{360} = \tfrac{1}{6}$) the yield would be $10.00, $\tfrac{1}{6}$ of $60.00. $10.00 is 1% of $1000 and the same figure could have been determined by merely moving the decimal point two places to the left, $10.00.

Thus we evolve the rule that: To find interest at six per cent for sixty days, move the decimal point two places to the left.

ILLUSTRATION: How much interest must be paid on $1768.47 for 60 days with interest at the rate of 6%?

Interest on $1768.47 for 60 days at 6% = $17.6848 or $17.68.

This was determined merely by moving the decimal point in $1768.47 two places to the left, the result being $17.68.47

Interest for Other Terms.—Interest for terms other than 60 days may be calculated by applying the principle of aliquot parts. The common aliquot parts of 60 are: 30 (1/2), 20 (1/3), 15 (1/4), 12 (1/5), 10 (1/6), 6 (1/10), 5 (1/12), 4 (1/15). Interest is first determined for 60 days and then the proper fractional part or combination of fractional parts is determined.

ILLUSTRATION: $875.00 is borrowed for 30 days with interest at the rate of 6% per annum. What is the amount of the interest?

Interest on $875.00 for 60 days at 6%   = $8.75
30 days equals ½ of 60 days.
Interest on $875.00 for 30 days (½ of $8.75) = $4.38   (Ans.)

Interest at Other Rates.—Quite frequently the rate of interest is not 6% but some other rate agreed upon by the parties involved. One method of calculating this interest is by applying the principle of aliquot parts. The aliquot parts of six are 3 (1/2), 2 (1/3), 1½ (1/4), 1 (1/6), ½ (1/12). In calculating interest at a rate other than 6%, the interest is first calculated at 6% by the 60-day method and then the proper fractional part is determined from that.

ILLUSTRATION: $1000 was borrowed for 30 days at 8%. What is the amount of the interest?

Interest on $1000 @ 6% for 60 days = $10.00

Interest on $1000 @ 6% for 30 days =   5.00
Interest on $1000 @ 2% (1/3 of 6%) =   1.67

Interest on $1000 @ 8%   = $ 6.67   (Ans.)

Interest Tables.—If much of a firm's business involves interest, precomputed tables are used to avoid the necessity of calculating the interest for every transaction. Table II shows simple interest on amounts from $1.00 to $9.00 for various periods of time and at various rates. In using this table to find the interest on a given principal at a given rate, one should

a. Run down the side of the table until he comes to the given rate.

b. If the principal in question is divisible to one figure by 10 or a multiple of 10, use the resulting quotient as a basic principal, that is, for $900 use 9, for $60 use 6, for $8000 use 8. If the principal is an odd number use one.

c. After selecting the basic principal in the correct interest rate group, follow along the line to the left until you reach the column headed by the number of days or months for which you are computing the interest.

d. Multiply the figure thus found by the true principal if you are using one for a base principal or move the decimal point to right the correct number of times if you are using a one-figure quotient determined by dividing by 10, or a multiple of 10.

ILLUSTRATION: $500 is borrowed for 20 days with interest at 5%. What amount of interest will have to be paid?

Using the interest table:

a. Run down the side of the table to the 5% section.

b. As $500 divided by 100 equals 5, use $5 as a basic principal.

c. Following along the $5 line to the 20-day column, it will be noted that the interest on $5 at 5% for 20 days equals 0.01388.

d. Moving the decimal point two places to the right to multiply by 100, it will be found that interest on $500 at 5% for 20 days equals $1.38. (Ans.)

ILLUSTRATION: $463.75 was borrowed for 3 months with interest at 7%. What amount of interest will have to be paid?

TABLE 2

SIMPLE INTEREST

Rate	Principal	Time.									
		1 Year	6 Mo.	5 Mo.	4 Mo.	3 Mo.	2 Mo.	1 Mo.	20 d.	10 d.	1 d.
	$1	.040	.0200	.01666$^{\vee}$6	.013$^{\vee}$3	.01000	.0066$^{\vee}$6	.00333$^{\vee}$3	.0022$^{\vee}$2	.00111$^{\vee}$1	.000111$^{\vee}$1
	2	.080	.0400	.03333$^{\vee}$3	.026$^{\vee}$6	.02000	.0133$^{\vee}$3	.00666$^{\vee}$6	.0044$^{\vee}$4	.00222$^{\vee}$2	.000222$^{\vee}$2
	3	.120	.0600	.05000	.040	.03000	.0200	.01000	.0066$^{\vee}$6	.00333$^{\vee}$3	.000333$^{\vee}$3
	4	.160	.0800	.06666$^{\vee}$6	.053$^{\vee}$3	.04000	.0266$^{\vee}$6	.01333$^{\vee}$3	.0038$^{\vee}$8	.00444$^{\vee}$4	.000444$^{\vee}$4
4%	5	.200	.1000	.08333$^{\vee}$3	.066$^{\vee}$6	.05000	.0333$^{\vee}$3	.01666$^{\vee}$6	.0111$^{\vee}$1	.00555$^{\vee}$5	.000555$^{\vee}$5
	6	.240	.1200	.10000	.080	.06000	.0400	.02000	.0133$^{\vee}$3	.00666$^{\vee}$6	.000666$^{\vee}$6
	7	.280	.1400	.11666$^{\vee}$6	.093$^{\vee}$3	.07000	.0466$^{\vee}$6	.02333$^{\vee}$3	.0155$^{\vee}$5	.00777$^{\vee}$7	.000777$^{\vee}$7
	8	.320	.1600	.13333$^{\vee}$3	.106$^{\vee}$6	.08000	.0533$^{\vee}$3	.02666$^{\vee}$6	.0177$^{\vee}$7	.00888$^{\vee}$8	.000888$^{\vee}$8
	9	.360	.1800	.15000	.120	.09000	.0600	.03000	.0200	.01000	.001000
	$1	.045	.0225	.01875	015	.01125	.0075	.00375	.0025	.00125	.000125
	2	.090	.0450	.03750	.030	.02250	.0150	.00750	.0050	.00250	.000250
	3	.135	.0675	.05625	.045	.03375	.0225	.01125	.0075	.00375	.000375
	4	.180	.0900	.07500	.060	.04500	.0300	.01500	.0100	.00500	.000500
4½%	5	.225	.1125	.09375	.075	.05625	.0375	.01875	.0125	.00625	.000625
	6	.270	.1350	.11250	.090	.06750	.0450	.02250	.0150	.00750	.000750
	7	.315	.1575	.13125	.105	.07875	.0525	.02625	.0175	.00875	.000875
	8	.360	.1800	.15000	.120	.09000	.0600	.03000	.03000	.01000	.001000
	9	.405	2025	.16875	.135	.10125	0675	.03375	.0225	01125	.001125
	$1	.050	.0250	.02083$^{\vee}$3	.016$^{\vee}$6	.01250	.0083$^{\vee}$3	.00416$^{\vee}$6	.0027$^{\vee}$7	.00138$^{\vee}$8	.000138$^{\vee}$8
	2	.100	.0500	.04166$^{\vee}$6	.033$^{\vee}$3	.02500	.0166$^{\vee}$6	.00833$^{\vee}$3	.0055$^{\vee}$5	.00277$^{\vee}$7	.000277$^{\vee}$7
	3	.150	.0750	.06250	.050	.03750	.0250	.01250	.0083$^{\vee}$3	.00416$^{\vee}$6	.000416$^{\vee}$6
	4	.200	.1000	.08333$^{\vee}$3	.066$^{\vee}$6	.05000	.0333$^{\vee}$3	.01666$^{\vee}$6	.0111$^{\vee}$1	.00555$^{\vee}$5	.000555$^{\vee}$5
5%	5	.250	.1250	.10416$^{\vee}$6	.083$^{\vee}$3	.06250	.0416$^{\vee}$6	.02083$^{\vee}$3	.0138$^{\vee}$8	.00694$^{\vee}$4	.000694$^{\vee}$4
	6	.300	.1500	.12500	.100	.07500	.0500	.02500	.0166$^{\vee}$6	.00833$^{\vee}$3	.000833$^{\vee}$3
	7	.350	.1750	.14583$^{\vee}$3	.116$^{\vee}$6	.08750	.0583$^{\vee}$3	.02916$^{\vee}$6	.0194$^{\vee}$4	.00972$^{\vee}$2	.000972$^{\vee}$2
	8	.400	.2000	.16666$^{\vee}$6	.133$^{\vee}$3	.10000	.0666$^{\vee}$6	.03333$^{\vee}$3	0222$^{\vee}$2	.01111$^{\vee}$1	.001111$^{\vee}$1
	9	.450	.2250	.18750	.150	.11250	.0750	.03750	.0250	.01250	.001250
	$1	.060	.0300	.02500	.020	.01500	.0100	.00500	.0033$^{\vee}$3	.00166$^{\vee}$6	.000166$^{\vee}$6
	2	.120	.0600	.05000	.040	.03000	.0200	.01000	.0066$^{\vee}$6	.00333$^{\vee}$3	.000333$^{\vee}$3
	3	.180	.0900	.07500	.060	.04500	.0300	.01500	.0100	.00500	.000500
	4	.240	.1200	.10000	.080	.06000	.0400	.02000	.0133$^{\vee}$3	.00666$^{\vee}$6	.000666$^{\vee}$6
6%	5	.300	.1500	.12500	.100	.07500	.0500	.02500	.0166$^{\vee}$6	.00833$^{\vee}$3	.000833$^{\vee}$3
	6	.360	.1800	.15000	.120	.09000	.0600	.03000	.0200	.01000	.001000
	7	.420	.2100	.17500	.140	.10500	.0700	.03500	.0233$^{\vee}$3	.01166$^{\vee}$6	.001166$^{\vee}$6
	8	.480	.2400	.20000	.160	.12000	.0800	.04000	.0266$^{\vee}$6	.01333$^{\vee}$3	.001333$^{\vee}$3
	9	.540	.2700	.22500	.180	.13500	.0900	.04500	.0300	.01500	.001500
	$1	.070	.0350	.02916$^{\vee}$6	.023$^{\vee}$3	.01750	.0116$^{\vee}$6	.00583$^{\vee}$3	.0038$^{\vee}$8	.00194$^{\vee}$4	.000194$^{\vee}$4
	2	.140	.0700	.05833$^{\vee}$3	.046$^{\vee}$6	.03750	.0233$^{\vee}$3	.01166$^{\vee}$6	.0077$^{\vee}$7	.00388$^{\vee}$8	.000388$^{\vee}$8
	3	.210	.1050	.08750	.070	.05250	.0350	.01750	.0116$^{\vee}$6	.00583$^{\vee}$3	.000583$^{\vee}$3
	4	.280	.1400	.11666$^{\vee}$6	.093$^{\vee}$3	.07000	.0466$^{\vee}$6	.02333$^{\vee}$3	.0155$^{\vee}$5	.00777$^{\vee}$7	.000777$^{\vee}$7
7%	5	.350	.1750	.14583$^{\vee}$3	.116$^{\vee}$6	.08750	.0583$^{\vee}$3	.02916$^{\vee}$6	.0194$^{\vee}$4	.00972$^{\vee}$2	.000972$^{\vee}$2
	6	.420	.2100	.17500	.140	.10500	.0700	.03500	.0233$^{\vee}$3	.01166$^{\vee}$6	.001166$^{\vee}$6
	7	.490	.2450	.20416$^{\vee}$6	.163$^{\vee}$6	.12250	.0816$^{\vee}$6	.04083$^{\vee}$3	.0272$^{\vee}$2	.01361$^{\vee}$1	.001361$^{\vee}$1
	8	.560	.2800	.23333$^{\vee}$3	.186$^{\vee}$6	.14000	.0933$^{\vee}$3	.04666$^{\vee}$6	.0311$^{\vee}$1	.01555$^{\vee}$5	.001555$^{\vee}$5
	9	.630	.3150	.26250	.210	.15750	.1050	.05250	0350	.01750	.001750

* Note that *all* repeating decimals may be extended indefinitely. Thus, the interest on $1.00 at 4% for 4 months is given as .013$^{\vee}$3 or 1⅓ cents, because the decimal .013$^{\vee}$3 = .01333333. ∴.; hence the interest on $1,000,-000, at the same rate and for the same time, is $13,333.33⅓. Decimals which are not repeating decimals are *exact*.

Using the interest table:

a. Run down the side of the table to the 7% section.

b. As this principal may not be reduced to a single figure by dividing by 10 or a multiple of 10, use $1.00 as a basic principle.

c. Following along the $1.00 line to the 3 months column, it will be noted that interest on $1.00 at 7% for 3 months equals 0.01750.

d. Multiplying this amount by $463.75, the true principal, it will be found that interest on $463.75 at 7% for 3 months equals $8.115625 or $8.12.   (Ans.)

If the number of days in a given problem does not appear in the table, the amount of interest for various numbers of days may be combined; thus, interest for 70 days may be determined by adding together the interest for 60 days (2 months) and the interest for 10 days.

ILLUSTRATION: $5000 was borrowed for 80 days with interest at 6%.   What amount of interest must be paid when the obligation is due?

Using the interest table:

a. Run down the side of the table to the 6% section.

b. Eliminate the zeros by pointing off 3 places and thus adopt $5 as the basic principal.

c Follow along the $5 line to the 2 months (60-day) column and note that

Interest on $5 at 6%		= 0.0500

also that in the 20-day column

Interest on $5 at 6%		= 0.016666

Therefore   Interest on $5 @ 6% for 80 days = 0.066666

d. Move the decimal point 3 places to the right to multiply by 1000, and thus we find that

Interest on $5000 @ 6% for 80 days = $66.6666, or $66.67 (Ans.)

If the number of days does not readily lend itself to such combinations, it is frequently more simple to find the interest for one day and then multiply by the number of days.

ILLUSTRATION: $750 was borrowed for 17 days with interest @ 7%. What amount of interest must be paid when the obligation is due?

Using the table:

a. Run down the column to the 7% section.

b. As the principal cannot be reduced to one figure, use $1 as the basic principal.

c. Follow along the $1 line to the 1 day column and note that

Interest on $1 at 7% for 1 day $= 0.0001944$

Therefore

Interest on $1 @ 7% for 15 days

(0.0001944×15) $= 0.0029160$

d. The interest on $750 at 7% for 15 days equals $750 \times 0.0029160 = 2.187 or $2.19 (Ans.)

It may be noted in the foregoing illustrations in which the interest table was used, the calculations in some instances were rather awkward. This is due to the fact that the particular table being used is not necessarily the best for all interest computations. Firms making use of precomputed interest tables will usually have those that best fit their particular needs.

Legal and Lawful Rates of Interest.—The legal rate of interest is the rate that may legally be charged in the absence of any definite agreement between the parties. This is particularly true of judgments and overdue accounts where interest is to be charged but it may also apply in other situations where interest is applicable but where no specific rate has been agreed upon.

The lawful rate (sometime called the contract rate) is the maximum rate that can be charged when a definite agreement has been made. In some states the legal and the lawful rates are the same. In other states they vary widely, while in still other states certain conditions are attached to the contract rate. The charging of a rate of interest above the lawful or contract rate is known as "usury," which in some states is a crime, in others a misdemeanor. In either case, it is punishable by a variety of penalties. New York and Maryland do not permit corporations to plead "usury" as a defense.

Automobile Loans.—As in the case of smaller consumer loans, the borrower of money toward the purchase of an automobile is not charged an interest rate computed for each payment period on the unpaid balance at the beginning of that period. Instead, the amount he is to pay over the period is taken from tables, such as Table 2A, and he makes a uniform payment at each period as is shown in the second column of this table. These tables are computed for various rates of interest. This particular table is based on the very low return to the bank of 5½ percent. However, by reference to the figures in the table, such as the finance charge of $55.24 per year for $1000 to be paid during the year, it can be computed that the actual interest rate is between 10 and 11 percent. This does not mean, however, that the lending institution earns that percentage on the money loaned. The overhead costs of these loans include the costs of preparing and sending periodic bills, the cost of bookkeeping for periodical payments and part of the costs of repossession and resale of automobiles on which the payments are defaulted. Thus, of the $55.24 annual charge the interest earned by the lending institution is only 5.5 percent, even on the unpaid balances, a sum which can be calculated to amount to $28.13. The difference between the $55.24 and this figure, which is $27.11, represents these costs of making and collecting the automobile loans of the institution. Note that in this table one of these charges, the life insurance premium on the owner of the car, which must be carried by the finance institution, is shown separately, but all the other costs, including the insurance policy, which the bank must carry on the automobile itself, are contained in the charge of $27.11.

Personal Finance.—Of course, smaller loans, especially those made on articles which deteriorate more rapidly than automobiles, carry higher overhead charges. Another reason for this is the fact that the smaller the loan the higher the proportion

TABLE 2A

SAMPLE AUTOMOBILE LOAN SCHEDULE

Un-paid Bal.	12 Months				18 Months			
	Amt. Per Mo.	Amt. of Note	Life Prem.	Finance Charge	Amt. Per Mo.	Amt. of Note	Life Prem.	Finance Charge
1	.08	.96	.00	⁻.04	.06	1.08	.01	.07
2	.17	2.04	.01	.03	.12	2.16	.02	.14
3	.26	3.12	.02	.10	.18	3.24	.03	.21
4	.35	4.20	.03	.17	.24	4.32	.04	.28
5	.44	5.28	.03	.25	.30	5.40	.04	.36
10	.88	10.56	.06	.50	.60	10.80	.08	.72
15	1.32	15.84	.08	.76	.90	16.20	.12	1.08
20	1.76	21.12	.11	1.01	1.21	21.78	.16	1.62
25	2.20	26.40	.13	1.27	1.51	27.18	.20	1.98
30	2.65	31.80	.16	1.64	1.81	32.58	.24	2.34
35	3.09	37.08	.19	1.89	2.12	38.16	.28	2.88
40	3.53	42.36	.21	2.15	2.42	43.56	.31	3.25
45	3.97	47.64	.24	2.40	2.72	48.96	.35	3.61
50	4.41	52.92	.26	2.66	3.03	54.54	.39	4.15
55	4.86	58.32	.29	3.03	3.33	59.94	.43	4.51
60	5.30	63.60	.32	3.28	3.63	65.34	.47	4.87
65	5.74	68.88	.34	3.54	3.93	70.74	.51	5.23
70	6.18	74.16	.37	3.79	4.24	76.32	.55	5.77
75	6.62	79.44	.39	4.05	4.54	81.72	.59	6.13
80	7.06	84.72	.42	4.30	4.84	87.12	.62	6.50
85	7.51	90.12	.45	4.67	5.15	92.70	.66	7.04
90	7.95	95.40	.47	4.93	5.45	98.10	.70	7.40
95	8.39	100.68	.50	5.18	5.75	103.50	.74	7.76
1,000	88.37	1,060.44	5.20	55.24	60.60	1,090.80	7.75	83.05
1,100	97.21	1,166.52	5.72	60.80	66.66	1,199.88	8.52	91.36
1,200	106.04	1,272.48	6.24	66.24	72.72	1,308.96	9.30	99.66
1,300	114.88	1,378.56	6.76	71.80	78.78	1,418.04	10.07	107.97
1,400	123.72	1,484.64	7.28	77.36	84.84	1,527.12	10.85	116.27
1,500	132.56	1,590.72	7.80	82.92	90.90	1,636.20	11.62	124.58
1,600	141.39	1,696.68	8.32	88.36	96.96	1,745.28	12.40	132.88
1,700	150.23	1,802.76	8.84	93.92	103.02	1,854.36	13.17	141.19
1,800	159.07	1,908.84	9.36	99.48	109.08	1,963.44	13.95	149.49
1,900	167.90	2,014.80	9.88	104.92	115.14	2,072.52	14.72	157.80

TABLE 2A (Cont.)

Un-paid Bal.	12 Months				18 Months			
	Amt. Per Mo.	Amt. of Note	Life Prem.	Finance Charge	Amt. Per Mo.	Amt. of Note	Life Prem.	Finance Charge
2,000	176.74	2,120.88	10.40	110.48	121.20	2,181.60	15.49	166.11
2,100	185.58	2,226.96	10.92	116.04	127.26	2,290.68	16.27	174.41
2,200	194.42	2,333.04	11.44	121.60	133.33	2,399.94	17.04	182.90
2,300	203.25	2,439.00	11.96	127.04	139.39	2,509.02	17.82	191.20
2,400	212.09	2,545.08	12.48	132.60	145.45	2,618.10	18.59	199.51
2,500	220.93	2,651.16	13.00	138.16	151.51	2,727.18	19.37	207.81
2,600	229.77	2,757.24	13.52	143.72	157.57	2,836.26	20.14	216.12
2,700	238.60	2,863.20	14.03	149.17	163.63	2,945.34	20.92	224.42
2,800	247.44	2,969.28	14.55	154.73	169.69	3,054.42	21.69	232.73
2,900	256.28	3,075.36	15.07	160.29	175.75	3,163.50	22.47	241.03
3,000	265.12	3,181.44	15.59	165.85	181.81	3,272.58	23.24	249.34
3,100	273.95	3,287.40	16.11	171.29	187.87	3,381.66	24.01	257.65
3,200	282.79	3,393.48	16.63	176.85	193.93	3,490.74	24.79	265.95
3,300	291.63	3,499.56	17.15	182.41	199.99	3,599.82	25.56	274.26
3,400	300.46	3,605.52	17.67	187.85	206.05	3,708.90	26.34	282.56
3,500	309.30	3,711.60	18.19	193.41	212.11	3,817.98	27.11	290.87
3,600	318.14	3,817.68	18.71	198.97	218.17	3,927.06	27.89	299.17
3,700	326.98	3,923.76	19.23	204.53	224.23	4,036.14	28.66	307.48
3,800	335.81	4,029.72	19.75	209.97	230.29	4,145.22	29.44	315.78
3,900	344.65	4,135.80	20.27	215.53	236.35	4,254.30	30.21	324.09
4,000	353.49	4,241.88	20.79	221.09	242.41	4,363.38	30.98	332.40

formed by these costs. This situation can be shown clearly by an illustration.

ILLUSTRATION: A purchase is made of furniture totalling $580.50. The agreement is that one-third of the total is to be paid at the time of purchase and the remainder is to be paid off in monthly installments. The rate on the unpaid balance is 8% and it is to be handled by a finance company. One-third of $580.50 ($193.50) was paid when the contract was executed, leaving a balance of $387 to be paid in monthly installments. How much must be paid monthly?

The true rate of interest (i.e., the rate without other charges) can be computed approximately by assuming payments of $\frac{1}{12}$ of $387 ($32.25) plus accrued interest each month for the year. This would work out as follows:

	Principal		*Interest*		
1st	$32.25	+	$0.22	=	$32.47
2nd	32.25	+	0.43	=	32.68
3rd	32.25	+	0.65	=	32.90
4th	32.25	+	0.86	=	33.11
5th	32.25	+	1.08	=	33.33
6th	32.25	+	1.29	=	33.54
7th	32.25	+	1.51	=	33.76
8th	32.25	+	1.72	=	33.97
9th	32.25	+	1.94	=	34.19
10th	32.25	+	2.15	=	34.40
11th	32.25	+	2.37	=	34.62
12th	32.25	+	2.58	=	34.83
	$387.00		$16.80		$403.80

In the standard method of computing finance charges, however, a flat 8% is charged on the entire balance of $387. This amounts to $30.96 in interest so that the customer pays this sum instead of $16.80, a difference of $14.16 which represents the expenses and profit to the lending institution. Compare this figure of $14.16 excess on a $387 advance for one year against the $27.11 charge by the bank for nearly three times as much money for the same period, and it will be seen how much greater the charges are on the smaller loans and the more perishable security. Such credit loans, by the way, are not usually handled by banks.

Small Loans.—While in the two examples cited the excess finance charge over normal interest, which represents the expenses of the lending institution, were not excessive, it is obvious and is often true that small loan rates have been excessive in the past and are often so at present. For that reason, a number of states in the United States and provinces of Canada have established regulations governing the amount which may be charged on various types of small loans. These regulations have been summarized for a few states in Table 2B.

TABLE 2B

SMALL LOAN REGULATIONS SUMMARIZED FOR FOUR STATES

MASSACHUSETTS—6%/no limit

Tender Act: A right to prepay with int. at 18% p.a., plus $5.00.
Ceiling $1,000

Loan & Investment Cos.... Morris or similar plan companies.

Less than $500,   12% discount.
More than $500,   9% discount.

Retail Sales Finance....... Disclosure.
Refund: Rule of 78ths, after deducting acq. chg.
  Motor Vehicles (other than those subject to motor vehicle sales finance statute)—$12.50
Other goods—$5.00

Motor Vehicle Sales
   Finance............... Retail Instalment Sales of Motor Vehicles

N. & U. C., model year of year of
  sale or prior $8/$100
N. & U. C., not in above and not
  more than 2 yrs. old $10/$100
O. U. C. $12/$100
Del. chg: $5 or 5%
Refund: Rule of 78ths, after deducting acq. chg.
  of $12.50

NEW YORK—6%/6%

Bank Instalment Loans

$6/$100 discount.   Ceiling $5,000.
Max. 25 mos. if $1,200 or less.
37 mos. max. over $1,200, not in excess of $5,000.
37 mos. max. for R. P. improvement, min. $10.
Del. chg: 5¢/$1, max. $5 or an aggregate of 2% of loan, but not in excess of $25.
Refund: Rule of 78ths, subject to $10 min. chg.

Loan & Investment Cos.... Industrial Banks

$6/$100 discount.   Ceiling $5,000.
25 mos. max., if less than $1,200.
37 mos. if more than $1,200 or if R. P.
Min. chg: $10.
Fee: $1/$50 up to $250.   Max: $5.
More than $250: $5 plus 1% of excess above $250.   Max: $20.
Del. chg: 5¢/$1.   Max: $5 per instalment.
  Total Max: 2% or $25.
Refund: Pro-rata as of following date, subject to $10 min. chg.

TABLE 2B (Cont.)

Retail Sales Finance....... Retail Instalment Sales Act
(All Goods Act)

Unpaid principal balance $500 $10/$100 p.a.
Unpaid principal balance over $500 8/$100 p.a.
Revolving credit: on that amount not in excess
  of $500—1½% per mo.
Del. chg: 5% or $5.
Refund: Rule of 78ths
Merchandise certificate refund: pro-rata.

Motor Vehicle Sales
Finance...............

N. C., current model $7/$100
N. & U. C., not more than 2 yrs. old $10/$100
O. U. C. $13/$100
Insurance premium $7/$100
Del. chg: 5% or $5.
Ext. or Refinance chg: service fee not to exceed
  $5, plus total additional chg. not to exceed
  1% per mo.
Refund: Rule of 78ths, after deducting acq.
  chg. $15.
Refinance chg: 1% per mo. or refund and charge
  as per contrat on unpaid balance.

NOTE: In this table, p.a. means per annum, N.C. means new car, and
U.C. means used car.

Commercial Partial Payments.—A partial payment is an amount that is not sufficient to liquidate an indebtedness. The finance plan for financing installment sales and the small loan payments already discussed are merely partial payment plans that apply in personal financing. Where business organizations borrow and make partial payments, other methods are applied. When such partial payments are made on interest-bearing items, a problem arises as to the amount due at the time of final settlement. There are two rules that are commonly followed: (1) the *Merchants' Rule,* and (2) the *United States Rule.*

Merchants' Rule.—Under this rule interest is charged on the principal for the full time and is credited on the payments from the date of each payment to the date of final payment. The interest on the principal less the interest credited on the periodic payments equals the interest charged.

ILLUSTRATION: On May 1, a man borrowed $5000 to be paid back at the rate of $1000 each month. The interest rate is 6%.

He pays $1000 on the first of June, July, August, and September. Applying the Merchants' Rule, how much must be paid on October 1 to settle the account?

Interest on $5000 for 5 months, May 1 to Oct. 1, @ 6% = $125.00

Interest credited as follows:

On $1000 for 4 months (from June 1 to Oct. 1) @ 6%, $20.00
On $1000 for 3 months (from July 1 to Oct. 1) @ 6%,   15.00
On $1000 for 2 months (from Aug. 1 to Oct. 1) @ 6%,   10.00
On $1000 for 1 month (from Sept. 1 to Oct. 1) @ 6%,   5.00

Total interest credit......................   50.00

Interest due October 1.....................   75.00
Unpaid principal October 1.................   1000.00

$1075.00

United States Rule.—Under this rule, all interest accrued on the unpaid balance is deducted from the payment before the remainder is deducted from the principal or that part of the principal that is still unpaid at the time payment was made.

ILLUSTRATION: Applying the United States Rule to the problem cited in the Illustration under Merchants' Rule, how much would the man have to pay on October 1 to settle his account:

Original Principal............................. $5000.00
Payment on June 1............................. $1000.00
Less int. on $5000 @ 6% for 1 mo.............   25.00   975.00

                                                                    4025.00
Payment on July 1............................. $1000.00
Less int. on $4025 @ 6% for 1 mo.............   20.13   979.87

                                                                    3045.13
Payment on August 1........................... $1000.00
Less int. on $3045.13 @ 6% for 1 mo...........   15.23   984.77

                                                                    2060.36
Payment on September 1........................ $1000.00
Less int. on $2060.36 @ 6% for 1 mo...........   10.30   989.70

                                                                    1070.66
Interest on $1070.66 @ 6% for 1 mo............   5.35

Amount due October 1............... $1076.01

TABLE 3

This table shows nine methods of borrowing $1000, showing the sources, arrangements made, amount at borrower's disposal, total finance charge, which includes interest, apparent and real interest rate paid, and disadvantages of each main method of borrowing.

Sources	Terms	Total Finance Charge You Pay	Interest Rate		Advantages	Disadvantages
			Real	Apparent		
1. Personal Loan Company	Loan must be repaid in monthly installments. Usually, borrower must be employed, or have collateral—such as a car or jewelry—worth the value of the loan. $500 is top amount you can borrow in many states.	$127.00	26.2%	12.7%	Informal atmosphere. Usually easy to get new loans once your credit has been established by promptly paying back one loan.	High interest rate; large amount of interest. By repaying in monthly installments, you steadily cut down the amount of loan money you can use.
2. Life Insurance Policy (Excluding term insurance)	Amount you can borrow grows larger each year you keep policy in force. This "Loan Value" is specified in policy—as is interest rate. You can repay whenever you wish—and in part or whole amount.	60.00	6.0%	6.0%	Low rate and mount of interest; easy availability; no limit on length of loan; defaulted interest payments merely added to amount of loan.	It takes 10 years before the average ordinary life policy taken out at age 25 will have a loan value of $1000.
3. Automobile Installment Finance Company	Available to help finance the purchase of new or used autos —or to pay for repairs or accessories.	92.31	18.7%	9.2%	Readily available, at time of car purchase. Gives dealer an added reason to give you good service.	Moderately high interest rate and amount of interest. By repaying in monthly installments you steadily cut down the loan money you can use.

TABLE 3 (Cont.)

Sources	Terms	Total Finance Charge You Pay	Interest Rate		Advantages	Disadvantages
			Real	Apparent		
4. Bank Personal Loan Dept. (Monthly repayment—collateral)	Borrower turns over good collateral—such as stocks, bonds, or chattel mortgage on car. Loan must be repaid in monthly installments. Typical face value of loan is $43.72, but the $43.72 is deducted in advance.	43.72	6.0%	8.9%	Moderate rate and total interest for this type of loan. Can be used to finance purchase of new or used car.	Ties up your stocks or bonds in bank's hands. By repaying in monthly installments, you steadily cut down amount of loan money you can use.
5. Bank Personal Loan Dept. (Lump sum repayment—collateral	Borrower turns over good collateral—such as stocks or bonds. Lump sum repayment at end of year permitted.	76.50	7.65%	7.65%	Low interest rate; low total amount of interest. Whole amount of loan available to you for whole year.	Your collateral tied up in bank and may have to be sold at an unfavorable time—if loan comes due and you have no other way to repay it.
6. Mortgage on House	Your house is security for loan, which must be repaid in monthly installments.	36.40	3.6%	7.5%	Low interest rate and total amount of interest.	By repaying in monthly installments you steadily cut down the amount of loan money you can use.
7. Credit Union	Membership in the credit union is necessary. Loan must be repaid in monthly installments. Typical face value of loan is $1,000, but $40 is deducted in advance.	50.00	4.0%	8.1%	Easy availability to members of the group.	Moderately high cost. By repaying in monthly installments you steadily cut down the amount of loan money you can use.

Mortgages.—A mortgage is a loan on property where conditional title to the property is given as a pledge of repayment. The payments on the mortgage and the interest charges are fixed according to the interest rate and the period of payment agreed upon with the lender. Most modern mortgages are based on a constant monthly payment plan (see Table 5, Plan III).

In addition to the mortgage payments it is usually possible to make other payments such as taxes, assessments, insurance, and upkeep, part of the same monthly payment plan. The bank or savings and loan association holds this money in escrow (the money is accumulated in a special fund). When the taxes or insurance come due they are paid by the bank for the homeowner.

Table 4 shows the major expenditures in financing a home after down payments and preliminary costs are paid. Expenses are assumed to be $18 for taxes and assessments, $3 for insurance, and $20 for upkeep for every $1000 loaned on the house.

TABLE 4

ANNUAL HOME OWNERSHIP COST FOR EACH $1000 BORROWED *

Payment Period	4% interest			5% interest			6% interest		
	10 years	15 years	20 years	10 years	15 years	20 years	10 years	15 years	20 years
Interest and payment on each $1000 of loan per year † (based on a systematic loan reduction plan).................	$122	$ 89	$ 73	$127	$ 95	$ 79	$133	$101	$ 86
Taxes and assessments....	18	18	18	18	18	18	18	18	18
Insurance...............	3	3	3	3	3	3	3	3	3
Upkeep.................	20	20	20	20	20	20	20	20	20
Total annual outlay on each $1000 borrowed...	$163	$130	$114	$168	$136	$120	$174	$142	$127

* University of Ill. Small Homes Council Bulletin A1.3.

† To find the total payment on interest and principal to maturity for each $1000 of loan, multiply the annual payment by the number of years in the payment period. Subtract the $1000 principal from the total payment to find the total interest to maturity for each $1000 of loan.

TABLE 5

FOUR PLANS FOR REPAYMENT OF $10,000 IN 10 YEARS WITH INTEREST AT 6% *

	End of Year	Interest Due (6% of money owed at start of year)	Total Money Owed Before Year-End Payment	Year-End Payment	Money Owed After Year-End Payment
	0				$10,000
	1	$600	$10,600	$ 600	10,000
	2	600	10,600	600	10,000
	3	600	10,600	600	10,000
	4	600	10,600	600	10,000
Plan I	5	600	10,600	600	10,000
	6	600	10,600	600	10,000
	7	600	10,600	600	10,000
	8	600	10,600	600	10,000
	9	600	10,600	600	10,000
	10	600	10,600	10,600	0
	0				$10,000
	1	$600	$10,600	$1,600	9,000
	2	540	9,540	1,540	8,000
	3	480	8,480	1,480	7,000
	4	420	7,420	1,420	6,000
Plan II	5	360	6,360	1,360	5,000
	6	300	5,300	1,300	4,000
	7	240	4,240	1,240	3,000
	8	180	3,180	1,180	2,000
	9	120	2,120	1,120	1,000
	10	60	1,060	1,060	0
	0				$10,000.00
	1	$600.00	$10,600.00	$1,358.68	9,241.32
	2	554.48	9,795.80	1,358.68	8,437.12
	3	506.23	8,943.35	1,358.68	7,584.67
	4	455.08	8,039.75	1,358.68	6,681.07
Plan III	5	400.86	7,081.93	1,358.68	5,723.25
	6	343.40	6,066.65	1,358.68	4,707.98
	7	282.48	4,990.45	1,358.68	3,631.77
	8	217.91	3,849.68	1,358.68	2,491.00
	9	149.46	2,640.46	1,358.68	1,281.78
	10	76.90	1,358.68	1,358.68	0.00
	0				$10,000.00
	1	$ 600.00	$10,600.00	$ 0.00	10,600.00
	2	636.00	11,236.00	0.00	11,236.00
	3	674.16	11,910.16	0.00	11,910.16
	4	714.61	12,624.77	0.00	12,624.77
Plan IV	5	757.49	13,382.26	0.00	13,382.26
	6	802.94	14,185.20	0.00	14,185.20
	7	851.11	15,036.31	0.00	15,036.31
	8	902.18	15,938.49	0.00	15,938.49
	9	956.31	16,894.80	0.00	16,894.80
	10	1,013.69	17,908.49	17,908.49	0.00

* Grant, E. L., *Principles of Engineering Economy,* Revised Edition, The Ronald Press.

New Note Method.—Some banks in handling this problem avoid some of the involved calculation by having the debtor pay his thousand dollars each month plus accrued interest and give a new note for the balance. This greatly simplifies the problem for both the bank and the borrower.

ILLUSTRATION: Using the same problem, find the amount to be paid and the amount of the new note to be given at the end of each month.

		Principal	Interest	Amount Paid	New Note
May 1.............	Borrowed	$5000			
June 1.............	Paid		1000 + $25.00 =	$1025	$4000
July 1.............	Paid		1000 + 20.00 =	1020	3000
August 1...........	Paid		1000 + 15.00 =	1015	2000
September 1........	Paid		1000 + 10.00 =	1010	1000
October 1..........	Paid		1000 + 5.00 =	1005	0

Total Interest.................... $75.00

Series of Notes.—Still another method of handling this matter is by having the borrower make out five $1000 notes bearing interest at 6%, one due each month. This procedure is even more simple than the new note plan.

ILLUSTRATION: Still using the same problem, assume that the borrower of the $5000 was asked to make out a series of five $1000 notes each bearing interest at 6%. How much must be paid when each note is due?

			Interest	Amount
May 1............	Borrowed	$5000		
June 1............	Paid	1000	$5.00	$1005
July 1............	Paid	1000	10.00	1010
August 1..........	Paid	1000	15.00	1015
September 1.......	Paid	1000	20.00	1020
October 1.........	Paid	1000	25.00	1025
			$75.00	$1075

Relative Merits.—There is relatively little difference among the methods treated. The United States Method gives a slightly larger interest return to the lender than does the Merchants' Method; in the problem used to illustrate these various methods, this difference amounted to $1.01. Because of this, the United States Method is usually used where large sums are involved. In considering the relative merits of the *new note* and the *series of notes* plans, it should be noted that while the interest paid under the two methods is $75.00 in each case, the same as under the Merchants' Rule, if one considers the present worth of the interest in relation to the final due date of the obligation, one perceives that the "series of notes" plan tends in the direction of the Merchants' Rule, while the New Note Method plan approximately equals the United States Rule.

Negotiable Instruments.—Because interest is so closely associated with certain negotiable instruments, it seems advisable to give them some brief attention at this point. A negotiable instrument is usually defined as being an instrument the legal title of which may be passed from one party to another by endorsement and delivery or merely by delivery. According to the New York Negotiable Instruments Law which is standard, basic factors that make a business paper negotiable are: (1) It must be in writing signed by the one who is to pay, (2) It must contain an unconditional promise or order to pay a certain sum in money, (3) It must be payable on demand or at a fixed or determinable future time, (4) It must be payable to order or to bearer, (5) Where the instrument is addressed to a drawee, he must be named or otherwise indicated therein with reasonable certainty.

Negotiable Instruments differ from other contracts in two rather vital respects: (1) as to quality of the title, and (2) as to consideration. When a person receives title to a negotiable instrument in the absence of any knowledge of any infirmity in the title of the person delivering that title to him, he receives a good valid title. In ordinary contracts, the title passes by assignment and the assignee becomes subject to all the defenses that may exist between the original parties.

All contracts must have consideration, but in the case of negotiable instruments this quality is conclusively presumed between all others than the original parties.

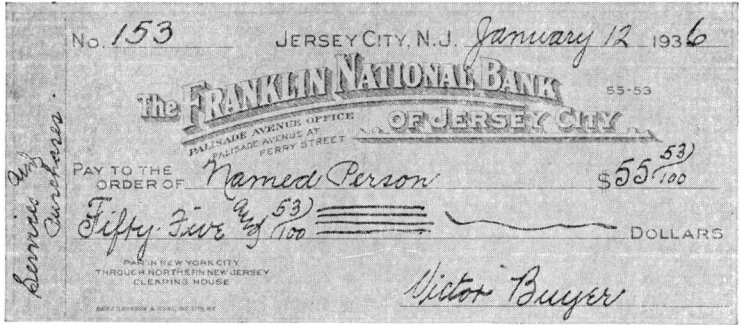

Fig. 3.—Check

Instruments of Exchange.—Broadly speaking, negotiable instruments fall into two classifications: (1) Instruments of Exchange, and (2) Instruments of Credit. An instrument of exchange

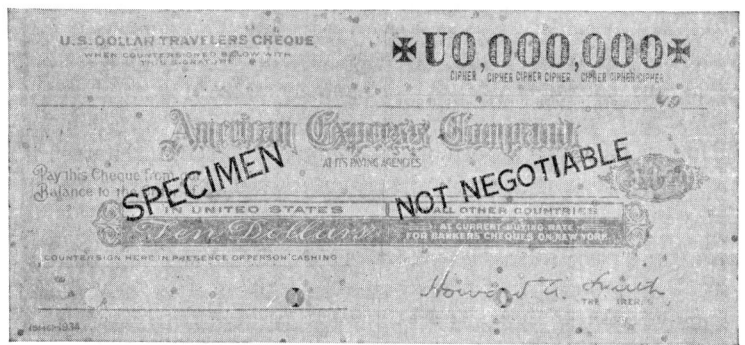

Fig. 4.—Travelers' Check

is an unconditional order in writing to pay to the order of a specified person or to bearer a certain sum of money. An instrument of exchange is used to transfer money without actually exchanging the cash, and bears no interest. The most commonly used negotiable instruments that fall into this category are: (1) check, (2) cashier's check, (3) certified check, (4) bank draft, (5) Post Office money order, (6) Express money order, and (7) travelers' check.

Instruments of Credit.—This type of instrument may be defined as being an agreement to pay at a later date a fixed sum of money to the order of a specified person or to bearer. It must be in writing and must be signed by the person who is to pay. This type of instrument is used in connection with various types of deferred payments and frequently, although not always, bears

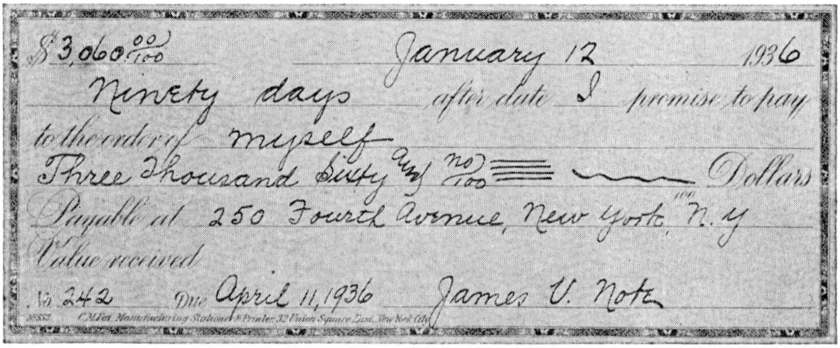
$3,060 00/100 January 12 1936
Ninety days after date I promise to pay
to the order of myself
Three Thousand Sixty no/100 Dollars
Payable at 250 Fourth Avenue, New York, N.Y.
Value received
No. 242 Due April 11, 1936 James V. Note

FIG. 5.—Promissory Note

interest. The most commonly used negotiable instruments that fall in this category are: (1) promissory notes, (2) commercial drafts, (3) trade acceptances, and (4) bonds. Interest on notes and other forms of negotiable instruments that bear interest is calculated the same as any ordinary interest, usually by using the 60-day method or by using an interest table.

ILLUSTRATION: On August 15, $2500 is borrowed on a 90-day note bearing interest at 6%. What amount must be paid when the note is due:

Principal (Face of the Note)		$2500.00
Interest on $2500 @ 6% for 60 days	$25.00	
Interest on $2500 @ 6% for 30 days	12.50	
90		37.50
Total Amount to be Paid when Note is Due		$2537.50 (Ans.)

Bank Discount.—Promissory notes and other forms of instruments of exchange are used in connection with credit operations (1) between merchandising and industrial organizations as well as between persons, (2) between such individuals and business organizations and banks. If a person or firm receives a note, draft, or trade acceptance from another, he may hold it until it is due and then collect the face plus the interest if it happened to be an interest-bearing draft. If, however, he would like to have the money before it is due he may take it to the bank and receive an amount equal to its present value. This is known as discounting the paper at the bank.

Discounting a Non-Interest-Bearing Note.—The process of bank discount involves five steps (1) determining the value of the paper at maturity (when it is due), (2) determining the date of maturity, (3) counting the exact number of days between the day that the paper is taken over by the bank (called the day of discount) and the date of maturity. This is known as the term of discount. (4) Calculating the discount (really interest) for the term of discount based on the value at maturity, (5) determining the Net Proceeds by deducting the discount from the value at maturity.

ILLUSTRATION: On May 2, Harold Jones receives a 60-day non-interest-bearing note for $750 from one of his customers. He holds it until May 17 and then takes it to the bank and discounts it. The rate of discount at the bank is 6%. What is the net proceeds?

The five steps are as follows:

(1) Value at Maturity. In the case of a non-interest-bearing note, only the face of the note is due at maturity. In this case, the value at maturity is $750.

(2) Date of Maturity is the due date of the note. This note is due 60 days after May 2. There are 29 more days in May. Twenty-nine plus 30, in June, makes 59. Fifty-nine plus one in July makes 60. Therefore, the date of maturity is July 1. It might be noted here that when a note reads days, days are counted,

if it reads months, months are counted. If this had read "two months" the due date would be two months after May 2, or July 2. As it read 60 days, the due date is July 1.

(3) Term of Discount is unexpired time, the exact number of days between the date of discount and the date of maturity. As this note was discounted on May 17, there are 14 more days in May, 30 in June, and one in July, a total of 45 days. This could have been readily ascertained by consulting Table 6, a table for finding the number of days between dates.

TABLE 6

For Finding Number of Days Between Any Two Dates in Two Consecutive Years.*

	First Year.													Second Year.											
Day Mo.	Jan.	Feb.	March.	April.	May.	June.	July.	Aug.	Sept.	Oct.	Nov.	Dec.	Day Mo.	Jan.	Feb.	March.	April.	May.	June.	July.	Aug.	Sept.	Oct.	Nov.	Dec.
1	1	32	60	91	121	152	182	213	244	274	305	335	1	366	397	425	456	486	517	547	578	609	639	670	700
2	2	33	61	92	122	153	183	214	245	275	306	336	2	367	398	426	457	487	518	548	579	610	640	671	701
3	3	34	62	93	123	154	184	215	246	276	307	337	3	368	399	427	458	488	519	549	580	611	641	672	702
4	4	35	63	94	124	155	185	216	247	277	308	338	4	369	400	428	459	489	520	550	581	612	642	673	703
5	5	36	64	95	125	156	186	217	248	278	309	339	5	370	401	429	460	490	521	551	582	613	643	674	704
6	6	37	65	96	126	157	187	218	249	279	310	340	6	371	402	430	461	491	522	552	583	614	644	675	705
7	7	38	66	97	127	158	188	219	250	280	311	341	7	372	403	431	462	492	523	553	584	615	645	676	706
8	8	39	67	98	128	159	189	220	251	281	312	342	8	373	404	432	463	493	524	554	585	616	646	677	707
9	9	40	68	99	129	160	190	221	252	282	313	343	9	374	405	433	464	494	525	555	586	617	647	678	708
10	10	41	69	100	130	161	191	222	253	283	314	344	10	375	406	434	465	495	526	556	587	618	648	679	709
11	11	42	70	101	131	162	192	223	254	284	315	345	11	376	407	435	466	496	527	557	588	619	649	680	710
12	12	43	71	102	132	163	193	224	255	285	316	346	12	377	408	436	467	497	528	558	589	620	650	681	711
13	13	44	72	103	133	164	194	225	256	286	317	347	13	378	409	437	468	498	529	559	590	621	651	682	712
14	14	45	73	104	134	165	195	226	257	287	318	348	14	379	410	438	469	499	530	560	591	622	652	683	713
15	15	46	74	105	135	166	196	227	258	288	319	349	15	380	411	439	470	500	531	561	592	623	653	684	714
16	16	47	75	106	136	167	197	228	259	289	320	350	16	381	412	440	471	501	532	562	593	624	654	685	715
17	17	48	76	107	137	168	198	229	260	290	321	351	17	382	413	441	472	502	533	563	594	625	655	686	716
18	18	49	77	108	138	169	199	230	261	291	322	352	18	383	414	442	473	503	534	564	595	626	656	687	717
19	19	50	78	109	139	170	200	231	262	292	323	353	19	384	415	443	474	504	535	565	596	627	657	688	718
20	20	51	79	110	140	171	201	232	263	293	324	354	20	385	416	444	475	505	536	566	597	628	658	689	719
21	21	52	80	111	141	172	202	233	264	294	325	355	21	386	417	445	476	506	537	567	598	629	659	690	720
22	22	53	81	112	142	173	203	234	265	295	326	356	22	387	418	446	477	507	538	568	599	630	660	691	721
23	23	54	82	113	143	174	204	235	266	296	327	357	23	388	419	447	478	508	539	569	600	631	661	692	722
24	24	55	83	114	144	175	205	236	267	297	328	358	24	389	420	448	479	509	540	570	601	632	662	693	723
25	25	56	84	115	145	176	206	237	268	298	329	359	25	390	421	449	480	510	541	571	602	633	663	694	724
26	26	57	85	116	146	177	207	238	269	299	330	360	26	391	422	450	481	511	542	572	603	634	664	695	725
27	27	58	86	117	147	178	208	239	270	300	331	361	27	392	423	451	482	512	543	573	604	635	665	696	726
28	28	59	87	118	148	179	209	240	271	301	332	362	28	393	424	452	483	513	544	574	605	636	666	697	727
29	29	...	88	119	149	180	210	241	272	302	333	363	29	394	...	453	484	514	545	575	606	637	667	698	728
30	30	...	89	120	150	181	211	242	273	303	334	364	30	395	...	454	485	515	546	576	607	638	668	699	729
31	31	...	90	...	151	...	212	243	...	304	...	365	31	396	...	455	...	516	...	577	608	...	669	...	730

* Subtract the number opposite the first date from the number opposite the last. If the 29th of February is included, add one day.

(4) Discount is really interest based on the value at maturity for the terms of discount. Bank discount is calculated precisely the same as is interest:

Interest on $750 @ 6% for 60 days............ $7.50
Interest on $750 @ 6% for 15 days............. 1.875

Interest on $750 @ 6% for 45 days............. $5.625 or $5.63

(5) Net Proceeds is the amount due after the discount has been deducted from the value at maturity. In this case, $750 less $5.63, or $744.37 (Ans.)

Discounting an Interest-Bearing Note.—The only difference between discounting an interest-bearing note and a non-interest bearing note is in the value at maturity. In a non-interest bearing note, the value at maturity is face only, in an interest bearing note the value at maturity is the face plus interest for the full life of the note.

ILLUSTRATION: On April 15, the Jones Manufacturing Company received a 90-day note from one of its customers. The note was for $1200 with interest at 6%. On May 1, the Jones Company discounted it at the bank. What is the Net Proceeds?

The five steps are as follows:

1. Value at Maturity: Interest on $1200 for 90 days is $18.00. The value at maturity is $1200 plus $18.00, $1218.00.

2. Date of maturity:

April	15 more days
May	31 more days
June 30	30 more days
	76
July	14 due date
	90

3. Terms of discount:

By actual count

May 1–31	30 days
June	30
July	14

$\overline{74}$ days

By using Table 6

July 14	195
May 1	121

$\overline{74}$ days

4. Discount:

Interest on \$1218 @ 6% for 60 days....\$12.18
Interest on \$1218 @ 6% for 12 days.... 2.436
Interest on \$1218 @ 6% for 2 days....   .406

$\overline{}$
\$15.022 = \$15.02

5. Proceeds:

Value at maturity.......	\$1218.00
Less Discount..........	15.02
Net Proceeds.....	\$1202.98 (Ans.)

Exact Interest.—Various financial organizations when dealing with each other and governments as a general rule use the exact or "accurate" method of calculating interest. In this method the 365 day year (in leap year 366) is used as the time basis rather than the 360-day year so-called bankers' time. When large financial transactions are involved, the slight five- or six-day inaccuracy of bankers' time makes a decided difference. The amount of interest is determined by finding the exact number of days that the obligation remained unpaid and then multiplying the principal by the exact number of days over 365 by the rate of interest expressed in fractional form. Cancellation may be applied if possible.

ILLUSTRATION: The state and county taxes of the City of Willbum amounting to $347,689 were due and payable on June 30. The city was unable to meet this obligation until October 1, at which time payment was made in full, plus accrued interest at the rate of 6%. Find: (a) the amount of exact interest on the obligation; (b) the amount of interest if it were calculated on the basis of bankers' time (360-day year); (c) which is greater, and by how much:

(a) The obligation was due June 30 and paid October 1. Using Table 6 it may be noted that the exact time between these two dates is

$$
\begin{array}{ll}
\text{October 1} & 274 \\
\text{June 30} & 181 \\
\hline
& 93 \text{ days}
\end{array}
$$

The exact interest equals

$$\$347{,}689 \times \frac{93}{365} \times \frac{6}{100} = \$5315.36 \quad (\text{Ans.})$$

(b) If bankers' time had been used, this would have been calculated as follows:

Interest @ 6% for 60 days....... $3476.89
Interest @ 6% for 30 days....... 1738.445
Interest @ 6% for 3 days....... 173.8445

$5389.1795 = $5389.18   (Ans.)

(c) The difference in the interest figured by the two methods equals:

Interest calculated on Bankers' Time.......... $5389.18
Interest calculated on Exact Time............. 5315.36

Interest calculated on the basis of Bankers'
Time greater than interest calculated on basis
of Exact Time by........................ $73.82   (Ans.)

Compound Interest.—Interest that is earned on other interest earned in previous periods and added to the principal is called compound interest. Interest may be compounded annually, semi-annually, quarterly, or at even more frequent intervals.

ILLUSTRATION: A man deposits $500 on January 2, 1954, in a savings bank which pays interest at the rate of 4% per annum, compounded quarterly. Assume that the quarters correspond with the calendar year and that interest is credited to accounts as of March 31, June 30, September 30, and December 31. If the account was allowed to stand for two years, how much would be on deposit at the end of that time? This would work out as follows:

	Principal	Interest	Amount
January 1, 1954, Deposit..	$500.00	. . . . .	. . . .
March 31, 1954...........	500.00	$5.00	$505.00
June 30, 1954............	505.00	5.05	510.05
September 30, 1954.......	510.05	5.10	515.15
December 31, 1954........	515.15	5.15	520.30
March 31, 1955...........	520.30	5.20	525.50
June 30, 1955............	525.50	5.25	530.75
September 30, 1955.......	530.75	5.30	536.05
December 31, 1955.......	536.05	5.36	541.41 (Ans.)

Compound interest earned over period of two years equals $41.41.

Pre-computing Compound Interest.—At times, an individual is interested for one reason or another in knowing how much a given sum of money might build up to if left at interest for a period of years. This may be calculated by

(1) Adding the interest rate per interest period to $1.00 and multiplying this by itself as many times as there are interest periods in the whole term of years.

(2) Multiplying this product by the amount to be deposited in the first place, the original principal, to ascertain the new amount. Because such problems usually involve a large number

of interest periods, compound interest tables are generally used. Such tables give the amount that $1.00 will amount to at compound interest for any given number of periods at various periodic rates. Table 7 is a compound interest table. To use it, determine the number of interest periods, (a) follow down the left column until that figure is reached, (b) follow the line across to the column headed by the periodic rate, (c) multiply the number thus determined by the principal.

ILLUSTRATION: A man deposits $1200 in a bank which pays interest at the rate of 4% per annum compounded semi-annually. If the deposit is allowed to remain in the bank, how much will have accumulated at the end of 15 years?

(a) If interest is paid semi-annually at the rate of 4% per annum, the semi-annual rate or periodic rate is 2%.

(b) If interest is paid semi-annually, there are two interest periods per year. In fifteen years, there are thirty interest periods.

(c) Turning to Table 7, it will be noted that interest on $1.00 compounded for 30 periods at 2% = 1.81134.

(d) Principal $1200.00
$\times$ 1.81134
————————
$2173.60800 = $2173.61 (Ans.)

Calculated by Logarithms.—Compound interest may also be computed by using logarithms. This method is frequently used when compound interest tables are not available or when the periodic interest rate is now shown in tables that are available.

The formula followed when using logarithms is:

Sum = Amount Deposited $\times$ (1 + periodic interest rate)
number of periods

That is: $S = x(1 + i)^n$

TABLE 7

Compound Interest Table

Amount of $1 at compound interest for periods 1 to 50 at various *periodic rates.

Periods. n.	*Periodic Rate							
	2%	3%	3½%	4%	4½%	5%	6%	7%
1	1.02000	1.03000	1.03500	1.04000	1.04500	1.05000	1.06000	1.07000
2	1.04040	1.06090	1.07123	1.08160	1.09203	1.10250	1.12360	1.14490
3	1.06121	1.09273	1.10872	1.12486	1.14117	1.15763	1.19102	1.22504
4	1.08243	1.12551	1.14752	1.16986	1.19252	1.21551	1.26248	1.31080
5	1.10408	1.15927	1.18769	1.21665	1.24618	1.27628	1.33823	1.40255
6	1.12616	1.19405	1.22926	1.26532	1.30226	1.34010	1.41852	1.50073
7	1.14869	1.22987	1.27228	1.31593	1.36086	1.40710	1.50363	1.60578
8	1.17166	1.26677	1.31681	1.36857	1.42210	1.47746	1.59385	1.71819
9	1.19509	1.30477	1.36290	1.42331	1.48610	1.55133	1.68948	1.83846
10	1.21899	1.34392	1.41060	1.48024	1.55297	1.62889	1.79085	1.96715
11	1.24337	1.38423	1.45997	1.53945	1.62285	1.71034	1.89830	2.10485
12	1.26824	1.42576	1.51107	1.60103	1.69588	1.79586	2.01220	2.25219
13	1.29361	1.46853	1.56396	1.66507	1.77220	1.88565	2.13293	2.40985
14	1.31948	1.51259	1.61870	1.73168	1.85194	1.97993	2.26090	2.57853
15	1.34587	1.55797	1.67535	1.80094	1.93528	2.07893	2.39656	2.75903
16	1.37279	1.60471	1.73399	1.87298	2.02237	2.18287	2.54035	2.95216
17	1.40024	1.65285	1.79468	1.94790	2.11338	2.29202	2.69277	3.15882
18	1.42825	1.70243	1.85749	2.02582	2.20848	2.40662	2.85434	3.37993
19	1.45681	1.75351	1.92250	2.10685	2.30786	2.52695	3.02560	3.61653
20	1.48595	1.80611	1.98979	2.19112	2.41171	2.65330	3.20714	3.86968
21	1.51567	1.86029	2.05943	2.27876	2.52024	2.78596	3.39957	4.14057
22	1.54598	1.91610	2.13151	2.36991	2.63365	2.92523	3.60354	4.43041
23	1.57690	1.97358	2.20611	2.46471	2.75217	3.07152	3.81976	4.74054
24	1.60844	2.03279	2.28332	2.56330	2.87602	3.22510	4.04894	5.07237
25	1.64061	2.09378	2.36324	2.66583	3.00544	3.38635	4.29188	5.42744
26	1.67342	2.15659	2.44595	2.77246	3.14068	3.55567	4.54939	5.80736
27	1.70689	2.22129	2.53156	2.88336	3.28201	3.73346	4.82224	6.21388
28	1.74103	2.28792	2.62016	2.99870	3.42970	3.92013	5.11170	6.64885
29	1.77585	2.35656	2.71187	3.11864	3.58406	4.11614	5.41840	7.11427
30	1.81134	2.42726	2.80672	3.24339	3.74532	4.32194	5.74351	7.61227
31	1.84759	2.50003	2.90501	3.37312	3.91386	4.53804	6.08812	8.14513
32	1.88454	2.57508	3.00670	3.50805	4.08998	4.76494	6.45340	8.71529
33	1.92224	2.65233	3.11193	3.64837	4.27403	5.00319	6.84061	9.32536
34	1.96068	2.73190	3.22085	3.79430	4.46637	5.25335	7.25115	9.97813
35	1.99989	2.81386	3.33358	3.94608	4.66735	5.51600	7.68611	10.6766
36	2.03989	2.89827	3.45025	4.10392	4.87738	5.79182	8.14728	11.4240
37	2.08069	2.98518	3.57101	4.26806	5.09686	6.08141	8.63611	12.2236
38	2.12230	3.07478	3.69599	4.43880	5.32618	6.38548	9.15428	13.0793
39	2.16475	3.16702	3.82535	4.61635	5.56590	6.70475	9.70354	13.9948
40	2.20804	3.26203	3.95924	4.80100	5.81637	7.03999	10.2855	14.9745
41	2.25221	3.35989	4.09781	4.99306	6.07811	7.39199	10.9029	16.0227
42	2.29725	3.46069	4.24124	5.19276	6.35162	7.76159	11.5571	17.1443
43	2.34320	3.56451	4.38968	5.40047	6.63744	8.14967	12.2505	18.3444
44	2.39006	3.67144	4.54332	5.61649	6.93613	8.55715	12.9855	19.6285
45	2.43786	3.78159	4.70233	5.84115	7.24826	8.98504	13.7647	21.0025
46	2.48662	3.89503	4.86692	6.07480	7.57443	9.43426	14.5906	22.4727
47	2.53635	4.01188	5.03726	6.31779	7.91528	9.90597	15.4660	24.0458
48	2.58708	4.13224	5.21356	6.57050	8.27146	10.4013	16.3939	25.7290
49	2.63882	4.25621	5.39604	6.83330	8.64368	10.9213	17.3776	27.5300
50	2.69160	4.38389	5.58491	7.10665	9.03265	11.4674	18.4202	29.4571

* Periods may be annual, semi-annual or quarterly, etc. Periodic rates are proportioned to the length of the period. Thus, 4% annual = 2% semi-annual rate.

ILLUSTRATION: $642.80 was to be left on deposit for 12 years at a bank paying interest at the rate of 3½% compounded semi-annually. What amount will be on deposit at the end of 12 years?

Interest for 12 years at 3½% compounded semi-annually means that there will be 24 interest periods at the rate of 1¾% per period; therefore, the amount at maturity (S) will be

$$S = 642.80 \times (1.0175)^{24} = \log 642.80 + 24 \times \log 1.0175$$

$$\log 642.80 = 3.808076$$
$$12 \log\ 1.0175 = 0.180840$$
$$\log S = 3.988916$$

$$\log 1.0175 = 0.007535$$
$$\times\quad 24$$
$$0.030140$$
$$0.015070$$
$$0.180840$$

$$S = \$974.80 \quad (\text{Ans.})$$

Interest on Bank Deposits.—It should be noted that there probably would be a slight discrepancy between the amount as worked out in the preceding solution and the amount as built up by the bank over the years. This would be due to the fact that banks usually ignore cents in the principal in calculating the interest at the end of each period.

Some other factors pertaining to interest in bank deposits that might be noted here are:

(1) While most interest is earned in savings accounts only, some banks pay interest on checking accounts. This practice varies widely, it usually being paid only when a reasonably good daily balance is maintained varying in different banks from $500 to $5000. The rate is usually 4% per annum.

(2) Savings banks usually have rules whereby money deposited on or before a specified day in the month, as the 5th or 10th, shall draw interest from the first of the month. Deposits made after that date will draw interest from the first of the following month.

(3) Money usually has to be on deposit for a minimum of three months before any interest is credited. If withdrawals are

made during an interest period, the withdrawal is usually deducted from money on deposit at the beginning of the period, and no interest is paid on such funds.

ILLUSTRATION: A man withdraws $1000 from a savings account 15 days before the end of the interest period. How much interest does he lose?

He loses all interest accrued on this sum for 2½ months—about $8.33 if the rate is 4% per annum compounded quarterly.

This and various other restrictive rules tend to reduce the actual rate of interest paid, especially if one makes deposits and withdrawals with any degree of frequency.

Some circumvent the above loss of interest by borrowing from the bank, using the savings account for security for the time that must elapse between the day the money is needed and the day the interest is due to be credited.

ILLUSTRATION: If the man mentioned in the previous illustration had followed this practice, how much of his interest would he have saved?

Interest on $1000 @ 4% for 3 months.....	$10.00
Interest on $1000 @ 6% for 15 days......	2.50
Net interest saved.............	$ 7.50 (Ans.)

Service Charges.—Many banks now make a charge for servicing checking accounts when an adequate balance is not maintained by the depositor. Here again practice varies in different banks, the balance to be maintained varying from $50 to $500 and the service charge ranging from 50¢ to $2.00. Some banks charge so much a check. Others permit the depositor to draw a minimum number of checks without making a charge, while still others use combinations of these various conditions.

PROFIT AND LOSS

Almost all business is organized for the purpose of making a profit. The profit (or the loss) for a fiscal period is usually shown in a statement prepared by the bookkeeper or accountant which is known as a Profit and Loss Statement. While the form of this statement will vary somewhat in terms of the specific business for which it is drawn up, it will fundamentally include sections which will set forth some analysis of (1) operating income, (2) operating costs, (3) non-operating income, and (4) non-operating cost. The net result of the statement will be the net profit for the period in question. It might be well to point out that the terms "income," "profit," "revenue," and "earnings" are used more or less synonymously by accountants in the preparation of profit and loss statements and that the term "fiscal period" means a financial period of any length of time. A few firms prefer to calculate their profit every week. Many calculate it once a month. Some use an arbitrarily adopted financial period of 4 or 5 weeks. Others use a fiscal period of 2 months, 3 months, 6 months, or a year.

Frequency in calculating profits or losses is a great aid to proper management. As a basic rule, profits or losses should be calculated as frequently as is commensurate with the value of such calculations to the management with due consideration given to the cost involved. In addition to having profits and losses calculated at frequent intervals, most firms have a definite summary of their financial affairs prepared at the end of their fiscal year and on the basis of this report they pay income taxes, divide profits, and make plans for the future. The fiscal year is a twelve-month period and may or may not coincide with the calendar year. Because of income tax and other reports that must be made, many firms have their fiscal year coincide with the calender year, but many others prefer to have the fiscal year end at a dull season when final inventory and other work necessary at the close of a fiscal year may be performed with the least possible disturbance to the business. The following is a profit and loss statement of a retail grocery store for the month ending January 31, 19—.

EDWIN S. HELLER

PROFIT AND LOSS STATEMENT FOR PERIOD EXTENDING FROM JANUARY 1 TO JANUARY 31, 19—

Income from Sales—
Sales.................................... $24,276.50
  Less Returns & Allowances................. 341.25 $23,935.25

Cost of Goods Sold—
Mdse Inventory Jan. 1....... $6,842.67
  Purchases....... $18,482.20
  Less Ret. & All.. 331.61

Frgt. & Cartage In........... 18,150.59
Less Inventory Jan. 31....... 141.17 $25,144.43
      6,497.60

Net Cost of Goods Sold 18,646.83

Gross Profit... $5,288.42

Operating Expenses—
  Selling Expenses—
    Salaries of Sales Force........ $1,575.00
    Advertising.................. 360.00
    Store Supplies............... 175.65
    Rent of Store................ 400.00
    Delivery Expenses........... 640.75
    Insurance on Stock.......... 45.15
    Taxes...................... 15.65
    Light, Heat & Power......... 75.20
    Repairs to Store Equipment.. 41.20
    Depr. on Store Equipment.... 27.49
    Depr. on Delivery Equipment. 18.20

      Total Selling Expenses... .... $3,374.29
  General Administrative Expenses—
    Management & Off. Salaries.... 525.00
    Office Supplies & Postage....... 162.50
    Rent of Office................ 100.00
    Depr. on Office Equipment..... 15.20

     Total Adm. Exp........................ 802.70

     Total Operating Expenses...........................   4,176.99
                $1,111.43

Add: Other Income:
  Discount on Purchases...................... $201.76
  Interest on Notes Receivable................. 22.16
  Interest on Bank Deposits................... 14.20

   Total Extraneous Income............................   238.12

   Total Income...................................... $1,349.55

Deduct: Other Costs:
  Discount on Sales........................... $321.60
  Interest on Notes Payable................... 41.16

   Total Extraneous Cost..............................   362.76

   Net Profit.. $986.79

Percentage of Profit.—When talking about the percentage of profit, one must be sure to know what is being used as a base. If a man buys an article for $100 and sells it for $150, it is obvious that he made a profit of $50, but what was the percentage of profit? There is much controversy as to what should be used as the base, the cost or the selling price. If we use the cost, $100, we would immediately determine that the rate of profit was 50% ($\frac{50}{100}$). If we use the selling price as a base, we then would find that the rate of profit is $33\frac{1}{3}$ ($\frac{50}{150}$). Technically, the use of the selling price as basis for calculating profits is not correct because the selling price includes profit which will cause the base to vary. On the other hand, however, the selling price as above affords the business man an opportunity to calculate not only gross and net profits, but also many other relationships on the same base.

ILLUSTRATION: The Profit and Loss Statement of the business of Edwin S. Heller is shown above. (a) What percent of the sales represents Net Profit? (b) Cost of Goods Sold? (c) Gross Profit? (d) Operating expense? (e) Operating Profit? (f) Non Operating Income? (g) Non-Operating Cost?

Each of these percentages will be determined by using the net sales as a base (letting it equal 100%) and dividing it into the item in question. Thus the percent of (a) net profit based on the sales equals

$986.79 \div $23,935.25 = 4.12\%$ (Ans.)

All the other percentages in question are determined in the same way. Thus we find that

(b) Cost of Goods Sold equals   77.91% of sales ($18,646.83 ÷ $23,935.25)
(c) Gross Profit equals   22.09% of sales (5,288.42 ÷ 23,935.25)
(d) Operating Expense equals   17.45% of sales (4,176.99 ÷ 23,935.25)
(e) Operating Profit equals   4.64% of sales (1,111.43 ÷ 23,935.25)
(f) Non-operating Income equals   0.99% of sales (238.12 ÷ 23,935.25)
(g) Non-operating Cost equals   1.52% of sales (362.76 ÷ 23,935.25)

Price Fixing.—In determining the price at which a commodity may be sold, the business man must keep in mind the cost to produce or procure that commodity, the cost of doing business, and a fair margin of profit.

Experience will usually show a man approximately what these percentages are and he may guide himself accordingly. If he finds that, for example, 28¢ of every dollar of sales must be used to pay the running expenses of the business and that 2¢ of every dollar of sales must be used to give him a fair return on his investment, he knows that 30¢ of every sales dollar must represent gross profit. He, therefore, in setting his selling price will let the cost of the article represent 70% of his selling price.

ILLUSTRATION: A shoe retailer can buy shoes at $2.45 per pair from the manufacturer and he must make a gross profit of 30% on the selling price. At what price should he sell the shoes?

The Cost $2.45 = 70\%$ or $\frac{7}{10}$ of the selling price

$.35 = \frac{1}{10}$ of the selling price

$3.50 = \frac{10}{10}$ or 100% of the selling price   (Ans.)

In some lines of business it is possible to follow this rule and apply it to all commodities sold. However, a number of factors will frequently require the business man to vary this procedure. Competition in some lines may require him to cut his margin of gross profit, while the very nature of other lines may permit him to charge more.

If several lines of commodities are carried, as in a department store, the cost of operating each department should be calculated and price ratios adjusted accordingly. Fast moving commodities in departments which do not cost much to operate may be sold at a relatively low margin of gross profit, while slower moving commodities in more expensive departments will have to be sold at a higher margin of gross profit. Thus groceries may conceivably be sold at a mark-up of 15 to 25%, while furniture may require a mark-up of 30 to 40%.

Leaders.—Many business organizations, particularly retail stores, sell certain articles at cost or even below cost in order to attract customers with the hope that once these customers buy that particular article, they will also purchase some other regularly priced commodities. Such articles are called "leaders" and their prices are fixed in terms of their cost, price asked at other places, and the probability of a given price attracting profitable customers.

Need for Records.—Records of sales and cost of sales should be carefully kept in order that a business man may know how the business is progressing. Too frequently, the inclination is to watch the volume of sales and not pay enough attention to the cost. Carefully kept records will frequently assist in the adjustment of costs and selling prices so that business may be done most profitably and at the same time competition will be adequately met.

Price Marking.—Most stores find it advisable to mark the selling price of each article on the article itself or on a tag or label attached to the article. This reduces the number of errors in quoting prices to customers, it means that sales people do not have to depend so much upon their memories, and it makes it possible to shift sales people from one counter to another without fear that they will sell goods at incorrect prices.

Very often the tag or label contains not only the selling price but also the cost price, the latter usually in code. Such a procedure facilitates the work at inventory time and·at the same time keeps the cost a secret from both the customer and the sales person. It also makes it possible for the manager to adjust intelligently prices downward on a commodity that is moving slowly.

The code used for marking the cost price usually consists of a word or group of words which contain ten different letters, each representing the figures from zero to nine. So that the secrecy of the code may be more completely preserved, extra letters such as x or y are usually used to represent digits that are repeated one or more times in the price. "Brown Chest," or "White Cloud," are words that may be used as codes.

ILLUSTRATION: What are two word groups that may be used as codes?

B R O W N   C H E S T   or   W H I T E   C L O U D
1 2 3 4 5   6 7 8 9 0   1 2 3 4 5   6 7 8 9 0

They may also be used in reverse.

ILLUSTRATION: A retailer bought shoes at \$3.30 per pair and had an established mark-up of 25% based on the selling price. How would the price tag read if "Brown Chest" with x as a repeater were used as a code for the cost?

\$3.30 equals 75% or $\frac{3}{4}$ of the selling price. Then the selling price will be \$4.40 per pair. The price tag would read as follows:

$$\begin{array}{ll} \text{Cost} & oxt \\ \text{Selling Price} & 4.40 \quad \text{(Ans.)} \end{array}$$

The words cost and selling price do not usually appear on the tag, the code for the cost price usually appearing above the line and the selling price listed below the line. The selling price may also be coded, but this is usually not done because there is no particular need for secrecy and there are fewer chances for error if the price is plainly marked.

Selling Price Based on Cost.—Some firms still base their percentage of mark-up on cost rather than selling price. When this is done, the percentage of mark-up is determined by noting the percentage the gross profit bears or must bear to the cost of goods sold. If this established percentage must be $33\frac{1}{3}$%, then that percent of the cost is calculated and added to it to determine the selling price.

ILLUSTRATION: A hat costs \$1.65 and the mark-up is $33\frac{1}{3}$% based on the cost. What is the selling price:

$$\begin{array}{ll} \$1.65 = 100\% \text{ cost} \\ \underline{.55 = 33\frac{1}{3}\% \text{ of cost (gross profit)}} \\ \$2.20 = 133\frac{1}{3}\% \text{ of cost (selling price)} \quad \text{(Ans.)} \end{array}$$

Odd Figures.—Many stores prefer not to quote prices at odd figures. To take care of this problem, they frequently make a rule that articles will be priced at the next figure divisible by five or ten above the one actually determined by calculations.

ILLUSTRATION: A store established a rule that prices should be fixed at the next figure divisible by 5 or 10 above the one actually determined by calculation. At what price will the following goods be marked?

Unit Cost	Mark-Up Based on Selling Price
0.47	25%
2.25	35%
6.48	$33\frac{1}{3}\%$

Unit Cost	Mark-Up Based on Selling Price	Mark-Up	Actually Calculated Selling Price	Fixed Price 5 and 10 Rule
0.47	25%	0.16	0.63	0.65
2.25	30%	0.66	2.91	2.95
6.48	$33\frac{1}{3}\%$	3.24	9.72	9.75

Instead of using figures divisible by five or ten, business organizations sometimes use figures that are supposed to have a good psychological effect on the buying public such as 39¢, 49¢, 69¢, 98¢, etc. The calculations are made the same but the special price scale is applied.

Manufacturing Cost.—Manufacturing Costs are usually divided into three major items, (1) raw materials, (2) direct labor, and (3) expenses applied to production called overhead burden, or indirect costs. This last item would include expenses of supervision, light, heat, power, depreciation, factory supplies, taxes, rentals, etc.

In preparing a statement showing the cost of goods manufactured, the problem is relatively simple. One simply lists from the bookkeeping records the cost of all materials used in production, the cost of all labor directly applied to production, and the indirect costs such as those listed. This information, along with the proper adding in of old inventories of goods in process, and deducting new inventories, will give one the cost of goods manufactured for a given period.

ILLUSTRATION: Make up a statement showing the cost of goods manufactured by the Warren Shoe Company during the month of June, 19—.

WARREN SHOE COMPANY

Cost of Goods Manufactured June 1–June 30, 19—

Materials—
Upper leather used.........................	\$4561.75	
Sole leather used...........................	1321.73	
Lining material used.......................	298.21	
Findings material used.....................	327.62	
Cost of raw materials used...............		\$6,509.31
Direct Labor................................		5,981.27
		\$12,490.58

Manufacturing Expenses—
Salaries and wages.........................	\$1327.61	
Rent......................................	350.00	
Rentals and Royalties......................	157.62	
Depreciation on Lasts, Dies & Patterns......	275.62	
Light, heat, & power.......................	76.21	
Taxes.....................................	27.25	
Depreciation on Machine Equipment.........	42.57	
Total Manufacturing Expenses........................		\$2,256.88
Total Cost of Manufacturing......................		\$14,747.46
Add: Goods in Process, Inv. June 1.........................		2,321.65
		\$17,069.11
Deduct: Goods in Process, Inv. June 30......................		2,576.21
Total Cost of Goods Manufactured................		\$14,492.90

Estimating Cost.—The real problem in dealing with manufacturing cost is not that of looking back over records to find what goods did cost, but rather looking ahead and estimating what they are going to cost. Every manufacturer has to quote prices, frequently in advance of actually making the goods, and the price he quotes must be low enough to help him to compete favorably with other manufacturers and at the same time be high enough to cover the cost of the goods along with giving him a fair margin of profit.

Estimating the cost of materials and the cost of direct labor is relatively simple. A manufacturer can usually tell about how much the material going into a product will cost, and about how much the labor directly applied to the product will cost. The allowance for overhead, however, is quite a different problem because the volume of production causes the cost of producing any particular unit to vary. Overhead costs (rent, superintendence, depreciation, etc.) are about the same whether the factory is almost idle or running at capacity production, and will jump up perceptibly only when it is necessary to enlarge quarters, add to equipment, etc.

There are various ways of estimating the overhead to be added in as part of the estimated cost of a unit. One very popular method is that of determining by experience that ratio that has existed in the past between the prime cost (raw materials plus direct labor) and the factory expenses. By referring to the statement of the cost of goods manufactured by Warren Shoe Company shown previously, you will notice that this ratio is about one to six; in other words, the manufacturing expenses amount to a figure that is about one-sixth of the prime cost or about one-seventh of the total cost of manufacturing. If experience has shown that approximately this ratio has existed each month, it may be used as the standard and may be applied when estimating the cost of goods to be produced.

ILLUSTRATION: A manufacturer desires to fix a selling price on shoes he is planning to make. The raw materials going into the shoes (upper leather, sole leather, trimmings, linings, findings, etc.) are estimated to cost $1.40 per pair. The direct labor required on the shoe (cutting, stitching, stock fitting, lasting, etc.) is estimated to cost $1.25 per pair. $16\frac{2}{3}\%$ of the prime cost has been established as the standard factory overhead charge. In addition, a standard mark-up of 20% based on the selling price is applied to cover the cost of selling, office administration and other general overhead costs. What is the cost to manufacture the shoes, and at what may they be sold?

Raw materials...........................	$1.40
Direct labor............................	1.25
Prime Cost............................	$2.65
Factory overhead (16⅔% of $2.65)...........	0.44
Cost to Manufacture....................	$3.09
Mark-up to cover general overhead (20% of Selling Price)............................	0.77
Calculated Selling Price.................	$3.86

NOTE: This price of $3.86 would probably be rounded off to $3.90 or $3.95 or if competition was particularly keen, it might be fixed at $3.85.

There is real danger in too much dependence on overhead standard rates that have been established solely on the basis of experience. Instead of accepting the figures as such, one should look behind the figures to determine why such a ratio exists, if it can be justified, and what improvements can be made to lower the relative cost of overhead. Are the factory costs too high? Can efficiency methods be adopted that will tend to reduce these costs or speed up production without necessitating expansion of the plant? These and many similar questions should be carefully thought of before one adheres too closely to overhead ratios and percentages established solely on the experiences of the past.

COMMISSIONS AND BROKERAGE

Agents are frequently used by growers and manufacturers who for some reason or other do not choose to undertake to market some or all of their goods themselves; or such agents are used when people desiring to procure certain merchandise find it inconvenient for them to do the buying themselves. These agents or factors are usually called *commission merchants*, their commission usually being a certain percent of the selling or buying price, or sometimes a flat rate per unit (bu, bbl, bale, ton, etc.) bought or sold.

When using the services of a commission merchant to market his goods, the grower or manufacturer simply consigns the goods to the merchant who receives them, pays any unpaid freight charges, has them hauled to his place of business, frequently insures them and pays other expenses incidental to handling them, and sells them at the best price he can get. Sometimes the selling is of the direct sale type where the agent contacts his customer or vice-versa, and a sale is consummated if the price and terms are agreeable to both; in other lines, the goods are sold at auction to the highest bidder.

When the goods are finally sold, the commission merchant renders an "Account Sales" upon which he lists the number of units sold at given prices and these are extended and totalled, the total thus determined is called the gross proceeds.

Also on the Account Sales are listed the various incidentally incurred expenses along with the commission which is usually 8 to 10 percent of the gross proceeds. The total of these charges is deducted from the gross proceeds to determine the net proceeds. The amount of the net proceeds is usually remitted with the account sales.

ILLUSTRATION: A commission merchant receives a shipment of 50 cases of eggs, each case containing 30 dozen. He sold 40 cases (1200 dozen) at 18¢ per dozen and the remaining 10 cases (300 dozen) at 17¢ per dozen. He paid freight and cartage $15.27 and insurance $2.32. Commission was charged at the rate of 10% on the gross sales. How would the Account Sales appear?

The Account Sales would appear as follows:

40 cases Eggs, 1200 dozen @ 18¢ per dozen..........		$216.00
10 cases Eggs, 300 dozen @ 17¢ per dozen..........		51.00
		$267.00
Charges—		
Freight.........................	$15.27	
Insurance......................	2.32	
Commission, 10%...............	26.70	44.29
Net Proceeds............................		$222.71

Southern Specialty Fruit
Produce

JAMES WILLIAMS   Shipping No. 39

COMMISSION MERCHANT

2 WASHINGTON STREET

NEW YORK,_____ Dec. 21, _____19____

Sold for

Account of   *William Adams*

Morgantown

West Virginia

References:

CHASE FRANKLIN NATIONAL BANK

HANOVER CENTRAL BANK
& TRUST CO., of N. Y.

4 20	50 Yams	3		1.40		4 20		
		3		1.35		4 05		
		38		1.25		47 50		
		6	Lost Repacking					
		50					55 75	
			Icing					
			Loading					
			Assorting					
			Express					
			Freight					
			Cartage					
			Commission		5 58		5 58	
			Net Proceeds				50 17	

E. & O. E.

TO AVOID ERRORS AND DELAYS, ALWAYS MAIL US INVOICE OF WHAT YOU SHIP

FIG. 6.—Account Sales

Account Purchase.—When a commission merchant is commissioned to buy merchandise for a client, the procedure is just the reverse. He buys it, sometimes at auction, sometimes through private purchase, and pays whatever expenses are necessary, such as insurance, freight, etc., in transferring the goods to the principal.

The report of the purchase is called an "Account Purchase" and lists the number or articles or units bought with the unit price paid, the extension and total, known as the "Prime Cost." To the prime cost are added the various costs involved in making the purchase including the commission. This final amount is called the "gross cost."

ILLUSTRATION: A commission merchant buys 1000 lbs. of raw silk at $1.39 per pound for a client, pays freight and cartage $27.62, and charges a 5% commission. How would the account purchases appear?

The account purchases would appear as follows·

July 22—1000# Raw Silk, $1.39..................		$1390.00
Prime Cost.................................		$1390.00
Charges—		
Freight and Drayage.............	$27.62	
Commission....................	69.50	
Total Charges..........................		97.12
		$1487.12

Salesmen's Commissions and Bonus.—There are a variety of systems used in paying salesmen, perhaps the most common of which are: (a) straight salary, (b) salary plus commission on all sales, (c) salary plus commission on certain items or groups of items, (d) salary plus commission on sales above a certain predetermined quota, (e) straight commission.

Straight Salary.—Many firms pay their salesmen on a straight salary basis feeling that their salesmen will work well without special commission or bonus incentives. This plan almost entirely eliminates a certain ruthless or high pressure type of salesmenship that so frequently destroys good will. The salary is almost always reasonably substantial. Sales work is looked upon as being the life-blood of any industry and successful salesmen are usually well paid.

Salary Plus Commission on All Sales.—Some firms prefer to have the salesmen have an opportunity to earn more if they can sell more, but at the same time like to give them the security of a regular salary regardless of business conditions. The salary as such is usually relatively small, set on what might be called a subsistence level and the rate of commission is set so that with normal effort a man should be able to earn a fairly substantial income.

ILLUSTRATION: A man receives a salary of $100 and a 5% commission on all sales. His sales for the month of October were $5000. What are his earnings for the month:

$$\begin{array}{ll}
\text{Salary} \dots\dots\dots\dots\dots\dots\dots & \$100.00 \\
\text{Commission, } 5\% \text{ of } \$5000 \dots & 250.00 \\
\hline
& \$350.00 \quad \text{(Ans.)}
\end{array}$$

Salary Plus Commission on Certain Items.—Some firms have a fundamental policy of paying a straight salary but use the commission as a special incentive to have salesmen sell certain items or groups of items. This may be and frequently is only a temporary arrangement and is used to move a special lot of slow-moving merchandise, to introduce a new item or line, to make salesmen "selling conscious" of articles that they have been neglecting or for other reasons.

ILLUSTRATION: A paint company noticed that its line of lacquers was not selling well and offered its salesmen a special commission of 5% on all sales in that line. One salesman whose salary was $275 per month sold $425.00 worth of lacquers during the month of May. What was his earning during the month?

$$\begin{array}{ll}
\text{Salary} \dots\dots\dots\dots\dots\dots\dots\dots & \$275.00 \\
\text{Commission, } 5\% \text{ of } \$425.00 \dots\dots & 21.25 \\
\hline
& \$296.25
\end{array}$$

Salary Plus Commission.—A salesman's quota may be set in various ways, but is usually determined by experience in the past. One favorite method for establishing a quota is to average the three or four best months that the salesman is paid a commission (sometimes called bonus in such cases) on all sales above the established figure.

ILLUSTRATION: A company pays a commission (or bonus) of 1% on all sales above the salesmen's quota. The quota is established by averaging the total sales made by each salesman in the best four months that each had in the preceding year. A salesman whose best months in the preceding year were January $18,750.00, March $17,925.00, September $19,256.00, and December $16,225.00, who was paid a monthly salary of $325.00, made sales totalling $19,475.00 in a given month. What was his total earnings?

Quota equals $18,750.00
      17,925.00
      19,256.00
      16,225.00

    $72,156.00 ÷ 4 = $18,039.00, or, in round numbers,
             $18,000.00, established monthly quota for
               new year.

Earnings for the Month:

Salary..................................... $325.00
Commission of 1% on ($19,475 − $18,000).....   14.75

               $339.75 (Ans.)

Straight Commission.—Under this plan, the salesman receives no salary as such but is entirely dependent upon his commission. A straight commission usually means that a flat rate of commission is paid on all articles sold. In some cases, however, a difference of commission is paid on different lines of goods sold by the firm.

ILLUSTRATION: A firm handles office equipment and supplies. Salesmen are paid no salaries but receive a commission of 10% on all equipment sold and 15% on all supplies sold. In one month, a salesman sold equipment totalling \$2257.65 and supplies totalling \$926.18. What were his monthly earnings?

Commission on equipment	= 10% of \$2257.65	\$225.77
Commission on supplies	= 15% of 926.18	138.93
		\$364.70 (Ans.)

PAYROLLS

A list of employees and the amount to be paid to each for a specific time is called a *payroll*. When pay is calculated on a time basis, the number of hours worked and the rate per hour is usually included. When the employee is paid on a piece-work basis, the number of pieces completed and the rate per piece is frequently included. The total of the payroll is the amount to be paid to all employees for the time specified, which is usually a week but may be a longer period. When the total is determined, a check is made out payable to the order of payroll. If the company pays each employee by check, the payroll check which covers the whole payroll is usually deposited in a special payroll account maintained at the bank for the purpose and special individual checks for each employee are drawn against this particular account and are distributed to the employees.

If the company pays each employee in cash, it is necessary to prepare a currency memorandum in order to know just how many bills and various coins will be needed and from it a Payroll Currency slip which is taken to the bank with the payroll check so that the bank will know how many bills and coins of various denominations to give to the paymaster or his representative when cashing the check. This currency and change is then distributed among the various pay employees and these in turn are passed out at a given time to the employees. Most companies usually require a receipt from the employee when he is given his pay envelope.

Time Basis.—Many firms pay on a time basis which is usually fixed at so much per hour for so many hours per day and so many days per week. A very common time schedule is 7 hours per day and 5 days per week, making a total of 35 hours per week. While

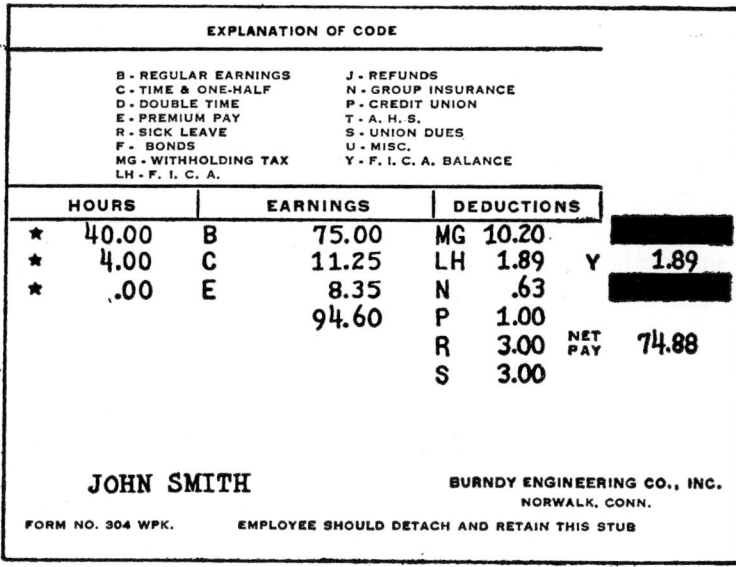

EXPLANATION OF CODE	
B - REGULAR EARNINGS	J - REFUNDS
C - TIME & ONE-HALF	N - GROUP INSURANCE
D - DOUBLE TIME	P - CREDIT UNION
E - PREMIUM PAY	T - A. H. S.
R - SICK LEAVE	S - UNION DUES
F - BONDS	U - MISC.
MG - WITHHOLDING TAX	Y - F. I. C. A. BALANCE
LH - F. I. C. A.	

HOURS		EARNINGS		DEDUCTIONS			
★ 40.00	B	75.00	MG	10.20			
★ 4.00	C	11.25	LH	1.89	Y	1.89	
★ .00	E	8.35	N	.63			
		94.60	P	1.00			
			R	3.00	NET PAY	74.88	
			S	3.00			

JOHN SMITH

BURNDY ENGINEERING CO., INC.
NORWALK, CONN.

FORM NO. 304 WPK.   EMPLOYEE SHOULD DETACH AND RETAIN THIS STUB

FIG. 7.—A Typical Check or Payroll Envelope Stub Showing a Detailed Statement of Income and Deductions for an Employee in a Medium-sized Plant. Most of the code letters are self-explanatory. "Premium pay" (E) is a bonus paid the employee for exceeding the production quota established for his job. The amount varies, depending upon how much the quota is exceeded. "F.I.C.A." (LH) is Federal Insurance Contributions Act or Federal Old-Age and Survivors Insurance or the deduction for "Social Security." "AHS" (T) stands for Associated Hospital Service ("Blue Cross," "Blue Shield," etc.).

some firms have a standard week of more than 35 hours, many others, particularly during recent years, have tended to reduce the number of hours per week so that now we find plants operating on a basis of 30, 32, 36, and 40 hours per week, variously distributed among the days with a tendency toward no work on Saturday. Time worked in excess of the standard number of hours per day is called overtime; and while some firms pay merely the regular rate per hour for this overtime, most firms pay extra, usually at the rate of time-and-one-quarter or time-and-one-half and, especially if it is on Sunday, double time.

Where employees work by the hour, the hours per day are usually calculated from a time card which the employee is required to punch on a time clock every time he comes into or leaves the plant.

ILLUSTRATION: Harry Allen is employed by the Standard Manufacturing Company as a machinist at the rate of $2.00 per hour. The plant operates on the basis of 8 hours per day for five days a week, a 40-hour week, and pays time-and-one-half for overtime. How would Mr. Allen's time card with pay calculated appear?

No. 91

HARRY ALLEN

Employed as Machinist at 2.00 per hour, Week Ending Tuesday, May 21, 19—

Day	A.M.		P.M.		Overtime	Total Hours
	In	Out	In	Out		
Wednesday...	7:55	12:05	1:00	5:04	......	8
Thursday.....	7:56	12:01	12:55	5:02	......	8
Friday........	7:59	12:10	12:57	5:10	5:30–8:30	11
Saturday.....						
Sunday.......						
Monday......	7:51	12:05	12:58	5:03	......	8
Tuesday......	8:00	12:02	12:55	5:07	......	8

Payroll credit:

Regular 40 hours at $2.00 per hour............. $80.00

Overtime 3 hours at $3.00 per hour............. 9.00

Total pay $89.00

The regular payroll for a company operating on a time basis is simply a summary of these individual time cards.

ILLUSTRATION: The Standard Manufacturing Company had working for it during the week ending May 21, 19— the employees listed below along with the hours per day each worked and his hourly rate. The company pays time-and-one-half for overtime. What is the total payroll for the week?

STANDARD MANUFACTURING COMPANY

Payroll for Week Ending May 21, 19—

Employee		W	Th	F	S	S	M	T	Rate per hour	Reg. time	Over- time	Wages		
No.	Name											Reg.	Over.	Total
91	Allen, Harry....	8	8	11	0	0	8	8	$2.00	40	3	$ 80.00	$ 9.00	$ 89.00
92	Moulton, James	8	8	8	0	0	7	8	$2.10	39	0	81.90	......	81.90
93	Brown, Edward.	8	10	11	0	0	10	8	$1.75	40	7	70.00	18.38	88.38
94	Paul, Samuel....	8	7	7	0	0	10	8	$2.10	38	2	79.80	6.30	86.10
95	Garvis, John....	8	8	8	0	0	8	8	$1.85	40	0	76.00	......	76.00
96	Young, James...	8	10	8	0	0	9	8	$2.00	40	3	80.00	9.00	89.00
										Totals,........		467.70	42.68	510.38

(Ans.)

It should be noted that in calculating overtime hours, they are calculated in terms of day rather than week, and it is perfectly possible as in the case of Samuel Paul for a man not to have a full week of regular time to his credit and yet have overtime credit. It should further be noted referring back to the time card of Harry Allen that odd minutes are not counted unless they are more than fifteen. Also that most firms do not count time before the regular starting time in the morning or during the noon hour unless the employee has been specifically asked to work at either or both of these times.

Change Memorandum.—This memorandum is really an analysis of payroll made so that the paymaster will know how many bills and coins of various denominations will be needed for the pay envelopes.

ILLUSTRATION: Prepare a change memorandum for the payroll of the Standard Manufacturing Company for the week ending May 21, 19—.

CHANGE MEMORANDUM

For Payroll of Week Ending May 21, 19—

No. of Employee	Total Wages	Currency					Coins				
		20	10	5	2	1	50	25	10	5	1
91	$89.00	4	0	1	0	4	0	0	0	0	0
92	81.90	4	0	0	0	1	1	1	1	1	0
93	88.38	4	0	1	0	3	0	0	3	1	3
94	86.10	4	0	1	0	1	0	0	1	0	0
95	76.00	3	1	1	0	1	0	0	0	0	0
96	89.00	4	0	1	0	4	0	0	0	0	0
		23	1	5	0	14	1	1	5	2	3
Totals..	$510.38	460	10	25	0	14	.50	.25	.50	.10	.03

Payroll Currency Slip.—This slip is prepared so that it may accompany the payroll check to the bank. It is a summary of the change memorandum and lets the paying teller of the bank know how much money is wanted in various denominations of bills and coins.

THE MERCHANT'S BANK

Orange, N. J.

Payroll Currency Slip

Depositor: Standard Mfg. Co.

Date: May 24, 19—

Bills	Dollars	Cents
20	460	0
10	10	0
5	25	0
2	0	0
1	14	0
Coins		
Halves.........	0	50
Quarters........	0	25
Dimes..........	0	50
Nickels.........	0	10
Cents..........	0	3
Total........	509	1.38

It should be noted that there is a discrepancy between the date on the payroll and the date on the currency slip. This exists because most firms have their week end sometime during the week, as on Monday, Tuesday, or Wednesday, and they pay on the following Friday or Saturday. This is done in order to give the clerical staff in the office adequate time in which to make up the time cards and properly prepare the payroll records.

Piece Work.—Instead of paying by the hour, day, or week, many firms pay many of their employees so much for each unit of work they complete. The piece rate is usually established by determining how many pieces the average man can do in a given length of time. This means that a fast worker will usually be rewarded by being able to earn more than his slower co-workers.

The piece-work production is usually reported at the office on a slip. In some plants the worker keeps a record of the number of units he has completed during the day and has the foreman of the room counter-sign his slip; in other plants the worker holds the work he completes until a checker counts it and gives the worker credit, releasing the units so that they may go on to the next operation. In still other plants, the work goes through the plant in numbered cases or job lots and the worker simply records in his book the job number of each lot on which he performs his operation and reports this number together with the price at the end of each day. A checker then takes these slips and enters them in a book especially prepared for the purpose, checks the completion of the operation against the case or lot number. This record is usually made by placing the date and number or initial of the employee in the proper column in his checker's book. It thus prevents two people from being paid for doing the same work or one person from being paid twice for the job. It also helps to keep employees from claiming pay for work that they have not actually completed.

In preparing the payroll when the plant is on a piece-work basis, some organizations record on the payroll sheet the number of pieces or units completed each day by the worker, add the total for the week, multiply this total by the price paid per unit, and thus get the total amount to be paid to the worker.

ILLUSTRATION: Mr. L. Cook, employee number 61, completed 23 units of work on Wednesday, 24 Thursday, 20 Friday, 11 Saturday, 21 Monday, and 20 Tuesday. He is paid $37\frac{1}{2}$¢ per unit. What is the amount of his earnings for the week?

		W	Th	F	S	M	T	Total Units	Price per Unit	Total Pay
61	Cook, L.	23	24	20	11	21	20	119	.37½	44.63 (Ans.)

In other plants such a system is not possible. Any one worker may work on different operations and thus complicate the problem or may work on much the same type of operation but may get a different price because he is working types of materials. Such plants will usually have some form upon which are calculated the daily earnings of the worker and these daily earnings are entered on the payroll sheet. At the end of the week, the daily earnings are totalled to determine the workers' pay for the week.

ILLUSTRATION: The work slips of Mr. H. Sailer, employee number 48, indicate that he earned $6.45 Wednesday, $7.78 Thursday, $5.01 Friday, $3.78 Saturday, $6.51 Monday, and $7.02 Tuesday. What is the amount of his earnings for the week?

		W	Th	F	S	M	Tu	Total Pay
48	Sailer, H.	6.45	7.78	5.01	3.78	6.51	7.02	36.55 (Ans.)

Other Pay Bases.—While time and piece-work are by far the most widely used bases for paying workers, because of the many obvious possibilities for injustices to both the employer and the employee, several special systems have come into vogue in recent years. These systems are fundamentally time or piece-work systems or a combination of the two, but have special features that set them apart and which make them, at least in some places, much more satisfactory than the basic systems. They are generally known as incentive wage plans. One particularly interested in these special wage plans should make himself acquainted with the Halsey Premium Plan, the Taylor Differential Piece-Work Plan, the Gantt Task and Bonus Plan, along with various others.

INSURANCE

The basis of all insurance is risk. Insurance is an agreement between a professional risk taker (insurance company) and an individual or firm, whereby the insurance company agrees under certain specified conditions to indemnify an individual or his heirs in case of some certain type of loss. The principal risks that a man faces have to do with his life, his property, and his liability for losses caused to other persons through some legal fault of his.

Life Insurance.—The average man or woman carries life insurance for two reasons: first, to take care of the payment of expenses incurred in connection with his last illness and burial, and secondly, to leave at least some funds to his dependents so that they may not suffer too much of an economic strain in the event of his death. The contract between the insurance company and the insured is called a *policy*, and the fee paid to the insurance company is known as a *premium*. The person who receives the face of the policy upon the death of the insured is called the *beneficiary*.

Premium.—The amount of the premium of life insurance depends upon the type of insurance and the age of the person being insured. The actual payment made to the company by the insured is technically called the gross or office premium. The cost of the policy as such is based on the death rate given in the mortality table and at a given interest rate. This basic cost of the policy is called the net premium. The gross premium is determined by adding an amount to cover expenses of all kinds and profit to the net premium. This is called loading. Sometimes the premium of a policy is paid in one sum, this being called the net single premium. Usually it is calculated on an annual basis and is called the net annual premium. While many people pay their insurance premiums annually, many pay a slight extra charge for the privilege of paying semi-annually, quarterly or monthly.

Insurance Tables.—The actual mathematics of determining the various premiums is somewhat involved and as a result tables have been prepared that are used to calculate these premiums. Insurance companies usually issue a table which simply lists their rates by age per thousand dollars of face value of policy. Naturally, a separate table or a separate column in a composite table is devoted to each type of insurance. Table 8 is a composite table showing the rates per $1000 on various types of insurance.

TABLE 8

A Typical Table Showing Rates Per Thousand for Different Types of Insurance Without Accidental Death Benefit *

Age	Life Pd. up at 85	Life Pd. up at 65	20 Pay Life	"Mod. Life 3" 1st 3 Yrs.	There-after	End't at 65	End't at 60	20 Yr. End't	10 Yr. End't	Term to 65	10 Yr. Term	5 Yr. Term	Age
15	17.08	18.07	28.67	13.46	15.84	19.79	21.57	51.05	105.68	....	....	....	15
16	17.45	18.50	29.17	13.79	16.22	20.29	22.16	51.15	105.82	....	....	....	16
17	17.84	18.97	29.69	14.14	16.64	20.80	22.79	51.24	105.94	12.21	7.28	6.96	17
18	18.23	19.43	30.19	14.49	17.05	21.35	23.43	51.33	106.04	12.47	7.48	7.22	18
19	18.62	19.90	30.68	14.85	17.47	21.90	24.09	51.40	106.12	12.73	7.66	7.45	19
20	19.02	20.39	31.19	15.21	17.89	22.47	24.79	51.44	106.17	12.99	7.77	7.60	20
21	19.55	20.99	31.79	15.58	18.33	23.15	25.60	51.50	106.23	13.26	7.89	7.75	21
22	20.07	21.62	32.39	15.97	18.79	23.87	26.46	51.54	106.27	13.54	8.00	7.88	22
23	20.62	22.28	33.01	16.37	19.26	24.62	27.36	51.59	106.30	13.82	8.10	7.99	23
24	21.19	22.97	33.63	16.80	19.76	25.40	28.32	51.65	106.33	14.10	8.19	8.08	24
25	21.78	23.68	34.26	17.24	20.28	26.22	29.31	51.70	106.36	14.40	8.29	8.18	25
26	22.38	24.43	34.92	17.70	20.82	27.08	30.36	51.76	106.38	14.72	8.38	8.26	26
27	23.02	25.21	35.57	18.17	21.38	27.98	31.47	51.85	106.40	15.05	8.52	8.35	27
28	23.69	26.03	36.27	18.68	21.98	28.92	32.65	51.92	106.44	15.38	8.63	8.41	28
29	24.37	26.90	36.98	19.22	22.61	29.94	33.91	52.03	106.48	15.76	8.80	8.52	29
30	25.09	27.82	37.71	19.78	23.27	30.99	35.25	52.15	106.52	16.13	8.97	8.61	30
31	25.74	28.70	38.31	20.38	23.98	32.01	36.58	52.29	106.58	16.54	9.20	8.74	31
32	26.43	29.61	38.95	21.02	24.73	33.11	38.02	52.45	106.65	16.96	9.46	8.91	32
33	27.17	30.62	39.62	21.67	25.49	34.27	39.58	52.65	106.74	17.42	9.76	9.11	33
34	27.94	31.69	40.31	22.38	26.33	35.52	41.24	52.87	106.86	17.90	10.09	9.34	34
35	28.75	32.82	41.04	23.13	27.21	36.86	43.06	53.11	106.98	18.40	10.49	9.62	35
36	29.61	34.04	41.81	23.91	28.13	38.30	45.03	53.40	107.15	18.93	10.94	9.94	36
37	30.51	35.36	42.60	24.74	29.11	39.84	47.16	53.71	107.32	19.50	11.45	10.31	37
38	31.49	36.78	43.44	25.61	30.13	41.51	49.52	54.07	107.52	20.09	11.99	10.72	38
39	32.49	38.28	44.32	26.54	31.22	43.31	52.08	54.46	107.77	20.71	12.63	11.18	39
40	33.57	39.92	45.25	27.51	32.36	45.24	54.91	54.91	108.03	21.38	13.32	11.71	40
41	34.68	41.70	46.19	28.55	33.59	47.34	58.03	55.38	108.35	22.08	14.10	12.32	41
42	35.87	43.62	47.20	29.63	34.86	49.62	61.48	55.92	108.69	22.80	14.96	12.98	42
43	37.12	45.72	48.25	30.78	36.21	52.10	65.33	56.49	109.06	23.59	15.92	13.72	43
44	38.44	47.98	49.33	32.00	37.65	54.80	69.66	57.10	109.49	24.38	16.96	14.53	44
45	39.84	50.47	50.47	33.28	39.15	57.77	74.55	57.77	109.96	25.24	18.12	15.45	45
46	41.31	53.21	51.66	34.63	40.74	61.05	80.11	58.51	110.49	....	19.38	16.46	46
47	42.86	56.21	52.91	36.07	42.44	64.64	86.52	59.30	111.05	....	20.79	17.56	47
48	44.51	59.54	54.22	37.59	44.22	68.66	93.94	60.17	111.68	....	22.32	18.78	48
49	46.24	63.24	55.59	39.18	46.09	73.15	102.69	61.11	112.38	....	24.00	20.12	49
50	48.08	67.42	57.02	40.88	48.09	78.19	113.15	62.13	113.15	....	25.82	21.59	50
51	50.01	72.12	58.56	42.65	50.18	83.92	....	63.24	113.93	....	27.56	23.19	51
52	52.06	77.47	60.15	44.53	52.39	90.45	....	64.46	114.74	....	29.44	24.97	52
53	54.20	83.64	61.83	46.52	54.73	98.01	....	65.77	115.57	....	31.49	26.92	53
54	56.48	90.83	63.63	48.62	57.20	106.86	....	67.18	116.43	....	33.69	29.02	54
55	58.89	99.30	65.50	50.83	59.80	117.33	....	68.71	117.33	....	36.08	31.35	55
56	61.42	..	67.49	53.15	62.53	....	....	70.37	118.25	....	....	33.25	56
57	64.09	....	69.58	55.60	65.41	....	....	72.14	119.23	....	....	35.33	57
58	66.89	...	71.78	58.17	68.44	....	....	74.06	120.24	....	....	37.52	58
59	69.84	....	74.11	60.88	71.62	....	....	76.10	121.27	....	....	39.90	59
60	72.93	....	76.54	63.70	74.94	....	....	78.29	122.34	....	...	42.39	60
61	76.50	...	79.47	66.94	78.75	....	....	80.97	124.07				61
62	80.32	....	82.60	70.39	82.81	....	....	83.90	125.98		Minimum $5000		62
63	84.42	....	85.99	74.06	87.13	....	....	87.08	128.06		Convertible		63
64	88.80	....	89.60	77.98	91.74	....	....	90.53	130.33	Prior to 60	During Term Period		64
65	93.49	....		82.16	96.66	....	....	94.27	132.81				65
66	98.56	...	97.69	86.61	101.89	....	....	98.32	135.52				66
67	104.02	..	....	....	...			..	..				67
68	109.92	..	....	...	...			..	..				68
69	116.31	...	....	..	..			...	..				69
70	123.29	....		....	...			..	..				70

* *Courtesy The Prudential Life Insurance Co. of America.*

Types of Life Insurance.—Insurance companies issue several types of life insurance policies, the most commonly known of which are: (1) ordinary life, (2) endowment, (3) paid-up, and (4) term.

Ordinary life insurance is that type of policy which remains in force during the life of the insured, and upon death the face of the policy is paid to the beneficiary. Usually premiums are paid annually from the time the policy is taken out until death. (It should be noted that many companies do not require premiums to be paid on ordinary life policies after the age of 85.) To find the cost of such insurance, one should refer to the table of rates, run down the age column to the age of the person buying the insurance, and multiply the rate given in the ordinary life column by the the number of thousand dollars worth of insurance the purchaser desires to take. The amount thus determined is the rate to be paid annually from that date until the death of the individual.

ILLUSTRATION: A man, age 35, desires to take out a $5000 ordinary life insurance policy. What annual premium must he pay?

The rate for ordinary life insurance (paid up at 85) at the age of 35 (according to Table 8) is $28.75 per thousand. The rate for $5000 will be

$$\$28.75 \times 5 = \$143.75 \quad \text{(Ans.)}$$

Endowment policies call for the payment of a regular premium for a given period of years—usually twenty but may be fewer or more—at the end of which time the face of the policy becomes due and payable to the insured. In the event of the death of the insured at any time during which the policy is in force, the face of the policy is paid to the beneficiary. The procedure for calculating the premium is similar to that used in the preceding illustration, the only exception being that the rate is selected from the proper endowment column.

ILLUSTRATION: A man, age 25, desires to take out a policy that will pay him $2500 at age 45, providing he lives, or in the event of his death before that time, will pay a like sum to his beneficiary. How much must he pay annually?

The rate for 20-year endowment insurance at age 25 is $51.70 per $1000. The rate for $2500 will be

$$51.70 \times \tfrac{2500}{1000} = \$129.25 \quad \text{(Ans.)}$$

Paid-up policies are a compromise between the endowment policy and whole life policies. Under the paid-up policy the insured pays a regular premium for a given period of years (usually 10, 15, or 20 years) and pays no more, but the face of the policy is not payable to him as in the case of endowment policies but rather is payable to the beneficiary upon the death of the insured regardless of whether that death occurs before or after the premiums have all been paid. The value of this type of insurance is that a man may pay up all the premium charges during his healthiest and most productive years and still remain insured during his entire life. The procedure for calculating the premium is similar to that used in the preceding illustration.

ILLUSTRATION: A man, age 30, desires to take out a $10,000 policy, the premiums of which will be paid up when he is 50. The rate for insurance requiring 20 annual premiums at age 30 is $37.71 per $1000. The annual rate for $10,000 will be

$$\$37.71 \times 10 = \$377.10$$

Term insurance is purchased for only a given length of time (usually one, five, or ten years) but may be kept in force indefinitely by the regular payment of the premium. If the insured dies while the policy is in force, the face of the policy is paid to the beneficiary, but if the insured survives that given length of time and fails to renew his policy, he is no longer protected. This type of insurance is frequently carried during that period of a man's life when his family is most dependent upon him.

ILLUSTRATION: A man, age 34, has three small children and feels that he should carry additional protection, realizing that should he die within the next few years his family might be in unusually straitened circumstances. He decides to take out a ten year term policy for $5,000 to supplement other insurance which he is carrying. What will it cost him per year?

The rate for ten year term insurance at age 34 is $10.09. The cost of a $5000 policy will be

$$\$10.09 \times 5 = \$50.45$$

Cash Surrender.—In all other forms of insurance except term insurance (whole life, endowment, and paid-up) the insured is to some extent protected even though he fails to continue to pay his premiums and thus lets his policy lapse. If this takes place, he may do one of three things: he may ask the insurance company to pay him an amount in money that is equal to the present value of the policy. This is called the cash surrender value and is really that portion of the premium that has been set aside by the company as a reserve out of which to pay the policy, plus accumulated interest.

Loans.—If the insured chooses to do so, he may borrow at a fixed interest rate, the cash surrender value of his policy rather than accept the cash settlement and surrender his policy. The value of the loan plan is that the insured may get his money and at the same time keep his policy in full force. Should he die before the loan has been repaid, the beneficiary will receive the face of the policy less the unpaid loan and interest accrued thereon.

Extended Insurance.—If, in the case of a lapsed policy, the insured does not choose to apply for and receive the cash surrender value of his policy, he may simply receive extended insurance. This means that the cash surrender value will be used to keep his policy in force on a term basis until such value is exhausted. Should the insured during this period die, the beneficiary will receive the face value of the policy. Should he survive this period, however, the insurance company is no longer liable.

Paid-up Insurance.—The holder of a lapsed policy may, however, choose to accept paid-up insurance rather than extended insurance. Under such a plan, the cash surrender value will be used to buy insurance that will remain in force during life without his paying any more premiums. Naturally, the face of the new policy will be less than that of the older policy, the face being the amount of paid-up insurance that could ordinarily be bought for the cash surrender value.

Special Benefits.—Many life insurance policies carry clauses which cover special risks such as general accident, travel, accident, disability, or some combination of these. An additional fee is charged for these special coverages.

Other Personal Policies.—Practically all other policies covering risks to the person such as accident insurance, disability insurance, health insurance, are also based on more or less standard experience tables and the annual fees are calculated accordingly. As a general rule, such policies are on a term basis (usually one year) and expire at the time unless they are renewed. Naturally, they have no cash surrender value except a refund that may be claimed if the policy is cancelled during the year.

In quoting rates for this type of insurance, many companies classify people in terms of their occupation and give more favorable rates to those in the fields that are least hazardous.

HOME INSURANCE

Types of Policies.—The earliest peril against which homes were insured was fire. In the course of time, however, the extended coverage fire policy developed, providing coverage against damage by such other agencies, natural and otherwise, as for example, lightning (since it can cause structural damage as well as fire damage), smoke, wind, hail, explosion, falling aircraft, civil commotion, vandalism and malicious mischief. In the past an additional policy was often written to cover losses incurred of the contents of the home (personal property) by these perils, and still another to cover personal property losses by theft. Then a fourth type of policy would be written to insure the homeowner against the cost of defending, and settling, personal liability lawsuits by outsiders for injuries sustained on the property.

In the more recent past, there has been a major trend toward the writing of policies that combine these various types of coverage into single policies, which are usually called Homeowners Policies. There are three varieties in common use which are in order of increasing coverage and cost, the Standard Plan (also called the A policy), the Broad-Form Plan (also called the B policy) and the Combination Plan (also called the C policy). These policies differ, not only in the amount of coverage they provide for other risks than that on the house itself, but also in the number and kind of the perils and risks covered. Therefore, they are discussed here in order in the following paragraphs, in which the coverage of each is discussed in some detail. In order to arrive at comparative figures the rates of each variety of policy will be given for a homeowner living in a Class A section of Nassau County, New York, one who has a masonry house and the other having a frame house. It is necessary to assume specific localities because the insurance rate books divide homes into 18 premium groups, of which Group 1 pays the lowest rate and Group 18 pays the highest. The assignment of a home to these groups depends upon (1) the zone in which the home is located; thus Nassau County is zone 2 in New York State; (2) the fire protection facilities available in the neighborhood in which the home is located; all of Nassau County is listed as Class A in this matter of protection. It is then found from the group chart that a masonry home so located belongs in premium group 7, while a frame home similarly located belongs in premium group 9. It is also assumed that the insurable value of the home is $20,000. One should note that insurable value means cost of replacement of the home less a factor which depends upon its present state of depreciation. Thus a newly constructed house would have an insurable value of its full cost of replacement, while that figure would be decreased for an older home according to its condition. In other words, the insurable value has no direct relation to selling price, which includes the cost of land and is subject to many other elements than construction costs.

Coverage and Costs.—All three of these homeowners policies provide insurance for specified perils to the (1) dwelling and (2) out buildings, plus specified insurance coverage of (3) household goods on the premises, (4) household goods away from the premises and (5) additional living expenses incurred while damages to the home are being repaired. All three of the homeowners policies protect against lawsuits resulting from personal injuries of outsiders, the Standard Plan and the Broad Form Plan up to $10,000 and the Combination Plan up to $25,000. Moreover, these injuries include not only such property connected accidents as tripping on the stairs or slipping on the sidewalk, but also the acts of children or pets.

It is, however, on the types of perils for which buildings are insured that the risk covered by the three policies varies so widely. The Standard Plan insures against damages due to fire, lightning, wind storms, hail, explosion, smoke, aircraft (or vehicles), civil commotion, vandalism, malicious mischief, glass breakage, and theft. The Broad-Form Plan adds to these perils such others as damage due to collapse of building, destruction by weight of ice, snow and sleet, falling trees and other falling objects besides aircraft and rupture of steam or hot water heating systems.

The Combination Plan includes certain other damages, especially those which result from obvious carelessness on the part of the occupants of the home, such as rain or snow damage due to windows left open and damage due to objects dropped within the house, which would even include spillage of paint and similar material.

On a home with an insurable value of $20,000, and located in Nassau County the Standard Plan (Plan A) would provide coverage of $20,000 for the dwelling, $2,000 for the out buildings, $8,000 for household goods on the premises, $1,000 for household goods off the premises, and $2,000 for additional living expenses incurred during the time needed to repair the home damaged.

The premium for this coverage, on a three year prepaid basis, would be $138 per year for the masonry home, and for the frame home it would be $147 per year.

The Broad-Form Plan (Plan B) would provide coverage for these homes of $20,000 for the dwelling, $2,000 for the out buildings, $8,000 for household goods on the premises, $1,000 for household goods off the premises, and $4,000 for additional living expenses. The cost of this coverage would be $177 for the masonry home and $192 for the frame home.

The Combination Plan (Plan C) would provide $20,000 for the dwelling, $2,000 for the out buildings, $10,000 for personal property, and $4,000 for additional living expenses. The rate for this coverage on the masonry home is $348 and on the frame home, $369, on a $100 deductible basis for all perils except fire and lightning. The reason for the major difference in rates between the Broad-Form Plan and the Combination Plan is due to the additional perils and articles insured. It is to be noted that this policy covers personal property instead of household goods. Personal property includes personal articles on the premises as well as household goods. Even this policy, however, does not include all the coverage available to the homeowner. Many other items can be added by the payment of increased premiums, so that it has become more necessary than ever to base the policy as finally written with specific reference to the home that is insured.

AUTOMOBILE INSURANCE

There are three major types of automobile insurance coverage: (1) that covering the liability of the owner of the vehicle for bodily injury or property damage caused by his vehicle; (2) that covering the cost of repairing damages which his own vehicle may sustain due to accidents; and (3) damages which his vehicle may incur or losses which he may sustain due to other hazards, such as fire, theft and glass breakage. Of these, the first is by far the most possible.

The variables which govern automobile liability (casualty) insurance rates depend upon a number of considerations. The first of these is the territory in which the automobile is domiciled, and individual rate tables are prepared for each territory. The United States is divided into many territories, some of which may be quite small in area. Thus New York City is divided into eight territories: Bronx County North; Bronx County South; Kings County; Manhattan; Queens County; Queens County Suburban; Staten Island; and New York City Suburban (the only area within New York City limits that belongs in this territory is Governor's Island). Then each territorial table has columns for each amount of insurance, the amount being stated at the top of the column as two figures, one for the amount applying to injuries to one person, and the other for more than one person. The usual standard amounts so provided are: $10/20; $20/20; $20/40; $25/50; $50/100; and $100/300.) These figures mean, of course, the number of thousands of dollars of insurance. There are two columns for property damage, one up to $5,000 and another, on which the premiums are slightly greater, up to $10,000. The important part of the table, however, is the first column which lists 28 classes of risks, and the importance of these classes can be judged from the fact that the lowest premium in Manhattan, which is the annual rate for Class 1A-O EX for $50/100 coverage is $145.20, while the highest rate shown for this amount of coverage in Manhattan is for Class 2C-2X and amounts to $555.39. The method of determining these classifications is critically important in determining the premium which the owner of the private passenger automobile must pay. These classes are as follows:

CLASS 1—Owned by an individual or jointly by two or more relatives who are residents in the same household. None-Business Use—No Operator under 25.

Class 1A—Not used * to and from work.

Class 1B—Used * to or from work less than 10 road miles one way.

Class 1C—Used * to or from work 10 or more road miles one way.

* Used—Means customarily used in driving to and from work, including car pools, share the ride arrangements, and driving to and from railroad or bus station, whether or not parked all day—Class 1B or 1C used to and from work.

CLASS 2—Owned by an individual or jointly by two or more relatives who are residents in the same household. Business and None-Business Use—Operators under 25.

Class 2A—Male operator under 25 not owner or principal operator

OR

Male owner or operator *under* 25—married.

CLASS 2B—Male operator under 25 not owner or principal operator and a resident student at a school, college or educational institution over 100 road-miles from the place of principal garaging.

CLASS 3A—Customarily used in business—individually owned. No male operator or owner under 25.

Reference to the table, however, discloses that in addition to the entries of these primary classes of uses and drivers there are two other items in the class symbol. One of these is the use, following the classes 1A, 2B, etc., of a number which may be 0, 1 or 2. These numbers are the driving records subclassification and are determined from driving record (demerit) points assigned to the automobile. One point is assigned for each automobile accident involving the applicant, or any operator of the automobile resident in his household for every automobile accident within four years resulting in property damage above $50 or in bodily injury or death. Points are also assigned, of course, for such matters as driving while intoxicated, leaving the scene of an accident, and speeding which results in accidents or revocation of license.

The third item in the class designation consists of the letters EX or IN, the former denoting an experienced driver licensed for three years or more and the latter the inexperienced driver licensed for less than three years.

In connection with automobile liability rates it should also be noted that there has been made by many companies an extensive revision of this classification system. The classes mentioned have been replaced by some 260 "primary classifications" based on 52 separate "driver" categories—i.e., categories based on the age, sex and marital status of operators of the insured automobile —combined with five categories of car use: cars used for pleasure only, driven less than 10 miles to work, one-way; cars driven 10 miles or more to work, one-way; cars used for business; and cars used by farmers. The method of applying this method can be illustrated by means of the tables. Table 8A gives base rates for private passenger automobiles in various territories. Note that these rates cover not only bodily injury, but also property damage and medical payments. These rates are to be multiplied by two factors, the first of which depends upon the kind of driver and the second upon whether there is one or more cars in the family and whether the car is a compact or not. The factors for the drivers are given in Table 8B. While the factors for the cars are given in Table 8C, one determines the basic rate from Table 9, multiplies it by the driver factor from Table 8B, and then adds or deducts from it according to the factor in Table 8C. Then the final result is increased by a percentage according to the accident record of the driver on the point system described earlier in this section.

Risks in which the owner or principal operator of the automobile has been licensed less than three years are currently charged the base rate if no points have been assigned for accidents or convictions during the experience period. Under the proposed revisions, one driving record point will be charged to such risks. This one point will result in a surcharge of 30% of the "base premium." The "base" premium is the premium charged for subclass 0—that is, for a risk without driving record points. The surcharges for two, three or four or more points are 70%, 120% and 180% respectively, of the base premium.

TABLE 8A

SAMPLE BASE PREMIUM PAGE FOR THE
AUTOMOBILE CASUALTY MANUAL

NAME OF STATE
(State Code XX)

BASE PREMIUMS FOR
PRIVATE PASSENGER AUTOMOBILES

Bodily Injury		Territory Schedule							
Limits	Code	01	02	03	04	05	06	07	08
10/20	2	$46	$25	$33	$37	$31	$19	$29	$47
25/50	5	55	30	39	44	37	23	35	56
50/100	6	60	33	43	48	40	25	38	61
100/300	8	65	35	47	52	44	27	41	66

Property Damage		Territory Schedule							
Limits	Code	01	02	03	04	05	06	07	08
5,000	1	$22	$20	$19	$23	$20	$20	$20	$24
10,000	2	23	21	20	24	21	21	21	25
25,000	4	24	22	21	25	22	22	22	26

Medical Payments		Territory Schedule							
Limits	Code	01	02	03	04	05	06	07	08
500	1	$ 5	$ 4	$ 5	$ 5	$ 5	$ 4	$ 4	$ 5
750	2	6	5	6	6	6	5	5	6
1,000	3	7	6	7	7	7	6	6	7
2,000	4	9	8	9	9	9	8	8	9
5,000	6	12	11	12	12	12	11	11	12

Refer to company for the Base Premiums for limits other than those shown above.

Age, Sex and Marital Status	Pleasure Use	Work Less Than 10 Miles	Work 10 or More Miles	Business Use	Farm Use
NO YOUTHFUL OPERATOR					
Only Operator in Household is a Female Age 30-64	.90	1.00	1.30	1.40	.65
One or More Operators Age 65 or Over	1.00	1.10	1.40	1.50	.75
All Other	1.00	1.10	1.40	1.50	.75
YOUTHFUL UNMARRIED FEMALE OPERATOR —WITHOUT DRIVER TRAINING (No Other Youthful Operator)					
Age of Youngest Female Operator is					
17 or under	1.55	1.65	1.95	2.05	1.30
18	1.40	1.50	1.80	1.90	1.15
19	1.25	1.35	1.65	1.75	1.00
20	1.10	1.20	1.50	1.60	.85
YOUTHFUL UNMARRIED FEMALE OPERATOR —WITH DRIVER TRAINING (No Other Youthful Operator)					
Age of Youngest Female Operator is					
17 or under	1.40	1.50	1.80	1.90	1.15
18	1.25	1.35	1.65	1.75	1.00
19	1.15	1.25	1.55	1.65	.90
20	1.05	1.15	1.45	1.55	.80
YOUTHFUL MARRIED MALE OPERATOR Without Driver Training					
Age of Youngest Male Operator is					
17 or under	1.80	1.90	2.20	2.30	1.55
18	1.70	1.80	2.10	2.20	1.45
19	1.60	1.70	2.00	2.10	1.35
20	1.50	1.60	1.90	2.00	1.25
With Driver Training					
Age of Youngest Male Operator is					
17 or under	1.60	1.70	2.00	2.10	1.35
18	1.55	1.65	1.95	2.05	1.30
19	1.50	1.60	1.90	2.00	1.25
20	1.45	1.55	1.85	1.95	1.20
With or Without Driver Training					
21	1.40	1.50	1.80	1.90	1.15
22	1.30	1.40	1.70	1.80	1.05
23	1.20	1.30	1.60	1.70	.95
24	1.10	1.20	1.50	1.60	.85
25					
26					
27					
28					
29					

Age, Sex and Marital Status	Pleasure Use	Work Less Than 10 Miles	Work 10 or More Miles	Business Use	Farm Use
YOUTHFUL UNMARRIED MALE OPERATOR (Not Owner or Principal Operator) Without Driver Training					
Age of Youngest Male Operator is					
17 or under	2.30	2.40	2.70	2.80	2.05
18	2.10	2.20	2.50	2.60	1.85
19	1.90	2.00	2.30	2.40	1.65
20	1.70	1.80	2.10	2.20	1.45
With Driver Training					
Age of Youngest Male Operator is					
17 or under	2.05	2.15	2.45	2.55	1.80
18	1.90	2.00	2.30	2.40	1.65
19	1.75	1.85	2.15	2.25	1.50
20	1.60	1.70	2.00	2.10	1.35
With or Without Driver Training					
21	1.55	1.65	1.95	2.05	1.30
22	1.40	1.50	1.80	1.90	1.15
23	1.25	1.35	1.65	1.75	1.00
24	1.10	1.20	1.50	1.60	.85
25					
26					
27					
28					
29					
YOUTHFUL UNMARRIED MALE OWNER OR PRINCIPAL OPERATOR Without Driver Training					
Age of Youngest Male Operator is					
17 or under	3.30	3.40	3.70	3.80	3.05
18	3.10	3.20	3.50	3.60	2.85
19	2.90	3.00	3.30	3.40	2.65
20	2.70	2.80	3.10	3.20	2.45
With Driver Training					
Age of Youngest Male Operator is					
17 or under	2.70	2.80	3.10	3.20	2.45
18	2.65	2.75	3.05	3.15	2.40
19	2.60	2.70	3.00	3.10	2.35
20	2.55	2.65	2.95	3.05	2.30
With or Without Driver Training					
21	2.50	2.60	2.90	3.00	2.25
22	2.30	2.40	2.70	2.80	2.05
23	2.10	2.20	2.50	2.60	1.85
24	1.90	2.00	2.30	2.40	1.65
25	1.70	1.80	2.10	2.20	1.45
26	1.50	1.60	1.90	2.00	1.25
27	1.35	1.45	1.75	1.85	1.10
28	1.20	1.30	1.60	1.70	.95
29	1.10	1.20	1.50	1.60	.85

TABLE 8C

		Sub-Class				
		0	1	2	3	4
		Single Car				
Non-Compact	Factor	+0.00	+0.30	+0.70	+1.20	+1.80
Compact	Factor	−0.10	+0.20	+0.60	+1.10	+1.70
		Multi-Car				
Non-Compact	Factor	−0.15	+0.00	+0.20	+0.45	+0.75
Compact	Factor	−0.25	−0.10	+0.10	+0.35	+0.65

Example. Assume that the named insured is a married male, 42 years old, with a daughter, age 17, who has successfully completed a driver training course, and a son, age 19, who has not had driver training. Neither child is married. The insured has one automobile which he uses to drive to and from work, six miles each way. The insured has incurred two driving points under the Safe Driver plan.

For this insured the primary rating classification factor is 2.00—the factor for a youthful unmarried male operator, not the owner or principal operator, age 19, without driver training, works less than 10 miles. To this primary factor is added the secondary classification factor of 0.70. (This is the factor for a single automobile, non-compact, under Safe Driver subclass 2.)

Thus, the final rating factor for this automobile is 2.70. The base premium for all coverages desired would be multiplied by this factor to determine the final premium.

Automobile Collision and Comprehensive Insurance.—Just as liability insurance is computed, as described above, by tables and factors, similar methods are used for computing collision and comprehensive insurance. The latter depends, of course, upon the car itself as well as territorial factors and are necessary for the cars of all manufacturers, all models and ages. For example, a collision policy on a two year old Chevrolet Impala in Manhattan would have the following premiums for a $50 deductible policy: $119 if the operator being over 25 (male) and if the operator had a four year accident free record, provided he used the car for pleasure only. If he had had one accident the rate would increase to $145; if he used the car for business, but had had no accidents the rate would increase to $165 (if there was a male operator under 25 in the family the accident free rate would increase to $224. In this case it might be desirable to carry the $100 deductible rate, which would only be $146. The comprehensive policy on insurance of the vehicle covering fire, theft and glass breakage, etc., for this two year old Chevrolet Impala would be $17 on a $50 deductible basis of $51 for full coverage.

Stock

Stock is the term applied to shares which represent ownership in a corporation. A corporation is an intangible person created by the state upon the request of individuals interested in organizing it. It operates under a charter granted by the state which, among other things, specifies the amount of stock that the corporation is authorized to issue, and the par value, if any.

Par Value.—The par value of stock is the face value, the value at which it must be originally issued. Most states now permit stock to be issued without par value. Such stock is known as no-par-value stock, each share merely representing a fractional part of the total ownership in the business.

Capital Stock.—The capital stock of a corporation represents the amount authorized and paid into the corporation as capital for conducting the business. The capital stock is divided into a certain number of shares which may have a par value.

Dividends.—When profit is made by the corporation, the board of directors may decide to retain some or all of it in the business for working capital or pay some or all of it to the stockholders as their share of the profit. This share of profit is known as a dividend. They are usually declared and paid quarterly, and are stated and paid in terms of a certain percent of the par value of the stock, as a 2% quarterly dividend, or they may be quoted as so much a share, as $2.00 a share quarterly dividend.

Common Stock.—The regular stock of a corporation is known as common stock. It carries with it no special preferences with regard to the distribution of profits or of assets. Whatever profits are left after dividends on preferred stocks are paid may be distributed among the common stockholders.

Preferred Stock.—Some stock, in order to attract the more conservative investor, carries a guarantee to pay dividends before any profits are distributed to the common stockholders. It also may carry preference with regard to the distribution of assets in case the business is dissolved. Some preferred stock carries the provision that, in case the dividends are not paid when they should be, the unpaid dividends will be allowed to accumulate and will be paid before any dividend is paid on common stock. This is known as cumulative preferred stock.

Market Value.—For those interested in buying or selling stocks, the market value is more important than is the par value. The market value is the price at which it may be bought or sold and depends upon a number of factors including whether or not it pays a good dividend. A stock exchange is a place where stocks are bought and sold, the members of the exchange acting as brokers for those who are the actual buyers or sellers.

Round Lots and Odd Lots.—Most trading is done in units of 100 shares, such lots being known as Round Lots. When fewer than 100 shares is traded, or if a total number of shares traded in a transaction is not divisible by 100, the fractional part of 100 shares is known as an odd lot. Federal and New York State taxes are charged to sellers in all transactions and the buyers of odd lots only.

Brokerage Charges.—The members of the stock exchange, acting as brokers for those trading stock, charge a fee known as "brokerage." The fees for stocks bought or sold on the New York Stock Exchange and the American Stock Exchange are:

On stocks selling at $1.00 per share and above, commission

rates are based upon the amount of money involved in a single transaction and are not less than the following:

On Round Lots (each single transaction not exceeding 100 shares, in a unit of trading; a combination of units of trading; or a combination of a unit or units, plus an odd lot):

Money Involved	Commission
Under $100	As mutually agreed
$100 to $399	2% plus $3
$400 to $2,399	1% plus $7
$2,400 to $4,999	$\frac{1}{2}$% plus $19
$5,000 and over	$\frac{1}{10}$% plus $39

A table of rates based on the above formulas for selected key round-dollar stock prices is given below.

For Odd Lot Transactions (sales of less than a round lot) the rates are the same as above, less $2.00. On odd lots, you also pay an extra differential for the odd lot dealer who handles the sale for your broker. On the vast majority of stocks, for which a round lot is 100 shares, the odd lot differential is $\frac{1}{8}$ of a point (or $12\frac{1}{2}$¢) per share for stock selling for less than $40 in round lots; and $\frac{1}{4}$ of a point (or 25¢) for stocks selling at $40 or more in round lots.

Minimum—When the amount involved is less than $100, commission is as mutually agreed; $100 or more, not more than $1.50 per share or $75 per single transaction, but in no case less than $6.

On stocks, rights, and warrants selling below $1 a share, commissions are on a per share basis and range from 10 cents per 100 shares for stocks selling at $\frac{1}{256}$ to $5.25 per 100 shares for stocks selling for $\frac{7}{8}$ or more but less than $1.

Minimum—Less than $100, as mutually agreed, an amount of $100 or more, not less than $6 or the rate per share, whichever is greater.

Commodities.—The unit trading of grains is 5,000 bushels on which the brokerage varies somewhat, being twenty-two dollars for wheat. On the other hand, for sugar the brokerage is twenty-five dollars for 112,000 pounds.

Commission Rate Table
(Selected Round-Dollar Stock Prices)

Price	Rate per 100 Shares*	Price	Rate per 100 Shares*
$ 1	$ 6.00	19	$26.00
2	7.00	20	27.00
3	9.00	21	28.00
4	11.00	22	29.00
5	12.00	23	30.00
6	13.00	24	31.00
7	14.00	25	31.50
8	15.00	30	34.00
9	16.00	35	36.50
10	17.00	40	39.00
11	18.00	45	41.50
12	19.00	50	44.00
13	20.00	60	45.00
14	21.00	70	46.00
15	22.00	80	47.00
16	23.00	90	48.00
17	24.00	100	49.00
18	25.00	125	51.50

*Round lots only.

Stock Transfer Tax.—In addition to paying the brokerage charges, the seller, and sometimes the buyer, of stocks must pay a Federal tax and a state tax.

Federal Transfer Taxes

For stocks, warrants, and rights to subscribe to stocks: $0.04 on each $100 (or major fraction thereof) of the actual sales price of the securities, with a maximum of $0.08 per share for stocks selling at more than $200 per share.

For bonds and debentures: $0.05 for each $100 (or fraction thereof) of par or face value, regardless of the actual selling price (equivalent to $0.50 per bond of $1,000 par value). There is no Federal transfer tax on the sale or transfer of rights to subscribe to bonds or debentures.

STATE TRANSFER TAXES

New York—$0.01 per share under $5; $0.02 per share between $5.00 and $10; $0.03 per share between $10 and $20; $0.04 per share of $20 or more. New York State does not tax the sale or transfer of bonds, debentures, rights, or warrants.

Florida—$0.15 per $100 par value or per share on no par stock, regardless of selling price.

South Car.—$0.04 per $100 par value or per share on no par stock, regardless of selling price.

Texas—$0.033 per $100 par value or per share of no par stock, regardless of selling price.

Computing the Cost of Stocks.—There are four steps in computing the cost of a lot of stock. They are: Base Cost, Brokerage, Taxes (if any), and Total Cost.

ILLUSTRATION: A man bought 225 shares of a stock, the market price of which was 135. What was the total cost?

The cost would be calculated as follows:

1. 225 shares are 2 round lots and one odd lot of 25.
2. Base cost = 135×100 (one round lot) = $13,500.

Cost of 2 round lots = $13,500 \times 2$ = ...	$27,000.00
135×25 (odd lot) = ...	3,375.00
Total Base Cost	$30,375.00

3. Brokerage:

   Round lot brokerage = $(\frac{1}{10}\% + \$39) \times$ No. of round lots

$(\frac{1}{10}\% \times \$13,500 + \$39) \times 2$ =	
($13.50 + $39) \times 2$	$105.00
Odd lot brokerage = $(\frac{1}{10}\% + \$39)$ less $2.	
$(\frac{1}{10}\% \times \$3375 + \$39) - \$2$ =	
($3.38 + $39) − 2 = ...	$ 40.38
Total Brokerage	$145.38

4. Tax:

0.04% of $30,375 = $12.15	
0.04% × 225 = $9.00	
Total Tax	$ 21.15

5. Total Charges	$ 166.53
6. Total Cost	$30,541.53

It should be noted that, if this transaction had been for a round lot, the number of shares divisible by 100, no tax would have had to be paid.

Computing the Proceeds of a Sale of Stock.—There are also four steps in computing the proceeds from a sale of stock. They are:

1. Total proceeds. Multiply the selling price per share by the number of shares.

2. Brokerage. Multiply the number of shares by the brokerage rate as shown in the illustration below.

3. Taxes. Calculate stock transfer taxes on all shares sold.

4. Net Proceeds. Add the brokerage to the taxes to determine the total charges and deduct the total charges from the gross proceeds.

ILLUSTRATION: If a man sold 500 shares of a stock, the market value of which was $42\frac{3}{8}$ per share, how much did he receive?

The proceeds would be calculated as follows:

1. Total proceeds:
    $42.375 \times 100 \times 5 = 4237.50 \times 5 = $ $21,187.50
2. Round lot brokerage $= (\frac{1}{2}\% + \$19)$
    $\times$ No. round lots ($\frac{1}{2}\% \times 4237.50$
    $+ 19) \times 5 = (21.19 + 19) \times 5 =$
    $\$40.19 \times 5 = $ .... \$ 200.95
3. Taxes:
    0.04% of \$21,187.50 $= 84.75$
    $500 \times 0.04 = $ N.Y.S. Tax
        Total Tax .... \$ 104.75
        Total Charges .... -305.70
                                              \$20,881.80   (Ans.)

Listed and Unlisted Stocks.—Leading stock exchanges do not permit the trading of all stocks on the floor of the exchange, but rather restrict the trading to those that are listed by the exchange after an investigation into the corporation issuing the stock. These stocks are known as listed stocks. Other stock may be bought and sold with or without the services of a brokerage, but may not be traded on the floor of an exchange not listing them. These are known as unlisted stocks.

Short Selling.—The practice of selling stocks before the seller actually owns such stock is known as "short selling." The short seller hopes to make delivery of stock sold short by buying the stocks at a price lower than that at which he sold such stock. Because he is interested in having the price of the stock go down he is known as a "bear." The trader who buys first is said to be "long" on stock and hopes to sell at a price higher than he paid. As he is interested in the market going up, he is known as a "bull."

Bonds.—Governments and private corporations, when they borrow money for long periods, issue certificates of indebtedness known as bonds. When issued by private corporations, these bonds are usually secured by a mortgage on real estate, movable property, or by all the assets of the corporation. Bonds differ from stocks in that stocks represent a share of ownership in the corporation, while bonds represent indebtedness of the corporation, a liability.

There are many classes of bonds and various bases of classification. Most bonds carry a pledge of payment at a specified date and of a specified amount, the redemption date and redemption price, respectively. However, some bonds are subject to call at the demand of the bondholder, usually at specified dates before maturity. Serial bonds provide for redemption at a number of dates, instead of a single one. Annuity bonds provide for the repayment of both principal and interest, in equal installments at uniform periods.

Bonds are usually issued with a face value of $1000, although they sometimes are issued in smaller units such as $500.00, $100.00, or even $50.00. They bear interest which is usually payable semi-annually; also they have a due date. Some bonds have attached to them interest coupons, one for every six-month interest period during the life of the bond. These may be clipped every six months and cashed at any bank. This type of bond is known as a *coupon bond.*

Brokerage.—Bonds are bought and sold through brokers much the same as are stocks. Commission rates for the New York Stock Exchange are:

BOND COMMISSION RATES

(Except Government, Short Term or Called Bonds)

	Minimum Commission Rate per $1000 (of Principal) Bond			
Price per $1000 (of Principal) Bond	On orders of 1 or 2 bonds	On orders of 3 bonds	On orders of 4 bonds	On orders of 5 bonds or more
Selling at less than $10 (1%)................	$1.50	$1.20	$0.90	$0.75
Selling at $10 (1%) and above but under $100 (10%)	2.50	2.00	1.50	1.25
Selling at $100 (10%) and above...........	5.00	4.00	3.00	2.50

A Federal Bond Transfer tax of 50¢ per $1000 of par value is charged the seller of all bonds. United States Government, state, municipal, and foreign government bonds are exempt from any tax.

Accrued Interest.—Bonds may be sold on the date that interest is due and paid or may be and frequently are sold at other times. When sold between interest dates, the seller of the bond is entitled to the interest that has been earned since the last interest date, but which is not yet paid. This is known as accrued interest.

Calculating the Proceeds of a Sale.—There are four or five steps in calculating the proceeds of a sale of bonds. They are:

1. Market Value. Multiply the quoted price by the par value of the bonds being sold.
2. Calculate the accrued interest and add to the market value.
3. Calculate the brokerage charges.
4. Calculate the Federal Tax and add it to the brokerage to determine the total charges.
5. Deduct the total charges from the market value plus accrued interest. If the bonds are sold on an interest date, the second step is eliminated as there would be no accrued interest.

ILLUSTRATION: Ten $1000 Railroad $4\frac{1}{2}\%$ bonds due in 1984 with interest, payable on April 1 and October 1, were sold on September 1, 1954, at $107\frac{3}{4}$. What were the proceeds?

1. Market Value
     $10,000 \times 1.0775$............................... $10,775.00
2. Accrued Interest
     Interest for 5 months @ $4\frac{1}{2}$ on $10,000 par
     value $10,000 \times \frac{4.5}{100} \times \frac{150}{360}$...................   187.50

     Present Value.............................. $10,962.50
3. Brokerage
     $1.25 per $1000 of par value equals 1.25×10. $12.50
4. Transfer Tax
     50¢ per 1000 of par value equals 50¢ $\times$ 10.... $5.00

     Total Charges............................... -17.50

     Net Proceeds................................ $10,945.00

Calculating the Cost of Bonds.—There are three or four steps (depending upon whether or not the bonds are bought on interest dates) that must be followed in calculating the cost of bonds.

1. Calculate the market value.
2. Calculate the accrued interest (if bond is bought between interest dates) and add to market value.
3. Calculate brokerage charges.
4. Add brokerage fees to present value to determine total cost.

ILLUSTRATION: One $1000 5% bond due on June 1, 1959, was bought on September 1, 1954, at the rate of 105, the cost would be calculated as follows:

1. Market Value
     1000×1.05.................................... $1050.00
2. Accrued Interest, 5% on $1000 for 3 months
     $1000 \times \frac{5}{100} \times \frac{90}{360}$................................   12.50

     Present Value................................. $1062.50
3. Brokerage, $2.50 per $1000 par value.....................   2.50

4. Total Cost.. $1065.00

ANNUITIES

Broadly speaking, an *annuity* is a number of equal payments made at equal periods of time. While technically the term annuity means "annual," practically the term annuity is applied whether the payments are made annually or at other intervals such as semi-annually, quarterly, or monthly.

TABLE 9A

MONTHLY LIFE INCOME FROM RETIREMENT ANNUITY PER $100
ANNUAL PAYMENT TAKEN OUT BY MAN AGED 40

| Retirement Age | Monthly Income | | |
	Whole Life Annuity[1]	Installment-Refund Annuity[2]	Life Annuity Income Guaranteed for 10 years[3]
50	$ 5.25	$ 4.65	$ 5.16
55	8.76	7.58	8.41
60	13.92	11.27	12.56
65	19.94	15.72	17.65
70	27.48	20.50	24.17

[1] Terminates with death of annuitant. No further payment made by company.

[2] If annuitant dies before payments exceed purchase price, the difference is paid to beneficiary. If annuitant dies after payments exceed purchase price, beneficiary receives nothing.

[3] Payments guaranteed for life. Death of annuitant before 10 years results in payment to beneficiary until end of 10 year period.

Contingent Annuities: Basically there are two kinds of annuities: contingent annuities, and certain annuities. Contingent annuities are annuities in which the date of the last payment or the first payment or both cannot be foretold. Old age pension plans are contingent annuities.

Annuities Certain: Annuities certain are annuities in which the dates involved, beginning and ending, may be definitely established. There are many types of annuities that are certain, one of the most common being the type in which an individual is interested in investing or depositing a fixed sum annually (or at other intervals) and desires to know how much the money will be worth at the end of a given number of years.

ILLUSTRATION: Beginning on July 1, 1954, a man deposits $500 a year at a place where interest is paid annually at the rate of 5%. How much of a fund will he have built up after he has made his fifth deposit? The deposit date and the interest date are the same.

Calculated by Simple Arithmetic: There are several methods of determining the amount that will be on deposit when the last payment is made, the most cumbersome being the procedure whereby one calculates the interest at the end of the first period, adds it to the principal, adds the new deposit, calculates the interest on the total at the end of the second period, adds it to the balance, adds the new deposit, etc. Follow this procedure for each year.

This would work out as follows:

July 1, 1954	First Deposit..................................	$ 500.00
July 1, 1955	Interest at 5% on $500........................	25.00
	Second Deposit...............................	500.00
		$1025.00
July 1, 1956	Interest at 5% on $1025.......................	51.25
	Third Deposit................................	500.00
		$1576.25
July 1, 1957	Interest at 5% on $1576.25....................	78.81
	Fourth Deposit...............................	500.00
		$2155.06
July 1, 1958	Interest at 5% on $2155.06....................	107.75
	Fifth Deposit................................	500.00
	Total.....................................	$2762.81

Calculated by Logarithms: Naturally this method takes too long for practical purposes, especially if the number of deposits is high. If this type of problem is encountered frequently, it is best to be provided with annuity tables. These are pre-computed tables indicating how much $1.00 accumulates to if deposited at regular periods of time at regular interest rates. The amount given in the table is then multiplied by the amount of the regular deposit. If annuity tables are not available, this problem may be calculated by the use of logarithms. The formula is:

$$\frac{(1 + i)^n - 1}{i} \times \text{regular payment} = \text{Amount of Annuity}$$

which, in this particular problem, would be:

$$\frac{(1 + .05)^5 - 1}{0.05} \times 500 = \text{Amount of Deposit at end of 5th year.}$$

The solution by logarithms is as follows:

$$\text{Log} = 1.05 \quad = 0.021189$$
$$\times 5$$
$$\overline{0.105945} = 1.2762$$

$$\therefore \quad (1.05)^5 = 1.2762, \quad (1.05)^5 - 1 = 0.2762$$

Amount = $\underline{500 \times 0.2762}$

$$0.05$$

Log   500 = 2.698970
Log   0.2763 = 9.441381 − 10
Co-log 0.05   = 1.301030

$$\overline{3.441381} = \text{Total } \$2763 \quad \text{(Ans.)}$$

Note.—This total is 17¢ greater than the one determined by arithmetic. This is due to the slight inaccuracies resulting from the use of six-place logarithms.

Amount of Annual Payment: Somewhat the reverse of the problem just presented is that in which an individual desires to know how much of an annual deposit he must make at a given rate of interest in order to build up a given sum. This may be determined by calculating the amount of annuity of $1.00 for the given number of periods and dividing this into the amount desired. If annuity tables are available, this amount of an annuity of $1.00

at a given interest rate for a given number of periods may be readily determined. In the absence of such tables, the problem may be worked out by logarithms on the following formula:

$$\text{Total amount} \div \frac{(1 + i)^n - 1}{i}$$

ILLUSTRATION: A man chooses to build up a sum of \$5000 over a period of 10 years by making an annual deposit. The interest rate is 4% and it may be assumed that the annual deposit will be made on the same date that interest is credited. How much must be paid annually? The formula applied to this problem will read:

$$\$5000 \text{ divided by } \frac{(1 + .04)^{10} - 1}{0.04}$$

$$\begin{aligned} \text{Log } 1.04 &= .017033 \\ &\quad \times 10 \\ \hline .170330 &= 1.4802 \end{aligned}$$

$$\therefore \quad (1.04)^{10} = 1.4802; \quad (1.04)^{10} - 1 = 0.4802$$

$$\begin{aligned} \text{Log} \quad 0.4802 &= 9.681422 - 10 \\ \text{Co-log } 0.04 \quad\quad & 1.397940 \\ \hline 1.079362 &= 12.005 \end{aligned}$$ (Amount of annuity of \$1.00 will equal at end of 10-year period)

$$\begin{aligned} \text{Log } \$5{,}000 &= 3.698970 \\ \text{Less} \quad & 1.079369 \\ \hline 2.619608 &= \$416.48 \end{aligned}$$ Amount of Annual Payment

(Ans.)

Sinking Fund: A sinking fund is a fund established for the purpose of paying off a debt or of making some other necessary payment. Many industrial bond issues, and some others, are paid off on their due dates from a sinking fund; in fact, the terms of the bond quite frequently require this procedure. The sinking fund is established by placing in the fund each year an amount in cash that if invested immediately at a given rate of interest, will accumulate to a sum equal to the total indebtedness on the date that the obligation must be paid. The amount of the annual payment is determined in exactly the same manner as it was in the preceding problem.

ILLUSTRATION: A corporation issued $500,000 worth of bonds due in 20 years and desired or was required to set up a sinking fund, the amount of the annual payment, assuming interest at $4\frac{1}{2}\%$, would be determined by using the following formula:

$$\$500,000 \text{ divided by } \frac{(1.045)^{20} - 1}{0.045}$$

Present Value of an Annuity: The term annuity means annual payment. Another basic problem in dealing with annuities is that of determining the sum which, placed at a given rate of interest, will make it possible to pay out a given amount each year for a given number of years. This is known as the Present Value of an annuity.

ILLUSTRATION: A man chooses to place at interest a sum that will permit him to pay out $1200 per year for a period of four years, the first withdrawal to be made one year after the fund is established. The interest rate is 5%. How much must he deposit?

This, in reality, is a problem of calculating the compound discount on the sums to be paid. If annuity tables are available it will be found that an annuity of $1.00 for 4 periods at 5% is 3.545950. If an annuity of $1200 is to be available, 3.545950 × $1200 = $4255.14, the amount that must be deposited to establish the fund.

If the annuity tables are not available, this problem may be worked out with the use of logarithms by following the following formula:

$$\frac{1 - \dfrac{1}{(1 + i)^n}}{i} \times 1200$$

In terms of this problem, the formula would read:

$$1 - \frac{1}{(1 + .05)^n} \times \$1200$$

$$\text{Log } 1.05 = 0.021189$$
$$\times 4$$
$$\overline{0.084756}$$

Subtracted from log $1 = 1.0$

$$9.915244 - 10 = 0.82270$$
$$1 - 0.82270 = 0.17730$$

Log $0.17730 = 9.248709 - 10$
Co-log $0.05 \quad = 1.301030$

$0.549739 = 3.546$, Amount necessary for annuity of $1

Log $\$1200 = 3.079181$

$3.628920 = \$4255.20$, the amount of the fund to be established.

Deferred Annuities: This is an annuity, the payments on which do not begin for some time after the fund is established. Such an annuity might be established when a child is very young with a view toward paying his expenses through college.

ILLUSTRATION: Let us assume that in the previous problem, the fund was to be established 12 years earlier; or in other words, that payments were to be begun 13 years after the fund was originally set up. How much must be deposited if interest at the rate of 5% per annum will be earned?

This problem simply involves the calculation of present worth of the fund to be set up as of 12 years in advance. If the fund normally to be established was $4255.20, a somewhat smaller sum will suffice if the deposit is to be made 12 years in advance and allowed to accumulate at compound interest (we will assume 5%) for all that time. This problem may be solved by determining the present value of $1.00 and multiplying at the present worth of the ordinary annuity which in this case is $4255.20.

$\dfrac{1}{(1 + i)^n}$ = present worth of $1.00, therefore the formula as applied to this problem is:

$$\dfrac{1}{(1 + .05)^{12}} \times \$4255.20$$

$$\text{Log } 1.05 = 0.0211809$$
$$\times\ 12$$

$$\overline{0.254268}$$

$$\text{Subtracted from log } 1 = 1.0$$

$$\overline{9.745732} - 10 = 0.55684$$
$$\text{Log } \$4255.20 = 3.628920$$

$$\overline{3.374652} = \$2369.50\text{, the amount that}$$
must be deposited if the annuity is to be established 12 years in advance.

Amortization: Strictly speaking, to amortize means to extinguish or liquidate a debt. Actually, however, there are two methods of disposing of debts: One, by the sinking fund method, which has already been described, and the other by the method known as amortization which, in common practice, means to liquidate the debt by making a series of equal periodic payments which include a part payment on the principal as well as interest on the principal outstanding.

Actually, this problem is another annuity problem in which the amount of the periodic payment is to be determined. It may be solved simply by dividing the present worth of an annuity of $1.00 at the given rate of interest for the given number of years into the full amount of the debt to be amortized. If annuity tables are not available, this problem which is called that of determining the amount of an annuity, may be solved by using logarithms, applying the following formula:

$$\dfrac{i}{1 - \dfrac{1}{(1 + i)^n}} \times \text{debt}$$

ILLUSTRATION: A debt of $12,000 is to be amortized over a period of 15 years, the interest rate being 5%. How much must be paid each year?

In terms of this problem, the formula would read:

$$\frac{0.05}{1 - \dfrac{1}{(1.05)^{15}}} \times \$12,000$$

$$
\begin{aligned}
\text{Log of } 1.05 &= 0.021189 \\
&\ \underline{\times\ 15} \\
&\ 0.317835
\end{aligned}
$$

$$
\begin{aligned}
\text{Log of } 1 &= \overline{1.0} \\
\text{Less } &\ \underline{.317835} \\
&\ 0.682165 = 0.48102
\end{aligned}
$$

$$
\begin{aligned}
1 - 0.48102 &= 0.51898 \\
\text{Log } 0.05 &= 8.698970 - 10 \\
\text{Co-log } 0.51898 &= \underline{.284848} \\
\text{Log } 12,000 &= \underline{4.079181} \\
&\ 3.062999 = \$1156.10, \text{ the amount that}
\end{aligned}
$$

must be paid annually
(Ans.)

Depreciation.—There are a number of methods of computing depreciation charges. The method used in a particular instance is often a matter of the practice of the particular business, or for the particular type of asset. The choice of a method is not necessarily a unilateral decision of the individual business or accountant; the policy of governmental taxing agencies must frequently be considered, in view of the close relationship between depreciation charges and earnings. It is the policy of the U. S. Government to permit the annual depreciation deduction to be computed in any manner that is consistent with recognized trade practices. Three methods were specifically listed, and therefore these methods are discussed in this chapter. They are the straight-line method, the declining-balance method, and the sum of the years-digits method. There are, however, certain other general methods of computing depreciation charges, which are permitted if they are so used as to give results consistent with those obtained by the listed methods. Two of these methods—the unit of production method, the sinking fund method and a combination method are also included in this section.

Before discussing any of these methods, however, it is to be understood that their use in U. S. practice is subject to the general limitations upon permissible depreciation which are set forth in Publication 311 of the U. S. Treasury Department of Internal Revenue, which is part of the Code of Federal Regulations. That publication sets forth general regulations on depreciation which stipulate the general character of property which may be depreciated, and the general rule that experience with similar equipment in the same industry be the basis for the assumed life of equipment. There is also, however, Internal Revenue Service Publication #173 of the U. S. Treasury Department which is called Bulletin F, and which gives information and tables of the average useful life of depreciable equipment. The data in this bulletin are arranged by industries, and vary greatly in the detail with which individual industries are treated.

The Straight Line Method.—The straight line method is one of the most simple methods of computing depreciation charges. It consists of dividing the total depreciation expense (which is initial cost of the asset minus its scrap value) by the number of periods of useful life. The quotient is depreciation expense per period, which by this method is the same for each period. For example, an asset having an initial cost of \$1100.00; a scrap value of \$100.00; and a useful life of 10 years would have a depreciation expense of $\dfrac{\$1100-\$100}{10} = \$100.00$ per year. Since by this method this figure is the same for each year of the life, the book value (which is initial cost − depreciation fund) decreases at the same rate from year to year, and its graph is a straight line, which gives the name to the method.

As another example, find the annual depreciation charge of a machine costing \$865.00, having a scrap value of \$61.25 at the end of a probable life of 5 years. Also find the book value at the end of 3 years.

$$\text{Depreciation expense} = \$865 - \$61.25 = \$803.75$$
$$\text{Depreciation charge per year} = \frac{\$803.75}{5} = \$160.75$$
$$\text{Depreciation fund after 3 years} = 3 \times \$160.75 = \$482.25$$
$$\text{Book value after 3 years} = \$865.00 - \$482.25 = \$372.75$$

It is to be noted that the straight-line method does not require the existance of a scrap value at the end of the period of depreciation, if the nature of the asset is such that its terminal value is then zero. In that case, the annual depreciation charge is the initial cost divided by the probable life in years.

However, by U. S. tax regulations if an asset is sold at the end of a depreciation period for more than the assumed scrap value, or if it is sold during the period for more than its current book value, the difference must be taken as a capital gain. On the other hand, failure to deduct the full amount of depreciation allowable each year does not prevent such amount from reducing the adjusted basis, nor does it entitle the taxpayer to a greater deduction in a subsequent year.

The Declining-Balance Method.—The declining-balance method is also designated by the more descriptive name of constant-percentage method. It is also referred to as a liberalized method, since it may be so applied as to give depreciation charges during the early years that exceed those calculated by the straight-line method. However, the extent of that excess is limited by U. S. Government regulations (generally to twice the depreciation charges for the first year calculated by the straight-line method). Also, the type of property to which the liberalized declining method may be applied is restricted; it cannot be applied to leases, patents and other intangible assets.

The basis of the declining-balance method is to depreciate the asset each year by a constant percentage of its book value.

This percentage may be assumed arbitrarily, or computed by the formula,

$$\text{Percentage Depreciation } d = 1 - \sqrt[n]{\frac{S.V.}{C.}} \tag{1}$$

Where n is the period of depreciation in years, $S.V.$ is the Scrap Value and C is the original cost.

For example, an asset costing $4,000.00 on January 1, 1961 estimated to have a scrap value of $800.00 in ten years, would have an annual percentage depreciation:

$$d = 1 - \sqrt[10]{\frac{\$800}{\$4000}} = 1 - \sqrt[10]{0.2}$$

The figure $\sqrt[10]{0.2}$ is evaluated by logarithms to be .85.
So that $d = 1 - .85 = .15 = 15\%$

Then applying this annual depreciation rate, the asset would have the following depreciation charges and book value for the first 5 years:

Depreciation Rate 15%

(Values of Asset costing $4,000.00 at Jan. 1 in various years)

Year	1961	1962	1963	1964	1965
Book Value	$4,000.00	$3,400.00	$2,890.00	$2,456.50	$2,088.02
Depreciation Charge		600.00	510.00	433.50	368.48
Depreciation Fund		600.00	1,110.00	1,543.50	1,911.98

Since the asset was purchased on January 1, 1961, it has undergone one year's depreciation by January 1, 1962. The charge for this depreciation is 15% of its book value at the beginning of 1961, which is 15% of $4,000.00, the cost. Therefore, the new book value as of January 1, 1962 is book value at beginning of year minus the depreciation charge, which is $4,000.00 − $600.00 = $4,400.00, the $600.00 going into the depreciation fund.

On January 1, 1963, the depreciation charge for 1962 is 15% of $3,400.00 (book value at beginning of 1962) = $510.00. Then the new book value as of January 1, 1963 is that at beginning of year minus the depreciation charge, which is \$3,400.00 minus \$510.00 = \$2,890.00, the \$510.00 going into the depreciation fund.

Thus, for each year, the depreciation charge is found by multiplying the book value at the beginning of the year by the depreciation rate, while the book value at the end of the year is the book value at beginning of the year less the depreciation charge. The depreciation fund is the total of the depreciation charges.

As explained there, machinery and other depreciable assets are often carried on the books and reported on the balance sheet at their initial cost, while the depreciation charges that have been made against these assets over the years are reported as a total "fund" or "reserve" which is subtracted from the initial cost to show the present value of the particular class of assets. It follows that a depreciation fund is merely a total of charges that have been made. It is thus not a "fund" at all in the ordinary everyday use of that term to denote a sum of money. The latter type of fund is called a sinking fund, and is treated in the last section of this chapter.

To derive general formulas, write these terms in symbols, using C for cost, BV for book value, D for depreciation charge, d for depreciation rate (decimal value), and $\sum D$ (the Greek letter $\sum$, sigma, means "sum of") for depreciation fund.

Then at the end of one year:

Depreciation charge for the year $= D_1 = Cd$

Book value $= (BV)_1 = C - Cd = C(1 - d)$

Depreciation fund $= \sum_{}^{1} D_i = D_1 = Cd$

Then at the end of the second year:

Depreciation charge for the second year $= D_2 = (BV)_1 d = C(1 - d)d$

Book value at end of second year $= (BV)_2 = (BV)_1 - D_2 = C(1 - d) - C(1 - d)d = C(1 - d)^2$

Depreciation fund at end of second year $= \sum_{}^{2} D_i = C - (BV)_2$

At the end of the third year;

Depreciation charge for the third year $= D_3 = (BV)_2 d = C(1 - d)^2 d$

Book value at end of third year $= (BV)_3 = (BV)_2 - D_3 = C(1 - d)^2 - C(1 - d)^2 d = C(1 - d)^3$

Depreciation fund at end of third year $= \sum^{3} D_i = C - (BV)_3$

Then at the end of the nth year

Depreciation charge for the nth year $= D_n = C(1 - d)^{n-1} d$ (2)

Book Value at end of nth year $= (BV)_n = C(1 - d)^n$ (3)

Depreciation fund at end of nth year $= \sum^{n} D_i = C - (BV)_n$ (4)

To apply these general formulas, solve the problem: An asset is purchased for \$8,000.00 on January 1, 1955. If it is depreciated by the declining-balance method at 14% annually, what is its book value on January 1, 1965, what is its depreciation fund on that date, and what is the depreciation charge from January 1, 1964 to January 1, 1965 (tenth year)?

Then by the Formula (3)

$$(BV)_{10} = C(1 - d)^{10} = (\$8,000)(1 - .14)^{10}$$
$$= (\$8,000)(.86)^{10}$$

This computation is readily made by logarithms to give \$1,770.39, which is the book value of the asset after depreciation at 14% for 10 years.

Then by Formula (4), depreciation fund = cost − book value = \$8,000.00 − \$1,770.39 = \$6,229.61, and by Formula (2) the depreciation charge during the tenth year $= C(1 - d)^{n-1} d = (\$8,000.00)(1 - .14)^9 (.14)$. This computation is readily made by logarithms, as was that of the book value above, to give \$288.20, which is the depreciation during the tenth year.

Note that by the declining-balance method an asset depreciated at an annual rate of 14% had a book value after 10 years of \$1,770.39 or about 22% of its initial cost, whereas if the straight-line method had been used on the basis of a 10-year life and no scrap value, its first year depreciation would have been only 10% of the cost, but the book value would have been reduced to 0 in 10 years. Thus, the declining balance method gives more rapid initial depreciation, but slower depreciation later and never depreciates completely.

To compensate partly for this effect, the U. S. Government allows higher initial depreciation rates where the declining-balance method is used, extending up to twice the straight line rate, that is, an asset with an estimated life of ten years may be depreciated as much as 20% annually, under specified restrictions.

The advantage of using the accelerated declining-balance method, or other accelerated methods, is felt particularly in instances where depreciation is hastened by obsolescence. For example, the useful life of a machine may be terminated, not by wear to a point where maintenance becomes excessive, but by the introduction of new machines so much more efficient that the old machine can no longer operate competitively. In industries undergoing rapid development, therefore, methods of accelerated depreciation are highly advantageous.

It should be added here that the U. S. Tax Code states that the annual depreciation deduction could be increased by an allowance for extraordinary obsolescence, starting from the year in which it was reasonably certain that the property was affected by revolutionary inventions, abnormal growth, or radical economic changes, which cause its abandonment prior to the end of its normal useful life.

The Sum of Years—Digits Method.—This method is allowable, under U. S. Tax Regulations, for property similar to that for which the liberalized declining-balance method may be used.

In this method the depreciation in any year is found by multiplying the cost of the asset less its scrap value by a fraction, whose numerator is the number of the year in question (counting from date on which the asset is to be scrapped) and whose denominator is the values of all the years added together.

For example, an asset costing $6,000.00 purchased January

1, 1960, with a useful life of 6 years, and a scrap value of $460.00, would have:

The number 6 assigned to the first year of its life.
The number 5 assigned to the second year of its life.
The number 4 assigned to the third year of its life.
The number 3 assigned to the fourth year of its life.
The number 2 assigned to the fifth year of its life.
The number 1 assigned to the sixth and last year of its life.
The number 21 as the total of its year numbers.

Then the depreciation charge during the first year
$$= \tfrac{6}{21} \text{ (Cost } - \text{ Scrap Value)} = \tfrac{6}{21} \text{ (\$6,000.00} - 460.)$$
$$= \tfrac{2}{7} \text{ (\$5540.00)}$$
$$= \$1,582.86.$$

The depreciation charge during the second year
$$= \tfrac{5}{21} \text{ (\$5,540.00)} = \$1,319.05$$

The depreciation charge during the third year
$$= \tfrac{4}{21} \text{ (\$5,540.00)} = \$1,055.23$$

The depreciation charge during the fourth year
$$= \tfrac{3}{21} \text{ (\$5,540.00)} = \$791.43$$

The depreciation charge during the fifth year
$$= \tfrac{2}{21} \text{ (\$5,540.00)} = \$527.62$$

The depreciation charge during the sixth and last year
$$= \tfrac{1}{21} \text{ (\$5,540.00)} = \$263.81$$

Note that the depreciation fund at the end of any period is found by adding the depreciation charges, and that the book value at the end of any period is the difference between initial cost and depreciation fund.

For example, the depreciation fund at the end of six years is

$$\$1582.86, D_1$$
$$1319.05, D_2$$
$$1055.23, D_3$$
$$791.43, D_4$$
$$527.62, D_5$$
$$263.81, D_6$$

Depreciation or fund at end of 6 years $= \sum_{}^{6} D_i = \$5540.00$

which checks the difference between cost and scrap value, as it should, since the calculation above was made on the basis of a useful life of 6 years.

Unit of Production Methods.—The basis of these unit of production methods, which apply most directly to machinery, is production. If a machine fabricates one article, or a limited number of kinds of articles, its output is most conveniently measured in terms of the number of articles produced. On the other hand, if the machine is used for working on a relatively small number of large articles, especially if several machines of the same type are so used, then the output is probably better measured in machine hours.

Whichever of these methods are used, they require the maintenance of records showing the number of hours for which the machine is operated daily, or the number of units which it produces daily, so that its annual output in machine hours or production units can be computed. There are also required estimates of the useful life of the machine in the same terms, that is, in terms of machine hours or units produced. In addition, of course, the initial cost and estimated salvage value of the machine are also required.

For example, a machine costing \$6,000.00 has an estimated salvage value of \$400.00, and an estimated production during its useful life of 70,000 units. What is its depreciation charge, depreciation fund and book value during its first four years of operation if its production is: First year, 10,000 units, second year, 6,000 units, third year, 8,000 units, fourth year, 7,000 units?

Then, first year; Depreciation charge $= D_1 = \left(\dfrac{10,000}{70,000} \right)$

(\$6000.00 $-$ \$400.00) $=$ \$800.00; Depreciation fund $= \sum^{1} D_i =$ \$800.00; Book value $=$ \$6000.00 $-$ \$800.00 $=$ \$5200.00.

Second year. Depreciation charge $= D_2 = \left(\dfrac{6,000}{70,000} \right)$

(\$6000.00 $-$ \$400.00) $=$ \$480.00; Depreciation fund $= \sum^{2} D_i =$ $D_1 + D_2 =$ \$1,280.00. Book value $=$ \$6000.00 $-$ \$1,280.00 $=$ \$4720.00.

Third year. Depreciation charge $= D_3 = \left(\dfrac{8,000}{70,000}\right)$ ($6000.00

$- \$400.00) = \$640.00$; Depreciation fund $= \sum_{}^{3} D_i = D_1 + D_2 + D_3 = \$1920.00$. Book value $= \$6000.00 - \$1920.00 = \$4080.00$.

Fourth year. Depreciation charge $= D_4 = \left(\dfrac{7,000}{70,000}\right)$ ($6000.00 $- \$400.00) = \$560.00$; Depreciation fund $= \sum_{}^{4} D_i = D_1 + D_2 + D_3 + D_4 = \$2480.00$; Book value $= \$6000.00 - \$2480.00 = \$3520.00$.

Note that the general formula for depreciation charge, as is apparent from the foregoing examples, can be written as

$$D_i = \left(\frac{N_i}{N_T}\right)(C - S.V.) \tag{5}$$

where D_i is the depreciation charge in any one year; N_i is the number of units produced in that year; N_T is the estimated total number of units produced during the useful life of the machine; and C and $S.V.$ are the initial cost of the machine and its estimated scrap value at the end of its useful life, respectively.

Depreciation calculations based upon machine hours are made in essentially the same way as those just described for the unit production basis. The annual depreciation charge can be computed by the relation $(C - S.V.)$ where N_i is the number of hours the machine is operated in a given year, N_T is the estimated total number of hours which the machine can operate in its useful life, C is initial cost and $S.V.$ is estimated scrap value at the end of useful life. Also, the unit cost method can be used for machine-hour calculations of depreciation in the form, $N_i = \left(\dfrac{C - S.V.}{N_T}\right)$ where N_i is again the number of hours the machine is operated in a given year, and the expression in parentheses is the depreciation charge per hour operated.

Sinking Fund Method.—The depreciation calculations made up to this point have been based upon the general usage of the term depreciation. In other words, the depreciation charges have been treated from the point of view of deductions from earnings to offset the gradual loss of assets by depreciation, which indeed they are. However, if sums of money are segregated to constitute a sinking fund to replace the depreciated assets, a somewhat infrequent practice, then the depreciation charges can be reduced by the interest which they will earn from the date of deposit to the date of retirement of the depreciated asset. The actual amount of the sinking fund can be computed by the method for finding the value of the rent of an ordinary annuity, as explained in Chapter 14.

As an example of the application of this method, find the depreciation charge for each year, and the value of the sinking fund at the end of each year for a machine costing $865.00, having a scrap value of $61.25 at the end of a probable life of 5 years, if interest is credited at 4%.

The total depreciation cost for the 5 years, which is the sum the sinking fund must then total, is Cost − Scrap Value = $865.00 − $61.25 = $803.75.

The annual payment into the sinking fund is found by the formula for the periodic reserve payment into an annuity, which is

$$R = P \frac{i}{1 - \dfrac{1}{(1 - i)^n}} \tag{6}$$

where P is the principle, i is the decimal interest rate per period, and n is the number of periods. Substituting

$$R = (\$803.75) \frac{.04}{1 - \dfrac{1}{(1 - .04)^5}}$$

Solving by logarithms, we find that R is $148.39 annually.

Then at the end of the first year, the depreciation sinking fund is $148.39.

At the end of the second year, the amount of the depreciation sinking fund is ($148.39) (1.04) + $148.39 = $154.32 + $148.39 = $302.71.

At the end of the third year, the amount of the depreciation sinking fund is ($302.71) (1.04) + $148.39 = $314.81 + $148.39 = $463.30.

At the end of the fourth year, the amount of the depreciation sinking fund is ($463.30) (1.04) + $148.39 = $481.82 + $148.39 = $630.21.

At the end of the fifth year, the amount of the depreciation sinking fund is ($630.21) (1.04) + $148.39 = $655.42 + $148.39 = $803.81, which is six cents greater (due to the fractional cent rounding-off) than the required amount.

Note that the annual depreciation charges computed by this method resemble those by the straight-line method in that they do not change throughout the depreciation period. They differ, however, in that they are not only charges, but sums of money, and are therefore credited with interest, which reduces their amount. To gain an idea of the effect of this reduction, compare the annual charges on a $865.00 asset, with a scrap value of $61.25 and a life of 5 years. When this computation was made by the sinking fund method above, the annual depreciation charge and deposit was found to be $148.39, against an annual charge of $160.75 for the same asset when computed earlier in this chapter by the straight line method. The figure of $148.39 was based upon 4% interest rate on the sinking fund; higher rates would increase the difference, and lower rates would decrease it.

Combination Methods.—In recent years the United States Internal Revenue Service has been willing to accept in many instances a combination of the straight line and the double declining balance methods. This combination can best be illustrated by an example. Consider an asset with an accepted life of 10 years, costing $100. By the accepted method one would depreciate it by the straight line amount of $10 for the first year. In later years, he could choose whichever of these two methods gave the highest depreciation. Thus the second year one would use the $9.00 figure of the declining balance method multiplied by 2, which is $18, leaving a value of $72. Likewise, on the third year he would use twice the declining balance amount of $7.20, which is $14.40, leaving a balance of $57.60. Simply the fourth year one would use a figure of $11.52, leaving a balance of $46.08. Then thereafter one could revert to the $10 annual basis of the straight line method, which gives greater deductions than the double declining balance method.

FOREIGN EXCHANGE

In its conventional sense, the term Foreign Exchange means the commercial paper and instruments used in foreign trade and the problems attending the settling of them. The most commonly used items are *Bills of Exchange*.

Bill of Exchange.—A draft drawn by one party ordering a second party to pay to the first party or to a third party a given sum of money is a bill of exchange. When such drafts are used for domestic exchange, they are usually known as checks, or drafts. When they are used in foreign exchange, they are called foreign drafts or, more frequently, foreign bills of exchange. Bills of exchange may be payable on demand or they may be payable at a later date, the most common times being 30, 60, or 90 days after date.

Cables.—Orders for the transfer of money abroad that are transmitted by wire are known as telegraphic transfers or, more frequently in the United States, as cables. They are not bills of exchange in the strict sense of the term because they are not written bills and not negotiable instruments, but they are used extensively to take the place of bills of exchange because of the speed with which a transaction may be completed. If a regular demand draft is drawn and mailed, several days will elapse between the time it is mailed in this country and the time it is delivered in a European country or almost any other foreign country. The same transaction may be completed in a few minutes by using a "cable" instead of a regular draft.

Buying and Selling Foreign Exchange.—Buying and selling foreign exchange is handled through banks who sell exchange at whatever it will bring at a given time and who buy it for whatever it is worth in terms of current values. In many countries, there are two or more rates, only one of which applies to ordinary commercial transactions. It is therefore important when using listed exchange figures to verify with the bank that the rate used applies to the transaction in question. Because of the widespread use of cables, the basic rate is usually quoted in terms of the rate for cables with other forms of exchange (demand 30-, 60-, or 90-day drafts) being usually worth a little less. Newspapers such as the *New York Times* and the *Journal of Commerce* carry tables of quotations listing current foreign exchange values.

United States Money.—The dollar was established as the unit of United States money by an Act of Congress on August 8, 1786, and the subdivisions and multiples of this unit as then established are:

> 10 mills make 1 cent
> 10 cents make 1 dime
> 10 dimes make 1 dollar

English Money.—The pound sterling is the unit of English money. While the smaller denominations have always been computed on an arbitrary system as shown by the table below, the British Parliament has considered, in recent years, a proposal to change their currency to a decimal system based on the pound. The present system is as follows:

English Money

> 4 farthings = 1 pence (d)
> 12 pence   = 1 shilling (s)
> 20 shillings = 1 pound Sterling (£)

To find out how much English money will be exchanged for a given quantity of American money, it is necessary to divide the number of dollars by the value of the pound. If there is a decimal in the result equate it into shillings and pence.

ILLUSTRATION: A man wants to exchange $500 for as much English money as he can get for it. Assume that the English pound is quoted at $2.80. How many pounds, shillings, and pence will he receive?

$$500 \div 280 = 178.57 = £178, 11s, 5d. \quad \text{(Ans.)}$$

Other Foreign Money.—Most other foreign countries use a decimal system as a basis for their monetary systems. The following is a list of them:

Country	Principal Unit	Its One-Hundredth Part
France	franc	centime
Belgium	franc	centime
Switzerland	franc	centime
Italy	lira	centesimo
Greece	drachma	lepta
Spain	peseta	centimo
Finland	mark	penni
Germany	mark	pfennig
Brazil	cruzeiro	reis
Yugoslavia	dinar	para
Venezuela	bolivar	centimo
Argentina	peso	centavo

To find out how much foreign money will be exchanged for a given quantity of American money, it is necessary to divide the number of dollars by the value of the foreign unit. When dealing with the above-listed monies, the decimal equals the number of coins of smaller denomination.

ILLUSTRATION: A man wants to exchange $50 for as much French money as he can get. Assume that the franc is quoted at 0.2025. How many francs and centimes will he receive?

$$50 \div 0.2025 = 246.85$$
.246 francs, 85 centimes   (Ans.)

Method of Quoting.—Exchange may be quoted (1) on a premium and discount basis, (2) on a direct price basis, or (3) on an indirect price basis.

Domestic exchange is always quoted on a premium and discount basis, and some foreign monies of the same denominations, such as the Canadian dollar, are usually quoted that way.

ILLUSTRATION: If Canadian dollars are quoted at 7.5% discount, how much would be paid for $1,000 exchange to Canada?

$$\$1000 - 7.5\% \text{ per dollar}$$
$$= \$1000 - 0.075\,(\$1000)$$
$$= \$1000 - \$75 = \$925.00$$

Direct Price Basis.—This method means that exchange is quoted in terms of how many cents or dollars must be paid per unit or per 100 units of a foreign money. English money has always been quoted this way in the United States and at present practically all foreign monies are quoted this way. If the quotation for the English pound is 2.80, it means that two dollars and eighty cents must be paid for every English pound sterling one wants to buy or that that amount will be paid for any which a customer has to sell.

ILLUSTRATION: A man owes a bill of 627 pounds in England and must buy sterling with which to pay the bill. How much must he pay if the pound is quoted at $2.80?

He will multiply the current price ($2.80) by 627 in order to determine how much he will have to pay. In this case, it would be $2.80 \times 627 = \$1753.60$ (Ans.)

ILLUSTRATION: Another man sold goods of value $7827.62 to an English firm. How many pounds did he receive?

The English firm would divide the $7827.62 by 2.80 to determine the number of pounds sterling it would have to pay to settle. In this case, the answer would be:

$$\$7827.62 \div 2.80 = 2795.07 \text{ pounds} = £2795, 1s, 5d.$$

Indirect Price.—Many foreign monies are, or used to be, quoted indirectly, especially when the unit is less in value than the dollar. When the quotation states the number of foreign units that can be bought for a dollar, it is called the indirect price.   England quotes all of its exchange on this indirect basis.

Methods of Payment.—Most foreign exchange transactions arise from transactions involving the sale of goods by an individual or firm in one country to an individual or firm in another country.   There are three methods for arranging for the payment of such goods: (1) payment with the order, (2) establishing credit to be drawn on when goods are ready to be shipped, and (3) draft drawn at time of shipment.   A fourth type, open account, could be mentioned, but it is used rarely and then by firms who have been doing business for extended periods.

Payment with Order.—Under this plan of payment, the purchaser sends exchange with his order.   It really amounts to payment in advance and is seriously objected to by most purchasers. It is not generally used, but at times it seems justifiable.   If an American firm were buying goods in Germany on such terms, it would be necessary for the American firm to buy a foreign draft (bill of exchange) payable on demand and attach it to the order for the goods.   The draft would be for the total purchase price of the goods calculated in the foreign money.

ILLUSTRATION: An American firm orders goods from Germany.   The goods were of such nature that the German firm demanded payment with the order.   The order would amount to 325 marks.   How much must be sent if the marks cost 25.00 cents?

$$325 \times 25 \text{ cents} =$$
$$325 \times 0.25 = \$81.25 \quad \text{(Ans.)}$$

Establishing Credit.—Establishing credit to be drawn against is required when a firm selling goods to a foreign customer desires to be certain that the goods will be paid for when they are ready to ship.   This is used when the purchasing firm is not well known, and particularly when the goods must be made to order.

ILLUSTRATION: If a French firm ordered goods to cost approximately $1200 under such terms, it would be necessary for the French firm to have its bank establish the necessary credit in a New York bank against which the American firm may draw when the goods are ready for shipment. When the goods are ready for shipment, the selling firm draws a draft against the established credit. The draft the seller draws must be accompanied by the shipping documents.

How much will the necessary credit cost if the franc is worth $0.2025?

$1200 ÷ $0.2025 = 5925 francs, 92 centimes. (Ans.)

Draft with Shipment.—The draft drawn at the time of shipment is the procedure in most general use. The draft is drawn on the foreign customer when the goods are ready to be shipped. The draft usually has all shipping documents attached. These documents are (1) Commercial Invoice, (2) Consular Invoice (not required in some countries), (3) Export Declaration, (4) Ocean Bill of Lading, (5) Marine Insurance Certificate or Policy. These drafts may be payable at sight or 30 days, 60 days, or 90 days after sight (or less frequently, at other periods of time). Sight drafts (or on demand) are usually marked D/P, meaning documents are endorsed and delivered at the time the buyer accepts the draft. The purchaser cannot claim the goods until these documents have been endorsed and delivered to him.

ILLUSTRATION: A firm in Italy sold an American firm an invoice of goods amounting to 172,600 lira. Upon shipment of the goods, a 60-day draft was drawn. How much did the American firm have to pay if the rate of exchange was $0.0016?

172,600 × 0.0016 = 276.16 (Ans.)

Posted or Nominal Rates.—Foreign exchange quotations show what are known as "Actual" rates and are used by bankers, regular traders in foreign exchange, and dealers in large transactions. Small letters of credit and small checks are sold at a slightly higher rate than the "actual." This is called the Posted or Nominal Rate.

Travelers' Money.—In order to safeguard his funds, the average traveler carries travelers' checks rather than regular cash. These checks are issued by the American Express, by some banks, and by some tourist companies. They are sold to prospective travelers in denominations from $10.00 up and are charged for at the rate of 75¢ per hundred dollars plus the face. Identification is established by having the traveler sign all checks in the presence of the person cashing them. These checks are issued in dollar denominations and may be cashed anywhere in the world at whatever may be the current rate at the time of cashing.

ILLUSTRATION: A traveler cashed a $75.00 travelers' check in England. The rate was $2.80. How much did he receive?

$75 ÷ 2.80 or 26.79 pounds, which equals £26, 15s, 10d.   (Ans.)

FREIGHT SHIPMENTS

Freight is one of the least expensive methods of shipping goods and is used in the shipping of all sorts of commodities, the speed of delivery of which is not vitally important.

The principal document used in freight shipments is the bill of lading. It is issued in triplicate and contains (1) the name of the consignor (the one shipping the goods), (2) the consignee (the one to whom the goods are shipped), (3) weight, description, and other essential factors about the goods being shipped, and (4) directions pertaining to the route over which the goods should travel. It is the agreement between the shipper and the carrier whereby the carrier agrees to deliver the goods to the consignee.

The three copies of the bill of lading are called (1) original, (2) shipping order, and (3) memorandum copy. After being duly signed by both the shipper and the carrier, the original is forwarded to the consignee, the shipping order is retained by the carrier, and the memorandum copy is retained by the shipper for his files.

There are two kinds of bills of lading: (1) straight and (2) order. The straight bill of lading is the most commonly used and is not negotiable. It is used when the terms of payment have been satisfactorily adjusted between the consignor and consignee, and where delivery of goods is not contingent upon payment for them.

The order bill of lading is negotiable and is used when the consignee is to pay for the goods before they are delivered to him. When it is used, the original (instead of being sent direct to the consignee) is forwarded (usually with sight draft attached) through banks to his bank where he is asked to pay the draft. When this is done, the bill of lading is endorsed to him and with it he may claim his goods at the freight depot. This form of bill of lading is also used in connection with reconsignment and divergence procedures.

Freight Rates.—The charges made by the carrier are based on a hundred pound minimum, by less than carload shipments (l.c.l.) and per car rates for carload shipments. In boat freight the

(Uniform Domestic Straight Bill of Lading adopted by Carriers in Official, Southern and Western Classification territories, March 15, 1922, as amended August 1, 1930.)

THIS SHIPPING ORDER must be legibly filled in, in Ink, in Indelible Pencil, or in Carbon, and retained by the Agent.

Shipper's No. ______

Agent's No. ______

LACKAWANNA CENTRAL RAILROAD **Company**

RECEIVED, Subject to the classifications and tariffs in effect on the date of the issue of this Shipping Order,

AT ______, 19

FROM ______

the property described below, in apparent good order, except as noted (contents and condition of contents of packages unknown), marked, consigned, and destined as indicated below, which said company (the word company being understood throughout this contract as meaning any person or corporation in possession of the property under the contract) agrees to carry to its usual place of delivery at said destination, if on its own road or its own water line, otherwise to deliver to another carrier on the route to said destination. It is mutually agreed, as to each carrier of all or any of said property over all or any portion of said route to destination, and as to each party at any time interested in all or any of said property, that every service to be performed hereunder shall be subject to all the conditions not prohibited by law, whether printed or written, herein contained, including the conditions on back hereof, which are hereby agreed to by the shipper and accepted for himself and his assigns.

(Mail or street address of consignee—For purposes of notification only.)

Consigned to Ewill Receive

Destination Central Point State of Illinois County of Iowa

Route ______

______ (Delivering Carrier) Car Initial ______ Car No. ______

NO. PACKAGES	DESCRIPTION OF ARTICLES, SPECIAL MARKS, AND EXCEPTIONS	*WEIGHT (SUBJECT TO CORRECTION)	CLASS OR RATE	CHECK COL.
6	cases specialized books	860	1st	
	25-30-17-28-44-19			
5	bdls sundry sheet stock	265	1st	

If this shipment is to be delivered to the consignee without recourse on the consignor, the consignor shall sign the following statement:

The carrier shall not make delivery of this shipment without payment of freight and all other lawful charges. (*See Section 7 of conditions.*)

______ (Signature of Consignor.)

If charges are to be prepaid, write or stamp here, "To be Prepaid."

Received $ ______ to apply in prepayment of the charges on the property described hereon.

______ *Agent or Cashier.*

Per ______ (The signature here acknowledges only the amount prepaid.)

Charges advanced:

$ ______

*If the shipment moves between two ports by a carrier by water, the law requires that the bill of lading shall state whether it is "carrier's or shipper's weight."

NOTE.—Where the rate is dependent on value, shippers are required to state specifically in writing the agreed or declared value of the property.

The agreed or declared value of the property is hereby specifically stated by the shipper to be not exceeding

Van Mac Pub. Co. per V.P.P. $1250.00

Will Van Mac Pub Co Shipper: Willard B. Jones Agent

Per V. J. P. Per Paul V. Cronk

Permanent Post-office address of Shipper New York City

The fibre boxes used for this shipment conform to the specifications set forth in the box-maker's certificate thereon, and all other requirements of Rule 41 Consolidated Freight Classifications.

FIG. 10.—Bill of Lading.

charges are based on weight or cubic space occupied, depending on which is the most advantageous. As it is impossible for a railroad to know just how much it costs to transport goods, freight charges are established by endeavoring to determine how much it is worth to the customer to have the goods transported. Hence it costs much more to transport a carload of silk than it does to transport a carload of sand. There are two kinds of rates, commodity rates and class rates. Commodity rates are charged when goods are shipped in large quantities. Class rates are charged on all types of articles which are classified into several different groups.

Freight Tariff Book.—This gives the classification of all articles, the names of all railroad depots, and the rate to be charged between them.

To use this book, one must look up the classification of his article, then look up the cost of transporting that article from the shipping point to the destination.

ILLUSTRATION: A manufacturer of shoes finds that the rate to ship shoes from his point to the point of destination is $2.75 per cwt. How much will it cost him to ship 10 cases weighing as follows: 141, 142, 141, 143, 145, 135, 135, 137, 138, 132 lbs.?

This will be calculated as follows:

$$
\begin{array}{r}
10 \text{ cases weigh } 141 \\
142 \\
141 \\
143 \\
145 \\
135 \\
135 \\
137 \\
138 \\
132 \\
\hline
\text{Total weight } 1389
\end{array}
$$

$1389 \times 2.75 \div 100 = \$38.19$ Total Charges

Express Shipments

The cost of express shipments is calculated in about the same manner. The American Railway Express classifies goods as first, second, or third class, and has about 300 different scales of rates, each designated by a block number. The scale of rates applies to any particular commodity and depends upon distance, weight, size, value, and whether or not it is fragile or perishable. In truck shipment, rates are not so well standardized and depend on many factors, frequently competition having much to do with them. They are usually quoted in terms of weight, however, and are calculated just as are the other types of shipments.

Typical express rates are listed below. In this instance the minimum charge for any express shipment is $3.60. When figuring charges add 3% Federal tax for each shipment. For instance, if your order weighs 30 pounds and you live about 200 miles from shipping point, the charge would be $4.00. Add 3% ($0.12) Federal Tax, making the total charge $4.12.

Weight	100 Miles	200 Miles	300 Miles	400 Miles	500 Miles
5 Pounds	$3.60	$3.60	$3.60	$3.60	$3.60
10 Pounds	3.60	3.60	3.70	3.80	3.90
15 Pounds	3.70	3.80	3.95	4.15	4.35
20 Pounds	3.80	3.95	4.20	4.50	4.80
30 Pounds	3.85	4.00	4.30	4.65	4.95
40 Pounds	3.90	4.10	4.45	4.80	5.20
50 Pounds	3.95	4.25	4.60	5.00	5.60
75 Pounds	4.20	4.70	5.30	5.95	6.95
100 Pounds	4.40	5.20	6.05	7.15	8.50

Taxes

The amount of tax to be collected is determined by preparing a budget in the preparation of which each department or division estimates the amount it will expend during the next fiscal year. The various departmental estimates are then grouped in order to determine the budget for that governmental unit.

ILLUSTRATION: The officials of a small New England town estimated that the budgetary needs for the next year would be those listed below. They also estimated the income that would probably be received from various sources other than from real estate and personal property. How much must be raised from these sources?

Estimated Expenditures:

State Tax	$ 3,133.00
County Tax	6,102.03
Town Charges	3,700.00
Town Maintenance	9,000.00
State Aid Construction	1,758.00
Public Health Nurse	1,000.00
Interest	400.00
Libraries	515.00
Street Lighting	2,750.00
Memorial Day	50.00
Schools	12,068.07
Abatements	100.00
Police Department	300.00
Elections and Registration	200.00

$41,076.10

Less:

Bank Stock	92.75

92.75

$40,983.35

Less Estimated Income from Other Sources:

Auto Tax	$1,300.00
Interest and Dividends	700.00
R. R. Tax	2,000.00
Savings Bank Tax	2,000.00

6,000.00

$34,983.35

Plus Overlay 1,319.90

$36,303.25

Less Local Exemption 72.22

Total to Raise $36,231.03

Direct Taxes.—There are two forms of taxes, direct and indirect. Direct taxes are those paid by the individual as taxes. Most of these are paid by the individual directly to the governmental unit, such as poll taxes, property taxes, income taxes, and various license fees; while some direct taxes, such as taxes on gasoline, theatre tickets, and most sales taxes, are paid as tax by the purchaser but are collected by the seller and then turned over to the government levying the tax.

Indirect Taxes.—Indirect taxes are those that are paid at the source by importers of various merchandise as a "duty" or by manufacturers of various merchandise such as cigars, cigarettes, and alcoholic beverages as an "excise tax." While these taxes are ultimately passed along to the purchaser in the form of increased price, he is never as conscious of paying them as he is of paying direct taxes.

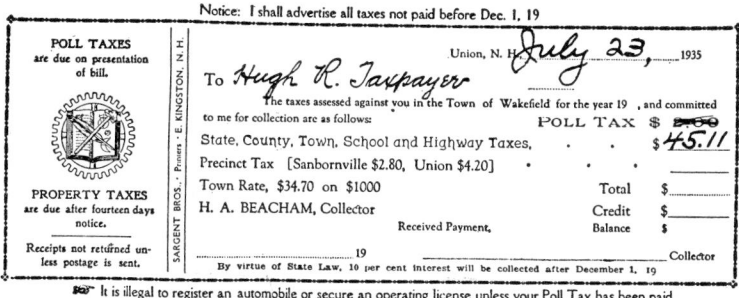

Fig. 11.—Tax Bill.

Property Taxes.—Taxes on real estate or personal property, called property taxes, are assessed by cities and towns, and paid by people owning such items. Some cities and towns pay relatively little attention to personal property, others endeavor to assess it as carefully as they assess real estate.

The property tax is the principal source of income for cities and towns, and the amount to be collected is determined by the amount required by the budget after deducting the estimated income from other sources such as poll taxes and license fees of one sort or another.

Tax Rate.—The rate at which property taxes are to be collected is usually quoted as so many mills on a dollar valuation, or so many dollars on a hundred or thousand dollars of property valuation. This is called the Tax Rate. Property values in a town or city is determined by officials known as assessors whose duty it is to place annually a value on all real estate and on taxable personal property. When the total value of taxable property in the community is determined, that total is divided into the total estimated budget (after the estimated income from other sources has been deducted) in order to determine the tax rate.

ILLUSTRATION: In a small town in New Hampshire the total assessed value of real estate and personal property was $1,163,636. The total budget, including taxes to be paid to the county and state by the town, called for expenditures of $41,076.00. After deducting estimated income from other sources such as auto tax, interest on investments, the amount to be raised was $33,601.06. What is the tax rate?

The tax rate is determined by dividing $33,601.06 by $1,163,-636 (assessed value of real and personal property) and the result is $28.88 per thousand.

Tax Payment.—After the tax rate has been determined, the rate is applied to all assessed valuation and bills are sent to property owners. The amount is calculated by multiplying the assessed valuation by the rate. If there is a poll tax in force, this is included in the same bill.

ILLUSTRATION: A man owns real estate valued at $1400, and taxable personal property valued at $850. The property tax rate is 29.73 and there is a two-dollar poll tax in the state. What is his total tax?

The tax bill should be calculated as follows:

$$\$1400 \times \frac{29.73}{1000} = \$41.62$$

$$850 \times \frac{29.73}{1000} = 25.27$$

$$\text{Poll Tax} \quad \underline{2.00}$$

$$\text{Total Tax} \quad \$68.89$$

Income Taxes.—A tax on incomes is levied by the Federal Government and also by several state governments. All persons living in the United States whose income is in excess of certain minimums, and all corporations making a profit are required to pay income taxes. Income Tax Returns, which is a report of income for the year, must be filed with the Collector of Internal Revenue for the United States Government on or before April 15, and with tax laws at whatever date is specified by that law.

In preparing his return, the taxpayer should realize that in some matters he is permitted a choice of more than one method of procedure, so that it is clearly to his advantage to select the one that requires payment of the least tax. An obvious example is the question of whether or not to itemize deductions. If this is not done, the taxpayer is permitted to take a lump sum deduction (if the income reported exceeds $5000.) up to a maximum of $1000. of either 10% of his total income or (if married) $200. plus $100. for each exemption.

Another permissible choice is that permitted to married couples who are living together, of filing joint or separate returns. In general a joint return is the most advantageous. Under the following conditions it might pay to file separate returns:

1. If both husband and wife have capital losses totalling more than $1000. On the joint return only $1000. is deductible, which on the separate returns, both may deduct $1000. Note, however, that joint losses may be carried forward as deductions in future years.

2. If both husband and wife had income, but one of them had high medical expenses. Since 3% of the adjusted gross income must be deducted from the medical expense to find the allowable deduction, the deduction would be 3% of only the income of the spouse having the high medical expenses.

Of course, these two advantages of single returns may be outweighed by the advantages of joint returns, which include (1) lower rate schedule on joint returns than separate ones; (2) child care expenses usually not deductible from separate returns; (3) income of one spouse so low that his or her dependency deductions or other deductions exceed his or her income so that the full deductions can only be taken on a joint return.

The only way to prove which type of return is most advantageous in a given case is to compute the taxes by both methods.

Withholding Tax.—The United States Government and some state governments require employers to withhold a portion of the individual's income tax from his wages and pay it periodically to the government in question. Self-employed persons prepay their estimated taxes to the United States Government quarterly through the year, making a final adjustment payment

TABLE 10

WITHHOLDING RATES

Weekly Pay	Single Person	Married Person	
		No Dependents	Two Children
$75	$9.30	$6.40	$2.40
100	14.10	10.60	6.50
125	19.10	14.60	10.30
150	24.60	19.30	14.70
200	35.80	28.10	23.20
250	49.80	38.10	32.70
300	64.80	48.10	42.70
350	79.80	58.10	52.70
400	93.30	70.00	63.20
500	123.30	97.70	89.60
600	153.30	126.20	118.10

by April 15 of the following year. This practice is also followed, in addition to the withholding method, by employed people whose incomes exceed a specified amount. The United States withholding tax for employees is based upon a graduated scale as shown in Table 10.

Social Security.—Under the Federal Insurance Compensation Act 4.2% of the first $600 earned per year by an employee is paid by the employer to the Federal Government, along with an equal amount contributed by the employer. This money provides Social Security insurance.

Capital Stock Tax.—The corporation, in addition to the regular income tax, is required to make an additional report and pay an additional tax on the declared value of their stock.

Inter-corporate Dividends.—To assist in controlling holding companies, the Federal government has made 10% of dividends paid from one corporation to another taxable at the corporation income tax rate.

Estate Taxes.—The Federal government and several states levy taxes on estates of deceased persons, providing those estates are in excess of certain amounts. The government (largely to protect the estate tax) also levies taxes on sizable gifts. In the tax law, levies begin at two percent on that part of the estate that is in excess of $40,000 and range up to 70% of that part in excess of $50,000,000. These levies apply to the entire estate left by an individual, regardless of how many persons or institutions may inherit parts of it. New gift taxes are approximately three-fourths as high as the estate levies.

Sales Taxes.—Several states now levy a sales tax of one sort or another, the most common being a tax on retail sales. The tax is most manageable when there are few exempt items. The taxes are collected by the seller on fixed scales and paid to the government levying the tax as a certain percent of gross sales.

Duties on Imports.—Taxes are levied by the Federal government upon commodities imported. These taxes are called Duties, or Customs. There are two kinds, *ad valorem* and *specific*. Ad valorem duties are levied as a certain percent of the value. The value may be fixed by appraisal by U. S. Customs officials or may be determined by the invoice price in the country from which the article is imported.

Specific duty is a certain amount levied on certain articles. It may be per ton or per pound, bushel, yard, gallon, quart, or other unit measure. A duty may be either ad valorem or specific, but in some cases both types are levied on the same article. There are two forms of entry for imported goods: consumptive entry and warehouse entry. Under consumptive entry, the duty is paid on the goods at the time they come in. Under warehouse entry, the goods are placed in bonded warehouses and the duty must be paid when the goods are removed therefrom unless they are subjected to other regulations. Usually they must be removed within three years.

ILLUSTRATION: Assume that the duty on printing paper is 10% ad valorem, and ¼¢ per pound specific. A man bought 10,000 lbs. in England and paid the equivalent of $350. How much tax must be paid?

The duty would be as follows:

Ad valorem, 10% of $350.00 .... $35.00
Specific ¼¢ × 10,000 .... 25.00
Total Duty .... $60.00   (Ans.)

XXIV

Accounting

Purpose of Accounting.—Accounting is the maintenance of a systematic record of business transactions, and its use in preparing control reports or summaries. This information is very important, and often obligatory, in many of the external relations of any business. Every application for a bank-loan, and most efforts to secure additional capital or credit must be accompanied by accounting figures. Any claim for property loss made upon an insurance company must be based upon an evaluation of the assets lost or damaged. Whenever a business is sold, the accounting records play a significant and often a critical part in determining the total price. Last, but by no means least, these records are required by various tax agencies, including the U. S. Internal Revenue Service.

While accounting figures are so necessary in many of the outside relations of a business, they are, if anything, even more valuable in its internal operation. For their control summaries furnish the most practical means for (1) planning and keeping its operations profitable, and for (2) sustaining a favorable ratio of current assets to current liabilities. Both of these conditions must be maintained if the business is to continue in existence.

Accounting records are kept in "books," which may be bound or loose-leaf, or not books at all, but filing cabinets containing separate sheets. There are two kinds of books, those of first entry which are called journals and contain descriptions of the transactions; and those in which the journal entries are copied (posted) to record the effect upon the assets and liabilities of the business of the particular transaction; these books of later entry are called ledgers.

How to Start and Keep Books.—In a small business the book-keeping system can be started with a single journal, which would then be called a day-book, and a single ledger, which would have separate pages for each account. The various transactions with their descriptions are recorded in the journal. However, transactions of the same kind are often, for practical reasons, grouped into a single entry—thus the cash sales of a retail store are commonly entered as a total figure for each day. As the business grows, some types of transactions become so numerous that time can be saved by having for them a separate journal, or a different place in the same journal or the filing cabinet. Among such frequently occurring transactions are sales, purchases, cash transactions, etc.

There is, however, one type of transaction for which a separate journal should be started at once. It is wages or salaries. As soon as the business grows to the point where employees are hired, a Payroll Record should be separated from the journal, or day-book. In this Payroll Record should be entered under the name of each employee his gross wages or salary, the amount of all deductions, including Social Security deductions and Withholding Tax deductions, and the net payments, which are the wages actually paid to the employee.

There are five kinds of ledger accounts: the two kinds of operating accounts are income entries, representing what is taken in, and expense entries, or what is paid out. The other three kinds of accounts are asset accounts (what is owned), liability accounts (what is owed), and capital accounts (amounting to the difference of what is owned and what is owed). Thus Sales is an income account; Purchases, an expense account; Accounts Receivable, an asset account; Accounts Payable, a liability account; and the Owner's Investment Account, a capital account.

The basic idea underlying accounting is expressed by the word balance, as used in the sense of balancing the books. Every transaction is regarded as a transfer from one account to another, in other words, an "addition" to one account is balanced by a "subtraction" from another, and since the two operations are theoretically done at the same time, the books are always in balance if no error has been made. However, instead of the words "addition" and "subtraction," which apply here only in the algebraic sense, the accountant uses the words debit and credit, and in the ledger, which has two columns for the purpose, debits are entered on the left and credits on the right. The meaning of these terms depends upon the nature of the account—for example, cash sales are debits to the cash account and credits to the sales account, payment of rent is a credit to the cash account and a debit to the expense account, etc. While the cash account entries are comparatively easy to reason out (receipts being debits and expenditures credits), this process is not nearly so easy for some of the other accounts. The best way to gain this knowledge is by following through a few steps in the accounting of a small business.

To start such a system, buy two books or types of forms—one for the single journal, or day-book, which would appear as shown in Figure 1; and the other for the ledger, which would appear as shown in Figure 2. Then take a group of representative transactions and record them in these books, beginning, of course, with the journal. The following list gives a description of a series of simple transactions:

October 1. Robert Jones invested in his new business $4,800.00 in cash and $2,000.00 in merchandise.

October 2. He paid $100.00 for rent, and $65.80 for stationery, account books and other office supplies.

October 3. He bought merchandise from K. Smith and Co. for $1945.52 on account, and he sold merchandise for $441.92 in cash.

October 4. He bought a safe for $160.00.

October 5. He sold merchandise to Walter Mitchell for $620.10, receiving $300.00 in cash.

October 6. He sold $460.48 merchandise to John Peters, receiving $100.00 cash, and Peters' 60-day note for the balance.

October 8. He bought merchandise in the amount of $958.20 from United Metals Co., paying $500.00 in cash.

October 9. Walter Mitchell returned $20.50 merchandise bought on October 5.

October 10. He paid postage $10.00; he bought merchandise at auction for cash $56.75; and he drew from business for personal use $50.00.

These transactions would be recorded in a single journal as shown in Figures 1a and 1b.

Note how these entries are made. The capital investment of $4800.00 is an acquisition of cash by the business; therefore, it is entered in the journal as a debit to Cash (right-hand column) and a credit (left-hand column) to the owner's Investment Account. Similarly, his merchandise investment of $2000.00 is a debit to Merchandise Inventory and a credit to his Investment Account. Rent paid is an expenditure of cash, and is therefore credited to Cash and debited to a General Expense account. Office supplies purchased are entered in the same way, except that a special account, separate from General Expense, called Expense Supplies, is commonly used for them. Purchases on account are obviously a credit to the supplier, entered to his account, and a debit to the Purchases of the business. Conversely, sales are a credit to the Sales account of the business, and if made for cash, as was the entry of $441.92, they are a debit to the Cash account.

Continuing with these transactions, the acquisition of a safe for $160.00 in cash is a credit to the Cash account and a debit to Furniture and Fixtures. The sale to Walter Mitchell for part cash is a credit to Sales, balanced by a debit to Cash and by another debit to Mitchell's account for the unpaid balance. Similarly, the sale to John Peters for part cash and a note for the balance is a credit to Sales, balanced by a debit to Cash and another debit to Notes Payable. The name of the maker of the note, its date and maturity are noted in the journal, as indeed is any fact of major importance about an entry. The purchase for part cash from the United Metals Co., is a debit to Purchases, balanced by a credit to Cash and another credit to the account of the United Metals Co. The merchandise returned by Walter

Oct.	1	Cash	2		4	8	0	0	00									
		Robert Jones, Investment	1								4	8	0	0	00			
Oct.	1.	Merchandise Inventory	3		2	0	0	0	00									
		Robert Jones, Investment	1								2	0	0	0	00			
Oct.	2	General Expense	4		1	0	0	0	00									
		Cash	2								1	0	0	00				
		Rent for October																	
Oct.	2	Expense Supplies	5			6	5	8	0									
		Cash	2									6	5	80				
Oct.	3	Purchases	6		1	9	4	5	52									
		K. Smith & Co.	7								1	9	4	5	52			
Oct.	3	Cash	2		4	4	4	1	92									
		Sales	8								4	4	4	1	92			
Oct.	4	Furniture & Fixtures	9		1	6	0	0	00									
		Cash	2								1	6	0	00				
		Cost of safe																	
Oct.	5	Cash	2		3	0	0	0	00									
		Walter Mitchell	10		3	2	0	1	0									
		Sales										6	2	0	10			
Oct.	6	Cash	2		1	0	0	0	00									
		Notes Receivable	11		3	6	0	4	8									
		Sales	8								4	6	0	48				
		Note of John Peters																	
		for 60 days dated 10/6																	
Oct.	8	Purchases	6		9	5	8	20										
		Cash	2								5	0	0	00				
		United Metals Co.	12								4	5	8	20				

FIG. 1a.

Oct.	9	Sales Returns	13		2 0 50					
		Walter Mitchell	10				2 0 50		
		Returned mdse.							
		claimed defective							
Oct.	10	Expense Supplies	5		1 0 00				
		Cash	2				1 0 00		
		Postage							
Oct.	10	Purchases	6		5 6 75				
		Cash	2				5 6 75		
Oct.	10	Robert Jones, Personal	14		5 0 00				
		Cash	2				5 0 00		

Fig. 1b.

Mitchell is obviously entered as a credit to his account, and a debit to Sales Returns. (The debit could be made to sales, but this is contrary to accounting practice, and less informative.) Stamps bought for $10.00 on October 10 are a credit to Cash, and a debit to the Expense Supplies account. Similarly, merchandise bought for cash, $56.75, is a credit to Cash and a debit to the Purchases account. Finally, the withdrawal of $50.00 on October 10 by the proprietor of the business is a credit to the Cash account, and a debit to his Personal Account. Note that two accounts are kept for the proprietor: his Investment Account (the first of the above entries) and his Personal Account. This also is proper accounting practice, important in the preparation of statements and tax returns.

By following through the foregoing simple transactions, item by item, and comparing them with the journal entries shown in Figures 1a and 1b, one can easily gain insight into the handling of accounting transactions. Its value becomes clear at once from a consideration of the next step, which is the posting of these journal entries into the ledger.

The first step in starting a ledger is to begin accounts, each on a separate page, for every account that appears in the journal. These will include every customer who owes money to the business, every creditor to whom the business owes money, and every internal account of the business. Posting the foregoing transactions from the journal will yield the ledger accounts and entries which are arranged in Figures 2 to 15, in the order they would appear by posting them just as they occur in the journal.

By comparing these ledger accounts, one by one, with the journal entries (Figures 1a and 1b) one can see how the posting is done. The order of the ledger accounts here, which results from taking them, for purposes of demonstration, in the journal order, is not one that an accountant would follow. However, this matter of order is not important, because the page number of the ledger account is shown on the journal entry. It appears in a column between the title of the entry and its amount, as can be seen by referring to Figures 1a and 1b.

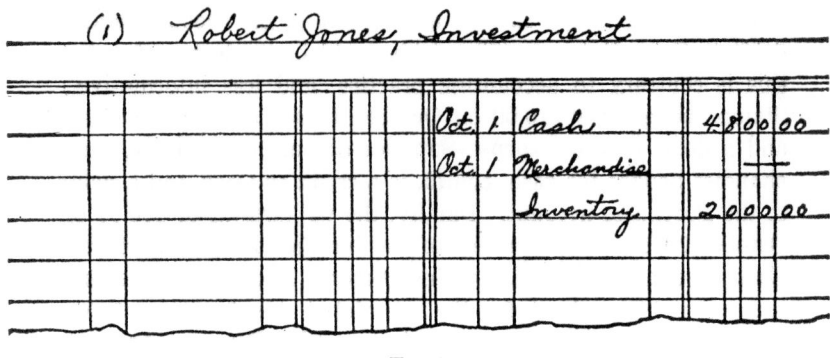

(1) Robert Jones, Investment

								Oct 1	Cash		4 8 0 0	00
								Oct 1	Merchandise		—	
									Inventory		2 0 0 0	00

Fig. 2.

(2) Cash

Oct.	1	Robert Jones, Investment	4 8 0 0	00	Oct	2	Gen Expense		1 0 0	00
Oct.	3	Sales	4 4 1	92	Oct	2	Expense Supplies		6 5	80
Oct.	5	Sales	3 0 0	00	Oct	4	Furn. + Fixt.		1 6 0	00
Oct.	6	Sales	1 0 0	00	Oct	8	Purchases		5 0 0	00
					Oct	10	Expense Supplies		1 0	00
					Oct	10	Purchases		5 6	75
					Oct	10	Robert Jones, Personal		5 0	00

Fig. 3.

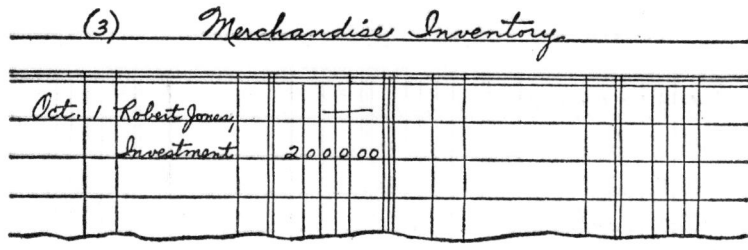

FIG. 4.

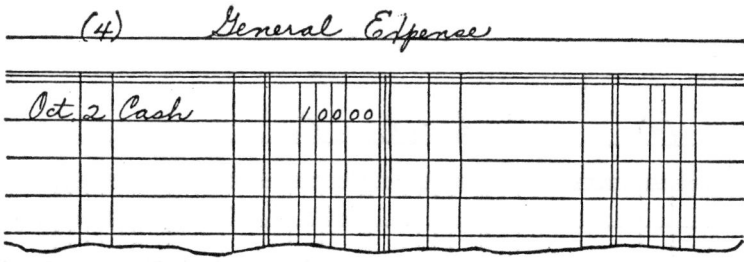

FIG. 5.

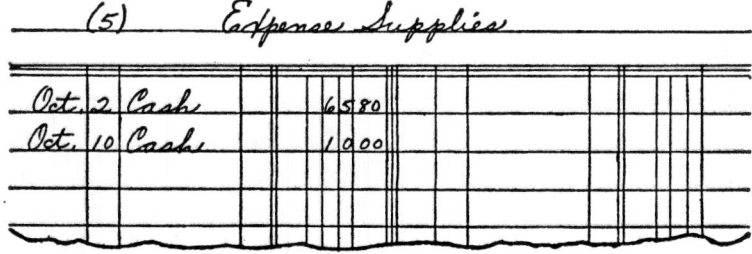

FIG. 6.

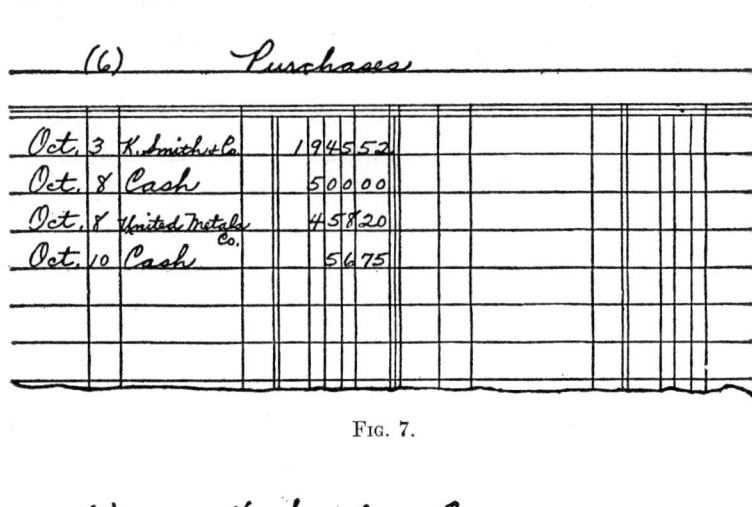

(6)   Purchases

Oct. 3	K. Smith & Co.	1945 52			
Oct. 8	Cash	500 00			
Oct. 8	United Metals Co.	458 20			
Oct. 10	Cash	56 75			

FIG. 7.

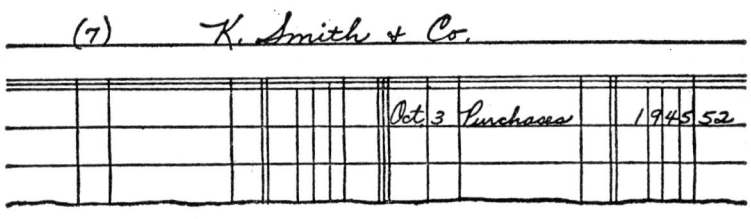

(7)   K. Smith & Co.

			Oct. 3	Purchases	1945 52

FIG. 8.

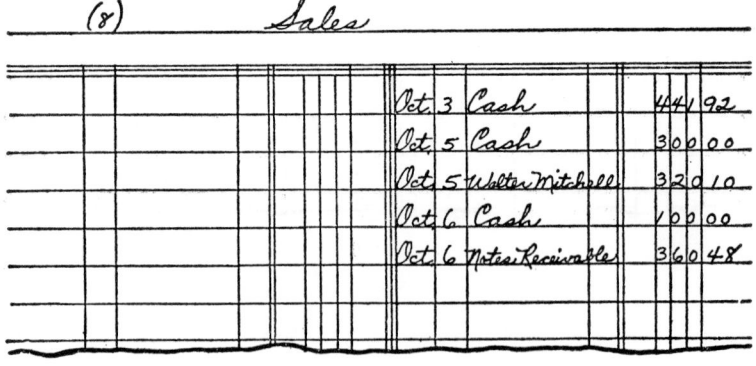

(8)   Sales

		Oct. 3	Cash	444 92
		Oct. 5	Cash	300 00
		Oct. 5	Walter Mitchell	320 10
		Oct. 6	Cash	100 00
		Oct. 6	Notes Receivable	360 48

FIG. 9.

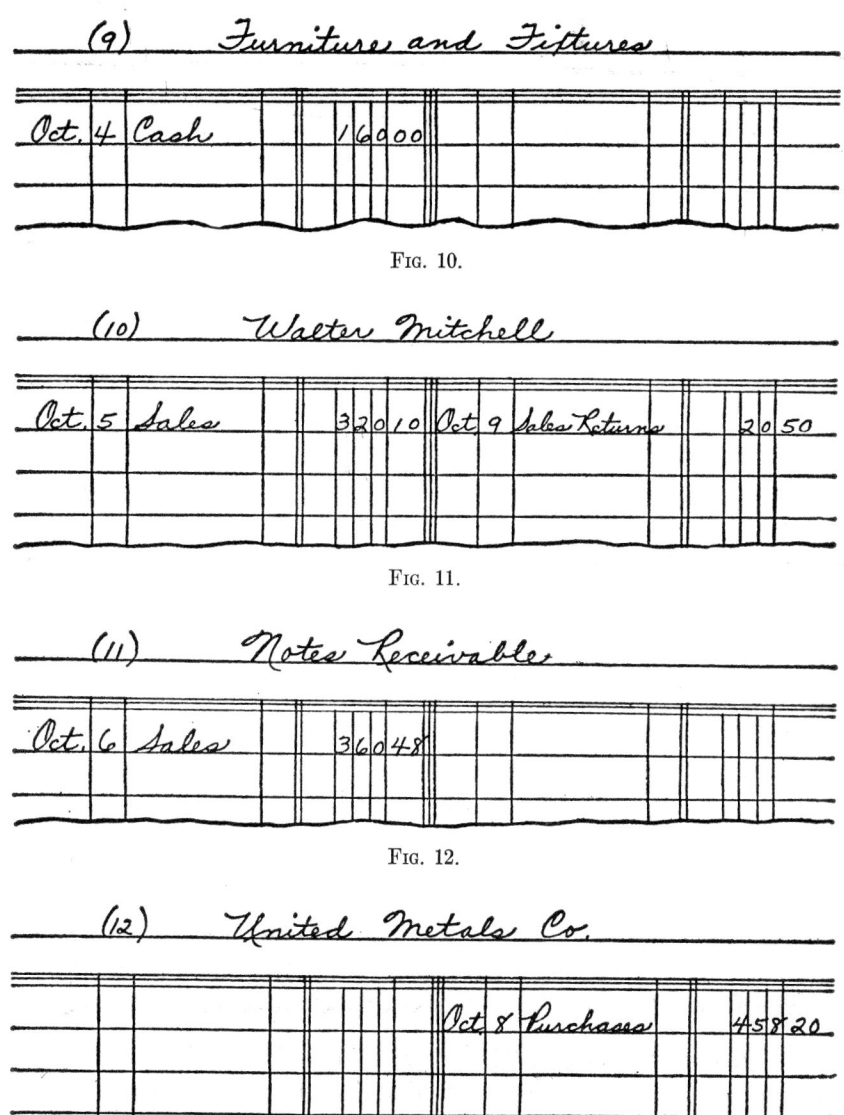

(9)   Furniture and Fixtures

| Oct. 4 | Cash | | 160 00 | | | | |

FIG. 10.

(10)   Walter Mitchell

| Oct. 5 | Sales | | 320 10 | Oct 9 | Sales Returns | | 20 50 |

FIG. 11.

(11)   Notes Receivable

| Oct. 6 | Sales | | 360 48 | | | | |

FIG. 12.

(12)   United Metals Co.

| | | | | Oct 8 | Purchases | | 458 20 |

FIG. 13.

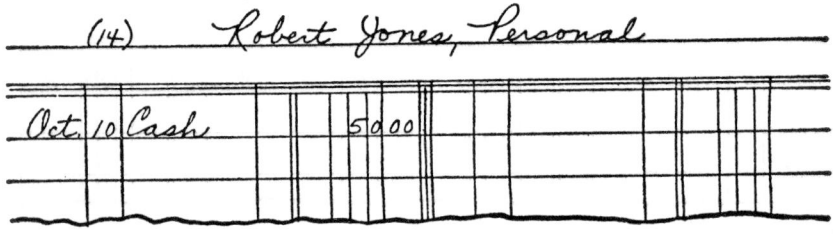

(13)   *Sales Returns*

| Oct. | 9 | Walter Mitchell | | 20 | 50 | | | | | | | |

F𝐼𝐺. 14.

(14)   *Robert Jones, Personal*

| Oct. | 10 | Cash | | | 50 | 00 | | | | | | |

F𝐼𝐺. 15.

Closing the Books.—In order to illustrate the steps involved in adjusting and closing the books, and preparing statements, these operations are carried out for the business of Robert Jones as of the close of the day of October 10. Note that as a consequence of this short period of operation, both the number of transactions and the number of accounts are very small. Note also that reports for tax purposes must be made on calendar year basis, unless permission is obtained from the local U. S. Internal Revenue office to use some other fiscal period. This change is often worth making when the season of slow business and low inventory regularly occurs at some other time than the start of the year.

The Trial Balance.—The first step in closing the books is to total each ledger account, subtracting the sum of the credits from the sum of the debits, or vice versa if the total credits exceed the total debits, ruling off the accounts, and entering the net balance of each account in the debit or credit column, as the case may be, of the ledger. If one were to perform these operations for the fourteen accounts that were prepared for the business of Robert Jones through October 10, the following Trial Balance would be obtained:

TRIAL BALANCE, October 10

(1)	Robert Jones, Investment		$6,800.00
(2)	Cash	$4,699.37	
(3)	Merchandise Inventory (Start of Period)	2,000.00	
(4)	General Expense	100.00	
(5)	Expense Supplies	75.80	
(6)	Purchases	2,960.47	
(7)	K. Smith & Co.		1,945.52
(8)	Sales		1,522.50
(9)	Furniture and Fixtures	160.00	
(10)	Walter Mitchell	299.60	
(11)	Notes Receivable	360.48	
(12)	United Metals Co.		458.20
(13)	Sales Returns	20.50	
(14)	Robert Jones, Personal	$10,726.22	$10,726.22

The trial balance "balances" as indeed it must in a double-entry system unless there has been an error in posting or in arithmetic.

Adjustment Entries.—The next step in closing the books is to make the adjustment entries. They are required in accounting for various purposes, for example, to take care of expenses incurred but not payable, such as accrued wages, or of the unused period of prepaid expenses, such as unexpired insurance. The only adjustment information required in the business situation under discussion is the change in merchandise inventory. This is determined by taking a new inventory as of the close of business on October 10. This inventory is found to have a value of $3,974.60, which is posted in the Merchandise Inventory Account as a credit. Then the purchases are also posted to this account as a credit, and the difference between the sum of the beginning inventory and purchases, less the final inventory, is the cost of goods sold. Figure 16 shows how these entries appear on

(3)   *Merchandise Inventory*

Oct 1	Robert Jones, Investment	1	2000 00	Oct 10	Inventory		3974 60
Oct 10	Purchases	L6	2960 47	Oct 10	Cost of Goods Sold (To Profit + Loss)		985 87
			4960 47				4960 47
Oct 10	Inventory		3974 60				

FIG. 16.

the Merchandise Inventory Account, and it also shows the method of ruling-off an account and entering its new balance, which you should do at the close of an accounting period for all asset, liability and capital accounts. Another operation in closing these books is to post sales returns to the Sales Account, and to enter the difference as net sales.

(15)   *Profit and Loss*

Oct 10	Cost of Goods Sold L3 (From Mdse Inventory)	985 87	Oct 10	Net Sales	L8	1502 00
Oct 10	General Expense L4	100 00				
Oct 10	Expense Supplies L5	75 80				
Oct 10	Robert Jones, Investment L1	340 33				
		1502 00				1502 00

FIG. 17.

(1) Robert Jones, Investment

Oct	10	Robert Jones Personal	14	50 00	Oct	1	Cash	2	4800 00
Oct	10	Capital + Profit		7090 33	Oct	1	Merchandise Inventory	3	2000 00
				7140 33	Oct	10	From Profit + Loss	15	340 33
									7140 33
					Oct	10	Capital + Profit		7090 33

FIG. 18.

Profit and Loss Account.—The next step in closing the books is to set up a new account in the ledger, the Profit and Loss Account. To it is transferred the cost-of-goods-sold adjustment entry from the Merchandise Inventory Account, the net sales entry from the Sales Account, and the various expense accounts. The difference found in this account, that is, profit or loss, is balanced by a transfer to the proprietor's Investment Account. This particular Profit and Loss Account is shown in Figure 17, and the proprietor's Investment Account as of October 10 (including the transfers from his Personal Account and from Profit and Loss) is shown in Figure 18.

Profit and Loss Statement.—The Profit and Loss Account which has just been posted is useful, not only in closing the books at the end of a period, but also in preparing the Profit and Loss Statement for that period. This statement has great value, as stated at the beginning of the chapter, not only in the external relations of the business, as with banks and taxing agencies, but also in the internal management. For that reason, it is presented in sufficient detail to show figures important to efficient control.

A Profit and Loss Statement for the Robert Jones business would be prepared as shown on following page.

Financial Statement.—The other important statement is the Financial Statement, which is also called the Balance Sheet. It shows the financial condition of the business. The information may be arranged in various ways, one of the most common shown on next page, in which the liabilities (and proprietorship entries) are grouped below the assets, or on a facing page.

Simplified Methods for Small Business.—To save time and money in keeping books, there are a number of short-cuts which accountants have approved for small business. Some are limited to particular circumstances or kinds of business, and those must be developed with an accountant. There is one short-cut method that is often found useful. A simple file is used for accounts payable, placing in it all the invoices grouped according to your creditors' names. This file saves the work of entering each invoice because one is, in effect, using the file as an accounts-payable journal.

Related to this method, there is another policy that should always be followed as a matter of sound accounting practice. It is this: to make the bank account of the business an accurate income-record by depositing in full each day's income. If wages, invoices or other expenses are paid in cash, checks are drawn for them to cash, the purpose of the payment being noted in the checkbook, and they are then endorsed and deposited in the bank with the remainder of that day's income. If this procedure is followed the cancelled checks will give all in one place, "receipts" or vouchers for every expenditure; while the deposits will constitute a complete record of income. These records will be of the greatest value, not only in any possible discussion with tax examiners or collectors, but also for future uses in obtaining bank loans or even proving volume of sales to a prospective purchaser of your business.

Accounting for Representative Types of Business.—The next step in the explanation of accounting methods is to review actual practices followed in specific types of enterprise. Therefore, to cover the needs of the small business as broadly as possible, ac-

ROBERT JONES

PROFIT AND LOSS STATEMENT, October 10

Sales	$1,522.50	
Less Sales Returns	20.50	
Net Sales		$1,502.00
Merchandise Inventory, Oct. 1	$2,000.00	
Purchases	2,960.47	
	4,960.47	
Merchandise Inventory	3,974.60	
Cost of Goods Sold		985.87
Gross Profit		$ 516.13
Expenses:		
General Expense	100.00	
Expense Supplies	75.80	
		175.80
Net Profit for the Period		$ 340.33

ROBERT JONES

FINANCIAL STATEMENT, October 10

Assets

Cash		$4,699.37
Notes Receivable		360.48
Accounts Receivable		299.60
Merchandise Inventory		3,974.60
Furniture and Fixtures		160.00
Total Assets		$9,494.05

Liabilities

Accounts Payable	$2,403.72	
Total Liabilities		$2,403.72
Net Worth		$7,090.33

Proprietorship

Robert Jones, Investment, Oct. 1		6,800.00
Profit for the Period	$340.33	
Less Withdrawals	50.00	
	290.33	
New Worth		$7,090.33

counting systems will be discussed for (1) the mail-order business; (2) the retail business; (3) the small manufacturing business; and (4) the service business.

The Mail-Order Business.—In considering the accounting of a mail-order business, one will discover at once that two types of records are necessary. The first, which is the overall accounting of the business, is designed to produce the same statements and records already discussed. The second is the accounting of the individual sales operations; that is, each mailing, each newspaper or magazine advertisement, and as far as possible each radio or television broadcast or each short-term series of broadcasts. Such records are indispensable to the profitable operation of a mail-order business. Therefore, examples are now given of an accounting record for each of these three types of operation. See Examples 1, 2 and 3.

These three examples cited are given to show how to keep the accounting records of individual mail-order advertisements. The process can be shortened somewhat with accumulated experience with some of these operations. For example, the shipping cost entries in all three cases are repetitive and can be combined, because the chief purpose of these figures is to determine whether these advertisements are profitable. For that reason also it was assumed that the mailing and shipping operations were done by outside agencies, for otherwise the labor charges for them would have to be found by analysis of the operating expenses of the mail-order company.

The critical question is that of whether the remaining operating expenses and administrative expenses are covered by the percentages of profit shown. These expenses include Clerical Salaries, Administrative Salaries, Rent, Insurance, Taxes on Payrolls, Other Taxes, Transportation on Goods Received. Moreover, if the product is manufactured on the premises and not purchased ready to ship, its unit cost, which was taken at $1.84, must include its share of all general and administrative expenses, as well as those directly chargeable. The answer to the question of whether the profit from the operation covers the overhead de-

EXAMPLE 1
MAILING TO CUSTOMER LIST A
Mailing Period: March 12-15

Nature of List:	Purchasers of Cake Assortment A
Nature of Offer:	Autumn-Leaf Maple Candy Assortment ($4.95) with premium offer of maple syrup to be sent C.O.D. (or cash with order prepaid)
Nature of Mailing:	Two color circular, 17″ × 22″ in size; multigraphed letter on lithographed letterhead; business reply post-card; all in a #10 envelope, marked for "Bulk Mailing" under Section 34.66 Postal Laws & Regulations. (Such mailings pay $1\frac{1}{2}$¢ postage up to the weight limit. They require a permit from the local postoffice, and must be sorted as directed.)

Quantity of Mailing: 16,430

I. *Cost of Mailing:

Circulars and Letters:

Delivered cost of circulars, including paper and printing	$191.30
Cost of lithographing and paper for letterheads	53.35
Cost of multigraphing letters	44.62
Cost of paper-stock and printing of postcards	32.90
Total cost of circulars and letters	$322.17

Mailing Expense.

†Addressing Envelopes from List	65.72	
†Folding, inserting, sealing and stamping	49.29	
†Sorting Mail	16.43	
Postage	246.50	
Postage due on orders received	13.65	
Total mailing expense	$391.30	
Total Cost of Mailing		$713.47

* These costs are shown more in detail than is consistent with the other accounting schedules in this chapter. This variation has been made to show you some of the specialized operations in direct-mail work.

† The labor operations are charged as if they were done by an outside mailing service. If the work is performed in the mail-order business, then these charges are not made directly; but there is a wage cost to pay the workers. Note also that this addressing charge is for labor only; if an outside list is used it will be larger to include list rental.

(Continued)

EXAMPLE 1 (Cont'd)

II. Income from Mailing
    Orders Received   525
    Shipments Not Accepted   66
    Sales   459
    Gross Income from Sales $2,272.05

III. Shipping Expenses:
    C.O.D. charges on prepaid orders $ 35.17
    Express charges on shipments not accepted   26.52
    C.O.D. collection charges   150.30
    Shipping cartons   60.65
    Shipping labor (see ** above)   50.14
        Total shipping expenses   323.05

IV. Cost of Goods Sold:
    459 assortments @ $1.84   844.56
    12 assortments returned in unsaleable condition   22.08
        Total Cost of Goods Sold   866.64

Recapitulation:
    Gross Income from Sales   2,272.05
    Total cost of mailing   713.47
    Total shipping expenses   323.05
    Total cost of goods sold   866.64
    Total costs   1,903.16
    Profit before overhead $ 368.89

EXAMPLE 2
MAGAZINE ADVERTISEMENT

Medium: Cosmopolitan
Date of Issue: March, 19——
Space Used: 53 lines in outside column on Page 11.
Nature of Offer: Autumn-Leaf Maple Candy Assortment ($4.95) with premium offer of maple syrup to be sent C.O.D. (or cash with order prepaid)

I. Cost of Advertisement
    Space $355.00
    Art-work   24.00
    Halftones   33.60
        Total Cost of Advertisement   412.60

(Continued)

EXAMPLE 2 (Cont'd)

II. Income from Advertisement
Orders Received	322	
Shipments not accepted	53	
Sales	269	
Gross Income From Sales		$1,331.55

III. Shipping Expenses
C.O.D. charges on prepaid orders	17.22	
Expenses charged on shipments not accepted	21.20	
C.O.D. collection charges	80.70	
Shipping cartons	38.10	
*Shipping labor	30.90	

(*See note in report of MAILING TO CUSTOMER LIST A)

Total Shipping Expenses		188.12

IV. Cost of Goods Sold
269 Assortments $1.84	$495.96	
10 Assortments Returned in Unsaleable Condition	18.40	
Total Cost of Goods Sold		514.36

Recapitulation
Gross Income From Sales		1,331.55
Total Cost of Advertisement	426.00	
Total Shipping Expenses	188.12	
Total Cost of Goods Sold	514.36	
Total Costs		1,128.48
Profit before Overhead		$ 203.07

EXAMPLE 3

RADIO ADVERTISEMENT

Station: WINS in New York City

Dates of Broadcast: Five consecutive days, beginning March 3 and ending March 7.

Time Used: One-minute spot announcements each day at 3:30 P.M.

Nature of Offer: Autumn-Leaf Maple Candy Assortment ($4.95) with premium offer of maple syrup to be sent C.O.D. (or cash with order prepaid)

I. Cost of Broadcasting: Five announcements at $55.00 each, giving a total of $275.00

(Continued)

EXAMPLE 3 (Cont'd)

II. Income from Broadcasting

Orders Received	214	
Shipments Not Accepted	39	
Sales	175	
Gross Income From Sales		$866.25

III. Shipping Expenses

C.O.D. Charges on Prepaid Orders	$10.54	
Express Charges on Shipments Not Accepted	15.60	
C.O.D. Collection Charges	52.50	
Shipping Cartons	25.30	
*Shipping Labor	20.40	
(* See note in report of MAILING TO CUSTOMER LIST)		124.34

IV. Cost of Goods Sold

175 Assortments @ $1.84	322.00	
8 Assortments Returned in Unsaleable Condition	14.72	
Total Cost of Goods Sold		336.72

Recapitulation

Gross Income From Sales		$866.25
Total Cost of Broadcasting	$275.00	
Total Shipping Expenses	124.34	
Total Cost of Goods Sold	336.72	
Total Costs		736.06
Profit Before Overhead		$130.19

pends upon its amount, and shows why the overhead of a small-scale mail-order business must be held to an absolute minimum. As the business expands, the savings made possible by larger printings of letters and circulars, and the increased proportion of orders from the use of larger units of advertising space, can be hoped to take care of increased overhead, and even provide increased net profit.

It is to be emphasized that the three examples shown were purposely chosen as hypothetical cases, with correspondingly hypothetical figures. They are included to show accounting methods—not relative returns from different types of media.

Retail Store Accounting.—For the purposes of this treatment, a representative retail store has been taken as a general bookstore, that is, a bookstore carrying current fiction and nonfiction, without specializing in any particular kind of books, and with certain other merchandise, such as stationery, greeting cards, pens and pencils, etc. A special feature of bookstore accounting arises from the practice of carrying books on consignment, which should be shown in a separate inventory (not included in the regular inventory as part of the assets) and as these books are sold they should be credited to the account of the company to whom they belong, in a special Consignment Account Payable.

The balance sheet accounts to be kept in a retail bookstore would then be classified as follows:

ASSETS

Current Assets
1. Cash in Bank
2. Cash on Hand
3. Accounts Receivable (one for each account)
4. Notes Receivable
5. Bad Debt Reserve
6. Merchandise Inventory
7. Rental Library Stock
8. Other current assets

Fixed and Other Assets
9. Real Property
10. Furniture and Fixtures
11. Depreciation Reserve
12. Prepaid Expenses

LIABILITY AND CAPITAL ACCOUNTS

Current Liabilities
20. Accounts Payable (one for each account)
21. Notes Payable
22. Library Deposits
23. Expenses Accrued but Not Due

Deferred Liabilities
24. Long Term Debts

Capital Accounts
30. Invested Capital
31. Surplus
32. Profit and Loss

The balance sheet of a bookstore is prepared by summarizing the foregoing accounts.

The operating accounts of a bookstore would consist of the following:

OPERATING ACCOUNTS

Income Accounts
- 40. Book Sales
- 41. Returned Books
- 42. Other Merchandise Sales
- 43. Returns of Other Merchandise
- 44. Income from Library Rentals
- 45. Income from Library Sales

Expense Accounts
- 50. Book Purchases
- 51. Other Merchandise Purchases
- 52. Library Purchases
- 53. Salaries
- 54. Rent
- 55. Telephone
- 56. Light and Heat
- 57. Taxes
- 58. Office Supplies and Office Postage
- 59. Shipping and Packaging Supplies and Shipping Charges
- 60. Charges on Incoming Shipments
- 61. Insurance
- 62. Bad Debts

The operating statement is prepared in the usual way, as explained in the early part of this chapter, by adding the income accounts (less returns), and subtracting the cost of goods sold to determine gross profit, and then subtracting total expenses to determine net profit. The cost of goods sold is determined for each of the three categories of income by adding purchases and charges on incoming shipments to inventory at beginning of the period, and subtracting inventory at end of period.

For the control interpretation of these figures in bookstore operation, the reader is referred to HANDBOOK OF ACCOUNTING METHODS, Second Edition, by J. K. Lasser. There will be seen the application of various ratios in the control of a business, particularly in detecting certain unsound conditions before they result in a net loss, rather than a net profit. For the retail bookstore business, for example, the suggested control figures are: (1) the ratio of total payroll (including owners' drawings) to sales, which should not exceed 13–14%; the ratio of total payroll to gross profit, which should not exceed 40–50%; (2) the ratio of rent to sales, which should not exceed 6–8%; and (3) the ratio of advertising to sales, which should rarely exceed 2%.

Of course, the figures given above may vary somewhat according to the special circumstances of a particular bookstore. They should, however, be watched closely because these three items, payroll, rent and advertising are usually the largest expenses of the bookstore. In exercising accounting control over other types of business, the same principle—that of watching ratios of the largest expense items to sales, or to gross profits— should be followed. Moreover, one other accounting figure that is so important to the bookstore is also very significant in many other types of retail business. That figure is inventory. In arriving at its true value, one needs to recognize differences in rate of depreciation or obsolescence of different classes of merchandise. For example, in a bookstore, the current fiction titles in general lose saleability more rapidly than the non-fiction titles, and therefore as books grow older the inventory-value of the former should be "written-down" more rapidly. The same principle should be applied in the accounting of other retailing businesses handling timely or seasonal products.

Accounting for the Small Manufacturing Business.—The business of food packing includes a wide range of possible enterprises for small-scale operation. Such enterprises are not, of course, necessarily small, but may vary in size from the home-kitchen production of jams and jellies to the giant cannery. The accounting methods suggested are, therefore, sufficiently flexible to meet the needs of the somewhat larger business, but not, of course, the truly large enterprise.

There is, however, one peculiarity of all enterprises of this nature, large and small. That is the limited seasonal availability of the raw material, which requires a short but intensive period of manufacturing activity which, if not planned carefully in advance, may leave in its wake an inventory of high-cost product that cannot be marketed profitably.

Holding this major consideration in mind, one can then set up the balance sheet accounts to be kept in a food packing business as follows:

Assets

Current Assets
1. Cash in Bank
2. Cash on Hand
3. Accounts Receivable
4. Notes Receivable
5. Marketable Investments
6. Bad Debt Reserve
7. Inventories of Raw Materials and Finished Products

This account includes a considerable number of items, such as delivered cost of raw materials, less discounts and allowances, and all the other costs, including manufacturing labor, which enter into the total cost of production of the finished product. This inventory also includes various supplies which are used in the manufacturing process, but are not considered to be raw materials.

8. Other Current Assets

Fixed Assets
9. Real Property
10. Machinery
11. Furniture and Fixtures
12. Depreciation Reserve
13. Prepaid Expenses

Liability and Capital Accounts

Current Liabilities
20. Accounts Payable
21. Notes Payable
22. Expenses Accrued but Not Due

Deferred Liabilities
23. Long Term Debt

Capital Accounts
30. Invested Capital
31. Profit and Loss

125. Collection Expense
126. Depreciation
    (a) Building*
    (b) Operating Equipment
    (c) Office and General
127. Donations
128. Insurance
    (a) Building*
    (b) Operating Equipment
    (c) Employer's Liability—Operating
    (d) Office and General
129. Legal and Accounting
130. Light and Heat
131. Maintenance and Repairs
    (a) Building*
    (b) Office and General
132. Miscellaneous
133. Office Expense
134. Postage
135. Rent
136. Salaries—Office and General
137. Stationery and Supplies
138. Taxes
    (a) Land and Building
    (b) Operating Equipment
    (c) General
    (d) Social Security
139. Telephone and Telegraph
140. Travel and Entertainment
150. Distribution (Cr.)

Nonoperating Accounts

161. Purchase Discounts
162. Interest Income
163. Miscellaneous Income
171. Sales Discounts
172. Interest Expense
173. Miscellaneous Charges
181. Employees' Bonuses—Profit Sharing
182. Income Taxes

* If all property is rented, this account may not be required.

The method of allocation of these overhead expenses among departments is as follows:

1. Building Expense is allocated by using the floor space occupied as the basis.

2. Fixed Charges are based on book value of fixed assets, including depreciation computed upon the basis of a property ledger or analysis of fixed assets.

3. Payroll Expense is computed upon compensation paid at applicable rates.

4. General Expense includes all other overhead expenses which cannot be applied on a more direct basis, and which are therefore charged to operating departments upon the basis of sales or gross income.

Also needed for the preparation of separate accounts for the different departments is a departmental breakdown of the income and current expense accounts, which, together with the distributed overhead, yields the various accounts to be kept.

Index of Tables

Index

The balance sheet of a food packing business is prepaid by summarizing the foregoing accounts.

The operating accounts of a food packing business would consist of the following:

OPERATING ACCOUNTS

Income Accounts
    40. Sales
    41. Sales Allowances
    42. Trade Discounts
    43. Returned Goods
Expense Accounts
   I. Direct Manufacturing Costs
     50. Raw Materials

> In addition to showing the cost of raw materials, plus delivery cost, less discounts and allowances, a further breakdown if often made, for better cost control, into cost of raw product, cost of additives (sugar, seasonings, flavors, etc.), cost of containers, etc. This type of breakdown, which is essentially for cost accounting purposes, is important in the control of costs.

     51. Manufacturing Labor, Including Labor Taxes and Insurance
     52. Rent on Machinery Not Owned
     53. Fuel
     54. Manufacturing Supplies
     55. Packing and Shipping Labor
  II. Indirect Manufacturing Costs
     60. Bookkeeping and Other Manufacturing Office Salaries
     61. Indirect Labor (loading trucks, etc.)
     62. Light, Heat and Power
     63. Water
     64. Repairs
     65. Insurance
     66. Property Taxes
     67. Depreciation
     68. Miscellaneous
 III. Sales Costs
     70. Salesmen's Salaries
     71. Salesmen's Expense Accounts
     72. Commissions and Brokerage
     73. Advertising
 IV. General and Administrative Expenses
     80. Executive Salaries

81. Office Salaries (other than charged in account #60 above)
82. Expense Accounts (other than charged in account #71 above)
83. Office Expense
84. Insurance
85. Rent
86. Taxes (other than charged in accounts #51 and 61 above)
87. Depreciation (other than charged to Manufacturing in account #67 above)

A special record of great usefulness in a manufacturing business is the sales-production statement. It helps to avoid overproduction by showing the following groups of figures: (1) orders booked during period, orders shipped, balance unshipped orders; and (2) the inventory at beginning of period, the production during the period, and the orders shipped. From these two sets of figures, two figures are obtained for the balance of unshipped orders, which should check if this important figure has been computed correctly.

Accounting for the Small Service Business.—From the accounting standpoint, a garage is an interesting type of service business, because however small it may be, it requires departmentalized accounting methods. Only by computing separately the results of the various services and types of products sold can you know which of them are most profitable, or what is more serious, whether some of them are actually losing money.

The method by which this separate accounting is accomplished is by summarizing first the overhead expenses, which are then allocated among departments, each of which has its own operating statement showing income, expenses and net profit. One can see how this work is done by reviewing actual lists of accounts for the garage as a whole, and then for its individual departments.

OVERHEAD EXPENSES FOR GARAGE

121. Advertising
122. Association Dues and Assessments
123. Auto Expense
124. Bad Debts and Allowances